BUSINESS/SCIENCE/TECHNOLOGY DIVISION
CHICAGO PUBLIC LIBRARY.
400 SOUTH STATE STREET
CHICAGO, IL 60605

REF
TS
191.8
.H36
1999

HWLCTC

Handbook of industrial robotics.

Chicago Public Library

REFERENCE

Form 178 rev. 1-94

BUSINESS/SCIENCE/TECHNOLOGY DIVISION
CHICAGO PUBLIC LIBRARY,
400 SOUTH STATE STREET,
CHICAGO, IL 60605

HANDBOOK OF
INDUSTRIAL ROBOTICS

EDITORIAL BOARD

T. Arai
University of Tokyo

J. J. DiPonio
Ford Motor Co.

Y. Hasegawa
Waseda University

S. W. Holland
General Motors Corp.

S. Inaba
FANUC Inc.

A. C. Kak
Purdue University

S. K. Kim
Samsung Electronics

J.-C. Latombe
Stanford University

E. Lenz
*Technion—Israel Institute of
  Technology*

H. Makino
Yamanashi University

G. Salvendy
Purdue University

G. Seliger
IPK/IFW Berlin

K. Tanie
MEL Tsukuba

K. Trostmann
Technical University Denmark

Y. Umetani
Toyota Technological Institute

H. Van Brussel
Catholic University of Leuven

H.-J. Warnecke
Fraunhofer Institute

R. H. Weston
*Loughborough University of
  Technology*

D. Whitney
Massachusetts Institute of Technology

HANDBOOK OF INDUSTRIAL ROBOTICS

Second Edition

edited by
Shimon Y. Nof

JOHN WILEY & SONS, INC.
New York, Chichester, Weinheim, Brisbane, Singapore, Toronto

This book is printed on acid-free paper. ⊚

Copyright © 1999 by John Wiley & Sons, Inc. All rights reserved.

Published simultaneously in Canada.

No part of this publication may be reproduced, stored in a retrieval system or transmitted in any form or by any means, electronic, mechanical, photocopying, recording, scanning or otherwise, except as permitted under Section 107 or 108 of the 1976 United States Copyright Act, without either the prior written permission of the Publisher, or authorization through payment of the appropriate per-copy fee to the Copyright Clearance Center, 222 Rosewood Drive, Danvers, MA 01923, (978) 750-8400, fax (978) 750-4744. Requests to the Publisher for permission should be addressed to the Permissions Department, John Wiley & Sons, Inc., 605 Third Avenue, New York, NY 10158-0012, (212) 850-6011, fax (212) 850-6008, E-Mail: PERMREQ@WILEY.COM.

This publication is designed to provide accurate and authoritative information in regard to the subject matter covered. It is sold with the understanding that the publisher is not engaged in rendering professional services. If professional advice or other expert assistance is required, the services of a competent professional person should be sought.

Library of Congress Cataloging-in-Publication Data
Handbook of industrial robotics / edited by Shimon Y. Nof. — 2nd ed.
     p.   cm.
   Includes index.
   ISBN 0-471-17783-0 (alk. paper)
   1. Robots, Industrial—Handbooks, manuals, etc.   I. Nof, Shimon Y.
   TS191.8.H36   1999
   670.42′72—dc21   98-8017

Printed in the United States of America

10 9 8 7 6 5 4 3 2 1

CHICAGO PUBLIC LIBRARY
BUSINESS / SCIENCE / TECHNOLOGY
400 S. STATE ST.   60605

This handbook is dedicated
to all of us who believe in
the wonders of human ingenuity
and robot servitude for the
betterment of our life

CHICAGO PUBLIC LIBRARY
BUSINESS / SCIENCE / TECHNOLOGY
400 S. STATE ST.
60605

CONTENTS

PART 9   ROBOTICS IN OPERATIONS

PART 10   ROBOTICS IN VARIOUS APPLICATIONS

PART 11   ROBOTICS AROUND THE WORLD

PART 12   ROBOTICS TERMINOLOGY

FOREWORD

LOOKING AHEAD

In 1939, when I was 19 years old, I began to write a series of science fiction stories about robots. At the time, the word *robot* had been in existence for only 18 years; Karel Capek's play, *R.U.R.*, in which the word had been coined, having been performed for the first time in Europe in 1921. The concept, however, that of machines that could perform tasks with the apparent "intelligence" of human beings, had been in existence for thousands of years.

Through all those years, however, robots in myth, legend, and literature had been designed only to point a moral. Generally, they were treated as examples of overweening pride on the part of the human designer; an effort to accomplish something that was reserved to God alone. And, inevitably, this overweening pride was overtaken by Nemesis (as it always is in morality tales), so that the designer was destroyed, usually by that which he had created.

I grew tired of these myriad-told tales, and decided I would tell of robots that were carefully designed to perform certain tasks, but with *safeguards built in;* robots that might conceivably be dangerous, as any machine might be, but no more so.

In telling these tales, I worked out, perforce, certain rules of conduct that guided the robots; rules that I dealt with in a more and more refined manner over the next 44 years (my most recent robot novel, *The Robots of Dawn,* was published in October, 1983). These rules were first put into words in a story called "Runaround," which appeared in the March, 1942, issue of *Astounding Science Fiction.*

In that issue, on page 100, one of my characters says, "Now, look, let's start with the three fundamental Rules of Robotics . . ." and he proceeds to recite them. (In later stories, I took to referring to them as "the Three Laws of Robotics" and other people generally say "Asimov's Three Laws of Robotics.")

I am carefully specific about this point because that line on that page in that story was, as far as I know, the very first time and place that the word *robotics* had ever appeared in print.

I did not deliberately make up the word. Since *physics* and most of its subdivisions routinely have the "-ics" suffix, I assumed that "robotics" was the proper scientific term for the systematic study of robots, of their construction, maintenance, and behavior, and that it was used as such. It was only decades later that I became aware of the fact that the word was in no dictionary, general or scientific, and that I had coined it.

Possibly every person has a chance at good fortune in his life, but there can't be very many people who have had the incredible luck to live to see their fantasies begin to turn into reality.

I think sadly, for instance, of a good friend of mine who did not. He was Willy Ley who, for all his adult life was wedded to rocketry and to the dream of reaching the moon; who in his early twenties helped found rocket research in Germany; who, year after year wrote popular books on the subject; who, in 1969, was preparing to witness the launch of the first rocket intended to land on the moon; and who then died six weeks before that launch took place.

Such a tragedy did not overtake me. I lived to see the transistor invented, and solid-state devices undergo rapid development until the microchip became a reality. I lived to see Joseph Engelberger (with his interest sparked by my stories, actually) found Uni-mation, Inc., and then keep it going, with determination and foresight, until it actually constructed and installed industrial robots and grew enormously profitable. His devices were not quite the humanoid robots of my stories, but in many respects they were far more sophisticated than anything I had ever been equipped to imagine. Nor is there any doubt that the development of robots more like mine, with the capacities to see and to talk, for instance, are very far off.

I lived to see my Three Laws of Robotics taken seriously and routinely referred to in articles on robotics, written by real roboticists, as in a couple of cases in this volume. I lived to see them referred to familiarly, even in the popular press, and identified with my name, so that I can see I have secured for myself (all unknowingly, I must admit) a secure footnote in the history of science.

I even lived to see myself regarded with a certain amount of esteem by legitimate people in the field of robotics, as a kind of grandfather of them all, even though, in actual fact, I am merely a chemist by training and a science-fiction writer by choice—and know virtually nothing about the nuts and bolts of robotics; or of computers, for that matter.

But even after I thought I had grown accustomed to all of this, and had ceased marveling over this amazing turn of the wheel of fortune, and was certain that there was nothing left in this situation that had the capacity to surprise me, I found I was wrong.

Let me explain . . .

In 1950 nine of my stories of robots were put together into a volume entitled *I, Robot* (the volume, as it happens, that was to inspire Mr. Engelberger).

On the page before the table of contents, there are inscribed, in lonely splendor *The Three Laws of Robotics:*

1. A robot may not injure a human being, or, through inaction, allow a human being to come to harm.

2. A robot must obey the orders given it by human beings except where such orders would conflict with the First Law.

3. A robot must protect its own existence as long as such protection does not conflict with the First or Second Law.

And underneath, I give my source. It is *Handbook of Robotics, 56th Edition, 2058 A.D.*

Unbelievable. Never, until it actually happened, did I ever believe that I would *really* live to see robots, *really* live to see my three laws quoted everywhere. And certainly I never actually believed that I would ever *really* live to see the first edition of that handbook published.

To be sure, it is *Handbook of Industrial Robotics,* for that is where the emphasis is now, in the early days of robotics—but I am certain that, with the development of robots for the office and the home, future editions will need the more general title. I also feel that so rapidly does the field develop, there will be new editions at short intervals. And if there are new editions every 15 months on the average, we will have the fifty-sixth edition in 2058 A.D.

But matters don't stop here. Having foreseen so much, let me look still further into the future. I see robots rapidly growing incredibly more complex, versatile, and useful than they are now. I see them taking over all work that is too simple, too repetitive, too stultifying for the human brain to be subjected to. I see robots leaving human beings free to develop creativity, and I see humanity astonished at finding that almost everyone *can* be creative in one way or another. (Just as it turned out, astonishingly, once public education became a matter of course, that reading and writing was not an elite activity but could be engaged in by almost everyone.)

I see the world, and the human outposts on other worlds and in space, filled with cousin-intelligences of two entirely different types. I see silicon-intelligence (robots) that can manipulate numbers with incredible speed and precision and that can perform operations tirelessly and with perfect reproducibility; and I see carbon-intelligence (human beings) that can apply intuition, insight, and imagination to the solution of problems on the basis of what would seem insufficient data to a robot. I see the former building the foundations of a new, and unimaginably better society than any we have ever experienced; and I see the latter building the superstructure, with a creative fantasy we dare not picture now.

I see the two together advancing far more rapidly than either could alone. And though this, alas, I will not live to see, I am confident our children and grandchildren will, and that future editions of this handbook will detail the process.

Isaac Asimov
(1920–1992)

New York, New York
January 1985

GUEST FOREWORDS

THE EVOLUTION CONTINUES

When the first *Handbook of Industrial Robotics* was published, robots were relatively new. Research projects were more prevalent than industrial usage. Implementation success was less than certain, and those applying robotics had to have a wide variety of special skills. Skeptics almost always outnumbered believers.

Time has substantiated the vision of those early practitioners. Robotics technologies, including the associated control systems and sensors, have evolved over the intervening years. Today robots are an accepted element in the portfolio of manufacturing equipment. Many industries have proven the reliability, ease of use, and excellent productivity of robots within their operations. Research projects continue, but they are building on the strong base of established implementations to open up even more areas for productive application.

Motorola has been an active participant throughout this evolution. Our interest in robotics began back in the early 1980s, not as a research project, but as a solution to a serious business issue. We needed a better way to manufacture pagers to stay competitive. The Bandit program, which vowed to "steal every good idea we could find," identified industrial robots as a major component of the solution. The rest, as they say, is history.

As robots have evolved during the past decade, so has the world around them. All companies now operate in a global environment, where products from far-distant competitors are accessible from international sales forces or simply over the Internet; competition has never been stronger. At the same time, customers are increasing their demands, expecting better prices, wider variety, improved performance, and shorter delivery times. These are the serious business issues that are facing us today.

Once again, robots play an important role in providing solutions at Motorola and many other companies. Their exceptional speed, repeatability, and tirelessness improve productivity and hold down costs. They can work with tiny parts or challenging materials to produce otherwise unattainable leading-edge products. With appropriate programming and sensor input, they can move smoothly from assembling one product model to the next. Once again, the *Handbook of Industrial Robotics*, in its second edition, explains the good ideas and knowledge that are needed for solutions.

What will the future bring? Will nanorobots become indispensable as our semiconductor products shrink smaller and smaller? At the other end of the scale, will space robots eventually help us maintain our satellite-based telecommunications systems? The work of robot researchers to make advances in surgery, agriculture, and many other diverse fields continues to open up new and exciting possibilities. The evolution continues.

CHRISTOPHER B. GALVIN

Chief Executive Officer,
Motorola, Inc.
Schaumburg, Illinois

NEW MISSIONS FOR ROBOTICS

The first edition of the *Handbook of Industrial Robotics* was published in the United States in 1985. The Handbook was large, almost 1,400 pages, and covered a wide variety of topics regarding industrial robotics. The Handbook was a significant resource for the robot specialists concerned with education globally. That time was also the golden age of factory robots. The annual worldwide robot population statistics surveyed by the International Federation of Robotics disclosed rapid expansion of robots in manufacturing industries. In particular, the electric and electronics and the automotive industries were the leading robot users. As a result, these industries have enjoyed the fruits of robotization and have succeeded in supplying their quality products to society at a lower price. The progress of robot technology has enabled robots to work not only in factories. Their field of productive applications has expanded to construction, agriculture, fishery, mining, ocean development, and so forth. In the service area robots are now deployed in medical, restaurant, care of the elderly and disabled persons, amusement, and other assignments.

During the initial period of new robot applications, several challenging problems and research themes were defined. Extensive research and development projects in robotics followed. As a result, some of the newly developed robots have gradually become justified for the new missions.

The outcome is the birth of new industrial robots. Presently the field of industrial robotics is expanding rapidly. The second edition of *The Handbook of Industrial Robotics* will contribute to this new robot age. The material covered in this Handbook reflects the new generation of robotics developments. It is a powerful educational resource for students, engineers, and managers, written by a leading team of robotics experts.

YUKIO HASEGAWA

Professor Emeritus
Waseda University
Tokyo, Japan

CONTINUATION IS A SOURCE OF POWER

From ancient times only human beings have known how to manufacture products. With this ability we have created cultures. Although finance and consumer services tend to be the main business topics in recent years, we ought not to forget the importance of manufacturing. Manufacturing will always continue to be the biggest creator of wealth and indispensable to economic growth. Toyota Motor Corporation has inherited the essence of manufacturing matters from its experience with automatic loom manufacturing. Toyota has considered the importance of manufacturing not only from the aspect of products, such as automobiles, but also from the aspect of processes to manufacture them. Toyota also has emphasized the importance of related matters in the education of employees, the development of technology, and the establishment of effective organizations. TPS (Toyota Production System), also called Lean Production, is the collected study of these efforts. JIDOKA (Autonomation) is one of the two main pillars of TPS, together with JIT (Just-in-Time). Industrial robots have played a significant role in enabling JIDOKA.

Robots have unique characteristics. Needless to say, robots are manufacturing machines which represent the state of the art in mechanics, electronics, and control theory. They are attractive in a certain sense, and offer us visions of the future world. For robotics researchers and engineers, the most important aspect is probably the fact that robots are the embodiment of their design concepts and philosophy.

In an automated production line, one can observe its underlying concept and level of automation. At present, the typical level of industrial robots can be considered as "making what we can." It has not yet reached the level of "making what we want." Specifically, robots are being called upon to work outside of factories. They are being expected to coexist and cowork with humans. We still have many hurdles ahead of us in realizing these expectations. How can we overcome these hurdles? People from various disciplines not only have to cooperate and study industrial robotics from the technological aspect, but also have to consider ethics, philosophy, and design concepts. And the most important consideration is the need to move ahead. We must never forget how difficult it is to recover after stopping progress and the powerful ability of continuous effort.

The second edition of *The Handbook of Industrial Robotics* organizes and systematizes the current expertise of industrial robotics and its forthcoming capabilities. These efforts are critical for solving the underlying problems of industry. This continuation is a source of power. I believe this Handbook will stimulate those who are concerned with industrial robots and motivate them to be great contributors to the progress of industrial robotics. I am most grateful for the efforts of those who have contributed to the development of this second edition and hopeful that our visions will come true in the near future.

HIROSHI OKUDA

President
Toyota Motor Corporation
Tokyo, Japan

CHANGING OUR WORLD

We all know "the machine that changed the world": the car. Now let me tell you about another machine that is changing the world, the robot. Today companies in virtually every industry are achieving significant improvements in productivity and quality by taking advantage of automation technologies. I am happy to report that robotics is one of the key technologies leading mankind into the twenty-first century. Robotics is now hailed for its reliability and accepted by today's workforce, and is benefiting small, medium-sized, and large enterprises. This Handbook describes very well the available and emerging robotics capabilities. It is a most comprehensive guide, including valuable information for both the providers and consumers of creative robotics applications. The attached CD-ROM vividly illustrates robots in action. Nearly 40 years ago Joe Engelberger and other pioneers envisioned a day when robots would perform the dangerous and dull jobs, enabling people to lead a more creative life. The vision developed slowly, as robot technology had to mature, workers had to be convinced that robots were not a threat to their jobs, and managers had to be shown that robots can indeed help their companies thrive as global competitors. In 1974, when our trade association was founded, American companies remained unconvinced about the benefits of robotics. But Japanese companies had started to embrace robots, a key factor in the emergence of Japan as a global manufacturing power.

By the early 1980s robotics was hailed as the "next industrial revolution." Robots were expected not only to solve all our manufacturing problems, but to cook our food, clean our homes, and care for the elderly and disabled. Robotics companies posted record new orders in 1985 as companies rushed to buy robots, often without fully understanding what would be required to implement them effectively.

In the mid-1980s many robotics companies exited the field because robotics products were considered high-priced and ineffective. But then, from 1987–1992, the robot manufacturers worked hard to improve their products and decreased their dependence on the automotive industry. They engineered new products with better intelligence and control, better and simpler vision systems and interfaces, and greater service. Exciting nonmanufacturing applications started to become viable in areas such as security, health care, environmental clean-up, and space and undersea exploration. Research and development advances in robot control, graphic simulation, and off-line programming made sophisticated robots easier to program and simpler to deploy and maintain.

During the last five years robotics companies have posted gains in new orders of 131%. A total of 12,149 robots valued at over $1 billion were ordered in 1997, a new record. Shipments also topped $1 billion for the first time. The big surge in robot use in the United States made it again one of the hottest markets for robotics. Companies that gave up on robotics long ago are now taking a fresh look and discovering that robotics can provide the solutions they need. Predicting the future of the robotics industry can be left to the futurists. However, I think it is safe to say that this industry has a solid foundation and is well positioned for the twenty-first century. The automotive market is still the largest, but applications in the electronics industry should grow at an average rate of 35% a year; in the food and beverage industry robot installations for packaging, palletizing, and filling are expected to grow by 25–30% annually for the next few years. These industries are likely to represent the largest markets for the near term, but we also anticipate growth in the aerospace, appliance, and nonmanufacturing markets.

After a quarter-century of being involved with robotics, I have concluded that the robotics industry is here to stay. And robotics does not stop here. Sojourner is the first, but certainly not the last, intelligent robot sent by humans to operate on another planet, Mars. Robotics, robots, and their peripheral equipment will respond well to the challenges of space construction, assembly, and communications; new applications in agriculture, agri-industries, and chemical industries; work in recycling, cleaning, and hazardous waste disposal to protect our environment and the quality of our air and water; safe, reliable, and fast transportation relying on robotics in flight and on intelligent highways. Robotics prospered in the 1990s; it will thrive and proliferate in the twenty-first century.

DONALD A. VINCENT

Executive Vice President
Robotic Industries Association
Ann Arbor, Michigan

PREFACE

The Handbook of Industrial Robotics was published first in 1985, translated to Russian in 1989–90, and republished in 1992. The second edition is warranted by the relative settling of this young professional field since the 1980s in some of its subareas; the emergence of completely new areas; and the considerable developments in the artificial intelligence aspects of robotics. We continue to use the term *industrial robotics* (henceforth, robotics) to distinguish robots working in service and production from science-fiction and purely software "robots."

In the early days of robotics, the 1920s, the physical fear of monstrous, human-like machines prevailed. In the 1960s, after pioneering robot applications in industry, there was skepticism, sometimes mixed with ridicule, as to whether robots were at all practical. In the 1980s, with increasing robot deployment and proven success, the main concern was whether robots were going to replace us all. Indeed, in the first edition we added to Asimov's original (1940) Three Laws of Robotics the Three Laws of Robotics Applications:

1. Robots must continue to replace people on dangerous jobs. (This benefits all.)
2. Robots must continue to replace people on jobs people do not want to do. (This also benefits all.)
3. Robots should replace people on jobs robots do more economically. (This will initially disadvantage many, but inevitably will benefit all as in the first and second laws.)

By now robotics has become accepted as an essential and useful technology, still exciting our imagination with feats such as the Sojourner's Mars exploration, the emergence of nanorobotics, and the advances in medical robotics. Interestingly, researchers are now even seeking to develop human-like robots with which people would feel more comfortable to work and live.

After consultation with my colleagues, the members of the Handbook's Editorial Board, the table of contents was revised to reflect the above trends. 120 leading experts from 12 countries participated in creating the new edition. I am grateful to all of them for their excellent contributions. All the original chapters were revised, updated, and consolidated. Of the 66 chapters in the new edition, 33 are new, covering important new topics in the theory, design, control, and applications of robotics. A larger robotics terminology with over 800 terms (300 more than in the original) was compiled from this Handbook material. A CD-ROM was added to convey to our readers the colorful motions and intelligence of robotics.

Each chapter was reviewed by two independent peer reviewers and by myself, and revised based on the reviewers' comments. In addition to the Editorial Board members, I wish to thank the following reviewers:

Arvin Agah	S. Krimi	Jacob Rubinovitz
George A. Bekey	C. S. George Lee	Hagen Schempf
Johann J. Borenstein	Masayuki Matsui	Michael J. P. Shaw
Yavuz A. Bozer	John J. Mills	Shraga Shoval
Grigore C. Burdea	Colin L. Moodie	Jose M. A. Tanchoco
Nicholas Dagalakis	Gordon R. Pennock	Tibor Vamos
Yael Edan	Venkat N. Rajan	L. Vlacic
Thomas D. Jerney	Aristides A. Requicha	Daniel A. Whitney
Chang-Ouk Kim	Elon Rimon	Eyal Zussman

I also thank the professionals from John Wiley and Sons, the Pro-Image production team, and the talented editorial and research help from my graduate students: José Ceroni, Chin-Yin Huang, Marco Lara, Jeff Liberski, NaRaye Williams, Jianhao Chen, Jorge Avila, Nitin Khanna, Keyvan Esfarjani, and Jose Peralta.
Finally, I thank my wife Nava, my parents Yaffa and the late Dr. Jacob Nowomiast, and our daughters, Moriah and Jasmin, for all their valuable advice and support.

SHIMON Y. NOF (NOWOMIAST)

West Lafayette, Indiana

CONTRIBUTORS

Arvin Agah, Department of Electrical Engineering and Computer Science, The University of Kansas, Lawrence, Kansas

Narendra Ahuja, Beckman Institute for Advanced Science and Technology, University of Illinois at Urbana-Champaign, Urbana, Illinois

Hadi Abu-Akeel, FANUC Robotics North America, Inc., Rochester Hills, Michigan

Fumihito Arai, Department of Micro System Engineering, Nagoya University, Nagoya, Japan

Ronald C. Arkin, College of Computing, Georgia Institute of Technology, Atlanta, Georgia

C. R. Asfahl, University of Arkansas, Fayetteville, Arkansas

O. Barth, Fraunhofer Institute for Manufacturing Engineering and Automation (IPA), Stuttgart, Germany

George A. Bekey, Institute for Robotics and Intelligent Systems, Department of Computer Science, University of Southern California, Los Angeles, California

J. T. Black, Auburn University, Auburn, Alabama

Karl F. Böhringer, University of California, Berkeley, California

Valerie Bolhouse, Ford Motor Company, Redform, Michigan

M. C. Bonney, University of Nottingham, University Park, United Kingdom

Wayne Book, Georgia Institute of Technology, Atlanta, Georgia

Yavuz A. Bozer, Department of Industrial and Operations Engineering, The University of Michigan, Ann Arbor, Michigan

H.-J. Bullinger, Institute of Work Organization, Stuttgart, Germany

Grigore C. Burdea, Rutgers—The State University of New Jersey, Piscataway, New Jersey

José A. Ceroni, School of Industrial Engineering, Purdue University, West Lafayette, Indiana

S. J. Childe, Manufacturing and Business Systems, University of Plymouth, Plymouth, United Kingdom

Yee-Yin Choong, GE Information Services, Inc., Rockville, Maryland

David R. Clark, Kettering University, Flint, Michigan

Philippe Coiffet, Laboratoire de Robotique de Paris, Velizy, France

Nicholas G. Dagalakis, National Institute of Standards and Technology, Intelligent Systems Division, Gaithersburg, Maryland

Brian Daugherty, Ford Motor Company, Redform, Michigan

Michael P. Deisenroth, Department of Industrial and Systems Engineering, Virginia Polytechnic Institute and State University, Blacksburg, Virginia

John V. Draper, Robotics and Process Systems Division, Oak Ridge National Laboratory, Oak Ridge, Tennessee

Charlie Duncheon, P.E., Adept Technology, Inc., San José, California

Yael Edan, Department of Industrial Engineering and Management, Ben-Gurion University of the Negev, Beer Sheva, Israel

Mohammad R. Emami, University of Toronto, Toronto, Ontario

Gay Engelberger, HelpMate Robotics Inc., Danbury, Connecticut

Joseph F. Engelberger, HelpMate Robotics Inc., Danbury, Connecticut

Ronald S. Fearing, University of California, Berkeley, California

K. Feldmann, Institute for Manufacturing Automatization and Production Systematics (FAPS), University of Erlangen-Nuremberg, Erlangen, Germany

Toshio Fukuda, Center for Cooperative Research in Advanced Science and Technology, Nagoya University, Nagoya, Japan

J. D. Gascoigne, Department of Manufacturing Engineering, Loughborough University of Technology, Loughborough, Leicestershire, United Kingdom

Ken Y. Goldberg, University of California, Berkeley, California

Andrew A. Goldenberg, University of Toronto, Toronto, Ontario

K. Götz, Institute for Manufacturing Automatization and Production Systematics (FAPS), University of Erlangen-Nuremberg, Erlangen, Germany

Allen C. Grzebyk, FANUC Robotics North America, Inc., Rochester Hills, Michigan

A. Gunasekaran, University of Massachusetts, Dartmouth, Massachusetts

M. Hägele, Fraunhofer Institute for Manufacturing Engineering and Automation (IPA), Stuttgart, Germany

Nicholas G. Hall, The Ohio State University, Columbus, Ohio

Lane A. Hautau, FANUC Robotics North America, Inc., Rochester Hills, Michigan

Steven W. Holland, General Motors Corp., Warren, Michigan

Thomas Hörz, Fraunhofer Institute for Manufacturing Engineering and Automation (IPA), Stuttgart, Germany

Chin-Yin Huang, School of Industrial Engineering, Purdue University, West Lafayette, Indiana

Brian Huff, The Automation & Robotics Research Institute, The University of Texas at Arlington, Fort Worth, Texas

Seth Hutchinson, Beckman Institute for Advanced Science and Technology, University of Illinois at Urbana-Champaign, Urbana, Illinois

Yukio Iwasa, Nomura Research Institute, Ltd., Tokyo, Japan

Thomas D. Jerney, Adept Technology, Inc., San José, California

A. C. Kak, Robot Vision Laboratory, School of Electrical and Computer Engineering, Purdue University, West Lafayette, Indiana

Sungkwun Kim, Samsung Electronics, Kyunggi-Do, Korea

Akio Kosaka, Robot Vision Laboratory, Purdue University, West Lafayette, Indiana

George L. Kovács, CIM Research Laboratory, Computer and Automation Research Institute, Hungarian Academy of Sciences, Budapest, Hungary

S. Krimi, Institute for Manufacturing Automatization and Production Systematics (FAPS), University of Erlangen-Nuremberg, Erlangen, Germany

Krishna K. Krishnan, Department of Industrial and Manufacturing Engineering, Wichita State University, Wichita, Kansas

Vijay Kumar, General Robotics and Active Sensory Perception (GRASP) Laboratory, University of Pennsylvania, Philadelphia, Pennsylvania

Marco A. Lara, School of Industrial Engineering, Purdue University, West Lafayette, Indiana

C. S. George Lee, School of Electrical and Computer Engineering, Purdue University, West Lafayette, Indiana

Mark R. Lehto, School of Industrial Engineering, Purdue University, West Lafayette, Indiana

Lonnie Love, Oak Ridge National Laboratory, Robotics and Process Systems Division, Oak Ridge, Tennessee

R. S. Maull, School of Business and Management, University of Exeter, Exeter, United Kingdom

J. B. Mills, Manufacturing and Business Systems, University of Plymouth, Plymouth, United Kingdom

John J. Mills, The Automation & Robotics Research Institute, The University of Texas at Arlington, Fort Worth, Texas

Laxmi P. Musunur, FANUC Robotics North America, Inc., Rochester Hills, Michigan

Doug Niebruegge, ABB Flexible Automation, New Berlin, Wisconsin

Erich Niedermayr, Corporate Technology Department—Production Engineering, Siemens AG, Munich, Germany

Shimon Y. Nof, School of Industrial Engineering, Purdue University, West Lafayette, Indiana

James P. Ostrowski, General Robotics and Active Sensory Perception (GRASP) Laboratory, University of Pennsylvania, Philadelphia, Pennsylvania

J. Pack, Institute of Work Organization, Stuttgart, Germany

Lynne E. Parker, Center for Engineering Systems Advanced Research, Oak Ridge National Laboratory, Oak Ridge, Tennessee

Joseph Pössinger, Corporate Technology Department—Production Engineering, Siemens AG, Munich, Germany

Adrien Presley, Division of Business and Accountancy, Truman State University, Kirksville, Missouri

Venkat N. Rajan, i2 Technologies, Chicago, Illinois

M. Reichenberger, Institute for Manufacturing Automatization and Production Systematics (FAPS), University of Erlangen-Nuremberg, Erlangen, Germany

Aristides A. G. Requicha, Laboratory for Molecular Robotics and Computer Science Department, University of Southern California, Los Angeles, California

Charles A. Rosen, Machine Intelligence Corp., Sunnyvale, California

Norbert Roth, Corporate Technology Department—Production Engineering, Siemens AG, Munich, Germany

Jacob Rubinovitz, The Davidson Faculty of Industrial Engineering and Management, Technion, Israel Institute of Technology, Haifa, Israel

Gavriel Salvendy, School of Industrial Engineering, Purdue University, West Lafayette, Indiana

Hagen Schempf, Robotics Institute, Carnegie Mellon University, Pittsburgh, Pennsylvania

G. Schmierer, Fraunhofer Institute for Manufacturing Engineering and Automation (IPA), Stuttgart, Germany

Claus Scholpp, Fraunhofer Institute for Manufacturing Engineering and Automation (IPA), Stuttgart, Germany

R. D. Schraft, Fraunhofer Institute for Manufacturing Engineering and Automation (IPA), Stuttgart, Germany

Klaus Schröer, Fraunhofer-IPK, Berlin, Germany

Manfred Schweizer, Fraunhofer Institute for Manufacturing Engineering and Automation (IPA), Stuttgart, Germany

Mario Sciaky, Sciaky S. A., Vitry-Sur-Seine, France

G. Seliger, IPK/IFW, Technical University of Berlin, Berlin, Germany

Michael J. Shaw, Beckman Institute for Advanced Science and Technology, University of Illinois at Urbana-Champaign, Urbana, Illinois

P. A. Smart, Manufacturing and Business Systems, University of Plymouth, Plymouth, United Kingdom

Geary V. Soska, The Goodyear Tire and Rubber Co., Akron, Ohio

G. T. Stevens, The Automation & Robotics Research Institute, The University of Texas at Arlington, Fort Worth, Texas

K. Sugimoto, Hitachi Limited, Yokohama, Japan

Kinya Tamaki, School of Business Administration, Aoyama Gakuin University, Tokyo, Japan

Kazuo Tanie, Mechanical Engineering Laboratory, AIST-MITI, Tsukuba, Japan

William R. Tanner, Productivity Systems, Inc., Farmington, Michigan

Russell H. Taylor, Department of Computer Science, The Johns Hopkins University, Baltimore, Maryland

Elpida S. Tzafestas, Intelligent Robotics and Automation Laboratory, Department of Electrical and Computer Engineering, National Technical University of Athens, Zographou, Athens, Greece

Spyros G. Tzafestas, Intelligent Robotics and Automation Laboratory, Department of Electrical and Computer Engineering, National Technical University of Athens, Zographou, Athens, Greece

Yoji Umetani, Toyota Technological Institute, Nagoya, Japan

P. Valckenaers, Catholic University of Leuven (KUL), Leuven, Belgium

H. Van Brussel, Catholic University of Leuven (KUL), Leuven, Belgium

Donald A. Vincent, Robotic Industries Association, Ann Arbor, Michigan

H.-J. Warnecke, Fraunhofer Institute, Munich, Germany

Atsushi Watanabe, FANUC Ltd., Robot Laboratory, Yamanashi, Japan

John G. Webster, University of Wisconsin, Madison, Wisconsin

R. H. Weston, Department of Manufacturing Engineering, Loughborough University of Technology, Loughborough, Leicestershire, United Kingdom

Daniel E. Whitney, Massachusetts Institute of Technology, Cambridge, Massachusetts

NaRaye P. Williams, School of Industrial Engineering, Purdue University, West Lafayette, Indiana

Y. F. Yong, BYG Systems Ltd., Nottingham, United Kingdom

Miloš Žefran, General Robotics and Active Sensory Perception (GRASP) Laboratory, University of Pennsylvania, Philadelphia, Pennsylvania

E. Zussman, Department of Mechanical Engineering, Technion—Israel Institute of Technology, Haifa, Israel

PART 1
DEVELOPMENT OF INDUSTRIAL ROBOTICS

CHAPTER 1

HISTORICAL PERSPECTIVE AND ROLE IN AUTOMATION

Joseph F. Engelberger
HelpMate Robotics Inc.
Danbury, Connecticut

It does not behoove a commentator on a history that he participated in making to repudiate past observations. Therefore this overview appears as it did in the first edition of the handbook and is followed by an update that picks up on the history of industrial robotics as the industry evolved after 1985.

Any historical perspective on robotics should at the outset pay proper homage to science fiction. After all, the very words *robot* and *robotics* were coined by science fiction writers. Karel Capek gave us *robot* in his 1922 play *Rossum's Universal Robots* (RUR), and Isaac Asimov coined the word *robotics* in the early 1940s to describe the art and science in which we roboticists are engaged today.

There is an important distinction between these two science fiction writers. Capek decided that robots would ultimately become malevolent and take over the world, while Asimov from the outset built circuits into his robots to assure mankind that robots would always be benevolent. A handbook on industrial robotics must surely defend the Asimov view. That defense begins with the history of industrial robotics—a history that overwhelmingly finds benefits exceeding costs and portends ever-rising benefits.

Science fiction aside, a good place to start the history is in 1956. At that time George C. Devol had marshalled his thoughts regarding rote activities in the factory and his understanding of available technology that might be applied to the development of a robot. His patent application for a programmable manipulator was made in 1954, and it issued as patent number 2,988,237 in 1961. This original patent was destined to be followed by a range of others that would flesh out the principles to be used in the first industrial robot.

Also in 1956, Devol and Joseph Engelberger met at a fortuitous cocktail party. Thus began an enduring relationship that saw the formation and growth of Unimation Inc. The first market study for robotics was also started in 1956 with field trips to some 15 automotive assembly plants and some 20 other diverse manufacturing operations. Figure 1 is a reproduction of an actual data sheet prepared during this first market study.

Giving a fairly tight specification regarding what was needed to do simple but heavy and distasteful tasks in industry, the original design team set to work. First came appropriate components and then a working robot in 1959. Shortly thereafter Devol and Engelberger celebrated again—we see them in Figure 2 being served their cocktails, this time by a prototype Unimate industrial robot.

By 1961 the prototype work had progressed far enough to let an industrial robot venture forth. Figure 3 shows the first successful robot installation: a die casting machine is tended in a General Motors plant.

At this juncture it may be well to step back and retrospectively evaluate whether or not robotics should have become a successful innovation. A useful vantage point is provided by a 1968 Air Force sponsored study called Project Hindsight. The objective of the study was to determine what circumstances are necessary for an innovation to become successful. Project Hindsight concluded that there were three essential prerequisites for success:

Handbook of Industrial Robotics, Second Edition, Edited by Shimon Y. Nof
ISBN 0-471-17783-0 © 1999 John Wiley & Sons, Inc.

CONSOLIDATED CONTROLS CORPORATION

UNIMATION SURVEY DATA SHEET

DATE: 5-14-56

OBSERVER: MJD

LOCATION:

TYPE OF WORK PERFORMED:

Press blanking of side panels from sheet stock.

SEQUENCE OF PRESENT OPERATION:

Sheet steel put into die against 3 locating pins. Trim
drops through bed, stamped part withdrawn and stacked.

APPROXIMATE CYCLES PER MINUTE: 3

MAXIMUM NO. OF SEQUENCES:

Horizontal	Vertical	Rotary
8	6	4

HORIZONTAL TRAVERSE, ACCURACY, AND MAX. SPEED: 3 FT; $\pm$1/8"; 6 IN/SEC
VERTICAL   "   "   "   "   "   : 4 FT; $\pm$1/8"; 6 IN/SEC
ROTARY   "   "   "   "   "   : 270°; $\pm$12 MIN; 60°/SEC

HAND ACTION REQUIRED:

Suction cup

APPROXIMATE WEIGHT OF PART: 2 LBS

NO. OF OPERATORS: 18 per shift - 2 shifts

PROCESS MODIFICATION REQUIRED:

None

AVAILABLE AREA: 4 FT x 5 FT

/lb
8-23-56

Figure 1   Reproduction of actual data sheet used in first field market study.

4

Figure 2   Devol and Engelberger being served cocktails by a prototype Unimate robot.

1. There must be a perceived need.
2. Appropriate technology and competent practitioners must be available.
3. There must be adequate financial support.

For robotics there was a perceived need, certainly in the eyes of Devol and Engelberger, although it would be many years before this perception would be broadly shared. Appropriate technology was available, and very competent practitioners could be drawn from aerospace and electronic industries. Finally, a venturesome financial support was brought to bear from such companies as Condec Corporation and Pullman Inc.

Back in 1922, and still in 1940, it was quite possible for Capek and Asimov to perceive a need for robots. There were certainly many heinous jobs that created subhuman working conditions. It might also have been possible in those times to gather financial support (witness all of the harebrained schemes that Mark Twain innocently sponsored); however, the technology simply was not at hand.

Three technologies born during or after World War II are crucial to successful robotics. First, servo-mechanisms theory was unknown before World War II. Second, digital computation came into its own after World War II. Finally, solid state electronics made it all economically feasible.

It is interesting to look at what has happened to the cost of electronic computation since the first tentative steps were made to produce a control system for an industrial robot. Figure 4 is a semilog plot of the cost of a typical computation function versus time. What in 1955 might have cost \$14 by 1982 would cost seven cents. That is a 200-fold reduction in cost. It allows today's roboticist to luxuriate in computational hardware and make his heavy investments in the software. In 1956 one of the design challenges was to accomplish necessary functions with frugal use of electronics hardware. One of Unimation's triumphant decisions was to go solid state in its controller design at a time when vacuum tube controller execution would have been substantially cheaper. At that time a five-axis controller for a numerically controlled machine tool could have been acquired at an Original Equipment Manufacturer (OEM) discount price of about \$35,000.

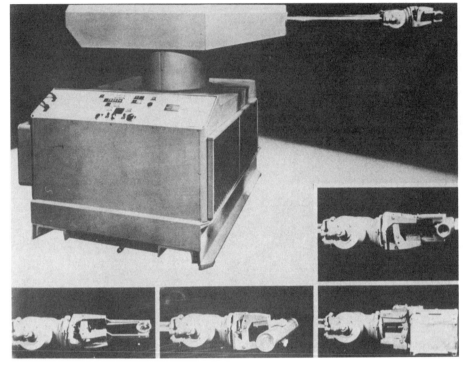

Figure 3 First robot installation (die casting machine in a General Motors plant).

Unimation engineers prided themselves on a purpose-built design that could be achieved in 1959 for $7000.

For the first robot the cost to manufacture was 75% electronic and 25% hydrome-chanical. Today that cost ratio is just reversed.

One should note that automation was already flourishing before World War II. There were many high-volume products that were made in large quantities by what today is called "hard automation." Charlie Chaplin in his 1936 movie *Modern Times* was able to

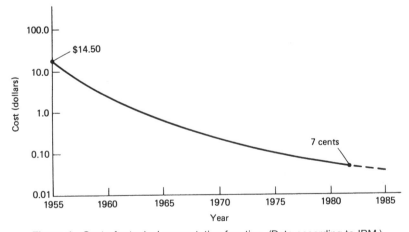

Figure 4 Cost of a typical computation function. (Data according to IBM.)

decry satirically the role of a human floundering in an automated manufacturing scene. However, all of that automation used mechanics that today we archly call "bang-bang." It is ironic that the word *robot* has become so glamorous that some companies, and even some countries, include mechanisms using this "bang-bang" technology in their categorization of robotics. The Japanese recognize "limited sequence" robots, which are conceptually turn-of-the-century technology, as being the single largest segment of the Japanese robot population (more about the Japanese role in this historic perspective shortly).

In 1961 the first industrial robot went to work. Unimation's founder and president proved just how clouded his crystal ball was by going from 1961 until 1975 before his company was able to show a profit. The publicity was great; it attracted many abortive competitive efforts. But those who provided that third ingredient, money, were sorely disappointed in the slow progress. Just consider: The first robot worked quite well! It is now in the Smithsonian Institute. Some of its brethren are still functioning today. Many of the earliest robots built have accumulated more than 100,000 hours of field operation, more than 50 man-years of working. The concept was viable, the product was viable. Why was it taking so long to gain sufficient acceptance to make robot manufacture economically viable?

In retrospect we recognize that manufacturing is an extremely conservative activity. It does not take to change lightly. Worse than that, no one really needs a robot! Anything a robot can do, a human can also do. The only justification for hiring a robot is that a robot will work at a cheaper rate. Even that justification is not convincing if one's competitors are not making extensive use of robots. The institutional load was a formidable one, and at times it seemed almost insurmountable.

Enter the Japanese. In the early 1960s Japanese visitors to Unimation increased in frequency, and by 1967 Engelberger was an invited speaker in Tokyo. In the United States it was difficult to gain the attention of industrial executives, but the Japanese filled the hall with 700 manufacturing and engineering executives who were keenly interested in the robot concept. They followed the formal presentation with three hours of enthusiastic questioning. In 1968 Kawasaki Heavy Industries took a license under all of Unimation Inc.'s technology, and by 1971 the fever had spread and the world's first robot association was formed—not in the United States but in Japan. The Japan Industrial Robot Association (JIRA) started out with an opening membership of 46 companies and with representatives having personal clout in the industrial community. The first president of JIRA was Mr. Ando, the Executive Vice President of Kawasaki Heavy Industries, a three-billion-dollar company.

Thereafter the rest of the industrial world slowly began to awaken. The Robot Institute of America was founded in 1975, well after the first International Symposium on Industrial Robotics (ISIR) was held in Chicago in 1970. That first ISIR attracted 125 attendees despite being held during a crippling snowstorm. Before this handbook is published the 13th ISIR will also be history, and all indications are that 1200 will attend the conference itself and the industrial exhibition will attract some 25,000 visitors.

Perhaps the institutional job has finally been accomplished. Look at the industrial giants who are attempting to stake out claims in the robotics arena. Beyond the biggest in Japan who are already well represented, we have such companies as General Motors, General Electric, Westinghouse, IBM, and United Technologies in the United States, and major European industrialists such as G.E.C. in England, Siemens in Germany, Renault in France, Fiat in Italy. Add to these a legion of smaller companies who fragment the market and make their mark in specialized robots, robot peripherals, or consulting and robotic system design.

The government of virtually every major industrial country in the world, capitalist or communist, have declared robotics to be an arena of intense national interest worthy of support from public coffers. So obviously robotics has arrived, hasn't it? Or, really, has it? We have a plethora of robot manufacturers, very few of whom are profitable. There is a shakeout under way unique in industrial history. It is occurring before any robot manufacturer has achieved great financial success.

The commercially available technology is not remarkably different from what existed 20 years ago. Moreover, none of the obvious applications is even close to saturation. Figure 5 lists applications that have been proven both technically and economically and still represent great robotic opportunities. There is little imagination necessary to go beyond the current level of commercially available technology to the addition of rudimentary

CURRENT APPLICATIONS

Die Casting	Machine Loading
Spot Welding	Stamping
Arc Welding	Plastic Molding
Glass Handling	Investment Casting
Heat Treatment	Conveyor Transfer
Forging	Palletizing
Paint Spraying	Inspection

Figure 5 Current robot applications.

vision or modest tactile sensing ability to accomplish another broad spectrum of jobs such as those listed in Figure 6. Further, jobs outside of the industrial robot stamping ground are already on the technically visible horizon. Some of these are listed in Figure 7.

What wonderful good luck to have founded a company, nay, even an industry, when one is young enough to participate during the industry's adolescence and to speculate on the tremendous technical excitement ahead as robotics reaches its majority. A handbook on industrial robotics will need be a living document for at least the balance of this century to keep up with the inevitable expansion of the technology. From the historical perspective one wishes the editors good health, long life, and a proclivity for conscientious reporting.

AND THEN THEY WERE 750,000 STRONG!

At the end of 1996 the industrial robot industry worldwide was running at an annual rate of $7 billion and projecting $8 billion for 1997 with a population increase of 103,000.

What Else?

The "shake-out" predicted in 1985 has taken place. Five giant entrants in the USA— General Motors, General Electric, Westinghouse, IBM, and United Technologies—have all either sold or folded their robot manufacturing operations. But one major U.S. independent manufacturer is left, Adept Technology Inc.

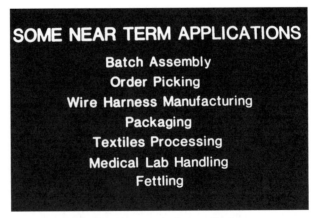

SOME NEAR TERM APPLICATIONS

Batch Assembly
Order Picking
Wire Harness Manufacturing
Packaging
Textiles Processing
Medical Lab Handling
Fettling

Figure 6 Near-term robot applications.

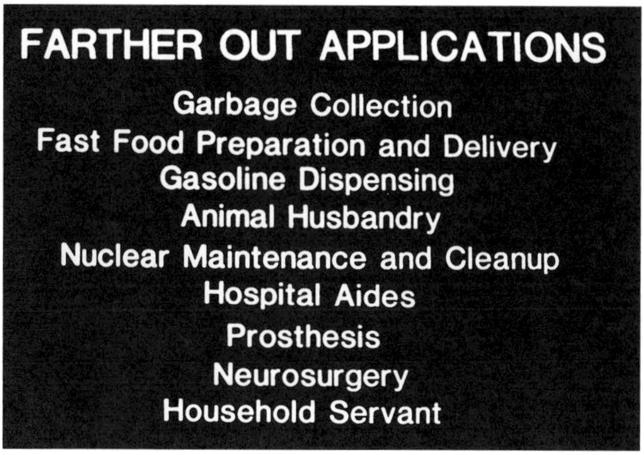

FARTHER OUT APPLICATIONS

Garbage Collection
Fast Food Preparation and Delivery
Gasoline Dispensing
Animal Husbandry
Nuclear Maintenance and Cleanup
Hospital Aides
Prosthesis
Neurosurgery
Household Servant

Figure 7 Long-range applications.

Meanwhile, with the U.S. market approaching $1 billion, foreign manufacturers have entered with both distribution and manufacturing facilities. Prominent among them are ABB Robotics, Yaskawa Electric Corporation, Kawasaki Heavy Industries, and Fanuc.

The cost of industrial robots has continued to come down at the same time as capabilities have improved. An average price in 1991 was $110,000. By 1996 it was $82,000. In constant dollars that is a drop of approximately 50%.

The cost of a computation function has continued to follow the trend line of Figure 4. Now it is down to a fraction of one cent. It is literally possible to match the 1983 tongue-in-cheek Engelberger offer, "I will *sell* you the mechanical arm, *give* you the electronics and *rent* you the software."

The technology spectrum available to roboticists has blossomed. Table 1 lists the modalities that give robots greatly enhanced performance. Sensory perception allows accommodation to workpiece variables, speed has become as fast as most hard automation systems, and reliability regularly exceeds 10,000 hours MTFB.

The original glamor of robotics has paled for industrial robotics. At the 27th ISIR in Milan both the conference attendance, 300, and the number of visitors to the industrial robot booths, 10,000, were down from 1200 and 25,000 respectively in 1985. Why? Because industrial robots have become commodities in the eyes of customers and particularly of the general public. There is no science fiction excitement left. There are no fathers with wide-eyed children on their shoulders!

But maybe that will change when the ISIR becomes the ISR, International Symposium on Robotics. Dropping the "industrial" from the symposium title is an admission that robotics is not limited to factory operations. In *Robotics in Service* this author discussed a range of service jobs that would become eligible given the growth of technology. Table 2 lists the potential applications. In Chapter 64 of this handbook, Gay Engelberger, past president of ISRA, the International Service Robot Association, speaks to the state of the embryonic service robot industry.

Industrial robot installations have become more complex, involving multiple robots, intricate peripherals, moving targets, and multiproduct assembly lines, all of which put great demands on industrial engineers and require long trial runs.

Surcease comes from simulation software. Companies like Deneb and Silma allow complete systems to be simulated prior to installation. Simulation companies have libraries on commercial robot kinematics, and these permit off-line 3D evaluations of any application. Before an installation goes live, robot programs have been created, interferences eliminated, and "speeds and feeds" demonstrated. Costly field hardware and software modifications are obviated.

In 1985 we reported on the growth of interest in robotics worldwide. In 1996 this interest was powerfully evidenced by the existence of the IFR, International Federation of Robotics. (Note the absence of the adjective *industrial*.) The IFR has 26 member organizations representing as many countries. It serves the interests of all members and of the industry. Member countries appeal to the IFR for the prestigious right to host the

Table 1   Robotics Toolchest

Electronics
- Low-cost, high-speed microprocessors
- Vast memories, negligible cost

Servos
- DC
- AC
- Stepper
- Hydraulic

Controllers
- Point-to-point
- Continuous path
- Sensor-driven

Application Software
- VAL
- KAREL
- RCCL
- Others

Position and Motion Sensors
- Encoders
- Resolvers
- Compasses
- Passive beacons
- Active beacons
- Ceiling vision
- Inertial (Gyro)
- Clinometer
- GPS

Range Scanning
- Ultrasound
- Light triangulation
- LIDAR
- Optical flow
- Capacitive
- Inductive

Vision
- Structured light
- Stereo
- Scene analysis
- Template matching
- Colorimeter
- Bar code readers

Tactility
- Wrist force sensing
- Torque sensing
- Fingertip arrays
- Limit switches
- Contact bumpers

Voice Communication
- Synthesis
- Recognition

Artificial Intelligence
- Expert systems
- Sensory fusion
- Fuzzy logic
- Semantic networks

Table 2   Service Robot Applications

- Hospital Porter
- Commercial Cleaning
- Guard Service
- Nuclear Power Maintenance
- Parapharmacist
- Underwater Maintenance
- Parasurgeon
- Farming
- Gas Station Attendant
- Hotel Bellboy
- Space Vehicle Assembly
- Military Combat
- Companion for Infirm and Handicapped

International Symposium on Robotics. A Secretariat is located in Sweden (Björn Weich-brodt, International Federation of Robotics, PO Box 5510, S-114 85, Stockholm, Sweden).

All in all, one must conclude that the industrial robotics industry is mature but still learning new tricks and growing. No single application is saturated. Spot welding, the first major application, still commands the largest market share. The automobile industry is still the largest user. Japan is still the dominant builder and user of industrial robots, boasting 60% to 70% of the installations and of the annual market.

When the future becomes history, we shall see if Japan remains dominant as robots take on service jobs—including even personal service, as projected by Asimov. Damon Runyon once said, "The race is not always to the swift, nor the battle to the strong; but, brother, that is the way to bet."

Watch for the third edition of this handbook to give the historical perspective on robotics migrating from industrial to service applications.

CHAPTER 2
ROBOTICS IN JAPAN: EMERGING TRENDS AND CHALLENGES

Yoji Umetani
Toyota Technological Institute
Nagoya, Japan

Yukio Iwasa
Nomura Research Institute, Ltd.
Tokyo, Japan

1 INTRODUCTION

Industrial robots and their application systems were put into full-fledged practical use in the 1980s, and the demand for such robots and systems has increased steadily ever since. During the bubble economy period in Japan in the latter half of the 1980s, particularly, every industry was keen to invest in plant and equipment. In 1990 the quantity of shipments (total quantity of shipments for manufacturing industry in Japan and for export) of industrial robots reached a peak of 79,782 units, and in 1991 the monetary value of shipments reached a peak of 588,274 million yen.

However, in line with the burst of the bubble economy in the 1990s, the desire for investment in plant and equipment has fallen rapidly in every industry, and consequently the demand for industrial robots has greatly decreased.

The economy now seems to be in a slow upturn trend, and the question of whether the future demand for industrial robots will increase is attracting much attention.

Based on the background stated above, the authors conducted research about industrial robots on the following three questions:

1. Transition of demand and the number of robots in operation
2. Future demand environment
3. Demand forecast until 2010

The direction and content of the research were studied by the Special Committee on Long-Term Demand Forecast for Industrial Robots (chaired by Professor Yoji Umetani of the Toyota Technological Institute), established by the Japan Robot Association (JARA). The research work and preparation of reports were conducted by Nomura Research Institute, Ltd.

Reports are discussed here only in outline. Further details can be found in the reports themselves.

2 TRANSITION OF DEMAND AND THE NUMBER OF ROBOTS IN OPERATION

2.1 Transition of Demand

Only a part of the results obtained from the analysis of the transition of demand for industrial robots will be discussed here. This analysis is based on the Fact-Finding Survey of Business Entities Related to Manipulators and Robots conducted by JARA.

Figure 1 shows the shipment transition of industrial robots (total quantity of shipments for manufacturing industry in Japan and for export).

Handbook of Industrial Robotics, Second Edition, Edited by Shimon Y. Nof
ISBN 0-471-17783-0 © 1999 John Wiley & Sons, Inc.

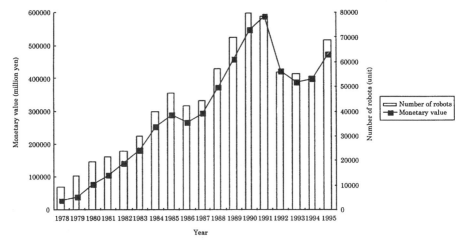

Figure 1   Shipment transition of industrial robots (total quantity of shipments for manufacturing industry in Japan and for export).

Full-scale shipment of industrial robots started in the 1980s. The total quantity of shipments for manufacturing industry in Japan and for export reached a peak of 79,782 units in 1990, and the monetary value of shipment reached a peak of ¥588,274 million in 1991.

However, after the burst of the bubble economy, shipments fell drastically in terms of both quantity and monetary value. In 1992 the quantity of shipments was 28.9% lower and the monetary value 28.3% lower than in the previous year. The quantity of shipments kept declining until 1994, the monetary value until 1993. However, in 1995 both the quantity and the monetary value of shipments shifted upward, which could be a sign of a full-scale recovery in demand.

Table 1 shows the transition of shipments, classified by field of demand (on the basis of monetary value of shipments).

In the classification by fields of demand, the monetary value of robots for export accounted for the largest share in every year but 1990. Among manufacturing industries in Japan, the automobile industry took up the largest share from 1991 through 1994, and radio/TV/communication devices the largest in 1995.

In the classification by usage, mounting accounted for the largest share every year. Recently bonding has become higher in ranking, reflecting the increase in demand for robots in assembling electronic circuit boards.

In the classification by type, though numerically controlled robots accounted for the largest share in 1991, playback robots had the largest share in 1992 and intelligent robots the largest in 1993.

2.2   Number of Robots in Operation

Figure 2 shows the component percentage of industrial robots in operation in the manufacturing industry in Japan.

The number of industrial robots in operation in the manufacturing industry in Japan in 1996 is estimated to be 464,367 units, based on the Fact-Finding Survey of Business Entities Related to Manipulators and Robots conducted by JARA and on the assumption that the replacement cycle is 10 years (as indicated in questionnaires answered by users).

In the classification by industry, the number of robots in operation is largest in the electric machinery, automobile/motorcycle, and plastic product manufacturing industries, respectively. These three industries use 74.5% of the robots in operation.

In the classification by process, the number of robots in operation is greatest in assembling (25.6%), followed by resin molding, inserting/installing/bonding, arc welding, and machining/grinding.

In the classification by type, the greatest number are playback robots, followed by numerically controlled robots, fixed sequence robots, and sequence robots.

Classification	Ranking	1988	1989	1990	1991	1992	1993	1994	1995
	1	Export	Export	Electric machines	Export	Export	Export	Export	Export
	2	Electric machines	Automobiles	Export	Automobiles	Automobiles	Automobiles	Automobiles	Radio/TV/communication devices
Classification by industry	3	Automobiles	Electric machines	Automobiles	Electric machines	Electric machines	Radio/TV/communication devices	Radio/TV/communication devices	Automobiles
	4	Radio/TV/communication devices	Radio/TV/communication devices	Radio/TV/communication devices	Home electric appliances	Metal products	Computers	Electric machines	Computers
	5	Plastic products	Plastic products	Metal products	Radio/TV/communication devices	Radio/TV/communication devices	Metal products	Plastic products	Plastic products
	1	Mounting	Mounting	Mounting	Mounting	Mounting	Mounting	Mounting	Mounting
	2	Spot welding	Spot welding	Arc welding	Arc welding	Spot welding	Spot welding	Spot welding	Bonding
Classification by usage	3	Inserting	Arc welding	Inserting	Spot welding	Arc welding	Arc welding	Bonding	Spot welding
	4	Arc welding	Inserting	Spot welding	Inserting	Loading/unloading	Bonding	Arc welding	Arc welding
	5	Bonding	Assembling in general	Assembling in general	Loading/unloading	Inserting	Assembling in general	Assembling in general	Assembling in general
	1	Numerically controlled robots	Numerically controlled robots	Numerically controlled robots	Numerically controlled robots	Numerically controlled robots	Intelligent robots	Intelligent robots	Intelligent robots
	2	Playback robots	Playback robots	Playback robots	Playback robots	Intelligent robots	Playback robots	Numerically controlled robots	Numerically controlled robots
Classification by type	3	Intelligent robots	Intelligent robots	Intelligent robots	Intelligent robots	Playback robots	Numerically controlled robots	Playback robots	Playback robots
	4	Sequence robots	Sequence robots	Sequence robots	Sequence robots	Sequence robots	Sequence robots	Fixed sequence manipulators	Fixed sequence manipulators
	5	Fixed sequence manipulators	Fixed sequence manipulators	Fixed sequence manipulators	Fixed sequence manipulators	Fixed sequence manipulators	Fixed sequence manipulators	Sequence robots	Sequence robots

Classification by industries

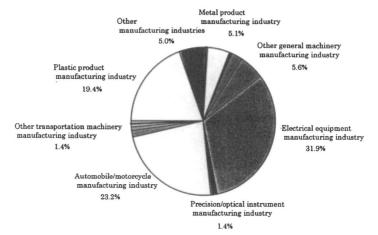

Classification by processes

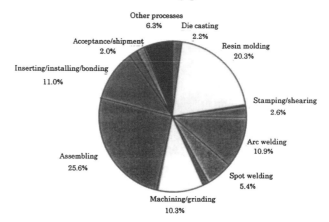

Classification by types

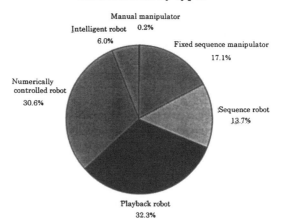

Figure 2 Component percentage of industrial robots in operation in the manufacturing industry in Japan (number of robots in operation in 1996: 464,367 units).

3 DEMAND ENVIRONMENT

The authors administered a questionnaire asking users (and potential users) about the future demand environment and demand forecast. Partial results follow.

3.1 Environmental Changes That Will Promote Demand for Robots

Most of the respondents said that the number one factor in the environmental changes that would promote demand for robots in the 10 years to come was "Necessity of improving manufacturing processes will become higher as the average age of operators gets higher," followed by, in order, "Since manufacturing industry is moving in the direction of multi-variety, small-quantity production, the need for production facilities capable of coping with versatile products will increase," "Improved working ratio of facilities will become important in terms of lower manufacturing cost," "Need for improved product quality and yield will become higher," "The sum of investment in plant and equipment will become higher because of higher labor cost," and "Responding to the requirement of shorter working hours will become more necessary."

3.2 Trend of Investment in Plant and Equipment

As for the forecasts on the investment in plant and equipment in the one or two years to come (1997 and 1998), many respondents believed that "We will increase investment in plant and equipment to promote automated production" or "We will make investment in plant and equipment as in other years." This suggests recovery of investment in plant and equipment after overcoming a prolonged slump in business. Also, many respondents thought that the rate of investment in robots in plant and equipment would increase in the future.

3.3 Stage of Robotization

Many respondents were at the stage of "Introducing robots is under study in spite of our desire to introduce robots because there remain some problems in terms of economy and performance." Therefore, reduced price and improved performance of robots will create room for greatly increasing the demand for robots.

3.4 Potential Rate of Robotization

The potential rate of robotization for the works currently being conducted manually in the manufacturing industry is 35.7%.

3.5 Change in Robotized Operations

The rate of robotized measurement is on the increase. This shows the trend toward more robots being used for measuring, though most of the industrial robots are being used for handling or working at present.

3.6 Future Tasks for Improving Robots

Many respondents were desirous of providing robots with "fine and precise control," "recognizing function," and "low cost" in all manufacturing processes as future tasks to be achieved to further promote introduction of robots in the future.

4 DEMAND FORECAST

Future demand for industrial robots was estimated in the research based on the following two methods:

- Micro demand estimate, based on questionnaires
- Macro demand estimate, based on regression analysis of macroeconomic indicator (nominal sum of investment in plant and equipment by private businesses) and robot shipments

Table 2 shows the results. Demand (total quantity of shipments for manufacturing industry in Japan and for export) for industrial robots will be ¥645 to ¥808 billion in the year 2000, ¥710 to ¥891 billion in 2005, and ¥920 to ¥1,152 billion in 2010.

5 CONCLUSION

As indicated by the research, future demand for industrial robots will show the following features:

Table 2　Result of Demand Estimate for Industrial Robots

(unit: ¥million)

		1995 year (actual value)	1997 year	2000 year	2005 year	2010 year
Classification by industry	Iron and steel industry	1,801	1,760	4,780 ~ 5,970	4,850 ~ 6,060	6,670 ~ 8,340
	Nonferrous metal industry	901	940	2,800 ~ 3,500	3,060 ~ 3,830	4,090 ~ 5,110
	Metal product manufacturing industry	7,758	8,090	22,400 ~ 28,180	21,150 ~ 26,430	26,690 ~ 33,370
	Metal processing machinery manufacturing industry	2,852	3,010	2,790 ~ 3,490	2,850 ~ 3,560	3,090 ~ 3,860
	Other general machinery manufacturing industry — Boiler/prime mover manufacturing industry	256	290	17,320 ~ 21,650	17,750 ~ 22,190	23,310 ~ 29,140
	Other general machinery manufacturing industry — Civil engineering machinery manufacturing industry	722	840			
	Other general machinery manufacturing industry — Other general machinery manufacturing industry	6,615	7,390	175,320 ~ 219,150	178,430 ~ 223,040	218,350 ~ 272,940
	Electrical equipment manufacturing industry — Computer manufacturing industry	17,462	18,630			
	Electrical equipment manufacturing industry — Home electric appliances manufacturing industry	7,261	7,490			
	Electrical equipment manufacturing industry — Electrical machinery/equipment manufacturing industry	14,168	14,770			
	Electrical equipment manufacturing industry — Radio/TV/communication device manufacturing industry	42,404	45,050			
	Electrical equipment manufacturing industry — Other electrical equipment manufacturing industry	18,248	18,750			
	Precision/optical instrument manufacturing industry	937	1,010	1,800 ~ 2,250	1,420 ~ 1,780	1,870 ~ 2,340
	Automobile/motorcycle manufacturing industry	35,435	36,280	60,320 ~ 75,400	64,090 ~ 80,120	77,160 ~ 96,450
	Other transportation machinery manufacturing industry — Shipbuilding/ship repair industry	959	950	5,380 ~ 6,720	5,640 ~ 7,050	6,840 ~ 8,550
	Other transportation machinery manufacturing industry — Other transportation machinery manufacturing industry	1,897	1,980			
	Food/beverage/tobacco manufacturing industry	4,808	5,310	3,150 ~ 3,940	4,980 ~ 6,230	5,510 ~ 6,890
	Fabric/clothing/leather product manufacturing industry	262	272	480 ~ 600	1,800 ~ 2,250	2,220 ~ 2,780
	Lumber/wooden product/cork manufacturing industry	181	200	1,280 ~ 1,610	660 ~ 830	850 ~ 1,060
	Paper/paper product/printing/publishing industry	867	950	1,620 ~ 2,020	2,180 ~ 2,730	2,670 ~ 3,340
	Chemical substance/chemical product manufacturing industry	1,433	1,520	3,080 ~ 3,860	3,820 ~ 4,780	5,020 ~ 6,270

	Plastic product manufacturing industry	15,648	17,110	26,810 ~ 33,520	28,490 ~ 35,620	35,670 ~ 44,590
	Ceramic/earth and rock product manufacturing industry	1,013	1,490	2,210 ~ 2,760	2,540 ~ 3,170	3,150 ~ 3,940
	Other manufacturing industries	10,626	12,740	6,430 ~ 8,040	10,220 ~ 12,780	10,900 ~ 13,630
Classification by process	Casting	407	420	990 ~ 1,240	1,010 ~ 1,260	1,250 ~ 1,560
	Die casting	1,483	1,490	4,710 ~ 5,880	4,550 ~ 5,690	5,770 ~ 7,220
	Resin molding	15,330	16,350	29,260 ~ 36,580	29,810 ~ 37,260	36,770 ~ 45,970
	Thermal treatment	86	100	120 ~ 150	130 ~ 160	160 ~ 200
	Forging	102	130	240 ~ 290	230 ~ 290	330 ~ 410
	Stamping/shearing	5,438	5,930	15,460 ~ 19,320	15,230 ~ 19,040	18,980 ~ 23,720
	Arc welding	15,749	16,160	31,970 ~ 39,970	31,620 ~ 39,530	37,530 ~ 46,920
	Spot welding	8,178	8,510	16,360 ~ 20,450	14,230 ~ 17,790	16,150 ~ 20,190
	Other welding	458	680	670 ~ 840	740 ~ 930	870 ~ 1,090
	Painting	2,910	3,050	7,820 ~ 9,770	7,000 ~ 8,750	8,570 ~ 10,720
	Machining/grinding	15,517	15,910	29,650 ~ 37,060	32,380 ~ 40,470	40,070 ~ 50,080
	Assembling	20,662	21,970	38,530 ~ 48,170	38,280 ~ 47,840	45,200 ~ 56,500
	Inserting/installing/bonding	77,775	78,890	135,280 ~ 169,100	141,670 ~ 177,080	182,080 ~ 277,600
	Acceptance/shipment	7,216	7,880	5,890 ~ 7,360	6,870 ~ 8,580	6,970 ~ 8,720
	Inspection/measurement	4,998	5,490	6,320 ~ 7,900	8,830 ~ 11,040	10,450 ~ 13,060
	Other processes	18,490	24,250	15,150 ~ 18,930	21,660 ~ 27,070	23,270 ~ 29,090
Classification by type	Manual manipulator	413	460	1,810 ~ 2,270	780 ~ 970	1,130 ~ 1,420
	Fixed sequence manipulator	11,770	11,270	14,960 ~ 18,700	10,080 ~ 12,600	12,020 ~ 15,020
	Remote controlled robot	553	860	150 ~ 190	230 ~ 280	170 ~ 210
	Sequence robot	9,771	10,460	22,850 ~ 28,570	13,570 ~ 16,960	15,830 ~ 19,790
	Playback robot	62,834	65,060	63,210 ~ 79,010	47,680 ~ 59,600	49,030 ~ 61,280
	Numerically controlled robot	62,304	66,230	63,030 ~ 78,780	50,970 ~ 63,720	53,090 ~ 66,360
	Intelligent robot	47,154	52,850	172,390 ~ 215,480	230,910 ~ 288,640	303,160 ~ 378,950
	Total of robots proper	194,799	207,190	338,390 ~ 422,990	354,220 ~ 442,780	434,420 ~ 543,020
	Peripheral device/accessory	46,374	52,140	97,120 ~ 121,400	125,750 ~ 157,190	186,180 ~ 232,720
	Total sum of demand for manufacturing industry in Japan	241,173	259,330	435,510 ~ 544,390	479,970 ~ 599,970	620,600 ~ 775,750
	Export	232,019	262,330	209,980 ~ 263,820	231,420 ~ 290,750	299,220 ~ 375,940
	Grand total	473,192	521,650	645,490 ~ 808,210	711,390 ~ 890,720	919,810 ~ 1,151,680

1. The demand for industrial robots in Japan, which showed signs of recovery in 1995, will continue to recover and will reach the record level of 1991.

2. The demand will stagnate temporarily somewhere between 2000 and 2010 due to decrease in replacement demand.

3. The demand in Japan will stagnate somewhere between 2000 and 2010 but will shift upward again toward 2010.

4. Total sum of shipments for demand in Japan and for export will reach ¥645 to ¥808 billion in the year 2000. This increase will slow down slightly somewhere between 2000 and 2010 due to the stagnated demand in Japan. After that period, however, the growth rate will turn upward again, reaching the level of ¥920 to ¥1,152 billion in 2010.

In order to cope with the various user needs expected to emerge in the future, robot manufacturers should develop the following types of robots:

- Highly flexible robots capable of coping with multivariety low-quantity production systems
- Safe and easy-to-use robots capable of responding to elderly operators
- Low-priced robots that will provide a high advantage of investment to users compared with alternative labor
- High-performance robots capable of expanding the range of further introduction of robots

ACKNOWLEDGMENT

The authors are deeply grateful to those who cooperated with the questionnaire and interviews for this research.

REFERENCES

Economic Planning Agency, Government of Japan. 1977. *Economic Survey of Japan (1995–1996)*. Tokyo: Printing Bureau, Ministry of Finance.

International Federation of Robotics. 1996. *World Industrial Robots 1996 Statistics, 1983–1996 and Forecasts to 1999*. Geneva: United Nations.

Japan Robots Association. 1997. *Report of Long Term Forecast on Demands of Industrial Robots in Manufacturing*. (March).

CHAPTER 3
ROBOTS AND MACHINE INTELLIGENCE

Charles A. Rosen
Machine Intelligence Corp.
Sunnyvale, California

Section 1 was written over 15 years ago. It presents this author's assessment at that time of the state of the art in robotics and related machine intelligence and his forecast of future developments in robotic systems. At the request of the editor it is included herein for historical reasons and because most of the projected developments have been realized and are in current practice. In Section 2 a new forecast is attempted of developments in the next generation of robots based on the present status and explosive growth of computer-related technologies.

1 ROBOTS AND MACHINE INTELLIGENCE (1982)

1.1 Introduction

The factory of the far future will be composed of a complex array of computer-controlled processes, programmable machine tools, and adaptive, sensor-mediated fixed and mobile industrial robots. These systems will be operated and maintained by a small cadre of skilled technicians and supervised by a small group of highly professional engineers, computer scientists, and business people. Planning, design, production, distribution, and marketing of products will depend critically on computers, used as information and knowledge-processing tools by the staff and as autonomous controllers (in the general sense) for each manufacturing process. Systems of such complexity must of necessity evolve since, at present, major components of these systems are not yet capable of performing required functions or are not cost-effective when they can. Even when such subsystems have attained acceptable performance, the difficult and laborious problems of standardization, interfacing, and integration into smoothly operating factory systems will remain.

What is the difference between a so-called "intelligent" computer system and all other computer systems? The criteria for "intelligence" vary with time. In less than 30 years the explosive growth of available computer science and technology has provided us with the means for supplementing and supplanting human intellectual functions far beyond our present capabilities for exploitation. At an early date arithmetic computation or "number crunching" was considered a function performable only by intelligent natural species. In a remarkably short time (as measured on an evolutionary scale) early pioneers realized the potential of the symbol-processing capabilities of the digital computer, a revolutionary advance in abstraction rivaled by few historical events. The encoding, manipulation, and transformation of symbols, representing objects of the world, action, induction and deduction processes, natural laws, theories and hypotheses, cause and effect, are intellectual functions that are now being performed with increasing sophistication by computers.

It is now commonplace to consider important computer applications such as storage and retrieval, data management systems, modeling, word processing, graphics, process controllers, computer games, and many others as merely information-processing techniques devoid of intelligence. Somewhat higher in abstraction, pattern recognition systems, initiated by the development of optical character recognition techniques, have led in theory and practice to explosive growth involving the extraction and classification of relevant information from complex signals of every type. Many of these applications have

Handbook of Industrial Robotics, Second Edition, Edited by Shimon Y. Nof
ISBN 0-471-17783-0 © 1999 John Wiley & Sons, Inc.

become commonplace and are no longer considered "intelligent" systems. At present it is acceptable to label programs as part of "machine intelligence" or "artificial intelligence" when they are concerned with studies of perception and interpretation, natural language understanding, common sense reasoning and problem solving, learning and knowledge representation, and utilization (expert systems). After 20 years of primarily empirical development, including conceptualization, debugging, and analysis of computer programs, only a few implementations of this technology are now being introduced into industry, and doubtless they are already considered as "mechanistic" rather than "intelligent" systems. In the following sections the current thrust toward implementing robot systems (that progressively become more and more "intelligent") is explored.

1.2 Available Robot Systems (1982)

1.2.1 First-Generation Robot Systems

The majority of robots in use today are first-generation robots with little, if any, computer power. Their only "intelligent" functions consist of "learning" a sequence of manipulative actions choreographed by a human operator using a "teach-box." These robots are "deaf, dumb, and blind." The factory world around them must be prearranged to accommodate their actions. Necessary constraints include precise workplace positioning, care in specifying spatial relationships with other machines, and safety for nearby humans and equipment. In many instances costs incurred by these constraints have been fully warranted by increases in productivity and quality of product and work life. The majority of future applications in material handling, quality control, and assembly will require more "intelligent" behavior for robot systems based on both cost and performance criteria.

1.2.2 Second-Generation Robot Systems

The addition of a relatively inexpensive computer processor to the robot controller led to a second generation of robots with enhanced capabilities. It now became possible to perform in real time, the calculations required to control the motions of each degree of freedom in a cooperative manner to effect smooth motions of the end-effector along predetermined paths—for example, along a straight line in space. Operations by these robots on workpieces in motion along an assembly line could be accommodated. Some simple sensors, such as force, torque, and proximity, could be integrated into the robot system, providing some degree of adaptability to the robot's environment.

Major applications of second-generation robots include spot welding, paint spraying, arc welding, and some assembly—all operations that are part of automated manufacturing. Perhaps the most important consequence has been the growing realization that even more adaptability is highly desirable and could be incorporated by full use of available sensors and more computer power.

1.2.3 Third-Generation Robot Systems

Third-generation robot systems have been introduced in the past few years, but their full potential will not be realized and exploited for many years. They are characterized by the incorporation of multiple computer processors, each operating asynchronously to perform specific functions. A typical third-generation robot system includes a separate low-level processor for each degree of freedom and a master computer supervising and coordinating these processors and providing higher-level function.

Each low-level processor receives internal sensory signals (such as position and velocity) and is part of the servo-system controlling that degree-of-freedom. The master computer coordinates the actions of each degree-of-freedom and can perform coordinate transformation calculations to accommodate different frames of reference; interface with external sensors, other robots, and machines; store programs; and communicate with other computer systems. Although it is possible to perform all the functions listed with a single computer, the major trend in design appears to favor distributed hierarchical processing, the resulting flexibility and ease of modification justifying the small incremental costs incurred by use of multiple processors.

1.3 Intelligent Robot Systems (1982)

1.3.1 Adaptive, Communicating Robot Systems

A third-generation robot equipped with one or more advanced external sensors, interfaced with other machines, and communicating with other computers could be considered to exhibit some important aspects of intelligent behavior. Interfaced with available machine vision, proximity, and other sensor systems (e.g., tactile, force, torque), the robot would acquire randomly positioned and oriented workpieces; inspect them for gross defects; transport them to assigned positions in relation to other workpieces; do insertions or other mating functions while correcting its actions mediated by signals from force, torque, and proximity sensors; perform fastening operations; and finally verify acceptable completion of these intermediate assembly processes. Its computer would compile statistics of throughput and inspection failures and would communicate status with neighboring systems and to the master factory system computer. The foregoing scenario is just one of many feasible today. The major functional elements of such an intelligent system are the following:

1. The capability of a robot system to adapt to its immediate environment by sensing changes or difference from some prespecified standard conditions and computing, in real time, the necessary corrections for trajectories and/or manipulative actions
2. The capability of interacting and communicating with associated devices (such as feeders and other robots) and with other computers so that a smoothly integrated manufacturing system can be implemented, incorporating fail-safe procedures and alternative courses of action to maintain production continuity

Clearly the degree of intelligence exhibited by such systems depends critically on the complexity of the assigned sequence of operations and how well the system performs without failure. At present the state of the art in available machine vision and other sensory systems requires considerable constraints to be engineered into the system and therefore limits applications to relatively simple manufacturing processes. However, rapid progress in developing far more sophisticated machine vision, tactile, and other sensory systems can be expected, with consequent significant increases in adaptability in the next two to five years. The level of "intelligence," however, will reside primarily in the overall system design, which is quite dependent on the sophistication of the master program that orchestrates and controls the individual actions of the adaptive robots and other subsystems.

1.3.2 Programming the Adaptive Robot

Programming the adaptive robot consists of roughly two parts:

1. A program that controls the sequence(s) of manipulative actions, specifying motions, paths, speed, tool manipulation, and so on. Several different sequences may be "taught" and stored and called up as required by some external sensory input or as a result of a conditional test.

The programming of these sequences has traditionally been implemented by using a "teach-box" for first- and most second-generation robot systems. This method of on-line programming is very attractive, being readily learned by factory personnel who are not trained software specialists.

2. A program that controls the remainder of the adaptive robot's functions, such as sensory data acquisition, coordinate transformations, conditional tests, and communications with other devices and computers. Programming this part is off-line and does require a professional programmer.

Some form of "teach-box" programming will likely be retained for many years, even for third-generation robots, because the replacement of off-line programming would require the development of a complex model of the robot and its total immediate environment, including dynamic as well as static characteristics. The master program that controls adaptive behavior and communications will call up, as subroutines, the manipulative sequences taught on-line as described.

Machine intelligence research has for many years included the development of high-level programming languages designed specifically for robotic assembly (see Chapters 18, 19). An appropriate language would permit off-line programming of complex assembly operations, with perhaps some calls to special on-line routines.

An ultimate goal is the use of a natural language to direct the robot. This may develop at increasing levels of abstraction. For example, an instruction as part of a program controlling an assembly task might be:

PICK UP THE BOLT FROM BIN NUMBER ONE

At this level the "smart" system must interpret the sentence, be able to recognize the object (bolt) using a vision sensor, plan a trajectory to bin number one (the position of which is modeled in its memory), acquire one bolt (after determining its position and orientation using a vision sensor), check by tactile sensing that the bolt has been acquired, and then await the next high-level instruction.

In a more advanced system operating at a higher level of abstraction, the instruction might be:

FASTEN PART A TO PART B WITH A QUARTER-INCH BOLT

This presumes that previous robot actions had brought parts A and B together in correct mating positions. The "smarter" system here hypothesized would know where all sizes of bolts are kept and carry out the implied command of picking a quarter-inch bolt from its bin, aligning it properly for insertion, inserting it after visually determining the precise position of the mating holes, and checking it after insertion for proper seating. A high-level planning program, having interpreted the instruction, would invoke the proper sequence of subroutines.

It is apparent that one can increase robotic intelligent behavior indefinitely by storing more and more knowledge of the world in computer memory, together with programmed sequences of operations required to make use of the stored knowledge. Such levels of intelligent behavior, while attainable, are still at the research stage but can probably be demonstrated in laboratories within five years.

1.3.3 Future Developments in Intelligent Robotic Systems

The development of intelligent robot systems is truly in its earliest stages. The rapid growth of inexpensive computer hardware and increases in software sophistication are stimulating developments in machine intelligence, especially those to be applied usefully in commerce and industry. General acceptance of third-generation adaptive robot systems will lead to the widespread belief that much more intelligence in our machines is not only possible but also highly desirable. It is equally probable that expectations will be quite unrealistic and exceed capabilities.

The following sections examine some interesting aspects of machine (or "artificial") intelligence (AI) research relevant to future robotics systems.

Sensors

Sensing and interpreting the environment are key elements in intelligent adaptive robotic behavior (as in human behavior). Physicists, chemists, and engineers have provided us with a treasure of sensing devices, many of which perform only in laboratories. With modern solid-state techniques in packaging, enhancement of ruggedness, and miniaturization, these sensors can be adapted for robot use in factories.

Extracting relevant information from sensor signals and subsequent interpretation will be the function of inexpensive, high-performance computer processors. With these advanced sensors a robot will in time have the capability to detect, measure, and analyze data about its environment considerably beyond unaided human capabilities, using both passive and active means for interaction. Sensory data will include many types of signals from the whole electromagnetic spectrum, from static magnetic fields to X-rays; acoustic signals ranging from subsonic to ultrasonic; measurements of temperature, pressure, humidity; measurements of physical and chemical properties of materials using many available spectroscopic techniques; detection of low-concentration contaminants; electrical signals derived from testing procedures (including physiological); and many more sensory modalities.

One may expect such sensors with their integrated computer processors to be made available in modular form with standardized computer interfaces to be selected as optional equipment for robotic systems.

Knowledge-Based (Expert) Systems

The technology of knowledge-based (expert) systems, typified by Stanford's "Dendral" and "Mycin" and SRI's "Prospector" systems, has been developed sufficiently for near-term implementation in factories. In such systems carefully selected facts and relations about a large body of specialized information and knowledge in a well-defined restricted domain have been encoded with the aid of one or more high-level human experts in that domain. A trained practitioner (but not necessarily an expert in that domain) can access the encoded expertise in an interactive give-and-take interchange with the computer program. The program can include empirical rules, laws of physics, models of processes, tabled values, and databases of many types.

Expert systems are expected to be highly useful for the factory at many levels of the production process. In CAD/CAM (computer-aided design and computer-aided manufacturing), expert systems can aid the designer in selection of materials, presentation of available purchased parts, selection of mechanisms, analysis of stress and temperature distributions, methods of assembly, and details of many other manufacturing processes. In particular, an expert system could be developed to aid the designer and manufacturing engineer in the design of parts and assemblies destined to be produced by robots, not by humans. Relaxation of requirements for manual dexterity and for visual scene analysis, for example, would greatly enhance the performance of existing robot and sensor systems, with their severely constrained sensory and manipulative capabilities. Design of workpieces that are easy to identify, inspect, handle, mate with other parts, and assemble requires a new form of expertise, which, when acquired, can be transferred to a computer-based expert system.

There are many other uses for expert systems in the total manufacturing process, such as in purchasing, marketing, inventory control, line balancing, quality control, and distribution and logistics. It is expected that many proprietary programs will be developed and be offered to the manufacturing community.

Continuous Speech Recognition and Understanding

Word and phrase recognition systems with limited vocabularies are available commercially today. A few systems can handle a few simple sentences. In most instances prior "training" of the system is required for each user. Research is proceeding to develop a continuous speech recognition system with an extended vocabulary and, if possible, speaker independence. Such a system would depend heavily on concurrent research in natural language processing systems.

Even with only a moderately advanced phrase or sentence recognition system, an attractive on-line programming method for "teaching" a robot system is suggested. The manipulative part will be modified and debugged on-line. No computer modeling is required—the real world is its own model. Future developments will enable special computer programs to be written that will optimize the performance of the robot system, given as input the relatively crude program generated on-line by the human.

The programming system outlined may have other attractive uses. Any advanced robot system could be operated in teleoperator mode—that is, under continuous operator control or under semiautonomous control, in which the operator sets up the robot system for some repetitive task and a subroutine then takes over to complete the assigned task. In this mode a human can be time-shared, using the robot system as a "slave" to do the dangerous or less intellectually demanding parts of a task.

Other Aspects of Machine Intelligence

The subfields of problem solving, planning, automatic programming and verification, learning, and, in general, common sense reasoning are all in the very early stages of development. They cannot now be considered as viable near-term options for at least five years. Incremental advances to enhance the intelligent behavior of our robotic systems will be incorporated at an accelerated pace when a large number of third-generation adaptive systems are in place and functioning cost-effectively. At that time the conservative manufacturing community will have become accustomed to the notion of machines that can adapt behavior according to conditions that cannot be precisely predetermined. They will then be prepared to accept additional intelligent functions certain to result from accelerating machine intelligence research programs now under way worldwide.

Hybrid Teleoperator/Robot Systems for Services and Homes

In the foregoing sections developments and applications of robot systems were considered primarily for the manufacturing industries. In the United States only 20% of the working population of approximately 100 million people are engaged in manufacturing, about 4% in agriculture, about 5% in the mining and extraction industries, and the remainder in the so-called service industries. The services include military, construction, education, health and social, transportation and distribution, sales, firefighting, public order and security, financial, recreation, and white-collar support. It is the author's considered opinion that adaptations and extensions of present robot/teleoperator systems will be developed for service use within the next generation that will exceed in number and total value all the installations in factories. Further, one can also predict a mass market for robot/teleoperator systems developed for use in the home. These will serve primarily as servants or aids to the aged and physically disabled who have limited physical capabilities for activities such as lifting, carrying, cleaning, and other household chores. By the turn of this century it is estimated that well over 15% of the total U.S. population will be in this class (approximately 35 million individuals).

At present, fully autonomous robots cannot cope with the more difficult environmental conditions in the relatively unstructured home or outdoors. Common sense reasoning capabilities would have to be developed for a "self-acting" robot. However, a partially controlled robot (hybrid teleoperator/robot) can be developed within the present state of the art that would be economically viable for the majority of these service tasks. Most of the intelligence would be supplied by the human operator with the most modern user-friendly interfaces to control the physical motion and manipulation of the robot, using switches, joysticks, and spoken word and phrase input devices. Subroutines controlling often used manipulative procedures would be called up by the operator and implemented autonomously by the robot/teleoperator which will be fitted with available sensors. Seldom occurring tasks would be performed by the operator effecting step-by-step control of the robot system in the same way that present industrial robots are "trained" using a teach-box. A specialized procedure could be stored and called up as needed. In short, dangerous, arduous, and repetitive physical manipulation of objects and control of simple manipulative actions will be performed by our new "slaves"; our goal in developing these "slaves" will be progressively to minimize human detailed control as we learn to improve our robot systems.

One can debate the social desirability of making possible the cost-effective elimination of many manual tasks. There is little doubt that in our free-market system the development of such systems cannot be prevented, only slowed down. The thrust toward the implementation of these technologies is worldwide, and international competition will guarantee that these new systems will be made available.

When the market for specialized robot systems approaches the size of the automotive industry, the price for a teleoperator/robot system will be comparable to (or less than) that of a car, which has evolved into a comparatively far more complicated system demonstrating the successful integration of electronic, mechanical, thermodynamic, and many other technologies together with effective user-friendly control.

For the home we can visualize a small mobile vehicle fitted with a relatively slow-moving arm and hand and visual and force/tactile sensors, controlled by joysticks and speech, with a number of accessories specialized for carrying objects, cleaning, and other manipulative tasks. High speed and precision would not be necessary. It would be all-electric, clean, and safe to operate. Its on-board minicomputer could be used for purposes other than controlling the robot/teleoperator—for example, recreation, recordkeeping, and security. It would be particularly useful for the aged and handicapped but would not be limited to these groups. A versatile assistant for those with reduced strength and other physical disabilities appears to be an effective substitute for expensive live-in or visiting household help and care.

There are many opportunities for introducing teleoperator/robot systems for military and commercial services. The U.S. Army has initiated a program for developing material-handling systems for support services and is studying their potential use under battle conditions. Both wheeled and legged mobile robot systems are under development. Teleoperator/robot systems could be effectively used in the construction and agriculture industries. Applications in space, deep seas, mining, and the Arctic are bring explored. A host of other commercial applications appear feasible, including loading and unloading trucks, firefighting, handling dangerous and noxious chemicals, painting and cleaning structures and buildings (outdoors), and road maintenance. These will require more expensive and sophisticated machines, probably specialized for the particular applications, and ranging widely in size, load-handling capacity, precision, and speed. Finally, more imaginative applications will be addressed, including the hobby market, game playing, dynamic shop-window displays, science fiction movies, and choreographed robots for modern dance.

1.4   Summary (1982)

After 25 years of laboratory research and development, machine intelligence technology is being exploited to a small but rapidly growing degree in manufacturing, primarily in improving the adaptability of robots through the use of sensors. Expert systems, planning, and advanced programming languages (including natural language) will provide significant improvements within the next 5 to 10 years. Hybrid teleoperator/robot systems will be developed for the service industries and ultimately may constitute the largest market for "smart" robots. We are now in a transition between the solid-state revolution and the intelligent/mechanical slave revolution.

2   ROBOTS AND MACHINE INTELLIGENCE (1998)

2.1   Emerging Technologies

In the past 15 years we have experienced revolutionary advances in computer hardware and software and significant improvements in sensors and electromechanical systems. Industrial robots, especially sensor-mediated robots, are now mainstream components in factory automation. We are now on the threshold of extending the use of robots and telerobots far more generally into the service, transportation, military, health, entertainment, and other segments of our economy. It is highly likely that far more robots and telerobots will be deployed in these fields than for factory use. In the following paragraphs some major developments in emerging technologies are presented that may be exploited in developing whole new classes of robots. This list is far from complete, as we cannot possibly predict discoveries of new physical effects, materials, and processes, nor the geometrically paced improvements in computer components.

2.1.1   Computers and Computer Components

Today one can purchase for less than $1000 a personal computer equipped with several processors, multi-megabyte random access memory, several-megabyte mass storage, complex operating systems, communication systems, interfaces with graphics, speech, video and external devices, and a large selection of specialized software programs. Computations are performed at speeds of several hundreds of megahertz. These computers are composed of inexpensive mass-produced modular components readily available for embedment into robot systems.

In the next 15 years we can anticipate the development of enormously powerful fixed and portable computers that will rival the performance of our present supercomputers. With relatively unlimited mass and random access memory, operating speeds in the gigahertz range, and greatly improved speech, gesture, multimedia, and wireless interfaces, it will be possible to input and process multiple sensor data, permit control of many more degrees-of-freedom of robot "arms" and "hands," and access worldwide data banks and programs. Multiprocessors and reprogrammable arrays will be available for embedment into special-purpose machines. Upgrading and enhancement of performance will be effected by software transmitted via local or world networks. In short, these projected computer developments are the most important factors in providing the "smarts" for the emergence of truly autonomous robots.

2.1.2   Sensors

Many "dumb" repetitive robots have been upgraded for complex applications, such as assembly, by the addition of sensory control. The major sensors in use are force, torque, and vision. In the near future more sophisticated machine vision sensors will include the use of multispectral color and depth sensing to attain true "scene analysis" capabilities—a long-sought goal in machine intelligence research. High-resolution digital cameras and algorithms utilizing depth data are becoming available now. At long last there will be the capability to recognize workpieces in a cluttered environment and accurately acquire and manipulate randomly oriented parts in assembly and material handling. Further, in-process inspection will be an invaluable byproduct of these technologies.

In the next generation new sensors based on known physical effects will empower robots with more than human capabilities to interface dynamically with the environment for testing, monitoring, accurately measuring important features useful for robot autonomy, and other scientific purposes. These sensors will enable a universal expansion of robot systems for nonfactory domains. The whole electromagnetic spectrum is available, taking advantage of inexpensive embedded processors, detectors, and filters. Included are microwave, thermal radiation, infrared, ultraviolet, X-ray, nuclear, radar, and laser measurements. The whole acoustic spectrum is similarly available, including ultrasonics. Mass-spectrometers now only accessible in laboratories can be miniaturized and effect measurement of physical and chemical properties of liquids and gases. High-temperature superconductors may produce the intense magnetic fields required to miniaturize an affordable nuclear magnetic resonance scanner. Such a scanner would permit viewing of the interior of opaque objects, especially the interior of parts of human bodies in locations other than hospitals. In all the above applications such sensing will be effected in real time, capturing dynamic data.

2.1.3 Energy Storage Systems for Robots

At present the lack of suitable energy storage to power mobile robots has been a major deterrent to the widespread use of nonfactory robots. Gasoline-powered engines are noisy, polluting, and hard to maintain, especially remotely. New methods of generating and storing electric energy to replace lead–acid and nickel–cadmium batteries are critically needed. There are worldwide research and development programs, primarily driven by the growing need for electrically powered automobiles and rechargeable batteries for portable tools, telephones, and computers.

In the next two to three years it is expected that the nickel–metal–hydride battery will be mass-produced and improve energy storage per unit of weight by a factor of two to three. An order-of-magnitude improvement is needed. On the horizon are promising developments of fuel cells, the electrochemical converters of hydrocarbon into electric energy. Fuel cells may use gasoline, methanol, ethanol, liquefied gas, or hydrogen derivatives to efficiently produce electric energy with no pollutants, noise, or moving parts. The intrinsic high energy of hydrocarbons will provide the order-of-magnitude increase in energy storage required for mobile robots and other uses.

Competing storage systems being developed include mechanical flywheels, superconductive solenoids, and solar-powered cells. These appear less promising, but at the very least each of these systems may be suited for specialized applications. Finally, completely new battery systems may be invented, driven by the enormous market for such devices. Less probable, but still possible, are the discovery of new energy sources based on exotic nuclear quantum effects.

2.1.4 Networks

The World Wide Web running on the Internet has made universally available multimedia information and computer programs. Fixed and mobile robots equipped with small radio receiver/transmitters will readily access these sources and be remotely controllable in real time via satellite systems now being deployed, communicating across the whole world. The Global Positioning System, now freely available worldwide, can accurately locate a mobile robot to within several feet and, together with detailed maps downloaded from the Web, greatly simplify robot guidance. Using the same technology, a local positioning network in a building or factory can position a robot with inch accuracy and ultimately position a robot end-effector to within a small fraction of an inch, thus greatly improving robot autonomy.

Some nodes on the Web will provide two-way communication with "super"-supercomputers, accessing specialized data banks and large-scale computer facilities to conduct high-speed simulations and problem solving and planning routines, reducing the computer hardware required to be carried on board the robot. A separate Intranet may be deployed for just this purpose. Full autonomy requires access to an enormous body of interrelated facts of the world to implement "common sense" reasoning—one of the major goals of machine intelligence.

2.1.5 Learning Machines, Neural Nets, and Genetic Programming

In the late 1950s and early 1960s a new class of computing machine, the Perceptron, was introduced. It emulated the learning and classification functions of intelligent organisms. Hardware and software networks were implemented, composed of large numbers of neuron-like logical elements interacting in parallel that successfully demonstrated supervised learning by example, especially in the field of pattern recognition. In essence, a neural net "learns" to perform specified functions without requiring formal computer programming, automatically altering the internal connectivity of its neural elements, when presented with a set of input examples and desired outputs, iterating this set many times until the machine has converged to an acceptable level. The past 15 years have seen explosive growth in both theory and practice. At present, neural net processing is contributing in a major way to machine intelligence, with thousands of practitioners. The standard symbol-processing digital computer performing sequential serial or parallel operations can, under program control, "learn" the same kinds of functions and more. Considerable unnecessary competition between these two schools of thought in machine intelligence circles exists, and it is highly likely that both methods will be combined for advanced applications.

Many neural net machines are now in use, either simulated on standard digital computers or composed of specialized processor chips. They perform tasks such as character recognition, speech recognition, modeling of nonlinear systems, stock prices prediction, expert systems, and multisensory fusion. Huge neural-net machines are being developed in research establishments, implemented by millions of neuron-like elements fabricated on computer chips.

An important, relatively new branch in this field has lately become prominent—genetic programming. In emulation of the principles of biological natural selection, algorithms have been developed and incorporated into neural nets to effect the progressive alteration of their internal architecture, automatically optimizing performance. When optimized, the new structures can be realized in special hardware.

In the future, fully autonomous robots will incorporate these powerful technologies and be capable of learning to adapt behavior when coping with new or changing environments and improving their performance in an evolutionary manner.

2.1.6   Superconductivity

Ceramic materials exhibiting lossless electrical conductivity at liquid nitrogen temperatures have been discovered, and vigorous research is being conducted to attain room-temperature compounds—a possible but improbable objective. However, extracting and liquefying nitrogen from the air is not difficult. With appropriate thermal insulation, small computer chips can be cooled using advanced solid-state "Peltier Effect" technology, reducing electric resistance of interconnections and thus reducing heat losses and increasing processing speed. New refrigerating methods being developed may be able to provide the liquid nitrogen to cool the windings of small electric motors. Eliminating heat losses and greatly increasing the magnetic field would result in a significant reduction in size and increase in mechanical torque. Small, light, efficient motor drives for robots would become available, especially for mini- and microrobots. It may also become possible to store electric energy in superconductive solenoids with a single central refrigerating unit supplying the cooling required for motors, general wiring, and computer chips.

2.2   New Robot Systems—Our Future Roboslaves

The following (necessarily incomplete) list of projected robot systems is based on the new technologies described above. Initial versions will be semiautonomous—i.e., they will incorporate computer subroutines, such as sophisticated vision and other sensor systems—but will basically be controlled by humans. They will slowly be upgraded with more autonomous features, but fully autonomous robots operating in complex environments are probably at least a generation in the future.

2.2.1   Service Robots

It is the author's considered opinion that the most fertile field for exploiting robots is in the service sector of the economy. Some applications follow:

Domestic Telerobots

It is estimated that in the U.S. at least 15% of the population will be aged 65 years or older by the year 2025. Together with millions of younger disabled people, they will total more than 35 million people, a substantial proportion of whom could afford a household aid—a telerobot mounted on an electric wheelchair or walker that would supplement the user's ability to perform simple functions such as picking up and transporting objects, opening doors, and many other mundane but useful daily tasks. Supplementing the several-degree-of-freedom (d.o.f.) chair, the telerobot would be fitted with a three-d.o.f. arm, grasping end-effector, laser pointer, vision system, and speech control. In mass production such a system could be priced at less than $10,000 in today's money, less than the price of a small automobile. It will be highly desirable to fit the home with "markers"—insignia such as coded color patches and bar codes placed strategically on walls, floors, and cabinets and appliances, simplifying guidance and recognition of objects using the vision sensor. Such aids for the aged, infirm, and disabled will become mandatory to reduce the growing costs of human helpers.

Robots for Dangerous, Dirty, and Undesirable Jobs

Many robots will become economically justified for firefighting, security, and crime control. Many other dirty or dangerous tasks, such as digging, drain cleaning, bomb disposal, and handling noxious wastes and chemicals, could readily be performed by telerobots. Similarly, in gardening, mowing, cultivation, pruning, and waste handling, telerobots will become welcome aids for many prosperous consumers.

Robots for Transportation Systems

Today automobiles have many telerobotic characteristics. They incorporate driver-controlled automatic transmissions, powered brakes and steering, cruise control, and many computer-controlled processes. We can anticipate major robotic improvements in the all-electric or hybrid-electric car of the future. The Global Position System, radar, and optical obstacle avoidance will greatly improve safety and traffic flow, especially on fully instrumented roadways. Ultimately, riderless autonomous vehicles will transport freight safely at high speeds on these roadways. In the far future autogyro telerobots, safely guided by sensors and signals on the ground, may populate the precious airspace above our major roadways, further assisting traffic flow.

Robots for the Health Services

Telerobots for surgical tasks are now beginning to be used in joint replacement and the microrealm, in which precision manipulation with micron resolution is enhanced. Pick-and-place robots are automating material-handling operations in drug tests and drug-discovery research. Semiautonomous robots are delivering necessities to patients in hospitals.

Rising personnel costs in hospitals and nursing homes will force the development of many fixed and mobile telerobots similiar to those described above for domestic use. These can be far more costly and therefore can incorporate many more functions, such as specialized end-effectors for feeding, sensors for monitoring vital functions, and drug delivery apparatus.

Emerging technologies will enable the development of very tiny robots the size of beetles or even ants. Fitted with microsensors, radio, and cutting or laser-burning tools, they could be introduced into the bloodstream and guided wirelessly to major organs, where they could present views, emplace drugs or monitoring devices, or perform surgical procedures such as clearing blocked arteries. External electric fields will provide the electric power. Development of microrobots will be accelerated by their planned use by the military.

2.2.2 Robots for Military Services

Sensor-guided missiles using the Global Position System, spy satellites, unmanned flying drones, and unmanned armored vehicles are already dominant armaments of modern military powers. Although many of these robots are semiautonomous, the trend is toward complete autonomy.

Planners for future military systems envision a robotized army with many foot soldiers replaced by robots and telerobots supported by aircraft drones and a wide information command and control network. A reduced number of skilled foot soldiers fitted with small portable supercomputers will control groups of robots guided to dangerous monitoring positions and control them in some offensive and defensive operations. In essence, robots will keep soldiers out of harm's way. In a separate development, "exoskeleton" enclosures for individual soldiers are being implemented. The soldier will be protected from biological or chemical attack and also be physically assisted, enhancing his load-carrying capacity and mobility. Thus protected, he will be an ideal human controller of many telerobots.

Miniature fixed and mobile telerobots, insect-sized and somewhat larger, will be developed primarily for spying and monitoring purposes. Clusters of cooperating robots fitted with infrared, vision, radar, biochemical and other sensors could form a dynamic "electronic" fence deployed at enemy borders. Tiny "throwaway" mobile robots could be dropped behind enemy lines to spy or conduct disruptive operations. With appropriate sensors they could find specific targets, attach themselves, and serve as markers to guide offensive missiles.

Many other robotic systems will be developed for less exotic but important tasks such as material handling of munitions and supplies, operating unmanned vehicles and tanks, detecting and destroying mines, and operating unmanned batteries of rocket launchers.

More funds will be made available for developing robots for the military than for any other purpose. As in the past, via technology transfer the fruits of these developments will ultimately "trickle down" and be made available for general use.

2.2.3 Miscellaneous Robotic Systems

Robots will be engaged in possibly large numbers in many other commercial fields. Agricultural robots will cultivate and pick fruit and vegetables, spray chemicals, and spread fertilizer. Small mobile robots will become a new class of toys engaged in mock battles and interactive computer games. Mobile robots may become members of teams actively participating in spectator sports performed in stadiums, satisfying the human desire to view violent gaming behavior. Exploration in space, the deep seas and, the arctic wastelands will need robots. Establishing space stations on the Moon or on Mars and maintaining them cannot be done without robots. In the arts, extending beyond Disney's animated figures, it will be possible to develop esthetically pleasing robots performing dances and even ballet with superhuman grace and agility.

3 SUMMARY

The preceding is an overview of emerging technologies and their potential use in future advanced robot and telerobotic systems. Within several generations such systems will be ubiquitous, doing much of our dirty work, improving our quality of life, and providing us with new forms of entertainment. They will have become our everyday helpers, our roboslaves.

ADDITIONAL READING

Future Developments in Robotics

Forbes, N. "Life After Silicon—Ultrascale Computing." *The Industrial Physicist* (December 1997). Research into quantum effects that could lead to new methods of computation.

Hively, W. "The Incredible Shrinking Finger Factory." *Discovery* (March, 1988). Article about Peter Will's development of microrobots.

National Academy. *Star21: Strategic Technologies for the Army of the 21st Century.* Washington: National Academy Press, 1992. A report by a committee assembled by the National Academy. An intensive study by over a hundred experts recommending needed developments in many technological fields for the U.S. Army of the next generation.

"The Next Fifty Years." *Communications of the ACM* (February 1997). A special issue with contributions by many authors from academia and business, covering many topics in computer-related subjects.

Stork, D. G., ed. *Hal's Legacy: 2001's Computer as Dream and Reality.* Cambridge: MIT Press, 1997. Essays by a number of experts commenting on the movie *2001,* its assumption of future machine intelligence, and their own projections.

Emerging Technologies

"Ballistic Missile Defense." *Spectrum Magazine* (September 1997). Special issue: articles regarding satellites for tracking missiles and missile defense systems.

Dario, P., et al. "Robotics for Medical Applications." *IEEE Robotics and Automation Magazine* (September, 1996). Review of robotics and automation in medical practice.

"The Future of Transporation." *Scientific American* (October 1997). Special issue: describes electric and hybrid-electric vehicles, automated highways, microsubs, flywheels, etc.

Goldberg, D. E. *Genetic Algorithms in Search, Optimization and Machine Learning.* Reading: Addison-Wesley, 1989. Introduction to genetic programming and applications.

Jochem, T., and D. Pomerleau. "Life in the Fast Lane." *AI Magazine* (Summer 1996). Article on ALVINN (Autonomous Land Vehicle in Neural Net). Adaptive control system using supercomputer implementing neural net that "learns" how to drive vehicle.

Moore, K., J. Frigo, and M. Tilden. "Little Robots in Space." *American Scientist* (March/April 1998). Describes development of mini-robot—"satbot"—100 gm, 2 cm across, at Los Alamos National Laboratory.

Riezenman, M. J. "The Search for Better Batteries." *IEEE Spectrum* (May 1995). Review of new storage batteries.

Scigliano, E. "Rover." *M.I.T. Technology Review* (January/February 1998). Describes mobile robot Sojourner on Mars and new developments in planetary exploration.

CHAPTER 4
EMERGING TRENDS AND INDUSTRY NEEDS

Steven W. Holland
General Motors Corp.
Warren, Michigan

Shimon Y. Nof
Purdue University
West Lafayette, Indiana

1 FUTURE PERSPECTIVE

In summer 1997 the Robotics Industries Association celebrated Joe Engelberger's special birthday and honored him as the Father of Robotics. The award presenters reminded the audience that the first robot was installed in 1961, less than 40 years ago. In honoring Joe, the RIA officials reflected on how the inventions of the electric lightbulb, the airplane, and other technological advancements affected daily life (see Table 1). Who, looking at the airplane in 1938, could have ever dreamed that jets and space ships would follow in the next 35 or so years? Who, looking at the electric lightbulb in 1914, would ever have imagined where that technology would be 35 years later? Observing other important technological innovations from which the world has benefitted, one can extrapolate the wonders that robotics will deliver to mankind over the next 40 years. It will truly surprise the most creative thinkers.

1.1 Deployment of Robotics

Deployment of robots and related intelligent technologies throughout the world is documented in the International Federation of Robotics publications, which are available from the RIA (Robotic Industries Association) (IFR, 1997). Based on the internationally accepted definition of what constitutes a robot, Japanese industry has deployed significantly more robots than anyone else (see Figure 1). The numbers are somewhat biased in that Japan deploys many of the simpler robots, while the United States tends to undercount these robots. Nevertheless, the Japanese still have deployed robots to a greater extent than any other country, for a variety of business and social reasons. Robotics as a technology is advancing, however, and it will be necessary to broaden our perspective beyond the "robot arm," which requires just three degrees of freedom and reprogrammability for multiple application types. In the mid-1990s 75,000 robots were deployed annually at an average price per robot of about $82,000 (compared to $108,000 in 1991). The world's total robot market amounted to almost $6 billion, led by Japan, with a 45% market share; Germany, Italy, France, and the UK, 20% (together); and the United States, 15%. In the 1980s the robotics industry was dominated by many small companies. In the 1990s, three of the world's largest robot producers combined for about one third of the world total robot sales. It is forecast that toward the end of the twentieth century about 145,000 robots will be deployed annually. The engineering industries, including fabricated metal products, machinery, electrical machinery, transport equipment, and precision instruments, account for 65% to 80% of all deployed robots.

Handbook of Industrial Robotics, Second Edition, Edited by Shimon Y. Nof
ISBN 0-471-17783-0 © 1999 John Wiley & Sons, Inc.

Table 1 Perspective: The Influence of Inventions on Daily Life[a]

Invention	When Invented?	How Many Years Ago?	How Are People Affected in the 1990s?
Electric lightbulb	1879	120	Can anyone imagine life without electric light bulb?
Automobile	c1890	Over 100	137,000 vehicles are produced daily worldwide.
Airplane	1903	Almost 100	Every day about 3.5 million people fly on 45,000 scheduled flights.
Industrial robot	1961	Almost 40	About 1 million robots are deployed, mostly in manufacturing, increasingly in service, agriculture, medical operations, construction, other useful tasks.

[a] From R. Ash, *The World in One Day*. New York, DK, 1997; R. Famighetti, ed., *The World Almanac and Book of Facts*. Mahawa: K-III Reference Corp., 1998.

1.2 Robotics Is a Critical Technology

"Robotics is a critical technology" (Thurow, 1996). "No other technology promises to transform so wide a span of the industrial base as robotization. Although use of robots is in a preliminary stage, unprecedented applications lie ahead for these marvelous (soon to be even more marvelous) machines" (Coats, 1994). These quotes come from studies concluding that robotics is indeed one of the critical technologies at the conclusion of the twentieth century and the dawn of the twenty-first.

In a 1997 congressional presentation on robotics and intelligent machines (RIMCC, 1997), Pat Eicker of Sandia Labs stated:

Imagine a world where smart cars avoid collisions; where surgeons guide molecularly-precise instruments instead of hand-held scalpels; where satellites the size of marbles monitor rogue nations; where grasshopper-size sensors watch a battlefield. A world that tackles problems from the galactic to the microscopic—with adaptive machine intelligence. Imagine assembly workers and machines working in concert to dramatically increase productivity and competitiveness . . . Intelligent machine technology is poised, right now, to offer applications so profound that they will fundamentally transform many aspects of our everyday life.

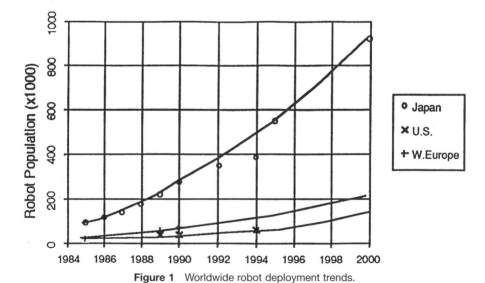

Figure 1 Worldwide robot deployment trends.

Let us briefly describe the development of robotics science and technology from its inception to where it is evolving.

1.3   Can R&D Ignite Industrial Applications?

Most of the exciting advancements during the few decades of robotics stemmed from the creative work in the R&D arena. Significant developments such as electric robots began in university labs; so did most of the impressive work in assembly robotics, sophisticated path planning, off-line programming, and more.

A pattern of a "robotics food chain" can be observed, starting with ideas developed and validated at universities and national labs, followed by robotics manufacturers and system integrators, and ultimately implemented by the final users. In turn, reactions and observation of technological barriers are fed back from final users to researchers. Collaborative efforts among several "food chain" parties have often shortened the time delay and enhanced concurrent development and application.

2   THE DAWN OF ROBOTICS IN INDUSTRY

Robotics history in General Motors is depicted in Figure 2. The earliest robots (early 1960s) were introduced to the public by Joe Engelberger on the Johnny Carson TV show. Joe was "hawking" his new robots and demonstrated their abilities to the audience (Figure 3). The exponential growth of the GM robot population continued in two major periods: one ending around 1987 and the other still under way as of this writing (Figure 4).

In the 1980s industrial robots were in their first phase of rapid diffusion, with worldwide shipments annual growth rates of up to 30%, averaging 25% per year. In the early 1990s, largely because of a slump in the Japanese economy, the annual growth rate diminished to an average of just above 8%. Since 1995 (with 26% annual increase) there has been a hefty recovery, with an annual increase of 15–20% in worldwide robot shipments. About 75,000 units were shipped in 1995 and 90,000 in 1996, and 145,000 are estimated for 1999. A large share of new robots replace older ones that have been removed from operation. It has been suggested that in the late 1990s two thirds of the robots supplied in Japan are for replacement of older models. This has been driven by falling prices and rapidly improving technology, making replacement a very cost-effective option.

The majority of the robots have been deployed by the automotive sector. When other sectors of the economy invest as heavily in robotics, it is predicted that robots' diffusion will increase even more. In 1981 GM's Dick Beecher made what many regarded as an outrageous prediction: "We expect to have 14,000 robots in GM by 1990!" His prediction was fulfilled just a few years later, in 1995. The delay was caused by overestimation of how soon robotics would be able to penetrate assembly and painting. On the other

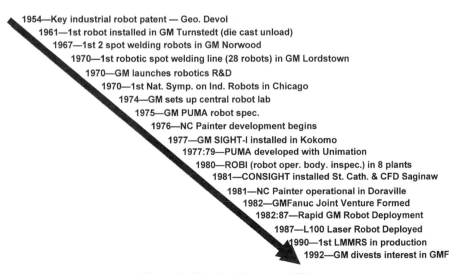

1954—Key industrial robot patent — Geo. Devol
1961—1st robot installed in GM Turnstedt (die cast unload)
1967—1st 2 spot welding robots in GM Norwood
1970—1st robotic spot welding line (28 robots) in GM Lordstown
1970—GM launches robotics R&D
1970—1st Nat. Symp. on Ind. Robots in Chicago
1974—GM sets up central robot lab
1975—GM PUMA robot spec.
1976—NC Painter development begins
1977—GM SIGHT-I installed in Kokomo
1977:79—PUMA developed with Unimation
1980—ROBI (robot oper. body. inspec.) in 8 plants
1981—CONSIGHT installed St. Cath. & CFD Saginaw
1981—NC Painter operational in Doraville
1982—GMFanuc Joint Venture Formed
1982:87—Rapid GM Robot Deployment
1987—L100 Laser Robot Deployed
1990—1st LMMRS in production
1992—GM divests interest in GMF

Figure 2   Robotics history and GM.

- circa 1961—GM Turnstedt Div., Trenton, NJ
  – unload die cast machine
- "Rent-a-Robot"...

Figure 3 "Prehistory": the dawn of robotics.

hand, the success of robotics applications for spot welding was underestimated. By now the sources of value generated by robotics application are much better understood.

Robotics in other segments of manufacturing, mainly electronics, microelectronics, food, apparel, and the process industry, are also maturing. In nonmanufacturing industries there are significant applications in the service sector, such as hospital, healthcare, security, maintenance, and recycling. There are also important, often amazing, applications in surgery, space, construction, transporation, shipbuilding, and agriculture (see Section X on Robotics in Various Industries.) Thus, the impact of industrial robotics and its integration in daily life are increasing significantly.

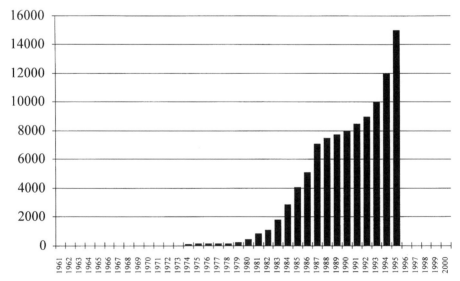

Figure 4 GM robot population.

2.1 Trends in Industry

Lean production, total quality and quality standards, agility, global manufacturing, the virtual enterprise, and service orientation are some of the key "drivers" of industry in the 1990s. Robotics should be examined in the context of what is driving manufacturing in general. In a report compiled by representatives from over 100 companies, universities, and government agencies (Agility Forum, 1997), key drivers and attributes of the "next generation manufacturing" were suggested and recommendations were derived. The global drivers identified were: ubiquitous availability and distribution of information; accelerating pace of change in technology; rapidly expanding technology access; globalization of markets and business competition; global wage and job-skill shifts; environmental responsibility and resource limitations; increasing customer expectations.

The corresponding attributes that a next-generation company or an enterprise must possess to be competitive have also been identified. These attributes include the company's responsiveness not only to customers and global markets, but to its human resources, and the responsiveness of the company's physical plants and equipment. Implications and recommendations about enabling imperatives for overcoming barriers to competitiveness include, for example, workforce and workplace flexibility; rapid product and process realization; adaptive, responsive information systems; management of change; extended enterprise collaboration and integration; and two enabling imperatives of particular interest to robotics: knowledge supply chains and next-generation equipment and processes.

2.1.1 Knowledge Supply Chains

Knowledge supply chains are integrated processes that use knowledge resources in industry and academia to enhance an enterprise's competitiveness by providing the enterprise with timely talent and knowledge. Knowledge is considered a commodity that can be transferred globally. Lack of the right knowledge can render a company noncompetitive. Typically, focused studies of *material* supply chains have reduced by 20% the cost of material acquisition and by 50% the acquisition cycle time. Similarly, exploiting "pull-type" *knowledge* supply chains can impact significantly the knowledge-oriented global competition. In terms of robotics, R&D knowledge that has already ignited industry applications can continue to drive robotics solutions. Knowledge supply chains within enterprises can deliver timely details about proven robotics applications. These success cases can then be repeated with minimum new effort, and information about them can be disseminated as part of a systematic organizational learning process.

2.1.2 Next-Generation Equipment and Processes

For a company to be responsive and able to adapt to specific production needs, its equipment and processes must be reconfigurable, scaleable, and cost-effective. In terms of robotics, this implies modular tools and controls. Modularity allows rapid specification and configuration of production facilities from open-architecture, reusable processing machinery and controllers. When production runs are complete, the facility can be disassembled and its components stored until needed again. Therefore, key enabling technologies include flexible, modular, and in-line cellular processes and equipment; high-speed processing machines; macro-, micro-, and nanotechnologies; and rapid prototyping.

2.2 Impacts on Industry

It is impossible to imagine producing an automobile anywhere in the world anymore without robots. Even if labor were available at no cost, robots would still be the best option for achieving the level of quality that customers demand. At GM, as at other companies, people are the most important asset, and robotics must support the people. Support is considered at two main levels: supporting people to perform their job more efficiently and supporting people by human-oriented interfaces. This approach corresponds well with that of the unions. UAW, for example, has traditionally supported technological progress. Don Ephelin, former Vice President of UAW, stated that robotics applications must be sensitive to workers and workers must have a long-term share in the benefits that stem from robotics. This approach is evident in the implementation of robotics in the automotive industry.

The largest-volume application of robotics at GM, as it typically is at other companies, has been spot welding (Figure 5), followed by painting and coating, machine load/unload, dispensing, part transfer, assembly, and arc welding. Although spot welding robotics in

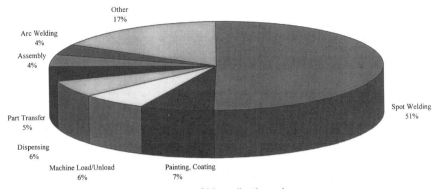

Figure 5 GM application mix.

the 1960s was based on 1940s welding technology, today the situation is reversed: robotics is the foundation of modern welding systems. Throughout its development, spot welding robotics has exploited primarily two basic robot work abilities: repeatability and programmability. The key drivers for robotics in GM have been:

1. Ergonomic solutions by taking over stressful, hazardous, and injury-prone tasks
2. Quality through consistent processing
3. Lean, improved productivity
4. Flexibility in product mix and product changeover
5. Repeat applications in common processes

Robotics technology impacts on GM have been both dramatic and positive. Robots are displacing hard automation in lean body shops. Commonality of application embraces robotics and accelerates their deployment, for instance, in standard paint shops, common door assembly, and other areas. Simulation technology is fundamentally changing the way tooling projects are executed. A sign of maturity is the fact that most used, nonhydraulic robots are being redeployed. In short, a quiet revolution is taking place as robotics technology lives up to its potential.

3 THE CHANGING NATURE OF ROBOTICS

The awesome advancements of computers are the key to the progress of robotics. Borrowing from John Hopcroft of Cornell University, *robotics is the science of representing and reasoning about the motion and action of physical objects in the real world.* The changing nature of robotics can be briefly summarized as shown in Figures 6 and 7. The drivers of change in robotics parallel the imperatives listed above for next-generation technology. They are:

1. Computer-based controller advances and cost reduction
2. Sophisticated motion planners optimized for process constraints, e.g., in arc welding, painting, coating, sealing
3. Highly evolved, robust hardware
4. Innovative new robot kinematics
5. Powerful simulation and analysis capability, e.g., RRS (realistic robot simulators), virtual factories, and rapid deployment automation

Other chapters of this handbook cover these topics in more detail.

3.1 Problems for Industry in Applying Robotics

The five drivers listed above for the changing nature of robotics also represent several barriers for industry applications. Further development in these areas is necessary. Part of the problem is the difficulty in translating research results promptly and effectively into a competitive advantage. Knowledge supply chains can offer a means for more timely

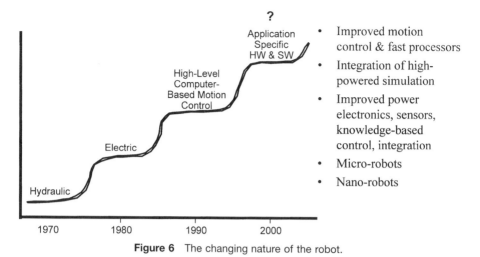

Figure 6   The changing nature of the robot.

transition. There are also challenges in turning a virtual factory into a working reality. Problems with complex integration of distributed components and information are another barrier.

3.2   Emerging Trends

Several emerging trends and technologies can influence cost reduction of robotics, according to the International Federation of Robotics (1997) and other sources. Seven general trends are:

1. *Standard robot assembly cells* are developed by a group of integrators and manufacturers (Light Flexible Mechanical Assembly Task Force). Modular conveyors, multipurpose grippers, machine vision, flexible feeders, and standardized peripherals will minimize the cost of customizing applications for given cases and offer inherent adaptability to market changes.

Generation	Typical Example	Control
1	Unimate 2000, circa 1968	Cam Switch
2	Cincinnati Milacron T3	Mini-computer
3	CM T3 w/V4 Control	μ-computer
4	CM 786 (electric)	"
5	[GM]Fanuc 360, circa 1984	Improved μ-computer
6	[GM]Fanuc 400	Remote I/O
7	[GM]Fanuc 420, circa 1992	"
8	ABB 6000	Motion Envelope with PC I/O
9	???	Multi-motion control, Improved motors, controller & power
10	Parallel Robots	
11	Interactive Assist devices	

Figure 7   Robot generations.

2. *Standardized application software,* which emerges for standardized solutions such as the standard assembly cell, can reduce the cost and effort associated with customized software development.

3. *Robot controllers* continue to decrease in cost and size with the continuing improvement in cost/performance and miniaturization of computers and semiconductor technology.

4. Implementation time is reduced by use of *graphical, interactive simulation* packages. In addition to off-line programming and cell-design analysis and automation, the graphic simulation can serve well for displaying the design to parties responsible for the solution but less involved in every technical detail.

5. Emerging *libraries of robot equipment and peripherals,* such as conveyors and feeders, which are currently available in 2D static form, are being converted to 3D solid models. Libraries of former facility robotics designs can also be reused to configure future applications quickly.

6. *Flexible feeding* of parts by programmable feeders that rely on software and machine vision, instead of mechanical hardware, can simplify robot integration.

7. *Design for flexible manufacturing and assembly* continues to be developed, with major impact on product realization. Machine vision can already compensate for small component tolerances in surface mount placement and is expected to have a similar contribution in part positioning and assembly robotics.

The above seven trends are accompanied with the emergence of three new classes of robots. These new robots will be highly useful in the automotive and other manufacturing industries and will also have significant impact on nonmanufacturing applications. They are:

1. *Intelligent assist devices,* which are based on current robotics technology but are intrinsically safe and designed to work in conjunction with people.

2. *Flexible fixturing robots,* which are designed to hold and position parts on which other robots, people, or automation can work. This new class of robots (also called *parallel robots*) may rival the installed base of all current robots and promises to make a significant dent in the expensive, slow to manufacture, inflexible tooling that everyone has to design and build to support today's automation systems. Fixturing robots contrast greatly with traditional robot arms. They tend to have much shorter duty cycles and much smaller work envelopes, need stiffness as opposed to speed, and provide absolute position accuracy as opposed to repeatability. (See chapter 47 Flexible Fixturing.)

3. *Force-controlled assembly robots,* as opposed to position-controlled robots, are emerging. Together with the emerging trends mentioned above, which are related to assembly robotics, these new, force-controlled robots will change significantly the way we assemble products.

3.3 Growth Areas and Research Needs

Future emerging growth areas for robotics applications in automotive production that imply opportunities for new research include:

1. Body shop framing and re-spot expanding into subassemblies
2. Increasing applications of robotics in glass preparation and installation
3. Sealing and dispensing with robotics
4. Flexible painting and coating
5. Assembly and disassembly for recycling
6. Die finishing, e.g., by stoning
7. Material handling as added functions for robots in their "spare time"
8. Handling of nonrigid materials, e.g., moldings, upholstery, carpeting, hoses, wiring
9. Laser welding
10. Laser and fine plasma cutting
11. Human–robot interface, including safe calibration, setup, maintenance, repair, and error recovery

Completely new types of robots and robotics are also emerging, such as microrobots, nanorobots, and virtual reality interfaces. Their potential impact on industry is still under investigation. All these future growth areas and challenges will also influence the usefulness and benefits of robotics in many other industries beyond the automotive industry. The benefits are not limited to the current regions of robotics deployment, but are already integrated into the worldwide global networks of enterprises.

In several recent surveys of experts from industry and academia about general production research needs and priorities (Nof, 1998), six problem areas were studied. These areas are listed below. For each problem area, examples are given of its priority research needs that are directly or indirectly related to robotics systems design and implementation. (The needs established in these surveys are consistent with and complement observations made earlier in this chapter from other sources.)

1. *Human–machine interface:* graphic simulation, virtual reality, methods and systems for interactive collaboration, conflict prevention and resolution
2. *Flexible integration framework:* reference languages for design integration, open software tools integration, modular-configurable machines
3. *Facility design and layout:* integration of supply chains, knowledge-based planning, neural network and other learning methods
4. *Implementation and economics:* design simulation-to-specification, management simulation-to-specification, benchmarking of incremental implementation
5. *Enterprise strategy:* models to facilitate negotiations, product design adequacy prediction, supplier-producer relations models
6. *Total quality:* total quality management models, measures of individual contributors to quality, and interdisciplinary team-building for quality.

When experts from specific industries were surveyed about their particular research needs, the first three priorities overall were found to be:

1. Design for manufacturing and assembly
2. Statistical process control
3. Quality management

However, for mostly electrical or electrical/mechanical manufacturers, soldering, material flow management, and surface mount technology were the three leading research priorities.

For the semiconductor industry, following the research priority of design for manufacturing and assembly are the areas of lead-bond interconnect and die attachment. In all these areas the contribution of future robotics can be revolutionary.

Table 2   Expected Lead Time Reductions with Future Robotics Technology

Robotics System	Minutes	Hours	Days	Weeks	Months	Years
• Specification (Initial)				*	O	
• Specification (changeover)		*			O	
• Configuration and programming			*		O	
• Debug and calibration		*		O		
• Mean time to failure					O	*
• Diagnostics and mean time to repair	*		O			

O current lead time
* Expected lead time

4 SUMMARY

Robotics has evolved rapidly over the past 40 years, having already had an enormous impact on industry. Its potential for future development is promising further astonishing solutions to the emerging problems of next-generation companies and enterprises competing in the global market. What industry expects from robotics technology in the twenty-first century can be summarized as follows: It must be implemented in a shorter time relative to current facilities and provide cost-effective and reliable production systems that drive quality, price, and lead-times to the highest level of customer satisfaction. Table 2 illustrates the time implications of this emerging vision.

REFERENCES

Agility Forum. 1997. *Next Generation Manufacturing Report.* Bethlehem.
Ash, R. 1997. *The World in One Day.* New York: DK.
Coats, J. F. 1994. *Research and Technology Management* (March/April).
Famighetti, R., ed. 1998. *The World Almanac and Book of Facts.* Mahawa: K-III Reference Corp.
International Federation of Robotics. 1997. *Industrial Robots 1997.* Geneva: United Nations.
Nof, S. Y. 1998. "Next Generation of Production Research: Wisdom, Collaboration, and Society." *International Journal of Production Economics.*
Robotics and Intelligent Machines Cooperative Council. 1997. *Highlights of the Congressional Exposition: We Must Cooperate to Compete* (Washington, September 30).
Thurow, L. C. 1996. *The Future of Capitalism: How Today's Economic Forces Shape Tomorrow's World.* New York: William Morrow.

PART 2
MECHANICAL DESIGN

CHAPTER 5
MANIPULATOR DESIGN

H.-J. Warnecke
R. D. Schraft
M. Hägele
O. Barth
G. Schmierer
**Fraunhofer Institute for Manufacturing
Engineering and Automation (IPA),
Stuttgart, Germany**

1 INTRODUCTION

In 1996 some 680,000 industrial robots were at work and major industrial countries reported growth rates in robot installation of more than 20% compared to the previous year (see Figure 1). The automotive, electric, and electronic industries have been the largest robot users, their predominant applications being welding, assembly, material handling, and dispensing. The flexibility and versatility of industrial robot technology have been strongly driven by the needs of these industries, which account for more than 75% of the world's installation numbers. Still, the motor vehicle industry accounts for 33% of the total robot investment worldwide (IFR, 1997).

In their main application, robots have become a mature product exposed to enormous competition from internationally operating robot manufacturers, resulting in dramatically falling unit costs. A complete six-axis robot with a load capacity of some 10 kg was offered at less than $60,000 in 1997. In this context it should be noted that the robot unit price accounts for only 30% of the total system cost. However, for many standard applications in such areas as welding, assembly, palletizing, and packaging, preconfigured, highly flexible robot workcells are offered by robot manufacturers, thus providing cost-effective automation, especially for small and medium-sized productions.

Robots are considered a typical representative of mechatronics, which integrates aspects of manipulation, sensing, control, and communication. Rarely are a comparable variety of technologies and scientific disciplines focused on the functionality and per-

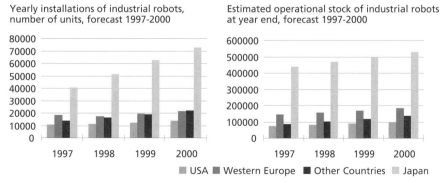

Yearly installations of industrial robots, number of units, forecast 1997-2000

Estimated operational stock of industrial robots at year end, forecast 1997-2000

⬛ USA ⬛ Western Europe ⬛ Other Countries ▦ Japan

Figure 1   Yearly installation numbers and operational stock of industrial robots worldwide (source: World Industrial Robots 1997, United Nations, and IFR).

Handbook of Industrial Robotics, Second Edition, Edited by Shimon Y. Nof
ISBN 0-471-17783-0   © 1999 John Wiley & Sons, Inc.

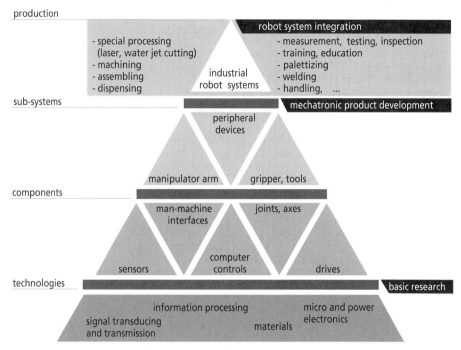

Figure 2   Robotics and mechatronics.

formance of a system as in robot development and application. Robotics integrates the state of the art of many frontrunning technologies, as depicted in Figure 2.

A sound background in mechatronics and the use of adequate design methods forms the basis of creative, time-efficient, effective robot development. Sections 2 through 9 give an overview of the basics of robot technology and adequate methods with their tools for arriving at a functional and cost-effective design.

2   PRINCIPLES OF ROBOT SYSTEMS ENGINEERING

The planning and development of robots is, as for any other product, a stepwise process with typical tasks, methods, and tools for each phase (see Figure 3). *Product planning*

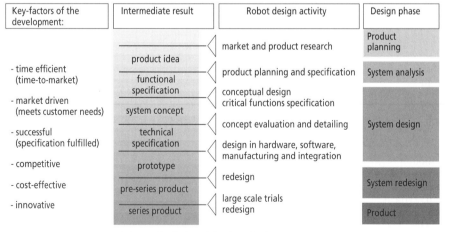

Figure 3   Industrial robot design process.

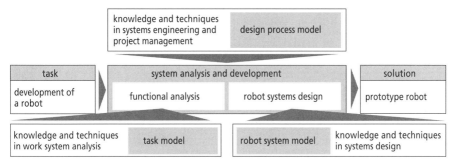

Figure 4   General system engineering method for robot development.

includes all steps from an initial product idea to the specification of a product profile, covering all information on major performance data, anticipated product cost, quantities, distribution channels, and projected development time and cost.

The subsequent *systems analysis and design phase* includes the robot's functional specification, conceptual design, and detailing in hardware and software, as well as a complete documentation of all information relevant to manufacturing, assembly, operation, and maintenance of the robot. The design phase ends with the experimental verification of the robot prototype.

The *system redesign phase* covers all activities toward improving the robot system on the grounds of detected deficits, quality, performance and cost potentials, requested modifications, and planned product variants.

Figure 4 shows a general engineering model of the design phase in accordance to the systems engineering method by (Daenzer and Huber, 1994).

Systems analysis and development consists here of two methods and three models. It supports the engineer during both the functional analysis and the robot system design.

Functional analysis extracts from a task all functions and performance requirements that will be automated and specifies the environment of the task execution. All subsequent design work and the competitive posture of the robot rely on the functional analysis.

Due to their importance in the subsequent development process, the functional specifications of robots have been investigated in depth. Herrmann (1976) showed the functional relationship between manual tasks and basic performance data of a robot. Another approach is to review existing robots in terms of cost, performance, etc., so that a competitive market position for a new design can be determined (Furgaç, 1986).

Many *robot system design* methods and tools have been discussed in the past. Most stress the kinematic layout and its optimization as the central design property. Some methods represent an integrated approach resolving the interdependencies between design tasks such as kinematics and transmission (Klafter, Chmielewski, and Negin, 1989; Wanner, 1988; Inoue et al., 1993; Nakamura, 1991).

2.1   Design Tasks

Specialization of robot design has a direct impact on the overall R&D goals and thus the robot's general appearance (see Figure 5).

In the past, the number of multipurpose or universal robot designs was overwhelming. However, many applications have occurred frequently enough that robot designs that take up specific process requirements could emerge. Examples of the different designs and their specific requirements are given in Figure 6.

2.2   Functional Analysis

Although many requirements may be obvious or are already defined, some central design parameters must be deduced from a work systems analysis. In order to arrive at the determination of geometric, kinematic, and mechanical performance data of a robot design, typical work systems that should be automated must be selected, observed, and described.

This analysis usually focuses on properties of objects that are handled, transported, machined, or worked on. In a "top-down" approach a given task can be broken down

Specialization of robots

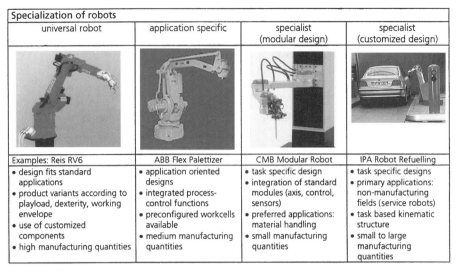

universal robot	application specific	specialist (modular design)	specialist (customized design)
Examples: Reis RV6	ABB Flex Palettizer	CMB Modular Robot	IPA Robot Refuelling
• design fits standard applications • product variants according to playload, dexterity, working envelope • use of customized components • high manufacturing quantities	• application oriented designs • integrated process-control functions • preconfigured workcells available • medium manufacturing quantities	• task specific design • integration of standard modules (axis, control, sensors) • preferred applications: material handling • small manufacturing quantities	• task specific designs • primary applications: non-manufacturing fields (service robots) • task based kinematic structure • small to large manufacturing quantities

Figure 5   Specialization of robot designs, with examples (courtesy Reis Robotics, ABB Flexible Automation, CMB Automation).

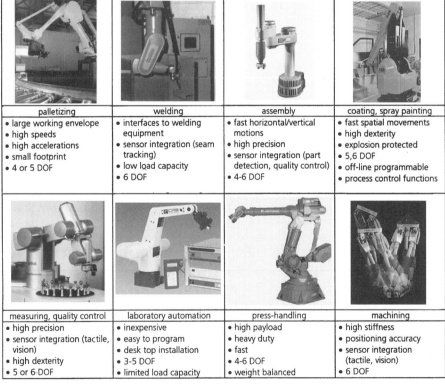

palletizing	welding	assembly	coating, spray painting
• large working envelope • high speeds • high accelerations • small footprint • 4 or 5 DOF	• interfaces to welding equipment • sensor integration (seam tracking) • low load capacity • 6 DOF	• fast horizontal/vertical motions • high precision • sensor integration (part detection, quality control) • 4-6 DOF	• fast spatial movements • high dexterity • explosion protected • 5,6 DOF • off-line programmable • process control functions
measuring, quality control	laboratory automation	press-handling	machining
• high precision • sensor integration (tactile, vision) • high dexterity • 5 or 6-DOF	• inexpensive • easy to program • desk top installation • 3-5 DOF • limited load capacity	• high payload • heavy duty • fast • 4-6 DOF • weight balanced	• high stiffness • positioning accuracy • sensor integration (tactile, vision) • 6 DOF

Figure 6   Application-specific designs of robots and their major functional requirements (courtesy FANUC Robotics, CLOOS, Adept Technology, ABB Flexible Automation, Fenoptik, CRS Robotics, Motoman Robotec).

into task elements, as in classic work analysis methods such as methods time measurements (MTM) or work factor (WF). However, the focus of the functional description of the considered tasks is on the analysis of geometries, motions, and exerted forces of object, tools or peripheral devices and with their possible dependence upon observation and supervision by sensors. Object or tool motions are divided into elementary motions without sensor guidance and control and sensorimotor primitives, defined as an encapsulation of perception and motion (Morrow and Khoshla, 1997; Canny and Goldberg, 1995).

Quantification of these motion elements led to the definition of the geometric, kinematic, and dynamic properties that play a central role in the subsequent robot design. Figure 7 gives a model of the breakdown of the tasks into motion and perceptive elements.

Typical performance and functional criteria are given in Table 1. The parameter with the strongest impact on the robot's complexity, cost, and appearance is its number of independent axis, i.e., degrees of freedom (d.o.f.). These are given by predefined goal frames and spatial trajectories that the robot endeffector must meet.

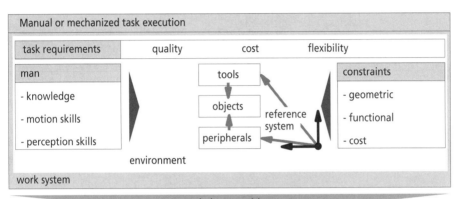

Figure 7   Task model for the functional analysis of robots.

Table 1   Basic Performance and Functional Criteria of a Robot

Criteria	Characterization
Load capacity	• Weight, inertia of handled object and endeffector • External movements/forces on endeffector or axes • Load history (RMS): static, periodic, stochastic, etc.
Degrees of freedom	• Required dexterity of endeffector • Number of degrees of freedom of peripherals (turn table etc.)
Handled object, tools	• Dimensions, size of objects/parts, • Kind of tools (torch, gripper, grinder, etc.) • Interfaces to robot
Task characteristics	• Changes from gripping to machining • Object/part presentation • Accessibility of objects/parts • Tolerances (parts, part presentation) • Fixing and positioning • Speed, acceleration
Accuracy	• Positioning accuracy • Repeatability • Path accuracy
Path control	• Point-to-point (PTP) • Continuous path (CP), motion profile • Rounding
Environmental conditions	• Quantifiable parameters (noise, vibration, temperature, etc.) • Not quantifiable parameters
Economical criteria	• Manufacturing cost, development cost • Break-even point, tradeoffs • Delivery time • Quality • Capacity (typical cycle times, throughput, etc.) • Point of sales (robot; workcell; production line)
Repair, maintenance	• Installation • Programming (on-line, off-line) • Remote servicing • Maintenance tasks and maintenance cycles • Changeability of modules and parts
Flexibility	• Workcell, CIM integration (logical and geometric interfaces) • Error handling, diagnosis • Cooperation with peripheral devices such as turntables, material handling equipment, other robots

The result of the functional analysis is a requirement list that quantifies the robot's functionality and performance data according to:

• Fixed criteria that must be met
• Target criteria that should be meet
• Wishes ("nice to have" criteria)

2.3   Robot Systems Design

In the robot design, solutions have to be developed on the basis of the robot's functional specifications. Finding solutions for given functional and performance requirements is both an intuitive and a systematic process (Pahl and Beitz, 1995). A stepwise process helps in gradually approaching optimal solutions. The subdivision of the robot system into subsystems or modules supports a systematic and complete search for solution principles. Often so-called morphological tables are used, which list solution principles for each subsystem or module. Compatible solution principles can be combined with system solutions that are evaluated against each other.

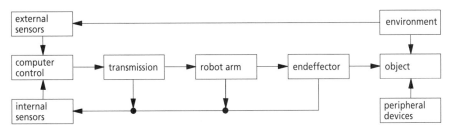

Figure 8 Robot system model.

The modular structure of a robot is obvious (Stadler, 1995; see Figure 8). For each subsystem possible solution principles are given in catalogues. Some of these are discussed in Sections 4 to 8.

The robot design process can be subdivided into three phases (Table 2, Figure 3), with increasing levels of detail:

1. The robot concept phase follows a top-down approach—a gradual development from the system's outer structure (kinematics) to its inner structure (transmissions).

Table 2 Robot Design Process

I Robot Concept	
Process Step	Result
Selection of kinematic structure	Kinematic structure
Estimate of link, joint parameters	Kinematic model (DH-Parameters), joint travels
Selection of transmission principle	Structure of joint driving system
Selection of transmission components	Geometrical, performance data, and interfaces of selected components

II Robot Structural Design and Optimization	
Process Step	Optimization Criteria
Optimization of robot link and joint parameters (DH parameters)	Minimum number d.o.f., kinematic dexterity, footprint Maximum workspace
Optimization of kinetic performance	Minimum motion times Minimum joint accelerations
Selection of motors, gears, bearings, couplings	Maximum torques Minimum torque peaks (uniform torque profiles) Maximum heat dissipation
Cabling	Minimum test and bend, space occupancy
Selection of materials	Minimum weight, machining, corrosion Maximum stiffness
Dimensioning of axes, housing, base, tool flange	Minimum machining, weight, part numbers, assembly

III Robot Detail Design	
Process Step	Result
Part design	Part drawings
Robot system assembly	Bill of materials, assembly, calibration instructions
Electrical, electronical design (switching cabinet)	Electric circuit layout bill of materials
Documentation	Operation manual, servicing instructions

Table 3   Categories of Tools for Robot Design

Tool Category	Purpose
Symbolic equation solver	Generation of mathematical robot models, analysis, optimization of robot kinematic, kinetic, and dynamic parameters
Multibody simulation packages	Modeling, analysis, and synthesis of dynamic multibody structures
Finite element packages (FEM)	Structural, modal and vibration analysis of bodies and multibody system
Computer-aided design (CAD) packages	Representation, dimensioning, documentation of parts, structures, and systems
Computer-aided engineering (CAE) packages	Versatile tools covering all steps in the product development process from product specification to manufacturing

2. The robot structural design and optimization phase refines the chosen concept.
3. The robot detail design further details the optimized design to produce all documents for manufacturing, assembly, operation and maintenance of the robot.

Suitable tools for robot design help the engineer to find successful solutions quickly at the various levels of the design process. Design variations can easily be carried out and evaluated, as well as complex computations for robot kinematic, kinetic, dynamic and structural analysis and optimization. For this purpose many program packages are available, as listed in Table 3.

Special computer-aided engineering (CAE) tools for design simulation, optimization, and off-line programming for robots are widely used for their kinematic and transmission design. These tools offer both computer-aided design (CAD) and multibody simulation features. Various robot CAE packages have been presented, of which IGRIP (DENEB Robotics) and ROBCAD (Tecnomatix) have found wide acceptance. An example of their use in robot design is given in Figure 9.

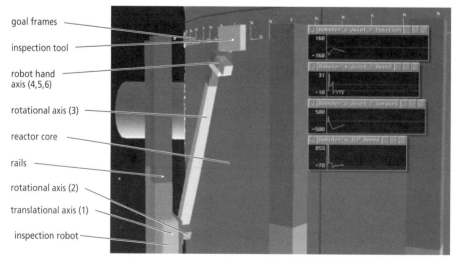

goal frames
inspection tool
robot hand axis (4,5,6)
rotational axis (3)
reactor core
rails
rotational axis (2)
translational axis (1)
inspection robot

Figure 9   Application of IGRIP in the design of a reactor outer core inspection robot. In this example the robot's joint 2 position, speed, torque, and tool center point Cartesian speed are displayed given a specific motion.

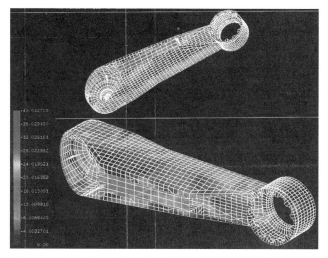

Figure 10   Finite element model of the Reis RV6 robot arm (axis 2) with stress pattern (courtesy Reis Robotics), see also Fig. 5 for the robot's structure.

Finite element methods (FEMs) are widely used in robot mechanical design. A large number of well-established FEM programs offer valuable support in the robot's structural analysis and optimization (Burnett, 1987):

- Structural deflections and failure due to external loads
- Vibration models of both forced and unforced vibrations
- Structural stress patterns
- Structural heat flow (i.e., from servo-motors or gears).

FEMs represent complex 3D objects as a finite assembly of continuous geometric primitives with appropriate constraints (Figure 10). Today most packages interactively guide the user through the three major steps of a finite element calculation (Roy and Whitcomb, 1997):

1. *Preprocessing:* the 3D shape of the candidate structure is discretized into a finite number of primitive continuous geometric shapes with common vertices. Many programs perform automatic mesh generation.
2. *Solution generation:* the model response for given loads, material data, boundary conditions, etc. is computed. Many packages offer optimization features, i.e., the variation of shapes within specified boundaries for homogeneous stress distribution.
3. *Postprocessing:* besides tabulated data, results are usually displayed graphically.

2.4   Design Evaluation and Quality Control

World-class manufacturers are better able than competitors to align their products and services with customer requirements and needs. Achieving this goal requires an ongoing interchange between customer and manufacturer, designer and supplier. Quality function deployment (QFD) is a suitable method for managing functional and performance inputs through the entire development process (Clausing, 1994).

The "house of quality" is a tool that records and communicates product information to all groups involved in the robot design process. It contains all relevant requirements for evaluating the design (Akao and Asaka, 1990; Ragsdell, 1994). The first task to be completed in any QFD undertaking is to clarify interdependencies in the design process that would otherwise be hidden or neglected:

- Systematic listing and evaluation of requirements
- Identification of critical functions and performance requirements
- Quality and cost-oriented requirements
- Evaluation of critical parts and production processes

In order to trace requirements from the beginning of product planning to the most detailed instructions at the operating level, QFD uses an interlocking structure to link ends and means at each development stage (Figures 11, 12).

Numerous QFD tools, such as QFD/CAPTURE and HyperQFD, support the engineer in the specification process and the quality control of the system development and its production.

3 ROBOT KINEMATIC DESIGN

The task of an industrial robot in general is to move a body (workpiece or tool) with a maximal of six degrees of freedom (three translations, three rotations) into another point and orientation within the workspace. The complexity of the task determines the robot's required kinematic configuration.

Industrial robots, according to ISO 8373, are kinematic chains with at least three links and joints. The number of degrees of freedom of the system determines how many independently driven and controlled axes are needed to move a body in a defined way in space. In the kinematic description of a robot we distinguish between:

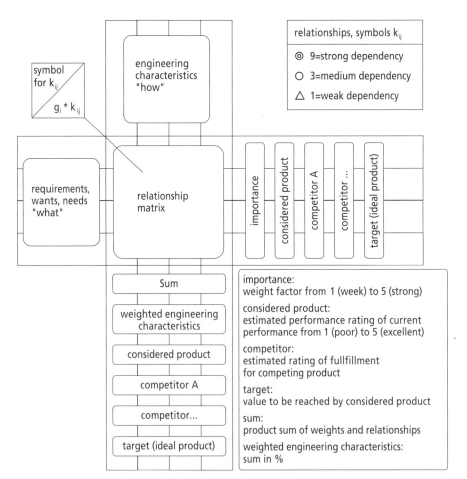

Figure 11   House of quality: QFD evaluation model for product design.

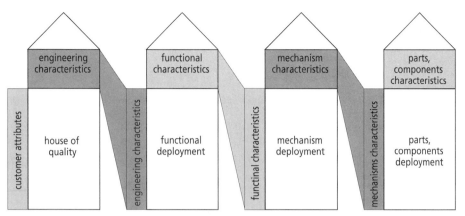

Figure 12 Deployment of function and performance requirements in the robot design process.

Arm: an interconnected set of links and powered joints that support or move a wrist and hand or endeffector.

Wrist: A set of joints between the arm and the hand that allows the hand to be oriented to the workpiece. The wrist is for orientation and small changes in position.

Figure 13 illustrates the following definitions:

- The *reference system* defines the base of the robot and also, in most cases, the zero position of the axes and the wrist.
- The *tool system* describes the position of a workpiece or tool with 6 d.o.f. (x_k, y_k, z_k and A, B, C)
- The robot (*arm* and *wrist*) is the link between the reference system and tool system.

3.1 Robot Kinematic Configuration

Axes are distinguished as follows:

Rotary axis: an assembly connecting two rigid members that enables one to rotate in relation to the other around a fixed axis

Translatory axis: an assembly between two rigid members enabling one to have linear motion in contact with the other

Complex joint: an assembly between two closely related rigid members enabling one to rotate in relation to the other about a mobile axis.

Figure 14 gives an overview of the symbols used in VDI guideline 2861 and in this chapter. The kinematic chain can be combined by translatory and rotatory axes.

The manifold of possible variations of an industrial robot structure can be determined as follows:

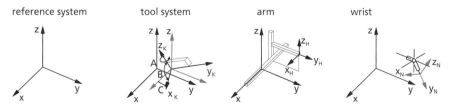

Figure 13 Definition of coordinate systems for the handling task and the robot.

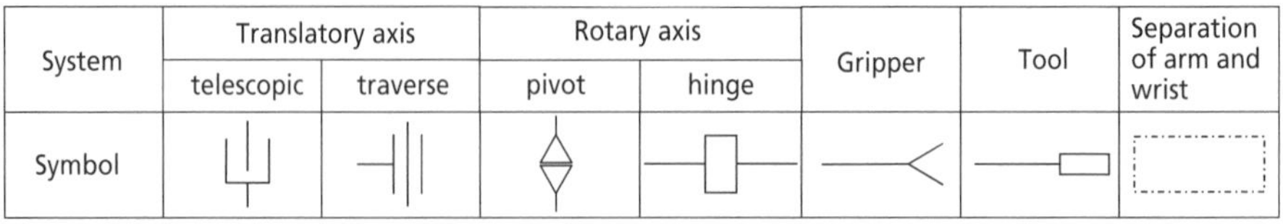

System	Translatory axis		Rotary axis		Gripper	Tool	Separation of arm and wrist
	telescopic	traverse	pivot	hinge			
Symbol							

Figure 14 Symbols for the kinematic structure description of industrial robots according to VDI Guideline 2681.

$$V = 6^{\text{d.o.f.}}$$

where V = number of variations

d.o.f. = number of degrees of freedom

These considerations show that a very large number of different chains can be built; for example, for six axes 46,656 different chains are possible. However, a large number are inappropriate for kinematic reasons (Angeles, 1997):

- Positioning accuracy decreases with the number of axes.
- The kinetostatic performance depends directly on the choice of the robot kinematic configuration and its link and joint parameters.
- Power transmission becomes more difficult as the number of axes increases.

Industrial robots normally have up to four principal arm axes and three wrist axes. Figure 15 shows the most important kinematic chains of today. While many of the existing robot

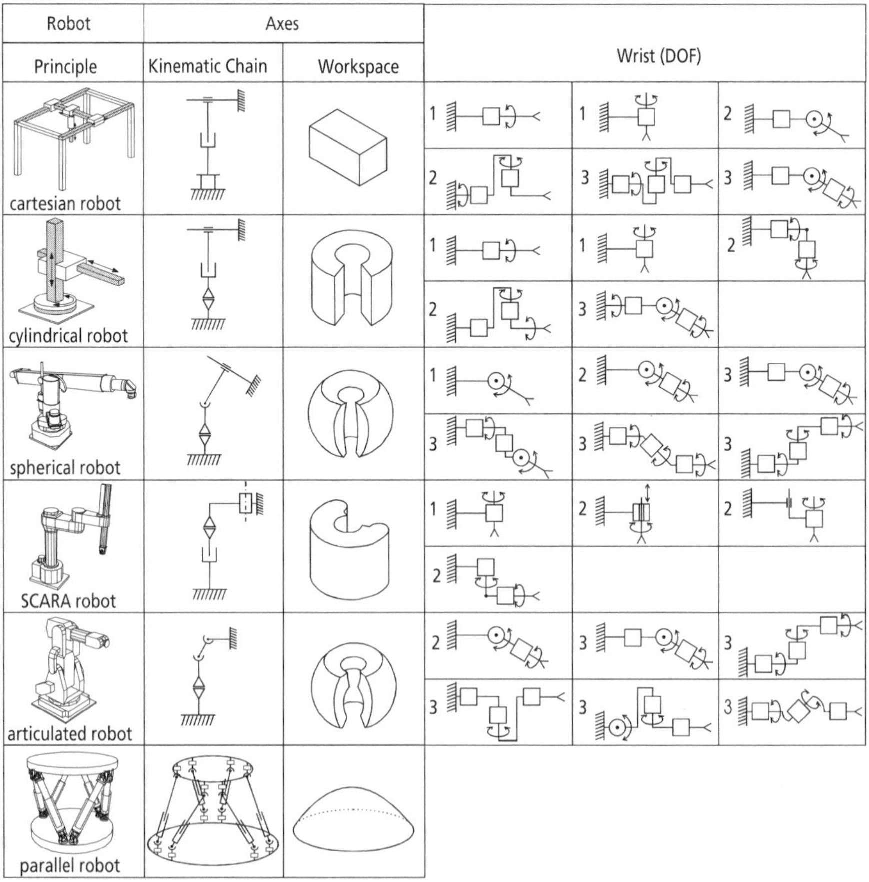

Figure 15 Typical arm and wrist configurations of industrial robots.

Figure 16   Floor and overhead installations of a 6-d.o.f. industrial robot on a translational axis, representing a kinematically redundant robot system (courtesy KUKA).

structures use serial kinematic chains (with the exception of closed chains for weight compensation and motion transmission), some parallel kinematic structures have been adopted for a variety of tasks. Most closed-loop kinematics are based on the so-called hexapod principle (Steward platform), which represents a mechanically simple and efficient design. The structure is stiff and allows excellent positioning accuracy and high speeds, but shows only very limited working volume.

If the number of independent robot axes (arm and wrist) is greater than six, we speak of kinematically redundant arms. Because there are more joints than the minimum number required, internal motions may exist that allow the manipulator to move while keeping the position of the end-effector fixed (Murray, Li, and Gastry, 1993). The improved kinematic dexterity may be useful for tasks taking place under severe kinematic constraints. Figure 27 shows a prominent example of an 11-degree-of-freedom redundant arm kinematics installed on a mobile base for aircraft cleaning. Other redundant configurations, such as a six-axis articulated robot installed on a linear axis (Figure 16) or even a mobile robot (automated guided vehicle), are quite common and are used as measures to increase the working volume of a robot.

3.1.1   Cartesian Robot

Cartesian robots have three prismatic joints, whose axes are coincident with a Cartesian coordinate system. Most Cartesian robots come as gantries, which are distinguished by a framed structure supporting the linear axes. Gantry robots are widely used for handling tasks such as palletizing, warehousing, and order picking or special machining tasks such as water jet or laser cutting where robot motions cover large surfaces.

Most gantry robot designs follow a modular system built up. Their principal axes can be arranged and dimensioned according to the given tasks. Wrists can be attached to the gantry's z-axis for end-effector orientation (Figure 17). Furthermore, a large variety of linear axes can be combined with gantry robots. Numerous component manufacturers offer complete programs of different-sized axes, drives, computer controls, cable carriers, grippers, etc.

3.1.2   Cylindrical and Spherical Robots

Cylindrical and spherical robots have an arm with two rotary joints and one prismatic joint. A cylindrical robot's arm forms a cylindrical coordinate system, and a spherical robot's arm forms a spherical coordinate system. Today these robot types play only a

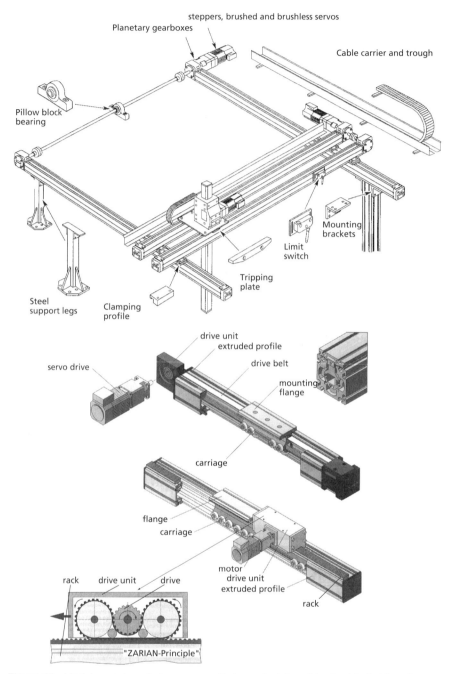

Figure 17 Modular gantry robot program with two principles of toothed belt-driven linear axes (courtesy Parker Hannifin, Hauser Division).

minor role and are preferably used for palettizing, loading, and unloading of machines. See Figure 18.

3.1.3 SCARA-Type Robots

As a subclass of cylindrical robots, the Selective Compliant Articulated Robot for Assembly (SCARA) consists of two parallel rotary joints to provide selective compliance

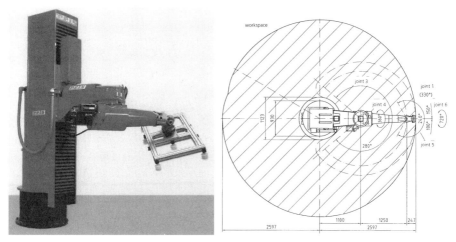

Figure 18   Five-d.o.f. cylindrical robot with depiction of its workspace (in millimeters) (courtesy Reis Robotics).

in a plane, which is produced by its mechanical configuration. The SCARA was introduced in Japan in 1979 and has since been adopted by numerous manufacturers. The SCARA is stiff in its vertical direction but, due to its parallel arranged axes, shows compliance in its horizontal working plane, thus facilitating insertion processes typical in assembly tasks. Furthermore, the lateral compliance can be adjusted by setting appropriate force feedback gains. In SCARAs direct drive technology can bring in all potentials: high positioning accuracy for precise assembly, fast but vibration-free motion for short cycle times, and advanced control for path precision and controlled compliance. Figure 19 shows the principle of a direct-drive SCARA.

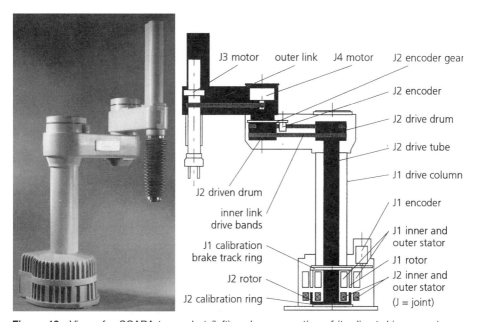

Figure 19   View of a SCARA-type robot (left) and cross section of its direct-driven arm transmission for joint 2 (courtesy Adept Technology).

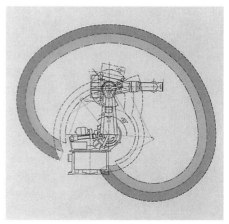

Figure 20 Articulated robot (KUKA) and ist workspace. Note the gas spring acting as a counterbalance to the weight produced by axis 2 (courtesy KUKA).

3.1.4 Articulated Robots

The articulated robot arm, the most common kinematic configuration, consists by definition of at least three rotary joints. High motor torques produced by the axes' own weight and relatively long reach can be counterbalanced by weights or springs. Figure 20 displays a typical robot design.

3.1.5 Modular Robots

For many applications the range of tasks that can be performed by commercially available robots may be limited by the robots' mechanical structure. It might therefore be advantageous to deploy a modular robotic system that can be reassembled for other applications. A vigorous modular concept has been proposed that allows universal kinematic configurations:

- Each module with common geometric interfaces houses power and control electronics, an AC servo-drive, and a harmonic drive reduction gear.
- Only one cable, which integrates DC power and field bus fibers, connects the modules.
- The control software is configured for the specific kinematic configuration using a development tool.
- A simple power supply and a PC with appropriate field bus interfaces replace a switching cabinet.

Figure 21 illustrates the philosophy of this system and gives an example.

3.1.6 Parallel Robots

Parallel robots are distinguished by concurrent prismatic or rotary joints. Of the many proposed parallel robot configurations, two kinematic designs have become popular:

1. The tripod with three translatory axes connecting end-effector, plate and base plate, and a 2 or 3-d.o.f. wrist
2. The hexapod with six translatory axes for full spatial motion

At the extremities of the link are a universal joint and a ball-and-socket joint. Due to the interconnected links, the kinematic structure generally has many advantages, such as high stiffness, accuracy, load capacity, and damping (Masory, Wang, and Zhuang, 1993; Wang and Masory, 1993). However, kinematic dexterity is usually limited.

Figure 21   Modular robot system consisting of rotary and translatory axis modules, grippers, and configurable control software (courtesy amtec).

Parallel robots have opened up many new applications where conventional serial chain robots have shown their limits, such as in machining, deburring, and parts joining, where high process forces at high motion accuracy are overwhelming.

Parallel robots can be quite simple in design and often rely on readily available precision translatory axes, either electrically or hydraulically powered (Merlet, 1995). Figure 22 gives examples of tripod and hexapod platforms.

Although parallel manipulators have been introduced quite recently and their design is quite different from that of most classical manipulators, their advantage for many

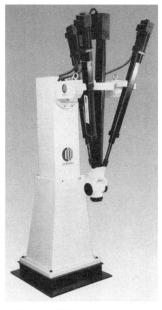
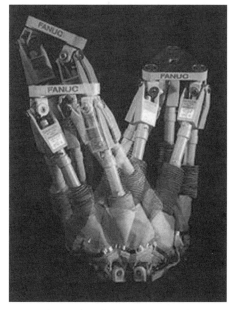

Figure 22   The COMAU Tricept, a 6-d.o.f. tripod, and the FANUC FlexTool Steward platform with six servo-spindle modules connecting the bottom and moving plate (courtesy COMAU, FANUC Robotics).

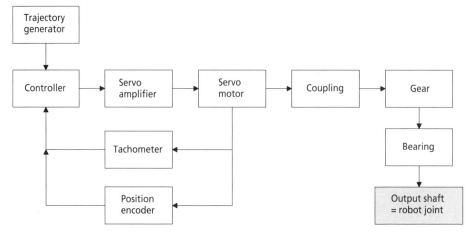

Figure 23   Drive chain of industrial robots.

robotics tasks is obvious, so they will probably become indispensable in the near future. (See also Chapter 47.)

4 DRIVE CHAIN

The drive chain of common industrial robots is shown in Figure 23. The control system is fed by the trajectory generator with desired joint values and delivers control commands to the servo-amplifier, which passes on the amplified control signal to the servo-motor. The tachometer and position encoder are typically located on the motor shaft and close the velocity and position loop. The torque produced by the servo-motor is transferred via coupling, gear, and bearing to the output shaft of the robot.

4.1 Computation of Drive Chain

Computation of drive chain begins after kinematics, motor, and gear performance have been set (Figure 24). The desired performance of the manipulator is then defined in terms of working cycles. This leads to motion profiles for each joint, where position, velocity, and acceleration are defined. Transition to joint torques is done by the dynamic model of the manipulator which should include gravitational and frictional forces/moments. In general, the equation system of joint torques and joint position, velocity, and acceleration is highly coupled, so special simulation software is used for this task. Peak torque and an equivalent torque number, calculated as a function of joint torque, time proportions of working cycle, and mean input speed, are typical numbers for gear validation, where motor validation is done by continuous and peak torque reduced by the gear. If the preselections of motor and gear performance do not match the desired manipulator performance, there are two ways of adaptation:

1. Increase motor and gear performance.
2. Decrease manipulator performance.

After this, calculation begins again until the deviation between required and available performance is within a tolerated error range.

5 SERVO-DRIVES

Table 4 sets out the principles for servo-drives.

In general, servo-drives for robotics applications convert the command signal of the controller into movements of the robot arm. Working cycles of industrial robots are characterized by fast acceleration and deceleration and therefore by short cycle times in the application process. Thus, servo-motors for industrial robots have to fulfill a high standard concerning dynamics.

Because of their good controllability and large speed and power range, only electrical servo-drives find practical use in common industrial robots. Direct electrical drives are

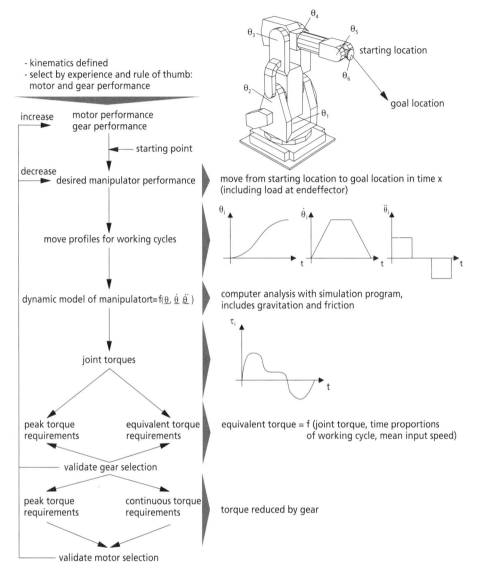

Figure 24   Computation of drive chain torques.

Table 4   Principles for Servo-drives

Driving Principle	Field of Application	Advantages	Disadvantages
Hydraulic	Manipulators with very high loads and/ or large workspace	• High dynamics • Very high power to weight ratio	• Requires equipment: pumps, hoses, servo-valves • "Dirty" • Require maintenance • Low efficiency
Electric	Standard for industrial robotics	• High dynamics • Very good controllability • Large power range • Large speed range	• Reduction gear necessary • Heating

found only in a few SCARA-type robots (Figure 19). In all others the high rated speed of 1000 rpm is lowered by reduction gears. Hydraulic drives are used only at very high loads and/or in large workspaces because of their high power-to-weight ratio.

5.1 Electrical Motor Drives

A wide variety of electrical motors are available, each with its own special features.

5.1.1 Stepper Motors

Stepper motors usually provide open loop position and velocity control. They are relatively low in cost and interface easily with electronic drive circuits. Developments in control systems have permitted each stepper motor step to be divided into many incremental microsteps. As many as 10,000 or more microsteps per revolution can be obtained. Motor magnetic stiffness, however, is lower at these microstepping positions. Typically, stepper motors are run in an open-loop configuration. In this mode they are underdamped systems and are prone to vibration, which can be damped either mechanically or through application of closed-loop control algorithms. Power-to-weight ratios are generally lower for stepper motors than for any other types of electric motors.

5.1.2 DC Motor Drives

DC motor drives are characterized by nearly constant speed under varying load and easy control of speed via armature current. Consequently, the direct current drive had a dominant role at the beginning of electric servo-driving technique for robot systems. A disadvantage, however, is wear-ridden mechanic commutation by brushes and commutators, which also limits the maximally transferring current. DC motor drives are more and more replaced by drives with electronic commutation, although recent developments in commutator technique have raised service life of the brushes to up to 30,000 hours.

5.1.3 Alternating Current Drives

Synchronous Motors

Synchronous alternating current drives have been made more powerful by the replacement of conventional oxide ceramic magnets by rare earth magnets (somarium cobalt, neodymium ferrite boron), which have much higher power density. Newly developed power semiconductor technology (e.g., IGBT, insulated gate bipolar transistor) has improved dynamics and controllability. Synchronous motors are used with power up to 10–20 kW and rotational speeds up to 3000 rpm. Because the rotating field is commutated contactless, the drive is nearly maintenance-free.

Table 5 Types of Servomotors

Type of Servomotor	Maximum Output Power	Specific Properties
Stepper motor	1 kW	• Running in open servo-loop • Heating in stalling • Poor dynamics
DC-brush motor	5 kW	• Good controllability via armature current • High starting torque • Brushes subject to wear
DC-brushless motor	10 kW	• Maintenance-free • Commutation by resolver or Hall-effect or optical sensor • High power density with rare earth magnets
AC-synchronous motor	20 kW	
AC-asynchronous motor	80 kW	• Maintenance-free • Very robust motor • High speed range • Expensive to control

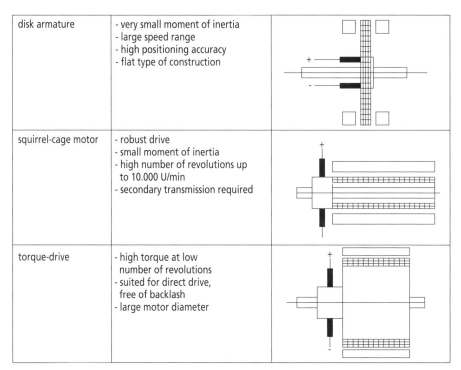

disk armature	- very small moment of inertia - large speed range - high positioning accuracy - flat type of construction	
squirrel-cage motor	- robust drive - small moment of inertia - high number of revolutions up to 10.000 U/min - secondary transmission required	

torque-drive	- high torque at low
	  number of revolutions
	- suited for direct drive,
	  free of backlash
	- large motor diameter	

Figure 25   Construction types of direct current drives.

Asynchronous Motors

Asynchronous drives are more robust than synchronous drives and have a higher density of power. They are used with power up to 80 kW and rotational speeds up to 10,000 rpm. By weakening the magnetic field, asynchronous drives can operate in a large field of constant power output beyond the nominal number of revolutions, which offers a great advantage over synchronous motors.

5.1.4   Linear Drives

If a rotary electric drive is cut open at the circumference and evenly unrolled, the result, in principle, is a linear drive (see Figure 26). For Cartesian robot transmissions, linear drives, for example, can be used for highly dynamic pick-and-place tasks. Compared with spindle drives, the most important advantages of the linear drives, which are mostly realized as synchronous and asynchronous drives for high loads, are high velocities (up to 3 m/s) and high accelerations (up to 10 g). Linear drives will probably replace some spindle drives in high-speed applications in the near future.

5.2   Electrohydraulic Servo-drives

Electrohydraulic servo-drives consist of a hydromotor or a hydrocylinder and a servo-valve for controlling the motor. Combined with a positional encoder, electrohydraulic servo-drives exhibit very good dynamics in closed loop control. Disadvantages are the need for much equipment, need for maintenance, and low efficiency. Figure 27 shows a hydraulic-driven manipulator.

6   GEARS

6.1   Introduction

Electric servomotors generate acceptable efficiency (up to 95%) only at relatively high numbers of revolutions (1000 rpm). For this reason they have not to this point been

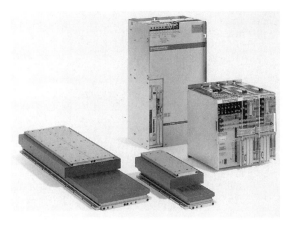

Figure 26   Linear electrical drive (courtesy Indramat).

well suited for direct drives, with the exception of some high-performance SCARA-type robots.

Robot gears convert high input numbers of revolutions and low torque into lower output numbers of revolutions and high torque (see Figure 28). At the same time they reduce the moment of inertia of the robot arm for the servo-drive.

Besides the conversion of torque and number of revolutions, robot gears induce some disadvantages in the drive chain:

- No gear is backlash-free. This is a problem in position accuracy and control.
- Gears can induce torsional oscillation because they act as elastic elements in the drive chain.

The requirements in Table 6 cannot be fulfilled separately. For example, the minimization of backlash requires pretensioned gear teeth, which on the other hand leads to a extension of friction and as a consequence to reduced efficiency.

Figure 27   Manipulator "Skywash" with hydraulic drives (courtesy Putzmeister).

Figure 28   Conversion of speed and torque by robot gear.

6.2   Gears for Linear Movements

Linear robot axes can be found in

- Cartesian robots for pick-and-place tasks
- Vertical axes for SCARA-type robots
- Gantry robots
- Peripheral axes, mostly under the base of an articulated robot in order to enlarge workspace

Different gears are used according to the dimension of joint range, demanded accuracy, and load, as shown in Table 7.

Figures 29 and 30 show a rack-and-pinion and a ball bearing spindle.

The maximum length of toothed belt and ball bearing spindle is limited by bending and torsional resonance frequencies.

Table 6   Quality Requirements for Robot Gears

Requirements	Typical Values
Very small backlash	Few arcmin
High efficiency	80–95%
Large reduction in few steps	100–320
Low inertia, low friction	Dependent on gear size
High torsional stiffness	Dependent on gear size
High power density, low weight	

Table 7

Naming	Common Joint Ranges	Accuracy	Load
Toothed belt	<10 m	Max. 0.1 mm	Small to medium
Ball bearing spindle	<5 m	Max. 0.001 mm	Small to medium
Rack-and-pinion	Theoretically unlimited	Max. 0.01 mm	Medium to high

Figure 29   Worm gear and rack-and-pinion (courtesy Atlanta).

Figure 30 *x, y*-axis with ball bearing spindle.

Usually the servo-motor is not joined directly with the linear gear but with an intermediate gear (transmission, bevel, or worm gear).

6.3 Gears for Rotary Movements

Gears for rotary movements can be found in all articulated robots and SCARA-type robots. They are usually offered as compact, pot-shaped kits. Depending on the gear type, output bearings up to the main axes are integrated.

Table 8 shows different types of common rotary drive gears.

Figure 31 gives a survey of construction schemes.

Examples of robot gears for rotary joints are shown in Figures 32 to 35.

7 COUPLINGS

Couplings are installed between servo-motor and gear and servo-motor and encoder to balance alignment faults between shafts. They also transfer torque and rotating movements. The coupling influences the performance of the whole drive chain. To apply couplings the following quantities must be considered:

- Torsional stiffness of coupling
- Damping of coupling

Figure 36 shows some typical couplings for robotic systems.

Following recent developments in articulated robots, the servo-motor–gear connection is executed without coupling. Instead, the servo-motor is directly flanged to the solid-constructed gear aggregate.

Table 8 Properties of Different Rotor Gears

Type of Gear	Typical Properties	Application
Planetary gear	Very high efficiency	Base axes
Harmonic drive gear	Very compact Very high reduction in one-gear stage Small to medium torques	Wrist axes
Cycloid gear	High efficiency High torque	Base axes
Involute gear	Very compact Very high reduction in one-gear stage Small to medium torques	Wrist axes

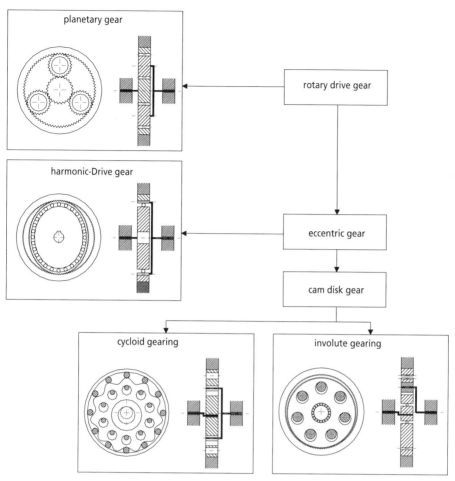

Figure 31 Survey of rotary drive gears.

Figure 32 KUKA robot with Harmonic Drive gears (courtesy Harmonic Drive).

Figure 33   Fanuc robot with cycloid gears (courtesy Teijin Seiki).

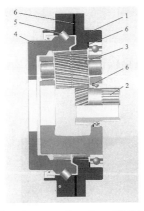

Figure 34   Planetary gear (courtesy ZF Friedrichshafen).

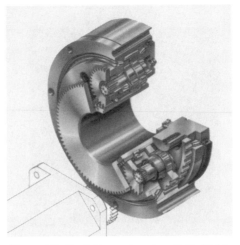

Figure 35   Cycloid gear (courtesy Teijin-Seiki).

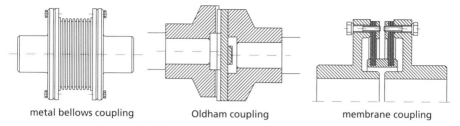

metal bellows coupling   Oldham coupling   membrane coupling

Figure 36   Different types of couplings used in robotics.

8   MEASURING SYSTEMS

Measuring systems feed back the positional signal from the robot joint to the controller. The resolution of the position sensor is a limiting factor for the precision of the robot system. Three types of encoders are commonly used in industrial robotics:

8.1   Optical Absolute Encoder

The electrical output signal of an optical absolute encoder provides a continuous digital signal that holds the information about the absolute position of the robot joint (see Figure

Figure 37   Optical measuring systems.

Table 9

Measuring System	Principle	Output Signal	Common Resolutions	Typical Properties
Optical absolute encoder	Linear/rotary	Digital absolute	12–16 bit	• No position loss during power down • Multiturn possible
Optical incremental encoder	Linear/rotary	Digital relative	5,000–100,000 pulses/rev	• Position loss during power down • High resolution
Resolver	Rotary	Analog absolute	12–16 bit	• No position loss during power down • Robust • Inexpensive

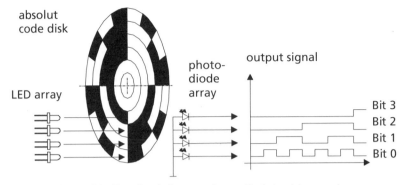

Figure 38 Functional diagram of an optical absolute encoder.

38). Sampling of joint position works contact-free using optical devices. Light from an LED array passes through a code disk fixed to the robot joint, which carries parallel tracks of binary code patterns of black and transparent segments. The optical signal is transformed into an electrical signal by an ASIC chip, which consists of a photovoltaic diode array, comparator, memory, code inverter, and driver. Typical resolutions of optical absolute encoders are 12–16 bits.

The most common code used to represent a numerical value in absolute encoders is the Graycode. When there is a transition from one numerical value to the next, only one bit changes. Imaging binary code, transition from 2 to 3 will change bit numbers 1, 2, and 3 at the same time. If there are only small tolerances on the code disk between dark and transparent segments of different bit-tracks, these bits will not change at exactly the same time and the output signal will not be well defined at this moment.

8.2 Incremental Optical Encoders

Compared to optical absolute encoders, incremental encoders provide only a relative signal, which means that pulses are counted in relation to a fixed reference point (see Figure 39). Loss of the reference point, as when the power supply is switched off for a short time, means that the position is lost and must be referenced again by searching the reference point. This is a big disadvantage in robotic systems, because up to six axes have to be referenced again, which is possible only in a special joint position configuration.

The direction of movement of the code disk is evaluated by comparing the signal of two photovoltaic diodes arranged so that they receive the signal from the LED 90° degree phase-shifted. Signal conditioning is done afterwards by a Schmitt trigger and an up/down counter.

As opposed to absolute encoders, typical incremental encoder design has a resolution of up to 5,000 pulses per revolution. By evaluation of the signal edges (or transitions),

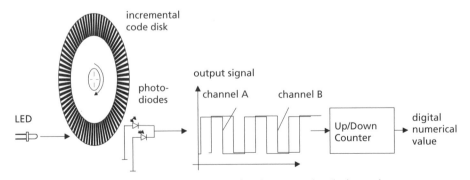

Figure 39 Functional diagram of an incremental optical encoder.

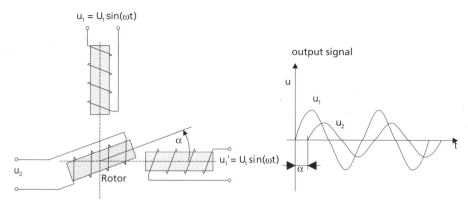

Figure 40   Functional principle of the resolver.

the resolution is increased by factor 4. If the encoder provides a sine output signal, by interpolation of the signal the resolution can be increased by factor 10.

Thus, in ultra-precision linear applications where resolution of some μm or better is required, only incremental encoders can be used. These applications usually occur in linear Cartesian systems for measurement or precision assembly.

8.3   Resolver

A resolver is an analogous measuring system giving out an absolute angular signal via one rotation (see Figure 40). Multiturn resolvers incorporate a reduction gear with a counter of revolutions to give out a continuous signal for more than one revolution.

Immediately after starting, the resolver indicates the exact position. The axis does not have to be moved to make new references, which provides a decisive advantage. Via a rotary transformer (brushless version), alternating voltage is fed into the rotor. The stator is composed of two windings into which this signal is induced. Both stator windings are mechanically arranged to distribute their signals by 90°. The computation of the combined output signals is done by an ASIC. Common resolutions of resolvers are 12–16 bits.

Resolvers are high-resolution, robust, and inexpensive position sensors that can easily be designed with a hollow shaft. For that reason, resolvers increasingly replace optical sensor systems in rotational robot joints.

9   ROBOT CALIBRATION

9.1   Introduction

A general task for a robot is to move its TCP coordinate system to a specific orientation and a certain point within its workspace. In this context, accuracy is the ability of the robot to move its end-effector to this pose and repeatability is the robot's ability to return to a previously achieved pose. In other words, in considering repeatability, the desired pose has been previously attained and therefore the necessary joint values are known. In considering accuracy, however, the pose is specified in task space and the desired set of joint displacements that corresponds to this pose must be determined. Typically, industrial robots have much better repeatability than accuracy.

For the robot's end-effector to be moved into desired pose in task space for the first time, the predefined coordinates of the desired pose must be translated into the robot's joint values. This conversion from the robot's task space to the robot's joint space is accomplished using a mathematical model that represents the manipulator's kinematics. This model relates the joint displacements to the TCP pose and vice versa and is used by the robot controller. If the mathematical model is incorrect, the manipulator will not have sufficient accuracy. However, the repeatability is not affected by this error. Additionally, the repeatability of a manipulator is relatively constant across the workspace, while the accuracy can vary significantly and is normally much worse than the repeatability. The mathematical model therefore contains a certain amount of parameters that, if correctly identified, allow the model to match the actual robot and enhance the robot's accuracy within its workspace. Equation (1) shows the relationship between the robot's task space and joint space:

$$P = f(\theta, c) \tag{1}$$

P is the (6×1) vector that describes the pose in task space (see Figure 41). Its six components represent three translational and three rotational degrees of freedom. θ is the $(n \times 1)$ vector that describes the pose in joint space where n is equivalent to the amount of robot axes. c represents the set of parameters used in the model, and f is the function that transforms the joint space coordinates θ into task space coordinates P with respect to the set of parameters c.

Robot calibration requires a process of defining an appropriate mathematical model and then determining and implementing the various model parameters that make the model match the robot as closely as possible (Mooring, Roth, and Driels, 1991). For costs for calibration to be reduced, the mathematical model must be as complex as necessary but as simple as possible. (See also Chapter 39.)

9.2 Steps in Robot Calibration

Figure 42 depicts the steps in robot calibration.

9.2.1 Modeling

A detailed discussion of fundamentals of kinematic modeling may be found in the literature (Asada and Slotine, 1986; Craig, 1986; Fu, Gonzalez, and Lee, 1987; Payonnet, Aldon, and Liegeois, 1985; Nikravesh, 1988).

Depending on the complexity and scope of the calibration task, Mooring, Roth, and Driels (1991) suggest three different levels of calibration procedures:

Level 1 (joint level) calibration has as its goal the determination of the relationship between the actual joint displacement and the signal generated by the joint displacement transducer. See Figure 42.

Level 2 calibration covers the entire robot model calibration. In addition to the correct joint angle relationships, the basic kinematic geometry of the robot must be identified.

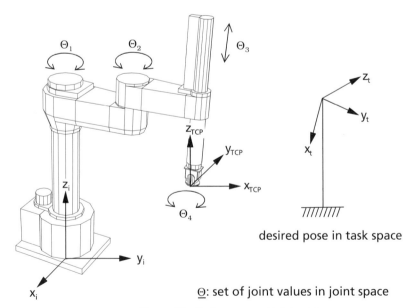

desired pose in task space

Θ: set of joint values in joint space

Figure 41   Task space and joint space.

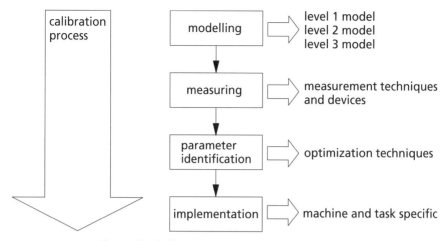

Figure 42   Calibration process and problems to be solved.

Level 3 (nonkinematic) calibration deals with nonkinematic errors in positioning of the robot's TCP that are due to effects such as joint compliance, friction, and clearance.

Each level implies a different mathematical model, and that means a different functional form for Equation (1) of the considered robotic system. While level 1 and 2 models correspond to practical needs, level 3 models are to be examined by scientists and have less practical relevance. Figure 43 shows a very simple kinematic model of a robotic device for a level 1 (= joint level) calibration.

Simulation systems such as IGRIP (Deneb Robotics Inc.) or ROBCAD (Tecnomatics) provide calibration tools that contain standard kinematics of industrial robots and algorithms that compute the forward kinematics and solve the inverse kinematics numerically.

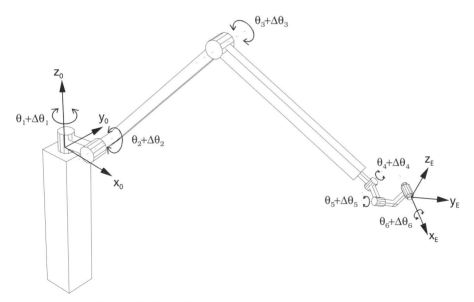

Figure 43   Kinematic schema of a six-d.o.f. industrial robot.

The following paragraphs will explain the next steps in manipulator calibration for a level 1 mathematical model. More complex procedures can be found in Mooring, Roth, and Driels (1991).

9.2.2 Measurement Devices and Techniques

Measurement devices and techniques differ depending on the mathematical model's set of parameters that have to be identified. Measurement devices can be divided into two groups:

Internal measurement devices are sensors that belong to the robotic system and are attached to the robot permanently: the joint displacement transducers. The most commonly used joint displacement transducers can be listed as follows:

- The *encoder* is a device for measuring linear or angular displacement. It produces a digital signal that can be easily interfaced to the robot control. The achievable accuracy depends on the encoder's resolution and can be high.
- The *resolver* is a device for measuring displacement of a revolute joint. Its output is a digital signal representing the error between the desired and the actual shaft angle. Its accuracy is high, but it is significantly more expensive than an encoder package with comparable resolution.

Internal measurement devices, such as encoders, are discussed above in Section 8. A more detailed introduction to internal measurement devices can be found in Doebelin (1983). (See also Chapters 12, 13.)

External measurement devices are sensors that belong to the measurement setup needed for the actual calibration procedure and are either attached to the robotic system temporarily or installed close to the robot. External sensors are used for such applications as measuring the pose of the robot's end-effector. See Figures 44 and 45.

The goal of the measurement process for a level 1 (= joint level) mathematical model is to accurately determine the robot's end-effector pose with a set of external measurement devices for a given set of joint displacements recorded by internal measurement devices.

This task can be completed with a 3D laser tracking system. An example of such a system is the LEICA LT/LTD 500 system from LEICA AG, Switzerland. It contains a freely movable target reflector, which must be attached to the robot's end-effector, and a laser tracker, which consists of a mirror system with two motors and two encoders, a laser interferometer, and a precision distance sensor.

A PC is used as a control computer for the measurement system. The system is capable of recording the target reflector with end-effector velocities of more than 4 m/s and end-effector accelerations of more than 2 g. To receive position information, one target reflector is sufficient. In order to measure the complete end-effector pose, at least three targets at the end-effector have to be recorded simultaneously.

9.2.3 Parameter Identification

Parameter identification is the mathematical process of using the data internally and externally measured to identify the coefficients of the mathematical model. Parameter identification describes the process of finding the optimal set of parameters for given sets of internal and corresponding external measurements. In other words, for a level 1 mathe-

Figure 44 External 1D rope sensors for long-distance measurements (courtesy ASM, Germany).

Figure 45 LEICA LTD 500 Laser Tracking System (courtesy Leica AG, Switzerland).

matical model of a manipulator, the internal joint displacement transducers' readings plus the identified joint offset parameters should form the correct input for the forward kinematics, so that the computed end-effector pose and the externally measured end-effector pose match as closely as possible.

The identification techniques therefore attempt to determine the parameter set that will optimize some performance index. Various approaches to achieve this task relate to such fields as control theory and dynamic systems modeling. The difference in these approaches comes, for example, in the type of model that is used or in the assumptions that are made about noise. Mooring, Roth, and Driels (1991) suggest the following classification:

1. *Deterministic vs. stochastic:* depending on whether or not probabilistic models for process and measurement noise are utilized.

2. *Recursive vs. nonrecursive:* depending on whether the whole set of observed data is saved and processed in its entirety or used sequentially.

3. *Linear vs. nonlinear:* depending on the type of mathematical model that is used.

Linear and nonlinear least square fittings are the most straightforward approaches for estimating the unknown model parameters from measured data. For very complex nonlinear target functions, genetic algorithms can serve as optimization techniques. Gerald and Wheatley (1990) give an overview of applied numerical analysis.

9.2.4 Implementation

The implementation procedure is the process of incorporating the improved version of the robot's mathematical model—in other words, the mathematical model plus the identified parameters of this model—into the controller so that the correct relationship between workspace coordinates and joint transducer values is achieved. This is seldom a simple process; the forward kinematic model and the inverse kinematic model have to be implemented in the controller. For simple kinematic models the inverse equations can be solved explicitly. For more complex models iterative methods have to be established.

The theory and examples of inverse kinematics can be found in the literature. See, e.g., Raghavan and Roth (1993); Zhiming and Leu (1990). (See also Chapter 6.)

10 PERFORMANCE TESTING

ISO 9283 is the International Standard that describes methods of specifying and testing the following performance characteristics of manipulating industrial robots:

- Unidirectional pose accuracy and pose repeatability
- Multidirectional pose accuracy variation
- Distance accuracy and distance repeatability
- Pose stabilization time
- Pose overshoot
- Drift of pose characteristics
- Path accuracy and path repeatability
- Cornering deviations

Criteria to be tested	Clause in ISO 9283	Applications								
		Spot welding 1	Handling/ loading/ unloading 1	Assembly 1	Assembly 2	Inspection 1	Inspection 2	Machining deburring polishing cutting 2	Spray painting 2	Arc welding 2	Adhesive sealing 2
Unidirectional pose accuracy	7.2.1	✔	✔	✔	✔	✔	✔			✔	
Unidirectional pose repeatability	7.2.2	✔	✔	✔	✔	✔	✔			✔	
Multi-directional pose accuracy variation	7.2.2		✔		✔	✔	✔				
Distance accuracy	7.3.2	✔ 3	✔ 3	✔ 3	✔ 3	✔ 3	✔ 3				
Distance repeatability	7.3.3	✔ 3	✔ 3	✔ 3	✔ 3	✔ 3	✔ 3				
Pose stabilization time	7.4	✔	✔	✔	✔	✔	✔				
Pose overshoot	7.5	✔	✔	✔	✔	✔	✔			✔	
Drift of pose characteristics	7.6	✔	✔	✔	✔	✔	✔			✔	
Path accuracy	8.2			✔		✔	✔	✔	✔	✔	✔
Path repeatability	8.3			✔		✔	✔	✔	✔	✔	✔
Corner deviations	8.4			✔		✔	✔	✔	✔	✔	✔
Stabilization path length	8.4.2			✔		✔	✔	✔	✔	✔	✔
Path velocity accuracy	8.5.2							✔	✔	✔	✔
Path velocity repeatability	8.5.3							✔	✔	✔	✔
Path velocity fluctuation	8.5.4							✔	✔	✔	✔
Minimum positioning time	9	✔	✔	✔	✔					✔	
Static compliance	10	✔	✔	✔	✔			✔			

Notes: 1 Application where pose-to-pose control is normally used
2 Application where continuous path control is normally used
3 Only in case of explicit programming

Figure 46 Performance testing methods versus applications according to ISO 9283.

- Path velocity characteristics
- Minimum positioning time
- Static compliance

Figure 46 is a guide to selecting performance criteria for typical applications. Many manufacturers that use industrial robots, such as General Motors and Mercedes Benz, have also developed their own standards (Schröer, 1998). Copies of these performance testing criteria can be obtained from the companies.

REFERENCES

Akao, Y., and T. Asaka. 1990. *Quality Function Deployment: Integrating Customer Requirements into Product Design.* Cambridge and Norwalk: Productivity Press.

Angeles, J. 1997. *Fundamentals of Robotic Mechanical Systems—Theory, Methods and Algorithms.* New York: Springer-Verlag.

Asada, H., and J.-J. Slotine. 1986. *Robot Analysis and Control.* New York: John Wiley & Sons.

Burnett, D. S. 1987. *Finite Element Analysis.* Reading: Addison-Wesley.

Canny, J. F., and K. Y. Goldberg. 1995. "A RISC-Based Approach to Sensing and Manipulation." *Journal of Robotic Systems* 6, 351–363.

Clausing, D. P. 1994. "Total Quality Development." *Manufacturing Review* 7(2), 108–119.

Craig, John J. 1986. *Introduction to Robotics Mechanics and Control.* Reading: Addison-Wesley.

Daenzer, W. F., and F. Huber, eds. 1994. *Systems Engineering: Methodik und Praxis.* 8. verbesserte Auflage. Zürich: Verlag Industrielle Organisation.

Doebelin, E. 1983. *Measurement Systems: Application and Design.* New York: McGraw-Hill.

DRAFT AMENDMENT ISO 9283: 1990/DAM 1. 1990. Manipulating Industrial Robots—Performance Criteria and Related Test Methods. Amendment 1: Guide for Selection of Performance Criteria for Typical Applications. Geneva: International Organization for Standardization.

Fu, K. S., R. C. Gonzalez, and C. S. G. Lee. 1987. *Robotics: Control, Vision, and Sensing.* New York: McGraw-Hill.

Furgaç, I. 1986. "Aufgabenbezogene Auslegung von Robotersystemen." Dissertation, Universität Berlin.

Gerald, C. F., and P. O. Wheatley. 1990. *Applied Numerical Analysis.* 4th ed. Reading: Addison-Wesley.

Hayward, V., and R. Kurtz. 1989. "Preliminary Study of Serial–Parallel Redundant Manipulator." In *NASA Conference on Space Telerobotics,* Pasadena, January 31. 39–48.

Herrmann, G. 1976. "Analyse von Handhabungsvorgängen im Hinblick auf deren Anforderungen an programmierbare Handhabungsgeräte in der Teilefertigung." Dissertation, Universität Stuttgart.

Inoue et al. 1993. "Study on Total Computer-Aided Design System for Robot Manipulators." In *24th ISIR,* November 4–6, Tokyo. 729–736.

International Federation of Robotics (IFR). 1997. *World Industrial Robots 1997—Statistics, Analysis and Forecasts to 2000.*

ISO 9283 First Edition 1990-12-15. 1990. Manipulating Industrial Robots—Performance Criteria and Related Test Methods. Reference number ISO 9283: 1990(E). Geneva: International Organization for Standardization.

Klafter, R. D., T. A. Chmielewski, and M. Negin. 1989. *Robot Engineering: An Integrated Approach.* Englewood Cliffs: Prentice-Hall International.

Masory, O., J. Wang, and H. Zhang. 1993. "On the Accuracy of a Stewart Platform. Part II: Kinematic Calibration and Compensation." In *Proceedings of the IEEE International Conference on Robotics and Automation,* Atlanta, May 2–6.

Merlet, J.-P. 1995. "Designing a Parallel Robot for a Specific Workspace." Research Report 2527 INRIA, April.

Mooring, B. W., Z. S. Roth, and M. R. Driels. 1991. *Fundamentals of Manipulator Calibration.* New York: John Wiley & Sons.

Morrow, J. D., and P. K. Khosla. 1997. "Manipulation Task Primitives for Composing Robot Skills." In *Proceedings of the 1997 IEEE International Conference on Robotics and Automation,* Albuquerque, April. 3454–3459.

Murray, R. M., Z. Li, and S. S. Sastry. 1993. *A Mathematical Introduction to Robotic Manipulation.* Boca Raton: CRC Press.

Nakamura, Y. 1991. *Advanced Robotics Redundancy and Optimization.* Reading: Addison-Wesley.

Nikravesh, Parviz E. 1988. *Computer-Aided Analysis of Mechanical Systems.* Englewood Cliffs: Prentice-Hall.

Pahl, G., and W. Beitz. 1995. *Engineering Design: A Systematic Approach.* 2nd ed. Trans. K. Wallace and L. Blessing. New York: Springer-Verlag.

Paredis, C., and P. Khoshla. 1997. "Agent-Based Design of Fault Tolerant Manipulators for Satellite Docking." In *Proceedings of the 1997 IEEE International Conference on Robotics and Automation,* Albuquerque, April. 3473–3480.

Payannet, D., M. J. Aldon, and A. Liegeois. 1985. "Identification and Compensation of Mechanical Errors for Industrial Robots." In *Proceedings of the 15th International Symposium on Industrial Robots,* Tokyo.

Raghavan, M., and B. Roth. 1993. "Inverse Kinematics of the General 6R Manipulator and Related Linkages." *Transactions of the ASME* **115**(3), 502–508.

Ragsdell, K. M. 1994. "Total Quality Management." *Manufacturing Review* **3,** 194–204.

Roy, J., and L. Whitcomb. 1997. "Structural Design Optimization and Comparative Analysis of a New High-Performance Robot Arm via Finite Element Analysis." In *Proceedings of the 1997 IEEE International Conference on Robotics and Automation,* Albuquerque, April. 2190–2197.

Schröer, Klaus (Editor); Commission of the European Communities, Directorate-General XII: Programme on Standards, Measurements and Testing: Handbook on Robot Performance Testing and Calibration: Improvement of Robot Industrial Standardisation IRIS. Stuttgart: Fraunhofer IRB Verlag, 1998.

Stadler, W. 1995. *Analytical Robotics and Mechatronics.* McGraw-Hill Series in Electrical and Computer Engineering. New York: McGraw-Hill.

Wang, J., and O. Masory. 1993. "On the Accuracy of a Stewart Platform. Part I: The Effect of Manufacturing Tolerances." In *Proceedings of the IEEE International Conference on Robotics and Automation,* Atlanta, May 2–6. 114–120.

Wanner, M. C. 1988. "Rechengestützte Verfahren zur Auslegung der Mechanik von Industrierobotern." Dissertation, Universität Stuttgart.

Zhiming, Z., and M. C. Leu. 1990. "Inverse Kinematics of Calibrated Robots." *Journal of Robotic Systems* **7**(5), 675–687.

CHAPTER 6
KINEMATICS AND DYNAMICS OF ROBOT MANIPULATORS

Andrew A. Goldenberg
Mohammad R. Emami
University of Toronto
Toronto, Ontario

1 INTRODUCTION

This chapter reviews current practical methodologies for kinematics and dynamics modeling and calculations. A kinematics model is a representation of the motion of the robot manipulator without considering masses and moments of inertia; a dynamics model is a representation of the balancing of external and internal loads acting on the manipulator whether it is stationary or moving. Both models are used widely in design, simulation, and, more recently, real-time control.

These topics are considered fundamental to the study and use of robotics. In the early development of this branch of science and engineering, kinematics and dynamics modeling were the main topics treated in the literature. Over the years, kinematics and dynamics modeling have generated the greatest number of publications related to robotics. This chapter attempts to extract some of the most relevant issues, but does not provide a summary of all the published work. Its aim is to present the standard tools for kinematics and dynamics modeling without prerequisites. The reader is also referred to the corresponding chapter in the first edition of this handbook (Walker, 1985).

TERMINOLOGY

n	number of degrees of freedom of the manipulator
$\mathbf{q}$	vector of joint variable displacements
$\dot{\mathbf{q}}$	vector of joint variable velocities
$\ddot{\mathbf{q}}$	vector of joint variable accelerations
$\mathcal{Q}$	joint space
$\mathcal{T}$	task space
$\mathbf{F}_i$	coordinate frame attached to link i
$R_{i-1,i}$	3×3 rotation matrix between frames $(i-1)$ and i
$\mathbf{p}_{i-1,i}$	position vector of the origin of frame i with respect to frame $(i-1)$ expressed in frame $(i-1)$
$H_{i-1,i}$	4×4 homogeneous transformation matrix between frames $(i-1)$ and i
ϵ	axis of general rigid body rotation
ν	axis of general rigid body translation
Ψ	4×4 twist matrix
$W(P_w)$	wrist workspace
$W(P_e)$	end-effector workspace
$\boldsymbol{\omega}_i$	angular velocity vector of frame i
$\dot{\boldsymbol{\omega}}_i$	angular acceleration vector of frame i
$\boldsymbol{\nu}_i$	linear velocity vector of the origin of frame i
$\dot{\boldsymbol{\nu}}_i$	linear acceleration vector of the origin of frame i
$\mathbf{v}_{ci}$	linear velocity vector of the center of mass of link i
$\dot{\boldsymbol{\nu}}_{ci}$	linear acceleration vector of the center of mass of link i
$\mathbf{p}_{ci}$	position vector of the center of mass of link i with respect to frame i

Handbook of Industrial Robotics, Second Edition, Edited by Shimon Y. Nof
ISBN 0-471-17783-0   © 1999 John Wiley & Sons, Inc.

I_{ci} 3 × 3 moment of inertia matrix of link i about its center of mass expressed in
          frame i
m_i mass of link i
J 6 × n Jacobian matrix
τ vector of joint torques and forces
$\mathbf{f}_e$ vector of resulting reaction forces at the end-effector
$\mathbf{g}_e$ vector of resulting reaction moments at the end-effector
$\mathbf{f}_{ext}$ vector of external forces acting at the end-effector
$\mathbf{g}_{ext}$ vector of external moments acting at the end-effector
$\mathbf{G}_e$ wrench vector
M $n \times n$ manipulator inertia matrix
$\mathbf{g}^r$ gravity acceleration vector

2   KINEMATICS

This section considers the motion of the robot manipulator irrespective of inertial and external forces. A study of the geometry of motion is essential in manipulator design and control in order to obtain the mapping between the end-effector location (position and orientation) and the movement of manipulator links, as well as the mapping between the end-effector velocity and the speed of manipulator links. The final goal is to use these mappings to relate the end-effector (or gripper, or tool mounted on the end-effector) motion to joint displacements (generalized coordinates) and velocities.

2.1   Forward Kinematics

The objective of forward kinematics is to determine the location of the end-effector with respect to a reference coordinate frame as a result of the relative motion of each pair of adjacent links. Attention is restricted to the case of an *open-chain* manipulator, a serial link of rigid bodies connected in pairs by *revolute* and/or *prismatic* joints, for relative rotation and relative translation, respectively.

2.1.1   Different Configuration Spaces for Robot Manipulators

The configuration of a robot manipulator can be specified using either of the following algebraic spaces:

1. The *joint space* $\mathcal{Q}$ is the set of all possible vectors of joint variables. The dimension of the joint vector is equal to the number of joints (or degrees of freedom), i.e., $\mathcal{Q} \subset \mathbb{R}^n$. Each joint variable is defined as an angle $\theta \in \mathcal{S} = [0, 2\pi)$ for a revolute joint, or a linear translation $d \in \mathbb{R}$ for a prismatic joint. Let $\mathbf{q} \in \mathcal{Q}$ denote the vector of generalized coordinates.

2. The *task space* $\mathcal{T}$ is the set of pairs $(\mathbf{p}, R)$, where $\mathbf{p} \in \mathbb{R}^3$ is the position vector of the origin of link coordinate frame and $R \in SO(3)$ represents the orientation of the link frame, both with respect to a general reference frame. Here, $SO(3)$ denotes the group of 3 × 3 proper rotation matrices. Thus, the task space is a *Special Euclidean* group $SE(3)$, defined as follows:

$$SE(3) = \{(\mathbf{p}, R): \mathbf{p} \in \mathbb{R}^3, R \in SO(3)\} = \mathbb{R}^3 \times SO(3). \tag{1}$$

Using the above notation, the forward kinematics is a mapping H, defined as follows:

$$H: \mathcal{Q} \rightarrow SE(3). \tag{2}$$

This mapping can be represented by a 4 × 4 *homogeneous transformation* matrix, defined as

$$\begin{bmatrix} R & \mathbf{p} \\ \hline [0] & 1 \end{bmatrix}.$$

2.1.2   The End-Effector Position and Orientation

In order to obtain the forward kinematics mapping H, suitable coordinate frames should be assigned to the manipulator base, end-effector, and intermediate links. One standard method attributed to Denavit and Hartenberg (1965) is based on the homogeneous transformation H. The Denavit–Hartenberg (DH) convention uses the minimum number of parameters to completely describe the geometric relationship between adjacent robot links.

Each link and joint of the manipulator is numbered, as illustrated in Figure 1. The frame $\mathbf{F}_i$ attached to link i is defined with the $\mathbf{z}_i$ along the axis of joint $(i + 1)$; the origin is located at the intersection of $\mathbf{z}_i$ and the common normal to $\mathbf{z}_{i-1}$ and $\mathbf{z}_i$, and $\mathbf{x}_i$ is along the common normal, as illustrated in Figure 1. The homogeneous transformation matrix between links i and $(i - 1)$ is then expressed as (Walker, 1985):

$$
H_{i-1,i} = \left[\begin{array}{c|c} R_{i-1,i} & \mathbf{p}_{i-1,i} \\ \hline [0] & 1 \end{array}\right] = \left[\begin{array}{cccc} \cos\theta_i & -\sin\theta_i\cos\alpha_i & \sin\theta_i\sin\alpha_i & a_i\cos\theta_i \\ \sin\theta_i & \cos\theta_i\cos\alpha_i & -\cos\theta_i\cos\alpha_i & a_i\sin\theta_i \\ 0 & \sin\alpha_i & \cos\alpha_i & d_i \\ 0 & 0 & 0 & 1 \end{array}\right] \quad (3)
$$

where

$R_{i-1,i}$ = relative rotation of frame $\mathbf{F}_i$ with respect to $\mathbf{F}_{i-1}$;
$\mathbf{p}_{i-1,i}$ = position vector of the origin of $\mathbf{F}_i$ with respect to $\mathbf{F}_{i-1}$, expressed in $\mathbf{F}_{i-1}$;
$[0]$ = 1×3 null matrix.

The three link parameters a_i, α_i, d_i and one joint variable θ_i required to specify the transformation (3) are defined as follows:

$a_i \equiv$ the length of the common normal between $\mathbf{z}_{i-1}$ and $\mathbf{z}_i$ (link length)
$\alpha_i \equiv$ the angle between $\mathbf{z}_{i-1}$ and $\mathbf{z}_i$ measured about $\mathbf{x}_i$ (twist angle)
$d_i \equiv$ the distance from $\mathbf{x}_{i-1}$ to $\mathbf{x}_i$ measured along $\mathbf{z}_{i-1}$ (link offset or distance)
$\theta_i \equiv$ the angle between $\mathbf{x}_{i-1}$ and $\mathbf{x}_i$ measured about $\mathbf{z}_{i-1}$.

The homogeneous transformation between the base frame $\mathbf{F}_0$ and the end-effector frame $\mathbf{F}_n$ (for an n-d.o.f. manipulator) can then be systematically determined by successive multiplication of the intermediate transformations, namely:

$$
H_{0n} = H_{01}H_{12} \cdots H_{i-1,i} \cdots H_{n-2,n-1}H_{n-1,n} \quad (4)
$$

The matrix H_{0n} contains the rotation matrix between frames $\mathbf{F}_0$ and $\mathbf{F}_n$ (R_{0n}), and the location of the origin of $\mathbf{F}_n$ with respect to $\mathbf{F}_0$, expressed in $\mathbf{F}_0$:

$$
H_{0n} = \left[\begin{array}{c|c} R_{0n} & \mathbf{p}_{0n} \\ \hline [0] & 1 \end{array}\right] = \left[\begin{array}{ccc|c} n_{0n}^x & o_{0n}^x & a_{0n}^x & p_{0n}^x \\ n_{0n}^y & o_{0n}^y & a_{0n}^y & p_{0n}^y \\ n_{0n}^z & o_{0n}^z & a_{0n}^z & p_{0n}^z \\ \hline 0 & 0 & 0 & 1 \end{array}\right] \quad (5)
$$

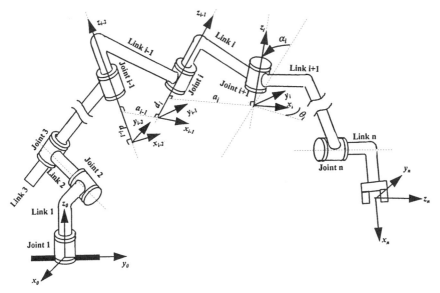

Figure 1   DH link frames and parameters.

where vectors $\mathbf{n}_{0n}$, $\mathbf{o}_{0n}$ and $\mathbf{a}_{0n}$ specify the orientation of the $\mathbf{x}_n$, $\mathbf{y}_n$ and $\mathbf{z}_n$ axes, respectively, of frame $\mathbf{F}_n$ with respect to frame $\mathbf{F}_0$.

2.1.3 Standard Coordinate Frames

Figure 2 shows some of the standard frames commonly used in industrial applications. The position of the origin and the orientation of each frame with respect to the base frame is obtained by successive multiplications of the intermediate homogeneous transformation matrices. For example, the representation of the tool frame with respect to the base frame is determined by

$$H_{0t} = H_{0n}H_{nt} \tag{6}$$

where H_{nt} and H_{0n} are the homogeneous transformation matrices between the end-effector and the tool frames and between the end-effector and the base frames, respectively.

2.1.4 Computational Considerations and Symbolic Formulation

In practical applications it is always effective to minimize the computational time required to perform the kinematic calculations. The calculations can be performed recursively because the open-chain manipulator can be seen as being constructed by adding a link to the previous links. This reduces the number of multiplications and additions at the cost of creating local variables in order to avoid the use of common terms throughout the computation. Algorithm 1 illustrates a backward recursive formulation for calculating the forward kinematics (Hoy and Sriwattanathamma, 1989).

Symbolic kinematic equations that describe the end-effector (or tool) position and orientation as explicit functions of joint coordinates can be derived in advance of real-time computation. If suitable trigonometric simplifications are implemented, symbolic representation helps to reduce the number of arithmetic operations. Either general-purpose (such as MATHEMATICA and MAPLE) or special-purpose (such as SD/FAST (West-macott, 1993)) symbolic modeling software can replace manual derivation to generate symbolic kinematic relationships automatically (Vukobratovic, 1986).

Transcendental functions are a major computational expense in forward kinematics calculations when standard software is used. Instead, lookup table implementations of these functions may reduce the required calculation time by a factor of two to three, or more (Ruoff, 1981). Moreover, using fixed-point instead of floating-point representation can speed up the operations. A 24-bit representation of joint variables is adequate due to the typically small dynamic range of these variables (Turner, Craig, and Gruver, 1984).

2.1.5 Manipulator Workspace

Evaluation of the manipulator workspace is a subject of interest for purposes of both analysis and synthesis. The workspace of a manipulator is defined as the set of all end-effector locations (positions and orientations of the end-effector frame) that can be reached by arbitrary choices of joint variables within the corresponding ranges. If both end-effector

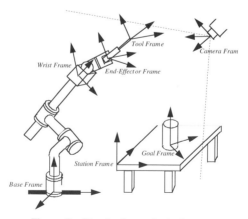

Figure 2 Standard coordinate frames.

LOOP: FOR $i = n - 1$ to 1

$$1.\ \text{SET: } R_{i,n} = \begin{bmatrix} n^x_{i,n} & o^x_{i,n} & a^x_{i,n} \\ \text{-----} \\ n^y_{i,n} & o^y_{i,n} & a^y_{i,n} \\ \text{-----} \\ n^z_{i,n} & o^z_{i,n} & a^z_{i,n} \end{bmatrix} = \begin{bmatrix} [R^x_{i,n}] \\ [R^y_{i,n}] \\ [R^z_{i,n}] \end{bmatrix}$$

$$2.\ \text{CALCULATE: } \begin{cases} [M_{i,n}] = \cos \alpha_i [R^y_{i,n}] - \sin \alpha_i [R^z_{i,n}] \\ r_{i,n} = \cos \alpha_i p^y_{i,n} - \sin \alpha_i p^z_{i,n} \\ s_{i,n} = \begin{cases} p^x_{i,n} + a_i : \text{if joint } i \text{ is revolute} \\ p^x_{i,n} \quad\ \ : \text{if joint } i \text{ is prismatic} \end{cases} \end{cases}$$

$$3.\ \text{CALCULATE: } \begin{cases} [R^x_{i-1,n}] = \cos \theta_i [R^x_{i,n}] - \sin \theta_i [M_{i,n}] \\ [R^y_{i-1,n}] = \sin \theta_i [R^x_{i,n}] + \cos \theta_i [M_{i,n}] \\ [R^z_{i-1,n}] = \sin \alpha_i [R^y_{i,n}] + \cos \alpha_i [R^z_{i,n}] \\ p^x_{i-1,n} = \cos \theta_i s_{i,n} - \sin \theta_i r_{i,n} \\ p^y_{i-1,n} = \sin \theta_i s_{i,n} + \cos \theta_i r_{i,n} \\ p^z_{i-1,n} = \sin \alpha_i p^y_{i,n} + \cos \alpha_i p^z_{i,n} + d_i \end{cases}$$

NEXT i

Algorithm 1   Backward recursive formulation of the forward kinematics problem.

position and orientation are considered, the workspace is the *complete workspace;* disregarding the orientation of the end-effector gives the *reachable workspace.* The subset of the reachable workspace that can be attained with arbitrary orientations of the end-effector is the *dexterous workspace.* Most industrial manipulators have spherical wrists; therefore, for a 6-d.o.f. manipulator, the wrist is positioned using the first three joints. If the wrist point P_w and the end-effector point of interest P_e are different, then, after the workspace of the wrist point $W(P_w)$ is determined, a sphere of radius $\overline{P_w P_e}$ is moved so that its center is on the boundary of the wrist workspace $W(P_w)$. The inner and outer envelopes are the boundaries of the dexterous and reachable workspaces, respectively. Nevertheless, due to machining tolerances, assembly errors, and other limitations, it is impossible to build a perfect wrist with three orthogonal revolute axes intersecting at one point. Thus a general methodology is required for determining the manipulator workspace (Ceccarelli, 1994).

2.1.6   The Product of Exponentials (PE) Formula

A geometric description of the robot kinematics can be obtained based on classical *screw theory* (Yuan, 1971). The fundamental fact is that any arbitrary rigid body motion is equivalent to a rotation θ about a certain line ε ($\|\varepsilon\| = 1$) combined with a translation l parallel to that line. The homogeneous transformation of the rigid body motion with respect to a reference frame can then be represented as

$$H(\theta, l) = e^{\Psi \theta} H(0, 0) \tag{7}$$

where $H(0, 0)$ is the initial homogeneous representation of the rigid body with respect to the same reference frame when $\theta = 0$ and $l = 0$. In Equation (7), Ψ is a 4×4 matrix called the *twist* and is defined as

$$\Psi = \begin{bmatrix} E & -E\mathbf{p} + (l/\theta)\epsilon \\ [0] & 0 \end{bmatrix} \tag{8}$$

where E is a 3×3 skew-symmetric matrix of the rotation axis $\epsilon = [\epsilon_x\ \epsilon_y\ \epsilon_z]^T$ such that

$$E = \begin{bmatrix} 0 & -\epsilon_z & \epsilon_y \\ \epsilon_z & 0 & -\epsilon_x \\ -\epsilon_y & \epsilon_x & 0 \end{bmatrix} \tag{9}$$

and $\mathbf{p} = [p_x\ p_y\ p_z]^T \in \mathbb{R}^3$ is the position vector of an arbitrary point located on the rotation axis ϵ and expressed in the same reference frame. The matrix exponential mapping $e^{\Psi\theta}$ is based on the following formulation (Park and Okamura, 1994).

$$e^{\Psi\theta} = \left[\begin{array}{c|c} e^{E\theta} & A(-E\mathbf{p} + (l/\theta)\epsilon) \\ \hline [0] & 1 \end{array} \right] \tag{10}$$

where
$$e^{E\theta} = [I] + E \sin \theta + E^2(1 - \cos \theta) \tag{11}$$

and
$$A = [I]\theta + E(1 - \cos \theta) + E^2(\theta - \sin \theta) \tag{12}$$

For a pure rigid body translation ($\theta = 0$) along some axis v ($\|v\| = 1$), the twist is defined as

$$\Psi_{\text{trans}} = \left[\begin{array}{c|c} [[0]] & lv \\ \hline [0] & 0 \end{array} \right] \tag{13}$$

where $[[0]]$ and $[0]$ are 3×3 and 1×3 null matrices, respectively. In this case the matrix exponential mapping becomes

$$e^{\Psi_{\text{trans}}l} = \left[\begin{array}{c|c} [I] & lv \\ \hline [0] & 1 \end{array} \right] \tag{14}$$

with $[I]$ as the 3×3 identity matrix.

Based on the above formulation of rigid body motion, for an open-chain manipulator, the homogeneous transformation of each link i *with respect to the base frame* is obtained by an exponential mapping

$$H_{0i}(q_i) = e^{\Psi_i q_i} H_{0i}(0) \tag{15}$$

where

q_i = joint variable;
Ψ_i = twist of link i.

The homogeneous representation of the end-effector with respect to the base frame is obtained by combining a sequence of mappings into the so-called *product of exponentials* (PE) formula (Brockett, 1983).

$$H_{0n}(q_1, q_2, \ldots, q_n) = e^{\Psi_1 q_1} e^{\Psi_2 q_2} \cdots e^{\Psi_n q_n} H_{0n}(0, 0, \ldots, 0). \tag{16}$$

The matrix $H_{0n}(0, 0, \ldots, 0)$ represents the homogeneous transformation of the end-effector frame with respect to the base frame when the manipulator is in its *reference configuration,* i.e., all joint variables are zero. The twist Ψ_i corresponds to the screw motion of the ith link as a result of moving joint i with all other joint variables held fixed at $q_j = 0$ ($j \neq i$).

One of the features of the PE formula is that, in contrast to the DH representation, there is no need to attach a frame to each link: once the base, end-effector, and a reference configuration frame have been chosen, a unique set of link twists is obtained that describes the forward kinematics of the robot. This property and the geometric representation make the PE formula a superior alternative to the DH convention.

2.2 Inverse Kinematics

Inverse kinematics is used to find the values of the joint variables that will place the end-effector at a desired location, i.e., desired position and orientation relative to the base, given the manipulator geometry (link lengths, offsets, twist angles, and the location of the base). Formally, for an n-d.o.f. manipulator, given the homogeneous matrix H_{0n} (5), the values of $q_1, q_2, \ldots, q_n$ are calculated. In general, the matrix equation (4) corresponds to 12 scalar equations; because the rotation matrix R_{0n} is orthonormal, only 6 of the 12 equations are independent. Therefore, the problem of inverse kinematics of a general 6-d.o.f. manipulator corresponds to solving a set of six nonlinear, transcendental equations with six unknowns (joint variables). There may be no solution, a unique solution, or multiple solutions to the inverse kinematics problem.

2.2.1   Solvability and Number of Solutions

A general approach for systematically solving the inverse kinematics problem (Lee and Liang, 1988) is to consider the set of nonlinear equations as a set of multivariate polynomials in $s_i \equiv \sin \theta_i$ and $c_i \equiv \cos \theta_i$ for $i = 1, 2, \ldots, n$. This is possible since the entries of each homogeneous transformation matrix (3) are unary (i.e., of degree one or less) in s_i and c_i. Then, by elimination of variables in a systematic way (Salmon, 1964), $n - 1$ variables are eliminated in a system of n polynomials in n variables, and a single polynomial in one variable is obtained. This method is called *dyalitic elimination,* and the resultant polynomial is called the *characteristic polynomial.* Once the roots of this polynomial are found, the eliminated variables can be determined from a set of linear equations. This general algorithm is presented in the next subsection, which addresses the existence and number of solutions of the inverse kinematics problem.

Generally, at least six joints are required to attain arbitrary three-dimensional task positions and orientations. The necessary condition for the existence of a solution is that the desired end-effector location lie in the reachable workspace. If the desired location is inside the workspace, then the existence of at least one solution is guaranteed. The existence of an analytical, closed-form solution to the inverse kinematics problem depends on the order of the characteristic polynomial. If the characteristic polynomial is of order 4 or less, since the roots can be obtained as algebraic functions of the polynomial coefficients, the corresponding inverse kinematics problem can be solved analytically. Otherwise, iterative numerical methods must be relied upon to obtain the roots of the polynomial. In this case, the problem is considered numerically solvable when: 1) an upper bound on the number of solutions exists. 2) an efficient algorithm for computing all solutions is available. Based on recent results in kinematics (Selfridge, 1989), all 6-d.o.f. open-chain manipulators with revolute and prismatic joints are solvable. The number of solutions depends on the number of prismatic joints and kinematic parameters. For the general case of six revolute joints (6R manipulator) or one prismatic and five revolute joints (5R1P manipulator), there are at most 16 different configurations for each end-effector location. For 4R2P manipulators the number of possible configurations drops to 8, and for the 3R3P the number is 2. These numbers are independent of the physical order of revolute and prismatic joints in the chain. In all of the above cases, the number of *real* roots of the characteristic polynomials (and hence the number of real configurations) may be less than the numbers cited above by any multiple of 2. Certain values of the kinematic parameters may also reduce the number of possible configurations. A detailed investigation can be found in Mavroidis and Roth (1994). As an example, a 6R manipulator with three consecutive joint axes intersecting in a common point (Pieper and Roth, 1969) or with three parallel joint axes (Duffy, 1980) has at most 8 configurations, and the characteristic polynomial is of order 4 with repeated roots; therefore analytical solutions exist. A 6R manipulator with a spherical wrist is very common in industry. The analytical technique for this case is first to solve for the first three joint variables to satisfy the desired wrist point location and then to find the last three joint variables to achieve the required hand orientation (Pieper and Roth, 1969).

2.2.2   A General Solution for Six-Degree-of-Freedom Manipulators

A systematic method of solving the inverse kinematics of 6-d.o.f. manipulators is to arrange the set of nonlinear equations as a set of multivariate polynomials in s_i and c_i and then eliminate all variables except θ_3, thus obtaining a polynomial of order 16 in $\tan(\theta_3/2)$ such that the joint angle θ_3 can be computed as its roots. The remaining joint variables are obtained by substituting and solving for some intermediate equations. In this section, the procedure is presented for general 6R manipulators. The extension to manipulators with prismatic joints is also discussed. The following algorithm is a summary of the algorithm presented in Raghavan and Roth (1993).

Step 1

Determine the DH parameters and homogeneous transformation matrices $H_{i-1,i}$ and then rewrite the forward kinematics matrix equation in the following form:

$$H_{23}H_{34}H_{45} = H_{12}^{-1}H_{01}^{-1}H_{06}H_{56}^{-1} \tag{17}$$

Step 2

Equate each of the first three elements of the third and fourth columns of both sides of Equation (17). This gives two sets of three scalar equations, from which all the other equations are formed. These sets are written as two three-dimensional vector equations, denoted P (corresponding to the third column) and Q (corresponding to the fourth column):

$$P \equiv \begin{bmatrix} P_{1l} = P_{1r} \\ P_{2l} = P_{2r} \\ P_{3l} = P_{3r} \end{bmatrix}; \qquad Q \equiv \begin{bmatrix} Q_{1l} = Q_{1r} \\ Q_{2l} = Q_{2r} \\ Q_{3l} = Q_{3r} \end{bmatrix} \tag{18}$$

where P_{il} and Q_{il} refer to the left-hand side and P_{ir} and Q_{ir} refer to the right-hand side of the equations.

The set of all six equations can be written in the following matrix form:

$$A\mathbf{X}_1 = B\mathbf{Y} \tag{19}$$

For a 6R manipulator, A is a 6×9 matrix whose elements are linear combinations of s_3, c_3, B is a 6×8 matrix with constant elements, and $\mathbf{X}_1$ and $\mathbf{Y}$ are 9×1 and 8×1 matrices, respectively, defined as

$$\mathbf{X}_1 = [s_4 s_5 \quad s_4 c_5 \quad c_4 s_5 \quad c_4 c_5 \quad s_4 \quad c_4 \quad s_5 \quad c_5 \quad 1]^T \tag{20}$$

$$\mathbf{Y} = [s_1 s_2 \quad s_1 c_2 \quad c_1 s_2 \quad c_1 c_2 \quad s_1 \quad c_1 \quad s_2 \quad c_2]^T \tag{21}$$

Step 3

Construct the following scalar and vector equations to obtain eight new scalar equations:

$$Q \cdot Q \equiv [Q_{1l}^2 + Q_{2l}^2 + Q_{3l}^2 = Q_{1r}^2 + Q_{2r}^2 + Q_{3r}^2] \tag{22}$$

$$P \cdot Q \equiv [P_{1l}Q_{1l} + P_{2l}Q_{2l} + P_{3l}Q_{3l} = P_{1r}Q_{1r} + P_{2r}Q_{2r} + P_{3r}Q_{3r}] \tag{23}$$

$$P \times Q \equiv \begin{bmatrix} P_{2l}Q_{3l} - P_{3l}Q_{2l} = P_{2r}Q_{3r} - P_{3r}Q_{2r} \\ P_{3l}Q_{1l} - P_{1l}Q_{3l} = P_{3r}Q_{1r} - P_{1r}Q_{3r} \\ P_{1l}Q_{2l} - P_{2l}Q_{1l} = P_{1r}Q_{2r} - P_{2r}Q_{1r} \end{bmatrix} \tag{24}$$

$$P(Q \cdot Q) - 2Q(P \cdot Q) \equiv \begin{bmatrix} P_{1l} \sum_{i=1}^{3} Q_{il}^2 - 2Q_{1l} \sum_{i=1}^{3} P_{il}Q_{il} = P_{1r} \sum_{i=1}^{3} Q_{ir}^2 - 2Q_{1r} \sum_{i=1}^{3} P_{ir}Q_{ir} \\ P_{2l} \sum_{i=1}^{3} Q_{il}^2 - 2Q_{2l} \sum_{i=1}^{3} P_{il}Q_{il} = P_{2r} \sum_{i=1}^{3} Q_{ir}^2 - 2Q_{2r} \sum_{i=1}^{3} P_{ir}Q_{ir} \\ P_{3l} \sum_{i=1}^{3} Q_{il}^2 - 2Q_{3l} \sum_{i=1}^{3} P_{il}Q_{il} = P_{3r} \sum_{i=1}^{3} Q_{ir}^2 - 2Q_{3r} \sum_{i=1}^{3} P_{ir}Q_{ir} \end{bmatrix} \tag{25}$$

These eight equations have the same functional form as P and Q. Therefore, combining all the equations generates a set of 14 nonlinear equations of the form

$$\overline{A}\mathbf{X}_1 = \overline{B}\mathbf{Y} \tag{26}$$

where $\overline{A}$ is a 14×9 matrix whose elements are linear combinations of s_3, c_3, and $\overline{B}$ is a 14×8 constant matrix.

Step 4

Use any 8 of the 14 equations in (26) to solve for $\mathbf{Y}$ in terms of $\mathbf{X}_1$. The resulting system of 6 equations takes the form

$$\Gamma_1 \mathbf{X}_1 = 0 \tag{27}$$

where Γ_1 is a 6×9 matrix. As a result, joint variables θ_1 and θ_2 are eliminated from the set of equations.

Step 5

Change Equation (27) into polynomial form by the following substitutions:

$$s_i = \frac{2x_i}{1 + x_i^2}; \qquad c_i = \frac{1 - x_i^2}{1 + x_i^2} \qquad \text{for} \quad i = 3, 4, 5 \tag{28}$$

where $x_i = \tan(\theta_i/2)$. Then multiply each equation by $(1 + x_4^2)$ and $(1 + x_5^2)$ to clear the denominators, and multiply the first four equations by $(1 + x_3^2)$ to obtain the following form:

$$\Gamma_2 \mathbf{X}_2 = 0 \tag{29}$$

where Γ_2 is a 6×9 matrix whose entries are linear combinations of s_3, c_3. For a general 6R manipulator, the vector $\mathbf{X}_2$ is

$$\mathbf{X}_2 = [x_4^2 x_5^2 \quad x_4^2 x_5 \quad x_4^2 \quad x_4 x_5^2 \quad x_4 x_5 \quad x_5^2 \quad x_5 \quad 1]^T \tag{30}$$

Step 6

Multiply the 6 equations in Equation (29) by x_4 to obtain 6 more equations. The set of all 12 equations forms the following homogeneous system:

$$\Gamma \mathbf{X} = \begin{bmatrix} \Gamma_2 & \vert & [0] \\ -- & \vert & -- \\ [0] & \vert & \Gamma_2 \end{bmatrix} \mathbf{X} = 0 \tag{31}$$

where $[0]$ is the 6×9 null matrix and $\mathbf{X}$ is the vector of power products, which for a 6R manipulator is obtained as follows:

$$\mathbf{X} = [x_4^3 x_5^2 \quad x_4^3 x_5 \quad x_4^3 \quad x_4^2 x_5^2 \quad x_4^2 x_5 \quad x_4^2 \quad x_4 x_5^2 \quad x_4 x_5 \quad x_4 \quad x_5^2 \quad x_5 \quad 1]^T \tag{32}$$

Step 7

Apply the condition of having nontrivial solutions to the homogeneous system in order to obtain the characteristic equation in x_3, i.e.,

$$\det(\Gamma) = 0 \tag{33}$$

which is a polynomial of order 16 in the case of a general 6R manipulator.

Step 8

Obtain the roots of the characteristic polynomial by numerical methods. The real roots correspond to the real configurations of the inverse kinematics problem. For each value of x_3 thus obtained, the corresponding joint variable θ_3 may be computed from the formula $\theta_3 = 2 \tan^{-1}(x_3)$.

Step 9

Substitute each real value of x_3 into the coefficient matrix of Equation (31), and then solve for X to obtain unique values for x_4 and x_5, and hence θ_4 and θ_5, using $\theta_i = 2 \tan^{-1}(x_i)$.

Step 10

Substitute the values of θ_3, θ_4, and θ_5 into Equation (26) and use a subset of 8 equations to solve for Y; then use the numerical values of s_1, c_1 and s_2, c_2 to obtain θ_1 and θ_2, respectively.

Step 11

Substitute values of θ_1, θ_2, θ_3, θ_4, and θ_5 into the first and second entries of the first column of the following kinematics relationship:

$$H_{56} = H_{45}^{-1} H_{34}^{-1} H_{23}^{-1} H_{12}^{-1} H_{01}^{-1} H_{06} \tag{34}$$

to obtain two linear equations in s_6 and c_6 from which a unique value for θ_6 is obtained.

For 5R1P manipulators, the above algorithm remains unchanged. However for a prismatic joint k, $\sin \theta_k$ is replaced by d_k, and $\cos \theta_k$ is replaced by $(d_k)^2$. For 4R2P manipulators, there are fewer power products and therefore fewer equations are required, leading to a characteristic polynomial of order 8. However, the procedure is essentially the same as above (Kohli and Osvatic, 1992). For 3R3P manipulators, the procedure simplifies considerably, leading to a characteristic polynomial of order 2 (Kohli and Osvatic, 1992).

2.2.3 Repeatability, Accuracy, and Computational Considerations

Industrial robots are rated on the basis of their ability to return to the same location when repetitive motion is required. The locations attained for any number of repeated motions may not be identical. The *repeatability* of the manipulator is the expected maximum error at any attained location with respect to an *average* attained location of repeated motions. Manipulators with low joint backlash, friction, and flexibility usually have high repeatability.

Another robot rating criterion is the precision with which a desired location can be attained. This is called the *accuracy* of the manipulator, and it is the error between the desired and attained locations. The accuracy can be enhanced if two major obstacles are overcome. First, due to manufacturing errors in the machining and assembly of manipulators, the physical kinematics parameters may differ from the design parameters, which can produce significant errors between the actual and predicted locations of the end-effector. To solve this problem, calibration techniques are devised to improve accuracy through estimation of the individual kinematics parameters (Hollerbach, 1989). The second difficulty is that the numerical algorithms for solving the inverse kinematics problem are not efficient for practical implementations. For instance, for a general 6R manipulator the algorithm illustrated in Section 2.2.2 takes on average 10 seconds of CPU on an IBM 370-3090 using double precision arithmetic (Wampler and Morgan, 1991), while a speed on the order of milliseconds would be required in real-time applications. Furthermore, the problem of computing the roots of a characteristic polynomial of degree 16 can be ill-conditioned (Wilkinson, 1959). Closed-form solutions are quite efficient, but they exist only for a few special manipulators. This is one reason most industrial manipulators are limited to simple configurations and inverse kinematics calculations are not performed on-line. Recently some efficient algorithms have been suggested for the general solution of the inverse kinematics problem for 6-d.o.f. manipulators (Manocha and Canny, 1994). These algorithms require that the following operations be performed.

Off-line Symbolic Formulation and Numeric Substitution

For any class of manipulators, symbolic preprocessing can be performed to obtain the entries of matrices $\bar{A}$, $\bar{B}$, and Γ in Equations (26) and (31) as functions of kinematic parameters and elements of the end-effector homogeneous matrix H_{06}. The symbolic derivation and simplification can be performed using MATHEMATICA or MAPLE. Then, the numerical values of the kinematic parameters of a particular manipulator can be substituted in advance into the functions representing the elements of matrices $\bar{A}$, $\bar{B}$, and Γ. Given the desired location of the end-effector, the remaining numerical substitutions are performed on-line.

Changing the Problem of Finding the Roots of the Characteristic Polynomial to an Eigenvalue Problem

A more efficient method of finding nontrivial solutions for the matrix Equation (31) is to set up an equivalent eigenvalue problem (Ghasvini, 1993), rather than expanding the determinant of the coefficient matrix. The matrix Γ in Equation (31) can be expressed as (Ghasvini, 1993)

$$\Gamma = Lx_3^2 + Nx_3 + K \qquad (35)$$

where L, N, and K are 12×12 matrices. For a nonsingular matrix L, the following 24×24 matrix is constructed:

$$\Pi = \begin{bmatrix} [0] & | & [I] \\ \text{---} & | & \text{---} \\ -L^{-1}K & | & -L^{-1}N \end{bmatrix} \qquad (36)$$

where $[0]$ and $[I]$ are 12×12 null and identity matrices, respectively. The eigenvalues of Π are equal to the roots of Equation (33). Moreover, the eigenvector of Π corresponding to the eigenvalue x_3 has the structure

$$V = \begin{bmatrix} \mathbf{X} \\ x_3\mathbf{X} \end{bmatrix} \qquad (37)$$

which can be used to compute the solutions of Equation (27). If the matrix L is singular, then the following two matrices are constructed:

$$\Pi_1 = \begin{bmatrix} [I] & [0] \\ [0] & L \end{bmatrix}; \qquad \Pi_2 = \begin{bmatrix} [0] & [I] \\ -K & -N \end{bmatrix} \qquad (38)$$

The roots of Equation (33) are equal to the eigenvalues of the generalized eigenvalue problem $(\Pi_1 - x_3\Pi_2)$, which has the same eigenvalues as Equation (36).

2.3 Velocity Kinematics

2.3.1 Link Velocity, Jacobian

In addition to the kinematic relationship between joint displacements and the end-effector location, the relationship between the joint velocity vector and end-effector linear and angular velocities is also useful. The absolute velocities (i.e., relative to the base coordinate frame) of the link i coordinate frame are computed from the absolute velocities of the link $i - 1$ frame as follows:

If the ith joint is revolute, i.e., $\dot{q}_i = \dot{\theta}_i$,

$$\boldsymbol{v}_i = \boldsymbol{v}_{i-1} + \omega_i \times \mathbf{p}_{i-1,i}$$
$$\omega_i = \omega_{i-1} + \mathbf{z}_{i-1}\dot{q}_i$$

(39)

and if the ith joint is prismatic, i.e., $\dot{q}_i = \dot{d}_i$,

$$\boldsymbol{v}_i = \boldsymbol{v}_{i-1} + \omega_i \times \mathbf{p}_{i-1,i} + \mathbf{z}_{i-1}\dot{q}_i$$
$$\omega_i = \omega_{i-1}$$

(40)

where ω_i is the angular velocity vector of frame F_i (attached to link i), $\boldsymbol{v}_i$ is the linear velocity vector of its origin, and $\mathbf{z}_i$ is the unit vector along the axis of joint i.

Combining Equations (39) and (40), the relationship between the link frame and joint velocities is obtained as follows (Whitney, 1969):

$$\mathbf{V}_i = \begin{bmatrix} \boldsymbol{v}_i \\ \omega_i \end{bmatrix} = J_i(q_1, q_2, \ldots, q_{i-1})\dot{\mathbf{q}} \tag{41}$$

The $n \times 1$ matrix $\dot{\mathbf{q}} = [\dot{q}_1 \ \dot{q}_2 \cdots \dot{q}_n]^T$ consists of joint velocities, and J_i is a $6 \times n$ "Jacobian" of link i defined as follows:

$$J_i(q_1, q_2, \ldots, q_{i-1}) = \begin{bmatrix} \mathbf{t}_0 \ \mathbf{t}_1 \cdots \mathbf{t}_{i-1} & \underbrace{[0] \ [0] \cdots [0]}_{(n - i) \text{ columns}} \end{bmatrix} \tag{42}$$

where $[0]$ is the 6×1 null matrix and other columns of the Jacobian are of the form

$$\begin{cases} \mathbf{t}_j = \begin{bmatrix} \mathbf{z}_j \times \mathbf{p}_{j,i} \\ \mathbf{z}_j \end{bmatrix} & \text{for joint } j \text{ revolute} \\ \\ \mathbf{t}_j = \begin{bmatrix} \mathbf{z}_j \\ [0] \end{bmatrix} & \text{for joint } j \text{ prismatic} \end{cases} ; \quad j = 0, 1, 2, \ldots, i - 1 \tag{43}$$

with $[0]$ as the 3×1 null matrix.

The robot Jacobian J_n (or simply the Jacobian J) is obtained from Equations (42) and (43) with i equal to n (the total number of joints):

$$J(\mathbf{q}) = [\mathbf{t}_0 \ \mathbf{t}_1 \cdots \mathbf{t}_{n-2} \ \mathbf{t}_{n-1}] \tag{44}$$

and therefore the end-effector linear and angular velocities are obtained from the linear mapping

$$\mathbf{V} = J(\mathbf{q})\dot{\mathbf{q}} \tag{45}$$

Numerical computation of the Jacobian depends on the frame in which the vectors in Equation (43) are expressed. If all vectors are taken in the base frame, the resulting Jacobian matrix denoted by J^0 is represented in the base frame, and so are the absolute velocities of the end-effector from Equation (45). A recursive formulation of J^0 is readily obtained from the recursive formulation of the forward kinematics by adding the following step to the backward loop of Algorithm 1:

$$
\textbf{4. CALCULATE:}\quad \mathbf{t}_{i-1} =
\begin{cases}
\begin{bmatrix}
a^y_{i-1,n}P^z_{i-1,n} - a^z_{i-1,n}P^y_{i-1,n} \\
a^z_{i-1,n}P^x_{i-1,n} - a^x_{i-1,n}P^z_{i-1,n} \\
a^x_{i-1,n}P^y_{i-1,n} - a^y_{i-1,n}P^x_{i-1,n} \\
a^x_{i-1,n} \\
a^y_{i-1,n} \\
a^z_{i-1,n}
\end{bmatrix}
& \text{if joint } i - 1 \text{ is revolute} \\[2em]
\begin{bmatrix}
a^x_{i-1,n} \\
a^y_{i-1,n} \\
a^z_{i-1,n} \\
0 \\
0 \\
0
\end{bmatrix}
& \text{if joint } i - 1 \text{ is prismatic}
\end{cases}
$$

$$(46)$$

The absolute velocities of the end-effector can also be expressed in its own frame, with the Jacobian being computed in this frame (denoted by J^n). The Jacobian matrix J^n can be directly obtained from the following formulation:

$$
J^n =
\begin{bmatrix}
R^T_{0n} & | & [0] \\
\hline
[0] & | & R^T_{0n}
\end{bmatrix}
J^0
\tag{47}
$$

As in the forward kinematics problem, a symbolic formulation is recommended for real-time tasks.

2.3.2 Link Acceleration

Link accelerations are computed the same way as link velocities. The absolute linear and angular accelerations (i.e., relative to the base coordinate frame) of the link i coordinate frame are obtained from the absolute accelerations of the link $i - 1$ frame as follows:

for joint i revolute, i.e., $\dot{q}_i = \dot{\theta}_i$ and $\ddot{q}_i = \ddot{\theta}_i$,

$$
\dot{\boldsymbol{v}}_i = \dot{\boldsymbol{v}}_{i-1} + \dot{\boldsymbol{\omega}}_i \times \mathbf{p}_{i-1,i} + \boldsymbol{\omega}_i \times (\boldsymbol{\omega}_i \times \mathbf{p}_{i-1,i})
$$
$$
\dot{\boldsymbol{\omega}}_i = \dot{\boldsymbol{\omega}}_{i-1} + \mathbf{z}_{i-1}\ddot{q}_i + \boldsymbol{\omega}_i \times (\mathbf{z}_{i-1}\dot{q}_i)
\tag{48}
$$

and for joint i prismatic, i.e., $\dot{q}_i = \dot{d}_i$ and $\ddot{q}_i = \ddot{d}_i$,

$$
\dot{\boldsymbol{v}}_i = \dot{\boldsymbol{v}}_{i-1} + \dot{\boldsymbol{\omega}}_i \times \mathbf{p}_{i-1,i} + \boldsymbol{\omega}_i \times (\boldsymbol{\omega}_i \times \mathbf{p}_{i-1,i}) + \mathbf{z}_{i-1}\ddot{q}_i + 2\boldsymbol{\omega}_i \times (\mathbf{z}_{i-1}\dot{q}_i)
$$
$$
\dot{\boldsymbol{\omega}}_i = \dot{\boldsymbol{\omega}}_{i-1}
\tag{49}
$$

The above equations can be written in compact form as

$$
\dot{\mathbf{V}}_i =
\begin{bmatrix}
\dot{\boldsymbol{v}}_i \\
\boldsymbol{\omega}_i
\end{bmatrix}
= J_i(q_1, q_2, \ldots, q_{i-1})\ddot{\mathbf{q}} + \dot{J}_i(q_1, \ldots, q_{i-1}, \dot{q}_1, \ldots, \dot{q}_{i-1})\dot{\mathbf{q}}
\tag{50}
$$

2.3.3 Singularity

Generally, the link frames velocities corresponding to a particular set of end-effector linear and angular velocities can be obtained from the inverse of the linear mapping (45) as

$$
\dot{\mathbf{q}} = J^{-1}(\mathbf{q})\mathbf{V}
\tag{51}
$$

The linear mapping (51) exists only for configurations at which the inverse of the Jacobian matrix exists, i.e., *nonsingular* configurations. In a *singular configuration,* the end-effector cannot move in certain direction(s); thus the manipulator loses one or more degrees of freedom. In a singular configuration, the Jacobian rank decreases, i.e., two or more columns of J become linearly dependent; thus the determinant of the Jacobian becomes zero. This is a computational test for the existence of singular configurations. Singular configurations should usually be avoided because most of the manipulators are designed for tasks in which all degrees of freedom are required. Furthermore, near singular configurations the joint velocities required to maintain the desired end-effector velocity in certain directions may become extremely large. The most common singular configurations for 6-d.o.f. manipulators are listed below (Murray, Li, and Sastry, 1994):

1. *Two collinear revolute joint axes* occur in spherical wrist assemblies that have three mutually perpendicular axes intersecting at one point. Rotating the second joint may align the first and third joints, and then the Jacobian will have two linearly dependent columns. Mechanical restrictions are usually imposed on the wrist design to prevent the wrist axes from generating a singularity of the wrist.

2. *Three parallel coplanar revolute joint axes* occur, for instance, in an elbow manipulator (Murray, Li, and Sastry, 1994, p. 90) that consists of a 3-d.o.f. manipulator with a spherical wrist when it is fully extended or fully retracted.

3. *Four revolute joint axes intersecting in one point.*

4. *Four coplanar revolute joints.*

5. *Six revolute joints intersecting along a line.*

6. *A prismatic joint axis perpendicular to two parallel coplanar revolute joints.*

In addition to the Jacobian singularities, the motion of the manipulator is restricted if the joint variables are constrained to a certain interval. In this case, a reduction in the number of degrees of freedom may occur when one or more joints reach the limit of their allowed motion.

2.3.4   Redundant Manipulators and Multiarm Robots

A *kinematically redundant* manipulator is one that has more than the minimum number of degrees of freedom required to attain a desired location. In this case, an infinite number of configurations can be obtained for a desired end-effector location. Multi-arm robots are a special class of redundant manipulators. When two or more robot arms are used to perform a certain task cooperatively, an increased load-carrying capacity and manipulation capability are achieved. For a general redundant manipulator with n degrees of freedom $(n > 6)$, the Jacobian is not square, and there are only $(n - 6)$ arbitrary variables in the general solution of mapping (51), assuming that the Jacobian is full rank. Additional constraints are needed to limit the solution to a unique one. Therefore, redundancy provides the opportunity for choice or decision. It is typically used to optimize some secondary criteria while achieving the primary goal of following a specified end-effector trajectory. The secondary criteria considered so far in the literature are robot singularity and obstacle avoidance, minimization of joint velocity and torque, increasing the system precision by an optimal distribution of arm compliance, and improving the load-carrying capacity by optimizing the transmission ratio between the input torque and output forces. Hayward (1988) gives a review of the different criteria. As an example, one common approach is to choose the minimum (in the least squares sense) joint velocities that provide the desired end-effector motion. This is achieved by choosing (Hollerbach, 1984)

$$\dot{\mathbf{q}} = J*(\mathbf{q})\mathbf{V} \tag{52}$$

where
$$J*(\mathbf{q}) = J^T(JJ^T)^{-1} \tag{53}$$

is the *pseudo-inverse* of the Jacobian matrix.

2.4   Parallel Manipulators

Parallel manipulators are mechanisms that contain two or more serial chains connecting the end-effector to the base. Generally, parallel manipulators can offer more accuracy in positioning and orienting objects than open-chain manipulators. They can also possess a high payload/weight ratio, and they are easily adaptable to position and force control. On the other hand, the workspace of parallel manipulators is usually smaller than that of serial manipulators. A typical configuration of parallel manipulators is the so-called *in-parallel* manipulators, in which each serial chain has the same structure and one joint is actuated in each chain (Duffy, 1996). A popular example of this type is the Stewart platform, shown in Figure 3. The mechanism consists of a rigid plate connected to the base by a set of prismatic joints, each of which is connected to the plate and the base by spherical joints, allowing complete rotational motion. Only the prismatic joints are actuated.

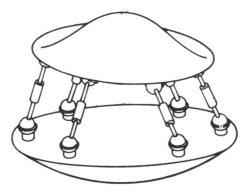

Figure 3   A general form of the Stewart platform.

The forward kinematics of a parallel manipulator can be expressed by equating the end-effector location to each individual chain. Consider a parallel manipulator with m chains so that each chain contains n_i ($i = 1, 2, \ldots, m$) joints. The forward kinematics model can then be described as

$$H_{0e} = H_{01}H_{12}^1 \cdots H_{n_1-1,n_1}^1 = H_{01}^2 H_{12}^2 \cdots H_{n_2-1,n_2}^2 = \cdots = H_{01}^m H_{12}^m \cdots H_{n_m-1,n_m}^m \quad (54)$$

where H_{ij}^k is the homogeneous transformation matrix between joints i and j in the chain k. All quantities are specified with respect to a unique base and end-effector coordinate frame. Equation (54), called the *structure equation,* introduces constraints between the joint displacements of the manipulator. As a result, unlike for serial manipulators, the joint space for a parallel manipulator is not the Cartesian product of the individual joint spaces but a subset of it that satisfies Equation (54). In a parallel manipulator, if N and L are the number of joints and links, respectively, and l_i is the number of degrees of freedom of the ith joint, then the number of degrees of freedom of the manipulator can be obtained by taking the total number of degrees of freedom for all links and subtracting the number of constraints imposed by the joints attached to the links. For the specific case where all joints apply independent constraints, the number of degrees of freedom F can be calculated as

$$F = 6L - \sum_{i=1}^{N} (6 - l_i) = 6(L - N) + \sum_{i=1}^{N} l_i \quad (55)$$

The inverse kinematics problem is no more difficult for a parallel manipulator than for the open-chain case, as each chain can be analyzed separately.

The end-effector velocity of a parallel manipulator can be obtained from each chain equivalently:

$$\mathbf{V} = J_1 \dot{\mathbf{q}}_1 = J_2 \dot{\mathbf{q}}_2 = \cdots = J_m \dot{\mathbf{q}}_m \quad (56)$$

Obviously not all joint velocities can be specified independently. The relationship between joint torques and end-effector forces is more complex in a parallel manipulator than in a serial manipulator, as there are internal interactions between forces produced by the different chains in a parallel manipulator. The reader is referred to more detailed sources, such as Duffy (1996), and open-chain manipulators will be considered in the sequel.

3   STATICS

This section discusses the relationship between the forces and moments that act on the manipulator when it is at rest. In open-chain manipulators, each joint is usually driven by an individual actuator. The corresponding input joint torque (for revolute joints) or force (for prismatic joints) is transmitted through the manipulator arm linkages to the end-effector, where the resultant force and moment balance an external load. The relationship between the actuator drive torques (or forces) and the end-effector resultant force and moment is determined using the manipulator Jacobian (Asada and Slotine, 1986):

$$\tau = (J^e)^T \begin{bmatrix} \mathbf{f}_e \\ \mathbf{g}_e \end{bmatrix} = (J^e)^T \mathbf{G}_e \qquad (57)$$

where $\tau \in \mathbb{R}^n$ is the vector of joint torques (and forces) and $\mathbf{f}_e \in \mathbb{R}^3$ and $\mathbf{g}_e \in \mathbb{R}^3$ are vectors of the resulting reaction force and moment, respectively, of the external loads acting on the end-effector and expressed in the end-effector frame:

$$\mathbf{f}_e = -\mathbf{f}_{ext}; \qquad \mathbf{g}_e = -\mathbf{g}_{ext} \qquad (58)$$

The *generalized force* $\mathbf{G}_e \in \mathbb{R}^6$, which consists of the force/moment pair, is called the *wrench* vector.

3.1   Duality Between the Velocity Kinematics and Statics

The statics (57) is closely related to the velocity kinematics (45) because the manipulator Jacobian is being used for both mappings. In a specific configuration, the kinematics and statics linear mappings can be represented by the diagram shown in Figure 4 (Asada and Slotine, 1986). For the velocity kinematics, the Jacobian is a linear mapping from the n-dimensional vector space $\mathbb{R}^n$ to the six-dimensional space $\mathbb{R}^6$. Note that n (number of joints) can be more than six in the case of redundant manipulators. The range subspace $R(J)$ represents all possible end-effector velocities that can be generated by the n joint velocities in the present configuration. If $n > 6$, there exists a null space $N(J)$ of the Jacobian mapping that corresponds to all joint velocity vectors $\dot{\mathbf{q}}$ that produce no net velocity at the end-effector. In a singular configuration, J is not full rank, and the subspace $R(J)$ does not cover the entire vector space $\mathbb{R}^6$, i.e., there exists at least one direction in which the end-effector can not be moved.

The statics relationship is also a linear mapping from $\mathbb{R}^6$ to $\mathbb{R}^n$ provided by the transpose of the Jacobian. The range subspace $R(J^T)$ and null subspace $N(J^T)$ can be identified from the Jacobian mapping. The null subspace $N(J^T)$ corresponds to the end-effector wrenches that can be balanced without input torques or forces at the joints, as the load is borne entirely by the structure of the arm linkages. For a redundant manipulator, the range subspace $R(J^T)$ does not cover the entire space $\mathbb{R}^n$, and there are some sets of input joint torques or forces that cannot be balanced by any end-effector wrench. These configurations correspond to the null space of the kinematics mapping $N(J)$ that contains the joint velocity vectors that produce no end-effector motion.

The velocity kinematics and statics are dual concepts that can be stated as follows:

1.   In a singular configuration, there exists at least one direction in which the end-effector can not be moved. In this direction, the end-effector wrench is entirely balanced by the manipulator structure and does not require any input joint torque or force.

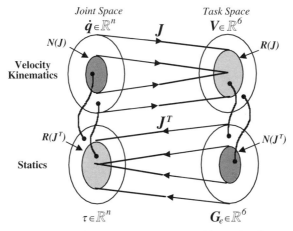

Figure 4   The duality relation between statics and velocity kinematics linear mappings (from H. Asada and J. J. E. Slotine, *Robot Analysis and Control*. New York: John Wiley & Sons, Inc., 1986).

2. In each configuration of a redundant manipulator, there is at least one direction in which joint velocities produce no end-effector velocity. In this direction, the joint torques and forces cannot be balanced by any end-effector wrench. Therefore, in order to maintain a stationary arm configuration, no input joint torque or force that generates end-effector wrench should be applied.

3. For a general manipulator, in each configuration the directions of possible motion of the end-effector also define the directions in which wrenches that are applied to the end-effector can be entirely balanced by the joint torques and forces.

4   DYNAMICS

The dynamics model describes the balance between internal and external loads applied to the manipulator. The input joint torques and forces generated by the actuators balance the other external forces and moments and the internal loads of the manipulator. The dynamics model is fundamental in mechanical and control system design and simulation of motion and also for real-time control (model-based control). The dynamics of a manipulator with n degrees of freedom can be expressed using the following equation of motion of each link i (Spong and Vidyasagar, 1989):

$$\tau_i + [J^T \, \mathbf{G}_{ext}]_i - N_i(\mathbf{q}, \dot{\mathbf{q}}) = \sum_{j=1}^{n} M_{ij}(\mathbf{q})\ddot{q}_j + \sum_{k=1}^{n} \sum_{j=1}^{n} h_{ijk}(\mathbf{q})\dot{q}_j\dot{q}_k; \qquad i = 1, 2, \ldots, n \quad (59)$$

The left-hand side of Equation (59) consists of all external forces and moments applied to the ith link, which are decomposed into three parts. The first part τ_i is the input torque (or force) applied by the actuator of joint i. The second part $[J^T\mathbf{G}_{ext}]_i$ is the mapping of the external wrench $\mathbf{G}_{ext}$ to the joint space, in particular the effect of the external wrench on link i. The third part $-N_i(\mathbf{q}, \dot{\mathbf{q}})$ represents any other external force and moment that act on the ith link, including gravity torque, friction, etc.

The right-hand side of (59) contains the reaction (internal) loads of the manipulator. The first term represents the inertial load on the ith link, while the second term accounts for the Coriolis effect (for $j \neq k$) and centrifugal effect (for $j = k$). The entries $[M_{ij}]$ of the manipulator inertia matrix are configuration-dependent, and the off-diagonal elements ($i \neq j$) generate the coupling inertial loads. Coriolis and centrifugal terms $h_{ijk}(\mathbf{q})$ also depend on configuration and introduce further interaction effects in the manipulator dynamics. The components of $h_{ijk}(\mathbf{q})$ are defined as follows (Spong and Vidyasagar, 1989):

$$h_{ijk} = \frac{\partial M_{ij}}{\partial q_k} - \frac{1}{2} \frac{\partial M_{jk}}{\partial q_i}. \qquad (60)$$

Equation (59) can be rewritten in compact form as

$$\tau + J^T \, \mathbf{G}_{ext} = M(\mathbf{q})\ddot{\mathbf{q}} + C(\mathbf{q}, \dot{\mathbf{q}})\dot{\mathbf{q}} + N(\mathbf{q}, \dot{\mathbf{q}}) \qquad (61)$$

where the elements of matrix C are defined as

$$C_{ij}(\mathbf{q}, \dot{\mathbf{q}}) = \sum_{k=1}^{n} h_{ijk}\dot{q}_k \qquad i, j = 1, 2, \ldots, n \qquad (62)$$

The matrices M and C express the inertial characteristics of the manipulator and have the following important properties (Spong and Vidyasagar, 1989):

1. $M(\mathbf{q})$ is symmetric and bounded positive definite.
2. $(\dot{M} - 2C) \in \mathbb{R}^{n \times n}$ is a skew-symmetric matrix.

These properties are useful in reducing the number of operations in the calculation of the dynamics model, and they are also used in proofs of stability of many control laws for robot manipulators.

The manipulator dynamics model can be used in two modes:

1. To simulate the motion of the manipulator (joint displacements, velocities, and accelerations) for given input joint torques and forces; this is referred to as the *forward dynamics* problem

   2. To calculate the joint torques and forces required to attain specific joint displacements, with given velocities and accelerations; this is called the *inverse dynamics* problem.

The computational algorithms of the above dynamics problems and related considerations are discussed in the sequel.

4.1   Inverse Dynamics

4.1.1   The Recursive Algorithm and Computational Considerations

Given the joint displacements, velocities, and accelerations, the joint torques can be directly computed from Equation (61). However, since at each trajectory point the configuration-dependent coefficients M, C, and N must be computed, the amount of computation required is extremely high, and it increases very rapidly as the number of degrees of freedom n increases. Calculating Equation (61) requires $(32n^4 + 86n^3 + 171n^2 + 53n - 128)$ multiplications and $(25n^4 + 66n^3 + 129n^2 + 42n - 96)$ additions, or more than 100,000 arithmetic operations for a 6-d.o.f. manipulator. This heavy computational load is a bottleneck for the use of the inverse dynamics model in real-time control, especially since the calculation would have to be repeated at the rate of 60 Hz or higher.

   If the dynamics equations are formulated in a *recursive* form, the computational complexity can be significantly reduced from $O(n^4)$ to $O(n)$ so that the required number of operations would vary linearly with the number of degrees of freedom. Most of the fast algorithms for the inverse dynamics problem are based on the Newton–Euler approach, and consist of two main steps. In the first step, the angular velocity and acceleration of each link of the manipulator and the linear velocity and acceleration of its center of mass are calculated starting with the first link and continuing through the last. In the second step, the force and moment exerted on each link are calculated backward starting from the last link. Furthermore, if the dynamic equations are expressed in a tensor form and tensor properties are used, the computational complexity of the inverse dynamics algorithm can be further reduced. Algorithm 2 presents an efficient inverse dynamics formulation that requires $(104n - 77)$ multiplications and $(92n - 70)$ additions for computing the joint torques or forces for each trajectory point. The basic formulation of this algorithm was originally developed in Balafoutis, Patel, and Misra (1988). Although this algorithm is not the fastest available formulation, it is simple and effective for real-time implementations. See He and Goldenberg (1990) for more details on computational considerations.

   While the current recursive formulations are quite efficient, symbolic closed-form equations derived for a particular manipulator are likely to be the most efficient formulation (Burdick, 1986). If symbolic operations are used simplified formulations can easily be obtained for real-time applications.

4.2   Forward Dynamics

In forward dynamics, joint displacements and velocities are obtained as a result of applied joint torques (forces) and other external forces and moments. From Equation (61), given the current joint positions and velocities and the current external wrenches, the vector of joint accelerations can be calculated. From this calculation the corresponding joint velocities and positions are computed by subsequent integrations. In this way, the problem of forward dynamics is divided into two phases:

   Phase 1 Obtain joint accelerations from Equation (61).
   Phase 2 Integrate the joint accelerations and obtain the new joint velocities; then integrate again and obtain the new joint displacements.

In Phase 2 any suitable numerical integration method can be used. In Phase 1 the following form of Equation (61) is used:

$$M(\mathbf{q})\ddot{\mathbf{q}} = [C(\mathbf{q}, \dot{\mathbf{q}})\dot{\mathbf{q}} + N(\mathbf{q}, \dot{\mathbf{q}}) - J^T\,\mathbf{G}_{ext}] - \tau \qquad (63)$$

Phase 1 can be performed by completing the following three steps:

   Step 1.1. Calculate the elements of $[C(\mathbf{q}, \dot{\mathbf{q}})\dot{\mathbf{q}} + N(\mathbf{q}, \dot{\mathbf{q}}) - J^T\mathbf{G}_{ext}]$
   Step 1.2. Calculate the elements of the inertia matrix $M(\mathbf{q})$.

Step 1: Forward Kinematics
OBTAIN link rotation matrices $R_{i-1,i}$ for $i = 1, 2, \ldots, n$
SET $R_{n,n+1} = [I]_{3\times 3}$

Step 2: Initialization

SET $\begin{cases} \boldsymbol{\omega}_0 = \dot{\boldsymbol{\omega}}_0 = \mathbf{v}_0 = [0 \quad 0 \quad 0]^T; \quad \dot{\mathbf{v}}_0 = -\mathbf{g}; \quad \tilde{\omega}_0 = [0]_{3\times 3} \\ \mathbf{f}_{n+1} = -\mathbf{f}_{ext}; \quad \mathbf{g}_{n+1} = -\mathbf{g}_{ext}; \quad m_i; \quad I_{ci}; \quad \mathbf{p}_{ci} \end{cases}$

Step 3: Forward Loop
FOR $i = 1$ to n

CALCULATE: $\begin{cases} \begin{cases} \boldsymbol{\omega}_i = R_{i-1,i}^T(\boldsymbol{\omega}_{i-1} + [0 \quad 0 \quad \dot{q}_i]^T) & \text{if joint } i \text{ is revolute} \\ \dot{\boldsymbol{\omega}}_i = R_{i-1,i}^T(\dot{\boldsymbol{\omega}}_{i-1} + \tilde{\omega}_{i-1}[0 \quad 0 \quad \dot{q}_i]^T + [0 \quad 0 \quad \ddot{q}_i]^T) \end{cases} \\ \begin{cases} \boldsymbol{\omega}_i = R_{i-1,i}^T \boldsymbol{\omega}_{i-1} \\ \dot{\boldsymbol{\omega}}_i = R_{i-1,i}^T \dot{\boldsymbol{\omega}}_{i-1} \end{cases} & \text{if joint } i \text{ is prismatic} \end{cases}$

SET: $\tilde{\omega}_i = \begin{bmatrix} 0 & -\omega_i^z & \omega_i^y \\ \omega_i^z & 0 & -\omega_i^x \\ -\omega_i^y & \omega_i^x & 0 \end{bmatrix}$; $\tilde{\dot{\omega}}_i = \begin{bmatrix} 0 & -\dot{\omega}_i^z & \dot{\omega}_i^y \\ \dot{\omega}_i^z & 0 & -\dot{\omega}_i^x \\ -\dot{\omega}_i^y & \dot{\omega}_i^x & 0 \end{bmatrix}$

CALCULATE: $\begin{cases} \Omega_i = \tilde{\dot{\omega}}_i + \tilde{\omega}_i \tilde{\omega}_i \\ \begin{cases} \dot{\mathbf{v}}_i = R_{i-1,i}^T(\dot{\mathbf{v}}_{i-1} + \Omega_i \mathbf{p}_{i-1,i}) & \text{if joint } i \text{ is revolute} \\ \dot{\mathbf{v}}_i = R_{i-1,i}^T(\dot{\mathbf{v}}_{i-1} + \Omega_i \mathbf{p}_{i-1,i} + 2\tilde{\omega}_i[0 \quad 0 \quad \dot{q}_i]^T + [0 \quad 0 \quad \ddot{q}_i]^T) & \text{if joint } i \text{ is prismatic} \end{cases} \\ \dot{\mathbf{v}}_{ci} = \dot{\mathbf{v}}_i + \Omega_i \mathbf{p}_{ci} \\ \tilde{g}_{ci} = -(\Omega_i I_{ci}) + (\Omega_i I_{ci})^T \\ \mathbf{f}_{ci} = m_i \dot{\mathbf{v}}_{ci} \end{cases}$

SET: $\mathbf{g}_{ci} = [-\tilde{g}_{ci}(1, 2) \quad \tilde{g}_{ci}(1, 3) \quad -\tilde{g}_{ci}(2, 3)]^T$

NEXT i

Step 4: Backward Loop
FOR $i = n$ to 1

SET: $\bar{P}_{i-1,i} = \begin{bmatrix} 0 & -p_{i-1,i}^z & p_{i-1,i}^y \\ p_{i-1,i}^z & 0 & -p_{i-1,i}^x \\ -p_{i-1,i}^y & p_{i-1,i}^x & 0 \end{bmatrix}$; $\tilde{P}_{ci} = \begin{bmatrix} 0 & -p_{ci}^z - p_{i-1,i}^z & p_{ci}^y + p_{i-1,i}^y \\ p_{ci}^z + p_{i-1,i}^z & 0 & -p_{ci}^x - p_{i-1,i}^x \\ -p_{ci}^y - p_{i-1,i}^y & p_{ci}^x + p_{i-1,i}^x & 0 \end{bmatrix}$

CALCULATE: $\begin{cases} \mathbf{f}_i = R_{i,i+1} \mathbf{f}_{i+1} + \mathbf{f}_{ci} \\ \mathbf{g}_i = R_{i,i+1} \mathbf{g}_{i+1} + \mathbf{g}_{ci} + \tilde{P}_{ci} \mathbf{f}_{ci} + \tilde{P}_{i+1,i} \mathbf{f}_{i+1} \end{cases}$

CALCULATE: $\begin{cases} \tau_i = [0 \quad 0 \quad 1]R_{i-1,i} \mathbf{g}_i & \text{if joint } i \text{ is revolute} \\ \tau_i = [0 \quad 0 \quad 1]R_{i-1,i} \mathbf{f}_i & \text{if joint } i \text{ is prismatic} \end{cases}$

NEXT i

Algorithm 2 Tensor recursive formulation of the inverse dynamics problem.

Step 1.3. Solve the set of simultaneous linear equations (63) for $\ddot{\mathbf{q}}$.

Step 1.1 can be directly computed using an inverse dynamics algorithm with the joint accelerations set to zero. If the inverse dynamics algorithm is represented (Walker, 1985) as the function

$$\tau = INVDYN(\mathbf{q}, \dot{\mathbf{q}}, \ddot{\mathbf{q}}, \dot{\mathbf{v}}_0, \mathbf{G}_{ext}) \tag{64}$$

then Step 1.1 can be performed as

$$[C(\mathbf{q}, \dot{\mathbf{q}})\dot{\mathbf{q}} + N(\mathbf{q}, \dot{\mathbf{q}}) - J^T \mathbf{G}_{ext}] = INVDYN(\mathbf{q}, \dot{\mathbf{q}}, 0, \dot{\mathbf{v}}_0, \mathbf{G}_{ext}) \tag{65}$$

Using the inverse dynamics algorithm, the complexity of this step is $O(n)$.

The computation of the matrix M can also be performed using the inverse dynamics formulation. First $\dot{\mathbf{q}}$, $\dot{\mathbf{v}}_0$ and $\mathbf{G}_{ext}$ are set to be zero. Then the ith column of M, denoted by $[M]_i$, can be obtained by setting the joint acceleration in the inverse dynamics algorithm to be $\boldsymbol{\delta}_i$ which is a vector with all elements equal to zero except for the ith element being equal to one. Thus

$$[M]_i = INVDYN(\mathbf{q}, 0, \boldsymbol{\delta}_i, 0, 0) \qquad i = 1, 2, \ldots, n \qquad (66)$$

By repeating this procedure for n columns of M, the entire inertia matrix is obtained. This step requires n applications of the inverse dynamics with the order of complexity $O(n)$.

M and the right-hand side of Equation (63) having been computed, the final step is to solve the set of simultaneous linear equations for $\ddot{\mathbf{q}}$. The order of complexity of this step is $O(n^3)$, thus making the order of complexity of the overall forward dynamics algorithm $O(n^3)$. The symmetry and positive definiteness of M can be exploited in Step 1.2 to calculate only the lower triangle of the matrix and in Step 1.3 to use specialized factorization techniques in order to improve the computational efficiency of the algorithm (Featherstone, 1987). New and efficient algorithms have recently been suggested (Lilly, 1993).

5 CONCLUSIONS

This chapter presents basic kinematics, statics, and dynamics modeling algorithms. These questions have been addressed extensively in the literature, but the approach taken here attempts to extract only the fundamentals needed by a reader wishing to obtain a complete set of tools to generate models of robot manipulators. The tools suggested could be computerized without great effort, and they could be useful in the design of new manipulators or workcells, the analysis of existing manipulators, simulations, and real-time control. The procedures and algorithms suggested here are considered the most computationally effective for fulfilling the basic tasks of modeling.

Some of the issues discussed are not commonly encountered in the literature. For example, the use of screw theory and products of exponentials in kinematics modeling, and even symbolic formulations in kinematics and dynamics modeling, have limited applicability, but can be very effective.

Further study is needed to investigate the numerical accuracy and robustness of inverse kinematics and forward dynamics calculations. In particular, there are strong indications that model-based control is more effective; thus incorporation of kinematics and dynamics models into real-time control is recommended. The computational overhead resulting from this methodology requires extensive work to minimize the number of arithmetic operations and to exploit parallel processing. In addition, the accuracy of on-line calculations must be addressed. In a different direction, special-configuration manipulators not describable as standard open-loop kinematic chains must also be investigated and specific models generated for them.

REFERENCES

Asada, H., and J. J. E. Slotine. 1986. *Robot Analysis and Control.* New York: John Wiley & Sons.

Balafoutis, C. A., R. V. Patel, and P. Misra. 1988. "Efficient Modeling and Computation of Manipulator Dynamics Using Orthogonal Cartesian Tensors." *IEEE Journal of Robotics and Automation* 4(6), 665–676.

Brockett, R. W. 1983. "Robotic Manipulators and Product of Exponentials Formula." In *Proceedings of International Symposium of Mathematical Theory of Networks and Systems,* Beer Sheba, Israel.

Burdick, J. W. 1986. "An Algorithm for Generation of Efficient Manipulator Dynamic Equations." In *Proceedings of the 1986 IEEE International Conference on Robotics and Automation,* San Francisco. 212–218.

Ceccarelli, M. 1994. "Determination of the Workspace Boundary of a General n-Revolute Manipulator." In *Advances in Robot Kinematics and Computational Geometry.* Ed. A. J. Lenarcic and B. B. Ravani. Dordrecht: Kluwer Academic. 39–48.

Denavit, J., and R. S. Hartenberg. 1965. "A Kinematic Notation for Low-Pair Mechanisms Based on Matrices." *ASME Journal of Applied Mechanics* (June), 215–221.

Duffy, J. 1980. *Analysis of Mechanisms and Manipulators.* New York: John Wiley & Sons.

———. 1996. *Statics and Kinematics with Applications to Robotics.* Cambridge: Cambridge University Press.

Featherstone, R. 1987. *Robot Dynamics Algorithms.* Dordrecht: Kluwer Academic.

Ghasvini, M. 1993. "Reducing the Inverse Kinematics of Manipulators to the Solution of a Generalized Eigenproblem." In *Computational Kinematics*. Ed. J. Angeles, G. Hommel, and P. Kovacs. Dordrecht: Kluwer Academic. 15–26.

Hayward, V. 1988. "An Analysis of Redundant Manipulators from Several Viewpoints." In *Proceedings of the NATO Advanced Research Workshop on Robots with Redundancy: Design Sensing and Control*, Salo, Italy, June 27–July 1.

He, X., and A. A. Goldenberg. 1990. "An Algorithm for Efficient Manipulator Dynamic Equations." *Journal of Robotic Systems* **7**, 689–702.

Ho, C. Y., and J. Sriwattanathamma. 1989. "Symbolically Automated Direct Kinematics Equations Solver for Robotic Manipulators." *Robotica* **7**, 243–254.

Hollerbach, J. M. 1984. "Optimal Kinematic Design for a Seven Degree of Freedom Manipulator." Paper presented at International Symposium of Robotics Research, Kyoto.

———. 1989. "A Survey of Kinematic Calibration." In *The Robotics Review I*. Ed. I. Khatib, J. J. Craig, and T. Lozano-Pérez. Cambridge: MIT Press.

Kohli, D., and M. Osvatic. 1992. "Inverse Kinetics of General 4R2P, 3R3P, 4R1c, 4R2C, and 3C Manipulators." *Proceedings of 22nd ASME Mechanisms Conference*, Scottsdale, DE. Vol. 45. 129–137

Lee, H. Y., and C. G. Liang. 1988. "Displacement Analysis of the General Spatial 7-link 7R Mechanism." *Mechanism and Machine Theory* **23**, 219–226.

Lilly, K. W. 1993. *Efficient Dynamic Simulation of Robotic Mechanisms*. Dordrecht: Kluwer Academic.

Manocha, D., and J. F. Canny. 1994. "Efficient Inverse Kinematics for General 6R Manipulators." *IEEE Transactions on Robotics and Automation* **10**(5), 648–657.

Mavroidis, C., and B. Roth. 1994. "Structural Parameters Which Reduce the Number of Manipulator Configurations." *Journal of Mechanical Design* **116**, 3–10.

Murray, R. M., Z. Li, and S. S. Sastry. 1994. *A Mathematical Introduction to Robotic Manipulation*. London: CRC Press. Chapter 3.

Park, F. C., and K. Okamura. 1994. "Kinematic Calibration and the Product of Exponentials Formula." In *Advances in Robot Kinematics and Computational Geometry*. Ed. A. J. Lenarcic and B. B. Ravani. Dordrecht: Kluwer Academic. 119–128.

Pieper, D. L., and B. Roth. 1969. "The Kinematics of Manipulators under Computer Control." In *Proceedings of 2nd International Congress for the Theory of Machines and Mechanisms*, Zakopane, Poland. Vol. 2. 159–168.

Raghavan, M., and B. Roth. 1993. "Inverse Kinematics of the General 6R Manipulator and Related Linkages." *Transactions of the ASME* **115**, 502–508.

Ruoff, C. 1981. "Fast Trigonometric Functions for Robot Control." *Robotics Age* (November).

Salmon, G. 1964. *Lessons Introductory to the Modern Higher Algebra*. New York: Chelsea.

Selfridge, R. G. 1989. "Analysis of 6-Link Revolute Arms." *Mechanism and Machine Theory* **24**(1), 1–8.

Spong, M. W., and M. Vidyasagar. 1989. *Robot Dynamics and Control*. New York: John Wiley & Sons.

Turner, T., J. Craig, and W. Gruver. 1984. "A Microprocessor Architecture for Advanced Robot Control." Paper presented at 14th ISIR, Stockholm, October.

Vukobratovic, M., and M. Kircanski. 1986. *Kinematics and Trajectory Synthesis of Manipulation Robot*. New York: Springer-Verlag. 53–77.

Walker, M. W. 1985. "Kinematics and Dynamics." In *Handbook of Industrial Robots*. 1st ed. Ed. S. Y. Nof. New York: John Wiley & Sons. Chapter 6.

Wampler, C., and A. P. Morgan. 1991. "Solving the 6R Inverse Position Problem Using a Generic-Case Solution Methodology." *Mechanisms and Machine Theory* **26**(1), 91–106.

Westmacott, G. D. 1993. "The Application of Symbolic Equation Manipulation to High Performance Multibody Simulation." In *Robotics: Applied Mathematics and Computational Aspects*. Ed. Kevin Warwick. Oxford: Oxford University Press. 527–540.

Whitney, D. E. 1969. "Resolved Motion Rate Control of Manipulators and Human Prosthesis." *IEEE Transactions on Man–Machine Systems* **MMS-10**(2), 47–53.

Wilkinson, J. H. 1959. "The Evaluation of the Zeros of Ill-Conditioned Polynomials, Parts I and II." *Numerical Methods* **1**, 150–180.

Yuan, M. S. C., and F. Freudenstein. 1971. "Kinematic Analysis of Spatial Mechanisms by Means of Screw Coordinates, Part 1—Screw Coordinates." *Transactions of the ASME* **93**, 61–66.

CHAPTER 7
ROBOT HANDS AND END-EFFECTORS

Kazuo Tanie
Mechanical Engineering Laboratory
AIST-MITI
Tsukuba, Japan

1 INTRODUCTION

The robot hand is a tool that enables robots to interact with environments. It is expected to perform tasks like a human hand. The objective of the interaction is generally to execute a required task through applying actions to a task environment. Because it can be considered a tool for creating an effect on the environment and is generally located at an end of a robotic arm, it is often called an *end-effector*. One of the best ways of making a good robot end-effector would be to develop a hand like a human hand, which has five fingers that can move dexterously and perform various complex tasks. However, such a hand requires complex mechanisms and control algorithms, causing development difficulties and increases in cost. Robots in industrial applications are often required to handle only limited shapes of objects or parts and do limited kinds of tasks. For these requirements, human-like hands are not economical. Various types of single-purpose hands with simple grasping function or task-oriented tools are commonly used instead. A device to handle only limited shapes of objects is sometimes called a *gripper* or *gripping device*. In some applications simplification of gripping function is more important than the limited versatility and dexterity of such devices. In other applications handling and manipulating many different objects of varying weights, shapes, and materials is required, and an end-effector that has complex functions will be more suitable. This type of end-effector is called a *universal hand*. Extensive research is currently being carried out on the design and manipulation of universal hands, but to date few practical versions exist.

This chapter describes several kinds of end-effectors in practical use and discusses recent developments in universal hands and their control algorithms. The chapter also explains the practical implementation and design criteria of end-effectors. Gripper selection is explained in Chapter 48.

2 CLASSIFICATION OF END-EFFECTORS

End-effectors have several variations, depending on the many ways in which they are required to interact with environments in carrying out required tasks. This brings about an increasing number of mechanisms used to design end-effectors and the functions installed in them. End-effectors can also be designed using several kinds of actuators if actively controlled elements are needed. This causes more variations to be developed. This section discusses the classifications of end-effectors based on the functions installed. The next section describes drive systems.

In general, end-effectors can be classified by function as follows:

1. Mechanical hands
2. Special tools
3. Universal hands

The most common tasks given to end-effectors involve grasping functions. The tool for grasping—that is, a hand—is thus the most commonly used. *Class 1* refers to hands with fingers whose shape cannot be actively controlled. The fingers in such a hand are generally

Handbook of Industrial Robotics, Second Edition, Edited by Shimon Y. Nof
ISBN 0-471-17783-0 © 1999 John Wiley & Sons, Inc.

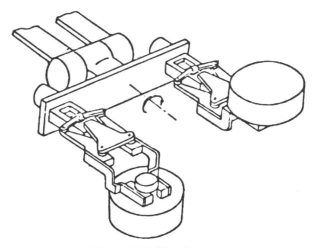

Figure 1 Multihand system.

unjointed and have a fixed shape designed specifically for the object to be handled. Class 1 devices are less versatile and less dexterous than class 3 devices, but more economical to produce. Finer classifications of class 1 can be made, using, for example, the number of fingers. Two-, three-, and five-finger types exist. For industrial applications the two-finger hand is the most popular. Three- and five-finger hands, with some exceptions, are customarily used for prosthetic hands for amputees. Another classification is by the number of hands—single or multiple—mounted on the wrist of the robot arm. Multihand systems (Figure 1) enable effective simultaneous execution of more than two different jobs. Design methods for each individual hand in a multihand system are subject to those for single hands. Another classification, by mode of grabbing, is between external and internal systems. An external gripper (Figure 2) grasps the exterior surface of objects with closed fingers, whereas an internal gripper (Figure 3) grips the internal surface of objects with open fingers. There are two finger-movement classifications: translational finger hands and swinging finger hands. Translational hands can move their own fingers, keeping them parallel. Swinging hands employ a swinging motion of the fingers. Another classification is by the number of degrees of freedom (d.o.f.) implemented in hand structures. Typical mechanical hands are classified as 1 d.o.f. A few hands have more than 2 d.o.f.

Class 2 refers to special-purpose devices for specific tasks. Vacuum cups and electro-magnets are typical devices in this class. In some applications the objects to be handled may be too large or thin for a hand to grasp them. In such cases a special tool suitable for holding the object has a great advantage over the other types of device. Also, in some applications it will be more efficient to install a specific tool according to the required

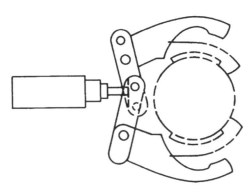

Figure 2 External gripper.

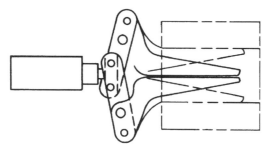

Figure 3   Internal gripper.

task. Spot welders for spot welding, power drills for drilling, paint sprayers for spray-painting, and so on are popularly used as end-effectors for industrial robots.

Class 3 is composed of multipurpose hands with usually more than three fingers and /or more than one joint on each finger, providing the capability of performing a wide variety of grasping and manipulative assignments. To develop this type of hand, however, many problems must be solved in the design of mechanisms and controls. Currently several mechanical designs have been proposed for universal hands, some of which are commercially available, and several efficient control algorithms for stable grasp and manipulation have been proposed, though not many applications have been found.

The following sections describe the three end-effector classes.

3 ACTUATOR SYSTEM FOR END-EFFECTORS

In robot systems three kinds of drive methods are practically available: pneumatic, electric, and hydraulic. Actuators are mainly used in hand-type end-effectors.

Pneumatic drive can be found in end-effectors of almost all industrial robots. The main actuator systems in pneumatic drive are the cylinder and motor. They are usually connected to on/off solenoid valves that control their direction of movement by electric signal. For adjusting the speed of actuator motion, airflow regulation valves are needed. A compressor is used to supply air (maximum working pressure 10 kg/cm²) to actuators through valves. The pneumatic system has the advantage of being less expensive than other methods, which is the main reason that many industrial robots use it. Another advantage of the pneumatic system derives from the low stiffness of the air-drive system. This feature of the pneumatic system can be used effectively to achieve compliant grasping, which is one of the most important functions of the hand: it refers to grasping objects with delicate surfaces carefully. On the other hand, the relatively limited stiffness of the system makes precise position control difficult. Air servo-valves are being developed for this purpose. However, they increase the cost and result in some loss of economic advantage of the air drive.

The electric system is also popular. There are typically three kinds of actuators: DC motor, AC motors, and stepping motors. AC motors are becoming more popular because of brushless structure, which reduces the maintenance cost and makes the system applicable to tasks in combustible environments. In general, each motor requires appropriate reduction gear systems to provide proper output force or torque. Direct-drive torque motors (DDMs) are commercially available, but their large size makes designing compact systems difficult. Few robot hands use DDM. In the electric system a power amplifier is also needed to provide a complete actuation system. Electric drive has several benefits:

1. A wide variety of products are commercially available.
2. Constructing flexible signal-processing and control systems becomes very easy because they can be controlled by electric signals, enabling the use of computer systems as control devices.
3. Electric drive can be used for both force and position control.

Drawbacks of the electric system are that it is somewhat more expensive than the pneumatic system and has less power generation and less stiffness than the hydraulic system.

The hydraulic drives used in robot systems are electrohydraulic drive systems. A typical hydraulic drive system consists of actuators, control valves, and power units. There are three kinds of actuators in the system: piston cylinder, swing motor, and hydraulic motor. To achieve position control using electric signals, electrohydraulic conversion devices are available. For this purpose electromagnetic or electrohydraulic servo-valves are used. The former provides on/off motion control, and the latter provides continuous position control. Hydraulic drive gives accurate position control and load-invariant control because of the high degree of stiffness of the system. On the other hand, it makes force control difficult because high stiffness causes high pressure gain, which has a tendency to make the force control system unstable. Another claimed advantage of hydraulic systems is that the ratio of the output power per unit weight can be higher than in other systems if high pressure is supplied. This drive system has been shown to provide an effective way of constructing a compact high-power system.

Other than the preceding three types of drive, there are a few other drive methods for hand type end-effector. One method uses a springlike elastic element. A spring is commonly used to guarantee automatic release action of gripping mechanisms driven by pneumatic or hydraulic systems. Figure 4 shows a spring-loaded linkage gripping mechanism using a pneumatic cylinder (Sheldon, 1972). Gripping action is performed by means of one-directional pneumatic action, while the spring force is used for automatic release of the fingers. This method considerably simplifies the design of the pneumatic or hydraulic circuit and its associated control system. The spring force can also be used for grasping action. In this case, because the grasping force is influenced by the spring force, to produce a strong grasping force a high-stiffness spring is necessary. This usually causes the undesirable requirement of high-power actuators for the release action of the fingers. Therefore the use of spring force for grasping action is limited to low-grasping-force gripping mechanisms for handling small machine parts such as pins, nuts, and bolts. The spring force can be used for a one-directional motion of the pneumatic and the hydraulic actuator because the piston can be moved easily by the force applied to the output axis (piston rod). The combination of a spring and electric motor is not viable because normal electric motors include a high reduction gear system which makes it difficult to transmit the force inversely from the output axis.

Another interesting method uses electromagnets. The electromagnet actuator consists of a magnetic head constructed with a ferromagnetic core, conducting coil, and actuator rod made of ferrous materials. When the coil is activated, the magnetic head attracts the actuator rod, and the actuator displacement is locked at a specified position. When the coil is not activated, the actuator rod can be moved freely. This type of actuator is usually employed with a spring and produces two output control positions. Figure 5 shows a hand using the electromagnetic drive. The electromagnetic actuator 1 produces the linear motion to the left along the L–L line. The motion is converted to grasping action through the cam 2. The releasing action is performed by the spring 3. The actuator displacement that this kind of actuator can make is commonly limited to a small range because the

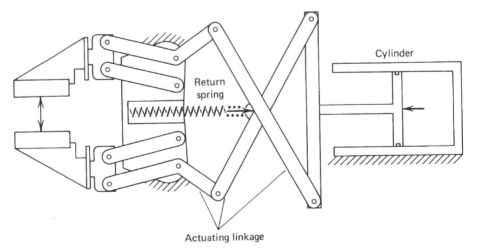

Figure 4   Spring-loaded linkage gripper (courtesy Dow Chemical Co., Ltd.).

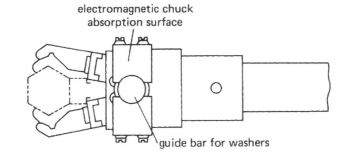

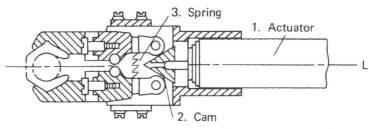

Figure 5   Hand using an electromagnetic drive (courtesy Seiko-seiki Co., Ltd.).

force produced by the magnetic head decreases according to the increase of the actuator displacement. This drive method can therefore be effectively used only for gripping small workpieces.

In the design of effective robotic end-effectors, the selection of drive system is very important. Selection depends on the jobs required of the robot. Briefly, if an end-effector has some joints that need positional control, the electric motor is recommended. If force-control function is needed at some joints—for example, to control grasping force—electric or pneumatic systems are recommended. If high power is needed, the hydraulic drive is recommended.

4   MECHANICAL HANDS

Human-hand grasping is divided into six different types of prehension (Schlesinger, 1919): palmar, lateral, cylindrical, spherical, tip, and hook. See Figure 6. Crossley and Umholts (1977) classified manipulation functions by the human hand into nine types: trigger grip, flipping a switch, transfer pipe to grip, use cutters, pen screw, cigarette roll, pen transfer, typewrite, and pen write. Human hand can perform various grasping functions using five fingers with joints. In designing a hand, a subset of the types of human grasp and hand manipulation is considered, according to the required task. To achieve minimum grasping function, a hand needs two fingers connected to each other using a joint with 1 d.o.f. for its open–close motion. If the hand has two rigid fingers, it has the capability of grasping only objects of limited shapes and is not able to enclose objects of various shapes. Also, this type of hand cannot have manipulation function because all degrees of freedom are used to maintain prehension. There are two ways to improve the capability for accommodating the change of object shapes. One solution is to put joints on each finger that can move according to the shape of the grasped object. The other is to increase the number of fingers. The manipulation function also emerges from this design. To manipulate objects the hand must usually have more fingers and joints driven externally and independently than does the hand designed for grasping objects. The more fingers, joints, and degrees of freedom the hand has, the more versatile and dexterous it can be. With this principle in mind, the type of hand can be selected and the mechanical design considered.

Several kinds of grasping functions described above can be realized using various mechanisms. From observation of the usable pair elements in gripping devices, the following kinds, among others, have been identified (Chen, 1982): linkage, gear-and-rack,

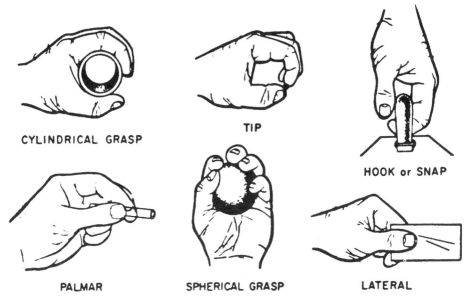

CYLINDRICAL GRASP

TIP

HOOK or SNAP

PALMAR   SPHERICAL GRASP   LATERAL

Figure 6   The various types of hand prehension. (From G. Schlesinger, *Der Mechanische Aufbau der künstlichen Glieder*, Part 2 of *Ersatzglieder und Arbeitshilfen*. Berlin: Springer, 1919.)

cam, screw, and cable and pulley. The selection of these mechanisms is affected by the kind of actuators and grasping modality to be used. Many hand mechanisms have been proposed. The following sections explain the hand mechanisms in practical use. Chen (1982) lists other possible hand mechanisms.

4.1   Mechanical Hands with Two Fingers

4.1.1   Swinging Gripping Mechanisms

One of the most popular mechanical hands uses the swinging gripping mechanism. The hand has the swing motion mechanism in each finger. It is useful for grasping objects of limited shape, especially cylindrical workpieces. Figure 7 shows a typical example of such a hand. This type of hand is driven by linear motion actuator, such as a pneumatic cylinder, because it uses a slider-crank mechanism. The design of the mechanism that makes the finger swing motion will depend on the type of actuator used. If actuators are used that produce linear movement, such as pneumatic or hydraulic piston cylinders, the device contains a pair of slider-crank mechanisms. Figure 8 shows a simplified drawing of a pair of slider-crank mechanisms commonly used in hands with pneumatic and hydraulic piston cylinder drive. When the piston 1 is pushed by hydraulic or pneumatic pressure to the right, the elements in the cranks 2 and 3 rotate counterclockwise with the fulcrum A1 and clockwise with the fulcrum A2, respectively, when γ is less than 180°. These rotations make the grasping action at the extended end of the crank elements 2 and 3. The releasing action can be obtained by moving the piston to the left. The motion of this mechanism has a dwell position at $\gamma = 180°$. For effective grasping action, $\gamma = 180°$ must be avoided. An angle γ ranging from 160° to 170° is commonly used. Figure 9 shows another example of a mechanism for a hand with swing finger motion that uses a piston cylinder and two swing-block mechanisms. The sliding rod 1, actuated by a pneumatic or a hydraulic piston, transmits motion by way of the two symmetrically arranged swinging-block linkages 1-2-3-4 and 1-2-3'-4' to grasp or release the object by means of the subsequent swinging motions of links 4 and 4' at their pivots A1 and A2.

   Figure 10 shows a typical example of a hand with swing motion fingers using a rotary actuator in which an actuator is placed at the cross point of the two fingers. Each finger is connected to the rotor and the housing of the actuator, respectively. The actuator movement directly produces grasping and releasing actions. This type of hand includes a revolutionary pair mechanism, which is schematically illustrated in Figure 11a. According

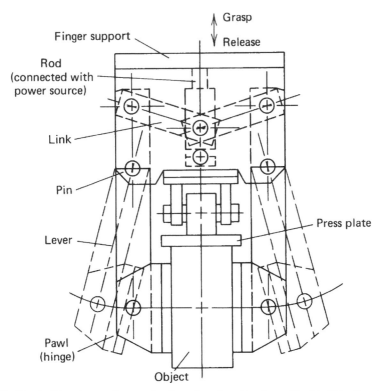

Figure 7   Hand with swing motion mechanisms (courtesy Yasukawa Electric Mfg. Co., Ltd.).

to the different types of actuator used instead of the rotary actuator, there are several variations in this type of hand. Two are shown in Figure 11*b* and *c*. Figure 11*b* uses a cylinder–piston actuator instead of a rotary actuator. Figure 11*c* shows a hand that uses a cam to convert the linear piston motion to the grasping–releasing action. An application example of this mechanism has been shown in Fig. 5. Figure 12*a* shows a cross-four-bar link mechanism with two fulcrums, A and B. This mechanism is sometimes used to make a finger-bending motion. Figure 12*b* shows a typical example of a finger constructed with the cross-four-bar link. There are two ways of activating this mechanism. First, a rotary actuator can be used at point A or B to rotate the element AD or BC. The actuator movement produces the rotation of the link element CD, which produces a bending motion of the finger. Second, a slider-crank mechanism activated by a cylinder piston can

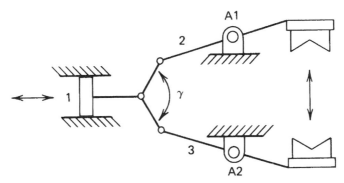

Figure 8   Schematic of a pair of slider-crank mechanisms.

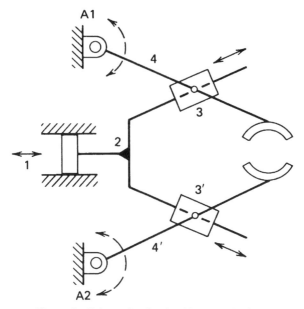

Figure 9 Schematic of swing-block mechanism.

be used to rotate the element AD or BC. The lower illustration in Figure 12*b* depicts this type of drive. The finger-bending motion can be obtained in the same way as with rotary actuators. The use of a cross-four-bar link offers the capability of enclosing the object with the finger. This mechanism can be used for a hand that has more than two fingers.

4.1.2   Translational Hand Mechanisms

The translational mechanism is another popular gripping mechanism widely used in hands for industrial robots. It enables the finger to be closed and opened without the attitude being changed. The mechanism is usually somewhat more complex than the swinging type. The simplest translational hand uses the direct motion of the piston–cylinder type actuator. Figure 13 shows such a hand using a hydraulic or pneumatic piston cylinder.

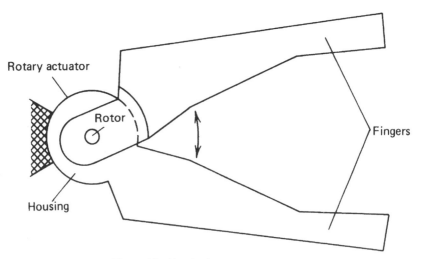

Figure 10   Hand using a rotary actuator.

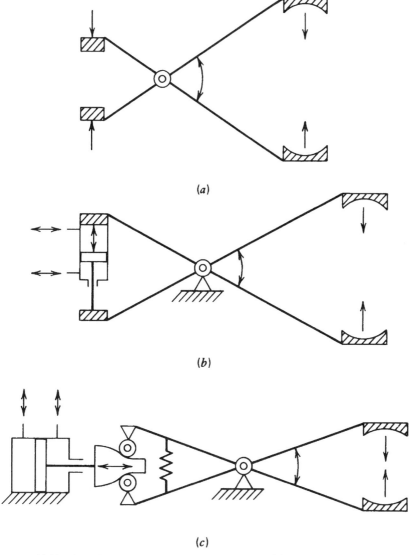

Figure 11   (a) Hand mechanism that includes a revolute pair. (b) Hand using a revolute pair and a piston cylinder. (c) Hand using a cam and a piston cylinder.

As depicted in the figure, the finger motion corresponds to the piston motion without any connecting mechanisms between them. The drawback of this approach is that the actuator size determines the size of the hand, which sometimes makes it difficult to design the desired size of hand, though it is suitable for the design of wide-opening translational hands. Figure 14 shows a translational hand using a pneumatic or hydraulic piston cylinder, which includes a dual-rack gear mechanism and two pairs of the symmetrically arranged parallel-closing linkages. This is a widely used translational hand mechanism. The pinion gears are connected to the elements A and A′, respectively. When a piston rod moves toward the left, the translation of the rack causes the two pinions to rotate clockwise and counterclockwise, respectively, and produces the release action, keeping each finger direction constant. The grasping action occurs when the piston rod moves to the right in the same way. There is another way to move the parallel-closing linkages in Figure 14. Figure 15 shows a mechanism using a rotary actuator and gears in lieu of the

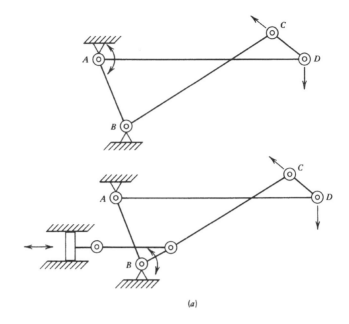

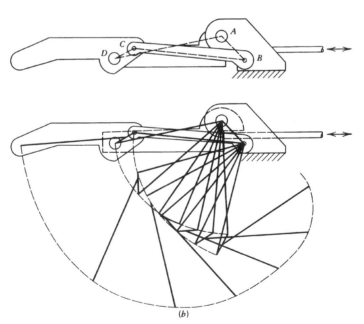

Figure 12   (a) Schematic of cross-four-bar link mechanism. (b) A finger using cross-four-bar link mechanism.

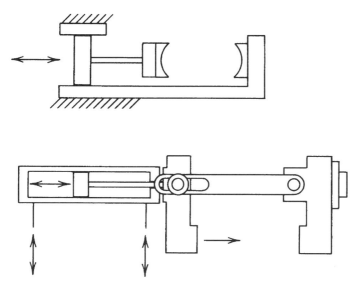

Figure 13   Tanslational hand using a cylinder piston.

combination of a piston cylinder and two racks in Figure 14. Figure 16 shows two other examples of translational hand mechanism using rotary actuators. Figure 16a consists of an actuator and rack-and-pinion mechanism. The advantage of this hand is that it can accommodate a wide range of dimensional variations. Figure 16b includes two sets of ball–screw mechanisms and an actuator. This type of hand enables accurate control of finger positions.

4.1.3   Single-Action Hand

Where fewer actuators are installed in the hand than the hand has fingers, each finger cannot be moved independently. This is true of most of the hands reviewed above. For

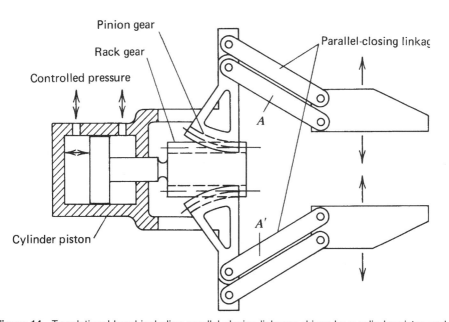

Figure 14   Translational hand including parallel-closing linkages driven by a cylinder piston and a dual-rack gear.

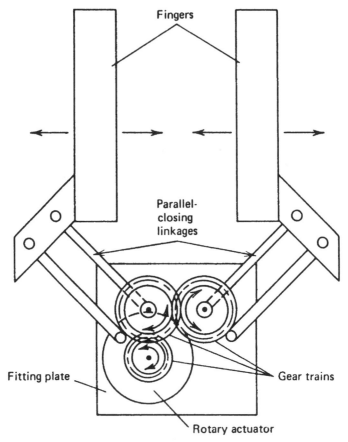

Figure 15 Translational hand including parallel-closing linkages driven by a rotary actuator and gears.

example, the mechanical hand shown in Figure 8 has only one actuator used to move two fingers. Figure 13 shows a hand with one stationary finger or fixed finger and one moving finger, commonly referred to as a *single-action* hand. When this kind of hand picks up an object rigidly fixed on the base, the hand must be exactly position-controlled to the position that will allow the two fingers to touch the surface of the part to be grasped with one actuator motion. Otherwise, coordinated motion of robot arm and fingers will be needed. The use of a hand in which each finger can be driven by an independent actuator will avert such a situation, but such a hand increases the cost. To reach a compromise solution, some precautions will generally be required in the task environment. A method of making the part to be handled free to move during pickup or release is often introduced.

4.1.4 Consideration of Finger Configuration

Design of Object-Shaped Cavity

When rigid fingers are used, the finger configuration must be contrived to accommodate the shape of the object to be handled. For grasping the object tightly, object-shaped cavities on the contact surface of the finger, as shown in Figure 17a, are effective. A cavity is designed to conform to the periphery of the object of a specified shape. For example, if cylindrical workpieces are handled, a cylindrical cavity should be made. A finger with this type of cavity can grasp single-size workpieces more tightly with a wider contact surface. However, such a structure will limit the hand's ability to accommodate change in the dimension of an object to be handled. Versatility can be slightly improved

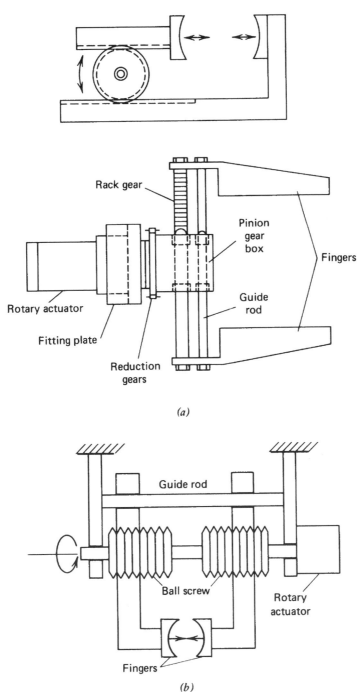

Figure 16   (a) A translational hand operated by a rotary actuator with rack-and-pinion mechanism (courtesy Tokiko Co., Ltd.). (b) A translational hand using a rotary actuator and ball-screw mechanism.

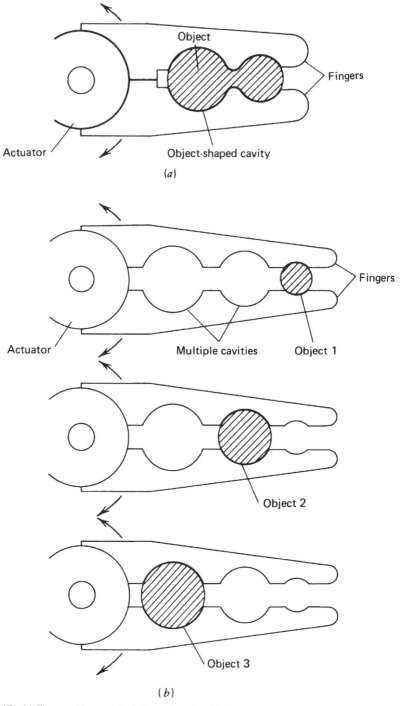

Figure 17   (a) Finger with an object-shaped cavity. (b) Finger with multiple object-shaped cavities.

by the use of a finger with multiple cavities for objects of differing size and shape. Figure 17*b* shows examples of fingers with multiple cavities.

Design of V-Shaped Cavity

In manufacturing, many tasks involve cylindrical workpieces. For handling cylindrical objects, a finger with a V-shaped cavity instead of an object-shaped cavity may be adopted. Each finger contacts the object at two spots on the contact surface of the cavity during the gripping operation. The two-spot contact applies a larger grasping force to a limited surface of the grasped object and may sometimes distort or scratch the object. However, in many tasks this problem is not significant, and this device has great advantages over a hand with object-shaped cavities. One advantage is that it can accommodate a wide range of diameter variations in the cylindrical workpiece, allowing the shape of the cavity to be designed independent of the dimensions of the cylindrical object. Another advantage is that it is easier to make, resulting in reduced machinery costs. Figure 18*a* shows a typical geometrical configuration of a grasping system for a cylindrical object using a V-shaped cavity finger hand. Some relation exists between the configuration pa-

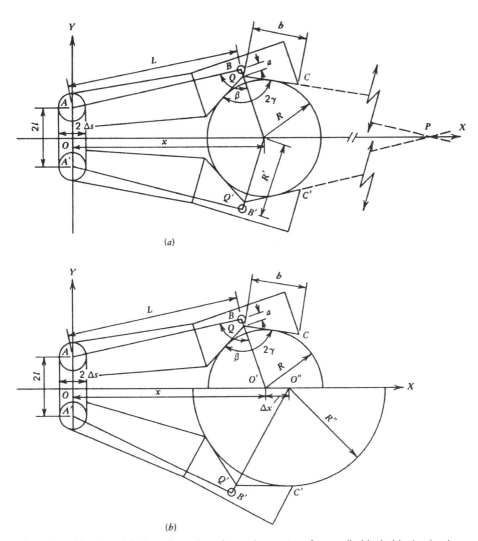

Figure 18 (a) Geometrical configuration of grasping system for a cylindrical object using two-finger hand with a V-shaped cavity. (b) The deviation of center position of grasped objects.

rameters of the V-shaped cavity and the diameter of possible cylindrical workpieces to be grasped (Osaki and Kuroiwa, 1969). Suppose that parameters of the grasping system γ, β, R, R', L, l, a, and b, symbols B, C, Q, B', C', Q', and O', and the coordinate system $O-xy$ are defined as shown in the figure. Since the cylindrical workpiece grasped and the hand construction cannot intersect, the following inequality is obtained:

$$x - R < \Delta s \tag{1}$$

where Δs is the ½ width of the hand element and x is the distance between the center of cylindrical workpiece and the origin O, as shown in Figure 18a. The distance x is expressed by the following equation:

$$x = \sqrt{\left[L^2 + \left(\frac{R}{\sin \gamma} + a \right)^2 - 2L \left(\frac{R}{\sin \gamma} + a \right) \cos \beta \right] - l^2} \tag{2}$$

In the swinging-type hand, the hand will be often designed so that β is a constant value. If $\beta = 90°$, as it generally is, Equation (2) becomes

$$x = \sqrt{L^2 + \left(\frac{R}{\sin \gamma} + a \right)^2 - l^2} \tag{3}$$

In a translational hand in which each cavity block is kept parallel to every other is used, i.e., $\angle BO'O = 90°$, the following equation can be obtained:

$$\cos \beta = \frac{R' - l}{L} \qquad \left(R' = \frac{R}{\sin \gamma} + a \right) \tag{4}$$

After substitution of Equation (4) into Equation (2), x can be expressed by the following:

$$x = \sqrt{L^2 - \left(\frac{R}{\sin \gamma} + a - l \right)^2} \tag{5}$$

For the object to contact each finger at the two spots on the surface of the cavity, D, γ, and b must satisfy the following inequality:

$$D < 2b \tan \gamma \tag{6}$$

where D is the diameter of the cylindrical object and equals $2R$. If the swinging gripper is assumed, another condition must be considered because the longitudinal direction of each finger varies with objects of differing size. To hold an object safely in the hand without slippage, the extended line QC must cross the extended line $Q'C'$ at point P in front of the hand. From this constraint, the upper diameter limit of the object that can be grasped by the hand is expressed with some parameters of the cavity as follows:

$$D < 2 \sin \gamma \cdot \left[L \cdot \tan \left(\frac{\pi}{2} - \gamma \right) + \frac{l}{\tan(\pi/2 - \gamma)} \right] \tag{7}$$

In the special case in which $l = 0$, as in Figure 10, Equation (7) becomes

$$D < 2 \sin \gamma \cdot L \tan \left(\frac{\pi}{2} - \gamma \right) \tag{8}$$

If the translational hand is used, inequalities (7) or (8) can be ignored because the attitudes of the finger and the V-shaped cavity are kept constant. Equations (1), (6), (7), and (8) provide the basic parameter relations to be considered when the V-shape cavity is designed. The other consideration in designing the V-shape cavity is that the x coordinate of the center of the cylindrical object varies with the diameter. This can be recognized from Equations (2), (3), and (5). Figure 18b explains the deviation ($O'O'' = \Delta x$) of the center positions of two different-sized objects grasped by a hand. The deviation

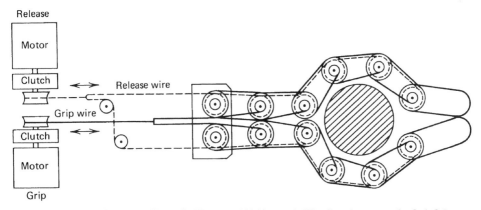

Figure 19   The soft gripper. (From S. Hirose and Y. Umetani, "The Development of a Soft Gripper for the Versatile Robot Hand," in *Proceedings of the Seventh International Symposium on Industrial Robots,* Tokyo, 1977, 353–360.)

will be undesirable for some tasks. The reduction must be considered (Osaki and Kuroiwa, 1969). The deviation is generally smaller if a translational hand is used than if a swinging hand is used, if fingers of the same size are employed. A well-designed translational hand using a rack-and-pinion or ball–screw mechanism, as shown in Figure 16, can make the deviation almost zero. The hand shown in Figure 13 yields the same effect. For a swinging hand, longer fingers will help to reduce the deviation.

Soft-Gripping Mechanism

To provide the capability to completely conform to the periphery of objects of any shape, a soft-gripping mechanism has been proposed. See Figure 19 (Hirose and Umetani, 1977). The segmental mechanism is schematically illustrated in Figure 20. The adjacent links and pulleys are connected with a spindle and are free to rotate around it. This mechanism is manipulated by a pair of wires, each of which is driven by an electric motor with gear reduction and clutch. One wire, called the *grip wire,* produces the gripping movement. The other, the *release wire,* pulls antagonistically and produces the release movement from the gripping position. When the grip wire is pulled against the release wire, the

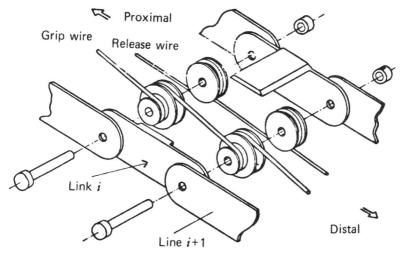

Figure 20   Segmental mechanism of the soft gripper. (From S. Hirose and Y. Umetani, "The Development of a Safe Gripper for the Versatile Robot Hand," in Proceedings of the Seventh International Symposium on Industrial Robots, Tokyo, 1977, 353–360.)

finger makes a bending motion from the base segment. During this process of wire traction, the disposition of each finger's link is determined by the mechanical contact with an object. When the link i makes contact with an object and further movement is hindered, the next link, ($i + 1$), begins to rotate toward the object until it makes contact with the object. This results in a finger motion conforming to the peripheral shape of the object. In this system it is reported that the proper selection of pulleys enables the finger to grasp the object with uniform grasping pressure.

4.1.5 Calculation of Grasping Force or Torque

The maximum grasping force or grasping torque is also an important specification in hand design. The actuator output force or torque must be designed to satisfy the conditions the task will require. How hard the hand should grasp the object depends on the weight of the object, the friction between the object and the fingers, how fast the hand is expected to move, and the relation between the direction of the hand movement and the attitude of the object grasped. In the worst case, the direction of the gravity and the acceleration force vectors applied to the grasped object are parallel to the grasp surface of the fingers. Then friction alone must hold the object. This situation is thus assumed in evaluating the maximum grasping force or grasping torque. After the maximum grasping force or torque has been determined, the force or torque that the actuator must generate can be considered. The calculation of those values requires the conversion of the actuator output force or torque to the grasping force or torque, which depends on the kind of actuators and mechanisms employed. Table 1 shows the relation between the actuator output force or torque and the grasping force for the hand with various kinds of mechanisms and actuators.

4.2 Mechanical Hands with Three or Five Fingers

4.2.1 Three-Finger Hand

Increasing the number of fingers and degrees of freedom will greatly improve the versatility of the hand, but will also complicate the design process. The design methods for three-finger hands are not yet well established from the practical point of view. However, several examples have been developed, mainly for experimental or research uses. The simplest example is a hand with three fingers and one joint driven by an appropriate actuator system.

The main reason for using the three-finger hand is its capability of grasping the object in three spots, enabling both a tighter grip and the holding of spherical objects of differing size while keeping the center of the object at a specified position. Three-point chuck mechanisms are typically used for this purpose (see Figure 21). Each finger motion is performed using a ball–screw mechanism. Electric motor output is transmitted to screws attached to each finger through bevel gear trains that rotate the screws. When each screw is rotated clockwise or counterclockwise, the translational motion of each finger will be produced, resulting in the grasping–releasing action. Figure 22 shows another three-finger hand with the grasping-mode switching system (Skinner, 1974). This includes four electric motors and three fingers and can have four grasping modes, as shown in Figure 23, each of which can be achieved by the finger-turning mechanism. All fingers can be bent by motor-driven cross-four-bar link mechanisms, and each finger has one motor. The finger-turning mechanism is called a *double-dwell* mechanism, as shown in Figure 24 (Skinner, 1974). Gears that rotate the fingers are shown, and double-headed arrows indicate the top edge of the finger's bending planes for each prehensile mode. This mechanism transfers the state of the hand progressively from three-jaw to wrap to spread to tip prehension. The gears for fingers 2 and 3 are connected to the motor-driven gear directly, whereas the gear for finger 1 is connected to the motor-driven gear through a coupler link. Rotating the motor-driven gear in three-jaw position, finger 1 rotates, passes through a dwell position, and then counterrotates to reach the wrap position. Similarly, finger 1 is rotated out of its spread position but is returned as the mechanism assumes tip prehension. Finger 2 is rotated counterclockwise 60° from its three-jaw position to the wrap position, then counterclockwise 120° into the spread position, then counterclockwise 150° into the tip position. Finger 3 rotates through identical angles but in a clockwise direction. A multi-prehension system of this type is effective for picking up various-shaped objects.

4.2.2 Five-Finger Hand

A small number of five-finger hands have been developed, with only a few for industrial use. Almost all are prosthetic hands for amputees. In the development of prosthetic arms, cosmetic aspects are more important to the mental state of the disabled than functions.

Table 1   Relations Between the Actuator Output Force or Torque and The Grasping Force

Type of Gripper Mechanism and Configuration Parameters	Relations between Actuator Output and Grasping force
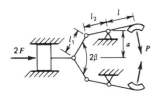	$$IP = l_2\left[\tan\beta\sqrt{1 - \left(\frac{l_1\sin\beta - a}{l}\right)^2} - \frac{l_1\sin\beta - a}{l}\right]\cdot F$$
	$$IP = l_1\frac{\cos(\phi - \theta)}{\cos\theta}\cdot F$$
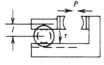	$$IP = \tau$$
	$$P = \frac{\tau}{l}$$
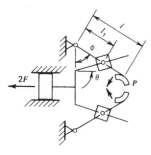 r: the pitch radius of the thread α: pitch angle of the thread (rectangular)	$$P = \frac{\tau}{r\tan\alpha}$$
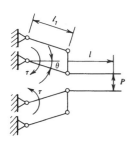	$$P = \frac{\tau}{l_1\cos\theta}$$

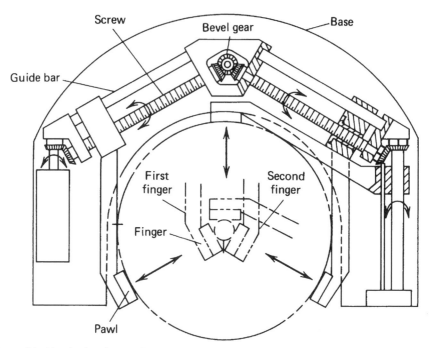

Figure 21 Hand using three-point chuck mechanism (courtesy Yamatake Honeywell Co., Ltd.).

Anthropomorphism is thus required in their design. For industrial use, function is more important than cosmetic aspects, and anthropomorphism is beyond consideration. This is why there are few five-finger industrial hands. Nevertheless, an example of a five-finger hand developed for prosthetic use is described below because such hands include many mechanisms that will be effective in the design of industrial hands. A disabled person must produce control signals for operation of a prosthetic arm. The number of independent control signals available determines how many degrees of freedom the prosthetic device can have. Typical models of a five-finger hand for prostheses have only 1 d.o.f. Each finger is connected to a motor by appropriate mechanisms. Figure 25 shows an example, called the WIME Hand (Kato et al., 1969). Each finger is constructed using a cross-four-bar link mechanism that gives the finger proper bending motion. One element of each of the five sets of cross-four-bar links includes a crank rod. All crank rods are connected to the spring-loaded plate 1, which is moved translationally by an electric motor drive-screw mechanism 2. When the motor rotates clockwise or counterclockwise, the plate 1 moves toward the left or the right, respectively, and activates the cross-four-bar link of each finger to bend the finger and produce the grasping operation. To ensure that the hand holds the object with the equilibrium of the forces between the fingers and the object, the arrangement of fingers must be carefully considered. In typical five-finger hands, the thumb faces the other four fingers and is placed equidistant from the index finger and middle finger so the tips of the fingers can meet at a point when each finger is bent (see Figure 26). If each finger connects to the drive system rigidly, finger movements are determined by the motion of the drive system. The finger configuration cannot accommodate the shape change of grasped objects. To remedy this problem, the motor output can be transmitted to each finger through flexible elements.

4.3 Mechanical Hands for Precision Tasks

For tasks that require manipulation of microobjects, such as microassembly in microindustry or manipulating cells in bioindustry, hands that can manipulate very precise objects are needed. The technology for developing such hands has drawn much attention since the beginning of the 1990s. With the advancement of microelectronics, manufacturing technologies for microobjects are becoming popular. Several very small microgripping devices have been developed. These are discussed in Chapter 10 and 11. This section

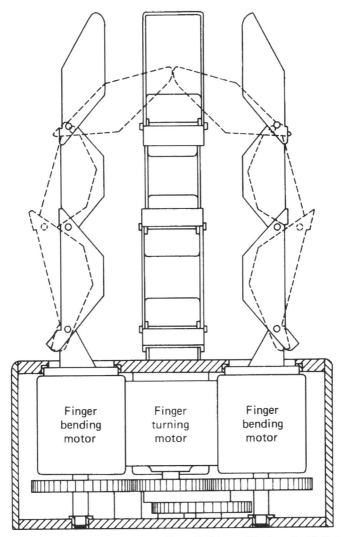

Figure 22 Multiple prehension hand system. (From F. Skinner, "Design of a Multiple Prehension Manipulator System," ASME Paper 74-det-25, 1974.)

discusses gripping devices that, though not small themselves, can precisely manipulate very small objects.

There are several ways of designing such hands. One way employs a flexure structure using thin metal plate. See Figures 27 and 28 (Ando et al., 1990; Kimura et al., 1995). For the actuator, a piezoelectric stack element (PZT) is used. It is inserted in the central section of the hand. The PZT deforms when voltage is applied, and the deformation is amplified through the structure, composed of several arms and hinges that are spring joints, causing grasping motion at the fingertip. Opening motion is caused by the spring effect in the structure when the applied voltage is turned off to zero. At the hinge near the tip, a strain gauge is installed for grasp force detection. It is also used to compensate for the hysteresis that the PZT actuator generally has. The hand shown in Figure 28 can grasp a load less than 0.8 mm in diameter. The maximum grasping force is about 40 mN.

Another efficient design for gripping devices for microobject handling uses a parallel mechanism. The parallel arm, like the Stewart platform-type arm, provides the benefit of precise positioning functions at the fingertip, since the joints relating to the fingertip motion are arranged in parallel, with the result that the positioning error at each joint will

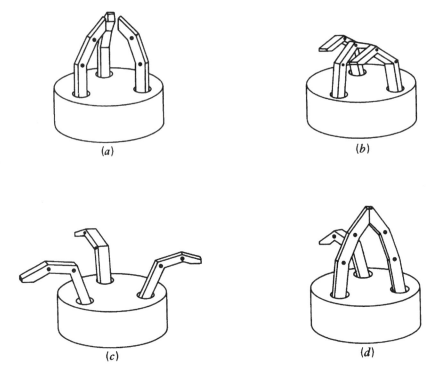

Figure 23 Mechanical equivalent prehensile modes. (a) Three-jaw position; (b) wrap position; (c) spread position; (d) tip position. (From F. Skinner, "Design of a Multiple Prehension Manipulator System," ASME Paper 74-det-25, 1974.)

not be stacked at the fingertip. Figure 29 shows a finger for such a gripping device using a parallel link mechanism (Arai, Larsonneur, and Jaya, 1993; Tanikawa, Arai, and Matsuda, 1996). A piezoelectric actuator is installed on each link located between the base plate and the upper plate attached to the end-effector (finger). Both ends of each link are connected to the base through a ball joint. The deformation of each link that is made by the actuator produces 6-d.o.f. fingertip motion. Thus, using two fingers, microobjects can be grasped and manipulated in three-dimensional space (Figure 30). Positioning two fingertips at exactly the same location, as is necessary for grasping very small objects easily, is difficult. To address this problem, a two-finger microhand, in which the link for a finger is fixed on the other finger's upper base, has been developed, as shown in Figures 31 and 32. It is called the *chopstick-type microgripper*. Because the finger set installed on the upper plate of one finger is positioned relative to the other, the hand can be controlled more simply and with higher reliability in grasping an object. Using such a hand, successful manipulation of a 2-mm glass ball has been reported.

5 SPECIAL TOOLS

As mentioned above in Section 2, there are two functional classifications of end-effector for robots: gripping devices for handling parts, and tools for doing work on parts. In some tasks, especially handling large objects, designing gripping mechanisms with fingers is difficult. Also, in some applications dealing with only a specific object, a special tool for the required task will be much easier and more economical to design than the gripping devices. As mentioned above, specific tools such as drills and spot welders are popularly used. Details are discussed in other chapters of this handbook. The following sections describe special tools especially for part handling used in industrial robots instead of mechanical hands.

5.1 Attachment Devices

Attachment devices are simply mounting plates with brackets for securing tools to the tool mounting plate of the robot arm. In some cases attachment devices may also be

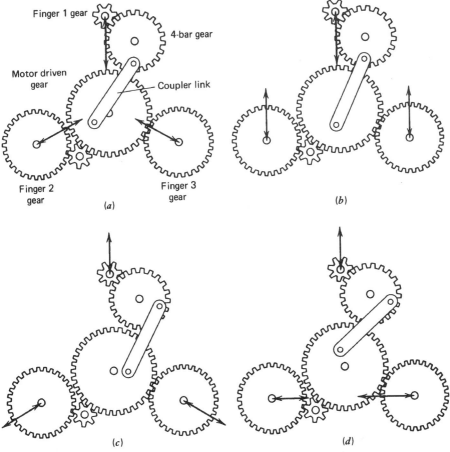

Figure 24 The double-dwell finger-turning mechanism. (From F. Skinner, "Design of a Multiple Prehension Manipulator System," ASME Paper 74-det-25, 1974.)

designed to secure a workpiece to the robot tool mounting plate, as with a robot manipulating a part against a stationary tool where the cycle time is relatively long. In this case the part is manually secured and removed from the robot tool mounting plate for part retention.

5.2   Support and Containment Devices

Support and containment devices include lifting forks, hooks, scoops, and ladles. See Figure 33. No power is required to use this type of end-effector; the robot simply moves to a position beneath a part to be transferred, lifts to support and contain the part or material, and performs its transfer process.

5.3   Pneumatic Pickup Devices

The most common pneumatic pickup device is a vacuum cup (Figure 34), which attaches to parts to be transferred by suction or vacuum pressure created by a Venturi transducer or a vacuum pump. Typically used on parts with a smooth surface finish, vacuum cups are available in a wide range of sizes, shapes, and materials. Parts with nonsmooth surface finishes can be picked up by a vacuum system if a ring of closed-cell foam rubber is bonded to the surface of the vacuum cup. The ring conforms to the surface of the part and creates the seal required for vacuum transfer. Venturi vacuum transducers, which are relatively inexpensive, are used for handling small, lightweight parts where a low vacuum flow is required. Vacuum pumps, which are quieter and more expensive, generate greater vacuum flow rates and can be used to handle heavier parts.

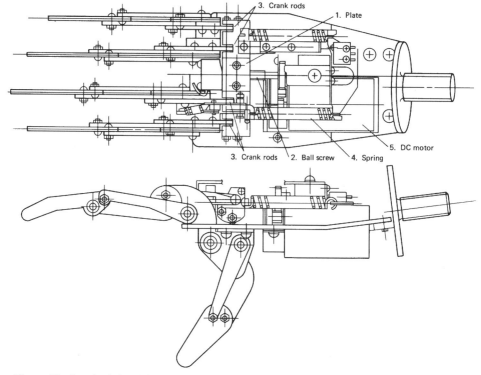

Figure 25 Prosthetic hand for forearm amputee, WIME hand (courtesy Imasen Engineering Co., Ltd.).

With any vacuum system, the quality of the surface finish of the part being handled is important. If parts are oily or wet, they will tend to slide on the vacuum cups. Therefore some additional type of containment structure should be used to enclose the part and prevent it from sliding on the cups. In some applications a vacuum cup with no power source can be utilized. Pressing the cup onto the part and evacuating the air between the cup and part creates a suction capable of lifting the part. However, a stripping device or valve is required to separate the part from the cup during part release. When a Venturi or vacuum pump is used, a positive air blow-off may be used to separate the part from the vacuum cup. Vacuum cups have temperature limitations and cannot be used to pick up relatively hot parts.

Another example of a pneumatic pickup device is a pressurized bladder, which is generally specially designed to conform to the shape of the part. A vacuum system is

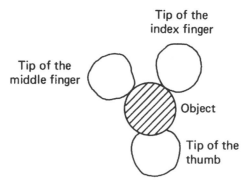

Figure 26 Three-point pinch.

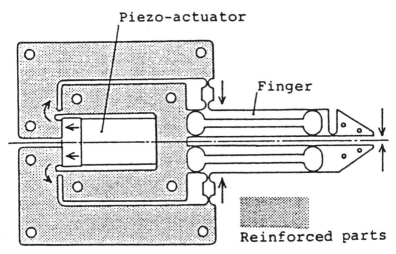

Figure 27   Microgripper using thin flexure structure. (Courtesy Mechanical Engineering Laboratory, AIST-MITI, Japan.)

used to evacuate air from the inside of the bladder so that it forms a thin profile for clearance in entering the tooling into a cavity or around the outside surface of a part. When the tooling is in place inside or around the part, pressurized air causes the bladder to expand, contact the part, and conform to the surface of the part with equal pressure exerted on all points of the contacted surface. Pneumatic bladders are particularly useful where irregular or inconsistent parts must be handled by the end-effector.

Pressurized fingers (see Figure 35) are another type of pneumatic pickup device. Similar to a bladder, pneumatic fingers are more rigidly structured. They contain one straight half, which contacts the part to be handled; one ribbed half; and a cavity for pressurized

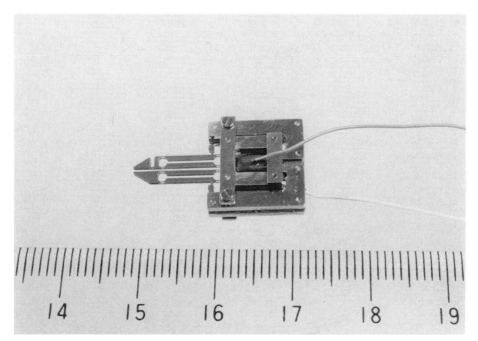

Figure 28   An example of microgripper using thin flexure structure. (Courtesy Mechanical Engineering Laboratory, AIST-MITI, Japan.)

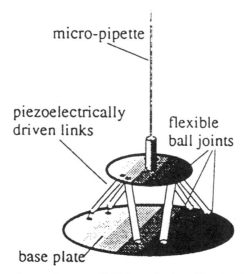

Figure 29 Microgripper finger using a parallel link mechanism. (Courtesy Mechanical Engineering Laboratory, AIST-MITI, Japan.)

air between the two halves. Air pressure filling the cavity causes the ribbed half to expand and "wrap" the straight side around a part. With two fingers per end-effector, a part can thus be gripped by the two fingers wrapping around the outside of the part. These devices can also conform to various shape parts and do not require a vacuum source to return to their unpressurized position.

5.4 Magnetic Pickup Devices

Magnetic pickup devices are the fourth type of end-effector. They can be considered when the part to be handled is of ferrous content. Either permanent magnets or electromagnets are used, with permanent magnets requiring a stripping device to separate the part from the magnet during part release. Magnets normally contain a flat part-contact surface but can be adapted with a plate to fit a specific part contour. A recent innovation in magnetic pickup devices uses an electromagnet fitted with a flexible bladder containing iron filings, which conforms to an irregular surface on a part to be picked up. As with vacuum pickup

Figure 30 Microgripper using two parallel link fingers. (Courtesy Mechanical Engineering Laboratory, AIST-MITI, Japan.)

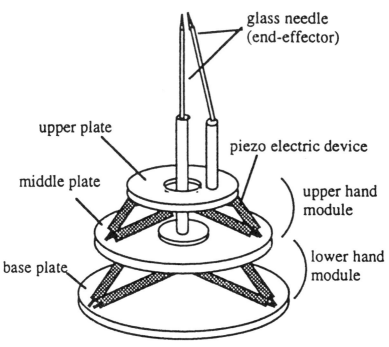

Figure 31   "Chopstick-type" microgripper. (Courtesy Mechanical Engineering Laboratory, AIST-MITI, Japan.)

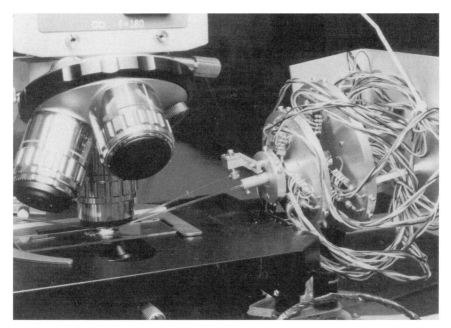

Figure 32   Microobject manipulation by chopstick-type microgripper. (Courtesy Mechanical Engineering Laboratory, AIST-MITI, Japan.)

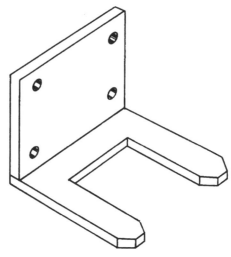

Figure 33 Support and containment device.

devices, oily or wet part surfaces may cause the part to slide on the magnet during transfer. Therefore containment structures should be used in addition to the magnet to enclose the part and prevent it from slipping.

Three additional concerns arise in handling parts with magnets. First, if a metal-removal process is involved in the application, metal chips may also be picked up by the magnet. Provisions must be made to wipe the surface of the magnet in this event. Next, residual magnetism may be imparted to the workpiece during pickup and transfer by the magnetic end-effector. If this is detrimental to the finished part, a demagnetizing operation may be required after transfer. Finally, if an electromagnet is used, a power failure will cause the part to be dropped immediately, which may produce an unsafe condition. Although electromagnets provide easier control and faster pickup and release of parts, permanent magnets can be used in hazardous environments requiring explosion-proof electrical equipment. Normal magnets can handle temperatures up to 60°C (140°F), but magnets can be designed for service in temperatures up to 150°C (300°F).

6 UNIVERSAL HANDS

6.1 Design of Universal Hands

Universal hands have many degrees of freedom, like the human hand. They have been researched by several investigators. Figure 36 shows the joint structure in a typical uni-

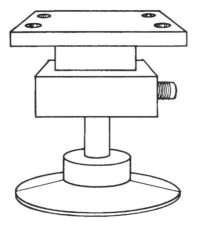

Figure 34 Vacuum cup pickup device.

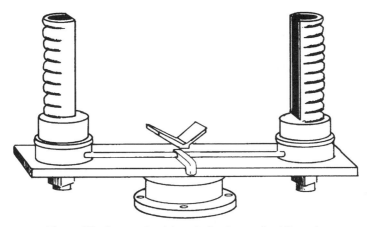

Figure 35   Pneumatic pickup device (pressurized fingers).

versal hand. The example uses 4-d.o.f. fingers, but 3-d.o.f. fingers are often used. In such fingers the joint nearest to the fingertip will be removed. The number of fingers is three or four, though three fingers usually provide enough functions for universal hands.

Increasing the number of degrees of freedom causes several problems. One difficult problem is how to install in the hand the actuators necessary to activate all degrees of freedom. This requires miniature actuators that can produce enough power to drive the hand joint. A few universal hands have finger joints directly driven by small, specially designed actuators. See Figure 37 (Umetsu and Oniki, 1993). The fingers in this hand each have 3 d.o.f. A miniature electric motor with a high reduction gear to produce sufficient torque is directly installed at each joint. The direct drive joint is very effective for simplifying the hand structure. However, suitable actuators are not always commercially available. An actuator with sufficient power generation will usually be too large to attach at each joint of the finger. The most frequent solution is to use a tendon drive

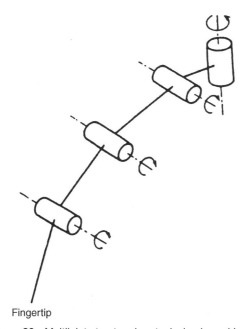

Fingertip

Figure 36   Multijoint structure in a typical universal hand.

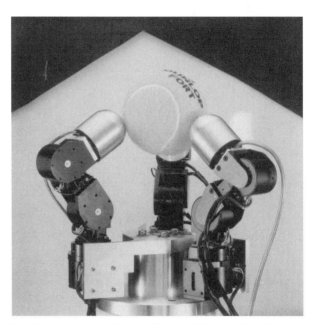

Figure 37   Yaskawa universal hand. (Courtesy Yaskawa Electric Corporation.)

mechanism that enables actuators to be placed at an appropriate position away from the joint. There are two tendon drive mechanisms. One uses tendon-pulley mechanisms. Figure 38 shows the Utah/MIT Hand, which has four fingers (Jacobsen et al., 1981). Tendon-pulley mechanisms are shown in Figure 39. One end of the tendon is fixed at a hinge located on a finger segment to be moved. The other end is connected to a pneumatic actuator (cylinder). Between the ends, the tendon is guided through several guided pulleys, which can be passively rotated by tendon motion. Two tendons, agonistic and antagonistic, are used to drive each joint. A tendon tension sensor is attached to each actuator to control tendon tension and interaction among the tendons. This sensor is used to control the tendon tension and the interaction among the tendons (Jacobsen et al., 1981). The other tendon drive mechanism uses a combination of pulleys and hoses to guide the tendons. This mechanism has the advantage of reducing the complexity of the guidance mechanisms. The drawback is that it increases the friction in a power transmission system. Figure 40 shows a hand using this drive mechanism. The hand has three fingers: a thumb, index finger, and middle finger (Okada, 1982). Each finger contains two or three segments made of 17-mm brass rods bored to be cylindrical. The tip of each segment is truncated at a slope of 30° so that the finger can be bent at a maximum angle of 45° at each joint, both inward and outward. The workspace of the finger is thus more extensive than that of the human finger. The thumb has three joints and the index and middle fingers each have four joints. Each joint has 1 d.o.f. and is driven using a tendon-pulley mechanism and electric motors. A pulley is placed at each joint, around which two tendons are wound after an end of each tendon is fixed to the pulley. The tendon is guided through coil-like hoses so that it cannot interfere with the complicated finger motion. Using coil-like hoses protects the tendon and makes possible the elimination of relaying points for guiding tendons. To make the motions of the fingers flexible and the hand system more compact, the tendons and hoses are installed through the finger tubes. The section drawing of the hand system in Figure 41 explains the joint drive mechanisms. Motors for driving the respective joints are located together within a trunk separated from the hand system.

   One solution proposed to solve the problem of friction between hoses and wires is to support the hoses with flexible elements such as springs (Sugano et al., 1984). Figure 42 shows the construction of a tendon-pulley drive system with tendon-guiding hoses supported by springs. Bending a hose whose ends are rigidly fixed will extend the hose. This reduces the cross-section area of the hose, which increases the friction. In the system

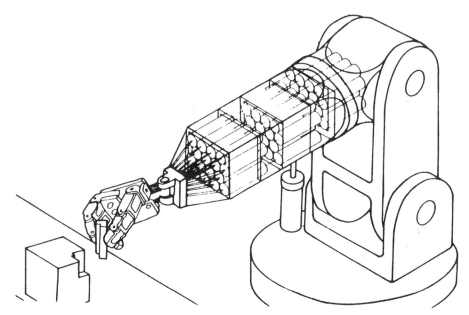

Figure 38   Utah–MIT hand. (From M. Brady and R. Paul, *Robotics Research.* Cambridge: MIT Press, 1984, p. 602.)

shown in Figure 42, the spring, rather than the hose, will be extended when the hose is bent, and the cross-section area of the hose will not be affected by the change of hose configuration. This prevents friction from increasing.

Another problem with the tendon drive method is controlling the tendon tension properly during motion. Tension control using sensors is used to avoid relaxing of tension. Coupled tendon drive mechanisms are also effective. These mechanisms use a network of tendons, each of which interferes with the others. Figure 43 shows such a tendon drive system for a finger with three joints. The hand using it, the Stanford/JPL Hand (Figure 44) (Mason and Salisbury, 1985), uses four tendons and electric motors to drive three joints. Each motor is installed so that it interferes with the others through tendons. The tendon tension can thus be adjusted by controlling each motor cooperatively. This structure raises control complexity, but is useful for controlling tendon tension.

6.2   Control of Universal Hands

To grasp an object stably and manipulate it dexterously using universal hands, each finger joint in the hand must be moved cooperatively. For this purpose a stable grasp control algorithm has been proposed (Li and Sastry, 1990; Yoshikawa, 1990). From the principle of statics, the requirements of fingertip force for each finger are as follows:

1. Fingertip force should be applied to the grasped object within the friction cone.
2. For the object to be grasped statically, the sum of fingertip forces and their moments around a specific point must be zero. Also proper internal forces at each fingertip should be generated to support the gravity force of the grasped object when the object is grasped statically.
3. For the grasped object to be manipulated, proper additional forces at each fingertip must be generated to produce the desired motion of the object, keeping the above conditions.

To meet the above requirements, each finger in the universal hand will be required to have force control functions as well as position control. Proper compliance at the fingertip also plays an important role in achieving robust stable grasp. Almost all universal hands developed so far therefore have force and position control functions.

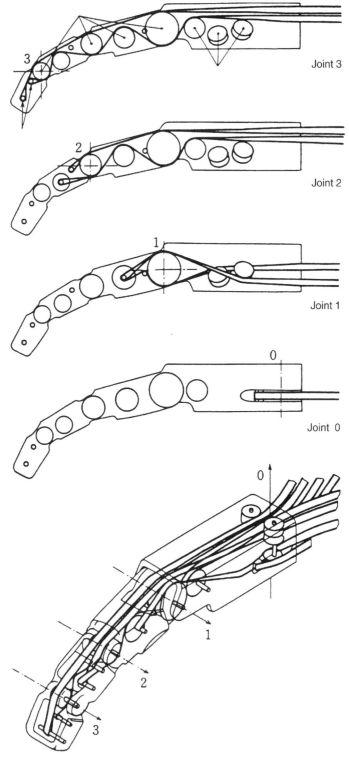

Figure 39 Tendon-Pulley Mechanisms in a finger of Utah–MIT hand. (From M. Brady and R. Paul, Robotics Research, Cambridge: MIT Press, 1984, p. 606–607.)

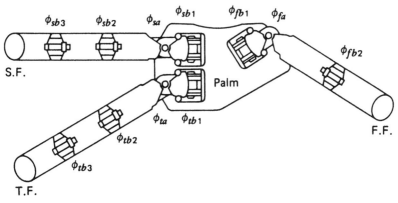

Figure 40 Versatile hand system. (From T. Okada, "Computer Control of Multijointed Finger System for Precise Object-Handling." *IEEE Transactions on Systems, Man and Cybernetics* **SMC-12**(3), 1982, 289–299.)

Humans can do various complex tasks using dexterous finger motions. Sometimes they will manipulate the object by using slip motion of the grasped object on the finger surface. For a universal hand to perform such skilled operations, more complex requirements should be satisfied in addition to those listed above. For dexterous universal hand control to be achieved, many problems must be solved relating to grasp planning, grasp location finding, adaptive grasp force control, and so on. For details about grasp and manipulation control using universal hands, see Murray, Li, and Sastry (1994) and Maekawa, Tanie, and Komoriya (1997).

7 PRACTICAL IMPLEMENTATION OF ROBOT END-EFFECTORS

The end-effector practically used in industrial robots is commonly made up of four distinct elements (see Figure 45):

1. Attachment of the hand or tool to the robot end-effector mounting plate
2. Power for actuation of end-effector tool motions
3. Mechanical linkages
4. Sensors integrated into the end-effector

The role of the element and how to implement power and sensors are discussed below from the practical point of view.

7.1 Mounting Plate

The means of attaching the end-effector to an industrial robot are provided by a robot end-effector mounting plate located at the end of the last axis of motion on the robot arm. This mounting plate contains either threaded or clearance holes arranged in a pattern for attaching a hand or tools. For a fixed mounting of a hand or tool, an adapter plate with a hole pattern matching the robot end-effector mounting plate can be provided. The remainder of the adapter plate provides a mounting surface for the hand or tool at the proper distance and orientation from the robot end-effector mounting plate. If the task of the robot requires it to automatically interchange hands or tools, a coupling device can be provided. An adapter plate is thus attached to each of the hands or tools to be used, with a common lock-in position for pickup by the coupling device. The coupling device may also contain the power source for the hands or tools and automatically connect the power when it picks up the end-effector. Figures 46, 47, and 48 illustrate this power connection end-effector tool change application. An alternative to this approach is for each tool to have its own power line permanently connected and the robot simply to pick up the various tools mounted to adapter plates with common lock-in points.

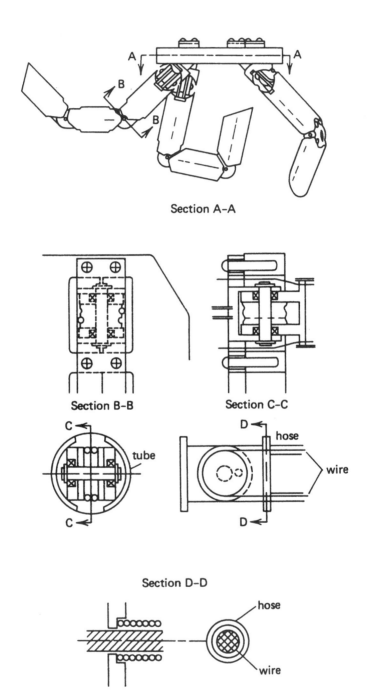

Section A–A

Section B–B   Section C–C

Section D–D

Figure 41 Joint drive mechanism of the versatile hand. (From T. Okada, "Computer Control of Multijointed Finger System for Precise Object-Handling," *IEEE Transactions on Systems, Man, and Cybernetics* **SMC-12**(3), 1982, 289–299.)

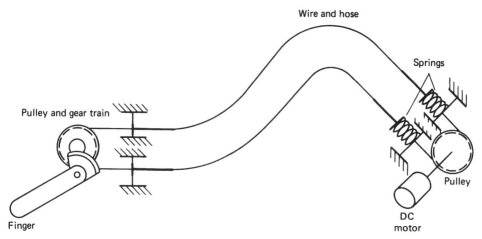

Figure 42   Tendon-pulley drive system with tendon-guiding hoses supported by springs. (From S. Sugano et al., "The Keyboard Playing by an Anthropomorphic Robot," in *Preprints of Fifth ISM-IFToMM Symposium on Theory and Practice of Robots and Manipulators, Udine, 1984*.)

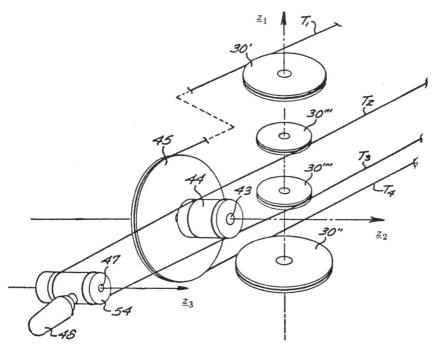

Figure 43   Coupled tendon-pulley drive mechanisms in Stanford/JPL hand. (From M. T. Mason and J. K. Salisbury, *Robot Hands and the Mechanics of Manipulation.* Cambridge: MIT Press, 1985, p. 80.)

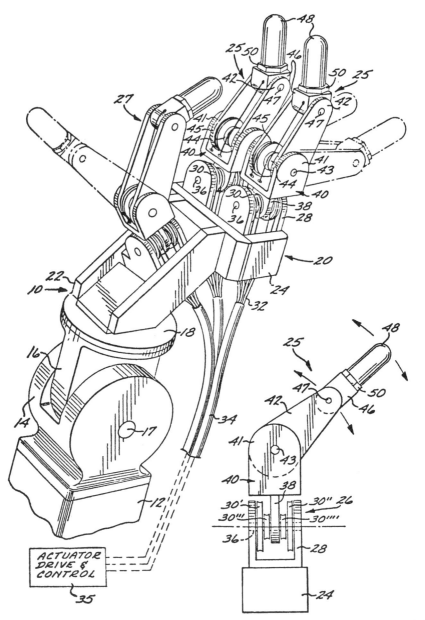

Figure 44 Stanford/JPL hand. (From M. T. Mason and J. K. Salisbury, *Robot Hands and the Mechanics of Manipulation.* Cambridge: MIT Press, 1985, p. 81.)

7.2 Power Implementation

Power for actuation of end-effector motions can be pneumatic, hydraulic, or electrical, as mentioned above, or the end-effector may not require power, as in the case of hooks or scoops. Generally, pneumatic power is used wherever possible because of its ease of installation and maintenance, low cost, and light weight. Higher-pressure hydraulic power is used where greater forces are required in the tooling motions. However, contamination of parts due to leakage of hydraulic fluid often restricts its application as a power source for tooling. Although it is quieter, electrical power is used less frequently for tooling power, especially in part-handling applications, because of its lower applied force. Several

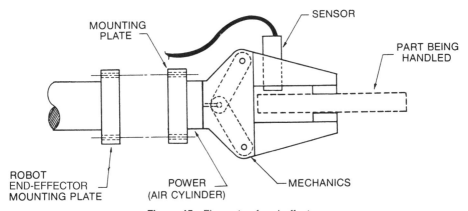

Figure 45   Elements of end-effector.

light payload assembly robots utilize electrical tooling power because of its control capability. In matching a robot to an end-effector, consideration should be given to the power source provided with the robot. Some robots have provisions for tooling power, especially in part-handling robots, and it is an easy task to tap into this source for actuation of tool functions. As previously mentioned, many robots are provided with a pneumatic power source for tool actuation and control.

7.3   Mechanical Linkages

End-effectors for robots may be designed with a direct coupling between the actuator and workpiece, as in the case of an air cylinder that moves a drill through a workpiece, or may use indirect couplings or linkages to gain mechanical advantage, as in the case of a pivot-type gripping device. A gripper-type hand may also have provisions for mounting interchangeable fingers to conform to various part sizes and configurations. In turn, fingers attached to hands may have provisions for interchangeable inserts to conform to various part configurations.

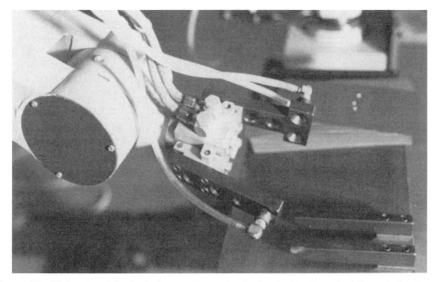

Figure 46   Pickup hand for tool change. Power for tool actuation is ported through the fingers for connection to the various tools to be picked up. (From R. D. Potter, "End-of-Arm Tooling," in *Handbook of Industrial Robotics,* ed. Shimon Y. Nof. New York: John Wiley & Sons, 1985.)

Figure 47 Tool is rack-ready to be picked up by robot. Note cone-shaped power connection ports in tool mounting block. (From R. D. Potter, "End-of-Arm Tooling," in *Handbook of Industrial Robotics,* ed. Shimon Y. Nof. New York: John Wiley & Sons, 1985.)

Figure 48 Another tool with power connection block ready for robot pickup. (From R. D. Potter, "End-of-Arm Tooling," in *Handbook of Industrial Robotics,* ed. Shimon Y. Nof. New York: John Wiley & Sons, 1985.)

7.4 Sensors

Sensors are incorporated in end-effectors to detect various conditions. For safety consid-erations sensors are normally designed into tooling to detect workpiece or tool retention by the robot during the robot operation. Sensors are also built into end-effectors to monitor the condition of the workpiece or tool during an operation, as in the case of a torque sensor mounted on a drill to detect when a drill bit is dull or broken. Sensors are also used in tooling to verify that a process is completed satisfactorily, such as wire-feed detectors in arc welding torches and flow meters in dispensing heads. More recently, robots specially designed for assembly tasks contain force sensors (strain gauges) and dimensional gauging sensors in the end-effector.

8 PRACTICAL DESIGN FOR ROBOT END-EFFECTORS

This section first explains general criteria for robot end-effector design, including the importance of analyzing the object to be handled or worked on. The analysis relates to preengineering and data collection. The development of end-effector concept and design will proceed only after the analysis is completed. Second, the preengineering phase, in-volving analysis of the workpiece and process, with emphasis on productivity consider-ations, will be explained. Third, some important guidelines for actual end-effector design will be summarized.

8.1 General End-Effector Design Criteria

Although robot end-effectors vary widely in function, complexity, and application area, certain design criteria pertain to almost all of them. First, the end-effector should be as lightweight as possible. Weight affects the performance of the robot. The rated load-carrying capacity of the robot, or the amount of weight that can be attached to the robot end-effector mounting plate, includes the weight of the end-effector and of the part being carried. The load that the robot is carrying also affects the speed of its motions. Robots can move faster carrying lighter loads. Therefore, for cycle time considerations, the lighter the tool, the faster the robot is capable of moving. The use of lightening holes, whenever possible, and lightweight materials such as aluminum or magnesium for hand components are common solutions for weight reduction. Figure 49 shows a double-action pickup hand with lightening holes.

Second, the end-effector should be as small in physical size as possible. Minimizing the dimensions of the tool also helps to minimize weight. Minimizing the size of the

Figure 49   Double-action pickup hand showing interchangeable fingers with lightening holes and V-block locating features. (From R. D. Potter, "End-of-Arm Tooling," in *Handbook of Industrial Robotics,* ed. Shimon Y. Nof. New York: John Wiley & Sons, 1985.)

tooling provides for better clearances in workstations in the system. Load-carrying capacities of robots are usually based on moment of inertia calculations of the last axis of motion; that is, a given load capacity at a given distance from the tool mounting plate surface. Minimizing the size of the end-effector and the distance from the end-effector mounting plate to the center of gravity of the end-effector thus enhances robot performance. At the same time, it is desirable to handle the widest possible range of parts with the robot end-effector. This minimizes changeover requirements and reduces costs for multiple tools. Although minimizing the size of the end-effector somewhat limits the range of parts that can be handled, there are techniques for accomplishing both goals. Adjustable motions may be designed into the end-effector so that it can be quickly and easily manually changed to accommodate different-sized parts. Interchangeable inserts may be put in the end-effector to change the hand from parts of one size or shape to another. The robot may also automatically interchange hands or tools to work on a range of parts, with each set of end-effectors designed to handle a certain portion of the entire range of parts. This addresses weight and size considerations and reduces the total number of tools required. Figure 50 shows a standard dual part-handling end-effector for gripping parts varying within a certain size range.

Maximizing rigidity is another criterion that should be designed into the end-effector. Again, this relates to the task performance of the robot. Robots have specified repeatabilities and accuracies in handling a part. If the end-effector is not rigid, this positioning accuracy will not be as good and, depending on part clearances and tolerances, problems in the application may result. Excessive vibration may also be produced when a nonrigid or flimsy end-effector is attached to the end-effector mounting plate. Because robots can move the end-effector at very high rates of speed, this vibration may cause breakage or damage to the end-effector. Attaching a rigid end-effector eliminates these vibrations.

The maximum applied holding force should be designed into the end-effector. This is especially important for safety reasons. Robots are dynamic machines that can move parts at high velocities at the end of the arm, with only the clamp force and frictional force holding the parts in the hand. Because robots typically rotate the part about a fixed robot

Figure 50   Standard tooling for parts-handling applications, with such features as parallel motion fingers, dual part-handling capability, and compliance in hand. (From R. D. Potter, "End-of-Arm Tooling," in *Handbook of Industrial Robotics,* ed. Shimon Y. Nof. New York: John Wiley & Sons, 1985.)

base centerline, centrifugal forces are produced. Acceleration and deceleration forces also result when the robot moves from one point to another. The effect of these forces acting on the part makes it critical to design in an applied holding force with a safety factor great enough to ensure that the part is safely retained in the hand during transfer and not thrown as a projectile, with the potential of causing injury or death to personnel in the area or damage to periphery equipment. On the other hand, the applied holding force should not be so great that it actually causes damage to a more fragile part being handled. Another important consideration in parts transfer relating to applied holding force is the orientation of the part in the hand during transfer. If the part is transferred with the hand axis parallel to the floor, the part, retained only by the frictional force between the fingers and part, may have a tendency to slip in the hand, especially at programmed stop points. By turning the hand axis perpendicular to the floor during part transfer, the required holding force may be decreased, and the robot may be able to move at higher speed because the hand itself acts as a physical stop for the part.

Maintenance and changeover considerations should be designed into the tooling. Perishable or wear details should be designed to be easily accessible for quick change. Change details such as inserts or fingers should also be easily and quickly interchangeable. The same type of fastener should be used wherever possible in the hand assembly, thereby minimizing the number of maintenance tools required.

8.2   Preengineering and Data Collection

8.2.1   Workpiece Analysis

The part being transferred or worked on must be analyzed to determine critical parameters to be designed into the end-effector. The dimensions and tolerances of the workpiece must be analyzed to determine their effect on end-effector design. The dimensions of the workpiece will determine the size and weight of the end-effector required to handle the part. It will also determine whether one tool can automatically handle the range of part dimensions required, whether interchangeable fingers or inserts are required, or whether tool change is required. The tolerances of the workpieces will determine the need for compliance in the end-effector. Compliance allows for mechanical "float" in the end-effector in relation to the robot end-effector mounting plate to correct misalignment errors encountered when parts are mated during assembly operations or loaded into tight-fitting fixtures or periphery equipment. If the part tolerances vary so that the fit of the part in fixture is less than the repeatability of the robot, a compliance device may have to be designed into the end-effector. Passive compliance devices such as springs may be incorporated into the end-effector to allow it to float to accommodate very tight tolerances. This reduces the rigidity of the end-effector. Other passive compliance devices, such as remote center compliance (RCC) units, are commercially available. These are mounted between the robot end-effector mounting plate and the end-effector to provide a multiaxis float. RCC devices, primarily designed for assembly tasks, allow robots to assemble parts with mating fits much tighter than the repeatability that the robot can achieve. Active compliance devices with sensory feedback can also be used to accommodate tolerance requirements.

The *material and physical properties* of the workpiece must be analyzed to determine their effect on end-effector design. The best method of handling the part, whether by vacuum, magnetic, or mechanical hand, can be determined. The maximum permissible grip forces and contact points on the part can be determined, as well as the number of contact points to ensure part retention during transfer. Based on the physical properties of the material, the need for controlling the applied force through sensors can also be resolved.

The *weight and balance* (center of gravity) of the workpiece should be analyzed to determine the number and location of grip contact points to ensure proper part transfer. This will also resolve the need for counterbalance or support points on the part in addition to the grip contact points. The static and dynamic loads and moments of inertia of the part and end-effector about the robot end-effector mounting plate can be analyzed to verify that they are within the safe operating parameters of the robot.

The *surface finish and contour* (shape) of the workpiece should be studied to determine the method and location of part grasp (i.e., vacuum for smooth, flat surfaces, mechanical hands for round parts, etc.). If the contour of the part is such that two or more independent grasp means must be applied, this can be accomplished by mounting separate gripping devices and/or special tools at different locations on the end-effector, each gripping or attaching to a different section of the part. This may be a combination of vacuum cups, magnets, and/or mechanical hands. Special linkages may also be used to tie together two different grasp devices powered by one common actuator.

Part modifications should be analyzed to determine whether minor part changes that do not affect the functions of the part can be made to reduce the cost and complexity of the end-effector. Often, simple part changes, such as holes or tabs in parts, can significantly reduce the end-effector design complexity and build effort in the design of new component parts for automation and assembly by robots.

Part inconsistencies should be analyzed to determine the need for provision of out-of-tolerance sensors or compensating tooling to accommodate these conditions.

For *tool handling,* the tool should be analyzed to determine the characteristics of the end-effector required. This is especially true for the incorporation of protective sensors in the tool and end-effector to deal with part inconsistencies.

8.2.2 Process Analysis

In addition to a thorough analysis of the workpiece, an analysis of the application process should be made to determine the optimum parameters for the end-effector.

The *process method* itself should be analyzed, especially in terms of manual versus robot operation. In many cases physical limitations dictate that a person perform a task in a certain manner where a robot without these constraints may perform the task in a more efficient but different manner. An example of this involves the alternative of picking up a tool and doing work on a part or instead picking up the part and taking it to the tool. In many cases the size and weight-carrying capability of a person is limited, forcing the person to handle the smaller and lighter weight of the part or the tool. A robot, with its greater size and payload capabilities, does not have this restriction. Thus it may be used to take a large part to a stationary tool or take multiple tools to perform work on a part. This may increase the efficiency of the operation by reducing cycle time, improving quality, and increasing productivity. Therefore, in process analysis, consider the alternatives of having the robot take a part to a tool or a tool to a part and decide which approach is most efficient. When a robot is handling a part, rather than a tool, power-line connections to the tool, which experience less flexure and are less prone to problems when stationary than moving, are less of a concern.

Because of its increased payload capability, a robot may also be equipped with a multifunctional end-effector. This can simultaneously or sequentially perform work on a part that previously required a person to pick up one tool at a time to perform the operation, resulting in lower productivity. For example, the end-effector in a die-casting machine unloading application may not only unload the part, but also spray a die lubricant on the face of the dies.

The *range and quantity of parts or tools* in the application process should be analyzed to determine the performance requirements for the end-effector. This will dictate the number of hands or end-effectors that are required. The end-effector must be designed to accommodate the range of part sizes, whether automatically in the tool for the end-effector, through automatic tool change, or through manual changeover. Manual changeover could involve adjusting the tool to handle a different range of parts or interchanging fingers, inserts, or tools on a common hand. To reduce the manual changeover time, quick disconnect capabilities and positive alignment features such as dowel pins or locating holes should be provided. For automatic tool change applications, mechanical registration provisions, such as tapered pins and bushings, ensure proper alignment of tools. Verification sensors should also be incorporated in automatic tool change applications.

Presentation and disposition of the workpiece within the robot system affect the design of the end-effector. The position and orientation of the workpiece at either the pickup or release stations will determine the possible contact points on the part, the dimensional clearances required in the end-effector to avoid interferences, the manipulative requirements of the end-effector, the forces and moments of the end-effector and part in relation to the robot end-effector mounting plate, the need for sensors in the end-effector to detect part position or orientation, and the complexity of the end-effector.

The *sequence of events and cycle time requirements* of the process have a direct bearing on tooling design complexity. Establishing the cycle time for the operation will determine how many tools (or hands) are needed to meet the requirements. Hands with multiple parts-handling functions often allow the robot to increase the productivity of the operation by handling more parts per cycle than can be handled manually. The sequence of events may also dictate the use of multifunctional end-effectors that must perform several operations during the robot cycle. For example, in machine unloading, mentioned above, the end-effector not only grasps the part, but sprays a lubricant on the molds or dies of the machine. Similarly, a robot end-effector could also handle a part and perform work on it at the same time, such as automatic gauging and drilling a hole. The sequence of events in going from one operation to another may cause the design of the end-effector to include some extra motions not available in the robot by adding extra axes of motion in the end-effector to accommodate the sequence of operations between various system elements.

In-process inspection requirements will affect the design of the end-effector. The manipulative requirements of the end-effector, the design of sensors or gauging into the end-effector, and the contact position of the end-effector on the part are all affected by the part-inspection requirements. Precision in positioning the workpiece is another consideration for meeting inspection requirements.

The *conditional processing* of the part will determine the need for sensors integrated into the end-effector, as well as the need for independent action by multiple-gripper systems.

The *environment* must be considered in designing the end-effector. The effects of temperature, moisture, airborne contaminants, corrosive or caustic materials, and vibration and shock must be evaluated, as will the material selection, power selection, sensors, mechanics, and provision for protective devices in the end-effector.

8.3   Guidelines for End-Effector Design

Guidelines for end-effector design to best meet the criteria discussed above are as follows:

1. Design for quick removal or interchange of the end-effector by requiring a small number of tools (wrenches, screwdrivers, etc.) to be used. Use the same fasteners wherever possible.
2. Provide locating dowels, key slots, or scribe lines for quick interchange, accuracy registration, and alignment.
3. Break all sharp corners to protect hoses and lines from rubbing and cutting and maintenance personnel from possible injury.
4. Allow for full flexure of lines and hoses to extremes of axes of motion.
5. To reduce weight, use lightening holes or lightweight materials wherever possible.
6. For wear considerations, hardcoat lightweight materials and put hardened, threaded inserts in soft materials.
7. Conceptualize and evaluate several alternatives in the end-effector.
8. Do not be "penny-wise and pound-foolish" in designing the end-effector; make sure enough effort and cost is spent to produce a production-worthy, reliable end-effector and not a prototype.
9. Design in extra motions in the end-effector to assist the robot in its task.
10. Design in sensors to detect part presence during transfer (limit switch, proximity, air jet, etc.).
11. For safety in part-handling applications, consider what effect a loss of power to the end-effector will have. Use toggle lock gripping devices or detented valves to promote safety.
12. Put shear pins or areas in end-effector to protect more expensive components and reduce downtime.
13. When handling tools with robot, build in tool inspection capabilities, either in the end-effector or peripheral equipment.
14. Design multiple functions into the end-effector.
15. Provide accessibility for maintenance and quick change of wear parts.
16. Use sealed bearings for the end-effector.
17. Provide interchangeable inserts or fingers for part changeover.
18. When handling hot parts, provide heat sink or shield to protect the end-effector and the robot.
19. Mount actuators and valves for the end-effector on the robot forearm.
20. Build in compliance in the end-effector fixture where required.

21. Design action sensors in the end-effector to detect open/close or other motion conditions.

22. Analyze inertia requirements, center of gravity of payload, centrifugal force, and other dynamic considerations in designing the end-effector.

23. Look at motion requirements for the gripping device in picking up parts (single-action hand must be able to move part during pickup; double-action hand centers part in one direction; three or four fingers center part in more than one direction).

24. When using an electromagnetic pickup hand, consider residual magnetism on parts and possible chip pickup.

25. When using vacuum cup pickup on oily parts, also use a positive blow-off.

26. Look at insertion forces of robot in using end-effector in assembly tasks.

27. Maintain orientation of the part in the end-effector by force and coefficient of friction or locating features.

9 SUMMARY

To date, most of the applications of industrial robots have involved a specially designed end-effector or simple gripping devices. In most practical applications the use of simple end-effector is the best solution from the point of view of productivity and economy. If more complex tasks will be required of the robot, universal hands with more dexterity will be needed. Several important contributions have been made so far in the theory of multifingered hand control, but the technology of designing multifingered hand hardware has not yet matured. Research is ongoing to develop more flexible general-purpose universal hand hardware that can adapt to a variety of sizes and shapes of parts. The advancement of this technology to make possible more dexterous robotic manipulation can be expected.

ACKNOWLEDGMENTS

This chapter is revised from Chapter 8 of the first edition of this handbook, with some materials added from Chapter 37, "End-of-Arm Tooling," by R. D. Potter.

REFERENCES

Ando, Y., et al. 1990. "Development of Micro Grippers." In *Micro System Technologies 90.* Ed. H. Reichl. Berlin: Springer-Verlag. 855–849.

Arai, T., R. Larsonneur, and Y. M. Jaya. 1993. "Basic Motion of a Micro Hand Module." In *Proceedings of the 1993 JSME International Conference on Advanced Mechatronics,* Tokyo. 92–97.

Chen, F. Y. 1982. "Gripping Mechanisms for Industrial Robots." *Mechanism and Machine Theory* **17**(5), 299–311.

Crossley, F. R. E., and F. G. Umholts. 1977. "Design for a Three-Fingered Hand." In *Robot and Manipulator Systems.* Ed. E. Heer. Oxford: Pergamon Press. 85–93.

Hirose, S., and Y. Umetani. 1977. "The Development of a Soft Gripper for the Versatile Robot Hand." In *Proceedings of the Seventh International Symposium on Industrial Robots,* Tokyo. 353–360.

Jacobsen, S. C., et al. 1981. "The Utah/MIT Dextrous Hand: Work in Progress." In *Robotics Research.* Ed. M. Brady and R. Paul. Cambridge: MIT Press. 601–653.

Kato, I., et al. 1969. "Multi-functional Myoelectric Hand Prosthesis with Pressure Sensory Feedback System-Waseda Hand 4P." In *Proceedings of the Third International Symposium on External Control of Human Extremities,* Belgrade.

Kimura, M., et al. 1995. "Force Control of Small Gripper Using Piezoelectric Actuator." *Journal of Advanced Automation Technology* **7**(5), 317–322.

Li, Z., and S. Sastry. 1990. "Issues in Dextrous Robot Hands." In *Dextrous Robot Hands.* Ed. S. T. Venkataraman and T. Iberall. New York: Springer-Verlag. 154–186.

Maekawa, H., K. Tanie, and K. Komoriya. 1997. "Tactile Feedback for Multifingered Dynamic Grasping." *Control Systems* **17**(1), 63–71.

Mason, M.T., and J. K. Salisbury. 1985. *Robot Hands and the Mechanics of Manipulation.* Cambridge: MIT Press. 80–81.

Murray, R. M., Z. Li, and S. S. Sastry. 1994. *A Mathematical Introduction to Robotic Manipulation.* London: CRC Press. 382–387.

Okada, T. 1982. "Computer Control of Multijointed Finger System for Precise Object-Handling." *IEEE Transactions on Systems, Man and Cybernetics* **SMC-12**(3), 289–299.

Osaki, S., and Y. Kuroiwa. 1969. "Machine Hand for Bar Works." *Journal of Mechanical Engineering Laboratory* **23**(4). [In Japanese.]

Schlesinger, G. 1919. *Der Mechanische Aufbau der künstlichen Glieder.* Part 2 of *Ersatzglieder und Arbeitshilfen.* Berlin: Springer.

Sheldon, O. L. 1972. "Robots and Remote Handling Methods for Radioactive Materials." In *Proceedings of the Second International Symposium on Industrial Robots,* Chicago. 235–256.

Skinner, F. 1974. "Design of a Multiple Prehension Manipulator System." ASME Paper 74-det-25.

Sugano, S., et al. 1984. "The Keyboard Playing by an Anthropomorphic Robot." In *Preprints of Fifth CISM-IFToMM Symposium on Theory and Practice of Robots and Manipulators,* Udine. 113–123.

Tanikawa, T., T. Arai, and T. Matsuda. 1996. "Development of Micro Manipulation System with Two-Finger Micro Hand." In *Proceedings of the 1996 IEEE/RSJ International Conference on Robots and Intelligent Systems,* Osaka. 850–855.

Umetsu, M., and K. Oniki. 1993. "Compliant Motion Control of Arm-Hand System." In *Proceedings of the 1993 JSME International Conference on Advanced Mechatronics,* Tokyo. 429–432.

Yoshikawa, T., and K. Nagai. 1990. "Analysis of Multi-fingered Grasping and Manipulation." In *Dextrous Robot Hands.* Ed. S. T. Venkataraman and T. Iberall. New York: Springer-Verlag. 187–208.

ADDITIONAL READING

Engelberger, J. F. *Robotics in Practice.* London: Kogan Page, 1980.

Kato, I., ed. *Mechanical Hand Illustrated.* Tokyo: Survey Japan, 1982.

CHAPTER **8**

MOBILE ROBOTS AND WALKING MACHINES

Hagen Schempf
Carnegie Mellon University
Pittsburgh, Pennsylvania

1 BACKGROUND

The notion of a mobile robot, as perceived by the general public, is aligned with the creatures that science fiction has brought to the big screen and the media have popularized.[1] The mobile robots in use today are actually of a different morphology due to the implementation constraints imposed by the laws of nature. The design of these systems is always a compromise between many variables that the different engineering disciplines must optimize in order to develop a mission-capable, cost-effective, and reliable remote device or automaton. To better understand the scope of this section, one should understand the topic of design of a mobile/legged machine in terms of its place and applications impact and the true practical and theoretical problems challenging the evolution of robotics morphologies and technologies.

1.1 Introduction

In order to develop any robotic system, whether teleoperated or automated, one must realize that robotics represents a truly multidisciplinary field where the different areas of design (mechanical, electrical, and software) must be successfully integrated into a complex system. As such, mechanical design cannot be thought of as a totally separate, dominant, or standalone discipline, but rather as an activity closely linked with the others. In practical terms, this implies that decisions on the robot configuration (locomotor type, passive/active structure, etc.) are not purely mechanical, but heavily influenced and sometimes even dominated by software or electrical requirements.

That said, in reality the performance specifications imposed by the task that the robot has to accomplish usually dictate the importance of each discipline. Thus, if a machine is to accomplish a task, such as accessing an area through a constrained space, the mechanical design might drive the packaging of all other components; if the sensing requirements are stringent, the type of sensor and support electronics/computers to run the system might dominate the design; if remoteness and duration of autonomous operation are crucial, the type and size of power source and electronic power-management hardware might dominate the overall design.

This chapter will focus on the mechanical aspects of mobile robot design, without explaining the lengthy and rich process of overall robot design, by concentrating on different mechanical design issues related to generic subsystems found on mobile/legged robots.

1.2 Overall Scope, Nomenclature, and Definition

The term *mobile robot* can be thought of as the most generic description of a robotic system that is able to carry out a set of generic tasks in different locations without being constrained to work from a fixed location, as is the case in industrial robotic manipulator arms operating within workcells. Mobility can be achieved in a variety of ways through movable platforms using different forms of locomotors, such as wheels, tracks, and legs.

---

[1]Whether science fiction writers influence robot designers or vice-versa is left to the reader to ponder.

Handbook of Industrial Robotics, Second Edition, Edited by Shimon Y. Nof
ISBN 0-471-17783-0   © 1999 John Wiley & Sons, Inc.

145

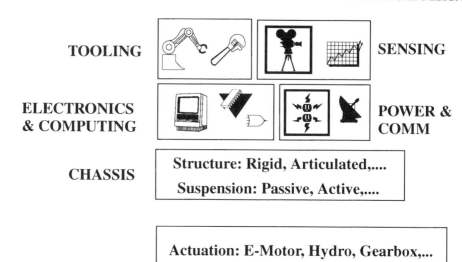

TOOLING   SENSING

ELECTRONICS   POWER &
& COMPUTING   COMM

CHASSIS

Structure: Rigid, Articulated,....

Suspension: Passive, Active,....

LOCOMOTOR

Actuation: E-Motor, Hydro, Gearbox,...

Modality: Legs, Wheels, Track,......

Figure 1 Diagrammatic view of a mobile robot system.

Hence a mobile robot can be thought of as a platform with a certain characteristic configuration, and able to transport sensors, tools, and/or manipulators to any accessible and desirable location to perform a desired task. The basic diagram depicted in Figure 1 thus attempts to capture the generic nature of a mobile robot, on which legs are just one of the many possible forms of locomotors.

The definition of a mobile robot should certainly include the ability of the machine to operate with some form of autonomy, based on on-board decision-making without requiring a human operator in the loop. On the other hand, since this chapter is limited to the more mechanical aspects, it will not solely investigate truly autonomous platform designs, but rather any and all remote (whether teleoperated, supervisory-operated, or autonomous) platforms, since mechanical design transcends definitions of autonomy. A well-designed mobile platform can be robotized and even made autonomous, and the goal of this chapter is to illustrate the most important and interesting aspects of mobile robot/platform design. (See also Chapters 9, 14, 17.)

This chapter will be limited to exploring the theoretical and practical mechanical design aspects of the chassis and locomotor components of a mobile robot and will not dwell on the details of the power and communications, electronics and computing, and tooling and sensing systems. The discussion will be limited to those components that are responsible for the mobility of a robotic system and the different configurations developed for specific environments and applications, as manifested in the chassis or structure of the system that houses the remaining components of any robotic system.

Typical examples of different mobile robot locomotor modalities including legs, wheels, tracks, and hybrids (locomotor with other or combinatorial mobility features) are illustrated in Figure 2. Legged vehicles such as monoped, biped, and quadruped or insect-like walkers usually have one, two, or more legs (usually 1, 2, 4, 6, or 8). Wheeled vehicles come in configurations of single-, three-, and multiwheeled systems, with two or more wheels actively driven and the others idling, and steered using angular heading changes (Ackerman steering) or skid-steer driving (like the steering in a tank). Tracked vehicles usually have a two-sided continuous belt/chain loop wrapped over idler and driver pulleys, allowing the vehicle to drive in a skid-steer fashion while using a large contact area on the ground for support and traction.[2] Hybrid locomotors come in many

---

[2]Due to the size of its contact area, the track can be thought of as a very large wheel.

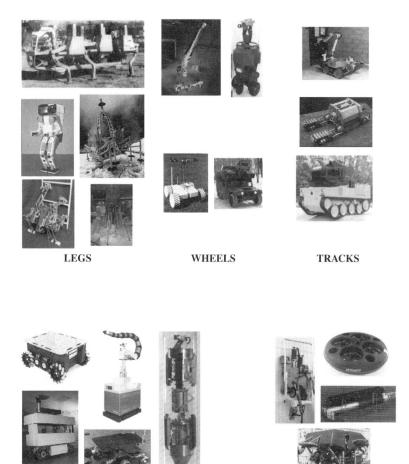

LEGS   WHEELS   TRACKS

HYBRIDS   OTHER

Figure 2   Locomotion modalities found in mobile robots (courtesy Ohio State University, Carnegie Mellon University and RedZone Robotics, Inc., Kraft Telerobotics, Inc., Jet Propulsion Laboratory (NASA) and REMOTEC, Inc., MAK GmbH, Sandia National Laboratories, Honda Motors, Inc., Odetics, Inc., TOSHIBA, Inc., NASA AND Jet Propulsion Laboratory, Aerobot, Inc., Foster-Miller, Inc., Woods Hole Oceanographic Institution, VIT, Inc.).

varieties, including vehicles with omnidirectional wheels, combined wheels and legs, and other omnidirectional wheels, whether fixed or articulated (i.e. a tired-walker, on which the feet on a legged robot are replaced by drivable wheels).

1.3   Locomotor Choice

The choice of locomotor depends primarily on the type of environment that one will be operating in. A fairly generic decision-tree approach follows.

1.3.1   Air and Sea

The only choices if the robotic system is to be deployed in the air are exclusively reaction-based systems such as jets, propellers, and aerodynamic wings (gliders or self-propelled). The name of the game here is to generate airflow with reaction forces, whether helicopter rotors or wings with which jets can glide and fly.

In the case of water-based systems, choices include floating and propeller- or water-jet based propulsion systems with steering-angle control of lift surfaces (rudder) or control of the propulsion-system-orientation (overboard motor or jet-nozzle) itself. Underwater the choices are similar to those with floating systems in that ducted propellers, dubbed thrusters, or even jets (though less used due to lower efficiencies) are used to propel the vehicle. Flotation provides for neutral buoyancy in many cases, while control surfaces can be used to steer the system in all directions (such as in a torpedo or submarine). Ground-dwelling robots crawling along the ocean floor might be built using legs, with the advantage that gravity-loads can be modified through the use of flotation systems, giving designers more freedom.

1.3.2 Land

The choice of locomotor on land is more complex due to the wide spectrum of environments one can encounter. Available choices typically include wheel(s), tracks, legs, and hybrids. The choice of locomotor thus depends completely on the type of environment and other drivers. The guidelines below discuss each of these locomotors.

Wheels are most useful in terrain where obstacles are few, not taller than 40% of the wheel height (to get over them) and smaller than the undercarriage height of the vehicle. For the generation of sufficient flotation and traction, designers might choose to add more than three or four wheels to the systems and then control the size of the wheel based on the necessary support requirements (reduce ground pressure). Steering mechanisms can be avoided if the wheels are sideways-coupled and driven differentially to create forward/reverse motion with differential speeds, resulting in skidding and thus steering (as in a tank). Wheel compliance and active/passive suspensions take care of smoothing out the ride at higher speeds. Wheeled vehicles are capable of much higher speeds than wheeled or tracked systems and are thus used in more benign environments, such as indoors and on roads, and in somewhat smooth outdoor areas.

Tracked vehicles can be thought of vehicles with very large-diameter wheels, with the contact patch of a depressed wheel equivalent to that of the track. Unless the track is articulated, most tracked vehicles are not able to get over obstacles higher than 40% to 50% of their largest/highest sprocket diameter. Shaping and articulation of the tread-drives can increase this height limitation, however. Tracks are used mostly in outdoor rough terrain situations where an optimum of flotation, traction, speed, and obstacle handling is sought.

Legs are typically useful for environments where the obstacle size is equal to or greater than half the maximum underbody clearance height of the machine or the largest dimension of the locomotor height. A legged system is probably preferable for a boulder-field with boulders large enough that a reasonably sized wheeled vehicle could not climb over; interboulder spacing close enough to foil a vehicle of a reasonable size trying to get around them; or slopes steep and loose enough that a tracked or wheeled vehicle would slip hopelessly. Legs are a viable option if power consumption and speed constraints inherent in legged vehicles are not the main consideration and the obstacle size and spatial distribution demand a different type of locomotor. Another advantage of legs is that their infinite adjustability allows for stable platform motions during locomotion—that is, the main body of the vehicle can maintain a stable posture despite the roughness of the terrain at any system speed; when this is not the case, as for other locomotors, sensory and navigation accuracy are greatly affected. Legs are thus typically used in extreme terrain such as that encountered in logging, space exploration, or seabed work (mine detection, pylon inspection, etc.).

Hybrid locomotion systems can vary greatly. They are typically extensions on the systems described above, meant to get around some of the drawbacks of each. For instance, an articulated tracked vehicle can emulate a large-height track and thus climb onto obstacles almost as high as the frontally articulated track can. A legged robot with wheels at the end of each leg rather than a footpad can walk over large obstacles yet also achieve higher traverse speeds over smoother terrain. Robots or remote platforms operating in more constrained environments, such as pipes or vertical surfaces (windows, steel ship-/tank sides), might use inchworm rolling or clamping locomotors or possibly even (electro-)magnetic or vacuum cup feet to hold themselves against a vertical surface to avoid slippage due to gravity loads.

1.4 Sensory Systems

A mobile or teleoperated robot system is usually a sensor-studded platform, with different sensors performing different functions. Sensory systems in this context can be thought of in terms of a hierarchical structure, as depicted in Table 1. The types of sensors are arranged in terms of increasing importance (downwards) for use in autonomous systems,

Table 1   Hierarchical Sensory Systems Structure

Health and safety monitoring
Safeguarding
Position estimation and navigation
Modeling and mapping
Other sensory modes

and data use in increasingly complex tasks necessary for autonomy. A simplistic overview of each is provided below to impart a better understanding of these individual categories.

1.4.1   Health and Safety Monitoring

Sensory systems are typically included aboard any remote system to allow for self-monitoring of essential parameters to ensure the proper operation of the system. The monitoring is typically performed by the on-board computer system at low and high bandwidths, some of it in analog form (circuit-logic without any software control) and some in digital form (computer-software monitored). Such variables might include pressure, temperature, flow rates and pressures, electrical motor-currents, and power-supply voltages. The monitoring bandwidth depends solely on the criticality and characteristic response time associated with potential failure modes in case any of these variables should stray from their operating range.

1.4.2   Safeguarding

Safeguarding of the mobile robot system is implemented so that the remotely operational system does not harm its environment and/or itself to a point where continued operation would become too risky or impossible. As such, there are many styles of noncontact (more warning time and typically more costly) and contact sensors (last line of defense and typically cheaper) already in use by the community. Such noncontact sensors might include acoustics-based sonar sensors or optical systems such as LED-based transmission-/reflection-based proximity sensors. Contact sensors encompass such systems as simple trigger switches, pressure mats, and force/torque monitoring devices (typically on manipulators), which can be used to monitor excessive force/torque thresholds, or even for active control tasks where the control of the force/torque exerted on the environment is an issue.

One interesting feature more and more common in real operational systems is a "heartbeat." Such a system is centered around a small microprocessor system that checks the operation of the computing systems and the communications networks and other key parameters and issues a "ping" to an on- and off-board monitoring system. The monitoring system decides whether to shut down or keep the system operational, based on whether it receives the "ping" at the requested intervals. This is an interesting approach implemented solely in hardware to monitor the complex interlinked systems aboard a mobile robot.

1.4.3   Position Estimation and Navigation

Most mobile robots need to know where they are located in the world, whether in a relative or absolute sense, and how best to proceed autonomously to the next waypoint or end goal. Most sensors used for that purpose are either relative or absolute, their measurements being based respectively on a previous position or data point (relative) or a absolute reference point (magnetic north, earth's rotation, and latitude and longitude). Much activity has revolved around improving positioning capabilities by integrating relative and absolute sensors in a sensory fusion process, typically accomplished with Kalman filtering techniques, in the attempt to yield a better complete estimate of the state of the platform (XYZ position, yaw/pitch/roll orientation, and the associated velocities).

Relative sensors in terrestrial use include accelerometers and inclinometers that can measure planar and vertical rates of velocity change and inclination with regard to gravity vectors, and tachometers and encoders (optical or resistive) to measure rotational speed and angular positions). Abovewater and underwater positioning systems can be accomplished using radio beacon triangulation (LORAN-C) or underwater sonar ranging and triangulation to position a vehicle accurately within an ocean bed-deployed acoustic net. Another method uses a platform-mounted camera coupled with vision software to extract key features (lines on the road, corridors, doors, etc.) for relative positioning and heading control. Laser-based navigation systems relying on time-of-flight measurements can actually provide a locally accurate position estimate and data of value for absolute positioning and modeling.

Absolute sensors such as laser/mechanical gyros and compasses are fairly standard, and, with the advent of GPS (Global Positioning System) and the more accurate differential GPS, absolute position, heading, and rates are available to robot designers. Laser-based navigation systems such as retroreflective laser scanner systems used for AGVs (Automated Guided Vehicles) can generate an absolute position estimate after the barcode targets have been surveyed. The same is true for an underwater acoustic positioning net. The net-transponders can be located accurately with regard to the surface ship deploying an ROV (Remotely Operated Vehicle) and the ship then positioned via GPS, allowing the actual ROV position to be calculated in terms of latitude and longitude.

Integration of sensors is typically accomplished via odometry (counting wheel rotations and accounting for heading changes) coupled with inertial navigation systems (combination of accelerometers, compasses, gyros, and inclinometers). The sensory fusion is typically performed by some nontrivial (possibly) Kalman filtering method, depending on the desired accuracy of the state estimate.

1.4.4 Modeling and Mapping

The modeling and mapping of the world surrounding an autonomous robot is of great utility in developing an awareness of the surrounding world with sufficient accuracy and resolution to allow higher speed and safer autonomous traverse through/over the terrain in a post- or real-time situation. The goal is to collect sufficient and accurate data on the location of all features in the scene, assign to the features an absolute size/shape/location with regard to the vehicle, and use that information to build a three-dimensional map by which to navigate and image the surrounding world for postmission tasks.

Typically the sensors used for such tasks are (in increasing order of resolution, decreasing computational load, and increased implementation cost) acoustic sonar rings, stereo cameras, and (at least) two-dimensional laser scanners. Acoustic sensors are cheap and simple to integrate, but developing a real-time, accurate, high-density, and more than cross-sectional map of the world continues to be a research challenge. Camera-based three-dimensional mapping systems (binocular and trinocular stereo) continue to be a field of research and implementation examples. Many systems are available or in research use today, with differing levels of accuracy and bandwidth. Their main use is for obstacle detection and avoidance. Although the spatial resolution of a stereo pair is better than for acoustic sensors, the use of stereo imagery for digital map-building is still limited. The most costly and accurate solution to three-dimensional modeling and mapping continues to be the laser scanner system. Such a system is able to measure time of flight of laser beams illuminating the environment through a two- to three-dimensional scanning mechanism, allowing a computer to build up accurate high-resolution imagery of the surrounding environment. The challenge has been and continues to be to build accurate three-dimensional models of computationally tractable elements so that in real time such a three-dimensional environmental model can be used for navigation and sometimes even for postprocessing applications (such as built mapping of facilities, reconnaissance).

1.4.5 Other Sensory Systems

Other sensory systems in use today, most of them in the area of manipulation are centered around devices able to measure presence/absence of contact and small contact forces and torques (such as infrared sensory skins and resistive piezoelectric skins attached to robot arms). In addition, researchers have embarked on providing sensory perception beyond the visual (video), auditory (sound), and motion (vibration) cues to now include smell.

1.5 Applications Evolution

Mobile robots can be found in an ever-increasing number of application arenas, whether on land, at or under the sea, and even in the air. Certain areas pioneered the introduction of mobile systems, driven mainly by the commercial impetus and application viability. A somewhat representative, yet far from exhaustive, application family tree for mobile robots is depicted in Table 2.

Table 2   Application Family Tree for Mobile Robots

Mobile Robots		
Air	Sea	Land
UAVs	ROVs	Military
Military	Exploration	Hazardous/nuclear
	Oil rig service	Medical
		Planetary exploration
		Construction
		Agriculture
		Mining
		Service
		Manufacturing/warehousing
		Remote inspection
		Environmental remediation
		Entertainment
		Mass transporation

In the air, UAVs (unmanned air vehicles) are used mostly by the military in reconnaissance missions. At sea and underwater, ROVs (remotely operated vehicles) are used for underwater exploration and search and rescue as well as offshore oil rig servicing and maintenance. By far the largest applications arena is in land-based systems. The military is developing remote scout vehicles for battleline reconnaissance, and the nuclear industry uses inspection robot systems for its power plants. The medical industry is developing surgery robots and assistive displays, and planetary exploration is proceeding with remote rovers. The construction, mining, and agricultural industries are automating their equipment to allow for automated field operations, such as excavation, harvesting, and mineral haulage. The service sector has several automated systems for cleaning and mail/hospital delivery, and the manufacturing/warehousing industries use AGVs (automated guided vehicles) to transfer materials/products on site. Several businesses have remote inspection robots for tanks and gas/sewer pipelines, and environmental cleanup depends on remote robotic systems to map and dismantle contaminated sites. The entertainment industry is using automatons and automated figures in theme parks and movie sets, and mass transportation industries are in the alpha/beta testing phases for automating transportation beyond people-movers to actual buses and private vehicles. The above list will continue to grow rapidly and reflects just a subset of possible arenas for mobile robot applications.

1.6   Theoretical Aspects and Research Topics

The areas that always seem to generate a lot of interest in the field of mobile robotics are locomotion and chassis/structural configuration. In general terms, which type of locomotor is needed for a certain application depends largely on what type of terrain that the mobile robot will be expected to operate in. *Terrainability* is a term coined in the research community. Its various definitions generically describe the ability of the robot to handle various terrains in terms of their ground support, obstacle sizes and spacing, passive/dynamic stability, etc. The overall configuration of the entire system can be developed to suit particular needs of the application, such as reconfigurability for ease of transport or access, suspension for high-speed operation, or even articulated frameworks and suspensions to increase the machine's terrainability.

Locomotion is probably the area in mechanical mobile robot design most heavily studied in theoretical and experimental terms. Important issues in this area are ground support (contact pressure), traction, and obstacle-handling capabilities. More typical issues in locomotor design are the details of actuator (motor, gearbox) design, suspension/steering methods, and ground-contactor design (i.e., feet on legged machines, magnetic or vacuum tracks/feet, and even wheel/track tread/grouser systems).

Robot system configuration is an area where the designer's systems integration experience, innovativeness, and inventiveness play a big role. Simple guidelines, such as having a framework of three or four wheels for a statically stable wheeled mobile system or choosing a single, dual-, or four- or six-legged walking system, can be implemented in different configurations to suit the task. Designers usually use such key criteria as static stability or power consumption to make these choices during the design process. Other criteria that affect the system configuration include:

1. Desired system rigidity (selection and configuration of structural members)
2. Ground clearance (size of locomotor and underhang of structural "belly")
3. Overall weight (material selection and on/off-board power source)
4. Reconfigurability (collapsible locomotor or actuated frame structure)

It would be a very one-sided and incomplete endeavor to try to capture this part of the design process numerically, since it largely depends on the application at hand, the particularity of the desired performance specifications, and the inventiveness of the development team member(s). Suffice it to say that a large variety of capable and innovative robot systems have been developed with only wheels or legs, suited to different applications, and this innovative trend can be expected to continue.

2 THEORY

2.1 Terrainability—Sinkage and Traction

Outdoor off-road mobile robots must be able to locomote on and over their deployment terrain. The two most important criteria are sinkage and traction. Both depend highly on soil properties such as granularity and moisture, as well as the weight and contact area of the locomotors for the robot and the surface properties of the locomotors for proper traction. A typical set of figures used to explain sinkage via contact pressure and terrain characteristics is shown in Figure 3.

In order to properly size a track system, for instance, theories and data developed by Bekker (1958, 1960) represent the best source of such information. In order to define such parameters as the track width b, the track height D, track length l, grouser depth h and the robot weight W, use can be made of the well-known Bekker parameters (internal friction angle ϕ, cohesion coefficient c, density ρ, deformation moduli k_c and k_ϕ, power ratio n, and derived parameters N_c, N_q and N_γ) and their use in several key equations derived by Bekker and shown below:

Sinkage depth z vs. vehicle weight and soil parameters

$$z = \left[\frac{\dfrac{W}{A}}{\dfrac{k_c}{b} + k\phi} \right]^{1/n} \tag{1}$$

Drawbar pull $DP = H \cdot R_c$ [Soil traction (shear + friction)]–{Soil compaction resistance}

$$DP = H - R_c = [Ac + H' + w \tan \phi] - \left\{ 2\left[(n + 1)(k_c + bk_\phi)^{1/n} \right]^{-1} \left[\frac{W}{l} \right]^{(n+1)/n} \right\} \tag{2}$$

An important rule of thumb is that the track sinkage should be less than 40% of the track height or the tracks will not be able to move through the medium. This requires as large a contact area as possible. Track width should also be maximized to reduce sinkage

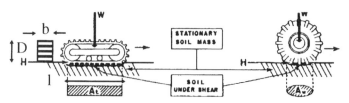

Figure 3 Traction diagram to evaluate locomotor terrainability.

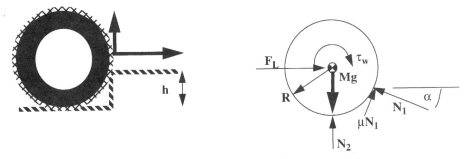

Figure 4   Idealized step-climbing wheeled geometry and forces.

in cohesive materials (loam and clay). To increase traction while reducing compaction and bulldozing resistance, the track length should be optimized to allow for good traction (in addition to sizing and shaping the grousers) and reduced sinkage, without affecting the turning ability of the robot (expressed as the ratio of length to width, ideally between 1.0 and 1.8).

2.2   Terrainability—Obstacle-Handling: Wheeled Step-Climbing

The step-climbing problem for wheeled vehicles has been and remains one of the most illustrative examples of locomotor design and optimization. The typical "ideal" geometry and free-body diagrams used to illustrate the theory behind wheel-size selection and actuator power specification are shown in Figure 4.

To determine how much torque (τ_w) would be required for a front-wheel-driven ($F_L = 0$) wheeled robot to make it over a certain-sized step (h) using a certain-sized wheel (R), given a certain distributed wheel-load (Mg), it can be shown that the equation can be reduced to the following (note $N_2 = 0$ for lift-off and that no slippage on the step-edge occurs):

$$\tau_w \geq MgR \cos \alpha = MgR \cos\left[\operatorname{atan}\left\{ \frac{R - h}{\sqrt{R^2 - (R - h)^2}} \right\} \right] \tag{3}$$

Note that the required torque becomes smaller if the mobile robot has all-wheel drive because the rear wheels provide additional push ($F_L > 0$), where the magnitude of said push depends on the net traction of the rear wheels (depending on soil properties and tire geometry). Note further that as $h > R$, the necessary motor torque rapidly approaches infinity, which implies that smooth tires cannot overcome a step height equal to their wheel radius, even if rear-wheel drive assist is present.

In order to reduce the reduced terrain capability of wheeled vehicles, tires (and even caterpillar treads) can be designed with grousers to allow the climbing of obstacles where $R = h$. The fact that tires are not always perfectly rigid cylinders, but rather compliant geometries, helps in step-climbing, as depicted in Figure 5. Notice that a grouser provides an effective leverage point, allowing the wheel to lift itself onto the step, assuming enough

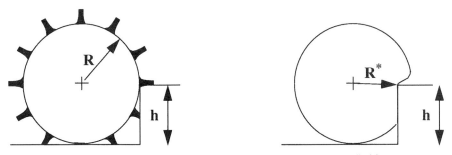

Figure 5   Impact of wheel grousers and compliance on step-climbing.

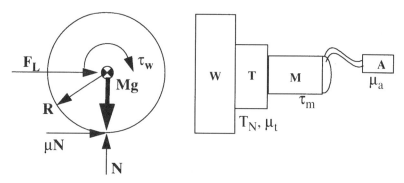

Figure 6   Idealized rendition of a wheel and its drivetrain.

torque is available—and note that a pure rear-wheel-drive vehicle could not overstep an obstacle where $h = R$. In the case of the compliant tire contacting the step, one can see that $R^* \leq R$, reducing the required torque τ_w, while also providing an effective grouse-like overhang that would tend to allow climbing obstacles where h can even be slightly larger than R, assuming no slippage occurs.

2.3   Locomotor Actuation—Wheeled Actuator Sizing

Another recurring requirement revolves around the sizing of the drive system to allow a wheeled vehicle to negotiate inclined slopes. Typically, in the case of electrical wheel-driven system, a generic front and side view of a wheel and its actuator system can be diagrammatically simplified, as shown in Figure 6. Depicted are a wheel (W) of radius R and mass M, coupled to a transmission (T) with a transmission ratio T_N and efficiency μ_t, a motor (M) outputting torque τ_m driven by an amplifier (A) with efficiency μ_a. The net forward propulsive force F_L required to keep the vehicle at a certain speed (remember that efficiencies refer to a measure of dynamic frictional effects and *not* static conditions) on certain terrain can be derived from the equation

$$F_L = \frac{\tau_w}{R} = \frac{\tau_m T_N}{R} \mu_t \mu_a \tag{4}$$

If one defines the terrain to be a slope of angle θ and the vehicle to have a mass M, as shown in Figure 7, the required motor torque (τ_m) for a four-wheel-driven mobile robot vehicle can be calculated as:

$$4 \times F_L \geq Mg \sin \theta \leq 4 \times \frac{\tau_m T_N}{R} \mu_t \mu_a \quad \text{or in rearranged form:} \quad \tau_m \geq \frac{MgR}{4T_N} \times \frac{\sin \theta}{\mu_t \mu_a} \tag{5}$$

Notice that the assumption here again is that none of the wheels slip, implying that the

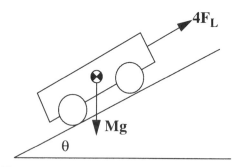

Figure 7   Inclined slope-climbing vehicle condition.

frictional forces at each wheel (μN), which differ due to the uneven weight distribution, are larger than the applied force F_L (i.e., $\mu N \geq F_L$).

2.4   Static Stability—Legged Configurations and Gait Selection

In configuring legged robots, one important decision is whether to design a dynamically or statically stable system. A good example of a dynamically stable legged robot would be that of a monoped, and possibly even a biped or quadruped, though these are more quasistatically stable when at rest (such as humans or animals when standing). In the case of a passively stable legged walker, most systems usually have six or eight legs. Guaranteed stability can best be illustrated by the illustration of an eight-image mosaic of a six-legged walker with a specific gait shown in Figure 8. The weight-supporting legs (those in contact with the ground and sharing in the load supporting the walker) are shaded, while the recovering legs (those off the ground and moved forward for the next step) are shown in solid white. The support polygon is shown as an enclosed area underneath the walker.

As the walker traverses a certain distance of terrain, its legs go through a sequence of steps. During any phase of the gait, the center of gravity of the walker, denoted by the CG symbol (⊗—Location of center of gravity) *must* always lie within the shaded support polygon described by the polygon drawn through all ground-contacting feet of the walker—that is, the polygon's vertices. As can be seen in Figure 8, the legged geometry of this six-legged walker is acceptable in terms of guaranteeing stability for flat-floor walking. The same method can be used for incline walking simply by using the projection location of the center-of-gravity vector for the walker with the plane described by the feet of all ground-contacting legs.

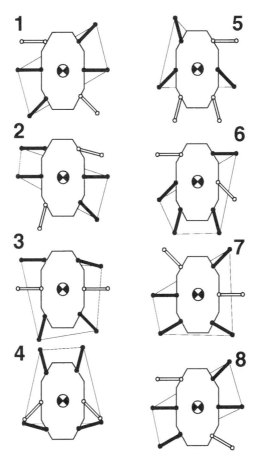

Figure 8   Gait of a six-legged walker.

2.5 Locomotion Power—Legged Propulsion Power

A legged walker typically needs to clear obstacles of a certain size as it crosses rugged terrain, requiring the legs to jointly lift the body of the vehicle to specific heights. In general, the total energy consumed for the execution of such motions is dependent on the mass of the vehicle (M), the characteristic height to which it needs to be lifted (H_c), and the frequency with which these motions occur (N_r). Power consumption can be computed by determining the characteristic raise time (Δt) for the system. The graphic expressing this power consumption is depicted in Figure 9, where the robot crosses an obstacle of height H_c. Assuming that the walker should keep as close to the ground as possible, in order to maximize stability and reduce power consumption due to gravity-load cancellations in articulated legs, the net power consumption due to raising and lowering the body P_b can be expressed as:

$$P_b = 2\,\frac{MgH_c}{N_r\Delta t} \tag{6}$$

Note that power is also consumed when the body propels itself forward or in reverse, with load-bearing legs propelling the body and recovering legs moving their own weight. A walker is hence always consuming power, even going downhill, because to date no simple method of energy recovery exists for legged machines as it does in wheeled vehicles (regenerative braking).

3 LEARNING BY CASE STUDIES

This section illustrates the application of the more theoretical tools in a number of practical mechanical design areas in mobile and legged robotics. The examples have been grouped into locomotor systems and overall system configurations.

3.1 Locomotor Systems

3.1.1 Drivetrain Design

A drivetrain was designed for a mobile robotic system intended to inject a toxic re-waterproofing chemical into each of the more than 15,000 heat-resistant tiles on the underside of NASA's space shuttles while they were serviced at Cape Canaveral. The robot system, the meccanum wheel used for the system, and a cross section of the drivetrain design are illustrated in Figure 10.

The wheeled mobile base carries a planar manipulator system with a vertical wrist and a sensing and tooling plate. The wheels carry on the inside a compact brushless DC motor coupled with a cycloidal planetary reducer, controlled via a motor-mounted resolver, and load-isolated using a pair of angular contact bearings. A novel feature is the coupling of a spline hub from a Toyota four-wheeler to allow the decoupling of the nonbackdriveable actuator from the wheel, allowing operators to move the robot around without electric power and without having to backdrive the transmission. The torque and speed-curve performance were developed around two set points. The one of driving over a two-inch obstacle at very low speeds (for stall conditions see the step-climbing equations), and maximum speed of the robot during locomotion (for motor torques at speed see the drivetrain efficiency calculations).

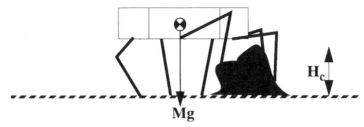

Figure 9 Legged power consumption for rugged terrain traverse (courtesy of Carnegie Mellon University).

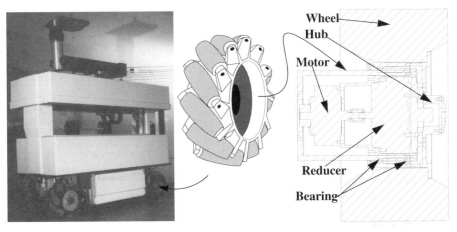

Figure 10 Tessellator robot system, meccanum wheel, and drivetrain cross section (courtesy of ADIDAS GmbH, Bally Design, Inc. and Hagen Schempf).

3.1.2 Impulse Leg Actuator

An impulse actuator, such as the ones used in a single-legged soccer ball-kicking robot, can deliver much larger amounts of power over shorter time periods than the typical continuous active power systems found in standard actuation systems, consisting of a motor (position/torque/velocity controlled), transmission, and output coupling. The system dubbed *RoboLeg* was developed by Bally Design and Automatika, Inc. for Adidas AG in order to rapidly and reliably test new soccer shoe and soccer ball designs for better spin control and hang time.

As shown in Figure 11, the system uses an anthropomorphic leg geometry in which the hip and knee joints are powered using impact actuators. The entire kick is program-

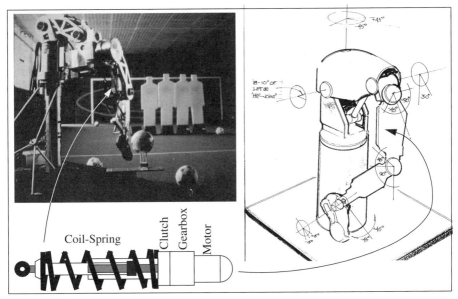

Figure 11 RoboLeg robotic soccer-ball-kicking leg using impulse actuators to simulate human double-pendulum kick in the game of soccer (courtesy of Carnegie Mellon University and Department of Energy).

mable and is finished in less than 1/10 of a second. The impact actuator consists of a high-speed heavily geared motor transmission system that preloads an energy storage device such as a coil spring (linear actuator—shown above) or spring washers (rotary actuator). A simple clutch on the output of the transmission box allows the actuator to be connected to the motor (loading phase) or the system to be free-wheeled so that the energy stored in the spring can be released while minimizing friction and without back-driving the actuator.

3.1.3 Legged Clamper Design

Another interesting mechanical design is a clamper–inchworm system, which locomotes along a cylindrical pipe section. It is used to propel along the pipe a tooling head that uses mechanical and water cutters to abate asbestos insulation from the pipe. The overall system view is shown in Figure 12, as well as the two positions of the clamping mechanism used to clamp onto the pipe. The system architecture is a simple push–pull linkage mechanism driven by a pinion/sector gear arrangement (crank–slider linkage arrangement) and an idler and follower linkage that eventually clamps onto the pipe using sharpened wheels (for self-alignment). Using the principle of mechanical advantage, the clamping forces can be very high for a certain range of motion of the linkage. By resizing the linkage lengths, the system can be tailored to clamp onto pipes of different diameters.

3.1.4 Legged System Design

Several mechanical designs for legs and their attachment frames currently exist. Those developed for some of the better-known robot systems, such as the ASV Walker, the Ambler, PLUSTECH's legged logger, and the NASA Dante II robot, are shown in Figure 13. In the Ambler the legged configuration is an orthogonal leg. The ASV Walker has an articulated pantograph mechanism, Dante II uses a pantograph leg in a sideways arrangement. PLUSTECH uses a rotary-jointed leg configuration.

3.1.5 Steering System

The Nomad, designed to perform 200+ km autonomous desert traverses on earth in preparation for planetary exploration, uses an interesting linkage to allow for the compact storage of the drivewheels and drivetrain, as well as the combined steering of the two wheels on each side, in order to achieve smoother turning radii while minimizing the slippage typical for skid-steer-driven wheeled vehicles. See Figure 14.

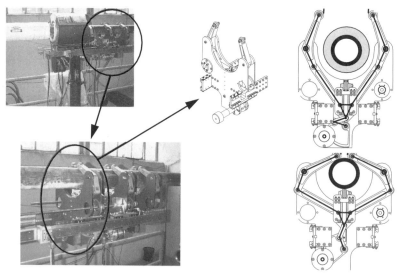

Figure 12 BOA's inchworm locomotor and detailed view of open/close clamper positions (courtesy Plustech, Finnland, Ohio State University and Kenneth Waldron, Carnegie Mellon University and NASA).

PLUSTECH Logger

ASV Walker

Dante II

AMBLER

Figure 13 Articulated leg designs: ASV, PLUSTECH, Ambler, and Dante II (courtesy of Carnegie Mellon University and NASA).

3.2 System Configurations

3.2.1 Collapsible Frame Structure

A mobile robot system that uses almost its entire structural system in an active fashion to reconfigure itself is Houdini™ (see Figure 15), a robot developed for the Department of Energy. The robot was developed as an internal tank cleanup system that would remotely access hazardous (or petrochemical) waste-storage tanks through a 24-inch-diameter opening in the collapsed position and then open up to a full 8′ × 6′ footprint and carry a manipulator arm with exchangeable end-effector tooling and articulated plow.

The "benzene ring"-shaped frame members actuate the two tread drives sideways and interlock at their end of travel, with the plow opened through passively coupled linkages onto the frame members.

3.2.2 Articulated Frame-, Suspension-, and Obstacle-Handling System

Some interesting research and commercial examples of articulated wheeled and tracked designs are currently in use as research, exploratory, and remote reconnaissance systems. Two noteworthy systems are NASA's latest Rocky 7 wheeled robot, whose planetary incarnation, Sojourner, explored the surface of Mars in 1997 for over 80 days as part of NASA's Pathfinder mission, and the Andros family of robots from REMOTEC, Inc., which are used in hazardous waste cleanup and law enforcement. See Figure 16. In Rocky a passive rocker–bogey linkage arrangement allows for the handling of larger obstacles than the wheel size would allow, thereby reducing the required wheel size and suspension

Figure 14 The articulated steering and stowage feature of the Nomad autonomous rover (courtesy of Carnegie Mellon University and Department of Energy).

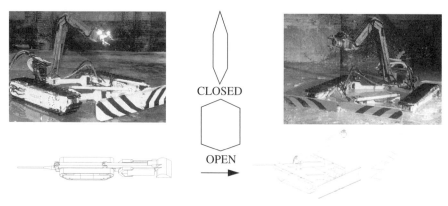

Figure 15 Houdini's reconfigurable structural frame design for access into tight spaces (courtesy of NASA, Jet Propulsion Laboratory, and REMOTEC, Inc.).

system. The Andros Mark V robot system uses a set of articulated rear- and front-driven treads to allow it to climb over obstacles and up stairs more readily than a tracked vehicle of similar size.

3.2.3 Underwater Robot Frame System

Underwater robots are typically built quite differently from land-based systems because they can take advantage of buoyancy (corollary to zero-gravity space systems). Hence they need only a minimal framework to retain a desirable configuration and provide for tie-downs for equipment, and need to resist only thruster and hydrodynamic forces, not gravitational forces. Typical ROVs, used in the exploration of the *Titanic* and the discovery of underwater volcanoes and the recovery of Roman amphoras from Mediterranean shipwrecks, are the Jason Jr. and Jason ROVs built at the Woods Hole Oceanographic Institution. These vehicles vary in size by up to a factor of 5, but in each case the framework consists of simple tubular aluminum (speed-rail) or flat-stock painted steel used to tie down the pressure housings that contain all power systems, electronics, and computing. The remainder of the framework is tailored to fasten to the flotation system, which for high pressures is usually syntactic foam (miniature glass spheres in an epoxy resin matrix). Overall views for both ROVs and a frame view of Jason Jr. are shown in Figure 17.

3.2.4 MonoCoque Frame Design

Another type of frame design can almost be termed a frameless design, in that the actual robot structural elements are functional components of the robot and are not by themselves attached to another framing system. A good example of this approach is the Neptune oil

Rocky 7   Andros Mark V

Figure 16 Articulated wheeled and tread-driven mobile robot systems.

JASON *JASON JR.*

Figure 17 Jason Jr. & Jason ROVs with typical ROV frame structure (courtesy of Carnegie Mellon University and US Army Corps of Engineers).

storage tank-inspection robot system, built for the U.S. Army Corps of Engineers and Raytheon Engineers, Inc. (see Figure 18). The robot was designed to be used in explosive environments, specifically light-crude storage tanks, where a remote internal visual and ultrasonic inspection of the bottom-plates of the tank without the need for draining and cleaning the tank and providing access to humans was desirable. This system too was designed to fit through a 24-inch-diameter opening. It is made solely of internally pressurized equipment enclosures housing electronics and power systems, which are tied together in the rear by a hollow backplate for cable-routing. Not a single frame member is used to hold any of the enclosures or the tread locomotors.

3.3 Configuration Analysis

It is always tempting in case studies to provide purely informational reviews of robot designs that have been generated over the years. To be complete, such a review would be a directory in itself, since there are a large number of differing systems out in the world and in operation. It is almost always difficult to determine why the configuration of a specific robot was chosen, what others were considered, and what rationale was used to decide upon a specific configuration, locomotor, power, computing, and sensory system. Pontificating after the fact and analyzing designs based on one's personal perspective is thus not necessarily an objective or fair undertaking. This dilemma of trying to provide a generic overview and guidelines while working without the benefit of detailed insight into each of the robot systems, suggests an important point about mobile robot configuration: The experience and backgrounds of a multidisciplinary team strongly influence a robot configuration, subject to the availability and maturity of existing technologies, the constraints of the application, and the environment. Useful design guidelines, combined with a rigorous approach to evaluating alternative locomotion, actuation, power, com-

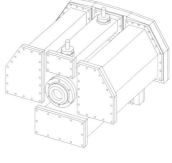

Figure 18 Neptune aboveground oil storage tank-inspection robot.

puting, and sensory systems, are not treated or summarized in a single place, but are sparsely distributed. It would seem high time to generate such a documentary volume to guide the design of future mobile robots—a task far from trivial.

4   EMERGING TRENDS AND OPEN CHALLENGES IN ROBOT DESIGN

In overall mobile system design, and mechanical design in particular, work continues and is still needed in the following areas:

1. System configuration
2. Overall power and computing/communications autonomy
3. Wheeled/tracked locomotion
4. Legged locomotion.

4.1   System Configuration

There is no steadfast set of rules when it comes to system configuration design. The trend in industry and in research is towards the development of task-specific machines rather than generic systems that are usable in a wide array of environments. Like industrial robot arms, which are targeted at highly dexterous assembly tasks, high-precision trajectory following for spraypainting, or point-to-point heavy-load transfer, mobile and legged robots have to be designed for a specific task set. A pipe-crawling robot, for example, can be used for sewer-line inspection, but not necessarily for large tank inspections. Because mobile and legged systems work in a world that is more unstructured (except for AGVs in factory/warehousing operations) and much more expansive, without controllable conditions, they are usually designed for a well-formulated task to reduce complexity and cost and increase reliability and overall performance. The notion of a generic mobile robot akin to an android or Pegasus-like device might remain on the drawing boards of science fiction writers and cartoonists for a while to come unless better software, higher-density compact power supplies, and compact actuators and sensors become a technical reality.

4.2   Autonomy—Power and Computing/Communications

Designers of mobile and legged robot systems, whether autonomous or teleoperated, all continue to grapple with the fact that there are limits to the levels of available power, computing, and communications. Any mobile system is usually a compromise in terms of these major drivers because, due to the range or working environment, the system might have to carry an umbilical power cable. On the other hand, due to size and processing constraints, the necessary computing power might have to reside off-board rather than on-board and rely upon a high-bandwidth data link to receive data and return commands. This constraint is the one most rapidly being eroded, since computing power is still on a geometric growth rate while power and space requirements are continually decreasing (though only linearly at best). Once on-board computing is implemented, the communications link requirements for real-time/telepresence experiences can become rather daunting. Given satellite communications and RF spread-spectrum technologies, as well as fiberoptic telecommunications, we have begun tackling this bottleneck with new applications continually demanding faster, cleaner, and better links, thereby pushing technology developments. Communications bottlenecks will in the foreseeable future continue to drive designers into compromises in overall system design.

4.3   Wheeled/Tracked Locomotion

A need will remain for standard wheeled and tracked mobile robot systems in existing applications where existing equipment or related configurations will require automation through retrofit. As developments in the past 5 to 10 years have shown, incremental innovations in terms of omnidirectional wheels and articulated track systems will enable designers to develop systems for particular applications, blazing the trail for the use of such locomotors in applications with similar environments/terrain. The use of articulated-frame and suspension systems to increase the terrainability of wheeled systems and the design of traction features for increased locomotion capability will benefit from continued work in the research and commercial world. Tracked systems will become more capable through the use of articulated and steerable chain designs, enabling smaller and simpler system designs to carry out novel tasks. The only limit in this field will be the innovative spirit and inventive capabilities of designers and inventors!

4.4 Legged Locomotion

The field of legged locomotion will continue to be a primary target of curiosity and one-off system developments. Most major multilegged system developments have been in the area of research and exploration. Notable systems have been developed for research and potential planetary exploration use, but none has really been a commercial success. To date the only legged systems that have, to the author's knowledge, seen any real-world use are drag-line open-pit mine-excavator systems and possibly rough-terrain logging systems in Finland (though the commercial success of the latter is questionable at this time). In order for such systems to become successful, their control and computing complexity must be reduced through better software and more capable sensing and computing, and the ultimate operational Achilles heel of any system, parts count and complexity (resulting in increased failure modes and reduced reliability), must be reduced to allow operational availability (mean time between failures, or MTBF) comparable to that of wheeled and tracked systems.

One continued area of research limited to university-level and internal company research efforts (mostly in Japan) is biped locomotion, specifically the development of anthropomorphic systems with locomotion capabilities akin to those of humans and state-of-the-art sensing and computing systems. The immediate goals of studying overall system capability and long-term potential are obvious, yet the advent of a commercial bipedal system with a clearly defined task in a commercial setting has still escaped developers. Research will and should continue in this area, but overall system capabilities as expressed in terms of power, sensory, computing, and communications autonomy have to see improvements of one to two orders of magnitude before bipeds can become viable automatons. It might only be a matter of time before dramatic new technology developments help bring that vision closer to reality.

4.5 System Integration

Retrofitting and integrating components to achieve a working, modular, maintainable system is not an area for research or scientific pursuit, but a completely experiential artform, in that it has no rigid rules and depends on the individuals in charge. In the aerospace industry, the most comparable job description is that of the packaging engineer. In robotic systems, all engineering disciplines should be trained in the pitfalls and dos-and-don'ts of systems integration. No authoritative publication on mobile robot design and integration has been published yet, and such a work is long overdue.

4.6 System Retrofitting

The use of mobile robots is also expanding into the more traditional fields of construction, agriculture and mining. Researchers and industrial outfits are working towards retrofitting their existing equipment with autonomy modules (sensing, computing, actuation, etc.) to allow their manually operated equipment not only to operate remotely, but to operate autonomously and under minimal supervision by an operator. This area will continue growing to the point where integrated autonomous or intelligent platforms built today by such household names as John Deere, Caterpillar, Joy Mining Co., Toro, and Hoover will eventually offer systems of substantial automation and autonomy that will challenge all fields, including electronics, computing, and mechanical design, to develop a fully integrated, working, robust, and long-lived maintainable product. Again, though not a pure science, this field of applied research and engineering will continue to expand and possibly dominate the field of robotics in the future.

REFERENCES

Terrainability

Bekker, M. G. 1958. *Theory of Land Locomotion*. Ann Arbor: University of Michigan Press.
———. 1960. *Off-the-Road Locomotion*. Ann Arbor: University of Michigan Press.
Caterpillar, Inc. 1991. *Caterpillar Performance Handbook*. 22nd ed. Peoria: Caterpillar, Inc.
Schempf, H., et al. 1995. "Houdini: Site- and Locomotion Analysis-Driven Design of an In-tank Mobile Cleanup Robot." Paper delivered at American Nuclear Society Winter Meeting, San Francisco, November).
Wong, J. Y. 1989. *Terramechanics and Off-Road Vehicles*. Dordrecht: Kluwer Scientific.

Legged Locomotion

Bares, J., and W. L. Whittaker. 1993. "Configuration of Autonomous Walkers for Extreme Terrain." *International Journal of Robotics Research* **12**(6), 535–559.

Chun, W. "The Walking Beam: A Planetary Rover." *Martin Marietta Astronautics Group Journal.* **1**, 32–39.

Hirose, S., et al. 1991. "Design of Prismatic Quadruped Walking Vehicle TITAN VI." In *Proceedings of 1991 International Conference on Advanced Robotics,* Pisa.

McGhee, R. B., "Vehicular Legged Locomotion." *Advanced Automation Robot* **1**, 259–284.

Raibert, M. H. 1986. *Legged Robots That Balance.* Cambridge: MIT Press.

Schempf, H., et al. 1995. "RoboLeg: A Robotic Soccer-Ball Kicking Leg." Paper delivered at Seventh International Conference on Advanced Robotics, Japan.

Todd, D. J. 1985. *Walking Machines.* 1985. New York: Chapman & Hall.

Waldron, K. J., and S.-M. Song. *Machines That Walk.* Cambridge: MIT Press.

Mobile Robots

Bekker, M. G. 1969. *Introduction to Terrain-Vehicle Systems.* Ann Arbor: University of Michigan Press.

Jones, J. L., and A. M. Flynn. 1993. *Mobile Robots.* Wellesley: A.K. Peters.

Schempf, H., et al. 1995. "NEPTUNE: Above-Ground Storage Tank Inspection Robot System." *IEEE Robotics and Automation Magazine.* (June) 9–15.

Schempf, H., et al. 1997. "BOA II and PipTaz: Robotic Pipe-Asbestos Insulation Abatement Systems." Paper delivered at International Conference on Robotics and Automation, Albuquerque, April.

ADDITIONAL READING

Asada, H., and J.-J. Slotine. *Robot Analysis and Control.* New York: John Wiley & Sons, 1986.

Asada, H., and K. Youcef-Toumi. *Direct-Drive Robots: Theory and Practice.* Cambridge: MIT Press, 1987.

Ayers, R. U., and S. M. Miller. *Robotics Applications and Social Implications.* Cambridge: Ballinger, 1983.

Borenstein, J., H. R. Everett, and L. Feng. *Navigating Mobile Robots: Systems and Techniques.* Wellesley: A.K. Peters, 1996.

Brady, M., ed. *Robotics Science.* Cambridge: MIT Press, 1989.

Brady et al., eds. *Robot Motion Planning and Control.* Cambridge: MIT Press, 1983.

Busby, F. *Undersea Vehicles Directory.* Arlington: Busby Associates.

Craig, J. J. *Introduction to Robotics: Mechanics and Control.* Reading: Addison-Wesley, 1988.

Crimson, F. W. *U.S. Military Tracked Vehicles.* Osceola: Motorbooks International, 1997.

Dodd, G. G., and L. Roscol, eds. *Computer-Vision and Sensor-Based Robots.* New York. Plenum Press, 1979.

Engelberger, J. F. *Robotics in Practice.* New York: AMACOM, 1980.

———. *Robotics in Service.* Cambridge: MIT Press, 1989.

Groen, F. C., S. Hirose, and C. E. Thorpe, eds. *Intelligent Autonomous Systems.* Washington: IOS Press, 1993.

Heiserman, D. L. *Robot Intelligence with Experiments.* Ridge Summit: TAB Books, 1981.

———. *How to Design and Build Your Own Custom Robot.* Blue Ridge Summit: TAB Books, 1981. [This is geared mostly toward the hobbyist].

Helmers, C. T. *Robotics Age: In the Beginning.* Hasbrouck Heights: Hayden Book Co., 1983.

Hirose, S. *Biologically Inspired Robots.* Oxford: Oxford University Press, 1993.

Irvin, E. I. *Mechanical Design of Robots.* New York: McGraw Hill, 1988.

JTEC Report. *Space Robotics in Japan.* Baltimore: Loyola College, 1991.

Jones, L. J., and A. M. Flynn. *Mobile Robots: Inspiration to Implementation.* Wellesley: A.K. Peters, 1993. [This is geared mostly toward the hobbyist.]

Khatib, O., J. J. Craig, and T. Lozano-Pérez, eds. *The Robotics Review.* Cambridge: MIT Press, 1989.

Leonard, J. J., and H. T. Durrant-Whyte. *Directed Sonar Sensing for Mobile Robot Navigation.* Boston: Kluwer Academic, 1992.

Mason, M. T., and J. K. Salisbury. *Robot Hands and the Mechanics of Manipulation.* Cambridge: MIT Press, 1985.

Moravec, H. *Mind Children: The Future of Robot and Human Intelligence.* Cambridge: Harvard University Press, 1988.

Paul, R. P. *Robot Manipulators: Mathematics, Programming and Control.* Cambridge: MIT Press, 1981.

Raibert, M. H. *Legged Robots That Balance.* Cambridge: MIT Press, 1986.

Rosheim, M. E. *Robot Wrist Actuators.* New York: John Wiley & Sons, 1985.

———. *Robot Evolution: The Development of Anthrobotics.* New York: John Wiley & Sons, 1994.

Schrait, R. D., Schmierer, G. "Serviceroboter," Springer Verlag, Heidelberg, Germany, 1998 (German & English edition).

Shahinpoor, M. *A Robot Engineering Textbook.* Cambridge: Harper and Row, 1987.

Skomal, E. M. *Automatic Vehicle Locating Systems.* New York: Van Nostrand Reinhold, 1981.

Song, S.-M., and K. J. Waldron. *Machines That Walk: The Adaptive Suspension Vehicle.* Cambridge: MIT Press, 1989.

Society of Manufacturing Engineers. *Teleoperated Robotics in Hostile Environments.* 1st ed. Dearborn: Society of Manufacturing Engineers, 1985.

Thorpe, C. E., ed. *Vision and Navigation: The Carnegie Mellon Navlab.* Boston: Kluwer Academic, 1990.

Thring, M. W. *Robots and Telechirs.* Chichester: Ellis Horwood, 1983.

Todd, D. J. *Walking Machines: An Introduction to Legged Robots.* New York: Chapman & Hall, 1985.

Wolowich, W. A. *Basic Analysis and Design.* New York: HRW Press, 1987.

CHAPTER 9

TELEOPERATION, TELEROBOTICS, AND TELEPRESENCE

Wayne Book
Georgia Institute of Technology
Atlanta, Georgia

Lonnie Love
Oak Ridge National Laboratory
Oak Ridge, Tennessee

1 INTRODUCTION

A *teleoperator* is a manipulator that requires the command or supervision of a remote human operator. The manipulator arm for a teleoperator has many design problems in common with the arm for an autonomous robot. Unlike the robot, however, the teleoperator has the human involved with each execution of the task. As a consequence, the human interface of the teleoperator is more critical than for most autonomous robots. *Telepresence* refers to the extreme enhancement of sensory feedback for teleoperation, providing realism to the operator that approaches the ideal of "presence at a distance." The operator can exercise judgment and skill in completing the task, even in the face of unforeseen circumstances. The distinction between the robot and the teleoperator is blurred when the operator only supervises the operation of the teleoperator or when a robot is being led through a motion by its human programmer. *Telerobotics* further extends the realm of manipulation into shared control by machine and human.

2 EMERGING APPLICATIONS

Industrial applications of teleoperators are numerous. They typically involve work conditions inappropriate for humans. The environment may be hazardous or unpleasant, or the required forces and reach may be greater or smaller and more accurate than the human can directly provide. If the task is predictable and repetitious, an autonomous robot is appropriate. If the judgment and skill of a human are needed, or if the task is one of a kind, use of a teleoperator should be considered. Telerobotics tends to address weaknesses in either extreme of autonomy with sharing of control as appropriate to the application. The following sections illustrate the use of teleoperated systems in a variety of applications.

2.1 Nuclear Waste

One of the largest problems facing the U.S. Department of Energy consists of removing liquid, solid, and mixed waste from over 200 underground storage facilities (Zorpette, 1996). Technological challenges include operations in confined and highly radioactive areas and manipulation of materials that may be of a solid, liquid, or muddy consistency. Furthermore, the only access into the tanks is through risers that range from 16 to 48 inches in diameter.

Oak Ridge National Laboratory (ORNL), teaming with Spar Aerospace, Redzone, Pacific Northwest, and Sandia National Laboratories, has assembled a teleoperation system designed to tackle this problem (Burks et al., 1997). The system consists of a waste dislodging and conveyance (WD&C) system, a long-reach manipulator, and a remotely operated vehicle. For deployment, each of the teleoperated systems must fit through the

Handbook of Industrial Robotics, Second Edition, Edited by Shimon Y. Nof
ISBN 0-471-17783-0   © 1999 John Wiley & Sons, Inc.

Figure 1 MLDUA and HMA at ORNL cold test facility (courtesy Oak Ridge National Laboratory, managed by Lockheed Martin Energy Research Corp. for the U.S. Department of Energy under contract number DE-AC05-96OR224).

limited space of the risers. This constraint forces the use of long, slender robots. Spar Aerospace designed and built a long-reach manipulator, the Modified Light Duty Utility Arm (MLDUA), for ORNL (Kruse and Cunliffe, 1997). This teleoperated system has a reach of approximately 17 feet, payload capacity of 200 pounds, and maximum profile diameter of 9 inches during insertion into a tank. Another long-reach manipulator is the Hose Management Arm (HMA). This arm has four degrees of freedom and a confined sluicing end-effector (CSEE). The CSEE not only sucks waste out of the tank, but has a series of high-pressure cutting jets for breaking up solid waste. Figure 1 shows the MLDUA grasping an end-effector on the HMA. Mining strategies consist of using either the MLDUA or the vehicle to move the CSEE through the waste.

2.2 Teleoperation in Space

Human operations in space or planetary exploration result in high costs due to life support and hazards in the space environment. The primary example of a space-based teleoperation system is the Remote Manipulation System (RMS) on the fleet of U.S. space shuttles. These manipulator arms, located in the main shuttle bay, are controlled directly by a human operator viewing the system through a window. Two three-axis variable-rate command joysticks provide velocity input commands from the operator to the robot's controller. Applications of the RMS include deployment and capture of satellite systems, space-based assembly, and construction of the International Space Station (ISS). The Space Station RMS (SSRMS) will become a permanent part of the ISS with the task of assembly and maintenance of ISS.

An emerging application of telerobotics is planetary exploration. One recent example is the Mars Pathfinder Sojourner Rover (see Figure 2), developed by the Jet Propulsion Laboratory (JPL) (Bekey, 1997). This remotely operated vehicle transmits visual and sensory information from the Mars surface to operators in California. With only a single uplink opportunity per day and with speed-of-light time delays from Earth to Mars, the rover is required to perform its daily operations autonomously. These operations include terrain navigation and response to unexpected contingencies.

2.3 Undersea Applications

Many undersea operations require teleoperated systems. The oil industry is one of the primary users of undersea teleoperated systems. Remotely operated vehicles (ROVs) in-

Figure 2   JPL's Sojourner (courtesy U.S. National Aeronautics and Space Administration).

spect and maintain underwater equipment thousands of feet below the surface. Tasks include construction of offshore structures and the inspection, maintenance, and repair of offshore oil rigs. Oceanographic research and exploration use teleoperated systems to explore the depths of the ocean. One of the leading manufacturers of underwater telerobotic systems is Schilling Robotic Systems, which has outfitted many underwater ROVs with hydraulic teleoperated manipulators. Figure 3 shows a Schilling Dual Titan 7F Integrated Work System installed on the U.S. Navy Deep Drone IV ROV.

Figure 3   Deep Drone IV (courtesy Schilling Robotic Systems, Inc.).

2.4 Medical Applications

One of the emerging areas of teleoperation and telerobotic applications is the medical field. Minimally invasive medical procedures reduce recovery time and subsequently reduce medical costs. Ironically, laparoscopic and endoscopic tools are mechanically similar to the first teleoperation systems. Hill et al. (1994) developed a teleoperated endoscopic system with integrated 3D stereo viewing, force-reflection, and aural feedback. Arai et al. (1996) developed a teleoperated catheter and demonstrated teleoperation over a distance of approximately 350 km. The operator can command movement of a two-degree-of-freedom catheter and feel reaction forces. Communication and visual feedback are provided through a high-speed optical fiber network. The work by Hunter et al. (1993) shows the potential for scaled surgical teleoperation. With force amplification provided from the scalpel, the surgeon can feel tissue-cutting forces that would normally be imperceptible. Visual cues to the surgeon, transmitted directly or merged with virtual environments, are displayed on a helmet-mounted display or a projection screen.

2.5 Firefighting and Rescue

Intense heat, heavy smoke, potential explosions, and pollutants are just some of the dangers facing today's firefighters. Fire departments are turning to teleoperated systems to ensure the safety of personnel and enable them to combat today's disasters effectively. Teleoperated firefighting equipment must operate in extremely hostile environments featuring intense heat, dense smoke, and toxic fumes. Not only must the robot combat fires, but it may retrieve victims who cannot be rescued by firefighters (Waddington and Kawada, 1997).

One example of a firefighting teleoperation system is the Rainbow 5, developed by USAR. This remotely operated vehicle has been in use since 1986. It is primarily used where firefighters would have difficulty in approaching, such as oil and petroleum fires and large-scale fires with intense heat, and/or where there is a danger of explosion. The Rainbow 5 carries four TV cameras and a gas detector. Water, drive, and obstruction-clearance equipment are available for use on the vehicle. Control is via a wire or radio. The operator can operate at a distance up to 100 meters. The remotely operated nozzles receive water or foam via a 75-mm hose and can pump up to 5000 liters of water per minute.

A related application for teleoperators on vehicles is bomb disposal. Extensive experience with bomb disposal using teleoperated vehicles and manipulators resides in several companies, such as Remotec Inc.

3 SUBSYSTEMS AND TERMINOLOGY

The terminology used in this chapter is introduced in Figure 4. A teleoperator system consists of a *remote unit,* which carries out the remote manipulation, a *command unit* for input of the operator's commands, and a *communications channel* for linking the command unit and remote unit.

The remote unit generally consists of the manipulator arm, or multiple arms, with an end-effector for grasping or otherwise engaging the work piece and special tools. Each articulation or joint of the manipulator provides a degree of freedom (d.o.f) to the manipulator. Commonly used joints are rotational or hinge joints providing one axis of rotation and prismatic or sliding joints with one direction of translation. The motion of the joints determines the motion of the end-effector. A minimum of six degrees of freedom is necessary for the end-effector to be arbitrarily positioned in the workspace with arbitrary orientation. Manipulators with more than the minimum number of degrees of freedom required for a task are *redundant.* Usually one joint actuator is provided per joint. Common actuator types are electromagnetic, hydraulic, and pneumatic. The motion of the joint actuator determines the position of the joint and is often *servocontrolled* based on the error between the desired position or velocity and the measured position or velocity of the joint. Servocontrol is also called *feedback* or *closed-loop control. Open-loop control* requires no measurement of the actual position for feedback to the controller. Electric stepping motors and hydraulic actuators can provide holding torque without feedback and can be controlled open loop. Not explicitly discussed here are possible mobile bases that extend the range of arm operation and incorporate functions similar to those found in the manipulator arm.

The command unit must contain an input interface unit for the operator to input the desired activities of the remote unit. For symbolic input this may be a button box or

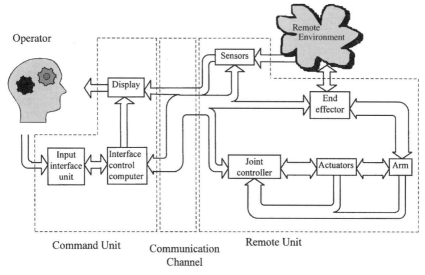

Figure 4   Teleoperator terminology displayed in conceptual block diagram.

keyboard. For analog input this may be a hand controller, joystick, glove device, or master unit. By *master unit* we refer to a device kinematically equivalent to the manipulator or slave unit, although kinematic equivalence is by no means generally necessary. The command unit must also provide some way, such as the interface control computer, to transform the operator inputs to commands compatible for transmission to the remote unit via the communications channel. This computer also controls a portion of the information displayed to the operator through a variety of modalities via the display unit. Local models of geometry and physical behavior (e.g., impedance) may be maintained by the interface control computer to supplement direct display of sensory information. It is in this area that rapid changes have occurred due to extraordinary progress in computing technology.

The communications channel may consist of individual wires carrying analog signals to command each joint, or the signals may be digitized and multiplexed for efficient use of the channel capacity. The physical medium of the communications channel can be a coaxial cable, an optical fiber, a radio link, or a sonar link, depending on the environment through which one must communicate. The Internet provides an interesting channel for some forms of teleoperation, but the significant and variable time delays and latency are still obstacles to be overcome.

3.1   Major Categories of Teleoperators

One way to categorize teleoperators is by their source of power. The earliest teleoperators were powered by the operator through a direct mechanical connection, either metal cables, tapes, or rods. The separation distance of the remote unit from the operator was limited, as were the speeds and forces of operation. The operator directly supplied and controlled the power to the various degrees of freedom. An advantage of the direct drive was the direct force feedback to the operator. Externally powered teleoperators opened up a new range of opportunities and problems. The operator now inputs signals to a control system that modulates the power to the remote unit. The remote unit is typically powered by electric motors or, when higher forces are demanded, hydraulic actuators. The power the operator must provide is greatly reduced, but the operator's effectiveness may be diminished unless the forces experienced by the remote unit can be displayed to the operator in a natural manner.

Another means of categorizing teleoperators is by the human interface, in particular the input interface unit, arranged as follows roughly in order of increasing sophistication:

1. *On–off control* permits joint actuators to be turned on or off in each direction at fixed velocity. This is the most primitive interface available and the simplest to

implement. Simultaneous joint motion may be possible, but true joint coordination is not.

2. *Joint rate control* requires the operator to specify the velocity of each separate joint and thus mentally transforms the coordinates of the task into arm joint coordinates. Many industrial devices (backhoes, cranes, forklift trucks) use joint rate control because of the simple hardware required to implement it.

3. *Master–slave control* allows the operator to specify the end position of the slave (remote) end-effector by specifying the position of a master unit. Commands are resolved into the separate joint actuators either by the kinematic similarity of the master and slave units or mathematically by a control unit performing a transformation of coordinates.

4. *Master–slave control with force reflection* incorporates the features of simple master–slave control and provides to the operator resistance to motions of the master unit that correspond to the resistance experienced by the slave unit. This interface is also called *bilateral master–slave control.*

5. *Resolved motion rate control* allows the operator to specify the velocity of the end-effector. The commands are resolved into the remote axes mathematically. The interface for the human may be a button box or joystick.

6. *Supervisory control* takes many forms, but in general allows the operator to specify some of the desired motion symbolically instead of analogically. The computer interpreting the symbolic commands then issues the joint commands. This mode of operator interface results in a hybrid between teleoperator and robot.

3.2 Major Categories of Telerobotics

Telerobotic systems generally consist of a human operator and an autonomous path planner. An important issue in telerobotic systems is the combination of human and machine intelligence. The primary objective is for the telerobotic system to perform tasks that cannot be done by either human or autonomous control. For a telerobotic system, the task is divided into subtasks that are easily automated (robotic) or require direct human interaction (teleoperated).

3.2.1 Superposition of Robotic and Teleoperated Velocity Commands

A number of strategies are available for combining robotic and teleoperated commands. Function Based Sharing Control (FBSC), by Tarn and coworkers (Guo et al., 1995), redefines the basic strategy for defining a task. Most path-planning strategies use a time index to define a desired position and/or velocity of the robot's joints or end-effector. Tarn's approach constructs a path in terms of a position-dependent path planner. There are two advantages to this strategy. First, if the robot collides with an unforeseen obstacle, the joint error remains fixed of integrated with time. Second, the planned motion of the robot can have additional stimuli superimposed on the planned path. These stimuli can include additional sensors or teleoperation input devices.

Figure 5 illustrates the basic concept. The task plan develops an original autonomous plan. The slave robot follows this plan autonomously while the operator passively ob-

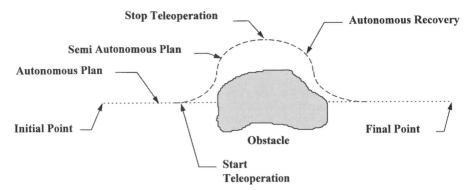

Figure 5 Obstacle avoidance and autonomous recovery using Function-Based Sharing Control.

serves the task execution. Upon observing an obstacle, he may intervene by grasping the input interface unit, such as a joystick, to provide a perturbation, defined as the semiautonomous plan, of the robot's path from the original autonomous plan. After clearing the obstacle, the operator can release the joystick and the planning controller generates an autonomous recovery path.

There are presently four modes of operation for the FBSC. The operator input to the path planner, through the joystick, can speed up, slow down, stop, or provide orthogonal perturbations to the path planner. This approach does not presently provide a method of providing any force reflection to the operator. Tarn (Wu, Tarn, and Xi, 1995) addressed this problem by augmenting the FBSC with a hybrid control strategy. This approach effectively provides force control along constrained degrees of freedom and FBSC along the unconstrained degrees of freedom.

3.2.2 Superposition of Robotic and Teleoperated Position Commands

Love and Magee (1996) developed a strategy that superimposed teleoperated position, instead of velocity, commands on the original autonomous path. Their master input device, which was under impedance control, used a decaying potential field to keep the master robot at a stable equilibrium point. When an operator perturbs the master robot from its equilibrium position, the time index on the remote robot's path planner is suspended. The command tip position of the slave robot is the summation of the master robot's position with the last position on the slave robot's autonomous path planner. This approach, illustrated in Figure 6, is beneficial for teleoperation systems in which there is a large scale difference between the workspaces of the master and slave robot. In addition, since the master robot provides position commands to the slave robot and the master is under impedance control, addition of remote force reflection is straightforward.

3.3 Major Categories of Master Input Devices

The physical interface to the human may include a master input device. There are two general types of interfaces: unilateral and bilateral. Two of the key parameters in selecting a master input device are the number of degrees of freedom and whether the system is actuated.

3.3.1 Unilateral Master Input Devices

Unilateral and bilateral master input devices have a number of common specifications. First is the number of controlled degrees of freedom. A mouse can provide only two

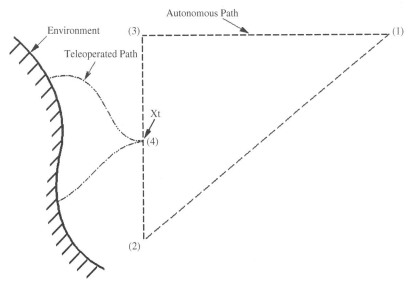

Figure 6 Telerobotic position path: perturbation from planned straight line path occurs at point (4).

degrees of freedom. More sophisticated joysticks, such as the Logitech Spaceball, provide up to six degrees of freedom from a single input device. While the spaceball provides the capability for a single input device to command six independent degrees of freedom, it is generally difficult to keep all of the commands decoupled. One alternative approach is to have two three-degree-of-freedom joysticks, such as the FlyBox by B&G Systems. One joystick provides translational commands while the second joystick provides orientation. While there is no constraint on the interpretation of these commands, the general approach is to convert displacement of the joystick to joint or end-effector velocity commands.

A second form of passive input device, generally used in virtual environment applications, are six-degree-of-freedom position-tracking systems. Magnetic tracking systems available from Polhemus and Ascension Technology use sets of pulsed coils to produce magnetic fields. The magnetic sensors determine the strength and angles of the fields. Tracking systems based upon magnetic technologies are generally inexpensive and accurate and have relatively large ranges of motion. Limitations of these trackers are high latency and interference from ferrous materials within the fields. Table 1 provides a list of commercially available tracking systems showing the range in performance as well as price at time of writing.

3.3.2 Bilateral Master Input Devices

Many conventional teleoperation systems provide some form of force reflection. Jacobsen et al. (1990) have implemented force feedback of this type in control of a 10-degree-of-freedom teleoperator. Colgate (1991) and Kazerooni (Bobgan and Kazerooni, 1991) have investigated the impact of impedance on the master robot. Many researchers have recently pursued techniques to augment or completely replace direct force reflection with virtual or computer generated forces (Hannaford, 1989; Morikawa and Takanashi, 1996). Rosenberg (1993) uses virtual fixtures to provide a local tactile representation of the remote environment. Love and Book (1996) use virtual fixtures, illustrated in Figure 7, to constrain the motion of the operator when the teleoperator has dissimilar kinematics between the master and remote unit. Furthermore, they adapt the impedance of a master robot based upon on-line identification of the slave robot's remote environment. The impedance of the master robot adjusts to the impedance of the slave robot, ensuring stability as well as reduction in the energy applied by the human during task execution.

Also commercially available are force-reflecting master units, also called *haptic displays*. The PHANToM is a desk-grounded pen-based mechanism designed for virtual force feedback. The system, shown in Figure 8, consists of a serial feedback arm that ends with a fingertip thimble-gimbaled support. Of the six degrees of freedom of the arm, the three translational axes are active while the gimbaled orientation is passive. Each interface has an $8 \times 17 \times 25$-cm workspace. The peak output force is 10 N, with a maximum continuous output force of 1.5 N. Table 2 provides a list of commercially available force feedback master devices.

While force reflection has definite advantages in performing remote operations, there are situations in which force-reflective teleoperation is presently not practical. Teleoperation over large distances is seriously impeded by signal transmission delays between the remote and command units. Round trip delays between ground and low earth orbit are minimally 0.4 seconds; for vehicles near the moon the delays are typically 3 seconds. Likewise, remote operation in the deep ocean can typically impose a 2-second round trip delay over a sonar link (Sheridan, 1993). These delays not only affect visual cues, but can induce instability when using force-reflective teleoperation. Ferrel (1966) provided early experiments showing how time delays limit bilateral teleoperation performance. With excessive delays, the force feedback no longer serves as a source of information, but acts as a disturbance. Vertut (Vertut, 1981) addressed this problem in terms of stability and proposed limiting the velocities and reducing the bandwidth of the system. Anderson and Spong (1989) use scattering theory to formally express this instability and define a new stable teleoperation control strategy. Niemeyer and Slotine (1991) have shown that placing energy-dissipating elements in the communication link guarantees stability in spite of time delays. This topic continues to be an area of active research and development.

3.3.3 Controlled Passive Devices

Passive actuation of an operator interface uses actuators that cannot impart power to the device but can control the way the device responds to externally applied power. For

Table 1 Commercial Passive Tracking Devices[a]

Product	Vendor	DOF	Frequency	Latency	Resolution	Working Volume	Cost
Fastrak	Polhemus	6	120 Hz	8.5 msec	0.0002 in./in. 0.025°	10–30 ft	$6050
PC/BIRD	Ascension Technology	6	144 Hz	10 msec	0.08 in., 0.15°	4 ft	$2475
CyberMaxx	Victor Maxx Technologies	3	75 Hz	29.6 msec	0.1° vertical, 0.1° horiz	360° horiz, ±45° tilt	$799
SELSPOT II	Selcom AB	6	10 kHz	NA	0.025% of millirad	200 m	$29,980
ADL-1	Shooting Star Technology	6	240 Hz	0.35–1.8 msec	0.25 in., 0.15–0.3°	35 in. diameter, 18 in. high	$1299
GyroPoint	Gyration, Inc.	3	NA	NA	0.2°	75 ft	$299
BOOM3C	FakeSpace, Inc.	6	>70 Hz	200 msec	0.1°	6 ft diameter, 2.5 ft height	$95,000

[a] From C. Youngblut et al., "Review of Virtual Environmental Interface Technology," IDA Paper P-3186, Institute for Defense Analyses, March 1996. © 1996 Institute for Defense Analyses.

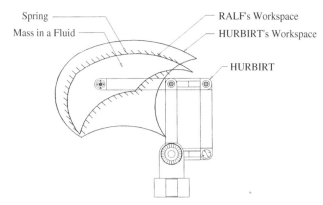

Figure 7 Virtual walls with crosshatch representing slave's workspace. (From L. Love and W. Book, "Adaptive Impedance Control for Bilateral Teleoperation of Long Reach Flexible Manipulators," in *Proceedings of the International Federation of Automatic Controls,* San Francisco, 1996.)

teleoperator input devices the input power is supplied by the human operator. By controlled passive actuation, the response of the device can provide to the human a haptic display of conditions present at the remote manipulator. If the operator is viewed as a force source, the velocity response of the master is controlled to give the operator a feeling of resistance or constraint. Passively controlled interfaces have potential advantages of inherent safety and low energy requirements for a corresponding high force capability. By removing the possibility of active motions, a compromise has been made in the total capability of the device that may be appropriate depending on the circumstances of the application. Medical, rehabilitation, virtual reality, and industrial applications are among the candidates for this new approach to operator interfaces with force feedback or haptic display.

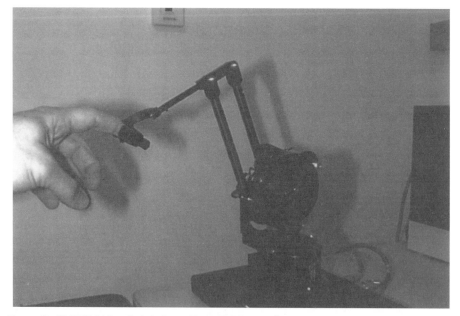

Figure 8 PHANToM haptic inferface. (From T. Massie, "Design of a Three Degree of Freedom Force-Reflecting Haptic Interface," S.B. Thesis, Department of Electrical Engineering and Computer Science, Massachusetts Institute of Technology. © Oak Ridge National Laboratory.)

Table 2 Commercial Force Feedback Devices[a]

Device	Vendor	Device Type	Force Provided to	DOF	Force Resolution	Applied Force
4-d.o.f. Master Feedback Master	EXOS, Inc.	desktop	hand via joystick	4	NA	12 oz-in. cont., 59 oz-in. peak
Force Exoskeleton ArmMaster	EXOS, Inc.	exoskeleton	shoulder and elbow	5	NA	56.6 in.-lb cont., 489 in.-lb peak
Impulse Engine 3000	Immersion Corp.	desktop	hand via joystick	3	0.00435 N	8.9 N cont.
HapticMaster	Nissho Electronics Corp.	desktop	hand via knob	6	2.85 gf	5.6 kgf/cm cont., 1.8 kgf peak
Hand Exoskeleton Haptic Display (HEHD)	EXOS, Inc.	exoskeleton	thumb & index finger joints	4	NA	1–5 lb peak
PER-Force 3DOF	Cybernet Systems Corp.	desktop	hand via joystick	3	0.035 oz	1 lb cont., 9 lb peak
PHANToM	Sens Able Devices, Inc.	desktop	fingertip via thimble	3	12 bit	1.5 N cont., 10 N peak
SAFiRE	EXOS, Inc.	exoskeleton	wrist, thumb, & index finger	8	NA	2 lb peak

[a] From C. Youngblut et al., "Review of Virtual Environmental Interface Technology," IDA Paper P-3186, Institute for Defense Analyses, March 1996. © 1996 Institute for Defense Analyses.

Two distinct classes of devices have been built and tested. Book et al. (1996) have used clutches and brakes to control the coupling between axes of a planar haptic display. Four clutches are used. Two clutches (configured as brakes) couple their respective axes to ground. A third clutch couples axes directly together, and the fourth couples axes with the opposite direction of motion. By modulation of the actuation of the devices, programmable constraints (like walls) can be imposed on the motion or resistance to motion can be adjusted as required for display of the remote environments. Colgate, Peshkin, and Wannasuphoprasit (1996) have used continuously variable transmissions to couple the axes of their passive interface. This coupling allows motion in one direction that is selectable by picking the variable ratio. The Colgate approach results in less dissipation and is well adapted to constraining motion.

The use of passive controlled interfaces may provide the designer of specialized teleoperation systems new options for providing force feedback to the human operator.

3.4   Major Categories of Visual Displays

Many teleoperation tasks provide direct viewing of the remote hazardous environment. Leaded glass provides safe viewing of radioactive objects. Likewise, portholes and pressurized vehicles provide safe, direct viewing of remote environments. However, not all teleoperated tasks provide a means for direct viewing. Studies reveal the effect visual cues have in the performance of telemanipulated tasks (Massimino and Sheridan, 1989). Factors such as control (pan, tilt, and zoom), frame rate, resolution, and color (grayscale) dramatically affect task performance. Often the visual display of the remote environment is provided by monoimage television, which provides the user with one view to both eyes. In order to enhance depth perception, stereoscopic television is often suggested as a desirable alternative. Accounts differ on the rated performance of these systems. Some studies show significant improvement in task execution (Cook, 1995), while others do not (Draper et al., 1991). Many of the present problems are associated with the technology. Studies show that adverse affects, such as eyestrain, headaches, and disorientation, limit the immersion time to under 30 minutes (Ebenholtz, 1992). Some studies show that many of these problems may be reduced through careful design and engineering (Rushton, Mon-Williams, and Wann, 1994). This area of research is expected to expand with the growing interest in the fields of virtual reality and virtual environments.

3.4.1   Visual Displays for Telepresence

Telepresence may be defined as a mental state entered when sensory feedback has sufficient scope and fidelity to convince the user that he or she is physically present at the remote site (Draper, 1995). It is widely accepted in the telerobotics community that telepresence is beneficial. However, as Sheridan commented, "It has yet to be shown how important is the sense of [telepresence] per se as compared to simply having . . . good sensory feedback" (Sheridan, 1992).

One common comment in the design of teleoperation systems is that one cannot have enough camera views. Traditional camera controls provide analog knobs that control a camera's pan and tilt motion. Using eye movement tracking, head-mounted displays control the motion of remote cameras by inspection of the operator's eye movement (Yoshidaa, Rolland, and Reif, 1995). More recently, systems have become available that combine computer-generated environments or virtual reality displays with given camera images (Bejczy, 1995). The goal is to provide additional information not available from the independent camera information. However, there are still technical issues, as discussed in the previous section, related to fatigue when attempting to immerse an operator in a virtual environment.

One example of the utility of virtual reality in teleoperation is point-and-direct telerobotics (McDonald et al., 1997). Virtual tools enable an operator to "reach into" a remote video scene and direct a robot to execute high-level tasks. The fusion of video and 3D graphics using the TELEGRIP simulation package extends the viewing and interaction capabilities between the operator and the display.

4   THE DESIGN CONTEXT

The design of a teleoperator ultimately involves the specification of components and parameters. These components and parameters result in characteristics of the manipulator behavior. Knowing the characteristics of the teleoperator and the characteristics of the task, one can determine the relative performance of that teleoperator by some performance index. Optimization of a design requires that the penalties associated with the cost and reliability of the components be considered as well as their performance. This chapter will consider only the relationship between performance and characteristics unique to teleoperators. For a wide range of opinions on performance evaluation, see the report of a workshop on the subject sponsored by the National Bureau of Standards (Sheridan, 1976).

The most relevant quantifiable measures of performance for teleoperators are based on task completion time. Measures considered here are task total time, time effectiveness ratio (time relative to the unencumbered hand), and unit task time (time for elemental task components). Information-based performance measures encapsulate time and accuracy of movement. Operator fatigue, success ratio, and satisfaction are hard to quantify but nonetheless important. Quantifying these performance measures requires the task or a range of tasks be specified. They are task-dependent measures to some extent. The most relevant tasks to be specified are the tasks for which the teleoperator will be used. Unfortunately, the tasks are not often known in advance with great certainty due to the general purpose nature of teleoperators.

Other performance measures are highly relevant for some applications. The interaction force between manipulator and manipulated object is highly relevant if physical damage is of concern. See, for example, Hannaford et al. (1991) and Love and Book (1996). Likewise, a measure of power and forces provided by the human during task execution is relevant when considering human factors such as fatigue and strength (Love, 1995).

4.1 Information-Based Performance Measures

One successful measure of performance is the information transmission rate achieved by the teleoperator. This is not totally task-independent, but has been correlated with simple characterizations of the task. The information transmitted by the operator is equal to the reduction in uncertainty in the relative position of the end-effector and the target, usually measured in bits. The time required to transmit the information determines the information transfer rate. The experimental determination of these correlations is based on measuring the task completion time, calculating an index of difficulty, and then normalizing the result. The index of difficulty I_d proposed by Fitts (1954) for use in direct manual positioning tasks (unencumbered hand) is:

$$I_d = \log_2 \frac{2A}{B}$$

where

A = the distance between targets in a repetitive motion
B = the width of the target

This index of difficulty and its variations have been applied to teleoperators by Book and Hannema (1980) for a simple manipulator of programmable dynamics and by McGovern (1974) for more complex manipulators with fixed dynamics. Hill (1976) combined this measure with the unit task concept, described below, to predict task times. He and his coworkers document claims that only the fine motion or final positioning phase is governed by information transmission. The gross motion phase of manipulation is governed by arm dynamics, both human and manipulator. These results are collected in summary form in Book (1985).

4.2 Time Effectiveness Ratio

One popular and easily understood measure of performance is the task time multiplier or time effectiveness ratio. When multiplied by the task time for the unencumbered hand, it yields the task time for the teleoperator. This is a useful rough estimate of performance, but varies significantly with the task type. There is some indication that it is approximately constant for similar manipulators and for narrow variation of the task difficulty.

4.3 Unit Task Times

Perhaps the most practical method for predicting the performance of teleoperators at a specific task is based on completion times for component subtasks or unit tasks. The time required for units such as *move, grasp, apply pressure,* and *release,* some of which have parameters, can be approximately predicted for a given existing manipulator. Unfortunately, they have not been related to the manipulator characteristics or the design parameters in any methodical way.

4.4 Task Characterization

Because all tasks that a teleoperator might be called on to perform cannot be included in a test of performance of a manipulator, important features of tasks must be characterized to evaluate its performance. More important, for design where no teleoperator yet exists, a general, if approximate, characterization is needed on which to base predictions. The simple Fitts's Law type of characterization requires only the ratio of distance moved to tolerance of the target position, i.e., A/B. More complex measures of the task consider the degrees of freedom constrained when the end-effector is at the final position. Positioning within a given tolerance in one dimension leaves the five remaining dimensions free, hence one degree of constraint. Inserting a sphere in a circular hole leaves the remaining four dimensions free, hence two degrees of constraint. A rectangular peg placed in a rectangular hole requires that three positions and two orientations be constrained, leaving only one degree of freedom. The tolerance in each degree of constraint can be used to specify the difficulty of the final positioning for insertions. The self-guiding feature of insertions with chamfers, for example, is difficult to account for and depends on manipulator characteristics such as compliance and force reflection.

As mentioned above, a task can also be described by its component subtasks. The unit task concept for industrial workers was developed in the 1880s by Taylor (1947) and later refined by Gilbreth (1911). Hill (1979) proposed a modified list of unit tasks for manipulators, as shown in Table 3. For design optimization one would like unit task times as a function of the manipulator characteristics. (See also the RTM method in Chapter 32.)

4.5 Teleoperator Characteristics

The important characteristics of a teleoperator in predicting performance should be traceable to the design decisions that resulted in those characteristics. Given the wide variety of teleoperators of interest, it is more productive to work with abstract characteristics that will be relevant to most if not all designs. Relating these characteristics to parameters

Table 3   Unit Task Descriptions for Manipulators

Unit	Description
Move (d)	Transport end-effector a distance d mm. Enhanced by greater arm speed. Hill (1977) documents a linear dependence on d.
Turn (a)	Rotate end-effector about the long axis of the forearm a degrees. Enhanced by force feedback (50% to 100%) or passive compliance. Degraded by friction and backlash.
Apply pressure	Apply force to overcome resistance with negligible motion. Enhanced by force feedback. Degraded by friction.
Grasp	Close end-effector to secure object. Enhanced by gripper force feedback and gripper speed.
Release	Open end-effector to release object. Enhanced by gripper force feedback and speed.
Pre-position (t)	Align and orient object in end-effector within a tolerance of t mm. Degraded by friction, backlash, and low bandwidth.
Insert (t)	Engage two objects along a trajectory with tolerance t mm. Enhanced by force feedback. Degraded by friction, backlash, and low bandwidth. Nonlinear with tolerance.
Disengage (t)	Reverse Insert. Greatly enhanced by force feedback at small t (factor of 2). Degraded by friction, backlash, and low bandwidth.
Crank	Follow a constrained circular path, pivoting at the "elbow." Greatly enhanced by force feedback (50% to 100%) or passive compliance.
Contact	Insure contact with a surface, switch, etc. Enhanced by force feedback.

and components for teleoperators is similar for robots and other systems. Some of the characteristics can be presented quantitatively, while others can only be discussed.

Characteristics considered by various researchers, designers, and authors include:

- Reach and workspace shape and volume
- Space occupied by the remote unit and command unit and the scaling of force and displacement between them
- Dexterity
- Degrees of freedom
- Velocity, acceleration, and force obtainable with and without payload
- Accuracy, repeatability, and resolution of position, force, and velocity
- Backlash, dead band, or free motion between the input and response
- Coulomb or dry friction and viscous friction of command and remote unit
- Bandwidth or frequency response of the remote unit to small amplitude inputs
- Time delay between issuance of a command and initiation of the resulting action by the remote unit as well as delay in transmission of sensory feedback
- Rigidity or compliance of the remote unit to externally applied forces
- Inertia of the remote unit and of the command unit
- Static loads that must be counteracted by the actuators of the remote unit or the operator

4.6 Performance Predictions

Many researchers have predicted performance of an existing manipulator for hypothetical tasks to be remotely performed. For design purposes one must predict the performance of a manipulator that does not yet exist so that design decisions can be made to achieve specified or desired performance. The limited existing results will be presented in the context of design, and qualitative discussion of other characteristics will be given.

4.7 Task Time Based on Index of Difficulty

Perhaps the most methodical results obtained applicable to design are due to Book and Hannema (1980). The results are based on experiments using a simple two-degree-of-freedom manipulator with programmable characteristics. The characteristics considered are among the most important for determining manipulator performance: arm servobandwidth, backlash (lost motion), and coulomb or dry friction. The task considered was a simple repetitive positioning task that involved moving to and tapping within a tolerance band.

4.8 Results for Multiple Degrees of Constraint

By using the concept of degree of constraint, it is possible to extend the index of difficulty to more complex tasks. This has been done for two manipulators with quite different characteristics and results presented (Hill and Matthews, 1976). The total task time was assumed to be the sum of travel, positioning, and insertion times. The positioning and insertion times were related to the degrees of constraint, with the index of difficulty being the sum of the indices for each constraint taken separately.

One of the useful concepts of this work was that the positioning and insertion task may involve one, two, or three phases. If the index of difficulty is low enough (less than 5 bits), it may be accomplished with only the open-loop travel phase dependent only on gross motion characteristics. If the index of difficulty is between roughly 5 and 10 bits, it will be completed within a fine motion phase. Fine motion characteristics influence both the time required and the upper limit on bits for completion in this phase. If the task tolerances are very high relative to the fine motion capabilities of the manipulator, a third phase may be required. In this phase a random variation of end position occurs about the nominal target. The probability of completing the task in any given time interval is constant and dependent on the standard deviation of the end point error. Only the second phase is limited by information channel capacity.

The fine motion phase begins with an uncertainty that depends on the accuracy of the open-loop move of the first phase. If the open-loop move results in a Gaussian distribution of final positions with standard deviation s, the reduction in uncertainty (information transmitted in the fine motion phase) results from truncating the tails of that Gaussian distribution to lie within $\pm x/2$. Precise tasks with low tolerance relative to the open-loop accuracy (represented by s) require a large reduction of uncertainty in the fine motion phase. Coarse tasks require little reduction in uncertainty after the open-loop move and in fact are probably completed without a fine motion phase.

4.9   Time Effectiveness Ratio Versus Manipulator and Task Type

Work by Vertut (1975) complements the work described above in that commercial manipulators with various control schemes were used. More complicated tasks were considered, although this prevented some of the generalization possible with simpler component tasks. Operator fatigue was also considered.

Vertut (1975) presented his results in terms of time effectiveness ratio, i.e., compared to the times of the unencumbered human hand. The general time effectiveness ratio predictions are shown in Figure 9 for four of the six manipulators types identified below. The types are generally the same as listed under categories above, with the abbreviations as follows:

- Light-duty master–slave—bilateral (LD)
- Heavy-duty master–slave—bilateral (HD)
- Position control—unilateral master–slave (PC)
- Resolved motion rate control (RMRC)
- Rate control (RC)
- On–off control (OOC)

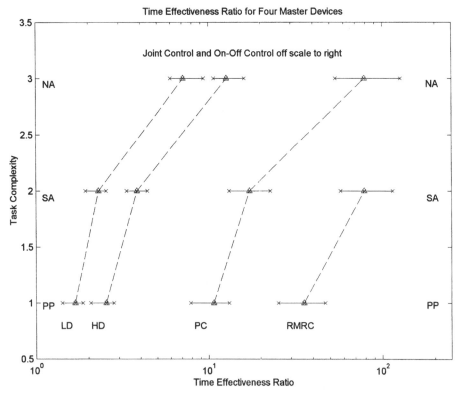

Figure 9   General chart for time effectiveness ratio. See text for nomenclature. (From W. J. Book, "Teleoperator Arm Design," in *Handbook of Industrial Robotics,* ed. S. Nof. New York: John Wiley & Sons, 1985. Chapter 9.)

The task types are described below.

- Pick-and-place (PP). A positioning task plus a grasp and release task. It consists of simply picking up an object and placing it in a new, specified location.
- Simple assembly (SA). A removal and insertion task. A simple insertion of a peg in a hole is an example.
- Normal assembly (NA). Involves insertion and turning, for example.

Figure 9 shows the significance of the human interface in determining the overall task time and the relative difficulty of the various task types. Time effectiveness ratios vary by a factor of over 100 for the different types of interfaces. As is always the case with comparisons of complete systems, one cannot attribute with certainty all the effects to one characteristic.

McGovern (1974) also considered the time effectiveness ratio as a means for performance prediction. A concise comparison of the work by Hill and McGovern is found in Book (1985).

4.10 Unit Task Times for Teleoperators

Prediction of total task times from the sum of unit task times has been effectively demonstrated for existing manipulators. Experiments performed to determine the unit task times for the specific combination of operator and manipulator have then been applied to other tasks, with predictions accurate to within 20%. The variability of the operators and the limited data for correlation account for a large fraction of this variation.

An example of decomposing a knob-turning task into motion elements is given by Hill (1979). In the task the operator touches a plate (signaling the start of the task), grasps the vertical handle of a rotary switch, turns it 90° clockwise, then 180° counterclockwise, and finally 90° clockwise, returning it to vertical. He releases the knob and touches the plate signaling the end of the task. The motion elements are as follows, with unit parameters in parenthesis:

1. Contact
2. Move (250 mm)
3. Grasp knob turn ($+90°$)
4. Dwell time
5. Turn ($-180°$)
6. Dwell time
7. Turn ($+90°$)
8. Release move (250)
9. Preposition (76.2)

5 SUMMARY

This chapter focused on recent developments in teleoperation, fundamental terminology, and a framework for making the decisions unique to teleoperator design. New application areas are opening, and the technology that supports teleoperation is progressing rapidly, especially computer, sensor, and display technology. The fundamentals of completing a manipulation task remain the same, however. Expected future challenges include many of the issues that have been pervasive in the past. Time delay in force-reflective teleoperation remains a difficult problem. Operator fatigue and "cybersickness" limit the effectiveness of task execution of immersive environments.

The following additional reading materials complement this chapter:

1. Sheridan (1992). Provides theoretical background to telerobotics and supervisory control.
2. Youngblut et al. (1996). Provides useful descriptions of present virtual environment hardware and approaches to virtual environment synthesis.
3. *Presence: Teleoperators and Virtual Environments,* a journal from MIT Press. Provides recent research articles related to teleoperation and virtual environments.
4. Burdea (1996).

Other chapters in this handbook will also add to the reader's understanding of this chapter. Consider specifically Chapter 13 on sensors, Chapter 14 on stereo-vision systems, Chapter 16 on control of mobility, and Chapter 17 on virtual reality and robotics.

REFERENCES

Anderson, R., and M. Spong. 1989. "Bilateral Control of Teleoperators with Time Delay." *IEEE Transactions on Automatic Control* **34**(5), 494–501.

Arai, F., et al. 1996. "Multimedia Tele-surgery Using High Speed Optical Fiber Network and Its Application to Intravascular Neurosurgery." In *Proceedings of the 1996 International Conference on Robotics and Automation,* Minneapolis, April. 878–883.

Bejczy, A. 1995. "Virtual Reality in Telerobotics." In *Proceedings of the International Conference on Automation and Robotics,* Barcelona, September.

Bekey, G. "On Space Robotics and Sojourner Truth." *IEEE Robotics and Automation Magazine* **4**(3), 3–4.

Bobgan, P., and H. Kazerooni. 1991. "Achievable Dynamic Performance in Telerobotic Systems." In *Proceedings of the 1991 IEEE International Conference on Robotics and Automation,* Sacramento, April. 2040–2046.

Book, W. J. 1985. "Teleoperator Arm Design." In *Handbook of Industrial Robotics.* Ed. S. Nof. New York: John Wiley & Sons. Chapter 9.

Book, W., and D. Hannema. 1980. "Master–Slave Manipulator Performance for Various Dynamic Characteristics and Positioning Task Parameters." *IEEE Transactions on Systems, Man, and Cybernetics* **SMC-10**(11), 764–771.

Book, W. J., et al. 1996. "The Concept and Implementation of a Passive Trajectory Enhancing Robot." In *1996 International Mechanical Engineering Congress and Exposition* (*IMECE*), November 17–22, Atlanta. 633–638.

Burdea, Grigore C. 1996. *Force and Touch Feedback for Virtual Reality.* New York: John Wiley & Sons.

Burks, B., et al. 1997. "A Remotely Operated Tank Waste Retrieval System for ORNL." *RadWaste Magazine* (March), 10–16.

Colgate, E. 1991. "Power and Impedance Scaling in Bilateral Manipulation." In *Proceedings of the IEEE International Conference on Robotics and Automation,* Sacramento, April. Vol. 3. 2292–2297.

Colgate, J. E., M. A. Peshkin, and W. Wannasuphoprasit. 1996. "Nonholonomic Haptic Display." In *Proceedings of the IEEE International Conference on Robotics and Automation,* April. 539–544.

Cook, T. 1995. "Teleoperation of Heavy Equipment: Derivation of Stereo Vision Requirements; A Case Study." In *Proceedings of the 25th International Conference on Environmental Systems,* San Diego, July.

Draper, J. 1995. "Teleoperators for Advanced Manufacturing: Applications and Human Factors Challenges." *International Journal of Human Factors in Manufacturing* **5**, 53–85.

Draper, J., et al. 1991. "Three Experiments with Stereoscopic Television: When It Works and Why." In *Proceedings of the 1991 IEEE International Conference on Systems, Man, and Cybernetics,* Charlottesville VA, October.

Ebenholtz, S. 1992. "Motion Sickness and Oculomotor Systems in Virtual Environments." *Presence: Teleoperators and Virtual Environments* **1**(3), 302–305.

Ferrell, W. 1966. "Delayed Force Feedback." *Human Factors* (October), 449–455.

Fitts, P. M. 1954. "The Information Capacity of Human Motor System in Controlling the Amplitude of Movement." *Journal of Experimental Psychology* **47**, 381–391.

Gilbreth, F. B. 1911. *Motion Study, a Method for Increasing the Efficiency of the Workman.* New York: D. Van Nostrand.

Guo, C., et al. 1995. "Fusion of Human and Machine Intelligence for Telerobotic Systems." In *Proceedings of the IEEE International Conference on Robotics and Automation,* Nagoya, May. 3110–3115.

Hannaford, B. 1989. "A Design Framework for Teleoperators with Kinesthetic Feedback," *IEEE Transactions on Robotics and Automation* **5**(4), 426–434.

Hannaford, B., et al. 1991. "Performance Evaluation of a Six-Axis Generalized Force-Reflecting Teleoperator." *IEEE Transactions on Systems, Man, and Cybernetics* **21**(3), 620–633.

Hill, J. W. 1976. "Study to Design and Develop Remote Manipulator Systems." Stanford Research Institute contract report on contract NAS2-8652, July.

Hill, J. W. 1977. "Two Measures of Performance in a Peg-in-Hole Manipulation Task with Force Feedback." Paper delivered at Thirteenth Annual Conference on Manual Control, Massachusetts Institute of Technology, June.

Hill, J. W. 1979. "Study of Modeling and Evaluation of Remote Manipulation Tasks with Force Feedback." Final Report, SRI International Project 7696, March.

Hill, J. W., and S. J. Matthews. 1976. "Modeling a Manipulation Task of Variable Difficulty." Paper delivered at Twelfth Annual Conference on Manual Control, University of Illinois, May.

Hill, J., et al. 1994. "Telepresence Surgery Demonstration System." In *Proceedings of the 1994 IEEE International Conference on Robotics and Automation,* San Diego, May. 2302–2307.

Hunter, I., et al. 1993. "A Teleoperated Microsurgical Robot and Associated Virtual Environment for Eye Surgery." *Presence: Teleoperators and Virtual Environments* **2**(4), 265–280.

Jacobsen, S., et al. 1990. "High Performance, High Dexterity, Force Reflective Teleoperator." In *Proceedings of the Conference on Remote Systems Technology.* Washington.

Kruse, P., and G. Cunliffe. 1997. "Light Duty Utility Arm for Underground Storage Tank Inspection and Characterization." In *Proceedings of the ANS 7th Topical Meeting on Robotics and Remote Systems,* April. 833–839.

Love, L. 1995. "Adaptive Impedance Control." Ph.D. Thesis, Georgia Institute of Technology, September.

Love, L., and W. Book. 1996. "Adaptive Impedance Control for Bilateral Teleoperation of Long Reach Flexible Manipulators." In *1996 International Federation of Automatic Controls* (*IFAC*) *World Congress,* San Francisco, June 1–5. Vol. A, 211–216.

Love, L., and D. Magee. 1996. "Command Filtering and Path Planning for Remote Manipulation of a Long Reach Flexible Robot." In *Proceedings of the IEEE International Conference on Robotics and Automation,* Minneapolis. 810–815.

Massie, T. "Design of a Three Degree of Freedom Force-Reflecting Haptic Interface." S.B. Thesis, Department of Electrical Engineering and Computer Science, Massachusetts Institute of Technology.

Massimino, M., and T. Sheridan. 1989. "Variable Force and Visual Feedback Effects on Teleoperator Man-Machine Performance." In *Proceedings of the NASA Conference on Space Telerobotics,* Pasadena.

McDonald, M., et al. 1997. "Virtual Collaborative Control to Improve Intelligent Robotic System Efficiency and Quality." In *Proceedings of the IEEE International Conference on Robotics and Automation,* Albuquerque.

McGovern, D. E. 1974. "Factors Affecting Control Allocation for Augmented Remote Manipulation." Ph.D. Thesis, Stanford University, November.

Morikawa, H., and N. Takanashi. 1996. "Ground Experiment System for Space Robots Based on Predictive Bilateral Control." In *Proceedings of the 1996 IEEE International Conference on Robotics and Automation,* Minneapolis, April. 64–69.

Niemeyer, G., and J. Slotine. 1991. "Stable Adaptive Teleoperation." *IEEE Journal of Oceanographic Engineering* **16**(1).

Rosenberg, L. 1993. "The Use of Virtual Fixtures to Enhance Telemanipulation with Time Delays." In *Proceedings of the ASME Winter Annual Meeting, DSC—Vol. 49,* New Orleans. 29–36.

Rushton, S., M. Mon-Williams, and J. Wann. 1994. "Binocular Vision in a Bi-Ocular World: New-Generation Head-Mounted Displays Avoid Causing Visual Deficit." *Displays* **15**(4), 255–260.

Sheridan, T. 1992. *Telerobotics, Automation, and Human Supervisory Control.* Cambridge: MIT Press.

———. 1993. "Space Teleoperation Through Time Delay: Review and Prognosis." *IEEE Transactions on Robotics and Automation* **9**(5), 592–606.

Sheridan, T. B., ed. 1976. *Performance Evaluation of Robots and Manipulators.* U.S. Department of Commerce, National Bureau of Standards Special Publication 459, October.

Taylor, F. W. 1947. *Scientific Management.* New York: Harper & Brothers.

Vertut, J. 1975. "Experience and Remarks on Manipulator Evaluation." In *Performance Evaluation of Programmable Robots and Manipulators.* Ed. T. B. Sheridan. U.S. Department of Commerce, NBS SP-459, October. 97–112.

Vertut, J., et al. 1981. "Short Transmission Delay in a Force Reflective Bilateral Manipulation." In *Proceedings of the 4th Symposium on Theory and Practice of Robots and Manipulators,* Poland. 269–285.

Waddington, S., and T. Kawada. 1997. "Taking the Heat." *Unmanned Vehicles* **2**(1), 10–13.

Wu, Y., T. Tarn, and N. Xi. 1995. "Force and Transition Control with Environmental Uncertainties." In *Proceedings of the IEEE International Conference on Robotics and Automation,* Nagoya, May. 899–904.

Yoshidaa, A., J. Rolland, and J. Reif. 1995. "Design and Applications of a High-Resolution Insert Head-Mounted-Display." In *Proceedings of Virtual Reality Annual International Symposium.* 84–93.

Youngblut, C., et al. 1996. "Review of Virtual Environment Interface Technology." IDA Paper P-3186, Institute for Defense Analyses, March.

Zorpette, G. 1996. "Hanford's Nuclear Wasteland." *Scientific American* (May), 88–97.

CHAPTER **10**

MICROROBOTICS

Toshio Fukuda
Fumihito Arai
Nagoya University
Nagoya, Japan

1 INTRODUCTION

Microsystems have recently been actively researched. They are used in a variety of fields, including maintenance and inspection in industry, microsurgery in medicine, and micromanipulation in biology (Fukuda and Arai, 1993; Fukuda and Ueyama, 1994; Irwin, 1997; Ishihara, Arai, and Fukuda, 1996). Micromachines have the scale advantage to reduce the size of components. Miniaturization is key to carrying out tasks in narrow spaces. Micromanipulators are used to manipulate minute objects with high precision. Since the microworld in which micromachines and micromanipulators are used is not subject to the same physical laws as the macroworld in which human operators work, it is necessary to translate sensing signals from that world to ours. To operate micromachines and micromanipulators, it is necessary to incorporate the human–machine interface into the total system. In this chapter, therefore, we consider the microrobot system, including micromachines, micromanipulators, and the human–machine interface, all together. Design of microrobot systems is closely related to the issues scaling problems and control methods.

The main characteristics of microrobot systems are as follows:

1. Simplicity
2. Preassembly
3. Functional integration
4. Multitude
5. Autonomous decentralization

These characteristics are closely related.

Simplicity results from the production methods and scale restrictions of microrobots. When given tasks cannot be executed by an individual robot due to the simplicity of its functions, the functions of multiple robots can be integrated or fused. The discussion here is limited to the functions of individual robots.

The configuration of multiple microrobots requires a new control architecture, such as decentralized autonomous control (Fukuda and Ueyama, 1994; Mitsumota et al., 1996). Decentralized microrobot systems depend on cooperation and coordination, which make possible the complementing of functions to carry out various tasks. These characteristics are similar to those of robot systems designed according to the concept of the cellular robotic system (Fukuda and Ueyama, 1994; Fukuda and Nakagawa, 1987), which indicates a new methodology for designing and constructing microrobot systems.

2 FUNDAMENTAL TECHNOLOGY FOR MICROROBOT SYSTEMS

Micromachine technology is necessary for making microrobot systems. The word *micro* is associated with machines under a millimeter in size, but machines do not necessarily have to be that small to be considered micromachines. From the viewpoint of miniaturization of a machine, the size of micromachine is relatively argued. On that account, it

Handbook of Industrial Robotics, Second Edition, Edited by Shimon Y. Nof
ISBN 0-471-17783-0   © 1999 John Wiley & Sons, Inc.

is a problem whether the physical size is extremely small in comparison with that of present machines. The microrobot is not only a miniaturized machine, but also an integrated system consisting of sensors, actuators, and a logic circuit (Figure 1). In a sense of conventional machine, these functional parts are miniaturized and integrated by wire. This is the first step in manufacturing a microrobot. The microrobot's unified structure makes possible mass production by a batch process, a new concept in manufacturing robot systems.

Will it be possible to make such microrobots? As an elementary technique of the micromachine, in recent years, micromotors and microsensors have been developed. Microsensors made by semiconductor technique have been in use since the 1960s. In this case, a logic circuit is also formed by silicon base. It was subsequently proposed that silicon be used as the mechanical material of micromachines, and research into microactuators began in the 1980s. At the University of California at Berkeley and the Massachusetts Institute of Technology (Mehregany et al., 1990; Behi, Mehregany, and Gabriel, 1990), micromotors about 100 microns in rotor diameter were developed based on the silicon process. The semiconductor technique of the silicon process has brought about the ability to make microactuators and microsensors as functional elements besides the manufacturing of microcomputer devices. This research has influenced the reality of microrobots. In addition to the microrobots that are fabricated by the conventional semiconductor techniques, an idea for a molecular machine has been presented (Drexler, 1986).

3 RESEARCH ISSUES ON MICROROBOT SYSTEMS

Many problems pointed out early in the research into microrobot systems later became concrete. The following research topics are important in the development of microrobotics:

1. Scaling issues
2. Microfabrication
3. Micromanipulation
4. Measurement methods
5. Actuation methods

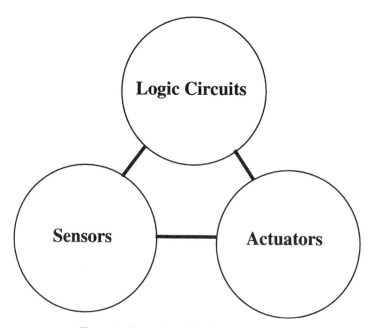

Figure 1 Integration of the key components.

6. Control methods
7. Communication
8. Energy supply methods
9. Human–machine interface

3.1 Scaling Issues

In the world of micromachines and micromanipulators, it is necessary to consider scaling issues on design, control, and measurement of microrobot systems. The difference in dimensions between the macroworld and microworld causes a difference in balance between magnitudes of forces. Magnitude of force is influenced by miniaturization to a degree that differs depending an object's physical property. For example, the viscosity force becomes more influential on the motion of the small object than with the inertial force (Ishihara, Arai, and Fukuda, 1996). Such influences can be roughly evaluated by the scaling law. Table 1 shows the scaling effect for several forces.

In microrobot systems common sense cannot be applied to predict motion. The influence of friction cannot be ignored. Friction force becomes larger than inertia force, making lubrication more important. The abrasion phenomenon greatly affects the preservation of the micromachines and micromanipulators. The Reynolds number is so small that water behaves as a viscous fluid from the viewpoint of common sense. Miniaturization intensifies the surface effect, as explained in Section 3.3.

3.2 Microfabrication

A photofabrication technique is commonly used in fabricating micromechanical structures (Irwin, 1997). The advantage of this technique is that the microactuators, microsensors, and microprocessor are easily united. The microfabrication method based on the photofabrication technique is suitable for two-dimensional structures, but producing three-dimensional structures by conventional photolithography is difficult. Several different approaches to realizing three-dimensional microstructures have therefore been proposed (Irwin, 1997). For example, the LIGA (Lithographie Galvanoformung, Abformung) process is one approach to making a microstructure with a high aspect ratio.

In order to design the microrobots and optimize the manufacturing process, it is necessary to introduce CAD/CAM systems for micromanufacturing systems. In microfabrication preassembly design should be adopted for constructing microrobots, and manipulation of tiny parts is required. An understanding of the mechanical property of silicon and the other functional materials, such as gallium and arsenic or polymer materials, is also necessary.

3.3 Micromanipulation

3.3.1 Basic Motion and Adhesive Forces

As research on micromachines has progressed, basic elements, such as microsensors, microactuators, and micromechanisms, have been developed. Yet most of the microfabri-

Table 1 Scaling Effect of Forces

Force	Equation	Scaling Effect	
electrostatic force	$eS/2 \cdot V^2/d^2$	L^0	e = permittivity, S = surface area, V = applied voltage, d = gap between electrodes
electromagnetic force	$B/(2m) \cdot Sm$	L^2	m = permeability, B = magnetic field density, Sm = area of cross section of coil
inertial force	$M \cdot d^2x/dt^2$	L^4	M = mass, t = time, $\times$ = displacement
viscosity force	$cS/L \cdot dx/dt$	L^2	c = viscosity coefficient, S = surface area, L = length, t = time, $\times$ = displacement
elastic force	$ES \cdot \Delta L/L$	L^2	E = Young's modulus, S = cross section area, ΔL = strain, L = length

cation method is still based on conventional lithography or other two-dimensional fabrication methods. Three-dimensional microstructures are generally difficult to manufacture. New fabrication methods have been proposed, such as the LIGA process and stereo lithography, but integrating sensing, actuating, and mechanical parts remains a problem. An assembly process will be required to some extent, as well as modification and maintenance of the microsystem. From this point of view, micromanipulation is a fundamental technology for microsystems.

Micromanipulation is broadly classified into the *contact* type and the *noncontact* type. This section focuses on the contact type, especially in the air (Fukuda and Tanaka, 1990; Fukuda et al., 1991a; Fukuda et al., 1991b; Arai et al., 1995; Arai et al., 1996a; Arai et al., 1996b; Arai et al., 1997; Arai and Fukuda, 1997; Fearing, 1995). Noncontact manipulation is also important and is mentioned in Section 3.8.3. The following are the typical contact micromanipulation movements in the air:

1. Lift up (hold, attach, stick, vacuum, etc.)
2. Place (release, attach, stick, vacuum, etc.)
3. Arrange (lift up and place, slide, rotate, etc.)
4. Push (hold, twist, cramp, deform, etc.)

Manipulation objects are supposed to be of the micron order. Adhesive forces thus have more effect on manipulation of the microobjects than does the gravitational force (Arai et al., 1995; Fearing, 1995).

Research has been carried out on the effects of the following adhesive forces, among others:

- The van der Waals force
- Electrostatic forces
- Surface tension force
- Hydrogen bonding force

These forces are calculated for the sphere that contacts the wall, as shown in Figure 2 (Arai et al., 1995; Arai et al., 1996b). The sphere is made of SiO_2, and the wall is made of Si. As noted above, adhesive forces are more influential than gravitational force in the microworld. The balance of these forces depends on environmental conditions, such as humidity, temperature, surrounding medium, surface condition, material, and relative motion. Micromanipulation is quite different from macroobject manipulation. Manipulating a microobject of less than 100 microns is quite difficult and calls for manipulation skills quite different from those required for macroobject manipulation. Physical phenomena in the microworld must be considered. Thermal, optical, electrical, and magnetic effects will change or become dominant when objects are miniaturized. Conventional studies on micromanipulation have not dealt much with physical phenomena in the microworld. Micromanipulation requires a new approach based on physics in the microworld (Arai et al., 1995; Arai et al., 1996b; Arai and Fukuda, 1997; Fearing, 1995).

3.3.2 Strategy for Micromanipulation

The environment for micromanipulation must be clean. Dust and small particles must be removed up to the appropriate class. Dust should not be generated in the micromanipulation process. To this end, abrasion should be avoided and most of the micromanipulation task based on pick-and-place motion and deformation by pushing. Adhesive forces affect micromanipulation tasks, and reducing them will improve the manipulability of the microobject. However, these forces can be utilized actively. If the adhesive force between the manipulation object and the place where the object is supposed to be located is greater than that between the gripper and the object, the object adheres to that place. It is then easy to remove the end-effector from the object. However, it is difficult to remove the object again. If we wish to rearrange the object without generating friction, we must consider adhesive force control and reduction of the adhesive forces.

If the above considerations are observed, rearrangement of the object can become very easy. It is possible to avoid contact rotation and contact slide motion of the manipulation object and to prevent dust generation and contamination based on microtribological phenomena. To control the stick-and-remove motion actively or reduce the adhesive forces,

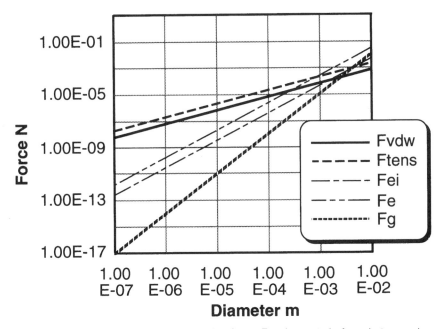

F_{vdw}: van der Waals force; F_{tens}: surface tension force; F_{ei}: electrostatic force between charged body and uncharged body; F_e: electrostatic force between the charged bodies.

Figure 2 Comparison of gravitational force and adhesive forces.

a special micro-end-effector must be developed. Several approaches to designing a micro-end-effector can be considered, including the following:

1. Modification of surface configuration (for example, by coating/treatment, roughness change (Arai, Andou, and Fukuda, 1995), or micropyramid (Arai et al., 1997*b*), with or without vibration)
2. Electrostatic force change (Arai et al., 1997*b*)
3. Temperature change (moisture control with microheater) (Arai and Fukuda, 1997)
4. Pressure change (Arai and Fukuda, 1997; Sato et al., 1993)

These special devices can be made using microfabrication methods. Strategy for micro-manipulation is summarized in Figure 3 (Arai, Andou, and Fukuda, 1995).

3.3.3 Possible Force Controller for Micromanipulation

Force interaction occurs between the object and the end-effector in contact-type micro-manipulation. The force sensor is important for assembly of the microparts or microma-nipulation of biological objects. Measuring the force applied at the microobject directly is difficult. If the microgripper holds the object, the force applied at the finger of the gripper can be measured.

Several kinds of force sensors for micromanipulation and atomic force microscopes have been measured (Arai et al., 1996*a*; Arai et al., 1996*b*; Kimura, Fujiyoshi, and Fu-kuda, 1991; Tortonese et al., 1991; Akiyama, Blanc, and de Rooij, 1996). We designed a force sensor based on the piezoresistivity effect because it has good linearity (Arai et al., 1996*b*). We made a micro-end-effector with the piezoresistive force sensor and the micro pyramids on the <100> Si cantilever. Because the micro-end-effector is made by Si surface micromachining, it is suitable for mass production. The assembly process for the endeffector is simplified. There is no need to set a semiconductor strain gauge by hand, which will cause inconsistency in the products. Resolution of this force sensor was less than 1 micron N (Arai et al., 1996*b*).

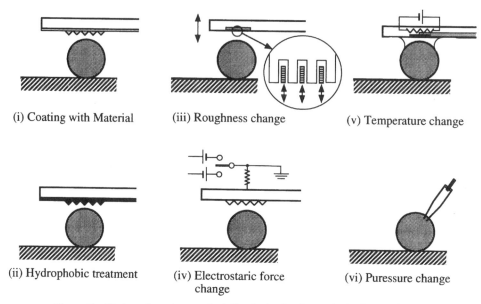

(i) Coating with Material   (iii) Roughness change   (v) Temperature change

(ii) Hydrophobic treatment   (iv) Electrostaric force   (vi) Puressure change
                                 change

Figure 3   Strategy for micromanipulation (reduction/control of adhesive forces).

Table 2 shows the classification of the force measurement methods. A strain gauge-type force sensor including the piezoresistive force sensor is suitable for real-time force feedback control. However, local stress around a few micrometer area is impossible to measure. If the structure is miniaturized to less than a few micrometers, the piezoresistive force sensor is difficult to make inside this area. On the other hand, the Laser Raman Spectrophotometer can measure local stress without contact. Raman microspectroscopy has been found capable of measuring stress over areas as small as 0.8 micrometers (Arai et. al., 1996a). However, real-time force control is difficult. At present, it takes a couple of seconds to measure the stress, and accuracy is very sensitive to temperature change. The Laser Raman Spectrophotometer is suitable for stress evaluation of miniaturized structures (Arai et al., 1996a). Vision system is also applicable for measuring deformation of the microstructure. The optical interference fringe can be used for high resolution. The force applied can thus be measured precisely. Realizing fast data sampling is still difficult because it is based on image processing. Force sampling speed is restricted by the frequency of the image signal. On the other hand, in the biological field optical tweezers are used to measure the small force of pN order acting on beads with diameters of a few microns based on the displacement of the target.

Table 2   Force Measurement Methods

	Strain Gauge (Piezoresistive)	LRS (Laser Raman)	Optical Interference	Deformation
measurement quantity	resistance change	Raman shift (stress)	space between slit	displacement
noncontact sensing	no	yes	yes	yes
measurement speed	real time	2–90 seconds	slow	real time
localized information	no	yes	yes	no
accuracy	good	good	very good	very good
rigidity	high	high	low	low

Rigidity of the microgripper is important for manipulation of a microobject. Based on the scaling effect, when the structure is miniaturized, the stress per unit force is more suitable for force measurement than the deformation per unit force (Arai et al., 1996a). We thought of measuring stress of the microgripper to measure the force applied at the microobject. Then we applied the Laser Raman Spectrophotometer for force measurement. We could make a relatively stiff microfinger with high accuracy (Arai et al., 1996a).

Alternatively, we proposed a variable stiffness microgripper. We used ER (electroreological) fluid to realize this unique function (Arai et al., 1997). ER fluids are functional fluids whose mechanical property such as viscosity is changed (1) in high speed, (2) reversibly by the applied voltage, and (3) greatly. ER fluids can be classified roughly into suspension (Bingham fluid) and homogeneous (liquid crystal). Shear stress property against shearing deformation speed differs between these fluids. When the voltage is applied, the Bingham fluid acts like Coulomb friction and the liquid crystal acts like a damper and generates a damping force proportional to the shearing velocity. These effects emerge according to the strength of the applied voltage, and mechanical property is controlled easily and quickly. From these characteristics the ER fluid is widely applied to the mechatronics devices. An ER valve, ER damper, ER clutch, and so on have been proposed. We have proposed an ER joint that can change the damping property of the joint by changing the applied voltage between the electrode. Here the space between the electrodes is filled up with the ER fluid. We applied this idea to (1) the safety mechanism with passive spring and (2) variable stiffness-type force sensor. The ER joint can change the damping property in a few milliseconds, so it can be used to protect the fragile part of the end-effector or the object. On the other hand, the ER joint can be used for vibration isolation. Using this function, we can improve work efficiency in transportation of the microobject.

3.4 Measurement Methods

In order to control micromachines and micromanipulators, it is necessary to measure or sense environments and objects. Development of microsensors is one of the main research topics in microrobot systems. Microsensors have been applied in industrial fields, including acceleration sensors and pressure sensors (Irwin, 1997). Sensing and measuring abilities are directly related to the performance of the controlled system. Sensing ability influences the control of micromachines and micromanipulators and organization of the architecture of microrobot systems, including human–machine interface.

3.5 Actuation Methods

Many different types of microactuators have been proposed, including the following:

- Electrostatic
- Electromagnetic
- Piezoelectric
- Optical
- Shape memory alloy (SMA)
- Giant magnetostrictive alloy
- Polymer

Each actuator has each characteristic. Property and application examples are given in Irwin (1997) and Ishihara, Arai, and Fukuda (1996).

3.6 Control Methods

Control issues with microrobots are different than with conventional robots due to the change in environment and configuration. As noted above, physical phenomena in the microworld are different than in the macroworld. Microrobot systems are highly nonlinear. The mechanical system control method is based mostly on the model-based approach. Precise modeling of the system is thus needed for fine motion control.

In addition to control methods, the control architecture for microrobot should be discussed. As an example of cooperative behavior, ant-like robots (Denouborg, 1990) can be regarded as a model for microrobots. The control architecture should be organized on the basis of decentralized autonomous control architecture, since controlling each microrobot from the central controller is too difficult and complex. The limited capacity of a microrobot makes it difficult to provide complete intelligence or knowledge, meaning that the organization of the control architecture must be constructed hierarchically. From this derives the idea of collective or swarm intelligence, which is actively studied in the field of multiple-robot systems (Fukuda and Ueyama, 1994; Mitsumoto et al., 1996).

3.7 Communication

For microrobot systems that include multiple micromachines or micromanipulators with the human–machine interface, communication ability is very important for carrying out given tasks. Multiple microrobots must communicate with each other to cooperate, coordinate, and synchronize. Mutual communication is a key issue for knowledge acquisition and knowledge exchange in microrobot systems. For micromanipulation, communication is also required among the micromanipulator, the mobile microrobots, and the operator.

3.8 Energy Supply Methods

3.8.1 Classification of Energy Supply Methods

One of the ultimate goals of microrobotics is to make an intelligent ant-like mobile microrobot able to perform given tasks. Most microrobots at present are supplied energy through a cable. But the cable, through friction, disturbs the motion of small robots. The energy supply method of the microactuator thus becomes important (Irwin, 1997). Energy supply methods are classified into *internal* (internal energy sources) and *external* (external energy supply to the system without a cable).

3.8.2 Internal Energy Supply

The energy source is contained within the moving body. Electric energy is frequently used. Batteries and condensers have been used. Batteries are good in terms of output and durability, but are difficult to miniaturize. Bates et al. (1993) made a micro lithium battery with thickness on the micron order and electricity density of 60 mA/cm^2. It was rechargeable for 3.6 V–1.5 V. As for condensers, an autonomous mobile robot 1 cm^3 in volume was developed in 1992 by Seiko Epson based on conventional watch production technology. A high-capacity condenser of 6 mm diameter, 2 mm thickness, and 0.33 F electric capacity was used as an energy source. The electric capacity of the condenser was small compared with that of a secondary battery. This microrobot used two stepping motors with current control of pulse width modulation. It could move for about 5 minutes on a charge of only 30 seconds.

3.8.3 External Energy Supply

Energy is given to the object from outside. Here we include noncontact-type micromanipulation of the small object. External energy can be classified as follows:

1. Optical energy
2. Magnetic energy
3. Electric energy
4. Ultrasonic energy
5. Mechanical vibration

Optical energy can be classified as follows:

1.1 Optical pressure by irradiating laser beam
1.2 Optical energy to strain conversion by photostrictive phenomenon
1.3 Optical energy to heat conversion

As an example of 1.1, remote operation of the microobject by a single laser beam as a tweezer has been proposed (Ashkin, 1970). As an example of 1.2, optical piezoelectric actuators such as PLZT (Fukuda et al., 1995) have been developed. As an example of 1.3, low-boiling-point liquid material has also been used with the optical heat conversion material. Utilization of the pyloelectric effect has been proposed to supply the energy from outside. For an energy transformation from optical energy to electric energy, PLZT has been employed (Ishihara and Fukuda, 1993). Generally, this can be substituted by the other pyloelectric element which can generate pyloelectric current by the temperature change.

As an example of magnetic energy, microwave, which has been used for noncontact energy transmission to airplanes and a solar energy-generation satellite, has been considered, but there are few reports. GMA (giant magnetostrictive alloy) can be considered for noncontact energy transmission. We applied the GMA for an actuator of an in-pipe mobile robot (Fukuda et al., 1991c).

As an example of electric energy, dielectrophoretic force can be used for noncontact-type micromanipulation of small objects (e.g., DNA, cells, glass beads) (Arai et al., 1997; Fuhr, 1997; Morishima, 1997).

As an example of ultrasonic energy, radiation pressure of ultrasonic waves can be used for noncontact-type manipulation of small objects. Alumina particles with an average diameter of 16 microns were trapped at the nodes of the ultrasonic standing wave field in water. The position of the nodes can be controlled by frequency change and switching of the electrodes. The position of the small object can be controlled this way (Kozuka et al., 1997).

As an example of mechanical vibration, selective energy transmission to the elastic object on the vibrating plate has been proposed. Another example is hydrodynamic flow. External force through the external medium can be used to move, for example, a maintenance pig robot in a pipeline filled with liquid.

3.9 Human–Machine Interface

Human–machine interface should be taken account in the system because of the difference of the physical phenomena in our common sense as described above. The difference causes the necessity of translation of signals between microrobots and human operators. For the development of human–machine interface, communication issues between human beings and microrobots, control issues of sensitivity and time delay, and differences in degree of freedom must be considered. Virtual reality (VR) can be applied in microrobot systems. VR technology is helpful in translating the world of micromachines and micromanipulators into the world of human beings.

3.10 Examples of Mobile Microrobots

Several mobile microrobots have been made so far (Mitsumoto et al., 1995; Stefanini, Carrozza, and Dario, 1996; Yano et al., 1996; Nicoud and Matthey, 1997). Figure 4 shows the Programmable Micro Autonomous Robotic System (P-MARS), developed by our

Figure 4   Programmable Micro Autonomous Robotic System (P-MARS).

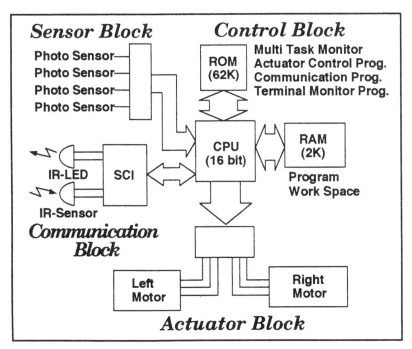

Figure 5 System configuration of P-MARS.

group in 1994 (Mitsumoto et al., 1995). This autonomous mobile robot is within 20 ×
20 × 20 mm and has photosensors, two stepping motors, CPU (16 bit), 8 batteries, and
an IR communication device. The system configuration is shown in Figure 5. The robot
has communication ability with the host computer and the other microrobots. The program
can be downloaded and rewritten by communication. Research works on distributed control of multiple microrobots based on the immune network architecture (Mitsumoto et al.,
1996) are on going.

4 SUMMARY

This chapter introduced and described the research issues in microrobotics. The scaling
issue indicates the necessity for considering the influence of the physical laws dominating
in the microworld. This point should be considered in designing and controlling microrobot systems in order to further extend the applications of microrobot systems.

REFERENCES

Akiyama, T., N. Blanc, and N. F. de Rooij. 1996. "A Force Sensor Using a CMOS Inverter in View
    of Its Application in Scanning Force Microscopy." In *Proceedings of IEEE Micro Electro Me-
    chanical Systems (MEMS)*. 447–450.
Arai, F., and T. Fukuda. 1997. "A New Pick up and Release Method by Heating for Micromanipu-
    lation." In *Proceedings of IEEE Micro Electro Mechanical Systems*. 383–388.
Arai, F., et al. 1995. "Micro Manipulation Based on Micro Physics—Strategy Based on Attractive
    Force Reduction and Stress Measurement." In *Proceedings of International Conference on In-
    telligent Robotics and Systems (IROS)*. Vol. 2. 236–241.
Arai, F., et al. 1996a. "New Force Measurement and Micro Grasping Method Using Laser Raman
    Spectrophotometer." In *Proceedings of IEEE International Conference on Robotics and Auto-
    mation.* Vol. 3. 2220–2225.
Arai, F., et al. 1996b. "Micro Endeffector with Micro Pyramids and Integrated Piezoresistive Force
    Sensor." In *Proceedings of International Conference on Intelligent Robotics and Systems (IROS)*.
    Vol. 2. 842–849.
Arai, F., et al. 1997. "Bio-Micro-Manipulation (New Direction for Operation Improvement)." In
    Proceedings of International Conference on Intelligent Robotics and Systems (IROS). Vol. 3.
    1300–1305.

Ashkin, A., 1970. "Acceleration and Trapping of Particles by Radiation Pressure." *Physical Review Letters* **24**(4), 156.

Bates, J.B., et al. 1993. "Rechargeable Solid State Lithium Microbatteries." In *Proceedings of Micro Electro Mechanical Systems.* 82–86.

Behi, F., M. Mehregany, and K. Gabriel. 1990. "A Microfabricated Three-Degree-of-Freedom Parallel Mechanism." In *Proceedings of Micro Electro Mechanical Systems.* 159–165.

Denoubourg, J. L., et al. 1990. "The Dynamics of Collective Sorting Robot-Like Ants and Ant-Like Robots." In *From Animal to Animates I.* Cambridge: MIT Press. 365–363.

Drexler, K. E. 1986. *Engines of Creation.* New York: Anchor Books.

Fearing, Ronald S. 1995. "Survey of Sticking Effects for Micro Parts Handling." In *Proceedings of 1995 IEEE/RSJ International Conference on Intelligent Robotics and Systems.* Vol. 2. 212–217.

Fuhr, G. 1997. "From Micro Field Cages for Living Cells to Brownian Pumps for Submicron Particles." In *Proceedings of 1997 International Symposium on Micromechatronics and Human Science.* 1–4.

Fukuda, T., and F. Arai. 1993. "Microrobotics—On the Highway to Nanotechnology." *IEEE Industrial Electronics Society Newsletter* (December) 4–5.

Fukuda, T., and S. Nakagawa. 1987. "Approach to the Dynamically Reconfigurable Robotic System (Concept of a System and Optimal Configuration)." In *Proceedings of IECON '87.* 588–595.

Fukuda, T., and T. Tanaka. 1990. "Micro Electro Static Actuator with Three Degrees of Freedom." In *Proceedings of IEEE International Workshop on Micro Electro Mechanical Systems* (*MEMS*). 153–158.

Fukuda, T., and T. Ueyama. 1994. *Cellular Robotics and Micro Robotic Systems.* Vol. 10 of World Scientific Series in Robotics and Automated Systems. World Scientific.

Fukuda, T., et al. 1991*a*. "Design and Dexterous Control of Micromanipulator with 6 D.O.F." In *Proceedings of the 1991 IEEE International Conference on Robotics and Automation,* Sacramento, April. 1628–1633.

Fukuda, T., et al. 1991*b*. "Electrostatic Micro Manipulator with 6 D.O.F." In *Proceedings of IEEE International Workshop on Intelligent Robots and Systems '91* (*IROS '91*). 1169–1174.

Fukuda, T., et al. 1991*c*. "Giant Magnetostrictive Alloy (GMA) Applications to Micro Mobile Robot as a Micro Actuator Without Power Supply Cables." In *Proceedings of IEEE Micro Electro Mechanical Systems.* 210–215.

Fukuda, T., et al. 1995. "Performance Improvement of Optical Actuator by Double Side Irradiation." *IEEE Transactions on Industrial Electronics* **42**(5), 455–461.

Irwin, J. D. 1997. *The Industrial Electronics Handbook.* Boca Raton: CRC Press. Section X, Emerging Technologies. 1468–1591.

Ishihara, H., and T. Fukuda. 1993. "Micro Optical Robotic System (MORS)." In *Proceedings of Fourth International Symposium on Micro Machine and Human Science.* 105–110.

Ishihara, H., F. Arai, and T. Fukuda. 1996. "Micro Mechatronics and Micro Actuators." *IEEE/ASME Transactions on Mechatronics* **1**(1), 68–79.

Kimura, M., M. Fujiyoshi, and T. Fukuda. 1991. "Force Measurement and Control of Electrostatic Micro-actuator." In *Proceedings of Second International Conference on Micro Machine and Human Science* (*MHS*). 119–124.

Kozuka, T., et al. 1997. "Two-Dimensional Acoustic Micromanipulation Using a Line-Focused Transducer." In *Proceedings of 1997 International Symposium on Micromechatronics and Human Science.* 161–168.

Mehregany, M., et al. 1990. "Operation of Microfabricated Harmonic and Ordinary Side-Drive Motors." In *Proceedings of Micro Electro Mechanical Systems.* 1–8.

Mitsumoto, N., et al. 1995. "Emergent Type of Micro Swarm Robotic System (Proposition of Micro Swarm Robotic System and Design of Programmable Micro Autonomous Robotic System (P-MARS))." *Transactions of Japan Society of Mechanical Engineers* **61-592C**(12), 4666–4673. [In Japanese.]

Mitsumoto, N., et al. 1996. "Self-Organizing Multiple Robotic System (A Population Control through Biologically Inspired Immune Network Architecture)." In *Proceedings of the IEEE International Conference on Robotics and Automation.* Vol. 2. 1614–1619.

Morishima, K., et al. 1997. "Screening of Single Escherichia Coli by Electric Field and Laser Tweezer." In *Proceedings of 1997 International Symposium on Micromechatronics and Human Science.* 155–160.

Nicoud, J.-D., and O. Matthey. 1997. "Developing Intelligent Micro-Mechanisms." In *Proceedings of the 1997 International Symposium on Micromechatronics and Human Science* (MHS '97). 119–124.

Sato, T., et al. 1993. "Novel Manipulator for Micro Object Handlings as Interface between Micro and Human Worlds." In *Proceedings of 1993 IEEE/RSJ International Conference on Intelligent Robots and Systems.* Vol. 3 1674–1681.

Stefanini, C., M. C. Carrozza, and P. Dario. 1996. "A Mobile Microrobot Driven by a New Type of Electromagnetic Micromotor." In *Proceedings of Seventh International Symposium on Micro Machine and Human Science* (*MHS '96*). 195–201.

Tortonese, M., et al. 1991. "Atomic Force Microscopy Using a Piezoresistive Cantilever." In *Proceedings of 1991 International Conference on Solid-State Sensors and Actuators* (*Transducers '91*). 448–451.

Yano, S., et al. 1996. "A Study of Maze Searching with Multiple Robots System." In *Proceedings of Seventh International Symposium on Micro Machine and Human Science* (*MHS '96*). 207–211.

CHAPTER 11

NANOROBOTICS

Aristides A. G. Requicha
University of Southern California
Los Angeles, California

1 NANOTECHNOLOGY

Nanorobotics is an emerging field that deals with the controlled manipulation of objects with nanometer-scale dimensions. Typically, an atom has a diameter of a few ångstroms (1 Å $= 0.1$ nm $= 10^{-10}$ m), a molecule's size is a few nm, and clusters or nanoparticles formed by hundreds or thousands of atoms have sizes of tens of nm. Therefore, nanorobotics is concerned with interactions with atomic- and molecular-sized objects and indeed is sometimes called *molecular robotics*. These two expressions, plus *nanomanipulation,* are used synonymously in this chapter.

Molecular robotics falls within the purview of nanotechnology, which is the study of phenomena and structures with characteristic dimensions in the nanometer range. The birth of nanotechnology is usually associated with a talk by Nobel Prize winner Richard Feynman entitled "There Is Plenty of Room at the Bottom," the text of which may be found in Crandall and Lewis (1992). Nanotechnology has the potential for major scientific and practical breakthroughs. Future applications ranging from very fast computers to self-replicating robots are described in Drexler's seminal book (Drexler, 1986). In a less futuristic vein, the following potential applications were suggested by well-known experimental scientists at the Nano4 conference held in Palo Alto in November 1995:

- Cell probes with dimensions $\sim 1/1000$ of the cell's size
- Space applications, e.g., hardware to fly on satellites
- Computer memory
- Near field optics, with characteristic dimensions ~ 20 nm
- X-ray fabrication, systems that use X-ray photons
- Genome applications, reading and manipulating DNA
- Optical antennas

Nanotechnology is being pursued along two converging paths. From the top down, semiconductor fabrication techniques are producing smaller and smaller structures—see, e.g., Colton and Marrian (1995) for recent work. For example, the line width of the original Pentium chip is 350 nm. Current optical lithography techniques have obvious resolution limitations because of the wavelength of visible light, which is on the order of 500 nm. X-ray and electron-beam lithography will push sizes further down, but with a great increase in complexity and cost of fabrication. These top-down techniques do not seem promising for building nanomachines that require precise positioning of atoms or molecules.

Alternatively, one can proceed from the bottom up, by assembling atoms and molecules into functional components and systems. There are two main approaches for building useful devices from nanoscale components. The first, based on self-assembly, is a natural evolution of traditional chemistry and bulk processing. See, e.g., Gómez-López, Preece, and Stoddart (1996). The other is based on controlled positioning of nanoscale objects,

The Laboratory for Molecular Robotics at the University of Southern California is supported primarily by the Z. A. Kaprielian Technology Innovation Fund.

Handbook of Industrial Robotics, Second Edition, Edited by Shimon Y. Nof
ISBN 0-471-17783-0 © 1999 John Wiley & Sons, Inc.

direct application of forces, electric fields, and so on. The self-assembly approach is being pursued at many laboratories. Despite all the current activity, self-assembly has severe limitations because the structures produced tend to be highly symmetric, and the most versatile self-assembled systems are organic and therefore generally lack robustness. The second approach involves nanomanipulation and is being studied by a small number of researchers, who are focusing on techniques based on scanning probe microscopy (SPM), which is described below.

A top-down technique closely related to nanomanipulation involves removing or depositing small amounts of material by using an SPM. This approach falls within what is usually called *nanolithography*. SPM-based nanolithography is akin to machining or to rapid prototyping techniques such as stereolithography. For example, one can remove a row or two of hydrogen atoms on a silicon substrate that has been passivated with hydrogen by moving the tip of an SPM in a straight line over the substrate and applying a suitable voltage. The removed atoms are "lost" to the environment, much like metal chips in a machining operation. Lines with widths from a few to 100 nm have been written by these techniques. See Wiesendanger (1994) and Stalling (1996) for surveys of some of this work. This chapter focuses on nanomanipulation proper, which is akin to assembly in the macroworld.

Nanorobotics research has proceeded along two lines. The first is devoted to the design and computational simulation of robots with nanoscale dimensions. See Drexler (1992) for the design of robots that resemble their macroscopic counterparts. Drexler's nanorobot uses various mechanical components such as nanogears built primarily with carbon atoms in a diamondoid structure. A major issue is how to build these devices, and little experimental progress has been made towards their construction.

The second area of nanorobotics research involves manipulation of nanoscale objects with macroscopic instruments. Experimental work has been focused on this area, especially through the use of SPMs as robots. The remainder of this chapter describes SPM principles, surveys SPM use in nanomanipulation, looks at the SPM as a robot, and concludes with a discussion of some of the challenges that face nanorobotics research.

2 SCANNING PROBE MICROSCOPES

The scanning tunneling microscope (STM) was invented by Binnig and Rohrer at the IBM Zurich laboratory in the early 1980s and won them a Nobel Prize four years later. The principles of the instrument can be summarized with the help of Figure 1. (Numeric values in the figure indicate only the orders of magnitude of the relevant parameters.)

A sharp conducting probe, typically made from tungsten, is placed very close to a sample, which must also be a conductor. The tip is biased with respect to the sample, as shown in the figure. The tip can be moved towards or away from the sample, i.e., in the $-z$ or $+z$ directions in the figure, by means of piezoelectric actuators. At ångstrom-scale distances, a quantum mechanical effect called *tunneling* causes electrons to flow across the tip/sample gap, and a current can be detected. To first approximation, the tunneling current depends exponentially on the distance between tip and sample. This current is kept constant by a feedback circuit that controls the piezoelectric actuators. Because of the current/distance relationship, the distance is also kept constant, and the z accuracy is very high because any small z variation causes an exponential error in the current. Additional piezo motors drive the tip in a xy scanning motion. Since the tip/sample gap is kept constant by the feedback, the scanning tip traverses a surface parallel to the sample surface. The result of a scan is a $z(x, y)$ terrain map with enough resolution to detect atomic-scale features of the sample, as indicated diagrammatically in Figure 1.

Various instruments analogous to the STM have been built. They exploit physical properties other than the tunneling effect on which the STM is based. The most common of these instruments is the *atomic force microscope* (AFM), which is based on interatomic forces. All of these instruments are collectively known as *scanning probe microscopes* (SPMs). The principles of operation of the AFM are shown in Figure 2. The forces between atoms in the tip and sample cause a deflection of the cantilever that carries the tip. The amount of deflection is measured by means of a laser beam bouncing off the top of the cantilever. (There are other schemes for measuring deflection.) The force depends on the tip/sample gap, and therefore servoing on the force ensures that the distance is kept constant during scanning, as in the STM. The AFM does not require conducting tips and samples and therefore has wider applicability than the STM.

If the tip of the AFM is brought very close to the sample, at distances of a few Å, repulsive forces prevent the tip from penetrating the sample. This mode of operation is

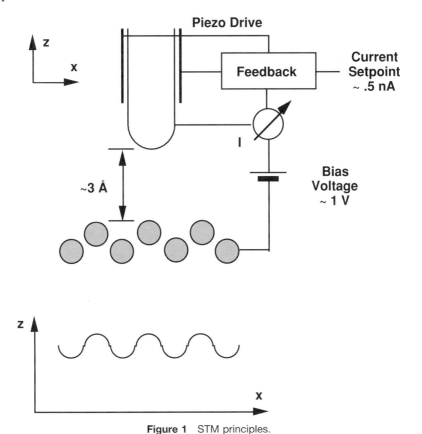

Figure 1   STM principles.

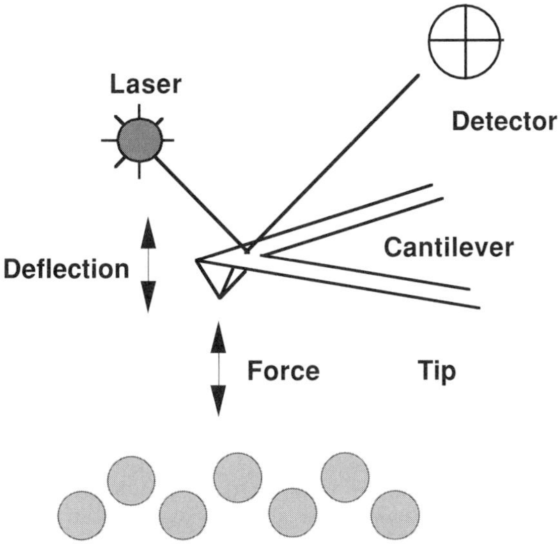

Figure 2   AFM principles.

called *contact* mode. It provides good resolution but cannot be used with delicate samples, e.g., biomaterials, which are damaged by the contact forces. Alternatively, the AFM tip can be placed at distances in the order of several nm or tens of nm, where the interatomic forces between tip and sample are attractive. The tip is vibrated at a frequency near the resonance frequency of the cantilever, in the kHz range. The tip/sample force is equivalent to a change in the spring constant of the cantilever and causes a change in its resonance frequency. This change can be used as the error signal in the feedback circuit that controls the tip. (There are alternative detection schemes.) This mode of operation is called *non-contact* mode. It tends to have poorer resolution than contact mode but can be used with delicate samples.

For more information on SPM technology see, for example, Wiesendanger (1994). Although the SPM is not even 20 years old, it has had a large scientific impact. There is a voluminous literature on SPM applications, scattered through many journals such as *Science* and the *Journal of Vacuum Science and Technology,* and proceedings of meetings such as the biennial conference on Scanning Tunneling Microscopy.

3   NANOMANIPULATION WITH THE SPM

Since the early days of the SPM it has been known that tip/sample interaction could produce changes in both tip and sample. Often these were undesirable, such as a blunt probe due to a crash into the sample. But it soon became clear that one could produce new and desirable features on a sample by using the tip in a suitable manner. One of the first demonstrations was done by Becker and coworkers at Bell Labs, who managed to create nanometer-scale germanium structures on a germanium surface by raising the voltage bias of an STM tip (Becker, Golovchenko, and Swartzentruber, 1987). Much of the subsequent work falls under the category of nanolithography and will not be discussed here. The following subsections survey nanomanipulation research involving the SPM.

3.1   Pushing and Pulling

Pushing and pulling operations are not widely used in macrorobotics, although interesting work on orienting parts by pushing has been done by Matt Mason and his students at CMU, Ken Goldberg at USC, and others (Erdmann and Mason, 1988; Goldberg, 1993). The techniques seem suitable for constructing 2D structures.

Interatomic attractive forces were used by Eigler et al. at IBM Almadén to precisely position xenon atoms on nickel, iron atoms on copper, platinum atoms on platinum, and carbon monoxide molecules on platinum (Stroscio and Eigler, 1991). The atoms are moved much as one displaces a small metallic object on a table by moving a magnet under the table. The STM tip is placed sufficiently close to an atom for the attractive force to be larger than the resistance to lateral movement. The atom is then pulled or dragged along the trajectory of the tip. Eigler's experiments were done in ultrahigh vacuum (UHV) at very low temperature (4K). Low temperature seems essential for stable operation. Thermal noise destroys the generated patterns at higher temperatures. Meyer's group at the Free University of Berlin dragged lead atoms and dimers and carbon monoxide (CO) and ethene (C_2H_4) molecules on a copper surface at temperatures of tens of K by using similar STM techniques (Meyer and Rieder, 1997).

Lateral repulsive forces were used by Güntherodt's group at the University of Basel to push fullerene (C_{60}) islands of ~ 50 nm size on flat terraces of a sodium chloride surface in UHV at room temperature with a modified AFM (Lüthi et al., 1994). The ability to move the islands depends strongly on species/substrate interaction; e.g., C_{60} does not move on gold, and motion on graphite destroys the islands. Lateral forces opposing the motion are analogous to friction in the macroworld, but cannot be modeled simply by coulomb friction or similar approaches that are used in macrorobotics.

Mo at IBM Yorktown rotated pairs of antimonium atoms between two stable orientations 90° apart (Mo, 1993). This was done in UHV at room temperature on a silicon substrate by scanning with an STM tip with a higher voltage than required for imaging. The rotation was reversible, although several scans were sometimes necessary to induce the desired motion.

Samuelson's group at the University of Lund succeeded in pushing gallium arsenide (GaAs) nanoparticles of sizes in the order of 30 nm on a GaAs substrate at room temperature in air (Junno et al., 1995). The sample is first imaged in noncontact AFM mode. Then the tip is brought close to a nanoparticle, the feedback is turned off, and the tip is moved against the nanoparticle. Schaefer et al. (1995) at Purdue University push 10–20-nm gold clusters with an AFM in a nitrogen environment at room temperature. They first image the clusters in noncontact mode, then remove the tip oscillation voltage and sweep the tip across the particle in contact with the surface and with the feedback disabled.

At USC's Laboratory for Molecular Robotics we have pushed colloidal gold nanoparticles with diameters of 5–30 nm at room temperature, in ambient air, and on mica or silicon substrates (Baur et al., 1997). We image in noncontact mode, then disable the feedback and push by moving the tip in a single line scan, without removing the tip oscillation voltage. Alternatively, we can also push by maintaining the feedback on and changing the setpoint so as to approach the sample. Figure 3 shows on the right a "USC" pattern written with gold nanoparticles. The z coordinate is encoded as brightness in this figure. On the left is the original random pattern before manipulation with the AFM. Figure 4 shows the same "USC" pattern displayed in perspective. A 3D VRML (Virtual Reality Modeling Language) file obtained from the terrain map shown in Figure 3 is available for browser viewing at http://www-lmr.usc.edu/~lmr.

Smaller objects have been arranged into prescribed patterns at room temperature by Gimzewski's group at IBM's Zurich laboratory. They push molecules at room temperature in UHV by using an STM. They have succeeded in pushing porphyrin molecules on copper (Jung et al., 1996), and more recently they have arranged bucky balls (i.e., C_{60}) in a linear pattern using an atomic step in the copper substrate as a guide (Cuberes, Schlittler, and Gimzewski, 1996).

C_{60} molecules on silicon also have been pushed with an STM in UHV at room temperature by Maruno et al. in Japan (Maruno, Inanaga, and Isu, 1993) and Beton et al. in the U.K. (Beton, Dunn, and Moriarty, 1995). In Maruno's approach the STM tip is brought closer to the surface than in normal imaging mode, then scanned across a rectangular region with the feedback turned off. This causes many probe crashes. In Beton's approach the tip also is brought close to the surface, but the sweep is done with the feedback on and a high value for the tunneling current. Their success rate is in the order of only 1 in 10 trials. Recently, carbon nanotubes have been manipulated by several groups. See, e.g., Falvo et al. (1997).

Single-stranded DNA on a copper substrate has been manipulated at Osaka University (Tanaka and Kawai, 1997) by using STM techniques similar to those described by Gimzewski's group, cited above.

3.2 Picking and Placing

Much of industrial macrorobotics is concerned with pick-and-place operations, which typically do not require very precise positioning or fine control. There are a few examples

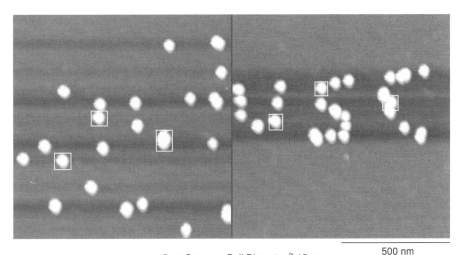

5μm Scanner, Ball Diameter ~ 15nm

500 nm
□ - stationary balls

Figure 3 The initial pattern of 15-nm Au balls (left) and the "USC" pattern obtained by nano-manipulation (right). (Reprinted by permission of the American Vacuum Society from C. Baur et al., "Robotic Nanomanipulation with a Scanning Probe Microscope in a Networked Computing Environment," *Journal of Vacuum Science and Technology B* **15**(4), 1577–1580, 1977.)

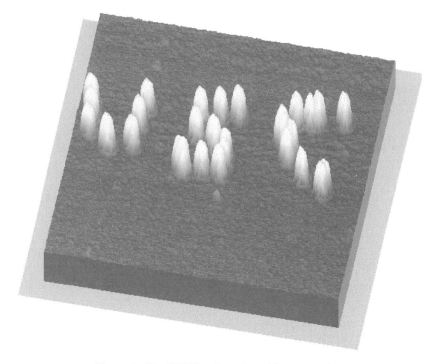

Figure 4 The "USC" pattern viewed in perspective.

of experiments in which atoms or molecules are transferred to SPM tips, which are moved and the atoms transferred back to the surfaces.

Eigler et al. succeeded in transferring xenon atoms from platinum or nickel surfaces to an STM tip by moving the tip sufficiently close for the adsorption barriers of surface and tip to be comparable (Stroscio and Eigler, 1991). An atom may leave the surface and become adsorbed to the tip, or vice versa. Benzene molecules also have been transferred to and from tips.

Eigler's group also has been able to transfer xenon atoms between an STM tip and a nickel surface by applying voltage pulses to the tip (Stroscio and Eigler, 1991). This is attributed to electromigration caused by the electric current flowing in the tunneling junction. All of Eigler's work has been done in UHV at 4K.

By using techniques similar to Eigler's pulse-induced electromigration, Meyer's group at the Free University of Berlin has succeeded in transferring xenon atoms and propene molecules between a copper substrate and an STM tip at temperatures of 15K (Meyer and Rieder, 1997).

Avouris's group at IBM Yorktown and the Aono group in Japan have transferred silicon atoms between a tungsten tip and a silicon surface in UHV at room temperature by applying voltage pulses to the tip (Lyo and Avouris, 1991; Uchida et al., 1993). The mechanism for the transfer is believed to be field-induced evaporation, perhaps aided by chemical phenomena at the tip/surface interface in Avouris's work.

3.3 Compliant Motion

Compliant motion is the most sophisticated form of macrorobotic motion. It involves fine motions, as in a peg-in-hole assembly, in which there is accommodation and often force control, for example to ensure contact between two surfaces as they slide past each other. Compliance is crucial for successful assembly operations in the presence of spatial and other uncertainties, which are unavoidable.

The study of the nanoscale analog of compliant motion seems to be virgin territory. We speculate that the analog of compliance is chemical affinity between atoms and molecules and suspect that such "chemical compliance" may prove essential for nanoassembly operations at room temperature in the presence of thermal noise. It seems likely that successful assembly of nanoscale components will require a combination of precise positioning and chemical compliance. Therefore, work on self-assembling structures is relevant.

4 THE SPM AS A ROBOT

4.1 Motion

To first approximation, an SPM is a three-degree-of-freedom robot. It can move in x, y, and z, but cannot orient its tip, which is the analog of a macrorobotic hand. (How a six-degree-of-freedom SPM could be built, and what could be done with it, are interesting issues.) The vertical displacement is controlled very accurately by a feedback loop involving tunneling current or force (or other quantities for less common SPMs). But nanoscale x, y motion over small regions (e.g., within a 5-micron square) is primarily open-loop because of a lack of suitable sensors and the difficulty of reducing noise in a feedback scheme. Accurate horizontal motion relies on calibration of the piezoelectric actuators, which are known to suffer from a variety of problems such as creep and hysteresis. In addition, thermal drift of the instrument is very significant. At room temperature a drift of one atomic diameter per second is common, which means that manipulation of atomic objects on a surface is not unlike picking parts from a conveyor belt. Thermal drift is negligible if the SPM is operated at very low temperatures, and experiments in atomic-precision manipulation are typically done at 4K. This involves complex technology and is clearly undesirable. For room-temperature nanomanipulation, drift, creep and hysteresis must be taken into account. Ideally, compensation should be automatic. Research is under way at USC on how to move accurately an SPM tip in the presence of all these sources of error.

To complicate matters further, the SPM often must operate in a liquid environment, especially if the objects to be manipulated are biological. Little is known about nanomanipulation in liquids.

4.2 Sensing

The SPM functions both as a manipulator and a sensing device. The lack of a direct and independent means of establishing "ground truth" while navigating the tip causes delicate problems. The SPM commands are issued in instrument or *robot* coordinates, and sensory data also are acquired in robot coordinates, whereas manipulation tasks are expressed in sample or *task* coordinates. These two coordinate systems do not coincide, and indeed are in relative motion due to drift. Accurate motion in task coordinates may be achieved by a form of visual servoing, tracking features of the sample and moving relative to them.

A simple example of a practical implementation of these ideas is illustrated in Figure 5. To construct the pattern shown in Figures 3 and 4, we move the tip to the approximate position of a nanoparticle to be pushed and then search for it through single-line scans.

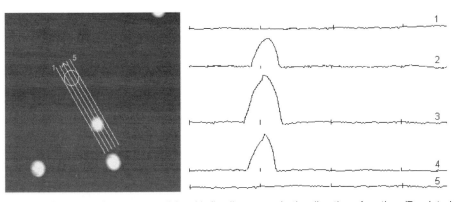

Figure 5 Searching for a nanoparticle with five line scans in the direction of motion. (Reprinted by permission of the American Vacuum Society from C. Baur et al., "Robotic Nanomanipulation with a Scanning Probe Microscope in a Networked Computing Environment," *Journal of Vacuum Science and Technology B* **15**(4), 1577–1580, 1977.)

Feature tracking assumes that features are stationary in task coordinates. This assumption may fail at room temperature if the features are sufficiently small, because of thermal agitation. Hence, atomic manipulation at room temperature may require artificially introduced features that can be tracked to establish a task coordinate system. It must also deal with the spatial uncertainty associated with the thermal motion of the atoms to be moved. Larger objects such as molecules and clusters have lower spatial uncertainty and should be easier to handle.

The SPM output signal depends not only on the topography of the sample but also on the shape of the tip and on other characteristics of the tip/sample interaction. For example, the tunneling current in an STM depends on the electronic wavefunctions of sample and tip. To first approximation one may assume that the tip and sample are in contact. Under this assumption one can use configuration-space techniques to study the motion of the tip (Latombe, 1991). Figure 6 illustrates the procedure in 2D. On the top of the figure we consider a tip with a triangular end and a square protrusion in the sample. On the bottom we consider a tip with a semicircular end and the same protrusion. We choose as reference point for the tip its apex and reflect the tip about the reference point. The configuration-space obstacle that corresponds to the real-space protrusion obstacle is obtained by sweeping the inverted tip over the protrusion so that the apex remains inside the obstacle. Mathematically, we are calculating the Minkowski sum of the inverted tip and the protrusion. The path of the tip in its motion in contact with the obstacle is the detected topographical signal. As shown in the figure, the sensed topography has been broadened by the dimensions of the tip. Note, however, that the detected height is correct. (Minkowski operations and related mathematical morphology tools were introduced in the SPM literature only recently (Villarrubia, 1994).) Tip effects are sometimes called *convolution,* by analogy with the broadening of an impulse passing through a linear system, and one talks of *deconvolving* the image to remove tip effects.

The configuration-space analysis outlined above is purely geometric and provides only a coarse approximation to the SPM output signal. More precise calculations may be

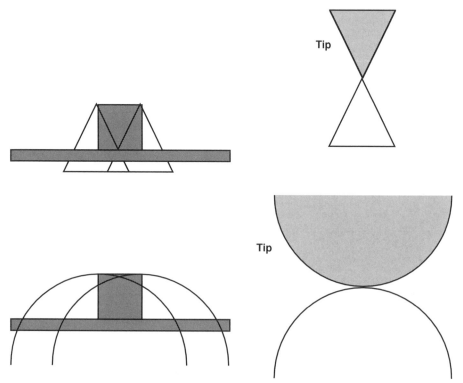

Figure 6   Configuration space obstacles for a triangular tip (top) and for a semicircular tip (bottom).

performed numerically. For example, in contact AFM we can assume specific atomic distributions for the tip and sample and a specific form for the interatomic forces and compute the resulting tip/sample force. Work along these lines has been reported for STMs by Aono's group in Japan (Watanabe, Aono, and Tsukada, 1992).

A major issue in tip-effect compensation is that the shape of the probe is not known and indeed may vary during operation. For example, atoms may be adsorbed on the tip or lost because of the contact with the sample. The most promising approach for dealing with tip effects consists of estimating the tip shape by using it to image known features. If necessary, artificial features may be introduced into the scene for tip estimation purposes. The estimated tip shape can then be used to remove (at least in part) the tip effects from the image by using Minkowski operations (Villarrubia, 1994). Removal procedures that take into account more sophisticated, nongeometric effects do not appear to be known.

Sensor fusion techniques may be used, at least in principle, for increasing the quality of the sensory data, since it is possible to access several signals during an SPM scan. For example, vertical and lateral force (akin to friction) can be recorded simultaneously in typical AFMs. To our knowledge, sensor fusion has not been attempted in SPM technology.

It is clear that faithful sensory data should facilitate manipulation tasks. What is not clear is whether clever manipulation strategies can compensate for the imperfections of SPM data.

4.3 End-Effectors

The SPM tip is the primary end-effector in nanomanipulation. A plain, sharp tip seems to be adequate for most pushing operations. In some cases it may also suffice for picking and depositing objects, especially in conjunction with electrostatic forces generated by applying a suitable bias to the tip. This requires both a conducting tip and a conducting substrate. Pick-and-place tasks, however, usually require the nanoscale analog of a gripper. Very little is known about molecular grippers.

One can think of a nanogripper as a molecule or cluster of molecules that are attached to a tip and capable of picking up and transporting other molecules or particles. (Tips with attached molecules are said to be *functionalized*.) Ideally, these grippers should be switchable, so as to pick and release objects on command. Candidates for grippers are certain molecules such as cyclodextrins, which have large cavities that can carry other molecules. Coating a tip with strands of DNA may also permit picking up objects that are coated with complementary strands. In both of these examples, switching the gripper on and off is not a solved problem.

Techniques for changing SPM tips automatically do not exist. Changes must be done manually, and it is very difficult or impossible to maintain sample registration, i.e., to return to the same position on the sample after tip replacement. This implies that one often must image using a tip with an attached gripper. Again, little is known about imaging with such functionalized tips.

4.4 Fixtures

In the macroworld, fixtures are often necessary to hold mechanical parts during assembly and other manufacturing operations. In micromechanics, sacrificial layers are used as scaffolding to fabricate certain microelectromechanical systems (MEMS). At the nanoscale, the analogs of fixtures are substrates that ensure that certain objects remain fixed during manipulation while others are allowed to move. Substrate selection seems to be highly dependent on the chemistry of the objects being manipulated.

4.5 Manipulation Processes

In macrorobotics the physical processes involved are mechanical and relatively well understood. At the nanoscale the processes are chemical and physical and are still an area of active research. In addition, nanomanipulation takes place in several different environments, such as liquids, air, or UHV. The environment has a strong influence on the physics and chemistry of the processes. Nanomanipulation is not restricted to mechanical interactions. For example, light, electrostatic fields, and the pH of a liquid all are candidates for controlled interaction with nanoparticles.

4.6 User Interfaces, Programming, and Planning

Graphic user interfaces for SPMs are provided by all the instrument vendors. Typically, they run in the Microsoft Windows environment on PCs and have the familiar look and feel of modern mouse-driven interfaces. Much more sophisticated, immersive interfaces for STMs are being developed by the nanomanipulator group at the University of North Carolina (UNC) at Chapel Hill (Taylor et al., 1993). The UNC system uses a virtual reality display with tactile feedback. A user can literally feel a surface's atoms, fly along the surface, and modify it by applying voltage pulses in real time.

High-level programming and planning systems are highly desirable, indeed essential, for assembling complex structures. One must begin with relatively low-level programming primitives and build upon them constructs at a higher level of abstraction. High-level commands must be compiled into low-level primitives. This compilation may involve sophisticated computations; for example, to ensure collision-free paths in an environment with large spatial uncertainty. What are the relevant high-level manipulation tasks? For example, what is the nanoscale equivalent of a peg-in-hole insertion? In short, we may need to adapt much of what is known about macrorobotics to the nanoworld. It is likely that new concepts will also be needed, because the physics and chemistry of the phenomena and objects are quite different in these two worlds.

4.7 Assembly Applications

What hardware primitives are suitable as building blocks? The nanotechnology literature suggests hardware primitives based on DNA structures such as those built in Seeman's lab (Seeman et al., 1993), proteins, and diamondoid structures (Drexler, 1992). Biomaterials such as DNA and proteins may be too flimsy, whereas diamondoid structures are expected to be very strong. No experiments have yet been reported in which any of these components are successfully assembled into a composite structure.

Which tasks should one attempt first? What should one try to build? The main options are in the realms of electronics, photonics, mechanics, or biomaterials.

5   CHALLENGES

Nanorobotics manipulation with SPMs is a promising field that can lead to revolutionary new science and technology. But it is clearly in its infancy.

Typical nanomanipulation experiments reported in the literature involve a team of very skilled Ph.D.-level researchers working for many hours in a tightly controlled environment (typically in ultrahigh vacuum and at low temperature, often 4K) to build a pattern with tens of nanoparticles. It still takes the best groups in the world some 10 hours to assemble a structure with about 50 components. This is simply too long—changes will occur in many systems, e.g., contamination or oxidation of the components, on a time scale that will constrain the maximum time available for nanomanipulation. Requiring all operations to take place at 4K and in UHV also is not practical for widespread use of nanomanipulation. In short, nanomanipulation today is more of an experimental tour de force than a technique that can be routinely used. It is clear that complex tasks cannot be accomplished unless the SPM is commanded at a higher level of abstraction. Compensation for instrument inaccuracies should be automatic, and the user should be relieved from many low-level details.

Building a high-level programming system for nanomanipulation is a daunting task. The various component technologies needed for nanomanipulation must be developed and integrated. These technologies include:

- Substrates that serve as nanofixtures or nanoworkbenches on which to place the objects to be manipulated
- Tips, probes, and molecules that serve as grippers or end-effectors
- Chemical and physical nanoassembly processes
- Primitive nanoassembly operations that play a role analogous to macroassembly tasks such as peg-in-hole insertion
- Methods for exploiting self-assembly to combat spatial uncertainty in a role analogous to mechanical compliance in the macroworld
- Suitable hardware primitives for building nanostructures
- Algorithms and software for sensory interpretation, motion planning, and driving the SPM

This is a tall order; it requires an interdisciplinary approach that combines synergistically the knowledge and talents of roboticists and computer scientists with those of physicists, chemists, materials scientists, and biologists.

SPM-based assembly methods face a major scale-up challenge. Building complex structures one atom (or even one nanoparticle) at a time is very time-consuming. Almost surely, SPMs will have applications in the exploration of new structures, which may later be mass produced by other means. This is the nanoworld analog of rapid protyping technologies such as stereolithography, which are becoming popular at the macroscale.

There are at least two approaches for fighting the serial nature of SPM manipulation. The first involves the use of large arrays of SPMs on a chip. These chips are being developed by Cornell's nanofabrication group under Noel Macdonald, by Calvin Quate's group at Stanford, and probably elsewhere. Programming such arrays for coordinated assembly tasks poses interesting problems. The second approach is subtler. It consists of using the SPM to construct structures that are capable of self-replication or serving as blueprints for fabrication. The best-known such structures involve DNA, but other systems also exist. Self-replication is inherently an exponential process.

In summary, nanomanipulation with SPMs may have a revolutionary impact on science, technology, and the way we live. To fully exploit its potential, we will have to develop powerful systems for programming nanorobotic tasks. Much of what is known in macrorobotics is likely to be relevant, but may have to be adapted to the nanoworld, where phenomena and structures are quite different from their macroscopic counterparts. Research at USC and elsewhere is progressing, with promising results.

Nanomanipulation, perhaps coupled with self-assembly, is expected to succeed eventually in building true nanorobots, i.e., devices with overall dimensions in the nanometer range capable of sensing, "thinking," and acting. Complex tasks are likely to require a group of nanorobots working cooperatively. This raises interesting issues of control, communications, and programming of robot "societies."

REFERENCES

Baur C., et al. 1997. "Robotic Nanomanipulation with a Scanning Probe Microscope in a Networked Computing Environment." *Journal Vacuum Science and Technology B* **15**(4), 1577–1580; in *Proceedings of the 4th International Conference on Nanometer-Scale Science and Technology*, Beijing, September 8–12, 1996.

Becker R. S., J. A. Golovchenko, and B. S. Swartzentruber. 1987. "Atomic-Scale Surface Modifications Using a Tunneling Microscope." *Nature* **325,** 419–421.

Beton P. H., A. W. Dunn, and P. Moriarty. 1995. "Manipulation of C_{60} Molecules on a Si Surface." *Applied Physics Letters* **67**(8), 1075–1077.

Colton, R. J., and C. R. K. Marrian, eds. 1995. *Proceedings of the 3rd International Conference on Nanometer-Scale Science and Technology*, Denver, October 24–28; *Journal of Vacuum Science and Technology B* **13**(3).

Crandall, B. C., and J. Lewis. 1992. *Nanotechnology*. Cambridge: MIT Press. 347–363.

Cuberes, M. T., R. R. Schlittler, and J. K. Gimzewski. 1996. "Room-Temperature Repositioning of Individual C_{60} Molecules at Cu Steps: Operation of a Molecular Counting Device." *Applied Physics Letters* **69**, 3016–3018.

Drexler, K. E. 1986. *The Engines of Creation*. New York: Anchor Books.

———. 1992. *Nanosystems*. New York: John Wiley & Sons.

Erdmann, M. A., and M. T. Mason. 1988. "An Exploration of Sensorless Manipulation." *IEEE Journal of Robotics and Automation* **4**(4), 369–379.

Falvo, M. R., et al. 1997. "Bending and Buckling of Carbon Nanotubes under Large Strain." *Nature* **389**, 582–584.

Goldberg, K. 1993. "Orienting Polygonal Parts Without Sensors." *Algorithmica* **10**(2), 201–225.

Gómez-López, M., J. A. Preece, and J. F. Stoddart. 1996. "The Art and Science of Self-Assembling Molecular Machines." *Nanotechnology* **7**(3), 183–192.

Jung, T. A., et al. 1996. "Controlled Room-Temperature Positioning of Individual Molecules: Molecular Flexure and Motion." *Science* **271**, 181–184.

Junno, T., et al. 1995. "Controlled Manipulation of Nanoparticles with an Atomic Force Microscope." *Applied Physics Letters* **66**, 3627–3629.

Latombe, J.-C. 1991. *Robot Motion Planning*. Boston: Kluwer.

Lüthi, R., et al. 1994. "Sled-Type Motion on the Nanometer Scale: Determination of Dissipation and Cohesive Energies of C_{60}." *Science* **266**, 1979–1981.

Lyo, I.-W., and P. Avouris. 1991. "Field-Induced Nanometer- to Atomic-Scale Manipulation of Silicon Surfaces with the STM." *Science* **253**, 173–176.

Maruno, S., K. Inanaga, and T. Isu. 1993. "Threshold Height for Movement of Molecules on Si(111)-7×7 with a scanning tunneling microscope." *Applied Physics Letters* **63**(10), 1339–1341.

Meyer, G., and K.-H. Rieder. 1997. "Controlled Manipulation of Single Atoms and Small Molecules with the Scanning Tunneling Microscope." *Surface Science* **377/379**, 1087–1093.

Mo, Y. W. 1993. "Reversible Rotation of Antimony Dimers on the Silicon (001) Surface with a Scanning Tunneling Microscope." *Science* **261**, 886–888.

Schaefer, D. M., et al. 1995. "Fabrication of Two-Dimensional Arrays of Nanometer-Size Clusters with the Atomic Force Microscope." *Applied Physics Letters* **66**(8), 1012–1014.

Seeman, N. C., et al. 1993. "Synthetic DNA Knots and Catenanes." *New Journal of Chemistry* **17**(10/11), 739–755.

Stalling, C. T. 1996. "Direct Patterning of Si(001) Surfaces by Atomic Manipulation." *Journal of Vacuum Science and Technology B* **14**(2), 1322–1326.

Stroscio, J. A., and D. M. Eigler. 1991. "Atomic and Molecular Manipulation with the Scanning Tunneling Microscope." *Science* **254**, 1319–1326.

Tanaka, H., and T. Kawai. 1997. "Scanning Tunneling Microscopy Imaging and Manipulation of DNA Oligomer Adsorbed on Cu(111) Surfaces by a Pulse Injection Method." *Journal of Vacuum Science and Technology B* **15**(3), 602–604.

Taylor, R. M., et al. 1993. "The Nanomanipulator: A Virtual Reality Interface for a Scanning Tunneling Microscope." *Proceedings of ACM SIGGRAPH '93*, Anaheim, August 1–6. 127–134.

Uchida, H., et al. 1993. "Single Atom Manipulation on the Si(111)7×7 Surface by the Scanning Tunneling Microscope (STM)." *Surface Science* **287/288**, 1056–1061.

Villarrubia, J. S. 1994. "Morphological Estimation of Tip Geometry for Scanned Probe Microscopy." *Surface Science* **321**(3), 287–300.

Watanabe, S., M. Aono, and M. Tsukada. 1992. "Theoretical Calculations of the Scanning Tunneling Microscope Images of the Si(111)$\sqrt{3} \times \sqrt{3}$–Ag Surface: Effects of Tip Shape." *Applied Surface Science* **60/61**, 437–442.

Wiesendanger, R. 1994. *Scanning Probe Microscopy and Spectroscopy*. Cambridge: Cambridge University Press.

PART 3
CONTROL AND INTELLIGENCE

CHAPTER 12

DESIGN OF ROBOT CONTROLLERS

Hadi A. Akeel
FANUC Robotics North America, Inc.
Rochester Hills, Michigan

Atsushi Watanabe
FANUC Ltd.
Yamanashi, Japan

1 ROBOT CONTROLLER DESIGN

A robot controller is usually a microprocessor-based electronic device that can be programmed with instructions to describe and control the operation of the robot. A robot controller can also house signal amplifiers that help drive the robot actuators or motors in accordance with command signals responsive to the programmed instructions. Figure 1 shows an industrial robot controller.

1.1 Role of the Robot Controller

The robot controller drives the motors attached to each robot axis and coordinates the motion of each axis to control the tool center point (TCP) at the end of the robot arm. It can control the input and output (I/O) signals (digital and analog) to control external devices, such as a gripper or a welding gun, based on a sequence synchronized with robot motion. The robot controller communicates with other controllers, personal computers (PCs), or a host computer and uses sensors to obtain information on the robot environment so it can modify the robot tasks accordingly.

1.1.1 Controller Fundamentals

This section describes the fundamental action of the controller for a trajectory-controlled robot, also known as a *playback* or *continuous path* robot. This is the most popular kind of robot used for industrial applications. The controller controls the path of the TCP along a smooth trajectory in space based on the coordinates of a preprogrammed set of discrete locations along the desired path. It controls the robot axes and peripheral equipment based on a robot program written in a robot language by the operator. The robot program consists of two major kinds of instructions: motion and nonmotion. Motion instructions move the robot axes along a designated path, and nonmotion instructions control the I/O signals or the execution of the sequence of instructions in the robot program.

To teach the positions where the robot axes should move, the operator jogs the robot, using the teach pendant keys to move the robot axes along a desired path. The operator can also use the teach pendant to record the desired discrete path positions in the robot program. Figure 2 shows a teach pendant with jog keys and a visual display screen.

The operator then modifies the motion instructions in the program to specify the speed of the robot motion or the connection type between the path segments. I/O instructions can also be inserted in the program to control the peripheral equipment or respond to sensor signals at the desired position.

The robot controller executes the robot program by reading the instructions from a section of the information to the motion control section. The motion control section plans the path to each designated point by using the specified motion type and speed, then interpolates the path, using an acceleration and deceleration algorithm. The servo-control

Handbook of Industrial Robotics, Second Edition, Edited by Shimon Y. Nof
ISBN 0-471-17783-0 © 1999 John Wiley & Sons, Inc.

Figure 1   Industrial robot controller (courtesy Fanuc Ltd., Yamanashi, Japan).

section controls the servo-motors by considering the dynamics of the robot axes in real time (Craig, 1986).

1.2   Robot Controller Configuration

Figure 3 shows the configuration of a robot controller from a functional perspective. The controller consists of a user interface, a robot program and control variable storage section, and a program sequence execution section, also known as the *interpreter.* The controller provides motion, I/O, and sensor control sections as well as network communication, and communicates with a programmable controller and other peripheral equipment.

1.2.1   User Interface

The user interface section of the controller is the interface between the robot and the operator through devices such as a teach pendant or PC. It provides the operator with the means to create programs, jog the robot, teach positions, perform diagnostics, and so forth.

1.2.2   Robot Program and Control Variable Storage

The robot program and the control variables used to control the robot axes are stored in the robot program and control variable storage section of the controller. Integrated circuit (IC) memory, such as DRAM or CMOS RAM, or a hard disk is normally used. This section is accessed not only from the user interface section, but also from the program sequence execution section. Other external controllers, through the network communication section, also use this section.

Figure 2   Teach pendant (courtesy Fanuc Ltd., Yamanashi, Japan).

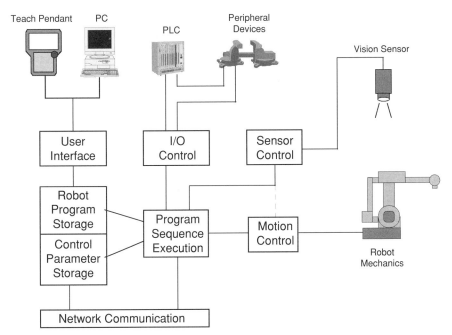

Figure 3   Functional configuration of a robot controller.

1.2.3 Program Sequence Execution (Interpreter)

The program sequence execution section executes the robot program created by inter-preting the user interface section and stores it in the robot program storage section. It interprets the instructions in the program and sends the information to the motion control section to move the robot axes. It also controls the program execution sequence by com-municating with the I/O control and sensor control sections.

1.2.4 Motion Control

The motion control section controls the motion of the robot axes according to the infor-mation passed from the program sequence execution section. It includes three smaller components: a path planning module, a path interpolation module, and a servo-control module (Lewis, 1993). It also communicates with the sensor control section and modifies the path based on the information received from the sensors. Section 1.6 provides more information on motion control.

1.2.5 I/O Control

The I/O control section of the controller controls the end-effector, such as a hand or a spot gun, and the peripheral devices and communicates with external controllers such as a programmable controller. It is usually necessary when the robot performs a productive operation to synchronize some peripheral devices with the robot motion and to receive signals from a host programmable controller to synchronize the robot operations with the production line. I/O signals include analog I/O, digital I/O (which shows one of two states: *on* or *off*), and group I/O (which is a combination of multiple digital I/O signals).

1.2.6 Sensor Control

Sensors are connected to the robot controller and can be utilized to modify the destination point or the path of the robot to accommodate the requirements of designated workpieces or changes in the robot environment as measured by the sensors. The sensor control section of the controller communicates the sensor signals to the robot controller. Sensor signals can represent positional or path offset data, speed adjustment data, or process modification data. The sensor control section generally uses serial communication or a bus interface for communication because the amount of data communicated is usually more than a simple digital signal (Qu and Dawson, 1995).

1.2.7 Network Communication

The network communication section of the controller controls the communication with the peripheral equipment, programmable controller, other controllers, host computers, and so forth. The communicated data include simple I/O signals, diagnostic data represen-tative of the controller's status, alarm information, robot programs, and control variable files. Section 1.9 describes network communication in more detail.

1.2.8 Programmable Controller

The programmable controller has two major roles. First, it connects the robot controller and the peripheral devices under the robot control. In this case the robot controller is the *master* and the programmable controller, when it receives a command from the robot controller, sends multiple commands to the peripheral devices and reports the status of the multiple peripheral devices to the robot. Second, the programmable controller might connect one or several robot controllers and a host line controller. In this case the robot controller is the *slave* and the programmable controller sends commands for program selection or execution to the robot controller.

1.2.9 Peripheral Equipment

Peripheral equipment includes elements such as the end-effector, which is usually attached to the robot wrist, jigs such as clamps and sliders, and process-specific controllers such as for welding, dispensing, and painting. They are often controlled directly by the robot controller or through a programmable controller using digital signals. For example, an arc welding robot, as shown in Figure 4, might control a workpiece positioner with multiple servo-driven axes, seam-tracking sensors, and part presence sensors, and it might control as well the welding process controller.

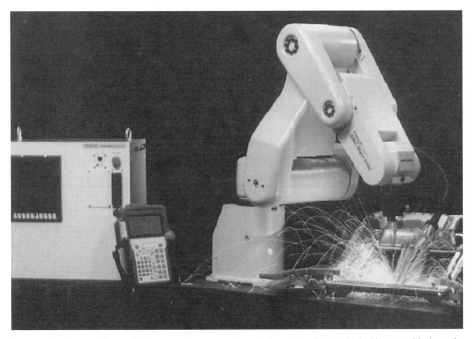

Figure 4   Arc welding robot with process equipment (courtesy Fanuc Ltd., Yamanashi, Japan).

1.3   Basic Architecture of the Robot Controller

The robot controller normally contains at least one digital processor with data storage memory, a set of servo-amplifiers, user interface hardware, and communication electronics (Proctor et al., 1993; Anderson, Cole, and Holland, 1993). The processor performs program interpretation, program sequence control, and motion control to generate motion commands, which are then communicated to a set of servo-amplifiers to regulate the electric current that drives the servo-motors. Figure 5 shows the general hardware architecture of the robot controller.

1.3.1   Processor

Microprocessors are used in robot controllers. The recent progress in digital microprocessor technology has drastically improved the performance of robot controllers. Digital servo-control technology has also progressed. The servo-system is controlled by a digital microprocessor to include position feedback control and current feedback control. In this case the main CPU performs program interpretation, sequence control, and motion planning. A digital signal processor (DSP) is often used for servo-control.

1.3.2   Memory

Memory is the component that stores the system software to be executed by the central processor. Memory also stores user programs and program execution parameters and is used as a working medium for CPU data processing. Nonvolatile memory such as Flash ROM or EPROM is used to store system software. CMOS RAM is used to store user programs and parameters because of its fast access speed. However, CMOS RAM memory requires battery backup and regular exchange of batteries. Volatile memory such as DRAM with fast access speed is used as the working medium for CPU. Floppy disks, hard disks, or PCMCIA memory cards are also used as external memory devices.

1.3.3   Servo-amplifier

The servo-amplifier increases the servo-control system drive voltage from a low-level to a high-level voltage suitable for driving the motor. The servo-amplifier can utilize a DC,

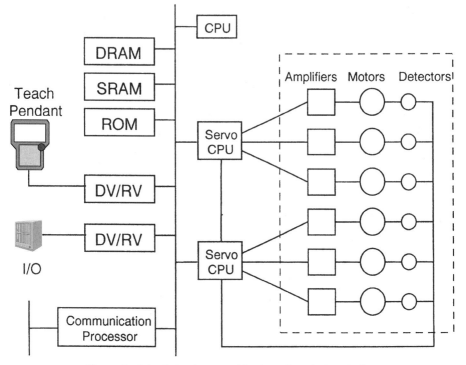

Figure 5   A basic hardware architecture of a robot controller.

AC, or pulse width modulated (PWM) signal. The pulse width modulated motor drive is predominant in modern robotic systems and is the drive discussed here.

The servo-control system provides a low-level PWM signal that is amplified by the servo-amplifier. The output is connected to the motor. In order to rotate the motor, a three-phase signal is used. The three-phase signal is varied in frequency to adjust motor speed and varied in amplitude to adjust the available torque. This is described in Section 1.6.9 under the heading *Current Control Loop*.

A PWM signal is used to create this variable frequency–variable amplitude signal. To create this signal, shown in Figure 6, three voltage commands, one for each phase, are generated by the servo-control system. These voltage commands are fed into a set of voltage comparators in addition to a triangle waveform. The output of the comparator changes state when one of these input signals exceeds the other. Thus, whenever the voltage command is higher than the triangle waveform, the output will be high. When the triangle waveform is higher, the output will be low. This makes a series of variable width on or off pulses in step with the voltage command. Six such pulse streams are created.

These low-level signals are sent to the servo-amplifier, where they are connected to the high-power transistors that drive the motor. These transistors are shown as "Circuit 2 Inverter" in Figure 6. The DC power for the output transistors is derived from a three-phase, full-wave bridge rectifier shown as "Circuit 1 Converter" in Figure 6. The purpose of this converter section is to change incoming fixed frequency (either 50 or 60 Hz) alternating current into direct current. The direct current is inverted back to a variable frequency–variable amplitude three-phase PWM drive for the motor (Andeen, 1988).

1.3.4   Motor

Two kinds of motors are commonly used: DC and AC servo-drive motors. DC motors require mechanical brushes, which are subject to wear, and therefore have a limited life and reliability. AC servo-drive motors have no such problem and are generally preferred.

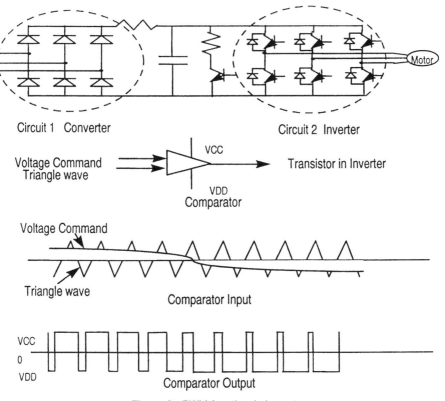

Figure 6 PWM functional elements.

Figure 7 shows the construction of a typical motor. The rotor, equipped with permanent magnetic poles, is in the center of the motor. The stator, with wire-wound poles, surrounds the rotor.

Figure 8 is a representation of a permanent magnet AC motor constructed with a three-phase winding in which phases are spaced 120° apart. When electric current is passed through the winding coils in the stationary part of the motor, the stator, a rotating magnetic field (vector F) is generated. The permanent magnets, equally spaced about the periphery of the rotating part of the motor (or the rotor), interact with the magnetic field of the stator to generate a force that causes the rotor to rotate.

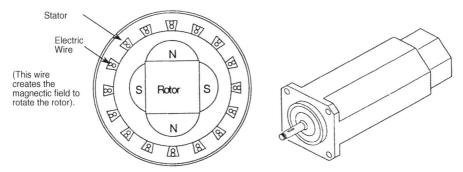

Figure 7 AC motor construction.

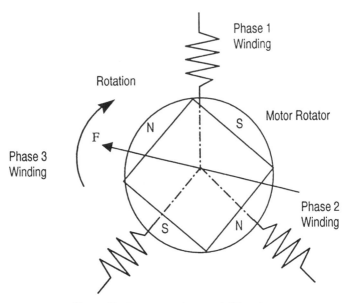

Figure 8 A permanent magnet AC motor.

1.3.5   Position Sensor

The motor contains a position sensor that detects the position of the rotor relative to the stator. The rotor speed is derived from the position information by differentiation with respect to time. The servo-system uses sensor data to control the position as well as the speed of the motor. In general, one of two sensor types is used: the *pulse coder* (also known as *digital encoder*) type, or the *resolver* type.

A pulse coder generates intermittent digital signals indicative of incremental rotor angular rotations. The signals can be generated optically or magnetically. An optical encoder includes a slitted disk mounted on the rotor, an LED light source fixed relative to the stator, and an optical receiver such as a photo diode. As the disk rotates, the receiver generates digital signals in response to the light pulsing through the slits, as shown in Figure 9. Magnetic encoders work on a similar principle but utilize magnetic sensors and ferretic rotor teeth.

A resolver is constructed with a rotating magnet and a stationary coil; the phase difference between the current generated through the coil (the output) and a reference signal (the input) is indicative of the rotor position in Figure 9.

1.4   User Interface

The most standard and distinctive user interface for the robot controller is the teach pendant (Rahimi and Karwowski, 1992). It consists of a display unit typically containing a liquid crystal display (LCD), keys for data input, keys for jogging, an EMERGENCY STOP button, an on/off switch, and a DEADMAN switch for safety. The teach pendant is enabled when the on/off switch is set to *on* and the DEADMAN switch is pressed. While the teach pendant is enabled, all operations by either the operator panel or external signals are disabled for the operator's safety. In most robot controllers all robot operations, including jogging, programming, and diagnostic functions, can be performed using the teach pendant.

A PC can be used as a front-end interface for robot controllers. The PC is a more intelligent user interface; it can utilize graphics and automation to assist the operator's job. The following operations are typically done at the user interface. Figure 10 shows a robot with a PC user interface also used for off-line robot programming.

1.4.1   Create Program

The program can be created by means of the user interface on-line when the robot is jogged and positions are recorded using the teach pendant. The program can also be

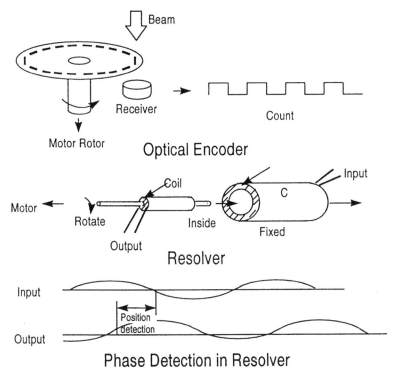

Figure 9   Position sensing with an optical encoder.

created off-line, away from the robot, using a PC. Section 1.8 provides more information on off-line programming.

1.4.2   Jogging and Position Recording

During jogging, the operator can move the robot in the required direction by pushing the JOG keys on the teach pendant (shown in Figure 2). There are x, y, and z keys to perform linear motion of the TCP, and another set of x, y, and z keys to perform a rotation about the robot axes in Cartesian coordinates. The frame that defines the direction of x, y, and z can be selected from the world frame, work frame, or tool frame. The world frame represents a fixed reference. The work frame is referenced to a workpiece. The tool frame is referenced to a tool attached to the end of the robot at the TCP. Each jog key can be also assigned to each robot axis and the operator can move each robot axis directly. Moreover, according to the robot axis configuration, the robot can be jogged along a cylindrical frame or a polar frame.

The jogging speed is selected by using an override or jog speed select key on the teach pendant. Usually, when the robot is jogged to retreat from a destination position, a high speed is used for robot motion. When the robot approaches the destination position and needs a more precise positioning, a low speed is used. Some robot systems have a step feed function that can move the robot a small amount each time the jog key is pressed.

The position data are recorded after the robot reaches a required position. When the position record key is pressed, the current TCP location and orientation of a tool indicating vector is recorded in the robot program as the position data. Position data can be recorded by Cartesian frame representation of the TCP location and orientation or by joint offset representation of each robot axis position. When the current position is recorded, some robot systems can record motion instructions at the same time, including motion type (joint, linear, or circular), motion speed, and termination type (fine or continuous). A fine termination assures that the TCP traces the recorded position, and a continuous termination assures a smooth transition between adjacent path segments.

Figure 10 Robot with PC user interface, Teach pendant, and off-line Programming System (courtesy FANUC Robotics N. A. Inc., Rochester Hills, Michigan).

1.4.3 Edit Program

The instructions in a program can be modified to include additional instructions, such as output signal instructions, and await instructions for input signals or conditional branch instructions based on the value of an input signal. When program editing is complete, the robot can then operate while checking the current status and perform the corresponding work taught by the program. During operation, the controller continuously checks the status of all I/O and executes any necessary interrupt instructions.

1.4.4 Test Run

When program editing is finished, a test run should be done to check that the robot program is correct. In a test run the robot might move unexpectedly because of a programming mistake or incorrect I/O signals. The use of a low override speed allows the robot to move at a speed slower than the programmed speed. When the step mode is used, the program is executed step by step to confirm the flow of program execution. After robot motion is confirmed with a low override, robot motion and program execution flow must be checked again by increasing the speed incrementally by small amounts.

Usually, when the override is changed, the robot path might also change. The robot path must be confirmed by gradually increasing the override. If the robot system has a function that maintains constant path execution, the robot can move on the same path with any override and the effort and time to perform a test run are greatly reduced.

When all confirmation of robot motion and interface signals between the robot and external devices is complete, the program can be executed in production mode. In production mode the robot usually executes the program by an external start signal from a programmable controller that can also control other devices in the production line.

1.4.5   Error Recovery

When an alarm occurs in production, the robot will stop. The operator must go to the location where the alarm has occurred, investigate the problem, determine the cause of the alarm, and move the robot to a safe position. If the robot has stopped in the middle of a program, the operator resumes the program by using the teach pendant again with a low velocity override and returns the robot to the wait position. After the operator removes the cause of the alarm and resets the system, production can be restarted. It is very important for system design to minimize the down time of a system when an alarm occurs.

1.5   Robot Program

The basic concept in robot program control is the teaching playback method (Pullmann, 1996; Samson, Espian, and Le Borgne, 1991). Motion instructions are taught in order of the desired motion. Also, application instructions or I/O instructions are taught in the order of program execution. Two kinds of programming languages describe robot behavior: the interpreter language, which interprets a program line by line and usually can be programmed easily by using menus, and the compiler language, which converts a program written in a high-level programming language such as PASCAL or C into intermediate codes for easy execution. The robot program is typically written in a high-level robot language with instructions for motion, I/O, program sequence, event control, and application specifics, as diagrammed in Figure 11.

1.5.1   Motion Instruction

A motion instruction can include the following elements:

- Position data
- Interpolation type
- Speed
- Termination type

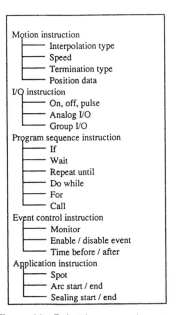

Figure 11   Robot language elements.

Position data specify a position to which the robot will move. Representation of position is in either Cartesian or joint offsets. In Cartesian representation the location and orientation of the TCP is stored. Configuration data, which describe the robot position when the robot can reach the same position by more than one set of joint offsets, are also included. The reference to the coordinate system on which the TCP position is based can also be included. In joint offset representation the offset values from the zero position are stored.

The *interpolation type* specifies how to move to the specified position. In general, either joint, linear, or circular interpolation is chosen. In joint interpolation all joints are equally interpolated from the current position to the next, or target, position. In linear interpolation the TCP moves along a straight line from the current position to the target position. In circular interpolation the TCP moves along a circular arc. In circular interpolation not only the target position but the intermediate position or radius is usually specified.

Speed specifies the speed of the robot to the target position. In joint interpolation a ratio to the maximum speed of the joint is specified. In linear or circular interpolation, the speed of the TCP is specified. The time to the target position can sometimes be specified.

The *termination type* specifies how to move through the target position. It might be specified to stop at the target position or to move near to the target position without stopping. When specifying to move through the target position, the operator can specify how near the robot should get to the target position. In addition to these elements, sequence instructions related to robot motion timing for spot welding or arc welding can also be attached to a motion instruction.

1.5.2   I/O Instruction

I/O instructions are instructions to control I/O signals. For input signals, some instructions read an input signal line and store the value into memory, and some control the program sequence or pause the execution with sequence control instructions. For output signals, some instructions turn on/off or send pulses to the specified signal lines. Some instructions read memory and send them to output lines.

1.5.3   Program Sequence Instruction

Program sequence instructions can include branch instructions, wait instructions, call instructions, or return instructions. Branch instructions can be nonconditional branches that jump to a specified line without condition or conditional branches that jump to a specified line according to whether the specified condition is true or false.

Wait instructions are instructions used to wait for a specified time or wait until a specified condition is satisfied. In the latter case a timeout in case the condition is not satisfied can be specified.

Call instructions call a specified subprogram. Arguments or a return value might be available. The call instruction might be able to call a subprogram indirectly. A called subprogram returns control to the caller program at the return instruction.

1.5.4   Event Control Instructions

A sequence of instructions can be executed while pausing or concurrently with the current execution of a program when a specified condition is satisfied. This instruction is useful when an event occurs that cannot be predefined, such as monitoring input from an operator or a dropped workpiece during program execution.

1.5.5   Application Instructions

Application instructions are used to execute an application-specific motion or sequence. Examples include spot welding instructions for spot welding applications, arc start and end instructions for arc welding applications, sealing start and end instructions for sealing applications, and hand open and close instructions for handling applications. For all cases, application instructions are designed to make it easy to specify the application-specific and complicated sequence of motion or I/O.

1.6   Motion Control

1.6.1   Motion Control Overview

Motion control involves several steps: path planning, interpolating, path filtering, and servo control, as shown in Figure 12 (Samson, 1991; Sciavicco, 1996).

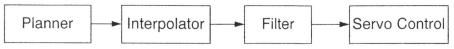

Figure 12   The main steps in motion control.

Path Planning

The path planner prepares the data for path interpolation. For example, in linear motion the direction vector from the start position to the destination position is calculated and the move distance at each interpolation interval (ITP) is calculated according to the specified speed.

Interpolation

At each interpolation interval, based on the information provided by the path planner, an intermediate position is computed (between the start position and destination position) by the path interpolator. The eventual output of the path interpolator at each interpolation interval is a set of joint angles that forms the input to the servo-control loop to command the robot to move along the path.

Filter

Filtering is a method of controlling the acceleration and deceleration of each path segment. It is applied to the output of the path interpolator in order to provide smooth acceleration/deceleration and keep the motor torque within the capabilities of the servo-motor.

Servo Control

Interpolated data are finally converted to the motor command data. The servo-control loop generates command signals in accordance with proven servo-control methods to maintain close correspondence between the command position and the active motor position as detected by the position sensor feedback.

1.6.2   Motion Command Types

The most common types of motion commands are (Shpitalni, Koren, and Lo, 1994):

- Joint motion
- Linear motion
- Circular motion

Joint Motion

With a joint motion command, each robot axis is moved from its start angle to its destination angle, as shown in Figure 13. This type of motion is simple and is commonly used in situations where there is no requirement to control the TCP (tool center point) path. The planner computes the overall travel time T for synchronizing all joint axes as follows. For each axis i, calculate the traveling angle, ΔJi = destination angle i – start angle i, and based on taught speed V compute its corresponding traveling time, $Ti = \Delta Ji/V$. To synchronize all joint axes to start and stop at the same time, the overall segment time is chosen as the maximum of all Ti's, i.e., $T = \max(Ti, i = 1 \ldots n)$. For each interpolation interval ITP, the joint increment for axis i is given by $\Delta Ji * ITP/T$.

Linear Motion

With a linear motion command the robot TCP is moved along a straight line at the programmed speed, as shown in Figure 14. This is accomplished by computing intermediate positions along a straight line. The planner computes the difference location vector (Δx, Δy, Δz) and difference orientation vector (Δw, Δp, Δr) in going from the start Cartesian position to the destination Cartesian position. By interpolating along this difference of location and orientation vectors and converting each interpolated Cartesian position using inverse kinematics to joint angle commands for servo-control, the robot TCP can follow a straight line path (Brady et al., 1982; Hartenberg and Denavit, 1964).

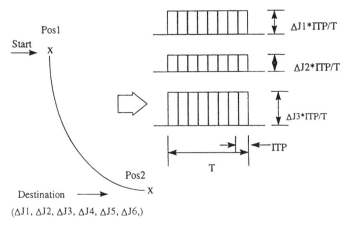

Figure 13 Joint motion command increments.

Circular Motion

With a circular motion command the robot TCP is moved along a circular arc at the programmed speed, as shown in Figure 15. A circular motion requires three Cartesian positions as inputs: start, via, and destination positions (three points define a circular arc). The planner fits a circular arc passing through the start, via, and destination locations. By computing intermediate positions along this arc, the robot TCP can follow a circular path. Similar to linear motion, orientation difference vectors are computed going from start to via and from via to destination. Each interpolated Cartesian position is then converted into its respective joint angle commands with inverse kinematics.

1.6.3 Forward/Inverse Kinematics

Given the mechanical structure of a robot, *forward kinematics* refers to the process of computing the Cartesian position (x, y, z, w, p, r) of the robot TCP given its joint angles. *Inverse kinematics* refers to the process of computing the robot's joint angles given its Cartesian TCP position (Lenarcic and Ravani, 1994).

1.6.4 Tool Center Point (TCP)

In describing a robot path, reference is usually made to the tool center point (TCP). The location is defined from the faceplate of the robot to the end of the tool. As mentioned

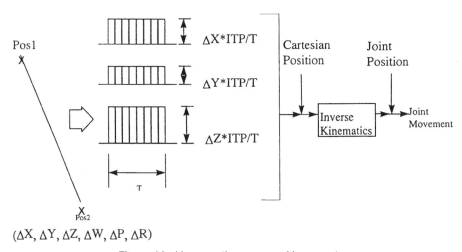

Figure 14 Linear motion command increments.

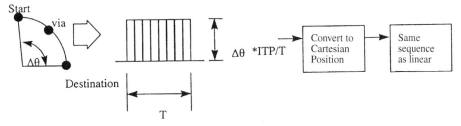

Figure 15   Circular motion command increments.

above, sometimes the tool or robot can be controlled along the linear path. During path execution the robot's TCP is instructed to move along the path.

1.6.5   Frame

A frame is a set of three planes at right angles to each other. The point where all three planes intersect is the *origin* of the frame. This set of planes is called a *Cartesian coordinate system.* In the robot system, the intersecting edges of the planes are the x, y, and z axes of the frame.

Frames are used to describe the location and orientation of a position. The location is the distance in the x, y, and z directions from the origin of the reference frame. The orientation is the rotation about the x, y, and z axes of the reference frame. When a position is recorded, its location and orientation are automatically recorded as x, y, z, w, p, and r relative to the origin of the frame it uses as a reference.

1.6.6   Mastering

The mastering operation calibrates the relationship between the position sensor attached to each axis motor and each axis angle defined for each robot. The robot can determine each axis angle from the position sensor data, using the data generated by the mastering operation. Mastering also allows for the definition of geometric parameters used in describing the analytic parameters of a robot geometric model to correct for discrepancies between design parameters and actual manufactured geometrics.

1.6.7   Acceleration and Deceleration

The planned speed in the previous section is assumed to be constant between the start position and the destination position. Realistically, a speed transition period is necessary near the start and destination positions to avoid theoretically infinite accelerations and decelerations. Accordingly, an acceleration algorithm is implemented to attain accelerations, thereby maintaining the drive torque requirements within the limits of the drive motors. During path planning, acceleration and deceleration times are computed based on robot inertia and drive motors' maximum output torques. Variable acceleration algorithms are often utilized to attain smooth velocity variations and overcome stability problems associated with nonlinearities and disturbances in the robot dynamic system. Figure 16 shows linear speed transitions with constant acceleration and deceleration.

1.6.8   Continuous Motion

In applications, a common requirement is not to decelerate to rest at every taught position, but rather to maintain continuous motion around them. One way to achieve this is by blending two consecutive path segments. Each path segment is planned with an acceleration and deceleration profile as though the robot stops at its destination position (as described in the previous section). Path blending can be accomplished by overlapping consecutive segment profiles, whereby the next segment starts accelerating while the current segment starts decelerating. This produces a continuous motion path with smooth corner rounding, as shown in Figure 17.

1.6.9   Servo Control

The robot has complex dynamics with substantial inertia variation, coriolis forces, and external disturbances such as friction and backlash. Consequently, controlling the servomotor of each robot axis to follow the joint angle commands from path planner is a challenging task. The following is a brief description of the traditional PID control and

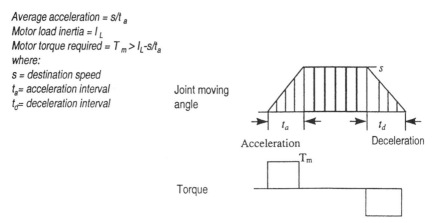

Average acceleration = s/t $_a$
Motor load inertia = I $_L$
Motor torque required = T $_m$ > I$_L$-s/t$_a$
where:
s = destination speed
t$_a$= acceleration interval
t$_d$= deceleration interval

Joint moving
angle

Acceleration   Deceleration

Torque

Figure 16   Linear speed transitions.

feedforward control and the application of these concepts to the control of individual robot joints (Spong and Vidyasagar, 1989; Slotine, 1992).

Preliminary: PID Control/Feedforward Control

Figure 18 shows a typical control system represented by a block diagram of its elements. For a robot the control object is a robot joint driven by an actuator such as an electric motor. When the motor receives a control signal u, it exerts a torque that results in a motion x of the joint. The output x is subtracted from a command signal c, and the error e is modified by a compensator to provide the signal u. When e is positive, the output x

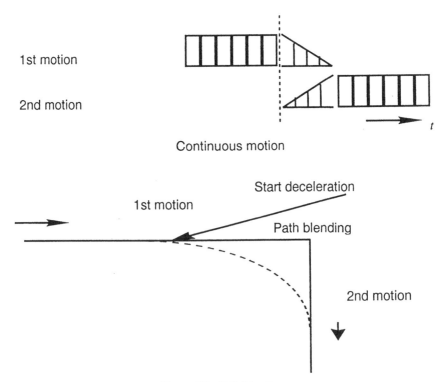

1st motion

2nd motion

Continuous motion

1st motion   Start deceleration

Path blending

2nd motion

Figure 17   Path blending.

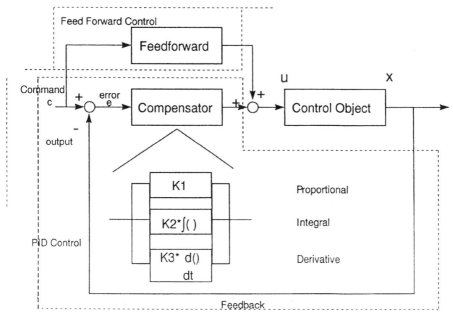

Figure 18   Typical control system.

lags the command c and vice versa. e is the position *following error*. When $u = 0$, the motor exerts no torque on the joint.

The objective of the control system is to make the output x follow the command c. How well this can be achieved depends on the control scheme used. Proportional, integral, derivative (PID) control achieves this objective by assigning specific functions to the compensator. For proportional control the compensator is a constant gain function, i.e., the error is multiplied by a constant $K1$, hence $u = K1*e$. The magnitude of $K1$ determines how fast x catches up with the command c. For simple servo control the proportional control scheme is adequate by itself, especially when the following error e can be tolerated.

However, to improve performance further compensation is necessary. For example, to hold a load against gravity the motor must exert certain torque, hence u must be non-zero, and a steady state following error e will always be present. To eliminate the steady state error, the integral control scheme is utilized. With integral control the error e is integrated over a certain time period and multiplied by a suitable gain $K2$. The output of the compensator is $u = K2*\int e$. dt can be non-zero even if the error e is zero, hence no steady state or following error.

To achieve higher performance we need to anticipate how the command signal c is changing. That leads to the use of derivative control. By differentiating e with respect to time, we can predict the immediate future value of e, which we can use in the control loop by multiplying de/dt with the derivative gain $K3$ (i.e., $u = K3*de/dt$) to provide active damping and prevent an overshoot (i.e., output x is ahead of command c for a certain period of time).

In PID control the proportional scheme is always required, but the use of the integral and/or derivative schemes is optional and depends on the control performance required. However, even when all schemes are used, it is well known that a PID controller introduces servo-delays because the control action is generated by the error e, i.e., it requires a non-zero e to initiate its control action. To reduce this servo-delay, feedforward control is very effective. The idea is simple but requires good knowledge of the control object dynamics. Given u and the control object dynamics, we can compute x. Inversely, if we can compute the inverse dynamics of the control object (which is the difficult part), we can multiply the inverse dynamics function with the command c to find out what the input u should be in order to produce x equal to c, hence no error. In practice, feedforward control is always used in conjunction with feedback control to compensate for unmodeled dynamics.

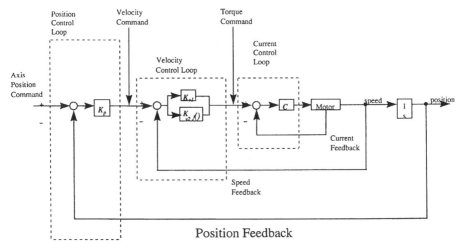

Figure 19   Industrial robot axis control loop (typical).

Robot Axis Control Loop

A common method in robot servo control is to control each robot's motor independently. Figure 19 illustrates a typical industrial control scheme utilizing several control loops. The overall objective is position control, that is, to move the motor to the desired joint angle command. The control scheme consists of three loops, with current control as the innermost loop, followed by a velocity control loop, then the outermost position control loop. Each of these subloops utilizes variations of the basic PID control scheme of the previous section.

Current Control Loop

A typical AC motor has three armature coils separated 120° apart in phase. There are three identical current control loops, one for each phase. Figure 20 shows the PI control scheme used for current control. The overall input is the torque command required. This torque command is multiplied by the phase information of the individual armature to generate its respective current command. For each current control loop, servo control is achieved by measuring the armature current, generating the current error signal by subtracting the feedback current from the command current, and applying PI control to this error signal. As explained above, I is used to eliminate the steady-state error.

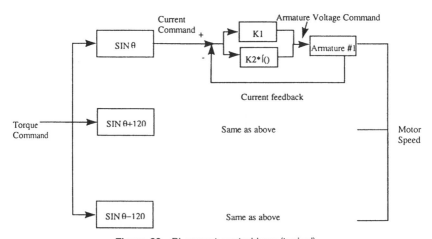

Figure 20   PI current control loop (typical).

Position and Velocity Control Loop

The velocity loop is constructed around the inner current loop. A PI control method is used for this velocity loop, as shown in Figure 19, where the error signal is generated by subtracting the feedback motor speed from the command speed. Finally, a position loop is constructed on top of this velocity loop, completing the control of a single robot joint. In Figure 19 only P control is used for the position loop, and the error signal is generated by subtracting the feedback motor position from the command position. The combined effect of position, velocity, and current loops provides a good responsive control of the motor position to follow the desire joint command input with zero steady state error and zero overshoot. Though not shown in Figure 19, servo-delay can be eliminated by feed-forward control, as explained in the previous section.

1.7 Sensor Control

1.7.1 Overview

Industrial robots can accurately move to taught points and move along taught paths. However, if the workpiece is moved slightly, the robot will perform the application poorly or even fail. It is necessary for the workpiece to be placed in the same position every time. Sometimes the location of the workpiece is not accurately known or moves during processing, or the type of workpiece is not known in advance. To overcome these problems, a sensor should be connected to the robot system to obtain visual, force, or position information from the environment. The robot can then work in these situations by modifying its taught points and paths according to the input from the sensor (Hall, 1992).

1.7.2 Sensor Types

Two types of sensors can be used with robots: single measurement and continuous measurement. Single measurement sensors measure position deviation and are used to adjust robot taught points or paths one time. Vision sensors of this type can provide information to adjust the robot points or paths in two dimensions (three degrees of freedom) or three dimensions (six degrees of freedom). Continuous sensors measure the robot path deviation continuously. Such tracking sensors are used by the robot to adjust the robot path continuously in real time while it is moving. In both cases the sensor information must be translated from the sensor coordinate frame to the robot's coordinate frame. In addition, with vision sensors the reference position of the workpiece where the offset should be zero needs to be taught (Zhuang and Roth, 1996).

   Another kind of sensor is the force sensor, which provides the sense of touch to the robot for fine robot arm control. Sensor information allows robots to adapt to variations in their environment. Figure 21 shows the interaction between sensors and other robot components.

1.7.3 Vision Sensor

Purpose and Configuration of Vision Sensor

A vision sensor uses a television camera to provide a sense of sight to the robot (Zuech and Miller, 1987; Horn, 1986). Vision sensors can be used in the production environment

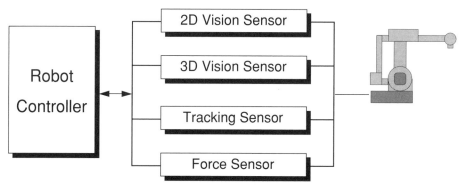

Figure 21 Controller–sensor interactions.

for position detection, inspection, measurement, and recognition. The most common use of vision sensors with robots is for position detection to compensate for the robot's position.

The interaction between the vision sensor and the robot controller in these types of applications typically proceeds as follows. The details vary from application to application and are controlled by customized application software.

The controller is notified that the workpiece is present by a digital input signal from either the cell controller PLC or a simple photocell sensor. The controller then sends a request to the vision sensor to take a picture and identify, locate, or inspect the workpiece. Because the sensor might take up to several seconds to perform its task, the controller might move the robot arm to an approach position after the picture is taken but while it is waiting for the results to be calculated by the vision sensor.

The vision sensor then returns its results and the robot controller adjusts its program according to the workpiece's identity, position, or condition as measured by the vision sensor. The adjustment could consist of selecting a different grip position or drop-off point depending on the workpiece identity, adjusting the pick-up position or working path to compensate for a shift in the position or orientation of the workpiece, or not picking up the workpiece and signaling to the cell controller PLC that the part has failed an inspection.

Depending on the robot's task, various camera configurations can be used (see Figure 22):

- Single camera view to detect the two-dimensional location and orientation of a small object
- Two camera views to detect the two-dimensional location and orientation of a large object that cannot be imaged with one camera with sufficient resolution
- Two overlapping camera views to detect the three-dimensional location of a small object by triangulation
- Three or more camera views to detect the three-dimensional location and orientation of a large object

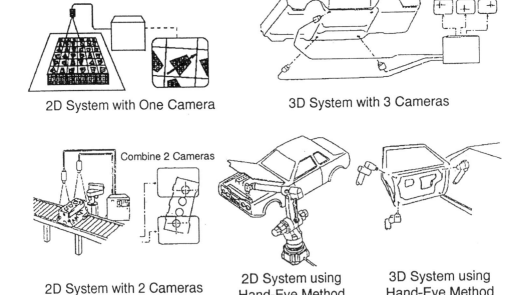

2D System with One Camera   3D System with 3 Cameras

2D System with 2 Cameras   2D System using Hand-Eye Method   3D System using Hand-Eye Method

Figure 22   Modes of using vision cameras.

Any of the above configurations can be achieved using multiple fixed cameras or a single movable camera attached to the robot gripper.

Image Processing

Flow of Image Processing.   The image of the workpiece is formed on the sensor array of the camera through the lens, as shown in Figure 23. The sensor array typically comprises 487 rows and 512 columns of discrete sensor elements. Each of these elements generates an electrical voltage proportional to the intensity of the incident light. These voltages are then converted into a television video signal, which is sent to the vision system. At the vision system the incoming video signal is digitized into a special computer memory area. The light intensity at each point is represented by a number in this image memory (Sood and Wechsler, 1993).

Image processing applies various computations to the digitized image to accomplish the desired vision task (position detection, inspection, measurement, or recognition). In the case of position detection the result is the location of the workpiece or desired portion of the workpiece in the image memory. By using the camera calibration information, this location can be converted to a location in the robot's coordinate frame so that the robot can make use of it (Ma, 1996).

Binary Image Processing and Grayscale Image Processing.   Image processing for industrial use is classified into two categories: binary and grayscale. Binary images look like silhouettes and grayscale images look like black and white television images (Fu, Gonzalez, and Lee, 1987).

Binary Image Processing.   Binary images have only two levels of brightness. A special brightness level called the *threshold,* as shown in Figure 24, is selected. All points with brightness at that level or above are set to white and represented in the image memory as 1's. All points with lower brightness are set to black and represented as 0's. Binary images can be processed quickly because the amount of memory required to store an image is very small. When vision sensors first became available for practical industrial use, they depended on binary image processing because processor speed and memory size at the time were limited.

Determining the proper threshold to separate black from white is a difficult problem in binary image processing. If the threshold is not correct, objects in the image will become smaller or larger than the vision system expects them to be. Even if the threshold is determined correctly for one workpiece, it might not work for another workpiece with

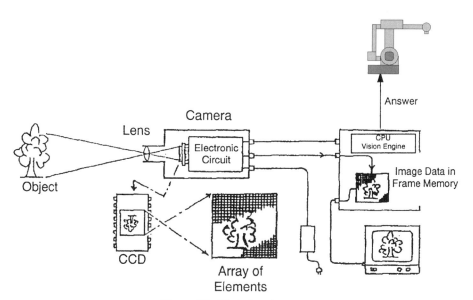

Figure 23   Workpiece images.

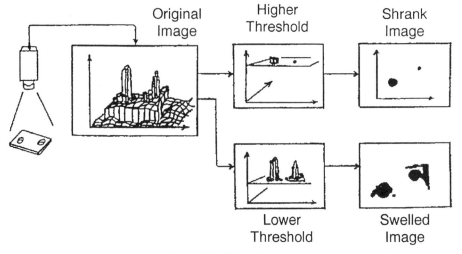

Figure 24 Binary images.

a small discoloration. A threshold that works correctly on a sunny day might not work at all on a rainy day.

Because binary image processing is very sensitive to environmental conditions, when binary image processing sensors are used, special measures must be taken to ensure that a consistent high-contrast image can be obtained. Back lighting is often used to generate a silhouette image. Other methods include very bright lighting or blackout curtains around the object to eliminate uncontrollable outside light.

Grayscale Image Processing. In grayscale image processing, each element in the image memory is represented by a number with a range of more than just 0–1, typically 0–255. Grayscale image processing has become practical as computer processor speed and memory capacity have increased. It is still slower than binary image processing, but it has significant advantages in reliability and is fast enough for many applications.

Grayscale image processing is much less sensitive than binary image processing to changes in the light or coloration of the workpieces. Because of this, the required lighting equipment is relatively simple and inexpensive. In binary image processing the edge of a workpiece is found at the border between white and black regions of the image as determined by the selected threshold. In grayscale image processing, on the other hand, the edge of the same workpiece is found at the point where the brightness changes the most rapidly between the object and the background. This is a more accurate and reliable method for finding the location of the edge, as shown in Figure 25.

1.7.4 Tracking Sensor

Principle of Tracking Sensor

Tracking sensors are typically attached to the robot end-effector to provide tool path position sensing relative to a part (Castano and Hutchinson, 1994). A typical tracking sensor application is arc welding, which requires continuous path motion control and a high degree of fidelity between the tool path and the part.

In arc welding applications the sensor is mounted to the weld torch assembly so that the sensing region is just ahead of the welding arc along the path of the weld joint. The sensor is designed to project a line of laser illumination across the weld joint and the reflected laser light is imaged on a change coupled device (CCD) array, shown in Figure 26. The position of the reflected light on the sensor array is proportional to the weld joint position, so that any error between the anticipated tool path and the actual path generates an error quantity that is communicated to the robot motion control system for correction.

The laser light is generated by a laser diode. There are two common methods for creating a line across the part surface. One method uses a cylindrical lens to diffuse the laser beam into a line. This method is relatively inexpensive, but the incident radiation

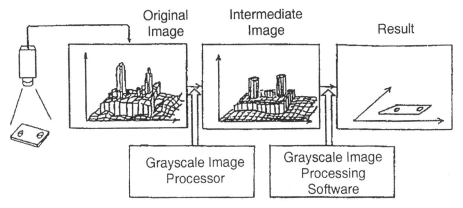

Figure 25   Grayscale image processing.

level on the part surface is low. The second method consists of scanning the laser beam across the part surface by oscillating a mirror mounted to a galvonometer or motor shaft. Scanning rates of 10–20 Hz are typical. The scanned technique results in more robust tracking data because the full power of the laser is concentrated on the part surface. In arc welding the light emission of the arc presents a significant level of noise that can interfere with the tracking sensor performance. In order to eliminate the majority of this noise, an optical interference filter is typically used in front of the sensor CCD array that passes only the wavelength of the laser light. Figure 26 illustrates the geometric relationship between the laser light projector and the CCD (detector) and a typical lap joint for arc welding. Due to the angle between the projector and the detector, the workpiece surface geometry can be determined by triangulation in two dimensions (Y–Z). The third dimension (X) is derived by traversing the sensor along the part surface.

Adaptive Arc Welding Robot System with Tracking Sensor

The tracking sensor provides path position and part shape (weld joint geometry) feedback to the robot control. In most cases the sensor controller operates by attempting to match the measured profile to a template that matches the idealized shape of the part surface. The high quantity of sensed position data is reduced to key points that are reported to the robot control for both path program adjustments and process modifications. In the case of arc welding, the reported data from the sensor include the offset position in the

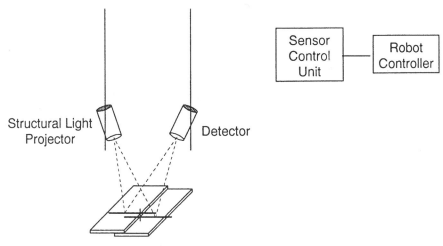

Figure 26   Tracking sensor.

sensor coordinate frame from the ideal position as well as geometric variations of the weld joint. Weld process parameters, including weld deposition rate and heat input, can be modified based on the measured weld joint geometry. A typical case where process modification is required arises when the weld joint gap increases. If the weld procedure has been optimized for a weld joint with no gap in the weld seam and the actual part has a different condition, unacceptable welding can result. The control system can be programmed to modify the output process parameters to accommodate the actual weld seam geometry, as illustrated in Figure 27 (Paul, 1981).

1.7.5 Force Sensor

Figure 28 illustrates the configuration of the force control robot system. The six-axis force sensor, which uses strain gauges, is mounted between the robot wrist flange and the hand to detect force and torque at the end of the robot arm in real time while the robot is working. It can detect six elements, forces along the x, y, and z axes of the sensor coordinate, and torques around those axes. The robot controller performs force control processing using detected information from the force sensor (Allen et al., 1996).

Force Control

Figure 29 shows a force control method called *impedance control*, which controls the robot so that an imaginary system consisting of mass, spring, and damper is realized at the end of the robot arm. Parameters for this mass, spring, and damper can be set to any values by the software and various characteristics can then be realized. For example, very soft compliance can be set along one direction of the robot hand while the other directions can be set rigid. With the combination of these parameters and the setting of the target force, for example, tracing a plane while pressing at a constant force can be realized.

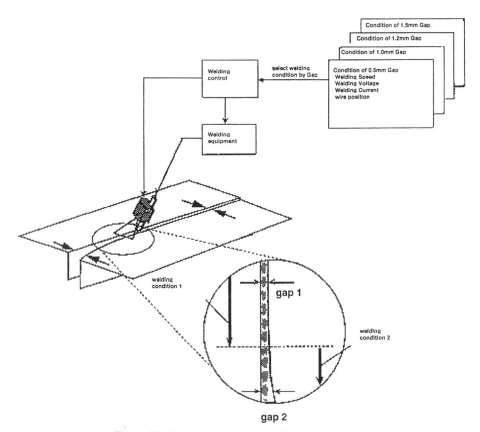

Figure 27 Adaptive arc welding with tracking sensor.

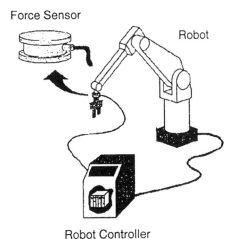

Figure 28 Robot with force sensor.

This can be adapted to grinding and polishing tasks. In another application, by setting the spring parameter to zero and the mass parameter to a small value, direct operation of the robot can be realized, whereby the operator can move the robot by grabbing the end-effector. This allows the operator to move and interact with the robot directly without a teach pendant, significantly improving the robot teaching effort (Yoshikawa and Sudou, 1993; Bruyninckx and De Schutter, 1996).

Inserting Task with Force Control

Force control technology can be applied to precise insertion of mechanical parts, as shown in Figure 30. For example, insertion of a machined shaft into a bearing is very precise, with clearance of a few millimeters for part diameters of several tens of millimeters. If the insertion direction is slightly inclined, binding might occur, preventing completion of the insertion task. Such tasks usually require skilled workers but can be automated with advanced force control technology (Nelson and Khosla, 1996).

This force control approach modifies the error of insertion angle based on the measured moment, which is generated at the end of the inserted part. When the angle modification

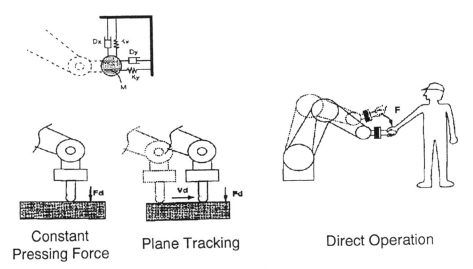

Figure 29 Impedance force control.

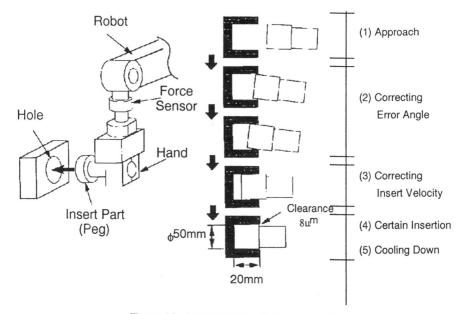

Figure 30   Insertion task with force control.

is done, insertion is performed by controlling target force with set impedance. When the predefined insertion depth is reached and the predefined force is detected, insertion is successfully completed.

In a conventional insertion task, insertion direction has to be the same as direction of gravity to remove the effect of gravity. In force control with the gravity compensation algorithm, insertion can be done in any direction.

1.8   Off-Line Programming System

1.8.1   Relationship of Off-Line System and Robots

The interrelationships of the components of an off-line programming system are shown in Figure 31 and explained in the following sections (Megahed, 1993; Zomaya, 1992). (See also Chapter 19 on off-line programming.)

Off-Line Programming System

Off-line programming systems present the behavior of robots, workpieces, or peripheral devices without the cost or danger of an actual robot. For example, an operator can investigate the planned layout of components or motion of robots and try different approaches much faster than with an actual robot.

An operator can also develop robot programs with off-line systems using libraries built over time. Automatic systems can write programs directly from manufactured product specifications. Multiple robots on the screen can be executed simultaneously in order to examine the timing of signal communication. Programs can be edited, updated, archived, and ported between robots without taking a robot out of service.

CAD Systems

Computer-aided design (CAD) systems are used to develop or modify the graphical representation of workpieces or peripheral devices and provide them to an offline system.

Motion Simulator

The RRS motion simulator simulates the internal behavior of the robot to realize a more accurate simulation (RRS Interface Specification, 1996).

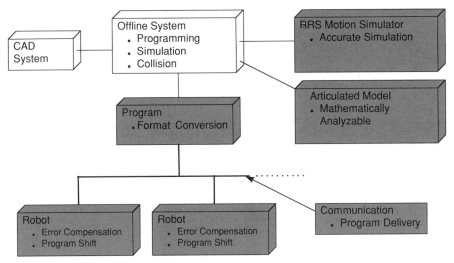

Figure 31   Components of off-line robot programming.

Articulated Model

Modeling a robot as a set of rigid bodies allows mathematical analysis and optimization (Slotine, 1996).

Program Conversion

Robot programs in the off-line system format are converted to and from actual robot format.

Robot

Robots interpret programs and execute processes. Calibrated robots compensate for the error between the off-line system model and actual environment in order to match the simulation precisely.

1.8.2   Robot Model and Environment Model

To simulate robots, workpieces, or peripheral devices off-line, numeric models are necessary. Geometric models are necessary for internal collision checks and a mechanism model is necessary for performance prediction.

1.8.3   Reachability Analysis and Collision Analysis

In a reachability analysis internal mechanism models predict whether robots can locate end-effectors at the positions needed for the desired process. In a collision analysis geometric models predict whether any components will collide. In a near-miss or clearance analysis geometric models predict the closest approach or component clearance.

1.8.4   Programming

In an off-line system a robot can run programs before they are tried on the actual robot. Reachability and collision between programmed points can be checked while robots are running. Not only path, but cycle time and I/O timing between robot and peripheral devices or between a pair of robots can be programmed. Processes with multiple components including robots can be programmed. Process details can be observed and validated that would otherwise be difficult or dangerous with an actual robot.

1.8.5   Motion Simulation

Programs for robots are interpreted by off-line systems that often display the robot graphically. However, since the actual robot motion algorithm reflects their mechanical complexity, off-line systems must trade off behavior accuracy and mathematical precision.

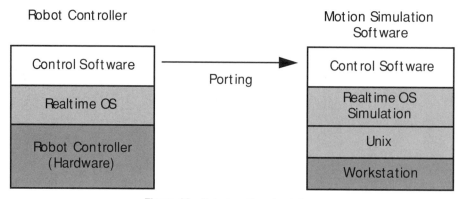

Figure 32   Robot motion simulation.

Structure of Motion Simulator

In order to realize the most accurate simulation, the motion control software on the robot controller should be ported to a computer for off-line system. To do that completely, the simulation must include the hardware and operating system of the robot controllers, as shown in Figure 32.

Accuracy of Motion Simulator

Motion algorithms are too complex for off-line systems to simulate without an internal mechanism simulation. Precision performance prediction will probably remain elusive since wear and load will impact the motion itself. Articulated models provide precise mathematical forecasts and represent the opposite end of the accuracy-versus-precision tradeoff.

Figure 33 shows the accuracy of a motion simulator for typical motions.

1.9   Robot Cell Architecture

Industrial robots frequently operate with other robots or other factory automation equipment as part of a robot application. The key to operation is being able to exchange information between the devices that are part of the factory automation cell, as shown in Figure 34 (Hoshizaki and Bopp, 1990; Nnaji, 1993).

Typically, communication between parts of the robot cell can be divided into one of three classes:

- Computer network communication
- Communication between controllers
- Serial I/O communication (Snyder, 1985; Dorf, 1983)

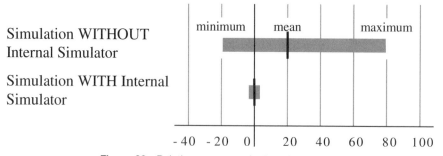

Figure 33   Relative accuracy of robot simulations.

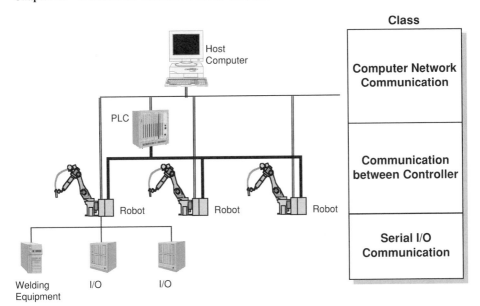

Figure 34   Robot cell control architecture.

1.9.1   Computer Network Communication

This consists of communications, usually as data files or programs, between a host computer system and the robot, programmable controller, or other intelligent device in the cell. The host computer normally is responsible for monitoring overall operation of the robot cell and sending and backing up application-specific data (such as program to be executed) to the devices in the cell. In most cases the network used to connect the host computer and the cell device is a standard network such as Ethernet.

1.9.2   Communication Between Controllers

This center communication level within the cell is used to transfer I/O data (digital input and output points or values) and messages between robots or between a robot and a programmable controller. The I/O data are usually transmitted between robots using a industrial field I/O network supported by the devices in the cell. Typical industrial field I/O networks currently in use include:

- Profibus
- ME-Net
- Interbus-S
- DeviceNet

The field I/O network used is typically high-speed and deterministic, meaning that the time to communicate between devices is the same under a wide variety of conditions.

1.9.3   Local Serial I/O Communication

Between the robot and other peripheral automation devices within the cell, industrial networks are also used to transfer the value of digital I/O points to and from the robot. If the number of digital I/O points communicated between the robot and other I/O devices within the cell is small, direct wiring (I/O point to I/O point) between the robot controller and the cell device is often used.

1.10   Development Trends in Robot Controllers

1.10.1   Enhanced Robot Accuracy

The increased validity of robot performance based on off-line simulations requires increased accuracy from robot mechanical units. Historically, robots have been very repeatable, with typical errors of about a tenth of a millimeter. However, the absolute accuracy over the work envelope can have errors as large as tens of millimeters. Reliance on off-line programming requires robots to have better calibration and more complete modeling to reduce the absolute accuracy error to about 1 mm (or less, depending on robot design). Robot controllers will include accuracy-enhancing modules through the use of sensors and improved calibration algorithms.

1.10.2 Increased Use of Sensors

Ease of robot position teaching and touchup has always been an area of concern for end users. Sensor technology will allow less skilled operators to teach robot positions for both initial programming and position touchup after initial programming. The sensor technology must be low-cost, reliable, safe, and easy to integrate into robot systems to gain widespread acceptance.

The integration of multiple sensory data, or *sensor fusion,* is becoming increasingly important as sensor technology becomes more reliable and affordable. The cross-coupling of sensory feedback makes this a difficult area for reliable industrial closed-loop control and is a topic of much research. Robot controllers should see appreciable enhancement in their ability to integrate and disseminate sensor inputs.

1.10.3 Open Architecture

The increased capability and preponderance of PC-based application software and the emergence of reliable operating systems are fueling the demand for open architecture in robot controllers. The controller will have in the future a greater level of architecture openness providing interfaces to a selection of third party applications. However, because of the intricate interaction between the robot's servo-system, the mechanical structure, and the systems that guard for human safety, the real-time servo-control functions of the robot controller are likely to remain proprietary to the robot manufacturer. The likely implementation for the near future may well be a PC front end connected to a proprietary real-time system.

ACKNOWLEDGMENTS

The editors wish to acknowledge the support of Dr. Seiuemon Inaba, Chairman and CEO of FANUC Ltd., Yamanashi, Japan, and Mr. Eric Mittelstadt, Chairman and CEO of Fanuc Robotics North America, Inc., Rochester Hills, Michigan; and the fundamental contributions of Mr. R. Hara of Fanuc Ltd., Messrs. Gary Rutledge, Claude Dinsmoor, Jason Tsai, Dean McGee, Sai-Kai Cheng, Bruce Coldren, Robert Meier, Peter Levick, and Ms. Debra Basso, all of the FANUC Robotics Development Staff, who made this work possible.

REFERENCES

Allen, P. K., et al. 1996. In *Integration of Vision and Force Sensors for Grasping.* IEEE/SICE/RSJ International Conference on Multisensor Fusion and Integration for Intelligent Systems, December 8–11, Washington. 349–356.

Anderson, B. M., J. R. Cole, and R. G. Holland. 1993. "An Open Standard for Industrial Controllers." *Manufacturing Review* **6**(3), 180–191.

Andeen, G. B. 1988. *Robot Design Handbook.* New York: McGraw-Hill.

Brady, J. M., et al. *Robot Motion: Planning and Control.* Cambridge: MIT Press.

Bruyninckx, H., and J. De Schutter. 1996. "Specification of Force-Controlled Actions in the 'Task Frame Formalism'—A Synthesis." *IEEE Transactions on Robotics and Automation* **12**(4), 581–589.

Castano, A., and S. Hutchinson. 1994. "Visual Compliance: Task-Directed Visual Servo Control." *IEEE Transactions on Robotics and Automation* **10**(3), 334–342.

Craig, J. J. 1986. *Introduction to Robotics: Mechanics and Control.* Reading: Addison-Wesley.

Dorf, R. 1983. *Robotics and Automated Manufacturing.* Reston: Reston Publishing.

Fu, K. S., R. C. Gonzalez, and C. S. G. Lee. 1987. *Robotics: Control, Sensing, Vision, and Intelligence.* New York: McGraw-Hill.

Hall, D. L. 1992. *Mathematical Techniques in Multisensor Data Fusion.* Boston: Artech House.

Hartenberg, R. S., and J. Denavit. 1964. *Kinematic Synthesis of Linkages.* New York: McGraw-Hill.

Hoshizaki, J., and E. Bopp. 1990. *Robot Applications Design Manual.* New York: John Wiley & Sons.

Horn, B. K. P. 1996. *Robot Vision.* Cambridge: MIT Press.

Hutchinson, S., G. D. Hager, and P. I. Corke. "A Tutorial on Visual Servo Control." *IEEE Transactions on Robotics and Automation* **12**(5), 651–670.

Lenarcic, J., and B. Ravani, eds. 1994. *Advances in Robot Kinematics and Computational Geometry.* Boston: Kluwer Academic.

Lewis, F. L. 1993. *Control of Robot Manipulators.* New York: Macmillan.

Ma, S. D. 1996. "A Self-Calibration Technique for Active Vision Systems." *IEEE Transactions on Robotics and Automation* **12**(1), 114–120.

Megahed, S. 1993. *Principles of Robot Modelling and Simulation.* New York: John Wiley & Sons.

Nelson, B. J., and P. K. Khosla. 1996. "Force and Vision Resolvability for Assimilating Disparate Sensory Feedback." *IEEE Transactions on Robotics and Automation* **12**(5), 714–731.

Nnaji, B. O. 1993. *Theory of Automatic Robot Assembly and Programming.* 1st ed. London: Chapman & Hall.

Paul, R. P. 1981. *Robot Manipulator: Mathematics,. Programming and Control.* Cambridge: MIT Press.

Pollmann, A. 1996. *Logic/Object-Oriented Concurrent Robot Programming and Performance Aspects.* Berlin: Walter De Gruyter.

Proctor, F., et al. 1993. *Open Architectures for Machine Control.* Gaithersburg: National Institute of Standards & Technology.

Qu, Z., and D. M. Dawson. 1995. *Robot Tracking Control of Robot Manipulators.* Institute of Electrical and Electronics Engineers.

Rahimi, M., and W. Karwowski. 1992. *Human–Robot Interaction.* London: Taylor & Francis.

Realistic Robot Simulation (RRS) Interface Specification RRS-Owners. 1996. IPK-Berlin, 49-30/39006-0.

Samson, C. 1991. *Robot Control: The Task Function Approach.* Oxford: Clarendon Press.

Samson, C., B. Espiau, and M. Le Borgne, *Robot Control: The Task Function Approach.* Oxford: Oxford University Press.

Sciavicco, L. 1996. *Modeling and Control of Robot Manipulators.* New York: McGraw-Hill.

Shpitalni, M., Y. Koren, and C. C. Lo. 1994. *Real-Time Curve Interpolators.* Computer-Aided Design.

Slotine, J.-J. E. 1986. *Robot Analysis and Control.* Haruhiko Asada.

———. 1992. *Robot Analysis and Control.* New York: John Wiley & Sons.

Snyder, W. 1985. *Industrial Robots: Computer Interfacing and Control.* Englewood Cliffs: Prentice-Hall.

Sood, A. K., and H. Wechsler. 1993. *Active Perception and Robot Vision.* New York: Springer-Verlag.

Spong, M. W., and M. Vidyasagar. 1989. *Robot Dynamics and Control.* New York: John Wiley & Sons.

Yoshikawa, T., and A. Sudou. 1993. "Dynamic Hybrid Position/Force Control of Robot Manipulators–On-Line Estimation." *IEEE Transactions on Robotics and Automation* **9**(2), 220–226.

Zhuang, H., and Z. S. Roth. 1996. *Camera-Aided Robot Calibration.* Boca Raton: CRC Press.

Zomaya, A. Y. 1992. *Modelling and Simulation of Robot Manipulators: A Parallel Processing Approach.* River Edge: World Scientific.

Zuech, N., and R. K. Miller. 1987. *Machine Vision.* Lilburn: Fairmont Press.

ADDITIONAL READING

Chen, C. H., L. F. Pau, and P. S. P. Wang, eds. *The Handbook of Pattern Recognition and Computer Vision.* River Edge: World Scientific, 1993.

Davies, E. R. *Machine Vision: Theory, Algorithms, Practicalities.* New York: Academic Press, 1990.

Dougherty, E., ed. *Digital Image Processing: Fundamentals and Applications.* New York: Marcel Dekker.

Everett, H. R. *Sensors for Mobile Robots: Theory and Applications.* A. K. Peters, 1995.

Gonzalez, R., and R. Woods. *Digital Image Processing.* Reading: Addison-Wesley, 1992.

Haralick, R., and L. Shapiro. *Computer and Robot Vision.* Vol. 1 New York: Addison-Wesley, 1992.

Jones, J. L., and A. M. Flynn. 1993. *Mobile Robots.* Wellesley: A. K. Peters, 1993.

Rembold, U., B. O. Nnaji, and A. Storr. *Computer Integrated Manufacturing and Engineering.* Reading: Addison-Wesley, 1993.

Wechsler, H. *Computational Vision.* Boston: Academic Press, 1991.

CHAPTER 13

SENSORS FOR ROBOTICS

C. R. Asfahl
University of Arizona
Fayetteville, Arizona

1 INTRODUCTION

Early industrial robots had little or no sensory capability. Even as late as the 1980s, feedback was limited to information about joint positions, combined with a few interlock and timing signals (Kak and Albus, 1985). The concept in those early days of the age of robotics was that the environment in which the robot would work would be strictly controlled and objects would be of known geometry, location, and orientation. Further, it was assumed that a high level of consistency would always exist with respect to part geometries, location, and orientation. Sometimes such assumptions made sense, as in the unloading of die-casting and injection-molding machines, for example, and in spot welding of rigidly fixtured automobile bodies. The ready availability of these ideal applications made them natural targets for implementation of the blind no-see, no-feel robots of the 1980s. However, toward the end of the decade robot application began to run aground as the industry began reaching around for opportunities in the not-so-controlled industrial environments, especially in smaller-volume industries.

In the 1990s it became widely recognized that industrial robots should have an awareness of their work environment, including the workpiece, its location and orientation, and the intrusion of unexpected objects or persons within their workspaces. This awareness has both practical and safety considerations.

The subject of safety will be addressed first. In the early 1980s the capability of an industrial robot to kill a human being was the subject of theoretical discussion. From 1985 to 1995 repeated human fatalities from encounters with industrial robots became reality. The incident referred to in the headline that follows (Park, 1994) was not a freak accident; rather, unfortunately, it is but a sample of several such accidents that have occurred around the world, primarily in the United States and Japan.

"Man Hurt by Robot Dies"
   The 22-year-old worker was fatally wounded after a robotic arm at Superior Industries Inc. pinned him against a machine. OSHA is investigating the accident.

Robots, perhaps more than any other industrial machine, suffer from a lack of public tolerance for accidental and dangerous contacts with humans. Even when used in the safest possible way, robots are considered by many to be sinister, uncaring job-destroyers to be feared even when they are safe. When the possibility of personal injury or even fatalities inflicted by industrial robots is added, there is little public sympathy for the robot or its designers and builders. Lawyers know this, of course, and a personal injury lawsuit is almost a certainty whenever a worker is badly injured or killed by an industrial robot. Court juries are a barometer of public sentiment and accordingly should not be expected to be forgiving when it comes to robots.

Although worker fatalities at the hands of robots have been rare, there have been thousands of robot accidents, resulting in either property damage, worker injury, or both. A defining feature of an industrial robot is its programmability, and ironically this feature is what makes the robot useful, but at the same time dangerous. The program for an industrial robot can be intricate and adaptable, allowing for reaction to varying demands and irregular actions or motions. But these irregular actions are often unsuspected by the

Handbook of Industrial Robotics, Second Edition, Edited by Shimon Y. Nof
ISBN 0-471-17783-0   © 1999 John Wiley & Sons, Inc.

unwary operator, and the result is an accident. While recognizing the hazards of robot movements, we should also recognize that the industrial robot has played a significant role in removing personnel exposure from dangerous workstations and unhealthful industrial environments. The key to keeping the ledger on the positive side with respect to robot safety is to design robotic systems that control or prevent exposure to hazards. Sensors, the subject of this chapter, can play a large role in such control and prevention systems.

The other reason for the recent trend toward sensory capability for robots is a practical one. If robots are able to "see" the orientation of a workpiece, they can be programmed to correct the problem. If the workpiece has a defect in its geometry that can be sensed by the robot, the robot can remove that workpiece for subsequent scrap or rework, or perhaps even rework the workpiece itself. If an object or a worker intrudes upon the robot's workspace, the robot can be programmed to stop, sound an alarm, or even remove the obstruction, or do all three. These capabilities are opening up new application areas for industrial robots in environments having less than ideal control.

It is one thing to sense something unexpected in the environment; it is another to understand what it is and to have a strategy for dealing with it. Robots that have this understanding and capability are called *intelligent* robots. Of course, there is a broad spectrum of what can be classified as intelligence. Chapter 14 of this handbook discusses robot vision systems and reveals that there is a lot more intelligence in vision systems, both robot and natural, than most people imagine.

This chapter focuses upon the most basic levels of robot sensing, including touch sensors, photoelectrics, and noncontact proximity sensors. Robot vision can be considered an extension of robot sensing and is discussed at the conclusion of this chapter. This chapter considers only the environment external to the robot. Robots, and automated machines in general, also need sensors and feedback systems to monitor and deal with their own internal states, but these topics have been reviewed in many other publications (see Chironis, 1966; Doeblin, 1966; Harris and Crede, 1961; Lion, 1959) and are not covered here.

The chapter begins with the simplest forms of sensing and roughly proceeds to the most complex. Remote center compliance (RCC) is probably the simplest form of sensing; indeed, it is hardly considered sensing at all. RCC is a means for robot tooling to adapt to slightly misaligned workpieces or components for assembly. It is an important link between the ideal world of the robot and the real world of robot applications. So important is RCC that it is considered in depth elsewhere in this handbook (see Chapter 50). Recent developments have increased the sophistication of RCC by applying sensors in a technology referred to as *instrumented remote center compliance* (IRCC). IRCC allows robot programs to follow contours, perform insertions, and incorporate rudimentary touch programming into the control system (Nevins et al. 1980).

After RCC, the next simplest form of adapting to the robot's environment is mechanical limit switches, probably the most widely used sensor in robotics and automation systems worldwide. Photoelectric sensing devices and their close relatives, infrared devices, come in many varieties and range selectivities. Other applications require proximity sensors, which sense objects without the benefit of light beams but also do not touch the object. The next topic is force and torque sensors, a type that is very important to the robot application of automatic assembly. A more sophisticated form of force sensing is tactile sensing, in which the robot senses several features and surface qualities of an object, much as humans do with their sense of touch. Miscellaneous sensors for robots include ultrasonic ranging systems for navigation and obstacle avoidance. A SensorGlove is equipped with instruments to measure human gestures and other hand movements with surprising precision. To increase the reliability and sophistication of sensor systems, multiple sensors are integrated using sensor fusion and the principles of sensor economy and selection procedures. Perhaps the newest development in the field of robot sensing is the integration of the sensing system into a networked communication system using bus technology. Recognizing the advantages and disadvantages of each type of sensor, the chapter closes with a decision diagram for assisting in the optimum selection of robotic sensors to match their intended applications.

2 MECHANICAL LIMIT SWITCHES

Remote center compliance, covered in Chapter 50, is a method of reacting to an assembly situation, but it really does not constitute an actual sensing of an object in the robot's environment. The most basic and widely used device for sensing objects and their position is the mechanical limit switch, although many system designers do not think of a limit switch as a sensor, either.

A mechanical limit switch contains a spring-activated snap mechanism to close or open a set of physical contacts. An actuator is physically deflected to trigger the mechanism and operate the switch. Actuators for these switches come in many varieties, as indicated in Figure 1.

Mechanical limit switches work best in robotics applications that permit physical contact between the switch actuator and the object to be sensed. The physical contact must be of sufficient force to trip the snap action of the switch mechanism. However, very sensitive "cat whisker" actuators are available for delicate robotic sensing applications. Mechanical limit switches are known for their rugged construction and excel in harsh industrial environments. Explosion-proof models can even be subjected to exposure to hazardous (flammable or combustible) atmospheres.

Mechanical limit switches may be connected to the input module of a commercially available industrial robot, but more often they are connected in quantity to a programmable logic controller (PLC). The PLC either acts as a control system for a robot or works as a supervisory unit controlling several robots or a system of robots and other automatic equipment.

Robotic system designers can benefit from some design tips to enhance the reliability and function of mechanical limit switch applications. If the limit switch is actuated by a cam or dog, a transistional slope on both the actuate and release sides of the cam can prevent sudden shocks to the mechanical actuator. Also, any mechanical switch requires a finite hold time to effect the switchover, so cams or other actuating objects should be designed to hold the switch in the actuated position long enough to do the job. Sometimes the actuator deflection is too small to trip the switch or, conversely, the deflection is too great and ruins the switch. Either problem can be remedied by careful kinematic design or by taking advantage of the adjustability of the switch actuator. In applications in which the object to be sensed moves in a direction perpendicular to the lever, a rotary actuator

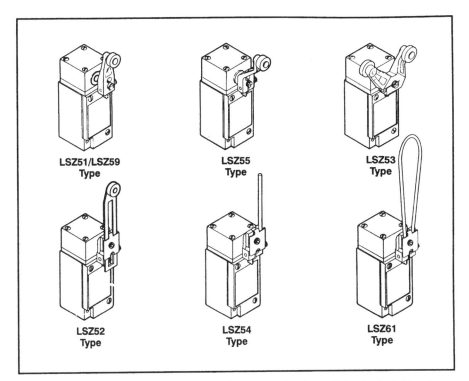

Figure 1 Rotary actuated limit switches can be equipped with a variety of adjustable actuators to suit a particular robotics application (courtesy Honeywell's MICRO SWITCH Division).

should be used with a lever arm. If the object moves in a straight line directly toward the switch, a pushrod actuator can be utilized to trip the switch. Attention to design of limits of travel is particularly important for pushrod actuators. The robotics designer should remember that the adjustability of the lever-type actuators has limits. Attempts to custom rig and attach a special arm to a limit switch actuator may result in unreliable operation of the switch. Finally, care should be taken to locate the switch in a position that will not expose it unnecessarily to excessive heat or the possibility of accidental tripping by coming into contact with personnel, materials, product, or scrap.

3   PHOTOELECTRIC DEVICES

If mechanical limit switches are the most widely used sensors in robotics today, then photoelectric sensors are without doubt second, and their variety is even more complex. Like mechanical limit switches, photoelectrics are often connected to the input modules of the industrial robot or to a programmable logic controller.

To most people the term *photoelectrics* means a conventional light beam sensed by a light-sensitive switch that relays the on/off signal to a controller. This understanding is correct, but limited in perspective. Although early photoelectrics used conventional incandescent light, modern systems almost exclusively use light sources from light-emitting diodes (LEDs). The advent of low-cost, efficient LEDs in the 1970s made possible tremendous advances in photoelectrics in the decades that have followed. The 1990s saw the recognition of the potential utility of photoelectrics as sensors for robots.

Among the many advantages of LEDs, perhaps the greatest is the capability of modulating or pulsing the light source. The sensor, in turn, can be tuned to recognize the light source at its prescribed modulation frequency and screen out all other ambient (and thus irrelevant) light of various wavelengths and modulation frequencies. Because the sensor is tuned to look for light only at this very special frequency, it can greatly amplify the signal and still not get false readings from ambient light. Therefore, even a tiny LED light source, as in a robotics application, can be recognized by a sensor over a much larger distance than would be possible for an equivalent incandescent.

Other advantages of LEDs make them superior for robotics applications. LEDs are extremely reliable and their solid-state construction has no moving parts and no fragile, hot filaments to burn out or become limp and droop away from the focal point. In the industrial environment to which many robots are subjected, it is comforting to have a photoelectric that is not sensitive to shock and vibration nor to wide-ranging temperature extremes. All of these reliability factors add another advantage: because the LED should last for the lifetime of the entire sensor, the circuitry of the sensor can be encapsulated, resulting in a rugged, sealed unit ideal for use on industrial robots subjected to harsh environments.

The introduction to this chapter referred to infrared sensors as close relatives of photoelectrics. The only difference between infrared and visible light is wavelength (or inversely, frequency). With the older, incandescent systems, infrared was a specialty application where it was desired to make the light beam invisible or perhaps to avoid false sensor signals from ambient light in the visible range. But with LED technology has come increased use of infrared for several reasons. Infrared radiation (wavelength greater than 800 nm) has a physical advantage in having greater optical penetrating power than does visible light. Thus, in an industrial environment with considerable smoke, fog, or dust, a robot equipped with infrared sensing systems may be able to "see" better than a comparable system using light in the visible spectrum. Also, the physical response of the phototransistor when subjected to infrared is superior to the response to visible light (Figure 2).

The simplest mode of operation for photoelectrics is *opposed-mode* (Figure 3), in which a beam is broken by an opaque object that is positioned between the source and sensor. Robots can use opposed-mode photoelectrics to detect intrusions into the work envelope by objects or personnel. Human fatalities from contact with industrial robots could have been prevented by opposed-mode photoelectrics. Mirrors can enhance the capability of opposed-mode photoelectrics to protect the entire perimeter of a robotic workstation. Figure 4 illustrates a typical perimeter guard system using opposed-mode photoelectrics with a rated beam range of 45 feet. The system shown in Figure 4 uses selective blanking to mute the sensitivity of the system in prescribed areas in which it is necessary to violate the workstation perimeter (as in the input conveyor shown in the figure).

A real breakthrough for industrial robotics applications is the concept of "floating blanking," which permits small objects (up to one inch in cross section) to violate the

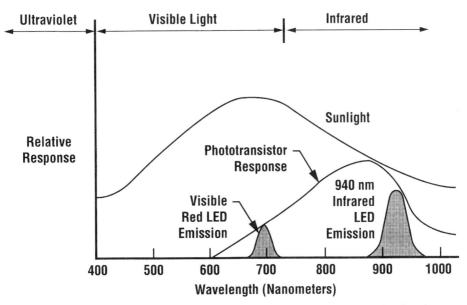

Figure 2 Spectral response of phototransistor (courtesy Banner Engineering Corp.).

perimeter without tripping the sensor. Floating blanking permits small workpieces to be fed to the robotics workstation, passing *through* the perimeter protection system.

Industrial robot gripper systems can also use opposed-mode photoelectrics to detect presence of objects or even features of objects. However, robot grippers are often small and unsuitable for the mounting of photoelectric emitters and receivers. One solution to this problem is to use fiber optics to pipe the beam to and from the application from a

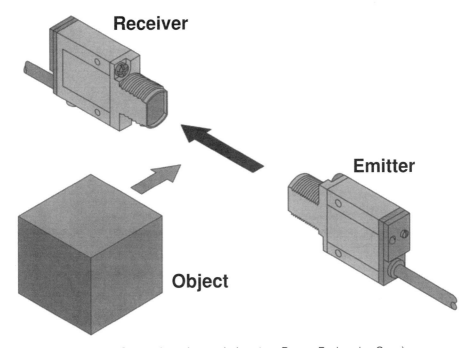

Figure 3 Opposed sensing mode (courtesy Banner Engineering Corp.).

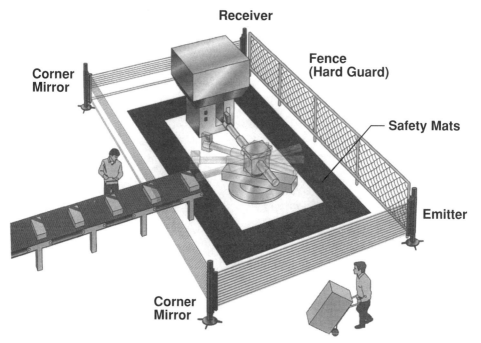

Figure 4 Typical perimeter guard system application using corner mirrors to protect a robotic workstation (courtesy Banner Engineering Corp.).

remote emitter and to a remote sensor, respectively. Fiber optics are bundles of flexible glass or single-strand plastic fibers that are capable of conducting the light beam into very intricate positions, as might be found around the gripper area of an industrial robot. Specially designed "bifurcated" fiber optic cables permit both the *send* and *receive* optical bundles to be combined into a single cable assembly at the robot gripper or other point of application (Figure 5).

It is possible to consolidate the photoelectric source and sensor into a single unit without the use of fiberoptics. One way is to use the *retroreflective* sensing mode, as shown in Figure 6. The retro target resembles the safety reflector on the rear of a bicycle, except that the reflector is white instead of red. Despite the retroreflective mode's advantage of permitting scanning from one side only, many robotics applications will favor the more conventional opposed-mode photoelectrics because of their superior range and performance. Also, space on an industrial robot or within an intricate workspace may make the placement of a retro target impractical.

If the surface of the object to be sensed has a moderate degree of light reflectivity, a *diffuse* mode photoelectric can be used (Figure 7). It is not necessary for the surface to

Figure 5 Bifurcated fiber optic assembly (courtesy Banner Engineering Corp.).

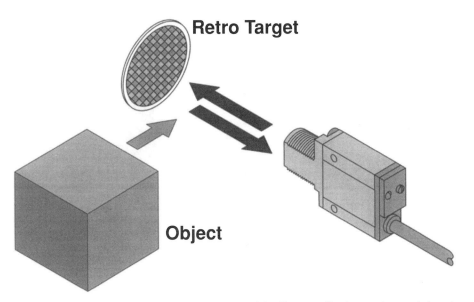

Figure 6   Photoelectric source and sensor are combined in retroreflective sensing mode (courtesy Banner Engineering Corp.).

have mirror-like qualities, because only a fraction of the emitted light is required to trigger the sensor. Even a white or light surface reflects a significant amount of diffuse light. As might be expected, the diffuse mode is more limited in range than opposed-mode photoelectrics. Also, note that in diffuse mode the sensed object *makes* the beam, whereas in opposed-mode or retroreflective mode the sensed object *breaks* the beam.

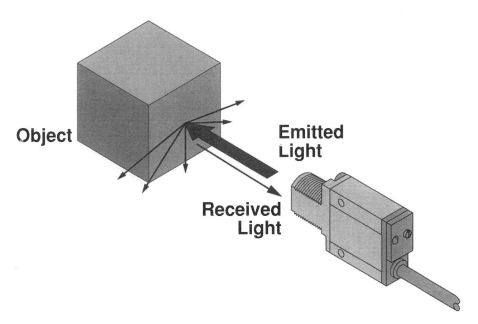

Figure 7   Diffuse sensing mode combines photoelectric source and sensor into a single unit (courtesy Banner Engineering Corp.).

4 PROXIMITY DEVICES

Mechanical limit switches and photoelectrics comprise the majority of robot sensing systems, but some robotics applications present problems that cannot be solved using mechanical switches or even photoelectrics. Suppose the object to be sensed is located behind a barrier, such as when it is inside a paper carton or on the other side of a glass or plastic shield. Or suppose the object to be sensed by the robot is too soft to deflect a mechanical limit switch and at the same time transparent to light or infrared radiation. Another problem is presented by objects too tiny or lightweight to be detected by the most sensitive and miniature mechanical limit switch or too tiny to either break or make a photoelectric beam and thus be "seen" by the robot. These types of robotics applications may benefit from proximity sensors.

With proximity sensors robots can sense objects that no human can, e.g., they can sense objects that cannot be seen or touched or sensed in any way by a human's five senses. The proximity sensor takes advantage of various properties of the material that permit the material to alter an electric or magnetic field. This disturbance is detected by the sensor and consequently a presence is indicated.

Many robots operate in hostile environments involving vibration, temperature extremes, and/or corrosive fluids or mists. In such an environment mechanical switches with moving parts may jam. Even photoelectrics may be degraded by the environment. Proximity switches offer the advantage of high-quality sealing and extra-long life and are also excellent for high-speed switching applications, such as counting objects on a production line.

Figure 8 illustrates a typical proximity sensor. These units are so well concealed within a sealed, threaded rod that many practitioners do not realize that the threaded rod actually contains a sophisticated sensor within. It is easy to see how proximity sensors can be mounted in a variety of ways wherever the system presents a plate or sheet to which the sensor can be mounted using the clamping nuts shown. Note in the diagram that the unit is equipped with onboard LEDs, useful for both monitoring the operation of the sensor and troubleshooting.

Many proximity sensors can detect any metal object because they are triggered by the conductive properties of the material being sensed. Others can detect only ferrous objects because they are triggered by the magnetic properties of the material being sensed. The latter method might seem to be a limitation, but the ferrous-only sensors can be used to even greater advantage than the other type in some robotics applications. Picture a situation in which the object to be sensed is ferrous but is located behind an aluminum barrier. The robot can then "see through" the aluminum barrier by employing a ferrous-only proximity switch that ignores the presence of the aluminum barrier.

Still another variety, the *capacitive* proximity switch, utilizes the dielectric property of the material to be sensed to alter the capacitance field set up by the sensor. In practical

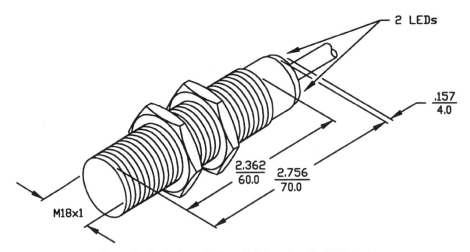

Figure 8 Typical proximity switch (courtesy TURCK, Inc.).

terms this means that the sensor can be used to detect objects constructed of *non*conductive material. Thus, the robot can detect the presence or absence of objects made of wood, paper, or plastic. Even liquids can be detected, meaning that capacitive proximity sensors can be used to detect liquid levels in a tank. In a robotics application the robot could detect the point at which it dips its gripper into a cooling bath. Of course, an ordinary temperature sensor might be used in such an application as well.

If all else fails, a robotics application may call for an ultrasonic proximity sensor, a device that operates in principle like a submarine sonar detector. A cone-shaped beam pulsed at ultrasonic frequencies is transmitted out toward a target object and "echoed" back to reflect not only the presence of the object but also the distance to it. The advantage is that transparent materials, such as glass, are detected as easily as opaque materials.

An advantage of ultrasonic proximity sensors is that they can operate over a comparatively long range. Because ultrasonic sensors operate on sound waves, one might suspect the disadvantage of noisy environments possibly causing false signals and erroneous tripping of the sensor. However, as was seen in the previous section with photoelectrics, the manufacturers of ultrasonic proximity sensors have done a good job of designing sensors that will recognize only the design oscillation frequency of the devices so that random ambient noise is not a problem. Finally, because ultrasonic proximity devices are like sonar, one might think that they would work well underwater, but robotics applications that result in direct wetting of the sensor may cause false outputs and should be avoided.

In summary, it can be seen that proximity sensors, subject to a few limitations, can use a variety of physical concepts that are simply not functional using human capabilities alone. Thus, proximity sensing devices open up new robotics applications that solve problems that may be impossible for humans to solve by themselves.

5   FORCE AND TORQUE SENSORS

The previous two sections, dealing with photoelectrics and proximity sensors, applied principally to the problem of dealing with objects external to the robot, either detecting the presence of these objects or determining their location or orientation. A more intimate mode of sensing deals with the force with which the robot comes into contact with objects. The success of a robotics application often hinges on the robot's ability to detect resistance from the object it is squeezing or attempting to displace. The real world is full of unexpected variation, and the robot often must adjust its movements to adapt to the position of objects or to the arrangement of its environment. The solution to this problem often involves a force sensor. If the resistance encountered is an angular rotation, as in a twisting robot hand, the appropriate choice is a torque sensor.

A common term for a force sensor is *strain gauge*. The mechanical deflection of a solid material is measurable and can be translated into the amount of force applied by direct proportion, taking advantage of Hooke's law. Thus the strain can be gauged and an applied force implied or sensed.

Roboticists sometimes fall into the trap of assuming that force sensors can act effectively as collision-avoidance systems for robots, but this usually does not work. The force sensor detects the deformation after the collision has occurred and the damage is already done. This is analogous to expecting a human to move about without vision or other assistance, blindly bumping into walls and other obstructions and then reacting to the force of the collision. Force sensors have been successful as robotic collision avoidance systems in those applications in which the robot uses one or more probes that encounter the external object before the robot body or hand encounters it.

Perhaps a more appropriate challenge for force and torque sensors is to detect assembly resistance and permit the robot to adjust accordingly. Indeed, force and torque sensors have been very important to the development of assembly applications for industrial robots. Thus, if two mating parts are misaligned, an unusual resistance will be detected and the robot can back off, adjust, and try again at a slightly different position or angle. As mentioned above, the RCC devices described in Chapter 50 might solve this problem without force sensors being resorted to.

If it is determined that a force sensor is indeed needed, there are some characteristics or features to look for. An important consideration is repeatability, i.e., does the sensor register the same each time when identical force is applied? Another consideration is overload protection; the sensor should be equipped with a rigid mechanical stop to prevent a force beyond the sensor's design limits from ruining the sensor. As was seen earlier with photoelectrics and proximity sensors, extraneous force "noise" may need damping to eliminate false signals due to vibrations caused by, say, a collision or initial impact of the robot gripper upon an object.

With force sensors, the objective is often to detect a certain *difference* between forces before and after an operation is applied. In these situations force is applied against the sensor in both states; it is the *difference* in force between the two states that is of interest. In such situations it may be necessary to zero-out the force in a given state as irrelevant to the problem. This is sometimes accomplished by means of a *tare button,* which zeroes out the *tare* force so that it can concentrate on the difference between gross and tare.

One difficulty with force sensors is that the measurable deflection caused by the applied force distorts the robot hand and interferes with its grasping function. The NASA Space Telerobotics Program (Bejczy) claims to have overcome this problem by designing the force sensor in such a way that the grasping force does not disturb the angular alignment of the claws. Although some deflection does occur, both claws translate by the same amount, thereby maintaining angular alignment.

Force sensors for robots can take many forms. The Fukuda Laboratory of Nagoya University, Japan ("Micro Force Sensor," 1997) has announced a micro force sensor less than 2 cm in length for use in intravascular surgery. The sensor, illustrated in Figure 9, is used to detect the force between catheter and blood vessel.

Photoelectrics has been shown to be an important medium of robotic sensing in its own right, but it can also act as the basis for force sensor devices. The University of Arkansas Electro-Optic Sensor (U.S. Patent No. 4,814,562) uses photoelectrics to detect the reduction in the cross-sectional area of one or more holes drilled through flexible gripper contact materials mounted on the robot hand. As force is applied, the holes through the gripper contact material are squeezed so that LED light passing through the hole is reduced. A surprisingly linear relationship is shown to exist between applied force and photoelectric sensor output voltage.

The forces and torques encountered by a robot arm during assembly can be measured directly by using a wrist force sensor, which basically consists of a structure with some compliant sections and transducers that measure the deflections of the compliant sections. Figure 10 shows a strain-gauge-type wrist force sensor built at SRI (Rosen and Nitzan,

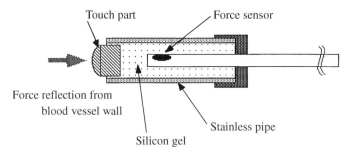

Structure of Micro force sensor

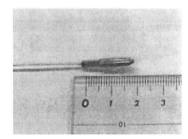

Micro force sensor

Figure 9 Micro Force Sensor for Intravascular Surgery. (Used with permission from *Proceedings of 1997 IEEE International Conference on Robotics and Automation.* © 1997 Fukuda Laboratories.)

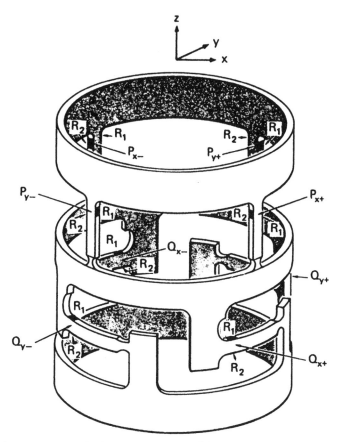

Figure 10   A strain-gauge wrist force sensor built at SRI. It is built from a milled 3-in.-diameter aluminum tube having eight narrow elastic beams with no hysteresis. The neck at one end of each beam transmits a bending torque, which increases the strain at the other end, where it is measured by two foil strain gauges. A potentiometer circuit connected to the two strain gauges produces an output that is proportional to the force component normal to the strain-gauge plates. (Courtesy Rosen and Nitzan.)

1977). It is built from a milled 3-in.-diameter aluminum tube having eight narrow elastic beams with no hysteresis. The neck at one end of each beam transmits a bending torque that increases the strain at the other end where it is measured by two foil strain gauges. A potentiometer circuit connected to the two strain gauges produces an output that is proportional to the force component normal to the strain-gauge planes. This arrangement also automatically compensates for variations in the temperature. This sensor is capable of measuring all three components of force and the three components of torque. Other wrist sensors have been designed by Watson and Drake (1975) and Goto, Inoyama, and Takeyasu (1974). Forces and torques can also be measured with pedestal-mounted sensors, as demonstrated by Watson and Drake (1975).

Forces and torques can also be sensed indirectly by measuring the forces acting on the joints of a manipulator. For joints driven by DC electric motors, the force is directly proportional to the armature current; for joints driven by hydraulic motors, it is proportional to back pressure. Inoue (1971) and Paul and Shimano (1982) have demonstrated assembly using such indirect force sensing. Whereas Inoue programmed a manipulator to insert a shaft into a hole, Paul's work accomplished the assembly of a water pump consisting of a base, a gasket, a top, and six screws. Paul computed the joint forces by measuring the motor current, and his program included compensation for gravity and inertial forces.

6 TACTILE SENSING

Simple force and torque sensing, described in the previous section, are a far cry from the type of sensing we call human touch. The human hand, especially the fingers, together with the analytical abilities of the brain and nervous system, are capable of determining a great deal of detail associated with objects and the associated environment. Determining the shape of an object, for instance, without vision requires a sophisticated sensing system that gathers a large array of force datapoints and draws conclusions from the distribution of these datapoints.

To give robots tactile sensing capability, scientists have relied upon a wide variety of physical principles to accomplish the gathering of large quantities of force sensor array data. Russell (1990) has compiled an excellent review of the various devices for this purpose along with diagrams of the principles of their operation.

Most tactile sensor devices generate a set of binary data for a two-dimensional array. Such devices are capable of registering only an open/closed switching position for any given array position. Although a simple open/closed datapoint does not register quantity of force, except that it exceeds a certain threshold, the arrangement of datapoints in the array conveys a great deal of information about the size, shape, and orientation of an object as held by a robot gripper or lying on a surface at a pick-up point for the robot.

Perhaps the most forbidding challenge in the design of a binary array tactile sensor is to find a switching mechanism tiny enough to fit into a single elemental (pixel) position of the two-dimensional array. Commercially available limit switches, tiny as they have become, are usually too large and too expensive except for tactile arrays of very large dimension and low resolution.

One strategy for achieving the elemental binary switching is to form a metal sheet with an array of bumps, each of which is capable of being depressed with a click at a given threshold force. Because metal is a conductor, each bump so depressed can close a circuit by contacting a terminal point at the corresponding position in a substrate beneath the metal sheet. The force threshold required to collapse the bumps is a design specification that depends upon the sheet material and the bump geometry. Care must be taken, of course, to design the bump geometry such that the bump will snap back to its original convex shape once the force is removed. Another way to alter the force threshold is to apply liquid or gas pressure between the metal bump sheet and its contact substrate, i.e., to inflate the sandwich, so to speak, so that a greater force is required to snap the bump into its deformed, or circuit-closed, state (Garrison and Wang, 1973).

Natural rubber and silicone-based elastomers can be made conductive by imbedding them with silver particles, carbon dust, or other conducting material. When pressure is applied, these flexible materials can close tiny circuits at elemental points distributed about a two-dimensional array. An added advantage of these materials is that they represent a good nonslip gripper surface material that is somewhat tolerant of slight mislocation of the object to be handled by the robot.

Polyvinylidene fluoride (PVF_2) is a piezoelectric material, i.e., it emits electric signals when deformed. Dario (1983) reports a development by Hillis and Hollerback that uses PVF_2 in a tactile sensor illustrated in Figure 11. A practical solution to the problem of dividing the two-dimensional array into tiny discrete elements was achieved by inserting a layer of nylon stocking material between the PVF_2 sheet and the contact substrate.

The subject of tactile sensing is not limited to the two-dimensional array system for describing two-dimensional shapes. The human tactile sense can work with three dimensions and further is sensitive to texture, temperature, weight, rigidity, and density, to name a few. Each of these properties of the object to be sensed is a different problem in terms of the science and technology of robot sensing. The problem of determining texture, for instance, is more one of feeling the vibrations while brushing one's hand across a surface feature. In the case of robotic sensing, the vibrations must be picked up as the robotic sensor is brushed across the target surface. This chapter has only introduced the highlights of tactile sensing and attempted to convey the magnitude and complexity of the problem. More detailed examination of various tactile sensing strategies is suggested in Russell's excellent text on the subject (1990).

7 MISCELLANEOUS SENSORS

This chapter has addressed the mainstream of sensors used in robotics and industrial automation applications. However, other sensors should be mentioned as possibilities for unusual applications of robotics. Mobile robots, for example, are increasing in importance,

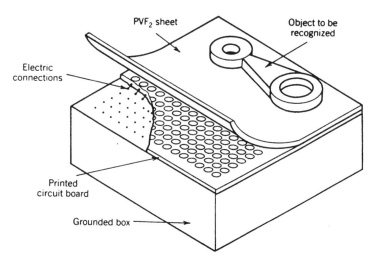

Figure 11 Electrical contact grid used in the design of a tactile sensor. The polyvinylidene fluoride (PVF$_2$) sheet has piezoelectric qualities. (From P. Dario et al., "Piezoelectric Polymers: New Sensor Materials for Robotic Applications," in *Proceedings,* 13th International Symposium on Industrial Robots and Robots 7. Dearborn: Society of Manufacturing Engineers, 1983. Reprinted courtesy of P. Dario and the Society of Manufacturing Engineers, Dearborn, Michigan. © 1983 Society of Manufacturing Engineers.)

though they are not as popular in industrial applications as their fixed-base cousins. For mobile robots, ultrasonic sensors are a convenient means of ranging and positioning as well as avoiding obstacles in the path of the robot (Figueroa and Lamancusa, 1986). Another application for ultrasonic sensors is in seam-tracking for robots used in industrial welding, illustrated in Figure 12 (Figueroa and Sharma, 1991; Mahajan and Figueroa, 1997).

An unusual application of robot sensor technology is the SensorGlove, illustrated in Figure 13, developed and patented by a team at the Technical University of Berlin (Hommel, Hofmann, and Henz, 1994). The sensor incorporates 12 position sensors on the back of the hand and 12 pressure sensors on the palm. The sensors are surprisingly accurate, with resolutions of less than 1° for the position sensors. The objective is sensor-based gesture recognition. To accomplish this, the quantity of precision data made available by the system of sensors is analyzed using various algorithms and computational methods, such as classical statistical techniques, neural networks, genetic algorithms, and fuzzy logic. A remarkable and ambitious application being attempted for this technology is automatic recognition of human sign languages, leading ultimately to the development of a "gesture telephone" for the hearing-impaired. In medicine the SensorGlove can be envisioned as a potential tool for precise telecontrol of surgical robots operating in spaces too small for invasion of the human hand. In dangerous industrial environments or in outer space, the SensorGlove could be used to remotely control robotic-type hands in real time.

8 BUS TECHNOLOGY FOR SENSORS

One of the most exciting developments in technology to appear toward the end of the twentieth century was interconnectivity. Interconnectivity made possible local area networks (LANs) and even the Internet, which connects hundreds of thousands of individual computers around the world, all at the same time. The basis for interconnectivity lies in the concept of bus technology, which uses multiplexing devices to permit a wide variety of communication signals to be transmitted and received on a single pair of conductors. By the use of a variety of coding protocols, communication packets can be addressed to various receiving points and synchronized in such a way as to employ a common conductor and eliminate thousands of point-to-point direct, parallel connections between the devices.

Front View   Side View

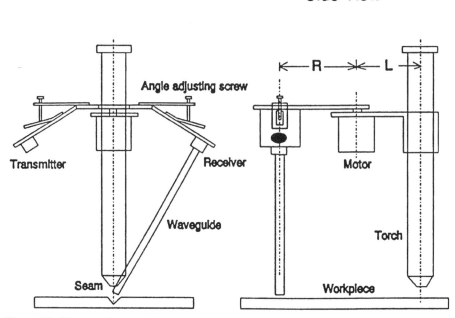

Figure 12 Diagram of setup of an ultrasonic seam-tracking system for industrial welding. (Used with permission from A. Mahajan and F. Figueroa, "Intelligent Seam Tracking Using Ultrasonic Sensors for Robotic Welding," *Robotica* **15**, 275–281. © 1997 Cambridge University Press.) (Figueroa and Sharma, 1991.)

At the very end of the century, bus technology, which had been used to connect computers, began to be seen as a way to interconnect all kinds of control devices, such as switches, robotic sensors, and other industrial automated system field devices. The conventional way of controlling these devices had been to use parallel interconnections between each device and a central controller, usually a PLC. The PLC typically has banks of input/output terminals for independent addressing of each sensor input and actuator output. The PLC can serve as the robot controller, sensing inputs and controlling joint movements as outputs. In another configuration the PLC acts as a supervisory controller for the robot and other associated processing equipment in a workstation or even for banks of robots within a system.

The numbers of interconnectors in a conventional parallel connection scheme can be staggering. It is estimated (Svacina, 1996) that a single automotive assembly line could have as many as 250,000 termination points for connecting field devices to their controllers. As additional robotics and automatic designers dream up new ways for sensors to be connected to controllers and industrial robots, the number of termination points will increase. With this increase, connectivity for sensors and other automation field devices will be effected using bus technology in a fashion similar to the development of area networks for computers.

At the time of this writing (near the end of the twentieth century), it is too early to predict the significance of this development or where it will carry the field of industrial robotics. Svacina (1996) has hailed it as the "Fifth Stage of the Industrial Revolution," behind water, steam, electricity, and programmable controllers, respectively. It is certain that bus technology will make robotic sensors more economically feasible, easier to implement, more flexible, and easier to replace if something goes wrong.

9 SENSOR FUSION

This chapter would not be complete without recognizing research on the integration of sensors into a synergistic intelligent system that permits the robotic system to understand

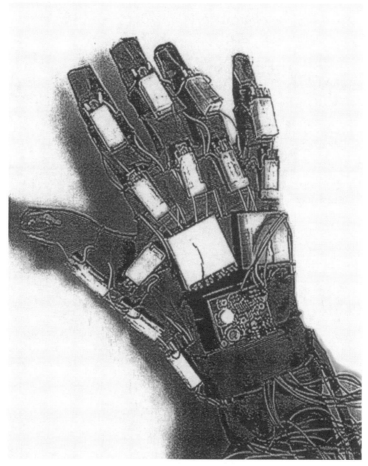

Figure 13   High-precision SensorGlove. (© 1997 Frank Hofmann, Department of Computer Science, Technical University of Berlin; used with permission.) (Hommel, Hofmann, and Jurgen, 1994.)

its environment in much the same way that humans do. This area of study is generally called *sensor fusion,* which has been defined as "the coordination of data from diverse sources to produce a usable perspective" (Lawrence, 1997). The term *diverse sources* can be interpreted to mean not only single-dimensional points of data, such as temperature or pressure, but also multidimensional sources, such as pixel arrays in various media, exemplified by ordinary television images or perhaps infrared pixel arrays.

A familiar area in which various sensor media have been used to interpret an image is the field of astronomy. Casting the image in various media perspectives provides a clearer understanding of the object under observation. Another application area is using spatial sound and color information to track human beings on video camera and automatically point the camera at the person talking (Goodridge, 1997).

For an industrial robot to make practical use of various sensor media, the fusion of the various media must be performed in real time, especially when the robot is using sensor fusion for navigation. Ghosh and Holmberg (1990) report a multisensor system for robot navigation that uses neural networks to perform sensor data fusion. Their research investigated several types of neural networks for this application and compared speed of learning and effectiveness under varying input conditions. Another architecture for intelligent sensing for mobile robots was developed by Mahajan and Figueroa (1997). The system was designed for a four-legged robot, but the intelligent integration of sensors could work for a wheeled robot as well. The scheme of this architecture is illustrated in Figure 14.

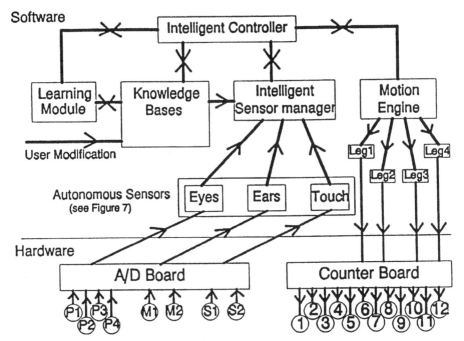

Figure 14 System architecture for integration of sensor data for a mobile robot. (Used with permission from A. Mahajan and F. Figueroa, "Four-Legged Intelligent Mobile Autonomous Robot" *Robotics and Computer-Integrated Manufacturing* **13**(1), 51–61, 1997. © 1997 Elsevier Science Ltd.)

General guidelines or principles, known as *sensor economy principles,* have been developed by Edan and Nof (1995) to facilitate the effective integration of multiple sensors into manufacturing systems, including industrial robots. Multiple sensors make possible redundancy, which in turn leads to increased reliability of the system. The redundancy goes beyond the simple backup of failed sensor components. Sensors are subject to noise and measurement errors, depending upon their type and position in an application. Applying systematic analysis using design principles and evaluating the tradeoffs using cost/benefit analyses lead to better utilization of sensor technology and higher-performance, more cost-effective systems.

10   SENSOR SELECTION FOR APPLICATIONS

An objective of this handbook is to present not only theory but applications for robotics. Toward this end, this section attempts to assist the reader in selecting the appropriate robotic sensor for a given application. In making such an attempt, one must acknowledge that the roboticist has many feasible alternatives to consider in sensor selection. Only by fully understanding the various strengths and weaknesses of the various devices examined in this chapter, along with a review of current prices and costs of alternatives, can the roboticist optimize the selection of the ideal sensor for a given robotics application.

Recognizing the limitation of absolutes, we present a decision tool in Figure 15 for assisting the reader in placing in perspective the various sensors discussed in this chapter. It should be remembered that a combination of available sensors may represent the best solution. With the use of sensor fusion, a combination of sensors can represent an intelligent synergism that is more effective than a simple association of various components.

11   MACHINE VISION

The field of machine vision can be considered an extension of robot sensing. There are many misconceptions about machine vision, the principal one being that machine vision is and should be capable of duplicating human vision. Another basic misconception sur-

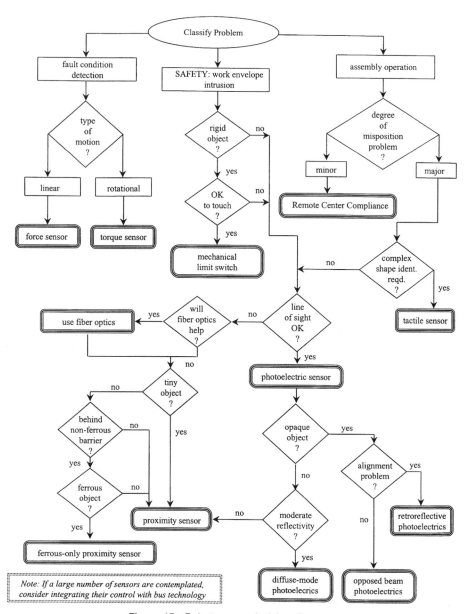

Figure 15 Robot sensor decision diagram.

rounds the word *vision* itself. Vision is about the production of images, but the vision process involves the mind and reasoning processes as much as it involves producing an image. Applications of machine vision should be developed in much the same way as applications of more basic forms of robot sensing, by establishing objectives and then designing the system to accomplish those objectives and no more.

The Optimaster Machine Vision system, shown in Figure 16, illustrates the basic components of a machine vision system and their interaction. A camera is typically used to gather data by scanning a pixel at a time into an image array. The task is to assign a digital, more specifically binary, value to each pixel. The array need not be two-dimensional. A single scan line across a field of view may yield the essential information

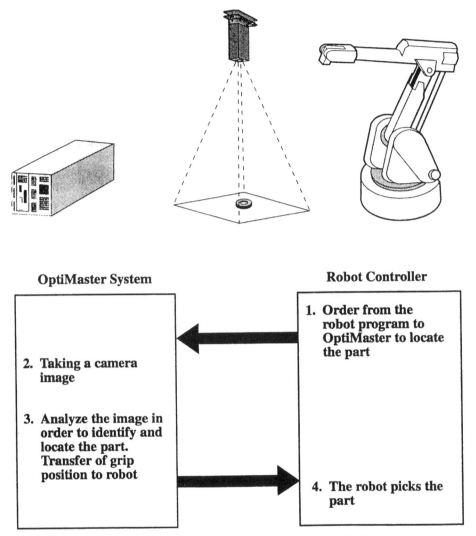

Figure 16   OptiMaster Machine Vision System (courtesy OptiMaster).

desired in the robotics application. If the objective of the system can be accomplished with a one-dimensional scan, there is no need to use additional memory to scan a two-dimensional image.

The amount of data contained in the individual pixel is a function of the number of bits of computer storage assigned to each element in the array. For example, a black and white image can be stored in an array consisting of only one binary bit for each element in the array. Various shades of light content, called *grayscale,* can be distinguished by allocating several binary digits to each pixel. An array using 6 binary digits per pixel can represent $2^6 = 64$ levels of gray scale.

Lighting is an important consideration when generating a machine vision image. A key to the success of the system is the consistency of the lighting. Naturally, if the illumination source is variable, the computer controller monitoring the image can become confused about the content of the image. Dedicated lighting is usually preferable to ambient light in this respect. Another consideration is whether to use direct lighting or backlighting. The former is preferred for distinguishing features on an object and the latter is better for shape identification.

Once the image has been stored in a digital array, analysis can be used to extract the desired data from the image. This is the area in which many advances in software have greatly increased the functionality of machine vision systems in recent decades. One of the first steps in analysis is to use *windowing* to select a small area of the image in which to focus the analysis. The area of interest may be identified by its position in the array if the object is rigid and fixtured during the image acquisition. This is called *fixed* windowing, as contrasted with *adaptive* windowing, in which the window is selected by its relationship to other features in the image. One or more fiducial points may be used to establish the position and orientation of the image before identifying the window of interest.

Once a window of interest has been identified, *thresholding* can be used to enhance the features or add contrast. Thresholding actually reduces the grayscale discrimination in a prescribed way. One advantage of thresholding is its ability to screen out the operational variability of the scanning system. The effectiveness of the thresholding scheme

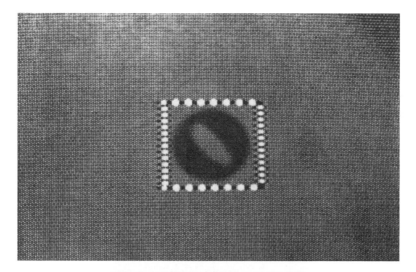

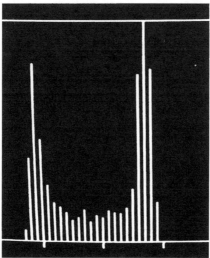

Figure 17   Bimodal histogram of a dark object on a light background. (From C. Ray Asfahl, *Robots and Manufacturing Automation*, rev. ed. New York: John Wiley & Sons, 1992. Used by permission of John Wiley & Sons.)

is dependent upon the selection of the threshold level for discrimination. The thresholding process can be rendered useless if the threshold is too low, as all pixels will be designated as light, or too high, as all pixels will be designated dark. Some applications may call for multiple thresholds, especially if the background is of variable grayscale.

Another method of analysis readily accomplished by computer is *histogramming*. Ignoring the position of the pixels within the array, histogramming tabulates frequencies for various grayscale levels found throughout the array. Histogramming is especially useful for identification. A sample histogram is illustrated in Figure 17. The left peak in the histogram corresponds to the dark image and the right peak corresponds to the lighter background.

Shape identification is accomplished by a variety of analysis tricks, such as run-lengths, template matching, and edge detection. The run-length method uses a variety of line-scans through the image to count run lengths of black versus white pixels. Depending upon the shape of the object being scanned, the run-lengths can switch from black to white and vice versa several times. The machine vision analysis system can count these switches back and forth to identify a characteristic shape. Run-length analysis is particularly effective for finding the orientation of an object of known geometry. The process is illustrated in Figure 18.

Template matching uses a known pixel array pattern to compare against a captured image of the object to be identified. The comparison is pixel-by-pixel, so the process can be time-consuming, even by high-speed computer. One trick to speed up the process is

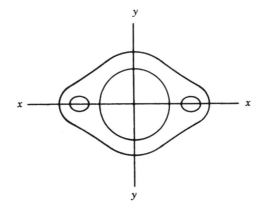

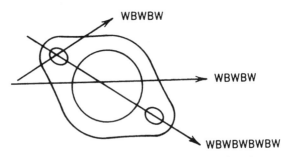

Figure 18 Using a shape identification algorithm to select and properly orient gaskets on a conveyor line. (From C. Ray Asfahl, *Robots and Manufacturing Automation,* rev. ed. New York: John Wiley & Sons, 1992. Used by permission of John Wiley & Sons.)

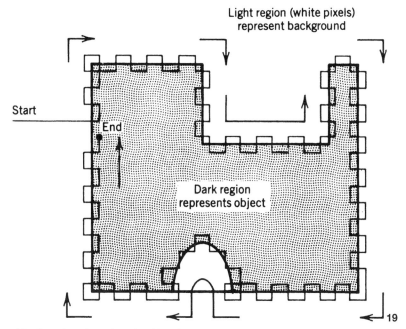

Light region (white pixels)
represent background

Dark region
represents object

Figure 19   An edge detection algorithm finds and then, by continually crossing it, follows the edge completely around the object. (From C. Ray Asfahl, *Robots and Manufacturing Automation,* rev. ed. New York: John Wiley & Sons, 1992. Used by permission of John Wiley & Sons.)

to focus on a unique and identifying feature and limit the template matching process to this area alone. In other words, the windowing process can be used together with template matching to make a much more efficient search than by using template matching alone.

Algorithms have been devised for creeping around the perimeter of an object by advancing a pixel at a time and performing a test to determine whether the new pixel has changed from dark to light or vice versa. If the new pixel is dark, for instance, the algorithm can direct the search to turn left, and if the new pixel is light, the algorithm can direct the search to turn right. In this fashion, program control can keep following close to the edge, even turning corners as they are encountered. The process is illustrated in Figure 19.

Toward the end of the twentieth century, machine vision systems have become much more affordable without sacrificing capability. Earlier systems employed dedicated computers with rigid, proprietary software. The newer systems use a more open architecture, allowing the system to be based on an ordinary personal computer (PC). Machine vision requires computer operating cycles too fast for older-model PCs, but newer models can accommodate the demands of machine vision. The open architecture is more amenable to assembling systems from a variety of peripherals, keeping the costs down. The resulting systems are more flexible and have the advantage of being more readily upgraded, component by component, as needed. Gregory (1997) sees a trend in this development. The speed of PC processors will continue to improve at a rapid pace, while the speed requirements of machine vision applications will be constrained by practical considerations to slower speeds. The result will be an increasing demand for PC-based machine vision systems as each new application moves into the realm of economic feasibility.

REFERENCES

Agin, G. J., 1985. "Vision Systems." In *Handbook of Industrial Robotics.* Ed. S. Y. Nof. New York: John Wiley & Sons. 231–261.

Asfahl, C. R. 1992. *Robots and Manufacturing Automation.* Rev. ed. New York: John Wiley & Sons.

Bejczy, A. "Grasping-Force Sensor for Robot Hand." NASA Space Telerobotics Program. [Internet address: ⟨http://ranier.hq.nasa.gov/telerobotics_page/Technologies/0203.html⟩]

Chironis, N. P. 1966. *Machine Devices and Instrumentations.* New York: McGraw-Hill.

Dario, P., et al. 1983. "Piezoelectric Polymers: New Sensor Materials for Robotic Applications." In *Proceedings, 13th International Symposium on Industrial Robots and Robots 7.* Vol. 2. Dearborn: Society of Manufacturing Engineers.

Doeblin, E. O. 1966. *Measurement System: Application and Design.* New York: McGraw-Hill.

Drake, S. N., P. C. Watson, and S. N. Simunovic. 1977. "High Speed Assembly of Precision Parts Using Compliance Instead of Sensory Feedback." In *Proceedings of 7th International Symposium of Industrial Robots.* 87–98.

Edan, Y., and S. Nof. 1995. "Sensor Selection Procedures and Economy Principles." In *Proceedings of the 13th International Conference on Production Research,* Jerusalem, August 6–10. London: Freund.

Ernst, H. A. 1962. "MH-1, a Computer-Operated Mechanical Hand." In *AFIPS Conference Proceedings, SJCC,* 39–51.

Everett, H. R. 1995. *Sensors for Mobile Robots.* Wellesley: A. K. Peters.

"Factory Robot Stabs Worker." 1981. *Athens [Ohio] Messenger,* December 8.

Figueroa, F., and J. S. Lamancusa. 1986. "An Ultrasonic Ranging System for Robot Position Sensing." *Journal of the Acoustical Society of America* **80**.

Figueroa, J. F., and S. Sharma. 1991. "Robotic Seam Tracking with Ultrasonic Sensors." *International Journal of Robotics and Automation* **6**(1), 35–40.

Garrison, R. L., and S. S. M. Wang. 1973. "Pneumatic Touch Sensor." *IBM Technical Disclosure Bulletin* **16**(6), 2037–2040.

Ghosh, J., and R. Holmberg. 1990. "Multisensor Fusion using Neural Networks." In *Proceedings of the Second IEEE Symposium on Parallel and Distributed Computing,* Dallas, December. 812–815.

Goodridge, S. G. 1997. "Multimedia Sensor Fusion for Intelligent Camera Control and Human-Computer Interaction." Ph.D. Dissertation, North Carolina State University.

Goto, T., T. Inoyama, and K. Takeyasu. 1974. "Precise Insert Operation by Tactile Controlled Robot HI-T-HAND Expert-2." In *Proceedings of 4th International Symposium of Industrial Robots.* 209–218.

Gregory, R. 1997. "The Rise of PC-Based Machine Vision." *Robotics World* (Winter).

Harris, C. M., and C. E. Crede. 1961. *Shock and Vibration Handbook.* New York: McGraw-Hill.

Hill, J. W., and A. J. Sword. 1973. "Manipulation Based on Sensor-Directed Control: An Integrated End Effector and Touch Sensing System. In *Proceedings of 17th Annual Human Factor Society Convention,* Washington, October.

Hommel, G., F. G. Hofmann, and J. Henz. 1994. "The TU Berlin High-Precision SensorGlove." In *Proceedings of the WWDU '94, Fourth International Scientific Conference,* University of Milan, Milan. Vol. 2. F47–F49.

Industrial Switches and Sensors. Freeport: MICRO SWITCH, Honeywell, Inc., August.

Inoue, H. 1971. "Computer Controlled Bilateral Manipulator." *Bulletin of the Japanese Society of Mechanical Engineering* **14**, 199–207.

Kak, A. C., and James S. Albus. 1985. "Sensors for Intelligent Robots." In *Handbook of Industrial Robotics.* Ed. S. Y. Nof. New York: John Wiley & Sons. 214–230.

Lawrence, P. N. 1997. "Sensor Fusion." Austin: Lawrence Technologies. [Internet address: ⟨http://www.khoros.unm.edu/khoros/submissions/Lawrence/lawrence.html⟩]

Lion, K. S. 1959. *Instrumentation in Scientific Research.* New York: McGraw-Hill.

Mahajan, A., and F. Figueroa. 1997. "Four-Legged Intelligent Mobile Autonomous Robot." *Robotics and Computer-Integrated Manufacturing* **13**(1), 51–61.

———. 1997. "Intelligent Seam Tracking Using Ultrasonic Sensors for Robotic Welding." *Robotica* **15**, 275–281.

"Micro Force Sensor for Intravascular Surgery." In *Proceedings of 1997 IEEE International Conference on Robotics and Automation,* Albuquerque, April 1997. 1561.

Nevins, J. L., et al. *Exploratory Research in Industrial Assembly and Part Mating.* Report No. R-1276, March 1980. Cambridge: Mass.: Charles Stark Draper Laboratory.

OptiMaster Product brochure, February 1996.

Parks, M. 1994. "Man Hurt by Robot Dies." Fayetteville, AR: *Northwest Arkansas Times,* October.

Paul, R., and B. Shimano. 1982. "Compliance and Control." In *Robot Motion.* Ed. M. Brady et al. Cambridge: MIT Press. 405–417.

Photoelectric Controls. Minneapolis: Banner Engineering Corp., 1996.

Proximity Sensors. Minneapolis: TURCK, Inc., 1994.

Rosen, C. A., and D. Nitzan. 1977. "Use of Sensors in Programmable Automation." *Computer* (December), 12–23.

Russell, R. A. 1990. *Robot Tactile Sensing.* Upper Saddle River: Prentice-Hall.

Svacina, B. 1996. *Understanding Device Level Buses.* Minneapolis: TURCK.

Takeda, S. 1974. "Study of Artificial Tactile Sensors for Shape Recognition Algorithm for Tactile Data Input." In *Proceedings of 4th International Symposium on Industrial Robots.* 199–208.

Tanimoto, M., et al. 1997. "Micro Force Sensor for Intravascular Surgery." In *Proceedings of 1997 IEEE International Conference on Robotics and Automation,* Albquerque, April. 1561.

Watson, P. C., and S. H. Drake. 1975. "Pedestal and Wrist Force Sensors for Automatic Assembly." In *Proceedings of 5th International Symposium on Industrial Robots.* 501–511.

CHAPTER 14

STEREO VISION FOR INDUSTRIAL APPLICATIONS

Akio Kosaka
A. C. Kak
Purdue University
West Lafayette, Indiana

1 INTRODUCTION TO STEREO VISION

Depth information is vital for enhancing the level of automation of present-day program-mable manipulators. Depth perception is, for example, needed for bin-picking.[1] After all, who can question the fact that a robust and affordable solution to the problem of automatic bin-picking would obviate the need for expensive fixturing and significantly reduce the complexity of parts feeders?

Depth information can be extracted with a number of competing technologies. Some of the most impressive results to date have been obtained with structured light imaging (Grewe and Kak, 1995; Kim and Kak, 1991; Chen and Kak, 1989). This approach consists of scanning a scene with a laser stripe, capturing the image of the stripe with an off-set camera, and then, through triangulation, calculating the 3D coordinates of each of the illuminated points in the scene. Figure 1 shows an illustration of 3D vision with this approach.

While systems based on the structured-light principle are inexpensive and highly ro-bust, their principal shortcoming is that they can be used only when the ambient light does not interfere excessively with the laser illumination. Also, some object surfaces with mirror finishes cannot be imaged by laser stripes in this manner, since such surfaces send practically all of the incident light in one direction, which may not be toward the camera. For these reasons, it is generally believed that the technology of the future for depth perception will be based on stereo or multicamera imaging in which no laser sources are needed to illuminate a scene. For this reason, this chapter will focus on the art and science of stereo imaging.

The basic principle underlying depth perception via stereo is simple (Kak, 1985), but its robust implementation has proved problematic. All that is needed for stereo is a pair of corresponding pixels in two different cameras looking at the same scene. *Correspond-ing pixels* refers to two pixels in two different cameras being the images of the same object point in the scene. Given the image coordinates of two corresponding pixels, the coordinates of the object point can be calculated straightforwardly. Exploitation of this principle is challenging because finding which pixel pairs in the two images are corre-sponding pixels is a nontrivial task. Finding which pixels in the two images go together, in the sense that they are the projections of the same object point, is called the *corre-spondence problem.*

Any solution to the correspondence problem is facilitated by the realization that a corresponding pair of pixels in the two images must lie on what are known as *epipolar lines.* Using the concept of epipolar lines when searching in the right image for the corresponding pixel of a left-image pixel, one can limit the search to a single line in the

---

[1]By *bin-picking* we mean the ability to pick up a part from a random orientation and position within the work environment of the manipulator. The part may present itself to the manipulator as a single object against a uniform background, or, in more complex scenarios, as one of many objects in a bin.

Handbook of Industrial Robotics, Second Edition, Edited by Shimon Y. Nof
ISBN 0-471-17783-0 © 1999 John Wiley & Sons, Inc.

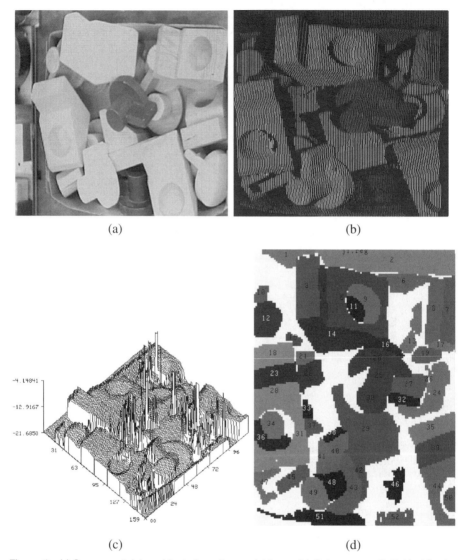

Figure 1 (a) Scene containing objects from the model-base. (b) Color-composite light-stripe image of the scene. (c) Three-dimensional plot of the points detected. (d) The result of the segmentation process.

right image. But even if one takes advantage of epipolar lines, the correspondence problem remains intractable for many practical applications for the simple reason that for a given pixel in, say, the left image there are often too many contenders on the epipolar line in the right image. Over the years various strategies have been proposed to further constrain the search for the corresponding pixel. Some of the better-known strategies are the *ordering constraint* and the *disparity smoothness constraint* (Faugeras, 1993; Grewe and Kak, 1994). The results obtained from these strategies have been at best mixed. We believe that better results have been obtained by using more than two viewpoints in what is known as *multicamera stereo*. For example, if we use three cameras instead of just two, we can solve the correspondence problem more robustly as follows: For a given pixel in camera 1, we can test the suitability of a candidate pixel in camera 2 by checking whether or not there exists in camera 3 a pixel at the intersection of the two epipolar lines, one associated with the pixel in camera 1 and the other associated with the pixel in camera 2 (Faugeras, 1993). Another approach to multicamera stereo consists of keeping one camera fixed and moving a second camera in small steps along a straight line while recoding images for each position of the second camera (Okutomi and Kanade, 1993). When the two cameras are close together, solving the correspondence problem is relatively easy because the corresponding pixels will be at approximately the same coordinates in the two images. (But it can be shown, unfortunately, that the accuracy of depth calculation suffers when the two cameras are close together.) For subsequent positions of the second camera, the solution to the correspondence problem is aided by the solutions discovered for the near positions.

This chapter focuses on what has proved so far to be the most robust approach to stereo vision for industrial applications: model-based stereo. This approach is made robust by the use of object-level knowledge in the stereo fusion process. More specifically, the stereo system uses object knowledge to extract from each of the images higher-level pixel groupings that correspond to discernible features on one or more surfaces of the object, then tries to solve the correspondence problem using these features. Because these features tend to be distinctive, the problem of contention during the establishment of correspondence is minimized. More specifically, the approach discussed in this chapter is characterized by the following:

- It is not necessary that 3D objects be represented by their full 3D CAD models for model-based stereo vision. We will show how much-simplified models suffice even for complicated-looking industrial objects. An object model need contain only those features that would be relevant to the stereo matching process.

- The object features retained in the model should allow the calculation of the 3D pose of the object. Suppose the most prominent features on the object were three bright red dots that happened to be collinear. Since their collinearity would make mathematically impossible the calculation of the object pose, such features would not suffice.

- An advantage of this approach is that the vision system needs to extract from the images only those features that can potentially be the camera projections of the object features retained in the model.

- The features extracted from each of the images can be fused to construct scene features, which can then be matched to the model features for the generation of pose hypotheses for the object.

- The pose hypotheses for the objects can be verified by the finding of supporting model features.

Model-based stereo based on the above rationale would possess the following operational advantages:

- Because the stereo process would be driven by a small number of simple object features, there would be no need to employ the full 3D CAD models of objects. Therefore, the overall complexity of the system would be much reduced by the use of simple visual cues that can be used for recognition and localization of the objects.

- Compared to model-based monocular vision, this approach permits robust estimation of object pose even for relatively small objects.

- Because the approach does not depend on rich 3D CAD descriptions of objects, it is more robust with respect to occlusion. Because only a small number of the object features are included in its model, the system should work as long as the cameras can see a reasonable subset of these features.

- The 3D geometric constraints that are invoked when comparing scene features with model features reduce the possibility of misregistration of image features with model features.

Section 2 presents mathematical preliminaries dealing with camera calibration and 3D depth estimation. Section 3 presents an overall architecture for a model-based stereo vision system. Sections 4 and 5 present two examples of model-based stereo vision for industrial applications.

2 MATHEMATICAL PRELIMINARIES

2.1 Camera Calibration

Camera calibration is always a tedious task in computer vision. If the goal were merely to recognize objects, then, depending on the methodology used, it would be possible to devise vision systems that did not require that the cameras be calibrated. But for many important cases of geometric reasoning, especially when it is necessary for the robot to interact with the scene, it is difficult to circumvent calibration of the cameras. This subsection presents mathematical preliminaries necessary for understanding camera calibration. A description of homogeneous transformations between coordinate frames will be followed by a description of a camera model often used in the robot vision community.

2.1.1 Homogeneous Coordinate Frame

Let us assume that we have two coordinate frames W and C whose origins are represented by O_w and O_c, as shown in Figure 2. The three orthogonal axes of the coordinate frame W are represented by $\mathbf{i}_w$, $\mathbf{j}_w$, and $\mathbf{k}_w$, whereas those of C are represented by $\mathbf{i}_c$, $\mathbf{j}_c$, and $\mathbf{k}_c$. (Note that these three axis vectors in each case exist in some underlying reference frame.) If point (x_c, y_c, z_c) in the coordinate frame C corresponds to point (x_w, y_w, z_w) in W, then we can represent the relationship between the two coordinates by the following equation (Fu, Gonzalez, and Lee, 1987; McKerrow, 1990):

$$\begin{bmatrix} x_w \\ y_w \\ z_w \\ 1 \end{bmatrix} = \begin{bmatrix} \mathbf{i}_w^T \mathbf{i}_c & \mathbf{i}_w^T \mathbf{j}_c & \mathbf{i}_w^T \mathbf{k}_c & p_x \\ \mathbf{j}_w^T \mathbf{i}_c & \mathbf{j}_w^T \mathbf{j}_c & \mathbf{j}_w^T \mathbf{k}_c & p_y \\ \mathbf{k}_w^T \mathbf{i}_c & \mathbf{k}_w^T \mathbf{j}_c & \mathbf{k}_w^T \mathbf{k}_c & p_z \\ 0 & 0 & 0 & 1 \end{bmatrix} \begin{bmatrix} x_c \\ y_c \\ z_c \\ 1 \end{bmatrix} \tag{1}$$

where (p_x, p_y, p_z) represents the x, y, z coordinates of the origin of C measured in the coordinate frame W and the matrix shown in the equation, represented by $_wH_c$ is called the *homogeneous transformation matrix* from the C coordinate frame to the W coordinate frame.

The homogeneous transformation matrix shown in Equation (1) is actually decomposed into two different terms—a rotation matrix R and a translation vector $\mathbf{t}$—and can be rewritten as

$$_wH_c = \begin{bmatrix} R & \mathbf{t} \\ 0 & 1 \end{bmatrix} \tag{2}$$

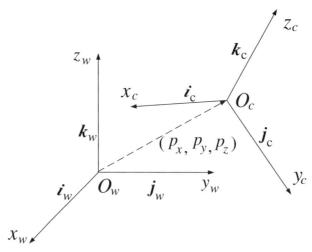

Figure 2 Relationship between two coordinate frames.

Let us consider an object-centered coordinate frame M corresponding to an object located somewhere in the world coordinate frame. The object pose estimation problem is then defined as follows: Given the world coordinate frame W and the camera coordinate frame C, estimate the transformation $_wH_m$ from the object coordinate frame to the world coordinate frame using visual measurement from the camera coordinate frame C to the object coordinate frame M.

2.1.2   Pinhole Camera Model

Let us assume that a camera is located in the world coordinate frame W and the camera has its own coordinate frame C whose origin is located on the camera focal point and the camera optic axis coincides with the z axis of the camera coordinate as shown in Figure 3.

Now we apply a pinhole camera model to express the relationship between a 3D world point (x_w, y_w, z_w) and its projected camera image point (u, v). Assume that the world point (x_w, y_w, z_w) is represented by (x_c, y_c, z_c) in the camera coordinate frame. We can write:

$$
\begin{bmatrix} uw \\ vw \\ w \end{bmatrix} = \begin{bmatrix} \alpha_u & 0 & u_0 & 0 \\ 0 & \alpha_v & v_0 & 0 \\ 0 & 0 & 1 & 0 \end{bmatrix} \begin{bmatrix} x_c \\ y_c \\ z_c \\ 1 \end{bmatrix} \tag{3}
$$

where α_u and α_v represent the scaling factors of the image in u and v directions, and (u_0, v_0) represents the center of the camera image plane. The 3×4 matrix in the equation is called the *intrinsic camera calibration matrix*. In terms of the world point (x_w, y_w, z_w) in the world coordinate frame, the transformation can be represented in the following form:

$$
\begin{bmatrix} uw \\ vw \\ w \end{bmatrix} = \begin{bmatrix} \alpha_u & 0 & u_0 & 0 \\ 0 & \alpha_v & v_0 & 0 \\ 0 & 0 & 1 & 0 \end{bmatrix} \begin{bmatrix} h_{11} & h_{12} & h_{13} & h_{14} \\ h_{21} & h_{22} & h_{23} & h_{24} \\ h_{31} & h_{32} & h_{33} & h_{34} \\ h_{41} & h_{42} & h_{43} & h_{44} \end{bmatrix} \begin{bmatrix} x_w \\ y_w \\ z_w \\ 1 \end{bmatrix} \tag{4}
$$

In this equation, the homogeneous transformation matrix consisting of the elements h_{ij}'s will be denoted $_cH_w$. This matrix represents the transformation from the world coordinate frame to the camera coordinate frame. The 3×4 matrix C

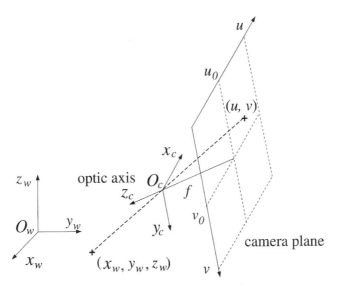

Figure 3   Pinhole camera model and coordinate frames.

$$
C = \begin{bmatrix} \alpha_u & 0 & u_0 & 0 \\ 0 & \alpha_v & v_0 & 0 \\ 0 & 0 & 1 & 0 \end{bmatrix} \begin{bmatrix} h_{11} & h_{12} & h_{13} & h_{14} \\ h_{21} & h_{22} & h_{23} & h_{24} \\ h_{31} & h_{32} & h_{33} & h_{34} \\ h_{41} & h_{42} & h_{43} & h_{44} \end{bmatrix} \tag{5}
$$

is called a *camera calibration matrix.*

The problem in camera calibration is to obtain the homogeneous transformation matrix $_cH_w$ and the intrinsic camera parameters α_u, α_v, u_0, and v_0 from a set of 3D landmark points and their projected image points in the camera image plane. Many techniques have been presented in the literature for this purpose (Tsai, 1988; Faugeras, 1993; Weng, Cohen, and Herniou, 1992).

2.2 Three-Dimensional Depth Estimation from Stereo Correspondence

We will now discuss how we can obtain the 3D depth information from a stereo correspondence. Let us assume that two cameras (left and right cameras) are calibrated by a pinhole model. Let $C = (c_{ij})$ and $D = (d_{ij})$ be the camera calibration matrices for the left and right cameras. Recall that these matrices include the homogeneous transformations from the world frame to the camera frames.

Let $P(x, y, z)$ be a 3D point in the world coordinate frame and $Q_a(u_a, v_a)$ and $Q_b(u_b, v_b)$ be the 2D points projected onto the left and right camera image planes.

From Equation (4), we can express the relationship between the 3D point P and the projected points Q_a and Q_b as follows:

$$
\begin{bmatrix} u_q w_a \\ v_a w_a \\ w_a \end{bmatrix} = C \begin{bmatrix} x \\ y \\ z \\ 1 \end{bmatrix} \tag{6}
$$

$$
\begin{bmatrix} u_b w_b \\ v_b w_b \\ w_b \end{bmatrix} = D \begin{bmatrix} x \\ y \\ z \\ 1 \end{bmatrix} \tag{7}
$$

Conversely, let us consider the case that we are given a 2D point $Q_a(u_a, v_a)$ in the left image. Let $P(x, y, z)$ be the 3D point that is projected to the point Q_a. Then the 3D point $P(x, y, z)$ must lie on a 3D line in the world coordinate frame, called the *line of sight*. Let α be this line. The projection of this line α onto the right camera image plane forms a line, called the *epipolar line* with respect to the point Q_a. We will now derive the mathematical representation of this epipolar line of Q_a in the right image. Figure 4 shows the epipolar line associated with the point $Q_a(u_a, v_a)$.

Let us assume that arbitrary point on the epipolar line is represented by $Q_b(u_b, v_b)$ in the right image. Then the following two equations must be satisfied:

$$
\begin{bmatrix} x \\ y \\ z \end{bmatrix} = \begin{bmatrix} c_{11} & c_{12} & c_{13} \\ c_{21} & c_{22} & c_{23} \\ c_{31} & c_{32} & c_{33} \end{bmatrix}^{-1} \begin{bmatrix} u_a w_a - c_{14} \\ v_a w_a - c_{24} \\ w_a - c_{34} \end{bmatrix} \tag{8}
$$

$$
\begin{bmatrix} u_b w_b \\ u_b w_b \\ w_b \end{bmatrix} = \begin{bmatrix} d_{11} & d_{12} & d_{13} \\ d_{21} & d_{22} & d_{23} \\ d_{31} & d_{32} & d_{33} \end{bmatrix} \begin{bmatrix} c_{11} & c_{12} & c_{13} \\ c_{21} & c_{22} & c_{23} \\ c_{31} & c_{32} & c_{33} \end{bmatrix}^{-1} \begin{bmatrix} u_a w_a - c_{14} \\ v_a w_a - c_{24} \\ w_a - c_{34} \end{bmatrix} + \begin{bmatrix} d_{14} \\ d_{24} \\ d_{34} \end{bmatrix} \tag{9}
$$

Let

$$
A = \begin{bmatrix} d_{11} & d_{12} & d_{13} \\ d_{21} & d_{22} & d_{23} \\ d_{31} & d_{32} & d_{33} \end{bmatrix} \begin{bmatrix} c_{11} & c_{12} & c_{13} \\ c_{21} & c_{22} & c_{23} \\ c_{31} & c_{32} & c_{33} \end{bmatrix}^{-1} \qquad \mathbf{b} = A \begin{bmatrix} -c_{14} \\ -c_{24} \\ -c_{34} \end{bmatrix} + \begin{bmatrix} d_{14} \\ d_{24} \\ d_{34} \end{bmatrix} \tag{10}
$$

Then

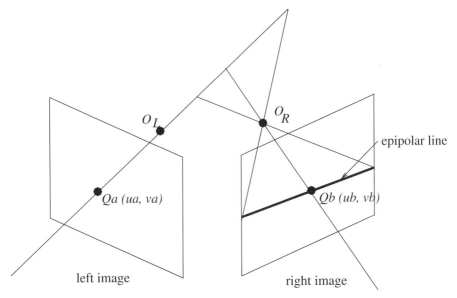

Figure 4   The epipolar line for point $Q_a(u_a, v_a)$ appears in the right image.

$$\begin{bmatrix} u_b w_b \\ v_b w_b \\ w_b \end{bmatrix} = A \begin{bmatrix} u_a w_a \\ v_a w_a \\ w_a \end{bmatrix} + \mathbf{b} \tag{11}$$

We define another vector $\mathbf{a}$ as

$$\mathbf{a} = A \begin{bmatrix} u_a \\ v_a \\ 1 \end{bmatrix} \tag{12}$$

Then

$$\begin{bmatrix} a_1 & b_1 & u_b \\ a_2 & b_2 & v_b \\ a_3 & b_3 & 1 \end{bmatrix} \begin{bmatrix} w_a \\ 1 \\ -w_b \end{bmatrix} = 0 \tag{13}$$

The determinant of the above matrix should be 0. Therefore,

$$(a_2 b_3 - a_3 b_2)u_b + (a_3 b_1 - a_1 b_3)v_b + (a_1 b_1 - a_1 b_0) = 0 \tag{14}$$

This equation represents a single line in the right camera image—the epipolar line. Once we obtain the epipolar line, we can use this line to verify whether or not left point Q_a is geometrically matched with right point Q_b, since the location of right point Q_b is constrained to be on the line. This constraint, called *epipolar line constraint*, is discussed below.

For depth perception, a stereo correspondence of the left and right image points like $Q_a(u_a, v_a)$ and $Q_b(u_b, v_b)$ leads to the estimation of the 3D coordinates of the original point $P(x, y, z)$. We will now describe how we can estimate these coordinates.

Because Equations (7) are linear in terms of (x, y, z), we obtain the following form:

$$A \begin{bmatrix} x \\ y \\ z \end{bmatrix} = \mathbf{b} \tag{15}$$

where A, different from its definition displayed in Equation (10), is given by

$$A = \begin{bmatrix} c_{11} - u_a c_{31} & c_{12} - u_a c_{32} & c_{13} - u_a c_{33} \\ c_{21} - v_a c_{31} & c_{22} - v_a c_{32} & c_{23} - v_a c_{33} \\ d_{21} - u_b d_{31} & d_{12} - u_b d_{32} & d_{13} - u_b d_{33} \\ d_{21} - v_b d_{31} & d_{22} - v_b d_{32} & d_{23} - v_b d_{33} \end{bmatrix} \qquad \mathbf{b} = \begin{bmatrix} -c_{14} + u_a c_{34} \\ -c_{24} + v_a c_{34} \\ -d_{14} + u_b d_{34} \\ -d_{24} + v_b d_{34} \end{bmatrix} \tag{16}$$

The least-square estimate of (x, y, z) for the above equation is given by

$$\begin{bmatrix} x \\ y \\ z \end{bmatrix} = (A^T A)^{-1} A^T \mathbf{b} \tag{17}$$

It is important to notice that the matrix A is overconstrained. This is because the constraint includes the epipolar line constraint.

So far we have discussed how we can obtain the 3D depth information from a pair of points matched by stereo correspondence. We will now discuss the case for lines. Lines are often used to model features in model-based vision techniques. In most cases such lines are extracted in individual images as line segments with end points. Due to the measurement error associated with image segmentation, however, the end points in the left images are not precisely matched with the end points in the right images. This section deals with the 3D reconstruction of lines that can handle the ambiguity in the locations of the extracted end points.

Let Q_1 and Q_2 be the end points for the line in the left image and Q_3 and Q_4 be the end points for the line in the right images. We will take the following steps to obtain the 3D line description:

1. Compute the epipolar lines for Q_1 and Q_2 in the right image. Let α_1 and α_2 be the epipolar lines for Q_1 and Q_2. Compute the intersections with α_1 and α_2 and the line $Q_3 Q_4$ in the right image. Let Q_1' and Q_2' be these intersections. Reconstruct the 3D point P_1 and P_2 from the pairs (Q_1, Q_1') and (Q_2, Q_2').
2. Conversely, compute the epipolar lines for Q_3 and Q_4 in the left image. Let α_3 and α_4 be the epipolar lines for Q_3 and Q_4. Compute the intersections with α_3 and α_4 and the line $Q_1 Q_2$ in the left image. Let Q_3' and Q_4' be these intersections. Reconstruct the 3D point P_3 and P_4 from the pairs (Q_3, Q_3') and (Q_4, Q_4').
3. Now we have four 3D points P_1, P_2, P_3, and P_4. Look for the farthest two points among these four points, say P_a and P_b. These two 3D points will be registered for the end points of the reconstructed 3D line.

Figure 5 shows how to generate the points P_1, P_2, P_3, and P_4.

Here in Step 3 we have taken the two farthest points for the end points of the line. This is due to the fact that the line segment may be occluded in either of the images and therefore the largest extension of the line segment in the 3D space will be the most appropriate choice for the line registration.

2.3 Pose Estimation from Three-Dimensional Model-to-Scene Feature Correspondence

In this subsection we will discuss how we can estimate the pose of an object given a set of correspondences between a certain number of model features and features extracted from the camera images. This problem is well researched. For example, Faugeras and Herbert (1986), Chen and Kak (1989), and Kim and Kak (1988) discuss pose estimation after a correspondence is established between vertices and surface normals on a model object and those extracted from a range map.

This section focuses on the problem of pose estimation given that the scene features— in particular the 3D line features and the 3D points extracted from the stereo images—

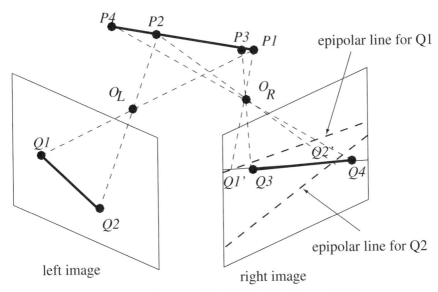

Figure 5   3D line reconstruction from two line segments.

have been paired up with model features. The pose will consist of a homogeneous transformation that will take the object from its standard pose to the pose it actually occupies in the scene. As we should expect, the transformation matrix will consist of two components—the rotation matrix R and the translation vector $\mathbf{t}$.

Case 1   Line Features

Although a 3D line segment is generally defined by the two end points of the line segment, it is sometimes useful to represent the line segment by its orientation vector and the centroid of the line segment. Let $\mathbf{a}$ be the orientation vector of the line where the vector $\mathbf{a}$ is normalized and $\mathbf{p}$ be the centroid of the line segment. Let us assume here that model line features, each specified by a pair of the orientation vector $\mathbf{a}_i^M$ and the centroid $\mathbf{p}_i^M$ for $i = 1, 2, \ldots, n$, are matched to the scene line features, each specified by a pair of the orientation vector $\mathbf{a}_i^S$ and the centroid $\mathbf{p}_i^S$ for $i = 1, 2, \ldots, n$.

Then, if we consider a transformation from the model frame to the scene frame as specified by the rotation matrix R and the translation vector $\mathbf{t}$, for each line correspondence $(\mathbf{a}_i^M, \mathbf{p}_i^M) \rightarrow (\mathbf{a}_i^S, \mathbf{p}_i^S)$, we can write

$$R\mathbf{p}_i^M + \mathbf{t} = \mathbf{p}_i^S \tag{18}$$

$$R\mathbf{a}_i^M = \mathbf{a}_i^S \tag{19}$$

However, in reality a scene line feature obtained from images may involve unknown artifacts and be contaminated with noise. Therefore, the centroid $\mathbf{p}_i^S$ and the orientation vector $\mathbf{a}_i^S$ may not be precisely obtained. This problem can be handled in the following manner.

Let us assume that a scene line feature is represented by two end points $\hat{\mathbf{q}}_i^S$ and $\hat{\mathbf{r}}_i^S$ and the orientation vector $\hat{\mathbf{a}}_i^S$ obtained from actual measurements. We can then represent the relationship among $\mathbf{p}_i^M, \mathbf{a}_i^S, \hat{\mathbf{q}}_i^S, \hat{\mathbf{r}}_i^S$, and $\hat{\mathbf{a}}_i^S$ using the parametric description of the line as follows:

$$R\mathbf{p}_i^M + \mathbf{t} = \alpha_i(\hat{\mathbf{r}}_i^S - \hat{\mathbf{q}}_i^S) + \hat{\mathbf{q}}_i^S + \boldsymbol{\epsilon}_i \tag{20}$$

$$R\mathbf{a}_i^M = \hat{\mathbf{a}}_i^S + \boldsymbol{\eta}_i \tag{21}$$

where

$$\hat{\mathbf{a}}_i^S = \frac{\hat{\mathbf{r}}_i^S - \hat{\mathbf{q}}_i^S}{|\hat{\mathbf{r}}_i^S - \hat{\mathbf{q}}_i^S|} \tag{22}$$

α_i is used to represent a point on the line defined by the two end points, and ϵ_i and η_i represent the noise components in the measurements.

We will now present a two-stage algorithm that estimates first R and then $\mathbf{t}$.

Step 1   Estimation of R

From Equation (21) we can solve for the rotation matrix R by solving the following least-squares minimization problem:

$$\min E_R = \min_R \sum_{i=1}^{n} |\hat{\mathbf{a}}_i^S - R\mathbf{a}_i^M|^2 \tag{23}$$

As explained in the tutorial appendix of Kim and Kak (1991), this minimization can be carried out readily by using the quaternion approach (Faugeras and Herbert, 1986). This approach yields a solution $\tilde{R}$ for R in terms of the components of an eigenvector of the following matrix:

$$B = \sum_{i=1}^{n} B_i^{\mathrm{T}} B_i \tag{24}$$

where

$$B_i = \begin{bmatrix} 0 & (\mathbf{a}_i^M - \hat{\mathbf{a}}_i^S) \\ (\hat{\mathbf{a}}_i^S - \mathbf{a}_i^M)^{\mathrm{T}} & [\hat{\mathbf{a}}_i^S + \mathbf{a}_i^M]_\times \end{bmatrix} \tag{25}$$

and $[\bullet]_\times$ is a mapping from a 3D vector to a 3×3 matrix:

$$[(x, y, z)]_\times = \begin{bmatrix} 0 & -z & y \\ z & 0 & -x \\ -y & x & 0 \end{bmatrix} \tag{26}$$

Let $q = [q_0, q_1, q_2, q_3]^{\mathrm{T}}$ be a unit eigenvector of B associated with the smallest eigenvalue of B. The solution for the rotation matrix R in Equation (23) is then given by

$$\tilde{R} = \begin{bmatrix} q_0^2 + q_1^2 - q_2^2 - q_3^2 & 2(q_1q_2 - q_0q_3) & 2(q_1q_3 + q_0q_2) \\ 2(q_2q_1 + q_0q_3) & q_0^2 - q_1^2 + q_2^2 - q_3^2 & 2(q_2q_3 - q_0q_1) \\ 2(q_3q_1 - q_0q_2) & 2(q_3q_2 + q_0q_1) & q_0^2 - q_1^2 - q_2^2 + q_3^2 \end{bmatrix} \tag{27}$$

Step 2   Estimation of t

Once $\tilde{R}$ is obtained, it is relatively easy to obtain $\mathbf{t}$ from Equation (20). We now minimize

$$\min E_t = \min_{\mathbf{t}, \alpha_i} |\tilde{R}\mathbf{p}_i^M + \mathbf{t} - \alpha_i(\hat{\mathbf{r}}_i^S - \hat{\mathbf{q}}_i^S) - \hat{\mathbf{q}}_i^S|^2 \tag{28}$$

Setting the derivatives $\delta E_t / \delta \mathbf{t}$ and $\delta E_t / \delta \alpha_i$ to zero leads to a linear equation for $\mathbf{t}$ and $\alpha_1, \ldots, \alpha_n$. Therefore, the optimal solution of $\mathbf{t}$ can be easily obtained.

Note that we need model-to-scene correspondences for at least two nonparallel lines for this procedure to yield estimates for the rotation matrix R and the translation vector $\mathbf{t}$.

Case 2   Point Features

It is also possible to estimate the pose of an object by setting up correspondences between designated points on a model object and the points in a camera image of the object. If we denote the corresponding model and scene points by $\mathbf{p}_i^M$ and $\mathbf{p}_i^S$, respectively, we have

$$R\mathbf{p}_i^M + \mathbf{t} = \mathbf{p}_i^S \tag{29}$$

We do not now, as we did for the line features, have a separate equation that could be used for estimating the rotation matrix R. To circumvent this difficulty, we construct a virtual line for each pair of the point features i and j:

$$R(\mathbf{p}_i^M - \mathbf{p}_j^M) = (\mathbf{p}_i^S - \mathbf{p}_j^S) \tag{30}$$

This being a relationship on the rotations of the virtual lines, it is also true for their normalized versions:

$$R \frac{\mathbf{p}_i^M - \mathbf{p}_j^M}{|\mathbf{p}_i^M - \mathbf{p}_j^M|} = \frac{\mathbf{p}_i^S - \mathbf{p}_j^S}{|\mathbf{p}_i^S - \mathbf{p}_j^S|} \tag{31}$$

In the presence of measurement noise, the above two equations will not be satisfied exactly. We now include the measurement noise factors ϵ_{ij} and η_i as follows:

$$R \frac{\mathbf{p}_i^M - \mathbf{p}_j^M}{|\mathbf{p}_i^M - \mathbf{p}_j^M|} = \frac{\hat{\mathbf{p}}_i^S - \hat{\mathbf{p}}_j^S}{|\hat{\mathbf{p}}_i^S - \hat{\mathbf{p}}_j^S|} + \epsilon_{ij} \tag{32}$$

$$R\mathbf{p}_i^M + \mathbf{t} = \hat{\mathbf{p}}_i^S + \eta_i \tag{33}$$

As before, we can again use a two-stage algorithm to obtain the estimates for the rotation matrix R and the translation vector $\mathbf{t}$. For estimating R, we define

$$\mathbf{a}_{ij}^M = \frac{\mathbf{p}_i^M - \mathbf{p}_j^M}{|\mathbf{p}_i^M - \mathbf{p}_j^M|} \qquad \hat{\mathbf{a}}_{ij}^S = \frac{\hat{\mathbf{p}}_i^S - \hat{\mathbf{p}}_j^S}{|\hat{\mathbf{p}}_i^S - \hat{\mathbf{p}}_j^S|} \tag{34}$$

to obtain

$$R\mathbf{a}_{ij}^M = \hat{\mathbf{a}}_{ij}^S + \epsilon_{ij} \tag{35}$$

R is now a solution to the following minimization problem:

$$\min E_R = \min_R |\hat{\mathbf{a}}_{ij}^S - R\mathbf{a}_{ij}^M|^2 \tag{36}$$

This minimization is solved using exactly the same method as before for the calculation of the pose from model-to-scene line correspondences. Once optimal $\tilde{R}$ is obtained, obtaining $\mathbf{t}$ from Equation (29) is an easy task. We now minimize

$$\min E_t = \min_{\mathbf{t}} \sum_{i=1}^{n} |\hat{\mathbf{p}}_i^S - \tilde{R}\mathbf{p}_i^M - \mathbf{t}|^2$$

Setting the derivative $\delta E_t / \delta \mathbf{t}$ to zero leads to the following equation:

$$\tilde{\mathbf{t}} = \sum_{i=1}^{n} \hat{\mathbf{p}}_i^S - \tilde{R}\left(\sum_{i=1}^{n} \mathbf{p}_i^M\right) \tag{37}$$

This actually represents the translation from the centroid of the points in the model frame to the centroid of the corresponding points of the scene frame. For the point features, we require at least three noncollinear points to estimate the transformation.

Case 3   Combination of Line and Point Features

The above two cases dealt with the cases of exclusively the line features and next exclusively the point features in setting up model-to-scene correspondences. As should be obvious by now, it is also possible to use a combination of line and point features in such correspondences. The mathematics remains the same.

3   BASIC STRATEGY FOR MODEL-BASED STEREO VISION

We will now explain our basic strategy for model-based stereo vision. A typical flow of control is shown in Figure 6. The *3D object modeling module* can take a 3D CAD model

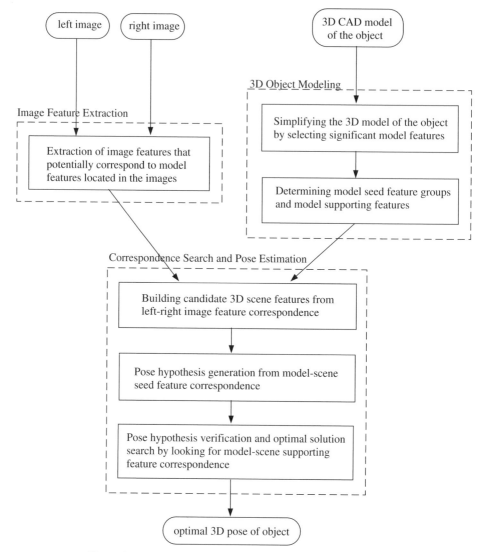

Figure 6   A typical flow of control for model-based stereo vision.

of the object and create from it a much simplified model consisting of only those features that would later be used for stereo vision-based recognition and localization of the object. After a simplified 3D model of the object is created, this module extracts from the simplified model what we call *seed feature groups* (SFG). These features can be used for pose estimation. A group may, for example, consist of two nonparallel line features on the simplified 3D model or, say, three noncollinear point features.

The *image feature extraction module* extracts image features that may potentially correspond to the projections of 3D model features. For example, if we deal with 3D lines for 3D model features, this module will extract 2D lines in the images.

The *correspondence search and pose estimation module* has the overall responsibility for estimating the pose of the object using the image features extracted by the image feature extraction module. This module consists of three main submodules. The *module for building candidate 3D scene features* receives 2D image features from the stereo pair of images and then generates 3D features that may potentially correspond to 3D model features. For example, if 3D lines are selected for 3D model features, then this module generates scene candidates for the 3D line segments. We will call these generated line segments *candidate scene features.*

The *pose hypothesis generation module* receives the 3D model seed features and the 3D candidate scene features generated by the module for building candidate scene features. For each 3D model seed feature, it first collects all 3D candidate scene features that are potentially matchable with the 3D model seed feature. A 3D model seed feature is paired up with a candidate 3D scene feature by checking whether or not the pairing satisfies all the implied geometric constraints. The paired-up model seed features and candidate scene features are used for estimating the pose of the object.

For each hypothesized pose of the object, the *pose hypothesis verification and optimal solution module* then generates a 3D expectation map of the model features in the scene space, using the hypothesized pose. This 3D expectation map includes the expected scene features and therefore the module compares the expected scene features and the actually generated scene features to verify whether the hypothesized pose will be supported by other matches of the model features and the scene features. More specifically, the module counts the number of such supporting model features matched and computes the overall 3D fitting error to search for an optimal correspondence between the model features and the scene features.

The following subsections explain the details of each module with the help of two examples involving two different types of industrial objects.

3.1   Modeling Three-Dimensional Objects

Three-dimensional CAD models are now available for most industrial objects. As mentioned above, using the rich descriptions of CAD models for driving the scene interpretation algorithms only adds complexity to the entire vision process. Vision systems become much more efficient if in the 3D model we retain only that information which is necessary and sufficient for recognition and localization.

Figures 7 and 8 show two examples of objects that can be recognized and localized efficiently and robustly using the methods presented here. In Figure 7, the goal is to recognize and localize the dumpster of the truck. The dumpster may be partially occluded by rocks in an outdoor environment. Figure 8 shows a bin of objects known as *alternator covers*. The goal here is for a robot to recognize the parts automatically and pick as many of them as possible.

Figures 9 and 10 depict the simplified 3D models used for these two classes of objects. For the dump truck example shown in Figure 9, only five line segments are needed for recognition and localization despite the occlusions that can be expected for such scenes. The simplified model for the case of alternator covers is shown in Figure 10. The model uses the large center hole that is present in all objects of this type and the four smaller screw holes surrounding the large hole. (Note that the large hole in the alternator covers can be used for grasping these objects with specially designed fingers on a robotic arm.)

Figure 7   Dump truck recognition: original left and right images taken in a complex environment.

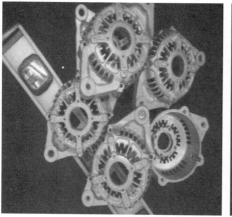

Figure 8 Alternator cover recognition: original stereo views of the workspace.

As can be surmised from Figures 9 and 10, the features incorporated in the simplified 3D models are of the following types:

1. 3D point features
2. 3D line features
3. Planar region features constituting simple shapes such as circles and polygons

In the memory of the computer, associated with each such model feature is a set of 3D geometric properties. For example, 3D line features are represented by their two end points in terms of the 3D coordinates of the points in an object-centered coordinate frame. Planar circle features are represented by the centroid of the circle, the orientation of the circle in the object coordinate frame, and the radius of the circle.

After computation of the relevant geometric properties of each of the retained model features, the modeling module next computes the attributes of the relations between each pair of model features. These attributes can, for example, be the minimum distance between the two or the orientation of the minimum distance vector.

The last task of the modeling module is to propose model seed feature groups. Because pose is estimated using minimization of error technique, it is better to use more features than necessary. In other words, while theoretically it is sufficient to use just one pair of nonparallel lines on a model object and their correspondents in a scene for estimating the

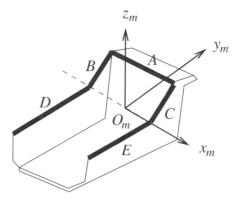

Figure 9 Five model lines are used to recognize a dump truck. {A, B, C}, {A, B, D}, {A, B, E}, {A, C, D}, {A, C, E}, {A, D, E}, {B, C, D}, {B, C, E}, {B, D, E}, {C, D, E} are registered as model seed features in the experiments described in the experimental results section.

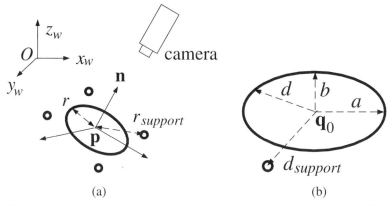

Figure 10   (a) The sufficient 3D model/representation of the alternator cover. (b) The 2D appearance analysis of a seed feature and its position relative to a supporting feature.

pose of an object, better results can be expected if we use more than two nonparallel lines in such a calculation. For this reason, we will use at least three line features in a seed group on objects like the one shown in Figure 7. For that specific object type, referring to the line feature labels shown in Figure 9, we have used the following seed groups: $\{A, B, C\}$, $\{A, B, D\}$, $\{A, B, E\}$, $\{A, C, D\}$, $\{A, C, E\}$, $\{A, D, E\}$, $\{B, C, D\}$, $\{B, C, E\}$, $\{B, D, E\}$, $\{C, D, E\}$. On the other hand, for the alternator cover shown in Figure 10, it is sufficient to use the single large hole as a model seed feature if we ignore the rotation of the object around an axis perpendicular to the hole.

3.2   Feature Extraction from Images

As mentioned above, this module extracts from the two stereo images the features that are subsequently used for establishing stereo correspondences. While the extraction of image features corresponding to 3D object points and straight lines is fairly straightforward, since object points project into image points and object straight edges project into image straight lines, the same cannot be said for the image features that correspond to the planar regions on an object. The image feature that corresponds to a planar face, real or virtual, of an object depends highly on the shape of that face. In the case of alternator covers, for example, circular planar features on the object, such as those corresponding to the large hole in the center, project into ellipses in the images for a majority of the viewpoints. In order to extract these features, the system needs an efficient algorithm for extracting ellipses. Fortunately, over the years a large number of tools have been developed for the extraction of such and other features from images (Ballard, 1981; Horowitz and Pavlidis, 1976; Rosenfeld and Kak, 1982). Later in this chapter we will briefly mention the specific tools that have proved successful for a fairly wide variety of industrial objects in our laboratory.

3.3   Generating Potential Three-Dimensional Scene Features from Image Feature Stereo Correspondence

After extraction of the image features, the system next tries to establish stereo correspondences between the image features pulled from the left and the right images. The pairing up of a feature from the left image with a feature from the right image creates a 3D scene feature in the world coordinate frame that can then be tested against a 3D model feature.

Although the correspondence problem is not as difficult now as it is when one tries to pair up pixels directly in the two images, there are still contentions to be resolved when establishing correspondences between the features extracted from the two images. The following constraints can be invoked if for a given feature in the left image we have more than one candidate in the right image:

1.   *Epipolar constraint.* The point features must obviously satisfy the epipolar constraint described earlier. In some cases, depending on the level of occlusion and

obscuration in scenes, this constraint can also be invoked for point-like attributes of other types of image features. For example, the centroid is a point attribute of line segments and projections of planar shapes.

2. *Image similarity constraint.* If the left and the right viewpoints are relatively close together, the image features observed in the left and the right images should possess nearly similar attributes if they are projections of the same model feature. This similarity constraint can be invoked for both geometric attributes, such as 2D orientation, 2D length, 2D distance, and 2D area, and nongeometric attributes, such as color, average gray levels, and contrast.

3. *Workspace constraint.* If we know in advance that the object is confined to a known part of the workspace in the world coordinate frame, then 3D scene features generated from a stereo correspondence must lie within this space.

In general, the image similarity constraint may be difficult to apply. Depending on how the vision system is set up, the similarity factors may be sensitive to the viewpoints chosen, the lighting condition in the workspace, and any differences in the left and right camera characteristics.

We will now describe the specifics of how correspondence is established between the image features from the two images. However, it is important to point out beforehand that any attempt at establishing correspondence for, say, a left-image feature must allow for *nil-match* (Boyer and Kak, 1988; Haralick and Shapiro, 1979), which means that due to obscuration an image feature in one image may not possess a correspondent in the other image. This can also happen if the feature-extraction routines fail to extract a feature for one reason or another. Taking the possibility of a nil-match into account, here are the two main steps for establishing image feature correspondences:

1. List all image features extracted from the left and the right images that may potentially be projections of the features incorporated in the simplified 3D models of the objects.

2. For each image feature in the left image and each image feature in the right image, check whether these image features can be matched in terms of the epipolar constraint, the image similarity constraint, and the workspace constraint. If two image features from the two images satisfy these constraints, fuse the two image features into a single 3D scene feature.

In general, the above three constraints—or an applicable subset thereof—will not be sufficient for resolving all contentions. In other words, at the end of the above two steps, an image feature i in one image may be matched with multiple image features c_{ij} in the other image ($j = 1, 2, \ldots, n_i$). Pairing up the ith image feature from the first image with each of the n_i candidates will generate n_i candidate 3D scene features that can subsequently be compared with the 3D model features for final resolution of the contentions. This final resolution is achieved with the help of the module described next.

3.4 Pose Hypothesis Generation and Pose Verification/Optimal Solution Search

As might have been surmised by this time, a major advantage of model-based stereo vision is that the availability of 3D scene features in the world frame allows for a direct comparison of such features with the model features in the 3D world frame. These comparisons can be used to reject untenable pairings of the image features from the two images. This is achieved by hypothesizing the pose of the object for a given set of scene-to-model correspondences and then searching for supporting evidence in the images. If the hypothesized pose cannot be verified in this manner, then an error was made in the correspondences established between the scene features in the two images.

The final pose that is accepted for each object is optimal in the sense that of all the pose hypotheses that can be generated, the finally accepted object pose maximizes the number of matchings between the 3D scene features and the 3D model features while minimizing the pose estimation error. The following two steps describe this process in more detail:

1. For each 3D model seed feature grouping, search for 3D scene features that potentially correspond to the model features contained in the grouping. Make sure

that the geometric constraints applicable to the seed features are satisfied by the 3D scene features.

2. By matching a 3D model seed feature grouping with candidate 3D scene features, generate a pose hypothesis of the object. Let k be the index of the hypothesis for $k = 1, 2, \ldots, K$. For each hypothesis indexed k, generate a 3D expectation map of all the model features that should be visible in the images. Such model features in the 3D expectation map will be called *expected scene features*. Let $M = \{m_1, m_2, \ldots, m_n\}$ be the set of expected scene features and $S = \{s_1, s_2, \ldots, s_m\}$ be the set of 3D scene features actually generated in the world frame by the matching of the image features. Associated with each expected scene feature m_i in M are certain geometric parameters that must agree with the supporting feature selected from the 3D scene features. We will explain this point further for the case of line features. In Figure 11 the solid line segments are the renditions of the 3D scene line segments obtained by fusing line features from the left and the right images. By matching some model seed feature grouping with some set of 3D scene features in this case, a pose hypothesis was constructed. The dashed lines show the 3D model features in the expectation map for this pose hypothesis. We must now seek image support for those model features that did not play a role in the formation of the pose hypothesis. The dashed lines then are the expected scene features denoted m_i above and the solid lines the 3D scene features denoted s_j. Each 3D model feature has associated with it certain geometric parameters. For a line feature these would be the two end point coordinates $\mathbf{p}_1$, $\mathbf{p}_2$, the orientation vector $\mathbf{a}$, the centroid $\mathbf{p}_c$. The hypothesized pose transformation is applied to these parameters for each expected scene feature m_i located in the world coordinate frames. More specifically, if R and $\mathbf{t}$ are the rotational and the translational components of a hypothesized pose, then

$$\mathbf{p}_1' = R\mathbf{p}_1 + \mathbf{t} \qquad \mathbf{p}_2' = R\mathbf{p}_2 + \mathbf{t} \qquad \mathbf{a}' = R\mathbf{a} \qquad \mathbf{p}_c' = R\mathbf{p}_c + \mathbf{t} \qquad (38)$$

where $\mathbf{p}_1'$, etc. represent the transformed parameters. Now let $\hat{\mathbf{p}}_1$, $\hat{\mathbf{p}}_2$, $\hat{\mathbf{a}}$, $\hat{\mathbf{p}}_c$ be the geometric parameters for an actual scene feature located in the world frame. Then, for determining whether a 3D scene feature s_j supports an expectation map feature m_i, we compare their geometric parameters as follows:

> Because the actual scene feature s_j located in the world frame may possess a shorter length due to the occlusion, we do not want to directly compare the end point coordinates with the expected scene feature m_i. Instead, we compute the distance from each of the two end points of the actual scene feature s_j to the expected scene feature m_i. By this distance, we mean here the shortest distance from the point to the line segment, the line segment

3-D expectation map of model features

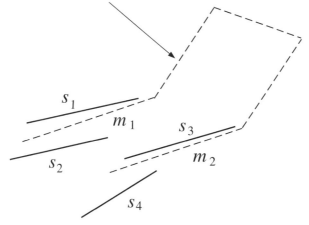

Figure 11 3D expectation map.

being limited by the end points, as shown in Figure 12. As for the orientation, because both s_j and m_i are defined in the same world coordinate frame, we simply compute the angle between them. We say that an expected scene feature m_i is compatible with an actual 3D scene feature s_j if (1) the distances from s_j to m_i do not exceed some threshold and (2) the angle between s_j and m_i does not exceed some threshold.

For the point features we simply compute the 3D distance between the point coordinates of the expected scene feature m_i and the actual scene feature s_j. If this distance does not exceed some threshold, we will say that these two features are compatible. For a hypothesized pose indexed k, if a 3D scene feature s_j is found to be compatible with a 3D model feature m_i, we form the pair (m_i, s_j) for recording this fact.

3. For each pose hypothesis k, count the number of m_i such that for some j there exists a marked pair (m_i, s_j). Let $n(k)$ be the total count thus obtained. Collect all hypotheses that have the maximal $n(k)$. For these hypotheses, compute the fitting error of the model-to-scene correspondence. We accept that pose which yields the smallest fitting error.

The overall model-to-scene fitting error, denoted OFE, is defined by

$$OPE = \frac{1}{n}(E_{R_{\min}} + \beta E_{t_{\min}}) \tag{39}$$

where n represents the number of features used in the equation. $E_{R_{\min}}$, defined previously in Equations (23) and (36), represents the fitting error associated with the estimation of the rotation matrix R, and $E_{t_{\min}}$, defined previously in Equations (28) and (37), the fitting error associated with the translation vector $\mathbf{t}$. β is a user-specified factor to weight the two components of the error unequally if so desired.

In order to further elucidate the practical considerations that go into designing such vision systems, we will now discuss in greater detail the two cases of model-based stereo vision, one for the case of dump truck-like objects and the other for the case of alternator covers.

4 STEREO VISION FOR DUMP TRUCK RECOGNITION AND LOCALIZATION

The goal of this vision system is to determine the position and orientation of a dump truck using stereo vision techniques. This task was recently deemed to be important in the context of automation of some of the more labor-intensive aspects of construction. It is envisioned that at some point in the future a team of robots would work in conjunction with minimally supervised dump trucks and excavators for preparing construction sites and clearing debris from volcanic or other hazardous areas. Sensors, such as cameras during daylight conditions, would be used for automated localization of the dump truck so that its position would become available to the excavator.

A stereo pair of images of a dump truck in a simulated dirt environment, as shown in Figure 7, is obtained with the help of a gripper-mounted camera on a robot by changing the viewpoint of the camera. The left and the right viewpoint images are in Figures 7a and b, respectively.

Line features are extracted from the stereo images by first applying the Canny's edge operator (Canny, 1986) to the images for the detection of the edge pixels. Subsequently, a thinning operator is applied to the edge pixels to yield a thinned edge image. The feature extraction algorithm then tracks the edge pixels from the end point pixels or the junction pixels and finds smooth curves that correspond to line segments in the images. Figure 13 shows the results of line segment extraction for the left and right images.

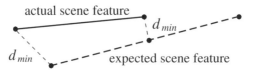

Figure 12 The distance used to compare the expected scene feature (dashed) and the actual scene feature (bold).

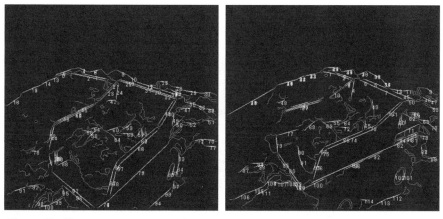

Figure 13   Line segments extracted by tracking edge pixels generated from Canny's operator.

Next, for each line segment in the left image the system constructs a list of line segments from the right image as potential candidate matches, taking into account the various constraints mentioned in the previous section. In accordance with our earlier discussion, the candidate pairings lead to the formation of 3D scene features. Matching candidate 3D scene features with the model seed feature groups leads to pose hypotheses. Eventually an optimal pose for the 3D object is calculated by maximizing the number of matches between the 3D model object and 3D scene features and minimizing the pose estimation error. The final model-to-scene matches obtained in this manner are illustrated in Figure 14. In this figure the 3D scene line segments participating in the optimal correspondence with the model are superimposed on the original left and right images. Figure 15 shows the pose estimate result, in which the detailed wireframe of the dump truck is reprojected to the original stereo images using the estimated pose and therefore displays the accuracy of the pose estimation.

This system was evaluated on more than 40 sets of images in the Robot Vision Lab. The average positional error in the calculated location of the different points on the truck was between 0.5 cm and 1.0 cm. Compare this to the overall size of the truck: 40 cm (L) × 20 cm (W) × 25 cm (H). The average processing time was 30 seconds on a Sun Workstation Sparc-1000. Note that no special image processing hardware was used in this system.

Figure 14   The line segments participating in the optimal correspondence are superimposed onto the left and right camera images.

Figure 15 The detailed wireframe of the dump truck is reprojected to the original stereo images using the estimated pose. This figure therefore verifies the accuracy of the pose estimation.

5 MODEL-BASED STEREO VISION FOR RECOGNITION AND LOCALIZATION OF ALTERNATOR COVERS

As object complexity increases, so does the difficulty of vision-based approaches to robotic bin-picking. This section provides further details of a model-based stereo vision system for industrial objects of fairly high complexity. Because of the model-based approach, the stereo vision system itself is simple; it is made all the simpler by the simplified 3D models of the objects used, in which only that information about the objects that would be relevant to vision tasks is retained. To demonstrate at the outset how powerful this model-based approach is, shown in Figure 16 is a depiction of the pose-estimation results obtained. The poses estimated for each of the objects that was successfully recognized from the stereo images of Figure 8 are used. The bright pixels in the images are reprojections into the two camera images of the large hole in the middle of the model object and the four smaller holes surrounding the large hole. The large cross hairs in the middle of each large hole are a reprojection into the camera image of the center of each hole in the model object. Any visible offset between the large cross hairs and the center of the large hole is a measure of the pose estimation error for that object.

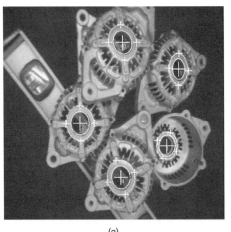
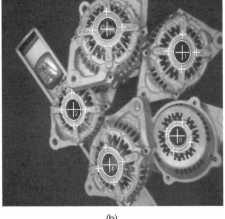

(a)   (b)

Figure 16 The result of recognition and pose estimation on both left (a) and right (b) images. The estimated 3D poses of the manipulation landmarks are reprojected as the dotted ellipses to verify the accuracy of pose estimation.

Given the complicated shape of alternator covers, it should be obvious it would not be possible to base the stereo vision approach on straight line segments extracted from the images. What are needed are the image features that would be camera projections of the circular features shown in the simplified 3D model presented in Figure 10. Therefore, the image feature-extraction module in this case needs to be very different from the one presented for the case of dump trucks. Now needed is a region-based segmentation of the camera images that would be capable of pulling out the ellipses—if not the full ellipses, then at least its parts—that are the camera projections of the circular features. The image feature-extraction module in this case, therefore, uses a specially developed region-based segmenter (Rahardja and Kosaka, 1996) based on the split-and-merge approach (Horowitz and Pavlidis, 1976). The segmenter uses the following steps:

1. Canny's edge detector (Canny, 1986), which has proven to be superb in preserving the localization of edges, is first applied to the images for detecting the edge pixels.
2. Because the edge detector does not produce continuous edge contours, an edge-linking routine is applied to extend the dangling edges. The Canny edge operator is particularly susceptible to producing edge breaks near T-junctions because of the uniqueness of response criterion used in deciding whether to accept a pixel as an edge pixel.
3. The edges thus extracted are incorporated in a split-and-merge-based region segmenter. In the split-and-merge segmenter, a region is considered homogeneous if edge pixels do not exist within its interior.

The following properties are measured for each of the regions output by the segmenter:

- Shape complexity = $\text{perimeter}^2/\text{area}$
- Region area
- Grayscale mean value.

The images in Figure 17 show a sample result of the region extraction algorithm applied to the stereo images of Figure 8. The output of the Canny edge operator is shown in Figures 17a and b for the two stereo images. The outputs of the split-and-merge segmentation algorithm taking into account the Canny edges are depicted in Figures 17c and d. Figures 17e and f show the candidate regions accepted for the left and the right images by placing user-specified thresholds on the three properties listed above.

Note that in Figures 17c, d, e, and f different regions are colored differently for the purpose of visualization. The number of regions obtained as candidate regions in e and f obviously depends on what thresholds are specified for the three region properties listed above. In practice, one would want to specify these thresholds in such a manner that the number of regions extracted exceeds the number of object features of interest in the images. In other words, it is better to err on the side of high false alarm at this stage than to reduce the probability of detection of the desired regions.

Each candidate region output by the segmenter is subject to the *Ellipse Verification Test* (EVT). This test consists of the following steps:

1. Compute the moment of inertia of a region and denote it by M. Also, let $\mathbf{q}_0$ denote the center of mass of the region. If $\mathbf{q}$ denotes an arbitrary point on the boundary of the region, then for a truly elliptical boundary the following distance measure

$$d = \sqrt{(\mathbf{q} - \mathbf{q}_0)^\mathrm{T} M^{-1}(\mathbf{q} - \mathbf{q}_0)} \tag{40}$$

will be constant for all boundary points. This fact can be used to advantage in the following manner. As depicted in Figure 18a, we traverse the boundary of a region and compute d for each pixel on the boundary. Because the segmented region is unlikely to be a perfect ellipse, d will not be absolutely constant as we traverse the boundary. We compute the standard deviation σ_d of d and compare it to a decision threshold

$$\sigma_d \leq \sigma_{\text{threshold}} \tag{41}$$

If this condition is satisfied, we accept the region as an ellipse, provided the second condition listed below is also satisfied. If the threshold test fails, the region is

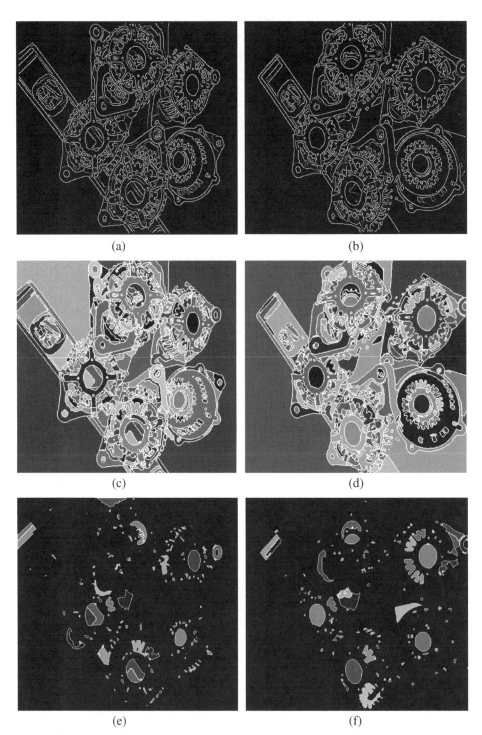

(a)   (b)

(c)   (d)

(e)   (f)

Figure 17 Result of the feature extraction. (*a*), (*b*) Canny's edge detection applied to the stereo pair images shown in Figure 8. (*c*), (*d*) Results of region extraction by taking into account extracted edges shown in (*a*) and (*b*). (*e*), (*f*) Candidate seed features and supporting features are extracted from the segmented regions shown in (*c*) and (*d*) by utilizing the parametric constraints of the simplified object model.

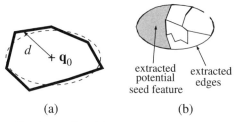

(a)   (b)

Figure 18   (a) Feature verification by EVT. (b) Region growing assisted by the edge contour that surrounds the seed feature candidate.

discarded for stereo matching. We have found empirically that the normalized $\sigma_{\text{threshold}} \leq 0.1$ performs satisfactorily.

2.   From the moment of inertia computation, the lengths of the principal axes obtained for a candidate region should also be within certain predesignated bounds that depend on the size of the circular feature on the object model.

It is not uncommon that the segmented regions by themselves may not satisfy the criteria for acceptance as ellipses, but if the contiguous regions were patched together, the composite would look more like an ellipse. This idea is shown schematically in Figure 18b. In the output of the segmenter, the camera projection of a circular object feature is so fragmented that none of the fragments shown there would pass our test for acceptance as ellipses. However, if we could merge the fragments together, we would indeed have a viable candidate for stereo matching. An algorithm that can achieve such merging is as follows:

1.   Choose a region that has failed EVT to serve as a seed in a region-growing process. Surround this seed region with a circular search area of radius d_{max} whose value, to be determined by trial and error at system design time, depends on the feature size on the object.
2.   Within the search area, find a convex enclosure of the edge pixels output previously by the Canny's operator.
3.   Count the number of pixels in the intersection of the convex enclosure and the set of edge pixels. If the pixel count exceeds a threshold, consider the composite region obtained by merging together the regions inside the convex enclosure as a metaregion that could be an ellipse.
4.   Apply EVT to the metaregion.

The regions thus extracted by the feature-extraction module from the left and the right images must then be fused for the creation of 3D scene features. Each pairing generates a pose hypothesis for one of the alternator covers. Of course, as mentioned earlier, the various constraints must be satisfied, such as the epipolar constraint and the workspace constraint. Here the epipolar constraint would be applied to the centroids of the paired up regions and the workspace constraint to the centroid of the 3D entity obtained by fusing two regions from the two stereo images.

The pose hypotheses are verified as before by determining whether there exists in the images sufficient support for the model feature not used for pose hypothesis formation. Pose hypotheses are formed by using the camera projections of the large hole in the center of the simplified 3D model shown in Figure 10, leaving the small holes for pose verification. This creates two difficulties:

1.   Any pose hypothesis generated by matching the two camera projections of the large hole will have a rotational ambiguity around the normal to the hole.
2.   The small size of the four additional holes shown in Figure 10 means that there will usually be a large number of image-generated candidates for them.

This latter point is illustrated with the help of Figure 19. In each image there, the large cross hairs correspond to the centroids of the those regions in the outputs of the region-

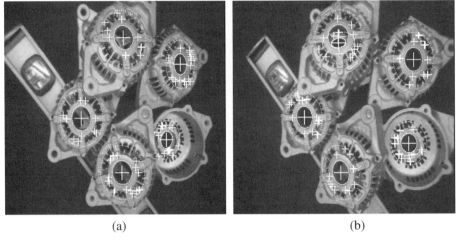

(a)   (b)

Figure 19 The result of EVT for (a) the left viewpoint and (b) the right viewpoint. Dotted ellipses are the accepted seed features. Potential supporting features for each associated seed feature candidates are highlighted with crosses.

based segmenter that satisfied the ellipticity condition with regard to both the shape and the lengths of the principal axes. The small cross hairs correspond to the other regions in the vicinity of the large regions that also satisfied the ellipticity condition and had the requisite principal axes. The regions in the output of the segmenter corresponding to the small cross hairs are candidates for the small holes shown in Figure 10.

The aforementioned rotational ambiguity in pose is resolved in the following manner. Let us say that during the verification of a pose hypothesis we have chosen in the left image, in the vicinity of the large elliptical region that was used for forming the pose hypothesis, an appropriately small elliptical region, whose centroid is at $\mathbf{q}_1^{\text{left}}$, as a candidate for verification. By invoking the epipolar constraint, let us further say that we have identified in the right image a candidate, whose centroid is at $\mathbf{q}_1^{\text{right}}$, for matching with the left image region centered at $\mathbf{q}_1^{\text{left}}$. Searching around the large region in the left image, we can continue in this manner until we have compiled a set A of four pairs of matches shown below

$$A = \{(\mathbf{q}_1^{\text{left}}, \mathbf{q}_1^{\text{right}}), (\mathbf{q}_2^{\text{left}}, \mathbf{q}_2^{\text{right}}), (\mathbf{q}_3^{\text{left}}, \mathbf{q}_3^{\text{right}}), (\mathbf{q}_4^{\text{left}}, \mathbf{q}_4^{\text{right}})\}$$

Let $\mathbf{p}_i$ be the centroid of the 3D supporting feature estimated from those in the images at $\mathbf{q}_i^{\text{left}}$ and $\mathbf{q}_i^{\text{right}}$. Then an optimal value for the pose can be obtained by minimizing the following objective function over all possible A sets:

$$F_A = \sum_{i,j=1,2,3,4} (\|\mathbf{p}_i - \mathbf{p}_j\| - C_{ij})^2 \tag{42}$$

Figure 20 Typical error of the pose estimation. These images were taken at the estimated distances of (a, d) 300 mm, (b, e) 150 mm and (c, f) 100 mm away from the estimated centroid **p** of the object as the camera was driven along the estimated normal vector **n** of the seed feature of the object.

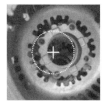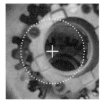

Figure 21 Worst case of the pose estimation. As opposed to the previous figures, the estimation error is relatively large. Our algorithm, however, estimates the 3D object pose accurately enough for the manipulator to grasp the object.

where C_{ij} is the 3D distance between the centroids of two supporting features i and j in the 3D model of the object. To find an optimal pose, the algorithm performs an exhaustive search over all sets of possible sets A.

We will now present the results of an experimental evaluation of the system. As for the dimensions of these objects, the radius of the bearing hole—that is, the large hole in the center—is 15 mm. The distance from the center of the bearing hole to each of the four neighboring screw holes is 22.5 mm. The stereo images were taken with a gripper-mounted camera on a PUMA 700 manipulator. Two gripper/camera positions were chosen as the left and right viewpoints, separated by 243.5 mm and a vergence angle of 20°. The approximate distance from the camera locations to the objects is 350 mm.

The stereo images for each trial were digitized using a 512×480 matrix. The processing time on a SUN SPARC 1000 machine was roughly 1.5 minutes. Of course, it would be possible to reduce this time by orders of magnitude by using specialized image-processing hardware.

To verify the accuracy of pose estimation, the robot arm with the camera in its gripper, instead of actually grasping the recognized objects, is made to zoom in on each object along the estimated normal vector passing through the center of the large hole. Watching the movement of the robot arm and the location of the large cross hairs gives a sense of how accurate the system is with regard to pose calculation. As the robot arm zooms in, the images seen though the camera are continuously displayed on a monitor. Figures 20 and 21 show three such images as the camera is zooming in; these images were taken at distances of (*a*) 300 mm, (*b*) 150 mm, and (*c*) 100 mm from the center of the large bearing hole. In these figures, the white cross hair shows the estimated location of the hole center via pose estimation and the white dotted circle the estimated perimeter of the circle that should pass through the screw holes. The extent to which the white cross hair is displaced from the actual center of the large hole in the camera image is a direct measure of the error in pose estimation.

Thirty-five experiments with an average of six objects in the workspace were performed for performance evaluation. Typical 3D positional and orientational errors are less than 7 mm and less than 10°, respectively.

6 CONCLUSIONS

This chapter presented what we believe is the best approach today for recognition and localization of 3D industrial objects using off-the-shelf cameras. The approach presented is robust and works well for even complex objects. It derives its power from the fact that object knowledge is used for solving the most difficult problem in stereo vision—the correspondence problem. The approach presented combines the best of bottom-up and top-down processing to achieve a happy medium that can easily be adapted for different classes of objects.

REFERENCES

Ballard, D. H. 1981. "Generalizing the Hough Transform to Detect Arbitrary Shapes." *Pattern Recognition* **13**(2), 111–122.

Besl, P. J. 1988. *Surfaces in Range Image Understanding.* New York: Springer-Verlag.

Bolles, R. C., and R. A. Cain. 1982. "Recognizing and Locating Partially Visible Objects: The Local-Feature-Focus Method." *International Journal of Robotics Research* **1**(3), 57–82.

Bolles, R. C., and P. Horaud. 1986. "3DPO: A Three-Dimensional Part Orientation System." *International Journal of Robotics Research* **5**(3), 3–26.

Boyer, K. L., and A. C. Kak. 1988. "Structural Stereopsis for 3-D Vision." *IEEE Transactions on Pattern Analysis and Machine Intelligence* **PAMI-10**(2), 144–166.

Canny, J. 1986. "A Computational Approach to Edge Detection." *IEEE Transactions on Pattern Analysis and Machine Intelligence* **PAMI-8**(6), 679–698.

Chen, C. H., and A. C. Kak. 1989. "A Robot Vision System for Recognizing 3-D Objects in Low-Polynomial Time." *IEEE Transactions on Systems, Man, and Cybernetics* **19**(6), 1535–1563.

Faugeras, O. D. 1993. *Three-Dimensional Computer Vision*. Cambridge: MIT Press.

Faugeras, O. D., and M. Herbert. 1986. "Representation, Recognition, and Locating of 3-D Objects." *International Journal of Robotics Research* **5**(3), 27–52.

Fu, K. S., R. C. Gonzalez, and C. S. G. Lee. 1987. *Robotics: Control, Sensing, Vision, and Intelligence*. New York: McGraw-Hill.

Grewe, L., and A. C. Kak. 1994. "Stereo Vision." In *Handbook of Pattern Recognition and Image Processing: Computer Vision*. New York: Academic Press. Chapter 8, 239–317.

———. "Interactive Learning of a Multiple-Attribute Hash Table Classifier for Fast Object Recognition." *Computer Vision and Image Understanding* **61**(3), 387–416.

Haralick, R. M., and L. G. Shapiro. 1979. "The Consistent Labeling Problem: Part I." *IEEE Transactions on Pattern Analysis and Machine Intelligence* **PAMI-1**(2).

Horiwitz, S. L., and T. Pavlidis. 1976. "Picture Segmentation by a Tree Traversal Algorithm." *Journal of ACM* **23**(2), 368–388.

Kak, A. C. 1985. "Depth Perception for Robots." In *Handbook of Industrial Robotics*. Ed. S. Y. Nof. New York: John Wiley & Sons. 272–319.

Kim, W. K., and A. C. Kak. 1991. "3-D Object Recognition Using Bipartite Matching Embedded in Discrete Relaxation." *IEEE Transactions on Pattern Analysis and Machine Intelligence* **13**(3), 224–251.

McKerrow, P. J. 1990. *Introduction to Robotics*. Reading: Addison-Wesley.

Okutomi, M., and T. Kanade. 1993. "A Multiple-Baseline Stereo." *IEEE Transactions on Patern Analysis and Machine Intelligence* **15**(4), 353–363.

Rahardja, K., and A. Kosaka. 1996. "Vision-Based Bin-Picking: Recognition and Localization of Multiple Complex Objects Using Simple Visual Cues." Proceedings of the IEEE/RSJ International Conference on Intelligent Robots and Systems vol. 3, pp. 1448–1457.

Rosenfeld, A., and A. C. Kak. 1982. *Digital Picture Processing*. 2nd ed., Vol. 2. New York: Academic Press.

Tsai, R. Y. 1988. "A Versatile Camera Calibration Technique for High Accuracy 3D Machine Vision Metrology Using Off-the-Shelf TV Cameras and Lenses." *IEEE Journal of Robotics and Automation* **RA-3**(4), 323–344.

Weng, J., P. Cohen, and M. Herniou. 1992. "Camera Calibration with Distortion Models and Accuracy Evaluation." *IEEE Transactions on Pattern Analysis and Machine Intelligence* **14**(10), 965–980.

CHAPTER 15

MOTION PLANNING AND CONTROL OF ROBOTS

Vijay Kumar
Miloš Žefran
James P. Ostrowski
University of Pennsylvania
Philadelphia, Pennsylvania

1 INTRODUCTION

In any robot task the robot has to move from a given initial configuration to a desired final configuration. Except for some degenerate cases, there are infinitely many motions for performing the task. Even in complex tasks, where the interactions of the robot with the environment may impose additional constraints on the motion, the set of all possible motions is still typically very large. Further, for a given motion there might be many different inputs (generally actuator forces or torques) that will produce the desired motion. Motion planning is the process of selecting a motion and the associated set of input forces and torques from the set of all possible motions and inputs while ensuring that all constraints are satisfied.

Motion planning can be viewed as a set of computations that provide subgoals or set points for robot control. The computations and the resulting motion plans are based on a suitable model of the robot and its environment. The task of getting the robot to follow the planned motion is called *control* and is addressed in Chapter 12. Robot control involves taking the planned motion as the desired motion and the planned actuator inputs (when provided) as the nominal inputs, measuring (sensing) the actual motion of the system, and controlling the actuators to produce the appropriate input forces or torques[1] to follow the desired motion. Generally, the inputs required for robot control will deviate from the nominal inputs to compensate for modeling errors and non-ideal actuators and sensors.

This chapter provides a review of the state of the art in motion planning. Because planning is viewed as a process that generates set points for control, often the distinction between planning and control blurs. Thus, in this chapter we will be forced to consider some issues pertaining to control. Further, since robots are dynamic systems that are governed by the laws of physics and are essentially continuous in behavior, our discussion will be biased toward methods that generate continuous motion plans. Discrete methods, in contrast, address the problem of obtaining motions via points or intermediate positions and orientations through which a lower-level controller can interpolate. Finally, we will not be worried about the specifics of a particular robot system. The ideas in this chapter are general and are applicable to different types of robot systems, including articulated, serial-chain robot arms, in-parallel, platform-like manipulators, and mobile robots.

2 MOTION PLANNING

2.1 The Task, Joint, and Actuator Spaces

The motion of a robot system can be described in three different spaces. First, the task is specified in the so-called *task space* or *Cartesian space*. It is customary to use Cartesian

---

[1]Generally, it is possible to design controllers that will control each actuator to produce a desired force (for a linear actuator) or torque (for a rotary actuator).

Handbook of Industrial Robotics, Second Edition, Edited by Shimon Y. Nof
ISBN 0-471-17783-0   © 1999 John Wiley & Sons, Inc.

x, y, and z coordinates to specify the position of a reference point on, say, the end-effector of a robot with respect to some absolute coordinate system and some form of Euler angles to specify the orientation. However, for a multi-degree-of-freedom robot, specifying the position of the end-effector may not specify the position of all the robot links.[2] For an n-degree-of-freedom robot, it may be necessary to specify the robot motion in the *joint space* by specifying the motion of n independent robot joints. The joint space is the Cartesian product of the intervals describing the allowable range of motion for each degree of freedom. Often it is simply a subset of $\mathbb{R}^n$. Finally, because a robot is a mechanical system governed by the equations of motion derived from physics, for each motion that is consistent with the kinematic and dynamic constraints, there must exist at least one set of input forces and torques that produce the motion. The actuator forces or torques that achieve all possible motions define the *actuator space*. Since actuator forces or torques are real-valued, the actuator space is a subset of $\mathbb{R}^m$, where m is the number of actuators.

Based on this classification, it is easy to see that it is possible to define a motion planning problem in the task space, joint space, or actuator space. In order for a robot to perform the task of, say, welding, it may be sufficient to plan the end-effector trajectory. One requires a motion plan in the task space. Such a motion plan may be satisfactory if there are no obstacles in the environment and the dynamics of the robot do not play an important role. However, if there are obstacles or if the robot arm has more than six degrees of freedom, it may be necessary to plan the motion of the arm in the joint space. Such a motion plan would guide the robot around obstacles while guaranteeing the desired end-effector trajectory for the welding task. It is easy to see that motion planning in the joint space can take constraints in the task space into account, but not the other way around. In other words, given a motion plan in the task space, it is possible to solve for a compatible motion plan in the joint space that accommodates additional constraints, particularly for a kinematically redundant robot where $n > 6$. However, a motion plan in the joint space automatically specifies the task space trajectory. Finally, if the task involves such dynamic constraints as actuator force/torque limits or limits on the contact forces in an assembly task, or if the task involves controlling a robot with torque-controlled motors, one may be interested in the history of actuator forces or torques for a specific task. Now the motion planning must be done in the actuator space, and the resulting plan will include a joint space trajectory and an actuator space trajectory.

The maps between the different spaces are shown in Figure 1. Because the direct kinematics and the forward dynamics maps always exist (see Chapters 5, 6), a motion plan in the actuator space (joint space) entails a motion plan in the joint space (task space). However, in redundant systems the reverse may not be true. It may not be possible to find a unique inverse kinematics in kinematically redundant systems. Similarly, in statically indeterminate systems (for example, multifingered grippers or multiarm sys-

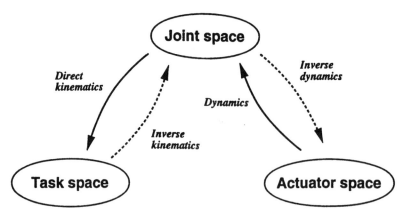

Figure 1   Spaces in which motion can be observed and mappings between them.

---

[2]Refer to Chapter 5, 6 for a discussion of ambiguities in the inverse kinematics map and the implications for kinematically redundant manipulators.

tems), where $m > n$, the inverse dynamics map may not exist. Finally, underactuated systems characterized by $m < n$ are typically subject to nonintegrable kinematic constraints called *nonholonomic constraints*. In such systems, given a feasible task space trajectory, finding the joint space trajectory can be very difficult. Further, not all joint space trajectories may be realizable with m actuators. Such systems are discussed in Section 4.3.

It is worth noting that the level at which motion planning is performed may be dictated by the type of robot controller. For example, if the joints are controlled by position control servos, there may be little point to determining a motion plan in the actuator space. In such a situation a joint space motion plan will generally suffice. However, if the controllers allow joint torque or actuator force control, it is meaningful to find a motion plan in the actuator space.

2.2 Obstacles and Configuration Space

A key theme in motion planning is the presence of obstacles and the need to find a path that is free of collisions. We will discuss the modeling of obstacles with the help of a simple example.

Consider a mobile robot in the horizontal plane. The task space is the group of all possible orientations and translations in the plane. This space is the Lie group $SE(2)$, the special Euclidean group[3] in two dimensions. The presence of obstacles creates holes or voids in the task space. Figure 2 shows the representation of a polygonal obstacle in the task space of a rectangular robot. The x and y axes in the plot in Figure 2b show the possible positions of a reference point on the mobile robot, labeled (x, y) in Figure 2a, and the third (θ) axis shows rotations in the plane. Each point in the three-dimensional plot in Figure 2b shows a possible position and orientation of the mobile robot. Note that $\theta = 0°$ must be identified with $\theta = 360°$ and therefore it is necessary to view the top face and bottom face of the plot as being connected. For each orientation θ, the set of all (x, y) that result in the robot touching the obstacle is a closed polygon. This polygon changes as a function of θ and generates the solid in Figure 2b. Points outside the solid are valid interference-free configurations, while other points represent intersections with the obstacle. The top view of the polygons that are cross sections of the solid is shown in Figure 2c.

The representation in Figure 2 is often called the *configuration space* or *C-space* representation of the obstacle. In the spirit of the definition of the configuration space that is used in classical mechanics, the figure simply shows the admissible values that the coordinates (x, y, θ) can have.

The generalization of this basic idea is as follows. The configuration of any robot system can be represented by a point in the configuration space (Lozano-Pérez and Wesley, 1979). In the case of an articulated robot arm, the configuration space is the joint space, while in the example we just considered, the configuration space is the Cartesian space. In the configuration space, an obstacle Obs_i maps to a set of configurations C_{Obs_i} in which the robot touches or penetrates the obstacle. Finding a collision-free path for a robot thus corresponds to finding a trajectory in the configuration space that does not intersect any of the sets C_{Obs_i}. Even for more complicated situations, such as a moving rigid body in three-dimensional space with obstacles, a C-space representation of each obstacle is possible. Although the computation of sets C_{Obs_i} is very time-consuming, it is usually only performed once, and a path between any arbitrary two configurations can be computed very efficiently afterwards. However, even for very simple robots it is often very difficult to visualize the obstacles C_{Obs_i} or depict them on paper.

2.3 Classification of Motion-Planning Algorithms

Until now we have treated motion planning as a set of computations that produce a set of subgoals or set points for a lower-level controller. The results involve explicit computations of the trajectory of the system, and in some cases the associated inputs, in the appropriate space. Such motion plans are called *explicit*.

In contrast to explicit motion plans, *implicit motion plans* provide trajectories (and actuator inputs) that are not explicitly computed before the motion occurs. Instead, the

---

[3]A good introduction to Lie groups with applications to robotics is provided in Murray, Li, and Sastry (1994).

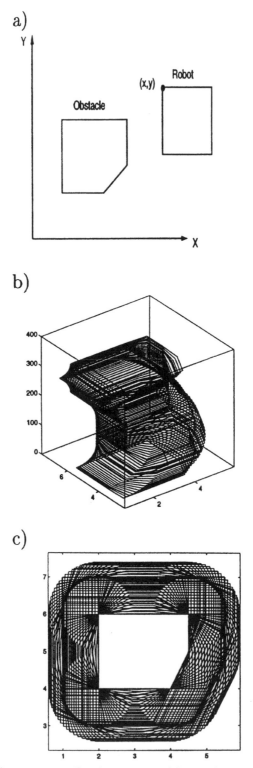

Figure 2 (a) The polygons representing the geometry of the mobile robot and the obstacle. (b) The representation of the obstacle in the configuration space ($SE(2)$). (c) The top view of the configuration space representation of the obstacle (projection on the x–y plane).

motion plan specifies how the robot interacts with the environment and how it responds to sensory information. An example of such an implicit scheme is the potential field algorithm developed by Khatib (1986). The key idea is to construct an artificial potential field according to which the robot is attracted to the goal while at the same time being repelled by obstacles in the environment. The motion of the robot is determined by the artificial potential field. However, this motion is never precomputed.

Explicit methods can either be *discrete* or *continuous.* Discrete approaches (Brooks, 1983; Canny, 1988; Latombe, 1991) focus on geometric constraints, generally introduced by the presence of obstacles, and the problem of finding a set of discrete configurations between the start and the goal configuration that are free from collisions. In contrast, explicit, continuous methods attempt to find feedforward or open-loop control laws for the control of the dynamic system (Bobrow, Dubowsky, and Gibson, 1985; Flash and Hogan, 1985; Shiller and Dubowsky, 1991). In fact, explicit, continuous motion plans can be viewed as open-loop control laws. They consist of the reference trajectory for the system and the associated actuator forces, and they are completely devoid of sensory feedback.

In contrast to open-loop control laws, closed-loop control laws can be viewed as implicit plans that characterize the deviation of the dynamic system from the nominal trajectory, which is specified, possibly by an explicit motion plan. Generally, such implicit methods are continuous in flavor.[4]

3 IMPLICIT METHODS

Implicit motion plans prescribe a strategy for controlling the robot in the presence of obstacles in the environment. In other words, they specify the desired dynamic behavior for the robot system. The motion plan is an algorithm that tells the robot how to move given its current state and its current knowledge.

In the potential field approach (Khatib, 1986) a potential function is defined in configuration space such that it has a minimum at the goal configuration. If the configuration space is a subset of $\mathbb{R}^n$, this is not difficult to do. If the coordinates of the robot system, for example the independent joint variables, are designated by $q \in \mathbb{R}^n$ and the goal configuration is $q_0 \in \mathbb{R}^n$, the potential function $\Phi_g(q) = (q - q_0)^T K(q - q_0)$ can be chosen so that K is a positive definite, $n \times n$, matrix. If a velocity controller is used to drive each joint with a velocity given by:

$$\dot{q} = -\nabla\Phi(q) \tag{1}$$

where $\Phi(q) = \Phi_g$, the robot is asymptotically driven to the goal configuration q_0.

If obstacles are present in the environment, this controller can be modified so that the velocity of the robot $\dot{q}$ points away from an obstacle when it gets near the obstacle. Let $d(q, Obs_i)$ be a function that returns the distance between the point on the robot that is closest (in $\mathbb{R}^3$) to the obstacle Obs_i and the point on Obs_i that is closest to the robot. One can construct a function, $\Phi_{Obs_i}(q) = -k_i/d(q, Obs_i)^l$, where l is a suitably chosen integer and k_i is a positive constant. If we choose $\Phi(q) = \Phi_g + \Phi_{Obs_i}(q)$, the velocity controller moves the robot toward the goal while moving away from Obs_i. We can add a potential function $\Phi_{Obs_i}(q)$ for each obstacle. Further, we can modify $\Phi_{Obs_i}(q)$ so that the range of the repulsive potential is limited. This way, only the obstacles that are close to the robot will affect its motion and the robot only needs local information about the environment to compute its direction according to Equation (1).

This approach is very attractive from a computational point of view because no processing is required prior to the motion. Further, it is easy to specify a dynamic behavior that tends to avoid obstacles. In fact, one school of thought believes human motor control and planning is organized in a similar fashion (Feldman, 1974).

The main drawback of the potential field approach is that when obstacles are present, the potential function, $\Phi(q)$, may not be convex and local minima (Koditschek, 1987) may exist that can "trap" the robot at points away from the goal. A second disadvantage

---

[4]Although closed-loop control laws can be specified as discrete control laws for a sampled data system, this discretization is not in the same spirit as in discrete methods for motion planning. The discretization is done in the time scale, while the configuration space, for example, is generally never discretized.

of this approach lies in the difficulty in predicting the actual trajectory. In practice, one has to carefully pick the constants in the potential function in order to guarantee obstacle avoidance. Further, the generated trajectories are usually far from being optimal by any measure. Finally, it is difficult to incorporate nonintegrable kinematic (nonholonomic) constraints and dynamic constraints into this approach.

The implicit planning strategy discussed thus far is a local strategy. In other words, it is based only on the knowledge of the immediate environment. The pitfalls associated with the local minima may be circumvented by a "global" strategy that is based on the complete knowledge of the environment. In many special, but important, cases, it is possible to construct a potential function (called the *navigation function*) (Rimon and Koditschek, 1992) that has a global minimum at the goal configuration, such that all other equilibrium points are saddle points (unstable equilibria) that lie in a set of measure zero. These special cases correspond to "sphere-like" models and diffeomorphisms of such models. However, constructing such a navigation function requires complete knowledge of the topology of the environment, and the simplicity and the computational advantage of the original potential field approach are lost. A review and algorithms of sensor-based, implicit motion planning for mobile robots can be found in Rimon, Kamon, and Canny (1997).

4 EXPLICIT METHODS

An explicit motion plan consists of the trajectory of the robot between the start and the goal configuration and, in some cases, the associated inputs in the appropriate space. We first discuss discrete methods and then continuous methods for generating explicit motion plans. The general class of continuous, explicit methods based on the variational approach is discussed in greater detail in Section 5.

4.1 Discrete Methods

Discrete planning algorithms consist of essentially two classes of algorithms: road map and cell decomposition methods (Latombe, 1991). Road map methods (Brooks, 1983; Canny, 1988; Ó'Dúnlaing, Sharir, and Yap, 1983) construct a set of curves, called a *road map,* that "sufficiently connects" the space. A path between two arbitrary points is found by choosing a curve on the road map and connecting each of the two points to this curve with a simple arc.

Cell decomposition methods (Schwartz and Sharir, 1983), on the other hand, divide the configuration space into nonoverlapping cells and construct a connectivity graph expressing the neighborhood relations between the cells. The cells are chosen so that a path between any two points in the cell is easily found. To find a trajectory between two points in the configuration space, a "corridor" is first identified by finding a path between the cells containing the two points in the connectivity graph. Subsequently, a path in the configuration space is obtained by appropriately connecting the cells that form the corridor.

The most general versions of road map and cell decomposition methods work for environments in which the obstacles in the configuration space can be described as semi-algebraic sets. In both types of methods the configuration space is discretized and represented by a graph. Thus, the problem of trajectory planning is eventually reduced to finding a path in this graph.

Such discrete motion planning methods are useful in environments that are cluttered with obstacles and in tasks in which the main challenge is simply to find a feasible collision-free path from the start configuration to the goal configuration. They generate a set of intermediate configurations so that the path between adjacent configurations can be easily determined. The excellent book by Latombe (1991) discusses motion planning methods for mobile robots and serial chain manipulators with many examples. While the methods accommodate geometric constraints as well as holonomic kinematic constraints, they do not lend themselves to nonholonomic and dynamic constraints. However, it is possible to refine a path that satisfies all geometric constraints so that it is consistent with nonholonomic constraints (Laumond et al., 1994). Thus, it is not difficult to imagine a two-step planning process in which one first finds a suitable initial plan in the configuration space (joint space) without worrying about nonholonomic kinematic constraints and the dynamics of the system, then develops a more realistic plan that satisfies all the constraints.

4.2 Continuous Methods

As we have said before, a continuous explicit method is essentially an open-loop control law. When a trajectory that is consistent with a set of constraints is selected, the trajectory and its derivatives constitute the feedforward component of the control law. Depending on the level at which the motion is planned, the motion planning is called *kinematic* or *dynamic* motion planning.

We will consider two types of continuous methods for motion planning in the presence of constraints. The first involves using tools from the nonholonomic motion planning literature to generate control laws consistent with the constraints. These methods have been applied to a special but important class of robotic systems that includes wheeled mobile robots with trailers, autonomous underwater vehicles, reorientable satellites, and unmanned aerial vehicles. These control laws, which are used to drive the system along the desired path, are discussed next in Section 4.3. Although these methods are very specialized, they are based on controllability tests, which are a necessary component for control and planning for nonlinear systems.

The second set of methods involves finding optimal trajectories that are consistent with geometric, kinematic, and dynamic constraints. This can be done in a continuous setting by solving an optimal control problem. Equivalently, the problem of minimizing a functional subject to constraints that may be differential or algebraic equations can be cast in the framework of variational calculus. Optimal control methods are treated in much greater detail in Section 5, and a discussion of how to incorporate modeling uncertainties is given in Section 6.

4.3 Underactuated Systems

It is often the case in robotics, and in particular in robotic motion planning, that the system of interest has fewer actuators, say m, than degrees of freedom, n. These systems are traditionally called *underactuated* and have recently received a great deal of attention in the literature. This is partly due to the fact that traditional models of control do not work for this case, since the linearization will always fail to be controllable. The most often studied examples of underactuated systems arise when certain directions of motion are forbidden due to no-slip constraints, such as those found in wheeled mobile robots. These constraints, called *nonholonomic kinematic* constraints, are nonintegrable constraint equations involving the joint variables and their first derivatives.

A typical example of such a system is seen in a wheeled platform made by Transition Research Corporation. The platform is driven by two coaxial powered wheels and has four passive castors for support. A schematic is shown in Figure 3. The robot's position, $(x, y, \theta) \in SE(2)$, is measured via a frame located at the center of the wheelbase. The robot's motion is driven by the movements of its two wheels, whose angles, (ϕ_1, ϕ_2), are measured relative to vertical. If the wheels do not slip with respect to the ground, we get three independent constraints:

$$-\dot{x} \sin \theta + \dot{y} \cos \theta = 0 \tag{2}$$

$$\dot{x} \cos \theta + \dot{y} \sin \theta + w\dot{\theta} = \rho\dot{\phi}_1 \tag{3}$$

$$\dot{x} \cos \theta + \dot{y} \sin \theta - w\dot{\theta} = \rho\dot{\phi}_2 \tag{4}$$

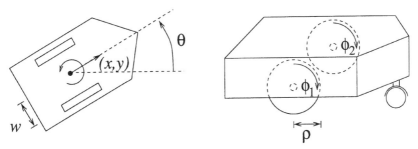

Figure 3 Two-wheeled planar mobile robot.

These constraints must be satisfied at all times and so partially determine the motion of the system in terms of three first-order ODEs in (x, y, θ):

$$
\begin{bmatrix} \dot{x} \\ \dot{y} \\ \dot{\theta} \end{bmatrix} = \begin{bmatrix} \cos\theta & -\sin\theta & 0 \\ \sin\theta & \cos\theta & 0 \\ 0 & 0 & 1 \end{bmatrix} \begin{bmatrix} \dfrac{\rho}{2}(\dot{\phi}_1 + \dot{\phi}_2) \\ 0 \\ \dfrac{\rho}{2w}(\dot{\phi}_1 - \dot{\phi}_2) \end{bmatrix} = \dfrac{\rho}{2} \begin{bmatrix} \cos\theta \\ \sin\theta \\ \dfrac{1}{w} \end{bmatrix} u_1 + \dfrac{\rho}{2} \begin{bmatrix} \cos\theta \\ \sin\theta \\ -\dfrac{1}{w} \end{bmatrix} u_2 \qquad (5)
$$

This is a nonintegrable set of equations and therefore represents a set of nonholonomic constraints. This system is underactuated—it is possible to move the vehicle to any point in $SE(2)$ (a three-dimensional task space) with only two inputs $u_1 = \dot{\phi}_1$, $u_2 = \dot{\phi}_2$ ($m = 2$). However, not all trajectories in $SE(2)$ can be followed by this vehicle. Thus, a kinematic motion plan must incorporate the constraint (5).

We treat the problem of generating kinematic motion plans for underactuated systems in this section and leave the treatment for general systems for the next section. We will first start with a statement for the kinematic motion planning problem for nonholonomic systems that are *driftless*.

Problem 1 (Nonholonomic Motion Planning Problem)

Let the robot be described in the state space with the equations:

$$
\dot{q} = f_1(q)u_1 + \ldots + f_m(q)u_m \qquad (6)
$$

where $q \in Q \subset \mathbb{R}_n$ is the vector of configuration space variables and $u \in \mathbb{R}_m$ is the vector of inputs. Let the initial and final configurations be given by

$$
q(t = t_0) = q_0 \qquad (7)
$$

and

$$
q(t = t_1) = q_1 \qquad (8)
$$

The motion planning problem is to find a piecewise continuous, bounded input vector $u(t)$ so that Equation (8) is satisfied.

Standard results from nonlinear control theory provide a means for testing whether a given system is controllable—that is, whether a particular set of inputs can be chosen to move the system between any two desired points. The results allow us to test for the existence of a solution to the motion planning problem above. These tests are based on the use of the *Lie bracket*, $[f_i, f_j]$, defined as

$$
[f_i, f_j](q) = \frac{\partial f_j}{\partial q} f_i - \frac{\partial f_i}{\partial q} f_j \qquad (9)
$$

where f_i and f_j ($i, j \in 1, \ldots, m$) are vector fields on the configuration space, Q. The Lie bracket is a skew-symmetric operator that returns a vector field and provides a measure of how two flows (corresponding to the two vector fields f_i and f_j) commute. Roughly speaking, the Lie bracket corresponds to the infinitesimal motions that result when one flows forward by f_i and f_j in succession and then flows backward again by $-f_i$ and $-f_j$ in succession.

In linear systems this sequence of inputs would obviously lead to no net motion. For nonlinear systems, however, one can generate new directions of motion simply by commuting vector fields using the Lie bracket. This is done by recursively bracketing each of the new directions of motion with those that were used to generate them. We will denote by Δ_0 the span of the original input vector fields, $f_1, \ldots, f_m$, and denote the successive bracketing of input vector fields by

$$
\Delta_i = \Delta_{i-1} + \text{span}\{[\alpha, \beta] : \alpha \in \Delta_0, \beta \in \Delta_{i-1}\} \qquad (10)
$$

Then the test for controllability is given by Chow's theorem, which states that the system is controllable if and only if $\Delta_k = \mathbb{R}^n$ for some k.[5]

In the above example of a mobile robot (see Equation (5)), we would label $f_1 = [\rho/2 \cos \theta, \rho/2 \sin \theta, \rho/2w]^T$ and $f_2 = [\rho/2 \cos \theta, \rho/2 \sin \theta, -\rho/2w]^T$. The Lie bracket of these two inputs is then just $f_3 = [f_1, f_2] = [-\rho^2/2w \sin \theta, \rho^2/2w \cos \theta, 0]^T$. Since these vector fields span the allowable directions of motion in $SE(2)$, the system is controllable.

The controllability tests used in Chow's theorem also provide a constructive means for generating desired motions and hence for controlling a nonholonomically constrained robot along a motion plan. These techniques, usually known as *steering* methods, grew from research in nonlinear control (Brockett, 1981; Sussmann, 1983) and have been very successfully applied to nonholonomic systems (Li and Canny, 1993). A large class of nonholonomic systems can be steered to a desired configuration using sinusoids (Murray and Sastry, 1993), and extensions to this basic idea are derived in Sussmann and Liu (1993). Recent advances in modeling of systems with symmetry and modeling of nonholonomic mechanical systems (Bloch et al., 1994; Ostrowski, 1995) make some of these ideas applicable to dynamic motion planning.

To utilize this theory, one begins by assuming that there are m directions of motion that can be controlled directly by the m inputs. The remaining $n - m$ directions of motion must be controlled by appropriately combining these inputs. It has been shown (Murray and Sastry, 1993; Sussmann and Liu, 1993) that the remaining directions, which are generated using the iterated Lie brackets, can be controlled by pairing sinusoidal inputs whose frequencies are appropriately matched according to the Lie brackets that generate them.

To be more specific, consider the goal of producing a motion associated with a first-order Lie bracket of the form $[f_i, f_j]$. The method of steering using highly oscillatory inputs states that motion in the $[f_i, f_j]$ direction can be achieved by using inputs of the form $u_i = \eta_i(t) + \sqrt{\omega}\eta_{ij}(t) \sin \omega t$ and $u_j = \eta_j(t) + \sqrt{\omega}\eta_{ji}(t) \cos \omega t$. In the limit $\omega \to \infty$, the motion of the system will then follow

$$\dot{q} = f_i(q)\eta_i(t) + f_j(q)\eta_j(t) + \frac{1}{2} \eta_{ij}(t)\eta_{ji}(t)[f_j, f_i](q) \tag{11}$$

In other words, using inputs of the form of u_i and u_j allows one to follow the Lie bracket direction as if it were one of the original controlled directions. Similar methods can be used to generate motions in directions of higher-order Lie brackets by appropriately modifying the frequency of the inputs. For example, in order to generate a direction of motion corresponding to the Lie bracket $[f_i, [f_i, f_j]]$, one would choose sinusoids of frequency ωt and $2\omega t$ respectively. Details for constructing such motion plans are discussed in Murray and Sastry (1993).

The use of high-frequency oscillations, however, may not always be appropriate. We note that other techniques similar to this exist, such as (near) nilpotent approximation (Lafferriere and Sussmann, 1993). These methods rely on similar notions of determining appropriate input patterns based on the vector fields generated by the Lie bracket tests used in Chow's theorem. The ability to direct the movements of an underactuated system in all the desired directions is very important for the motion planning problem. It allows the division of the problem into (1) the generation of a feasible path under the assumption that there are no constraints (or the system is fully actuated), and (2) the steering control of the robot about this nominal path using physically allowable inputs.

5 OPTIMAL CONTROL AND VARIATIONAL APPROACHES

The problem of finding an optimal trajectory with an associated set of inputs can be viewed as a constrained minimization problem. The problem of minimizing a functional on the trajectories of a dynamical system described with a set of differential equations is traditionally studied in optimal control. The basic problem is formulated as follows.

---

[5]Under certain regularity assumptions, this sequence of Δ_i's will always terminate at $k \leq n$, where $n = \dim(Q)$ (Murray and Sastry, 1993).

Problem 2 (General Motion Planning Problem)

Let the robot be described in the state space with the equations:

$$\dot{x} = f(x, u, t) \tag{12}$$

where x is the vector of state variables and u is the vector of inputs. Suppose that during the motion the robot must satisfy a set of equality and inequality constraints:

$$g_i(x, u, t) = 0 \qquad i = 1, \ldots, k \tag{13}$$

and

$$h_i(x, u, t) \leq 0 \quad i = 1, \ldots, l \tag{14}$$

and let the initial and final configurations be given by

$$\alpha(x, t)\big|_{t_0} = 0 \tag{15}$$

and

$$\beta(x, t)\big|_{t_1} = 0 \tag{16}$$

Choose a cost functional:

$$J = \Psi(x(t_1), t_1) + \int_{t_0}^{t_1} L(x(t), u(t), t) \, dt \tag{17}$$

The functions f, g, h, and L are assumed to be of class C^2 in all variables, while the functions α, β, and Ψ are assumed to be of class C^1.

The motion planning problem is to find (a piecewise smooth) state vector $x^*(t)$ and (a piecewise continuous, bounded) input vector $u^*(t)$ that satisfy Equations (12)–(16) and minimize the cost functional (Laumond et al., 1994).

The function f describes the dynamics of the system and g the kinematic constraints on the system. The function h describes the unilateral constraints due to the presence of obstacles and limits on joint positions, velocities, actuator inputs, and contact (including frictional) forces. α describes the starting configuration and β the goal configuration. The cost functional J consists of a penalty on the end configuration and an integral. The function L may be chosen so that a meaningful quantity such as a distance, time, or energy is minimized.

5.1 Kinematic Motion Planning

We consider the kinematic motion planning problem when no obstacles are present. Here we are simply interested in generating a smooth path or motion from the initial to the final position and orientation, while perhaps interpolating through intermediate points. This problem has important applications in robot trajectory generation and computer graphics. We pose it as a trajectory generation problem on a Riemannian manifold as follows.

Problem 3 (Kinematic Motion Planning)

Let the task space be described with a manifold Σ and take a Riemannian metric $\langle \cdot, \cdot \rangle$ on Σ. Choose the desired initial and final configurations p_0 and p_1 for the robot in the task space ($p_0, p_1 \in \Sigma$) and let C_0 and C_1 be the additional conditions that have to be satisfied at these two points. Take the family $\mathcal{G}$ of all smooth curves $\gamma : [0, 1] \to \Sigma$ that start at p_0, end at p_1, and satisfy C_0 and C_1: $\mathcal{G} = \{\gamma \mid \gamma(0) = p_0, \gamma(1) = p_1, C_0(\gamma), C_1(\gamma)\}$. Define a function $L : \mathcal{G} \to T\Sigma$ and a functional $J: \mathcal{G} \to \mathbb{R}$ of the form:

$$J(\gamma) = \int_0^1 \left\langle L\left(\gamma, \frac{d\gamma}{dt}\right), L\left(\gamma, \frac{d\gamma}{dt}\right) \right\rangle dt \tag{18}$$

The kinematic motion planning problem on Σ is to find a curve $\gamma \in \mathcal{G}$ that minimizes the functional, Equation (18).

We consider, for the purposes of illustration, reaching from one point to another in the plane, where the task space or the manifold Σ is simply the vector space $\mathbb{R}^2$. In order to achieve a smooth motion, it is meaningful to choose a cost function that is the integral of the square norm of an appropriate time derivative of the position vector. For example, if the motion starts from rest with zero acceleration and ends with a prescribed velocity and acceleration, it is necessary to satisfy boundary conditions on the velocity and acceleration at the start and goal points (conditions C_0 and C_1 in Problem 3). A smooth motion satisfying these conditions can be achieved by minimizing the integral of the square norm of jerk (jerk is the derivative of acceleration):

$$J = \frac{1}{2} \int_{t_0}^{t_1} (\ddot{v}_x^2 + \ddot{v}_y^2) \, dt \tag{19}$$

where $v_x = \dot{x}$ and $v_y = \dot{y}$ are the velocity components. The resulting trajectories are given by the Euler–Lagrange equations:

$$\frac{d^5 v_x}{dt^5} = 0, \qquad \frac{d^5 v_y}{dt^5} = 0 \tag{20}$$

This model for trajectory generation has also been used to model human performance in voluntary, planar, reaching tasks (Flash and Hogan, 1985).

Now consider manipulation tasks that include positioning and orienting an object in the plane. The task space Σ is now the Lie group $SE(2)$. We are interested in planning smooth motions on $SE(2)$ while guaranteeing that the motions are independent of the choice of the inertial reference frame (world coordinate system) and the end-effector reference frame. In order to formalize these ideas, it is convenient to formulate kinematic motion planning in the framework of differential geometry and Lie groups (Boothby, 1986; Murray, Li, and Sastry, 1994). Using purely geometric ideas also makes the planning method independent of the description (parameterization) of the space.[6] Geometric analysis reveals that it is necessary to define the concept of distance (a Riemannian metric) and a method of differentiation (an affine connection) on $SE(2)$ before the notion of smoothness can be defined (Žefran and Kumar, 1997). Since a Riemannian metric naturally leads to an affine connection, once a metric is chosen, trajectories with different smoothness properties can be generated. Further, by properly choosing a metric we can obtain trajectories with desired invariance properties.

In the framework of Problem 3, L is a vector-valued function that is a local measure of smoothness of the trajectory and usually depends on the affine connection corresponding to the chosen metric. For example, the general expression for the minimum-jerk cost functional is:

$$J_{\text{jerk}} = \int_a^b \langle \nabla_V \nabla_V V, \nabla_V \nabla_V V \rangle \, dt \tag{21}$$

Contrast this equation with Equation (19). In Equation (21) $V = d\gamma/dt$ is the velocity vector field and ∇ is the affine connection obtained from the chosen Riemannian metric. $\nabla_V V$ is the Cartesian acceleration and $\nabla_V \nabla_V V$ is the Cartesian jerk. The trajectories that minimize (21) are given by the Euler–Lagrange equation:

$$\nabla_V^5 V + R(V, \nabla_V^3 V)V - R(\nabla_V V, \nabla_V^2 V) = 0 \tag{22}$$

where R denotes the metric-dependent tensor describing the curvature properties of the space. In the special case where the task space is $\mathbb{R}^2$, this equation reduces to Equation (20).

Figure 4 shows some examples of trajectories that are generated using this approach. Figure 4a shows the trajectories for the case where the distance between two points is minimized and the cost functional is

---

[6]This is particularly important in the context of rotations in three dimensions.

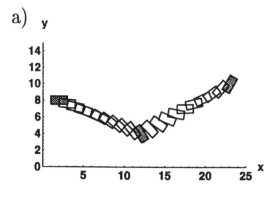

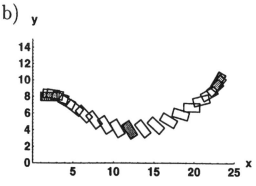

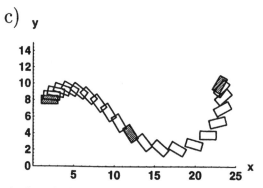

Figure 4 Examples of trajectories between specified configurations (shown shaded). (*a*) Shortest distance trajectories (geodesics). (*b*) Maximally smooth interpolant computed from a positive definite left-invariant metric. (*c*) Maximally smooth interpolant computed from an indefinite bi-invariant metric.

$$J_{\text{vel}} = \int_a^b \langle V, V \rangle \, dt$$

and the Riemannian metric is chosen to be such that the square of the length of a velocity vector consisting of an angular velocity ω and a linear velocity v expressed in the end-effector frame is given by:

$$\langle V, V \rangle = D \times \omega^2 + v^2$$

where D is a suitably chosen length scale. This metric is independent of the choice of the inertial frame but depends on the position and orientation of the end-effector frame and the length scale D. The trajectory in Figure 4a turns out to be independent of D.

We can also obtain generalized spline motions using this framework. For example, a maximally smooth trajectory that is C^1 continuous (i.e., it can satisfy arbitrary boundary conditions on the velocities) is a generalization of a cubic spline and can be obtained by minimizing the integral of the acceleration along the trajectory:

$$J_{acc} = \int_a^b \langle \nabla_V V, \nabla_V V \rangle \, dt$$

Figures 4b and c show generalized cubic splines that satisfy boundary conditions on positions and velocities and pass through a given intermediate configuration for two different choices of the metric for the space. The metric in Figure 4b is the scale-dependent metric discussed above. The resulting trajectory depends on the choice of the end-effector reference frame. The trajectory in Figure 4b is derived from a metric that is independent of both the end-effector and inertial reference frames. However, this metric is indefinite. See Žefran and Kumar (1997) for more details on the choice of metrics and the implications for robot motion planning.

While kinematic motion plans may not be adequate for some applications, they have the advantage that they can be easily computed. Some important problems even have explicit, closed-form solutions (Žefran and Kumar, 1997). If a detailed dynamic model of a mobile robot system is not available, it may be desirable to simply determine the smoothest trajectory that satisfies the required boundary conditions. In such a situation, left-invariance (invariance with respect to the choice of the inertial frame) is desirable. If dynamic motion plans are desired, the kinematic motion plans can be used as the first step of a two-step planning process. However, the framework of Problem 3 is not particularly useful in the presence of constraints. Even if geometric constraints such as those due to obstacles are present (Buckley, 1985), it is necessary to resort to the motion-planning problem in Problem 2.

5.2 Dynamic Motion Planning

5.2.1 Time-Optimal Motion Plans

Optimal control methods have been extensively used in robotics for time-optimal control. The objective is to minimize the duration of motion between the initial and goal configurations while satisfying geometric, kinematic, and dynamic constraints (Kahn and Roth, 1971). There are two approaches to time-optimal control. The first set of methods addresses the general problem of finding the trajectories and the associated actuator inputs, as specified in Problem 2. The second set of methods address the simpler problem in which a path that satisfies geometric constraints is specified and the goal is to obtain a set of actuator inputs that achieves time-optimal motion along the path while satisfying dynamic constraints.

Time-optimal control along a specified path is equivalent to the time-optimal control of a double integrator system. The optimal solution will therefore be "bang-bang." In other words, each input is always at its upper or lower limit, and it instantaneously switches between the limits finitely many times (Bobrow, Dubowsky, and Gibson, 1985; Žefran, Kumar, and Yun, 1994). Based on the definition of a *velocity limit curve* in the two-dimensional state space for the specified path, it is possible to compute the switching points (Bobrow, Dubowsky, and Gibson, 1985; Slotine and Yang, 1989). There is extensive literature on this subject including algorithms for contact tasks such as contour following (Huang and McClamroch, 1988) and for the so-called singular trajectories (Shiller, 1992).

Time-optimal solutions are often not practical because they require discontinuities (jumps) in the actuator forces. One approach to overcoming this practical difficulty is to minimize a suitable combination of time duration and the energy required to perform the motion (Singh and Leu, 1989). Clearly, a minimum time motion will require a lot of energy and there is an obvious tradeoff between the two criteria.

The problem of solving Problem 2 for time-optimal trajectories is significantly more complicated. One approach to this problem is suggested by Shiller and Dubowsky (1991) for computing time-optimal inputs and trajectories for a robot moving among obstacles.

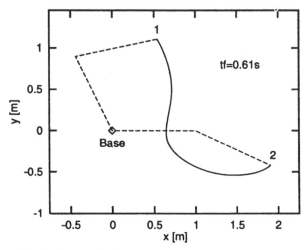

Figure 5 Time-optimal trajectory of the manipulator going from configuration 1 to configuration 2.

The trajectories are approximated with B-splines and the algorithm from Bobrow, Dubowsky, and Gibson (1985) is used in conjunction with graph-search techniques to find the time-optimal trajectory. Alternatively, it is possible to discretize the trajectory and use finite-difference methods to solve the optimization problem (Žefran, Kumar, and Yun, 1994). An example of this is shown in Figure 5 for a planar manipulator with two revolute joints. Each link is 1 m long with a homogeneous mass distribution of 1 kg. The joints are actuated by motors with torque limits of 20 Newton-meter and 10 Newton-meter respectively. The torque histories are shown in Figure 6. The solution for the torques can be seen to be bang-bang, and there are three switches.

The key to evaluating such solution techniques is the computational time. With a mesh size of 200 time steps this process takes roughly 1 minute on a 100-MHz machine. This is acceptable as an off-line tool for designing trajectories in a workcell. However, the time required can be expected to be significantly higher for more complicated robot geometries and in the presence of obstacles.

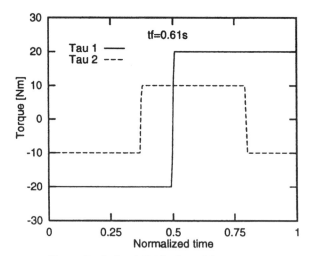

Figure 6 Optimal distribution of the torques.

5.2.2   Other Cost Functionals

In applications where time and limits on the actuator forces are not critical, other cost functionals can be used to obtain smooth motion plans. For example, energy-optimal plans are satisfying from at least an intellectual standpoint (Vukobratović and Kirćanski, 1982), although they may not be very useful for industrial applications. The minimization of the integral of the square norm of the vector of joint torques has also been proposed for obtaining optimal trajectories (Suh and Hollerbach, 1987). Because the energy consumed by an actuator (for example, a DC motor) often goes as the square of the torque, the minimum torque trajectories are also the minimum energy trajectories.

The optimal trajectories for minimum time or energy may result in actuator inputs that exhibit rapid rates of change. Because actuators are themselves dynamic systems with finite response times, they may not be able to follow the optimal input histories. One solution to this problem is to try to minimize the rate of change of actuator inputs (Uno, Kawato, and Suzuki, 1989). One possible approach is use the cost function:

$$J = \int_{t_0}^{t_1} (\dot{\tau}_1^2 + \dot{\tau}_2^2 + \ldots + \dot{\tau}_m^2) \, dt$$

Figure 7 shows an example of the motion plan for two mobile manipulators negotiating the free space between two obstacles, generated by minimizing the integral of the rate of change of actuator torques. In this problem we have constraints due to the dynamics, the obstacles, and the nonholonomic nature of the mobile platforms. It is interesting to note that the system shows some apparently discrete behaviors—the two mobile manipulators reconfigure from a "march-abreast" formation to a "follow-the-leader" formation, then follow a straight line to the closest obstacle and pursue a trajectory that hugs that obstacle until they are beyond the constriction before reconfiguring back to the "march-abreast" formation and taking the unconstrained straight-line path to the goal (Žefran, Desai, and Kumar, 1996).

5.2.3   Numerical Methods for Optimal Trajectories

Numerical methods for solving optimal control problems can be divided into four categories:

1. Indirect methods
2. Direct methods
3. Discretization methods
4. Dynamic programming methods

In indirect methods the solution that is an extremal is computed from the minimum principle (first-order necessary conditions) by solving a two-point boundary value prob-

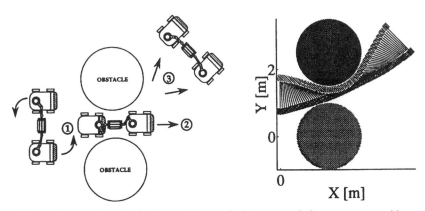

Figure 7   A motion plan for two mobile manipulators negotiating a narrow corridor.

lem. The most popular technique for solving boundary value problems is to guess the missing initial values and improve the guess until the boundary values are matched. If the initial guess is close to the exact solution, this method can be shown to converge. For obvious reasons, the technique is also called the *method of neighboring extremals* (Bryson and Ho, 1975). The main drawback of the method is that it is very sensitive to the initial guess.

In contrast, direct methods are designed to minimize the cost functional with respect to u. Gradient-based methods, for example, rely on the computation of $\partial J / \partial u$. In analogy with the minimization of functions in finite dimensions, first-order and second-order gradient methods can be formulated, as well as conjugate gradient algorithms (Jacobson, 1968; Lasdon, Mitter, and Warren, 1967; Ma and Levine, 1993; Mayne and Polak, 1987). They generally deal with constraints by adjoining them to the cost functional in the form of *penalty functions*. Compared to indirect methods, direct methods prove to be quite robust and globally convergent, but computationally more expensive.

If the continuous problem is discretized in time, the optimization problem reduces to a finite-dimensional problem. In discretization techniques, methods of finite-dimensional mathematical programming are used to find a solution. Some methods only discretize the vector of controls (Teo, Goh, and Wong, 1991), while the others discretize both the state and controls (Pytlak and Vinter, 1995). Another possibility is to interpolate the unknown functions with a set of basis functions (collocation methods) (Neuman and Sen, 1973). As with the direct methods, the constraints are often adjoined to the cost functional with the penalty functions. The main advantage is that approximate solutions can be obtained with coarse mesh sizes very quickly and then refined with finer mesh sizes to obtain more accurate solutions.

An efficient numerical technique for solving variational problems based on finite-element techniques and finite-dimensional optimization is described in Žefran (1996). Applications of this framework to several practical examples, such as two-arm coordinated manipulation with frictional constraints, multifingered grasping, and coordination of multiple mobile manipulators, are presented in Žefran, Desai, and Kumar (1996). The typical run time for a system with a 20-dimensional state space on a 200-MHz machine (for example, the system shown in Figure 7) is roughly two minutes.

In contrast to previous methods, dynamic programming relies on discretizing the state space. The basic idea is to compute the optimal value of the cost functional for each grid point taken as the initial state for the optimization. This value is called the *optimal return function* (Bryson and Ho, 1975) and is computed by a recursive algorithm (Bellman, 1957). Given the values of the optimal return function, it is easy to construct an extremal starting at an arbitrary initial point. Dynamic programming always produces a global minimum, but the computation becomes prohibitively expensive as the dimension of the state space increases.

6 MODELING ERRORS AND UNCERTAINTY

There are several properties that a dynamic motion planning method should possess:

1. The method must account for the dynamics of the system and provide the task space trajectory, the joint space trajectory, and the actuator forces.
2. It is desirable that trajectory generation and resolution of kinematic and actuator redundancies be performed within the same framework.
3. The method must be capable of dealing with additional equality and inequality constraints, such as kinematic closure equations, nonholonomic constraints, joint or actuator limits, and constraints due to obstacles.
4. Because there are usually one or more natural measures of performance for the task, it is desirable to find a motion with the best performance.
5. One would like to develop plans that are robust with respect to modeling errors.
6. The robot's ability to use sensory information for error correction should be explicitly incorporated.

While the discussion in Section 5 mostly focused on 1–4, issues pertaining to modeling errors and poor sensory information have so far been neglected. This section addresses the motion planning problem when only a nominal, parametric model of the environment is available and the model may be subject to errors.

Game theory provides the natural framework for solving optimization problems with uncertainty. The motion-planning problem can be formulated as a two-person zero-sum game (Basar and Olsder, 1982) in which the robot is a player and the obstacles and the other robots are the adversary. The goal is to find the control law that yields the best performance in the face of the worst possible uncertainty (a saddle-point strategy). The motion planning problem can be solved with open-loop control laws (as we have considered thus far) or with closed-loop control laws or feedback policies. Rather than develop the notation and the theory that is required for the framework of game theory, we present representative examples and discuss optimal open-loop and closed-loop control laws.

In order to explore how the framework of the previous section may be extended, consider the situation where we have a deterministic model of the robot and exact state information but there is uncertainty in the environment. Except in very structured environments, it is dangerous to assume a prior distribution for the noise or uncertainty in the environment. It is more realistic to assume a sensor system and estimation algorithms that return, along with the measurement of each parameter in the model, a confidence interval for each parameter in which the true value lies. Examples of such confidence set-based estimation algorithms are discussed in Kamberova, Mandlebaum, and Mintz (1997).

6.1   Open-Loop Motion Plans

The approach in the previous sections for generating open-loop trajectories for deterministic systems can be extended in an obvious fashion to problems with uncertainty. The uncertainty in the environment is incorporated through conservative bounds on the feasible regions in the state space. This effectively amounts to making the obstacles bigger, reflecting the uncertainty in the position and the geometry of the obstacles.[7] With the additional sensory information that becomes available during the execution of the plan, the bounds on the feasible regions can be made less conservative and the open-loop plan can be refined accordingly. This method is attractive because our numerical method lends itself to efficient remeshing and refining and the computational cost of refining an open-loop plan, when the changes in the model remain small, is an order of magnitude less than the cost of generating an initial open-loop plan. Thus, open-loop plans may be recursively refined (Zhang, Kumar, and Ostrowski, 1988).

An example of this approach is demonstrated in Figure 8, where two nonholonomic robot vehicles of the type considered in Section 4, Robot 1 and Robot 2, are to interchange their positions while moving through a narrow corridor formed by two obstacles. Each

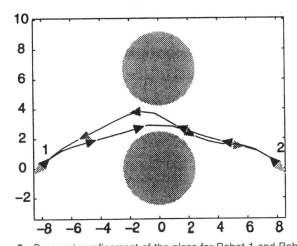

Figure 8   Successive refinement of the plans for Robot 1 and Robot 2.

---

[7]The resulting motion plan can be shown to be a saddle-point solution in the set of all open-loop strategies.

robot replans (refines) the initial open-loop trajectory at the points shown by the markers. The shading of each robot in this diagram represents the time elapsed, moving from an initial dark shading (at $t = 0$) to a light shading (at $t = 1$). Neither robot knows the other's task or planned route. Each robot determines its open-loop control based only on its estimate of the current position and orientation of the other robot and the obstacle. Thus, the robots change their motion plans only when the two robots are about to collide. While the refinement of the plans is locally optimal, it is clearly not globally optimal, and the resulting trajectories are more expensive than needed. In this simulation, Robot 1 is given a priority over Robot 2 and so follows a path that is closer to being optimal.

6.2 Closed-Loop Motion Plans

While it is possible to incorporate modeling uncertainties using such approaches, they invariably lead to suboptimal paths. Further, these paths are designed to stay away from areas that have even a very small probability of being occupied by an obstacle. There is clearly a tradeoff between the conservative strategy of skirting the uncertain boundaries of the obstacle and the more aggressive strategy that incorporates better sensory information about the obstacle as it gets closer to it. Such an aggressive strategy requires feedback control, suggesting that the motion planning should be reformulated as a search for the optimal feedback control law. In this formulation it is necessary to concurrently consider the dynamics of the system and the problem of estimating the geometry of the environment.

A simplified but realistic problem that is mathematically tractable is discussed below. We assume a single robot and an obstacle (or obstacles) that can be observed with a sensor. The sensor estimates the position of the obstacle with some uncertainty bounds depending on the distance between the robot and the obstacle. We consider a simple model for the robot dynamics:

$$\dot{x} = u \tag{23}$$

where the vector x is the position of the robot and u is the vector of inputs. The obstacles (including other robots) define a set of points in R^2 parameterized by a vector $y \in \mathbb{R}^k$. The initial model of the environment is denoted by y_0. $d(x, y)$ is a distance function whose value is the Euclidean distance between the nearest pair of points, one on an obstacle and the other on the robot. We use $\hat{y} \in R^k$ to denote the estimated obstacle(s). The basic idea is that $\hat{y} \rightarrow y$ as $d(x, y) \rightarrow 0$. An example is provided by a simple sensor model:

$$\hat{y}_i = y_i + (y_{0,i} - y_i)e^{-\beta d(x,y)^{-1}} \tag{24}$$

where the exponential law is scaled by the parameter β so that the effect is approximately linear in an appropriate neighborhood of the obstacle and levels off to the initial (worst-case) value further away.

For this problem, we are interested in obtaining a (static) feedback control law[8] $u^* = u(x, y)$ that will minimize the traveled distance as well as ensure that the robot avoids the obstacle and reaches the goal. We can allow the robot to come arbitrarily close to the obstacle, but we want to prevent collisions. Thus the allowable feedback policies are restricted to ones for which $d(x(t), \hat{y}(t)) \geq 0$, through the time interval $[0, T]$.

In general it is difficult to find even a sufficing feedback strategy $u(x, y)$ for the above problem (Rimon and Koditschek, 1992). One way to simplify the computation is to parameterize the control law and find the optimal values for the parameters. For example, we can try to find the optimal linear feedback control law:

$$u(x, y) = A(x^d - x) + B(x - y) \tag{25}$$

The task then becomes to find the optimal values for the matrices A and B. For even simple problems it is difficult to find a feasible linear feedback law. It is more practical to consider the set of all piecewise linear feedback laws. In order to find the optimal feedback policy, we can divide the path into discrete intervals, in each of which the

---

[8]Strictly speaking, the u is a function of x and $\hat{y}$.

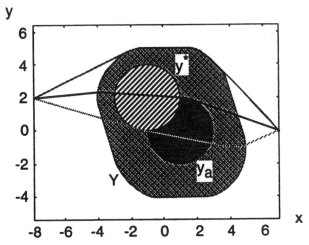

Figure 9   A comparison of the worst-case path with the optimal feedback law with that generated by the open-loop control law.

feedback parameters A and B are held constant. The task now is to determine the values of A and B in each time interval.

In Figure 9 we show the motion plan when an obstacle is known to belong to a compact subset Y in R^m, but the exact location is unknown. The set Y is shown shaded, the worst obstacle location y^* is shown hatched, and the actual object location $y_a \in R^m$ is shown in black. The figure shows three trajectories. The longest path (shown gray) is the most conservative one that would have to be taken using a purely open-loop approach. The intermediate path (shown dotted) is the worst-case path that could possibly arise with the optimal feedback law. In other words, this is the path given by u^* for the worst-case object y^*. Finally, the shortest path (shown solid) is the path followed for the obstacle y_a under the optimal feedback law u^*.

This approach can be used to solve more complicated min–max (or inf–sup) motion planning problems. However, while the simplified model (23–24) may guarantee the existence of a saddle point (Basar and Olsder, 1982), this is not the case in more complicated situations. Even if there are saddle-point solutions, there may be many such solutions. Finally, the computational cost of generating a min–max solution is an order of magnitude higher than solving the open-loop problem (essentially a single-person game).

7   CONCLUDING REMARKS

In this chapter we have presented a summary of the state of the art in motion planning and control. Because our motivation for planning motions comes from the need to control a robot system, this discussion was biased toward continuous methods in motion planning. In particular, we went to considerable length to explain how tools from optimal control can be used to obtain a kinematic or dynamic motion plan. Many tools are now available to solve such problems. However, motion planning and control for systems with uncertainty in sensing and modeling is still an ongoing research area. In the continuous setting, motion planning in the presence of uncertainty can be formulated as a two-person game and results from game theory can be used to solve such problems. However, the problems are not only intractable from an analytical viewpoint, but computationally very difficult.

REFERENCES

Basar, T., and G. J. Olsder. 1982. *Dynamic Noncooperative Game Theory.* London: Academic Press.

Bellman, R. *Dynamic Programming.* 1957. Princeton: Princeton University Press.

Bloch, A. M., et al. 1994. "Nonholonomic Mechanical Systems and Symmetry." Tech. Rep. CDS 94-013, California Institute of Technology.

Bobrow, J., S. Dubowsky, and J. Gibson. 1985. "Time-Optimal Control of Robotic Manipulators Along Specified Paths." *International Journal of Robotic Research* **4**(3), 3–17.

Boothby, W. M. 1986. *An Introduction to Differentiable Manifolds and Riemannian Geometry.* Orlando: Academic Press.

Brockett, R. W. 1981. "Control Theory and Singular Riemannian Geometry." In *New Directions in Applied Mathematics.* Ed. P. Hilton and G. Young. New York: Springer-Verlag. 11–27.

Brooks, R. A. 1983. "Solving the Find-Path Problem by Good Representation of Free Space." *IEEE Transactions on Systems, Man, and Cybernetics* **SMC-13**(3), 190–197.

Bryson, A. E., and Y.-C. Ho. 1975. *Applied Optimal Control.* New York: Hemisphere.

Buckley, C. E. 1985. "The Application of Continuum Methods to Path Planning." Ph.D. Thesis, Stanford University.

Canny, J. F. 1988. *The Complexity of Robot Motion Planning.* Cambridge: MIT Press.

Feldman, A. G. 1974. "Change in the Length of the Muscle as a Consequence of a Shift in Equilibrium in the Muscle-Load System." *Biofizika* **19,** 534–538.

Flash, T., and N. Hogan. "The Coordination of Arm Movements: An Experimentally Confirmed Mathematical Model." *The Journal of Neuroscience* **5**(7), 1688–1703.

Huang, H.-P., and N. H. McClamroch. 1988. "Time-Optimal Control for a Robot Contour Following Problem." *IEEE Transactions on Robotics and Automation* **4**(2), 140–149.

Jacobson, D. H. "Differential Dynamic Programming Methods for Solving Bang-Bang Control Problems." *IEEE Transactions on Automatic Control* **AC-13**(6), 661–675.

Kahn, M. E., and B. Roth. 1971. "The Near-Minimum-Time Control of Open-Loop Articulated Kinematic Chains." *ASME Journal of Dynamic Systems, Measurement, and Control* **93,** 164–172.

Kamberova, G., R. Mandelbaum, and M. Mintz. 1997. "Statistical Decision Theory for Mobile Robots." Tech. Rep., GRASP Laboratory, University of Pennsylvania.

Khatib, O. 1986. "Real-Time Obstacle Avoidance for Manipulators and Mobile Robots." *International Journal of Robotics Research* **5**(1), 90–98.

Koditschek, D. E. 1987. "Exact Robot Navigation by Means of Potential Functions: Some Topological Considerations." In *Proceedings of 1987 International Conference on Robotics and Automation,* Raleigh. 1–6.

Lafferriere, G., and H. J. Sussmann. 1993. "A Differential Geometric Approach to Motion Planning." In *Nonholonomic Motion Planning.* Ed. Z. Li and J. F. Canny. Boston: Kluwer Academic. 235–270.

Lasdon, L. S., S. K. Mitter, and A. D. Warren. 1967. "The Conjugate Gradient Method for Optimal Control Problems." *IEEE Transactions on Automatic Control* **AC-12**(2), 132–138.

Latombe, J.-C. 1991. *Robot Motion Planning.* Boston: Kluwer Academic.

Laumond, J.-P., et al. 1994. "A Motion Planner for Non-holonomic Mobile Robots." *IEEE Transactions on Robotics and Automation* **10**(5), 577–593.

Li, Z., and J. F. Canny, eds. 1993. *Nonholonomic Motion Planning.* Boston: Kluwer Academic.

Lozano-Pérez, T., and M. A. Wesley. 1979. "An Algorithm for Planning Collision-Free Paths Among Polyhedral Obstacles." *Communications of the ACM* **22**(10), 560–570.

Ma, B., and W. S. Levine. "An Algorithm for Solving Control Constrained Optimal Control Problems." In *Proceedings of the 32nd IEEE Conference on Decision and Control,* San Antonio. 3784–3790.

Mayne, D. Q., and E. Polak. 1987. "An Exact Penalty Function Algorithm for Control Problems with State and Control Constraints." *IEEE Transactions on Automatic Control* **AC-32**(5), 380–387.

Murray, R. M., Z. Li, and S. S. Sastry. 1994. *A Mathematical Introduction to Robotic Manipulation.* Boca Raton: CRC Press.

Murray, R. M., and S. S. Sastry. 1993. "Nonholonomic Motion Planning: Steering Using Sinusoids." *IEEE Transactions on Automatic Control* **38**(5), 700–716.

Neuman, C. P., and A. Sen. 1973. "A Suboptimal Control Algorithm for Constrained Problems Using Cubic Splines." *Automatica* **9,** 601–613.

Ó'Dúnlaing, C., M. Sharir, and C. K. Yap. 1983. "Retraction: A New Approach to Motion Planning." In *Proceedings of the 15th ACM Symposium on the Theory of Computing,* Boston. 207–220.

Ostrowski, J. P. 1995. "The Mechanics and Control of Undulatory Locomotion." Ph.D. Thesis, California Institute of Technology.

Pytlak, R., and R. B. Vinter. 1995. "Second-Order Method for Optimal Control Problems with State Constraints and Piecewise-Constant Controls." In *Proceedings of the 34th IEEE Conference on Decision and Control,* New Orleans.

Rimon, E., I. Kamon, and J. F. Canny. 1997. "Local and Global Planning in Sensor Based Navigation of Mobile Robots." Paper presented at the 8th International Symposium of Robotics Research, Hayama, Japan, October 3–7.

Rimon, E., and D. E. Koditschek. 1992. "Exact Robot Navigation Using Artificial Potential Functions" *IEEE Transactions on Robotics and Automation* **8**(5), 501–518.

Schwartz, J. T., and M. Sharir. 1983. "On the 'Piano Movers' Problem: 1. The Case of Two-Dimensional Rigid Polygonal Body Moving Amidst Polygonal Barriers." *Communications on Pure and Applied Mathematics* **36,** 345–398; "On the 'Piano Movers' Problem: 2. General

Techniques for Computing Topological Properties of Real Algebraic Manifolds." *Advances in Applied Mathematics* **4,** 298–351.

Shiller, Z. 1992. "On Singular Points and Arcs in Path Constrained Time Optimal Motions." *ASME, Dyn. Syst. Contr. Division DSC-42* **42,** 141–147.

Shiller, Z., and S. Dubowsky. 1991. "On Computing the Global Time-Optimal Motions of Robotic Manipulators in the Presence of Obstacles." *IEEE Transactions on Robotics and Automation* **7,** 785–797.

Shin, K. G., and N. D. McKay. 1985. "Minimum-Time Control of Robotic Manipulators with Geometric Constraints." *IEEE Transactions on Automatic Control* **AC-30**(6), 531–541.

Singh, S., and M. C. Leu. "Optimal Trajectory Generation for Robotic Manipulators Using Dynamic Programming." *ASME Journal of Dynamic Systems, Measurement, and Control* **109.**

Slotine, J.-J. E., and H. S. Yang. 1989. "Improving the Efficiency of Time-Optimal Path-Following Algorithms." *IEEE Transactions on Robotics and Automation* **5,** 118–124.

Suh, K. C., and J. M. Hollerbach. 1987. "Local Versus Global Torque Optimization of Redundant Manipulators." In *Proceedings of 1987 International Conference on Robotics and Automation,* Raleigh.

Sussmann, H. J. 1983. "Lie Brackets, Real Analyticity and Geometric Control." In *Differential Geometric Control Theory.* Ed. R. Millman, R. W. Brockett, and H. J. Sussmann. Boston: Birkhauser. 1–116.

Sussmann, H. J., and W. Liu. 1993. "Lie Bracket Extensions and Averaging: The Single-Bracket Case." In *Nonholonomic Motion Planning.* Ed. Z. Li and J. F. Canny. Boston: Kluwer Academic. 109–148.

Teo, K. L., C. J. Goh, and K. H. Wong. 1991. *A Unified Computational Approach to Optimal Control Problems.* Burnt Mill, Harlow: Longman Scientific & Technical.

Uno, Y., M. Kawato, and R. Suzuki. 1989. "Formation and Control of Optimal Trajectory in Human Multijoint Arm Movement." *Biological Cybernetics* **61,** 89–101.

Vukobratović,, M., and M. Kirćanski. 1982. "A Method for Optimal Synthesis of Manipulation Robot Trajectories." *ASME Journal of Dynamic Systems, Measurement, and Control* **104.**

Žefran, M. 1996. "Continuous Methods for Motion Planning." PhD Thesis, University of Pennsylvania.

Žefran, M., and V. Kumar. 1997. "Rigid Body Motion Interpolation." *Computer Aided Design* **30**(3), 179–189.

Žefran, M., J. Desai, and V. Kumar. 1996. "Continuous Motion Plans for Robotic Systems with Changing Dynamic Behavior." In *Proceedings of 2nd International Workshop on Algorithmic Foundations of Robotics,* Toulouse, France.

Žefran, M., V. Kumar, and X. Yun. 1994. "Optimal Trajectories and Force Distribution for Cooperating Arms." In *Proceedings of 1994 International Conference on Robotics and Automation,* San Diego. 874–879.

Zhang, H., V. Kumar, and J. Ostrowski. 1998. "Motion Planning under Uncertainty." In *Proceedings of 1998 IEEE International Conference on Robotics and Automation,* Leuven, Belgium. 638–643.

CHAPTER 16

INTELLIGENT CONTROL OF ROBOT MOBILITY

Ronald C. Arkin
Georgia Institute of Technology
Atlanta, Georgia

1 INTRODUCTION

Mobile robot systems are typically concerned with material-handling problems in industrial settings. These problems generally require determining where to go to acquire the material, moving to that location, acquiring the load, then transporting it to the destination and offloading it. While in many factories automated guided vehicles (AGVs) handle these sorts of operations, autonomous robotic systems can potentially provide extended capabilities (e.g., sophisticated obstacle avoidance) that AGVs currently do not possess. In addition, new service functions for mobile robot systems become addressable, such as surveillance, cleaning, and operation among free-ranging humans. While AGVs may be the method of choice for highly structured environments, autonomous mobile robots play an ever-increasing role when the world becomes less predictable and more dynamic.

This article focuses on what is required to produce intelligent mobility in domains that go beyond current material-handling factory floor scenarios. Often the environment cannot be reengineered easily to fit the needs of the robot, and thus the robot must be provided with additional means to cope with the inherent uncertainty in these worlds. But where does this uncertainty come from?

- *Sensing:* Sensor interpretation is an inherently error-prone process. The difficulty begins with noise in the incoming signal itself and is further compounded by confounding factors in the environment, such as poor or uneven lighting, shadows, and obscurations. Different sensors have different problems; for example, ultrasonic sensors, which are very cheap and commonly used for mobile systems, are notorious for their poor resolution and specular reflections. Vision systems are sensitive to lighting conditions and have to infer three-dimensional structure from two-dimensional imagery.

- *Action:* Mobile robots do not permit the use of inverse kinematics to recover their position, as is normally the case with a robotic arm. Commanding a mobile robot to turn and move in a particular direction will typically result in an action that is only an approximation of that command. The quality of the approximation is dependent on a wide range of factors, including traction and mechanical accuracy. A slippery surface will result in much greater uncertainty. But how can the robot know without sensing when it is slipping?

- *World modeling:* Often models such as blueprints, maps, CAD models, and the like can be made available for navigation or recognition purposes. There are inevitably errors in these models, some perhaps subtle: they may have been inaccurate initially, the world may have changed since their creation, or they may be incomplete. How can a robot function with potentially inaccurate or outright incorrect models?

These factors, among others, have led many researchers to utilize techniques from the artificial intelligence community to provide robots with intelligent and flexible control systems so that they may operate in dynamic and uncertain environments. A look at the

Handbook of Industrial Robotics, Second Edition, Edited by Shimon Y. Nof
ISBN 0-471-17783-0   © 1999 John Wiley & Sons, Inc.

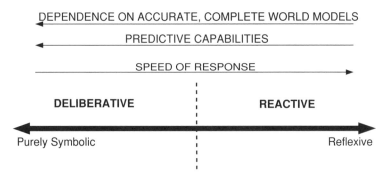

Figure 1   Robot control system spectrum.

current state of the art shows that a range of intelligent control systems exist that have differing utility in various circumstances. Figure 1 depicts this robot control spectrum.

At the left side of this spectrum is a class of planning systems that are deliberative in nature. Basically these fit into the mold of first planning, then acting, and then, since plans often fail, compensating for errors by creating a new plan. On the other end of this spectrum are reactive systems, which act with virtually no planning; they are reflexive systems that tightly couple sensors and actuators through behaviors. In addition, hybrid control systems exist that extract capabilities from both ends of the spectrum and integrate them on an as-needed basis.

This chapter surveys the various approaches currently being used to control autonomous robotic systems. In the last decade a major paradigm shift has occurred with the creation of reactive behavior-based systems. Considerable emphasis is therefore placed on that area, as well as on the even more recent ascent of hybrid systems. Figure 2 shows a wide range of autonomous robotic systems currently in use at the Georgia Tech Mobile Robot Laboratory, ranging from relatively small (the three blizzard robots in the front row, which won the 1994 AAAI Mobile Robot Competition (Balch et al., 1995)) to much larger (the GT Hummer, similar to those used in DARPA's Unmanned Ground Vehicle Program (Arkin and Balch, 1997*b*). See Chapter 8 on the design of mobile robots.

Figure 2   Robots of Georgia Tech Mobile Robot Laboratory. In the front row are three blizzard robots: Io, Callisto, and Ganymede. In the second row are two Nomad robots: Shannon and Sally. Behind them are three Denning robots: George, Ren, and Stimpy. To the rear is the GT Hummer.

2 A HISTORICAL PERSPECTIVE

A brief look at four earlier efforts in autonomous robots will help give a better understanding of how we arrived at our current state.

1. *Shakey:* Stanford Research Institute's Shakey, one of the earliest mobile robots (Nilsson, 1984), was an experiment in the integration of perception, reasoning, and action. Shakey used computer vision to recognize objects in an artificial blocks world environment; used formal logical reasoning to determine what to do and when to do it; and carried out its motor actions in a step-by-step process. This early work provided planning software (STRIPS) that profoundly influenced the direction of robot planning systems for at least a decade. Some (Brooks, 1991) considered these early results misleading because they relied heavily on world models constructed from perceptual experience that are inherently prone to error and uncertainty.

2. *Stanford Cart:* Moravec, first at Stanford and later at Carnegie-Mellon (Moravec), created an indoor mobile vehicle that could successfully detect obstacles using stereo vision and navigate around them. A metric spatial representation was used (tangent space) that stored the location of these obstacles in a map of the world as they were detected.

3. *Hilare:* Researchers at LAAS in France extended the role of representational knowledge and freed it from either purely logical forms or purely metric ones (those that attempt to reason over an accurate spatial map). In particular, they introduced the notion of a multilevel representation that tied together different perspectives on the world, including a topological map that contained information regarding connectivity of navigable regions.

4. *Grey Walter's Tortoise:* Walter, a cyberneticist, developed a small automaton in the 1950s based on vacuum tube technology and analog circuitry. This small robot used no map, but rather responded directly to stimuli in the environment (lights and bumps). Its task was basically to explore the environment and recharge itself as its power supply dwindled, which it successfully accomplished.

These early efforts led to the development of the principal schools of thought prevalent today regarding mobile robot control. Hierarchical control relies heavily upon planning and maps of the world. Reactive control avoids the use of maps and instead responds behaviorally to various stimuli presented within the environment (à la Walter). Hybrid systems exploit the benefits of both of these approaches. We now discuss each of these in turn.

3 HIERARCHICAL PLANNING AND CONTROL

The hierarchical planning approach has several distinguishing characteristics. A clearly identifiable subdivision of functionality exists for each level in the hierarchy. A hierarchical planning system for a mobile robot might typically include a high-level *mission planner,* which establishes objectives, determines optimality criteria, and invokes spatial or common sense reasoning; a *navigator,* which develops a global path plan subject to the mission planner's designated constraints; and a *pilot,* which implements the navigator's global plan in a step-by-step process while avoiding detected obstacles. There is often an associated *cartographer* process for independently maintaining the world model necessary for the reasoning of the planning components. The higher levels of the system architecture establish subgoals for lower levels. Predictable and predetermined communication routes are present between hierarchical modules. The planning requirements regarding time and space change during descent in the planning/control hierarchy: real-time responses become more critical at lower levels, global spatial concerns at higher levels. These systems generally are well suited for structured environments with limited uncertainty, such as a factory shop floor.

This approach dominated in the 1980s but has yielded somewhat in favor of both reactive and hybrid systems. Some representative architectural examples of hierarchical control systems include:

- *Albus's theory of intelligence:* Albus's work on hierarchical control systems at the National Bureau of Standards (NBS) in the United States led to the formulation of a robotic architecture that embodied his theoretical committments on the nature of intelligence in both biological and robotic systems (Albus, 1991). A multilevel temporal and spatial hierarchy for planning is advocated, with each level consisting of four main components: task decomposition, world modeling, sensory processing, and value judgment. A global memory cuts through each of the levels and permits communication and the sharing of knowledge between them. A telling aspect of this work is its perspective on perception: perception's role is declared to be the establishment and maintenance of correspondence between the internal world model and the external real world (Albus, 1991). This architecture resulted in the generation of an early standard for robotic architectures, the NASA/NBS Standard Reference Model for Telerobot Control System Architecture (NASREM), which has waxed, then waned, in popularity since its inception in the mid-1980s.

- *Nested hierarchical information model:* Meystel, at Drexel University, developed a formal method for the expression of hierarchical systems that correlates human teams and control structures (1986). This approach makes the claim that any autonomous control system can be organized as a team of hierarchically related decision makers. It includes an assumption that tasks can result in structured subtasks that can be delegated to lower levels within the hierarchy. The model incorporates a generalized controller whereby hierarchies can be generated via recursion. Preconditions are established at each level of recursion to insure proper execution.

- *Intelligent control:* Saridis, at Rensselaer Polytechnic Institute, created a three-level architecture that implemented his principle of "increasing precision with decreasing intelligence" as the hierarchy is traversed (Saridis, 1983). It employs a three-level control structure consisting of the organization level (possessing high intelligence and low precision), the coordination level (medium intelligence and precision), and the execution level (low intelligence, high precision).

4   REACTIVE/BEHAVIOR-BASED CONTROL SYSTEMS

An alternative to hierarchical planning and control, reactive systems emphasize the importance of tightly coupling sensing and action without the use of intervening representations or world models. When individual parallel modules that map sensing stimuli onto robotic responses are used to describe the overall architecture, the system is referred to as *behavior-based*. These primitive modules (behaviors) are the building blocks for these symbolic representation-free systems. At one level they can be viewed as a collection of asynchronous parallel feedback control systems whose outputs are coordinated by a specialized operator, such as an arbiter that selects one action to undertake.

Reactive systems rely on the immediacy of sensor data as opposed to world models. They are particularly well suited for dynamic and unstructured environments, where accurate world models are potentially difficult to construct, untimely, or errorful. The notion that "the world is its own best model" (Brooks, 1991) epitomizes the spirit of this approach.

A simple behavioral controller appears in Figure 3. This system has three primitive behaviors: avoiding obstacles, moving towards a perceived goal, and remaining on a path. The coordinating process determines the aggregate overt action of the robot by selecting or combining the individual behavioral outputs in a meaningful way.

A wide range of approaches to the design of behavior-based systems exist (Arkin, 1998). These methods differ in their means of behavioral expression, integration, selection, and control. Several representative examples follow:

- *Subsumption architecture:* The subsumption architecture, developed by Brooks at MIT (Brooks, 1986), was the first well-established behavior-based robotic architecture. It involves a layered, but nonhierarchical, approach where higher-level behaviors "subsume" the activity of low-level ones. While there exist many behaviors running in parallel, a priority-based arbitration mechanism is used to select which one is in control at any given time. No internal world model is maintained—the system relies completely on incoming sensor data. The arbitration is implemented using suppression and inhibition connections between behavioral components. In subsumption-style robots individual behavioral layers work on individual goals independently. More complex actions are layered on top of simpler actions. No centralized control or central sensor representations exist. The topology

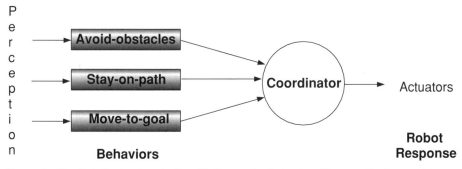

Figure 3 Simple behavioral controller with three active behaviors. The coordination module can take many forms, including priority-based arbitration, dynamic action-selection, and behavioral fusion.

of the behavioral set is fixed by hierarchy. Lower levels are unaware of higher levels.

The subsumption architecture has been demonstrated on many robots at the MIT AI Lab, ranging from small hexapods to soda can-collecting robots to robotic tour guides (Brooks, 1990). Several commercial mobile robots are available utilizing this architecture, including those made by companies such as IS Robotics in Somerville, Massachusetts, and RWI Inc. in Jaffrey, New Hampshire.

- *Reactive circuits:* Another reactive architectural design was developed at SRI (Kaelbling). This approach utilizes hierarchies of competence (subsumption-like increasing levels of functionality) plus abstraction (the aggregation of primitive behaviors into abstract groups of behaviors). The REX and GAPPS languages were developed using formal methods to express particular design implementations. This added the possibility of establishing provable properties regarding the performance of these systems. The architecture was developed and tested on Shakey's successor at SRI, Flakey.

- *Schema-based reactive control:* Arkin, at the Georgia Institute of Technology, developed an alternative paradigm for behavior-based control (Arkin, 1989). It exploited schema theory (Arbib, 1995) as a means for modularizing and organizing both behavioral and perceptual activity. No hard-wired layering is present, in contrast to subsumption, but rather a dynamic network of behaviors is provided for. High-level planning selects and parameterizes the behaviors based on the mission, environment, and internal state of the robot. Potential field techniques (Khatib, 1985) are used to generate the response of each individual motor behavior. Instead of arbitration, weighted vector summation of each behavior's contribution produces the overt behavioral response of the robot. Each schema (behavior) is an independent computing agent. Perceptual strategies are embedded within the motor schema to provide the necessary sensor information on a need-to-know basis. A wide range of robots was deployed using this architectural approach, from small multiagent systems (Balch et al., 1995) to a large all-terrain Hummer (Arkin and Balch, 1997) (Figure 2).

- *Other reactive behavior-based architectures:* Many other architectures followed these earlier examples. One method, developed by Slack (1990), involved the use of navigational templates to generate flow fields produced by rotation around objects. It was developed to address the problem of local minima and maxima commonly found in reactive control systems, where robots could become trapped. The Distributed Architecture for Mobile Navigation (DAMN) architecture, developed by Rosenblatt at Carnegie-Mellon (Rosenblatt, 1995), can be considered a fine-grained alternative to subsumption. It utilizes multiple arbiters and permits behaviors to vote for a particular response. The arbiter selects the action to undertake, based upon the one receiving the largest number of votes. This approach is related to earlier work by Maes using action selection (Maes, 1989). Here each behavior maintains an activation level that is related to its current situational relevance. The arbiter selects the most relevant action, based on these activation levels, instead of enforcing a prewired prioritized layering as was originally found in subsumption.

Many other behavior-based architectures exist. Arkin (1998) gives additional information.

5 HYBRID ARCHITECTURES

Subsequent to the revolution caused by behavior-based control, it was recognized that there might be advantages to be gained by combining the benefits of both hierarchical planning and reactive control techniques. Purely reactive control regimes have several potential shortcomings: available world knowledge, which may have potential value, is ignored; many reactive systems are inflexible; and they can suffer from shortsightedness (local failures) due to their heavy reliance on immediate sensing.

The advantages of hierarchical planning were rediscovered: they can readily integrate world knowledge, and navigational planning in the context of a broader scope is straightforward (i.e., it is easy to see the "big picture"). But reactive systems have distinct advantages too: they produce robust, demonstrable navigational successes, afford modular development and incremental growth, and are perfectly situated for arriving sensor data.

An effort was made by several roboticists to reap the benefits of both of these techniques without introducing their frailties. The goal was to yield robust plan execution while exploiting a high-level understanding of the robot's world. These new architectures embodied a hybrid philosophy that recognized that reactive control and hierarchical planning systems address different parts of the same problem: plan execution is best associated with reactive control, while plan formulation is better served by a hierarchical, deliberative process. Thus reactive control and hierarchical planning are deemed to be fully compatible.

We now examine several of these hybrid robot control architectures.

- *AuRA:* Autonomous Robot Architecture (AuRA), developed by Arkin (Arkin, 1987; Arkin and Balch, 1997*a*), was the first architecture to merge classical artificial intelligence planning techniques with behavior-based control for robot navigation. AuRA partitions the navigational task into two phases, planning and execution (Figure 4). A hierarchical planner is used to develop the behavioral configuration for execution by the reactive controller. In its initial form it used the A* search algorithm over a specialized free-space world representation to compute the sequence of actions that a robot needed to undertake for a given mission (Arkin, 1986). Each leg of the mission was then translated into a behavioral configuration and sent to the motor schema-based controller for execution. Should difficulties arise, the hierarchical planner is reinvoked to address them, resulting in a new configuration for the controller. Motor schemas, as described in the previous section, are used for the actual conduct of the robot's mission. Later versions of AuRA (Arkin and Balch, 1997*b*) incorporate qualitative navigational planning and extensions to domains such as manufacturing, military operations, and multiagent robotics. Various aspects of the approach were tested on a wide range of robots (Figure 2).

- *Atlantis:* Atlantis, developed at NASA's Jet Propulsion Lab by Gat (1992), consists of a three-layer architecture. The reactive layer is a behavior-based system that uses the ALFA language. ALFA resembles the REX language used in the reactive circuits architecture described earlier. At the other end of this hybrid system is a deliberative layer that plans and maintains world models. The intervening interface layer, the *sequencer,* controls sequences of reactivity and deliberation and is used to handle cognizant failures, such as self-recognition when deliberation is necessary due to a failure to achieve the systems goals by reactive methods alone. Atlantis was successfully tested in difficult outdoor navigation on the JPL Rover (a prototype Mars Rover) in a California arroyo.

- *PRS:* PRS (Procedural Reasoning System), developed by Georgeff and Lansky at SRI (Georgeff, 1987), is not a behavior-based system per se. Instead it continuously creates and executes plans but abandons them as the situation dictates. PRS plans are expanded dynamically and incrementally on an as-needed basis. Thus the system reacts to changing situations by changing plans as rapidly as possible should the need arise. The system was fielded on the robot Flakey at SRI. A later version,

User Input

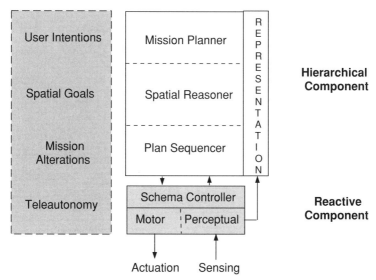

Figure 4 AuRA, a hybrid reactive/hierarchical architecture. It consists of a three-layer deliberative planner interfaced to a schema-based reactive behavioral controller. Representational knowledge is generated only in support of higher-level planning.

developed at the University of Michigan (UM-PRS) (Lee et al., 1994), was geared towards navigation for battlefield robot applications.

- *TheoAgent:* TheoAgent, developed by Mitchell at Carnegie-Mellon (1990), embodies the philosophy "React when it can, plan when it must." Rule-based stimulus-response rules encode the reactivity. The method for selecting the rules constitutes the planning component. TheoAgent is primarily concerned with learning new rules; in particular, to learn to act more correctly, learn to be more reactive (reduce time), and learn to perceive better (distinguish salient features). These goals were tested on a Hero 2000 mobile robot that was tasked to locate garbage cans.

6   SUMMARY AND OPEN QUESTIONS

Autonomous robotic mobility has been explored through the development of various supporting architectures. Currently there are three major schools of thought:

1. The hierarchical planning approach, well suited for relatively stable and predictable environments, uses planning and projection to determine suitable courses of action for a robot.
2. The behavior-based reactive approach, best suited for dynamic and unpredictable worlds, tightly ties sensing to action in lieu of constructing symbolic artifacts to reason over.
3. Hybrid architectures attempt to span the gap between planning and reactive systems and hence be more generalizable.

Considerable research is progressing in other areas directly related to intelligent mobility (Arkin, 1998). Learning, for example, can be implemented at many levels. It can provide knowledge of the structure of the world gained over time, or it can allow a robot to tune its behavioral responses as it experiences the results of its actions. The performance of teams of robots working together is now also being explored in a meaningful way. More exotic notions such as imagination and survivability are just beginning to be explored. Other open research questions include making robots more accessible to the average user; improving perceptual capabilities; fully exploiting the notions of expectation and attention during planning and execution; and providing effective means for evaluating autonomous systems. This brief discussion merely touches the surface of this large body of research.

REFERENCES

Albus, J. 1991. "Outline for a Theory of Intelligence." *IEEE Transactions on Systems, Man, and Cybernetics* **21**(3), 473–509.

Arbib, M. 1995. "Schema Theory." In *The Handbook of Brain Theory and Neural Networks*. ed. M. Arbib. Cambridge: MIT Press, 830–834.

Arkin, R. C. 1986. "Path Planning for a Vision-Based Autonomous Robot." In *Proceedings of SPIE Conference on Mobile Robots,* Cambridge, Mass. 240–249.

———. 1987. "Towards Cosmopolitan Robots: Intelligent Navigation in Extended Man-made Environments." Ph.D. Thesis, COINS Technical Report 87–80, University of Massachusetts, Department of Computer and Information Science.

———. 1989. "Motor Schema-Based Mobile Robot Navigation." *International Journal of Robotics Research* **8**(4), 92–112.

———. 1998. *Behavior Based Robotics,* Cambridge: MIT Press.

Arkin, R. C., and T. Balch, 1997a. "Cooperative Multiagent Robotic Systems." In *AI-Based Mobile Robots: Case Studies of Successful Robot Systems.* Ed. D. Kortenkamp, R.P. Bonasso, and R. Murphy. Cambridge: AAAI/MIT Press.

———. 1997b. "AuRA: Principles and Practice in Review." *Journal of Experimental and Theoretical Artificial Intelligence* **9**(2), 175–189.

Balch, T., et al. "Io, Ganymede, and Callisto—A Multiagent Robot Trash-Collecting Team." *AI Magazine* **16**(2), 39–51.

Brooks, R. 1986. "A Robust Layered Control System for a Mobile Robot." *IEEE Journal of Robotics and Automation* **RA-2**(1), 14–23.

———. 1990. "Elephants Don't Play Chess." In *Designing Autonomous Agents.* Ed. P. Maes. Cambridge: MIT Press. 3–15.

———. 1991. "Intelligence Without Representation." *Artificial Intelligence* **47**, 139–160.

Gat, E. 1992. "Integrating Planning and Reaction in a Heterogeneous Asynchronous Architecture for Controlling Real-World Mobile Robots." In *Proceedings of 10th National Conference on Artificial Intelligence (AAAI-92).* 809–815.

Georgeff, M., and A. Lansky. "Reactive Reasoning and Planning." In *Proceedings of the AAAI-87.* 677–682.

Kaelbling, L. 1986. "An Architecture for Intelligent Reactive Systems." SRI International Technical Note 400, Menlo Park, October.

Khatib, O. 1985. "Real-Time Obstacle Avoidance for Manipulators and Mobile Robots." In *Proceedings of the IEEE International Conference Robotics and Automation,* St. Louis. 500–505.

Lee, J. L., et al. 1994. "UM-PRS: An Implementation of the Procedural Reasoning System for Multirobot Applications." In *Proceedings of the Conference on Intelligent Robotics in Field, Factory, and Space (CIRFFSS '94),* March. 842–849.

Maes, P. 1989. "The Dynamics of Action Selection." In *Proceedings of the Eleventh International Joint Conference on Artificial Intelligence (IJCAI-89),* Detroit. 991–997.

Meystel, A. 1986. "Planning in a Hierarchical Nested Controller for Autonomous Robots." In *Proceedings of the 25th Conference on Decision and Control,* Athens. 1237–1249.

Mitchell, T. 1990. "Becoming Increasingly Reactive." In *Proceedings of the 8th National Conference on Artificial Intelligence (AAAI-90),* Boston. 1051–1058.

Moravec, H. "The Stanford Cart and the CMU Rover." In *Proceedings of the IEEE* **71**(7), 872–884.

Rosenblatt, J. 1995. "DAMN: A Distributed Architecture for Mobile Navigation." Paper delivered at Working Notes AAAI 1995 Spring Symposium on Lessons Learned for Implemented Software Architectures for Physical Agents, Palo Alto, March.

Nilsson, N. 1984. "Shakey the Robot." Technical Note 323, Artificial Intelligence Center, SRI International, Menlo Park.

Saridis, G. 1983. "Intelligent Robotic Control." *IEEE Transactions on Automatic Control* **28**(5), 547–556.

Slack, M. 1990. "Situationally Driven Local Navigation for Mobile Robots." JPL Publication No. 90-17, NASA, Jet Propulsion Laboratory, Pasadena, April.

CHAPTER 17

VIRTUAL REALITY AND ROBOTICS

Grigore C. Burdea
Rutgers—The State University of New Jersey
Piscataway, New Jersey

Philippe Coiffet
Laboratoire de Robotique de Paris
Velizy, France

1 INTRODUCTION

Readers of this handbook are by now familiar with the reliance of robotics on allied technologies such as multimodal sensing or neural networks, which are becoming part of modern robotic systems. The need to incorporate such technologies stems from the ever-increasing expectations for robot performance in both industrial and service applications.

Virtual reality (VR) is a high-end human–computer interface that allows users to interact with simulated environments in real time and through multiple sensorial channels (Burdea and Coiffet, 1994). Sensorial communication with the simulation is accomplished through vision, sound, touch, even smell and taste. Due to this increased interaction (compared to standard CAD or even multimedia applications), the user feels "immersed" in the simulation and can perform tasks that are otherwise impossible in the real world. Virtual reality is related to, but different from, telepresence, where the user (operator) interacts with and feels immersed in a distant but *real* environment.

The benefits of VR technology have been recognized by many scientists and engineers, and today virtual reality is being applied to areas such as architectural modeling, manufacturing, training in servicing complex equipment, and design of new plants. In robotics, VR is being used in design of manipulators, programming of complex tasks, and tele-operation over large distances.

After a description of VR system architecture and dedicated interfaces, this chapter describes three areas of synergy between robotics and virtual reality: programming, teleoperation, and force/touch feedback interfaces. A summary of future research challenges concludes the chapter.

2 VIRTUAL REALITY SYSTEM ARCHITECTURE

Figure 1 illustrates a typical single-user VR system (Burdea and Coiffet, 1994). Its main component is the *VR engine,* a multiprocessor real-time graphics workstation. The real-time graphics are produced by dedicated boards called *graphics accelerators,* which have speeds ranging from tens of thousands to millions of triangles displayed every second. It is important to display (or refresh) the virtual scene some 20–30 times every second in order to maintain visual continuity.

Interactions between the user and the VR engine are mediated by I/O devices that read user's input and feedback simulation results. A simple keyboard, joystick, or 3D position sensor (so-called *tracker* (Krieg, 1993)) can be used in simple tasks. More dexterous simulations where the user's hand gestures need to be measured call for more complex interfaces such as sensing gloves (Kramer, Lindener, and George, 1991). Feedback from the VR engine typically is through mono or stereo graphics, which the user sees through *head-mounted displays* (HMDs) connected to the computer (Roehl, 1995). Other feedback modalities are spatially registered sound (also called *3D sound*) (Wenzel, 1992) and force/touch feedback (Burdea, 1996).

Handbook of Industrial Robotics, Second Edition, Edited by Shimon Y. Nof
ISBN 0-471-17783-0   © 1999 John Wiley & Sons, Inc.

VR SYSTEM ARCHITECTURE

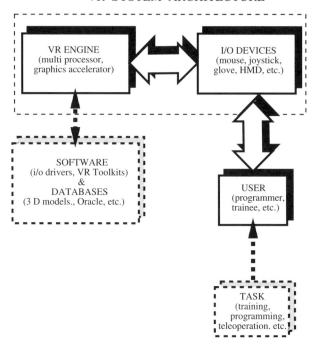

Figure 1   VR system block diagram. (From G. Burdea and P. Coiffet, *Virtual Reality Technology.* New York: John Wiley & Sons, 1994. © 1994 Editions Hermes. Reprinted by permission.)

The VR engine responds to user's commands by changing the view to, or composition of, the virtual world. This synthetic world is modeled off-line using dedicated software libraries such as WorldToolKit (Sense8 Co., 1995) and databases of preexisting models such as Viewpoint Dataset Catalog (Viewpoint Datalabs, 1994). Such data sets exist for objects ranging from human organs to airplanes and cars to household items and animals. Each model typically is available at different detail levels (and different costs). Because these models populate the virtual world, they must be rendered in real time (preferably at 30 frames/sec), limiting the geometrical complexity of the virtual world. Otherwise the scene takes too long to be displayed and the feeling of immersion is lost.

Another computation load for the VR engine is related to object dynamics and contact forces present during interaction between the virtual objects. Collisions must be detected automatically, contact forces calculated and (sometimes) displayed to the user, and resulting surface deformation calculated at a rate that should at least keep up with the graphics changes. These computations compound such that they may be too much to be handled by a single computer. In that case several computers are networked, each having in charge some aspect of the simulation (Richard et al.,1996). Thus the VR system becomes a single-user distributed one over a communication medium such as the Ethernet. Several users can share the same virtual world in a multiuser distributed VR environment (Mark et al.,1996).

3 ROBOT PROGRAMMING USING VR

One of the most tedious tasks in industrial robotics is off-line programming. While in the past debugging a new program required the robot to be taken off-line for about 30 hours, modern approaches such as CAD and computer graphics have reduced this time to 6 hours. This is still unacceptable for industries using a large number of robots, and in fact the debugging may fail if the model of the robot and its sensors and surroundings are not perfect.

One project attempting to advance the state of the art in robot programming is ongoing at IPA in Stuttgart, Germany (Strommer, Neugebauer, and Flaig, 1993). As illustrated in

Figure 2, the programming is done on a virtual robot and a virtual environment, with the programmer interaction being mediated by VR I/O devices. The programmer, who wears a sensing glove and HMD and feels immersed in the application, can navigate using a trackball and look at the scene from any angle, seeing details that may not be visible in real life. Once the code is debugged, it is downloaded to a real robot connected to the same VR engine and the same task is executed. Feedback from the sensors on the real robot is then used to fine-tune the program.

The manipulator may become unstable if there is contact between the robot and its environment and the data on object hardness are missing or proximity sensor data are unreliable or noisy. Instabilities may also occur when the robot is very long and flexible or has a large inertia. This is the case for underground storage tank cleanup, where long-reach manipulators are used. Researchers at Sandia National Laboratory have developed a VR simulation in which virtual force fields are placed around objects to prevent collisions. The cleaning robot and surrounding tank were modeled with the Sequential Modular Architecture for Robotics and Teleoperation (SMART) software developed at Sandia (Anderson and Davies, 1994). Virtual spring–damper combinations surround the robot

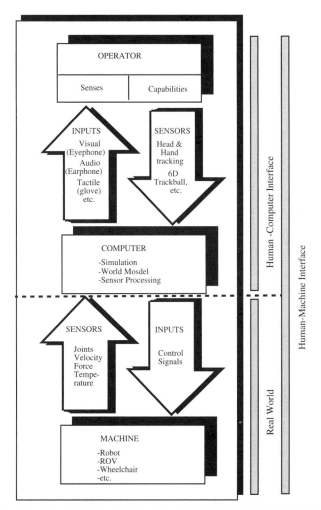

Figure 2   Block diagram of the IPA robotics VR workstation. (Adapted from Strommer, Neugebauer, and Flaig, "Transputer-Based Virtual Reality Workstation as Implemented for the Example of Industrial Robot Control," in *Procedings of Interface to Real and Virtual Worlds Conference,* Montpellier, France, March 1993. © 1994 Editions Hermes. Reprinted by permission.)

tool and the tank model is decomposed in graphics primitives that are used to detect collisions. In such cases repelling forces are automatically generated.

It is not expected that VR programming will replace the current off-line programming approach in the short run. Rather, it is an extension of robotic simulation libraries allowing a better (and sometimes more realistic) view of the robot interaction. As CAD packages evolve, VR will gain more acceptance in the programming arena, but off-line programming may still exist in some form. In nonindustrial applications VR programming will be even more complex, owing to the complexity of the world it aims to replicate. Cost concerns should restrict usage to certain areas, but education and training should accept VR due to its flexibility and economy of scale.

4   VR AIDS TO TELEOPERATION

The limited intelligence and adaptability of today's robots are insufficient in cases where the task is complex or when the environment is *unstructured,* meaning it can change in an unpredictable fashion. In such cases it is necessary to keep the operator in the loop if the task is to succeed. Thus the robot slave is controlled at a distance (teleoperated) by the user through a *master* arm. Other situations where teleoperation is necessary involve adverse environments such as nuclear plant servicing (or decommissioning), undersea operations, space robotics, and explosive environments. In such cases the robot performs the task for the human operator, and protects the operator from harm.

4.1   Teleoperation with Poor Visual Feedback

Unfortunately, teleoperation suffers from a number of drawbacks. One is the degradation of sensorial information about the remote environment. Controlling something you cannot see, as when the scene is visually occluded or dark, is difficult. Here VR can help by overimposing a virtual model of the object of interest over the video signal from the remote camera(s). This "virtual camera" allows the operator to see portions of the object that are not within direct view. Oyama and his colleagues at the University of Tokyo developed a teleoperation system for fire rescue operations where the visual feedback is degraded by smoke (Oyama et al., 1993). They used a virtual slave robot and surrounding scene that were overimposed on the real-time video from the smoke-filled environment. Key to successful teleoperation under these adverse conditions is good calibration between the virtual and real robots. The authors used a manual calibration scheme that is good only for simple tasks.

Visual feedback from the remote scene can be affected by inadequate level of detail. In the extreme case of teleoperation in microworlds, direct visual feedback cannot give the necessary molecular level detail. An example is the *nanomanipulator* project at the University of North Carolina at Chapel Hill (Grant, Helser, and Taylor, 1998). The tele-manipulation system consists of a PHANToM master arm controlling the head of a scanning tunneling microscope. As illustrated in Figure 3, the computer renders a virtual image of the atomic surface, which is then felt haptically by the operator. The user

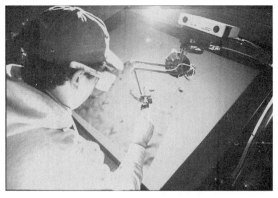

Figure 3   VR-aided teleoperation at a nanometer scale (courtesy Computer Science Department, University of North Carolina at Chapel Hill). Photo by Todd Gaul.

squeezes a trigger on the master arm, firing electrical pulses that allow modification of the surface geometry at a nanometer scale.

4.2   Teleoperation with Time Delay

Another drawback in teleoperation is the time delay that occurs whenever communication between the robot master and slave takes place over long distances. One such case is teleoperation in space, where delays on the order of one second or more can occur between the time the operator moves the master arm and the time the video image shows the remote slave responding. Thus task completion time in teleoperation is always larger than the time needed to perform the same task locally. One approach to alleviate this problem is the superposition of a virtual or *phantom* robot over the video image of the remote robot (Bejczy, Kim, and Venema, 1990). When no command is issued, the two images overlap. Once the operator moves the master arm, the phantom robot responds immediately, allowing the operator to precisely and rapidly complete the command without the intermediate stops associated with the usual "move-and-wait" teleoperation mode.

Time delays are detrimental not only to the task completion time, but also to system stability whenever interaction forces are fed back to the operator. Normally force feedback is beneficial in teleoperation, but if this information comes too late (when interaction has already ended), it can be detrimental. Kotoku at MITI, Japan (1992), attempted to solve this problem by calculating the contact forces between the virtual robot and its virtual environment and then feeding these forces back to the user instead of the actual forces coming from the remote robot slave. Because the virtual robot responds immediately to operator's input, there is no delay in the force feedback loop. For a very simple planar task, tests showed that for a 0.5-second time delay the virtual slave produced more stable control and motion that was three times faster than otherwise possible.

Observing that the above approach will fail in unstructured remote environments and for more complex tasks, Rosenberg at Stanford University (1993) proposed an alternative method of solving the problem of teleoperation with time delay. He introduced the notion of *virtual fixtures,* which refers to abstract sensorial data overlaid on top of the remote space and is used to guide the operator in the performance of the task. Tests showed that when the time delay was 0.45 seconds, the task completion time was 45% larger when no virtual fixtures were used. The enhanced "localization" (guidance) provided by virtual fixtures reduced this time difference to only 3%.

Another approach to reducing the teleoperation time delay problem is to send high-level commands from the master whenever the remote slave is capable of sufficient autonomy. This approach, called *teleprogramming,* reduces the bandwidth of the communication required. It was developed by, among others, Grialt and his colleagues at LAAS, France, and Sayers and Paul (1994) at the University of Pennsylvania. Virtual fixtures such as *point fixture* and *face fixture* are automatically activated by the system and displayed on the operator workstation. These fixtures are color-coded to signify desired versus undesired locations (or actions) and have a force feedback component that the operator feels through the master arm. Thus, desired locations that increase the chance of succeeding in the task have an attractive force response, while locations of increased uncertainty trigger a repelling force that increases as the user approaches that location. The VR simulation thus takes a very active role in assisting the operator in the performance of the remote task.

4.3   Multiplexed Teleoperation

VR can be beneficial to teleoperation whenever a single master station is attempting to control, and multiplex between, several (dissimilar) slave robots. Researchers at the Laboratoire de Robotique de Paris, France, were able to control four robots in France and Japan (Kheddar et al., 1997) using a LRP Master (Bouzit, Richard, and Coiffet, 1993). The key was the use of a *functional-equivalent* simulation in which the slave and the master were replaced by a virtual hand, as shown in Figure 4. A range of commands (some at high level) were sent to the slave robots, which then relied on local sensing to perform the remote task.

Conversely, several operators (masters) can be multiplexed to control a single slave robot. Such an approach is warranted when a team of operators, each with a different field of expertise, collaborates to solve a remote application, such as disposal of nuclear waste (Cannon and Thomas, 1997). All operators see the same video scene, and each controls a virtual hand icon in order to manipulate icons in that scene. Once consensus

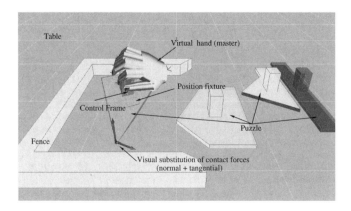

(a)

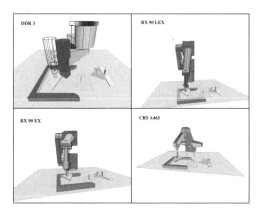

(b)

Figure 4 VR-aided teleoperation of multiple dissimilar robots. (a) Functional-equivalent master. (b) Task execution by four robots. (From A. Kheddar et al., "Parallel Multi-Robot Long Distance Teleoperation," in *Proceedings of ICAR '97,* Monterey. Copyright IEEE 1997.)

is reached, actions are executed by the remote robot under input from a single (supervisory) operator.

5 ROBOTS AS VR HAPTIC INTERFACES

The above discussion outlined some of the ways VR simulations can help robotics. Robotics can in turn be beneficial to VR simulations. A case in point is the use of robotic manipulators as force feedback interfaces in various VR simulations, the *robotic graphics* concept introduced by McNeely at Boeing Co. (McNeely, 1993). The robot carries a turret with various shapes designed to replicate those of virtual objects in the simulation. By tracking the motion of the user's hand, the robot is able to orient itself to provide "just-in-time" contact force feedback. When the user pushes against a hard and immobile virtual object, such as a virtual wall, the robot locks its brakes and replicates the correspondingly large forces at the point of contact.

Older position-controlled robots, such as the one used by McNeely, raise safety concerns because the user "wears" the robot and could be injured accidentally. This problem has been addressed in newer special-purpose manipulators designed from the outset as haptic interfaces. An example is the PHANToM arm developed at MIT and illustrated in

Figure 5 (Massie and Salisbury, 1994). This relatively small arm has a work envelope accommodating the user's wrist motion in a desktop arrangement. It also has optical position encoders to read the gimbal position, as well as three DC actuators to provide up to 10 N translating forces to the user's fingertip.

Unlike position-controlled manipulators, the PHANToM is fully backdrivable, allowing the user to push back the robot without opposition. Indeed, the user will not feel any forces as long as there is no interaction in the virtual world. Further, the low inertia and friction and the gravity-compensated arm result in very crisp, high-quality haptic feedback. The high mechanical bandwidth of the interface (800 Hz) allows it to feed back small vibrations such as those associated with contact with rough surfaces. Thus it is possible to map surface mechanical textures over the various virtual objects being manipulated and then distinguish them based on how they feel (hard–soft, smooth–rough, sticky–slippery, etc).

6   CONCLUSIONS

The above discussion is by necessity limited, but the reader should be able to see by now the great potential and mutual benefits that robotics and virtual reality offer each other. These benefits, as summarized in Table 1, range from aid to robot programming, to teleoperation over long distances and between a single master station and multiple dissimilar robot slaves.

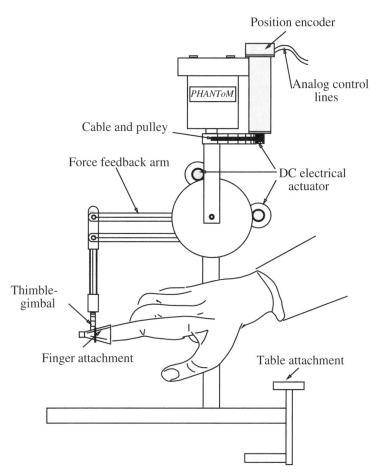

Figure 5   The PHANToM master. (Adapted from T. Massie and K. Salisbury, "The PHANToM Haptic Interface: A Device for Probing Virtual Objects," in *Proceedings of ASME Winter Annual Meeting,* 1994. DSC-Vol. 55-1. © 1994 ASME. Reprinted by permission.)

Table 1 VR and Robotics Synergy

Application	Traditional Approach	Virtual Reality
Robot Programming	Tedious, knowledge of specific robotic language	User-friendly, high-level programming
Teleoperation	Poor sensorial feedback Impossible with large time delays Single-user	Enhanced sensorial feedback Possible with large time delays Multiplexed
Haptic feedback	Adequate in some cases Expensive Human-factors expertise	Special-purpose interfaces Safety issues Improved simulation realism

Because virtual reality research and development is a more recent activity than robotics, its benefits will take time to be recognized and used by the robotics community. Furthermore, full implementation will require that present technical limitations be overcome. Specifically, there is a need for faster graphics and faster communication links. Also needed is a sustained effort to include physical modeling (object weight, hardness, surface friction, temperature, etc.) to produce realistic simulations. Ergonomics and safety studies should quantify the real (as opposed to perceived) advantages and disadvantages of VR integration within robotic applications.

Finally, as VR technology matures and the newer force/touch feedback interfaces become more prevalent, existing robotic technology and know-how will be invaluable to advance the state of the art of virtual reality in general. It is important to realize, however, that virtual interactions result in *real* forces being applied to the user (operator). Situations where the user is in close proximity to the robot producing the force feedback sensation call for drastic safety measures. New actuator technology capable of high power/weight and power/volume ratios need to be integrated in new-generation robots. Costs must also come down in order to allow widespread usage.

REFERENCES

Anderson, R., and B. Davies. 1994. "Using Virtual Objects to Aid Underground Storage Tank Teleoperation." In *Procedings of 1994 IEEE International Conference on Robotics and Automation,* San Diego, May. 1421–1426.

Bejczy, A., W. Kim, and S. Venema. 1990. "The Phantom Robot: Predictive Displays for Teleoperation with Time Delay." In *Proceedings of 1990 IEEE International Conference on Robotics and Automation,* Cincinnati, May. 546–551.

Bouzit, M., P. Richard, and P. Coiffet. 1993. "LRP Dextrous Hand Master Control System." Technical Report, Laboratoire de Robotique de Paris, 21 pp.

Burdea, G. 1996. *Force and Touch Feedback for Virtual Reality.* New York: John Wiley & Sons.

Burdea, G., and P. Coiffet. 1994. *Virtual Reality Technology.* New York: John Wiley & Sons.

Cannon, D., and G. Thomas. 1997. "Virtual Tools for Supervisory and Collaborative Control of Robots." *Presence* 6(1), 1–28.

Grant, B., A. Helser, and R. Taylor II. 1998. "Adding Force Display to a Stereoscopic Head-Tracked Projection Display," in *Proceedings of IEEE VRAIS '98,* Atlanta, March. 81–88.

Kheddar, A., et al. 1997. "Parallel Multi-Robot Long Distance Teleoperation." in *Proceedings of IGAR '97,* Monterey. 1007–1012.

Kotoku, T. 1992. "A Predictive Display with Force Feedback and Its Application to Remote Manipulation System with Transmission Time Delay." In *Proceedings of the 1992 IEEE/RSJ International Conference on Intelligent Robots and Systems,* Raleigh. 239–246.

Kramer, J., P. Lindener, and W. George. 1991. "Communication System for Deaf, Deaf-Blind, or Non-vocal Individuals Using Instrumented Glove." U.S. Patent 5,047,952.

Krieg, J. 1993. "Motion Tracking: Polhemus Technology." *Virtual Reality Systems* 1(1), 32–36.

Mark, W., et al. 1996. "UNC-CH Force Feedback Library." UNC-CH, TR. 96-012, 47 pp.

Massie, T., and K. Salisbury. 1994. "The PHANToM Haptic Interface: A Device for Probing Virtual Objects." In *Proceedings of ASME Winter Annual Meeting.* DSC-Vol. 55-1. 295–300.

McNeely, W. 1993. "Robotic Graphics: A New Approach to Force Feedback for Virtual Reality." In *Proceedings of IEEE VRAIS '93 Symposium,* Seattle. 336–341.

Oyama, E., et al. 1993. "Experimental Study on Remote Manipulation Using Virtual Reality." *Presence* 2(2), 112–124.

Richard, P., et al. 1996. "Effect of Frame Rate and Force Feedback on Virtual Object Manipulation." *Presence* 5(1), 1–14.

Roehl, B. 1995. "The Virtual I/O 'I-glasses!' HMD." *VR World* May–June, 66–67.

Rosenberg, L. 1993. "The Use of Virtual Fixtures to Enhance Telemanipulation with Time Delay." In *Proceedings of ASME Winter Annual Meeting.* DSC-Vol. 49. 29–36.

Sayers, C., and R. Paul. 1994. "An Operator Interface for Teleprogramming Employing Synthetic Fixtures." *Presence* **3**(4), 309–320.

Sense8 Co. 1995. "WorldToolKit Reference Manual Version 2.lb." Sausalito.

Strommer, W., J. Neugebauer, and T. Flaig. 1993. "Transputer-Based Virtual Reality Workstation as Implemented for the Example of Industrial Robot Control." In *Proceedings of Interface to Real and Virtual Worlds Conference,* Montpellier, France, March. 137–146.

Viewpoint Datalabs. 1994. *Viewpoint Dataset Catalog.* 3rd ed. Orem.

Wenzel, E. 1992. "Localization in Virtual Acoustic Display." *Presence* **1**(1), 80–107.

PART 4

PROGRAMMING AND INTELLIGENCE

CHAPTER **18**

ON-LINE PROGRAMMING

Michael P. Deisenroth
Virginia Polytechnic Institute and State University
Blacksburg, Virginia

Krishna K. Krishnan
Wichita State University
Wichita, Kansas

1 INTRODUCTION

With increased competition and need for enhancing productivity, robots have become an important part of manufacturing and assembly operations. They are used in a variety of functions, such as pick-and-place parts, spot and seam weld, spray paint, and assembly. The industrial robot of today is an automated mechanism designed to move parts or tools through some desired sequence of motions or operations. As the robot proceeds from one cycle of a work task to the next, the sequence of robot operations may vary to allow the robot to perform other tasks based on changes in external conditions. Additionally, the same type of robot, or even the same robot, may be required to perform a completely different set of motions or operations if the workcell is revised or the desired tasks are changed. The robot control program must be able to accommodate a variety of application tasks, and it must also be flexible within a given task to permit a dynamic sequence of operations (Dorf, 1984; Engelberger, 1980; Warnecke and Schraft, 1982). The flexibility of the robot is governed to a large extent by the types of motions and operations that can be programmed into the control unit and the ease with which that program can be entered and/or modified. This programming can be achieved either off-line or on-line. Off-line programming can be achieved through specially devised programming languages such as VAL, WAVE, AML, MCL, and SIGLA. These programs are usually debugged using simulation. On-line programming on the other hand is usually achieved using teach programming methods through either teach pendant or lead-through programming methods.

1.1 CONCEPTS OF ON-LINE PROGRAMMING

On-line programming is accomplished through teach programming methods. Teach programming is a means of entering a desired control program into the robot controller. The robot is manually led through a desired sequence of motions by an operator who is observing the robot and robot motions as well as other equipment within the workcell. The teach process involves the teaching, editing, and replay of the desired path. The movement information and other necessary data are recorded by the robot controller as the robot is guided through the desired path during the teach process. At specific points in the motion path the operator may also position or sequence related equipment within the work envelope of the robot. Program editing is used to add supplemental data to the motion control program for automatic operation of the robot or the associated production equipment. Additionally, teach-program editing provides a means of correcting or modifying an existing control program to change an incorrect point or compensate for a change in the task to be performed. During the teach process the operator may desire to replay various segments of the program for visual verification of the motion or operations. Teach

Handbook of Industrial Robotics, Second Edition, Edited by Shimon Y. Nof
ISBN 0-471-17783-0   © 1999 John Wiley & Sons, Inc.

replay features may include forward and backward replay, single-step operations, and operator-selectable replay motion speeds.

The approach taken in teach programming is somewhat dependent on the control algorithm used to move the robot through a desired path. Three basic control algorithms may be used: point-to-point, continuous, and controlled path (Agrawal, 1997; LaValle et al., 1997).

Robots with *point-to-point* control move from one position to the next with no consideration of the path taken by the manipulator. Generally, each axis runs at its maximum or limited rate until it reaches the desired position. Although all axes will begin motion simultaneously, they will not necessarily complete their movements together. Figure 1*a* illustrates the trajectory taken by a robot moving with point-to-point motion.

Continuous path control involves the replay of closely spaced points that were recorded as the robot was guided along a desired path. The position of each axis was recorded by the control unit on a constant-time basis by scanning axes encoders during the robot motion. The replay algorithm attempts to duplicate that motion. Figure 1*b* illustrates a continuous path motion.

Controlled path motion involves the coordinated control of all joint motion to achieve a desired path between two programmed points. In this method of control, each axis moves smoothly and proportionally to provide a predictable, controlled path motion. In Figure 1*c* the path of the end-effector can be seen to follow a straight line between the two points of the program.

In on-line programming two basic approaches are taken to teach the robot a desired path: teach pendant programming and lead-through programming. Teach pendant programming involves the use of a portable, hand-held programming unit referred to as a teach pendant. Teach pendant programming is normally associated with point-to-point motion and controlled path motion robots. When a continuous path has to be achieved, lead-through programming is used for teaching the desired path. The operator grasps a handle secured to the arm and guides the robot through the task or motions. Lead-through programming is used for operations such as spraypainting or arc welding. Lead-through programming can also be utilized for point-to-point motion programming. Whether the robot is taught by the teach pendant method or the lead-through method, the programming task involves the integration of three basic factors:

1. The coordinates of the points of motion must be identified and stored in the control unit. The points may be stored as individual joint axis coordinates, or geometric coordinates of the tip of the robot may be stored.

2. The functions to be performed at specific points must be identified and recorded. Functional data can be path oriented, that is, path speed or seam weaving or point oriented. For example, paint spray on or wait for signal.

3. The point and functional data are organized into logical path sequences and subsequences. This includes establishing what paths should be taken under specific conditions and when various status checks should be made.

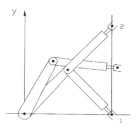
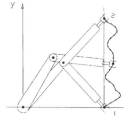
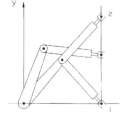

a) Point-to-Point Path   b) Continuous Path   c) Line Tracking Motion
    Motion   Motion

Figure 1 Motion control algorithms determine the actual path taken when the program is replayed in the automatic mode.

These three factors are integrated into the teaching process and do not exist as separate programming steps. During teach pendant programming the operator steps through the program point by point, recording point coordinate and functional data for each point. Path sequence is a direct consequence of the order in which the points are taught.

Teach programming is the most natural way to program an industrial robot. It is on-line programming in a truly interactive environment. During the programming task the operator is free to move around the workcell to obtain the best position to view the operation. The robot controller limits the speed of the robot, making programming safer. The operator is able to coordinate the robot motion with the other pieces of equipment with which the robot must interact. The operator can edit previously taught program replay and can test the path sequence and then continue to program until the complete program is stored in the memory of the control unit.

Teach programming does not place a significant burden on the operator. It is easy to learn and requires no special technical skill or education. While learning to program the robot, the operator is not learning a programming language in the traditional sense. This allows the use of operators who are most familiar with the operations with which the robot will interact. For example, painting and welding robots are best taught by individuals skilled in painting and welding, and parts handling and other pick-and-place tasks can be taught by material handlers or other semi-skilled labor.

Another advantage of teach programming is the ease and speed with which programs can be entered and edited. Advanced programming features permit easy entry of some common tasks, such as parts stacking, weaving, or line tracking. Stored programs can be reloaded and modified to accommodate changes in part geometry, operating sequence, or equipment location. Commonly used program sequences can be stored and recalled for future applications when needed.

The disadvantage of teach programming stems from the on-line nature and the point-by-point mode of programming. Because the robot is programmed in the actual workcell, valuable production time is lost during the teaching process. For short programs this may be of little consequence, but if the operations to be programmed involve many points and are complex in nature, having a number of branches and decision points, significant equipment downtime may result. In such cases off-line programming techniques may be more appropriate.

Teach programming has been a basic form of on-line programming for a number of years, but it is still the only effective means of on-line programming. It is the most widely used method of programming industrial robots and continues to play an important role in robot programming. The increased sophistication that has been added to the teach program capabilities has created a low-cost, effective means of generating a robot program.

2   TEACH PENDANT PROGRAMMING

Teach pendant programs involve the use of a portable, hand-held programming unit that directs the controller in positioning the robot at desired points. This is best illustrated by the example shown in Figure 2. In this example a robot is required to pick up incoming parts from the conveyor on the left, place them into the machining center, and then carry them to the finished parts conveyor on the right. Twin grippers on the end of the arm allow unloading of a finished part followed immediately by loading of a new part, thus reducing wasted motion and overall cycle time. The robot is interfaced to the machining center and to both part conveyors. An operator will lead the robot step by step through one cycle of the operation and record each move in the robot controller. Additionally, functional data and motion parameters will be entered as the points are programmed. The teach pendant is used to position the robot, whereas the controller keyboard may be required for specific data entry. Once the setup is completed the programming process can begin as follows.

2.1   Teaching Point Coordinates

1. Move the robot arm until the left gripper is just above the part at the end of the input conveyor, then open the left gripper.
2. Align the gripper axes with the part to be picked up.
3. Store this program by pressing the *record* or *program* button on the teach pendant.

Machining Center

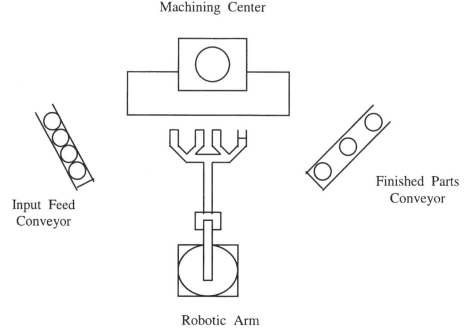

Input Feed
Conveyor

Finished Parts
Conveyor

Robotic Arm

Figure 2   Teach pendant programming example of a robotic workcell.

4.   Lower the arm until the left gripper is centered on the part to be grasped. Store this point.

5.   Close the left gripper so that the part can be lifted up. Store this point.

6.   Raise the arm so that the part is clear of the conveyor and at a desired level to rotate toward the machining center. Store this point.

7.   Move the arm so that the right gripper is in front of the finished part on the machining center, then open the right gripper.

8.   Align the gripper axes with the finished part. Store this point.

9.   Extend the arm until the right gripper is centered on the finished part. Store this point.

10. Close the right gripper so that the part can be retracted. Store this point.

11. Raise the arm so that the part clears the machining center table slightly. Store this point.

12. Retract the arm so that the finished part is clear of the machine and the unfinished part can be positioned at the desired approach point. Store this point.

13. Rotate the arm so that the unfinished part is positioned in front of the machine and slightly above the table surface. Store this point.

14. Extend the arm so that the unfinished part is in the desired positron in the machining center. Store this point.

15. Lower the arm until the unfinished part is on the table. Store this point.

16. Open the left gripper to release the unfinished part. Store this point.

17. Retract the arm so that it is clear of the machining center and ready to rotate to the finished parts conveyor. Store this point.

18. Move the arm until the finished part is above the finished parts conveyor. Store this point.

19. Lower the finished part onto the conveyor. Store this point.

20. Open the right gripper and release the finished part. Store this point.

21. Raise the arm so that it is clear of the conveyor and the finished part. Store this point.

22. Return the robot to the initial position and indicate end of cycle.

The preceding 22 steps provide the controller with the desired program point coordinates. The operator will also enter functional data associated with a number of the programmed points. For example, the robot must be made to wait at the points programmed in steps 3 and 8 until external signals indicate that it is clear to continue; an output signal must be activated at the point programmed at step 17 so that the machining center can begin operations. Additionally, the robot must be taught how to process the first and last part of the whole batch. The program sequence for these parts differs slightly from the sequence given.

2.2 Teaching Functional Data

Functional data such as desired velocity can be programmed during the normal teach process. The operator first positions the robot by depressing the appropriate buttons on the teach pendant. Functional data can then be specified by keyboard entry or on the pendant. The record button is then pressed and all relevant data are stored in the robot control memory. The operator then continues until the sequence is complete. If a program branch sequence is to be programmed, the operator must replay through the mainline to the point at which the new sequence is to be entered. The control unit can then be altered and the branch sequence programmed point by point.

2.3 Robot Control Unit and Teach Pendant

The keyboard and display of the robot control unit are often used in conjunction with the standard teach pendant while in the teach mode. In general there are three types of teach pendants:

1. Generic teach pendant (Figure 3a)
2. Teach pendant with graphical control buttons (Figure 3b)
3. Teach pendant with display (Figure 3c)

The ABB robot teach pendant in Figure 4a is a typical modern teach pendant. Motion control is achieved through a joystick with keypads for function and data entry and a display for operator messages (see Figure 4b for detailed components). The joystick can be programmed by keypad entries to control robot positioning or tool orientation. The keys on the keypad may be used to define a fixed function, or they may be defined by the operating system software to execute other functions. Operations controlled by these soft keys are displayed to the operator in the display screen adjacent to the keys. The menu-driven prompting of the operator facilitates rapid learning of the system and eliminates costly errors. The keyboard panel is used during programming to enter functional data and facilitate the editing process. The teach pendant is connected to the control unit by a long cable that permits the operator to program from a position that provides good visibility of the actions being performed. It is used to direct robot motion and to cause points to be stored. Another example of robot teach pendant is Sankyo's LCD Touch Screen Pendant (Figure 5). Operations are controlled by touching the functions on the LCD Screen. In addition, STAUBLI's robot control unit (Figure 6) is a typical modern control unit for robots. Figure 6 shows that a teach pendant is a part of the robot control unit.

Because the purpose of a robot is normally to transport some workpiece or tool through a desired path or to a specific point, the operator is usually more concerned with the location of the tool or gripper than with any other point of the arm. Most robot control systems track the *tool center point* (TCP) instead of specific joint angles. As points are stored into the memory of the control, the coordinates of the TCP with respect to the robot origin are calculated and stored. This includes both positional data and orientation data. When the robot is in the automatic mode, these TCP data values must be translated into joint position values so that the control algorithms can be executed. This added computational burden increases the complexity of the control program, and many robot controllers simply store the joint angles. This specific factor is of great importance when considering the type of motions available to the operator during the teach process and the range of functions available within the program.

While robot motion involves individual or simultaneous motion of the joints and linkages, the robot programmer is primarily concerned with the motion of the tool or gripper. Early teach pendants restricted robot motion during teaching to axis-by-axis moves. The feature of teaching the desired motion of the TCP greatly simplifies teach programming of many parts-handling operations.

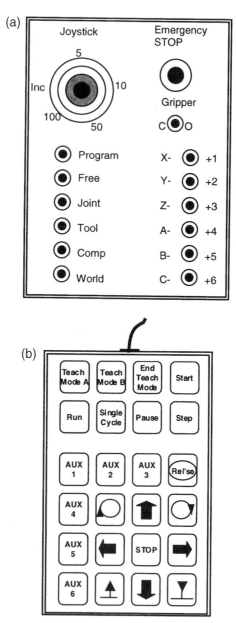

Figure 3 (a) Generic teach pendant. (b) Teach pendant with graphic control buttons. (c) Teach pendant with display.

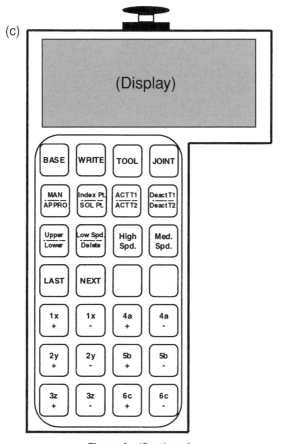

Figure 3   (Continued)

2.4   Teaching in Different Coordinate Systems

To further enhance the teach process, robot controllers allow positioning of the arms using different coordinate systems, such as rectangular, spherical and cylindrical coordinates. In the rectangular system the motions of right and left are in the Y plane, in and out in the X plane, and up and down in the Z plane. In the spherical system the motions are rotation about the base, angular elevation of a ray beginning at the base, and in and out along the ray. The cylindrical motion is defined by rotation of the base, in and out motion in a horizontal, radial direction, and up and down motion in a vertical plane. A fourth programming positioning motion that is extremely useful is a wrist-oriented rectangular coordinate system. Here the robot positioning motion is taken in a Cartesian coordinate system that is aligned with the wrist orientation.

Positioning of the TCP during programming does not affect tool orientation. Pitch, roll, and yaw rotations are normally controlled by a separate set of axis buttons. When the TCP is positioned by one of the coordinate motions described earlier, it is translated through space. When the orientation of the TCP is changed, the TCP remains fixed and the robot actuates the motion of the axes to rotate around the specified point. The path taken under program control in the automatic mode is independent of the path taken while teaching, since only end-point data are stored.

Axis-by-axis positioning is acceptable when the end-point data are all that is desirable. If the application requires precise path trajectories, both during programming and replay, coordinated joint motion is highly desirable to maximize operator productivity. This added

(a)

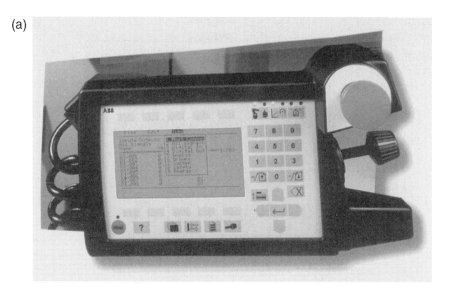

(b)

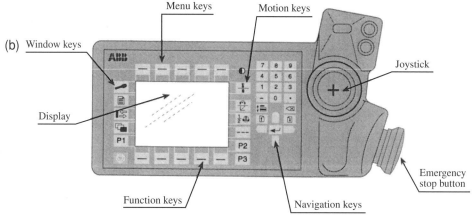

Figure 4 (a) Modern robot teach pendant (courtesy ABB Robotics, Inc.). (b) Components of modern robot teach pendant (courtesy ABB Robotics, Inc.).

complexity in the control of the robot during programming requires that the TCP positional data be stored and that the control algorithm be of the continuous path type.

2.5 Teach versus Replay Motion

The relationship between teach path motion control and the real-time replay motion control algorithm is not always fully understood. A robot can be programmed with a joint coordinated motion scheme and replayed in a point-to-point mode. This may be done to minimize the demands placed on the axis servo-systems or reduce computational burden in the control unit. Robots programmed by axis-by-axis motion may be replayed point-to-point or by controlled path motion, depending on the results desired. Many robots are limited to point-to-point path motion or a course-controlled path motion. Early robots that offered sophisticated path control during programming were often limited to controlled path motion with straight-line motion between the programmed points. Currently, control units provide a point-to-point replay motion after teaching a joint coordinated motion.

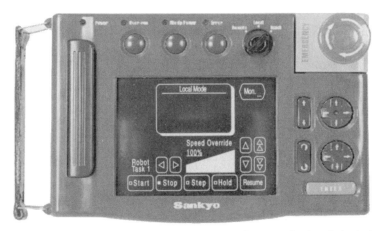

Figure 5 Teach pendant with LCD touch screen (courtesy Sankyo Robotics).

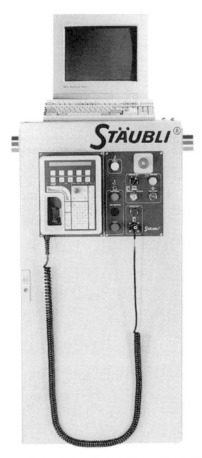

Figure 6 Teach pendant within a robot control unit (courtesy Staubli, Inc.).

2.6   Teaching Interface Signals

Most industrial robots perform production operations that require interconnections to the surrounding equipment. In the example presented in Figure 2, the robot must be interfaced to both conveyors and the machining center. Input signals are required from the feed conveyor to indicate that a part is ready for pickup and from the machining center to indicate that the machine is done and a finished part is ready for pickup. It is also desirable to arrange some sort of sensor to insure that the finished-parts conveyor is clear before unloading and placing the next finished part on the unit (Jacak et al., 1995). The robot must output a signal to the machining center to initiate the start of the machining cycle after the robot has loaded a new part and withdrawn a safe distance. Such types of input and output signals are included in the path program through the use of specific functional data associated with specific programmed points. Figure 7 illustrates the control communication diagram for the ABB System 4 industrial robot controller.

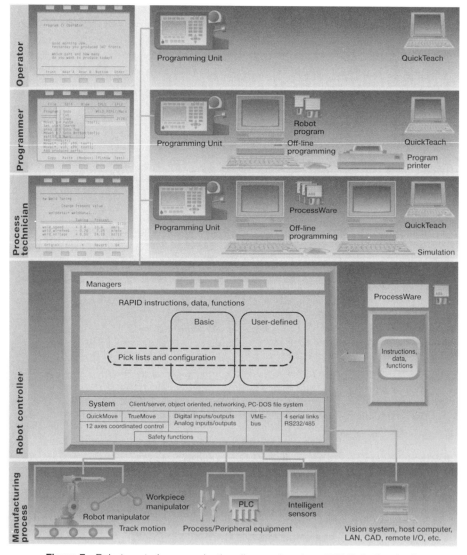

Figure 7   Robot control communication diagram (courtesy ABB Robotics, Inc.).

2.7   Teaching Program Branching

These input and output signals may be digital, such as limit switches and relay contacts, or analog, such as furnace temperature and arc voltage. Most industrial robot controllers include standard interfaces for programmable controller, valve packs and peripheral control for process interconnection. In addition to these signals, the robot must also respond to interrupt service routines that are activated by external input signals in the cycle. Interrupt signals are constantly monitored by the robot controller, and the robot immediately abandons the current operation and performs the interrupt routine.

Decision-making allows a robot system to deviate from its normal path program and perform other tasks based on changes in external events. Teach pendant programming systems permit a variety of capabilities in robot decision making and path modification. The specific features available on a given robot are a function of the sophistication of the control system and the imagination of the control system's programmers. Advanced decision-making concepts are discussed in Section 3.

2.7.1   Standard Branching

A standard branch exists when the robot is programmed to check a single input signal at a given path point and choose the next destination based on that signal. The robot control unit is not constantly monitoring the status of the input signal. The signal is checked only when the robot reaches the branch point. Thus an external condition that is momentary in nature will not cause branching if it goes undetected. The path sequence associated with a branch may return to the point where it was initiated or it may close on some other motion point. In an actual application a number of standard branches may be included in the path program. It should also be noted that a number of branches may be initiated from the same physical location. This would permit the controller to examine a number of input signals one at a time and branch when the first signal is found at the desired level.

2.7.2   Conditional Branching

Conditional branching is an extension of the standard branch concept. Instead of a single input signal being examined to test for a desired state, a number of input signals are examined at the same time to see if they are all in their desired state. These two concepts have been presented separately, however, to emphasize the difference in their nature. If a robot controller is limited to standard branching, the path program may become excessively long or the number of input signals required may be large. Functionally they are externally equivalent and have the same appearance.

2.7.3   Offset Branching

Offset branching provides the programmer with the means of establishing a path sequence that is relative to the physical location of the robot arm when the branch is requested. When the branch is initiated, each point in the path sequence is offset or translated relative to the current robot position and/or the wrist orientation. Offset branch programming is primarily associated with robot control systems that utilize TCP coordinates as stored data. Offset branches are used when a specific sequence of motions is repeated at a number of points in a path program. An example of offset programming is the insertion of screws or fasteners during an assembly operation.

3   LEAD-THROUGH TEACH PROGRAMMING

Lead-through teach programming is used primarily for continuous-path motion-controlled robots. During the lead-through teach process an operator grasps a removable teach handle, which is attached directly to the arm, and leads the robot through a desired sequence of operations to define the path and relative velocity of the arm. While the operator is teaching the robot, position transducers measure the joint angles associated with each robot axis, and the robot control unit records this information for use during production replay. Automatic replay of the digitized data provides the continuous path motion necessary for precise duplication of the motion and technique utilized by the human operator. Lead-through teach programming and continuous path motion-controlled robots are well suited for operations such as spraypainting, sealant application, and arc welding (Akeel, 1982; Green, 1980).

Lead-through teach programming can also be used to program a robot for point-to-point motion control. In this mode of lead-through teach programming, only the end points of each path segment are recorded for replay. Path velocity is normally recorded in this mode as a function entry associated with each path segment. Additionally, path replay patterns can be incorporated for use in a specific application such as seam weaving for arc welding. Arc welding robots offer both continuous path lead-through teaching with the arc on and point-to-point teaching with the arc off. A lead-through teach material-handling robot will utilize the point-to-point mode to program part loading and transfer operations.

The major advantage of lead-through teach programming over teach-pendant programming is ease of programming. Lead-through teach programming is easy to learn and can be performed by an operator who is directly associated with the production task (painting or welding). To further facilitate programming operations, the robotic arm is kinematically balanced to allow easy movement by the operator during lead-through programming (Colbaugh et al., 1996; McCarragher et al., 1997). The counterbalanced arm permits the operator to manipulate the end-of-arm tooling manually in a way similar to manual operations. Lead-through teach-programmed robots often have dual handles, one to support the arm and the second to guide the tooling.

3.1 Lead-Through with a Programming Arm

The robot arm when performing lead-through is often difficult to maneuver. In the past, mechanisms for counterbalancing the weight have been used. Currently a number of robot vendors market lightweight programming arms that are kinematically identical to the actual manipulator. The operator manipulates the programming arm and joint position data are recorded. The stored data path is then used for replay with the actual production robot. Because the programming arm is used in parallel to the actual manipulator, a second robot control unit will be necessary to digitize joint position data or the programming arm must be connected to the control unit of the actual manipulator.

Another concept found in lead-through teach programming is a teaching system to program multiple robots of the same type. The programming facility is either a lightweight programming arm or a production robot. The teaching system is led through the desired sequence of operations and the path program is created. The program is then passed to the control unit on which the program is to be automatically replayed. The nature of the application areas associated with lead-through robots can accommodate the minor differences between different arms and the resulting differences in replay paths.

3.2 Software for Lead-Through Programming

Lead-through teach-programming software systems incorporate basic program editing features. Often the program is short enough that the entire program can be deleted and a new program recorded. Alternatively, the desired path sequence can be broken down into a number of path segments and each segment taught separately. The segments can then be merged into the desired sequence to create the path program. Because the robot controller maintains each segment as a separate entity, the user can delete an undesired segment and reprogram the proper path. The new segment can then be merged into the path program for the desired result. If the program has been created in the continuous path mode, a program segment is the smallest path unit that can be deleted or replaced.

Robot vendors use a number of different schemes in recording the joint-position data associated with lead-through programming. The simplest and most direct scheme is to digitize the data at some present rate and record the value as obtained. The robot controller will then attempt to drive the robot through these points during replay. A second scheme involves digitizing joint-position data at some multiple of the desired data recording rate. The digitized values can further be mathematically smoothed by a moving-average technique to obtain the desired positional data for recording. This technique can be helpful in removing random fluctuations in the programmed path caused by operator actions. The path can also be determined by fitting a spline through the data gathered and then redistributing the points along the mathematical function to insure a more uniform velocity. Irrespective of the scheme chosen for storing joint-position data, velocity control is normally limited to some fractional multiple of the programmed speed. The replay velocity of each program segment can be edited to fine-tune the program. However, the replay motion will contain the acceleration and deceleration patterns exhibited by the programmer.

4   ADVANCED PROGRAMMING CONCEPTS

A number of special features or options in teach programming have been created that increase the ease of programming or add flexibility and versatility to a part program. Some of these features are presented in the following sections to illustrate the extent to which teach-pendant programming has developed over the years.

Branch and interrupt structures have been called subroutines by a number of robot manufacturers and could have been presented as such in the preceding section. They somewhat lack all of the features normally associated with high-level language subprograms, but they do serve a similar purpose. Subroutines, or subprogram blocks, can be stored on a floppy disk in the form of a common library. These subroutines can then be available for future use when the same or similar sequence of motions is desired. As control path programs are developed, frequently used routines can be added to the library. Such a library will continue to increase in size and value. This programming feature has greatly reduced programming time and cost.

4.1   Role of Sensors in On-Line Programming

Sensors play a vital role in on-line programming. They ease the programming task and increase the flexibility of the path motion. The functions that can be programmed include search, contour following, and speed control. The search function is normally associated with a stacking operation. Contour following permits the robot to move along a desired surface that is not defined completely. It can be used to guide the robot motion during automatic replay or to teach the robot a specific part-oriented path for subsequent operations. The speed control function can be used to adjust the velocity of the robot as it moves along a given path. The sensors used with these functions vary from simple on/off digital sensors to fully analog sensors. Multiple sensors can be used simultaneously during the search and contour functions. This allows the functions to be performed in multiple dimensions.

4.2   Teaching and Supervisory Control

A communications link between the robot controller and an external computer system is provided to store taught programs and download them for use at some future time. Various approaches have been developed, such as natural language (Knoll et al., 1997) and single demonstration (Atkeson and Schaal, 1997), but a more interesting use of this link can be found in branch modification. The data comprising a complete branch sequence in the robot controller can be transmitted to the external system. This system can then modify the data values and transmit the branch back to the robot controller. When a branch is executed, new data points will be the points of motion of the robot and not the points taught earlier. A higher level of robot decision-making is therefore possible with the coupling of the robot with a supervisory system or some more complex sensor systems employing vision, force, or tactile transducers, e.g., tactile Jacobian (Chen et al., 1997). The supervisory control feature has also been useful in widely diversified applications. For instance, the branch modification scheme has been mainly responsible for the application success of vision systems used for part location tasks with respect to taught robots. A dummy branch is created to accommodate part pickup and then the branch is transmitted to the vision control system. This system establishes the robot path that must be taken to pick the part up and revises the point data in the dummy path. The branch is then downloaded and the robot proceeds according to the new data. Seam tracking for arc welding operations has been another application area for branch modification. A vision system or laser-scanning device is used to locate the seam to be welded. Positional data and weld parameters can then be downloaded into the control unit to execute the desired weld.

4.3   Teaching and Line Tracking

Many manufacturing operations are carried out on continuously moving production lines. Line tracking involves the ability of the robot to work in conjunction with such lines. There are two basic ways to accomplish line tracking with a teach-pendant robot. *Moving-base* line tracking requires that the robot be mounted on some form of transport system that will cause the robot to move parallel to the production line. *Stationary-base* line tracking combines the sophistication of the control unit with sensor data to modify dynamically the robot motion commands to permit work to be performed on the moving line by a stationary robot.

Both moving-base and stationary-base line tracking allow the user to program the robot in a stationary position. The teach pendant is used to position the robot at desired points that are programmed in the normal manner. During automatic replay a synchronization signal ensures that the stationary-base line-tracking robot origin is in the same position relative to the moving line during program replay. Because the points are programmed while the two objects are in the same relative position, the path will be performed as desired. The stationary-base line-tracking system utilizes a position sensory device that sends signals to the control unit indicating the position of the part relative to a fixed origin at all times. The signal is then used to produce a shift in the zero position value of the taught points. Because the joint axis motions must be calculated dynamically during replay anyway for robots that store TCP, the tracking feature only slightly complicates the computational task.

Moving-base line-tracking robots are easily programmed but require expensive installation of the transport system and may create interference between adjacent stations. Stationary-base line-tracking robots, however, increase the complexity of programming as it flows through the work envelope. Because the work moves relative to the robot, the robot may be able to work on different sides of the work. Although the robot may have expanded the possible working surfaces, each surface is available only during a fraction of the overall design cycle time. The operator must be concerned with the work window from both time and space viewpoints. Accessibility of each surface with respect to time will have to be kept in mind during programming.

4.4 Abort and Utility Branches

Finally, the operator must consider abort and utility branches. Abort branches represent pretaught safe paths by which the robot can exit a work area if the taught points are no longer feasible. Utility branches are similar to the interrupt service routines discussed above, but they must be performed as the tool is in motion.

4.5 Process Functions

A number of programming features are available that adapt general purpose industrial robots to various processes. Process specific add-on software modules are available, which provide dedicated application robots. For example, in welding robots repeatability, lift capacity, and speed of robots, coupled with proper weld parameters, can produce welds that are consistent and of high quality. Additional programming functions allow the operator to enter the wire feed, speed, and arc voltage into a table called the *weld schedule*. As the operator teaches a given weld path, an index to this schedule is attached to program points to set the desired process parameters. When the path is repeated in the automatic replay mode, the robot control unit will assess these functional values and output analog control set points to the weld controller. Because the weld parameters are part of the stored path program, weld quality is reproducible from part to part. A standard velocity function allows various weld segments to be programmed at different velocities to produce different weld depths. If the operator desires to change the speed of welding over a given set of points, the teach pendant can be used to edit the necessary stored data. Another standard feature that can be used to ease weld programming is offset branching. A root path of a multi-pass weld can be programmed once and then offset as desired to obtain multiple passes.

Similar software is also available for painting robots. Examples of parameters that can be specified include spray velocity and robot arm velocity.

5 CURRENT AND FUTURE TRENDS IN ON-LINE PROGRAMMING

Teach programming has established itself as the primary mode of on-line programming. Improvement in on-line programming has been in the development of more user-friendly software and customization of software for various types of application. With the advent of more powerful computers, the algorithms used for analysis of the path traced during teach mode and its regeneration have become more sophisticated. The development of dedicated robotic systems for activities such as painting and welding has also resulted in customized software. One advance has been in the development of windows-style graphic user interfaces with function keys and pull down menus, which allow the functional parameters of the manufacturing process to be programmed easily. Additionally, the functional process parameter inputs were separated from the teach pendant, allowing expert manufacturing people to input the parameters. In earlier systems the emphasis was on using a lightweight robot arm for the development of the trajectory and ease of programming. The trend has reversed, and robots today find applications where teach programming is done on dummy facilities. Typical examples of this are found in medical surgery applications and hazardous waste material handling. The future of robot programming may be in the use of virtual reality tools. Computer-aided designs and virtual reality tools (Ferrarini et al., 1977; Rizzi and Koditschek, 1996; Natonek et al., 1995) are already being used in performing the analysis of assembly and manufacturing processes. This can be easily applied to the programming of robots. Meanwhile, robot calibration (Hollerbach and Wampler, 1996; Lu and Lin, 1997) is also an important research issue in robotic on-line programming.

On-line programming using the teach method is flexible, versatile, and easily accomplished. Although off-line programming has been developing at a rapid pace easier-to-program languages, and tools for off-line debugging have been introduced, the programming mode has not yet been simplified to the same level as that of on-line systems. As a result, on-line programming is continuing the success it has enjoyed in the past and will remain for some time the most widely used method of programming computer-controlled industrial robots.

ACKNOWLEDGMENT

The authors wish to thank Mr. Chin-Yin Huang from Purdue University for helping in the revision of this chapter.

REFERENCES

Agrawal, S. K., and T. Veeraklaew. 1997. "Designing Robots for Optimal Performance During Repetitive Motion." In *Proceedings of the 1997 IEEE International Conference on Robotics and Automation,* April. 2178–2183.

Akeel, H. A. 1982. Expanding the Capabilities of Spray Painting Robots." *Robots Today* **4**(2), 50–53.

Atkeson, C. G., and S. Schaal. "Learning Tasks from a Single Demonstration" In *Proceedings of the 1997 IEEE International Conference on Robotics and Automation.* 1706–1712

Chen, N. N., H. Zhang, and R. E. Rink. 1997. "Touch-Driven Robot Control Using a Tactile Jacobian." In *Proceedings of the 1997 IEEE International Conference on Robotics and Automation.* 1737–1742.

Colbaugh, R., K. Glass, and H. Seraji. 1996. "Adaptive Tracking Control of Manipulators: Theory and Experiments." *Robotics and Computer-Integrated Manufacturing* **12**, 209–216.

Dorf, R. C. 1984. *Robotics and Automated Manufacturing.* Reston: Reston.

Engelberger, J. F. 1980. *Robotics in Practice.* American Management Association.

Ferrarini, L., G. Ferretti, C. Maffezzoni, and G. Magnani. 1997. "Hybrid Modeling and Simulation for the Design of an Advanced Industrial Robot Controller." *IEEE Robotics and Automation Magazine* **4**(2), 27–51.

Green, R. H. 1980. "Welding Auto Bodies with Traversing Line-Tracking Robots." *Robotics Today* (Spring), 23–29.

Hollerbach, J. M., and C. W. Wampler. 1996. "The Calibration Index and Taxonomy for Robot Kinematic Calibration Methods." *International Journal of Robotics Research* **15**, 573–591.

Jacak, W., B. Lysakowaska, and I. Sierocki. 1995. "Planning Collision-Free Movements of Robots." *Robotica* **13**(2), 297–304.

Knoll, A., B. Hildebrandt, and J. Zhang. 1997. "Instructing Cooperating Assembly Robots Through Situated Dialogues in Natural Language." In *Proceedings of the 1997 IEEE International Conference on Robotics and Automation.* 888–894.

LaValle, S. M., H. H. Gonzalez-Banos, C. Becker, and J. C. Latombe. 1997. "Motion Strategies for Maintaining Visibility of a Moving Target." In *Proceedings of the 1997 IEEE International Conference on Robotics and Automation,* April. 731–736.

Lu, T., and G. C. I. Lin. 1997. "An On-Line Relative Position and Orientation Error Calibration Methodology for Workcell Robot Operations." *Robotics and Computer-Integrated Manufacturing* **13**(2), 88–99.

McCarragher, B. J., G. Hovland, P. Sikka, P. Aigner, and D. Austin. 1997. "Hybrid Dynamic Modeling and Control of Constrained Manipulation Systems." *IEEE Robotics and Automation Magazine* **4**(2), 27–44.

Natonek, E., L. Flueckiger, T. Zimmerman, and C. Baur. 1995. "Virtual Reality: An Intuitive Approach to Robotics." *Proceedings of SPIE* **2351,** 260–270.

Rizzi, A. A., and D. E. Koditschek. 1996. "An Active Visual Estimator for Dexterous Manipulation." *IEEE Transactions on Robotics and Automation* **12**(5), 697–713.

Warnecke, H. J., and R. D. Schraft. 1982. *Industrial Robots: Application Experience.* Kempston, Bedford: IFS Publications.

CHAPTER 19

OFF-LINE PROGRAMMING

Y. F. Yong
BYG Systems Ltd.
Nottingham, United Kingdom

M. C. Bonney
University of Nottingham
University Park, United Kingdom

1 INTRODUCTION

1.1 What Is Off-Line Programming?

Present *teach* methods of programming industrial robots have proved to be satisfactory where the proportion of teaching time to production time is small, and also when the complexity of the application is not too demanding. They involve either driving a robot to required positions with a teach pendant or physically positioning the robot, usually by means of a teach arm. Teach methods as such necessitate the use of the actual robot for programming.

Off-line programming may be considered as the process by which robot programs are developed, partially or completely, without requiring the use of the robot itself. This includes generating point coordinate data, function data, and cycle logic. Developments in robot technology, both hardware and software, are making off-line programming techniques more feasible. These developments include greater sophistication in robot controllers, improved positional accuracy, and the adoption of sensor technology. There is currently considerable activity in off-line programming methods, and these techniques are employed in manufacturing industries.

1.2 Why Should Off-Line Programming Be Used?

Programming a robot by *teaching* can be time-consuming—the time taken quite often rises disproportionately with increasing complexity of the task. As the robot remains out of production, teach programming can substantially reduce the utility of the robot, sometimes to the extent that the economic viability of its introduction is questioned.

Many early robot applications involved mass production processes, such as spot welding in automobile lines, where the reprogramming time required was either absent or minimal. However, for robot applications to be feasible in the field of small and medium batch production, where the programming times can be substantial, an off-line programming system is essential. The increasing complexity of robot applications, particularly with regard to assembly work, makes the advantages associated with off-line programming even more attractive. These advantages may be summarized as follows:

1. *Reduction of robot downtime.* The robot can still be in production while its next task is being programmed. This enables the flexibility of the robot to be utilized more effectively (Roos and Behrens, 1997).

2. *Removal of programmer from potentially hazardous environments.* As more of the program development is done away from the robot this reduces the time during which the programmer is at risk from aberrant robot behavior.

3. *Single programming system.* The off-line system can be used to program a variety of robots without the need to know the idiosyncracies of each robot controller.

Handbook of Industrial Robotics, Second Edition, Edited by Shimon Y. Nof
ISBN 0-471-17783-0 © 1999 John Wiley & Sons, Inc.

These are taken care of by appropriate postprocessors, minimizing the amount of retraining necessary for robot programmers.

4. *Integration with existing CAD/CAM systems.* This enables interfaces to access standard part data bases, limiting the amount of data capture needed by the off-line system. Centralization of the robot programs within the CAD/CAM system enables them to be accessed by other manufacturing functions such as planning and control.

5. *Simplification of complex tasks.* The utilization of a high-level computer programming language for the off-line system facilitates the robot programming of more complex tasks.

6. *Verification of robot programs.* Existing CAD/CAM systems, or the off-line system itself, can be used to produce a solid world model of the robot and installation. Suitable simulation software can then be used to prove out collision-free tasks prior to generation of the robot program.

2   DEVELOPMENT OF OFF-LINE PROGRAMMING

2.1   Some Parallels with NC

The technique of off-line programming has been employed in the field of numerical controlled (NC) machining for some considerable time, and a significant body of knowledge has been built up in this area. Although there are fundamental differences between the programming/controlling of NC machines and industrial robots, there are similarities in the problems and phases of development for off-line systems. It is therefore instructive to draw some parallels with NC in the course of charting the developments of industrial robot programming methods.

Early NC controllers were programmed directly using codes G, F, S, . . . and so on, with appropriate X, Y, and Z positioning coordinates. These specified certain parameters such as tool movements, feed rates, and spindle speeds, and were specific to individual controllers.

The majority of present-day industrial robots have controllers more or less equivalent to this, that is, they are programmed directly using robot-specific functions. The programming is done entirely on-line, and coordinate positions are taught manually and recorded in the computer memory of the controller. Program sequences can then be replayed.

The next development phase for programming NC controllers utilized high-level computer languages to provide greater facilities for the NC programmer. Programming became truly off-line, with program development being performed on computers remote from the shop floor. High-level textual NC programming languages such as APT and EXAPT developed. Tools and parts could be described geometrically using points, lines, surfaces, and so on, and these were used to generate cutter path data. Appropriate interfaces were required between the languages and specific NC controllers.

Just as NC programming progressed toward integration with CAD systems, off-line programming of industrial robots will develop in a similar fashion. Ultimately, systems incorporating direct robot control will be developed to cater to multirobot installations.

2.2   Levels of Programming

A useful indicator of the sophistication of a robot-programming system is the level of control of which it is capable. Four levels can be classified as follows (Thangaraj and Doelfs, 1991):

1. *Joint level.* Requires the individual programming of each joint of the robot structure to achieve the required overall positions.

2. *Manipulator level.* Involves specifying the robot movements in terms of world positions of the manipulator attached to the robot structure. Mathematical techniques are used to determine the individual joint values for these positions.

3. *Object level.* Requires specification of the task in terms of the movements and positioning of objects within the robot installation. This implies the existence of a world model of the installation from which information can be extracted to determine the necessary manipulator positions.

4. *Objective level.* Specifies the task in the most general form; for example, "spray interior of car door." This requires a comprehensive database containing not only

a world model but also knowledge of the application techniques. In the case of this example, data on optimum spraying conditions and methods would be necessary. Algorithms with what might be termed "intelligence" would be required to interpret the instructions and apply them to the knowledge base to produce optimized, collision-free robot programs.

Programming at object, and particularly objective, levels requires the incorporation of programming constructs to cater for sensor inputs. This is necessary to generate cycle logic. For example, the "IF, THEN, ELSE" construct could be employed in the following manner:

$$\text{IF (SENSOR = value)} \quad \begin{array}{ll} \text{THEN} & \text{action 1} \\ \text{ELSE} & \text{action 2} \end{array}$$

Most present-day systems, on-line and off-line, provide manipulator-level control. Language systems currently under development are aiming toward the object level of programming, with objective level being a future goal.

3   GENERAL REQUIREMENTS FOR AN OFF-LINE SYSTEM

In essence, off-line programming provides an essential link for CAD/CAM. Success in its development would result in a more widespread use of multiaxis robots and also accelerate the implementation of flexible manufacturing systems (FMS) in industry.

As indicated in the preceding section, off-line programming can be affected at different levels of control. Different systems employ different approaches to the programming method.

Yet, despite their differences, they contain certain common features essential for off-line programming. This following list gives the requirements that have been identified to be important for a successful off-line programming system:

1. A three-dimensional world model, that is, data on the geometric descriptions of components and their relationships within the workplace.
2. Knowledge of the process or task to be programmed.
3. Knowledge of robot geometry, kinematics (including joint constraints and velocity profiles), and dynamics.
4. A computer-based system or method for programming the robots, utilizing data from 1, 2, and 3. Such a system could be graphically or textually based.
5. Verifications of programs produced by 4. For example, checking for robot joint constraint violations and collision detection within the workplace.
6. Appropriate interfacing to allow communication of control data from the off-line system to various robot controllers. The choice of a robot with a suitable controller (i.e., one that is able to accept data generated off-line) will facilitate interfacing.
7. Effective man–machine interface. Implicit in off-line programming is the removal of the programmer from the robot. To allow the effective transfer of his skills to a computer-based off-line system, it is crucial that a user-friendly programming interface be incorporated.

4.   PROBLEMS IN OFF-LINE PROGRAMMING

4.1   Overview

$$\text{THEORETICAL} \qquad \text{ROBOT} \qquad \text{REAL}$$
$$\text{MODEL} \; \underline{} \; \text{PROGRAM} \; \underline{} \; \text{WORLD}$$

Off-line programming requires the existence of a theoretical model of the robot and its environment; the objective is to use this model to simulate the way in which the robot would behave in real life. Using the model, programs can be constructed that, after suitable interfacing, are used to drive the robot.

The implementation of off-line programming encounters problems in three major areas. First, there are difficulties in developing a generalized programming system that is independent of both robots and robot applications. Second, to reduce incompatibility between robots and programming systems, standards need to be defined for interfaces. Third, off-line programs must account for errors and inaccuracies that exist in the real world. The following sections provide a more detailed discussion of these problem areas.

4.2   Modeling and Programming

The modeling and programming system for off-line work can be categorized into three areas: the geometric modeler, the programming system, and the programming method. Each has its own inherent difficulties, but the major problems arise when attempts are made to generalize functional features. Although generalization (to cater for different types of robots and applications) is necessary to make the system more effective, it is also important to ensure that corresponding increases in complexity do not inhibit the functional use of the system.

4.2.1   Geometric Modeler

A problem with any geometric modeler is the input of geometrical data to allow models to be constructed. In a manual mode this process is time-consuming and error-prone. Sufficient attention must be given to improving methods of data capture. One way is to utilize data stored in existing CAD systems. Even so, this necessitates the writing of appropriate interfaces.

The data structure used must be capable not only of representing relationships between objects in the world but also of updating these relationships to reflect any subsequent changes. It must also allow for the incorporation of algorithms used by the robot modeler. To accomplish these requirements efficiently can be a difficult task.

4.2.2   Robot Modeler

An off-line programming system must be able to model the properties of jointed mechanisms. There are several levels at which this may be attempted.

The first level is to develop a robot-specific system, that is, for use with only one type of robot. While this greatly simplifies the implementation, it also limits the scope of application of the system.

A second-level approach is to generalize to a limited class of structures. For example, most commercial robots consist of a hierarchical arrangement of independently controlled joints. These joints usually allow only either rotational or translational motion. Standard techniques exist for modeling such manipulators. (See Chapter 57 for a detailed description of simulators/emulators of robotic systems.) Even at this level, controlling the robot in terms of the path to be followed by the mounted tool is not easy. There is no general solution that covers all possibilities. The structures must be subclassified into groups for which appropriate control algorithms can be developed.

The third level is to attempt to handle complex manipulator structures. Some robots have interconnected joints that move as a group and are therefore not mathematically independent. The mathematics of complex mechanisms such as these is not understood in any general sense, although particular examples may be analyzed.

In the near future the kinematics of generalized off-line systems will probably cope with a restricted set of mechanisms as described in the second level. The incorporation of dynamic modeling (to simulate real motion effects such as overshoot and oscillation) is considered too complex to accomplish in a generalized fashion.

4.2.3   Programming Method

The geometric and robot modelers provide the capability of controlling robot structures within a world model. A programming method is required to enable control sequences, that is, robot movement sequences, to be defined and stored in a logical manner. The method should allow robot commands, robot functions, and cycle logic to be incorporated within these sequences to enable complete robot programs to be specified.

The latter requirements cause complications when the system is to be applied to different application areas. For example, the functional requirements and robot techniques for arc welding are significantly different from those involved in spraypainting. Modularization of the programming method into application areas should ease these complications and produce a more efficient overall system.

The implementation of programming can be at one of several levels, joint, manipulator, object, or objective level, as discussed in Section 2.2. Most systems under development are aimed at object level, which requires programmer interaction to specify collision-free, optimized paths. This interaction is considerably simplified if graphics are used within the programming system. This enables the programmer to visualize the problem areas and to obtain immediate feedback from any programmed moves.

Generalization of the programming method to cater for multirobot installations creates many difficulties. The incorporation of time-based programming and methods of communication between separate programs is necessary. Future systems will have greater need for these enhanced facilities as the trend toward FMS increases.

4.3   Interfacing

An off-line programming system defines and stores the description of a robot program in a specific internal format. In general, this is significantly different from the format employed by a robot controller for the equivalent program. Hence it is necessary to have a form of interfacing to convert the program description from an off-line system format to a controller format.

One of the major problems lies in the existence of a wide range of different robot controllers together with a variety of programming systems, each employing a different format of program description. To avoid a multiplicity of interfaces between specific systems and controllers, standards need to be defined and adopted.

Standardization could be employed in one or more of the following areas:

1. *Programming system.* The adoption of a standard system for off-line programming would considerably reduce interfacing efforts.
2. *Control system.* A standardized robot control system would have similar beneficial effects to standardizing the programming system. Commercial and practical considerations make adoption of this approach an unlikely occurrence.
3. *Program format.* The definition of a standard format for robot program descriptions would also reduce interfacing problems. Such a format would of necessity, be independent of the programming systems and controllers. The adoption of CLDATA in numerical control (NC) is a useful precedent for this approach.

4.4   Real-World Errors and Inaccuracies

Owing to implicit differences between an idealized theoretical model and the inherent variabilities of the real world, simulated sequences generally cannot achieve the objective of driving the robot without errors. In practice, the robot does not go to the place predicted by the model, or the workpiece is not precisely at the location as defined in the model. These discrepancies can be attributed to the following components:

1. *The robot*
   (a) Insufficiently tight tolerances used in the manufacture of robot linkages, giving rise to variations in joint offsets. Small errors in the structure can compound to produce quite large errors at the tool.
   (b) Lack of rigidity of the robot structure. This can cause serious errors under heavy loading conditions.
   (c) Incompatibility between robots. No two robots of identical make and model will perform the same off-line program without small deviations. This is caused by a combination of control system calibration and the tolerancing problems outlined.
2. *The robot controller*
   (a) Insufficient resolution of the controller. The resolution specifies the smallest increment of motion achievable by the controller.
   (b) Numerical accuracy of the controller. This is affected by both the word length of the microprocessor (a larger word length results in greater accuracy) and the efficiency of the algorithms used for control purposes.
3. *The workplace*
   (a) The difficulty in determining precise locations of objects (robots, machines, workpiece) with reference to a datum within the workplace.

  (b)   Environmental effects, such as temperature, can adversely affect the perform-
        ance of the robot.
4.   *The modeling and programming system*
  (a)   Numerical accuracy of the programming system computer—effects such as
        outlined under 2(b).
  (b)   The quality of the real-world model data. This determines the final accuracy
        of the off-line program.

The compounding effects of these errors across the whole off-line programming system
can lead to discrepancies of a significant magnitude. For off-line programming to become
a practical tool, this magnitude must be reduced to a level whereby final positioning
adjustments can be accomplished automatically. (See Chapter 39.)

To achieve this a combination of efforts will be required. First, the positional accuracy[1]
of the robot must be improved. Positional accuracy is affected by factors such as the
accuracy of the arm, the resolution of the controller, and the numeric accuracy of the
microprocessor. Second, more reliable methods for determining locations of objects within
a workplace must be applied. Third, the incorporation of sensor technology should cater
to remaining discrepancies within a system. Improvements in tolerances of the components on which the robot has to work will aid the overall performance of the system.

5   TASK-LEVEL MANIPULATOR PROGRAMMING

Robots are useful in industrial applications primarily because they can be applied to a
large variety of tasks. The robot's versatility derives from the generality of its physical
structure and sensory capabilities. However, this generality can be exploited only if the
robot's controller can be easily, and hence cost-effectively, programmed.

Three methods of robot programming can be identified; the following list reflects their
order of development:

1.   Programming by guiding (a teach method)
2.   Programming in an explicit robot-level computer language (joint level and manip-
        ulator level control as defined in 1.2)
3.   Programming by specifying a task-level sequence of states or operations (normally
        object level control as defined in 1.2 but could be objective level)

5.1   Programming by Guiding

The earliest and most widespread method of programming robots involves manually moving the robot to each desired position and recording the internal joint coordinates corresponding to that position. In addition, operations such as closing the gripper or activating
a welding gun are specified at some of these positions. The resulting *program* is a sequence of vectors of joint coordinates plus activation signals for external equipment. Such
a program is executed by moving the robot through the specified sequence of joint coordinates and issuing the indicated signals. This method of robot programming is known
as *teaching by showing* or *guiding*.

Robot guiding is a programming method that is simple to use and to implement.
Because guiding can be implemented without a general-purpose computer, it was in widespread use for many years before it was cost-effective to incorporate computers into
industrial robots. Programming by guiding has some important limitations, however, particularly regarding the use of sensors. During guiding the programmer specifies a single
execution sequence for the robot; there are no loops, conditionals, or computations. This
is adequate for some applications, such as spot welding, painting, and simple materials
handling. In other applications, however, such as mechanical assembly and inspection,
one needs to specify the desired action of the robot in response to sensory input, data
retrieval, or computation. In these cases robot programming requires the capabilities of a
general-purpose computer programming language.

---

[1]Positional accuracy of the robot is its ability to achieve a commanded world position. This is distinct
from *repeatability*, which relates to the variation in position when it repeats a taught move.

5.2   Robot-Level Computer Languages

Some robot systems provide computer programming languages with commands to access sensors and specify robot motions. The key advantage of these *explicit* or *robot-level* languages is that they enable the data from external sensors, such as vision and force, to be used in modifying the robot's motions. Through sensing, robots can cope with a greater degree of uncertainty in the position of external objects, thereby increasing their range of application.

The key drawback of robot-level programming languages, relative to guiding, is that they require the robot programmer to be expert in computer programming and design of sensor-based motion strategies. Robot-level languages are certainly not accessible to the typical worker on the factory floor. Programming at this level can be extremely difficult, especially for tasks requiring complex three-dimensional motions coordinated by sensory feedback. Even when the tasks are relatively simple, as today's industrial robot tasks are, the cost of programming a single robot application may be comparable to the cost of the robot itself. This is consistent with trends in the cost of software development versus the cost of computer hardware. Faced with this situation, it is natural to look for ways of simplifying programming.

5.3   Task-Level Programming

Many recent approaches to robot programming seek to provide the power of robot-level languages without requiring programming expertise. One approach is to extend the basic philosophy of guiding to include decision-making based on sensing. Another approach, known as *task-level* programming, requires specifying goals for the positions of objects, rather than the motions of the robot needed to achieve those goals. In particular, a task-level specification is meant to be completely robot-independent; no positions or paths that depend on the robot geometry or kinematics are specified by the user. Task-level programming systems require complete geometric models of the environment and of the robot as input; for this reason, they are also referred to as *world-modeling* systems. These approaches are not as developed as the guiding and robot-level programming approaches, however.

In a task-level language robot, actions are specified only by their effects on objects. For example, users would specify that a pin should be placed in a hole rather than specifying the sequence of manipulator motions needed to perform the insertion. A *task planner* would transform the task-level specifications into robot-level specifications. To do this transformation, the task planner must have a description of the objects being manipulated, the task environment, the robot carrying out the task, the initial state of the environment, and the desired final state. The output of the task planner would be a robot program to achieve the desired final state when executed in the specified initial state. If the synthesized program is to achieve its goal reliably, the planner must take advantage of any capabilities for compliant motion, guarded motion, and error checking. We assume that the planner will not be available when the program is executed. Hence the task planner must synthesize a robot-level program that includes commands to access and use sensory information.

5.4   Components of Task-Level Programming

The world model for a task must contain the following:

Geometric descriptions of all objects and robots in the task environment

Physical descriptions of all objects; for example, mass and angular moments and surface texture

Kinematic descriptions of all linkages

Descriptions of the robot system characteristics; for example, joint limits, acceleration bounds, control errors, and sensor error behavior

A world model for a practical task-level planner also must include explicit specification of the amount of uncertainty there is in model parameters, such as object sizes, positions, and orientations.

Task planning includes the following:

Gross-motion planning

Grasping planning

Fine-motion planning

5.4.1 Gross-Motion Planning

The center of focus in a task-level program is the object being manipulated. However, to move it from one place to another, the whole multilinked robot manipulator must move. Thus it is necessary to plan the global motion of the robot and the object to ensure that no collisions will occur with objects in the workspace whose shape and position are known from the world model.

5.4.2 Grasping Planning

Grasping is a key operation in manipulator programs because it affects all subsequent motions. The grasp planner must choose where to grasp objects so that no collisions will result when grasping or moving them. In addition, the grasp planner must choose grasp configurations so that the grasped objects are stable in the gripper.

5.4.3 Fine-Motion Planning

The presence of uncertainty in the world model affects the kind of motions that a robot may safely execute. In particular, positioning motions are not sufficient for all tasks; guarded motions are required when approaching a surface, and compliant motions are required when in contact with a surface. A task planner must therefore be able to synthesize specifications for these motions on the basis of task descriptions.

The preceding three types of planning produce many interactions. Constraints forced by one aspect of the plan must be propagated throughout the plan. Many of these constraints are related to uncertainties, both initial uncertainties in the world model and the uncertainties propagated by actions of the robot.

5.5 Toward Fully Automated Robot Programming

As indicated in Section 4.2 most off-line systems are developing at the object level. These task level programs could be sequences of relatively simple high-level instructions describing what needs to be done to carry out an assembly task.

In the longer term this task can also be automated. A system called an *assembly planner* will be given an even higher-level command concerning production requirements. It will examine the CAD database and produce a task-level program. The task planner described will then produce a robot-level program as before.

The system developed by Kawasaki Heavy Industries and Matsushita Electric Industrial Co., Ltd. for the Japan Robot Association program "Robot Programming Simplification Project" (Arai et al., 1997) presents a novel approach to robot-task teaching. The target users are workers in small to medium-size companies and not familiar with industrial robots in small batch sized production facilities. The objective of the system is the generation of robot programs by the workers by using a rich graphical user interface on a personal computer-based platform. Key features of the approach, presented in Figure 1, are:

- It enables users to teach the robot behavior at a task-instruction level instead of a motion-instruction level.
- It generates automatically the motion program from a task-level program and a knowledge database for robot programming and task accomplishment.

The difference from other systems proposed in the literature (Weck and Dammertz, 1995; Eriksson and Moore, 1995; Douss et al., 1995) is in the reduction of the teaching process by enhancing the programming language level to a task-instruction level. The system is configured into a personal computer platform as shown in Figure 2. The system is connected to a robot simulator or a 3D CAD system via a local area network and to the robot controller via a serial communication port for transfer of the programs. An initial welding application was implemented and tried in the system. For this task the teaching flow is as shown in Figure 3. The task-level program for the application is created from the specification of the welding parts and the condition data for the welding process. The created task includes all the necessary robot operations, such as detection of work deviation, torch approach at a welding start point, and torch removal at a welding end point. The task created is checked for its procedure by using a task simulator. Modifi-

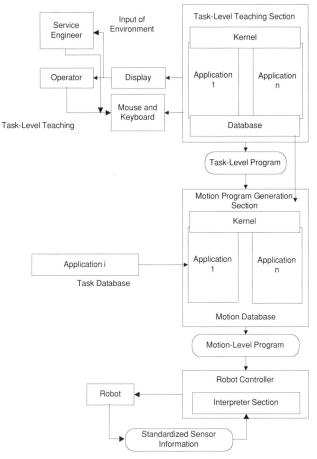

Figure 1   Off-line robot programming system.

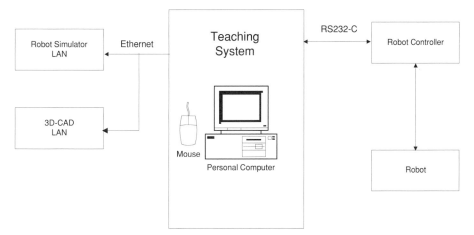

Figure 2   System configuration.

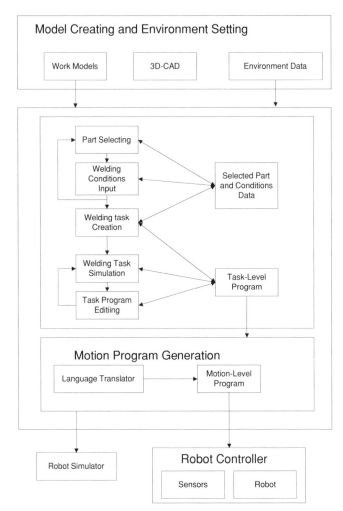

Figure 3 Task-level instruction flow.

cations to the task are introduced and the program modified accordingly. Once a correct procedure is obtained, the task-level program is converted into a motion-level program and transferred to the robot controller for its execution test. The window structure for the interface with the worker is as shown in Figure 4.

The system has been tested since 1996 by the Japan Robot Association, with positive results. The results for the corrugation welding process show that using the system can reduce the teaching time to about one-sixth. This is due to the familiarity of the user interface to users of personal computers, the intuitive ease in understanding the instruction task through the graphical display, and the minimal number of instructions guided by the system for creating the robot language program.

6 MODELS

The role of models in the synthesis of robot programs is discussed in the remainder of this chapter. First, however, we explore the nature of each of the models needed for a task and how they may be obtained.

6.1 The World Model

The geometric description of objects is the principal component of the world model. The major sources of geometric models are CAD systems, although computer vision may

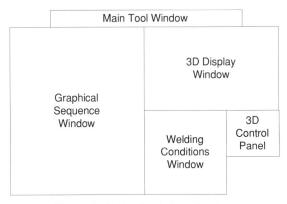

Figure 4   System's window structure.

eventually become a major source of models. There are three major types of commercial CAD systems, differing in their representations of solid objects as follows:

1. *Line:* objects are represented as lines, and curves are needed to draw them.
2. *Surface:* objects are represented as a set of surfaces.
3. *Solid:* objects are represented as combinations of primitive solids.

Line systems and some surface systems do not represent all the geometric information needed for task planning. A list of edge descriptions (known as a *wireframe*), for example, is not sufficient to describe a unique polyhedron. We assume instead that a system based on solid modeling is used.

In these systems, models are commonly constructed by performing set operations on a few types of primitive volumes. Figure 5 shows an object constructed by taking the union of three volumes and then subtracting a fourth. Besides the volumetric primitives that make a particular object, it is necessary to specify their spatial interrelationships—usually in terms of coordinate transforms between the local coordinate systems of the primitives.

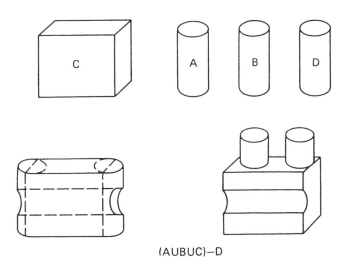

(AUBUC)−D

Figure 5   The two objects in the bottom row are obtained from set operations on primitive volumes A, B, C, D. The different results are obtained from different relative positions of the primitives.

The descriptions of the primitive and compound objects vary greatly among existing systems. Some of the representational methods currently in use are polyhedra, generalized cylinders, and constructive solid geometry (CSG) (Requicha, 1996).

The legal motions of an object are constrained by the presence of other objects in the environment, and the form of the constraints depends in detail on the shapes of the objects. This is the fundamental reason that a task planner needs geometric descriptions of objects. Additional constraints on motion are imposed by the kinematic structure of the robot itself. If the robot is turning a crank or opening a valve, then the kinematics of the crank and the valve impose additional restrictions on the robot's motion. The kinematic models provide the task planner with the information required to plan manipulator motions that are consistent with external constraints.

The major part of the information in the world model remains unchanged throughout the execution of the task. The kinematic descriptions of linkages are an exception, however. As a result of the robot's operation, new linkages may be created and old linkages destroyed. For example, inserting a pin into a hole creates a new linkage with one rotational and one translational degree of freedom. Similarly, the effect of inserting the pin might be to restrict the motion of one plate relative to another, thus removing one degree of freedom from a previously existing linkage. The task planner must be appraised of these changes, either by having the user specify linkage changes with each new task state or by having the planner deduce the new linkages from the task state or operations descriptions.

In the planning of robot operations, many of the physical characteristics of objects play important roles. The mass and inertia of parts, for example, determine how fast they can be moved or how much force can be applied to them before they fall over. Similarly, the coefficient of friction between a peg and a hole affects the jamming conditions during insertion. Likewise, the physical constants of the robot links are used in the dynamics computation and in the control of the robot.

The feasible operations of a robot are not sufficiently characterized by its geometrical, kinematical, and physical descriptions. One important additional aspect of a robot system is its sensing capabilities: touch, force, and vision sensing. For task-planning purposes, vision enables obtaining the configuration of an object to some specified accuracy at execution time; force sensing allows the use of compliant motions; touch information could serve in both capacities.

In addition to sensing, there are many individual characteristics of manipulators that must be described, such as velocity and acceleration bounds, positioning accuracy of each of the joints, and workspace bounds.

Much of the complexity in the world model arises from modeling the robot. Fortunately, it need be done only once.

6.2   The Task Model

A model state is given by the configurations of all the objects in the environment; tasks are actually defined by sequences of states of the world model or transformations of the states. The level of detail in the sequence needed to specify a task fully depends on the capabilities of the task planner.

The configurations of objects needed to specify a model state can be provided explicitly—for example, as offsets and Euler angles of rigid bodies and as jont parameters for linkages—but this type of specification is cumbersome and error-prone. Three alternative methods for specifying configurations have been developed:

1. Use a CAD system to position models of the objects at the desired configurations.
2. Use the robot itself to specify robot configurations and to locate features of the objects.
3. Use symbolic spatial relationships among object features to constrain the configurations of objects. For example, $Face_1$ *AGAINST FACE$_2$*.

Using the CAD model is the obvious approach but positioning objects and the robot accurately in a CAD workplace using only graphical means is not easy. A method is needed which is easy to interpret and modify. One way is to describe a configuration by a set of symbolic spatial relationships that are required to hold between objects in that configuration. For example, in Figure 6 the position of block 1 relative to block 2 is specified by the relations f_2 *AGAINST* f_1 and f_4 *AGAINST* f_2.

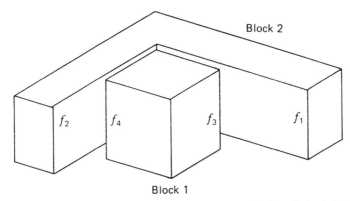

Figure 6   The position of block 1 relative to block 2 can be specified symbolically by f_3 AGAINST f_1 and f_4 AGAINST f_2.

The CAD model enables one to see the proposed positions. More importantly, if the CAD model is part of a simulator and the task is described by means of a work task language which can be interpreted as spatial relationships, then the graphical simulator enables one to see the effects of the specified task. In particular, the simulation indicates whether the problems discussed later in this section e.g., collision avoidance are likely to be difficult. An illustration of off-line programming of a robot arc-welding application, using GRASP-VRI which can be run on a low cost PC platform or on Silicon Graphics workstations is described in greater detail in Sorenti, 1997.

One advantage of using symbolic spatial relationships is that the configurations they denote are not limited to the accuracy of a light pen or a manipulator. Another advantage of this method is that families of configurations such as those on a surface or along an edge can be expressed. The relationships, furthermore, are easy to interpret by a human and therefore easy to specify and modify. The principal disadvantage of using symbolic spatial relationships is that they do not specify configurations directly; they must be converted into numbers or equations before they can be used.

Model states are simply sets of configurations. If task specifications were simply sequences of models, then, given a method such as symbolic spatial relationships for specifying configurations, we should be able to specify tasks. This approach has several important limitations, however. One is that a set of configurations may overspecify a state. A classic example of this difficulty arises with symmetric objects, such as a round peg in a round hole. The specific orientation of the peg around its axis given in a model is irrelevant to the task. This problem can be solved by treating the symbolic spatial relationships themselves as specifying the state, since these relationships can express families of configurations. A more fundamental limitation is that geometric and kinematic models of an operation's final state are not always a complete specification of the desired operation. One example of this is the need to specify how hard to tighten a bolt during an assembly. In general, a complete description of a task may need to include parameters of the operations used to reach one task state from another.

The alternative to task specification by a sequence of model states is specification by a sequence of operations, or more abstractly, transformations on model states. Thus, instead of building a model of an object in its desired configuration, we can describe the operation by which it can be achieved. The description should still be object-oriented, not robot-oriented; for example, the target torque for tightening a bolt should be specified relative to the bolt and not the manipulator. Most operations also include a goal statement involving spatial relationships between objects. The spatial relationships given in the goal not only specify configurations, but also indicate the physical relationships between objects that should be achieved by the operation. Specifying that two surfaces are *against* each other, for example, should produce a compliant motion that moves until the contact is actually detected, not a motion to the configuration where contact is supposed to occur.

6.3   Gross Motion

Gross robot motions are transfer movements for which the only constraint is that the robot and whatever it is carrying should not collide with objects in the environment. Therefore, an ability to plan motions that avoid obstacles is essential to a task planner. The most straightforward way to do this is to guide the robot through the CAD model at the terminal. However, to do this automatically, several obstacle-avoidance algorithms have been proposed in different domains. This section briefly reviews those algorithms that deal with robot manipulator obstacle avoidance in three dimensions, which can be grouped into the following classes:

1. Hypothesize and test
2. Penalty function
3. Explicit free space

The hypothesize and test method was the earliest proposal for robot obstacle avoidance. The basic method consists of three steps: first, hypothesize a candidate path between the initial and final configuration of the manipulator; second, test a selected set of configurations along the path for possible collisions; third, if a possible collision is found, propose an avoidance motion by examining the obstacle(s) causing the collision. The entire process is repeated for the modified motion.

The main advantage of hypothesize and test is its simplicity. The method's basic computational operations are detecting potential collisions and modifying proposed paths to avoid collisions. The first operation, detecting potential collisions, amounts to the ability to detect non-null geometric intersections between the manipulator and obstacle models. This capability is part of the repertoire of most geometric modeling systems. However, the second operation, modifying a proposed path, can be very difficult. Typical proposals for path modification rely on drastic approximations of the obstacles, such as enclosing spheres. These methods work fairly well when the obstacles are sparsely distributed so that they can be dealt with one at a time. When the space is cluttered, however, attempts to avoid a collision with one obstacle will typically lead to another collision with a different obstacle.

The second class of algorithms for obstacle avoidance is based on defining a penalty function on manipulator configurations that encodes the presence of objects. In general, the penalty is infinite for configurations that cause collisions and drops off sharply with distance from obstacles. The total penalty function is computed by adding the penalties from individual obstacles and, possibly, adding a penalty term for deviations from the shortest path. At any configuration we can compute the value of the penalty function and estimate its partial derivatives with respect to the configuration parameters. On the basis of this local information, the path search function must decide which sequence of configurations to follow. The decision can be made so as to follow local minima in the penalty function. These minima represent a compromise between increasing path length and approaching too close to obstacles (Minami et al., 1996).

The penalty function methods are attractive because they seem to provide a simple way of combining the constraints from multiple objects. This simplicity, however, is achieved only by assuming a circular or spherical robot; only in this case will the penalty function be a simple transformation of the obstacle shape. For more realistic robots the penalty function must be much more complex. Otherwise, motions of the robot that reduce the value of the penalty function will not necessarily be safe.

Penalty functions are more suitable for applications that require only small modifications to a known path. In these applications search is not as central as it is in the synthesis of robot programs.

The third class of obstacle-avoidance algorithms builds explicit representations of subsets of robot configurations that are free of collisions, the *free space*. Obstacle avoidance is then the problem of finding a path within these subsets that connects the initial and final configurations. The proposals differ primarily on the basis of the particular subsets of free space that they represent and in the representation of these subsets (see Figure 7).

The advantage of free-space methods is that by explicitly characterizing the free space they can guarantee to find a path if one exists within the known subset of free space. This does not guarantee, however, that the methods will always find a path when one exists because they only compute subsets of the free space. Moreover, it is feasible to search for short paths, rather than simply finding the first path that is safe. The disadvantage is that the computation of the free space may be expensive. In particular, the other methods may be more efficient for uncluttered spaces. However, in relatively cluttered

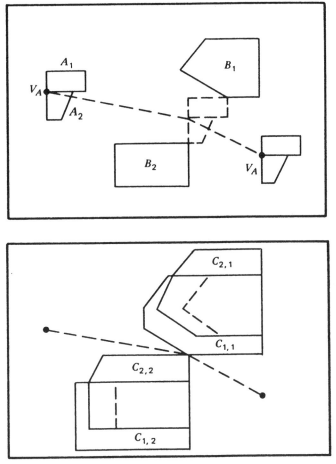

Figure 7   The problem of moving $A = A_1 \cup A_2$ between the obstacles B_1 and B_2 is equivalent to the problem of moving V_A among the modified obstacles $C_{i,j}$ (assuming A cannot rotate).

spaces the other methods will either fail or expend an undue amount of effort in path searching.

6.4   Grasping

A typical robot operation begins with the robot grasping an object; the rest of the operation is deeply influenced by choices made during grasping. Several proposals for choosing collision-free grasp configurations on objects exist, but other aspects of the general problem of planning grasp motions have received little attention.

In this section, *target object* refers to the object to be grasped. The surfaces on the robot used for grasping, such as the insides of the fingers, are *gripping surfaces*. The manipulator configuration that grasps the target object at that object's initial configuration is the *initial grasp configuration*. The manipulator configuration that places the target object at its destination is the *final grasp configuration*.

There are three principal considerations in choosing a grasp configuration for objects whose configuration is known.

1. *Safety.*   The robot must be safe at the initial and final grasp configurations.
2. *Reachability.*   The robot must be able to reach the initial grasp configuration and, with the object in the hand, find a collision-free path to the final grasp configuration.

3. *Stability.* The grasp should be stable in the presence of forces exerted on the grasped object during transfer motions and parts-mating operations.

If the initial configuration of the target object is subject to substantial uncertainty, an additional consideration in grasping is *certainty:* the grasp motion should reduce the uncertainty in the target object's configuration.

Choosing grasp configurations that are safe and reachable is related to obstacle avoidance; there are significant differences, however. First, the goal of grasp planning is to identify a single configuration, not a path. Second, grasp planning must consider the detailed interaction of the manipulator's shape and that of the target object. Note that candidate grasp configurations are those having the gripping surfaces in contact with the target object while avoiding collisions between the manipulator and other objects. Third, grasp planning must deal with the interaction of the choice of grasp configuration and the constraints imposed by subsequent operations involving the grasped object. Because of these differences, most existing proposals for grasp planning treat it independently of obstacle avoidance.

Most approaches to choosing safe grasps consist of three steps: choose a set of candidate grasp configurations, prune those that are not reachable by the robot or that lead to collisions, then choose the optimal (in some sense, grasp among those that remain).

The initial choice of candidate grasp configurations can be based on considerations of object geometry, stability, or uncertainty reduction. For parallel-jaw grippers, a common choice is grasp configurations that place the grippers in contact with a pair of parallel surfaces of the target object. An additional consideration in choosing the surfaces is to minimize the torques about the axis between the grippers.

The other aspects of the grasping problem, stability and uncertainty, have received even less attention. The stability condition used in several of the proposals amounts to checking that the center of mass of the target object is on or near the axis between the gripper jaws. This condition minimizes the torques on the grip surfaces.

In many applications enough uncertainty is present that a grasping strategy must be designed to reduce the uncertainty in the configuration of the target object relative to the gripper. Often such a strategy must involve sensing. Vision or touch may be used to identify the configuration of an object on a conveyor belt. Many tricks for grasping under uncertainty have been developed for particular applications, but a general theory for synthesizing grasping strategies does not exist. One common class of grasping strategies relies on touch sensing to achieve compliance of the robot gripper to the configuration of the target object.

6.5   Fine Motion

The presence of uncertainty in the world model and the inherent inaccuracy of the robot affect the kind of motions that a robot may safely execute. In particular, positioning motions are not sufficient for all tasks; guarded motions are required when approaching a surface, and compliant motions are required when in contact with a surface. A task planner must therefore be able to synthesize specifications for these motions on the basis of task descriptions.

Lynch and Mason (1995) describe a method for specifying compliant motions based on kinematic models of the manipulator and the task. The method requires as input a nominal path for the manipulator. Planning the nominal path is akin to the obstacle-avoidance problem discussed before: achieving the specified task while avoiding collisions with nearby objects. There is a significant difference, however. In the compliant task the legal configurations of the robot are constrained by the kinematics of the task as well as the kinematics of the robot. The task kinematics constrain the legal robot configurations to lie on a surface in the robot's configuration space. This type of surface is called a *C-surface.* The obstacle-avoidance algorithm used in planning the nominal path must incorporate this additional constraint. In practice, both the C-surface and the configuration-space obstacles may be difficult to compute exactly.

The foregoing discussion suggests that task planning, for tasks requiring compliant motions, may be done by first finding a collision-free path on the model C-surface for the task, from the initial to the goal configuration, and then deriving a force-control strategy that guarantees that the path of the robot stays on the actual C-surface, close to the desired path. This two-step procedure assumes that the robot is already on the desired C-surface, that is, in contact with some object in the task. The problem remains of achieving this contact; this is the role played by guarded motions.

The goal of a guarded motion is achieving a manipulator configuration on an actual C-surface while avoiding excessive forces. The applied forces must be limited to avoid damaging the robot or the objects in the task and to avoid moving objects in known configurations. Because the exact configurations of the objects and the robot are not known, it is necessary to rely on sensors, force or touch, to guarantee that contact is achieved and no excessive forces are exerted. In addition, the speed of the approach must be relatively low so that the robot can be stopped before large forces are generated.

The task planner must specify the approach path to the C-surface and the maximum force that may be exerted for each guarded motion. The approach path may be computed by finding a path that avoids collisions with nearby objects while guaranteeing a collision with the desired C-surface. This is a special case of the obstacle-avoidance problem discussed before, one requiring motion close to obstacles. Computing the force threshold for a guarded move is a very difficult problem since the threshold depends on the strength and stability of the objects giving rise to the C-surface. One can always use the smallest detectable force as a threshold, but this choice constrains the manipulator to approach the C-surface very slowly, thus wasting valuable time. It is possible that bounds on these thresholds may be obtained from simple models of the task.

When the robot's configuration relative to the task is uncertain, it may not be possible to identify which C-surface in the world model corresponds to the C-surface located by a guarded move. In such situations it is not enough to plan the nominal path; the task planner must generate a sensor-based fine-motion strategy that will guarantee that the desired C-surface is reached.

6.6   Uncertainty Planning

All the planning modules described above had to deal with uncertainty in the world model. Uncertainties in a model of the initial state of the world may be amplified as each action takes place.

6.6.1   Physical Uncertainty

There are three main sources of physical uncertainty.

Mechanical Complexity

Robot manipulators are complex mechanical devices. There are upper bounds on speed and payload and limits to accuracy and repeatability. The absolute positional accuracy of a manipulator is the error that results when it is instructed to position its end-effector at a specified position and orientation in space. This may depend on temperature, load, speed of movement, and the particular position within its work area. Furthermore, in a situation where these parameters are all fixed, a manipulator will not in general return to precisely the same location and orientation when commanded to repeat an operation over and over. The error measures the positional repeatability. There can be contributions from stochastic effects and from long-term drift effects, which can be corrected by calibration. The positional and repeatability errors of current manipulators are sufficiently large to cause problems in carrying out a large class of planned tasks in the absence of feedback during plan execution. Manipulators can be made more accurate by machining their parts more accurately and increasing their structural stiffness. There is, however, a trade-off between cost and performance of manipulators, so that there is a point of diminishing returns in trying to build ever more accurate mechanisms.

Parts Variability

To make matters worse for the task planner, multiple copies of a mechanical part are never identical in all their dimensions. It is impossible to manufacture parts with exact specifications. Instead, designers specify tolerances for lengths, diameters, and angles. Parts made from the design can take on any physical values for the parameters that fall within the designed tolerances. The effects of these variations might be large enough by themselves to be a significant factor in the success or failure of a planned robot manipulator task. When many parts are assembled into a whole, the individual small variations can combine and become large.

Uncertainty of Initial Position and Orientation

Often the most significant source of uncertainty is the position and orientation of a workpiece when it is first introduced into the task. Mechanical feeders sometimes deliver parts with large uncertainties in position and orientation, sometimes on the order of 50% of the size of the part. Conveyor belts deliver parts with even larger uncertainties. A task planner often includes actions in the plan that are aimed at significantly reducing these initial uncertainties—for instance, the grasp strategies and guarded moves described previously.

Besides physical uncertainty, there will always be uncertainty in the runtime system's knowledge of the state of the world, as all sensors are inaccurate. Usually the maximum inaccuracy can be characterized as a function of sensor reading.

6.6.2 Error Propagation

The effects of actions on uncertainties can be modeled in the world model, and so the task planner can propagate uncertainties throughout the sequence of tasks. There have been two approaches to error propagation.

Numeric Error Propagation

Numeric bounds are estimated for initial errors, the propagation functions are linearized, and linear programming techniques are used to estimate the resultant errors. If the errors are too large for the next task to handle, then deliberate error-reduction strategies are introduced, such as sensing or one of the C-surface methods described above.

Symbolic Error Propagation

Unforced decisions, such as workplace location, tool parameters, and compliant motion travel, are represented as symbolic variables. Uncertainties are functions of these variables. Errors are propagated symbolically, and the resultant uncertainties are also symbolic expressions. Constraints on uncertainties necessary for the success of subsequent tasks provide symbolic constraints on the as yet unforced decisions. If the constraints are too severe, sensors must be introduced to the task plan. They too are analyzed symbolically, and constraints on sensor choice are generated. Relaxation methods over the task sequence are applied to satisfying the constraints. The result is that the requirements of tasks late in the sequence can generate preparatory actions early in the sequence.

7   SUMMARY

Off-line programming is a necessary step toward increasing the productivity of industrial robots in industry. The most important reason for using an off-line programming system is to reduce the robotics system's set-up time by overlapping the programming time with the robot's normal operation. The available off-line programming systems range from textual programming systems (text editor and a program compiler) to highly sophisticated programming and simulations systems, allowing interactive task-level programming, its simulation, and ultimately the generation of the robot program code. Systems such as IGRIP and ROBCAD (Roos, 1997) are widely used in the automotive industry, one of the largest robotics systems consumer worldwide. However, there are still challenges to overcome. Distribution of operations under the same controller, real-time operation, and the autonomy of the robotics system are aspects imposing harsh requirements on the off-line programming system. These requirements have been solved by employing techniques such as virtual reality (Schraft, Strommer, and Neugebauer, 1992; Sorenti, 1997), neural networks (Chan, Tam, and Leung, 1994), and fuzzy logic (Lee, 1997). Research performed and undergoing will result in a new generation of off-line programming systems in which autonomy will be one of the key characteristics.

ACKNOWLEDGMENT

The authors wish to thank J. A. Ceroni of Purdue University for major help in revising and updating this chapter. Sections 5 and 6 of this chapter are revised with permission from Chapter 22 of the first edition of this handbook, "Task-Level Manipulator Programming," by T. Lozano-Pérez and R. Brooks.

REFERENCES

Arai, T., T. Itoko, and H. Yago. 1997. "A Graphical Robot Language Developed in Japan." *Robotica* **15,** 99–103.

Chan, R. H., P. K. Tam, and D. N. Leung. 1994. "Solving the Motion Planning Problem by Using Neural Networks." *Robotica* **12,** 323–333.

Douss, M., A. Genay, and P. Andre. 1995. "Off-Line Programming Robots from a CAD Database. Example of an Industrial Application." In *Proceedings of the International Conference on CAD/CAM, Robotics and Factories of the Future,* Pereira, Colombia. Vol. 8. 845–850.

Eriksson, P., and P. Moore. 1995. "An Environment for Off-Line Programming and Simulation of Sensors for Event Driven Robot Programs." In *Proceedings of the International Conference on Recent Advances in Mechatronics,* Istanbul. Vol. 2. 1104–1108.

Lee, S. B. 1997. "Industrial Robotics Systems with Fuzzy Logic Controller and Neural Network." In *Proceedings of the International Conference on Knowledge-Based Intelligent Electronic Systems,* Adelaide, Australia.

Lynch, K. M., and M. T. Mason. 1995. "Pulling by Pushing, Slip with Infinite Friction, and Perfectly Rough Surfaces." *International Journal of Robotics Research* **14**(2), 174–183.

Mayr, H., and S. Stifler. 1989. "Off-Line Generation of Error-Free Robot/NC Code Using Simulation and Automatic Programming Techniques." In *Proceedings of the IFIP TC5/WG 5.3 International Conference,* Jerusalem.

Minami, M., Y. Nomura, and T. Asakura. 1996. "Trajectory Tracking and Obstacle Avoidance Control to Unknown Objects for Redundant Manipulators Utilizing Preview Control." *Nippon Kikai Gakkai Ronbunshu* **62**, 3543–3550.

Requicha, A. 1996. "Geometric Reasoning for Intelligent Manufacturing." *Communications of the ACM* **39**(2), 71–76.

Roos, E., and A. Behrens. 1997. "Off-Line Programming of Industrial Robots—Adaptation of Simulated User Programs to the Real Environment." *Computers in Industry* **33**(1), 139–150.

Schraft, R. D., W. M. Strommer, and J. G. Neugebauer. 1992. "Virtual Reality Applied to Industrial Robots." In *Proceedings of the International Conference Interface to Real and Virtual Worlds 1992,* Montpellier, France.

Sorenti, P. 1997. "Efficient Robotic Welding for Shipyards—Virtual Reality Simulation Holds the Key." *Industrial Robot* **24**(4), 278–281.

Thangaraj, A. R., and M. Doelfs. 1991. "Reduce Downtime with Off-Line Programming." *Robotics Today* **4**(2), 1–3.

Weck, M., and R. Dammertz. 1995. "OPERA—A New Approach to Robot Programming." *Annals of the CIRP* **44**, 389–392.

CHAPTER 20

LEARNING, REASONING, AND PROBLEM SOLVING IN ROBOTICS

Spyros G. Tzafestas
Elpida S. Tzafestas
National Technical University of Athens
Zographou, Athens

1 INTRODUCTION

The robotic manipulator whose operation could be programmed was developed in the mid-1950s by Devol. This manipulator was capable of following a sequence of motion steps determined by instructions in the program (Hunt, 1983; Scott, 1984). Based on Devol's robot concept, the first industrial robot was put in operation by Unimation in 1959. Our major interest is in programming industrial and other robots to do useful work. The basic hardware components needed for constructing any robotic system are a control computer, a set of sensors for acquiring information about the state of the environment, and a set of appropriate actuators for influencing the state of the environment. However, the current trend is not toward the special-purpose robots that characterized automation in the 1960s, but toward general-purpose robots able to perform a wide repertory of industrial tasks with no change in hardware and only minor changes in programming (Tzafestas, 1991). To this end, not only must the hardware be general purpose, but the software must be general enough to simplify the task of reprogramming the robot to perform a new job. In practice it is very difficult, if not impossible, to anticipate and list all situations that the robot may face in its future work. Therefore, the designer of the robot software program must specify general classes of situations and equip the robot with sufficient intelligence and problem-solving abilities to be able to cope with any situation belonging to one of the taught classes. Sometimes the situation faced by the robot is ambiguous and uncertain and the robot must be able to evaluate different possible courses of action. If the robot's environment is not varying, the robot is given a model of its environment that enables it to predict the results of its actions. If the environment is varying, the robot must be able to learn it. These and other requirements call for the development and embedding in the robot system of artificial intelligence (AI) capabilities, including learning, reasoning, and problem solving.

This chapter reviews the available techniques for performing those three operations, with particular focus on robot task planning and robot path planning problem solving. Due to space limitations the theoretical details are not given, but they can be found in the references below which also include many case studies and application examples. Section 2 deals with robot learning and knowledge acquisition. Section 3 reviews the general AI reasoning methodologies, most of which have found wide application in robotics. Section 4 discusses the general problem-solving issues and investigates in more detail the major techniques available for robot task planning and robot path planning, including the newer concept of reactive algorithm. Finally, Section 5 gives some concluding remarks and mentions a number of successful applications.

2 ROBOT LEARNING AND KNOWLEDGE ACQUISITION

Knowledge differs from information in that it is structured in long-term memory and is the outcome of learning. We can say that learning takes place whenever a permanent

Handbook of Industrial Robotics, Second Edition, Edited by Shimon Y. Nof
ISBN 0-471-17783-0 © 1999 John Wiley & Sons, Inc.

change in behavior results from experience. In other words, a basic characteristic of learning is permanent change in behavior. Learning can be characterized from either a logical or a psychological point of view. The knowledge that a human acquires through learning includes *facts* (about events or entities) and *rules* or *generalizations.* The learning of rules and generalizations can be accomplished in the following ways (Michalski, Carbonell, and Mitchell, 1983):

- *Simple memorization* of facts through programming a direct implementation of new knowledge. This learning style (also called *rote learning*) is not useful, since all the responsibility is left to the teacher instead of the learner. Applied to machine learning, it is nothing more than programming.
- *Learning by analogy:* Here the learner must draw more inferences on the incoming knowledge. This learning style is based on the identification of similarity with some knowledge structure existing in memory and employs the operations of abstraction and specialization.
- *Learning from instruction:* Here the knowledge is obtained from a teacher, a book, or some other source and is usually expressed in a form or language not coinciding with the internal representation language of the learner.
- *Learning from examples:* Here the learner has to generalize from examples, typically preselected by a teacher. This learning style, also called *inductive learning,* is the one mostly used for learning by machines and robots.
- *Learning by discovery:* This is a more general kind of learning from examples, where the observations or examples are not preselected by a teacher but autonomously selected by the learner.
- *Other learning styles* include learning through vision and through tactile or other sensors.

The simplest, most straightforward way to teach a robot something is by showing it a well-selected sequence of examples. For example, we may want the robot to learn rules from examples. Other examples of robot inductive learning include learning through vision or learning by showing. Programs for computer induction are available for several purposes, such as object recognition, concept discrimination, and hierarchical taxonomy. For object recognition a training set of individual instances of the object is provided, each with a description; i.e., with a conjunction of predicate-type propositions. Here the goal of the learner is to establish a maximally specific generalization of concept. A bottom-up inductive algorithm for learning a concept description involves the following steps:

1. Consider a specific instance and assume that its description is the target class description.
2. Consider other instances. If they do not match the current class description, modify appropriately the current description so as the new description is valid for the latest instance as well as the previous ones. Some methods of description revision are condition dropping, disjunction introduction, universal generalization, and exception introduction.

Learning algorithms (bottom-up or top-down) vary according to the particular generalizing and specializing rules used and the rule selection criteria employed (Cornuejols and Moulet, 1997; Tzafestas, Ma, and Tzafestas, 1997). Some learning algorithms include negative instances in the training set. A representative set of learning programs is reviewed by Michalski (1983).

A popular inductive learning program available commercially is ACLS (Analog Concept Learning System) (Paterson and Niblet, 1982). This system generates classification and action rules by induction from examples. The training examples are inputed to ACLS as a list of records. The last field of the record is entered with the name of the class to which the record is to be attributed, such as *circle, ellipse,* or *square.* All the other fields are entered with the values of the primitive attributes; i.e., the attributes that the user considers relevant to the classification task. ACLS uses these examples to derive a classification rule in the form of a *decision tree* that branches according to the values of attributes, and then generates a PASCAL conditional expression logically equivalent to the decision tree. This PASCAL code can be run to classify new examples. Whenever a new example is found that does not match the current rule, ACLS can be asked to revise the rule so as to cover the new instance and to display, store, or output it as before. After the presentation of a sufficient number of examples, ACLS can be asked to induce a rule that matches and explains all of them. ACLS has been used in several applications by the Edinburgh University group (A. Blake, P. Mowforth, and B. Shepherd) (Michie, 1985).

Robot learning may correspond to various degrees of changes to internal structure and external behavior. The following cases of robot learning may occur:

- Learning parameter changes of feedback controls
- Learning operational architectural properties (concurrency, coordination, etc.)
- Learning and building topological models
- Learning sophisticated perceptual or motor skills
- Learning successful operational associations between sensing and acting
- Learning new concepts

These cases are instances of a continuum of possibilities between two extremes. At the one extreme, the robot has full knowledge of the domain and thus no learning is necessary. Near this extreme is the case where the robot has some parameters missing. Here adaptive and neural learning is typically employed (Ahmad and Guez, 1997; Tzafestas, 1995; Omatu, Khalid, and Yusof, 1996; Harris, 1994). The other extreme corresponds to the case where no domain knowledge is available. This case can be viewed as impractical. Clearly, to construct a successful robotic system, learning must be restricted to the aspects of the system where the human's knowledge is not sufficient to engineer a solution. Learning from scratch is inefficient for practical problems of unusual complexity. A set of contributions dealing with several learning issues lying between the above two extremes was presented in the ROBOLEARN '96 Workshop (Hexmoor and Meeden, 1996). These studies include the application of Q-learning to navigate a robot among obstacles, neural network learning to adapt preplanned motion sequences of a walking biped robot, reinforcement learning to refine primitive actions via repetitive execution in a structured environment, evolutionary (genetic algorithm) learning, topological map learning, and robot localization learning.

We will now look at the specific control activity of a robot system designed for industrial production in an assembly line (Tzafestas and Stamou, 1997; Mason, 1981). This activity typically involves two layers. The higher level establishes the fundamental cycle executed by the robot and can be realized by a conventional robotic program or as a dispatcher acting as a plan execution subsystem. The lower level takes care of the compliant motion of the robot and actually completes the loop between sensing and action. The compliant motion process involves the tasks for which the robot must establish or maintain contact with its environment such that the capability of continuously adapting the action of the robot to its perception is required. To achieve compliant motion in the presence of uncertainty, one can use either *passive compliance* via elastic joints or *active* compliance via appropriate controllers (e.g., PID controllers) (Mason, 1981). A very good approach to achieving active compliance, which is a complicated nonlinear adaptive control problem, is to use empirical (non-model-based) controllers such as fuzzy or neural controllers (Tzafestas, 1995; Omatu, Khalid, and Yusof, 1996; Tzafestas and Venetsanopoulos, 1994). Neural control is primarily based on *supervised learning,* which is a kind of learning from examples (including bad examples). However, it must be remarked that learning only from examples provided by a teacher allows the design of controllers acting at most as good as the teacher. This limitation can be greatly reduced if noise is filtered from the examples and the available domain theory is used. A way for getting even better results is to enhance the learning controller by other learning paradigms such as reinforcement learning or incremental learning (Berenji, 1993; Barto, Sutton, and Anderson, 1983). In reinforcement learning no supervisor is available to critically judge the selected control action at each time step. The effect of control action is learned by the system indirectly. Reinforcement control is usually based on the *credit assignment* principle, where, given the results (performance) of a system, reward or blame is distributed to the individual elements that contributed to the achievement of that performance (Barto, Sutton, and Anderson, 1983).

We now turn to the knowledge acquisition (KA) process, which is a prerequisite for the design of intelligent and expert systems in specific domains like robotics (Tzafestas and Tzafestas, 1997). The acquisition of knowledge from human experts (and other sources) requires special skills and techniques. The KA process is not monolithic. Like all software engineering, it makes use of information in several forms, such as specifications, experience, principles, laws, and observation, recorded in a variety of other media. Three factors critical to the development of large knowledge bases are specificity, explicitness, and broadness. *Specificity* implies that a system exhibits intelligent understanding and action of high-level competence because of the specific knowledge (facts, models, heuristics, metaphors, etc.). *Explicitness* of the knowledge is a must for an intelligent system, although compiled forms of knowledge may also be present. Finally, *broadness* is a breadth hypothesis that poses that intelligent performance often requires the problem solver to fall back on increasingly general knowledge and/or to analogize about specific knowledge obtained from diverse domains. Presently there are a variety of KA methods, ranging from traditional psychological methods (interview, observation, protocol analysis methods) to more recent, perhaps nontraditional, ones (Tzafestas and Tzafestas, 1997). Of particular interest is KA through concept learning. Concept learning techniques are distinguished, on the basis of the amount of the explicitly represented background knowledge, into two classes:

1. *Knowledge-sparse techniques,* which are used when little or no domain knowledge is explicitly available during learning (and involve parameter learning, similarity-based learning, and hierarchical learning)
2. *Knowledge-rich techniques,* which need considerable domain knowledge and are divided into explanation-based learning and learning by discovery

A widely used knowledge representation approach in robotics is in terms of an analytical language (Fischhoff, 1989). The basic steps in any analytical language are:

● Identification of the key structural elements of the problem
● Characterization of the interrelationships among these elements
● Estimation of the quantitative parameters of the model
● Evaluation of the overall quality of the representation, iterated if necessary

A method worth mentioning for the acquisition of the knowledge required for robot path planning and navigation uses *attributed graphs* (Bourbakis, 1995). Graph knowledge representation forms are processed and combined appropriately by producing a scheme that acquires the extracted knowledge during navigation. The graph forms represent the extracted shape of each current free navigation space as well as its properties and relations. The syntactic and semantic information regarding the shape is represented on a directed graph with attributes. This KA methodology is applicable in 2D robot vision and image representation applications.

3 REASONING IN AI AND ROBOTICS

Reasoning is the process of drawing conclusions (inferences or judgments) from known or assumed facts. In artificial intelligence this definition covers mainly common sense reasoning, i.e., human-like reasoning based on common sense knowledge about the world. Many problems in robotics require for their solution a considerable amount of common sense reasoning from a computer or a human. Besides common sense reasoning, reasoning by computers includes all activities required for the problem solution: organization and analysis of facts and data association of a sequence of simple inferences, that prove hypotheses or draw conclusions. These activities must be embodied and carried out automatically by the intelligent system.

Early automated reasoning systems include the Intelligent Machine of Samuel (1959) and the General Problem Solver (GPS) of Newell and Simon (1963), which was capable of simulating human reasoning. GPS was based mainly on means–ends analysis, where a sequence of operators are generated that reduce the difference between the current solution (state) and the desired solution (goal state). The field of automated (or computer-aided) reasoning is now at a very mature state of development, and a large repertoire of special reasoning techniques (or strategies) is available for robotic and other engineering or real-life problems. These techniques differ in *deepness, generality, precision,* etc. Formally, reasoning is the process of deriving new facts from given ones. It includes elements for identifying the domain knowledge that is relevant to the problem at hand, resolving conflicts, activating rules, and interpreting the results. Reasoning is implemented via suitable computer programs called *inference engines* (*mechanisms*). The primary reasoning methods of AI are:

1. Forward reasoning (chaining)
2. Backward reasoning
3. Forward/backward reasoning

Logical (or predicate logic-based) reasoning is the most straightforward formalism for automated reasoning. It is distinguished into *deductive* and *inductive* reasoning (Mortimer, 1998). The principal tool of logical reasoning is Robinson's resolution principle (1965) (Charniak and McDermot, 1985; Nilsson, 1971), which provides a unique general scheme for drawing conclusions. Resolution is mathematically expressed as:

$$\text{IF } (A \lor B) \text{ AND } (\sim B \lor C) \qquad \text{(parent clauses)}$$
$$\text{THEN } A \lor C \qquad \text{(resolvent clause)}$$

Closely related to inductive reasoning (induction) is *abductive* reasoning (abduction) which states that "IF B is true AND A $\rightarrow$ B THEN A must be true." Both induction and abduction are reliable only to a certain degree and can be employed with certainty only when there is no doubt that the resulting conclusion might be wrong. Depending on the knowledge representation method, reasoning includes reasoning with frames, reasoning with semantic nets, reasoning with production rules, and reasoning with scripts (Charniak and McDermot, 1985). Rule-based reasoning can follow the forward or the backward chaining scheme. Other automated reasoning types, besides common sense reasoning, include causal reasoning, nonmonotonic reasoning, default reasoning, reasoning under uncertainty, and temporal reasoning. Causal reasoning can reconstruct past states (events) or future states and is applicable when the knowledge about the world is stored as causal knowledge in the form of a causal model. Causal reasoning has been applied successfully in medical diagnosis (CASNET and INTERNIST systems) and natural language understanding problems.

When the world of the problem changes (i.e., when the knowledge stored in the knowledge base (KB) changes during the process of reasoning), one must apply *nonmonotonic* reasoning. *Monotonic* reasoning produces new knowledge (facts, statements, etc.) that never makes the available knowledge (already stored in the KB) invalid. It presumes that the world is completely known and nothing will ever change about it. This is not always true in practical problems, like the ones encountered in robotics, where some contradictory evidence may be observed and can be dealt with by the deletion of some knowledge elements already stored in the KB. Nonmonotonic reasoning is based on inferences made through nonmonotonic logic. One of the available nonmonotonic reasoning systems is the Truth Maintenance System (TMS) (Antoniou, 1997).

Default reasoning employs default assumptions (or values) whenever the problem information is uncertain or incomplete. These are the "best suitable" assumptions (values) made (selected) among the possible alternatives. Default reasoning leads to conclusions that are not accessible through monotonic reasoning. A default value D is typically assigned as: "If A is not available or A is unknown, then take D as the value of A," and it can be permanent or temporary.

Reasoning under uncertainty (plausible reasoning) is a type of inexact reasoning that derives conclusions on the basis of inaccurate, imperfect, or incomplete premises. Its usual form is abduction. Uncertainty can originate from several sources and appears in many different forms, such as random event, uncertainty in judgment, and lack of evidence or lack of certainty in evidence. The principal numerical techniques of reasoning under uncertainty are probabilistic (Bayesian) technique, confirmation theory, evidential reasoning, and fuzzy reasoning. The one most used in robotics, as in path and task planning in uncertain and unknown environments, is fuzzy reasoning (Tzafestas and Venetsanopoulous, 1994; Tzafestas and Stamou, 1994; Tzafestas et al., 1994; Tzafestas, Hatzivassiliou, and Kaltsounis, 1994; Tzafestas, Zikidis, and Stamou, 1996; Tzafestas, 1994).

Temporal reasoning, or reasoning with (or about) time, is needed in many problems concerned with a changing environment, such as system monitoring/diagnosis and robotic planning where tasks have to be executed in time and collision avoidance with other objects and robots must occur at all times (McDermot, 1982).

Finally, another type of reasoning that finds extensive applications in robotics is *spatial* reasoning, or reasoning about the visual world (Winston, 1984). For example, when a robot must find its way through the world among stationary or moving obstacles, it must be able to understand the world on the basis of

1. General spatial knowledge
2. Perception
3. Spatial reasoning

Spatial reasoning includes a variety of subproblems: visual world recognition, spatial planning (i.e., pathfinding under obstacle avoidance), image and scene analysis, and physical (mechanical) reasoning. All these problems need for their solution appropriate spatial knowledge acquisition and representation schemes. Spatial knowledge (the robot's surrounding world) can be efficiently represented (mapped) into the KB in several ways, usually referred to the representations of shape, volume, boundaries, etc. of the object in a two- or three-dimensional physical world. Objects are represented through their components and the interrelations/interactions among them. Alternatively, objects can be represented as a combination of different volumes that correspond to the object modules. Objects can also be described through a classification of their bounding surfaces or an approximation of such surfaces by planes. Robot task planning and path planning belong to the class of spatial planning. The path planning (or find path) methods are broadly classified into:

1. Path-based methods based on maps of objects and routes
2. Shape-based methods relying on information about obstacles as geometric objects

Finally, a considerable part of spatial robotic reasoning is related to the problem of image and scene analysis (Fisher, 1989). *Image analysis* (or early vision) is passive, data-driven, and domain-independent. *Scene analysis* (or high-level vision) is concerned with the recognition of visual objects, their shapes, and so on. Both image and scene analysis are prerequisites for the design and implementation of any intelligent and autonomous robotic system. More details on the path/task problem-solving process are given in Sections 4.2 and 4.3.

4 PROBLEM SOLVING

4.1 General Issues

Problem solving is a well-established area of AI research dealing with knowledge-based methods for solving a wide spectrum of problems. In the 1960s problem-solving methods were studied in connection to unreal situations, such as classical puzzles like the Missionaries and Cannibals problem or the Towers of Hanoi problem. Problem solving by planning is now an area of practical value, especially applied to robotics, where automated problem solving leads to more intelligent and powerful robots than those based on traditional preprogramming. Expert systems are useful for problem solving in domains where rich knowledge is available, whereas planning systems are usually applied to problems that have well-defined logical structures.

Problem solving is an intelligent process is based on three elements:

1. *General knowledge* (common to all problems under solution)
2. *Specific knowledge* (about the selected specific problems to be solved)
3. *Reasoning knowledge* (appropriate for the selected problem domain)

Newell and Simon (1972) published one of the first books dedicated to problem-solving methods. It includes several problem-solving systems, such as General Problem Solver and the Logic Theorist. The Logic Theorist, which was later evolved into a real problem-solver, starts with five basic axioms and generates (using deduction rules) more and more general theorems until one of the theorems contained in the *Principia* of Whitehead and Russell (1950) is obtained.

A General Problem Solver (Newell and Simon, 1963; Mortimer, 1998; Charniak and McDermot, 1985; Nilsson, 1971) is a search program where the problem to be solved is specified through a given initial state and a desired final (goal) state. Then GPS applies the best sequence of legal operations or moves for leading as quickly as possible to the goal state. This is done by evaluating the current state and by operator ordering, i.e., by applying the operations that promise the best movement in the search space. The initial, final, and intermediate states can all be represented by logic clauses or equivalent graphs. The problem-solving process is then simply a search procedure on the graph of the planning problem, from an initial to a final state. A successful plan is either a single operation that leads from an initial to a final state or a single step and another plan; that is, planning is actually a recursively defined search policy. Examples of problem-solving cases, including path optimization and robot motion-planning problems, can be found in Newell and Simon (1963); Charniak (1985); Nilsson (1971); Winston (1984); Newell and Simon (1972); Patterson (1990).

To solve a problem by a computer problem-solving technique, the problem must be adequately formulated and represented. To this end, a model analogous to the problem is usually employed and stored in the computer. Then the problem solution reduces simply to the simulation of the model. The solution of a problem in a specific knowledge domain requires both the *domain specification* and the *problem formulation*. The domain specification involves the description of a specific area (engineering, medicine, etc.) where problems can be solved only by skilled experts. The *domain knowledge* is an aggregation of basic knowledge about the facts and their relations, the objects, the possible goals, the relevant restrictions, the procedures, and the heuristics of the domain at hand. The problem formulation (declarative or procedural) is the description of the initial state of the problem and the goals to be achieved by solving it. *Problem reduction* (or decomposition) is the process of splitting up the initial complex problem into a set of (primitive) subproblems the solution of which is known. The most suitable way to represent a decomposed problem is to use *AND/OR* graphs, which are solution graphs that contain *OR* nodes and *AND* nodes. An *AND/OR* graph is an *AND/OR* tree with the root node as the initial state of the problem.

The standard available problem-solving methods are (Charniak and McDermot, 1985; Nilsson, 1971; Patterson, 1990; Popovic and Bhatkar, 1994):

- *Formal logic,* based on logic inference rules such as *modus ponens, modus tollens,* and Robinson's resolution principle.
- *Constraint propagation,* the generation of new constraints from old ones. Constraints (numeric or symbolic) express the limits of the object's implementation or the interfaces between the objects.
- *Means–ends analysis,* based on the minimization (at every step of the search) of the difference between the current state and the goal state of the problem. The calculation of this difference is performed using a suitable function depending on the domain of interest. The difference is used for selecting from a look-up table the most appropriate rules (operations) to minimize the calculated difference.
- *Generate-and-test,* based on the repeated generation of possible problem solutions and testing of their effectiveness and acceptability. This method is particularly useful for causal reasoning and problem reduction. Clearly, two modules are needed: a *generator* (producing enumerated possible solutions) and a *tester* (evaluating each generated solution and accepting or rejecting it).
- *Search methods,* based on searching the solution space for possible solution routes. The search problem is equivalent to the route optimization in the solution space (graph). AI search methods are more intelligent than standard system optimization methods because they are knowledge-based.

The problem-solving methods are distinguished into:

1. *Brute-force search methods,* which are uninformed methods with a component for selecting the next state for consideration along the solution route
2. *Heuristic search methods,* which are informed methods that use suitable heuristic rules of thumb to help the solution-finding process

The first class includes depth-first search, breadth-first search, depth-first iterative deeping, progressive deeping, and bidirectional search. The second class contains hill climbing, best–first search, branch and bound search, A*, AO*, and B* search.

The blackboard (BB) method (Nii, 1986; Hayes-Roth, 1985) is a distinct problem-solving method applicable to robotic and other applications. The BB method employs three elements:

- *Knowledge sources* (KS): The knowledge about the problem is divided into independent but complementary subsets, the knowledge sources.

- *BB data structure* is a central data structure where all the knowledge sources have exclusive access for retrieval and storage or modification of information. Those modifications in the state of the BB drive the system opportunistically and incrementally to a solution (not necessarily unique) of the problem at hand. The interaction/communication of the knowledge sources is possible only implicitly through the BB. The stored data may contain partial solutions; constraints, properties, and description of the problem world; target limitations; etc.

- *Control mechanism:* There is no explicit control mechanism; rather, the KS's respond dynamically and opportunistically to the changes of the BB and are self-activating, in the sense that they know when and how to contribute to the solution of the problem. A central monitor chooses a KS to execute next, and the process is repeated until a solution is found or a deadlock is traced. After each cycle of execution, all KS's examine their condition parts to see if they fire, i.e., if they are ready to execute. Since the monitor operates on the data of the higher hierarchical levels, it may be viewed as a metaknowledge control mechanism.

The BB solving method has been applied to many problems, such as robot control, diagnostics, pattern recognition, multiple task planning, intelligent control, and manufacturing systems (Pang, 1989; O'Grady and Lee, 1990; Marik and Lhotska, 1989; Tzafestas and Tzafestas, 1991).

4.2   Robot Task Planning

Two classical generic AI tools suitable for robot task planning are GPS (Newell and Simon, 1963; Newell and Simon, 1972) and STRIPS (Fikes and Nilsson, 1971). GPS employs the means–ends search strategy, and STRIPS uses the backward planning strategy. The basic assumption in STRIPS is that conjunctive goals are independent. Given a set of goals $g = g_1 \wedge g_2 \wedge \ldots \wedge g_n$ and the current state s, STRIPS first develops a plan, say P_1, for a goal, say g_1. Let $P_1 \cdot s_1$ be s_2. The plan is generated backwards, i.e., it finds an operator whose effect can accomplish g_1. Let P_1 be a sequence of operations that can solve g_1 from s_1. Once s_1 is solved and the state becomes s_2, STRIPS tries to solve another goal. Subsequently, a plan P_2 is generated and applied to the end of P_1. The state at this point becomes, say, s_3. This process is repeated until all g_i, $i = 1, 2, \ldots, n$ are solved. An enhancement of STRIPS, obtained by differentiating those operators that are critical to the desired goals from those that are essential only for details, is ABSTRIPS. This is done in a hierarchical way (Sacerdoti, 1974).

Robot task planning is performed in three closely related and computationally dependent stages:

1. World modeling
2. Task specification
3. Robot program synthesis

The world model of a task involves a geometric description of the objects and robots in the workspace, a physical description of the objects (mass, inertia, elastic parameters, etc.), a kinematic description of the robot linkages, and finally a description of the robot characteristics (joint limits, velocity bounds, sensor features, etc.). The tasks are provided to the task planner as a sequence of models of the world state at several steps during the execution of tasks. A popular and easy-to-use method for task specification is through the use of symbolic spatial relationships among object configurations. Robot program synthesis is the most crucial stage of robot task planning. It involves *grasp planning, motion planning,* and *plan evaluation* (Figure 1). The output of the program synthesis stage is a program made of grasp commands, motion specifications, sensor commands, and error tests. Thus the synthesized program is actually a robot-level programming language for the robot at hand.

In contrast to traditional programming languages, robot programming languages have to be easy to use for ordinary users and enable complex tasks to be described and realized. The current trend is toward producing automatic robot programming languages with the aid of AI. The available robot programming languages are distinguished into

Task-level commands

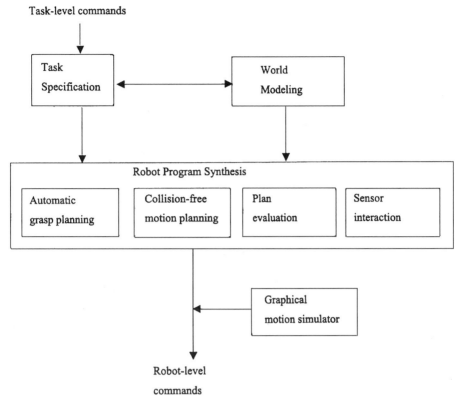

Figure 1 General architecture of a robotic task planner.

1. Robot-oriented programming languages
2. Task-oriented programming languages
3. Goal-oriented programming languages

Class 1 languages belong to the lowest level of languages. The user must specify in detail the actions that the robot must take. Class 2 languages allow the user to determine high-level tasks directly rather than having to specify the details of every action the robot will take. Task-oriented programming languages are capable of performing many planning tasks automatically. Class 3 languages allow the user to command desired subgoals of the tasks directly. A goal-oriented programming language employs a model state given by the configurations of all the objects in the environment and performs the tasks as sequences of states of the world model. Thus automated AI problem-solving techniques must be used and implemented. This means that a goal-oriented programming system must produce the sequence of high-level robot operations the execution of which can lead to the goal state. The transformation of a goal-level task description to a task-level description is called *goal-level planning problem,* and the transformation of a task-level description to a robot-level program is called *task-level planning problem.*

Robot world modeling deals with the geometric description (modeling) of robots, objects, and obstacles. All of them are actually solid geometric objects that have to be modeled, i.e., represented or described. The major solid object representation schemes for robot planning are *cell decomposition* (including spatial occupancy enumeration), *sweep representation, constructive solid geometry,* and *boundary representation* (Winston, 1984; Fisher, 1989; Sheu and Xue, 1993). Two standard methods for cell decomposition in robot task planning are *spatial occupancy* and *octree representation.* In spatial occupancy enumeration schemes all of the cells in the scheme are cubic and lie in a fixed spatial grid. The use of spatial occupancy enumerations can be made more effective by using quadtrees in the two-dimensional case or octrees in the three-dimensional case.

In sweep/representation a set of points moving in space may sweep a one-, two-, or three-dimensional object, which can be represented by the moving set plus the trajectory. Constructive solid geometry is a method representing a complex solid as a composition of simpler (primitive) objects, where composition is implemented through Boolean operators. Finally, in the boundary representation method the surface of a solid object is segmented in nonoverlapping faces, each face being modeled by its bounding edges and vertices. Mobile robots and obstacles can also be described by any one of the above methods, but the resulting representations are very complex. Therefore, in the literature mobile robots or obstacles are usually approximately modeled by fixed-shape objects (e.g., a circle or a polygon), though in some cases (e.g., multilimbed walking robots) a fixed-shape object is not appropriate (Xue and Sheu, 1990).

Most of the currently available robot languages (e.g., AL, AML, RAIL, RPL) lie at the robot programming level. In the robot-level languages the positions and orientations of the objects (manipulators, obstacles, workspace, holes, etc.) have to be specified explicitly by coordinate frames. A robot motion is specified as a sequence of positional goals, namely a start position, a goal position, and a sequence of intermediate points through which the robot has to pass in order to avoid the obstacles.

Since the introduction of Smalltalk in the early 1980s it has been recognized that object-oriented programming languages offer several attractive features for software development. In an object-oriented language data can be naturally expressed as a set of objects that covers any kind of conceptual abstraction and organization. Two representative object-oriented languages are ROSS (McArthur and Klahr, 1982) and OSIS (Yoo and Sheu, 1993). Let us demonstrate the application of ROSS to a robotic system consisting of many manipulators that have to cooperate in the task of simple assembly using polyhedral objects. The given task is first inputted to the scheduler (Figure 2) which selects an initial set of basic actions to be performed. This set of actions is then distributed to manipulators in a contract-negotiation way. After the receipt of a task by a manipulator, the manipulator's processor proposes some actions that are transmitted in a conversation-like manner to the world model. These actions are simulated to evaluate their consequences and see if they are acceptable. Then the manipulator makes necessary modifications of the actions. Each adopted action is then transmitted to all the other manipulators. The above task planning scheme is much more understandable and convenient for the human user than STRIPS (Fikes and Nilsson, 1971) or the Graph Traverser (best next graph search) (Doran and Michie, 1966) programs, since it resembles in many respects the human plan generation mode.

The world model in ROSS is of the *perceptual cause and effect* type (Fahlman, 1974). All processing in ROSS is based on message-passing among a collection of actors and objects. ROSS is appropriate for modeling and understanding complex real-world robotic and other systems that cannot be treated by more analytical tools. This system can explore problems that may appear in parallel systems, such as deadlocks.

The basic general concepts of object-oriented programming are:

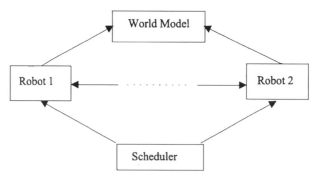

Figure 2   A cooperating multimanipulator system assembly.

1. Object and class
2. Method and message
3. Integrity constraints
4. Deductive laws

Each object has an identifier (e.g., *car3* identifies a Renault, *Ann-Marie* identifies a woman). A class is a group of similar objects (e.g., the class of robots or airplanes). Associated with each class is a set of attributes. Objects and classes are used to model the structural issues of real-world situations. A complex object is an object made up from a set of other objects. A hierarchy can therefore be derived for a set of nested complex objects. This is called *inheritance hierarchy.* In connection with the concept of data types, there is an additional concept, called *methods* or *operations* (e.g., comparisons with the class of integers). Every object-oriented system has a set of rules that designates the control and knowledge associated with the object. These are divided into *integrity constraints* (conditions about the object) and *deductive laws* (conditions deducing a higher-order fact about the object on the basis of lower-order facts).

Using OSIS, one can program a robot at all levels (robot-level, task-level, goal-level). To program a robot at the robot level, one expresses an operation as a query for which the details of the operation are provided at the qualification part of the query. Three robot-planning examples in OSIS follow (Sheu and Xue, 1993; Xue and Sheu, 1990):

- **Robot-level**

  ```
  var r is robot arm
  var v is vector
  sas ("arml", r.id)
  vector-assign ((0 0 1), v)
  move (r, v)
  while (r.gripper.sensed_force<=10)
  ```

  Here a robot moves gently and stops when the sensed force is greater than 10 ounces.

- **Task-level**

  ```
  range of t is solid
  range of s is solid
  range of r is robot
  var d is location
  move (r, t, t.location, d)
  where seq(t.id, "a") and seq(r.id, "pumal") and seq(s.id, "b") such that against
  (t.f1, s.g1) and against (t.f2, s.g2).
  ```

  Here the task-level request is to move a solid object a to a position d such that (a) its face feature f_1 is against the face feature g_1 of another object b and (b) its face feature f_2 is against the face feature g_2 of b.

- **Goal-level**

  ```
  range of t is T
  range of s is T
  range of r is robot
  var d is location
  var e is location
  move (r, t, t.location, d)
  where on-top (t, s) such that against (t.f_1, s.f_1) and against (t.f_2, s.f_2)
  ```

  Here the task is to move all pairs of objects s and t on the same type T, where t is on top of s to a position such that (a) the face feature f_1 of s is against the face feature f_1 of t and (b) the face feature f_2 of t is against the face feature f_2 of s.

Other well-known early AI systems for robot task planning are BUILD (Fahlman, 1974), AUTOPASS (Lieberman and Wesley, 1977), and RAPT (Popplestone, 1980). In the AUTOPASS system only the component relationships (*part of, attachment, constraint, assembly*) are specified. Thus, on the basis of the description a database is constructed in which components and assemblies are represented by nodes in a graph structure. RAPT (Edinburgh University) is a system for transforming symbol specifications of geometric goals. It is very suitable for assembly problems where parts are recognized using vision technology. Finally, a newer high-level system developed at Imperial College (London) is AUTO-GRASP (Besant, 1989), which is implemented in LISP and is capable to program a robot using object-level functions down to robot (joint) level.

Robot task planning and other knowledge-based intelligent functions of robots are frequently supported by *expert systems* (ES), which embody in the robot's computer the expertise of human experts or knowledge acquired from other sources (Puppe, 1993). Small-sized expert systems can be built from scratch using almost any language (procedural or symbolic), but large-sized ones need the use of some expert system language or tool. Expert system languages and tools can be used for the following:

1. To get knowledge into the robot
2. To test it
3. To fill gaps
4. To extend it
5. To modify it

Sometimes the knowledge can be put back into the human world in unrecognizably improved shape. A system is said to be an expert system if it performs at or near the level of human experts. Usually an ES is a system that is applied to difficult problems for which we need human experts as consultants. Expert systems have been applied in almost all scientific, engineering, and real-life domains (Tzafestas, 1993). Early examples of ES's are MYCIN (for diagnosis and treatment of infectious diseases), PROSPECTOR (for exploration geology problems), XCON(R1) (for configuring computer systems to customer needs), and DENDRAL (for mass spectra interpretation). The details of the structural and operational features of expert systems vary from one system to another, depending on the application requirements and the approach taken by the developers. The features common to the ES's are:

1. Explicit representation of the domain knowledge
2. A general-purpose interface mechanism providing control
3. Provision for reasoning under uncertainty (i.e., with uncertain evidence and knowledge)
4. Provision of justification, explanation and other run-time user support

The principal methods for knowledge representation used in expert systems are predicate logic, semantic networks, frames, scripts, objects, and production rules. The most popular is the production rule representation, where knowledge is represented by *IF–THEN* rules of the type *IF A THEN B,* where A is the antecedent (premise) part and B is the consequent (conclusion, action) part. The separation of knowledge and control permits the run-time facilities to provide, on demand, explanation and justification. Expert systems contain from about 100 to several thousand rules. The principal task in a consultation is to apply all those rules that are relevant to the problem at hand and avoid those that are not. The knowledge base contains no information about how to find the rules that apply and how to evaluate them. To do this a general-purpose *control-structure* or *inference engine* is needed. A rule is said to "fire" when it is selected for evaluation. Once its premises have all been evaluated, the truths of its conclusions can be determined. The selection of the rules to fire and their order can be done following either the forward or the backward chaining or the combination of both. Expert systems have been used with success in many robotic areas, such as intelligent task planning, assembly sequence finding, path planning, on-line scheduling, intelligent welding, and intelligent supervision and monitoring (Tzafestas, 1991; Jordanides and Torby, 1991; Adeli, 1990).

4.3 Robot Path Planning

Robot path-planning techniques are divided into *model-based* and *nonmodel-based.* The first class assumes that all the knowledge about the robot's workspace is available (learned) a priori and used to construct geometric models of parts and their relations (known navigation map). The second class assumes that some or all of the knowledge is obtained through sensors. According to the type of obstacles (static or moving) we have *path planning among stationary obstacles* and *path planning among moving obstacles.*

To solve the path-planning problem among moving obstacles the problem is usually decomposed into two subproblems:

1. Plan a path to avoid collision with stationary obstacles
2. Plan the velocity along the path to avoid collision with moving obstacles

Robot path planning supports robot task planning by determining and planning collision-free paths for the completion of the task. A solution path in a known environment must be collision-free, must be the shortest one, and must have the greatest minimum clearance among the paths along the collision-free path. Three kinds of robots have been considered in the literature: robot manipulators, mobile robots, and mobile robots with manipulators (briefly, mobile manipulators). Path planning for mobile robots is divided into *global* and *local* path planning. In global path planning the robot is fixed (e.g., it is considered to be a circle or polygon) and the environment is assumed to be arbitrary. In local path planning the problem is to move an arbitrary-shaped robot to pass a low clearance locality such as a door, corner, or corridor. The effectiveness of both global and local path planning is measured on the basis of their complexity and their convergence rate. A robot manipulator is usually represented in its joint space, where each manipulator configuration is represented by a point. The sets of points in the joint space that correspond to the configurations of the robotic manipulator that collide with obstacles are called *joint space obstacles*. Once the joint space obstacles are determined, the path-planning problem can be transformed to the problem of finding a path for a point from the initial to the goal configuration in the joint space that does not intersect with the joint space obstacles. Using this formulation, one can apply mobile robot path-planning techniques to manipulator path planning. The problem is much more difficult for mobile manipulators since they have more degrees of freedom, which of course gives greater maneuverability and higher flexibility in avoiding collisions.

The first accurate (exact, nonheuristic, and nonapproximate) solution to the global path-planning problem for mobile robots was given for the "piano movers" problem by Schwartz and Sharir (1983*a*, 1983*b*, 1983*c*; Schwartz and Ariel-Sheffi, 1984). This problem was shown to have a solution in polynomial time with respect to the smooth surfaces of the walls (boundaries and surfaces), the bodies, and the maximal number of equations describing them, but the complexity is exponential with respect to the degrees of freedom (d.o.f.) of the system (Schwartz and Sharir, 1983*b*). The problem was then reduced from three to two dimensions, and the solution was derived by studying the space of all collision-free positions of the robot (with n d.o.f.). This algorithm has complexity $O(m^5)$ where m is the number of distinct convex walls involved in the original problem (Schwartz and Sharir, 1983*a*). Another version of the piano mover's problem was treated via the *connectivity graph* approach (Schwartz and Sharir, 1983*c*). This problem was to find a continuous motion connecting two given configurations of m 2D circular bodies in a region bounded by a collection of walls, or else to establish that no such motion exists. The complexity of the solution with respect to the number of walls for a fixed number of moving circles is polynomial. The motion planning of 2D robotic systems involving multiple arms, all joined at a common end point and free to rotate for passing each other, was solved in Schwartz and Sharir (1983*c*) with an algorithm of complexity $O(n^{k+4})$ where k is the number of arms. Finally, the same authors have solved the problem for the case of moving in 3D space amidst polyhedral obstacles (Schwartz and Sharir, 1984).

Lozano-Pérez presents a method for finding the shortest path via the *configuration space* (C-space) concept (1983, 1987). Conceptually, the C-space shrinks a moving object (robot) to a point and expands the obstacles to the space of the moving object. The crucial step is to compute the C-space obstacles for each obstacle in the workspace. This technique was applied to polygonal (2D) and polyhedral (3D) objects. The shortest path from the initial to the final position among the obstacles is the *visibility graph* (V-graph). The C-space approach works well for translational object motions but needs special care for rotational motions (Moravec, 1980; Gouzenez, 1984; Hatzivassiliou and Tzafestas, 1992).

Other approaches for global path planning include the following:

- Free space decomposition approach (Brooks, 1983)
- Octree representation approach (Herman, 1986)
- Potential field approach (Andrews and Hogan, 1983; Okutomi and Mori, 1986)

The free space is described by generalized cones, inside which the search attempts to find a collision-free path for the moving robot. The algorithm starts by computing the generalized cones that describe the space. Then the cones are pairwise examined to see if and where their spines intersect. Finally, a path is found by the A* search algorithm from the initial to the goal position, following the spine intersection. Bourbakis (1989) defines the navigation path using the shapes of the current free spaces, selecting low-cost corridors in an unknown environment.

An octree is a hierarchical decomposition of a cubic space into subcubes where initially the root node is used to represent the workspace. If the whole workspace is completely filled by objects or completely empty, it is decomposed into eight equal subcubes (octants), which become the children of the root. The procedure is repeated until all nodes are completely filled or until some resolution limit is reached. The search through an octree can be carried out by a combination of various search methods, such as hypothesize-and-test, hill climbing, and A*.

Potential field methods are not based on graph search, but they use some artificial potential field functions for the obstacles and the goal positions to control the robot path. A first way to determine the free-to-move space is through the use of repulsive and attractive poles (Andrews and Hogan, 1983). The repulsive poles cause a robot to move away from an object, whereas attractive poles work in the opposite direction. The big problem in potential field methods is the *trapping at local minima.* Naturally, the planner prefers the low-potential areas, and so the robot may reach a *potential basin* (which is an equilibrium state) and become trapped. An early attempt to overcome the trapping at a potential basin was made by Okutomi and Mori (1986). A new potential-field method based on the vector-field histogram of Borenstein and Koren (19991a, 1991b) was proposed by Katevas and Tzafestas (1997, 1998). This method relies on one-dimensional environment representations named *active kinematic histograms,* which are derived by the available information for the obstacle space and the robot. A histogram shifting procedure is applied to determine the kinematic feasibility and result on the robot's steering limits. This active kinematic histogram method is suitable for fast, collision-free, and kinematically feasible path planning for nonpoint, nonholonomically constrained mobile robots. The method was applied successfully on a wheelchair (the SENARIO chair) (Katevas et al., 1997), where, for overcoming the local minima traps, a local minima trap detector was developed and used. Newer solutions to the global path planning problem involve the topology-based discretization method (Lück, 1997) and the state-space augmentation method (Zhou and Nguyen, 1997).

As already mentioned, in local path planning the problem is to move a robot that is arbitrarily specified through a fixed environment. Thus a local path planner that is specialized for a fixed environment is a *local expert.* A solution to the local path-planning problem for a rectangular robot that has to pass a corner in a corridor was given by Lee (1990). If the robot is longer than, say, "l," then it would not be able to pass the corner. The value of "l" is determined by the so-called *critical configuration algorithm.* The problem of moving a chair through a door was solved by Yap (1987), by representing the chair by a polygon. The smallest door allowing the chair to pass through it was rigorously determined by an algorithm of complexity $O(n^2)$.

The path-planning problem for the robot manipulators can be solved by the C-space technique. The basic issues here are how to map obstacles from the Cartesian space to the joint space and how to plan a collision-free path for manipulators in a high-dimensional joint space. The first issue was treated by Lumelsky (1987) and Hwang (1990). The second issue can be dealt with via the potential field approach (Warren, 1989) by splitting the potential field into two paths: one for points inside a C-space obstacle and one for points outside.

Path planning for mobile manipulators and shape-changeable robots needs more sophisticated algorithms. As we saw, a mobile robot is usually approximated by a fixed-shape object (a circle or a polygon), whereas a robot manipulator is represented by points in the joint space. To solve the path-planning problem for mobile manipulators and other shape-changeable robots (e.g., multilinked walking robots), a good model for robot and environment representation should be found. The collision-free path-planning problem in a two-dimensional environment for a mobile robot possessing two arms was considered and solved by Xue and Sheu (1989). The planner derived involves an algorithm that finds the *collision-free feasible configurations* for the reconfigurable object once the position and orientation of the carried object are specified and fixed (CFFC algorithm) and an algorithm that selects the best candidate path (on the basis of an objective criterion). Then all the collision-free feasible configurations along the best candidate path are found with the CFFC algorithm and a collision-free path is searched in these collision-free configurations. Other path-planning solutions in this area were proposed by Mayorga and Wong (1997), Xue et al. (1996), and Lee and Cho (1997).

We close our discussion on robot path planning with a look at the case where the robot environment is unknown. No attempt is made here to minimize the robot path as in the known environment case. The robot must be equipped with suitable sensors and may or may not have memory. When no memory is used, the path planning is exclusively based on proximity information provided by the sensors, and use is made of algorithmic strategies for moving the robot through the unknown environment. When the system has memory, some kind of learning is incorporated for incrementally constructing a visibility graph of the environment. Learning is accomplished as the robot visits a number of goal points. In both cases (memory, no memory) the solution is of the heuristic type (Tzafestas and Stamou, 1994; Lumelsky, 1984; Lumelsky and Stepanov, 1986). The critical issues in multirobot navigation (in the same dynamic environment) are:

1. Detection of other moving robots
2. Traffic priority rules
3. Communication among robots

For issue 1 the basic questions are how to detect other moving robots in the same space, what the size of the moving object is, what the motion direction of the moving object is, and how many objects exist in the robot area. The first and second questions are answered using visual and sonar sensors, the third question is dealt with by registering the direction of the changes that occur in the shape of the navigation space, and for the final question each robot acquires and defines the changes in different locations of the shape of the navigation space.

For issue 2 we distinguish two cases: objects of the same velocity and objects moving in different nonparallel directions. In the first case, if the objects move in the same parallel direction they have equal traffic priority. If the objects possess opposite parallel motion they are given the same priority if the moving lanes are not identical, and if they move on the same lane they are shifted to the right a proper distance if space is available; otherwise one of the objects has to retract for a while until the open navigation lane allows a parallel-shifting function. In the second case (nonparallel directions) the robots estimate the distance they have to travel until they reach the "anticipated" intersection point. If this distance is large enough to secure no-collision, they continue with the same velocity, but if this is not so, then the robot with the free space has the higher priority, while the other robot has to wait or reduce velocity or change direction.

For issue 3 the robots communicate to exchange information/knowledge about the structure of the common navigation environment in order to be able to improve their future navigation paths. In many cases man–robot interaction exists; i.e., the human makes decisions that affect the operation of the robot and the robot provides information to the human that affects these decisions. A complete navigation system for a mobile robot that involves a local navigator, a global navigator, and a map recalling subsystem is presented in Tzafestas and Zikidis (1997). The local navigation (obstacle avoidance and goal seeking) is performed using ultrasonic sensor data fed to an actor–critic-type reinforcement-learning neurofuzzy controller. The global navigation is accomplished using a topologically orderer Hopfield neural network that works with an ultrasonic sensor-built environment map represented by the so-called *eligibility factor matrix* (Elfes, 1987).

4.4   Reactive Robots

Since the mid-1980s, in an attempt to provide an efficient alternative to classical planning methods (a good collection of papers appears in Allen, Hendler, and Tate, 1990), the paradigm of reactive systems has emerged. Whereas planning relies on the existence and updating of a perfect global world model, reactive systems have adopted, at the other extreme, a view of total absence of any internal model. Instead, every action is in fact a direct reaction to perceived stimulus (stimulus-response (S-R) scheme). However simple in itself this might appear, its power grows exponentially when several reactive components are connected and used within the same system. Interactions between such reflexes may result in what appears to an observer concrete activity or even purposeful reasoning. For the case of a single robot, if S stands for its perceptions and R for the commands finally fed to the actuators, it appears as though something exists between (S-x-R). This something is nothing else but a *reactive algorithm* (RA): *reactive* because as explained it does not maintain an explicit model of the world or of the global task (although it does have some knowledge about its conditions of operation and about its operation itself), and *algorithm* because it appears as such and because it was so designed (S-RA-R scheme).

This reactive approach was initiated by Brooks and has been found to work best in partially unknown environments, where sensor readings are noisy or unreliable, and operation must be real-time and extremely responsive to unexpected events (Brooks, 1991a, 1991b). Those robots are equipped with a set of competences that they use in order to "survive" in the environment they are put into. Completion of the designer's task occurs as a side effect of the robot's struggle for survival. In this context, reactive algorithms precisely minimize processing of sensor data and accelerate it by exploiting parallelism on fixed topology networks. Another feature of reactive algorithms is that sensor readings or other "stimuli" may arrive at any point inside it, what may serve both to monitor command output and to ease design. Other advantages of the use of reactive algorithms include low implementation cost, design simplicity, enhanced robustness (due to minimal processing and crude data representations), and potential adaptivity, since a designer may decide to adapt one of the parameters the algorithm uses (there are not too many of them, and they are usually visible because the RAs are rather primitive).

Other desirable properties of the reactive, behavior-based paradigm are (Brooks, 1991a, 1991b): no central world representation bottlenecks, minimization of time lags between perception and actuator level command output, and fault tolerance. The power of the behavior-based paradigm lies in its simplicity. It is its interaction with the world that gives rise to potentially complex phenomena. By exploiting implicit knowledge about the world and the task, we can significantly simplify our life as designers and, by using such reactive autonomous robots, we can build low-cost, robust, fault-tolerant systems that demonstrate conceptual as well as pragmatic ease of design.

Actual use of reactive robots to industrial contexts has already been sketched (see, for instance, Arkin and Murphy (1990) and Doty and Van Aken (1993)). Many recently appeared service robots (Schraft, Degenhart, and Hägele, 1993) that comply with flexibility and adaptivity constraints have come closer to such reactive robots. An application of reactive robots for production management is discussed by Tzafestas (1995) where the primitive reactive model is enhanced by motivational autonomy and operational coupling features.

5   SUMMARY AND CONCLUDING REMARKS

Most present-day robots are far from human-like, both in appearance and behavior. They are typically one-armed robots positioned at fixed locations on the shop floor. Modern applications, such as underwater operations, space teleoperations, dexterous manipulations in nuclear and other toxic environments, and precise industrial operations, require further enhancement of the mechanical, electronic, and intelligence capabilities of present-day robots. Modern-life applications call for the development and use of fully autonomous multimanipulator mobile robots able to react to human voice commands and rapid environmental changes and to receive, translate, and execute general instructions. They must have the ability to sense, learn, acquire knowledge, perceive, make decisions, reason under uncertainty, and compute and execute adaptive/knowledge-based control actions. These intelligence features can be embedded into the robotic machines through AI methods and tools. Three areas of primary importance in the design and implementation of autonomous intelligent robots of this kind are learning, reasoning, and problem solving. These areas, their capabilities, and limitations have been described in this chapter.

Of the various learning algorithms, the one most frequently employed in robotics is learning from examples (inductive learning) that are usually obtained through vision (robot training by showing). For inductive learning many commercial software tools are available that can be integrated into robotic systems. Here, learning via a neural network (supervised or nonsupervised) can be used for several purposes, especially classification and control (Franklin, Mitchell, and Thrun, 1996). In changing and uncertain environments the robot uses some type of plausible reasoning. The typical reasoning paradigms in robotics are spatial reasoning and temporal reasoning. For problem solving the most common techniques used in robotics problems are means–ends reasoning, heuristic search, and the blackboard model. The tools for robot problem solving discussed in this chapter are GPS, STRIPS, ABSTRIPS, ROSS, OSIS, and AUTO-GRASP. Other systems have been proposed by Chang and Wee (1988) and Kak et al. (1986) for robotic assembly tasks.

To conclude, one can say that the field of intelligent robotics has arrived at an advanced state, but much remains to be done to improve both the methodological issues (learning, reasoning, and problem solving) and the machine hardware issues (sensors, actuators, drivers, end-effectors, etc.). Many aspects of knowledge representation are crucial for overall system design, while the perception process remains the major bottleneck for real-time autonomous navigation of mobile robots. Some sensor types, such as sonar, offer rapid but coarse environmental feedback, and computer vision holds great promise for robot navigation but demands specialized and costly hardware and software even to approach real-time navigation. The principal perceptual tasks for a mobile robot seem to be path following, goal identification, obstacle avoidance, and localization. Today more robust algorithms are needed to secure reliable and versatile performance of the robot. The potential applications of intelligent mobile robots are diverse, including service robots in warehouses and hospitals, domestic robots for home and office tasks, planetary rovers, and robots for supervision and repair functions in dangerous environments (nuclear reactors, mines, or military).

REFERENCES

Adeli, H., 1990. *Knowledge Engineering.* Vol. 2. New York: McGraw-Hill.

Ahmad, Z., and A. Guez. 1997. "Adaptive and Learning Control of Robotic Manipulators." In *Methods and Applications of Intelligent Control.* Ed. S. G. Tzafestas. Boston: Kluwer. 401–422.

Allen, J. F., J. Hendler, and A. Tate, eds. 1990. *Readings in Planning.* San Mateo: Morgan Kaufmann.

Andrews, J. R., and N. Hogan. 1983. "Impedance Control as a Framework for Implementing Obstacle Avoidance in a Manipulator." In *Proceedings of ASME Winter Annual Meeting on Control of Manufacturing Processes and Robotic Systems,* Boston. 243–251.

Antoniou, G. 1997. *Nonmonotonic Reasoning.* Cambridge: MIT Press.

Arkin, R. C., and R. R. Murphy. 1990. "Autonomous Navigation in a Manufacturing Environment." *IEEE Transactions on Robotics and Automation* **6**(4), 445–454.

Barto, A. G., R. S. Sutton, and C. W. Anderson. 1983. "Neuronlike Elements That Can Solve Difficult Learning Control Problems." *IEEE Transactions on Systems, Man, and Cybernetics* **13,** 835–846.

Berenji, H. R. 1993. "Fuzzy Neural Control." In *An Introduction to Intelligent and Autonomous Control.* Ed. P. J. Antsaklis and K. M. Passino. Boston: Kluwer. 215–236.

Besant, C. B. 1989. "The Application of Artificial Intelligence to Robotics." In *Parallel Processing and Artificial Intelligence.* Ed. M. Reeve and S. E. Zenith. New York: John Wiley & Sons. 175–191.

Borenstein, J., and Y. Koren. 1991*a*. "The Vector Field Histogram—Fast Obstacle Avoidance for Mobile Robots." *IEEE Transactions on Robotics and Automation* **7**(3), 278–288.

———. 1991*b*. "Histogramic In-Motion Mapping for Mobile Robot Obstacle Avoidance." *IEEE Transactions on Robotics and Automation* **7**(4), 535–539.

Bourbakis, N. G. 1989. "A Heuristic Collision Free Path Planning for an Autonomous Navigation Platform." *Journal of Intelligent and Robotic Systems* **1**(4), 375–387.

———. 1995. "Knowledge Extraction and Acquisition During Real-Time Navigation in Unknown Environments." *International Journal of Pattern Recognition and Artificial Intelligence.* **9**(1), 83–99.

Brooks, R. 1983. "Solving the Find-Path Problem by Good Representation of the Free Space." *IEEE Transactions on Systems, Man, and Cybernetics* **13**(3), 190–197.

Brooks, R. A. 1991*a*. "Intelligence Without Representation." *Artificial Intelligence* **47**(1–3), 139–159.

———. 1991*b*. "Intelligence Without Reason." MIT AI Memo. No. 1293, April; Computers and Thought Workshop, IJCAI-91.

Chang, K. H., and W. G. Wee. 1988. "A Knowledge-Based Planning System for Mechanical Assembly Using Robots." *IEEE Expert* (Spring), 18–30.

Charniak, E., and D. McDermot. 1985. *Introduction to Artificial Intelligence.* Reading: Addison-Wesley.

Cornuejols, A., and M. Moulet. 1997. "Machine Learning: A Survey." In *Knowledge-Based Systems: Advanced Concepts, Techniques and Applications.* Ed. S. G. Tzafestas. Singapore: World Scientific.

Doran, J. E., and D. Michie. 1966. "Experiments with the Graph Traverser Program." *Proceedings of the Royal Society A* **294,** 235–259.

Doty, K. L., and R. E. Van Aken. 1993. "Swarm Robot Materials Handling Paradigm for a Manufacturing Workcell." In *Proceedings of 1993 International Conference on Robotics and Automation,* Atlanta, May. 778–782.

Elfes, A., 1987. "Sonar-Based Real-World Mapping and Navigation." *IEEE Journal of Robotics and Automation* **3**(3), 249–265.

Fahlman, S. E. 1974. "A Planning System for Robot Construct Tasks." *Artificial Intelligence* **5**(1).

Fikes, R. E., and N. J. Nilsson. 1971. "STRIPS: A New Approach to the Application of Theorem Proving to Problem Solving." *Artificial Intelligence* **2**.

Fischhoff, B. 1989. "Eliciting Knowledge for Analytical Representation." *IEEE Transactions on Systems, Man, and Cybernetics.* **19**(3), 448–461.

Fisher, R. B. 1989. *From Surfaces to Objects: Computer Vision and Three Dimensional Scene Analysis.* New York: John Wiley & Sons.

Franklin, J. A., T. M. Mitchell, and S. Thrun, eds. 1996. *Recent Advances in Robot Learning,* Boston: Kluwer.

Gouzenez, L. 1984. "Strategies for Solving Collision-Free Trajectories Problems for Mobile and Manipulator Robots." *International Journal of Robotics Research* **3**(4), 51–65.

Harris, C. J., ed. 1994. *Advances in Intelligent Control.* London: Taylor & Francis.

Hatzivassiliou, F. V., and S. G. Tzafestas. 1992. "A Path Planning Method for Mobile Robots in a Structured Environment." In *Robotic Systems: Advanced Techniques and Applications.* Ed. S. G. Tzafestas. Dordrecht: Kluwer. 261–270.

Hayes-Roth, B. 1985. "A Blackboard Architecture for Control." *Journal of Artificial Intelligence.* **26**, 251–321.

Herman, M. 1986. "Fast 3-Dimensional Collision-Free Motion Planning." In *Proceedings of IEEE International Conference on Robotics and Automation,* San Francisco. 1056–1063.

Hexmoor, H., and L. Meeden, eds. 1996. *ROBOLEARN '96: International Workshop on Learning for Autonomous Agents.* Technical Report-96-11. Computer Science, State University of New York, Buffalo.

Hunt, V. D. 1983. *Industrial Robotics Handbook.* New York: Industrial Press.

Hwang, Y. K. 1990. "Boundary Equation of Configuration Obstacles for Manipulators." In *Proceedings of 1990 IEEE International Conference on Robotics and Automation.* 298–303.

Jordanides, T., and B. Torby, eds. 1991. *Expert Systems and Robotics.* Series F, Vol. 71. Berlin: Springer-Verlag.

Kak, A. C., et al. 1986. "A Knowledge Based Robotic Assembly Cell." *IEEE Expert,* 64–83.

Katevas, N. I., et al. 1997. "The Autonomous Mobile Robot SENARIO: A Sensor-Aided Intelligent Navigation System for Powered Wheelchair." *IEEE Robotics and Automation Magazine* **4**(4), 60–70.

Katevas, N. I., and S. G. Tzafestas. "Local Mobile Robot Path Planning: A New Technique Based on Active Kinematic Histogram." In *Proceedings of 1st Mobile Robotics Technology for Health Care Services Symposium (MOBINET '97),* Athens, Greece, May. 195–204.

———. 1998. "The Active Kinematic Histogram Method for Path Planning of Non-Point, Non-Holonomically Constrained Mobile Robots." *Journal of Advanced Robotics* (in press).

Lee, C.-T. 1990. "Critical Configuration Path Planning and Knowledge-Based Task Planning for Robot Systems." Ph.D. Thesis, School of Electrical Engineering, Purdue University.

Lee, J. K., and H. S. Cho. 1997. "Mobile Manipulator Motion Planning for Multiple Tasks Using Global Optimization Approach." *Journal of Intelligent and Robotic Systems* **18**(2), 169–190.

Lieberman, L. I., and M. A. Wesley. 1977. "AUTOPASS: An Automatic Programming System for Computer Controlled Mechanical Assembly." *IBM Journal of Research and Development* (July).

Lozano-Pérez, T. 1983. "Spatial Planning: A Configuration Space Approach." *IEEE Transactions on Computers* **32**(2), 108–120.

———. 1987. "A Simple Motion-Planning Algorithm for General Robot Manipulators." *IEEE Journal of Robotics and Automation* **3**(3), 224–238.

Lück, C. L. 1997. "Self-Motion Representation and Global Path Planning Optimization for Redundant Manipulators Through Topology-Based Discretization." *Journal of Intelligent and Robotic Systems* **19**(1), 23–38.

Lumelsky, V. J. 1987. "Dynamic Path Planning for a Planar Articulated Robot Arm Moving Amidst Unknown Obstacles." *Automatica* **23**(5), 551–570.

Lumelsky, V. J., and A. A. Stepanov. 1986. "Dynamic Path Planning for a Mobile Automaton with Limited Information on the Environment." *IEEE Transactions on Automatic Control* **31**(11), 1058–1063.

Marik, V., and L. Lhotska. 1989. "Some Aspects of the Blackboard Control in the Diagnosis FEL Expert Shell." In *Artificial Intelligence and Information-Control Systems of Robots.* Ed. D. Plander. Amsterdam: North-Holland.

Mason, M. 1981. "Compliance and Force Control for Computer Controlled Manipulators." *IEEE Transactions on Systems, Man, and Cybernetics* **11**.

Mayorga, R. V., and A. K. C. Wong. 1997. "A Robust Method for the Concurrent Motion Planning of Multi-Manipulator Systems." *Journal of Intelligent and Robotic Systems* **19**(1), 73–88.

McArthur, D., and P. Klahr. 1982. *The ROSS Language Manual.* N-1854-AF. Santa Monica: The Rand Corp.

McDermot, D. 1982 "A Temporal Logic for Reasoning About Processes and Plants." *Cognitive Science* **6**, 101–155.

Michalski, R. S., J. C. Carbonell, and T. M. Mitchell. 1983. *Machine Learning.* Palo Alto: Tioga.

Michie, D. 1985. "Expert Systems and Robotics." In *Handbook of Industrial Robotics.* Ed. S.Y. Nof. New York: John Wiley & Sons. 419–436.

Moravec, H. P. 1980. "Obstacle Avoidance and Navigation in the Real World by a Seeing Robot." Technical Report AIM-340, Stanford University.

Mortimer H. 1988. *The Logic of Induction.* Chichester: Ellis Horwood.

Newell, A., and H. Simon. 1963 "GPS: A Program That Simulates Human Thought." In *Computers and Thought.* Ed. E. A. Feigenbaum and J. Feldman. New York: McGraw-Hill.

———. 1972. *Human Problem Solving.* Englewood Cliffs: Prentice-Hall.

Nii, H. P. 1986. "Blackboard Systems." Stanford University, Report KSL 86-18; *AI Magazine* **7**(2), (3).

Nilsson, N. J. 1971. *Problem Solving Methods in Artificial Intelligence.* New York: McGraw-Hill.

O'Grady, P. O., and K. H. Lee. 1990. "A Hybrid Actor and Blackboard Approach to Manufacturing Cell Control." *Journal of Intelligent and Robotic Systems* **3**, 67–72.

Okutomi, M., and M. Mori. 1986. "Decision and Robot Movement by Means of a Potential Field." *Advanced Robotics* **1**(2), 131–141.

Omatu, S., M. Khalid, and R. Yusof. 1996. *Neuro-Control and Its Applications.* Berlin: Springer-Verlag.

Pang, G. K. H. 1989. "A Blackboard System for the Off-Line Programming of Robots." *Journal of Intelligent and Robotic Systems* **2**, 425–444.

Paterson, A., and T. Niblett. 1982. *ACLS Manual.* Oxford: Intelligent Terminals.

Patterson, P. W. 1990. *Introduction to Artificial Intelligence and Expert Systems.* Englewood Cliffs: Prentice-Hall.

Popovic, D., and V. P. Bhatkar. 1994. *Methods and Tools for Applied Artificial Intelligence.* New York: Marcel Dekker.

Popplestone, R. J. 1980. "An Interpreter for a Language for Describing Assemblies." *Artificial Intelligence* **14**.

Puppe, F. 1993. *Systematic Introduction to Expert Systems: Knowledge Representations and Problem Solving Methods.* New York: Springer-Verlag.

Sacerdoti, E. D. 1974. "Planning in a Hierarchy of Abstraction Spaces." *Artificial Intelligence* **5**(2), 115–135.

Samuel, A. L. 1959. "Some Studies in Machine Learning Using the Game of Checkers." *IBM Journal of Research and Development* **3**, 221–229.

Schraft, R. D., E. Degenhart, and M. Hägele. "Service Robots: The Appropriate Level of Automation and the Role of Operators in the Task Execution." In *Proceedings of 1993 IEEE Systems, Man, and Cybernetics Conference.* 163–169.

Schwartz, J. T., and E. Ariel-Sheffi. 1984. "On the Piano Mover's Problem: IV—Various Decomposable 2-Dimensional Motion Planning Problems." *Communications on Pure and Applied Mathematics* **37**, 479–493.

Schwartz, J. T., and M. Sharir. 1983a. "On the Piano Movers Problem: I—The Case of 2-Dimensional Rigid Polygonal Body Moving Amidst Polygonal Barriers." *Communications on Pure and Applied Mathematics* **36**, 345–398.

———. 1983b. "On the Piano Movers Problem: II—General Techniques Topological of Real Algebraic Manifolds." *Advances in Applied Mathematics* **4**, 298–351.

———. 1983c. "On the Piano Movers Problem: III—Coordinating the Motion of Several Independent Bodies: The Special Case of Circular Bodies Moving Amidst Polygonal Barriers." *International Journal of Robotics Research* **2**(3), 46–75.

———. 1984. "On the Piano Movers Problem: V—The Case of Rod Moving in 3-Dimensional Space Amidst Polyhedral Obstacles." *Communications on Pure and Applied Mathematics* **37**, 815–848.

Scott, P. P. 1984. *The Robotics Revolution.* New York: Basil Blackwell.

Sheu, P., and Q. Xue. 1993. *Intelligent Robotic Planning Systems.* Singapore: World Scientific.

Tzafestas, E. S. 1995. "Reactive Robots in the Service of Production Management." In *Proceedings of 4th International Conference on Intelligent Autonomous Systems (IAS-4),* Karlsruhe, March. 449–456; *Journal of Intelligent and Robotic Systems* **21**(2), 172–191.

Tzafestas, S. G., ed. 1991. *Intelligent Robotic Systems.* New York: Marcel Dekker

———, ed. 1993. *Expert Systems in Engineering Applications.* New York: Springer-Verlag.

———. 1994. "Fuzzy Systems and Fuzzy Expert Control: An Overview." *Knowledge Engineering Review* **9**(3), 229–268.

———. 1995. "Neural Networks in Robot Control." In *Artificial Intelligence in Industrial Decision Making, Control and Automation.* Ed. S. G. Tzafestas and H. B. Verbruggen. Boston: Kluwer. 327–387.

Tzafestas, S. G., and G. B. Stamou. 1994. "A Fuzzy Path Planning Algorithm for Autonomous Robots Moving in an Unknown and Uncertain Environment." In *Proceedings of EURISCON '94: The 2nd European Robotics and Intelligent Systems Conference,* Malaga, Spain, August. Vol. 1. 140–149.

———. 1997. "Concerning Automated Assembly: Knowledge-Based Issues and a Fuzzy System for Assembly Under Uncertainty." *Computer Integrated Manufacturing Systems* **10**(3), 183–192.

Tzafestas, S. G., and E. S. Tzafestas. 1991. "The Blackboard Architecture in Knowledge-Based Robotic Systems." In *Expert Systems and Robotics.* Ed. T. Jordanides and B. Torby. New York: Springer. 285–317.

———. 1997. "Recent Advances in Knowledge Acquisition Methodology." In *Knowledge-Based Systems: Advanced Concepts, Techniques and Applications.* Ed. S. G. Tzafestas. Singapore: World Scientific. 3–59.

Tzafestas, S. G., and A. N. Venetsanopoulos, ed. 1994. *Fuzzy Reasoning in Information, Decision and Control Systems.* Boston: Kluwer.

Tzafestas, S. G., and K. C. Zikidis. 1997. "Complete Mobile Robot Navigation via Neural and Neurofuzzy Control." In *Proceedings of First Mobile Robotics Technology for Health Care Services Symposium (MOBINET '97).* Athens, Greece, May. 205–217.

Tzafestas, S. G., F. V. Hatzivassiliou, and S. K. Kaltsounis. 1994. "Fuzzy Logic Design of a Nondestructive Robotic Fruit Collector." In *Fuzzy Reasoning in Information, Decision and Control Systems.* Ed. S. G. Tzafestas and H. Verbruggen. Boston: Kluwer. 243–256.

Tzafestas, S. G., Z. Ma, and E. S. Tzafestas. 1997. "Some Results in Inductive Learning: Version Space, Explanation-Based and Genetic Algorithm Approaches." In *Knowledge-Based Systems: Advanced Concepts, Techniques and Applications.* Singapore: World Scientific. 87–124.

Tzafestas, S. G., K. C. Zikidis, and G. B. Stamou. 1996. "A Neural Network System with Reinforcement Learning Driven by a Fuzzy Knowledge-Based System: Application to Autonomous Robots." In *Proceedings of SOCO '96: International ICSC Symposium on Soft Computing: Fuzzy Logic, Artificial Neural Networks and Genetic Algorithms,* Reading, March.

Tzafestas, S. G., et al. 1994. "An Expert System for Flexible Robotic Assembly under Uncertainty Based on Fuzzy Inference. *Studies in Informatics and Control* **5**(4), 401–407.

Warren, C. W. 1989. "Global Planning Using Artificial Potential Field." In *Proceedings of 1989 IEEE International Conference on Robotics and Automation.* 316–321.

Winston, P. H. 1984. *Artificial Intelligence.* Reading: Addison-Wesley.

Xue, Q., and P. Sheu. 1989. "Collision-Free Path Planning for Reconfigurable Robot." In *Proceedings of 4th International Conference on Advanced Robotics,* Columbus, Ohio, June.

———. 1990. "Path Planning for Shape-Changeable Robot in 3-D Environment." In *Proceedings of IEEE International Conference on Tools for Artificial Intelligence,* Washington, November. 9–15.

Xue, Q., et al. 1996. "Planning Collision-Free Paths for Reconfigurable Dual Manipulator Equipped Mobile Robot." *Journal of Intelligent and Robotic Systems* **17**(1), 223–242.

Yap, C.-K. 1987. "How to Move a Chair Through a Door." *IEEE Journal of Robotics and Automation* **3**(3), 172–181.

Yoo, S. B., and P. Sheu. 1993. "Evaluation and Optimization of Database Programs in an Object-Oriented Symbolic Information Environment." *IEEE Transactions on Knowledge and Data Engineering* **5**(3), 479–495.

Zhou, Z.-L., and C. C. Nguyen. 1997. "Globally Optimal Trajectory Planning for Redundant Manipulators Using State Space Augmentation Method." *Journal of Intelligent and Robotic Systems* **19**(1), 105–117.

CHAPTER 21
NEURO-FUZZY SYSTEMS

C. S. George Lee
Purdue University
West Lafayette, Indiana

1 INTRODUCTION

Fuzzy logic systems (FLS) and neural networks (NN) have increasingly attracted the interest of researchers, scientists, engineers, and practitioners in various scientific and engineering areas. The number and variety of applications of fuzzy logic and neural networks have been growing, ranging from consumer products and industrial process control to medical instrumentation, information systems and decision analysis, and robotics and automation.

Fuzzy sets and fuzzy logic are based on the way the human brain deals with inexact information. They were developed as a means for representing, manipulating, and utilizing uncertain information and to provide a framework for handling uncertainties and imprecision in real-world applications. Neural networks (or artificial neural networks) are modeled after the physical architecture of the brain. They were developed to provide computational power, fault tolerance, and learning capability to the systems. Although the fundamental inspirations for these two fields are quite different, there are a number of parallels that point out their similarities. This chapter examines the basic concepts of fuzzy logic control systems and neuro-fuzzy systems.

Fuzzy logic systems and neural networks are both numerical model-free estimators and dynamical systems, and they share the ability to improve the intelligence of the systems working in an uncertain, imprecise, and noisy environment. They have an advantage over traditional statistical-estimation and adaptive-control approaches to function estimation. Both estimate a function without requiring a mathematical description of how the output functionally depends on the input; that is, they learn from numerical examples. Basically, fuzzy logic systems estimate functions with *fuzzy-set* samples, while neural networks use *numerical-point* samples. Both have been shown to have the capability of modeling complex nonlinear processes to arbitrary degrees of accuracy.

In this chapter we shall examine the basic concepts of fuzzy logic control systems and neuro-fuzzy systems. Section 2 covers fundamental concepts and operations of fuzzy logic control systems, including fuzzy sets and relations, linguistic variables, and approximate reasoning. Section 3 focuses on the utilization of various neural network architectures and their associated learning algorithms to realize the basic elements and functions of traditional fuzzy logic control systems. Three neuro-fuzzy systems will be closely examined to illustrate their learning ability: the fuzzy adaptive learning control network (FALCON), the fuzzy basis function network (FBFN), and the adaptive-network-based fuzzy inference system (ANFIS). Section 4 summarizes the results with a discussion on applications.

2 FUZZY LOGIC CONTROL SYSTEMS

During the past decade fuzzy logic control (FLC), initiated by the pioneer work of Mamdani and Assilian (1975), has emerged as one of the most active and fruitful areas for research in the application of fuzzy set theory, fuzzy logic, and fuzzy reasoning. Its application ranges from industrial process control to medical diagnosis and securities trading. Many industrial and consumer products using this technology have been built, especially in Japan, where FLC has achieved considerable success. In contrast to con-

Handbook of Industrial Robotics, Second Edition, Edited by Shimon Y. Nof
ISBN 0-471-17783-0 © 1999 John Wiley & Sons, Inc.

ventional control techniques, FLC is best utilized in complex, ill-defined processes that can be controlled by a skilled human operator without much knowledge of their underlying dynamics.

The basic idea behind the FLC is to incorporate the "expert experience" of a human operator in the design of the controller in controlling a process whose input/output relationship is described by a collection of fuzzy control rules (e.g., *IF–THEN* rules) involving linguistic variables rather than a complicated dynamic model. This utilization of linguistic variables, fuzzy control rules, and approximate reasoning provides a means to incorporate human expert experience in designing the controller.

In this section we shall introduce the basic architecture and operation of fuzzy logic control systems. We will find that FLC is strongly based on the concepts of fuzzy sets and relations, linguistic variables, and approximate reasoning. These are covered in the next subsection.

2.1 Fuzzy Sets and Relations, Linguistic Variables, and Approximate Reasoning

In this subsection basic concepts that are central to the development and the understanding of fuzzy logic control and decision systems are reviewed. More detailed treatments of these subject areas can be found in these excellent textbooks: Klir and Folger (1988); Lin and Lee (1996); Dubois and Prade (1980); Zimmermann (1991); Kaufmann (1975).

A classical (*crisp*) set is a collection of distinct objects. It is defined in such a way as to dichotomize the elements of a given universe of discourse into two groups: members and nonmembers. This dichotomization is defined by a *characteristic function*. Let U be a universe of discourse that could be discrete or continuous. The characteristic function $\mu_A(x)$ of a crisp set A in U takes its values in $\{0, 1\}$ and is defined as

$$\mu_A(x) = \begin{cases} 1 & \text{if and only if } x \in A \\ 0 & \text{if and only if } x \notin A \end{cases} \tag{1}$$

Note that (i) the boundary of set A is rigid and sharp and it performs a two-class dichotomization (i.e., $x \in A$ or $x \notin A$), and (ii) the universe of discourse, U, is a crisp set.

A fuzzy set, on the other hand, introduces vagueness by eliminating the sharp boundary that divides members from nonmembers in the group. Thus, the transition between the full membership and nonmembership is *gradual* rather than abrupt, and this is realized by a membership function. Hence fuzzy sets may be viewed as an extension and generalization of the basic concepts of crisp sets; however, some theories are unique to the fuzzy set framework.

A fuzzy set A in the universe of discourse U can be defined as a set of ordered pairs,

$$A = \{(x, \mu_A(x)) \,|\, x \in U\} \tag{2}$$

where $\mu_A(\cdot)$ is called the *membership function* of A, and $\mu_A(x)$ is the *grade* (or degree) of membership of x in A, which indicates the degree that x belongs to A. The membership function $\mu_A(\cdot)$ maps U to the membership space M, that is, $\mu_A: U \to M$. When $M = \{0, 1\}$, set A is nonfuzzy and $\mu_A(\cdot)$ is the characteristic function of the crisp set A. For fuzzy sets the *range* of the membership function (i.e., M) is a subset of the nonnegative real numbers whose supremum is finite. In most general cases M is set to the unit interval $[0, 1]$.

Consider an example where U be the real line $\Re$ and let crisp set A represent "real numbers which are greater than or equal to 5." Then we have

$$A = \{(x, \mu_A(x)) \,|\, x \in U\}$$

where the characteristic function is

$$\mu_A(x) = \begin{cases} 0 & x < 5 \\ 1 & x \geq 5 \end{cases}$$

which is shown in Figure 1*a*. Now let fuzzy set A represent "real numbers which are close to 5." Then we have

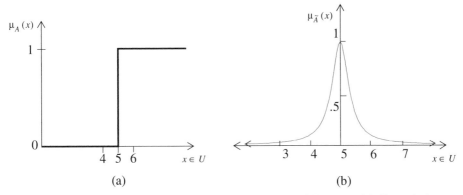

Figure 1   Characteristic functions of (a) crisp set A and (b) fuzzy set A in Example 1.

$$A = \{(x, \mu_A(x)) \mid x \in U\}$$

where the membership function is

$$\mu_A(x) = \frac{1}{1 + 10(x - 5)^2}$$

which is shown in Figure 1b. Obviously, one can select another membership function such as $\mu_A(x) = 1/[1 + (x - 5)^2]$ to represent the fuzzy set A. Hence, the assignment of the membership function of a fuzzy set is *subjective* in nature, although it cannot be assigned arbitrarily. A qualitative estimation reflecting a given ordering of the elements in A may be sufficient. This example shows that fuzziness is a type of imprecision that stems from a grouping of elements into classes that do not have sharply defined boundaries; it is not the lack of knowledge about the elements in the classes (e.g., a particular parametric value). Note that $\mu_A(x) \in [0, 1]$ indicates the membership grade of an element $x \in U$ in fuzzy set A, which reflects an ordering of the objects in fuzzy set A, and it is not a probability because $\Sigma \, \mu_A(x) \neq 1$.

Besides the ordered-pair representation in Equation (2), an alternative way of representing a fuzzy set is

$$A = \begin{cases} \sum_{x_i \in U} \mu_A(x_i)/x_i, & \text{if } U \text{ is discrete} \\ \int_U \mu_A(x)/x, & \text{if } U \text{ is continuous} \end{cases} \tag{3}$$

where the summation and the integral signs denote the union of the elements with their associated membership grades.

The above discussion shows that the construction of a fuzzy set depends on the specification of an appropriate membership function and the universe of discourse. The most commonly used membership functions (MFs) are: triangular MFs, trapezoidal MFs, Gaussian MFs, generalized bell-shaped MFs, and sigmoid MFs. They are listed in Table 1 and shown in Figure 2.

With the understanding of fuzzy sets and MFs, one can define set-theoretic operations on fuzzy sets. These operators are defined via their membership functions. Let A and B be two fuzzy sets in U with MFs $\mu_A(x)$ and $\mu_B(x)$, respectively. Several useful set-theoretic operators are defined in Table 2 for compactness.

The intersection and union operations of fuzzy sets in Table 2 are often referred to as *triangular norms* (*t*-norms) and *triangular conorms* (*t*-conorms), respectively (Dubois and Prade, 1980, 1985b). *t*-norms are a class of intersection functions and can be described as a two-parameter function of the form

$$t : \quad [0, 1] \times [0, 1] \rightarrow [0, 1] \tag{4}$$

Table 1 Membership Functions

Membership Functions	Equations
Triangular	$\Delta(x; a, b, c) = \max\left(\min\left(\dfrac{x-a}{b-a}, \dfrac{c-x}{c-b}\right), 0\right)$
Trapezoidal	$Trap(x; a, b, c, d) = \max\left(\min\left(\dfrac{x-a}{b-a}, 1, \dfrac{d-x}{d-c}\right), 0\right)$
Gaussian	$G(x; m, \sigma) = \exp(-[(x-m)/\sigma]^2)$
Generalized bell-shaped	$Bell(x; a, b, c) = 1/(1 + |(x-c)/a|^{2b})$
Sigmoid	$s(x; a, c) = 1/(1 + \exp[-a(x-c)])$

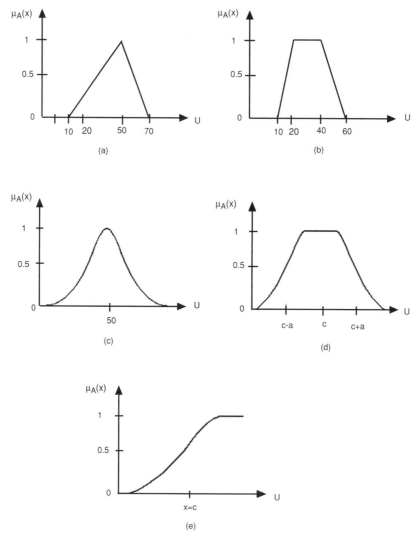

Figure 2 Various membership functions. (a) Triangular MF, $\Delta(x; 10, 50, 70)$. (b) Trapezoidal MF, $Trap(x; 10, 20, 40, 60)$. (c) Gaussian MF, $G(x; 50, 20)$. (d) Generalized bell-shaped MF, $Bell(x; a, b, c)$. (e) Sigmoidal MF, $s(x; 0.1, c)$.

Table 2   Set-Theoretic Operators

Operations	Equations
Complement, $\bar{A}$	$\mu_{\bar{A}}(x) = 1 - \mu_A(x), \quad \forall x \in U$
Intersection, $A \cap B$	$\mu_{A \cap B}(x) = \mu_A(x) \wedge \mu_B(x), \quad \forall x \in U$ $= \min[\mu_A(x), \mu_B(x)]$
Union, $A \cup B$	$\mu_{A \cup B}(x) = \mu_A(x) \vee \mu_B(x), \quad \forall x \in U$ $= \max[\mu_A(x), \mu_B(x)]$
Equality, $A = B$	$\mu_A(x) = \mu_B(x), \quad \forall x \in U$
Subset, $A \subseteq B$,	$\mu_A(x) \leq \mu_B(x), \quad \forall x \in U$
Cartesian product, $A_1 \times A_2 \times \cdots \times A_n$	$\mu_{A_1 \times A_2 \times \cdots \times A_n}(x_1, x_2, \ldots, x_n) = \min[\mu_{A_1}(x_1), \mu_{A_2}(x_2), \ldots, \mu_{A_n}(x_n)]$ where $x_i \in U_i$ and $U = U_1 \times U_2 \times \cdots \times U_n$
Algebraic sum, $A + B$	$\mu_{A+B}(x) = \mu_A(x) + \mu_B(x) - \mu_A(x) \cdot \mu_B(x)$
Algebraic product, $A \cdot B$	$\mu_{A \cdot B}(x) = \mu_A(x) \cdot \mu_B(x)$
Bonded sum, $A \oplus B$	$\mu_{A \oplus B}(x) = \min\{1, \mu_A(x) + \mu_B(x)\}$
Bounded difference, $A \ominus B$	$\mu_{A \ominus B}(x) = \max\{0, \mu_A(x) - \mu_B(x)\}$

such that $\mu_{A \cap B}(x) = t[\mu_A(x), \mu_B(x)]$, where the function $t(\cdot, \cdot)$ satisfies the boundary, commutativity, monotonicity, and associativity conditions (see Lin and Lee (1996) for details). Similarly, t-conorms (or s-norms) are a class of union functions and can be described as

$$s : \quad [0, 1] \times [0, 1] \rightarrow [0, 1] \qquad (5)$$

such that $\mu_{A \cup B}(x) = s[\mu_A(x), \mu_B(x)]$, where the function $s(\cdot, \cdot)$ satisfies the similar commutativity, monotonicity, and associativity conditions, but different boundary conditions.

The t-norms and t-conorms are often used to define other operations, such as fuzzy conjunction, fuzzy disjunction, and fuzzy implication. Typical t-norms and t-conorms that are used in FLC are listed in Table 3, where for simplification of notation we use $a \equiv \mu_A(x)$ and $b \equiv \mu_B(x)$.

It can be easily proved that the t-norms and t-conorms in Table 3 are bounded by the following inequalities:

$$t_{dp}(a, b) \leq t(a, b) \leq \min(a, b) \qquad (6)$$

$$\max(a, b) \leq s(a, b) \leq s_{ds}(a, b) \qquad (7)$$

where $t_{dp}(a, b)$ is the drastic product and $s_{ds}(a, b)$ is the drastic sum in Table 3. Hence the standard min and max operations are, respectively, the upper bound of t-norms and the lower bound of t-conorms. This provides a wide range of t-norms and t-conorms for the user's applications.

Fuzzy relations are a generalization of classical relations to allow for various degrees of association between elements. An n-ary fuzzy relation is a fuzzy set in $U_1 \times U_2 \times \cdots \times U_n$ and can be expressed using the fuzzy set representation as

Table 3   Typical Examples of Nonparametric t-norms and t-conorms

	t-norms		t-conorms (s-norm)
intersection	$a \wedge b = \min(a, b)$	union	$a \vee b = \max(a, b)$
algebraic product	$a \cdot b = ab$	algebraic sum	$a \hat{+} b = a + b - ab$
bounded product	$a \odot b = \max(0, a + b - 1)$	bounded sum	$a \oplus b = \min(1, a + b)$
drastic product	$a \hat{\cdot} b = \begin{cases} a, & b = 1 \\ b, & a = 1 \\ 0, & a, b < 1 \end{cases}$	drastic sum	$a \hat{\vee} b = \begin{cases} a, & b = 0 \\ b, & a = 0 \\ 1, & a, b > 0 \end{cases}$

$$R_{U_1 \times U_2 \times \cdots \times U_n} = \{[(x_1, x_2, \ldots, x_n), \mu_R(x_1, x_2, \ldots, x_n)]$$
$$| (x_1, x_2, \ldots, x_n) \in U_1 \times U_2 \times \cdots \times U_n\}. \tag{8}$$

An important operation of fuzzy relations is the composition of fuzzy relations. There are two types of composition operators: *max–min* and *min–max* compositions, and these compositions can be applied to both *relation–relation* compositions and *set–relation* compositions. The max–min composition is commonly used in FLCS. Let $P(X, Y)$ and $Q(Y, Z)$ be two fuzzy relations on $X \times Y$ and $Y \times Z$, respectively. Then the max–min composition of $P(X, Y)$ and $Q(Y, Z)$, denoted as $P(X, Y) \circ Q(Y, Z)$, is defined by

$$\mu_{P \circ Q}(x, z) \triangleq \max_{y \in Y} \min[\mu_P(x, y), \mu_Q(y, z)], \qquad \text{for all } x \in X, z \in Z \tag{9}$$

Equation (9) indicates that the max–min composition for fuzzy relations can be interpreted as indicating the strength of the existence of a relational chain between the elements of X and Z. The max–min composition can be generalized to other compositions by replacing the min-operator by any t-norm operator. Especially when the algebraic product in Table 3 is adopted, we have the *max-product* composition, denoted as $P(X, Y) \odot Q(Y, Z)$ and defined by

$$\mu_{P \odot Q}(x, z) \triangleq \max_{y \in Y}[\mu_P(x, y) \cdot \mu_Q(y, z)], \qquad \text{for all } x \in X, z \in Z \tag{10}$$

Analogous to the relation–relation compositions, we have the max–min composition for set–relation compositions. Let A be a fuzzy set on X and $R(X, Y)$ by a fuzzy relation on $X \times Y$. The max–min composition of A and $R(X, Y)$, denoted as $A \circ R(X, Y)$, is defined by

$$\mu_{A \circ R}(y) \triangleq \max_{x \in X} \min[\mu_A(x), \mu_R(x, y)], \qquad \text{for all } y \in Y \tag{11}$$

Again, the min-operator of the max–min composition can be replaced by any t-norm.

The concept of *linguistic variables* was introduced by Zadeh (1975) to provide a means of approximate characterization of phenomena that are too complex or too ill-defined to be amenable to description in conventional quantitative terms, and it plays a key role in fuzzy logic control. Basically, a linguistic variable is a variable whose values are words or sentences in a natural or artificial language. For example, *speed* is a linguistic variable if it takes the values such as *slow, moderate,* and *fast.*

A linguistic variable is characterized by a quintuple $(x, T(x), U, G, M)$ in which x is the name of the variable; $T(x)$ is the term set of x, that is, the set of names of linguistic values of x with each value being a fuzzy number (or fuzzy set) defined on U; G is a syntactic rule for generating the names of values of x; and M is a semantic rule for associating with each value of x its meaning. For example, if *speed* is interpreted as a linguistic variable with $U = [0, 100]$, that is $x = $ "*speed*," then its term set $T(speed)$ could be $T(speed) = \{slow, moderate, fast\}$. Here the syntactic rule, G, for generating the names (or the labels) of the elements in $T(speed)$ is quite intuitive. The semantic rule M could be defined as $M(slow)$ is the fuzzy set for "a speed below about 40 miles per hour (mph)" with membership function μ_{Slow}, $M(moderate)$ is the fuzzy set for "a speed close to 55 mph" with membership function $\mu_{Moderate}$, and $M(fast)$ is the fuzzy set for "a speed above about 70 mph" with membership function μ_{Fast}. These terms can be characterized by fuzzy sets whose membership functions are shown in Figure 3.

In approximate reasoning the reasoning procedure is performed through the generalization of the *modus ponens* inference rule in classical logic. The *modus ponens* inference rule states that $A \wedge (A \rightarrow B)) \rightarrow B$, which is interpreted as: If the proposition $(x$ is $A)$ is true and if the proposition "If $(x$ is $A)$ is true then $(y$ is $B)$ is true" is also true, then the proposition $(y$ is $B)$ is true. The *modus ponens* is closely related to the forward data-driven inference, which, when extended to fuzzy logic, is particularly useful for fuzzy logic control. Consider the following extension of the *modus ponens* inference rule,

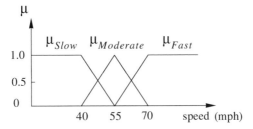

Figure 3   The terms of linguistic variable *speed*.

$$A' \wedge (A \to B)) \to B' \tag{12}$$

which can be rewritten in the form of:

premise 1: x is A'

premise 2:   IF x is A, THEN y is B (13)

conclusion: y is B'

If $A' = A$ and $B' = B$, then Equation (12) reduces to the *modus ponens* inference rule, and if $A' \neq A$ and $B' \neq B$, then Equation (12) is called the *generalized modus ponens* (GMP). Equation (12) raises an interesting question of how to deduce the the consequence (or conclusion) from a set of premises (or antecedents). If we model the IF–THEN rule of Equation (13) as a fuzzy implication, then we need to develop a systematic method of "combining" premise 1 with premise 2 to derive the conclusion "y is B'." Recognizing that A' in premise 1 is a fuzzy set/relation and premise 2 is a fuzzy relation (or fuzzy implication), the max–min set–relation composition in Equation (9) can be utilized to derive the conclusion

$$B' = A' \circ R \tag{14}$$

whose membership function is

$$\mu_{A' \circ R}(y) = \max_{y \in Y} \min(\mu_{A'}(x), \mu_R(x, y)), \qquad \text{for all } y \in Y \tag{15}$$

This is known as the *compositional rule of inference*. Thus, the generalized *modus ponens* is a special case of the compositional rule of inference. Unlike the *modus ponens* in classical logic, the generalized *modus ponens* does not require that the precondition "x is A'" must be identical with the premise "x is A." Hence the generalized *modus ponens* is related to the *interpolation rule* for solving the important problem of *partial* matching of preconditions of the *IF–THEN* rules, which arises in the operation of any FLCS.

2.2   Basic Structure and Operation of Fuzzy Logic Control Systems

The typical architecture of a FLC is shown in Figure 4, which is composed of four principal components: a *fuzzifier*, a *fuzzy rule base*, an *inference engine*, and a *defuzzifier*. If the output from the defuzzifier is not a control action for a plant, then the system is a fuzzy logic *decision* system. The fuzzifier has the effect of transforming crisp measured data (for example, speed is 10 miles per hour) into suitable linguistic values (i.e., fuzzy sets, for example, speed is *too slow*). The fuzzy rule base stores the empirical knowledge of the operation of the process of the domain experts. The inference engine is the kernel of a FLC, and it has the capability of simulating human decision-making by performing approximate reasoning to achieve a desired control strategy. The defuzzifier is utilized to yield a nonfuzzy decision or control action from an inferred fuzzy control action by the inference engine. More details about the operations of these components are described next.

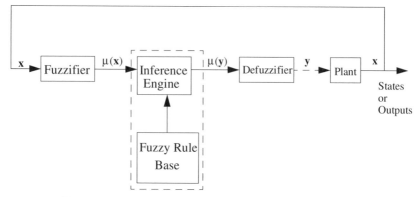

Figure 4   Basic architecture of the fuzzy logic controller (FLC).

2.2.1   Input/Output Spaces

The purpose of fuzzy logic controllers is to compute values of control (or action) variables from the observation or measurement of state variables of the controlled process such that a desired system performance is achieved. Thus a proper choice of process state variables and control variables is essential to the characterization of the operation of a fuzzy logic control system (FLCS) and has a substantial effect on the performance of a FLC. Typically, the input variables in a FLC are the state, state error, state error derivative, state error integral, etc. Following the definition of linguistic variables, the input vector **x**, which includes the input state linguistic variables x_i, and the output state vector **y**, which includes the output state (or control) linguistic variables y_i in Figure 4, can be defined, respectively, as

$$\mathbf{x} = \{(x_i, U_i, \{T_{x_i}^1, T_{x_i}^2, \ldots, T_{x_i}^{k_i}\}, \{\mu_{x_i}^1, \mu_{x_i}^2, \ldots, \mu_{x_i}^{k_i}\})|_{i=1,\ldots,n}\} \tag{16}$$

$$\mathbf{y} = \{(y_i, V_i, \{T_{y_i}^1, T_{y_i}^2, \ldots, T_{y_i}^{l_i}\}, \{\mu_{y_i}^1, \mu_{y_i}^2, \ldots, \mu_{y_i}^{l_i}\})|_{i=1,\ldots,m}\} \tag{17}$$

where the input linguistic variables, x_i, form a fuzzy input space, $U = U_1 \times U_2 \times \cdots \times U_n$, and the output linguistic variables, y_i, form a fuzzy output space, $V = V_1 \times V_2 \times \cdots \times V_m$. From Equations (16) and (17), we observe that an input linguistic variable x_i in a universe of discourse U_i is characterized by $T(x_i) = \{T_{x_i}^1, T_{x_i}^2, \ldots, T_{x_i}^{k_i}\}$ and $\mu(x_i) = \{\mu_{x_i}^1, \mu_{x_i}^2, \ldots, \mu_{x_i}^{k_i}\}$ where $T(x_i)$ is the *term set* of x_i; that is, the set of names of linguistic values of x_i with each value $T_{x_i}^{k_i}$ being a fuzzy set (or a fuzzy number) with membership function $\mu_{x_i}^{k_i}$ defined on U_i. So $\mu(x_i)$ is a semantic rule for associating with each value its meaning. For example, if x_i indicates speed, then $T(x_i) = \{T_{x_i}^1, T_{x_i}^2, T_{x_i}^3\}$ may be "slow," "medium," and "fast." Similarly, an output linguistic variable, y_i, is associated with a term set, $T(y_i) = \{T_{y_i}^1, T_{y_i}^2, \ldots, T_{y_i}^{l_i}\}$, and $\mu(y_i) = \{\mu_{y_i}^1, \mu_{y_i}^2, \ldots, \mu_{y_i}^{l_i}\}$. The size (or cardinality) of a term set, $|T(x_i)| = k_i$, is called the *fuzzy partition* of x_i. Fuzzy partition determines the granularity of the control obtainable from a FLC. Figure 5a depicts two fuzzy partitions in the same normalized universe $[-1, +1]$. For a two-input FLC, the fuzzy input space is divided into many overlapping grids (see Figure 5b). Furthermore, the fuzzy partitions in a fuzzy input space determine the *maximum* number of fuzzy control rules in a FLCS. For example, in the case of a two-input–one-output fuzzy logic control system, if $|T(x_1)| = 3$ and $|T(x_2)| = 7$, then the maximum number of fuzzy control rules is $|T(x_1)| \times |T(x_2)| = 21$. The input membership functions, $\mu_{x_i}^k$, $k = 1, 2, \ldots, k_i$, and the output membership functions, $\mu_{y_i}^l$, $l = 1, 2, \ldots, l_i$, used in a FLC are usually parametric functions (see Table 1).

Proper fuzzy partitioning of input and output spaces and a correct choice of membership functions play an essential role in achieving the success of a FLC design. Unfortunately, they are not deterministic and have no unique solutions. Traditionally, a heuristic trial-and-error procedure is usually used to determine an optimal fuzzy partition. Furthermore, the choice of input and output membership functions is based on subjective decision criteria and relies heavily on time-consuming trial and error. A promising ap-

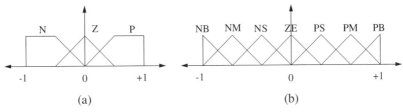

Figure 5   Diagrammatic representation of fuzzy partition. (a) Coarse fuzzy partition with three terms: N, negative; ZE, zero; and P, positive. (b) Finer fuzzy partition with seven terms: NB, negative big; NM, negative medium; NS, negative small; ZE, zero; PS, positive small; PM, positive medium; and PB, positive big.

proach to automate and speed up these design choices is to provide a FLC with learning abilities to learn its input and output membership functions and the fuzzy control rules. This will be explored later in Section 3.

2.2.2   Fuzzifier

A fuzzifier performs the function of fuzzification which is a subjective valuation to transform a measurement data into a valuation of a subjective value. Hence it can be defined as a mapping from an observed input space to labels of fuzzy sets in a specified input universe of discourse. Since the data manipulation in a FLC is based on fuzzy set theory, fuzzification is necessary and desirable at an early stage. In fuzzy control applications the observed data are usually crisp (though they may be corrupted by noise). A natural and simple fuzzification approach is to convert a crisp value, x_0, into a fuzzy singleton, A, within the specified universe of discourse. That is, the membership function of A, $\mu_A(x)$, is equal to 1 at the point x_0, and zero at other places. In this case, for a specific value $x_i(t)$ at time t, it is mapped to the fuzzy set $T_{x_i}^1$ with degree $\mu_{x_i}^1(x_i(t))$ and to the fuzzy set $T_{x_i}^2$ with degree $\mu_{x_i}^2(x_i(t))$, and so on. This approach is widely used in FLC applications because it greatly simplifies the fuzzy reasoning process. In a more complex case where observed data are disturbed by random noise, a fuzzifier should convert the probabilistic data into fuzzy numbers. For this, Dubois and Prade (1985) defined a bijective transformation that transforms a probability measure into a possibility measure by using the concept of the degree of necessity.

2.2.3   Fuzzy Rule Base

Fuzzy control rules are characterized by a collection of fuzzy IF–THEN rules in which the preconditions and consequents involve linguistic variables. This collection of fuzzy control rules characterizes the simple input/output relation of the system. For a multiinput/multioutput (MIMO) system,

$$R = \{R_{\text{MIMO}}^1, R_{\text{MIMO}}^2, \dots, R_{\text{MIMO}}^q\} \tag{18}$$

where the general form of the ith fuzzy logic rule is

$$R_{\text{MIMO}}^i: \quad \text{IF}(x_1 \text{ is } T_{x_1} \text{ and } \cdots \text{ and } x_n \text{ is } T_{x_n}),$$
$$\text{THEN } (y_1 \text{ is } T_{y_1} \text{ and } \cdots \text{ and } y_m \text{ is } T_{y_m}), \quad i = 1, \dots, q \tag{19}$$

The preconditions of R_{MIMO}^i form a fuzzy set $T_{x_1} \times \cdots \times T_{x_n}$ and the consequence of R_{MIMO}^i is the union of m independent outputs. So the rule can be represented by a fuzzy implication:

$$R_{\text{MIMO}}^i: \quad (T_{x_1} \times \cdots \times T_{x_n}) \to (T_{y_1} + \cdots + T_{y_m}), \quad i = 1, \dots, q \tag{20}$$

where "$+$" represents the union of independent variables. Since the outputs of an MIMO rule are independent, the general rule structure of an MIMO fuzzy system can be represented as a collection of m multiinput/single-output (MISO) fuzzy systems by decomposing the above rule into m subrules with T_{y_i} as the single consequence of ith subrule.

For clarity, the general form of the fuzzy control rules in the case of an MISO system is:

$$R^i : \quad \text{IF } x \text{ is } A_i, \ldots, \text{AND } y \text{ is } B_i, \text{ THEN } z = C_i, \quad i = 1, 2, \ldots, q \quad (21)$$

where $x, \ldots, y$, and z are linguistic variables representing the process state variables and the control variable, respectively, $A_i, \ldots, B_i$, and C_i are the linguistic values of the linguistic variables $x, \ldots, y$, and z in the universes of discourse $U, \ldots, V$, and W, respectively. For example,

> IF the speed is SLOW and the acceleration is DECELERATING,
>
> THEN INCREASE POWER GREATLY

The preconditions and consequent of Equation (21) are linguistic values. A variant of this type is to represent the consequent as a function of the process state variables, $x, \ldots, y$, that is,

$$R_i : \quad \text{IF } x \text{ is } A_i, \ldots, \text{AND } y \text{ is } B_i, \text{ THEN } z = f_i(x, \ldots, y) \quad (22)$$

where $f_i(x, \ldots, y)$ is a function of the process state variables, $x, \ldots, y$. Fuzzy control rules of Equations (21) and (22) evaluate the process state (i.e., state, state error, state error integral, etc.) at time t and compute and decide the control actions as a function of the state variables $(x, \ldots, y)$. It is worthwhile to point out that both fuzzy control rules have linguistic values as inputs and either linguistic values (as in Equation (21)) or crisp values (as in Equation (22)) as outputs.

2.2.4 Inference Engine

This is the kernel of the FLC in modeling human decision-making within the conceptual framework of fuzzy logic and approximate reasoning. The generalized *modus ponens* in Equations (12) and (13), which is the special case of the compositional rule of inference, plays an important role in the partial matching of the preconditions of the IF–THEN rules. In general, the compositional rule of inference can be realized by four types of compositional operator. These correspond to the four operations associated with the max–t-norms, namely, max–min composition (Zadeh, 1973), max–product composition (Kaufmann, 1975), max–bounded-product composition (Mizumoto, 1981), and max–drastic-product composition (Mizumoto, 1981). In FLC applications the max–min and the max–product compositional operators are the most commonly and frequently used due to their computational simplicity and efficiency. Let max—$\star$ represent any one of the above four composition operations. Then Equation (14) becomes

$$B' = A' \star R = A' \star (A \rightarrow B), \quad (23)$$

$$\mu_{B'}(y) = \max_{y \in Y} \{\mu_{A'}(x) \star \mu_{A \rightarrow B}(x, y)\}$$

where $\star$ denotes the t-norm operations, such as mim, product, bounded product, and drastic product operations. As for the fuzzy implication, $A \rightarrow B$, there are nearly 40 distinct fuzzy implication functions described in the existing literature. Some of the most commonly used fuzzy implications are listed in Table 4.

The inference engine in Figure 4 is to match the preconditions of rules in the fuzzy rule base with the input state linguistic terms and performs implication. For example, if there were two rules:

$$R1: \quad \text{IF } x_1 \text{ is } T_{x_1}^1 \text{ and } x_2 \text{ is } T_{x_2}^1, \text{ THEN } y \text{ is } T_y^1$$

$$R2: \quad \text{IF } x_1 \text{ is } T_{x_1}^2 \text{ and } x_2 \text{ is } T_{x_2}^2, \text{ THEN } y \text{ is } T_y^2$$

then the firing strengths of rules R1 and R2 are defined as α_1 and α_2, respectively. Here α_1 is defined as

$$\alpha_1 = \mu_{x_1}^1(x_1) \wedge \mu_{x_2}^1(x_2) \quad (24)$$

Table 4   Various Fuzzy Implication Rules

Rule of Fuzzy Implication	Implication Formulae	Fuzzy Implication: $\mu_{A \to B}(u, v)$
R_c: min-operation (Mamdani)	$a \to b = a \wedge b$	$= \mu_A(u) \wedge \mu_B(v)$
R_p: product operation (Larsen)	$a \to b = a \cdot b$	$= \mu_A(u) \cdot \mu_B(v)$
R_{bp}: bounded product	$a \to b = 0 \vee (a + b - 1)$	$= 0 \vee [\mu_A(u) + \mu_B(v) - 1]$
R_{dp}: drastic product	$a \to b = \begin{cases} a, & b = 1 \\ b, & a = 1 \\ 0, & a, b < 1 \end{cases}$	$= \begin{cases} \mu_A(u), & \mu_B(v) = 1 \\ \mu_B(v), & \mu_A(u) = 1 \\ 0, & \mu_A(u), \mu_B(v) < 1 \end{cases}$
R_a: arithmetic rule (Zadeh)	$a \to b = 1 \wedge (1 - a + b)$	$= 1 \wedge (1 - \mu_A(u) + \mu_B(v))$
R_m: max–min rule (Zadeh)	$a \to b = (a \wedge b) \vee (1 - a)$	$= (\mu_A(u) \wedge \mu_B(v)) \vee (1 - \mu_A(u))$
R_s: standard sequence	$a \to b = \begin{cases} 1, & a \le b \\ 0, & a > b \end{cases}$	$= \begin{cases} 1, & \mu_A(u) \le \mu_B(v) \\ 0, & \mu_A(u) > \mu_B(v) \end{cases}$
R_b: Boolean fuzzy implication	$a \to b = (1 - a) \vee b$	$= (1 - \mu_A(u)) \vee \mu_B(v)$
R_g: Gödelian logic	$a \to b = \begin{cases} 1, & a \le b \\ b, & a > b \end{cases}$	$= \begin{cases} 1, & \mu_A(u) \le \mu_B(v) \\ \mu_B(v), & \mu_A(u) > \mu_B(v) \end{cases}$
R_Δ: Goguen's fuzzy implication	$a \to b = \begin{cases} 1, & a \le b \\ \dfrac{b}{a}, & a > b \end{cases}$	$= \begin{cases} 1, & \mu_A(u) \le \mu_B(v) \\ \dfrac{\mu_B(v)}{\mu_A(u)}, & \mu_A(u) > \mu_B(v) \end{cases}$

where $\wedge$ is the *fuzzy AND* operation. The most commonly used fuzzy AND operations (i.e., *t*-norms) are intersection and algebraic product. Hence α_i, $i = 1, 2$, becomes

$$\alpha_i = \mu_{x_1}^i(x_1) \wedge \mu_{x_2}^i(x_2) = \begin{cases} \min(\mu_{x_1}^i(x_1), \mu_{x_2}^i(x_2)) \\ \text{or} \\ \mu_{x_1}^i(x_1)\mu_{x_2}^i(x_2) \end{cases} \quad (25)$$

The above two rules, R1 and R2, lead to the corresponding decision with the membership function, $\hat{\mu}_y^i(w)$, $i = 1, 2$, which is defined as

$$\hat{\mu}_y^i(w) = \alpha_i \wedge \mu_y^i(w) = \begin{cases} \min(\alpha_i, \mu_y^i(w)) \\ \text{or} \\ \alpha_i \mu_y^i(w) \end{cases} \quad (26)$$

where w is the variable that represents the support values of the membership function. Combining the above two decisions, we obtain the output decision

$$\hat{\mu}_y(w) = \hat{\mu}_y^1(w) \vee \hat{\mu}_y^2(w) \quad (27)$$

where $\vee$ is the *fuzzy OR* operation. The most commonly used fuzzy OR operations (i.e., *t*-conorms) are union and bounded sum. Hence the output decision becomes

$$\hat{\mu}_y(w) = \hat{\mu}_y^1(w) \vee \hat{\mu}_y^2(w) = \begin{cases} \max(\hat{\mu}_y^1(w), \hat{\mu}_y^2(w)) \\ \text{or} \\ \min(1, \hat{\mu}_y^1(w) + \hat{\mu}_y^2(w)) \end{cases} \quad (28)$$

Notice that Equation (28) is a membership function. Before feeding the signal to the plant, we need a defuzzification process to obtain a crisp decision, and the defuzzifier block in Figure 4 serves this purpose. Among the commonly used defuzzification strategies, the *center of area* method yields a superior result. Let w_j be the support value at which the membership function, $\hat{\mu}_y^j(w)$, reaches the maximum value $\hat{\mu}_y^j(w)|_{w=w_j}$, then the defuzzification output is

$$y = \frac{\sum_j \hat{\mu}_y^j(w_j)w_j}{\sum_j \hat{\mu}_y^j(w_j)} \quad (29)$$

The above describes the standard function operations in a traditional fuzzy logic control/decision system. There are some alternatives for fuzzy OR, fuzzy AND, and reasoning operations (Lee, 1990). For example, Sugeno (1985) proposed that the consequent of a rule is a function of input linguistic variables (see Equation (22)). In this case the defuzzification process is unnecessary and there are no exact membership functions for output linguistic variables. Currently, in a traditional fuzzy logic control system most system designers choose their membership functions empirically and construct the fuzzy logical rules from experts. Bringing learning abilities to the fuzzy logic system is an important issue. This learning/adapting procedure finds suitable fuzzy logic rules and adapts the fuzzifier and the defuzzifier to find the proper shapes and overlaps of membership functions by learning the desired output. Three different feedforward neural networks will be discussed in the next section to provide this learning/adapting ability. These neural-network-based architectures eliminate the rule-matching process and distributively stores the control/decision knowledge in the connections and link weights.

3 NEURO-FUZZY CONTROL SYSTEMS

This section focuses on the utilization of various neural network architectures and their associated learning algorithms to realize the basic elements and functions of traditional fuzzy logic control systems. These so-called *neuro-fuzzy control systems* (NFCS) are connectionist models, and they apply neural learning techniques to determine and tune the parameters and/or structure of the connectionist model. Neuro-fuzzy control systems require two major types of tuning: *structural* tuning and *parametric* tuning. Structural tuning concerns the tuning of the structure of fuzzy logic rules, such as the number of variables to account for and for each input or output variable the partition of the universe of discourse; the number of rules and the conjunctions that constitute them; etc. Once a satisfactory structure of the rules is obtained, the NFCS needs to perform parametric learning. In this parameter-learning phase the possible parameters to be tuned include those associated with membership functions such as centers, widths, and slopes; the parameters of the parameterized fuzzy connectives; and the weights of the fuzzy logic rules.

Among the three classes of learning schemes of neural networks, the unsupervised learning, which constructs internal models that capture regularities in their input vectors without receiving any additional information, is suitable for structure learning in order to find clusters of data indicating the presence of fuzzy logic rules. Supervised learning, which requires a teacher to specify the desired output vector, and reinforcement learning, which requires only a single scalar evaluation of the output, are suitable for parameter learning to adjust the parameters of fuzzy logic rules and/or membership functions for the desired output in NFCS. In cases where the membership functions are differentiable, gradient-decent-based learning methods (e.g., backpropagation algorithm) for parameter learning can be easily derived.

In this section we shall introduce three different connectionist models for realizing a traditional fuzzy logic control system. The models have both structure learning and parameter learning capabilities. A basic concept of these neuro-fuzzy control models is first to use structure learning algorithms to find appropriate fuzzy logic rules and then to use parameter learning algorithms to fine-tune the membership functions and other parameters. Due to limitations of space and the vast amount of literature available on neural networks, introductory materials to neural networks are not discussed here. More detailed information on neural networks can be found in the following excellent textbooks: Hertz, Krogh, and Palmer (1991), Lin and Lee (1996), and Zurada (1992).

3.1 Fuzzy Adaptive Learning Control Network (FALCON)

In this section we shall introduce a general connectionist model, the fuzzy adaptive learning control network (FALCON), proposed by Lin and Lee (1991) to study hybrid structure/parameter learning strategies. The FALCON is a feedforward multilayer network that integrates the basic elements and functions of a traditional fuzzy logic controller into a connectionist structure that has distributed learning abilities. In this connectionist structure the input and output nodes represent the input states and output control/decision signals, respectively, and in the hidden layers there are nodes functioning as membership functions and rules. The FALCON can be contrasted with a traditional fuzzy logic control and decision system in terms of its network structure and learning abilities. Such fuzzy control/decision networks can be constructed from training examples by neural learning techniques, and the connectionist structure can be trained to develop fuzzy logic rules and determine proper input/output membership functions. This connectionist model also provides human-understandable meaning to the normal feedforward multilayer neural network in which the internal units are always opaque to the users. So, if necessary, expert knowledge can be easily incorporated into the FALCON. The connectionist structure also avoids the rule-matching time of the inference engine in the traditional fuzzy control system. We shall first introduce the structure and function of the FALCON and then consider its learning scheme.

Figure 6 shows the structure of the FALCON. The system has a total of five layers. Nodes in layer 1 are input nodes (*linguistic nodes*) that represent input linguistic variables, and layer 5 is the output layer. There are two linguistic nodes for each output variable. One is for training data (desired output) to feed into the network, and the other is for decision signals (actual output) to be pumped out of the network. Nodes in layers 2 and 4 are *term nodes,* which act as membership functions representing the terms of the respective linguistic variables. Actually, a layer-2 node can either be a single node that performs a simple membership function (e.g., a triangle-shaped or bell-shaped function) or composed of multilayer nodes (a subneural net) that performs a complex membership function. So the total number of layers in this connectionist model can be more than 5. Each node in layer 3 is a rule node that represents one fuzzy rule. Thus, all the layer-3 nodes form a fuzzy rule base. Links in layers 3 and 4 function as a *connectionist inference engine,* which avoids the rule-matching process. Layer-3 links define the preconditions of the rule nodes, and layer-4 links define the consequents of the rule nodes. Therefore, for each rule node, there is at most one link (maybe none) from some term node of a linguistic node. This is true both for precondition links (links in layer 3) and consequent links (links in layer 4). The links in layers 2 and 5 are fully connected between linguistic nodes and their corresponding term nodes. The arrow on the link indicates the normal signal flow direction when this network is in use after it has been built and trained. We shall later indicate the signal propagation, layer by layer, according to the arrow direction. Signals may flow in the reverse direction in the learning or training process.

With this five-layered structure of the FALCON, we shall then define the basic functions of a node. The FALCON consists of a node which has some finite fan-in of connections represented by weight values from other nodes and a fan-out of connections to other nodes. Associated with the fan-in of a node is an integration function f, which serves to combine information, activation, or evidence from other nodes. This function provides the net input to this node

$$\text{net}_i = f(u_1^{(k)}, u_2^{(k)}, \ldots, u_p^{(k)}; w_1^{(k)}, w_1^{(k)}, w_2^{(k)}, \ldots, w_p^{(k)}) \qquad (30)$$

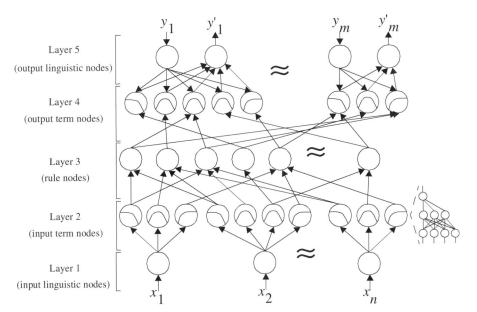

Figure 6 The structure of the FALCON.

where $u_1^{(k)}$, $u_2^{(k)}$, ..., $u_p^{(k)}$ are inputs to this node and $w_1^{(k)}$, $w_2^{(k)}$, ..., $w_p^{(k)}$ are the associated link weights. The superscript (k) in the above equation indicates the layer number. A second action of each node is to output an activation value as a function of its net input,

$$\text{output} = o_i^{(k)} = a(\text{net}_i) = a(f) \tag{31}$$

where $a(\cdot)$ denotes the activation function.

We shall next describe the functions of the nodes in each of the five layers of the FALCON.

1. **Layer 1.** The nodes in this layer only transmit input values to the next layer directly. That is,

$$f = u_i^{(1)} \quad \text{and} \quad a = f \tag{32}$$

From this equation, the link weight at layer 1 $(w_i^{(1)})$ is unity.

2. **Layer 2.** If we use a single node to perform a simple membership function, then the output function of this node should be this membership function. For example, for a bell-shaped function we have

$$f = \mu_{x_i}^j(m_{ij}, \sigma_{ij}) = -\frac{(u_i^{(2)} - m_{ij})^2}{\sigma_{ij}^2} \quad \text{and} \quad a = e^f \tag{33}$$

where m_{ij} and σ_{ij} are, respectively, the center (or mean) and the width (or variance) of the bell-shaped function of the jth term of the ith input linguistic variable x_i. Hence the link weight at layer 2 $(w_{ij}^{(2)})$ can be interpreted as m_{ij}. If we use a set of nodes to perform a membership function, then the function of each node can be just in the standard form (e.g., the sigmoid function), and the whole subnet is trained off-line to perform the desired membership function by a standard learning algorithm (e.g., the backpropagation algorithm).

3. **Layer 3.** The links in this layer are used to perform precondition matching of fuzzy logic rules. Hence the rule nodes should perform the fuzzy AND operation,

$$f = \min(u_1^{(3)}, u_2^{(3)}, \ldots, u_p^{(3)}) \quad \text{and} \quad a = f \tag{34}$$

The link weight in layer 3 $(w_i^{(3)})$ is then unity.

4. **Layer 4.** The nodes in this layer have two operation modes: *down–up* transmission and *up–down* transmission modes. In the down–up transmission mode the links in layer 4 should perform the fuzzy OR operation to integrate the fired rules which have the same consequent:

$$f = \sum_i u_i^{(4)} \quad \text{and} \quad a = \min(1, f) \tag{35}$$

Hence the link weight $w_i^{(4)} = 1$. In the up–down transmission mode the nodes in this layer and the links in layer 5 function exactly the same as those in layer 2 except that only a single node is used to perform a membership function for output linguistic variables.

5. **Layer 5.** There are two kinds of nodes in this layer. The first kind of node performs up–down transmission for training data being fed into the network. For this kind of node

$$f = y_i \quad \text{and} \quad a = f \tag{36}$$

The second kind of node performs down–up transmission for the decision signal output. These nodes and the layer-5 links attached to them act as the defuzzifier. If m_{ij} and σ_{ij} are, respectively, the center and the width of the membership function of the jth term of the ith output linguistic variable, then the following functions can be used to simulate the *center of area* defuzzification method in Equation (29):

$$f = \sum_j w_{ij}^{(5)} u_{ij}^{(5)} = \sum_j (m_{ij}\sigma_{ij})u_{ij}^{(5)} \quad \text{and} \quad a = \frac{f}{\sum_j \sigma_{ij}u_{ij}^{(5)}} \tag{37}$$

Here the link weight in layer 5 ($w_{ij}^{(5)}$) is $m_{ij}\sigma_{ij}$.

Based on the above connectionist structure, a two-phase hybrid learning algorithm must be developed to determine proper centers (m_{ij}) and widths (σ_{ij}) of the term nodes in layers 2 and 4. Also, it will learn fuzzy logic rules by deciding the existence and connection types of the links at layers 3 and 4; that is, the precondition links and consequent links of the rule nodes.

We shall now present a hybrid learning algorithm to set up the FALCON from a set of supervised training data. The hybrid learning algorithm consists of two separate stages of learning strategy which combine unsupervised learning and supervised gradient-descent learning procedures to build the rule nodes and train the membership functions. The performance of this hybrid learning algorithm is superior to that of the purely supervised learning algorithm (e.g., the backpropagation learning algorithm) because of the *a priori* classification of training data through an overlapping receptive field before the supervised learning. The learning rate of the original backpropagation learning algorithm is limited by the fact that all layers of weights in the network are determined by the minimization of an error signal that is specified only as a function of the output, and a substantial fraction of the learning time is spent in the discovery of internal representation.

In phase 1 of the hybrid learning algorithm a self-organized learning scheme (i.e., unsupervised learning) is used to locate initial membership functions and to detect the presence of fuzzy rules. In phase 2 a supervised learning scheme is used to optimally adjust the parameters of the membership functions for desired outputs. To initiate the learning scheme, training data and the desired or guessed coarse of fuzzy partition (i.e., the size of the term set of each input/output linguistic variable) must be provided by an expert.

Before this network is trained, an initial structure of the network is first constructed. Then, during the learning process, some nodes and links of this initial network are deleted or combined to form the final structure. In its initial form (see Figure 6) there are $\Pi_i|T(x_i)|$ rule nodes with the inputs of each rule node coming from one possible combination of the terms of input linguistic variables under the constraint that only one term in a term set can be a rule node's input. Here $|T(x_i)|$ denotes the number of x_i terms (i.e., the number of fuzzy partitions of the input linguistic variable x_i). So the input state space is initially divided into $|T(x_1)| \times |T(x_2)| \times \cdots \times |T(x_n)|$ linguistically defined nodes (or fuzzy cells) which represent the preconditions of fuzzy rules. Also, the links between the rule nodes and the output term nodes are initially fully connected, meaning that the consequents of the rule nodes are not yet decided. Only a suitable term in each output linguistic variable's term set will be chosen after the learning process.

3.1.1 Self-Organized Learning Phase

The problem for the self-organized learning can be stated as: Given the training input data $x_i(t)$, $i = 1, 2, \ldots, n$, the corresponding desired output value $y_i^d(t)$, $i = 1, 2, \ldots, m$, the fuzzy partitions $|T(x_i)|$ and $|T(y_i)|$, and the desired shapes of membership functions, we want to locate the membership functions and find the fuzzy logic rules. In this phase the network works in a two-sided manner; that is, the nodes and links in layer 4 are in the up–down transmission mode so that the training input and output data are fed into the FALCON from both sides.

First, the centers (or means) and the widths (or variances) of the membership functions are determined by self-organized learning techniques analogous to statistical clustering technique. This serves to allocate network resources efficiently by placing the domains of membership functions covering only those regions of the input/output space where data are present. Kohonen's feature-maps algorithm is adopted here to find the center m_i of the ith membership function of x, where x represents any one of the input and output linguistic variables, $x_1, \ldots, x_n, y_1, \ldots, y_m$:

$$\|x(t) - m_{\text{closest}}(t)\| = \min_{1 \le i \le k} \{\|x(t) - m_i(t)\|\}, \tag{38}$$

$$m_{\text{closest}}(t + 1) = m_{\text{closest}}(t + \alpha(t)[x(t) - m_{\text{closest}}(t)], \tag{39}$$

$$m_i(t + 1) = m_i(t), \quad \text{for } m_i \ne m_{\text{closest}} \tag{40}$$

where $\alpha(t)$ is a monotonically decreasing scalar learning rate and $k = |T(x)|$. This adaptive formulation runs independently for each input and output linguistic variable. The determination of which of the m_i is m_{closest} can be accomplished in constant time via a winner-take-all circuit. Once the centers of the membership functions are found, their widths can be determined by the *N-nearest-neighbors* heuristic by minimizing the following objective function with respect to the widths σ_i:

$$E = \frac{1}{2} \sum_{i=1}^{N} \left[\sum_{j \in N_{\text{nearest}}} \left(\frac{m_i - m_j}{\sigma_i} \right)^2 - r \right]^2 \tag{41}$$

where r is an overlap parameter. Since the second learning phase will optimally adjust the centers and the widths of the membership functions, the widths can be simply determined by the first-nearest-neighbor heuristic at this stage as

$$\sigma_i = \frac{|m_i - m_{\text{closest}}|}{r} \tag{42}$$

where the user initially sets r to an appropriate value.

After the parameters of the membership functions have been found, the signals from both external sides can reach the output points of term nodes in layers 2 and 4 (see Figure 6). Furthermore, the outputs of term nodes in layer 2 can be transmitted to rule nodes through the initial connection of layer-3 links. So we can get the firing strength of each rule node. Based on these rule-firing strengths (denoted as $o_i^{(3)}(t)$) and the outputs of term nodes in layer 4 (denoted as $o_j^{(4)}(t)$), we want to determine the correct consequent links (layer 4 links) of each rule node to find the existing fuzzy logic rule by *competitive learning* algorithms. As stated before, the links in layer 4 are initially fully connected. We denote the weight of the link between the ith rule node and the jth output term node as w_{ji}. The following competitive learning law is used to update these weights for each training data set,

$$\dot{w}_{ji}(t) = o_j^{(4)}(-w_{ji} + o_i^{(3)}) \tag{43}$$

where $o_j^{(4)}$ serves as a win–loss index of the jth term node in layer 4. The theme of this law is that *learn if win*. In the extreme case, if $o_j^{(4)}$ is a 0–1 threshold function, then this law says *learn only if win*.

After competitive learning involving the whole training data set, the link weights in layer 4 represent the strength of the existence of the corresponding rule consequent. Among the links which connect a rule node and the term nodes of an output linguistic node, at most one link with maximum weight is chosen and the others are deleted. Hence, only one term in an output linguistic variable's term set can become one of the consequents of a fuzzy logic rule. If all the link weights between a rule node and the term nodes of an output linguistic node are very small, then all the corresponding links are deleted, meaning that this rule node has little or no relation to this output linguistic variable. If all the links between a rule node and the layer-4 nodes are deleted, then this rule node can be eliminated since it does not affect the outputs.

After the consequents of rule nodes are determined, a rule combination is performed to reduce the number of rules. The criteria for combining a set of rule nodes into a single rule node are: (1) They have exactly the same consequents, (2) some preconditions are common to all the rule nodes in the set, and (3) the union of other preconditions of these rule nodes comprises the whole term set of some input linguistic variables. If a set of nodes meets these criteria, a new rule node with only the common preconditions can replace this set of rule nodes. An example is illustrated in Figure 7.

3.1.2 Supervised Learning Phase

After the fuzzy logic rules have been determined, the whole network structure is established. The network then enters the second learning phase to adjust the parameters of the input and output membership functions optimally. The problem for the supervised learning can be stated as: Given the training input data $x_i(t)$, $i = 1, 2, \ldots, n$, the corresponding desired output value $y_i^d(t)$, $i = 1, 2, \ldots, m$, the fuzzy partitions $|T(x_i)|$ and $|T(y_i)|$, and the fuzzy logic rules, adjust the parameters of the input and output membership functions optimally. The fuzzy logic rules of this network were determined in the first-phase learn-

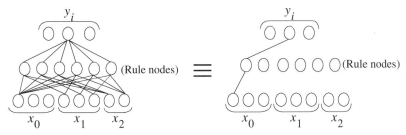

Figure 7 Example of combination of rule nodes in the hybrid learning algorithm for the FALCON.

ing or, in some application domains, they can be given by experts. In the second-phase learning the network works in the feedforward manner; that is, the nodes and the links in layers 4 and 5 are in the down–up transmission mode. The backpropagation algorithm is used for this supervised learning. Considering a single-output case for clarity, the goal is to minimize the error function

$$E = \frac{1}{2}(y^d(t) - y(t))^2 \tag{44}$$

where $y^d(t)$ is the desired output and $y(t)$ is the current output. For each training data set, starting at the input nodes, a forward pass is used to compute the activity levels of all the nodes in the network to obtain the current output $y(t)$. Then, starting at the output nodes, a backward pass is used to compute $\partial E/\partial w$ for all the hidden nodes. Assuming that w is the adjustable parameter in a node (e.g., m_{ij} and σ_{ij} in our case), the general learning rule used is

$$\Delta w \propto -\frac{\partial E}{\partial w}, \tag{45}$$

$$w(t + 1) = w(t) + \eta\left(-\frac{\partial E}{\partial w}\right) \tag{46}$$

where η is the learning rate and

$$\frac{\partial E}{\partial w} = \frac{\partial E}{\partial(\text{activation function})} \frac{\partial(\text{activation function})}{\partial w} = \frac{\partial E}{\partial a}\frac{\partial a}{\partial w} \tag{47}$$

To illustrate the learning rule for each parameter, we shall show the computations of $\partial E/\partial w$, layer by layer, starting at the output nodes, and we will use the bell-shaped membership functions with centers m_i and widths σ_i (single-output case) as the adjustable parameters for these computations.

1. **Layer 5.** Using Equations (47) and (37), the adaptive rule of the center m_i is derived as:

$$\frac{\partial E}{\partial m_i} = \frac{\partial E}{\partial a^{(5)}} \frac{\partial a^{(5)}}{\partial m_i} = -[y^d(t) - y(t)]\frac{\sigma_i u_i^{(5)}}{\sum_i \sigma_i u_i^{(5)}} \tag{48}$$

where $a^{(5)}$ is the network output $y(t)$. Hence the center parameter is updated by

$$m_i(t + 1) = m_i(t) + \eta[y^d(t) - y(t)]\frac{\sigma_i u_i^{(5)}}{\sum_i \sigma_i u_i^{(5)}} \tag{49}$$

Similarly, using Equations (47) and (37), the adaptive rule of the width σ_i is derived as:

$$\frac{\partial E}{\partial \sigma_i} = \frac{\partial E}{\partial a^{(5)}} \frac{\partial a^{(5)}}{\partial \sigma_i} = -[y^d(t) - y(t)] \frac{m_i u_i^{(5)}(\Sigma_i \ \sigma_i u_i^{(5)}) - (\Sigma_i \ m_i \sigma_i u_i^{(5)}) u_i^{(5)}}{(\Sigma_i \ \sigma_i u_i^{(5)})^2} \quad (50)$$

Hence the width parameter is updated by

$$\sigma_i(t + 1) = \sigma_i(t) + \eta[y^d(t) - y(t)] \frac{m_i u_i^{(5)}(\Sigma_i \ \sigma_i u_i^{(5)}) - (\Sigma_i \ m_i \sigma_i u_i^{(5)}) u_i^{(5)}}{(\Sigma_i \ \sigma_i u_i^{(5)})^2} \quad (51)$$

The error to be propagated to the preceding layer is

$$\delta^{(5)} = -\frac{\partial E}{\partial a^{(5)}} = -\frac{\partial E}{\partial y} = y^d(t) - y(t) \quad (52)$$

2. **Layer 4.** In the down–up transmission mode there is no parameter to be adjusted in this layer. Only the error signals ($\delta_i^{(4)}$) need to be computed and propagated. The error signal $\delta_i^{(4)}$ is derived as

$$\delta_i^{(4)} = -\frac{\partial E}{\partial a_i^{(4)}} = -\frac{\partial E}{\partial u_i^{(5)}} = -\frac{\partial E}{\partial a^{(5)}} \frac{\partial a^{(5)}}{\partial u_i^{(5)}} \quad (53)$$

where (from Equation (37))

$$\frac{\partial a^{(5)}}{\partial u_i^{(5)}} = \frac{m_i \sigma_i(\Sigma_i \ \sigma_i u_i^{(5)}) - (\Sigma_i \ m_i \ \sigma_i u_i^{(5)})\sigma_i}{(\Sigma_i \ \sigma_i u_i^{(5)})^2} \quad (54)$$

and from Equation (52),

$$-\frac{\partial E}{\partial a^{(5)}} = \delta^{(5)} = y^d(t) - y(t) \quad (55)$$

Hence the error signal is

$$\delta_i^{(4)}(t) = [y^d(t) - y(t)] \frac{m_i \sigma_i(\Sigma_i \ \sigma_i u_i^{(5)}) - (\Sigma_i \ m_i \sigma_i u_i^{(5)})\sigma_i}{(\Sigma_i \ \sigma_i u_i^{(5)})^2} \quad (56)$$

In the multiple-output case the computations in layers 4 and 5 are exactly the same as the above and proceed independently for each output linguistic variable.

3. **Layer 3.** As in layer 4, only the error signals need to be computed. According to Equation (35), this error signal can be derived as:

$$\delta_i^{(3)} = -\frac{\partial E}{\partial a_i^{(3)}} = -\frac{\partial E}{\partial u_i^{(4)}} = -\frac{\partial E}{\partial a_i^{(4)}} = -\frac{\partial E}{\partial a_i^{(4)}} \frac{\partial a_i^{(4)}}{\partial u_i^{(4)}} = -\frac{\partial E}{\partial a_i^{(4)}} = \delta_i^{(4)} \quad (57)$$

Hence the error signal is $\delta_i^{(3)} = \delta_i^{(4)}$. If there are multiple outputs, then the error signal becomes $\delta_i^{(3)} = \Sigma_k \ \delta_k^{(4)}$, where the summation is performed over the consequents of a rule node; that is, the error of a rule node is the summation of the errors of its consequents.

4. **Layer 2.** Using Equations (47) and (33), the adaptive rule of m_{ij} (multiinput case) is derived as

$$\frac{\partial E}{\partial m_{ij}} = \frac{\partial E}{\partial a_i^{(2)}} \frac{\partial a_i^{(2)}}{\partial m_{ij}} = \frac{\partial E}{\partial a_i^{(2)}} \ e^{f_i} \ \frac{2(u_i^{(2)} - m_{ij})}{\sigma_{ij}^2} \quad (58)$$

where

$$\frac{\partial E}{\partial a_i^{(2)}} = \frac{\partial E}{\partial u_i^{(3)}} = \frac{\partial E}{\partial a_i^{(3)}} \frac{\partial a_i^{(3)}}{\partial u_i^{(3)}} \quad (59)$$

where (from Equation (57))

$$\frac{\partial E}{\partial a_i^{(3)}} = -\delta_i^{(3)} \tag{60}$$

and from Equation (34),

$$\frac{\partial a_i^{(3)}}{\partial u_i^{(3)}} = \begin{cases} 1, & \text{if } u_i^{(3)} = \min(\text{inputs of rule node } i), \\ 0, & \text{otherwise} \end{cases} \tag{61}$$

Hence

$$\frac{\partial E}{\partial a_i^{(2)}} \equiv -\delta_i^{(2)} = \sum_k q_k \tag{62}$$

where the summation is performed over the rule nodes that $a_i^{(2)}$ feeds into and

$$q_k = \begin{cases} -\delta_k^{(3)}, & \text{if } a_i^{(2)} \text{ is minimum in } k\text{th rule node's inputs}, \\ 0, & \text{otherwise} \end{cases} \tag{63}$$

So the adaptive rule of m_{ij} is

$$m_{ij}(t+1) = m_{ij}(t) + \eta \delta_i^{(2)} e^{f_i} \frac{2(u_i^{(2)} - m_{ij})}{\sigma_{ij}^2} \tag{64}$$

Similarly, using Equations (47), (33), and (59)–(63), the adaptive rule of σ_{ij} is derived as:

$$\frac{\partial E}{\partial \sigma_{ij}} = \frac{\partial E}{\partial a_i^{(2)}} \frac{\partial a_i^{(2)}}{\partial \sigma_{ij}} = \frac{\partial E}{\partial a_i^{(2)}} e^{f_i} \frac{2(u_i^{(2)} - m_{ij})^2}{\sigma_{ij}^3} \tag{65}$$

Hence the adaptive rule of σ_{ij} becomes

$$\sigma_{ij}(t+1) = \sigma_{ij}(t) + \eta \delta_i^{(2)} e^{f_i} \frac{2(u_i^{(2)} - m_{ij})^2}{\sigma_{ij}^3} \tag{66}$$

The convergence speed of the above algorithm is found superior to that of the normal backpropagation scheme because the self-organized learning process in phase one has done much of the learning work in advance. Finally, it should be noted that this back-propagation algorithm can be easily extended to train the membership function that is implemented by a subneural net instead of a single-term node in layer 2, since, from the above analysis, the error signal can be propagated to the output node of the subneural net. Then, by using a similar backpropagation rule in this subneural net, the parameters in this subneural net can be adjusted.

The FALCON has been used to simulate the control of a fuzzy car discussed by Sugeno and Nishida (1985). Other applications, such as modeling and controlling underwater robots, developing an empirical model for predicting earthquake-induced liquefaction hazards, and plasma etch process modeling, are currently being pursued by the author.

3.2   Fuzzy Basis Function Network (FBFN)

Another NFC is the *fuzzy basis function network* (FBFN) proposed by Wang and Mendel (1992). In the FBFN a fuzzy system is represented as a series expansion of fuzzy basis functions (FBFs), which are algebraic superpositions of membership functions. Each FBF corresponds to one fuzzy logic rule. Based on the FBF representation, an orthogonal least-squares (OLS) learning algorithm can be utilized to determine the significant FBFs (i.e., fuzzy logic rules (structure learning) and associated parameters (parameter learning)) from input/output training pairs. Since a linguistic fuzzy IF–THEN rule from human experts can be directly interpreted into an FBF, the FBFN provides a framework for combining both numerical and linguistic information in a uniform manner.

Consider a multiinput–single-output fuzzy control system: $X \subset \mathcal{R}^n \to Y \subset \mathcal{R}$. Suppose that we have M fuzzy logic rules in the following form:

$$R^j : \text{IF } x_1 \text{ is } A_1^j \text{ AND } x_2 \text{ is } A_2^j \text{ AND} \cdots \text{AND } x_n \text{ is } A_n^j, \text{ THEN } y \text{ is } B^j \qquad (67)$$

where x_i, $i = 1, 2, \ldots, n$, are the input variables of the fuzzy system, y is the output variable of the fuzzy system, A_i^j and B^j are linguistic terms characterized by fuzzy membership functions $\mu_{A_i^j}(x_i)$ and $\mu_{B^j}(y)$, respectively, and $j = 1, 2, \ldots, M$. We also assume *singleton fuzzifier, product inference, centroid defuzzifier,* and *Gaussian membership functions* are used. Then for a given input (crisp) vector $\mathbf{x} = (x_1, x_2, \ldots, x_n)^T \in X$, we have the following defuzzified inferenced output:

$$y = f(\mathbf{x}) = \frac{\sum_{j=1}^{M} \bar{y}^j (\prod_{i=1}^{n} \mu_{A_i^j}(x_i))}{\sum_{j=1}^{M} (\prod_{i=1}^{n} \mu_{A_i^j}(x_i))} \qquad (68)$$

where $f : X \subset \mathcal{R}^n \to \mathcal{R}$, $\bar{y}^j$ is the point in the output space Y at which $\mu_{B^j}(\bar{y}^j)$ achieves its maximum value, and $\mu_{A_i^j}(x_i)$ is the *Gaussian membership function* defined by

$$\mu_{A_i^j}(x_i) = a_i^j \exp\left[-\frac{1}{2} \left(\frac{x_i - m_i^j}{\sigma_i^j} \right)^2 \right] \qquad (69)$$

where a_i^j, m_i^j, and σ_i^j are real-valued parameters with $0 < a_i^j \leq 1$.

According to Equation (68), we can define *fuzzy basis functions* (FBFs) as follows

$$p_j(\mathbf{x}) = \frac{\prod_{i=1}^{n} \mu_{A_i^j}(x_i)}{\sum_{j=1}^{M} \prod_{i=1}^{n} \mu_{A_i^j}(x_i)}, \qquad j = 1, 2, \ldots, M \qquad (70)$$

where $\mu_{A_i^j}(x_i)$ are the Gaussian membership functions defined in Equation (69). Then the fuzzy control system described by Equation (68) is equivalent to an *FBF expansion* or a *FBF network* (FBFN):

$$f(\mathbf{x}) = \sum_{j=1}^{M} p_j(\mathbf{x})\theta^j \qquad (71)$$

where $\theta^j = \bar{y}^j \in \mathcal{R}$ are constants. In other words, a fuzzy control system described by Equation (68) can be viewed as a linear combination of the FBFs.

There are two points of view for training the FBFNs. First, if we view all the parameters a_i^j, m_i^j, and σ_i^j in $p_j(\mathbf{x})$ as free design parameters, then the FBFN is nonlinear in the parameters. In this case we need nonlinear optimization techniques such as the backpropagation algorithm to train the FBFN. This is similar to what is used for the FALCON in the last section. On the other hand, if all the parameters in $p_j(\mathbf{x})$ are fixed at the very beginning of the FBFN design procedure, then the only free design parameters are θ^j. In this case, $f(\mathbf{x})$ in Equation (71) is linear in the parameter and some very efficient linear parameter estimation methods can be used to train or "design" the FBFN. We shall focus on the utilization of the Gram–Schmidt orthogonal least squares (OLS) algorithm (Chen, Cowan, and Grant, 1991) to design the FBFN. This OLS learning strategy can determine the proper number of significant FBFs automatically. It is a kind of hybrid structure–parameter learning method; that is, it can find the number of significant FBFs and its associated θ^j for the FBFN.

Suppose we are given N input/output training pairs: $(\mathbf{x}(t), d(t))$, $t = 1, 2, \ldots, N$. The task is to design a FBFN $f(\mathbf{x})$ such that

$$d(t) = f(\mathbf{x}(t)) + e(t) = \sum_{j=1}^{M} p_j(\mathbf{x}(t))\theta^j + e(t) \qquad (72)$$

where the approximation error function, $e(t)$, between $f(\mathbf{x}(t))$ and $d(t)$ is minimized. In order to present the OLS algorithm, Equation (72) is arranged from $t = 1$ to N in a matrix form:

$$\mathbf{d} = \mathbf{P}\theta + \mathbf{e} \qquad (73)$$

where $\mathbf{d} = [d(1), \ldots, e(N)]^T$, $\mathbf{P} = [\mathbf{p}_1, \ldots, \mathbf{p}_M]$ with $\mathbf{p}_i = [p_i(1), \ldots, p_i(N)]^T$, $\theta = [\theta_1, \ldots, \theta_M]^T$, and $\mathbf{e} = [e(1), \ldots, e(N)]^T$. To determine θ in Equation (73), the OLS algorithm is utilized. The purpose of the original Gram–Schmidt OLS algorithm is to perform an orthogonal decomposition for $\mathbf{P}$ such that $\mathbf{P} = \mathbf{WA}$, where $\mathbf{W}$ is an orthogonal matrix and $\mathbf{A}$ is an upper-triangular matrix with unity diagonal elements. Substituting $\mathbf{P} = \mathbf{WA}$ into Equation (73), we have

$$\mathbf{d} = \mathbf{P\theta} + \mathbf{e} = \mathbf{WA\theta} + \mathbf{e} = \mathbf{Wg} + \mathbf{e} \tag{74}$$

where

$$\mathbf{g} = \mathbf{A\theta} \tag{75}$$

Using the OLS algorithm, we can select some dominant columns from $\mathbf{P}$ (i.e., some significant FBFs) and find the associated θ^j values to form the final FBFN. In other words, the OLS algorithm transforms the set of $\mathbf{p}_i$ into a set of orthogonal basis vectors and uses only the FBFs that correspond to the significant orthogonal basis vectors to form the final FBFN.

Since matrix $\mathbf{A}$ in Equation (75) is known from the OLS algorithm, if we can find an estimate of $\mathbf{g}$, $\hat{\mathbf{g}}$, then $\mathbf{A\theta} = \hat{\mathbf{g}}$ can be solved for $\hat{\mathbf{\theta}} = \mathbf{A}^{-1}\hat{\mathbf{g}}$, which is a solution for $\mathbf{\theta}$. To find $\hat{\mathbf{g}}$, from Equation (73) we have

$$\mathbf{e} = \mathbf{d} - \mathbf{P\theta} = \mathbf{d} - \mathbf{Wg} \tag{76}$$

Taking the derivative of $\mathbf{e}^T\mathbf{e}$ with respect to $\mathbf{g}$ to determine $\hat{\mathbf{g}}$, we have

$$\hat{\mathbf{g}} = [\mathbf{W}^T\mathbf{W}]^{-1}\mathbf{W}^T\,\mathbf{d} \tag{77}$$

or each element of $\hat{\mathbf{g}}$ can be expressed as

$$\hat{g}_i = \frac{\mathbf{w}_i^T\,\mathbf{d}}{\mathbf{w}_i^T\mathbf{w}_i}, \qquad i = 1, 2, \ldots, M \tag{78}$$

Once $\hat{\mathbf{g}}$ is found, we solve Equation (75) for $\mathbf{\theta}$, $\hat{\mathbf{\theta}} = \mathbf{A}^{-1}\hat{\mathbf{g}}$. Then the final FBFN using all the M FBFs is

$$f(\mathbf{x}) = \sum_{j=1}^{M} p_j(\mathbf{x})\hat{\theta}^j \tag{79}$$

Now if M is huge, then it is advantageous to use less number of FBFs, say $M_s \ll M$. That is, we want to select the M_s *significant* FBFs out of the M FBFs such that the approximation error $\mathbf{e} = \mathbf{d} - \mathbf{P\theta}$ is minimized. This can be accomplished by using the OLS algorithm and selecting each significant FBF from 1 to M_s such that the approximation error is *maximally* reduced in each selection. To understand how this can be accomplished, let us take a look at the error function in Equation (76) again; we have

$$\mathbf{e}^T\mathbf{e} = \mathbf{d}^T\mathbf{d} - 2(\mathbf{d} - \mathbf{Wg})^T\mathbf{Wg} - \mathbf{g}^T\mathbf{W}^T\mathbf{Wg} \tag{80}$$

Since $(\mathbf{d} - \mathbf{Wg})^T\mathbf{Wg} = 0$, we have

$$\mathbf{e}^T\mathbf{e} = \mathbf{d}^T\mathbf{d} - \mathbf{g}^T\mathbf{W}^T\mathbf{Wg}$$

$$= \mathbf{d}^T\mathbf{d} - \sum_{i=1}^{M} g_i^2\mathbf{w}_i^T\mathbf{w}_i \tag{81}$$

or

$$\frac{\mathbf{e}^T\mathbf{e}}{\mathbf{d}^T\mathbf{d}} = 1 - \sum_{i=1}^{M} \frac{g_i^2\mathbf{w}_i^T\mathbf{w}_i}{\mathbf{d}^T\mathbf{d}} \tag{82}$$

Thus, from the OLS algorithm, we can select the orthogonal basis vector $\mathbf{w}_i$, $i = 1, 2, \ldots, M_s$, such that the term $g_i^2\mathbf{w}_i^T\mathbf{w}_i/\mathbf{d}^T\mathbf{d}$ is maximized to maximally reduce the error function in Equation (82). This error reduction term is the same as $[err]_k^{(i)}$ in Equations (86) and (92), which represents the error reduction ratio caused by $\mathbf{w}_k^{(i)}$ as discussed in the following OLS algorithm.

3.2.1 Algorithm OLS: Orthogonal Least-Squares (OLS) Method for FBFN

This algorithm selects M_s ($M_s \ll N$) significant FBFs from the N input/output training pairs $(\mathbf{x}(t), d(t))$, $t = 1, 2, \ldots, N$, where $\mathbf{x}(t) = (x_1(t), \ldots, x_n(t))^T$.

- *Input:* N input/output training pairs $(\mathbf{x}(t), d(t))$, $t = 1, 2, \ldots, N$, and an integer $M_s \ll N$, which is the desired number of significant FBFs to be selected.
- *Output:* M_s significant FBFs will be selected from the N FBFs formed from the N input/output training pairs $(\mathbf{x}(t), d(t))$, $t = 1, 2, \ldots, N$.
- *Step 0* (initial FBF determination): Form N initial FBFs, $p_j(\mathbf{x})$'s, in the form of Equation (70) from the training data (the M in Equation (70) equals N), with the parameters determined as follows:

$$a_i^j = 1, \qquad m_i^j = x_i(j),$$

$$\sigma_i^j = \left[\max_{j=1\ldots N} x_i(j) - \min_{j=1\ldots N} x_i(j) \right] \Big/ M_s \tag{83}$$

  where $i = 1, 2, \ldots, n$, and $j = 1, 2, \ldots, N$.
- *Step 1* (initial step): For $1 \le i \le N$, compute

$$\mathbf{w}_1^{(i)} = \mathbf{p}_i \tag{84}$$

$$g_1^{(i)} = \frac{(\mathbf{w}_1^{(i)})^T \mathbf{d}}{(\mathbf{w}_1^{(i)})^T \mathbf{w}_1^{(i)}} = \frac{\mathbf{p}_i^T \mathbf{d}}{\mathbf{p}_i^T \mathbf{p}_i}, \tag{85}$$

$$[err]_1^{(i)} = \frac{(g_1^{(i)})^2 (\mathbf{w}_1^{(i)})^T \mathbf{w}_1^{(i)}}{\mathbf{d}^T \mathbf{d}} = \frac{(g_1^{(i)})^2 \mathbf{p}_i^T \mathbf{p}_i}{\mathbf{d}^T \mathbf{d}} \tag{86}$$

  where $\mathbf{p}_i = [p_i(\mathbf{x}(1)), \ldots, p_i(\mathbf{x}(N))]^T$, and $p_i(\mathbf{x}(t))$ are obtained in Step 0. Then find $i_1 \in \{1, 2, \ldots, N\}$ such that

$$[err]_1^{(i_1)} = \max_{i=1,\ldots,N} [err]_1^{(i)} \tag{87}$$

  and set

$$\mathbf{w}_1 = \mathbf{w}_1^{(i_1)} = \mathbf{p}_{i_1}, \quad g_1 = g_1^{(i_1)} \tag{88}$$

  The vector $\mathbf{w}_1$ and value g_1 in Equation (88) are, respectively, the selected first column of $\mathbf{W}$ and the first element of $\mathbf{g}$ in Equation (74).
- *Step k* ($k = 2, 3, \ldots, M_s$): For $1 \le i \le N$, $i \ne i_1, \ldots, i \ne i_{k-1}$, compute the following equations:

$$\alpha_{jk}^{(i)} = \frac{\mathbf{w}_j^T \mathbf{p}_i}{\mathbf{w}_j^T \mathbf{w}_j}, \quad 1 \le j < k \tag{89}$$

$$\mathbf{w}_k^i = \mathbf{p}_i - \sum_{j=1}^{k-1} \alpha_{jk}^{(i)} \mathbf{w}_j \tag{90}$$

$$g_k^{(i)} = \frac{(\mathbf{w}_k^{(i)})^T d}{(\mathbf{w}_k^{(i)})^T \mathbf{w}_k^{(i)}} \tag{91}$$

$$[err]_k^{(i)} = \frac{(g_k^{(i)})^2 (\mathbf{w}_k^{(i)})^T \mathbf{w}_k^{(i)}}{\mathbf{d}^T \mathbf{d}} \tag{92}$$

Then find $i_k \in \{1, 2, \ldots, N\}$ such that

$$[err]_k^{(i_k)} = \max_{\substack{i=1,\ldots,N \\ i \neq i_1, \ldots, i \neq i_{k-1}}} ([err]_k^{(i)}) \tag{93}$$

and select

$$\mathbf{w}_k = \mathbf{w}_k^{(i_k)}, \quad g_k = g_k^{(i_k)} \tag{94}$$

as the kth column of $\mathbf{W}$ and the kth element of $\mathbf{g}$ in Equation (74).

- *Step* ($M_s + 1$) (final step): Solve the following triangular system for the coefficient vector for $\theta^{(M_s)}$:

$$\mathbf{A}^{(M_s)} \theta^{(M_s)} = \mathbf{g}^{(M_s)} \tag{95}$$

where

$$\mathbf{A}^{(M_s)} = \begin{bmatrix} 1 & \alpha_{12}^{(i_2)} & \alpha_{13}^{(i_3)} & \cdots & \alpha_{1M_s}^{(i_{M_s})} \\ 0 & 1 & \alpha_{23}^{(i_3)} & \cdots & \alpha_{2M_s}^{(i_{M_s})} \\ \cdots & \cdots & \cdots & \cdots & \cdots \\ 0 & 0 & \cdots & 1 & \alpha_{M_s-1,M_s}^{(i_{M_s})} \\ 0 & 0 & 0 & \cdots & 1 \end{bmatrix} \tag{96}$$

$$\mathbf{g}^{(M_s)} = [g_1, \ldots, g_{M_s}]^T, \quad \theta^{(M_s)} = [\theta_1^{(M_s)}, \ldots, \theta_{M_s}^{(M_s)}]^T \tag{97}$$

Then the final FBFN is

$$f(\mathbf{x}) = \sum_{j=1}^{M_s} p_{i_j}(\mathbf{x}) \theta_j^{(M_s)} \tag{98}$$

where $p_{i_j}(\mathbf{x})$ make up the subset of the FBFs determined in Step 0 with i_j determined in the above steps.

The $[err]_k^{(i)}$ in Equations (86) and (92) represents the error-reduction ratio caused by $\mathbf{w}_k^{(i)}$. Hence the OLS algorithm selects significant FBFs based on their error-reduction ratio; that is, the FBFs with the largest error-reduction ratios are retained in the final FBFN. According to the error-reduction ratios, we can set a threshold value to decide how many FBFs, or equivalently, how many fuzzy logic rules, M_s, are to be used. The guideline is that only those FBFs whose error-reduction ratios are greater than the threshold value are chosen. Wang and Mendel (1992) illustrated the use of FBFNs in the control of a nonlinear ball and beam system.

3.3   Neural Fuzzy Controllers with TSK Fuzzy Rules

In this section we consider the neural fuzzy control systems with TSK fuzzy rules whose consequents are linear combination of their preconditions (see Equation (22)). The TSK fuzzy rules are in the following forms:

$$R^j: \quad \text{IF } x_i \text{ is } A_i^j \text{ AND } x_2 \text{ is } A_2^j \text{ AND } \cdots \text{ AND } x_n \text{ is } A_n^j,$$

$$\text{THEN } y = f_j = a_0^j + a_1^j x_1 + a_2^j x_2 + \cdots + a_n^j x_n \tag{99}$$

where

$\quad x_i$ = an input variable
$\quad y$ = the output variable
$\quad A_i^j$ = linguistic terms of the precondition part with membership functions $\mu_{A_i^j}(x_i)$
$a_i^j \in \Re$ = coefficients of linear equations $f_j(x_1, x_2, \ldots, x_n)$
$\quad j = 1, 2, \ldots, M, i = 1, 2, \ldots, n$

To simplify our discussion, we shall focus on a specific NFC of this type called an *adaptive-network-based fuzzy inference system* (ANFIS) (Jang, 1992).

For simplicity, assume that the fuzzy control system under consideration has two inputs x_1 and x_2, one output y, and the rule base contains two TSK fuzzy rules as follows:

$$R^1 : \text{IF } x_1 \text{ is } A_1^1 \text{ AND } x_2 \text{ is } A_2^1, \text{ THEN } y = f_1 = a_0^1 + a_1^1 x_1 + a_2^1 x_2 \qquad (100)$$

$$R^2 : \text{IF } x_1 \text{ is } A_1^2 \text{ AND } x_2 \text{ is } A_2^2, \text{ THEN } y = f_2 = a_0^2 + a_1^2 x_1 + a_2^2 x_2 \qquad (101)$$

For given input values x_1 and x_2, the inferred output y^* is calculated by

$$y^* = \frac{\mu_1 f_1 + \mu_2 f_2}{\mu_1 + \mu_2} \qquad (102)$$

where μ_j's are firing strengths of R^j, $j = 1, 2$, and are given by

$$\mu_j = \mu_{A_1^j}(x_1) \cdot \mu_{A_2^j}(x_2), \qquad j = 1, 2 \qquad (103)$$

if product inference is used. The corresponding ANFIS architecture is shown in Figure 8, where node functions in the same layers are of the same type as described below.

1. **Layer 1.** Every node in this layer is an input node that just passes external signals to the next layer.
2. **Layer 2.** Every node in this layer acts as a membership function, $\mu_{A_j}(x_i)$, and its output specifies the degree to which the given x_i satisfies the quantifier A_i^j. Usually we choose $\mu_{A_j}(x_i)$ to be bell-shaped with maximum equal to 1 and minimum equal to 0, such as

$$\mu_{A_j}(x_i) = \frac{1}{1 + \left[\left(\dfrac{x_i - m_i^j}{\sigma_i^j}\right)^2\right]^{b_i^j}} \qquad (104)$$

or

$$\mu_{A_j^i}(x_i) = \exp\left\{-\left[\left(\frac{x_i - m_i^j}{\sigma_i^j}\right)^2\right]^{b_i^j}\right\} \qquad (105)$$

where $\{m_i^j, \sigma_i^j, b_i^j\}$ is the parameter set to be tuned. In fact, any continuous and piecewise-differentiable functions, such as commonly used trapezoidal or triangular-shaped membership functions, are also qualified candidates for node functions in this layer. Parameters in this layer are referred to as *precondition parameters*.
3. **Layer 3.** Every node in this layer is labeled Π, which multiplies the incoming signals, $\mu_j = \mu_{A_1^j}(x_i) \cdot \mu_{A_2^j}(x_2)$, and sends the product out. Each node output represents the firing strength of a rule. In fact, other *t*-norm operators can be used as the node function for the generalized AND function.
4. **Layer 4.** Every node in this layer is labeled N, which calculates the *normalized firing strength* of a rule. That is, the *j*th node calculates the ratio of the *j*th rule's firing strength to the sum of all rules' firing strengths, $\overline{\mu}_j = \mu_j/(\mu_{A_1^j}(x_1) + \mu_{A_2^j}(x_2))$.
5. **Layer 5.** Every node j in this layer calculates the weighted consequent value, $\overline{\mu}_j(a_0^j + a_1^j x_1 + a_2^j x_2)$, where $\overline{\mu}_j$ is the output of layer 4 and $\{a_0^j, a_1^j, a_2^j\}$ is the parameter set to be tuned. Parameters in this layer are referred to as *consequent parameters*.
6. **Layer 6.** The only node in this layer is labeled Σ, which sums all incoming signals to obtain the final inferred result of the whole system.

Thus, we have constructed an ANFIS that is functionally equivalent to a fuzzy control system with TSK rules. This ANFIS architecture can then update its parameters according to the backpropagation algorithm. We can derive the update rule for each parameter in the ANFIS as we did for the FALCON in Section 3.1. Moreover, the Kalman filter algorithm introduced in Section 3.2 can also be used to find the consequent parameters of an ANFIS. This is achieved by arranging all the consequent parameters in one vector $(a_0^1, a_1^1, a_2^1, a_0^2, a_1^2, a_2^2)^T$ and using the Kalman filter algorithm to solve the following overconstrained simultaneous linear equations:

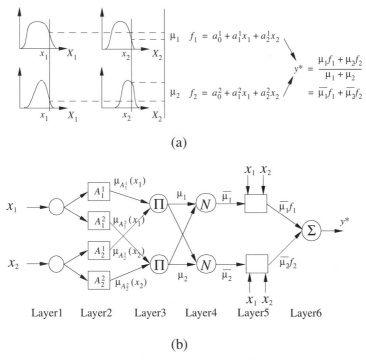

(a)

(b)

Figure 8 The structure of an ANFIS. (*a*) A fuzzy inference/control system. (*b*) Equivalent ANFIS.

$$
\begin{bmatrix}
\overline{\mu}_1^{(1)} & \overline{\mu}_1^{(1)}x_1^{(1)} & \overline{\mu}_1^{(1)}x_2^{(1)} & \overline{\mu}_2^{(1)} & \overline{\mu}_2^{(1)}x_1^{(1)} & \overline{\mu}_2^{(1)}x_2^{(1)} \\
\overline{\mu}_1^{(2)} & \overline{\mu}_1^{(2)}x_1^{(2)} & \overline{\mu}_1^{(2)}x_2^{(2)} & \overline{\mu}_2^{(2)} & \overline{\mu}_2^{(2)}x_1^{(2)} & \overline{\mu}_2^{(2)}x_2^{(2)} \\
\vdots & \vdots & \vdots & \vdots & \vdots & \vdots \\
\overline{\mu}_1^{(p)} & \overline{\mu}_1^{(p)}x_1^{(p)} & \overline{\mu}_1^{(p)}x_2^{(p)} & \overline{\mu}_2^{(p)} & \overline{\mu}_2^{(p)}x_1^{(p)} & \overline{\mu}_2^{(p)}x_2^{(p)}
\end{bmatrix}
\begin{bmatrix}
a_0^1 \\ a_1^1 \\ a_2^1 \\ a_0^2 \\ a_1^2 \\ a_2^2
\end{bmatrix}
=
\begin{bmatrix}
d^{(1)} \\ d^{(2)} \\ \vdots \\ d^{(p)}
\end{bmatrix}
\tag{106}
$$

where $[(x_1^{(k)}, x_2^{(k)}), d^{(k)}]$ are the kth training pair, $k = 1, 2, \ldots, p$, and $\overline{\mu}_1^{(k)}$ and $\overline{\mu}_2^{(k)}$ are the third layer outputs associated with the input $(x_1^{(k)}, x_2^{(k)})$.

4 CONCLUDING REMARKS

The synergism of combining fuzzy logic and neural networks into a learning system has been briefly discussed and explored. We first started with basic structure and operations of a traditional fuzzy logic control system, then discussed the necessary background materials in fuzzy sets and relations, linguistic variables, and approximate reasoning.

At present there is a lack of systematic procedure for the design of fuzzy logic systems. The most straightforward approach is to define membership functions and fuzzy logic rules subjectively by studying a human-operated system or an existing controller and then testing the design for the proper output. If the design fails the test, then the membership functions and/or fuzzy logic rules are adjusted. Again, there is no procedure or method to guide the adjustment. Thus, providing learning ability to fuzzy logic control systems via neural learning technique is a significant contribution.

Various neural learning techniques are explored to determine the structure and fine tune the parameters of NFCS. As a result, various learning algorithms are developed to construct different NFCS from input/output training data. The hybrid structure–parameter learning algorithms for the FALCON are found most useful when the input–output training data are available for off-line training. Details of other FALCON systems can be found in Lin and Lee (1996). The *fuzzy Hebbian learning law* proposed by Wang and Vachtsevanos (1992) can also be used to find fuzzy rules in the structure learning phase of the FALCON. For the FBFNs, Wang and Mendel (1992) have shown that the FBFN is able to uniformly approximate any real continuous function on a compact set to arbitrary accuracy. This result is based on the Stone–Weierstrass theorem. They have also derived a backpropagation-based learning strategy for FBFNs in addition to the OLS algorithm. The ANFIS is complementary to the FALCON because the consequents of its fuzzy logic rules are expressed as a function of the process state variables. Although this eliminates the defuzzification process, the approximate reasoning power is not as rich as the FALCON's. Jang (1992) also proposed a generalized control scheme, based on the ANFIs, that can construct a fuzzy controller through *temporal backpropagation* such that the state variables can follow a given desired trajectory as closely as possible.

The notion of stability, controllability, and observability is well established in modern control theory. However, due to the complexity of mathematical analysis for fuzzy logic controllers, stability analysis requires further study and controllability and observability issues need to be defined and studied for fuzzy control systems. Hence there is no theoretical guarantee that a general fuzzy control system will not go chaotic, although such a possibility appears to be extremely slim from practical experience. Currently much attention is being focused on the stability analysis of fuzzy systems. Some recent results can be found in Wang (1993, 1994) and Kang (1993).

Over the past decade we have witnessed a very significant increase in the number of applications of fuzzy logic-based techniques and neuro-fuzzy-based techniques to various commercial and industrial products/systems. In many applications, especially to control nonlinear, time-varying, ill-defined systems and manage complex systems with multiple independent decision-making processes, the FLC-based systems have proved to be superior in performance when compared to conventional control systems.

Notable applications of FLC include:

- Steam engine (Mamdani and Assilian, 1975; Ray and Majumder, 1985)
- Warm water process (Kickert and Van Nauta Lemke, 1976)
- Heat exchange (Ostergaard, 1977)
- Activated sludge wastewater treatment (Tong, Beck, and Latten, 1980; Itoh et al., 1987; Yu, Cau, and Kandel, 1990)
- Traffic junction control (Pappis and Mamdani, 1977)
- Cement kiln (Larsen, 1980; Umbers and King, 1980)
- Aircraft flight control (Larkin, 1985; Chiu et al., 1991)
- Autonomous orbital operations (Lea and Jani, 1992)
- Turning process (Sakai, 1985)
- Robot control (Uragami, Mizumoto, and Tananka, 1976; Scharf and Mandic, 1985; Tanscheit and Scharf, 1988; Ciliz et al., 1987; Isik, 1987; Palm, 1989)
- Model car parking and turning (Sugeno and Murakami, 1984,1985; Sugeno and Nishida, 1985; Sugeno et al., 1989)
- Automobile speed control (Murakami, 1983; Murakami and Maeda, 1985)
- Elevator control (Fujitec, 1988)
- Automobile transmission and braking control (Kasai and Morimoto, 1988)
- Power systems and nuclear reactor control (Bernard, 1988; Kinoshita et al., 1988)
- Arc welding (Murakami et al., 1989; Langari and Tomizuka, 1990)
- Adaptive control (Graham and Newell, 1989)
- Walking machine (DeYong et al., 1992; Yoshida and Wakabayashi, 1992)
- Highway incident detection (Hsiao, Lin, and Cassidy, 1993)
- Fuzzy computers (Yamakawa, 1987)

Fuzzy logic control systems have also found applications in household appliances, such as air conditioners (Mitsubishi); washing machines (Matsushita, Hitachi); video recorders (VCR) (Sanyo, Matsushita); television autocontrast and brightness control cameras (Canon), autofocusing and jitter control (Shingu and Nishimori, 1989; Egusa et al., 1992); vacuum cleaners (Matsushita); microwave ovens (Toshiba); palmtop computers (Sony); and many others.

An interesting application of fuzzy control is the camera tracking control system at the Software Technology Laboratory, NASA/Johnson Space Center, used to investigate fuzzy logic approaches in autonomous orbital operations (Lea and Jani, 1992). Other recent applications of fuzzy control in various disciplines can be found in Sugeno (1985*b*) and Marks (1994).

For the past several years neuro-fuzzy systems have gradually replaced traditional fuzzy logic control systems in household appliances. This is because neural network hardware can be retrained if there is a change in the working environment. Due to their learning ability, neuro-fuzzy systems are also finding potential applications in controlling underwater robots and multilimb walking robots, developing an empirical model for predicting earthquake-induced liquefaction hazards, and performing plasma etch process modeling. As we continue to search for new techniques in neural learning, the application domain of neuro-fuzzy systems will continue to expand to other areas that were once thought to be unsolvable.

ACKNOWLEDGMENT

This work was supported in part by the National Science Foundation under Grant BES-9701850 and in part by the Sze Tsao Chang Memorial Engineering Fund.

REFERENCES

Bernard, J. A. 1988. "Use of Rule-Based System for Process Control." *IEEE Control Systems Magazine* **8**(5), 3–13.

Chen, S., C. F. N. Cowan, and P. M. Grant. 1991. "Orthogonal Least Squares Learning Algorithm for Radial Basis Function Networks." *IEEE Transactions on Neural Networks* **2**(2), 302–309.

Chiu, S., et al. 1991. "Fuzzy Logic for Control of Roll and Moment for a Flexible Wing Aircraft." *IEEE Control Systems Magazine* **11**(4), 42–48.

Ciliz, K., et al. 1987. "Practical Aspects of the Knowledge-Based Control of a Mobile Robot Motion." In *Proceedings of 30th Midwest Symposium on Circuits and Systems,* Syracuse.

DeYong, M., et al. 1992. "Fuzzy and Adaptive Control Simulations for a Walking Machine." *IEEE Control Systems Magazine* **12**(3), 43–49.

Dubois, D., and H. Prade. 1980. *Fuzzy Sets and Systems: Theory and Applications.* New York: Academic Press.

———. 1985*a*. "Unfair Coins and Necessity Measures: Toward a Possibilistic Interpretation of Histograms." *Fuzzy Sets and Systems* **10**(1), 15–20.

———. 1985*b*. "A Review of Fuzzy Set Aggregation Connectives." *Information Science* **36, 85–** 121.

Egusa, Y., et al. 1992. "An Electronic Video Camera Image Stabilizer Operated on Fuzzy Theory." In *Proceedings of IEEE International Conference on Fuzzy Systems,* San Diego. 851–858.

Fujitec, F. 1988. *Flex-8800 Series Elevator Group Control Systems.* Technical Report, Osaka, Japan.

Graham, B. P., and R. B. Newell. 1989. "Fuzzy Adaptive Control of a First Order Process." *Fuzzy Sets and Systems* **31**(1), 47–65.

Hertz, J., A. Krogh, and R. G. Palmer. 1991. *Introduction to the Theory of Neural Computation.* New York: Addison-Wesley.

Hsiao, C. H., C. T. Lin, and M. Cassidy. 1993. "Application of Fuzzy Logic and Neural Networks to Incident Detection." *Journal of Transportation Engineering, ASCE* **120**(5), 753.

Isik, C. 1987. "Identification and Fuzzy Rule-Based Control of a Mobile Robot Motion." In *Proceedings of International Symposium on IEEE Intelligent Control,* Philadelphia.

Itoh, O., et al. 1987. "Application of Fuzzy Control to Activated Sludge Process." In *Proceedings of 2nd IFSA Congress,* Tokyo. 282–285.

Jang, J. S. R. 1992. "Self Learning Fuzzy Controllers Based on Temporal Back Propagation." *IEEE Transactions on Neural Networks* **3**(5), 714–723.

Jang, J. S. R., and C. T. Sun. 1993. "Functional Equivalence Between Radial Basis Function Networks and Fuzzy Inference Systems." *IEEE Transactions on Neural Networks* **4**(1), 156–159.

Jang, J. S. R., C. T. Sun, and E. Mizutani. 1996. *Neuro-Fuzzy and Soft Computing.* Englewood Cliffs: Prentice-Hall.

Kasai, Y., and Y. Morimoto. 1988. "Electronically Controlled Continuously Variable Transmission." In *Proceedings of International Congress on Transportation Electronics,* Dearborn.

Kaufmann, A. 1975. *Introduction to the Theory of Fuzzy Subsets.* New York: Academic Press.

Kickert, W. J. M., and H. R. Van Nauta Lemke. 1976. "Application of a Fuzzy Logic Controller in a Warm Water Plant." *Automatica* **12**(4), 301–308.

Kinoshita, M., et al. 1988. "An Automatic Operation Method for Control Rods in BWR Plants." In *Proceedings of Specialists' Meeting on In-Core Instrumentation and Reactor Core Assessment,* Cadarache, France.

Klir, G. J., and T. A. Folger. 1988. *Fuzzy Sets, Uncertainty, and Information.* Englewood Cliffs: Prentice-Hall.

Langari, G., and M. Tomizuka. 1990. "Self Organizing Fuzzy Linguistic Control Systems with Application to Arc Welding." In *Proceedings of IEEE Workshop on Intelligent Robots and Systems (IROS),* Tsuchiura, Japan.

Larkin, L. I. 1985. "A Fuzzy Logic Controller for Aircraft Flight Control." In *Industrial Applications of Fuzzy Control.* Ed. M. Sugeno. Amsterdam: North-Holland. 87–104.

Larsen, P. M. 1980. "Industrial Applications of Fuzzy Logic Control." *International Journal of Man–Machine Studies* 12(1), 3–10.

Lea, R. N., and Y. Jani. 1992. "Fuzzy Logic in Autonomous Orbital Operations." *International Journal of Approximate Reasoning* 6(2), 151–184.

Lee, C. C. 1990. "Fuzzy Logic in Control Systems: Fuzzy Logic Controller, Parts I and II." *IEEE Transactions on Systems, Man, and Cybernetics* 20(2), 404–435.

Lin, C. T., and C. S. G. Lee. 1991. "Neural-Network-Based Fuzzy Logic Control and Decision System." *IEEE Transactions on Computers* 40(12), 1320–1336.

———. 1994. "Reinforcement Structure/Parameter Learning for Neural-Network-Based Fuzzy Logic Control Systems." *IEEE Transactions on Fuzzy Systems* 2(1), 46–63.

———. 1996. *Neural Fuzzy Systems: A Neuro-Fuzzy Synergism to Intelligent Systems.* Englewood Cliffs: Prentice-Hall.

Mamdani, E. H., and S. Assilian. 1975. "An Experiment in Linguistic Synthesis with a Fuzzy Logic Controller." *International Journal of Man–Machine Studies* 7(1), 1–13.

Marks II, R. J., ed. 1994. *Fuzzy Logic Technology and Applications,* New York: IEEE Press.

Mizumoto, M. 1981. "Fuzzy Sets and Their Operations." *Information and Control* 48, 30–48.

Murakami, S. 1983. "Application of Fuzzy Controller to Automobile Speed Control System." In *Proceedings of the IFAC Symposium on Fuzzy Information, Knowledge Representation and Decision Analysis,* Marseille. 43–48.

Murakami, S., and M. Maeda. 1985. "Application of Fuzzy Controller to Automobile Speed Control System." In *Industrial Applications of Fuzzy Control.* Ed. M. Sugeno. Amsterdam: North-Holland. 105–124.

Murakami, S., et al. 1989. "Weld-Line Tracking Control of Arc Welding Robot Using Fuzzy Logic Controller." *Fuzzy Sets and Systems* 32, 221–237.

Ostergaard, J. J. 1977. "Fuzzy Logic Control of a Heat Exchange Process." In *Fuzzy Automata and Decision Processes,* Ed. M. M. Gupta, G. N. Sardis, and B. R. Gaines. Amsterdam: North-Holland. 285–320.

Palm, R. 1989. "Fuzzy Controller for a Sensor Guided Manipulator." *Fuzzy Sets and Systems* 31, 133–149.

Pappis, C. P., and E. H. Mamdani. 1977. "A Fuzzy Logic Controller for a Traffic Junction." *IEEE Transactions on Systems, Man, and Cybernetics* 7(10), 707–717.

Ray, K. S., and D. D. Majumder. 1985. "Structure of an Intelligent Fuzzy Logic Controller and Its Behavior." In *Approximate Reasoning in Expert Systems.* Ed. M. M. Gupta et al. North-Holland: Elsevier Science.

Sakai, Y. 1985. "A Fuzzy Controller in Turning Process Automation." In *Industrial Applications of Fuzzy Control.* Ed. M. Sugeno. Amsterdam: North-Holland. 139–152.

Scharf, E. M., and N. J. Mandic. 1985. "The Application of a Fuzzy Controller to the Control of a Multi-Degree-Freedom Robot Arm." In *Industrial Applications of Fuzzy Control.* Ed. M. Sugeno. Amsterdam: North-Holland. 41–62.

Shingu, T., and E. Nishimori. 1989. "Fuzzy Based Automatic Focusing System for Compact Camera." In *Proceedings of IFSA89.* 436–439.

Sugeno, M. 1985a. "An Introductory Survey of Fuzzy Control." *Information Science* 36, 59–83.

———, ed. 1985b. *Industrial Applications of Fuzzy Control.* Amsterdam: North-Holland.

Sugeno, M., and K. Murakami. 1984. "Fuzzy Parking Control of Model Car." In *Proceedings of 23rd IEEE Conference on Decision and Control,* Las Vegas.

———. 1985. "An Experiment Study on Fuzzy Parking Control Using a Model Car." In *Industrial Applications of Fuzzy Control.* Amsterdam: North-Holland. 125–138.

Sugeno, M., and M. Nishida. 1985. "Fuzzy Control of Model Car." *Fuzzy Sets and Systems* 16, 103–113.

Sugeno, M., et al. 1989. "Fuzzy Algorithmic Control of a Model Car by Oral Instructions." *Fuzzy Sets and Systems* 32, 207–219.

Tanscheit, R., and E. M. Scharf. 1988. "Experiments with the Use of a Rule-Based Self-Organizing Controller for Robotics Applications." *Fuzzy Sets and Systems* 26, 195–214.

Tong, R. M., M. B. Beck, and A. Latten. 1980. "Fuzzy Control of the Activated Sludge Wastewater Treatment Process." *Automatica* 16(6), 695–701.

Umbers, I. G., and P. J. King. 1980. "An Analysis of Human Decision-Making in Cement Kiln Control and the Implications for Automation." *International Journal of Man–Machine Studies* 12(1), 11–23.

Uragami, M., M. Mizumoto, and K. Tananka. 1976. "Fuzzy Robot Controls." *Journal of Cybernetics* 6, 39–64.

Wang, L. X. 1993. "Stable Adaptive Fuzzy Control of Nonlinear Systems." *IEEE Transactions on Fuzzy Systems* 1(2), 146–155.

———. 1994. *Adaptive Fuzzy Systems and Control.* Englewood Cliffs: Prentice-Hall.

Wang, L. X., and J. M. Mendel. 1992. "Fuzzy Basis Functions, Universal Approximation, and Orthogonal Least-Squares Learning." *IEEE Transactions on Neural Networks* **3**(5), 807–814.

Yamakawa, T. 1987. "A Simple Fuzzy Computer Hardware System Employing Min and Max Operations—A Challenge to 6th Generation Computer." In *Proceedings of 2nd IFSA Congress, Tokyo.*

Yoshida, S., and N. Wakabayashi. 1992. "Fuzzy and Adaptive Control Simulations for a Walking Machine." *IEEE Control Systems Magazine* **12**(3), 65–70.

Yu, C., Z. Cao, and A. Kandel. 1990. "Application of Fuzzy Reasoning to the Control of an Activated Sludge Plant." *Fuzzy Sets and Systems* **38**, 1–14.

Zadeh, L. A. 1973. "Outline of a New Approach to the Analysis of Complex Systems and Decision Processes." *IEEE Transactions on Systems, Man, and Cybernetics* **SMC-1**, 28–44.

———. 1975. "The Concept of a Linguistic Variable and Its Application to Approximate Reasoning." Part I, *Information Sciences* **8**, 199–249; Part II, **8**, 301–357.

Zimmermann, H.-J. 1991. *Fuzzy Set Theory and Its Applications.* 2nd rev. ed. Boston: Kluwer Academic.

Zurada, J. M. 1992. *Introduction to Artificial Neural Systems.* Singapore: Information Access & Distribution.

CHAPTER 22

COORDINATION, COLLABORATION, AND CONTROL OF MULTIROBOT SYSTEMS

Michael J. Shaw
Narendra Ahuja
Seth Hutchinson
University of Illinois at Urbana-Champaign
Urbana, Illinois

1 INTRODUCTION

Modern automated manufacturing systems increasingly consist of computer-controlled machine tools that have rapid changeover capability to ensure flexibility in making a variety of product lines. Robots are used in these fully automated systems to put the workpiece in place or to perform part of the manufacturing operations. Material-handling systems, which can be robotic systems themselves, are used to move workpieces around the shop floor.

The control system of a *multirobot system* (MRS) is the unit responsible for the coordination of physical flows as well as information flows. Instead of a central hierarchical controller, the networking environment and the increasingly powerful standalone controllers used by the individual robots and machine tools have lead to the use of distributed control. Moreover, because of the large amount of information flowing around the shop floor, distributed control makes real-time adaptive control much more feasible. To that end, the control of MRS must be able to coordinate the control units of the robots and machine tools in the systems, effectively assign jobs to the most capable machine tools or robots, and resolve any conflicts between various robots' activities—and it must do this all while achieving maximum efficiency. A host of methodologies for intelligent cooperative control based on the *multiagent system* (MAS) paradigm for cooperative problem solving has been developed. This chapter will review the coordination and integration of robotic systems, especially of MRSs, on three levels. Section 2 discusses the coordination on the system level; Section 3 addresses the coordination issues concerning motion planning; and Section 4 describes the coordination issues that involve sensory data.

2 MULTIROBOT CONTROL AND COORDINATION

2.1 Intelligent Systems for Cooperative Control

Cooperative problem solving has been extensively and independently studied in the *distributed artificial intelligence* community (Durfee, 1991; Gasser and Huhns, 1989). A number of models for cooperative problem solving have been developed based on the blackboard paradigm (Lesser, 1991), the scientific community model (Kornfeld and Hewitt), and the multiagent model (Huhns and Briddeland, 1991; Guha and Lenat, 1994; Shaw and Fox, 1993) surveyed the applications of DAI techniques for group decision support. Four different types of multiagent problem-solving systems have been developed:

1. *Distributed problem-solving systems.* In these systems the overall problem to be solved is decomposed into subproblems assigned to the agents. Each agent asyn-

Handbook of Industrial Robotics, Second Edition, Edited by Shimon Y. Nof
ISBN 0-471-17783-0   © 1999 John Wiley & Sons, Inc.

chronously plans its own actions and turns in its solutions to be synthesized with the solutions of other agents. The agents in these systems use either task sharing (Smith, 1980) or data sharing (Lesser and Corkill, 1981) to cooperate with other agents. The office information system described in Woo and Lochovsky (1986) and the scheduling system described in Shaw and Whinston (1989) are examples of these systems.

2. *Collaborative reasoning systems.* The agents in this type of system would solve the same problem collaboratively. The main issue here is not the decomposition into subproblems assigned to the agents but the mechanism for coordinating the interactions among the participating agents so that the problem can be solved jointly by the group simultaneously. The focus of this type of MAS is on fully utilizing and integrating the different perspectives and specializations of the agents (Tan, Hayes, and Shaw, 1995). The multiagent system described in Tan, Hayes, and Shaw (1995) falls into this category.

3. *Connectionist systems.* This type of multiagent system uses agents as the basic computational units without much intelligence. But collectively they can solve complicated problems. An example is the society-of-mind model described in Minsky (1985). The agents learn to solve problems more effectively by adjusting their connections with each other.

4. *Distributed information systems.* This type of multiagent system has become especially important in view of the recent developments in on-line information retrieval technology, as typified by Internet and World Wide Web services. Agents are embedded in the system to assist users in such tasks as information searches, retrievals, and filtering. Agents in such a web environment mostly serve as the intelligent user interface for more effective communications and information processing (Maes, 1994).

Although the agents, the problems to be solved, and the strategies used are different in the aforementioned systems, it is nevertheless possible to construct a general framework to describe the strategic factors shared by these systems. These factors are summarized below.

1. *Goal identification and task assignment.* Two different types of problem-solving processes can be used by the group of agents. In the first type the problem is presented to the whole group and the agents collectively carry out the deliberation process, which usually consists of a series of steps including issues identification, proposing solutions, discussion, prioritizing, and finalizing the group solutions. For the second type of group problem solving the tasks involved in having the problem solved are structured and known and the common strategy used consists of four steps: problem decomposition, task assignment, local problem solving, and solution synthesis.

2. *Distribution of knowledge.* In the design of a distributed knowledge-based system, the set of domain knowledge possessed by the group of agents can be distributed in different fashions. The consideration of knowledge distribution is similar to the design consideration involved in placing multiple copies of data files in distributed databases. That is, the different areas of knowledge and their duplicate copies should be placed according to the level of demands by the agents.

3. *Organization of the agents.* The control structure is determined by the organization of the agents. The group, for example, can have a central coordinator that gives commands to the other agents in a hierarchical fashion. The agents alternatively can have a flat structure, sometimes with the flexibility of dynamically forming subteams. The organizational structure incorporated in MAS would dictate how the group of agents coordinate (Jin and Levitt, 1993).

4. *Coordination mechanisms.* Coordination is necessary in MAS for resolving conflicts, allocating limited resources, reconciling different preferences, and searching in a global space for solutions based on local information (Papastarrou and Nof, 1992). Coordination mechanisms can be based on a variety of forms of information exchanged, such as data, new facts just generated, partial solutions/plans, preferences, and constraints. A number of coordination mechanisms have been developed for MAS, each with its unique protocol for determining the timing of

activities, triggering of events, action sequences, and message content in the co-ordination process.

5. *Learning schemes.* Learning should be an integral part of problem solving for improving strategies, knowledge, and skills. There are several learning processes that can be incorporated in multiagent systems. Learning can be triggered by data exchange, knowledge transfer, or heuristics migration, where the mechanisms involved are relatively simple. It can also be done by extending single-agent learning to the multiagent systems, where one agent can learn by observing other agents. Of particular interest is the use of group processes often used in human decision-making situations, such as group induction, nominal group techniques, or brain-storming, for achieving the learning effects among the whole group. These group processes use structured information exchanges to develop new concepts, which leads to solutions not attainable by any of the agents alone.

In implementing a multiagent problem-solving system, these five strategic considerations can be implemented by group metaknowledge. Directly affecting how the multiagent system operates, this group metaknowledge can be embedded in the network through which the agents communicate. Smith (1980) described the use of a "problem-solving layer" for incorporating such group metaknowledge in the layered network architecture. It can also be viewed as a "protocol" (Finin, Fritzson, and McKay, 1992) that every agent has to observe for the group to operate efficiently. Depending on the organization of the system and the knowledge distribution, group metaknowledge would be stored in the agents' knowledge bases. It is used by the multiagent system to direct and coordinate the group problem-solving sessions.

2.2 The Role of Coordination

Coordination is the key design component for MAS. Since each problem-solving agent possesses only local view and incomplete information, it must coordinate with other agents to achieve globally coherent and efficient solutions. The design of coordination can be viewed from three different perspectives: information content, exercise of control, and coordination mechanisms. Coordination can be achieved based on different types of information among the agents. The initiative to coordinate may result from a variety of means of control: it may be self-directed, externally directed, mutually directed, or a combination (Lesser and Corkill, 1981).

Many coordination mechanisms have been developed for MAS. Some representative ones follow:

- *Coordination by revising actions.* One of the primary factors that necessitates co-ordination is the potential conflicts between the actions taken by the agents. The objective of coordination, then, is to derive a plan of actions for the group such that all the conflicts are avoided. Cammarata (Cammarata, McArthur, and Steeb, 1983) used the collision-avoidance problem in air traffic control to illustrate co-ordination mechanisms based on passing the information constantly among prob-lem solvers, each of which guides the course of an aircraft. The flight plan would be revised if there were a risk of conflict with the plan of other aircraft.

- *Coordination by synchronization.* In a multiagent system the action of an agent usually affects some other agents. The objective of coordination in a way is to regulate the interactions among agents and to control the timing and sequence of these interactions. Corkill (1979) and Georgeff (1983) incorporated communication primitives, similar to the ones used in distributed operating systems, for synchro-nizing the problem-solving process of the agents.

- *Coordination by negotiation.* Negotiation is a widely used form for mutually di-rected coordination: the process involves two-way communication to reach a mu-tually agreed upon course of actions. The contract-net mechanism described in Smith (1980) used negotiation to coordinate the sharing of tasks among agents, in which the negotiation process is done by contract bidding. Croft and Lefkowitz (1988) adopted the negotiation process to maintain the consistency of different agents' plans. Cammarata, McArthur, and Steeb (1983) also used negotiation for finalizing revision of the agents' plans. Sathi and Fox (1989) developed a con-straint-directed negotiation approach for coordinating the agents for resource real-location.

- *Coordination by opportunistic goal satisfaction.* The blackboard model for problem solving (Nii, Aiello, and Rice, 1989) has been used extensively in the multiagent environments (Lesser and Corkill, 1981; Durfee, Lesser, and Corkill, 1987). It provides a paradigm for coordinating multiple problem-solving agents (i.e., knowledge sources) to contribute to the group solution process opportunistically.

- *Coordination by exchanging preferences.* In studying how the group of autonomous, self-interested agents should interact with each other in achieving globally satisfactory solutions, of one another. Using a game-theoretic approach, Rosenschein (Rosenschein and Genersereth, 1985) described a coordination scheme based on the exchange of the payoff matrices.

- *Coordination by message passing.* This coordination method is used by the Actor system (Agha, 1986, 1989), where the coordination can be incorporated by sending messages from one agent to the other. While expressing coordination strategies in terms of low-level message is tedious, error-prone, and inflexible, interaction patterns can be specified by abstractions, which allow dynamic installation and removal of coordination constraints on groups of agents. KQML, which is a language for knowledge and information exchange among agents (Finn, Fritzson, and Mc-Kay, 1992) emphatically allows agents to interpret the messages they exchange by sharing a framework of knowledge, i.e., shared ontology.

2.3 Real-Time Control Using Bidding and Negotiation Schemes

Using the multiagent problem-solving paradigm, a variety of bidding and negotiating schemes have been developed in the literature for real-time distributed control. Farber and Larson (1972) described a bidding scheme for allocating computing resources in distributed computing environment. Smith (1980) developed the idea further and incorporated the negotiation procedure into their *contract-net* mechanism for coordinating a distributed set of artificial intelligence agents. Contract-net was one of the earliest distributed artificial intelligence (DAI) systems and has influenced a whole generation of research work in the DAI area. Shaw and Whinston (1989) applied the contract-net mechanism to the shop-floor control of flexible manufacturing systems (FMS) and demonstrated the benefits of distributed control in automated manufacturing. Upton (1989), Lin and Solberg (1992), Wang and Veeramani (1992), and Veeramani and Wang (1997) also applied a variety of realtime bidding schemes to distributed manufacturing systems with multiple robots and machine tools. The generic bidding scheme consists of the following procedure:

1. A manufacturing job arrives at the MRS.
2. A request-for-bid message is broadcast to the robots and machine tools (i.e., agents).
3. Each agent evaluates the job requirements and its capabilities, job queues, and other real-time manufacturing information.
4. Each agent submits a bid within a biding duration (Tb) enclosing a "price" determined by its own bidding calculation.
5. All the bids submitted are computed and compared.
6. The agent submitting the best bid is selected to get the job.

Bidding schemes are inherently suboptimal because of the lack of consideration of future jobs entering the systems. To resolve that problem while retaining the real-time control capability, modifications can be incorporated, especially in two dimensions. First, Tb can be made longer to consider more system information on a longer term when deciding the assignment, and second, the bidding scheme can be applied to multiple jobs so that agents can be assigned to the most beneficial job among the group entered. On the other hand, there is a clear tradeoff between the performance improvement of the bidding scheme and the response time for making the assignment determined by Tb.

The real-time control and coordination of multiple robot systems can be extended to that of multiple automated manufacturing cells, each consisting of fully automated machining centers and material handling with robots and automated guided vehicles. An example of this is illustrated in Figure 1.

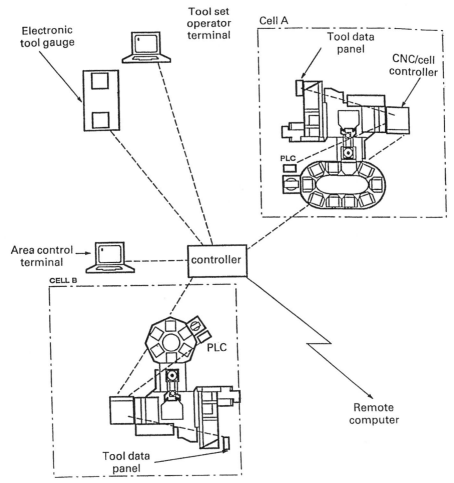

Figure 1 Distributed control and coordination in multiple manufacturing cell systems. (From M. Shaw and G. Wiegand, "Intelligent Information Processing in Flexible Manufacturing Systems," *FMS Magazine* **6**(3), 137–140, 1988).

3 MOTION PLANNING

3.1 Motion Planning for Multirobot Systems

In the case of multiple robots the geometric motion-planning problem corresponds to finding trajectories for each robot such that at no time do any two robots collide and at no time does any robot collide with a fixed obstacle. More formally, assume that there are p robots, denoted $A_1 \ldots A_p$, with the configuration space of the ith robot denoted by C_i. We denote by W the workspace inhabited by the robots, and by $A_i(\mathbf{q})$ the subset of W occupied by the ith robot when it is at configuration $\mathbf{q}$. The set of points in W occupied by static obstacles is denoted by B. The *free* configuration space for the ith robot (i.e., the set of configurations for which the ith robot does not collide with any obstacle) is denoted by C_i^{free}, and is defined by $C_i^{\text{free}} = C_i - \{\mathbf{q} \in C_i | A_i(\mathbf{q}) \cap B \neq \emptyset\}$. Using this notation, the goal of robot motion planning is to find a trajectory $\tau_i: [0, 1] \rightarrow C_i^{\text{free}}$ such that $A_i(\tau_i(s)) \cap A_j(\tau_j(s)) = \emptyset$ for all $s \in [0, 1]$, and such that $\tau_i(0) = \mathbf{q}_i^{\text{init}}$ and $\tau_i(1) = \mathbf{q}_i^{\text{goal}}$ for each i.

In *centralized* approaches we treat the p robots as though they comprise a single robot whose configuration space is given by $C = C_1 \times \cdots \times C_p$. We define the free configuration space by

$$C^{\text{free}} = C - \{\mathbf{q} \in C \mid A_i(\mathbf{q}_i) \cap B \neq \emptyset\} - \{\mathbf{q} \in C \mid \exists_{i,j} A_i(\mathbf{q}_i) \cap A_j(\mathbf{q}_j) \neq \emptyset\}$$

where $\mathbf{q}_i$ is understood to be the projection of the configuration $\mathbf{q}$ onto C_i. We define the initial and goal configurations of this composite robot by $\mathbf{q}^{\text{init}} = (\mathbf{q}_1^{\text{init}} \ldots \mathbf{q}_p^{\text{init}})$ and $\mathbf{q}^{\text{goal}} = (\mathbf{q}_1^{\text{goal}} \ldots \mathbf{q}_p^{\text{goal}})$. With this approach, the motion-planning problem is to find a trajectory $\tau:[0, 1] \rightarrow C^{\text{free}}$ such that $\tau(0) = \mathbf{q}^{\text{init}}$ and $\tau(1) = \mathbf{q}^{\text{goal}}$. Under this formalism, in principle any of the methods commonly used for robot motion planning may be applied to the multiple robot problem. However, in practice, because robot motion-planning algorithms have complexities that increase exponentially with the dimension of the configuration space, such approaches are not generally feasible. Descriptions of most existing methods for robot motion planning may be found in Latombe (1991) and Hwong and Ahuja (1992).

A few specific examples of the centralized approach for special cases have been reported. The problem of planning the motion of N disks in the plane was addressed by Schwartz and Sharir (1983b), where they extend their earlier work using critical curve analysis (1983a). Their approach proceeds by performing a critical curve analysis on the configuration spaces of the individual disks and then planning a collision-free path through free cells in the composite configuration space. For the case of two constrained robot manipulators, Ardema and Skowronski (1991) report a technique that proceeds by modeling the problem as a noncooperative game.

The most successful of the centralized planning approaches is the randomized planner reported in Barraquand, Langlois, and Latombe (1992). Originally this approach was developed as a way to cope with the local minima that plague artificial potential fields methods for motion planning. Motion planners that use artificial potential fields typically proceed by defining an artificial repulsive field (which acts to steer the robot away from obstacles) and an artificial attractive field (which acts to pull the robot toward the goal configuration). These methods are often thwarted in their attempt to find solutions by the presence of local minima in the combined potential field (Hwang and Ahuja, 1992b; Rimon and Koditschek, 1992). To cope with this problem, Barraquand and Latombe developed RPP, a planner that uses randomized motions to escape local minima when they are encountered during the course of a gradient descent search (1991). Because the volumes of the attractive basins of the local minima tend to decrease relative to the volume of the configuration space as the dimension of the configuration space increases, this method has shown great promise for high dimensional configuration spaces. Thus it is a good candidate for a centralized motion-planning algorithm for the case of multiple robots.

3.2   Decoupled Approaches to Multirobot Motion Planning

Decoupled approaches construct plans for the individual robots and then combine these plans while considering the interactions of the robots. Two approaches have been considered: prioritized planning and path coordination. We will briefly describe each. The advantage of a decoupled approach is that planning for a single robot is a much easier computational problem than planning simultaneously for multiple robots (due to the exponential complexities of robot motion-planning algorithms, discussed above). The disadvantage of a decoupled approach is that completeness is lost; there is no guarantee that a successful plan that coordinates the motions of all robots can be generated by combining plans for individual robots. Figure 2 shows an example of a problem that cannot be solved by a decoupled algorithm. If the robots are treated completely independently (as with the path-coordination method), then each robot will be assigned a straight-line motion to its goal. There is no way to coordinate the resulting plans to achieve the goals while avoiding collision. If the robots are treated according to priority, then whichever robot is assigned the highest priority will be assigned the straight-line motion to the goal. The robot with less priority will be unable to find any plan to achieve its goal in light of this choice by the first robot.

3.2.1   Prioritized Planning

Prioritized planning is an iterative approach (Erdmann and Lozano-Pérez, 1986; Warren, 1990). Without loss of generality, assume that the robots are indexed in order of decreasing priority. At the ith iteration of the planning process, a path is planned for A_i, while treating robots $A_1 \ldots A_{i-1}$ as moving obstacles with known trajectories. Thus at the first iteration a plan is created for the robot with the highest priority, A_1, ignoring robots A_2

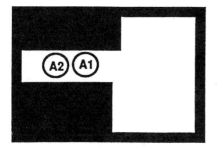 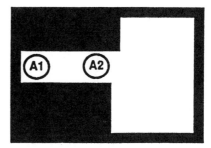

Initial Configuration   **Goal Configuration**

Figure 2   A problem that cannot be solved by decoupled planning.

. . . A_p. At the second iteration a plan is created for A_2, treating A_1 as a moving obstacle, and so on. The problem of motion planning in the presence of moving obstacles with known trajectories is beyond the scope of the present chapter, but examples can be found in Fujimura and Samet (1993); Reif and Sharir (1985); Shih, Lee, and Gruver (1990); Spence and Hutchinson (1995).

Priorities could be assigned randomly, but in general it is more productive to derive priorities from knowledge of constraints on the motions of the robots. For example, in Buckley (1989) priorities are assigned by attempting to maximize the number of robots that can move to the goal using a straight-line motion. Other prioritization schemes include assigning higher priority to larger robots (since, intuitively at least, bigger robots will encounter greater difficulty navigating among obstacles) or assigning higher priorities to robots that are more distant from their goal configurations.

3.2.2   Path Coordination

The method of path coordination was introduced in O'Donnell and Lozano-Pérez (1989) and has been applied to the problem of planning the motions for two robots. The method proceeds by first deriving paths for each of the two robots individually, then coordinating the two paths. Specifically, the first step is to determine $\tau_i : s_i \in [0, 1] \rightarrow \tau_i(s_i) \in C_i^{\text{free}}$ for $i = 1, 2$. Then a schedule is constructed in $s_1 \times s_2$ space (i.e., the unit square). To create this schedule, we define the set of positions along the two paths at which the robots collide:

$$SB = \{(s_1, s_2) \in [0, 1] \times [0, 1] \mid A_1(\tau_1(s_1)) \cap A_2(\tau_2(s_2)) \neq \emptyset \}$$

The space $[0, 1] \times [0, 1] - SB$ is then searched for a continuous path from $(0, 0)$ to $(1, 1)$. Such a path is called a *free schedule,* and it defines a parameterization for the two paths, τ_1 and τ_2, that bring the robots to their goal configurations while avoiding collisions. For computational reasons, the unit square is discretized, and this discrete representation is actually searched for the free schedule. In the literature this discretized representation is known as the *coordination diagram.* The coordination space for two manipulators is analyzed in Bien and Lee (1992) and Chang, Chung, and Lee (1994). Scheduling issues are studied through the use of coordination diagrams in Lin and Tsai, (1990), and additional multiple-robot scheduling issues are presented in Kountouris and Stephanou (1991).

3.3   A Game-Theoretic Approach

Recently a new approach to the multiple robot motion-planning problem has been reported by LaValle and Hutchinson. By considering independent performance criteria for each robot, they introduce a form of optimality that is consistent with concepts from multiobjective optimization and game theory research. Algorithms derived from the principle of optimality presented for three problem classes along the spectrum between centralized and decoupled planning:

1. Coordination along fixed, independent paths
2. Coordination along independent road maps

3. General, unconstrained motion planning

Previous multiple-robot motion-planning approaches that consider optimality have combined several individual criteria into a single, scalar criterion and as a result can fail to find many potentially useful motion plans. For example, in Bien and Lee (1992) and Shin and Zheng (1992) the objective is to minimize the maximum time taken by any robot to reach the goal. In Wang and Lever (1994) the scalar objective criterion is merely the aggregate of the individual objectives. In general, when individual objectives are combined, certain information about potential solutions and alternatives is lost (for general discussions see, e.g., Hipel, Radford, and Fang (1993); Sawaragi, Nakayama, and Tanino (1985); and Zionts (1985)). For example, the amount of sacrifice that each robot makes individually to accomplish its goals is not usually taken into account. It might be that one robot's goal is nearby, while the other robot has a distant goal. Combining the objectives by scalarization might produce a good plan for the robot that has the distant goal; however, the execution cost for the other robot would hardly be considered.

In LaValle and Hutchinson it is shown that, given a vector of independent objective functionals, there exists a natural partial ordering on the space of motion plans, yielding a search for the set of minimal motion plans. For any other motion plan that can be considered, there will exist at least one minimal plan that is clearly better or equivalent, and the set of all minimal motion plans is typically small. Hence their method filters out large sets of motion plans that are not worth considering and presents a small set of reasonable alternatives. Within this framework, additional criteria, such as priority or the amount of sacrifice one robot makes, can be applied to automatically select a particular motion plan. If the same tasks are repeated and priorities change, then a different minimal plan would be selected, as opposed to reexploring the space of motion plans. These minimal strategies are consistent with certain optimality concepts from multiobjective optimization (e.g., Hipel, Radford, and Fang (1993); Sawaragi, Nakayama, and Tanino (1985); and Zionts (1985)) and dynamic game theory (e.g., Başar and Olsder (1982) and Owen (1982)) literature.

4 SENSOR BASED COORDINATION AMONG MULTIPLE ROBOTS

4.1 Sensors

A number of sensors have been used to estimate the environmental parameters, e.g., distance, orientation, shape, motion, and material of objects. The sensors may involve contact with the object, or they may function remotely. Contact sensors respond to touch-based signals such as force and torque and are typically useful in grasping and other manipulatory operations to avoid slipping and crushing of the grasped object. Noncontact sensors detect variations in some type of radiation, e.g., sound and light, and provide a means of navigation for manipulation. A summary of these sensors can be found in Fu, Gonzalez, and Lee (1987).

4.1.1 Contact Sensors

Force and torque sensors are useful for making measurements at joints, between the robot base and the surface on which they are mounted, and between the tip of the arm and end-effector. Strain gauges are a commonly used sensor of this type. The distribution of forces along a surface is sensed by using arrays of binary or analog touch sensors. A single binary touch sensor detects whether an object is pressing against the sensor or not, e.g., through the use of a microswitch. A distribution of touch and no-touch values across the sensor array provides a coarse measurement of the spatial distribution of force. The individual binary sensor may also be replaced by an analog transducer that converts the force into a physical displacement or electric current that is easily measured. A common class of such analog arrays is *artificial skins,* across which the resistance of a compliant conductor varies as a function of compression due to the local force. Sensors also exist to detect relative motion between two sliding surfaces.

4.1.2 Noncontact Sensors

Noncontact sensors are useful for the detection of objects and their properties from a distance. A commonly used class is *proximity sensors*, which detect the presence of magnetic material in the vicinity of the sensor by recording the interference the material causes with the field generated by the sensor. This interaction can also be used to estimate object motion characteristics. Other sensors detect object interference with optical and infrared beams. Most of these sensors directly exploit analog physical processes that are calibrated with the measurements of interest. Many other sensors, such as range sensors, involve explicit computation of properties of interest. Energy (ultrasound, laser) is transmitted and the times and phases of the echos are analyzed to estimate the locations of objects. The design of the transmitted signal and analysis of the received signal are important parts of the sensor development.

Visual sensing is the most pervasive type of noncontact sensing. The contents of the visual signal or images depend on the the the nature of illumination and scene and the type of light sensors. The illumination characteristics depend on the number, shape, locations, spectral properties, and directional characteristics of the light sources. The nature of the illumination is further complicated by the indirect path the light may take in reaching the object after it leaves the source, due to reflections from the many objects that may be present in the scene. Some of the light reaching each object is absorbed and the rest is reflected, in general, in an anisotropic manner, with a part going in the direction of specular (mirror) reflection and the rest forming a lobe around that direction. Good reflectors such as metals return most energy in the direction of reflection, while the returns from other surfaces may be more diffused. Finally, the optical sensor elements located at the cameras themselves have their positions, orientations, and spectral sensitivities, which determine the recorded response to the total irradiation.

In controlled environments, such as on a factory floor, the illumination can be and often is controlled by using special types and arrangements of light sources. For example, point and planar light stripes are moved across the scene to create a dynamic illumination pattern. This, in conjunction with suitably positioned detectors, is used to estimate shapes and layout of objects. Many times the beam of light emanating from the source is spatially coded such that the resulting image pattern captures the object shapes. However, more challenging aspects of coordination arise in the uncontrolled environments where the scene must be understood under complex lighting with no *a priori* knowledge of the types or configurations of objects.

4.2   Coordination of Visual Cues

An image is a two-dimensional function that captures a complex interaction among the three components of image formation: illumination, scene, and sensor. The goal of computer vision includes estimation of these three components. This is an ill-posed problem in the sense that the formation of an image involves loss of information. For example, images are two-dimensional, whereas a scene is three-dimensional. Fundamentally, the information lost is unrecoverable. However, the three components and the interactions among them are not arbitrary; they are bound by some basic laws that restrict the resulting images to being "well formed." Rectilinear propagation of light, the reflectance characteristics of a surface, the piecewise contiguous nature of matter, and coherent movement of objects are examples of the laws that allow only certain "legal" images ever to be encountered. These lead to diverse characteristics defined in terms of specific images and imaging properties, which reveal particular aspects of the three unknown components. Such characteristics, often called *visual cues* in psychology, can be exploited to computationally recover the three components, which are otherwise unrecoverable.

Visual cues may be spatial or temporal, may be defined for an image taken by one camera or multiple spatially dispersed cameras, and may apply to a single image taken at one time instant or a sequence of images acquired over an entire time interval. Examples include the following. Nearby scene points appear nearby in the image because the depth of objects is a piecewise smooth function of viewing direction. Uniform surface features contain spatial gradients in the image space due to perspective shifts. Occlusions among objects gives rise to discontinuities in image properties. The spatial relationships among images' features of the same scene taken from multiple viewpoints contain information about scene geometry. Objects moving at different distances have different image speeds. The silhouette of an object's image captures the three-dimensional shape of the object. The changes in a moving object's silhouette more vividly convey the object geometry. Directions of image motion provide good partition of scene into distinctly moving objects. Dynamic control of imaging parameters such as direction, and focus, zoom, and aperture setting of the cameras, result in active visual cues.

Each visual cue represents and depends on a specific set of scene (i.e., objects, layout, illumination, and sensor) characteristics, which in turn determine the relevance and reliability of the cue. To make possible a comprehensive understanding of the scene, the relevant cues must be identified dynamically and the information they contain about the scene must be extracted along with the reliability of the information, and a composite description of the scene must be cumulatively constructed from the pieces provided by the cues. Thus, both the selection and use of the cues must be continuously coordinated for effective use of imaging and image interpretation processes. The following paragraphs summarize some examples of such coordination.

4.2.1 Texture

The importance of various texture gradients has been studied extensively in the psychology literature (see, for example, Braunstein and Payne, 1969; Phillips, 1970; Rosinski, 1974; Rosinski and Levine, 1976). Vickers (1971) was among the first to advocate an approach involving accumulation of evidence from multiple texture gradients. Cutting and Millard (1984) attempt to quantify the relative importance of various texture gradients. They test human subjects on synthetically generated textures, which are designed to contain only a subset of the normally occurring texture gradients. Experimental results show that for slant judgments of flat surfaces the perspective and density gradients are more important than the aspect–ratio gradient, whereas in the perception of curved surfaces the aspect–ratio gradient is dominant, with perspective and density gradients having little impact.

Most shape-from-texture algorithms use indirect methods to estimate texel features by making some assumptions about the nature of texture elements. For example, texel density may be estimated by measuring edge density, under the assumption that all detected edges correspond to the borders of texture elements (Aloimonos and Swain, 1985; Aloimonos, 1986; Kanatani and Chan, 1986; Rosenfeld, 1975). Alternatively, texture elements may be assumed to have uniform edge direction histograms; surface orientation can then be estimated from any deviations from isotropy observed in the distribution of edge directions (Davis, Janos, and Dunn, 1983; Kanatoni, 1984; Witkin, 1981). However, the directional-isotropy assumption is very restrictive; for example, it does not hold true in images containing elongated texels such as waves. Texture coarseness and directionality may be characterized using Fourier domain measurements (Bajcsy, 1976), ignoring the effect of super- and subtextures. Various research (Ikeuchi, 1980; Kender, 1980; Nakatani et al., 1980) has developed algorithms to analyze textures containing parallel and perpendicular lines. Most natural textures are too irregular to be analyzed in this way.

All of these methods may encounter problems when applied to complex natural textures seen under natural lighting conditions. Since texels are not identified and explicitly dealt with, it becomes difficult to distinguish between responses due to texels and those due to other image features, such as subtexture. It appears to be necessary to recognize the texture elements before the various measures can be computed as intended. Blostein and Ahuja (1989) argue that correct weak segmentation is not possible without simultaneously performing a strong segmentation: in order to eliminate all edges except those that arise from texel boundaries, one has to in effect identify the texels. Explicit texel identification also provides a unifying framework for examination of the various texture gradients (such as gradients of apparent texel area, aspect ratio, density, etc.) that may be present in an image. A given image may exhibit a combination of texture gradients. In general, the accuracy of the surface information obtainable from these gradients varies from image to image. Since it is not known in advance which texture gradients are useful for determining the three-dimensional orientation of surfaces, a shape-from-texture system should evaluate the information content of different types of gradients in a given image and use an appropriate mix of these gradients for surface estimation.

4.2.2 Silhouettes

Structure and motion estimation for objects with smooth surfaces and little texture is an important but difficult problem. Silhouettes are the dominant image features of such objects. Several methods have been proposed for structure estimation from silhouettes under known camera motion (Blake and Cippola, 1992; Giblin and Weiss, 1987; Szeliski and Weiss, 1993; Vaillant and Faugeros, 1992; Zheng, 1994). These approaches have demonstrated that a given set of three or more nearby views of a smooth object, the structure of the object up to second order can be obtained along its silhouette. The recovery of structure and motion from a monocular sequence of silhouettes has been investigated by Giblin, Rycroft, and Pollick (1993). Joshi, Ahuja, and Ponce (1995) address the problem of estimating the structure and motion of a smooth object undergoing arbitrary unknown rigid motion from its silhouettes observed over time by a trinocular stereo rig. This technique is useful when the viewer is a passive observer and has no knowledge or control of the object's motion in the scene (see Kutulakos and Dyer (1994) for a complementary approach, where a viewer plans its motion for building a global model of an object). Another application is model construction for object recognition (Joshi et al., 1994): due to self-occlusion, a simple motion, such as rotation on a turntable, may not reveal all the interesting parts of a complex object. It is desirable to move the object arbitrarily and still be able to construct the model completely. To coordinate the information in the silhouettes observed in successive images, a model of the local structure is constructed that can be used to estimate the motion. Trinocular imagery is useful because three images allow recovery of model parameters of the local structure (up to second order) (Blake and Cippola, 1992; Szeliski and Weiss, 1993; Vaillant and Faugeras, 1992) and three frames are sufficient to differentiate between viewpoint-dependent and viewpoint-independent edges in the image (Vaillant and Faugeras, 1992).

4.2.3   Dynamic Features and Texture

The identification and analysis of the relative strengths of different cues, and intercue coordination, constitute a research problem. In general, the available cues, and sometimes even their relative merits, depend upon the scene under consideration. Sull and Ahuja (1994) address the problem of coordination among cues for the case of an observer moving above a planar, textured surface, such as while in an aircraft that is landing or taking off. The goal is to recover the translational and rotational motion of the observer and the orientation of the plane as a function of time, from multiple attributes present in the sequence of images of the plane acquired during the motion. Their approach allows the use of the following image cues: point features, optical flow, regions, lines, texture gradient, and vanishing line. These cues have complementary strengths and shortcomings. For example, region correspondences are easier to find because regions have nonzero sizes, but they give coarse estimates of motion; on the other hand, points and point correspondences are harder to find but have better positional accuracy and hence give more accurate estimates. Further, certain features may have special significance and utility for a specific type of scene. For example, for images taken from a flying aircraft, the vanishing line is an important cue since it carries information about aircraft orientation with respect to the ground plane. (The vanishing line is defined as the intersection of the image plane with a plane that includes the camera center and is parallel to the object plane.) The cues used by Sull and Ahuja could be adapted to achieve increased robustness in other scenarios while still following their basic approach to coordinated interpretation.

4.2.4   Active Stereo

Active acquisition and interpretation of visual data is another theme that involves coordination as a central feature. This has been facilitated in recent years with the availability of sophisticated hardware for controlling imaging elements (Ballard, 1989; Burt, 1988; Clark and Ferrier, 1988; Krotkov, 1989). For example, incorporation of anthropomorphic features, such as spatially varying sensors (Bajcsy, 1992; Tistarelli and Sandini, 1990), can reduce irrelevant sensory information. In their analysis of surface reconstruction from stereo images, Marr and Poggio (1979) point out the role of eye movements in providing large relative image shifts for matching stereo images having large disparities, thus implying the need for active data acquisition. Aloimonos, Weiss, and Bandyopadhyay (1987) show that active control of imaging parameters leads to simpler formulations of many vision problems that are not well behaved in passive vision. Ballard and Ozcandarli (1988) also point out that the incorporation of eye movements radically changes (simplifies) many vision computations; for example, the computation of depth near the point of fixation becomes much easier. Geiger and Yuille (1987) describe a framework for using small vergence changes to help disambiguate stereo correspondences.

Ahuja and Abbott (1993) and Das and Ahuja (1996, 1995) have developed computational approaches to surface estimation from stereo images, especially in a mutually cooperative mode such as discussed in Bajcsy (1988), Krotkov (1989), and Sperling (1970). Ahuja and Abbott (1993) have argued that to reconstruct surfaces for large scenes having large depth ranges, it is necessary to integrate the use of camera focus, camera vergence, and stereo disparity. Surface estimation must be performed over a scene in a piecewise fashion, and the local surface estimates must be combined to build a global description.

In Ahuja and Abbott (1993) an algorithm is outlined to achieve such integration through iteration of visual target selection, fixation, and stereo reconstruction. The scope of that algorithm is limited to the reconstruction of a single continuous surface, e.g., one object. When the entire surface of the fixated object has been scanned, the acquired surface map does not smoothly extend and therefore surface reconstruction must be resumed by fixating on a new object (Das and Ahuja, 1996). Since surface reconstruction from stereo requires coarse initial estimates, such estimates must be obtained for the new object before reconstruction can continue. Other efforts have concentrated on modeling biological mechanisms of interactions among vergence, accommodation, and stereopsis. Erkelens and Collewijn (1985) discuss interactions between vergence and stereo for biological systems. Sperling (1970) presents a model for the interaction of vergence, accommodation (focus), and binocular fusion in human vision.

5 SUMMARY

This chapter addresses the issue of coordination and integration of robotic systems, especially multiple robot systems, on three levels: the system level, the motion planning level, and the sensory information-processing level. They share the common need in robotic control to integrate multiple sources of information input. Furthermore, it is important on all three levels to apply real-time coordination. Through a discussion of task allocation on the system level, path coordination on the motion planning level, and data integration on the sensory information processing level, the main purpose of this chapter is to emphasize the importance of distributed and cooperative computing for robotics.

REFERENCES

Agha, G. 1986. *Actors: A Model of Concurrent Computation in Distributed Systems.* Cambridge: MIT Press.

———. 1989 "Supporting Multi-paradigm Programming on Actor Architectures." In *Proceedings of Parallel Architecture and Languages Europe.* Vol. 2. Berlin: Springer-Verlag. 1–19.

Ahuja, N., and A. L. Abbott. 1993. "Active Stereo: Integrating Disparity, Vergence, Focus, Aperture, and Calibration for Surface Estimation." *IEEE Transactions on Pattern Analysis and Machine Intelligence* 15, 1007–1029.

Aloimonos, J. 1986. "Detection of Surface Orientation from Texturei: The Case of Planes." In *Proceedings of IEEE Conference on Computer Vision and Pattern Recognition.* 584–593.

Aloimonos, J., and M. Swain. 1985. "Shape from Texture." In *Proceedings of 9th International Joint Conference on AI.* 926–931.

Aloimonos, Y., I. Weiss, and A. Bandyopadhyay. 1987. "Active Vision." In *Proceedings of First International Conference on Computer Vision,* London. 35–54.

Ardema, M. D, and J. M. Skowronski. 1991. "Dynamic Game Applied to Coordination Control of Two Arm Robotic System." In *Differential Games: Developments in Modelling and Computation.* Ed. R. P. Hämäläinen and H. K. Ehtamo. Berlin: Springer-Verlag. 118–130.

Bajcsy, R., 1988. "Active Perception." *Proceedings of IEEE* 76, 996–1005.

———. 1992. "Active Observer." In *Proceedings of DARPA Image Understanding Workshop,* San Diego. 137–147.

Bajcsy, R., and L. Lieberman. 1976. "Texture Gradient as a Depth Cue." *Computer Graphics Image Processing* 5, 52–67.

Ballard, D. H. 1989. "Reference Frames for Animate Vision." In *Proceedings of IJCAI-89,* Detroit. 1635–1641.

Ballard, D. H., and A. Ozcandarli. 1988. "Eye Fixation and Early Vision: Kinetic Depth." In *Proceedings of Second International Conference on Computer Vision,* Tarpon Springs, Florida. 524–531.

Barraquand, J., and J.-C. Latombe. 1991. "Robot Motion Planning: A Distributed Representation Approach." *International Journal of Robotics Research* 10(6), 628–649.

Barraquand, J., B. Langlois, and J. C. Latombe. 1992. "Numerical Potential Field Techniques for Robot Path Planning." *IEEE Transactions on Systems, Man, and Cybernetics* 22(2), 224–241.

Başar, T., and G. J. Olsder. 1982. *Dynamic Noncooperative Game Theory.* London: Academic Press.

Bien, Z., and J. Lee. 1992. "A Minimum-Time Trajectory Planning Method for Two Robots." *IEEE Transactions on Robotics and Automation* 8(3), 414–418.

Blake, A., and R. Cippola. 1992. "Surface Shape from the Deformation of Apparent Contours." *International Journal of Computer Vision* 9(2).

Blostein, D., and N. Ahuja. 1989. "A Multiscale Region Detector." *Computer Vision, Graphics and Image Processing* (January), 22–41.

Braunstein, M. L., and J. W. Payne. 1969. "Perspective and Form Ratio as Determinants of Relative Slant Judgments." *Journal of Experimental Psychology* 81(3), 584–590.

Buckley, S. J., "Fast Motion Planning for Multiple Moving Robots." In *IEEE International Conference on Robotics and Automation.* 322–326.

Burt, P. J. 1988." Algorithms and Architectures for Smart Sensing." In *Proceedings of DARPA Image Understanding Workshop,* Cambridge, Mass. 139–153.

Cammarata, S., D. McArthur, and R. Steeb. 1983. "Strategies of Cooperation in Distributed Problem Solving." In *Proceedings of 8th International Joint Conference on Artificial Intelligence.* 767–770.

Chang, C., M. J. Chung, and B. H. Lee. 1994. "Collision Avoidance of Two Robot Manipulators by Minimum Delay Time." *IEEE Transactions on Systems, Man, and Cybernetics* 24(3), 517–522.

Clark, J. J., and N. J. Ferrier. 1988. "Modal Control of an Attentive Vision System." In *Proceedings of Second International Conference on Computer Vision,* Tarpon Springs, Florida. 514–523.

Corkill, D. 1979. "Hierarchical Planning in a Distributed Environment." In *Proceedings of Sixth International Joint Conference on Artificial Intelligence,* Tokyo, August. 168–175.

Croft, W. B., and Lefkowitz. 1988. "Knowledge-Based Support of Cooperative Activities." In *Proceedings of 21st Annual Hawaii International Conference on System Sciences.* Vol. 3. 312–318.

Cutting, J. E., and R. T. Millard. 1984. "Three Gradients and the Perception of Flat and Curved Surfaces." *Journal of Experimental Psychology: General* 113(2), 198–216.

Das, S., and N. Ahuja. 1995. "Performance Analysis of Stereo, Vergence and Focus as Depth Cues for Active Vision." *IEEE Transactions on Pattern Analysis and Machine Intelligence* 17(12), 1213–1219.

———. 1996. "Active Surface Estimation: Integrating Coarse-to-Fine Image Acquisition and Estimation from Multiple Cues." *Artificial Intelligence* 83, 241–266.

Davis, L., L. Janos, and S. Dunn. 1983. "Efficient Recovery of Shape from Texture." *IEEE Transactions on Pattern Analysis and Machine Intelligence* PAMI-5(5), 485–492.

Durfee, E. H., 1991. "Introduction to Special Issue." *IEEE Transactions on Systems, Man, and Cybernetics* 21(6).

Durfee, E. H., V. R. Lesser, and D. D. Corkill. 1987. "Cooperation Through Communication in a Distributed Problem Solving Network." *Distributed Artificial Intelligence.* 29–58.

Erdmann, M., and T. Lozano-Pérez. 1986. "On Multiple Moving Objects." In *IEEE International Conference on Robotics and Automation.* 1419–1424.

Erkelens, C. J., and H. Collewijn. 1985. "Eye Movements and Stereopsis During Dichoptic Viewing of Moving Random-Dot Stereograms." *Vision Research* 25, 1689–1700.

Farber, D., and K. Larson. 1972. "The Structure of a Distributed Processing Systems Software." In *Proceedings of Symposium on Computer Communications Networks and Teletraffic,* Poletechnica Institute of Brooklyn, April.

Finin, T., R. Fritzson, and D. McKay. 1992. "A Language and Protocol to Support Intelligent Agent Interoperability." In *Proceedings CE&CALS Washington Conference,* Washington, D.C.

Fu, K. S., R. C. Gonzalez, and C. S. G. Lee. 1987. *Robotics: Control, Sensing, Vision, and Intelligence.* New York: McGraw-Hill.

Fujimura, K., and H. Samet. 1993. "Planning a Time-Minimal Motion Among Moving Obstacles." *Algorithmica* 10, 41–63.

Gasser, L., and M. Hahns. 1989. *Distributed Artificial Intelligence.* Vol. 2. London: Pitman.

Geiger, D., and A. Yuille. 1987. "Stereopsis and Eye-Movement." In *Proceedings of First International Conference on Computer Vision,* London. 360–374.

Georgeff, M. 1983. "Communication and Interaction in Multi-agent Planning." In *AAAI-83.*

Giblin, P., and R. Weiss. 1987. "Reconstruction of Surfaces from Profiles." In *IEEE Conference on Computer Vision and Pattern Recognition.*

Giblin, P., J. Rycroft, and F. Pollick. 1993. "Moving Surfaces." In *Mathematics of Surfaces V.* Cambridge: Cambridge University Press.

Guha, R., and D. Lenat. 1994. "Enabling Agents to Work Together." *Communications of the ACM* 37(7), 126–142.

Hipel, K. W., K. J. Radford, and L. Fang. 1993. "Multiple Participant-Multiple Criteria Decision Making." *IEEE Transactions on Systems, Man, and Cybernetics* 23(4), 1184–1189.

Huhns, M. N., and D. M. Bridgeland. 1991. "Multiagent Truth Maintenance." *IEEE Transactions on Systems, Man, and Cybernetics* 21(6), 1437–1445.

Hwang, Y. K., and N. Ahuja. 1992a. "Gross Motion Planning—A Survey." *ACM Computing Surveys* 24(3), 219–291.

———. 1992b. "A Potential Field Approach to Path Planning." *IEEE Transactions on Robotics and Automation* 8(1), 23–32.

Ikeuchi, K. 1980. "Shape from Regular Patterns (An Example of Constraint Propagation in Vision)." A. I. Memo 567, Massachusetts Institute of Technology.

Jin, Y., and R. E. Levitt. 1993. "i-agents: Modeling Organizational Problem Solving in Multi-agent Teams." *Intelligent Systems in Accounting, Finance, and Management* 2, 247–270.

Joshi, T. N., Ahuja, and J. Ponce. 1995. "Structure and Motion Estimation from Dynamic Silhouettes under Perspective Projection." In *Proceedings of 5th International Conference on Computer Vision,* Cambridge, June. 290–295.

Joshi, T., et al. 1994. "Hot Curves for Modelling and Recognition of Smooth Curved 3d Shapes." In *Proceedings of IEEE Conference on Computer Vision and Pattern Recognition.*

Kanatani, K. 1984. "Detection of Surface Orientation and Motion from Texture by a Stereological Technique." *Artificial Intelligence* **23,** 213–237.

Kanatani, K., and T. Chou. 1986. "Shape from Texture: General Principle." In *Proceedings of IEEE Conference on Computer Vision and Pattern Recognition,* Miami, June. 578–583.

Kender, J. 1980. "Shape from Texture." Ph.D. Thesis, Carnegie-Mellon University. CMU-CS-81-102.

Kornfeld, W., and C. Hewitt. 1981. "The Scientific Community Metaphor." *IEEE Transactions on Systems, Man, and Cybernetics* **11,** 24–33.

Kountouris, V. G., and H. E. Stephanou. 1991. "Dynamic Modularization and Synchronization for Intelligent Robot Coordination: The Concept of Functional Time-Dependency." In *IEEE International Conference on Robotics and Automation,* Sacramento, April. 508–513.

Krotkov, E. P. 1989. *Active Computer Vision by Cooperative Focus and Stereo.* New York: Springer-Verlag.

Kutulakos, K., and C. Dyer. 1994. "Global Surface Reconstruction by Purposive Control of Observer Motion." In *Proceedings of IEEE Conference on Computer Vision and Pattern Recognition.*

Latombe, J.-C. 1991. *Proceedings of Robot Motion Planning.* Boston: Kluwer Academic.

LaValle, S., and S. Hutchinson. 1991. "Multiple-Robot Motion Planning under Independent Objectives." *IEEE Transactions on Robotics and Automation.*

Lesser, V. R. 1991. "A Retrospective View of fa/c Distributed Problem Solving." *IEEE Transactions on Systems, Man, and Cybernetics* **21**(6), 1347–1362.

Lesser, V. R., and D. D. Corkill. 1981. "Functionally Accurate, Cooperative Distributed Systems." *IEEE Transactions on Systems, Man, and Cybernetics* **11**(1), 81–96.

Lin, C.-F., and W.-H. Tsai. 1990. "Motion Planning for Multiple Robots with Multi-mode Operations via Disjunctive Graphs." *Robotica* **9,** 393–408.

Lin, G., and J. Solberg. 1992. "Integrated Shop-Floor Control Using Autonomous Agents." *IIE Transactions on Design and Manufacturing* **24**(3), 57–71.

Maes, P. 1994. "Agents That Reduce Work and Information Overload." *Communications of the ACM* **37**(7), 37–40.

Marr, D., and T. Poggio. 1979. "A Computational Theory of Human Stereo Vision." *Royal Society of London B* **204,** 301–328.

Minsky, M. 1985. *The Society of Mind.* New York: Simon & Schuster.

Nakatani, H., et al. 1980. "Extraction of Vanishing Point and Its Application to Scene Analysis Based on Image Sequence." In *Proceedings of International Conference on Pattern Recognition.* 370–372.

Nii, H. P., N. Aiello, and J. Rice. 1989. "Experiments on Cage and Poligon: Measuring the Performance of Parallel Blackboard Systems." *Distributed Artificial Intelligence* **2,** 319–384.

O'Donnell, P. A., and T. Lozano-Pérez. 1989. "Deadlock-Free and Collision-Free Coordination of Two Robot Manipulators." In *IEEE International Conference on Robotics and Automation.* 484–489.

Owen, G. 1982. *Game Theory.* New York: Academic Press.

Papastavrou, J., and S. Nof. 1992. "Decision Integration Fundamentals in Distributed Manufacturing Topologies." *IIE Transactions* **24**(3), 27–42.

Phillips, R. J. 1970. "Stationary Visual Texture and the Estimation of Slant Angle." *Quarterly Journal of Psychology* **22,** 389–397.

Reif, J. H., and M. Sharir. 1985. "Motion Planning in the Presence of Moving Obstacles." In *Proceedings of IEEE Symposium on Foundations of Computer Science.* 144–154.

Rimon, E., and D. E. Koditschek. 1992. "Exact Robot Navigation Using Artificial Potential Fields." *IEEE Transactions on Robotics and Automation* **8**(5), 501–518.

Rosenfeld, A. 1975. "A Note on Automatic Detection of Texture Gradients." *IEEE Transactions on Computers* **C-24,** 988–991.

Rosenschein, J. S., and M. R. Genersereth. 1985. "Deals Among Rational Agents." In *Proceedings of 9th International Joint Conference on Artificial Intelligence.* 91–99.

Rosinski, R. R. 1974. "On the Ambiguity of Visual Stimulation: A Reply to Eriksson." *Perception Psychophysics* **16**(2), 259–263.

Rosinski, R., and N. Levine. 1976. "Texture Gradient Effectiveness in the Perception of Surface Slant." *Journal of Experimental Child Psychology* **22,** 261–271.

Sathi, A and M. Fox. 1989. "Constraint-Directed Negotiation of Resource Reallocations." *Distributed Artificial Intelligence* **2,** 163–194.

Sawaragi, Y., H. Nakayama, and T. Tanino. 1985. *Theory of Multiobjective Optimization.* New York: Academic Press.

Schwartz, J. T., and M. Sharir. 1983a. "On the Piano Mover's Problem: I. The Case of a Two-Dimensional Rigid Polygonal Body Moving Amidst Polygonal Barriers. *Communications on Pure and Applied Mathematics* **36**, 345–398.

———. 1983b. "On the Piano Mover's Problem: III. Coordinating the Motion of Several Independent Bodies." *International Journal of Robotics Research* **2**(3), 97–140.

Shaw, M., and M. Fox. 1993. "Distributed Artificial Intelligence for Group Decision Support: Integration of Problem Solving, Coordination, and Learning." *Decision Support Systems* **9**, 349–367.

Shaw, M., and A. B. Whinston. 1989. "Learning and Adaptation in Distributed Artificial Intelligence Systems." *Distributed Artificial Intelligence* **2**, 413–430.

Shaw, M., and G. Wiegand. 1988. "Intelligent Information Processing in Flexible Manufacturing Systems." *FMS Magazine* **6**(3), 137–140.

Shih, C. L., T.-T. Lee, and W. A. Gruver. 1990. "A Unified Approach for Robot Motion Planning with Moving Polyhedral Obstacles. *IEEE Transactions on Systems, Man, and Cybernetics* **20**, 903–915.

Shin, K. G., and Q. Zheng. 1992. "Minimum-Time Collision-Free Trajectory Planning for Dual-Robot Systems." *IEEE Transactions on Robotics and Automation* **8**(5), 641–644.

Smith, R. G. 1980. "The Contract Net Protocol: High Level Communication and Control in a Distribute Problem Solver." *IEEE Transactions on Computers* **29**, 1104–1113.

Spence, R., and S. A. Hutchinson. 1995. "An Integrated Architecture for Robot Motion Planning and Control in the Presence of Obstacles with Unknown Trajectories." *IEEE Transactions on Systems, Man, and Cybernetics* **25**(1), 100–110.

Sperling, G. "Binocular Vision: A Physical and a Neural Theory." *American Journal of Psychology* **83**, 467–534.

Sull, S., and N. Ahuja. 1994. "Integrated 3-D Analysis and Analysis Guilded Synthesis of Flight Image Sequences." *IEEE Transactions on Pattern Analysis and Machine Intelligence* **16**(4), 357–372.

Szelinski, R., and R. Weiss. 1993. "Robust Shape Recovery from Occluding Contours Using a Linear Smoother." In *Proceedings of IEEE Conference on Computer Vision and Pattern Recognition.*

Tan, G. W., C. Hayes, and M. Shaw. 1995. "An Intelligent Agent Framework for Concurrent Product Design and Planning." Technical Report AI-DSS-95-03, Beckman Institute.

Tistarelli, M., and G. Sandini. 1990. "Robot Navigation Using an Anthropomorphic Visual Sensor." In *Proceedings of IEEE Conference on Robotics and Automation,* Cincinnati. 374–381.

Upton, D. 1989. "*The Operation of Large Computer Controlled Manufacturing Systems.*" Ph.D. Thesis, Purdue University.

Vaillant, R., and O. Faugeras. 1992. "Using Extremal Boundaries for 3-D Object Modeling." *IEEE Transactions, Pattern Analysis and Machine Intelligence* **14**(2), 157–173.

Veeramani, D., and K. J. Wang. 1997. "Performance Analysis of Auction-based Distributed Shop-floor Control Scheme from the Perspective of the Communication System." *International Journal of Flexible Manufacturing Systems* **9**(2), 121–143.

Vickers, D. 1971. "Perceptual Economy and the Impression of Visual Depth." *Perception and Psychophysics* **10**(1), 23–27.

Wang, F.-Y and P. J. A. Lever. 1994. "A Cell Mapping Method for General Optimum Trajectory Planning of Multiple Robotic Arms." *Robots and Autonomous Systems* **12**, 15–27.

Wang, K. J., and D. Veeramani. 1994. "Comparison of Distributed Control Schemes from the Perspective of the Communication System." In *Proceedings of IIE Research Conference,* Atlanta, May. 394–399.

Warren, C. W. 1990. "Multiple Robot Path Coordination Using Artificial Potential Fields." In *IEEE International Conference on Robotics and Automation.* 500–505.

Witkin, A. P. 1981. "Recovering Surface Shape and Orientation from Texture." *Artificial Intelligence* **17**, 17–45.

Woo, C. C., and F. H. Lochovsky. 1986. "Supporting Distributed Office Problem Solving in Organizations." *ACM Transactions on Office Information Systems,* 185–204.

Zheng, J. Y. 1994. "Acquiring 3-D Models from Sequences of Contours." *IEEE Transactions on Pattern Analysis and Machine Intelligence* **16**(2), 163–178.

Zionts, S. 1985. "Multiple Criteria Mathematical Programming: An Overview and Several Approaches." In *Mathematics of Multi-Objective Optimization.* Ed. P. Serafini. Berlin: Springer-Verlag. 227–273.

CHAPTER 23

GROUP BEHAVIOR OF ROBOTS

George A. Bekey
University of Southern California
Los Angeles, California

Arvin Agah
The University of Kansas
Lawrence, Kansas

1 INTRODUCTION

"Is more better?" This is the fundamental question that has inspired one of the important recent trends in robotics. Research performed under such titles as *multirobot systems, distributed robotic systems, swarm robotics, sociorobotics, decentralized robotics,* and *cellular robotics* has focused on the investigation of issues and applications of systems composed of groups of robots. This area of research has gained enough momentum to have resulted in an international conference, Distributed Autonomous Robotics Systems (Asama et al., 1996*a*), devoted solely to this field. The general idea is that teams of robots deployed to achieve a common goal are not only able to perform tasks that a single robot is unable to, but also can outperform systems of individual robots in terms of efficiency and quality. In addition, groups of robots provide a level of robustness, fault tolerance, and flexibility, since the failure of one robot does not result in the mission being unsuccessful as long as the remaining robots share the tasks of the failed robot. Examples of tasks appropriate for robot teams are large area surveillance, environmental monitoring, autonomous reconnaissance, large object transportation, planetary exploration, and hazardous waste cleanup.

The common impression of group behavior of robots is generally of a small number of robots communicating via radio frequency links to synchronize their efforts in carrying a large object. However, researchers in the relatively new field of multirobot systems have been experimenting with and producing unconventionally novel systems. One currently available system is a group of robots that read and write into small data units that can be carried and deposited at suitable places (Arai et al., 1996). The idea behind these intelligent data carriers is inspired by insects leaving chemical trails for others in the colony. In terms of locomotion, not only can a group of robots cover a large area, but it is also shown in hardware that robots can help lift each other over large obstacles (Asama et al., 1996*b*)! This is achieved by equipping all robots with forklifts capable of helping lift the robot itself (by pushing onto the ground) or another robot. Thus it is clear that group behavior of robots promises to be not only an area of robotics research with applications, but a very interesting area as well.

The research in the area of multirobot systems is illustrated within the robotics and artificial intelligence *research space* (as opposed to the *search space*) in Figure 1. The brain-versus-body graph is shown, including systems with no body and one brain, no body and multiple brains, one body with no brains (machines), one body with one brain (traditional robot), tens of bodies with brains (multirobot systems or distributed robotic systems), and hundreds of bodies with brains (swarm systems). The groups of robots discussed in this chapter cover the area of distributed robotic systems and swarm robotics.

Handbook of Industrial Robotics, Second Edition, Edited by Shimon Y. Nof
ISBN 0-471-17783-0 © 1999 John Wiley & Sons, Inc.

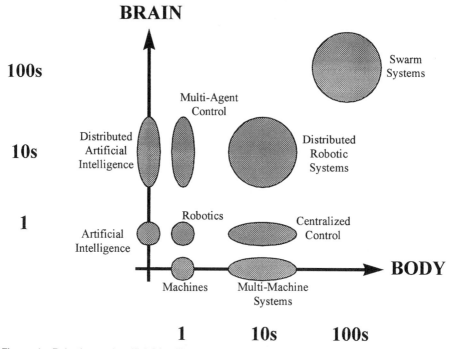

Figure 1 Robotics and artificial intelligence research space. This research space is the brain-versus-body graph. Artificial intelligence deals with study of intelligence not embedded within a body, while distributed artificial intelligence considers multiple brains. In multiagent control multiple brains are used to control one body, while in centralized control one brain is used to control many bodies. Robotics traditionally has been the investigation of one body with one brain, although that is changing. The distinction between distributed robotic systems and swarm systems is that the former is concerned with systems composed of tens of bodies with brains, while the latter deals with systems with hundreds or more components.

2 HISTORY AND RECENT RESEARCH

One of the first known studies of a multirobot system dates back to 1950. It is described in (Walter, 1950). Two electromechanical tortoises of the mock-biological class *Machina Speculatrix* were built to help in investigating the behavioral complexity and independence of autonomous mobile robots consisting of miniature radio tubes, light and touch sensors, crawling and steering motors, and batteries. The robots explored the environment when no light source was present. If a light source was present, it attracted the machines until they were too close, in which case they steered away. Placing of lights on the robots allowed for the investigation of self-recognition using a mirror and the study of group behavior of robots.

This research was ahead of its time, and there was little activity in the study of multirobot systems until the late 1980s and early 1990s, when a relatively large number of researchers began to investigate and build systems of multiple robots. A selected number of works that are representative of the research activities are described in this article, and more are listed at the end under Additional Reading.

A cellular robotic system is a dynamically reconfigurable robotic system, adaptable to the task and environment, composed of diverse robotic cells combined in different configurations (Fukuda, Ueyama, and Arai, 1992). Each cell is autonomous and mobile, able to seek and physically join other appropriate cells. Investigations have included self-organization and self-evaluation of robot cells, communication among robots, and control strategies for group behavior (Ueyama, Fukuda, and Arai, 1992). Another example of multirobot systems composed of different types of robots is an autonomous decentralized robotics system developed for use in complex tasks such as nuclear power plant maintenance (Suzuki et al., 1996).

Investigation of cooperative behavior of multirobot systems composed of a small number of identical robots (from 2 to 20) has focused on the issues of team organization and task performance. Twenty wheeled mobile robots equipped with grippers and their own radio transmitter and receiver have been used to study group behavior of robots (Mataric, 1992). Using three levels of control strategies, *ignorant coexistence, informed coexistence,* and *intelligent coexistence,* each robot used its measures of local population density and population gradient to balance its behavior between collision and isolation. The absolute positions and the headings of the robots were transmitted via the radios, and different strategies were used to test the flocking behavior of the robots. In another research project cooperative behavior of multirobot systems was studied, employing a group of five robots without explicit communication in a collective box pushing task (Kube and Zhang, 1994). The effects of population size on the box-pushing task were studied. Cooperation of a group of robots without communications was also studied in (Arkin, 1992), where the behavior of a robot was composed of a collection of building blocks (motor schemas). The robots could switch to a number of states, including forage, acquire, and deliver, iteratively in order to perform the task of finding certain objects and delivering them.

Development of a specific theory of interactions and learning among multiple robots performing certain tasks has been studied in (Agah and Bekey, 1997). A primary objective of this research was the study of the feasibility of a robot colony in achieving global objectives when each individual robot was provided only with local goals and local information. In order to achieve this objective, a novel cognitive architecture for the individual behavior of robots in a colony was introduced. Experimental investigation of the properties of the colony demonstrates the group's ability to achieve global goals, such as the gathering of objects, and to improve its performance as a result of learning, without explicit instructions for cooperation. A photograph of the robot group is shown in Figure 2. Since this architecture was based on representation of the "likes" and "dislikes" of

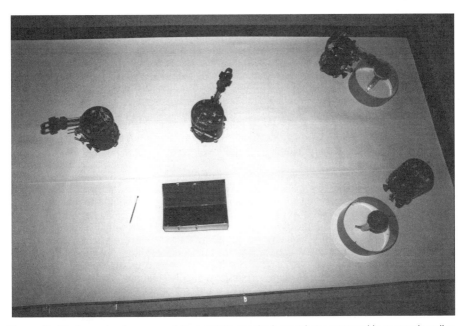

Figure 2   Photograph of a group of four mobile-manipulator robots engaged in a sample collection and retrieval task. Each robot is autonomously controlled by an on-board microprocessor and powered by rechargeable batteries. Actuation is achieved through independently controlled wheels for motion and a manipulator and a gripper for object manipulation. Robot sensors include optical sensors and touch sensors. The body of each robot is made from ABS plastic, is cylindrical, and is approximately 16 cm in diameter, and each robot's height is 32 cm. The robots were designed and built in order to test and validate a number of group behavior concepts initially developed through simulation experiments (Agah and Bekey, 1997).

the robots, it was called the Tropism System Cognitive Architecture (Agah and Bekey, 1997).

Systems that are composed of a relatively large number of simple members (hundreds) are referred to as *swarm systems*. These systems exhibit collectively intelligent behavior (Beni and Hackwood, 1990). Robots of different types, each executing different protocols, would be arranged in a variety of patterns to form structures such as a measuring swarm, a distributed sensor where each member of the swarm is sensitive to a specific value. The objective was then to determine optimal structures of the swarm (Hackwood and Beni, 1992).

Another trend in the study of group behavior of robots has been the investigation of insect-like behaviors in simulated robots. A distributed sorting algorithm, similar to brood sorting of ant colonies, was implemented by Deneubourg et al., (1991), suggesting that the communication of robots was indirect through their effects on the environment. Chain-making behavior of robots was studied by Goss and Deneubourg, (1992), where robots spread themselves out in the environment while remaining in contact with each other. The robots in the area used a central beacon with a limited signal range. Once a robot reached the limit of the beacon signal, and no other beacon signals were available, the robot would switch to the beacon mode. Assigning a unique identification to each beacon signal, with increasing numeric values, allowed the return of the robots by the process of following the chain in decreasing order of beacon identification numbers.

Other representative examples of research on design, control, and applications of multirobot systems include multirobot team design for specific tasks (Parker, 1996), coordination of autonomous and centralized decision-making (Yuta and Premvuti, 1992), and consistent ordering of discrete events in distributed robotics (Wang, 1993), among many others. In addition, a number of review articles and collections of papers have appeared in the literature, providing a good starting point for those interested in further exploration of the field of multirobot systems (Asama et al., 1996a; Cao et al., 1995).

3 CONCEPTS AND TRENDS

The most significant concept in group behavior of robots is cooperation. Only through cooperative task performance can the superiority of robot groups be demonstrated. The cooperation of robots in a group can be classified into two categories: implicit cooperation and explicit cooperation. In implicit cooperation each robot performs individual tasks, while the collection of these tasks is toward a unified mission. For example, when multiple robots are engaged in collecting rock samples and returning them to a common place, the team is accomplishing a global mission while cooperating implicitly. This type of group behavior is also called *asynchronous cooperation,* as it requires no synchronization in time or space. In explicit cooperation robots in a team work synchronously with respect to time or space in order to achieve a goal. One example is transportation of heavy objects by multiple robots, each having to contribute to the lifting and moving of the object. This task requires the robots to be positioned suitably with respect to each other and to function simultaneously.

Group behavior of explicit cooperation requires communication. The concept of communication for enhancing group behavior has taken several forms. In its simplest form the communication is achieved through changes in the environment. Moving an object could communicate a message to the next robot, which senses the change in the object location. In broadcast communication systems messages are transmitted in a broad range to whichever robot is within a receiving range. In contrast, in the peer-to-peer format messages are sent by one robot specifically to a designated robot. In a more novel approach to communication, researchers have achieved accurate multirobot navigation and self-localization using intelligent data carriers, small self-contained units that are moved around (Arai et al., 1996).

Based on approach and methodology, the research efforts on the group behavior of robots can be categorized according to a number of distinguishing factors:

- Inspirations from distributed artificial intelligence versus artificial life versus sociobiology versus distributed computing versus traditional robotics
- Homogeneous robot groups (identical robots) versus heterogeneous groups (variety of robots
- Communicating robots versus multirobot systems without communications
- Preassigned task allocation versus distributed dynamic task allocation

- Theoretical foundations (analytical studies and formalisms) versus applied approach (engineering).
- Mobile robots (group navigation, local interactions, explicit cooperation) versus manipulators (optimal force distribution, collision-free path planning, optimal motion planning) versus swarm robots (distributed sensing, distributed action, distributed intelligence, self-organization)
- Centralized systems (center for intelligence) versus decentralized (distributed intelligence)
- Unit being complete robots versus modular units combining to form a robot

The overall trend of research in multirobot systems will continue to be toward having the merits of robustness, fault tolerance, and flexibility outweigh the demerits of group interaction and coordination complexity (harmonious, compatible task performance without deadlocks and duplications).

4 APPLICATIONS

Applications of robot teams are in four basic areas:

1. Large objects must be handled.
2. Large areas must be covered.
3. Iterative tasks must be performed.
4. Robustness and fault tolerance is required.

Large-material handling and transportation can benefit from the use of multiple robots, since one very large robot is not energy-efficient, is too expensive to build and maintain, requires too much maneuvering space, and is not adaptable for large objects of different shapes. Covering of large areas such as surveillance, map generation, or sample collection can be accomplished in a more timely manner when multiple robots are employed. Iterative tasks such as cleaning jobs or transport of hundreds of items from one location to another can also benefit in terms of time when many robots are used. In applications where the system is expected to be robust and fault-tolerant, a system composed of a group of robots has the advantage that it can recover from failure of one robot, if it is assumed that the other team members can share the tasks of the failed robot.

Although there are relatively few real-world applications of multirobot systems (perhaps due to the young age of the field), the applications are growing in number. Actual applications of robot teams include mobile robots for military surveillance and safe travel of groups through dangerous areas (Parker, 1996). Teams of small underwater robots have been used for a distributed measurement system in a pulp beach process for quality control of a number of bleach towers and algae removal (Vainio, 1996). In potentially hazardous environments robot groups have been deployed for plant maintenance and operation of a pressurized water reactor nuclear power plant through a human operator interfacing with the team of robots (Suzuki et al., 1996), and surveillance and characterization of radiologically contaminated facilities prior to and during cleanup activities (Anderson, McKay, and Richardson, 1996). In addition to these implemented applications, there are numerous potential applications of robot groups that are currently under investigation by researchers in this field, including multirobot systems for assembly and manufacturing, coordinated movements of automated vehicles, and robot teams for planetary exploration. Swarm robot systems are also studied in terms of their applications to measurements and sensing of environments where large robots are not capable of monitoring. It is through the collective sensing and proper integration of the swarm robots that pooled capabilities are achieved.

5 ISSUES AND CHALLENGES

The current trend in multirobot group research follows two correlated tracks:

1. Study of issues in multirobot systems in hardware with small population sizes (under 20)
2. Study of issues in multiagent systems in simulation with large population sizes

Construction, maintenance, and utilization of large groups of robots have proven to be infeasible due to time and budget requirements. It is too difficult to build a team of 100 robots, make sure that all are functioning, and perform experiments with them. Instead researchers have been conducting hardware experiments with only a few robots and then augmenting their hardware studies with computer modeling and simulation of robot groups with large populations. It should be noted that the effects of team size and its scaling are integral issues in robot group studies and the reliability of the simulation results remains to be seen.

Although some important results have been obtained by researchers in the area of multirobot systems, a great deal of work remains to be done in order for the group behavior of robots to be fully understood and utilized in real-world applications. This includes a number of issues that require investigations, as they are open challenges facing researchers attempting to understand the group behavior of robots. Following are some of these issues in the form of questions:

- What task features would necessitate a robot group?
- What environmental factors influence the robot group?
- What is a good robot group size?
- Should there be leadership in a robot group, and if so, in what form?
- How should the robot group be structured?
- Should the robot members be specialized or multitasked?
- What rules govern the group behavior of the robots?
- Should the robot group behavior be self-organizing?
- How can the remaining robots detect and recover from the failure of one or more robots?
- How can the robot group's performance be enhanced?
- Is communication necessary for task performance, and what format should it have?
- What real-world applications can truly benefit from teams of robots?
- How can robot groups be designed and fabricated for suitable real-world applications?

It is by thorough investigation of the above issues that the group behavior of robots can be understood, yielding potential benefits in real-world applications.

REFERENCES

Agah, A., and G. A. Bekey. 1997. "Phylogenetic and Ontogenetic Learning in a Colony of Interacting Robots." *Autonomous Robots Journal* **4,** 85–100.

Anderson, M. O., M. D. McKay, and B. S. Richardson. 1996. "Multi-robot Automated Indoor Floor Characterization Team." In *Proceedings of the IEEE International Conference on Robotics and Automation,* Minneapolis, April. 1750–1753.

Arai, Y., et al. 1996. "Self-localization of Autonomous Mobile Robots Using Intelligent Data Carriers." In *Distributed Autonomous Robotic Systems 2* Ed. H. Asama et al. Tokyo: Springer-Verlag. 401–410.

Arkin, R. C. 1992. "Cooperation Without Communication: Multiagent Schema-Based Robot Navigation." *Journal of Robotic Systems* **9,** 351–364.

Asama, H., et al. 1994. "Collaborative Team Organization Using Communication in a Decentralized Robotic System." In *Proceedings of the IEEE/RSJ International Conference on Intelligent Robots and Systems,* Munich, September 1994. 816–823.

Asama, H., et al. 1996*b.* "Mutual Transportation of Cooperative Mobile Robots Using Forklift Mechanisms." In *Proceedings of the IEEE International Conference on Robotics and Automation,* Minneapolis, April. 1754–1759.

Beni, G., and S. Hackwood. 1990. "The Maximum Entropy Principle and Sensing in Swarm Intelligence." In *Toward a Practice of Autonomous Systems.* Ed. F. J. Varela and P. Bourgine. Cambridge: MIT Press. 153–160.

Cao, Y. U., et al. 1995. "Cooperative Mobile Robotics: Antecedents and Directions." In *Proceedings of the IEEE/RSJ International Conference on Intelligent Robots and Systems,* Pittsburgh, August. Vol. 1. 226–234.

Deneubourg, J. L., et al. 1991. "The Dynamics of Collective Sorting Robot-Like Ants and Ant-Like Robots." In *From Animals to Animats.* Ed. J.-A. Meyer and S. W. Wilson. Cambridge: MIT Press. 356–363.

Fukuda, T., T. Ueyama, and F. Arai. 1992. "Control Strategies for Cellular Robotic Network." In *Distributed Intelligence Systems.* Ed. A. H. Lewis and H. E. Stephanou. Oxford: Pergamon Press. 177–182.

Goss, S., and J. L. Deneubourg. 1990. "Harvesting by a Group of Robots." In *Toward a Practice of Autonomous Systems*. Ed. F. J. Varela and P. Bourgine. Cambridge: MIT Press. 195–204.

Hackwood, S., and G. Beni. 1992. "Self-Organization of Sensors for Swarm Intelligence." In *Proceedings of the IEEE International Conference on Robotics and Automation,* Nice, France, May. 819–829.

Kube, C. R., and H. Zhang. 1994. "Collective Robotics: From Social Insects to Robots." *Adaptive Behavior* **2,** 189–218.

Mataric, M. J. 1992. "Minimizing Complexity in Controlling a Mobile Robot Population." In *Proceedings of the IEEE International Conference on Robotics and Automation,* Nice, France, May. 830–835.

Parker, L. 1996. "Multi-robot Team Design for Real-World Applications." In *Distributed Autonomous Robotic Systems 2*. Ed. H. Asama et al. Tokyo: Springer-Verlag. 91–102.

Suzuki, T., et al. 1996. "Cooperation Between a Human Operator and Multiple Robots for Maintenance Tasks at a Distance." In *Distributed Autonomous Robotic Systems 2*. Ed. H. Asama et al. Tokyo: Springer-Verlag. 50–59.

Ueyama, T., T. Fukuda, and F. Arai. 1992. "Configuration of Communication Structure for Distributed Intelligent Robotic System." In *Proceedings of the IEEE International Conference on Robotics and Automation*, Nice, France, May. 807–812.

Vainio, M., et al. 1996. "An Application Concept of an Underwater Robot Society." In *Distributed Autonomous Robotic Systems 2*. Ed. H. Asama et al. Tokyo: Springer-Verlag. 103–114.

Wang, J. 1993. "Establish a Globally Consistent Order of Discrete Events in Distributed Robotic Systems." In *Proceedings of the IEEE International Conference on Robotics and Automation,* Atlanta, May. Vol. 3. 853–858.

Walter, W. G. 1950. "An Imitation of Life." *Scientific American* **182,** 42–45.

Yuta, S., and S. Premvuti. 1992. "Coordinating Autonomous and Centralized Decision Making to Achieve Cooperative Behaviors Between Multiple Mobile Robots." In *Proceedings of the IEEE/RSJ International Conference on Intelligent Robots and Systems,* Raleigh, North Carolina, July. 1566–1574.

ADDITIONAL READING

Asama, H., et al. eds. *Distributed Autonomous Robotic Systems 2*. Tokyo: Springer-Verlag, 1996*a*.

PART 5

ORGANIZATIONAL AND ECONOMIC ASPECTS

CHAPTER 24

INDUSTRIAL ROBOTICS STANDARDS

Nicholas G. Dagalakis
National Institute of Standards and Technology
Gaithersburg, Maryland

1 INTRODUCTION

A standard is defined in "National Policy on Standards" (1978) as:

A prescribed set of rules, conditions, or requirements concerning definition of terms; classification of components; specification of materials, performance, or operations; delineation of procedures; or measurement of quantity and quality in describing materials, products, systems, services, or practices.

A good standard under the proper market conditions can help to increase competition, reduce the cost of products and services, break trade barriers, and expand markets. The phenomenal success of the personal computer (PC) market is the best example of good architecture and interface standards. An additional benefit is the countless lives saved and accidents prevented by health and safety standards.

Due to space limitations, only a brief review of the subject of industrial robot standards can be provided here. The objective is to cover the following three subjects:

1. Provide a brief general description of the U.S. standards setting process.
2. Describe a few of the most important standards.
3. Provide as many relevant references known to the author as possible.

In the United States the only organization that is active in writing Industrial Robotics Standards is the Robotic Industries Association (RIA) (see Appendix B). RIA is a member of the American National Standards Institute (ANSI) (see Appendix B), which has generated a legal framework within which the various member associations can write standards. The RIA committees and subcommittees have to follow the rules set by ANSI for standards writing so that the standards conform to its legal framework. As member of the International Standardization Organization (ISO) (see Appendix B), ANSI has designated RIA as the U.S. representative to ISO on matters relating to international standards in the field of industrial robots. In order to fulfill its role as the ISO representative, RIA has set up Technical Advisory Groups (TAGs) of experts to advise the RIA standards manager on matters of international standards.

The authority to initiate a standards-writing effort has been given to the RIA Executive Committee for Standards Development, R15, and the RIA Board of Directors. Any interested party can apply to R15 and petition the creation of a standards subcommittee to write a standard on a certain subject. If the petition is accepted, a subcommittee is set up and is assigned the responsibility of preparing the standard. The subcommittee has to follow the ANSI rules of balance and voting (ANSI Board Standards Review) to achieve this goal. The procedure is intentionally long to achieve consensus, and it could last for many years. All proposed standards must be approved by the RIA Committee for Standards Approval before they become public.

Appendix A lists all the U.S. and international standards on industrial robots that are known to the author. Also listed in Appendix A are the committee drafts. Although the drafts are not standards, they have been written by field experts and contain useful in-

Handbook of Industrial Robotics, Second Edition, Edited by Shimon Y. Nof
ISBN 0-471-17783-0 © 1999 John Wiley & Sons, Inc.

formation. The most important and controversial of the standards have been those related to safety and performance. These two standards will be discussed in the next section.

2 SIGNIFICANT STANDARDS ACTIVITIES

2.1 U.S. Robot Performance Standard

This standard consists of two volumes. R15.05-1 covers the point-to-point and static performance characteristics (see Appendix A, U.S.A. Standards, 4), and R15.05-2 covers the path-related and dynamics performance characteristics (see Appendix A, U.S.A. Standards, 6).

The philosophy of the U.S. subcommittee on robot performance standards R15.05 is to write standards that are useful to buyers in helping them select the best robot for their specific applications.

The main tools used to force comparability of test results are a standard test path and standard test loads. The test loads are limited to the ones shown in Table 1. Thus, if two different robots from two different vendors are being considered for an application, and one has a payload capacity of 45 kg and the other 50 kg, they will both be tested under a standard load of 40 kg. The center of gravity of the 40-kg load with its associated support brackets shall have an axial CG offset of 12 cm and a radial CG offset of 6 cm from the mechanical interface coordinate system (mechanical flange and end-effector interface). The standard test path is located on the standard test plane and lies along a reference center line, as shown in Figure 1. The standard test path segments can only assume three lengths: 200 mm, 500 mm, or 1000 mm. Detailed instructions to determine the position and orientation of the standard test plane and the reference center line are given in the standard. The rules to select the proper size segment for a particular size robot are also given in the standard. Thus, two comparably sized robots from two different vendors will probably have the same-size test path.

The performance characteristics used by R15.05-1 are accuracy, repeatability, cycle time, overshoot, settling time, and compliance. This standard allows the vendor to tune operating parameters to optimize the values of desired performance characteristics. For example, they can maximize repeatability at the expense of cycle time. To identify the type of characteristic that is being optimized during a particular test, the standard establishes four performance classes. If class II testing is performed, the robot operates under optimum cycle time conditions. If class III testing is performed, the robot operates under optimum repeatability conditions. Due to the importance of the position repeatability *figure of merit* (FOM), it should be mentioned here that its mathematical definition in this standard is different from that of the ISO standard. These two definitions will be discussed and compared in the next section. Class I testing requires no specific parameter optimization. Class IV testing allows optimizing of robot performance characteristics not covered by classes II and III.

Table 1 Standard Test Load Categories[a]

Test Load Category	Mass (kg$_j$)	Axial CG Offset (mm)	Radial CG & TP Offset (mm)	Axial TP Offset (mm)
1	1.0	20	10	40
2	2.0	40	20	80
3	5.0	60	30	120
4	10.0	80	40	160
5	20.0	100	50	200
6	40.0	120	60	240
7	60.0	140	70	280
8	80.0	160	80	320
9	100.0	180	90	360
10	120.0	200	100	400
11	140.0	220	110	440
12			OPTIONAL	

CG—Center of Gravity

TP—Test Point

[a] See Appendix A, U.S.A. Standards, 6.

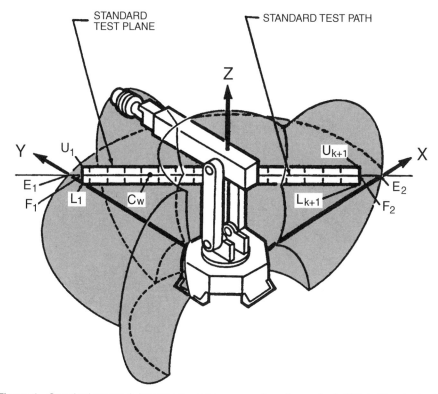

Figure 1   Standard test path location in working space (see Appendix A, U.S.A. Standards, 4).

The R15.05-2 standard defines the fundamental path-related performance characteristics and dynamic performance characteristics. Again, to assure comparability of test results, the standard specifies the use of the standard loads shown in Table 1 and the physical point on the end-effector where the path is measured, called the *test point* (TP). The standard test path is composed of a rectangle and a circle that are located on the standard test plane. Their dimensions are functions of the three standard lengths 200 mm, 500 mm, or 1000 mm, which assures that robots of approximately the same size will probably be tested on the same-size paths. The measurement of the path-performance characteristics is done on the evaluation planes. The concept of the evaluation planes was established by the committee to eliminate serious sources of inaccuracies in path metrology. All modern metrology systems use a digitizer that records data at discrete instances of time. A path-tracking test is repeated several times. The measurement data from each test with the same sequence number are grouped together to calculate the performance characteristics. Unfortunately, since it is impossible to perfectly synchronize the metrology system with the robot controller, the grouping of the measurement data points is in error. The metrology system data from each test are shifted with respect to the robot motion by the amount of time the metrology system and the robot controller are out of phase. This out-of-phase amount of time varies from test to test. The use of the evaluation planes eliminates the effect of time synchronization. Figure 2 shows the evaluation planes for the rectangular reference path. Evaluation planes are aligned normal to the standard test plane and the corresponding test path. The intersections between the attained paths and the evaluation planes define the points that will be used for the path-related figures of merit calculations. Linear interpolation shall be used when an attained (measured) point does not lie on the evaluation plane.

Again, to identify the type of characteristic that is being optimized during a particular test, the standard establishes three performance classes. If class I testing is performed, the robot operates under optimum path following conditions. The performance characteristics used to evaluate path following are *path accuracy* and *path repeatability*. It is interesting to note that the committee has defined two types of accuracy, *relative* and

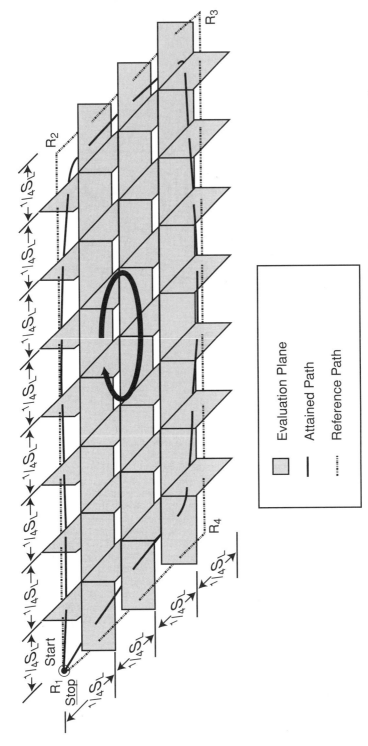

Figure 2 Rectangular reference and attained paths showing evaluation planes (see Appendix A, U.S.A. Standards, 6).

452

absolute. The relative path accuracy uses a previously measured path as the reference, while the absolute path accuracy, which is at the present optional, uses a mathematically defined path as the reference. The origin and coordinate system of this path are defined through manual teach programming. The objective of the committee is to use this technique to evaluate the manual teach off-line programming ability of the robot. This is a common method of programming robots today. It involves teaching a few points of a part or fixture and then making off-line programmed vector moves from those taught locations.

2.2   ISO Robot Performance Standard

As the scope section of this standard states, the specified tests are primarily intended to develop and verify individual robot specifications, prototype testing, or acceptance testing. The philosophy of the ISO subcommittee that developed this standard was not to use it to compare the performance of similar capacity and size robots as did the U.S. R15.05-1 and R15.05-2. The first version of this standard, ISO 9283:1990, did not even specify standard test paths and test loads. The lengths of the paths and size of the test loads were specified as a percentage of the robot workspace and rated load. Since no two robots have the same workspace and rated load, it was not possible for them to be tested under the same conditions, thus making comparisons very difficult. The U.S. subcommittee complained to ISO about this, and ISO corrected the second version of this standard, which now contains an annex listing standard test path lengths, loads (same as R15.05-2), and velocities. The use of these standards is optional, however.

The test planes and test paths of this standard are defined with respect to a cube located inside the workspace of the robot. Various diagonal planes of this cube are used to locate the test planes, paths, and points. This standard specifies tests for the measurement of 14 performance characteristics. The most commonly used characteristics are those of accuracy and repeatability.

Figure 3 shows the results from a set of robot position performance test data. The robot was commanded to move to the origin of the coordinate frame (rectangle), but instead attained all the positions marked by the triangles. The centroid of these positions, called the *barycenter* by this standard, is marked by the cross. The cloud of attained positions usually forms an ellipsoid. The lengths and orientations of the principal axes of this ellipsoid provide significant information about the performance of the robot at this position of its workspace. To average the results of this test over a significant portion of the workspace, both the U.S. and ISO standards require that this test be performed at several locations on the test plane and that the data are mixed together. Since the orientation of the ellipsoid is different at each location, the mixing of the data gives a cloud that can be approximated by a sphere. The distance between the commanded position and the barycenter represents the systematic part of the error (bias), and its main contributor is kinematic model errors. The cloud represents the random part of the error caused by electronic noise and friction. The mathematical description of the results of this test is different whether one uses the mathematical formulas of this or the other standard. Table 2 lists the formulas used to calculate positioning accuracy and repeatability by the two standards. For the same position data, the R15.05-1 calculated accuracy d_{PA} is always greater than the ISO calculated accuracy AP_p. For the same position data, the R15.05-1 calculated repeatability r_{REP} is always smaller than the ISO calculated repeatability RP_1.

The path accuracy and repeatability tests specified by this standard created problems from the beginning. The first version of this standard, ISO 9283:1990, did require that a path tracking test be repeated several times and the measurement data from each test with the same sequence number be grouped together for the calculation of the performance characteristics. The U.S. and Chinese robot performance subcommittees complained that grouping the measurement data this way can introduce errors, as is explained in Section 2.1. The U.S. subcommittee offered the use of the evaluation planes as a solution to this metrological problem. This was partially accepted. The new version of the ISO standard does require the use of calculation planes that are perpendicular to the commanded path. Unfortunately, the new method leaves the number and location of the planes up to the discretion of those who perform the test. Thus, a creative vendor can locate the calculation planes at those locations where the path errors are small and make the robot appear to perform better than it actually does. Comparing path-related performance of robots from different robot manufacturers is impossible based on the present test specifications.

In the early 1990s several proposals to develop application-specific performance standards were submitted to ISO. The most notable were one for arc welding, another for

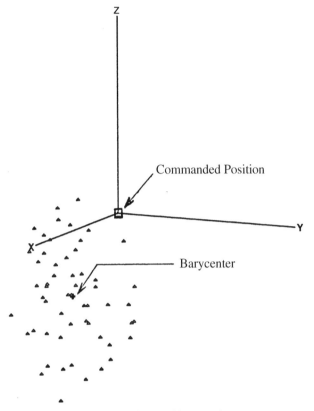

Figure 3 Robot position test data.

Table 2 Accuracy and Repeatability Definitions

Attained Position i: X_{ai}, Y_{ai}, Z_{ai} Commanded Position: X_c, Y_c, Z_c

Mean Attained Position:

$$\bar{X} = \frac{1}{N}\sum_{i=1}^{N} X_{ai}, \quad \bar{Y} = \frac{1}{N}\sum_{i=1}^{N} Y_{ai}, \quad \bar{Z} = \frac{1}{N}\sum_{i=1}^{N} Z_{ai}$$

$$1_i = \sqrt{(X_{ai} - \bar{X})^2 + (Y_{ai} - \bar{Y})^2 + (Z_{ai} - \bar{Z})^2}, \quad \bar{1} = \frac{1}{N}\sum_{i=1}^{N} 1_i, \quad S_1 = \sqrt{\frac{\sum_{i=1}^{N}(1_i - \bar{1})^2}{N - 1}}$$

ISO 9283   ANSI/RIA R15.05-1

Position Accuracy: Mean Position Accuracy:

$$AP_p = \sqrt{(\bar{X} - X_c)^2 + (\bar{Y} - Y_c)^2 + (\bar{Z} - Z_c)^2} \quad \bar{d}_{PA}$$
$$= \frac{1}{N}\sum_{i=1}^{N}\{\sqrt{(X_{ai} - X_c)^2 + (Y_{ai} - Y_c)^2 + (Z_{ai} - Z_c)^2}\}$$

Positioning Repeatability: Positional Repeatability:
$RP_1 = \bar{1} + 3S_1$ $\bar{r}_{REP} = \bar{1}$

spot welding, and a third for sealing/adhesive. None of them moved beyond the draft status. At some point the decision was made not to have separate application-specific standards but to make them appendices to ISO 9283. That has not happened yet.

2.3   U.S. Robot Safety Standard

Because of safety concerns, this is probably the most popular industrial robots standard in the United States (see Appendix A, U.S.A. Standards, 7). There are also European Union ("Manipulating Industrial Robots," 1992), Japanese, and ISO standards (see Appendix A, International Standards, 4). The U.S. standard is currently under review. The new version will become available soon. The discussion here will concentrate on the new version of the standard.

One of the fastest-growing industrial robot markets is for used robots. Large industrial users are modernizing their fleet of robots and selling the old used ones. An industry of robot remanufacturers has developed to rebuild and resell these robots. The RIA subcommittee responsible for the safety standard (R15.06) decided to strengthen the section on "remanufacture and rebuild of robots" to address the safety concerns of users who want to buy these robots. This section contains a detailed list of requirements that must be met by any robot that changes ownership (ensuring the healthy growth of this market).

A robot component that will see significant changes after this standard becomes effective is the teach pendant. Ordinary teach pendants are required to be equipped with an enabling device, which is usually a spring-loaded switch that must be kept pressed in order to enable any machine motion to take place. Most people call this device a *dead man's switch* because it will deactivate when the operator drops the teach pendant in an emergency. Recent research has revealed that some people in a panic state freeze and hold onto an emergency device instead of releasing it. This has prompted the safety committee to require that all teach pendants be equipped with an enabling device that, upon release or maximum compression, stops any robot motion and all associated equipment that may present a hazard.

The practice of placing the robot controller console anywhere it is convenient on the plant floor will not be acceptable anymore. The location of the operator controls shall be constructed to provide clear visibility of the area where work is performed. The controller and all equipment requiring access during automatic operation shall be located outside the safeguarded space of the robot. This will reduce the likelihood of equipment and machinery being operated when another person is in a hazardous position. Restricted space is the volume of space to which a robot, including the end-effector, workpiece, and attachments, is restricted by limiting devices. The safeguarded space is defined by safeguards. Safeguards are positioned so as to prevent access to a hazardous location in the workspace.

For personnel safety the new standard allows the implementation of firm safeguarding procedures or a comprehensive risk-assessment study and then installation of the safeguards determined to be appropriate. This study shall be prepared by the user or supplier during the design of the robot workcell and revised and updated any time there is a change that can affect safety. Based on this study, the minimum required safeguard devices and their location shall be determined. Table 3 (given in the standard) clearly illustrates the steps to determine the types of safeguards needed. In this table *PSSD* stands for *presence sensing and safeguarding device*. The standard provides two very informative tables on safeguarding devices and expected typical performance.

The manual teaching operation brings the teacher into close proximity with the moving robot and all its associated moving equipment, thus increasing the possibility of an accident. The standard provides a long list of safety rules that must be followed during this type of operation. Only trained personnel are allowed to perform this operation. Before teaching commences, all safety devices must be tested. The teacher is allowed to enter the safeguarded space, but only under slow speed control mode. This is a required control mode for all controllers that provide for pendant control. The speed specified by the standard is 250 mm/sec (approximately 10 in./sec), and it is measured at the tool center point (TCP). The objective of this requirement is to allow the operator sufficient time to react in an emergency during manual teaching. If additional personnel are allowed into the safeguarded space, they must be furnished with enabling devices, which give them the ability to stop motion independently.

A special mode of operation designed to confirm that a robot's programmed path and process performance are consistent with expectations is defined by the new standard. Called the *attended program verification* (APV) mode, it allows personnel inside the

Table 3 Safeguard Selection Decision Matrix[a]

Severity of Injury	Exposure	Avoidance	Safeguard Category and Selection (Performance)	Examples
S2 SERIOUS INJURY Normally irreversible	**F2** Frequent access, long duration	**P2** Scarcely possible	Category 4 Safeguards — Category 3 requirements plus: Follow prescribed safeguarding procedures contained in 6.3, Protection inside safeguarded space, Eliminate trapping / pinch hazards, Category 4 stopping circuits	PSSD's at pinch points, Limiting devices, Dual channel control circuitry with full fault detection and output testing
		P1 Possible	Category 3 Safeguards — Category 2 requirements plus: Minimize the restricted space, Elimination of hazards, Category 3 stopping circuits	Limiting devices, Design out hazard, Dual channel control circuitry with practical fault detection
	F1 Selom access, short duration	**P2** Scarcely Possible		
		P1 Possible	Category 2 Safeguards — Category 1 requirements plus: Perimeter guarding, Point of operation guarding, Passive safeguarding devices, as appropriate Category 2 stopping circuits	PSSD or fixed barrier, Single channel circuitry with periodic testing
S1 SLIGHT INJURY Normally reversible	Robots rated at <2 Kg and <250 mm/sec		Category 1 Safeguards — Awareness means, Training, Administrative procedures	Awareness: Barriers, Signs, Signals, Single channel stopping circuit, Well tried components, Well tried safety principles

[a] (See Appendix A, U.S.A. Standards, 7 Revised, Draft 14, 1997-09-30. See the standard for the final version of this matrix. Paragraph 6.3 and Category 2, 3, and 4

safeguarded space. This is a testing mode when the robot is allowed to move at full programmed speed, which presumably exceeds the slow speed velocity limits. The reason this is allowed is that some operations, such as welding and laying adhesive, depend on speed and require close observation to identify trouble spots. When it is impractical to remotely observe the operation, APV is permitted. A long list of robot safety requirements and user safety rules shall be obeyed during APV.

The work of the operator and maintenance and repair personnel is safeguarded with rules similar to those specified for the teacher. The main difference is that in this case no human body parts are allowed inside the safeguarded space while a robot is in automatic operation. To assist the release of trapped colleagues, clear directions shall be provided for the emergency movement without drive power of the robot mechanisms.

CONCLUSIONS

Most of the industrial applications of robots today do not require high accuracy and repeatability. Furthermore, manual teach programming is used for most of these applications, which eliminates the effect of the kinematic mechanism model errors. For that reason there has not been a great demand for standard performance test results from the robot manufacturers. The situation is changing, however, as more and more people realize the economic benefits of off-line programming and a hybrid manual-teach–off-line programming technique is growing in popularity. New robotic applications in arc welding, optoelectronic devices assembly, etc. have high performance requirements. These new developments might revive the interest in performance testing. Most users would like to have application-specific performance test results. Right now this luxury is available only for a few big buyers.

It has been relatively easy to comply with the first generation of safety standards. With the passage of time, and after numerous industrial accidents involving robots, the situation is changing. The new version of the robot safety standard is far more stringent, requiring much more effort and expense to achieve compliance. An unexpected outcome of this stringent standard will probably be the complete replacement of the present generation of robots, which could bring a technological renewal to the industry.

APPENDIX A

U.S.A. Standards

See Appendix B for organization addresses.

1. "Standard Guide for Classifying Industrial Robots," American Society for Testing and Materials (ASTM), Designation: F 1034–86.

    This standard defines methods that may be used to classify industrial robots.

2. "American National Standard for Industrial Robots and Robot Systems—Common Identification Methods for Signal- and Power-Carrying Conductors," American National Standards Institute (ANSI), ANSI/RIA R15.01-1-1990.

    This standard defines common identification methods for signal- and power-carrying conductors applicable to industrial robots and robot systems.

3. "American National Standard for Industrial Robots and Robot Systems—Hand-Held Robot Control Pendants—Human Engineering Design Criteria," American National Standards Institute (ANSI), ANSI/RIA R15.02/1-1990.

    This standard defines human factors characteristics for hand-held control devices that accompany industrial robots and industrial robot systems.

4. "American National Standard for Industrial Robots and Robot Systems—Point-to-Point and Static Performance Characteristics—Evaluation," American National Standards Institute (ANSI), ANSI/RIA R15.05-1-1990.

    This standard defines methods for the static performance evaluation of industrial robots. Its main objective is to facilitate comparison based on performance.

5. "American National Standard for Industrial Robots and Robot Systems—Infant Mortality Life Test," Robotic Industries Association (RIA), BSR/RIA R15.05-3-1991.

    This standard defines the minimum testing requirements that will qualify a newly manufactured or rebuilt robot to be placed into use without additional testing.

6. "American National Standard for Industrial Robots and Robot Systems—Path-Related and Dynamic Performance Characteristics—Evaluation," American Na-

tional Standards Institute (ANSI), ANSI/RIA R15.05-2-1992.

This standard defines methods for the dynamic performance evaluation of industrial robots. Its main objective is to facilitate comparison based on performance.

7. "American National Standard for Industrial Robots and Robot Systems-Safety Requirements," American National Standards Institute (ANSI), ANSI/RIA R15.06-1992.

This was the first version of the Industrial Robots safety standard. A revision of this standard was initiated a few years ago that resulted in significant changes of the original standard. The new version will probably become available to the public sometime in 1998.

International Standards

1. "Manipulating Industrial Robots—Coordinate Systems and Motions," International Standardization Organization (ISO), 9787, first edition 1990-12-01.

This standard defines and specifies three robot coordinate systems and also gives the axis nomenclature. A revision of this standard was initiated a few years ago. The new version will probably become available to the public sometime in 1997.

2. "Manipulating Industrial Robots—Performance Criteria and Related Test Methods," International Standardization Organization (ISO), 9283, first edition 1990-12-15.

This standard describes methods of specifying and testing several performance characteristics of manipulating industrial robots. A revision of this standard was initiated a few years ago that has resulted in significant changes of the original standard. The new version will probably become available to the public sometime in 1998.

3. "Manipulating Industrial Robots—Presentation of Characteristics," International Standardization Organization (ISO), 9946, first edition 1991-02-15.

This standard specifies requirements for how characteristics of robots shall be presented by the manufacturer.

4. "Manipulating Industrial Robots—Safety," International Standardization Organization (ISO), 10218, first edition 1992-01-15.

This standard provides guidance on the safety considerations for the design, construction, programming, operation, use, repair, and maintenance of manipulating industrial robots and robot systems.

Committee Drafts

1. "Manipulating Industrial Robots—Vocabulary," Revision of ISO/TR 8373:1988, Draft International Standard ISO/DIS 8373, 1993.

Provides a list of terms most commonly used for industrial robots. The terms are briefly defined or explained.

2. "Manipulating Industrial Robots—Mechanical Interfaces—Part 1: Circular (Form A)," Revision of ISO 9409-1:1988, Committee Draft ISO/CD 9409-1, 1992-09-21.

This Committee Draft (CD) defines the main dimensions, designation, and markings for the circular mechanical interface of the manipulator end-effector.

3. "Manipulating Industrial Robots—Mechanical Interfaces—Part 2: Cylindrical Shafts," Committee Draft ISO/CD 9409-2, 1992-09-21.

This Committee Draft (CD) defines the main dimensions, designation, and markings for the shaft mechanical interface of the manipulator end-effector.

4. "Manipulating Industrial Robots—Automatic End-effector Exchange Systems— Vocabulary and Presentation of Characteristics," Committee Draft ISO/CD 11-593, 1992-06.

This Committee Draft (CD) defines the terms that are necessary to describe automatic end-effector exchange systems.

5. "Manipulating Industrial Robots—An Overview of Test Equipment and Metrology Methods for Robot Performance Evaluation in Accordance with ISO 9283," Committee Draft Technical Report ISO/DTR 13309, 1994-03.

This Committee Draft Technical Report (DTR) provides information on the state-of-the-art metrology instruments for the testing and calibration of industrial robots.

6. "Manipulating Industrial Robots—Vocabulary of Object Handling with End-Effectors and of Characteristics of Grasp-Type Grippers," Committee Draft ISO/CD 14539, 1996.

   This Committee Draft (CD) defines the terms which are necessary to describe object handling.

APPENDIX B: ORGANIZATIONS

American National Standards Institute, 11 West 42nd St., New York, NY 10036.
American Society for Testing and Materials (ASTM), 1916 Race St., Philadelphia, PA 19103.
International Organization for Standardization, Case postale 56. CH-1211 Geneve 20. Switzerland.
Robotic Industries Association, 900 Victors Way/P.O. Box 3724, Ann Arbor, MI 48106.

REFERENCES

"National Policy on Standards for the United States and a Recommended Implementation Plan." 1978. National Standards Policy Advisory Committee, Washington, D.C., December, p. 6.
"Manipulating Industrial Robots—Safety." 1992. European Standard, EN 775:1992, October.
ANSI Board Standards Review. "Procedures for the Development and Coordination of American National Standards."

CHAPTER 25

ORGANIZATION AND AUTOMATION IMPACTS ON PRODUCTION WORKERS QUALIFICATION (EUROPEAN EXPERIENCE)

H.-J. Bullinger
J. Pack
Institute of Work Organization
Stuttgart, Germany

1 INTRODUCTION

A report of the IRDAC (Industrial Research and Development Advisory Committee of the European Commission) (IRDAC, 1994) shows that an adequate response to industrial and market change can only be achieved through enhanced collaboration between educational institutions and industry. The report recommends that it is necessary to:

- Develop total competence in people
- Prepare people and society for a lifetime of learning
- Stimulate a learning culture in companies

The whole discussion on learning organizations and lifelong learning is based on the presence of learning incentives at the place of work or in the process of work. But are they present? Our long-term analysis of job designs and requirements for qualification shows that many deficits, such as missing learning incentives and even requirements, exist, leading to a permanent reduction of qualification and intellectual flexibility, i.e., a psychic deformation (Kohn and Schooler, 1982).

2 THE INSTRUMENT FOR MEASUREMENT OF QUALIFICATION REQUIREMENTS

The work analyses have been classified using a work evaluation system (TBS) (Hacker et al., 1995) containing 52 ordinal rating scales, rated by specially trained work psychologists. The instrument is valid and reliable related to the test criteria of psychological tests. The 52 scales were composed of five dimensions:

1. The completeness of the job versus executing and releasing only portions of the whole job
2. The cooperation and communication requirements (and thus the chances of implementing group work or teamwork)
3. The responsibility for the work result
4. The required intellectual and problem-solving processes
5. The learning incentives and requirements posed by the work task

---

Handbook of Industrial Robotics, Second Edition, Edited by Shimon Y. Nof
ISBN 0-471-17783-0   © 1999 John Wiley & Sons, Inc.

Within the scope of this instrument, minimum requirements (based on expert validation and cross validation with other instruments) stand for qualification-suited work contents oriented to the average qualification potential of regular employees in Germany. Nowadays even the average serial assembly worker in Germany has a vocational and professional qualification, as does the average gainfully employed worker (Schuler, 1992). If this minimum level is not reached, it means that the average qualification existing in production systems is not fully utilized through the requirements of the work. It also means that performing these activities over a long period of time might lead to a permanent reduction of qualification and intellectual flexibility, i.e., a psychic deformation (Kohn and Schooler, 1982).

In the following illustrations the average levels-of-requirement dimensions have been related to the minimum level in percentage. Levels under 100% depict the extent of personnel resources not utilized, or the *psychic deformation potential*. This percentage can also be regarded as a personnel-related *LEAN-Management indicator*. It has to be taken into consideration, however, that meeting the minimum levels (100%) is not sufficient, since this would only lead to a static state, i.e., the mere maintenance of existing qualifications. Only if the minimum values are exceeded can task-related learning incentives be created, and thus the skeleton conditions for a dynamic qualification development. This is because the values represent the degree of qualification and learning conduciveness, i.e., they mark the employees' chances for further personal development while learning during the actual work process and hence the potential for learning in organizations.

3 RESULTS OF A REPRESENTATIVE STUDY IN GERMAN SERIAL ASSEMBLY SYSTEMS

Within the scope of a project to evaluate flexible automated assembly systems and additional research in other projects, 21[1] assembly systems were investigated with regard to their work tasks and qualification requirements. Work analyses were performed for 140 different work tasks executed at 464 workplaces. The assembly systems investigated spanned a wide spectrum of products, from printed-board assemblies to watches to household appliances to engines and automobiles.

The main goal of the requirement analyses was to find out:

- What qualification requirements occur in flexible automated assembly systems
- To what extent the tasks are designed to be conducive to qualification
- What deficits are related to task design

As Figure 1 shows, the minimum values for qualification-suited work contents are not reached in any of the five dimensions in the average serial assembly. Since these minimum values are oriented to the medium qualification potential of gainfully employed workers, i.e., this qualification exists in serial assembly, the following statement can be made: *The current design of work tasks in serial assembly prevents the complete utilization of existing qualification resources and constitutes an obstacle to a dynamic qualification extension and the development of learning processes during the actual work process.* The main deficits lie in the fact that the intellectual requirements and learning incentives are far too low. The existing potentials of the employees are not being utilized. Contradicting these low requirements is the far higher requirement for responsibility for this work (73% of the minimum value). This contradiction points to a dilemma we often find in serial assemblies, namely the low intellectual challenge and low chances of influencing the work situation versus the excessive quantities that have to be produced and the relatively high responsibility for work quality.

Due to the division of labor, however, the scope of responsibility is restricted to partial results of the overall task and contradicts the adverse possibilities of meeting this responsibility. On the one hand, responsibility for the work result is required, but at the same time, no range in relation to time or content is granted to meet this requirement.

The cooperation and communication requirements are also far too low, and so are the objective, basic conditions for the implementation of group work. It can be said that the current work task design in the average serial assembly hardly allows the implementation

---

[1] We have actually gathered data about 52 assembly systems. The analysis shows the same trends, but the evaluation is not yet completed.

Job Requirement Levels:

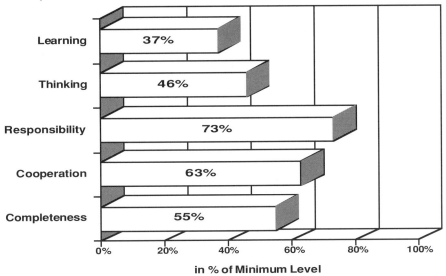

Figure 1 Average of five job analysis dimensions (representing 464 workplaces in 21 serial assembly systems).

of group work. This means that the implementation of group work can only be accomplished through a redefinition, a rearrangement, of the work tasks. In the long run, nonutilization of existing qualifications leads to their decline.

4 CORRELATION BETWEEN QUALIFICATION REQUIREMENTS AND AUTOMATION DEGREE

When we analyze the correlation between requirement level and automation level (Figure 2), a totally different picture appears. Thus it can be realized that qualification requirements in highly automated and therefore mostly flexible and nonsynchronous systems are highly evident, since in these systems the share of the remaining nonautomated assembly functions consists mostly of functions such as controlling, removal of faults, retooling, and maintenance. Even work in such assembly systems does not reach the minimum value of 100% with the exception of the cooperation requirements. This can be explained by the fact that a division of labor exists here, such as job rotation on the level of plant operators, and more complex tasks, such as the removal of more serious faults, maintenance, and system control are carried out by suppliers and maintenance staff based on the division of labor.

The workplace requirements in purely manual assembly are relatively higher but with a similar structure. In comparison, the most partial work with high requirements is found in the first-mentioned distribution, but in contrast to the other automation levels, manual assemblies have a similar structure regarding a rather medium or low division of labor. Manual assemblies comprise 37% of all the studied workplaces, have the highest share of work cycles longer than 12 minutes, and are mostly nonsynchronous and highly flexible regarding product variants and types. This characterization explains why workplaces with manual systems come directly after flexibly automated assemblies in average value of requirements. (That does not necessarily mean that a spot check would find no manual assemblies with low requirements.)

At workplaces where there are "single-purpose" automated assemblies with residual work, employees' qualifications are always required at a very low level. Those systems almost exclusively demand activities with a cycle time of less than three minutes (97%), with some work synchronization and without system flexibility. These kinds of work are all set according to the division of labor. Between the mentioned extreme values lie those average values of requirements in hybrid assemblies with automated "island" solutions and those that have equally distributed manual and automated areas. Statements regarding

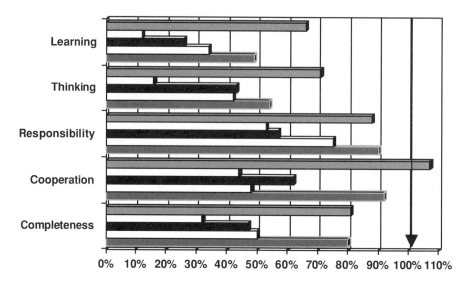

Figure 2   Requirement level and automation degree.

the requirement structure of these systems can hardly be made at this point, since the simultanuously existing manual and automated areas blur the differences of qualification requirements between and within these areas with respect to the average value. Conclusions can only be made relative to the constructing types of assembly work.

5   CORRELATION BETWEEN QUALIFICATION REQUIREMENTS AND SYSTEM FLEXIBILITY

The level of qualification requirements is clearly influenced by the flexibility of assemblies in terms of an assembly-relevant variety of types and variants (Figure 3).

In assemblies where almost the same products are produced without assembly-relevant variants, workplaces offer low incentives for learning, seldom require the ability to think and cooperate, and are limited to a rather incomplete performance of single, nonisolated partial tasks. Except for responsibility, the qualification requirements rise linearly with the transition to product variants that change work flow and content (medium flexibility) up to different assembly-relevant product types (high flexibility). With higher flexibility in an assembly system the employees' qualification potential can be better used, but it reaches on average the minimum value only conditionally. The rise in requirements can be explained by an increasing variety of work caused by changes of variants and types, by greater flexibility, requiring changes in work organization, and by higher frequency of long-cycle work, lower level of work synchronization, and lower division of labor.

6   CORRELATION BETWEEN REQUIREMENTS AND WORK SYNCHRONIZATION

The correlation between qualification requirements and synchronization is not so clear (Figure 4).

Nonsynchronous workplaces generally have more complete work contents, require a higher level of intelligence, and offer more incentives for learning. In this regard differences hardly exist between nonsynchronous and partially synchronous workplaces. This can be explained by the usable leeway of activity available to each employee, but only in fully nonsynchronous systems. This leeway of activity influences the requirements. Thinking and learning processes need usable time that is free from repetitive cycles in order to materialize.

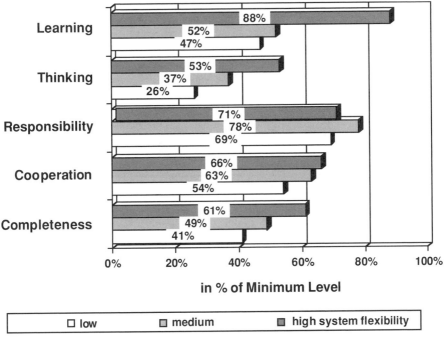

in % of Minimum Level

□ low ▨ medium ▨ high system flexibility

Figure 3 Requirement level and system flexibility.

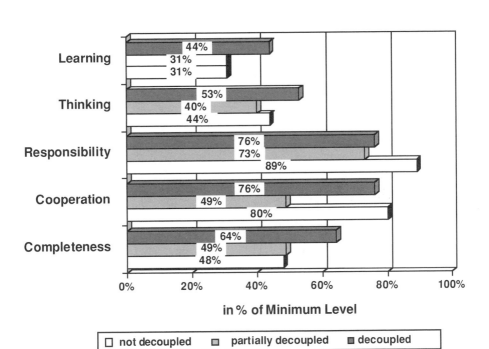

in % of Minimum Level

□ not decoupled ▨ partially decoupled ▨ decoupled

Figure 4 Requirement level and work synchronization.

The level of requirements regarding responsibility and cooperation is highest where work is synchronized. This is a consequence of the fact that assembly faults cannot be corrected when there is synchronization and thus continue and accumulate in the entire system. Since the systems that are synchronized are mainly production line systems, the cooperation requirements are relatively higher because of the mutual dependency.

7 REQUIREMENTS AND DIVISION OF LABOR

The fact that the withdrawal from Taylor's division of labor is associated with an immediate increase in qualification requirements is obvious. Worth mentioning is the fact that, in contrast to all the other correlations mentioned, with a medium level of division of labor (teamwork with the integration of indirect functions) the minimum requirements regarding qualification-adequate work design are met for the first time (Figure 5).

It is interesting that a medium level of division of labor, which in this case is especially characterized by job rotation, leads to a clear approach to the minimum value only in the area of cooperation and responsibility. The structuring measures of job rotation do not automatically lead to complete work with thinking requirements suitable for qualification and learning incentives, since those measures are limited to a change between similar activities at an executing level (for example, pure assembling). That does not mean that those measures would not be useful. They may contribute to physical and psychological changes of strain, apart from the above-mentioned effects. Additionally, this job rotation allows the employee to gain an overview of the different kinds of work that are subject to change and therefore provides a complete idea of the assembly processes and the correlation between cause and effect. This complete "physical" representation (brain model) of processes is especially useful as an orientation in new situations and generally useful as an aid to decision-making in cases of uncertainty. Decisions can then be derived logically and causally and lead to an optimized use of human intelligence, creativity, and flexibility, as far as the task allows it. In order to obtain qualification-suitable work, the job rotation needs to be supplemented through the integration of partial work that is different in its requirements and relevant for learning.

8 QUALIFICATION REQUIREMENTS AND CYCLE TIME

A similar but even clearer pattern regarding the distribution of requirements is evident in relation to work repetitiveness.

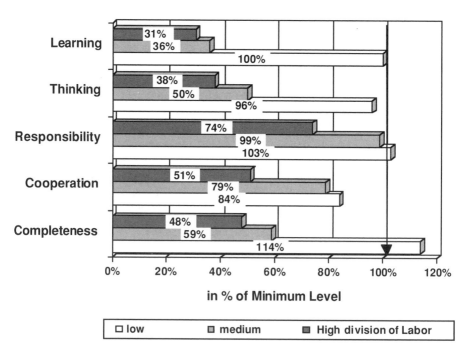

Figure 5 Requirement level and division of labor.

Activities that last three minutes or less can be characterized as absolutely unsuitable for qualification and as insufficient for the general existing human basic qualification. These activities hardly require thinking operations, do not enable learning processes, and offer purely space for cooperation with colleagues. These activities are, moreover, extremely fragmented (Figure 6).

This is understandable, since activities that are repeated more than a hundred times per day do not offer any learning incentives or requirements for thinking. Indeed, where, in such short-time activities, could the time be found for learning, thinking and cooperating? There are severe restrictions on the integration of partial work with different requirements—for example, material disposal, controlling, and rework—since for these kinds of work the employee would have only 18 seconds for each working step within a cycle time of 3 minutes if this partial work is to have a 20% share. The already mentioned low requirements stand out in contradiction to the higher requirements regarding responsibility for this kind of work (70% of the minimum value). This contradiction illustrates a dilemma occurring mainly in serial assemblies: low mental requirements and low influence on the working situation, and at the same time a quantitative excessive requirement for relatively high responsibility for assembly quality. Much confusion exists about the decision whether the requirements regarding order quantity or regarding quality have to be fulfilled. This dilemma is often solved by choosing items in order to fulfil the production plan, leaving no time for quality responsibility.

Workplaces with significant repetitive work extent from more than 3 minutes up to 12 minutes show a considerable improvement in the requirement level but do not reach at all the minimum values necessary for activities suitable for qualification. This is because these kinds of work mainly consist of similar partial activities and partial activities that are different in their requirements are rarely used in work design (industrial engineering).

In the case of activities lasting more than 12 minutes, it is evident that once the repetitive short cycles are dropped in all dimensions, the minimum values regarding activities suitable for qualification are more than fulfilled. This shows very clearly that the temporal work content for the design of work contents has a very strong influence, especially in serial assemblies. A sufficiently long work cycle (circa 10 minutes) is a basic

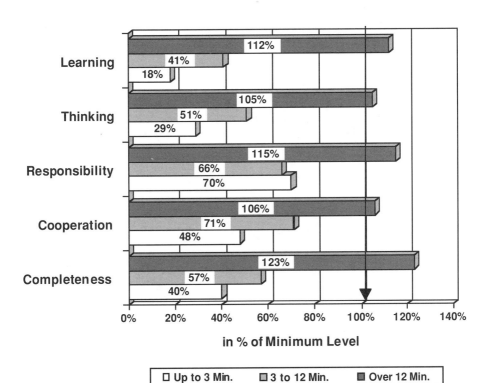

Figure 6 Requirement level and repetitive work extent.

condition for a work design suitable for qualification, but does not guarantee it, since without an additional concern regarding its content, only low effects can be expected for activities with different requirements. The simple increase of the cycle time through addition of similar partial work leads to a zero-sum game (0 + 0 = 0), at least regarding qualification requirements. This simple temporal increase has shown considerable positive effects in other areas. The employees are able to work out usable and temporal breaks, which have stress-reducing effects, coupled with psychological relaxation.

9 QUALIFICATION REQUIREMENTS IN MANUFACTURING SYSTEMS— RESULTS OF CASE STUDIES

With the same set of instruments as used in the assembly study, 32 cases of manufacturing systems were studied. Since the higher variability of differences of manufacturing systems does not offer generalized conclusions, the results of two case studies will be presented here.

9.1 Highly Automated System for Cast-Iron Casings

The manufacturing system consists of seven linked transfer lines in a U-shaped layout. There is no rigid linkage because there are several ways of pulling casings from the lines with handling equipment. Several handling robots, automated measuring machines, and a reworking station are integrated in the lines. The transfer lines are decoupled by buffers of 30 minutes.

A group of 10 skilled workers is responsible for the whole unit's output and quality, including retooling, presetting of tools, repairs of machinery, changing of program parameters, and maintenance. Electrical or electronic repairs are allowed only for certified group members. All other work is carried out by all group members in rotation or as necessary. Being responsible for personnel employment, vacation planning, and free shifts, the group is nearly independent in its work. Since the tasks of all group members have the same contents and through planned or unplanned (by absenteeism) job rotation everyone does the same work, only one work analysis for the whole system was necessary. The different parallel activities of any group member were analyzed as partial activities and combined in one analysis based on time-weighting (Figure 7).

The qualification requirements for working in this system greatly extend beyond the minimum level in each of the five dimensions. This means that working in the system gives the chance not only to conserve qualification, but also for further development of

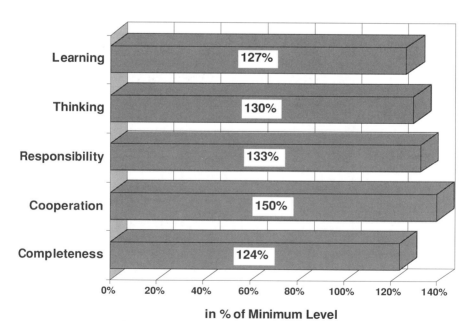

in % of Minimum Level

Figure 7   Requirements for qualification in a system for cast-iron casings.

knowledge, skills, and abilities through integrated learning incentives or the development of important dimensions of human personality.

9.2 Fully Automated Molding and Box-Casting Unit for Large-Size Spheroidal Graphite Parts

The unit consists of a fully automated box-filling machine for the upper and lower box, with two robots for cleaning the mold and spraying the mold release agent, an automated system for transferring and feeding the cores to the boxes, a machine for turning and closing the upper and lower box, an automated casting machine with parameter control, an automated unpacking and sand separation unit, and a box-cleaning buffer. The whole unit is rigidly linked by automated transport systems, and in a one-minute cycle time one box is molded with nearly one ton of spheroidal graphite iron. If only one part of machinery stops, the whole unit will not produce.

The unit's main electronic and electrical parameters are supervised and steered in an operation center that also controls parallel units. For direct operation in the unit, four workers are needed: one for sand core control, one for controlling the filled and formed boxes and supervising the filling machinery and the two robots, one molder (founder) for supervising the molding machines shot parameters and quality control, and one for supervising the unpacking station and preventing stops, e.g., by cleaning light sensors.

Only the molder is nearly desynchronized from the unit's cycle time. The other workers have to stay steadily at their workstations, even at the unpacking station, because if it stops the whole unit will stop too. There is no job rotation between these workers.

The requirements for qualification in this manufacturing system are very different, depending on the different tasks.

The requirement for qualification for supervising the automated molding and controlling parameters and the iron alloy are the highest and include learning incentives (Figure 8).

Controlling formed boxes and supervising filling machinery and robots reaches nearly the minimum level of qualification requirements. This means that the mean qualification of workers at that task will be used and there is no danger of dequalifying (Figure 9).

The two other tasks, sand core controlling and unpacking, underuse the mean qualification of workers. Doing this job for a longer time will result in losses of qualification, cognitive flexibility, and the ability to learn (Figures 10 and 11).

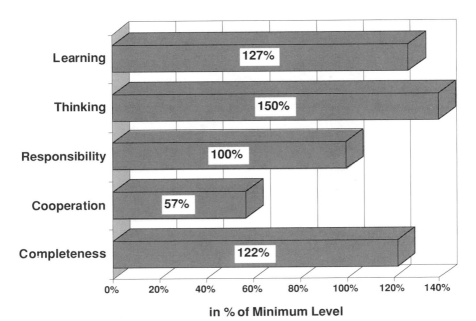

Figure 8 Requirements for qualification: supervising an automated molding system.

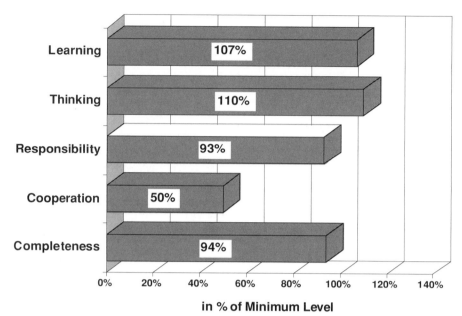

Figure 9   Requirements for qualification: controlling formed boxes and supervising robots.

This is the result of technical planning of the automated system that filled an auto-mation gap with residual manual work strictly synchronized with the cycle time. To avoid such results it would be necessary to design a technical decoupling by buffers and, as an organizational measure, install periodic job rotation with full team responsibility for the system.

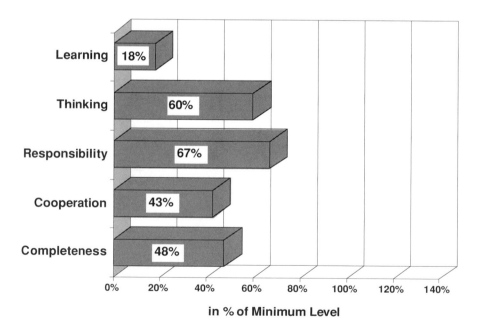

Figure 10   Requirements for qualification: controlling sand cores.

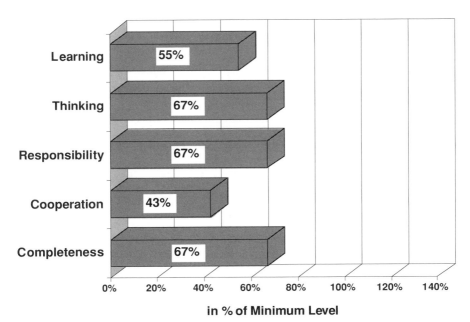

in % of Minimum Level

Figure 11   Requirements for qualification: supervising automated unpacking station.

REFERENCES

IRDAC. 1994. *Quality and Relevance—The Challenge to European Education—Unlocking Europe.* Brussels: Human Potential.

Kohn, M. L., and C. Schooler. 1982. "Job Conditions and Intellectual Flexibility: A Longitudinal Assessment of Their Reciprocal Effects." *American Journal of Sociology* **87**(6), 1257–1286.

———. 1983. *Work and Personality: An Inquiry into the Impact of Social Stratification.* Norwood: Ablex.

Hacker, W., et al. 1995. *Tätigkeits-Bewertungssystem.* Zürich, Stuttgart.

Schuler, M. 1992. "Qualifizierung angelernter IndustiearbeiterInnen in der Montage." *WSI-Mitteilungen* (June), 367.

ADDITIONAL READING

Bullinger, H.-J., R. Ilg, and S. Zinser. "Structures and Processes in Learning Organizations: European Experiences. In *Manufacturing Agility and Hybrid Automation I.* Louisville: IEA Press, 1996.

Bullinger, H.-J., C. Lott, and A. Korge. "Target Management: Customer Oriented Reorganization of Manufacturing." In *Manufacturing Engineering: 2000 and Beyond.* Ed. I. D. Marinescu. London: Freund, 1996. 3–15.

———. "Cognitive Resources and Leadership Performance: What about the Task Requirements?" *Applied Psychology* **44,** 48–50, 1995.

Hacker, W. "Action Regulation Theory and Occupational Psychology: Review of German Empirical Research since 1987." *German Journal of Psychology* **18**(2), 91–120, 1994.

———. "Complete vs. Incomplete Working Tasks—A Concept and 1st Verification." In *The Psychology of Work and Organization.* Ed. G. Debus and H. W. Schroiff. Amsterdam: North-Holland, 1985.

———. "Objective Work Environment: Analysis and Evaluation of Objective Work Characteristics." In *A Healthier Work Environment—Basic Concepts and Methods of Measurement.* Ed. WHO Regional Office for Europe. Copenhagen, 1993. 42–57.

———. "Occupational Psychology Between Basic and Applied Orientation—Some Methodological Issues." *Le Travail Humain* **56**(2, 3), 157–169, 1993.

———. "On Some Fundamentals of Action Regulation." In *Discovery Strategies in the Psychology of Action.* Ed. G. P. Ginsburg, J. Brenner, and M. von Cranach. European Monographs in Social Psychology. 35. London: Academy Press, 63–84, 1985.

Pack, J. "The Use of Human Resources in West German Serial Assembly Systems." In *Human Factors in Organizational Design and Management IV.* Ed. G. E. Bradley and H. W. Hendrick. Amsterdam: North-Holland, Elsevier Science, 1994.

CHAPTER 26

MANAGEMENT POLICIES OF COMPUTER-INTEGRATED MANUFACTURING/ROBOTICS

A. Gunasekaran
University of Massachusetts
Dartmouth, Massachusetts

1 INTRODUCTION

This chapter deals with management policies with regard to CIM/robotics, including strategic, organizational, operational, technological, and human issues. The role of automation such as CAD, CAE, CIM, and robotics in improving the productivity of manufacturing has been given due attention for the implementation and hence the competitiveness of manufacturing. The objective of this chapter is to discuss the various management policies on the integration and adaptability of CIM/robotics. The installation of CIM/robotics obviously replaces workers, but not entirely; therefore, there is a need to discuss the role of human issues in the operations of CIM/robotics in manufacturing. Also, CIM/robotics requires a different environment to function with the objective of improving productivity and quality. With the role of CIM/robotics in manufacturing and management policies for integration and adaptability in mind, an attempt is made in this chapter to discuss these issues and offer solutions to overcome the problems of integration and adaptability and enhance competitiveness.

2 COMPUTER-INTEGRATED MANUFACTURING/ROBOTICS

CIM is the architecture for integrating engineering, marketing, and manufacturing functions through information technologies. In the broad sense CIM involves the integration of all the business processes from supplier to end consumer. CIM can be used as a strategy for enterprise resource planning for business-wide integration. This indicates the relationship between business process reengineering and computer-integrated manufacturing with the objective of achieving the enterprise integration and management for improving productivity and quality. Robotics has been used as part of automation in manufacturing/service industries to improve the system. The motivation for CIM/robotics has been based on the perceived need for the manufacturing industry to respond to changes more rapidly than in the past. CIM/robotics promises many benefits, including increased machine utilization, reduced work-in-process inventory, increased productivity of working capital, reduced number of machine tools, reduced labor costs, reduced lead times, more consistent product quality, less floor space, and reduced set-up costs.

Most of the published reports concentrate on the technological and operational issues at too early a stage of the development of CIM/robotics. However, the role of strategic, organizational, and behavior issues needs to be given due consideration in improving the integration and adaptability of CIM/robotics. The implementation of CIM/robotics is thus discussed in this chapter with reference to strategic, organizational, technological, behavioral, and operational issues. Furthermore, there is no systematic framework available for improving the integration and adaptability of CIM/robotics taking into account different managerial, technological, and operational issues. This chapter examines one aspect of human interaction with computer-integrated systems, that of fault diagnoses or troubleshooting. The complexity (and attendant unreliability) of the new manufacturing systems

Handbook of Industrial Robotics, Second Edition, Edited by Shimon Y. Nof
ISBN 0-471-17783-0 © 1999 John Wiley & Sons, Inc.

has meant that fault diagnosis has become an increasing proportion and an integral part of operators' jobs. Establishing and maintaining high levels of diagnostic accuracy and efficiency is important for a variety of reasons. One is that equipment downtime is expensive; another is that diagnostic errors might, under some circumstances, be a threat to both safety and quality. As yet there has been little research into human fault diagnosis in CIM/robotics. The main issue here is to identify critical human–machine interaction design principles and highlight research questions yet to be addressed in contemporary manufacturing environments, with respect to the diagnosis of system failures (Morrison and Upton, 1994).

2.1 Importance of Integration Strategies and Technologies

The integration and adaptability issues are generally affected by factors such as business strategy, manufacturing strategy, availability of knowledge workers, software professionals, complexity of material flow, information flow pattern and decision-making processes, product and process complexities, supplier/purchasing activities, and behavioral issues. The major components of CIM are CAD and CAM technologies, computer numerical control (CNC) equipment, robots, and FMS technology (Groover, 1987). The computer system is used to integrate the design and manufacturing process and other production-planning and control systems (such as inventories, materials, and schedules) and the integration of manufacturing activities with both vendors and suppliers (Levary, 1992).

2.2 Role of Human Resources in Integration

Human-centered systems are being implemented in increasing numbers in various European countries, industries, and firms. The key issue in designing human-centered systems is finding the balance between technological components and human skills and abilities. Another important issue is the evaluating systems in terms of their aims. Cooperative evaluation is a procedure for obtaining data about problems experienced when working with a prototype for a software product, so that changes can be made to improve it. What distinguishes cooperative evaluation is the collaboration that occurs as users and designers evaluate the system together. The participatory design and joint application can be used to build human-centered systems (Uden, 1995).

Introduction of CIM/robotics as a catalyst for human resource and organizational change within manufacturing companies has received attention from researchers. The strategic benefits expected to accrue from full CIM/robotics implementation have been reviewed, and it has been shown why these are frequently too important to be ignored by production organizations (Hassard and Forrester, 1997). However, the adoption of CIM/robotics brings with it new demands for interdisciplinary working across traditional functional boundaries. It is frequently argued that since humans are error-prone, it is necessary to limit their influence in manufacturing by using CIM/robotics technology to automate physical/cognitive tasks. However, many researchers assert that workers are essential to manufacturing and CIM/robotics technology should be used to assist workers in making the best product possible for the marketplace. In recent years the possibility of achieving workerless manufacturing, a situation that would exist in a completely automated manufacturing system, has been discussed. Human aspects will remain an integral part of current manufacturing for technical, economic, and cybernetics reasons. The outline of a procedure for systematically allocating functions to humans and machines in a purely manufacturing context has significant influence on the success of integration and adaptability of CIM/robotics (Mital, 1997).

2.3 Impact of CIM/Robotics on Productivity and Quality

CIM/robotics has focused and continues to focus mainly on the premises of the computer. CIM is the integration of the total manufacturing enterprise through the use of integrated systems and data communications coupled with new managerial philosophies to improve organizational and personal efficiency. In this environment CIM/robotics was put into use as a production system. It was closely linked to the market via sales and distribution departments in order to avoid wastage of personnel money or goods. These systems have two aims:

1. To optimize business income by integrating the entire company without compromising the flexibility of the individual departments
2. To build a large-scale system using information network technology (Gunter, 1990)

Table 1   CIM/Robotics Achievement[a]

Question	Percentage Range	Average Change
Increase in manufacturing productivity?	20–200%	120%
Increase in product quality?	60–200%	140%
Decrease in lead time from design to sale?	30–100%	60%
Decrease in lead time from order to shipment?	30–50%	45%
Increase in capital equipment utilization?	20–1500%	340%
Decrease in WIP inventory?	30–100%	75%

[a] From L. Gunter, "Strategic Options for CIM Integration," in *New Technology and Manufacturing Management,* ed. M. Warner. New York: John Wiley & Sons, 1990. ©1990 John Wiley & Sons, Inc.

Computer and information network technologies have been developed to support the growth of CIM/robotics. Table 1 shows what some experts feel CIM/robotics can achieve.

3   INTEGRATION AND ADAPTABILITY ISSUES IN THE IMPLEMENTATION OF CIM/ROBOTICS

Integration and adaptability are key issues in the implementation of CIM/robotics. Therefore, it is appropriate to discuss these issues and how they should be taken into account in the implementation of CIM/robotics (Gunasekaran and Nof, 1997).

3.1   Important Issues of Integration in CIM/Robotics

The integration of systems is frequently hindered by resistance to converging the activities of different functions within the business. Organizational integration and the elimination of departmental barriers are proving to be difficult to achieve in practice and will in turn hinder the technical development of the "seamless" integration required. The integration of CAD/CAE, CNC machines, AGVs, etc. had a huge impact on the development of CIM/robotics. In support of the critical roles that humans play in the success of CIM/robotics, the most common recommendation is the dire need for education and training in relation to the adoption of CIM/robotics. It could even mean a redefinition of responsibilities from the top to the bottom of the organization. Research in CIM/robotics design and implementation has mainly been in the area of production. However, the major issues in CIM/robotics are directly related to information systems such as CAD, CAE, and MRPII (Gowan and Mathieu, 1994).

Because manufacturers face shorter contract terms, greater market volatility, and a higher risk of equipment becoming technologically obsolete, they must examine how best to invest in robotics to maximize their strengths while minimizing risks and costs. Developing an aggressive, integrated long-term asset management strategy can help. The three key elements of this strategy are acquisition, maintenance, and disposition. The goal is to optimize the productivity and flexibility of an entire manufacturing system throughout its life cycle while minimizing overall costs (Carpenter, 1997).

3.2   A Conceptual Model for Integration Issues of CIM/Robotics

A conceptual model illustrating the integration and adaptability issues of implementing CIM/robotics is presented in Figure 1. The organization has to develop the strategy that best fits the environment in which it operates. The model explains the importance of the alignment between various implementation strategies for improving integration and adaptability of CIM/robotics. For instance, strategic-level issues, such as alignment between business and manufacturing strategies, require suitable organizational structure, technology, employee involvement, and the nature of production planning and control systems. This relationship is represented by the closed loop, as shown in Figure 1, to explain the interaction and dependency between managerial, technological, and operational-level issues. Technological issues, such as CNC machines, AGVs, and computers, and operational issues, such as production planning, scheduling, and control, should be given due attention in the implementation of CIM/robotics. There are a number of organizational issues that companies consider when analyzing, designing, and managing the implementation of CIM/robotics systems. Organizations appear to have paid only limited attention to finding ways of managing design projects that are conducive to multidisciplinary and innovative

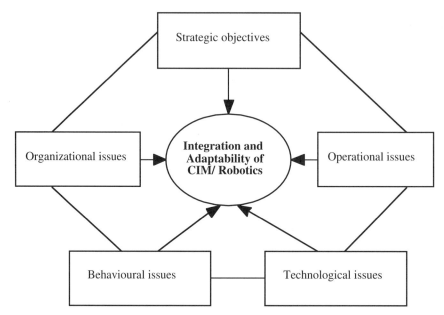

Figure 1   Integration and adaptability issues in the implementation of CIM/robotics.

teamwork. The organizational structure should support the kind of communication and decision-making together with the responsibilities to facilitate the implementation of CIM/robotics, and vice versa. The behavioral aspects should be given due attention, since they are related to human relations and human–machine interface, which play a significant role in obtaining cooperative supported work, not just for the implementation of CIM/robotics, but for the operations of CIM/robotics.

In the next section implementation of CIM/robotics is discussed, with the objective of providing further insights into integration and adaptability issues.

4   IMPLEMENTATION ISSUES FOR CIM/ROBOTICS

In this section an attempt is made to provide insight into the implementation issues of CIM/robotics, integration and adaptability being key issues. These issues are discussed on the basis of major issues such as strategic and technological, operational, behavioral, and organizational issues. The objective is to identify the most critical and pressing issues in the implementation of CIM/robotics.

4.1   Strategic Issues in the Implementation of CIM/Robotics

The manufacturing strategy should fully support the objectives of the business strategy. Moreover, there is a need for alignment between manufacturing strategy, such as CIM/robotics implementation, and business strategy to achieve the full potential and benefits of implementing CIM/robotics. The characteristics of a company in terms of capital, knowledge workers, complexity of the material flow, layout types, etc. should be considered in designing and implementing CIM/robotics.

While CIM/robotics integrates the system components, it does not necessarily introduce flexibility into the system. The focus should not be on integration alone, but on the simultaneous introduction of flexibility as part and parcel of the integration process (Babber and Rai, 1990). Computer-integrated flexible manufacturing (CIFM) or robotics is a long-term proposition and entails significant capital investment. Such a top-down strategy should receive strong support from upper management, and team effort should be emphasized. However, they seem not to have offered a strategic framework to improve flexibility in the implementation of CIM/robotics, in particular where automation warrants a kind of standardization of products and processes.

4.1.1   Justification for the Implementation of CIM/Robotics

Traditional cost-accounting systems generally fail to detect the many benefits of automation, such as CIM and robotics. Inability to financially justify an investment in CIM/

robotics often has been cited as the reason for not proceeding with an automation project (Bolland and Goodwin, 1988). However, automation's benefits can be more favorably highlighted by use of a discounted cash flow method such as net present value. Discounted cash flow produces a single figure so the total of cash inflows and outflows over the equipment's lifetime can be determined. Jain (1989) writes that effective CIM/robotics requires that every employee understand, accept, and endorse the need for attaining a competitive advantage. The importance of quality, customer service, product innovation, quick response to market needs, and competitive prices must be part of the organization's collective objectives.

The justification of a CIM/robotics system is one of the most important steps in implementing CIM/robotics. Factors such as reduced costs of material, direct and indirect labor, and reduced scrap, rework, and inventory would justify the scope of CIM/robotics implementation, together with organizational and strategic benefits. Nevertheless, performance measures such as reduced cycle time, increased capacity, and improved quality of products and customer services are less obvious but very important items (Kaltwasser, 1990). CIM/robotics as a strategy for integration must physically link parts of the facility and handle the flow of information, especially in the context of improving the speed of the material flow. According to O'Hara (1990), many firms begin a CIM/robotics project without conducting a proper analysis of the firm's management of information. To ensure that goals are attained and that the new system is compatible with a firm's strategic direction, an information flow profile needs to be created. However, these authors have not taken into account various cost tradeoffs in discussing the availability of computers and their integration.

The important part of the justification process is determining whether CIM/robotics fits in with cooperate strategy and can achieve performance goals. Strategy justification evaluates the ability of CIM/robotics to meet corporate goals, maintain or increase market demand, and reduce production cost. Strategy justification also identifies other relevant long-term alternatives. For strategy justification three types of evaluation are required: strategy planning, market assessment, and functional analysis of operation. Traditional cost-accounting systems generally fail to detect the many benefits of automation. The inability to financially justify an investment in CIM/robotics has often been cited as the reason for not proceeding with the automation project (Bolland and Goodwin, 1990).

4.1.2   CIM/Robotics in Manufacturing

Sound strategic decisions are at the heart of any successful industrial organization. Most manufacturing businesses have long-term survival as a basic objective. However, to stay in business, more new technologies such as CIM/robotics are needed. The characteristics of a company in terms of customer satisfaction, market share, product quality, etc. should be considered in designing and implementing CIM/robotics. A strategic impact study evaluates how CIM/robotics will fit within the overall cooperate strategy. The strategy plan ideally should be developed in an interactive workshop setting. Participation by top management from all functions allows for integration of goals and consensus on future direction. The objectives of CIM/robotics-oriented strategy planning are to:

1. Evaluate corporate strength and weakness.
2. Define the company's mission and long-term goals.
3. Develop manufacturing performance objectives.
4. Evaluate CIM/robotics' capability to meet goals and objectives.

Toyota Motor Corporation has employed many robot systems in automobile body assembly lines all over the world since 1970. They use robot systems mainly in the task areas of spot welding, arc welding, sealing, and inspection. Robot applications are increasing. These robot systems were mainly implemented from 1980 to 1990 to develop and achieve a new integrated body assembly line called *Flexible Body Line* (FBL). Automation for manpower savings was an extremely important technical issue in that period. However, in the current period, in which FBL development has nearly been completed, the installation and use of robots seem to be reaching the saturation point. Nowadays the primary issues are the important technical problems of reducing lead time and creating both new manufacturing systems and new automobile products to meet the demands of the market, such as recreational vehicles (RVs) and multipurpose vehicles (MPVs). Taking such situations into consideration, their technical approaches for actual lines are as follows: technical control and management of robot systems in practical lines space saving robot systems and CAE systems for preparation of body assembly engineering. Sakamoto, Suita, and Shibata (1997) discuss and explain simple maintenance programs in spot welding robot systems to improve and maintain robot reliability. A special spot-welding gun system employing an integrated servo-motor with high control abilities to reduce running cost, space-saving robot systems with wider ranges of application to reduce lead time has been reported.

The concept of adaptive robotized multifunction assembly (ARMA) cell, as used at CTRI Robotics, a division of Dassault Aviation, evolved in response to the demand for flexibility in automation system design, the inherent limitations of overspecialized equipment having been recognized. da Costa (1996) defines the mobility and placement corrective systems, the assembly process, information system, programming system, quality system, cell pilot, process control, and man–machine interface. The author concludes that the ARMA concept was implemented to provide the capabilities necessary to assemble the Rafale fighter automatically but, more importantly, is flexible enough also to assemble other and future aircraft (da Costa, 1996).

4.1.3 Strategy Development for CIM/Robotics

Strategy should be understood in terms of the interaction among product and process strategies, critical success factors, and product life cycle. *Implementation* stresses an iterative incremental process based on strategy, user involvement, and tolerance. *Innovation* is the result of a successfully implemented strategy. It consists of both organizational learning and change (Fjermestad and Chakrabarti, 1993). While many large companies have implemented CIM/robotics, small firms have rarely achieved company-wide integration of CIM/robotics. This is because small firms often feel that they lack the in-house expertise to implement new computer technologies and that hiring a consulting firm to assist will be too expensive. A number of companies are slow in adopting CIM/robotics in their organizations. This is especially problematic in today's consumer-oriented marketplace, where global competitive pressures are forcing firms to develop highly integrated information systems. The CIM/robotics system relates primarily to two areas of operation—logistics planning and human competence engineering.

4.1.4 Experiences with the Implementation Strategies for CIM/Robotics

Automation and robots are frequently mentioned as solutions to industry-wide problems of increasing costs, declining productivity, skilled-labor shortages, safety, and quality control. Despite numerous attempts to develop automation and robotics for construction field operations, few practical applications can be found on construction sites today. The promises of robotics remain unfulfilled, and attempts to transfer automation technology from manufacturing have not been optimal. Identification of opportunities for automation requires analysis of construction work at the appropriate level. A hierarchical taxonomy that divides construction field operations into several levels can be a more useful approach. The basic-task level is the appropriate level for construction automation. A set of basic tasks that describe construction field work facilitates automation in construction. Comparison of construction to highly repetitive manufacturing operations will provide insight into the relationships among product design, process design, and fabrication. In manufacturing, product and process design are closely interrelated. In construction, process design is completely separated from product design but is intimately related to fabrication. Until construction product and process design become more highly integrated, automation must occur at the basic-task level. Advances in construction automation will continue to be characterized by a machine performing physically intensive basic tasks, operated by a human craftsperson performing the information-intensive basic tasks (Everett and Slocum, 1994).

The many companies working with the University of Wisconsin—Madison's Center for Quick Response Manufacturing are proof that U.S. companies still have a long way to go in understanding and implementing speed-based strategies. Moreover, the Center's partnership with these companies illustrates that universities can provide industry with practical knowledge and assist them in implementing solutions to their current problems. One such company is ABB Flexible Automation, a world leader in the design and manufacture of robot-based automation for a variety of manufacturing industries. ABB's Industrial Automation Division diligently pursues customer-focused improvement efforts, including improving their responsiveness. This desire to improve the company's customer focus motivated ABB to become a member of the Center (Suri, Veeramani, and Church, 1995).

Lockheed Missile and Space Company Inc., Austin (Texas) Division, is one of the companies that has implemented an MRP II system as the key component of CIM/ robotics. The implementation took more than four years and required a $5 million investment in software and hardware. Lockheed's approach to assembling a project team is a good model for this first step in implementation. The major responsibility for defining detailed system specifications, developing user manuals, and the like fell upon representatives from manufacturing, material, and product assurance. The company's experience shows that successful implementation requires a well-informed and actively involved management group, education and training, a strong project manager with the skills and authority to make things happen, psychometric profiles to master project team members and adequate staffing of the project team with committed and capable personnel (Howery, Bennett, and Reed, 1991).

The synergic interaction between mechanical and electrical engineering and computer science, characterized by the synonym *mechatronics,* is a relatively new approach in modern engineering. Mechatronics is most apparent in the field of robotics. The demands for mechatronic systems are considerable in many key industrial areas. Therefore a survey of mechatronic approaches in the research and development of high-technology products and manufacturing processes and their application in various fields of industry can help in the integration and adaptability of CIM/robotics. The approaches presented have been developed in the context of the introduction of advanced robotics and information technology into modern production automation (Drews and Starke, 1993).

4.1.5 CIM/Robotics and Their Implementation Strategies

CIM/robotics is a manufacturing strategy for improving productivity and quality. It is essential that the organizational characteristics in terms of infrastructure and nature of the products manufactured be taken into account in selecting CIM/robotics. The implementation of CIM/robotics requires a clear, precise corporate strategy, the success of which will depend upon careful planning of several logical steps:

- Priming the corporate culture for change
- Clearly defining expectations
- Appointing a champion for CIM/robotics design and implementation
- Establishing a project team
- Performing a comprehensive environmental analysis
- Identifying the technology the strategy requires
- Formalizing operating policies
- Establishing working partnerships with suppliers and vendors
- Tracking and reporting progress

To summarize the strategic aspects of implementing CIM/robotics, there is a need to develop a framework for alignment between business and CIM/robotics as a strategy. There are a number of issues relating to poor planning and implementation of Advanced Manufacturing Technologies (AMTs):

- Firms were taking a long time to decide whether to proceed with completing implementation of AMT including robotics. On average, firms took nearly two years from idea generation to full implementation.
- All available sources of information were not being used in the decision-making process.
- Inadequate financial evaluation of the proposed investment was carried out.
- Many firms did not universally involve people from a variety of functions in the planning process (Sohal, Samson, and Weill, 1991).

A multicriterion-based decision model, consisting of 20 criteria structured across two hierarchical levels, is discussed here. The first step in the development of an appropriate set of criteria for a multiattribute-based decision model requires a definition of the enterprise's major objectives. These objectives provide the basis for the first-level criteria used by the model. First-level criteria need to be further subdivided into lower-order criteria.

At the lowest level of criteria the model must define how each criterion is to be measured. The provision of a detailed examination of lowest-order criteria, as well as a discussion on how criteria should be measured, are beyond the scope of this chapter. The model presented in Table 2 ensures that the second-level criteria are properly defined and form measurable entities. Therefore, each set of second-level criteria shown in Table 2 represents a logical breakdown of their associated first-level criteria. Thus, economic impact, for example, is a composite criterion consisting of profitability, operational risk, and net present value. This means that an evaluation of the economic impact of CIM/robotics is performed using measurement of profitability, operational risk, and the net present value (Nakamura, Vlacic, and Ogiwara, 1996).

The following are major issues with the regard to the success of AMT implementation (Sohal, 1997):

- Gaining competitive advantage
- Nature of AMT investment
- Decrease in the level of investment activity
- Cross-functional participation
- Level of management involvement
- Training of AMT
- Flexibility and responsiveness
- Need for post-implementation review.

The major obstacles for the implementation of CIM/robotics are shown in Table 3.

The integration of various activities at strategic, tactical, and operational levels primarily involves design of hardware such as CNC machines, robotics, AGVs, and software such as MRP II and Kanban. There is a need to integrate them with business processes for improving quality and productivity and hence competitiveness. However, there is also a need to examine manufacturing organizations to recognize the wide range of technological options for transmitting information, including fax machines, electronic mail systems, and even telephones. Lack of integration is not, therefore, the result of a lack of information-transmission technology or even a lack of available information, but rather of a lack of motivation on the part of individuals to use the available options.

Table 2   Decision Model

Main Criteria	Basis for Evaluation of the Main Criteria
Economic impact	profitability operations risk net present value
Strategic impact	customer satisfaction (service) reduced lead time improved quality of products larger portion of market share
Social impact	labor loading labor productivity training requirement motivation
Operational impact	delivery schedule performance productivity inventory maintainability flexibility quality control
Organizational impact	decision making efficiency of integration organizational structure

Table 3 Major Obstacles to the Implementation of CIM/Robotics[a]

Obstacle	Mean (rating)	Standard Deviation
lack of top management support and commitment	5.18	0.95
inadequate leadership	5.18	0.83
inadequate planning	4.93	1.01
lack of functional manager support and commitment	4.91	0.98
inadequate analysis of user needs	4.82	0.90
inadequate funding	4.78	1.06
failure to understand CIM/robotics and its potential	4.77	0.92
inadequate communications	4.71	0.98
insufficient education and training of managers and workers	4.68	1.11
inadequate system design	4.63	1.07
lack of people with technical expertise	4.61	0.95
corporate culture not right for CIM/robotics	4.60	1.07
management averse to risk of investing in new technology	4.59	1.16
human resistance to change	4.51	0.85
cost justifying with conventional methods	4.51	1.15
unrealistic expectations	4.45	1.23
high cost of CIM/robotics	4.36	0.96
inadequate organizational structure	4.35	1.14
key people are usually overcommitted	4.30	1.32
inadequate equipment	4.29	1.18
system incompatibility	4.24	1.19

[a] From R. E. McGaughey and C. A. Snyder, "The Obstacles to Successful CIM," *International Journal of Production Economics,* 37, 1994, 247–258.

4.2 Organizational Issues

Organizational structure can be seen as the assignment of people with different tasks and responsibilities. Interdepartmental integration can be achieved by authority, centralization, and regulation. An ideal organization should embrace cultural diversity in terms of education, skills, gender, race, ethnicity, and nationality, and have managers to oversee the change process. Characteristics of an organization such as full structural integration, unimpeded interpersonal communication, absence of prejudice, low levels of conflict with users and vendors, and pluralism will resolve the conflict between *management information system* (MIS) and manufacturing. It is important that people be trained in the organizational changes that CIM/robotics will introduce in the factory. Central to this discussion will be cross-functional training, where MIS learns manufacturing concepts while manufacturing learns about methods for *information system* (IS) analysis and design. In a CIM/robotics system there is a need for correlation and integration of data across the planning, design, implementation, and operation phases.

4.2.1 Functional Integration and CIM/Robotics

The implementation of computer-aided production management system (CAPM) depends upon the complexity of the overall manufacturing systems. The implementation can be made easy if the overall manufacturing is first simplified. Also, any methodology must include a software specification, since most companies will require a computerized solution. Finally, the overall performance of the system may be enhanced by suitable changes in the infrastructure that supports the software and integration.

It has been shown that a number of organizational and strategic imperatives are present that affect, and are affected by, the CIM/robotics technology, particularly when one considers the CAE–CAPM interface. The strategic management and investment appraisal literature can provide a starting point for evaluating the cost and benefits of adopting CIM/robotics. The main issue arising from Hassard and Forrester (1997) from an industrial viewpoint is how a company, once it has adopted CIM/robotics as a goal, can develop its own CIM/robotics system. The key to this is a business-oriented justification for CIM/robotics and top management support for any such project. The evolution of the true CIM/robotics goal, whether this is thought of in terms of the *computer-integrated business* (CIB) (Harrison, 1990) or, more radically, the *computer-integrated manufacturing enterprise* (CIM/robotics-E), will not be realized within a company unless there is knowledgeable perception of the practical issues of CIM/robotics development, greater understanding of management decision processes, and use of objective and appropriate investment procedures.

CIM/robotics technology implies the overall integration of managerial functions such as marketing, design, engineering, accounting, personnel, and finance. The higher degree of cross-functional integration demands strong infrastructural support for the efficient operation of manufacturing systems. When implementing the technology, management must be sure that the whole organization, including its structure, strategy, people, and power and authority distribution, is compatible with CIM/robotics. As many authors have pointed out, organizational issues play a predominant role in accepting new technologies. However, it is a two-way process: from the technological perspectives such as the suitability of CIM/robotics and from organizational perspectives such as infrastructure, business process characteristics, and skills available (Zhao and Steier, 1993).

Differences in functional goals produce a large number of incompatible activities, which is a major source of variability for the production system. Production and administrative throughput times can be reduced only if functional goals are aligned. The most effective means of aligning them is redesigning the organization structure so that members of different functional groups work together more closely. True organizational integration amounts to correcting the problems that have created poorly integrated, loosely coupled organizational systems in the first place. If these problems—which are essentially organizational rather than technological—are ignored, CIM/robotics implements run the risk of institutionalizing ineffective organizational procedures and communications linkages by automating them rather than correcting them.

The adoption of CIM/robotics brings with it new demands for interdisciplinary working and the sharing of information. However, in most traditionally organized companies this in itself yields a host of difficulties, not least of all in trying to break the barriers that exist between different parts of the business: that is, trying to effect organizational as well as technical, integration. Design engineering modules within CIM/robotics encompass those activities traditionally performed by designers and engineers, while computer-aided production management often replaces the manual activities of production scheduling, control, and operation, which have historically been within the domain of production management. The integration of systems in many organizations is currently being hindered by resistance to converging these parts of the business. Organizational integration and the elimination of departmental barriers are proving to be more difficult to achieve in practice and will in turn hinder the technical development of the "seamless" integration required.

It appears that for truly effective CIM/robotics to work in practice there needs to be a detailed appraisal of organizational structures and process. The goal must be the appropriate development and orientation of the organization's structure and the attitudes of people within the company. Preparing the organization and people in this way for the introduction of CIM/robotics must be seen as a prerequisite to enabling the technology to work for the maximum benefit of the business (Hassard and Forrester, 1997).

4.2.2 Top Management Support for the Implementation of CIM/Robotics

Attempting to install CIM/robotics without a total management system driving the company will lead to failure. The decision to install such an all-embracing system cannot be summarily turned over to the data processing staff. On the other hand, CIM/robotics is not workable without highly skilled workers, the production and plant managers, supervisors, and operators, who often know more about how things should and should not work than anyone else in the company. However, the most important reason for a company's failure at implementation is lack of understanding of what is really needed for any of these techniques to be successful (Hazeltine, 1990). Thus the main approach used in the CIM/robotics system for dealing with organizational variability should be to increase the level of flexibility in order to handle variability at the point of impact (within manufacturing). However, both integration and variability handling within the CIM/robotics system are purely technological issues rather than organizational issues (Duimering, Safayeni, and Purdy, 1993).

To be successful, an initiative must have the direct involvement of top management. Top management must not only commit itself, invest company resources, and accept long-term results, but must eventually modify the company organization as required. Those employees whose normal work tasks are touched by the project must be trained, involved, and motivated. Although performance measurement is necessary, it must be remembered that measurement is a tool, not an end in itself. If normal company operating procedures must be bypassed to obtain timely results, then the company bureaucracy is the real problem and a reorganization with trim procedures should be the real objective (Levetto, 1992). However, no framework has been developed on how to motivate top management in the implementation of CIM/robotics. The implementation of advanced technology requires an interactive incremental process based on implementation, user involvement, and tolerance. The implementation of CIM/robotics needs both organizational learning and change (Fjermestad and Chakrabarti, 1993). *Business process reengineering* (BPR) can play a tremendous role in the implementation of CIM/robotics.

4.3   Behavioral Issues in the Implementation of CIM/Robotics

CIM/robotics requires a teamwork approach in which every member has a key role to play. One of the biggest obstacles to CIM/robotics implementation is that systems are conceived by corporate headquarters and then pushed down onto a plant manager. Training and involvement help to minimize worker resistance. However, it ultimately comes down to corporate culture. The effectiveness of the training should be evaluated and based on learning objectives. Some of the types of training that can be used for CIM/robotics are classroom instruction, computer-based training, workshops, and videotapes. As mentioned earlier, top management support is essential for the successful implementation of CIM/robotics.

4.3.1   Human Factors in the Integration and Adaptability of CIM/Robotics

Human factors should be considered at the earliest stages of the planning and implementation of CIM/robotics systems. If not, a CIM/robotics project may fail as workers struggle to operate and maintain a system superficially designed to prevent their efforts. Human factors are important in areas such as installation, operation and maintenance, and safety. Installation requires workers well trained in automation principles. Operation and maintenance requirements include workstations and computer interfaces designed according to established human factor principles and work environments that provide human interaction during the job performance and scheduled breaks in order to prevent feelings of isolation. In addition, safety enablers should include minimizing contact between humans and automated equipment, installing sensors and intruder alarms that stop equipment when a human enters the workcell or crosses the path of an automated vehicle, and installing of panic buttons accessible from anywhere in the workcell.

An unmanned or unattended factory with a manned control room can be a solution to the ever-declining number of workers in factories. The control room should be kept clean and comfortable for supervisors, who would mostly be engaged in planning and programming the manufacturing process of the unmanned factory. The future plant requires computer training and training in self-management and conflict management. The Monsanto Chemical Company's experience in this matter involved employees participation in extracurricular activities, such as design committees, quality circles and improvement teams. This gave the workers' ideas validation and routes to implementation. However, the major benefits envisioned for CIM/robotics will not come about if the importance of human resources is not taken into consideration or strategic planning is ignored. Furthermore, the process of creating change and implementing new systems must be understood for the benefits of CIM/robotics to be achieved (Yoshikawa, 1987).

4.3.2   Human Resource Management in CIM/Robotics Environment

Firms should implement action-oriented operation strategies that stress integration of human values, taking into account the critical factors, such as workers' pride and positive attitudes, that contribute to any successful operations strategy. CIM/robotics' initial implementation requires long-term strategies, a great amount of research and development, and possibly the forgoing of immediate financial benefits. The most essential element in such a strategy is the preparation of the workforce for the impending changes. This requires consultation at all levels and a systematic training effort. Moreover, since CIM/

robotics systems are tailor-made for specific companies, such training must be carried out mainly by the companies themselves in cooperation with systems suppliers. Since the need for human work is reduced in the automated functions, training and job enrichment for the existing people are a necessity (Ebel, 1989).

4.4 Technologies Issues for CIM/Robotics

The key elements in any CIM/robotics application are:

1. Information management and communication systems—the computers that link the various developments and decision-making activities to each other
2. Material management and control systems involving material flow control
3. Process management and control, which includes down loading the appropriate inputs or process instructions and monitoring the process itself
4. Integration using computers

If a CIM/robotics solution is practical, there must be common operating plans for all affected departments, with an integrated system using a common database. However, the aspects of human–machine interface are to be considered in designing key elements of the CIM/robotics system.

4.4.1 Role of Information Technology in CIM/Robotics

The programmable automation, including Internet, Multimedia, flexible manufacturing systems (FMS), robots, AGVs, and EDI in business transactions, can make possible tremendous improvements in manufacturing. The reluctance of many companies to adopt these new technologies may reflect gaps in their capital budgeting processes due to lack of understanding. The companies that are able to exploit the hidden capabilities of a new technology most effectively are generally those that adopt it early, continually experiment with it, and keep upgrading their skills and equipment as the technology evolves. The fundamental problems in the development and implementation of CIM/robotics include lack of integration, islands of automation, suboptimization of resources, and inability to migrate to future technology.

A graphic simulation system for robotics operations has been developed in conjunction with a kinematic error model to assist management in making investment decisions. Taguchi-type experimental design can be used to predict robot process capability. Two sets of charts (histograms) have been used by the system to illustrate the repeatability and accuracy of a robot performing a given operation. Central to the system are three characteristics: a description of the geometric error of the kinematic error, a function with which to request and display data on a robot moving from one position to the next, and a family of charts showing the capability of robots for implementing a required task. The results provided by the system can help management to analyze robotics projects and make decisions concerning robot selection and implementation in a more systematic manner (Kamrani, Wiebe, and Wei, 1995).

Through the elimination of manually produced paper systems and keying in of data, the *radio frequency data communications* (RFDC) system has brought the benefits of improved efficiency to the everyday working environment. RFDC has become an operational tool for providing the timely, accurate data necessary for the implementation of CIM/robotics. For example, RFDC-based warehousing and manufacturing operations are required to conform to the longer-term business IT strategy. In addition, an open system helps to maximize existing investment in system development and improve the performance of material-handling systems (Hiscox, 1994). The application of MIS-oriented software design tools and methodologies to manufacturing applications has generally resulted in poor performance in manufacturing systems. However, the problems with software arise from the complexities not only of driving and controlling various elements of the factory, but also of interfacing manufacturing software with both engineering and business software (Gowan and Mathieu, 1994). The integration and adaptability problems arise in manufacturing due to lack of integration, islands of automation, suboptimization of resources, and the inability to accept new technologies. These problems can be addressed through the object-oriented approach to modelling of manufacturing enterprises.

4.4.2 CIM/Robotics and Other Associated Technology Integration

Organizations using CIM/robotics technology gain the capability of concurrent engineering—that is, design and manufacturing of desired products can take place simultaneously. However, the connecting mechanisms with customers, such as understanding of customer demands, forecasting of market segment change and quality improvement of products have to be built into the production process by management. In order for the competitive advantage through implementing the CIM/robotics technology to be regained, concurrent engineering technique and philosophy must also be used simultaneously.

Many companies have implemented CIM/robotics and are successful in achieving integration of various operations. An "enterprise engineering methodology" for the strategic management of technologies in CIM/robotics is presented by Sarkis, Presley, and Liles (1995). This methodology is based on preparing manufacturing system that will accept CIM/robotics effectively. The design of automated equipment such as robots, computers, and CNC machines must conform to high-level standards of communication protocols to develop an intelligent manufacturing cell. The following are the reasons why companies elect not to choose a CIM/robotics environment (Boubekri, Dedeoglu, and Eldeeb, 1995):

- The amount of investment required
- Insufficient skills
- Difficulty of implementing computerized systems
- Limiting or multiplying synergy
- Different support infrastructure
- Lack of a unique set of standards that fulfills all the requirements of a system

4.5   Operational Issues

Day-to-day tasks, such as scheduling, are performed at the operational level. To reduce administrative throughput time, the focus of improvement must be placed on the information-processing portion of the equation. Different technologies have applications in different areas to reduce processing times, and there is a need to combine the technologies to utilize their best features in CIM/robotics applications. The use of software packages such as SIMAN and CINEMA is very important for modeling the behaviors of the CIM/robotics system and building a simulation-driver animation of the manual manufacturing system and the proposed CIM/robotics system. CIM/robotics requests the reorganization of the production planning and control system with the objective of simplifying the material and information flows. Manufacturing concepts such as JIT and MRPII and technologies such as CE and AGVs provide the base for easy implementation of CIM/robotics and improvement of integration and adaptability (Gunasekaran, 1997).

4.5.1   Simulation in the Integration and Adaptability of CIM/Robotics

To reduce administrative throughput times, the focus of improvement must be placed on the information-processing portion of the CIM/robotics. To reduce processing times, the actual activities performed by various functions must be coordinated. The use of software packages such as SIMAN and CINEMA is very important for modelling the behavior of the CIM/robotics system and building a simulation-driven animation of the manual manufacturing system and the proposed CIM/robotics system. For example, to achieve competitive advantage through implementing the CIM/robotics technology, concurrent engineering technique and philosophy can be used simultaneously. Moreover, the success of CIM/robotics system depends upon effective scheduling and control.

Automatic data extraction, processing, and transfer interfacing between the design and construction phases are needed as a key element in CIM/robotics. The CAD/CAM interface eliminates the need for manual interfacing between these two computer-aided phases. The CAD/CAM interface improves the process of the constructed facility realization by eliminating manual data processing, thereby reducing many sources of errors. It also makes the process more cost-effective because it reduces labor inputs, especially those presently invested in robot programming. A model for automatic data extraction, processing, and transfer has been proposed for the tile-setting paradigm. This model can generate construction management data as well as data needed for automatic on-site construction (robotics). The model can be implemented in the AutoCAD and AutoLISP environments. The model and the implementation system can be tested in laboratories with a scaled robot adapted to perform interior finishing tasks (Navon, 1997).

4.5.2   Techniques Available for Operations Control in CIM/Robotics

CIM/robotics design should view the development process as a sequence of activities that follows the traditional sequential waterfall model of software engineering. In the design and implementation of CIM/robotics of the Novatel company, management approached the problems first by identifying the limits of CIM/robotics with the objective of achieving a fully computerized network organization. Improving product and service quality is important for a manufacturing company. Process or product redesign may be necessary to effect significant improvement in quality. The need to respond to swift and unpredicted market changes requires flexibility in the production process and system (Groves, 1990).

An autonomous distributed control system to improve flexibility and speed will be useful in the implementation of CIM/robotics. In order to realize the autonomous distributed control necessary for intelligent manufacturing systems (IMS), a model and a method can be used. The model consists of numerous autonomous agents and a field where these agents can exchange information for cooperation. The importance of the protocol needed for cooperation of the individual agents should also be addressed. The autonomous distributed control system model can be used in the study of an automated guided vehicle (AGV) system. Here, numerical simulations using the Petri net model can be performed to evaluate the effectiveness of the model (Nagao et al., 1994).

The basis of the CIM/robotics concept is the integration of various technologies and functional areas of the organization to produce an entirely integrated manufacturing organization. However, the networking that is essential to CIM/robotics implementation is difficult to achieve. The CIM/robotics literature cites several problems related to CIM/robotics networking, including multiple vendor installations, scarcity of software applications, immaturity of connectivity products, and network management. CIM/robotics should be looked upon not as a panacea, but as one viable tool for companies to use to stay competitive in the international market. Therefore, operational refinements should be carefully reviewed with the intention of providing for what the Japanese call *kaizen*, or continuous improvement. A study by Snyder (1991) examines continuous improvement in processes (CIP) as a preimplementation strategy for CIM/robotics. Furthermore, a systems development life cycle approach can be used to identify the activities that should be initiated before a CIM/robotics strategy (Snyder, 1991).

5 HUMAN RESOURCE MANAGEMENT AND THE ROLE OF UNIONS

There can be no doubt that many of the activities and functions currently being performed by humans will gradually be taken over by automation. Indeed, in many cases this is to be welcomed (in general these functions are fairly well proceduralized, require little creative input, and permit algorithmic analysis). The functions that will not be automated, at least in the near future, are those requiring cognitive skills of a high order: design, planning, monitoring, exception handling, and so on. While it is easy to understand the need for large capital investment associated with the CIM/robotics, the technical reasons are not always explicit.

5.1 The Reasons for CIM/Robotics

The prime requirements for automating any function are the availability of a model of the activities necessary for that function, the ability to quantify that model, and a clear understanding of the associated information and control requirements. One cannot automate function simply by copying what people do (Mital, 1997). There are several reasons why automation purely or largely based on what people do does not work well:

1. In many cases we really do not know exactly what people are doing.
2. We are limited by technology in building machines that have capabilities comparable to those of humans, particularly flexibility and decision-making, and that mimic human techniques.
3. People have unique capabilities, such as sensing.
4. People are often innovative or resourceful and improve processes in ways that machines cannot—for instance, in mating parts that do not quite meet the tolerances or finding ways to prevent errors or inefficiencies in process.

Clearly some functions should be performed by machines, for the following reasons (Mital, 1997):

1. Design accuracy and tolerance requirements
2. The nature of the activity being such that it cannot be performed by humans (e.g., water jet cutting, laser drilling).

3. Speed and high-production volume requirements
4. Size, force, weight, and volume requirements (e.g., materials handling)
5. Hazardous nature of the work (e.g., welding, painting)
6. Special requirements (e.g., preventing contamination)

5.2 Activities Performed by Humans in CIM/Robotics Environment

Equally, some functions should be performed by humans, for the following reasons:

1. Information-acquisition and decision-making needs (e.g., supervision, some forms of inspection)
2. Higher-level skill needs (e.g., programming)
3. Specialized manipulation, dexterity, and sensing needs (e.g., maintenance)
4. Space limitations (e.g., work that must be done in narrow and confined spaces)
5. Poor-reliability equipment is involved or equipment failure could be catastrophic
6. Technology is lacking (e.g., in soil remediation)

The assignment of tasks to humans and machines must be based on in-depth analysis of:

1. Information and competence requirements of the activity
2. Capabilities and limitations of humans and the automated equipment
3. Availability of technology
4. Safety and comfort
5. The design principles for good jobs
6. Economics

5.3 Education and Training for CIM/Robotics

5.3.1 Human Issues
1. Determining what kind of assistive devices are needed to enhance human performance (inspection, assembly, machine loading, etc.) and how to design them
2. Determining supervisory needs
3. Determining human considerations to account for during manufacturing planning

5.3.2 Training Issues
1. Identifying the kind of training—cross training versus retraining
2. Training workers and supervisors to adapt to automation
3. Measuring training performance and effectiveness
4. Identifying the role of workers in the production process
5. Determining the necessary training support
6. Developing training documentation
7. Determining documentation, aids, etc. needs
8. Preparing workers for change

5.4 Implications of CIM/Robotics for the Relationship Between Unions and Management

The pressure to introduce robots or automation into a factory usually comes from the marketplace. Goods have to be sold at an economical price, and if profit margins are to be adequate, manufacturing efficiency must be high. In some areas of the world labor is so cheap that there is little incentive to automate. But in the developed countries labor is expensive and the solution to manufacturing efficiency is found in the utilization of capital equipment (Morgan, 1984). The importance of workers having the right attitude towards robots and robot systems cannot be overemphasized. Managers need to take as much trouble in launching a new robot system in the factory as they would for launching a new product on the market. Management must use any and every technique available to ensure that their workforce is "sold," and remains sold, on what the company is trying to do. The range of skills required will vary greatly depending on who is involved. By and large the greatest demand in the new systems is for operators, foremen, and supervisors. At the operator level the skills required to use the robot system are evident (Ayres and Miller, 1995).

Operational skills will be greatly enhanced through practice, and any training course must have plenty of hands-on experience. Usually initial training is done off-site, and it is well to allow adequate time after on-site installation for more practice before serious production begins. It is wise to ensure that as many operators as possible are trained to use the robot system. This reduces the company's vulnerability to sickness and labor turnover (Ayres and Miller, 1995). The level of management immediately above the operators must be as thoroughly familiar with the robot system as the operators are. As with the operators, the training of foremen or supervisors should be a combination of on-the-job and off-the-job experience. The training required of middle and senior managers is of a more conceptual nature. The relationship between management and trade unions, often called *industrial relations,* is a notoriously tricky area to write about, but in the final analysis managers control the working environment and they get the industrial relations they deserve. Poor planning, bad selling, bad negotiations, and inadequate training can all cause poor robot systems performance.

Although 14,000 robots were in use in Japan toward the end of 1982, no worker had really been displaced because of robotization. This is because:

1. Robots were used to meet labor shortages, particularly in dirty, dangerous, and heavy and/or highly repetitive work.
2. Many robots were used in industries that were expanding production.

This has also been supported by statistics from the United States. From this perspective, the effect of robotization on employment seems to be negligible. However, robotization does have some impact on employment opportunities, in particular where the process is labor-intensive, as in the metalworking industry (Ayres and Miller, 1995). The following can be said about the future of robotization in terms of employment and union relations:

1. Nearly half of all the unskilled and semiskilled "operative" workers—those who could be replaced by robots—are concentrated in the metalworking industry.
2. Older, established workers are generally protected by union seniority rules, except in cases where the whole plant closes. However, robotization will lead to unemployment among young workers in labor-intensive industries. This would lead to the creation of a class of insecure and marginal workers, providing a potential source of social problems and political dissension.
3. A migration of workers from the regions with the most job losses would occur. In the regions the workers migrated they would accept lower-paying jobs and hence join the "underclass" of insecure marginal workers who never became established with a stable employer.
4. There would likely be a disproportionate impact on racial minorities and women. Women employed in semiskilled and unskilled manufacturing jobs are less likely to be represented by labor organizations than their male counterparts and hence economic discrimination will surface.
5. Unions representing the affected workers will probably experience sharp declines in membership and political/economic clout.

Unions can be expected to favor two approaches in dealing with job losses for their members threatened by robotization. The first is to transfer and retrain the displaced employees into other jobs that have been created by attrition or by growth. This type of remedy is likely to be least costly to employers and to constitute a minimum barrier to the introduction of robots. The limitation of this approach is that it assumes a pace of robotization consistent with the number of suitable job openings created within the same plant. The other approach, which is favored by many unions, is to attempt to protect threatened jobs by raising the cost of introducing robotization and, in this way, transferring part of the productivity gain from employers to employees. Policies in this case include implementing restrictive work rules, shortening the work week, lengthening paid vacations, and adding paid personal holidays. They also include employment guarantees and employer-financed pensions for older employees who retire early (Ayres and Miller, 1985).

The most remarkable thing about the job displacement and job creation impact of industrial robots is the clear distinction between the jobs eliminated and the jobs created. The jobs eliminated are semiskilled, while the jobs created require significant technical background. The traditional skills of craftsmen are being made unnecessary by machines or diluted into simpler skills. On the other hand, new types of skills have emerged. Major characteristics of these new types of skills are profound knowledge concerning complicated machinery and its functioning, programming ability, and perspectives on the total machine system. The transformation makes it more and more difficult for older workers to follow and adapt to the new technology.

6   MAJOR INTEGRATION AND ADAPTABILITY ISSUES IN THE IMPLEMENTATION OF CIM/ROBOTICS

CIM is an important emerging issue for both practitioners and researchers as firms have increasingly adopted it as a competitive weapon. The issues addressed by McGaughey and Roach (1997) are noteworthy for both practitioners and researchers. The longer a company has used CIM, the less likely participants are to perceive commitment of resources and strategic concerns as important obstacles. The longer a company has used CIM/robotics, the more likely participants are to perceive the CIM/robotics planning effort to have been successful. There are significant positive correlations among commitment of resources, strategic concerns, and organizational receptivity. The following discussion offers several insights gained on integration and adaptability issues of CIM/robotics implementation in practice:

1. CIM/robotics should be implemented only after the basic foundations have been put in place in the company. It may be more productive to redesign the organizational structure before implementing available technology than to hope the technology will bring about manufacturing effectiveness. Simplification of information flow and material flow will establish a solid foundation for adopting CIM/robotics technology.

2. Integration and adaptability issues of CIM/robotics should be evaluated keeping in mind the lack of knowledge about CIM/robotics and its potential strategic implications for longer-term planning, the effect of delaying CIM/robotics implementation on company competitiveness, and the effect of operations integration.

3. The integration and adaptability issues for CIM/robotics are influenced by factors such as the required hardware platform, integration requirements, and data processing skills. These factors must be considered in implementing CIM/robotics. Knowledge workers, such as computer operators and software engineers, and a multifunctional workforce are essential to improving integration and adaptation in the implementation of CIM/robotics.

4. Human workers play a significant role in influencing the integration and adaptability issues of CIM/robotics, especially through cooperative supported work. This indicates the importance of providing comprehensive training to equip workers with knowledge of automation, computer technologies, and manufacturing processes.

5. Despite the arguments regarding flexibility of CIM/robotics, the experience from practice is that automation is frequently too rigid to adapt to changing market needs and the production of new products. This indicates the importance of flexibility of CIM/robotics in designing the system and reorganizing the production planning and control system.

6. There is a need for a unique set of standards that satisfies all the requirements of a CIM/robotics system.

7   A FRAMEWORK FOR THE IMPLEMENTATION OF CIM/ROBOTICS

The strategy for successful implementation of CIM/robotics should include the use of computers for integrating information and material flows, small-batch production with on-line production control system (e.g., FMS), and local area network (LAN) for integrating information flow within the organization. A conceptual framework is presented in Figure 2 to explain the main issues involved in improving the integration and adaptability aspects of CIM/robotics. The model presents a set of major elements of CIM/robotics implementation, including strategic, organizational, behavioral, technological, and operational

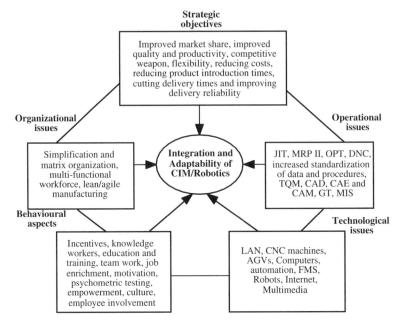

Figure 2   Integration and adaptability of CIM / robotics: a framework.

issues. Each of these elements is discussed here from the viewpoint of improving integration and adaptation in the implementation of CIM/robotics. The details follow.

7.1 Strategic Objectives

Top management selects CIM/robotics as a manufacturing strategy based on business strategy. Middle management then should work out the CIM/robotics development program. The workers, along with middle management staff, are responsible for the implementation of CIM/robotics. The implementation of CIM/robotics requires a clear, precise corporate strategy, the success of which will depend upon careful planning of several logical steps (Gunasekaran, 1997):

- Priming the corporate culture for change
- Clearly defining expectations
- Appointing a champion for CIM/robotics design and implementation
- Establishing a project team
- Performing a comprehensive environmental analysis
- Identifying the technology the strategy requires
- Formalizing operating policies
- Establishing a working partnership with suppliers and vendors
- Tracking and reporting progress

7.2 Organizational Issues

CIM/robotics requires cross-functional cooperation and high involvement of employees in the product-development process. For implementation of CIM/robotics to be successful, an initiative must have the direct involvement and commitment of top management. Top management also invests company resources and accepts long-term results, eventually modifying the company organization as required for successful CIM/robotics. Effective implementation of CIM/robotics requires a strong degree of communication and coordination among interdependent units in companies. Internal factors such as product and process characteristics, infrastructure, and skills available, and external factors such as market characteristics and government support and regulations, tremendously influence the implementation process of CIM/robotics. The process of integration using CIM/

robotics concepts becomes somewhat easier if the system has already a JIT production control system or manufacturing cell. Government support, whether financial or technical, will help minimize the risk of loss in production and business as a whole during implementation of CIM/robotics. For the financial constraints to be overcome, the level of integration should be focused on the critical areas that influence the performance of manufacturing.

Continuous organizational change driven by a need to be more competitive appears to be a major driving force for changing information systems. One should first study why flexibility is needed in CIM/robotics systems and what level of flexibility is required in software. Usually two types of flexibility requirements are identified, namely changes in operational procedures and changes in decision-making. The matrix organization will help to improve cross-functional coordination in the implementation of CIM/robotics. CIM/robotics technology may enhance competitive advantage, but it must be recognized that the integration of various computerized systems produced by different vendors often leads to technological difficulties. In addition, the capital investments needed for the development and implementation of CIM/robotics are substantial. Hence it is very important to provide a system after necessary changes in the organization have been made to facilitate the system for computer integration. CIM/robotics includes all the engineering and design functions of CAD/CAM, together with all the business functions such as sales, order entry, accounting, and distribution (Groover, 1987).

7.3   Behavioral Aspects

Cooperation among different levels of employees can be achieved in developing CIM/robotics through smoother communication systems. The workforce involved in the implementation and operation of CIM/robotics is composed of knowledge workers such as computer operators, software engineers, and network managers. Therefore, the type and level of training and education required should be determined taking into account infrastructure, integration, and adaptability issues. Effective teamwork (with empowerment and responsibility) has to be achieved for successful implementation of CIM/robotics. This can be done through a collective incentive scheme, training, and job enrichment.

7.4   Technological Aspects

The suitable configuration for CIM/robotics should be decided before the implementation process. This generally centers around the identification of tasks to computerize, selection of feasible software packages, and improvement of software compatibility. In order to include flexibility in CIM/robotics, manual policies, procedures, and practices should be established. The integration and adaptability of CIM/robotics can be made considerably easier with FMS, cellular manufacturing systems and JIT production systems. Technologies such as the Internet, multimedia, and LAN can be used to improve the integration of various business areas of manufacturing organizations. AGVs can play an important role in improving the integration of material flow within the production system using computers. Integration of operational activities with suppliers can be improved by on-line computer information systems such as an electronic data interchange (EDI). These can also play a vital role in an unmanned factory.

The integration of various functional areas (in terms of information and material flow) can be achieved using existing equipment and low-cost components such as contact sensors and relay systems. Hence the old equipment can be partially replaced without affecting the remaining components of the cell. The system's design is modular and the component programs are independent. The integration capability of the equipment is of utmost importance. Automation and technology have given factory management the tools they have always dreamed about. The integration of robotics into the welding process is a tradeoff of technology and quality, with cost being the primary factor. It requires strategic planning and careful execution. Via (1994) describes a simultaneous engineering approach to ensuring the proper integration of people, processes, and machines.

7.5   Operational Issues

CIM/robotics requires the reorganization of the production planning and control system, with the objective of simplifying the material and information flows. Manufacturing concepts such as JIT and MRP II and technologies such as CE and AGVs provide the basis for easy implementation of CIM/robotics to improve integration and adaptability. The essence of CE is the integration of product design and process planning into one common activity, CAD/CAE. Concurrent design helps to improve the quality of early design decisions and has a tremendous impact on the life cycle cost of the product. The implementation of CE will facilitate the integration and adaptation of CIM/robotics.

The major issue in the implementation of CIM/robotics is the reduction of the overall complexity of the manufacturing system. This may call for process simplification, such as through the use of flow lines or cells using group technology. One should recognize that because of data volumes, most companies will still need a computer-based production management system, so the approach must lead to a software specification. The overall performance of the system can be considerably enhanced through the infrastructure policies, procedures, and practices that control business processes, so any solution must be a balance between software and the infrastructure.

8 CONCLUSIONS

Management concerns with the integration and adaptability issues of CIM/robotics have been discussed in this chapter with reference to strategic, organizational, operational, technological, and human issues points of view. A conceptual model has been presented to illustrate the role of strategic, organizational, behavioral, technological, and operational issues in the management and implementation of CIM/robotics. Also, a framework has been provided to improve the integration and adaptability of CIM/robotics. In implementing CIM/robotics the danger of moving towards full integration with the wrong emphasis in undertaking the project should not be underestimated. Emphasis should be placed on information automation, organizational support, human resource involvement, and union relations to improve integration and adaptability in CIM/robotics. Nevertheless, investment justification for the implementation of CIM/robotics projects and the determination of the level of skills and training required to develop and maintain an effective CIM/robotics system need further attention from researchers and practitioners. Moreover, a functional approach should be developed for integrating the concept of operational safety during the specification phase of CIM/robotics systems.

ACKNOWLEDGMENTS

The author is grateful to Professor Shimon Y. Nof for his extremely useful and helpful comments on an earlier version of the manuscript. His comments helped to improve the presentation of this chapter. Also, the assistance provided by Mr. Hussain Marri is appreciated.

REFERENCES

Ayres, R. U., and S. M. Miller. 1985. "Socioeconomic Impacts of Industrial Robots: An Overview." In *Handbook of Industrial Robots*. Ed. S. Y. Nof. New York: John Wiley & Sons. 467–496.

Babber, S., and A. Rai. 1990. "Computer-Integrated Flexible Manufacturing: An Implementation Framework." *International Journal of Operations and Production Management* 10(1), 42–50.

Bolland, E., and S. L. Goodwin. 1990. "Corporate Accounting Practice Is Often Barrier to Implementation of Computer Integrated Manufacturing." *Industrial Engineering* 22(7), 24–26.

Boubekri, N., M. Dedeoglu, and H. Eldeeb. 1995. "Application of Standards in the Design of Computer-Integrated Manufacturing Systems." *Integrated Manufacturing Systems* 6(1), 27–34.

Carpenter, D. M. 1997. "Affording Robotics," *Robotics World* 15(1), 28–29.

da Costa, S. 1996. "Dassault Adaptive Cells." *Industrial Robot* 23(1), 34–40.

Drews, P., and G. Starke. 1993. "Mechatronics and Robotics in Europe." In *IECON Proceedings (Industrial Electronics Conference)*. Vol. 1. 7–13.

Duimering, P. R., F. Safayeni, and L. Purdy. 1993. "Integrated Manufacturing: Redesign the Organization Before Implementing Flexible Technology." *Sloan Management Review* 34(4), 47–56.

Ebel, K. H. 1989. "Manning the Unmanned Factory." *International Labour Review* 128(5), 535–551.

Everett, J. G., and A. H. Slocum. 1994. "Automation and Robotics Opportunities: Construction Versus Manufacturing." *Journal of Construction Engineering and Management* 120(2), 443–452.

Fjermestad, J. L., and A. K. Chakrabarti. 1993. "Survey of the Computer-Integrated Manufacturing Literature: A Framework of Strategy, Implementation, and Innovation." *Technology Analysis and Strategic Management* 5(3), 251–271.

Gowan, J. A., Jr., and R. G. Mathieu. 1994. "Resolving Conflict Between MIS and Manufacturing." *Industrial Management and Data Systems* 94(8), 21–29.

Groover, M. P. 1987. *Automation, Production Systems, and Computer Integrated Manufacturing*. Englewood Cliffs: Prentice-Hall.

Groves, C. 1990. "Hands-Off Manufacturing Transcends Limits of CIM." *Industrial Engineering* 22, 29–31.

Gunter, L. 1990. "Strategic Options for CIM Integration." In *New Technology and Manufacturing Management.* Ed. M. Warner. New York: John Wiley & Sons.

Gunasekaran, A. 1997. "Implementation of Computer-Integrated Manufacturing: A Survey of Integration and Adaptability Issues." *International Journal of Computer-Integrated Manufacturing* **10**(1–4), 266–281.

Gunasekaran, A., and S. Y. Nof. 1997. Editorial: Integration and Adaptability Issues. *Int. J. of CIM* **10**(1–4), 1–3.

Harrison, M. 1990. *Advanced Manufacturing Technology Management.* London: Pitman.

Hassard, J., and P. Forrester. 1997. "Strategic and Organizational Adaptation in CIM Systems Development." *International Journal of Computer-Integrated Manufacturing* **10**(1/4), 181–189.

Hazeltine, F. W. 1990. "The Key to Successful Implementations." *Production and Inventory Management Review and APICS News* **10**(11), 40–41, 45.

Hiscox, M. 1994. "The Management of RFDC Systems." *Sensor Review* **14**(2), 3–4.

Howery, C. K., E. D. Bennett, and S. Reed. 1991. "How Lockheed Implemented CIM." *Management Accounting* **73**(6), 22–28.

Jain, A. K. 1989. "Plan for Successful CIM Implementation." *Systems/3x and As World* **17**(10), 128–137.

Kaltwasser, C. 1990. "Know How to Choose the Right CIM Systems Integrator." *Industrial Engineering* **22**(7), 27–29.

Kamrani, A. K., H. A. Wiebe, and C. C. Wei. 1995. "Animated Simulation of Robot Process Capability." *Computers and Industrial Engineering* **28**(1), 23–41.

Levary, R. R. 1992. "Enhancing Competitive Advantage in Fast-Changing Manufacturing Environments." *Industrial Engineering* (December), 22–28.

Levetto, M. 1992. "Successful Implementation of Management Technique." *Industrial Management* **34**(5), 14–15.

McGaughey, R. E., and D. Roach. 1997. "Obstacles to Computer Integrated Manufacturing Success: A Case Study of Practitioner Perceptions." *International Journal of Computer-Integrated Manufacturing* **10**(1–4), 256–265.

McGaughey, R. E., and C. A. Snyder. 1994. "The Obstacles to Successful CIM." *International Journal of Production Economics* **37**, 247–258.

Mital, A. 1997. "What Role for Humans in Computer Integrated Manufacturing." *International Journal of Computer-Integrated Manufacturing* **10**(1–4), 190–198.

Morgan, C. 1984. *Robots: Planning and Implementation.* IFS.

Morrison, D. L., and D. M. Upton. 1994. "Fault Diagnosis and Computer Integrated Manufacturing Systems." *IEEE Transactions on Engineering Management* **41**(1), 69–83.

Nagao, Y., et al. 1994. "Net-Based Co-operative Control for Autonomous Distributed Systems." In *IEEE Symposium on Emerging Technologies and Factory Automation.* 350–357.

Nakamura, T., L. J. B. Vlacic, and Y. Ogiwara. 1996. "Multiattribute-Based CIE/CIM Implementation Decision Model." *Computer Integrated Manufacturing Systems* **9**, 73–89.

Navon, R. 1997. "COCSY II: CAD/CAM Integration for On-Site Robotics." *Journal of Computing in Civil Engineering* **2**(1), 17–25.

O'Hara, K. 1990. "Key to Successful CIM Implementation." *Production and Inventory Management Review and APICS News* **10**(4), 53–58.

Sakamoto, Y., K. Suita, and Y. Shibata. 1997. "Technical Approaches of Robotic Applications in Automobile Body Assembly Lines." *Robot* 114, 56–63.

Sarkis, J., A. Presley, and D. H. Liles. 1995. "The Management of Technology Within an Enterprise Engineering Framework." *Computers and Industrial Engineering* **28**(3), 497–511.

Sohal, A. S. 1997. "A Longitudinal Study of Planning and Implementation of Advanced Manufacturing Technologies." *International Journal of Computer-Integrated Manufacturing* **10**(1–4), 281–295.

Sohal, A. S., D. A. Samson, and P. Weill. 1991. "Manufacturing and Technology Strategy: a Survey of Planning for AMT." *Computer-Integrated Manufacturing Systems* **4**, 71–79.

Suri, R., R. Veeramani, and J. Church. 1995. "ABB Teams up with University to Drive Quick Response Manufacturing." *IIE Solutions* **27**(11), 26–30.

Uden, L. 1995. "Design and Evaluation of Human Centered CIM Systems." *Computer Integrated Manufacturing Systems* **8**(2), 83–92.

Via, T. H. 1994. "Design for Robotic and Automated Welding: A Systems Approach." Technical Paper, Society of Manufacturing Engineers.

Yoshikawa, H. 1987. "Computer Integrated Manufacturing: Visions and Realities about the Factory of the Future." *Computers in Industry* **8**, 181–196.

Zhao, B., and F. Steier. 1993. "Effective CIM Implementation Using Socio-technical Principles." *Industrial Management* **35**(3), 27–29.

ADDITIONAL READING

Ayres, R. U., and S. M. Miller. *Robotics: Applications and Social Implications.* Cambridge: Ballinger, 1983.

Gunasekaran, A., and S. Y. Nof (Eds.). Special Issue on Design and Implementation of CIM Systems, *Int J of CIM* **10**(1–4), 1997.

Hasegawa, Y. "How Society Should Accept the Full-Scale Introduction of Industrial Robots." Technical Report, Japan External Trade Organization, Machinery and Technology Department, 1982.

Hunt, H. A., and T. L. Hunt. *Human Resource Implications of Robotics.* Kalamazoo: W. E. Upjohn Institute for Employment Research, 1982.

Japan Industrial Robot Association. *The Robotics Industry of Japan: Today and Tomorrow.* Englewood Cliffs: Prentice-Hall, 1982.

Johnston, R. B., and M. Brennan. "Planning or Organizing: The Implications of Theories of Activity for Management of Operations." *Omega* **24**(4), 367–384, 1996.

Kuwahara, Y. "Living with New Technology—Japan's Experience with Robots." In *Highlights in Japanese Industrial Relations: A Selection of Articles for Japan Labour Institute.* Tokyo: Japan Institute of Labour, 1983. 75–79.

Motiwalla, L., L. Coudurier, and D. Shetty. 1995. "Intelligent Sensing for Real-Time Quality Control in Manufacturing." *Journal of Applied Manufacturing Systems* **7**(2), 33–41, 1995.

Yonemoto, K. "The Socio-economic Impacts of Industrial Robots in Japan." *Industrial Robot* **8**(4), 238–241, 1981.

CHAPTER 27

THE ROLE OF CIM AND ROBOTICS IN ENTERPRISE REENGINEERING

R. S. Maull
University of Exeter
Exeter, United Kingdom

S. J. Childe
J. B. Mills
P. A. Smart
University of Plymouth
Plymouth, United Kingdom

1 KEY POINTS

In this chapter we provide guidance for companies seeking help in the implementation of computer-integrated manufacturing/robotics (CIM/R) systems within enterprise reengineering projects. We will draw on our experiences in a U.K. company that has been on the leading edge in developing CIM/R technology over the last 15 years. Unusually for such illustrations, we will draw on a case that was, at best, only partially successful. We will draw out the lessons learned and point to a better, more integrated approach.

The main message to emerge from these experiences was the importance of integrating the CIM/R technology with other organizational processes in a systemic approach. The key to this approach lies in integrating activity and information models. Guidelines are provided in Section 4 on how to carry out this integration.

2 INTRODUCTION

We begin this chapter by providing a framework for enterprise reengineering. We illustrate the characteristics of each of the major stages and where CIM/R programs fit within the framework. We then move on to illustrate the application of CIM/R technologies within a major U.K. manufacturing company. The specific example describes the complex problem of the use of CIM/R for wiring loom assembly. Drawing from this and other similar projects, we have identified two crucial issues for enterprise reengineering. First, it is important to integrate the change program (in this case an integrated assembly system) with all the other activities of the business. This enables the company to simplify the nature of the integration. Second, after simplifying the processes the company can then integrate the CIM/R system with the other business processes in the enterprise by providing an integrated information model and database.

Substantial guidance is now available for companies on how to take a process perspective of their organization through the use of business process modeling as well as on how to use such models to simplify their business processes. We describe two aspects of business process modeling: the activity view and the information view. Activity views show how activities are interrelated, and information models show how data is used in a system. Drawing on our experiences, we provide guidance and examples on how to model using these methods.

Reconciling activity models with information models is difficult even within the same modeling methodology. Even more difficult is addressing the crucial issue of the changes in human and organizational issues required. There is therefore a need for integrated modeling approaches that help us to understand the human aspects. A number of enterprise integration (EI) reference architectures are now being developed that enable companies to identify where their CIM/R developments fit within a whole enterprise. We

Handbook of Industrial Robotics, Second Edition, Edited by Shimon Y. Nof
ISBN 0-471-17783-0 © 1999 John Wiley & Sons, Inc.

conclude this chapter by describing the issues addressed by the reference architectures and the role that detailed activity and information models might play in making these reference models more usable in industry.

3 ENTERPRISE REENGINEERING

A framework for analyzing types of business process reengineering (BPR) is proposed in Childe and Maull (1994). Figure 1 is an adaptation of that framework to account for later developments in EI and work on extended enterprise (EE).

The axes are:

- *Scope of change:* the extent to which the change program is radical or incremental in nature
- *Change team:* whether the change program is usually carried out by ad hoc teams or requires a full-time project team
- *Benefits and risk:* the benefits to be gained from a successful project and, inversely, the risks associated with a failed project
- *Focus:* where the major activity should be, from internal within the organization to external to include the links to other companies both up and down the supply chain

To understand the nature of the interventions we must consider a simple organizational diagram (see Figure 2). In this example a manufacturing company has four departments: marketing, design, manufacturing, and assembly. The flows of material and information are from marketing (which specifies the product and obtains orders) to design to manufacturing to assembly. The explosions in the diagram refer to major problem areas. These often occur where information and or material should be transferred from department to department. Typical problems include loss of material between departments and the re-keying of information to different departmental databases, which in turn can often lead to a loss in priority and urgency.

The thick line running through Figure 2 describes a complete process, which we define from an external customer perspective. It begins with a customer order and ends with a delivered product. Within this process a number of improvement initiatives may be made.

Personal or single-point improvement (depicted by a single person in the middle of the design function) is led by individuals improving the way they perform their day-to-day activities. The authors have been in companies where there are extensive personal improvement activities, often formalized through training and suggestions schemes. In such companies incremental change takes place on a regular basis, driven by individuals improving the way they work. Another way of bringing about such change is to replace individual machines within a process. Such changes often speed up machining or material

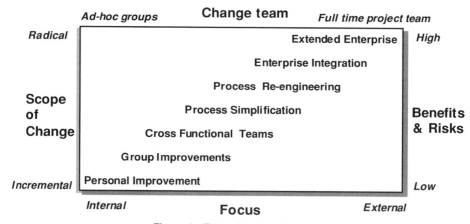

Figure 1   Enterprise reengineering.

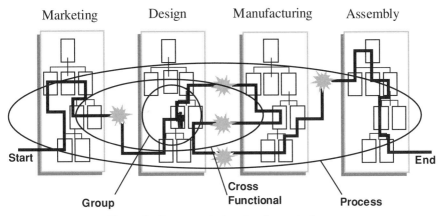

Figure 2 Processes in functional organizations.

handling. However, such change is often not focused on the whole process and has only limited improvement potential.

Group improvement is represented in Figure 2 by the circle within the design function. Here a group of individuals may come together to improve the way the members of the group work together, perhaps by sharing or exchanging responsibilities. Such change tends to be wider in scope than individual improvement, but it is still essentially operational in nature, entails minimal risk to the organization, and can be regarded as incremental change.

The work undertaken by cross-functional teams (CFTs) is depicted in Figure 2 by the small ellipse that extends beyond the localized small group improvement activity and into other functional areas of the firm. CFTs typically have the freedom to consider internal customer/supplier relationships as well as their own process. For example, the design groups might work with manufacturing to improve the manufacturability of their product. CFTs tend to be centered upon a particular business function.

Process simplification (PS) can be regarded as the first real type of process-based change. It focuses on the whole process. Often a team will be established whose job is to analyze the whole process for such non-value-added activities as storage and inspection. The team will be seeking to remove these activities. The benefits of this type of change are related to the price of non-conformance (the cost of scrap, rework, inspection, warranty departments, etc.) in the process. This typically would be around 20% of the cost of the process.

Process reengineering (PR) again focuses on the whole process but has a wider scope than the removal of waste. This equates with Hammer's (1990) approach to process reengineering, which questions whether the status quo is relevant to the future system. A reduced number of activities, organizational and job redesign, and new developments in information technology such as document image processing (DIP) or expert systems may be used. This type of change seeks to reduce the number of activities by up to 90% and addresses real strategic benefits.

EI (Vernadat, 1996) has its roots in CIM. It is an extension of BPR that focuses primarily on the complete integration of a set of processes within the organization. In many cases this is achieved through the use of automation and information; e.g., BAAN 4 (Perreault, 1997) and SAP R3 (Bancroft et al., 1996). Whereas process reengineering often focuses on improving specific processes and may be carried out without technology, technology is the key to EI and involves integration across all processes. As we will see in the case study for this type of change, complete business-wide activity and information models are required.

The concept of the extended enterprise (Childe, 1998) builds on EI and looks at the whole enterprise within the whole supply chain: a company's vertical relationships, its customers and suppliers. This type of change program looks at reducing the duplication of activities between firms so that the complete supply chain can be as efficient as possible. Change programs aimed at improving the EE are strategic in nature, high-risk, invariably involve a dedicated project team for two to three years' work, and are radical in nature, aimed at making major improvements in a company's cost and service levels.

3.1 Role of CIM/R Within Enterprise Reengineering

The framework proposed in Figure 1 can be useful for those developing CIM/R. It is important to bring about the CIM/R change at the correct level in the framework. Failure to do so can result in the CIM/R program failing to bring about the benefits expected. This is illustrated in the case study in the next section.

The authors have seen many examples of the use of complex robotics as part of single-point solutions. In these cases robots and automated machines are used to speed up individual process times. Historically these cases have resulted in "islands of automation" (Bradt, 1984). While technological developments such as postprocessors and the use of open systems have helped link together individual machines within an integrated solution, some systems problems do remain. Goldratt (1990) has clearly illustrated the problems associated with speeding up one part of a process without considering the impacts elsewhere and has pointed us to focusing on removing the constraints within the system. Even where such developments do speed the work through a bottleneck, a new constraint will often emerge.

Group improvements extend the scope of automation across a whole function. This may be through the use of cellular manufacturing, where a combination of robots and automation is used to complete, for example, the whole of manufacturing or assembly.

Cross-functional teams again extend the scope of the project. For example, in manufacturing, the cross-functional teams will be extended to look at the links to design and assembly. This will focus on CAD/CAM, process planning, manufacturing planning, and control systems.

CIM/R fits well within a PS or PR approach. As explained earlier, the essential difference between the two approaches is one of scope. PS aims at making incremental change across the whole process, whereas PR is more radical and seeks to make large-scale strategic changes. However, the focus of both these types of process improvement is on the whole process. The authors have defined four major operational processes in manufacturing: get order, develop product, fulfil order, and support product (Weaver, 1995). A useful structure for linking these processes is provided in Appendix A. A typical approach to PS or PR would focus on improving the end-to-end flow of material and information through one of these processes. So, for example, cellular manufacturing could be used to automate the activities associated with making the product. This could be linked to MRPII and material movement systems in a fully automated facility. If the change program were aimed at removing non-value-added activities, it would be an example of PS. If the change addressed fundamental issues such as whether this product should be manufactured or bought in, it would be an example of PR.

A program focusing on the EI level would introduce CIM/R as part of a total change program dedicated to integrating all the different aspects of the business. This would include substantial information systems integration and both horizontal and vertical integration. Horizontal integration is enabled by investigating the passing of material and information from one activity to another. This can be supported through activity modeling. Vertical integration is enabled by investigating the passing of information through a control system and an information systems architecture. This can be supported through information modeling and an architecture that identifies where various actions are taken. The complex issues associated with bringing together both these types of integration are discussed in Section 8.

Change programs aimed at bringing about the extended enterprise are highly strategic. Such programs link together a series of firms in a supply chain that serves the end customer. An extended enterprise change program seeks to identify and remove areas where tasks are duplicated within the supply chain. It often necessitates companies identifying core competencies. For example, a supply chain in the aerospace sector is usually made up of a prime contractor and a systems integrator in each of the three main areas: engines, avionics, and airframes. Identifying which activities are to be carried out by the systems integrator and which are to be carried out by the prime contractor clearly involves an identification of competencies and, in the medium term, where profits will lie. Such programs are an extension of the process concept through the supply chain.

The best means of illustrating these ideas is through a case study. The following is an account of a company that has moved through many of the stages of the framework in developing its CIM/R program. It is now engaged on a program of enterprise integration.

4   CASE STUDY

The company is one of the United Kingdom's most technologically advanced and has an extensive history of using CIM/R technologies. The company is a major U.K. aerospace business. It is both a manufacturer and a systems integrator. More than 1000 of the company's aircraft are in service in 19 countries. The company is a world leader in the design and manufacture of flight-critical structures and transmissions for civil and military aircraft.

Ten years ago the then Research and Technology (R&T) division of the company was given the task of developing novel CIM and robotics technologies across the range of the company's aerospace interests. The division's approach was to cooperate with production departments in the running of demonstration projects. On completion of the demonstration projects the production departments would be able to implement the new manufacturing technologies that had been proven in live situations. The main objectives were to achieve cost reduction and quality improvement and to develop new techniques to exploit materials such as composites. A major theme of this work was automation.

This example relates to the development of an alternative method of manufacturing for aircraft electrical looms. The process of loom building is composed of the following elements.

Cables are cut to length by hand. Each cable has an unique alphanumeric identification number, printed onto plastic shrink sleeves fitted onto each end of the cable. A kit of cut cables is issued to the loom builder together with additional hardware (pins, terminals, plugs, etc.) to complete the loom. The operator forms the loom on a loom board, collecting the cables into bunches, laying them into designated routes, adding and extracting from bunches, strapping the assemblies as they are completed and cutting them to length, stripping insulation from cable ends, and crimping terminals into place.

The company uses 70 different cable and terminal types, and many of the terminal types need dedicated hand tools. The necessary skill level for building the most complicated looms, of up to 2500 cables of 12 m (about 40 ft) in length, takes the average skilled electrician four years to acquire.

The R&T division established a system using CIM/R for the total automation of all the operations. The new system was a series of devices for each discrete manufacturing operation, such as measure and cut, print strip, etc. Most of the these were mechanized, but laser printing was used to print cable identifications continuously along the cable, and laser cable stripping was used to displace traditional mechanical cable strippers. Mechanical handling was developed to carry the work in progress between workstations. This demonstration phase concluded with the drafting of various schematic shop floor layouts using the hardware in various combinations with associated computer control systems.

The full implementation phase was carried out in three stages, moving from semi-automated to a completely automated CIM/R. See Figure 3.

In stage one instructions were sent via computer to a manual loom assembly. The process of operation was:

1. The operators made up a prototype loom.
2. They then used a digitizer to input cable length to the CAD system.
3. CAD instructions were then sent to a laser cutting/printing machine that automatically cut and printed the cable length, mirroring these features from the digitized prototype.

Stage two was the addition of automated cutting and crimping of the connectors.

Stage three contained the CIM/R automated loom assembly. Cables were now automatically cut to length and mechanically transferred via the laser stripper to the rapid wire-laying machine. An automated machine prepared the forming boards and strapped the bunches prior to laying. Future developments were to include integrating the loom assembly with CAD, configuration control, and MRP systems to improve the control of the process.

Using the ER framework, this company illustrates a case of a company developing an implementation plan that gradually moved upwards through the framework. At the outset it had a single-point solution, automating individual processes. It moved on to focus on group improvement, automating all of the processes within loom assembly. In the third stage it aimed to use cross-functional teams to integrate the assembly with other IT systems.

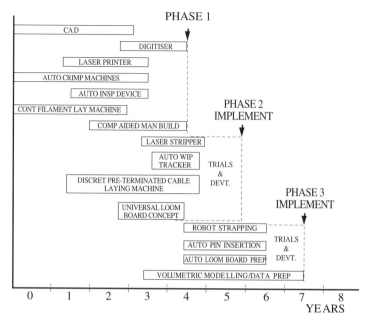

Figure 3 Project plan of the technology implementation.

The major problem the company was trying to solve with the CIM/R was that the workload in the shop was well in excess of the capacity available. Consequently, it was felt necessary to speed up the process to provide additional capacity by mechanizing all of the process. This complete mechanization of the process (stage three) was never achieved. The underlying problem was that the company could not integrate all the different elements of the process and the automated loom assembly process was not financially viable unless the whole process was mechanized. So why couldn't the company integrate all the different elements of the process?

The principal reason stems from the fact that the loom is the last part of the aircraft to be designed. In Figure 4 (an adaptation of Figure 2) the whole aircraft assembly begins with the commercial department contracting with customer agencies, such as governments, on aircraft specifications. The product specification is in turn negotiated with the design department on the exact specification, design modification level, etc. The product specification is passed on to manufacturing, to make the fuselage, blades, transmission, interior fittings, instrumentation, etc.

Individual customers would nearly always require modifications or additional electronic systems to be added to the standard aircraft design. Whenever additional systems

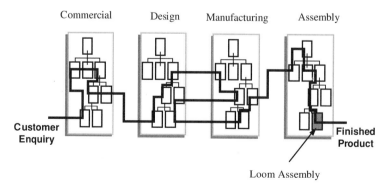

Figure 4 Aircraft processes.

are added and/or existing systems modified, they affect the wiring required and the placing of the wiring within the aircraft. As a result, the company is often still designing the loom in assembly—indeed, in some cases the loom is not formally designed. Due to the lateness in design and the complexity of the looms, the first off is often hand-assembled and then digitized.

A secondary problem relates to the substantial problem of crosstalk between wires. This cannot be identified and overcome until the whole loom has been defined. As a consequence of lateness, complexity, and potential crosstalk, the data required to enable the loom assembly to be completely automated are often not available.

Many of the problems associated with loom assembly were caused by product changes requested by the customer. These product changes led to late loom design, which made it very difficult to automate. The key would be to simplify the existing system and then apply automation. A critical issue would be ensuring data integrity so that the loom design could be accurately interpreted and an appropriate loom manufacturing process developed.

In this case the company attempted to analyze and improve its technical and organizational processes by moving up the enterprise reengineering framework. Initially it focused on the optimization of individual activities or subprocesses. Then it moved on to automating all of the process. Later still it needed to integrate the loom assembly with other IT systems. In our view this can be seen as a reductionist approach and is a typical failure mode (Hall, Rosenthal, and Wade, 1993; McHugh, Merli, and Wheeler, 1995). With the benefit of hindsight it was realized that the problems occurred in other activities in the complete design-to-manufacturing cycle. What was missing was a method for identifying how the loom assembly fitted into the whole business process of designing and making an aircraft—a systemic perspective. Once the company had a method for identifying how the loom assembly fitted into the whole process, it could begin to integrate the data requirements with those of the wider system. A systemic or whole-system perspective is often appropriate for the simplification stage.

Reflection on these and other similar issues arising from automation projects elsewhere in the organization has now moved the focus of the company's attention to an enterprise integration project that covers the reengineering of the whole enterprise. Automation projects still play a critical part in the reengineering initiative, but now they are part of an overall business process-based change program that identifies a business process view of the organization, within which it develops an enterprise-wide integrated information system.

Companies require two types of help. The first is recognition of the type of change they are seeking with their CIM/R projects. The ER framework should help pinpoint this. The second is help in integrating their CIM/R projects with the rest of the business. We believe that two aspects of integration are crucial for those companies developing automation projects: activity modeling (sometimes called *functional modeling*) and information modeling.

5 ACTIVITY MODELING

A wide variety of modeling methods support activity modeling. In a recent investigation Smart et al. (1995) found that IDEF0 was the most suitable for encouraging users to take a systemic perspective.

As part of a research program awarded by the U.K. government (Maull et al., 1997) the authors have used IDEF0 to develop a generic business process activity model. The generic model is based on the notions of architecture and hierarchy in business processes. This is more widely discussed by Maull et al. (1995). The provision of a hierarchical business process model both facilitates and simplifies the development of specific process models.

The research team at Plymouth has integrated its process architecture with some of the concepts from CIM–OSA (AMICE, 1993). The CIM–OSA standard provides a recognized framework around which to group processes. The first two types of processes are the *operate* processes, which add value, and the *support* processes, which enable the operate processes to function. Operate processes are viewed as those that are directly related to satisfying the requirements of the external customer; for example, the logistics supply chain from order to delivery. A detailed description of the operate dimension of the framework is discussed in Weaver (1995). The support processes include the financial, personnel, facilities management and information systems-provision activities. The third type, the *manage* processes, are those processes that develop a set of business objectives and a business strategy and manage the overall behavior of the organization. The grouping of the processes under *manage, operate, and support* provides a framework for the classification of process types.

The operate process has been the focus of our work. The four processes at the highest level of abstraction are get order, develop product, fulfil order, and support product. Appendix A provides an example of the integration of these four core processes at the highest level of abstraction. Each of the four core processes is further detailed in the model. For example, the generic order-fulfilment model currently includes over 110 activities at various levels of detail, which are integrated by the flows of physical and information entities.

It must be made clear, particularly in such a brief description, that although the operate processes can be identified as individual processes, it is the interaction of all the processes that results in the fulfilment of customer needs. Considering any of these processes in isolation without the links between each of them is likely to result in the suboptimization of the operate processes as a whole. Focusing on this high level of integration at the outset might have enabled the company to see the linkages between their loom assembly operations and the impact of changes in product design.

The authors (Maull et al., 1997) have identified six major advantages in using this type of business process model. Such a model:

1. Reduces the chance of the team reverting to a traditional functional view. The scope of the model ensures that the team focuses on the whole process.
2. Reduces the amount of modeling activity required.
3. Provides greater momentum than starting from a "blank sheet of paper."
4. Provides standard models that are not influenced by any political issues in the company.
5. Enables the team to criticize the standard models. Company members are encouraged to criticize and modify the models.
6. Supports the activity of comparison by encouraging communication and understanding through debate. By comparing current activities with the model, it encourages companies to think through why they carry out activities in that way.

For those wishing to know more details of the model and its use, the model is in the public domain and is available from the authors.

Despite these many benefits, there are a number of critical deficiencies with this type of modeling. Most importantly, it takes little account of the information integration within the process. While this is of little consequence if the objective is to describe how a process works and identify the knock on effects of changes to product designs, it is essential if the objective is to integrate the data entities within the whole business process. This sort of process modeling therefore needs to be accompanied by information modeling.

6 INFORMATION MODELING

At the company the loom assembly system shared data with a wide range of enterprise software systems, such as computer-aided design, computer-aided process planning, manufacturing resource planning (MRPII), product data management, and ultimately accountancy systems (for job costing). These systems need to be integrated so that changes in one system are quickly reflected in other systems. This would ensure that data integrity is maintained. For example, changes made using the electronic CAD system at the company need to be quickly fed through to the loom assembly system.

The traditional manner of carrying out this information integration activity is through entity-relationship modeling. Of the many different types of ER modeling methods available (Vernadat, 1996), the method we have most experience with is IDEF1x (Bruce, 1992).

IDEF1x facilitates the creation of a number of logical information models. Models created using the IDEF1x method are representations of things (entities), the properties of those things (attributes), and associations or links (relationships) to other entities in the model. An entity may be regarded as an abstraction of a real-world thing, an attribute is a specific data item of an entity, and relationships result from a business need to associate two entities.

Loomis (1986) characterizes the IDEF1x technique as follows. IDEF1x

- Produces diagrams that explicitly represent data semantics.
- Represents entities, relationships, attributes, and various assertions.

- Develops a data model by top-down analysis of entities and relationships, rather than by bottom-up synthesis of attributes.
- Represents a broad range of detail suitable for supporting the full process of developing information systems.
- Is independent of any database management system (DBMS) and can therefore be used to drive implementation in any environment.

The development of a data model is crucial in developing a CIM/R system that is integrated with the rest of the enterprise. An example of a data model for the purchasing activity of a large U.K. company is provided in Appendix B.

In developing this data model we had two major difficulties. First, many company personnel have difficulty in describing the supporting data structure of an activity. Typical responses were redescriptions of how information flowed through the business process. A substantial amount of expertise is required to extract information relating to the structure of data from the responses obtained. Second, even in a relatively limited model the volume of entities required to support the activity view of the process is very large (see Appendix B).

As a result of the difficulties in obtaining the data and the complexity of the models that tend to be developed, considerable expertise is needed from the analyst in building the entity relationships models.

7 GUIDELINES ON INTEGRATING ACTIVITY AND INFORMATION MODELS

We have developed an integrated approach to enterprise reengineering based around business process activity and information modeling. From our research we conclude that

1. Although benefits may be gained from the creation of both model views in parallel, company personnel may have difficulty in understanding information models. Prior construction of the activity model provides a common language that can be used in investigating and identifying the supporting data structure of a process.

2. Analysis of information flows at all levels in the hierarchy of the activity view should be conducted to establish potential entities for the information view.

3. The volume of entities required to support cross-functional processes is large and complex. The ability of company personnel who were not involved in the creation of the model to understand cross-functional information modeling is questionable.

4. If the creation of the model is purely for information system development, then only the affected entities may need to be included; that is, the areas of the process that will receive information system support.

5. The creation of an information view may uncover a number of key issues or rules governing company procedure. This knowledge may be used to further improve the redesigned process.

It is clear that for those companies wishing to integrate their CIM/R systems with other business processes support is available through a variety of activity and information modeling tools. However, the transition between the two views is still a "black art." Despite using two modeling techniques from the same toolset, IDEFO and IDEF1x, there is no simple means for their integration.

In addition, as critical as activity and information views of an enterprise are, they are missing the vital human and organizational dimension. Experience with the company has taught us that both modeling perspectives are absolutely essential, but much more help is required in integrating these views with the human aspects of enterprises. More help is therefore needed both to integrate modeling methods and to extend them to include human aspects.

8 REFERENCE ARCHITECTURES

A number of developments are currently taking place aimed at developing enterprise-wide integration models. These are often termed *reference architectures*. It is impossible to provide a detailed description of these extensive developments in the limited space available. However, most of the reference architectures have a number of common features, and these are detailed below. For a more detailed description of each individual reference architecture see Vernadat (1996), who has provided an excellent description of these approaches. He provides a synopsis of all the major international standard approaches to CIM.

The major reference architectures are ENV 40 003 (CEN, 1990), CIMOSA (AMICE, 1993), ARIS (Scheer, 1992), PERA (Williams, 1992, 1994), and GERAM (Williams, 1995; Bernus and Nemes, 1994).

CEN and ARIS have a substantial number of similarities with the CIMOSA framework. All three have the first two dimensions (although they sometimes use different terminology) and significant similarities in the third:

- *Stages of the systems life cycle.* This is associated with defining requirements, design, and implementation models
- *Different views.* This is the most relevant area for this chapter. It includes perspectives on models from
  - the function view, which provides both a dynamic and a static view of the activities of the enterprise
  - the information view, which provides the description of the enterprise objects identified in other views
  - the resource view—the resources required to execute enterprise activities (these are subsumed under *organization* in the ARIS architecture)
  - the organization view, a description of the organizational structure and responsibilities of individuals within the enterprise
  - a control view, included in ARIS alone, which federates the architecture and is related to the three other views of the ARIS architecture.
- *Genericity.* This includes the basic modeling constructs and sets of common models adapted to represent particular enterprises and industries. This is not explicitly dealt with by ARIS but is included as three levels within the ARIS process chains.

While CIM-OSA, CEN, and ARIS are complete reference architectures, they provide very little guidance on *how* such architectures should be implemented. Williams, Rothwell, and Li (1996) differentiate between Type 1 architectures, which present a pictorial model of the physical organization or structure, and Type 2 architectures, which describe the steps of the process of development of enterprise integration.

PERA is a Type 2 architecture that may be best described as a methodology. PERA defines all the stages of the manufacturing systems life cycle. The life cycle progresses through feasibility study and definition, enterprise definition, conceptual engineering, preliminary engineering, detailed engineering, construction, operation and maintenance, renovation, and enterprise dissolution. PERA is supported by an easy-to-understand workbook that takes the user through each of these stages. Each stage is informally described by a technical document as a set of procedures for leading a user through all the phases of an enterprise-integration program.

The logical step is the integration of the Type 1 architectures with the detail provided by PERA. This is currently taking place under the generalized enterprise reference architecture and methodology (GERAM). The purpose of GERAM is to serve as a reference for the whole EI community. It seeks to provide:

- Definitions on terminology
- A consistent modeling environment
- A detailed methodology
- Promotion of good practice for building reuseable standard models
- A unified perspective for products, processes, management, enterprise development, and strategic management.

This work is currently ongoing, with a number of publications and documents emerging.

The concept of EI is clearly a complex one and has attracted much academic interest and a wide range of alternative perspectives. Unfortunately, much of the work that has gone into developing these reference architectures is at the conceptual level. They are metalevel frameworks that enable users to evaluate and compare existing architectures and identify gaps. Their great strength is that they develop activity (function), information, and human/organizational views, all of which are capable of being integrated together. Their great weakness is that they are not equipped with models that users can quickly pick up and use. We would argue that what industry requires is partially defined reusable models, building blocks on which they can quickly develop integration models—for example, the generic business process models discussed above. Real benefits are expected with the integration of these examples within the all-embracing reference architectures. We await these developments with interest.

9   DISCUSSION AND CONCLUSIONS

This chapter has sought to take a very practical perspective on enterprise reengineering and the closely related topic of enterprise integration. From our experiences it is clear that those developing automated systems can no longer do so in isolation from the rest of the organization. Islands of automation are not an option. What is required is a systemic perspective that first aims to simplify the whole business process and then addresses automation and information integration. The company is now adopting just such an approach.

We have described two key approaches that support process integration: activity modeling and information modeling. Activity modeling may be approached by developing a model from the bottom up, but is perhaps best approached from the top down; that is, by using generic models and then amending them to reflect more accurately how a company works.

Information models are more complex. Generic models do exist, in that most major software products have data models embedded within them. Obtaining these software-based generic models is difficult because most software companies see them as an integral part of their intellectual property. Consequently, integrating the information requirements of a new automated system with existing software systems is often very difficult. Yet companies still need an understanding of the underlying data model to integrate their automated system, otherwise the data needed by the automated systems will be unreliable, inaccurate, and inconsistent. This will inevitably result in major implementation problems. A word of caution: developing information models for even relatively constrained areas such as purchasing is a large task. Developing an enterprise-wide information model would require the full-time effort of expert staff!

Integrating activity and information models is fraught with difficulty and often involves considerable time and resource expertise. In addition, these models fail to take any account of the human and organizational aspects of enterprise integration. New approaches based on reference architectures are currently being developed. These aim to provide a standard that will enable the integration of activity, information, and human modeling approaches. Work is even going on to standardize these architectures. The drawback of these architectures is that they focus on the overarching concepts. There are very few usable models available that can be picked up and used by companies to ensure the proper systemic integration of automation projects. Future developments in GERAM on the standards for developing these reusable models will draw together these conceptual frameworks to address the needs of industry.

REFERENCES

AMICE. 1993. *CIMOSA: Open System Architecture for CIM.* 2nd rev. ed. Berlin: Springer-Verlag.

Bernus, P., and L. Nemes. 1994. "A Framework to Define a Generic Enterprise Reference Architecture and Methodology." In *Proceedings of the Third International Conference on Automation, Robotics and Computer Vision (ICARCV '94)*, Singapore, November 8–11. 88–92.

Bradt, J. 1984. "Linking Islands of Automation." Technical Paper, Society of Manufacturing Engineers. Vol. 84 (January).

Bruce, T. A. 1992. *Designing Quality Databases with IDEF1x Information Models.* New York: Dorset House.

CEN. 1990. *ENV 40 003: Computer Integrated Manufacturing—Systems Architecture—Framework for Enterprise Modeling.* Brussels: CEN/CENELEC.

Childe, S. 1998. "The Extended Enterprise—A Concept of Co-operation." *Production Planning and Control* 9(4) 320–327.

Childe, S. J., R. S. Maull, and J. Bennett. 1994. "Frameworks for Understanding Business Process Re-engineering." International Journal of Operations and Production Management 14(2) 22–34.

Goldratt, E. M. 1990. *Theory of Constraints.* New York: North River Press.

Hall, E. A., J. Rosenthal, and J. Wade. 1993. "How to Make Reengineering Really Work." *Harvard Business Review* 71, 119–131.

Hammer, M. 1990. "Reengineering Work: Don't Automate, Obliterate." *Harvard Business Review* **68**, 104–112.

Loomis, M. E. S. 1986. "Data Modeling—The IDEF1x Technique." In *5th Annual International Phoenix Conference on Computers and Communications.* 146–151.

Maull, R. S., et al. 1995. "Different Types of Manufacturing Processes and IDEF0 Models Describing Standard Business Processes." Working Paper WP/GR/J95010-6, University of Plymouth.

———. 1997. "Using IDEF0 to Develop Generic Process Models." Paper presented at Computer Aided Production Engineering (CAPE '97), Detroit.

McHugh, P., G. Merli, and W. A. Wheeler. 1995. *Beyond Business Process Re-engineering.* New York: John Wiley & Sons.

Perreault, Y. 1997. *Using BAAN IV.* Que Corp.

Scheer, A. W., 1992. *Architecture of Integrated Information Systems—Foundations of Enterprise-Modeling.* Berlin: Springer-Verlag.

Smart, P. A., et al. 1995. *Report on Process Analysis Techniques.* University of Plymouth.

Vernadat, F. 1996. *Enterprise Modeling and Integration: Principles and Applications.* London: Chapman & Hall.

Weaver, A. M., 1995. "A Model Based Approach to the Design and Implementation of Computer Aided Production Management Systems." Ph.D. Thesis, University of Plymouth.

Williams, T. J., 1992. *The Purdue Enterprise Reference Architecture: A Technical Guide for CIM Planning and Implementation.* Instrument Society of America.

———. 1994. "The Purdue Enterprise Reference Architecture." *Computers in Industry* **24**(2, 3), 141–158.

———. 1995. "Development of GERAM, a Generic Enterprise Reference Architecture and Enterprise Integration Methodology." In *Integrated Manufacturing Systems Engineering.* Ed. P. Ladet and F. Vernadat. London: Chapman & Hall.

Williams, T. J., G. A. Rathwell, and H. Li. 1996. *A Handbook on Master Planning and Implementation for Enterprise Integration Programs.* Purdue Laboratory for Applied Industrial Control, Report No. 160.

Williams, T. J., et al. 1994. "Architectures for Integrating Manufacturing Activities and Enterprises." *Computers in Industry* **24**(2, 3), 111–139.

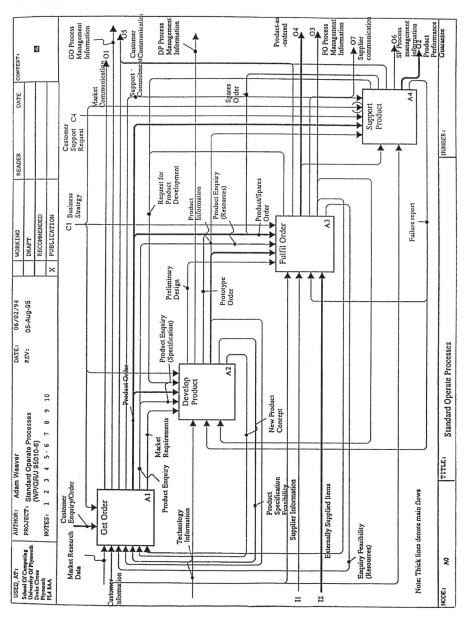

APPENDIX B

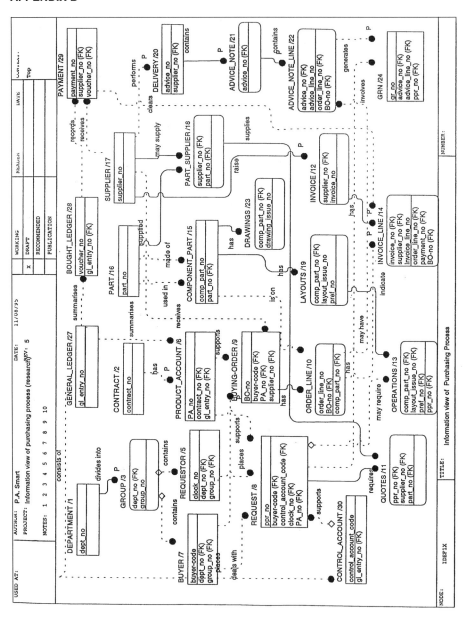

CHAPTER 28

ROBOT INTEGRATION WITHIN MANUFACTURING SYSTEMS

J. D. Gascoigne
R. H. Weston
Loughborough University of Technology
Loughborough, Leicestershire, United Kingdom

1 ROBOTS AS COMPONENTS OF MANUFACTURING SYSTEMS

To cope with the extremely high levels of complexity found within modern manufacturing companies, industry commonly deploys function decomposition, i.e., it breaks down the elements of an enterprise into blocks of functionality that have well-defined capabilities and qualities. (See Chapter 27.) Typical large-grained blocks of functionality might correspond to a marketing department, a sales department, an engineering department, a production department, and so forth. Intermediate grain-sized function blocks may include a paint shop, an assembly shop, a machining cell, an assembly cell, and so on. At a finer level of granularity will be *manufacturing components,* such as a foreman, cell supervisor, machine operator, conveyor system, CNC machine, or robot. Many machine and software systems can themselves be broken down into modular components that have more elemental functionality, such as a drive system, an intelligent sensor, a motion control software process, and so on.

Robots, like other machines, software processes, and people, can function as a class of *manufacturing component* (MC). Here the term *component*[1] is simply used to imply that robotic building blocks can be deployed as a part of a manufacturing system in different domains. Its use does not presuppose that current-generation robots can readily and generally be deployed as *reusable components* of manufacturing systems. Indeed, normally their use in a new application domain cannot be achieved without significant systems engineering effort. Also, it is understood that certain constraining concepts and qualities embedded into current-generation robots will mitigate against their use in certain application domains. Nonetheless, provided that we bear in mind practical constraints on the use of robots, we may beneficially consider their role as reusable components of manufacturing systems. The importance of taking such a perspective on robots has grown as the lifetimes of consumer products have fallen and the complexity of product markets has increased. As these trends continue, companies will increasingly need to deploy "agile" manufacturing systems (Nagel and Dove, 1992).

When engineering and evolving the capabilities of a manufacturing enterprise, it is evident that a clear understanding is required of the functional capabilities of available manufacturing components and of alternative ways in which MCs can be grouped together so that they operate in a structured way as part of a manufacturing system. Here it will be important to ensure that resultant manufacturing systems can operate effectively in conjunction with other manufacturing systems, thereby realizing the *manufacturing pro-*

---

[1] In this chapter the term *component* is used generally. It does not imply a focus on software issues (although such issues are of concern), nor does it assume use of distributed systems and object-oriented technology (although they may be deployed as required).

Handbook of Industrial Robotics, Second Edition, Edited by Shimon Y. Nof
ISBN 0-471-17783-0 © 1999 John Wiley & Sons, Inc.

cesses and other types of *business process*[2] needed to achieve the purpose(s) of an enterprise. Seldom will competitive behavior result if units of functionality (i.e., MCs) are
built into systems in an ad hoc way. In short, systems must be conceived and engineered
in an effective manner. When it is to be used in environments characterized by high levels
of complexity and/or uncertainty, such a system (be it people-centered, technology-
centered, or sociotechnical) should have an embedded *change ethic*. By definition, uncertainty can lead to unanticipated change. Therefore it will often not be possible to
include sufficient functionality to cope with unspecified forms of change. Rather it will
be more practical to design a system so that it inherently facilitates rapid and effective
change. By design it should enable agents responsible for change to evolve the system
as required. These agents could be external to the system in question (such as a system
architect, designer or engineer) or might be internal to the system (such as a decision-
making component capable of evolving the system behavior).

2   VIRTUAL MANUFACTURING COMPONENTS

We may conclude that future generations of manufacturing systems will need to be designed and engineered more scientifically and frequently. We may also conclude that the
architects, designers, engineers, builders, managers, and maintainers of manufacturing
systems will need appropriate representations (or models) of (1) manufacturing components, (2) process behaviors, and (3) alternative ways of rapidly and effectively integrating
components into cohesive systems. Regarding (1), *virtual manufacturing components*
(VMCs) that model real MCs are required, i.e., theoretical representations that characterize properties of MCs from the viewpoints or perspectives of the various people concerned
with a manufacturing system during its lifetime. Thereby these persons will separately
use their perception of an idealized component (characterized by a VMC) when performing a role as part of a team of people concerned with evolving the behavior of manufacturing systems and processes.

If individual VMC viewpoints are consistent with each other, then collective decision-
making should also be enabled. Generally speaking, VMCs will not represent all aspects
of a real MC. Indeed, in some cases a VMC may not model reality closely but may still
prove useful, such as in cases where high-level conceptual decisions are made and detailed
modeling is not practical or desirable.

Theoretically, models of VMCs can prove useful in transmitting specifications to manufacturing companies that describe the capabilities of real components and systems supplied by the vendors of sociotechnical systems. Similarly VMCs can transmit to vendors
generic descriptions of manufacturing requirements. The way in which a particular vendor
implements systems and components need not be overly constrained by models of VMCs.
For example, their use need not constrain the adoption of new technology as it comes on
stream. However, use of a VMC may well require implementers to abide by architectural
guides, such as those necessary to ensure that a change ethic is embedded into reusable
components or into an agile system.

3   EXAMPLE APPLICATION DOMAIN AND THE ROLE OF VMCs

3.1   Common Classes of Manufacturing Component

Consider a typical application domain in which robots may be deployed. A study of such
a domain will help to exemplify the importance of the VMC concept in respect to robotic
systems. Figure 1 illustrates the nature of interactions between common manufacturing
components deployed within manufacturing cells. The figure illustrates that certain man-

---

[2] Process decomposition represents an alternative approach to functional decomposition. Here focus
is on flows and relationships between activities (be they constituents of business or production processes) required to add value to inputs in an effective and efficient way. Rather than digress discussion
at this stage, we will assume that in general a process decomposition can prove very helpful to a
manufacturing company in aligning its elemental activities to business and production requirements,
whereas a fuctional decomposition will prove very helpful to system and component builders and
suppliers who service manufacturing companies.

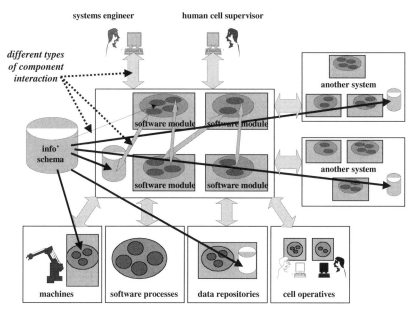

Figure 1   Examples of interaction in a manufacturing system.

ufacturing components (such as a human supervisor or a cell controller) will have a decision-making role, some (such as a robot) may assume an action-making role, whereas others (such as data repositories) will play a supportive service role. Also, some manufacturing components (such as cell operatives) may perform more than one role. Collectively these system elements provide the functional capabilities required to realize a specific set of production processes (Bauer, Bowden, and Browne, 1991; Jones and Saleh, 1991; Lartim, 1989). Hence, when designing a manufacturing cell, appropriate manufacturing components need to be selected and assembled together dependent on the nature of the production functions required. During cell operation the collective and individual functions carried out by selected MCs need to be organized and controlled. Generally speaking, manufacturing and control functions in this domain will be carried out in a semiautomated way; i.e., with a level of automation lying between the extremes of full automation and manual operation. Thus it is important that human, machine, and software components interoperate synergistically, irrespective of the degree of automation required in any specific case.

The nature of control tasks within a cell will vary considerably and be dependent upon the type of manufacture involved (e.g., small batch manufacture, mass production) and the way in which a selected set of MCs and their associated and/or embedded control system elements need to interact internally (locally) within a cell and (more globally) with manufacturing components and systems external to the cell. Thus under batch manufacturing conditions the elements of cell systems invariably will need to interact flexibly within their host environment in a manner that supports the manufacture of different products (Duffie and Prabhu, 1996).

A task will be done by one or more MCs, such as a CNC machine (for turning a part) or a human (to carry out a setting-up operation). A task concerned with transporting material within or across the boundary of a cell might be done by a combination of manufacturing components, such as a robot, conveyor, and tooling. Typically the choice of component types will depend largely upon static technical and business requirements in a given domain. Tasks provide the behavior needed to accomplish dynamic objectives and hence may need to be changed more regularly (i.e., may need to be programmable).

3.2   Use of the VMC Concept in System Design, Construction, and Evolution

Section 3.1 has outlined an example use of different types and groupings of MCs and the roles they play. In order to engineer complex systems the designers, builders, and developers require, at the very least, mental models of components, their relationships, and the way in which combinations of components behave dynamically to realize the various roles and tasks that need to be performed by the system. However, systems engineering will be facilitated if aspects of these mental models can be translated into suitable graphical representations (such as an activity flowchart or a timing diagram) so that they can be reused and communicated between the various "players" concerned with the lifetime of a system. Even greater benefit can accrue if the same, or possibly different, representations of these mental models can be translated into models that can be processed and executed by a computer. By producing executable models of components (i.e., VMCs), relationships between VMCs, and behavioral models related to combinations of VMCs, it becomes practical to design, synthesize, and analyze the operation of manufacturing systems, such as robotic cells, before the system is constructed. Alternatively, a possible system redesign, such as in response to changing production or human requirements, can be synthesized and analyzed before the time- and cost-consuming changes are made.

In modeling a manufacturing cell it is typically necessary to formalize the receiving of tasks and the assignment of these tasks to appropriate VMC combinations. This is necessary to define who carries out the tasks and how monitoring of the tasks will be carried out. Thereby use of VMCs can enable the system performance to be analyzed in the case of different options.

The foregoing description serves to illustrate that in seeking to produce more flexible, configurable, and extendible cells from reusable components, it is essential to find an effective way of defining and flexibly mapping tasks (and associated processes and activities) onto a selection of VMCs.

4 PROBLEM AND SOLUTION PERSPECTIVES

Models of components are used widely within contemporary approaches to designing manufacturing systems. However, often those models have been used in an ad hoc and nonstandard way and take the form of mental or possibly graphical models. The use of computer-executable VMCs is in its infancy, primarily because related research issues remain and comprehensive computational tools that support the concept have yet to become commercially available.

Whichever form (i.e., mental, graphical, or executable) of component model is used, the concept naturally supports the definition of a *solution structure*. The use of executable VMCs can also support the analysis of alternative solution structures, thus potentially generating better system designs. However, if a change ethic is to be embedded into a solution structure, it is evidently necessary to separate out the kind of things that are likely to change and, if practical, to represent these separately from, or as an organized division of, the structure itself. The chance of being able to reengineer systems quickly and effectively in the event of unforeseen changes is thereby much improved. In this context it is important to classify and characterize things that are likely to change during the system lifetime. For example, properties of the actual MCs deployed within system solutions are likely to change at different rates, e.g., the mechanical properties of a robot may not change as rapidly as its control system properties, which may not change as often as its tooling requirements, which may not change as frequently as the tasks it performs, and so on. Based on this kind of thinking, previous practical experience of the authors, their coresearchers, and industry partners has led them to definitively separate solution structure from VMC implementation and VMC interaction issues during the design and analysis of manufacturing systems. Table 1 outlines each of these issues and following sections describe how this division of issues can be separated further into more focused problem and solution perspectives to engender change ethic implemented systems.

Table 1 Top-Level Issues

Top-Level Issues	Interpretation
Solution structure	How VMCs are arranged and organized to meet a need
MC Implementation	How the constituent MCs are constructed
MC and VMC Interaction	How the necessary interactions are made possible

The rationale for separating VMC implementation issues from solution structure issues is similar to that which underpins concepts used widely in the field of software engineering (Graham, 1994). Significant benefit can accrue from separating conceptual design activities from the specifics of how solutions will be constrained by alternative implementation options. These specifics must be tackled (hence the need for a VMC implementation perspective), but if they are embedded inextricably into conceptual designs, then resultant systems will be incapable of supporting certain classes of change without major reengineering effort (Barber and Weston, 1998). Indeed, the reuse of manufacturing components can be promoted by separating implementation issues from solution structure issues. Also, the evolution of new generations of components by vendors should be promoted by more explicitly defining generic requirements of manufacturers.

Similarly, the separating out of VMC interaction issues can much improve the way in which systems are designed and implemented. Such a separation of issues is consistent with the notion of building systems from reusable components, since it has the potential to directly support the definition and realization of flexible links between components. We will also see later that it is consistent with the use of standard interaction protocols and integration services that can support the decomposition and distribution of present-generation manufacturing components into new, smaller-grained building blocks that can fit business needs more effectively than is currently possible.

5   SOLUTION STRUCTURE

The solution structure must be consistent with the requirements of its host environment. It should also arrange and organize the way in which solutions are engineered.

Generally it will be necessary to establish and maintain structural connections between subsolutions, which themselves meet a particular aspect of formalization of application requirements. These structural connections may be viewed as a metamodel connecting more focused representations of solutions, which themselves separate out and encode models of subsolutions. Table 2 defines generic solution structure perspectives that the authors and their co-researchers have found are needed when realizing and changing various types of manufacturing system. Such a separation of issues has proven to embed a change ethic into resultant systems in that the decomposition supports system design, build, and evolution in an effective and scalable way. The solution structure must be loosely linked to representations of VMC behavior and VMC interaction so that solution structure change is essentially independent of those issues.

Historically many manufacturing companies and systems builders have developed and used their own conventions and structural guidelines to facilitate the process of defining solution structures. However, no public domain, comprehensive, and formally defined methodology has been used industrially to support solution structure design and construction, although there has been an increasing use of formalisms when producing software to implement solution structures embedded into contemporary manufacturing systems. As a consequence, the issues listed in Table 2 are invariably inextricably tangled together, resulting in systems and machines that are inflexible (i.e., do not facilitate change and their inclusion into wider-scope solutions).

Table 2   Solution Structure

Subissue	Interpretation / Example
Selecting a primary architecture	Hierarchical–heterarchical, client–server, master–slave
Representing the process requirements	Capturing, analyzing, refining process definitions. Need tools, agreed formats, etc.
Representing business logic	Definition of functionality required to fulfill the requirement
Representing VMC behavior	How the behavior of candidate VMCs is made accessible to the design activity
Representing VMC interactions	How the potential interactions of candidate VMCs are made accessible to the design activity
Defining user–system interactions.	What interactions will be possible and when or in what order

During the last decade research and development initiatives world-wide on enterprise modeling and integration have promised to provide more complete and formal means of specifying solution structures and thereby specifying and analyzing the requirements of manufacturing systems and mapping these requirements onto organized sets of interoperating resources (i.e., VMCs and MCs). The following subsections illustrate how these developments may support the decomposition of solution structure issues in Table 1 and thereby lead more rapidly to agile and better designed systems.

5.1 Selecting a Primary Architecture

There has been much debate in the academic and research communities about the relative merits of so-called hierarchical and heterarchical system architectures, the degrees of sophistication in distributing the operational process decision-making, and so forth. In theory, the process definition in conjunction with the chosen parameters for flexibility and resolution of control of the process would be the determining factors in this choice.

In practice, however, the majority of implemented solutions have followed much simpler patterns. In general terms there are two common structures: viz.: (1) *client–server* (C–S) architecture and (2) *master–slave* (M–S) architecture. Both of these stem from the same basic concept of clustering a set of MEs around a larger ME that fulfills the role of co-ordinator or server, as appropriate. Examples are as follows:

1. *Cell/shop control:* Typically a controller (master) "pushes down" tasks to a cell of MEs (slaves). Associated activities include requesting task start/stop, tracking task progress, and collecting production information.
2. *Workpiece/quality tracking:* Typically a number of data entry devices (clients) "pushing up" blocks of data to a central collection point (server).
3. *Multiuser access to database systems:* Typically user interface devices (clients) "pulling down" and "pushing up" information to large centralized DBMS systems (servers).

Sometimes such systems involve multiple layers and possibly a mix of C–S or M–S operation for different layers or, occasionally, for different sets of interactions at the same layer. In the latter case one device might perform a cell control and workpiece tracking function for the same set of clustered devices. Similarly, most real systems will not be "pure" in that they will have a few "work arounds" built in. Even then the division between, and the type of, the various subsystems is generally readily discernible.

Apart from tending towards being implicitly hierarchical, another important characteristic of such solutions is that the C–S or M–S interactions are almost invariably synchronous in nature. That is, each interaction is basically a *request* from the client or the master followed by a *response* from the server or slave, respectively. Systems incorporating asynchronous interactions—for example, slave devices indicating error conditions or other events to a master without having been polled—are significantly more difficult to engineer and, more importantly, debug and subsequently maintain. This is due to the accompanying rapid increase in the number of possible system states, which becomes impractical to manage on any realistic scale outside the laboratory.

The above is not to say that more sophisticated structures would not be better. Indeed, various primary architectures suggested by the research community have been applied to a limited extent, such as the NIST model (Jones and McLean, 1986), the COSIMA model (Brown, Harhen, and Shiwan, 1988); PCF (Archimede, 1991) and the ESPRIT 809 model (Van Der Pluym, 1990). However, these cases are the exception, not the rule, in industry. Therefore either the current practice is lagging the theory or the theory is not sufficiently well developed for practical application.

Thus the necessary decisions on this issue are:

1. If the complexity requires, determining the appropriate levels of the solution and the division into subsystems at each level
2. Selecting the appropriate principle structure for each subsystem identified
3. Identifying (while minimizing) any necessary work arounds

5.2   Representing Process Requirements and Defining Process Definitions

Over the last decade many business process-modeling tools have become commercially available (Weston, 1996). Increasingly they have been used to formally describe the business processes of an enterprise with a view to:

1. Helping to generate a coherent view of what an enterprise needs to do to realize products and services in competitive ways
2. Allowing teams of people to view process models from their own perspectives, yet contribute to a unified whole
3. Begin to develop change options that quantify the benefits of reorganizing the tasks, operations, and activities carried out in a given company

However, seldom have business process descriptions been explicitly connected to models of systems and their interacting MCs. Rather, to date, the use of process models has been focused on "What an enterprise should do" rather than "how it can do the what." This may be largely because the latter is hard to do and to justify in business terms.

Nonetheless, with respect to IT systems design there is a growing need to connect models of business processes to models of software components. Indeed, this need is driving recent developments by vendors of ERP, MRP, and MES software packages in North America and Europe (Prins, 1996; Dirks, 1997; OMG, 1996). Here they are developing component-based and parameter-driven software applications, where values can be assigned to parameters to configure the application in alignment with specific process needs of different manufacturing companies.

Fairly comprehensive formal definitions of ways of connecting models of processes, systems, and components (such as software applications) are described as part of the CIMOSA specification (Kosanke and Klevers, 1990). Furthermore, other enterprise modeling methodologies and architectures, such as GRAI-GIM, PERA, ARIS, and TOVE, suggest alternative structural connections (Shorter, 1994). Figure 2 exemplifies the use of one of these architectures. It illustrates the use of an enterprise engineering workbench called *SEWOSA* (Aguiar, 1995; Aguiar, Coutts, and Weston, 1995), which developed and operationalized key facets of the CIMOSA open systems architecture. Thereby SEWOSA supports and connects the definition of a series of modeling views. This allows a team of system designers and implementers to define and realize suitable solution structures,

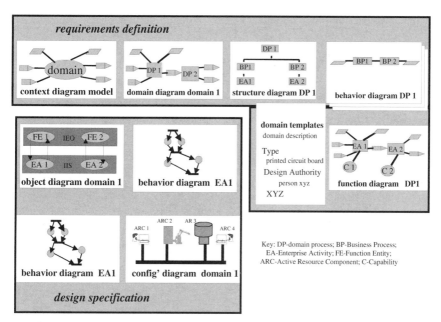

Figure 2   Some of the modeling views supported by the SEWOSA workbench.

including each of the elements defined by Table 2. A metamodel that conforms to the CIMOSA architecture is implemented and maintained by the SEWOSA workbench. This allows the separate modeling perspectives illustrated by Figure 2 to be developed by different designers and implementers, yet maintains consistency between the models so that complete solution structures are defined.

5.3   Representing Business Logic and System Behavior

The business logic (i.e., a set of logical functions that need to be included into a given system) will be application-dependent and (at least implicitly) will need to be linked to definitions of process requirements. There are many alternative ways of encoding business logic. Often, when defining and implementing function blocks of systems, general purpose programming languages are used, such as C++, C, PASCAL and VISUAL BASIC. However, with these languages the meaning (or semantics) of the business logic (which is some abstract representation of system functionality that can be understood and reused by a human designer) gets lost within the detailed system code. Therefore this knowledge is not retained so that it can be used to change the system design when and as new requirements emerge. Research is ongoing to address such problems (Coutts, 1998), but this situation makes large-scale systems programming costly and the solutions very inflexible. Potentially enterprise modeling frameworks and architectures such as CIMOSA and ARIS provide modeling constructs that can be used to support the definition of system functionality in a way that retains and promotes the reuse of design knowledge. Figure 2 also illustrates in outline how this can be done using the SEWOSA workbench. Here behavior models are separate from, but can be readily linked to, process and function models. Also, SEWOSA behavior models can be mapped readily onto lower-level interaction models, which are formally represented by Petri-nets and models of VMCs described in Estelle, UML, and EXPRESS (Murgatroyd, Edwards, and Weston, 1993; Weston and Gilders, 1995). Over the next 5 to 10 years we can expect to see many new systems engineering tools to emerge.

5.4   Representing VMC Behavior and VMC Interactions

As indicated in the previous section, various formal description techniques have emerged that are capable of representing the behavior of VMCs and the way in which VMCs interact. Means of representing behavior are included as part of popular object-oriented approaches to software and system design, e.g., as part of UML (Booch and Rumbaugh, 1995). Here state diagramming techniques are used. Also commonly used to model component behavior are extended versions of Petri net, since they can offer good visualization and simulation capabilities. A unification and extension of state diagramming and Petri net techniques is offered by using the binary transition language (BTL) developed by Coutts to execute behavior models over different integration infrastructures, such as CIM-BIOSYS, CORBA, and Internet (Coutts, 1998). Estelle and IDL are formal description languages developed to represent interaction and communication protocols (Weston and Gilders, 1995). Notwithstanding their technical capabilities, it is likely that the Interface Description Language (IDL) will be used most widely in view of its close connection to CORBA developments as part of standards initiatives by the Object Management Group (OMG, 1996). Historically the development of EXPRESS as a formal language has been linked to initiatives world-wide on information systems modeling. EXPRESS is now very widely used and has powerful description capabilities that can model information entities and their relationships as part of solution structure definitions.

Arguably, therefore, there are many descriptive techniques available to model VMCs and their behavior and interactions, although currently they are weakest in terms of describing concurrent behavior, particularly where tightly coupled synchronous behavior is needed (Nof, 1994). These formal descriptions of VMCs need to be a consistent and integral part of a complete solution structure. Shared modeling constructs are also needed to link formal descriptions of MC implementation and VMC interaction issues, as described in Sections 6 and 7 of this chapter. Because it implements the CIMOSA open systems architecture, the SEWOSA workbench can achieve such an integration whilst allowing VMC behavior and VMC interaction models to be developed and retained in largely independent ways. Again, this capability is depicted by Figure 2.

5.5   Defining User–System Interactions

If a change ethic is to be embedded into a system, then by definition it is necessary to consider the complete lifetime of a system. It follows that manufacturing system users include not only shop floor operatives, supervisors, and engineers, but also businessmen, managers, system architects, system designers, and so forth. It also follows that the process of defining human(user)–system interactions can be very complex. Invariably, using current approaches to designing and constructing systems, humans are treated essentially as a class of VMC, which accordingly will have a set of functional capabilities, qualities, and constraints (which unfortunately may be ill-defined). It is not obvious whether this situation will change significantly in the foreseeable future. However, if humans continue to be modeled rather mechanistically as VMCs it will be important to develop better understandings of the rules governing human–system interactions (such as allowable multimedia rules, profiles, permissions, sequences, etc.) and the constraints imposed on allowable interactions by current human computer interface technology.

6 MC IMPLEMENTATION

Previous experience of the authors when implementing robotic and other manufacturing systems has led them to seek to induce a change ethic into such systems by identifying three types of implementation issue, tabulated in Table 3.

6.1 Incorporation of the Real Elements

The actual MCs required to construct manufacturing systems may be broadly classified under the following headings.

1. *Humans* fulfill a range of roles, such as equipment operation, managerial, support, maintenance. Because humans cannot be directly interfaced to the rest of the system, they will be supported by a range of computer-based devices, from switches and lights to computer displays.
2. *Software applications* generally fall into the two subcategories of system and support functions. The former contribute directly to the process that the system instantiates and include planning, control, possibly manufacturing operations, data collection/storage, and so on. The latter include those functions that are indirectly necessary, such as postprocessing of CAD data to generate machine programs.
3. *Machines* will be computer-controlled to some degree and will generally be used for physical manufacturing operations, such as part handling, metal cutting, inspection.

Robots fall into the last category and in some ways are amongst the most flexible of manufacturing machinery. In practice they usually require a considerable amount of tooling, such as end-effectors or fixtures. Alternatively, they may be used to manipulate other equipment, such as welders. In either case flexibility is lost to a lesser or greater degree by the time the robot is installed and able to perform tasks. In some cases this occurs to the point that the robotic concept is lost and the robot and its tooling become effectively a dedicated machine of the same ilk as, say, a machining center. This does not negate the potential benefit of the robotic concept, since benefit will presumably have been gained by using the base robot as a component of the resulting machine instead of building from scratch. In other cases the installation does not constrain the flexibility of the robot so severely and subsequent reconfiguration of the solution in response to required changes in the process is not so difficult—at least in terms of physical change. However, if appropriate consideration has not been given to the issues discussed below, such reconfiguration will probably be limited due to cost and time constraints; pure ingenuity is not usually a problem. These issues principally revolve around questions of structure—of the component itself, of its interaction capabilities, and of the solution as a whole.

Table 3 VMC Implementation

Issue	Interpretation/Examples
Incorporation of the real elements	How physical device behavior is made externally available via requests and instructions
Definition and execution of behavior	Potential levels of granularity—commands, programs, abstract task performance
Access to element specific information	Purely local or externally available information; data formats and structures

Unfortunately, most manufacturing machines (and indeed software applications) are conceived as fundamentally standalone. Any external access (i.e., by other system components) to their internal function and data is normally an afterthought. Worse still, such access is often implemented using mechanisms and protocols completely different from those of similar equipment. Of course, people may also prefer individualism to a team culture.

The VMC concept is very useful in supporting an ability to design and specify solutions in abstract terms—that is, not dependent on the particular machines, interactions, etc. to be used. As the foregoing discussion illustrates, this is essential if the desired solution flexibility, cost-effectiveness and timeliness levels are to be achieved.

Unfortunately, a choice of available manufacturing machinery, applications, etc. may not exist that corresponds directly to component models (i.e., VMCs) used during system design. It is controversial when and indeed whether any such consistent set of real components will become available. However much the argument is put forward that the makers of such components will benefit from adopting a common model, the suspicion remains that many suppliers perceive their market edge as stemming from the very uniqueness of their product in operational detail as well as capability. This is in addition to suspicions about whether the "right" common model has or is being defined.

It therefore seems that solutions will be built for some considerable time using non-idealized components. If this is to be compatible with the development and adoption of abstract design techniques, then these components will need to be either modified or accommodated via additional interface functionality. Modification may involve the component's supplier and may be the route by which the latter eventually adopts and implements a common component model. The second method requires that the extra interface ability be generated either by the solution implementer or possibly by third party companies.

As discussed below, a gateway is often required to provide interconnection between a manufacturing component and a communication mechanism. It has been common for additional higher-level interface functions to be included to convert between interactions using the common model on the system side and interactions with specific machinery (and sometimes applications software) on the component side. Although somewhat inelegant, this approach does provide much greater flexibility by dissociating specific manufacturing hardware/software/humans from the rest of the system. The concept is well known to the "object-oriented" software community and is known as *wrapping*.

The downside of these approaches is, of course, increased cost. It would seem that end users are not currently sufficiently convinced of the benefits of enhanced system flexibility to go the extra mile to achieve it. This implies that the higher-level analysis and design tools are currently not capable of convincingly predicting the benefit (in the cases where it does exist, of course—it may not always). This may be due to their current state of development, but another factor is likely to be that the fact that they do not seamlessly—if at all—link to the business-oriented tools used higher up in the user organizations. Surveys of organizations closely involved in BPR activities have highlighted a gulf between the high-level design decision-making and the implementation phases (Barber et al., 1996). The revolution that has occurred in the former, particularly with respect to dealing with change, does not yet appear to have caused significant changes of practices in the latter.

6.2 Definition and Execution of Behavior

An increasing number of MCs are becoming more flexible by being programmable via stored data instruction sequences. Robots are a good example. What this programmability effectively means is that the way the behavior of a given MC is defined is split into a generic part and a specific part. The first includes, for example, axis control and transformations, while the latter constitutes a much higher-level definition of the what the robot will do. How high this level of behavior definition is determines how flexible an end solution can be. This is illustrated by three examples of interaction between a robot and another MC.

Some robots, such as the ASEA Irb6, can be remotely controlled by sending them a stream of spatial (or joint angle) positions. This has advantages when the robot movements must be very closely coupled in real time to another MC; an inspection device, for instance. Clearly such a coupling between the two MCs is very device-specific and could make system changes difficult. A design tradeoff must be made according to system requirements.

A more common method of defining robot behavior is via the execution of blocks of stored instructions or programs such as those defined by VAL II for Unimate and ADEPT robots. In operation the MC interactions will either select one of several available programs or possibly download new programs as necessary. With this scheme much greater system flexibility is potentially available—especially if robot manufacturers could be persuaded to use the same language! The impact of machine-specific issues can be contained at a much lower level in the system design and implementation process.

More sophistication still is possible whereby MC interactions are restricted to an essentially abstract level. Thus a command to specifically execute a given program would be substituted for by a request to perform a certain operation—i.e., "Load The Lathe" instead of "Run Program Load(Lathe)." The necessary behavior to achieve the operation would be determined locally by the target machine. In this way the details of the operation would be separated from the rest of the system and the same requests would be used whatever changes were made to the target MC.

The system builder does not currently have a wide choice in respect of these types of capabilities, and these will often take a (long way) back seat to more task-specific features (speed, accuracy, etc.). Suppliers will only respond to customer demand, and the philosophy behind most available devices may be summed up as primarily "stand-alone with the ability for some remote control" rather than "system component-oriented and flexible interaction-friendly."

6.3   Access to Element Specific Information

Many MCs maintain or generate information that would be useful to other system components. This information might relate to behavior definition, execution state, production progress or specific workpiece/tool data, and so on. In accord with the primarily stand-alone concept underlying most available MCs, such information is often maintained behind some form of supplied interface. Thus it is only available via certain mechanisms and in certain formats, both of which are generally highly device-specific.

Among others, the Manufacturing Message Specification (MMS) specification was developed as part of the MAP standardization initiative to standardize message protocol between system devices. Indeed, the ISO 9506 protocol standard (Aguirre Suarez, 1997) was developed to define *manufacturing system objects* (effectively a set of virtual components) and *message protocol between objects* (thereby standardizing aspects of message formats and semantics, i.e., to encode the meaning of specific messages). ISO 9506 was developed to include companion standards that define common types of message interaction between virtual manufacturing devices (VMDs), including robots, programmable logic controllers (PLCs), and CNC machines. Unfortunately, however, the industrial take-up of MMS has been limited. This may be because the architecture of the MMS/MAP approach is not ideal; because the use of common message formats and partially common semantics solves only a fraction of the general system integration problem (hence can only partially realize benefits from system agility); or because available supporting systems engineering tools do not facilitate the approach adequately.

Correspondingly even wider-scale initiatives linked to PDES/STEP and EXPRESS (Clements, Coutts, and Weston, 1993) have attempted to derive and increase the acceptance of data models, primarily for products but also for other system information such as process definitions.

The situation currently parallels that of external access to a components behavior where few available devices incorporate the concepts behind such models to any great degree. Thus in the short to medium term it seems likely that the translation of unified, abstract design time representations will require extra work at build time. It will require the incorporation of additional functionality around the chosen components to make them appear to fit the model as far as the rest of the system is concerned. Again, because the effort to do this is visibly significant and the potential future benefits difficult to quantify with current tools, the frequent result is a hard-wired information exchange structure between pairs or sets of components.

7   MC INTERACTION

Issues associated with MC interaction are related to the rationale behind the desire for the interactions. There are two discernible (though interrelated in practice) reasons:

1. Management of system function in support of the process. This naturally covers a whole spectrum of complexity. At its simplest it provides for coordination be-

tween two or more components—a robot unloads a machine when the operation is finished, for example. It may involve remote control of a set of components such as a metal cutting cell according to a predetermined or dynamically updated plan. This will automatically require operation progress monitoring, which may be extended to generate production data—either for quality management or improved production control or possibly to assist in redesigning the process involved.

2. Management of information (availability and usage) in support of the individual components. Again, a range of complexity prevails. The most basic example is probably the upload and download of programs or other machine-setting data. Increasing in sophistication are monitoring of component operations; the collection of production data for whatever reasons as above; and the access by system components to centralized information systems—part data, say.

It is important to note that neither of these principal activities implies any particular degree of automation of the system. For example, numerous benefits can be reaped just by supporting on-line transfer of program- and machine-setting data. Benefits over the off-line procedures have been shown to include significant improvements in set-up time (and hence system flexibility) and accuracy. Similarly, control of the system may involve supporting humans in terms of task sequence and progress monitoring.

In the case of robotic machinery, the most natural interactions involve, as might be expected, program transfer/selection/execution and associated status monitoring. These are therefore the more commonly encountered. As discussed below, however, much greater sophistication is possible. Exploiting this, while retaining a high level of system flexibility—i.e., being able to alter the system in response to new process requirements rather than just, say, minor part changes—requires commonality in several respects:

1. *Mechanisms:* moving data blocks from one device to another
2. *Structures and syntax:* how to interpret the data blocks
3. *Dialogue structure:* how to structure ongoing interactions
4. *Semantics:* what dialogues/messages to use to achieve a goal

Standardization—often at a detailed level—has often been seen as the way to make progress in these areas. Although much effort has been expended, a system implementer is still usually faced with the same problems of component incompatibility and a general inability to design solutions in the abstract rather than around a given set of particular components. Although it can be costly and time-consuming, it is not usually particularly difficult to integrate any given set of components to solve a given need. If the lifetime of the system is large and changes are few, such a "big machine" style of solution is effective.

However, achieving the agile systems perceived as necessary in response to contemporary business process concepts will require not detailed standardization alone but commonality of approach, both to individual component construction and to system structures. In this respect, previous experience of the authors and their colleagues has shown that significant benefit can be gained by separating the interaction issues listed in Table 4.

7.1 Data Exchange

The most basic facility underpinning any MC or VMC interaction is that of transferring blocks of data from one component to another (or several others). At its most basic it

Table 4 VMC Interaction

Issue	Interpretation/Examples
Data exchange	Common data exchange mechanisms Digital networks and services
Exchange of instructions and requests	Message sets, structures and formats Structured interaction dialogue
Information exchange	Common formats and structures
Special situations	Close coupled activities, highly constrained requirements

does not really look like data exchange. At this level a *transmitting* MC turns binary outputs on or off and (an)other MV(s) act on the changes of state. This method is very widely used in shop floor systems and is clearly particularly useful for interaction involving one or more devices of low processing capability, e.g., use of a switch to operate an actuator. It can, though, be usefully employed between more sophisticated devices such as a manufacturing machine and its attendant workpiece handling robot, for example.

The advantages are very low in cost and complexity and very high performance and reliability. These are always at a premium in real operating solutions. The disadvantages arise when more complex interactions are required, such as indicating to another MC which of a choice of parts should be loaded. Furthermore, this type of implementation is inherently localized, in terms both of geographical distribution and, more importantly, the specific machines or applications involved. It would be very difficult to achieve an agreed set of signals for use in certain situations or with a certain type of machine, apart from relatively trivial cases such as most manufacturing machines having a stop/start or program select button for interaction with humans.

Therefore for more complex interactions many MCs implement a more sophisticated data interchange mechanism, often taken from the computer industry. The typical solution involves an RS232 (or closely related) interface to get medium to large blocks of bytes from one MC to another. Unfortunately, presumably from a mix of parochial thinking and plain lack of proper understanding, there are probably almost as many unique implementations of RS232 interface as there are machines using it.

For example, one robot used by the author allegedly implemented a subset of DDCMP. The use of a subset in combination with mistakes in the implementation meant that it was impractical for the robot controller to communicate with any other device except one specially constructed for the task.

Thus the notion of commonality of interaction mechanisms supporting the potential "plug and play" of different devices is lost not only in terms of practicality on the shop floor, but also in genericity at the design level. Again, this type of mechanism (point-to-point links), although more flexible in physical terms (in the limit, though, the wiring alone becomes unmanageable), tends to localize and hard-wire the two devices and their interactions together.

The use of LAN technology greatly simplifies the physical distribution, and the available hardware has become standardized, relatively low cost, reliable, and widely available. It is used to connect PCs, etc. together in office environments. In fact, the growing use of common PC hardware as the basis of manufacturing machine controllers may point to a way forward. Apart from this, for reasons including cost, lack of demand, difficulty of use, and the more complex attendant issues covered below in Section 7.2, far from many MCs have hitherto possessed a direct LAN connection capability. However, in the wake of the MAP initiative an increased number of vendors of manufacturing components, such as CNC machines, PLCs, and SCADA packages, developed LAN interfaces to a selection of their products.

Generally the lack of LAN interfaces has hitherto been overcome by the use of gateways. These devices generally appear as a normal LAN interface on one side and a local machine-specific interface on the other. Thus the peculiarities of the local device can be isolated (to a greater or lesser degree) from the rest of the system. The local interface may range from binary signaling via custom RS232 through to a different LAN technology while the standard main LAN interface enables interaction with all the other MCs in the system that support use of the same types of interface. The gateway internals perform the necessary manipulation of the data blocks, such as buffering or retransmission on error. Much more functionality is usually incorporated, though, to fulfill the needs that follow hard on the heels of data interchange, as discussed above. See Section 6.1.

Whether a direct LAN interface or a gateway is used, the concept of protocols is vital. These protocols ensure not only that the "plugs fit the sockets," but that the data blocks can be transferred to a correct destination with an appropriate degree of confidence that no corruption has occurred. The protocol therefore covers matters of device addressing (e.g., Internet numbers or WWW URLs), splitting of long transfers into multiple short ones, detecting and correcting errors, and so forth. In the past these issues have been the subject of truly enormous debate and needed to be fairly well understood by a system builder. Fortunately, the maturing of the technologies now means that much of the technology is buried inside available products and, with one notable exception, the user need not be concerned with the details. The exception concerns the still relatively commonplace lack of interoperability between devices that ostensibly use the same set of protocols.

Indeed, in recent years there has been a significant focus of attention on so-called *middleware software* issues, where middleware software such as CORBA and Internet tools abstract the user further away from LAN and other implementation details by providing virtual models of data and message interchange (Aguirre, 1998).

7.2 Exchange of Instructions and Requests

Even when simple binary signaling is used, it is necessary to associate some form of protocol with the signals. At the simplest level this covers the linkage between the signal and the implied action. More complex signaling will require more protocol—an MC needing to signal back that action is now under way, for instance.

However, as soon as the more comprehensive mechanisms involving the exchange of blocks of data—or just sequences of numbers, as it in fact is—are adopted, much more comprehensive and sophisticated protocols are required. These are an extension of the protocols introduced above. While the data exchange protocols can be seen as enabling the transfer of data blocks, the messaging protocols enable the interpretation of the data when it has arrived. Thus a transferred block of data or message might be interpreted as an instruction to, for example, execute program "Load_Tool" with parameters "Tool_1" and "Holder_4." This transfer of meaning can only occur if the MCs involved both use the same protocol (or can negotiate a common protocol) for the message syntax. Thus the data block must be recognizable as an acceptable message, the numbers in the message must be of appropriate value and occur in an appropriate order, and so forth. Any errors that occur in this process will ideally be made known to the sending end, but this may not be possible if the data are complete garbage.

In fact, even more protocol than this is required, since the message must arrive at an appropriate time. That is to say, the recipient must be in an appropriate state to act on the message content or that the message be correctly positioned as part of some sequence. Thus a structure to a series of requests and responses is necessary. This structure will be very closely related to that in Section 5.1 above in terms of hierarchical relationships and synchronous/asynchronous interaction.

As a simple example, consider the case of a robot receiving a request to execute a certain instruction sequence or program when it is already executing another program. We could design the system so that this was never supposed to happen, although it is generally unwise to rely on this. At the other extreme, we could aim for great sophistication, so that the robot was capable of interrupting one task and then returning to it later. An intermediate approach would be to arrange for the robot to respond with a "request failed" indication and carry on. Each of these approaches has drawbacks, as exemplified in Table 5. Often the result is to move a decision one level up in the system.

The common practical solution is currently to keep the solutions simple and often fairly rigid, with all possible conditions preconsidered and allowed for. Unfortunately, this goes against the perceived need for highly agile configurable solutions composed of readily substitutable MCs forming ad hoc virtual subsystems on an as needed basis.

Thus it can be seen not only that a chosen structure is most important in interactions between MCs, but that it would be most advantageous if common structures were adopted

Table 5   Example of Interaction Strategies and Implications

Strategy	Implication
Conflicting request impossible	Implies that all possible requesting system elements are keeping track of the activities of the robot. Implies complex, static and, inflexible system design.
Conflicting request serviced by interrupting current task	Implies complex local optimization of task execution and management. Recovery from exceptions becomes very difficult because possible system states multiply.
Conflicting request rejected	Simple and flexible system, but what do requesting system elements do next? Keep trying? Forever? Try elsewhere? Reschedule a whole job?

in the construction of such solutions. This is true not only for individual messages, but also for longer-term interactions or *dialogues*. This is applicable across a whole industry sector, or at least just within a single enterprise. It does not necessarily mean that there is only one solution for a given situation or that the answer lies in detailed standardization of the low-level protocols. The key more likely lies in the direction of solution patterns and widely accepted common approaches. This has happened in certain well-defined, relatively localized situations where "typical" solutions have evolved over time. A good example would be the commercially available Manufacturing Execution System (MES) solutions for (principally) FMS-type systems.

7.3   Information Exchange

The basic level of MC interactions involves mainly control-oriented functions such as start/stop/status and so forth. Even when the upload and download of programs is incorporated, the data involved are usually specific to the target machine and opaque to the rest of the system. That is, it could not be recognized or interpreted as a set of instructions except by the particular device it is used by. Thus there is not really the concept of the exchange of information as opposed to data, and this again simplifies the system at the expense of making it highly specific.

However, to achieve high system flexibility it is necessary to move beyond this. For example, if a system incorporates robots whose programs definitions are specific to that particular brand and it was then found desirable to replace them with other robots, the programs would all have to be re-generated. Even worse, substitutions of manufacturing components by ones from a different class (e.g., human by robot, robot by dedicated machinery) would require wholesale reengineering of the system. Similarly, if the robot programs are inherently dependent on, say, a particular workpiece dimension, the impact of a design change will be large. Ideally MCs should be able to access and use product information during operation.

Thus it becomes highly desirable not only that common mechanisms be available for handling the information requirements of a system, but also that consideration be given to the format and content of the system-related information.

7.4   Special Situations

Despite the advantages of adopting commonized system structures, concepts, and mechanisms in enabling much greater abstraction of the design process, it seems certain that there will always be a need to allow for special cases. A good example of this is the use of programmable manipulators whose specific operations require to be modified at run time. This will typically be for two reasons.

The first is when, say, a robot must allow for normal variations in the manufacturing process. These variations may be in dimensional tolerance, part presentation, part recognition, etc., when the robot would typically operate in close cooperation with some form of sensing, such as mechanical or visual. Alternatively, a function such as in-process measurement might require modified path trajectories in real time.

The second general case is that where manipulators share a common workspace or need to cooperate to achieve a task. It may be possible to predetermine and preprogram for all the constraints involved, but this may drastically impair system flexibility.

In either case the interactions between the components are likely to require much greater performance than is available from either the mechanisms or protocols used for commonized interactions in the rest of the system. Specifically, the limitations will probably be related to timing and the highly specialized nature of the interaction content and the requirements to support synchronization between effectively parallel executing threads of behavior (Nof, 1994). Two problems arise from the specialized nature of the interactions or the high-performance mechanisms required to support them.

1. How are these represented in the abstract design process?
2. How is the impact of "oddball" elements localized in the system?

At present it is often simplest to treat the devices involved as one larger component as far as the design process is concerned and to treat their interactions as a separate design activity. This is more problematic if the devices interact with other system components as well as each other. In this case the design may be split into two (or more) parts and a different set of common concepts used within each part. In any circumstance it is important to recognize early and then separate and maintain the different domains as much as possible.

8 SIMPLE EXAMPLE

Due to the very nature of manufacturing systems integration, it is not possible to clearly illustrate the problems and potential solutions by way of simple examples. The bulk of the problems associated with integrating the activities of the constituent elements while supporting system reconfiguration without excessive reengineering arise only because the real examples are not simple. However, in order to relate some points of the above discussion to actual concrete parts of a system component, an example is presented below. This example is taken from a technology demonstration and proving cell constructed as part of system integration research. While the work is now several years old, the basic problems illustrated still have to be tackled—technology and standardization advances have not yet obviated them. In some areas new technology and approaches have been suggested to alleviate the problems, but many have not yet been proven on the shop floor.

The component under consideration was an Adept SCARA five-axis manipulator. The simple task was to assemble several switches in a switch plate according to commands received from a cell controller. The workpieces—switches and plate—were delivered via a conveyor system controller by an application running on a standard PC. Progress of the tasks was monitored by the cell controller. In fact, alternative configurations of switch plates could be assembled as determined by a production schedule.

In accord with the above discussion, the first requirement was for a data link. This was accomplished initially using Ethernet and eventually using a MAP specification broadband LAN. Since the Adept could not be connected directly to the network, a PC-based gateway was used. This contained both the necessary hardware and software to implement the seven-layer ISO conformant MAP protocol stack.

The robot had no remote interfacing ability at all beyond an RS232 port accessible via the VAL II programming language used to program all robot functions. It was therefore necessary to implement a point-to-point data link between the gateway and the robot. This was implemented in ordinary "C" code on the gateway and in the VAL II robot language itself. Thus the mechanisms for basic data exchange were achieved.

The specific behavior of the robot was defined in the VAL II language. In this case the only level of external control was to stop, start, and monitor the progress of the available VAL II programs. Again this facility was not supported by the robot and so had to be implemented by further VAL II software and "C" code in the gateway. A set of messages was used for interaction between remote devices and the robot. These messages conformed to the Manufacturing Message Specification (MMS) and were interpreted by the gateway, which then interacted as necessary with the robot using a local message protocol. This extra level of message encoding was appropriate because of the relatively high complexity of the processing required to implement MMS and the good but relatively low processing capabilities of the robot controller.

The information-oriented interactions with the robot were of two kinds: the download and upload of VALII programs and the read/write of VALII variables used by the programs. As with the behavioral interactions, MMS was used between the system and the gateway, which then performed the actions necessary. In this case, as with other machine programs in the implementation, the robot programs were treated by other system elements as opaque in that they were just blocks of bytes with no apparent meaning.

In term of system structure a master–slave relationship was used, with the ADEPT being one of the slave elements of a cell controller. The latter was itself a slave element in respect of its relationship with higher-level factory control. Thus all interactions between elements consisted of a request by the master and a response by the slave and no interactions were initiated by the slaves.

By the above means the robot was made capable of acting as part of a system and its limitations in this respect were overcome using a gateway-type solution. The use, thus enabled use of the standard data transfer and messaging services implied that substitution of the ADEPT by a different entity—dedicated, robotic, or even human—would require only localized reengineering and have minimal impact on the rest of the system.

9 CONCLUSIONS

We may conclude that theoretically recent advances in enterprise modeling, integration technology, and component technology should mutually induce a major step change in the way in which future manufacturing and machine systems will be engineered. Indeed, they promise an ability to configure more agile manufacturing enterprises in which the operation of systems remains closely aligned with business goals and customer needs. In practice, however, various technical and commercial obstacles remain that are likely to impede progress and result in incremental improvements over a significant time period.

Corresponding improvements are required in respect to the next generation of industrial robots. They should more readily facilitate configuration and engineering into agile manufacturing systems. Like other programmable machines, contemporary robots are themselves inherently configurable. However, they are invariably less so when viewed as an integral part of a system. We may conclude that robot manufacturers who rapidly embrace such concepts as business process and component modeling, semantic integration, and component decomposition and distribution are likely to realize a competitive edge. Furthermore, such developments should lead naturally to the deployment of reconfigurable and reusable robotic technology in new and advanced application areas.

REFERENCES

Aguiar, M. W. C. 1995. "Executing Manufacturing Models of Open Systems." Ph.D. Thesis, Loughborough University.

Aguiar, M. W. C., I. A. Coutts, and R. H. Weston. 1995. "Rapid Prototyping of Integrated Manufacturing Systems by Accomplishing Model-Enactment." In *Integrated Manufacturing Systems Engineering.* Ed. P. Ladet and F. Vernadat. London: Chapman & Hall. 62–83.

Aguirre Suarez, O. 1997. "Architecture and Methodology for the Development of Manufacturing Control Systems." Ph.D. Thesis, Universidad de Navarra.

Archimede, B. 1991. "Conception d'une architecture réactive, distribuée et hierarchisée pour le pilotage des systemes de production." Ph.D. Thesis, Université de Bordeaux.

Barber, M. I., and R. H. Weston. 1998. "Scoping Study on Business Process Reengineering: Towards Successful IT Application." *International Journal of Production Research* 36(3), 575–601.

Barber, M. I., et al. 1996. "A Study of Business Process Re-engineering Practice in the UK." MSI Publications, Loughborough University.

Bauer, A., J. Bowden, and J. Browne. 1991. *Shop Floor Control System: From Design to Implementation.* London: Chapman & Hall.

Booch, G., and J. Rumbaugh. 1995. "Unified Method for Object-Oriented Development Version 0.8." Rational Software Corp.

Brown, J., J. Harhen, and J. Shivan. 1988. *Production Management Systems: A CIM Perspective.* London: Addison-Wesley.

Clements, P., I. A. Coutts, and R. H. Weston. 1993. "A Life-Cycle Support Environment Comprising Open Systems Manufacturing Modeling Methods and the CIM-BIOSYS Infrastructural Tools." In *Proceedings of the Symposium on Manufacturing Application Programming Language Environment (MAPLE) Conference,* Ottawa, September. 181–195.

Coutts, I. A. 1998. "An Infrastructure to Support the Implementation of Distributed Software Systems." Ph.D. Thesis, Loughborough University.

Dirks, C. 1997. In *Enterprise Engineering and Integration: Building International Consensus: Proceedings of the ICEIMT '97 International Conference on Enterprise Integration and Modeling Technology,* October, Turin, Italy. Ed. K. Kosanke and J. G. Nell. Research Reports ESPRFT Project 21.859, EI-IC. Vol. 1. Berlin: Springer-Verlag.

Duffie, N. A., and V. V. Prabhu. 1996. "Heterarchical Control of Highly Distributed Manufacturing Systems." *International Journal of Computer Integrated Manufacturing* 9(4), 270–282.

Graham, I. 1994. *Migrating to Object Technology.* Reading: Addison-Wesley.

Jones, A. T., and C. R. McLean. 1986. "A Proposed Hierarchical Model for Automated Manufacturing Systems." *Journal of Manufacturing Systems* 5(1), 15–25.

Jones, A., and A. Saleh. 1991. *The Cell as Part of a Manufacturing System. Manufacturing Cells—Control, Programming and Integration.* B.H. Newnes. 37–61.

Kosanke, K., and T. Klevers. 1990. "CIMOSA: Architecture for Enterprise Integration." *Journal of Computer Integrated Manufacturing Systems* 3(1), 317–324.

Lartim, D. J. 1989. "Cell Control: What We Have, What We'll Need." *Manufacturing Engineering* (January), 41–48.

Murgatroyd, I. S., J. M. Edwards, and R. H. Weston. 1993. "Tools to Support the Design of Integrated Manufacturing Systems—An Object Oriented Approach." In *Proceedings of the International Conference on Industrial Engineering and Production Management,* Fucam (Catholic University of Mons), Mons, Belgium. 413–422.

Nagel, R., and R. Dove. 1992. *21st Century Manufacturing Enterprise Strategy.* Vol. 2. Bethlehem: Iacocca Institute, Lehigh University.

Nof, S. Y. 1994. "Critiquing the Potential of Object Orientation in Manufacturing." *International Journal of Computer Integrated Manufacturing* 7(1), 3–16.

Object Management Group OMG. 1996. "Product Data Management Enablers." Manufacturing DTF RFP-1, Version 1.2-MFG/96.05.01, Request for Proposal, May.

Prins, R. 1996. *Developing Business Objects: A Framework Driven Approach.* New York: McGraw-Hill. 294–295.

Shorter, D. N. 1994. "An Evaluation of CIM Modeling Constructs." Evaluation Report of Constructs and Views According to ENV40003. *Computers in Industry* **24,** 159–236.

Van Der Pluym, B. 1990. "Knowledge-Based Decision-Making for Job-Shop Scheduling." *International Journal of Computer Integrated Manufacturing* **3**(6), 354–363.

Weston, R. H. 1994. Model Driven Configuration of Manufacturing Systems in Support of the Dynamic, Virtual Enterprise." In *Advances in Concurrent Engineering-CE96, Proceedings of CE '96 Conference,* Toronto, August. Lancaster: Technomic.

Weston, R. H., and P. J. Gilders. 1995. "Enterprise Engineering Methods and Tools Which Facilitate Simulation, Emulation and Enactment via Formal Models." In *Working Conference on Models and Methodologies for Enterprise Integration (E195), IFIP TC5 Special Interest Group on Architectures for Enterprise Integration,* Heron Island, Australia, November 1–16.

PART 6

APPLICATIONS: PLANNING TECHNIQUES

CHAPTER 29

PRODUCT DESIGN AND PRODUCTION PLANNING

William R. Tanner
Productivity Systems, Inc.
Farmington, Michigan

1 INTRODUCTION

Industrial robots are generally applied under one of two sets of circumstances. The first is the situation involving a new facility, process, or product; here robots are incorporated into the initial plans and are implemented routinely along with other equipment and facilities. The second, more common, situation involves the application of robots to existing processes and operations, often in response to management direction or upon a suggestion from a supplier of robotic equipment. Here the robot must be integrated into ongoing operations, and changes to product, process, equipment, or facility which may be necessary are often difficult to accomplish.

To assure success in either case, the application of industrial robots must be approached in a systematic manner. Launching a robotic production system is best done in a multistep process that involves not only the robot, but also the product, production equipment, layout, scheduling, material flow, and a number of other related factors. Where robots are being integrated into existing operations, there are five discrete steps in this process (Tanner and Spiotta, 1980; Estes, 1979):

1. Initial survey
2. Qualification
3. Selection
4. Engineering
5. Implementation

Where robots are incorporated into a new operation, the first three steps are not specifically followed. This chapter addresses the first four of these steps. Emphasis is on practical, simple-to-use rules and principles. Readers are referred to other chapters in this section for more detailed techniques.

2 INITIAL SURVEY

The selection of the manufacturing process or processes to which a robot is to be applied should not be done arbitrarily. It requires the careful identification and consideration of all potential operations and begins with an initial survey of the entire manufacturing facility involved. The objective of the initial survey is to generate a "shopping list" of opportunities, that is, operations to which robots might be applied.

During the initial survey, one should look for tasks that meet the following criteria:

1. An operation under consideration must be physically possible for a robot.
2. An operation under consideration must not require judgment by a robot.
3. An operation under consideration must justify the use of a robot.

At this point in the process, a detailed analysis of each operation relative to these three criteria should not be undertaken; rather, a few simple rules of thumb should be applied.

Handbook of Industrial Robotics, Second Edition, Edited by Shimon Y. Nof
ISBN 0-471-17783-0   © 1999 John Wiley & Sons, Inc.

With regard to physical constraints, *generally avoid* operations where:

Cycle time is less than 5 sec.
Working volume exceeds 30 m³.
Load to be handled exceeds 500 kg.
Positioning precision must be better than ±0.1 mm.
"Randomness" in workpiece position and orientation cannot be eliminated.
"Randomness" in process cannot be eliminated.
Work lot size is typically less than 25 pieces.
Number of different workpieces per process is typically greater than 10.

With regard to judgment requirements, *generally avoid* operations where:

The robot's alternative actions cannot be readily identified.
The robot may be required to execute, at random, any one of more than five alternative actions.
Specific, quantified workpiece and process standards do not exist.
The critical properties of the workpiece and the process cannot be measured.
Workpiece identification, condition, and orientation may be ambiguous.

With regard to justification, the economic attractiveness of a potential robot application, as measured by return on capital or by payback period, is usually of primary importance. (See Ch. 34, Justification of Robotics Systems.) During the initial survey, value judgments regarding the relative merits of potential applications should be avoided. The purpose of this step is to develop objectively a list of opportunities that are technically and economically feasible and that will next be screened and prioritized.

3 QUALIFICATION

The second step in the launching of a robotic production system is the qualification of the operations identified in the first step as potential robot applications. Although some screening was done in that step, there are likely to be operations on the list that are not, upon further scrutiny, technically or economically feasible for robotics. Also, all operations on the list will not be of equal importance or complexity, nor can robot production systems be implemented on all of them simultaneously; thus the qualification step will also involve prioritizing the qualified applications.

Qualification and prioritization will be an iterative process. The first element of the process involves the review of each listed operation to answer the basic question, "Can I use a robot?" There are seven factors that should be considered at this time in deciding whether or not a potential exists to apply a robot on that particular operation:

1. Complexity of the operation
2. Degree of disorder
3. Production rate
4. Production volume
5. Justification
6. Long-term potential
7. Acceptance

For each of these factors a simple rule has been presented. Reviewing each operation on the list, one finds that the process will eliminate those where a robot should not be used unless the rules given can all be clearly applied to the operation in question.

Regarding complexity, although simple robots exist and are well suited to simple tasks, there are operations where a cylinder, a valve, and a couple of limit switches are sufficient. In other cases a gravity chute may suffice to transfer and even reorient a part from one location and attitude to another.

At the other end of the scale, operations that require judgment or qualitative evaluation should be avoided. Checking and accepting or rejecting parts on the basis of a measurable standard can be done with a robot. If, however, the only feasible measuring system is human sight or touch, then a robot is out of the question.

In the same vein, operations that involve a combination of sensory perception and manipulation should also be avoided. An example would be a machine tool loading operation that requires that the part being loaded into the chuck be rotated until engagement of a notch with a key is felt, after which the part is fully inserted. While the development of a "hand" for the robot capable of doing this is technically feasible, the complexity of the hand and its potential unreliability will certainly reduce the overall probability of success. Of even greater complexity are operations requiring visual determination of random spindle orientation and orienting the part to match.

The rule to apply here is:

Avoid both extremes of complexity.

Robots cannot operate effectively in a disorderly environment. Parts to be handled or worked on must be in a known place and have a known orientation. For a simple robot, this must be always the same position and attitude. For a more complex robot, parts might be presented in an array; however, the overall position and orientation of the array must always be the same. On a conveyor, part position and orientation must be the same, and conveyor speed must be known.

Sensor-equipped robots (vision, touch) can tolerate some degree of disorder; however, there are definite limitations to the adaptability of such robots today. A vision system, for example, enables a robot to locate a part on a conveyor belt and to position its arm and orient its hand to grasp the part properly. It will not, however, enable a robot to quickly remove a part, correctly oriented, from a bin of parts or from a group of overlapping parts on a conveyor belt.

A touch sensor enables a robot to find the top part on a stack. It does not, however, direct the robot to the same place on each part if the stack is not uniform or is not always in the same position relative to the robot.

The rule to apply here is:

Repeatability is necessary, disorder must be eliminated or made unambiguous.

When we consider production rate and cycle time, we find that small nonservo (pick-and-place) robots can operate at relatively high speeds. There is a limit, though, on the capability of even these devices. A typical pick-and-place cycle takes several seconds. A rate requiring pickup, transfer, and placement of a part in less than about 5 sec cannot be consistently supported by a robot.

Operations that require more complex manipulation or involve parts weighing pounds rather than ounces require even more time with a robot. The larger servo-controlled robots are able to move at speeds up to 1300 mm/sec; however, as speeds are increased, positioning repeatability tends to decrease. For a rough estimate of cycle time, 1 sec per move or major change of part orientation should be allowed. In addition, at least a half-second should be allowed at each end of the path to assure repeatable positioning. In handling a part, allow another half-second each time the gripper or handling device is actuated. (More accurate techniques are covered in Ch. 32, Robot Ergonomics: Optimizing Robot Work.)

The rule to apply here is:

Robots are generally no faster than people, as measured by cycle time; however, robots maintain the same pace whereas people do not.

There are two factors related to production volumes to consider. In batch manufacturing the typical batch size must be considered. In single-part volume manufacturing the overall length of the production run is important.

In small-batch manufacturing, changeover time is significant. Recalling that a robot needs an orderly environment, we find that part orienting and locating devices may need to be changed or adjusted before each new batch is run. A robot's end-of-arm tooling and program may also have to be changed for each new batch of parts. Generally, people do not require precise part location—their hands are instantly adaptable and their "reprogramming" is intuitive. A robot becomes impractical when its changeover time from batch to batch approaches 10% of the total time required to manufacture the batch of parts.

If a single part is to be manufactured at high annual volumes for a number of years, special-purpose automation should be considered as an alternative to robots. Per operation, a special-purpose device is probably less costly than the more flexible, programmable automation device, or robot. Single-function, special-purpose devices may also be faster and more accurate than robots. Where flexibility is required or obsolescence is likely, robots should be considered; where these are not factors, special-purpose automation may be more efficient and cost-effective.

The rule to apply here is:

For very short runs (about 25 pieces or less), use people; for very long runs (several million per year of a single part), use special-purpose automation; use robots in between.

The application of an industrial robot can represent a significant investment in capital and in effort. Economic justification must therefore be carefully considered: on the balance sheet, increased productivity; reduced scrap losses or rework costs; labor cost reduction; improved quality; improvement in working conditions; avoiding human exposure to hazardous, unhealthy, or unpleasant environments; and reduction of indirect costs are among the plus factors. Offsetting factors include capital investment; facility, tooling, and rearrangement costs; operating expenses and maintenance cost; special tools, test equipment, and spare parts; and cost of downtime or backup expense.

Ballpark estimates of the potential costs and savings should be made whether or not a reasonable return on the investment can be expected. The savings can be roughly estimated by multiplying the number of direct labor heads displaced per shift, times the number of production shifts per day, times the fully burdened annual wage rate. The costs can be roughly estimated by multiplying the basic cost of the robot planned for the operation times 2.5.

"Management direction," "following the crowd," and emotion are no substitutes for economic justification and, in the long run, will not support the application of robots. In some cases, safety or working conditions may override economics; however, these are usually exceptional circumstances.

The rule to apply here is:

If ballpark costs do not exceed ballpark savings by more than a factor of 2, the application can probably be economically justified.

Another consideration is the long-term potential for industrial robots in the particular facility. Both the number of potential applications and their expected duration must be taken into account.

Because of its flexibility, a robot can usually be used on a new application if the original operation is discontinued. Since the useful life of a robot may be as long as 10 years, several such reassignments may be made. Unless the first application of the robot is to be of relatively long duration, it is possible that reapplication must be considered. In the process of justifying the initial investment, the cost of reapplying the robot should also be included. If the initial application is of significantly shorter duration than the robot's useful life and no follow-on applications can be foreseen, it can seldom be justified.

As with any electromechanical device, an industrial robot requires some special knowledge and skills to program, operate, and maintain. An inventory of spare parts should be kept on hand. Auxiliary equipment for programming and maintenance or repair may also be required. Training of personnel, spare parts inventory, special tools, test equipment, and the like may represent a sizeable investment. The difference between the amount invested in these items to support a single robot or to support half a dozen or more robots is insignificant.

Maintenance and programming skills and reaction time in case of problems tend to deteriorate without use. Few opportunities will normally arise to exercise these skills in support of a single robot. Under these conditions, the abilities may eventually be lost and any serious difficulty with the robot may then result in its removal.

The rule to apply here is:

If there are no feasible opportunities for more than one robot installation, the single installation is seldom warranted; don't put just one robot into a plant.

Not everyone welcomes robots with open arms. Production workers are concerned with the possible loss of jobs. Factory management is concerned with the possible loss of production. Maintenance personnel are concerned with the new technology. Company management is concerned with effects on costs and profit. Collectively, all of these concerns may be reflected in a general attitude that "Robots are OK, but not here."

It is essential to know whether a robot will be given a fair chance. Reassignment of workers displaced by a robot can be disruptive. Training of personnel to program and maintain the robot can upset maintenance schedules and personnel assignments, and new skills may even have to be developed. The installation and startup can interrupt production schedules, as can occasional breakdowns of the robot or related equipment. Unless everyone involved is aware of these factors and is willing to accept them, the probability of success is poor.

The rule to apply here is:

A robot must be accepted by people, not only on general principles, but on the specific operation under consideration.

The foregoing screening process will, no doubt, eliminate a number of operations from the "shopping list." Those remaining should be operations that qualify as technically and economically feasible for the application of robots. These operations should now be prioritized, in preparation for the selection step. The prioritizing of operations and subsequent selection of an initial robot application can be facilitated by the use of an operation scoring system. The elements of the scoring system might include:

Complexity of the task

Complexity of end-of-arm tooling, part orienters, feeders, fixtures, and so on

Changes required to facilities and related equipment

Changes required to product and/or process

Frequency of changeovers, if any

Impact on related operations

Impact on workforce

Cost and savings potential

Anticipated duration of the operation

For each of the elements involved in the prioritization, a set of measures and a score range is established, with the more important elements having a higher range of points than the less important elements. A typical set of elements, measures, and score ranges is shown in Table 1.

Using this scoring system, each operation on the "shopping list" can be rated and prioritized; the operation with the highest score will be the prime candidate for the first application. Other factors, such as timing, management direction, experience, and human relations, might also be considered; however, subjectivity in establishing priorities should be minimized.

4   SELECTION

The third step in the launching of a robotic production system is the selection of the operation for which the robot will be implemented. If the initial survey was made with care, and if the qualification and prioritization have been objectively done, this step will be virtually automatic. If a scoring system is used in establishing priorities, then the selection should be made from among the two or three operations with the highest scores. Some consideration might be given to conditions and circumstances that were not measured in the second step; however, it is again important that subjectivity and arbitrariness be minimized.

It is imperative to review, on-site, the top few candidate operations before making the final selection to ascertain that they are, indeed, technically feasible and justifiable. For the first robot application, the cardinal rule is: *Keep it simple.* It is wise to forgo some degree of economic or other benefit for the sake of simplicity; even the most elementary of potential applications is likely to be more complex than anticipated. Avoid the temptation to solve a difficult technological problem; those opportunities will come with later installations. Resist the pressure of uninformed suggestions from managers or others trying to be helpful. Do not try to solve some production or manufacturing problem with the robot; if conventional approaches did not succeed, the robot is also likely to fail. Be aware of the robot's limitations; do not choose an application that requires 100% of some

Table 1   Scores for Ranking Robot Potential Applications

Element	Measured by	Score range
1. Complexity of task	Number of parts	1–10 to 5–5
	Number of operations	1–10 to 5–5
	Number of batches	1–5 to 5–1
2. Complexity of tooling and peripherals	Number of parts	1–10 to 5–5
	Part orientation at delivery	Single, oriented—10 Matrix—5 Bulk, random—0
	Ease of orienting parts	Easy—10 to difficult—5
3. Facility and equipment	Relocation required	No—2 Yes—0
	Utilities available	At site—5 Nearby—3 Not available—0
	Floor loading	Adequate—3 Need new—0
4. Product and/or process changes	Product changes required	No—10 to minor—5 to major-0
	Process changes required	No—10 to minor—5 to major—0
5. Impace on related operations	Synchronized with previous operation	No—3 Yes—0
	Synchronized with following operation	No—3 Yes—0
	"Bottleneck"	No—3 Yes—0
	Backup/buffer	Easy—3 Hard—1 No way—0
6. Impact on workforce	Monotonous, repetitious	Yes—3 No—0
	Bad environment	Yes—3 No—0
	Safety hazard	Yes—5 No—0
	Fast pace or heavy load	Yes—3 No—0
	Labor turnover	High—3 Low—1
7. Risk of unforeseen or random problems	Number of potential different occurrences	1–10 to 10–1
	Attitude/expectations of management	Understanding, reasonable—10 to tough, unrealistic—0
8. Potential benefits	Labor savings, per shift	One point per 0.1 person
	Production shifts per day	Five points per shift
	Quality improvement	Yes—5 No—0
	Productivity improvement	One point per percent increase
	Reduced repair and rework	Yes—5 No—0

capability, such as reach, load capacity, speed, or memory. Once the operation to which the robot will first be applied has been selected, the engineering of the application can begin. The prioritized, qualified "shopping list" should be retained as a source of further robot applications.

5 ENGINEERING

The fourth step in the launching of a robotic production system is the system engineering. There are a number of engineering activities involved, some of which must be done sequentially and some of which can be performed concurrently. The first activity is to return to the chosen workplace and thoroughly study the job to make sure that everything that must be done is identified and planned. During this study phase there are a number of considerations that must be addressed:

Alternatives to the robot—can the desired result be better accomplished by a special-purpose device, by basic facility changes, or by restructuring the operation?

Alternative robot attitudes—are there any advantages to mounting the robot in other than the usual feet-on-the-floor attitude, such as overhead?

Alternatives to existing process—are there advantages in reversing the usual "bring the tool to the work" approach and having the robot carry the work to the tool?

Backup—what arrangements, equipment, or actions will be taken to back up the robot during downtimes?

Environment—does the robot need special protection from excessive heat or cold, abrasive particulates, shock and vibration, fire, or explosion hazards, and so on?

Space—will the robot occupy significantly more space than the alternatives and, if so, what difficulties might this cause?

Layout—will the robot create problems with accessibility to it, other equipment, and the workplace for material handling, maintenance, inspection, and so on?

Safety—how will the installation be done so as to protect people from the robot and vice versa? Unusual, intermittent, random occurrences—have all the things that could possibly go wrong with the operation been anticipated, and have contingency plans been made for each?

The detailed study of the operation chosen has as its objective a thorough familiarity with the operational requirements, sequence, and pace, as well as with the occasional random disruptions that seem to occur in any process. Because the robot is not adept at handling disruptions or disturbances in its normal routine, approaches must be developed to minimize their occurrence and impact, to prevent damage when they happen, and to recover rapidly afterward. Significant data to be gathered about the operation include the following:

Number and description of elemental steps in the operation

Size, shape, weight, and the like of parts or tools handled

Part orientation at delivery, acquisition, in-process, and at disposal

Method and frequency of delivery and removal of parts

If batch production, lot sizes, characteristics of all parts in family, frequency of changes, and changeover time for related equipment

Production requirements per hour, day, and so on

Cycle times—floor to floor and elemental

Inspection requirements, defect disposition

At this point a layout drawing of the installation is made. Typically this starts with a scale layout of the existing area, onto which the robot and its work envelope are superimposed. Locations for incoming and outgoing material, buffers, and intermediate positions of parts, if necessary, are determined. From this layout, potential interference points can be located and equipment relocations, if any, can be developed. Sources and routing for utilities, such as electrical power, compressed air, and cooling water, are also shown on the layout.

Simultaneously with preparation of the layout, a detailed description of the robot's task is written. This task description must be broken down to a level comparable to the individual steps of the robot's program; it will, in fact, become the basic documentation of that program. Elemental times are estimated for each step so that an approximate cycle time can be established for the entire task.

Working with the layout and the task description, the robot's program is optimized. The objectives are to minimize the number of program steps and robot moves to attain the shortest cycle time. Often, rearranging incoming and outgoing material locations or even the positions of the equipment in the work station can significantly affect the cycle time. Thus both the layout and the task description are necessary elements of this process. Product and/or process changes may also be necessary or desirable. If so, these should be identified and described at this time, with actual engineering to follow.

Up to this point the final selection of a specific model of robot should not be made. Ideally several robot models will be capable of performing this operation, and alternative layouts should be made for each. The task description is basic to the process and should be common for all robot models considered. The selection of the desired robot is now made, based upon best fit to the layout; performance advantages, if any; price, delivery, support, and other similar considerations.

Once the robot model has been chosen and the layout has been optimized, personnel and equipment access points are determined and hazard-guarding (safety barrier) locations are established. It is necessary that an area encompassing the robot's entire working volume be guarded against accidental intrusion by people. Although some installations use active intrusion devices such as light curtains or safety mats interlocked with the robot's control to stop the robot when a person enters the area, passive systems such as fences, walls, and guard rails are more dependable.

Working with the task description, the interlocks between the robot and related equipment are determined. The robot will not directly control the other equipment, that is, the robot's controller will not directly operate other machines in the work cell. The robot will, however, initiate other machine cycles, and its operation will, in turn, be initiated by other machines or devices. For each of the different inputs and outputs, an I/O port on the robot control must be hard-wired to or from some other device. In more complex operations the robot control may not have sufficient input/output capacity, and an external I/O device such as a programmable controller may be required. When determining inputs and outputs, it is also important to consider the backup method to be used. If manual backup is to be employed, then a manual control station may also have to be provided. This control station must have a "manual/automatic" mode selector, which should be a lockable selection switch.

Following the task description, robot selection, and layout finalization, there are several engineering tasks to be performed. These should all be undertaken simultaneously because they are very much interdependent. These engineering tasks include the following:

End-of-arm tooling design

Parts feeders, orienters, and positioners design

Equipment modifications

Part (product) redesign

Process revisions

5.1 End-of-Arm Tooling

Typically, a robot is purchased without the end-of-arm tooling unique to its intended task. The robot supplier may furnish a "standard gripper" actuating mechanism, or a suitable device may be obtained from another source; however, adaptation of a standard mechanism may still require some design effort. Likewise, a standard power tool such as a screwdriver or grinder or a spray gun or welding torch that is to be mounted on the end of the robot arm will require the design of mounting hardware, such as brackets, adaptors, and so on. The lack of standard robot/tooling mechanical interfaces means that little "off-the-shelf" hardware is available.

End-of-arm devices lack the dexterity of a human hand; thus, in the case of batch manufacturing, several interchangeable tools may be required. A multifunctional tool for such tasks must represent a practical compromise between simplicity, for reliability, and flexibility, to perform a number of functions or handle a number of different parts. Interchangeable tools should be designed for ease of removal and installation and for repeatable, precise location on the robot arm to avoid the necessity to reprogram the robot with each tool change. In some cases, automatic exchange of tools by the robot may be possible through the use of quick-disconnects, collet/drawbar arrangements similar to preset machine tool holders, tool racks, and the like. More discussion on end-of-arm tooling is in Ch. 48.

5.2 Parts Feeders

Another engineering requirement may be for parts feeders, orienters, and positioners, or other parts-acquisition systems. As noted earlier, today's robots require an ordered, repeatable environment and cannot easily acquire randomly oriented parts delivered in bulk. There are several solutions to this problem, including trays or dunnage that contain parts in positive locations and orientations; mechanical feeder/orienter devices; manual transfer of parts from bulk containers into feeder systems; and sensor-based acquisition systems, such as vision or tactile sensing. Table 2 (revised from Warnecke and Schraft, 1982) summarizes typical mechanisms for part feeding and their functions.

The mechanically simplest approach is to use parts containers that retain individual parts in specific locations. A robot with microprocessor or computer control and sufficient

Table 2   Typical Part-Feeding Devices and Their Functions

Feeding device	Feeding functions[a]						
	Transfer	Order	Orient	Position	Metering	Binning	Magazining
Bins or hoppers							
Feed hopper	•				•	•	
Tote bin						•	
Storage conveyor	•					•	
Magazines							
Pallet magazine			•				•
Drum magazine	•		•				•
Spiral magazine	•		•				•
Channel magazine	•		•				•
Transfer devices							
Belt/roller conveyor	•						•
Vibratory conveyor	•						
Walking beam conveyor	•						
Rotary index table	•			•	•		•
Pick-and-place			•	•	•		
Ordering devices							
Vibratory bowl feeder	•	•				•	
Rotary feeder	•	•				•	
Elevating feeder conveyor	•	•				•	
Centerboard hopper feeder	•	•				•	
Metering devices							
Pop-up pusher	•				•		
Screw feeder	•				•		
Escapement	•	•			•	•	

[a] Definition of functions: (1) Transfer: movement of parts to feed point. (2) Order: bringing random parts to a predetermined position *and* orientation. (3) Orient: bringing parts in a known orientation or direction. (4) Position: placing a part in an exact position. (5) Metering: physical separation of parts at feed point. (6) Binning: random storage of parts for feeding. (7) Magazining: storage of parts in a definite orientation.

memory can be programmed to move to each location in the container, in sequence, to acquire a part. Multiple layers of parts may be packed in this manner, with the robot also programmed to remove empty trays or separators between layers. The only requirement in the workplace is to provide locators for repeatably positioning the containers. The multiple pickup points, in addition to requiring a computer control or large memory capacity, may increase the average cycle time for the operation. And, if the robot lacks the capability to acquire parts from a matrix array, some other approach must be taken.

Another parts-presentation approach is to use mechanical feeder/orienters. These, for small parts, may be centrifugal or vibratory feeders which automatically orient parts in a feeder track. Larger parts may be handled with hoppers and gravity chutes or elevating conveyors and chutes, which also present parts in proper orientation at a specific pickup point. Usually these devices are adaptations of standard, commercially available equipment. Advantages of this approach are that the single acquisition point for each part minimizes nonproductive motions, and the orienters can often present the parts in attitudes that require little manipulation by the robot after pickup. Disadvantages are difficulties in orienting and feeding some parts, relatively high cost of mechanical feeders, lack of flexibility to handle a variety of parts (as in batch manufacturing), potential damage to delicate, fragile, highly finished, or high-accuracy parts, and inability to handle large, heavy, or awkward-shaped objects.

A third approach is the manual transfer of parts from bulk containers to mechanical feeders such as gravity chutes or indexing conveyors. An obvious disadvantage of this approach is the use of manual labor, especially to perform the very sort of routine, nonrewarding tasks to which robots should be applied. Advantages are relatively low capital investment requirements, ability to handle difficult or critical parts, and flexibility to accommodate a variety of similar parts, as in batch manufacturing.

A fourth approach is the use of sensors such as vision or tactile feedback devices to modify the robot's programmed motions, enabling it to acquire somewhat randomly oriented parts. Advantages are a reduction in the extent of mechanical orientation required and a potential to work with a variety of randomly mixed parts or to accommodate batch manufacturing lot changes with few or no physical changes required. Disadvantages are relatively high cost, compared to simple mechanical feeder/orienters; the possible need for special lighting; relatively slow processing time; and difficulty with touching or overlapping parts or with three-dimensional space (such as bins). The solutions to parts presentation for the robot often combine all of the approaches described, as well as others, such as automatic, single-part-at-a-time delivery from a previous operation by means of a conveyor or shuttle device.

5.3 Equipment Modification

A third engineering requirement may be the modification of existing equipment with which the robot will operate. Typical modifications may include adding a cylinder and solenoid valve to a machine-tool splash guard for automatic, rather than manual, opening and closing. Other changes may be made to guards and housings for improved access by the robot. Machine-tool chucks and collets may be modified to increase clearances or to provide leads or chamfers for easier insertion of parts. Powered clamping and shuttle devices may be substituted for manually actuated mechanisms. In machine-tool operations coolant/cutting fluid systems may be changed or chip blow-off systems added to automatically remove cuttings (chips) from the work and work holders. Assembly operations may require the development of simple jigs and fixtures in which to place parts during the process (remember that robots are generally single-handed devices and cannot hold something in one hand while adding components to it with the other). Likewise, manual tools such as screwdrivers and wrenches will have to be replaced with automatic power tools.

5.4 Part/Product Redesign

Part orienting and feeding and/or part handling by the robot's end-of-arm tooling may require some redesign of a product (Bailey, 1983). Ideally, the product should be designed so that it has only one steady-state orientation, that is, it should be self-orienting. As an alternative, the product should be designed so that its orientation for acquisition is not critical (for example, a flat disk or washer shape). A family of parts which are all to be handled by the robot should have some common feature by which they are grasped; this feature should be of the same size and in the same location on all products in the family.

Vacuum pickups are simple, fast, and inexpensive. Product designs that incorporate surfaces or features to which a vacuum pickup can be applied facilitate easier handling by the robot. Products should be designed so that the robot's task (such as load, unload, insert, and assemble) requires a minimum of discrete motions; complex motions, especially those that require the coordinated movements of two or more robot axes (such as a helical movement of the part), should be avoided. Tolerances should be "opened up" as much as possible. Chamfers should be provided on inserted parts to aid in alignment. Parts should be self-aligning or self-locating, if possible. Parts that are to be mechanically or gravity oriented and fed to the robot should be designed so that they do not jam, tangle, or overlap.

Because product redesign is costly and time-consuming, it should not be undertaken lightly, but should be considered only when its potential benefits significantly outweigh its cost. In the design of new products, however, incorporation of features that facilitate the use of robots should add little or nothing to the cost and should, thus, be encouraged. In Table 3 (developed from material in Boothroyd et al., 1994, and in Nof et al., 1997) and Figure 1 rules and principles are provided for design of parts and products for robotic assembly. Such rules can guide designers in the design of parts for robotic handling.

5.5 Process Revisions

Another engineering requirement may be the modification of the process with which the robot is involved. Process revision may include changing an operational sequence so that critical part orientation is not required. Process revision may involve moving several machines into an area and setting up a machining cell to take advantage of initial part orienting and to increase the robot's utilization. It may involve linking of operations with conveyors so as to retain part orientation for the robot, or the incorporation of compartmentalized pallets or dunnage to retain orientation between operations. Process revision may involve the rescheduling of batch operations to increase lot size or to minimize changeover between batches.

Process revisions can often be accomplished at minimal cost, particularly those that involve only scheduling, and can sometimes significantly increase the efficiency of the robotic production system. Like product changes, process revisions should not be undertaken lightly, however, but should be carefully examined for cost-effectiveness. In Table 4 guidelines are provided for revisions in assembly process for robotics applications.

6 SUMMARY AND EMERGING TRENDS

The first four steps of launching a robotic production system, *initial survey, qualification, selection,* and *engineering,* which have been described in this chapter, should, if followed

---

Table 3 Part/Product Redesign for Robotic Assembly

A. *Rules for Product Design*

  1. Minimize the number of parts.
  2. Product must have suitable base part on which to build.
  3. Base part should ensure precise, stable positioning in horizontal plane.
  4. If possible, assembly should be done in layers, from above.
  5. Provide chamfers or tapers where possible to aid correct guidance and positioning.
  6. Avoid expensive and time-consuming fastening operations, e.g., screwing, soldering.

B. *Rules for Part Design*

  1. Avoid projections, holes, or slots that may cause entanglement in feeders.
  2. Symmetrical parts are preferred because less orientation is needed and feeding is more efficient.
  3. If symmetry cannot be achieved, design appropriate asymetrical features to aid part orienting.

C. *Design Features That Determine Cost-Effectiveness of Robotic Assembly*

  1. Frequency of simultaneous operations.
  2. Orienting efficiency.
  3. Feeder required.
  4. Maximum feed rate possible.
  5. Difficulty rating for automatic handling.
  6. Difficulty rating for insertions required.
  7. Assembly operations required (number and type).
  8. Total number of component parts per assembly.

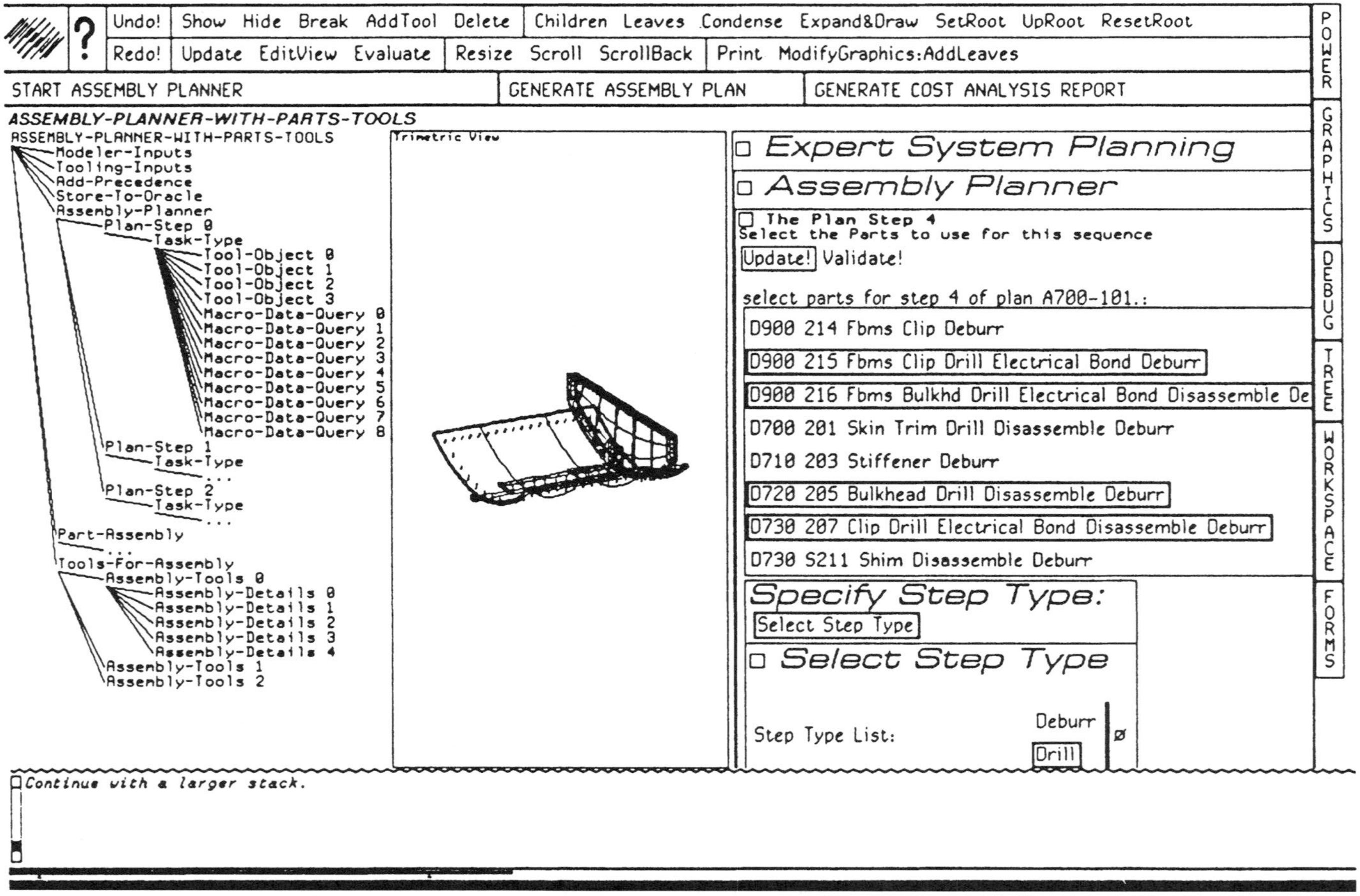

Figure 1 Product design and production planning can be done with knowledge-based systems, as in the following example: The automated assembly planner for airframe subassemblies includes CAD analysis, cost and time evaluation, assembly instruction generation and down-loading capability to shop machines (courtesy of Northrop Corp.).

Table 4   Process Revision for Robotic Assembly

Issue	Guidelines
1. Part handling	Provide component features, such as tapers and chamfers, lips, leads that make parts self-guide, self-align and self-locate readily, so that less accuracy is required during part handling.
2. Self-locating	Ensure that parts are self-locating if they are not secured immediately after insertion.
3. Motion economy	Apply robot motion economy principles, including:
(a) grippers	Minimize the number of grippers needed, and consider adding a simple feature for easier, more secure gripping.
(b) simpler robots	Minimize the complexity of required robots. **Reduce** (1) number of arms and joints (determined by the number of necessary orientations); (2) dimensions (determined by where robots must reach); (3) loads to be carried. **Result:** simpler robots, less energy use, simpler maintenance, smaller workspace.
(c) motion path	Simplify required motion paths; point-to-point motions require simpler position and velocity control compared with continuous path control.
(d) sensors	Do not apply sensors, but if needed, simpler touch and force sensors are preferred to vision systems; minimize the number and complexity of sensors to reduce hardware and operation cost.
(e) given abilities	Utilize robot abilities already determined as justified. Example: a sensor justified for one function can improve other tasks.

[a] From S. Y. Nof, W. E. Wilhelm, and H. J. Warnecke, *Industrial Assembly.* London: Chapman & Hall, 1997.

carefully and thoroughly, make the fifth and last step, *implementation,* relatively easy and trouble-free.

Computer-aided design tools and knowledge-based process planning (Chang, 1990; Swift, 1987) offer designers more advanced techniques to accomplish the basic steps described in this chapter. Advanced applications for product design and production planning for robotics include operations scheduling (Jiang, Senevirante, and Earles, 1997), and design is based on process modeling (Guvenc and Srinivasan, 1997).

REFERENCES

Bailey. J R. 1983. "Product Design for Robotic Assembly." In *Proceedings of the 13th International Symposium on Industrial Robots.* Chicago, April. 1144–1157.

Boothroyd, G., P. Dewhurst, and W. Knight. 1994. *Product Design for Manufacture and Assembly.* New York: Marcel Dekker.

Chang, T. C. 1990. *Expert Process Planning for Manufacturing,* Reading: Addison-Wesley.

Estes, V. E. 1979. "An Organized Approach to Implementing Robots." In *Proceedings of the 16th Annual Meeting of the Numerical Control Society,* Los Angeles, March. 287–307. [Describes the approach taken by General Electric Consulting Services to implement robotics successfully in the company.]

Guvenc, L., and K. Srinivasan. 1997. "An Overview of Robot-Assisted Die and Mold Polishing with Emphasis on Process Modeling." *Journal of Manufacturing Systems* 16(1), 48–58.

Jiang, K., L. D. Senevirante, and S. W. E. Earles. 1997. "Assembly Scheduling for an Integrated Two-Robot Workcell." *Journal of Robotics and CIM* 13(2), 131–143.

Nof, S. Y., W. E. Wilhelm, and H. J. Warnecke. 1997. *Industrial Assembly.* London: Chapman & Hall.

Swift, K. 1987. *Knowledge-Based Design for Manufacture.* London: Kogan Page.

Tanner, W. R., and R. H. Spiotta. 1980. "Industrial Robots Today." *Machine Tool Blue Book* **75**(3), 58–75. [Analyzes applications in which robots can be efficient and economical.]

Warnecke, H. J., and R. D. Schraft. 1982. *Industrial Robots Application Experience.* IFS Publications.

CHAPTER 30

OPERATIONS RESEARCH TECHNIQUES FOR ROBOTICS SYSTEMS

Nicholas G. Hall
The Ohio State University
Columbus, Ohio

1 INTRODUCTION

1.1 Scope of the Chapter

The scientific field of *operations research,* also known as *management science,* has much to offer towards the solution of complex planning problems such as those that arise in robotic systems. This chapter discusses a variety of operations research techniques, with the aim of enabling both industrial and academic readers to establish connections between those techniques and robotic systems applications. An overview of the most important techniques is provided, along with enough details to enable the reader to assess the potential relevance and applicability of the technique in each case. Many related references that give further details are provided.

The chapter is written with the expectation that the reader is either a system manager or controller involved in implementing or operating a robotic system, or a researcher familiar with operations research techniques and hoping to apply them to a robotic system in a way that would be potentially valuable to the manager or controller.

The use of operations research techniques for the planning, design, and operation of robotic systems is valuable because a robotic facility is a complex system. Even a small facility typically consists of many interconnected hardware and software components and involves the use of various limited resources such as buffer space, part feeders or orienters, and material-handling equipment. The design and operation of such a system are difficult tasks because of the many available choices and also because of the interactions between the different components. Complex choices of this type can often be evaluated and resolved more effectively by using the powerful techniques of operations research than by any other strategy. This suggests the importance of operations research as a topic in a study of industrial robotics.

Space considerations dictate the extent to which each technique can be discussed. In general, widely used operations research techniques are discussed briefly, and their use for robotic systems is illustrated with supporting references. However, where the study of robotic systems has generated important new problems or techniques, a more detailed discussion is provided.

1.2 Operations Research

Examples of the use of a scientific approach to the management of organizations can be found as far back in time as the late nineteenth century. However, it is generally agreed that the origins of operations research as a coherent body of knowledge can be traced to military support activities during World War II. A lack of critical resources during wartime, and the urgency of the war effort, motivated both U.K. and U.S. military leaders to encourage the development and use of scientific management techniques for solving complex logistical problems. To this end, teams of scientists in both countries were formed

Handbook of Industrial Robotics, Second Edition, Edited by Shimon Y. Nof
ISBN 0-471-17783-0 © 1999 John Wiley & Sons, Inc.

to study strategic and tactical operations and to improve their efficiency. Operations research has been credited with making major contributions to the war effort.

After the war ended, operations research spread quickly into the private sector, where it was applied with considerable success to complex planning problems. The usefulness of operations research was greatly enhanced by significant scientific advances (for example, the development of the simplex method for linear programming by G. Dantzig in 1947). This usefulness was further enhanced by the development of high-speed digital computing, which made possible the implementation of operations research techniques for much larger and more complex planning problems in industry. By the early 1950s the use of operations research by technical specialists could be frequently observed in manufacturing and service industries as well as in government planning problems. Within universities both business and engineering schools developed programs in operations research to support the growing interest in industry. Related programs also developed in mathematics and computer science departments. Finally, the growth of personal computing and the widespread development of affordable and easy-to-use software that occurred in the 1980s and 1990s have given access to operations research techniques, not just to technical specialists but to the entire industrial and commercial communities.

A major source of information about operations research is the continuing series *Handbooks in Operations Research* (1989–). A comprehensive introductory text on operations research is provided by Hillier and Lieberman (1995). This book also provides references to classic works in each general topic area within operations research. An interesting history of the important topic area of *mathematical programming* is provided by Lenstra, Rinnooy Kan, and Schrijver (1991). For a comprehensive treatment of the operations research techniques underlying *production and manufacturing* decisions, the text by Nahmias (1993) is highly recommended. The professional society that promotes operations research in the United States is INFORMS, which has over 11,000 members and a useful homepage at "www.informs.org". A valuable compendium of related information appears in the 1997 OR/MS Resource Directory. Finally, the definitive on-line source for operations research information, including people, educational programs, research groups, companies, and many electronic links, is the homepage of M. Trick at "http://mat.gsia.cmu.edu".

1.3 Phases of the Robotic System Design Process

In the planning, implementation, and operation of a robotic system, an organization either explicitly or implicitly goes through all the following distinct phases of activity:

1. *System planning.* This involves a feasibility study and preliminary system design, usually with the consideration of several alternatives, as well as preliminary economic evaluations.

2. *System design.* This involves deciding which operations will be performed by the system, selecting equipment, and deciding how it is to be configured. This stage is critical to the eventual performance of the system, since the choices made here limit the options that are available under the later system operation phase.

3. *System operation.* The system should be operated in an efficient way that satisfies production targets and other management criteria (for example, reduction of work-in-process). It should be maintained with minimum disruption to ongoing processes. Also, the performance of the system should be regularly evaluated, and potential modifications, upgrades, and expansion should be considered.

In each of these activity phases appropriately used operations research techniques can greatly assist the decision-maker. Details of the system planning, design, and operation issues, as well as suitable operations research techniques for evaluating them, are discussed in Sections 2 through 4, respectively, of this chapter.

1.4 Choosing an Operations Research Technique

Operations research has developed and now offers a wide variety of techniques. These techniques differ considerably in their attributes and in the type of evaluative and prescriptive information they provide. Most operations research techniques construct an abstract representation, or *model,* of a decision problem, which is then solved using computer software. There is an important general tradeoff, discussed later in this subsection, with respect to the solvability and realism of a model.

To give the reader an overview of the various available operations research techniques, we make use of a simple classification scheme based on two attributes of those techniques. The first attribute describes whether the technique is based on a *deterministic* or *probabilistic* model. In deterministic models the assumption is made that all relevant data are known with certainty. However, deterministic models can be used to develop good approximations for problems in which uncertainty is present in a limited way. For example, some redundancy or estimated "safety factor" can be built into the model to allow for uncertainty. By contrast, in probabilistic models uncertainties that arise in a problem are explicitly incorporated into a mathematical model that represents that problem. The second attribute describes whether the technique uses a *static* or *dynamic* model. Static models look at a snapshot of a system's performance or at an aggregate measure that summarizes performance over a period of time. By contrast, dynamic techniques explicitly study the behavior of the system at various points in time. In the case of probabilistic models that use an aggregate measure of long-run system performance, the term *static* is replaced by *steady-state*.

Given these attributes, we can classify operations research techniques as shown in Table 1. In Table 1 the reader can find any technique discussed in the remainder of this chapter as well as a reference to the section where it is discussed. It is thus easy to understand how the important issues of uncertainty and time horizon can be modeled by that technique. Some techniques may be used with various attributes, in which case they are located in Table 1 according to their most typical use. However, in the case of two techniques a * denotes frequent use under two different sets of attributes.

In choosing an operations research technique an important criterion is finding the right level of complexity. More complex models have a greater ability to capture the full details of a decision problem, but they tend to be harder to solve than simpler models. The choice of an appropriate level of complexity may also depend upon what phase of the design process is being considered. For example, early stages of the design process may require only approximate comparisons of the performance of different designs. However, in planning detailed operations it may be useful to *optimize* the system, which requires a model that captures all or almost all of the complexities of the real-world problem. Such a model may have many variables and constraints, and finding optimal solutions to the model may present a difficult computational challenge that requires expensive hardware and software. Moreover, gathering data and formatting the input information may be time-consuming and costly. Therefore, the costs involved in the optimization effort should be

Table 1 Classification of Operations Research Techniques

Attribute 1	Attribute 2	
	Static / Steady-State	Dynamic
Deterministic	integer and mixed integer programming (2.3.1)	location and layout* (2.3.5)
	linear programming (2.3.2)	scheduling* (4.3.2)
	analytic hierarchy process (2.3.3)	
	heuristics (2.3.4)	
	location and layout* (2.3.5)	
	group technology and cell formation (3.3.1)	
	batching (4.3.1)	
	scheduling* (4.3.2)	
Probabilistic	decision analysis (2.3)	strategic planning (2.3)
	queueing models (3.3.2)	forecasting (2.3)
	queueing networks (3.3.3)	simulation (3.3.4)
		reliability and maintenance (3.3.5)
		hierarchical production planning (4.3.3)

weighed against the potential benefits to be gained from the use of improved operating procedures that may be identified by the model.

1.5   Structure of the Chapter

The chapter is organized according to the phases of system planning, design, and operation identified in Section 1.3. Each of the next three Sections (2 through 4) considers one of those phases. For each phase we begin with a discussion of the most important issues and requirements that need to be considered. This discussion is followed by a summary of several performance measures that might be appropriate for that phase of the overall process. Then an overview is presented of the operations research techniques that are potentially useful for problems arising in that phase, along with many references giving examples and additional information. Finally, Section 5 places the various uses of operations research techniques for robotic systems planning, design, and operation within an overall model of organizational decision-making and discusses the importance of decision support systems.

2   SYSTEM PLANNING

This section considers the preliminary system planning phase. Here the concern is to evaluate the feasibility of a robotic system by developing initial estimates of cost, staffing and training requirements, floor space requirement, return on investment, and other measures of importance to management.

2.1   Issues and Requirements

In order to establish guidelines for the scope of the planning study, the issues and requirements that are most important to management should be identified before the study is initiated. Typically those issues and requirements include the following:

1. *Management objectives.* The purposes of considering the adoption of a robotic system need to be identified. The expected benefits should be listed. It is important to estimate the costs of disruption to existing processes resulting from the new system. All parties likely to be affected by the new system must be consulted. The expected benefits of a robotic system typically include improved productivity, reduced labor costs, improved worker safety, improved product quality, and increased product flexibility. However, in any given application there may be some benefits that are predictable in advance and others that can only be evaluated once the system is fully operational. Estimates for the latter should be included in preliminary evaluation models.

2. *Location and layout.* The distinction between location and layout issues is that location relates to the placement of the plant itself, whereas layout relates to the positioning of a robotic system within an existing plant. In location problems proximity to sources and to markets is important, as are cost factors. The layout decision attempts to make material flow as efficient as possible. If a high priority placement is given to the robotic system, it is important to ensure that the flow of material to other systems within the plant is not significantly disrupted.

3. *Operations and equipment to be automated.* The operations that will be considered for automation should be identified. The use of group technology and cellular manufacturing principles should be evaluated. Robot-served cells can be organized by product (i.e., all machines that are used to produce a given product or family of products) or by process (i.e., all products processed on a given machine). While it is not necessary at the planning stage to finalize these decisions in detail, group technology principles (see Section 3.3.1) are available to develop approximate scenarios.

4. *Resource requirements.* The major resource requirements to be considered are floor space, material flow capacity, capital cost, and adequately trained personnel. In order to assess the availability of these resources, estimates of the required resources should be prepared.

The purpose of identifying these issues and requirements is to focus the preliminary design study on the decisions and options that are most important. If this is done, the study becomes much simpler and consequently the variety of operations research techniques that can be applied increases. Since different techniques provide substantially different evaluative and prescriptive information, it is useful to have a range of these techniques available.

Given these general guidelines, specific requirements of this preliminary phase are typically to identify the following:

1. One or more alternative systems that can meet management objectives
2. For each alternative, estimates of performance indicators
3. For each alternative, the equipment and the operations to be automated

2.2   Performance Measures

In order to make relevant comparisons between current technology and available alternatives for robotic systems, several performance measures are needed. The following are typically useful:

1. *Amount of investment.* The total capital cost and annual incremental operating cost of the new system must be estimated.
2. *Return on investment.* The return on investment, the net present value, and the payback period must be estimated. These measures and a variety of others are discussed by Canada and Sullivan (1989).
3. *Flexibility.* It is important to assess how easily short-term changes in production requirements can be handled by the new alternatives. The potential for long-run capacity expansion and for adapting to fundamental shifts in product design are relevant. Also, the impact of significant reductions in product life cycles on the system is useful to know.

It should be emphasized that these questions are relative and not absolute. That is, they should be used only in the context of a comparison between the current and future alternatives. Useful references in the area of capacity expansion include the books by Manne (1967) and Freidenfelds (1981). Models for technology acquisition are described by McCardle (1985), Monahan and Smunt (1989), and Fine and Freund (1990). A sophisticated multiple objective decision model for evaluating advanced manufacturing systems is provided by Demmel and Askin (1992). An extensive review of the literature related to the selection of industrial robots, and an overview of various modeling approaches to the selection problem, are provided by Khouja and Offodile (1994). A decision support system that combines optimization and simulation tools to evaluate a computer-integrated manufacturing system is described by Kassicieh, Ravinder, and Yourstone (1997).

2.3   Operations Research Techniques for System Planning

We begin with three examples of general planning techniques that can be applied to robotic systems. Since applications of these techniques to robotic systems are similar to well-documented applications elsewhere, we provide only brief comments and several references containing additional information. We then discuss in greater detail several techniques that can have a major impact on the planning of a robotic system.

1. *Strategic Planning.* Strategic planning is usually associated with long-term decisions. The purchase and implementation of an expensive new robotic system is a long-term commitment. Valuable references on the topic of strategic planning generally include Porter (1980), Hayes and Wheelwright (1984), and Cohen and Zysman (1987). As noted by Skinner (1974), changes in the manufacturing task can be a major source of inconsistency between company policies in different functional areas. Goldhar and Jelinek (1983) discuss arguments in favor of new manufacturing technology.
2. *Forecasting.* A natural use of forecasting to aid in planning a robotic system is to predict future demand for the products being considered for automated production. This is important because the benefits of a robotic system will depend upon the production requirements for its products and because of the considerable cost of such a system. A variety of time series and other forecasting techniques are available. Useful general references for forecasting include Box and Jenkins (1970), Montgomery and Johnson (1976), and Gross and Peterson (1983). An interesting review of forecasting practices in industry is provided by Sanders and Manrodt

(1994). Yurkiewicz (1996) presents a thorough survey of commercially available forecasting software.

3. *Decision Analysis.* Decision analysis (which includes decision trees) is a technique that has direct application to robotic systems. Uncertainties involving, for example, market strength or competitors' reactions can be included in a model to evaluate which of several robotic systems to build. Other relevant uses of decision analysis involve planning where to locate a new robotic facility when the location of future demands is unknown. Good general references for decision analysis include Keeney and Raiffa (1976), Bunn (1984), and Goodwin and Wright (1991). Fishburn (1989) provides an interesting perspective on the historical development of decision theory and related topics. Mehrez, Offodile, and Ahn (1995) use decision analysis to study the impact of a robot's *repeatability,* i.e., physical precision over time, on output rate and quality. Buede (1996) presents a thorough survey of commercially available decision analysis software.

2.3.1 Integer and Mixed Integer Programming

Integer programming (Nemhauser and Wolsey, 1988; Salkin and Mathur, 1989) is an optimization technique that aids in complex decision-making. The problem-solving power of this approach is demonstrated by Johnson, Kostreva, and Suhl (1985). We illustrate the use of this technique by developing a model that will maximize the value of the output of a robotic system over a given time horizon, T. The decisions to be made include which operations to include in the robotic system and which robot-served workstations to purchase, subject to a capital investment budget. We use the following notation for input data.

$O_1, \ldots, O_m$ = the set of possible operations that can be included in the robotic system
$W_1, \ldots, W_n$ = the set of workstations that can be purchased to process the operations
c_j = the purchase cost of workstation W_j
B = the capital investment budget
$O(j)$ = the index set of operations that can be performed by workstation W_j
$W(i)$ = the index set of workstations that can perform operation O_i
t_{ij} = the time for workstation W_j to perform operation O_i
h_j = the number of available hours at workstation W_j within time horizon T, allowing for maintenance and other unavoidable downtime
r_i = the number of times operation O_i is required within time horizon T
v_{ij} = the net value, i.e., revenue less cost, produced by operation O_i at workstation W_j

We also define the following decision variables:

$$Y_j = \begin{cases} 1 \text{ if workstation } W_j \text{ is purchased} \\ 0 \text{ otherwise} \end{cases}$$

X_{ij} = the number of times operation O_i is performed at workstation W_j within time horizon T

With these data and variable definitions, the preliminary design problem of maximizing the value of operations performed in the robotic system can be written as follows.

$$\text{maximize} \quad \sum_{i=1}^{m} \sum_{j \in W(i)} v_{ij} X_{ij} \tag{1}$$

$$\text{subject to} \quad \sum_{j=1}^{n} c_j Y_j \leq B \tag{2}$$

$$\sum_{j \in W(i)} X_{ij} \leq r_i, \quad i = 1, \ldots, m \tag{3}$$

$$\sum_{i \in O(j)} t_{ij} X_{ij} \leq h_j Y_j, \quad j = 1, \ldots, n \tag{4}$$

$$X_{ij} \geq 0 \quad \text{and integer}, \quad i = 1, \ldots, m, \quad j = 1, \ldots, n \tag{5}$$

$$Y_j \in \{0, 1\}, \quad j = 1, \ldots, n \tag{6}$$

The objective (1) represents the total value produced. It sums the value produced by operation O_i on workstation W_j over all operations and workstations. Constraint (2) ensures that the total capital investment does not exceed the budget, B. Constraints (3) ensure that the number of times an operation is performed does not exceed its production requirement. Constraints (4) ensure that the total workload assigned to workstation W_j does not exceed the available hours there, which are h_j if W_j is purchased and 0 otherwise. Constraints (5) ensure that the number of times operation O_i is assigned to workstation W_j is a nonnegative integer. Finally, constraints (6) enforce the choice either to purchase workstation W_j or not to do so. This model is useful in that it can evaluate, for example, the tradeoff between expensive workstations that are capable of performing many operations and cheaper but less flexible workstations.

The above integer program may be difficult to solve for large data sets, either by using standard software or with a customized technique. The reason is that the computational difficulty involved in solving this model increases very rapidly with the number of integer variables. The reader may refer to the introductory chapters of Garey and Johnson (1979) for a more detailed explanation. Note that the above formulation has $mn + n$ integer variables.

To circumvent this computational difficulty the following approximation is often useful. We remove the integer requirement from Constraints (5), thus requiring only that $X_{ij} \geq 0$. This greatly reduces the number of integer variables to n, and consequently also the difficulty of solution. For example, if $m = n = 30$, solving the original integer program with 930 variables is a major challenge that requires expensive software and probably a long running time on a mainframe computer, whereas the new *mixed integer* program has only 30 integer variables and can easily be solved on a personal computer. Moreover, it is frequently the case that few X_{ij} variables in the mixed integer program have fractional values (e.g., $X_{11} = 2.4$), and that these can be rounded to integer values (e.g., $X_{11} = 2$ or 3) without serious detriment to the quality of the overall solution.

Applications of integer programming to flexible manufacturing systems are discussed by Stecke (1983) and Berrada and Stecke (1986). Commercially available software packages for integer and mixed integer programming are reviewed by Saltzman (1994).

2.3.2 Linear Programming

Perhaps the most important contribution made by operations research to decision-making is the technique of *linear programming*. Bradley, Hax, and Magnanti (1977), Schrijver (1986), and Bazaraa, Jarvis, and Sherali (1990) provide very detailed treatments of the theoretical background of linear programming. As an example of the power of this technique, Bixby et al. (1992) describe the solution of an extremely large problem with 12,753,313 decision variables and 837 constraints. Some exciting recent developments in this area are reviewed and evaluated by Lustig, Marsten, and Shanno (1994). For a very readable discussion about how to build practical models that access the power of this approach, the reader may refer to Williams (1990).

Because of the existence of highly efficient solution methods, which can be implemented in inexpensive commercial software, linear programs with many thousands of variables and constraints are readily solvable. Linear programming offers less modeling flexibility than mixed integer programming, in that all the variables are considered divisible. However, there are many applications where this limitation is not very problematic.

To illustrate the use of linear programming, we consider a subproblem of the problem from Section 2.3.1. In this subproblem the choice of workstations has already been made. It only remains to decide how to allocate the available operations to those workstations.

Using the notation of Section 2.3.1, the value maximization problem can be written as follows: maximize (1), subject to (3), (4′) and (5′). Constraints (4′) are formed from Constraints (4) by setting $Y_j = 1$ if workstation j was purchased, and $X_{ij} = 0$ for $i = 1$, $\ldots$, m if workstation j was not purchased. Constraints (5′) are formed from Constraints (5) by removing the integer requirement on the X_{ij} variables, thus requiring only that $X_{ij} \geq 0$. Note that the integer Y_j variables and all the constraints in which they appear are no longer present. Thus the problem can be solved as a linear program, with much less time and cost than the formulation in Section 2.3.1, subject to the same issues about fractional X_{ij} values as were discussed there.

Closely related to linear programming is *goal programming* (Charnes, Cooper, and Ferguson, 1955; Ignizio, 1976), which allows the modeling of conflicting goals or objectives. Imany and Schlesinger (1989) demonstrate the use of goal programming to optimize the robot selection process taking into account requirement priorities.

Schrage (1991) discusses one very popular linear programming software package in detail. Fourer (1997) presents a comprehensive review of commercially available software packages for linear programming. A valuable source of on-line information about linear programming, including free software and discussions about solvability, is the Frequently Asked Questions Homepage maintained by J. Gregory at "http://www.mcs.anl.gov/home/otc/Guide/faq".

2.3.3 Analytic Hierarchy Process

The *analytic hierarchy process,* developed by Saaty (1980), is a technique for making comparisons between several options across multiple factors. The difficulty in such comparisons arises when the relative performance of the options is inconsistent across the various factors. Since the number of robotic systems being compared is likely to be quite small, and since that comparison involves several factors such as cost, space utilization, and flexibility (see Section 2.1), the analytic hierarchy process is ideally suited to help with the choice of which system to purchase. Saaty (1986) presents an in-depth discussion of the theoretical foundations of the analytic hierarchy process and addresses some concerns of earlier authors about its correctness. Zahedi (1986) and Golden, Wasil, and Levy (1989) give bibliographies of its applications.

As an illustration, let us assume that we wish to compare three robotic systems (A, B, and C) using the three factors mentioned above. This creates a decision-making hierarchy in which the choice of which system to purchase is on the top level, the three factors mentioned are on the second level, and the individual systems are on the third level. This hierarchy is illustrated in Figure 1.

The analytic hierarchy process is implemented as follows. The decision-maker provides various *pairwise comparisons* between factors, indicating which is more important and how much more, using a discrete scale. Also, within each factor similar pairwise comparisons are made regarding how well each system performs for that factor. The outcome of the analysis is a useful overall ranking of Systems A, B, and C. This technique has the advantage that responses on a pairwise comparison point scale are often more reliable than those on an abstract interval scale. A further advantage is that the technique identifies inconsistent input data, thereby giving the decision-maker a chance to review and correct

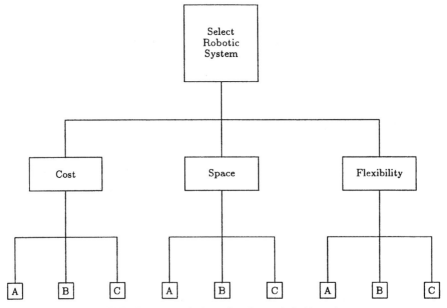

Figure 1 Robotic system choice using the analytic hierarchy process.

it. The analytic hierarchy process is less suitable than mixed integer or linear programming for making comparisons between thousands of alternatives that differ only slightly, or for dealing with constraints, such as might be necessary in detailed production planning. However, it can be a valuable aid to comparing several options across several factors.

Applications of the analytic hierarchy process to manufacturing systems are described by Arbel and Seidmann (1984), Varney, Sullivan, and Cochran (1985), Bard (1986), and Sullivan (1986). Seidmann, Arbel, and Shapira (1984) use the analytic hierarchy process to assist in robotic system design. Frazelle (1985) presents an application to choices among material-handling alternatives. Shang and Sueyoshi (1995) use data envelopment analysis (Farrell, 1957; Charnes, Cooper, and Rhodes, 1978; Sengupta, 1995) to unify an analytic hierarchy process model and a simulation model (see Section 3.3.4) for the problem of selecting a flexible manufacturing system. An interesting new variant of the analytic hierarchy process is described by Lipovetsky (1996). A popular analytic hierarchy process software package is EXPERT CHOICE.

2.3.4   Heuristics

Models of the type described in Section 2.3.1 may be computationally hard to solve to optimality, as discussed by Garey and Johnson (1979). Since optimality is hard to achieve, a useful "second-best" approach is to solve those problems *heuristically*. A heuristic is a simple problem-solving approach that may not give an optimal solution. Many computer codes for integer and mixed integer programming problems allow the user to terminate the search process when it becomes very lengthy and to identify the best solution found so far. Typically the heuristic solution found in this way will be within a few percentage points of optimal, which for many practical purposes may be good enough. Examples of heuristic design and evaluation are provided by Talbot and Gehrlein (1986) and Adams, Balas, and Zawack (1988). The theoretical state of the art in this area is comprehensively examined in Hochbaum (1997).

It is also possible to design heuristics directly for decision problems. As an example, consider the problem of choosing which of n workstations to buy. Here the net present value (respectively, capital cost) of workstation W_j is denoted by b_j (respectively, c_j). The total capital cost must not exceed a given budget, B. We can model the problem of maximizing the total net present value using the following integer program.

$$\text{maximize} \quad \sum_{j=1}^{n} b_j Y_j$$

$$\text{subject to} \quad \sum_{j=1}^{n} c_j Y_j \leq B$$

$$Y_j \in \{0, 1\}, \quad j = 1, \ldots, n$$

In this model $Y_j = 1$ if workstation W_j is purchased, and $Y_j = 0$ otherwise. This model is known as a 0–1 knapsack problem (Balas, 1980), which may be hard to solve computationally if n is large. Moreover, additional constraints or complications may arise in practical decision problems involving robotic facilities. Examples include space restrictions, material flow capacities, or the opportunity for discounts with multiple purchases. However, the following well-known heuristic often provides solutions that are close to optimal for the above formulation.

Greedy Heuristic

Step 1.   Order the workstations such that $b_1/c_1 \geq \cdots \geq b_n/c_n$.

Step 2.   Consider the next workstation in index order. If enough of the budget remains to purchase it, then do so.

Step 3.   If not all workstations have been considered, then go to Step 2. Otherwise, stop.

The Greedy Heuristic is intuitive in that it first considers workstations with larger ratios of net present value to capital cost, that is, the most attractive workstations.

Hochbaum and Maass (1984) describe a variety of heuristic approaches to robotics and related problems. Heuristics for layout problems in automated manufacturing are provided by Heragu and Kusiak (1988) and Kouvelis and Kim (1992). Ball and Magazine

(1988), Ahmadi and Tang (1991), Lofgren, McGinnis, and Tovey (1991), and Lane and Sidney (1993) describe and evaluate heuristics for automated manufacturing systems. Sethi et al. (1990) discuss heuristics for the selection and ordering of part-orienting devices in flexible manufacturing systems.

2.3.5 Location and Layout

The main *location* decision is where to build a new robotic facility. The following factors are particularly important in such decisions:

1. Transport costs from/to other facilities
2. Fixed costs
3. Production capacity, as determined by site characteristics

A typical facility location problem is illustrated in Figure 2. The objective is to locate a single new robotic facility as close as possible to three markets and three suppliers. However, only sites A, B, C, and D are available, and the choice between them involves a comparison of costs and travel distances and times.

The operations research literature includes a large number of studies on location problems. An overview is provided by Francis, McGinnis, and White (1983). Schmenner (1982) discusses how large companies make location decisions and concludes that key factors are labor costs, unionization, closeness to markets and suppliers and other company facilities, and quality of life.

An important choice in modeling location problems is how to measure distance, using either a straight-line or a rectilinear measure. The latter measure is based on the assumption that travel is only possible in the north–south or east–west directions, as might be the case in urban planning problems. Assuming that location at any point on a plane is possible typically makes location problems easier to solve, as for example in a marketing application by Gavish, Horsky, and Srikanth (1983). However, for many applications, only a discrete set of possible locations is available. The intermediate and broadly applicable case of location on networks is discussed by Handler and Mirchandani (1979). Among the best known studies of location problems are those of Erlenkotter (1978) and Ross and Soland (1978). Jungthirapanich and Benjamin (1995) develop a knowledge-based decision support system for facility location. The issues arising in the location of automated facilities are not very different from those arising in facility location generally, and therefore the reader should find much of the literature on this topic to be directly useful.

The main *layout* decisions are where, within an existing plant, to locate a robot-served workstation and its peripheral equipment such as storage devices. In this context the layout

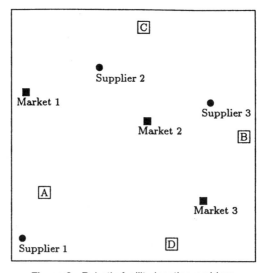

Figure 2 Robotic facility location problem.

problem is very similar to the location problem shown in Figure 2. Proximity to other workstations is in general desirable, but space and material-handling limitations may limit the available options. Relevant factors include:

1. Transport time from/to other workstations within the plant
2. Dimensions of the workstation
3. Limited reach of transporters, e.g., robot arms
4. Production capacity, as determined by material flow bottlenecks

Francis, McGinnis, and White (1992) provide a comprehensive discussion of both location and layout problems. They describe several basic horizontal flow patterns (for use within a single floor) as well as vertical flow patterns (for use across several floors). Layout analysis (Apple, 1977; Tompkins et al., 1996; Heragu, 1997) often begins by establishing a relationship chart that indicates the desirability (or occasionally, undesirability) of locating two operations close to one another. A from–to chart is used to identify the trip frequencies between workstations. These two charts can be used as input information for a variety of techniques, many of which are implemented in commercial software. Among the better known packages are CRAFT 3-D (Cinar, 1975), SPACECRAFT (Johnson, 1982), BLOCPLAN (Donaghey and Pire, 1991) and MULTIPLE (Bozer, Meller, and Erlebacher, 1994). Francis et al. (1994), extending earlier work by Drezner and Nof (1984), consider the problem of finding the component bin locations and part sequence that will minimize total time in a robotic assembly problem. Genetic search (Michalewicz, 1996) is a heuristic search procedure based on an analogy with a natural evolutionary process in which intermediate solutions are allowed to "reproduce" new and hopefully improved solutions. Tate and Smith (1995) use genetic search techniques to develop a facility layout model. Dynamic layout problems, in which changes in the data motivate changes in the layout over time, are considered by Rosenblatt (1986) and Urban (1993). Neural networks (Aleksander and Morton, 1990; Warwick, 1995) are also based on a biological analogy, and perform searches by adaptive parallel processing of information. Tsuchiya, Bharitkar, and Takefuji (1996) present and test a heuristic neural network procedure for facility layout. A design approach that features parallel and continual interaction between product, process and system design is recommended by Rampersad (1995) for a robotic assembly system layout problem. An interesting survey that highlights research trends and gives an overview of several useful solution procedures for layout problems is provided by Meller and Gau (1996).

Scriabin and Vergin (1975) sparked an interesting debate about the relative effectiveness of human layout planners and computer packages. They conclude that, in large problems, the superior pattern recognition abilities of a human planner lead to more effective layouts. However, further studies (Block, 1977) and subsequent developments in artificial intelligence (Brown and White, 1990) argue against such a conclusion.

3   SYSTEM DESIGN

In this section we assume that, as a result of decisions made in the system planning phase, we have already identified a small number of options for the set of operations to be automated and the set of robot-served workstations to be purchased. Whereas the data used at the system planning stage may have been aggregate and approximate, in the system design phase both more detail and more accuracy are required. In particular, the characteristics of individual operations need to be considered so that the related material handling requirements can be met within the system that is eventually adopted.

3.1   Issues and Requirements

The main issues and management requirements that need to be considered at the system design phase are as follows.

1. *Equipment layout.* A good layout should be found for automated workstations, storage facilities, and personnel, taking into account patterns of material flow within the robotic system and within other systems that interface with it. Related operations research techniques are discussed in Section 2.3.5.
2. *Cell design.* If a cellular layout is to be used, a decision is needed about whether cells should be organized by product or by process. Procedures for forming groups of operations should be developed.

3. *Material-handling equipment.* A decision is needed about what material-handling equipment should be purchased. A similar decision is needed about what storage equipment, such as an automated storage and retrieval system, to be purchased.

4. *Accessories.* It is necessary to decide the number and types of peripheral equipment, including part feeders and part orienters, magazines, and tool holders, to be purchased.

3.2 Performance Measures

The performance measures that are useful at the system design phase fall into two types. First, they include refinements of the measures suggested for the system planning phase. Second, they include new measures that can now be considered because of the greater detail in the available data.

1. *Throughput rate.* For each cell or for the overall system a natural objective is to maximize the throughput rate. In modern manufacturing systems it is common to use a *minimal part set,* which is the smallest possible set of parts having the same proportions as the overall production target. Minimizing the cycle time to produce a minimal part set repetitively is equivalent to maximizing the throughput rate.

2. *Turnaround time.* This is the elapsed time from the entry of a job into the robotic system until it leaves, and is a measure of system responsiveness.

3. *Work-in-process inventory.* Since holding cost may be a significant part of overall manufacturing cost, there is a need to control work-in-process inventory. This relates closely to buffer storage design also. Further, a result to be discussed in Section 3.3.2 suggests that in a steady state the average turnaround time is proportional to the average work-in-process inventory.

4. *Equipment utilization.* A traditional measure of manufacturing effectiveness is the minimization of equipment idle time. A more modern view, however, suggests that this is less useful than the other measures above.

5. *System reliability.* It is important to understand how the reliability of the different components of the robotic system will interact in various configurations, and thus to assess the reliability of the overall system.

3.3 Operations Research Techniques for System Design

Because of the availability of more accurate data, the operations research techniques discussed here are capable of greater accuracy than those used at the system planning phase. However, estimates may still be needed for the requirements of individual operations, for example. Therefore, it is important to find ways to perform sensitivity analysis or investigate "what if" scenarios for the information which is output by those techniques.

3.3.1 Group Technology and Cell Formation

In *group technology* (Mitrofanov, 1966; Burbidge, 1975) a major decision is how to decompose the manufacturing system into *cells.* Families of component parts are identified and each family is fully or predominantly processed within a machine group. The most obvious advantage of group technology is setup time reduction arising from the similarity of the parts within the same cell. Other advantages include improved productivity and quality, as well as reduced inventories and tool requirements. In the remainder of this subsection we summarize the main approaches to cell formation. Singh (1993) provides a useful review of the literature on this topic.

Coding and classification schemes work by coding parts according to their attributes. Bedworth, Henderson, and Wolfe (1991) provide a summary of such schemes. However, there is relatively little work that has integrated such schemes with more formal operations research techniques. Kusiak (1985) uses a model from location theory to achieve this integration.

In machine-component grouping analysis methods, cells are formed by algebraic operations on a 0–1 machine-component index. Chu and Pan (1988) provide a review of these methods, which lack the ability to deal effectively with bottleneck machines, and do not consider operating details such as processing times. Ng (1996) shows how to separate machines and components into independent families when such an ideal separation is possible, and then uses this information to design a cell formation procedure that performs very well on standard test problems.

Similarity-based clustering techniques design measures of similarity and then cluster jobs in various ways. For example, Rajagopalan and Batra (1975) cluster jobs using graph theory. Other mathematical approaches that eschew the use of similarity coefficients include a decomposition scheme and heuristics (Askin and Chiu, 1990) and a network approach (Vohra et al., 1990). These approaches often suffer from the problem with bottleneck machines mentioned in the previous paragraph.

Other potentially valuable techniques include a knowledge-based system (Kusiak, 1988) and a neural network approach (Kaparthi, Suresh, and Cerveny, 1993). Burke and Kamal (1995) demonstrate the use of the fuzzy adaptive resonance theory neural network. Xie (1993) focuses on the minimization of intercell traffic, a problem for which three heuristics are designed and tested. Methods for reconfiguring existing cells are discussed by Mohamed (1996).

3.3.2   Queueing Models

As mentioned in Section 3.2, an important performance measure for robotic systems is work-in-process inventory. This varies constantly while the system is operating, due to the random processes of jobs arriving and jobs being completed. However, *queueing models* can take this random behavior into account and provide useful performance measures. When used in this way, a queueing model is a probabilistic, steady-state model. Jobs waiting to be processed are essentially "customers" waiting in line while other jobs are being processed. General introductions to queueing models are provided by Kleinrock (1975, 1976) and Gross and Harris (1985).

Queueing models can help in evaluating how much service should be provided. Within the context of robotic systems, this means how many workstations and with what capabilities (e.g., material-handling capacities and processing speeds) and how many operators. A service level that is too high incurs unnecessary capital and operating costs, but one that is too low creates expensive work-in-process inventories. Queueing models do not optimize this design question directly. However, given a particular system configuration, they can provide several steady-state measures of its performance.

In order to present some useful results for queueing models, we define the following notation:

ρ = utilization factor of the system
λ = average arrival rate of the jobs
L_q = average number of jobs waiting to be processed
L = average number of jobs waiting or in process
W = average total of waiting and processing time of a job
μ = average service rate per server
m = number of servers

The utilization factor is defined to be the average arrival rate divided by the total average service rate, or $\rho = \lambda/m\mu$. A queueing system is described using the three field notation $A|S|m$, where A describes the arrival process, S describes the service process, and m is the number of servers. Some common types of arrival or service process appearing in the first two fields include: M = Markovian (times are independent and exponentially distributed), G = general (times follow an arbitrary distribution), and D = deterministic (all times are the same).

The most important result in queueing models, initially proved by Little (1961) for a simple system and then extended by Stidham (1974) to more general systems, is:

$$L = \lambda W \tag{7}$$

In the context of robotic systems, Equation (7) implies that the work-in-process inventory is proportional to the average amount of time a job spends in the system. This relationship emphasizes the importance of timely throughput of work-in-process.

Single-Robot System

We can consider a single-robot system as a queueing model with only one server. Figure 3 illustrates a system of this type. Jobs arrive from other workstations independently and according to an exponential distribution. For a general service time distribution this system is denoted $M|G|1$.

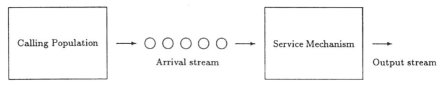

Figure 3   Single-server queueing system.

The following formula gives the average number of jobs waiting in an $M|G|1$ system (Kleinrock, 1975):

$$L_q = \rho^2(1 + C_s^2)/[2(1 - \rho)]$$

where C_s denotes the coefficient of variation (i.e., the standard deviation divided by the mean) of the service times. However, a common occurrence in robotic systems is that all jobs require the same operation by the robot. In this case we have an $M|D|1$ queueing model where $C_s = 0$, and thus

$$L_q = \rho^2/[2(1 - \rho)] \tag{8}$$

It is important to note that Equation (8) is nonlinear. In fact, if ρ starts to increase from a low level such as 0.1, L_q increases slowly at first but then more and more rapidly. Table 2 shows the values of L_q associated with various values of ρ in an $M|D|1$ system.

If the jobs arrive from other workstations, then an approximation that is usually satisfactory is that both the times between job arrivals and the service times are exponentially distributed. This situation is known as an $M|M|1$ system, for which it can be shown (Trivedi, 1982) that

$$L_q = \rho^2/(1 - \rho) \tag{9}$$

Multiple Robot System

The results from the single robot system can be extended to a system with m identical processors, where each job can be performed by any processor. A system of this type is illustrated in Figure 4. Following the above assumption about exponentially distributed interarrival and service times gives an $M|M|m$ system, for which it can be shown (Trivedi, 1982) that

$$L_q = \rho(m\rho)^m P_0/[(1 - \rho)^2 m!] \tag{10}$$

where P_0 is the long-run average proportion of time for which all m servers are idle, and where P_0 has a closed form expression.

Pooling of Facilities

A very interesting issue in queueing models is whether to pool facilities. Suppose we wish to process two types of jobs using two robot-served workstations. A choice between two alternative designs presents itself. In the first design one workstation is dedicated to each type of job. In the second design the workstations are pooled and each is used for both types of jobs. The first (dedicated workstation) design is likely to be less expensive in capital cost. This is because the second design may need duplicated equipment such as part orienters and specialized gripping devices to handle both job types. However, it can be shown from Equations (9) and (10) that the first design has more jobs waiting on average. For example, if $\rho = .75$ and the two types of jobs arrive at the same rate, there

Table 2   Average Number of Jobs Waiting in an $M|D|1$ System

ρ	.15	.25	.35	.45	.55	.65	.75	.85	.95
L_q	.0132	.0417	.0942	.1841	.3361	.6036	1.125	2.4083	9.025

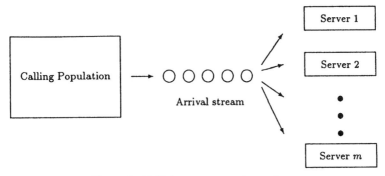

Figure 4   Multiple-server queueing system.

are on average 2.25 extra jobs (4.5 vs. 2.25) waiting in the first design, creating an increase of 100% in work-in-process.

Despite this significant mathematical advantage of a combined service system, Roth-kopf and Rech (1987) make a persuasive behavioral case for the first design above in queueing systems where the customers are human, rather than jobs. Some of their arguments, moreover, such as the relative proximity of the waiting and service areas in the first design, can also be applied to robotic systems.

Ram and Viswanadham (1990) and Viswanadham and Narahari (1992) discuss the use of queueing models to evaluate the performance of a flexible machine center problem.

3.3.3   Queueing Networks

Queueing networks extend the methodology of queueing models discussed in Section 3.3.2. In doing so they provide a much greater ability to model the complexities of real automated manufacturing systems. Jobs that in general may have different routings through the available workstations are transported, by a robot or other automated material handler, from one workstation to another. Comprehensive introductions to queueing networks are provided by Walrand (1988), Viswanadham and Narahari (1992), Buzacott and Shanthikumar (1993), and Gershwin (1994). Extensive surveys are provided by Bitran and Dasu (1992) and Buzacott and Shanthikumar (1992).

There are two basic types of queueing networks: open and closed. An open queueing network, illustrated in Figure 5, has one or more points at which customers enter the system and one or more points at which they leave the system after processing. By contrast, in a closed queueing network, as illustrated in Figure 6, customers neither enter nor leave the system. In a manufacturing context such customers might represent parts in an open network. In a closed network the customers might represent pallets that accompany parts through the network. Little's (1961) result in Equation (7) extends to queueing networks and implies that the fraction of waiting time at any workstation is equal to the fraction of work-in-process at that workstation.

Jackson (1957, 1963) gives an important result about open queueing networks in which an arriving job enters the network at workstation i (which has m_i processors) with probability q_i, for $i = 1, \ldots, m$. After processing, the job is routed to workstation j with probability p_{ij}. Let $P(k_1, \ldots, k_m)$ denote the steady-state probability that there are k_i jobs at workstation i for $i = 1, \ldots, m$, and let $P_i(k_i)$ denote the probability that there are k_i jobs in a one workstation $M|M|m_i$ queueing system. Then Jackson's theorem states that

$$P(k_1, \ldots, k_m) = \prod_{i=1}^{m} P_i(k_i)$$

The importance of this result is that it enables the use of standard queueing theory (Section 3.3.2) to derive performance measures for queueing networks.

Analogous results for closed queueing networks are due to Gordon and Newell (1967). Both types of model are known as *product form* queueing networks. This class was extended and presented in a unified way by Baskett et al. (1975), and within the extended

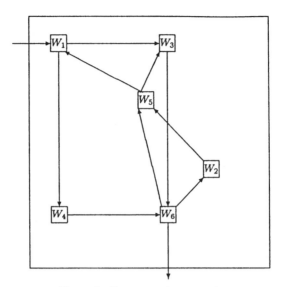

Figure 5 Open queueing network.

class many measures of system performance can easily be obtained. However, a number of features of automated manufacturing systems may violate the product form requirement and thus require the use of approximations. These features include nonexponential service times, nonstandard customer service orders, production deadlines, assembly product structures, breakdowns, the use of dynamic rerouting of jobs to avoid bottlenecks, and resource constraints.

The following types of queueing networks have been studied extensively for both open and closed cases: tandem queues (where machines are in series), networks with feedback (such as items returned for rework after inspection), networks with blocking (where a job waits at a workstation until the next workstation is ready), open central server networks, and product form networks. Additionally, product form networks allow the analysis of problems with multiple job classes.

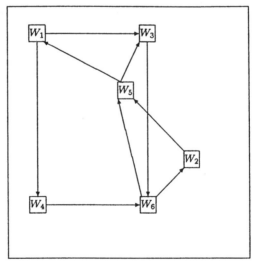

Figure 6 Closed queueing network.

Solberg (1977), Suri and Hildebrandt (1984), and Yao and Buzacott (1987) provide examples of the application of queueing networks to automated manufacturing systems. Chen et al. (1988) demonstrate the accuracy of performance predictions from a queueing network model of a semiconductor wafer fabrication facility. A survey of sensitivity analysis results is provided by Liu and Nain (1989). Bitran and Tirupati (1989) analyze tradeoff relationships between different performance measures in manufacturing queueing networks. Dallery and Stecke (1990) and Calabrese (1992) address the issue of how to allocate workloads and servers in similar environments. Software packages that are available to analyze queueing networks include CAN-Q (Solberg, 1980) and MANUPLAN (Suri and Diehl, 1985). Seidmann, Schweitzer, and Shalev-Oren (1987) perform a comparative evaluation of four queueing network software packages. A variety of methods for analyzing queueing networks are described by Robertazzi (1994).

3.3.4   Simulation

Simulation is a controlled statistical sampling technique for evaluating the performance of complex probabilistic manufacturing and other systems. General introductions to simulation techniques can be found in Law (1986), Law and Kelton (1991), and Schriber (1991).

With simulation, the performance of a real system is *imitated* by a computer model that uses preestablished probability distributions to generate random events (such as demand rates, processing times, and machine breakdowns) that affect the performance of the system. The outcome of a simulation study is statistical information about the performance of the system. When applied to large manufacturing systems, simulation tends to generate and process a huge amount of data, which can be time-consuming and expensive compared to the use of most other operations research techniques.

In order to use simulation to assess the performance of a robotic system, it is first necessary to collect information about the various hardware and software components of the system. The information to be collected includes both deterministic information (such as known processing times for standard parts) and probabilistic information (such as demand rates for new products). For the probabilistic information it is necessary to fit the data collected to a probability distribution using standard statistical techniques (Kleijnen, 1987). These distributions are used to generate specific events, the effects of which are simulated by a computer model that captures as many of the intricacies of the real operating system as possible. The tradeoff between realism and solvability is less problematic in simulation than in the optimization approaches discussed in Sections 2.3.1 and 2.3.2 because simulation possesses much greater flexibility and can therefore be extended, without significant loss of solvability, to allow for complex realities.

There are essentially three approaches to simulating a robotic system:

1. *Simulation package.* This approach (Runner and Leimkuhler, 1978; Phillips and Handwerker, 1979) involves the purchase of a manufacturing system simulation package, which can then be modified to approximate the real system more accurately. While this approach is likely to be relatively quick and simple, it may not be possible to introduce all the complexities of a specific real-world system into a particular package.

2. *Simulation language.* This relatively sophisticated approach (Gordon, 1975; Pritsker and Pegden, 1979) requires knowledge of a simulation language. Also, there may be interface problems with system subroutines that are written in a standard programming language such as Fortran. The main advantage of this approach, however, is that it offers almost unlimited modeling flexibility. Thus, a real-world system can be modeled as accurately as desired.

3. *Standard programming language.* This approach takes advantage of already existing knowledge of a standard programming language such as Pascal, which is used to write a simulation code for the system. Interfacing with system subroutines is easy and the simulation package is universally portable. The modeling flexibility of this approach is similar to that of using a simulation language. However, all the basic components must be written by the user, and the overall programming task is likely to be much greater.

Further discussion of these tradeoffs appears in Bevans (1982). The use of simulation to evaluate the performance of manufacturing systems is discussed by Halder and Banks

(1986), Carrie (1988), and Trunk (1989). An example of the use of simulation to plan cellular layouts is provided by Morris and Tersine (1990). De Lurgio, Foster, and Dickerson (1997) use simulation to assist in the selection of equipment for integrated circuit manufacturing. A useful survey of simulation practice in industry is provided by Christy and Watson (1993). Hlupic and Paul (1995) conduct a thorough case study evaluation of four manufacturing system simulation packages: WITNESS, EXCELL+, SIMFACTORY II.5, and ProModel PC. None of the packages dominates the others with respect to all performance criteria, but they discuss the relative advantages of each. Kleijnen (1995) and Kleijnen and Rubinstein (1996) discuss methods for validating a simulation analysis and performing sensitivity analysis on the results. A comprehensive survey of commercially available simulation software packages appears in Swain (1995). Banks (1996) provides useful advice about how to interpret checklists of features when buying a simulation software package. A survey of simulation packages, including some that are designed specifically for robotic systems, appears in the May 1997 issue of *IIE Solutions*. Application-specific packages, such as IGRIP and TELEGRIP, may capture more details of robotic systems, or may require less programming to do so, than general-purpose manufacturing simulation packages.

3.3.5   Reliability and Maintenance

The *reliability* of a robotic system is critical because the substantial investment in capital cost needs to be repaid through consistent output of high-quality products. Thus, in evaluating a robotic system, it is important to be able to estimate how frequently and for how long it will be unavailable due to component failures. Methods for making such estimates fall within the topic of reliability theory. Classic introductory texts in this area are provided by Barlow and Proschan (1965, 1975).

Single-Component Systems

In order to illustrate the use of reliability theory, we define the following notation. Let T be a random variable that represents the lifetime of a component. Let $F(t)$ denote the cumulative distribution function of T, i.e., $F(t) = P\{T \leq t\}$. Assuming that T is differentiable, let $f(t)$ denote its density function, i.e., $f(t) = dF(t)/dt$. A valuable concept is the failure rate function $r(t)$, which is defined by

$$r(t) = f(t)/[1 - F(t)]$$

For small values of Δ the probability that a component which has survived to time t fails by time $t + \Delta$ is given by $r(t)\Delta$. The failure rate function can be increasing, constant, or decreasing over time.

It is also possible to derive closed form expressions for the mean T_a and the variance V_a of the time for which the system is available during a time horizon of length T. Let f (respectively, s_f) denote the mean (respectively, standard deviation) of the time from the last repair to the next failure. Let r (respectively, s_r) denote the mean (respectively, standard deviation) of the time from the last failure to the next repair. Then we have

$$T_a = fT/(f + r) \tag{11}$$

$$\text{and} \quad V_a = [(fs_r)^2 + (rs_f)^2]/(f + r)^3$$

Equation (11) is highly intuitive, in that the long-run proportion of time in which the system is available is $f/(f + r)$, i.e., the mean time from a repair to the next failure divided by the mean time between two successive repairs.

Multicomponent Systems

Reliability theory also helps us to evaluate the performance of systems with multiple components. Suppose that there are m components, where R_i is defined as the proportion of time in which the ith component is available. Note that R_i represents the reliability of the ith component. The following analysis assumes independence of the availability of the components. This assumption greatly simplifies the mathematics involved in estimating the reliability of the system, which might otherwise defy exact solution procedures. It should be pointed out that the independence assumption is critical and, if made erroneously, can lead to poor predictions. For example, Nahmias (1993) identifies this as a

Figure 7 Series structure.

source of error in a U.S. Nuclear Regulatory Commission (1975) study of reactor safety prior to the 1979 Three Mile Island accident.

Three possible designs for multicomponent systems are (a) a series structure, which is the least reliable, (b) a parallel structure, which is the most reliable, and (c) a hybrid structure.

A series structure is illustrated in Figure 7. Here the system fails unless all components are available. The reliability of the system, R_s, is thus given by

$$R_s = \prod_{i=1}^{m} R_i$$

For example, if $m = 6$ and $R_1 = \ldots = R_6 = .9$, then $R_s = 0.531441$, which is much too low for most applications.

A parallel structure is illustrated in Figure 8. Here the system fails only if all components fail. Thus, the reliability of the system is given by

$$R_s = 1 - \prod_{i=1}^{m} (1 - R_i)$$

In the above example we obtain $R_s = 0.999999$, a very high level of reliability.

One example of a hybrid structure is illustrated in Figure 9. This system fails unless at least one out of each pair of components $(1, 2), (3, 4), \ldots, (m - 1, m)$ is available. The reliability of this structure is given by

$$R_s = \prod_{i=1}^{m/2} [1 - (1 - R_{2i-1})(1 - R_{2i})]$$

In the above example we obtain $R_s = 0.970299$.

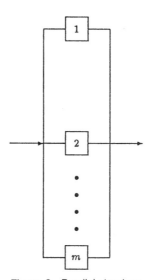

Figure 8 Parallel structure.

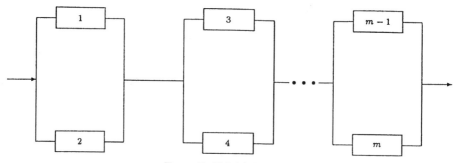

Figure 9 Hybrid structure.

Another common type of hybrid structure fails if fewer than any k out of the m components are available. Here, assuming that $R_1 = \ldots = R_m = R$, the reliability of the system is given by

$$R_s = \sum_{j=k}^{m} m! R^j (1 - R)^{m-j} / j! (m - j)!$$

In the above example for the case $k = 3$, we obtain $R_s = 0.99873$.

It should be noted that the reliability of both hybrid structures is close to that of the parallel structure. However, because of the multiple ways in which a hybrid structure can be designed, it is often more flexible and cheaper than a parallel structure.

Prasad, Nair, and Aneja (1991) discuss how to assign components in multicomponent systems to maximize reliability. Hwang and Rothblum (1994) consider the problem of assigning interchangeable parts of many types to modules with a series structure and prove that the best parts of each type should be assigned to the same module. Chung and Flynn (1995) discuss how to identify optimal replacement policies in hybrid structures.

Maintenance and Planned Replacement

A survey of *maintenance* models is provided by Pierskalla and Voelker (1976). Vanneste and Van Wassenhove (1995) review recent developments in maintenance theory and practice, and propose an integrated approach that combines some of the best features of other methods. Feldman and Chen (1996) perform some postoptimality analyses for various policies. A computational comparison of different maintenance policies is performed by Kelly, Mosier, and Mahmood (1997).

Planned replacement is a type of preventive maintenance. Its purpose is to reduce the likelihood of a highly disruptive and expensive failure during operation by investing in a cheaper replacement during scheduled downtime. It is possible to show that, if the failure rate function is constant or decreasing, it is never worthwhile to perform planned replacement. However, if the failure rate function is increasing, as in most practical systems, then there exists a closed form expression for the expected total cost of planned and unplanned replacements. By using spreadsheet analysis, this expression can be minimized to find the optimal replacement time.

The examples discussed here represent only a brief introduction to reliability and maintenance theory. Generalizations that are relevant to robotic systems include multicomponent systems with various other configurations and models that allow various levels of repair. Details can be found in Trivedi (1982). In many cases the analysis is computationally difficult, which has led to the development of reliability analysis software, of which CARE (Stiffler et al., 1979) is an early example. Smith (1993) and Ushakov (1994) each provide a list of commercially available software packages in an appendix. The periodicals *Data Sources* and *Maintenance Technology* provide current information about reliability and maintenance software packages, respectively.

4 SYSTEM OPERATION

In discussing the system operation phase we assume that a robotic system has been chosen and purchased. We further assume that it has been through testing and final installation and has been declared operational. The next priority is to ensure that the system operates as efficiently as possible. Therefore the focus is on day-to-day operating decisions.

4.1 Issues and Requirements

The overall objectives are to meet production targets and to operate the system smoothly and efficiently. In order to achieve these objectives, several issues and requirements demand the attention of management.

1. *Batch sizes.* Appropriate batch or lot sizes may be determined by setup times, system capacity, and material availability. Decisions need to be made about the size, content, and timing of each batch.
2. *Scheduling.* For each workstation the detailed sequence and timing of jobs must be determined. One of the special features of scheduling in robotic systems is the need to coordinate scheduling and automated material-handling decisions.
3. *Disruptions.* Occasional major disruptions in system operation may occur due to equipment failure, nonavailability of material, or sudden changes in production requirements. It is important to have in place operating procedures for dealing with such situations. Operations research techniques for addressing these problems usually resolve a batching and/or scheduling problem.
4. *Maintenance.* Operating procedures for both scheduled and emergency maintenance should be in place. Some issues in connection with maintenance policies are discussed in Section 3.3.5.

4.2 Performance Measures

The measures used to evaluate system performance at the operation phase are similar to those used at the system design phase. Here, however, they are used with greater detail and precision. This is because the removal of the system design issues makes room for additional considerations within a solvable operations research model, and also because an actual robotic system with known configuration and operating characteristics is being used. Another difference is the time horizon over which performance is being assessed. At the system design phase the only reasonable time horizon is long-term because of the capital cost and disruption involved in changing the system configuration. At the operation phase, however, short-term operating decisions can be made and implemented. The planning process can be repeated when disruptions occur or when production requirements or other parameters change.

4.3 Operations Research Techniques for System Operation

Because short-term operating problems tend to be well defined and present a single objective, they are ideally suited to modeling and solution using operations research techniques. The issues of manufacturing efficiency are not unique to robotic systems. However, there are some characteristics of such systems that distinguish the problems that need to be solved. The following sections emphasize the novel features of using operations research techniques in this environment.

4.3.1 Batching

In traditional manufacturing, *batching* of jobs is frequently necessary because if jobs are performed in the order in which they arrive, there may be many lengthy and expensive setups between them. In batch processing, jobs that are similar are processed together, and within a batch the setup time and cost tend to be small or zero. Only a relatively small number of large setups are necessary when changing batches. Forming a given set of jobs into appropriate batches is a problem that has received considerable attention from manufacturing researchers. Standard references include Wagner and Whitin (1958), Crowston and Wagner (1973), Elmaghraby (1978), Peterson and Silver (1979), Carlson, Jucker, and Kropp (1983), and Afentakis, Gavish, and Karmarkar (1984). Recent studies that integrate batching and scheduling issues include those of Monma and Potts (1989), Albers and Brucker (1993), and Webster and Baker (1995).

However, there are some special characteristics of robotic systems that affect the batching issue. First, to the extent that manufacturing cells are formed around part families of very similar jobs, as discussed in Section 3.3.1, both setup time and cost will typically be much less than in a traditional manufacturing environment. This characteristic tends to reduce the importance of effective batching. Second, there may be unusual constraints that arise from limited magazine capacities for parts or from the need for different jobs to share specialized tools. These constraints tend to reduce the effectiveness of traditional batching rules in the context of a robotic system. Batching problems that are closely related to robotic systems are discussed by Agnetis, Rossi, and Gristina (1996) and Agnetis, Pacciarelli, and Rossi (1997).

4.3.2 Scheduling

Scheduling concerns the allocation of limited resources to jobs over time. A very readable introduction to this topic is provided by Pinedo (1995). Melnyk (1997) provides a comprehensive survey of scheduling software products used in industry and indicates the suitability of each package for a variety of manufacturing environments. We focus on work directly related to robotic and similar systems. Because this work is extensive, we consider it in two parts, those studies related to robotic cells and those related to other robot-served configurations, respectively. For this purpose, a robotic cell is a robot-served cluster of workstations that contains no internal buffers for work-in-process, and in which only a single family of parts is produced. Thus there are no setup times between parts. The importance of robotic cells is discussed by Levner, Kogan, and Levin (1995), who describe them as the simplest automated blocks, or "islands of automation" in modern computer-integrated manufacturing. Comprehensive surveys of recent work on related scheduling problems are provided by Crama (1997) and Lee, Lei, and Pinedo (1997). Hall and Sriskandarajah (1996) discuss a large variety of scheduling problems with close mathematical connections to robotic systems.

Robotic Cells

An example of a robotic cell is illustrated in Figure 10. An extensive review of the early literature on the scheduling of robotic cells appears in Sethi et al. (1992). More recent references can be found in Hall, Kamoun, and Sriskandarajah (1997). Two types of decisions need to be made in the scheduling of a robotic cell. First, the operations of the material-handling robot must be planned. There are many possible robot moves within the cell. A sequence of moves that returns the cell to its original state forms a *robot move cycle*. As a practical example of a robot move cycle, consider the problem of inserting components into printed circuit boards (see Li and Kumar (1997) for a history of this problem and an alternative solution procedure). In order to identify a robot move cycle

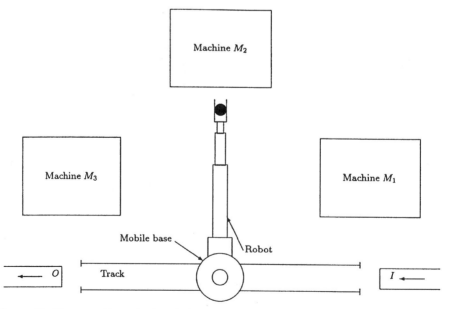

Figure 10 Three-machine robotic cell. (From C. Sriskandarajah, N. G. Hall, and H. Kamoun, "Scheduling Large Robotic Cells Without Buffers," *Annals of Operations Research* **76**, 1998, 287–321.)

it is first necessary to compute the times between each pair of insertion points. The time used if point j follows point i is computed as the travel time for the robot arm moving from point i to pick up a component, plus the travel time to point j, plus the insertion time at point j. A matrix of these total times can be used to develop an efficient sequence of insertions, which defines a robot move cycle. In robotic cells where several parts may be in a cell simultaneously, it is important to eliminate unnecessary or wasted moves that may make a cycle suboptimal. Sethi et al. (1992) show that in an m machine cell there are $m!$ potentially optimal robot move cycles.

Second, a sequence must be determined for the parts that need to be processed in the cell. Because in general the processing times vary between the parts, the robot move cycle and the part sequence must be determined jointly. Hall et al. (1997) provide an efficient solution procedure that minimizes the time to produce a minimal part set in a two-machine cell. Their procedure has been implemented in a fully equipped two-machine cell at the Manufacturing Systems Laboratory of the University of Toronto. Hall, Kamoun, and Sriskandarajah (1998) and Sriskandarajah, Hall, and Kamoun (1998) identify necessary and sufficient conditions under which an efficient solution procedure exists for larger cells. They show that if these conditions are not satisfied, then the part sequencing problem is intractable (Garey and Johnson, 1979). Variations of this problem are discussed by Kats and Levner (1996).

Robotic cell problems in which all parts are identical have also received the attention of researchers. Sethi et al. (1992) suggest that this problem can be solved by the repetition of a very simple robot move cycle. Hall, Kamoun, and Sriskandarajah (1997) partially prove this result, but do not consider some of the more complex possibilities in their analysis. Crama and van de Klundert (1997) demonstrate that repetition of a simple cycle does provide optimal schedules in three-machine cells. Brauner and Finke (1997) show that a similar result holds in four machine cells with two-part cycles. Furthermore, they show that it does not hold if either the number of machines or the number of parts in a cycle is increased. Levner, Kats, and Levit (1997) and Crama and van de Klundert (1998) provide procedures to minimize cycle time. A variation of this problem is discussed by Levner, Kats, and Sriskandarajah (1996).

Kamoun, Hall, and Sriskandarajah (1998) describe and test heuristic solution procedures for scheduling problems in robotic cells. Levner, Kogan, and Levin (1995) consider a robotic cell consisting of two tandem machines with limited intermediate storage, and provide an efficient procedure for minimizing the schedule length. Levner, Kogan, and Maimon (1995) take into account the effect of transportation time and setups in a similar problem.

Other Configurations

The problems discussed here fall outside our definition of a robotic cell, but nevertheless involve automated material-handling operations that can be performed by robots. As in robotic cells, a major issue is the *coordination* of the material-handling and scheduling functions. However, there may be other relevant issues, such as setup times for jobs, transportation times, time window constraints, and the need to minimize work-in-process. Crane scheduling problems that allow substantial intermediate buffers are considered by Park and Posner (1994, 1996).

Lei and Wang (1991) consider a system involving several workstations in series and two material-handling hoists that move jobs between stations. They develop a procedure for finding the best material-handling cycle that is common to the two hoists. Related work appears in Armstrong, Lei, and Gu (1994). Similar environments in which the objective is to minimize the number of transporters (e.g., robots) are studied by Lei, Armstrong, and Gu (1993) and Armstrong, Gu, and Lei (1996).

Other problems of interest arise in a manufacturing system consisting of several machines, where a job can be processed by any machine. However, job setup on the chosen machine must be performed by a robot. Since there is only one robot, its activity must be coordinated with the schedules of the jobs on the machines. This problem is considered by Glass, Shafransky, and Strusevich (1996) where jobs are dedicated to particular machines and by Hall, Potts, and Sriskandarajah (1998) where jobs can be scheduled on any machine. In both these papers the results include solution procedures, proofs of intractability, and the analysis of heuristics.

Geiger, Kempf, and Uzsoy (1997) describe an interesting application of scheduling methodology to the wet etching process in semiconductor manufacturing. Karuno, Nagamochi, and Ibaraki (1997) consider various scheduling problems that arise in a tree network served by a single material handler (e.g., a robot). They provide efficient solution procedures for several cases of these problems and distinguish other cases by proving that they are intractable.

4.3.3 Hierarchical Production Planning

The idea of *hierarchical production planning* originated with Hax and Meal (1975). Early studies in this area include those of Bitran, Haas, and Hax (1981) and Graves (1982). The logic behind hierarchical production planning is as follows. Instead of one large planning problem, a number of smaller subproblems are solved, which are typically much easier. Then a (usually approximate) solution to the original problem is constructed from the subproblem solutions. A comprehensive review of the hierarchical production planning literature is provided by Libosvar (1988). Gershwin (1994) gives a very readable introduction to this topic.

Much of the methodology that underlies hierarchical production planning originates with *optimal control theory* (Sethi and Thompson, 1981). For example, Caramanis and Liberopoulos (1992) and van Ryzin, Lou, and Gershwin (1993) use control theory to provide approximate feedback controls for probabilistic manufacturing systems.

However, more recent methodology known as the *hierarchical controls approach* (Lehoczky et al., 1991; Sethi and Zhang, 1994b) may provide an improvement. It has been demonstrated in several probabilistic planning systems that, by using this approach, it is possible to identify solutions that become closer to optimal as the rates of fluctuation of the individual processes within the system increase. The definitive text in this rapidly developing area is by Sethi and Zhang (1994a). Samaratunga, Sethi, and Zhou (1997) provide a computational comparison between a hierarchical controls approach and other approaches by earlier authors to planning a probabilistic manufacturing system. Although the hierarchical controls approach is not designed specifically for automated manufacturing systems, it appears to have the potential to improve their operations.

5   ORGANIZATIONAL DECISION-MAKING AND DECISION SUPPORT SYSTEMS

The preceding sections of this chapter contain a discussion of many operations research techniques that can enhance the design and operation of robotic systems significantly. To use these techniques effectively it is important to understand them in the context of overall *organizational decision-making*. A comprehensive study of organization design theory and practice is provided by Robey (1991). Some interesting perspectives about the subjective implications of organizational structures are offered by Banner and Gagné (1995). Operations research techniques often demonstrate remarkable success at solving well-defined local problems. However, the overall performance of a robotic facility depends upon successful decision-making over different time horizons and at different organizational levels. The means by which the improvements offered by operations research techniques can be translated into overall system performance is a *decision support system*. This system facilitates communication and coordination within the organization. A valuable reference source for background on decision support systems for manufacturing is Parsaei, Kolli, and Hanley (1997). Suri and Whitney (1984) and Lee and Stecke (1996) discuss decision support for automated manufacturing systems. In this section we first discuss the integration of operations research techniques into the decision-making structure of a robotic facility, then provide an overview of a decision support system for such a structure.

An overview of a typical decision-making structure for a robotic facility appears in Table 3. The structure divides into three time horizons, for long-term, medium-term, and short-term decisions, respectively. It should be noted that Sections 2 through 4 above are organized by phase during the *adoption* of a robotic system, whereas all the three time horizons (and corresponding managerial levels) discussed below relate to the *continuing* management and control of a robotic facility. Therefore, there is no one-to-one correspondence between those three sections and the three time horizons of decision-making discussed below.

The first time horizon of decision-making is long-term decisions. These are typically performed by upper-level management over a time horizon of several months or years. Typical long-term decisions include system modification or expansion, major changes in product mix, and the setting of general economic targets. Several relevant and valuable operations research techniques are summarized in the first row, last column of Table 3. Software to support these techniques can be implemented on a workstation computer, for example. Further guidance can be offered by extended part-programming and program-

Table 3 Decision-Making Structure in a Robotic Facility

Time Horizon	Management Level	Typical Tasks	Typical Operations Research Techniques
long-term (months/years)	upper management	system modification or expansion	decision analysis, analytic hierarchy process, queueing models
		product mix changes	linear programming, simulation
		setting of economic targets	forecasting
medium-term (days/weeks)	robotic facility manager	batching tasks	batching
		capacity planning	mixed integer and integer programming, linear programming
		responding to higher-level changes	simulation
short-term (minutes/hours)	robotic facility line supervisor or computer	scheduling operations	scheduling and sequencing
		material-handling and tooling assignments	scheduling and sequencing, heuristics
		problem resolution	simulation

verification tools. Also, the performance of the facility will need to be evaluated, and plans for implementing improvements will need to be developed. Coordination between functional areas of business, e.g., between marketing and product mix decisions, is critical.

The second time horizon is medium-term decisions. These are typically made by a robotic system line manager over a time horizon of several days or weeks. Medium-term decisions include initial batching of production, and planning the availability and effective utilization of resources such as production and material-handling capacity. It may also be necessary to respond to changes in long-term production plans or in material availability. Appropriate operations research techniques to assist in these decisions are shown in the second row, last column of Table 3. Software to support them can be implemented, for example, on a personal computer that is networked to other control functions within the plant.

The third time horizon is short-term decisions. These are typically made by a production system line controller over a time horizon of several minutes or hours. They can also be made automatically as a result of system monitoring by a computer. Short-term decisions include scheduling operations, determining material-handling assignments, resolving temporary bottlenecks and breakdowns, and managing tools and other peripheral equipment. Useful operations research techniques to assist in these decisions are shown in the third row, last column of Table 3. Supporting software can be implemented on a networked personal computer.

As important as decision-making is *within* each level of the organization, so also are the issues of *communication* and *coordination* between levels. Figure 11 illustrates the flow of information between management levels within the organization. Information has to be transferred efficiently from each level to the others, and detailed procedures for such transfers should be established and reviewed regularly. An effective decision support system, as discussed below, is needed to achieve this.

The term decision support system refers to the interaction between human and machine in reaching decisions (Hussain and Hussain, 1981). As discussed by Davis and Hamilton (1993), a decision support system is not intended to replace human decision-makers, but rather to enhance human judgment. Nof (1983) and Lockamy and Cox (1997) provide concise introductions to decision support systems for manufacturing. Within the context of a robotic facility, a decision support system provides a link between the strategic and tactical planning processes illustrated in Figure 11, and supports decision-making for managerial and operational control. The types of information handled by a decision support system in a robotic facility include product and process specifications, order requirements, availability of material, capacity requirements and capabilities, and production

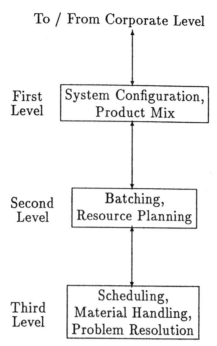

Figure 11   Integration of organizational decision-making in a robotic facility.

schedule details. This information supports strategic decisions about the conversion of resources into outputs and tactical decisions to control the flow of materials. Another important use of this information, as discussed by Hayes, Wheelwright, and Clark (1988), is to provide performance feedback to help with ongoing operational improvements.

An overview of a decision support system for a robotic facility appears in Figure 12. Levary (1997) provides a detailed discussion of a similar system for computer-integrated manufacturing. Some comments about the components of that system now follow. Kaplan (1990) suggests that traditional accounting systems are often not modified sufficiently to allocate costs in a meaningful way in automated manufacturing systems and discusses several alternative approaches. Deterministic and probabilistic operations research techniques are represented as two separate inputs because their role in decision support is conceptually different. Deterministic techniques typically offer a direct (optimal or heuristic) prescription regarding decisions. Probabilistic techniques, by contrast, typically perform an evaluation of a given decision or set of decisions, which can then be compared with similar evaluations for alternative decisions. The former approach offers a much more direct way of identifying good decisions. However, it is not always possible to model a problem deterministically. Because both types of operations research techniques are essentially quantitative and implemented through software, which makes the inclusion of subjective information difficult, that information is represented separately.

There are two distinct databases, an engineering database and a business database. The engineering database includes the characteristics of the product, the robotic facility, and the manufacturing process. Information about the current technology and opportunities for system enhancement is also valuable. The business database includes information on inventory levels, resource availability and production capacity, cash flow, and sales and distribution characteristics. In addition, the accounting system needs to be maintained in the business database. Information from external sources includes, for example, changes in the regulatory environment.

To conclude, appropriate hardware and software components need to be integrated into the organizational decision-making structure in order to maximize the return of a robotic facility. The resulting decision support system should be considered an integral part of the facility, and operation of the facility should explicitly provide for all the components of such a decision support system. It is an effective decision support system that permits

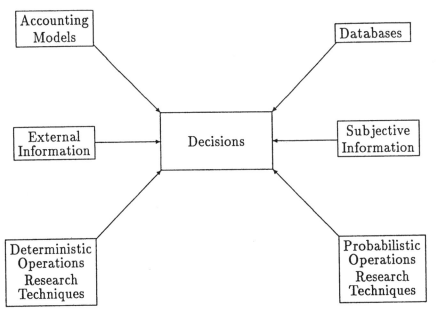

Figure 12 Decision support system in a robotic facility.

the sophisticated techniques of operations research to be fully realized in the performance of a world-class robotic facility.

ACKNOWLEDGMENTS

The author thanks Professor Rajan Suri (University of Wisconsin–Madison) for permission to use materials from Chapter 31 of the 1st edition of this handbook. This project is supported by the Summer Fellowship Program, Fisher College of Business, The Ohio State University. The author thanks Professor Marc Posner (The Ohio State University) for detailed comments on an earlier draft of this chapter.

REFERENCES

Introduction to Operations Research

Handbooks in Operations Research and Management Science. Amsterdam: North-Holland, 1989.

Hillier, F. S., and G. J. Lieberman. 1995. *Introduction to Operations Research.* 6th ed. New York: McGraw-Hill.

Lenstra, J. K., A. H. G. Rinnooy Kan, and A. Schrijver. 1991. *History of Mathematical Programming.* Amsterdam: North-Holland.

Nahmias, S. 1993. *Production and Operations Analysis.* 2nd ed. Homewood: Irwin.

OR/MS Resource Directory. *OR/MS Today* (December), 61–75, 1997.

Capacity Expansion and Technology Acquisition

Canada, J. R., and W. G. Sullivan. 1989. *Economic and Multiattribute Evaluation of Advanced Manufacturing Systems.* Englewood Cliffs: Prentice-Hall.

Demmel, J. G., and R. G. Askin. 1992. "A Multi-Objective Decision Model for the Evaluation of Advanced Manufacturing System Technologies." *Journal of Manufacturing Systems* **11,** 179–194.

Fine, C. H., and R. Freund. 1990. "Optimal Investment in Product-Flexible Manufacturing Capacity." *Management Science* **36,** 449–466.

Freidenfelds, J. 1981. *Capacity Expansion: Analysis of Simple Models with Applications.* New York: North-Holland.

Kassicieh, S. K., H. V. Ravinder, and S. A. Yourstone. 1997. "A Decision Support System for the Justification of Computer-Integrated Manufacturing." In *Manufacturing Decision Support Systems.* Ed. H. R. Parsaei, S. Kolli, and T. R. Hanley. London: Chapman & Hall.

Khouja, M., and O. F. Offodile. 1994. "The Industrial Robots Selection Problem: Literature Review and Directions for Future Research." *IIE Transactions* **26**(4), 50–61.

Manne, A. S., ed. 1967. *Investments for Capacity Expansion: Size, Location and Time Phasing.* Cambridge: MIT Press.

McCardle, K. F. 1985. "Information Acquisition and the Adoption of New Technology." *Management Science* **31**, 1372–1389.

Monahan, G. E., and T. L. Smunt. 1989. "Optimal Acquisition of Automated Flexible Manufacturing Processes." *Operations Research* **37**, 288–300.

Strategic Planning

Cohen, J. S., and J. Zysman. 1987. *Manufacturing Matters: The Myth of the Post-Industrial Economy.* New York: Basic Books.

Goldhar, J. P., and M. Jelinek. 1983. "Plan for Economies of Scope." *Harvard Business Review* **61**, 141–148.

Hayes, R. H., and S. C. Wheelwright. 1984. *Restoring Our Competitive Edge: Competing Through Manufacturing.* New York: John Wiley & Sons.

Porter, M. E. 1980. *Competitive Strategy: Techniques for Analyzing Industries and Competitors.* New York: The Free Press.

Skinner, W. 1974. "The Focused Factory." *Harvard Business Review* **52**, 113–121.

Forecasting

Box, G. E. P., and G. M. Jenkins. 1970. *Time Series Analysis, Forecasting and Control.* San Francisco: Holden Day.

Gross, C. W., and R. T. Peterson. 1983. *Business Forecasting.* 2nd ed. New York: John Wiley & Sons.

Montgomery, D. C., and L. A. Johnson. 1976. *Forecasting and Time Series Analysis.* New York: McGraw-Hill.

Sanders, N. R., and K. B. Manrodt. 1994. "Forecasting Practices in U.S. Corporations: Survey Results." *Interfaces* **24**, 92–100.

Yurkiewicz, J. 1996. "Forecasting Software Survey." *OR/MS Today* (December), 70–75.

Decision Analysis

Buede, D. 1996. "Aiding Insight III." *OR/MS Today* (August), 73–79.

Bunn, D. W. 1984. *Applied Decision Analysis.* New York: McGraw-Hill.

Fishburn, P. C. 1989. "Foundations of Decision Analysis: Along the Way." *Management Science* **35**, 387–405.

Goodwin, P., and G. Wright. 1991. *Decision Making under Uncertainty.* Englewood Cliffs: Prentice-Hall.

Keeney, R. L., and H. Raiffa. 1976. *Decisions with Multiple Objectives.* New York: John Wiley & Sons.

Mehrez, A., O. F. Offodile and B.-H. Ahn. 1995. "A Decision Analysis View of the Effect of Robot Repeatability on Profit." *IIE Transactions* **27**, 60–71.

Integer and Mixed Integer Programming

Berrada, M., and K. E. Stecke. 1986. "A Branch and Bound Approach for Machine Load Balancing in Flexible Manufacturing Systems." *Management Science* **32**, 1316–1335.

Garey, M. R., and D. S. Johnson. 1979. *Computers and Intractability: A Guide to the Theory of NP-Completeness.* San Francisco: Freeman.

Johnson, E. L., M. M. Kostreva, and U. H. Suhl. 1985. "Solving 0–1 Integer Programming Problems Arising From Large-Scale Planning Models." *Operations Research* **33**, 803–819.

Nemhauser, G. L., and L. A. Wolsey. 1988. *Integer Programming and Combinatorial Optimization.* New York: John Wiley & Sons.

Salkin, H., and K. Mathur. 1989. *Foundations of Integer Programming.* New York: North-Holland.

Saltzman, M. J. 1994. "Broad Selection of Software Packages Available." *OR/MS Today* (April), 42–51.

Stecke, K. E. 1983. "Formulation and Solution of Nonlinear Integer Production Planning Problems for Flexible Manufacturing Systems." *Management Science* **29**, 273–288.

Linear Programming

Bazaraa, M. S., J. J. Jarvis, and H. D. Sherali. 1990. *Linear Programming and Network Flows.* 2nd ed. New York: John Wiley & Sons.

Bixby, R. E., et al. 1992. "Very Large-Scale Linear Programming: a Case Study in Combining Interior Point and Simplex Methods." *Operations Research* **40**, 885–897.

Bradley, S. P., A. C. Hax, and T. L. Magnanti. 1977. *Applied Mathematical Programming.* Reading: Addison-Wesley.

Charnes, A., W. W. Cooper, and R. O. Ferguson. 1955. "Optimal Estimation of Executive Compensation by Linear Programming." *Management Science* **1,** 138–151.

Fourer, R. 1997. "Software Survey: Linear Programming." *OR/MS Today* (April), 54–63.

Ignizio, J. P. 1976. *Goal Programming and Extensions.* Lexington: Heath.

Imany, M. M., and R. J. Schlesinger. 1989. "Decision Models for Robot Selection: A Comparison of Ordinary Least Squares and Linear Goal Programming Methods." *Decision Sciences* **20,** 40–53.

Lustig, I. J., R. E. Marsten, and D. F. Shanno. 1994. "Interior-Point Methods for Linear Programming: Computational State of the Art." *ORSA Journal on Computing* **6,** 1–14.

Schrage, L. 1991. *LINDO: An Optimization Modeling System, Text and Software.* 4th ed. Danvers: Boyd & Fraser.

Schrijver, A. 1986. *Theory of Linear and Integer Programming.* New York: John Wiley & Sons.

Williams, H. P. 1990. *Model Building in Mathematical Programming.* 3rd ed. New York: John Wiley & Sons.

Analytic Hierarchy Process

Arbel, A., and A. Seidmann. 1984. "Performance Evaluation of Flexible Manufacturing Systems." *IEEE Transactions on Systems, Man, and Cybernetics* **SMC-14**(4), 606–617.

Bard, J. F. 1986. "A Multiobjective Methodology for Selecting Subsystem Automation Options." *Management Science* **32,** 1628–1641.

Charnes, A., W. W. Cooper, and E. Rhodes. 1978. "Measuring the Efficiency of Decision Making Units." *European Journal of Operational Research* **2,** 429–444.

Farrell, M. J. 1957. "The Measurement of Productive Efficiency." *Journal of The Royal Statistical Society, Series A* **120,** 253–290.

Frazelle, E. 1985. "Suggested Techniques Enable Multi-Criteria Evaluation of Material Handling Alternatives." *Industrial Engineering* **17,** 42–48.

Golden, B. L., E. A. Wasil, and D. E. Levy. 1989. "Applications of the Analytical Hierarchy Process: A Categorized, Annotated Bibliography." In *The Analytic Hierarchy Process: Applications and Studies.* Ed. B. L. Golden, E. A. Wasil, and P. T. Harker. Berlin: Springer-Verlag. 37–58.

Lipovetsky, S. 1996. "The Synthetic Hierarchy Method: an Optimizing Approach to Obtaining Priorities in the AHP." *European Journal of Operational Research* **93,** 550–564.

Saaty, T. L. 1980. *The Analytic Hierarchy Process.* New York: McGraw-Hill.

———. 1986. "Axiomatic Foundation of the Analytic Hierarchy Process." *Management Science* **32,** 841–855.

Seidmann, A., A. Arbel, and R. Shapira. 1984. "A Two-Phase Analytic Approach to Robotic System Design." *Robotics and Computer-Integrated Manufacturing* **1,** 181–190.

Sengupta, J. 1995. *Dynamics of Data Envelopment Analysis: Theory of Systems Efficiency.* Norwell: Kluwer.

Shang, J., and T. Sueyoshi. 1995. "A Unified Framework for the Selection of a Flexible Manufacturing System." *European Journal of Operational Research* **85,** 297–315.

Sullivan, W. G. 1986. "Models IEs Can Use to Include Strategic, Non-Monetary Factors in Automation Decisions." *Industrial Engineering* **18,** 42–50.

Varney, M. S., W. G. Sullivan, and J. K. Cochran. 1985. "Justification of Flexible Manufacturing Systems with the Analytical Hierarchy Process." In *Focus on the Future: Proceedings of the 1985 Industrial Engineering Show.* Ed. C. Savage. Atlanta: Industrial Engineering & Management Press.

Zahedi, F. 1986. "The Analytic Hierarchy Process—A Survey of the Method and Its Applications." *Interfaces* **16,** 96–108.

Heuristics

Adams, J., E. Balas, and D. Zawack. 1988. "The Shifting Bottleneck Procedure for Job Shop Scheduling." *Management Science* **34,** 391–401.

Ahmadi, R. H., and C. S. Tang. 1991. "An Operation Partitioning Problem for Automated Assembly System Design." *Operations Research* **39,** 824–835.

Balas, E. 1980. "An Algorithm for Large Zero–One Knapsack Problems." *Operations Research* **28,** 1130–1154.

Ball, M. O., and M. J. Magazine. 1988. "Sequencing of Insertions in Printed Circuit Board Assembly." *Operations Research* **36,** 192–201.

Heragu, S. S., and A. Kusiak. 1988. "Machine Layout Problem in Flexible Manufacturing Systems." *Operations Research* **36,** 258–268.

Hochbaum, D. S. Ed. 1997. *Approximation Algorithms for NP-Hard Problems.* Boston: P.W.S. Kent.

Hochbaum, D. S., and W. Maass. 1984. "Approximation Schemes for Packing and Covering Problems in Robotics and VLSI." In *Lecture Notes in Computer Science.* Berlin: Springer-Verlag.

Kouvelis, P., and M. W. Kim. 1992. "Unidirectional Loop Network Layout Problem in Automated Manufacturing Systems." *Operations Research* **40,** 1107–1125.

Lane, D. E., and J. B. Sidney. 1993. "Batching and Scheduling in FMS Hubs: Flow Time Considerations." *Operations Research* **41,** 1091–1103.

Lofgren, C. B., L. F. McGinnis, and C. A. Tovey. 1991. "Routing Printed Circuit Cards Through an Assembly Cell." *Operations Research* **39,** 992–1004.

Sethi, S. P., et al. 1990. "Heuristic Methods for Selection and Ordering of Part-Orienting Devices." *Operations Research* **38,** 84–98.

Talbot, F. B., and W. V. Gehrlein. 1986. "A Comparative Evaluation of Heuristic Line Balancing Techniques." *Management Science* **32,** 430–454.

Location and Layout

Aleksander, I., and H. Morton. 1990. *An Introduction to Neural Computing.* London: Chapman & Hall.

Apple, J. M. 1977. *Plant Layout and Material Handling.* 3rd ed. New York: John Wiley & Sons.

Block, T. E. 1977. "A Note on 'Comparison of Computer Algorithms and Visual Based Methods for Plant Layout' by M. Scriabin and R. C. Vergin." *Management Science* **24,** 235–237.

Bozer, Y. A., R. D. Meller, and S. J. Erlebacher. 1994. "An Improvement-Type Layout Algorithm for Single and Multiple-Floor Facilities." *Management Science* **40,** 918–932.

Brown, D. E., and C. C. White, III. 1990. *Operations Research and Artificial Intelligence: The Integration of Problem-Solving Strategies.* Norwell: Kluwer.

Cinar, U. 1975. "Facilities Planning: A Systems Analysis and Space Allocation Approach." In *Spatial Synthesis in Computer-Aided Building Design.* Ed. C. M. Eastman. New York: John Wiley & Sons.

Donaghey, C. E., and V. F. Pire. 1991. *BLOCPLAN-90 User's Manual.* Industrial Engineering Department, University of Houston.

Drezner, Z., and S. Y. Nof. 1984. "On Optimizing Bin Picking and Insertion Plans for Assembly Robots." *IIE Transactions* **16,** 262–270.

Erlenkotter, D. 1978. "A Dual-Based Procedure for Uncapacitated Facility Location." *Operations Research* **26,** 992–1009.

Francis, R. L., L. F. McGinnis, Jr., and J. A. White. 1983. "Locational Analysis." *European Journal of Operational Research* **12,** 220–252.

———. 1992. *Facility Layout and Location: An Analytical Approach.* 2nd ed. Englewood Cliffs: Prentice-Hall.

Francis, R. L., et al. 1994. "Finding Placement Sequences and Bin Locations for Cartesian Robots." *IIE Transactions* **26,** 47–59.

Gavish, B., D. Horsky and K. Srikanth. 1983. "An Approach to the Optimal Positioning of a New Product." *Management Science* **29,** 1277–1297.

Handler, G. Y., and P. B. Mirchandani. 1979. *Location on Networks.* Cambridge: M.I.T. Press.

Heragu, S. 1997. *Facilities Design.* Boston: PWS.

Johnson, R. V. 1982. "SPACECRAFT for Multi-Floor Layout Planning." *Management Science* **28,** 407–417.

Jungthirapanich, C., and C. O. Benjamin. 1995. "A Knowledge-Based Decision Support System for Locating a Manufacturing Facility." *IIE Transactions* **27,** 789–799.

Meller, R. D., and K.-Y. Gau. 1996. "The Facility Layout Problem: Recent and Emerging Trends and Perspectives." *Journal of Manufacturing Systems* **15,** 351–366.

Michalewicz, Z. 1996. *Genetic Algorithms + Data Structures = Evolution Programs.* 3rd ed. Berlin: Springer-Verlag.

Rampersad, H. K. 1995. "Concentric Design of Robotic Assembly Systems." *Journal of Manufacturing Systems* **14,** 230–243.

Rosenblatt, M. 1986. "The Dynamics of Plant Layout." *Management Science* **32,** 323–332.

Ross, G. T., and R. M. Soland. 1978. "Modeling Facility Location Problems as Generalized Assignment Problems." *Management Science* **24,** 345–357.

Schmenner, R. W. 1982. *Making Business Location Decisions.* Englewood Cliffs: Prentice-Hall.

Scriabin, M., and R. C. Vergin. 1975. "Comparison of Computer Algorithms and Visual Based Methods for Plant Layout." *Management Science* **22,** 172–181.

Tate, D. M., and A. E. Smith. 1995. "Unequal-Area Facility Layout by Genetic Search." *IIE Transactions* **27,** 465–472.

Tompkins, J. A., et al. 1996. *Facilities Planning.* 2nd ed. New York: John Wiley & Sons.

Tsuchiya, K., S. Bharitkar, and Y. Takefuji. 1996. "A Neural Network Approach to Facility Layout Problem." *European Journal of Operational Research* **89,** 556–563.

Urban, T. L. 1993. "A Heuristic for the Dynamic Facility Layout Problem." *IIE Transactions* **25,** 57–63.

Warwick, K. 1995. "Neural Networks: An Introduction." In *Neural Network Applications in Control.* Ed. G. W. Irwin, K. Warwick, and K. J. Hunt. Stevenage: The Institution of Electrical Engineers. 1–16.

Group Technology and Cell Formation

Askin, R. G., and K. S. Chiu. 1990. "A Graph Partitioning Procedure for Machine Assignment and Cell Formation in Group Technology." *International Journal of Production Research* **28,** 1555–1572.

Bedworth, D. D., M. R. Henderson, and P. M. Wolfe. 1991. *Computer Integrated Design and Manufacturing.* New York: McGraw-Hill.

Burbidge, J. L. 1975. *The Introduction of Group Technology.* New York: John Wiley & Sons.

Burke, L., and S. Kamal. 1995. "Neural Networks and the Part Family/Machine Group Formation Problem in Cellular Manufacturing: A Framework Using Fuzzy ART." *Journal of Manufacturing Systems* **14,** 148–159.

Chu, C. H., and P. Pan. 1988. "The Use of Clustering Techniques in Manufacturing Cellular Formation." In *Proceedings, International Industrial Engineering Conference,* Orlando. 495–500.

Kaparthi, S., N. C. Suresh, and R. P. Cerveny. 1993. "An Improved Neural Network Leader Algorithm for Part-Machine Grouping in Group Technology." *European Journal of Operational Research* **69,** 342–356.

Kusiak, A. 1985. "The Parts Families Problem in FMSs." *Annals of Operations Research* **3,** 279–300.

———. 1988. "EXGT-S: A Knowledge Based System for Group Technology." *International Journal of Production Research* **26,** 887–904.

Mitrofanov, S. P. 1966. *The Scientific Principles of Group Technology.* Boston Spa, Yorkshire: National Lending Library Translation.

Mohamed, Z. M. 1996. "A Flexible Approach to (Re)Configure Flexible Manufacturing Cells." *European Journal of Operational Research* **95,** 566–576.

Ng, S. M. 1996. "On the Characterization and Measure of Machine Cells in Group Technology." *Operations Research* **44,** 735–744.

Rajagopalan, R., and J. L. Batra. 1975. "Design of Cellular Production Systems—A Graph Theoretic Approach." *International Journal of Production Research* **13,** 567–579.

Singh, N. 1993. "Design of Cellular Manufacturing Systems: An Invited Review." *European Journal of Operational Research* **69,** 284–291.

Vohra, T., et al. 1990. "A Network Approach to Cell Formation in Cellular Manufacturing." *International Journal of Production Research* **28,** 2075–2084.

Xie, X. 1993. "Manufacturing Cell Formation Under Capacity Constraints." *Applied Stochastic Models and Data Analysis* **9,** 87–96.

Queueing Models

Gross, D., and C. M. Harris. 1985. *Fundamentals of Queueing Theory.* 2nd ed. New York: John Wiley & Sons.

Kleinrock, L. 1975. *Theory.* Vol. 1 of *Queueing Systems.* New York: John Wiley & Sons.

———. 1976. *Computer Applications.* Vol. 2 of *Queueing Systems.* New York: John Wiley & Sons.

Little, J. D. C. 1961. "A Proof of the Queueing Formula $L = \lambda W$." *Operations Research* **9,** 383–387.

Ram, R., and N. Viswanadham. 1990. "Stochastic Analysis of Versatile Workcenters." *Sadhana, Indian Academy Proceedings in Engineering Sciences* **15,** 301–317.

Rothkopf, M. H., and P. Rech. 1987. "Perspectives on Queues: Combining Queues Is Not Always Beneficial." *Operations Research* **35,** 906–909.

Stidham, S. 1974. "A Last Word on $L = \lambda W$." *Operations Research* **22,** 417–421.

Trivedi, K. S. 1982. *Probability and Statistics with Reliability, Queueing and Computer Science Applications.* Englewood Cliffs: Prentice-Hall.

Viswanadham, N., and Y. Narahari. 1992. *Performance Modeling of Automated Manufacturing Systems.* Englewood Cliffs: Prentice-Hall.

Queueing Networks

Baskett, F., et al. 1975. "Open, Closed and Mixed Networks of Queues with Different Classes of Customers." *Journal of the Association for Computing Machinery* **22,** 248–260.

Bitran, G. R., and S. Dasu. 1992. "A Review of Open Queueing Network Models of Manufacturing Systems." *Queueing Systems: Theory and Applications* [Special Issue on Queueing Models of Manufacturing Systems] **12,** 95–133.

Bitran, G. R., and D. Tirupati. 1989. "Tradeoff Curves, Targeting and Balancing in Manufacturing Queueing Networks." *Operations Research* **37,** 547–564.

Buzacott, J. A., and J. G. Shanthikumar. 1992. "Design of Manufacturing Systems Using Queueing Models." *Queueing Systems: Theory and Applications* [Special Issue on Queueing Models of Manufacturing Systems] **12,** 135–213.

———. 1993. *Stochastic Models of Manufacturing Systems.* Englewood Cliffs: Prentice-Hall.

Calabrese, J. M. 1992. "Optimal Workload Allocation in Open Networks of Multiserver Queues." *Management Science* **38,** 1792–1802.

Chen, H., et al. 1988. "Empirical Evaluation of a Queueing Network Model for Semiconductor Wafer Fabrication." *Operations Research* **36**, 202–215.

Dallery, Y., and K. E. Stecke. 1990. "On the Optimal Allocation of Servers and Workloads in Closed Queueing Networks." *Operations Research* **38**, 694–703.

Gershwin, S. B. 1994. *Manufacturing Systems Engineering.* Englewood Cliffs: Prentice-Hall.

Gordon, W. J., and G. F. Newell. 1967. "Closed Queuing Systems with Exponential Servers." *Operations Research* **15**, 254–265.

Jackson, J. R. 1957. "Networks of Waiting Lines." *Operations Research* **5**, 518–527.

———. 1963. "Jobshop-Like Queueing Systems." *Management Science* **10**, 131–142.

Liu, Z., and P. Nain. 1989. "Sensitivity Results in Open, Closed and Mixed Product Form Queuing Networks." INRIA Research Report, No. 1144, INRIA, Le Chesnay, France.

Robertazzi, T. G. 1994. *Computer Networks and Systems: Queueing Theory and Performance Evaluation.* 2nd ed. New York: Springer-Verlag.

Seidmann, A., P. J. Schweitzer, and S. Shalev-Oren. 1987. "Computerized Closed Queueing Network Models of Flexible Manufacturing Systems: A Comparative Evaluation." *Large Scale Systems* **12**, 91–107.

Solberg, J. J. 1977. "A Mathematical Model of Computerized Manufacturing Systems." In *Proceedings of 4th International Conference on Production Research,* Tokyo. 22–30.

———. 1980. *CAN-Q User's Guide,* Report No. 9 (revised), School of Industrial Engineering, Purdue University.

Suri, R., and G. W. Diehl. 1985. "MANUPLAN: A Precursor to Simulation for Complex Manufacturing Systems." In *Proceedings of 1985 Winter Simulation Conference.* 411–420.

Suri, R., and R. R. Hildebrandt. 1984. "Modeling Flexible Manufacturing Systems Using Mean Value Analysis." *Journal of Manufacturing Systems* **3**, 27–38.

Walrand, J. 1988. *An Introduction to Queueing Networks.* Englewood Cliffs: Prentice-Hall.

Yao, D. D., and J. A. Buzacott. 1987. "Modeling a Class of FMSs with Reversible Routing." *Operations Research* **35**, 87–92.

Simulation

Banks, J. 1996. "Interpreting Simulation Software Checklists." *OR/MS Today* (June), 74–78.

Bevans, J. P. 1982. "First Choose an FMS Simulator." *American Machinist* (May), 143–145.

Carrie, A. 1988. *Simulation of Manufacturing Facilities.* New York: John Wiley & Sons.

Christy, D. P., and H. J. Watson. 1993. "The Application of Simulation: A Survey of Industry Practice." *Interfaces* **13**, 47–52.

De Lurgio, S. A., Sr., S. T. Foster, Jr., and G. Dickerson. 1997. "Utilizing Simulation to Develop Economic Equipment Selection and Sampling Plans for Integrated Circuit Manufacturing." *International Journal of Production Research* **35**, 137–155.

Gordon, G. 1975. *The Application of GPSS V to Discrete System Simulation.* Englewood Cliffs: Prentice-Hall.

Halder, S. W., and J. Banks. 1986. "Simulation Software Products for Analyzing Manufacturing Systems." *Industrial Engineering* (July), 98–103.

Hlupic, V., and R. J. Paul. 1995. "A Critical Evaluation of Four Manufacturing Simulators." *International Journal of Production Research* **33**, 2757–2766.

Kleijnen, J. P. C. 1987. *Statistical Tools for Simulation Practitioners.* New York: Marcel Dekker.

———. 1995. "Verification and Validation of Simulation Models." *European Journal of Operational Research* **82**, 145–162.

Kleijnen, J. P. C., and R. Y. Rubinstein. 1996. "Optimization and Sensitivity Analysis of Computer Simulation Models by the Score Function Method." *European Journal of Operational Research* **88**, 413–427.

Law, A. M. 1986. "Introduction to Simulation: A Powerful Tool for Complex Manufacturing Systems." *Industrial Engineering* (May), 46–63.

Law, A. M., and W. D. Kelton. 1991. *Simulation Modeling and Analysis.* 2nd ed. New York: McGraw-Hill.

Morris, J. S., and R. J. Tersine. 1990. "A Simulation Analysis of Factors Influencing the Attractiveness of Group Technology Cellular Layouts." *Management Science* **36**, 1567–1578.

Phillips, D. T., and M. Handwerker. 1979. "GEMS: A Generalized Manufacturing Simulator." In *Proceedings of the 12th International Conference on Systems Science,* Honolulu, January.

Pritsker, A. A. B., and C. D. Pegden. 1979. *Introduction to Simulation and SLAM.* New York: John Wiley & Sons.

Runner, J. A., and E. F. Leimkuhler. 1978. "CAMSAM: A Simulation Analysis Model for Computer-Aided Manufacturing Systems." Report 13, School of Industrial Engineering, Purdue University.

Schriber, T. J. 1991. *An Introduction to Simulation.* New York: John Wiley & Sons.

"Simulation Software Buyer's Guide." *IIE Solutions* (May), 64–77, 1997.

Swain, J. J. 1995. "Simulation Survey: Tools for Process Understanding and Improvement." *OR/MS Today* (August), 64–72.

Trunk, C. 1989. "Simulation for Success in the Automated Factory." *Materials Handling Engineering* (May), 64–76.

Reliability and Maintenance

Barlow, R. E., and F. Proschan. 1965. *Mathematical Theory of Reliability.* New York: John Wiley & Sons.

———. 1975. *Statistical Theory of Reliability and Life Testing.* New York: Holt, Rinehart & Winston.

Chung, C. S., and J. Flynn. 1995. "A Branch-and-Bound Algorithm for Computing Optimal Replacement Policies in K-Out-Of-N Systems." *Operations Research* **43**, 826–837.

Feldman, R. M., and M. Chen. 1996. "Strategic and Tactical Analyses for Optimal Replacement Policies," *IIE Transactions* **28**, 987–993.

Hwang, F. K., and U. G. Rothblum. 1994. "Optimality of Monotone Assemblies for Coherent Systems Composed of Series Modules." *Operations Research* **42**, 709–720.

Kelly, C. M., C. T. Mosier, and F. Mahmood. 1997. "Impact of Maintenance Policies on the Performance of Manufacturing Cells." *International Journal of Production Research* **35**, 767–787.

Pierskalla, W. P., and J. A. Voelker. 1976. "A Survey of Maintenance Models: The Control and Surveillance of Deteriorating Systems." *Naval Research Logistics Quarterly* **23**, 353–388.

Prasad, V. R., K. P. K. Nair, and Y. P. Aneja. 1991. "Optimal Assignment of Components to Parallel-Series and Series-Parallel Systems." *Operations Research* **39**, 407–414.

Smith, D. J. 1993. *Reliability, Maintainability and Risk: Practical Methods for Engineers.* 4th ed. Oxford: Butterworth Heinemann.

Stiffler, J. J., et al. 1979. "CARE III Final Report, Phase I." NASA Contractor Report 159122.

U.S. Nuclear Regulatory Commission. 1975. *Reactor Safety Study.*

Ushakov, I. A., ed. 1994. *Handbook of Reliability Engineering.* New York: John Wiley & Sons.

Vanneste, S. G., and L. N. Van Wassenhove. 1995. "An Integrated and Structured Approach to Improve Maintenance." *European Journal of Operational Research* **82**, 241–257.

Batching

Afentakis, P., B. Gavish, and U. Karmarkar. 1984. "Computationally Efficient Optimal Solutions to the Lot-Sizing Problem in Multistage Assembly Systems." *Management Science* **30**, 222–239.

Agnetis, A., D. Pacciarelli, and F. Rossi. 1997. "Batch Scheduling in a Two-Machine Flow Shop with Limited Buffer." *Discrete Applied Mathematics* **72**, 243–260.

Agnetis, A., F. Rossi, and G. Gristina. 1996. "An Exact Algorithm for the Batch Sequencing Problem in a Two-Machine Flowshop with Limited Buffer." Working Paper, Dipartimento di Informatica e Sistemistica, Universita degli Studi di Roma "La Sapienza."

Albers, S., and P. Brucker. 1993. "The Complexity of One-Machine Batching Problems." *Discrete Applied Mathematics* **47**, 87–107.

Carlson, R. C., J. V. Jucker, and D. H. Kropp. 1983. "Heuristic Lot Sizing Approaches for Dealing with MRP System Nervousness." *Decision Sciences* **14**, 156–169.

Crowston, W. B., and H. M. Wagner. 1973. "Dynamic Lot Size Models for Multi-Stage Assembly Systems." *Management Science* **20**, 14–21.

Elmaghraby, S. E. 1978. "The Economic Lot Scheduling Problem (ELSP): Review and Extensions." *Management Science* **24**, 587–598.

Monma, C. L., and C. N. Potts. 1989. "On the Complexity of Scheduling with Batch Setup Times." *Operations Research* **37**, 798–804.

Peterson, R., and E. A. Silver. 1979. *Decision Systems for Inventory Management and Production Planning.* New York: John Wiley & Sons.

Wagner, H. M., and T. Whitin. 1958. "Dynamic Version of the Economic Lot Scheduling Problem." *Management Science* **5**, 89–96.

Webster, S., and K. R. Baker. 1995. "Scheduling Groups of Jobs on a Single Machine." *Operations Research* **43**, 692–703.

Scheduling

Armstrong, R., S. Gu, and L. Lei. 1996. "A Greedy Algorithm to Determine the Number of Transporters in a Cyclic Electroplating Process." *IIE Transactions* **28**, 347–355.

Armstrong, R., L. Lei, and S. Gu. 1994. "A Bounding Scheme for Deriving the Minimal Cycle Time of a Single-Transporter N-Stage Process with Time Window Constraints." *European Journal of Operational Research* **78**, 130–140.

Brauner, N., and G. Finke. 1997. "Final Results on the One-Cycle Conjecture in Robotic Cells." Working Paper, Laboratoire LEIBNIZ, Institut IMAG, Grenoble, France.

Crama, Y. 1997. "Combinatorial Optimization Models for Production Scheduling in Automated Manufacturing Systems." *European Journal of Operational Research* **99**, 136–153.

Crama, Y., and J. van de Klundert. 1997. "Cyclic Scheduling in 3-Machine Robotic Flow Shops." Working Paper, École d'Administration des Affaires, Université de Liège, Belgium.

————. 1998. "Cyclic Scheduling of Identical Parts in a Robotic Cell." *Operations Research* **45,** 952–965.

Geiger, C. D., K. G. Kempf, and R. Uzsoy. 1997. "A Tabu Search Approach to Scheduling an Automated Wet Etch Station." *Journal of Manufacturing Systems* **16,** 102–115.

Glass, C. A., Y. M. Shafransky, and V. A. Strusevich. 1996. "Scheduling for Parallel Dedicated Machines with a Single Server." Working Paper, Faculty of Mathematical Studies, University of Southampton.

Hall, N. G., H. Kamoun, and C. Sriskandarajah. 1997. "Scheduling in Robotic Cells: Classification, Two and Three Machine Cells." *Operations Research* **45,** 421–439.

Hall, N. G., and C. Sriskandarajah. 1996. "A Survey of Machine Scheduling Problems with Blocking and No-Wait in Process." *Operations Research* **44,** 510–525.

————. 1998. "Scheduling in Robotic Cells: Complexity and Steady State Analysis." *European Journal of Operational Research* **109,** 43–65.

Hall, N. G., C. Potts, and C. Sriskandarajah. 1998. "Parallel Machine Scheduling with a Common Server." To appear in *Discrete Applied Mathematics.*

Kamoun, H., N. G. Hall, and C. Sriskandarajah. 1998. "Scheduling in Robotic Cells: Heuristics and Cell Design." To appear in *Operations Research.*

Karuno, Y., H. Nagamochi, and T. Ibaraki. 1997. "Vehicle Scheduling on a Tree with Release and Handling Times." *Annals of Operations Research* **69,** 193–207.

Kats, V., and E. Levner. 1996. "Polynomial Algorithms for Scheduling of Robots." In *Intelligent Scheduling of Robots and Flexible Manufacturing Systems.* Ed. E. Levner. Holon, Israel: CTEH Press. 77–100.

Lee, C.-Y., L. Lei, and M. Pinedo. 1997. "Current Trends in Deterministic Scheduling." *Annals of Operations Research* **70,** 1–42.

Lei, L., R. Armstrong, and S. Gu. 1993. "Minimizing the Fleet Size with Dependent Time Window and Single-Track Constraints." *Operations Research Letters* **14,** 91–98.

Lei, L., and T.-J. Wang. 1991. "The Minimum Common-Cycle Algorithm for Cyclic Scheduling of Two Hoists With Time-Window Constraints." *Management Science* **37,** 1629–1639.

Levner, E., V. Kats, and V. E. Levit. 1997. "An Improved Algorithm for Cyclic Flowshop Scheduling in a Robotic Cell." *European Journal of Operational Research* **97,** 500–508.

Levner, E., V. Kats, and C. Sriskandarajah. 1996. "A Geometric Algorithm for Finding Two-Unit Cyclic Schedules in a No-Wait Robotic Flowshop." In *Intelligent Scheduling of Robots and Flexible Manufacturing Systems.* Ed. E. Levner. Holon, Israel: CTEH Press. 101–112.

Levner, E., K. Kogan, and I. Levin. 1995. "Scheduling a Two-Machine Robotic Cell: A Solvable Case." *Annals of Operations Research* **57,** 217–232.

Levner, E., K. Kogan, and O. Maimon. 1995. "Flowshop Scheduling of Robotic Cells with Job-Dependent Transportation and Setup Effects." *Journal of the Operational Research Society* **46,** 1447–1455.

Li, H., and R. Kumar. 1997. "Optimizing Assembly Time for Printed Circuit Board Assembly." In *Manufacturing Decision Support Systems.* Ed. H. R. Parsaei, S. Kolli, and T. R. Hanley. London: Chapman & Hall. 217–237.

Melnyk, S. 1997. "1997 Finite Capacity Scheduling." *APICS—The Performance Advantage* (August).

Park, K.-S., and M. E. Posner. 1994. "Single Crane Scheduling Problems." Working Paper, Department of Industrial and Systems Engineering, The Ohio State University.

————. 1996. "Scheduling with a Material Handler." Working Paper, Department of Industrial and Systems Engineering, The Ohio State University.

Pinedo, M. 1995. *Scheduling: Theory, Algorithms and Systems.* Englewood Cliffs: Prentice-Hall.

Sethi, S. P., et al. 1992. "Sequencing of Parts and Robot Moves in a Robotic Cell." *International Journal of Flexible Manufacturing Systems* **4,** 331–358.

Sriskandarajah, C., N. G. Hall, and H. Kamoun. 1998. "Scheduling Large Robotic Cells Without Buffers." *Annals of Operations Research* **76,** 287–321.

Hierarchical Production Planning

Bitran, G. R., E. A. Haas, and A. C. Hax. 1981. "Hierarchical Production Planning: A Single-Stage System." *Operations Research* **29,** 717–743.

Caramanis, M., and G. Liberopoulos. 1992. "Perturbation Analysis for the Design of Flexible Manufacturing System Flow Controllers." *Operations Research* **40,** 1107–1125.

Graves, S. C. 1982. "Using Lagrangean Techniques to Solve Hierarchical Production Planning Problems." *Management Science* **28,** 260–275.

Hax, A. C., and H. C. Meal. 1975. "Hierarchical Integration of Production Planning and Scheduling." In *Logistics.* Vol. 1 of *Studies in the Management Sciences.* Amsterdam: North-Holland/TIMS.

Lehoczky, J., et al. 1991. "An Asymptotic Analysis of Hierarchical Control of Manufacturing Systems under Uncertainty." *Mathematics of Operations Research* **16,** 596–608.

Libosvar, C. 1988. "Hierarchies in Production Management and Control: A Survey." M.I.T. Laboratory for Information and Decision Sciences, LIDS-P-1734.

Samaratunga, C., S. P. Sethi, and X. Y. Zhou. 1997. "Computational Evaluation of Hierarchical Production Control Policies for Stochastic Manufacturing Systems." *Operations Research* **45**, 258–274.

Sethi, S. P., and G. L. Thompson. 1981. *Applied Optimal Control and Management Science.* Boston: Martinus Nijhoff.

Sethi, S. P., and Q. Zhang. 1994*a. Hierarchical Decision Making in Stochastic Manufacturing Systems.* Cambridge: Birkhäuser.

———. 1994*b.* "Hierarchical Production Planning in Dynamic Stochastic Manufacturing Systems: Asymptotic Optimality and Error Bounds." *Journal of Mathematical Analysis and Applications* **181**, 285–319.

van Ryzin, G., S. X. C. Lou, and S. B. Gershwin. 1993. "Production Control for a Two-Machine System." *IIE Transactions* **25**, 5–20.

Organizational Decision-Making and Decision Support Systems

Banner, D. K., and T. E. Gagné. 1995. *Designing Effective Organizations: Traditional and Transformational Views.* Thousand Oaks: Sage.

Davis, G. B., and S. Hamilton. 1993. *Managing Information: How Information Systems Impact Organizational Strategy.* Homewood: Business One Irwin.

Hayes, R. H., S. C. Wheelwright, and K. B. Clark. 1988. *Dynamic Manufacturing: Creating the Learning Organization.* New York: The Free Press.

Hussain, D., and K. M. Hussain. 1981. *Information Processing Systems for Management.* Homewood: Irwin.

Kaplan, S. 1990. "The Four-Stage Model of Cost Systems Design." *Management Accounting* **71**, 22–26.

Lee, H. F., and K. E. Stecke. 1996. "An Integrated Decision Support Method for Flexible Assembly Systems." *Journal of Manufacturing Systems* **15**, 13–32.

Levary, R. R. 1997. "Computer-Integrated Manufacturing: A Complex Information System." In *Manufacturing Decision Support Systems.* Ed. H. R. Parsaei, S. Kolli, and T. R. Hanley. London: Chapman & Hall. 280–291.

Lockamy, A., III, and J. F. Cox, III. 1997. "Linking Strategies to Actions: Integrated Performance Measurement Systems for Competitive Advantage." In *Manufacturing Decision Support Systems.* Ed. H. R. Parsaei, S. Kolli, and T. R. Hanley. London: Chapman & Hall. 40–54.

Nof, S. Y. 1983. "Theory and Practice in Decision Support for Manufacturing." In *Data Base Management: Theory and Applications.* Ed. C. W. Holsapple and A. B. Whinston. Dordrecht: Reidel. 325–348.

Parsaei, H. R., S. Kolli, and T. R. Hanley. Ed. 1997. *Manufacturing Decision Support Systems.* London: Chapman & Hall.

Robey, D. 1991. *Designing Organizations.* 3rd ed. Homewood: Irwin.

Suri, R., and C. K. Whitney. 1984. "Decision Support Requirements in Flexible Manufacturing." *Journal of Manufacturing Systems* **3**, 64–69.

CHAPTER **31**

COMPUTATION, AI, AND MULTIAGENT TECHNIQUES FOR PLANNING ROBOTIC OPERATIONS

Venkat N. Rajan
i2 Technologies
Chicago, Illinois

Shimon Y. Nof
Purdue University
West Lafayette, Indiana

1 INTRODUCTION

Planning of robotic operations in discrete product manufacturing involves two stages: (1) a system-independent identification of one or more ways in which the product is put together (*process planning*), and (2) a system-dependent determination of which machine performs which task (*execution planning,* or *sequencing and scheduling*). This chapter deals in general with integrated strategies for process and execution planning. The process planning approaches focus on assembly planning techniques. The execution planning strategies are based on computational techniques, artificial intelligence-based search methods, and game-theoretic multiagent cooperation planning techniques.

2 ASSEMBLY PLANNING

Assembly planning is defined as the process of interpreting design information into assembly process information. The various steps that constitute the assembly planning process include (Rajan and Nof, 1996a):

1. Interpretation of assembly design data
2. Selection of assembly processes
3. Selection of assembly process parameters
4. Selection of assembly equipment/system
5. Selection of auxiliary equipment
6. Assembly sequencing
7. Selection of inspection devices
8. Calculation of assembly times

2.1 Interpretation of Assembly Data

Assembly data consist of structure information, geometry information, and process information. The assembly structure information is generally provided in the form of a bill of material or an indented parts list. These structures generally represent one possible decomposition of the assembly product, the one envisioned by the engineering group. The assembly geometry information may be provided in the form of an exploded assembly drawing, assembly mating detail, or assembly CAD models. In most cases the main

Handbook of Industrial Robotics, Second Edition, Edited by Shimon Y. Nof
ISBN 0-471-17783-0 © 1999 John Wiley & Sons, Inc.

process associated with the assembly joint is also predefined by the selection of the type of joint (welded, bolted, snap fit, bonded, etc.). For some specific types of joints the process parameters may also be specified (such as the size of the bolt or rivet).

The interpretation of the assembly data involves identification of the joints that need to be created, the types of fits, the datums against which the measurements need to be made, and the features involved in the joining process. The initial and final configuration of mating parts is determined from this analysis. A preliminary operation process sheet is prepared that describes the various steps involved in creating the various joints. The operation process sheet consists of a high-level description of the assembly steps. A taxonomy of generic assembly tasks is given by Csakvary (1985), Kondelon (1985), and Owen (1985). An updated list of tasks is shown in Table 1.

2.2 Selection of Assembly Processes

Assembly processes are classified into four categories: welding, brazing/soldering, bonding, and mechanical fastening. Within these categories subclasses are defined and specific assembly processes are identified. A taxonomy of assembly processes is given by Schey (1987). While the broad assembly process category may be dictated by engineering, further selection of the specific assembly process may be left to the planner. For example, the design engineer may specify that a fillet weld needs to be created but not specify whether the weld is to be created using gas metal arc welding (GMAW) or laser welding. In these cases, depending on the materials involved, required strength, finish, material condition, and other factors, the particular process needs to be selected. In other cases, such as mechanical fastening, the entire process may be specified by the design engineer. For example, the design engineer may specify that a bolted joint needs to be created and specify the type of bolt to be used. In these cases further process selection may not be necessary. The operation process chart is updated by including the selected assembly process. In some cases the selected assembly process may introduce additional pre- and postprocessing steps. For example, many welding processes require surface preparation prior to welding. They may also require a variety of processes after welding, such as heat treatment process to relieve residual stresses and machining to improve appearance and surface quality. These additional activities are also included in the operation process chart. Kalpakjian (1992) states that the main considerations in the selection of a joint and welding process in addition to material characteristics are:

1. Configuration of the parts or structure to be welded and their thickness and size
2. The methods used to manufacture component parts
3. Service requirements, such as the type of loading and stresses generated
4. Location, accessibility, and ease of welding
5. Effect of distortion and discoloration
6. Appearance

Table 1   List of Common Assembly Tasks

Transport (single robot, parallel and sequential cooperation between robots)
Insert peg and retainer
Provide temporary support (flexible fixturing)
Press (one robot holds component while the other performs the press operation)
Form (cut, pinch, crimp bend: some robots perform operation, while others provide support)
Gauge
Simple peg insertion
Push and twist
Fastening (screw, rivet)
Remove location peg or temporary support (in general, remove fixture)
Orient (including flipping parts over)
Weld or solder
Inject adhesive or sealant
Multiple peg insertion

7. Costs involved in edge preparation, welding, and postprocessing of the weld, including machining and finishing operations

2.3 Selection of Assembly Process Parameters

Once the assembly process has been selected, the process parameters need to be specified. Some common parameters for some of the joining processes are shown in Table 2. The proper selection of the joining process parameters is essential to obtain the desired performance from the joint.

2.4 Selection of Assembly Equipment/System

The selection of the assembly equipment/system is primarily an economic issue. For large-volume production of assemblies with minimum variability, hard automation is preferred. For medium volume production with medium variety, robotic assembly is used. For very low-volume, high-variety production, manual assembly is used. Because manual assembly is fault-tolerant, it is also used when problems exist with the design of automated and robotic systems. Examples include high percentage of defective components, tight fits between components, etc.

Boothroyd and Dewhurst (1983) classify assembly systems into three major categories: manual assembly, robotic assembly, and automated assembly. They also provide a methodology for selection of the assembly system based on the product characteristics. The steps of this methodology are described in detail in Boothroyd and Dewhurst (1983). The methodology uses product-related information such as annual production volume per shift, number of parts in an assembly, whether the system is to produce a single product or multiple products, number of parts required for multiple styles of the product, number of major design changes expected during product life, and company policy on investment in labor-saving machinery, to identify the most suitable types of assembly systems.

2.5 Selection of Auxiliary Equipment

Common auxiliary equipment required in an assembly system includes part feeders, material-handling equipment, worktables/benches, part bins, work carriers, and fixtures. Part feeders are used to make parts available, one at a time, in the same position and orientation. These devices are used to increase the production rate in manual assembly and simplify robotic assembly and are mandatory in automatic assembly (if some form of magazining is not used). The selection of part feeders is based on the characteristics of the part, such as symmetry, nesting/tangling features, stickiness, and size.

Material-handling equipment, commonly conveyors, is used to transport work carriers from one assembly station to another. Again, the selection of the equipment is based on production rate, part characteristics, and process requirements. When the production volume is relatively low or when the stations are operated asynchronously, worktables or assembly benches may be utilized. Generally such benches are also equipped with part bins in the case of small parts assembly. The operator is required to retrieve parts from

Table 2 Important Process Parameters for Various Joining Processes

Welding	Adhesive Bonding	Brazing/Soldering	Mechanical Fastening
Electrode material	Adhesive material	Filler material	Surface preparation
Surface preparation	Surface preparation	Surface preparation	Fastener type, size, and
Weld pitch	Pressure	Temperature	number
Pressure	Temperature	Voltage	Postprocesses
Temperature	Postprocesses	Amperage	Process duration
Voltage	Process duration	Speed	Force and torque
Amperage	Size—length, width,	Postprocesses	Tool and fixture
RPM, Speed	area, volume	Process duration	
Postprocesses		Size—length, volume	
Process duration			
Size—length, volume			
Polarity			

the bins and assemble them with the subassembly on a work carrier or fixture. The bins should be selected to accommodate a batch of parts, a quantity that will allow uninterrupted operation of the workstation. For small parts and standard purchased parts this may represent a quantity that is consumed over a number of shifts, while for larger-sized or expensive parts the quantity may be the production batch size, where two to three batches may be consumed in one shift. The workbench needs to provide sufficient space to store the work carrier, bins, handtools, etc. However, the overall size and arrangement should be such that all the parts and equipment are within easy reach of the operator/assembly device. Thus ergonomic and safety considerations must be taken into account in the actual layout of the workbench.

The primary function of work carriers is to transport the subassemblies from one workstation to another. Each work carrier has a stable base that allows transportation on a conveyor system without dislodging parts from the subassembly. A work carrier may also function as an assembly or inspection fixture. Fixtures are used to locate and clamp parts during assembly operations. They provide repeatable relative position and orientation for the parts. Assembly fixtures may be commonplace devices such as vises that clamp parts but do not provide any locating capabilities. Dedicated and modular fixtures may also be designed to include location and clamping features.

A generic methodology for assembly fixture design is as follows:

1. Determine parts required to build the assembly.
2. Determine the base part to be located on the fixture.
3. Choose locating features for each part. If no apparent locating features are present, locating tabs and other features may be added.
4. Design fixture that guarantees that the parts are properly clamped, deflections are minimized to ensure that tolerances are met, setup changes are minimized, and accessibility and ergonomic issues are considered (Rajan, Sivasubramanian, and Fernandez, 1997). Accessibility includes ability to load/unload parts and ability to access operation surfaces using tools.

2.6 Assembly Sequencing

Assembly sequencing involves the determination of one or more sequences in which the parts can be put together to guarantee a feasible assembly. In order to guarantee the maximum possible flexibility during the operation-planning process, the set of all feasible sequences must be identified. Assembly constraints provide an implicit and compact representation of all feasible assembly sequences. Assembly constraints are of two types: those that arise from geometry, called *geometric precedence constraints,* and those that arise from process-related issues, called *process constraints.*

A large body of literature exists on methodologies for determining assembly constraints (De Fazio and Whitney, 1987; Homem de Mello and Sanderson, 1991; Huang and Lee, 1988; Lee, 1994; Henrioud and Bourjault, 1991; Rajan, 1993; Nof and Rajan, 1993; Wolter, 1991; Nnaji, Chu, and Akrep, 1988; Shpitalni, 1989; Woo, 1987). Geometric precedence constraints focus on how the presence of some assembly components affect the assembly of other components. The derived constraints should be such that partial assembly states that prevent the assembly of a component are eliminated from further consideration. Process constraints are created when certain joining process-related issues determine the order in which components are assembled. Examples of such issues include heat or pressure inputs to the assembly that may damage parts, and the need for inspection of subassemblies before they are installed in the final assembly.

Here we will present the Minimal Precedence Constraint (MPC) method (Rajan, 1993; Nof and Rajan, 1993) for the identification of geometric precedence constraints that implicitly represent all geometrically feasible assembly sequences. The query-based approach, an alternative methodology for the explicit identification of all feasible sequences, is presented in Chapter 50. A mathematical approach has been devised for determining completely constrained components and their precedence constraints. While the methodology is generally applicable, for this discussion we will assume that only single straight-line translations are valid for assembly/disassembly motions. This assumption helps to simplify the explanation and, more importantly, to reduce the computational complexity.

The minimal precedence constraint for an assembly component is defined as the alternative minimal assembly states that will prevent the assembly of the component. An assembly state is said to be minimal if all of its subsets do not prevent the assembly of the given component. In contrast to the query-based approach, the MPC approach focuses attention on the component rather than the liaison. While the identification of alternative minimal assembly states that prevent the assembly of a component may require more effort, the number of times this process need to be repeated is equal to N, the number of assembly components. Because it may be possible to add many of the assembly components in the final step of the process, these components do not have any partial assembly states that prevent their assembly. Therefore only components that cannot be added in the final assembly step (called *completely constrained* components) need to be considered for the generation of the MPCs, and the actual number of times the process needs to be repeated may be considerably less than N.

The set of directions along which a component can be moved to its final assembly location (called the location direction set, $LD_{i,\varnothing}$) from outside the assembly can be represented by a circle in 2D and a sphere in 3D (because of the single straight-line translation assumption). The directions are in fact lines going from the periphery of the circle or sphere to its center. If there are no other components present in the assembly state, i.e., nothing has been assembled as yet, any direction is available for locating an assembly component. As other components are assembled, the available set of directions along which the component can be located becomes smaller. For a completely constrained component each of the location directions is constrained by one or more of the other components in the assembly.

For example, consider the component A in the 2D example assembly shown in Figure 1. If we represent the constraint direction set (set of directions along which component i constrains the assembly of component j) as CD_{ij}, then the sets of directions along which components B, C, D, and E constrain the assembly of component A can be mapped to the circle representing all the location directions, as shown in Figure 2a (empty circles at the ends of the CDs indicate that component A is not constrained along that direction, i.e., it can slide along that direction). Since the union of the constraint direction sets is the same as the location direction set $LD_{i,\varnothing}$ (i.e., assembly of A along each of the location directions is obstructed by one or more of the other components), component A is completely constrained. The circle can then be partitioned into regions such that each region is constrained by a unique set of components, as shown in Figure 2b. The conjunction of these sets represents the precedence constraint for component A:

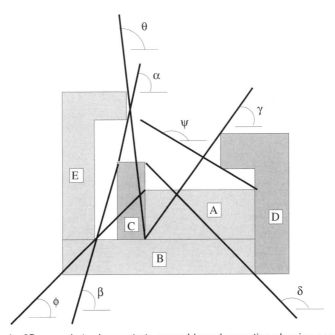

Figure 1   2D example to demonstrate assembly and execution planning concepts.

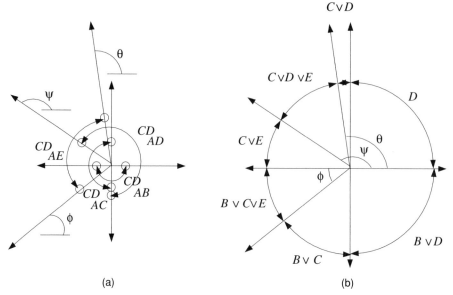

Figure 2 Union of the constraint direction sets of component A caused by components B, C, D, and E in 2D example. The overlaid constraint direction sets are shown in (a) and the overlap between the sets is shown in (b).

$$PR_A = (B \lor D) \land (B) \land (B \lor C) \land (B \lor C \lor E) \land (C \lor E)$$
$$\land (C \lor D \lor E) \land (C \lor D) \land (D)$$

By converting this expression to the disjunctive form and eliminating subsumed expressions, the minimal precedence constraint for the component is determined. For the 2D example, the completely constrained components are A and C (C becomes completely constrained because of the single translational assembly motion assumption). Their MPCs are as follows:

$$PR_A = (B \land C \land D) \lor (B \land D \land E)$$
$$PR_C = (A \land B \land E)$$

The precedence relation of A indicates that any assembly state consisting of the components B, D, and at least one of the components C and E, and not consisting of component A, is infeasible. In other words, component A has to be in place before B, D, and at least one of the components C and E have all been added to the assembly. Note that as each of these components is added to the assembly, the set of location directions becomes correspondingly smaller. This indicates that depending on the versatility of the robot used to perform the assembly of component A, the actual sequence will have to be determined. For example, if the robot can assemble component A only in the vertical downward direction, then the assembly sequence mandates that component A be assembled before component D. The sequence of assembly of the remaining components is essentially irrelevant. This issue will be addressed in more detail later.

2.7 Selection of Inspection Devices

Inspection of assemblies in many ways parallels inspection of individual components. Dimensions such as overall length and gap size and surface forms such as parallelism and perpendicularity may need to be inspected for assemblies. Inspections of individual joining process results may also be performed. For example, the surface finish of a weld or bond may be inspected or its strength tested. Finally, the functionality of the product or subassembly may be tested to ensure that it performs according to specifications.

Process inspections are used to evaluate dimension, form, surface finish, etc. created by various joining processes. A variety of inspection methods and tools such as visual inspection, linear and angular measurement devices, gauges, coordinate measuring machines, etc. may be used for these purposes. Additional description of common inspection methods and tools is given in Kennedy, Hoffman, and Bond (1987). Final inspection or functional testing of subassemblies and products may require specialized equipment depending on the function being evaluated. Joint range of motion, strength, deflection, and conductivity are examples of functional characteristics that may be tested to verify that the product or subassembly performs according to specifications. Generally two classes of testing procedures are defined: destructive testing methods such as tension, fracture toughness, corrosion, creep, and peel tests, and nondestructive testing methods such as visual, radiographic, magnetic particle, eddy current, liquid die penetrant, and ultrasonic tests.

2.8   Calculation of Assembly Times

Standard setup and operation times need to be estimated for the purpose of execution planning and monitoring of operations. For manual assembly, standard techniques such as the methods time measurement (MTM) or work factor (WF) (Mundel and Danner, 1994) can be used to estimate setup and operation times. For robotic assembly the robot time and motion (RTM) system, discussed in detail in Chapter 32, can be used to obtain good estimates of setup and run times.

3   EXECUTION PLANNING

Execution planning is the process of allocating tasks to the machines in a system. In the context of robotic systems it involves:

1. An identification of which tasks can be performed by the robots in the system
2. An assignment of tasks to robots to optimize a given performance measure

3.1   Matching Task Requirements to System Capabilities

The assembly task requirements need to be mapped to the robot and system capabilities to determine whether and how the assembly tasks can be successfully completed. Geometric task requirements include precedence constraints and motion requirements (gross and fine motion trajectories). These geometric requirements, along with the task physical requirements and the assembly tasks operational plan, are used during the matching process. The task physical requirements relate to the physical properties of the objects to be handled during the execution of the task, and the assembly tasks operational plan is a description of the tool and fixture requirements of the assembly tasks.

Similar to the assembly task requirements description, the robot and system capability specification consists of three elements: geometric, physical, and operational. The geometric capability information specifies the system layout and motion capabilities of the robots. The physical capability information is related to the physical capabilities of the robots, such as payload. The operational capability information specifies the tool- and fixture-handling capabilities of the robots and system cooperation information relating to parallel and sequential operational capabilities of the robots.

The matching process needs to determine the assembly task cooperation requirements in terms of three cooperation modes: mandatory, optional, and concurrent (Nof and Hanna, 1989). These modes are defined as follows:

1. *Mandatory task cooperation:* The successful completion of a task requiring this cooperation mode involves the mandatory participation of two or more robots. Two main subclasses are *sequential,* where the participation of the robots is in some specified sequence, and *parallel,* where the robots participate simultaneously in performing the task. This mode captures the limitations in processing capability of an individual robot and exploits the capabilities of cooperating robots.

2. *Optional task cooperation:* This cooperation mode encompasses those tasks that can be independently completed by more than one robot in the system. The two main classes of this cooperation mode are *redundancy within a robot type* and *redundancy between robot types.* This mode captures the overlap in capabilities of the robots in the system.

3. *Concurrent task cooperation:* This cooperation mode represents a subset of optional cooperation mode tasks, where use of additional robots in cooperation will

lead to improvement in performance. This mode captures the flexibility available in the system due to cooperation and its benefits.

First, the assembly task requirements should be matched with the corresponding capabilities of individual robots in the system to determine optional task cooperation requirements. For any tasks that cannot be satisfied by optional task cooperation, mandatory task cooperation requirements should be generated by matching the capabilities of cooperating robot sets to satisfy the task requirements. Only feasible robot sets should be considered, based on the cooperation constraints specified. If one or more assembly tasks exist that cannot be satisfied either by optional or mandatory cooperation, then the given assembly cannot be successfully completed by the given system and the planning process is aborted. Otherwise, for tasks satisfied optionally, concurrent task cooperation requirements should be generated.

It must be noted that while we consider all three cooperation modes, a given robotic system may exhibit only a subset of these capabilities. Any robotic system consisting of two or more robots will exhibit optional task cooperation. If there is overlap between the robot workspaces, mandatory and concurrent cooperation capabilities may be realized.

Mandatory cooperation is invoked for tasks that cannot be performed using optional cooperation, i.e., multiple robots are required to satisfy the task needs. The failure of optional cooperation occurs due to the inability of any individual robot's capability to match one or more requirements:

1. *Fixture requirement:* If the required fixture is not available to any robot in the system, then flexible fixturing can be considered, leading to mandatory parallel cooperation. In this case a robot acts as the fixture for a component to be fixtured while other robots assemble components with it.

2. *Size requirement:* Mandatory task cooperation may be required when the component being handled is large and requires multiple supports during transport and assembly. Typically this would lead to mandatory parallel cooperation.

3. *Load requirement:* In this case the component weight is greater than the payload of a single robot and therefore requires multiple supports during transport and assembly. This leads to a mandatory parallel cooperation requirement among multiple member robot sets.

4. *Reach requirement:* The successful completion of an assembly task requires the ability of the robot to reach the initial and final locations of the component. Both these locations may not be reachable by a single robot. Therefore the component needs to be transferred from a robot that can access the input location to a robot that can perform the assembly task through intermediate robots and locations. This implies mandatory sequential cooperation.

5. *Trajectory requirement:* When no single robot can perform the specified trajectory, mandatory cooperation is not necessarily the solution. If the trajectory is modified, some robot may be able to perform the task. This should be handled in the optional cooperation requirements-generation process. If no such modification provides the desired solution, then mandatory sequential cooperation can be used to split the trajectory and use different robots to execute different segments of the trajectory.

6. *Linkage assembly:* When an assembly consists of a linkage as a component that requires multiple points to be located simultaneously for successful assembly, mandatory parallel cooperation is desired. Although this is not a failure of optional cooperation requirements, it demonstrates an important use of mandatory cooperation requirements.

It is important to note that these are not the only cases that lead to mandatory cooperation requirements. Adding other considerations in the satisfaction of the various requirements may cause additional situations to arise. Only a representative and important set of mandatory cooperation requirements is considered here.

The output of the matching process consists of capability information of robot sets to perform the assembly tasks. A task plan represents the sequence of robot actions needed to successfully complete a given task. For each action the robot or robots required to perform it are specified. Since the assembly tasks can generally be considered to be pick-and-place operations, a standard task plan is used (Rajan, 1993). Each robot set is assumed to start from its safe position, move to the pick location of the object, take the object, move to the drop location of the object, leave the object, and move back to the safe position. Under optional cooperation all these actions for all the objects in a task are performed by a single robot. Under mandatory or concurrent cooperation some actions are performed independently while others are performed sequentially when sequential cooperation is used, or performed in parallel when parallel cooperation is used.

Depending on the redundant action capabilities, alternative task plans may exist for the same robot set to perform a given task. Additionally, for a given task plan, since the precedence constraint analysis showed that alternate assembly directions may exist for assembling a given component, alternative trajectories exist for the robot assigned the *leave* assembly action. As the task plan remains the same, the different trajectories provide different costs for performing the task. These alternative costs should be generated by considering the ability of the robot sets to perform the *leave* action along all the directions of the location direction set. The subset of this set that are successfully executed by the robot set is defined as the *robot location direction set* for the given robot set–task–task plan combination, i.e., RLD_{ijk}, where i is the robot set, j is the task, and k is the particular task plan. A cost is associated with each member of the robot location direction set that specifies the cost of executing the complete task when the particular trajectory is used.

For a given robot set–task pair, the lowest cost obtained over all the task plans is specified as the capability measure of the robot set for the given task. The capability measures of the robot sets for the various tasks form the entries of the *cooperation requirements matrix*. Therefore the entries in the cooperation requirements matrix represent the lowest cost that can be realized by using a given robot set for a given task. Let $N(R)$ be the set of robots in the system and $N(T)$ the set of assembly tasks. Then the cooperation requirements matrix can be expressed as a mapping as follows:

$$C: S \times N(T) \rightarrow \Re^+$$

where S is the set of allowed robot sets and $C(i, j)$ represents the capability measure of robot set i for task j. Let R be the number of elements in $N(R)$. Then S is a subset of the power set of $N(R)$, i.e., the maximum number of elements in S will be 2^R-1. In general, due to cooperation limitations imposed externally, or due to absence of a motion space (Li et al., 1989) between cooperating robots, the set S is much smaller than the power set. When a robot set i cannot perform a task j, the corresponding capability measure $C(i, j) = \infty$.

To illustrate the execution planning concepts, two examples are presented. The first example (we will refer to it as the *blocks* example) illustrates optional cooperation among the robots to complete a given assembly. The assembly, shown in Figure 3, consists of seven components. The components are labeled *base, cent, tops, plsx, mnsx, plsy,* and *mnsy.* Components *tops* and *base* are of the same type, and components *plsx, mnsx, plsy,* and *mnsy* are also of a single part type. All the components consist of only planar mating features. It must be noted that, in general, assemblies consist of various components with cylindrical mating features. The assembly shown in this case study represents an extreme case where no such features exist. This assembly allows us to illustrate the mating and constraint generation processes, since they are more difficult for planar mating features and are computationally more expensive relative to cylindrical features. The planar mating features also provide more robot location directions, since the mating direction set for the component is in general more than a single direction. Thus, it allows us to demonstrate the flexibility in the capabilities of the various robots.

The assembly consists of a single completely constrained component, *cent,* which is constrained by the assembly state consisting of all the other components. The only minimal precedence constraint is:

$$cent = base \; plsx \; mnsx \; plsy \; mnsy \; tops$$

The workcell used for the first example is shown in Figure 4. The cell consists of three robots: an ADEPT 1 (*adept*) four-axis SCARA robot, a PUMA 550 (*puma5*) five-axis articulated robot, and a PUMA 560 (*puma6*) six-axis articulated robot. All the robots are mounted with the same two-finger gripper with a motion range of 0 to −27°. The standard cell also consists of a centrally located worktable that is common to the three robots and is used as the location for building the assemblies. The components are input to the cell on three component input tables. Input tables 1 and 2 are common to the PUMA 550 and PUMA 560 robots, while table 3 is common to the ADEPT 1 and PUMA

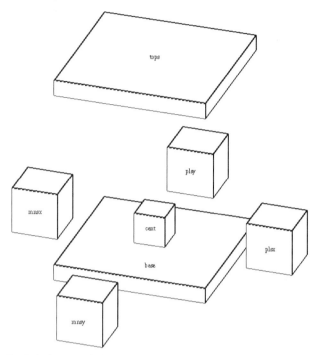

Figure 3 Exploded assembly for the *blocks* example showing the seven components.

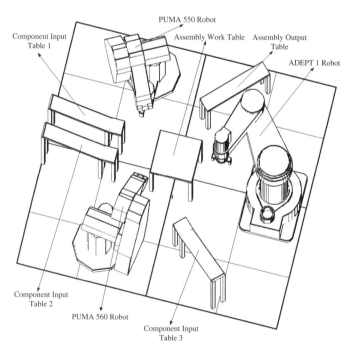

Figure 4 Workcell used for the *blocks* example. The cell consists of three robots, part input tables, assembly worktable, and assembly output table.

Table 3   Cooperation Requirements Matrix for Tasks of the *Blocks* Example

Cooperation Requirements Matrix (all task times are in seconds):

	Task 0	Task 1	Task 2	Task 3
Set 0:	∞	∞	5.764	5.751
Set 1:	5.425	5.720	∞	∞
Set 2:	5.551	5.297	5.451	5.551
Set 3:	∞	∞	∞	∞
Set 4:	∞	∞	∞	∞
Set 5:	∞	∞	∞	∞
Set 6:	∞	∞	∞	∞

	Task 4	Task 5	Task 6	Task 7
Set 0:	5.844	5.855	∞	5.714
Set 1:	∞	∞	5.412	5.800
Set 2:	5.615	5.519	5.550	∞
Set 3:	∞	∞	∞	∞
Set 4:	∞	∞	∞	∞
Set 5:	∞	∞	∞	∞
Set 6:	∞	∞	∞	∞

Robot Sets:

Set 0:	ADEPT 1
Set 1:	PUMA 550
Set 2:	PUMA 560
Set 3:	ADEPT 1 & PUMA 550
Set 4:	ADEPT 1 & PUMA 560
Set 5:	PUMA 550 & PUMA 560

560 robots. When the assembly is completed, it is removed from the assembly worktable to an output table that is accessible to the PUMA 550 and ADEPT 1 robots. A coarse cell space partitioning is used to describe the shared regions. The partitioning is essentially the shared workcell equipment, i.e., the assembly worktable, input tables, and output table are defined as shared regions for the robots that have these workcell elements in common as described above.

To generate the cooperation requirements the cell shown in Figure 4 is used with input location table 1 for the *tops* and *base* components, table 2 for the *cent* component, and table 3 for the *plsx, mnsx, plsy,* and *mnsy* components. The task plan specifies eight tasks: seven tasks to assemble the components and the last task to remove the assembly from the cell. In order to ensure that *base* is the first component to be assembled, a precedence relation is specified such that task 6 precedes all other tasks. Similarly, task 0 for assembling *tops* succeeds tasks 1–6, and task 7 succeeds task 0. Based on the matching of the physical, operational, and geometric requirements of the tasks to the corresponding capabilities of the robots in the cell, all tasks require optional cooperation. Due to the motion limitations of the robots, the PUMA 550 robot can perform tasks 0, 1, 6, and 7, the PUMA 560 robot can perform tasks 0–6, and the ADEPT 1 robot can perform tasks 2, 3, 4, 5, and 7. The cooperation requirements matrix is shown in Table 3.

The exploded assembly for the second example (we will refer to it as the *linkage* example) is shown in Figure 5. The assembly consists of five components, *base, blk1, blk2, llnk,* and *rlnk*. The component *blk2* is fixed to the *base* by a press fit, while *blk1* slides along the *base*. Rotational joints are defined between *blk1* and *llnk, llnk* and *rlnk*, and *rlnk* and *blk2*, thus making the assembly a kinematic device.

The standard cell shown in Figure 4 is modified to accommodate an additional component input table due to the presence of four different part types in the assembly. Also, a fixture, *blk_fix*, is introduced in the cell located on a table near the assembly worktable.

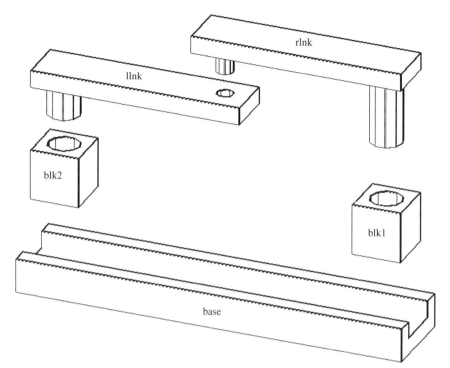

Figure 5 Exploded assembly for the *linkage* example showing the five components.

All the robots in the cell are specified to have the operational and physical capability for the handling of the fixture, but geometric motion constraints limit the actual system capability for its handling to the PUMA 550 and PUMA 560 robots.

The task plan consists of six tasks:

1. Placing the *base* on the worktable
2. Assembling of *blk1* using a press operation
3. Assembling *blk2*
4. Assembling the *llnk* component by fixturing *blk1* using the fixture
5. Assembling the component *rlnk* by fixturing *llnk* to maintain its orientation
6. Finally, removing the assembly from the cell to the output table

Task 1 precedes all tasks, tasks 2 and 3 precede tasks 4 and 5, while task 6 succeeds all tasks.

The optional cooperation requirements analysis is applied to each of the tasks. Task 1 fails the physical requirement because the force exerted by the *base* on the tools of the PUMA 550 and 560 robots is greater than their static payload. However, the operational and geometric requirements are satisfied. Therefore mandatory parallel cooperation is required for task 1. Task 2 can be performed optionally by either the ADEPT 1 or the PUMA 560 robots because the required press force is less than their static payload. Task 3 can also be performed optionally by either robot. Task 4 can be performed optionally by the PUMA 550 and 560 robots because they can both handle the required fixture and satisfy the required motions for locating the fixture and assembling *llnk*. Task 5 fails the optional cooperation requirements because the required fixture is not available. Therefore mandatory parallel cooperation is required. Task 6 fails the physical requirement, and therefore mandatory parallel cooperation is required.

The mandatory cooperation requirements are generated for tasks 1, 5, and 6. The physical requirement for task 1 is satisfied for the PUMA 550–PUMA 560 robot combination by distributing the weight of the base equally between the two robots. The start of the task execution by the two robots is shown in Figure 6. All the other robot sets fail to satisfy the geometric task requirements because of the motion limitation of the ADEPT 1 robot. Task 5 can be satisfied by mandatory parallel cooperation between the PUMA 550–PUMA 560 robot set and the ADEPT 1–PUMA 560 robot set. In both cases the PUMA 560 performs the assembly of the *rlnk* component, while the other robot provides the fixture support on the *llnk* component. Figure 7 shows a snapshot taken during the execution of task 5 when the PUMA 550–PUMA 560 robot set is used. Although the PUMA 550–PUMA 560 robot combination can perform the task by reversing the roles of the two robots, i.e., using PUMA 560 to provide fixture support, while PUMA 550 performs the *rlnk* assembly, the geometric requirement fails due to intersecting links of the robots. Task 6 can be performed successfully by the robot set ADEPT 1–PUMA 550 in two different ways, depending on which pick location is assigned to which robot. But only one of the plans is feasible, since the robot links intersect at the beginning of the other path.

Concurrent task cooperation requirement for tasks 2, 3, and 4 are evaluated. Tasks 2 and 3 cannot be decomposed any further, and there is no advantage in splitting the motion because the speeds of the robots are the same. Task 4, however, can be decomposed into three parts: location of the fixture, assembly of the component *llnk,* and removal of the fixture. The operational, physical, and geometric requirements of the three subtasks are satisfied by the PUMA 550 and the PUMA 560 robots. Therefore six task plans can be generated for the concurrent cooperation mode execution of task 4.

The cooperation matrix is shown in Table 4. For task 4 only the best plan value for the concurrent cooperation mode is shown. In addition to the task precedence constraints, the motion and resource constraints are required to generate the global plan. The motion constraints are generated as the robot location direction sets, while the resource constraints are the *blk_fix* fixture and the spatial resource constraints. The fixture table is specified as a shared area. All the other spatial resource constraints are the same as those used in the *blocks* example.

3.2   Global Execution Plan Generation

The purpose of execution planning is to assign tasks to the robot sets in the system based on their capabilities and generate a set of individual robot plans that are consistent and coordinated. The interactions that exist between individual robot plans arise due to as-

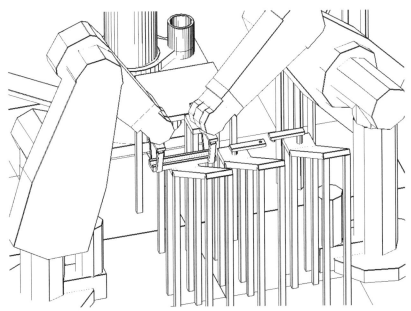

Figure 6   Start of the assembly task for the *base* component using mandatory parallel cooperation between the PUMA 550 and PUMA 560 robots for the *linkage* example.

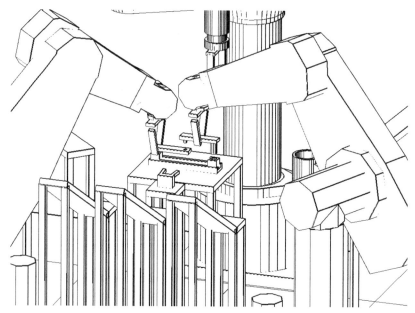

Figure 7   Snapshot taken during the execution of task 5 involving the assembly of the *rlnk* component using mandatory parallel cooperation between the PUMA 550 robot acting as a flexible fixture and the PUMA 560 robot performing the assembly of the component for the *linkage* example.

Table 4   Cooperation Requirements Matrix for Tasks of the *Linkage* Example

Cooperation Requirements Matrix (all task times are in seconds):

	Task 1	Task 2	Task 3	Task 4
Set 0:	∞	5.688	5.955	∞
Set 1:	∞	∞	∞	12.446
Set 2:	∞	5.267	5.803	12.384
Set 3:	∞	∞	∞	∞
Set 4:	∞	∞	∞	∞
Set 5:	5.509	∞	∞	11.329
Set 6:	∞	∞	∞	∞

	Task 5	Task 6
Set 0:	∞	∞
Set 1:	∞	∞
Set 2:	∞	∞
Set 3:	∞	6.073
Set 4:	5.577	∞
Set 5:	5.603	∞
Set 6:	∞	∞

Robot Sets:

Set 0:	ADEPT 1
Set 1:	PUMA 550
Set 2:	PUMA 560
Set 3:	ADEPT 1 & PUMA 550
Set 4:	ADEPT 1 & PUMA 560
Set 5:	PUMA 550 & PUMA 560
Set 6:	ADEPT 1 & PUMA 550 & PUMA 560

sembly, robot, and system constraints. To ensure successful completion of the assigned assembly tasks, these constraints need to be satisfied in the final consistent plans. Also, for mandatory and concurrent parallel cooperation modes the motions of the cooperating robots have to be coordinated to ensure that all cooperative tasks are successfully completed.

The task-assignment and execution-planning process can also be viewed as a multiple-goal planning problem (Yang, Nau, and Hendler, 1991). In this case each of the tasks represents an individual goal that needs to be accomplished. In the assignment and execution planning process we wish to identify a plan for each task (goal), resolve interactions between plans and goals, and ensure coordination within a plan when required. We will define the execution planning problem to be one of selecting a plan for each goal and resolving interactions between goals and plans and ensuring coordination within goals such that the cost is minimized.

The complexity of the planning problem is determined by the strategy used. The two strategies that can be used are as follows:

1. *Simultaneous plan generation strategy:* The plan selection and constraint satisfaction can be performed simultaneously. In this case all the available plans for all the tasks (goals) are considered, and a global plan is constructed by considering alternate combinations of goal plans.

2. *Sequential plan generation strategy:* The plan selection can be initially performed. This is based solely on the cooperation requirements matrix information. For a given set of plans, one for each goal, constraint satisfaction is performed.

The first method will generate an optimal global execution plan as a search over the entire state-space is performed. However, the complexity of the search is very high. In the latter case, decoupling the plan selection and constraint satisfaction processes considerably reduces the search complexity. However, the constraint satisfaction of the set of selected plans does not guarantee the globally optimal solution, nor is a feasible solution guaranteed. It is possible that a given assignment will not provide a feasible execution plan because it violates one or more constraints.

3.2.1 Simultaneous Plan Generation Strategy

When plan selection and constraint satisfaction are performed simultaneously, a best-first search algorithm (Yang, Nau, and Hendler, 1991) is used to generate the optimal global plan solution.

Search Algorithm

The search is performed over a state space where each state represents the partial global plan for a collection of tasks (goals). A state at the ith level represents a plan for i tasks (goals). The O level is the empty state. An open list is maintained for the states yet to be expanded. The states in the open list are ordered in the best-first fashion. The first state in the list is selected for expansion at the beginning of each iteration. If the state contains all the assembly tasks, then the optimal global plan has been obtained and the search stops. Otherwise the state is expanded by merging the plans for an additional task. The set of tasks to be included is formed by selecting feasible tasks that do not violate fixed assembly and robot constraints. The successor states are formed by merging the various plans for a selected task with the partial global plan. Let *OPEN* represent the open list of states and *COST*( ) the cost function.

Feasible Tasks

For each state to be expanded, the set of feasible tasks needs to be identified. Let F be the set of tasks already in the state *FIRST*. The available set of tasks is then $[N(T) - F]$. For each task t, $t \in \{N(T) - F\}$, if adding the task to the current state will not lead to violation of fixed assembly constraints of any other task u, $u \in N(T) - F$, then t is a feasible task.

Plan Merging

The current state is expanded by merging the various plans of the various feasible tasks to generate the successor states. To merge a given plan into the current global plan, one must determine the feasibility of the plan. A plan is feasible if its robot constraint is not violated by the global plan. In other words, if the current global plan does not allow for any location motion of the given plan to be executed, then the given plan is infeasible. For a feasible plan the types of merging actions performed will be dependent on the interactions between the partial global plan tasks and the tasks in the global plan. Five types of action interactions arise: action combination, precedence relation, resource sharing, cooperative action, and independent action. The definitions of these interactions and the merging process are described in detail in Rajan and Nof (1996*b*). The merging process enforces assembly and resource constraints, both secondary and spatial. It eliminates unnecessary actions caused by the same resources being used in consecutive tasks. Finally, when interactions are minimal, it increases the parallelism in task execution. In general, combinations of the first four interactions occur. Because the precedence constraints between tasks need to be satisfied to generate a consistent global plan, they have to be satisfied before other interactions can be considered. After the precedence relations are satisfied, resource sharing and cooperative action are considered, followed by action combination interactions. The cooperative action interaction is similar to the resource-sharing interaction because the robots involved in the cooperative action become resource constraints on the task plan.

The assignment of tasks in the optimal global plan and the interactions between the task plans for the *blocks* example are shown in Figure 8. Only the action combination and resource interactions are shown. The optimal global plan is shown in Figure 9. As shown, only certain action combination interactions were resolved, while others were precluded by the spatial resource interactions. The total cost (execution time) for the optimal global plan is 29.791 seconds. The global execution plan for the *linkage* assembly can be similarly generated and its total execution time is 32.393 seconds.

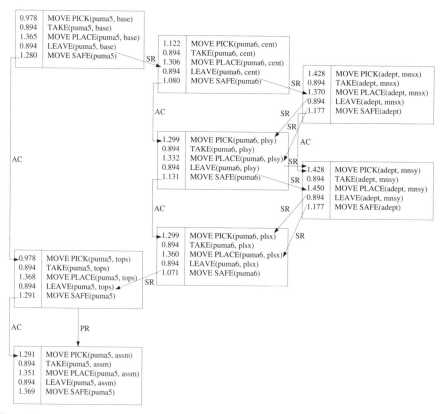

Figure 8 Task assignment and interactions for the optimal global plan for the *blocks* example. The interactions shown are: action combination (AC), shared resource (SR), and task precedence relation (PR).

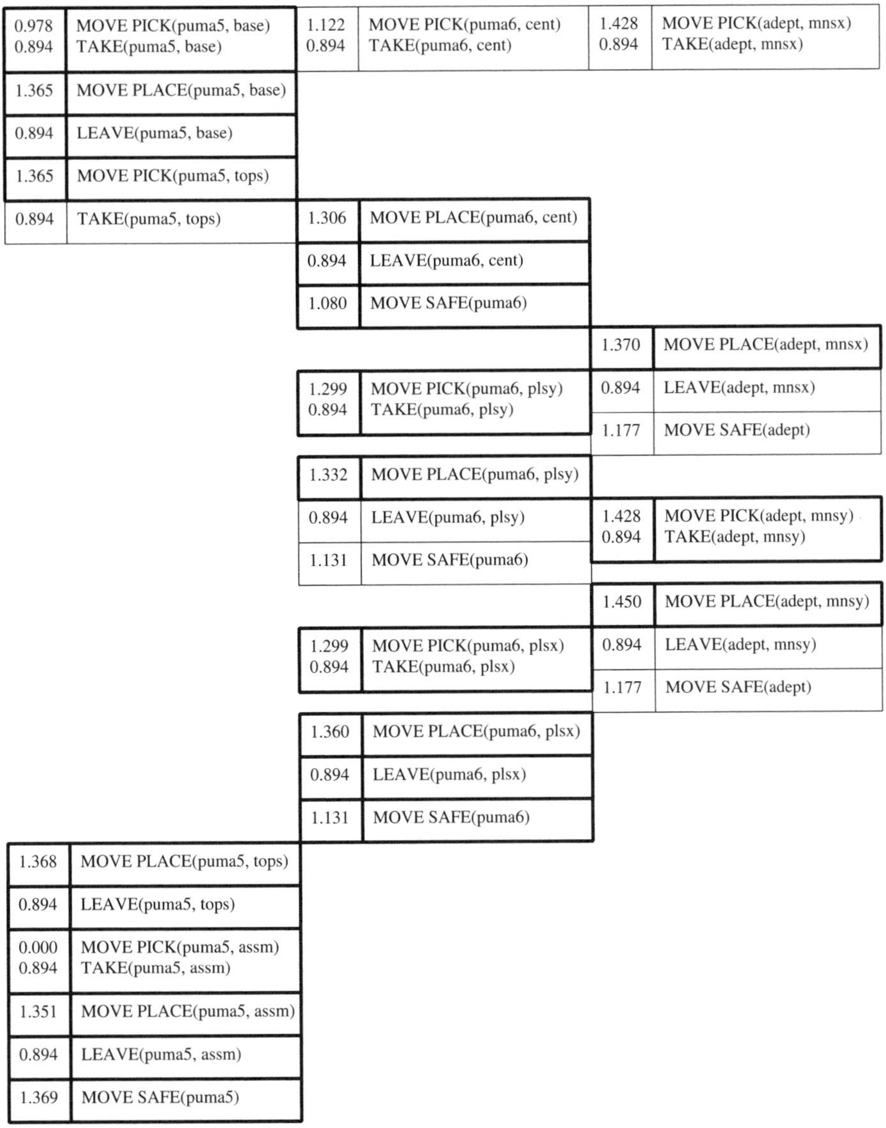

| 0.978 | MOVE PICK(puma5, base) | 1.122 | MOVE PICK(puma6, cent) | 1.428 | MOVE PICK(adept, mnsx) |
| 0.894 | TAKE(puma5, base) | 0.894 | TAKE(puma6, cent) | 0.894 | TAKE(adept, mnsx) |

1.365	MOVE PLACE(puma5, base)

0.894	LEAVE(puma5, base)

1.365	MOVE PICK(puma5, tops)

| 0.894 | TAKE(puma5, tops) | 1.306 | MOVE PLACE(puma6, cent) |

0.894	LEAVE(puma6, cent)

1.080	MOVE SAFE(puma6)

1.370	MOVE PLACE(adept, mnsx)

| 1.299 | MOVE PICK(puma6, plsy) | 0.894 | LEAVE(adept, mnsx) |
| 0.894 | TAKE(puma6, plsy) | 1.177 | MOVE SAFE(adept) |

1.332	MOVE PLACE(puma6, plsy)

0.894	LEAVE(puma6, plsy)	1.428	MOVE PICK(adept, mnsy)
		0.894	TAKE(adept, mnsy)
1.131	MOVE SAFE(puma6)		

1.450	MOVE PLACE(adept, mnsy)

| 1.299 | MOVE PICK(puma6, plsx) | 0.894 | LEAVE(adept, mnsy) |
| 0.894 | TAKE(puma6, plsx) | 1.177 | MOVE SAFE(adept) |

1.360	MOVE PLACE(puma6, plsx)

0.894	LEAVE(puma6, plsx)

1.131	MOVE SAFE(puma6)

1.368	MOVE PLACE(puma5, tops)

0.894	LEAVE(puma5, tops)

| 0.000 | MOVE PICK(puma5, assm) |
| 0.894 | TAKE(puma5, assm) |

1.351	MOVE PLACE(puma5, assm)

0.894	LEAVE(puma5, assm)

1.369	MOVE SAFE(puma5)

Figure 9 Optimal global plan for the *blocks* example. The highlighted action sequence indicates the critical path.

3.2.2 Sequential Plan Generation Strategy

When plan selection and constraint satisfaction are performed sequentially, the planning process is considerably less complex. The task-assignment process generates a particular assignment based on the cooperation requirements matrix. This step does not attempt to satisfy any of the constraints. Its main objective is to optimize the given performance measure. It is possible to identify a variety of performance measures for performing task assignment. In the following sections we will discuss two of the important assignment methods we have identified. The first finds the optimal assignment based solely on the cooperation requirements matrix. The second determines the minimum robot set based on the cooperation requirements matrix. The global plan-generation method in the case when the task assignment is specified is then discussed.

Minimum Total Cost Assignment

If we use the performance goal to be minimum total cost (using operation times for the $C(i, j)$ entries, the measure is makespan), the assignment problem can be solved easily. For each task j we need to select the row i that provides the minimum cost. Because we are not imposing any constraints on the assignment except that a task cannot be assigned to more than one robot set, the total cost is minimized by this approach. Therefore this scheme provides the optimal assignment of tasks to robot sets based only on the cooperation requirements matrix. This solution is the locally optimal solution, in that the best possible robot set for each task is being selected. However, because it imposes the various constraints described previously, this is not necessarily a feasible global plan, and even if feasible, it is not necessarily the optimal global plan.

Minimum Robot Set Assignment

The simultaneous plan-generation strategy and the minimum cost-assignment scheme shown above do not restrict the number of robots used in performing the tasks. Therefore, even if a smaller robot set is available that can successfully perform all the tasks, these methods may not necessarily generate the minimum set to be the final assignment. By restricting the number of robots participating in the task execution process, two main advantages can be realized:

1. The number of interactions between robots is reduced. Because fewer robots participate in the execution of the tasks, the interactions are fewer.

2. Robots that are not selected to participate in the execution of a particular set of tasks can be used to complete other jobs in parallel.

The disadvantage of using the minimum robot set is that it may not be the optimal solution for the task assignment. This, combined with constraint satisfaction, could lead to a very poor solution. The minimum robot set problem can be formulated as an integer program (Rajan, 1993). The assignment process provides a single task plan for each task. However, due to interactions between the tasks and to robot location directions, various global plans can be constructed using the same set of individual task plans. The generation of the global plan given a set of individual task plans is discussed below.

Global Plan Generation

Given one plan for each task, the generation of the global plan is considerably simplified. The best-first search strategy specified in Section 3.2.1 can be used, with one task plan per goal, to generate the global plan. However, by performing the assignment and constraint satisfaction sequentially, we can no longer be certain that a global plan exists for the particular assignment. Therefore a check is required during the search process to ensure that at each iteration at least one feasible partial global plan exists. Since partial global plan states are placed in the *OPEN* list, if at the start of any iteration the *OPEN* list is empty, then no feasible global plan exists and therefore the search process can be terminated with failure.

The minimum total cost assignment and the corresponding global plan for the *blocks* example is shown in Figure 10. If an assignment is generated based solely on the cooperation requirements matrix, the solution obtained is considerably worse than the optimal global plan. The total cost of the plan is 35.987 seconds. The minimum robot set assignment-generation process is shown in Table 5. The optional cooperation requirements matrix shows that no single robot can execute all the tasks. However, any two-member robot set is capable of completing all the tasks successfully. The minimum cost, minimum robot set solution is the PUMA 550–PUMA 560 combination.

For the *linkage* example the minimum time solution assigns task 1 to the PUMA 550–PUMA 560 robot set, tasks 2 and 3 to the PUMA 560 robot, task 4 to the PUMA 550–PUMA 560 robot set, task 5 to the ADEPT 1–PUMA 560 robot set, and task 6 to the ADEPT 1–PUMA 550 robot set. Based on the cooperation requirements matrix, the total time for this solution is 39.558 seconds. The minimum robot set solution includes all three robots, as no smaller subset can successfully perform all the assigned tasks. The global plan generated for this solution will be the same as the optimal global plan solution.

3.2.3 Game-Theoretic Methods

The game-theoretic paradigm (Owen, 1982) is very relevant to the assignment and execution planning problems from a multi-agent perspective. The robots in the system are

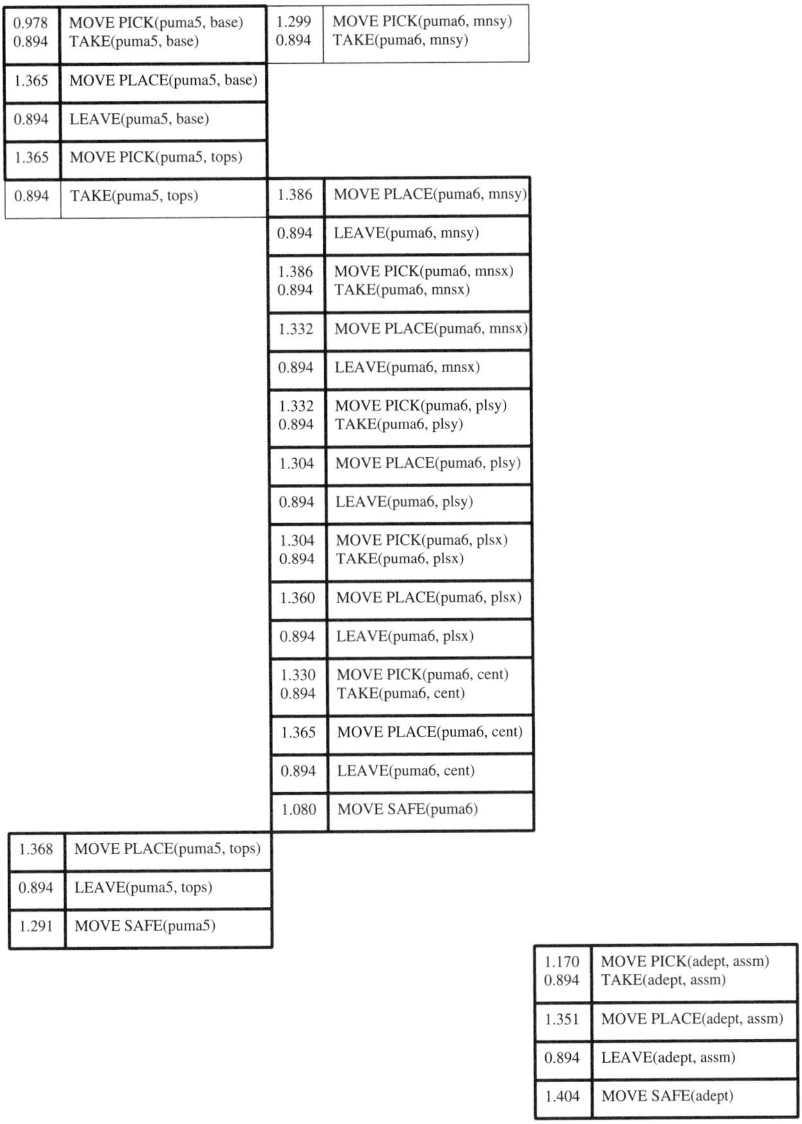

0.978	MOVE PICK(puma5, base)	1.299	MOVE PICK(puma6, mnsy)
0.894	TAKE(puma5, base)	0.894	TAKE(puma6, mnsy)
1.365	MOVE PLACE(puma5, base)		
0.894	LEAVE(puma5, base)		
1.365	MOVE PICK(puma5, tops)		
0.894	TAKE(puma5, tops)	1.386	MOVE PLACE(puma6, mnsy)
		0.894	LEAVE(puma6, mnsy)
		1.386	MOVE PICK(puma6, mnsx)
		0.894	TAKE(puma6, mnsx)
		1.332	MOVE PLACE(puma6, mnsx)
		0.894	LEAVE(puma6, mnsx)
		1.332	MOVE PICK(puma6, plsy)
		0.894	TAKE(puma6, plsy)
		1.304	MOVE PLACE(puma6, plsy)
		0.894	LEAVE(puma6, plsy)
		1.304	MOVE PICK(puma6, plsx)
		0.894	TAKE(puma6, plsx)
		1.360	MOVE PLACE(puma6, plsx)
		0.894	LEAVE(puma6, plsx)
		1.330	MOVE PICK(puma6, cent)
		0.894	TAKE(puma6, cent)
		1.365	MOVE PLACE(puma6, cent)
		0.894	LEAVE(puma6, cent)
		1.080	MOVE SAFE(puma6)

1.368	MOVE PLACE(puma5, tops)
0.894	LEAVE(puma5, tops)
1.291	MOVE SAFE(puma5)

1.170	MOVE PICK(adept, assm)
0.894	TAKE(adept, assm)
1.351	MOVE PLACE(adept, assm)
0.894	LEAVE(adept, assm)
1.404	MOVE SAFE(adept)

Figure 10 Optimal minimum cost task assignment and the corresponding global plan for the *blocks* example. The highlighted action sequence indicates the critical path.

viewed as the players of the game. The global objective of these robots is to perform the jobs assigned to the system such that some global objective function is optimized. Additionally, each robot perceives the local objectives of completing the tasks assigned to it in the best possible manner, i.e., to optimize some local objective function. Depending on the situation, the local or global objective may not exist. When both local and global objectives exist, the interactions between the robots have to be resolved such that the combination of the local and global objectives is optimized. We will refer to this situation as the *coalition* mode. The solution for this mode is based on *n*-person cooperative game theory. Some solution schemes are discussed below.

The cooperation assignment-planning problem is one of determining which robot set is selected to perform the tasks and which tasks are performed by which robot in the set. The robots as players of the game are attempting to maximize their gain. The value of the game to any individual robot is generally 0 because no single robot is capable of

Table 5   Minimum Robot Set Solution Generation Using the Enumeration Scheme for the *Blocks* Example

0–1 Matrix:	Task 0	Task 1	Task 2	Task 3	Task 4	Task 5	Task 6	Task 7
Set 0:	0	0	1	1	1	1	0	1
Set 1:	1	1	0	0	0	0	1	1
Set 2:	1	1	1	1	1	1	1	0
Set 3:	0	0	0	0	0	0	0	0
Set 4:	0	0	0	0	0	0	0	0
Set 5:	0	0	0	0	0	0	0	0
Set 6:	0	0	0	0	0	0	0	0

1-Member Sets:
Set 0: $0 \wedge 0 \wedge 1 \wedge 1 \wedge 1 \wedge 1 \wedge 0 \wedge 1 = 0$
Set 1: $1 \wedge 1 \wedge 0 \wedge 0 \wedge 0 \wedge 0 \wedge 1 \wedge 1 = 0$
Set 2: $1 \wedge 1 \wedge 1 \wedge 1 \wedge 1 \wedge 1 \wedge 1 \wedge 0 = 0$

2-Member Sets:
Set 3: $(0 \vee 1 \vee 0) \wedge (0 \vee 1 \vee 0) \wedge (1 \vee 0 \vee 0) \wedge (1 \vee 0 \vee 0) \wedge (1 \vee 0 \vee 0) \wedge (1 \vee 0 \vee 0) \wedge$
$(0 \vee 1 \vee 0) \wedge (1 \vee 1 \vee 0)$
$= 1$ (minimum cost = 45.485)
Set 4: $(0 \vee 1 \vee 0) \wedge (0 \vee 1 \vee 0) \wedge (1 \vee 1 \vee 0) \wedge (1 \vee 1 \vee 0) \wedge (1 \vee 1 \vee 0) \wedge (1 \vee 1 \vee 0) \wedge$
$(0 \vee 1 \vee 0) \wedge (1 \vee 0 \vee 0)$
$= 1$ (minimum cost = 44.249)
Set 5: $(1 \vee 1 \vee 0) \wedge (1 \vee 1 \vee 0) \wedge (0 \vee 1 \vee 0) \wedge (0 \vee 1 \vee 0) \wedge (0 \vee 1 \vee 0) \wedge (0 \vee 1 \vee 0) \wedge$
$(1 \vee 1 \vee 0) \wedge (1 \vee 0 \vee 0)$
$= 1$ (minimum cost = 44.071)

Minimum Robot Set Solution:
Set 5: PUMA 550 and PUMA 560 with cost of 44.071 seconds

---

performing all the assigned tasks. If a global plan exists, then the set of all robots, the grand coalition, is capable of performing all the assigned tasks. There might also exist many other robot sets, coalitions, that can also successfully complete the assigned tasks. Robot sets that can perform all the assigned tasks are considered *feasible coalitions* and their value is greater than 0. All other robot sets have values of 0. With this structure, in the following sections we will show the application of the bargaining set and the Caplow Power Index to the cooperation assignment planning problem.

Bargaining Set Solution

To use the bargaining set (Kahan and Rapoport, 1984) as a solution concept, we need to define the *characteristic function* and the concept of *value* to each of the robots. In the previous sections we have specified a cost for each robot set to perform each task given by the cooperation requirements matrix, $C(i, j)$. Because the cost is to be minimized, we will define the value function as:

$$v(i, j) = \Gamma - C(i, j), \text{ if } C(i, j) < \infty,$$
$$= 0, \text{ otherwise,}$$

$i = 1, \ldots, N(R), j = 1, \ldots, N(T)$, where Γ is some large positive number

Thus the robots will attempt to form coalitions such that their value is maximized. As mentioned earlier, the value for a robot acting alone is normally 0 because a job may consist of mandatory cooperation tasks that cannot be performed by a single robot. Even if the job contains all optional cooperation tasks that can be performed by a single robot, using concurrent cooperation may provide a higher value than acting alone.

We enforce the global objective requirement that all assigned tasks should be performed successfully by considering only those partitions of the set of robots that can successfully perform all the assigned tasks. In such partitions robot coalitions may exist for performing mandatory and concurrent cooperation tasks. Thus, the value for such sets is greater than 0, while the value for those that do not belong to this set is 0. Within a feasible coalition a variety of task assignments to robots are possible to complete the assigned tasks. However, the set of all possible assignments within a particular coalition is finite. A payoff vector for a coalition is defined as the value vector obtained by a particular assignment within the coalition. Since the value to an individual robot is 0, any payoff vector satisfies individual rationality as a robot gets a minimum of 0 if no tasks are assigned to it. The combination of coalition structures and payoff vectors provide the individually rational payoff configurations. Since the bargaining set consists of individually rational payoff configurations, this condition is satisfied. The bargaining set can then be computed by finding those payoff configurations that satisfy the condition that no member of a coalition has a justified objection against any other member of the coalition (Kahan and Rapoport, 1984).

The bargaining set solution for the *blocks* example is shown in Table 6. Only the two-member coalitions are considered. For each feasible coalition structure a variety of individually rational payoff configurations exist, based on the assignment of tasks to robots. Since each robot has a value of 0 by acting independently, a positive payoff is obtained only by participating in a coalition. As shown, the bargaining set consists of six payoff vectors, any of which can be chosen as the final assignment. The bargaining set solutions provide an equitable distribution of the tasks between the cooperating robots within the

Table 6   Bargaining Set Solution by Considering Only the Two-Member Coalitions for the *Blocks* Example. Γ is Some Large Number Such That $\Gamma - * > 0$

Individually Rational Payoff Configurations:
(Γ − 28.928, Γ − 16.557, 0.000; (adept[2, 3, 4, 5, 7], puma5[0, 1, 6]), puma6);
(Γ − 23.214, Γ − 22.357, 0.000; (adept[2, 3, 4, 5], puma5[0, 1, 6, 7]), puma6);
(Γ − 28.928, 0.000, Γ − 16.398; (adept[2, 3, 4, 5, 7], puma6[0, 1, 6]), puma5);
(Γ − 23.164, 0.000, Γ − 21.850; (adept[3, 4, 5, 7], puma6[0, 1, 2, 6]), puma 5);
(Γ − 23.177, 0.000, Γ − 21.949; (adept[2, 4, 5, 7], puma6[0, 1, 3, 6]), puma5);
(Γ − 23.084, 0.000, Γ − 22.013; (adept[2, 3, 5, 7], puma6[0, 1, 4, 6]), puma5);
(Γ − 23.073, 0.000, Γ − 21.917; (adept[2, 3, 4, 7], puma6[0, 1, 5, 6]), puma5);
(Γ − 17.413, 0.000, Γ − 27.401; (adept[4, 5, 7], puma6[0, 1, 2, 3, 6]), puma5);
(Γ − 17.320, 0.000, Γ − 27.465; (adept[3, 5, 7], puma6[0, 1, 2, 4, 6]), puma5);
(Γ − 17.309, 0.000, Γ − 27.369; (adept[3, 4, 7], puma6[0, 1, 2, 5, 6]), puma5);
(Γ − 17.333, 0.000, Γ − 27.564; (adept[2, 5, 7], puma6[0, 1, 3, 4, 6]), puma5);
(Γ − 17.322, 0.000, Γ − 27.468; (adept[2, 4, 7], puma6[0, 1, 3, 5, 6]), puma5);
(Γ − 17.229, 0.000, Γ − 27.532; (adept[2, 3. 7], puma6[0, 1, 4, 5, 6]), puma5);
(Γ − 11.569, 0.000, Γ − 33.016; (adept[5, 7], puma6[0, 1, 2, 3, 4, 6]), puma5);
(Γ − 11.558, 0.000, Γ − 32.920; (adept[4, 7], puma6[0, 1, 2, 3, 5, 6]), puma5);
(Γ − 11.465, 0.000, Γ − 32.984; (adept[3, 7], puma6[0, 1, 2, 4, 5, 6]), puma5);
(Γ − 11.478, 0.000, Γ − 33.083; (adept[2, 7], puma6[0, 1, 3, 4, 5, 6]), puma5);
(Γ − 5.714, 0.000, Γ − 38.535; (adept[7], puma6[0, 1, 2, 3, 4, 5, 6]), puma5);
(0.000, Γ − 22.357, Γ − 22.137; (adept, puma5[0, 1, 6, 7]), puma6[2, 3, 4, 5]);
(0.000, Γ − 16.938, Γ − 27.688; (adept, puma5[1, 6, 7]), puma6[0, 2, 3, 4, 5]);
(0.000, Γ − 16.637, Γ − 27.434; (adept, puma5[0, 6, 7]), puma6[1, 2, 3, 4, 5]);
(0.000, Γ − 16.945, Γ − 27.687; (adept, puma5[0, 1, 7]), puma6[2, 3, 4, 5, 6]);
(0.000, Γ − 11.212, Γ − 32.985; (adept, puma5[6, 7]), puma6[0, 1, 2, 3,4, 5]);
(0.000, Γ − 11.520, Γ − 33.238; (adept, puma5[1, 7]), puma6[0, 2, 3, 4, 5, 6]);
(0.000, Γ − 11.225, Γ − 32.984; (adept, puma5[0, 7]), puma6[1, 2, 3, 4, 5, 6]);
(0.000, Γ − 5.800, Γ − 38.535; (adept, puma5[7]), puma6[0, 1, 2, 3, 4, 5, 6]);
Bargaining Set:
(Γ − 28.928, Γ − 16.557, 0.000; (adept[2, 3, 4, 5, 7], puma5[0, 1, 6]), puma6);
(Γ − 23.214, Γ − 22.357, 0.000; (adept[2, 3, 4, 5], puma5[0, 1, 6, 7]), puma6);
(Γ − 23.164, 0.000, Γ − 21.850; (adept[3, 4, 5, 7], puma6[0, 1, 2, 6]), puma5);
(Γ − 23.177, 0.000, Γ − 21.949; (adept[2, 4, 5, 7], puma6[0, 1, 3, 6]), puma5);
(Γ − 23.084, 0.000, Γ − 22.013; (adept[2, 3, 5, 7], puma6[0, 1, 4, 6]), puma5);
(Γ − 23.073, 0.000, Γ − 21.917; (adept[2, 3, 4, 7], puma6[0, 1, 5, 6]), puma5);

constraints of the various payoff vectors that can occur. If the grand coalition is considered, some of the solutions generated for the two-member coalitions may be invalidated, since the robots in the two-member solutions may be able to raise objections without receiving any counterobjections.

Caplow Power Index Solution

The set of winning coalitions for a multirobot system is the set of robot sets that can successfully complete all the assigned tasks. The weight assigned to each robot is the number of tasks that the robot can successfully perform by independent operation. The winning coalition definition satisfies the global objective, while the assignment of weights satisfies the local objectives. Based on these definitions, the power indices for each member of the winning coalitions can be computed and the undominated winning coalitions identified. The robot sets that are undominated provide the solution for the game. The best possible assignment within each set is then computed, and the assignment that provides the lowest cost is selected to perform the assigned tasks.

The Caplow Power Index (Shenoy, 1978) solution for the *blocks* example is shown in Table 7. All the two-member sets and the grand coalition are feasible coalitions because they can successfully complete all the assigned tasks. The weights for the robots are assigned based on the number of tasks they can successfully complete. The two-member coalition ADEPT 1–PUMA 560 and the grand coalition are dominated by the other two-member coalitions. Therefore, the final coalition selected to perform the task can be either the ADEPT 1–PUMA 550 robot set or the PUMA 550–PUMA 560 robot set.

The bargaining set and Caplow Power Index solutions for the *linkage* example are the grand coalition, since this is the only feasible coalition.

4 SUMMARY AND CONCLUSIONS

In this chapter we have discussed concepts and techniques for planning robotic operations with emphasis on integrated assembly and execution planning. Robotic operation planning is a bridge between product design and process execution, the first part of which is planning that converts design data into operational process information. In assembly, as in other operations, it consists of interpretation of design data, selection of processes and parameters, selection of assembly system and auxiliary equipment, operations sequencing, selection of inspection devices, and calculation of task times. The second part of robotic operation planning is execution planning, which consists of identifying which tasks can be performed by which robot sets, assigning tasks to robot sets, and generating a consistent and coordinated global execution plan. While the selection of an operation sequence can be performed independently of the actual robotic system used for the process, it is recommended that the sequence selection and execution planning be performed concurrently to ensure feasible and optimal task assignment and execution. The minimal precedence constraint concept has been presented for the identification of geometric precedence constraints of the assembly components. Simultaneous and sequential task assignment execution planning strategies have also been presented. The application of

---

Table 7   The Caplow Power Index Solution for the *Blocks* Example

Weights:
   adept: 5; puma5: 4; puma6: 7

Winning Coalitions:
   adept/puma5, adept/puma6, puma5/puma6, adept/puma5/
     puma6

Caplow Power Indices:

Coalition Structure	Index Value (adept, puma5, puma6)
(adept/puma5, puma6)	(2/3, 1/3, 0)
(adept/puma6, puma5)	(1./3, 0, 2/3)
(adept, puma5/puma6)	(0, 1/3, 2/3)
(adept/puma5/puma6)	(1/3, 0, 2/3)

Undominated Coalition Structures:
   (adept/puma5, puma6), (adept, puma5/puma6)

two game-theory concepts, bargaining sets and the Caplow Power Index, has also been discussed.

Operations and assembly planning provide the constraints that are imposed on the execution process by the product, robots, and system. If the minimal set of such constraints is identified, significant flexibility in planning of robotic operations can be realized. Proper task assignment and execution planning that concurrently consider the product and robot/system constraints will ensure that feasible and optimal task execution occur. However, it must be noted that these activities demonstrate complicated interactions that need to be carefully resolved.

REFERENCES

Boothroyd, G., and P. Dewhurst. 1983. "Design for Assembly: Selecting the Right Method." *Machine Design* (November 10), 94–98.

Csakvary, T. 1985. "Planning Robot Applications in Assembly." In *Handbook of Industrial Robotics.* 1st ed. Ed. S. Y. Nof. New York: John Wiley & Sons.

De Fazio, T. L., and D. E. Whitney. 1987. "Simplified Generation of All Mechanical Assembly Sequences." *IEEE Journal of Robotics and Automation* **RA-3**(6), 640–658.

Henrioud, J.-M., and A. Bourjault. 1991. "LEGA: A Computer-Aided Generator of Assembly Plans." In *Computer-Aided Mechanical Assembly Planning.* Ed. L. S. Homem de Mello and S. Lee. Norwell: Kluwer Academic. 191–215.

Homem de Mello, L. S., and A. C. Sanderson. 1991. "Representation of Mechanical Assembly Sequences." *IEEE Transactions on Robotics and Automation* **7**(2), 211–227.

Huang, Y. F., and C. S. G. Lee. 1988. "Precedence Knowledge in Feature Mating Operation Assembly Planning." Technical Report No. TR-ERC 88-22, Engineering Research Center for Intelligent Manufacturing Systems, Schools of Engineering, Purdue University.

Kahan, J. P., and A. Rapoport. 1984. *Theories of Coalition Formation.* Hillsdale: Lawrence Erlbaum Associates.

Kalpakjian, S. 1992. *Manufacturing Processes for Engineering Materials.* 2nd ed. Reading: Addison-Wesley.

Kennedy, C. W., E. G. Hoffman, and S. D. Bond. 1987. *Inspection and Gaging.* 6th ed. New York: Industrial Press.

Kondelon, A. S. 1985. "Application of Technology-Economic Model of Assembly Techniques and Programmable Assembly Machine Configurations." M.S. Thesis, Massachusetts Institute of Technology.

Lee, S. 1994. "Subassembly Identification and Evaluation for Assembly Planning." *IEEE Transactions on Systems, Man, and Cybernetics* **24**(3), 493–503.

Li, Z., et al. 1989. "Motion Space Analysis of an Object Handled by Two Robot Arms." In *Proceedings of the 28th Conference on Decision and Control,* Tampa, December 13–15. 2487–2493.

Mundel, M. E., and D. L. Danner. 1994. *Motion and Time Study.* 7th ed. Englewood Cliffs: Prentice-Hall.

Nnaji, B. O., J.-Y. Chu, and M. Akrep. 1988. "A Schema for CAD-Based Robot Assembly Task Planning for CSG-Modeled Objects." *Journal of Manufacturing Systems* **7**(2), 131–145.

Nof, S. Y., and D. Hanna. 1989. "Operational Characteristics of Multi-Robot Systems with Cooperation." *International Journal of Production Research* **27**(3), 477–492.

Nof, S. Y., and V. N. Rajan. 1993. "Automatic Generation of Assembly Constraints and Cooperation Task Planning." *Annals of the CIRP* **42**(1), 13–16.

Owen, G. 1982. *Game Theory.* New York: Academic Press.

Owen, T. 1985. *Assembly with Robots.* Englewood Cliffs: Prentice-Hall.

Rajan, V. N. 1993. "Cooperation Requirement Planning for Multi-Robot Assembly Cells." Ph.D. Dissertation, Purdue University.

Rajan, V. N., and S. Y. Nof. 1996a. "Minimal Precedence Constraints for Integrated Assembly and Execution Planning." *IEEE Transactions on Robotics and Automation* [Special Issue on Assembly Task Planning for Manufacturing] 175–186.

———. 1996b. "Cooperation Requirements Planning (CRP) for Multiprocessors: Optimal Assignment and Execution Planning." *Journal of Intelligent and Robotic Systems* **15**, 419–435.

Rajan, V. N., K. Sivasubramanian, and J. E. Fernandez. 1997. "Accessibility and Ergonomic Analysis of Assembly Product and Jig Designs." to appear in *International Journal of Industrial Ergonomics.*

Schey, J. A. 1987. *Introduction to Manufacturing Processes.* 2nd ed. New York: McGraw-Hill.

Shenoy, P. P. 1978. "On Coalition Formation in Simple Games: A Mathematical Analysis of Caplow's and Gamson's Theories." *Journal of Mathematical Psychology* **18**, 177–194.

Shpitalni, M., G. Elber, and E. Lenz. "Automatic Assembly of Three-Dimensional Structures via Connectivity Graphs." *Annals of the CIRP* **38**(1), 25–28.

Wolter, J. D. 1991. "On the Automatic Generation of Assembly Plans." In *Computer-Aided Mechanical Assembly Planning.* Ed. L. S. Homem de Mello and S. Lee. Boston: Kluwer Academic. 263–288.

Woo, T. C. 1987. "Automatic Disassembly and Total Ordering in Three Dimensions." Technical Report No. 87-9, Department of Industrial and Operations Engineering, University of Michigan, Ann Arbor.

Yang, Q., D. S. Nau, and J. Hendler, "Merging Separately Generated Plans with Restricted Interactions." Technical Report No. UMIACS-TR-91-73, University of Maryland Institute for Advanced Computer Studies, University of Maryland, College Park.

CHAPTER 32

ROBOT ERGONOMICS: OPTIMIZING ROBOT WORK

Shimon Y. Nof
Purdue University
West Lafayette, Indiana

1 THE ROLE OF ROBOT ERGONOMICS

The word *ergonomics,* in Greek, means "the natural laws of work." Traditionally, it has meant the study of the anatomical, physiological, and psychological aspects of humans in working environments for the purpose of optimizing efficiency, health, safety, and comfort associated with work systems. Correct and effective introduction of robots to industrial work requires use of ergonomics. However, planning the work of robots themselves in industry brings about a completely new dimension.

For the first time, as far as we can tell, we have the ability to design not only the work system, its components and environment, but also the structure and capabilities of the *operator*—the robot.

The purpose of this chapter is to explain a number of techniques that have been developed and applied in recent years for planning various applications of robots in industry. Some of the techniques bear strong similarity to the original, human-oriented ergonomics techniques. But all these techniques have been developed or adapted for the unique requirements of industrial robots. Their common objective: to provide tools for the study of relevant aspects of robots in working environments for the purpose of optimizing overall performance of the work system. Specifically, robot work should be optimized to: (1) minimize the time per unit of work produced; (2) minimize the amount of effort and energy expended by operators; (3) minimize the amount of waste, scrap, and rework; (4) optimize quality of work produced; and (5) maximize safety.

A general ergonomics procedure for optimizing industrial robot work is depicted in Figure 1. For given job requirements, it entails the analysis and evaluation of whether a human or a robot should be employed for the job. If a robot, the best combination of robot models and work method, implying also the best workplace, should be selected. In integrated human and robot systems the best combination must be designed. The subsequent sections cover the ergonomics techniques that are useful to follow the foregoing procedure in practice.

The following topics are covered (see Figure 2):

Understanding the job requirements
Analysis of work characteristics
Work methods planning
Motion economy principles (MEPs)
Performance measurement
Workplace design
Integrated human and robot ergonomics

2 UNDERSTANDING THE JOB REQUIREMENTS

There are specific considerations related to industrial robot work. The following considerations are based on surveys and case studies (Nof and Rajan, 1994).

Handbook of Industrial Robotics, Second Edition, Edited by Shimon Y. Nof
ISBN 0-471-17783-0 © 1999 John Wiley & Sons, Inc.

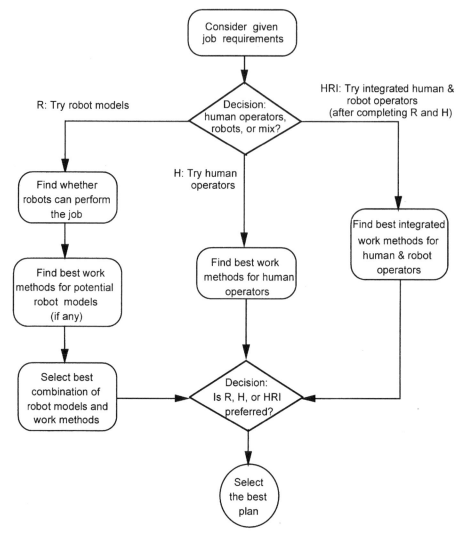

Figure 1 Ergonomics procedure for optimizing robot, human, or integrated work.

2.1 Hazardous Environment Operations

A variety of applications involve human activity under hazardous conditions. Such applications provide a primary motivation for the use of robots instead of humans to improve working conditions.

Die casting. Eliminate hand degating and handling of gates, storage of parts and trimmings, material handling.

Forging. Eliminate human operation in hot, dirty, noxious environments.

Injection molding. Eliminate human operation in hot, noxious environments and ensure compliance with OSHA regulations involving dies.

Stamping. Eliminate hand-in-die hazard.

Machine loading. Eliminate human operation in environments with cutting oil fumes, chips, and noise.

Material handling. Eliminate tedious and heavy tasks.

Welding. Eliminate handling of heavy weld guns, heat exposure, and eyesight damage.

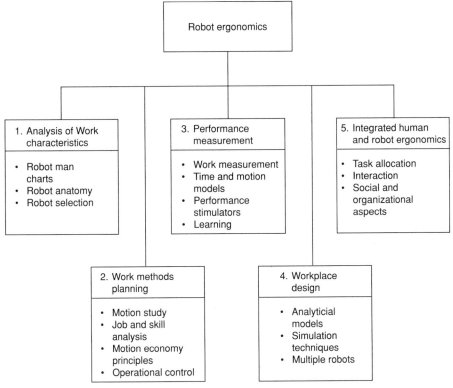

Figure 2 General scope and functions of robot ergonomics.

Finishing and painting. Eliminate paint spray hazard and exposure to flammable paint material.

2.2 Improved Quality

Use of robots can improve quality of a process due to higher accuracy and repeatability or by performing the process under conditions that lead to higher quality but are not conducive for human operation.

Die casting. Reduced scrap due to consistent cycle times, less damage from dies and tie bars, better quality on long and flat parts by controlling quench sequence

Investment casting. Uniform shell and better castings, consistent timing, draining, faster spinning

Forging. Hot trim instead of cold trim improves part quality, consistent part placement, properly timed operations

Die forging. Consistent lubrication

Injection molding. No damage or contamination; stable cycle times, especially on more than one unit

Welding. Consistent arc welding, reduction of weld spots in spot welding due to consistency

Finishing and painting. Consistent film build and coverage

Assembly. No missing parts, no misassembled parts, no visibly defective parts used

Inspection. Consistent results due to reduced reading and recording errors

2.3 Reduced Cost

Costs can be reduced due to savings in labor cost and reduced use of some process and safety equipment.

Die casting. Eliminate labor cost; reduced remelt cost; avoid cost of safety equipment, as much as $20,000/machine; typical payout period is from 9 months to 2.5 years, depending on heads, shifts, etc.

Investment casting. Labor savings

Forging. Labor savings, no relief required

Injection molding. Labor savings

Stamping. Labor savings, particularly with large parts

Machine loading. Labor savings; reduce in-process handling by arranging machines in "cell"; however, relocation of machines may be a major expense

Material handling. Labor savings; savings due to elimination of hoists and other mechanical aids for handling heavy parts; may require part orientation, which may be costly

Welding. Labor savings on multiple-shift operations; savings in energy consumption and electrode-weld tip replacement due to consistent spot welding; can reduce administrative and training costs for replacement of welders in arc welding operations where labor turnover tends to be high; may require precise fixturing, indexing, or line-tracking ability that add to costs

Finishing and painting. Material savings; overspray, trigger control, and removing humans from the booth reduces costs by reducing makeup air requirements and allows concentration of solvents for paint recovery or emissions control; labor savings

Assembly. Labor savings, requires capital expenditures to avoid obsolescence

Inspection. Visual inspection can often be combined with handling-transfer operation for savings, avoid cost of expensive custom-made gauging devices

2.4 Improved Performance

Robots can perform tedious, repetitive tasks at greater speeds and with continuous operation.

Die casting. Faster production rate, run through breaks, and lunch; more parts/day; run multiple machines served by single robot

Investment casting. Faster than humans; can handle heavier trees, e.g., wax cluster with 100 turbine blades in same time as 50-blade cluster, whereas a human could not handle a 100-blade tree

Forging. Higher production, especially on heavy parts for which a human requires help of a second human and/or hoist

Injection molding. Faster production on large parts, especially on large machines or on molds with more than two cavities; completion of secondary operations such as trimming, palletizing, and packing during machine cycle

Stamping. Reduced changeover time for different parts compared with usual press automation; avoid purchase of safety equipment required with manual press operations, but could lead to expensive interlocks. A recent installation, e.g., had two lines of three presses, two robots per line transferring parts from press 1 to press 2 and from 2 to 3, a third line with two presses and one robot—five robots in all at about $35,000 each, interlocks between robots and presses cost about $100,000, which is about 45% of total installation cost

Machine loading. Ability to handle heavy parts

Material handling. More uniform packing, higher density, ability to handle heavy parts, faster than people on long reach, no walking

Finishing and painting. Better overspray, trigger control, less booth cleanup from reduced overspray

Assembly. Higher productivity, especially in systems with robots and people; if robots are set up to pace operations, can often combine assembly and test or in-process inspection with 100% assurance of test performance, people tend to skip such steps because it is not obvious when not performed

Inspection. Higher productivity due to higher speeds, can provide 100% inspection on operations for which sampling was formerly done, can inspect moving parts

3   ANALYSIS OF WORK CHARACTERISTICS

To implement effectively an ergonomics procedure (following Figure 1) for given job requirements, a general list of considerations in planning robot work can be prepared. In addition to such a general analysis, it is also necessary to know the detailed characteristics and skills of today's industrial robots as well as those of humans. A series of the robot–human charts can serve this purpose well.

3.1   The Robot–Human Charts

Robot–human charts, originally prepared by Nof, Knight, and Salvendy (1980), are developed with two functions in mind, namely (1) to aid engineers in determining whether a robot can perform a job and (2) to serve as a guideline and reference for robot specifications. Table 1 presents the robot–human charts for comparison of physical skills of humans and robots. The robot–human charts can also be useful in job design of combined systems, which integrate both robots and human operators. They contain three principal types of work characteristics.

1. *Physical skills and characteristics,* including manipulation, body dimensions, strength and power, consistency, overload, underload performance, and environ-

Table 1   Robot–Human Charts[a]

Characteristics	Robot	Human
	a. Comparison of Robot and Human Physical Skills and Characteristics	
	Manipulation	

Body	a. One of four types:
	  1. Uni- or multiprismatic
	  2. Uni- or multirevolute
	  3. Combined revolute–prismatic
	  4. Mobile
	b. Typical maximum movement and velocity capabilities:
	  Right–left traverse
	    5–18 m at 500–1200 mm/s
	  Out–in traverse
	    3–15 m
	    500–1200 mm/s
		a. A mobile carrier (feet) combined with 3-d.o.f. wrist like (roll, pitch, yaw) capability at waist
		b. Examples of waist movement:[b]
		  Roll: $\simeq 180°$
		  Pitch: $\simeq 150°$
		  Yaw: $\simeq 90°$

Arm	a. One of four primary types:
	  1. Rectangular
	  2. Cylindrical
	  3. Spherical
	  4. Articulated
	b. One or more arms, with incremental usefulness per each additional arm
	c. Typical maximum movement and velocity capabilities:
	  Out–in traverse
	    300–3000 mm
	    100–4500 mm/s
	  Right–left traverse
	    100–6000 mm
	    100–1500 mm/s
	  Up–down traverse
	    50–4800 mm
	    50–5000 mm/s
	  Right–left rotation
	    50–380°[c]
	    5–240°/s
	  Up–down rotation
	    25–330°
	    10–170°/s
		a. Articulated arm composed of shoulder and elbow revolute joints
		b. Two arms, cannot operate independently (at least not totally)
		c. Examples of typical movement and velocity parameters:
		  Maximum velocity:
		    1500 mm/s in linear movement.
		  Average standing lateral reach:
		    625 mm
		  Right–left traverse range:
		    432–876 mm
		  Up–down traverse range:
		    1016–1828 mm
		  Right–left rotation range:
		  (horizontal arm)
		    165–225°
		  Average up–down rotation:
		    249°

Table 1 (Continued)

Characteristics	Robot	Human
Wrist	a. One of three types: 1. Prismatic 2. Revolute 3. Combined prismatic/revolute Commonly, wrists have 1–3 rotational d.o.f.: roll, pitch, yaw, however, an example of right–left and up–down traverse was observed	a. Consists of three rotational degrees of freedom: roll, pitch, yaw.
	b. Typical maximum movement and velocity capabilities: *Roll* 100–575°^{d} 35–600°/s *Pitch* 40–360° 30–320°/s *Yaw* 100–530° 30–300°/s *Right–left traverse* (*uncommon*) 1000 mm 4800 mm/s *Up–down traverse* (*uncommon*) 150 mm 400 mm/s	b. Examples of movement capabilities Roll: ≈ 180° Pitch: ≈ 180° Yaw: ≈ 90°
End-effector	a. The robot is affixed with either a hand or a tool at the end of the wrist. The end-effector can be complex enough to be considered a small manipulator in itself	a. Consists of essentially 4 d.o.f. in an articulated configuration. Five fingers per arm each have three pitch revolute and one yaw revolute joints
	b. Can be designed to various dimensions	b. Typical hand dimensions: Length: 163–208 mm Breadth: 68–97 mm (at thumb) Depth: 20–33 mm (at metacarpal)
Body dimensions	a. Main body: Height: 0.10–2.0 m Length (arm) 0.2–2.0 m Width: 0.1–1.5 m Weight: 5–8000 kg	a. Main body (typical adult): Height: I.5–1.9 m Length (arm): 754–947 mm Width: 478–579 mm Weight: 45–100 kg
	b. Floor area required: from none for ceiling-mounted models to several square meters for large models	b. Typically about 1 m² working radius
Strength and power	a. 0.1–1000 kg of useful load during operation at normal speed: reduced at above normal speeds	a. Maximum arm load: <30 kg; varies drastically with type of movement, direction of load, etc.
	b. Power relative to useful load	b. Power: 2 hp ≈ 10 s 0.5 hp ≈ 120 s 0.2 hp ≈ continuous 5 kc/min Subject to fatigue: may differ between static and dynamic conditions

Table 1   (Continued)

Characteristics	Robot	Human
Consistency	Absolute consistency if no malfunctions	a. Low b. May improve with practice and redundant knowledge of results c. Subject to fatigue: physiological and psychological d. May require external monitoring of performance
Overload–underload performance	a. Constant performance up to a designed limit, and then a drastic failure b. No underload effects on performance	a. Performance declines smoothly under a failure b. Boredom under local effects is significant
Environmental constraints	a. Ambient temperature from $-10°$ to $60°C$ b. Relative humidity up to 90% c. Can be fitted to hostile environments	a. Ambient temperature range $15–30°C$ b. Humidity effects are weak c. Sensitive to various noxious stimuli and toxins, altitude, and air flow

b. Comparison of Robot and Human Mental and Communicative Skills

1. Computational capability	a. Fast, e.g., up to 10 Kbits/sec for a small minicomputer control b. Not affected by meaning and connotation of signals c. No evaluation of quality of information unless provided by program d. Error detection depends on program e. Very good computational and algorithmic capability by computer f. Negligible time lag g. Ability to accept information is very high, limited only by the channel rate h. Good ability to select and execute responses i. No compatibility limitations j. If programmable—not difficult to reprogram k. Random program selection can be provided l. Command repertoire limited by computer compiler or control scheme	a. Slow—5 bits/sec. b. Affected by meaning and connotation of signals c. Evaluates reliability of information d. Good error detection correction at cost of redundancy e. Heuristic rather than algorithmic f. Time lags increased, 1–3 sec g. Limited ability to accept information (10–20 bits/sec) h. Very limited response selection/execution (1/sec); responses may be "grouped" with practice i. Subject to various compatibility effects j. Difficult to program k. Various sequence/transfer effects l. Command repertoire limited to experience and training
2. Memory	a. Memory capability from 20 commands to 2000 commands, and can be extended by secondary memory such as cassettes b. Memory partitioning can be used to improve efficiency c. Can forget completely but only on command d. "Skills" must be specified in programs	a. No indication of capacity limitations b. Not applicable c. Directed forgetting very limited d. Memory contains basic skills accumulated by experience e. Slow storage access/retrieval f. Very limited working register: ≈ 5 items

Table 1 (Continued)

Characteristics	Robot	Human
3. Intelligence	a. No judgment ability of unanticipated events b. Decision-making limited by computer program	a. Can use judgment to deal with unpredicted problems b. Can anticipate problems
4. Reasoning	a. Good deductive capability, poor inductive capability b. Limited to the programming ability of the human programmer	a. Inductive b. Not applicable
5. Signal processing	a. Up to 24 input/output channels, and can be increased, multitasking can be provided b. Limited by refractory period (recovery from signal interrupt)	a. Single channel, can switch between tasks b. Refractory period up to 0.3 sec
6. Brain–muscle combination	a. Combinations of large, medium, and small "muscles" with various size memory, velocity and path control, and computer control can be designed	a. Fixed arrangement
7. Training	a. Requires training through teaching and programming by an experienced human b. Training doesn't have to be individualized c. No need to retrain once the program taught is correct d. Immediate transfer of skills ("zeroing") can be provided	a. Requires human teacher or materials developed by humans b. Usually individualized is best c. Retraining often needed owing to forgetting d. Zeroing usually not possible
8. Social and psychological needs	a. None	a. Emotional sensitivity to task structure—simplified/ enriched; whole/part b. Social value effects

9. Sensing	a. Limited range can be optimized over the relevant needs
	b. Can be designed to be relatively constant over the designed range
		a. Very wide range of operation (10^{12} units)
		b. Logarithmic:
		  1. visual angle threshold— 0.7 min
		  2. brightness threshold— 4.1 $\mu\mu l$
		  3. response rate for successive stimuli $\simeq$ 0.1 sec
		  audition:
		  1. threshold—0.002 dynes/m^2
		  tactile:
		  1. threshold—3 g/mm^2

	c. The set of sensed characteristics can be selected. Main senses are vision and tactile (touch) d. Signal interference ("noise") may create a problem e. Very good absolute judgment can be applied f. Comparative judgment limited by program	c. Limited set of characteristics can be sensed d. Good noise immunity (built-in filters) e. Very poor absolute judgment (5–10 items) f. Very good comparative judgment
10. Interoperator communication	Very efficient and fast intermachine communication can be provided	Sensitive to many problems, e.g., misunderstanding
11. Reaction speed	Ranges from long to negligible delay from receipt of signal to start of movement	Reaction speed $\frac{1}{4}$–$\frac{1}{3}$ sec

Table 1 (Continued)

Characteristics	Robot	Human
12. Self-diagnosis	Self-diagnosis for adjustment and maintenance can be provided	Self-diagnosis may know when efficiency is low
13. Individual differences	Only if designed to be different	100–150% variation may be expected
c. Comparison of Robot and Human Energy Considerations		
1. Power requirements	Power source 220/440 V, 3 phase, 50/60 Hz, 0.5–30 KVA. Limited portability	Power (energy) source is food
2. Utilities	Hydraulic pressure: 30–200 kg/cm² Compressed air: 4–6 kg/cm²	Air: Oxygen consumption 2–9 liters/min
3. Fatigue, downtime, and life expectancy	a. No fatigue during periods between maintenance	a. Within power ratings, primarily cognitive fatigue (20% in first 2 hr; logarithmic decline)
	b. Preventive maintenance required periodically	b. Needs daily rest, vacation
	c. Expected usefulness of 40,000 hr (about 20 one-shift years)	c. Requires work breaks
	d. No personal requirements	d. Various personal problems (absenteeism, injuries, health)
4. Energy efficiency	a. Relatively high, e.g. (120–135 kg)/(2.5–3.0 KVA)	a. Relatively low, 10–25%
	b. Relatively constant regardless of workload	b. Improves if work is distributed rather than massed

[a] Revised from S. Y. Nof, J. L. Knight, and G. Salvendy, "Effective Use of Industrial Robots—A Job and Skills Analysis Approach," *AIEE Transactions* **12**(3), 216–225, 1980.

[b] Where possible, 5th and 95th percentile figures are used to represent minimum and maximum values. Otherwise, a general average value is given.

[c] A continuous right–left rotation is available.

[d] A continuous roll movement is available.

mental constraints. Table 1 provides details of this category. Typical ranges of maximum motion capabilities (TRMM) are given for several categories of body movement and speed and arm and wrist motions. To clarify the meaning of TRMM, consider the following example. For robot arm right–left traverse, the table lists a maximum movement range of 100 to 6000 mm at a maximum velocity range of 100 to 1500 mm/s. This means that for the surveyed population of robot models, it was found that a maximum arm right–left linear motion is typically between 100 mm (for some models) and up to 6000 mm (for some other models). The maximum velocity values for right–left travel were found to be from 100 up to 1500 mm/s.

2. *Mental and communicative characteristics.* The robot–human charts contain mental and communicative system attributes for robots and humans.

3. *Energy considerations.* A comparison of representative values of energy-related characteristics, such as power requirements, and energy efficiency for robots and humans.

Certain significant features that distinguish robots and human operators, such as the following, can effectively be used to select jobs that robots can do well.

1. The more similar two jobs are, the easier it is to transfer either robot or human from one job to the other. For humans such transfer is almost entirely a question of learning or retraining. For robots, however, as job similarity decreases, robot reconfiguration and reprogramming (retraining) become necessary for economical interjob transfer.

2. Humans possess a set of basic skills and experience accumulated over the years and, therefore, may require less detail in their job description. Today's robots, on

the other hand, perform each new task essentially from scratch and require a high degree of detail for every micromotion and microactivity.

3. Robots do not have any significant individual differences within a given model. Thus an optimized job method may have more generality with robot operators than with human operators.

Robot sensing, manipulative, and decision-making abilities can be designed for a given specialized task to a much greater degree than can a human's abilities. Of course, this specialization may entail the cost of decreased transferability from one task to another.

Robots are unaffected by social and psychological effects (such as boredom) that often impose constraints on the engineer attempting to design a job for a human operator.

3.2 Job Selection for Robots

In planning work in industry two decisions must be made.

1. *Selection.* Who should perform a given task or set of tasks—a human operator or a robot?
2. *Specification.* What are the specifications of the job and the skills? If a robot was indicated in the first decision, complete robot specifications are also sought.

Usually three cases can be identified:

1. A human operator must perform the job because the task is too complex to be performed economically by any available robot.
2. A robot must perform the job because of safety reasons, space limitations, or special accuracy requirements.
3. A robot can replace a human operator on an existing job, and the shift to robot operation could result in improvements such as higher consistency, better quality, lower costs, and so forth. Labor shortages in certain types of jobs may also result in robot assignments.

In the first two cases the selection is clear. In the third case the main concern is whether a robot can at all perform a given task. The robot–human charts provide a means of identifying job activities that can or cannot be done by robots or humans. Another approach for assessing different dimensions in the problem is a systematic comparison between robot time and motion (RTM) task method elements for a robot and methods time measurement (MTM) elements for a human operator. A simple comparison in terms of reaching ability and motions is illustrated in Figures 3 and 4. Additional information for this decision can be obtained from a database of robot work abilities.

3.3 Robot Anatomy and Distribution of Work Abilities

A thorough examination of industrial robots and their controls provides an anatomy of the basic structure and controls of robots and reveals their resulting limitations, particularly in the area of sensor ability and task interactions. To determine the current abilities of industrial robots, literature describing numerous models was summarized and analyzed in two forms (Nof, 1985; Pennington, Fisher, and Nof, 1982): characteristic frequency distributions and motion–velocity graphs. Such a survey can provide the job analyst with detailed and specific work ability to determine if a robot can perform a job and which robot is preferred.

4 PLANNING WORK METHODS AND PROCESSES

An effective work method determines how well limited resources, such as time, energy, and materials, are being used and has a major influence on the quality of the product or output. A strategy for designing new methods or for improving existing methods is composed of seven steps (Nadler, 1981):

1. Determine the purpose of the method.
2. Conceptualize ideal methods.
3. Identify constraints and regularity.
4. Outline several practical methods.

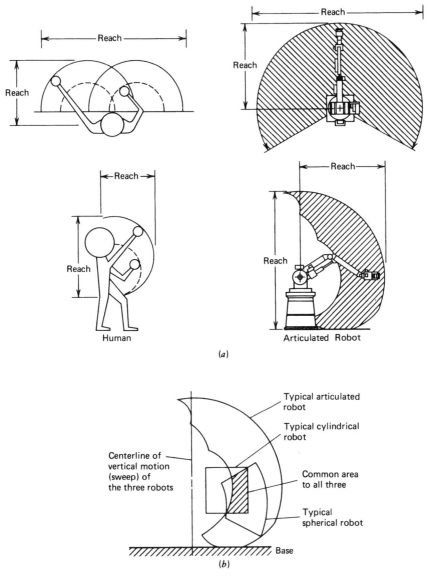

Figure 3 Work envelope comparisons. (a) Human vs. robot. (b) Different robot models.

5. Select the best method outline by evaluating the alternatives and using criteria such as hazard elimination, economics, and control.
6. Formulate the details of the selected method outline.
7. Analyze the proposed method for further improvement.

Work methods must be documented for records, ongoing improvement, time study, and training. Several tools are available for methods documentation as well as for gathering and analyzing information about work methods (Clark and Close, 1992), e.g., process charts, workplace charts, multiple activity charts, and product flow sequence charts.

4.1 Motion Study

Work performance is usually accomplished by a combination of motions, and this is certainly true for robot work. The effectiveness of the movement can be measured in

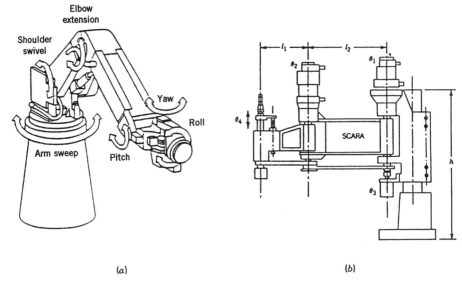

Figure 4 Typical motions of industrial robots. (*a*) vertically articulated robot. (*b*) SCARA, horizontally articulated robot.

terms of time and accuracy. Motion study applies various techniques to examine thoroughly all of the motions that comprise a work method. Based on this examination, alternative methods can be evaluated, compared, and improved. One of the most useful practices in motion study is the use of video cameras. While the various charts identified earlier are commonly used in motion study, the use of videotape or film has several advantages: it provides data acquisition and records facts that are unobtainable by other means, it can provide quick study results, it allows individual analysis of simultaneous activities, it allows detailed analysis away from the actual work area, and it provides a permanent record for further reference and training. One disadvantage, however, is that this technique can be applied only if there is access to an existing system.

4.2 Job and Skills Analysis

Job and skills analysis methods have been used for cost-reduction programs in human–human and conventional human–machine work environments and for the effective selection and training of personnel. The job analysis focuses on what to do, while the skills analysis focuses on how. The method, as it has been traditionally applied for human work, includes three major stages:

1. Examine the task to be analyzed to understand what and how the operator must do the job. If it is not possible to observe an existing task, assess how the task would be performed.
2. Using a table, document the what and the how of task performance.
3. From the documentation, examine in a systematic way the possibilities of performing the task in different ways, including the following guidelines that are relevant for human operators:
    (a) Use of kinesthetic (sense of direction) instead of visual senses
    (b) Reduction of factors complicating the task performance ("noise")
    (c) Reduction or use of perceptual load associated with task performance
    (d) Use of principles of motion economy and workplace design to simplify the task
    (e) Resequence, eliminate, or combine elements

Alternative ways are compared by the time it takes to perform them, complexity, quality, error probability, cost, safety, training and skill requirements, and so forth. Based on such analysis, the best method to perform a task can be selected and implemented.

The features that distinguish robot and human operators, as discussed previously, necessitate modification of the original human job and skills analysis method (Nof, Knight, and Salvendy, 1980). As before, a task is broken down to elements that are specified with their time and requirements. However, the columns are modified as follows:

Limbs column. The human left-hand, right-hand column is replaced by a limbs column, because the robot may be designed to have any number and variety of limbs (e.g., arms, grippers, special tool hands, etc.).

Memory and program column. A column for memory and decision details is added for the analysis of robot performance. These details will determine what type of computer memory and processing capability is needed, if any. Humans have their own memory, but decisions that are self-explanatory and often trivial for humans must be completely specified for robots. Robots may be designed, as explained earlier, with no programmability (i.e., with a fixed sequence), a variable sequence that is fixed differently from time to time, or with computer and feedback mechanisms that control the robot operations. The cost of a robot increases, of course, with the degree of programmability and sophistication of computer control: higher levels of robot control require additional investment in hardware. Furthermore, limited controllers may be based on pneumatic control that requires relatively simple human skills to resequence. More complex operations may require computer control with associated system software and control programmers. Therefore, a work method that permits a fixed sequence will need a simple, cheaper robot and be preferred to one that needs periodic resequencing or complex programming.

Comment column. The comment column is for details about additional requirements such as position tolerances and utilities. Special precautions that are typical in the human-oriented analysis are probably not necessary here because they should appear in written decision logic information.

As indicated previously, robots possess no basic knowledge or experience and, therefore, necessitate much detail in the task specifications. Thus elements will most commonly specify micromotions, with their time measured in seconds (or minute/100). Once a task is specified, its analysis basically follows the three stages described in the human-oriented method: (1) examine task elements, (2) document the what and how of all elements, and (3) systematically examine and evaluate alternative ways. However, because it is possible to select a robot and design its capabilities to best suit the task requirements, the performance evaluation in the last stage should be expanded as follows.

From the documentation, examine systematically, using robot motion economy principles, the possibilities of performing the task in different ways and of using different robots.

5   ROBOT MOTION ECONOMY PRINCIPLES (MEPs)

Principles of motion economy have traditionally guided the development, troubleshooting, and improvement of work methods and workplaces, with special attention to human operators. In developing such principles for robot work, some of the traditional principles can be adopted, but new principles unique to robots must also be added. A MEP can be given as a design rule, a decision algorithm or heuristic, or a mathematical expression.

The MEPs can be applied as guidelines for the initial planning of new robotic installations, where the planned tasks have to satisfy, as much as possible, the recommendations made by the guideline. The MEPs can also serve in analyzing existing robotic facilities, particularly to check whether under changes made by new requirements the robotic operations still follow the MEPs objectives.

Following is a list of MEPs for human operators and their relevance for robot work (Nof and Robinson, 1986). Recent research (Nof, 1991; Edan and Nof, 1995*a*, 1996; Shoval, Rubinovitz, and Nof, 1998) has added MEPs specifically for the design and evaluation of robots' work, and also sensor economy principles (SEPs) for sensor selection (Edan and Nof, 1995*b*, 1998).

5.1   Principles Related to Operator

1. *Hands and arms should follow smooth, continuous motion patterns.* This principle may be useful for robot work when process quality considerations require it (e.g., in painting to eliminate jerky strokes). See principle 2 under "Principles Concerning Robots."

2. *Human hands should move simultaneously and symmetrically, beginning and ending their motions together.* This principle is irrelevant for robots because coordinated robot arms can operate well under distributed computer controls.

3. *Minimize motion distance, within practical limits.* To reduce motion time and improve accuracy by not operating overextended limbs. Relevant for both humans and robots.

4. *Both hands should be used for productive work.* This principle is irrelevant for robots because the number of arms can be chosen to maximize utilization.

5. *Work requiring the use of eyes should be limited to the field of normal vision.* Directly relevant to robots whenever vision is required.

6. *Actions should be distributed and assigned to body muscles according to their abilities.* This principle is only partially useful for robots—the robot work abilities should be specified according to the precise task requirements.

7. *Muscular force required for motions should be minimized.* On the one hand, body momentum should be used to advantage; on the other hand, momentum should be minimized to reduce the force required to overcome it (e.g., it is better to slide parts than to carry them).

5.2 Principles Related to the Work Environment

1. *Workplace should be designed for motion economy* (i.e., tools, parts and materials should be placed in fixed, reachable positions, and in the sequence in which they are used by a robot). In addition, the height of a work surface should be designed within the robot work envelope.

2. *Tools and equipment should be designed and selected with ergonomics guidelines.* This principle is concerned with the dimension and shape of items handled by end-of-arm tooling, with safety and effectiveness in mind.

3. *Materials handling equipment should be selected to minimize handling time and weight.* For instance, parts should be brought as close as possible to the point of use; fixed items position and orientation simplify pickup and delivery.

5.3 Principles Concerning Time Conservation

1. *All hesitations, temporary delays, or stops should be questioned.* Regular, unavoidable delays should be used to schedule additional work. For instance, in a foundry, delays caused by metal cooling-off period can be used by robots for gate removal; in machine tending, when one machine is performing a process the robot can load–unload another machine.

2. *The number of motions should be minimized.* Elimination and combination of work elements are the two most common methods to achieve this goal.

3. *Working on more than one part at a time should be attempted.* This principle can be generalized to multiarm robots. Again, the number of robot arms or hands can be chosen for the most effective work method.

5.4 Principles Concerning Robots

These new principles are based mainly on the fact that robot work abilities can be designed and optimized to best fit the task objectives.

1. *Reduce the robot's structural complexity* (i.e., minimize the number of arms and arm joints that are determined by the number of hand orientations; reduce the robot dimensions that are determined by the distance the robot has to reach; reduce the load that the robot has to carry). These will result in a requirement for a cheaper robot, lower energy consumption, simpler maintenance, smaller workspace.

2. *Simplify the necessary motion path.* Point-to-point motion requires simpler control of positioning and velocity compared with continuous path motion.

3. *Minimize the number of sensors,* because each sensor adds to installation and operating costs by additional hardware, information processing, and repairs. Use of robot with no senses is preferred to one sense, one sense is preferred to two, and so on.

4. *Use local feedback if at all necessary, instead of nonlocal feedback:* for example, use of aligning pins for compliance provides quick local feedback. Use of touch or force sensors requires wiring and processing by a control system. Local feedback may add operations and may increase the overall process time, but usually needs no wiring and no information processing.

5. *Use touch or force sensors, if at all necessary, instead of vision systems,* because the cost of the latter is significantly higher and requires more complex computer support. However, when vision is necessary as part of an operation, attempt to minimize the number of lenses that are required, as this will simplify information processing requirements and shorten the visual sensing time.

6. *Take advantage of robot abilities that have already been determined as required* to reduce the cost and time of the job method. In other words, if a sensor or a computer must be provided for a certain task function, use it to improve the performance of other functions.

Five main measures have been applied in experimental research of robot MEPs:

1. *Reachability:* The number of positions or cells that a particular robot succeeds in reaching during its task execution;

2. *Cycle times:* The total time it takes a particular robot to accomplish its task. When the experiments involve repeated cycles of task motions, the total time is divided by the number of reachable task positions.

3. *Arm joint utilization* (AJU): This measure is calculated based on the total movement of all joints during the task performance and is based on the kinematics of robot motions. The minimization of the arm joint utilization was defined originally as a MEP by Nof and Robinson (1986). This principle is concerned with the design of tasks by alternative robot models to make greater use of the robot arm joints whose motions require less effort on the drive system. In general, the joints nearer the end-effector are preferred because the load is smaller and smaller moments have to be applied. As a result, wear is reduced, less maintenance is needed, and operations are therefore more accurate and consistent.

For different types of robot kinematic chains, task motions will require moving a different set of joints. To calculate the AJU of a given robot performing a given task, it is assumed for simplicity that a unit move is equivalent to both $1°$ and 1 mm. A measure of drive system effort is generated by a weighted mean of the average AJU at each joint. The weight is calculated proportionally based on the distance of the joint from the base drive. For instance, for the first three joints of a robot, joint one (base) proportional effort is considered three times larger than the proportional effort of joint three, joint two is twice that of joint three, and so on. Thus the equation for the weighted utilization is:

$$\text{AJU Measure} = \sum_{j=1}^{N} (0.5d\theta_1 + 0.33d\theta_2 + 0.17d\theta_3)$$

where

N = number of motions by the robot
$d\theta_i = \theta_{ij} - \theta_{i(j-1)}$
i = joint number, $i = 1, 2, 3$
j = position $j, j = 1, \ldots, N$

Multiplying each mean absolute joint variation by its respective proportional weight multiplier and summing up the products yields a comparative estimate of the total effort required of the drive system while performing a task and of the wear of the joints. Because the motions of each robot model are different, the AJU measure estimates for any given task are different. A relatively lower AJU would be more desirable.

The two following measures consider both the kinematics and dynamics of robot work and were defined by Shoval, Rubinovitz, and Nof (1998).

4. *Arm joint accuracy* (AJA): The AJA measures the accuracy of the robot arm during motion. In point-to-point tasks the accuracy is important at the end of the motion,

while in continuous path tasks accuracy is an important measure throughout the motion. Calculation of AJA is as follows:

$$\text{AJA Measure} = \sum_{j=1}^{N} \int_{t_{j-1}}^{t_j} (k_1 \varepsilon_1 + k_2 \varepsilon_2 + k_3 \varepsilon_3) \, dt$$

where

N = number of motions by the robot
k_i = joint i kinematic coefficient, which also depends on link geometry
ε_i = the difference between the reference and actual position of joint I
$t_{j-1} - t_j$ = the time during which the robot moves from position $j - 1$ to position j

The reference position is calculated by the robot controller; the actual position is calculated by the joint encoder. A lower-value AJA implies a relatively better accuracy estimate and provides another way to compare the performance of alternative robot models.

5. *Arm joint load* (AJL): AJL estimates the load to which the arm is subjected during task motions. It is calculated as follows:

$$\text{AJL Measure} = \sum_{j=1}^{N} \int_{t_{j-1}}^{t_j} (T_1 + T_2 + T_3) \, dt$$

where

N = number of motions by the robot
$t_{j-1} - t_j$ = the time during which the robot moves from position $j - 1$ to position j
T_i = the moment applied on joint i during motion

Both AJA and AJL can also be calculated based on the absolute values of inaccuracy and load moments. Based on the five measures, a number of new MEPs have been developed experimentally for kitting and similar point-to-point tasks. Subject to the experimental assumptions, these MEPs can be generalized as follows (see Figures 5 and 6).

1. For flat work surfaces vertically articulated robots have the best overall reachability, reaching the maximum number of positions and orientations, compared with cylindrical, spherical, and SCARA robots.
2. Despite the relatively large work volume of each robot, the work area which is reachable is limited. Therefore, the selection of the work-surface location and orientation (see Figure 6) has significant influence on the robots' performance.
3. The work position relative to the robot base affects robot performance in terms of motion, accuracy, and load. For a small, vertically articulated robot the height level of the work area has little impact, while the distance from the base does: when this distance increases, the effort (AJU, AJL) increases, but the accuracy (AJA) improves.
4. For the same task, the SCARA robot performs the fastest overall; the cylindrical robot performs the slowest.
5. For most work positions faster performance is achieved by widthwise travel through sequential work points, for vertically articulated and SCARA robots; there is no significant difference between widthwise and lengthwise motions for cylindrical robots.
6. The vertically articulated robot has the best (minimum) arm joint utilization when comparing specific work positions; the cylindrical robot has the most wasteful (maximum) joint utilization; the SCARA robot has the best overall joint utilization for all its reachable work positions, followed by the vertically articulated robot.
7. Horizontal travel through a work position, e.g., through a bin, is significantly more efficient than vertical travel for the majority of work positions for small and medium vertically articulated robots, medium and large spherical robots, and the SCARA robot.

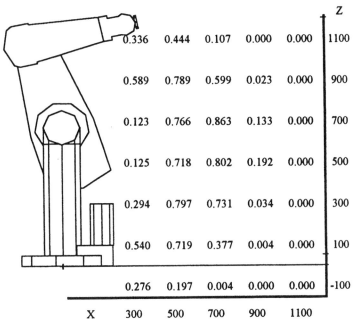

					Z
0.336	0.444	0.107	0.000	0.000	1100
0.589	0.789	0.599	0.023	0.000	900
0.123	0.766	0.863	0.133	0.000	700
0.125	0.718	0.802	0.192	0.000	500
0.294	0.797	0.731	0.034	0.000	300
0.540	0.719	0.377	0.004	0.000	100
0.276	0.197	0.004	0.000	0.000	-100
X 300	500	700	900	1100	

Figure 5 Composite ranking of reachable points in the workspace of a given assembly robot (PUMA 560) indicates the optimal work surface and position when ranking is 0.863. (From J. Nof, J. P. Witzerman, and S. Y. Nof, "Reachability Analysis with a Robot Simulator–Emulator Worksta-tion," Research Memo 96-J, School of Industrial Engineering, Purdue Univeristy, 1996.)

6 PERFORMANCE MEASUREMENT

Performance measurement, including work measurement, performance prediction, and performance evaluation, accompanies a work system throughout its life cycle. In the planning and design stages performance prediction is required to evaluate the technical and economic feasibility of a proposed job plan and to compare and select the best out of a set of feasible alternatives. During the development and installation stages, perform-ance measurement provides a yardstick for progress toward effective job implementation. In the regular, ongoing operations, performance evaluation serves to set and revise work standards, troubleshoot bottlenecks and conflicts, train workers, and estimate cost and duration of new work orders. Another vital function of performance measurement at this stage is the examination of new work methods, technologies, and equipment that can be used to upgrade, expand, and modernize the existing operations on the job.

6.1 Work Measurement and Time-and-Motion Studies

Robot designers have long been concerned with planning robot motions in an optimal way (i.e., accurately follow specified trajectories while avoiding collisions and moving at the minimum amount of time (Brady et al., 1982; Sabe, Hashimoto, and Harashima, 1996). From a work method point of view, cycle times can be reduced by carefully following the MEPs described above (e.g., Kalley and McDermott, 1989).

6.2 RTM Method

The RTM method (Nof, 1985) for predetermined robot cycle times is based on standard elements of fundamental robot work motions. RTM is analogous to the MTM technique, which has long been in use for human work analysis. Both methods enable users to estimate the cycle time for given work methods without having to first implement the work method and measure its performance. Therefore these methods can be highly useful for selection of equipment as well as work methods without having to purchase and commit to any equipment. MTM users, however, must consider human individual varia-bility and allow for pacing effects. RTM, on the other hand, can rely on the consistency

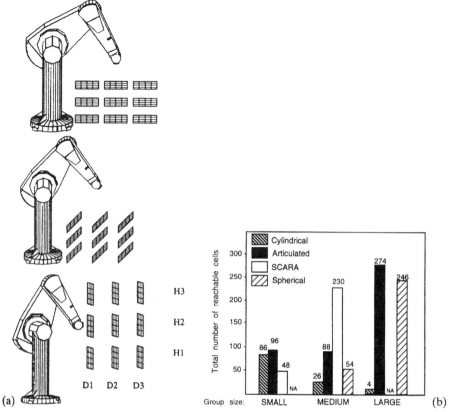

Figure 6 Robot motion economy analysis for ten robot types. (*a*) Twenty-seven alternative positions for bin or work surface. (*b*) reachable cells (out of 432 possible cells). (From Y. Edan and S. Y. Nof, 1995a.)

of robots and apply computational models based on physical parameters of each particular robot model.

The RTM methodology provides a high-level, user-friendly technique with the following capabilities:

1. Systematically specifies a work method for a given robot in a simple, straightforward manner.

2. Applies computer aids to evaluate a specified method by time to perform, number of steps, positioning tolerances, and other requirements so that alternative work methods can be compared.

3. Repeats methods evaluation for alternative robot models until the best combination is established.

The RTM system is composed of three major components: RTM elements, robot performance models, and an RTM analyzer. The system has been implemented, verified and applied with a variety of robot models. Several companies adopted and applied the RTM methodology.

The RTM user can apply 11 general work elements to specify any robot work by breaking the method down to its basic steps. The RTM elements are shown in Table 2. They are divided into five major groups:

1. *RTM Group 1.* Movement elements—REACH, MOVE, and ORIENT

2. *RTM Group 2.* Sensing elements—STOP-ON-ERROR, STOP-ON-FORCE, TOUCH, and VISION

Table 2   RTM Symbols and Elements

Element Number	Symbol	Definition of Element	Element Parameters
1	Rn	*n-Segment reach:* Move unloaded manipulator along a path composed of *n* segments	Displacement (linear or angular) and velocity or
2	Mn	*n-Segment move:* Move object along path composed of *n* segments	Path geometry and velocity
3	ORn	*n-Segment orientation:* Move manipulator mainly to reorient	
4		*Stop on position error*	
4.1	SEi	Bring the manipulator to rest immediately without waiting to null out joints errors	Error bound
4.2	SE2	Bring the manipulator to rest within a specified position error tolerance	
5		*Stop on force or moment*	
5.1	SFi	Stop the manipulator when force conditions are met	Force, torque, and touch values
5.2	SF1	Stop the manipulator when torque conditions are met	
5.3	SF2	Stop the manipulator when either torque or force conditions are met	
5.4	SF3	Stop the manipulator when touch conditions are met	
	SF4		
6	VI	Vision operation	Time function
7	GRi	*Grasp an object*	
7.1	GR1	Simple grasp of object by closing fingers	Distance to close–open fingers
7.2	GR2	Grasp object while centering hand over it	
7.3	GR3	Grasp object by closing one finger at a time	
8	RE	Release object by opening fingers	
9	T	Process time delay when the robot is part of the process	Time function
10	D	Time delay when the robot is waiting for a process completion	Time function
11	MB	*Mobility elements*	
11.1	SMF	Straight movement forward	Parameters: velocity and path functions
11.2	SMS	Straight movement sideways	
11.3	STS	Spin turn on the spot	
11.4	CAO	Curve with a 90° angle with change of orientation	
11.5	DMC	Diagonal movement with a constant orientation	

3. *RTM Group 3.* Gripper or tool elements—GRASP and RELEASE
4. *RTM Group 4.* Delay elements—PROCESS-TIME-DELAY and TIME-DELAY
5. *RTM Group 5.* Mobility elements—straight movements, spin turn, curve, and diagonal movement

By application of the RTM elements with the parameters shown in Table 2, alternative robot work methods can be analyzed, evaluated, compared, and improved.

6.2.1 RTM Models and Analyzers

Four main approaches for modeling RTM elements have been developed since the original RTM was introduced in 1978: empirical models expressed as element-tables (e.g., Table 3) or as regression equations; models based on the robot velocity control (e.g., Figure 7); equations based on the kinematic solution of the robot. Examples of such models include the ARM method for the SCARA robot (Wang and Deutsch, 1992) and the ROMUM method (Kwon and Kim, 1997). While rough estimates by table values may be needed initially, with estimation accuracy of 2–12% more accurate equations are applied by most commercial robot simulators for detailed cycle time and motion analysis. For instance, Terada (1995) describes an off-line simulator of Fanuc robots with 1–5% accuracy in estimating cycle times.

The RTM analyzer has been developed (Nof, 1985) to provide a means of systematically specifying robot work methods with direct computation of performance measures. The input to the RTM analyzer includes control data (e.g., task and subtask titles, type of robot, and type of RTM model to apply) and operation statements. The statements specify robot operations, each represented by RTM element and its parameters; and control logic that provides capabilities of REPEAT blocks, PARALLEL blocks, conditional branching, and probabilistic generation of conditions.

A simplified RTM was developed for point operations such as spot welding, drilling, and riveting. In point operations the robot carries a tool or a part during the whole operation; the same operation type is performed repeatedly along one or more paths, and the execution time of the tool at each point is well known and can be calculated rather than estimated. Taking advantage of these characteristics, the specification and computation can be simplified with little loss of accuracy.

6.3 Performance Evaluation and Measurement by Robot Graphic Simulators/Emulators

Another type of robot performance analysis is by graphic simulation. The main objective of such simulations is to observe the geometric and spatial aspects of the robot workplace and work methods. A number of such simulators are available, as described in Chapter 37. Some simulators include accurate mathematical models of kinematics and dynamics and can provide emulation of the actual, physical behavior of robots and machines. This emulation can yield relatively accurate estimates of work performance.

General robot graphic simulators include libraries of numerous robot models from various robot makers and have the capability to translate simulation programs to various robot languages. Kaneshima (1997) describes the realistic robot simulation project, started

Table 3 RTM Tables for Estimated REACH/MOVE and ORIENT Elements by Some Vertically Articulated Robots

Distance to Move (cm)	1. Reach (R1) or Move (M1) Time (s) at Velocity (cm/s)				
	5.0	12.5	25.0	50.0	100.0
1	0.4	0.4	0.4	0.4	0.4
30	6.4	2.8	1.6	1.0	0.8
100	21.3	8.7	4.5	2.4	1.4

Angle to Move (deg)	2. ORIENT (OR1) Time (s) at Velocity (cm/s)				
	5.0	12.5	25.0	50.0	100.0
15	3.0	1.4	0.8	0.6	0.6
60	10.8	4.6	2.5	1.4	0.9
120	21.3	8.7	4.6	2.5	1.4

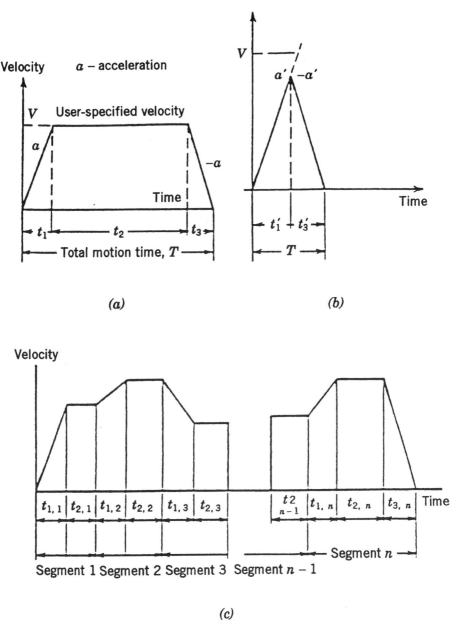

Figure 7 Velocity patterns. (a) Velocity pattern of regular one-segment motions. (b) Velocity pattern of short one-segment motions. (c) Velocity pattern of general multisegment motions.

in 1992, in which suppliers of robot controllers and simulators cooperate to develop a common interface leading to more accurate robot motion simulation.

7   ROBOT WORKPLACE DESIGN

The best workplace is the workplace that supports and enables the most effective accomplishment of the chosen work method. The layout of a workplace determines, in general, four main work characteristics:

1. The distance of movements and the motions that must be carried out to fulfill the task.

2. The amount of storage space.
3. The delays caused by interference with various components operating in the workplace.
4. The feelings and attitudes of operators toward their work. The latter, obviously, is not a concern for robots.

Typical configurations of robotic workplaces are shown in Figures 8 and 9. In traditional ergonomics a workplace is designed for human operators. Hence there are anthropometric, biomechanical, and other human factors considered. In analogy, in regard to robotic workplaces, robot dimensions and other physical properties such as reachability, accuracy, gripper size, and orientation will determine the design of the workplace (Rodrigues, 1996; Craig, 1997). In addition, requirements such as those shown in Table 4 must be considered: whether tasks are variable and/or are handled by one or more robots; whether workplace resources are shared by the robots; the nature and size of components stationed in the workplace; and the characteristics of parts flow into, within, and outside of the workplace. Several researchers have studied specific layout models for robot systems (Tamaki and Nof, 1991; Shiller, 1997). Because all robot operations are controlled by a computer, several robots can interact while concurrently performing a common task and even share resources (Rajan and Nof, 1990; Nof and Hanna, 1989)—a difficult requirement when people must interact under a tight-sequencing control. Thus a robotic workplace can be optimized with regard to the layout and also planned for effective control of operations.

Figures 10 and 11 illustrate the use of a facility design language (FDL) (Witzerman and Nof, 1995a, 1995b), based on the ROBCAD graphic simulator, to design robotic cells.

7.1 Multirobot Applications

Uses of multiple robots with various levels of interactions are emerging applications in industry. Multirobot systems can mainly be classified into two groups: *autonomous* and *cooperative*. Under autonomous operation each robot performs a prespecified set of tasks and does not interact with the other robots in the system. Such a system can be viewed as a collection of single robot workcells organized as a multiple robot system to perform a collection of tasks.

A cooperative multirobot system exploits the ability of the robots to perform tasks independently or by cooperation. Thus it provides the flexibility of single robot cells by having robots perform tasks independently along with the added dimension of task performance obtained by using the cooperative capabilities of robot sets (Rajan and Nof, 1990; Nof and Hanna, 1989). A given cooperative multirobot system is capable of performing a wide variety of tasks. In addition, such systems have enhanced reliability due to the ability to perform tasks assigned to failed robots by using other single robots or cooperative robot sets with the required capabilities. Some of the types of tasks that can be performed by cooperating robot sets are shown in Figure 12. While autonomous multirobot systems can be used to mimic high production volume and hard automation lines with greater flexibility, cooperative multirobot systems are mainly useful for small-volume, large-variety part production (Figure 12h, i).

Some research issues related to multirobot systems are kinematics–dynamics of cooperating robots (Rodriguez, 1989), motion planning and collision avoidance in shared workspaces (Roach and Boaz, 1987), multirobot cooperation activity control (Maimon and Nof, 1986), and cooperation requirement planning for multirobot assembly (Rajan and Nof, 1996).

7.2 Classification and Measurement of Cooperation among Robots

We divide the modes of cooperation possible among robots into task cooperation and resource sharing. Task cooperation itself can be of three types: mandatory, optional, and concurrent.

7.2.1 Mandatory Task Cooperation

This is a cooperation mode that is needed when certain operations must be carried out by two or more robots simultaneously. In general, designing a workcell with such a cooperation mode should be avoided as much as possible because the productivity per each individual robot here is diminished. S_a, measure of the level of mandatory cooper-

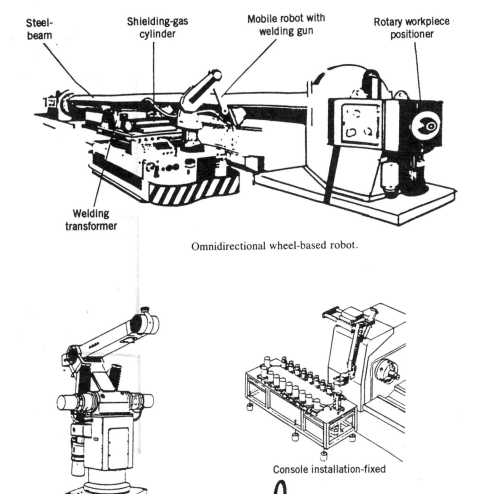

Omnidirectional wheel-based robot.

Figure 8 Types of installations of industrial robots in the workplace.

a. Random load-unload station

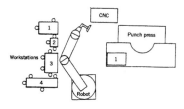

b. Wrap-around station

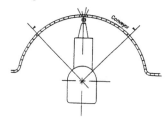

c. Load-unload with pallets (or AGV)

• at least three axes of motion required

• program must consider parts organization

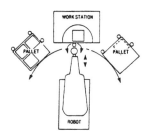

d. Load-unload with conveyors (or feeders)

• at least two axes of motion requires

• known parts location simplifies program

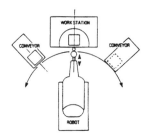

e. Safety considerations

• Overpowering restriction posts, barriers, interlock gates, footpads protect humans

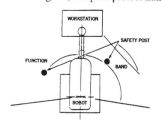

f. Same work-height

• less motion axes, simpler program

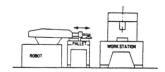

Figure 9   Elementary ergonomics considerations in workplace design.

ation, is defined based on the fraction of such operations and the number of robots that must cooperate.

Given a cell with N robots and K tasks that are assigned to the cell, define

$$S_a = \left(\sum_{i=1}^{K} \left(\sum_{j=1}^{N} a_{ij} - 1 \right) \right) \Big/ (n-1)K \tag{1}$$

where the mandatory cooperation matrix, A, is defined by:

$$a_{ij} = \begin{cases} 1 & \text{when robot } R_j \text{ must participate in task } i \ (j = 1, \ldots, N, \ i = 1, \ldots, K) \\ 0 & \text{otherwise} \end{cases}$$

A typical example of the mandatory task cooperation mode is where a number of arms must lift together a cumbersome structure while maintaining its position and orientation over a period of time.

Table 4   Classification of Robotic Workplaces

Robots	Tasks	Stations Inside Workplace: Machines, Feeders, Other Equipment	Input/Output Flow
Simple robot	Single or multiple tasks Fixed or variable motion sequence	Uniform or different station area Fixed or variable equipment	Single or multiple entry, exit One or multidirection of flow
Multiple robots	Single or multiple tasks per robot per team With or without task interaction	With or without equipment sharing (for multiple robots)	With or without part grouping (binning, kitting, magazining, palletizing)
Multiple robot cells	With or without cell overlap		
Human–robot cells	Mutually exclusive or interactive; human as backup	Teleoperation, exchange, or overlap	Shared, or exclusive

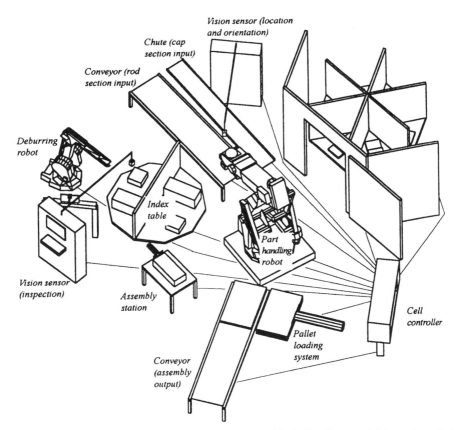

Figure 10   An integrated CAD for an engine rod assembly facility (the top-right area is a shot-pinning cell) layout and control. (From Nof and Huang (1998).)

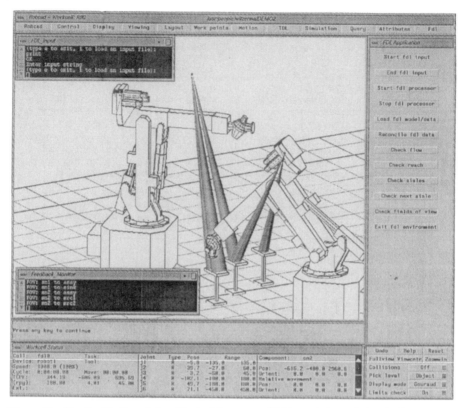

Figure 11 The Facility Description (Design) Language (FDL) can help optimize the location and position of sensors and controls. (From J. P. Witzerman and S. Y. Nof, "FDL: Facility Description Language," in *Proceedings of IIE Research Conference,* Nashville, May 1995, 449–455; id., "Integration of Cellular Control Modeling with a Graphic Simulator/Emulator Workstation," *International Journal of Production Research* **33**, 3193–3206, 1995.)

7.2.2 Optional Task Cooperation

This cooperation mode is designed for certain operations that can be carried out by one of several robots. In general, this cooperation may increase the productivity of the whole cell, as robots can share in performing a larger variety of tasks as soon as they become available. S_o, measure of the level of optional cooperation, is defined based on the fraction of such operations and the number of robots that can (but do not have to) participate. Define the optional cooperation matrix, P, as follows:

$$p_{ij} = \begin{cases} 1 & \text{when robot } R_j \text{ can perform task } i \\ 0 & \text{otherwise} \end{cases}$$

then define

$$S_o = \left(\sum_{i=1}^{K} \left(\sum_{j=1}^{N} p_{ij} - 1 \right) \right) \bigg/ (N-1)K \qquad (2)$$

According to this definition, $S_o = 0$ implies that there is no optional sharing of tasks and $S_o = 1$ implies full optional operation by any one of the cell robots on any of the tasks performed in the cell. To illustrate a situation with partial optional cooperation consider the case $N = 3$ robots, $K = 4$ tasks. Suppose R_1 or R_3 can perform task 1; R_1 or R_2 or R_3 can perform task 2; R_2 or R_3 can perform task 3; and only R_3 can perform task 4. The optional cooperation matrix in this case is

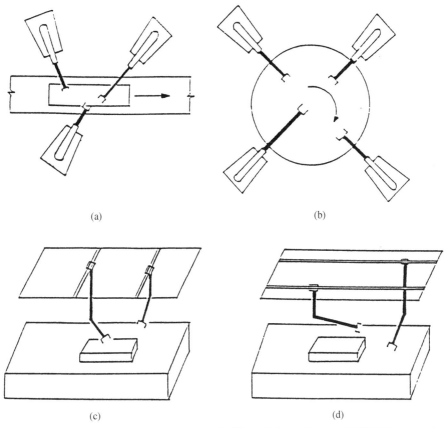

Figure 12 Typical cases of multiple robot work. (From Nof and Drezner (1993).) (a) alongside a conveyor. (b) around a circular station. (c) gantry arms with partial longitudinal motions. (d) gantry arms with full longitudinal motions.

$$P = \begin{bmatrix} 1 & 0 & 1 \\ 1 & 1 & 1 \\ 0 & 1 & 1 \\ 0 & 0 & 1 \end{bmatrix}$$

and $S_o = 4/((3-1)\cdot 4) = 0\cdot 5$.

7.2.3 Concurrent Task Cooperation

This is another mode of cooperation by design, which may increase the cell productivity. It is found when several robots can carry out simultaneously, in parallel, several operations—for example, on a number of product facets, thus cutting the total completion time. Unlike the two previous cooperation modes, this one is characterized both by the ability of different robots to cooperate and by the relative reduction in task time that depends on the number of cooperating robots. Thus S_c, measure of the level of concurrent cooperation, is defined by the fraction of such operations and by the subsequent reduction in their completion time.

Define:

$$S_c = \sum_{i=1}^{K} (1 - t_{i|n}/t_{i|1})/K \tag{3}$$

where $t_{i|n}$ are elements of the concurrent cooperation time matrix, $T_{i|n}$, and $t_{i|n}$ is the task

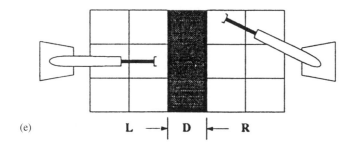

(e)   **L** →| **D** |← **R**

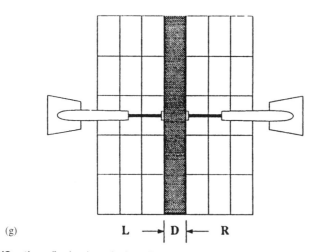

(f)   **L** →| **D** |← **R**

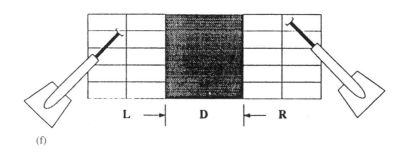

(g)   **L** → **D** ← **R**

Figure 12 (Continued) (e–g) exclusive allocation of bins and assembly places to "left-right" robots. (L—clearly "left" robot; R—clearly "right" robot; D—to be determined by optimization.)

i time given n participating robots, $n \le N$. For instance, suppose $N = 3$, $K = 3$, and conditional task times in minutes are $t_{1|n} = (10, 5, —)$; $t_{2|n} = (15, —, 6)$; $t_{3|n} = (12, —, —)$. Then according to (3)

$$S_c = [(1 - 5/10) + (1 - 6/15) + (1 - 12/12)]/3 = 0 \cdot 37$$

Note that by definition, if there is no possibility of concurrent cooperation, $S_c = 0$. When for a task i there are several alternative levels of concurrent cooperation (as, for instance, if in this example $t_{2|n}$ had been (15,12,6)), then $t_{i|n}$ in (3) must be calculated first

Figure 12 (Continued)   (*h*) Ceiling-mounted robot arms load/unload machining centers. Independently, each robot serves exclusively its assigned center; collaboratively, robots can exchange and transfer parts between centers (courtesy ABB Robotics Co.).

as the mean of the task time distribution (in the latter example, for $t_{2|n}$ it would be $(1 - (12 + 6)/30)$.

7.2.4   Level of Resource Sharing

This level, S_r, is defined based on the number of resources that are shared by more than one robot and the number of robots that can share in each. To compute S_r a resource sharing matrix, RS, is defined first for each type of cell resource as follows:

$$RS_l(k,j) = \{m_{l,k,j}\} \qquad \text{for} \quad \begin{aligned} l &= 1, 2, \ldots, L \\ k &= 1, 2, \ldots, M_l \\ j &= 1, 2, \ldots, N \end{aligned} \qquad (4)$$

where

L = number of resource types
M_l = number of resource units of type
$m_{l,k,j}$ = $\begin{cases} 1 & \text{if robot } R_j \text{ can utilize unit } k \text{ of resource type } l \\ 0 & \text{otherwise} \end{cases}$

Next define the sharing ratio of resource type l, $\beta_l = S_l/M_l$, where S_l is the number of units of resource type l that are shared by more than one robot. Then:

$$S_r = \sum_{l=1}^{L} \left[\left(\sum_{k=1}^{M_l} \sum_{j=1}^{N} m_{l,k,j} \right) \cdot \beta_l / (N \cdot M_l) \right] \Big/ L \qquad (5)$$

For example, suppose a cell with $N = 3$ robots has $L = 3$ resource types, including (*a*) three workbenches, one for each robot ($l = 1$); (*b*) two drills of the same type ($l = 2$), one usable by R_1 and R_2, the other just by R_3; and (*c*) a common in-process buffer ($l = 3$). In this case, $M_l = (3, 2, 1)$, $\beta_l = (0, \frac{1}{2}, 1)$, and

Figure 12 (Continued) (*i*) Robots insert keys in engine-cylinders. Independently, each robot is assigned its exclusive engine; cooperatively, robots can reach neighbor engines for mutual backup and error-recovery (courtesy ABB Robotics Co.).

$$RS_1 = \begin{bmatrix} 1 & 0 & 0 \\ 0 & 1 & 0 \\ 0 & 0 & 1 \end{bmatrix}; \quad RS_2 = \begin{bmatrix} 1 & 1 & 0 \\ 0 & 0 & 1 \end{bmatrix}; \quad RS_3 = (1, 1, 1)$$

Now, according to (5),

$$S_r = (3 \cdot 0/9 + 3 \cdot \tfrac{1}{2}/6 + 3 \cdot 1/3)/3 = 0 \cdot 4166$$

Note that for simplicity reasons, the definition of S_r in Eq. (5) does not attempt to quantify the specific amount of resource usage by each sharing robot, which is difficult to estimate a priori. Also, for convenience we opt to use β_l to consider the relative sharing of each resource type, rather than measure S_r by each individual resource separately.

The different cooperation modes and their measures are summarized in Table 5. Other types of cooperation modes, relatively more complex, exist too. For instance, complex conditional combinations of mandatory, optional, and concurrent cooperation may be found.

A workplace design study (Hanna and Nof, 1989) has included a number of probabilistic and simulation analyses of the various cooperation modes. The main design questions that were studied were:

Table 5 Terms and Measures of Multirobot Cooperation Modes

	Task cooperation
1. *Mandatory*	
Definition	Two or more robots must carry out the task simultaneously.
Example	One robot serves part of the time as a programmable fixutre for other robots.
Specification	Mandatory cooperation matrix, A (task i × robot j).
Measure	Measure of the level of manadatory cooperation, S_a (%).
2. *Optional*	
Definition	Any one of several robots can carry out the task, and only one is required.
Example	Three robots can unload parts from an input conveyor, but just one is needed for each part.
Specification	Optional cooperation matrix, P (task i × robot j).
Measure	Measure of the level of optional cooperation, S_o (%).
3. *Concurrent*	Two or more robots can carry out concurrently portions of the same task.
Definition	
Example	Several robots can reach and drill holes in a casting simultaneously rather than sequentially.
Specification	Concurrent cooperation time matrix $T = \{t(\text{task } i \text{ given } n \text{ robots})\}$.
Measure	Measure of the level of concurrent cooperation, S_c (%).
	Resource sharing
Definition	Two or more robots can utilize sequentially the same unit of resource-type.
Example	A single straightening device for circuit-board pins can be used by several robots.
Specification	Resource sharing matrix for each resource-type I, RS_I (unit k × robot j).
Measures	Sharing ratio of resource-type I, β_I; Measure of the level of resource sharing, S_r (%).

How much concurrent, optional, and combination of concurrent and optional cooperation should be designed into the system?

How do different resource sharing and task cooperation combinations influence the workstation performance?

The above questions were analyzed from both the production performance point of view, and the cost-effectiveness point of view (see Figure 13).

7.2.5 Design Recommendation for Multirobot Systems with Cooperation

Mandatory Task Cooperation

This type of task cooperation may save cell resources, as in the case of a robot taking the role of fixtures.

In other cases it may be required by the nature of the task, as in handling accurately a cumbersome component by several robots.

In all mandatory cooperation cases the result is a relative loss in production rate per robot (see Figure 13c). Therefore, careful cost-effectiveness analysis must be performed before mandatory cooperation is allowed.

Optional Task Cooperation

This type of task cooperation can result in significant improvement in performance. Because of imbalance in allocating shared operations among a finite, discrete number of multiple robots, however, there may be inconsistency in performance improvement. In other words, it may decrease for certain levels of optional sharing even if it shows significant improvement for S_o levels that are lower and higher.

Optional cooperation may require special design efforts with regard to robot selection, e.g., in terms of reachability. Therefore, the lowest level of optional cooperation, S_o,

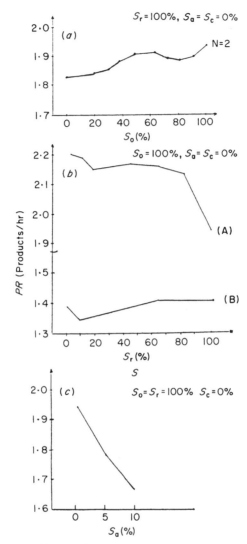

Figure 13 Production rate of final products vs. different levels of task cooperation and resource sharing, in a cell with $N = 2$ robots. (a) Production rate, PR, vs. level of optional task cooperation, S_o (b) Production rate, PR, vs. level of resource sharing, S_r in a two-robot cell. Part interarrival time is distributed $u(4, 5)$ min (A) or $u(14, 15)$ min (B). (c) Production rate, PR, vs. mandatory cooperation level, S_a. (From S. Y. Nof and D. Hanna, "Operational Characteristics of Multi-robot Systems with Cooperation," *International Journal of Operations Research* **27**(3), 477–492.)

should be designed if there is no significant difference in performance for higher levels of S_o. (In the experiment $S_o = 62\%$ is the preferred level as shown in Figure 13a.)

On the other hand, issues of backup and reliability provided by optional cooperation can justify higher levels of S_o.

Concurrent Task Cooperation

Concurrent task cooperation leads to shorter production cycle-times, and is highly desirable. On the other hand, like optional cooperation, it is bounded by the amount of conflict resolution.

Resource Sharing

In general, a higher level of S_r implies resource savings and often also space savings. Therefore it is desirable to design the highest level of S_r as long as there is no significant decline in performance relative to the case of $S_r = 0\%$ (no resource sharing). In the experiments, this level is $S_r = 64\%$ for the case of relatively high part arrival rate and $S_r = 100\%$ for the lower part arrival rate (Figure 13b).

Imbalance can occur in allocating resources among the sharing robots, similar to the imbalance in optional task cooperation. As a result, performance may decrease for certain regions of S_r, and should not be assumed to always increase continuously. Therefore, wherever possible, balanced resource allocation should be attempted.

Number of Cooperating Robots

More cooperating robots in a cell will provide a higher service level, but unless all robot motions are predetermined and preprogrammed, this is at a cost of increasing conflicts. The conflicts result from shared space and, possibly, shared resources. As a result, conflict resolution time has to be considered. In a study (Nof and Hanna, 1989), the optimal number of cooperating robots was found to be between three and four in terms of production rate.

Cost-effectiveness analysis of different multirobot configurations has to consider savings due to shared resources and reduced, common space versus increase in conflict resolution time, e.g., waiting for a shared resource to become available. In this study such analysis indicated an economic advantage to workstations with up to $N = 4$ cooperating robots.

In planning multirobot workstations it is necessary to consider these recommendations and determine specifically what cooperation levels are desirable and what cooperation combinations are preferred.

8   INTEGRATED HUMAN AND ROBOT ERGONOMICS

A vital area in robotic job design is the integration of humans and robots in work systems. While industry has tended to separate employees from robot activities as much as possible, mainly for safety reasons, there are several important issues to consider.

8.1   Roles of Human Operators in Robotic Systems

Except for unmanned production facilities, people will always work together with robots, with varying degrees of participation (Figure 14). Parsons and Kearsley (1982) offer the acronym SIMBIOSIS to the roles of humans in robotic systems:

Surveillance—monitoring of robots

Intervention—setup, startup, shutdown, programming, and correcting

Maintenance

Backup—substituting manual work for robotic at breakdown or changeover

Input and output—manual handling before and after the robotic operation

Supervision—management, planning, and exception handling

Inspection—quality control and assurance beyond automatic inspection

Synergy—combination of humans and robots in operation (e.g., assembly or supervisory control of robots by humans)

In all of these roles the objective is to optimize the overall work performance. The idea is to plan a robotic system with the required degree of integration to use best the respective advantages of humans and robots working in concert. Human factors considerations in planning robot systems for these roles include job and workplace design, work environment, training, safety, pacing, and planning of supervisory control.

8.2   Learning to Work with Robots

Both individuals and organizations using robots must learn to work effectively with the robots. Aoki (1980) describes the progress of his company in using robots. Robot performance in die casting was improved over time, following the learning curve model. The workers planning and operating the robot first improved its work method and program; then they improved the program (and method) further and introduced improvements in hardware. This is basically a process of organizational learning that is based on learning by a group of individuals to better use robots. Many other companies report similar progress in adopting robot operators, similar to adopting any new technology.

In an early study (Argote, Goodman, and Schkade, 1983) the objective was to investigate how employees, as individuals, perceive and accept a new robot, the first in their

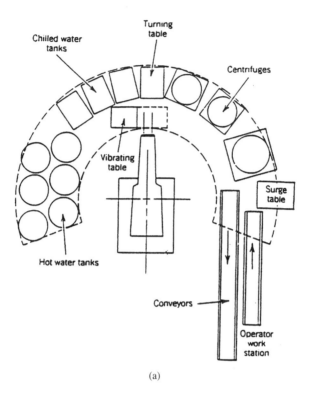

(a)

(b)

Figure 14 Robotics applications that integrate human operators for the operation completion. (a) A large robot carries out the air separation from mercury in clinical thermometer manufacturing. Human role: control the flow of baskets on conveyors in/out to prevent glass breakage. (b) A cell designed for assembly of lighting fixtures (courtesy Adept Technology, Inc.). Human role: replenish components; final assembly inspection; partial backup.

company. Workers were interviewed 2.5 months before and 2.5 months after the robot introduction. As can be expected, with time and experience workers increased their understanding about what a robot really is. However, with time, workers' beliefs about robots (e.g., the potential hazards associated with robots) became more complex and pessimistic. In addition, an increase in stress was indicated among workers interacting directly with the robot. Further research is needed on this problem; however, the researchers saw their findings as another indication that effective strategies for correct introduction of robots to the factory are vital to the success of robot implementations.

8.3 Human–Robot Interaction

Human–robot interaction (HRI) has emerged as an important area of study and application during the last decade (see also Chapter 33). HRI includes the design of real and virtual interfaces, communication and off-line programming, adaptive and social behavior, anthropomorphic interfaces and robot devices, collaboration, human-friendly interactions, and more. The design and analysis of interactions among distributed humans, robots and computers integrates computer science and human–machine interactions (Anzai, 1994). For instance, integrated topics include hardware architectures of autonomous mobile robots, robot operating systems, wireless communication, robot groupware, reactive planning and task scheduling, and mixed-initiative interactions.

8.3.1 Interaction and Interface Design

Robot ergonomics plays a central role in HRI. At the center of the interaction are the transfer of skills from human operators and designers to robotic devices, and interactive feedback from robots to humans during teaching/programming and during operation (Okada, Yamamoto, and Anzai, 1993; Kaiser, 1997; see also chapters 18 on On-Line Programming and 19 on Off-Line Programming). The most common interface is the teach pendant; however, there is little standardization in teach pendant design. Joysticks are easier to operate and therefore are preferred for disabled users, but relatively more training is necessary with a joystick interface (Dario et al., 1996). Keyboards have also become popular for more complex, computer-supported interactions. Particularly in remote manipulation under human supervision, sophisticated computer support is necessary. Sensor-based interaction has also become practical. Machine vision and speech recognition interfaces have been developed for interaction support functions. Acoustic sensors coupled with knowledge-based control are integrated with the robot system for safety monitoring and collision avoidance during interaction (Imai, Hiraki, and Anzai, 1995).

Ergonomic design of a robot teach pendant involves considerations of interaction effectiveness, task allocation, and safety (Podgorski and Beleslawski, 1990; Morley and Syan, 1995; Kwon, 1996). Tasks have to be allocated between interacting humans and robots in order to improve performance. A fuzzy relation model for such allocation has been developed (Bing, Halme, and Helbo, 1996). A task allocation model can be useful for planning HRI work and organizing it methodically (Martensson, 1987).

8.3.2 Knowledge-Based Interaction

Besides applications of a teach-pendant interface, improvement of interaction has also been attempted by developing perception interface systems (Osborn, 1996), by providing communication devices (Klingspor, Demiris, and Kaiser, 1997), and by implementing interactive programming by demonstration (Friedrich, Hoffmann, and Dillmann, 1997). Knowledge-based interaction methods and knowledge-based control have been developed for interactive robotic devices (Nof, 1989; Widmer and Nof, 1992) and for mobile robot learning by dynamic feature extraction (Hiraki and Anzai, 1996). The next frontier in investigating knowledge-based interaction is the development of social interaction abilities (Brooks, 1997).

8.4 Human–Robot Collaboration

Computer-supported collaboration between robots and humans is also associated with robot ergonomics. According to *the integration and collaboration taxonomy* (Nof, 1994; see Figure 15) human–robot collaboration usually falls under the human–machine (H–M) classification. When several robots are involved, both the H–M and M–M collaboration schemes may be involved, assuming the robots can interact among themselves and with the human operator. Furthermore, when several people operate a multirobot system, all three types of collaboration, H–H, H–M, and M–M, may be included.

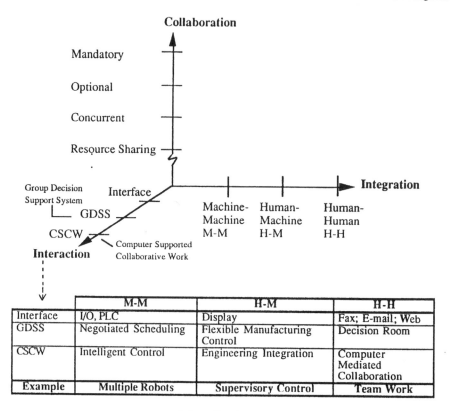

Integration taxonomy

	Integration Problem	Integration Example *		Collaboration Type
1	Processing of task without splitting and with sequential processing	Concept design followed by physical design	H-H	Mandatory, sequential
2	Processing of task with splitting and parallel processing	Multi-robot assembly	M-M	Mandatory, parallel
3	Processing of task without splitting. Very specific operation.	Single human planner (particular)	H-M	Optional, similar processors
4	Processing of task without splitting. General task.	Single human planner (out of a team)	H-M	Optional, different processor types
5	Processing of task, can have splitting	Engineering team design	H-H	Concurrent
6	Resource allocation	Job-machine assignment	M-M	Competitive
7	Machine substitution	Data base backups	M-M	Cooperative

Figure 15 Integration and collaboration taxonomy. (From S. Y. Nof, ed., *Information and Collaboration Models of Integration.* Dordrecht: Kluwer Academic, 1994.)

Several groups have begun to design anthropomorphic manipulators that can interact and collaborate more naturally and safely with their human users and consider the human interaction constraints (Morita and Sugano, 1997). For instance, real-time visual recognition by robots has been developed as part of the human–robot interface (Kaneko et al., 1996) to augment the responsive capacity of the interaction channels. Cooperation by a manipulator in a virtual reality (Luecke and Edwards, 1996) can provide a haptic interface device. The need for anthropomorphic robot design has been suggested (Roth, 1995). It has been motivated by the design principle that "form follows function." This design approach has led to the development of robots that look like humans. In addition to the idea that anthropomorphic robots may better and more gracefully perform human-oriented tasks, such as running, climbing, and walking down stairs, there is the assumption that humans will feel more comfortable collaborating with them.

8.4.1   Cobots and Humanoids

A *cobot* is a robot designed specifically to collaborate with humans. It is typically a safe, mechanically passive robotic device that is intended for direct physical contact with a human operator (Wannasuphoprasit et al., 1997). The operator supplies the power for motions, while the cobot responds with software-controlled movements. Because of the mobility that is often expected for collaboration, biped robots have been designed (Qinghua, Takanashi, and Kato, 1993; Seward, Bradshaw, and Margrave, 1996; Yamaguchi and Takanichi, 1997). Robots that are designed to resemble human physical characteristics are called *humanoids*.

Humanoids have been developed for fast motions (Inaba et al., 1997), better, human-like task execution (Hwang et al., 1997), and operations within human environments (Yamaguchi et al., 1996). The humanoid development trend assumes that from an ergonomics consideration a human-like machine may be more suitable for performance of tasks in the human world. In a nonindustrial environment, it is hypothesized, people would feel more comfortable collaborating with a human-like machine. This hypothesis requires further study.

On the issue of intelligent behavior and control, the humanoid robot developers seek to apply capabilities that comprise artificial life systems. These capabilities include cognitive functions, speech, eye-coordination of manipulation, and self-organization of behavior (Dennett, 1994; Rutkowska, 1995); behavior-based interaction (Brooks, 1996); and collaborative–interactive robot organisms that are controlled by artificial life, massively parallel computers (Kitano, 1994). The anthropomorphic robot and humanoid design thus creates a new level of robot ergonomics challenge: optimizing these machines not only for their functional work ability, but also for their ability to collaborate and communicate effectively.

9   ROBOT ERGONOMICS IN PRACTICE

Since the early studies of industrial robot ergonomics in the 1970s, models, methods, and tools of robot ergonomics have been widely practiced for the design and optimization of a variety of robot jobs and work assignments. Some applications occurred in R&D laboratories, while many others were developed by industry practitioners. A representative sample of typical applications is given in Table 6. For each work or operation illustrated, a specific application is described and the particular ergonomics function that was practiced for planning this application is listed. The sample covers all the main areas of robot ergonomics that were described in this chapter (Figure 2).

10   SUMMARY AND EMERGING TRENDS

Techniques and principles developed by robot companies, robot users, and researchers have been identified and described as necessary for planning effective industrial robot work. Traditional engineering tools have been extended to design robotic and human–robot work systems. New techniques have evolved from traditional methods for the design of robots for specific planned work. Basically, robots offer a new dimension in work design by allowing creation of the operator (i.e., the robot) itself. Interestingly, the emerging design of robots that would resemble humans (humanoids) adds new dimensions to human–robot interactions. Table 7 lists available and emerging capabilities and trends related to robot ergonomics. It is expected that further research and study will lead to the development and refinement of additional, innovative robot ergonomics knowledge that corresponds with new types of robot capabilities.

ACKNOWLEDGMENT

The education I received from Professor H. Gershoni, who taught me systems integration through methods engineering at the Technion, stimulated my interest in the general area that I have termed *Robot Ergonomics*. For this inspiration I am grateful to him. Thanks are also due to the many students who worked with me enthusiastically in this area over the years: Hannan Lechtman, Andy Robinson, Ed Fisher, Richard Pennington, Oded Mai-

Table 6 Robot Ergonomics in Practice—An Application Sample

Work/Operation	Application	Ergonomics Function	Example References
Production line	Scheduling operations in a robotic cell	Cycle-time minimization	Levner, Kats, and Levit, 1997
Material handling	Design of robot-conveyor facility	Method planning: Time/energy optimization	Song and Liu, 1993
Dies and molds finishing	In-process inspection of robotic finishing operation	Performance measurement	Schmidt and Schauer, 1994
Assembly	Robot design for repetitive assembly motions	Motion optimization	Agrawal and Veeraklaew, 1997
Assembly redesign	Design of flexible feeder for variable assembly	Cycle-time estimation	Goldberg, Craig and Carlisle, 1997
Kitting	Design of robotic kitting facilities and methods for flexible assembly	Workplace design; Methods planning	Tamaki and Nof, 1991
Assembly cell design	Integration of sensors, task plans, and assembly execution	Sensor and task planning	Tung and Kak, 1996
Mobile robots	Path planning to avoid obstacles in an uncertain environment	Method planning: Time/motion optimization	Kimmel, Kiryati, and Bruckstein, 1994
Interactive mobile robots	A language for supervisory control of a mobile robot fleet	Human–robot interaction	Causse and Crowley, 1993
Electric power generation	Design of manipulator work in an overhead power generation facility	Parallel motion planning	Tanaka et al., 1996
Telemanipulation	Design of telepresence controllers	Human–robot interface	Suss, 1991; Kazerooni and Moore, 1997
Space telerobotics	Work method planning for antenna assembly in space	Motion planning and simulation	Morikawa et al, 1990
Space shuttle manipulator	Simulation of interaction between human intuitive controller and remote manipulator	Human–robot interaction	Clover and Cruz, 1997
Safety system	Design of computer vision and speech recognition for safety control	Safety; Interaction	Baerveldt, 1992
Hospital robot	Design of fetch and carry tasks, navigation in crowded hallways	Interface design	Krishnamurthy and Evans, 1992
Service robot	Design of robot for handing objects to humans	Human–robot interaction	Agah and Tanie, 1997

Table 7   Emerging Capabilities and Trends Related to Ergonomics of Industrial Robotics Systems

1. Knowledge-based robot programming languages
2. Human–robot interaction by natural-language communication
3. Self-diagnostic fault tracing and self-repair
4. Inherent safety
5. Neuro-fuzzy controllers for enhanced robot mobility, learning and autonomy
6. Telepresence and teleoperation by virtual reality
7. Multirobot collaboration
8. Microrobotics and microassembly
9. Nanorobotics
10. Human–robot physical collaboration
11. Humanoids, bipeds with specialized intelligence and superior physical work skills
12. Robots with social interaction skills

mon, Bob Wilhelm, Cristy Sellers, Daud Hanna, Neal Widmer, Bernhard Rembold, Nitin Khanna, Venkat Rajan, Jim Witzerman, Yael Edan, and the students who worked on robot ergonomics projects in my class, IE574; also to my colleagues, Zvi Drezner, Jim Knight, Gave Salvendy, Kinya Tamaki, Wil Wilhelm, Jacob Rubinovitz, Shraga Shoval, and Eyal Zussman. Their contributions were valuable to the development of this chapter. Research reported here was supported in part by General Motors, the Alcoa Foundation, DARPA, and NSF.

REFERENCES

Agah, A., and K. Tanie. 1997. "Human Interaction with a Service Robot: Mobile-Manipulator Handing over an Object to a Human." In *Proceedings of IEEE International Conference on Robotics and Automation.* Vol. 1. 575–580.

Agrawal, S. K., and T. Veeraklaew. 1997. "Designing Robots for Optimal Performance Motion." In *Proceedings of IEEE International Conference on Robotics and Automation,* Albuquerque. Vol. 3. 2178–2183.

Anzai, Y. 1994. "Human–Robot–Computer Interaction: A New Paradigm of Research in Robotics," *Advanced Robotics* **8**(4), 357–369.

Aoki, K. 1980. "High-Speed and Automated Assembly Line—Why Has Automation Successfully Advanced in Japan?" Paper presented at the 4th International Conference on Production Research, Tokyo.

Argote, L., P. S. Goodman, and D. Schkade. 1983. "The Human Side of Robotics: A Study on How Workers React to a Robot." Tech. Rep. CMU-RI-TR-83-11, Carnegie-Mellon University.

Baerveldt, A. J. 1992. "Cooperation Between Man and Robot: Interface and Safety." In *Proceedings of IEEE International Workshop on Robot and Human Communication,* Tokyo.

Bing X., A. Halme, and J. Helbo. 1996. "Interaction Method in Human–Robot System." In *Proceedings of the Human-Oriented Design of Advanced Robotics Systems,* IFAC. 93–98.

Brady, M., et al. 1982. *Robot Motion Planning and Control.* Cambridge: MIT Press.

Brooks, R. A. 1996. "Behavior-Based Humanoid Robotics." In *Proceedings of the IEEE/RSJ International Conference on Intelligent Robots and Systems,* Osaka, November. Vol. 1. 1–8.

———. 1997. "From Earwigs to Humans." *Robots and Autonomous Systems* **20**(2–4), 291–304.

Causse, O., and J. L. Crowley. 1993. "A Man Machine Interface for a Mobile Robot." *IEEE/RSJ International Conference on Intelligent Robots and Systems.* Vol. 1. 327–335.

Clark, D. O., and G. C. Close. 1992. "Motion Study." In *Handbook of Industrial Engineering.* 2nd ed. Ed. G. Salvendy. New York: John Wiley & Sons. Chapter 3.2.

Clover, C. L., and N. C. Cruz. 1997. "A High Performance Computing Framework for Human–robot Interaction in a Synthetic Environment." In *Proceedings of Simulation MultiConference on High Performance Computing.* Ed. A. Tentner. 313–318.

Craig, J. J. 1997. "Simulation-Based Robot Cell Design in AdeptRapid." In *Proceedings of IEEE International Conference on Robotics and Automation,* Albuquerque, April. Vol. 4. 3214–3219.

Dario, P., et al. 1996. "Robot Assistants: Applications and Evolution." *Robotics and Autonomous Systems* **18**, 225–234.

Dennett, D. C. 1994 "The Practical Requirement for Making a Conscious Robot." *Philosophical Transactions of the Royal Society A, Phys. Sci. Eng.* **349**, 133–146.

Edan, Y., and S. Y. Nof. 1995*a*. "Motion Economy Analysis for Robotic Kitting Tasks." *International Journal of Production Research* **33**, 1213–1227.

———. 1995*b*. "Economics of Multi-sensor Systems." In *Proceedings of International Conference on Production Research-13,* Jerusalem, August.

————, 1996. "Graphic-Based Analysis of Robot Motion Economy Principles." *Robotics and Computer-Integrated Manufacturing* **12**(2), 185–193.

————. 1998. "A Method for Sensor Selection Based on Economy Principles." Submitted to *IIE Transactions on Design and Manufacturing.*

Friedrich, H., H. Hoffmann, and R. Dillmann. 1997. "3d-Icon Based User Interaction for Robot Programming by Demonstration." In *Proceedings of IEEE International Symposium on Computational Intelligence in Robotics and Automation, CIRA 97.* 240–245.

Goldberg, K., J. Craig, and B. Carlisle. 1995. "Estimating Throughput for a Flexible Part Feeder." In *Proceedings of 4th International Symposium on Experimental Robotics,* Stanford, June. 486–497.

Hiraki, K., and Y. Anzai. "Sharing Knowledge with Robots." *International Journal of Human–Computer Interaction* **8**(3), 325–342.

Hwang, Y. K., et al. 1997. "Motion Planning of Eye, Hand and Body of Humanoid Robot." In *IEEE International Symposium on Assembly and Task Planning,* Marina del Rey, August. 231–236.

Imai, M., K. Hiraki, and Y. Anzai. 1995. "Human–Robot Interaction with Attention." *Systems and Computers in Japan* **26**(12), 83–95.

Inaba, M., et al. 1997. "A Remote-Brained Full-Body Humanoid with Multisensor Imaging System of Binocular View, Ears, Wrist Force and Tactile Sensor Suit." In *Proceedings of IEEE International Conference on Robotics and Automation,* Albuquerque, April. Vol. 3. 2497–2502.

Kaiser, M. 1997. "Transfer of Elementary Skills via Human–Robot Interaction." *Adaptive Behavior* **5**(3–4), 249–280.

Kalley, G. S., and K. J. McDermott. 1989. "Using Expert Systems to Incorporate the Principles of Motion Economy as a Means of Improving Robot Manipulator Path Generation." *Computers and Industrial Engineering* **16**(2), 207–213.

Kaneko, S., et al. 1996. "Real Time Visual Recognition in Handling Interface for Human–Robot Collaboration." In *Proceedings of Symposium on Robotics and Cybernetics, IMACS Computer Engineering in Systems Applications,* Lille, France, June. 874–879.

Kaneshima, S. 1997. "Realistic Robot Simulation." *Robot* 114, 71–76.

Kazerooni, H., and C. L. Moore. 1997. "An Approach to Telerobotic Manipulations." *Journal of Dynamic Systems, Measurement and Control* **119**(3), 431–438.

Kimmel, R., N. Kiryati, and A. M. Bruckstein. 1994. "Using Multi-layer Distance Maps for Motion Planning on Surfaces with Moving Obstacles." In *Proceedings of IAPR International Conference on Pattern Recognition,* Israel. Vol. 1. 367–372.

Kitano, H. 1994. "Genesis Machine: Artificial Life Based Massively Parallel Computing System." *Journal of the Institute of Electronics, Informaiton and Communication Engineers* **77**(2), 150–154.

Klingspor, V., J. Demiris, and M. Kaiser. 1997. "Human–Robot Communication and Machine Learning." *Applied Artificial Intelligence* **11**(7–8), 719–746.

Krishnamurthy, B., and J. Evans. 1992. "HelpMate: A Robotic Courier for Hospital Use." In *Proceedings of IEEE International Conference on Systems, Man and Cybernetics,* Chicago, October. Vol. 2. 1630–1634.

Kwon, K. S. 1996. "Optimum Design for Emergency Stop Button on Robot Teach Pendants." *International Journal of Occupational Safety and Ergonomics* **2**(3), 212–217.

Kwon, K. S., and J. S. Kim. 1997. "Analysis of Robot Work by Modeling the Unit Motion of Robots." In *Proceedings of the International Conference on Manufacturing Automation (ICMA),* Hong Kong, April. Vol. 2. 1051–1056.

Levner, E., V. Kats, and V. E. Levit. 1997. "An Improved Algorithm for Cyclic Flowshop Scheduling in a Robotic Cell. *European Journal of Operations Research* **97**(3), 500–508.

Luecke, G. R., and J. C. Edwards. 1996. "Virtual Cooperating Manipulators as a Virtual Reality Haptic Interface." In *Proceedings of Third Annual Symposium on Human Interaction with Complex Systems.* 133–140.

Maimon, O., and S. Y. Nof. "Analysis of Multi-robot Systems." *IIE Transactions* **18**(9), 226–234.

Martensson, L. 1987. "Interaction Between Man and Robots." *Robotics* **3**(1), 47–52.

Morikawa, H., et al. 1990. "A Prototype Space Telerobotic System." In *Proceedings of IROS IEEE International Workshop on Intelligent Robots and Systems,* Ibaraki, Japan.

Morita, T., and S. Sugano. 1997. "Development of an Anthropomorphic Force-Controlled Manipulator WAM-10." In *Proceedings of IEEE International Conference on Advanced Robotics,* Monterey, June. 701–706.

Morley, E. C., and C. S. Syan. 1995. "Teach Pendants: How Are They for You?" *Industrial Robot* **22**(4), 18–22.

Nadler, G. 1991. *The Planning and Design Professions: An Operational Theory.* New York: John Wiley & Sons.

Nof, J., J. P. Witzerman, and S. Y. Nof. 1996. "Reachability Analysis with a Robot Simulator–Emulator Workstation." Research Memo 96-J, School of Industrial Engineering, Purdue University.

Nof, S. Y. 1985. "Robot Ergonomics—Optimizing Robot Work." In *Handbook of Industrial Robotics.* Ed. S. Y. Nof. New York: John Wiley & Sons.

———. 1989. "Knowledge-Based Real-Time Machine Interaction." In *Proceedings of 22nd Hawaii International Conference on System Sciences,* Kailua-Kona, Hawaii. Vol. 3. 192–199.

———. 1991. "Advances in Robot Motion Economy Principles." In *Proceedings of ISIR,* Detroit. Vol. 22. 2213–2221.

———, ed. 1994. *Information and Collaboration Models of Integration.* Dordrecht: KIuwer Academic.

Nof, S. Y. and Drezuer, Z. 1993. "The Multiple-Robot Assembly Plan Problem." *Journal of Intelligent and Robotic Systems,* 5, 57–71.

Nof, S. Y., and D. Hanna. 1989. "Operational Characteristics of Multi-robot Systems with Cooperation." *International Journal of Production Research* 27(3), 477–492.

Nof, S. Y. and Huang, C.-Y. 1998. "The Production Robotics and Integration Software for Manufacturing (PRISM): An overview." Research Memo 98-3. School of Industrial Engineering, Purdue University.

Nof, S. Y., J. L. Knight, and G. Salvendy. 1980. "Effective Utilization of Industrial Robots—A Job and Skill Analysis Approach." *AIE Transactions* 12(3), 216–225.

Nof, S. Y., and V. N. Rajan. 1994. "Robotics." In *Handbook of Design, Manufacturing and Automation.* Ed. R. Dorf and A. Kusiak. New York: Wiley Interscience.

Nof, S. Y., and P. Robinson. 1986. "Analysis of Two Robot Motion Economy Principles." *Israel Journal of Technology* 23, 125–128.

Okada, T., Y. Yamamoto, and Y. Anzai. 1993. "The Active Interface for Human–Robot Interaction." In *Proceedings of the Fifth International Conference on HCI.* Vol. 1. 231–236.

Osborn, J. 1996. "A Perception System for Advanced Human Robot Interaction." In *Proceedings of American Nuclear Society Meeting on Control and Human–Machine Interface Technologies.* Vol. 2. 1529.

Parsons, H. M., and G. P. Kearsley. 1982. "Robotics and Human Factors: Current Status and Future Prospects." *Human Factors* 24(5), 535–552.

Pennington, R. A., E. L. Fisher, and S. Y. Nof. 1982. "Analysis of Robot Work Characteristics." *Industrial Robot.* 166–171.

Podgorski, D., and S. Boleslawski. 1990. "Ergonomic Aspects of Robot Teach Pendant Design." In *Proceedings of the Third National Conference on Robotics,* Wroclaw, Poland, September Technical Polytechnic of Wroclaw. No. 37. 175–182. [In Polish.]

Qinghua, L., A. Takanishi, and I. Kato. 1993. "Learning of Robot Biped Walking with the Cooperation of a Human." In *Proceedings of 2nd. IEEE International Workshop in Robot and Human Communication,* Tokyo, November. 393–397.

Rajan, V. N., and S. Y. Nof. 1990. "A Game-Theoretic Approach for Cooperation Control." *International Journal of Computer Integrated Manufacturing* 3(1), 47–59.

———. 1996. "Cooperation Requirements Planning (CRP) for Multiprocessors: Optimal Assignment and Execution Planning." *Journal of Intelligent and Robotic Systems* 15, 419–435.

Roach, J. W., and M. N. Boaz. 1987. "Coordinating the Motions of Robot Arms in Common Workspace." *IEEE Journal of Robotics and Automation* 3(5), 437–444.

Rodrigues, J. 1996. "How to Plan a Robot-Based Automated Molding Cell." *Robotics Today* 9(3), 1–6.

Rodriguez, G. 1989. "Recursive Forward Dynamics for Multiple Robot Arms Moving a Common Task Object." *IEEE Transactions on Robotics and Automation* 5(4), 510–521.

Roth, B. 1995. "How Can Robots Look Like Human Beings," In *Proceedings of IEEE International Conference on Robots and Automation,* Nagoya, Japan, May. Vol. 3. 3164.

Rutkowska, J. C. 1995. "Can Development Be Designed? What We May Learn from the Cog Project." In *Proceedings of the Third European Conference on Artificial Life,* Granada, Spain, June. 383–395.

Sabe, K., H. Hashimoto, and F. Harashima. 1996. "Analysis of Real Time Motion Estimation with Active Stereo Camera." In *Proceedings of the 4th International Workshop on Advanced Motion Control,* Mie, Japan, March. Vol. 1, 365–370.

Schmidt, J., and U. Schauer. 1994. "Finishing of Dies and Molds: An Approach to Quality-Oriented Automation with the Help of Industrial Robots." *Industrial Robot* 21(1), 28–31.

Seward, D. W., A. Bradshaw, and F. Margrave. 1996. "The Anatomy of a Humanoid Robot." *Robotica* 14, 437–443.

Shiller, Z. 1997. "Optimal Robot Motion Planning and Work-Cell Layout Design." *Robotica* 15, 31–40.

Shoval, S., J. Rubinovitz, and S. Y. Nof. 1998. "Analysis of Robot Motion Performance and Implications to Economy Principles," In *Proceedings of the IEEE International Conference on Robotics and Automation,* Leuven, Belgium.

Song, Y. Y., and L. Lin. 1993. "Optimization of Cycle Time and Kinematic Energy in a Robot/Conveyor System with Variable Pick-up Locations." *International Journal of Production Research* 31(7), 1541–1556.

Suss, U. 1991. "M/sup 2/IDI: A Man–Machine Interface for the Digital Telemanipulator Control System DISTEL." *Robotersysteme* 7(3), 164–168.

Tamaki, K., and S. Y. Nof. 1991. "Design Method of Robot Kitting System for Flexible Assembly." *Robotics and Autonomous Systems* **8**(4), 255–273.

Tanaka, S., et al. 1996. "Work Automation with the Hot-line Work Robot System Phase II." In *Proceedings of IEEE International Conference on Robotics and Automation,* Minneapolis. Vol. 2. 1261–1267.

Terada, T. 1995. "Motion Simulation in Robot Offline System." *Robot* 104, 21–26.

Tung, C.-P., and A. C. Kak. 1995. "Integrating Sensing, Task Planning, and Execution for Robotic Assembly." In *IEEE Transactions on Robotics and Automation* **12**(2), 187–201.

Wang, C. C., and S. J. Deutsch. "Accurate Robot Motion Time (Arm-Time) Prediction" In *Proceedings of the Japan-USA Symposium on Flexible Automation,* ASME, New York. 83–89.

Wannasuphoprasit, W., et al. 1997. "Cobot Control." In *Proceedings of IEEE International Conference on Robotics and Automation,* Albuquerque, April. Vol. 4. 20–25.

Widmer, N. S., and S. Y. Nof. "Design of Knowledge-Based Performance Progress Monitor." *Computers and Industrial Engineering* **22**(2), 101–114.

Witzerman, J. P., and S. Y. Nof. 1995a. "FDL: Facility Description Language." In *Proceedings of IIE Research Conference,* Nashville, May. 449–455.

———. 1995b. "Integration of Cellular Control Modeling with a Graphic Simulator/Emulator Workstation." *International Journal of Production Research* **33**, 3193–3206.

Yamaguchi, J., and A. Takanishi. 1997. "Development of a Biped Walking Robot Having Antagonistic Driven Joints Using Nonlinear Spring Mechanism." In *Proceedings of IEEE International Conference on Robotics and Automation,* Albuquerque, April. Vol. 1. 185–192.

Yamaguchi, J., et al. 1996. "Development of a Dynamic Biped Walking System for Humanoid–Development of a Biped Walking Robot Adapting to the Humans' Living Floor." In *Proceedings of IEEE International Conference on Robotics and Automation,* Minneapolis, April. Vol. 1. 232–239.

ADDITIONAL READING

Helander, M. G. *A Guide to the Ergonomics of Manufacturing.* London: Taylor & Francis, 1995.

Konz, S. *Work Design: Industrial Ergonomics.* 4th ed. Scottsdale: Publishing Horizons, 1995.

Nof, S. Y., W. E. Wilhelm, and H. J. Warnecke. *Industrial Assembly.* London: Chapman & Hall, 1997.

Proceedings of the IEEE International Workshop on Robot and Human Communication, RO-MAN Tsukuba, Japan, November 1996.

Proceedings of the International Workshop on Biorobotics: Human–Robot Symbiosis. Robots and Autonomous Systems **18**(1–2), 1996.

Rahimi, M., and W. Karwowski, eds. *Human–Robot Interaction.* London: Taylor & Francis, 1992.

CHAPTER 33

HUMAN FACTORS IN PLANNING ROBOTICS SYSTEMS

Yee-Yin Choong
GE Information Services, Inc.
Rockville, Maryland

Gavriel Salvendy
Purdue University
West Lafayette, Indiana

1 OVERVIEW

The purpose of this chapter is to demonstrate the nature and characteristics of the human factors disciplines that will impact effective planning, design, control, and operation of industrial robotics systems. A good understanding of the human element in planning robotics systems contributes to greater acceptance and adoption and more effective utilization of industrial robots. Human factors issues—which impact effective implementation and utilization of industrial robotics systems that are economically viable, humanly acceptable, and result in increased productivity and quality of life—must be considered in the analysis, design, implementation, control, and operation of robotics systems. Those issues include social, human performance, supervisory control, and human–computer interaction. Although these issues are conceptually integrated, for operational purposes, each is discussed in separate sections in this chapter.

2 SOCIAL ISSUES

There are at least three major social factors that impact the effective and widespread utilization of industrial robotics systems: worker displacement, worker retraining, and function allocation of workers. The possible social impacts of robot diffusion are also discussed at the end of this section.

2.1 Worker Displacement

The extent of worker displacement due to automation is difficult to ascertain. Technological change does not necessarily create new jobs or avoid job displacements (Senker, 1979). Senker states that there are two phases in major technological revolutions: in the initial phase new technology primarily generates employment, and in the latter (mature) phase it tends to displace labor. Senker asserts that the mature phase has been reached in the "electronics technological revolution." The extension of this, as it applies to industrial robots, is that the low cost and high reliability of microprocessors result in decreasing robotics costs and concurrently enlarge their range of applicability. Thus the expansion of production does not always imply a proportionate increase in employment (Senker, 1979).

To demonstrate a possible net decrease in workforce as a result of robotization of manual production operations, a simple material-handling operation is presented. If these manual material-handling systems are required to feed three numerical control (NC) machines, a great deal of cost is incurred for manual labor and indirect costs due to manual labor. A worker may incur a total first-shift cost of nearly $30 per hour. Tote bins typically cost $125 to $150 (Mangold, 1981). Other expenses may include forklift operators to move pallets of tote bins, and so on.

Handbook of Industrial Robotics, Second Edition, Edited by Shimon Y. Nof
ISBN 0-471-17783-0 © 1999 John Wiley & Sons, Inc.

Suppose these NC machines were arranged in a manufacturing cell; based on the electronics and software capabilities currently available, one robot is capable of tending each machine even though each may perform a different operation. It must be realized that some type of material-feeding system must be in effect or designed for the robot. It may be possible to eliminate the tote bins, which may assume production space. Forklifts and their operators may be modified or eliminated, and the material-handling personnel may also be displaced. Over an extended time period humans can work only one shift per day. Since a robot is capable of working more than one shift, it may replace more than one worker (Mangold, 1981). In this particular situation it can be seen that direct and indirect cost savings may be quite significant, and robots can contribute to disproportionate displacement per job.

The impact of similar situations for unskilled and semiskilled workers is fewer hours of work, which translates to fewer jobs and less job security. This has spurred organizations such as the United Auto Workers (UAW) to develop positions on integrated automation, chiefly concerning industrial robots. Precarious as it is to permit one specific organization to speak for all production employees, the UAW does encompass a larger proportion of employees in an industry that utilizes the greatest number of industrial robots. The UAW does not place a specific emphasis on robots but instead considers them another technological advancement it must consider (Weekley, 1979). The union also recognizes that enhanced productivity is necessary for long-term economic viability (Mangold, 1981; Weekley, 1979). However, it is aware of possible detrimental impacts upon its membership, primarily due to job insecurity. The union believes that technological advancement is acceptable and encourages it as long as the current workforce retains job security (Mangold, 1981).

Quality of work life (QWL) is also affected by the introduction of robotics. In reference to job security, it is possible that robots may ultimately improve job security. Robot adaptability enables robots to be assigned and tooled to many production tasks; the degree to which those tasks are similar to tasks performed by humans increases the likelihood that a human and robot are interchangeable in task performance (Nof, Knight, and Salvendy, 1980). In the case of consumer items, where market fluctuations may be quite drastic, robots can offer a distinct advantage in production assemblies (Sugarman, 1980). In QWL terms, a company may initiate "robot layoffs" due to downward fluctuation in demand and temporarily assign humans to the assembly line, thus minimizing the displacement effects of a market downturn.

It is often argued that technological change (i.e., industrial robots) will create jobs. There is a wide range of estimates on the extent to which displaced jobs will be compensated by newly developed jobs due to increases and complex robot utilization (Albus, 1982). There is a significant demand for highly trained personnel in computer programming, mechanical engineering, electronic design, and so forth, all highly skilled positions to implement, utilize, and/or maintain industrial robots (Albus, 1982). In all probability the workers to be displaced will be those in unskilled and semiskilled jobs. A large void in skill level exists between the jobs eliminated and the jobs created.

2.2 Worker Retraining

One of the most acute problems associated with introducing and utilizing flexible manufacturing systems (FMS) and industrial robots is that the skill requirements in these new technologies do not capitalize and build on the skills, perception, and knowledge accumulated by the industrial workers (Rosenbrock, 1982). This implies that acquired industrial skills, which were widely utilized in the premicroelectronics–automation era (Salvendy and Seymour, 1973), are completely lost and have become redundant for the industrial robot revolution era.

This has two major implications. First, it must be assessed who can be retrained for the new skills. This can be achieved by analyzing skills and knowledge requirements for robotics jobs. Secondly, from this analysis either work samples or tests that simulate the job can be developed. After the assessment of the reliability and validity of these tests, the samples can be administered to displaced workers to assess the likelihood of their success in mastering new skills through retraining (Borman and Peterson, 1982). Based on this evaluation and on the nature of human abilities, it may be estimated that more than half of these displaced workers will not possess adequate skills for the new robot-oriented and computer-based manufacturing work environment. To eliminate or reduce this situation industrial robotics systems must be so designed, developed, and operated to capitalize on (as far as possible) acquired and used human skills.

2.3   Function Allocation of Workers

The development and introduction of industrial robots has helped to improve the precision and economy of operations. At the same time, however, a considerable number of unanticipated problems and failures are incurred, for the most part due to breakdowns in the interaction between human operators and computers controlling the robotics systems. This can result in job unsatisfaction and frustration of human operators. A number of critical questions for the design of new organizational structures arises, such as:

- What is the optimal allocation of functions between human supervisory control and the computer?
- What is the relationship between the number of machines controlled by one supervisor and the productivity of the overall systems? What is the optimal number of machines that a supervisor should control?
- What is the impact of work isolation of the supervisor in a computer-controlled work environment on the quality of working life and mental health of the operator?

In allocating functions between computer and humans, emphasis must be placed on optimizing human arousal, job satisfaction, and productivity.

Evidence pertaining to job design (Table 1) indicates that the numbers of people who prefer to work and are more satisfied and productive in performing the task in simplified mode are equal to those who prefer enriched jobs, but that 9% of the labor force does not like work of any type (Salvendy, 1978). When 270 shop floor workers performed their work in both enriched and simplified modes, it was evident that the numbers of people who preferred simplified jobs and those who preferred enriched jobs were equal. It is typical that the older workers (past 45 years of age) prefer simplified jobs and the younger workers prefer enriched jobs.

In the simplified job design the operator performs only very small components of the total job and does not have decision latitude about task performance. These simplified jobs can be enlarged either vertically or horizontally. Thus either the operator may do more of the same thing, thereby enlarging the job vertically, or additional tasks may be added for the task performance, thereby enlarging the task horizontally, which results in job enrichment. As noted above, typically older workers prefer to work at simplified jobs and younger workers prefer to work at, and are both more satisfied and productive in, enriched jobs. The overwhelming majority of computer-based supervisory control tasks are manned by younger operators. Hence in allocating the function between human and computer the division should be made such that the task content of the human is sufficiently enriched to provide for psychological growth of the individual.

The introduction of automation technology was expected to result in reduced workload. It turned out, however, that automation does not have a uniform effect on workload (Wiener, 1982). The operator's task has shifted from active control to supervisory control. Humans no longer continuously control a process themselves (although sometimes they still need to revert to manual control), but instead they monitor the performance of highly

Table 1   Worker Job Satisfaction in Relation to Job Simplicity[a]

Variable Measured	Satisfied with Enriched Jobs (Dissatisfied with Simplified Jobs)	Dissatisfied with Enriched Jobs (Satisfied with Simplified Jobs)	Do Not Like Work of Any Type
1. Percent of labor force	47	44	9
2. Percent of labor force in category 1 who are dissatisfied	4	87	100
3. Productivity of the labor force in category 1	91	92	84
4. Percent of labor force over 45 years of age	11	82	50

[a] From G. Salvendy, "An Industrial Dilemma: Simplified Versus Enriched Jobs," in *Production and Industrial Systems*, ed. R. Murumatsu and N. A. Dudley. London: Taylor & Francis, 1978.

autonomous machine agents (robots). This imposes new demands on attention, and it requires that the operator knows more about systems in order to be able to understand, predict, and manipulate their behavior. In making the allocation of functions, it should be noted that an optimal arousal level exists for maximizing productivity and job satisfaction: when arousal level is too low, boredom sets in; when arousal is too high, mental overload occurs.

2.3 Social Impacts

According to the U.S. Congress Office of Technology Assessment (Office of Technology Report, 1982), there are a number of institutional and organizational barriers to the use of informational technology; this also has bearing on the use of industrial robots. The barriers include high initial cost, the lack of high-quality programming, and the dearth of local personnel with adequate training. In this connection Rosenbrock (1982) addresses two vital behavioral and social issues:

1. The skills that robots call for will usually be new, yet there is no reason why they should not be based on older skills and developed from them.
2. Industrial robots will aid us to carry on rapidly the process of breaking jobs into their fragments and performing some of these fragments by machines, leaving other fragments to be done by humans.

This has broad implications for the training and retraining of personnel and the design of the psychological contents of jobs. Many of these implications associated with the introduction of industrial robots have effectively been managed in Japan (Hasegawa, 1982). Seven positive and seven negative aspects of the social impacts of robots are summarized in Table 2.

3 HUMAN INDUSTRIAL WORK PERFORMANCE

A number of human performance capabilities impact the effective design and operation of robotics systems: human abilities and limitations in work performance, information-processing and decision-making capabilities, and design of controls to fit work requirements. Although these variables are conceptually integrated, for purposes of presentation each of them is discussed separately.

3.1 Human Abilities and Variations

Variation in human performance levels occurs both among different operators and within a single operator over a period of time. This variation arises from the following three general classes of operator characteristics:

1. Experience and training
2. Enduring mental and physical characteristics
3. Transitory mental and physical characteristics prevailing at the time of task performance.

Table 2 Social Impacts of Robot Diffusion

Positive Impacts	Negative Impacts
1. Promotion of worker's welfare	1. Unemployment problems
2. Improvement of productivity	2. Elimination of pride in old skills
3. Increase in safety of workers	3. Shortage of engineers and newly trained skilled workers
4. Release of workers from time restrictions	4. Production capacity nonproportional to the size of the labor force
5. Ease in maintaining quality standard	5. Decrease in flow of labor force from under-developed to developed countries
6. Ease of production scheduling	6. Safety and psychological problems of robot interaction with human
7. Creation of new high-level jobs	7. Great movement of labor population from the second to the third sector of industry

Human performance is also influenced by task characteristics such as equipment variability, defects and malfunctions, and, especially among different operators, the methods employed by operators to perform their tasks.

The combined impact of these various factors on the performance variability of an individual operator (i.e., within-operator variability) has been documented among blue-collar workers in manufacturing industries. These studies (Salvendy and Seymour, 1974) indicate that reliability[1] of production output varies from 0.7 to 0.9, with a mean of 0.8. This implies that about 64% (i.e., $0.8^2 \times 100\%$) of an operator's performance in one week can be predicted by his or her performance observed during a prior week. Conversely, 36% of the operator's performance cannot be explained in this manner, but can be apparently explained by such factors as those previously listed.

It should be noted that individual variability within a working day is markedly smaller than between working days. Furthermore, performance variability within a workday is smallest from midmorning to early afternoon. These patterns of within-operator variability, as well as warmup and slowdown at the beginning and end of the workday, must be accounted for in the design, control, and operation of robotics systems in which the human is a part.

Based on many studies, it is well known that human performance variability among operators is much larger than that observed within the same operator over successive observations. Generally a performance range of 2 to 1 encompasses 95% of the working populations (Wechsler, 1952). However, in practical work situations the range encountered is likely to be much smaller than this because of preemployment selection, attrition of some low-performance operators, and peer pressures that may limit the output of high-ability operators. The recognition of this range of performance levels is critical to the design of robotics systems and to the development of effective production planning and control techniques.

3.2   Human Information-Processing and Decision-Making

3.2.1   Human Information-Processing

The operator's ability to perform the crucial mental activities and to perform tasks effectively rests upon fundamental cognitive processes and functions. These basic mental functions and processes (or stages) appear in Figure 1, which represents an information-processing model of the human operator. In the model of Figure 1 three major informa-

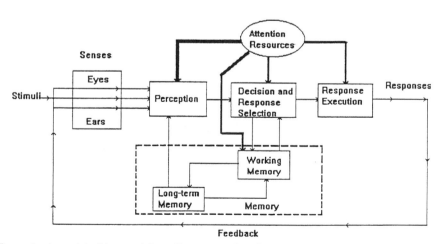

Figure 1   A model of human information processing. (From *Handbook of Human Factors and Ergonomics*, 2nd ed., ed. G. Salvendy. New York: John Wiley & Sons. Chapter 4, p. 1. © 1997. Reprinted by permission of John Wiley & Sons, Inc.)

---

[1] The reliability coefficient is a measure of consistency determined by the extent to which two successive samples of same-task performance provide similar results. Thus, for example, reliability of performance may be obtained by correlating one week's performance with another week's.

tion-processing stages are shown: perception, decision, and response execution. Also shown are three memory systems (sensory, short-term, and long-term), which depend upon, and are limited by, the information-processing capacities of these three major stages and the storage characteristics of the three memory systems.

Limits of human performance arise from two characteristics of the major information-processing stages: (1) they require a minimum time in which to perform their functions, and (2) they have limits as to the amount of information they can process per unit time. If information arrives too rapidly, a stage may become overloaded and unable to process information effectively. This limit to the rate at which a stage can handle (i.e., transmit) information is its channel capacity.

The information-processing model contains three memory systems. These systems contribute several essential functions in psychomotor performance. They act as buffers to store temporarily (from 1 to 2 sec) rapidly arriving sensory information (sensory memory). They temporarily store up to seven "chunks" (words, names, digits, etc.) of information (short-term memory). Finally, they provide long-term storage that underlies learning and improvement in psychomotor performance (long-term memory).

Industrial work often requires the operator to time-share, or simultaneously perform, several separate subtasks. This time-sharing demand occurs on three levels. First, even in simple tasks the operator must receive information, make decisions, and control response movements. Efficient performance may require that these activities occur parallel to one another. Second, more complex tasks often require the operator to make several separate responses simultaneously—for example, concurrent, but separate, hand motions. Third, the operator may be asked to perform two quite separate tasks at once. How efficiently can activities at each of these levels overlap?

At the first level, evidence suggests that information reception can efficiently overlap both decision-making and response control. However, the initiation and correction of movements interfere with decision-making. These response functions occur primarily in the second phase of movement control. Hence performance can be enhanced by terminating movements with mechanical stops rather than with closed-loop, operator guidance.

Time-sharing at the second level can be enhanced if the same mental function, information reception, decision-making, or response control is not needed simultaneously by both activities. The refractory period[2] of the central decision-making stage requires that successive inputs to this process be separated by at least 300 msec. For example, if the operator is required to identify and respond to two successive signals, those signals should not occur within 300 msec of each other.

The processes involved in closed-loop movement control, including monitoring, selecting an appropriate corrective response, and initiating the correction, impose particularly high information-processing demands. Hence when these processes are required by two simultaneous subtasks (e.g., independently moving each gripper), information overload and consequent interference between the subtasks can be expected.

Another critical factor in time-sharing efficiency is response–response compatibility. Some combinations of responses can be performed more easily than others. For example, in executing simultaneous movements performance is best when the hands (or feet) move in the same direction (e.g., both forward). Similarly, responses that start at the same time are easier to time-share than those that do not. Selecting, initiating, or monitoring parallel (or successive) responses that have similar characteristics apparently requires less information processing than occurs in the case of unrelated movements.

Time-sharing efficiency will be greatly enhanced when highly compatible stimulus–response (S–R) relationships are used. High S–R compatibility reduces the load on the decision-making stage responsible for selecting responses. For example, a vibrating machine control provides a highly compatible signal for the response of grasping the control more firmly. The operator may do this almost immediately with no disruption of other movement activities.

Finally, at the most complicated level, performing two separate tasks at once, performance depends on a wide variety of factors, including the priorities that the operator attaches to the competing tasks. Typically, when an easy task was combined with a more difficult one, a greater percentage decline in performance was found for the easier task (Kantowitz and Knight, 1976).

---

[2] The period during which the operator is unable to process any new information.

Time-sharing efficiency improves with task experience for a variety of reasons. First, there is evidence that time-sharing is a general ability that can be enhanced by training. Operators who efficiently time-share one pair of tasks often are superior at time-sharing other task pairs. Second, as operators become well trained, tasks impose lower information-processing loads and even appear to become "automatic." Several reasons for this have been considered, including the following:

- An internal task model frees the operator from processing redundant information.
- Kinesthetic information, which may be processed faster than visual information (Table 3) and which often is highly S–R compatible, is gradually substituted for visual information.
- Certain information-processing steps (e.g., "check" operation) may be minimized or deleted entirely. More efficient movement sequences involving less second-phase, close-loop control are developed.

Because the "automatic" time-shared tasks each impose lower information-processing demands upon the operator, there is less likelihood of overload, and efficient time-sharing is possible.

3.2.2   Decision-Making

Decision-making refers to the processes whereby operators evaluate information made available by the initial perceptual processing. Decision-making results in the selection of an intended course of action. Two decision-making characteristics are especially important: how much time decision-making requires and how accurate decisions are.

Decision delays stem from two sources: capacity limitations and refractory limitations. Capacity limitations arise because decision-making stages can process information at only a limited rate. The amount of information transmission involved in a decision increases logarithmically with the number of possible stimuli that might be presented and the number of alternative responses from which the operator might select. In general, doubling the number of possible stimuli and responses increases the information transmitted in the decision by one bit.

Hick (1952) showed that the rate of information flow per unit time remains constant at about 1 bit/220 msec. However, if the operator exceeds these margins by trying to go too fast, accuracy drops very rapidly and the rate of information transmission will fall. This occurs when the operator tries to increase the speed more than about 20%.

The other source of decision-making delay is a fixed delay of about 300 msec that must separate successive decisions. This is the so-called *psychological refractory period.* If information is presented to the decision-making stage within 300 msec of a previous decision, decision-making will be delayed until the psychological refractory period has elapsed. This refractory delay does not decline with practice.

3.3   Design of Controls

Two major concerns must be practiced in designing monitoring systems for robots that are compatible with human performance capabilities, namely, which control is best to use for which purpose and, given the selection of a certain control, the determination of the appropriate and applicable range for size, displacement, and resistance for each control. Furthermore, the human parameters that vary from one individual to the next must also

Table 3   Minimum Reaction Times (Kp) for Various Stimulation Modalities

Stimulation Modality	Reaction Time (msec)
Visual	150–225
Auditory	20–185
Tactual	115–190

be accounted for in the ergonomic design and arrangement of the controls. The values for these parameters are presented in Figure 2.

4 SUPERVISORY CONTROL OF ROBOTICS SYSTEMS

The industrial robotics system are typically jointly supervised by computer and human. One or more human operators are setting initial conditions for robots, intermittently adjusting, and receiving information displayed on a computer with data from robots and the task environment.

4.1 Supervisory Control Concept

The human should feel in control of the plant and thus the computer software should be in a position to help the human operator perform the tasks efficiently rather than in a position to take over control. The human operator should have the option to override the computer if he or she feels that it is necessary. This is because humans are much more flexible to novel situations than the computer. An important point is that the human's role should be coherent. This coherence of the human's role must be assured in the initial stages of the design process when system tasks are allocated between man and computer.

If the operator supervising an FMS through the computer is given too little to do, boredom results, which leads to degraded performance and less productivity. On the other hand, if the operator is given too much to do, mental overload occurs, which also leads to decreased performance and less productivity.

In assigning functions to humans in a FMS, many decision-making responsibilities correspond to an enriched job, whereas low arousal levels and minimum decision-making for the operator correspond to a simplified job. Thus in allocating responsibilities between human and computer in FMS, one must be aware of levels of arousal for the human and the degree of decision-making allocated between human and computer.

4.2 Supervisory Control Models

The operator in FMS may shift attention among many machines, rendering to each in turn as much attention as is necessary to service it properly or keep it under control. The human tends to have more responsibility for multiple and diverse tasks. It is appropriate to view the human as a time-shared computer with various distributions of processing times and a priority structure that allows preemption of tasks. This can be done by queueing theory formulation (Rouse, 1977; Chu and Rouse, 1979; Walden and Rouse, 1978).

Tulga and Sheridan (1980) developed a multitask dynamic decision paradigm with such parameters as interarrival rate of task, time before tasks hit the deadline, task duration, the productivity of the human for performing tasks, and task value densities. In this experiment a number of task-sharing finite completion times and different payoffs appear on a screen. The subject must decide which task to perform at various times to maximize the payoff. When the human performs a variety of tasks by the aid of computer, allocation of responsibility for different tasks is important for optimum performance.

Several investigators (Rouse, 1977; Chu and Rouse, 1979) have studied multitask decision-making where the human is required to allocate attention between control tasks and discrete tasks. In a queueing theory formulation the human "server" serviced various tasks, arriving at exponentially distributed interarrival times. The growing model predicted the mean waiting time for each task as well as the mean fraction of attention devoted to each task.

5 HUMAN–COMPUTER INTERACTION

As mentioned previously, the industrial robotics system is supervised by human operators through the assistance of computers. Success in making the interaction between human and computer smooth and seamless will assure the effective implementation and utilization of industrial robots and the increased productivity and quality of human life. At least four major issues need to be addressed in the realm of human–computer interaction: the design of the computer terminal workstation, the design of the computer–user interface, the evaluation of the interaction, and networking and the rapidly growing World Wide Web.

All under "Path of C. motion" = **Turning movement**

Control	Dimension (mm)	D / R	M
Handwheel 	D : 160 - 800 d : 30 - 40	160 - 200 mm 200 - 250 mm (D)	2 - 40 Nm 4 - 60 Nm
Crank 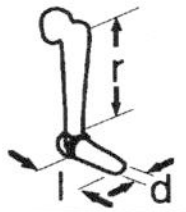	Hand (Finger) r : < 250 (<100) l : 100 (30) d : 32 (16)	<100 mm 100 - 250 mm (R)	0,6 - 3,0 Nm 5 - 14 Nm
Rotary knob 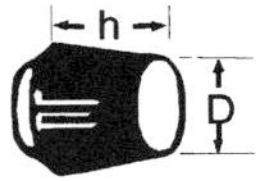	Hand (Finger) D: 25-100 (15-25) h: >20 (>15)	15 - 25 mm 25 - 100 mm (D)	0,02 - 0,05 Nm 0,3 - 0,7 Nm
Rotary selector switch 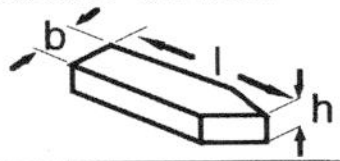	l: 30 - 70 h: > 20 b: 10 - 25	30 mm 30 - 70 mm (D)	0,1 - 0,3 Nm 0,3 - 0,6 Nm
Thumbwheel 	b: > 8	0,4 - 5 N	
Rollball 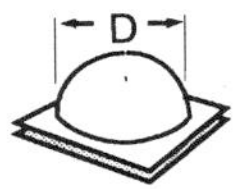	D: 60 -120	0,4 - 5 N	

Operational characteristics columns (across the top): 2 positions, >2 positions, Continuous adjustment, Precise adjustment, Quick adjustment, Large force application, Tactile feedback, Setting visible, Accidental actuation.

Legend: ○ Not suitable   ◐ Acceptable   ● Recommended

Figure 2 Hand- and foot-operated control devices and their operational characteristics and control functions. (From H.-J. Bullinger, P. Kern, and M. Brown, "Controls," in *Handbook of Human Factors and Ergonomics*, 2nd ed., ed. G. Salvendy. New York: John Wiley & Sons. Chapter 21, pp. 702–704. © 1997. Reprinted by permission of John Wiley & Sons, Inc.)

Turning movement (Path of C. motion)

Control	Dimension (mm)	Force F (N) / Moment M (Nm)	2 positions	>2 positions	Continous adjustment	Precise adjustment	Quick adjustment	Large force application	Tactile feedback	Setting visible	Accidental actuation
Lever	d : 30 - 40 l : 100 - 120	10 - 200 N	●	●	●	◐	◐	●	◐	◐	○
Joystick	s : 30 - 150 d : 10 - 20	5 - 50 N	●	◐	●	●	◐	◐	◐	◐	○
Toggle switch	b : >10 l : >15	2 - 10 N	●	◐	○	○	●	○	●	●	○
Rocker switch	b : >10 l : >15	2 - 8 N	●	○	○	○	●	○	●	●	◐
Rotary disk	d : 12 - 15 D : 50 - 80	1 - 7 N	●	◐	◐	○	◐	○	○	○	◐
Pedal	b : 50 - 100 l : 200 - 300 l : 50 - 100 (Forefoot)	Sitting : 16 - 100 N Standing : 80 - 250 N	◐	◐	◐	◐	◐	●	◐	○	○

Legend: ○ Not suitable　◐ Acceptable　● Recommended

Figure 2 (Continued)

654

Figure 2 (Continued)

Path of C. motion	Control	Dimension (mm)	Force F (N) Moment M (Nm)	2 positions	>2 positions	Continuous adjustment	Precise adjustment	Quick adjustment	Large force application	Tactile feedback	Setting visible	Accidental actuation
Turning movement	Handle (Slide) 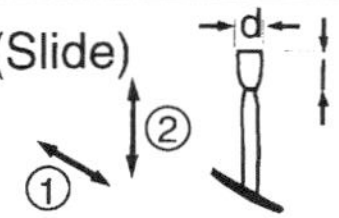	d : 30 - 40 l : 100 - 120	F_1 : 10 - 200 N F_2 : 7 - 140 N	●	●	●	◐	◐	●	◐	◐	○
	D-Handle 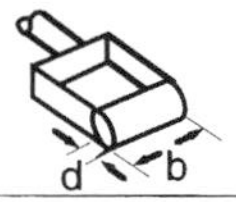	d : 30 - 40 b : 110 - 130	10 - 200 N	●	●	●	◐	◐	●	◐	◐	○
	Push button 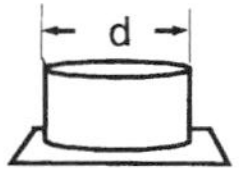	Finger : d > 15 Hand : d > 50 Foot : d > 50	Finger : F = 1 - 8 N Hand : F = 4 - 16 N Foot : F = 15 - 90 N	●	○	○	○	●	○●●	○	○	●○
	Slide 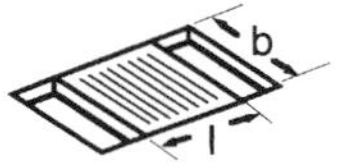	l : >15 b : >15	1 - 5 N (Touch grip)	●	◐	◐	◐	◐	○	○	●	●
	Slide 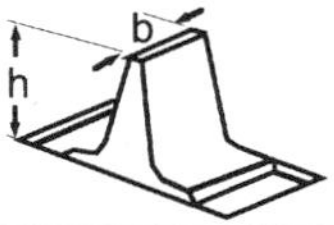	b : > 10 h : > 15	1 - 10 N (Thumb - finger grip)	●	◐	◐	◐	◐	◐	○	●	◐
	Sensor key 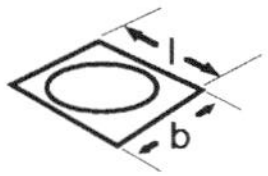	l : > 14 b : > 14		●	○	○	○	●	○	○	○	◐

Legend: ○ Not suitable   ◐ Acceptable   ● Recommended

655

5.1 Computer Terminal Workstation Design

5.1.1 Components of the Work System

From an ergonomic point of view the different components of the work system (e.g., environment, robot, work tasks, and people) interact dynamically with each other and function as a total system, as shown in Figure 3. In an ergonomic approach the person is the central focus and the critical factors of the work system are designed to help the person be effective, motivated, and comfortable. The physical, physiological, psychological, and social needs of the person are considered to ensure the best possible workplace design for productive and healthy human–computer interaction. Table 4 shows ergonomic recommendations for video display terminals (VDTs) that will improve the human interface characteristics.

5.2 User Interface Design

The human supervisor gives directives and receives feedback/information through a computer interface in communicating with the industrial robots. The design of a computer–user interface shares many of the general user interface design objectives, such as that it be easy to learn, easy to use, effective, efficient, comfortable, and safe to use. A well-designed user interface that meets these objectives can improve the quality of work, increase the satisfaction of users, improve the productivity of the workforce, and enhance the safety of the system that the software program controls.

The content of an interface should abstract the critical features of the target system (Norman, 1983; Rasmussen, Pejtersen, and Goodstein, 1994; Vicente and Rasmussen, 1992); the form of interface presentation should be consistent with human perceptual and cognitive characteristics (Wickens, 1992; Carroll, Mack, and Kellogg, 1988; Tulla, 1988); the means for interface manipulation should be compatible with human response tendencies (Myers, 1991; Shneiderman, 1983); and interface actions should help users effectively achieve their intended impact upon the target system (Sheridan, 1984; Shneiderman, 1992).

From the user's point of view, a specific goal of computer–user interface design is to reduce the user's cognitive loads and stresses. Williges, Williges, and Elkerton (1987) examined the design objectives from the user's perspective and, based on a review of

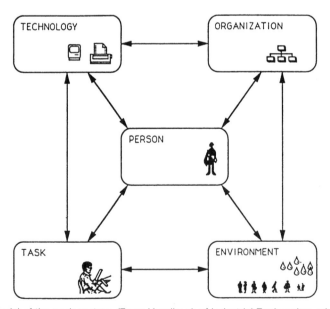

Figure 3   Model of the work system. (From *Handbook of Industrial Engineering*, ed. G. Salvendy. New York: John Wiley & Sons. Chapter 42, p. 1122. © 1992. Reprinted by permission of John Wiley & Sons. Inc.)

Table 4   Ergonomic Recommendations for the VDT Technology, Work Environment, and Workstation[a]

Ergonomic Consideration	Recommendation
1. Viewing screen	7:1
a. Character /screen contrast	height = 20–22 min of visual arc
b. Screen character size	Width = 70–80% of height
c. Viewing distance	Usually 50 cm or less
d. Line refresh rate	70 hertz
e. Eye viewing angle from horizon	10°–40°
2. Illumination	
a. No hardcopy	300 lux
b. With normal hard copy	500 lux
c. With poor hard copy	700 lux
d. Environmental luminance contrast	
Near objects	1:3
Far objects	1:10
e. Reflectance from surfaces	
Working surface	40–60%
Floor	30%
Ceiling	80–90%
Walls	40–60%
3. HVAC	
a. Temperature—winter	20°–24°C (68°–75°F)
b. Temperature—summer	23°–27°C (73°–81°F)
c. Humidity	50–60%
d. Airflow	0.15–0.25 m/sec
4. Keyboard	
a. Slope	0°–15°
b. Key top area	200 mm²
c. Key top horizontal width	12 mm (minimum)
d. Horizontal key spacing	18–19 mm
e. Vertical key spacing	18–20 mm
f. Key force	$0.25N$–$1.5N$ (0.5–$0.6N$ preferred)
5. Workstation	51 cm minimum
a. leg clerance	(61 cm preferred minimum)
b. Leg depth	38 cm minimum
c. Leg depth with leg extension	60 cm minimum
d. Work surface height—nonadjustable	70 cm
e. Work surface height—adjustable for one surface	70–80 cm
f. Work surface height—adjustable for two surfaces	Keyboard surface 59–71 cm Screen surface 70–80 cm
6. Chair	
a. Seat pan width	45 cm minimum
b. Seat pan depth	38–43 cm minimum
c. Seat front tilt	5° forward to 7° backward
d. Seat back inclination	110°–130°
e. Backrest height	45–51 cm

[a] From M. J. Smith et al., "Human–Computer Interaction," in *Handbook of Industrial Engineering*, 2nd ed., ed. G. Salvendy. New York: John Wiley & Sons, 1992. 1107–1144.

psychological concepts and research results, summarized a list of seven general design principles as a basis for developing specific design objectives: compatibility, consistency, memory, structure, feedback, workload, and individualization.

From the interface engineering point of view, Shneiderman (1992) discussed in detail the following four classes of engineering goals of interface design:

1. A software interface must supply the necessary functionality to ensure all required tasks can be carried out.

2. The system must ensure reliability, availability, security, and data integrity.
3. Interface features and styles should be standardized and portable across multiple applications.
4. Interface development should be on schedule and within budget.

It should be noted that a computer interface should not only help users understand the target system, but also assist them to achieve their intended impact or influences upon it. In supervisory control applications, users need to exert authoritative, immediate, and effective control over the target system (Sheridan, 1984). A goal of interface design is certainly to help users accomplish their task of controlling or influencing a target system, in addition to understanding it.

5.2.1 Process of Interface Design

Interface design is a creative, iterative, and cooperative activity. The design of high-quality interfaces requires scientific reasoning, artistic imagination, and aesthetic judgment; it involves many rounds of revisions; and it usually takes place in a multidisciplinary team environment.

Many theories and ideas have been proposed to characterize the design process. Almost all of them describe design as a process composed of a number of stages, including identification of goals and constraints, requirement analysis and task analysis, design generation and synthesis, prototyping and iterative redesign, and testing and evaluation. These stages are shown in Figure 4.

A number of techniques and tools that might be used in each stage of the interface design process are summarized in Table 5.

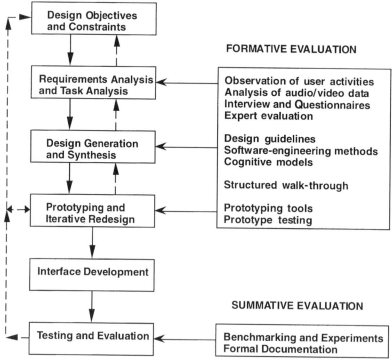

Figure 4 Process of interface design and human factors evaluation methods. (From Y. Liu, "Software–User Interface Design," in *Handbook of Human Factors and Ergonomics*, 2nd ed., ed. G. Salvendy. New York: John Wiley & Sons. Chapter 51, p. 1693. © 1997. Reprinted by permission of John Wiley & Sons, Inc.)

Table 5   Techniques and Tools for Stages of the Interface Design Process

Stage in Process	Techniques and Tools
Goal identification	Prioritize various design goals
Requirements analysis	Observations Interviews Focus groups Questionnaires
Task analysis	Cognitive complexity theory (CCT) (Kieras and Polson, 1985) Cognitive grammar (Reisner, 1981) Command language grammar (Moran, 1981) GOMS (goals, operators, methods, and selections) models (Card, Moran, and Newell, 1983) Natural GOMS language (NGOMSL) (Kieras, 1988) TAG (task-action grammar) (Payne and Green, 1986)
Prototyping	XWindows: Motif Macintosh: HyperCard and MacApp IBM-PC: Toolbox and Visual Basic Web-based applications: HTML
Testing and evaluation	Usability testing Summative evaluations

5.2.2   Computer–User Interface Design Issues

The design of different computer–user interfaces requires consideration of different issues throughout the design process, including interface capabilities, user characteristics, the target system, interface representation of the target system, presentation of information to the user, and user manipulation of the interface.

Interface Capabilities

One of the important issues that interface designers must consider is the available interface technology. Interface hardware and software are changing on a daily basis; interface designers must keep track of the changes and stay informed about the options available so that they can select the best options to serve their needs.

Hardware devices are the physical input/output devices used for human–computer interaction. Although the most commonly used hardware devices are keyboard, mouse, and CRT display, other input devices such as touchscreen, stylus, light pen, trackball, trackpad, and touch-sensitive panel are replacing or augmenting the role of conventional input devices. The rapid development of computer technology and the increased concern for human factors have led to the appearance of a large variety of new devices. Input devices for tracking eye movement, hand gestures, or whole-body movement and for speech recognition provide novel ways of sending information to the computer. Output devices such as high-resolution graphic displays and multimedia displays with sophisticated video and audio capabilities are also becoming more and more prevalent in workplaces, and they provide more options for displaying information to the users.

In parallel with the rapid development of hardware capabilities, a fundamental change has occurred in the way software interfaces are constructed and used. The current style of interface design is to design interfaces that are composed of a number of windows, each of which occupies a section of the computer screen and is responsible for a specific part of a user's task by communicating with the user within the space provided by the window. A window can be built with a window widget contained in the widget set of a programming toolkit. In order to facilitate the communication with users, a number of widget items can also be constructed and attached to a window. The most commonly used widgets include labels, forms, lists, scroll bars, pushbuttons, toggle switches, radio boxes, windows, dialog boxes, text widgets, menu bars, and pulldown menus. For a general discussion of interface widgets, see Eberts (1994).

Whereas the interface widgets help establish the control structure and "open the windows" of an interface, graphics libraries help construct and render images in the space provided by the windows. To a great extent, we can say that widgets are for making frames and graphics libraries are for painting the pictures.

A graphics library contains a collection of subroutines for drawing and animating two-dimensional (2D) and three-dimensional (3D) color graphics images. More specifically, the subroutines can be called from a program to display characters that form texts, draw 2D and 3D geometric primitives that form complex objects, show color and lighting effects, create animation, provide coordinate transformations such as viewing and projection transformations, render curves and surfaces, create and edit objects, provide depth cueing and hidden-surface removal for 3D scenes, collect user inputs, and so on.

Multimedia interfaces and virtual reality (VR) are newcomers in the field of interface technologies. Multimedia, as its name suggests, attempts to improve the "look and feel" of interfaces and change the mode of user–computer interaction by providing high-resolution computer graphics, digital full-motion interactive video, stereo sound, and CD-ROM, all on a personal computer. Multimedia technology distinguishes itself from current audio and video technologies such as television by the interactivity it provides users. However, current technological trends suggest that the distinction between the two technologies will undoubtedly become blurred as they merge with each other (Hoffert and Gretsch, 1991). VR is an emerging computer technology that attempts to eradicate the barriers between users and computers by creating a "virtual" environment where users "live" and "navigate" while performing their tasks. Head-mounted displays, 3D goggles, and data-gloves are some of the enabling technologies for virtual reality, and new concepts and devices are emerging rapidly.

User Characteristics

No matter how sophisticated computer interface technology may become, the ultimate judges of the technology are the human users with their information-processing capabilities and limitations. Interface designers should take into account the characteristics of user behavior and preferences and be knowledgeable about related psychological theories and explanations.

According to Norman (1983, 1986), when using a device or working with a physical system, users tend to form their views or *mental models* (also called *user models*) of how the device or system works. Users construct mental models by relating the system to similar systems they are already familiar with; thus past experience and training play an important role in the formation of mental models.

Although users tend to form mental models of devices or systems, there is no guarantee that these models are always adequate. In fact, people's mental models are often deficient in a number of ways. People tend to forget the details of the system they are using, confuse similar systems, and maintain erroneous behavior or action patterns if they help reduce their mental efforts.

An important task of an interface designer is to help users form adequate mental models, which need not be technically accurate but must be functional to allow users to accomplish their tasks effectively and efficiently. To do so designers need to understand the relationship among the designer's conceptual model of the system, the system image presented to the user, and the user's mental model formed on the basis of that image (Norman, 1983). The designer's conceptual model is an accurate, consistent, and complete abstraction of the system, expressed in designer's terminology or some formal modeling language, used in designing and developing the system. The system image is what can be seen by the user about the system. The designer expects that the user's mental model matches the designer's conceptual model, but the system image is the only channel through which the designer can communicate with the user. Therefore the designer should carefully construct the system image so as to minimize potential differences between their model of the system and the user's mental model, as shown in Figure 5.

The Target System

The computer interface plays an intermediary role between the user and the target system because, oftentimes users use the interface to influence or control the target system (e.g., the human operator supervises robots through the computer interface). The effective design of such interfaces requires not only the consideration of available interface technology and user characteristics, but also a clear understanding of the target system, including its purpose, its function, the concrete and abstract objects that comprise the system, the relationships among these objects, the system variables that define the state of the system, and the factors that determine the values of these variables.

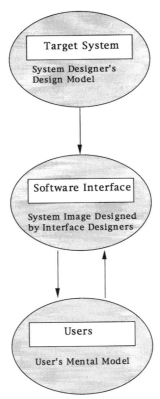

Figure 5   Distinction between design model, system image, and user's mental model. Interface designers need to use well-designed system image to bridge the gap between system designer's design model and user's mental model of the same system. (From Y. Liu, "Software–User Interface Design," in *Handbook of Human Factors and Ergonomics,* ed. G. Salvendy. New York: John Wiley & Sons. Chapter 51, p. 1703. © 1997. Reprinted by permission of John Wiley & Sons, Inc.)

A target system can be analyzed and described in different ways and at different levels of detail and rigor. One way to describe a system is to categorize it into a particular class according to one of its attributes. Another commonly used method for system analysis is modeling, which allows designers to analyze and describe a system at a desired level of detail by selecting a proper modeling methodology.

The target system and the software–user interface are usually designed by separate individuals, who are often from different groups or organizations. The two types of design activities may take place at different locations or at different times. It should be noted that establishing effective communication between the interface designers and the system designers may not always be an easy task. The two design teams may have different objectives and constraints and may encounter conceptual and "language" barriers in the process of communication because of possible differences in their conceptual bases and their use of terminology. Furthermore, if the interface designers are asked to add a new interface to an existing system, the time gap between the new interface design and the target system design may add more difficulties to the establishment of effective communication. Interface designers must realize and overcome these difficulties so that they can successfully bridge the gap between the user's mental model and the system designer's conceptual model of the same target system.

Interface Representation of the Target System

Much of interface design is concerned with constructing abstract models of systems and presenting them to the users. In other words, interface designers need to solve both a representation problem and a presentation problem to decide both the content and the form of the interface. This section discusses three major methods that might be used to represent a target system.

Abstraction and Selective Representation. Different aspects of a target system are not of equal importance to the users; some are essential, some are optional, and others can be distracting to the users in carrying out their tasks. An interface should represent the essential aspects of the target system and filter out the potentially distracting aspects.

Different users have different needs in their interactions with different target systems, and apparently they should be provided with different ways of abstraction and selective representation. A map of city roads and highways is useful for automobile drivers; an aviation map is useful for aircraft pilots; however, a map showing both the road and the air navigation information can be distracting to both types of users. The goals and the requirements of the user and the target system should determine the particular form of abstraction and selective representation for a specific application.

Structural Decomposition. One of the methods for developing representations of target systems is structural decomposition, which represents a system by decomposing it into structural components and elements. The behavior of the system is described as a causal chaining of events occurring at the component level.

Structural decomposition is a useful method because humans often need to know the parts and elements that comprise a system, the functions and characteristics of these components, and the relationships among them. Structural information is often particularly important for the training of new users and the diagnosis of system failures.

The Abstraction Hierarchy. Although the structural decomposition approach is an important tool for modeling human interaction with complex systems, the approach is limited in its ability to model high-level, goal-oriented behavior (Rasmussen, Pejtersen, and Goodstein, 1994). Rasmussen (1985) proposed a theoretical framework called *abstraction hierarchy* that complements the structural approach and emphasizes the importance of looking at the entire system and representing it by abstraction to a proper functional level and separation of relevant functional relations. Furthermore, the abstraction hierarchy emphasizes the need to include several levels of abstraction and decomposition in representing a work system.

Murray and Liu (1995) applied the abstraction hierarchy in their analysis of operator behavior in a highway traffic control center. They found that the more experienced operators tended to move up and down through the various levels of the hierarchy more fluidly and rapidly in their way of seeking information and reasoning. The novice operators, on the other hand, were more inclined to focus their attention on a tighter range at the lower part of the abstraction hierarchy. This finding suggests that an interface represented in the form of the abstraction hierarchy may not only enhance operator performance, but also can serve as a useful device for user training and testing.

Presentation of Information to the User

The previous section discussed representational issues of interface design for a target system. This section is concerned with the style or the format of presenting information to the users. It should be noted that information presentation and user manipulation are conceptually integrated in user activities and are two interdependent aspects of the interface design problem, though the topic of user manipulation of the interface is discussed in a separate section.

It is impossible to offer a comprehensive review in this section of the vast body of literature on how to present information effectively through a software interface. Four topics are selected for discussion to demonstrate the presentation issues in interface design, each involving an interface technique and a fundamental psychological or human factors concept.

It should be emphasized that effective interface design requires the consideration of a large number of issues that are not completely covered in this section. Some of these issues are summarized in Table 6.

Screen Design. When computer screens are cluttered with homogeneous items, users often need to spend a long time to find the item they are looking for. The psychological process involved in finding a target in a crowded visual field is called *visual search* or *visual selective attention*. An item is distinct if it differs significantly from the other items in at least one physical attribute—size, brightness, color (more precisely, hue), rate of blinking, and so on.

Table 6 Major Display Design Issues[a]

Screen design—Color, highlighting, layout, and information density (Tullis, 1988; Shneiderman, 1992; Marcus, 1992; Tullis, 1981)
Typography (Baecker, 1990; Sassoon, 1993)
Data display format (Tullis, 1988; Smith and Mosier, 1986)
Object displays (Chernoff, 1973; Coekin, 1969; Woods, Wise, and Hanes, 1981)
Information visualization (Pickover and Tewksbury, 1994; Rosenblum, 1994; Tufte, 1983, 1990)
Three-dimensional displays (Ellis, 1991; Okoshi, 1980; Wickens, Todd, and Seidler, 1989)
Multiwindow strategy (Marcus, 1992; Billingsley, 1988; Henderson and Card, 1986)
Selection of display modalities (Wickens, Sandry, and Vidulich, 1983; Wickens and Liu, 1988)
Hypertext and hypermedia (Nielsen, 1990; Rivlin, Botafogo, and Shneiderman, 1994; Rizk, Streitz, and Andre, 1990; Shneiderman and Kearsley, 1989)
Design of voice interfaces (Baber and Noyes, 1993; Strathmeyer, 1990)
Design of error messages and online help (Shneiderman, 1992; Isa et al., 1983; Kearsley, 1988)
Documentation (Boehm-Davis and Fregly, 1985; Carroll, 1984; Crown, 1992)

[a] From Y. Liu, "Software–User Interface Design," in *Handbook of Human Factors and Ergonomics,* ed. G. Salvendy. New York: John Wiley & Sons. Chapter 51, p. 1703 © 1997. Reprinted by permission of John Wiley & Sons, Inc.

Christ (1975) reviewed the results of 42 studies on the effects of color on visual search and showed that color aids searching if the color of the target is unique and known to the users in advance. Shneiderman (1992) discussed 14 guidelines for the effective use of color in computer user interfaces. Other authors have also provided important suggestions about how to create effective color displays (Marcus, 1992; Galitz, 1989; Thorell and Smith, 1990). For example, it is suggested that color can be used to group related items; changes in color can be used to indicate status changes. However, designers should avoid using too many colors on the same screen and should be consistent in color coding by using the same color-coding rules throughout the application.

Tullis (1988) developed a comprehensive list of guidelines for screen design, which includes specific guidelines with regard to the amount of information on a screen, grouping of information, display of text, use of highlighting, standardization of displays, and graphic representations. For example, it is suggested that designers minimize the amount of information on a screen by displaying only the necessary information and using concise wording. It is also suggested that a consistent format be used for all the screens of the same application so that the users will know where a given piece of information is located.

Object Displays. The idea of object displays was developed to help users process multiple pieces of information in parallel. The idea of object displays is based on psychological theories of preattentive perceptual processes in processing visual information. A large number of studies have shown that visual processing of the multiple dimensions or attributes of a single object is a parallel and obligatory process. It is parallel because people pay attention to all the dimensions or attributes at once, rather than one of them at a time. It is obligatory because people have no control over this process—one cannot notice one dimension or attribute of an object but ignore the others.

When users need to process and integrate multidimensional information in order to accomplish their tasks, it is desirable for user interface designers to capitalize on this human perceptual characteristic by creating multidimensional object displays. In fact, many examples of object displays have appeared, including Chernoff's faces (1973), polygon-polar diagrams (Coetin, 1969; Woods, Wise, and Hanes, 1981), box displays (Barnett and Wickens, 1988), and other multidimensional object displays (Carswell and Wickens, 1987; Wickens and Andre, 1990).

The expression of a human face, the shape of a polygon, and the area of a rectangle are all called *emergent features* in the psychology literature—global perceptual features that are created via interactions among individual parts or local features (Pomerantz, 1981). As pointed out by Bennett, Toms, and Woods (1993), it is not an easy task to represent variables as emergent features that highlight the critical data relationships. Bennett, Toms, and Woods (1993) suggested that designers need to decide which variables should be included in the graphic form and how they should be assigned to the dimensions of the object. Furthermore, decisions must also be made about whether all variables should be converted to a common scale and how to represent the task context.

Information Visualization.   Graphical presentation of information has been used in diverse areas of application and has demonstrated its value as a method of presenting information (Tufte, 1983, 1990). The advancement of computer graphics technology is providing more options for portraying abstract concepts and data relations in various graphics forms. Visualization has emerged as a major field of computer graphics research, and the techniques have been successfully applied in diverse disciplines (Pickover and Tewksbury, 1994; Rosenblum, 1994).

Proper use of metaphors is critical for creating effective graphical representation of abstract concepts and variables. Metaphor is an important emerging concept in user interface design; interfaces incorporating proper metaphors can make the strange familiar, the invisible visible, and the abstract concrete.

Metaphors have been used extensively in user–interface design, and many of these applications are summarized in Carroll, Mack, and Kellogg (1988) and Eberts (1994). For instance, the desktop metaphor represents files and folders as file- and folder-like icons, lists of choices as menu icons, and file deletion as the action of moving a file icon to a wastebasket icon.

Metaphors and icons have their limitations, too. Icons and graphic representations take up excessive screen space (Shneiderman, 1992). A poorly selected metaphor or icon could be confusing if it does not convey its meaning clearly, and it could be misleading if it supports alternative interpretations. Interface designers need to be clear about the benefits and costs of using metaphors for their specific application.

Three-Dimensional Displays.   The development of computer graphics technology has made it possible to develop 3D computer graphic displays that provide the viewers a sense of three dimensionality—"realistic" and "natural" viewing conditions that closely resemble perceiving 3D objects in a 3D space.

As summarized by Wickens, Todd, and Seidler (1989), there are both benefits and costs associated with using 3D displays. One benefit is that a 3D visual scene is a more natural, "ecological," or compatible representation of a 3D world than that provided by 2D displays. Another benefit is that a single integrated display of a 3D object or scene reduces the need for mental integration of two or three 2D displays. A cost of using 3D displays is that any projection of a 3D world inevitably produces an inherent perceptual ambiguity, in addition to the potential costs associated with additional hardware, software, and programming requirements for implementing 3D displays.

User Manipulation of the Interface

Based on the way that users provide inputs to the computer, computer user interfaces can be classified into four major classes: *command language-based, menu-selection, direct manipulation,* and *anthropomorphic.*

Command language-based interfaces are the oldest and often the most difficult to use. Users are required to remember and type command sequences that are often cryptic and composed by complex syntax. User frustration in front of a command-based interface is often observed because of the great memory load and typing demands and the low tolerance for errors shown by this type of interface. Important issues that have been considered by command language researchers include command abbreviation behavior (Benbasat and Wand, 1984), command organization strategies (Carroll and Thomas, 1982), and selection of command names (Black and Moran, 1982; Laudauer, Calotti, and Hartwell, 1983).

Menu selection avoids many of the problems associated with command language interfaces. The interface displays the choices as menu items and users can select an item easily by pointing and clicking or with one or two keypresses. Menu selection can thus greatly reduce the need for memorization and typing of complex command sequences. The design of effective menu selection interfaces requires the consideration of numerous issues, including menu structure, sequence and phrasing of menu items, shortcuts through the menus for frequent users, menu layout and graphical menu features, display rates and response time, and selection mechanisms. Detailed discussions of these issues can be found in Norman (1991) and Shneiderman (1992).

Direct manipulation interfaces (DMIs) are based on the idea that users should be allowed to manipulate computer interfaces in a way that is analogous to the way they manipulate objects in space.

Examples of DMI can be found in many areas of applications, including word processing, desktop publishing, computer-aided design (CAD), flight simulations, and video games. These DMI word processors in use are also called *WYSIWYG* (what you see is what you get) word processors. DMI word processors display the document in the form in which it will appear when printing is done, allow users to make changes directly to the displayed document, and display the changes immediately. DMI employs familiar icons and associated actions to help users accomplish their tasks without the need to remember abstract commands. For example, a user may delete a file by dragging the corresponding file icon into a wastebasket icon.

Anthropomorphic interfaces interact with users in a way that is analogous to the way humans interact with each other. Natural language interfaces and interfaces that recognize gestures, facial expressions, or eye movements all belong to this type of interface. The design and development of anthropomorphic interfaces require not only hardware and software supports but also a clear understanding of how humans communicate with each other with natural languages and through gestures, facial expressions, and eye contact. A number of books and articles have been published covering related topics, such as Eberts (1994), Fels and Hinton (1995), Schmandt (1994), and Walker, Sproull, and Subramani (1994).

5.3 Evaluation

Computer–user interface evaluation includes formative evaluation and summative evaluation.

5.3.1 Formative Evaluation

Formative evaluation is conducted before an interface is constructed to help improve the interface as part of an iterative design process. The main goal of formative evaluation is thus to learn which detailed aspects of the interface are good and bad and how the design can be improved. Activities of formative evaluation include user observation, interview, focus group, survey, cognitive modeling, and "cognitive walk-through."

5.3.2 Summative Evaluation

Summative evaluation aims at assessing the overall quality of an interface—for example, for use in deciding between two alternatives or as a part of competitive analysis to learn how good the competition really is. Summative evaluation is usually conducted after an interface or a prototype of the target system has been developed to test its usability.

User testing with real users is the most fundamental usability method and is in some sense irreplaceable, since it provides direct information about how people interact with computers and what problems they have during the interaction. Even so, other usability engineering methods (Nielsen, 1994*a*) can serve as good supplements to gather additional information or to gain usability insights at a lower cost. These methods include heuristic evaluation, expert review, and checklists.

A usability test usually consists of several components: a test plan, a pilot test, test users, test experimenter, test tasks, the actual test, test data, and a laboratory where the test will be conducted.

Test Plan

A test plan should be written down before the start of the test and should address the following issues (Nielsen, 1994*b*):

- The goal of the test: What do you want to achieve?
- Where and when will the test take place?
- How long is each test session expected to take?
- What computer support will be needed for the test?
- What software needs to be ready for the test?
- What should the state of the system be at the start of the test?
- What should the system or network load and response times be? If possible, the system should not be unrealistically slow, but neither should it be unrealistically fast because the experimental system or network has no other users. One may have to slow down the system artificially to simulate realistic response times.
- Who will serve as experimenters for the test?

- Who are the test users going to be, and how are you going to get hold of them?
- How many test users are needed?
- What test tasks will the users be asked to perform?
- What criteria will be used to determine when the users have finished each of the test tasks correctly?
- What user aids (manuals, online help, etc.) will be made available to the test users?
- To what extent will the experimenter be allowed to help the users during the test?
- What data are going to be collected, and how will they be analyzed once they have been collected?
- What will the criterion be for pronouncing the interface a success? Often this will be the "planned" level for previously specified usability goals, but it could also be a looser criterion such as "no new usability problems found with severity higher than 3."

Pilot Test

No usability testing should be performed without the test procedure first having been tried out on a few pilot subjects.

During pilot testing the instructions for some of the test tasks will typically be incomprehensible to the users or misinterpreted by them. Similarly, any questionnaires used for subjective satisfaction rating or other debriefing will often need to be changed based on pilot testing. Also, a mismatch often occurs between the test tasks and the time planned for each test session. Most commonly, the tasks are more difficult than one expected, but of course it may also be the case that some tasks are too easy. Depending on the circumstances of the individual project, one will either have to revise the tasks or make more time available for each test session.

Pilot testing can also be used to refine the experimental procedure and clarify the definitions of various things that are to be measured.

Test Users

The main rule regarding test users is that they should be as representative as possible of the intended users of the target system.

One of the main distinctions between categories of users is that between novice and expert. Almost all user interfaces need to be tested with novice users, and many systems should also be tested with expert users. Typically these two groups should be tested in separate tests with some of the same and some different test tasks.

Sometimes one will have to train the users with respect to those aspects of a user interface that are unfamiliar to them but not relevant for the main usability test. This is typically necessary during the transition from one interface generation to the next, where users will have experience with the old interaction techniques but will be completely baffled by the new ones unless they are given some training.

Test Experimenters

A test experimenter is needed to be in charge of running the test, no matter what test method is chosen. In general it is preferable to use good experimenters who have previous experience in using the chosen method.

In addition to knowledge of the test method, the experimenter must have extensive knowledge of the application and its user interface. System knowledge is necessary for the experimenter to understand what the users are doing as they perform tasks with the system and to make reasonable inferences about the users' probable intentions at various stages of the dialogue.

Test Tasks

The test tasks should be chosen to be as representative as possible of the uses to which the system will eventually be put in the field. Also, the tasks should provide reasonable coverage of the most important parts of the user interface. The test tasks can be designed based on a task analysis or on a product identity statement listing the intended uses for the product.

The tasks need to be small enough to be completed within the time limits of the user test, but they should not be so small that they become trivial. The test tasks should specify precisely what result the user is being asked to produce, since the process of using a computer to achieve a goal is considerably different from just playing around. Test tasks should normally be given to the users in writing. Not only does this ensure that all users get the tasks described the same way, but it allows users to refer to the task description during the experiment instead of having to remember all the details of the task. After the user has been given the task and has had a chance to read it, the experimenter, in order to minimize the risk that the user has misinterpreted the task, should allow the user to ask questions about the task description.

Test tasks should never be frivolous, humorous, or offensive. Instead, all test tasks should be business-oriented and as realistic as possible. To increase both the users' understanding of the tasks and their sense of using the software in a realistic way, the tasks can be related to an overall scenario. For example, the scenario could be that the user has just been hired as a human supervisor to control a robot using a computer in an FMS environment.

The test tasks can also be used to increase the user's confidence. The very first test task should always be extremely simple in order to guarantee the user an early success experience to boost morale. Similarly, the last test task should be designed to make users feel that they have accomplished something.

The Test

A usability test typically has four stages: preparation, introduction, the test itself, and debriefing.

In preparation for the experiment, the experimenter should make sure that the test room is ready for the experiment, that the computer system is in the start state that was specified in the test plan, and that all test materials, instructions, and questionnaires are available. To minimize the user's discomfort and confusion this preparation should be completed before the arrival of the user.

During the introduction the experimenter welcomes the test user and gives a brief explanation of the purpose of the test. The experimenter then proceeds with introducing the test procedure. The experimenter should emphasize that the purpose of the test is to evaluate the software, not the user, and should encourage the user to speak freely. Confidentiality of the test results should be assured to the user. If video- or audiorecording will be taking place, the user needs to be informed.

During the test itself the experimenter should normally refrain from interacting with the user and should certainly not express any personal opinions or indicate whether the user is doing well or poorly. Also, the experimenter should avoid helping the test user, even if the user gets into quite severe difficulties. The main exception to this rule is when the user is clearly stuck and is getting frustrated with the situation. The experimenter may also decide to help a user who is encountering a problem that has occurred several times before with previous test users, but should only do so if it is clear beyond any doubt from the previous tests what the problem is and what different kinds of subsequent problems users may encounter as a result of the problem in question. It is tempting to help too early and too much, so experimenters should exercise caution in deciding when to help. Sometimes subjective satisfaction opinions are desired from the user for each task or for the whole test. The user could be asked to fill in subject satisfaction questionnaires after performing each task or after completing the whole test.

After the test the user is debriefed. During debriefing users are asked for any comments they might have about the system and any suggestions they may have for improvement. Such suggestions may not always lead to specific design changes, and one will often find that different users make completely contradictory suggestions, but this type of user suggestion can serve as a rich source of additional information to consider in the redesign.

The experimenter can also use the debriefing to ask users for further comments about events during the test that were hard for the experimenter to understand. Even though users may not always remember why they did certain things, they are sometimes able to clarify some of their presumptions and goals.

Test Data

Typically test data collected in a usability test are quantitative measurements. Common quantifiable usability measurements include (Nielsen, 1997):

- Time to complete a specific task
- Number of tasks completed within a given time limit
- Ratio between successful interactions and errors
- Time spent recovering from errors

- Number of user errors
- Number of immediately subsequent erroneous actions
- Number of commands or other features utilized by the user
- Number of commands or other features never used by the user
- Number of system features remembered during a debriefing after the test
- Frequency/time of use of the manuals and/or the help system
- How frequently the manual and/or help system solved the user's problem
- Proportion of user statements that were positive versus critical toward the system
- Number of times the user expresses clear frustration (or clear joy)
- Proportion of users who say that they would prefer using the system over some specified competitor
- Number of times the user had to work around an unsolvable problem
- Proportion of users using efficient working strategies compared to the users who use inefficient strategies
- The amount of "dead" time when the user is not interacting with the system (the system can be instrumented to distinguish between two kinds of dead time: response-time delays where the user is waiting for the system and thinking-time delays where the system is waiting for the user)
- Number of times the user is sidetracked from focusing on the real task

Clearly only a subset of these quantifiable measurements would be collected during any particular measurement study.

Other than quantitative measurements, many usability studies in industry use qualitative methods instead of the exact measurements. Often the main focus of a test is collecting information about what aspects of a design seem to work and what aspects cause usability problems, even if no exact measures are available to discriminate between the two.

Laboratory

Many user tests take place in specially equipped usability laboratories (Nielsen, 1994*b*). Usually a usability laboratory consists of two rooms, one for experiment and the other for observation. Typically there will be soundproof one-way mirrors separating the observation room from the test room to allow the experimenters, other usability specialists, and the developers to discuss user actions without disturbing the test users.

Typically a usability laboratory is equipped with several video cameras (normally two) under remote control from the observation room. These cameras can be used to show an overview of the test situation and to focus in on the user's face, the keyboard, the manual and the documentation, and the screen. A producer in the observation room then typically mixes the signal from these cameras to a single video stream that is recorded and possibly timestamped for later synchronization with an observation log entered into a computer during the experiment. Such synchronization makes it possible to find the video segment corresponding to a certain interesting user event later without having to review the entire videotape.

In addition to permanent usability laboratories, it is possible to use portable usability laboratories for more flexible testing and field studies. With a portable usability laboratory any office can be rapidly converted to a test room and user testing can be conducted where the users are rather than the users having to be brought to a fixed location. Typical equipment includes a laptop computer, a camcorder, and a lavaliere microphone (two microphones are preferred so that the experimenter can also get one). The regular directional microphone built into many camcorders is normally not sufficient because of the noise of the computer. Also, a tripod helps steady the image and carry the camera during the test sessions.

5.4 Networking and the World Wide Web

Industrial robots are parts of computer-integrated FMS systems, which require computer networking to facilitate communication among groups of people and between robots and human operators. Networking is simply a way for computing systems to communicate— to "socialize"—with each other. Hence networking is also called *social computing.*

Social computing can be defined as the use of computers and computing systems for the purpose of interaction between people for work, entertainment, or even simple communication. Social computing thrives because almost every organization is composed of divisions or departments where people work together in groups toward a common goal. The need to bring together the process of collaboration and computing has led to the development of various collaborative systems.

The development of various types of collaborative systems can be best described along a continuum of technological advancements over time: group decision support systems (GDSS), computer-supported cooperative work (CSCW), Groupware, and collaborations based on the Internet's World Wide Web (WWW).

5.4.1   Group Decision Support Systems (GDSS)

The earliest objective of group decision support systems was to aid professionals to conduct group meetings that would otherwise have been handled using paper and pencil. Since decision-makers were typically confronted with unstructured problems (DeSanctis and Gallupe, 1985) and were often viewed as rushing to a decision before adequately defining a problem (Lewis and Keleman, 1990), early GDSS focused on problem definition and documentation.

5.4.2   Computer-Supported Cooperative Work (CSCW)

The single most important technological advancement fueling the growth of collaborative systems is undoubtedly the computer network. The concept of connecting computers together is analogous to social networking or grouping. It stands to reason that since the aim of collaborative systems is to bring together groups working toward a common goal, the tools used by the individuals in these groups must also have the ability to communicate with each other. In fact, CSCW has been defined as "the use of computer and electronic communication tools as a media for communicating" (Johnson et al., 1986).

5.4.3   Groupware

Definitions of *groupware* range from considering it essentially an umbrella term for "the technologies that support person-to-person collaboration," including anything from "E-mail to Electronic Meeting Systems to workflow" (Coleman, 1995), to tying it in somewhat more specifically with communication, document management, and database access. In essence, groupware can be thought of as a superset of desktop environments and messaging systems on local area networks (LANs). User interfaces, group processes, and concurrency control are key features of groupware products that aid group editing, distributing, and sharing of documents (Ellis, Gibbs, and Rein, 1991). All this places high demands for robustness and responsiveness as well as data replication on groupware products.

5.4.4   World Wide Web (WWW)

The Internet is a platform that is ideally suited for collaboration not only because of its geographical expanse but also because of its open nature, which allows users from around the globe to communicate and interact with each other without the need to have specific, proprietary end-user applications (Barua, Chellappa, and Whinston, 1997). This is made possible by the TCP/IP[3] protocol and a suite of other communication protocols built on top of this stack. This holds true for the Internet's WWW as well.

In addition to open access, the Web provides an extremely high degree of media richness. It is the first of the Internet applications to offer multimedia and hypertext capability on a seamless basis, made possible by its client–server architecture. The "document" in this case includes hypertext links to other information resources and in-line images that are part of the document itself.

The advancements the Web includes—global geographical scope, high degree of media richness, nonproprietary open system, client–server architecture—all combine to bring great opportunities for collaboration on a global basis. The Web takes a revolutionary step in the direction of the ultimate goal of worldwide information dissemination and interaction with the ability to link information resources anywhere in the world. Although we are closer than ever to this goal with the help of WWW, two major issues still need

---

[3] A protocol refers to "language" that computers use to speak to each other. Transmission control protocol/Internet protocol (TCP/IP) defines a set of rules or guidelines that allow machines with a unique IP (Internet protocol) number to recognize each other. Any application that requires communication with other machines on such a network needs to have this protocol implemented in it.

to be resolved: interactivity and information organization. For the WWW to be an electronic forum for productive interaction, interactive capabilities will need to be integrated with the information repositories distributed over the Internet. Caution also needs to be taken that the isolated and distinctive developments of Web servers will quickly lead to information overload, inefficiency, and chaos. Both increased interactive ability and improved search capabilities will be the focus of innumerable development efforts that will undoubtedly make the Web even more conducive to successful group collaboration.

The Web makes it possible to customize a system to meet specific organizational requirements or business needs without the restraints of proprietary standards and protocols. Unlike a proprietary system or application, which provides only a few customization choices to users through limited application programming interfaces, an open system such as the Internet provides a wide range of design choices, all of which require careful consideration on the part of the developers.

REFERENCES

Albus, J. 1982. "Industrial Robot Technology and Productivity Improvement." In *Exploration Workshop on the Social Impact of Robotics,* OTA. U.S. Congress Number 90-240 0-82-2.

Baber, C., and J. M. Noyes, eds. 1993. *Interactive Speech Technology: Human Factors Issues in the Application of Speech Input/Output to Computers.* Bristol: Taylor & Francis.

Baecker, R. M. 1990. *Human Factors and Typography for More Readable Program.* Reading: Addison-Wesley.

Barnett, B. J., and C. D. Wickens. 1988. "Display Proximity in Multicue Information Integration: The Benefit of Boxes." *Human Factors* **30**, 15–24.

Barua, A., R. Chellappa, and A. B. Whinston. 1997. "Social Computing: Computer Supported Cooperative Work and Groupware." In *Handbook of Human Factors and Ergonomics.* Ed. G. Salvendy. New York: John Wiley & Sons. 1760–1782.

Benbasat, I., and Y. Wand. 1984. "Command Abbreviation Behavior in Human–Computer Interaction." *Communications of the ACM* **27**, 376–383.

Bennett, K. B., M. L. Toms, and D. D. Woods. 1993. "Emergent Features and Graphical Elements: Designing More Effective Configural Displays." *Human Factors* **35**, 71–97.

Billingsley, P. A. 1988. "Taking Panes: Issues in the Design of Windowing Systems." In *Handbook of Human–Computer Interaction.* Ed. M. Helander. Amsterdam: Elsevier Science.

Black, J., and T. Moran, 1982. "Learning and Remembering Command Names." In *Proceedings of the Conference on Computer Human Interaction (CHI '82).* New York: ACM. 8–11.

Boehm-Davis, D. A., and A. Fregly. 1985. "Documentation of Concurrent Programs." *Human Factors* **27**(4), 423–432.

Borman, W. C., and N. G. Peterson. 1982. "Selection and Training of Personnel." In *Handbook of Industrial Engineering.* Ed. G. Salvendy. New York: John Wiley & Sons.

Bullinger, H., P. Kern, and M. Braun. 1996. "Controls." In *Handbook of Human Factors and Ergonomics.* Ed. G. Salvendy. New York: John Wiley & Sons. 697–728.

Card, S. K., T. P. Moran, and A. Newell. 1983. *The Psychology of Human–Computer Interaction.* Hillsdale: Erlbaum.

Carroll, J. M. 1984. "Minimalist Training." *Datamation* **30**, 125–136.

Carroll, J. M., and J. Thomas. 1982. "Metaphor and the Cognitive Representation of Computing Sytems." *IEEE Transactions on Systems, Man, and Cybernetics* **SMC-12**, 107–115.

Carroll, J. M., R. L. Mack, and W. A. Kellogg. 1988. "Interface Metaphors and the User Interface Design." In *Handbook of Human–Computer Interaction.* Ed. M. Helander. Amsterdam: Elsevier. 67–85.

Carswell, C. M., and C. D. Wickens. 1987. "Information Integration and the Object Display." *Ergonomics* **30**, 511–527.

Chernoff, H. 1973. "The Use of Faces to Represent Points in k-Dimensional Space Graphically." *Journal of the American Statistical Association* **68**, 361–368.

Christ, R. E. 1975. "Review and Analysis of Color Coding Research for Visual Displays." *Human Factors* **17**, 542–570.

Chu, Y. Y., and W. B. Rouse. 1979. "Adaptive Allocation of Decision Making Responsibility Between Human and Computer in Multitask Situations." *IEEE Transactions on Systems, Man, and Cybernetics* **SMC-9**(12), 769–778.

Coekin, J. A. 1969. "A Versatile Presentation of Parameters for Rapid Recognition of System State." In *International Symposium on Man–Machine Systems,* IEE Conference Record No. 69, 58-MMS, Vol. 4. Cambridge: IEE.

Coleman, D. 1995. "An Overview of Groupware." In *Groupware: Technology and Applications.* Ed. D. Coleman and R. Khanna. Englewood Cliffs: Prentice-Hall.

Crown, J. 1992. *Effective Computer User Documentation.* New York: Van Nostrand Reinhold.

DeSanctis, G. L., and R. B. Gallupe. 1985. "Group Decision Support Systems: A New Frontier." *Data Base* **16**, 3–9.

Eberts, R. 1994. *User Interface Design*. Englewood Cliffs: Prentice-Hall.

Ellis, C., S. Gibbs, and G. Rein. 1991. "Groupware: Some Issues and Experiences. (Using Computers to Facilitate Human Interaction)." *Communications of the ACM* **34**(1), 38–58.

Ellis, S. R., ed. 1991. *Pictorial Communication in Virtual and Real Environments*. London: Taylor & Francis.

Fels, S., and G. Hinton. 1995. "Glove-Talk II: An Adaptive Gesture-to-Formant Interface." In *Proceedings of the Conference on Computer Human Interaction* (CHI '95). New York: ACM. 456–463.

Galitz, W. O. 1989. *Handbook of Screen Format Design*. 3rd ed. Wellesley: Q.E.D. Information Sciences.

Hasegawa, Y. 1982. "How Robots Have Been Introduced into the Japanese Society." Paper presented at the Micro-electronics International Symposium, Osaka, Japan, August 17–19.

Henderson, A., and S. K. Card. 1986. "Rooms: The Use of Multiple Virtual Workspaces to Reduce Space Contention in a Window-Based Graphical User Interface." *ACM Transactions on Graphics* **5**(3), 211–243.

Hick, W. E. 1952. "On the Rate of Gain of Information." *Quarterly Journal of Experimental Psychology* **4**, 11–26.

Hoffert, E. M., and G. Gretsch. 1991. "The Digital News System at EDUCOM: A Convergence of Interactive Computing, Newspapers, Television and High-Speed Networks." *Communications of the ACM* **34**, 113–116.

Isa, B. S., et al. 1983. "A Methodology for Objectively Evaluating Error Messages." In *Proceedings of the Conference on Computer Human Interaction (CHI '83)*. New York: ACM. 68–71.

Johnson, B., et al. 1986. "Using a Computer-Based Tool to Support Collaboration: A Field Experiment." In *Proceedings of the Conference on Computer-Supported Cooperative Work (CSCW '86)*. 343–353.

Kantowitz, B. H., and J. L. Knight. 1976. "Testing Tapping Time-Sharing: II. Auditory Secondary Task." *Acta Psychologica* **40**, 343–362.

Kearsley, G. 1988. *Online Help Systems: Design and Implementation*. Norwood: Ablex.

Kieras, D. E. 1988. "Towards a Practical GOMS Model Methodology for User Interface Design." In *Handbook of Human–Computer Interaction*. M. Helander. Amsterdam: Elsevier Science. 135–157.

Kieras, D. E., and P. G. Polson. 1985. "An Approach to the Formal Analysis of User Complexity." *International Journal of Man–Machine Studies* **22**, 365–394.

Laudauer, T. K., K. M. Calotti, and S. Hartwell. 1983. "Natural Command Names and Initial Learning." *Communications of the ACM* **23**, 556–563.

Lewis, L., and K. Keleman. 1990. "Experiences with GDSS Development: Lab and Field Studies." *Journal of Information Science* **16**, 195–205.

Liu, Y. 1997. "Software-User Interface Design." In *Handbook of Human Factors and Ergonomics*. Ed. G. Salvendy. New York: John Wiley & Sons. 1689–1724.

Mangold, V. 1981. "The Industrial Robot as Transfer Device." *Robotics Age* (July/August), 20–26.

Marcus, A. 1992. *Graphic Design for Electronic Documents and User Interfaces*. New York: ACM Press.

Moran, T. P. 1981. "The Command Language Grammar." *International Journal of Man–Machine Studies* **15**, 3–50.

Murray, J., and Y. Liu. 1995. "Towards a Distributed Intelligent Agent Architecture for Human, Machine Systems in Hortatory Operations." Tech. Rep. 95-17. Ann Arbor: Department of Industrial and Operations Engineering, University of Michigan.

Myers, B. A. 1991. "Demonstrational Programming: A Step Beyond Direct Manipulation." In *People and Computers* VI. Ed. D. Draper and N. Hammond. Cambridge: Cambridge University Press. 11–30.

Nielsen, J. 1990. *Hypertext and Hypermedia*. New York: Academic Press.

———. 1994a. *Usability Engineering*. Boston: AP Professional.

———. 1994b. "Usability Laboratories." *Behaviour and Information Technology* **13**(1, 2), 3–8.

———. 1997. "Usability Testing." In *Handbook of Human Factors*. New York: John Wiley & Sons. 1543–1568.

Nof, S. Y., J. L. Knight, and G. Salvendy. 1980. "Effective Utilization of Industrial Robots—A Job and Skills Analysis Approach." *AIIE Transactions* **12**(3), 216–225.

Norman, D. A. 1983. "Some Observations on Mental Modes." In *Mental Models*. Ed. D. Gentner and A. L. Stevens. Hillsdale: Erlbaum.

———. 1986. "Cognitive Engineering." In *User Centered System Design*. Ed. D. A. Norman and S. W. Draper. Hillsdale: Erlbaum.

Norman, K. L. 1991. *The Psychology of Menu Selection: Designing Cognitive Control of the Human–Computer Interface*. Norwood: Ablex.

Office of Technology Report. 1982. *Information Technology and Its Impact on American Education.* U.S. Government Printing Office, GPO Stock No. 052-003-00888-2.

Okoshi, T. 1980. "Three-Dimensional Displays." *Proceedings of the IEEE* **68**(5), 548–564.

Payne, S. J., and T. R. G. Green. 1986. "Task-Action Grammars: A Model of the Mental Representation of Task Languages." *Human–Computer Interaction* **2**(2), 93–133.

Pickover, C. A., and S. K. Tewksbury. 1994. *Frontiers of Scientific Visualization.* New York: John Wiley & Sons.

Pomerantz, J. R. 1981. "Perceptual Organization in Information Processing." In *Perceptual Organization.* Ed. M. Kubovy and J. R. Pomerantz. Hillsdale: Erlbaum. 141–180.

Rasmussen, J. 1985. "The Role of Hierarchical Knowledge Representation in Decision Making and System Management." *IEEE Transactions on Systems, Man, and Cybernetics* **SMC-15**(2), 234–243.

Rasmussen, J., A. M. Pejtersen, and L. P. Goodstein. 1994. *Cognitive Systems Engineering.* New York: John Wiley & Sons.

Reisner, P. 1981. "Formal Grammar and Human Factors Design of an Interactive System." *IEEE Transactions on Software Engineering* **SE-7**(2), 229–240.

Rivlin, E., R. Botafogo, and B. Shneiderman. 1994. "Navigating in Hyperspace: Designing a Structure-Based Toolbox." *Communications of the ACM* **37**(2), 87–96.

Rizk, A., N. Streitz, and J. Andre, eds. 1990. *Hypertext: Concepts, Systems and Applications.* Cambridge: Cambridge University Press.

Rosenblum, L. J., ed. 1994. *Scientific Visualization: Advances and Challenges.* London: Academic Press.

Rosenbrock, H. H. 1982. "Robots and People." *Measurement and Control* **15,** 105–112.

Rouse, W. B. 1977. "Human–Computer Interaction in Multitask Situations." *IEEE Transactions on Systems, Man, and Cybernetics* **SMC-7**(5), 384–392.

Salvendy, G. 1978. "An Industrial Dilemma: Simplified Versus Enlarged Jobs." In *Production and Industrial Systems.* Ed. R. Murumatsu and N. A. Dudley. London: Taylor & Francis.

Salvendy, G., and W. D. Seymour. 1973. *Prediction and Development of Industrial Work Performance.* New York: John Wiley & Sons. 105–125.

Sassoon, R. 1993. *Computers and Typography.* Oxford: Intellect.

Schmandt, C. 1994. *Voice Communication with Computers: Conversational Systems.* New York: Van Nostrand Reinhold.

Senker, P. 1979. "Social Implications of Automation." *The Industrial Robot* **6**(2), 59–61.

Sheridan, T. 1984. "Supervisory Control of Remote Manipulations, Vehicles and Dynamic Processes." In *Advances in Man–Machine Systems Research.* Vol. 1. Ed. W. B. Rouse. New York: JAI Press. 49–137.

Shneiderman, B. 1983. "Direct Manipulation: A Step Beyond Programming Languages." *IEEE Computer* **16**(8), 57–69.

———. 1992. *Designing the User Interface: Strategies for Effective Human-Computer Interaction.* 2nd ed. Reading: Addison-Wesley.

Shneiderman, B., and G. Kearsley, 1989. *Hypertext Hands-On! An Introduction to a New Way of Organizing and Accessing Information.* Reading: Addison-Wesley.

Smith, M. J., and P. C. Sainfort. 1989. "A Balance Theory of Job Design for Stress Reduction." *International Journal of Industrial Ergonomics* **4**, 67–79.

Smith, M. J., et al. 1992. "Human–Computer Interaction." In *Handbook of Industrial Engineering.* 2nd ed. Ed. G. Salvendy. New York: John Wiley & Sons. 1107–1144.

Smith, S. L., and J. N. Mosier. 1986. "Guidelines for Designing User Interface Software." Report ESDTR-86-278. Bedford, Electronic Systems Division, the MITRE Corp. Available from National Technical Information Service, Springfield, Virginia.

Strathmeyer, C. R. 1990. Voice in Computing: An Overview of Available Technologies. *IEEE Computer* **23**(8), 10–16.

Sugarman, R. 1980. "The Blue Collar Robot." *IEEE Spectrum* **17**(9), 52–57.

Thorell, L. G., and W. J. Smith. 1990. *Using Computer Color Effectively.* Englewood Cliffs: Prentice-Hall.

Tufte, E. R. 1983. *The Visual Display of Quantitative Information.* Cheshire: Graphics Press.

———. 1990. *Envisioning Information.* Cheshire: Graphics Press.

Tulga, M. K., and T. B. Sheridan. 1980. "Dynamic Decisions and Work Load in Multitask Supervisory Control." *IEEE Transactions on Systems, Man, and Cybernetics* **SMC-10**(5), 217–232.

Tullis, T. S. 1981. "An Evaluation of Alphanumeric, Graphic and Color Information Displays." *Human Factors* **23**, 541–550.

———. 1988. "Screen Design." In *Handbook of Human–Computer Interaction.* Ed. M. Helander. Amsterdam: Elsevier.

Vicente, K. J., and J. Rasmussen. 1992. "Ecological Interface Design: Theoretical Foundations." *IEEE Transactions on Systems, Man, and Cybernetics* **SMC-22**(4), 589–606.

Walden, R. S., and W. B. Rouse. 1978. "A Queueing Model of Pilot Decision Making in a Multitask Flight Management Task." *IEEE Transactions on Systems, Man, and Cybernetics* **SMC-8**(12), 867–875.

Walker, J., L. Sproull, and R. Subramani. 1994. "Using a Human Face in an Interface." In *Proceedings of the Conference on Computer Human Interaction (CHI '94)*. New York: ACM. 85–91.

Wechsler, D. 1952. *The Range of Human Capabilities.* 2nd ed. Baltimore: Williams & Wilkins.

Weekley, T. L. 1979. "A View of the United Automobile Aerospace and Agricultural Implement Workers of America (UAW) Stand on Industrial Robots." SME Technical Paper MS79-776, Dearborn, Michigan.

Wickens, C. D. 1992. *Engineering Psychology and Human Performance.* 2nd ed. New York: HarperCollins.

Wickens, C. D., and A. D. Andre. 1990. "Proximity Compatibility and Information Display: Effects of Color, Space, and Objectness on Information Integration." *Human Factors* **32**, 61–78.

Wickens, C., and Y. Liu. 1988. Codes and Modalities in Multiple Resources: A Success and a Qualification." *Human Factors* **30**(5), 599–616.

Wickens, C. D., D. Sandry, and M. Vidulich. 1983. "Compatibility and Resource Competition Between Modalities of Input, Output, and Central Processing." *Human Factors* **25**, 227–248.

Wickens, C. D., S. Todd, and K. Seidler. 1989. "Three-Dimensional Displays: Perception, Implementation, and Applications." CSERIAC Rep. CSERIAC-SOAR-89-001, CSERIAC AAMRL, Wright-Patterson Air Force Base, Ohio.

Wiener, E. L. 1982. "Human Factors of Advanced Technology." (*"Glass Cockpit"*) *Transport Aircraft* (NASA Contractor Report No. 177528), Moffett Field: NASA–Ames Research Center.

Williges, R. C., B. H. Williges, and J. Elkerton. 1987. "Software Interface Design." In *Handbook of Human Factors*. New York: John Wiley & Sons. 1416–1449.

Woods, D. D., J. Wise, and L. Hanes. 1981. "An Evaluation of Nuclear Power Plant Safety Parameter Display System." In *Proceedings of the 25th Annual Meeting of the Human Factors Society*. Santa Monica: Human Factors Society.

CHAPTER **34**

JUSTIFICATION OF ROBOTICS SYSTEMS

John J. Mills
G. T. Stevens
Brian Huff
The University of Texas at Arlington
Fort Worth, Texas

Adrien Presley
Truman State University
Kirksville, Missouri

1 INTRODUCTION

The adoption of advanced technologies such as robotics and flexible manufacturing systems is widely seen as a key to continued competitiveness in the world market (Schonberger, 1990). When properly selected and defined, such technologies offer substantial potential for cost savings, flexibility, product consistency, and improved throughput. However, justification of these technologies is difficult when they are judged solely on traditional economic criteria. The lack of consideration of their strategic and long-term benefits has led to the failure of manufacturers to adopt these technologies (Kaplan, 1986; Zald, 1994; Davale, 1995). Too often these benefits are ignored, leading to poor decisions on the technologies to implement. The long-range cost of not automating can be much greater than the short-term cost of acquiring an automation technology (Badiru, 1991).

An important point that needs to be made early in the justification of any robotics system is that the objective of any robotics system project is not to emulate existing methods and systems, simply replacing humans with robots, but to develop a new, integrated system providing financial, operational, and strategic benefits. Financial benefits include decreased labor costs. Although investment in a robotics system project is similar to other capitalized equipment projects, four major differences stand out.

1. Robots can provide such flexibility in production capability that the capacity of a company to respond effectively to future market changes has a clear economic value, though this capacity is usually difficult to measure.

2. As components of computerized production systems, robots force their users to rethink and systematically define and integrate the functions of their operation. This in itself carries major economic benefits and frequently lets a company "clean up its act."

3. A robot is by definition reprogrammable and reusable and has a useful life that can often be longer than the life of a planned production facility.

4. In addition to reducing direct labor and benefit costs, using robots can also significantly reduce requirements for employee services and facilities.

These differences lead to operational benefits that include:

Handbook of Industrial Robotics, Second Edition, Edited by Shimon Y. Nof
ISBN 0-471-17783-0 © 1999 John Wiley & Sons, Inc.

1. Flexibility
2. Increased productivity
3. Reduced operating costs
4. Increased product quality
5. Elimination of health and safety hazards
6. Higher precision
7. Ability to run longer shifts
8. Reduced floor space

Strategic benefits are derived from new, emerging strategic philosophies that change the way companies approach the marketplace. The majority of automated production systems deployed over the last three decades have been designed to meet the needs of a specific product or a narrowly defined family of products. The economic justification for the implementation of these systems has been based on the anticipated cost savings attributable to the use of automated production methods versus traditional manual production techniques. Over the past decade, however, major changes have occurred in the consumer goods marketplace. Time-based competition and mass customization are emerging as key competitive strategies (Goldman). The average life of a product in the marketplace has fallen from several years to as little as six months for products based on rapidly evolving technology.

Product–centric system development strategies are failing to develop automated production systems that truly have the agility required to support the production needs of a continuously evolving family of products. To create these agile automated systems a new system development strategy is required that incorporates a more systemic view of the automated production environment. This new strategy focuses on the development of common automated manufacturing processes and a modular set of system building blocks that can be rapidly deployed in different combinations and permutations. The class of system that emerges from this new system development strategy is referred to as *reconfigurable automation*. Reconfigurable systems built from these core automation and process technologies have demonstrated their ability to support new production requirements rapidly.[1]

Justification of this new class of automation must be determined from a strategic perspective. Many automated systems developed under the old product–centric paradigm have failed economically due to the fact that there was not enough demand for the product(s) the system was designed to produce. These product-specific systems have generally lacked the ability to be reconfigured at a cost that would allow them to meet the needs of additional products. As a consequence, these systems are typically decommissioned well before their capital cost can be recovered and are held idle in storage until they are fully depreciated for tax purposes. Once the full tax depreciation allowance has been taken, they are sold as salvage at a small fraction of their original development cost. It can be argued that if the systems were designed to be reconfigurable, their economic life could be extended over multiple product life cycles. This argument, however, must be tempered by the fact that there is a price associated with providing system reconfigurability. To support rapid product changeover, quick change tooling is generally required. Module interfaces must be designed and implemented. Generic system capabilities must be built into the common system building blocks, which may not provide any additional benefit to a specific product. It is estimated that these features can increase the cost of reconfigurable system hardware by as much as 25% over that of a comparable dedicated system. In addition, the software required to configure and run this class of automation can be much more expensive to develop than simple part-specific programs.

Traditional economic analysis tools have hindered the adoption of this new reconfigurable systems technology because they fail to explicitly consider the economic benefit of reusing capital over multiple projects and do not recognize the strategic benefits of the technology. Because we have traditionally justified the purchase of robotics on a product-by-product basis, the less expensive dedicated robotic systems have obviously emerged as the more attractive options. This failure to explicitly consider the economic benefit of

---

[1]See Sony Factory Automation, the Smart Cell Technology on the World Wide Web, URL = ⟨http://www.sony.co.jp/ProductsPark/Professional/FA/⟩.

capital reuse has become a barrier to the adoption of reconfigurable automation. Modifications and extensions to traditional cost comparison techniques are required to help recognize the short-term economic and long-term strategic value of investing in agile robotic technologies that can be reused to support the production requirements of an evolving family of products.

In this chapter we first address the traditional economic justification approach to robotic system justification. We then address new approaches to robotic manufacturing system justification based on strategic considerations. Finally we discuss some of the justification approaches currently being researched for justifying reconfigurable systems.

2 GENERAL PROCEDURE FOR EVALUATING TRADITIONAL ROBOT APPLICATIONS

Economic evaluation provides the decision framework to compare the financial benefits of automation through robotics with the present system and with other alternatives. The economic justification is based on the comparison between the capital cost and operating expenses of the robot installation being considered and the cash flow benefits projected. The purpose of this section is to describe a procedure for economic evaluation and justification of proposed robotic system projects.

A primary decision issue is whether a robot is indeed the best solution for a particular application. The justification used by companies that have applied robots generally follows the eight benefit areas listed above. The results of recent surveys on such justification factors are found in Table 1.

2.1 General Procedure and Robotic Systems Project Evaluation and Justification—Justifing Industrial Robot Applications

Figure 1 summarizes the series of steps used to evaluate fully the economics of a robotic system project. The step numbers presented in the figure correspond to the detailed discussion that follows. The evaluation and analysis of a robotic system project can be divided into two phases: precost and cost analysis. An example is provided in the cost analysis phase.

2.2 Precost Planning Analysis Phase

Prior to any thorough economic evaluation of capitalized equipment, several initial considerations and planning studies must be carried out. The first six steps include determining the best manufacturing method, selecting the best jobs to automate with robots, and the feasibility of these options. Noneconomic considerations must be studied as well as pertinent data collected concerning product volumes and operation times. The precost phase focuses basically on the feasibility of a proposed robotic system project: feasibility in terms of technical capability to perform the necessary job and in terms of production capacity and utilization relative to predicted production schedules. This phase follows the first six steps outlined in Figure 1.

Table 1   Justification for Using Robots as Found in Companies That Are Using Robots, Ranked According to Priority[a]

Rank	U.S. Companies	Japanese Companies[b]
1	Reduced labor cost	Economic advantage
2	Improved product quality	Increased worker safety
3	Elimination of dangerous jobs	Universalization of production system
4	Increased output rate	Stable product quality
5	Increased product flexibility	Labor shortage
6	Reduced material waste	
7	Compliance with OSHA regulations	
8	Reduced labor turnover	
9	Reduced capital cost	

[a] From S. Y. Nof, "Decision Aids for Planning Industrial Robot Operations," in *Proceedings of the 1987 Annual Industrial Engineering Conference, Institute of Industrial Engineers.*
[b] Survey in Japan consisted of only five categories.

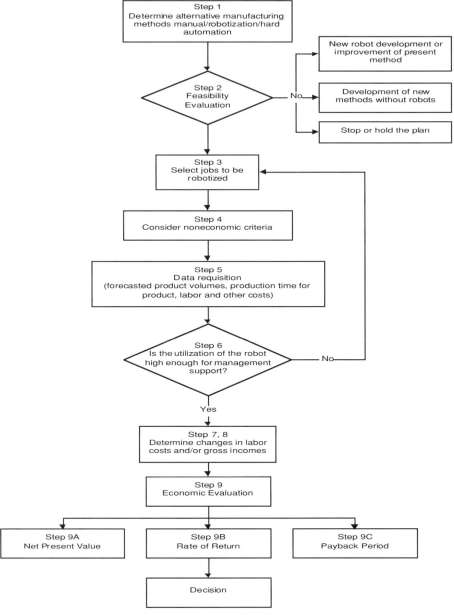

Figure 1 Flowcart of project economic evaluation procedure. (From Y. Hasegawa, *Handbook of Industrial Robotics.* New York: John Wiley & Sons, 1985. Chapter 33.)

Step 1 Determination of Afternative Manufacturing Methods

The three main alternative manufacturing methods, manual labor, flexible automation and robots, and hard automation, are compared in general in Figure 2 (Hasegawa, 1985). These alternatives are economically compared by their production unit cost at varying production volumes. For low volumes manual labor is usually most cost-effective, but new approaches to reconfigurable assembly are slowly changing this situation. Flexible, programmable automation and robots are most effective for medium production volumes.

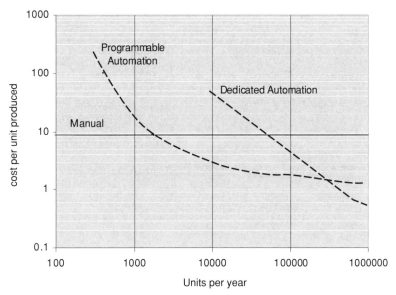

Figure 2 Comparison of manufacturing methods for different production volumes. (From Y. Hasegawa, *Handbook of Industrial Robotics.* New York: John Wiley & Sons, 1985. Chapter 33.)

These medium volumes can range, depending on the particular products, from a few tens or hundreds of products per year per part type to hundreds of thousands of products per year. For annual volumes of 500,000 and above, hard automation is usually preferred.

In the area of assembly Boothroyd, Poli, and Murch (1982) have derived specific formulas for the assembly cost (Table 2). Using these formulas they have compared several alternative assembly systems, including an operator assembly line, an assembly center with two arms, a universal assembly center, a free-transfer machine with programmable workheads, and a dedicated machine. The free-transfer machine with programmable workheads, the assembly center with two arms, and the universal assembly system are robotic systems. Effects of the number of parts per assembly, annual volume, and style/design variations have been studied. They have derived formulas for each of these types of machines and assumed certain costs. The generic formula derived by them is:

$$C_{pr} = t_{pr}(W_t + WM'/SQ)$$

The parameters in this expression have different relationships, depending on the type of assembly machine.

As an example, Figures 3 and 4 compare the assembly costs for these systems versus volume for one product with 50 parts against those for 20 products, each with 50 parts. Note that the "universal assembly machine" is a hypothetical one used by the authors as an example of what might be.

Note that in this example the addition of multiple products has increased the costs of the assembly center with two arms and the free transfer machine with programmable workheads by ~100% and the dedicated hybrid machine by 1000%.

Step 2 Technical Feasibility Study

It is important to check the feasibility of the robotic system plan carefully. There have been some cases that have passed (by mistake) the economic justification but still had a problem of feasibility. A possible reason that there have been many failures in robotic system plans is that the robotic system system design includes many complicated conditions. These conditions are usually more complex than the conventional production systems design and are not yet fully understood.

In a feasibility study of the alternatives the following items must be considered:

Table 2　Summary of Equations Used in the Comparison of Assembly Systems Costs

Assembly System	t_{pr}	W_t	M'
1. Operator assembly line no feeders	$kt_0(1 + x)$	nW/k	$(n/k)(2C_B + N_pC_c)$
2. Operator assembly line with feeders	$kt_0'(1 + x)$	nW/k	$(n/k)(2C_B + N_pC_c) + N_p(ny + N_dC_F)$
3. Dedicated hybrid machine	$t + xiT$	$3W$	$[nyC_T + (T/t)(ny/i)C_B] + N_p\{(ny + N_D)(C_F + C_W) + [ny + (T/2t)(ny/i)]C_C\}$
4. Free-transfer machine with programmable workheads	$k(t + xT)$	$3W$	$(n/k)[C_{dA} + (T/t + 1)C_B + N_p[(ny + N_D)C_M + nC_g + (n/k)(T/2t + 0.5)C_c]$
5. Assembly center with two arms	$n(t/2 + xT)$	$3W$	$2C_{6A} + N_p[C_c + nC_g + (ny + N_d)C_M]$
6. Universal assembly center	$n(t/2 + xT)$	$3W$	$(2C_{3A} + nyC_{PF} + 2C_{ug}) + N_pC_c$

C_{pr} = product unit assembly cost
S = number of shifts
Q = equivalent cost of operator in terms of capital equivalent
W = operator's rate in dollars per sec
d = degree of freedom
k = number of parts assembled by each operator or programmable workhead
C_{dA} = cost of a programmable robot or workhead
C_B = cost of transfer device per workstation
C_c = cost of a work carrier
C_F = cost of an automatic feeding device
C_g = cost of a gripper per part
C_M = cost of a manually loaded magazine
C_{PF} = cost of a programmable feeder
C_s = cost of a workstation for a single station assembly
C_T = cost of a transfer device per workstation
C_{ug} = cost of a universal gripper
C_W = cost of a dedicated workhead
T = machine downtime for defective parts
t_0 or t_0' = machine downtime due to defective parts
t = mean time of assembly for one part
x = ratio of faulty parts to acceptable parts

1. Is the product designed for robot-handling and automated assembly?
2. Is it possible to do the job with the planned procedure?
3. Is it possible to do the job within the given cycle time?
4. Is it possible to ensure reliability as a component of the total system?
5. Is the system sufficiently staffed and operated by assigned engineers and operators?
6. Is it possible to maintain safety?
7. Is it possible to keep the designed quality level?
8. Can inventory be reduced?
9. Can material handling be reduced in the plant?
10. Are the material-handling systems adequate?
11. Can the product be routed in a smooth batch-lot flow operation?

The alternatives that have passed the feasibility screening are moved to the next level of evaluation. But if the plan is not passed, as shown in Figure 1, there should be a search

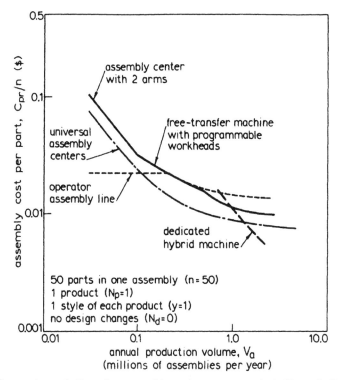

Figure 3 Comparison of alternative assembly systems, one product. (From G. Boothroyd, C. Poli, and L. E. March, *Automatic Assembly.* New York: Marcel Dekker, 1982.)

for other solutions, such as developing a new robot, improving the proposed robot, and developing other alternatives without robots.

Step 3   Select Which Job to Automate with Robots

Job selection for a single robot or a group of robots is a difficult task. In general the following five job grouping strategies can be used to determine feasible job assignments.

1. Component products belonging to the same product family
2. Products presently being manufactured near each other
3. Products that consist of similar components and could share part-feeding devices
4. Products that are of similar size, dimensions, weight, and number of components
5. Products with a rather simple design that can be manufactured within a short cycle time

Step 4   Noneconomic and Intangible Considerations

Several noneconomic issues should be addressed in regard to specific company characteristics, company policy, social responsibility, and management's direction. While a methodology to account for these is presented in Section 3, they are important even if that methodology is not used. Justification of robotic systems needs to consider such questions as:

1. Will the robotic system meet the general direction of the company's automation?
2. Will the robotic system meet the fundamental policy of the standardization of equipment and facilities?
3. Will the plan be able to meet future product model change or production plan change?

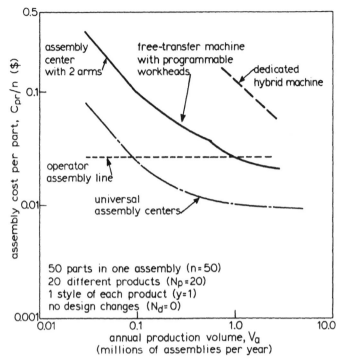

Figure 4 Comparison of alternative assembly systems, 20 products. (From G. Boothroyd, C. Poli, and L. E. March, *Automatic Assembly*. New York: Marcel Dekker, 1982.)

4. Will the plan promote improved quality of working life for and raise the morale of workers?
5. Will the plan influence a good company reputation?
6. Will the plan promote technical progress of the company?

Table 3 lists several other difficult-to-quantify benefits associated with robot applications. Considerations of such intangibles can only be done with a different methodology than that presented here. Such a strategic methodology is presented in Section 3.

Several special differences between robots and other capitalization equipment also provide numerous intangible benefits.

1. Robots are reusable.
2. Unlike hard automation, robots are multipurpose and can be reprogrammed for many different tasks.
3. Because of reprogrammability, the useful life of the robotic system can often be three or more times longer than that of fixed (hard) automation devices.
4. Tooling costs for robotic systems also tend to be lower owing to the programming capability around certain physical constraints.
5. Production operations can often be started up much sooner because of the lesser construction and tooling constraints.
6. Modernization in the plant can be implemented by eliminating discontinued automation systems.

Questions often arise concerning long-range unmeasurable effects of robotic system on economic issues. A few such issues include:

1. Will the robotic system raise product value and price?

Table 3 Difficult-to-Qualify Benefits of Robotization[a]

Robotization Can Improve	Robotization Can Reduce or Eliminate
Flexibility	Hazardous, tedious jobs
Plant modernization	Safety violations and accidents
Labor skills of employees	Personnel costs for training
Job satisfaction	Clerical costs
Methods and operations	Cafeteria costs
Manufacturing productivity capacity	Need for restrooms, need for parking spaces
Reaction to market fluctuations	Burden, direct, and other overhead costs
Product quality	Manual material handling
Business opportunities	Inventory levels
Share of market	Scrap and errors
Profitability	New product launch time
Competitive position	
Growth opportunities	
Handling of short product life cycles	
Handling of potential labor shortages	
Space utility of plant	
Level of management	

[a] Analyzing the amount of change in each of these categories in response of robotization and assigning quantitative values to the intangible factors is necessary if they are to be included in the financial analysis (see Section 2). Otherwise they can only be used as weighting factors when determining the best alternative.

2. Will the robotic system expand the sales volume?
3. Will the plan decrease the production cost?
4. Will the robotic system decrease the initial investment amount?
5. Will the robotic system reduce lead time for products?
6. Can manufacturing costs be reduced?
7. Can inventory costs be reduced?
8. Will robotic system reduce direct and indirect labor costs or just shift workers' skills?
9. Can the burden (overhead) rate be reduced?
10. Will the robot be fully utilized?
11. Will setup time and costs be reduced?
12. Can material-handling costs be reduced?
13. Will damage and scrap costs be reduced?

Step 5 Determination of Costs And Benefits

To best proceed further with the methodology of determining costs and benefits, we have provided an example using realistic numbers. In order to determine the economic desirability of a robot installation, the following data are required:

1. The cost of the robot installation (capital investment)
2. Estimated (calculated) changes in gross incomes (revenue, sales, savings) and costs resulting from the robot installation

For this example the installed cost operation costs and salvage are given in Table 4. While the numbers in Table 4 are rather low, they are solely for illustration of the method and should not be taken as typical of a robotic system's cost.

Step 6 Decisions Concerning Future Applications

Underutilized robots cannot be cost-justified, owing to the high initial startup expenses and low labor savings. Additional applications or planned future growth are required to

Table 4   Data for Numerical Example

Project Costs	
Purchase price of robot	$40,000
Cost of end-effector	$10,000
Software integration	$20,000
Cost of part feeders	$10,000
Installation cost	$ 4,000
Total	$84,000
Actual realizable salvage	$ 4,000

drive the potential cost-effectiveness up; however, there is also an increase in tooling and feeder costs associated with each new application.

2.3   Cost Analysis Phase

Step 7   Period Evaluation, Depreciation, and Tax Data Requirements

Before the economic evaluation can be made, determination of the evaluation period, tax rates, and tax depreciation method must be made. In this example the evaluation period is six years. This example uses the United States Internal Revenue Service's MACRS (five years). Therefore the MACRS percent ages used to determine the yearly tax depreciation amounts are those given in Table 5.

The tax rate used in this example is 40%. These values are, of course, not fixed and can be changed for a particular application if deemed appropriate.

Table 5   MACRS Percentages

Year	Percentage
1	20.00
2	32.00
3	19.20
4	11.52
5	11.52
6	5.76

Step 8   Project Cost Analysis

This phase focuses on detailed cost analysis for investment justification and includes five general steps.

The project cost was given in Table 4 ($84,000). It is now necessary to determine (estimate) the yearly changes in operations cost and cost savings (benefits). For this example the changes are shown in Table 6.

Table 6   Costs and Savings ($/Year)

Year	1	2	3	4	5	6
Labor savings	40,000	40,000	50,000	50,000	50,000	50,000
Quality savings	9,000	9,000	10,000	10,000	10,000	10,000
Operating costs (increase)	(20,000)	(20,000)	(15,000)	(10,000)	(10,000)	(10,000)

Step 9   Evaluation Techniques

The techniques used for the economic analysis of robotic applications are similar to those for any manufacturing equipment purchase and use present value rate of return or payback (payout) methods. All of these methods require the determination of the yearly net cash flows, which are defined as

$$X_j = (G-C)_j - (G-C-D)_j(T) - K + L_j$$

where

X_j = net cash flow in years j
G_j = gross income (savings, revenues) for year j
C_j = the total costs for year j exclusive of book (company depreciation and debt interest)
D_j = tax depreciation for year j
T = tax rate (assumed constant)
K = total installed cost of the project (capital expenditure)
L_j = salvage value in year j

The net cash flows are given in Table 7. Some sample calculations follow:

$$X_0 = -84,000$$

$$X_1 = (49,000 - 20,000) - (49,000 - 20,000 - 16,800)*(.4)$$

$$= \$24,120$$

$$X_6 = (60,000 - 10,000) - (60,000 - 10,000 - 4838)*(.4) + 4,000$$

$$= \$35,935$$

Table 7 Net Cash Flow

End of Year	K&L	Total G[a]	C	D[b]	X
0	84,000	—	—	—	-84,000
1		49,000	20,000	16,800	24,120
2		49,000	20,000	26,880	28,152
3		60,000	15,000	16,128	33,451
4		60,000	10,000	9,677	33,871
5		60,000	10,000	9,677	33,871
6	L = 4,000	60,000	10,000	4,838	35,935

[a] These are the sums of labor and quality savings (See Table 6).
[b] D_1 = 0.2* (84,000) = 16,800
 D_2 = 0.32* (84,000) = 26,880, etc.

2.3.1 Net Present Value (NPV)

Once the cash flows have been determined, the net present value (NPV) can be determined using the equation

$$NPV = \sum_{j=0}^{n} \frac{X_j}{(1 + k)^j} = \sum_{j=0}^{n} X_j(P/Fk, j)$$

where

X_j = net cash flow for year j
n = number of years of cash flow
k = minimum acceptable (attractive) rate of return referred to as a MARR

$\dfrac{1}{(1 + k)^j}$ = discount factor, usually designated as $P/Fk, j$

Using the cash flows in Table 7 and a k value of 25%, the NPV is

$$NPV = -84,000 + 24,120*P/F25, 1 + 28,152*P/F25, 2 + \ldots + 35,935*P/F25, 6$$
$$= \$4,836$$

The project is economically acceptable if the NPV is positive. Also, a positive NPV indicates the rate of return is greater than k.

2.3.2 Rate of Return

The rate of return is the interest rate that makes the NPV ≥ 0. It is sometimes referred to as the *internal rate of return* (IRR). Mathematically, it is defined as

$$0 = \sum_{j=0}^{n} \frac{X_j}{(1 + i)^j} = \sum_{j=0}^{n} X_j(P/Fi, j)$$

where $i = IRR = $ internal rate of return.

For this example the rate of return is determined from

$$0 = -84{,}000 + 24{,}120 P/Fi,\ 1 + 28{,}152 P/Fi,\ 2 + \ldots + 35{,}935 P/Fi,\ 6$$

To solve for i a trial and error approach is needed. Assuming 25%, the right-hand side gives \$4,833 (see NPV calculations) and with 30% it is \$–5,137. Therefore the rate of return using linear interpolation is

25%	4,836
i	0
30%	–5,137

$$i = 25 + \frac{4{,}836*(5)}{9973} = 27.42\%$$

This rate of return (27.42%) is now compared to a minimum acceptable rate of return (MARR). This is the MARR used in calculating the NPV. If IRR $\geq$ MARR the project is acceptable. Otherwise it is unacceptable. Consequently the NPV and the rate of return methods will give the same decision regarding the economic desirability of a project (investment). It is pointed out that the definitions of cash flow and MARR are not independent. Also, the omission of debt interest in the cash flow equation does not necessarily imply that the initial project cost (capital expenditure) is not being financed by some combination of debt and equity capital. When total cash flows are used, the debt interest is included (approximated) in the definition of MARR as shown in the equation

$$\text{MARR} = k_e(1-c) + k_d(1-T)c$$

where

k_e = required return for equity capital
k_d = required return for dept capital
T = tax rate
c = debt ratio of the pool of capital used for current capital investments

In practice it is not uncommon to adjust (increase) k_e and k_d for project risk and uncertainties in the economic climate. There are other definitions of cash flow definitions (equity and operating), with corresponding MARR definitions. A complete discussion of the relationship between cash and flow and MARR definitions is given in Stevens (1994).

2.3.3 The Payback (Payout) Period

A method that is sometimes used for the economic evaluation of a project is the payback period (or payout period). The payback period is the number of years required for incoming cash flows to balance the cash outflows. Expressed mathematically, the payback period, p, is $0 = \sum_{j=0}^{p} X_j$.

This is one definition of the payback period. While there is another definition that employs a discounting procedure, this is the definition most often used in practice. Using the cash flow given in Table 7, the payback equations for two years gives

$$-84,000 + 24,120 + 28,152 = \$-31,728$$

and for three years is

$$-84,000 + 24,120 + 28,152 + 33,451 = \$1723$$

Therefore, using linear interpolation, the payback period is

$$p = 2 + \frac{31,728}{33451} = 2.95 \text{ years}$$

Step 10 Additional Economic Considerations

The following points should be remembered in using the evaluation techniques given in this section.

1. The values for the components in the cash flow equation are incremental values. They are increases or decreases resulting directly from the project (investment) under consideration.
2. The higher the NPV and rate of return, the better and the lower the payback period.
3. The use of the payback period as a primary criterion is questionable. It does not consider the cash flows after the payback period.
4. In the case of evaluating mutually exclusive alternatives, select the alternative with the highest NPV. Selection of the alternative with the highest rate of return is incorrect. This point is made clear in many references (see Stevens (1994), Blank (1989), and Thuesen and Fabrycky (1989)).
5. In selecting a subset of projects from a larger group of independent projects due to some constraint (restriction), the objective is to maximize the NPV of the subset of projects subject to the constraint(s).

3 NEW APPROACHES TO THE JUSTIFICATION OF ROBOTICS APPLICATIONS

3.1 Issues in Strategic Justification of Advanced Technologies

The analysis typically performed in justifing advanced technology outlined in Section 2 is financial and short-term in nature. This has caused difficulty in adopting systems and technology that have both strategic implications and a number of intangible benefits that usually cannot be captured by traditional justification approaches. The literature has identified a number of other issues that have made the justification and adoption process for strategic technologies difficult, including high capital costs and risks, myopic approaches to justification, difficulty in quantifying indirect and intangible benefits, inappropriate capital budgeting procedures, and technological uncertainties (Sarkis et al., 1994).

Another complicating factor in the justification and adoption of integrated technologies is the cultural and organizational issues involved. The impact of implementing, for example, a flexible manufacturing system crosses many organizational boundaries. The success or failure of this implementation depends on the buy-in of all organizations and individuals involved. Traditional methods of justification often do not consider these organizational impacts and are not designed for group consensus building.

The literature discusses many of the intangible and nonquantifiable benefits of implementing robot systems. The most often mentioned is flexibility. Zald (1994) discusses four kinds of flexibility provided by robotic systems:

Mix flexibility	the ability to have multiple products in the same product process at the same time
Volume flexibility	the ability to change the process so that additional or less throughput is achieved
Multifunction flexibility	the ability to have the same device do different tasks by changing tools on the device
New product flexibility	the ability to change and reprogram the process as the manufactured product changes

Other frequently mentioned benefits include improved product quality, better customer service, improved response time, improved product consistency, reduction in inventories, improved safety, better employee morale, improved management and operation of processes, shorter cycle times and setups, and support for continuous improvement and *just in time* (JIT) efforts (Badiru, 1991; Davale, 1995).

In Step 4 above several questions concerning the noneconomic and intangible considerations of justifying robotic systems were provided. This section expands on these and presents a methodology that was developed to overcome many of the shortcomings of traditional economic justification methods. An illustrative example of justifying a robotic system is used to present the methodology.

3.2 Methodology Overview

In response to the needs outlined above, the Automation and Robotics Research Institute, with funding from the National Center for Manufacturing Sciences, has developed a pervasive activity-based business case tool for the justification of advanced technologies such as robotic systems (National Center for Manufacturing Sciences, 1994). This methodology considers both financial and strategic benefits of the technology. The methodology consists of five integrated phases:

1. Identify system impact.
2. Identify transition impact.
3. Estimate costs and benefits.
4. Perform decision analysis.
5. Audit Decision (National Center for Manufacturing Sciences, 1994).

The analysis in this methodology is accomplished through a series of documents and matrices. The documentation data include information about the enterprise, such as strategies and objectives, as well as information about the system being justified. The detailed requirements for this documentation, as well as the details of the methodology, can be found in National Center for Manufacturing Sciences (1994). The data in the documents are used to create the matrices. The matrices help identify relationships among the strategic and activity-based impacts of the technology. Five major *linkage matrices* are used to link objectives, strategies, attributes, activities, strategic metrics, and operational metrics. Four *analysis matrices* are used to measure the actual impact of the adoption of a technology or business practice.

3.3 Illustrative Example

This section describes the methodology through the use of an illustrative example involving the justification of a robotic system. It uses benefits and costs related to implementation of robot systems found in literature and practice. To maintain conciseness in this discussion, not all forms and matrices will be shown.

In the *identify system impact* and *identify transition impact* phases the technology and the plan for its implementation are linked to the enterprise. This linkage uses an activity-based approach to estimate financial costs and benefits. The impacts of the new technology on the strategies of the enterprise are also identified. The impact of transition (implementation)-specific issues are explicitly considered. For example, a phased implementation of a flexible machining system would have a different impact, both in terms of magnitude and timing, than if the system were implemented in a so-called "crash" or overnight implementation.

In these phases many of the documentation forms required by the methodology are completed. The matrices are created from the documentation in preparation for further analysis. An understanding of the enterprise's strategic direction is an outcome of the initial analysis. The vision, strategy, and objectives are documented in appropriate forms. Based on this information, a matrix (the *strategies to objectives* matrix) showing the linkages is created. For this example several strategies impacted by the technology have been identified, including those related to *financial performance, use of flexible processes and technologies, improved communications, customer service,* and *improved workplace conditions.*

Other linkage matrices (*activities to components, strategies to strategic attributes, strategies to traditional metrics,* and *strategies to strategic metrics*) are used as understanding and selection tools to help discriminate which individual items will be included in the analysis. The activities to components matrix is used to link the system operationally to the activities of the enterprise. Components are simply a convenient way of breaking what could be a large system (e.g., a flexible assembly system) into smaller units (e.g., the physical processing equipment, the control software, etc.) to increase the rigor and precision of the analysis. Likewise, the strategies to strategic attributes ties the system to the strategies of the enterprise. Attributes are equivalent to components in that they are a way to break down the system. In this case we are searching for factors about the system that do not directly affect any single activity but in some way impact the strategies. Example attributes include flexibility and extensibility of a system.

The methodology uses two types of metrics in its analysis: *cost driver* and *strategic*. Cost driver metrics are used to help in defining activity-based effects of the system and are used in the activity analysis phase. Examples include *reduced WIP* and *reduced expediting*. Knowing the magnitude of these metrics allows for a more accurate estimation of activity-based cost savings.

Strategic metrics are used to identify and estimate the impact of the technology to the strategies of the firm. Three types of strategic metrics are recommended:

1. *Financial* metrics, such as the two used in this example, *net present value* (NPV) and *payback period*, incorporate the cost impacts of the investment as identified in the traditional analysis with noncost impacts. These are discussed in Section 30.2.

2. *Quantitative* metrics, such as lead time and throughput, refer to those metrics for which numerical estimates can be obtained but are difficult or impossible to put in dollar terms. In the example three metrics related to the benefits of robotic systems described in literature, *cycle time reduction, quality improvement,* and *work in process level* in inventory, are used.

3. *Qualitative* metrics are used to measure the impacts that are difficult to measure in dollar terms. These are often expressed in terms of categorical values such as better/same/worse and A/B/C/D. In the example the investment's effect on consistency of product quality, ability to respond to changing market demands, and safety will be rated on poor/bad/neutral/good/excellent scales.

The strategies to strategic metrics matrix (Figure 5) allows the magnitude of the relationships of matrix elements to be identified. It is assumed that not all strategies are

Alternative 1	Strategy Weight	Financial		Quantitative			Qualitative		
		NPV	Payback	% Cycle Time Reduction	% Quality Improvement	WIP	Improved Information Flow	Response to Market Demands	Safety
Strategies									
Reconfiguarable Processes/Techn	0.1			0.2	0.2	0.2		0.4	
Communications	0.1						0.5	0.5	
Workplace Issues	0.1								1
Customer Service	0.2			0.3	0.2	0.2		0.3	
Financial Performance	0.5	0.8	0.2						
Strategic Metric Weight		0.40	0.10	0.08	0.06	0.06	0.05	0.15	0.10
Target		7.5	1	50	50	1000	5	5	5
Upper		7.5	5	50	50	1000	5	5	5
Lower		0	1	0	0	0	0	0	0
Type		I	D	I	I	D	I	I	I

Figure 5 Strategies to strategic metrics matrix.

equally important to an enterprise at a certain point in time. Likewise, not all strategic metrics are equally adept at measuring the impact against a particular strategy. To account for this, weights are calculated for each strategic metric. Two factors are considered in assigning weights: the relative importance of each strategy to the overall objectives of the enterprise and the relative ability of each metric to measure the realization of each strategy. The importance of each metric is indicated by a numerical weight between 0 and 1. The sum of the strategic weights and the sum of the metrics weights for a strategy are normalized to 1.

In the example in Figure 5 the financial performance strategy has been judged to be the most important, with a weight of 0.5. Two metrics, NPV and payback, are to be used to measure this strategy, with weights of 0.8 and 0.2, respectively, consistent with the discussion in Section 2. The overall weight for each strategic metric is calculated by summing the product of strategy weight and individual metric weight for each strategic metric. Since the summation of all the metric weights and strategy weights is equal to one, this will guarantee that the summation of the strategic metric weights is also equal to one. Since financial performance is the only strategy these metrics will be used for, the weights for the metrics are 0.4 and 0.1, calculated as 0.5 × .8 and 0.5 × .2, respectively. Similarly, the weight for the % cycle time reduction metric is calculated as (0.1 × 0.2) + (0.2 × 0.3) for a value of 0.08. The strategies to strategic metrics matrix essentially defines the decision model of the decision-maker.

The final output of the analysis is a *score* for the alternative based on a summation of its score for each metric. Because the metrics may all be expressed in different units (dollars, percent, counts, grades, etc.), it is necessary to normalize the values into a consistent scale. This is accomplished through utility function conversions. The functions convert estimated values provided for each metric into a value between 0 and 5 (5 being optimal) based on upper, lower, and target values. The last four rows of the strategies to strategic metrics matrix in Figure 5 are used to keep track of various utility functions. The target row defines a value to which the decision-maker aspires. In the case of NPV a value greater than 7.5 is desired. The upper and lower rows define the bounds of the values for the utility function calculation as well as what may be viewed as the *ideal* and *bottom* (best and worst) points as viewed by the decision-maker. It is assumed that any value beyond the best and worst is a constant utility value. The type row specifies the type of utility function is to be used. Three basic types of linear utility functions for this methodology:

1. Increasing (I) functions are used when a higher value is preferred, as with NPV. It is assumed that anything equal to or less than the lower value receives a normalized utility value of 0, anything equal to or greater than the upper value receives a normalized utility value of 5. Values in between receive a proportionate score.

2. Decreasing (D) functions are essentially similar but with lower values preferred, as with a metric such as initial investment.

3. Peaked (P) utility functions (which we do not show here) aim for a value at a specified target, with deviation in either direction resulting in a lesser score. An example is delivery date, where being early and late are both undesirable.

3.3.1 Strategic Analysis Matrices

Four analysis matrices are used in the methodology. The first, the *activity analysis* matrix (Figure 6), is used to derive the activity-based costs and benefits of the new technology. Along the vertical axis are listed the activities that the analysis team has determined are affected by the system under consideration. The components (breakdown of the system into logical or physical units), cost driver metrics, and cost categories are listed along horizontal axis. Cost categories are the components that form the total cost of performing a particular activity. In this example the two cost savings categories identified in Table 6, labor and quality, are shown. The values listed represent savings. Note that for some of the cost categories negative values may result if there is an increase in the cost of performing the activity as a result of implementing the system under consideration. Total savings are calculated for each of the analysis matrices (an activity analysis matrix will be created for each year). The total savings by year are then carried over into a *cash flow analysis* matrix (not shown) for use in performing analysis on standard financial criteria such as NPV. A column in the cash flow analysis would contain the operating costs shown

Alternative 1 / Year 1	Components			Cost Drivers			Cost Categories		
Activities	Physical Robot System	Monitoring/Data Collection	Control Software	Reduced WIP	Reduced Expediting	Training Time	Labor	Quality	Activity Total
Schedule Shop Floor		x	x	x	x	x	$3.5	$.0	$3.5
Supervise Shop Floor	x	x	x	x	x	x	$2.0	$.0	$2.0
Perform Manufacturing Operations	x	x	x	x	x	x	$28.5	$9.0	$37.5
Collect Data/Monitor Progress		x	x	x	x	x	$3.5	$.0	$3.5
Perform Control Actions	x	x		x	x	x	$2.5	$.0	$2.5
									$49.0

Figure 6 Activity analysis matrix.

in Table 4. The calculations of these financial criteria are conducted as detailed in Section 2.

The values for the criteria are then transferred into the financial metrics portion of the next matrix, the *strategic analysis* matrix (Figure 7). Note that the utility function information and weighting information have been transferred directly from the strategies to strategic metrics matrix in Figure 5. Again, the values for the financial metrics are calculated from the activity analysis matrix. The estimated or calculated values are entered in the observed value column. The utility function is then applied to this value to arrive at the value in the normalized value column. For example, the value for NPV as calculated in Section 2, $4.8k, is normalized to a value of 3.2. The observed value for payback,

Alternative 1	Strategic Attributes					Utility Functions				Values		Calculations	
Metrics	Support JIT & Cellular	Integration	Information	Dependability	Expandability	Target	Upper	Lower	Type (I, D, P)	Observed Value	Normalized Value	Weight	Strategic Metric Total
Financial Quantitative													
NPV	x	x	x	x	x	$0	$7.5	$0	I	$4.8	3.2	0.40	1.28
Payback	x	x	x	x	x	1	5	1	D	2.95	2.6	0.10	0.26
Quantitative													
% Cycle Time Reduction	x	x	x	x		50	50	0	I	47	4.7	0.08	0.38
% Quality Improvement						50	50	0	I	48	4.8	0.06	0.29
WIP	x		x			0	1000	0	D	220	3.9	0.06	0.23
Qualitative													
Improved Information Flow	x	x	x	x	x	5	5	0	I	5	5	0.05	0.25
Response to Market Demands						5	5	0	I	4	4	0.15	0.60
Safety		x	x	x	x	5	5	0	I	4	4	0.10	0.40
											Total		3.69

Figure 7 Strategic analysis matrix.

2.95 years, lies between the upper and lower values of its utility function. Rounding off, the system receives a value of 2.6 for this metric.

The next step is to apply the weights for each metric that were derived in the strategies to strategic metrics matrix. Again taking the NPV example, the normalized value of 3.2 is multiplied by the weight of the metric (0.4) to obtain the score for this metric of 1.28. Likewise, payback is calculated as 0.1×2.6 for a score of 0.26. Finally, the scores for all metrics are added to arrive at a total score for this system of 3.69 (out of a maximum of 5).

Alone this value does not mean much, but it should be used as a comparative value with other alternatives. To compare alternatives the *alternatives comparison* matrix (Figure 8) is used. This matrix allows the scores of several alternatives to be displayed on one matrix for easy comparison. In this example the comparison is to a baseline alternative. We see that even though the system under consideration does not perform as well on the financial metrics, its strong performance on the quantitative and qualitative metrics allows it to outscore the baseline overall by a score of 3.69 to 3.06. This example points out the importance of considering all benefits, not simply financial benefits, in deciding on alternatives.

Metrics	Weight	Alternative 1		Baseline
Financial Quantitative				
NPV	0.40	1.28		1.60
Payback	0.10	0.26		0.50
Quantitative				
% Cycle Time Reduction	0.08	0.38		0.16
% Quality Improvement	0.06	0.29		0.14
WIP	0.06	0.23		0.07
Qualitative				
Improved Information Flow	0.05	0.25		0.10
Response to Market Demands	0.15	0.60		0.30
Safety	0.10	0.40		0.20
Weighted Total		3.69		3.06

Figure 8 Alternatives comparison matrix.

4 PROPOSED MODIFICATIONS AND EXTENSIONS TO TRADITIONAL COST COMPARISON TECHNIQUES

4.1 The Adjustment of the Minimum Acceptable Rate of Return in Proportion to Perceived Risk

Not all capital investments carry equal risk. It is also reasonable to expect higher potential returns from projects with higher perceived risk. It is common practice for management to increase the MARR for high-risk projects. This has the effect of forcing the proposed projects with higher risk to demonstrate that they are capable of generating greater return on their capital investment.

A similar strategy is proposed to recognize the fact that capital equipment that can be reused for additional projects is less likely to see a rapid drop in its value due to unforeseen reductions in a given product's demand and market life. Product-specific capital, however, would be susceptible to such risks. Traditionally, more stringent MARR requirements would be applied to the entire capital investment. It is proposed that only the portion of the capital investment that is at risk due to sudden changes in market demand should be forced to meet these more stringent rate of return requirements. This approach to assigning risk will explicitly recognize and promote the development and reuse of modular automation by compensating for its higher initial development and implementation costs through lower rate of return requirements.

4.2   Depreciation and Realizable Salvage Value Profiles

Capital depreciation mechanisms and projected project can also have a significant effect on the financial justification of automation. Two types of depreciation must be considered: tax depreciation and book depreciation. Tax depreciation methods provide a systematic mechanism for recognizing the reduction in value of a capital asset over time. Since depreciation is tax-deductible, it is generally in the best interest of the company to depreciate the asset as quickly as possible. Allowable depreciation schedules are determined by the tax code. Tax depreciation can become a factor when comparing product-specific automation to reusable systems when the projected product life is less than the legislated tax deprecation life. Under these circumstances the product-specific systems can be sold on the open market and the remaining tax depreciation would be forfeited. A second alternative is that the system would be decommissioned and stored by the company until it is fully depreciated and then sold for salvage. In theory it would be much more unlikely that the life of a reusable system would be shorter than the mandated tax depreciation period due to the fact that it can be redeployed on an additional project.

It is important to note that tax depreciation schedules are determined by tax and accounting conventions rather than the actual anticipated useful life of the asset. Book depreciation schedules are therefore developed to predict the realizable salvage value of the capital asset over its useful life. Two separate viewpoints can be considered when determining realizable salvage value: the value of the asset when used to support an additional project within the same organization and the asset's value when it is sold to an external buyer for salvage. A used piece of industrial automation that can be profitably redeployed within the same company is clearly of more value to that organization than it will be to an equipment reseller. This is particularly the case if the system has been constructed from a set of modular components specifically designed to enhance reuse.

Traditionally, when product-specific automation is sold off for scrap, the potential buyer will most likely only be interested in key system components like the manipulator and the controller. All of the product-specific and process-specific tooling will add very little additional value in the eyes of an equipment reseller. All the custom engineering and software development costs required to field the system will also be lost. If, on the other hand, the system has been constructed from modular automation components that are well understood by the manufacturing organization, it is conceivable that the used system may have more value than a completely new set of system components. The redeployed system may lower the time and cost required to provide useful automation resources. Investments made in the development of process technology may be of value in subsequent projects. Application software, if developed in a modular fashion, also has the potential for reuse.

5   SUMMARY

While the calculation of the economics of robotic systems is often straightforward, their justification is often fraught with difficulty and many opportunities for long-range improvement are lost because the purely economic evaluation apparently showed no direct economic benefit. Modern methods are emerging to avoid such mistakes, which can be costly in the long run. In this chapter, in addition to providing the traditional economic justification methodology we have also provided a methodology based on strategic considerations and presented a discussion on a move away from the product-oriented production of today toward the more process-oriented production of tomorrow.

6   ACKNOWLEDGMENTS

The authors gratefully acknowledge the significant and valuable contribution of Yukio Hasegawa, on whose chapter in the 1st edition of this handbook we have based this chapter. A large portion of the first part of this chapter is adapted from his. We have taken his overall approach, simplified it, added new material about noneconomic justification, and updated the references where appropriate. The authors also acknowledge the contribution of Professor Nof to Professor Hasegawa's chapter. The section on strategic justification in this chapter would not have been possible without financial support from the National Center for Manufacturing Sciences, which is the owner of the copyright on the methodology described therein. Finally, the assistance of Ms. Sharon Peterson in the typing and editing of the chapter is gratefully recognized.

REFERENCES

Badiru, A. B. 1991. "A Management Guide to Automation Cost Justification." *Industrial Engineering* **22**(2), 26–29.

Blank, L. T., and A. J. Tarquin. 1989. *Engineering Economy.* 3rd ed. New York: McGraw-Hill.

Boothroyd, G., C. Poli, and L. E. Murch. 1982. *Automatic Assembly.* New York: Marcel Dekker.

Davale, D. 1995. "Justifying Manufacturing Cells." *Manufacturing Engineering* **115**(6), 31–37.

Goldman, S. L. "Agile Manufacturing: A New Paradigm for Society." White Paper, Agile Manufacturing Enterprise Forum, The Iacocca Institute, Lehigh University.

Hasegawa, Y. 1985. *Handbook of Industrial Robotics.* Ed. S. Y. Nof. New York: John Wiley & Sons. 665–687.

Kaplan, R. S. 1986. "Must CIM Be Justified by Faith Alone?" *Harvard Business Review* **64**(2), 87–93.

National Center for Manufacturing Sciences. 1994. *A Methodology for the Strategic Justification of Integrated Enterprise Technologies.* Ann Arbor: NCMS.

Nof, S. Y. 1982. "Decision Aids for Planning Industrial Robot Operations." In *Proceedings of the 1982 Annual Industrial Engineering Conference, Institute of Industrial Engineers.*

Sarkis, J., et al. 1994. "Development of the Requirements for a Strategic Justification Methodology." *Intelligent Automation and Soft Computing: Proceedings of the First World Automation Congress.* Vol. 1. Ed. M. Jamshidi et al. Albuquerque: TSI Press. 87–91.

Schonberger, R. J. 1990. *World Class Manufacturing.* New York: The Free Press.

Stevens, G. T., Jr. *The Economic Analysis of Capital Expenditures for Managers and Engineers.* Needham Heights: Ginn Press.

Thuesen, G. J., and W. J. Fabrycky. 1989. *Engineering Economy.* 7th ed. Englewood Cliffs: Prentice-Hall.

Zald, R. 1994. "Using Flexibility to Justify Robotics Automation Costs." *Industrial Management* **36**(6), 8–9.

PART 7

APPLICATIONS: DESIGN AND INTEGRATION

CHAPTER 35

ROBOTIC MANUFACTURING CELLS

J. T. Black
Auburn University
Auburn, Alabama

Laxmi P. Musunur
FANUC Robotics North America, Inc.
Rochester Hills, Michigan

1 INTRODUCTION

In a factory with a future the cell will be the key manufacturing system design element in developing a truly integrated manufacturing system. Most cells in place today are manned, but they are evolving toward unmanned systems that are totally computerized, robotized, and automated for autonomous operation. For O.R. models of robotics systems design see Chapter 30.

The factory with a future will have manufacturing and assembly cells linked by a pull (kanban-type) system for material and information control. Initially the cells should be manned by standing, walking workers. The cell should have the attributes given in Table 1. When justified, the cells can be partially or completely unmanned or robotic.

In manufacturing and assembly cells machines and operations are placed next to each other to permit one-part-at-a-time movement within the cell. Decouplers (Black and Schroer, 1988) are placed between the machines in the cell, improving the cell flexibility. The decoupler is so named because it reduces (i.e., decouples) the time and/or functional dependency of one machine on another within the cell. Moreover, it frees (decouples) the worker (or robot) from the machines so that the worker (or robot) can move to different machines in the cell. In fact, the workers can move in the opposite direction to the movement (flow) of the parts within the cell. The decoupler also carries production control information through the cell. Decouplers usually hold one part and are not storage queues. Decouplers in robotic cells have many additional functions, including inspection, part manipulation, intercell transportation, and process delay.

2 MANNED CELLS

While much has been written about flexible manufacturing systems (FMSs), fewer than 5000 FMS installations exist worldwide. FMSs are expensive, complex systems that are not very flexible and cover only a small segment of the component part spectrum. As shown in Figure 1, based on 1980–89 factory data, manned and unmanned manufacturing cells span the wide region between job shop and dedicated flow lines (Black, 1988a). Over the past 15 years linked-cell manufacturing systems (L-CMS) have been systematically replacing the classical job shop/flow shop manufacturing systems. The L-CMS has had many names (see Table 2), the most recent one being *lean production*. The batch and queue job shop with elements of flow is being eliminated and replaced with a system of linked manufacturing and assembly cells. The L-CMS, shown in Figure 2, has manned manufacturing and assembly cells linked with a kanban-type production control system, permitting the material to be pulled to final assembly. Downstream usage (by the customer) dictates upstream production quantities. Information flows in the opposite direction to material. The heart of the lean manufacturing system is the U-shaped manned manufacturing cell, shown in Figure 3. The worker moves from machine to machine unloading, checking, and loading parts. Each time the worker makes a loop around the cell, a part

Handbook of Industrial Robotics, Second Edition, Edited by Shimon Y. Nof
ISBN 0-471-17783-0   © 1999 John Wiley & Sons, Inc.

Table 1   Cell Design Attributes—Manned Cell

⇒ Standing, walking worker (90% utilized)
⇒ U-shape with one-piece flow and decouplers
⇒ "Buffers" to compensate for long changeovers between the cells
⇒ Every operation 100% mistake proofed, ergo integrated quality
⇒ Machines built with narrow footprint—4' wide
⇒ Material loading from rear of cell (along aisle)
⇒ Chips fed to rear of cell (along aisle)
⇒ 3 hours inventory between machining and assembly in link
⇒ 1 to 2 days inventory between component plant and assembly
⇒ Volume flexible (accommodate changes in demand or demand fluctuations)

is produced. The cell is designed to make a family of parts, usually a product family, requiring the same set of processes in the same order. Suppose the worker can make the loop in two minutes. This is the cycle time (CT) for the cell. The cell produces a part for the subassembly cell, which in turn produces subassemblies for the final assembly line. Say that final assembly needs 480 parts per day. The cell needs to run two shifts per day to meet this daily requirement. Each process in the cell will typically have single-cycle automatic capability. The machine can be loaded and started and will complete the machining cycle automatically. Suppose the machining times (MT) for the seven processes are 0.5, 0.4, 0.4, 0.3, 0.4, 0.3, and 0.5 respectively going from saw to grinder. Note that all MTs are less than the CT. Thus CT can be changed without having to alter any of the processes. This is how flexibility is obtained. The output can be doubled by adding a second worker to the cell, as shown in Figure 4. Each worker has a loop time of about one minute and we are still okay with respect to MT < CT.

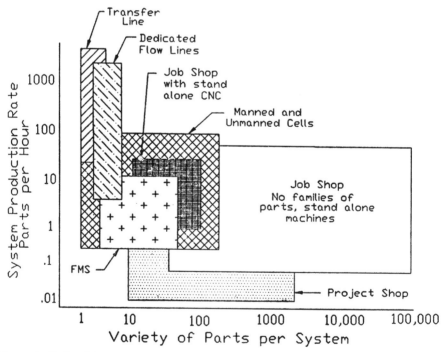

Figure 1   Part variety versus production rates for various manufacturing systems. (From E. P. DeGarmo, J. T. Black, and R. A. Kohser, *Materials and Processes in Manufacturing*, 8th ed. Upper Saddle River: Prentice-Hall International, 1997. 1133.)

Table 2   Names for the Linked-Cell Manufacturing System

- Lean Production (MIT group of authors, researchers)
- Toyota Production System (Toyota Motor Company)
- Ohno System (Taiichi Ohno, Inventor of TPS)
- Integrated Pull Manufacturing System (AT&T)
- Minimum Inventory Production System (Westinghouse)
- Material as Needed—Man (Harley Davidson)
- Just-in-Time/Total Quality Control (JIT/TQC) (Schonberger)
- World Class Manufacturing (Schonberger)
- Zero Inventory Production Systems—ZIPS (Omark and Bob Hall)
- Quick Response or Modular Manufacturing (Apparel Industry)
- Stockless Production (Hewlett-Packard)
- Kanban System (many companies in Japan)
- Continuous Flow Manufacturing (CFM) (Chrysler)
- One Piece Flow (Sekine)

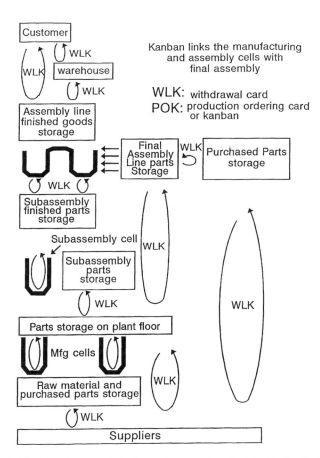

Figure 2   The L-CMS has the manufacturing and assembly cells linked to the final assembly area with kanban inventory links.

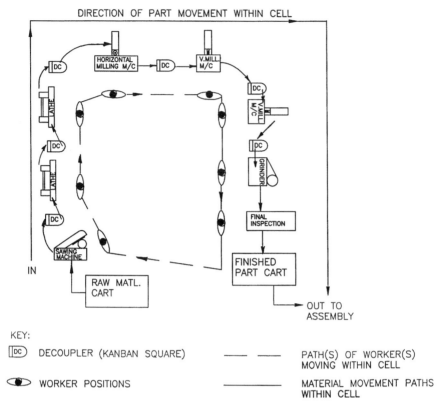

KEY:

DC	DECOUPLER (KANBAN SQUARE)	——— ——— PATH(S) OF WORKER(S) MOVING WITHIN CELL
WORKER POSITIONS		——————— MATERIAL MOVEMENT PATHS WITHIN CELL

Figure 3 Manned manufacturing cell, made up from existing equipment, using one standing, walking operator.

The system design permits the functions of quality control, inventory control, and production control, which have resided classically in the production system, to be integrated into the cells. This integration requires that the cell be flexibly designed. That is, the methodology to design manned cells to be flexible requires the elimination of setup, the development of multifunctional workers who walk from operation to operation within the cells, the U-shaped layout, small lots moving between cells with one-piece part movement within the cell, and rapid quality feedback. In addition, the machines are given regular maintenance and do not usually break down during the shift.

The decouplers between each machine are used to connect the flow when using multiple workers, as shown in Figure 4. Worker 1 unloads machine 2 (lathe) and puts the part in decoupler 2 between the machines, then steps across the aisle to the grinder, unloads the grinder, and loads it with a part from decoupler 6. As shown in Figure 5, the workers can easily vary the loops and change the point of handoff to anywhere in the cell that has a decoupler.

In the same way subassembly and assembly lines can be redesigned to develop one-piece flow, as shown in Figure 6. In this figure (from Sekine, 1990) the process flowchart and existing assembly process are shown for a system using two conveyors and 10 operators. The assembly cell uses 8 operators and increases the output while decreasing the stock on hand. The area with the dotted line represents a "rabbit chase" where two workers (m and n) move in a loop, one behind the other, going from operation ① to ⑪ and ⑫ and back across the aisle to ①. Operation ⑦, caulking, is done some of the time by operator x and some of the time by operator y. The main difference is that the operations in assembly tend to be manned and the equipment is manned single-cycle automatic.

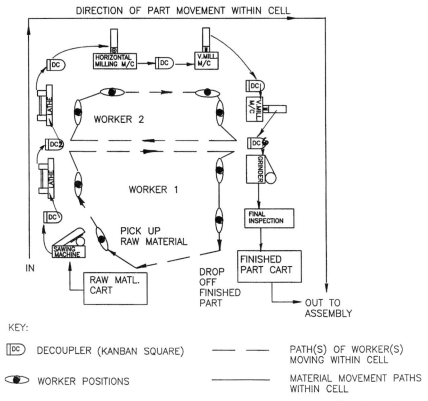

Figure 4   Manned manufacturing cell with two operators connected by decouplers.

3   UNMANNED MANUFACTURING CELLS

The flexibility problem, however, becomes much more difficult in unmanned cells (see Figure 7), for example where the most flexible element in the cell (the workers) has been removed.

Unmanned cells must be able to operate autonomously, replacing the human's decision-making capability (thinking) and superior sensory system, while producing superior-quality parts (zero defects) without any machine tool or cutting tool failures to disrupt the system. The cell must be able to react to changes in demand for the parts and therefore must be flexible in capacity (must easily change cycle time). The cells must be able to adapt to change in the product mix and the cell must be able to accommodate changes in the design of existing parts for which the cell was initially designed (a family of parts). The latter capability requires the CAD/CAM link to include a cell-based computer-aided process planning (CAPP) because the new part is being made by a group of machines. The machines must be modified for robotic loading. The decouplers permit the robot to move in either direction and may also serve to check parts (automatic inspection) and reorient parts for loading into the next machine. Decouplers will be discussed more in Section 5.

4   A FACTORY OF LINKED CELLS

The redesign of the factory into manufacturing and assembly cells streamlines and simplifies the manufacturing system for the component parts, making the system's component part-producing portion compatible with the subassembly and final assembly lines. The next step is integrating quality control and machine tool maintenance and production control functions into the cell. As this integrated system evolves, computerization, automation, and robotization elements will be introduced to solve production flow and ca-

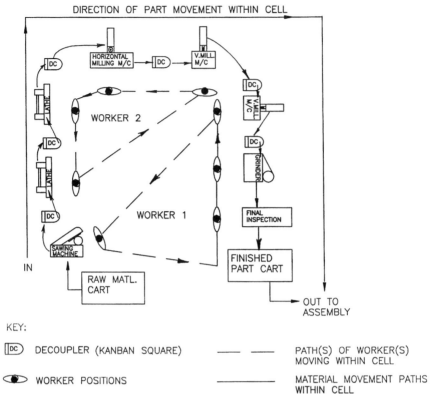

Figure 5   The loops for the operators are variable. The handoff points can change as needed.

pacity problems. Production problems pull the automation solutions. When the cells arrive at the condition where no defective parts are produced, machines do not break down, and setups are virtually instantaneous, then the inventory between the manufacturing and assembly areas can be greatly reduced and fully robotic cells can be a reality.

4.1   Five Basic Concepts

Five basic concepts are embodied in the cellular structure.

1. *The way to superior-quality products is through defect prevention.* Producing defects costs money. Superior-quality products are reliable products. To get superior quality, quality control must be integrated into the cells. Workers are given the responsibility and authority to make the parts right the first time. The idea is to inspect to prevent defects rather than to find defects later. The cells operate on a MO, CO, MOO policy: make one, check one, move one on. When the quality is very good, automation can be considered. However, remember that when the operator is removed, inspection falls to in-process probes or to decouplers that can inspect the part and provide feedback to the upstream machine to make corrections, if necessary. The products and components must be designed for quality and easy inspection. Total preventive maintenance of the machines and decouplers must be implemented.

2. *Excessive inventory is a liability.* Inventory can be used as an independent system control variable rather than as a dependent variable. Techniques required include reducing setup time with the cells and the flow lines, reducing lot sizes, and connecting the elements (cells, subassemblies, final assembly) with a pull system for inventory control (kanban).

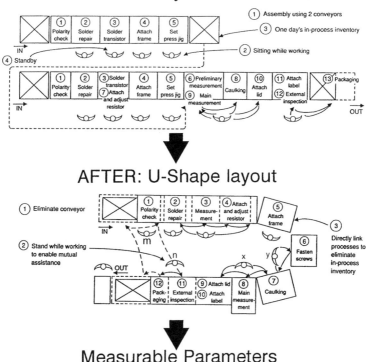

Figure 6   Assembly cell redesigned into U-shaped cell with standing, walking workers. From Sekine (1990).

3. *Manufacturing systems must be designed and constructed to be flexible.* Here is where decouplers come into the design picture. To be flexible a manufacturing system must be able to react quickly to changes in design (handle different products or product design changes easily) and to changes in demand (handle quantity increases and decreases). Manned cells use multifunctional workers, have quick setups and small lot sizes, and are U-shaped. The decouplers permit the easy increase or decrease of the number of workers within the cell. In unmanned cells retaining flexibility requires a tradeoff in machine utilization. Decouplers help to replace some functions of the worker.

4. *People who run the process are the company's most valuable asset.* In reality these workers are often the company's most poorly used asset. Workers must be given respect and the opportunities to improve themselves and their working environment and contribute to the continuous improvement of the processes and the system. Remember that people can think and computers cannot. Let the people do what people do best and machines, robots, and computers do what they do best, which brings us to concept 5.

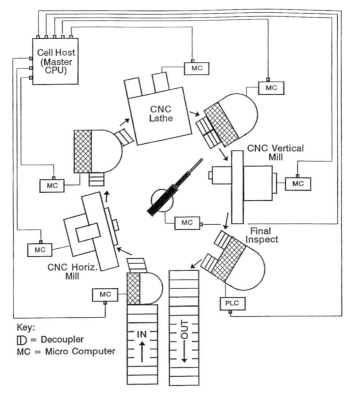

Figure 7 Unmanned robotic cell for gear manufacturing, having CNC machine tools, robot for material handling, and decouplers for flexibility and capability.

5. *Integrate first, then computerize.* The foundation for robotics, automation, and computerization of a manufacturing system is lean production (Ayres and Butcher, 1995), that is, the L-CMS. Perhaps we should write *IM,C* instead of *CIM*.

4.2 Advantages of Manufacturing Cells

Assembly and manufacturing cells are characterized by grouping machines, operations, and people to fabricate groups of parts (parts having a common set of processes and operations).

Cellular manufacturing offers these advantages:

- Quality feedback between manufacturing operations processes is immediate.
- Material handling is markedly reduced.
- Setup time is reduced or even eliminated.
- In-process monitoring, feedback, and control of inventory and quality are greatly improved.
- A smoother, faster flow of products through the manufacturing operations is achieved (shorter throughput time).
- Cycle time variability and line-balancing constraints are reduced.
- Implementation of automation of manufacturing operations is easier.
- Process capability and reliability are improved.

When machines and processes are placed in a cell, the processes become dependent on each other. However, the cell design decouples the worker (or robot) from a *specific machine*. The worker becomes multiprocess, having the ability to run many processes within the cell. The decoupler breaks the dependency of one machine or process on the next and permits the worker to move in the opposite direction to the flow of the parts.

4.3   Unmanned Cells Evolve from Manned Cells

The initial designs of unmanned cells must include the long-term goals of *flexibility, reliability, process capability,* and *autonomy.* As the unmanned design evolves from the manned design, the cell becomes a group of computer-controlled machines (including material-handling equipment) that is managed by a supervisory computer called the *cell host.* The cells act independently but are connected hierarchically to a production system-level computer, forming a factory of "linked, computerized cells." In converting cells from manned to unmanned, one of the primary constraints is robot process capability, discussed in Chapter 42.

5   DECOUPLERS IN CELLS

Decouplers are used between the processes within a cell (manned or unmanned). The simplest decoupler is a kanban square. In the Hewlett-Packard electronics assembly cell shown in Figure 8 the workers used tape to make a kanban square on each table between each assembly station. The removal of a unit from the square signaled the upstream

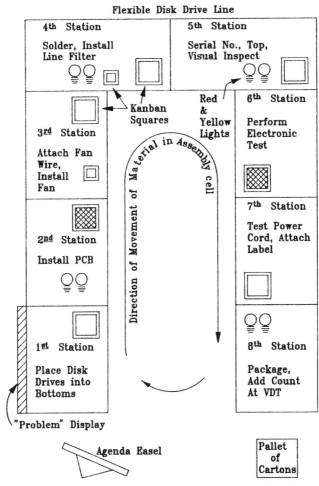

Figure 8 An assembly cell for disk drives, designed by workers at Hewlett-Packard, Greeley Division, with kanban squares.

worker to start the operations to make a unit to replace the one removed. This decoupler controlled production (material movement) within the cell, pulling the product from station to station. The presence of one unit in the kanban square effectively decouples each station while providing the vehicle for information to travel upstream within the cell.

The cell can be operated by one worker or multiple workers, depending on the needed output. This design allows the workers to be on flextime, arriving for work anytime between 6:30 a.m. and 8:30 a.m. This would not work if the cell had to deliver disk drives directly to another assembly area on a just-in-time basis, because the output of the cell will vary with the number of workers. However, the inventory link between the subassembly cell and final assembly takes care of the problem of variable output rates. (Yes, Virginia, there is inventory in a lean manufacturing system—it is in the links between the cells.)

5.1   Decoupler Functions

Decouplers typically reside between the processes within the cell and hold one part. For unmanned cells the decoupler must have specific input and output points. Decouplers can perform the following functions.

5.1.1   Functionally Decouple the Process

Decouplers decouple the processes from the cycle times. Figure 9 shows a manned cell with seven machines, decouplers, and two workers. Varying the number of workers changes the production rate to meet changes in demand. Two or three workers could be put in the cell as easily as one. Remember that adding workers to or subtracting them from the line changes the balance. The addition of decouplers relaxes the need for precise line balancing. The workers, foremen, and supervisors on the plant floor rebalance the work. Decouplers relax the need for immediate fine-tuning. The problem of cycle time variability in the line is overcome. Flexibility is gained. Figure 10 shows a robotic cell with two robots. Here the decouplers again serve to connect the flow of the parts between the machines. Suppose the part going from machine 4 to machine 5 needs to be reoriented as well as transported. Robot 2 drops the part in the decoupler shown in Figure 11. The part is reoriented and waiting on the output side for robot 1 to pick it up and deliver it to machine 6.

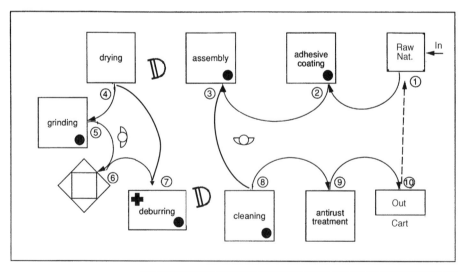

Figure 9   Manned manufacturing cell with two workers.

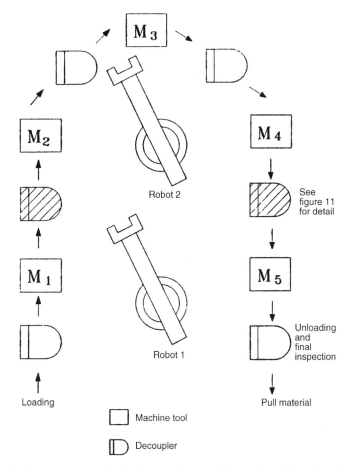

Figure 10   U-shaped unmanned cell with two robots and decouplers integrating the processes.

5.1.2   Poka-yoke Inspection

Decouplers can inspect the critical dimensions of parts so that only good parts are pulled to the next station. Such inspections can be done automatically in a decoupler with a feedback signal to the process, okaying the next part for manufacture. This is an example of poka-yoke (defect-prevention) technique for 100% inspection. The purpose of such techniques is to improve quality by preventing defects from occurring.

5.1.3   Freedom in Worker Movement

Decouplers provide flexibility in worker (or robot) movement. The expected direction for worker movement is downsteam, in the direction of parts movement. However, putting decouplers between the machines enables the worker (or robot) to move upstream, opposite to part flow.

5.1.4   Control of Cell Inventory Stock on Hand

Decouplers control the inventory level within the cell. The ideal situation is to have one-part movement from machine to machine. The inventory within the cell is called the *stock on hand*. There is usually only one part per decoupler. Decouplers operate within the cells and manage the stock on hand.

5.1.5   Interprocess Part Transporation

Decouplers can transport parts from process to process. Gravity slides or roller chutes work well here, getting the part transferred independently of the worker or robot. This

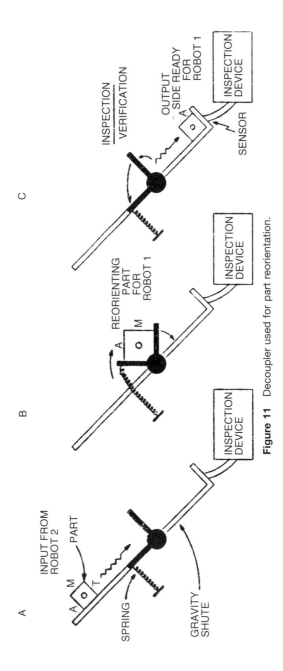

Figure 11 Decoupler used for part reorientation.

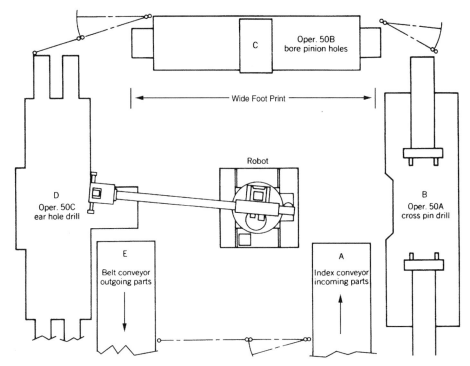

Figure 12   Machines with wide footprints restrict the number of machines that can be serviced by a robot.

transfer can be a serious problem in cells using machines built for stand-alone operations in the job shop. Figure 12 shows a robotic application for loading and unloading three drilling machines that process 20-lb castings. The robot gripper as well as the three drilling machines are designed to handle two parts at a time using a double-handed gripper. This design highlights the most serious problem with cell design. The machine tools built for stand-alone operations have very wide footprints. This reduces the number of machines that can be clustered around the robot to typically three to five machines and requires a robot with a long reach. Usually robots have their poorest capability at the extremes of their work envelopes, so they have both accuracy and precision problems. Decouplers can be used to alleviate some of these problems.

5.1.6   Branching

Decouplers can branch or combine part flows within the cell. Parts can be metered as needed at the junctions. This situation occurs when a process must be duplicated to handle a capacity problem (i.e., $MT_i > CT$). One machine is feeding two machines (or more) and two machines are feeding one machine. An example is where the manned portion of a cell is making left- and right-hand components (for a car) and the robotic portion has a robot station for left-hand parts and another station for right-hand parts.

5.1.7   Part Reorientation, Reregistration

In unmanned cells utilizing robots, decouplers can handle a family of parts, locating a part for the robot gripper on the output side. Often the part the robot wants to insert into the next machine must be manipulated or reoriented as well as inspected. The robot can place the part into the decoupler and pick it up later at a precise location on the output side, restoring precise part-to-gripper registration in a new orientation. Reorienting parts is a simple human task but very difficult to accomplish in unmanned (robotic) environments. Precisely locating the part on the output side eliminates the need to locate the part precisely on the input side (during robotic pickup). Gravity chutes move the parts. Part manipulation (reorientation) may require a turned part, for example, to be flipped end over end. Thus, decouplers assist robot handling and restore flexibility to unmanned cells. See Figure 11.

Decouplers can undertake secondary functions within the cell, jobs typically done by the operator. For example, the use of decouplers allows the worker (or robot) to delay the process sequence as needed to change the product's state—a heat treatment or temperature increase or decrease before the next process or a curing time for an adhesive, etc. Other secondary uses of decoupler may be degreasing or cleaning the parts to remove chips or oil before inserting the part into the next machine.

6 DEVELOPING A ROBOTIC CELL

Figure 13 shows a schematic of a robotic manufacturing cell for performing a series of metal working operations on a workpiece. The cell consists of an entry and exit conveyor and four CNC machines to perform the material-removal operations.

The process involves a robot picking up a part from the entry conveyor with a hand that has two sets of part grippers. The robot then unloads the finished part from the turning center, indexes the hand, and loads a new part. The robot then moves to the grinder, unloads the finished ground part, indexes the hand, and loads the turned part. Then the robot moves to drilling machine A and unloads the drilled part, indexes the hand, loads the ground part, and performs the same procedure on drilling machine B. Finally the robot deposits the finished part from drilling machine B onto the exit conveyor, picks up a new part from the entry conveyor, and repeats the entire procedure in a clockwise sequence. Note that the processing time for each of the machines should be less than the time required for the robot to make a complete loop around the cell. Suppose the robot can do all the loading/unloading tasks in two minutes and that the turning center time cycle is two minutes and other machine cycles are one minute; then the grinder and drilling machines will be sitting idle for about one minute each cycle. However, in the majority of the cases the allowable cycle time is already dictated by the downstream production requirements of the customer. For example, suppose the customer requires 30 parts per hour at 100% operating efficiency—allowance having been made in this rate for downtime required to perform adjustments, service, maintenance, etc. In this case the mandatory maximum cycle is two minutes, or 120 seconds. This means that the machining cycle time of two minutes must be able to include the robot load/unload time and the robot must be capable of loading and unloading all four machines and the conveyors in two minutes.

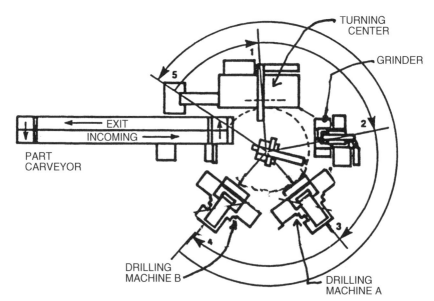

Figure 13 Robotic work envelope for cell (decouplers not shown) (courtesy FANUC Robotics North America, Inc.).

Next the total time required to load and unload all the equipment involved is examined to see if it can be accomplished in two minutes. Figure 13 indicates that loading and unloading the turning center will require more time than the other machines since it involves extra steps to pick up the part from the entry conveyor and drop the finished part in the exit conveyor. At each of the other machines the raw part is always in the robot's hand, ready to be loaded. The load/unload cycle for drilling machine B will also be longer since the robot has to rotate about 300° counterclockwise to return to the conveyor and deposit the finished part. Table 3 shows the breakdown of the two cycles. Since the two longest load/unload cycles fall well within the allotted time of approximately 30 seconds each and the two remaining machine cycles are even shorter, the assumption that the robot can load and unload all machines, and the conveyor, within two minutes is justified.

Next the MT of the slowest machine, the turning center, is examined. Assume that the metal-cutting portion of the machine cycle takes 88 seconds. Table 4 shows a breakdown of this cycle. Based on all these checks, there is more than adequate time to meet the production requirements. This is important because sometimes it may allow the use of a

Table 3   Cycle Time Breakdowns[a]

A. Turning Center Load/Unload Breakdown	
Lower gripper over part on conveyor	1.0 sec
Grip part	1.0 sec
Raise part off conveyor	1.0 sec
Retract robot arm	2.0 sec
Rotate robot arm to turning center	2.0 sec
Reach into machine	2.0 sec
Move empty hand over part	1.0 sec
Grip finished part	1.0 sec
Remove finished part from chuck	1.0 sec
Partial retract for index clearance	1.5 sec
Index hand 180° to present raw part	1.0 sec
Reach into machine again	1.5 sec
Place raw part in chuck	1.0 sec
Release raw part	1.0 sec
Retract hand to clear chuck	1.0 sec
Retract arm to clear machine	2.0 sec
	21.0 sec
B. Drilling Machine B Load/Unload Breakdown	
Rotate arm to machine B	2.0 sec
Reach into machine	2.0 sec
Move empty hand over part	1.0 sec
Grip finished part	1.0 sec
Remove finished part from chuck	1.0 sec
Partial retract for index clearance	1.5 sec
Index hand 180° to present raw part	1.0 sec
Reach into machine again	1.5 sec
Place raw part in chuck	1.0 sec
Release raw part	1.0 sec
Retract hand to clear chuck	1.0 sec
Retract arm to clear machine	2.0 sec
Rotate 300° back to conveyor	5.0 sec
Deposit part on conveyor	1.0 sec
Raise to clear part	1.0 sec
	23.0 sec

[a] Courtesy FANUC Robotics North America, Inc.

Table 4 Turning Center Cycle Breakdown[a]

Stop spindle and retract tools	2.0 sec
Open door	2.0 sec
Robot reach into machine	2.0 sec
Move empty hand over part	1.0 sec
Grip finished part	1.0 sec
Remove finished part from chuck	1.0 sec
Partial retract for index clearance	1.5 sec
Index hand 180°	1.0 sec
Reach into machine again	1.5 sec
Move raw part into chuck	1.0 sec
Release raw part	1.0 sec
Retract hand to clear chuck	1.0 sec
Retract arm to clear machine	2.0 sec
Close door	2.0 sec
Start spindle and advance tools	2.0 sec
	22.0 sec
Total time available	120.0 sec
Less non-cutting time	22.0 sec
Time available for cutting cycle	98.0 sec

[a] Courtesy FANUC Robotics North America, Inc.

less expensive robot to take advantage of the extra time available in both the robot load/unload cycle and the machine cycle.

7 INTERFACING

To interface machine tools with the robot and the decouplers, using a PLC in the cell may be necessary. Based on the complexity of interface requirements, it is possible to eliminate the PLC and directly use the robot controllers to provide the safety and sequence interlocks between the robot and the equipment. The following sensors are recommended as a minimum.

A *part present* limit switch signal will be required at the decoupler at the end of the input conveyor, telling the robot that there is a part waiting. This decoupler will present the part to the robot in the correct orientation and exact location. A *part present* signal from the decoupler should be present after drilling machine B and before the exit conveyor, saying that there is an empty space and the robot can deposit the finished part. It is also wise to equip the robot with *a part present* switch to detect parts dropped during robot handling and movement. Sensors can be used in the machines to verify the correct location and orientation of the part. The decoupler can be equipped with inspection devices, part-washing capability, deburring, or whatever function is necessary. Each machine must have the capability to send status signals to the cell controller.

Cells designed with new machines are capable of supplying these signals. In the case of older machines and situations where robots are being retrofitted to existing equipment, revisions have to be made to the electrical panel and control computer of the machines.

Table 5 Cell Design Attributes—Robotic Cell

Stationary Robot
U-shaped with one-piece flow
Flexibility—handle a family of parts
Reliable—does not breakdown
Quality control built into system to prevent defects

8 MISALIGNMENT/COMPLIANCE

When a part is being loaded by the robot into a machine tool, there may be minor misalignment of the part in the workholding device due to forces on the part from the robot. When the chuck is closed on the part, excessive force may be exerted on the robot arm. In these situations some robot manufacturers provide programmable compliance within the robot. FANUC, for example, offers a software feature called *SoftFloat* that allows the robot to become compliant while it is stationary and waiting for the chuck to close. Any misalignment will be absorbed by the compliant action of the robot. Once the workholding device is closed and the part located, the robot can open its gripper and move away under normal position control.

9 PART STORAGE BETWEEN OPERATION—DECOUPLERS

When machines require frequent adjustment, holding parts between machines using decouplers at points A, B, and C, as shown in Figure 14, may provide an advantage. Using decouplers offers several advantages, such as the ability to reverse the robot movements and the ability to shut down a machine for adjustment. For example, if the grinder were shut down, the robot would deposit the part removed from the turning center into decoupler A, bypass the grinder, pick up a part from decoupler B, and load it into the next machine. Branch routines would have to be included in the robot program to allow it to make these decisions. While service is being performed on the grinder, an electrically interlocked physical barrier guard must be in place to protect service personnel from accidents.

There are different possible designs of decouplers. A simple method is to provide a length of downward sloping roller conveyor capable of storing a sufficient number of parts to keep the system running for the length of time required for average maintenance time. The difficulty is in designing a decoupler that can handle a family of parts flexibly. A part present sensor would also be required at the pickup point. Based on the previous example of 30 parts/hour, storing 10 parts would accommodate a 20-minute shutdown without disrupting the output of the system. A simpler design for a decoupler would be a kanban square with multiple locations. This adds considerable complexity to robot programming, but a greater number of parts can usually be stored.

10 INCREASING PRODUCTION RATE

To explore another system-planning possibility, go back to the earlier situation where the turning center machining cycle time was twice as long as that for the grinder and two drilling machines. Recall that these later machines would be sitting idle half the time waiting for parts from the turning center. For these machines the cutting speeds or feed rates may be decreased. This of course increases the machining cycle time. However, small decreases in cutting speed can greatly improve the tool life (tools may last the entire shift). Decreasing the feed may improve the surface finish.

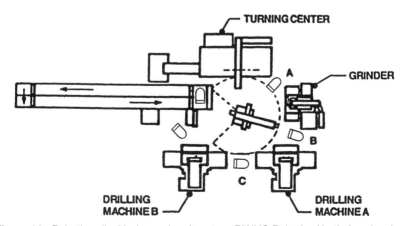

Figure 14   Robotic cell with decouplers (courtesy FANUC Robotics North America, Inc.).

Alternatively, production could be doubled by adding a second turning center to the cell. However, this may also necessitate adding a second robot and rearranging the equipment and the robot paths. As complex as this revision may seem, it does enable a 100% production increase by adding only two more pieces of equipment and some decouplers.

11 ADVANCED FUNCTIONS

Robot manufacturers offer a variety of software/hardware features that simplify the design and building of robotic cells. Customers should verify that the required features are available for the robot that they are planning on using in a manufacturing cell.

11.1 Space Check

For some robots the space check function, incorporated into a robot, monitors the predetermined interference area. When another robot or a peripheral unit is located within that area, the function stops the operation of the robot if a move command specifying movement into that area is issued to the robot. The function automatically releases the robot to operate when the other robot or peripheral has moved out of the interference area. This check is suitable for a multirobot cell, as shown in Figure 10.

11.2 Reference Position

This feature defines position limits within which an output signal will turn on. To use it the position of the robot is specified, an output signal is defined, and the tolerance range for each axis joint position is optionally specified. This feature can be used to enhance the safety and functionality of a robotic cell.

11.3 Line and Circular Tracking

Tracking enables a robot to treat a moving workpiece as a stationary object. Unlike the application shown in Figure 9, this feature is used in conveyor applications where the robot must perform tasks on moving workpieces without stopping the assembly line or rotary table.

There are some special considerations when selecting a tracking robot. For example, when a conveyor is stopped it may oscillate back and forth before it is completely stopped. However, the robot comes to a complete stop sooner. This may pose a problem if the robot is working close to the part on the conveyor. For a six-axis articulated robot it may be helpful to utilize a robot that has brakes only on axis 2 and 3, so that when the conveyor rocks back and forth the robot will move with the conveyor if the part hits the end-effector. For example, FANUC M-6 is a 6-kg payload robot that comes with two different brake options. One model comes with joint 2 and joint 3 brakes only and the other has all-axis brakes. The model with joint 2 and joint 3 brakes only may be preferable in this type of tracking application. Typically this option is not available on robots with higher payload capacity.

12 EXAMPLE: SIMPLE ROBOTIC MANUFACTURING CELL

Figure 15 shows a system in operation at Custom Products Corporation, Oconomowoc, Wisconsin. In this cell a FANUC LR Mate robot is mounted directly on the machine structure, eliminating the need for a mounting structure. LR Mate is a 5 or 6 axis robot weighing about 32 kg (70 lb). This robot is also designed to withstand constant splashing of the coolant, as is typical of machining centers. Die-cast parts are manually loaded into a vibratory bowl feeder located close to the LR Mate robot. The LR Mate robot picks a part from the feeder track, places it into the machining center, and returns to the feeder for another part. As parts are machined they are automatically released from the machine and transferred out of the cell via an exit conveyor. The entire load/unload sequence is completed in less than seven seconds. This robotic material handling system has allowed Custom Products to increase its efficiency by 30%.

13 EXAMPLE: LARGE ROBOTIC MANUFACTURING CELL

The robotic system shown in Figure 16 uses two six-axis S-420F robots and six four-axis A-510 robots, each equipped with custom end-of-arm tooling, to automatically load antilock brake components in and out of three types of milling machines. The entire eight-robot system was installed as an integrated modular cell.

In operation the S-420 robot, which is able to lift up to 264 pounds (120 kg), automatically picks up four parts from an inbound conveyor and loads them into one of four

Figure 15 Mounted robot on the machine tool (courtesy FANUC Robotics North America, Inc.).

Figure 16 Large conveyor system with robots and pallets (courtesy FANUC Robotics North America, Inc.).

vertical machining centers. After processing, the robot unloads the parts and returns them to the conveyor.

Next the system uses six vertical machining centers, each matched with an A-510 robot, which can handle loads weighing as much as 44 pounds. The robots load and unload the vertical mills, then return finished brake components to a conveyor that transfers parts for final machining.

During the final processing stage parts are loaded into one of four horizontal machining centers, again using an S-420F robot. The parts are unloaded from the horizontal mills and returned to an outbound conveyor.

The above system is in operation at Remmele Engineering, Big Lake, Minnesota. According to Remmele, this cell has allowed them to meet their customer's just-in-time delivery requirements.

14 SUMMARY

In contrast to the early 1980s, the robotics industry has been growing very steadily in the 1990s. This is due to several technological advances that have resulted in increased reliability of robots and due to the systematic method of building the robotic manufacturing cells. Today robot manufacturers offer robots with 30,000–50,000 hours of mean time between failure (MTBF). The trend toward increased use of robotic manufacturing cells will continue into the twenty-first century, and we will see applications in new areas.

REFERENCES

Ayres, R. U., and D. C. Butcher. 1993. "The Flexible Factory Revisited." *American Scientist* **81,** 448–459.
Black, J. T. "The Design of Manufacturing Cells." In *Proceedings of Manufacturing International '88,* Atlanta. Vol. 3. 143.
Black, J. T., and B. J. Schroer. 1988. "Decouplers in Integrated Cellular Manufacturing Systems." *Journal of Engineering for Industry* **110,** 77.
De Garmo, E. P., J. T. Black, and R. A. Kohser. 1997. *Materials and Processes in Manufacturing.* 8th ed. Upper Saddle River: Prentice-Hall International.

ADDITIONAL READING

Black, J. T. "Cellular Manufacturing Systems, An Overview." *Industrial Engineering* **15**(11), 36–48, 1983.
————. "IMPS—Integrated Manufacturing Production Systems." In *Proceedings of the IIE 1988 Integrated Systems Conference.* 125, 1988.
————. *The Design of the Factory with a Future.* New York: McGraw-Hill, 1990.
Burbidge, J. L. "Group Technology in the Engineering Industry." London: *Mechanical Engineering Publications, LTD,* 1979.
GMFanuc Robotics, "Developing a Workcell Application." Handout.
Ham, I., K. Hitomi, and T. Yoshida. *Group Technology—Applications to Production Management.* Dordrecht: Kluwer-Nijhoff, 1985.
Jiang, B. C., et al. "Determining Robot Process Capability Using Taguchi Methods." *Robotics and CIM* **6**(1), 55–66, 1989.
Monden, Y. *Toyota Production System: Practical Approach to Production Management.* Norcross: IIE Press, 1983.
Nakajima, S. *Introduction to TPM.* Cambridge: Productivity Press, 1988.
Schonberger, R. J. *Japanese Manufacturing Techniques.* New York: The Free Press, 1982.
Shingo, S. *A Revolution in Manufacturing: The S.M.E.D. System.* Cambridge: Productivity Press, 1985.
————. *Study of Toyota Production System from Industrial Engineering Viewpoint.* Cambridge: Productivity Press, 1989.
————. *Zero Quality Control: Source Inspection and the Poka-Yoke System.* Cambridge: Productivity Press, 1986.
Spur, G., G. Seliger, and B. Vichweger. "Cell Concepts for Flexible Automated Manufacturing." In *Proceedings of CASA/SME Autofact Europe Conference,* September 1983. 197.
Suh, N. P. *The Principles of Design.* Oxford: Oxford University Press, 1990.
Suh, N. P., A. C. Bell, and D. C. Gossard. "On an Axiomatic Approach to Manufacturing and Manufacturing Systems." *Journal of Engineering for Industry* **100**(2), 127–130, 1978.
Zelenovic, D. M. "Flexibility—A Condition for Effective Production Systems." *International Journal of Production Research* **10**(3), 319, 1982.

CHAPTER 36

RELIABILITY, MAINTENANCE, AND SAFETY OF ROBOTS

David R. Clark
Kettering University
Flint, Michigan

Mark R. Lehto
Purdue University
West Lafayette, Indiana

1 INTRODUCTION

Reliability, maintainability, and safety are closely related issues that greatly influence the effectiveness of robot applications. Attaining highly reliable, easily maintained, and safe applications of robotics can be difficult. Robots are complex computer systems that sense information from their environment, autonomously make decisions, and manipulate the environment with effectors. The complexity of robotics applications increases the chance that malfunctions and errors will occur and can make it difficult to detect and correct them before they result in system failures or accidents. The most easily analyzed failures and errors can be directly mapped to particular elements of the robotics system. More complex error modes involve an interaction between system elements, and error modes may also be interrelated, leading to difficult tradeoffs. For example, performing maintenance tasks may increase reliability but at the same time may expose operators to additional hazards.

Solutions to reliability, maintenance, and safety problems can be found at each stage of the robotics system's life cycle. Implementing solutions in the design and development stages increases the chance of completely eliminating problems and can help prevent design defects. Design solutions include the provision of redundancy, fault detection and error correction algorithms, and many different types of safety guarding. Solutions applicable in the production phase are drawn from the field of quality assurance (Hancock, Sathe, and Edosomwan, 1992) and focus on preventing manufacturing defects. Such solutions include establishment of specifications and tolerances, use of high-reliability components, on-line inspection and control procedures, testing and component burn-in, and employee training and motivation programs. Solutions applicable in the use stage include maintenance and many robot safeguarding approaches.

The following discussion will separately discuss the reliability, maintainability, and safety of robot systems. Although these topics will be addressed in separate sections, an effort will be made to show the strong linkage between these topics throughout this discussion.

2 RELIABILITY

The *reliability* of a system is normally defined as the probability that the system will give satisfactory performance (i.e., not fail) over a given time interval. The frequency at which failures occur over time is the failure rate, $\lambda(t)$, and is measured as the number of failures occurring per unit time. The mean time between failures (MTBF) is another commonly used measure of reliability. For a constant failure rate, λ, the MTBF is equal to $1/\lambda$.

The reliability of a system generally changes over its life cycle. Figures 1*a* and 1*b* illustrate the typical "bathtub" curve, showing how the failure rate for most systems

Handbook of Industrial Robotics, Second Edition, Edited by Shimon Y. Nof
ISBN 0-471-17783-0   © 1999 John Wiley & Sons, Inc.

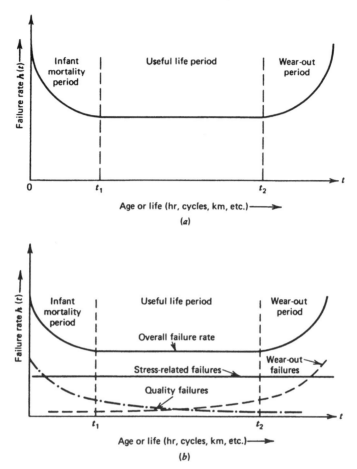

Figure 1   (a) Failure rate-life characteristic curve. (b) Failure rate based on components of failure. (From K. C. Kapur, "Reliability and Maintainability," in *Handbook of Industrial Engineering,* ed. G. Salvendy. New York: John Wiley & Sons, 1992. Chapter 89.)

changes over time. When a system is first put into service, the failure rate is often elevated. This so-called "infant mortality" phase often reflects quality problems or manufacturing defects (Figure 1b). Toward the end of a system's useful life, the failure rate again increases as components of the system begin to wear out.

It is important to realize that in practice, a robot may not be operational even if the robot itself has not failed. Shipping, handling, storage, installation, operation, maintenance, and other activities can all reduce the proportion of time a robot is operational. Operational *readiness* is the probability that a system is operating or can be operated satisfactorily and takes into account the effects of such factors on the robot system.

Availability is the probability at any point in time that the system will be ready to operate at a specified level of performance. Availability is calculated using the MTBF and the mean time to restore (MTTR) or repair the system after a failure occurs. The relationship is expressed as:

$$\text{Availability} = MTBF/(MTBF + MTTR) = 1/(1+MTTR/MTBF) \qquad (1)$$

If a system is required to have a certain availability factor and its MTBF is known, then the required restoration or repair time can be determined. Note, however, that the availability factor alone may be an insufficient criterion. In production circumstances, equipment having an MTBF of 500 hr and an MTTR of 4 hours (MTTR/MTBF ratio of 0.008) may be much more acceptable than equipment having a 5000-hour MTBF but a 40-hour MTTR (also a ratio of 0.008).

The MTTR is, then, a measure of overall *maintainability*. The MTTR will depend on how quickly, easily, and accurately a malfunction can be diagnosed and corrected and the system restored to normal operation.

2.1 Reliability Data

Little has been published on the actual level of reliability achieved by robots. Robots have been estimated to have a useful life of at least 40,000 hours and a MTBF of at least 400 hours (Dhillon, 1991). Robot availability has been estimated to be at least 97%, and higher estimates have been reported (Munson, 1985). However, these estimates of robot reliability can be somewhat misleading. Part of the issue is that the MTBF varies greatly between applications. Dhillon (1991) cites data showing that about 30% of robot applications have a MTBF of 100 hours or less; 60% have a MTBF of 500 hours or less; 75% have a MTBF of 1000 hours or less; 85% have a MTBF of 1500 hours or less; and 90% have a MTBF of 2000 hours or less.

It also must be recognized that the robot itself is not necessarily the main cause of robot system downtime. This point was emphasized in a study of 37 robot systems used in three companies covering over 20,000 hours of production (Jones and Dawson, 1985). The data showed that problems not directly attributable to robot units greatly reduced the availability of the robot system. The overall availability of the robot systems was only 75% of production time, despite the fact that the robot-related downtime averaged across the three companies was only 5.3%. The robot-related problems therefore accounted for only about 20% of the overall downtime, suggesting to Jones and Dawson "that it is the systems as a whole and not the robots alone which are the main cause of 'unavailability.' "

Robot systems can be nonoperational for many different reasons (Table 1). Some data are available documenting the prevalence of particular types of robot system failures or failure modes and their impact on downtime (Jones and Dawson, 1985). That study showed that the prevalence of particular failure modes depends greatly on the type of application. Erratic or out-of-tolerance movements were a fairly frequent problem for all of the companies surveyed. Tool failures were a particularly prevalent problem for spot welding applications. Controller failures and robot arm failures were a particularly prevalent problem for arc welding applications. Average downtime for the latter types of failures was quite long; for controller failures one company reported an average of 6 hours of downtime; for robot arm failures one company reported an average of 3.6 hours of downtime. Average downtime for tool failures ranged from 15.8 to 17.7 minutes; average downtime for erratic or out-of-tolerance movements ranged from 13.1 to 77.5 minutes.

This variation in the prevalence and influence of particular failure modes on the operational reliability of robot systems, along with the substantial influence of system factors, suggest that there is a great need to carefully analyze reliability for any particular application from an overall system perspective. Along these lines, Figure 2 presents a structured method for achieving reliability targets. As indicated by the figure, inspections and tests conducted at each reliability control point in the robot equipment's life cycle, beginning with initial system configuration and change control activities and ending with field maintenance and surveillance, ensure that appropriate system specifications and standards are both established and met. This process requires substantial interaction and cooperation between all parties involved, including suppliers, vendors, the robot manufacturer, and the customer who is purchasing the robot system.

At a more analytical level, a number of tools have been developed to help designers focus on the critical failure modes of a robot system, including human error. Such approaches guide the systematic identification of reliability problems and the development of countermeasures. Following such procedures documents what has been done, helps prevent critical omissions, and in some cases provides quantitative estimates of system reliability that can be used to measure the expected costs and benefits of particular countermeasures. It should also be emphasized that these approaches can be applied to evaluate reliability, safety, and productivity alike. The first step is to identify failures modes and their effects. Analytical tools can then be applied that quantitatively document both the criticality of particular failure modes and the effectiveness of proposed solutions.

Table 1   Categories of Robot System Failure[a]

Robot-Related	Nonrobot-Related
Component Failures	*Component Failures*
Failure of robot arm	Failure of other production equip-
joint sensor	ment
joint motor	Failure of safety equipment
other	interlocked gates, fences, guards
Tool failure	other
Cable/transmission problem	Failure of services
Controller failure	electricity
software error	water
hardware error	compressed air
Power supply fault	other
power surge	Parts out of tolerance
blown fuses	
loss of power	*Process Failures*
Teach pendant fault	Sequence faults
faulty button	System failure
other	Human error
Sensor fault	Defective process
Hydraulics failure	
	Other Delays
Process Failures	System checks
Out of synchronization	Part checks
Emergency stops	Maintenance
Controlled stops	Repair
Robot collision	
Stiffness in robot arm	
Dropped/damaged part	
Robot frozen	
Erratic movements	
Movements out of tolerance	

[a] Adapted from R. Jones and S. Dawson, "People and Robots—Their Safety and Reliability," in *Robot Safety,* ed. M. C. Bonney and Y. F. Yong. New York: Springer-Verlag, 1985.

2.2   Failure Modes and Effects Analysis

Failure modes and effects analysis (FMEA) is a systematic procedure for documenting the effects of system malfunctions on reliability and safety (Hammer, 1993). Variants of this approach include preliminary hazard analysis (PHA) and failure modes effects and criticality analysis (FMECA). In all of these approaches, worksheets are prepared that list the components of a system, their potential failure modes, the likelihood and effects of each failure, and both the implemented and the potential countermeasures that might be taken to prevent the failure or its effects. Each failure may have multiple effects. More than one countermeasure may also be relevant for each failure.

Identification of failure modes and potential countermeasures is guided by past experience, standards, checklists, and other sources. This process can be further organized by separately considering each step in the production of a product performed by the robot. For example, the effects of a power supply failure for a welding robot might be separately considered for each step performed in the welding process.

FMEA is clearly a useful tool for documenting sources of reliability and safety problems and also helps organize efforts to control these problems. The primary shortcoming of the approach is that the large number of components in robot systems imposes a significant practical limitation on the analysis. That is, the number of event combinations that might occur is an exponential function of the large number of components. Applications in the field of robotics consequently are normally confined to the analysis of single-event failures (Visinsky, Cavallaro, and Walker, 1994*a*).

2.3   Quantitative Measures

Quantitative approaches focus on describing the reliability of a system in terms of the component probabilities, $p_1(t), \ldots p_n(t)$ that the n components of the system are func-

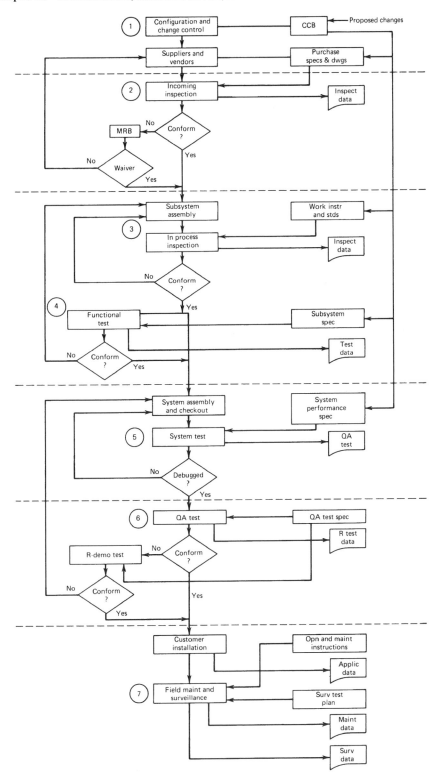

Figure 2   Reliability control points in the robot system's life cycle. (From G. E. Munson, "Industrial Robots: Reliability, Maintenance, and Safety," in *Handbook of Industrial Robotics,* ed. S. Y. Nof. New York: John Wiley & Sons, 1985. 722–758. Photo courtesy Unimation, Inc.)

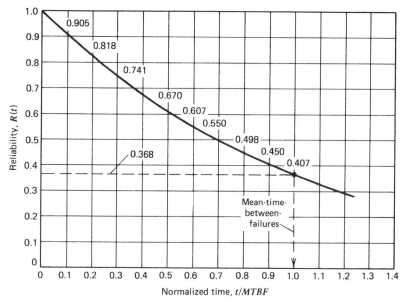

Figure 3 Exponential reliability function. (From G. E. Munson, "Industrial Robots: Reliability, Maintenance, and Safety," In *Handbook of Industrial Robotics,* ed. S. Y. Nof. New York: John Wiley & Sons, 1985. Photo courtesy NAVWEPS, Naval Weapons Center 00-65-502.)

tioning at some time, t. The reliability function $R(p)$ specifies the probability the system is functioning in terms of these component probabilities.

When the failure rate, $\lambda(t)$, of a system component is constant, the probability of failure, $f(t)$, at some time t is exponentially distributed.[1] That is,

$$f(t) = \lambda e^{-\lambda t} \qquad \text{where } t \geq 0, \ \lambda > 0 \tag{2}$$

and λ is the constant failure rate. The probability, $F(t)$, that the system component fails between time 0 and time t is obtained as:

$$F(t) = \int_0^t f(t) \, dt \tag{3}$$

$$= \int_0^t \lambda e^{-\lambda t} \, dt \qquad \text{where } t \geq 0, \ \lambda > 0 \tag{4}$$

$$= 1 - e^{-\lambda t} \tag{5}$$

and

$$p(t) = 1 - F(t) = e^{-\lambda t} \tag{6}$$

$$= e^{-t/\text{MTBF}} \tag{7}$$

Equation (7) shows that the probability any particular system component is functioning goes to zero as t gets large. For a given time interval, the probability of functioning increases when λ decreases, and decreases when the MTBF increases. Figure 3 illustrates this relationship in terms of t/MTBF, or the proportion of the MTBF that the system has been running.

---

[1] The normal, lognormal, Weibull, and Gamma distributions are also used in reliability theory to describe the probability of component failure over time (Kapur, 1992).

Table 2   Unimate System Reliability Estimate[a]

Failure classification	Failure rate (x 10⁻⁶)	MTBF (hours)
Part failures only:		
Electronic/Electrical	555	1800
Mechanical/Hydraulic	673	1485
Non-part failures:		
Electronic/Electrical	267	3745
Mechanical/Hydraulic	475	2100
System failures:		
Parts only	1228	815
Non-tolerance	742	1350
Combined	1970	508

Estimated reliability feasibility, Unimate 2000: MTBF = 500 Hours

[a] Photo courtesy Unimation, Inc.

If the system consists of a single component, it holds that the reliability function

$$R(p) = p(t) = e^{-t/\text{MTBF}} \qquad (8)$$

In many cases, however, the analyst is interested in predicting the system reliability from the reliability of its components. Dhillon (1991) provides a comprehensive overview of documents, data banks, and organizations for obtaining failure data to use in robot reliability analysis. Component reliabilities used in such analysis can be obtained from sources such as handbooks (MIL-HBDK-217F: *Reliability of Electronic Equipment*), data provided by manufacturers (Table 2), or past experience. Limited data are also available that document error rates of personnel performing reliability-related tasks, such as maintenance (Table 3). Methods for estimating human error rates have been developed (Gertman and Blackman, 1994), such as THERP (Swain and Guttman, 1983) and SLIM-

Table 3   HEPs Associated with Maintenance Activities[a]

Task Name	Individual HEP	Redundant HEP
Prepare for pressure-decay check for turbopump gearbox	6.8×10^{-2}	4.7×10^{-2}
Prepare for electrical check	6.7×10^{-2}	4.8×10^{-2}
Visually inspect work area after test, ensure integrity and clean tools	5.3×10^{-2}	4.0×10^{-2}
Service oil sump of turbo gearbox	4.6×10^{-2}	3.2×10^{-2}
Prepare for turbopump torque test	4.3×10^{-2}	2.8×10^{-2}
Perform turbopump torque check of subassembly	4.0×10^{-2}	2.5×10^{-2}
Prepare for installation of gearbox pressurization kit	4.0×10^{-2}	2.6×10^{-2}
Prepare for subassembly turbopump torque check	1.7×10^{-2}	9.0×10^{-3}
Perform leak check of turbopump fuel pump seal	2.4×10^{-1}	1.9×10^{-1}
Perform electrical check	2.2×10^{-1}	1.7×10^{-1}
Service oil sump of gearbox	4.6×10^{-2}	3.2×10^{-2}

[a] From D. L. Gertman and H. S. Blackman, *Human Reliability and Safety Analysis Data Handbook.* New York: John Wiley & Sons, 1994.

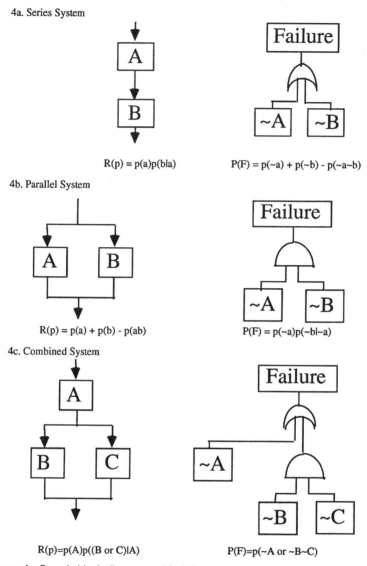

Figure 4   Generic block diagram and fault tree representations of simple systems.

MAUD (Embrey, 1984). THERP follows an approach analogous to fault tree analysis, as discussed below in Section 4.2, to estimate human error probabilities (HEPs). In SLIM-MAUD expert ratings are used to estimate HEPs as a function of *performance-shaping factors* (PSFs).

Given that system component failure rates or probabilities are known, the next step in quantitative analysis is to develop a model of the system showing how system reliability is functionally determined by each component. The two most commonly used models are (1) the systems block diagram and (2) the system fault tree. Fault trees and system block diagrams are both useful for describing the effect of component configurations on system reliability. The most commonly considered configurations in such analysis are (1) serial systems, (2) parallel systems, and (3) mixed serial and parallel systems.

In a purely serial system (Figure 4*a*) each component of the system must function, or else the system will fail. This tendency is shown in the block diagram on the left side of the figure and the corresponding fault tree on the right side of the figure. The block and fault tree representations differ in that the block diagram shows an operational system, while the fault tree shows failures of the system. Note that:

$$R(p) = p(a)p(b|a) \tag{9}$$

where $p(a)$ is the probability that component a is operational, at time t, and $p(b|a)$ is the conditional probability that component b is functional, at time t, given that a is operational. Also, note that

$$R(p) = 1 - P(F) \tag{10}$$

where $P(F)$ is the probability of system failure given by the system fault tree. Boolean algebra can be used to analyze the cut-sets or failure paths. For a serial system each component failure results in system failure. Consequently, the number of ways the system can fail equals the number of components, leading to the result:

$$P(F) = p(\sim a) + p(\sim b) - p(\sim a \sim b) \tag{11}$$

$$= p(\sim a) + p(\sim b) - p(\sim a)p(\sim b|\sim a) \tag{12}$$

where

$$
\begin{aligned}
p(\sim a) &= \text{probability that component } a \text{ fails} \\
p(\sim b) &= \text{the probability that component } b \text{ fails} \\
p(\sim a \sim b) &= \text{the probability that both } a \text{ and } b \text{ fail} \\
p(\sim b|\sim a) &= \text{the conditional probability that component } b \text{ fails, given that component} \\
&\quad\; a \text{ has also failed}
\end{aligned}
$$

Also, note that:

$$p(\sim a) = 1 - p(a) \tag{13}$$

If the component failures are independent, then:

$$p(b|a) = p(b), \quad \text{and} \tag{14}$$

$$R(p) = p(a)p(b) \tag{15}$$

or more generally,

$$R(p) = \prod_{i=1}^{m} p_i(t) \tag{16}$$

where m equals the number of serial components.

In a parallel system (Figure 4b) every component of the system must fail for the system to fail. Consequently,

$$R(p) = p(a) + p(b) - p(ab) \tag{17}$$

using the equations

$$R(p) = 1 - P(F), \quad \text{and} \tag{18}$$

$$P(F) = p(\sim a)p(\sim b|\sim a) \tag{19}$$

we get the result

$$R(p) = 1 - p(\sim a)p(\sim b|\sim a) \tag{20}$$

If the component failures are independent, we then obtain the general equation

$$R(p) = 1 - \prod_{i=1}^{m} (1 - p_i(t)) \tag{21}$$

where m equals the number of serial components.

A mixed series–parallel system is shown in Figure 4c. As shown in the figure, system reliability can be calculated from component reliabilities following a similar approach to that outlined above for the series and parallel systems. By application of this approach, the probability of each failure path or cut-set can be calculated for the system. These probabilities can then be combined with cost data to evaluate the cost-effectiveness of particular countermeasures (Lehto, 1997). The above approach can be extended to include robot repair rates in Markov chain analysis (Dhillon, 1991). Analytical solutions are fairly straightforward when failure rates and repair rates are assumed to be constant and independently distributed (see Ross, 1970).

Simulation is a useful approach that applies elements of the above approach, but allows more complex models to be developed by making fewer simplifying assumptions about the system (Pritsker, 1992). Workcell simulations of how a robot interacts with other equipment are commonly used to improve layouts, maintenance and production schedules, and other elements of processes that impact system reliability. Note that tasks performed by people, as well as errors and violations, can be included in the simulation models.

2.3 Design for Reliability

A number of generic design strategies can be followed to increase the reliability of robot systems. These strategies can be roughly classified into three categories: (1) fault avoidance, (2) fault tolerance, and (3) fault detection.

Methods of fault avoidance focus on preventing or reducing the occurrence of faults. One common strategy is to select highly reliable components. Safety margins can be provided by designing components to withstand stresses well above those expected in normal operation. Probability theory can be applied to predict the probability of component failure for particular safety margins and conditions (Kapur, 1992). In this approach probability density functions $f(x)$ and $g(y)$ are used to, respectively, describe the stresses incurred, x, and the ability to withstand particular levels of stress, y. The probability of component failure is then calculated as:

$$P(F) = P(y > x) \tag{22}$$

$$= \int_{-\infty}^{\infty} g(y) \left[\int_{-\infty}^{y} f(x)\, dx \right] dy \tag{23}$$

Other methods of fault avoidance include burn-in, timed replacement, component screening, software screening and debugging, and accelerated life testing (Hammer, 1993). Burn-in methods involve subjecting the component to expected stress levels for a limited time prior to installation. This approach is useful when a component has a high "infant mortality" failure rate. Timed replacement involves the replacement of components prior to entering the "wear-out period," in which failure rates are increased (see Figure 1b). Component screening involves the establishing of tolerances, or upper and lower limits for component attributes. Components outside of the tolerable range are rejected. Software screening and debugging involves testing the software under a wide variety of foreseeable conditions as well as applying methods used in software engineering to eliminate errors in code. Accelerated life testing involves testing components under more adverse conditions than those normally present.

Fault tolerance is a second commonly applied approach. Ideally, when components fail, robot systems should lose their functionality gradually rather than catastrophically. Fault tolerance is a necessary strategy because highly reliable components can be very expensive. Furthermore, the complexity and large number of components within robot systems make hardware and software failures inevitable. Methods of fault tolerance include system redundancy, error correction, and error recovery.

Three forms of redundancy commonly employed in robot systems are: (1) structural (hardware), (2) kinematic (adding extra degrees of freedom), and (3) functional redundancy (Visinsky, Cavallaro, and Walker, 1994a). Structural redundancy is present when a function of the robot system is concurrently performed by two or more system components. In such configurations system failure can theoretically occur only when all of the parallel components fail. The use of standby systems illustrates a similar form of hardware redundancy. Kinematic redundancy entails the provision of extra degrees of freedom for the robot's effectors. Doing so, allows fault tolerant algorithms to calculate alternative movement paths when joints in effectors fail. Functional redundancy involves the use of functionally equivalent data from dissimilar components.   Fault-tolerant computing schemes that provide for error correction and recovery are used extensively in robot systems. Such methods include parallel processing of information, use of modular code, voting schemes, and a variety of system rollback schemes. Parallel processing and the use of modular code are both forms of redundancy. Voting schemes and probabilistic methods provide a means of error correction by comparing information received from multiple redundant sources. System rollback schemes store system state information. This theoretically allows the system to return to an earlier "correct" state and resume operation after an error occurs. In the simplest implementations the system will retry the operation after a brief pause. More complex schemes will respond to the error by appropriately reconfiguring the system before attempting the operation again. Expert systems provide another means of implementing fault tolerance in robot systems (Visinsky, Cavallaro, and Walker, 1992, 1994*b*).

Fault detection plays a general role in both improving and maintaining the reliability of robot systems and is a critical element of real-time fault-tolerant computing. A wide variety of fault-detection algorithms have been developed for this purpose (Visinsky, Cavallaro, and Walker, 1994*a*). Analytical redundancy is a concept often exploited by these algorithms. That is, knowledge of the structural and functional redundancy of the robot system is used to guide the comparison of data obtained from different sensors. Faulty sensors provide data that deviate by more than the fault detection threshold from values calculated or obtained from other sources. Determining the appropriate fault-detection threshold can be difficult because both sensor noise and the criticality of particular errors change, depending on the operation performed, component failures, and other factors.

3   MAINTENANCE

Proper maintenance is essential to attaining high levels of robot reliability and safety. Modern robot-maintenance programs focus on predicting and preventing failures rather than simply fixing them after they occur (Dhillon, 1991). Well in advance of robot installation and start-up, the maintenance manager will want to establish a program of preventive maintenance, coordinate it with the production manager, and provide for its support. Formally establishing a maintenance program is an essential first step in making sure a robot system will be properly maintained. A maintenance program will typically:

1. Specify, assign, and coordinate maintenance activities
2. Establish maintenance priorities and standards
3. Set maintenance schedules
4. Specify diagnosis and monitoring methods
5. Provide special tools, diagnostic equipment, and service kits
6. Provide control of spare parts inventory
7. Provide training to maintenance personnel and operators
8. Measure and document program effectiveness

These elements of a maintenance program are briefly discussed below.

3.1   Robot Maintenance Activities

Maintenance activities or tasks include (Dhillon, 1991):

1. Repair
2. Service
3. Inspections
4. Calibration
5. Overhauls
6. Testing
7. Adjusting
8. Aligning

9. Replacing components

Maintenance and production have traditionally been viewed as independent activities. However, coordination of maintenance activities with production is essential, as is developing a mutual understanding of the needs and purposes. Production must allot time for maintenance. If a one-shift operation is involved, scheduling should be no problem. If it is a three-shift operation, then either weekend maintenance must be planned for or downtime scheduled on one of the shifts.

Some users service robot systems with several different classes of skilled people (e.g., electricians, millwrights, electronic technicians, hydraulic trades persons) having separate and divided responsibilities, while others use personnel with integrated skills. There are merits to both approaches, and the solution is what works best for the specific installation. However, for the benefit of efficiency, proficiency, and fast response time, the trend is toward integrated skills and responsibilities. The latter trend, inspired much by developments in Japan, both emphasizes the critical role the operator must play in maintenance and focuses on cooperation between production and maintenance departments as the key to increasing equipment reliability (Nakajima et al., 1992).

Rather than all responsibility for maintenance being assigned to the maintenance department, operators are now being assigned many routine maintenance tasks. Tasks assigned to operators include (Nakajima et al., 1992):

1. Maintaining basic equipment conditions (cleaning and lubricating equipment; tightening bolts)
2. Maintaining operating conditions through proper operation and inspection of equipment
3. Detecting deterioration of equipment through visual inspection and early identification of signs indicating abnormalities during equipment operation

Maintenance personnel are expected to concentrate their efforts on less routine tasks. Such tasks include:

1. Providing technical support for the production department's maintenance activities
2. Restoring deterioration of equipment through inspections, condition monitoring, and repair activity
3. Improving operating standards by tracing design weaknesses and making appropriate modifications

Inspection is a particularly critical maintenance task that is also often difficult and tends to be error-prone (Drury, 1992). Interventions to improve the performance of human inspectors, such as the development of inspection procedures, employee training, and specification and control of task and environmental conditions, are likely to improve the reliability of inspection in robot applications.

3.2 Setting Maintenance Priorities and Standards

Maintenance efforts must be prioritized because maintenance programs are expensive and the resources allocated to maintenance must consequently be allocated efficiently. The annual cost of robot maintenance has been estimated to be about 10% of the initial procurement cost (Dhillon, 1991), but obviously these costs will vary depending upon what is done.

Maintenance priorities must reflect the cost to the company of equipment failures. FMEA, discussed in Section 2.2, provides data useful in setting maintenance priorities. Table 4 provides an example (taken from Nakajima et al., 1992) illustrating how maintenance priorities can be set as a function of the effect of failures. As indicated in the table, the priority assigned to particular equipment should reflect the degree that failures:

1. Disrupt production
2. Waste expensive materials
3. Reduce quality
4. Increase downtime
5. Impact safety

Table 4 TPM Priority Management Table for Rating Equipment[a]

Area	Item	Evaluation			Evaluation Standard
Production	1. How often is the equipment used?	4	2	1	80% or above: 4 59% or below: 1
	2. Is there backup equipment?	5	4	2	1	No (or) yes, but it takes too many person-hours: 5 Available at other plants: 4 Covered by stock: 2 Backup equipment exists: 1
	3. How high is the dedication? (the proportion of products of a similar type produced by the equipment)	4	2	1	100–75%: 4 0–35: 1 35–75%: 2
	4. To what extent will a failure effect other processes?	5	4	2	1	Affects the entire plant: 5 Strongly effects other processes: 4 Only effects this machining center: 1
Quality	5. Value of monthly scrap losses (burned rubber, wasted cloth, wasted production)	4	2	1	(Burnt rubber) (Wasted cloth) (Wasted production) Over $1000 Over$5000 Over $1000 $500–$1000 $2000–5000 $5000–1000 Under $500 Under $2000 Under $500
	6. How will the process run on this equipment affect the quality of the finished product?	5	4	2	1	Decisively: 5,4 Not significantly: 1 Somewhat: 2
Maintenance	7. Frequency of failures in terms of cost of monthly repairs?	4	2	1	Over $5000: 4 Under $3000: 1 $3000–5000: 2
	8. Mean time to repair (MTTR)	4	2	1	MTTR over 3 hr: 4 Under 1 hr: 1 1–3 hr: 2
Safety	9. To what extent does a failure affect the work environment? (noise, etc.)	5	4	2	1	Can be life threatening: 5 No significant effect: 1 Stops work: 4

A (priority-ranked equipment)(30 or more points); B (20–29 points); C (19 points or less).

[a] From S. Nakajima et al., "Maintenance Management and Control," in *Handbook of Industrial Engineering*, 2nd ed., ed. G. Salvendy. New York: John Wiley & Sons, 1992. Chapter 73.

Once maintenance priorities are developed, maintenance standards and procedures, as well as maintenance plans, should then be established that reflect these priorities. Maintenance priorities should also be reviewed on a periodic basis because they are likely to change over time for a variety of reasons. Typical high-priority activities include inspection and test procedures, lubricating effectors, changing internal filters in hydraulic and pneumatic robots, testing hydraulic fluids for contamination, checking water separators in pneumatic robots, changing air filters for cooling systems, and performing functional checks of controls and control settings. Maintenance standards and procedures will include specifications, tolerances, and methods to be followed when performing maintenance tasks. Checklists (Figure 5), maintenance manuals, and training programs are common means of making sure this information is made available to maintenance personnel.

3.3 Maintenance Schedules

Maintenance activity is generally scheduled to take place at regular intervals, after breakdowns, or when certain conditions are met. Time-based maintenance schedules are typically suggested by the robot's manufacturer (Table 5). Inspections are especially likely to be performed on a periodic basis. Daily inspections are normally performed by the operator prior to beginning operations. Less routine inspections are often performed on weekly, monthly, biannual, and annual schedules by maintenance personnel. After sufficient running time and performance results have been accumulated, users of robot systems may choose to deviate from the original schedule provided by the manufacturer. Replace-

Step Description	Checked	Corrective Action

Power Distribution and Interlock:

6. Check remote STOP function if used, by opening NC switch between A9 and B9 on customer access panel.

7. Check remote HOLD if used, by opening NC switch between A8 and B5 on customer access panel.

8. Check door interlock switch(es) if installed.

9. Check servo power relay by removing 4CR from relay bank. Unimate will be in HOLD.

ELECTRONIC (Power ON)

1. 918 W Board Lights.
   (a) Check that the encoder lamp monitor is functioning with Cycle Start out and Mode switch in REPEAT by pulling an encoder bulb.
   (b) Parity Error light should remain OFF in REPEAT with CYCLE START ON.
   (c) Check IT$_2$ light is ON at all times *except* when in REPEAT, AUTO with CYCLE START ON, and *not* in HOLD.
   (d) Check DC Indicator light is ON as long as AC power is being supplied to the power supply.

2. 918 T Board Light
   (a) Check Total Position Coincidence light will light when all motions reach TOTAL (position) Coincidence.

3. 918 D/C Board Light
   (a) Check True Total Coincidence light. (And so on.)

Figure 5 Sample page. Preventive maintenance checklist. (From G. E. Munson, "Industrial Robots: Reliability, Maintenance, and Safety," in *Handbook of Industrial Robotics,* ed. S. Y. Nof. New York: John Wiley & Sons, 1985. 722–758. Photo courtesy Unimation, Inc.)

Table 5   Sample of a Long-Term Parts-Replacement Schedule[a]

Part No.	Description	INTERVALS AT WHICH PARTS ARE TO BE REPLACED																			
		1000	2000	3000	4000	5000	6000	7000	8000	9000	10,000	11,000	12,000	13,000	14,000	15,000	16,000	17,000	18,000	19,000	20,000
403BD1	Prev-main. kit	X	X	X	X	X	X	X	X	X	X	X	X	X	X	X	X	X	X	X	X
377H3B	Encoder lamp										X										X
922N1	Hand gear train										X										X
121AH2 121AH3	U cup feedthru							X							X						
922W1	Rear drive bend										X										X
922V1	Rear drive yaw										X										X
127A1	Servo valves										X										X
313BH	Relief valve										X										X
825H1	Accumulator										X										X
1912AG5	Encoder cable										X										X
825F1 *	Scavenge pump										X										X
403BA1	Hyd. pump overhaul kit										X										X
182AU1	Press switch															X					
318L2	Air regulator															X					
318L3	Air splenoid															X					
313BD1	Unload valve																				X

[a] From G. E. Munson, "Industrial Robots: Reliability, Maintenance, and Safety," in *Handbook of Industrial Robotics*, ed. S. Y. Nof. New York: John Wiley & Sons, 1985. 722–758. Photo courtesy Unimation, Inc.

ment analysis provides one means of formally analyzing this problem, by comparing the expected value of replacing versus not replacing equipment at particular points in time.

Breakdown-based maintenance policies schedule the replacement of parts and other repair activity after failures occur. This approach can maximize the life of equipment components that might otherwise be replaced prior to completing their useful life. On the other hand, a breakdown-based policy can result in disruptive interruptions of production and serious damage to equipment that far outweigh the cost of replacing components before they begin to wear out.

Condition-based, or predictive, maintenance policies schedule part replacements and repairs after the equipment shows signs of deterioration (Figure 6). This approach theoretically results in less frequent maintenance activity than interval-based policies and fewer disruptions of production than a breakdown-based policy.

3.4 Diagnosis and Monitoring Methods

Many different approaches are used to monitor the status of robotics equipment, diagnose their condition, and then trigger maintenance activities, such as part replacement, repair, and overhauls, when warranted. Condition-based maintenance policies typically schedule inspections at regular intervals by the operator or maintenance personnel. Control charts and other methods used in the field of statistical quality control (Speitel, 1992; Brumbaugh and Heikes, 1992) are also used to determine deterioration over time in production quality and a consequent need for equipment maintenance activity, including nonroutine inspections, adjustments, recalibration, part replacement, or other repairs.

Other commonly applied diagnosis techniques include vibration analysis, wear debris (contaminant) monitoring of lubricants and hydraulic fluids, behavior monitoring, and structural integrity monitoring (Nakajima et al., 1992). Industrial robots also contain internal diagnostic functions, which are useful during troubleshooting and signal the need for conducting maintenance on the machine.

3.5 Tools and Diagnostic Equipment

Robot maintenance normally requires that maintenance personnel be provided special tools and diagnostic equipment. Robot manufacturers can provide help and guidance regarding this issue and generally provide such information as part of the robot's documentation package. Ordinarily the investment in special tools will be minimal, but the cost without them can be significant. Among the most common special tools are circuit card pullers, torque wrenches, seal compressors, accumulator charging adapters, and alignment fixtures. The list will be to some extent peculiar to the specific equipment involved.

In addition to special tools, the maintenance department must be equipped with proper diagnostic equipment. In some cases this will only involve oscilloscopes, multimeters,

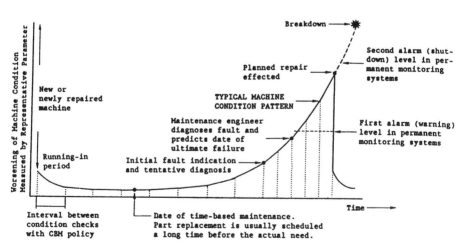

Figure 6 Procedure of condition-based maintenance policy. (From S. Nakajma et al., "Maintenance Management and Control," in *Handbook of Industrial Engineering,* 2nd ed., ed. G. Salvendy. New York: John Wiley & Sons, 1992. Chapter 73.)

gauges, and similar devices probably already available in the department. In other cases special diagnostic tools must be purchased.

3.6   Inventory Control

An adequate selection of spare parts can greatly reduce robot downtime after failures occur. Most manufacturers will provide a list of recommended inventory based on the number of machines in the facility and the statistical likelihood of need. The typical cost of this inventory may be 12% of the robot cost when only 1 or 2 robots are purchased, but may drop to 5% or less for 10 or more robots (Munson, 1985). Manufacturers will also often provide preventive maintenance kits. When components are supplied and stocked as kits, a long list of items need not be pulled from inventory to be taken to the job site.

Spare parts must be in secure areas and stocked in an organized manner to reduce the time required to find them. A well-designed inventory control system will also specify reorder points and order quantities that appropriately balance the cost of keeping parts in inventory against the expected cost of running out of needed parts. Inexpensive parts, or parts that frequently fail, normally should be kept in stock. On the other hand, it may be cost-effective to return expensive parts, or parts that rarely fail, to the supplier for repair or replacement rather than keeping them in inventory.

3.7   Training for Maintenance and Repair

A skilled workforce is a prerequisite of any well-planned and executed preventive maintenance program. Personnel need to be trained how to maintain, troubleshoot, and repair robots. They also need to know how the robot operates and how to operate it. Although manufacturers will usually have a field service organization of highly trained technicians, it is essential that these skills reside in-house so that the user can be virtually self-supporting and defer to others only when confronted with unusual problems. The most timely use of manufacturers' technicians is during installation and start-up phases, but even then it should be a team effort during which the customer personnel receive on-the-job training.

Although manufacturers will normally provide training to their customers, there still may be reasons to establish in-house maintenance training courses. Prior to starting this effort, management must understand the need for training and be committed to meeting that need. Once management is committed, the next step is to meet the people who will be directly involved to describe and discuss what is being planned and why it is needed. A series of "working" meetings should then follow that involve personnel at all levels. During these meetings the concerned parties need to express their concerns, discuss potential problems, and develop plans of action. In a cooperative working environment, the often-neglected maintenance department will ideally become an integral part of the team, ready and able to fulfill its function when and as needed.

During this process, it is essential that the group determine what skills are needed, what skills are lacking, who will be assigned what tasks, and how they will be trained. Table 6 presents a typical curriculum of a maintenance course. After the initial training program, it may be useful to train and upgrade personnel on a continuing basis to maintain skills and make sure that personnel learn how to maintain newly acquired equipment and perform new maintenance procedures and other important tasks.

3.8   Recordkeeping and Measurement of Maintenance Program Effectiveness

A final consideration is that the effectiveness of the maintenance program should be periodically reviewed. This of course means that all maintenance work must be carefully recorded. The cost of the maintenance program and its influence on system performance also must be documented. Accomplishing this latter goal requires that a failure-reporting and documentation system be established. Robot inspection and repair records must also be kept (Dhillon, 1991). At a minimum it is necessary to describe the failure and the repair, dates and times, operating time since the last repair, designation of the robot, repair person's name, place of repair, serial numbers, repair starting date and time, type of inspection or event that triggered the repair, and model numbers (Dhillon, 1991).

In the short term, documentation of what has been done provides invaluable information for people working on different shifts. In the long term, documentation of what has been done, and of the effectiveness of doing it, provides a database for establishing trends and fine-tuning the program. It can also provide the manufacturer with a history,

Table 6 Maintenance Course Curriculum[a]

Unit	Goals	First Day	Second Day	Comments
1: Dismantling Dismantling, cleaning, and inspection	Learn the structure and functions of machinery Learn to measure deterioration for more effective preventive maintenance	1. Orientation 2. Create a work schedule 3. Create an inspection checklist 4. Inspection (dynamic and static) 5. Dismantling 6. Cleaning 7. Summary	1. Create sketch 2. Create parts dimension table 3. Specifications and standards 4. Measuring and inspecting parts 5. Ordering new parts 6. Summary	Limited to two groups (3 to 4 persons) Daily reports required
2: Modification Modify and reassemble	Learn to reassemble machinery Learn important points in reassembly processes Learn to make improvements that extend equipment life	1. Parts modification 2. Investigating causes of deterioration 3. Troubleshooting procedures 4. Minor assembly 5. Summary	1. Assembly 2. Improvements for longer life 3. Sealing 4. Lubrication 5. Summary	Instructor adds comments
3: Test Run Trial operation and summary	Acquire hands-on experience in manual repairs and adjustments Confirm results through dynamic inspections Learn the daily maintenance procedures and as maintenance planning	1. Inspection while equipment shutdown 2. Manual repairs 3. Logging repairs properly 4. Maintenance planning table 5. Trial operation 6. Summary	1. Creating cutaway models 2. Procedures manual and summary	

[a] From S. Nakajima et al., "Maintenance Management and Control," in *Handbook of Industrial Engineering*, 2nd ed., ed. G. Salvendy. New York: John Wiley & Sons, 1992. Chapter 73.

should there be a chronic problem. To help ensure that accurate and complete data are collected, repair personnel and operators must be made aware that documenting failures is as important as repairing them. Forms used to report failures and record inspections and repairs must be made conveniently available, simple, and clear. Simply put, filling out the necessary documentation forms should be quick and easy.

4 SAFETY

Robots are often viewed as an ideal solution to many industrial safety and health problems (Willson, 1982) because they can perform tasks in dangerous work environments that previously were performed by humans. Robot welders and painters eliminate the need to expose workers to toxic fumes and vapors. Robots load power presses. They operate die-casting and injection-molding equipment, manipulate objects in outer space, handle hazardous waste, and can work in radioactive environments and underground storage tanks (Special Issue, 1992). At the same time, as the use of robot and other automated systems increases, it is becoming clear that such systems pose unique and significant safety problems.

Part of the problem is that robots are powerful devices that move quickly over considerable distances. People often have difficulty predicting robot motions and consequently accidents occur when people enter the robot's work area. In response to the growing recognition that robot systems pose a safety problem, governmental and consensus organizations have developed several robot safety standards. Many companies have also developed in-house safety standards and programs. Concurrently, many methods of robot safeguarding have been developed by manufacturers and industrial users of robots. Such methods include:

1. A variety of designed-in robot safety features
2. Perimeter safeguards
3. Warning signs, markings, and signals
4. Intelligent collision-avoidance and warning systems
5. Personnel selection, training, and supervision
6. Workcell design for safety

4.1 Robot Accident Data

Currently, limited data on robot-related accidents are available. There is some evidence that robot-related accident rates (number of accidents per hours of exposure) may be quite high (Helander, 1990; Carlsson, Harms-Ringdahl, and Kjellen, 1983), but it is difficult to draw conclusions given the small sample sizes studied. Some additional insight can be drawn from a recent study showing that 12–17% of the accidents in industries employing advanced manufacturing technology were associated with automated production equipment (Backström and Döös, 1995). The latter study also found that such accidents were more severe than those associated with other machines.

A recent review of studies that investigated robot-related accidents provides some insight into why such accidents occur (Beauchamp and Stobbe, 1995). These studies showed that the majority of robot accidents occurred after the victim entered the robot's work area to perform programming or maintenance tasks, and often involved unexpected movement of the robot. The unexpected movements were, in most cases, caused by equipment failures or human errors. In a surprisingly large number of instances, operators were attempting to make adjustments on robots that were still operating or still in operable condition (i.e., had not been isolated).

4.2 Robot Risk Assessment

Risk assessment is perhaps the most fundamental safety and health activity. Risk assessment makes use of the same tools used to evaluate reliability that were discussed earlier in Sections 2.2 and 2.3. The first step in risk assessment is to identify hazards. Each hazard is then evaluated in terms of its severity and probability, and countermeasures are identified. In the final step of this process the cost of control measures is then compared to expected loss reduction. A framework for robot risk assessment developed by the Machine Tool Trades Association in the United Kingdom is given in Table 7 (*Safeguarding Industrial Robots,* 1982). Note that the first four items in the table correspond to hazard identification and evaluation. Items 5 through 12 emphasize the evaluation of possible countermeasures. Item 13 emphasizes the need for documenting the process.

Table 7   Framework for Robot Risk Assessment[a]

1. Determine the mode of operation—that is, normal working, programming, maintenance.
2. Carry out a hazard analysis to determine potential areas of doing harm.
3. Determine whether "designed" or "aberrant" behavior is to be considered.
4. Determine if hazards are liable to lead to injury.
5. If so, then consider whether there are any recognized methods of guarding the particular machine concerned. At present such standards may well be available for the associated machinery, but probably not for the robot.
6. Consider whether such standards are appropriate, particularly in the context of machines being used in conjunction with robots. For example, the risk assessment could be different for a machine with a human operator than for one that is associated with a robot. One factor that will affect this risk is whether or not the human operator will take over from the robot during, for example, robot failure.
7. If no standard is available, consider what logical steps should be taken to establish a reasonable standard for the particular application.
8. Determine whether a fixed guard can be used.
9. If fixed guards cannot be used, then consider the use of interlocked guards. Determine the type of interlocking system appropriate to the circumstances. The interlocking system should give a reasonable level of integrity appropriate to the risk in question and should enable regular effective maintenance check to be made. Any "monitoring" system should be carefully examined to assess its effectiveness.
10. When analyzing the system under "aberrant" behavior, a similar process of examination of the hazards and then a risk assessment is carried out. The hazards may be more difficult to determine because they may only exist on failure of part of the machine system, such as a control system malfunction on the robot. An alternative to guarding in such circumstances might be to improve the safety integrity of the control system in question, retaining the interlocking safeguards proposed for "designed" behavior.
11. After particular measures have been taken, a reassessment of the system integrity will be necessary. If the hazards are minimized/prevented, reassessment of risks will also be necessary. In most cases, particularly where the risks are high, it is preferable to assume "worst case" when designing the safeguarding system and work on the premise that specific malfunctions will occur. It is not prudent to rely solely on the digital programmable electronic system of, say, the robot for all safeguarding features unless a very detailed assessment has been carried out, which may be beyond the competence of the average user.
12. After the analysis has been carried out for normal working, programming, and maintenance—any safeguarding interlocks considered necessary for any one of these modes must be compatible with the requirements of the other from both a functional and a safety integrity point of view—consideration should also be given to emergency stop controls and whether adequate integrity is achieved.
13. The need cannot be overemphasized for documentation concerned with the analysis, decisions, and systems of work, and so on relating to hazards analysis, risk assessment, safety integrity assessments, maintenance requirements, and so on.

[a] Adapted from *Safeguarding Industrial Robots, Part 1, Basic Principles.* The Machine Tools Trade Assn., 1982.

Hazard identification is guided by past experience, codes and regulations, checklists, and other sources. This process can be organized by separately considering systems and subsystems and potential malfunctions at each stage in their life cycles. Numerous complementary hazard-analysis methods, which also guide the process of hazard identification, are available (Lehto and Salvendy, 1991; Hammer, 1993). Commonly used methods include failure modes and effects analysis, work safety analysis, human error analysis, and fault tree analysis.

Failure mode and effects analysis is a systematic procedure for documenting the effects of malfunctions discussed earlier in Section 2.2. Work safety analysis (WSA) (Soukas and Rouhiainen, 1984) and human error analysis (HEA) (Soukas and Pyy, 1988) are related approaches that organize the analysis around tasks rather than system components. This process involves the initial division of tasks into subtasks. For each subtask, potential effects of product malfunctions and human errors are then documented, along with the implemented and the potential countermeasures. In robot applications, the tasks that would be analyzed fall into the categories of normal operation, programming, and maintenance.

Fault tree analysis (FTA) is a commonly used approach used to develop fault trees. The approach is top-down in that the analysis begins with a malfunction or accident and works downwards to basic events at the bottom of the tree (Hammer, 1993). Computer tools that calculate minimal cut-sets and failure probabilities and also perform sensitivity analysis (Contini, 1988) make such analysis more convenient. Certain programs help analysts draw fault trees (Ruthberg, 1985; Knochenhauer, 1988). Human reliability analysis (HRA) event trees are a classic example of this approach (Figure 7).

4.3   Robot Safety Standards

Numerous standards, codes, and regulations are potentially applicable to robot systems. The best-known standard in the United States that addresses robot safety is ANSI/RIA R15.06. This standard was first published in 1986 by the Robotics Industries Association (RIA) and the American National Standards Institute (ANSI) as ANSI/RIA R15.06, the *American National Standard for Industrial Robots and Robot Systems—Safety Requirements*. A revised version of the standard was published in 1992, and the standard is currently undergoing revisions once again. (See also Chapter 24, Industrial Robotics Standards.)

The checklists in Table 8 summarize the contents of the ANSI/RIA R15.06 standard pertaining to (1) manufacture, remanufacture, and rebuilding of robots, (2) Installation of robots and robot systems, and (3) safeguarding of personnel. Note that the material in Tables 8*a* and 8*b* is based on the recommended manufacturer and user requirements in the 1992 revision and the March 5, 1997, unapproved draft of the next revised version. Table 8*c* on safeguarding personnel is based on only the currently applicable 1992 version because of the extensive nature of the changes proposed, especially with respect to the implementation of safeguards through a procedure of hazard identification and control and the installation of specified or appropriate safeguards.

Several other standards developed by ANSI are potentially important in robot applications and address a wide variety of topics, such as machine tool safety, machine guarding, lockout/tagout procedures, mechanical power transmission, chemical labeling, material safety data sheets, personal protective equipment, safety markings, workplace signs, and product labels. For example, safety signs and labels are addressed by the ANSI Z535 series:

1. ANSI Z535.1 Safety Color Code
2. ANSI Z535.2 Environmental and Facility Safety Signs
3. ANSI Z535.3 Criteria for Safety Symbols
4. ANSI Z535.4 Product Safety Signs and Labels
5. ANSI Z535.5 Accident Prevention Tags

Other potentially relevant standards developed by nongovernmental groups include the National Electric Code, the Life Safety Code, and the proposed UL1740—Safety Standard for industrial robots and robotics equipment. Literally thousands of consensus standards contain safety provisions.

The best-known governmental standards in the United States applicable to robot applications are the general industry standards specified by the Occupational Safety and

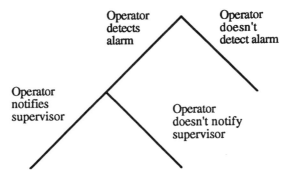

Figure 7   HRA event tree. (Adapted from D. I. Gertman and H. S. Blackman, *Human Reliability and Safety Analysis Data Handbook.* New York: John Wiley & Sons, 1994.)

Table 8 ANSI/RIA R15.06 Robot Safety Checklist
(a) *Manufacture, Remanufacture, and Rebuild of Robots*

Safety Provision	1992 Section	1997 Changes
Robot manufactured, remanufactured, or rebuilt according to standard	4.1	
Moving part hazards to be eliminated, preferentially, by (1) design or (2) protection; otherwise (3) warning must be provided	4.2.1	
Design to minimize hazards associated with breaking/loosening components or released energy	4.2.2	4.2.4
Energy source(s) disconnect, with lockout/tagout capability	4.2.3	4.2.5
Means for controlled stored energy release; identifying labeling	4.2.4	4.2.6
Design to prevent hazardous motion due to electromagnetic interference, radio frequency interference, and electro-static discharge	4.2.5	4.2.7
Each axis of robot movable without drive power	4.3	
Actuating controls designed/located to prevent inadvertent activation	4.4.1	
Labels and status indicators for all actuating controls	4.4.2–.3	
Means to disable controls designed to operate robot from remote location	4.4.4	
Hardware-based emergency stop circuit to override all other controls and stop all motion	4.5.1	
Emergency stop device for each operator station	4.5.2	
Provision within emergency stop circuit for additional emergency stop devices and hardware-based output signals	4.5.3–.4	
Pendant used within restricted envelope cannot initiate automatic operation, can only have momentary-contact motion controls, must have an emergency stop circuit, must have exclusive control of motion, and can only initiate slow speed motion unless through means of constant actuation	4.6.1–.5	
Attended continuous operation initiated only outside restricted envelope, requires continuous activation from a single exclusive control and, initiates slow speed only	4.7.1–.4	omitted
In slow speed, tool center or joints not to exceed 250 mm/sec	4.8	4.7
Controls designed so single-point failures (1) shall not prevent stopping, (2) will initiate stopping of system and prevent initiation of subsequent cycle		4.8
Mechanical stop capable of stopping motion at rated load, maximum speed and extension	4.9	
Load-rated lifting means for entire robot and associated components	4.10	
Electrical connector matching and securing means	4.11	
Secure/protect hoses subject to failure, causing hazard	4.12	
Loss or change of power not to result in hazard	4.13	4.2.3
Single-point failures not to result in hazard	4.14	4.13
Documentation of all control functions/locations, range and load capacity, lifting procedures, precautions, operating instructions, maintenance/repair information, installation information, precautions for electromagnetic interference, radio frequency interference, and electro-static discharge, electrical requirements [new version may add: limiting devices, hard steps, copy of ANSI/RIA R15.06, lockout, failure modes, hazard analysis training, testing, attended program verification, enabling devices, standards met, emergency recovery, movement without power]	4.15	4.14

[a] Based on 1992 version and possible changes based on March 5, 1997, draft.

Table 8 (Continued)
(*b*) *Installation of Robots and Robot Systems*

Safety Provision	Section[a]
Installed according to manufacturer specifications	5.1
Electrical grounding according to manufacturer specifications, applicable codes, or both	5.2
Power sources according to manufacturer specifications, applicable codes, or both	5.3
Controls/equipment outside restricted envelope if they must be accessible during automatic operation [new version may add: controls protected against inadvertent actuation]	5.4
Installed to avoid creating trapping or pinch points with buildings, structures, utilities, other machines, equipment [new version may add specific distances]	5.5, 6.2
Power disconnect, with lockout/tagout capability, outside restricted envelope	5.6
Limiting devices used to establish restricted envelopes shall not create additional hazards	5.7
Robot-compatible environmental conditions, including but not limited to explosive mixtures, corrosive conditions, humidity, dust, temperature, electromagnetic interference, radio frequency interference, and electro-static discharge	5.8
Shutdown of equipment associated with robot shall not create hazard	5.9
Additional precautionary notice provided where manufacturer-provided precautionary labeling obscured by application or process	5.10
Signs, barriers, floor line markings, railing, or equivalent to conspicuously identify restricted envelope	5.11
Dynamic restricted envelope tied to safeguarding interlocking logic	5.12
Hardware-based emergency stop circuit to override all other controls and stop all motion	5.13
Emergency stop device for each operator station	5.13.1
Emergency stop pushbuttons to be unguarded red, palm/mushroom-types requiring manual reset	5.13.2
Red, palm/mushroom-type pushbuttons limited to emergency stop use	5.13.3
Restarting automatic operation after an emergency stop to require deliberate action to follow startup procedure outside the restricted envelope	5.13.4
Common emergency stops for overlapping restricted spaces	5.13.5[b]
Design or safeguarding of end-effectors to not cause hazard due to loss or change of power	5.14
Directions for emergency movement without drive power	5.15[b]

[a] 1992 version unless otherwise noted.
[b] Proposed in March 5, 1997, draft.

Health Administration (OSHA). OSHA also published *Guidelines for Robotics Safety* (OSHA, 1987) and includes a section on industrial robot safety in the *Occupational Safety and Health Technical Manual* (OSHA, 1996). Other potentially relevant standards in robot applications include those specified by the Environmental Protection Agency (EPA) on disposal and cleanup of hazardous materials and the Nuclear Regulatory Commission's (NRC) standards regarding employee's working with radioactive materials.

Also, many companies that use or manufacture robot systems develop their own guidelines (Blache, 1991). Companies often start with the ANSI/RIA R15.06 robot safety standard and then add detailed information that is relevant to their particular situation.

4.4 Robot Safety Features

A well-engineered robot will include, to the greatest extent possible, practical safety features that take into account all modes of operation—normal working, programming, and maintenance (Barrett, Bell, and Hudson, 1981). Some features are common to all robots; others are peculiar to the type of robot, particularly with regard to its motive power. A quick review of Table 8 reveals that there are many steps that can be taken to

Table 8 (Continued)
ANSI/RIA R15.06-1992 Robot Safety Checklist: Safeguarding Personnel

Safety Provision	Section
Provided safeguards used in accordance with risk assessment	6.1
Risk assessment based on capabilities or robot, application and process, operational tasks and their hazards, failure modes, probability/severity of injury, exposure rates, expertise of personnel	6.1.1
Safeguarding for all stages of robot/robot system development: integration, off-site testing, installation and on-site testing, production operation	6.1.2
Installed to avoid creating trapping or pinch points with buildings, structures, utilities, other machines, equipment [new version may add specific distances]	6.2, 5.5
Safeguarding devices (mechanical or nonmechanical devices to reduce area of maximum envelope, presence-sensing devices, and/or barriers) to comply with specific requirements	6.3, 4.9
Awareness devices (perimeter guarding, barriers and/or signals) not to be used as sole safeguarding means unless warranted by specific risk assessment	6.4, 6.1.1
Robot teacher to be trained	6.5.1
Before teaching, visual hazard and control function checks to be performed	6.5.2
Before entering restricted envelope, all safeguards to be in place and functioning	6.5.3
In teach mode: teacher has single and sole point of control; slow speed to be in effect; emergency stops functional; only teacher in restricted envelope	6.5.4
All personnel to clear restricted envelope before initiating automatic mode	6.5.5
Ensure safeguards for each robot-associated operation, especially when operator interaction required. Operators to be trained in proper operation, recognition of hazards, and response to such hazards	6.6, 9
Safeguarding during attended continuous operation, including: trained operator; slow speed control; safeguarding from adjacent robots or other industrial equipment; pendant with "dead man" and emergency-stop controls; single and sole point of control by operator; deliberate procedure for initiation of automatic mode	6.7, 4.6, 4.7, 9
Safeguarding maintenance or repair personnel: training; safeguarding from hazardous motion, lockout/tagout procedure unless alternate safeguarding used including visual inspection, functional control check, sole and single point of control by maintenance or repair personnel, placing of robot arm in predetermined position, use of blocking to limit robot motion, and/or monitoring by second person. All safeguards bypassed to be identified and restored to original effectiveness	6.8, 9
Establish effective inspection and maintenance program, considering manufacturer's recommendations	7
Interim safeguarding may be used prior to testing and startup	8.1
During testing and start-up, no personnel in restricted envelope	8.2
Manufacturer's recommendations to be followed during testing and startup	8.3
Before apply power, initial startup procedure checks to include: mechanical connections and stability; electrical, utility, and communication connections; peripheral equipment; maximum envelopoe limiting devices	8.4.1
After apply power, initial startup procedure checks to include: functional emergency stops, interlocks, safeguards, slow speed control; intended movement and restrictions, program execution	8.4.2, 4.8
Ensure those who program, teach, operate, maintain, or repair robot systems are trained and demonstrate competence	9.1
Training to include: standards; vendor recommendations; robot system purpose and operation; specific personnel responsibilities; emergency contact personnel; hazard identification; contingencies; function and limitations of safeguards	9.2
Retraining after system changes or upon belief that retraining is required	9.3

increase the safety of a robot system. Table 9 summarizes some common safety features of robots and their intended functions. Table 10, taken from a supplier catalog, schematically illustrates typical safety features of robot workcells. These features and their intended functions can be grouped into five generic categories:

1. Design for reliability features (both fault avoidance and fault tolerance)
2. Set-up and programming mode features
3. Power transmission and point-of-operation guards
4. Maintenance affordances
5. Emergency features

Fault avoidance and fault tolerance features increase system safety by increasing its reliability during normal operation. They play a particularly critical role in protecting equipment from damage. Such features prevent robots from making mistakes, such as attempting to reach into a press before it is open. Examples of robot features that improve safety by increasing fault avoidance and fault tolerance include:

1. The use of interlocks between the robot and the machinery with which it works that provide signals indicating whether the system is "ready" or safe for robot or other equipment to perform some step. Interrogation of these signals will be part of the robot's program and placed strategically in the proper sequence.

2. The use of sensors that indicate whether the system is "ready" or safe for robot or other equipment to perform some step.

3. Parity checks, checksums, cyclic redundancy, error detecting, and other ways of minimizing the effects of computer-related malfunctions.
4. "Software" stops, electrical stops, and "hard" stops.

Set-up and programming mode features address the fact that personnel will have to work within the robot's sphere of influence when they teach the robot. The standard

Table 9   Common Robot Safety Features

Feature	Function
power disconnect switch	removes all power at machine junction box
line indicator	indicates incoming power is connected at junction box
power-on button	energizes all machine power
control-power-only button	applies power to control section only
arm-power-only button	applies power to manipulator only
stop button	removes control and manipulator power
hold/run button	stops arm motion, but leaves power on
teach pendant trigger	must be held in by operator for arm power in teach mode
step button	permits program execution one step at a time
slow speed control	permits program execution at reduced speeds
teach/playback mode selector	provides operator with control over operating mode
program reset	drops system out of playback mode
condition indicators and messages	provides visual indication by lights or display screens of system condition
parity checks, error detecting, etc.	computer techniques for self-checking a variety of functions
servo-motor brake	maintains arm position at stand-still
hydraulic fuse	protects against high speed motion/force in teach mode
software stops	computer-controlled travel limit
hardware stops	absolute travel limit control
manual/automatic dump	provides means to relieve hydraulic/pneumatic pressure
remote connections	permits remoting of essential machine/safety functions

Table 10   Schematic of Robot Safety Features in a Workcell[a]

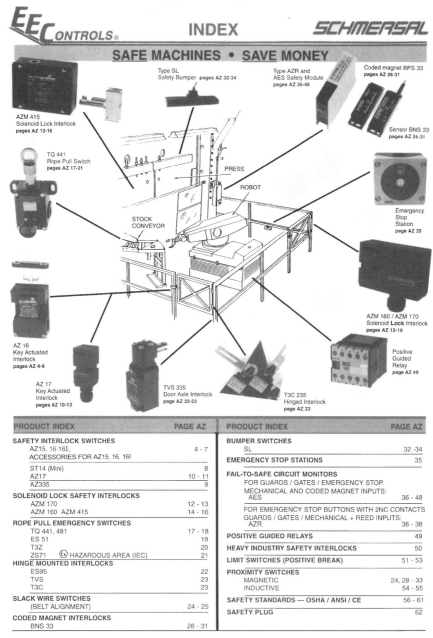

PRODUCT INDEX	PAGE AZ	PRODUCT INDEX	PAGE AZ
SAFETY INTERLOCK SWITCHES		**BUMPER SWITCHES**	
AZ15, 16 16I,	4 - 7	SL	32 -34
ACCESSORIES FOR AZ15, 16, 16I		**EMERGENCY STOP STATIONS**	35
ST14 (Mini)	8	**FAIL-TO-SAFE CIRCUIT MONITORS**	
AZ17	10 - 11	FOR GUARDS / GATES / EMERGENCY STOP.	
AZ335	9	MECHANICAL AND CODED MAGNET INPUTS:	
SOLENOID LOCK SAFETY INTERLOCKS		AES	36 - 48
AZM 170	12 - 13	FOR EMERGENCY STOP BUTTONS WITH 2NC CONTACTS	
AZM 160 AZM 415	14 - 16	GUARDS / GATES / MECHANICAL + REED INPUTS:	
ROPE PULL EMERGENCY SWITCHES		AZR	36 - 38
TQ 441, 481	17 - 18	**POSITIVE GUIDED RELAYS**	49
ES 51	19	**HEAVY INDUSTRY SAFETY INTERLOCKS**	50
T3Z	20		
ZS71 ⓔ HAZARDOUS AREA (IEC)	21	**LIMIT SWITCHES (POSITIVE BREAK)**	51 - 53
HINGE MOUNTED INTERLOCKS		**PROXIMITY SWITCHES**	
ES95	22	MAGNETIC	24, 28 - 33
TVS	23	INDUCTIVE	54 - 55
T3C	23	**SAFETY STANDARDS — OSHA / ANSI / CE**	56 - 61
SLACK WIRE SWITCHES		**SAFETY PLUG**	62
(BELT ALIGNMENT)	24 - 25		
CODED MAGNET INTERLOCKS			
BNS 33	26 - 31		

AZ 3

[a] Courtesy Schmersal Co. EEC Controls.

response to this situation is to provide a teach mode in which the robot's speed and power are reduced. For such an approach to be effective the speed of the robot arm must be reduced sufficiently to allow an operator to avoid being hit if the arm moves unexpectedly. The force of the arm also should be reduced sufficiently to prevent injury in case of entrapment. A number of recent studies have evaluated the ability of people to react appropriately to unexpected movement of the robot arm. Depending on the study, recommended speeds for when a person is in the robot's operating envelope ranged from 14–20 cm/s (Beauchamp and Stobbe, 1995). It is interesting to note that ANSI/RIA R15.06-1992 recommends a speed of no more than 25 cm/s.

The teach pendant is another robot component that plays a potentially critical role in safety (Figure 8). Programming errors during teaching are one of the reasons for unexpected robot movement (Carlsson, Harms-Ringdahl, and Kjellen, 1983). There is evidence that the design of the teach pendant may play a critical role in safety (Beauchamp and Stobbe, 1995). In their review article Beauchamp and Stobbe discuss an unpublished study in which the use of a joystick control resulted in significantly fewer errors than use of a pushbutton control. Other research showed that the size of the emergency button impacts reaction time. Beauchamp and Stobbe conclude that the teach pendant should have an emergency button on the front panel and that control buttons should be of the "dead man" type, to reduce the time needed to deactivate the control.

Other set-up and programming mode safety features focus on:

1. Eliminating or reducing the chance that failures of robot system components during teach mode result in full power being applied. In hydraulic robots these approaches include restricting the fluid flow in a fail-safe manner and, as an added precaution, incorporating a hydraulic "fuse" that will rapidly sense excess flow and shut the machine down.
2. Disarming, or preventing other machinery from being cycled when the robot is being programmed.

Power transmission and point-of-operation guarding becomes important when people enter the robot work zone or interact with other machinery used in the robot system. Numerous approaches are used to guard machinery in general. Enclosure guards and

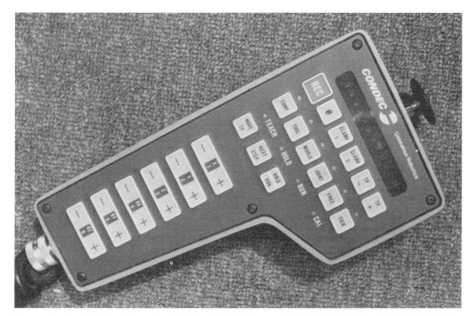

Figure 8 A robot teach pendant. (From G. E. Munson, "Industrial Robots: Reliability, Maintenance, and Safety," in *Handbook of Industrial Robotics,* ed. S. Y. Nof. New York: John Wiley & Sons, 1985. 722–758. Photo courtesy Unimation, Inc.)

interlocked guards (Brauer, 1990) are used in most robot systems for power transmission components. Point-of-operation guarding is less likely to be an important element of the robot, but plays a major role for other equipment commonly used in production processes that use robots.

Maintenance affordances are features designed into the robot specifically to facilitate maintenance. Such features include manual or automatic hydraulic or pneumatic pressure dumping, locking brakes on electric drive motors, and control system power-on only (i.e., no manipulator power). In at least one case of a servo-controlled robot the manufacturer provides a manual control pendant that permits moving of the manipulator articulations by an operator without power to the robot's control system. In addition to these built-in features, most robot manufacturers offer maintenance aids (tools and fixture) to facilitate maintenance safety.

Emergency features often provided with robots include providing a "HOLD" mode, in which all motion is stopped as soon as possible, a "STEP" mode in which it will complete the command currently being executed, but not proceed further, dead man's switches on the teach pendant, and remote emergency stop buttons.

4.5 Perimeter Safeguards

It has become clear that most robot accidents are caused by or in some way involve careless or intentional entry by people into the robot work envelope (Beauchamp and Stobbe, 1995). Such entry is inherently hazardous because the point of operation (the effector and its tooling) of a robot is essentially unguarded. Perimeter safeguards implement a number of different approaches intended to prevent entry into the work envelope. These approaches include:

1. Barriers and fencing
2. Presence sensing and interlocks
3. Warning signs, markings, signals, and lights

Barriers and fencing are commonly used methods of preventing entry into the robot's work envelope. A recent survey of robot applications revealed that 73% of the robot installations used barriers or fencing (Hirshfeld, Aghazadeh, and Chapleski, 1993). Barriers pose several potential advantages, in that:

- Their very existence (painted appropriately) is a warning even to someone unfamiliar with what they enclose and to the casual observer.
- Safety barriers preclude material-handling equipment and other vehicles in the plant from being inadvertently moved into a danger zone.
- Properly interlocked barrier gates tend to enforce procedural discipline when authorized personnel need to gain access to the work area. Tampering with the equipment when it is not operating is minimized.
- Properly designed, barriers will eliminate, or at least reduce, the potential hazard of objects (workpieces) rolling or even flying out of the work area.

Figures 9 and 10 show typical safety fence guard arrangements. Note in Figure 10 that the die-casting and extrusion machines form part of the barrier. Obviously it is essential that the fencing abut the machinery so that a person cannot squeeze through. Similar precautions must be observed in determining the height of the fencing and the gap at floor level. Note the control panel just outside the fenced area. These controls allow an operator to systematically shut down the machines before entering the work area, thereby protecting the equipment from itself as well as providing human safety.

Presence sensing and interlock devices are used to sense and react to potentially dangerous workplace conditions. Systems of this type include:

1. Pressure-sensitive floor mats
2. Light curtains
3. End-effector sensors
4. Ultrasound, capacitive, infrared, and microwave sensing systems
5. Computer vision

Floor mats and light curtains are used to determine whether someone has crossed the safety boundary surrounding the perimeter of the robot. Perimeter penetration will trigger

Reliable safety where you need it

The application possibilities of the FGS safety light curtain are universal. This allows for safe, flexible work place layouts. With reliable guarding and unrestricted access, material placement, work piece removal or maintenance is possible at all times. These are conditions for ergonomic improvements and reduction of manufacturing costs.

With its small dimensions and high resolution, the FGS can be installed closer to the hazardous zone. The FGS can operate with a smaller safety distance, saving valuable industrial space.

The FGS is flexible to accommodate other areas of application requiring additional functions. Some examples of these functions are *muting*, a by-pass function required to allow automatic feeding of material, and *blanking* of defined sections of the protective field to allow passage of material.

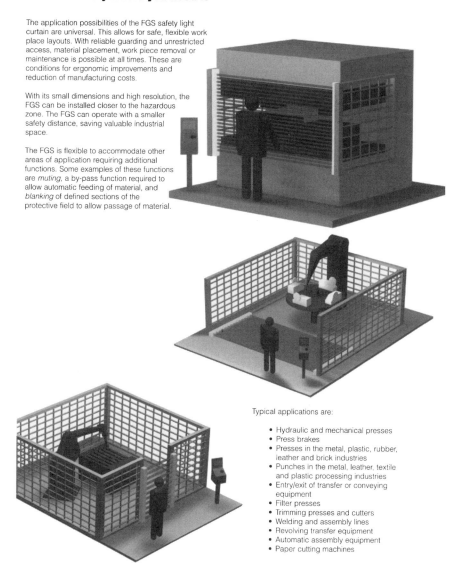

Typical applications are:

- Hydraulic and mechanical presses
- Press brakes
- Presses in the metal, plastic, rubber, leather and brick industries
- Punches in the metal, leather, textile and plastic processing industries
- Entry/exit of transfer or conveying equipment
- Filter presses
- Trimming presses and cutters
- Welding and assembly lines
- Revolving transfer equipment
- Automatic assembly equipment
- Paper cutting machines

Figure 9 Schematic illustrating installation of safety light curtain in robot workcell (courtesy of Sick Optic Electronic, Inc.).

a warning signal and in some cases cause the robot to stop. End-effector sensors detect the beginning of a collision and trigger emergency stops. Ultrasound, capacitive, infrared, and microwave sensing systems are used to detect intrusions. Computer vision theoretically can play a similar role in detecting safety problems.

Figure 11 illustrates how a presence-sensing system, in this case a safety light curtain device, might be installed for point of operation, area, perimeter, or entry/exit safeguarding. Figure 12 lists technical specifications for a light curtain system provided by the manufacturer. By reducing or eliminating the need for physical barriers, such systems make it easier to access the robot system during setup and maintenance. By providing

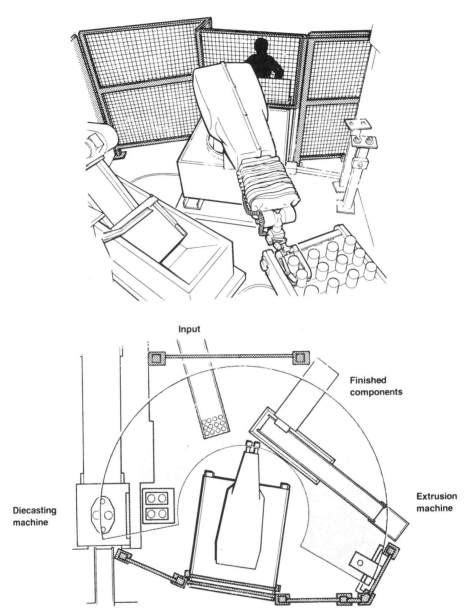

Figure 10 Safety enclosure formed, in part, by workcell machines. (From G. E. Munson, "Industrial Robots: Reliability, Maintenance, and Safety," In *Handbook of Industrial Robotics,* ed. S. Y. Nof. New York: John Wiley & Sons, 1985. Photo courtesy Machine Tool Trades Association.)

early warnings prior to entry of the operator into the safety zone, such systems can also reduce the prevalence of nuisance machine shutdowns (Figure 13).

Warning signs, markings, signals, and lights are commonly used safety measures in robot applications. A recent survey revealed that 73% of the robot applications used warning lights, 60% used warning signs, 33% used boundary chains, and 13% used painted boundaries (Hirshfeld, Aghazadeh, and Chapleski, 1993).

The most commonly accepted role of warnings is that of alerting operators to the presence of a hazard (Lehto and Miller, 1986). Other roles have, however, been proposed

Figure 11 Typical safety fence enclosure (courtesy of Sick Optic Electronic, Inc.).

for warnings, including those of informing, reminding, educating, and persuading (Lehto and Miller, 1986). It should be emphasized that product design features, instruction manuals, job performance aids, training, and supervision duplicate these latter roles and often perform them much more effectively than a warning (Lehto and Salvendy, 1995). Consequently, warnings should never replace these methods of attaining a safe workplace. In their proper role, however, warnings can be an effective supplement to the above approaches.

It is often blindly assumed that a comprehended warning message will be heeded. This assumption is unfortunate because many additional factors complicate the transition between understanding what should be done and actually behaving safely (Lehto and Miller, 1986; Lehto and Papastavrou, 1993). The perceived benefit of taking unsafe actions is of particular concern in robot safety. Simply put, operators may ignore warnings and intentionally enter robot work zones, without first isolating the robot, to save small amounts of time and to avoid what might seem minor inconveniences to outside observers. False alarms also reduce the effectiveness of warnings, for obvious reasons. Warning signals that alert workers when they *unintentionally* enter dangerous areas are much more likely to be effective than those that tell people what they already know or try to convince them to change an existing behavior pattern.

Lehto (1992) provides guidelines along these lines for the design of warnings that were inferred from models of human behavior and the available research on the effectiveness of warnings. These guidelines are:

1. Match the warning to the level of performance at which critical errors occur for the relevant population.
2. Integrate the warning into the task- and hazard-related context.
3. Be selective.
4. Make sure that the cost of compliance is within a reasonable level.
5. Make symbols and text as concrete as possible.
6. Simplify the syntax of text and combinations of symbols.
7. Make warning signs and labels conspicuous and each component legible.
8. Conform to standards and population stereotypes when possible.

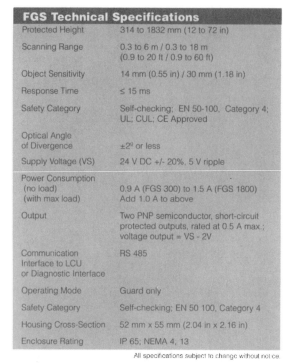

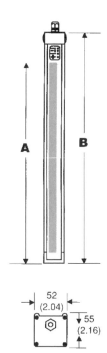

FGS Technical Specifications	
Protected Height	314 to 1832 mm (12 to 72 in)
Scanning Range	0.3 to 6 m / 0.3 to 18 m (0.9 to 20 ft / 0.9 to 60 ft)
Object Sensitivity	14 mm (0.55 in) / 30 mm (1.18 in)
Response Time	≤ 15 ms
Safety Category	Self-checking; EN 50-100, Category 4; UL; CUL; CE Approved
Optical Angle of Divergence	±2º or less
Supply Voltage (VS)	24 V DC +/- 20%, 5 V ripple
Power Consumption (no load) (with max load)	0.9 A (FGS 300) to 1.5 A (FGS 1800) Add 1.0 A to above
Output	Two PNP semiconductor, short-circuit protected outputs, rated at 0.5 A max.; voltage output = VS - 2V
Communication Interface to LCU or Diagnostic Interface	RS 485
Operating Mode	Guard only
Safety Category	Self-checking; EN 50 100, Category 4
Housing Cross-Section	52 mm x 55 mm (2.04 in x 2.16 in)
Enclosure Rating	IP 65; NEMA 4, 13

All specifications subject to change without notice.

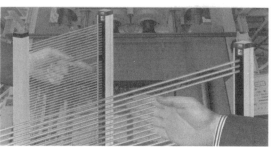

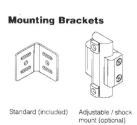

Mounting Brackets

Standard (included)   Adjustable / shock mount (optional)

Figure 12 Technical specifications of flexible guarding system safety light curtain (courtesy of Sick Optic Electronic, Inc.).

4.6 Intelligent Collision-Avoidance and Warning Systems

Intelligent collision-avoidance and warning systems process information from a wide variety of sensors to make real-time decisions. Traditionally, such systems make use of sophisticated ultrasound, capacitive, infrared, or microwave presence sensing systems and, in some cases, computer vision to detect the presence of obstacles and then react appropriately to their presence by stopping, replanning their motions, or giving alarm signals (Graham, Smith, and Kannan, 1991; Zurada and Graham, 1995).

While most such schemes have focused on intelligent control of a robot arm, such capabilities are especially useful for locomotion or autonomous navigation. Such capacities also might be combined with teleoperation of a robot in a complex spatial environment. In even more ambitious applications, an intelligent, interactive robot might work together in close physical proximity with people. For example, robots might assist disabled persons to perform physical tasks such as dressing themselves. In more mundane applications, a robot and human might work next to each other on an assembly line.

Although significant effort has been given to solving the difficult problems associated with real-time detection and prevention of potential collisions, little, if any, research has

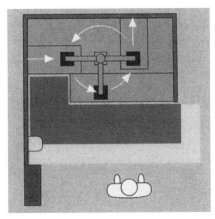

Figure 13 Schematic of proximity laser scanner robot warning system (courtesy of Sick Optic Electronic, Inc.).

addressed the potentially useful role of intelligent robot warning systems. This potential benefit follows, if an intelligent warning system can reduce false alarms (i.e., a warning is given but a collision is *not* imminent) and increase correct detections (i.e., a warning is given *and* a collision is imminent), and especially so, if it can adjust its behavior for different modes of robot operation. In traditional applications of industrial robots, this would include operation, programming, and maintenance tasks. It seems likely that nearly all employees involved in accidents occurring after they entered into the robot work envelope received warnings that were ignored because they did not provide information. In other words, the employees knew they were entering a hazardous area. What they didn't know, and needed to know, was probably much more specific. Simply put, they needed to know that a collision *was imminent* if they remained where they were.

An intelligent warning system, rather than indiscriminately providing warnings, will consider both system states and the past behavior of the target audience and adjust its behavior accordingly by appropriately balancing the expected cost of false alarms against the expected cost of correct detections. This problem can be formally analyzed from a team decision-making perspective to set the appropriate warning threshold (Lehto and Papastavrou, 1996; Papastavrou and Lehto, 1995). Current robot warning systems seem to take the intuitive, but misled, approach of "always warning." This approach maximizes correct detections, but also maximizes the number of false alarms. When warnings are always present they convey no information, and it is actually optimal for people to ignore the warning and base their decisions entirely on their own observations of the environment.

4.7   Employee Selection, Safety Training, and Supervision

Modification of the employee through employee selection, training, and supervision is necessary in almost all occupations (Lehto and Salvendy, 1995). As a consequence of selecting qualified personnel, training, and supervision, workers should know and follow safe procedures when working with robots. Particularly critical procedures include startup and shutdown procedures, set-up procedures, lockout and tagout procedures during maintenance, testing procedures, diagnosis procedures, programming and teaching procedures, and numerous procedures specific to particular applications, such as welding operations.

Simply put, employees must be capable of satisfying the physical and mental demands of their jobs. Selecting employees with the necessary physical and cognitive abilities is an essential first step. Given that employees have the necessary abilities, training them becomes an important part of the solution. Supervision, as the final element related to modification of the employee, should consistently enforce safe behavior patterns. Failure to enforce safety procedures is a very common cause of accidents.

Attaining the above elements of employee modification is a standard function of industrial and safety engineering in most plants. While each of these elements plays an important role, training is especially recognized as playing a critical role in the field of robot safety (Dhillon, 1991; Graham, 1991; ANSI/RIA R15.06, 1992; OSHA, 1996). A robot safety training program must, at the very least, teach workers what the hazards are and their severity, how to identify and avoid them, and what to do after exposure. A core focus of the course should be on teaching employees the safety procedures they need for their particular job and why the procedures need to be followed. Common mistakes leading to accidents should also be explained in detail. Particular focus should be placed on teaching employees why it is dangerous to enter the work envelope of a robot. Case studies of typical robot accidents should be covered, showing how common misconceptions regarding robot movements lead to accidents. Employees should also receive follow-up training through refresher courses. Robot manufacturers and vendors are often an excellent source of training materials, and in many cases will offer useful courses of their own.

4.8 Workcell Design for Safety

Safeguards should be considered an integral part of the workcell(s) design. They should be provided for at the planning stages. To do otherwise could incur extra expense later on and might compromise safety effectiveness. A generalized approach (Munson, 1985) is outlined below:

1. Develop an installation layout showing all equipment. Plan and elevation views will usually be required.

2. Lay out a safety enclosure (fencing) around the workcell to preclude human and machinery (e.g., forklift trucks) intrusion. Utilize nonmoving workcell machinery as part of the barrier where possible. Appraise the layout to see that electrical and control panels and consoles, especially start–stop controls, are strategically and conveniently located. Provide for access into the area as required for tool/setup changes, arrival and removal of pallets of workpieces, manual load/unload stations, and so on, and maintenance activities.

3. Evaluate whether restricted motion (hard stops) of the robot will facilitate or improve safety precautions or will conserve floor space. Adjust the layout accordingly.

4. Design the interlock/interface system between all elements of the workcell as required, including safety interlocks between the equipment and between the equipment and barrier-access gates. Evaluate the need for primary and secondary or redundant safety interlocks (e.g., in addition to interlocked safety gates designed to stop operations, trip devices such as optical sensors or electromagnetic shields). Also consider key locks on gates to limit access to authorized personnel only.

5. Review the design to be sure safety provisions are fail-safe.

6. Consider the use of visuals (signs, flashing lights) and audio devices to indicate condition of the operation and to sound alarms.

7. Review all aspects of the final design with sign-offs by manufacturing engineering, production, maintenance, safety officers, and any others deemed appropriate.

This is a suggested approach. Others can be developed with equal or better effectiveness. The important point is that a systematic process be followed, working with a checklist of items to be given attention and a constructive "what if" attitude from all departments involved.

In the final analysis, a safety engineered system will only be as safe as *people* permit it to be. Part of the commissioning of the installation should include a safety check of all of the built-in features—of the robot, the related machinery, the control and safety interlock system, and the barrier access gates and alarms. This should be a supervised evaluation following procedural documentation that is to be posted and always followed. Thereafter the safety system should be periodically tested for functionality to be sure that no aspect has been aborted, intentionally or unintentionally, with the passage of time.

5 SUMMARY

Reliability, maintenance, and safety of robot systems are closely related issues that require a substantial commitment from robot manufacturers and industrial users alike. On the part of the manufacturer, design solutions play a critical role in determining whether robot systems will be reliable, safe, and easily maintained. Industrial users, on the other hand, can greatly influence the degree to which robot systems meet these objectives by establishing and then following an effective preventive maintenance program and making sure that safe and effective procedures are followed throughout the robot system's life cycle by all personnel who come in contact with the system.

Design solutions include fault avoidance through various forms of redundancy, fault tolerance, on-line fault detection, design for maintainability, and the provision of numerous robot safety features. Maintenance and safety solutions include the development, documentation, and enforcement of appropriate maintenance and safety procedures, work-cell design, and worker selection, training, and supervision. These solutions include significant contributions from numerous fields, such as risk analysis, reliability theory, quality control, industrial and safety engineering, and human factors engineering. Governmental and consensus standard-making organizations also play an important role.

ACKNOWLEDGMENTS

The authors wish to express their thanks to George E. Munson, the author of the version of this chapter in the 1st edition of this handbook, and to Winston Erevelles of Kettering University for providing materials used in this chapter.

REFERENCES

American National Standard for Industrial Robots and Robot Systems—Safety Requirements. ANSI /RIA R15.06. Ann Arbor: Robotic Industries Assn.; New York: American National Standards Institute, 1992.

Backström, T., and M. Döös. 1995. "A Comparative Study of Occupational Accidents in Industries with Advanced Manufacturing Technology." *International Journal of Human Factors in Manufacturing* 5(3), 267–282.

Barrett, R. J., R. Bell, and P. H. Hudson. 1981. "Planning for Robot Installation and Maintenance: A Safety Framework." In *Proceedings of the 4th British Robot Association Annual Conference,* Brighton, May 18–21.

Beauchamp, Y., and T. J. Stobbe. 1995. "A Review of Experimental Studies on Human–Robot System Situations and Their Design Implications." *International Journal of Human Factors in Manufacturing* 5(3), 283–302.

Blache, K. M. 1991. "Industrial Practices for Robotic Safety." In *Safety, Reliability, and Human Factors in Robotic Systems.* Ed. J. H. Graham. New York: Van Nostrand Reinhold. Chapter 3.

Brauer, R. L. 1990. *Safety and Health for Engineers.* New York: Van Nostrand Reinhold.

Brumbaugh, P. S., and R. G. Heikes. 1992. "Statistical Quality Control." In *Handbook of Industrial Engineering.* 2nd ed. Ed. G. Salvendy. New York: John Wiley & Sons. Chapter 87.

Carlsson, J., L. Harms-Ringdahl, and U. Kjellen. 1983. "Industrial Robot Accidents at Work." Report # TRITA-AOG-0026, Royal Institute of Technology, Stockholm.

Contini, S. 1988. "Fault Tree and Event Tree Analysis." In *Advanced Informatic Tools for Safety and Reliability Analysis,*" Ispa, October 24–28.

Dhillon, B. S. 1991. *Robot Reliability and Safety.* New York: Springer-Verlag.

Drury, C. G. 1992. "Inspection Performance." In *Handbook of Industrial Engineering.* 2nd ed. Ed. G. Salvendy. New York: John Wiley & Sons. Chapter 88.

Embrey, D. E. 1984, *SLIM-MAUD: An Approach to Assessing Human Error Probabilities Using Structured Expert Judgment.* NUREG/CR-3518, Vols. 1 and 2. Washington: U.S. Nuclear Regulatory Commission.

Gertman, D. I., and H. S. Blackman. 1994. *Human Reliability and Safety Analysis Data Handbook.* New York: John Wiley & Sons.

Graham, J. H., ed. 1991. *Safety, Reliability, and Human Factors in Robotic Systems.* New York: Van Nostrand Reinhold.

Graham, J. H., P. E. Smith, and R. Kannan. 1991. "A Multilevel Robot Safety and Collision Avoidance System." In *Safety, Reliability, and Human Factors in Robotic Systems.* Ed. J. H. Graham. New York: Van Nostrand Reinhold.

Hammer, W. 1993. *Product Safety Management and Engineering.* 2nd ed. American Society of Safety Engineers (ASSE).

Hancock, W. M., P. Sathe, and J. Edosomwan. 1992. "Quality Assurance." In *Handbook of Industrial Engineering.* 2nd ed. Ed. G. Salvendy. New York: John Wiley & Sons. Chapter 85.

Helander, M. G. 1990. "Ergonomics and Safety Considerations in the Design of Robotics Workplaces: A Review and Some Priorities for Research." *International Journal of Industrial Ergonomics* 6, 127–149.

Hirshfeld, R. A., F. Aghazadeh, and R. C. Chapleski. 1993. "Survey of Robot Safety in Industry." *International Journal of Human Factors in Manufacturing* 3(4), 369–379.

Jones, R., and S. Dawson. 1985. "People and Robots—Their Safety and Reliability." In *Robot Safety.* Ed. M. C. Bonney and Y. F. Yong. New York: Springer-Verlag.

Kapur, K. C. 1992. "Reliability and Maintainability." In *Handbook of Industrial Engineering*. 2nd ed. Ed. G. Salvendy. New York: John Wiley & Sons. Chapter 89.

Knochenhauer, M. 1988. "ABB Atom's SUPER NET Programme Package for Reliability and Risk Analysis." In *Advanced Informatic Tools for Safety and Reliability Analysis*," Ispa, October 24–28.

———. 1997. "Decision Making." In *Handbook of Human Factors and Ergonomics*. 2nd ed. Ed. G. Salvendy. New York: John Wiley & Sons. 1201–1248.

Lehto, M. R., and J. M. Miller. 1986. *Warnings Volume I. Fundamentals, Design, and Evaluation Methodologies*. Ann Arbor: Fuller Technical Publications.

———. 1993. "Models of the Warning Process: Important Implications Towards Effectiveness." *Safety Science* 16, 569–595.

———. 1996. "Improving the Effectiveness of Warnings by Increasing the Appropriateness of Their Information Content: Some Hypotheses about Human Compliance." *Safety Science* 21, 175–189.

Lehto, M. R., and G. Salvendy. 1991. "Models of Accident Causation and Their Application: Review and Reappraisal." *Journal of Engineering and Technology Management* 8, 173–205.

———. 1995. "Warnings: A Supplement Not a Substitute for Other Approaches to Safety." *Ergonomics* 38, 2155–2163.

MIL-HBDK-217F: Reliability Prediction of Electronic Equipment, Department of Defense. Rome Laboratory, Griffiss Air Force Base, New York, January 1990.

Munson, G. E. 1985. "Industrial Robots: Reliability, Maintenance, and Safety." In *Handbook of Industrial Robotics*. Ed. S. Y. Nof. New York: John Wiley & Sons. 722–758.

Nakajima, S., et al. 1992. "Maintenance Management and Control." In *Handbook of Industrial Engineering*. 2nd ed. Ed. G. Salvendy. New York: John Wiley & Sons. Chapter 73.

OSHA. 1987. *Guidelines for Robotics Safety*. U.S. Department of Labor.

OSHA. 1996. "Industrial Robots and Robot System Safety." In *Occupational Safety and Health Technical Manual*. U.S. Department of Labor: Section III, Chapter 4.

Papastavrou, J., and M. R. Lehto. 1995. "A Distributed Signal Detection Theory Model: Implications for the Design of Warnings." *International Journal of Occupational Safety and Ergonomics*. 1(3), 215–234.

Pritsker, A. 1992. "Modeling for Simulation Analysis." In *Handbook of Industrial Engineering*. 2nd ed. New York: John Wiley & Sons. Chapter 100.

Ross, S. 1970. *Applied Probability Models with Optimization Applications*. San Francisco: Holden-Day.

Ruthberg, S. 1985. "DORISK—A System for Documentation and Analysis of Fault Trees." Paper delivered at SRE Symposium, Trondheim, Norway, September 30–October 2.

Safeguarding Industrial Robots, Part 1, Basic Principles. London: Machine Tool Trades Assn., 1982.

Soukas, J., and P. Pyy. 1988. "Evaluation of the Validity of Four Hazard Identification Methods with Event Descriptions." Research Report 516, Technical Research Center of Finland.

Soukas, J., and V. Rouhiainen. 1984. "Work Safety Analysis: Method Description and User's Guide." Research Report 314, Technical Research Center of Finland.

Special Issue on Robots in Hazardous Environments. *Journal of Robotic Systems* 9(2), 1992.

Speitel, K. F. 1992. "Measurement Assurance." In *Handbook of Industrial Engineering*. 2nd ed, Ed. G. Salvendy. New York: John Wiley & Sons. Chapter 86.

Swain, A. D., and H. Guttman. 1983. *Handbook for Human Reliability Analysis with Emphasis on Nuclear Power Plant Applications*, NUREG/CR-1278. Washington: U.S. Nuclear Regulatory Commission.

Visinsky, M. L., J. R. Cavallaro, and I. D. Walker. 1992. "Expert System Framework for Fault Detection and Fault Tolerance for Robotics." In *Robotics and Manufacturing: Recent Trends in Research, Education, and Applications. Proceedings Fourth International Symposium on Robotics and Manufacturing*, Sante Fe, November. New York: ASME Press. 793–800.

———. 1994a. "Robotic Fault Detection and Fault Tolerance: A Survey." *Reliability Engineering and System Safety* 46(2), 139–158.

———. 1994b. "Expert System Framework of Fault Detection and Fault Tolerance in Robotics." *International Journal of Computers and Electrical Engineering* 20(5), 421–436.

Willson, R. D. 1982. "How Robots Save Lives." MS82-130, Society of Manufacturing Engineers.

Zurada, J., and J. H. Graham. 1995. "Sensory Integration in a Neural Network-Based Robot Safety System." *International Journal of Human Factors in Manufacturing* 5(3), 325–340.

ADDITIONAL READING

Reliability

Engelberger, J. F. "Industrial Robots: Reliability and Serviceability." Paper presented at a conference on robots in Munich, Germany, November 1972.

————. "Designing Robots for Industrial Environment." Society of Manufacturing Engineers, MR76 600, 1976.

Kapur, K. C. "Reliability and Maintainability." In *Handbook of Industrial Engineering.* 2nd ed. Ed. G. Salvendy. New York: John Wiley & Sons, 1992. Chapter 89.

Pollard, B. W. " 'RAM' for Robots (Reliability, Availability, Maintainability)." Society of Manufacturing Engineers, MS80 692, 1980.

Maintenance

Howard, J. M., "Human Factors Issues in the Factory Integration of Robotics." Society of Manufacturing Engineers, MS528-127, 1982.

"Industry's Man in the Middle." *Iron Age* (January 21), 36–38, 1983.

Macri, F. C. "Analysis of First UTD (Universal Transfer Device) Installation Failures." Society of Manufacturing Engineers, MS77-735, 1977.

"Preventive Maintenance: An Essential Tool for Profit." *Production* (July), 83–87, 1979.

"The Race to the Automatic Factory." *Fortune* (February 21), 52–64, 1983.

Safety

Bonney, M. C., and Y. F. Yong, eds. *Robot Safety.* New York: Springer-Verlag, 1985.

Dhillon, B. S. *Robot Reliability and Safety.* New York: Springer-Verlag, 1991.

Graham, J. H., ed. *Safety, Reliability, and Human Factors in Robotic Systems.* New York: Van Nostrand Reinhold, 1991.

Hartmann, G. "Safety Features Illustrated in the Use of Industrial Robots Employed in Production in Precision Mechanical/Electronic Industries and Manufacture of Appliances." *Journal of Occupational Accidents* **8**(1), 91–98, 1986.

Jiang, B. C., and C. A. Gainer, Jr., "Cause-and-Effect Analysis of Robot Accidents." *Journal of Occupational Accidents* **9**(1), 27–45, 1987.

Lauck, K. E. "Standards for Industrial Robot Safety." *CIM Review* **2**(3), 60–68, 1986.

OSHA. "Industrial Robots and Robot System Safety." *Occupational Safety and Health Technical Manual.* U.S. Department of Labor, 1996. Section III, Chapter 4.

Robotics International. *Working Safely with Industrial Robots.* Society of Manufacturing Engineers, 1986.

CHAPTER 37

CAD AND GRAPHIC SIMULATORS/ EMULATORS OF ROBOTICS SYSTEMS

Jacob Rubinovitz
Technion, Israel Institute of Technology
Haifa, Israel

1 INTRODUCTION

The installation and operation of a robotic manufacturing system frequently proves to be a much more expensive venture than initially planned or imagined (Bock, 1987). Robotic systems are expensive to purchase, install, and operate. Unrealistic expectations based on technical equipment specifications may lead to errors in the robotic cell and application design. Such errors are very expensive and difficult to rectify once the robotic equipment is selected, purchased, and installed.

The operation of robots is also frequently less efficient and effective than expected. The main advantage of robots is their flexibility, which allows their implementation for manufacturing a variety of products in small and medium batch sizes. Flexibility is essential in modern manufacturing to respond to short product life cycles, varying demand, small production lots, and model changes. However, productivity is much easier to achieve in a dedicated mass production system, and few flexible systems are also highly productive. One of the key reasons for low productivity in robotic systems is the fact that robot programming requires the allocation of a considerable amount of robot production time, both for program development and for testing.

CAD-based graphic emulators and simulators of robotic systems can potentially help to avoid some of the roadblocks on the way to successful robot system installation and operation. Cell design, robot selection, verification of robot reach and of correct placement of the cell elements, off-line programming, and simulation of the robot task can all be done in a virtual CAD and simulation environment. Simulation models that accurately represent the proposed robot and cell geometry and the robot kinematic and dynamic performance are valuable tools for evaluating design alternatives, verifying feasibility, designing workcell layout, verifying robot programs, and evaluating cell performance.

In this chapter we will first review the development of graphical simulation tools for robot modeling, cell modeling, and off-line task programming. We will then discuss the different features that are required of such systems in order to deliver useful and realistic design and operation capabilities of robotic cells in a virtual environment. We will follow by presenting examples of implementation for robot design, cell design, and task-level programming in several application domains. Finally, we will review some recent work intended to narrow the gap between the virtual reality model and the real world in which robots and cells operate, and describe the leading commercial robotic systems simulators.

2 HISTORY AND BACKGROUND

The development of computer graphics and computer-aided design systems provided a powerful tool for the modeling and design of robotic cells and off-line programming of robots. The Robotic Industries Association (RIA) defines a robot as "a reprogrammable multi-functional manipulator designed to move material, parts, tools, or other specialized devices through variable programmed motions for the performance of a variety of tasks." This definition explains both the need for reprogramming and the fact that such programming involves motions and tasks performed in space by the manipulator arm. Thus,

Handbook of Industrial Robotics, Second Edition, Edited by Shimon Y. Nof
ISBN 0-471-17783-0 © 1999 John Wiley & Sons, Inc.

programming a robot is inherently different and more difficult than usual computer programming. It has to consider motion in space, potential collisions and ways to avoid them, the ability to reach certain locations in a required pose and orientation of the robot arm, and the dynamic effects during motion on grasped objects or delivered materials. It is only natural that the early robotic systems were programmed on site, using on-line methods either leading the arm physically or via a teach-pendant device through the required locations and motions in space.

However, such on-line programming methods consume a considerable amount of the robot production time. It is impossible to benefit from the robot flexibility and its ability to perform a variety of tasks if a heavy toll of reprogramming setup is required for every new task. The solution provided in early systems was the development of a textual robot programming language. In fact, multiple textual robot programming languages were developed by university researchers or industrial robot manufacturers. By the mid-1980s about 200 different robot programming languages existed, most of them robot-specific (Bonner and Shin, 1982; Ranky, 1984; Rembold and Dillmann, 1985). The textual off-line robot programming languages scene resembled the Tower of Babel. Most of the languages were tailored to support a single manufacturer's robot, and each had its own way of handling sensory input and robot programming functions. Portability of software from one robot to another was virtually nonexistent (Gruver et al., 1983; Soroka, 1983). On top of all these problems, the textual languages failed to deliver on their main promise, which was elimination of the costly setup involved in use of the robot for reprogramming. The textual program usually addressed, by symbolic names, physical locations in space. These locations had to be verified and taught on-line, using the actual robot, prior to program execution. Additional considerable on-line time was needed to fully test and debug such programs.

Thus it was only natural that the emerging 3D computer graphics and CAD modeling systems provided the much-needed environment for modeling the robot and the robotic cell and for programming it off-line. Early development attempts in this area combined the geometric model, which defined the robot environment and task, with extensive user interaction to specify robot paths in space and programs. In the Automatic Programming System for a Spray Robot, developed for the Citroen-Meudon FMS, the task specification consists of the part geometry described by a wire-frame model in a CAD system. The detailed path planning of the robot, including collision detection, is performed manually by the robot programmer (Bourez et al., 1983). The development of 3D solid modeling CAD systems provided the necessary tools for development of more advanced automated robotic tasks planners. The complete geometric representation provided by solid modeling paved the way for implementation of algorithms for path planning, collision detection, approach and grasp planning, and other functions necessary for an automated task-level robotic programming system.

The three key elements common to the CAD-based robot modeling and task programming systems are a world model, a task planner, and a robot interface.

- The *world model* is a geometric model of the cell composed of the robot and the work environment (tools, fixtures, pallets, machines, parts, etc.). The user specifies the required robot task using this world model.
- The *task planner* element of the system converts the user specification of a task into a robot-level program. It may use local and global planning and optimization algorithms to convert the task specification, within the constraints of the world model, into a detailed sequence of robot motion commands.
- The *robot interface* element of the system creates the link between the robot-level program and the robot controller.

Several works laid the foundation for today's CAD-based robotic cell simulation and task-level programming systems. A complete architecture for a task-level off-line robot programming system was proposed for ATLAS, the Automatic Task Level Assembly Synthesizer (Lozano-Pérez and Brooks, 1984). ATLAS is an architecture for a task-level system. It integrated previous research in the area of task planning and provided a framework for future research. Although not fully implemented in a working system, ATLAS defines the following three main elements:

1. The world model, which contains all the information on objects in the task environment and on the kinematic structure of the robot. This information defines the constraints on the motions to be planned by the task planner.

2. The task model, which describes the sequence of states or operations necessary to complete the task.

3. The task planner, which converts the user specification contained in the task model, within the constraints described by the world model, into a robot-level program to carry out the task. The task-planner itself contains several modules, which deal with planning subproblems such as gross motion planning, fine motion planning, and grasping.

A task-oriented robot programming system has been developed for an autonomous robot at the University of Karlsruhe (Rembold and Dillmann, 1985). The suggested structure for this system consists of eight basic parts: a CAD module, a world model, a planning module, an interactive programming and teach-in module, a robotic language compiler (SRL), an interpreter, a monitor, and a simulator.

The schemes suggested for the ATLAS programming system (Lozano-Pérez and Brooks, 1984) and the task-oriented programming system developed at the University of Karlsruhe (Rembold and Dillmann, 1985) include expert modules that use the geometric information in a CAD model to aid the user in task specification and accomplish partly automated robot program creation.

Another key development and implementation along these lines was RoboTeach, an off-line robot programming system developed by General Motors Research laboratories and based on GMSolid solid modeling package (Pickett, Tilove, and Shapiro, 1984). RoboTeach is built around a world modeling subsystem that contains the geometric representation of robots, robot tools, workpieces, and other objects in the robot workcell environment. This world model is designed to support subsystems for several functions, as follows:

- Robot workcell layout design
- Program generation and editing
- Program simulation
- Program translation and downloading

One of the first commercial CAD systems to include a complete robot simulation and kinematic modeling module was CATIA, from Dassault Systèmes (Borrel, Liegeois, and Dombre, 1983). An important feature included in CATIA was a user interface, CATIA CATGEO, that allows a bidirectional access to the geometric model through calls to FORTRAN library subroutine (CATIA, 1986). Such interface provides a development platform for robot task-planning and task-optimization algorithms.

A task-level off-line programming system for robotic arc welding that utilized the CATGEO interface of CATIA was developed in the Robotics Welding Lab at Pennsylvania State University (Rubinovitz and Wysk, 1988a). This system includes a global task-planning module that employs algorithms aimed to minimize the nonproductive time during travel between welds while selecting a welding sequence that minimizes thermal distortions in the welded area. The path planned for torch travel between the weld seams must not collide with the work environment. The task-planning algorithms use a model of the robot motion capabilities to assign optimal motion velocities to different travel segments and estimate time of motion. The final welding sequence is planned using a traveling salesman algorithm (TSA). The geometric information of the robotic welding cell and a specific welding task is created using CATIA and accessed by the task-planning algorithms through the CATGEO interface. The resulting optimal robot path is fed back through the same interface to the CATIA system for program simulation and verification prior to downloading to the robot controller. Figure 1a shows the CAD model of the welding cell that serves as the source for the automated task-planning algorithms. The resulting robot path is shown in Figure 1b.

There are several subproblems that need to be solved by robot cell and task-planning systems, each involving complex reasoning and strategic ramifications based on geometric data as represented by a valid and complete representation such as provided by solid modeling CAD systems. These planning subproblems include:

- Reachability within the robot work envelope. The solution of this problem is necessary for robot selection, optimal placement of the robot or other cell elements, and evaluation of task feasibility.

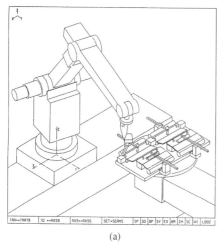 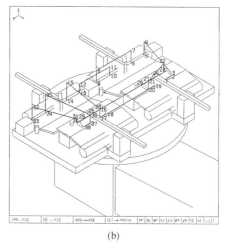

(a)   (b)

Figure 1 (*a*) CAD model of the welding cell. (*b*) The resulting robot TCP path.

- Path-finding and collision-avoidance problems, whose solution is necessary for almost all robotic task-planning domains, including assembly, material handling, welding, painting, and the navigation of autonomous vehicles.
- Grasping and part-mating problems, whose solution is required mainly for the task domains of robotic assembly and material handling.

The importance of specialized CAD and robot simulation systems for design, planning, and programming of robotic applications prompted the development of several such systems.

Some of these systems have been developed at university and research institution laboratories. These include IGRIP, SAMMIE, and GRASP, from the University of Nottingham; STAR, from the University of Wisconsin; and ROBOCELL, from the Centre for Flexible Manufacturing Research and Development of McMaster University. Commercial systems include CATIA, from Dassault Systèmes; ADJUST, from McDonnell-Douglas Automation Co.; ROBOCAM, from SILMA Inc.; ROBOT-SIM, from GE-Calma; ROBCAD, from Tecnomatix, Inc.; ROBOGRAPHIX, from Computervision; Ultra-GRIP, from Deneb Robotics Inc.; AdeptRapid, from the SILMA division of Adept Technology, Inc.; and WorkSpace from Robot Simulations Ltd.

3   OVERVIEW OF REQUIRED ELEMENTS FOR GRAPHIC SIMULATION OF ROBOTIC SYSTEMS

The main elements of a system, which allow the user to deal with a great number of robotics problems in a virtual environment of a graphic simulator, include the following (Rubinovitz and Wysk, 1988*a*; Owens, 1991; Zeghloul, Blanchard, and Ayrault, 1997):

- A CAD solid modeler, which allows the user to build a database with a valid and complete geometric description of the robot and its environment. This CAD database includes the models of the robot links and all the objects in the cell environment: machines, fixtures, feeders, grippers, parts, etc.
- Built-in libraries of commercially available industrial robots, common production equipment, and application-specific options that include tooling, such as a variety of commercially available grippers, spot weld guns, and arc weld guns. The libraries are an integrated part of the system and can be expanded by inclusion of elements from the CAD database created by the user.
- Data translators for standard data-exchange formats such as IGES, DXF, STEP, and VDAFS. These translators allow the importing of models of products, tools, and parts from other corporate CAD systems and support the rapid development of accurate simulation models. They also allow the exporting of model data to be used by other systems.

- Kinematics module, which allows the modeling of the robots and other mechanisms. This module includes direct and inverse kinematics algorithms, which are necessary to calculate the robot envelope, its reach, and the motion in space during simulation of the robot's movements.

- Algorithms for planning the robot placement and task accomplishment. These algorithms may vary in their complexity and level of task automation in different simulation systems. They include collision-detection algorithms, which are used to verify task accomplishment, optimal placement algorithms, which determine the relative position of the robot in the task environment, collision-free path-planning algorithms, which allow the system to generate trajectories in a cluttered environment, and cycle-time-estimation algorithms, which help to evaluate and optimize the time needed for a completion of a sequence of movements.

- Simulation and off-line programming of robots and industrial robot cells. This capability is built on the CAD modeling capabilities and computational algorithms mentioned above. It allows the user to interactively define robot tasks and create programs in the virtual cell environment and to verify the programs by a realistic simulation. The off-line programming can be achieved by three different methods:

  1. Interaction with the robots and cell elements in the virtual model
  2. A specific task definition and programming language
  3. Automatic task and process planning algorithms

  Combined use of all three methods is also possible. The simulation functions may include the concurrent movement of robots and of any other kind of kinematic devices; automatic interference checks and collision or near-miss detection; full trajectory trace of the robot tool-center point (TCP); management of virtual cameras; and inclusion of signals from virtual sensors. Verification functions of the created programs include collision and near-miss detection, checking the joint limits and speeds, dynamic performance analysis, and cycle time analysis.

- Calibration of the simulation model: calibration of the TCP, possible servo-axes and other auxiliary axes for specific robot models, and calibration of cell environment elements such as used workpieces or jigs. Some systems also offer functions for performing a signature analysis for an individual robot, which is based on direct data collection from the physical robot during motion.

- A postprocessor for the specific programming language of different industrial robots, and downloading capability for downloading the verified program to the robot controller. Uploading capability is also sometimes available, which allows the uploading and verifying of actual robot programs in the virtual environment.

- An open development environment, which allows the researcher or the advanced user to access all the geometric and kinematic models in the system and develop new applications, new types of analysis, and new planning and synthesis algorithms.

4 THE ROLE AND IMPORTANCE OF SIMULATORS OF ROBOTIC SYSTEMS

It is relatively easy to identify a need for a robotic system solution. But once you know you need a robotic system, how do you get it up and running efficiently on the factory floor? How do you select the right robot for the application, design the robotic workcell environment, and find optimal placement for the robot in the cell without costly errors? How do you estimate correctly the performance and throughput of your robotic system at the design stage? How do you resolve the flexibility–efficiency tradeoff if reprogramming, which is the key to robotic system flexibility, requires hours of on-line programming and debugging, thus rendering the robotic equipment inefficient for large-variety and small lot-size production?

Robotics simulation software systems provide some of the answers to these questions. They help manufacturers choose the right robots and make correct design decisions by simulating real-life work environments. The software packages help engineers build computer-generated virtual workspaces, then draw upon libraries of existing robots, complete with all of their specifications, to try out or test the feasibility of different robots for specific applications. Robots can be carefully calibrated in the simulation environment to make sure, for instance, that arms are able to reach where they are needed or that special end-of-arm tools and accessories can do their tasks with various parts and equipment types. Robot programs can be created and verified in the simulated environment by off-line programming. The simulated task program can be used for task performance estimation by predicting cycle time, the time it takes a robot to move through a defined series of motions. Subsequently, robot programs can be downloaded from the simulations directly to the robots themselves.

Some robotic simulation software vendors offer application-specific packages for processes such as arc welding, spot welding, painting, finishing, and material handling. These packages make the design, tool selection, and programming task for the specific application even more efficient, resolving favorably the flexibility–efficiency tradeoff equation.

In this section we will present examples of the contribution and importance of robot simulation systems to the solution of the following problems in different application domains:

- Robot design
- Robotic cell and system design
- Program testing and verification
- Task-level programming and task analysis and design

See also Chapter 32, Robot Ergonomical Optimizing Robot Work.

4.1 Design and Evaluation of Unique Robot Systems

Simulation tools play an important role in the design and evaluation of unique robot systems for applications where standard off-the shelf commercial robots are inadequate. Such applications range from space station construction to various construction tasks on earth.

MDSF, a generic manipulator development and simulation facility for flexible, complex robotic systems (Ma et al., 1997), was developed by Spar Aerospace Ltd. as a prime contractor to the Canadian Space Agency. It is an integral part of Canada's contribution to the International Space Station (ISS) program. MDSF is capable of simulating general robotic systems with complex and flexible structures undergoing all kinds of tasks, including payload handover and contact with the environment. Rigorous validation and configuration control have been imposed throughout the development course of the simulation facility. MDSF is currently applied for modeling and evaluating manipulators and tasks in the space station program.

The Robotics Technology Development Program (RTDP) was initiated by the Department of Energy (DOE) in 1994 to develop and use techniques for simulating a robot system using verified algorithms and models (Harrigan and Horschel, 1994; Griesmeyer and Oppel, 1996). The simulation tools are needed for the design and evaluation of unique and costly robot systems that will be used for maintenance, construction, and treatment activities across the DOE complex. Some of the robot systems needed are unique and are estimated to cost millions of dollars for the robot system alone. Design and analysis tools that can model structural performance, control behavior, suitability to accomplish desired tasks, and the interaction of end-effectors do not exist in one integrated package. Many key simulation parameters unique to these robots do not exist at all. The program will develop specific simulation modules and integrate them with validated existing commercial modules. The system would allow DOE and the national laboratories to simulate and evaluate design performance parameters of a robot system before it is purchased or committed to hardware. Unique advanced robot systems can be designed faster. The validation and simulation tools will assist in vendor proposals evaluation and ensure that the design that best fits the task requirements is selected and commissioned.

Further, RTDP is developing a test bed for arm-based retrieval technology development. This test bed will be used among other things, to validate simulations dealing with dynamic performance and adaptive control of a flexible manipulator. The simulation system will provide DOE researchers with tools that can be used to influence design and assess system performance before costly modifications are made or new systems are fabricated. Simulation modules will be validated against real hardware systems, such as the RTDP long-reach arm test bed, and then applied to conceptual and prototype designs.

A special purpose robot for interior-finishing works was designed and developed at the Technion, Israel Institute of Technology, using the commercial simulation system ROBCAD of Tecnomatix Technologies, Inc. (Warszawski and Navon, 1991). The simulation was used for the entire robot development process and was instrumental during one of the more important development stages, the selection of the configuration of the robot's arm. The development process of the robot included a number of interrelated studies: formulation of the robot performance specifications; the ensuing preliminary design; planning of robot activity in a building; analysis of the robot configuration; adaptation of building technology to the robot's constraints; and physical experiments with robotic performance of building tasks. The design and analysis stages in the development process were accomplished using the ROBCAD system. The simulation tools assisted in the selection of a preferred configuration for the robot arm, as required for optimal performance of the interior finishing tasks. Such simulation-assisted analysis helps to determine the values for several variables: the configuration of the joints, the reach of the arm, the length of its links, and the velocity attainable at the joints. The criteria employed in the selection of the preferred alternative were the general efficiency of operation, the productivity, and the cost of the arm.

4.2 Robotic Cell and System Design

The various commercially available robotic simulation packages provide the basic tools needed for robotic cells and systems design. These basic tools are essential in order to rapidly design and deploy automated manufacturing systems (Craig, 1997). Simulation-based robot cell design involves the integration of all cell elements and hence must also provide tools for the design and operation of various peripherals. AdeptRapid, a simulation-based design system developed by the SILMA division of Adept Technology, Inc., is using object-oriented techniques for cleanly implementing not only simulated robots but also libraries of cell peripherals. Another very important function in simulation-based design is the availability of optimization algorithms and accurate modeling of the space and time (performance) aspects of the cell elements. In order to predict the overall cycle time of a robotized cell within few percentage points of actual performance, not only robot motion times must be accurately estimated, but also timing associated with cell peripherals such as grippers, conveyors, and parts-feeding devices. With all these items accurately modeled, the system can perform optimizations to automatically search for superior designs. The use of a simulator to design and evaluate alternative concepts and communicate them to system integrators and end users can effectively reduce the overall time needed to install flexible automation.

The simulation tools for robot cell design and programming are accessible and available even on a standard low-cost PC-compatible computer (Owens, 1991). A software package called Workspace has been developed as a visualization aid for engineers and managers involved in the process of designing and debugging new or existing robot installations. Benefits of the use of such a system include better understanding of the robotic cell processes, which involve many moving parts and mechanisms; the ability to detect off-line collisions between robots and objects; and the ability to evaluate and optimize the time needed for a sequence of movements.

A major challenge in the design of robotic workcells using simulation is achieving a faithful representation of reality in the simulated model. Freund et al. (1994) present a new approach to realistic simulation of robotic workcells in industrial manufacturing processes. A workcell is composed of different classes of objects, such as robots, AGVs, conveyor belts, and sensors. The simulation system is used to design the workcell layout and assist in off-line programming of tasks. The approach described includes embedding an industrial robot controller into the simulation system by means of a versatile message-based interface and using a sophisticated environment model for intuitive simulation of grip-and-release operations, conveyor belts, and sensors. This is combined with a graphical 3D visualization of the workcell in real time. User interaction with the simulation system is based on a windows-oriented graphical user interface (GUI).

The design of a workcell involves evaluation of many alternatives in a search for an optimal solution. Commercial simulation systems help to model, create, and evaluate various alternative cell configurations, but seldom provide automated planning tools which optimize the design. The design engineer has to create a large number of different cell design configurations and try simultaneously to pursue a wide variety of often conflicting subobjectives. Automation of the search for optimal design is the goal of several research and development efforts of simulation systems. One such system is Universal Simulation System (USIS) (Woenckhaus et al., 1994), a graphic robot simulation package that implements largely automated procedures and search algorithms to support the design engineer during most of the workcell design stages.

An important aspect of robot cell design is how to guarantee operator safety. Various safeguards need to be included and planned for early in the design stages. Such safeguards include physical barriers, emergency stop buttons, and electronic sensor-based screens.

The increasing complexity of robotic workcells also makes it more difficult to ensure operator safety (Lapham, 1996). Simulation helps in evaluating safety hazards and planning appropriate solutions early in the design process, before it becomes too costly to make changes. Problems can be solved months in advance, and ideas can be tried in simulation before any tooling, safeguard, or positioners are constructed and safeguards or sensors purchased and placed in location. Lapham describes the role of simulation in providing safety-related information at the various stages of the design process. In the first stage of the process, the initial concept formulation, a better spatial idea of the issues involved is achieved by preparing a basic simulation. Safety considerations can be discussed early and problems visualized for each basic cell configuration concept.

The next step in the process is to determine exact equipment selection and placement. Robot simulation helps in envisioning the interaction of different devices in the cell and identifying potential safety hazards. In Workspace, a PC-based robot simulation program, the safety engineer can display the robot's work envelope to make sure that the guarding will fully enclose the robot. Another feature found in Workspace, near-miss and collision detection, can help an engineer to design a safer workcell. If a safety engineer states that a robot must stay at least six inches from any guarding, then this distance can be set, and any time the robot moves closer than six inches to the guarding a message will appear, helping in analyzing potential risks in the task design. Similar analysis tools are available in most of the commercial robot simulation packages.

Finally, simulation improves safety by making possible early evaluation of the task programs prior to actual downloading and execution in the cell. The safety inspector can verify that the program logic includes all the necessary safety precautions. The complete functioning of the task program within the cell can be simulated to make sure that everything is interacting correctly.

4.3 System Control, Robot Capability Modeling, and Program Verification

Verification of robot programs in a simulated, virtual reality environment requires an accurate model of the robot controller and of robot motion capabilities. The robot motion capabilities model should include kinematic and dynamic performance and is needed to predict performance times as well as trajectory errors during task execution. Such a graphic simulation system for robotics operations has been developed at the University of Michigan—Dearborn, in conjunction with a kinematic error model, to assist in predicting robot performance and capabilities (Kamrani, Wiebe, and Wei, 1995). The system uses a Taguchi-type experimental design to predict the robot process capability. Two sets of charts are created by the system to illustrate the repeatability and accuracy of a robot performing a given operation. These charts are histograms in the x, y, and z directions and scattered ellipses in the xy, yz, and xz planes. Three key functions in the system assist in mapping the robot capability: a description of the geometric error that is due to the kinematic error, a function with which to request and display data on a robot moving from one position to the next, and a family of charts showing the capability of a robot for implementing the required task. The results provided by the system can help management analyze robotics projects and make decisions concerning robot selection and implementation in a more systematic manner.

Graphic robot simulation can serve as a tool to derive motion economy principles (MEP) for various robot classes in different group sizes (Edan and Nof, 1996). If robot capability is mapped to utilize MEP, robot tasks can be planned in a way that will improve performance times while reducing joint wear. Edan and Nof have described the use of arm joint utilization (AJU) factor, which is continuously calculated during simulation based on the total movement of all joints during the task performance. In particular, they concentrated on the effect of bin orientation and location on robot performance and compared the performance of different robot classes for specific cell setup in a kitting process.

Graphic simulation is indispensable for verification and evaluation of novel robot control structures and unique task designs. A simulation system was used in such a manner at Purdue University to verify an intelligent control system for an agricultural robot that performs in an uncertain and unstructured environment (Edan, Engel, and Miles, 1993). The control system was modeled as a distributed, autonomous set of computing modules that communicated through globally accessible blackboard structures. This control architecture was implemented for a robotic harvester of melons. A CAD workstation was used to plan, model, simulate, and evaluate the robot and gripper motions using 3D, real-time animation. The intelligent control structure was verified by simulating the dynamic data flow scenarios of melon harvesting. Different control algorithms for improving harvest performance were evaluated in the simulated environment on measured melon locations. Picking time was reduced by 49% by applying the traveling salesman algorithm to define the picking sequence. Picking speeds can be increased by a continuous mode of operation. However, this decreases harvest efficiency. Therefore an algorithm was developed to attain 100% harvest efficiency by varying the vehicle's forward speed. By comparison of different motion control algorithms through animated visual simulation, the best control method was selected, leading to an improved performance.

4.4   Task-Level Programming and Task Analysis and Design

One of the most important uses of CAD-based simulation systems for robotics is in the area of task-level off-line programming of robots. By programming robots off-line, users can increase the system efficiency and be able to justify the use of robotic processes even for small and medium lot sizes. However, off-line programming of robots requires programming of complex motions in space and is almost an impossible task without visual aid and validation such as is provided by simulation systems.

The development of graphical simulation tools plays an increasingly important role for off-line programming (OLP) in the automotive industry (Rooks, 1997). Graphic simulation-based OLP systems use is increasing, and it is becoming an industry standard, particularly in the areas of robot welding and robot painting of vehicle bodies. Such systems have a major role in the success of "right first time" policies. Most of the users can easily measure the benefits of "right first time" task design and programming, which results in accurate programs, safe operation, short lead times, material savings, and significant man-hour and production line downtime savings. A brief summary of the advantages and savings realized by users of simulation and OLP systems in the automotive industry, as reported by Rooks, is provided below.

Comau, the Italian robot manufacturer and systems integrator for the automotive industry, managed to significantly reduce lead time in installing robotic applications by simulating and verifying the systems using Robcad from Tecnomatix Technologies and correcting the problems before installation of a line on the shop floor. According to one U.S. luxury car manufacturer, an extensive user of Deneb's IGRIP simulation and OLP software, the effort that goes into the simulation is fully rewarded by the resulting savings in equipment and man hours on-line. An estimated 1500 man hours have been saved by a contractor for Chrysler Corporation by the use of OLP and simulation to program an automotive line with 40 spot welding and materials-handling robots.

Painting is a particularly difficult process to program because of the many parameters to be controlled and the high degree of skill required to paint a vehicle body. Capponi Alesina, Fiat's largest painting subcontractor, is using Robcad simulation and OLP for the robot painting lines at Fiat's Pomigliano, Melfi, and Rivalta plants—in total, seven lines with 36 robots painting body interiors. The savings from OLP are not only in man hours and shorter lead-times, but also in the paint materials or products that are not needed for verification of programs in a virtual environment.

Applied Manufacturing Technologies (AMT), a U.S. manufacturing systems design house, discovered this when contracted by a major vehicle manufacturer to reprogram a robot line to accommodate a new heavy solids-content clearcoat. AMT estimates that programming off-line saved 20 scrap car bodies, thousands of gallons of clearcoat costing $1 million, and more than 1,000 man hours.

Navistar International Transportation Co. used IGRIP software from Deneb to program two paint-finishing lines for trucks in which 24 Graco painting robots are used. Model variety is much higher on a truck line, increasing the number of programs for each robot or paint machine. Each of these 24 robots requires 35 different motion programs, a total of 840 in all, to accommodate the various cab bodies, air shields, extenders, bonnets and other parts painted on the line. The model of the paint-finishing lines was created by importing existing Unigraphics CAD data. The robots and all auxiliary axes, including track motion and lift, were imported from the library built in to IGRIP, and paint deposition and other process parameters were defined. From the simulation, robot motion programs with the correct paint deposition parameters, such as atomization air pressure, fan width, and fluid delivery, were automatically generated. Robot signature analysis functions built in to IGRIP were used to identify the dimensional data of each real robot on the line. Then the software's integrated calibration functions used the real robots as the workcell measurement device. These data were uploaded into IGRIP and used to offset the programmed positions of the robots, parts, and fixtures. As a result, the simulation was "tuned" to reflect the real-world discrepancies so that after signature analysis and calibration the robot programs could be downloaded directly to the paint line. Navistar estimates that by using simulation it halved the time required to reprogram the paint lines, and all this without interrupting production or routine maintenance activities. However, even more significant than time savings and reduced production downtime were the materials savings achieved through simulation and OLP. The consistency of methods, syntax, and coating application resulted in smoother robot motions and better paint deposition. The switch to the new application process and the ability to control spray parameters better, which reduced overspray, waste, and solvent usage, gave savings in materials costs amounting to $11,000 per day.

Mercedes in Germany also faces a large number of models and variants at its line for robot welding truck bodies. It has been using IGRIP Spot, a special simulation program for spot welding, to off-line program the 80 robots it employs for spot and arc welding, stud welding, riveting, handling, and assembly. By using OLP, Mercedes is able to cut down the amount of programming in two ways. First, because the four welding cells used are identical, only one set of programs is required and, by accurate calibration of each cell, the specific off-sets for that cell can be calculated. Thus, rather than four individual cells being programmed, only one program has to be created and the other three are automatically generated using IGRIP's built-in calibration function. The second area of saving from OLP is in developing programs for part families. The program for the first-off member is produced and then the other members, or variants, are generated by the addition of modules that individualize the product—long and short wheelbase, left-hand and right-drive, and so on. Also, only one half of the body needs to be programmed and the other half is generated by mirror-imaging, which again is done automatically using a standard IGRIP function. Additional benefits of the system include:

- The ability to modify programs off-line without causing disruption to production, as required by the frequent introduction of new truck features.

- The ability to import and work directly on CAD product data, which improves the accuracy of the program. For instance, the weld gun can be positioned accurately relative to the truck body so that it is always normal to body surface. This ensures a weld of the highest quality and extends weld tip life.

- The ability to validate and thoroughly investigate programs before they are implemented on the shop floor using the simulation feature. As a result of the validation, collisions can be avoided between the simultaneously working four robots in a cell and between the robots and the fixturing, tooling, and components. This reduces the risk of equipment damage and line stoppage and increases safety on a welding line that operates with up-times of 97% and higher.

- The ability to design, try in a simulated environment, and modify if necessary new fixtures, tools, and welding guns so that they are correct when realized.

GRASP-VRI, a shipbuilding-specific software simulation and off-line programming tool, is being used to support the design and programming of large-scale robot arc welding cells in two of the world's largest and most modern shipyards (Sorenti, 1997). GRASP-VRI is based on GRASP, a virtual reality simulation system developed by BYG Systems Ltd., Nottingham, U.K. The system assists in increasing productivity while creating the arc welding programs for robots off-line. Major time savings are achieved by using the integrated program creation concept, which uses predefined, parametric libraries of generic arc welding tasks that were created and verified prior to and during cell commissioning. This approach is also applicable to other robotics applications, such as cutting, gluing, and inspection.

Automated task-planning strategies and algorithms are needed to further increase productivity in the different domains of robotic applications where OLP is used. Automated and optimized task planning is supported by the model in the simulation system and uses domain specific parameters for task planning. An example for this approach is an automated path-planning and off-line programming system of robots in spray-glazing operations (Bidanda, Narayanan, and Rubinovitz, 1993). The system creates and validates automated path-planning strategies and algorithms for robotic spray-glazing processes, using ROSE (ROBCAD Open System Environment) as a development and interface platform to a commercial simulation system. In order to optimize the spray-glazing process parameters, the system uses an analytical model of the spray-glazing process that was developed based on experimental data. Based on the process parameters, path-planning algorithms that ensure efficient glaze coverage are automatically generated. These algorithms are based on detection and identification of relevant features of the surface models of spray-glazed parts. The surface models are imported into ROBCAD, using an IGES interface, from a corporate CAD database created with the DUCT system. Finally, the spray-glazing programs are validated in the simulation system and downloaded into a robotic cell for spray glazing. Figure 2 shows the general structure of the system.

The system allows better integration of the design and production of spray-glazed parts. As a result, significant reduction is achieved in the lead time and work content needed to produce spray-glazed parts, along with material savings and better part quality. Specifically, the following improvements in the process are achieved:

- Higher quality and consistency of glaze spraying
- Reduced overspray and loss of glaze due to more accurate positioning of the spray gun
- Increased system productivity due to reduced robot programming time
- Reduced manufacturing lead time due to higher degree of integration between the design and manufacturing activities

An additional area where robot and cell simulation contributes to task planning and execution is in the creation of better operator interfaces in the virtual environment. A "teaching by showing" system using a graphic interface has been developed for robotic assembly operations at NTT Human Interface Lab in Tokyo (Ogata and Takahashi, 1994). This system provides a user interface with which an inexperienced operator can easily teach a task and a task execution system that can operate the task in a different environment. The operator shows an example of assembly task movements in the virtual environmental created in the computer. A finite automation defined task is used to interpret the movements as a sequence of high-level representations of operations. When the system is commanded to operate the learning task, the system observes the task environment, checks the geometrical feasibility of the task in the environment, and if necessary

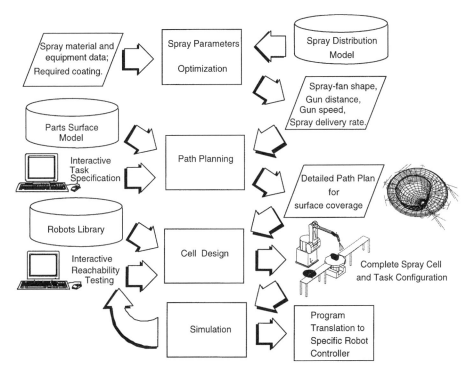

Figure 2   General structure of off-line programming and simulation system for robotic spray glazing.

replans the sequence of operations so that the robot can complete the task. Then the system uses task-dependent interpretation rules to translate the sequence of operations into manipulator-level commands and executes the task by replicating the operator's movement in the virtual environment.

Another novel man–machine interface that leads to task enhancement of an underwater robotic arm by graphical simulation was developed at Cranfield University (McMaster et al., 1994). The task of piloting an underwater robotic arm is assisted by using a computer-generated dynamic image of the marine work site. This technique will be used for deep-water welding repair schemes, and it is envisaged that such a system would operate without the assistance of divers. Ultimately the system would have the capability to carry out the complete intervention process, from detrenching a pipeline to final inspection and corrosion protection. With the use of knowledge of both the robotic arm and work site geometries a kinematic model can be built using a computer-based simulation and off-line programming system. This 3D solid modeling system is then used to provide the visual feedback to the operator of the underwater robot. With such a system the operator benefits from a higher degree of telepresence and improved task capability.

4.5   Robotic Systems Simulation in Education and Training

Robotic simulation systems are used to enrich the educational and research laboratory experience in many universities around the world. Nof gives an extensive report of such use in the development of Production Robotics and Integration Software for Manufacturing (PRISM) workstation at Purdue's School of Industrial Engineering (Nof, 1994).

The simulation systems allow experimentation and hands-on experience in the following areas that are difficult or infeasible to include in a traditional university laboratory:

- Demonstrations and design of various robot types and of kinematic principles of robot operation
- Concurrent design of robotic facility solutions
- Integration of robotic devices in CIM systems
- Off-line task- and application-oriented programming
- Computer-aided design of technologies such as welding, painting, clean-room assembly, and teleoperation
- Integrated manufacturing cell communication and control

Using the virtual reality manufacturing environment provided by the graphic simulation systems is an effective way to introduce realistic complex systems into the classroom. Using such systems, students can accomplish and demonstrate actual designs of robotic and manufacturing systems at different levels of complexity. They can design and assemble a new robot, program it, and evaluate its performance. They can also design and program a complete manufacturing cell, include various industrial robot models from an existing library, and then iteratively evaluate the performance of their cell design. Figure 3 shows a complete model of the CIM lab at the Faculty of Industrial Engineering and Management of the Technion. This model was created by Technion students as part of a final project and is currently used for validation of programs and designs prior to actual execution in the CIM lab.

The effective teaching of state-of-the-art courses in the area of industrial robotic applications and computer-integrated design and manufacturing is an ongoing challenge due to the hardware- and capital-intensive nature of the curriculum. An integral component of such courses are the laboratory assignments where students must design and program robotic applications. The high cost of robotic equipment has traditionally been resolved by using small table-top educational robots. These robots are limited in modeling of actual manufacturing applications. Inclusion of a single industrial robot in the laboratory setup solves the problem only partially due to both safety constraints and robot time availability when teaching a large number of students.

In addition, there are several key robotic manufacturing systems engineering principles and knowledge that must be taught and researched effectively by students and trainees but cannot be achieved in a traditional laboratory:

1. *Design of new robotic mechanisms and demonstration of their operation.* Within the simulated workstation environment it is possible to design, visualize, simulate, and evaluate new robotic devices, tooling, etc. without the prohibitive manufacturing time and cost.

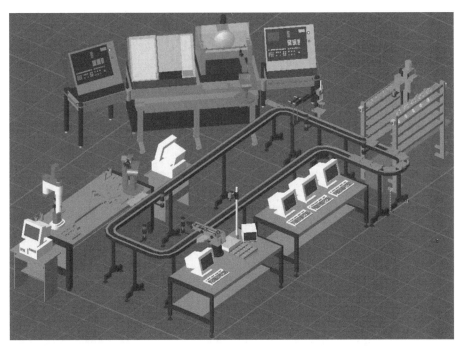

Figure 3 Simulation model of the CIM lab at the Faculty of Industrial Engineering and Management of the Technion (created with RobCad, courtesy Tecnomatix).

2. *Computer-integrated manufacturing.* Beyond the basic design knowledge about how industrial robots work, the students must learn to view robots as part of a CIM system. The workstation environment enables simulation and animation of the physical, spatial, geometric, and trajectory plans of the cell elements, as well as the operational planning aspects. This simulation is based on the kinematic and dynamic models of the integrated components in the CIM system.

3. *Robot database and communication.* The use of a database with extensive robot design and performance information and the use of communication between the robotic manufacturing system components are closely related to CIM and are vital tools for the engineering of effective robotic solutions.

4. *Off-line task-level programming.* Higher-level task- and application-oriented languages are fast becoming the most effective technique in programming robotic manufacturing equipment.

5. *Economic justification and performance measurement of robotic manufacturing systems.* The workstation environment enables experimentation with and evaluation of a large variety of robots. A graphic simulation and display of multiple robot models is needed for proper performance evaluations.

6. *Robotic manufacturing system design for industrial applications.* The workstation environment enables students to design or model a robotic application, such as welding, painting, processing, inspection, etc., and study its behavior using simulation. By this means they can provide realistic designs and suggest modifications for the various industrial applications that they will be required to implement in their professional careers.

7. *Concurrent engineering design and collaboration among engineers.* Once the basics of industrial robotics technology are understood through the lab work, the students can be taught collaborative design of effective robotic solutions. The workstation environment can demonstrate the principle of concurrent design using information from distributed sources, including a CAD system, combining product design, process and assembly plans, and quality/accuracy requirements with alternative robot systems from a vast library of industrial robots.

5 VIRTUAL REALITY AND THE REAL WORLD

A successful simulation system must be able to create a faithful, complete, exact, and precise model of the real-world environment. This is the major challenge in simulation of robotic systems and an ongoing area for further research. The-state-of-the-art CAD software provides tools for complete and exact representation of geometric features of the robot cell. The challenge lies in modeling the action within the cell: how well can the robot model be calibrated to reflect dynamic performance and specific response "signature" of an individual robot? How well modeled is the control behavior, including motion deviations due to specific computational methods of the real controller? How to model various sensors in the virtual cell environment and simulate and evaluate uncertainty?

This section will briefly overview the methods and research work aiming to reduce the gap between the virtual reality model and the real world of robots, cells, and tasks. (See also Chapter 17, Virtual Reality and Robotics.)

5.1 Calibration and Modeling of Specific Robots

A common reason for discrepancies between a robot model in a simulated cell and the actual robot on the factory floor is that there are inaccuracies in each specific working robot that have not been taken into account in the simulation model. The robot model in a simulation system is a perfect model made according to the manufacturer specifications and exact dimensions (assuming that no errors were made during the creation of robot models and libraries). The real robot on the factory floor is unique and different from the exact specifications and from other robots of the same type due to tolerances in the manufacturing process, backlash, and general wear and tear. The discrepancies between the model and the real robot can be quite significant. Programs created off-line in the simulation system frequently miss the path points in the real task by more than 10 mm (Bernhardt, 1997). In an arc welding task, for example, this may create a severe problem. The tolerance for the welding tip error from the seam path is ± 1 mm, and as a result the robot operator must reteach the entire program with a teach pendant, rendering futile the entire effort of off-line programming with a simulation system.

By calibrating the robot model it is possible to reduce the discrepancies with the actual robot. Calibration methods measure and map the performance and find the "signature" of the a particular robot, and these actual performance data are used to modify the kinematics of the robot model within the simulation. (See also Chapter 39, Precision and Calibration.)

A commercial system for robot calibration, the Robotrak system, is offered by Robot Simulations Ltd. It works in conjunction with the company's PC-based WorkSpace simulation software. Robotrak includes a robot performance-measuring device and data-collection software.

The Robotrak calibration system, shown in Figure 4, consists of the following components:

- Three incremental encoder measuring units
- A robot end-effector interface
- An ISA computer interface card for data acquisition
- A reference bar of known length
- A PC running the Robotrak software

Robotrak is capable of measuring static positions or motion paths to an accuracy of 0.2 mm (0.008 in.) in three dimensions. The calibration process is as follows.

The three high-resolution encoder units are placed anywhere within the working envelope of the robot, roughly at the corners of an equilateral triangle, and connected through the data acquisition card to a computer running the Robotrak software. A non-stretch cord with constant tension on it is then pulled from each of the encoder units and calibrated with the use of a reference bar. Once a conversion of encoder counts to length is established for each unit, the relative distance between the three encoder units is determined by pulling the cords from each unit to the adjacent unit. At this point the encoder units are ready to begin calibrating the robot.

Each cord is then pulled from the encoder and attached to the robot end-effector interface, which has already been attached to the robot faceplate or tool. Once the cords are in position, Robotrak uses the length of each cord to determine the x, y, and z coordinates of the robot end-effector. The robot is then moved through 50 points in space,

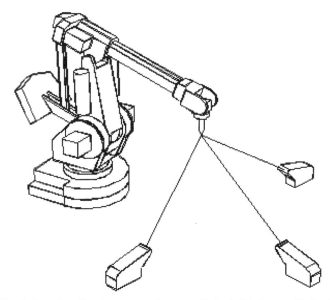

Figure 4 The Robotrak calibration system (courtesy Robotic Workspace Technologies, Inc.).

and at each point the Robotrak software records the robot's position through the three Robotrak encoders. Over the length of the test as many joints as possible should be moved.

Once the points have been recorded, the software will compare the measured positions (where the robot actually was) with the teach points (where the robot thought it was), and the errors associated with the kinematics of the robot are extracted and stored as a filter used for correction of the kinematic model in the WorkSpace system. Using this filter, precise programs for a particular robot on the shop floor can be created off-line in the simulation system.

The calibration process is usually a time-consuming procedure unless it is optimized by identifying the critical parameters and a minimal set of data points needed for the process. Graphical simulation of the robot calibration has been used to develop a systematic method for the evaluation of needed data points and calibration parameters (Zak, Fenton, and Benhabib, 1993; Pathre, 1994; Pathre and Driels, 1990). The graphical simulation system is also used for comparison of alternative calibration procedures/setups and for prediction of the expected accuracy of the robot after an actual calibration is performed. Simulated experiments can be conducted to demonstrate that an optimal calibration procedure was found.

Another advanced application of a virtual reality simulation for improving the calibration process for telerobotic servicing has been implemented at the California Institute of Technology (Kim, 1994). The calibration technique is based on matching a virtual environment of simulated graphics models, created using solid 3D geometry and perspective, with actual camera views of the remote site task environment. The result is a high-fidelity preview/predictive display in which the calibrated 3D graphics model is overlaid on live video. Reliable and accurate calibration is achieved by operator-interactive camera calibration and object localization. Both rely on linear and nonlinear least-squares algorithms. Since the object pose becomes known through object localization, the computer-generated trajectory mode can be effectively used in this approach. This calibration technique has been successfully utilized in Jet Propulsion Laboratory/Goddard Space Flight Center telerobotic servicing demonstration. The positioning alignment accuracy achieved by this technique from four camera views was 0.51 cm on average for a tool insertion in the servicing task.

5.2 Modeling the Actual Robot Control—Realistic Robot Simulation

One major problem in achieving exact simulation of the robotic cell control behavior is the availability of a model of the robot controller. The algorithms defining the robot's motion behavior are not publicly available. To overcome this problem a consortium of automotive companies, controller manufacturers, and simulation systems manufacturers initiated the realistic robot simulation (RRS) project (Bernhardt, Schreck, and Willnow, 1995; Bernhardt, 1997). It aimed at integrating original controller software (black box) into simulation and off-line programming systems via the specification of an adequate interface. The RRS project, managed by IPK-Berlin, started in January 1992. The RRS Interface Specification was released in January 1994.

The project goal was to improve the simulation accuracy of industrial robot simulation systems in order to achieve a more realistic simulation of robot controllers. The goal was achieved by the definition of a common RRS Interface for the integration of controller simulating modules into simulation systems. Using original controller software parts, controller manufacturers provide simulation modules for their latest controller types. Simulation system suppliers have implemented the RRS Interface in their software products.

The RRS interface has been tested on software and hardware platforms used for robotic simulation in the automotive industry and has demonstrated impressive results of accurate simulation of motion behavior, robot kinematics, and condition handling. It has been proven that the deviation between simulated and real joint values is less than 0.001 radians. Concerning task cycle times, a difference of less than 3% could be reached.

The partners in the RRS project are major automotive companies, robot controller suppliers, and robotic simulation systems developers. Automotive industry partners include Adam Opel AG, Audi AG, Bayerische Motoren Werke AG, Ford Werke AG, Mercedes-Benz AG, Peugeot S.A., Renault S.A., Volkswagen AG, and Volvo Car Corp. Robot controller suppliers include ABB-Robotics, ABB Trallfa Robot A/S, COMAU S.p.A., Fanuc Robotics Europa GmbH, KUKA Schweissanlagen und Roboter GmbH, Renault Automation S.A., Siemens AG, and Volkswagen AG Industrial Robots. Robot simulation system suppliers include Dassault Systems S.A., Deneb Deutschland GmbH, SILMA Inc., and Tecnomatix Ltd. The RRS, Interface Specification has been available to the public since January 1994, along with RRS, Interface Shell Package (RISP), a software interface designed to integrate original software from robot controllers into simulation systems. It is distributed by the Fraunhofer-Institute for Production Systems and Design Technology (IPK), Berlin, Germany.

5.3 Modeling Sensors and Simulating Uncertainty in the Virtual Cell

Most graphical simulation systems utilized in robotics are concerned with simulation of the robot and its environment without simulation of sensors and without modeling event-driven uncertainty that can be detected and reacted to based on sensory feedback. These systems have difficulty in handling robots that utilize sensory feedback for error recovery and adequate action planning in response to uncertainty. The inclusion of sensors and discrete-event simulation into the existing systems is a challenge that has to be resolved to further reduce the gap between the model and the reality.

Pioneering research and development of a simulator that integrates various sensors and can evaluate extreme conditions in the virtual mode was performed at the Computer Vision and Robotics Research Laboratory of The University of Tennessee, with support from the U.S. Department of Energy's University Program in Robotics for Advanced Reactors (Chen, Trivedi, and Bidlack, 1994). The system developed in this research can simulate the following sensing modalities: proximity, point laser range, laser range depth imagery, and vision intensity imagery. Simulation of contact sensors such as force/torque and tactile sensors is also being considered. The result is a system that can assist the robot user and designer to analyze and evaluate the structural feasibility of the mechanical devices before construction, the response of the robot under extreme situations, the correctness of the robot in performing a desired sequence of sensing and actions (flow of task operation), and the visualization of changing environment, status of the robot, and conditions of the effectors from various simulated sensory modalities.

The main conclusion of the above research is that the simulator in an integrated sensor-driven robotic system must incorporate simulation of sensory information feedback. Further work involves the enhancement of the virtual model by integration of voice-activated robot actions and by simulation of the dynamics of the environment. The simulation of the environment dynamics requires assigning physical properties such as weight, center of gravity, rigidity/deformability, surface friction parameters, etc. to each object involved in the manipulation actions. As a result, better determination of the consequence of an object's status as a result of the robot action will be possible, and it will depend on both geometrical and physical characteristics of the object.

6   SUMMARY

Graphic simulators and emulators of robotic systems are becoming an indispensable tool for companies designing, installing, and operating flexible automation and robots. The simulation and off-line programming systems are already an industry standard for the leading automotive manufacturing companies. Major breakthroughs have been achieved in bridging the gap between the virtual model in these systems and the real manufacturing cell. The advances in simulated model calibration and realistic modeling of the control performance enable effective and efficient use of such systems for off-line programming of robot tasks. While research is still needed to improve the modeling capabilities of uncertainty and sensory feedback, the existing systems provide major benefits, including better and right-on-target cell designs, accurate programs, safer operation, shorter lead times, and more productive use of the expensive robotics equipment.

REFERENCES

Bernhardt, R. 1997. "Approaches for Commissioning Time Reduction." *Industrial Robot* **24**(1), 62–71.
Bernhardt, R., G. Schreck, and C. Willnow. 1995. "Realistic Robot Simulation." *Computing and Control Engineering Journal* **6**(4), 174–176.
Bidanda, B., V. Narayanan, and J. Rubinovitz. 1993. "Computer-Aided-Design-Based Interactive Off-Line Programming of Spray Glazing Robots." *International Journal of Computer Integrated Manufacturing* **6**(6), 357–365.
Bock, G. 1987. "Limping Along in Robot Land." *Time* (July 13, 1987), 46, 67.
Bonner, S., and G. S. Kang. 1982. "A Comparative Study of Robot Languages." *Computer* **15**(12), 82–96.
Borrel, P., A. Liegeois, and E. Dombre. 1983. "The Robotics Facilities in the CAD-CAM CATIA System." In *Developments in Robotics 1983.* Kempston, Bedford: IFS.
Bourez, M., et al. 1983. "Automatic Programming of a Spray Robot." In *AUTOFACT Europe Conference Proceedings,* September. 126–134.
Boyse, J. W., and J. E. Gilchrist. 1982. "GMSolid: Interactive Modeling for Design and Analysis of Solids." *IEEE Computer Graphics and Applications* **2**(2).
CATIA User Manual. Program Number: 5668-760. IBM Corp. and Dassault Systèmes, 1986.
Chen, C., M. M. Trivedi, and C. R. Bidlack. 1994. "Simulation and Animation of Sensor-Driven Robots." *IEEE Transactions on Robotics and Automation* **10**(5), 684–704.
Craig, J. J. 1997. "Simulation-Based Robot Cell Design in AdeptRapid." In *Proceedings of the 1997 IEEE International Conference on Robotics and Automation, ICRA,* Albuquerque, April 20–25. Vol. 4. 3214–3219.
de Pennington, A., M. S. Bloor, and M. Ballia. 1983. "Geometric Modeling: A Contribution Towards Intelligent Robots." In *Proceedings of the 13th International Symposium on Industrial Robots,* Chicago, April. 35–54.
Derby, S. J. 1982. "Computer Graphics Robot Simulation Programs: A Comparison." In *ASME Robotics and Advanced Applications Conference Proceedings,* Winter Annual Meeting, November. 203–212.
Dillmann, R. 1983. "A Graphical Emulation System for Robot Design and Program Testing." In *Proceedings of the 13th International Symposium on Industrial Robots,* Chicago, April 11–15.
Edan, Y., and S. Nof. 1996. "Graphic-Based Analysis of Robot Motion Economy Principles." *Robotics and Computer Integrated Manufacturing* **12**(2), 185–193.
Edan, Y., B. A. Engel, and G. E. Miles. 1993. "Intelligent Control System Simulation of an Agricultural Robot." *Journal of Intelligent and Robotic Systems* **8**(2) 267–284.
ElMaraghy, H. A., L. Hamid, and W. E. ElMaraghy. 1987. "Robocad: A Computer Aided Robots Modelling and Workstation Layout System." *International Journal of Advanced Manufacturing Technology* **2**(2), 43–59.
Fougere, T. J., and J. J. Kanerva. 1986. " RobotSim—A CAD Based Workcell Design and Off-Line Programming System." In *SME Robots 10 Conference Proceedings,* Chicago, April.
Freund, E., et al. 1994. "Towards Realistic Simulation of Robotic Workcells." In *Proceedings of the IEEE/RSJ/GI International Conference on Intelligent Robots and Systems*, Munich. Vol. 1. 39–46.
Griesmeyer, J. M., and F. J. Oppel. 1996. "Process Subsystem Architecture for Virtual Manufacturing Validation." In *IEEE International Conference on Robotics and Automation*, Minneapolis, April. Vol. 3. 2371–2376
Gruver, W. A., et al. 1983. "Evaluation of Commercially Available Robot Programming Languages." In *Proceedings of the 13th International Symposium on Industrial Robots,* Chicago, April. 12158–12168.
Harrigan, R. W., and D. S. Horschel. 1994. "Robotics Technology Development Program Cross Cutting and Advanced Technology." In *Intelligent Automation and Soft Computing: Trends in*

Research, Development, and Applications: Proceedings of the First World Automation Congress (WAC '94) Held August 14–77, 1994 in Maui, Hawaii, U.S.A. Vol. 2 ed. M. Jamshidi, Albuquerque: TSI Press.

Kamrani, A. K., H. A. Wiebe, and C.-C. Wei. 1995. " Animated Simulation of Robot Process Capability." *Computers and Industrial Engineering* **28**(1), 23–41.

Kim, W. S. 1994. "Virtual Reality Calibration for Telerobotic Servicing." In *Proceedings of the 1994 IEEE International Conference on Robotics and Automation,* May 8–13. 2769–2775 .

Lapham, J. 1996. "Using Robot Simulation to Determine Safeguards." *Industrial Robot* **23**(6), 27–29.

Lozano-Pérez, T., and R. A. Brooks. 1984. "An Approach to Automatic Robot Programming." In *Solid Modeling by Computers: From Theory to Applications.* Ed. M. S. Pickett and J. W. Boyse. New York: Plenum Press. 293–327.

Ma, O., et al. 1997. "MDSF—A Generic Development and Simulation Facility for Flexible, Complex Robotic Systems." *Robotica* **15**(1), 49–62.

Martin, J. F., and C. C. Ruokangas. 1985. "An Architecture for an Adaptive Flexible Robotic Welding System." *SAMPE Journal* **21**(1), 20–24.

McMaster, R. S., et al. 1994. "Task Enhancement of an Underwater Robotic Arm by Graphical Simulation Techniques." In *Proceedings of the 1994 IEEE Oceans Conference,* Brest, France, September 13–16 1994. Vol. 2. 163–167.

Nof, S. Y. 1994. "The PRISM Work Station for Manufacturing Facilities Design Experimentation." Research Memorandum No. 94-9, School of Industrial Engineering, Purdue University.

Ogata, H., and T. Takahashi. 1994 "Robotic Assembly Operation Teaching in a Virtual Environment." *IEEE Transactions on Robotics and Automation* **10**(3), 391–399.

Owens, J. 1991. "Robot Simulation—Seeing the Whole Picture." *Industrial Robot* **18**(4), 10–12.

Pathre, U. 1994. "Simulation and Calibration for Off-Line Programming of Industrial Robots." In *Proceedings of the 5th World Conference on Robotics Research,* Cambridge, September 27–29. MS94-210-1-13.

Pathre, U., and M. R. Driels. 1990. "Simulation Experiments in Parameter Identification for Robot Calibration." *International Journal of Advanced Manufacturing Technology* **5**(1), 13–33.

Pickett, M. S., R. B. Tilove, and V. Shapiro. 1984. "Roboteach: An Off-Line Robot Programming System Based on GMSolid." In *Solid Modeling by Computers: From Theory to Applications.* Ed. M. S. Pickett and J. W. Boyse. New York: Plenum Press. 159–184.

Ranky, P. G. 1984. "Programming Industrial Robots in FMS." *Robotica* **2**, 87–92.

Rembold, U., and R. Dillmann. 1985. "Artificial Intelligence in Robotics." Report of Institute for Informatik III, Robotics Research Group, University of Karlsruhe, West Germany.

Rooks, B. W. 1997 "Off-Line Programming: A Success for the Automotive Industry." *Industrial Robot* **24**(1), 30–34.

Rubinovitz, J., and R. A. Wysk. 1988a. "Task-Level Off-Line Programming System for Robotic Arc Welding—An Overview." *Journal of Manufacturing Systems* **7**(4), 293–306.

———. 1988b. "Task Planning and Optimization for Robotic Arc Welding—An Algorithmic Approach." *Manufacturing Review* **1**(3).

Sjolund, P., and M. Donath. 1983. "Robot Task Planning: Programming Using Interactive Computer Graphics." In *Proceedings of the 13th International Symposium on Industrial Robots,* Chicago, April. 7122–7135.

Sorenti, P. 1997. "Simulation Holds the Key." *Industrial Robot* **24**(4), 278–281.

Soroka, B. L. 1983. "What Can't Robot Languages Do?" In *Proceedings of the 13th International Symposium on Industrial Robots,* Chicago, April. 1–8.

Warszawski, A., and R. Navon. 1991. "Robot for Interior-Finishing Works." *Journal of Construction Engineering and Management* **117**(3), 402–422.

Woenckhaus, C., et al. 1994. "USIS—An Integrated 3D-Tool for Planning Production Cells." In *Proceedings of the IEEE/RSJ/GI International Conference on Intelligent Robots and Systems,* Munich. Vol. 1. 31–38.

Zak, G., R. G. Fenton, and B. Benhabib. 1993. "Simulation Technique for the Improvement of Robot Calibration." *Journal of Mechanical Design, Transactions of the ASME* **115**(3), 674–679.

Zeghloul, S., B. Blanchard, and M. Ayrault. 1997. "SMAR: A Robot Modeling and Simulation System." *Robotica* **15**(1), 63–73.

CHAPTER 38

COMPUTATIONAL, AI, AND MULTIAGENT TECHNIQUES FOR DESIGN OF ROBOTICS SYSTEMS

George L. Kovács
Hungarian Academy of Sciences
Budapest, Hungary

1 INTRODUCTION

By 1989 both words in "intelligent robotics" were well understood by the academic community. In industry, too, robotics was widely accepted, but artificial intelligence (AI) applications were not yet common. The early appearance of Lee's book (Lee, 1989), together with its being intended to serve as a textbook for graduate and undergraduate students, explains its contents, with the following main items:

"Artificial Intelligence—Machine Intelligence." The author of this chapter notes that in the first decades of application of digital computers we claimed it as a basic advantage that they could complete tasks completely differently from humans (fast enumerations, exhaustive searches, etc.), while now we are trying to find human-like solutions.

"Intelligence and Flexibility." AI is concerned not only with the search for solutions to problems, but ultimately with the discovery of the nature of the problem itself. Different AI methods and tools are experiments to learn a little more about the problems tackled, since intelligence is too complex for us to expect one system to approach all aspects.

"Aspects of Intelligence." These are: memory, reasoning, learning (backward by expertise, forward by curiosity), understanding, perception, sensing, recognition, intuition, imagination, creativity, association, introspection, flexibility, common sense, coordination, communication, analysis, calculation, abstraction of concepts, organization of ideas, judgments and decisions, and planning and prediction. Intelligent robots are characterized by the abilities of sensing, thinking, and acting.

The basic knowledge representation schemes are the following:

Declarative: logic systems, networks
Imperative: procedures
Hybrid: frames, production systems

Thinking is believed to be substituted by searching for solutions over different graph structures and by different algorithms; e.g., and/or graphs representing (state space) or (goal, operators, states), searched by heuristics of hill climbing, depth-first, breadth-first, beam-search, best-first, min–max, etc.

Rule-based planning, blackboard systems, and pattern recognition in vision and speech processing are also important issues for AI applications in robotics.

Apart from the above topics (discussed in Lee, 1989), if we had to define *computational, AI, and multiagent techniques,* we would have no problems with *computational;* we could easily say what is meant by *multiagent;* and we would have much to say about

Handbook of Industrial Robotics, Second Edition, Edited by Shimon Y. Nof
ISBN 0-471-17783-0 © 1999 John Wiley & Sons, Inc.

AI, which refers to all artificial intelligence-related means and tools that can be used during the life cycle of a robotics system. As for *techniques,* the emphasis in this chapter is on design activities. Naturally one can consider all AI and multiagent techniques as part of computational techniques as soon as computers are applied. On the other hand, multiagent systems may be counted as a specific type of AI system. Finally, there are mathematical techniques, which can be used without computers. However, nowadays we include them with the computational techniques because they are much more effective with computer applications.

Robotics system may mean a variety of systems, from a simple single manipulator to totally equipped robotized CIM systems. Single manipulators and robots, systems of two or more robots, and flexible manufacturing and flexible assembly systems (FMS, FAS) with service robots are the most typical examples. Specific robotics applications, such as microrobots, surgery robots, space and underwater robots, walking and wall-climbing robots, and so on, should be considered within the scope of robotics systems.

Design (and/or *planning*) refers only to certain steps in the life cycle of a robot or robotics system. The main phases of a life cycle are *requirements specification, analysis, design, implementation, test, operation, maintenance, dismantling,* and *reuse and recycling* of the system or its parts. Design has to take into account all previous and later phases of the life cycle as well. Each of these steps consists of several different components that may differ in the case of different products/systems. For instance, design may be split *into conceptual design, preliminary design, detailed design,* and so on. Modeling and simulation, which were not mentioned separately, are generally important in the design and in the operation phase, although simulation of other steps may be useful too.

AI in design and planning of robotics systems means the application of multiagent systems (MAS), expert systems (ES), knowledge-based systems (KBS), artificial neural networks (ANN), Petri-nets, fuzzy systems (FS), genetic algorithms (GA), object-oriented technologies (OO), and several other techniques, including pattern recognition, vision, and voice and tactile systems and learning. These techniques may be used as standalone ones, or in almost any combinations of two, three, or more of them, depending on the given application advantages and requirements. Concurrent engineering (CE) requires even more interesting combinations of computational techniques with further possibilities for AI applications.

The above topics are covered abundantly in the literature (see Additional Reading). Most robotics literature deals with intelligent applications such as motion and path planning, collision avoidance, grasping, autonomous robots, mobile robots, walking robots, obstacle avoidance, jumping robots, navigation, humanoid robots, multirobot systems, telerobotics, teleoperation, telepresence, and so on. Only a small fraction of the publications deal with simulation and design issues of robotic systems. On the other hand, FMS and FMC design (using robots, too) are discussed often, with reference to the selection and layout, and integration of robots, machines and other equipment. Another common issue is the redesign or reconfiguration of systems in the case of major changes, such as breakdowns, decisive changes in production demand, and extensions or improvement of the system.

Certain relevant results reached by several authors are discussed in Section 2; the opinions and experience of the author are discussed in Sections 2 and 3. Conclusions and future trends are discussed in Section 4.

2 APPLICATIONS OF MATHEMATICAL, MULTIAGENT, AND AI METHODS AND TOOLS

The following robotics applications are selected from the literature according to some major features of the discussed systems, although some of them would fit into other domains, too. For example, simulation and design should not always be separated, and formal specifications should not necessarily belong to robot design.

This chapter is organized to follow some crucial steps and methods of the design/planning process. First, modeling and simulation are discussed as design assistance tools, then different design and implementation issues are detailed, with special emphasis on AI applications for component selection and layout design. These last two steps mean *configuration* and *reconfiguration,* which are extremely important for all production/manufacturing facilities. Different robots and manipulator design systems are then described, including the application of genetic algorithms, formal mathematical methods, multiagent systems, and design issues of intelligent, autonomous robots.

The targets of the different design systems, using different design methods and tools, are not always robots and manipulators in general; they are, in fact, rather diverse. Design issues of robots, robotics systems, assembly systems, sensors, trajectories, motion systems, etc. are discussed. Some techniques mentioned deal with general design issues, while others deal with well-defined, precise part problems. The common issue is that all deal with the problems that are the subject of this chapter. The author hopes that this diversity will allow readers to find what they need, while being able to skip the parts less relevant to them.

2.1   Modeling and Simulation in the Design of Robotics Systems

Modeling and model-based simulation are highly effective design tools. In a robot-rich environment (environment with robots) there are many robot-specific design tasks. These belong either directly to the robots or to the environment (e.g., an FMS), and their cooperation in the appropriate synergy is very important.

As simulation examples, multirobot systems, teleoperation, and a robot–human motion-planning system will be discussed.

Actor Based Robot and Equipment Synthetic System (ACTRESS) is an autonomous, decentralized system consisting of *robotors*. A robotor is a robot, a computing system, or any equipment (Habib, 1992). The key issue in this work is to simulate and analyze the main functions of each agent and design and analyze the actions to be carried out by the agents, analyzing real-time behavior in a variety of situations, considering their environment and how unexpected events are to be treated.

The main components of the system are the following:

- Autonomous mobile robots (geometry, local environment manager to communicate with a global manager, motion system, perceptor systems, local path planner, communication module, behavior description process, obstacle avoidance solver)
- Global environment manager with global path planner
- Task descriptor panel
- Human interface

Path planning is solved by means of rule-based expert systems. The design of an autonomous mobile robot system is presented based on simulation and analysis results. The system is a multiagent system.

Telemonitoring, a new type of teleoperation system using simulation, is discussed in Lee and Lee (1992). In designing and controlling a teleoperation system under time delay, it is necessary to incorporate human dynamic behavior into the control loop. Time delay between the human operator and the remote stations may degrade the fidelity and incur control instability. One solution is the application of feedback signals of force, velocity, and position. Another solution is an indirect control link using a delay-free task environment. In the latter case a model and a simulated environment are needed.

A dynamic model of the human is detailed. Normally the position and force control is done by telepresence, by which the human feels what is going on at the site. Telemonitoring is a new concept for enabling the operator to feel how the manipulator performs toward its control goals. The operator is provided with a new kinesthetic coupling (monitoring force feedback (MMF) is a linear combination of forces caused by the position error and force error) and an advanced control structure to monitor the well-being of the manipulator. A stability analysis is applied to evaluate the results.

Simulation for various time delays proved that in designing and evaluating a teleoperator control system with time delay it is imperative to have human dynamics to react to visual/kinesthetic stimuli in the control loop.

Real and simulated examples for telepresence in robotics in manufacturing applications are presented using multimedia communications (Haidegger and Kopácsi, 1996).

Simulation is safer and cheaper than building and testing when unknown obstacles may be in the environment of a moving robot (Skewis and Lumelsky, 1992). Modeling various robots, sensor systems, obstacles, testing several 2D and 3D motion-planning algorithms, testing human performance in motion planning, and real-time animation are the main issues in this paper.

Path planning with complete information is a typical off-line activity, while path planning with incomplete information or sensor-based motion planning is a real-time task. A 2D and then a 3D robot model were built to make simulation possible. The simulation is used for testing human performance in motion planning. Sometimes it is difficult to decide which systems are better, those with humans or those with automata, which are based on sensors only. Motion simulation provides some answers: first motion along

M-line, then motion along an obstacle boundary are solved by tactile boundary following. Then motion along an obstacle boundary is solved by means of proximity boundary following. The termination of boundary following, motion replay, and then motion used in the test of human motion planning are discussed.

Together with the motion problems the following main sensory types are simulated and discussed: tactile sensing, proximity sensing, visual sensing, and sensing for the test of human motion planning.

2.2 Design and Integration of Robotics Systems

A summary of design and integration issues can be found in an early reference (ESPRIT, 1992). This project was one of the early joint European projects. Its objective was to study the benefits of off-line robot programming and the problems of robot systems planning. These fields were identified and analyzed in the frame of ESPRIT Project 75: "Design Rules for the Integration of Industrial Robots into CIM Systems." Not only design, but integration, is discussed, since recent complex systems may have very serious integration problems. The integration of robots into manufacturing cells/systems requires the integration of information concerning product design, plant availability, and system layout as well.

The result of the study is a prototype planning system for robotized workcells. Integration of planning and off-line programming systems using simulation models is solved, too. The following main applications are discussed:

- Tasks with critical time specifications
- Complex robot operations with multipurpose end-effectors
- Complex kinematic structures with more than six joints

A computer-based system supports the entire planning process with the following objectives:

- Improvement of planning quality
- Reduction of planning costs
- Expansion of the spectrum of solutions to enable selection of the best choices
- Increase in creative planning
- Fast processing and preparation of planning data
- Improvement of information flow
- Enhancement of the transparency of planning
- Establishment of robot-integrated CIM systems

The planning process has the following phases: analysis, outline, documentation, design, evaluation, detailing, and installation.[1] The approach is successive and general. Each step (phase) consists of several modules and submodules.

Two types of workcells are considered:

1. Loose composition of components with low functional dependencies, with all components simply at the same location
2. Many functional dependencies among the components, where integration of the components and modeling are important

The authors claim that there is no systematic planning procedure for the implementation of robot-based systems, but there are planning aids for some simple tasks, such as loading/unloading machines only, and naturally generalization and extension need further study.

---

[1]These steps represent the phases of a traditional waterfall model design; however, they are used as steps for a life cycle engineering spiral model. For the spiral model generally the following phases are defined in an object-oriented approach, as in Jacobson et al. (1992): requirements specification, analysis, design, unit implementation, unit testing, integration, and maintenance. The mapping of the two sets of phases to each other is certainly possible.

The main application areas of integration are communication, networking, product description, manufacturing process control, CAD interfaces, graphics, database management systems, programming languages for operational control, and standard toolbox systems.

Design examples are given for the automotive, electric/electronic, and mechanical engineering industries in four application areas: handling, machining, assembly, and testing. Examples include planning and programming of a dot matrix printer assembly and programming of the Cranfield assembly benchmark.

The applications of the planning and interactive programming systems ROSI (Robot Simulator) and RIEP (Robot Interactive Explicit Programming) are given for spot welding of car doors, inspection of welding systems, assembly of matrix printers, space applications, sheet metal welding, laser cutting in 2D and, as a new technology, in 3D, assembly of bulky parts, hardfacing in nuclear industry, application in aerospace, palletizing of clutch cases, and seam welding of a motor carrier.

Another general important issue is the application of standards, where possible: IGES, STEP (product description), IRL (documentation purposes, readable, supports modularization), and ICR (simpler implementation with less memory and low-speed processors, faster execution, for robot control) interfaces.

A robot system design architecture was developed with blackboard-based, distributed, and object-oriented (OO) systems (Bagchi and Kawamura, 1992). Distributed and OO concepts are used to design and implement a single-manipulator robotic system, ISAC (Intelligent SoftArm Control), to assist the physically handicapped. An OO communication framework is presented, where intermodule communication is managed through remote method calls from client to server resource objects. The distributed modules receive or send tasks through a task blackboard. There is a multiagent, object-oriented task decomposition and execution in ISAC. C++ is used in a networked UNIX environment. Distributed systems are generally for multiple robots, but here the functions, such as manipulator controller, task planner, vision, and other sensors, are distributed among processes and processors. Distribution helps in integration of programs written by different groups or in adding new parts, and so on.

ISAC's goal is to feed soup to the user. Its tasks are voice communication with the user; spoon and soup bowl recognition; picking up the spoon and dipping it into the bowl; locating the mouth of the user; moving the spoon to the mouth, even if it moves; predicting sudden user moves; and avoiding collision.

The modules of ISAC are user interface, task decomposition, object recognition, face tracking, and motion execution. The system is able to learn new tasks by experience and observation.

The relatively simple blackboard provides methods to the client: add tasks to the blackboard, wait for a task to be serviced by another client, get tasks from the blackboard, update the blackboard with the status of a task that has been serviced. It is easy to add or remove modules. If a module is not in the blackboard, certain functions will not be executed. Low-level modules, such as move forward or backward, open and close gripper, are always present.

ANDECS is a computational environment for robot-dynamic design (Grübel et al., 1996). The system has OO environments for mechatronics and robot dynamics modeling, simulation of the robot dynamics (with event-driven differential-algebraic equations), multiobjective parameter and trajectory optimization, and multivariate result visualization. All parts of the system are coherently integrated using a common engineering database system.

Modeling is applied by the OO Dymola language and by the DSblock system developed for the ANDECS project. The run-time environment of ANDECS is DSSIM with three unique features:

- Close coupling with engineering database
- Powerful signal generator
- Design by optimization

Trajectories are optimized by parametrization with certain approximation schemes. Decision support is assisted by user-friendly interactions via graphic display. The computation can be stopped or interrupted at any time, providing the designer the following facilities:

- Selecting any design candidate by picking any point/curve
- Scanning the design history step by step and choosing the best fit solution
- Tradeoff analysis via the conflict editor
- Information zooming
- Interactive steering

Knowledge-based product design is a key issue associated with the application of robotic assembly systems (Gairola, 1987). Application of expert systems for design for assembly (DFA) means cost-saving, reducing rework, increasing quality, minimizing material cost, etc. It has the following major steps: simplify assembly task, improve assembly organization, and facilitate assembly execution.

2.2.1 Component Selection and Layout Design

Component selection in robotics and manufacturing systems is one of the tasks that calls for the application of expert, or knowledge-based (KB), systems, since all necessary inputs, requirements, possibilities, and results are relatively simple to define. As input, generally the tasks to be performed should be defined in the form of process plans or operation plans. The possibilities are defined by means of databases containing all available resources and KBs to specify all constraints, relationships, and specific requirements. The KBs also contain all applicable optimization criteria in the form of rules and facts (sometimes represented in frames). Even uncertain and incomplete knowledge can be managed without problems. The main task of the KB system is matching the goals to be achieved and the means available, taking into consideration all requirements and constraints and some optimization criteria. This component selection is called *configuration,* or *design* in several cases. Usually, design contains layout planning as well when the physical position of all equipment in a workshop is designed based on available space and the components selected. If a new set of components is selected for a given task, changes to an existing system are necessary, or different optimization criteria are used, reconfiguration or redesign takes place.

A knowledge-based method was developed to select and design the layout of robots and machine tools (Kopacek et al., 1995). The semiautomatic selection of assembly cell components with the ROBPLAN system contains four independent relational databases (manufacturing symbols, products, components, manufacturing planning) plus an expert system to assist in the selection. First the selection of robots, tools, grippers, etc. takes place, then layout planning, simulation, and final planning. Selection is based on the necessary assembly operations, which are defined as a set of elementary movements. First the subassemblies, then the assemblies are designed.

The robot selection is based on the maximum number of axes, type of the robot, number of robots, and maximum payloads. Grippers and fixtures are selected likewise.

A two-phase model can be used for robot selection (Fisher and Maimon, 1988). First a set of specifications, based on the analysis of the technological requirements of a specific set of robot-related tasks, is created to serve as functional specifications for robot design. Then the best-fit robots are selected from the set of available robots. The robot specifications contain about 25 different important parameters, including power requirements, control, accuracy, force, torque, gripper, integration possibilities, and price. In phase one a mathematical programming method and a rule-based expert system are presented to help in finding the best set of robotics technologies for the requirements of the set of considered tasks. In phase two the acceptable set of robots is chosen and the candidate list is prioritized using both methods above. Finally the combination of the mathematical tools (integer programming) and rule-based expert systems is suggested.

Another system for robot selection is ROBOSPEC (McGlennon et al., 1988), which is a rule-based, menu-driven, user-friendly system written in OPS5, one of the early dedicated expert system programming languages. The resulting robot specifications include kinematics, axes, drives, programming, and speed.

Several projects by this author and his team are concerned with the application of expert systems for robotized manufacturing system design (Kovács, Mezgár, and Kopácsi, 1991; Kovács and Mezgár, 1991; Kovács, Kopácsi, and Mezgár, 1992; Kovács and Kopácsi, 1992; Kovács and Létray, 1989; Kovács, 1989; Kovács, Létray, and Kopácsi, 1991; Molina, Mezgár, and Kovács, 1992). The basic idea is to match the requirements with the possibilities, similarly to the works discussed above. Different experimental (prototype) programs were written in PROLOG, CS-PROLOG, ALL-EX, and G2. A complete list of the available machine tools, robots, robot and machine tool controllers, fixtures, jigs, tools, etc. and their main technical and price parameters was produced first in a database. Because this list was rather restricted in Hungary in the 1980s, we were not concerned about the databases being too large. The input of the system was the description of the required production (process plans). This input was mapped by means of appropriate rules to technical and price parameters. The same parameters were used for the specification of the resulting system, which meant that, based on the optimization criteria set in the programs, the best-fit tools and equipment were chosen as an initial set of machines, robots, etc. to be used. Then a placement (layout) and scheduling system checked the chosen set of elements, and several refinements were made, all in simulation mode. If any problems occurred, a reconfiguration algorithm to find better solutions started. The same reconfiguration was used if changes, additions, breakdowns, etc. occurred or if the production requirements changed.

KBML, a rule-based expert system for equipment selection and layout, contains 60 complex rules (Kusiak and Heragu, 1988). The input of the system is the number of machines and robots, machine dimensions, connectivity matrix, location restrictions, type of possible layouts, type of material-handling systems, floor dimensions, etc., and the solution is given for the simplest single-row layout.

Another knowledge-based system for component selection is described by Paasiala et al. (1993). The main steps of the selection process are creating the database, eliminating unsuitable components by calculations, calculating quality functions for the candidate components, sorting the components according to the quality factors, and selecting components. The STEP standard is used for data transfer of components to an OO database and an expert system. The paradigm of concurrent engineering (CE) is used where applicable.

Knowledge acquisition is discussed widely. Although often neglected, it is a rather important, non-trivial requirement. There are two main types of knowledge acquisition methods, both of which can be found in the literature: product-oriented and component-oriented knowledge acquisition.

2.2.2   Robot and Manipulator Design

Several aspects of robot, manipulator, and robotics system design will be discussed in this section. Rather than basic design questions, interesting questions and views will be pointed out. Some general design aspects will be detailed, and then genetic algorithms, formal specifications and mathematical models, the multiagent approach, and intelligent autonomous robots will be discussed through application examples.

A basic reference model for Knowledge Integrated Manufacturing Systems (KIMS) is proposed by Putnik and de Silva (1995). This model is general and can be applied to robotic systems. The idea is to combine integration and knowledge, as is done in computer-integrated manufacturing (CIM) or intelligent manufacturing system (IMS). The suggested reference model is a $3 \times n \times 3$ cube, where:

- Along axis x are integration over knowledge, integration through knowledge, and integration by knowledge (3).
- Along axis y are CAD/CAM, CAPP, FMS, etc. Planning and Control techniques (n).
- Along axis z are knowledge representation, inference, and knowledge acquisition (3).

Basic entities such as fundamental block (FB), structural block (SB), and operating block (OB) are then defined. Two functions organize the knowledge: the coupling function (CF) and the output function (OF). A knowledge block is KE = (FB, SB, OB, CF, OF).

There are knowledge acceptors, knowledge transducers, and knowledge generators. All robots, manufacturing elements, and systems can be defined as classes and instances and can be connected as object-oriented diagrams if appropriate symbols are assigned to the instances and classes. Design can be done based on the class structures.

An integrated planning system is defined with two stages of planning (Filho and Carvalho, 1995):

- A stochastic model to provide aggregated planning and then decomposition for detailed planning. The two planning activities are then integrated. A stochastic optimization model and an equivalent deterministic model are given with detailed mathematics.

- Assignment, which has another well-manageable mathematical model.

A method has been suggested by which products and robotic assembly systems are designed at the same time; optimization and simulation steps are repeated. Optimal reliability is defined and calculated with precise mathematics (Gradetsky et al., 1995).

When two or more robots grasp the same object simultaneously, a closed kinematic chain is formed and there is a need for grasp planning (Rodriguez et al., 1995). The robots are characterized with a configuration parameter, which is a positive integer whose binary representation corresponds to configuration possibilities of the robot. There is a grasp configuration parameter, as well as a cost function to characterize the costs of a grasping for a manipulation. Optimization is done to find minimum total cost. Collision avoidance for the whole system is also taken into account.

Computer Aided Assembly Planning (CAAP) is a system based on the assembly CAD data. It is applied in the PRORA project, in which CAD models are transformed to assembly plans and then, based on the assembly planning, the task-level programming can be generated (Armilotta and Semerado, 1993).

Genetic Algorithms (GA)

Genetic algorithms are a kind of imitation of how living nature works. It is interesting how effectively such algorithms can be used to solve different design tasks by finding (close to) optimal solutions using the imitation.

Because of the nature of the performance of robots and manipulators, genetic algorithms can be suitable for characterizing their behaviors. Since one of the main design tasks is to seek performance optimality, the ability of genetic algorithms to reach (close to) optimal solutions is used to design robots and robotic systems. The following work provides a good example of the rather general statement above.

A detailed survey of genetic algorithms and their application is given in Kim and Khosla (1992). Multi-Population Genetic Algorithm (MPGA) is an efficient optimization technique for highly nonlinear problems. It is a parallel implementation of GA, which is a kind of adaptive search method, and is robust for the manipulator design to best fit to the tasks to be solved. The optimality is based on the use of a dexterity measure, here the relative manipulability. This is a typical task-based design (TBD) function. The number of optimization functions equals the number of task points. A progressive design framework is added to the system. Generally the optimization is based on an objective function and general constraints. This increases the complexity of the TBD exponentially with the increase in the number of task points, due to the huge search space and nonlinearity. If a separate optimization is ordered to consider all task points and these points are connected by connecting constraints, the complexity is almost constant. There is no requirement for continuity in the derivatives, so any fitting function can be applied. The design constraints typically include reachability, heuristics, joint limit, and joint angle change between two adjacent task points. These constraints are task-specific.

In the progressive design framework the design variables are dimension (link length), pose (joint angle), and the base position of the manipulator. Four steps are performed: kinematic design, prototyping, planning, and kinematic control. Design is performed progressively based on a coarse–fine approach.

This approach can be compared with the Simple Genetic Algorithm (SGA), where the operators are reproduction, crossover, and mutations. Starting with a random population, generating subsequent populations, and halting by some criteria, MPGA generates multiple populations, one for each optimizing function corresponding to a given task point.

Formal Specifications, Mathematical Models

Certain computational methods where the computations are based on given mathematical expressions are discussed in this section. These methods are the only applications in our discussion that do not require AI means and tools and more or less guarantee appropriate solutions. Most real-size problems require simplifications and abstractions in order to be solved by means of such methods because generally robotics system design problems are complex, nonlinear, dynamic, and heuristic. They can rarely be solved by means of a finite number of solvable equations and functions where the solution must be found in an acceptable time.

Modeling and simulation are applied in a multifaceted (multiple subdomains, multiple objectives) application such as robotics (Alagar and Periyasami, 1992). This approach helps in problem analysis and decision-making. The procedure models all subsystems separately and then combines them with proper interfaces.

Different formal modeling approaches are used: algebraic, VDM (Vienna Development Method), and Z are model-based, but they have no real-time capabilities, while CCS (Calculus of Communicating Systems) and CSP (Communicating Sequential Systems) satisfy concurrency. There are discrete and continuous models. Stochastic models can be built up from both. Physical models and symbolic models have to be distinguished, too. Physical models can be captured by using the semantic SCG tree approach (Alagar, Bui, and Periyasami, 1990). Mathematical models are symbolic. Set theory and first-order logic are used to construct formalisms in which models and their theory can be built. The aggregation of abstract models together with the mathematical goals governing their validity are collectively known as *formal systems*. In software engineering a formal system abstracting a particular software is called *formal specification* of that software. This work provides formal specifications (models and theory) applicable to robotics systems. Based on the specifications, the robot design can be performed if the task-level description of the robot is available.

A simulation built up based on a model should not be confused with a prototype. They may be similar, but their means and goals are completely different. Simulation is used to investigate possibilities based on models extracting some main features of a given system, while a rapid prototype is a programming technique that, based on a design, extracts main features to check the main functionalities of software.

As an example, a robotic assembly system is designed. The assembly is based on textual description of the tasks to be performed. The main domains of the design are physical, geometric, kinematics and inverse kinematics, dynamics, control, and sensors (environmental interface). All domains are analyzed and shown with the above methods. Finally, a VDM-based modular OO design is given. It can be implemented in Eiffel or C++.

Robot layouts can be designed to maximize the possibility of a group of robots grasping an object without difficulty (Sasaki et al., 1996). Based on quantitative relationships between the arrangement and the load, the unknown parameters of each robot can be calculated using matrix notations and functions. Simulations are applied to prove the correctness of calculations. This work provides a bridge to our next topics.

Multiagent Approach

The multiagent approach is used when a system is decomposed (or built up) in such a way that different cooperating subsystems work together with data and information exchange to solve problems. These problems may be "simple," as in the first reference below, or multiple objects may be involved. A natural problem is the solution of the design and operation of multiple (autonomous, moving) robots. Communicating agents having some intelligence are called *intelligent agents,* but even nonintelligent agents may have intelligent cooperation, giving intelligence to a whole system. On the other hand, nonintelligent cooperation of intelligent agents may result in a nonintelligent system. In the last few years multirobot systems and the multiagent approach have been much discussed in the literature, including the aspects of design.

Multiagent systems have been applied to robot trajectory design (Duhaut, 1993). To generate a robot trajectory, the inverse kinematics problem should be resolved by solving a set of equations that represents the geometric model of the robot. Generally a set of equations should be defined and solved, which is a tedious, time-consuming job. This work presents a multiagent approach to solving a reduced form of the inverse kinematics problem, with simple but still effective expressions.

Multiagent programming is analogous to a group of people working together to reach the solution of a problem. All links of a robot are taken into account as agents. ADA programs have been written for 2D and 3D solutions. The beginning of the link is the base of the agent and the end of the link is the head.

A limitation of the method is that the orientation of the effector is not taken into account and, if all positions of the links are coupled, a negotiation between the agents must be introduced to solve the interactions.

An advantage of the method is that because the solution is independent of the number of links, changing geometry can be taken into account. If a link is deleted, i.e., the system degrades, only the associated agent should be disabled.

A disadvantage of the method is that for some inputs oscillations may occur, which are possible to detect and then correct with a random move for a link. Another is that if the destination point is not in the reaching space of the robot, the program does not stop. The response time of the method cannot be bounded, i.e., the method cannot be used real-time, only off-line.

Agents with cognitive features using KB systems can be used in the design of autonomous mobile robots (Bussmann and Demazeau, 1994). An abstract model of an autonomous agent is presented. It integrates reactive and cognitive behavior. For such an integration the following requirements should be accomplished: reactivity, timely behavior, and symbolic representation. Knowledge representation, planning, scheduling and execution, plus the process of evaluation, are added. The evaluation module constantly supervises the environment and the agent's actions to ensure the agent's reactivity.

The reactive agent model specifies an agent by its behavior, i.e., how the agent reacts to certain stimuli from the environment. With every perceptual input or class of inputs certain actions are associated, which are performed immediately. Different solutions can be used, such as programmed, hard-wired, rule-based, or multilayered architecture. Conflicts between simultaneously firing rules are solved by priorities.

Cognitive (or intentional) agent models are motivated by the agent's knowledge. AI techniques lead to the introduction of beliefs and intentions into the reasoning process of the agents. Intentions enable the agent to reason about its internal state and that of the others. In the BDI architecture (Nilsson, Cohen, and Rosenschein, 1987), belief, desires, and intentions form the basis of all reasoning processes. Cognitive models are more complex than reactive ones, since more powerful and more general methods can be used. The integration of the two models requires the following:

- *Reactivity:* An agent should be able to react to unpredicted events by adapting its behavior.
- *Adaptation of behavioral speed:* An agent should be able to adapt its behavioral speed to the evolution of the system.
- *Symbolic representation and reasoning:* An agent should be able to manipulate explicitly its knowledge about the universe.

The agent model involves the following relationships, where the arrows show the directions of the activities and effects:

sensory input $\rightarrow$ perception $\rightarrow$ world model $\leftarrow$5— scheduler $\leftarrow$ planner

world model —1$\rightarrow$ evaluator $\leftarrow$ guards and triggers $\leftarrow$4— executor $\leftarrow$ scheduler

evaluator —2$\rightarrow$ goals $\rightarrow$ planner $\leftarrow$3— executor $\rightarrow$ actions

The numbers in the arrows refer to activities in the given transition:

1. Supervise
2. Create, suspend, kill
3. Replan
4. Set
5. Schedule

- The *world model* includes: Time and event model, plan schedule, and updating.
- The *planner* includes: Construct partial plans and complete them, choose prioritized goals first, plan dependencies and internal actions.
- The *scheduler and executor* includes: The plan is scheduled in time or violates a deadline, or the plan cannot be scheduled due to violation of environmental predicates, conflicts to other agents, or conflicts with other plans of the agent.
- The *evaluator* detects critical situations (guard, trigger). The evaluation includes:
  - probability of events: sure or probable or unprobable event
  - possibility of avoiding the situation; inevitable or avoidable by own means, or avoidable with the help of others

- the effect of the situation: threat to the existence of the agent or conflict with intended actions of the agent, change of the situational context of the agent's plans, or no effect.

The following decisions may be made: stop agent, ignore situation, continue to supervise the situation, avoid situation, take countermeasures, adopt the new situation. An important classification in the system is based on the priorities of the last three points.

Complete agents should be built to understand intelligence (Lambrinos and Schreier, 1996). A complete agent behaves autonomously in its environment without a human intermediary. A new control architecture (Extended Braitenberg Architecture) consisting of loosely coupled parallel running processes is used to control a mobile robot that collects objects and categorizes them. Detailed mathematical formulations are given for navigation, exploration, learning, etc.

A cellular machine is defined as a new type of intelligent autonomous system with three main features: fault tolerance, physical reconfigurability based on modularity, and autonomy (Sakao et al., 1996). The cellular machine should have *homogeneity,* since homogeneously composed identical cells realize redundancy, which is necessary for fault tolerance; modularity, which means that each cell is an independent, autonomous module; and *intelligence,* which means that each cell has its CPU and an appropriate code. As a result, such systems have self-organization capability, reconfigurability, self-maintenance, and adaptiveness to environmental changes and are capable of managing faults.

Another intelligent capability is cooperation among multiple robots. A generic scheme for multirobot cooperation can be based on an incremental and distributed plan-merging process (Alami, 1996). This scheme has coherence in all situations, including the propagation that may occur after a failure. It is able to detect deadlock situations and to call for a new task-distribution process if needed. The scheme can be used in a hierarchical manner and in a context where planning is performed in parallel with plan execution (concurrent approach).

The plan-merging operation (PMO) results in a coordination plan. Before executing any action, each robot has to ensure that the action is valid in the current multirobot context. All robots' plans are collected and the actual one is merged under protection by a mutual exclusion mechanism and without modifying the other robots' plans. All the components work as if there were a global plan.

There are two situations in which a PMO cannot be performed: where the goal can never be achieved, and where the robot can generate a plan, but it cannot be executed, i.e., another robot forbids the execution. Waiting for a while may help in the second case, as the other robot may change its mind.

Agents require agent-specific but flexible skills to cope with their tasks' and environment's variability. Elementary operations (EO) are presented to combine the agent-specific nature of skills with the requirements for a general action knowledge representation inherent to MASs (Friedrich, Rogalla, and Dillmann, 1997). The main advantages of MASs are modularity, robustness and fault tolerance, maintainability and extendibility. There are two models to describe a MAS:

- Single agent
- System, a collection of agents

Mathematical and set theoretical descriptions of MASs and of skills are given in detail. A manager distributes the tasks among agents. EO is a data structure with the encapsulation of skills and an interface between agent-specific skills and the agent's non-specific interagent communication. EO is a knowledge representation with the following requirements when a task should be solved by agents of the MAS:

- Negotiating tasks
- Solving unknown tasks by searching in subtasks set by planner agent: applicability conditions represented in symbolic way
- Supporting macrooperations to hide the agent-specific nature of skills
- Supporting the user–MAS interaction by skill representation via EOs

The EO structure includes:

1. A symbolic name to reflect semantics
2. A characteristic ID, a symbolic identifier for interagent communication
3. For planning and testing abilities, a precondition based on geometric relations, object features, and skill parameters
4. Post condition after skill application
5. Parameters: task-dependent, agent's non-specific, and agent-specific
6. The actual skill

All robots and robotics systems can be analyzed, simulated, and designed this way.

Navigation of autonomous robots, even in a 2D space, requires intelligent planning. The problem is to get the robots from an initial state—that is, location—to a goal state. During the motion the agents can be considered as moving obstacles to each other (Minagawa and Kakazu, 1996). Task-oriented ANN-based behavior functions with sensory inputs and motor outputs assigned to the agents, and the problem is solved by evolutionary programming (a real-valued mutation on network weights occurs according to Gaussian distribution). Deadlock and reactivation (with a simple energy transfer and sharing mechanism) problems are solved by local interactions between the agents. Optimal motion was not reached because the method of conflict resolution should be refined.

Intelligent Autonomous Robots

Aspects of intelligent autonomous robots, as agents, were discussed in the previous section. This section discusses several examples focused more on design.

How to design high flexibility and autonomy for robot systems, together with consistent, user-transparent guidelines for installation, programming, and operation in industrial working environments, is described by Freund and Rossmann (1994). The focus is on control (CIROS: multirobot controller) with modern human–machine interfaces for intuitive operation. Cell Oriented Simulation of Industrial Robots (COSIMIR) is the simulator of control and display of the user environment, the robot controller, and the application environment.

Teamwork is important in the design of distributed systems, and can be accomplished through the application of cooperating KBs (Schweiger, Ghandri, and Koller, 1994). Local and global knowledge of autonomous systems work in teams. There is a distributed, real-time, multiuser KB to facilitate teamwork in flexible manufacturing environments, resulting in cooperating knowledge bases. Every local KB is an object structure with demons, multiuser and real-time access. Each local KB has an active component that asynchronously passes changed data to interested application processes. If several autonomous systems cooperate to perform a certain task, the local KBs of these systems form a team to provide the knowledge needed by the autonomous systems. The local KBs cooperate by means of selective updating and location-transparent access to the global knowledge. The local KBs are responsible for task planning, motion control, path planning, and manipulation control.

An autonomous hexapod walking robot can be analyzed and designed with the MEL-MANTIS system (Koyachi et al., 1996). The basic concept is integration of locomotion and manipulation; control integration and mechanism integration are necessary for the practical working robot.

An example of the integrated limb mechanism is a six-bar linkage mechanism with four degrees of freedom. Kinematic analysis of the geometric and mechanical design suggests that the mechanism satisfies all contradictory conditions, not only for leg with slow and heavy motion, but also for arm with fast and light motion in upward posture. The workspace that two or three arms can reach is analyzed and evaluated by volume numerically for dual- and triple-arm work.

A prototype hexapod named *MELMANTIS* was designed and implemented to prove the design theory. The results show that a five-bar link mechanism is a good solution to concentrate the actuators near the body and provide large workspace and long reach with three degrees of freedom. A new leg mechanism is proposed that is transformable into an arm based on the six-link mechanism.

The forward kinematics is defined by the position vector of the foot P and the joint angle vector. The link dimensions are L1, L2, L3, L5 = Constant, L5 = L1, and L4 = kL3. The $k = 1$ and $k = 2$ cases were examined. $k = 2$ means faster arm than leg motion, which increases the velocity of the end-effector. All coordinates for the kinematics can be calculated by trigonometric functions.

The kinematic analysis is done by manipulability ellipsoids based on the Jacobi matrix for $k = 1$ and $k = 2$. Kinematic analysis was applied to estimate actual load. For this purpose the Jacobi matrix was applied again to relate the joint angle velocity to the velocity of a toe. Cartesian coordinates were used. The principle of virtual work was used. Finally, a kinematic analysis of two or three arms to grasp an object in 3D workspace was performed.

Intelligent sensors are central to the design and implementation of a four-legged walking robot, *FLIMAR* (Mahajan and Figueroa, 1997). The robot has intelligent sensing and decision-making capabilities. Multiple sensors with embedded incremental learning represent a new approach towards environmental perception and reaction. The sensors continuously monitor the environment as well as their own parameters. Each of the four legs has three degrees of freedom, i.e., there are 12 motors to move the robot. The robot responds to light, sound, and touch in different ways, depending on the environmental conditions. There are different degrees of intelligence at different levels of the OO architecture.

The basic movements, such as walking forward and backward and turning right and left, are defined by solving the inverse kinematics for each leg, where leg 1 and 4 and leg 2 and 3 are taken into account similarly, based on the simple geometrical data of the legs and the path on which to move.

The intelligent architecture means that all sensors are connected to a sensor manager, which is then connected to the appropriate knowledge bases, which are connected to the intelligent controller and the learning module. The learning module is between the KBs and the controller. The user interacts with the KBs. There is a priority controller to evaluate the environmental conditions.

Learning occurs at different levels: robot level (based on the environment and on preprogrammed maps), sensory level (there are autonomous agent sensors), quantitative parameters (how to tune the sensors), and qualitative parameters (to recognize new behaviors by the sensor).

2.3   Specific Applications

Environmental protection, maximum reuse of materials, and specific involvement of humans in robot design and control are discussed next as important topics of computational robotics systems design.

Design for environmentally optimal resource consumption and reusability is discussed by Shimojima and Fukuda (1994). The goal is to save and use resources effectively from the viewpoint of environmental protection. It may be hard to accept, but we should try to improve recycling instead of focusing only on improving productivity. This objective can be achieved by converting industrial waste into raw materials.

To define when waste will appear, error/failure/breakdown management is needed to:

- Predict the time of breakdown
- Predict the location of breakdown (which part)
- Recognize the cause of breakdown
- Plan how to fix or repair the breakdown or malfunction part
- Recognize that the product may change over time

The basic means and tools to be used are the appropriate sensors and sensory systems, together with AI technology for fault diagnosis in the running process and for defining how the repairing/fixing process will be carried out. A manipulator system with intelligent control methods is presented. A real example is described in the aircraft industry with 3D measuring system and integration of measurement values by fuzzy inference and adaptive sensing for fast and accurate sensing.

Human intelligence should be incorporated into the robot controller by means of a dual control architecture (Miyake, Kametani, and Sakairi, 1994). The architecture has specific software and hardware tools, although the application of specific hardware tools is against recent control trends toward placing more burden on PCs. Two layers are defined: the upper one is analogous to the human cerebrum (symbolic data processing regarding the global conditions of the robot environment), and the lower one is closer to the traditional controller. An intelligent system processor and motion-control processor work in synergy.

3 DESIGN AND IMPLEMENTATION OF AN INTELLIGENT ROBOTIC CELL

The CIM Research Laboratory of the Computer and Automation Research Institute of the Hungarian Academy of Sciences in Budapest (CARI HAS) started an FMS simulation–evaluation–scheduling–control project several years ago (Kovács, Mezgár, and Kopácsi, 1991; Kovács and Mezgár, 1991; Kovács, Kopácsi, and Mezgár, 1992; Kovács and Létray, 1989; Kovács, 1989; Kovács, Létray, and Kopácsi, 1991; Molina, Mezgár, and Kovács, 1992). The intelligent control part of the project aimed to design and implement a simple robotic demonstration cell. The purpose was to demonstrate the joint applicability of intelligent (KB) systems with standard manufacturing protocols for open communications (OSI). As a result of the intelligent design and implementation process of a real-time robot control system, the G2 intelligent, OO, real-time environment (G2 Reference Manual, 1990) and the MMS protocol (Brill and Gramm, 1991) were used together with VMD description of the target robot (Nagy and Haidegger, 1994) to solve simple, yet intelligent, control tasks. The developments are described in the following sections.

3.1 Problems of Real-Time, Intelligent Control

A simplified definition of intelligent control of a robotic system or an FMS is continuous or frequent observation and evaluation of the status and condition of the system performance, decision-making based on the evaluation results and predefined knowledge, and then operation according to the decisions. In the case of so-called normal operation, which runs according to the given schedules, there is no special need for intelligence or for interactions to modify the operation. The procedure suggested by Smith et al. (1992) may help if any disturbance or irregularity occurs, which is relatively common in case of highly sophisticated, complex systems.

Recent discrete robotic, manufacturing, and/or assembly systems often use MAP/MMS (Manufacturing Automation Protocol, Manufacturing Message Specification) (Brill and Gramm, 1991) because this technology is widely available from many vendors and offers a safe and open solution according to the requirements of OSI. Often users do not exactly know that they have such interconnections, yet they enjoy the useful features of MAP. It is noted that the MAP or related solutions are still relatively expensive.

MMS is a network protocol that provides a tool for defining the manufacturing-specific features of different industrial resources, such as CNC, robot, and PLC. At the same time MMS provides a modeling method of the resources from the communication point of view, and it also implements the communication messages. It is similar to the OO methodology in that different objects have different type of services.

On the other hand, intelligent control is coming into general demand. There is an active discussion among control engineers about the existence and need for intelligent control (Franklin et al., 1994–1995). Some experts claim that there is nothing that really could be called intelligent control. Most of them refer only to process control, not to discrete manufacturing (FMS) or robot systems control. The control tasks and problems of manufacturing systems are basically similar to those of batch-like process control. Despite this discussion, it is accepted that there is a need for intelligent control in terms of *knowledge-based control systems,* often called *expert systems.* Commercial expert systems provide effective problem description and software development tools; the programming itself is closer to the problem to be solved and to the user who has to support the control with limited real-time facilities (Laffey et al., 1988).

For real-time control one needs scheduling data, the start and finish dates of all operations of all equipment, and the possibility of downloading RC and CNC control programs to all equipment. Additionally, it is necessary to obtain and evaluate different signals from the associated equipment.

3.2 Real-Time Communication with Expert Systems

The practical problems of communication of expert systems in CIM applications can be divided into two parts: the hardware–software connection (physical) and the logical connection between the controller(s) and controlled devices. This decomposition was useful in both the design and implementation phases during the last projects of our CIM Laboratory. If this decomposition is not sharp, many problems may occur during the development and especially in subsequent maintenance.

There are relatively simple programming interfaces (e.g., C/C++) in most available expert system shells (ES). These interfaces provide data transfer and communication possibilities with external tasks, stations, etc. They support clear and simple programming to reach objects, call procedures, and set and get variables. The interfaces are dedicated to specific software tools of the ES and are general towards the external world without being able to take into account the requirements of the given application. Thus nearly every CIM (FMS) implementation requires special software development to cover this gap between the external world and the ES.

The communication functions depend on the capabilities of the expert system. The ways of learning and knowledge handling determine the logical levels of the communication. Three different types of working mode and different levels of the communication of an intelligent cell-controller in a CIM environment are shown in Figure 1. These levels are implemented within the same protocol. This picture explains the different possible meanings of given messages. The lowest level has the basic control and data acquisition-type messages. The other two levels have messages if and only if the intelligence of the cell controller is not hidden. *Hidden intelligence* means that the knowledge-based technology is applied only inside the cell controller and has no specific actions via the communication channel. A typical example of the hidden case occurs if a KB system is built up on top of a traditional control system using its original communication. The knowledge acquisition and the knowledge communications are separated. The first contains specific data for modifying or verifying the knowledge of the given controller. When a KB system shares its knowledge (new or modified), it belongs to the knowledge communication level (Buta and Springer, 1992). The communication messages of most real and pilot KB applications belong to the lowest or possibly the middle logical level.

3.3   Object-Oriented Communication Using MMS

In regard to the physical connection, there are several solution alternatives. Most controller and controlled device vendors offer good (proprietary) solutions to communicate, and vendor independent standards are also available.

In the robotics/CIM area there are also more accepted models or modeling tools to describe the objects of an FMS. In the communication point of view the most promising one is the object-oriented view of MMS. MMS gives a virtual manufacturing device (VMD) view of each resource of the FMS, including robots. Each resource in an MMS network is defined as a VMD. The VMD object itself has some general attributes, such as physical and logical status and vendor or model name. The VMD contains additional objects to define the resource-specific elements, such as the domains, program invocations, and variables, and such not widely used attributes as events, semaphores, and journals.

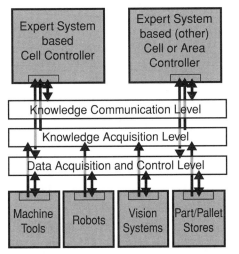

Type of knowledge processing within the cell-controller (according to the communication levels):

being modified (growing) knowledge base processing on dynamic data base and knowledge exchange

being modified (growing) knowledge base processing on dynamic data base

permanent (no change in rules) knowledge base processing on dynamic data base

Figure 1   Different logical communication levels of an intelligent cell controller. (From J. Nacsa and G. L. Kovács, "Communication Problems of Expert Systems in Manufacturing Environment," in *Preprints of the IFAC Symposium on Artificial Intelligence in Real-Time Control, AIRTC '94*, Valencia, Spain, 1994. Ed. A. Crespo i Lorente. 377–381.)

In a VMD it is possible to create, read, and write different types of variables, up- and download domains, which can be programs or machine data, start and stop different tasks (with program invocations), and handle events.

Figure 2 shows a simple but rather general type of VMD: a robot VMD. It was realized that this specification is usful for the higher level of the FMS (Nagy and Haidegger, 1994) to provide a communication-oriented view of the network elements and their resources.

The object-oriented view of MMS allows the VMD model of a certain device to be defined, and it is immediately possible to use this model as a specification of the communication where the services (what one can do with a given object of the VMD) are defined and are working in the MAP networks.

3.4 Case Study with an Intelligent Robot

The goal of this work has been to design a robotic cell with real-time intelligent control. One step was to apply all development results concerning scheduling with a hybrid, KB system and OSI communication with expert systems, as described in the previous sections. Next it was necessary to integrate the independently working and tested software–hardware modules.

As a first experimental setup (Figure 3), a simple configuration was chosen. From earlier projects a Mitsubishi robot was available with serial DNC connection to a PC that had a MAP interface. The game of tic-tac-toe was served by the robot as follows: A user could play against the G2 system. This well-known game can serve to demonstrate clearly the "intelligent" features of the control. On the other hand, the full capability of the MAP networking was utilized (Figure 3).

In the setup a special gateway was developed between the ES and the robot. It couples the G2 system interface (called *GSI*) to the MMS standard communication. This special gateway of the G2 allows one to develop any discrete manufacturing applications in G2, using MMS standard interface, without the need for interface modification. In the demo version the gateway supports the context management and some program invocation management services.

Two solutions were examined. One study (Tömösy, 1994) examined how the complete MMS object structure could be built up within the G2. The efforts resulted in a relatively complicated and not too useful structure. Because of the different services and their parameters, numerous external procedure calls were necessary. All implemented MMS services have their own GSI procedures, and all MMS object types have their internal G2 object descriptions. The complexity of the inner structure was the same as the complexity of the MMS itself.

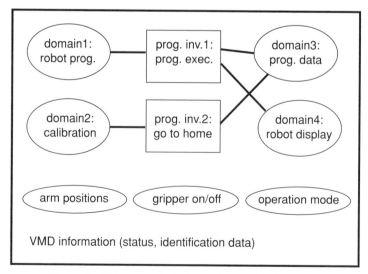

Figure 2 Virtual manufacturing device model of a robot.

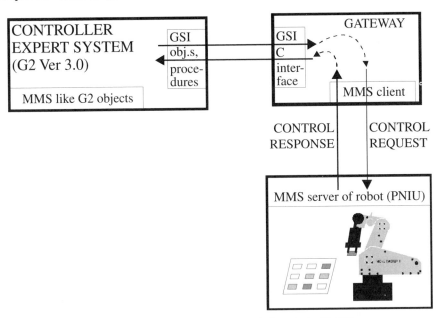

Figure 3 Knowledge-based control of a robot with OSI.

The second solution approach reduced the necessary objects and procedures to some general terms. It used limited number of external GSI procedure calls, but provided a relatively more general interface to the MMS, with fewer limitations.

The implementation was as follows: A special object hierarchy was built up within G2 to map the MMS objects in a more general way. We called it *MLGO* (MMS-like G2 objects). MLGO was coupled to the MMS network by means of special procedures via the GSI of G2. This way the user (who is the developer of the control system in G2) would be able to handle all remote resources via the MMS network from his/her G2 application using MLGO.

3.5 Findings from the Intelligent RS Study

Effective, close to optimal design and operation of complex, real-time stochastic systems such as robotized flexible manufacturing systems cannot be accomplished with only traditional (procedural) programming. The application of artificial intelligence means such as expert systems is required too, together with sophisticated modeling and simulation.

Several examples were investigated and some results were obtained in the CIMLab of CARI HAS during recent years. Investigations focused on combinations of a traditional simulation/animation package (SIMAN/Cinema), advisory expert systems (using ALL-EX shell), intelligent environment (G2), and C programming. The goal was to generate solutions to certain aspects of simulation, quality control, and scheduling and control of robots and flexible manufacturing systems.

The implementation of hybrid programs in SIMAN and G2 was fast enough. G2 was effective in running the different experimental programs. Data preparation, however, is a tedious and time-consuming activity.

Real-time intelligent control of FMS is considered the most challenging but most promising task to be solved within the area of AI in robotics and manufacturing. As an example, a robotics cell (where the robot is defined as a VMD) controlled by a real-time computer system has been studied. The control software was implemented using a real-time, OO, intelligent environment. Several research and development studies addressed MAP/MMS/VMD problems, and the applicability of expert systems was examined from the OSI communication point of view. An interfacing problem was solved when the G2 system and the MAP/MMS network were connected to reach real-time intelligent control.

These results should be useful and interesting in the future when they are applied to more complex and sophisticated large-scale robotics manufacturing and assembly systems. Such systems can hardly be designed and controlled without the combination of simulation and expert systems.

4 CONCLUSIONS; FURTHER TRENDS IN ROBOTIC DESIGN

It is clear that the application of computational, multiagent, and AI techniques and tools to design and operate robotics systems already has achieved positive concrete results worldwide. Most of these results are known and applied by the academic community, but industry needs more time to be convinced to use the most up-to-date techniques. Naturally, some of the results reported in the literature are tentative, experimental, or prototype solutions, and thus real industrial applications are yet to come. This is not a serious problem, since several years ago the same could be (and was) stated about CAD and CAM techniques and tools, and now these tools are widely used by industry.

Numerous different techniques, such as genetic algorithms, KB, fuzzy logic, ANN systems, Petri-nets, holonic systems, concurrent engineering, and multiagent techniques, can be combined with each other. Other AI tools are available, such as pattern recognition in the fields of vision, voice, and tactile applications. Hence there are different new possibilities for further study, research, and development, and problems that remain to be solved in this area.

There is a trend toward being more multidisciplinary, combining engineering tools and knowledge not only with computer techniques and knowledge, but also with psychological, behavioral, or cognitive results. These prospects offer future possibilities that are even more challenging and exciting.

Inoue (1996) has presented new ideas for the future, suggesting new research issues for robotics, including:

- Physical understanding of dexterity
- Real-world understanding through tightly coupled sensing and action
- Human–robot cooperative systems
- Biologically inspired autonomous systems

Hence, a triple shift of research directions is predicted:

1. From separate functions of vision/control/intelligence/design to perception–action coupling at all levels
2. From the motion of arms, legs, fingers, etc. to integrated whole-body motion of robots
3. From technologically oriented research to psychologically and biologically oriented research

Researchers are already working on certain aspects of these suggestions. Some aspects are in the understanding and comprehension phase; some of them will probably turn out to be false, which is natural to expect in the case of new research initiatives.

Nevertheless, several tangible results of the research have been applied already or will be applied soon in industry. For example, the worldwide IMS project has achieved significant improvements in holonic and lean manufacturing, concurrent engineering, and several other robotics-related topics. These and other results in autonomous, mobile, and other specific robotics are used in mining, space, and underwater applications. Finally, we can state that in spite of all difficulties, more intelligent, multiagent, and other systems are used for simulation, design, and control of robotics systems in manufacturing, assembly, and service applications.

ACKNOWLEDGMENT

The author wishes to thank Shimon Y. Nof for many comments that improved this chapter.

REFERENCES

Alagar, V. S., and K. Periyasami. 1992. "Formal Specifications Are Mathematical Models: An Example from Robotics." In *Proceedings of IROS '92* (*IEEE International Conference on Intelligent Robots and Systems*), Raleigh, July 7–10. 733–741.

Alagar, V. S., T. D. Bui, and K. Periyasami. 1990. "Semantic CSG Trees for Finite Element Analysis." *Computer Aided Design* **22**(4), 194–198.

Alami, R. 1996. "A Fleet of Autonomous and Cooperative Mobile Robots." In *Proceedings of IROS '96 (IEEE/RSJ International Conference on Intelligent Robots and Systems)*, Osaka, November 4–8. 1112–1117.

Armilotta, A., and Q. Semerado. 1993. "Assembly Planning in a Robot Cell Environment." In *Realising CIM's Industrial Potential*. Ed. C. Kooij, P. MacConaill, and J. Bastos. Amsterdam: IOS Press.

Bagchi, S., and K. Kawamura, 1992. "An Architecture of a Distributed Object-Oriented Robotic System." In *Proceedngs of IROS '92 (IEEE Industrial Conference on Intelligent Robots and Systems)*, Raleigh, July 7–10. 711–716.

Brill, M., and U. Gramm. 1991. "MMS: MAP Application Services for the Industry." *Computer Networks and ISDN Systems* **21**, 357–380.

Bussmann, S., and Y. Demazeau. 1994. "An Agent Model Combining Reactive and Cognitive Abilities." In *Proceedings of IROS '94 (IEEE/RSJ International Conference on Intelligent Robots and Systems)*, Munich, September 12–16. 2095–2102.

Buta, P., and S. Springer. 1992. Communicating the Knowledge in K-Based Systems." *Expert Systems with Applications* **5**, 389–394.

Duhaut, D. 1993. "Using a Multi-Agent Approach to Solve the Inverse Kinematics Program." In *Proceedings of IROS '93 (IEEE/RSJ International Conference on Intelligent Robots and Systems)*, Yokohama, July 26–30.

"ESPRIT Project 623 (1985–1990)." 1992. In *Integration of Robots into CIM*. Ed. R. Bernhardt et al. London: Chapman & Hall.

Filho, O. S., and M. F. Carvalho. 1995. "An Integrated Strategic Planning System." In *Proceedings of the 11th ISPE/IEE/IFAC International Conference on CAD/CAM. Robotics and Factories of the Future*, Pereira, Colombia, August 28–30. 209–216.

Fisher, E. L., and O. Z. Maimon. 1988. "Selection of Robots." In *Artificial Intelligence Implications for CIM*. Ed. A. Kusiak. Bedford: IFS. 172–187.

Franklin, G., et al. 1994–1995. Private Discussion by E-mail in Architectures for Intelligent Control Systems Discussion List, ⟨aics-1@ubvm.cc.buffalo.edu⟩.

Freund, E., and J. Rossmann. 1994. "Intelligent Autonomous Robots for Industrial Space Applications." In *Proceedings of IROS '94 (IEEE/RSJ International Conference on Intelligent Robots and Systems)*, Munich, September 12–16. 1072–1080.

Friedrich, H., O. Rogalla, and D. Dillmann. 1997. "Integrating Skills into Multi-agent Systems." In *Proceedings of the Second World Congress on Intelligent Manufacturing Processes and Systems*, Budapest, June 10–13.

Gairola, A. 1987. "Design for Assembly: A Challenge for Expert Systems." In *Artificial Intelligence in Manufacturing, Key to Integration? Proceedings of the Technology Assessment and Management Conference*, Zurich, November 1985. Ed. T. Bernold. Amsterdam: North-Holland. 199–212.

Gradetsky, V., et al. 1995. "Concurrent Design Methods for Assembly System and Reliability Problems in Design of Assembly Systems." In *Proceedings of the 11th ISPE/IEE/IFAC International Conference on CAD/CAM. Robotics and Factories of the Future*, Pereira, Colombia, August 28–30. 289–295.

Grübel, G., et al. 1996. "A Computation Environment for Robot-Dynamics Design Automation." In *Proceedings of IROS '96 (IEEE/RSJ International Conference on Intelligent Robots and Systems)*, Osaka, November 4–8. 1088–1093.

G2 Reference Manual, G2 Version 2.0. 1990. Gensym Corp., Cambridge.

Habib, M. K. 1992. "Simulation Environment for an Autonomous and Decentralized Multi-Agent Robotic System." In *Proceedings of IROS '92 (IEEE International Conference on Intelligent Robots and Systems)*, Raleigh, July 7–10. 1550–1557.

Haidegger, G., and S. Kopácsi. 1996. "Telepresence with Remote Vision in Robotic Applications." In *Proceedings of the International Workshop on Robotics in the Alpe–Adria–Danube Region*, Budapest, June 10–13. 557–562.

Inoue, H. 1996. "Whither Robotics: Key Issues, Approaches and Applications." In *Proceedings of IROS '96 (IEEE/RSJ International Conference on Intelligent Robots and Systems)*, Osaka, November 4–8. 9–14.

Jacobson, I., et al. 1992. *Object-Oriented Software Engineering: A Use Case Driven Approach.* Reading: Addison-Wesley.

Kim, J. O., and P. K. Khosla. 1992. "A Multi-Population Genetic Algorithm/MPGA/and Its Application to Design of Manipulators." In *Proceedngs of IROS '92 (IEEE International Conference on Intelligent Robots and Systems)*, Raleigh, July 7–10. 279–286.

Kopacek, P., et al. 1995. "Semiautomatic Knowledge Based Planning of Small Assembly Cells." In *Proceedings of the First World Congress on Intelligent Manufacturing, Processes and Systems*, University of Puerto Rico, Mayagüez, Puerto Rico, February. Ed. V. Milacic. 574–582.

Kovács, G. L. 1989. "Expert System for Manufacturing Cell Design." In *IFAC 6th Symposium on Automation in Mining, Mineral and Metalprocessing (MMM)*, Buenos Aires, Argentina. September 4–8. Preprints: Vol. 2. Session: Control 1, TS 14.1 324–329.

Kovács, G. L., and S. Kopácsi. 1992. "Cooperative Agents for the Simulation of FMC and FMS." Paper delivered at TIMS/ORSA Conference, Orlando, April.

Kovács, G. L., and Z. Létray. 1989. "Expert System for Work Cell Simulation and Design." In *PROLOG Language, the COOPERATOR; COMPCONTROL '89*, Bratislava, Czechoslovakia, September 12–14. Section 2.4.

Kovács, G. L., and I. Mezgár. 1991. "Expert Systems for Manufacturing Cell Simulation and Design." *Engineering Applications of Artificial Intelligence* 4(6), 417–424.

Kovács, G. L., S. Kopácsi, and I. Mezgár. 1992. "A PROLOG Based Manufacturing Cell Design System." In *Practical Application of Prolog Conference, Abstracts of the Poster Session*, London, April. The Institution of Civil Engineers. 3–4.

Kovács, G. L., Z. Létray, and S. Kopácsi. 1992. "Application of Cooperative Agents (Co-gents) for Simulation of Flexible Manufacturing Systems." In *IFIP International Workshop on Open Distributed Processing*, October 8–11, 1991. IFIP Transactions, C-1. Ed. J. De Meer, V. Heimer, and R. Roth. Amsterdam: North-Holland. 415–420.

Kovács, G. L., I. Mezgár, and S. Kopácsi. 1991. "Concurrent Design of Automated Manufacturing Systems Using Knowledge Processing Technology." *Computers in Industry* 17, 257–267.

Koyachi, N., et al., 1996. "Geometric Design of Hexopod with Integrated Limb Mechanism of Leg and Arm." In *Proceedings of the Japan–USA Symposium on Flexible Automation*, Boston, July 7–10.

Kusiak, A., and S. Heragu. 1988. "Rule Based Expert Systems for Equipment Selection and Layout." In *Artificial Intelligence, Manufacturing Theory and Practice*. Ed. S. T. Kumara, A. S. Soyster, and R. L. Kashyap. Norcross: Industrial Engineering & Management Press.

Laffey, T. J., et al. 1988. "Real-Time Knowledge Based Systems." *AI Magazine* 9(1), 27–45.

Lambrinos, D., and C. Schreier. 1996. "Building Complete Autonomous Agents: A Case Study on Categorization." In *Proceedings of IROS '96 (IEEE/RSJ International Conference on Intelligent Robots and Systems)*, Osaka, November 4–8. 170–177.

Lee, M. H. 1989. *Intelligent Robotics*. New York: Open University Press.

Lee, S., and H. S. Lee. 1992. "Modeling, Design and Evaluation for Advanced Teleoperator Control Systems." In *Proceedings of IROS '92 (IEEE International Conference on Intelligent Robots and Systems)*, Raleigh, July 7–10. 881–888.

Mahajan, A., and F. Figueroa. 1997. "Four-Legged, Intelligent Mobile Autonomous Robot." *Robotics and Computer-Integrated Manufacturing* 13(1), 51–61.

McGlennon, J. M. 1988. "ROBOSPEC—A Prototype Expert System for Robot Selection." In *Artificial Intelligence Implications for CIM*. Ed. A. Kusiak. Bedford: IFS.

Minagawa, M., and Y. Kakazu. 1996. "Multiple Robots Navigation in Cellular Space: Acquisition of Collective Behavior." In *Proceedings of the Japan–USA Symposium on Flexible Automation*, Boston, July 7–10. 1517–1523.

Miyake, N., Kametaní, and H. Sakairi. 1994. "Hardware and Software Architecture for Intelligent Robot Control—An Approach to Dual Control." In *Procedings of IROS '94 (IEEE/RSJ International Conference on Intelligent Robots and Systems)*, Munich, September 12–16. 2146–2151.

Molina, A., I. Mezgár, and G. Kovács. 1992. "Cooperative Hybrid Expert Systems for Concurrent Design of FMS." In *Preprints of the IFAC Workshop on Intelligent Manufacturing Systems*, Dearborn, Michigan, October 1–2. Ed. R. P. Judd and N. A. Kheir. New York: Pergamon Press. 51–56.

Nacsa, J., and G. L. Kovács. 1994. "Communication Problems of Expert Systems in Manufacturing Environment." In *Preprints of the IFAC Symposium on Artificial Intelligence in Real-Time Control, AIRTC '94*, Valencia, Spain. Ed. A. Crespo i Lorente. 377–381.

Nagy, G., and G. Haidegger. 1994. "Object-Oriented Approach for Analyzing, Designing and Controlling Manufacturing Cells." In *Proceedings of AUTOFACT '94*, Detroit. Ed. Moody. 1–10.

Nilsson, N. J., P. R. Cohen, and J. Rosenschein. 1987. "Intelligent Communication Agents." Technical Report, SRI International.

Paasiala, P., et al. 1993. "Automatic Component Selection." In *Realising CIM's Industrial Potential*. Ed. C. Kooij, P. MacConaill, and J. Bostos. Amsterdam: IOS Press. 303–312.

Putnik, G., and M. de Silva. 1995. "Knowledge Integrated Manufacturing." In *Proceedings of the First World Congress on Intelligent Manufacturing, Processes and Systems*, University of Puerto Rico, Mayagüez, Puerto Rico, February. Ed. V. Milacic. 688–698.

Rodriguez, C. F., et al. 1995. "Grasp Planning for Manipulation with Multirobot Systems." In *Proceedings of the 11th ISPE/IEE/IFAC International Conference on CAD/CAM. Robotics and Factories of the Future*, Pereira, Colombia, August 28–30. 922–925.

Sakao, T., et al. 1996. "The Development of a Cellular Automatic Warehouse." In *Proceedings of IROS '96 (IEEE/RSJ International Conference on Intelligent Robots and Systems)*, Osaka, November 4–8. 324–331.

Sasaki, J., et al. 1996. "Optimal Arrangement for Handling Unknown Objects by Cooperative Mobile Robots." In *Proceedings of IROS '96 (IEEE/RSJ International Conference on Intelligent Robots and Systems),* Osaka, November 4–8. 112–117.

Schweiger, J., K. Ghandri, and A. Koller. 1994. "Concepts of a Distributed Real-Time Knowledge Base for Teams of Autonomous Systems." In *Proceedings of IROS '94 (IEEE/RSJ International Conference on Intelligent Robots and Systems),* Munich, September 12–16. 1508–1515.

Shimojma, K., and T. Fukuda. 1994. "Intelligent Robot Reusable Manufacturing for Environment Conscious Design." In *Proceedings of IROS '94 (IEEE/RSJ Intrnational Conference on Intelligent Robots and Systems),* Munich, September 12–16. 1986–1991.

Skewis, T., and V. Lumelsky. 1992. "Simulation of Sensor-Based Robot and Human Motion Planning." In *Proceedings of IROS '92 (IEEE International Conference on Intelligent Robots and Systems),* Raleigh, July 7–10. 933–940.

Smith, P. E., et al. 1992. "The Use of Expert Systems for Decision Support in Manufacturing." *Expert Systems with Applications* **4,** 11–17.

Tömösy, G. 1994. "Implementation of MMS Services into an Expert System." Unpublished Diploma Work, MTA SZTAKI, Budapest. [In Hungarian.]

ADDITIONAL READING

Akman, V., P. J. W. ten Hagen, and P. J. Veerkamp, eds. *Intelligent CAD Systems II, Implementational Issues.* Berlin, Springer-Verlag, 1988.

Brown, D. C., M. B. Waldron, and H. Yoshikawa, eds. *Intelligent Computer Aided Design.* Proceedings of the IFIP WG 5.2 Conference on IntCAD '91, Columbus, September–October 1991. IFIP Transactions B-4. Amsterdam: North-Holland, 1992.

Chang, T. C. *Expert Process Planning for Manufacturing.* Reading: Addison-Wesley, 1990.

Crespo i Lorente, A., ed. *Preprints of the IFAC Symposium on Artificial Intelligence in Real-Time Control, AIRTC '94,* Valencia, Spain, 1994.

Famili, A., D. S. Nau, and S. H. Kim, eds. *Artificial Intelligence Applications in Manufacturing.* Menlo Park: AAAI Press/MIT Press, 1992.

Hallam, J., and C. Mellis, eds. *Advances in Artificial Intelligence, Proceedings of the 1987 AISB Conference,* Edinburgh, April 1987. New York: John Wiley & Sons, 1987.

Hatvany, J., ed. *Preprints of International Conference on Intelligent Manufacturing Systems,* Budapest, June 1986. [Organized by the *International Journal of Robotics and Computer Integrated Manufacturing.*]

Jorrand, P., and V. Sgurev, eds. *Artificial Intelligence II, Methodology, Systems, Applications, Proceedings of the AIMSA '86 Conference,* Varna, Bulgaria, September 1986. Amsterdam: North-Holland, 1987.

Kusiak, A. *Intelligent Manufacturing Systems.* Englewood Cliffs: Prentice-Hall, 1990.

Mezgár, I., and P. Bertók, eds. "KNOWHSEM: Knowledge Based Hybrid Systems." In *Proceedings of IFIP/IFAC Conference.* IFIP Transactions. Amsterdam: North-Holland, 1993.

Miller, R. K., and T. C. Walker. *Artificial Intelligence Applications in Manufacturing.* SEAI Technical and Fairmart Press, 1988.

Monostori, L., ed. *Preprints of the Second International Workshop on Learning in Intelligent Manufacturing Systems,* Budapest, April 1995. AKAPRINT, 1995.

Pham, D. T., ed. *Artificial Intelligence in Design.* Berlin: Springer-Verlag, 1991.

Proceedings of IFAC/IFIP/IFORS Workshops on Intelligent Manufacturing Systems, Dearborn, 1993, Vienna, 1994, Bucharest, 1995.

Proceedings of IROS '95 (IEEE/RSJ International Conference on Intelligent Robots and Systems), Pittsburgh, August 5–9, 1995.

Richer, M. H. *AI Tools and Techniques.* Norwood: Ablex, 1988.

Rosenberg, J. M. *Dictionary of Artificial Intelligence and Robotics.* New York: John Wiley & Sons, 1986.

Rychener, M. D., ed. *Expert Systems for Engineering Design.* Boston: Academic Press, 1988.

Rzevski, G., J. Pastor, and R. A. Adey, eds. *Proceedings of AIENG: Applications of Artificial Intelligence in Engineering VIII.* Southampton: Computational Mechanics, and London: Elsevier Applied Science, 1995.

ten Hagen, P. J. W., and T. Tomiyama, eds. *Intelligent CAD Systems I, Theoretical and Methodological Aspects, EUROGRAPHICS Workship.* Berlin: Springer-Verlag, 1987.

Widman, L. E., K. A. Loparo, and N. R. Nielsen, eds. *Artificial Intelligence, Simulation and Modeling.* New York: John Wiley & Sons, 1989.

Wright, P. K., and D. A. Bourne. *Manufacturing Intelligence.* Reading: Addison-Wesley, 1988.

Journals

Robotica, Cambridge University Press, U.K.
Data and Knowledge Engineering, North-Holland (Elsevier), Amsterdam.
Machine Learning, Kluwer Academic, Dordrecht.

Artificial Intelligence, Elsevier, Amsterdam.
Computers in Industry, Elsevier, Amsterdam.
Engineering Applications of Artificial Intelligence, Elsevier, Oxford, Pergamon Press.
Artificial Intelligence in Engineering, Elsevier, Oxford.
Applied Intelligence, Kluwer Academic, Dordrecht.
Journal of Intelligent Systems, Freund Publishing House, U.K.
International Journal of Intelligent Mechatronics, Design and Production, Middle East Technical University, Ankara, Turkey.
Journal of Artificial Intelligence Research, AI Access Foundation and Morgan Kaufmann.

CHAPTER 39

PRECISION AND CALIBRATION

Klaus Schröer
Fraunhofer-IPK
Berlin, Germany

1 SCOPE OF THE CHAPTER

Since the early days of robotics, the problem of investigating and improving a robot's precision has been of the utmost importance for robot manufacturers and even more for robot users. Investigating a robot's precision first leads to determining robot performance criteria through application of related test procedures. If a robot's basic precision or repeatability is to be improved, changes in construction design will usually be required. If a robot's dynamic characteristics (path following, cornering deviation, etc.) are to be improved, changes in control settings, parameters, or algorithms will usually be required. If a robot's absolute pose accuracy (i.e., its capability of precisely moving the tool center point (TCP) to a Cartesian position and orientation being defined in the robot base or a workcell coordinate frame) is to be improved, calibration techniques have to be applied. The same is true if the capability of a robot-based flexible manufacturing process (according to section 4.9 of ISO 9001) is to be determined and ensured based on tolerances to be met in the entire robot workspace.

Therefore this chapter begins with a short review of the history, status, and trends of definition and standardization of robot performance criteria. The subsequent sections are dedicated to calibration, first presenting background, scope, and components to be calibrated, continuing with sections on workcell and robot tool calibration and on robot calibration, and finally describing trends and requirements in industrial application.

In the calibration sections the intention is to provide a short technical and theoretical introduction to each topic, to give hints and advice concerning issues to be regarded, and finally to present references where more detailed information can be obtained.

Measurement systems and methods, although very important for practical implementation of calibration and performance testing, are not covered by this section. An overview of available measurement systems is included in the ISO document (ISO, 1994), which is a result of the revision process of ISO 9283.

2 PERFORMANCE TESTING

Robot performance testing is defined in international standard ISO 9283, which "is intended to facilitate understanding between users and manufacturers of robots and robot systems. It defines the important performance characteristics, describes how they shall be specified and recommends how they should be tested" (ISO 9283, Introduction). ISO 9283 defines the following characteristics: pose accuracy and pose repeatability, multidirectional pose accuracy variation, distance accuracy and distance repeatability, position stabilization time, position overshoot, drift of pose characteristics, exchangeability, path accuracy and path repeatability, path accuracy on reorientation, cornering deviations, path velocity characteristics, minimum posing time, static compliance, and weaving deviation.

ISO 9283 was first defined in 1990. A first revision was technically finished in 1996 and finally approved in 1998. In parallel, the American standard ANSI/RIA R15.05 was developed between 1990 and 1992. Part 1 of the ANSI standard defines point-to-point and static performance characteristics: accuracy, repeatability, cycle time, overshoot and settling time, and compliance. Part 2 defines path-related and dynamic performance characteristics: path accuracy, path repeatability, cornering deviations, and path speed characteristics. Part 3 defines guidelines for reliability acceptance testing: infant mortality life

Handbook of Industrial Robotics, Second Edition, Edited by Shimon Y. Nof
ISBN 0-471-17783-0   © 1999 John Wiley & Sons, Inc.

test, functional verification, 24-hour individual axes movement test, and 48-hour composite axes movement test.

Up to now, these standards have suffered from being hardly applied—particularly by robot users. A representative investigation conducted in 1996 in the framework of a European research project provided a survey on the current status of robot testing in Europe and showed (among other things) the following requirements and trends:

- Less than 10% of the companies experienced testing according to ISO 9283.
- Both robot manufacturers and users are prepared to conduct robot performance testing. They are convinced that it is essential and valuable.
- There is a strong demand from robot users for application-oriented testing.
- There are differing requirements on testing from different types of robot users and at different stages of a robot's life cycle.
- Robot users want to have test procedures that are simpler and easier to interpret and can be used for trouble-shooting in robot production.

3 BACKGROUND OF CALIBRATION AND BASIC DEFINITIONS

An industrial robot has to be used with regard to the following components of a robotic workcell, their coordinate frames, and the errors or inaccuracies introduced by the components (see Figure 1):

- The *robot itself* defines the pose (i.e., position and orientation) of the robot's tool-mounting flange (FL) with respect to the robot base frame (RB). In mathematical terms the robot defines a transformation between the coordinate frames associated with base and flange. This transformation is a function of the robot's joint positions, and its correct and precise computation requires knowledge of the entire electromechanical system, which can be obtained through calibrating the robot.
- The *robot tool* defines the transformation from the robot flange frame to the TCP frame. There may be changes due to wear or tool exchange during operation. The

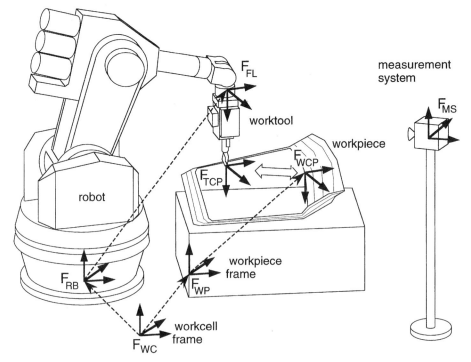

Figure 1 Workcell components.

robot operation is usually defined in terms of poses or paths that the TCP has to move to or follow.

- The *location of the robot in the real workcell* has of course to be determined in order to ensure precise operation. It is defined by a transformation between the RB and the workcell coordinate frame (WC). If the robot is mounted on a gantry or servo-track, the location of the robot base frame changes during operation.

- The *workpiece location* is defined by jigs, conveyors, or (turn)tables, thus defining the transformation between the workpiece reference frame (WP) and the workcell frame. During robot programming (particularly during off-line programming), the robot actions are frequently defined with respect to the workpiece reference frame because position and orientation of (for example) spot welding points must be derived from CAD data of the workpiece. Thus robot programming results in a set of work poses (plus related actions) being defined with respect to the WP. Each work pose is thus defined by a transformation between the workpiece reference frame and the actual work center point (WCP), and the robot operation is perfect if the WCP and the TCP of the robot coincide in each step of robot action.

- The *real workpiece itself* also introduces a significant amount of error or uncertainty due to workpiece variations or deflection. If this amount of error or uncertainty is higher than the tolerances of the process allow for, additional sensors have to be applied to overcome it.

- Finally, a *measurement system* frequently has to be used to calibrate or to orient the workcell components. This measurement system usually provides its measurements with respect to the measurement system coordinate frame (MS), and its location has to be determined with respect to the workcell frame.

Figure 1 illustrates the components and coordinate frames to be included in order to have precise robot operation. In mathematical terms the condition for precise robot operation is defined by the following equation:

$$\substack{WC\\RB}T \cdot \substack{RB\\FL}T(q) \cdot \substack{FL\\TCP}T = \substack{WC\\WP}T \cdot \substack{WP\\WCP}T \tag{1}$$

Here the transformations are denoted by T, and the argument q indicates that the transformation from robot base to robot flange depends on the actual joint positions q.

The transformations $\substack{WC\\RB}T$, $\substack{WC\\WP}T$, and also $\substack{MS\\WC}T$ are constant transformations that have to be determined after installation or change of the workcell. A procedure for their determination is called *workcell calibration*.

The transformation $\substack{FL\\TCP}T$ defines the geometry of the tool and has to be checked frequently if the tool is subject to wear or if it is exchanged. A procedure for its determination is called *tool calibration* and often differs from the procedures used for workcell calibration.

The transformation $\substack{RB\\FL}T(q)$ describes the robot itself. The value of $\substack{RB\\FL}T(q)$ computed by the robot control usually differs significantly from the pose of the flange which is actually attained. Causes for this deviation are:

- Differences between the controller model of the robot's kinematic geometry and the real kinematic structure
- Joint elasticity or backlash
- Transmission and coupling characteristics of gears
- Elasticity of links or of the robot mounting fixture on the floor (or ceiling)

These deviations are systematic errors that can be compensated for if their parameters are known through robot calibration procedures (thus improving a robot's absolute pose accuracy). Moreover, there are of course purely stochastic errors caused by joint encoder resolution or error motion of joint bearings that define a robot's basic system accuracy or repeatability.

Robot calibration is a term applied to the procedures used in determining actual values that describe the geometric dimensions and mechanical characteristics of a robot or multibody structure. A robot calibration system must consist of appropriate robot modeling techniques, accurate measurement equipment, and reliable model parameter determination methods. For practical improvement of a robot's absolute accuracy, error compensation methods are required that use calibration results.

Robot calibration packages are required in industry because of the demands large robot users have for calibrated robots, i.e., for robots showing a high absolute pose accuracy and capable of executing off-line programmed robot task programs without manual re-teaching. Certain calibration capabilities are offered by off-line programming system providers and by providers of measurement systems that can be used for robot calibration.

Many different solutions to calibrating a robot were developed in the past. Therefore it is necessary to discuss their capabilities and advantages as well as practical experiences with and requirements for applying robot calibration.

4   WORKCELL AND ROBOT TOOL CALIBRATION

Workcell calibration is required to determine the location of robot, workpiece, and measurement systems in the robot workcell, i.e., to determine the transformations: $^{WC}_{WP}T$, $^{WC}_{WP}T$, and $^{MS}_{WC}T$. The basic idea of this procedure is to select certain points with well-known 3D coordinates with respect to one coordinate frame, measure their 3D position with respect to the other coordinate frame, and compute the difference transformation from this information. Alternatives for computing this difference transformation are presented below after some examples.

4.1   Examples

1. To determine the location of a measurement system reference frame with respect to the workcell frame, select certain points in the workcell (e.g., holes, corners, spikes, or, even better, special measurement points already defined in the workcell design). Three-dimensional coordinates of these points in the CAD model of the workcell must be available. Measurement of these points with a 3D coordinate measurement system provides the coordinates of these points with respect to the measurement reference frame. The transformation $^{MS}_{WC}T$ can thus be computed. The transformation $^{MS}_{WP}T$ can be determined in the same way by selecting measurement points on the workpiece, its jig, conveyer, or table. With this information available, the location of the workpiece reference with respect to the workcell frame $^{WC}_{WP}T$ is also known:

$$^{WC}_{WP}T = {}^{WC}_{MS}T \cdot {}^{MS}_{WP}T; \quad {}^{WC}_{MS}T = {}^{MS}_{WC}T^{-1} \tag{2}$$

2. To determine the location of the robot base frame with respect to the workcell frame, basically the same operation could be applied if certain measurement points at the robot are available with well-known coordinate values with respect to the robot base frame. Alternatively, the robot's TCP can be measured for different robot poses. In this case the 3D coordinates of the measured TCP positions are not available from a CAD model, but have to be taken from the robot control display. Of course, the robot itself $^{RB}_{FL}T(q)$ and the location of the TCP with respect to the robot flange $^{FL}_{TCP}T$ must be known precisely in advance, i.e., the robot should be calibrated.

4.2   Determination of a Transformation

First, 3D coordinates of points to be measured have to be known with respect to a reference frame F_R. These points are then measured by a 3D coordinate measuring device with respect to this system's reference frame F_M. The objective is to compute the difference transformation T that maps points given in F_R to their coordinate representation in F_M.

Let a_i be the coordinates of one of the points in F_R and b_i be the same point with respect to F_M. Then the transformation $^M_R T$ is defined by the N equations:

$$^M_R T a_i = b_i, \quad i = 1, \ldots, N \tag{3}$$

Here the 4D representation of points using homogeneous coordinates are used here. Because embedding into 4D homogeneous coordinates is unique, the same symbols are used here for 3D coordinates and the related 4D representation. $^M_R T$ is a so-called 4 × 4 homogeneous matrix. It consists of a 3D translation and a 3 × 3 rotation matrix. It is unambiguously defined by 6 parameters. Usually 3 translations and 3 rotations are selected to describe the transformation. The functional relationship between the matrix $^M_R T$ and the angles of rotation defining its rotational part is nonlinear. See also Chapter 6, Kinematics and Dynamics of Robot Manipulators.

The problem of determining ${}_R^M T$ is therefore a nonlinear least-squares problem. There are iterative algorithms available, for example the Gauss–Newton method or the Levenberg–Marquardt method (see, e.g., Dennis and Schnabel, 1983; Press et al., 1994). A one-step algorithm (Umeyama, 1991) based on the singular value decomposition (SVD) of a matrix derived from the point sets $\{a_i\}$ and $\{b_i\}$ is highly recommended.

4.2.1 Number of Points

At least 3 different points ($N = 3$) are required to determine ${}_R^M T$. But these measurements usually include a significant amount of error. Therefore measuring at least 10 different points is recommended in order to have a real least-squares problem. Moreover, evaluating the residuals r_i after having determined ${}_R^M T$ is recommended:

$$r_i :: = {}_R^M T a_i - b_i \tag{4}$$

This evaluation will show whether there are measurement outliers included that should better be excluded from the process of determining ${}_R^M T$.

4.2.2 Location of Points

The points to be measured should be located such that they span a volume that includes the volume of interest where the transformation ${}_R^M T$ will be used. The transformation thus determined will be significantly less reliable outside the convex hull spanned by the measured points.

4.3 Workcell Calibration Without a Measurement System

Many robot users prefer to use the robot as a measurement system for workcell calibration, thus avoiding the use of a separate measurement system. In general this is possible, but this solution implicitly includes certain assumptions that must be regarded.

The robot TCP is manually (i.e., by using the teach pendant) moved to a number of target points on the workpiece, its jig, conveyor, or table, or in the workcell. The joint positions q_i are recorded after a visual (or sensor-supported) inspection has shown that the TCP coincides with the target point. The visual inspection is the weak point of the procedure because its accuracy cannot be controlled.

To describe the problem mathematically, now let b_i be the target point coordinates in the workpiece reference frame. The coordinates a_i of the same points with respect to the robot base frame are then computed by the robot control or an off-line programming system using the recorded joint positions q_i (6). Therefore the 3D coordinates of the TCP x_{TCP} with respect to the robot flange frame have to be known exactly (5).

$$x_{TCP} :: = {}_{TCP}^{FL} T \cdot \begin{pmatrix} 0 \\ 0 \\ 0 \\ 1 \end{pmatrix} \tag{5}$$

$$a_i = {}_{FL}^{RB} T(q_i) \cdot x_{TCP} \tag{6}$$

$$ {}_{RB}^{WP} T \cdot a_i = b_i, \quad i = 1, \dots, N \tag{7}$$

Finally, Equations (7) can be exploited as above to determine the transformation ${}_{RB}^{WP} T$ between workpiece reference frame and robot base. Of course, the values for ${}_{RB}^{WP} T$ are reliable only if the TCP coordinates x_{TCP} with respect to the robot flange frame are known precisely and the robot is calibrated. This means here that the transformation ${}_{FL}^{RB} T(q_i)$ in (6) exhibits the precise location of the robot flange (with respect to the robot base) as a function of the joint positions q_i.

If a noncalibrated robot is used, then the resulting transformation ${}_{RB}^{WP} T$ incorporates a local approximation of all the robot errors. Sometimes this is desired, because it may be sufficient for a certain robot task to execute within its tolerances, particularly if the robot model used in the robot control computations is sufficiently correct. But also in this case the TCP coordinates x_{TCP} with respect to the robot flange frame must be known from a tool calibration.

4.4   Tool Calibration

The procedure frequently used for tool calibration is of similar simplicity to workcell calibration. The robot TCP is manually (i.e., by using the teach pendant) moved to one and the same point, but with different orientations. The joint positions q_i are recorded after a visual (or sensor-supported) inspection has shown that the TCP coincides with the target point. There is no need to measure the 3D coordinates of the target point; only the joint positions need be recorded. With these data available, the transformation $^{FL}_{TCP}T$ can be determined by evaluation of the equations obtained by each two successive measurement positions i and $i + 1$:

$$
^{RB}_{FL}T(q_i) \cdot {}^{FL}_{TCP}T \cdot \begin{pmatrix} 0 \\ 0 \\ 0 \\ 1 \end{pmatrix} = {}^{RB}_{FL}T(q_{i+1}) \cdot {}^{FL}_{TCP}T \cdot \begin{pmatrix} 0 \\ 0 \\ 0 \\ 1 \end{pmatrix} \tag{8}
$$

Only the position part of the transformation $^{FL}_{TCP}T$ can be determined in this way. If the TCP frame is to have an orientation different from the flange frame's orientation, the required information can be taken from tool design information. For many applications this is sufficient, since the orientation accuracy requirements are often less tight than position accuracy requirements.

This procedure is supported by many off-line programming systems or provided as tool-calibration procedure by the robot manufacturer. Concerning the procedure's accuracy, the visual decision that the same TCP position is met is critical. This problem can be overcome through the use of a simple short-range 3D measurement system.

5   ROBOT CALIBRATION

5.1   Errors to Be Modeled

The transformation $^{RB}_{FL}T(q)$ describes the entire electromechanical multibody system. Mathematical models of all systematic errors must be included in this transformation if their numerical model parameter values are to be determined through the calibration procedure and later used for error compensation and improvement of the robot's absolute pose accuracy. Robot modeling for purposes of calibration has been investigated by several researchers. These investigations include the development of various robot modeling techniques for estimating the kinematic and sometimes the mechanical features of the static robot. This section surveys necessary or recommended model components.

Among the possible sources of systematic error listed in Section 3, the differences between the controller model of the robot's kinematic geometry and the real kinematic structure contribute the biggest share to the entire system error. With today's industrial robots, identifying the actual parameters of the kinematic model usually allows the reduction of the absolute position error from more than 10 mm to below 2 mm. A kinematic model of the robot is therefore the starting point for each calibration procedure.

While not all calibration procedures developed include the identification of a robot's static–mechanical features, experience has shown these features to have a significant impact on robot accuracy, particularly if accuracy requirements are very tight, i.e., below 1 mm. At least the elasticity of rotary joints (those with nonvertical rotation axis in particular) is therefore recommended to be included in the model used for calibration. In order to include elasticity, the reaction forces and torques induced by the robot's payload and its own body mass must then be computed. Examples of how to provide the calibration system user with flexible options to model and identify joint, beam, or plate elasticity are described, for example, in Schröer et al. (1994) and RoboCal (1995). Actuator parameters may be included if the complete system from motor encoders to TCP has to be identified or checked. Due to practical reasons and user needs, it is suggested that a "target" model (specifying the TCP with respect to the robot flange) be included in order to allow easy exchange between measurement target and different tools (e.g., for error compensation). A further practical reason is that integration of a target model allows for use of a calibration system with different types of measurement systems. This results in integrating the four models illustrated in Figure 2. If only kinematic calibration is required, the deformation model will be missing and information known from construction design of the actuating elements will be used to define the functional relation between motor encoder values and joint positions.

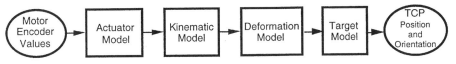

Figure 2   Robot model.

5.2   Calibration Procedures

For calibration this way of modeling requires so-called *kinematic-loop methods* (Hollerbach and Wampler, 1996), where the manipulator is placed into a number of poses providing joint and TCP measurement information and all model parameters are identified simultaneously by exploitation of the differentiability of the system model as a function of its parameters. These are the calibration methods this section refers to, and this approach was chosen by most research groups[1] and implemented by many commercial calibration packages.

Kinematic-loop methods can be applied with a variety of TCP measurement options: from full (i.e., 6D) TCP pose measurement to only 1D measurement, or using a ball bar with known length and unsensed spherical joints at each end (Driels, 1993). These different solutions were categorized by Hollerbach and Wampler (1996) by defining the calibration index. They also present a more theoretical survey on different calibration methods that is recommended as a complementary reading to this introduction and that includes more references than can be included here.

Quite different from kinematic-loop methods is the so-called *circle point analysis,* *screw axis measurement,* or *signature identification,* which can be used for kinematic calibration only. Each joint is identified as a screw, i.e., as a line in space. From knowledge of all of the joint screws, the kinematic parameters can be extracted straightforwardly. This method employs several steps of measurement and computation (Stone and Sanderson, 1987; Judd and Knasinski, 1990; Ziegert and Datseris, 1990), each step identifying a parameter subset based on geometric-conspicuous relations of successive joint axes (e.g., each step involves separate motion and measurement of one joint after the other). The basic assumption of circle point analysis is that separate motion of joints displays specific errors and that only those specific errors affect the measured target positions. However, this assumption is violated when elastic deformations are present.

The following sections introduce topics that a user of robot calibration should know about and that are relevant for deciding which capabilities a calibration system should have for a given application. These topics are:

- The *kinematic model* used to describe the robot structure
- Options for modeling the *elasticity*
- Models of *actuating elements,* particularly for robots with closed kinematic chains
- What *numerical procedures* for robot calibration do, noting that scaling is important
- Factors that influence the *reliability of calibration results,* in particular model parameter dependency, pose selection, rank deficiency, and condition number; also options for checking reliability.

5.3   Kinematic Model

Important to robot calibration methods is an accurate kinematic model that has identifiable parameters. Many different alternatives for modeling the kinematic geometry have been suggested, but it is generally accepted that parameter identification demands that three basic requirements of the kinematic model be met:

---

[1]See, e.g., Wu (1984), Sugimoto and Okada (1985), Hayati and Mirmirani (1985), Golikov (1986), Veitschegger and Wu (1986), Everett (1988), Driels (1993), Schröer (1993), and Stanton and Meyer (1993).

- *Completeness:* All spatial arrangements of successive joints in the kinematic chain must be describable.
- *Model continuity:* Small changes in the spatial geometry of joints must result in small changes in the describing parameters.
- *Minimality:* The kinematic model must include only a minimal number of parameters.

The requirements of completeness and model continuity are self-evident (Everett, Driels, and Mooring, 1987). Model continuity is a particularly important prerequisite for nonlinear optimization procedures to work reliably. The requirement of minimality is plausible because its violation results in redundancy of the model and rank deficiency during the numerical identification procedure.

Everett and Hsu (1988) have shown that a minimal and complete kinematic model has exactly

$$n = 4r + 2t + 6 \tag{9}$$

model parameters, where r is the number of rotary joints and t the number of prismatic joints. The number n defined by (9) is also the maximum number of independent model parameters of a complete kinematic model.

It must be recognized that mathematically precise statements about minimality can only be made for the geometric relations defined by the kinematic model. Statements about minimality of the model cannot be made for nongeometric parts of the robot model, such as the model of elastic deformations.

History of kinematic modeling of robots begins with the famous Denavit–Hartenberg convention, which was used and modified by many authors (Khalil and Kleinfinger, 1986):

$$T = R_z(\theta) \, T_z(d) \, T_x(a) \, R_x(\alpha), \qquad \theta, d, a, \alpha \in \Re \tag{10}$$

To overcome problems of model continuity with parallel joints it was modified by Hayati and Mirmirani (1985):

$$T = R_z(\theta) \, T_x(a) \, R_x(\alpha) \, R_y(\beta), \qquad \theta, a, \alpha, \beta \in \Re \tag{11}$$

Veitschegger and Wu (1986) suggested using a five-parameter model:

$$T = R_z(\theta) \, T_z(d) \, T_x(a) \, R_x(\alpha) \, R_y(\beta) \tag{12}$$

and Stone and Sanderson (1987) presented the S-model having six parameters:

$$T = R_z(\theta) \, T_z(d) \, T_x(a) \, R_x(\alpha) \, R_z(\gamma) \, T_z(b) \tag{13}$$

The CPC model developed by Zhuang, Roth, and Hamano (1990) has seven parameters and is different in as far as it directly uses the redundant representation of normalized 3D vectors to model a joint-axis direction:

$$T = \begin{pmatrix} 1 - \dfrac{b_x^2}{1+b_z} & \dfrac{-b_x b_y}{1+b_z} & b_x & 0 \\[2ex] \dfrac{b_x b_y}{1+b_z} & 1 - \dfrac{b_y^2}{1+b_z} & b_y & 0 \\[2ex] -b_x & -b_y & b_z & 0 \\[2ex] 0 & 0 & 0 & 1 \end{pmatrix} \cdot R_z(\beta) \cdot T_{xyz}(l_x, l_y, l_z) \tag{14}$$

One motivation for these investigations was the objective of finding a unique/single kinematic model uniformly applicable for calibration of all arrangements of subsequent joints. But several years ago it was proved that this single modeling convention cannot exist for fundamental topological reasons concerning mappings from Euclidean vector spaces to spheres (Gottlieb and Daniel, 1986; Baker and Wampler, 1988; Baker, 1990), known informally among topologists as "you can't comb the hair on a coconut."

A consequence of this result is that a list of alternative modeling conventions for different types and configurations of joints has to be defined. Implicitly, the above models having five or more parameters also are lists because some of the parameters are switched off for calibration depending on the geometric arrangement of subsequent joints. Therefore another solution (Schröer, 1993; Schröer, Albright, and Grethlein, 1997) directly provides a list of 17 parametrizations containing a suitable model (being complete, model-continuous, and minimal) for different joint types (rotary/prismatic) and different joint orientations (orthogonal/parallel) that also meets the following requirements on the kinematic model that arise from practical application:

1. If the current coordinate frame is not the robot base frame but a frame belonging to a joint, the parametrization must begin with an R_z or T_z transformation in order to integrate the joint motion.

2. The transformations from last joint to TCP and from robot base to first joint have to be parametrized in a way different from ordinary joint transformations.

3. Integration of an elastic deformation model shall be possible. This requires that the geometric locations of the joints described by the kinematic model coincide with their physical locations.

The user of a calibration package has to care mainly that the kinematic model used satisfies the completeness requirement. The user can then compute the parameters of each other complete kinematic model from the parameters obtained with the calibration package.

5.4 Elasticity Model

Concerning improvement of a robot's absolute pose accuracy, joint elasticity is the second factor (just behind the kinematic errors) to be considered.

Both joint elasticity and backlash can be modeled in the working direction of the joint. Joint elasticity can be modeled as a linear spring (Schröer, 1993). It is evident that joint elasticity and backlash can be individually identified only if varying reaction torques are produced by different robot poses. Although important for most robots, this identification is not possible for some robot axes, such as the primary axes of SCARA robots.

Links can be modeled as elastic beams with masses concentrated at the end points. Their torsion and bending can then be modeled (Schröer, 1993). The beam length needs to be included in the elasticity model in order to model the elasticity of links with prismatic joints (i.e., links of varying length).

Besides the elastic beam model, various other structural elasticity options to define the elastic deformation model should be offered to the user. Elastic plates can be specified for modeling deflections in the robot footing that frequently could be detected. A rigid element within a subsystem should be specified for a structure whose mass influences joint elasticity but not link elasticity. This flexibility will allow the definition of a model that is very close to reality.

5.5 Modeling the Actuating Elements

Integrating an actuator model into the robot model used for calibration is not necessary for practical error compensation because the parameters of gears, encoders, and motors usually are better known in advance than by indirect determination through robot calibration. But in many cases joint positions required for calibration are not directly available and thus motor encoder values must be used. To avoid a gap in the identification model it is therefore suggested that a model of the actuating elements be included.

Most actuating elements (particularly gears) have a linear transmission and coupling characteristic and therefore can simply be modelled by a matrix. But in many industrial applications robots can be found that use closed-loop mechanisms (KUKA IR663, FANUC, ABB). They offer the advantage of increased stiffness and carrying capacity. In regard to modeling, closed-loop mechanisms in robots present a challenge. They introduce complex nonlinear couplings between successive joints, and at times between actuating motors and joints, that do not exist in a purely open-loop manipulating robot. For more information on both types of closed-loop structures used with robots, see the papers mentioned below:

1. Fully parallel manipulators with multiple closed loops such as a Stewart platform or hexapod: Stoughton and Arai (1992); Wang and Masory (1993); Hollerbach and Lokhorst (1993); Wampler, Hollerbach, and Arai (1994).

2. Manipulators with closed-loop actuating elements such as four-bar linkages or slider-cranks that are used to actuate one or two rotary joints of a serial manipulator's open kinematic chain: Everett (1989); Schröer, Albright, and Lisounkin (1997). This is the most frequent way to utilize closed-loop mechanisms with industrial robots.

5.6 Numerical Procedures for Robot Calibration

Accuracy and reliability of calibration results strongly depend on the numerical methods employed for model parameter identification. This is particularly true if nonkinematic errors are modeled and identified.

For calibration purposes the robot is considered a stationary system described by the model function T. Having N joints, the input values are the joint encoder values $h \in \Re^N$ and the output values are the position and orientation of the TCP (e.g., expressed as a homogeneous matrix). Here only the TCP position is mentioned, but all results are valid as well if position and orientation are regarded.

Defining the model parameter vector as $p \in \Re^n$, the function describing the TCP position is then:

$$T: \Re^n \times \Re^N \longrightarrow \Re^3$$

$$(p, h) \longrightarrow T(p, h)$$
(15)

If the measured TCP position belonging to the robot pose defined by the encoder values h is denoted by $M(h)$, the stationary behavior of the system for k robot poses $h_i \in \Re^N$, $i = 1, \ldots, k$

$$\overline{h} :: = (h_1, \ldots, h_k)^T \in \Re^{kN}$$
(16)

is described by one function:

$$\overline{T}: \Re^n \times \Re^{kN} \longrightarrow \Re^{3k}$$

$$(p, \overline{h}) \longrightarrow \overline{T}(p, \overline{h})$$
(17)

$$\overline{T}(p, \overline{h}) :: = (T(p, h_1), \ldots, T(p, h_k))^T$$

and all measured positions are described by one $3k$-dimensional vector:

$$\overline{M}(\overline{h}) :: = (M(h_1), \ldots, M(h_k))^T \in \Re^{3k}$$
(18)

The identification itself is the computation of those model parameter values p^* that result in an optimal fit between the actual measured positions and those computed by the model. An investigation of various options for defining an objective function (Schröer, 1993) to numerically solve this optimization problem leads to the conclusion to use least squares (or modifications of the least-squares method effected by scaling and probabilistic considerations; see, e.g., Schröer et al., 1992; Everett, 1993; Schröer, 1993):

$$\min_{p \in \Re^n} \| \overline{M}(\overline{h}) - \overline{T}(p, \overline{h}) \|^2$$
(19)

Appropriate scaling of the calibration problem is necessary if reliable results are required. An overview of task variable scaling (i.e., line scaling) and parameter scaling (i.e., column scaling) techniques is presented by Hollerbach and Wampler (1996).

The least-squares problem is nonlinear, but is shown to be only mildly nonlinear (Deuflhard, 1992) if appropriate scaling techniques are applied; thus even the Gauss–Newton method can be used to solve it (Dennis and Schnabel, 1983). More powerful, damped Gauss–Newton methods (Dennis and Schnabel, 1983; Spellucci, 1983; Hollerbach and Wampler, 1996) like the Levenberg–Marquardt method can of course be applied too (Stanton and Meyer, 1993). All these require an iterative solution of the locally linearized problem:

$$\overline{M}(\overline{h}) - \overline{T}(p, \overline{h}) = D_p\overline{T}(p, \overline{h}) \cdot \Delta p \tag{20}$$

Here $D_p\overline{T}$ is the model function's Jacobian, containing the partial derivatives with respect to all model parameters p. The following notation is used for simplification purposes:

$$
\begin{aligned}
b &:: = \overline{M}(\overline{h}) - \overline{T}(p, \overline{h}) \in \mathcal{R}^m, \quad m = 3k, \\
A &:: = D_p\overline{T}(p, \overline{h}) \qquad \in \mathcal{R}^{m \times n}, \\
x &:: = \Delta p \qquad\qquad\quad \in \mathcal{R}^n, \\
r &:: = Ax - b \qquad\qquad \in \mathcal{R}^m
\end{aligned}
\tag{21}
$$

Algorithms used for solving (20) should apply any type of orthogonal matrix decomposition (because of error propagation), should have the ability to reach a solution even in the case of a rank deficient Jacobian A, and, in this case, should compute the uniquely defined minimal solution (Golub and vanLoan, 1983; Lawson and Hanson, 1974). This is accomplished, for example, by the algorithm HFTI, developed by Lawson and Hanson (1974). Also, the singular-value decomposition of matrix A can be applied to solve the least-squares problem (20), but this is more time-consuming without providing better results.

5.7   Reliability of Calibration Results

Reliability of calibration results is often investigated through evaluating the covariance matrix of the linearized problem (20). But the limitations of this approach for estimating reliability of calibration results should be carefully considered:

1.   The real model is far from being linear.
2.   The mathematical background of the covariance matrix cares only for TCP measurement errors.

But not only TCP measurement errors disturb the results. Moreover, the measured joint positions h are random variables as well, and much unknown, purely stochastic error is inherent in the mechanical system. The calibration problem is particularly characterized by the fact that distribution functions of these errors and their parameters are definitely unknown, but the uncertainty introduced by these errors is much more important than (usually well-known) TCP measurement error.

This means mathematically not only that TCP measurements included in vector b introduce errors, but also that matrix A in (21) is erroneous. This results in a disturbed linear least-squares problem. Its solution $\tilde{x}$:

$$(A + E)\tilde{x} = b + \Delta b \tag{22}$$

is compared to the solution x. Assuming $\|E\| \cdot \|A^+\| < 1$ (A^+ is referred to as the *pseudoinverse* of A), the following perturbation bounds can be proved for a full-rank Jacobian A (Lawson and Hanson, 1974):

$$\frac{\|\tilde{x} - x\|}{\|x\|} \le \frac{\kappa(A)}{1 - \|E\| \cdot \|A^+\|} \cdot \left\{ \left(1 + \kappa(A) \frac{\|r\|}{\|A\| \cdot \|x\|}\right) \frac{\|E\|}{\|A\|} + \frac{\|\Delta b\|}{\|b\|} \cdot \frac{\|b\|}{\|A\| \cdot \|x\|} \right\} \tag{23}$$

5.7.1   Rank Deficiency, Condition Number, Parameter Dependency, Pose Selection

The above inequality stresses the importance of the condition number $\kappa(A)$ of the matrix A, which is a quadratic amplification factor of the residuum norm. Rank deficiency and bad (i.e., high) condition number of the Jacobian A are caused by parameter dependency. If a minimal, complete, and model continuous kinematic model is used (see Section 5.3), rank deficiency or bad numerical condition can only be caused by nonkinematic model components (elasticity or gear model). Therefore parameter dependency has to be investigated and eliminated if reliable calibration results are required. Mathematical background and strategies for detecting and eliminating parameter dependencies are presented, for example, by Khalil and Kleinfinger (1991), Schröer et al. (1992), and Stanton and

Meyer (1993). They frequently evaluate the singular-value decomposition of matrix A. A robot calibration package should therefore always inform its user on rank deficiency and on the condition number $\kappa(A)$ of the model Jacobian.

Also, the selection of the robot pose set $h_i \in \Re^N$, $i = 1, \ldots, k$ measured for calibration affects the condition number $\kappa(A)$, i.e., the reliability of calibration results. A suitable pose selection strategy is of the utmost importance for the reliability of calibration results. Obviously, only through exploiting a robot's motion capability as far as possible can information for calibration be obtained. Restrictions often imposed by measurement system requirements (restricted measurement system workspace, restrictions in admissible TCP orientations, etc.) may cause pose sets of bad quality. Therefore it is highly recommended that an observability measure (i.e., a numerical value exhibiting the quality of a pose set) be computed for each pose set used for calibration and compared to observability measure values from pose sets that were generated without visibility or workspace restrictions. Observability measures are usually computed from the singular value decomposition of the model Jacobian A (see, e.g., Driels and Pathre, 1990; Borm and Menq, 1991; Nahvi, Hollerbach, and Hayward, 1994).

5.7.2 Checking Reliability

Many recent contributions in the field of robot calibration attack the problem of checking for purely stochastic error inherent in the mechanical system, thus checking reliability of calibration results. The first and simplest possibility is to check the residuals R at the end of the optimization process:

$$R = \overline{M(h)} - \overline{T}(p^*, \overline{h}) \tag{24}$$

If for one robot $\|R\|$ is significantly higher than for the majority of robots of the same type, this is a hint that there is a problem. This is not a very powerful tool, but a first indicator in many cases. Moreover, the residuals R after the optimization process should always be analyzed by a robot calibration package because measurement outliers can reliably be detected in this way and then be excluded from the numerical evaluation.

A more powerful method to check *a posteriori* the amount of unmodeled or purely stochastic error is also available (Schröer et al., 1992; Schröer, 1993). It does not require any *a priori* information on the error distribution and is based on a variation of the robot model that is used for calibration. The basic idea of this method is to compare different values obtained for the same model parameter when using slightly different models for parameter identification. This was demonstrated to be a reliable method to determine confidence intervals for all model parameters. From an analysis of these results, hints could be obtained on sources of unmodeled deterministic or purely stochastic error inherent in the mechanical system.

The implicit-loop method for numerically solving the calibration problem, which was developed by Wampler, Hollerbach, and Arai (1994) is another powerful tool for checking unmodeled or purely stochastic error. In this method (which is an interesting subject for future research in calibration) all measurements, whether from joint angle sensors h or from TCP measurements $M(h)$, are grouped together into a vector $Y = (h, M(h))$, and covariance matrices C_Y of Y and C_p of the model parameters p are assumed. Here they include the scaling information. Then the maximum likelihood estimate of Y and p is the minimizer of:

$$\chi^2 = \Delta Y^T \cdot C_Y^{-1} \cdot \Delta Y + \Delta p^T \cdot C_p^{-1} \cdot \Delta p \tag{25}$$

subject to the equality constraints defined by the model function itself:

$$\overline{M(h)} - \overline{T}(p, \overline{h}) = 0 \tag{26}$$

In this case the calibration problem numerically results in a quadratic minimization problem with nonlinear equality constraints which can be solved, for example, using one of the methods presented in Spellucci (1993, Chapter 3.3).

Another approach worth being a future subject for research originates from numerical mathematics: *generalized least squares* (Watson, 1985). In common with the implicit-loop method, it allows information to be obtained not only on the parameters of the calibration model, but also on measurement errors of joint positions and TCP poses.

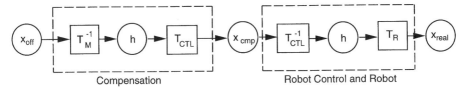

Figure 3   Off-line error compensation.

5.8 Error Compensation

Most available robot controls cannot yet make direct use of calibration data. A higher computational power is required if complete kinematic models (see Section 5.3), instead of the simple models with explicitly invertible kinematic equations, are used in the control algorithms. The same is true if elasticity is to be compensated for, since on-line computation of reaction forces and torques is then required by robot control. However, robot manufacturers plan to integrate this capability in their next control generation.

In order for calibration results for improving absolute accuracy of existing robots to be used, off-line error compensation methods can be applied.

Off-line compensation methods that use calibration results allow improvement in absolute pose accuracy with no changes in the robot control algorithms and parameters. Using the powerful robot model T_M from the calibration procedure and the model T_{CTL} on which the control algorithms are based, compensation poses are computed and transmitted to the robot control instead of the target poses taken from the off-line generated program (see Figure 3; x_{off}, off-line generated target pose; T_M, identified robot model; h, joint encoder values; T_{CTL}, controller model; x_{cmp}, compensated target pose; T_R, function representing real robot; x_{real}, real attained pose). A similar problem arises if existing robot application programs (perhaps generated by teach-in programming) have to be transferred to another robot and manual corrections (reteaching) have to be avoided. To solve this problem both robots have to be calibrated.

To describe the problem formally (see Figure 4): The robot represented by function T_{R1} with control T_{CTL1} executes the programmed task x_{prg} and thus attains pose x_{real} with joint encoder values h_1. With calibration results T_{M1} and T_{M2} for the source robot T_{R1} and the destination robot T_{R2}, the program is to be transferred to T_{R2}. The compensation problem can be solved as shown in Figure 5 (h_1, h_2, joint encoder values; x_{real}, attained target pose; T_{M1}, T_{M2}, identified robot models; T_{CTL2}, controller model; x_{cmp}, compensated target pose). The computed compensation poses x_{cmp} are then input to the destination system with robot control T_{CTL2} and robot T_{R2} as shown in Figure 3, and the robot pose x_{real} of the source system will also be attained by the destination system.

6 CONCLUSIONS AND TRENDS

Practical prerequisites for applying calibration techniques are availability of required measurement devices (i.e., their integration in the processes of production and production preparation) and a concept for handling measurement and calibration results. Both prerequisites are at the same time the most important obstacles to broader industrial use of calibration techniques.

The most important reason for this situation is that an investment in measurement technology and in organizational changes in the production process concerning the handling of measurement systems and the handling of data from measurement and calibration

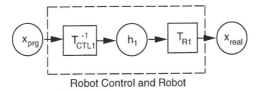

Figure 4   Robot performing a programmed task.

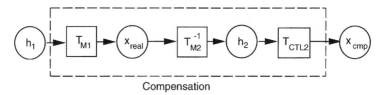

Compensation

Figure 5 Compensation for program transfer.

is initially required. For this to occur, decisions at the management level of a company must be made. It is evident that:

- Through ongoing changes a growing number of tasks in the field of planning, test, simulation, and preparation of the production process move to the computer.
- A growing need thus exists for closing the gap between computer model and reality through applying measurement and calibration techniques.
- Flexible, robot-based manufacturing processes require their process capability (according to section 4.9 of ISO 9001) to be based on tolerances that are met in the entire robot workspace.

Nevertheless, company management often avoids making the required decisions if the technical expert is not capable of guaranteeing return of investment within a quite short period. Therefore only a very small number of the large robot users have already decided to introduce measurement and calibration.

Among the industrial users of calibration a discussion on the right place to allocate calibration and related error compensation is still going on: Should error compensation be integrated into off-line programming (OLP) systems or into robot control?

For workcell calibration there are good arguments for feeding results back into the OLP system: this will allow for better and more realistic programming and optimization of robot tasks. There are also good arguments not to use robot-tool calibration results in OLP systems: the tool definition changes frequently, and these changes should be dealt with by the robot control only.

If robot calibration results are used in the OLP system, the off-line error compensation (see Section 5.8) is effective only in those robot poses that are explicitly part of the robot application program. The interpolation algorithms inside robot control will then continue using the less exact robot model. But for most applications this will be sufficient.

On the other side, integration of error compensation into robot control requires higher computing power (particularly if elasticity has to be compensated for). Large robot users in the automotive industry prefer integration into robot control and push robot manufacturers to provide robots with high absolute accuracy by doing all error compensation tasks internally.

To ease the process of implementing industrial use of measurement and calibration, the providers of measurement and calibration technology have to take account of industry's needs, including:

- Industry needs complete solutions for robot, (varying) tools, and workcell. Complete solutions contain procedures and measurement systems.
- Industry needs graduated calibration capabilities for accuracy check, partial recalibration, full calibration, and calibration for quality control in robot production. For partial recalibration and accuracy check, simple and inexpensive measurement devices should be applied (like light beams or camera systems) that can remain permanently installed inside the robot workcell.

A further measure that should be taken to ease introduction of calibration in industry is standardization of models and procedures.

REFERENCES

Baker, D. R. 1990. "Some Topological Problems in Robotics." *The Mathematical Intelligencer* **12**(1), 66–76.

Baker, D. R., and C. W. Wampler. 1988. "On the Inverse Kinematics of Redundant Manipulators." *International Journal of Robotics Research* **7**(2), 3–21.

Borm, J. H, and C. H. Menq. 1991. "Determination of Optimal Measurement Configurations for Robot Calibration Based on Observability Measure." *International Journal of Robotics Research* **10**(1), 51–63.

Dennis, J. E., and R. B. Schnabel. 1983. *Numerical Methods for Unconstrained Optimization and Nonlinear Equations.* Englewood Cliffs: Prentice-Hall.

Deuflhard, P. 1992. *Newton Techniques for Highly Nonlinear Problems—Theory, Algorithm, Codes.* New York: Academic Press.

Driels, M. R. 1993. "Using Passive End-Point Motion Constraints to Calibrate Robot Manipulators." *Journal of Dynamic Systems Measurement Control* **115**, 560–565.

Driels, M. R., and U. S. Pathre. 1990. "Significance of Oberservation Strategy on the Design of Robot Calibration Experiments." *Journal of Robotic Systems* **7**(2), 197–223.

Everett, L. J. 1989. "Forward Calibration of Closed-Loop Jointed Manipulators." *International Journal of Robotics Research* **8**(4), 85–91.

———. 1993. "Research Topics in Robot Calibration." In *Robot Calibration.* Ed. R. Bernhardt and S. Albright. London: Chapman & Hall. 19–36.

Everett, L. J., and T.-W. Hsu. 1988. "The Theory of Kinematic Parameter Identification for Industrial Robots." *Journal of Dynamic Systems, Measurement and Control* **110**, 96–100.

Everett, L. J., M. Driels, and B. W. Mooring. 1987. "Kinematic Modelling for Robot Calibration." In *Proceedings of 1987 IEEE International Conference on Robotics and Automation.* 183–189.

Golikov, N. Y., V. I. Kavinov, and A. E. Klepov. 1986. "Parameter Identification Algorithm of Kinematic Schemes of Robots." *Soviet Journal of Computer Systems Science* **24**(1), 117–120.

Golub, G. H., and C. F. vanLoan. 1983. *Matrix Computations.* Oxford: North Oxford Academic.

Gottlieb, D. H. 1986. "Robots and Topology." In *Proceedings of 1986 IEEE International Conference on Robotics and Automation.* 1689–1691.

Hayati, S. A., and M. Mirmirani. 1985. "Improving the Absolute Positioning Accuracy of Robot Manipulators." *Journal of Robotic Systems* **2**(4), 397–413.

Hollerbach, J. M., and D. M. Lokhorst. 1993. "Closed-Loop Kinematic Calibration of the RSI 6-DOF Hand Controller." In *Proceedings of 1993 IEEE International Conference on Robotics and Automation.* Vol. 2. 142–148.

Hollerbach, J. M., and C. W. Wampler. 1996. "A Taxonomy for Robot Kinematic Calibration Methods." *International Journal of Robotics Research* **15**(6), 573–591.

Ishii, M., et al. 1987. "A New Approach to Improve Absolute Positioning Accuracy of Robot Manipulators." *Journal of Robotic Systems* **4**(1), 1045–1056.

ISO/TC184/SC2/WG2 N291 Rev. 2. 1994. *Manipulating Industrial Robots. Overview of Test Equipment and Metrology Methods for Robot Performance Evaluation in Accordance with ISO 9283* (DTR 13309). November.

Judd, R. P., and A. B. Knasinski. 1990. "A Technique to Calibrate Industrial Robots with Experimental Verification." *IEEE Transactions on Robotics and Automation* **6**(1), 20–30.

Khalil, W., and J. F. Kleinfinger. 1986. "A New Geometric Notation for Open and Closed Loop Robots." In *Proceedings of 1986 IEEE International Conference on Robotics and Automation.* 1174–1180.

Khalil, W., M. Gautier, and C. Enguehard. 1991. "Identifiable Parameters and Optimum Configuration for Robots Calibration." *Robotica* **9**, 63–70.

Lawson, C. L., and R. J. Hanson. 1974. *Solving Least Squares Problems.* Englewood Cliffs: Prentice-Hall.

Nahvi, A., J. M. Hollerbach, and V. Hayward. 1994. "Closed-Loop Kinematic Calibration of a Parallel-Drive Shoulder Joint." In *Proceedings of IEEE International Conference on Robotics and Automation,* San Diego, May 8–13. 407–412.

Press, W. H., et al. 1994. *Numerical Recipes in C.* Cambridge: Cambridge University Press.

RoboCal—How to Calibrate a Robot with IPK Calibration Package. 1995. Berlin: Fraunhofer-IPK.

Schröer, Klaus. 1993. *Identifikation von Kalibrationsparametern kinematischer Ketten.* Produktionstechnik—Berlin, Bd. 126. Munich: Hanser, 1993.

———. 1994. *Calibration Applied to Quality Control and Maintenance in Robot Production—CAR 5220—Report on Project Results.* Berlin: Fraunhofer-IPK.

Schröer, K., S. L. Albright, and M. Grethlein. 1997. "Complete, Minimal and Model-Continuous Kinematic Models for Robot Calibration." *Robotics and Computer-Integrated Manufacturing* **13**(1), 73–85.

Schröer, K., S. L. Albright, and A. Lisounkin. 1997. "Modeling Closed-Loop Mechanisms in Robots for Purposes of Calibration." *IEEE Transactions on Robotics and Automation* **13**(2), 218–229.

Schröer, K., et al. 1992. "Ensuring Solvability and Analyzing Results of the Nonlinear Robot Calibration Problem." In *Proceedings of 2nd International Symposium on Measurement and Control in Robotics (ISMCR '92).* Tsukuba Science City, Japan. 851–858.

Spellucci, P. 1993. *Numerische Verfahren der nichtlinearen Optimierung.* Basel: Birkhäuser.

Stanton, D., and J. R. R. Meyer. 1993. "Robot Calibration within CIM-SEARCH/I." *Robot Calibration.* Ed. R. Bernhardt and S. Albright. London: Chapman & Hall. 57–76.

Stone, H. W., and A. C. Sanderson. 1987. "A Prototype Arm Signature Identification System." In *Proceedings of 1987 IEEE International Conference on Robotics and Automation.* 175–182.

Stoughton, R. S., and T. Arai. 1992. "A Modified Stewart Platform Manipulator with Improved Dexterity." In *Proceedings of 5th Annual Conference on Recent Advances in Robotics,* Florida Atlantic University, June 11–12. 165–178.

Sugimoto, K., and T. Okada. 1985. "Compensation of Positioning Errors Caused by Geometric Deviations in Robot System." In *Robotics Research: The Second International Symposium.* Ed. H. Hanafusa and H. Inoue: Cambridge: MIT Press. 231–236.

Umeyama, S. 1991. "Least-Squares Estimation of Transformation Parameters Between Two Point Patterns." *IEEE Transactions on Pattern Analysis and Machine Intelligence* **13**(4), 376–380.

Vaishnav, R. N., and E. B. Magrab. 1987. "A General Procedure to Evaluate Robot Positioning Errors." *International Journal of Robotics Research* **6**(1), 59–74.

Veitschegger, W. K., and C. H. Wu. 1986. "Robot Accuracy Analysis Based on Kinematics." *IEEE Journal of Robotics and Automation* **2**(3), 171–179.

Wampler, C. W., J. M. Hollerbach, and T. Arai. 1994. "An Implicit Loop Method for Kinematic Calibration and Its Application to Closed-Chain Mechanisms." GM Publication R&D-8188.

Wang, J., and O. Masory. 1993. "On the Accuracy of a Stewart Platform—Part I. The Effect of Manufacturing Tolerances." In *Proceedings of 1993 IEEE International Conference on Robotics and Automation.* Vol. 1. 114–120.

Watson, G. A. "Towards More Robust Methods of Data Fitting." *Bulletin of Institute of Mathematics and its Applications* **21**, 187–192.

Wu, C. H. 1984. "A Kinematic CAD Tool for the Design and Control of a Robot Manipulator." *International Journal of Robotics Research* **3**(1), 58–67.

Zhuang, H., Z. S. Roth, and F. Hamano. 1990. "A Complete and Parametrically Continuous Kinematic Model for Robot Manipulators." In *Proceedings of 1990 IEEE International Conference on Robotics and Automation.* 92–97.

Ziegert, J., and P. Datseris. 1990. "Robot Calibration Using Local Pose Measurements." *International Journal of Robotics and Automation* **5**(2), 68–76.

ADDITIONAL READING

Bernhardt, R., and S. Albright. *Robot Calibration.* London: Chapman & Hall, 1993.

Hollerbach, J. M., and C. W. Wampler. "A Taxonomy for Robot Kinematic Calibration Methods." *International Journal of Robotics Research* **15**(6), 573–591, 1996.

McCarthy, J. M. *Introduction to Theoretical Kinematics.* Cambridge: MIT Press, 1990.

Mooring, B. W., Z. S. Roth, and M. R. Driels. *Fundamentals of Manipulator Calibration.* New York: Wiley Interscience, 1991.

Schröer, K. *Handbook on Robot Performance Testing and Calibration.* Stuttgart: Fraunhofer IRB Verlag, 1998.

Stone, H. W. *Kinematic Modeling, Identification, and Control of Robotic Manipulators.* Boston: Kluwer Academic, 1987.

CHAPTER 40
ROBOTICS, FMS, AND CIM

H. Van Brussel
P. Valckenaers
Catholic University of Leuven (KUL)
Leuven, Belgium

1 INTRODUCTION

Manufacturing organizations increasingly face unpredictable demands for customized and complex products without defects that they have to produce in small quantities, under time-based competition, and at low cost. Moreover, technological innovations and developments constantly create new opportunities in this competitive environment.

Until very recently, CIM (computer-integrated manufacturing) was the magic formula for successful flexible automation of the world's factories. CIM aims at the comprehensive integration, by means of computers, of all stages of the manufacturing cycle. CIM is an applied enterprise-wide philosophy underlying the automation of the information flow, from the product order, via design, production, and delivery, up to maintenance and quality control.

An FMS (flexible manufacturing system) is only a part of the CIM concept. An FMS is a programmable production system consisting of a set of automatic workstations mutually connected by material-handling systems (transport systems) and governed by a—mostly hierarchical—control system. FMS can be considered the "hands" of the CIM system; it enables the realization of the CIM concept on the shop floor. As such, FMS is the least automated bottleneck between the highly automated information flows and the very flexible manufacturing processes.

Automation as such is not so difficult, but flexible automation certainly is. Flexibility of our manufacturing systems is the key issue in modern industry. In sharp contrast to its importance, flexibility is still an ill-defined concept, both in general and within manufacturing systems. This failure to understand the precise nature of flexibility makes it difficult to identify flexibility, to assess or quantify its value within the manufacturing environment, and to preserve useful flexibility without overdoing it. Extremely flexible but economically justified manufacturing systems sometimes look like the Holy Grail: very tempting, but impossible to find. That is perhaps why some companies are now retreating from offering a wide range of variants and are simplifying their product range.

As far as flexibility is concerned, transport forms one of the most notorious bottlenecks in typical FMS. This applies to both interworkstation transport and local transport at workstations (loading, unloading of parts and tools, assembly operations). High expectations were created by researchers: the flexibility problem in transport functions would be solved by the development of industrial robots, AGVs, and the like. The outcome of all those research efforts has been rather deceiving, however. Difficult interface problems, insurmountable so far, have prevented the smooth introduction of robotics into FMS. The most notorious problem lies in the interface of the robot with the parts to be handled. Universal grippers are still inaccessible, notwithstanding the considerable progress in research on artificial hands, grasp planning, etc. The uncertainty in the robot environment (inaccurately positioned parts) is another cause of problems, reducing the overall system flexibility. External sensors, such as vision, force, and tactile sensors, are used to resolve the problem, but the developed solutions are not generic so far. The way the transport flexibility problem is solved in modern FMS is by the use of standardized pallet systems. The parts are mounted manually on the pallets, eventually using standardized modular clamping tools. The pallets themselves have standardized mechanical interfaces so that

Handbook of Industrial Robotics, Second Edition, Edited by Shimon Y. Nof
ISBN 0-471-17783-0 © 1999 John Wiley & Sons, Inc.

they can be freely interchanged between different workstations by the use of AGVs or conveyor systems.

Besides the mechanical interface problems stated above, *information interfaces* are even more important, and often the only available alternative, for increasing the flexibility of FMSs. To execute a manufacturing task, the different components of an FMS must be able to exchange information smoothly and swiftly. This requires interoperable data interfaces and communication protocols. In this context, bandwidth and real-time response issues must be addressed almost on an equal level with data/information exchange.

In this chapter some ongoing developments in the quest for real flexibility in FMS, with special emphasis on—though the discussion is not restricted to—robots, will be explained. Particularly, the information technology aspects will be addressed, for the simple reason that information technology is the most rewarding and presently the most viable approach. The developed concepts constitute a radical departure from the typical CIM developments in industry. The reader is invited to compare these ideas with the mainstream developments encountered in present-day industry, and we hope that the potential of the proposed solutions will become clear.

2 DESIGN PRINCIPLES FOR FMS

In our quest for introducing genuine flexibility in manufacturing systems, we must identify those concepts and actions that really contribute to the flexibility of the system. In this section some fundamental design principles are outlined and situated in the context of designing FMSs. These principles are complementary to the basic design axioms as put forward, e.g., by Suh (1988). They bring the temporal dimension into those axioms.

The emergence and survival of large, complex systems such as flexible production systems requires the availability of suitable stable subsystems that are able to withstand external disturbances. For the developers of large complex systems this means that they need modules of appropriate size that permit the rapid development and adaptation of the overall system to the environment. In the domain of manufacturing systems subsystems exist that are sufficiently large but unsuitable for integration. Likewise, components exist that are suitable for integration but too small for the development of large complex systems in demanding environments. Typical examples of the former category are the well-known "islands of automation," whereas mass production systems are typically built with representatives of the latter category such as motors, pallets, and conveyors. Apparently designers fail to preserve the suitability for integration (or flexibility) during the development of members of large subsystems.

The emergence and survival of a CIM/E subsystem depend on the subsystem's ability to fit into a large number of manufacturing systems, at different places within the manufacturing system, and for a sufficiently long time during which neighboring components will be replaced. This goal has proved to be unattainable until today.

The classical top-down development paradigm needs to be refined for the development of large complex systems. It is still useful to identify the functionality that is required from subsystems. However, subsystem developers can no longer rely heavily on the specific context definitions provided by the approach. A top-down approach often is unnecessary to identify useful functionality, e.g., use of better and cheaper materials, actuators, and sensors. Therefore, bottom-up subsystem development is possible, useful, and probably necessary for the creation of complex artefacts. The two approaches complement each other. Top-down development corresponds to an evolutionary approach when designers build on the architectures of existing systems. Bottom-up development creates subsolutions that reflect new opportunities when designers propagate the possibilities offered by technological advances. Top-down development is appropriate in the final development phases, to develop end-user solutions in a short time and with little uncertainty about the outcome. Bottom-up development creates new building blocks for these end-user systems; it creates new opportunities. Without bottom-up development radically new systems cannot emerge.

In Van Brussel (1994) a design theory for flexible systems is developed. The most important design principles resulting from the theory are:

DP1.   Consider design decisions harmful until the opposite is proved.
DP2.   Look for design decisions that maintain flexibility.
DP3.   Avoid the buildup of inertia for any unsafe design decisions.

DP1 states that although it is necessary to prepare decisions, commitment should be avoided or postponed whenever possible. It is dangerous to commit to arbitrary choices that simply satisfy current requirements, since unpredictable integration requirements will emerge later and the arbitrary decisions probably will be inconsistent with these emerging requirements. The COMPLAN project (Kruth, 1996) provides an excellent example of a successful application of this design principle. Briefly, the COMPLAN process planning system supports multiple alternatives in its process plans, thus avoiding conflicts over manufacturing resources during production later on.

DP2 states that every design decision that brings a developer closer to a final solution without reducing flexibility significantly is a safe decision. Commitment to such decisions is unlikely to cause integration problems afterward, since they exclusively impose constraints that are already present in the environment. Therefore, developers must take the possible safe decisions in the beginning of the design sequence. Note that a system architecture design originates from early design decisions only when seen from the viewpoint of actual installations.

An example is the structure of most present-day trajectory generation systems in machine and robot controls. This structure, developed in the early days of robotics, is such that the trajectories are generated on-line. The design rationale behind this went as follows. At first, in laboratories, designers generated trajectories off-line, which allowed for more complex computations and for customized trajectory generation—more than one trajectory generator could produce trajectories for a single manipulator. However, the amount of computer memory that was required to store the trajectories exceeded by far the capacity of the typical 16-bit minicomputers that were in use for industrial controllers at the time; even today the memory requirements would be considered significant. In view of this situation, the developers of industrial robot systems made an unsafe design decision: they developed systems supporting exclusively on-line trajectory generation.

This is a classical example of how unsafe design choices sneak into a system design together with a perfectly sound/safe design decision. The safe decision was to use a compressed trajectory representation (like zipped computer files) that would be decompressed on-line to feed the trajectory control subsystem with the proper data. The unsafe design decisions resided in the representation of the compressed trajectories, which were simply the representations used for the early experiments. These compressed representations were based on geometric primitives (straight lines in joint space, straight lines in Cartesian space, and circular segments in Cartesian space for the more sophisticated systems), typically augmented by speed, acceleration, and deceleration settings. This design decision enormously reduced the set of available trajectories theoretically executable by the robot manipulator. If these trajectory specifications are interpreted in a strict sense, the set of trajectories that can be specified and the set of trajectories that can be executed intersect only when manipulator dynamics can be ignored (low velocity and acceleration). Industrial robots need to go fast.

A safe design decision tries to cover itself against later conflicts by hiding behind existing long-living constraints. In this example there is an abundance of such constraints: the robot manipulator is a mechanical structure with maximum velocity, acceleration, deceleration, and sensor resolution. For controlled movement the frequency of the first eigenmode imposes additional constraints. All these constraints provide ample opportunity to compress the trajectory representation such that all those trajectories that are theoretically executable by the robot manipulator are preserved. These compressed trajectories can be provided by whatever generator suits the problem/task. Moreover, the compression technique can rely on the required accuracy for the robot task. This latter point corresponds to a delayed decision (DP1).

To achieve maximum flexibility the trajectory decompression subsystem has to be able to support on-line as well as off-line generation. This simply means that the system has to support a mode with compression ratio 1 (no compression) and the data coming from another computer process. In other words, just pass the data to the low-level control system (trivial). This is necessary to support sensor-based trajectory generation (e.g., compliant motion).

DP3 states that earlier unsafe design choices are not on par with stable constraints in the environment; later decisions cannot be covered by an earlier (unsafe) design choice (covering by a safe choice implies that both choices use a common stable constraint in the environment). Violating this guideline will increase the inertia of the unsafe decision, making reversal more costly at later stages.

DP3 must also be seen in the perspective of the large differences in the reproduction cost of flexible, and therefore complex, subsystems in the different technological domains (brainware, software, firmware, hardware, and ironware). It is typically extremely expensive to safeguard flexibility in ironware. On the other hand, it is often trivial to develop software that is suitable for integration with an entire family of such ironware, where a data file defines the actual configuration. It is tempting to simplify the software development when the ironware configuration is known at design time. However, the small savings will prove very costly when, afterward, a design flaw is discovered in the ironware or when the ironware needs upgrading. Moreover, it will prevent reuse of the software for other systems, which implies that the development will not benefit from economics of scale and go up its learning curve at a slow pace.

3 INTRODUCING FLEXIBILITY IN FMS

It has been shown above that flexibility is introduced or killed in the very first stages of the design cycle. This is a way of saying that 80% of the cost of a system is allocated in the first 20% of the design cycle—another proof that top-down design is not the appropriate way to go when flexibility is of prime concern.

In this section the concept of device drivers is put forward as a solution to solve the flexibility problem when combining off-the-shelf automation components into flexible manufacturing systems. Further, the case study of a flexible odd-component inserter, successfully developed by the authors and implemented in a factory environment for many years now, is described. It shows in a convincing way the power and wide applicability of the above-defined concepts.

3.1 A New Use of an Old Concept: The Device Driver

More robot and machine tool programming languages exist than robot and machine manufacturers (Hocken, 1986). Every manufacturer has its own programming language(s). Moreover, what appears to be a large choice of robot programming languages is actually the reverse. The variety of requirements needed for, e.g., assembly systems denies the users the opportunity to choose robots with the desired programming language. Therefore a methodology is needed to solve the software development and maintenance problems. Traditionally, *postprocessing* is proposed: programs are written in a single robot-independent language, like APT, and automatically translated into the different languages of the robots or machines in the production system. The disadvantage of this approach is that it is technically unfeasible for both on-line and off-line (unless performance/speed is not critical) robot programming.

We propose the concept of device drivers as a flexible solution to the problem of programming heterogeneous sets of industrial robots (Valckenaers, 1993). Robot manipulators can be considered as input/output devices of a computer, like plotters, printers, or computer displays. The device driver is a set of subroutines that know how to drive the device (robot). Using these subroutines eases the programming of the device, without complete knowledge of the device hardware being required.

3.1.1 System Configuration for Device Driver Implementation

The hardware configuration consists of an off-line programmable robot system (robot and its controller) and an external computer platform (e.g., a transputer system or a PC). A communication link (e.g., a serial line) between the computer platform and the robot controller completes this hardware configuration.

The construction of a device driver consists of three tasks:

- Establishing a communication procedure
- Identifying the device functions
- Developing two small communicating programs: one in the robot controller, another in the computer platform

3.1.2 The Communication Procedure

This is partly a hardware task. The procedure must support the communication of rudimentary data types, like characters, integers, and floating point numbers. This task is the most difficult and time-consuming of the three device driver construction tasks. The uncertainty about the final outcome and the effort that will be required is also the largest. The task involves setting up the robot system and writing the first programs on the robot controller in its native language. Note that this task is necessary for every nontrivial application, even when no device driver is used.

3.1.3 Identification of the Device Functions

All the capabilities of the robot that cannot be executed directly on the external computer platform must be available as functions through the device driver. We can distinguish three classes of functions:

- Observation capabilities (e.g., asking the position or velocity of the robot)
- Action capabilities (e.g., motion commands, tool manipulations)
- Mode control functions (e.g., changing the velocity and the acceleration settings, enabling/disabling of continuous path mode)

Furthermore, all the capabilities that exist in the robot controller and that can be performed by the external computer platform (e.g., adding two numbers, editing a data file) must not be duplicated in the device driver. This avoids problems caused by the difference between the implementation of these functions on the external computer platform and on the robot controller (e.g., numerical accuracy).

The device driver provides the functions in their rudimentary form. Advanced commands (i.e., the elementary operations) must be developed with these functions in the application programs on the external computer platform.

3.1.4 Construction of Two Small Programs

The final task consists of the construction of two small communicating programs. The first one, in the robot controller, is basically an endless loop that reads a code from the communication link. This code selects a robot function. When additional data (e.g., coordinates of the target position for a motion command) are required, these data are read from the communication link. Then the program orders the robot to execute the task.

The second program, on the application computer, is a subroutine library that offers the robot functions to the application programmer as a set of basic procedures. Each procedure sends to the small program on the robot controller, via the communication link, the suitable code to activate the robot and the additional data. Figure 1 illustrates the device driver concept.

3.1.5 Remarks on the Device Driver Approach

The most important advantage of the device driver approach is the uncoupling of the robot system selection, which can therefore focus on manipulator capabilities (precision, speed, payload, etc.) and the application computer platform selection, which can focus on computing performance and support (choice of the hardware, of the language, of the environment, etc.)

When several kinds of equipment, with several languages, are used in a cell, the development of device drivers allows the entire cell to be programmed in the same application language. Moreover, when the computer platform is able to communicate with several external devices and run several processes at the same time, like transputer systems, the whole control system can be centralized on this platform. This solution that we use to control our assembly cell is very powerful for communicating processes, data management, software development, maintenance problems, and so on.

Nevertheless, the design of such an integrated control system must be done with method and homogeneity to obtain an effective integrated robust and powerful application. So, after the device driver development, the next step in the method is the creation of the elementary operations available in the cell.

Also, note that the existing robot systems are analogous to alphanumeric terminals. When robot systems provide more bandwidth between external computers and the manipulator control system, they become analogous to the graphic display subsystems in computer workstations. Such a change will have a profound effect on the nature of the commands provided by a device driver.

3.2 An Odd-Component Inserter Illustration

3.2.1 Introduction

The device driver methodology discussed above is used for an industrial application: robotic assembly of odd components on printed circuit boards (Figure 2). The automatic

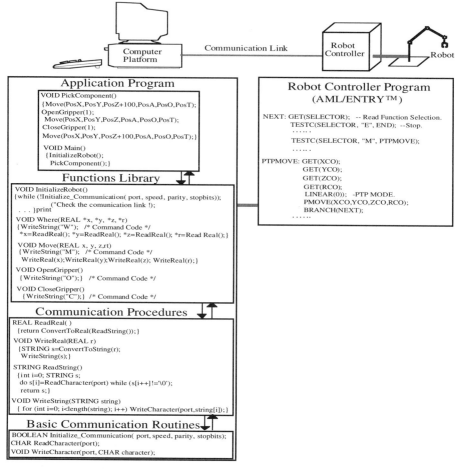

Figure 1   Illustration of the device driver concept.

assembly of standard electronic components by dedicated machines has been common practice for many years. However, this solution is not available for the nonstandard or odd components such as relays and hybrid circuits. Moreover, dedicated machines are only economical for standard components belonging to types that are assembled in sufficient quantities within the production unit. There is an undeniable need for automated assembly systems that can handle a large variety of components that cannot be assembled by dedicated machines for technical or economical reasons. The remainder of this subsection reports on a successful development that addresses this problem. Five robot stations are now operational at the plant of ALCATEL in Geel, Belgium, the industrial partner.

The case study concerns a somewhat soft target area in the domain of integrated automation. First, it is a so-called 2D application. Electronic components are placed next to each other on a single surface. Most of the existing automated and integrated batch-of-one systems belong to this category. Second, there exist dedicated process planning systems for this type of product (computer-aided electronic layout and routing systems). These generate the necessary process plans except for the final *process grounding* phase. Finally, the application is restricted to the automation of a stand-alone assembly station.

3.2.2 Accuracy

The application targets the automatic assembly of odd components on printed circuit boards, which are positioned in the robot station by means of pins fitting into reference

Figure 2   Picture of the odd-component inserter (courtesy ALCATEL).

holes on the boards. This results in an accuracy of about 0.02 mm in the plane of the board. The positioning margins for the component leads to fit in their mounting holes in the printed circuit board are about 0.1 mm with respect to the reference hole positions. This represents severe demands upon the robot and the component accuracy.

Dedicated assembly machines use mechanical methods to position the leads of the components accurately within the gripper that places the component on the printed circuit board. In other words, they grasp the components by their leads with mechanisms that correct small errors. Odd-component assembly systems cannot use this method because each component type requires a different mechanism; odd-component types lack the production quantities to recover the investment needed for the development of the mechanisms. Other factors against the mechanical solution are the development time for these mechanisms, the increased complexity of an assembly system that has to support a multitude of mechanisms, and the unsuitability of many odd components for mechanical solutions. The problem of removing the position uncertainty of the component leads is solved by the laser-based vision system described below.

The feeder systems exhibit considerable diversity. The development setup includes component trays and gravity-based feeders. The former present the components in an array of positions within a horizontal plane. The latter present the components at a single position under a fixed angle around a horizontal axis. An air-powered tilting mechanism between the gripper and the robot manipulator provides the additional binary degree of freedom needed to pick up components under this fixed angle. However, the software implementation is flexible with respect to the feeder type. In the actual production setup at least one additional type of feeder is already added to the system. This feeder type presents the components at a fixed location and requires a signal indicating that the robot gripper is holding the component so it can release the component.

A transportation belt moves the printed circuit boards in and out of the station. When an incoming printed circuit board runs against a stop, the positioning mechanism forces pins through the reference holes in the printed circuit board. Below the printed circuit board is an air-powered clinching mechanism. The mechanism moves a human-replaceable array of pins up and down. When the robot holds a component with its leads through the holes, this mechanism is activated to bend enough leads to secure the component. As long as there are no conflicts between the arrays needed by the different printed

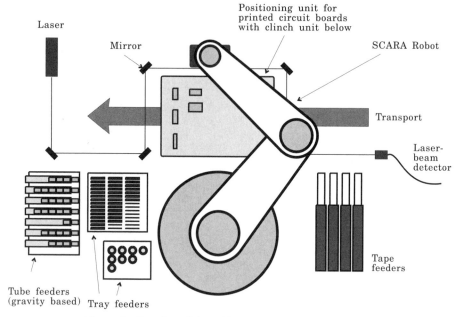

Figure 3 Overview of the odd-component assembly station.

circuit board types, a single array will suffice. Otherwise a human operator has to switch the arrays. Similar systems use an *xy*-table for this task to automate it completely.

The gripper has two parallel moving fingers that can be position- and force-controlled. The fingers are exchangeable, without human assistance, to accommodate a wide range of components. Gripper control is implemented on the application computer. Special care must be taken not to close the gripper at maximum force because this causes a jam. In retrospect, this mechanical design shortcoming justifies a dedicated microcontroller for the gripper. On top of the gripper is a one-degree-of-freedom force sensor based on a piezoelectric crystal. This force sensor is used to detect insertion failures.

3.2.3 The Laser-Based Vision System

The vision system is composed of three hardware components (Figure 4). The first component corresponds to an extremely benign sort of surgery on the robot system, which is about the furthest people in industry will go. In UNIX jargon, this component *tees* the signals from the robot position sensors. In other words, the signals from the robot position sensors are diverted into this component, which provides a copy for the robot controller and another copy for the second component of the vision system. The system is completely transparent to the original robot system. This component, called the *T-component* in the remainder of this text, is evidently robot system-dependent but simple to develop.

3.2.4 System Description

The system is composed of a robot, a gripper, feeders, a printed circuit board positioning and transportation system, the application computer and the vision system (Figure 3). The robot is an AdeptOne, a fast SCARA robot with sufficient accuracy for the application. It has three translational degrees of freedom and one rotational degree of freedom around the vertical axis. Its native programming system is a version of VAL-II. The application computer is a personal computer.

The second component receives the copy of the position sensor signals. It is a printed circuit board designed for the expansion bus in the application computer. It contains electronic circuits and counters to keep track of the robot position by means of the position sensor signals. This part of the system requires that the position sensor signals be quadrature pulses, which are typically produced by incremental optical encoders. The majority

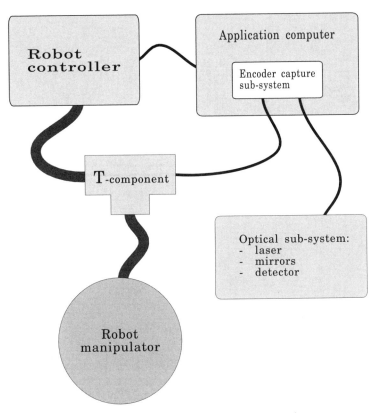

Figure 4   Schematic overview of the sensor system.

of industrial robots use this type of signals. This component is called the *encoder capture subsystem* in the remainder of this text.

The second component also receives a binary signal from the third component. In the latest version this is an optical signal entering through an optical fibre. This signal falls upon an optical sensor. When the signal passes a threshold, the robot position is copied, in less than 10 microseconds, into buffers and an interrupt is triggered on the application computer. The robot position can then be read by programs on the application computer. In addition, the application computer can perform other operations such as reading the robot position on its own initiative and initializing the position counters.

An optical subsystem constitutes the third component. It is composed of a low-power laser, mirrors to guide the laser beam, and an optical system to inject the beam into the optical fiber. When something intersects the laser beam, the robot position is recorded and the program on the application computer receives a notification of the event. This simple vision system enables the removal of uncertainty about the position of the component leads.

The robot moves the components such that their leads intersect the laser beam two times (Figure 5). This intersection of the laser beam can occur at any robot velocity. Each lead must intersect the laser beam separately and at two different angles. When the laser beam is intersected, the position of the robot gripper frame in the robot reference frame is recorded. The position of the laser beam in the same robot reference frame is also known. From this the position of the laser beam in the gripper frame at the instant of intersection can be derived. The intersecting component lead lies on this line. When this procedure is repeated at a different angle, a second line in the gripper frame on which the lead lies becomes known. Since these lines are not parallel, they intersect at the position of the lead in the gripper frame. Details are given in (Thielemans, 1984). As a result, the system can assemble any component it can grasp and make its leads intersect the laser beam one by one and at two different angles.

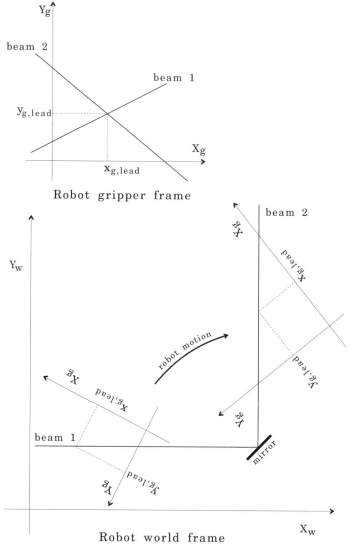

Figure 5 The laser-vision measurement principle.

3.2.5 Data-Driven

The off-line program that manages the odd-component assembly system is driven by a number of data files. One data file defines the workload; it defines how many printed circuit boards have to be made and where which components have to be placed on these printed circuit boards. The locations are given in the robot coordinate reference frame. In other words, this part of the process plan is *grounded*. The second data file defines the component parameters, such as the number of component leads, their relative position, the suitable finger-pair, and the grasping force for the gripper. A final data file defines the feeder setup. More details can be found in Valckenaers (1993).

The data-driven design provides the flexibility required by fast-changing demands from the environment. Examples are the introduction of new printed circuit boards and new components that forfeit hardware modifications. The data-driven design accommodates even certain changes in the hardware. Examples are the addition and removal of feeders and gripper fingers where the feeder types are already supported and the gripper fingers use the available exchange procedure. Note that the hardware variations that are accommodated through data-driven design have no impact on the *modus operandi*. However, the assembly system has already experienced one hardware variation that did change the *modus operandi*. In the production plant a new feeder type was introduced that required a signal when it must (actively) release a component. In practice it is impossible to provide for all conceivable procedure variations caused by hardware modifications in a data-driven manner.

3.2.6   Virtual Devices

The virtual device concept and modularity are more suited to handle hardware variations that involve the *modus operandi*. Consider the set of feeders, which is the most likely source for the latter type of variation. An interface or definition module defines the virtual device; it defines the elementary operations that are needed to fetch components from the feeder system. These operations include:

- A function returning the total number of feeders available
- A Boolean function testing whether a feeder delivers a certain component
- A procedure to fetch a component from a given feeder

The implementation module handles all the details. When a new type of feeder is introduced, only this implementation module and its data file must be adapted. The implementation module must be recompiled and the system relinked. Access to the source code of the other modules is not necessary; object code is sufficient. The gripper-finger exchange operations and the transportation unit operations were implemented in the same manner. Note that this solution cannot be implemented in existing robot programming languages, which do not support modularity or the distribution of object code instead of source code (for the protection of intellectual property).

3.2.7   Portability

The application software and the vision system were adapted for another SCARA robot, the IBM7575, programmed in AML/2. The hardware adaptation is restricted to the modification of the T-component. The software adaptation consists of the development of a new device driver and adaptation to the new robot environment. The environment changes include the disappearance of the force sensor, the tilting device, the clinch unit, the transportation system, and the finger exchange mechanism. The new mechanical gripper is air-powered and can only be opened and closed. This new gripper cannot sense the presence of a component nor control its opening width. A second port of the application uses the same robot and a similar setup, with the exception of the gripper, which is a vacuum gripper that senses the presence of a component.

The uncoupling of the application computer and the manipulator selection proved to be very helpful. Without it the software has to be rewritten completely in the other robot language and two of the three vision hardware components have to be redesigned. With the device drivers only the simplest component of the vision system must be adapted and only a small part of the software must be reimplemented.

The changes in the robot environment, although commonplace, forced software maintenance up to the level of the elementary operations. Examples of these operations are the fetching and inserting of components. The code for these elementary operations needs to be adapted because of its performance-critical nature. As a consequence, portability is not significantly affected by the differences in the set of functions offered by the device drivers. The modular implementation of the elementary operations is the critical factor. The application must be built upon a set of virtual devices that each provide a set of necessary elementary operations.

The main program must be protected against the specific requirements of these elementary operations. A superficial modular implementation is insufficient. An example is the multitude of parameters required by the electromechanical gripper that are not needed by the air-powered grippers: one grasping force and three gripper openings (before, during, and after grasping). It is unreasonable to assume that the main program provides these values. The main program provides component identifiers and the modular implementation of the elementary operations uses these identifiers to retrieve by itself the necessary parameter values. Modular design is not a syntactic matter. Much attention must be paid to avoid having the main software body depend on unstable properties of the robot environment.

4 CONCLUDING REMARKS

Off-the-shelf robots have not brought the ultimate solution for introducing flexibility in solving the material-handling/manipulating problems in FMS. The main reason is that in their present state of development robots are not stable intermediate components, suitable for ready integration in dynamically varying CIM environments.

Some generally valid design principles have been formulated to arrive at such autonomous, stable, plug-and-play compatible building blocks.

It has been proven, with the case study of an odd-component inserter, that the device driver principle is a valuable tool to ease the flexible integration of off-the-shelf automation components, with minor surgery, into highly flexible CIM systems.

4.1 Future Trends

Developments in information and communication technology (ICT) have a profound impact on the future of robotics, FMS, and CIM. Prominent among these are Web technologies and multiagent technologies; Internet/intranet solutions, implemented on multithreading computer platforms, will pervade the manufacturing world. Note that these technologies, for the first time, cause mainstream computing to catch up with the PLC (programmable logic controller): the mainstream computer transforms from a Turing machine into an interaction machine (Wegner, 1997).

For CIM architectures these ICT developments open up the path for the holonic manufacturing system, which uses cooperating autonomous agents to operate and control manufacturing systems (Van Brussel, 1994; Bongaerts et al., 1996; Valckenaers, 1997).

Closer to the robotics side, steady progress in sensor technology and advanced control over the manipulators—e.g., trajectories defined by B-splines allowing system dynamics and actuator limitations to be accounted for—will continuously increase the number of situations in which robotics can be used economically. The contribution of niche players and specialists—small companies optimizing trajectories or providing sensor applications—will increase.

REFERENCES

Bongaerts, L., et al. 1996. "Identification of Manufacturing Holons." In *Proceedings of European Workshop on Agent-Oriented Systems in Manufacturing,* Berlin.

Hocken, R., and G. Morris. 1986. "An Overview of Off-Line Robot Programming Systems." *CIRP Annals,* 495–503.

Kruth, J. P., et al. 1996. "Methods to Improve the Response Time of a CAPP System That Generates Non-linear Process Plans." *Advances in Engineering Software* 25, 9–17.

Suh, N. P. 1988. *The Principles of Design.* New York: Oxford University Press.

Thielemans, H., and H. Van Brussel. 1984. "Laser Guided Robotic Assembly of Printed Circuit Boards." Paper delivered at Fifth International Conference on Assembly Automation, Paris.

Valckenaers, P. 1993. "Flexibility for Integrated Production Automation." Ph.D. Thesis, K. U. Leuven.

Valckenaers, P., H. Van Brussel, and L. Bongaerts. 1995. "Programming, Scheduling and Control of Flexible Assembly Systems." *Computers in Industry* 26(3), 209–218.

Valckenaers, P., et al. 1997. "IMS Test Case 5: Holonic Manufacturing Systems." *Journal of Integrated Computer-Aided Engineering* 4(3), 191–201.

Van Brussel, H. 1994. "Holonic Manufacturing Systems, the Vision Matching the Problem." In *Proceedings of First European Conference on Holonic Manufacturing Systems,* Hannover.

Van Brussel, H., and P. Valckenaers. 1994. "A Theoretical Model to Preserve Flexibility in Flexible Manufacturing Systems." In NATO ASI Series E: *Information and Collaboration Models of Integration.* Ed. S. Nof. Dordrecht: Kluwer Academic. 89–104.

Wegner, P. 1997. "Why Interaction Is More Powerful than Algorithms?" *Communications of the ACM* 40(5), 80–91.

ADDITIONAL READING

Bedworth, D. D., P. M. Wolfe, and M. R. Henderson. *Computer-Integrated Design and Manufacturing.* McGraw-Hill Series in Industrial Engineering and Management Science. New York: McGraw-Hill, 1991.

Hannam, R. *Computer Integrated Manufacturing: From Concepts to Realisation.* Harlow: Addison-Wesley, 1997.

International Journal of Computer Integrated Manufacturing [Special Issue on Design and Implementation of CIM] **10**(1–4).

Rembold, U., A. Storr, and B. O. Nnaji. *Computer Integrated Manufacturing and Engineering.* Reading: Addison-Wesley, 1993.

Sanjay, B. J., and J. S. Smith, eds. *Computer Control of Flexible Manufacturing Systems: Research and Development.* London: Chapman & Hall, 1994.

Singh, N. *Systems Approach to Computer-Integrated Design and Manufacturing.* New York: John Wiley & Sons, 1996.

CHAPTER 41

A STRATEGY FOR IMPLEMENTATION OF ROBOTICS PROJECTS

Geary V. Soska
The Goodyear Tire and Rubber Company
Akron, Ohio

1 INTRODUCTION

Today more than ever, a company's survival depends on its ability to remain competitive in an ever-changing global economy.

Since the introduction of industrial robots in the early 1960s, many companies have benefited from their use. Reduced manufacturing costs, increased productivity, improvements in product quality, workplace safety, and the flexibility to meet customer needs have not only made significant contributions to the bottom lines of these companies, but have also helped enhance their competitive strength. However, benefits like these can only be realized if robotics projects are successful. Therefore, every project must be viewed as unique, and *all* aspects of the project must be carefully considered and analyzed to ensure its success. The key word here is **ALL**: not only the technical aspects, but also the nontechnical.

This chapter presents a robotics implementation strategy that, if adopted and adhered to, can ensure that success. However, trying to shortcut the strategy will result in failure. *Guaranteed!*

2 STRATEGY

The strategy is not rocket science. It is simple and straightforward and consists of the following four elements:

1. Paradigm shifts
2. Communications
3. Coordinated in-house effort
4. Partnerships

3 PARADIGM SHIFTS

Over the past 30 years the traditional school of thought held that you could purchase an $80,000 robot to eliminate direct labor costs while achieving a six-month payback. There are several fallacies in this line of thinking.

First, there is no such thing as an $80,000 robot. Sure, you can buy a robot that costs $80,000, but all you really have is a device that can be anchored to the factory floor. The paradigm shift that must occur is to think in terms of a *system*: a total solution that will help enhance your competitive strength. Those involved with the implementation of robotics projects must understand that, at best, an industrial robot represents only about 25% of the total project cost. The other 75% consists of all the things required to make the system work. It has been the failure to think in terms of a system that has caused hosts of highly viable industrial robotics projects never to progress beyond the budgeting phase. See Table 1, which illustrates how the various cost items associated with a robotics system can be broken out. The percentages shown in the table are fairly typical. However, they can and will vary depending on the complexity of a robotics project.

Handbook of Industrial Robotics, Second Edition, Edited by Shimon Y. Nof
ISBN 0-471-17783-0 © 1999 John Wiley & Sons, Inc.

Table 1   Cost Breakdown for a Fairly Complex Hypothetical Robotic System

Cost Element	% of Total Cost	Cost
Robot	25	$100,000
Systems engineering and documentation	15	$60,000
System tooling	10	$40,000
System peripherals	10	$40,000
Application software	10	$40,000
Project management	5	$20,000
Spare parts	5	$20,000
Training	5	$20,000
Installation/start-up	5	$20,000
Totals	**100**	**$400,000**

Second, using direct labor reduction as the sole means to justify an investment in robotic technology is ludicrous. It is nothing more than a cost accounting game, typically characterized by a shift in departmental overhead allocations. Common sense dictates that even if a labor unit is replaced by a robot, someone still has to operate, program, and maintain the robot. It is not going to do these things by itself. The paradigm shift that has to occur is to approach justification in terms of your long-term competitive advantage. Will the investment in an industrial robot help you to be more productive than your competitors? Will it allow you to offer a product with better quality than your competitors? Will a robot enable you to shorten the time it takes to get your product to market? Will it help to reduce or eliminate workers' compensation claims? Will a robot afford you the flexibility necessary to address changes in customer demands?

Third, a six-month or one-year payback is the exception, not the norm. Sure, you can purchase a robot for $80,000 and realize a payback in less than a year. It is a simple numbers game. Suppose you run a continuous four-shift operation with one labor unit per shift. If your labor rate plus benefits runs $50,000 per year, you can realize a payback in less than one year after taxes. However, you have only purchased a robot. You still have to put it into production. That is where a paradigm shift of thinking in terms of total system cost must occur. When you assess your payback in terms of total system cost, you are likely to find that your payback period is somewhere in the two- to three-year range. What's so bad about that? It depends whether you think in cost accounting terms or in terms of your long-term competitive advantage. This is a key reason why Japan has outpaced the United States in use of industrial robots.

4 COMMUNICATIONS

By the year 2250 the human race will have learned to read minds. People in all walks of life will be so well informed that there will be no need for written or verbal communications. No one will ever be in the dark for the lack of information. Until then, however, we will have to rely on conventional communication methods.

If a robotics project is to succeed, the rationale, or the why, behind the effort needs to be clearly communicated to all those involved. Of equal importance is the need to communicate with organized labor, as any form of automation is generally viewed as a threat to job security. It is human nature to resist change, and unless people understand why an industrial robot is being considered and how it will affect them, they will resist it.

Communications should also include any constraints that might put up potential roadblocks to implementation. These might include budget limitations, floorspace availability, a tight project-implementation schedule, limited human resources, an unfinalized product design, or the unavailability of parts when time comes for system acceptance trials. While none of these constraints is insurmountable, each needs to be communicated, discussed, and resolved.

5 COORDINATED IN-HOUSE EFFORT

No man or woman is an island. Since the introduction of industrial robots in the early 1960s, no single individual has successfully implemented an industrial robot project alone.

Successful projects have been achieved through a coordinated in-house effort. Every company in the world possesses islands of expertise. These islands can be found in both the corporate and plant environments. In the corporate environment one can find managers, purchasing agents, lawyers, project managers, and engineers. In the plant environment there are managers, purchasing agents, engineers, maintenance workers, and production workers. Each of these islands, whether in the corporate or plant environment, possesses a particular level of expertise that, if tapped, can and will ensure the successful implementation of a robotic project. Table 2 shows what contributions in expertise each of these islands can make.

6 PARTNERSHIPS

Partnerships are the last step in the strategy. This means establishing both an intimate and trusting relationship with suppliers. Remember, suppliers know their equipment and its capabilities better than anyone. Conversely, companies know their products and manufacturing processes better than anyone. Together you have a great marriage.

As with any marriage, there are the aspects of intimacy and trust. In a business relationship, intimacy and trust take the form of information sharing. This means:

- Granting suppliers access to your manufacturing facilities to get a first-hand look at your operations and manufacturing processes. Simply furnishing drawings and parts does little to ensure that suppliers understand the big picture.
- Allowing them to ask questions of plant personnel most closely involved with the process or operation you are considering for robotics. After all, these people probably know more about the process from a day-to-day standpoint than anyone else in the company.
- Explaining to them the intimate idiosyncrasies of your manufacturing process so they better understand what you really expect of their equipment, which will help them determine if their equipment can satisfy your needs.

Table 2   A Coordinated In-house Effort Provides Synergy

Island of Expertise	Contribution
Management (corporate and plants)	• Commitment and support • Overall strategy (the rationale or the why) • Identify and communicate the constraints • Communications with organized labor
Purchasing agents (corporate and plants)	• Relationships with suppliers • Bid requests • Cost-effective purchasing negotiations • Expediting orders
Lawyers	• Secrecy agreements with suppliers • Contracts • Patent applications
Project Managers (corporate)	• Schedule control • Budget control • Overall project coordination
Engineers (corporate and plants)	• Relationships with suppliers • Concept development • Specifications (functional and performance) • Technical expertise for system design • Manufacturing process knowledge • Installation and start-up • Troubleshooting skills
Maintenance workers (plants)	• Troubleshooting skills • Technical expertise (service and maintain equipment)
Production workers (plants)	• Intimate knowledge of manufacturing processes and their idiosyncrasies

- Providing them with parts and part drawings. Geometric and dimensional data is extremely important, especially if your supplier needs to develop a simulation to thoroughly analyze your application.

- Providing them with a central point of contact for any technical questions they might have. One readily available point of contact is worth more in terms of time and money than 50 contacts that can't be reached for information when it's needed.

- Informing them of any internal constraints you might have. Last-minute surprises have a way of spoiling everything, especially when a successful robot application is of the utmost importance.

- Trusting them with the information you provide. If husbands and wives can't trust each other, then the marriage is over. You might as well "go it alone" and take your chances.

There are a few other aspects of robotics implementation that must be stressed:

1. *Avoid the low-bid syndrome.* Buying low-bid is something you do with ash trays, paper clips, and restroom supplies. It is not something you do with high-technology capital equipment. Remember, you get what you pay for. Therefore it is imperative that you always consider *total life cycle costs* when evaluating bids. What may appear to be a bargain on the surface may later come back to bite you when you need parts or service or if you need to modify the scope of your project.

2. *Avoid mixing robots.* No two robots are programmed the same, nor do parts from robot A fit robot B. Mixing robots puts a tremendous cost burden on the end user in terms of support personnel and parts inventories. While there is nothing wrong with mixing robots throughout a company, avoid mixing them in a plant. The economics simply isn't there. You can be guaranteed that what you save in the initial purchase will end up costing you more in operating expenses over the long run. You are much further ahead if you narrow yourself to a single "best" supplier than if you "play the field."

3. *Don't expect automotive industry discounts.* Unless you purchase industrial robots in the same quantities as the automotive industry, don't expect to receive the same discount rates. If you are a company that buys one or two robots every three to five years, the chances of negotiating a significant discount on your purchases are very remote. If, on the other hand, you are a company that purchases robots in lots of 200 and up each year, your chances of negotiating significant discounts are excellent.

4. *Consignment parts.* Spare parts to support a robot installation can represent a significant investment. Consequently, it is highly desirable to negotiate a parts consignment program with your supplier. Again, unless you are buying in significant quantity, the probability of your being able to negotiate for consignment parts is extremely low.

7 CONCLUSION

Over the past 35 years industrial robots have proven their ability to create more efficient and cost-effective manufacturing environments that have ultimately enhanced the competitive strength of companies around the world. However, trying to implement them without a sound, value-based strategy can prove to be disastrous. It is only through paradigm shifts, communications, coordinated in-house efforts, and partnerships that companies can come to realize the true benefits of industrial robots.

ADDITIONAL READING

Soska, G. V. "How to Select a Robotics Supplier." *Robotics World* (September), 13–14, 1991.
———. "Budgeting for Industrial Robots." *Managing Automation* (June), 30–31, 1992.
———. "Fundamentals and Implementation of Industrial Robots." Tutorial: International Robotics and Vision Automation Show and Conference, April 5, 1993.
———. "Robotic Implementation . . . A Multi-Phase Systematic Approach." In *Proceedings: International Robotics and Vision Automation Show and Conference,* April 5, 1993. 2-1–1-10.

PART 8

ROBOTICS IN PROCESSES

CHAPTER 42
FABRICATION AND PROCESSING

J. T. Black
Auburn University
Auburn, Alabama

1 INTRODUCTION OF INDUSTRIAL ROBOT APPLICATIONS

The first commercial application of an industrial robot took place in 1961, when a robot was used to unload a die-casting machine, a particularly unpleasant task for humans. In fact, many early robot applications were tasks defined by the three H's: hot, heavy, hazardous. Thus jobs like welding, painting, and foundry operations were handed over to robots.

While still so used, in recent years robots have also been used in many applications where they outperform humans or have an economic advantage over humans or fixed automation systems.

At the same time, robots have gotten smaller, faster, and less expensive than they were 10 years ago. Robots that lift, transport, assemble, and insert components are widely used in assembly. The presence of robots that are smaller and therefore have better capabilities means that many repetitive manufacturing activities that previously would not have been considered candidates for a robot application are now legitimate robotics applications.

A good example is the use of robots in "clean rooms," which must be relatively free of contaminants and require human operators to wear special clothing, masks, and sometimes breathing apparatus. Here a robot may be cost-justified because in addition to human labor saving, the cost of the clean room and special clothing may be saved *and* the quality of the end product may be improved.

2 ELEMENTS OF ROBOTICS

Figure 1 illustrates the major elements in an industrial robot used in processing (Asfahl, 1991): the manipulator, or arm; the controller; the operator interface; and the end-effector (gripper), which is attached at the wrist on the arm.

The manipulator is the robot's most visible part. It moves under operator control or in accordance with a programmed sequence. Attached to the "wrist" is an end-effector, typically some type of tool or gripper for acting upon or manipulating parts or, more commonly in processing, holding the tool. The controller is used for storing programs, controlling the movements of the manipulator, communicating with other machines in the factory, and communicating with the operator. Operator communication can be through a variety of devices. The two most popular are a CRT terminal for entering commands and monitoring status and a teach pendant for moving the manipulator under operator control. To teach the robot a program, the operator usually moves the manipulator by using the teach pendant and records the desired positions into the robot's memory. When operating under automatic control, the robot "revisits" the taught positions according to the sequence defined by the control program. The control program or sequence can be entered using the CRT or special controls on the teach pendant. The end-effector on the wrist is generally custom-built for the particular processing task. The robot, gripper, product, and workholding devices must all be considered as a system when developing a robotic processing application.

3 ROBOT CLASSIFICATIONS

Industrial robots may be classified in a number of different ways: according to their application, size, control systems, power source, configuration, etc. In terms of processing, robots are classified according to their controls and configurations.

Handbook of Industrial Robotics, Second Edition, Edited by Shimon Y. Nof
ISBN 0-471-17783-0 © 1999 John Wiley & Sons, Inc.

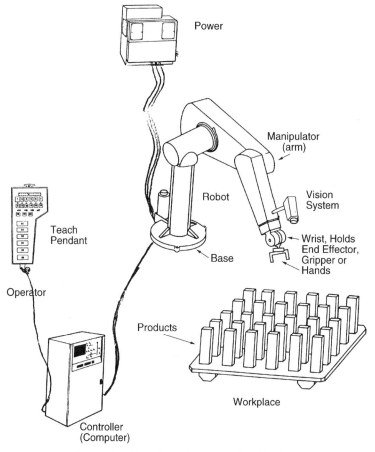

Figure 1   Major elements of an industrial robot.

3.1   Robot Controls

There are three basic types of robot controls, arranged from simple to complex: *nonservoed* (point-to-point), *servoed* (point-to-point), and *servoed continuous*.

Nonservoed robots are typically used for the simplest pick-and-place operations. The system is "programmed" by adjusting the positions of the limit stops and switches. The sequence of steps may be controlled by a variety of schemes—relay ladders, programmable controls, or some other type of logic. Typically the nonservoed robot moves a single joint until it reaches a limit. The next motion is selected by the logic. This sequence continues until the cycle repeats. This system has limited flexibility and must be physically reprogrammed whenever there is a change in the production operation.

In a servoed point-to-point robot, positions are recorded in some type of electronic memory. Joint positions are sensed using some type of feedback device. This allows the manipulator to reach any point (limited by the servo-system's resolution) within its reach, as long as that point has been programmed. This allows the servoed-robot to perform different operations on subsequent cycles, providing the appropriate positions have been recorded in the system. Another advantage of servoed-robots is their capability of following a prescribed motion path. The nonservoed robot is only capable of positioning each joint at one of two extremes of travel and is therefore incapable of following a desired curve. The servoed-robot can follow a series of points, which when taken together can define a curved path. This capability allows servoed-robots to be used for point-to-point (PTP) applications such as assembly or for continuous path (CP) applications such as welding, spraypainting, machining, and complex assembly.

Additional capabilities can be provided with an added level of control so the robot can move to positions that have not been defined in advance, but rather are determined based on some outside event or condition that is sensed by the equipment. For example, a robot fitted with a tactile sensor may be commanded to squeeze an object until some force conditions are met. The controller would autonomously determine where to move the manipulator or gripper in order to satisfy the desired conditions. Thus the advanced robot must have the capability to communicate with other machines, devices, and sensors and to act upon those external data, which means the robot controller must be capable of transforming the feedback information into commands that the manipulator can carry out. This requires the robot controller to perform some sophisticated calculations.

4 ROBOT CONFIGURATIONS

All robot manipulators are composed of combinations of joints and links. Joints allow relative motion between pairs of links, and links rigidly connect joints. Two joint types are typically used in robotics: *revolute joints,* which allow rotations between links, and *prismatic joints,* which allow linear motions.

Traditionally, manipulators fall into five categories (see Figure 2): polar or spherical, cylindrical, Cartesian, articulated, and selected compliance assembly robot arm (SCARA), a particular configuration of the articulated robot.

For processing, a robot with a total of six independent degrees of freedom (d.o.f.) is usually required to position the tool to a point in space with any orientation (see Figure 3). In the examples the manipulators have (typically) only three d.o.f. These three joints are adequate for positioning the tool, but not for orienting it. This is the job of the manipulator's wrist. Wrists, located at the end of the manipulator, typically contain two or three joints. At the other end of the wrist is a tooling plate for mounting a tool or gripper for the tool to the robot.

When the manufacturing engineer is selecting a machine tool (a lathe or a milling machine) for a processing job, the process capability of the machine must be considered (Degarmo, Black, and Kohser, 1997). To determine a machine's process capability, one measures the components produced by the machine. In the same light, when trying to determine whether or not a robot can do a processing task, one has to consider the process capability of the robot.

4.1 Robot Process Capability[1]

Robot process capability (RPC) is measured in terms of accuracy, repeatability, reproducibility and stability. The robot's capability depends very much on the task conditions (e.g., speed and load), as well as the design of the robot, the working envelope, the quality of the joints, and the type of control.

4.1.1 Definition of Terms for RPC

Robot process capability refers to the ability of a robot to consistently perform a job with a certain degree of accuracy, repeatability, reproducibility, and stability (see Figure 4). RPC is a function of task variables such as speed of movement, spatial position, and load. The elements of RPC may be described as follows:

L_o = target value
M_{ijk} = the ith measured value at starting joint j and at time period k, $i = 1$ to n, $j = 1$ to p
$\bar{M}_{i(jk)}$ = average of n values at the same starting point j, and the same time period k
S.D. $[M_{i(jk)}]$ = standard deviation of n values at the same starting point j and the same time period k
$\bar{M}_{i(jk)}$ = average of $n \times s$ values at the same time period k

---

[1] See Gilby and Parker (1982); Gilby, Mayer, and Parker (1984); Harley (1983); Hunt (1983); *Industrial Robots* (1983); Jiang, Black, and Duraisamy (1988); Jiang et al. (1989); Jiang et al. (1991); Lau and Hocken (1984).

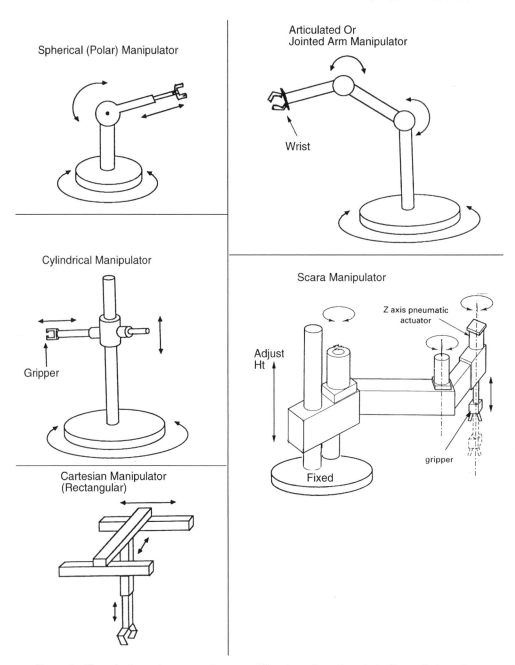

Figure 2 The robotic worker comes in many different configurations but will usually have three degrees of freedom.

S.D. $[M_{ij(k)}]$ = standard deviation of $n \times s$ at the same time period k

$\bar{M}_{ijk}$ = Average of $n \times s \times p$ values

- *Accuracy* refers to the ability to hit what is aimed at. It is the degree of agreement between independent measurements and the target value being measured.

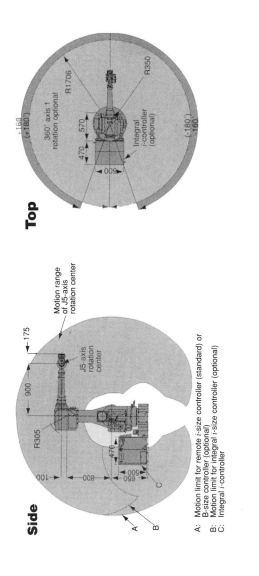

		Range	Speed
Motion	Axis 1	320°/360°	160°/sec
range	Axis 2	260°	120°/sec
and	Axis 3	420°	150°/sec
speed (1)	Axis 4	540°	240°/sec
	Axis 5	250°	240°/sec
	Axis 6	720°	340°/sec
Repeatability		±0.15mm (±0.006")	
Maximum load	Axis 3	10kg (22 lbs)	
capacity (2)	Axis 6	45kg (99 lbs)	
Maximum reach		1706mm (67.2")	
Maximum stroke		1304mm (51.3")	

A: Motion limit for remote *i*-size controller (standard) or
   B-size controller (optional)
B: Motion limit for integral *i*-size controller (optional)
C: Integral *i*-controller

Figure 3 The six axes of motion of a robot (including roll, pitch, and yaw movement of the wrist), the robotic work envelope, and some specifications.

835

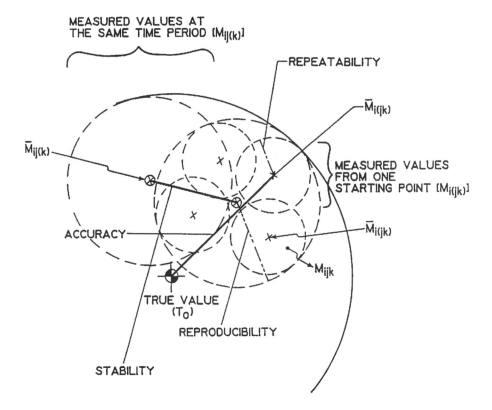

MEASURED VALUES AT
THE SAME TIME PERIOD [$M_{ij(k)}$]

REPEATABILITY

$\bar{M}_{i(jk)}$

$\bar{M}_{ij(k)}$

MEASURED VALUES
FROM ONE
STARTING POINT [$M_{i(jk)}$]

$\bar{M}_{i(jk)}$

ACCURACY

M_{ijk}

TRUE VALUE
(T_0)

REPRODUCIBILITY

STABILITY

Figure 4 Definition of robotic process capability (Jiang, Black, and Duraisamy, 1988; Jiang et al., 1989).

$$\text{Accuracy} = \bar{M}_{i(jk)} - T_o \text{ for } i = 1 \text{ to } n, \tag{1}$$

$$\text{where } \bar{M}_{i(jk)} = \Sigma \, M_{i(jk)}/n \tag{2}$$

- *Repeatability* (also called *precision*) refers to the process variability. This is the figure usually cited by the manufacturers in their sales brochures. It affects how well the robot returns to a taught position. It is the degree of agreement between independent measurements under specified conditions.

$$\text{Repeatability} = 3 \times \text{S.D.} \, [M_{i(jk)}], \text{ for } i = 1 \text{ to } n, \tag{3}$$

$$\text{where S.D.} \, [\bar{M}_{i(jk)}] = \{\Sigma \, [\bar{M}_{i(jk)} - M_{i(jk)}]/(n-1)\}^{0.5} \tag{4}$$

- *Reproducibility* refers to the deviation from or difference between the means of measurements taken from repeated tests that hit the same target but developed from different starting points.

$$\text{Reproducibility} = 3 \times \text{S.D.} \, [M_{i(jk)}], \text{ for } i = 1 \text{ to } n, j = 1 \text{ to } s \tag{5}$$

- *Stability* refers to the difference in the average of at least two sets of measurements obtained with the same target from the same starting point at different times. This is also called *drift*.

$$\text{Stability} = \bar{M}_{ij,k+1} - \bar{M}_{ijk}, k = 1 \text{ to } p - 1, \tag{6}$$

$$\text{where} \qquad \bar{M}_{ijk} - \Sigma \ \Sigma \ \Sigma \ (M_{ijk})/nsp \qquad\qquad (7)$$

Process capability refers to the uniformity of the process; the variability in the process is a measure of the uniformity of output. The process capability ratio (PCR) is a way to express this process capability, using the 6σ spread of the process (where σ is the standard deviation). For a quality characteristic with both upper and lower specification limits (USL and LSL, respectively), the ratio is PCR = USL $-$ LSL/6σ. Thus the PCR is the ratio of the difference between the specification limits and the natural spread or variability in the process ($\pm$ 3σ above and below the mean).

Most manufacturers calibrate their robots to determine their resolution. *Resolution* refers to the smallest incremental motion made at the end of the manipulator. This is determined by finding the resolution of each joint and then calculating the vector sum. This calculation is usually performed for the arm in the worst-case position, typically fully extended. If the mechanism can be assumed to be perfectly smooth (continuous), then the resolution is determined by the limitations of the control system.

The robot's *accuracy* is its capability to position the end of the manipulator at the desired target position. If everything were perfect, then the accuracy theoretically would therefore be one half the resolution. Unfortunately, many other factors (see Table 1) contribute to the accuracy error, including deflection of the arm under load, thermal expansion, drive and joint inaccuracies, and computational inaccuracies. Hence *the robot's true accuracy is much worse than one half the resolution* and much worse than the repeatability.

Repeatability, the most commonly quoted performance specification, describes the precision with which a manipulator can return to a point that it previously visited. If the

Table 1   Factors Associated with a Robot's Process Capability

Attributes	Factors
Robot Attributes	
Design	Type: servo-point-to-point, nonservo-point-to-point, servo-continuous-path
	Work envelope: cylindrical, spherical, articulated arm, rectangular
	Arm: number of joints, degree of freedom, weight, shape, length, material, tolerance
	End-effector: material, size, mobility, compliance, strength, sensor, shape
Control	Type: air logic, stepping drum
	Memory type and capacity: standard (single program, multiple program), optional expansion of standard different controller)
	Software: main program, subprogram, degree of intelligence.
	Power supply: electric, hydraulic, pneumatic
Specifications	Movement range, velocity range, repeatability, cycle time, reliability, number of years used, frequency of use, mobility, commonality, training and maintenance requirements
Other attributes	
Task attribute	Object: load, shape, size, material, load, distribution, accuracy requirements, repeatability requirements, force requirements, frequency of task performance requirements
	Task type: processing materials (painting, welding, grinding), material handling (pick up and place, insertion), registration, inspection (finding burrs)
Human attributes	Operator skill, programmer skill, operator variations. Psychological factors: mood, motivation, coordination
Environmental attributes	Temperature, air velocity, humidity, stability, dust, floor loading, peripheral equipment, safety, electrical noise

robot is taught using a teach–repeat technique, this is a valid measure of performance. Typically, repeatability is affected by the mechanical hysteresis in the robot system. Since the robot is moving to a point that was demonstrated to it physically, repeatability is not affected by the same factors as accuracy. Typically, a robot's repeatability is better than its resolution.

Most of the current specifications are for a static condition or a range of extreme conditions of performance, such as a weight-handling capacity of 0–20 kg, a velocity of movement of 1–20 cm/sec, and a positioning tolerance of $\pm$ 0.5 mm. When a robot performs a task during a process, however, neither the manufacturer nor the user knows its actual process capability characteristics. The manufacturer often lacks an acceptable methodology and measuring technique to determine this capability. Companies using robots often do not have the personnel or the measuring equipment needed to determine RPC. Unfortunately, many quality processing and control objectives cannot be met if the RPC is unknown.

In a process capability study of a machine tool, the parts machined on the machine are measured to determine the machine tool's process accuracy and capability. For example, in lathe turning, precision decreases (variability increases) as the diameter of the workpiece increases. If the robot carries the tool (drill or laser, for example), then the workpieces can be examined to determine the RPC. However, in robot material handling or processing activities like spraypainting or arc welding, the output cannot be examined (easily measured) directly to determine the process capability. Therefore a different method must be developed to determine a robot's process capability. To do that a means must be provided to independently measure the robot end-effector's spatial location three-dimensionally at any specific time.

4.1.2   How to Do the RPC Study

An RPC study can be done using Taguchi methods as an effective and economical way to determine the process capability of a robot (Jiang, Black, and Duraisamy, 1988; Jiang et al., 1989; Jiang et al., 1991). The significant factors and optimal treatment combinations of controllable factors can be determined using Taguchi experimental methods. The ultimate objective is to develop an effective method to enable the user to determine whether the robot is capable of performing its tasks before the robot is purchased or installed.

To meaningfully indicate a robot's process capability, the measurement data obtained must first be reliable. Therefore, for a useful evaluation, the testing method and conditions must be well designed. Also, for the measurement data to be meaningful, it must be analyzed and specified in a form that appropriately represents the data. There are two major concerns in determining the process capability of a robot: (1) specific measurement techniques based on the type of instrument used, and (2) testing methods, conditions, and specifications (TMCS).

4.1.3   Review of Measuring Techniques

Different devices for measuring a robot's spatial location have been developed (Lau and Hocken, 1984; Rembold, 1990; Warnecke, Schraft, and Wanner, 1985). Basically, a spatial location device should be able to spot and trace the path of the robot's end-effector with a specified accuracy, under various operating conditions. In addition, the device's repeatability should be 10 times greater (practically, at least 3 times greater) than that of the robot. This is commonly referred to as the *rule of 10*, meaning that the measuring precision should be an order of magnitude better than the precision of the device being measured (Degarmo, Black, and Kohser, 1997). The measuring device should also have a means for the storage and output of data so the data can be used for further analysis. The device should be simple and flexible and it should not affect the actual performance of the robot. There are various types of measurement devices that meet these requirements. Broadly speaking, the methods adapted for using these systems can be classified as *contact sensing* and *noncontact sensing*. Measurement techniques based on vision systems and laser systems are popular noncontact methods. The advantages and limitations of various techniques are summarized in Table 2. The theodolite system and the optical scanner system are briefly discussed. The laser ball bar is a new technique that can be applied to RPC studies (Ziegert and Datseris, 1990). See also Chapter 39, Precision and Calibration.

Table 2 Summary of Measuring Techniques

Method	Resolution	Accuracy	Repeatability	Advantages	Disadvantages
Contact sensing					
1. CMM $50,000 up	0.0005 mm	± 0.005 mm per axis	± 0.002 mm	a. simple in operation b. both positioning and path accuracy can be measured	a. high cost for DCC b. measuring speed is very slow
2. Latin square ball plate ($10,000)*	n/a	n/a	± 0.15 mm	a. experiment setup is defined by statistical procedure	a. slow speed b. for point-to-point (PTP) test only
3. LVDT sensor	0.0025 mm	n/a	n/a	a. both position and orientation can be measured with multiple LVDTs b. Low cost	a. slow speed b. for PTP only
Noncontact sensing					
4. Acoustic-based system	0.2 mm	n/a	n/a	a. use lightweight sensor b. can improve accuracy by using ultrasonic sensor	a. environmental effects are likely
5. Photogrammetry ($200,000)	n/a	n/a	n/a	a. same operation for a large number of markers (e.g., 200)	a. need to process films b. for PTP only c. not economical for a small number of markers (e.g., less than 10)
6. WATSMART system Positioning Sensor ($50,000 up)	12 bits (1:4096)	1 mm	point: 0.005% path: 0.01%	a. can be used for both PTP and continuous path (CP)	a. lighting environment should be controlled to avoid deflection b. LED position may cause inaccuracy c. high temperatures due to extensive usage may cause inaccuracy
7. SELSPOT system Positioning Sensor ($50,000 up)	12 bits (1:4096)	n/a	point: 0.005%	a. high speed b. good software capacity c. most widely used commercially available system	a. speed reduced by the number of markers used b. same lighting and temperature problems as WATSMART
8. Laser interferometer ($75,000 up)	1:100,000	25×10^{-6}	n/a	a. target velocities up to 25.4 m/min can be tracked b. highest accuracy system available c. can be used for both PTP and CP	a. high cost

Table 2 (Continued)

Method	Resolution	Accuracy	Repeatability	Advantages	Disadvantages
9. Bird system	n/a	n/a	n/a	a. can measure both position and orientation	a. high cost
10. Optical scanner system (34, 38) ($100,000)	n/a	0.05%	0.01%	a. good tracking speed (10 m/s) and accuracy (0.1 mm at 1 m distance) b. for both PTP and CP	a. not commercially available
11. ExpertVision ($50,000 up)	(depends 1:00	on the 1:4000	cameras used)	a. the motion of the object can be recorded and replayed b. the same hardware can be used for other vision application	a. will be commercially available b. high cost
12. Theodolite system ($50,000 up)	13×10^{-6}	n/a	n/a	a. simple algorithm and operation procedure b. system capability is good enough for industrial robots	a. speed is slower than LED-based system (max. 200 Hz) b. crossed field markers may cause confusion for the software c. calibration frame is not available
13. Proximeter system ($2,000)* ($20,000 up)	25×10^{-6}	n/a	n/a	a. can measure both position and orientation b. low cost	a. slow speed (about 1 point/min) b. manual operation could cause eye fatigue over a long period of operation

Theodolite System

The theodolite system is a noncontact measuring system employing a triangulation technique (Warnecke and Schraft, 1982). Two theodolites are used to determine the spatial location of the robot's end-effector, using a 3-D triangulation principle. The setup (Figure 5) consists of two theodolites, the respective positions of which are calibrated by means of landmarks obtained from the graduations of rules or the vertices of the calibrated trihedron. To show the position of the robot, a special target made of five arms (or other forms) ending in lightpoints is fixed to the robot. To measure the position of the lightpoint in the space of the robot, both theodolites are aimed at the same lightpoint. Then azimuth and elevation angles are recorded. This operation is repeated for three positions of the end-effector being measured. By processing of the collected data, the position and orientation of the end-effector are determined.

Comments:

1. The theodolite system can be accurate to 0.13 mm three-dimensionally.
2. Although the collected data are processed and analyzed by computer, the data collection is done manually and is time-consuming (approximately 1 point per minute).
3. The theodolite system cannot be used for a continuous-path test.

Optical Scanner System

This system was developed by Gilby et al. at the University of Surrey for testing the dynamic performance of a robot arm (Gilby and Parker, 1982; Gilby, Mayer, and Parker, 1984; Parker and Gilby, 1982–1983). The principle of operation of this system is to follow the target with two separate beams of light and record their positions with respect to their source location. The general layout of the system is shown in Figure 6. The instrument consists of a moving target that is rigidly fixed to the robot arm, and two similar static measurement units. Each of these units (called *subsystems*) consists of optical and me-

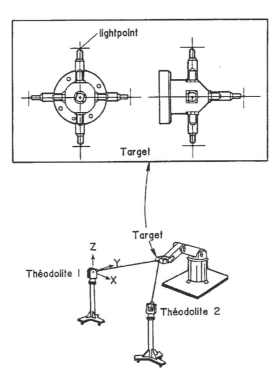

Figure 5 A typical theodolite system. (From Warnecke and Schiele, in *Proceedings of the 1st Robotics Europe Conference,* 1984. Berlin: Springer-Verlag, 1984.)

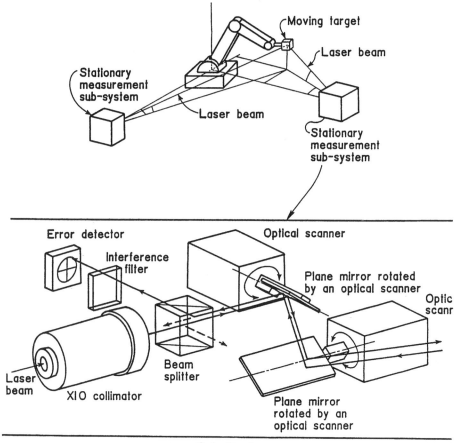

Sub-System

Figure 6 Optical scanner system. (From G. A. Parker and J. Gilby, "An Investigation of Robot Arm Position Measurement Using Laser Tracking Techniques." Department of Mechanical Engineering Report, University of Surrey. Reprinted by permission of Springer-Verlag Inc. from *Proceedings of the 1st Robotics Europe Conference, 1984.)*

chanical components and a control unit. Two separate beams of light (one from each subsystem) are made to follow the moving target, and the relative position of the target from the reference source is continuously recorded. To effect this tracking, the distances between the center of the target and the center lines of the beams are measured by light-sensitive detectors. Error signals from these detectors are used to correct the direction of the light beams so that they point directly at the optical center of the target. To enable higher tracking speed the instrument has been designed for low inertial.

The beam that is generated by the He–Ne laser is first expanded to about 8 mm in diameter and then passed through the beam-splitter and on to the center of a plane mirror attached to the optical scanner. This scanner is able to rotate the mirror rapidly through an angle of approximately 20° about an axis that is perpendicular to the beam's direction. The present tracking speed is about 2 m/s. A tracking speed of 10 m/s is under development. The beam reflected by this mirror impinges upon the axis of a second mirror also rotated by an optical scanner. The axis of this second mirror is parallel to the initial direction of the laser beam. After reflection by both mirrors, the beam emerges from the subsystem in a direction that is determined by the rotations of the mirrors. By suitable rotation of the shafts of the two optical scanners, the emergent beam is directed towards the target.

Comments:

1. This system has good tracking speed (2 m/s) and sampling at (selectable up to 250 Hz).
2. Measuring volume is 0.5 × 0.5 × 00.5, and up to 3 × 3 × 3 m.
3. Accuracy is 0.05% of range and repeatability is 50 m or 0.01% of range.

4.1.4 What We Know about RPC

An RPC study can be done on any robot to determine its accuracy, repeatability, reproducibility, and stability. These measures will vary for different robots with different operational parameters. The operational parameters shown in Table 3 are typical of those used in RPC studies for both continuous path and for PTP robots. Payload reflects the weight the robot is moving (the weight of the gripper and tooling). This may be fairly constant when the robot is carrying the tooling or may vary if the robot is carrying the workpieces rather than the tooling. The speed at which the end-effector is moving is always a significant factor. The shape of the working envelope is very important. This factor depends on the type of robot being used. The size of the working envelope (WE) reflects the reach distances, the direction reflects reaches from left to right in the working envelope, and the height reflects vertical extension. Orientation reflects rotations of the manipulator about the center of the WE. In processing there can be reaction forces on the robot (for example, an abrasive waterjet has a bit of a kick when turned on) and machining forces (cutting forces developed between the tool and workpiece) in processes like drilling, grinding, deburring, polishing, and so forth. See Sections 5.5 and 6.1.6. In an RPC study there will likely be interactions between factors (some engineers refer to this as *coupling*), so it is imperative that the RPC study be carried out using experimental design techniques. Taguchi experiments are particularly well suited for RPC studies because the necessary number of measurements is reduced and a set of operational parameters can be determined, which will bring the process capability close to its optimal value.

Every robot will have a different RPC depending on its design and application, which determine the operational parameters. Typically the robot will be six to eight times more precise (repeatable) than it is accurate. For a robot doing dispensing or material removal, repeatability of ± 0.006 in. are typical, while a robot doing painting would have a repeatability of ± 0.080 in. Spot welding robots have repeatabilities in the range of ± 0.020 in., while robots doing arc welding need to be ± 0.004 in.

Robots will perform best at speeds and loads about in the middle of the respective ranges and about in the middle of the WE. It is recommended that the RPC study be done before the robot is installed. Again, it is pointed out that an RPC is not a calibration study (which develops no load data). An RPC study tries to determine operational data. Calibration is discussed in Chapter 39.

Stability was not a problem for most robots, but most robots did not have good reproducibility (i.e., they were unable to get to the same spot from different starting points very well).

4.1.5 Robot Characteristics

In general, robots possess three important characteristics that make them useful in manufacturing operations: *transport, manipulation,* and *sensing.*

Table 3 Typical Operational Factors Used to Determine Robot Process Capability

Factor	Comments
Payload (weight)	0 to 95% of maximum pay load
Arm Speed	Low, medium, and high speeds
Shape of WE	Spherical, rectangular, etc.
Size of WE (reach)	Reflects maximum reach distance
Direction of WE (angular range)	Rotations left and right
Height of WE (Z axis)	Vertical extensions

Transport

One of the basic operations performed on workpieces as they pass through the manufacturing system is material handling or physical displacement. The workpieces are transported from one location to another, stored, machined, assembled, or packaged. In these transport operations the physical characteristics of the workpiece remain unchanged.

The robot's ability to pick up a workpiece, transport it through space, and release it makes it an ideal candidate for transport operations. Simple material-handling tasks, such as part transfer from one conveyor to another, may only require one- or two-dimensional movements. These types of operations are often performed by nonservo robots. Other parts-handling operations may be more complicated and require increased levels of manipulative ability and additional process capability in addition to the ability to transport. Examples of these more complex tasks include machine loading and unloading, palletizing, part sorting, and packaging. These operations are typically performed by servo-controlled point-to-point robots.

Manipulation

In addition to material handling, another basic operation performed on a workpiece as it is transformed from raw material to a finished product is processing, which generally requires some type of manipulation. That is, workpieces are inserted, oriented, or twisted to be in the proper position for machining, assembly, or some other operation. In many cases it is the tool that is manipulated rather than the workpiece being processed.

A robot's ability to manipulate either tooling or parts makes it very suitable for processing applications. Examples in this regard include robot-assisted machining; spot and arc welding; laser heat treating, joining, or machining; and spraypainting. More complex operations, such as assembly, also rely on the robot's manipulation abilities. In many cases the manipulations required in these processing and assembly operations are quite involved and therefore either a continuous-path or point-to-point robot with a large data storage capacity is required.

Sensing

In addition to transport and manipulation, a robot's ability to react to its environment by means of sensory feedback is also important, particularly in sophisticated processing applications. While a spot welding operation needs no feedback, an arc welding process requires a means to track the seam. These sensory inputs may come from a variety of sensor types, including proximity switches, force sensors, and laser and machine vision systems.

The steel collared worker is a handicapped worker and has relatively limited sensing capabilities compared to humans. For example, an operator in a manufacturing cell can easily check surface finish and check for burrs in the couple of seconds needed to unload a part from a machine. The robot will have difficulty doing this simple task due to the difficulty with which robots can be effectively interfaced with sensors and the lack of suitable low-cost sensing devices. As control capabilities continue to improve and sensor costs to decline, the use of sensory feedback in robotics applications will increase.

In each application one or more of the robot's capabilities of transport, manipulation, and sensing is employed. These capabilities, along with the robot's inherent reliability and endurance, make it ideal for many applications now performed manually, as well as in some applications now performed by traditional automated means.

5 APPLICATIONS OF ROBOTS

By the end of 1997 approximately 80,000 robots were installed in the United States (*Robotic Industries Association Quarterly Report,* 3rd Quarter, 1997). These installations are usually grouped into the application categories shown in Figure 7. This figure also shows the major robot capabilities and requirements for each application and the type of benefits obtained. A list of application examples by type is contained in Table 4. A brief description of each application category is contained in the following paragraphs.

5.1 Material Handling

In addition to tending die casting machines, early robots were also used for other material-handling applications. These applications make use of the robot's basic capability to transport objects, with manipulative skills being of less importance. Typically motion

Requirements for further Application of IR													
Requirements of ● major ◐ moderate ○ minor importance	Handling				Assembly			Processing					
	Paletizing	Machine loading	Interlinking	Other handling	Small part ass.	Spot welding	Arc welding	Milling, grinding	Forming	Spraypainting	Other coating	Inspection	All applications
Various features of the robot													
Low cost, reliable and effective vision sensing	◐	○	○	●	●	○	●	◐	◐	◐	◐	●	◐
Easier, standardized programming	◐	◐	◐	○	●	○	◐	●	◐	●	●	●	●
Improved gripper dexterity[a]	◐	◐	○	●	●	◐	○	◐	●	○	○	◐	◐
Greater flexibility for different applications	○	◐	◐	●	◐	○	○	◐	◐	○	○	○	○
Low cost, effective force sensing	○	○	○	○	◐	○	○	◐	●	○	○	◐	○
Lighter, smaller robots	○	◐	◐	◐	●	○	○	○	○	◐	◐	○	○
Improved control systems	○	○	○	○	◐	○	◐	○	◐	◐	◐	◐	○
Greater speed	●	●	●	●	●	◐	○	○	●	○	◐	○	◐
Process capability													
Improved positioning accuracy	○	◐	○	◐	●	◐	◐	◐	○	○	◐	●	○
Improved repeatability	◐	◐	◐	◐	●	◐	◐	◐	○	○	◐	●	◐
Improved reliability and stability	◐	●	◐	◐	●	◐	◐	◐	◐	◐	◐	◐	●
Planning needs													
Reduced robot costs	●	●	●	●	●	◐	◐	◐	●	◐	◐	◐	●
Improved ability to interface with existing equipment	○	◐	●	◐	◐	○	◐	○	◐	◐	○	○	○
Product design/redesign	○	◐	◐	◐	●	●	●	◐	◐	◐	◐	◐	○
Turnkey robotic cells & systems	◐	◐	●	◐	●	●	●	◐	●	◐	◐	◐	●
All needs	○	◐	◐	◐	●	○	◐	●	◐	○	◐	◐	

[a]Dexterity in grippers refers to their ability to imitate the versatility of the human hand.

Figure 7 Typical requirements for industrial robot applications in manufacturing.

takes place in two or three dimensions, with the robot mounted either stationary on the floor or on slides or rails that enable it to move from one workstation to another. Occasionally the robot may be mounted overhead, but this is rare. Robots used in purely material-handling operations are typically nonservo, or pick-and-place, robots.

The primary benefits of using robots for material handling are reduction of direct labor costs and removal of humans from tasks that may be hazardous, tedious, or fatiguing. Also, the use of robots typically results in less damage to parts during handling, a major reason for using robots for moving fragile objects. In many material-handling applications, however, other forms of automation may be more suitable if production volumes are large and no manipulation of the workpiece is required.

5.2   Machine Tending (Loading and Unloading)

In addition to unloading die casting machines, robots are also used extensively for other machine loading and unloading applications. Machine loading and unloading is generally considered a more sophisticated robot application than simple material handling. Robots can be used to grasp a workpiece from a supply point, transport it to a machine, orient it correctly, and then insert it into the workholder on the machine. This may require that the robot signal the machine tool when it (thinks it) has the workpiece in the right position in the workholder so that the latter can clamp the part in the right location. The robot then releases the part and withdraws the hand so that processing can begin. While this appears to be rather a simple set of tasks, it may not be so simple if the two machines cannot communicate with each other. Because it is rather unlikely that the robot and the machine tool are using the same software, some sort of communication link will have to be established in order to integrate these machines and coordinate the functions. After processing, the robot unloads the workpiece and transfers it to another machine or conveyor. The greatest efficiency is usually achieved when a single robot is used to service

Table 4 Examples of Robot Applications

Material Handling

Moving parts from warehouse to machines
Depalletizing wheel spindles into conveyors
Transporting explosive devices
Packaging toaster ovens
Stacking engine parts
Transfer of auto parts from machine to overhead conveyor
Transfer of turbine parts from one conveyor to another
Loading transmission cases from roller conveyor to another
Transfer of finished auto engines from assembly to hot test
Processing of thermometers
Bottle loading
Transfer of glass from rack to cutting line
Core handling

Machine Loading/Unloading

Loading auto parts for grinding
Loading auto components into test machines
Loading gears onto CNC milling machines
Orienting/loading transmission ring gears onto vertical lathes
Loading hot forging presses
Loading of electron beam welder
Loading cylinder heads onto transfer machines
Loading and unloading punch presses
Unloading die cast machine or plastic injection molding machine

Spraypainting, Dispensing, Dipping

Painting of aircraft parts on automated line
Painting of trucks and cars and agricultural equipment
Application of adhesives and sealants to car bodies
Application of thermal material to rockets
Painting of appliance components

Welding

Spot welding of auto bodies
Welding front end loader buckets
Arc welding hinge assemblies on agricultural equipment
Braze alloying of aircraft seams
Arc welding of tractor front weight supports, auto axles
Laser welding sheet metal parts

Material Removal

Laser drilling ceramic parts
Drilling aluminum panels on aircraft
Metal flash removal from castings, edging components
Finishing missile parts, deburring

Assembly

Assembly of aircraft parts (used with auto-rivet equipment)
Riveting small assemblies
Drilling and fastening metal panels
Assembling molds and cores for sand castings
Assembling appliance switches
Inserting and fastening screws
Electronics assembly
Application of two-part urethane gasket to auto part
Installation of a battery in an auto assembly line

Inspection and Other

Induction hardening
Inspection dimension on parts
Inspection of hole diameter and wall thickness

several machines, as in a robotic cell. Also, the single robot may be used to perform other operations while the machines are performing their primary functions. This may require that the robot be able to exchange grippers.

Another example of a machine loading and unloading application is the loading and unloading of hot billets into forging presses. This may necessitate the design of a special gripper that can accommodate the preforged billet of narrow cross section and later the forged, flattened billet of wider cross section. Loading and unloading machine tools, such as lathes and machining centers; stamping press loading and unloading; tending plastic injection molding machines; holding a part for a spot welding operation. Although robot hands are more durable and heat-resistant than human hands, robot hands can also get too hot. The robot can be programmed to dip its hand into a cooling bath at appropriate intervals.

Although adverse temperatures or atmospheres can make robots advantageous for machine loading and unloading, the primary motivation may still be to reduce direct labor costs. Overall productivity is also likely to increase because of the longer amount of time the robot can work compared to humans. In machine loading and unloading it is both the manipulative and the transport capabilities that make use of robots feasible. The task may require a high level of RPC.

Robot configurations have been developed specifically for tending tandem press lines for press-to-press transfer of sheet metal parts. See Figure 8. This is an excellent example of a robot for material handling and machine tending.

These robots require a variety of features. Configurations include swing-arm, track-mounted, and pendulum-type robots.

Pendulum robots are medium payload, overhead, or bridge-mounted four, five, or six-axis robots. This configuration allows for greater reach with offset tool (20–29 ft) and faster cycle time (420–550 pph). They have medium wrist moment load ratings (300–600 lb. ft) and can handle medium-size stamping that can be run on 108–144 in. press lines. The robot's reach with an offset tool can typically handle press centers from 16–22 ft. See Figure 9.

Some press-mounted solutions use two robots and a transfer station to transfer panels. Two robots are often used between the draw and trim presses to allow for part turnover requirements.

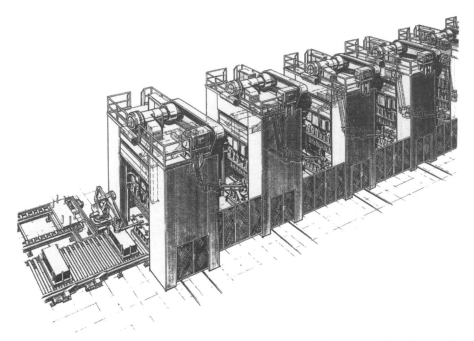

Figure 8 Typical pendulum robot press line. The high-speed robots are capable of processing up to 600 parts per hour.

Figure 9 This robotic transfer system features an overhead-mounted robot. The mounting method allows the robot, which is the tallest in the industry, to swing up and out of the way during die changes.

The six-axis robots have higher flexibility to handle parts that require nonlinear manipulations during load/unload operations. The configuration of the pendulum arm, along with a horizontal articulation, allows it to be overhead-mounted and offset to the side of the press line, allowing for die change clearance with die carts without moving the robot. The robot support structure can be floor- or press-mounted in various custom configurations.

Swing-arm robots are typically five- or six-axis, medium payload articulated-arm robots mounted on a six or seventh axis swing arm rotational base. This configuration allows for greater reach with offset tool (20–28 ft) and faster cycle time (420–550 pph). These robots are typically floor-mounted and sit between the presses. To allow for die change clearance, the robots may need to be mounted on slide bases to retract them out from between the presses.

Track-mounted robots are typically four- to six-axis, medium payload articulated-arm robots mounted on an auxiliary axis linear track that travels in the same direction as the flow of the press line. This configuration allows for greater reach with offset tool (22–26 ft) and slightly faster cycle time (360–500 pph). These robots are typically floor-mounted and sit between the presses. To allow for die change clearance, the robots may need to be mounted on slide bases to retract them out from between the presses.

Each of these robots has various qualities, advantages, benefits, costs, and limitations in press-to-press transfer applications. It is important that manufacturers evaluate and prioritize these features and benefits so the proper robot model or models can be selected. This will optimize the stamping process while supplying the highest quality parts at the highest yields with the lowest possible cost per part.

5.3 Spraying

Spraypainting is a natural application for industrial robots. Robots are able to achieve a level of consistency that it is difficult to expect human spraypainters to duplicate. Although a skilled human spraypainter is required to teach the robot a painting task initially, the robot, once taught, will do the spraypainting operation repeatedly with a consistency unattainable by the human who taught it. A comprehensive spraypainting line for automobile bodies uses robots not only for the spraypainting operation but also for opening the doors of the automobile bodies to facilitate the operation. Spraypainting robots are being utilized on a more modest scale in industries other than the automobile industry. In spraying applications the robot manipulates a spray gun to apply a material, such as paint, stain, adhesive, or plastic powder, to either a stationary or moving part. These coatings are applied to a wide variety of parts, including automotive body panels, appliances, and furniture. In those cases where the part being sprayed is on a moving conveyor line, the robot's sequence of spraying motions is coordinated with the motion of the conveyor. In a similar application robots apply resin and chopped glass fiber to molds for producing glass-reinforced plastic parts and spray epoxy resin between layers of graphite broadgoods in the production of advanced composites.

The manipulative ability of the robot is of prime importance in spraying applications. Accuracy is not a problem in spraying. A major benefit is higher product quality through more uniform application of material (repeatability). Other benefits include reduced labor costs and reduced waste of coating material and reduced exposure of humans to toxic materials. See Chapter 46.

5.4   Welding

The largest single application for robots at present is in spot welding automotive bodies. See Chapter 44. Spot welding is normally performed by a point-to-point servo-robot holding a welding gun. Arc welding can also be performed by robots using noncontact seam trackers, which have greatly increased the use of robots for arc welding. Today robotic arc welders are low-cost, easily programmed, and durable. Most of the robotic welding units have six axes: three in arm and three in wrist. The robot has a fixed load, with aluminum the material of choice (to reduce mass). Some welding robots demonstrate a repeatability of $\pm$ 0.004 inches anywhere in the WE at maximum speed. Simulation and off-line programming are being investigated to reduce the time needed to set up and run a welding project. The problem is that (1) off-line programming is still expensive and (2) the programs must be manually edited (to account for the factors discussed earlier under RPC, such as flexing, effect of weight, speed, and gear backlash). All these factors (and inactions of these factors) affect the accuracy. When you use the *teach* pendant (on-line programming), you are using the repeatability aspect of RPC rather than the accuracy.

5.5   Machining

In machining applications the robot typically holds a powered spindle and performs drilling, grinding, deburring routing, or other similar operations on the workpiece. A schematic of a robot performing a deburring operation is shown in Figure 10. The problem in deburring is that the burr varies in size (height and width) so the cutting forces also vary. Therefore burring tools are usually rotated at very high revolutions per minute (ca 20,000) and are equipped with a constant force compensation device (controls feed rate to maintain constant cutting force so as to minimize deflection of the cutting tool). Rotary files made of tungsten carbide, PCD, CBN, or TiN coated tool steels are used. In general deburring cutting forces are very small, so the deflections of the tooling are small and three-force accuracy is good.

In machining operations the workpiece can be placed in a fixture by a human or another robot. In some operations the robot moves the workpiece to a stationary powered spindle and tool, such as a buffing wheel, and manipulates the part against the tool in order to accomplish the processing. The general problem in machining is this: The process of chip formation creates cutting forces, which may be quite substantial in processes like drilling or milling. These forces can cause the tool to deflect and consequently a loss of accuracy since the tool does not take the desired path to achieve the desired part size or geometry. Robots in general do not have the rigidity typically found in machine tools.

Robot applications in machining are limited at present because of accuracy requirements, expensive tool designs, and lack of appropriate sensory feedback capabilities. Most machining processes require that the robot be accurate (able to go to the right place the first time and perhaps only once) and that it must usually be programmed off-line, which means that accuracy will be much poorer than repeatability.

5.6   Assembly

One area of current robot assembly operations includes the insertion of light bulbs into instrument panels, the assembly of typewriter ribbon cartridges, the insertion and place-

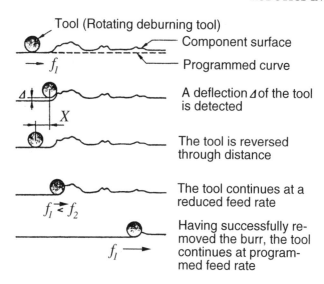

Tool (Rotating deburning tool)

Component surface

Programmed curve

f_l

A deflection Δ of the tool is detected

X

The tool is reversed through distance

$f_l \lesssim f_2$

The tool continues at a reduced feed rate

$f_l \longrightarrow$

Having successfully removed the burr, the tool continues at programmed feed rate

- Feed rate control functions which enable the robot itself to select appropriate feed rates for deburring or grinding.

Figure 10 Robotic deburring requires adaptive control functions that can adapt to the variable burr size.

ment of components onto printed circuit boards, and the automated assembly of small electric motors. More complex assembly tasks typically require improved sensory feedback, improved RPC (accuracy, repeatability, reproducibility), and better programming languages. See Chapters 50 and 51.

5.7 Inspection

A growing number of robot applications are in the area of inspection. Robots are used in conjunction with sensors, such as a vision system, lasers, or ultrasonic detector, to check part locations or identify defects. Such robots have been used to inspect valve cover assemblies for automotive engines, sort metal castings, and inspect the dimensional accuracy of openings in automotive bodies.

As in assembly and machining operations, a high degree of accuracy and extensive sensory abilities are required for inspection applications. In the future this is expected to be one of the high-growth application areas as low-cost sensors and improved positioning accuracy evolve. See Chapter 52.

5.8 User Industries

Not surprisingly, the industry with the most robots installed is the auto industry, with about 40% of the total U.S. robot population. Within the auto industry, welding is the most common robot application, using about 70% of all robots in that industry.

Although robots are used in almost every industry and type of application, the majority of installations are concentrated in a relatively few plants and types of applications. For example, it is estimated that just 10 plants contain nearly one-third of all robots installations and that the three categories of welding, material handling, and machine loading account for approximately 80% of all current applications. At the same time, it must be remembered that the market penetration of robots has been relatively limited in even the most common applications.

6 ROBOTS IN FABRICATION AND PROCESSING

6.1 Definition of Fabrication and Processing

The previous section provided an overview of all major types of industrial applications. Most of these application types are discussed in detail in subsequent chapters of this handbook, including chapters on welding; material handling and machine loading; assembly; inspection, quality control, and repair; and finishing, coating, and painting.

As a matter of convenience, in this chapter *fabrication and processing* is defined for those applications not covered in later handbook chapters. In some instances the applications described in the following sections reflect the use of robots for specific manufacturing functions, such as machining or heat treating, whereas in other cases they involve the use of robots in specific industries, such as plastics processing and glassmaking. In these latter cases there will be some obvious duplication with information presented in subsequent chapters.

6.1.1   Die Casting

As mentioned previously, die casting has historically been one of the major application areas for industrial robots. In fact, in 1961 the first commercial robot installation involved tending a die casting machine. Today die casting continues as one of the largest application areas in the United States.

In retrospect, die casting was an ideal area for the initial use of robots because this application possesses a number of important characteristics that make it amenable to robotization. Die cast parts are produced in relatively large volumes, but it is also important to keep the time required for equipment changeover to a minimum. The parts are precisely oriented when they are removed from the die casting machine, which makes them suitable for robot handling with standard or slightly modified grippers. Because of the maturity of the die casting process, there are few equipment or product design changes that would necessitate retooling. Furthermore, die casting operations are notoriously hot, dirty, and hazardous, providing a particularly unpleasant environment for human workers. And last, die casting is a relatively competitive industry that benefits from both the cost reduction provided by robots and the improved product quality resulting from their consistent performance.

Robots can be used to perform a number of functions in die casting. In simple installations the robot is used to remove the part from the die and place it on a conveyor. In more sophisticated applications the robot may perform a number of tasks, including part removal, quenching, trim press loading and unloading, and periodic die maintenance. The robot may also be used for insert placement and, in the case of aluminum die casting, loading the cold-shot chamber. In some cases a single robot can service two die casting machines. The specific functions performed depend on a number of factors, including casting cycle times, physical layout, and robot speed and type.

Although robots have been installed in a number of die casting plants, a number of important points must be considered in planning applications. Overall layout of the installation must be carefully thought out in those cases where the robot performs more than just part unloading or services more than one die casting machine. Similarly, the interfacing requirements between the robot and other equipment may become complex. Also, additional sensory inputs may be required to insure that all parts have been removed from the die and that the robot maintains its grip on the sprue.

Optimizing the cost-effectiveness of the installation may also be challenging. Although maximizing throughput of the die cast machine is clearly of primary importance, deciding which type of robot should be used and which functions should be performed by the robot is not so obvious. In some installations it is more appropriate to use other automated techniques for such functions as die lubrication and metal ladling. This would permit the use of a less costly nonservo robot. In other cases, however, the use of continuous-path servo-robots to perform these functions may more than justify the increased cost.

The benefits to be obtained by using robots in die casting have been well established. It is not uncommon to replace as many as two workers on a shift by one robot, and direct labor cost reductions of as much as 80% have been reported. Also, 20% increases in throughput are possible because of consistent cycle times and the elimination of rest periods and lunch breaks required by humans. Significant increases in product quality have also been obtained. This is primarily due to consistent cycle times and constant die temperatures that result in better-quality parts and less scrap. Net yield increases of 15% have been achieved. Other benefits of using robots in die casting include increased die life, reduced floor space for material handling and storage operations, and a significant decrease in the cost of safety equipment since human operators are not required. This added benefit of removing humans from tedious and unpleasant tasks associated with die casting should not be overlooked. See also Chapter 43.

6.1.2   Foundry Operations[2]

Although foundries represent one of the most difficult operation environments for human workers in industry today, the use of robots in this area has been relatively slow in materializing. This is probably primarily due to the diversity of castings typically encountered in most foundries and the relatively low-technology approach usually undertaken in such facilities. These factors notwithstanding, a considerable number of robots have been effectively employed in foundries, and additional installations are expected in the future.

Foundry applications of industrial robots have ranged from ladling of molten metal into molds to final cleaning of castings. Robots have been particularly useful in mold preparation, where they have been used for core handling and for spraying and baking of refractory washes on copes and drags. Robots have also been used for traditional material-handling operations, such as removing hot castings from shakeout conveyors and placing the casting in a trim press to remove gates and runners.

Another major use of robots is beginning to emerge in casting cleaning operations. When the casting is first removed from the mold it is still attached to gates and risers and is likely to have a considerable amount of flash that needs removing. Traditionally, removal of these unwanted appendages has been done manually and is an extremely unpleasant and costly task. Robots have met with some success in grinding flash and chipping and cutting away gates and risers.

Another well-established robot application in foundries is mold making for investment casting operation. In this application wax pattern trees are repeatedly dipped into a ceramic slurry and stucco sand to build up the mold shell (see Figure 11). As many as 12 coats may be required before the mold has reached its desired size. Following these repeated dipping and drying operations, the mold is heated so that the wax patterns are melted out. The mold is then fired and used for investment casting. The mold is destroyed when the casting is removed. A number of these mold-making operations are totally automated through the use of industrial robots.

The primary reasons for using robots in foundry operations are cost reduction and elimination of unpleasant or hazardous tasks for human workers. More important, significant improvements in product quality that come about through the consistency of robot operations are also a significant benefit in foundries. This latter benefit is particularly true in investment casting, where more uniform molds translate into much higher yield of good castings.

The use of robots in foundries also presents a number of challenges, particularly with respect to the operating environment and interfacing the robot with other equipment. The abrasive dust encountered in foundries may require the use of protective covers to prevent damage to the robot. In mold-making operations for investment casting, interfacing the robot with slurry mixtures, fluidized beds, conveyors, and drying ovens is usually a complex task because of the variation in processing requirements that may be reflected in typical product mixes. And last, implementation of casting cleaning operations may be difficult because of the sensory feedback normally required to deflash castings.

6.1.3   Forging[3]

Since 1974 a number of robots have been used in forging operations. These applications have ranged from loading and unloading of forging presses to the movement of workpieces from one die station to another. By far the largest category of applications in forging is material handling. Robots have been used to load furnaces, move heated billets between furnaces and drop hammers or forging presses, and move forged workpieces from presses to drawing benches, trim presses, conveyors, or pallets. Robots have also been used to apply lubricant to both workpieces and dies.

Robots have been used primarily in closed die forging and heading operations since these are relatively precise processes. To a lesser extent they have also been used in drop forging, upset forging, roll forging, and swaging, processes that have a relatively large degree of variability between workpieces.

---

[2]See Asfahl (1991); Degarmo, Black, and Kohser (1997); Harley (1983); Hunt (1983); Tanner (1981, 1983); and Warnecke, Schraft, and Wanner (1985).
[3]See Asfahl (1991); Degarmo, Black, and Kohser (1997); Hunt (1983); Tanner (1981, 1983); Warnecke, Schraft, and Wanner (1985).

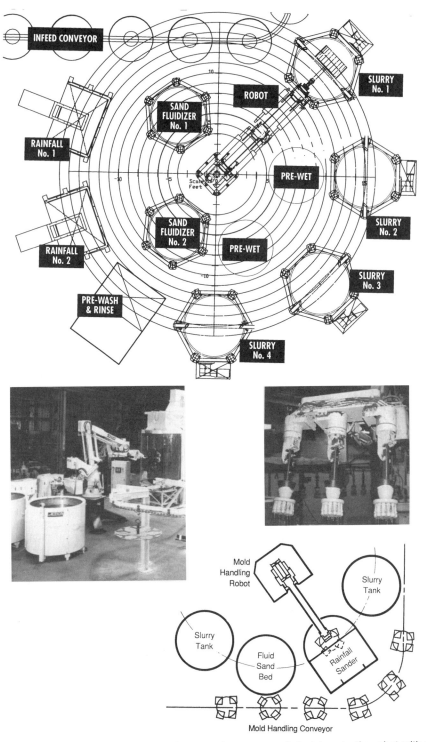

Figure 11 Dipping wax patterns requires a systems approach to integrate the robot with the conveyor and tanks.

Because of the harsh environment encountered in forging operations, the primary motivation for using robots is to eliminate unpleasant tasks for human workers, which also reduces direct labor costs. Another major reason for using robots in forging is to improve product quality, which can be accomplished through the robot's consistent operation. Furthermore, increases in throughput of up to 30% have been reported because of the robot's ability to work almost continuously.

Forging applications present unique challenges for robots. Handling of the hot billets may require water-cooled grippers or the periodic cooling of tooling by immersion in water baths. Similarly, the heavy shock loads produced by forging may necessitate the use of special grippers for isolation from these forces. Additional problems are created by time-varying changes in the process resulting from warpage and die wear. And last, it must be remembered that forging is still a relatively hazardous process characterized by extremely high forces and frequent abnormalities, such as workpieces not ejecting properly from dies. To overcome these difficulties, robots used in forging operations typically require a high level of interlocking with other equipment and sensory inputs to insure proper functioning of the process.

6.1.4 Heat Treatment[4]

As in applications such as die casting, foundry operations, and forging (see Chapter 39), the use of robots in heat treating primarily involves material-handling and machine-loading and unloading tasks. Robots typically are used to load and unload heat-treating furnaces, salt baths, and washing and drying stations.

The motivations for using robots in heat treating include elimination of unpleasant and hazardous tasks, cost reductions, improved product quality, and increased productivity. Generally, medium- and low-technology robots can be employed, and few unusual difficulties are encountered beyond protecting the robot from the high temperatures normally encountered in heat treating.

6.1.5 Forming and Stamping[5]

Press work, such as stamping, forming, and trimming, is another area where robots were applied early in their development. Such applications have ranged from feeding presses for stamping small parts to loading and unloading large presses for forming automotive body panels. Again, the applications are primarily machine loading and unloading and typically involve the use of medium- or low-technology pick-and-place robots.

Robots are usually used in press work applications for two reasons. First, press loading is considered a very dangerous task, and robots are used primarily for safety and to minimize hazard to human workers. Second, cost reduction can be achieved through both the elimination of human labor and increased productivity.

Although robots have been successfully applied in sheet metal forming, there are a number of cases when their use is not appropriate. Many stamping and forming operations are highly automated, particularly for long production runs, and robots have difficulty competing on an economic basis under these circumstances. Similarly, many press work applications are relatively high-speed operations, and robots are simply incapable of achieving the necessary operating speeds. These factors tend to limit robot applications to those situations where production quantities are moderate and manual techniques predominate, or where low-technology robots and special-purpose robot designs, such as two-armed robots, are appropriate.

6.1.6 Machining[6]

Machining as a process is usually used to achieve accurate and precise dimensions in the parts. As mentioned previously, machining applications for robots are limited by process capability requirements. Robots have been used to perform such tasks as drilling, routing, reaming, grinding, countersinking, and deburring. Many of these applications have been in the aerospace industry, with the exception of deburring, which has seen more widespread use. Almost all the machining applications where the tool contacts the work use

---

[4]See Tanner (1981).
[5]See Hunt (1983); Tanner (1981, 1983); Warnecke, Schraft, and Wanner (1985).
[6]See Asfahl (1991); Degarmo, Black, and Kohser (1997); Harley (1983); Molander (1983); Tanner (1981); Warnecke, Schraft, and Wanner (1985).

a rotating tool held in the hand of the robot. The rotating tool produces varying cutting forces, which can produce vibrations in the manipulator. Machine tools counter this problem with rigid construction of the machine, but this option is not usually available for robots. A long arm with two joints may have a lot of defection problems. In deburring, the cutting forces are small and steady (due to high rpm of the tool), and many deburring setups are equipped with constant force cutting algorithms. Since many machining applications tax the process capabilities of today's most sophisticated robots, it is not likely that robots will be widely used for this function. For most machining operations the robot's speed of end effect traverse may not be fast enough when compared to other processes and machine tools. Similarly, in tasks such as deburring, programming becomes a major problem because of the number of complex motions that must be executed by the robot.

Another major difficulty in using robots for machining operations is positioning accuracy. Present robots simply have difficulty achieving the necessary accuracy and repeatability (precision) needed to locate the tool with the work. Because of this, most robot machining operations rely on the extensive use of jigs and fixtures, which quickly erases any advantages the robot provides in terms of flexibility and low cost. Considerable research and development is being conducted in an attempt to improve positioning accuracy by means of sensory feedback. Although some progress has been made in this regard, the use of such sensors is rare, and improved solutions to the problem are still being sought. Some robotic machining operations use energy beams (like lasers) or an abrasive waterjet. In this case cutting forces are not a problem, but control of the beam during processing can be a problem (safety concerns).

6.1.7   Drilling[7]

Drilling is a simple machining process in that the feed rate is one-dimensional, usually the z axis. Drilling is a widely used fabrication process and should be considered a candidate application for robotics whenever the drilling is done by a handheld drill. Most drilling operations are done by fixed drill presses, but in the fabrication of large products such as aircraft, space vehicles, ships, cranes, railroad locomotives, and other transportation equipment, handheld drilling operations are standard procedure. The reason is that the workpieces are so large and unwieldy that it is more feasible to bring the tool and the jig to the workpiece than to fixture the workpiece in a drill press (see Figure 12).

An industrial robot equipped with a drill as its end-effector has many of the capabilities of a human operator with a handheld drill. It can move about and can position its work platform in an infinite number of planes. This positioning capability is essential in such tasks as drilling rivet holes in aircraft skin, the surface of which may be curved and therefore is in no consistent plane. The human operator with a handheld drill or an industrial robot can adjust the plane of the tool platform such that the direction of the feed is orthogonal to the surface of the aircraft skin. The human operator typically manually positions the handheld drill in an orthogonal orientation by visual determination. In such an operation the robot can be positioned more precisely than a human-held drill and thus quality is enhanced even when a template is used. Not only quality, but economics and safety are benefits to be gained from robotic drilling operations.

6.1.8   Plastics Processing[8]

In plastics processing the most common application of robots by far has been for unloading injection molding machines. Other applications have included unloading transfer molding presses and structural foam molding machines, handling large compression-molded parts, and loading inserts into molds. Robots have also been used for spraying and applying a variety of resins, as well as performing many secondary operations such as trimming, drilling, buffing, packaging, and palletizing of finished plastic products.

Estimates indicate that nearly 5% of all injection molding machines are now tended by robots, and in Japan this is one of the most common robot applications. The robot may be mounted on either the top or bottom of the molding machine, or it may be a stand-alone unit servicing more than one machine. Increasingly, special-purpose, low-cost robots are being employed for this function.

[7]See Asfahl (1991).
[8]See Hunt (1983); Meyer (1983); Tanner (1981, 1983); Warnecke, Schraft, and Wanner (1985).

Figure 12   Cincinnati Milacron T³776 robot drills rivet holes for vertical fin skins in F16 aircraft. (From C. R. Asfahl, *Robots and Manufacturing Automation,* 2nd ed. New York: John Wiley & Sons, 1992.)

The use of robots in plastics processing is no more difficult than other types of robot applications. The same level of integration with other equipment, such as conveyors and trim presses, is still required, and overall equipment layout is important. Gripper design may be a problem, however, because of workpiece size, quantity, and limpness (see Figure 13).

6.2   Other Areas

Robots are being used or considered for use in almost every conceivable industry. A brief sampling of some additional applications that might be considered as part of fabrication and processing follows:

1. *Electronics processing.* Electronics is starting to emerge as one of the important user industries for robotics. Applications include machine loading and unloading for parts fabrication and presses, placement of surface-mounted components for hybrid microcircuit assembly, component insertion for printed circuit board assembly, cable harness fabrication, and robot-assisted test and inspection.

2. *Glassmaking.* Robots are used in glassmaking because of their ability to withstand high temperatures and handle fragile workpieces, which eliminates unpleasant and hazardous tasks for humans and reduces overall costs. Robots have been used for charging molds with molten glass and for handling both sheet and contoured glass products.

3. *Primary metals.* Robots are being used in the production of primary metals such as steel and aluminum. In addition to traditional material-handling applications, robots have been used for such tasks as charging furnaces with ingots and furnace tapping.

4. *Textiles and clothing.* The use of robots in textile and clothing manufacturing presents unique problems because of the limp nature of the workpieces. However, robots are being used for such applications as material handling in spinning mills

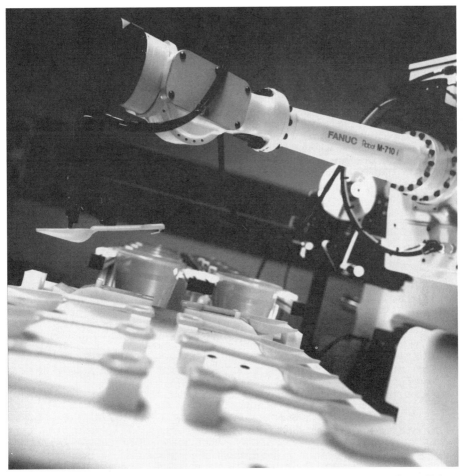

Figure 13   Robot with special gripper for handling plastic-molded parts (courtesy Fanuc Robotics).

and automatic placement in laser-cutting machines and sewing machines for clothing items.

5. *Food processing.* In addition to material-handling and packaging applications in food processing, robots have also been used for such tasks as decorating chocolates and actual food preparation.

6. *Chemical processing.* The chemical processing industry, which is normally thought of as a continuous-flow operation, is starting to use robots for a variety of applications, including material-handling and maintenance activities such as cleanup of chemical reactors.

7   FUTURE TRENDS IN MATERIALS AND PROCESSING

The use of industrial robots will continue to increase during the next decade. Robots will continue to be designed and manufactured with greater process capability. Many new jobs will be created for such positions as robot programmers, troubleshooters, and maintenance personnel.

Along with this increase in robot sales, a number of important technological and product developments are expected to take place. These include the development of smaller, faster, and lighter-weight robots, an increase in payload capacity relative to the weight of the robot, and dramatically improved grippers. In the sensor area major improvements are expected in machine vision systems, tactile sensing, and low-cost force sensors. Major developments are also anticipated in robot control and programming capabilities, including the use of hierarchical control concepts and off-line programming. At the same time, robot prices are expected to decline as production volumes increase.

These trends and anticipated developments can only help to accelerate the use of robots for fabrication and processing applications. Improved capabilities, particularly in control and sensory technology, coupled with declining costs, should make many currently difficult applications cost-effective realities in the future.

As robot sales continue to grow, new applications for robots are anticipated. Traditional robot applications such as spot welding, material handling, and painting may decline somewhat in terms of their respective market shares, while other emerging applications, such as assembly and machining, may increase significantly. Since machining, assembly, and similar uses of robots generally require more sophisticated equipment and interfacing, their relative impact on future manufacturing operations will be even more significant.

Some parallel-link robots are being developed for high-force and torque applications, but these robots are very complex to program and very expensive, so their future in commercial applications is limited.

REFERENCES

Asfahl, C. R. 1991. *Robots and Manufacturing Automation.* 2nd ed. New York: John Wiley & Sons.

Degarmo, E. P., J. T. Black, and R. A. Kohser. 1997. *Materials and Processes in Manufacturing.* 8th ed. Englewood Cliffs: Prentice-Hall.

Gilby, J. H., and G. A. Parker. 1982. "Laser Tracking System to Measure Robot Arm Performance." *Sensor Review,* 180–184.

Gilby, J. H., R. Mayer, and G. A. Parker. 1984."Dynamic Performance Measurement of Robot Arms." In *Proceedings of the 1st Robotics Europe Conference,* Brussels, Belgium. 31–44.

Harley, J. 1983. *Robots at Work: A Practical Guide for Engineers and Managers.* Bedford: IFS.

Hunt, V. D. 1983. *Industrial Robotics Handbook.* New York: Industrial Press.

Industrial Robots: A Summary and Forecast. Naperville, Illinois: Tech Tran Corp., 1983.

Jiang, B. C., J. T. Black, and R. Duraisamy. 1988. "A Review of Recent Developments in Robot Metrology." *Journal of Manufacturing Systems* 7(4), 339–357.

Jiang, B. C., et al. 1989. "Determining Robot Process Capability Using Taguchi Methods." *Robotics and Computer-Integrated Manufacturing* 6(1), 55–66.

Jiang, B. C., et al. 1991. "Taguchi-Based Methodology for Determining/Optimizing Robot Process Capability." *IIE Transactions* 23(2), 169.

Lau, K., and R. J. Hocken. 1984. "A Survey of Current Robot Metrology Methods." *Annals of the CIRP* 32(2), 485–488.

Meyer, J. D. 1983. "Industrial Robots in Plastics Manufacturing—Today and in the Future." In *Proceedings of the Regional Technical Conference on Automation, Tooling, and Thermosets,* Mississauga, Ontario, March. Greenwich, Conn.: Society of Plastics Engineers.

Molander, T. 1983. "Routing and Drilling with an Industrial Robot." In *Proceedings of the 13th International Symposium on Industrial Robots and Robot 7,* Chicago, April. Dearborn: Society of Manufacturing Engineers.

Parker, G. A., and J. Gilby. 1982–1983, "An Investigation of Robot Arm Position Measurement Using Laser Tracking Techniques." Department of Mechanical Engineering Report, University of Surrey.

Rembold, U. 1990. *Robot Technology and Applications.* New York: Marcel Dekker.

Tanner, W. R., ed. *Robotics Industry Association Quarterly Report,* 3rd Quarter, 1997. *Industrial Robots.* Vols. 1 and 2. Society of Manufacturing Engineers.

———. 1983. *Industrial Robots from A to Z: A Practical Guide to Successful Robot Applications.* Larchmont: MGI Management Institute.

Warnecke, H. J, and R. D. Schraft. 1982. *Industrial Robots: Application Experience.* Bedford: IFS.

Warnecke, H. J., R. D. Schraft, and M. C. Wanner. 1985. "Performance Testing." In *Handbook of Industrial Robotics.* Ed. S. Y. Nof. New York: John Wiley & Sons. 158–166.

Ziegert, J. and P. Datseris. 1990. "Robot Calibration Using Local Pose Measurements." *International Journal of Robotics and Automation* 5(2), 68–76.

CHAPTER 43

ROBOTICS IN FOUNDRIES

Doug Niebruegge
ABB Flexible Automation
New Berlin, Wisconsin

1 INTRODUCTION

The use of robots in the foundry is by no means a recent development. In fact, the history of robotics in die casting is a relatively long one. The first such recorded application dates back to 1961, when the Ford Motor Company installed a robot to tend a die casting machine at a foundry in the United States. The anticipated benefits of this initial application—namely, improved worker safety and more predictable production output—were soon realized. In addition, part quality also improved, thanks to the regularity of the robot operation, which maintained die temperatures at a constant level to improve the consistency of the castings.

These basic benefits of robotics still apply today. However, as robot technology has developed and industry's demands have changed, other advantages have emerged. So the true potential of the industrial robot is now being better realized.

The key reason a robot is more economical than a dedicated device lies in its flexibility—its ability to undertake a wide variety of other functions. It is the one capital asset that can be readily reconfigured to respond to changing demands and provide a practical way of automating an array of tasks. Cooling the casting by dipping or spraying it, placing the part into a trimming press, spraying the dies with a lubricant—these are just a few of the many added benefit functions that can be performed by CNC-controlled articulated-arm robots (see Figure 1).

With a life expectancy of at least five years and a typical payoff time of less than two years, robots are an economically viable alternative—especially when one considers that a robot can be reprogrammed and retooled for further projects with a minimum of investment.

2 ROBOTICS IN DIE CASTING

From the introduction of the die casting process, automation was recognized as a key factor in achieving consistent cycle times, which in turn had a significant impact on part quality and productivity.

The operation of the die casting machine was inherently automatic, so the need was to automate the material loading and part unloading operations. While the hazardous task of ladling the molten metal into the machine was soon automated, automating the process of removing or unloading the casting was harder to justify for two reasons. In the early days replacing relatively cheap operators with capital-intensive automation equipment was simply not economically viable. In addition, it soon became apparent that the operator did more than just remove the part. He visually inspected it, perhaps quenched it in a bath or a spray, possibly placed it in a trimming press, and then stacked or palletized it. This degree of flexibility could not be provided by the automatic take-out devices then being employed in industries such as automotive.

Today part unloading is one of the major applications for industrial robots in die casting. Part removal is relatively easy for a robot. It not only has the power to grip the part, but it also has the dexterity to reach in the dies and remove the part in the optimum direction without damaging the casting. Changeover to new parts is fast and accurate.

Handbook of Industrial Robotics, Second Edition, Edited by Shimon Y. Nof
ISBN 0-471-17783-0   © 1999 John Wiley & Sons, Inc.

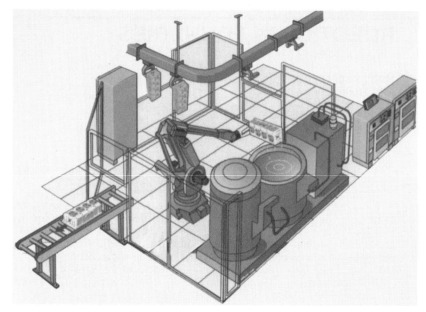

Figure 1 An integrated foundry cell automation (courtesy ABB Flexible Automation Inc.).

2.1 Auxiliary Foundry Operations

Robots are the solution to a wide range of other problems in the die casting foundry as well. They perform both pre- and postprocess operations with a precision and regularity not possible with even the most skilled operators—in one of the most hazardous of industrial environments, now often considered unfit for human occupancy. In fact, the real potential of robots in die casting did not begin to be realized until they were programmed to tackle some of the peripheral activities previously performed by human operators.

2.1.1 Parts Integrity Checking

The first of these activities was to check the integrity of parts. Robots can check for faults that might scrap a part by passing the new casting in front of an array of sensors programmed to inspect critical areas on the part. This task is greatly simplified by the robot's ability to position the part in three-dimensional space. A robotics system can also check the die for the presence of remaining injection material that could prove catastrophic if the next injection proceeded without the material being removed.

2.1.2 Cooling

Robotics systems soon demonstrated their worth in the next step in the die casting process, cooling the cast part. Robots are capable of dipping the part in a coolant tank, passing it under sprays or placing it onto a conveyor for air cooling, and then picking it up.

2.1.3 Trimming

Similarly, a robot can easily place or hold a part under a trimming press. This application was facilitated by the development of the "soft servo," which prevents the blow of the press from being transmitted to—and possibly damaging—the robot arm.

In some instances robotic deburring replaces the press trimming operation. This saves on the cost of press tools that need to be designed and changed for each casting. A new deburring program is simply loaded into the robot when a new part appears.

2.1.4 Storage

Further downstream, the robot has proven to be one of the most cost-effective and space-saving devices available for storing or palletizing finished castings. Its combination of flexibility and software allows it to create stacks and patterns of parts to meet virtually any requirement.

2.1.5   Pouring Metal

Robots also provide opportunities for operations other than the actual handling of parts within the die casting foundry. Pouring molten metal—a particularly hazardous operation requiring precise control to avoid spillage and ensure complete die filling—is now being done effectively by robots (Figure 2).

2.1.6   Die Insertion

Other preprocess operations, such as die insertion, are also well suited for arm robots, which can easily pick up inserts from indexing conveyors or carousels, or from magazines, totes, or boxes on a pallet.

2.1.7   Die Lubrication

Die lubrication is another process that is ideal for automation. A robot can be programmed to apply just the right amount of lubricant in the proper parts of a die. This minimizes material waste, optimizes lubrication for maximum part quality, and helps control die temperatures for further process improvements (Figure 3).

2.1.8   Integration

Automation is recognized as a key factor in achieving consistent cycle times, quality, and productivity in die casting processes. In fact, in many current installations the robot co-ordinates the entire operation of the die casting cell, serving as the principal instrument of cell automation. The progress of robots in die casting is a perfect testimony to their versatility in foundry operations.

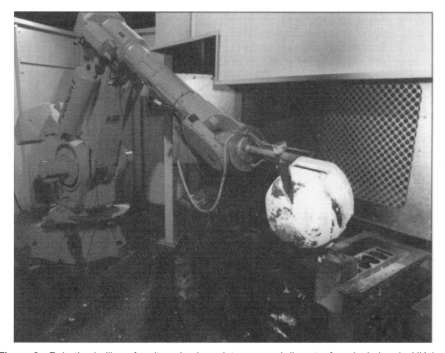

Figure 2 Robotics ladling of molten aluminum into carousel dies at a foundry in Leeds, UK, has the following features: avoids spillage by precise control and smooth movements; ensures complete die filling; conforms with IP67 Standards; has 40,000-hour mean time between failures (courtesy ABB Flexible Automation Inc.).

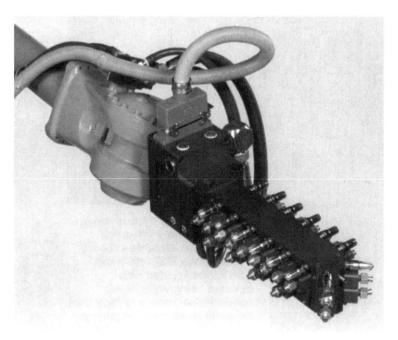

Figure 3   A spray tool mounted on a robot wrist can apply lubricant just where it is needed in a die to condition the die for maximum coating quality (courtesy ABB Flexible Automation Inc.).

3   ROBOTICS IN ALUMINUM CASTING

The automotive industry and other industries are switching from steel to aluminum parts due to their light weight and easier recycling. Aluminum is a readily available material that can be cast into shape with relative ease. However, special-purpose machines do not possess the accuracy and flexibility needed to cast aluminum wheels, engine and transmission components, and more complex aluminum parts with thinner profiles. Consequently, the die casting industry is investing heavily in advanced technology systems to meet this growing demand.

Articulated robots are used in the same application areas in modern aluminum foundries as they are in general die casting facilities. Here, too, robotics plays a critical role in improving part quality and consistency while lowering manufacturing costs (Figure 4).

4   ROBOTICS IN SAND CASTING

Until recently, few robots had been utilized in sand casting foundries. However, the considerable benefits provided by robots in die casting have spawned a growing interest in robotic benefits for other foundry processes.

One of the most notable developments has been the use of robots in the manufacture of sand cores. The industrial robot has the strength, precision, and dexterity to handle the relatively delicate but often very heavy sand cores. Because of the large number of sand cores that need to be produced, demand for this process is high and a number of successful robotic sand core systems have been installed.

Other sand casting tasks performed by robots are part insertion, core deburring, and gluing of multiple component cores. Robots are also being successfully installed for cleaning in sand core foundries, an application area with great future potential (Figure 5).

5   EXAMPLES OF APPLICATIONS

In today's foundry, robots have had a significant positive impact on manufacturing costs, productivity, uptime, product quality, scrap rates, changeover times, and working conditions. Current foundry robots have been developed to meet these many demands, along with the specialized demands placed on machinery that must operate reliably in a foundry

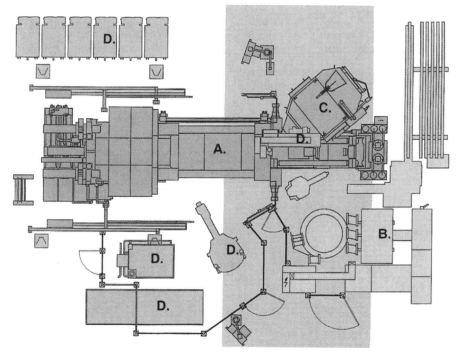

A. Die casting machine

B. Peripheral from SSM

C. Dosing oven for fluid die casting

D. Other peripherals

Figure 4 Automation of a combined casting machine. The shown die casting facility can cast conventional liquid-metal from both the dosing-and-feeding furnace (C) and semisolid materials (B). The real-time die casting machine has a dynamic casting capacity of 1,000 kN and a locking pressure of 12,000 kN. Three robots are used: (1) two-axis spraying robot; (2) six-axis robot for unloading and completion supervision; (3) robot for finishing in cooling basin and deburring (courtesy ABB Flexible Automation Inc.).

environment. Examples of the effective and innovative use of robots in foundries and die casting operations are many:

- The precise motion control achievable by robots is enabling foundries in Austria, Ireland, England, and Taiwan to pour molten metal with greater efficiency, less waste, and higher productivity (Figure 6).
- A robotic deburring cell installed as part of a machining line at a motorcycle factory in England has reduced the time needed to machine a crankcase from 18 minutes to 9 minutes.
- A robotic core-handling system at a major automaker has the flexibility to handle 14 different cylinder block core sets.
- Robotic dross skimming of aluminum ingots by an Australian smelter operates continuously 24 hours a day with increased levels of personnel safety.
- A pair of linked robot cells at a French plant produces consistently high-quality car door handle and steering lock parts, completely—from die casting through

(a)

(b)

Figure 5   Sand or permanent mold casting: (a) mold assembly; (b) mold loading; (c) robots vent molds (courtesy ABB Flexible Automation Inc.).

(c)

Figure 5 (Continued)

Figure 6 Robot fills ladles with molten brass in an Australian foundry producing bath taps. The mold and ladle must be tilted in synchronism to produce a high-quality casting. The robot exchanges contaminated ladles for new ones held in a magazine. One robot and one operator serve two casting machines (courtesy ABB Flexible Automation Inc.).

drilling and polishing to final palletization—with no human handling or intervention.

- An innovative engine block core robotic manufacturing center in Germany combines high output with the flexibility to accommodate design modifications and new models as they are introduced.

6 EMERGING TRENDS AND FUTURE CHALLENGES

There is a new mood in the foundry industry with respect to automation. Today the thousands of robots at work in foundries, primarily in die casting, bring their benefits of safety, productivity, and quality to hundreds of companies. Still, this is only the first wave.

A clear trend exists toward increased investment in automation, particularly robotics automation, in all segments of the industry. Global competitive pressures, along with increased demands from major customers, are the two primary factors fueling this trend.

Today's automotive industry, for example, is placing new demands on die casting suppliers to provide more than just castings—it now requires completely finished parts and even assemblies. The robot is the preferred vehicle to meet this demand by undertaking tasks such as placing inserts into the die casting machine, gate removing and deburring, and performing assembly operations on the casting.

This trend to add more value to a casting is being further amplified by the casting supply industry in response to the automotive industry's move to preferred or single-source suppliers. Aluminum die casters and material producers are increasingly being integrated into major suppliers with the resources to invest in the state-of-the-art systems needed to stay competitive. Robotics has been recognized by these large players as essential for winning business.

In the future it will no longer be enough for automotive casting suppliers simply to have the best price. Although cost will continue to be important, the winning supplier will need to add value to a part by performing assembly operations and through other means. Suppliers will also need to respond more quickly to changing customer needs.

Flexible automation is key to the future development of foundries and die casting facilities, with more operations being carried out by robots. New plants and large-scale modernization of existing plants will increasingly be driven by a high degree of robotic automation. As a result, the automated foundries of tomorrow will be outstanding in both the quality of their products and the cost-efficiency of their production.

ADDITIONAL READING

ABB Foundry Focus, published by ABB Robotics Products, Vasteras, Sweden, No. 3, 1997.

ABB Robotics Review, Foundry Edition, published by ABB Robotics Products, Vasteras, Sweden, No. 2, 1997.

Austin, Roger. "Robots in the Moulding Shop." *Industrial Robot* **23**(6), 24–26, 1996.

Barnett, S. "Automation in the Investment Casting Industry. From Wax Room to Finishing." *Foundry Trade Journal* **16**(1), 226–228, 1993.

Gomes, M. P. S. F., R. D. Hibberd, and B. L. Davies. "Robotic Preforming of Dry Fibre Reinforcements." *Plastics Rubber and Composites Processing and Applications* **19**(3), 131–136, 1993.

Hamura, Masayuki. "Automation of Robotic Deburring System for Iron Casting." *Robotics* (July), 98–104, 1995.

Ichikawa, T. "Application Example of Deburring Robot System." *Robot* (July), 60–65, 1996.

Kato, Mikio. "New (Large Size) Aluminum Pouring Robot." *Robot* (November), 33–40, 1996.

Kurtis, J., et al. "Tundish Working Lining Developments at Bethlehem's Burns Harbor Division." *Iron and Steelmaker* **21**(6), 13–19, 1994.

Mallon, J. M., IV. "Servo Robots Claim Savings for Injection Molders." *Modern Plastics* **74**(9), 115–116, 118, 1997.

"Robot Builders Enhance Controls, Look to Meet Downstream Needs." *Modern Plastics* **74**(2), 67–69, 1997.

Rodrigues, J. "How to Plan a Robot-Based Automation Cell Processing." In *Proceedings of Annual Technical Conference—ANTEC.* Vol. 1. Brookfield: Society of Plastics Engineers, 1997. 502–506.

Rooks, B. W. "Robots at the Core of Foundry Automation." *Industrial Robotics* **23**(6), 15–18, 1996.

———. "Software Synchronization for Radial Forging Machine Manipulators." *Industrial Robotics* **23**(6), 19–23, 1996.

Smock, D. "How VDO Yazaki Achieved a 96% Utilization Rate." *Plastics World* **54**(3), 64–65, 1996.

"Why Automate the Fettling Shop?" *Foundry Trade Journal* **170,** 1996.

CHAPTER 44

SPOT WELDING AND LASER WELDING

Mario Sciaky
Sciaky S. A.
Vitry-Sur-Seine, France

1 INTRODUCTION

One of the main applications of robots has been and still is in automotive spot welding. This proliferation is not because it is a simple process; on the contrary, it is a complex one. However, it is a good example of a cost-justified application that also relieves humans from a tedious and difficult job. Furthermore, the automotive industry around the world joined forces with robot makers to improve and advance this application area. J. Engelberger tells in his book (Engelberger, 1980) how it all happened. Traditionally, large, heavy car body parts had been held by clamping jigs, tacked together by operators using multiple welding guns, and then spot welded manually too.

In 1966 the first steps were taken to use a robot to guide the welding guns and combine the control of the robot with the control of the gun. In 1969 General Motors in the United States installed 26 Unimate robots on a car body spot welding line. Then in 1970 Daimler-Benz in Europe used Unimate robots for body-side spot welding.

Spot welding robots often work three shifts (with the third devoted partly to overall line maintenance), and produce 80 and sometimes well over 150 cars per hour. All major car manufacturers use robots for spot welding because of their speed, accuracy, and reliability. The repeatability and consistent positional accuracy that spot welding robots can achieve provide for a much better quality product than in manual operation. In some cases this can be done with fewer spots welded in locations that can be accurately and precisely controlled. The last section of this chapter reviews laser technology, specifically laser welding and laser sensing, in welding processes. Laser welding is the most promising of the advances in noncontact welding. However, its application has been slow to take off for robotics-based welding processes. Bulkiness of laser beam generation devices and difficulties with the beam driving along the robot's arm are among the barriers preventing the widespread implementation of robotics laser welding systems in industry. Recent advances in yttrium aluminum garnet (YAG) laser and fiber optics for laser beam conduction (Holt, 1996) are bringing industrial implementation of laser welding systems closer to reality. On the other hand, the utilization of laser-based vision system for arc welds profile sensing has been successfully applied (Boillot, 1997; Erickson, 1993). Factors such as light contamination and fumes from the welding process make the high intensity of the laser beam suitable for the task of gathering the profile information and feeding it back to the system's controller.

2 THE SPOT WELDING OPERATION

Welding is the process of joining metals by fusing them together. This process is distinguished from soldering and brazing, which join metals by adhesion. In spot welding, sheet metal sections are joined by a series of joint locations, *spots,* where heat generated by an immense electric current causes fusion to take place.

Three factors are critical to successful spot welding:

1. Pressure between electrodes
2. Level of current

Handbook of Industrial Robotics, Second Edition, Edited by Shimon Y. Nof
ISBN 0-471-17783-0   © 1999 John Wiley & Sons, Inc.

3. Welding time

The pressure exerted by the electrodes on the joined surfaces controls the resistance of the local column of material being welded. With too much pressure the air gap is minimized, resistance is reduced, and higher current is required to generate the heat for fusion. With insufficient pressure the resistance is higher and the spot may burn because of excessive heat. Thus there is an optimal relationship between the three factors, depending on the material of the workpieces and their thickness.

2.1 Welding Sequence

A typical spot welding sequence includes the following four steps:

1. Squeeze—hold the two surfaces between the electrodes.
2. Weld—turn current on for required duration; heat is generated at the spot.
3. Hold—keep the electrodes closed for the duration required to cool the spot; usually cooling water is circulated through the electrodes.
4. Release—release the electrodes' grip and rest.

The spot welding machine is automatically controlled to repeat this sequence accurately and repeatedly, and the control is adjustable for different welding conditions. The electrical components of a spot welding machine are depicted in Figure 1.

Materials most appropriate for spot welding are ferrous metals—they are electrical conductors and do not have low resistance (such as in aluminum and copper) that requires excessively large current. The most frequent workpieces are made of cold-rolled low-carbon steel, but high-strength steel and galvanized material are also common. Thickness ranges from 0.6 to 1.0 mm. Parts are usually stamped and blanked out in such a way as to facilitate the welding process. The most typical application areas for spot welding are, therefore, the manufacture of automobile bodies, domestic appliances, sheet metal furniture, and other sheet metal fabrications.

3 OVERVIEW OF SPOT WELDING METHODS

Resistance spot welding lines have gone through several transformations since inception. Early assembly lines consisted of a long conveyor on which the components were assembled by spot welding. The workers operated manual overhead welding sets, each performing his specific task. The system is still used today for low production rates and has the advantages of high flexibility and adaptability; model changes are made possible without undue expense.

The search for improved productivity brought about the use of multiple spot welding machines and transfer lines. The principle consists of moving the parts to be assembled through a series of automatic welding stations, each having one electrode for each spot weld. A line could have as many as several hundred electrodes. These automatic lines were highly inflexible, and a major drawback was the risk that it would be impossible to amortize them if the model for which they were designed did not sell well.

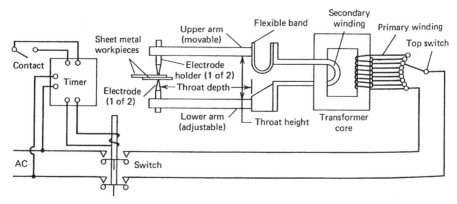

Figure 1 Electrical components of a spot welding tool.

With industrial computers and robots appearing on assembly lines, it is possible to return to the original flexibility offered by use of welding guns, but with the advantages of increased accuracy and production quality. Production facilities are able to accommodate three or four different body styles smoothly and in any order with multilevel computer management control coordinating all stages of production.

4   STRUCTURE AND COMPOSITION OF A ROBOTIC SPOT WELDER

A spot welding robot consists of three main parts:

1. A mechanical assembly comprising the body, arm, and wrist of the robot
2. A welding tool, generally a welding gun
3. A control unit

4.1   Mechanical Assembly

This is an articulated mechanical structure with the following functions:

To position the operational extremity of the robot—the tool it carries—at any point within its working volume

To orient this tool in any given direction so that it can perform the appropriate task

Table 1 gives numerical values for these and other criteria required of industrial spot welding robots.

4.2   The Welding Tool (Gun)

The welding tool is considered to be a resistance welding gun, composed of a transformer, a secondary circuit, and a pressure element.

When the spot weld distribution lines are straight and present no access problem, the attainable rate is 60 spots per minute. Welding two pieces of steel sheet metal each 1 mm thick at this rate requires a 10-cycle current pulse of 10 kA. The welding force to be applied by the electrodes to maintain pressure as explained before should be approximately 3000–3500 N. The power rating of the welding transformer and the secondary voltage will depend on the impedance of the secondary circuit.

These electrical considerations, along with the problem of accessibility, can be met with one of the three following transformer-mounting configurations (see Figure 2):

1. External transformers
2. On-board transformers
3. Built-in transformers

Table 2 gives numerical indications of the merits and inconveniences of each of these systems, as explained in the following sections.

4.2.1   Welding Guns with External Transformers

The welding transformer is suspended above the robot and is mounted on a track to follow the movements of the robot wrist without undue traction on the cables connecting the welding gun to the transformer. The length of these cables is sufficient to absorb the displacements and rotations of the wrist, and they are usually held by a balancing device that bears some of the weight. The movements of robots, especially the wrist rotations,

Table 1   Typical Ranges of Mechanical Requirements for Spot Welding Robots

Tool load	40–100 kg
Torque	120–240 Nm
Displacement speed	
Linear	0.5–1.5 m/sec
Angular	60–180 degree/second
Accuracy	± 0.5 or ± 1 mm
Repeatability	1.5 or 2 mm

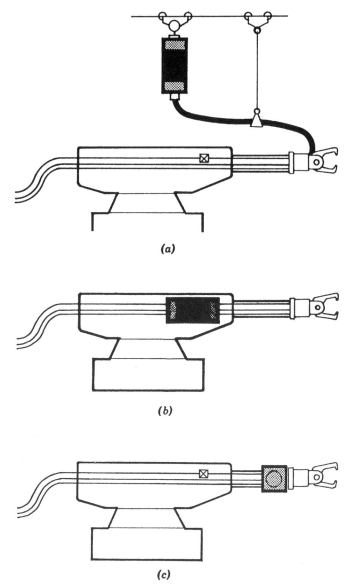

Figure 2 Three different methods of adapting resistance spot welding equipment to robots. (a) External overhead transformer with secondary cable. (b) On-board transformer with a short secondary cable. (c) Transformer integrated into welding gun.

make tensile and torsional stresses unavoidable on the welding gun; these stresses are transmitted to the robot wrist. Moreover, during welding the repeated application of 10 kA of welding current generates intense electrodynamic stresses, which are transferred to the gun and the wrist mechanism. Guns of this type are shown in Figure 3.

4.2.2 On-Board Transformers

An improvement on the preceding configuration consists of mounting the welding transformer on the robot, as close as possible to the welding gun. The design of the robot must be compatible with this approach; it is generally easy to achieve on Cartesian coordinate robots, and is possible on polar or spherical coordinate robots.

Table 2   Comparison of Various Spot Welding Equipment Suitable for Robots

Characteristic	External	On-Board	Incorporated into Gun
Secondary cable			
Length (m)	3.5	1.0	0
Cross section (mm^2)	150	150	—
Secondary voltage (V)	22.3	10.0	4.3
Maximum short-circuit current (kVA)	312	140	60
Power at 50% duty cycle (kVA)	160	70	30
Transformer mass (kg)	120	51	18

Figure 3   Spot welding operation with scissor-type welding guns on truck body (courtesy ABB Flexible Automation Inc.).

When the transformer can be mounted on the arm, the secondary cable length is significantly reduced. It is even possible to connect the transformer secondary outputs, by means of rigid conductors, close to the wrist. Such a transformer is heavy, and the overload makes necessary a reduction of the robot speed.

With some types of polar coordinate robots with articulated arms, the transformer is used as a counterweight to balance the weight of the gun and cable at the tip of the arm. There also exist certain robot designs where the transformer is housed inside the arm and the secondary cable is run inside.

4.2.3 Welding Guns with Built-in Transformers

An attractive solution, which cannot be applied in all cases, is to use a welding gun containing a built-in transformer specifically designed for use with robots. The main advantage is the elimination of the heavy secondary cables, replacing them with primary cables of smaller cross-sectional area. The secondary impedance becomes much lower since it is limited to the gap of the gun, and the size of the transformer is consequently reduced. For example, a 30 kVA transformer measures $325 \times 135 \times 125$ mm and has a mass of 18 kg. The transformer, fully integrated into the welding gun, is directly attached to the active face of the wrist so that its center of gravity is as close as possible to this face to minimize the moment of inertia

This system has two disadvantages:

1. The welding gun is bulkier. This can lead to penetration difficulties when carrying out certain welds where access is limited, for instance, in wheel wells.

2. A markedly greater mass; the transformer doubles the mass of the gun. A weight of 40 kg is acceptable; at values greater than 50 kg this can become a problem for most robots.

More than the static load, it is the stress generated on the active face of the robot wrist that can become excessive. Indeed, spot welding robots must allow for a torque of at least 120 Nm. This corresponds to a 50-kg gun whose center of gravity is 240 mm from the axis of the wrist. Many robots furnish torques exceeding 200 Nm and, as such, are better sited to handle welding guns with built-in transformers.

4.3 Control Systems

The process control in resistance welding involves an AC phase-shift controller and counting the number of periods of weld time. This can be accomplished by a conventional weld-control timer, but a more effective (and flexible) method is to integrate the process control into a centralized control that includes the robot positioning and the adaptive parameter information. This system makes efficient use of the real-time control system since the weld process takes place when the robot axes are static. Unification of all control functions eliminates the proliferation of interconnections resulting in systems composed of robot controllers, programmable logic controllers, weld timers, and communications links. Interfacing is greatly simplified.

The soundness of the spot welds can be monitored electrically by the dynamic resistance method. Information gathered serves two essential purposes. First, if the weld is determined to have insufficient penetration, respot robots further down the line add spots around the faulty point and the centralized control alerts personnel of the possibility of a malfunctioning gun or improper choice of parameters. Second, the dynamic resistance monitoring reveals if there was sufficient "splash" in the weld to cause the electrodes to stick together; the robot head cannot move. Concurrently, the opening and closing positions of the weld gun secondary circuit are monitored to prevent the axis controllers from indexing the gun to the next weld position. Immediate intervention is necessary.

Centralizing the weld process control and the axis control is necessary for adaptive parameter control. As various models come down the line, different weld distributions and sheet metal thicknesses are encountered having various heat-sinking capacities. The weld parameters must be programmed to accommodate for this.

Combining the overall line logic control with the weld control allows for power interlocking of the weld guns. In some cases power interlocking ensures significant savings in the power distribution system along a line. In other instances better-quality welds are produced because the network is capable of delivering the necessary current to each gun.

Computer-assisted maintenance is essential when many robots are involved. Electrode wear and replacement is a common problem. As the electrodes flatten with use, the current can be programmed to be gradually incremented to maintain a constant current density in the spot weld. Electrode lifetime is thus lengthened; the control system informs maintenance personnel when particular electrodes require replacement. Axis-displacement times are monitored and compared with predetermined values to warn of possible mechanical wear.

5 PLANNING ROBOTIC SPOT WELDING LINES

The planning of a robotic assembly spot welding line to meet a manufacturer's specific production needs requires detailed investigations to optimize the proposed solution. This section examines the particular aspect of spot welding by robots and their planning implications.

Design data from the product to be manufactured can be brought together to give a first estimate of the facility requirements. These include the following:

1. The parts to be assembled
2. The geometrical conformation of these parts and the corresponding number of stations required
3. The distribution of the spot welds and the number of robots required to weld them
4. The production rate and the number of lines required to meet the production needs
5. The desired degree of flexibility

To complete the design, additional information is needed:

6. The basic principles relating to the transfer and positioning of the assembly
7. The final selection of the robot, its equipment, and its installation
8. The environment and the available space

These decisions are based on the final design of the line and may be heavily influenced by political, social, and economic considerations. Normally, several designs are proposed; the final choice should reflect the best possible compromise between the technology and its cost.

5.1 Geometry of Parts to Assemble

The first step in a project is to undertake a detailed study of the parts to be assembled for purposes of classification. The operational procedure is determined by this classification.

In the automotive field there are numerous examples of various small reinforcing parts welded onto one main part. Such an assembly is shown in Figure 4: tunnel reinforcements, seat belt hooks, and cable brackets are welded onto the stamped panel. Extensive loading and small-part-handling equipment must be provided for in this type of assembly.

A second classification is the assembly of several parts of the same size. These include the front end, composed of baffles, apron, and radiator cross-members, and the rear floor frame, made up of longerons and cross-members.

A third classification is the "toy tab" method; approximate geometry is obtained by the mechanical interlocking of the subassemblies. An example is the "body-in-white," composed of the underbody, body sides, roof braces, and the roof.

A similar classification can be used on other fields such as household appliances and metal furniture.

The geometrical references are the significant zones of a part or subassembly that define its theoretical position in the X, Y, and Z dimensions. Compliance with the references of the component elements of a subassembly thus guarantees the geometry of the completed assembly.

Main references are those that must be maintained throughout the production process, from the stamping phase on. Secondary references are those that are used only when assembling the unit. Figure 5 shows a floor panel with its references. The references are maintained by mechanical elements (fixed or movable reference pins, locators, and clamps) for each assembly.

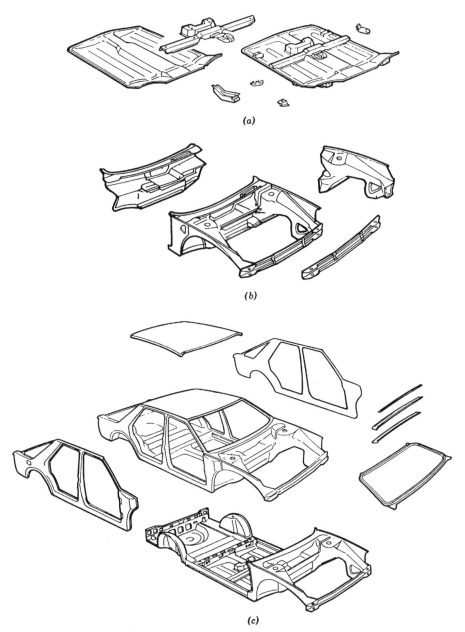

Figure 4 Examples of automobile sheet metal assemblies. (*a*) Stiffeners and accessories added to floor panel. (*b*) Front-end composition. (*c*) Final assembly of underbody, body sides, and roof.

The reference definitions must be preserved in each phase if the entire assembly is not completed in a single operation. The station or stations thus defined are *geometry conforming*. The location of these references makes it difficult for the robots to reach the welding areas. For the assembly of a number of small parts onto a main element (a main part on which are welded various smaller stiffeners), access may prove particularly troublesome.

5.2 Weld Distribution

In general, the parameters pertaining to the spot welds, such as quantity, location, and strength, are specified by the product designer. These data are drawn from research, design studies, and tests previously conducted during the product planning and design phase for composition, shapes, and assembly.

5.2.1   Selection of Tack Welding Points

In the geometrical conformation station(s) it is essential to carry out simultaneously designs concerning the number of points required for the assembly geometry and the relative positioning of the robot gun with respect to the reference elements (Figure 5).

5.2.2   Operational Procedure

The order in which the parts are loaded along the line must be carefully planned to allow for maximum accessibility of the welding tools until completion of the assembly (Figure 6).

5.2.3   Spot Weld Grouping

At this stage it is necessary to define the grouping of spot welds that can be welded by a single welding element. Similarly, a first approximation must be made of the "base times" required for welding the various weld groups as a function of their position in the assembly and of the existing layout elements.

5.2.4   Welding Gun Planning Phase

On the basis of the foregoing study, the minimum number and the configuration of the various welding elements necessary for these welds or weld groups can be established. If certain points are physically impossible to weld because of the welding method or the assembly itself, this will become apparent at this stage of the study.

5.3   Production Rate

Based on the production procedure, the production rate dictates the overall design and the components of an assembly line. In the automotive industry lines are usually moving continuously, and robots must track the lines to perform the spot welding operation "on the go."

5.3.1   Global Line Utilization Factor

This factor is a function of all of the elements entering into the makeup of the line, ranging from the supply of workpieces to be welded to the flow of finished products coming off the line. This factor depends on the design of the welding stations: it decreases if several stations are installed in succession; it increases when intermediate buffer stocks are provided.

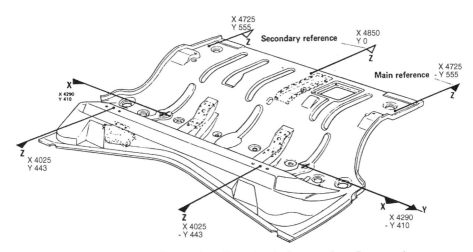

Figure 5   Main and secondary dimensional references for a floor panel.

Figure 6 Spot welding production line of truck floor panels (courtesy ABB Flexible Automation Inc.).

5.3.2 Welding Time

The desired rate, taking into account the line utilization factor initially selected, converts into the cycle time per part. After deduction of the product-handling and layout time, a preliminary estimate of the time available for the actual welding operation can be made. The importance of minimizing handling and layout time should be stressed (see Chapter 6).

5.3.3 Preplanning of the Assembly Line

The welding time thus established provides a preliminary basis for the choice of the type of assembly line to be implemented. In fact, when the calculated time is obviously insufficient (a large number of points to be welded in a minimum period of time), a first option concerns paralleling a number of welding lines for the same assembly.

5.3.4 Minimum Number of Robots

The welding and displacement times, taking into account accelerations and decelerations, make it possible to preplan the number of points that can be welded in one cycle by a single robot. Taking into account the spot weld grouping and the necessary or possible types of welding guns, it is possible to define the minimum number of robots theoretically necessary for the assembly. This figure, of course, must be subsequently verified.

5.4 Assembly Flexibility

The flexibility of an assembly line is defined by its degree of adaptability with respect to the various products that are, or may have to be, processed on that line, simultaneously or not. This flexibility is defined by the following options:

Suitability of the tooling for several different products
Capacity for adaption and the time needed for changeover to another product

5.4.1   Degrees of Flexibility

The user may request various degrees of flexibility for the lines, which may represent large investment costs.

A line may be initially designed to process only one type of product. Its overall design, however, makes possible partial or complete retooling at the end of the production run, and the replacement of the original product by a product of related design.

A line may be initially equipped for one part type, but its design provides for the accommodation of several variants by adding or adapting the appropriate tools. An alternative is to design the line capable of producing several versions of a single-base product model.

A totally flexible line is designed to produce without preliminary adjustments, in any order, a variety of different products and their variants. Of course, these different products must be similar and involve a comparable production technique.

5.4.2   Replacement Flexibility

This type of flexibility is defined by the ease with which a line can continue to operate at a reduced rate in the event of failure of one of its components (robot or tooling). The line-management programming may be such that the line robots automatically compensate for the missing work of a faulty robot, taking into account problems concerning the specific type of welding gun and the specific capacity of the robot. An alternative is to install respot robots at the end of the line to make welds that were missed.

5.5   Part Positioning and Transport

5.5.1   Position of the Part During Processing

Large-scale assemblies, such as floor pans and bodies-in-white, should be handled in their normal or "car" position. Subassemblies like body sides are handled in several positions, depending on the loading restrictions of the smaller parts and the location and accessibility to the weld points. The geometrical conformation tooling must be taken into account.

5.5.2   Part Transfer

In some plants floor pans and body-in-white assemblies are handled on skids or lorries. These handling devices are kept in all production stages as well as off the assembly line and are removed only after final assembly. These transfer units become an integral part of the assembly whether they are used as a transfer means or remain idle. For high production rates, skids cannot be used for part handling, but they remain fastened to the assembly when alternative transport means are employed.

Carriages are used when two different types of subassemblies are to be processed with geometrical conformation or when the subassembly is already assembled but its geometry must be sustained all along the line.

Automatic guided-vehicle systems are commonly used in flexible shops for the preparation of subassemblies such as doors, hoods, and dashboards.

5.6   Environment

5.6.1   Positioning of the Robot

The distribution and scatter of welding points on a given assembly or subassembly can determine the position of the workpiece with respect to the robot when this is physically possible or, if not, make it necessary to install the robot in a certain position. The robot can be installed on the floor, on a base, overhead, or inclined (see Figure 7). When the production line must be straddled, it is possible to install two, three, or even four robots on one gantry structure.

5.6.2   Floor Space

The space available around the workstation to be equipped may sometime be limited, and the general layout in this case calls for the selection of gantry robots. This type of robot configuration lends itself to more compact installations. A transfer line with gantry robots is typically 4–5 m wide, while the same line equipped with floor-mounted robots occupies 2–3 m, additional, on either side, for a total line width of 8–10 m.

6   SELECTION OF ROBOT MECHANICAL CONFIGURATION

Certain generalizations can be made when selecting a robot for a specific spot welding task. Each task is associated with specific geometrical criteria that favor particular me-

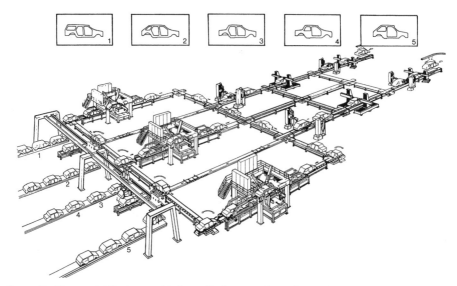

Figure 7 Five-model final assembly line with three conformation and tack welding stations each with six robots followed by two robotic finishing lines. Four hundred spot welds are made on each car at a production rate of 150 cars per hour.

chanical configurations (Jung, Kwon, and Kang, 1997). However, economic factors might modify this choice.

In automotive spot welding five typical spot welding tasks can be identified (Sakamoto, Suita, and Shibata, 1997). These tasks are explained next.

6.1 Geometrical Conformation and Tack Welding Stations

Because of its specific design, this task is one of the most difficult to deal with owing to the inherent shape of the body requiring geometrical conformation external to the volume. To weld the appropriate spots, it is thus necessary to pass through the surrounding conformation tooling. This type of station normally requires six robots: four positioned laterally on each side of the body and two at the front and the rear to weld the trunk and front end. (Three such parallel stations are shown in Figure 7.)

Regarding the lateral robots, those having cylindrical movements pose certain operational problems because the sweeping movements of the axes during penetration require large openings in the surrounding conformation tooling. For this reason, robots with polar movements and linear penetration are better suited to this type of situation and are thus frequently used. Nongantry Cartesian-coordinate robots are ideal for this application. The only factor limiting their widespread use is the large amount of floor space they require.

As for the two front and rear robots, it is almost impossible to use standard electric robots because of limitations on the positions they may occupy when suspended; they require special balancing adjustments for each configuration. Moreover, this also makes it difficult to install them in the middle of the line.

Hydraulic polar movement robots are the most suitable: they can be installed overhead, and their linear or articulated penetration permits welding at the bottom of the trunk. Their polar axis allows access to lateral points. Cartesian-coordinate gantry robots are also used, but their configuration is less favorable, the access to lateral points being critical. In all applications for this type of machine, it is essential that the robot head be as compact as possible and that the three wrist axes be arranged to minimize their sweep.

6.2 Underbody Lines

The main characteristic of this type of assembly is the need to perform all weld points from the outside (with some exceptions). In this kind of application, the robots must be chosen for their capacity to use large, heavy welding guns with long throat depths. The possible choices are thus restricted from the very start.

Cartesian-coordinate robots are the preferred solution to this problem. Indeed, this is one of the prime examples in which their application is economically attractive. In the gantry configuration, these robots can weld points symmetrically by lateral penetration; in some cases the robots are installed on pivoting arms on the sides of the gantry to increase their flexibility.

Articulated-arm robots and cylindrical-movement robots are possible second choices. Compared with linear-penetration robots, they possess the inherent advantage of not requiring installation on a raised base for accessibility.

6.3   Body Side or Floor Frame Lines

These parts, characterized by large openings, permit welding with lateral access as well as from the center. Cartesian-coordinate robots are by far the best choice for this particular task. Polar-movement robots with linear penetration are compatible with this application but have the disadvantage of requiring a raised base.

6.4   Lines with Vertical-Plane Welding

When side panels are oriented in their "road" position or slightly inclined, or for final body welding, gantry robots are difficult to use. Most other configurations are satisfactory, especially those with linear penetration to reach points inside the body.

6.5   Final Assembly Lines

All types of robots can be used for these tasks and are selected primarily on the basis of access. In the final assembly line shown in Figure 7 several different models are employed that provide access to the vehicle interior and enough flexibility to accommodate all five vehicle models.

7   EXAMPLES OF SPOT WELDING[1]

Figure 8 depicts a diagram for a spot welding application, augmented to operate with user-supplied equipment. The combination of equipment must be tightly integrated so that it is easy to install and to operate.

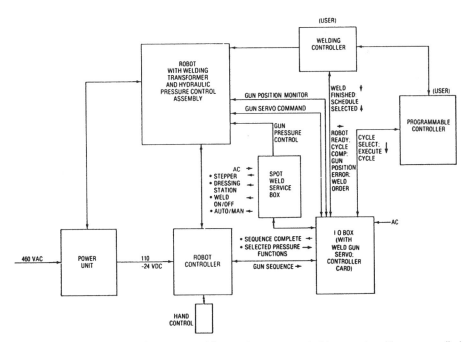

Figure 8   A block diagram for a spot welding system, augmented to operate with user-supplied equipment.

---

[1] Section 7 is revised with permission from Chapter 50, "The Operation of Robotic Welding," by P. G. Jones, J.-L. Barre, and D. Kedrowski in the 1st edition of this handbook.

In the automotive application described here, these are the typical parameters of the spot welding process:

Welds per minute: 50 spots/robot
Duration of welding cycle: ¼ to ⅓ sec (current flow) for each spot
Voltage: 6–8 V
Amperage: 20,000 A
Frequency: 60 cycles/sec

In existing applications robots make from slightly more than 37% to more than 75% of the total spot welds on a given vehicle.

The following description covers spot welding by a single robot on a single part. The objective is to illustrate how a robot system functions in this application.

Figure 9 depicts a spot welding cycle. Trajectory 1 is the front door frame, and trajectory 2 is the rear door frame. Trajectory 1 is comprised of numbers 1–14. Point 1 is the starting, or home, position for the robot. Points 2, 3, and 4 are passing points, through which the robot moves to reach the first working point, 5. Each of the working points, 5–14, must be precisely identified in terms of the welding parameters required. The points are so identified by weld identification numbers, shown on the outside of the trajectory. Each weld identification number carries with it the electrical and mechanical parameters for the weld at the specified point as specified by welding engineers responsible for setting welding parameters for every point to be welded: pressure to be exerted by the electrodes; duration of the welding impulse; shaping and development of the pulse. The parameters are implemented by the robot system through weld schedule numbers, transmitted by the controller to the spot weld service box and welding controller. The numbers in parentheses are weld schedule numbers. For each point to be welded, the spot weld service box transmits the required control signals to the hydraulic pressure control assembly to achieve the correct pressure and schedule of pressure development at the point during welding. The overall requirements vary according to the metals involved, the thicknesses, and the number of thicknesses; all variables must be predetermined and then programmed into the data for each trajectory that the robot is to execute. The welding controller delivers the required welding current, modulated to produce the appropriate heating cycle in the welded point.

Thus each point on the door frame to be welded has a welding identification number, supplied by the user. Each point is entered into a table by welding identification numbers, including the specified weld schedule. Using the robot controller's keyboard, the operator places this table into the controller's memory. In terms of the robot system's operation, the weld schedule number (1–6) associated with each point for the spot weld service box is the means for establishing the match-up between pressure identification number and

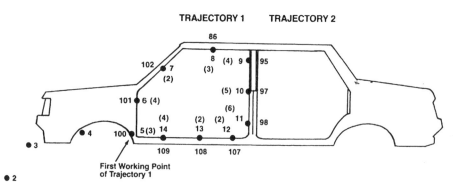

Figure 9 A spot welding cycle example.

required welding parameters. For example, pressure identification number 101 is point 6 on the robot's work trajectory and is assigned weld schedule number 4. Weld schedule 4 selects the specified pressure, controlled through the spot weld service box, and the welding controller provides preestablished electrical welding parameters associated with weld schedule 4.

Having entered the welding identification number and associated pressure schedule into the table in the controller's memory, the operator then teaches the robot the trajectory. At this time, or later, the operator can use the hand control's multiple-function capabilities to correlate every point on the trajectory with the specified weld schedule, assigning welding identification numbers and weld schedules for each point on the trajectory in accordance with the prescribed schedule.

The operator can also modify the welding identification number, using the editing capabilities integrated into the hand control's operation. The operator can check the exact parameters operative at any point along any of the spot welding trajectories, using status display information available through the robot controller's display screen. For each point the operator can check trajectory number, point number, robot speed assigned for that number, the welding identification number and weld schedule required for spot welding, and other pertinent data. In the execution phase the display screen shows a running account of progress through the cycle under way, point by point and trajectory by trajectory.

7.1 Welding Operational Sequence

The following is the operational sequence, from detection of the part in the robot's work envelope through completion of the cycle for that specific part (refer to Figure 8).

1. Based on the application system's detection of a given part, the programmable controller specifies a cycle number to the robot controller (cycle select). If the cycle agrees with a cycle in the controller's memory, cycle verification occurs, the controller sends ROBOT READY, and the programmable controller sends EXECUTE CYCLE. The robot controller sends the robot from its home position to the first working point of the cycle (point 5 in Figure 9).

2. As the robot approaches the first working point, the robot controller sends the weld schedule to the spot weld service box and the weld controller. The spot weld service box sends preestablished pressure-control voltage to the hydraulic pressure-control assembly, based on the pressure schedule transmitted by the robot controller, and the welding controller establishes the other parameters as specified.

3. The robot reaches the first working point:

   (a) The controller sends gun sequence to the weld gun servo-controller card in the I/O box.

      (i) The weld gun servo-controller card sends a servo-command to the hydraulic pressure-control assembly on the robot and monitors feedback circuitry to determine that the welding gun has closed to the correct position, which corresponds to the pressure specified for that point.

      (ii) As the weld gun servo-controller card receives confirmation that the gun is closed to the position prescribed in the gun sequence, it transmits the weld order to the programmable controller.

      (iii) The programmable controller enables the welding controller, causing the welding controller to generate the correct pulse (heating and duration) to spot weld the type of metal and number of thicknesses between the jaws of the welding gun at the present welding identification number.

      (iv) When it has finished welding at the point, the welding controller sends WELD FINISHED to the weld gun servo-controller card in the I/O interface box.

      (v) On receipt of WELD FINISHED, the weld gun servo-controller opens the welding gun by commands to the hydraulic pressure-control assembly servo-valve, and when the gun opens to a preset distance, as reported by feedback circuitry, the weld gun servo-controller card sends SEQUENCE COMPLETE to the robot controller.

   (b) The robot controller sends the robot to the next programmed welding point.

(c) When the robot's cycle ends, the robot controller sends CYCLE COMPLETE to the programmable controller. The programmable controller is tied to other controllers along the line and to the programmable controller (sequencer) that synchronizes operation of the entire line. When all robot operations, other machine operations, and any required manual operations on the line are in the appropriate condition, the sequencer operates the transfer line, moving parts as required to maintain flow along the line.

4. If the programmable controller requests a cycle that is not valid for the controller, or if the robot controller does not verify a valid cycle and send ROBOT READY, the selection process repeats. The robot does not move until all conditions are satisfied.

5. Control circuitry in the spot weld service box monitors electrode wear, sending a signal to the programmable controller to stop welding when electrodes are excessively worn (gun position error).

8 LASER WELDING

High-power CO_2 laser welding has been widely used in the industry as an alternative to both spot and arc welding due to its high productivity and excellent weld quality (Lankalapalli, 1996). Laser, a highly coherent beam of light, can be focused to a very small spot, giving rise to high power densities of over 106 W/cm^2. At these high power densities the substrate material evaporates because the beam delivers energy so rapidly that the mechanisms of conduction, convection, and radiation have no time to remove it. This leads to the formation of a keyhole (Figure 10). The keyhole can be considered as a cylindrical cavity filled with ionized metallic gas, which absorbs 95% of the power. The equilibrium temperatures reached are about 45,000°F. A curious phenomenon is that the heating occurs from the inside of the keyhole outward, not from the surface of the material down (as in conventional welding processes). This causes a pool of molten metal to form all around the keyhole, which hardens behind to it to form a weld as the beam moves along (Vasilash, 1988). Typical speed of the beam can reach hundreds of inches per minute. The keyhole mechanism provides high aspect ratio (depth-to-width ratio) welds in steels, where 10:1 ratios are common.

8.1 Advantages

The advantages of laser welding relative to spot welding are:

- Consistent weld integrity. Better control of the weld gun positioning and the consistency of welding parameters such as welding time, squeeze time, hold time, weld control settings, and weld tip conditions.
- Single-sided access.
- Reduced flange widths.
- Reduced heat-distortion zones. Narrow distortion zones with weld lines of approximately 1 mm with no mechanical distortion (noncontact welding process). Typical heat distortion zones in spot welding are equal to the size of the nugget and mechanical distortion of the material (wavy flanges) are likely to occur due to the pressure and heat effects.
- Increased structural strength. Average tensile strength of laser welding can be similar to or greater than that of the material being welded. Tensile and shear strength are greater than those generated by spot welding.
- High-speed processing. Operating speed of laser welding can reach in excess of 200 ipm. Spot welding, on the other hand, requires 3–4 sec per spot.
- Increased design flexibility. Narrow weld widths make possible new designs, improving the capabilities for styling, product design, and traditional production engineering processes.
- Noncontact process. Elimination of consumable items associated with spot welding and attendant maintenance results in process cost reduction.

The advantages of laser welding compared with traditional welding processes such as gas metal arc welding (GMAW) are that laser beam welding:

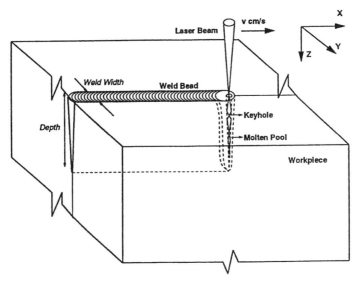

Figure 10   A schematic of the laser welding process.

- Is a low heat input process resulting in low distortion of the workpiece.
- Produces welds with high depth-to-width ratios and narrows the heat-affected zone, diminishing the alteration of the materials' mechanical properties.
- Does not require flux or filler material, leading to cleaner welds.
- Has the ability to weld dissimilar metals in intricate shapes.
- Yields high welding rates.

8.2 Disadvantages

The disadvantages of robotic laser welding systems have limited their application to stationary laser welding and cutting systems. The disadvantages include:

- Bulkiness of the laser beam generator and the need to mount it on the robot arm. This imposes additional requirements for beam accuracy.
- Workpiece fit-up and handling and beam manipulation accuracy requirements are relatively high.
- High capital investment.
- Hard and brittle welds may be produced in hardenable materials.

The disadvantages associated with the laser welding process have discouraged its implementation in robotics applications. Recent developments in YAG-based laser robotics welding systems are slowly overcoming these difficulties (Weckman, Kerr, and Liu, 1997). KUKA, for example, has developed a high-power YAG laser welding system (2kW) conducting the laser beam in a flexible optical fiber to the robot's tool (Holt, 1996). The reported benefit is welding speed similar to that of a 3kW CO_2 laser welding system employing automobile industry standard robots. Some earlier developments of laser welding robots include the L-100 by Fanuc Robotics, with an internal arrangement of mirrors, and the COBRA system by Sciaky, with an external arrangement of mirrors. The applications of laser welding were mainly in the automobile industry, also the largest consumer for welding technology. Examples include welding of drip moldings to roof panels (BMW); welding of two blanks into a single floor pan, which eliminates overlapping weld joint and sealing operation in the paint shop (Toyota, Audi); and laser welding to produce truck front end assemblies (Mercedes) (Vasilash, 1988). Research in the field of laser spot welding is still in progress (Yilbas and Kar, 1997), and few applications are reported in the literature (Dorn et al., 1996, Marley, 1996).

8.3 Laser Sensing in Arc Welding

The application of laser-based vision systems to welding processes results in quality assurance improvements, technical feasibility, increased welding speed, savings on consumable materials, and a solution to the shortage of trained workforce (Boillot, 1997). The objective of the sensing system is to provide information for motion control, torch position and orientation, and welding parameters such as arc current, arc voltage, arc length, and torch weaving. Laser-based vision systems provide also the capability of performing 100% inspection.

The largest number of laser vision systems are installed in arc welding robots and dedicated machines. The steps for performing arc welding are:

- Plan and program the welding sequence and set thermal and metallurgical parameters.
- Inspect joint fit-up.
- Initialize the position and orientation of the electrode and begin the actual welding operation.
- Dynamically control the electrode position and orientation and fine-tune the welding conditions. Modify the weld sequence in multipass welding.
- End operation.
- Inspect the result and eventually start the next weld bead for multipass welding.

Critical factors affecting the quality of the weld are:

- Distortion and shrinkage
- Surface condition of the base metal
- Temperature of the base metal
- Joint geometry
- Position and orientation of the torch
- Calibration of the welding power supply
- Wearout of the contact tube
- Composition and quality of the welding wire

Vision is the essential sense of the human welder. The requirements imposed on an automated welding control system for performing similar feedback control are:

- Optimization of welding parameters and welding sequences
- Selection of welding procedures
- Adaptive control based on joint geometry analysis
- Analysis of joint profile
- Rapid communication among sensors, robot, and operator
- Monitoring and control of operating conditions in real time.

Although several approaches to welding sensing have been proposed (Table 3), laser-based vision systems using triangulation have overcome the difficulties of a highly light- and gas-polluted environment. These systems provide high flexibility, low cost, and the ability to yield information useful for more than one adaptive function. In particular, laser vision systems are resistant to:

- Arc radiation
- Fumes
- Hot metallic spatters
- Vibrations
- Shocks
- High temperature

Laser vision systems provide:

Table 3 Sensor Characteristics for Adaptive Welding[a]

Types of Sensors	Feature Extraction	Functions	Comments
Tactile	Joint coordinates	Trajectory shift	Off welding time Time consuming Reduced duty cycle
Through the arc	Joint coordinates	Seam-tracking	On-line No added device Low cost Not all joints Not all models Lap thickness > 2.5–3.0 mm
Inductive	Joint coordinates	Seam-tracking	On-line Requires close proximity EM interference
Active vision	Joint coordinates Joint orientation Joint geometry Bead shape	Seam-tracking Filler metal control Bead inspection	On-line All joints All processes Very flexible High resolution Programmable field of view Makes use of intensity data
Passive vision	Pool top shape Vision of electrode	Seam-tracking Penetration control	Process dependent Limited applications

[a] Following J. P. Boillot, "New Laser Sensors Penetrate Robotics Welding Applications," *Robotics Today* **10**(2), 1–6, 1997.

- Insensitivity to metallic surface conditions of joints
- High performance
- Flexibility to cope with various joint geometries
- Compatibility with high-speed welding

Several steps are involved in guiding a robot welder and controlling the operating conditions of the process (Figure 11). The laser camera produces a profile of the joint calibrated in real time. The main features of the profiles are extracted and used to calculate the joint-geometry information (orientation, position, root gap, mismatch, and cross-sectional area). Torch position is also calculated from the joint profile information. The torch path is generated by a trajectory module considering position and orientation of the joint and robot wrist. Selection of the welding operational parameters is computed based on the robot speed, and tool position and orientation are computed via a model that links joint geometrical information to the controllable process variables.

Industrial implementation must consider factors such as the selection of the laser camera according to the specific process and type of joint to weld. Size and weight must be kept to a minimum to avoid obstruction in accessing the workpieces and overload of the robot. A typical industrial implementation consists of an articulated robot for welding car frames in the automotive industry. In this case the joint types are mainly lap and fillet, with stamped parts of curved shape that may affect considerably the fit-up of mating parts. Parts are preassembled in a jib and then welded by a six-axis articulated robot using laser vision system to track the weld joints. Precision ($\pm$ 0.2 mm), welding speed (1.25 m/min), and welding current (350 A) establish the conditions that make necessary the implementation of a highly efficient welding tracking system. Additionally, several car frame models are produced in the same workcell to achieve the company's desired production rate.

ACKNOWLEDGMENT

The author wishes to thank J. A. Ceroni of Purdue University for major contributions to this chapter.

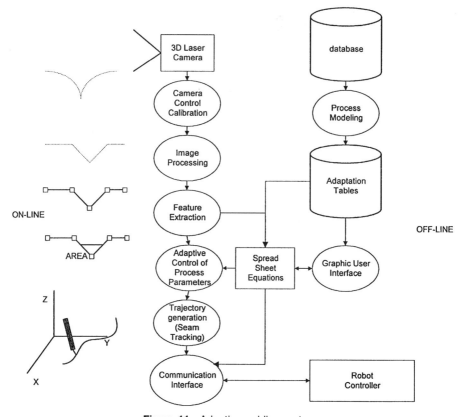

Figure 11   Adaptive welding system.

REFERENCES

Boillot, J. P. 1997. "New Laser Sensors Penetrate Robotics Welding Applications." *Robotics Today* **10**(2), 1–6.

Dorn, L., et al. 1996. "Laser Beam Spot Welding of Amorphous Foils." *Schweissen und Schneiden* **48**(9), E178–E179.

Engelberger, J. F. 1980. "Spot Welding Applications." In *Robotics in Practice.* American Management Association.

Erickson, M. 1993. "High Speed Arc Weld Seam Tracking." In *Proceedings of the 1993 International Robots and Vision Automation Show and Conference,* Detroit. 3–13.

Holt, T. 1996. "New Applications in High Power Laser Welding." *Welding Research Abroad* **42**(11), 4–6.

Jung, W. W., Y. D. Kwon, and S. S. Kang. 1997. "Selecting the Spot Welding Condition of Multi-Layer Vehicle Structure." In *Proceedings of the 1997 International Congress and Exposition SAE,* Detroit. 19–25.

Lankapalli, K. 1996. "Model Based Penetration Depth Estimation of Laser Welding Processes." Ph.D. Dissertation, Purdue University.

Marley, C. 1996. "Guide to Welding with Low Power YAG Lasers." *Welding Journal* **75**(11), 47–50.

Sakamoto, Y., K. Suita, and Y. Shibasta. 1997. "Technical Approaches of Robotics Applications in Automobile Body Assembly Lines." *Katakana,* 56–63. [In Japanese.]

Vasilash, G. S. 1988. "Lasers Heat up Robotics." *Production* **100**(11), 35–41.

Weckman, D. C., H. W. Kerr, and J. T. Liu. 1997. "Effects of Process Variables on Pulsed Nd:YAG Laser Spot Welds." Metallurgical and Materials Transactions **28**(4), 687–700.

Yilbas, B. S., and A. K. Kar. 1997. "Laser Spot Welding and Efficiency Consideration." *Journal of Materials Engineering and Performance* **6**(6), 766–770.

CHAPTER 45

ARC WELDING

José A. Ceroni
Purdue University
West Lafayette, Indiana

1 ARC WELDING TECHNOLOGY

Arc welding is a technique by which workpieces are joined with an airtight seal between their surfaces (Kannatey-Asibu, 1997). The principle of the process is the fusion of two metal surfaces by heat generated from an electric arc. During the welding process there is an ongoing electric discharge—thus the sparks between the welding electrode and the work (Figure 1). The resulting high temperatures of over 6000°F (3300°C) melt the metal in the vicinity of the arc. The molten material from the electrode is added to supplement the welding seam. Whereas spot welding is performed with alternating current, arc welding is performed with direct current, usually 100–200 A at 10–30 V.

Originally arc welding used carbon rods for electrodes. However, they did not add material to the weld, so metal filler rods had to be added. Modern methods essentially have replaced the carbon arc welding by providing quality solutions to the welding requirements (Huther et al., 1997). Some of these methods are listed in Table 1. In some methods, to prevent oxidation of the molten metal, electrodes are coated with flux material that melts during the welding process. Inert gas such as helium or argon serves also to prevent oxidation.

Automatic, electric arc welding equipment is comprised of a line through which continuous electrode wire is fed. The wire forms a tip that is graded along the desired weld trajectory. The following factors are critical to good welding results:

1. Correct feed rate of electrode wire
2. Optimal distance between the electrode and workpiece
3. Correct rate of advance of the electrode

A typical sequence in automatic arc welding operation is composed of five steps, as follows:

1. Turn on inert gas flow over work area.
2. Start weld cycle: begin wire feed, turn power on.
3. Stop wire feed.
4. Turn power off.
5. Turn gas flow off.

2 SELECTING A ROBOT FOR ARC WELDING

Robots are attractive for arc welding for several reasons (Berger, 1994; Teubel, 1994; Dilthe and Stein, 1993):

1. Robots replace human operators in the unpleasant, hazardous environment of radiation, smoke, and sparks from the arc welding.
2. Robots relieve human operators from carrying and guiding heavy welding guns, which frequently must be guided in uncomfortable positions.
3. Robots can consistently perform the precise welding motions.

Handbook of Industrial Robotics, Second Edition, Edited by Shimon Y. Nof
ISBN 0-471-17783-0 © 1999 John Wiley & Sons, Inc.

Table 1   Common Arc Welding Methods

Method		Comment
TIG	—Tungsten insert gas	Welding with a tungsten electrode
MIG	—Metal inert gas	Most common, using continuously fed metal electrode from a coil
GMAW	—Gas metal arc welding	
GTAW	—Hot-wire TIG	Welding processes with high weld-material deposition rate
SAW	—Submerged arc welding	
FCAW	—Flux cored arc welding	

Figure 1   Arc welding robotics station (courtesy ABB Flexible Automation Inc.).

The robot control system can easily communicate with the automatic control of the arc welder to synchronize the necessary robot motions with the welding sequence of steps (Nomura, 1994).

Usually, electrically driven robots are preferred for arc welding. Motion speeds required in arc welding are relatively slow, and the weight of the welding gun is relatively small too. When heavy welding guns, with a water cooling system, are applied, a hydraulic robot could be considered. Except for straight-line welding, continuous-path control is required. Interpolation is required to simplify accurate control of nonlinear welding lines.

3   THE ROBOTIC WELDING SYSTEM

A complete robotic welding system includes the robot, its controls, suitable grippers for the work and the welding equipment, one or more compatible welding positioners with controls, and a suitable welding process (with high-productivity filler metal to match). The system installation also requires correct safety barriers and screens, along with a plant materials-handling flow that can get parts into and out of the workstation on time.

3.1   Welding Positioners

The welding positioner is a critical and often overlooked part of that system. Robots are not compatible with perfectly good welding positioners designed for humans. The robot's position accuracy is useless if it is not matched by the positioner handling the weldment. Moreover, most human-oriented welding positioners are not designed for operation under microprocessor control. One solution to the control problem is to retrofit an existing positioner with new controls compatible with the robot. That still leaves the problem of positioning accuracy unsolved.

A human operator can tolerate much more gear backlash or small amounts of rational "table creep" during tilting than a robot can. The human will compensate for the variation without thinking about it. The robot expects the weld joint to be where its program thinks it is, not where the positioner actually puts it. A conventional welding positioner cannot be retrofitted at a reasonable cost for the precision required by a robot. The machine would have to be rebuilt.

Several major manufacturers of welding positioners already have models on the market designed specifically for use with robots.

Modern robotic positioners have backlash controlled to within $\pm$ 0.005 in./in. ($\pm$ 0.001 mm/mm) radius on all gears. Figure 2 shows some examples of part positioners for arc welding robots.

All the components of a robotic workstation must be interfaced so that each performs its function as initiated by the program. Typically a robotic positioner will include multiple-stop mechanical or solid-state limit switches on all axes for point-to-point positioning (open-loop) or with DC servo-drives for positioning with feedback to the control (closed-loop).

Speed is important in a robotic workstation. The full rotation of the worktable positioner on two different axes (while handling a full load) is typically between two to three times faster than a conventional geared positioner. A full rotation in both axes with simultaneous motion takes about 9 sec. A 90° table rotation may take 3 sec.

3.2   Robot Control

Programming modes for a welding controller (Sullivan and Rajaram, 1992) include speed and position settings, arc voltage, wire-feed speed and pulsing data, and other operating parameters. The starting and ending phases of each weld (with special parameters such as pre- and postflow of shielding gas, puddle formation, creep start, crater filling, and afterburn time) are programmed in separately stored routines (Harwig, 1997).

Movements of the welding robot are programmed independently by a separate control. Both control systems coordinate their functions on the handshaking principle (mutual computer affirmatives after each concluded clock cycle). The operator determines when the robot should switch from one workpiece to the next. The operator also can interrupt and adjust the program at will, even during the welding process.

Often there is a requirement for welding robot to move over a long distance to weld large workpieces—for example, in shipbuilding. A robotic linear track or gantry system can be used. For example, the system shown in Figure 3 uses an ABB IRB6 robot mounted on a gantry for arc welding of large parts in the shipbuilding industry.

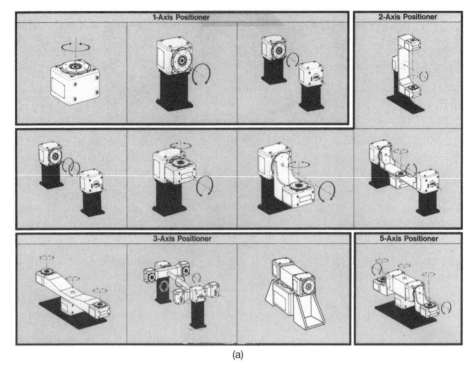

(a)

Figure 2   Examples of positioners for arc welding. (*a*) Different axis configurations for arc welding positioners. (*b*) Three configurations of positioners and their limit specifications. (Courtesy Matsushita Industrial Equipment Co., Ltd.)

3.3 Adaptive Welding

One of the recent advances in control systems of arc welding robotics processes is the development of adaptive welding systems. These systems consist of an off-line programming system and a real-time controller (Ferguson and Kline, 1997). While the off-line programming system allows the planning of the entire automated welded process, the real-time controller provides the system with the capability of implementing the welding plans and dealing with the anomalies that may arise during their execution. The off-line programming system provides several advantages to the welding process: development of welding data without stopping the operation of the robotics welding system, automatic collision detection and avoidance, and optimal part placement (see chapter 19 "Off-Line Programming" for further details). Key components of the off-line programming system are the *welding database,* the *motion planner module,* and the *graphical operation simulator.*

The purpose of the welding database is to maintain the welding process descriptions independent of path length or robot location. The definitions are developed to be unique to the significant welding variables (e.g., position, material, electrode). The database allows the automated selection of process parameters based on joint information of the job to be performed.

The motion planner module consists of a series of robotics motion planning algorithms necessary for the automated welding operations. It provides support in detection and avoidance of collisions, motion coordination of robots and positioners, checking of joint limits violations, and optimization of robot motion. The graphical simulation of the operation allows the user to review and modify, if necessary, the operation of the system interactively.

The real-time controller is that part of the system providing the adaptiveness for the welding process. The controller provides the interface to the process equipment via input/output devices, system input/output, and servo-motion boards. A required capability for this system component is the flexibility for controlling a variety of robotics equip-

One Axis
Positioner

Double-side Support
Rotating and Tilting
2 Axes Positioner

One-side Support
Rotating and Tilting
2 Axes Positioner

Specification	Min	Max
Maximum Payload	200 Kg	1,000 Kg
Rotating/Tilting Speed	4 rpm	250 rpm
Position Repeatability	± 0.1 mm	± 0.05 mm

(b)

Figure 2   (Continued)

Figure 3   ABB arc welding robot mounted on a gantry for welding of large parts in the shipbuilding industry (courtesy Kranendonk Factory Automation).

ment, welding processes, and sensors. The basic requirement for the controller is the capability to adapt in real time the programs supplied by the off-line programming system. The feedback information is gathered into the system through devices for touch sensing (through arc or touch probe for location of the welding joints) and welding seam tracking (using vision systems for tracking the arc welding seam) (Wu, Smith, and Lucas, 1996). The controller uses the joint location information for performing the seam tracking and part geometry information for adjusting in real time the welding parameters. Adaptability to changes in part geometry (joint gap) is one of the latest advances in adaptive welding (Ferguson and Kline, 1997).

4 ECONOMICS OF ARC WELDING

Without an analysis of total welding costs, manufacturing methods, present production requirements, and part design, a new robot could become a financial and production disaster. Conversely, with proper planning a user may increase productivity by 50–400%. Many users amortize their investment within one or two years (Knott, Bidanda, and Pennebaker, 1988).

The total capital cost for a complete robotics welding system including controls, materials-handling equipment, accessories, and tools can be double or triple the cost of the robot. The reason is mainly due to the cost of peripheral equipment that is designed for most effective robot operation. For instance, poor fit-up is a problem a skilled welder can usually adjust, but a robot cannot solve by itself. At best, poor fit-up requires lower welding speeds. At worst, it may mean incomplete welds.

Accessibility of the joint to the welding gun nozzle and electrode is always a design consideration in welding. Humans can compensate for bad design with exceptional skills. Robots cannot, but if joint design and part tolerances are correct, the work is properly positioned, and the program is well defined, robots can increase weld quality consistently over people.

The cost of bad welds can be excessive. Improving weld quality alone can be enough to pay for some robotic welding systems. For example, one weld repair on heavy code welded plate can cost from $400 to $4000 per 30 cm (1 ft) for initial welding, inspection, X-ray, tearing out the bad weld, rewelding, and reinspection. A bad weld can cost much more in money, and sometimes lives, if the weld causes a failure when the part is in service.

Welding positioners designed to work with robots are essential for parts with weld joints the robot cannot reach and for many joints the robots can get to. Simply reaching the joint is not enough to maximize the productivity of the fusion-welding process. The reason is that both oxy-fuel and arc welding speeds are strongly influenced by the position of the weld metal when it is being deposited.

Downhand welding often can increase weld-metal deposition rates by a factor of 10 or more compared with working out of position. What that means, using humans, is that one welding operator working downhand frequently can deposit as much weld metal as 10 operators working overhead. The same thing applies to a robot. Simply turning the work into the downhand position can increase productivity significantly. A robotic welding system can achieve maximum weld-metal deposition rates.

Other advantages of a robotic welding system are:

1. Arc-on time is maximized because the robot does not tire and take breaks.

2. Robots are easy to adapt to new work.

4.1   Weld Direction and Cost

Downhand welding costs less than vertical or overhead welding because of gravity. In flat downhand welding, gravity allows higher deposition rates and lower labor costs per pound of deposited weld because the molten weld metal is held in position by gravity.

The higher the deposition rate, the greater the chances are that the metal will sag and run out of the weld. Welding out of position causes problems whether a human or a robot deposits the weld metal. The preferred solution is to turn the work so that the weld is in the flat downhand position and to use a robot to do the work.

Operators using a welding-head manipulator, multiwire, multipass welding, and a high-production GMAW, FCAW, or subarc system do not need a welding robot, since they already employ a dedicated, automated welding system. Welding robots can be justified if the work falls between intermittent stick-electrode welding and highly automated production welding (Stenke, 1990).

4.2   Using Robots and Floor Space

Manufacturing floor space can be the most restrictive operating cost, and robotic welding systems make effective use of available floor space. Space comparisons with other materials-handling systems will generally favor a robotic welding system.

4.3   Using Multiple Positioners

The use of two positioners and one robot makes extra sense for another important reason. Since positioners are materials-handling devices specifically designed for handling and assembling weldments, the time used to load the work can be cut in half by keeping the workpiece on one positioner after it is welded and using the second positioner to keep the robot welding.

Some applications use workstations with three positioners and one robot. The first positioner is used for parts assembly and tack welding. The robot welds the work on the second positioner. The third positioner is used for parts inspection and nondestructive testing. A robot can be programmed to turn to any of the three positioners that is ready for welding. The labor time saved, alone, between stops for loading and unloading, has changed the entire workflow of some metal-fabricating plants.

5   WELDING COST CALCULATIONS

5.1   Cost of Filler Metals

No arc welding process is capable of converting all of the filler metal into weld metal. Some of the electrode's weight (unless it is an inert-gas-shielded solid wire) is converted into slag. A surprisingly large amount of most filler metals becomes spatter, and some of the electrode becomes fumes. If stick electrodes are used, a couple of inches of each electrode (typically one-sixth or 17% of it) also is thrown away as a stub. That immediately increases SMAW filler-metal costs by 20%.

Under ideal conditions the gas-metal-arc welding process (GMAW), using a solid wire, can be 99% efficient. At the other extreme, some shops deposit only 30% of the SMAW electrodes that they actually buy. The real cost per pound of filler metal is not that paid for the product in a carton or can, but the amount of electrode that can be converted into useful weld metal. Therefore weld-metal cost is calculated as:

$$\frac{\text{filler metal cost (\$/lb)}}{E_d}$$

where E_d is the fractional amount of the electrode deposited as weld metal (rather than the amount of the electrode that actually is consumed), that is, the electrode deposition efficiency.

The effects of differences in electrode deposition efficiencies among various electrodes were not as obvious before the advent of welding robots as they are now. Labor cost overwhelmed the total cost of welding, and filler-metal costs accounted for only a few percent (typically less than 5%) of total welding costs. With the increased productivity made possible by the use of a welding robot in a properly designed workstation, filler-metal costs become a significantly greater part of the total cost of depositing solid weld metal.

An analysis of filler-metal deposition efficiency and deposition rates can indicate that it may pay to use expensive wire to help the robot lower total costs. This could mean using high-deposition-rate processes such as flux-cored wire welding (FCAW), submerged-arc welding (SAW), or even hot-wire TIG (GTAW).

5.2 Deposition Efficiency

The influence of deposition efficiency on costs must not be taken lightly. Figure 4 shows how the relative cost of filler metal increases as deposition efficiency drops off. Deposition efficiency has an equivalent effect on labor costs (or robot amortization) discussed in Section 5.4.

5.3 Gas Costs

Most cored wires and all solid wires are designed to be used with shielding gases during actual welding. The contribution of the shielding gas cost to the overall weld-metal cost cannot be disregarded. Gas cost per pound of deposited weld metal can be calculated as

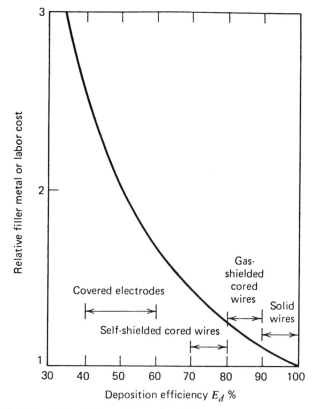

Figure 4 Effect of deposition efficiency E_d on the relative cost of the filler metal or labor. Base labor cost or the cost of the electrode is 1. Covered electrodes have the lowest deposition efficiency and thus the highest relative cost; solid wires boast the highest E_d and the lowest relative cost.

$$\frac{\text{gas price (\$/ft}^3) \times \text{gas flow rate (ft}^3/\text{h)}}{MR \times E_d}$$

where MR is instantaneous filler metal burn-off rate expressed in pounds per hour, and E_d is the electrode deposition efficiency expressed as a decimal.

Obviously this cost element can be kept to a minimum by selecting less expensive shielding-gas mixtures and setting the lowest possible flow rates, depending on the wire deposition rates and the arc power levels needed to obtain those rates. The welding equipment can have a significant effect on this cost because reduced flow rates are possible with well-designed nozzles.

On the other hand, lower gas-shielding flow rates too far increases the risk of producing weld porosity and other defects. In some wires the nature of the shielding gas or gas mixture directly influences the weld-metal transfer mode (for example, at least 85% argon in the shielding gas is required to produce spray-transfer welding instead of dip-transfer).

The operating characteristics of some welding wires are quite sensitive to even small changes in shielding-gas composition. A few percent change in a gas-mixture component can significantly change the shape of weld beds, the wetting characteristics of molten metal, and the depth of penetration into the base metal.

Shielding-gas composition also has an effect on the weld-metal deposition efficiency D_d because the amount of spatter, fume, and slag produced depends on how oxidizing the gas is. Straight CO_2 shielding, although low-cost, produces high spatter. This spatter reduces filler-metal recovery, increases maintenance costs and downtime. The E_d for solid mild-steel GMAW wires is less than pure CO_2 shielding (about 93% deposition efficiency) than with argon-rich gas mixtures such as 98% argon + 2% CO_2, or 95% argon + 5% oxygen, which produces deposition efficiencies of about 99% with properly deoxidized, solid mild-steel wires. The specially deoxidized wire costs a little more, as does the shielding gas, but that cost is more than made up by the high-deposition efficiency of the wire.

5.4   Labor Costs

Labor is the most important element in fabricating costs. It must even be included in robot welding, either directly as the cost of the robot tender or programmer, or indirectly as labor cost to load and unload the robot workstation. Besides depositing weld metal, welders and other workers must prepare joints for welding, position the components (or program the robotic positioner to do the job), fixture and tack weld parts, remove slag and spatter, and inspect the finished welds.

Shaving costs for welding filler metal, shielding gas, or equipment to save money will quickly be offset by increased labor cost to remove more spatter, repair bad welds, or for additional inspection required when less than optimum welding conditions are used.

When determining present cost of labor (in manual system) for depositing a given amount of weld metal, consider the fraction of each welder's day devoted to depositing weld metal (their duty cycle DC) as well as the rates at which the metal is deposited while the worker is welding (the actual deposition rate, $MR \times E_d$, where MR is the melt rate and E_d is the deposition efficiency). The labor cost per pound of deposited weld metal can be calculated as

$$\frac{\text{hourly wages + overhead}}{MR \times E_d \times DC}$$

The influence of deposition rate and duty cycle on the relative cost of labor and overhead is shown in Figure 5. A number of features are revealed here. First, labor costs are very high when deposition rates are low. Relatively small changes in the welding rate cause large changes in cost. Labor costs are relatively low when the welding processes allow high deposition rates to be used, and even small changes in those rates cause significant changes in labor cost. Finally, small changes in the operator's effectiveness as measured by the duty cycle (percent arc-on time) can have a large effect on total cost.

In other words, pay attention to increasing deposition rates ($MR \times E_d$, or melt rate times deposition efficiency) as well as arc-on-time, whether humans or robots are doing the welding. Increasing deposition rates from 2 to 4 lb/h (0.9–1.8 kg/h) can reduce relative costs from 50 to 25 units or 50%. The same 2 lb/h (0.9 kg/h) change at a 20 lb/h (9 kg/h) deposition rate still reduces costs from 5 to 4 units, or 20%.

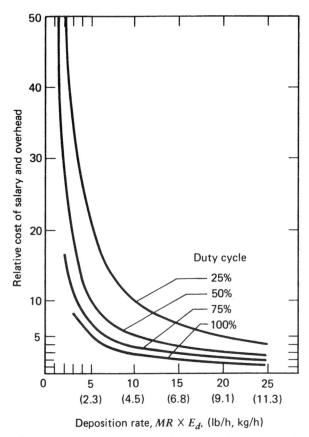

Figure 5 Effect of deposition rate (melting rate times deposition efficiency) and the welder's effectiveness (duty cycle) on labor costs and overhead.

Increasing the duty cycle (arc-on time) from 25 to 50% when depositing weld metal at 4 lb/h (1.8 kg/h) can reduce the per-pound cost for labor from 25 to 12.5 units. Or if the labor and overhead rate is $25/h for manual welding, the per-pound cost for the weld metal can be reduced by $12.50.

The same relative filler-metal cost reductions apply to robot welding, except that increasing a robot's productivity by 50% will cut in half the time to amortize the robotic workstation.

Again, the key to these savings is reduced labor costs if human welders are used. If robot are used, most of the labor costs are transferred into capital costs, which can be amortized, whereas labor costs for manual welding remain a constant part of the total welding costs as long as the work is done manually. Labor costs per hour for manual welding increase in proportion to electrode and gas costs. Labor costs per hour for robot welding remain low as long as the workstation is designed to keep up with the robot.

5.5 Joint Design Saves Money

Savings are also possible by reducing the amount of weld metal needed to fabricate a joint. The joint gap is designed to provide room for a welding electrode to reach into the joint, as well as to ensure complete fusion in the weld. Most welds designed for human operators are overwelded to get good average results. Just changing the included angle of a simple V-groove joint from 90° to 60° will reduce weld metal requirements by 40%. Several major filler-metal suppliers provide detailed tables for calculating the volumes of various weld joints. One of the better references is Lincoln Electric's *The Procedure Handbook for Arc Welding* (see the chapter entitled "Determining Welding Costs").

5.6   Fixture Design

Several general conclusions are relevant whenever a new weld fixturing is required.

Start with very simple fixtures, or clamp the part directly to the positioner table. Optimize the part location to provide the best accessibility and welding gun angle for the robot work envelope.

Use existing experience in manually welding a particular part to develop the best weld parameters for robotic welding.

When the part positioning problems have been ironed out and several test assemblies have been welded, only then consider going to more permanent fixturing.

To reduce setup and cycle time, multiple-step or full-component welding programs should be avoided. However, in some cases these more complex programs are necessary because of part size or to avoid complex fixturing and manufacturing changes.

Air- or water-cooled copper-backing blocks may be required to reduce burn-through on thin sections. These blocks can be built into the fixture or held in place by clamps.

5.7   Workpiece Size and Weight

The positioner and fixtures should be purchased or designed to be compatible with the part size and weight and the robot's tolerances. Close consideration should be given to the part's center of gravity when designing a system. A heavy part or fixture that is not balanced on the holding station can overload a positioner's tilt and rotation functions. Adequate clearance should be allowed for the part to be rotated and tilted in any direction. In conjunction with this requirement, the robot should initially be programmed in a safe location, or by off-line programming, to avoid accidental damage to the positioner while it is in operation.

5.7.1   Part-Location Positions and Clamping

When the fixture design has been finalized, make provisions to locate the part accurately and fasten it securely to the positioning station. Start with manual clamping prior to going to hydraulic, pneumatic, or electric systems. Use bolts, welding clamp pliers, or toggle clamps. More sophisticated systems can be considered after testing has established the proper clamp size and location and required clearances for robot and positioner motions.

5.7.2   Setup

Parts loading for a second assembly should take place simultaneously with robot welding on the first assembly. With this type of system, part-loading time is not critical as long as it is less than the robot welding program's cycle time. Provide a second MIG or TIG welding station in the robot enclosure for tack welding and touch-up work.

6   EXAMPLE: ARC WELDING OF AC MOTORS

The frames of small AC induction motors are made of cast iron, but the frames of medium-sized motors are made of steel. We show here the system by Hitachi for assembling and welding automatically the steel frame of a 55–132-kw AC motor (diameter: 345–553 mm; weight: 70–220 kg). It is common to use a welding robot after partial manual welding or welding with a fixture to keep the accurate position of workpieces during the welding operation. This requires a lot of manual work for preparation. Manual work is minimized by the system described here.

The components for welding are shown in Figure 6. For the steel-frame motor there are seven kinds of workpieces, such as a cylinder, fins, and stays. The total number of the components is 45.

6.1   The Welding System

The welding station consists of three subsystems (Figure 7): (1) two robots; (2) a positioner; (3) conveyors. The two articulated robots (Hitachi Process Robot) are electrically driven, with five degrees of freedom. One is for arc welding, the other is for supplying and positioning of components. The positioner tilts the motor cylinder so that the robot can reach the proper place of the workpiece. Indexing of the tilting by the positioner is at 15°. The conveyer system carries the cylinder from the previous station to the welding

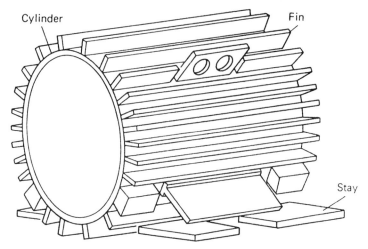

Figure 6   Components of steel-frame motor.

station. The hand for gripping the workpiece uses a magnetic chuck and an air-driven clamp.

The welding system is able to position and weld components within 0.5-mm accuracy. In addition to the welding robots themselves, other subsystems are also developed at Hitachi.

6.2   Evaluation

Since installation of the system, 75% of welding, measured in total arc welding length of the motor frame, has been automated. Also, a 20% speedup in production rate has been achieved. A total of 900 location points for parts handling and 1500 location points for welding are taught to the robots. An overview of the system is shown in Figure 8.

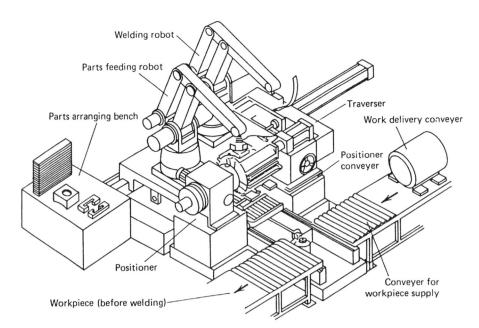

Figure 7   Schematic diagram of the welding system.

Figure 8   An overview of the system.

7   PROGRAMMING EXAMPLE: GMAW APPLICATION IN ARC WELDING

This section describes the operation of a robot system in gas metal arc welding (GMAW). The base robot system is augmented by equipment required for continuous arc welding, with seam tracking, and with adaptive feedback for error correction. This application integrates special software, available to the user through the MODE key switch on the robot controller.

Figure 9 is a block diagram of the complete arc welding system. The robot controller, hand control, I/O box, and power unit are standard components, with optional elements and features for both the hand control and robot controller. The robot is a standard six-axis robot. It is important to notice that the control is more than merely a device for teaching the robot. In this system the hand control gives the operator the ability to make real-time changes to individual parameters in a given weld table. The operator can also change entire tables during operation. Elements of the system are as follows:

Optional axes (positioning table). Controlled by an optional axis-control board in the robot controller, the positioning table is controlled by the same kind of signals from the robot controller. In the teaching phase, the operator maneuvers the table and robot as necessary to get the required relationships between the robot's welding torch and the welding seam of the part mounted on the table. Multiple-function capabilities of the hand-control manual control push buttons provide complete control of all eight axes involved. In program execution the robot controller coordinates all eight axes, presenting the seam to the robot's torch in the location and orientation taught, modified by adaptive feedback or changes inserted by the operator, dynamically, during execution.

Welding controller. A subassembly located in the welding power supply, the welding controller sets voltage and amperage as established by robot controller, holding at programmed levels during execution, unless signals are otherwise modified.

Welding power supply. This is the electrical power supply for arc welding. The robot controller controls amperage and voltage by welding controller interface.

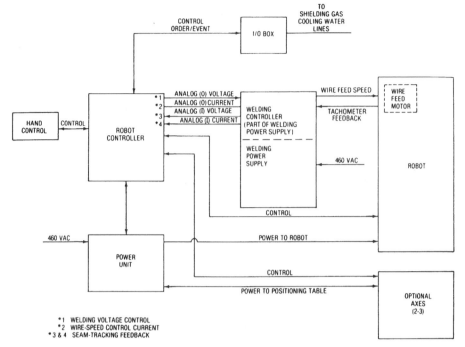

Figure 9 Block diagram of an arc welding robotics system.

Input/output interface box. A Standard I/O box, in this application it continuously links the robot controller to controls of cooling water and shielding gas.

Wire-feed motor. Installed on robot, controlled by input from robot controller, tachometer feedback to welding controller provides closed-loop control of wire-feed speed, which affects the characteristics of the welding arc. The wire provides filler for the welding seam; the wire chosen for specific applications must be compatible with the metals being welded.

7.1 Process Parameters

In operation, the welding system controls the following parameters, which collectively determine the quality of the welding. Before welding, the operator must enter these welding process variables into weld sequence parameter tables, each with an assigned weld sequence number. Thereafter the weld sequences can be recalled to apply specific combinations required for a given welding task. Also, the tables can be modified, either with the robot controller or with the hand control.

1. PREFLOW:	time prior to welding used to establish a flow of shielding gas to protect the arc from atmospheric gases.
2. RUN-IN TIME:	time required to establish an arc.
3. RUN-IN VOLTAGE:	voltage to establish the arc.
4. RUN-IN FEED:	feed rate of wire to establish the arc.
5. WELD VOLTAGE:	sustaining welding voltage, after run-in.
6. WELD FEED:	sustaining wire-feed rate.
7. CRATER TIME:	time required to fill and finish the crater at the end of welding.
8. CRATER VOLTAGE:	voltage required during crater finishing.
9. CRATER FEED:	wire-feed rate required during crater finishing.
10. BURNBACK TIME:	time required to burn the consumable electrode wire free of the welding surface.

11. BURNBACK VOLTAGE: voltage required for burnback.

12. POSTFLOW TIME: provides extra shielding gas to insure coverage of the crater as it cools.

13. WEAVE WIDTH: a lateral vector set at the robot controller, adjustable with hand control; determines the horizontal width of the welding seam.

14. LEFT DWELL: time robot holds torch on left edge of seam during seam tracking.

15. RIGHT DWELL: time robot holds torch on right edge of seam during seam tracking.

16. CROSS TIME: time robot takes to move torch laterally across the welding seam.

17. STICKOUT CURRENT: welding current level, which the RC-6 monitors to track a weld seam at a constant height (stickout).

18. TRAVEL SPEED: speed of the torch along the welding seam.

Variables 1–12 control the arc process; 13–16 are the basic parameters for weaving; 17, STICKOUT CURRENT, is the arc current that the robot controller maintains by moving the robot vertically with respect to the plane of the welding seam, either moving *in* to the seam, or moving *out* away from the seam. Variable 18, TRAVEL SPEED, sets the speed for torch movement along the seam.

7.2   Operation

The application sequence section of program software prompts the operator about programming a weld sequence while teaching a trajectory. Each of the preceding variables can be varied in terms of time, speed, voltage, current, or distance, depending on the variable itself and the limits established.

To weld, the operator teaches a trajectory. On a given point he or she can specify start or stop welding, the weld sequence number, and any other required parameters. After teaching the trajectory and linking it to a cycle, the operator can run the welding operation like any other cycle and can modify the major welding parameters from the hand control dynamically while welding: welding voltage, welding wire-feed rate, stickout tracking current, weave width, dwell time, and welding travel speed.

Electronics for seam tracking are integral to the robot controller, using data from the arc to correct both the cross-seam and stickout directions. The robot controller produces a weaving pattern that has programmable weave width, left-and-right dwell, and crossing time. Oscillation across the weld joint induces a change in the arc current. The seam-tracking electronics samples and analyzes the arc data at each sidewall and makes a proportional correction at the beginning of each weave cycle. Automatic current control maintains a constant stickout height above the workpiece. Figure 10, timing of arc welding sequence, displays the interaction of parameters and control signals during welding.

7.2.1   Weaving

During welding, *weaving* widens the welding seam and provides lateral torch motion, required for seam tracking. Weaving is controlled by a weaving vector, taught to the robot controller and inserted into the welding trajectory at the desired point, as are parameters such as speed. Figure 11 displays the weaving effect. In each case weaving moves the torch *across* the welding seam as the torch moves parallel to the seam centerline. This holds true for curvilinear paths as well. The operator can teach and store weaving vectors for recall.

Selection of WEAVE on the controller's MENU automatically places TOOL GEOMETRY in effect. In TOOL GEOMETRY the reference axes for torch movement are in a coordinate systems centered on the tip of the welding torch, shown in Figure 12 tool mode operation. The X-axis is the reference for STICKOUT. All of these directions are based on a perspective facing the end of the torch, in the plane of welding. Thus, positive STICKOUT ($+X$) moves the torch closer to the welder, increasing STICKOUT current; negative STICKOUT ($-X$) raises the torch from the weld, decreasing STICKOUT current. All commands issued by the robot controller as part of the program or as results of adaptive feedback for path control are referred to the TOOL GEOMETRY reference system.

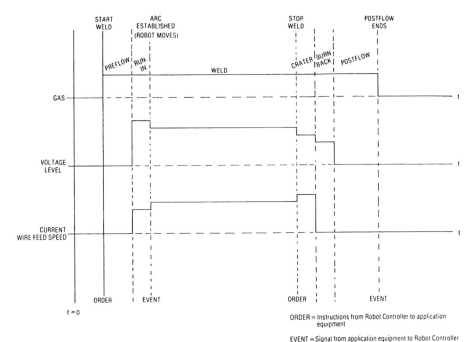

Figure 10   Timing of arc welding sequence.

Using a third function of the manual control, the hand control provides real-time modification of parameters during welding. In this application two buttons on the hand control reserved for user-defined functions are marked WELD CONTROL and WELD MODIFY. Pressing WELD CONTROL, the operator can use the six pairs of manual control buttons to increase/decrease welding voltage, wire-feed rate, stickout current, weave width, dwell times, and speed (torch travel speed). By pressing WELD MODIFY, the operator inserts the modifications into the currently used welding sequence table.

7.2.2   Adaptive Control

Adaptive feedback enables automatic error correction for stickout and weave. Required information flows from analog inputs from the welding controller through a serial link to the robot controller. The system extracts error-correction data for both stickout and weave centerline. TOOL GEOMETRY allows the robot controller to make changes directly into

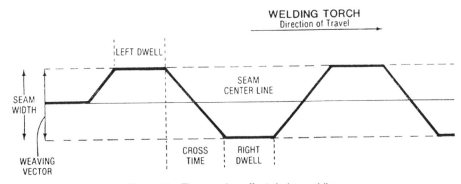

Figure 11   The weaving effect during welding.

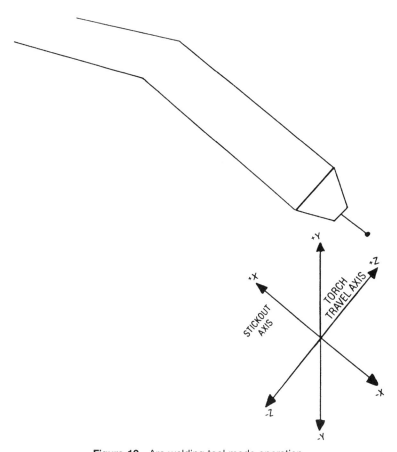

Figure 12   Arc welding tool mode operation.

the torch reference system. The origin of the coordinate system lies at the tip of the torch, and the X-axis is coincident with the centerline of the end portion of the torch.

Teaching trajectories and creating cycles with the welding application are essentially the same as with the standard software. Using the hand control and/or the robot controller, the operator starts or stops welding at any given point along a trajectory, modifies parameters and sequences during welding. When the operator stops welding (presses STOP WELD on the robot controller control screen), the system software automatically performs a controlled welding shutdown, filling the crater and burning back the electrode wire. Postflow shielding gas protects the crater from atmospheric gases until the welding cools.

A software feature called MULTIPASS allows the operator to rerun the same welding trajectory with multiple welding sequences, so that filling a very large weld seam can be done automatically, using the root pass, memorized by the controller and replayed with lateral bias as required to fill the seam.

Application software allows the operator to run a cycle or any portion of a cycle without arc to check the robot's accuracy. Status display screens (on the controller monitor) enable the operator to check the following parameters:

1. Speed of travel
2. Weaving
    (a) Width
    (b) Dwell time
    (c) Cross time

3. Weld sequence table number
  (a) Tracking: cross-seam only
  (b) Tracking: Stickout only
  (c) Tracking: both (cross-seam and stickout)

SUMMARY

The technological advances in the last 10 years have made arc welding the preferred joining technique for applications ranging from shipbuilding to space industry components. These advances provide the capability of welding more materials, including some previously considered impossible to weld (special alloys or different materials). Laser-welding systems also provide improved quality and efficient weld. However, laser-related implementation problems seem to constrain the range of laser-welding applications (see Chapter 44 Spot Welding and Laser Welding for a detailed discussion).

The field of welding robotics has been improved by increases in the efficiency and welding deposition rates of such systems. The developments in off-line programming and simulation methods of arc welding processes have contributed greatly to improving the actual performance of arc welding robotics systems.

The achievement of high-efficiency welding processes has also been supported by the development of new and more powerful control mechanisms. These controllers allow the welding robotics systems to perform on-line corrections to the process, along with the instantaneous inspection of the produced welds. Control systems based on laser vision systems have proved to be the most effective, given the environmental conditions in arc welding processes.

Control technologies are the future frontier for further developments. Research has already achieved efficient arc welding mechanisms. These advances have shifted the challenge to the control systems arena (Brantmark and Reyier, 1995). Adaptable arc welding robotics systems incorporating decision processes based on fuzzy logic and neural networks will become standard in the near future.

ACKNOWLEDGMENTS

This chapter has been updated and revised from its original version in the 1st edition of the *Handbook of Industrial Robotics*. Many thanks are due to the authors of the chapters in the 1st edition of this handbook whose contributions are part of this chapter. Sections 1–5 are based on Chapter 49, "Robots in Arc Welding," by Bruce S. Smith. Section 6 has been revised from Chapter 52, "Arc Welding of AC Motors," by Kenichi Isoda and Kazuhiko Kobayashi. The example presented in Section 7 has been revised from Chapter 50, "The Operation of Robotic Welding," by Peter G. Jones, Jean-Louis Barre, and Dan Kedrowski.

REFERENCES

Berger, H. 1994. *Automating the Welding Process: Successful Implementation of Automated Welding System.* New York: Industrial Press.

Brantmark, H, and I. Reyier. 1995. "Implementation of Application Oriented Function Packages in a Robot Control System." In *Proceedings of the 26th International Symposium on Industrial Robots,* Singapore, October.

Dilthey, U., and L. Stein. 1993. "Robot Systems for Arc Welding—Current Position and Future Trends." *Welding Research Abroad* **39**(12), 2–6.

Ferguson, C. R., and M. D. Kline. 1997. "Adaptive Welding for Shipyards." *Industrial Robot* **24**(5), 349–358.

Harwig, D. 1997. "Weld Parameter Development for Robot Welding." *Robotics Today* **10**(1), 1–7.

Huther, I., et al. 1997. "Analysis of Results on Improved Welded Joints." *Welding Research Abroad* **43**(8–9), 2–26.

Kannatey-Asibu, E. 1997. "Milestone Developments in Welding and Joining Processes." *Journal of Manufacturing Science and Engineering* **119**(4), 801–810.

Knott, K., B. Bidanda, and D. Pennebaker. 1988. "Economic Analysis of Robotic Arc Welding Operations." *International Journal of Production Research* **26**(1), 107–117.

Nomura, H., ed. 1994. *Sensors and Control Systems in Arc Welding.* London: Chapman & Hall.

Stenke, V. 1990. "Economic Aspects of Active Gas Metal Arc Welding." *Svetsaren, a Welding Review,* 11–13.

Sullivan, E. C., and N. S. Rajaram. 1992. "A Knowledge-Based Approach to Programming Welding Robots." *ISA Transactions* **31**(2), 115–148.

Teubel, G. 1994. "Experience and Application Up-date: Automation of Arc Welding Operations Using Robot Technology." In *Proceedings of the International Conference on Advanced Techniques and Low Cost Automation,* Beijing.

Wu, J., J. S. Smith, and J. Lucas. 1996. "Weld Bead Placement System for Multipass Welding." *IEE Proceedings Scientific Measurement Technology* **143**(2).

CHAPTER 46
PAINTING, COATING, AND SEALING

K. Sugimoto
Hitachi Limited
Yokohama, Japan

1 SPRAYPAINTING AND COATING ROBOT

1.1 Spraypainting Robot System

Industrial robots utilized in spraypainting applications are not quite similar in appearance and functions to robots in other application fields because of the environmental conditions in spraypainting. Conventional spraypainting robots are composed of three basic components: a control unit, a manipulator, and a hydraulic unit (Figure 1). Recently a manipulator driven by electric motors has become popular in this area.

Spraypainting areas of manufacturing facilities are normally the least up-to-date areas in the plant. In some industries spraypainting is still considered a human function and a skill or art rather than a science. In considering the implementation of a robot, it is advantageous not to rely on methods of the past as the only guidelines for the future.

Spraypainting robots in themselves should not be viewed as the complete solution to all painting problems. Rather, they should be recognized as integral parts of an automated system. And, like other forms of automation, spraypainting robots should possess the capabilities to complete their tasks and complement the total production operation.

Today's spraypainting robots can consistently duplicate the best work performed by skilled human painters, provided that sufficient time is allowed to adjust and refine the programming. Once the robot is programmed, it will repeat the exact motions of the sprayer/programmer and provide quality results, whether the application is a final top coat, primer, sealer, mold release, or almost any other material deposition.

The controller of the robot also provides the capability of interfacing with other equipment supportive of the spraypainting system, including color changers, conveyors, lift and transfer tables, and host computers. In fact, using the robot's controller capabilities can alleviate many of the design and operating concerns associated with present-day production spraypainting systems.

Today in the painting industry robots are applying automotive exterior top coat and underbody primer, stains on wood furniture, sound deadeners on appliances, porcelain coating on kitchen and bathroom fixtures and appliances, enamel on lighting fixtures, and even the exterior coating on the booster rockets that propelled the space shuttle into orbit (Figure 2).

1.2 System Components

1.2.1 The Robot Manipulator

Because many spraypainting robots are programmed or their programs are refined by a direct-teach method, workers play a key role in the design of a robotic spraypainting system. Close attention must be given to the interaction between the programmer, the robot, and the product. The programmer plays a vital role, and the physical limitations of both the programmer and the robot must be considered.

Manual spraypainting permits coating of remote and hard-to-reach areas because of the dexterity of the painter. Unless prior consideration is given to the size of the robot's arm and the spraypainting tool, the coating capability of the robot will be limited. Most spraypainting robots are equipped with three axes with actuators on the wrist, but this intricate painting capability may not be achieved by a robot with this type of arrangement.

Handbook of Industrial Robotics, Second Edition, Edited by Shimon Y. Nof
ISBN 0-471-17783-0 © 1999 John Wiley & Sons, Inc.

Figure 1 The main components of a spray finishing robot include a manipulator, control center, and hydraulic unit (photo courtesy DeVilbiss).

Limitations in the motion range of the wrist axes can seriously restrict the flexibility of the unit.

Ideally the robot arc should have at least as much freedom as the human wrist. The elephant nose-type wrist is sometimes utilized in a spraypainting robot. Figure 3 illustrates this type of wrist mechanism, named *flexi-arm,* manufactured by DeVilbiss. This design provides full arching without regard to the pitch-and-yaw axes.

Equipping the spray gun with a multiple-rotation capability permits rolling of the gun to coat interior surfaces. This is ideal for coating engine compartments, frame structures, undercarriages, and similar wall-type structures.

1.2.2 Programmable Positioning Equipment

The robot system provides more versatility and flexibility than manual or sophisticated automated systems. However, because robots are limited to their work envelopes, it is sometimes necessary to increase their spatial coverage. This is accomplished using auxiliary positioning axes, which can be programmed to reposition the robot manipulator to gain the optimum working position or for the robot to follow the conveyor's motion.

1.2.3 Fluid and Air Control

Control of fluid and air pressure is a major concern in any spraypainting system. The control of these functions is the key to obtaining a quality finish and efficient use of coating material.

A number of methods are available to achieve this control. Most automatic and manual spray guns are equipped with manual adjustments to perform these functions. However, these controls are normally preset by the spraypainter to coat the large surface areas of the product. They also are usually set to the level that requires the least amount of effort and skill by the worker. With feathering techniques an experienced worker may elect to minimize fluid flow for certain areas, but under most conditions this is not the case. Feathering features usually do not exist in automatic systems. Consequently the flow and pattern are set for maximum area coverage.

By taking advantage of the programmable output functions of the robot controller, it is possible to have fully automatic control over fluid flow, atomization, and pattern size. A standard automatic-type spray gun is modified to permit remote and separate control

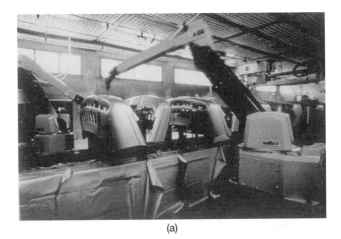

(a)

(b)

Figure 2   Painting and coating robotics examples. (*a*) Spraypainting of car-bumpers by two robots (photo courtesy ABB Flexible Automation Inc.). (*b*) Enamel powder coating of preheated (to 600°F) sinks (photo courtesy ABB Flexible Automation Inc.). (*c*) Multiple robots spraypaint a car body; doors, hood, and deck openers are coordinated in this integrated system (photo courtesy ABB Flexible Automation Inc.). (*d*) Coating wheels with multiple guns (photo courtesy ABB Flexible Automation Inc.). (*e*) Robotic painting facilities include down-draft and side-draft booths, drying ovens, pedestal- and rail-mounted robots (photo courtesy ABB Flexible Automation Inc.). (*f*) Inside a coating booth, robot arms spray-coat, while door openers are shown on left, right sides (photo courtesy Fanuc Robotics). (*g*) Dual nozzle for coating is shown in a coating line with multiple booths in series (photo courtesy Fanuc Robotics).

over atomization and pattern-forming air. Through a series of air-piloted and shuttle valves, remote-control regulators monitor fluid and air pressure to the spray gun. By using only two functions, it is possible to achieve three levels of controls, more than adequate for most applications.

The functions are programmed by the robot trainer during the teach cycle (Figure 4). The actuation of a miniature three-position toggle switch conveniently located on the teaching handle performs the programming and permits the trainer to vary pressure and spray pattern to achieve the desired results and compensate for variation in products and materials.

(c)

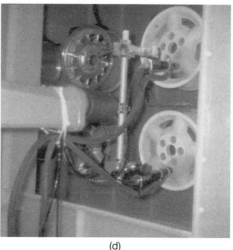

(d)

Figure 2 (Continued)

With this technique it is necessary to preset the desired levels of control prior to programming. Should this not be desirable, or should three levels of control be inadequate, analog control can provide the trainer with virtually infinite fingertip control at the teach handle.

1.2.4 Cleaning Spray Caps

A problem for most automated spraypainting systems is the requirement to clean spray guns to maintain the quality of the finish. This results in the additional cost of relief painters and cleanup crews in production downtime. Again, the programming output functions of the robot can solve this problem.

In most cases the cleanup required during production of an automated system or even on a hand gun consists of cleaning or changing the retaining ring and spray cap of the gun. If the gun is mounted on an automatic machine, the system may have to be shut down. But not with the robot. A spray-cleaning nozzle is mounted at a convenient location, and the robot is programmed to move the spray gun to the cleaning device and position the gun in front of the nozzle. At this time the programmable output function

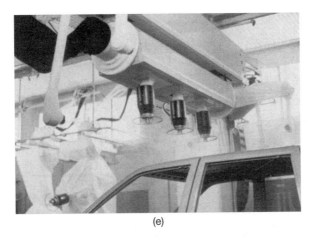

(e)

(f)

Figure 2 (Continued)

controls the spray-cleaning jets for the proper duration. Using a counter or timer and appropriate controls, it is possible to operate this cleaning program to meet the requirements of production and maintain clean air caps throughout the day. The cleaning cycle is determined by the frequency set on the counter or timer. With the use of proper solvents and nozzle design, the cleaning cycle can be completed in approximately 2–3 sec.

1.2.5 Electrostatic Control

The electrostatic application of coating is well known for its benefit of material savings. However, in many products electrostatic attraction, ionization, and the Faraday cage effect can pose problems. These effects can be compounded by a product design that requires deep-coating penetration or includes sharp corners, by insulated components within an assembly or other material, or by a problem with the substrate.

   With human sprayers and conventional automation, any one of these problems could prohibit electrostatic application. With robotic applications, programmable output functions can solve the problem. The solution is in programming one of the output functions to control the electrostatic power supply and cycle the voltage as required. This control can be built into the teach handle of the robot, enabling the programmer to control the electrostatics as required throughout the coating process. Thus the desired penetration can

(g)

Figure 2   (Continued)

be achieved in recessed areas by eliminating the Faraday cage effect created by the high voltage. It is also possible to achieve the benefits of electrostatics while, at the same time, selectively eliminating the detrimental effects associated with electrostatic coating.

In a spray application, more than in a pick-and-place or machine loading/unloading situation, ambient environmental conditions can play a significant role in the success or failure of a robot installation. It is essential that the characteristics of the coating material remain consistent in terms of viscosity, specific gravity, temperature, or pressure. If these items vary to any great degree, the finish will also vary. As a result, even though the

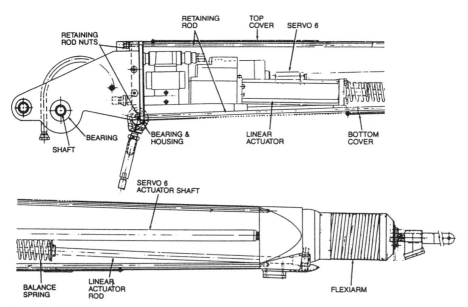

Figure 3   The flexi-arm, first manufactured by DeVilbiss, allows greater freedom of mobility during the coating process.

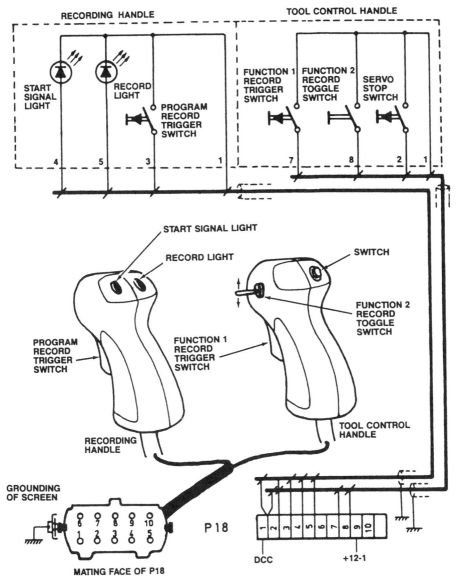

Figure 4 Wiring diagram of robot teach handles. Functions are programmed by a robot trainer during the teach cycle.

robot may be performing flawlessly, lack of control over these ambient conditions can greatly affect finish quality.

In many instances a robot is installed in a spraypainting operation where the success is dependent on factors other than the robot. When initial failure occurs, many times the robot is viewed as the cause when, in fact, problems have arisen with the material being sprayed, part orientation, or environmental factors in the booth.

1.3 Robot System Design

The controller acts as the brain of the robot system, providing data storage of position data, path generation and motion control, outside function control, and control interface with other programmable controllers or host computers. The memory medium for spraypainting robots may be solid-state, floppy disk, or any other type of mass memory.

The controller is normally designed to interface through input/output with other controls, such as programmable controllers, and other mechanisms, such as transfer tables, and spray gun triggering. It is necessary for this type of interface to provide a synchronous program between the robot movements and other components.

The second component, the robot manipulator, is the working end of the robot. This component is the actual spraypainter and must duplicate the movements of the human body. Arm movement and, most important, wrist movement must be duplicated for the robot to achieve a quality finish like the human spraypainter. A robot-finished job is only as good as the program taught to the robot. If the program is questionable, the robot-repeated work will be unacceptable.

1.3.1 Feature Capability

Most spraypainting robots are of the continuous-path variety, capable of multiple-program storage with random access. These two features allow for the smooth, intricate movements, duplicating the human wrist, that are needed to perform high-speed, efficient spraypainting. The random access of multiple programs is normally required on production lines that paint a variety of different style parts. Seldom will a particular line paint the same style and color of parts.

Many spraypainting robot installations have utilized lead-through teach-type robots that are taught by leading the robot manipulator, or a teach arm that simulates manipulation by a human operator, through movements and physically spraying the part. Programming the robot in this manner is simple. The operator attaches a teaching handle to the manipulator arm (Figure 5) and plugs it into a receptacle on the robot base. He then leads the arm through the designed program sequence to define the path and relative velocity of the arm and spray gun. After programming, the operator switches the control from "programming" to "repeat" and puts the robot into automatic mode. Robot applications have been used with a teach control pendant, but this type of teach method is usually successful only when the part is stationary. Off-line programming methods using a robot system model in a computer are becoming popular in many robot application fields. However, it is difficult to model the process of spraypainting, especially when a part has a complicated shape. The program made by off-line programming may need to be refined because of an insufficient process model and geometrical errors in the robot model. Therefore the lead-through teach method is still necessary for spraypainting robots.

Spraypainting robots require some additional features that are normally required for other types of robots. In most cases finishing robots must be equipped with noise filters to prevent interference of electrical noise from electrostatic spraying devices located on the end of the robot or in the near vicinity. Also usually required is an explosion-proof manipulator and an explosion-proof remote control operator's panel in the spray area.

Other features that enhance the spraypainting robot system are gun and cap cleaners and a cleaning receptacle into which the robot can submerge the spray gun so that the gun's exterior can be automatically cleaned after prolonged use or after color changes.

1.3.2 System Operation

Spraypainting robots must be designed and built for reliable operation in a dirty, solvent-filled atmosphere. In most cases the controller is in an area susceptible to paint, solvent, heat, and other elements harmful to a computer. The controller must be designed to operate easily for paint shop personnel who are not normally trained in sophisticated computer control machinery.

The equipment used in a spraypainting robot system must be properly maintained, especially since the manipulator end of the robot is subjected to solvent, mist, and accumulation of overspray. Robots will operate satisfactorily within a spray atmosphere for long periods; however, unmaintained equipment will eventually begin to cause production downtime and loss of productivity.

1.4 Application

The first continuous-path spraypainting robot system, at John Deere and Company, was installed as part of the chassis plant operation at the company's Tractor Works plant in Waterloo, Iowa. The criterion was to provide capacity to paint tractors at 92% efficiency in an 880-min day. In addition, there were eight basic tractor chassis models with a total of 36 paint programs to handle all variations. The system's goal was to paint 95% of

Figure 5 The teaching handle, attached to the manipulator arm, is used during programming of a lead-through-teach robot (photo courtesy DeVilbiss).

two-wheel-drive tractors and 90% of four-wheel-drive tractors. The first design consideration was the number of robots needed to store information about the different styles and variations of chassis and paint programs.

After initial testing, it was determined that robots could paint the tractor chassis to specifications. Figure 6 shows the cross section of the robots as they were installed in the spray booth relative to the tractor chassis. One robot was located on each side of the conveyor line, and the third was installed in a pit beneath the tractor chassis.

The robot in the pit sprayed the underside of the moving chassis and axles as well as the chassis sides. The chassis then proceeded into the next robot station, where the two robots sprayed the sides, top, and remaining areas. Owing to the size of the tractor, the system was designed so the robot moved in and out 18 in. perpendicular to conveyor travel. This allowed the robot to paint the end of the axle and the middle of the tractor chassis.

Robot programming was kept to a minimum to ensure a satisfactory start-up time for the new system. To reduce the number of programs, a feature termed *program linking* was used. Program linking incorporates the composite of three individual programs, one for the front end of the chassis, one for the center, and one for the rear. By the linking of these segments as required for each chassis model, the paint code was determined.

The benefits of program linking were easier and faster programming during installation and reduced memory storage. The latter benefit was illustrated with two front ends, which required painting variations on two-wheel-drive chassis in addition to the option of rockshaft or no rockshaft. Without program linking, each of the six variations would have

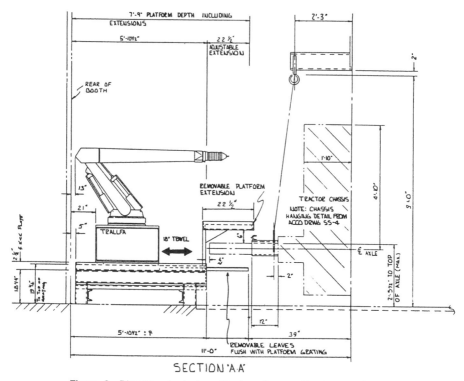

SECTION 'A·A'

Figure 6   Diagram of robot positioning when coating tractor chassis.

required 4.6 min per program, or 27 total minutes, to finish that model. Through program linking, each front end required only 1.5 min, or 4.5 total minutes. The middle section was the same for all the chassis of the model, requiring 1.5 min, and the rockshaft/no rockshaft added another 3 min, for a total 9 min, or two-thirds less time than if each model had a continuous program.

2 ROBOTS FOR SEALING AND ADHESIVE APPLICATIONS

2.1 Robotic Dispensing Systems

Industrial robots are increasingly utilized for material dispensing, especially in application of sealants and adhesives. The following major factors motivate this application area.

1. *Applying just the right amount of adhesive, sealant, or other materials.* Unlike manual application, where operators often dispense too much material that may cause poor quality, robots can accurately control the amount and flow of material dispensed. In addition to better quality, material cost savings can be very significant over long periods (up to about 30%).

2. *Consistent, uniform material dispensing.* Robots can maintain, with high repeatability, a consistent bead of material while laying it along accurate trajectories. Furthermore, where two or more components must be mixed while they are dispensed, as is the case, for example, in certain adhesives, robots can provide better control.

3. *Process flexibility.* As in other application areas, the robot can be used to dispense materials according to different programs, depending on the particular that is required. It can be easily reprogrammed when design changes occur.

4. *Improved safety.* The use of robots reduces the health hazards to workers from dispensed materials, including allergic reactions to epoxy resins and other substances, and potential long-term problems.

These four general factors are similar in many ways to those of spraypainting robot systems. Also similar to spraypainting, the major application of sealing and adhesive dispensing are in the automotive, appliance, aerospace, and furniture industries. In general, any process that requires joining of component parts in a variety of production is a potential candidate for robotic dispensing. In the area of adhesive application, one can certainly say that adhesive bonding is as important to the assembly of plastic parts as welding has always been to metal joining.

Typical examples of robotic dispensing are the following:

- Sealer to car underbody wheelhouse components
- Silicon on truck axle housing
- Urethane bead on windshield periphery before installation
- Two-component polyurethane or epoxy adhesive to automobile hoods made of sheet molding compound (SMC) between outer and inner shells
- Sealant application in the appliance industry, as described later in this section

2.2   System Components

The general structure of a robotic dispensing system is shown in Figure 7.

2.2.1   The Robot Manipulator

Typically a robot for accurate material dispensing is electrically actuated to achieve smooth motion and high repeatability. Five- to six-axis robots are required for flexible motions. Although speeds are usually on the order of 10–15 in. (25–38 mm) of bead per sec, the robot must be able to move at different speeds at different segments of the bead path. However, a key requirement is the ability to maintain a constant bead size along all contours. Robots can be floor or overhead mounted, depending on the particular dispensing orientation. A robot can be mounted on another translational axis or axes when material must be dispensed onto large parts, such as aircraft wings.

2.2.2   Type of Dispensed Material

In general, each material will have unique properties that must be considered in the application planning. A good summary of adhesives used in assembly and their properties can be found in Larson (1983). Two-part adhesives usually require static or dynamic mixing devices. When a mixer is attached to the robot manipulator, motion variations may cause inconsistencies in the bead. Hot-melt materials do not require mixing, but the hot temperatures (300–400°F or 150–200°C are typical ranges for sealants and adhesives) may present a problem for some robots.

2.2.3   Container, Pump, and Regulators

Pump selection depends on the properties of the material, the container size, and the dispensing rate. The dispensing system must track the level of material in the container and stop the automatic operation when the material is depleted. Another issue is the timing control. Certain materials dry out, harden, or solidify if not mixed or if left unused for a period of time. Some adhesives harden within a few minutes after mixing and must be

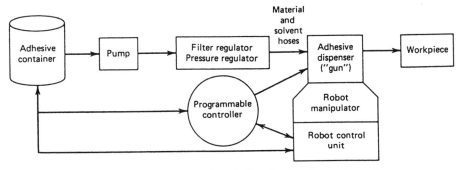

Figure 7   General structure of a robotic adhesive-dispensing system.

controlled very carefully. In such cases the system should have the capability to auto-matically purge spoiled material and clean the container and lines. Regulators of filters and line pressure are also essential accessories that are required in the dispensing system.

2.2.4 Programmable Controller

A programmable controller is used to supervise the overall dispensing option and com-municate between the robot, the container, and the dispenser ("gun"). Typically, it is responsible for the on/off activation of the dispenser in coordination with the robot motion and for the control of the material level.

2.2.5 Dispenser ("Gun")

Usually an automatic dispenser is attached to the robot wrist, together with an inlet hose for material and one for cleaning solvent. When a mixing unit must be used, as in two-component adhesives, the combined weight of dispenser, mixer, and three hoses neces-sitates mounting on the robot arm. In this case an additional tube is attached to the wrist and leads the mixed material from the mixer to the workpiece.

An important issue is control of bead integrity. An air jet sensor located on the dis-penser nozzle can be applied to detect any break in bead application. This type of sensor must be cleaned frequently to prevent any clogging. In more sophisticated application a vision system is used to monitor the bead integrity. When a break is detected, the robot is instructed to return and dispense material to the missing areas.

2.3 Robotic Dispensing of Foamed, Hot-Melt Sealant

In this section a particular robotic dispensing system of foamed, hot-melt sealant is ex-amined. This system was developed and installed for sealing perforations in multiple refrigerator cases at a General Electric (GE) refrigerator plant. It combines GE's 5-d.o.f. robot with Nordoson's FoamMelt System. Production results have been excellent and follow the list of potential benefits mentioned in Section 2.1.

The major appliance industry has experienced a problem in the manufacturing of re-frigerators that until recently has found only moderately acceptable solutions. The refrig-erator outer case contains holes and gaps necessary for fabrication and assembly. To prevent foam leaks and moisture migration, these perforations must be sealed prior to the injection of urethane foam insulation.

2.3.1 The Sealing Process

The process of sealing perforations in refrigerator cases with an adhesive appeared ini-tially to be relatively straightforward. The manufacturing parameters in GE's plant varied, however, and an extensive evaluation was required. These parameters were as follows:

- Two refrigerator models were to be sealed with provisions for future new models.
- These models included two different case sizes, 11 and 14 ft^3.
- The potential sealant application points varied from 15 to 20 points.
- Various hole sizes, contours, and locations were mandated by design in each case.
- Production rates varied from 15 to 18 sec.
- Cases were moving on a conveyer at the point of manufacturing, lying horizontally with the open side up.
- The substrates for application of sealant were both prepainted and galvanized steel.

Extensive planning and evaluation were required to demonstrate successfully the ele-ments of automation proposed for case sealing. Studies were performed to evaluate:

1. Foamed hot-melt adhesives as opposed to conventional sealants, and the suitability of those materials
2. The capabilities of robotic systems versus hard automation
3. The automatic material handling and fixturing

2.4 Automation Components of the Sealant Robotic Dispensing System

The system includes the following components:

1. Sealant
2. Robotic system

   3. Material-handling system

2.4.1   Foamed Hot-Melt Adhesive Sealant

In 1981 Nordson introduced a commercially available foamed hot-melt adhesive system. Early studies of sealing refrigerator cases with conventional hot melt were unsuccessful because of the sagging and running of the material. Foamed hot-melt adhesive was selected for trial in this application for two reasons. First, a foamed adhesive material is more cohesive than the same conventional hot-melt adhesive. This reduces material sag and run and enables holes to be bridged easily without penetration. Second, material savings are approximately 50% greater (for a material with 100% expansion) with a foamed rather than unfoamed adhesive.

Extensive development work was conducted in which sample refrigerators were manually sealed with a foamed adhesive. Holes were successfully bridged with the foamed sealant by raising one end of the case, applying the sealant above a hole, and allowing gravity to cause the material to flow over the holes. The sample refrigerators were then processed through a urethane foam injection system. The urethane foam provides the

Figure 8   Robot automatically dispenses foamed hot-melt adhesive extruded from a dispensing gun for sealing holes in refrigeration cases.

Figure 9 General view of robotic sealant dispensing cell. Hydraulic device (right) lifts and presents refrigerator cases properly fixtured and oriented to the robot. Each case is presented at 30° to enhance sealant flow coverage. Robot controller with programmable controller and foam container/extruder are shown left. The robot is mounted on a roller track for off-line programming and maintenance.

thermal insulation and structural rigidity between the liner and case and is produced by an exothermic reaction of two particular chemicals. Therefore the foamed hot-melt sealant must be able to withstand the urethane foam temperature and pressure. No urethane foam leaks were detected from any perforation manually sealed with the foamed adhesive sealant.

The adhesive application equipment for the final production installation consisted of Nordson's FoamMelt Unit, gun with a 12-in. (30-cm) heated extension, two 16-ft (5.3-m) hoses, and specific accessories. This system proved successful in actual production.

In the selection of the production sealant, several materials were tested. The results showed that a particular amorphous polypropylene-based sealant met the requirements of the application. This material was chosen for the following reasons:

- It foams with nitrogen and has excellent foamability, with an expansion of more than 100%.
- Its adhesion is excellent to galvanized and prepainted steel.
- Material setup is fast on steel substrates.
- It can withstand the temperatures and pressures encountered in the urethane foam-injection process.
- It foams and applies at moderate temperatures, which reduces energy consumption.
- It passes GE's odor and taste tests for major appliances.
- It meets the Federal Food and Drug Administration approved standards.
- No filler is added that might cause equipment wear.
- Previous studies demonstrated suitability for this application.

Figure 10 Robot is shown dispensing sealant along its programmed point-to-point path.

2.4.2 Robotic System

Sealing perforations in refrigerator cases with foamed hot-melt adhesives could be accomplished with either hard automation or a robotic system. Based on the manufacturing parameters of varied case sizes, hole sizes, contours, and locations, hard automation proved to be impractical. Criteria for a robotic system meeting the manufacturing parameters were established. Detailed feasibility studies and capability demonstrations were conducted involving several robot manufacturers.

The GE Model P5 Process Robot was selected and proved to be advantageous for this particular application. Tests confirmed the required cycle time and working envelope. Although selection of the robot involved many technical characteristics, three were essential:

1. The valid hole locations and contours demonstrated the need for point-to point programming with a teach pendant.
2. The manufacturing repeatability of hole locations and fixturing tolerance ($\pm$ 0.03 in., or $\pm$ 0.76 mm) demonstrates the need for high repeatable robot arm positioning.
3. The application required a six-axis robot in general. However, a five-axis robot can be used by considering the gun centerline axis as a sixth pseudoaxis by offsetting it with the fifth axis of the robot.

Figure 11 Close-up of sealant application. Gun tip must be positioned at proper orientation and distance from holes to provide successful sealing.

2.4.3 Automated Material Handling

The production process is limited without the benefits of automated material handling. In this system four constraints dictated the type of material-handling system needed for case sealing production:

1. The cases moved continuously side by side along a horizontal conveyor with the open side up.
2. The required maximum production cycle time was 15 sec.
3. Two different size cases were processed on the line.
4. A hydraulic tip-up station was required to raise one end of the case to allow foamed adhesive to flow by gravity over the holes.

With these constraints in mind, a material-handling system was conceived and designed (Figures 8–11). A hydraulic walking-beam-type transfer system was incorporated as the basis for handling. This was necessary because the time required to seal a case is about equal to the designated production cycle time; little time was left for a completed case to leave the sealing station and another to enter. To account for different size cases, adjustable pick-up pads were designed and built.

Figure 12 Multiple robots spray adhesive sealant on the interior seams of truck cabs (photo courtesy ABB Flexible Automation Inc.).

Finally, a hydraulic tip-up station was provided to allow the foamed sealant to flow by gravity over the holes in the case. This tip-up proved to be critical for the flow concept developed in early investigations and thus was incorporated into this equipment.

2.5   System Integration

Manufacturers today are producing individual cells of high technology. Often the marriage of two or more technologies is overlooked or simply not possible. The sealing of refrigerator cases discussed here is an example of where two high-technology components have been combined with the latest in automation techniques. The entire system was engineered for electrical control integration and interaction with mechanical components.

The electrical system was controlled by a programmable controller. It integrates the master control panel of the robot, FoamMelt Unit, hydraulic system, material-handling system, and input/output devices. Through ladder-diagram programming the controller commands the sequence of operations in both the manual and automatic modes. For example, the controller instructs the robot when to move through its motions. The robot then instructs the controller when to dispense adhesive. The controller further instructs the FoamMelt Unit to dispense a timed quantity of adhesive. Similarly, the programmable controller sequences the operation of other system devices.

Although many mechanical design features were incorporated into the system, a select few were subtle, yet very important. First, the dispensing gun was attached at an offset to the robot arm, enabling a five-axis robot to simulate a six-axis robot. Second, the suspension of the adhesive hoses removed restrictional forces from the robot arm. This proved necessary for consistent repeatability. Finally, the robot was placed on a roller track for off-line programming and maintenance. This allowed the walking beam to continue the transfer of parts while the robot was being maintained or reprogrammed with an auxiliary case. These features provided some additional support for avoiding problems.

The automatic sealing of refrigerator cases using foamed hot-melt adhesive and a robot has demonstrated a process of potential benefit to many industries today. Sealant applications are plentiful and possibly limited only by the adhesive chemistry and robot characteristics. For example, robots can be used to seal body panels in the automotive industry or provide a means for in-place gasketing in the appliance, automotive, or other industries (see Figures 12, 13). Manufacturers again have another tool for productivity and quality improvement.

Figure 13   A robot dispenses a urethane seal on a vehicle windshield (photo courtesy ABB Flexible Automation Inc.).

3   CONCLUSIONS

Robots have become an integral part of spraypainting, coating, and sealing applications in industry. Robots can work under harsh environmental conditions, eliminate safety and health hazards to workers, and help cut costs by reducing labor, energy, and material consumption. With improving design and off-line programming, these robotics application are becoming standard facilities, while maintaining high finish quality.

ACKNOWLEDGMENT

This chapter is a revision, consolidation, and update of two chapters from the 1st edition of this handbook: Chapter 76, "Robot Applications in Finishing and Painting" by Timothy J. Bublick, and Chapter 77, "Robots for Sealing and Adhesive Applications" by Patrick J. Bowles and L. Wayne Garrett.

REFERENCES

Dueweke, N. 1983. "Robotics and Adhesives—An Overview." *Adhesive Age* (April).
Larson, M. 1983. "Update on Adhesives." *Assembly Engineering* (June), 9–12.

ADDITIONAL READING

Robotic Coating, Painting, and Finishing

"Taking the Inside Track." *Finishing* **15**(11), 38–40, 1991.
Anand, S., and P. Egbelu. "On-Line Robotic Spray Painting Using Machine Vision." *International Journal of Industrial Engineering—Applications and Practice* **1**(1), 87–95, 1994.
Asakawa, N., and Y. Takeuchi. "Teachingless Spray-Painting of Sculptured Surface by an Industrial Robot." In *Proceedings—IEEE International Conference on Robotics and Automation*. Vol. 3. 1875–1879, 1993.
Graves, B. A. "Truck of the Year Shines with First-Class Finish." *Products Finishing* **59**(10), 48, 1995.
Grohmann, U. "Paint Robots in the Automotive Industry—Process and Cost Optimization." *The Industrial Robot* **23**(5), 11–16, 1996.
Hager, J. S., and N. J. Weigart. "Electrostatic Application of Waterborne Paint." *Automotive Engineering* **105**(6), 41, 1997.
Josefsson, E. "Powder Painting of Truck Cabs—Helping to Preserve the Environment." *ABB Review* 9–10, 38–42, 1996.
Loar, J. E. "Software Helps Speed Robot Adjustments." *Robotics World* **14**(1), 24–25, 1996.
Moon, S., and L. E. Bernold. "Vision-Based Interactive Path Planning for Robotic Bridge Paint Removal." *Journal of Computing in Civil Engineering* **11**(2), 113–120, 1997.
Owen, J. V. "Simulation: Art and Science." *Manufacturing Engineering* **114**(2), 61–63, 1995.
Yao, Y. L., and S. M. Wu. "Development of an Adaptive Force/Position Controller for Robot-Automated Composite Tape-Layering." *Journal of Engineering for Industry* **115**(3), 352–358, 1993.

Sealing Robotics

Haas, C. "Evolution of an Automated Crack Sealer: A Study in Construction Technology Development." *Automation in Construction* **4**(4), 293–305, 1996.
Kato, T. "Sealing Application with Work Handling by Remote TCP function." *Katakana/Robot* (September), 84–87, 1996.
Lasky, T. A., and B. Ravani. "Path Planning for Robotic Applications in Roadway Crack Sealing." In *Proceedings—IEEE International Conference on Robotics and Automation*. Vol. 3. Piscataway: IEEE, 1993. 859–866.
Sawano, S., and N. Utsumi. "Sealing." In *International Encyclopedia of Robotics*. Ed. R. Dorf and S. Nof. New York: John Wiley & Sons, 1988. 1455–1462.
Van Meesche, A., B. Bowe, and W. J. Cantillon. "Robotic Extrusion for Complex Profiles Using a Thermoplastic Vulcanite." In *New Plastics Applications for the Automotive Industry*. Warrendale: Society of Automotive Engineers, 1997. 127–129.

CHAPTER 47

FLEXIBLE FIXTURING

Lane A. Hautau
Allen C. Grzebyk
FANUC Robotics North America, Inc.
Rochester Hills, Michigan

José A. Ceroni
Purdue University
West Lafayette, Indiana

1 INTRODUCTION

Advances in robotics have contributed greatly to enabling manufacturers to respond quickly to market fluctuation and customer preferences. However, to become truly flexible the manufacturing system is required to provide for infinite product variety, profitable lots of size one, building different products on the same line, and achieving quick and low cost changeovers. These requirements coincide with Nissan Motor Company's definition of flexibility: "any product, any time, any place, any volume, by anybody" (Stuart, 1992). Robotics advances in fixturing are an important step allowing manufacturers to move to the next plateau in their quest to achieve the "five any's."

The concept of flexible fixturing has been around a long time. Systems employing variations of hard tooling have been used for years. Such systems often involve the utilization of one or more of the following approaches:

- Multiple dedicated lines
- Parallel processes
- Indexing hard tooling
- Tool trays
- Dedicated pallets
- Interchangeable tooling sections

Systems based on variations of hard tooling frequently suboptimize the product mix flexibility, convertibility, speed to market, floor space, and simplicity. However, through application of the proven benefits of robotics to fixturing, these shortcomings can be and are being addressed. In fact, many automotive and aircraft manufacturers have already implemented or are in the midst of developing reconfigurable fixturing systems.

The next generation of flexible fixturing systems has achieved its flexibility via servo-driven programmable positioners (Monashenko, 1992). Programmable positioners are mechanical devices that, through locators being moved and clamped to the appropriate orientation, allow manufacturing processes to be performed over a wide range of parts. Their reprogrammability provides the flexibility for fixturing different parts with minimal set-up time as the parts arrive at the manufacturing cell, as well as the autonomy for performing the adjustments automatically via a programmable controller. While programmable positioners possess the kinematics of a robot, they differ fundamentally in that positioners perform their value-added work standing still, while robots perform their value-added work mainly through motion. Programmable positioners are usually more compact, rigid, and repeatable than their robot predecessors. However, programmable positioners share robots' ability to reconfigure part to part, enabling manufacturers to

Handbook of Industrial Robotics, Second Edition, Edited by Shimon Y. Nof
ISBN 0-471-17783-0 © 1999 John Wiley & Sons, Inc.

build different products in the same fixture, with short changeovers that enable faster responses to market fluctuations. A comparison of programmable positioners versus hard tooling is shown in Table 1.

Programmable positioners offer processing options not readily available with the more conventional hard tooling systems. The programming capability provides the accommodation of operations requiring the sequential load of parts or the repositioning of clamps for providing access. Programmable positioners are also easier to set up, adjust, and tune in. Their programmability means that multiple fixturing scenarios can be analyzed, stored, and evaluated. Conventional tooling would require the costly, time-consuming addition, subtraction, or alteration of shim blocks. Programmable positioners can be shimmed in seconds by revising the mathematical data. Finally, programmable positioners permit better coping with design errors. In a conventional tooling system, if a bracket or locator is incorrectly designed, a lengthy delay can result from removing, reworking, and reinstalling the affected component. In a programmable positioner system a design error can often be compensated for by repositioning via controller software.

The ability of flexible fixturing systems to teach and shim location positions quickly allows plants to minimize downtime during startup and operation, resulting in operational savings (Brosilow, 1994; Lorincz, 1993). However, the initial capital investment for a flexible fixturing system can sometimes exceed the cost of a hard tooling variation. Other potential drawbacks occur where products are not designed for assembly or where one-year investment justification may hinder the adoption of flexible fixturing systems. Despite these problems, if changeovers occur the significant savings generated in capital investment can compensate largely for the initial investment. Instead of buying new fixturing being necessary, as it would be for conventional tooling systems, the clamps and locators of a programmable positioner system permit reusability of the system via reprogramming their positions. This flexibility feature not only reduces the capital investment for new fixtures, but also reduces the cost of process engineering, tool design, and lost production due to changeover downtime. Additional savings are also realized through having a standardized system. This standardization reduces the time required by maintenance and electrical personnel for familiarizing themselves with the maintenance and operation of new fixturing systems. Consequently, further savings can be expected from inventory reduction and simplification of spare fixtures.

2 INDUSTRIAL APPLICATIONS OF FLEXIBLE FIXTURING FOR ROBOTICS SYSTEMS

Although not extensively covering all the possible industry applications of flexible fixturing for robotics systems, the next examples illustrate the traditional application of robotic fixturing to arc welding and an innovative application developed recently.

Table 1 Comparison of Hard Tooling Variation and Programmable Positioner

Feature	Function	Hard Tool Variation	Programmable Positioner	Programmable Positioner Benefits
Flexibility	Multiple part run in same fixture	Multiple fixtures	Reposition servo-driven device	Equipment reduction
				Multiple styles on single line
				Floorspace savings
Convertibility	Product change	Scrap or rework tool	End-of-arm tooling position change	Lower time and cost to build new part
	Model change	Scrap entire cell	End-of-arm tooling position change	Reusable equipment
	Assembly added	Additional tooling or entire new fixture required	Tool position added	Equipment reduction
	Assembly removed	Scrap entire cell	Remove old positions, reprogram for new build	Reusable equipment

(*a*) *ESAB Orbit 500*

Workpiece:
   Max weight   1100 lb (500 kg)
   Max diameter   57 in. (1460 mm)
Indexing: 90°   360°
   Rotation   2.7 sec   7.2 sec
   Tilting   3.3 sec   8.9 sec

(*b*) *ESAB Orbit 160 R*

Workpiece:
   Max weight   352 lb (160 kg)
   Max diameter 45 in. (1150 mm)
Indexing: 90°   360°
   Rotation   2.3 sec   6.8 sec
Station interchange (180°): 4.5 sec

(*c*) *ESAB Orbit 160 RR*

Workpiece:
   Max weight   352 lb (160 kg)
   Max diameter   45 in. (1150 mm)
Indexing: 90°   360°
   Rotation 2.3 sec 6.8 sec
   Tilting   2.7 sec 7.2 sec

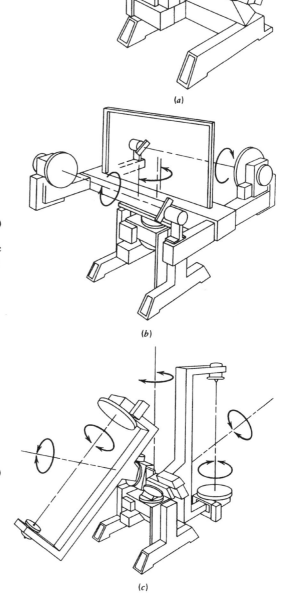

(*a*)

(*b*)

(*c*)

Figure 1 Five types of part positioners for arc welding robots. (From B. S. Smith, "Robots in Arc Welding," in *Handbook of Industrial Robotics,* ed. S. Y. Nof. New York: John Wiley & Sons, 1985. 913–929. Courtesy ESAB.)

(d) ESAB Orbit MHS 150

Workpiece:
  Max weight   330 lb (150 kg)
  Max diameter   45 in. (1150 mm)
Indexing: 90°   360°
  Rotation 6.0 sec 12 sec
  Tilting   4.0 sec   —
Station interchange (180°): 7.0 sec

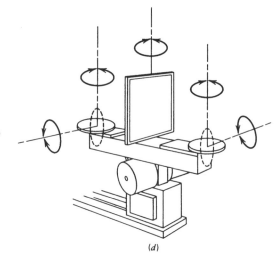

(d)

(e) ESAB Orbit MHS 500

Workpiece:
  Max weight   1100 lb (500 kg)
  Max diameter Any
Indexing: 90°   360°
  Rotation 11 sec 22 sec
  Tilting   5.0 sec   —

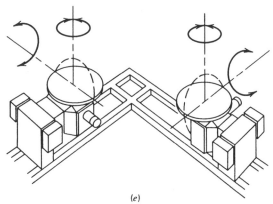

(e)

Figure 1   (Continued)

2.1   Robotic Arc Welding

The welding positioner is a critical and often overlooked part of the welding system. Robots are not compatible with perfectly good welding positioners designed for humans (Dilthey and Stein, 1993). The usually high robot accuracy is useless if not matched by the positioner handling the weldment. Also, most welding positioners are not designed for operation under microprocessor control. One solution to the control problem is to retrofit an existing positioner with new controls compatible with the robot. That solution still leaves the problem of positioning accuracy unsolved.

A human operator can tolerate much more gear backlash than a robot can, as well as a small amount of "table creep" during tilting. The human operator will compensate for the variation without thinking about it. The robot expects the weld joint to be where its program thinks it is, not where the positioner actually puts it. A conventional welding positioner cannot be retrofitted at a reasonable cost for the precision required by a robot. The machine would have to be rebuilt.

Modern robotics positioners have backlash controlled to within ± 0.005 in./in. (± 0.001 mm/mm) radius on all gears. Figure 1 shows different varieties of part positioners for arc welding robots. The specifications for the illustrated models are shown in the figure.

Figure 2   An arc welding robot station with programmable positioner system. (From B. S. Smith, "Robots in Arc Welding," in *Handbook of Industrial Robotics,* ed. S. Y. Nof. New York: John Wiley & Sons, 1985. 913–929.)

All the components of a robotic workstation must be interfaced so that each performs its function as initiated by the program. Typically a robotic positioner will include multiple-stop mechanical or solid-state limit switches on all axes for point-to-point positioning (open loop) or with DC servo-drives for positioning with feedback to the control (closed-loop). Figure 2 shows a complete robotic welding cell with two table positioners.

Speed is important in robotic workstations. The full rotation of the worktable positioner on two different axes (while handling a full load) is typically between two to three times faster than a conventional geared positioner. A full rotation in both axes with simultaneous motion takes about 9 sec. A 90° table rotation may take 3 sec.

2.2   Automotive Manufacturing and Assembly Applications

The increasing maturity of the robotics technology offers a new and promising opportunity for achieving a higher level of manufacturing flexibility. The availability of lower-cost and improved-drive mechanisms, motors, and motor controls make robotic fixturing applications feasible beyond the traditional positioners. FLEXTOOL from Fanuc (Brooke, 1994) is a system of robotic manipulators designed to locate and fixture automotive sheet metal for spot welding. The system provides increased product mix flexibility, increased speed and low cost to convert production lines to new products building, lower capital investment in process engineering and tooling design, and a lower floor space requirement. The FLEXTOOL system (Kosmala, 1995) consists of multiple servo-driven positioners utilized jointly with locators and clamps to fixture parts for assembly (Figure 3).

The FLEXTOOL system is able to perform value-added work between cycles by repositioning the parts, as well as during work cycles by locking up and holding the locator position. Figures 4 and 5 show the application of the system to the welding of car underbodies for a mix of small and large cars, both two- and four-door styles. Figure 4 shows the annual expected cash flow savings for the application, indicating when changes for the products are expected. From the graph insert of Figure 4, it is clear that the system presents its maximum savings capacity when changeovers take place. By integration of

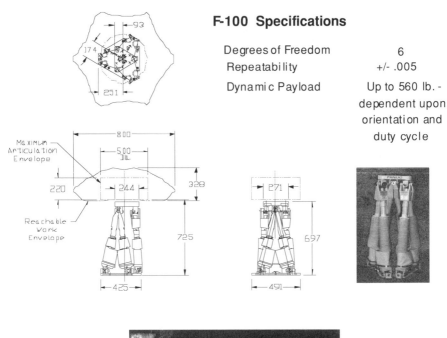

F-100 Specifications

Degrees of Freedom	6
Repeatability	+/- .005
Dynamic Payload	Up to 560 lb. - dependent upon orientation and duty cycle

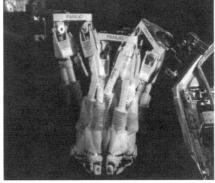

Figure 3 FLEXTOOL flexible fixturing system from FANUC (courtesy FANUC Robotics North America, Inc.).

the FLEXTOOL's positioner data into the CAD system in which the underbody is designed, the location and number of the FLEXTOOL positioners is determined and optimized (Figure 6). The main benefits of the system are its incorporation of lean tooling techniques and the elimination of many dump units, indexers, and other downtime-contributing devices currently present in a typical automotive body shop manufacturing system. The simplicity of the system's Windows-based interface, reprogrammable locators, and a reliable integration of servo-controls help manufacturers in achieving flexibility and gaining a competitive advantage in the agile manufacturing era.

3 SUMMARY AND EMERGING TRENDS

The global competitiveness that manufacturers face today has made mandatory the adoption of flexible manufacturing technologies. The flexibility in this case is directed toward the minimization of the nonproductive setup times when changeovers in the product line occur or when different products are mixed in the same production line. Incorporating programmability features into the fixture mechanisms brings the anticipated flexibility

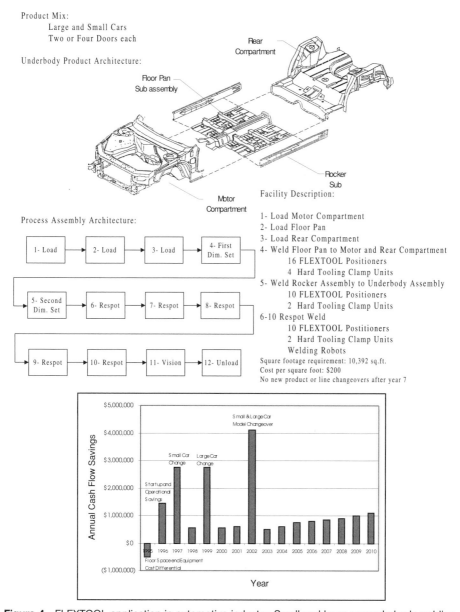

Product Mix:
    Large and Small Cars
    Two or Four Doors each

Underbody Product Architecture:

Process Assembly Architecture:

Facility Description:

1- Load Motor Compartment
2- Load Floor Pan
3- Load Rear Compartment
4- Weld Floor Pan to Motor and Rear Compartment
    16 FLEXTOOL Positioners
     4 Hard Tooling Clamp Units
5- Weld Rocker Assembly to Underbody Assembly
    10 FLEXTOOL Positioners
     2 Hard Tooling Clamp Units
6-10 Respot Weld
    10 FLEXTOOL Postitioners
     2 Hard Tooling Clamp Units
     Welding Robots
Square footage requirement: 10,392 sq.ft.
Cost per square foot: $200
No new product or line changeovers after year 7

Figure 4   FLEXTOOL application in automotive industry. Small and large car underbody welding in the same robotics welding cell (courtesy FANUC Robotics North America, Inc.).

closer to reality. Reprogrammable positioner systems allow the reusability of equipment, eliminating the necessity for tools specially designed for specific products that will be scrapped due to product specification changes. Systems of reprogrammable positioners result in increased flexibility at the same time that they generate important savings for the manufacturing operation. Clearly the trend toward increasing the flexibility and responsiveness of robotics fixturing systems must rely on the addition of programmability capabilities. These added capabilities will make possible the reduction of nonproductive times, improved time-to-market, and ultimately the survival of the organization.

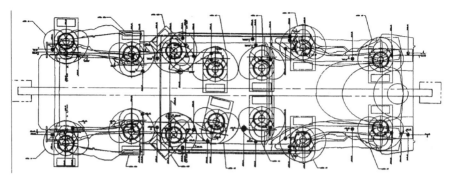

Figure 5 Overlay of the FLEXTOOL positioners clamping points for small and large cars (courtesy FANUC Robotics North America, Inc.).

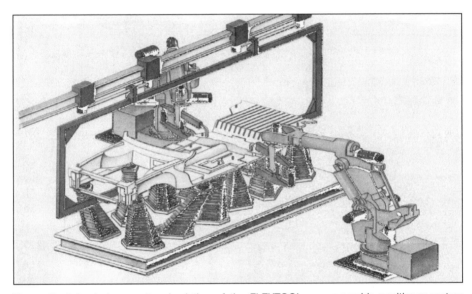

Figure 6 Animated computer simulation of the FLEXTOOL programmable positioner system (courtesy FANUC Robotics North America, Inc.).

REFERENCES

Brooke, L. 1994. "Flex Factor." *Chilton's Automotive Industries* **174**(11), 50–52.

Brosilow, R. 1994. "Need Flexible Scheduling? Robots Deliver." *Welding Design and Fabrication* **67**(12), 36–37.

Dilthey, U., and L. Stein. 1993. "Robot Systems for Arc Welding—Current Position and Future Trends." *Welding Research Abroad* **39**(12), 2–6.

Kosmala, J. B. 1995. "Time-Based Competition, the Product–Process Linkage, and FLEXTOOL Programmable Tooling." In *Proceedings of Body Assembly and Manufacturing,* International Body Engineering Conference. Vol. 15.

Lorincz, J. A. 1993. "How Much Automation Do You Need?" *Tooling and Production* **59**(9), 26–27

Monashenko, N. T. 1992. "Inverse Problem of Kinematics of Positioner with Linear Propeller." *Process Safety Progress* (November–December), 32–34.

Smith, B. S. 1985. "Robots in Arc Welding." in *Handbook of Industrial Robotics.* Ed. S. Y. Nof. New York: John Wiley & Sons. 913–929.

Stuart, T. 1992. "Brace for Japan's Hot New Strategy." *Fortune* (September 21), 62–74.

CHAPTER **48**

WORKPIECE HANDLING AND GRIPPER SELECTION

Hadi Abu-Akeel
FANUC Robotics North America, Inc.
Rochester Hills, Michigan

Atusushi Watanabe
FANUC Ltd.
Yamanashi, Japan

1 INTRODUCTION

This chapter surveys various applications of robots for workpiece handling and provides guidelines for gripper selection.

2 APPLICATIONS OF ROBOTS FOR HANDLING IN A MACHINING CELL

Robots used for handling in machining cells can either be mounted on the machine tool frame or free-standing (pedestal-mounted) separately from the machine tool.

Pedestal robots can be constructed with cylindrical coordinates, though articulated arm robots (Figures 1 and 2) are now commonly used. The load capacity at the wrist of such robots ranges from 40 to 120 kg. Robots that are machine frame-mounted can be constructed with articulated arms as shown in Figures 3 and 4, though robots with cylindrical or prismatic coordinates are also used; the maximum load capacity for such robots ranges from 5 to 60 kg.

3 EXAMPLES OF ROBOTIC WORKPIECE HANDLING

Figures 5 and 6 show examples of applications of self-standing robots to lathes. The robot grips workpieces from a rotary workpiece feeder one by one and loads them onto a lathe. The robot also grips machined workpieces from the lathe and puts them in a separate place on the rotary workpiece feeder. As a result, blanks on the rotary workpiece feeder are processed to machined parts after a certain time.

Figures 7 through 10 depict examples of applications of self-standing robots to machining centers. The robot sequentially picks up workpieces stacked on a table from the top and stacks machined workpieces in regular sequence.

Figures 11 and 12 show examples of applications of self-standing-type robots to grinding machines. The robot employs a hand suitable for handling cylindrical workpieces. An exclusive pallet is used for stacking workpieces (Figure 13).

4 WORKPIECE HANDLING BY A ROBOTIC VEHICLE

A machining cell is employed for machining, while an assembly cell is employed for assembling, both are major components of factory automation. In an automated factory workpieces are carried between an automatic warehouse and such cells by unmanned carriers. In machining cells, a robot then carries the workpieces from the workstation to NC machine tools. (See Figures 14 and 15.)

A pallet carried by an unmanned carrier is transferred to the loading station of the workpiece station. Then the pallet shifts to the active station. The self-standing robot sequentially grips workpieces on the pallet and sets them one by one onto the mounting jig on the machining center. After machining, the robot stacks machined workpieces onto

Handbook of Industrial Robotics, Second Edition, Edited by Shimon Y. Nof
ISBN 0-471-17783-0 © 1999 John Wiley & Sons, Inc.

the pallet on the active station. When all workpieces have been machined, the pallet shifts to the unloading station. The unmanned carrier carries the pallet on the unloading station to a specified position.

5 GRIPPER (HAND) SELECTION

Hands with two or three fingers are employed for gripping workpieces (see Figure 16). The servo-hand is a gripper with continuously controllable jaws for the smooth handling of delicate workpieces.

5.1 Hand with Three Fingers

A hand with three fingers can grip a cylindrical or rectangular workpiece.

5.1.1 Gripping of Cylindrical Workpieces

The weight and size of grippable workpieces and the gripping methods using fingers are determined according to the gripping conditions (see Figures 17, 18, and 19).

5.1.2 Gripping of Rectangular Workpieces

Rectangular workpieces are gripped by a hand with two or three fingers. The weight and sizes of grippable workpieces and gripping methods using fingers are determined according to the gripping conditions (see Figures 20 and 21).

5.1.3 Insertion of a Workpiece into Lathe Chuck

A workpiece is inserted into the chuck by a hand with the push buffer mechanism, as shown in Figures 22 and 23. One must be careful with the gripping positions of workpieces if they are particularly long.

5.2 Hand with Two Fingers

The hand with two fingers can grip cylindrical workpieces as depicted in Figure 24. The weight and size of grippable workpieces and the gripping methods of fingers are determined by the gripping conditions, as shown in Figure 25.

ACKNOWLEDGMENTS

This chapter is an updated version of Chapter 55 in the 1st edition of this handbook, "Workpiece Handling and Gripper Selection," by Seiuemon Inaba.

REFERENCES

Barash, M. M. 1976. "Integrating Machinery Systems with Workpiece Handling." *Industrial Robot* 3(2), 62–67.

Bey, I. 1982. "Automation of Workpiece Handling in Small Batch Production." In *Proceedings of the 1st International Conference on Flexible Manufacturing Systems,* Brighton, U.K., October.

Lian, D., S. Peterson, and M. Donath. 1982. "A Three-Fingered, Articulated Robotic Hand." In *Proceedings of the 13th I.S.I.R.,* Chicago, April. 18/91–101.

Lien, T. K. 1978. "Workpiece Handling by Robot in a Flexible Manufacturing Cell." In *Proceedings of the 8th I.S.I.R.,* Stuttgart, West Germany, June. 242–254.

Lundstrom, G., B. Glemme, and B. W. Rooks. 1977. *Industrial Robots Gripper Review.* Bedford: IFS. 1977.

Schafer, H. S., and E. Malstrom. 1983. "Evaluating the Effectiveness of Two-Fingered Parallel Jaw." In *Proceedings of the 13th I.S.I.R.,* Chicago, April. 18/112–121.

Schekulia, K. 1975. "Workpiece Handling on Machine Tools with Industrial Robots." In *Industrielle Fertigung* **65,** 685–86. [In German.]

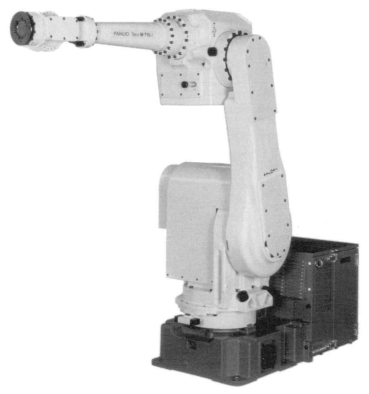

Figure 1   Overview of robot for machine loading.

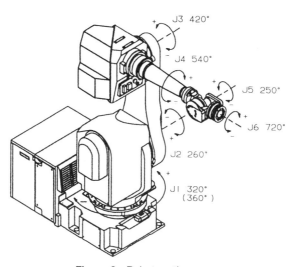

Figure 2   Robot motion axes.

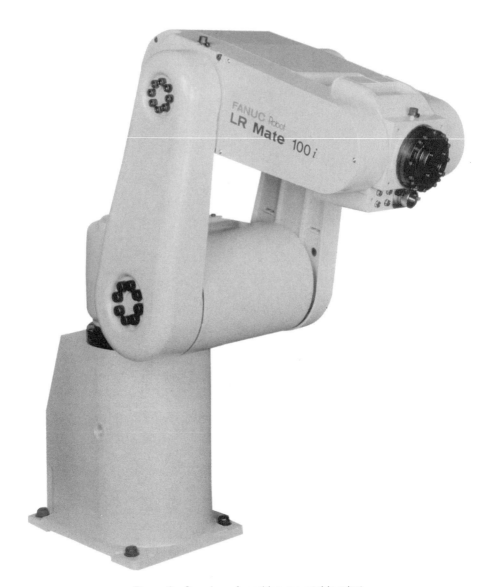

Figure 3 Overview of machine-mountable robot.

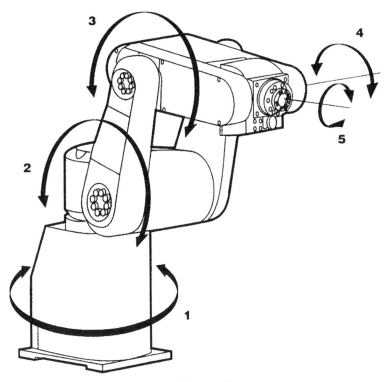

Figure 4   Robot motion axes.

Figure 5   Overview of an application to a lathe.

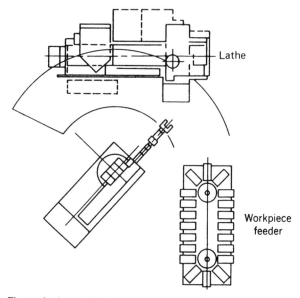

Lathe

Workpiece
feeder

Figure 6   Layout in the example of an application to a lathe.

FANUC ROBOT M

Figure 7   Overview of an application to a machining center.

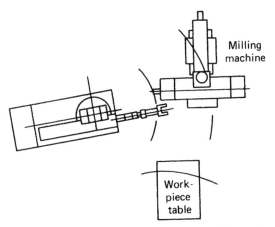

Figure 8 Layout of an application to a machining center.

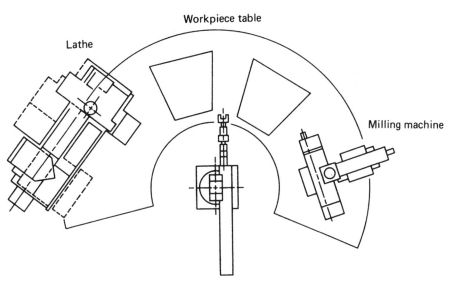

Figure 9 Layout for servicing two machine tools.

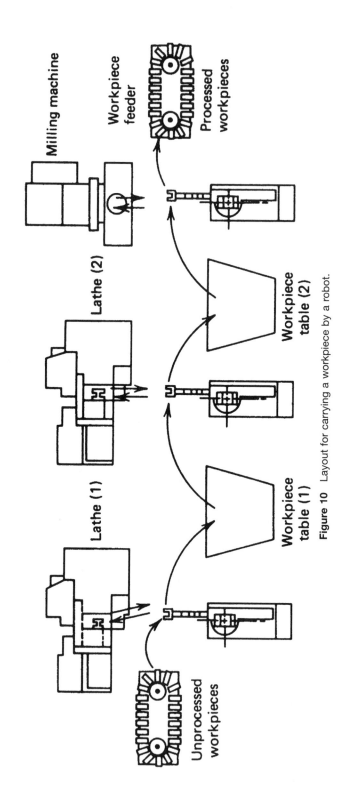

Figure 10 Layout for carrying a workpiece by a robot.

Milling machine

Workpiece feeder

Processed workpieces

Lathe (2)

Workpiece table (2)

Lathe (1)

Workpiece table (1)

Unprocessed workpieces

Figure 11   Overview of an application to a grinding machine.

Figure 12   Gripping a cylindrical workpiece.

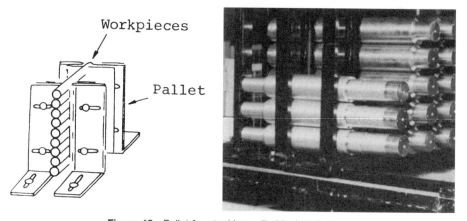

Figure 13   Pallet for stacking cylindrical workpieces.

Figure 14   Overview of unmanned carrier, workpiece station, and carriage of workpieces using robots.

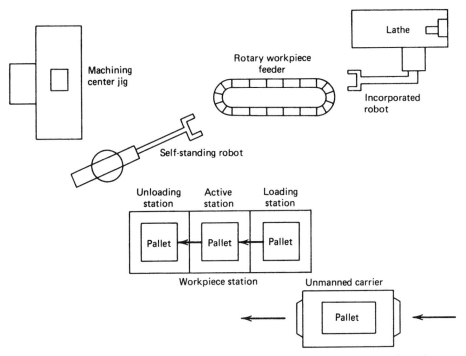

Figure 15   Unmanned carrier, workpiece station, and carriage of workpieces using robots.

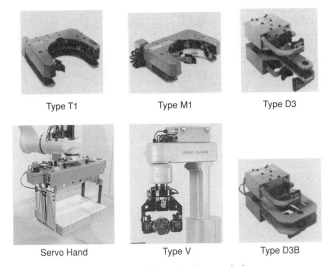

Figure 16   Hands for gripping workpieces.

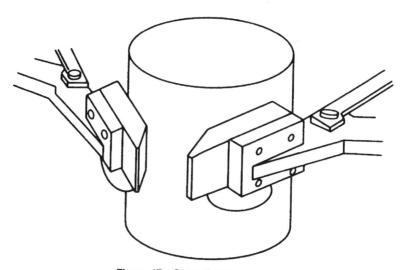

Figure 17   Outer diameter gripping.

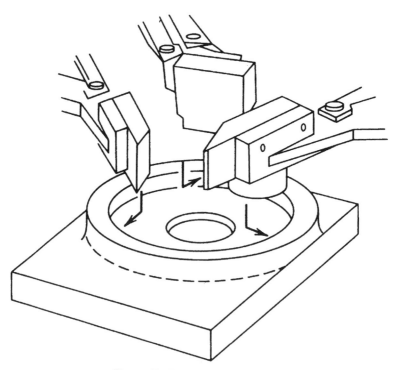

Figure 18   Inner diameter gripping.

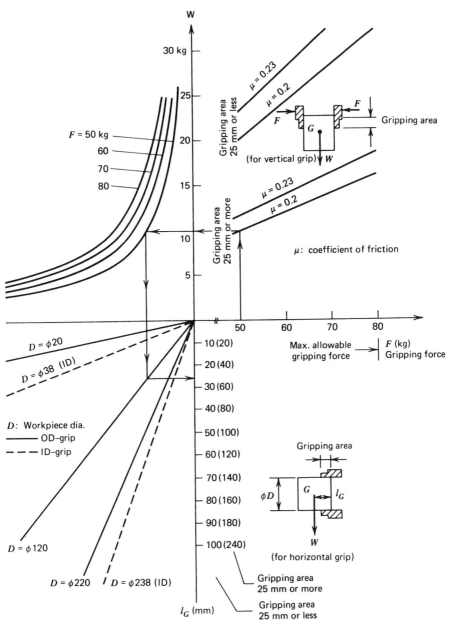

Figure 19 Gripping conditions using a hand with three fingers (cylindrical workpiece). For a vertical grip, when gripping force $F = 50$ kg, and gripping area is 25 mm or less, the maximum weight is 10 kg. For horizontal grip, when workpiece diameter is 120 mm, workpiece gravity center position l_G is 26 mm or less. The flexible push mechanism limits use of Hand T. Usually $\mu = 0.2$.

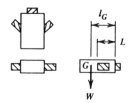

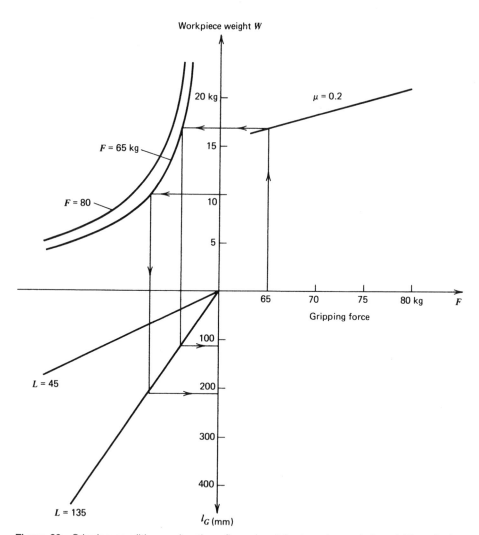

Figure 20 Gripping conditions using three-finger hand (rectangular workpieces). The gripping force when the width across two flats of workpiece, $B = 129$ mm. With the same service pressure, F varies with the workpiece width B. With the gripping force 65 kg, the maximum workpiece weight is 17 kg. When $L = 135$, l_G is 110 mm. When $F = 65$ kg, $W = 10$ kg, and $L = 135$ mm, l_G is 210 mm. The flexible push mechanism limits the use of the Hand T.

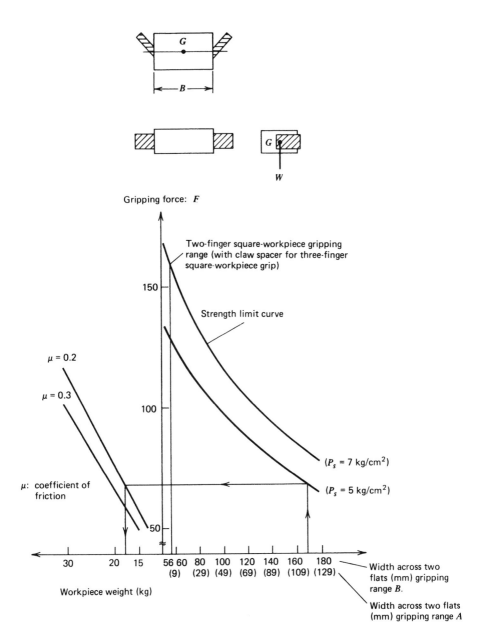

Figure 21   Gripping conditions using two-finger hand (rectangular workpieces). When the width across two flats of workpiece is 170 mm and no claw spaces are provided, the maximum workpiece weight is 18 kg. The hand material strength limits the maximum gripping force and workpiece weight. Always grip the workpiece at the center of gravity so that no moment is applied. The flexible push mechanism limits the use of Hand T.

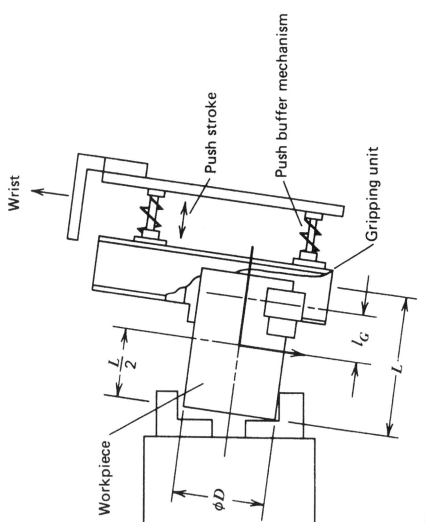

Figure 22 Insertion of workpieces into lathe chuck. L: length of workpiece. D: diameter of workpiece. l_G: eccentric gravity distance.

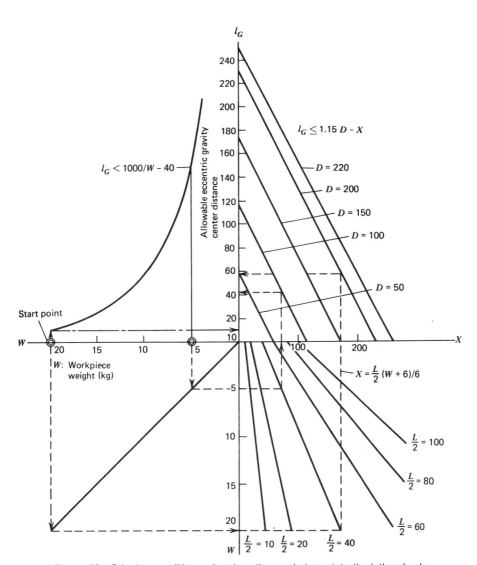

Figure 23   Gripping conditions when inserting workpieces into the lathe chuck.

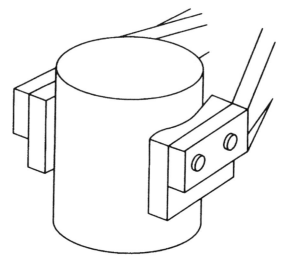

Figure 24 Outer diameter (OD) gripping.

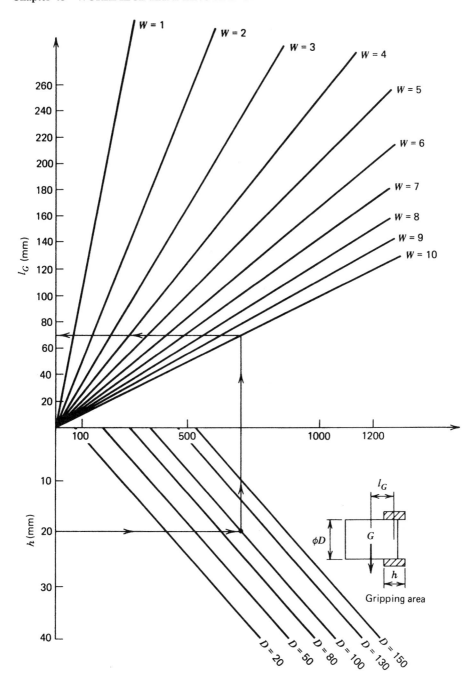

Figure 25   Gripping conditions of a two-finger hand. Example: When the gripping area is 20 mm, outer diameter 100 mm, and workpiece weight 10 kg, allowable eccentric center of gravity in horizontal grip is 70 mm.

PART 9

ROBOTICS IN OPERATIONS

CHAPTER 49

MATERIAL HANDLING AND WAREHOUSING

Yavuz A. Bozer
The University of Michigan
Ann Arbor, Michigan

1 INTRODUCTION

Robots have played an increasingly important role in material-handling systems; this is especially true for automated and computer-integrated material-handling systems. For many, the words "robot" and "material handling" will first bring to mind pick-and-place robots. While pick-and-place robots represent perhaps the most common and effective use of robots in material handling, in this chapter we take a wider view of robots and the role they play in material handling. In fact, here we take a systems view of material handling and treat systems based on automated, (re)programmable handling devices that can perform a variety of tasks as "robot-based" systems.

For example, we treat automated guided vehicles (AGVs) as "robots." Strictly speaking, some may argue that an AGV is not a robot, especially when one considers that a mobile robot (see Chapter 8) can be obtained by installing a robot on an AGV. However, with or without an on-board robot, the AGV itself represents an automated device that can be (re)programmed to perform a variety of handling tasks and, therefore, qualifies as a robot. In fact, the dividing line between mobile robots and automated (self-guided) vehicles that require no external guidance (such as a guidewire) has become quite fuzzy. Nevertheless, in this chapter we also treat more conventional robotics applications in material-handling systems, such as gantry robots.

2 ROBOTS IN MATERIAL HANDLING AND WAREHOUSING

Since robots are used extensively in material handling, we will not offer a comprehensive treatment of the topic. Rather, we will present a number of well-known automated, robot-based material-handling systems and show certain analytical models that can be used to design and analyze such systems, particularly from an industrial engineering viewpoint. The handling systems we present (such as AGV systems and gantry robots) are aimed largely at manufacturing applications. However, some of the systems we cover—including automated storage/retrieval (AS/R) systems and carousels—have been used successfully both in manufacturing and warehousing.

Let us start with a basic definition of material handling: Material handling is using the right *method* to provide the right *amount* of the right *material* in the right *orientation* and the right *condition* to the right *place,* at the right *time* and at the right *cost,* while maintaining a *safe* working environment. As the above "right definition" suggests, material handling, which is more than just handling material, is a key contributor to improved quality and efficiency (two prerequisites to lean manufacturing) as well as better safety—in terms of both accident prevention and avoiding long-term repetitive motion injuries. (For an extensive review of material-handling equipment and basic functional descriptions, including equipment illustrations, see Chapter 6 in Tompkins et al. (1996)).

Many tedious, repetitive (and sometimes potentially hazardous and/or strenuous) material-handling tasks have been and continue to be relegated to robots. Furthermore, robots can perform certain handling tasks with increased reliability, uniform speed, accuracy, and consistency compared to humans. In addition to pick-and-place robots, some

Handbook of Industrial Robotics, Second Edition, Edited by Shimon Y. Nof
ISBN 0-471-17783-0 © 1999 John Wiley & Sons, Inc.

of the successful robot applications in material handling include gantry robots, AGV systems, AS/R systems, and carousels, which we discuss in the next few sections. The models we show for the above systems, with few exceptions, apply generally to other robot-based handling systems that may have similar operating and/or hardware characteristics.

2.1 Pick-and-Place Robots

Pick-and-place robots have been used in a variety of material-handling applications, including palletizing (or depalletizing), case picking, bin picking, kitting, machine loading (unloading), parts feeding, and parts delivery. Such robots (or slightly modified versions) have also been used in storage/retrieval systems (which we discuss in the following sections) and case packing and sorting. Examples of pick-and-place robots are shown in Figures 1 and 2.

A typical configuration used with a pick-and-place robot applied in manufacturing is a circular one, as shown in Figure 3. In such cases the robot serves multiple functions by:

1. Acting as an interface between the conveyor and the machines
2. Loading and unloading the parts from the machines
3. Transferring the parts from one machine to the next

In designing such a system, one is typically concerned with the number of machines served by the robot, the cycle time at each machine, and the overall production rate of the system. In most cases where such a configuration is used, the cycle time at each machine is constant (or nearly constant) and all the parts follow the same production

Figure 1   A robot picks parts from a carousel parts feeder (courtesy FANUC Robotics North America, Inc.).

Figure 2   Gripping on the sides, a robot transfers furniture panels from a prepositioned pallet (courtesy Stäubli Unimation).

sequence (i.e., they visit the machines in the same sequence, starting with machine 1 and ending with machine 4). Therefore, to design such a system one typically needs to develop a "super cycle" during which one part is produced (out of machine 4) and one part is moved from each machine to the next with minimal waiting incurred by the machines. The length of the super cycle determines the output rate of the system.

In many robotic material-handling applications, especially those that involve pick-and-place robots, parts feeding plays an important role. To address the parts feeding problem, one needs to consider three subproblems (Swanson, Burridge, and Koditschek, 1998): singulation, orientation, and presentation. Singulation is concerned with separating a cluster of parts into individual parts; it can be very difficult if the parts have a tendency to nest within each other (like bottle caps) or become entangled (like paper clips). Orientation is concerned with (re)establishing the orientation of randomly oriented parts. Presentation is the action of "moving the singulated and oriented part to a known location, where a machine tool or robot can easily perform an operation on it" (Swanson, Burridge, Koditschek, 1998).

According to Swanson, Burridge, and Koditschek (1998), since "no automated technique of performing singulation for an arbitrary part exists," singulation strategies are often "custom designed through trial-and-error for each new type of part"; the orientation problem, on the other hand, "lies at the heart of the parts feeding problem" and "little is known about how to orient an arbitrary part beyond decades of craftsmanship and experience"; fnally, presentation is the "easiest part of the parts feeding problem, usually accomplished with a conveyor or buffer."

Various parts feeding equipment is available for performing one or more of the tasks of singulation, orientation, and presentation. Such equipment typically utilizes vibration, centrifugal force, and/or gravity. Examples include vibratory bowls, centrifugal or various rotating feeders, elevating feeders, magazines, jigs/fixtures, indexing conveyors, carousels, and vision-based systems. For example, the parts feeder shown in Figure 1 is based on a carousel, which can also be used as a storage/retrieval device, as we later describe in Section 2.5.

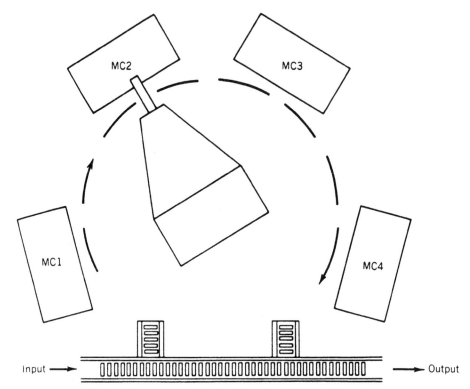

Figure 3 Manufacturing cell based on a pick-and-place robot. (From J. A. Tompkins and J. A. White, *Facilities Planning,* New York: John Wiley & Sons, 1984.)

2.2 Gantry Robots

Pick-and-place robots of the type discussed in Section 2.1 are typically stationary, pedestal-mounted robots. Therefore their work envelope is limited. To increase the work envelope and serve more machines, a (pick-and-place) robot can be installed on a gantry crane, resulting in a gantry robot as shown in Figure 4. (Strictly speaking, some gantry robots are installed on bridge cranes, but for convenience we will still refer to them as gantry robots). Although they require higher ceilings (at least 10 feet), gantry robots offer large work envelopes, heavy payloads, mobility, overhead mounting (which also leads to floor-space savings), and the capability to replace several stationary (floor- or pedestal-

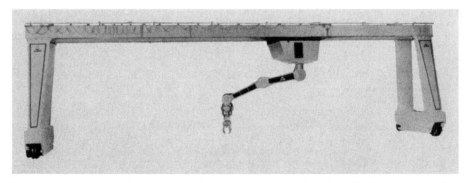

Figure 4 A gantry robot has a nearly unlimited work envelope (courtesy Air Technical Industries).

mounted) robots. Depending on the type of application, a gantry robot such as the one shown in Figure 4 or 5 can be used for material handling (moving parts between machines), machine loading/unloading, welding/cutting, or painting, among other tasks.

With reference to Figure 5, in a typical gantry robot installation a *superstructure* (A) supports the *runways* (B); the runways in turn support the *bridge* (C), and the bridge supports the *carriage* (D), to which the robot is attached. Travel in the X direction occurs when the bridge rolls on the runways, whereas travel in the Y direction occurs when the carriage rolls on the bridge. Travel in the Z direction is provided typically either by telescoping tubes (which work better in low-ceiling areas) or by a sliding mast. (In the latter case, sufficient clearance for the mast must be provided above the carriage.)

The gantry robot is composed of up to six axes, depending on the manufacturer and model (Ziskovsky, 1985). There are three linear axes (X, Y, and Z) and three rotational axes: α, β, and γ (see Figure 5). Typical distance ranges (in feet) for the linear axes are (Ziskovsky, 1985): $10 \leq X \leq 40$, $10 \leq Y \leq 20$, and $3 \leq Z \leq 10$; and for the rotational axes they are (in degrees): $\alpha \leq 330$, $\beta \leq 210$, and $\gamma \leq 330$. Note that, in many cases, depending on the exact location of the pick-up/deposit (P/D) points, the superstructure needs to be slightly larger (generally 2 ft or more in the X and Y directions) than the required work envelope. Also, superstructures that measure 500 ft (X) $\times$ 40 ft (Y) $\times$ 25 ft (Z), to handle loads of up to 15 tons, have been reported (Ziskovsky, 1985).

The typical velocity range for the gantry robot is 36 in./sec (180 fpm) to 39 in./sec (195 fpm) for the linear axes, and 60°/sec to 120°/sec for the rotational axes (Ziskovsky, 1985). Of course, depending on the load weight and the actual distances over which the robot travels, the effective velocity is likely to be less than the above maximum speeds. Also, the above velocities must be considered concurrently with the payload and maximum travel distances. Generally speaking, under constant repeatability (i.e., the accuracy with which a robot can return to a prespecified point), the larger the maximum travel distance in the Z axis, and the higher the velocity, the smaller the payload, and vice versa.

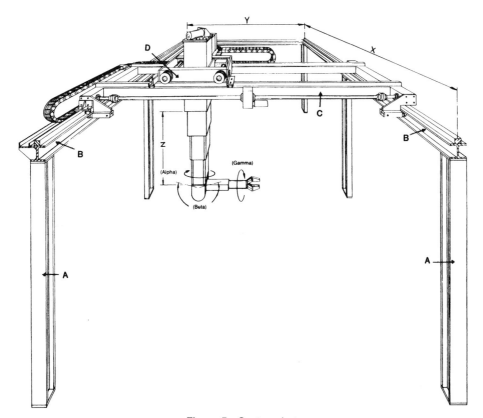

Figure 5   Gantry robot.

As we remarked earlier, a gantry robot can serve a larger number of machines than a pedestal robot. The increased flexibility offered by the gantry robot, however, also requires more careful planning in designing such a system. The number and location of the machines, the frequency with which parts are moved from one machine to the next, the load P/D times, and the crane travel velocities (and operating discipline) determine jointly whether the gantry robot will meet the workload imposed on it. This is especially important in single-device handling systems such as the gantry robot because the robot may become a bottleneck and adversely affect the overall productivity of the entire system. Also, preventive maintenance to minimize robot downtime is critical. The following simple analytical model determines the workload a gantry robot can handle.

Consider a gantry robot serving M stations with known locations. Each station is represented by a P/D point; each P/D point corresponds to either an input/output (I/O) point (where loads enter and exit the system) or a machine where processing takes place. A loaded trip occurs when the robot picks up a part at station i and delivers it to station j. An empty trip occurs when the robot has just delivered a part at station j and the next part it needs to move is at station k. (A part is defined as a single load that the robot moves in one trip; if multiple parts are handled together in one trip, we still treat them as one part for the purposes of the model. When a part is ready to be moved by the robot, we refer to it as a *move request*.)

Suppose the parts flow in the system (specified by the user) is expressed as a from–to chart, where f_{ij} denotes the rate at which parts must be moved from station i to station j. (We assume that $f_{ii} = 0$.) Let Λ_i be the rate at which parts must be delivered to station i (i.e., $\Lambda_i = \Sigma_j f_{ji}$); likewise, let λ_i be the rate at which parts must be picked up at station i (i.e., $\lambda_i = \Sigma_j f_{ij}$). We assume that parts flow is conserved at each machine (i.e., $\Lambda_i = \lambda_i$). The same is true for an I/O point if there is only one such point. Otherwise, flow is not necessarily conserved at each I/O point, since parts may enter the system from one I/O point and exit from another. Even with multiple I/O points, however, flow is conserved globally, i.e., $\Lambda_T = \Sigma_i \Lambda_i = \Sigma_i \lambda_i = \lambda_T = \Sigma_i \Sigma_j f_{ij}$.

Let τ_{ij} and σ_{ij} denote the loaded and empty travel time from station i to j, respectively. We stress that τ_{ij} includes the part pick-up time at station i and the part deposit time at station j. That is, τ_{ij} accounts for travel time along the linear axes (X, Y, and Z), plus any time incurred for rotational movement to pick up and later deposit the part. (The values of τ_{ij} and σ_{ij} are supplied by the user; they are computed based on the robot parameters, the load weight, and the given machine locations.)

As long as it is up and running, the robot is in one of three states at any given instant: (1) traveling loaded, (2) traveling empty, or (3) sitting idle at the last delivery point. (We will remark about machine buffers at the end of this section.) Let α_f and α_e denote the fraction of time the robot is traveling loaded and traveling empty, respectively. Let $\rho = \alpha_f + \alpha_e$ denote the utilization of the robot. Our goal is to estimate ρ for user-supplied values of f_{ij}, τ_{ij}, and σ_{ij}.

Computing α_f from the problem data is straightforward. That is,

$$\alpha_f = \sum_i \sum_j f_{ij} \tau_{ij} \tag{1}$$

Computing α_e is more difficult in general since its value depends on the operating discipline, which is also known as *empty device dispatching*. A dispatching rule used often in industry is the first-come-first-served (FCFS) rule, where the robot, upon completing a loaded trip and becoming empty, is assigned to the "oldest" move request in the system. If there are no move requests, the robot becomes idle at its last delivery point.

Modeling the robot as an M/G/1/FCFS queue, Chow (1986) first derives the "service time" (s_{kij}) for the robot to serve one move request. To serve a move request, the robot travels empty from station k to i, picks up the load, and travels to station j ($j \neq i$), where the load is deposited. (If $k = i$, empty travel time is 0.) The probability that the robot delivers a part to station k, say r_k, is equal to Λ_k/Λ_T. The probability that a given move request must be picked up at station i and delivered to j, say, r_{ij}, is equal to f_{ij}/λ_T. Hence the probability that the robot, starting empty at station k, performs a loaded trip from station i to j is given by $r_k r_{ij}$ (since the last delivery point of the robot and the origin of the next move request to be served are independent under the FCFS rule).

The expected service time, say, $\bar{s}$, is obtained as:

$$\bar{s} = \sum_k \sum_i \sum_{j\neq i} r_k r_{ij} s_{kij} = \sum_k \sum_i \sum_{j\neq i} \left(\frac{\Lambda_k}{\Lambda_T}\right)\left(\frac{f_{ij}}{\lambda_T}\right)(\sigma_{ki} + \tau_{ij}) \qquad (2)$$

time units/move request. Hence the service rate for the robot is equal to $1/\bar{s}$ (including empty travel), and we therefore obtain $\rho = \lambda_T/[1/\bar{s}]$. Note that, with the above approach, we could have obtained α_e through a slight modification of Equation (2); however, since we already obtained ρ, we can simply compute α_e as $\rho - \alpha_f$. The robot will satisfy the workload requirement if $\rho < 1$. (Of course, one can also derive the second moment of the service time and obtain additional results such as the expected waiting for a move request averaged across all the stations; Cho (1990) presents a more rigorous treatment of the FCFS rule and derives the expected waiting times at individual stations.) We will now demonstrate the model through a simple example.

Example 1: Consider a gantry robot that serves a manufacturing center with 11 stations. Suppose four job types (labeled A through D) are processed through the center. Assuming that stations 1 and 11 are the I/O stations, the average hourly production volume and the production route for each job type have been supplied as follows: Job A, 1.25 jobs/hour, 1-2-4-7-11; Job B, 1.00 jobs/hour, 11-10-7-5-3-1; Job C, 2.25 jobs/hour, 1-3-9-8-6-11; and Job D, 0.50 jobs/hour, 11-7-4-5-1. The layout of the center and the resulting from–to chart (including the Λ and λ values) are shown in Figure 6.

For simplicity, ignoring acceleration and deceleration, we assume the robot (empty or loaded) travels at an average speed of 80 fpm in the X and Y directions. The time required

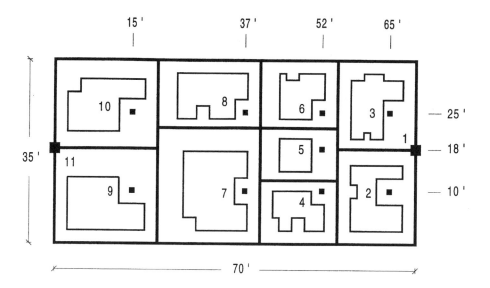

THE FROM-TO CHART (PARTS/HOUR)

	1	2	3	4	5	6	7	8	9	10	11	λ
1	.00	1.25	2.25	.00	.00	.00	.00	.00	.00	.00	.00	3.50
2	.00	.00	.00	1.25	.00	.00	.00	.00	.00	.00	.00	1.25
3	1.00	.00	.00	.00	.00	.00	.00	.00	2.25	.00	.00	3.25
4	.00	.00	.00	.00	.50	.00	1.25	.00	.00	.00	.00	1.75
5	.50	.00	1.00	.00	.00	.00	.00	.00	.00	.00	.00	1.50
6	.00	.00	.00	.00	.00	.00	.00	.00	.00	.00	2.25	2.25
7	.00	.00	.00	.50	1.00	.00	.00	.00	.00	.00	1.25	2.75
8	.00	.00	.00	.00	.00	2.25	.00	.00	.00	.00	.00	2.25
9	.00	.00	.00	.00	.00	.00	.00	2.25	.00	.00	.00	2.25
10	.00	.00	.00	.00	.00	.00	1.00	.00	.00	.00	.00	1.00
11	.00	.00	.00	.00	.00	.00	.50	.00	.00	1.00	.00	1.50
Λ	1.50	1.25	3.25	1.75	1.50	2.25	2.75	2.25	2.25	1.00	3.50	23.25

Figure 6 Layout and flow data from Example 1.

to pick up (and later deposit) a part, including Z travel and all rotational movements, is given as 1.5 min. Assuming *rectilinear travel* (i.e., the robot moves *only* in the X or Y direction at any given instant), it is straightforward to compute the σ_{ij} values. For example, $\sigma_{1,2} = 0.1625$ min (or 0.0027083 hr, since the production requirements are expressed on a per-hour basis). The τ_{ij} values are obtained simply by adding 1.5 min to the corresponding σ_{ij} values; for example, $\tau_{1,2} = 1.6625$ min (or 0.0277083 hr).

Using the above flow and travel time values in Equations (1) and (2), we obtain $\alpha_f = 0.7245$ and $\bar{s} = 0.03785$ hr/move request, respectively. Hence $\rho = 0.8800$ ($\alpha_e = 0.1556$), and we conclude that the robot meets the workload requirement, though an 88% utilization is likely to lead to long waiting times for the move requests (and more work-in-process inventory at the stations). Note that the robot would travel empty about 18% (0.1556/0.88) of the time that it is busy. (Typically the FCFS rule results in more empty travel; however, percentage of empty travel is fairly low in our example, primarily because the part P/D time is large compared to the loaded or empty travel times.)

In the example above, all the jobs, except for Job C, flow from one I/O point to the other with no backtracking. In the case of Job C, machining stations 6 and 9 are not located correctly. Swapping machines 6 and 9 should improve the layout without affecting the other jobs. Furthermore, we assumed rectilinear travel for the robot. Since the bridge and the carriage are usually powered independently, it is possible to achieve Chebyshev travel where the robot moves simultaneously in the X and Y directions; the travel time is determined by the maximum travel time in the X and Y direction.

Example 2: Given the data in Example 1, suppose the locations of machines 6 and 9 are swapped. Assuming Chebyshev travel, we again compute the σ_{ij} values. For example, $\sigma_{1,2} = 0.10$ min (or 0.001666 hr). The τ_{ij} values are obtained simply by adding 1.5 min to the corresponding σ_{ij} values; for example, $\tau_{1,2} = 1.60$ min (or 0.0266 hr).

From Equations (1) and (2), we obtain $\alpha_f = 0.6571$ and $\bar{s} = 0.03377$ hr/move request, respectively. Hence $\rho = 0.7853$ ($\alpha_e = 0.1281$) and of course the robot still meets the workload requirement. Swapping machines 6 and 9 reduced α_f by about 9%, and the robot utilization decreased from 88% to 78.5% (a reduction of about 11%), which should help reduce the work-in-process inventory at the stations. (The robot would travel empty about 16% of the time that it is busy; improving the layout further decreases the weight of travel times compared to part P/D times.)

The models we present here are based on the assumption that there is sufficient buffer space at each station that the robot does not get blocked when delivering a part, and that the robot does not wait for a machine to complete a part (i.e., a move request is placed only when a part is ready to be moved to the next machine). If these assumptions are relaxed, the possible states we defined earlier for the robot would have to be revised.

2.3 Automated Guided Vehicle Systems

An AGV is an automated handling device that can move unit loads from point A to point B with no human intervention and under computer control. Guidance is typically provided by a wire (embedded in the floor), and movement typically occurs on a horizontal plane. For a review of AGV types, guidance and control methods, see Koff (1987), among others. Here we focus on the unit load AGV (see Figure 7), which moves one container (i.e., one unit load) at a time on board the vehicle.

Unit load AGVs have been used quite successfully (especially in Japan and Europe) in manufacturing. They have also been the subject of extensive research, particularly from a systems design and analysis viewpoint. Due to limited space, we will show only one model for AGV systems. This model applies to *single-vehicle* systems, where one vehicle serves a set of machines. Such systems are obtained under the "tandem AGV" approach (Bozer and Srinivasan, 1991), where all the machines to be served by the AGVs are partitioned into nonoverlapping zones; one vehicle is assigned to each zone, and transfer points (in the form of short conveyors or other transfer mechanisms) are provided between adjacent zones. (For modeling multivehicle, conventional AGV systems, where all the machines are treated as one group and all the vehicles have access to all the machines, see Egbelu, 1987; Srinivasan, Bozer, and Cho, 1994; and Kim, 1995).

An AGV serves a set of machines and, as with gantry robots, the number and location of the machines, the frequency of parts movement among the machines, the load P/D times, and the vehicle travel speed (and operating discipline) determine jointly the workload on the AGV, which should not be a bottleneck.

Figure 7 Unit load AGV. (From J. A. Tompkins et al., *Facilities Planning,* 2nd ed. New York: John Wiley & Sons, 1984. Courtesy Jervis B. Webb Co.)

Although the FCFS operating discipline (i.e., dispatching rule) is straightforward to implement and relatively easy to model, it typically leads to unnecessary empty travel (because the empty vehicle location—relative to the move requests—is not considered when it is dispatched to its next assignment). To reduce empty travel (while maintaining some analytical tractability), Srinivasan, Bozer, and Cho (1994) proposed the modified FCFS (MOD FCFS) rule, where the vehicle, upon delivering a load at station i, first inspects station i. If it finds a move request at i, the vehicle serves that move request; otherwise it switches to the FCFS rule.

As before, our goal is to estimate the vehicle utilization (ρ). For the device to meet the workload under MOD FCFS, we need $\rho < 1$ (Srinivasan, Bozer, and Cho, 1994). The analytical model needed for the MOD FCFS dispatching rule (Srinivasan, Bozer, and Cho, 1994), however, is more complex than the models used for the FCFS rule (Cho, 1990; Chow, 1986). Given the parameters f_{ij}, Λ_i, λ_i, λ_T, and σ_{ij} (see Section 2.2), we first compute the following values:

$$\theta_i = \sum_j \lambda_j \sigma_{ij}, \tag{3}$$

$$\chi = \sum_i \lambda_i \theta_i \quad \text{and} \quad \hat{\chi} = \sum_i \Lambda_i \theta_i, \tag{4}$$

$$\phi_i = \frac{\hat{\chi}}{\lambda_T} - \frac{\Lambda_i \chi}{\lambda_i \lambda_T} \tag{5}$$

$$\Phi = \max_i \phi_i \tag{6}$$

Let q_i denote the probability that station i has no move requests at any given time. Assuming that the vehicle inspects station i at random points in time (an inspection occurs only when a load is delivered), Srinivasan, Bozer, and Cho (1994) show that the algorithm below converges to a unique ρ value that is less than one if $\alpha_f + \Phi < 1.0$; otherwise, the AGV cannot meet the workload requirement. (Recall that α_f is obtained from Equation (1).)

Algorithm to Obtain Approximate ρ Value

1. Set $n = 0$ and start with an initial value of ρ^n, say, $\rho^n = \alpha_f + \Phi$.
2. Compute the probability q_i for each station i:

$$q_i = \frac{\lambda_i}{\Lambda_i\chi}(\lambda_T - \lambda_i)(\rho - \alpha_f - \phi_i) + \frac{\lambda_i}{\lambda_T}(1 - \rho)$$

3. Set $\rho = 1 - \Pi_i\, q_i$.
4. Set $\rho^{n+1} = \rho^n + \Delta(\rho - \rho^n)$, where Δ is a sufficiently small step size. Let $n \leftarrow n + 1$.
5. Execute steps 2 through 4 until two successive values of ρ^n are reasonably close (i.e., the absolute difference between them is within a user-specified stopping tolerance). Set $\rho = \rho^n$ and **STOP**.

Provided that the AGV meets the workload, i.e., $\alpha_f + \Phi < 1.0$, the unique ρ value obtained from the algorithm above lies in the range $[\alpha_f + \Phi, 1.0)$. We now demonstrate the algorithm through an example.

Example 3: Consider a two-zone tandem AGV system with 14 stations. Stations 1 through 9 are in the first zone; the remaining stations are in the second zone. (For brevity, we will focus only on the first zone; similar results can be derived for the second zone.) Suppose three job types (labeled A through C) are processed through the first zone. Assuming that stations 1 and 6 are I/O stations, the average hourly production volume and the production route for each job type have been supplied as follows: Job A, 1.75 jobs/hour, 1-3-4-5-9-6; Job B, 1.00 jobs/hour, 6-2-7-8-6; and Job C, 2.25 jobs/hour, 6-4-5-2-8-1. The layout of the two zones, and the resulting from–to chart for the first zone (including the Λ and λ values), are shown in Figures 8 and 9, respectively.

The AGV travels in a bidirectional manner. The travel times (in min) between adjacent stations are shown in parentheses in Figure 8. (Travel time is not necessarily proportional to distance due to vehicle acceleration/deceleration and slower vehicle speed during turns). The time required to pick up (and later deposit) a load is given as 0.30 min. The empty travel time from station i to j is equal to the travel time via the shortest route from i to j. For example, $\sigma_{1,5} = \sigma_{5,1} = 2.00$ min. The loaded travel time is obtained by adding the load P/D time to the empty travel time. For example, $\tau_{1,5} = \tau_{5,1} = 2.30$ min.

Using the above flow and travel time values in Equations (3) through (6), we obtain $\chi = 9.828959$, $\hat{\chi} = 9.872709$, and $\Phi = 0.064829$. We next compute (from Equation (1)) $\alpha_f = 0.5550$. Since $\alpha_f + \Phi = 0.6198 < 1.0$, we conclude that the AGV in the first zone will meet the workload requirement and execute the aforementioned algorithm to compute the approximate vehicle utilization, ρ. Using a step size of $\Delta = 0.01$ and a stopping tolerance of 0.0001, the algorithm converges to $\rho = 0.9032$, which implies that $\alpha_e = 0.3482$. Note that the vehicle in the first zone is traveling empty about 39% of the time that it is busy.

For comparison purposes, if we assume FCFS dispatching for the empty AGV and apply the model shown in Section 2.2, we would obtain $\bar{s} = 0.04027$ hr/move request and $\rho = 0.9664$, which implies that $\alpha_e = 0.4114$. (Naturally, the α_f value is independent of the empty-vehicle dispatching rule.) Hence the AGV would still meet the workload under FCFS-dispatching, but the vehicle utilization would be too high and the fraction of empty vehicle travel would increase by about 18% relative to MOD FCFS.

2.4 Automated Storage/Retrieval Systems

AS/R systems play a significant role both in manufacturing and warehousing. Under complete computer control, AS/R systems can store, keep track of, and retrieve a large number of part types and containers. In manufacturing they have been used successfully for work-in-process storage on the factory floor. In warehousing they have been used in component/raw materials storage as well as order picking. AS/R systems tend to be capital-intensive, yet they offer long-term efficiency by reducing labor, energy, and land

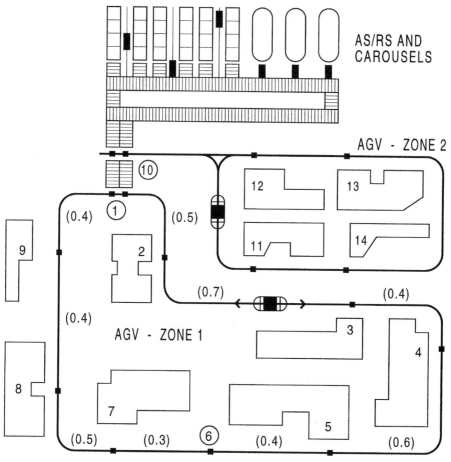

Figure 8   Layout for Example 3.

THE FROM-TO CHART (TRIPS/HOUR)

	1	2	3	4	5	6	7	8	9	λ
1	.00	.00	1.75	.00	.00	.00	.00	.00	.00	1.75
2	.00	.00	.00	.00	.00	.00	1.00	2.25	.00	3.25
3	.00	.00	.00	1.75	.00	.00	.00	.00	.00	1.75
4	.00	.00	.00	.00	4.00	.00	.00	.00	.00	4.00
5	.00	2.25	.00	.00	.00	.00	.00	.00	1.75	4.00
6	.00	1.00	.00	2.25	.00	.00	.00	.00	.00	3.25
7	.00	.00	.00	.00	.00	.00	.00	1.00	.00	1.00
8	2.25	.00	.00	.00	.00	1.00	.00	.00	.00	3.25
9	.00	.00	.00	.00	.00	1.75	.00	.00	.00	1.75
λ	2.25	3.25	1.75	4.00	4.00	2.75	1.00	3.25	1.75	24.00

Figure 9   Flow data for Example 3.

(or floor space) costs and by providing better parts visibility (i.e., knowing exactly where each part or subassembly is and how many are actually in stock), higher reliability, and reduced pilferage.

An AS/R system consists of one or more storage aisles and has four basic components:

1. Storage racks
2. Storage/retrieval (S/R) machine(s)
3. A handling system (such as conveyors) to serve as an interface
4. A computer to control the system

A *unit load* AS/R system stores and retrieves pallets. There are also *miniload* AS/R systems (for handling trays of small to medium parts) and *microload* AS/R systems (for handling tote boxes). See Tompkins (1996) for a more detailed description of the above and other types of AS/R systems.

A three-aisle unit load AS/R system is shown in Figure 10. Note that each storage aisle consists of a storage rack on either side, the aisle itself, and the S/R machine. Also, there is an input queue (for loads waiting to be stored) and output queue (for loads that have been retrieved) at each aisle; a conveyor is used as the interface. The input and output queue at each aisle represents the P/D point, which is generally modeled as a "point" located at the lower left-hand corner of the rack. (In some systems the P/D point is elevated by elevating the interface conveyors.) In most systems each S/R machine is aisle-captive. Otherwise, one or more transfer cars— to move the S/R machine(s) from one aisle to another—must be provided at the far end of the aisles. Some AS/R systems, especially those built for storing components and raw material, are housed in a stand-alone building erected for the system. Other systems, such as small, one- or two-aisle AS/R systems built for handling work-in-process inventory, are housed within the factory itself.

Since "lean manufacturing" dictates lower inventory levels, AS/R systems built today are smaller in size than those built 15–20 years ago. Nevertheless, manufacturing companies worldwide (including companies in the United States, Japan, Korea, and Europe) continue to use AS/R systems due to the advantages they offer. In fact, one may argue that the increased parts visibility and reliability alone offered by AS/R systems lead to reduced inventories, fewer errors, and fewer parts shortages. Minimizing errors and improving inventory accuracy are essential for running a manufacturing system with less inventories.

The storage/retrieval function in an AS/R system is performed by the S/R machine, which travels inside the aisle and automatically positions itself in front of the appropriate rack opening. With reference to Figure 11, a typical S/R machine consists of a (single or double) *mast* (A), a *top guide rail* (B), a *bottom guide rail* (C), and a *carriage* (D) that supports a *shuttle transfer mechanism* (E).

The S/R machine travels in the horizontal (X) direction when the mast, powered by a horizontal drive motor, moves up and down the aisle. Travel in the vertical (Y) direction occurs when the carriage, powered by a separate vertical drive motor, moves up and down the mast. The S/R machine can use the rack on either side of the aisle without having to turn inside the aisle; this is crucial for maintaining very narrow aisles and is accomplished by a telescoping shuttle mechanism mounted on the carriage. The shuttle mechanism is powered by a separate motor as well. Since the S/R machine is supported between two guide rails, it can reach fairly high speeds (typically, for unit load AS/R systems, 350–500 fpm in the X direction and 60–100 fpm in the Y direction), and it can travel in the X and Y directions simultaneously (which is known as *Chebyshev travel*).

One of the key design factors for an AS/R system is the system's throughput capacity. (Computing the storage capacity of a system is straightforward: We simply multiply the number of tiers by the number of columns in each rack, and then multiply the result by twice the number of aisles, since each aisle has two storage racks.) Suppose we define an *operation* as a storage or a retrieval performed by the S/R machine. The throughput capacity of an aisle is defined as the number of operations per hour performed by the S/R machine in that aisle. The throughput capacity of the system is equal to the throughput capacity of an aisle multiplied by the number of aisles (provided, of course, the S/R machines are aisle-captive and the handling system serving as an interface for the AS/R system does not become a bottleneck).

Under its basic mode of operation, to perform a storage operation the S/R machine picks up a load from the P/D point, travels to an empty opening in the rack, deposits the

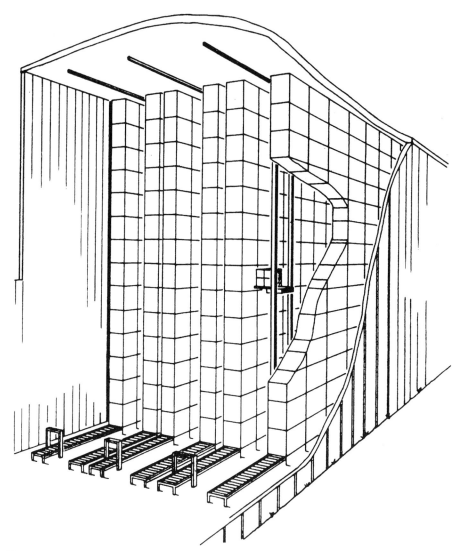

Figure 10 Unit load automated storage/retrieval system. (From *Warehouse Modernization and Layout Planning Guide,* Department of the Navy, NAVSUP Publication 529, 1985.)

load, and returns empty to the P/D point. To perform a retrieval operation, the S/R machine starts empty at the P/D point, travels to the appropriate rack opening, picks up the load, travels back to the P/D point, and deposits the load. Thus, each such trip—known as a *single command* cycle—consists of travel time (out to a rack opening and back) plus two load-handling operations (a pick-up and a deposit).

If a storage request (say, load s) and a retrieval request (say, load r) are both present, the S/R machine can perform a *dual command* cycle as follows: It picks up load s at the P/D point, travels to an empty opening, deposits load s, travels to the opening containing load r, picks up load r, travels back to the P/D point, and deposits load r.

Since the load pick-up and deposit times are constant, we will focus on the travel time of the S/R machine. Let:

$E(SC)$ = the expected travel time for a single command cycle
$E(TB)$ = the expected travel time from the storage location to the retrieval location during a dual command cycle (i.e., $E(TB)$ denotes the expected travel time between two points in the rack)

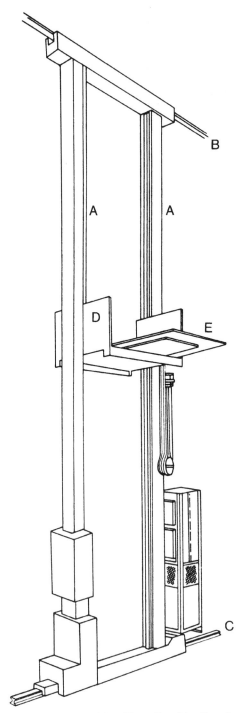

Figure 11   Unit load storage/retrieval machine. (From *Considerations for Planning an Automated Storage/Retrieval System*, 1982. Courtesy of Material Handling Industry.)

$E(DC)$ = the expected travel time for a dual command cycle

Assuming that the rack length, L (in feet), the rack height, H (in feet), the horizontal velocity of the S/R machine, h_v (in fpm), and the vertical velocity of the S/R machine, v_v (in fpm), are specified by the user, we first compute:

t_h = time required to travel horizontally from the P/D station to the furthest location in the aisle, ($t_h = L/h_v$ min), and

t_v = time required to travel vertically from the P/D station to the furthest location in the aisle, ($t_v = H/v_v$ min)

To determine compact expressions for $E(SC)$ and $E(DC)$, we then "normalize" the rack by dividing its shorter (in time) side by its longer (in time) side. That is, we let

$$T = \max(t_h, t_v), \tag{7}$$

$$b = \min(t_h/T, t_v/T) \tag{8}$$

where T (or the "scaling factor") designates the longer (in time) side of the rack, and b (or the "shape factor") designates the ratio of the shorter (in time) side to the longer (in time) side of the rack. The normalized rack is b time units long in one direction and one time unit long in the other direction. Also, $0 < b \le 1$; if $b = 1$, the rack would be *square-in-time* (SIT).

Given randomized storage (i.e., an operation is equally likely to involve any location in the rack), an end-of-aisle P/D point located at the lower left-hand corner of the rack, constant horizontal and vertical S/R machine velocities (i.e., no acceleration or deceleration is taken into account), and a continuous approximation of the storage rack (which is sufficiently accurate), the expected travel times for the S/R machine are derived as (Bozer and White, 1984):

$$E(SC) = T\left[1 + \frac{b^2}{3}\right], \tag{9}$$

$$E(TB) = \frac{T}{30}[10 + 5b^2 - b^3], \tag{10}$$

$$E(DC) = E(SC) + E(TB) = \frac{T}{30}[40 + 15b^2 - b^3] \tag{11}$$

Bozer and White (1984) show that $E(SC)$ and $E(DC)$ are minimized at $b = 1$; i.e., given a rack with a fixed area (but variable dimensions), the expected single- and dual-command travel times are minimized when the rack is SIT. (A SIT rack, however, may not be the lowest-cost rack, since rack cost is affected by the rack height.)

The expected *cycle time* for the S/R machine is obtained by adding the load P/D time to the expected travel time. That is, if T_{SC} denotes the expected single command cycle time, T_{DC} denotes the expected dual command cycle time, and $T_{P/D}$ (supplied by the user) denotes the time required to either pick up or deposit a load, we have

$$T_{SC} = E(SC) + 2T_{P/D}, \tag{12}$$

$$T_{DC} = E(DC) + 4T_{P/D} \tag{13}$$

Given the above cycle times, the throughput of an aisle (i.e., number of operations performed per hour) depends on how often the S/R machine performs a single- versus dual-command cycle. Note that in a single-command cycle the S/R machine completes one operation; in a dual-command cycle it completes two operations. With the following examples, we show how to compute the system throughput.

Example 4:   Consider an AS/R system rack that is 200 ft long and 45 ft tall. The P/D point is located at the lower left-hand corner of the rack. The S/R machine travels at an average speed of 60 fpm and 400 fpm in the vertical and horizontal direction, respectively.

Normalizing the rack, we obtain $T = 0.75$ min and $b = 0.6666$. Hence the single command expected travel time (from Equation (9)) is equal to 0.8611 min. The dual command expected travel time (from Equation (11)) is equal to 1.1593 min.

Example 5: Suppose the P/D point of the AS/R rack described in Example 4 is elevated by 15 ft. How would this affect the expected S/R machine travel times? Since elevating the P/D point reduces the vertical travel performed by the S/R machine, we would expect a reduction in $E(SC)$ and $E(DC)$.

To compute the values of $E(SC)$ and $E(DC)$, we view the rack in two sections: The first section measures 30 ft × 200 ft and represents the portion of rack that is *above* the P/D point; the second section measures 15 ft × 200 ft and falls *below* the P/D point. (Although the P/D point is at the *upper* left-hand corner of the second section, the results given by Equations (9)–(11) still hold due to symmetry.)

Normalizing the first section, we obtain $T_1 = 0.50$ and $b_1 = 1.0$. Therefore, $E(SC)_1 = 0.6666$ min. For the second section we obtain $T_2 = 0.50$ and $b_2 = 0.50$, for which $E(SC)_2 = 0.54166$ min. Since each rack opening is equally likely to be used, the probability of traveling to a particular section is proportional to the area of the section. Hence,

$$E(SC) = \left(\frac{30 \times 200}{45 \times 200}\right) E(SC)_1 + \left(\frac{15 \times 200}{45 \times 200}\right) E(SC)_2$$

or $E(SC) = 0.6250$ min, which represents a 27% reduction compared to Example 4.

To compute $E(DC)$ we first note that $E(TB)$ is independent of the P/D point location. That is, to compute $E(TB)$ we consider the entire rack (45 ft × 200 ft), for which $T = 0.75$ min and $b = 0.6666$. Hence, from Equation (10), we have $E(TB) = 0.2981$ min, and from Equation (11) we obtain $E(DC) = E(SC) + E(TB) = 0.6250 + 0.2981 = 0.9231$ min, which represents a 20% reduction compared to Example 4. Of course, an elevated P/D point reduces the expected S/R machine travel time, but one must also consider the cost associated with elevating the P/D point.

Example 6: Suppose the rack described in Example 5 is part of a 3-aisle AS/R system. Further, suppose that it takes 9 sec to pick up or deposit a load. During peak activity periods, if 70% of the operations are retrievals, and if 40% of the retrievals are performed as part of a dual command cycle, what throughput capacity would the system achieve, assuming that the S/R machine utilization cannot exceed 95%?

Using $T_{P/D} = 0.15$ min in Equations (12) and (13), respectively, we obtain $T_{SC} = 0.9250$ min and $T_{DC} = 1.5231$ min. Letting x denote the aisle throughput, we have $[0.28 \times (1.5231/2)]2x + [(0.42 + 0.02)(0.9250)]x = (60 \times 0.95) = 57$, or $x = 68.39$ opns/hr. Hence the throughput capacity of the system is equal to $68.39 \times 3 = 205.17$ opns/hr.

In this section we assumed that each trip starts and ends at the P/D point. One may implement alternative dwell point strategies for the S/R machine. For example, following a storage operation at opening i, the S/R machine may wait at opening i. If the next operation is a retrieval, the S/R machine can travel directly from opening i to the retrieval location. The reader may refer to Bozer and White (1984) for further details.

2.5 Carousels

A carousel, which can operate under complete computer control, is used typically for small to medium parts storage and retrieval. It consists of a set of floor-supported or suspended revolving bins that rotate to bring the containers to the P/D point. If rotation is horizontal (such as a carousel one might find at a dry cleaner), it is termed a *horizontal* carousel. If rotation is vertical (such as a carousel one might find at a hospital with many patient files), it is termed a *vertical* carousel.

Although storage/retrieval in a carousel-based system can be performed by humans, containers can also be moved automatically into and out of the carousel by a pick-and-place robot, known as a *robotic insertion/extraction* (I/E) device. A horizontal carousel with such a device is shown in Figure 12. The robotic I/E device, which may have grippers or utilize suction, is interfaced often with a take-away conveyor to handle the containers retrieved from the carousel. Containers that weigh up to 100 lb and measure up to 25 in. wide by 30 in. long are within the range of standard systems.

Equipped with robotic I/E devices, computer-controlled carousels can offer many of the same benefits as AS/R systems, and sometimes at a lower cost. Another automated system with a robotic I/E device that is used quite extensively for component and work-

Figure 12   Horizontal carousel with robotic insertion/extraction device (courtesy White Systems, Inc.).

in-process storage/retrieval (particularly in Japan) is the *rotary rack*. A rotary rack is similar to a horizontal carousel; however, each level of a rotary rack is powered individually. (A rotary rack that is *m* containers high functions like *m* horizontal carousels, each one only one container high and stacked on top of the other.) Although rotary racks are more costly, they achieve higher throughput rates when coupled with the right number of I/E devices. (One rotary rack supports more than one I/E device in general.)

3 SUMMARY

In this chapter we reviewed a number of robot-based material-handling systems that have been used successfully in manufacturing and warehousing. We also showed a few analytical models that can be used in the design and analysis of such systems. One may construct completely automated, computer-controlled, robot-based material-handling systems by integrating some of the devices we presented in this chapter. For example, one may combine an AS/R system, carousels, and robotic I/E devices with an AGV system to deliver raw materials and components from the AS/R system to the shop floor, store/retrieve work-in-process through the carousel, and move parts among the machines via the AGVs, with no human intervention (see Figure 8). A primary advantage of such integrated material-handling systems is the extraordinary flow control and high job visibility they offer. The robotic handling devices and the computer play a key role in such systems.

REFERENCES

Bozer, Y. A., and M. M. Srinivasan. 1991. "Tandem Configurations for Automated Guided Vehicle Systems and the Analysis of Single-Vehicle Loops." *IIE Transactions* **23**(1), 72–82.

Bozer, Y. A., and J. A. White. 1984. "Travel-Time Models for Automated Storage/Retrieval Systems." *IIE Transactions* **16**(4), 329–338.

Cho, M. S. 1990. "Design and Performance Analysis of Trip-Based Material Handling Systems in Manufacturing." Ph.D. Dissertation, Department of Industrial and Operations Engineering, University of Michigan.

Chow, W. M. 1986. "Design for Line Flexibility." *IIE Transactions* **18**(1), 95–103.

Egbelu, P. J. 1987. "The Use of Non-Simulation Approaches in Estimating Vehicle Requirements in an Automated Guided Vehicle Based Transport System." *Material Flow* **4**, 17–32.

Kim, J. 1995. "Transfer Batch Sizing in Trip-Based Material Handling Systems." Ph.D. Dissertation, Department of Industrial and Operations Engineering, University of Michigan.

Koff, G. A. 1987. "Automatic Guided Vehicle Systems: Applications, Controls and Planning." *Material Flow* **4**, 3–16.

Srinivasan, M. M., Y. A. Bozer, and M. S. Cho. 1994. "Trip-Based Material Handling Systems: Throughput Capacity Analysis." *IIE Transactions* **26**(1), 70–89.

Swanson, P. J., R. R. Burridge, and D. E. Koditschek. "Global Asymptotic Stability of a Passive Juggling Strategy: A Possible Parts Feeding Method." To appear in *Mathematical Problems in Engineering*.

Tompkins, J. A., et al. 1996. *Facilities Planning.* 2nd ed. New York: John Wiley & Sons.

White, J. A., and J. M. Apple. 1985. "Robots in Material Handling." In *Handbook of Industrial Robotics.* Ed. S. Y. Nof. New York: John Wiley & Sons. 955–970.

Ziskovsky, J. P. 1985. "Gantry Robots and Their Applications." In *Handbook of Industrial Robotics.* Ed. S. Y. Nof. New York: John Wiley & Sons. 1011–1022.

ADDITIONAL READING

Ackerman, K. B. *Practical Handbook of Warehousing.* 4th ed. New York: Chapman & Hall, 1997.

Allegri, T. H. *Materials Handling: Principles and Practice.* New York: Van Nostrand Reinhold, 1984.

Asfahl, C. R. *Robots and Manufacturing Automation.* 2nd ed. New York: John Wiley & Sons, 1992.

Hartley, J. R. *Robots at Work: A Practical Guide for Engineers and Managers.* Bedford: IFS, and Amsterdam: North-Holland, 1983.

Nof, S. Y. *Robotics and Material Flow,* Amsterdam: Elsevier Science Publishers, 1986.

Sciavicco, L., and B. Siciliano. *Modeling and Control of Robot Manipulators.* New York: McGraw-Hill, 1996.

Sheth, V. S. *Facilities Planning and Materials Handling: Methods and Requirements.* New York: Marcel Dekker, 1995.

CHAPTER 50

ASSEMBLY: MECHANICAL PRODUCTS

Daniel E. Whitney
Massachusetts Institute of Technology
Cambridge, Massachusetts

1 INTRODUCTION

The design of robotic assembly systems requires a combination of economic and technical considerations. As products become more complex, manufactured in many models that change design rapidly, and as manufacturers seek to be more responsive to changing demand and just-in-time manufacturing, it becomes necessary to design flexible assembly systems. The robotic assembly option must compete against the major alternatives—manual assembly, rigid automation, or some hybrid of these—on both technical and economic bases. People are the most flexible and the most dexterous assemblers, but their performance is variable, difficult to document and hold to a standard. Fixed automation is efficient and uniform in performance, but it is too expensive for small production runs because each assembly operation requires its own dedicated workstation. Robot assembly offers an alternative with some of the flexibility of people and the uniform performance of fixed automation. Detailed information about mechanical product assembly can be found in Boothroyd (1992) and Nof, Wilhelm, and Warnecke (1997).

In this chapter we will provide an overview of approaches for designing robot assembly systems for mechanical products. We will provide guidelines for analyzing products, creating concept system designs, and evaluating them economically against the basic alternatives. We will also describe the steps in detailed system design and give some case studies. Next we will provide a glimpse at the theory of part mating and the capabilities of the remote center compliance, a device that permits assembly of precise parts by imprecise equipment.

2 PLANNING AND STRATEGIC ISSUES, INCLUDING ECONOMICS

2.1 Strategic Considerations

The planning phase of any assembly system, including a robotic system, provides a unique opportunity to review the design of the product at a variety of levels. Assembly is the phase of production where the product first comes to life and can be tested for function. Assembly is also the phase where production directly interfaces with customer orders and warranty repairs. Thus assembly is more than putting parts together. Planning assembly is the opportunity to determine how to meet a variety of business needs, ranging from quality to logistics.

Anyone considering robot assembly must start by recognizing that robots are machines, not electric people. Thus their inherent strengths—repeatability, lack of fatigue, ability to switch programs with no learning effect, ability to link directly with factory information systems—must be balanced against their weaknesses—inability to make judgments or distinguish fine details in sensory domains—and one must be prepared to take advantage of the strengths without succumbing to the weaknesses.

It is often true that automated assembly is considered for products that are already being assembled manually. In such cases it is imperative that the entire assembly process be analyzed afresh because the strengths and weaknesses of people are totally different from those of machines. Often there are undocumented steps in the process that people have developed on their own to overcome hidden problems with parts or procedures. These steps must be found and eliminated by redesign because they will be difficult or

Handbook of Industrial Robotics, Second Edition, Edited by Shimon Y. Nof
ISBN 0-471-17783-0 © 1999 John Wiley & Sons, Inc.

impossible to automate. All steps where people use judgment, acute vision, or touch sensing, or where they are indirectly inspecting the parts, must be identified; ways must be found to perform these functions automatically, or else the need for them must be eliminated.

2.2 Basic Planning Issues

When a product is considered for robotic assembly, several specific analyses must be done:

- Examining the basic architecture of the product. The possibilities for different subassemblies should be identified because these provide opportunities for subassembly lines complete with their own final testing, as well as opportunities for creating different models of the product by mixing and matching different subassemblies. Architecture also includes determining whether all the assembly operations can be performed from above or whether the product must be reoriented during assembly. Reorientation takes time and involves design and purchase of additional grippers and fixtures.

- Investigating alternative assembly sequences (Baldwin, et al, 1991, Homem de Mello and Sanderson, 1991a, 1991b; Khosla and Mattikali, 1989; Nof and Rajan, 1993; Kim and Lee, 1983). This step is related to architecture considerations, and iteration between the two analyses may be needed. Different sequences involve different subassembly options, grouping of operations onto single stations, in-process testing opportunities, and requirements for reorientation of the product. Never use an existing manual assembly sequence without investigating alternatives.

- Identifying difficult assembly steps and redesigning the product to eliminate them. Such steps may involve maneuvering parts through tortuous paths, trial and error to select a part that fits, doing undefined tests like "check for looseness," and so on.

- Performing tolerance analyses of all assembly steps. Such analyses must include the entire tolerance chain, from the receiving part to the arriving part, including all tooling, grippers, feeders, and the robot arm itself. An assessment of assembly process capability should be performed, and six-sigma capability should be sought. Part mating theory (see list of additional reading) provides methods for calculating the success or failure of assembly operations based on geometric errors and friction. Improvements in capability can be obtained by adding chamfers to parts, properly tolerancing grip points in relation to mating points, and using error-absorbing tooling such as the remote center compliance, discussed later in this chapter.

- Identifying and evaluating methods of part presentation for speed, feasibility, and cost. This is one of the most troublesome aspects of machine assembly compared to manual. People can acquire, reorient, inspect, and assemble almost any parts that they can comfortably lift. Robots need to have parts presented to them arrayed, aligned, and inspected. In some instances randomly oriented parts can be presented, but a separate capability is needed to orient them, often relying on vision or a separate mechanism.

- An economic analysis must be performed (Milner, Graves, and Whitney, 1994). Such an analysis must be based on more than simple labor replacement and must include benefits and detriments like improved quality or yield, or time lost due to breakdowns. The proposed robot system must be compared to feasible alternatives that include people or fixed automation.

2.3 Design Process for Robot Assembly Systems

A programmable robot assembly system typically consists of one or more robot workstations and their associated grippers and parts presentation equipment (Feldman and Geyer, 1991; Ho and Moodie, 1994; Bard, Dar-El, and Shtub, 1992). These stations may be linked in a variety of ways, including a conventional conveyor. Alternatively, each robot can simply pass the work over to the next robot. System configurations include straight lines of simple workstations doing one task each, U-shaped setups where several robots share tools and parts feeders, and single workstations where one robot does all the tasks, changing tools as needed. These configurations typically are suitable for high-, medium-, and low-volume applications, respectively. The medium- and low-volume applications take advantage of the robot's ability to change tools and programs in order to perform a series of different assembly tasks.

The following steps are involved in planning a robot assembly system:

1. Project initiation or approval
2. Application survey for candidate products
3. Preliminary screening of candidates
4. Priority-setting of candidates
5. Selection of the project team
6. Data collection and documentation of the candidate product
7. Conceptual system configuration, in parallel with initial economic evaluation
8. Final system configuration and final economic evaluation
9. Detailed design of the system
10. Final design review

A company may carry out all, some, or none of the above steps, the remainder being done by outside consultants or assembly system vendors. If vendors are used, it is imperative that the strategic issues discussed in Section 2.1 be carried out by the company itself and not left to the vendor (Chen and Wilhelm, 1994; Lee and Johnson, 1991; Ghosh and Gagnon, 1989; Nof and Drezner, 1993; Park and Chung, 1993; Yano and Rachamadugu, 1991; Sellers and Nof, 1989; Van Brussel, 1990).

Following are some guidelines on several important issues from the list above. Other topics from this list are dealt with later in this chapter.

2.3.1 Criteria for Selecting a Candidate Product

Products whose production volume is less than about 100,000 per year rarely meet economic criteria for automated assembly of any kind, because labor is relatively low-cost and automation requires a large investment. Products that are small (less than 10 cm square), with unidirectional assembly comprising simple straight push motions, at most a few varieties, and production volumes in excess of 1,000,000 per year, are often suitable for automated assembly by dedicated fixed automation. Those with mid-range production volumes, complex assembly motions, and/or high variety are often suitable candidates for robot assembly. A product with many flexible components like wires or rubber gaskets can present challenges to robots, but these are increasingly being overcome with clever tooling and vision combined with skillful redesign.

Other considerations that encourage automatic assembly (not just robotic) include dangerous human environments, work that might cause repetitive strain injury, heavy parts, or the need to take special precautions for environmental reasons. Company politics or fads can also influence automation decisions. If at all possible, a rational screening and evaluation process should be used in order to avoid disappointments based on excessive hope or hype.

2.3.2 Selecting the Project Team

Assembly system design may occur in two circumstances. The most favorable is that where the system can be designed in conjunction with the design of the product. This is often called *concurrent engineering.* In this case the assembly project team is part of, or associated with, the overall product design team. Members of the team have expertise in all the relevant areas, such as product marketing (to determine required production volumes, customer options, quality requirements), product design (to determine product architecture, part count, materials, fastening methods), finance and accounting (to determine economic justification criteria), manufacturing (to determine part tolerances and costs), assembly and test (to determine assembly and quality assurance methods), and general management (to set project goals, budgets, and schedules, and to encourage team behavior). Vendors of parts, subassemblies, software, or fabrication or assembly equipment should also be represented and involved closely from the beginning.

When the product has already been designed, there is much less scope for optimization of the product–process combination. Team membership in this case may be restricted to people from fabrication, finance and assembly, and vendors of equipment.

A long-term strategic decision for a company is the extent to which it wants to have in-house expertise in automation. The advantages include the ability to integrate product and process design as well as to develop proprietary capabilities that reduce cost, improve responsiveness, or raise quality. The disadvantages include the cost of maintaining this capability between projects and the possibility that the group will become stale from lack of external stimulation, information, or competition. Many Japanese and European companies opt for in-house capability, while many U.S. firms opt for reliance on vendors.

2.3.3 System Concept Design

A piece of manufacturing equipment or a manufacturing system must do its job and must fit into the factory as a whole. That is, the system has global and local responsibilities. Correspondingly, the product to be fabricated or assembled presents both local and global characteristics that influence the system's design. The local issues are primarily technical, the traditional part mating or assembly sequence problems, whereas the global issues are primarily economic and concern management's objectives for the product. Such global issues as potential volume growth, number of models, frequency of design changes, and field repairability heavily influence both product and system design.

This morass of requirements is given some structure in Table 1.

2.3.4 Product Documentation and Data Collection

Information about the product includes drawings, bill of materials, assembly process instructions, varieties required, and any special information such as toxic chemicals, fragile parts, and so on. If possible, useful redesigns should be requested if they have not already been accommodated during concurrent engineering. These, too, may be grouped into global and local.

1. *Global:* Concentrate model differences in a few, perhaps complex, parts so that the rest of the parts and assembly operations will be common to all models. Reduce part count to keep the assembly system, parts feeding, and logistics systems simple. But do not pursue part count reduction to the point where parts become too complex, costly, or hard to design within the project's schedule. Remember, the cost of most products is concentrated in the parts themselves, with assembly typically being a small fraction.

2. *Local:* Design subassemblies for unidirectional assembly, so that reorientations, if needed, occur infrequently at the end of subassembly lines. Reduce fastener count and standardize on a few kinds and sizes of fasteners to permit common part feeding and tools. Take care to provide reference surfaces for fixturing or gripping that are carefully toleranced with respect to part mating surfaces. Figure 1 shows

Table 1 Structure of Assembly System Concept Design Issues

	Global: Management's Objectives	Local: The Parts and Assembly Operations
Product	Economics and market	Assembly sequences
	Volume growth	Types of operations
	Design volatility	Geometric constraints
	Quality, reliability, safety	Part size and weight
	Make–buy decisions	Shape and stiffness
	Build to order or stock	Tolerances and clearances
		Tests and inspections
Assembly system	Cost and productivity	System layout
	How it interfaces with the factory	Equipment choice
	Labor support needs	Assignment of operations to equipment
	Failure modes and uptime	Part feeding
	Space and energy needs	

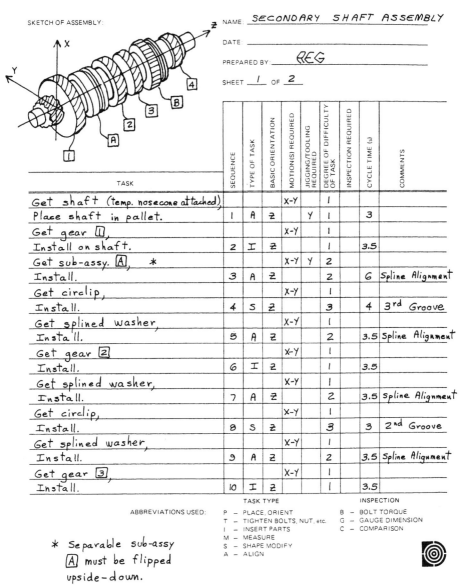

Figure 1 Assembly planning chart. This chart contains detailed information on each assembly step.

a chart on which assembly requirements information can be compiled for a sub-assembly.

System concept design also includes consideration of alternative assembly sequences. Computer algorithms exist that will generate all the feasible assembly sequences for a mechanical product. These algorithms require input information concerning relative approach directions between pairs of parts, as well as information on mutual interference between parts. This information can be put in manually, or in some cases it can be obtained from advanced computer-aided design (CAD) files or from a form like Figure 1. The algorithms sift through the approach direction and interference information and produce a list of precedence relations that constrain the possible order in which parts are added (Figure 2a). One way to represent the feasible sequences is with a network diagram like that in Figure 2b. All feasible sequences are represented, whether they are good or bad from a technical or economic standpoint.

Computer software also exists that will help designers select good sequences from those represented in the network. Since there is unlikely to be one optimal sequence, the software permits sequences to be edited from a variety of criteria (Baldwin, et al, 1991).

2.3.5 System Configuration Design

A variety of system configurations should be considered. The issues include how to handle product variety, how to provide space for backup workstations and access for repairs, and how to provide parts. Traditional straight lines are not the only possibility. Moreover, different methods have been used to handle model variety. One method uses coded pallets that tell a workstation whether its part should be installed. More complex stations use such information to decide which part to install. Some configurations provide a kit of parts that rides on the pallet, while one experimental system put the robot on the pallet and directed it to one station after another where it obtained parts or tools and performed operations.

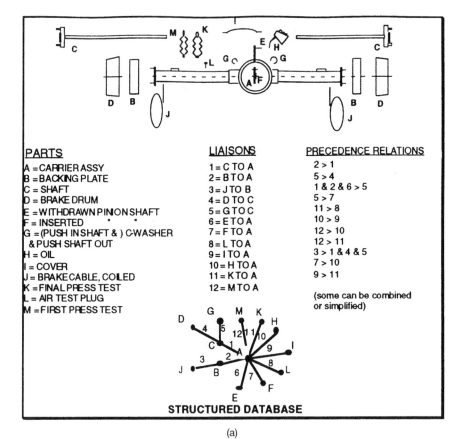

(a)

Figure 2 (a) Example product with assembly constraints. Each part is listed at the left. In the middle are listed all the "liaisons" or mates between parts. At the right is a list of the precedence relations derived by sequence analysis software. (b) Network of feasible assembly sequences for the product in Figure 2a. The empty box at the top represents a completely disassembled product, while the full box at the bottom represents the product fully assembled. Each box in between has interior segments blackened to show which liaisons have been completed. Any path from the top to the bottom obeys the precedence relations and can be read as a feasible assembly plan.

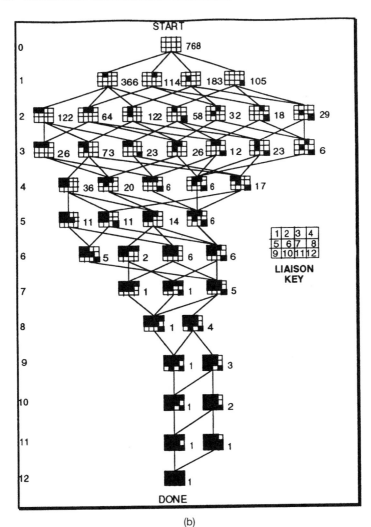

(b)

Figure 2   (Continued)

Figure 3 shows a product that was designed for simple production of a wide variety. The system is a simple line of pick-place robots or fixed cycle stations, but the parts feeders can be switched from one part to another within one assembly cycle.

2.3.6   Economic Justification

Clearly, advancing technology provides the system designer with more and varied alternatives. Inevitably, economics becomes the tool by which a choice is made. Cost models and rate of return calculations are customarily used to determine if an investment is attractive. Here we will deal only with simplified unit cost models.

In general, any automation system will include both labor and capital. Labor is accounted for via annual wages, but capital involves a one-time upfront investment. Most economic models seek to convert the one-time investment into an equivalent annual expense in order to permit it to be combined with the labor cost. This is done by using a fraction of the total investment, with the fraction represented by an annual cost factor based on desired rate of return.

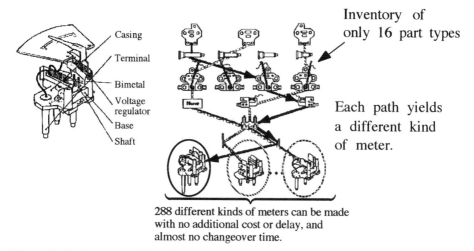

Inventory of only 16 part types

Each path yields a different kind of meter.

288 different kinds of meters can be made with no additional cost or delay, and almost no changeover time.

Figure 3 Panel meter produced by Denso, Ltd. A simple assembly machine with switchable parts feeders can select among a small inventory of parts and make a wide variety of end products.

The labor component of the cost of the system is represented by the number of operations done directly by people plus the number of support people. The capital component of cost is determined by calculating the number of machines necessary multiplied by the cost of each. The number of machines is equal to the number of operations to be done if the machines are inflexible. However, if the machines are flexible robots and can perform more than one operation, then the number of machines depends on the production rate required, the number of operations required per product, and the speed of the machines. In addition, there is the possibility that time will be lost for a variety of reasons. These include time lost while workers are on breaks, when the machines are broken or being repaired, and because the products produced were of unsatisfactory quality and had to be discarded or reworked.

A combined unit cost model that includes all these factors is given below in Table 2. For completeness this model includes a model for manual assembly and one for fixed automation.

3 PLANNING EXAMPLE

This section contains an example of assembly system concept design together with assembly process design. The product is a small gunpowder-filled ignitor. The ignitor (Figure 4) is about an inch long and has eight parts, counting the powder charge. The product is currently assembled manually with the aid of a press and simple hand tools. While production volume is expected to rise year by year, it is unlikely to become large enough to justify conventional fixed automation. Since labor costs are low, the robot system must be very low in cost. The challenge is to design a robot assembly system with inexpensive pick-place robots having three or four axes at most that move in simple arcs from one fixed stop to another. This system must not only stack some parts under a press (not too difficult), but must also thread a wire through them (difficult).

The proposed solution is shown in Figure 5. It is based on reversing the manual assembly sequence so that the wire can be attached to the housing and cap in a few simple twists and sideways moves by a fixed-stop robot and a wire-feeding station.

The final system concept is shown in Figure 6. It contains two fixed-stop robots with three degrees of freedom each (in/out, up/down, and a 180° flip about the in/out axis). Robot 1 performs the operations shown in Figure 5, while robot 2 puts the assembled unit in a tester and deposits in a "good" bin or a "bad" bin. This optional action can be done by this simple robot because it visits the "bad" bin first, then the "good" bin. If the unit is good, programmed action to open the gripper over the "bad" bin is skipped. The gripper always opens over the "good" bin.

Table 2

(a) *Unit Assembly Cost Model for Manual Assembly*

$$\text{Cost}_{\text{unit, manual}} = \frac{A\$ * \text{No\_People}}{Q * Y}, \$/\text{unit}$$

Q = annual production volume, units/yr

Y = yield, % of units made that are acceptable

$$\text{No\_People} = \left\lceil \frac{T * N * Q}{2000 * 3600} \right\rceil$$

[. . .] = round up contents to next larger integer

T = assembly time per part, seconds

N = number of parts per unit

$A\$$ = annual cost of a person, \$/yr

$A\$ = L_h * 2000$

L_h = labor cost, \$/hr

2000 = number of hours per shift—year

3600 = number of second per hour

(b) *Unit Assembly Cost Model for Fixed Automation*

$$\text{Cost}_{\text{unit, fixed}} = \frac{f_a * N * S\$}{Q * Y * E}$$

f_a = fraction of machine cost allocated per year

$S\$$ = average cost per station in the machine, assuming one station per part

E = efficiency of machine operation

E = 1—downtime fraction per shift, where downtime can occur for any reason, including employee breaks, regular maintenance, or machine breakdown and repair

(c) *Unit Assembly Cost Model for Flexible Automation*

$$\text{Cost}_{\text{unit, flex}} = \frac{f_a * I + L\$ * U}{Q * Y * E}$$

I = total investment in machines, tools, and fixtures

I = No_Machines * \$/machine + No_Tools * \$/tool + No_Fixtures * \$/fixture

$$\text{No\_Machines} = \left\lceil \frac{T * N * Q}{2000 * 3600} \right\rceil$$

$L\$$ = annual cost of workers associated with the system

$L\$$ = w * No_Machines * L_h * 2000

w = Number of workers per machine

This is not a very flexible system, except for the last step, but it meets the strategic and tactical needs for a low-cost system with good safety, uniformity of performance, and product quality. It also illustrates the possibilities of performing complex operations with simple robots.

4 DETAILED SYSTEM DESIGN

Detailed system design is composed of many complex steps, including detailed planning of each operation, preparation of process diagrams and flow charts, preparation of system operating software, creation of wiring diagrams and mechanical designs of grippers and fixtures, and many other detailed tasks. Of these, only a few can be discussed here, namely development of the process flow chart, the cycle time chart, and equipment selection and system layout. These steps provide the basic information for selecting robots and other equipment and guaranteeing that the system will meet its production rate requirements.

4.1 Process Flow Chart

The process flow chart is the graphic representation of the assembly tasks in sequence. It can be considered as the logic diagram for the conceptual system configuration design

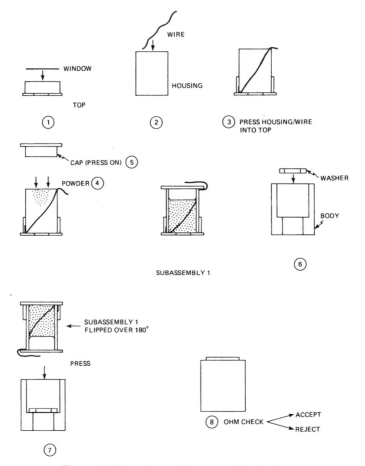

Figure 4   Manual assembly process for the ignitor.

work. The main advantage of developing a process flow chart is that it summarizes and presents all the assembly, orientation, and test operations together, forming the basis for both mechanical and software design. Figure 7 shows an example chart.

In this chart the principal direction of the process flow represents the sequential assembly tasks. Assembly tasks are represented by circles, inspection tasks by squares. Side lines come into the main line, representing subassemblies. Those on the left are members of every product unit, while those on the right are options included in only some units. Associated with each task are two numbers. The three-digit number is the task number. The two-digit number(s) call out the model numbers that utilize that part or subassembly. The product diagrammed in Figure 7 is made in three models.

4.2   Cycle Time Chart

In manufacturing, the single most significant factor affecting production volume is the cycle time. All system design parameters ultimately can be related to this parameter. Therefore it is desirable to define operation cycle time as early as possible in the system design process. Cycle time charts can be prepared using simple computer programs designed for project scheduling.

In developing a cycle time chart, the first question asked is how to determine the operation times when the equipment to perform the operations has not yet been selected. Usually one makes estimates based on robot manufacturers' data or makes use of commercial robot simulation software. For either of these purposes one estimates the size of the work area based on the size of the product being assembled, chooses a specific or

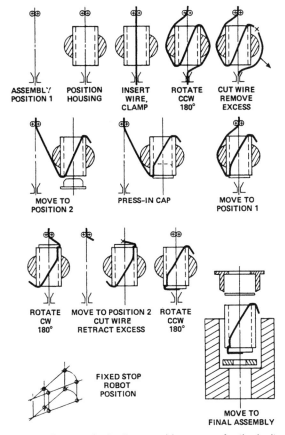

Figure 5   Proposed robotic assembly process for the ignitor.

generic robot with sufficient reach, and calculates or estimates motion times based on distance, acceleration/deceleration, and top speed. One then adds tool action times, short for grip, longer for screwdriving, and so on. Upon compiling all this information, one summarizes it in a chart similar to Figure 8. In this chart it is assumed that several operations can be done in parallel, using separate equipment. If only one robot is used, the cycle time will be 40 seconds.

The problem of dividing multiple tasks into groups and assigning them to workstations is a complex one because the operations typically take different lengths of time and different operating speeds can be obtained from different robots, with higher speed usually costing more. An optimum solution usually requires a computer search. What follows is a simplified process that results in an estimate.

Using the equation in Table 2c for calculating the required number of machines, we find that for an annual volume of 540,000 units, 14 operations, and a presumed average time per operation of $40/14 = 2.85$ seconds, we would need three robots per shift, working together. Each robot would do several of the operations. This is only an estimate, since we know that some tasks take longer than 2.85 seconds, and in fact we can do some operations in parallel. A detailed design would require trying different patterns of assigning the operations to the robots, as well as seeking faster robots for the tasks that currently take the longest. A robot simulation program would be used to check each trail design to assure that all the operations can be performed within the required time. If the production volume is so high or the operations take so long that the work cannot be done in one shift even if a separate robot does each operation, then two-shift or even three-shift operation will be necessary. More detailed analysis can apply line-balancing models (see Chapter 30).

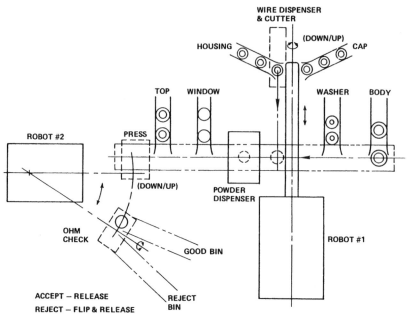

Figure 6 Ignitor assembly station including two robots.

4.3 Equipment Selection and System Layout

The previous step identified an approximate number of workstations and their required speed. Additionally, each station must have the required reach and payload capacity, and if necessary the ability to change tools. (Note that tool change time may use up part of the available cycle time. If this causes problems, possible remedies include choosing a different assembly sequence so that the same tool can be used for more operations in a row, redesigning the product to reduce the need for tool changes, and regrouping the operations onto different stations.)

Robots are specified according to several characteristics:

- *Work envelope*. This is probably the most important parameter affecting the selection of the robot. It is a good practice to choose a robot that can reach a little farther than one's first estimates of needed reach. Factors influencing reach requirement include size of the parts and the assembly, and the number of part feeders and tool magazines that must be reached.

- *Repeatability*. Robots usually repeat the same motions, so the ability to return to a preprogrammed point is the main capability of a robot. For maximum speed it is best that the robot use its internal position sensors to reach the required destination, rather than some kind of feedback such as vision. End-of-arm repeatability is a function of joint measurement resolution, arm length, gear backlash, and arm vibration caused by elasticity, among other factors. Repeatability should be good enough to get parts within the capture range of chamfers on the parts, assuming that an error-absorbing element such as a remote center compliance is available.

- *Accuracy*. Accuracy is important if the robot is programmed off-line from CAD data, but not if it is programmed simply to repeat an initial motion pattern put in by joystick. Accuracy can never be better than repeatability.

- *Payload*. Most robot assembly is done on parts that weigh less than 2 kg, and the large majority weigh only a few grams. The robot usually carries a tool that weighs much more, so the payload is usually dominated by tool weight.

- *Speed*. Robots are often advertised based on their top speed, but often robots never reach top speed. If a robot goes very fast, it may vibrate when stopping, adding time rather than saving it. Thus the time a robot needs to perform an operation

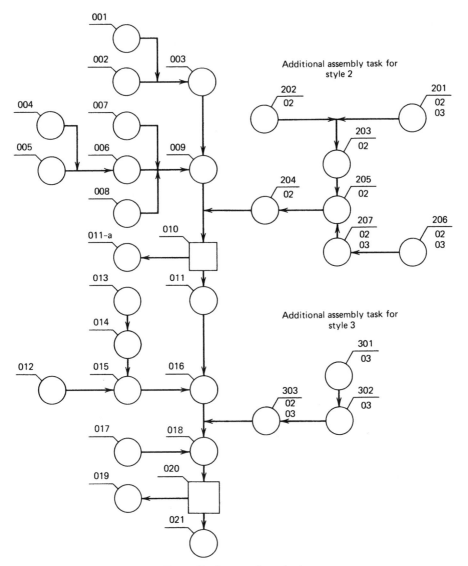

Figure 7 Process flow chart.

may depend more on careful design of the workstation so that the places the robot visits are near each other and each move is short. If time is tight, a robot simulation program should be used to be sure the operations can be done in the required time. Note that robots with longer reach are usually slower.

- *Degrees of freedom and type of drive and control.* The task or tasks at a station determine the number of degrees of freedom the robot needs. Careful attention to the product's architecture and the assembly sequence pays off here because robots with more axes cost proportionately more and move more slowly. There are a few standard configurations of robots, which are discussed in Chapter 5. Of these, the most common for assembly, applicable for assembly from above, is the SCARA type. Assembly robots are driven by electric actuators now, although hydraulics were more common 10 or 15 years ago. Hydraulics at one time were faster, but their tendency to leak made them unsuitable for many assembly environments. Both electrics and hydraulics can be servo-driven or controlled by simple fixed

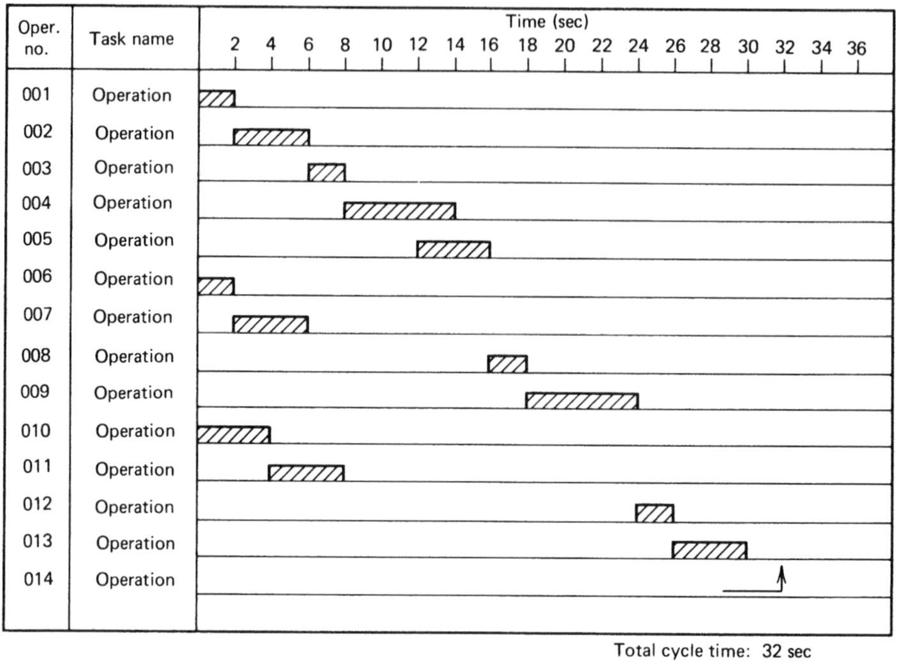

Oper. no.	Task name	Time (sec)
001	Operation	
002	Operation	
003	Operation	
004	Operation	
005	Operation	
006	Operation	
007	Operation	
008	Operation	
009	Operation	
010	Operation	
011	Operation	
012	Operation	
013	Operation	
014	Operation	

Total cycle time: 32 sec

Figure 8   Cycle time chart.

stops. The servo-driven kind is obviously better when optional motions or multiple stopping points are needed. In addition, direct control from CAD data and factory information systems is possible via servo-control.

5 CASE STUDIES

Case Study 1 (*See Figure 9*)

Product. Clutch-bearing assembly by Westinghouse
Yearly volume: 720,000 units
Manual assembly time: 38 sec
Number of product styles: 6 (3 types of bearing assemblies, 2 types of clutch assemblies)
Number of parts: 6 (two subassemblies and 4 individual parts)
Number of shifts: 1
Number of batches per month: 14
Average batch size: 4000 units
Weight of assembly: 0.55 kg
Average yield: 94%
Reason for automation: cost reduction
System description: Programmable assembly center based on overhead gantry robot configuration. The system includes a spiral storage feeder, a vibratory bowl feeder, two single and one dual parts magazines, fixtures, grippers, sensors, and control.
System operations:
Robot 1: Pick up a bearing housing from the spiral storage feeder and place it in the assembly fixture. Pick up a shaft and insert in the housing.
Robot 2: Pick up subassembly from position 1 and place in assembly fixture at position 2. Pick up dutch assembly and insert into bearing housing.

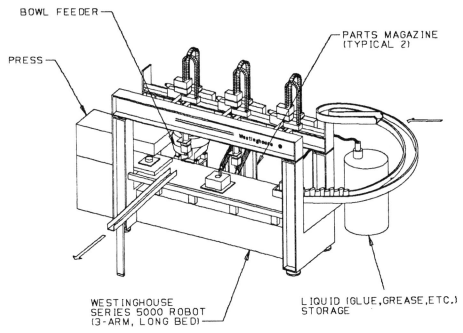

BOWL FEEDER

PARTS MAGAZINE (TYPICAL 2)

PRESS

Westinghouse

WESTINGHOUSE
SERIES 5000 ROBOT
(3-ARM, LONG BED)

LIQUID (GLUE,GREASE,ETC.)
STORAGE

Figure 9   Case Study 1.

Robot 3: A special apparatus slides two brackets in the fixture into the press. The robot picks up the subassembly from position 2 and places it in the press. The press joins all the components. The robot removes the assembly from the press and places it in the output chute.

Case Study 2: Video Tape Recorder Mechanism by Hitachi Ltd.
(See Figure 10)

The video tape recorder (VTR) mechanism is composed of pressed parts like the chassis and levers, plastic molded parts, electrical parts like motors and magnetic heads, soft or flexible parts like rubber belts and springs, and so on. In order to automate the assembly of this product, a great deal of redesign and simplification was required.

The assembly system is large and includes a series of base machines for assembly connected by conveyors (see Figure 11). Since the base machines are designed to be independently usable, a nonsynchronous assembly line of arbitrary length can be made by combining these base machines. Larger parts are placed in magazines and driven to the assembly stations by automatic guided vehicles. Some common small parts are fed from vibratory bowl feeders.

The assembly stations are composed of small-sized assembly robots, pick-and-place units, and single-purpose machines such as screwdrivers, lubricators, spring-fitting machines, and rubber belt-fitting machines. The workstation consists of a base machine, a robot, and a magazine supply unit. First a magazine is supplied and positioned at the predetermined location. Then the robot grips the parts one by one, starting from one end of the magazine, and fits them onto chassis positioned on the base machine.

6   PART MATING

Assembly in the small consists of physically mating individual parts. A large body of theory and experiment exists that can predict the likelihood of success of individual part mating events. There is insufficient space to deal with this theory in this handbook. The reader is referred to the references at the end of this chapter as well as to Chapter 64 of the 1st edition of this handbook.

A generic part mating event occurs in four stages, illustrated in Figure 12. These phases are approach, chamfer crossing, one-point contact, and two-point contact.

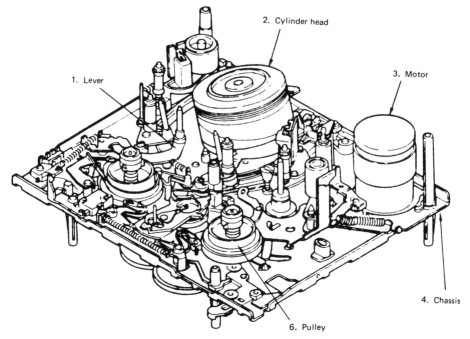

Figure 10   VTR mechanism.

Part mating theory will predict the success or failure mode of assembly of common simple geometries, such as round pegs and holes, screw threads, gears, and some simple prismatic part shapes. There are two common failure modes, both of which occur during two-point contact: *wedging* and *jamming.*

Wedging occurs when the parts are in a friction lock and there is energy stored between the contact points. There is no remedy for this condition once it occurs, except to pry the parts apart and start over. Dresser drawers often exhibit this assembly failure mode. Wedging can be avoided if the angular error between the arriving and receiving parts is kept smaller than the clearance ratio divided by the coefficient of friction. (For round parts the clearance ratio is the diametral clearance divided by the diameter.)

Jamming occurs when the applied insertion force is balanced by frictional reaction forces. In this case the remedy often is to direct the insertion force more parallel to the axis of the receiving part. Sensing of the jammed condition, followed by deliberate robot motion, is required. Jamming can also be avoided by using the remote center compliance. This device, described briefly below, permits the part to reorient itself in response to the two contact forces that arise during two-point contact so that the insertion force is directed more suitably.

7 REMOTE CENTER COMPLIANCE

The RCC is a unique device that aids the assembly of parts even when there are lateral or angular errors between them when chamfer crossing begins. The RCC supports the arriving part in such a way that it can move laterally without changing its angular orientation, or it can rotate about a point at or near its tip without translating. In this way lateral and angular errors can be absorbed independently.

Figures 13 and 14 show schematically how an RCC can be constructed from rods and hinges. Commercial RCCs are quite different but behave similarly.

8 EMERGING TRENDS AND OPEN ISSUES

Very high-speed and intricate robot assembly is possible today. End-of-arm tooling that includes some degrees of freedom of its own is often involved. Such tooling can attach springs, twist parts together, straighten out wires, and do other similarly dexterous tasks. SCARA robots with direct electric drives are capable of single task times of about two

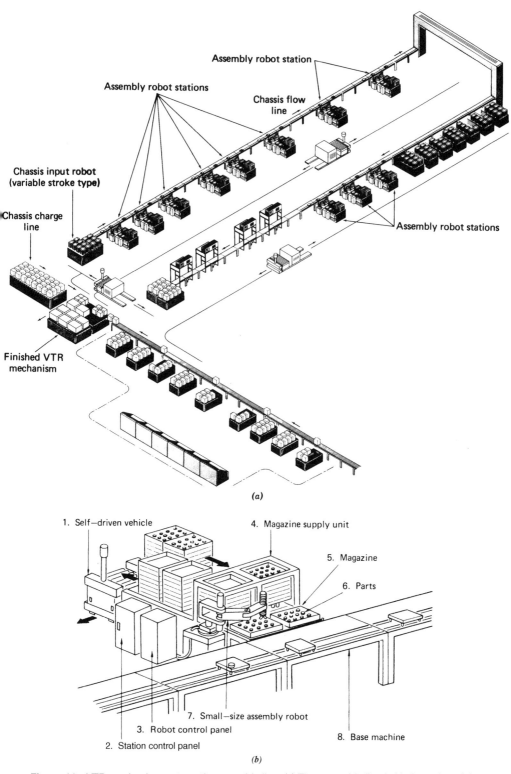

Figure 11   VTR mechanism automatic assembly line. (*a*) The assembly line is U-shaped, and the total length is 150 m. (*b*) The base machines adopt a nonsynchronous direct-feed system, which does not use pallets.

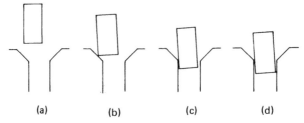

Figure 12   Four phases of assembly: (a) approach, (b) chamfer crossing, (c) one-point contact, (d) two-point contact.

seconds. Such systems usually do not use vision to achieve accuracy or dexterity but depend instead on accurate joint sensors. Vision is used to verify the presence of parts or to verify assembly actions. Examples include assembly of shutters, film transports, and other precision mechanisms inside Polaroid cameras, as well as assembly of multiple models of automobile instrument clusters, including handling of flexible printed circuits, lightbulbs, and compliant plastic parts. Current systems interact fully with factory information systems, reporting task times, type of product being assembled, failures of robot or product tests, and so on.

These systems still require most product architectures to permit assembly from one side, and parts must be placed in pallets or fed by bowl feeders to the correct position and orientation. Automatic pallet filling methods exist, making pallet feeding faster than bowl feeders for parts with complicated shapes. Robot assembly is also limited to products with small lightweight parts. The main economic barriers are still high cost, including part feeding cost, and low speed, relative to the cost of people. However, robots are preferred even when the cost is higher if the quality or yield is also higher (Martin-Vega et al., 1995; Makino and Arai, 1994; Kaplan, 1993, Van Brussel, 1990).

ACKNOWLEDGMENTS

This chapter contains material adapted from the following chapters in the *Handbook of Industrial Robots, First Edition:* "Planning Robot Applications in Assembly," by Tibor Csakvary, and "Assembly Cases in Production," by Kenichi Isoda and Michio Takahashi.

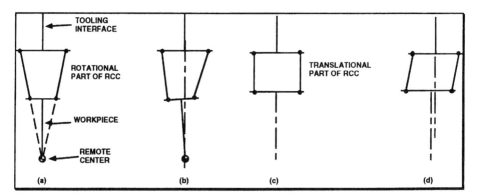

Figure 13   Elements of the RCC. (a) The rotational part of the RCC permits the workpiece to rotate about the remote center. (b) The rotational part in a rotated condition. (c) The translational part of the RCC permits the work piece to translate. (d) The translational part in a translated condition.

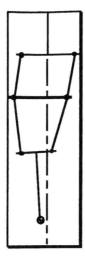

Figure 14   Complete RCC made by combining translational and rotational parts, capable of absorbing lateral and angular part mating errors.

REFERENCES

Bard, J., E. M. Dar-El, and A. Shtub. 1992. "An Analytic Framework for Sequencing Mixed Models Assembly Lines." *International Journal of Production Research* **30,** 35–48.

Boothroyd, G. 1992. *Assembly Automation and Product Design.* New York: Marcel Dekker.

Chen, J. F., and W. E. Wilhelm. 1994. "Optimizing the Allocation of Components to Kits in Small-Lot, Multiechelon Assembly Systems." *Naval Research Logistics* **41**(2), 229–256.

Feldman, K., and G. Geyer. 1991. "AI Programming Tools for Planning Assembly Systems." *Journal of Design and Manufacturing* **1**(1), 1–5.

Ghosh, S., and R. J. Gagnon. 1989. Review and Analysis of the Design, Balancing and Scheduling of Assembly Systems." *International Journal of Production Research* **27**(4), 637–670.

Ho, Y.-C., and C. L. Moodie. 1994. "A Heuristic Operation Sequence-Pattern Identification Method and Its Application in the Design of a Cellular Flexible Assembly System." *International Journal of CIM* **7**(3), 163–174.

Homem de Mello, L. S., and A. C. Sanderson. 1991*a.* "Representation of Mechanical Assembly Sequences." *IEEE Transactions on Robotics and Automation* **7**(2), 211–227.

———. 1991*b.* "A Correct and Complete Algorithm for the Generation of Mechanical Assembly Sequences." *IEEE Transactions on Robotics and Automation* **7**(2), 228–240.

Kaplan, G. 1993. "Manufacturing à la carte: Agile Assembly Lines, Faster Development Cycles." *IEEE Spectrum* **30**(9), pp. 24–26.

Khosla, P. K., and R. Mattikali. 1989. "Determining the Assembly Sequence from 3-D Model." *Journal of Mechanical Working Technology* **20,** 153–162.

Kim, S. H., and K. Lee. 1983. "An Assembly Modeling System for Dynamic and Kinematic Analysis." *Computer-Aided Design* **21**(1), 2–12.

Lee, H. F., and R. V. Johnson. 1991. "A Line Balancing Strategy for Designing Flexible Assembly Systems." *International Journal of Flexible Manufacturing Systems* **3,** 91–120.

Makino, H., and T. Arai. 1994. "New Developments in Assembly Systems." *Annals of CIRP* **43**(2), 501–512.

Martin-Vega, L. A., et al. 1995. "Industrial Perspective on Research Needs and Opportunities in Manufacturing Assembly." *Journal of Manufacturing Systems* **14**(1), 45–58.14.

Milner, J. M., S. C. Graves, and D. E. Whitney. 1994. "Using Simulated Annealing to Select Least-Cost Assembly Sequences." In *Proceedings of IEEE International Conference on Robotics and Automation,* San Diego, May. 2058–2063.

Nof, S. Y., and Z. Drezner. 1993. "The Multiple-Robot Assembly Plan Problem." *Journal of Intelligent and Robotic Systems* **5,** 57–71.

Nof, S. Y., and V. N. Rajan. 1993. "Automatic Generation of Assembly Constraints and Cooperation Task Planning." *Annals of CIRP* **42**(1), 13–16.

Nof, S. Y., W. E. Wilhelm, and H. J. Warnecke. 1997. *Industrial Assembly.* London: Chapman & Hall.

Park, J. H., and M. J. Chung. 1993. "Automatic Generation of Assembly Sequences for Multi-Robot Workcell." *Robotics and Computer-Integrated Manufacturing* **10,** 355–363.

Sellers, C. J., and S. Y. Nof. 1989. "Performance Analysis of Robotic Kitting Systems." *Robotics and Computer-Integrated Manufacturing* **6**(1), 15–23.

Van Brussel, H. 1990. "Planning and Scheduling of Assembly Systems." *Annals of CIRP* **39**(2), 637–644.

Yano, C. A., and R. Rachamadugu. 1991. "Sequencing to Minimize Work Overload in Assembly Lines with Product Options." *Management Sciences* **37**(5), 572–586.

ADDITIONAL READING

Design of Assembly Systems and Product Design for Assembly

Baumann, W., et al. "Operating and Idle Times for Cyclic Multi-Machine Serving." *Industrial Robot* **8**(1), 44–49, 1981.

Boothroyd, G., and A. H. Redford. *Mechanized Assembly.* New York: McGraw-Hill, 1968.

Boothroyd, G., P. Dewhurst, and W. Knight. *Product Design for Manufacture and Assembly.* New York: Marcel Dekker, 1994.

Graves, S. C., and B. W. Lamar. "A Mathematical Programming Procedure for Manufacturing System Design and Evaluation." In *Proceedings of the IEEE Conference on Circuits and Computers,* 1980.

Graves, S. C., and D. E. Whitney. "A Mathematical Programming Procedure for Equipment Selection and System Evaluation in Programmable Assembly." In *Proceedings of the IEEE Decision and Control Conference,* 1979.

JanJua, M. S. "Selection of a Manufacturing Process for Robots." *Industrial Robot* **9**(2), 97–101, 1982.

Lotter, B. *Manufacturing Assembly Handbook.* London: Butterworths, 1989.

Nevins, J. L., and D. E. Whitney, eds. *Concurrent Design of Products and Processes.* New York: McGraw-Hill, 1989.

O'Grady, P. and J. S. Oh. "A Review of Approaches to Design for Assembly." *Concurrent Engineering* **1**(3), 5–11, 1991.

Oh, J. S., P. O'Grady, and R. E. Young. "Constraint Network Approach to Design for Assembly." *IIE Transactions* **27**, 72–80, 1995.

Quinn, R. D., et al. "An Agile Manufacturing Workcell Design." *IIE Transactions* **29**(10), 901–909, 1997.

Warnecke, H. J., et al. "Assembly." In *Handbook of Industrial Engineering.* 2nd ed. Ed. G. Salvendy. New York: John Wiley & Sons, 1992. Chapter 19.

Wilhelm, W. E., and S. Ahmadi-Marandi. "A Methodology to Describe Operating Characteristics of Assembly Systems." *IIE Transactions* **14**(3), 204–213, 1982.

Wilhelm, W. E., M. V. Kalkunte, and C. Cash. "A Modeling Approach to Aid in Designing Robotized Manufacturing Cells." *Journal of Robotic Systems* **4**(1), 25–48, 1988.

Wilhelm, W. E., P. Som, and B. Carroll. "A Model for Implementing a Paradigm of Time-Managed, Material Flow Control in Certain Assembly Systems." *International Journal of Production Research* **30**(9), 2063–2086, 1992.

Assembly Sequence Analysis; Assembly Planning

Baldwin, D. F., et al. "An Integrated Computer Aid for Generating and Evaluating Assembly Sequences for Mechanical Products." *IEEE Journal of Robotics and Automation* **7**(1), 78–94, 1991.

DeFazio, T. L., and D. E. Whitney. "Simplified Generation of All Mechanical Assembly Sequences." *IEEE Journal of Robotics and Automation* **3**(6), 640–658, 1987.

Homem de Mello, L. S., and S. Lee, eds. *Computer-Aided Mechanical Assembly Planning.* Boston: KIuwer Academic, 1991. 263–288.

Rajan, V. N., and S. Y. Nof. "Minimal Precedence Constraints for Integrated Assembly and Execution Planning." *IEEE Transactions on Robotics and Automation* **12**(2), 175–186, 1996.

Wolter, J. D. "A Combination Analysis of Enumerative Data Structures for Assembly Planning." *Journal of Design and Manufacturing* **2**(2), 93–104, 1992.

Zussman, E., M. Shoham, and E. Lenz. "A Kinematic Approach to Automatic Assembly Planning." *Manufacturing Review* **5**(4), 293–304, 1992.

Part Mating Theory

Andreev, G. Y. "Assembling Cylindrical Press Fit Joints." *Russian Engineering Journal* **52**(7), 54, 1972.

Andreev, G. Y., and N. M. Laktionev. "Problems in Assembly of Large Parts." *Russian Engineering Journal* **46**(1), 60, 1966.

———. "Contact Stresses During Automatic Assembly." *Russian Engineering Journal* **49**(11), 57, 1969.

Changman, J., and G. Vachtsevanos. "A Fuzzy Intelligent Organizer for Control of Robotic Assembly Operations." In *Proceedings of 32nd IEEE Conference on Decision and Control,* 1993, Vol. 2. 1765–1765.

Drake, S. H. "Using Compliance in Lieu of Sensory Feedback for Automatic Assembly." Ph.D. Thesis, Massachusetts Institute of Technology, Mechanical Engineering Department, 1977.

———. "The Use of Compliance in a Robot Assembly System." Paper presented at the IFAC Symposium on Information and Control Problems in Manufacturing Technology, Tokyo, October 1977.

Gusev, A. S. "Automatic Assembly of Cylindrically Shaped Parts." *Russian Engineering Journal* **49**(11), 53, 1969.

Karelin, M. M., and A. M. Girel. "Accurate Alignment of Parts for Automatic Assembly." *Russian Engineering Journal* **47**(9), 73, 1967.

Laktionev, N. M., and G. Y. Andreev. "Automatic Assembly of Parts." *Russian Engineering Journal* **46**(8), 40, 1966.

McCallion, H., and P. C. Wong. "Some Thoughts on the Automatic Assembly of a Peg and a Hole." *Industrial Robot* **2**(4), 141–146, 1975.

Park, Y.-K., and H. S. Cho. "A Self-Learning Rule-Based Control Algorithm for Chamferless Part Mating." *Control Engineering Practice* **2**(5), 773–783, 1994.

Savischenko, V. M., and V. O. Bespalov. "Orientation or Components for Automatic Assembly." *Russian Engineering Journal* **45**(5), 50, 1965.

Sturges, R. H., and S. Laowattana. "Polygonal Peg Insertion with Orthogonal Compliance. In *Proceedings of 1994 Japan–U.S.A. Symposium on Flexible Automation—A Pacific Rim Conference,* Kobe, Japan, July 11–18. Vol. 2. 609–616.

———. "Passive Assembly of Non-axisymmetric Rigid Parts." In *Proceedings of the IEEE International Conference on Intelligent Robots and Systems,* 1994, Vol. 2. 1218–1225.

Whitney, D. E. "Quasi-Static Assembly of Compliantly Supported Rigid Parts." *ASME Journal of Dynamic Systems, Measurement and Control* **104,** 65–77, 1982.

———. "Part Mating in Assembly." In *Handbook of Industrial Robotics.* 1st ed. Ed. S. Y. Nof. New York: John Wiley & Sons, 1985. 1084–1116.

Remote Center Compliance

Ang, M. H. Jr., et al. "Passive Compliance from Robot Limbs and Its Usefulness in Robotic Automation." *Journal of Intelligent and Robotic Systems, Theory and Applications* **20**(1), 1–21, 1997.

De Fazio, T. L. "Displacement-State Monitoring for the Remote Center Compliance (RCC)— Realizations and Applications." In *Proceedings, 10th International Symposium on Industrial Robots,* Milan, March 1980.

Drake, S. H., and S. N. Simunovic "Compliant Assembly Device." U.S. Patent No. 4,155,169, May 22, 1979.

Leu, M. C., and Y. L. Jia. "Mating of Rigid Parts by a Manipulator with Its Own Compliance." *Journal of Engineering for Inustry* **117**(2), 240–247, 1995.

Nakao, M., et al. "A Micro-remote Centered Compliance Suspension for Contact Recording Head." *Journal of Applied Physics* **79**(8), 5797–5815, 1996.

Seltzer, D. S. "Use of Sensory Information for Improved Robot Learning." SME Paper MS79-799. Presented at Autofact, Detroit, November 1979.

Simunovic, S. "Force Information in Assembly Processes." Paper presented at 5th International Symposium on Industrial Robots, Chicago, September 1975. Proceedings published by Society of Manufacturing Engineers.

———. "An Information Approach to Part Mating." Ph.D. Thesis, Massachusetts Institute of Technology, 1979.

Sturges, R. H., and S. Laowattana. "Fine Motion Planning Through Constraint Network Analysis." In *Proceedings of IEEE International Symposium on Assembly and Task Planning,* 1995. 160–170.

Sturges, R. H., and K. Sathirakul. "Modeling Multiple Peg-in-Hole Insertion Tasks." In *Procedings of the Japan–USA Symposium on Flexible Automation,* 1996. Vol. 2. 819–821.

Watson, P. C. "A Multidimensional System Analysis or the Assembly Process as Performed by a Manipulator." Paper presented at 1st North American Robot Conference, Chicago, October 1976.

Watson, P. C. "Remote Center Compliance System." U.S. Patent No. 4,098,001, July 4, 1978.

Watson, P. C., and S. H. Drake. "Pedestal and Wrist Force Sensors for Automatic Assembly." CSDL Report No. P-176, June 1975, presented at 5th International Symposium on Industrial Robots, Chicago, September 1975.

———. "Methods and Apparatus for Six Degree or Freedom Force Sensing." U.S. Patent No. 4,094,192, June 13, 1978.

Whitney, D. E. "The Mathematics of Coordinated Control of Prosthetic Arms and Manipulators." *ASME Journal of Dynamic Systems, Measurement and Control* (December) 303–309, 1972.

———. "Force Feedback Control or Manipulator Fine Motions." *ASME Journal of Dynamic Systems, Measurement and Control* (June), 91–97, 1977.

Whitney, D. E., and J. L. Nevins. "Servo-Controlled Mobility Device." U.S. Patent No. 4,156,835, May 29, 1979.

Whitney, D. E., J. L. Nevins, and CSDL Staff. "What Is the Remote Center Compliance and What Can It Do?" CSDL Report P-728, November 1978. Presented at 9th International Symposium on Industrial Robots, Washington, D.C., 1979.

CHAPTER 51

ASSEMBLY: ELECTRONICS

K. Feldmann
S. Krimi
M. Reichenberger
K. Götz
University of Erlangen-Nuremberg
Erlangen, Germany

1 DEVELOPMENT OF COMPONENT PLACEMENT

The development of component placement depends on the electronics products that are built. Mobile telecommunication products and automotive electronics are regarded as the most important shares of the market. The market current shows an annual growth rate of about 50%. The requirements for those products and their assembly are:

- Integration of mechanics and electronics
- Miniaturization
- Function integration
- Reliability
- Design freedom
- Improved environmental compatibility
- Cost reduction

The main challenges are shown in Figure 1. These requirements influence component packaging and substrate technology, as discussed below. The properties of components and substrates themselves influence electronics assembly.

Miniaturization and reliability also influence quality assurance. In particular, high placement accuracy, with precision better than 20 μm with 6 σ, is a challenge in electronics assembly in the near future. This objective leads to new placement and soldering strategies, some of which are discussed below.

1.1 Trends in Component Packaging

Due to the requirements of electronics products, challenges in component packaging also include miniaturization and function integration. These requirements lead to new designs such as BGA, μBGA or μFC (Figure 2), thinner packaging such as TSOP or TQFP, and higher pin counts with smaller pitch.

As industrial roadmaps show, a pitch of less than 200 μm is expected to become a future standard. It is also expected that in the future the share of bare dies will grow significantly. The same roadmaps also predict the increasing share of BGA, μBGA, FC, and μFC.

In the future, assembly will have to handle at least three different types of components. This influences the specifications of production strategies, machines and shop floors.

While today most production systems use a convective soldering machine as the main reflow system, in the future, vapor phase for mass production and laser soldering or light soldering for selective soldering will also be used in one production line.

For that purpose and for higher placement accuracy, new machine concepts have been developed. In the ALERT project a placement machine was built that integrated a placement process and a soldering process. Intelligent scheduling of components allows rela-

Handbook of Industrial Robotics, Second Edition, Edited by Shimon Y. Nof
ISBN 0-471-17783-0 © 1999 John Wiley & Sons, Inc.

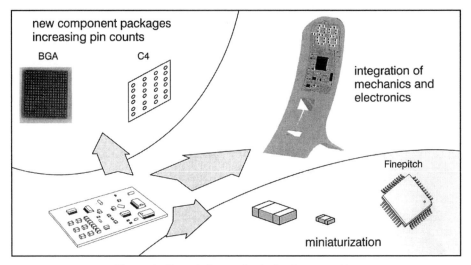

Figure 1   New challenges in electronics assembly.

tively high component density. It also allows components to be placed and soldered selectively by laser beam, although other components had already been placed and soldered by vapor phase or convective soldering. The main concepts in this system are listed in Figure 3 (Feldmann, Götz, and Sturm, 1994). While this machine system was developed for two-dimensional PCBs, in Sections 3.4.2 and 3.4.3 newer machine concepts for laser and light soldering with three-dimensional devices are discussed.

1.2   Trends in Substrate Technology and Function Integration

Some of the demands mentioned above are met by integration of mechanical and electrical functions in three-dimensional, molded circuit carriers: molded interconnect devices (MIDs), which are molded plastic parts with a partially metal-plated surface forming an electric circuit pattern (Figure 4).

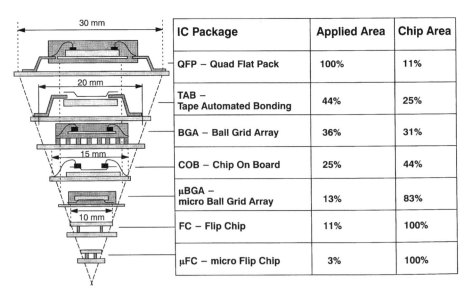

IC Package	Applied Area	Chip Area
QFP – Quad Flat Pack	100%	11%
TAB – Tape Automated Bonding	44%	25%
BGA – Ball Grid Array	36%	31%
COB – Chip On Board	25%	44%
μBGA – micro Ball Grid Array	13%	83%
FC – Flip Chip	11%	100%
μFC – micro Flip Chip	3%	100%

Figure 2   Trends in component packaging: Comparison of the areas applied by chip versus the areas applied by package (Philips EMT, Eindhoven, 1996).

ALERT: Advanced Laser Reflow Technology for S.M.T.

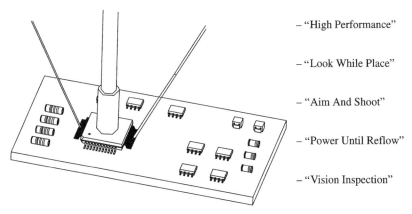

– "High Performance"

– "Look While Place"

– "Aim And Shoot"

– "Power Until Reflow"

– "Vision Inspection"

Figure 3 System concept for the laser mounting machine.

MIDs unite the functions of, for example, conventional circuit boards, casings, connectors, and cables. Some are also used for electronic components such as LEDs. MIDs however, do not normally replace conventional assemblies, but complement them. In Figure 6 integration potentials and markets for MIDs are shown. The most important among them are automotive electronics and telecommunication, especially mobile phones.

The benefits of MID can be stated in three categories: technology, economy, and ecology (Figure 5).

The main technological advantage of MIDs is that they integrate electronic and mechanical elements into circuit carriers having virtually any geometric shape. They have enabled entirely new functions and usually help in miniaturizing electronic products.

Through integration of mechanical components, substantial economic advantages can be achieved. Process chains are shortened, especially in assembly, and reliability is increased by the reduction of connections, thus lowering quality cost. Miniaturization can lead to reduced consumption of raw materials, and integration slims down inventories of components.

The thermoplastics normally used as substrate materials are inherently flame-retardant and can easily be recycled after plastic–metal separation. In addition, the number of different materials used in a product can usually be reduced.

The most important functions that can be integrated into an MID are classical electrical and mechanical functions, but new possibilities are also emerging. One of the most in-

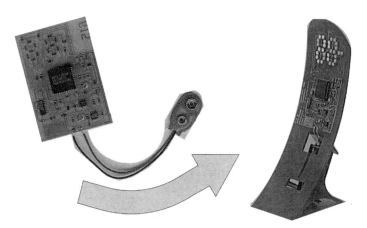

Figure 4 Transition from conventional printed circuit board to molded interconnect device.

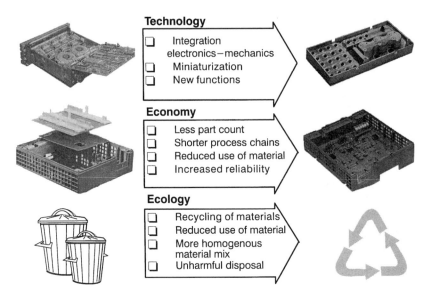

Figure 5 Main advantages of MIDs.

teresting possibilities is the integration of sockets, connectors, and other contact elements. These components are usually very expensive. Another interesting feature is integrated shielding areas.

Most applications realized with this innovative technology are devices without electronic parts. They are used as highly functional interconnections, mainly in automobile technology or telecommunications. But during the past few years the number of applications realized with the MID technology has grown rapidly. At the moment, several MIDs with electronic components have been introduced into the market. One of these applications is a part of an automotive ABS system ("MFD auf Basis von Polyamid," 1997).

Three-dimensional MIDs set special demands on electronics assembly. But although 3D MID is a growing market, other substrate materials, such as the well-known FR4 and ceramics, will share the market. For electronics assembly this means that not only a set of different component packages has to be handled but also a set of different substrates.

2 AVAILABLE PLACEMENT EQUIPMENT IN ELECTRONIC PRODUCTION

Printed wiring assembly is a production discipline characterized by high productivity, high quality, and fast innovation cycles. These characteristics result in high demands on

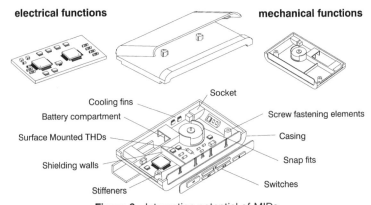

Figure 6 Integration potential of MIDs.

all the systems used during the production process in terms of relative accuracy, speed, and flexibility.

The introduction of surface mount technology favors automated assembly of components. In this case the components are placed flat on the surface of the printed circuit board (see Figure 7). Then guided tools for the placement of through-hole devices are dropped for bending and cutting the lead to size. Furthermore, the number of defective mounted components decreases. In the assembly of THTs (through hole devices) the pin can bend, making the whole component unusable. In addition, a considerably higher scale of integration is achieved by surface mount technology. Since the number of required tools (other than guided tools) is smaller, hence, less space is required for placement.

2.1   Placement Methods

Manual placement systems allow the hand-guided assembly of SMDs (surface mounted devices). They normally consist of a carousel or table with component feeders, a holder for the printed circuit board, and a hand-guided placement tool with a vacuum nozzle. To enhance placement accuracy, some systems are equipped with optical aids. Common application areas are prototypes, repair jobs, and very small series.

In automatic sequential placement (pick and place), a program-controlled machine individually removes the components from the feeders and individually places them on the printed circuit board (Figure 8).

The goal of automatic sequential/simultaneous placement is to achieve a high placement rate per cycle. Therefore specific placement heads must be developed that can pick a number of components simultaneously from the feeders and then place them simultaneously on the printed circuit board.

Simultaneous population results in a high placement rate of some 100,000 components per hour and therefore lower placement costs. However, this technique has several decisive disadvantages. Placement quality compared to a sequential insertion is considerably lower. Therefore, pure simultaneous placers (inserters) are used less frequently nowadays.

SMDs have made simultaneous placement technically possible. However, simultaneous placement is sensible only with high throughput and smaller variety of types in electronics manufacturing.

2.2   Kinematic Assembly Principles and Design Variants

In accordance with diverse assembly orders, a series of different machine types were developed. The kinematics of the different machine elements is an important characteristic.

Figure 9 depicts a broad division of the different assembly principles for automatic placement machines. However, an exact division of designs is not possible because of the diversity of placement systems.

Each of the different variants has advantages and disadvantages. For instance, a moving printed circuit board has the advantage that only a single-axis positioning system is required, but also the danger that components already placed can slip because of acceleration forces on the printed circuit board during movement.

2.3   Machine Components for Improvement in Assembly Quality

As in the case of machine tools and industrial robots, the market is imposing ever greater demands on the reliability of SMD placement systems. This can only be achieved through appropriate design measures, however, since high acceleration results in twisting at the machine tool table, drive systems, and placement head, placement accuracy is reduced.

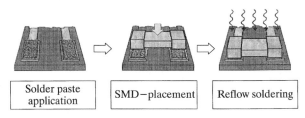

Solder paste application   SMD−placement   Reflow soldering

Figure 7   SMD production process.

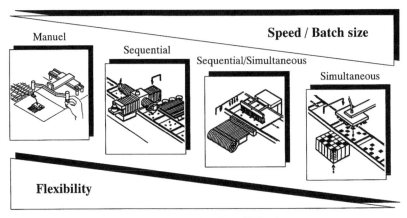

Figure 8 Placement methods of various SMD placement systems.

Figure 10 shows the machine components that are inserted for improving process ability and for component examination. The CRDL measuring instrument offers the possibility of testing the electric values of bipolar components during the placement cycles. The ends of the component are connected with two ties integrated at the placement head, and the permissible values of the component are checked (resistance, capacity, inductance, and polarity of diodes). If a faulty component is recognized, it is sorted into a drop tank.

The coplanarity of multicontact ICs is tested by means of specific remote sensing. This testing prevents a component from being wrongly soldered in spite of correct population when one or several pins protrude from the placement tier.

Optical centering of the components is used first to register the position of the component relative to the target position on the pipette, and relative to the shift and torsion. For measurement of the printed circuit board position in the working area of the placement system, a CCD camera mounted on the placement head sequentially records the position of the two or three fiducials on the PCB. Additional local fiducials are provided in the immediate vicinity of the placement positions in order to further increase the placement accuracy when fine pitch components are involved. This step provides the information needed to recognize and compensate for tolerances and nonlinear warping of the PCB during the placement process. The corrected placement position can be computed with the aid of the machine controller.

2.4 New Application for Assembly Systems in Electronic Assembly by 3D Molded Interconnect Devices

The new task of mounting SMD onto 3D circuit boards calls for new capabilities in the dispensing and mounting systems. Both processes work sequentially and are realized with

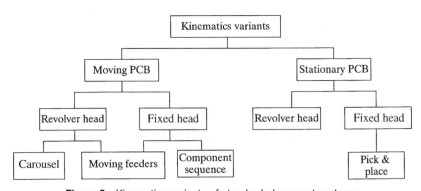

Figure 9 Kinematics variants of standard placement systems.

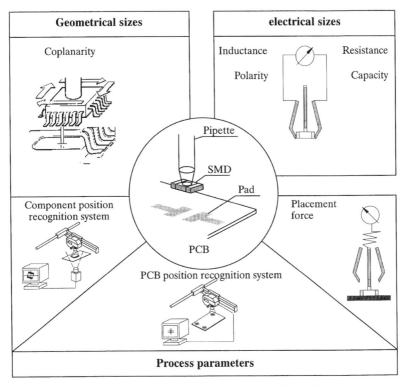

Figure 10   Machine components for assembly safety.

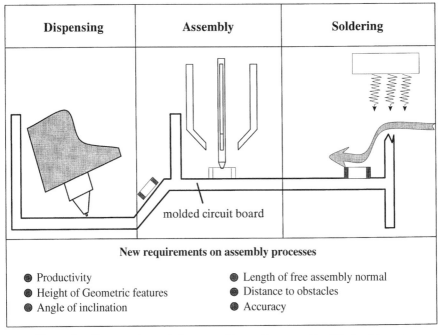

Figure 11   New requirements for assembly quality caused by 3D circuit board.

Cartesian handling systems for PCB production (Figure 11). SMD assembly into MID can be done with six-axis robots. Such robots are conceived as geometry-flexible handling systems for reaching each point in the working space with highly flexible orientation of the tool.

The available handling systems for SMD assembly into PCB are optimized for accuracy and speed. One alternative for working with these systems on MID is to move the workpiece during the process.

2.4.1 Six-Axis Robot Systems for SMD Assembly into MID

Robot placement systems are currently being used primarily to place so-called exotic THT (through hole technology) components such as coils and plugs. The reason is that their assembly is not possible with pick-and-place machines.

The robots can be equipped with pipettes that are built small enough to allow SMDs to be assembled very close to obstacles. If the length of the free assembly normal of an MID is limited, bent pipettes can be used without the help of inclined feeders or a centering station.

Available tool-changing systems enable the integration of additional processes such as the application of solder paste and reflow soldering into one assembly cell. A critical point in surface-mount technology is the necessary accuracy in centering the SMD to the pipette and placing components. This is realized with the use of two cameras, one mounted on joint five of a manipulator pointing down for fiducial registration, and the other mounted under the stage looking up for component registration.

Figure 12 shows the robot placement cell layout developed at the Institute for Manufacturing Automation and Production Systems at Friedrich-Alexander University Erlangen-Nuremberg. The structure of the robot cell is shown in Figure 13.

The robot cell layout was designed to minimize the total distance traveled by the robot to assemble with MID. Additionally, the feeder system is movable. As a result, the components can be placed at a distance-optimized position for the robot. With this feeder moving system a placement rate of 1800 components/hour can be achieved (up to 1200 components/hour with a stationary feeder).

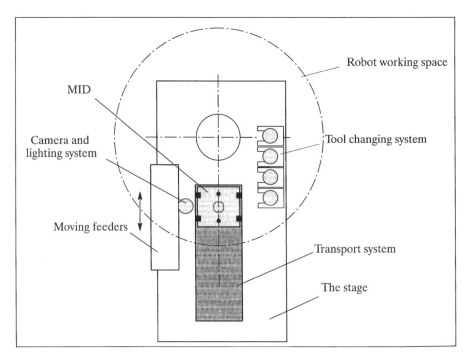

Figure 12 Layout of the robot placement system.

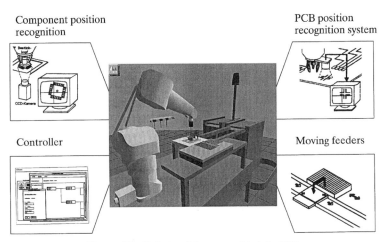

Figure 13 Robot cell for assembly into MID.

The geometric advantage and flexibility of the robot system come at the expense of lower productivity and accuracy compared to Cartesian systems. Since the applications for six-axis robots are widespread, the accuracy of available systems varies widely. Most precision robots fail in assembling fine pitch SMDs, but there are also robots with an accuracy of a few microns. The disadvantage of these systems, apart from the high price, is the volume of working space, which is too small for adding enough feeders for different SMDs. In regard to this problem, it must be noted that the use of bent pipettes requires complex movement. This constraint applies especially to SMDs, which should be placed in another orientation than allocated to their feeder. This means that the productivity decreases rapidly when bent pipettes are used.

Higher productivity for assembling into MIDs with a small length of free assembly normal can be achieved by using pipettes with an additional axis. This tool works like a straight pipette while getting the SMD from the feeder, and is bent while the robot is moving it to the requested x, y, and z distances. This integrated movement can be realized faster than the complex movement described above.

Since the use of tools with additional axes is recommended for productivity reasons and the placing accuracy of available six-axis robots is lower than that of Cartesian systems, the basis for a specialized MID assembly system should be Cartesian.

Commercially available standard robots offer a high flexibility because of their free programmability with regard to the part spectrum, but their placement accuracy is not sufficient for compliance with the small joining tolerances during the assembly of miniaturized products. There are some systems for optical laser measurements in modern, commercially available six-axis robots of load class 5–10 kg. Their orbital deviations total up to 0.6 mm amplitudes with a positioning inaccuracy of approximately 40 μm.

A new approach developed in a cooperative project between Munich and Erlangen universities should allow highly accurate positioning through the use of standard industrial robots. The approach of microsystem engineering offers new perspectives. This technology consists of a combination of technical and microtechnical components (hybrid systems) for fine work. One such system is the micropositioning module. (See also Chapter 55, Microassembly.)

The concept of robot accurate positioning provides a division of the working space in coarse (robot) and accurate (micropositioning module) domains. The micropositioning module is mounted at the flange between robot arm and gripper. During the transportation and assembly movements of the robot, precise manipulation results from the micropositioning system in the direct local area of the assembly space. Its total functionality includes the sensory recording of position deviations between the joining elements and compensation by generation of a contrary correction movement. The object is a three-dimensional tolerance compensation (robot inaccuracies, component and feeder tolerances) at degrees of freedom X, Y, and Z.

Figure 14 shows the accurate positioning module. Compensation movement is achieved with a miniaturized actuator system. The inertia of this system is very small compared to that of the robot. Therefore, it allows dynamic compensation of structure oscillations so that orbital deviations can already be reduced during the approach of the robot to the joining position.

2.4.2   MID Placement System

The next step was to design possibilities for extending the identified limits of conventional SMD assembly systems. The shortcomings of Cartesian systems lie in the possible height and angle of inclination of the MID and the fact that an unlimited length of free assembly normal is required. Therefore a concept was developed to extend these limits. A module for handling the MID in the working space of a SMD assembly system, as well as the use of a pipette with additional axes, are useful for the 3D assembly of components into MID. For the realization of both concepts it is necessary to develop hardware and control software of the PCB basic assembly system. Both have been realized with a widespread standard system at the FAPS institute. Figure 15 depicts the kinematics simulation of the 3D assembly system with the main modules for handling components and MID.

The system kinematics of standard placement systems can be extended by moving the workpiece during the assembly process. There are two ways to do this (Figure 16).

First, an additional movement in the direction of the z-axis extends the obstacle height that can be surmounted (six-axis pipette). Second, the MID can be inclined so that the process plane is oriented horizontally in the working space (MID handling system). The advantage is that the high accuracy and speed of a Cartesian system can be exploited. The tolerance chain is divided into two parts: the *SMD handling system* and the *MID moving system*. It is thus possible to compensate the possibly lower position accuracy of the MID moving system by correction of the handling system. For this purpose, the use

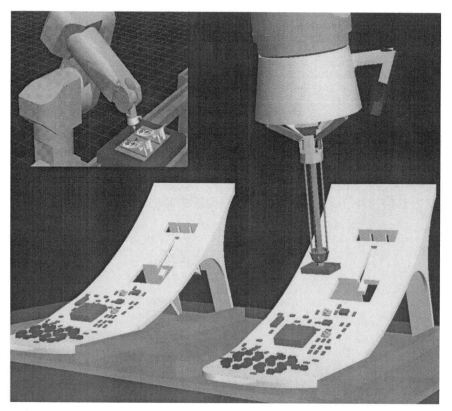

Figure 14   Improvement of placement accuracy with the accurate positioning module.

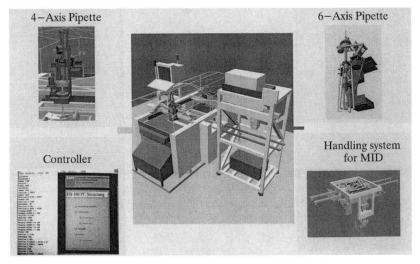

Figure 15   Modules of a kinematic simulation of the MID placement system.

of vision system to detect the position of the inclined MID is necessary. Here the already available position recognition systems for PCB production systems can be applied.

Inclination of the MID means that the process plane, which should be oriented horizontally, may be on another level than the usual. Its height must therefore be compensated by an additional z-axis. This means that both possibilities for extending with Cartesian assembly systems could be integrated into one lift and incline module. With this module integrated into the Cartesian SMD assembly system, it is possible to attach SMD components onto MID circuit planes with an angle of inclination up to 50°. The MID is fixed onto a carrier system that can be transported on the double belt conveyer. At the working position the carrier is locked automatically with the module, the connection to the conveyer is released, and the MID can be inclined and lifted into the required position. It is thus possible to extend the capability of the common Cartesian assembly system nearly without reduction of productivity.

3 NEW SOLDERING SYSTEMS USING INDUSTRIAL ROBOTS

3.1 Introduction

Soldering is a process in which two metals, each having a relatively high melting point, are joined together by means of a third metal or alloy having a low melting point. The molten solder material undergoes a chemical reaction with both base materials during the soldering process. To accomplish a proper solder connection a certain temperature has to be achieved.

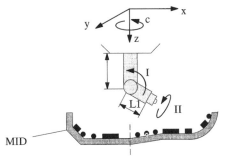

Figure 16   Kinematics of the six-axis pipette for SMD assembly into MIDs.

In electronic applications the main elements in the soldering process are PCBs as well as electronic components such as ICs, resistors, and capacitors. In most cases eutectic or near-eutectic alloys consisting of tin and lead with melting points around 180°C are used. Predominant in electronics manufacturing are the compositions 62 Sn:36 Pb:2 Ag, 63 Sn: 37 Pb, and 60 Sn:40 Pb. Other alloys with relatively high or low melting points are used only for special soldering situations; for example, in soldering of heat-sensitive components or sequential soldering. In recent years non-lead soldering materials were developed for health reasons.

Most solder connections manufactured in electronics are made with conventional *mass soldering* systems, in which many components are soldered simultaneously onto a printed circuit board. Two different mass soldering methods are used in electronics production. *Wave* or *flow* soldering is based on the principle of simultaneous supply of solder and soldering heat in one operation. Therefore the components on a PCB are moved through a wave of melted solder. In contrast, during the *reflow* soldering process solder preforms, solid solder deposits, or solder paste attached to the pads in a first operation are melted in a second step by transferred energy. Possible methods for heat transfer are by the use of convection, radiation, or condensation.

Despite the optimization of these mass soldering methods, an increasing number of solder joints are incompatible with conventional soldering methods and microsoldering or selective soldering methods have to be used. The main reasons for this development are (Figure 17):

- Rework systems for surface mount assemblies use selective soldering to heat the rework site.
- Mass soldering is uneconomical for applications with only a small number of soldering joints, such as smart cards.
- Heat-sensitive components such as sensors, fans, and connectors do not allow the heating of the whole device in a reflow furnace. Thus selective heating is necessary.
- Even though surface-mounted components are predominant in current applications, most electronic assemblies still require one or more through-hole mounted components. With the decreasing amount of THDs the conventional wave soldering process becomes uneconomical. In this case selective soldering can be an attractive alternative.
- Another possible application for selective soldering is the field of MID technology. Some of the thermoplastic materials used for MID are heat-sensitive and do not allow automatic soldering processes because of high temperature loads. Local heating of the solder joints reduces the thermal stress for the material.

Usually selective soldering is still done manually with a soldering iron, but many companies, for technical and economical reasons, are interested in automating the sol-

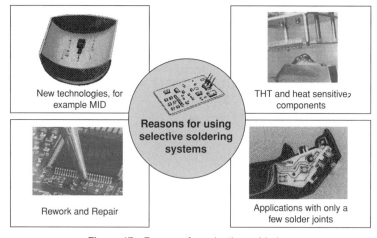

Figure 17   Reasons for selective soldering.

dering process. The drawbacks and limitations of the manual soldering process can be summarized as follows:

- Hand soldering is a low-volume process and soon becomes uneconomical with increasing production volume.
- Due to its operators' dependency, the hand soldering process is very expensive and uneconomical, especially in Western Europe, the United States, and Japan.
- For the same reason, the hand soldering process is a low-quality process without any reproducible process parameters. Because of the demands for high quality and precision, hand soldering is increasingly being eliminated in electronics manufacturing.
- Manual soldering often endangers workers' health due to the existence of solder smoke. Additionally, the hand soldering process is fatiguing and unpleasant for the operator.

3.2   Methods for Selective Soldering

For the last few years there has been a steady flow of requests for selective soldering. Several systems have been designed and introduced into the market. Depending on the kind of heat transfer, the following systems can be distinguished (Figure 18).

Among the microsoldering methods the *soldering iron* is the most frequently used. Although some soldering iron processes have been automated with robots, this process has various disadvantages. The main drawback of the iron is mechanical stressing of the joining components during soldering. On the other hand, contact between iron and component is a basic requirement for heat transfer into the solder joint when using a soldering iron. Furthermore, the process requires the skill of experienced operators for adjustment and extensive maintenance, for instance to clean or change the soldering tips.

Micro wave soldering (sometimes called *fountain soldering, micropoint soldering,* or *selective wave soldering*) is a selective process using liquid solder. The process is similar to the conventional wave soldering process, with the difference that only a few joints on a PCB are soldered simultaneously.

Selective heating of solder joints with hot air is a low-cost process, but is too slow for the poor heat convergence of air during the manufacturing process. For this reason hot air soldering is used mainly for repair.

The use of the *microflame* as a heat source for electronics manufacturing is restricted to only a few soldering tasks due to high temperatures of about 3000°C, which might damage adjacent components or the base material.

Another noncontact reflow soldering method is the use of *radiation* energy emitted by heat sources like lasers or special lamps. The focused light is directed onto the target and within a few seconds or less the connection is soldered. For satisfactory results the emitted wavelength should be in a range where it is well absorbed by the solder material.

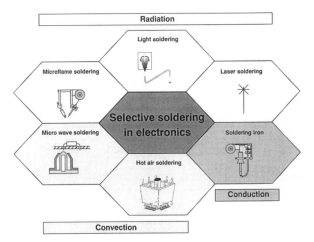

Figure 18   Methods of selective soldering.

3.3 Robotic Soldering

A complete robotic soldering system includes the robot, its control, the soldering equipment, and a plant materials-handling equipment that delivers electronic devices into and out of the workstation on time. Depending on the selective soldering system used, the robot has different functions:

- When working with a selective reflow system such as a soldering iron or laser, the robot has to handle the soldering tool. During soldering the heat source is moved over the PCB. In this case the robot is carrying out a productive operation.
- In selective wave soldering, in most cases the robot has to handle the PCB or the thermoplastic interconnection device and move it over the soldering wave.

The robot control is also used to program the soldering conditions, including wire-feed speed when using solder wire, soldering energy, and soldering time. Programming of position parameters can be made either by teach-in or CAD data of the PCB and can be supported by vision systems.

3.4 Special Soldering Situations Using Industrial Robots

Besides the well-known soldering iron, many alternative selective soldering systems are available on the market. Due to the reasons mentioned above, most of these systems are automated. The following solutions are discussed below:

- Micro wave soldering
- Light soldering using xenon or halogen lamps
- Laser soldering

3.4.1 Micro Wave Soldering

Micro wave soldering is a selective soldering process characterized by partial heating and solder supply. During soldering the solder plates on the PCB or the MID have to be moved over the small solder wave by a handling system. Melted solder is pumped from a heated temperature-controlled depot to the upper end of a nozzle into the soldering head. An adaptation to a wide range of soldering tasks can be achieved by the use of special nozzles and multinozzle heads. As a consequence, the productivity of micro wave soldering systems can be ensured. Depending on the special soldering tasks the handling of the PCBs or the MIDs can be done by different systems.

Applications with a low complexity, such as soldering of planar boards, in most cases need only a three- or four-axis Cartesian system, whereas soldering onto a three-dimensional interconnect device needs a six-axis robot due to the complicated geometric shape of the object (Figure 19). Selective wave soldering methods deliver high process capability with relatively short cycle times compared with other selective methods. On the other hand, good access to the soldering joints from below has to be ensured.

Figure 19   Micro wave soldering of a 3D MID using a six-axis industrial robot.

3.4.2 Light Soldering with Xenon or Halogen Lamps

The use of light energy in form of a focused xenon or halogen lamp is an example of selective noncontact reflow soldering (Figure 20). The emitted light has wavelengths of, depending on the heat source, between 300 and 1500 nanometers and covers the range of visible light. In most applications the light is bundled into an optical fiber, with a lens unit directed at the solder joint. The lens or the light soldering system itself can be integrated into a handling system or robot. To satisfy a wide range of soldering tasks, light soldering systems have focal diameters ranging from 1 to 5 mm. Main advantages over other soldering methods are the fine microstructure of the joints and the reduced creation of intermetallic phases due to very short soldering times of only a few seconds. In comparison to laser soldering systems the emitted light has a low energy density and therefore is less likely to damage the substrate material if the beam does not hit the target precisely. For this reason handling systems for light soldering do not need very high position accuracy or supporting systems such as vision systems. In most applications Cartesian systems are used for light soldering.

3.4.3 Laser Soldering

Soldering using laser beams is another noncontact heat source for selective soldering. Current systems use the radiation of Nd:YAG lasers for soldering. As in light soldering, the laser beam often is guided via optical fiber and lens. This heat source creates a powerful focused and coherent light of a single wavelength with high energy density and very small focus diameters. Therefore cycle times below one second are possible and a fine microstructure of the joints is achieved. On the other hand, Nd:YAG soldering systems have the disadvantage of producing light wavelengths that are mainly reflected by metal materials such as solder paste or metallizations of components or circuit boards. Due to the small focus diameter and the very high energy density, high position accuracy of the handling system is necessary. Therefore three- or four-axis high-precision Cartesian robots or six-axis robots and vision systems are necessary.

Recently laser diodes have been introduced that generate enough power to allow their utilization for soldering. They produce light frequencies (about 800 nanometers) that are better absorbed by metallic materials. In addition they are cheaper than existing Nd:YAG lasers and, as another advantage, the diodes can be mounted directly into the handling system without any optical fibers, so in the near future laser diodes with higher energy density may be a suitable solution for special soldering situations such as soldering on thermoplastic materials or other selective soldering tasks.

In order to improve the flexibility and performance of laser soldering, a combined mounting and laser soldering head has been developed at the University of Erlangen (Geiger and Feldmann, 1998). The head can be fixed to a high-precision robotic handling system. The components are mounted onto the PCB by a vacuum tool and simultaneously soldered by laser diodes that are focused in a line (Figure 21). With this simultaneous

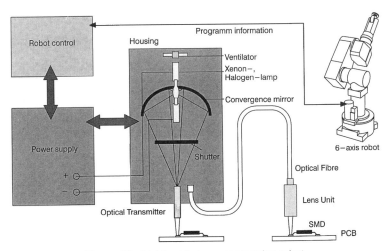

Figure 20 Light soldering system using robot.

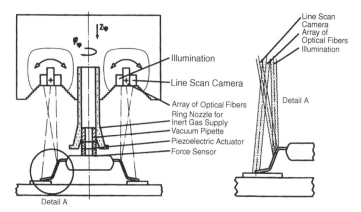

Detail A

Figure 21   Combined mounting and soldering head (From M. Geiger and K. Feldmann, *DFG-Antrag SFB 356 (1995–1998)*, Erlangen, 1995. 299–333).

mounting and soldering process the cycle time is reduced and the productivity of laser soldering can be increased.

REFERENCES

Feldmann, K., K. Götz, and J. Sturm. "CAD/CAM Process Planning for Laser Soldering in Electronics Assembly." In *Proceedings of LANE 94—Laser Assisted Net Shape Engineering.* Ed. M. Geiger and F. Vollertsen. Bamberg-Erlangen: Meisenbach GmbH. 715–724.
Geiger, M., and K. Feldmann. *DFG-Antrag SFB 356 (1995–1998).* Erlangen. 299–333.
"MID auf Basis von Polyamid als Teil eines ABS-Bremsmoduls." *Productronic* **10,** 1997.
Philips EMT, Eindhoven, 1996.

ADDITIONAL READING

Feldmann, K., and A. Brand. "Analytical and Experimental Research on Assembly Systems for Molded Interconnection Devices (3D-MID)." *Annals of the CIRP* **43**(1), 1994.
———. Molded Interconnection Device: A New Challenge for Industry Robots. Dearborn: Electronics Manufacturing Engineering, 1995.
Feldmann, K., A. Brand, and J. Franke. "Räumliche spritzgegossene Schaltungsträger helfen Kosten sparen." PRONIC **3,** 1993.
Feldmann, K., and J. Franke. "Three-Dimensional Circuit Carriers." *Kunststoffe* **4,** 1993.
Feldmann, K., K. Götz, and J. Sturm. "CAD/CAM Process Planning for Laser Soldering in Electronics Assembly." In *CIRP 95—Manufacturing Systems* **24**(5), 459–465, 1995.
Feldmann, K., and R. Luchs. "Micro Wave Soldering in Nitrogen Atmosphere." In *Proceedings of the Technical Program Nepcon West,* 1996. 299–306.
Feldmann, K., A. Rothhaupt, and J. Sturm. "Systematische Untersuchungen zum Lotpastenauftrag mittels Dispenstechnologie für unterschiedliche Rastermasse." In Tagungsband zum SMT/ASIC/-HYBRID-Kongress, 1993. Berlin: vde-Verlag, 1993.
IPC Molded Printed Board Subcommittee. *IPC-MB-380, Guidelines for Molded Interconnects.* Lincolnwood: Institute for Interconnecting and Packaging Electronic Circuits, 1989.
New Soft Beam Fiber Optic Light System. Panasonic, 1996.
Seliger, G., et al. "New Light Soldering-System Integrates the Assembly and Soldering of SMD-Components." In *Proceedings of 2nd International Congress Molded Interconnect Devices,* Bamberg, 1996. 303–311.
Die Welt der Surface Mount Technology. Siemens AG, Bereich Automatisierungstechnik, 1995.
Zaderej, V., et al. "New Developments in the Three Dimensional Molded Interconnection Industry." In *Proceedings of 1990 5th Printed Circuit World Convention 5,* Glasgow, 1990.

CHAPTER **52**

QUALITY ASSURANCE, INSPECTION, AND TESTING

Charlie Duncheon
Adept Technology, Inc.
San José, California

1 INTRODUCTION

Like so many robot applications, testing and inspection find robots replacing traditional testing methods which utilize manual labor. This chapter addresses applications that utilize robots at the core of the testing or inspection process. Dimensional testing by standard coordinate measuring machines and high-speed vision inspection by in-line dedicated vision systems will not be covered. Both on-line and off-line testing robot applications in various industries will be addressed. Because of the proprietary nature of some manufacturers' robot applications, some of the examples will be limited in detail.

2 APPLICATIONS OVERVIEW

2.1 Inspection

As stated in the introduction, in-line inspections with dedicated vision systems will not be addressed. However, many vision guided robot cells utilize inspection vision capability. This will be discussed in more detail in the application of robotic flexible feeding.

Only noncontact inspection will be addressed. Significant inspection is done on factory floors with three- to five-axis programmable coordinate measuring machines utilizing sensing probes. However, only inspection that uses both vision and robotic technology will be covered here.

2.2 Testing

Testing will be addressed as it relates to:

1. Robotic acquisition of component or product to be tested
2. Robotic loading of component or product into testing fixture
3. Robotic controller communication with testing controller
4. Robotic sorting of product or component based on test results
5. Surface and feature location (flexible gauging) verification utilizing robots and sensors

2.3 Inspection and Testing Application Examples

Examples of current applications using robotic technology are given in Table 1.

3 ROBOTIC TESTING OF MEDIA DISKS

The final step in the production of hard media disks is to test and certify each disk for several characteristics that affect the quality and performance of the disk once it is installed in a disk drive. To accomplish this process step, hard disk manufacturers use testing equipment that emulates the operation of a disk drive, including a spindle that spins the disk and a set of heads that certifies the disk by gliding over the entire surface of the disk.

Handbook of Industrial Robotics, Second Edition, Edited by Shimon Y. Nof
ISBN 0-471-17783-0 © 1999 John Wiley & Sons, Inc.

Table 1   Inspection and Testing Application Examples

Industry	Product/Component Tested	Notes
Automotive	Fuel gauges	Testing done after robotic assembly
	Speedometers	Vision used to verify pointer location
	Car bodies/frames	Flexible gauging
Electronics	Printed circuit board	Function and in-circuit tests
	Disk media	Testing disk "guide" and disk certification
Semiconductor	Integrated circuit	Testing robotic load, unload, accept/reject
Telecommunications	Cellular phones	Vision and force sensing

The disk certification system includes eight dedicated testers evenly spaced around the perimeter of the full circle workspace of a single AdeptOne direct drive robot. The robot is equipped with a sophisticated custom vacuum gripper with two independent disk paddles designed to contact only the outer edge of the disks. Each paddle has a 90° pitch motion, which is used to change the orientation of the disks from vertical in the cassettes to horizontal prior to placing onto the test spindles. The robot gripper also has a cassette gripper, which is used to handle entire cassettes of disks.

The workcell also includes six parallel cassette transfer conveyors. A single in-feed conveyor introduces cassettes of untested disks while five out-feed conveyors convey cassettes of certified disks. Disks are graded into five levels and segregated into cassettes. Each out-feed conveyor holds a cassette containing a different grade of disk (see Figure 1).

The role of the robot in the disk certification process is that of a high-speed, high-precision disk and cassette transfer mechanism. The process starts with a cassette of 25 untested disks, which is conveyed into the robot workcell. The AdeptOne robot picks up the entire cassette of disks off the in-feed conveyor and places it into a registration fixture, which precisely aligns the individual disks and raises their height slightly to accommodate the robot gripper.

The robot utilizes one of its vacuum paddles to pick up a vertically oriented disk out of the conveyor. Once it has removed the disk from the cassette, it moves over to the first available test spindle while keeping both paddles in the down position to minimize space requirements for the motion. Once the robot reaches the test spindle, it flips its disk paddles to horizontal. If there is a completed disk on the test spindle, the robot removes it with its free paddle. The gripper is then rolled 180° and the robot places the untested disk onto the test spindle and signals the tester to start the certification process. The robot rotates its paddles to the down position and quickly moves to the out-feed conveyors. The certified disk is then placed into the proper cassette.

The process is repeated for all eight test spindles until all the disks are removed from the first cassette. The empty cassette is then removed from the registration fixture and a new cassette is placed into the fixture. The empty cassette is placed into a buffer area to be used as an outbound cassette once one of the graded cassettes is full and conveyed out of the workcell.

The robot and testers operate asynchronously. The system software is sophisticated enough to allow the robot to service testers randomly as they require disks and also is smart enough to know which of the eight testers are on-line and which are off-line. This capability allows maintenance to be performed on individual testers while the remaining system continues to operate.

Disk certification is a critical step in the production of quality hard disks. The robot is used because of the speed, workspace, reliability, precision, and cleanliness of its direct drive design. Since every disk needs to be certified, and the certification process requires roughly one minute to complete, testing is an expensive bottleneck for hard disk manufacturers. Since this is one of the last steps in the process, quality is critical. The robot optimizes this process by maximizing the utilization of expensive testing equipment. The size of the AdeptOne's workspace allows eight test spindles to be placed around its perimeter, while the relatively high speed of the robot allows it to keep up with the throughput of the eight testers. This maximizes the throughput of the testing process by minimizing the amount of time testers remain idle during disk transfer. In addition, the performance of the system minimizes the amount of expensive clean room floor space

Figure 1 Disk certification test cell by Phase Metrics, Inc.

required by reducing the number of systems needed to meet the expected hard disk testing capacity.

Quality and throughput are also optimized due to the direct drive design of the robot. The elimination of gears from the robot's drive train allows the arm to easily achieve a class 10 clean room rating, ensuring that disks remain clean throughout the certification process. Also, since there are no harmonic drives to wear, the robot is able to maintain its .001-in. repeatability over its entire 40,000-hour design life ensuring precise disk placements and reducing the need for reteaching of the robot.

Finally, with downtime measured at over $2,000 per hour for a typical certification workcell, the robot and system design means exceptional robot reliability with minimal preventive maintenance downtime and a design life that translates into several product life cycles for the fast-changing hard disk industry.

Flexible automation such as this has allowed many hard disk manufacturers to meet the aggressive performance, quality, and cost requirements of the fast-paced computer industry while continuing to manufacture in the United States.

4 ROBOTIC TESTING OF CELLULAR PHONES

Many consumer and commercial electronic products require in-process testing to verify functions, electromagnetic compatibility (EMC) compliance, and visual appearance. This need especially applies to the cellular phone business. The test equipment often involves placing the assembled phone in a fixture or "nest," where electrical contacts are made and the test apparatus goes through a routine to check all parameters.

In testing of cellular phones the phones are placed into test fixtures and tested for some or all functions that are utilized by the consumer. The robot is programmed to contact certain keys. Integrated force sensing allows the robot to press the keys in the range of what would be human fingers. Once the keys are pressed, an integrated vision system using optical character recognition algorithms inspects for presence, intensity, and accuracy of the characters in the LCD displays.

The testing process is slower than the production rate on the assembly line, so multiple test stations, or nests, are required to achieve throughput on the line. The testing process may have unpredictable duration, based on variable warm-up routines and whether a marginal data point results in a retest, so the function of servicing the nest is random.

Random placing of cellular phones to multiple nonsynchronous test nests from an assembly line and removing and sorting the completed products is an ideal application for a SCARA robot. The SCARA robot can access the line to pick phones and then move to an available test nest, swapping the untested phone for a completed one at the nest. The robot then moves to place the completed phone to an "accept" or "reject" area, based on test results. "Two up" tooling can be used on the robot so it can carry an untested phone to the test fixture and rapidly interchange it with a completed phone, minimizing robot moves and loss of test time as the fixture waits to be reloaded.

Besides inspection, vision and other intelligence can increase flexibility. Robot guidance vision can recognize incoming phone type and location, reducing the need for line fixturing and operator set-up. Bar code tags can be read to select test routines and log product compliance. Robotic force sensing can also verify correct placement in the test fixture.

Critical robot features for phone testing are:

- Adequate reach to enable access to all required test stations
- Robot speed and settling time
- Load capacity, carried mass, and insertion force needs
- Precision for placement
- Sensory capability, vision, and force
- Networking to the line

The entire phone test and inspection cell can be simulated in virtual manufacturing environment to insure proper test cell throughput and robots' working envelope (Figure 2). Simulation can also insure robotic paths that avoid collisions. This simulation technology can significantly reduce the implementation time for robotic test cells.

5 ROBOTIC ASSEMBLY INTEGRATED WITH VISION-BASED INSPECTION

Machine vision can be utilized in robotic assembly cells for both robot guidance and inspection of either incoming parts or the completed assembly. Consider a robotic cell for an assembly consisting of two injection-molded parts that utilizes flexible feeding of randomly oriented parts for delivery to the robot. The example to be considered here uses an Adept robot with integrated vision, also manufactured by Adept (Figure 3). The vision system recognizes the incoming parts on the flexible feeder, the part orientation, and whether the part is accessible for robotic acquisition. Parts that are not accessible are bypassed and are recycled in the flexible feeder.

In addition to the vision-based guidance function of the vision system for flexible robotic assembly, critical inspection is performed. The inspection process is shown as one of the steps in the overall process in Figure 4. The vision system executes dimensional inspection to insure that the parts are dimensionally compatible with the inspection to be performed. If the part is deemed accessible by the robot gripper, but outside of the tolerance for inspection, the robot can pick up the part, discard it into a reject bin, and record the reject into the quality control database.

The key requirement in setting up a vision system is providing a good image that will simplify the vision analysis and yield better and more reliable results. To obtain a good

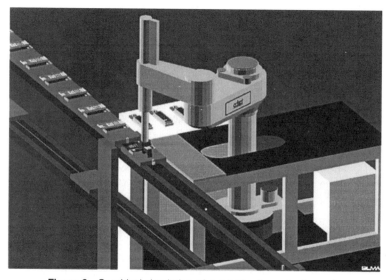

Figure 2 Graphical simulation of a robotic phone-testing cell.

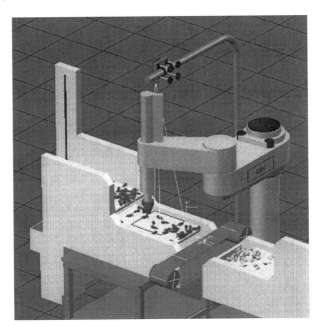

Figure 3   An AdeptOne robot with two Adept FlexFeeders.

image, proper lighting, optics, and visual sensing must be employed. In flexible feeding applications one can use backlighting, top lighting, or a combination of both. Backlighting can be used whenever parts can be uniquely identified from their silhouettes. It uses binary vision to extract part information from the image. Otherwise, top lighting becomes essential to identify internal part features necessary for recognition. Hence grayscale vision tools are needed to identify part features. No matter what lighting method is utilized, it is important that the light sources be at least an order of magnitude higher than the ambient light variations. Since the field of view is large, standard lenses will have considerable parallax distortion for parts that have considerable height (part height about 1% of the camera's distance from presentation surface). To maintain a general approach to the problem and keep the cost down, standard optics will be used. However, proper choice of the lens' focal length, spacer, and color filter is important. Color filtering is used most frequently with translucent parts on a backlit surface. More information on lighting, optics, and visual sensing for machine vision applications can be found in Cencik (1992), Dechow (1996), Bolhouse (1996), and Schmidt (1996). See also Chapter 13, Sensors for Robotics, and Chapter 14, Stereo-Vision for Industrial Applications.

In addition to the above requirements in setting up the vision system, special attention must be given to the large field of view. Since parts are randomly scattered onto a large, flat surface, the image resolution per part is limited. The downward-looking camera is set to identify primarily the various stable poses of the part. As a result of the large field of view of the camera, distortion resulting from slight misalignments of the camera to the perpendicular (perspective distortion) and from the part height (parallax distortion) have to be compensated for. While the former can be compensated for in a calibration routine, the latter cannot be easily compensated for when parts have a considerable height. In such cases when part resolution is low and image distortion is high, inspection must be carried out using a second camera. This camera can be mounted in an upward-looking position whereby the robot will fly over the camera, acquiring an image of the part in the gripper while the robot is in motion. This not only provides a larger image of the part, hence a higher pixel resolution per part and a better inspection, but also provides information about the error in the grip that can be compensated for in accurate part mating. Nesnas (1997) provides details analysis on setting up and programming vision systems for flexible part-feeding applications.

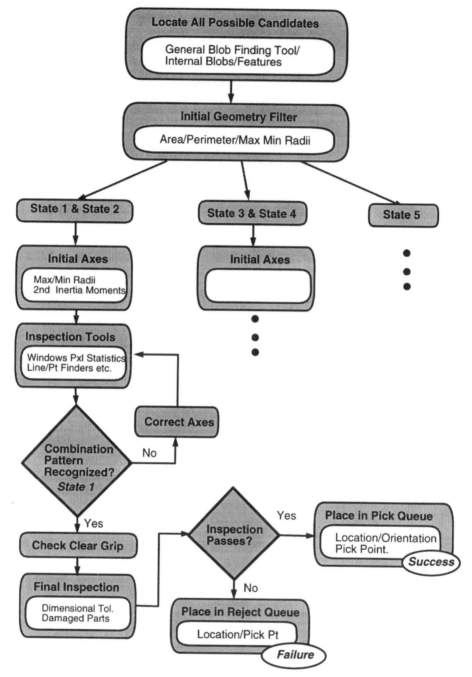

Figure 4 A general vision tree with part inspection.

6 PART RECOGNITION AND INSPECTION

Following is an outline of a general part feeding vision inspection application:

1. *Locate all possible candidates.* The first step in any object-finding operation for flexible feeding is to locate where possible candidates are, using any object finding

tool. The sophistication of the tool used determines how close the recognized object is to the final target. Let us assume that the tool used here is a blob finder that is not the most sophisticated from an object-recognition standpoint.

2. *Initial geometry filter.* Based on the results of the previous operation, blobs can be classified into different categories based on certain geometric properties of interest, such as area, perimeter, and maximum or minimum radii. More often than not, these properties do separate various stable states into several groups (Figure 4). If only one stable state is sought, then other groups can be ignored.

3. *Initial coordinate axes.* An initial orientation of the blob can be based on minimum/maximum radius, second moment axis, or hole/feature positions. The initial axis is needed as a reference frame for subsequent tool placements.

4. *Invariant inspections.* With the use of *invariant* properties with relative tools, this inspection process can be as simple as one inspection tool or as complex as a conditional tree. While redundant inspections increase the robustness of the vision algorithm, it is important to keep in mind the computational limitations of the current vision processor. If the inspection tree fails, the initial axes can be corrected and the tree reprocessed.

5. *Final coordinate axes.* This is perhaps the most difficult task since it affects the accuracy of the system. The final axes have to be accurately placed no matter where the part is in the field of view. It is important to avoid using features that are at different z-heights, since they can potentially reduce the accuracy of parts that are closer to the edges. It is important to use features/pixels that are at the same height at which the system was calibrated, since they have the most accurate pixel scaling. Using multiple features can have an averaging advantage over using a single edge or feature.

6. *Selecting a pick point.* When a pick point on the part is selected, care must be taken not to extrapolate too far from the most accurate point on the part. Extrapolation magnifies the positional errors. If the latter was the case, additional tools can be used to refine the accuracy of the pick point.

   Whenever possible, it is more reliable and efficient to pick parts using internal gripping as opposed to external gripping. Internal gripping does not usually require gripper clearance checking. Both internal and external grippers help align the part in at least one direction, unlike vacuum gripping which does not align parts. However, large part alignments can disturb neighboring parts.

7. *Clear grip inspection.* Using an external gripper, it is important to check that the fingertips have the necessary clearances for the specified pick location.

8. *Final inspection.* The vision system will carry out a final inspection of the part's tolerances and dimensions provided there is sufficient image resolution. The vision system can also identify damaged parts. If parts fail the final inspection, they are picked and placed in the rejection bin. Successful parts are placed in their assemblies/subassemblies. A log of the system's performance reflects the quality metric.

7   USE OF ROBOTICS FOR FLEXIBLE GAUGING

A robotic approach to automotive body-in-white gauging was chosen by BMW in Spartanburg, South Carolina, to satisfy a number of process requirements. These requirements were to:

1. Increase product quality by implementing trend analysis to modify frame production methods prior to product quality tolerances exceeding acceptable limits, and prior to body-in-white transfer to the painting operation.

2. Improve yields and reduce product rework costs by applying 100% inspection of critical fit-points, normally done by CMM machines on just one body per day.

3. Free up metrology lab to complete more product quality inspection tasks.

In addition to meeting the immediate needs listed above, BMW envisioned future benefits to robotic in-line gauging, including reduction of launch time for new models, correction of component quality problems at the supplier's site, and elimination of expensive checking fixtures.

The six-axis robot system with 3D laser sensor cameras was chosen over a traditional CMM approach because of its flexibility and speed. Six-axis robots mounted with laser sensor cameras were also more flexible and cost-effective than fixed camera solutions. Fixed camera solutions require one sensor for each characteristic inspected. BMW uses only four robots, each with a single sensor, to measure 137 characteristics.

The system specifications required that the measurements be made in line for ten different Z-3 BMW car models. The system repeatability had to be ±0.2 mm or better. It was decided to build a beta site first to verify system performance.

The system installed included the following major components (Figure 5).

- Four Staubli Unimation Model RX 130L six-axis robots
- Four Perceptron Tri-Cam 3D laser sensors
- One Perceptron P1000 controller
- One Lamb Technicon tooling base and cell control

The Siemens s3964r protocol was used.

The beta system was built, programmed, and tested, with the following results:

1. Programming was accomplished by operators with little or no previous robot programming experience.
2. Total system accuracy reached ±0.3 mm at six sigma, including equipment tolerances, mechanical thermal expansion, tooling tolerances, and operation within the nonoptimized robot work envelope.
3. ±0.2 mm at six sigma was achieved within the optimized robot work envelope.
4. Thermal expansion at 85% of maximum robot speed accounted for 0.1 mm or less of the total six-sigma deviation.
5. 137 measurements were made in 92 seconds with the robot running at 85% of the maximum speed. The approximate cycle for one-dimensional laser sensor mea-

Figure 5 Flexible robot gauging system at BMW.

surement was 0.4 sec. The cycle was approximately 0.8 sec for 3D laser sensor measurements. The average measurement cycle, including measurement time, robot motion, and serial communication, was 2.5 sec.

The planned system enhancements to allow the system to reach all original specifications are as follows:

1. Simulate the robot in the application to determine the ideal performance work envelope.
2. Implement a seventh axis rail to increase the effective robot work envelope as required.
3. Develop software to accommodate the thermal expansion effect of robot arms.

In summary, BMW successfully installed a flexible robot gauging solution that met all the existing production goals while offering the ability to decrease time to market and startup costs for future car models (see Lutz, 1997).

REFERENCES

Bolhouse, V. 1996. "Machine Vision's Role in Electronics Assembly." In Assembly Technology Expo, *AIA Conference Proceedings,* September.
Cencik, P. 1992. *Optics, Lighting, and Image Sensing for Industrial Machine Vision I.* San Jose: Adept Technology.
Dechow, D. 1996. "Practical Integration of Machine Vision for Assembly." In Assembly Technology Expo, *AIA Conference Proceedings,* September.
Lutz, C. 1997. "Robotic Gauging of Tomorrow: A BMW Case History." In *International Robots and Vision Conference Proceedings,* May.
Nesnas, I. A. D. 1997. "Computer Vision Strategies for Flexible Part Feeding." In *International Robots and Vision Conference Proceedings,* May.
Schmidt, P. E. 1996. "Machine Vision in Manufacturing—Simple or Difficult." In *Successfully Using Vision for Flexible Part Feeding Workshop*, RIA/AIA sponsored, October.

CHAPTER 53

MAINTENANCE AND REPAIR

Lynne E. Parker
John V. Draper
Oak Ridge National Laboratory
Oak Ridge, Tennessee

1 INTRODUCTION AND BACKGROUND

Maintenance is the process that preserves or restores a desired state of a system or facility. The maintenance process includes three major activities: inspection, planned maintenance, and disturbance handling (where *disturbances* are unplanned system states). Inspection is the activity in which information about state is monitored to allow prediction or early detection of disturbances. Planned maintenance is the activity in which elements of the system are modified or replaced according to a predetermined schedule, with the aim of avoiding or reducing the frequency of disturbances. Disturbance handling is the activity in which elements of the system are modified or replaced to restore the desired state following a disturbance.

Maintenance is a task that has some important differences from tasks commonly selected for industrial robots. First, maintenance often requires access to environments that are more dynamic and less predictable than is the case for many robotics applications. Second, tasks may be less predictable in maintenance, both in terms of the nature of tasks and the frequency of maintenance or latency between tasks. Finally, the cost benefits of robotic maintenance may be different from those of other robotic applications. In general, robots typically pay for themselves over their entire operational life; however, robots used in maintenance, particularly in the nuclear industry, may pay for themselves in a single application. This is particularly true if robots allow a facility to avoid shutdown and continue operating even during maintenance.

2 APPLICATION EXAMPLES AND TECHNIQUES

Robots, whether teleoperated, under supervisory control, or autonomous, have been used in a variety of applications in maintenance and repair. The following subsections describe many of these systems, focusing primarily on applications for which working robot prototypes have been developed.

2.1 Nuclear Industry

In the nuclear industry teleoperators have been well utilized in the maintenance role for more than four decades. Several features of maintenance make it a good application for teleoperators in this arena:

1. The low frequency of the operation calls for a general-purpose system capable of doing an array of maintenance tasks.
2. Maintenance and repair require high levels of dexterity.
3. The complexity of these tasks may be unpredictable because of the uncertain impact of a failure.

Handbook of Industrial Robotics, Second Edition, Edited by Shimon Y. Nof
ISBN 0-471-17783-0 © 1999 John Wiley & Sons, Inc.

For these reasons the choice for this role is often between a human and a teleoperator. Thus, when the environment is hazardous, a teleoperator is usually the best selection. If humans in protective clothing can perform the same job, the benefits of having teleoperators continuously at the work site need to be weighed against the cost of suiting up and transporting humans to and from the work site. While humans are likely to be able to complete tasks more quickly than teleoperators, using teleoperators can:

1. Shorten mean time to repair by reducing the response time to failures
2. Reduce health risks
3. Improve safety
4. Improve availability by allowing maintenance to take place during operations instead of halting operations

As an example of the importance of maintenance for nuclear industry robotics, the proceedings of the 1995 American Nuclear Society topical meeting on robotics and remote handling included 124 papers, nearly 25% of which were devoted to some aspect of maintenance. The 1997 meeting included 150 papers, more than 40% of which dealt with some aspect of maintenance. Furthermore, if environmental recovery operations are considered as a form of maintenance, then a much larger proportion of presentations at both meetings were maintenance-related.

Vertut and Coiffet (1985) list the following applications of teleoperated robots in nuclear maintenance:

1. Operation and maintenance of industrial nuclear facilities and laboratories
2. Maintenance in nuclear reactors
3. Decommissioning and dismantling nuclear facilities
4. Emergency intervention

Our exploration of robotics for nuclear-related maintenance will follow these categories.

2.1.1 Operation and Maintenance of Industrial Nuclear Facilities and Laboratories

Guidelines for applying teleoperators in the remote maintenance role exist (e.g., Burgess et al., 1988). Unfortunately, there is an historic tendency to ignore the importance of designing for robotic maintenance and lessons learned by past experiences with remote maintenance (Vertut and Coiffet, 1985). Nuclear facilities do not necessarily need to be designed for robotic operations for robots to be successfully deployed within them. However, designing a facility to accommodate robotic maintenance greatly improves efficiency. Chesser (1988) reports on an extensive maintenance demonstration carried out using a teleoperator to dismantle and reassemble components built for robotic maintenance. Reprocessing spent nuclear fuel requires a complex chemical process plant. To demonstrate the ability of a state-of-the-art teleoperator to replace equipment modules in such a plant, Chesser conducted a demonstration disassembly and reassembly of a prototype chemical process rack using the Oak Ridge National Laboratory's Advanced Servomanipulator (ASM). Figure 1 shows the ASM, which is also remotely maintainable because of its modular design. Using standard tools (impact wrenches, ratchet wrenches, and a torque wrench), ASM operators were able to dismantle the rack, including tubing jumpers, instruments, motors, tanks, etc. As an adjunct to this demonstration, another teleoperator was used to disassemble and reassemble the ASM to show its remote maintainability.

Particle accelerators are another type of nuclear facility sometimes maintained robotically. The CERN laboratory seems to have produced the largest body of experience in remote maintenance (Horne et al., 1991). That program features integrated use of a variety of remote devices, including inspection and surveillance systems, dexterous manipulators, and mobile robots. The CERN application has some characteristics common to nuclear applications:

1. Diverse array of maintenance tasks
2. Unpredictability of tasks and occurrences
3. Environment dangerous for humans

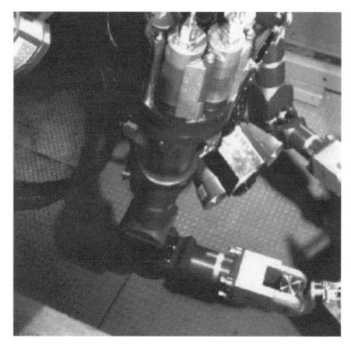

Figure 1 Advanced servo-manipulator (ASM) designed for maintaining chemical process racks and itself remotely maintainable.

4. Large facility size and hence a need for mobility

The evolving response to these factors depends on flexibility and versatility of robotic systems and their users. There is also a trend toward providing computer assistance during teleoperation to improve overall performance (Horne et al., 1991).

2.1.2 Maintenance in Nuclear Reactors

Nuclear reactors are even less likely to be designed for robotic maintenance than industrial nuclear facilities, thus leading to the frequent requirement for innovative approaches to gaining access to trouble spots. This, along with limitations on size imposed by reactor designs and the resulting long lead times necessary for purpose-built robots, has limited the use of robots in this arena in the past (Glass et al., 1996). However, increasingly stringent limits on worker exposure to ionizing radiation and exposure reductions possible with robotic maintenance may increase robot use in the future (Lovett, 1991). For experimental fusion reactors a remote maintenance philosophy seems to be an important part of design (see, e.g., MacDonald et al., 1991).

Several programs are addressing the difficulties of maintenance in nuclear reactors. For example, a preliminary analysis of tasks required in nuclear power plant maintenance and robot requirements for meeting these needs has been completed (Tsukune et al., 1994). A modular approach has been proposed to reduce costs incurred by customized robot designs (Glass et al., 1996), and problems in handling tools and fixtures designed solely for human use have been addressed by the development of more dexterous robot grippers (Ali, Puffer, and Roman, 1994).

2.1.3 Decommissioning and Dismantling Nuclear Facilities

In the United States this topic is currently receiving much attention within the Department of Energy community. At the time of this writing, the first completely robotic decommissioning effort is being conducted at the Argonne National Laboratory. Two systems—a pair of dexterous manipulator arms and a vehicle with a manipulator—are being used to dismantle the CP-5 (Chicago Pile number 5) reactor. Development work in support of this effort is described in Noakes, Haley, and Willis (1997). Hazardous waste site remediation is a related topic that is also the target of development work within the DOE. Underground storage tank remediation using robotics is currently being done by the Oak Ridge National Laboratory (Randolph et al., 1997) and Pacific Northwest Laboratories (Kiebel, Carteret, and Niebuhr, 1997). Figure 2 is a photograph of robotic devices deployed to retrieve waste at the Oak Ridge National Laboratory. Detailed task analyses of underground storage tank remediation illustrate the complexity of this task (Draper, 1993a, 1993b). In this application the unpredictability inherent in the mission makes it a prime candidate for the application of teleoperated robots, with automation generally limited to providing assistance to the operator under specific conditions. As an example of the unpredictability in this arena, a quarry remediation conducted at Oak Ridge started with an estimate that approximately 2500 objects would be retrieved; in fact, more than 17,000 items were removed from the quarry by robotic systems.

2.1.4 Emergency Intervention

The most famous case of emergency intervention in the nuclear arena in the United States is the accident at Three Mile Island (TMI-2). Merchant and Tarpinian (1985) and Bengal (1985) provide overviews of robotic programs developed for recovery operations at TMI-2. The accident is illustrative of principles of emergency intervention by robots. First, the event was unpredictable. Robotic systems were not available to perform reconnaissance or inspection for some time afterward. Robots on-site could have significantly reduced post-accident personnel exposure (Merchant and Tarpinian, 1985). Second, the accident created an environment very hostile to people and at the same time hostile to autonomous robots. Obstructions routinely present in buildings designed for human access are already difficult for robots to negotiate; the effects of the accident were to render access more difficult by introducing even less structured, unknown obstructions. Lesser-known interventions have also been carried out, Chester (1985) briefly describes some of them.

2.2 Highways

In the developed world, highways are a critical component of the transportation network. The volume of traffic on the roadways has been steadily increasing for many years as society becomes more and more mobile. However, the funding to maintain these roadways has not been keeping pace with the traffic volume. The result is deteriorating roadways

Figure 2   Robotic devices deployed to retrieve waste at Oak Ridge National Laboratory.

that cannot be adequately maintained. Conventional techniques to road repair lead to traffic congestion, delays, and dangers for workers and the motorists. Robotic solutions to highway maintenance applications are attractive due to their potential for increasing the safety of the highway worker, reducing delays in traffic flow, increasing productivity, reducing labor costs, and increasing the quality of the repairs.

Application areas to which robotics can be applied in this area include (Ravani and West, 1991):

- Highway integrity management (crack sealing, pothole repair)
- Highway marking management (pavement marker replacement, paint restriping)
- Highway debris management (litter bag pick-up, on-road refuse collection, hazardous spill cleanup, snow removal)
- Highway signing management (sign and guide marker washing, roadway advisory)
- Highway landscaping management (vegetation control, irrigation control)
- Highway work zone management (automatic warning system, lightweight movable barriers, automatic cone placement and retrieval)

Although relatively few implementations in highway maintenance and repair have been attempted, some successful prototypes have been developed (Zhou and West, 1991). The California Department of Transportation (Caltrans), together with the University of California at Davis (UC Davis), are developing a number of prototypes for highway maintenance under the Automated Highway Maintenance Technology (AHMT) program. Efforts are underway to develop systems for crack sealing, placement of raised highway pavement markers, paint striping, retrieving bagged garbage, pavement distress data collection, and cone dispensing. One result of this effort is a robotic system, ACSM, for automatic crack sealing along roadways (Winters et al., 1994). Shown in Figure 3, this machine senses, prepares, and seals cracks and joints along the highway. Sensing of cracks along the entire width of a lane is performed using two-line scan cameras at the front of the vehicle. Sealing operations occur at the rear of the vehicle using an inverted, slide-mounted SCARA robot. A laser range finder at the tooling verifies the presence of the cracks and provides guidance for the sealing operation. The vehicle is able to perform this operation moving at about 1.6–3.2 km/hr (1–2 mi/hr). Other crack sealing prototypes have been developed at Carnegie-Mellon University (Hendrickson et al., 1991) and the University of Texas at Austin (Haas, 1996).

Earlier prototypes for highway maintenance date back to the 1980s (Skibniewski and Hendrickson, 1990). Researchers at Tyndall AFB developed a rapid runway repair (RRR) telerobotic system for repairing craters in runways (Nease, 1991). The objective of this work was to safely restore pavement surfaces after enemy attack to ensure subsequent future successful aircraft operations. The system was based upon a John Deere multipur-

Figure 3 Automated crack sealing machine developed by UC Davis. (Photograph used with permission from Steve Velinsky, University of California at Davis.)

pose excavator enhanced for telerobotic operation. The system used a 4-d.o.f. joystick system using position and rate control, plus force feedback for human remote control. The resulting machine could dig, scrape, compact, break pavement, and change tools under preprogrammed, onboard control. Carnegie-Mellon University also developed a robotic excavator prototype called REX, which used topography and a computer-generated map of buried objects to generate and execute appropriate trajectories for the mission. The system used an elbow-type manipulator coupled with a master arm for manipulator setup and error recovery.

Automated pavement distress data-collection vehicles have also been developed (Zhou and West, 1991). The vehicle built by Komatsu Ltd. of Japan uses scanning lasers to examine a road surface's condition, measuring information regarding crack formation, wheel rutting, and longitudinal unevenness. The vehicle can travel at speeds up to 60 km/hr, evaluating roads up to 4 meters wide with ruts and potholes up to 0.25 meters deep.

The French Petroleum Studies Company developed an automatic cone dispenser that can dispense and remove up to two rows of warning cones, for a total of 240 cones. The system operates at about 15 km/hr. The Technique Special de Sécurité Company developed a mobile lane separator that can place and remove concrete road marker blocks at speeds up to 30 km/hr. Systems for automatic grading have been developed by Spectra-Physics of Dayton, Ohio, and Agtek Co. of California. These systems use laser guides to control the height of the grading blades, thus relieving the human operator of the need to perform manual positioning and control of the blades.

2.3 Railways

The railroad industry has recognized the economic benefits of automation, which has led to the development of a number of robotic solutions to maintenance and repair applications in the industry. The railway maintenance shops are the most common location of robots, which perform activities such as welding, grinding, cleaning, and painting (Martland, 1987). A Toronto Transit Commission project led to the design of an automated system for cleaning the undersides of subway cars (Wiercienski and Leek, 1990). Shown in Figure 4, this system involved the use of three industrial painting robots mounted on either side and under the subway vehicle being cleaned. An operator located remotely would begin the cleaning operations after preparing the vehicle in advance. The entire system would be controlled by a master computer that supervised the three individual robot controllers. A robot-mounted vision system would be used to correct the robots' positions along their tracks. This system is expected to yield dramatic improvements in working conditions and work quality over the previous human worker approach.

Another robotic system developed by the railway industry is the RMS-2 rail grinding system ("Toward the 'Smart Rail' Maintenance System," 1986). Developed by Speno Rail Services Co., this system has automated capabilities to sense the existing condition of the surface of the rails. Up to 99 patterns are stored that correspond to rail contour patterns. An integrated on-board computer system is used to generate on-board grinder controls for finishing the rail to the appropriate, predetermined rail contour. The RMS-2 has 120 stones for grinding spread along the underside of 5 of the 12 railway cars that make up the system. The system is capable of grinding rails at speeds up to six miles per hour.

2.4 Power Line Maintenance

Many common maintenance operations on overhead transmission lines are performed by human operators on live lines. Examples of these tasks include replacing ceramic insulators that support conductor wire and opening and reclosing the circuit between poles. These tasks are very dangerous for the human workers due to risks of falling from high places and electric shock. Obtaining skilled workers to perform these tasks is quite difficult due to the high training and labor requirements of the job. Performing the maintenance while the lines are deenergized would alleviate some of the risks, but would also create other problems in a society that demands interruption-free service from electric power companies.

Electric power companies have therefore been investigating the use of robotic systems for live-line power line maintenance since the mid-1980s. In particular, power companies in Japan, Spain, and the United States have developed teleoperated and semiautonomous approaches to this problem. One of the first systems developed was TOMCAT (Tele-operator for Operations, Maintenance, and Construction using Advanced Technology),

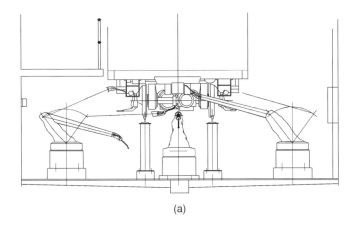

(a)

(b)

Figure 4 Robot system developed for the Toronto Transit Commission for cleaning the undersides of subway cars. (a) The schematic layout of the robots under the railcar. (b) The robot in operation (one robot manipulator can be seen at the right side of the photograph). (From W. Wiercienski and A. Leek, "Feasibility of Robotic Cleaning of Undersides of Toronto's Subway Cars," *Journal of Transportation Engineering* **116**(3), 272–279, 1990, by permission of the American Society of Civil Engineers.)

developed as part of an Electric Power Research Institute (Palo Alto, California) program. The basic TOMCAT concept was first demonstrated by Philadelphia Electric in 1979 (Dunlap, 1986), with subsequent development continuing in the 1980s, leading to a system prototype. The TOMCAT system consisted of an insulated bucket truck, a Kraft seven-function manipulator that was bolted to the end of the truck boom, a television viewing system for human supervisory control, and requisite control and power supplies. The operator control components were mounted on the back of the bucket truck and included a manipulator master with no force feedback.

A more recent robotic system for live-line maintenance has been developed by Kyushu Electric Power Co., Inc., in Fukuoka, Japan (Yano et al., 1995). The system configuration schematic for the dual-arm robot system they developed is shown in Figure 5. The earlier phases of this work involved the development of a two-manipulator telerobotic system; subsequent phases are incorporating more autonomy to evolve from a basic master–slave

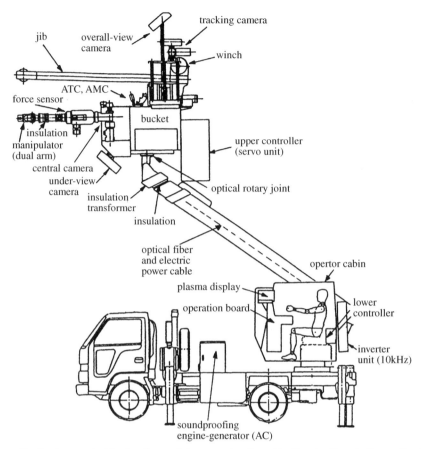

Figure 5 Dual-arm robot for live-line maintenance developed by Kyushu Electric Power Co., Inc. (From S. Nio and Y. Maruyama, "Remote-Operated Robotic System for Live-Line Maintenance Work," in *Proceedings of the 6th International Converence on Transmission and Distribution Construction and Live-Line Maintenance.* © 1993 IEEE.)

configuration to a human–robot cooperative system. Laser sensors are used on-board the robot to help with position control. In the current system the human works from a control station on the supporting truck, rather than on the elevated boom, which nearly eliminates the risk of injury due to falls.

Other related work in this area includes the ROBTET teleoperated system for live-line maintenance developed by researchers in Spain (Aracil et al., 1995) and the robot for automatic washing and brushing of polluted electric insulators (Yi and Jiansheng, 1993) developed in China.

2.5 Aircraft Servicing

Aircraft servicing applications may benefit from robotic maintenance in several areas. The size of modern multiengine jets makes inspection and coating removal and application particularly attractive in terms of improving quality and efficiency. As examples, Siegel, Kaufman, and Alberts (1993) describe concepts for automating skin inspections, and Birch and Trego (1995) and Baker et al. (1996) describe stripping and painting concepts. Automated stripping and painting systems are already in place at a few U.S. Air Force bases. A robotic assistant for rearming tactical fighter aircraft is being developed at the Oak Ridge National Laboratory for the U.S. Air Force.

2.6 Underwater Facilities

Teleoperated robots are widely used to maintain facilities beneath the surface of the ocean, mainly in service of the offshore oil industry. Specific applications include repairing communications cables, pipelines, well heads, and platforms. Teleoperators have also been deployed to clean marine growth from power plant cooling systems (Edahiro, 1985), inspect and clean steam generators (Trovato and Ruggieri, 1991), perform underwater construction (Yemington, 1991), and inspect and repair water conveyance tunnels (Heffron, 1990). While these efforts do not have the visibility of robotics work in industrial, space, or nuclear applications, this is an arena in which robotics and remote control technology are widely used. It is perhaps the most common venue for everyday use of teleoperated robots. One publication lists 63 companies involved in building remotely operated vehicles or manipulators for subsea work and 180 different commercially available, remotely controlled systems (Gallimore and Madsen, 1994). The remotely operated systems range from towed sensor arrays to submersibles with dexterous manipulators to large construction machinery.

2.7 Coke Ovens

Another example of robotics used in maintenance and repair operations is a robot developed by Sumitomo Metal Industries, Ltd., Japan, for repairing the chamber wall of a coke oven (Sakai et al., 1988). Damage to coke ovens occurs over years of operation due to repeated cycles of chamber door opening and coke pushing, which induce damaging changes in temperature. The result is cracks, joint separations, and chamber wall abrasion, which can lead to gas leakage, air pollution, and structural flaws in the ovens. Thus the effective repair of coke ovens is needed to extend the life of the ovens and allow for stable operation.

Especially challenging maintenance operations involve the repair of the central portion of the oven. This type of repair is very difficult due to the inaccessibility of the area, the high temperature, and the predominance of narrow cracks. Any technology for repair in this area must involve high heat-resistance components and mechanisms for external observation, resulting in repairs of high quality and durability. The solution to this repair problem was the development of a heat-resistant robot, shown in Figure 6 (Sakai et al., 1988), that can autonomously perform individual crack repair while being given high-level guidance through a human–machine interface. Of special benefit to the industry is the ability to perform these repairs without disturbing oven operation or incurring a large firebrick temperature drop.

2.8 Summary of Robotics Applications in Maintenance and Repair

In this chapter we have reviewed the primary application areas in which robotics is used for maintenance and repair. Some of these application areas have a significant ongoing effort in robotic development and usage, while others have received relatively little attention thus far. Table 1 provides a summary of these application areas, noting the importance

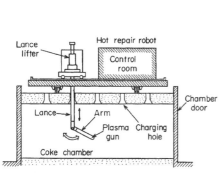

Figure 6 Hot repair robot developed by Sumitomo Metal Industries, Ltd., Japan, for repairing cracks and joints in the chamber wall of coke ovens during hot operation. (From T. Sakai et al., "Hot Repair for Coke Ovens," *Transactions of Iron and Steel Institute of Japan* **28,** 346–351. 1988. With permission from The Iron and Steel Institute of Japan.)

Table 1   Summary of Robotics Applications in Maintenance and Repair

Application Area	Maintenance Task		
	Inspection	Planned Maintenance	Disturbance Handling
Nuclear industry	Growing area, especially as new facility designs incorporate remote maintenance philosophy.	Well-established field, with several decades of successful robotic applications.	Much current activity related to decontamination, decommissioning, and dismantling.
Highways	Relatively new area with few current prototypes, except as packaged with crack sealing and pothole repair systems.	Relatively new area, with quickly growing interest and a huge potential impact. Several ongoing efforts should result in a number of new robot prototypes in the next 5 years.	Of significant interest, particularly for highway integrity management. A number of successful prototype systems are gradually making way into routine use. Several new efforts underway.
Railways	Few current systems, and little ongoing activity.	Most common area of railway robotics, but with little new activity.	Little current use.
Power line maintenance	Little current use.	Interest is increasing, especially for robotic techniques that work on live power lines.	Greatest area of current use, with much potential growth due to technology advances and need to remove humans from highly dangerous tasks.
Aircraft servicing	Steadily growing area, due to recent advances in automated inspection technologies.	Steadily growing area, especially for automated stripping and painting.	Little current use.
Underwater facilities	Steady progress over the last two decades, with continued advances.	Of increasing importance, with several new prototype systems under development.	Of increasing importance, with several new prototype systems under development.
Coke ovens	Little current use.	Little current use.	Fair amount of activity in late 1980s. Relatively little new work in this area.

of robotics to the maintenance tasks of inspection, planned maintenance, and disturbance handling.

3   DESIGN CONSIDERATIONS AND EMERGING TRENDS

Years ago Jordan (1963) observed that "[people] are flexible but cannot be depended upon to perform in a consistent manner whereas machines can be depended upon to perform consistently but they have no flexibility whatsoever." The three decades of development in robotics and artificial intelligence that have passed since he wrote have greatly improved the flexibility of machines but have not abrogated his observation. As Ruoff (1994) pointed out, "Robots have limited intelligence and ability to perceive. To compensate, applications have . . . relied either on human presence in the control loop or on the imposing of significant order on tasks." The applicability of autonomous robots, supervised robots, and teleoperated robots to maintenance and repair applications is especially dependent upon two aspects of the work environment—variability and accessibility. Autonomous robots still cannot function well in many dynamic, and thus variable, environments, so these applications tend to require either completely human or teleoperator solutions. Of course, environments with low human accessibility (because of physical constraints or danger) are usually good candidates for robotic solutions. Where accessibility is low but variability is high, teleoperators are usually best.

Until additional significant progress is made in autonomous systems, human involvement with robotic systems must increase with environmental variability. Figure 7 illustrates this principle: where variability is low, autonomous robots are efficient and human

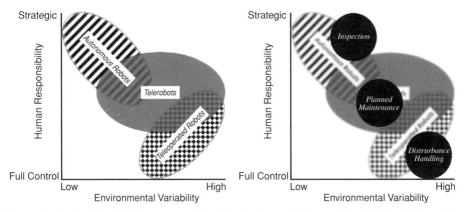

Figure 7 Application of autonomous robots, telerobots, and teleoperated robots to robotic maintenance and repair tasks.

involvement is at the level of strategic decision-making. Where variability is high, human sensing and decision-making are more important and the human user must take more responsibility. At the lowest level the human is responsible for executing movements; this is the region of true teleoperation. Telerobots, which combine autonomous subtask completion capabilities with human-in-the-loop control, may help teleoperators to be more efficient or autonomous robots to be more capable of dealing with variability. The left-hand side of Figure 7 illustrates the relationship between environmental variability and human input responsibility for various types of robotic systems. The right-hand side maps maintenance and repair tasks onto these ideas. Inspection tasks are most often carried out under routine conditions and may be performed with an autonomous robot for many applications. Planned maintenance may be somewhat less structured and thus requires more interaction with the environment in terms of manipulating task elements. Therefore it may be necessary to deploy a telerobot or teleoperator to complete these tasks. Disturbance handling is by nature unpredictable. Therefore, in general, this maintenance task is much more likely to require human-in-the-loop control.

From this survey of robotic systems for maintenance and repair, some trends in robotic maintenance in recent years can be identified. These trends include:

- *Computer control/monitoring/assistance:* increasing integration of computer control over subsystems and subtasks for monitoring user inputs and evaluating their appropriateness and for user assistance. As an example, Volpe (1994) describes using model-based artificial forces to enhance collision avoidance. By monitoring the position of a manipulator relative to a model of the remote work site, a repulsive force can be generated by the control system to resist movement into areas where collisions are possible or even to push the manipulator away from surfaces or features.

- *Virtual reality:* increasing integration of virtual reality interfaces into teleoperated systems. Virtual reality (VR) systems and teleoperators may be considered members of a superset called *synthetic environments technology* (Durlach and Mavor, 1995). They differ only in that a user interacts with a remote, *computer-generated* world in VR but a remote, *real* world in teleoperation. Therefore it should be no surprise that VR techniques are being adopted for teleoperation. However, application of VR to autonomous robotics is also likely to increase in the future for programming and monitoring robot operations. This may be limited to providing a virtual view of the workspace and manipulator or may be present in the form of graphical preview control. In the latter the user completes the task in the virtual world as a means of programming the robot. The latter is inherently less efficient than real-time control but proponents claim potentially greater safety and ease of use for graphical preview control (e.g., Small and McDonald, 1997), although there are no data supporting that opinion. Milgram, Yin, and Grodski (1997) provide examples of what may be termed an "augmented reality" interface for teleoperation, in which video from a remote site is combined with partial computer models of the site. See also Chapter 17, Virtual Reality and Robotics.

- *Increasing sensor integration:* use of larger numbers of sensors and multisensor integration, toward the goal of allowing autonomous robots to function more effectively in environments too variable for effective modeling. For example, Dudek and Freedman (1996) demonstrated how two different types of sensors, a sonar array and a laser range finder, can provide more accurate information about a work site than either type alone.
- *Responsiveness vs. payload:* sacrificing responsiveness (the ability of a manipulator to recreate a user's trajectories and impedance in time and space) for payload in manipulators. Telerobotic manipulators developed since about 1985 seem to be increasingly less responsive but more powerful.

Table 2 provides some comments on how these trends may affect each of the three maintenance tasks in future applications. These are necessarily speculative, but offer some guidance for the development of robotic technology in maintenance for the near and intermediate future.

Table 2   Emerging Trends and Their Potential Impact on the Robotics Activities of Inspection, Planned Maintenance, and Disturbance Handling

| | Maintenance Task | | |
Trend	Inspection	Planned Maintenance	Disturbance Handling
Computer control/ monitoring/ assistance	Will make inspections more autonomous and at the same time more effective. Robots are more capable of accurately monitoring position and measuring sensory input than humans are (although humans are better at pattern recognition).	Will assist in scheduling and reduce human involvement in tasks that can be sufficiently ordered. It is possible to remove much of the tedium from remote tasks by automating tasks or subtasks.	Most likely to improve recordkeeping and to help assure quality by assisting operators in quality-critical sub-tasks. Also may allow oversight of user actions, which may improve safety.
Virtual reality	Graphical overlays of expected task configuration will be helpful. Multiple sensor "views" can be integrated to provide more information.	May assist in robot navigation by providing displays to enhance user situation awareness. May be used to display robot progress through planned maintenance tasks without on-site video. May be used to display information about maintenance campaigns (as opposed to tasks).	Forms of computer-assisted teleoperation may be beneficial; integrating graphics and video could provide information about expected state, which could help with diagnosis. Could allow on-line guidance for operators.
Increasing sensor integration	Will make it possible to perform inspections in multisensor fashion, and develop a picture of inspected items that integrates several sensors, including energy not sensed by humans.	Arrays of sensors may provide more complete information about a component and aid in assuring the effectiveness of planned maintenance.	Arrays of sensors may provide more complete information about a disturbance and aid in diagnosis and remediation.
Responsiveness vs. payload	Not likely to have much impact.	May allow larger parts to be replaced but may make replacement more difficult under human control.	May allow application to a wider range of disturbances but may reduce efficiency.

REFERENCES

Ali, M., R. Puffer, and H. Roman. 1994. "Evaluation of a Multifingered Robot Hand for Nuclear Power Plant Operations and Maintenance Tasks." In *Proceedings of the 5th World Conference on Robotics Research*. Dearborn: Robotics International of the Society of Manufacturing Engineers. MS94-217-1-10.

Aracil, R., et al. 1995. "ROBTET: A New Teleoperated System for Live-Line Maintenance." In *Proceedings of the 7th International Conference on Transmission and Distribution Construction and Live Line Maintenance,* Columbus, Ohio. 205–211.

Baker, J. E., et al. 1996. "Conceptual Design of an Aircraft Automated Coating Removal System." In *Proceedings of ISRAM '96*, Sixth International Symposium on Robotics and Manufacturing, Montpellier, France, May 27–30.

Bengal, P. R. 1985. "The TMI-2 Remote Technology Program." In *Proceedings of the Workshop on Requirements of Mobile Teleoperators for Radiological Emergency Response and Recovery.* Argonne: Argonne National Laboratory. 49–60.

Birch, S., and L. E. Trego. 1995. "Aircraft Stripping and Painting." *Aerospace Engineering* **15,** 21–23.

Burgess, T. W., et al. 1988. *Design Guidelines for Remotely Maintained Equipment.* ORNL/TM-10864. Oak Ridge: Oak Ridge National Laboratory.

Chesser, J. B. 1988. *BRET Rack Remote Maintenance Demonstration Test Report* ORNL/TM-10875. Oak Ridge: Oak Ridge National Laboratory.

Chester, C. V. 1985. "Characterization of Radiological Emergencies." In *Proceedings of the Workshop on Requirements of Mobile Teleoperators for Radiological Emergency Response and Recovery.* Argonne: Argonne National Laboratory. 37–48.

Draper, J. V. 1993a. *Function Analysis for the Single-Shell Tank Waste Retrieval Manipulator System.* ORNL/TM-12417. Oak Ridge: Oak Ridge National Laboratory.

———. 1993b. *Task Analysis for the Single-Shell Tank Waste Retrieval Manipulator System.* ORNL/TM-12432. Oak Ridge: Oak Ridge National Laboratory.

Dudek, G., and P. Freedman. 1996. "Just-in-Time Sensing: Efficiently Combining Sonar and Laser Range Data for Exploring Unknown Worlds." In *Proceedings of the 1996 IEEE International Conference on Robotics and Automation.* Piscataway: The Institute of Electrical and Electronics Engineers. 667–672.

Dunlap, J. 1986. "Robotic Maintenance of Overhead Transmission Lines." *IEEE Transactions on Power Delivery* **PWRD-1**(3), 280–284.

Durlach, N. I., and A. S. Mavor, eds. 1995. *Virtual Reality: Scientific and Technological Challenges.* Washington: National Research Council, National Academy Press.

Edahiro, K. 1985. "Development of 'Underwater Robot' Cleaner for Marine Live Growth in Power Station." In *Teleoperated Robotics in Hostile Environments.* Ed. H. L. Martin and D. P. Kuban. Dearborn: Robotics International of the Society of Manufacturing Engineers. 108–118.

Gallimore, D., and A. Madsen. 1994. *Remotely Operated Vehicles of the World.* Ledbury, Herfordshire: Oilfield Publications.

Glass, S. W., et al. 1996. "Modular Robotic Applications in Nuclear Power Plant Maintenance." In *Proceedings of the 1996 58th American Power Conference.* Vol. 1. Chicago: Illinois Institute of Technology. 421–426.

Haas, C. 1996. "Evolution of an Automated Crack Sealer: A Study in Construction Technology Development." *Automation in Construction* **4,** 293–305.

Heffron, R. 1990. "The Use of Submersible ROVs for the Inspection and Repair of Water Conveyance Tunnels. In *Water Resources Infrastrucure: Needs, Economics.* 35–40.

Hendrickson, C., et al. 1991. "Perception and Control for Automated Pavement Crack Sealing." In *Proceedings of Applications of Advanced Technologies in Transportation Engineering* 66–70.

Horne, R. A., et al. 1991. "Extended Tele-robotic Activities at CERN." In *Robotics and Remote Systems: Proceedings of the Fourth ANS Topical Meeting on Robotics and Remote Systems.* Ed. M. Jamshidi and P. J. Eicker. Albuquerque: Sandia National Laboratory. 525–534.

Jordan, N. 1963. "Allocation of Functions between Man and Machines in Automated Systems." *Journal of Applied Psychology* **47**(3), 161–165.

Kiebel, G. R., B. A. Carteret, and D. P. Niebuhr. 1997. "Light Duty Utility Arm Deployment in Hanford Tank T-106." *ANS Proceedings of the 7th Topical Meeting on Robotics and Remote Systems* LaGrange Park: American Nuclear Society. 921–930.

Lovett, J. T. 1991. "Development of Robotic Maintenance Systems for Nuclear Power Plants." *Nuclear Plant Journal* **9,** 87–90.

Macdonald, D., et al. 1991. "Remote Replacement of TF and PF Coils for the Compact Ignition Tokamak." *Robotics and Remote Systems: Proceedings of the Fourth ANS Topical Meeting on Robotics and Remote Systems.* LaGrange Park: American Nuclear Society. 131–140.

Martland, C. 1987. "Analysis of the Potential Impacts of Automation and Robotics on Locomotive Rebuilding." *IEEE Transactions on Engineering Management* **EM-34**(2), 92–100.

Merchant, D. J., and J. E. Tarpinian. 1985. "Post-accident Recovery Operations at TMI-2." In *Proceedings of the Workshop on Requirements of Mobile Teleoperators for Radiological Emergency Response and Recovery.* Argonne: Argonne National Laboratory. 1–9.

Milgram, P., S. Yin, and J. J. Grodski. 1997. "An Augmented Reality Based Teleoperation Interface for Unstructured Environments." In *Proceedings of the Seventh ANS Topical Meeting on Robotics and Remote Systems.* LaGrange Park: American Nuclear Society. 966–973.

Nease, A. 1991. "Development of Rapid Runway Repair (RRR) Telerobotic Construction Equipment." In *Proceedings of IEEE National Telesystems Conference (NTC '91)*, Atlanta. 321–322.

Noakes, M. W., D. C. Haley, and W. D. Willis. 1997. "The Selective Equipment Removal System Dual Arm Work Module." In *ANS Proceedings of the 7th Topical Meeting on Robotics and Remote Systems* LaGrange Park: American Nuclear Society. 478–483.

Randolph, J. D., et al. 1997. "Development of Waste Dislodging and Retrieval System for Use in the Oak Ridge National Laboratory Gunnite Tanks." In *ANS Proceedings of the 7th Topical Meeting on Robotics and Remote Systems.* LaGrange Park: American Nuclear Society. 894–906.

Ravani, B., and T. West. 1991. "Applications of Robotics and Automation in Highway Maintenance Operations." In *Proceedings of 2nd International Conference on Applications of Advanced Technologies in Transportation Engineering.* 61–65.

Ruoff, C. F. 1994. "Overview of Space Telerobotics." In *Teleoperation and Robotics in Space.* Ed. S. B. Skaar and C. F. Ruoff. Washington: American Institute of Aeronautics & Astronautics. 3–22.

Sakai, T., et al. 1988. "Hot Repair Robot for Coke Ovens." *Transactions of Iron and Steel Institute of Japan* **28,** 346–351.

Siegel, M. W., W. M. Kaufman, and C. J. Alberts. 1993. "Mobile Robots for Difficult Measurements in Difficult Environments: Application to Aging Aircraft Inspection." *Robotics and Autonomous Systems* **11,** 187–194.

Skibniewski, M., and C. Hendrickson. 1990. "Automation and Robotics for Road Construction and Maintenance." *Journal of Transportation Engineering* **116**(3), 261–271.

Small, D. E. and M. J. McDonald. 1997. "Graphical Programming of Telerobotic Tasks." In *Proceedings of the American Nuclear Society 7th Topical Meeting on Robotics and Remote Systems.* LaGrange Park: American Nuclear Society. 3–7.

"Toward the 'Smart' Rail Maintenance System." *Railway Tract. Struct.* **82**(11), 21–24, 1986.

Trovato, S. A., and S. K. Ruggieri. 1991. "Design, Development and Field Testing of CECIL: A Steam Generator Secondary Side Maintenance Robot." In *Robotics and Remote Systems: Proceedings of the Fourth ANS Topical Meeting on Robotics and Remote Systems.* Ed. M. Jamshidi and P. J. Eicker. Albuquerque: Sandia National Laboratory. 121–130.

Tsukune, H., et al. 1994. "Research and Development of Advanced Robots for Nuclear Power Plants." *Bulletin of the Electrotechnical Laboratory* **58**(4), 51–65.

Vertut, J., and P. Coiffet. 1985. *Teleoperation and Robotics: Applications and Technology.* Englewood Cliffs: Prentice-Hall.

Volpe, R. 1994. "Techniques for Collision Prevention, Impact Stability, and Force Control by Space Manipulators." In *Teleoperation and Robotics in Space.* Eds. S. B. Skaar and C. F. Ruoff. Washington, D.C.: American Institute of Aeronautics and Astronautics. 175–212.

Wiercienski, W., and A. Leek. 1990. "Feasibility of Robotic Cleaning of Undersides of Toronto's Subway Cars." *Journal of Transportation Engineering* **116**(3), 272–279.

Winters, S., et al. 1994. "A New Robotics System Concept for Automating Highway Maintenance Operations." In *Proceedings of Robotics for Challenging Environments.* 374–382.

Yano, K., et al. 1995. "Development of the Semi-Automatic Hot-Line Work Robot System 'Phase II'." In *Proceedings of 7th International Conference on Transmission and Distribution Construction and Live Line Maintenance,* Columbus, Ohio. 212–218.

Yemington, C. 1991. "Telerobotics in the Offshore Oil Industry." In *Robotics and Remote Systems: Proceedings of the Fourth ANS Topical Meeting on Robotics and Remote Systems.* Ed. M. Jamshidi and P. J. Eicker. Albuquerque: Sandia National Laboratory. 441–450.

Yi, H., and C. Jiansheng. 1993. "The Research of the Automatic Washing–Brushing Robot of 500KV DC Insulator String." In *Proceedings of the 6th International Conference on Transmission and Distribution Construction and Live Line Maintenance,* Las Vegas. 411–424.

Zhou, T., and T. West. 1991. "Assessment of the State-of-the-Art of Robotics Applications in Highway Construction and Maintenance." In *Proceedings of 2nd International Conference on Applications of Advanced Technologies in Transportation Engineering.* 56–60.

ADDITIONAL READING

Draper, J. V. "Teleoperators for Advanced Manufacturing: Applications and Human Factors Challenges." *International Journal of Human Factors in Manufacturing* **5**(1), 53–85, 1995.

Martin, H. L., and D. P. Kuban, eds. *Teleoperated Robotics in Hostile Environments.* Dearborn: Robotics International of the Society of Manufacturing Engineers, 1985.

Najafi, F. T., and S. M. Naik. "Potential Applications of Robotics in Transportation Engineering." *Transportation Research Record* 1234, 64–73.

CHAPTER **54**
PRODUCT REMANUFACTURING

E. Zussman
Technion—Israel Institute of Technology
Haifa, Israel

G. Seliger
Technical University of Berlin
Berlin, Germany

1 INTRODUCTION

Remanufacturing is an economical form of reusing and recycling manufactured goods. It can be defined as a process in which worn-out products or parts are brought back to original specifications and conditions or converted into raw materials (Figure 1). Central to remanufacturing is the disassembly process, which is essential for material and component isolation, since the objectives of product remanufacturing are to maximize the parts obtained for repair and reuse and minimize the disposal quantities. Complementary processes to which products are subjected include refurbishment, replacement, repair, and testing. The disassembly process may be seen as the inverse of the assembly process, in which products are decomposed into parts and subassemblies. However, in product remanufacturing the disassembly path and the termination goal are not necessarily fixed, but rather are adapted according to the actual product condition.

Remanufacturing processes for obsolete products are technologically challenging due to their special characteristics, such as high variety of products, different product manufacturers, and uncertain product condition after usage. These conditions may affect the disassembly of fasteners or parts. The process is labor-intensive and there is a high probability of exposure to hazardous materials since such materials must be isolated from the rest of the product. Therefore automation is an adequate solution for increasing both productivity and safety (Seliger et al., 1996; Zussman et al., 1994).

In this chapter the current research and application of robots and automation in remanufacturing are reviewed. The review is limited to the remanufacturing of household electronics appliances such as TV sets, radios, and computers. The major technical and economic issues involve finding cost-effective ways to perform the remanufacturing operations in sequence and selecting the proper remanufacturing technology.

This chapter covers issues of remanufacturing process planning and plan adaptation due to actual product conditions. Two remanufacturing processes for worn out products, cathode ray tube (CRT) and computer keyboard, are presented and discussed.

2 REMANUFACTURING PLANNING

In dealing with worn-out products the aim is to maximize recycled resources and minimize possible damage by the leftovers that are disposed, while considering economic factors. We term this multipurpose goal *increasing the end-of-life (EOL) value.* Our focus is on *product remanufacturing,* namely how to achieve optimal recovery of obsolete product while maximizing the *EOL* value. For this task we introduce a representation that includes all feasible disassembly sequences. It is called a *recovery graph (RG).* It is a variant of an *and/or* graph, a directed graph, in which edges emanating from the same node are in either an *AND* or an *OR* relation with each other (Homem de Melo and

Handbook of Industrial Robotics, Second Edition, Edited by Shimon Y. Nof
ISBN 0-471-17783-0 © 1999 John Wiley & Sons, Inc.

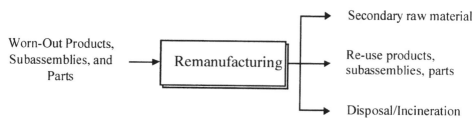

Figure 1   Material flow in a remanufacturing system.

Sanderson, 1990). In our context each node in the and/or graph of a product represents a possible subassembly. Edges in the graph emanating from the same node are partitioned via an *AND* relation, so that edges $\{(u, v_0), (u, v_1) \ldots (u, v_m)\}$ are all in an *AND* relation to each other if and only if subassembly u can be disassembled by a single operation into subassemblies $v_0, v_1, \ldots v_m$. (Equivalently, a single joint connects them to form u). An implicit *OR* relation exists between different *AND* groups emanating from the same node, e.g., if $\{(g, y), (g, z)\}$ and $\{(g, t), (g, t)\}$ are two such groups, then it is possible to disassemble g into either y and z or x and t (Figure 2).

It is easy to see that such an and/or graph is always acyclic and that each disassembly plan of the product corresponds to a subtree of this graph. The *recovery graph* of a product is its *and/or* graph, where with each node we associate an *EOL* value $c(v)$ incurred by reusing, refurbishing, utilizing, or dumping node v without further disassembly:

$$c(v) = \max\{c_{\text{reuse}}(v), c_{\text{refurbish}}(v), c_{\text{utilize}}(v), c_{\text{dispose}}(v)\}$$

For each group of *AND* edges, say $\{(u, v), (u, g)\}$, we associate a *disassembly cost* of disassembling the subassembly, represented by u, into the subassemblies represented by v and g. Once the various costs/benefits associated with parts/subassemblies are established, the remaining problem is concerned with the order and type of operations. Using the *RG*, we can define the disassembly process plan (DPP) as a 3 tupple: $P = \langle R, S, T \rangle$ where:

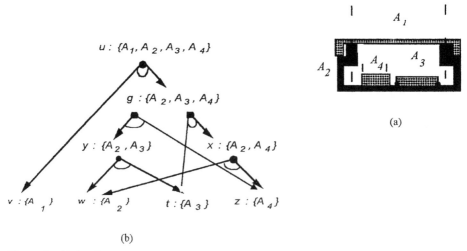

(b)

Figure 2   (a) An electronic packaging with four parts $\{A_1, A_2, A_3, A_4\}$; and (b) its respective recovery graph.

$R = \{n_1, n_2, \ldots n_K\}$—a set of nodes in a RG that are intermediate results of a disassembly plan, where n_1 is the entire product

$S = \{s_1, s_2, \ldots s_K\}$—a set of disassembly actions (disassembly operation plus machine), where each action s_k is related to node n_k and defines the corresponding AND group in node n_k $(k = 1, 2, \ldots K)$

$T = \{t_1, t_2, \ldots t_L\}$—a set of nodes that includes final results of a disassembly process

Let us consider a product and its respective RG with a set of nodes $\{1, 2, \ldots N\}$. Each node n has R_n AND groups. The rth AND group of node n is defined by disassembly action $s_{n,r}$, $r = 1, \ldots, R_n$, and edges $\{(n, m_1^r), \ldots, (n, m_{k_r}^r)\}$, where $m_1^r, \ldots, m_{k_r}^r$ are results of the disassembly action $s_{n,r}$. We associate with each node n an EOL value $c(n)$, and for each disassembly action $s_{n,r}$ (AND group) its disassembly cost $dis(n, s_{n,r})$.

The *remanufacturing value* of node n is defined recursively as:

$$d(n) = \max \left\{ c(n), \max_{1 \leq r \leq R_n} \left[\sum_{k=1}^{k_r} d(m_k^r) - dis(n, s_{n,r}) \right] \right\} \tag{1}$$

For nodes n, which represent components (leaves of the recovery graph), $d(n) = c(n)$, since n cannot be further disassembled. The remanufacturing value $d(n)$ presents the cost of the disassembly option for each node n. It can be seen that the value of the objective function of DPP $P = \langle R, S, T \rangle$ is $f(P) = d(n_1)$, where n_1 is the node representing the entire product (the root of the graph). Let $\Re$ be a set of all feasible disassembly process plans. Plan P_{opt} is called optimal if $f(P_{opt}) = \max_{P \in \Re} f(P)$. Hence, the disassembly planning problem is to find the best plan P_{opt}.

It is possible to prove that the objective function in the process plan can be described as (Zussman, 1995):

$$f(P) = \sum_{l=1}^{L} c(t_1) - \sum_{k=1}^{K} dis(n_k, s_k) \tag{2}$$

Once a disassembly plan is executed for a particular product an unexpected event can occur, either breakage of a machine or the impossibility of performing a disassembly operation due to the product condition. The disassembly process must then be adapted to select an alternative action, i.e., either traverse the RG or stop the process and request an external intervention.

3 REMANUFACTURING OF HOUSEHOLD ELECTRONICS

The following two case studies deal with remanufacturing process of cathode ray tube and computer keyboard. In both cases the remanufacturing plans are elaborate and further on are transferred into a remanufacturing system. The specific characteristics of the remanufacturing processs are detailed and integration of automation and robots is discussed and evaluated.

3.1 Remanufacturing of Cathode Ray Tubes

A CRT is composed of four parts: the glass panel (screen), a shadow mask, a glass funnel, and an electron gun (see Figure 3). The cone and the neck consist of glass with up to 20% lead oxide. The screen and glass funnel are connected using a glass frit solder. A CRT is assembled into a display unit (e.g., a TV set) that includes other parts such as plastic cabinet, electromagnetic shields, and printed circuit boards. We consider the CRT itself and not the other parts of the assembly. According to the EPA report (1997), the greatest part of the worn-out CRTs in the United States are currently disposed of, most of the rest are refurbished, and the rest are recycled. The main environmental concern with CRT is the large lead percentage in the funnel (28% lead) and screen (2%) glass as well as phosphors on the screen.

Because of the potential hazardous waste status and associated liability, and because of the inherent value in some components of CRTs and CRT products, various disassembly and recycling options have been initiated.

The CRT remanufacturing process is modeled below on a recovery graph that presents the recovery options. The process plan is transformed into a sequence that includes in-

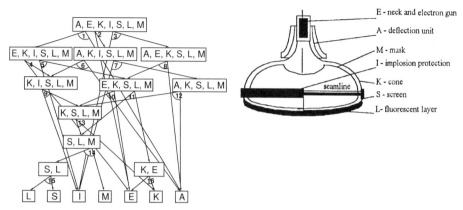

Figure 3 A CRT and its respective recovery graph. (From C. Hentschel, "Organization of Disassembly Systems," Ph.D. Thesis, Fraunhofer Gesellschaft/IPK, 1996.)

spection, fixing, and disassembly operations. The conventional process is mainly manual and is assisted by flexible fixtures, handling devices, and conveyor.

The recovery graph presents all feasible options for disassembling a color CRT (Figure 3). The remanufacturing goal is to maximize the EOL value of the CRT. Practically, a remanufacturing plan should lead to the separation of the hazardous materials from the rest of the product and enable the reutilization of some of the components. Given the product RG, one can determine a predictive plan. The predictive plan that was determined was to take off the deflection unit; next the neck and electron gun are disassembled, implosion protection is taken off, and then the cone (funnel) and the screen are disassembled. The CRT is positioned screen glass down so all the other parts can be easily separated.

Nevertheless, randomly arriving CRTs first have to be identified, and their process plan must be adapted to the product condition. A major concern is the vacuum condition in the CRTs. In certain cases it is uncertain whether the CRTs are already ventilated. Premature ventilation would cause encrustation of the fluorescent layer due to humidity reaching the tube. To reduce the risk of explosion during the process, the tube has to be ventilated as soon as possible.

A possible material flow in a remanufacturing system is presented in Figure 4. At the first stage the CRTs are identified and inspected manually. Following the inspection the deflection units are removed, the CRTs are ventilated, and their neck is removed. A bypass exists based on the CRT condition. Following the process plan is a linear; a heated wire is used to make the glass burst along a predefined line between the screen and the cone. The CRT is fixed by a vacuum fixture and its screen, facing downward, is then removed from its cone. The screen remains in the disassembly station, providing access to the shadow mask in the case of a color tube. With the mask taken out, the fluorescent material can be easily vacuumed off and stored as a hazardous material. To eliminate exposure to it automation is introduced, based on a manipulator with a vacuum gripper attached to its end-effector. The total duration of the disassembly is 240 sec. The system can handle various sizes of CRTs.

3.2 Remanufacturing of Computer Keyboard

The main components of a computer keyboard are housing polycarbonate (PC), key caps (polybutylene terephthalate (PBT)), and printed circuit board (PCB). Besides these are various parts made of rubber, which are not compatible with the plastic parts and cannot be recycled together with them. In this case study the keyboard was recovered by a fully automated disassembly system based on two robots and a conveyor (Figure 5).

The keyboards are manually loaded into the system, then transferred to station 2. Next an identification process is performed, where the parts are carried by the robot. The keyboards are then fixed in station 3 and disassembled. The disassembling steps are described in Table 1 with their respective time duration.

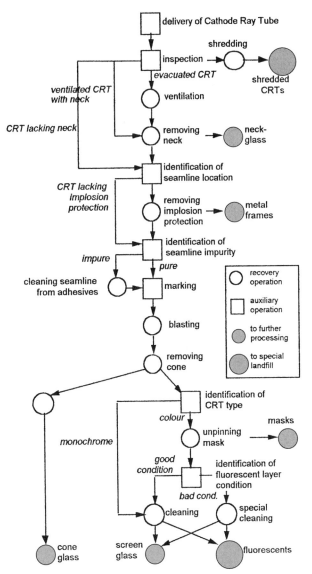

Figure 4   A remanufacturing system of CRTs. (From C. Hentschel, "Organization of Disassembly Systems," Ph.D. Thesis, Fraunhofer Gesellschaft/IPK, 1996.)

All the data described above were collected in the experimental disassembly system. The total disassembly time, as shown above, can be improved by fast inspection and identification. Nevertheless, it should be emphasized that such a disassembly system will have to deal with different keyboards from different manufacturers, types, and production years. Hence they must first be identified and the process plan of the disassembly process must be adapted to the product condition. Therefore a flexible automation and efficient information system is required.

Naturally, the return of investment and economic justification will be based on the quantities that arrive. Nevertheless, using of flexible automation will enable the disassembling of different parts with relatively small setting time. Based on the available separation technology, shredding the whole keyboard should also be taken into account.

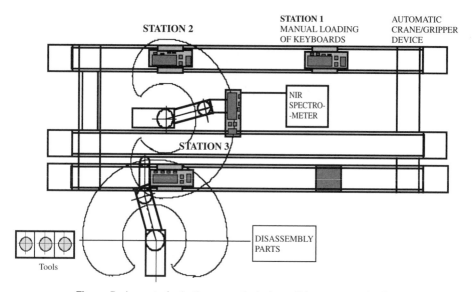

Figure 5   Layout of robotic remanufacturing cell for computer keyboards.

4   DISCUSSION

Utilizing modern production technologies and automation at the end of the product life cycle for remanufacturing seems to be a feasible approach for tackling waste problems efficiently. Combining the technology with a proper methodology for planning and control of the remanufacturing operations can guarantee cost minimization. No doubt the environmental legislation already enacted in Europe will give impetus to remanufacturing activities, especially when "take back policy" is in place.

Table 1   Summary of Automated Disassembling Operations[a]

Disassembling Step, Tool Description	Tool Changing Time (sec)	Working Time (sec)	Total Time (sec)	Weight (g)	Material
Fixation	0	6	6	0	
Key caps	21	73	94	174.3	PBT
Space bar	14.8	9	23.8	10.0	PBT
Milling of screws	11.7	106	117.7	0	
Screws	0	0	0	30	Fe
Removal of cover	10.8	14.5	25.3	272.5	PVC
Springs. mag. removal	5.7	27.3	33	30.77	
PCB	9.5	0	9.5	26.34	Div
Rubber mat	0	8	8	21.55	PS
Plastic sheets	0	8	8	38.28	Rubber
Rubber stop	0	7	7	0.65	Div
Remove base plate	0	3	3	329.1	PVC
Electric cable	0	3	3	50.0	PS
Total	73.5	261.8	338.3	983.4	

[a] From E. Langerak, "To Shred or to Disassemble: Recycling of Plastics in Mass Consumer Goods," in *Proceedings of the IEEE International Symposium on Electronics and the Environment*, May 1997, San Francisco.

Future production trends with emphasis on design for remanufacturing (McGlothlin and Kroll, 1995; Shu and Flowers, 1995) will enable efficient remanufacturing processes. Product joints and materials will be designed in such a way that material separation and product decomposition will become easier. Future remanufacturing systems will enable disassembly of a wide range of products, such as automobile parts, computers, and home appliances (Warnecke et al., 1994; Feldmann and Meedt, 1996).

REFERENCES

Feldmann, K., and O. Meedt. 1996. "Recycling and Disassembly of Electronic Devices." In *Life-Cycle Modeling for Innovative Products and Processes*. Eds. F.-L. Krause and H. Jansen. London: Chapman & Hall.

Hentschel, C. 1996. "Organization of Disassembly Systems." Ph.D. Thesis, Fraunhofer Gesellschaft/IPK.

Homem de Melo, L., and A. C. Sanderson. 1990. "AND/OR Graph Representation of Assembly Plans." *IEEE Transactions on Robotics and Automation* **6**, 188–199.

Langerak, E. 1997. "To Shred or to Disassemble: Recycling of Plastics in Mass Consumer Goods." In *Proceedings of the IEEE International Symposium on Electronics and the Environment*, San Francisco, May.

McGlothin, S., and E. Kroll. 1995. "Systematic Estimation of Disassembly Difficulties: Application to Computer Monitors." In *Proceedings of the IEEE International Symposium on Electronics and the Environment*, Orlando, May.

Overview of CRT Manufacturing and Management. 1997. Office of Solid Waste, U.S. Environmental Protection Agency (prepared by ICF Inc., Fairfax, Virginia). February.

Seliger, G., C. Hentschel, and M. Wagner. 1996. "Disassembly Factories for Recovery of Resources in Product and Material Cycles." In *Life-Cycle Modelling for Innovative Products and Processes*. Eds. F.-L. Krause and H. Jansen. London: Chapman & Hall.

Shu, L. H., and W. C. Flowers. 1995. "Considering Remanufacturing and Other End-of-Life Options in Selection of Fastening and Joining Methods." In *Proceedings of the IEEE International Symposium on Electronics and the Environment*, Orlando, May.

Warnecke, H.-J., M. Kahmeyer, and R. Rupprecht. 1994. "Automotive Disassembly—A New Strategic and Technologic Approach." In *Proceedings of the 2nd International Seminar on Life Cycle Engineering*, Erlangen, Germany.

Zussman E. 1995. "Planning of Disassembly Systems." *Assembly Automation* **15**(4), 20–23.

Zussman, E., A. Kriwet, and G. Seliger. 1994. "Disassembly-Oriented Assessment Methodology to Support Design for Recycling." *Annals of the CIRP* **43**(1), 9–14.

ADDITIONAL READING

Amezquita, T., and B. Bras. "Lean Remanufacturing of an Automotive Clutch." In *Proceedings of the 1st International Working Seminar on Reuse*. Eindhoven, The Netherlands, 1996.

Geiger, D., and E. Zussman. "Probabilistic Reactive Disassembly Planning." *Annals of the CIRP* **45**(1), 9–12, 1996.

Graedel, T. E., and B. R. Allenby. *Industrial Ecology.* Englewood Cliffs: Prentice-Hall, 1995.

Guide, V. D. R., Jr., and R. Srivastava. "Buffering from Material Recovery Uncertainty in a Recoverable Manufacturing Environment." *Journal of the Operational Research Society* **48**(5), 249–259, 1997.

Guide, V. D. R., Jr., M. E. Kraus, and R. Srivastava. "Scheduling Policies for Remanufacturing." *International Journal of Production Economics* **48**(2), 187–204, 1997.

Henstock, M. E. *Design for Recyclability.* London: Institute of Metals, 1988.

Pnueli, Y., and E. Zussman. "Evaluating Product End-of-Life Value and Improving It by Redesign." *International Journal of Production Research* **35**(4), 921–942, 1996.

Zeid, I., S. M. Gupta, and T. Bardasz. "A Case-Based Reasoning Approach to Planning for Disassembly." *Journal of Intelligent Manufacturing* **8**(2), 97–106, 1997.

CHAPTER 55
MICROASSEMBLY

Karl F. Böhringer
Ronald S. Fearing
Ken Y. Goldberg
University of California, Berkeley
Berkeley, California

1 INTRODUCTION

The trend toward miniaturization of mass-produced products such as disk drives, wireless communication devices, displays, and sensors will motivate fundamental innovation in design and production; microscopic parts cannot be fabricated or assembled in traditional ways. Some of the parts will be fabricated using processes similar to VLSI technology, which allows fast and inexpensive fabrication of thousands of components in parallel. Whereas a primary challenge in industrial robotics is how to securely grasp parts, at the micro scale, where electrostatic and van der Waals adhesion forces dominate, the challenge is how to release parts.

Microassembly lies between conventional (macro-scale) assembly (with part dimensions >1 mm) and the emerging field of nanoassembly (with part dimensions in the molecular scale, i.e., <1 μm). Currently microassembly is performed largely by humans with tweezers and microscopes or by high-precision pick-and-place robots. Both methods are inherently serial. Since individual parts are fabricated in parallel, it is intriguing to consider how they might be *assembled* in parallel (see Figure 1).

Microassembly poses new challenges and problems in design and control of hardware and software tools. These problems are discussed in the following section, which investigates the effects of downscaling on parts handling and gives a survey of sticking effects. Section 3 introduces a taxonomy of microassembly. Section 4 gives an overview of recent work on microassembly. Section 5 describes a new approach towards massively parallel, stochastic microassembly. The chapter concludes with a look at open problems and future trends.

Automated microassembly poses a list of new challenges to the robotics community. Conventional high-accuracy robots have a control error of 100 μm at best, which would translate to relative errors of 100% or more. Obtaining accurate sensor data is equally difficult. Sensors cannot easily be placed on tiny precision instruments without making them bulky or compromising their functionality. Image processing is one alternative, but it is still slow, costly, difficult to program, and susceptible to reflection and other noise. Moreover, the view may be obstructed by tools that are orders of magnitude larger than the parts being handled. Even when reliable images are obtained, one major challenge is how to coordinate and calibrate gross actuator motion with sensor data.

Models based on classical mechanics and geometry have been used to describe microassembly processes. However, due to scaling effects, forces that are insignificant at the macro scale become dominant at the micro scale (Fearing, 1995; Shimoyama, 1995). For example, when parts to be handled are less than one millimeter in size, adhesive forces between gripper and object can be significant compared to gravitational forces. These adhesive forces arise primarily from surface tension, van der Waals, and electrostatic attractions and can be a fundamental limitation to part handling. While it is possible to fabricate miniature versions of conventional robot grippers, for example from polysilicon (see Figure 2 (Keller and Howe, 1997) or Pister et al. (1992)), overcoming adhesion effects for the smallest parts will be difficult. Thus manipulation of parts on the order of

Handbook of Industrial Robotics, Second Edition, Edited by Shimon Y. Nof
ISBN 0-471-17783-0   © 1999 John Wiley & Sons, Inc.

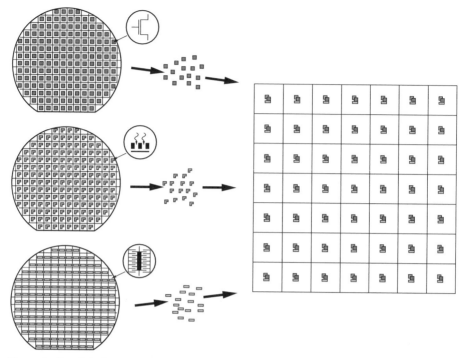

Figure 1 Parallel microassembly: Multiple micro-scale components (e.g., electronics, photonics, and MEMS) are built in parallel using standard manufacturing processes. They are positioned and combined with other components on a hybrid "pallet." Note that the fabrication density is very high, while the pallets may be of a larger size and have lower density.

10 microns or smaller may best be done in a fluid medium using techniques such as laser trapping or dielectrophoresis.

Interest in microassembly has been fueled by the availability of new components made from integrated circuits (ICs) and microelectromechanical systems (MEMS). The exponential increase in computing power of ICs has been made possible to a large extent by the dramatic advances in process miniaturization and integration. MEMS technology directly taps into this highly developed, sophisticated technology. MEMS and ICs share the efficient, highly automated fabrication processes using computer-aided design and analysis tools, lithographic pattern generation, and micromachining techniques such as thin film deposition and highly selective etching. Unlike ICs, MEMS include *mechanical* components, whose sizes typically range from about 10 to a few hundred μm, with smallest feature sizes of less than a micron and overall sizes of up to a millimeter or more. While recent years have brought an explosive growth in new MEMS devices ranging from accelerometers, oscillators, and micro optical components to microfluidic and biomedical devices, interest is now shifting towards complex microsystems that combine sensors, actuators, computation, and communication in a single micro device (Berlin and Gabriel, 1997). It is widely expected that these devices will lead to dramatic developments and a flurry of new consumer products, in analogy to the microelectronics revolution.

Current microsystems generally use *monolithic* designs, in which all components are fabricated in one (lengthy) sequential process. In contrast to the more standardized IC manufacture, a feature of this manufacturing technology is the wide variety of nonstandard processes and materials that may be incompatible with each other. These incompatibilities severely limit the manufacture of more complex devices. A possible solution to these problems is *microassembly,* the discipline of positioning, orienting, and assembling of micron-scale components into complex microsystems.

The goal of microassembly is to provide a means to achieve *hybrid* micro-scale devices of high complexity while maintaining high yield and low cost: various IC and MEMS

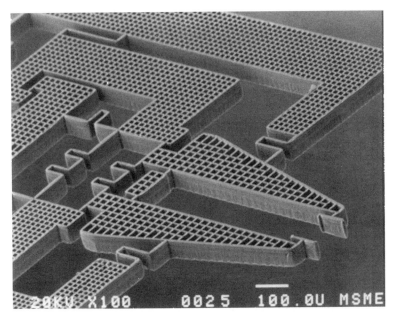

Figure 2 Microgripper made of high aspect ratio molded polycrystalline silicon. The white bar at the bottom of the picture represents 100 μm. The actuator is an electrically heated thermal expansion beam that causes the compound lever linkage to move the tips. (SEM photograph courtesy Chris Keller, Berkeley Sensor & Actuator Center.)

components are fabricated and tested individually before being assembled into complete microsystems.

2 STICKING EFFECTS FOR MICRO PARTS HANDLING

A typical robotic manipulation scenario is the sequence of operations pick, transport, and place. For parts with masses of several grams the gravitational force will usually dominate adhesive forces, and parts will drop when the gripper opens. For parts with size less than a millimeter (masses less than 10^{-6} kg), the gravitational and inertial forces may become insignificant compared to adhesive forces, which are generally proportional to surface area. When parts become very small, adhesive forces can prevent release of the part from the gripper. For example, a laser diode for an optical disk may be only 300 μm in size (Hara, 1993). Figure 4 illustrates some of the effects that can be seen in attempting to manipulate micro parts. As the gripper approaches the part, electrostatic attraction may cause the part to jump off the surface into the gripper, with an orientation dependent on initial charge distributions. When the part is placed to a desired location, it may adhere better to the gripper than to the substrate, preventing accurate placement.

Adhesion could be due to electrostatic forces, van der Waals forces, or surface tension. Electrostatic forces arise from charge generation (triboelectrification) or charge transfer during contact. Van der Waals forces are caused by instantaneous polarization of atoms and molecules due to quantum mechanical effects (Israelachvili, 1974). Surface tension effects arise from interactions of layers of adsorbed moisture on the two surfaces. This section of the chapter surveys causes of adhesion, provides estimates on the magnitude of their effect, and surveys methods for reducing the effect of adhesive forces.

For a simple numerical example to provide an idea of the scale of the adhesion forces, consider the force between a spherical object and a plane (such as one finger of the gripper in Figure 5). The approximate force between a charged sphere and a conducting plane is given by:

$$F_{\text{elec}} = \frac{q^2}{4\pi\epsilon(2r)^2} \tag{1}$$

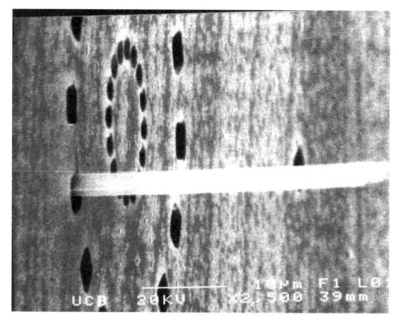

Figure 3 The peg-in-hole problem at micro scale. A $1 \times 4 \times 40$ μm silicon oxide peg was inserted into a $4 \times 4 \times 12$ μm square hole by using a microgripper as shown in Figure 2. (SEM photograph courtesy Chris Keller, Berkeley Sensor & Actuator Center.)

where q is charge, ϵ is the permittivity of the dielectric, and r is object radius. The assumed charge density is approximately 1.6×10^{-6} C m^{-2}. It is interesting to note that the contact of good insulators such as smooth silica and mica can result in charge density up to 10^{-2} C m^{-2} with pressures on the order of 10^{6} Pa at 1 μm distance (Horn and Smith, 1992).

The van der Waals force for a sphere and plane is given approximately by Bowling (1988) as:

$$F_{\text{vdw}} = \frac{hr}{8\pi z^2} \tag{2}$$

where h is the Lifshitz–van der Waals constant and z is the atomic separation between the surfaces. Of course, this formula assumes atomically smooth surfaces; severe corrections need to be made for rough surfaces since the van der Waals forces fall off very rapidly with distance. For a rough estimate, we will assume a true area of contact of 1% of apparent area, or estimated force 1% of maximum predicted with smooth surfaces.

In a high-humidity environment, or with hydrophilic surfaces, there may be a liquid film between the spherical object and planar surface contributing a large capillary force (Alley et al., 1992):

$$F_{\text{tens}} = \frac{\gamma(\cos\,\theta_1 + \cos\,\theta_2)A}{d} \tag{3}$$

where γ is the surface tension (73 mNm^{-1} for water), A is the shared area, d is the gap between surfaces, and θ_1, θ_2 are the contact angles between the liquid and the surfaces. Assuming hydrophilic surfaces and a separation distance much smaller than the object radius (Bowling, 1988; Sze, 1981):

$$F_{\text{tens}} = 4\pi r\gamma \tag{4}$$

where r is the object radius.

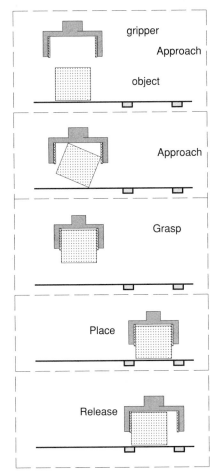

Figure 4   Pick-move-place operation with micro parts. Due to sticking effects, parts may be attracted to the gripper during the approach and release phase, causing inaccurate placement.

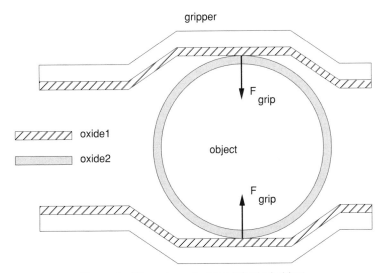

Figure 5   Microgripper holding spherical object.

For a spherical part of silicon the gravitational force is:

$$F_{grav} = \frac{4}{3}\pi r^3 \rho_{Si} g \tag{5}$$

where $\rho_{si} = 2300$ kg m^{-3} is the density of silicon. Figure 6 shows the comparison of forces. For accurate placement, adhesion forces should be an order of magnitude less than gravitational forces. Capillary forces dominate and must be prevented to allow accurate placement. Van der Waals forces can start to be significant (with smooth surfaces) at about 100 μm radius, and generated electric charges from contacts could prevent dry manipulation of parts less than 10 μm in size.

While Figure 6 shows electrostatic to be the least significant force except for gravity, it can be argued that it is actually the most significant force for grasping and manipulation of 10 μm to 1 mm parts. First, the van der Waals force is significant only for gaps less than about 100 nm (Scheeper et al., 1992; Israelachvili, 1974). Unless rigid objects are very smooth, the effective distance between the object and the gripper will be large except at a few points of contact. Second, actual contact with a fluid layer needs to be made for surface tension to be significant, and a dry or vacuum environment could be used to eliminate surface tension effects. Finally, the electrostatic forces can be active over ranges of the order of the object radius. Surface roughness is much less important for electrostatic forces than for van der Waals.

2.1 Literature on Adhesion

The adhesion of particles to substrates has received substantial study for problems such as particulate contamination in semiconductor manufacturing (Krupp, 1967; Zimon, 1969; Bowling, 1968; Hecht, 1990; Jones, 1995). The recent developments in MEMS, disk drives, and microassembly have stimulated the study of friction effects at the micro scale. The normal Coulomb friction effects seen at the macro scale are quite different at the micro-scale with large adhesive components. Several studies have examined surfaces using the atomic force microscope (Torii et al., 1994; Kaneko, 1991). A common problem in MEMS devices is that freestanding microstructures tend to stick to the substrate after being released during processing. The dominant mechanisms for sticking in these devices

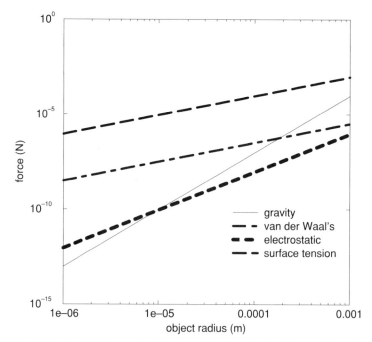

Figure 6 Gravitational, electric, van der Waals, and surface tension forces. Attractive force between sphere and plane.

(which are typically constructed as a cantilever plate suspended 1 or 2 μm above the substrate) appears to be surface tension pulling the plate down, followed by van der Waals bonding. Recent papers have studied this problem (Ando, Ogawa, and Ishikawa, 1991) and proposed solution methods of making the surfaces rough and hydrophobic (Legtenberg, 1994; Alley et al., 1992; Alley et al., 1994; Scheeper et al., 1992).

2.2 Adhesion Due to Electrostatic Forces

Ensuring that parts and grippers are electrically neutral is difficult (Harper, 1967). Significant amounts of charge may be generated by friction forces and differences in contact potentials. While grounded conductors will drain off charge, insulators can maintain very high surface charge distributions. The local field intensity near a surface charge distribution can be estimated using Gauss's law and Figure 7. Neglecting any interior field, the boundary conditions give

$$\hat{z} \cdot \epsilon_0 \vec{E}(z = 0) = \sigma_s \tag{6}$$

where $\vec{E}$ is the electric field, $\hat{z}$ is the surface normal, and σ_s is the surface charge density Cm^{-2}. The near field approximation for a surface charge is then:

$$\vec{E} \approx \frac{\sigma_s}{\epsilon_0} \hat{z} \tag{7}$$

The force per unit area for parallel plates is

$$P = \frac{1}{2} \epsilon_0 |E|^2 = \frac{\sigma_s^2}{2\epsilon_0} \tag{8}$$

where P is the pressure in Pascals.

At atmospheric pressure and centimeter size gaps, the breakdown strength of air (about 3×10^6 V m^{-1} (Lowell and Rose-Innes, 1980)) limits the maximum charge density to about 3×10^{-5} C m^{-2}, or peak pressures of about 50 Pa. Let l be the length of a side of a cube of silicon. Then the smallest cube which will not stick due to electrostatic force is:

$$l = \frac{\sigma_s^2}{2\epsilon_0 \rho_{Si} g} \tag{9}$$

or about 2 mm minimum size. Of course, a uniform charge distribution over such a large area is unlikely, although there could be local concentrations of charge of such magnitude. However, at very small gaps of the order of 1 μm (less than the mean free path of an electron in air), fields two orders of magnitude higher have been observed (Horn and Smith, 1992).

2.2.1 Contact Electrification

When two materials with different contact potentials are brought in contact, charge flows between them to equalize this potential. For metal–metal contact (Lowell and Rose-Innes, 1980; Krupp, 1967), a rough approximation to the surface charge density is:

$$\sigma_s = \frac{\epsilon_0 U}{z_0} \tag{10}$$

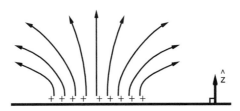

Figure 7 Field approximation near surface charge distribution.

Table 1   Charge from Contact Electrification[a]

Materials in Contact	Charge Density mCm⁻²	Electrostatic Pressure Nm⁻²	Condition	Reference
SiO_2–Al	2.0	2×10^5	1 mPa N_2 80 nm SiO_2	Lowell, 1990
soda glass–Al	0.13	10^3		
SiO_2–mica	5–20	1.4×10^6 to 20×10^6	N_2 at STP "atomically smooth"	Horn and Smith, 1992
epoxy–Cu	0.04	100	10^4 Pa air	Kwetkus, Gellert, and Sattler, 1991
glass–Au	4.2	10^6	air	Harper, 1967
nylon–steel	.0036	1	40–60% RH	
polystyrene	.0002	2×10^{-6}		

[a] Pressure is the effective pressure due to the generated charge.

where U is the contact potential difference, which is typically less than 0.5 V, and z_0 is the gap for tunneling, about 1 nm. Consider two metal spheres (insulated from their surroundings) brought into contact, then slowly separated. With a contact potential of 0.5 V, the initial charge density according to Equation (10) will be about 4 m cm⁻², with field strength 5×10^8 V m⁻¹. For small gaps (order 1 nm), electron tunneling and field emission (Lowell and Rose-Innes, 1967) will transfer charge, and then for larger gaps (order 1 μm) air breakdown can occur. In laboratory experiments, contact electrification has been shown to generate significant charge density, which could cause adhesion (see Table 1).

2.2.2   Charge Storage in Dielectrics

In principle, using conductive grippers can reduce static charging effects. However, the objects to be handled, such as silicon parts, can be covered with good insulator layers, such as native oxides. Up to 1 nm of native oxide is possible after several days in air at room temperature (Morita et al., 1990). This native oxide is a very good insulator and can withstand a maximum field strength of up to 3×10^9 V m⁻¹ (Sze, 1981). This implies that significant amounts of charge can be stored in the oxide. With the permittivity of silicon $\epsilon = 3.9\epsilon_0$, peak pressures according to Equation (8) are on the order of 10^8 Pa. With a contact area of only 10 (nm)², this would be a force of 10 nN, enough to support a 30 μm cube against gravity.

Consider an initially charged object grasped, as shown in Figure 5, by a grounded gripper. In regions where the two dielectrics are not in contact, charge will be induced on the opposite surface. As suggested in Figure 8, local regions of charge can remain in the dielectric layer in spite of "intimate" contact between two nominal conductors. The surface roughness can prevent charge neutralization through intimate contact of oppositely charged regions. The residual charge can cause adhesion.

It can be very difficult to remove stored charge in a dielectric layer. Consider a simplistic model for the electrical contact, with one capacitor representing the air gap and a second capacitor in series representing the dielectric layer, as shown in Figure 9. It is apparent that shorting the terminals will not instantaneously remove charge from both capacitors, hence there will be a residual attraction force between the gripper and the

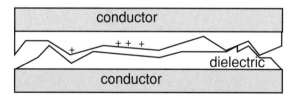

Figure 8   Physical model of contact with charge in oxide layer.

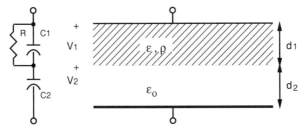

Figure 9 Equivalent circuit model of contact. ρ is the resisitivity of the dielectric.

object. The stored charge (and hence electric field) decays as a first-order exponential, with time constant:

$$\tau = \rho \left(\epsilon + \epsilon_0 \frac{d_1}{d_2} \right) \tag{11}$$

where ρ is the resistivity of the dielectric. For SiO_2 with resistivity $\rho = 10^{12}$ Ωm, dielectric thickness 10 nm, and air gap 20 nm, the time constant τ is about 40 seconds, significantly reducing cycle time. Charge storage in dielectric layers may result in undesired adhesions in electrostatic grippers (Nakusuji and Shimizu, 1992) and in electrostatic microactuators where contact is made with an insulating layer (Anderson and Colgate, 1991).

2.3 Summary

As we have seen, electrostatic, van der Waals, and surface tension forces can be significant compared to the weight of small parts. Conventional assembly methods such as pick-move-place do not scale well for submillimeter parts. One possible attractive alternative is assembly while immersed in a fluid, which eliminates electrostatic and surface tension effects (Yeh and Smith, 1994a).

Several design strategies can be used to reduce adhesive effects in microgrippers. Figure 10 compares fingertip shapes. Clearly the spherical fingertip has reduced surface contact area and better adhesive properties, unlike polysilicon microgrippers fabricated using planar surface micromachining.

Proper choice of gripper materials and geometry can be used to reduce adhesion:

1. Minimize contact electrification by using materials with a small contact potential difference for the gripper and object.
2. Use conductive materials that do not easily form highly insulating native oxides.
3. Gripper surfaces should be rough to minimize contact area.
4. The high contact pressure from van der Waals and electrostatic forces can cause local deformation at the contact site (Bowling, 1988). This deformation can increase the contact area and increase the net adhesive force. Hard materials are preferable to rubber or plastic.
5. A dry atmosphere can help to reduce surface tension effects. Surface tension can be used to help parts adhere better to the target location than the gripper.
6. Free charges such as in ionized air can combine with and neutralize exposed surface charges.

As discrete parts are designed continually smaller to make equipment smaller, more economical, and higher-performance, there will be a greater need for understanding how to manipulate and assemble microparts. Because of adhesive forces, grasping and particularly ungrasping of these parts can be complicated. Good models of surfaces and the physics of contact will be needed to implement reliable manipulation and assembly systems.

3 TECHNIQUES FOR MICROASSEMBLY

Current micromachined devices generally use monolithic designs in which all components are fabricated in one (lengthy) sequential process. Recently microassembly has been proposed as a means to achieve hybrid micro-scale devices of high complexity, while main-

Grasping with Spherical Fingertips

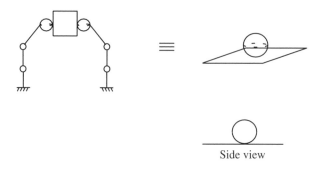

Side view

Grasping with Planar Fingertips

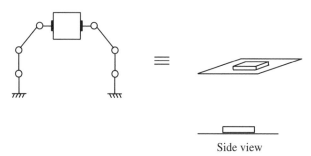

Side view

Figure 10 Comparison of finger types for grasping. A spherical fingertip will minimize electro-static and surface tension forces and can be roughened to minimize van der Waals forces.

taining high yield and low cost: various electronic and mechanical components are fabricated and tested individually before being assembled into complete systems. (See, e.g., Cohn, Kim, and Pisano, 1991; Cohn, 1992; Hosokawa, Shimoyama, and Miura, 1995; Yeh and Smith, 1994b, 1994c; Nelson and Vikramaditya, 1997; Böhringer et al., 1998.)

In this section we attempt to characterize the techniques currently in use for microas-sembly. Since microassembly is a new and very active area of research, this characteri-zation may not be complete, and other taxonomies are certainly possible.

3.1 A Taxonomy of Microassembly

1. *Serial microassembly:* parts are put together one by one according to the traditional pick-and-place paradigm. Serial microassembly includes the following techniques:
   - Manual assembly with tweezers and microscopes.
   - Visually based and teleoperated microassembly (Nelson and Vikramaditya, 1997; Feddema and Simon, 1998).
   - High-precision macroscopic robots: stepping motors and inertial drives are used for submicrometer motion resolution (see, e.g., Quaid and Hollis, 1976; Zesch, 1997; Danuser et al., 1997) or MRSI[1] assembly robots for surface-mount electronics components of submillimeter size).

---

[1]MRSI International, 25 Industrial Ave., Chelmsford, MA 01824.

- Microgrippers (Kim, Pisano, and Muller, 1992; Pister et al., 1992; Keller and Howe, 1995, 1997; see, e.g., Figure 2) with gripper sizes of 100 μm or less.

2. *Parallel microassembly:* multiple parts (of identical or different design) are assembled simultaneously. We distinguish two main categories:
  - *Deterministic:* The relationship between part and its destination is known in advance.
    - Flip-chip wafer-to-wafer transfer: a wafer with partially released components is carefully aligned and pressed against another substrate. When the wafers are separated again, the components remain bonded to the second substrate (Cohn and Howe, 1997; Singh et al., 1997; see, e.g., Figure 11).
    - Microgripper arrays (Keller and Howe, 1997) capable of massively parallel pick-and-place operations.
  - *Stochastic:* The relationship between the part and its destination is unknown or random. The parts "self-assemble" during stochastic processes, in analogy to annealing. The following effects can be used as motive forces for stochastic self-assembly:
    - Fluidic agitation and mating part shapes (Yeh, 1994*b*, 1994c; Whitesides, Mathias, and Seto, 1991).
    - Vibratory agitation and electrostatic force fields (Cohn, 1992; Cohn, Howe, and Pisano, 1995; Böhringer et al., 1995; Böhringer et al., 1998*b*).
    - Vibratory agitation and mating part shapes (Hosokawa, Shimoyama, and Miura, 1995).
    - Mating patterns of self-assembling monolayers (Srinivasan and Howe, 1997).

3.2 A Hierarchy of Assembly Forces

We noted in the introduction that for parts of dimensions 1 mm or less, surface adhesion forces may dominate "volume" forces such as gravity or inertia. Parts are trapped in locations where these adhesion forces are sufficiently strong, and may not be released (for example from a microgripper) even if traditional (macroscopic) dynamical analysis does not show force closure. Hence it is essential to have control over these forces during microassembly. One common technique to overcome adhesion is to employ vibration. Note, however, that since this technique relies on the inertia of the parts, vibration be-

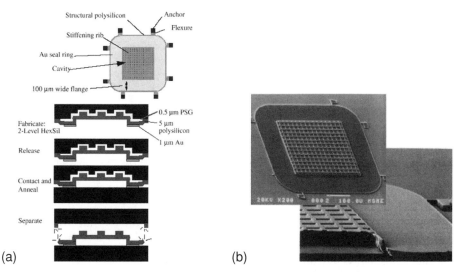

(a)   (b)

Figure 11 Microassembly by wafer-to-wafer transfer using breakaway tethers. (*a*) Process flow for a wafer-to-wafer transfer task. (*b*) Above: a micro shell structure after wafer-to-wafer transfer. Below: cleaved cross section of the micro shell. (From M. B. Cohn, "Assembly Techniques for Microelectromechanical Systems," Ph.D. Thesis, University of California at Berkeley, Department of Electrical Engineering and Computer Science, 1997.)

comes less efficient with decreasing part sizes, i.e., higher vibration amplitudes or frequencies are necessary for smaller parts.

As an example, consider the task of palletizing micro parts (as described in Figure 1 or Section 5). During the assembly process adhesion forces have to be overcome in the initial positioning phase as well as the final bonding phase. However, adhesion is used to keep the parts in place during the part transfer phase.

1. *Positioning:* Forces provided by vibration are larger than trapping forces (van der Waals, electrostatic, surface tension). The part can move freely and exhibits trajectories resembling Brownian motion.
2. *Annealing:* The vibration forces are gradually reduced until the trapping forces dominate. The part settles at a local potential minimum.
3. *Bonding:* Target spot adhesion (e.g., by indium soldering during wafer-to-wafer transfer) is greater than the trapping forces. Permanent bonds are created between part and target substrate.

3.3 Issues and Problems

Earlier work on microfabrication has looked almost exclusively at *in situ* batch fabrication, where (in accordance with the IC fabrication paradigm) all components are built in one fabrication process on a substrate (usually a silicon wafer). The shortcomings of this approach have already been outlined in the introduction. They include:

- *Incompatible materials:* For example, many opto-electronics require GaAs substrates that are incompatible with standard electronics.
- *Incompatible processes:* For example, processes requiring high temperatures destroy CMOS circuitry.
- *Exponential decline in yield:* Each step in a processing sequence has a nonzero failure probability associated with it. These probabilities multiply and hence dramatically reduce the yield for long processing sequences, prohibiting a complex process generated by simple concatenation of standard processes.

Microassembly overcomes these problems and makes possible hybrid devices with otherwise incompatible materials such as bipolar transistors, MOSFETs, photoelectronic components, and mechanical structures.

4 RECENT RESEARCH IN MICROASSEMBLY

Vibration has been widely in use in industrial parts feeders. A parts feeder is a machine that singulates, positions, and orients bulk parts before they are transferred to an assembly station. The most common type of parts feeder is the *vibratory bowl feeder,* in which parts in a bowl are vibrated using a rotary motion, so that they climb a helical track. As they climb, a sequence of baffles and cutouts in the track create a mechanical "filter" that causes parts in all but one orientation to fall back into the bowl for another attempt at running the gauntlet (Boothroyd, Poli, and Murch, 1982; Riley, 1983; Sandler, 1991). Sony's APOS parts feeder (Hitakawa, 1998) is another example of using vibration for parts handling. It uses an array of nests (silhouette traps) cut into a vibrating plate. The nests and the vibratory motion are designed so that the part will remain in the nest only in one particular orientation. Tilting the plate and letting parts flow across it allows the nests to fill up eventually with parts in the desired orientation. Although the vibratory motion is under software control, specialized mechanical nests must be designed for each part (Moncevicz, Jakiela, and Ulrich, 1991).

The term *self-assembly* has been applied to spontaneous ordering processes such as crystal and polymer growth. Recently the manufacture of systems incorporating large numbers of micro devices has been proposed. Positioning, orienting, and assembly are done open-loop, without sensor feedback. The principle underlying the APOS system (controlled vibration to provide stochastic motion, combined with gravity as a motive force for parallel, nonprehensile manipulation) is well suited for micro-self-assembly. Stochastic microassembly often encompasses vibration in combination with electrostatic, fluidic, and other forces that operate on singulated parts in various media (fluids, air, or vacuum).

Cohn, Kim, and Pisano (1991) reported on stochastic assembly experiments that use vibration and gravitational forces to assemble periodic lattices of up to 1000 silicon chiplets. Following work demonstrated the use of patterned electrodes to enable assembly of parts in arbitrary 2D patterns. In addition, hydrophobic–hydrophilic interactions in liquid media were employed for 3D self-assembly of millimeter-scale parts (Cohn, 1992). Electrostatic levitation traps were also described by Cohn, Howe, and Pisano (1995) with the aim of controlling friction. This work demonstrated a novel type of electrostatic interaction essentially unique to the micromechanical regime. Recent results by Böhringer, Goldberg, and the above authors have demonstrated the ability to break surface forces using ultra-low-amplitude vibration in vacuum ambient (Böhringer et al., 1997a). This promises to be an extremely sensitive technique for positioning parts as well as discriminating part orientation, shape, and other physical properties.

Yeh and Smith (1994b, 1995c) have demonstrated high-yield assembly of up to 10,000 parts using fluidic self-assembly. Work to date has focused on fabrication of parts and binding sites with desired trapezoidal profiles, surface treatments for control of surface forces, as well as mechanical and electrical interconnection of assembled parts. Parts have included both silicon and III–V devices. Semiconductor junction lasers were suspended in liquid and trapped in micromachined wells on a wafer by solvent–surface forces.

Hosokawa, Shimoyama, and Miura (1995) have analyzed the kinetics and dynamics of self-assembling systems by employing models similar to those used to describe chemical reactions. They performed assembly experiments with planar parts of various simple geometries at macro and micro scales (for example, assemblies of hexagons from isosceles triangles).

Deterministic parallel assembly techniques have been developed by Cohn and Howe for wafer-to-wafer transfer of MEMS and other microstructures (Cohn et al., 1996; Cohn and Howe, 1997). In their flip-chip process the finished microstructures are suspended on breakaway tethers on their substrate (Figure 11). The target wafer with indium solder bumps is precisely aligned and pressed against the substrate such that the microstructures are cold-welded onto the target wafer. This technique is well suited for fragile parts. However, it is not appropriate for large numbers of small parts since the tether suspensions and solder bumps require too much surface area. High-quality electrical, mechanical, and hermetic bonds have been demonstrated by Singh et al. (1997). Parts of 10–4000 μm size have been transferred, including functioning microresonators and actuators. High yield (>99%) as well as 0.1 μm precision have been demonstrated.

By downscaling and parallelizing the concept of a robot gripper, Keller and Howe (1995, 1997) have demonstrated microgrippers (Figure 2) and propose gripper arrays for parallel transfer of palletized micro parts. Arai and Fukuda (1997) built a manipulator array with heated micro holes. When the holes cool down, they act as suction cups whose lower pressure holds appropriately shaped objects in place. Heating of the cavities increases the pressure and causes the objects to detach from the manipulator.

Quaid and Hollis (1996) built extremely accurate systems for precision robotic assembly. Nelson and Vikramaditya (1997) used teleoperation and visual feedback for microassembly.

Several groups of MEMS researchers have designed and built actuator arrays for micromanipulation, which usually consist of a regular grid of *motion pixels*. Devices were built by, among others, Pister, Fearing, and Howe (1990), Fujita (1993), Böhringer et al. (1994; 1996), Kovacs et al. (Storment et al., 1994; Suh et al., 1996), and Will et al. (Liu et al., 1995; Liu and Will, 1995).

MEMS actuator arrays that implement *planar force fields* were proposed by Böhringer et al. (1994), who also built single-crystal silicon actuator arrays for micromanipulation tasks (1996). Micro cilia arrays fabricated by Suh (1996) were extensively used in their experiments, which successfully demonstrated strategies for parts translating, orienting, and centering (Böhringer et al., 1997b). This research in micromanipulation, which built on recent advances in sensorless manipulation (see Erdmann and Mason (1988), and Goldberg (1993)), again motivated various new macroscopic devices such as vibrating plates (see Böhringer, Bhatt, and Goldberg (1995); (Böhringer et al., 1998) and Reznik, Canny, and Goldberg (1997)) or a "virtual vehicle" consisting of a two-dimensional array of roller wheels that can generate motion fields in the plane (see Luntz and Messner (1997) and Luntz, Messner, and Choset (1997)). We expect this cross fertilization between macro and micro robotic systems to continue and expand in the future.

Research in MEMS and microassembly is substantially different from the rapidly growing research in nanotechnology, which endeavors to design and construct materials and devices at a molecular scale. Components typically consist of individual molecules or atoms, with dimensions in the nanometer range (1 nm = 10^{-9} m)—approximately three to four orders of magnitude below the range of microassembly. This chapter focuses solely on microtechnology. Microtechnology is generally seen as "top-down" discipline whose goal is to scale down traditional mechanisms. It draws from classical mechanics, robotics, and control theory. In contrast, nanotechnology constitutes a "bottom-up" approach from individual atoms and molecules to nanomachines. Its science base includes molecular chemistry and physics. Nevertheless, chemistry also serves as an inspiration and paradigm for microassembly, and models of chemical reactions can be used to analyze stochastic assembly of microcomponents (see Section 5). Recent groundbreaking work in nanoassembly has been performed by Whitesides et al. (Seto and Whitesides, 1991; Prime and Whitesides, 1991; Whitesides, Mathias, and Seto, 1991).

5 STOCHASTIC MICROASSEMBLY WITH ELECTROSTATIC FORCE FIELDS

Currently microassembly methods by humans or by high-precision pick-and-place robots are inherently serial. Since individual parts are fabricated in parallel, parallel assembly can be useful. In this section we propose a concept for massively parallel assembly.

The idea is to arrange microscopic parts on a reusable pallet and then press two pallets together, thereby assembling the entire array in parallel. We focus on how to position and align an initially random collection of identical parts. This approach builds on the planning philosophy of sensorless, nonprehensile manipulation pioneered by Erdmann and Mason (1988). To model electrostatic forces acting on parts moving on a planar surface, we use the *planar force field,* an abstraction defined with piecewise continuous functions on the plane that can be locally integrated to model the motion of parts (Böhringer et al., 1994).

Planar force fields, as defined by the magnitude and direction of force at each point, can be designed to position, align, and sort arrays of microscopic parts in parallel. Developing a science base for this approach requires research in device design, modeling, and algorithms.

As a feasibility study, we perform experiments to characterize the dynamic and frictional properties of microscopic parts when placed on a vibrating substrate and in electrostatic fields. We first demonstrate that ultrasonic vibration can be used to overcome friction and adhesion of small parts. In a second set of experiments we describe how parts are accurately positioned using electrostatic traps. We are also working to model part behavior as a first step toward the systematic design of planar force fields. The input is part geometry and desired final arrangement, and the output is an electrode pattern that produces the appropriate planar force field.

5.1 Experimental Apparatus

A piezoelectric actuator supports a vibratory table consisting of a rigid aluminum base, which has a flat glass plate (25 mm × 25 mm × 2 mm) attached to its top. A thin chrome gold layer (1000 Å) is evaporated onto the glass and patterned using photolithography. The signal from a function generator is amplified and transformed to supply the input voltage for the piezo transducer. The piezo is driven at ultrasonic frequencies in the 20 kHz range. At resonance we observe amplitudes of up to 500 nm (measured with laser interferometry), which correspond to accelerations of several hundred g's. Figure 12 shows a diagram of the experimental setup. The current experimental apparatus is shown in Figure 13. The apparatus can be operated in air or in a vacuum chamber.

Voltage is applied between the aluminum vibratory table and the chrome–gold electrode, which together act as a parallel plate capacitor. The applied voltage is limited by the breakdown voltage of air and glass and the path length (air: $3 \cdot 10^6$ V/m 1 cm = 30 kV; glass: 10^9 V/m 2 mm = 2 MV). The patterned top electrode creates fringing electrostatic fields. Its effect is a potential field whose minima lie at apertures in the top electrode. Parts are attracted to these electrostatic "traps."

The parts employed in our experiments are mainly surface-mount diodes and capacitors. They usually have rectangular shapes with dimensions between 0.75 mm and 2 mm. We also performed experiments with short pieces of gold wire (0.25 mm diameter).

5.2 Experimental Observations

5.2.1 Overcoming Friction and Adhesion

Small parts were randomly distributed on the substrate. When no signal is applied to the piezo, the parts tend to stick to the substrate and to each other due to static charges or

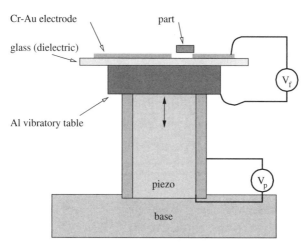

Figure 12 Experimental apparatus for self-assembly with electrostatic traps. A vibratory table with a gold-covered dielectric is attached to a piezoelectric actuator. The aperture in the upper electrode creates a fringing field that causes polarization in the part. The part is attracted to the aperture.

capillary or van der Waals forces. When sinusoidal signals of various frequencies and amplitude are applied, the parts break contact. This behavior was particularly pronounced at resonance frequencies (e.g., observed in the 20 kHz range). In this case the motion of the parts resembles liquid: tilting of the substrate surface by less than 0.2% was sufficient to influence their direction of (down-slope) motion. This implies a friction coefficient $\mu <$ 0.002.

When the substrate surface was leveled carefully, the parts exhibited random Brownian motion patterns until they settled in a regular grid pattern. This important observation is a strong indication that the system is sufficiently sensitive to react even to very small surface forces.

At high signal amplitudes the vibration induces random bouncing of the parts. Reducing the amplitude accomplishes an annealing effect; at lower amplitudes only in-plane translations and rotation occur. After such annealing sequences, surface mount diodes consistently settled with their solder bumps facing up. This observation suggests that even

Figure 13 Experimental apparatus for self-assembly experiments. A lithographically patterned electrode is attached to a piezoelectric actuator (vertical cylinder). Some parts can be seen in the lower left quadrant of the substrate.

very small asymmetries in part design can be exploited to influence its final rest position. Voltages of $V_{pp} = 2$ V were sufficient to sustain free motion of the parts. This corresponds to a vibration amplitude of approximately 30 nm.

5.2.2 Vacuum Experiments

These experiments were repeated both in air and in low vacuum (high mTorr range). First results indicate that the energy required to overcome adhesive forces decreases with pressure, probably due to squeeze film effects (Fearing, 1995a) and the vacuum created between the flat part bottom surface and the substrate when operated at ultrasonic frequencies. As a result, the atmospheric pressure acting on the top surface presses the part onto the surface. For example, simple calculations show that if a rectangular part with dimensions 1 mm × 1 mm × 0.1 mm and mass 0.1 mg were exposed to atmospheric pressure on one side and to vacuum on the other side, it would experience an acceleration of nearly 100,000 g.

5.2.3 Electrostatic Self-Assembly and Sorting

The electrode design represents a parallel-plate capacitor with apertures in the upper electrode. The resulting fringing fields induce polarization in neutral parts, so that they are attracted to the apertures and become trapped there. Once a part is trapped, it reduces the fringing field, which prevents attraction of more parts to this location. Figure 16 shows the positioning of four surface mount capacitors on four sites. The binding times for parts were automatically measured with an optical sensor and a recording oscilloscope. They exhibit an exponential distribution (Figure 14) with expected time of approximately 30 seconds.

5.2.4 Parts Sorting by Size

Large and small parts were mixed and placed randomly on a vibrating surface slightly tilted by ≈1°. Vibration caused a sorting effect such that parts were separated, with smaller parts settling at the lower end of the vibrating surface.

5.3 Modeling and Simulation

A variety of effects influence the behavior of the parts used in microassembly experiments including:

1. Electrostatic fields created by capacitor plates
2. Conductivity or dielectric constants of parts
3. Induced dipoles
4. Static charges on nonconductive and electrically isolated conductive parts

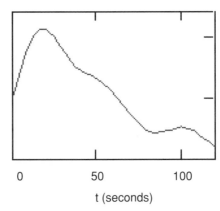

t (seconds)

Figure 14 Histogram of binding times for electrostatic trapping, from an experiment with a total of 70 sample runs. Data exhibit an exponential distribution.

Results from modeling based on a smooth approximation of the electrostatic potential are shown in Figure 15. The potential U is created by an electrode design as shown in Figure 16. The corresponding planar force field $F = \nabla U$ is shown in Figure 15b, together with a simulation of a part moving in the field. In this simulation the effective force on the part F_P was determined by integrating the force field over the part area $F_P = \int_P F\, da$ (a more accurate model will take into account the deformation of the field by the part, as well as, e.g., changes in its induced charge distribution). Then the force F_P is integrated over time to determine the part motion.

5.4   Algorithmic Issues for Massively Parallel Manipulation

As shown in the previous sections, planar force fields (PFFs) constitute a useful tool for modeling massively parallel, distributed manipulation based on geometric and physical reasoning. Applications such as parts feeding can be formulated in terms of the force fields required. Hence planar force fields act as an abstraction between applications requiring parallel manipulation and their implementation, e.g., with MEMS or vibratory

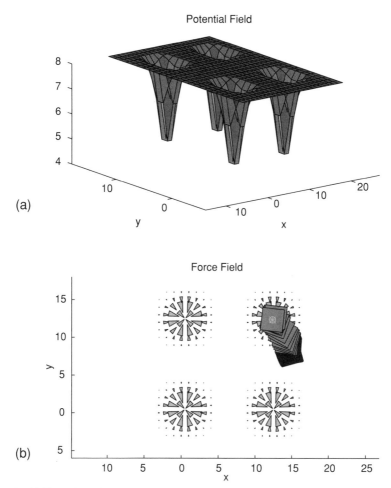

Figure 15 (a) Potential field created by an electrode with four small square-shaped apertures, as shown in the experimental setup in Figure 16. The four potential traps correspond to the four apertures. (b) Simulation of a square part moving in the corresponding force field (denoted by force vectors). The part translates and rotates until it reaches a local minimum in the potential field.

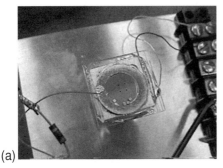
(a)

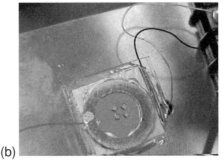
(b)

Figure 16 Parallel microassembly with electrostatic force fields. (a) Surface-mount capacitors are placed onto a glass substrate with a 100-nm thin patterned Cr–Au electrode. Frictional and adhesive forces are overcome by ultrasonic vibration. (b) Voltage applied to the electrode creates an electrostatic field. The parts are attracted to the apertures in the electrode (dark squares) and are trapped there.

devices. Such abstractions permit hierarchical design and allow application designs with greater independence from underlying device technology.

5.4.1 Recently Developed PFFs

Böhringer et al. (1997b) established the foundations of massively parallel manipulation with force fields. Among the PFFs developed in the past years the following have been thoroughly investigated:

> *Squeeze field:* Squeeze fields are fields with unit forces pointing perpendicularly towards a straight *squeeze line* (e.g., $\vec{F}(x, y) = (-\mathrm{sign}(x), 0)$). When placed in a squeeze field, every part reaches one out of a small number of possible equilibria.
>
> *Radial field:* A unit radial field is given by $\vec{F}(x, y) = -1/\sqrt{(x^2 + y^2)}\,(x, y)$, $(x, y) \neq 0$, and 0 otherwise. In a radial field, any polygonal part has a unique *pivot point*. The part is in a unique translational equilibrium if and only if its pivot point coincides with the center of the squeeze field.
>
> *Elliptic field:* The elliptic PFF (see Kavraki (1997)) is a continuous field of the form $\vec{F}(x, y) = (-\alpha x, -\beta y)$, where α and β are two distinct positive constants. The field poses and orients nonsymmetric parts into two stable equilibrium configurations.

5.4.2 Motion Planning with Artificial and Physical Potential Fields

Robotics motion planning is concerned with the problem of moving an object from an initial configuration q_i to a goal configuration q_g. In our case a manipulation plan consists of a sequence of planar force fields. A general question that arises in the context of PFFs is the following: *Which force fields are suitable for manipulation strategies?* That is: can we characterize all those force fields in which every part has stable equilibria? To answer these questions, we use recent results from the theory of *potential fields.* It can be shown that certain PFFs that implement potential fields have this property, whereas fields without potential do not induce stable equilibria on all parts. Previous work has developed control strategies with *artificial* potential fields (Khatib, 1986; Koditschek and Rimon, 1988; Rimon and Koditschek, 1992; Reif and Wang, 1995) and discrete approximations to physical potential fields (Böhringer et al., 1994; Böhringer et al., 1997b). The fields employed in this paper are nonartificial (i.e., *physical*). Artificial potential fields require a tight feedback loop, in which, at each clock tick, the robot senses its state and looks up a control (i.e., a vector) using a state-indexed navigation function (i.e., a vector field). In contrast, physical potential fields employ no sensing, and the motion of the manipulated object evolves open-loop (for example, like a body in a gravity field). Hence, for physical potential fields such as electrostatic fields, the motion planning problem has to be solved during device design. A design algorithm takes as input part geometry and desired goal configurations and returns an electrode geometry that creates the proper potential field.

During execution the system runs open-loop. This shift of complexity from run-time to design-time is crucial for efficient parallel microassembly methods.

5.5 Summary

Our experiments show that friction and adhesion between small parts can be overcome by ultrasonic vibration. We demonstrated that in such an effectively frictionless environment small parts can be accurately positioned in parallel with electrostatic traps. This research opens the door to parallelize the manufacture of a new generation of consumer and industrial products, such as hybrid IC/MEMS devices, flat panel displays, and VCSEL arrays.

The behavior of the parts on the substrate can be modeled using planar force fields, which describe the effective lateral force acting on the part (as a function of its location in configuration space). A key problem is to determine an electrode design that creates a specific planar force field such that parts are reliably positioned and oriented at desired locations. We attack this problem by the development of efficient models for manipulation in electrostatic force fields, and with new algorithms for motion planning with planar force fields.

Planar force fields have enormous potential for precise parallel assembly of small parts. The goal is to develop entirely new methodologies for precision part manipulation, algorithms, and high-performance devices. For updated information on this project refer to WWW pages at ⟨www.ee.washington.edu/faculty/karl/ElStatAssembly⟩. See also Chapters 10, 11 and 51.

6 CONCLUSION AND EMERGING TRENDS

Miroassembly is a challenging new area of research. This chapter constitutes an initial attempt to identify important issues in this field. We investigate surface sticking effects at the micro scale, give a taxonomy of microassembly techniques, discuss a hierarchy of assembly forces, and outline a brief summary of the current state of the art in microassembly. We also discuss a specific new technique for massively parallel assembly by employing vibration to overcome adhesion and electrostatic forces to position micro parts in parallel.

Advances in the field of microassembly can be expected to have an enormous impact on the development of future miniaturized consumer products such as data storage systems and wireless communication devices. Techniques known from robotics, such as vibratory parts feeding, teleoperation, and sensorless and planar force field manipulation, provide useful tools for research in microassembly. However, the possibly huge numbers of tiny parts employed in microassembly pose specific and unique challenges that will require innovative or unconventional solutions. These results may in turn inspire new approaches and techniques for macroscopic robots.

ACKNOWLEDGMENTS

The authors would like to thank John Canny, Michael Cohn, Bruce Donald, Anita Flynn, Hiroaki Furuichi, Roger Howe, Lydia Kavraki, Al Pisano, and Kris Pister for many fruitful discussions.

Work reported in this chapter has been supported in part by an NSF grant on Challenges in CISE: Planning and Control for Massively Parallel Manipulation (CDA-9726389), an NSF CISE Postdoctoral Associateship in Experimental Computer Science to Karl Böhringer (CDA-9705022), an NSF Presidential Faculty Fellowship to Ken Goldberg (IRI-9553197), an NSF Presidential Young Investigator grant to Ron Fearing (IRI-9157051), and NSF grant IRI-9531837.

REFERENCES

Alley, R. L., et al. 1992. "The Effect of Release-Etch Processing on Surface Microstructure Stiction." In *Proceedings of Solid State Sensor and Actuator Workshop,* Hilton Head Island, June. 202–207.

Alley, R. L., et al. 1994. "Surface Roughness Modification of Interfacial Contacts in Polysilicon." In *Transducers—Digest of International Conference on Solid-State Sensors and Actuators,* San Francisco, June. 288–291.

Anderson, K. M., and J. E. Colgate. 1991. "A Model of the Attachment/Detachment Cycle of Electrostatic Micro Actuators." In *Proceedings of ASME Micromechanical Sensors, Actuators, and Systems,* Atlanta, December. Vol. DSC 32. 255–268.

Ando, Y., H. Ogawa, and Y. Ishikawa. 1991. "Estimation of Attractive Force Between Approached Surfaces." In *Second International Symposium on Micro Machine and Human Science,* Nagoya. Japan, October. 133–138

Arai, F., and T. Fukuda. 1997. "A New Pick up and Release Method by Heating for Micromanipulation." In *Proceedings of IEEE Workshop on Micro Electro Mechanical Systems (MEMS),* Nagoya, Japan.

Berlin, A. A., and K. J. Gabriel. 1997. "Distributed MEMS: New Challenges for Computation." *IEEE Computational Science and Engineering* **4**(1), 12–16.

Böhringer, K.-F., V. Bhatt, and K. Y. Goldberg. "Sensorless Manipulation Using Transverse Vibrations of a Plate." In *Proceedings of IEEE International Conference on Robotics and Automation (ICRA),* Nagoya, Japan, May. 1989–1996.

Böhringer, K.-F., et al. 1994. "A Theory of Manipulation and Control for Microfabricated Actuator Arrays." In *Proceedings of IEEE Workshop on Micro Electro Mechanical Systems (MEMS),* Oiso, Japan, January. 102–107.

Böhringer, K.-F., et al. 1997a. "Electrostatic Self-Assembly Aided by Ultrasonic Vibration." In *AVS 44th National Symposium,* San Jose, CA, October.

Böhringer, K.-F., et al. 1997b. "Computational Methods for Design and Control of MEMS Micromanipulator Arrays." *IEEE Computational Science and Engineering* **4**(1), 17–29.

Böhringer, K.-F., et al. 1998a. "Sensorless Manipulation Using Transverse Vibrations of a Plate." *Algorithmica.* To appear in Special Issue on Algorithmic Foundations of Robotics.

Böhringer, K.-F., et al. 1998b. "Parallel Microassembly with Electrostatic Force Fields. In *Proceedings of IEEE International Conference on Robotics and Automation (ICRA),* May. Submitted for review.

Boothroyd, G., C. Poli, and L. E. Murch. 1982. *Automatic Assembly.* New York: Marcel Dekker.

Bowling, R. A. 1988. "A Theoretical Review of Particle Adhesion." In *Particles on Surfaces 1: Detection, Adhesion and Removal.* New York: Plenum Press. 129–155.

Cohn, M. B. 1992. "Self-Assembly of Microfabricated Devices." United States Patent 5 355 577, September.

———. 1997. "Assembly Techniques for Microelectromechanical Systems." Ph.D. Thesis, University of California at Berkeley, Department of Electrical Engineering and Computer Sciences.

Cohn, M. B., and R. T. Howe. 1997. "Wafer-to-Wafer Transfer of Microstructures Using Break-Away Tethers." United States Patent Application, May.

Cohn, M. B., R. T. Howe, and A. P. Pisano. 1995. "Self-Assembly of Microsystems Using Noncontact Electrostatic Traps." *ASME-IC.*

Cohn, M. B., C. J. Kim, and A. P. Pisano. 1991. "Self-Assembling Electrical Networks as Application of Micromachining Technology." In *Transducers—Digest of International Conference on Solid-State Sensors and Actuators,* San Francisco June.

Cohn, M. B., et al. 1996. "Wafer-to-Wafer Transfer of Microstructures for Vacuum Packaging. In *Proceedings of Solid State Sensor and Actuator Workshop,* Hilton Head, June.

Danuser, G., et al. 1977. "Manipulation of Microscopic Objects with Nanometer Precision: Potentials and Limitations in Nano-robot Design." *International Journal of Robotics Research.* Submitted for review.

Erdmann, M. A., and M. T. Mason. 1988. "An Exploration of Sensorless Manipulation." *IEEE Journal of Robotics and Automation* **4**(4), 369–379.

Fearing, R. S. 1996. "A Planar Milli-Robot System on an Air Bearing." *Robotics Research the 7th International Symposium,* edited by G. Giralt and G. Hirzinger, 570–581, London, Springer-Verlag.

———. 1995b. "Survey of Sticking Effects for Micro Parts Handling." In *IEEE/RSJ International Workshop on Intelligent Robots and Systems (IROS),* Pittsburgh.

Feddema, J. T., and R. W. Simon. 1998. "CAD-Driven Microassembly and Visual Servoing." In *Proceedings of IEEE International Conference on Robotics and Automation (ICRA),* May.

Fujita, H. 1993. "Group Work of Microactuators." In *International Advanced Robot Program Workshop on Micromachine Technologies and Systems,* Tokyo, October. 24–31.

Goldberg, K. Y. 1993. "Orienting Polygonal Parts Without Sensing." *Algorithmica* **10**(2–4), 201–225.

Hara, S., et al. 1993. "High Precision Bonding of Semiconductor Laser Diodes." *International Journal of Japan Society for Precision Engineering* **27**(1), 49–53.

Harper, W. R. 1967. *Contact and Frictional Electrification.* Oxford: Clarendon Press.

Hecht, L. 1990. "An Introductory Review of Particle Adhesion to Solid Surfaces." *Journal of the IES* **33**(2), 33–37.

Hitakawa, H. 1988. "Advanced Parts Orientation System Has Wide Application." *Assembly Automation* **8**(3), 147–150.

Horn, R. G., and D. T. Smith. 1992. "Contact Electrification and Adhesion Between Dissimilar Materials." *Science* **256**, 362–364.

Hosokawa, K., I. Shimoyama, and H. Miura. 1995. "Dynamics of Self-Assembling Systems— Analogy with Chemical Kinetics." In *Artificial Life IV: Proceedings of the Fourth International*

Workshop on the Synthesis and Simulation of Living Systems, Cambridge, Mass., July 6–8. 172–180.

Israelachvili, J. N. 1974. "The Nature of Van der Waals Forces." *Contemporary Physics* **15**(2), 159–177.

Jones, T. B. 1995. *Electromechanics of Particles.* Cambridge: Cambridge University Press.

Kaneko, R. 1991. "Microtribology Related to MEMS." In *Proceedings of IEEE Workshop on Micro Electro Mechanical Systems (MEMS),* Nara, Japan, January 30–February 2. 1–8.

Kavraki, L. 1997. "Part Orientation with Programmable Vector Fields: Two Stable Equilibria for Most Parts." In *Proceedings of IEEE International Conference on Robotics and Automation (ICRA),* Albuquerque, April.

Keller, C., and R. T. Howe. 1992. "Nickel-Filled Hexsil Thermally Actuated Tweezers. In *Transducers—Digest of International Conference on Solid-State Sensors and Actuators,* Stockholm, June.

————. 1997. "Hexsil Tweezers for Teleoperated Micro-assembly." In *Proceedings of IEEE Workshop on Micro Electro Mechanical Systems (MEMS),* Nagoya, Japan, January. 72–77.

Khatib, O. 1986. "Real Time Obstacle Avoidance for Manipulators and Mobile Robots." *International Journal of Robotics Research* **5**(1), 90–99.

Kim, C.-J., A. P. Pisano, and R. S. Muller. 1992. "Silicon-Processed Overhanging Microgripper. *Journal of Microelectromechanical Systems* **1**(1), 31–36.

Koditschek, D. E., and E. Rimon. 1988. "Robot Navigation Functions on Manifolds with Boundary." *Advances in Applied Mathematics.*

Krupp, H. 1967. "Particle Adhesion Theory and Experiment." *Advances in Colloid and Interface Science* **1,** 111–239.

Kwetkus, B. A., B. Gellert, and K. Sattler. 1991. "Discharge Phenomena in Contact Electrification." In *International Physics Conference Series No. 118: Section 4, Electrostatics.* 229–234.

Legtenberg, R., et al. 1994. "Stiction of Surface Micromachined Structures after Rinsing and Drying: Model and Investigation of Adhesion Mechanisms." *Sensors and Actuators* **43,** 230–238.

Liu, C., et al. 1995. "A Micro-machined Magnetic Actuator Array for Micro-robotics Assembly Systems." In *Transducers—Digest of International Conference on Solid-State Sensors and Actuators,* Stockholm, June.

Liu, W., and P. Will. 1995. "Parts Manipulation on an Intelligent Motion Surface." In *IEEE/RSJ International Workshop on Intelligent Robots and Systems (IROS),* Pittsburgh.

Lowell, J. 1990. "Contact Electrification of Silica and Soda Glass." *Journal of Physics D: Applied Physics* **23,** 1082–1091.

Lowell, J., and A. C. Rose-Innes. 1980. "Contact Electrification." *Advances in Physics* **29**(6), 947–1023.

Luntz, J. E., and W. Messner. 1997. "A Distributed Control System for Flexible Materials Handling." *IEEE Control Systems Magazine* **17**(1), 22–28.

Luntz, J. E., W. Messner, and H. Choset. 1997. "Parcel Manipulation and Dynamics with a Distributed Actuator Array: The Virtual Vehicle." In *Proceedings of IEEE International Conference on Robotics and Automation (ICRA),* Albuquerque, April. 1541–1546.

Moncevicz, P., M. Jakiela, and K. Ulrich. 1991. "Orientation and Insertion of Randomly Presented Parts Using Vibratory Agitation." In *ASME 3rd Conference on Flexible Assembly Systems,* September.

Morita, M., et al. 1990. "Growth of Native Oxide on a Silicon Surface." *Journal of Applied Physics* **68**(3), 1272–1281.

Nakasuji, M., and H. Shimizu. 1992. "Low Voltage and High Speed Operating Electrostatic Wafer Chuck." *Journal of Vacuum Science and Technology A (Vacuum, Surfaces, and Films)* **10**(6), 3573–3578.

Nelson, B., and B. Vikramaditya. 1997. "Visually Guided Microassembly Using Optical Microscopes and Active Vision Techniques." In *Proceedings of IEEE International Conference on Robotics and Automation (ICRA),* Albuquerque, April.

Pister, K. S. J., R. Fearing, and R. Howe. 1990. "A Planar Air Levitated Electrostatic Actuator system." In *Proceedings of IEEE Workshop on Micro Electro Mechanical Systems (MEMS),* Napa Valley, February. 67–71.

Pister, K. S. J., et al. 1992. "Microfabricated Hinges." *Sensors and Actuators A* **33**(3), 249–256.

Prime, K. L., and G. M. Whitesides. 1991. "Self-Assembled Organic Monolayers Are Good Model Systems for Studying Adsorption of Proteins at Surfaces." *Science* **252**.

Quaid, A. E., and R. L. Hollis. 1996. "Cooperative 2-dof Robots for Precision Assembly." In *Proceedings of IEEE International Conference on Robotics and Automation (ICRA),* Minneapolis, April.

Reif, J., and H. Wang. 1995. "Social Potential Fields: A Distributed Behavioral Control for Autonomous Robots." In *International Workshop on Algorithmic Foundations of Robotics (WAFR).* Ed. K. Goldberg et al. Wellesley: A. K. Peters. 431–459.

Reznik, D., J. F. Canny, and K. Y. Goldberg. 1997. "Analysis of Part Motion on a Longitudinally Vibrating Plate." In *IEEE/RSJ International Workshop on Intelligent Robots and Systems (IROS)*, Grenoble, France, September.

Riley, F. J., 1983. *Assembly Automation: A Management Handbook.* New York: Industrial Press.

Rimon, E., and D. Koditschek. 1992. "Exact Robot Navigation Using Artificial Potential Functions." *IEEE Transactions on Robotics and Automation* **8**(5), 501–518.

Sandler, B.-Z. 1991. *Robotics: Designing the Mechanisms for Automated Machinery.* Englewood Cliffs: Prentice-Hall.

Scheeper, P. R., et al. 1992. "Investigation of Attractive Forces Between PECVD Silicon Nitride Microstructures and an Oxidized Silicon Substrate." *Sensors and Actuators A (Physical)* **30,** 231–239.

Seto, C. T., and G. M. Whitesides. 1991. "Self-Assembly of a Hydrogen-Bonded 2 + 3 Supramolecular Complex." *Journal of American Chemistry Society* **113**.

Shimoyama, I. 1995. "Scaling in Microrobotics." In *IEEE/RSJ International Workshop on Intelligent Robots and Systems (IROS)*, Pittsburgh.

Singh, A., et al. 1997. "Batch Transfer of Microstructures Using Flip-chip Solder Bump Bonding. In *Transducers—Digest of International Conference on Solid-State Sensors and Actuators,* Chicago, June.

Srinivasan, U., and R. Howe. 1997. Personal communication.

Storment, C. W., et al. 1994. "Flexible, Dry-Released Process for Aluminum Electrostatic Actuators." *Journal of Microelectromechanical Systems* **3**(3), 90–96.

Suh, J. W., et al. 1996. "Combined Organic Thermal and Electrostatic Omnidirectional Ciliary Microactuator Array for Object Positioning and Inspection." In *Proceedings of Solid State Sensor and Actuator Workshop,* Hilton Head, June.

Sze, S. M. 1981. *Physics of Semiconductor Devices.* 2nd ed. New York: John Wiley & Sons.

Torii, A., et al. 1994. "Adhesive Force Distribution on Microstructures Investigated by an Atomic Force Microscope." *Sensors and Actuators A (Physical)* **44**(2), 153–158.

Whitesides, G. M., J. P. Mathias, and C. T. Seto. 1991. "Molecular Self-Assembly and Nanochemistry: A Chemical Strategy for the Synthesis of Nanostructures." *Science* **254**.

Yeh, H.-J., and J. S. Smith. 1994. "Fluidic Self-Assembly for the Integration of GaAs Light-Emitting Diodes on Si Substrates." *IEEE Photonics Technology Letters* **6**(6), 706–708.

———. 1994*b*. "Fluidic Self-Assembly of Microstructures and Its Application to the Integration of GaAs on Si." In *Proceedings of IEEE Workshop on Micro Electro Mechanical Systems (MEMS)*, Oiso, Japan, January. 279–284.

———. 1994*c*. Integration of GaAs Vertical Cavity Surface-Emitting Laser on Si by Substrate Removal." *Applied Physics Letters* **64**(12), 1466–1468.

Zesch, W. 1997. "Multi-Degree-of-Freedom Micropositioning Using Stepping Principles." Ph.D. Thesis, Swiss Federal Institute of Technology.

Zimon, A. D. 1969. *Adhesion of Dust and Powder.* New York: Plenum Press.

PART 10

ROBOTICS IN VARIOUS APPLICATIONS

CHAPTER 56

AUTOMOTIVE AND TRANSPORTATION APPLICATIONS

Valerie Bolhouse
Brian Daugherty
Ford Motor Company
Redform, Michigan

1 INTRODUCTION

The automotive industry was one of the early pioneers in the use of robotic automation and continues to install about 50% of the robots in the U.S. market. The motor vehicle industry accounted for 40% of the installed base of robot applications in 1995 in Spain, Germany, the United Kingdom, France, Taiwan, and Australia. If the suppliers to the automotive industry were taken into account, these numbers would be even higher. Characteristics of automotive manufacturing that have made it well suited for automation include high volume, capital-intensive plants, long product life cycle, an emphasis on quality, and a highly paid skilled workforce. The flexibility and cost performance of robots have made them an integral part of automation strategy in the automotive industry (Figure 1).

Automotive manufacturing encompasses a diverse range of manufacturing and assembly processes. Some vehicle assembly operations have a high level of automation (i.e., body shop and paint shop), while others, such as final trim, continue to use manual operations in all but a few applications. One of the biggest hurdles for robotic applications in automotive operations is the need for very robust processes. Due to the long serial production lines typical in automotive assembly, an uptime of 97%, which might sound high, is not acceptable for individual workcells. Many current operations can be automated, but not to the robust level required to support automotive production requirements. This is the primary reason that automation is so sparse in final assembly operations.

Components manufacturing for automotive encompasses an even more diverse range of processes: casting, machining, and assembly for powertrain, injection molding, painting, and assembly for interior and exterior trim, brazing and soldering in climate control, assembly and soldering in electronics, and so on. Examples of automation and the rationale for implementation will be examined for the major processes within automotive manufacturing. The automotive assembly process will first be described, along with a description of either the motivation to automate or the impediments to automation.

2 VEHICLE ASSEMBLY

Body and assembly (B & A) plants are the most recognizable use of robots within an automotive company (Figures 2–7). These plants typically receive stamped parts from internal stamping operations or outside suppliers and have hundreds of welding robots to perform the final welding sequences required to build a car or light truck body. The body is then E-coated and painted on a heavily automated line using both custom spray head automation and long-reach robot arms. While the body is in the paint area, other subassemblies are being completed on feeder lines throughout the plant. Everything from instrument panels to door assemblies or chassis systems may be fully or partially assembled depending on the plant. Finally, the painted body is transferred to the final assembly line, where, in a largely manual operation, all parts are added to the painted body to produce a vehicle that can be driven off the line.

Handbook of Industrial Robotics, Second Edition, Edited by Shimon Y. Nof
ISBN 0-471-17783-0   © 1999 John Wiley & Sons, Inc.

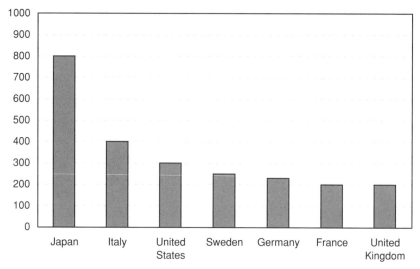

Figure 1   Robot population per 10,000 people employed in motor vehicle manufacturing (courtesy Robotic Industries Assn.).

Figure 2   Robotic assembly of decklid components.

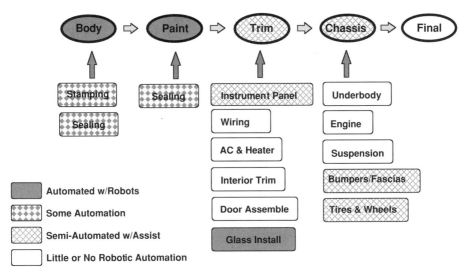

Figure 3 Automotive assembly: generic manufacturing sequence.

2.1 Body and Paint Shop

Most body shops are fully automated with a wide range of welding robots for handling the spot welding sequences and fixed automation for fixturing and postioning the sheet-metal pieces. Material-handling robots are often used to transport the sharp and heavy sheetmetal assemblies from one workcell location to the next. In addition to welding, robots are also used to apply adhesive and sealer during body assembly. Currently efforts are underway to increase the flexibility of the fixturing equipment as well. Many of these efforts are using robots to replace the traditional fixed clamping mechanisms for locating body parts. Increased flexibility is required to support the trend toward high mix/low volume niche market vehicles. In addition, automakers plan to reduce line development time and increase equipment reusability with flexible automation.

Figure 4 Automated welding operation in body shop.

Figure 5   Automated welding operation in body shop.

Painting operations in an automotive plant are also highly automated due to environmental and quality concerns. These issues and concerns about the overall quality of the work environment for paint shop workers have driven the development of extensive robotic paint lines (see Figure 6). Another factor that has contributed to the early use of robots in automotive painting is that the patterns are well defined and no contact is required between the robot and the vehicle body. The installation of robotic paint lines in modern vehicle paint shops has resulted in increased consistency and overall paint quality. It is also estimated that up to 50% less material is used due to the automated control of the systems. As in the body shop, robots are also used for waterproof sealant application in the paint area.

2.2   Final Assembly

The process of final assembly of the vehicle (including trim) has not changed radically since Henry Ford first implemented the assembly line. Shown in Figure 7 are the trim area from an assembly line in the early 1940s and from one in the 1990s. The operators select parts for assembly from a feeding area, and assemble them to the vehicle as it moves down the line. A build sheet specifies the unique option and color content. The vehicle position is not fixed to tight tolerances. Multiple operations are performed in one station. These characteristics make automation very difficult.

2.3   Robotic Applications in Final Assembly—Windshield Decking

One area within vehicle assembly that is now typically automated is windshield installation, or windshield decking, as it is known within the assembly plant. This task was previously very dangerous and labor-intensive, and involved manually unloading glass from a rack, manually dispensing sealant/adhesive, and manually aligning and placing the glass in the windshield opening.

A typical automation system is fairly complex and is designed as follows. First the vehicle body enters the assembly area stop station and the windshield opening is precisely located using a noncontact laser measurement system. At the same time, a robot dispenses adhesive on the windshield glass within a fixture that precisely locates the glass edges. A large six-degree-of-freedom robot with a series of vacuum cups picks up the windshield and inserts it into the opening on the vehicle body. The obvious area of difficulty within

Figure 6   Robotic painting of vehicle body.

this robot workcell is the sensing system for the windshield opening. Any misalignment or inaccuracy from the laser sensors obviously affects the performance of the overall workcell.

The quality of the automated assembly process is better and more consistent than the earlier manual method. Windshield decking is a great example of how a robot can be used to eliminate a dangerous manual operation while increasing quality at the same time.

(a)

Figure 7   Automotive final trim. (a) 1940s. (b) 1990s.

(b)

Figure 7   (Continued)

2.4 Robotic Applications in Powertrain Manufacturing

The manufacture of engines and transmissions is a complex process with many precision machining operations and intricate part assembly. The machining process is usually highly automated, with robots used for pick-and-place operations between machining transfer lines or workcells. The reliability and easy programmability of current industrial robots make them ideal for these types of operations rather than a custom-designed one-of-a-kind piece of hard automation.

The assembly of subsystem components for transmissions, including clutch assemblies, torque converters, and pressure-regulating valves, is typically done manually. One of the main reasons for this is the large number of component parts that are delivered in bulk packaging, such as springs, washers, and bearings. The tangled nature of the parts makes automated feeding a difficult option. The components could be delivered in semiconstrained dunnage, but this increases the cost of the components and defeats the cost efficiencies required for economic justification.

Automated assembly of these components represents an area in which robots and other automation devices have had very limited success. Usually several operations are done by each assembly worker. Even if one or two operations can be successfully automated, the remaining operations that cannot currently be automated will prevent the automation of the cell. Sometimes the assembly sequence can be modified to redistribute operations among robotic workcells and manual assembly stations. This is usually difficult in powertrain assembly due to the functional requirements of each part in the assembly sequence.

The final assembly areas within powertrain manufacturing have very little automation. This is due in part to the inherent limitations of point-to-point motion in current robots. Most of these assembly operations rely on people using tactile and force feedback to assembly the mating gears, spline shafts, wiring harnesses, and clutch assemblies. In addition, multiple parts are installed onto the engine or transmission within one workcell, thereby further increasing the difficulty of automation. Efforts are underway to automate the areas that have the greatest ergonomic concerns using force feedback control, but these applications require a great deal of customization.

3 AUTOMOTIVE COMPONENT MANUFACTURING

Component manufacturing has a different set of constraints than vehicle assembly, yet the application of robotics is just as sporadic throughout these operations. The component manufacturing plants do share some attributes with the assembly plants, including a work-force that is stable, highly paid, and skilled. The facilities required to manufacture are also highly capitalized, such that productivity and uptime are critical in order to maximize the return on assets.

Component products have cycle times ranging from 3 to 20 seconds. One component plant will manufacture products for several automotive assembly plants. The processes vary depending on what product is being manufactured, but could include assembly, welding, painting, machining, grinding, injection molding, and casting. We will describe several automotive component manufacturing examples to demonstrate the variety of applications.

3.1   Robotic Applications in Automotive Components Manufacturing— Electronic Modules

Each vehicle has multiple electronic modules—radio, door and seat modules, instrument cluster, and powertrain control modules, to name a few. The printed circuit board assembly for these modules is usually fully automated with dedicated equipment. There are typically some devices used for output power (such as to drive motors or other high current devices) that are considered nonstandard by the assembly equipment suppliers. These will either be assembled by hand or automated with customized automation equipment, including robots.

An automated assembly cell for electronics will typically use SCARA or Cartesian robots since the assembly process requires only four degrees of freedom. Custom parts feed tooling is developed to present the components to the robot for assembly. Tape and reel feeding is the predominant part-feeding method, although vibratory bowl feeding is also frequently used. Machine vision is not usually required to locate the device leads or the board for the assembly process. The board is located with tooling holes that are drilled at the same time as the device lead holes. The gripper will pick up the device such that the leads are in a known location. The robot then performs a pick-and-place operation, picking from the feeder and placing on the board (Figure 8).

Sensors can be used to improve the reliability of the operation. A force sensor can be used to measure the insertion force. If the insertion force is too high, it is assumed that the leads are not aligned with the holes in the board. The robot will then search for the hole location by moving slightly off-center and retrying the insertion. The robot is generally programmed to try several different locations about the trained path. If unsuccessful,

Figure 8   In-line Cartesian robots assembling printed circuit cards.

it will dispose of the device and try with a new part. Since the primary cause of misinserts is the device lead location, the second try is usually successful. After a preset number of failed attempts, the robot will shut down and alert maintenance of the issue.

It is not unusual for the total cost of the assembly cell to be two to four times the cost of the robot alone. A great deal of customization is usually required to fixture the board and feed the devices. This limits the flexibility of the cell since changes in the product form factor, component type, or volume requirements will generally require modifications in tooling. Because of the cost and complexity of the cell, this application is usually only automated in manufacturing sites with high labor costs. However, there are other benefits to automating besides labor that are also considered: ergonomics and product quality. Manual assembly can cause repetitive motion injury. Device leads cannot be clinched for manual drop-in, and the devices can pop out of the hole during material handling prior to solder.

Robots are used for other applications in automotive electronics manufacturing as well. There are a number of material-handling applications—loading and unloading testers or other processing equipment and packing completed product into shipping boxes. These applications do not require the amount of custom tooling that is needed for assembly, and are generally easier to cost-justify. In addition to the ergonomic advantage, they provide an improvement in productivity by keeping the test equipment supplied with product on a continuous basis.

3.2 Automotive Interiors

Robots are also used selectively for the manufacture of automotive interiors. The instrument panel consists of a structural cross beam member with a plastic/foam skin. Multiple assemblies are required to form the mounting locations for the wiring harness, body electrical system, instrument cluster, heater/AC controls, glove box, and radio and other driver information systems. The manufacturing processes required are stamping, welding, injection molding, and assembly.

The cross beam assembly consists of several component members (stampings) that are welded together to form a tightly toleranced structural member. A fixture is used to hold the components in location for welding. The station can be hard-tooled with individual weld guns for each weldment. Alternatively, it can be designed to be flexible with one weld gun on a robot arm, the robot being used to go to each of the locations to produce the weld. There are pros and cons to each method of implementation. The hard-tooled workcell can provide a higher throughput since all of the welds can happen simultaneously instead of the sequential process. However, it also requires the purchase of multiple guns (which can be quite expensive at $30,000 or more), as well as custom tooling that can be complex and difficult to maintain.

The robotically automated cell uses one robot and one gun. Hard-tooled fixtures are required for locating and locking the subassemblies in position for welding, but this tooling is considerably less complex than if a large weld gun also had to be accommodated. The maintenance operators also have only one gun to be concerned with, and if it starts to produce a bad weld, all of the welds will be bad and easily noticed. With multiple guns it is possible to miss one or two marginal welds because the overall product appears to be good.

The material-handling task to load the multiple stampings into the welding fixture is usually performed manually. The subassemblies are not easily located for automated handling. There are often 10 or more parts to be loaded into the fixture. An operator can pick from loosely packed containers, without the tolerance and clearance of robot tooling being required. A manual operation is more flexible and less costly to implement than an automated process. This rationale follows for many material-handling applications, which is why it is difficult to automate many of these tasks.

Material handling to remove parts from an injection molding machine will typically only be automated if the parts are heavy, the work area in and around the machine is hot, or the parts have to be removed from within the interior of the machine. Otherwise an operator is used to remove the parts and load them onto hooks or into other material-handling devices.

Throughout the other component manufacturing operations the applications follow a similar pattern. The dirty and difficult tasks in a plant are generally automated first. Repetitive motion injury and other work-related injuries are costly to manufacturers. Improved worker morale will mean a more productive workforce. The next operations to be automated are those where consistency is critical to the quality of the product. Consistent output is inherent to the robotic operations, provided the application is designed with sensors and feedback to accommodate the normal variations in input product attributes. Finally, applications will be automated for productivity. A robot can be faster and is more tireless than a human operator. However, this productivity improvement will only be achieved if the application is well implemented. A poorly designed robot cell could result in lower quality with considerable downtime.

4 ATTRIBUTES OF AUTOMOTIVE MANUFACTURING THAT DRIVE AUTOMATION

Vehicle assembly cycle times are relatively slow (45 to over 60 seconds per vehicle), but there are innumerable operations that must occur during that cycle time. Also, the product must continually flow through the process so as to not constrain an upstream process or starve the stations down-line. Any automation cell installed for vehicle assembly must be capable of meeting cycle time without line stoppages that will reduce the vehicle output.

Cycle times for automotive components manufacturing are typically shorter than vehicle assembly—10–20 seconds per part is common. Component manufacturing is inherently more batch oriented, with the option to buffer parts between stations. Line stoppages are less critical, but the shorter cycle times can be difficult to meet. The material movement into and out of the cell takes up a larger percentage of the overall cycle time.

At automotive cycle times even simple assembly or machine loading tasks can become grueling to an operator. Ergonomic design of the workcell becomes critical. If the parts are heavy or the reach long, the best solution might be automation. The dirty and difficult tasks in a plant are generally automated first. Repetitive motion injury and other work-related injuries are costly to manufacturers. Eliminating undesirable jobs improves worker morale, leading to a more productive workforce.

Plant productivity can be further increased by automating the constraint operation in the line. To maximize throughput the line needs to be configured so that the constraint machine can operate continuously. The constraint operation would not need to be stopped for breaks or lunch or for any other reason as long as the automation cell is provided with parts. Higher overall equipment efficiency is important for high-investment machines such as machining cells or functional testers.

Automation also provides for more consistent output quality. This can be attributed to the multiple sensors often required for the task, as well as the inherently repetitive actions performed by the robot. In addition, since automated operations are less tolerant to variations in the environment and materials than manual operations, care is taken by the process engineer to better control the tolerances of incoming materials. This discipline is required for high uptime in the cell and further contributes to more consistent output quality.

4.1 What Makes an Application Difficult

It is often difficult to cost-justify the simple replacement of manual labor with automation due to the relatively high capital investment required up front, and the ongoing costs to maintain the cell. Indirect labor costs will increase as the direct labor decreases, offsetting a portion of the cost savings. Complex cells can require considerable support for maintenance and programming, with the added need for a more highly skilled workforce. However, when quality, productivity, and ergonomics are also factored in, automation might become the best overall solution.

There are four features of an application that will impact the technical feasibility as well as the cost justification used to implement an automation workcell:

1. *Cell uptime.* Uptime becomes critical in a long serial line with minimal or no buffering between stations. A rough estimate of line throughput can be determined by the product of the uptime percentage of each individual station. A single station going down impacts the throughput of the entire line—therefore the uptime of an individual cell needs to be in the upper 90s. This is often not achievable with automation due to the wide variation seen in incoming materials.

2. *Cycle time.* The cycle time of the process needs to closely match the cycle time achievable with automation. If it is too short, the robot cannot perform the task within the period. However, if it is too long, the robot will be idle during the period. This is costly since the value-add time of the expensive automation is low. There is the tendency to increase the number of tasks the robot is programmed to

perform within the process cycle time, which will drive up the cell complexity and tooling cost.

3. *Positional accuracy of the workpiece.* Robots are best suited for point-to-point moves, or movement along a preprogrammed path. If the workpiece is naturally positioned and oriented throughout the process, installation of a robot to perform the task will require minimal tooling or rearranging. However, the workcell cost and complexity increase considerably if special means must be used to bring the workpiece into accurate and repeatable positions.

4. *Amount of tooling required to perform the task.* Parts feeding for assembly cells generally requires custom tooling and sometimes custom packaging for components. The custom packaging increases the variable cost of the product. The custom tooling is expensive, takes a long time to debug, and limits cell flexibility.

A manual line is inherently more flexible and costs less to implement than an automated solution. An operator can pull parts from pallets or bins, and can assemble them anywhere within his workspace. In addition, people can handle unexpected events or process problems that would cause an automated system to grind to a halt. If cycle times need to be decreased, more people can be assigned to the task. This is one of the tenets of "lean manufacturing." But people are not machines and are not well suited for all tasks. The optimal solution is usually a combination of the operator and some automation to handle those tasks that are undesirable for manual operations due to ergonomics, quality, or throughput requirements. Automotive manufacturers are not striving for the "lights out factory," but for the most cost-competitive processes to produce a quality product.

4.2 Future Trends in Automation

Over the last three decades robots have become faster, more accurate, easier to use, and less expensive. While it has become easier to automate applications, progress is still slow for a number of reasons:

1. The cost of the robot might have come down, but the peripheral hardware and integration required to complete the application is still very expensive. It is not unusual to have a cell where the cost of the robot hardware is 25% or less of the total cost of the cell. Material conveyance and product fixturing hardware (conveyors, transfer lines) are expensive for large parts such as vehicle bodies, engines, or transmissions.

2. The easy applications have been automated. The applications still to be done are those that require substantially more feedback (vision, force, torque), intelligence, or peripheral hardware. These requirements drive up the complexity and cost beyond what is viable.

3. Product life cycle is being reduced. Customers are demanding more frequent changes to products as the market. While the robots are fully reprogrammable and flexible, the peripheral custom tooling often is not flexible. A manual operation is inherently more adaptable since part location and part feeders are not critical. Manufacturers are reluctant to invest in automation that cannot be used for the next-generation product, knowing that the facilities might only be used for two or three years. Building flexibility into a cell often increases the cost and complexity of the automation (and manufacturers do not want to spend money today for features that might never be used).

Besides the evolutionary improvements in robotic technology, new technology is coming into the field that should increase the rate of application. Technology developed for space or military applications is being declassified and made available for commercial use. Universities and research laboratories are also performing research in machine perception, force-controlled robotics, and new architecture manipulators for assembly. As these technologies find their way into commercial products, applications that were previously not feasible or cost-justifiable will be implemented.

4.3 Trends in the Automotive Industry

There are also changes in the automotive industry itself that will impact robotic automation. One of the greatest is the trend towards modularity. Currently, most of the assembly of the vehicle performed in trim, chassis, and final occurs on the assembly line or on small feeder lines into the vehicle assembly line. As seen in Figure 9, the operator

Figure 9 Manual assembly of vehicle interior in trim.

must climb into the vehicle, where access is limited, to install console components. This would be a difficult application to automate because of the loss of locating tolerance of the console relative to the vehicle moving down the line and the restricted access with obstructions in the vehicle.

With the modularity concept, these component systems would be preassembled and delivered to the line for final assembly. Automotive suppliers will be providing full instrument panels, door cassettes, front end systems, and other parts to the vehicle assembly plants. The environment for the subassembly operation will not be constrained by the vehicle assembly line and can be designed for automated assembly. The subassembly can be mechanically fixtured with a tight tolerance and with good access to all sides of the product. The major impediments to automation will be removed.

Finally, the discipline of designing for automated assembly is now understood by the majority of automotive product design engineers. Companies are practicing concurrent or simultaneous engineering, and manufacturing feasibility studies are now done on component parts. Taken in combination with better automation technology, the future looks bright for the continued application of robotics in automotive applications.

4.4 Trends in Intelligent Transportation

Finally, an exciting trend in transportation is the development of Intelligent Transportation Systems (ITS), formerly known as Intelligent Vehicles Highway Systems (IVHS). Unlike applications of robotics in the automotive and transportation industry, ITS is virtually a combination of robotics *and* transportation, both in terms of robotic vehicles, and in terms of knowledge-based driving and transporation management. Vehicles have long been equipped with sensors and microcomputers to enhance safety and performance. Recent trends include the GPS, the global positioning systems, which enable navigation and tracking of fleets; in-vehicle information systems to support the driver and travelers; and instrumented highways in Europe, Japan, the United States, and other countries to provide better traffic planning and control, improved safety, and various commercial services. The reading list that follows can provide additional details. The frontier of intelligent transportation opens new vistas for the world of transportation robotics.

ADDITIONAL READING

Robotics and Automotive/Transportation Industry

Carmer, D. C., and L. M. Peterson. "Laser Radar in Robotics." In *Proceedings of the IEEE* **84**(2), 299–320, 1996.

Davies, H., M. Brunnsatter, and K. W. Ho. "Robots Used for Hong Kong Mass Transit Railway Repair." *Tunnels and Tunnelling* **27**(3), 28–30, 1995.

Fujiuchi, M., et al. "Development of a Robot Simulation and Off-line Programming System." In *Simultaneous Engineering in Automotive Development.* Warrendale: SAE, 1992. 69–77.

Grohmann, U. "Paint Robots in the Automotive Industry—Process and Cost Optimization." *ABB Review* **4**, 9–17, 1996.

Leavitt, W. "In the Drive to Maximize Productivity, Maybe We Could Use a Robotic Hand." *Trucking Technology* **2**(1), 32–39, 1995.

Otsuka, K. "Application of Robot in the Shipping Operations." *Katakana/Robot* (May), 55–59, 1994.

Rooks, B. W. "Off-line Programming: A Success for the Automotive Industry." *The Industrial Robot* **24**(1), 30–34, 1997.

Soga, T. "Automated Aircraft Washing System." *Katakana/Robot* (May), 16–23, 1994.

Tachi, Y., and N. Hagiwara. "Automation of Loading and Unloading Facilities at Depots." *Katakana/Robot* (May), 48–54, 1994.

Intelligent Transportation

Badcock, J. M., et al. "Autonomous Robot Navigation System—Integrating Environmental Mapping, Path Planning, Localization and Motion Control." *Robotica* **11**(2), 97–103, 1993.

Charkari, N. M., and H. Mori. "Visual Vehicle Detection and Tracking Based on the 'Sign Pattern.' " *Advanced Robotics* **9**(4). 367–382, 1995.

Harris, C. J., and D. Charnley. "Intelligent Autonomous Vehicles: Recent Progress and Central Research Issues." *Computing and Control Engineering Journal* **3**(4), 164–171, 1992.

Madden, M. G. M., and P. J. Nolan. "Application of AI Based Reinforcement Learning to Robot Vehicle Control." In *Applications of Artificial Intelligence in Engineering.* Billerica: Computational Mechanics, 1995. 437–445.

Peters, J., et al. "Estimate of Transportation Cost Savings from Using Intelligent Transportation System (ITS) Infrastructure." *ITE Journal* **67**(11), 42, 43, 45–47, 1997.

Rombaut, M., and N. Le Fort-Piat. "Driving Activity: How to Improve Knowledge of the Environment." *Journal of Intelligent and Robotic Systems* **18**(4), 399–408, 1997.

Sussman, J. M. "ITS: A Short History and Perspective on the Future." *Transportation Quarterly* **50**(4), 115–125, 1996.

Tank, T. and J.-P. M. G. Linnartz. "Vehicle-to-Vehicle Communications for AVCS Platooning." *IEEE Transactions on Vehicular Technology* **46**(2), 528–536, 1997.

CHAPTER **57**

ELECTRONICS, INSTRUMENTS, AND SEMICONDUCTOR INDUSTRY

Sungkwun Kim
Samsung Electronics
Kyunggi-Do, Korea

1 INTRODUCTION

1.1 Overview

Robotics technology has brought about a structural change in the manufacturing industry. Productivity, flexibility, quality, and reliability superior to that of conventional production methodology can be achieved. The ever-changing demand from the market has led to increased product variety and shorter product life cycles, and automating production has seemed to be the only solution to meet the demand. Many early automated production cells with robotic manipulators were plagued with insufficient user programming tools, positioning inaccuracy, poor reliability, primitive sensory interaction, and huge cost of capital investment for automation. The evolution of digital technology enabled robot controllers to take advantage of powerful microprocessors and digital signal processors (DSPs) that greatly enhanced performance while reducing cost.

The electronics industry, especially the consumer electronics and semiconductor industries, have benefitted from automation, since most production activity involved repetitive assembly and fabrication of an aggregate of parts in a large quantity. Dedicated machines were effective enough for certain simple and repetitive tasks because they delivered unequaled productivity to human workers, justifying the capital investment in a very short time span. The introduction of robot assembly offered more flexible and soft solutions to dedicated automation equipment. The notable attraction of robots as a major component in automation comes from their superb flexibility, which often results in a dramatic reduction in labor cost. Robots may replace human workers completely to a wide extent, liberating humans from hazardous, dirty, and fatiguing work environments. A well-designed automated assembly system may benefit from recent advances in production technology, including robotics, sensory devices, PLCs, and GUI-based production management software.

Past uses of robots in the electronics industry were usually confined to traditional pick-and-place operations like odd-shaped components insertion, transporting, palletizing, cut-and-clinching, and soldering. However, component miniaturization and complication in circuitry have brought us to the point where human workers are no longer capable of assembling or inspecting assembled products in terms of competitive productivity and quality. In the semiconductor industry automation cannot be overlooked. Even minor processes are considered for automation to achieve seamless and consistent fabrication of chips to maintain valid quality and economy.

In this chapter we review the current research and applications of robotic technology in the electronics industry, including the consumer electronics and semiconductor industries. In this chapter we will try to provide better understanding of underlying design concepts, practical issues involved, and operational problems of robotic manipulators of electronics industry in action. Lastly, we will present prospects for emerging manufacturing technology in the electronic industry. See also Chapters 51 and 55.

Handbook of Industrial Robotics, Second Edition, Edited by Shimon Y. Nof
ISBN 0-471-17783-0 © 1999 John Wiley & Sons, Inc.

Table 1   Technology Trend in the Electronics Industry[a]

Year	1997	2000	2005
Products			
Camcorder	380 g (MPEG, digital)	300 g (Full digital)	200 g
Cellular phone	80 g (digital)	50 g	30 g
New product	Multimedia, PCS, PDA	Universal and Intelligent	Laser TV
		ISDN	Portable video phone
Component technology			
Shaft size	0.15 mm	0.1 mm	0.1 mm
Assembly tolerance	2 μm	1 μm	0.5 μm
PCB	TAB	Multichip module	Bare chip mounting
	(0.25-mm pitch)	(0.2-mm pitch)	
	BGA	HGA	COG
Mount density	40 EA/cm^2	50 EA/cm^2	70 EA/cm^2
No. of QFP leads	500	1000	1500

[a] The forecast is based on a number of industrial reports, articles from magazines, and opinions from experts in the field. The weight of products exclude batteries, and the assembly tolerance includes that of the semiconductor packaging process.

1.2   Trends in the Electronics Industry

Electronics companies that have taken initiative have made and continue to make considerable investments in production engineering technologies and factory automation, mainly because of their desire to miniaturize and to maintain high product quality and competitive product cost. Manual assembly is no longer feasible as electronic components are reduced in size to as small as 1.0 by 0.5 mm (often referred as *1005* parts) and as component lead pitch decreases by as much as 0.2 mm.

As seen in Table 1, the electronic product technology is believed to be moving towards manufacturing miniaturized, personalized, and multifunctionalized products such as camcorders, cell phones, and personal digital assistance (PDA). The functionality of software is receiving more attention as products become intelligent and user-friendly. The notable progression of semiconductor products, including CPU, memory chips, and LCD, will continue.

1.3   Robots Used in the Electronics Industry

At the end of 1994 the accumulated worldwide supply of robots amounted to about 610,000 units. Japan accounted for more than half of the world robot population where a somewhat wider definition of industrial robot was used. In Japan the electrical machinery industry, with 51% of the 1994 market, led the rapid growth in robotics. In most countries welding is the predominant robotic application area, particularly for the dominant motor vehicle manufacturing countries. Machining is the second largest application area. Assembly is the largest application area in Japan, accounting for 41% of the total 1994 stock.

Generally robots with higher accuracy and rapid cycle time are considered primarily in the electronics industry. Payload is insignificant, and the reliability of the robot is absolutely crucial since the number of cycles the robot should perform is very high. A typical SCARA robot (Figure 1) has four-degree-of-freedom motions, very much like human arms. This device incorporates a shoulder and elbow joint and a wrist axis and

Table 2   Supply of Industrial Robots in 1994, Distributed by Relevant Application Areas to Electrical Industry and Countries (Number of Units)[a]

Application Area	U.S.A.	Japan	Germany	World
Assembly (mechanism, soldering, bonding)	651	4,427	723	6,199
Palletizing/packaging	N/A	575	209	929
Measuring, inspection, testing	N/A	522	44	639
Material handling	N/A	2,182	688	4,454

[a] From *World Industrial Robots 1995,* IFR.

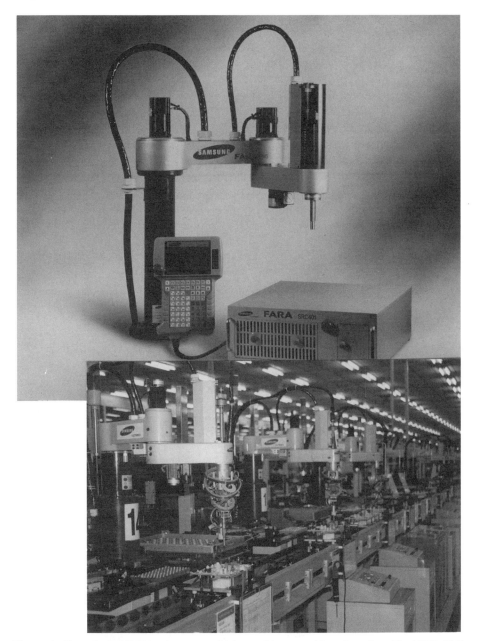

Figure 1   Four-d.o.f. SCARA robots are widely used in assembly automation. The FARA SM5 robot is a typical example of a SCARA-type manipulator with 0.7-sec cycle time, ±0.03-mm repeatability, and 2.5-Kg payload. It employs four AC servo-motors driven by full digital servo-drives based on DSP technology. (Photo courtesy Samsung Electronics, Inc.)

vertical motion. SCARA robots were invented in Japan in the early 1960s, and the design has long since been in the public domain. SCARA robots are ideal for a variety of general purpose applications that require fast, repeatable and articulated point-to-point movements such as pick-and-place, palletizing and depalletizing, loading and unloading, and assembly. Because of their unique elbow motions, they are also ideal for applications that require uniform acceleration through circular motions such as dispensing adhesives.

SCARA robots exhibit good performance in terms of repeatability and speed. Their simple construction results in simple kinematic solutions and superior vertical rigidity while keeping the cost down. Some SCARA robots are designed as direct-drive robots, eliminating the need for gears, which are the source of backlash and noise. A typical SCARA robot for assembly automation should exhibit ± 0.05-mm repeatability and less than 2 sec cycle time.

Due to the linear nature of their movements, Cartesian robots, often referred as *XY* robots (Figure 2), are intrinsically more accurate than rotary motion robots such as the SCARA. Typical Cartesian robots have offered useful repeatability and speed at lower cost. Simple yet robust construction of Cartesian robotic manipulators helped to gain wide usage in the assembly world, where performance in terms of accuracy and speed over a small area is critical. Chip mounters, semiconductor packaging machines, and automatic visual inspection machines are the robotic machines in which Cartesian manipulation is utilized at large.

A gantry robot is another form of a Cartesian robot. It combines the flexibility of an articulated robot with wide area coverage. Overhead-mounted, rectilinear robots transport heavy payloads in large workspaces with accuracy and flexibility. A basic gantry has three degrees of freedom to access any point within its *X*, *Y*, *Z* Cartesian workspace. An optional robotic wrist can provide additional degrees of freedom for applications where control of

Figure 2   The FARAMAN RC Series robot is a 4-d.o.f. Cartesian-type manipulator with a maximum speed of 1400 mm/sec and ± 0.03 mm repeatability that can handle up to 5 Kg of payload. A number of Cartesian robots are employed in an automated AV product assembly line. (Photo courtesy Samsung Electronics, Inc.)

the angle of approach and orientation of the payload are important considerations. Gantry robots are also widely used for material handling in the nuclear and automotive industries.

Most companies have experienced the drawbacks of manual pallet loading and unloading, which involves high injury potential and worker fatigue. Furthermore, the shortcomings of traditional palletizing automation—load-out inaccuracy and limited speed—are becoming more of a problem as companies put a higher demand on flexibility. An efficient robotic palletizing system can handle multiple products on a single line, or multiple lines of similar or different products. It can eliminate problems associated with repetitive tasks and injuries and can keep workers away from hazardous or uncomfortable environments. A palletizing robot is designed for heavy-duty palletizing, material-handling, packaging, and assembly tasks. The PL Series palletizing robot (Figure 3) is a high-payload, four-axis SCARA robot designed for heavy-duty palletizing, material handling, and packaging. With a reach of 1850 mm and working volume of 1200×1200 mm, the PL1 is designed to accommodate various standard pallet geometries while accessing conveyors, feeders, and other production equipment. A payload of 60 kg and a useful repeatability of ± 0.5 mm allow the PL1 to handle a wide variety of single or multiple package types, as well as heavy industrial components.

2   MANUFACTURING AUTOMATION IN ELECTRONICS INDUSTRY

2.1   Overview

Advanced manufacturing automation is based on a compound technology of mechanical, electrical, electronic, control, software, and manufacturing technology. The current requirements of automation equipment are to assemble and inspect multifunctional products with miniaturized components and high-precision parts. *Mechatronic* equipment, often classified as *robotic* equipment, should exhibit reliability, maintainability, productivity, and agility for new products. While demand from the industry has become extreme, the repeatability of assembly robots has improved from ± 0.05 to ± 0.01 mm over the past decade. Latest chip mounter incorporates 11 placement heads with ± 0.01-mm placement repeatability. Also, some high-speed robots now work at ± 0.012-mm repeatability. These levels of precision are considered well beyond human capabilities.

The process of manufacturing an electrical product, for example, starts with PCB assembly, where electrical components are mounted and soldered. In order to ensure prime quality, inspection and adjustment of the assembled PCB are performed before the final assembly, with its mechanism assembled separately. Finally, the assembled product is inspected and adjusted to best quality/performance before packaging (Figure 4).

Workpiece positioning is the most common problem in automating assembly. Precision positioning of parts and PCBs in consumer electronic products is especially troublesome. The use of jigs and machine vision is essential to pick and place parts accurately for an advanced assembly. Visual sensors can detect the positions of parts and also allow for mixed-flow production operations. Visual sensors can also identify parts' shapes and are therefore very useful in product quality control. The most recent application of CCD technology to visual sensors has achieved ± 0.02 mm positioning accuracy.

Flexible lines are required to cope with the demands of multiple-model, mixed-flow production. Movable jigs and sensors are used to adjust to changing part shapes. In mixed-flow assembly lines product model information must be controlled to match parts with the models on the line. Some companies have used memory cards on parts pallets to achieve this control. Integration of such part flows with a mixed-line assembly is based on sophisticated parts-feeding equipment, which may account for 80% of the automation success. It has set about improving the automation rate while developing ways to handle multiple-module, mixed-flow production. The mixed-flow production approach helps to hold down equipment cost and allows for flexibility in adjusting to demand fluctuations.

2.2   PCB Assembly

The PCB assembly includes placing active plastic and ceramic components as well as components like capacitors, resistors, and inductors. The assembly itself involves ball grid array (BGA), surface mount device (SMD), pin-throughhole components (PTHCs), and odd-shaped components. PTHCs are expensive, space-consuming, and heavy compared to SMDs. Generally, SMDs and small PTHCs can be placed with special-purpose mounters, whereas odd-shaped components such as connectors, coils, and switches are handled with robots. Automatic assembly of a typical PCB requires various kinds of mechatronic equipment, as follows.

Figure 3   The FARAMAN PL1 palletizing robot utilizes high-torque motors combined with precision harmonic drives that deliver production rates up to 900 cycles per hour and repeatability to ±0.5 mm. Brushless digital AC servo-technology delivers leading performance with low maintenance needs. A PL1 is in action at the color television manufacturing plant. Heavy CRTs are depalletized from pallets that are transported by conveyors from the warehouses. Each CRT is lowered to a product housing to begin the assembly of a television. (Photo courtesy Samsung Electronics, Inc.)

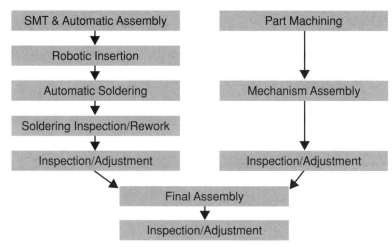

Figure 4   Assembly sequence of an electrical product. Product features are divided into a number of modules for design purposes. Each separate module may be assembled on a separate assembly line, and the assembled modules are collected for the final assembly.

2.2.1   Screen Printer

A screen printer (Figure 5) is used to apply solder paste and an adhesive dispenser is used to apply adhesives onto a PCB where surface mount-type components are to be placed. Robots with special tools such as end-effectors can be employed instead of the dedicated adhesive dispenser (Figure 6).

2.2.2   Chip Mounter

A chip mounter is employed for the high-speed automatic mounting of surface-mount type components. Modern high-speed chip mounters utilize high-speed vision assistance

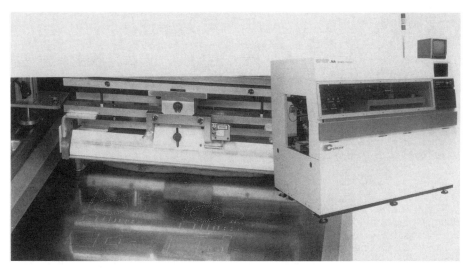

Figure 5   The Panasonic SP10P Screen Printer features automatic mask exchange and back surface cleaning. Board recognition time is 2.5 sec/2 dots at 20-sec printing cycle time. It can support up to 330 × 250-mm-sized PCB (photo courtesy Panasonic).

Figure 6 A Musashi Sigma series dispenser system can dispense a wide variety of liquid materials. Its application ranges from simple filling of epoxy resins to assemble LEDs to injecting liquid crystal to LCDs (photo courtesy Musashi Engineering Inc.).

achieving 0.15 sec/chip. Feedback from the vision data to the pick-up stage enables feeder table and pick-up nozzle correction when picking up chips. Chip mounters operating at higher speed have multiple pick-up nozzles to handle various components from 1005 chips to 25-mm QFP. Some larger chips, such as 52-mm QFPs and BGA components, are mounted with a multifunctional chip mounter (Figure 7) that can realize a high-speed mounting of 0.3 sec/chip for a 8–12-mm tape or 1.25 sec/chip for QFPs at ±0.05-mm accuracy, suitable for a stable mounting performance of 0.3-mm pitch QFP. A laser head recognition system detects lifted or skewed leads, remarkably improving the quality of assembled PCB.

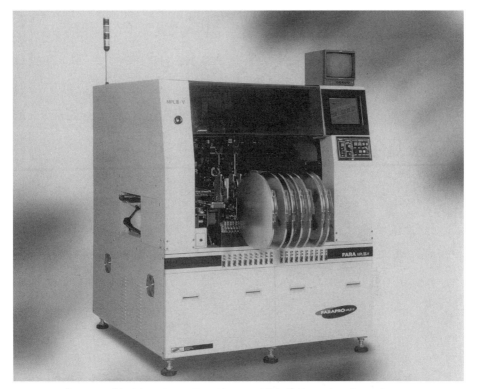

Figure 7 Samsung Multi-functional Chip Mounter MPL III-V can handle various components, such as 1005 chips, QFP, odd-form connector, BGA, and all types of SMDs. Maximum placing speed for standard chips and QFP placing speed with recognition are 0.6 sec/chip and 1.3 sec/QFP respectively. The use of five cameras ensures high-precision placement. Placement accuracy after compensation by a fixed camera recognition is better than ±0.05 mm. (Photo courtesy Samsung Electronics, Inc.)

2.2.3 Assembly Robot

Some odd-shaped components can be placed by a dedicated insertion machine, and some PTHCs are inserted by robots and soldered by robots (Figure 8). The utilization of robots to automate assembling PCBs has been growing steadily in recent years. The flexibility of a robot assembly cell allows effective assembly of odd-shaped hard-to-automate components.

2.2.4 Application Example: Automated Assembly of Printer PCB

Due to high volumes with stringent quality requirement, automating PCB assembly for the color ink-jet printer and laser beam printer (LBP) was inevitable. The assembly system replaced a number of manual assembly cells, namely odd-shaped component insertion, heat sink assembly, and ROM copying and labeling process. The automated line has produced a fourfold increase in productivity and reduced rejects from 4% to less than 1%. At the heart of the system are two robots operating per cell: a SCARA robot that inserts components while an XY robot is placed below the PCB surface performs cut-and-clinch operations of the extraneous leads. A three-axis XY robot with cut-and-clinch tools such as end-effectors performs cut-and-clinching of parts with lead pitch from 5 mm to 38 mm while manual insertion of odd-form components is done. It is a simple yet very effective companion to manual assembly workers in a small-batch production line where full automation is economically not feasible.

The assembly line is based on a nonsynchronous continuously moving conveyor belt system that carries pallets on which PCBs (around 80 × 200–330 × 250 mm, 1.2 t–1.6 t) are secured (Figure 9). The pallets are raised to a fixed position for an assembly by a pneumatic cylinder when a pallet enters a workcell. Then the PLC checks for pallet

Figure 8 A multipurpose assembly cell can handle a mixed flow of production by using a SCARA robot with multiple turret heads. In this model up to six different end-effectors can be attached to each head, making it very agile for the assembly of diverse odd-form components. (Photo courtesy Samsung Electronics, Inc.)

position, feeder status, robot status, and other signals before commanding the robot to execute assembly operations. The robot can insert about four parts in 20 seconds while using a force sensor at the wrist to detect an erroneous insertion. Various robot end-effectors, designed to grasp odd-shaped components, are used for each operation. A specially designed six-position turret (Figure 10) is employed for maximizing the number of operations performed in a cell. The XY robot does the cutting and clinching of the component legs as soon as the SCARA finishes insertions. Finished pallets are lowered again onto the conveyor belt to continue to a next cell, waiting in a buffer area if the next cell is occupied by another pallet. The effectiveness of this automatic assembly system lies in the effective combination of dedicated assembly machines (hard automation) and robots (soft automation).

2.3 Mechanism Assembly

Unlike electronic parts assembly, mechanism assembly involves dealing with nonuniform parts of various shapes and sizes. Fixing of parts is usually done by bolts, screws, and clips rather than by adhesives and soldering. Due to more complex and awkwardly shaped parts involved than in other assembly processes, more robots are required for the assembly and more tact time is required to complete each operation.

2.3.1 Application Example: Camcorder Mechanism Assembly—A Dual-Arm SCARA Application

The camcorder mechanism consists of molded plastic parts, motors, magnetic heads, belts, and pressed parts like chassis and brackets. Moreover, the camcorder mechanism assembly requires rubber belt application and lubrication of various moving parts.

Figure 9   A fully automated PCB assembly line consists of a number of robotic cells. Each cell is equipped with an assembly robot and peripherals performing odd-form component insertions, cut-and-clinch, or soldering. (Photo courtesy Samsung Electronics, Inc.)

A dual-armed SCARA robot (Figure 11) was employed to construct an automated 8-mm camcorder mechanism assembly line. The robots operate in cooperation, sharing workspace by continuously monitoring each other's position so that no physical encounter of the arms occurs, thanks to an advanced collision-avoidance algorithm. Considerable space and time savings can be achieved if robots can share work envelopes, operating as one tool executing several assembly sequences simultaneously in one cell. One robot may perform a series of insertions while the other loads components into its multiheaded turret. As soon as the robot finishes loading components from the feeding stations, it negotiates to enter the work envelope. If the targeted point of insertion is occupied by another robot, it will wait until the active robot finishes the operation and clears away from the area. The preparation for this robot motion is totally transparent to the users so that the users may program the robots separately. A fixed vision system seeks for a landmark on the pallet, and an offset is calculated for the robot to start placing parts with absolute confidence.

2.4   Inspection and Adjustment

The problems arising from inspection and adjustment come from increasing quantity requirements and decreasing geometry of products as found in assembly stages. Around 15% of the total resources consumed in manufacturing are taken by inspection and adjustment process in the electrical/electronics industry. The research and development of equipment for inspection and adjustment should be directed towards:

1.   Fully automated operation
2.   Simultaneous measurement and adjustment

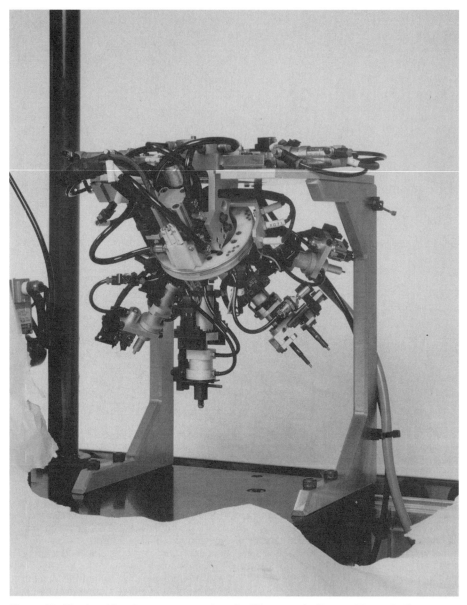

Figure 10 The turret head can accommodate six different end-effectors. The end-effectors can be grippers or special tools performing a push or a pull. The turret head rotates about the Z-axis of a robot by pneumatic power, selecting a suitable end-effector position for a given task. (Photo courtesy Samsung Electronics, Inc.)

Table 3   Examples of Typical Electrical Products That Require Automatic Mechanism Assembly

Categories	Products
Home appliances	VTR, camcorder, CD, CDP, DVD
Telecommunication	Cell phone, telephone, pager
Information technology	Hard disk drive, floppy disk drive, printer, scanner, fax, copier
Components	Electrical motor, compressor

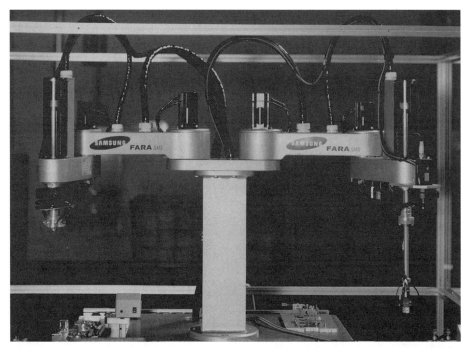

Figure 11   The SD1 dual-arm robot. The robot has two standard four-axis SCARA robots attached in a fixture, reducing the footprint to just 225 cm^2. The robot is controlled by a multitasking (Lynx Real-Time OS) controller based on a 66-MHz Intel 80486 CPU. Motion control is governed by a number of TMS320C31-based modular motion control boards plugged into the controller. (Photo courtesy Samsung Electronics, Inc.)

3. Compact design
4. Flexible system to handle varying product
5. CIM-ready standard interfaces, standard software

Automatic inspection and adjustment system technology is based on a compound science of software, measurement equipment, precision mechanism design, and control theory (Figure 13).

Assembled PCBs are tested for proper functioning of the circuitry as intended. An *in-line circuit tester* detects various defects of the PCB, including short/open circuit and missing, misplaced, poorly soldered, and damaged parts.

2.4.1   Application Example: CRT Inspection and Adjustment

An automatic CRT inspection and adjustment system (Figure 14) for the manufacture of television and computer monitors is devised. Fundamental characteristics such as focus, balance, brightness, and horizontal/vertical sizes of a screen are inspected and automatically adjusted until predefined quality criteria are met. The system consists mainly of two robots, a CRT color analyzer, a pattern generator, a PLC unit, an adjuster mechanism, and a personal computer.

The adjuster hand attached as the end-effector of the robot has eight motors to adjust the variable components when signaled by the main computer, which is also responsible for the whole operation and for the analysis of digital signals from the measuring device and the color analyzer. The measuring device is shifted around the CRT surface by a single-axis Cartesian robot. Units with unsuccessful results are marked as rejects on the pallet.

2.4.2   Cream Solder Inspection System

Machine vision-based systems have replaced the human vision inspection process completely. Human vision is no longer practical for most modern PCBs. Soldered joints,

Figure 12 A fully automatic VTR deck assembly line using 33 SCARA robots implemented by Samsung Electronics, Korea. The assembly sequence involves fitting 58 odd-form components, 19 screws, 11 washers, and 9 springs and greasing 19 points. This implementation is capable of producing 85,000 units per month with 18 seconds tact time. (Photo courtesy Samsung Electronics, Inc.)

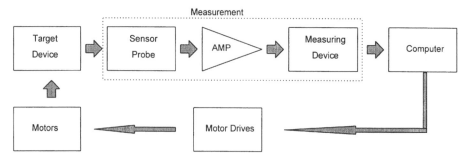

Figure 13 Simplified version of a typical automatic inspection and adjustment control block diagram. The sensor probe detects the signal from the target device and the signal is amplified for the measuring device to convert into useful data formats. The computer calculates the deviation and control quantity for a compensation. The motor drives are controlled by the computer to adjust the target devices until the deviation falls into a tolerable range (Kim, 1994; Stanley, 1984).

Figure 14   A three-axis Cartesian robot docks the adjuster hand to its adjusting points at the back of CRT by following the guidance of sensors as soon as a target CRT is locked up on a position by a pneumatic unit that lifts the pallet from the conveyor. (Photo courtesy Samsung Electronics, Inc.)

missing parts, shorted parts, dislocated parts, disoriented parts, and insecurely seated parts are identified in real time (Figure 15).

PCBs for SMDs require application of cream solder before placement of components with chip mounters. Implementing a vision system coupled with a Cartesian robot enabled the automatic inspection of all PCBs for proper cream solder application (Figure 16). The system visually examines the amount and location of cream solder applied by the screen printer or by the adhesive dispensing robot. It is normally a two-dimensional visual inspection, but the usage of an additional laser triangulation device enables the measurement of cream solder height as well. The monochrome vision system is based on a TMS320C30 running at 32 Mhz, capable of handling a maximum of two cameras that can provide four frames at 640×480 resolution. The Pentium-based host computer running MS Windows is responsible for the coordination of the whole inspection process and accommodates the vision board in one of its ISA slots. As many as 2000 locations for 200 models can be stored in the controller for inspection, and no less than two inspections are performed in a second.

2.5   Considerations for Clean-Room Applications

There are a number of robot types widely used in outdoor, hygienic, condensing environment and clean rooms. The external surfaces of such robots are protected by a corrosion-resistant coating, in compliance with IEC IP55 and NEMA type 4 requirements; all of its joints are sealed to keep moisture out of the robot, and an internal air circulation system prevents condensation. Most regular industrial robot designs can be adapted to devise a clean-room robot that functions satisfactorily under exceptional conditions such as higher atmospheric pressure, higher humidity, and infrared emission. There are a number of considerations for clean-room robot design and an anti-particle emission strategy.

1. *Design criteria.* In order to sustain the conditions of a clean room, the robots employed should exhibit a number of features: a carefully designed aerodynamic external shape to prevent particle deposition and prevent turbulence in air flow, a small footprint for minimum usage of costly clean-room space, durability against chemicals, and maintenance-free design of most mechanical parts.

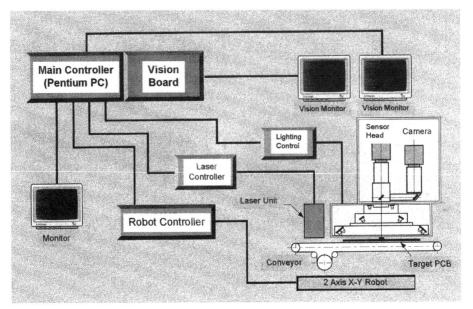

Figure 15 The VSS4 PCB inspection system utilizes a vision system based on a TMS34020 DSP. Cyber-Cage CG-2000-AP laser unit, Matrox vision board, and two monochrome FOV cameras. A Pentium PC system controls the inspection process, which performs at 30 msec/point.

2. *Source of particles* (see Table 4).
3. *Anti-particle emission measures.*

   (a) *Drive mechanism.* Relative movement of any driving mechanism results in wear and tear of surface, spatter/evaporation of lubricant and anticorrosive paint, and production of rust. An antiabrasive material such as special use stainless (SUS) should be used for linear guides after heat treatment to avoid corrosion, and use of maintenance-free lubricant will help to enhance surface hardness. Zero emission of particles cannot be achieved in any case, but brushless motors are the best.

   (b) *Transmission.* Again, the relative movement of any transmission mechanism results in wear and tear of surface where timing belts made of polyurethane are superior to gears or a rack-and-pinion system for durability and particle performance. Since the evaporation of lubricant results in the most particle formation as far as lubrication is concerned, grease for vacuum application should be used to minimize evaporation in high-speed operation.

   (c) *Material.* Material exhibiting no magnetism, electrostatical nature, and fine durability against fatigue, impact, and corrosion is preferred. Aluminum alloy, iron alloy, carbon steel, cast iron, and austenite/martensite stainless steel are widely used only when properly treated for clean-room application.

   (d) *Surface treatment.* Although aluminum alloy is an eligible material featuring lightness and good workability, treatment with alumite is required to improve stiffness. Teflon coating is effective to prevent rust when used in humid surroundings, and nonelectrolysis nickel plating is recommended to avoid contamination from surface oxidation when used with pneumatics. Iron alloy also requires a thick and uniform application of nonelectrolysis nickel plating for the same reason. Carbon steel, well known for its superior stiffness and low cost, is used with a coat of Teflon or a hard chromium plating applied to the surface, as it is prone to surface corrosion.

   (e) *Painting.* Three types of paints can be used: melanin resin, polyurethane, and epoxy + melanine resin.

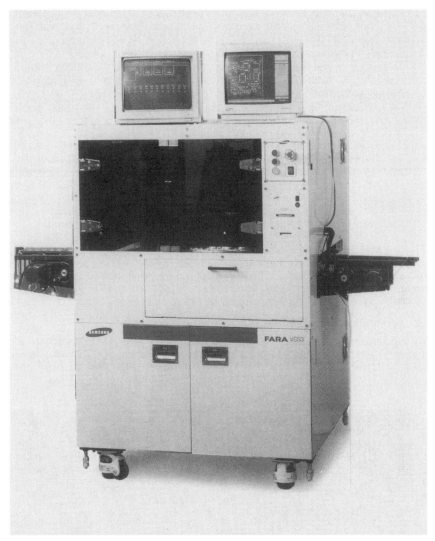

Figure 16 The system consists of a two-axis Cartesian robot, a host computer, a vision system including a laser head, and a PLC. The robot employed that carries the vision camera is a Cartesian type with 400 × 300-mm stroke and ±0.02-mm repeatability. The motion of the robot is handled by a pair of digital servo-drives are controlled by the PLC's position control unit. (Photo courtesy Samsung Electronics, Inc.)

Table 4   Sources of Particles in Robots

Category	Components
1. Drive mechanism	Motors, linear guide
2. Transmission	Gear box, ball screw, belt, pulley, oil seal, lubricant
3. Structural	Material, surface treatment, paint
4. Cabling	Cables, air plumbing

(f) *Cabling.* It is best to route cables inside robot arms to avoid incidental rubbing of cables against the robot or any other devices. Cables with polyurethane skin should be used.

4. *Particle isolation technology.* Even though all components used are considered for preventing contamination, one cannot expect zero particle generation from robots. A number of secondary anti-particle emission measures can be adopted to enhance particle performance of robots.

(a) *Vacuuming of particles.* The inner space of a clean robot may require to maintain vacuum to expel particles. Polyurethane or rubber gaskets, Gore-tex sealing, and labyrinth sealing are used to seal various moving parts from outside. Vacuum pressure of 200–1000 mmAq is maintained by using a ring blower or a helical roots blower that can handle at least 500 liters/min.

(b) *Magnetic fluid seal.* The magnetic fluid seal is used in many demanding spots where mechanical seals and oil seals are ineffective. Since it is a non-contact sealing device, there is no loss in torque and no particle is produced due to contact with rotating surface. However, it is not suitable for high pressure, high temperatures, and liquid sealing for a long period.

Figure 17 features a typical SCARA clean robot developed for implementing in clean-room manufacturing, such as automating hard disk drive (HDD), compact disk player (CDP)/digital video disk player (DVDP), pick-up production, that requires at least class 10 (meaning 10 counts of 0.3-μm diameter particle allowed in 1 ft^3; JIS B9992) clean-room capability.

2.5.1 Application Example: Hard Disk Drive Assembly

A cell-type assembly line is developed in which four clean-room SCARA robots, a class 10 Cartesian robot and a number of automatic screwdrivers are used to assemble various parts of HDDs (Figure 17). The assembly line consists of seven stations, where the first

Figure 17 An HDD assembly cell is realized by using an SC1 clean-room SCARA robot. It is one of the seven stations where disks are loaded onto the spindle of the drive. The SC1 clean-room SCARA robot meets the class 10 clean-room requirement. Fitting an optional exhaust pump will improve the cleanness performance to class 1. (Photo courtesy Samsung Electronics, Inc.)

station is operated as a preparation stage where the baseplates are fixed onto pallets. The line is flexible enough to handle a number of models of different storage capacity. Most of all, a baseplate fixed onto a pallet is accurately positioned into the workcell. The disks from the disk cassette or canister are retrieved and placed onto the spin motor. Screws and spacers are placed and fastened using the automatic screwdrivers. Vision-guided manipulation aids the assembly of the head by monitoring the location of the disks and the head. The assembly of the head requires extremely accurate yet swift manipulation and consistent force application that manual operation cannot achieve to keep up with the quantity targets. Several more screws are applied to fix the head subassembly before a cover is loaded on top of the baseplate. Finally, six more screws are applied and the assembled disk is transported to the next process, namely the *servo-track writing*. The cycle time for each station is 28 seconds, capable of producing 1,750 units per day (two shifts) when utilized for 85% of the total production time. The production volume increased from 1,400 to 1,800 after automation of the above processes.

3 MANUFACTURING AUTOMATION IN THE SEMICONDUCTOR INDUSTRY

3.1 Overview

The semiconductor industry has become very significant in the electrical/electronics industry, taking more attention in terms of investment and turnover than any other conventional fields of electronics business. The semiconductor industry has gone through a vast change in the last decade, generating consistent demand for new technologies to support the design and manufacture of fast-evolving silicon devices. The increasing complexity and condensing geometry of current submicron circuit designs are placing new demands on practically all characteristics of the semiconductor manufacturing process. Nowhere are these demands more intense than in the development of semiconductor manufacturing and inspection equipment. To meet these demands, knowledge of semiconductor fabrication and the desire to enhance manufacturing efficiency play a vital role and will become fundamental to designing and testing advanced automation platforms and advanced control strategy for the semiconductor industry.

The process of manufacturing semiconductors or integrated circuits (ICs) consists of complicated steps, producing several hundred copies of an integrated circuit on a single wafer. Generally the process involves the generation of 8 to 20 patterned layers on and into the substrate, ultimately forming the complete integrated circuit. This layering process creates electrically active regions in and on the semiconductor wafer surface. The whole process of manufacturing semiconductors can be divided into four major steps: circuit design, wafer fabrication, assembly, and inspection.

The semiconductor manufacturing begins with production of a single-crystalline ingot from a highly purified molten silicon. From the ingot the wafers—thin, round slices of semiconductor materials—are produced in various sizes. The wafers are mirror-polished on one side for the various processes of engraving circuitry on them. The wafers are then sent to the fabrication (FAB) area, where they are used as the starting material for manufacturing integrated circuits. By the help of CAD systems, the circuitry to be formed on the wafer is produced and a mask (reticle) is produced on glass.

3.2 Introduction to FAB (Fabrication) Process

The heart of semiconductor manufacturing is the FAB facility, where the integrated circuit is formed in and on the wafer. The FAB process, which takes place in a clean room, involves a series of principal steps, described below. Typically it takes from 10 to 30 days to complete the process.

3.2.1 Oxidation

The polished wafers are treated with oxygen or vapor in a high temperature (800°C–1200°C) to form a thin, uniform SiO_2 film on the polished surface.

3.2.2 Masking and Exposure

Masking is used to protect one area of the wafer while working on another. This process is referred to as *photolithography* or *photo-masking*. A photo aligner aligns the wafer with a mask and then projects an intense light through the mask and through a series of reducing lenses, exposing the photoresist with the mask pattern. Precise alignment of the wafer with the mask prior to exposure is critical. Most alignment tools are fully automatic.

3.2.3 Development

The wafer is then "developed" just like photographic film (the exposed photoresist is removed) and baked to harden the remaining photoresist pattern.

3.2.4 Etching and Other Processes

The wafer is then exposed to a chemical solution or plasma (gas discharge) so that areas not covered by the hardened photoresist are etched away. Further processes such as doping, dielectric deposition and metallization, and passivation are followed before assembly and testing of the fabricated chips.

3.3 Wafer Handling

Each process tool in a semiconductor plant must be supplied with a continuous flow of product in order to meet the fierce demand for production efficiency. Deep submicron processing, 0.35 microns or lower, uses wafers 8 inches or greater that cannot be handled by human operators because they cannot take any vibration or particles. Typically, process tools in the FAB are utilized to process wafers for only 40% of the total system uptime. The remaining 60% of the time those tools are idling, and 20% of that downtime is roughly divided between processes. The productivity of a facility can be improved dramatically with proper introduction of material transfer robots between tools. There are a number of considerations for such wafer transfer robots, as follows.

3.3.1 Contamination-Free

Relatively less familiar technology in robotics is introduced to ensure maximum control over the production and removal of dust particles produced by the robot itself and the cell. All robot arms are coated with polyacetar, for example, to minimize unwanted particle production. Only sealed clean-room motors, belts, and pulleys are used. Magnetic fluid seals are provided at the arm joints, and a fine mesh filter is used for exhaustion of the link enclosures when a more stringent cleanness requirement is stipulated, such as for the vacuum chamber application (should be prepared to withstand 10^{-8} Torr). Furthermore, vacuum devices rather than compressed air devices are used. All motor parts including the manipulator arms, are located well beneath the wafer surface to further reduce the chance of a wafer catching dust particles emitted from the robot.

3.3.2 Absolute Reliability

Reliability of robots is absolutely necessary because no interruption in production may be tolerated, and the cost of a wafer stack could exceed the cost of its production equipment. Most clean-room robots for wafer handling are designed with harmonic drives, resulting in backlash-free movement at 0.01-mm repeatability. The wafer loader's built-in sensor detects the presence of a wafer, tilt angle and cassette deformation, and so on. An optimum position for picking up or placing wafers is automatically determined while a contingency scheme is always active, for example, for dealing with missing wafers and cassette. The mechanics of the robot are simplified; only sealed clean-room motors, belts, and pulleys are to be used. A typical life expectancy should be 10 million cycles or more, with an MTBF of more than 35,000 hours.

3.3.3 Rapid Transfer

A typical SCARA-type loader (Figures 18, 19) employs two to four links to minimize wafer exchange time, allowing maximum agility in limited space. Thoughtful design of the manipulator arms is required to minimize the inertia, which will enhance the acceleration/deceleration characteristics, making it suitable for rapid wafer transfer. Low-inertia design allows rapid motion with no sensitivity loss within the closed-loop servo-system. A respectable speed of 1000 mm/sec in X-direction, 100 mm/sec in Z-direction, and 260°/sec in θ-direction can be achieved with such a robot that can handle up to a 12-in. wafer.

It is not only the robots that are essential in transporting wafers between stations. The mobile transport system handles semiconductor material between and/or within process bays to provide a streamlined production environment. A batch of wafers is usually transferred in boxes, cassettes, or pods suitable for loading and unloading operation of the vehicle. With automatic transfer systems, on-time material delivery and 24 hours of unmanned operation are attainable, allowing for optimum tool utilization.

Figure 18   The Daihen SPR-152 wafer transfer robot has two arms for shorter tact time. It incorporates magnetic sealing technology for particle-free operation, and the rigid arms ensure better horizontal accuracy (photo courtesy Daihen Co.).

3.3.4   Clean Mobile Robot

ICMR is an intelligent clean class 1 mobile robot (Figure 20) developed to transfer wafer cassettes between fabrication tools. It is constructed by combining a six-axis manipulator and an AGV that carries the robot on top. The AGV is powered by a number of on-board nickel–cadmium batteries that are constantly monitored and recharged by a voltage drop sensing module. The use of a couple of diagonally arranged steering motors, drive motors, and casters enables the AGV to be operated in a limited space since steering to an arbitrary direction can be achieved without rotating the vehicle. Unlike other mobile robots, the ICMR uses no predefined path but instead uses ultrasonic range finders to follow a wall. This feature makes retrofitting ICMRs into existing production lines effortless. A vision system attached to the robot end-effector is responsible for accurate pick-and-place operation as the vehicle comes to a load/unload position. The vision seeks a landmark for calibrating its end-effector with respect to the coordinates of the wafer cassette landing zone. Then the vehicle exchanges data with the target facility via eight-bit optical communication. Since the ICMR is heavy and friction with the floor changes abruptly with cruising speed in the clean-room environment, controlling the vehicle is no simple matter. The concept of membership function has been implemented to follow the wall, analyzing distance data from ultrasonic sensors and data from the encoders on the wheels. The target cruising speed is controlled by an open-loop control technique called the *dynamic scheduling tracking* method, where the control amount is the target speed and the control input is RPM of the driving wheel. The target location is reached by setting the target cruising speed, which is manipulated by dynamic scheduling utilizing the remaining distance and environmental information.

The robot attached to the AGV is a six-axis five bar-linked vertically articulated manipulator commonly found in industrial applications. An end-effector to handle the wafer cassettes (up to 8 kg) is fixed to the end of the five bar-link system, which is renowned for optimal balance effect to minimize the usage of battery power. The repeatability of

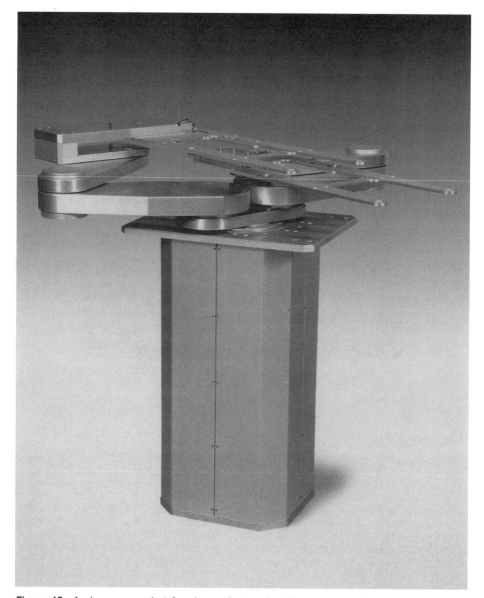

Figure 19 A clean-room robot for glass substrate-handling between cells or cluster tools. It exhibits the most repeatable positioning accuracy to ± 0.25 mm and class 1–10 particle performance to handle heavy glass substrates for FPD (up to 5 kg). (Photo courtesy Samsung Electronics, Inc.)

the robot is better than ± 0.05 mm and that of the vehicle is better than ± 0.1 mm. Most conceivable safety features can be found with the ICMR: ultrasonic sensors, proximity sensors, pressure sensors at the bumper, audio and visual warnings, and emergency stop buttons around the vehicle.

3.3.5 Monorail Transporter

Monorail transporters (Figure 21) are designed to operate at a much faster speed (up to 260 meters per minute) than conventional flat conveyor systems or mobile robots. Since it is built for midair operation to connect assembly cells, a rail system should be attached

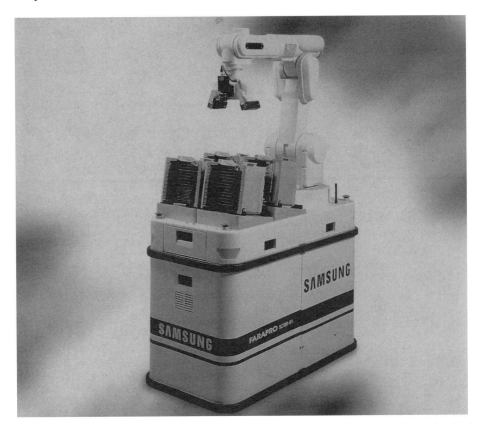

Figure 20   Intelligent clean-room mobile robot (ICMR). (Photo courtesy Samsung Electronics, Inc.)

to the ceiling or wall. Timely delivery of components to the point of need results in increased productivity by reducing operator time spent searching for materials while virtually eliminating stacking of excessive materials by the assembly cells or inside the production site. Some monorail transporters can climb slopes up to 60° along the railway while keeping the payload level at all times. The tracks can be arranged while avoiding pillars or equipment that may be in the way and directly connecting cells on the same floor or on different floors. The direction of travel can be changed easily by a wide variety of lane change devices (for parking, branching, merging, turning, etc.). Use of a monorail transporter improves facility use through increased storage density and reduced facility size due to less need for transport aisles.

Direct drive (DD) motors are employed for compact and lightweight design of a transporter. Characteristics such as high torque (torque/weight ratio of 6.0 Nm/kg or more) for climbing a slope and smooth acceleration and deceleration, which will protect valuable payloads in motion from vibration/physical shocks, are fundamental. DD motors are also maintenance-free, eliminating the need for lubrication and other maintenance activities. Monorail transporters receive a stable supply of electricity from trolley wires (feeder) made of a highly resistant material. The monorail transport system should be designed for a class 1 clean-room environment in order to be employed for semiconductor wafer fabrication.

3.4   Wafer Inspection

In the semiconductor manufacturing process an automatic, computer-driven electrical test system checks the functionality of each chip on the wafer. Chips that do not pass the test are marked with ink for rejection. As chip geometries become smaller and their circuitry grows more complex, manufacturers' need for productivity improvements through faster, more accurate wafer inspection continues to accelerate with probe systems.

PRI-AeroTrak Turntable

PRI-AeroTrak Vechicle at
a Simple Node

Horizontal Transfer Station

Figure 21 The PRI Aero-Trak monorail transporter travels at 80 fpm with a 25-lb payload. The rail system can be composed of a maximum of 255 vehicles. The rail system features queues, parking lots, turntables, elevators, and automatic path planning (photo courtesy PRI).

3.4.1 Probe System

A fully configured probing system can load, focus, align, theta-correct, position-correct (at start position and each subsequent step position), unload, and sort a number of cassettes of wafers. Figure 22 depicts a probing station. The main feature of the system is facilitation of the wafer to make fast and accurate contact with a probe pin. It consists of 11 axes for precise positioning and compensatory positioning of wafers. A linear positioning stage is employed to accurately position a wafer in the XY direction at 300 mm/sec. An additional Z-axis on top of the stage lifts the wafer for loading and inspection. Loading and unloading of the wafers is done by a dual-arm clean-room robot. Fine alignment of the wafer is performed with assistance from a vision system.

The linear positioning stage uses a number of high-accuracy brushless linear motors. Unlike mechanical drives incorporating ball screw or rack-and-pinion, the brushless linear motor offers highly reliable noncontact operation without backlash or component wear. The higher stiffness of linear motors results in optimal dynamic and settling time performance. Linear servo-motors also provide smooth closed-loop motion and are capable of generating the high peak output force required for fast acceleration and deceleration. In order to minimize the Abbe error, the linear motor and linear encoder are centered between two precision linear recirculating bearings. A total of four bearings provides the support for the user load as well as the moving coil and encoder reader head. An internal noncontact linear optical encoder provides high-resolution feedback (0.25 μm) for high positioning accuracy and velocity control. An integral linear servo-motor enables high servo-stiffness for fast move and settle performance (20 m/sec). With maximum velocities to 3 m/sec and acceleration of 4 g, the linear positioning stage is an ideal choice for the probing station.

3.5 Packaging

A diamond saw or laser beam typically slices the wafer into single chips. The inked chips from the EDS test are discarded and the remaining chips are visually inspected under a microscope or an automatic probing stations before packaging. The chip is then assembled into a package that provides the contact leads for the chip. The wire-bonding machine (Figure 24) then attaches wires (a fraction of the width of a human hair) to the leads of the package. Encapsulated with a plastic coating for protection, the chip is tested again prior to delivery to the customer.

Current chip assembly requirements are pushing wire bonding to its limits, and the use of TAB has been highly successful. The trend will continue to apply flip-chip with both solder and conductive adhesive technologies. In addition, pushing the chip on board will continue, not only by wire bonding the chip to the board, but also by TAB bonding to board and flip-chip solder bonding to organic board. Most large semiconductor companies typically use all three technologies: wire bond, TAB, and flip-chip. The development goal for wire bond by ball bond and wedge bond technologies is set at a pitch of 50 microns and for TAB at a pitch of 75 microns. Japan expects to push wire bond up to 1200 I/Os on a 20×20-mm chip using a staggered pin configuration.

3.5.1 Die Bonder

The die is the single square or rectangular piece of wafer into which an electrical circuit has been fabricated. Die bonding is the process of connecting wires from the package leads to the die (or chip) bonding pads. Alternatively, it is the process of securing a semiconductor die to a leadframe or package. There is an area on the periphery of a silicon die for making a connection to one of the package pins. A small-diameter gold or aluminum wire is bonded to the pad area by a combination of heat and ultrasonic energy. Die bonders attach chips to leadframes by means of specific bonding materials that are determined by the process and the machine type. The most commonly used bonding processes are epoxy, soft solder, LOC tape, and eutectic.

A die bonding machine consists of the following modules; an indexer, a die pick-up, a leadframe feeder, and a vision system for positioning wafers/dies and inspection (Figure 23). The bonding head utilizes a number of motors to transfer/position each die from a wafer to a sequence of leadframes. The programmable leadframe indexer provides the transport of leadframe. It is a recipe-driven system that utilizes a number of maintenance-free linear motors. The high-resolution vision system automatically checks the operating parameters in order to define chip position, identify faulty chips, and determine the process direction for chip pick-up.

Figure 22 The SWS1 probing station provides automatic prealigning, loading, and unloading of wafers in various diameters (6–12 in.) prior to inspections. The resolution of the system is 0.5 μm and the typical accuracy achieved is 6 μm. (Photo courtesy Samsung Electronics, Inc.)

Figure 23 The SDB-01 LOC die bonding machine utilizes a high-resolution video automatic die positioning system that checks the entered parameters in order to define chip position, identify bad chips (corner check, broken or damaged chip) and determine the process direction for chip pick-up. The system is capable of achieving bonding accuracy of ±25 μm, and the tact time of the system is 4 sec. (Photo courtesy Samsung Electronics, Inc.)

Communication with peripheral equipment could be helpful in realizing a high level of automation. The SECS I and II communication protocol ensures standard connectivity to other semiconductor industry standard equipment. It also helps to reduce operator handling by enabling remote control as well as alarm and status reporting.

3.5.2 Wire Bonder

The wire bonder (Figure 24) supplies the electrical interconnection between the chip and the external connectors on the leadframe and thus represents the interconnection of silicon circuitry to the world. A minute ball is formed on the end of a heated gold wire and is firmly bonded to a surface area on the chip of no more than 0.1 mm by a combination of thermopressure and ultrasonic technology.

Sliced wafers can be attached face-up into single-chip packages or multichip modules using eutectic solder reflow or epoxy die attachment methods, or attached face-down using flip-chip die attachment methods. For face-up mounting, the power, ground and signal pins on the die can be connected to the IC package or MCM using automatic wire bonding (both wedge and ball types). For face-down mounting, the bare die can be "bumped" with gold balls to facilitate attachment. The chip is now assembled into a package that provides the contact leads for the chip (Harris Corp., 1997). The wire-bonding machine is used to attach wires a fraction of the width of a human hair to the leads of the package.

3.5.3 TAB (Tape Automated Bonding)

TAB is a low-cost IC packaging method that attaches a die directly to a PCB using a thin film with conductive leads. TAB equipment (Figure 25) is being used in a wide range of applications, such as LCD drivers, IC cards, and other small systems that require thin, lightweight ICs. To increase performance of wire bonding, the wire is replaced with etched copper leads plated with gold. The leads are supported with polyimide plastic film. TAB is the dominant method of packaging chips for driving liquid crystal displays. It is

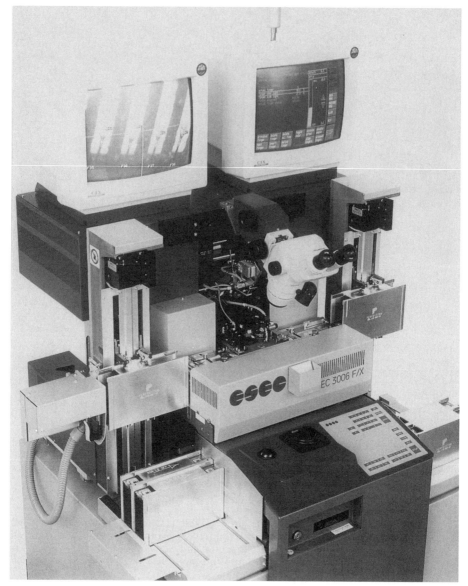

Figure 24   The ESEC 3006 wire bonder utilizes the Flying Bondhead, which is operated by frictionless air bearings and a direct linear drive (voice coil). It results in no wear, and high speed and accuracy are achieved. AI based on teach and search algorithms is used for the pattern recognition system (photo courtesy ESEC SA).

also used for some very cost-sensitive high-volume consumer electronics products, such as digital watches and calculators.

An assembly process was automated by using the TAB technology for 7–17-in. TFT LCD (t = 0.5–1.1 mm). The process involves aligning LCDs and press-bonding ICs from a tape onto the LCD panels. A number of kinds of equipment have been developed to implement various steps of applying the TAB technology in TFT LCD assembly: a cell loader, an anisotropic conductive adhesive films (ACF) applicator, a chip-mounter (prepress), a bonding machine (press), and finally a visual inspection system. The cell loader employs AC servo-motors and ball screws to accurately deliver LCDs into the assembly cells. An LCD panel is initially prepared by applying the ACF for the mounting of

Figure 25 The MOL1/OLB TAB system is suitable for the LCD outer lead bonding (OLB) process. LCD sizes ranging from 7–17 in. are processed to attach TAB IC components (min. 60 μm) to an LCD panel. Bonding accuracy is better than ±8 μm and it takes 2.2 sec per TAB operation. The vision system inspects the assembled LCD at 35 sec/2 sides at ±1-μm resolution (photo courtesy Panasonic and Samsung Electronics).

components of 55-μm lead pitch. The ACF can be used in a maximum pressure of 80 Kgf and a temperature range of 80°C–180°C with 0–20-sec pressurizing time. An ACF applied LCD panel is press-loaded with components by using the chip-mounter which can handle 2 types of ICs at a time with 0.1-mm placing accuracy. A vision system examines the subassembly for misplacement, missing chips, and particles. Then the pre-press mounted components are permanently bonded by applying appropriate temperature (30°C–350°C), pressure (15–400 Kgf) for a certain time (1–120 sec). Finally, the assembled LCD panel is inspected by the vision system of 1-μm resolution for defects. Most motion was dealt with linear motors to maximize accuracy and eliminate noise.

4 ROBOTS IN THE INSTRUMENT INDUSTRY

4.1 Coordinate Measuring Machine

In the 1950s the design of coordinate measuring machines (CMMs) (Figure 26) was known to industry. Diffraction gratings, which had mainly been used in spectroscopy until then, could be used to measure with great accuracy the movement of machine tables. The other development was the application of near-friction-free linear bearings for purposes of precise automatic control. This enabled the design of coordinate measuring machines with tables able to carry large workpieces for the purpose of measurement. The development of the touch trigger probe and the introduction of computers into the system greatly enhanced the performance and scope of coordinate measuring machines. The probe is a transducer, or sensor, that converts physical measurements into electrical signals, using various measuring systems within the probe structure.

Workpieces to be inspected are mounted on the CMM granite table. A Z-axis-mounted electronic touch probe (Figure 27) is then guided to the workpiece to check and record the accuracy of features such as planes, diameters, cones, spheres, cylinders, lines, slots, bolt hole patterns, and point-to-point measurements. Most modern CMMs can operate in either CNC or joystick and move smoothly on air bearings. Built-in glass linear scales provide a typical resolution of 0.0005 mm (0.00002 in.). Introduction of a special alloy in the axes ensures accuracy at varying room temperatures. Also, some advanced CMMs feature a fully digital servo-control system to ensure superior travel stability and vibration-free operation.

To enhance both repeatability and accuracy, CMMs usually incorporate a granite table on which to mount the workpiece. CMMs should be rigidly constructed, normally using a bridge-type configuration, mounted on vibration-proof mountings, and ideally housed

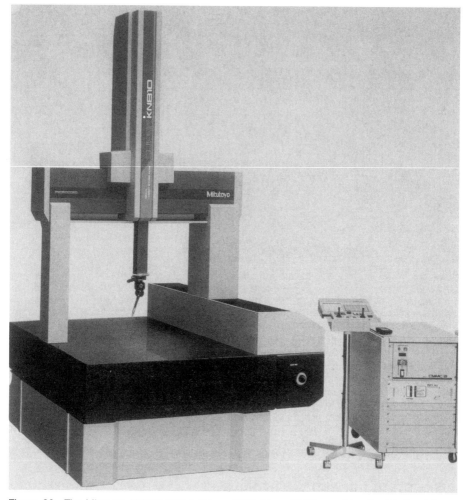

Figure 26 The Mitutoyo KN810 CMM. It is a CNC-driven CMM capable of achieving 0.5 μm measurement accuracy over a 850 × 1000 × 600-mm area. The KN810 can be driven at 140 mm/sec. The machine is used to inspect various features of manufactured components to ensure quality and conformance relative to an engineering drawing (photo courtesy Mitutoyo).

in a temperature- and humidity-controlled room. To ensure reference traceability back to national standards, they must be calibrated at regular intervals. Published accuracy data should be available on every CMM model. For added confidence, each CMM should be checked prior to shipment in accordance with ANSI-B-89.1.12 and either meet or exceed published accuracy levels. All equipment used to perform these tests should conform with the National Institute of Standards and Technology (NIST), and test results for all machines should be kept on file by the manufacturer.

4.2 3D Laser Tracking System

The measurement head of a laser tracking system consists of an interferometer, a tilting mirror combined with two or more optical angle encoders, motors, and a position detector (Figure 28). Coordinates are generated based on the method of polar joint determination. The direction of the laser beam in space and the distance between the measuring head and the reflector are measured. These measurements are taken and converted to Cartesian coordinates by a computer.

The laser beam is reflected back to the laser head directly along its transmission path as long as the reflector does not move. When the reflector is moved, the transmitter laser

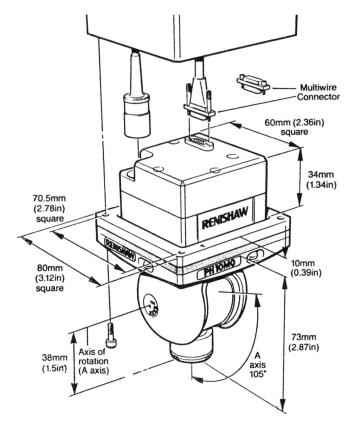

Figure 27   The Renishaw PH10M motorized probe head with ±0.5-μm (±0.00002-in.) repeatability. It has two axes, *A* and *B*, which turn 105°–0° in 7.5° steps and ± 180° in 7.5° steps respectively. It also features an autochange capability when used with an Autojoint probe system (photo courtesy Renishaw).

beam no longer hits the optical center of the reflector. Instead, the reflected beam follows a parallel path according to the rate of the shift. This parallel offset is determined at the position detector—a 2D position-sensitive photodiode within the measuring head—by the measurement of voltages and is factored into any distance measurements taken by the interferometer. The calculated angle corrections for the tilting mirror are converted to analog signals and sent to the motor amplifiers in order to point the laser beam back on the center of the reflector. This calculation is done many times per second, allowing continuous tracking of any path.

As with all interferometers, no absolute distances can be determined. It is possible only to determine changes in distance—that is, how much the reflector has moved towards or away from the measuring head. This is true because measurements are made by means of fringe counting. To achieve the absolute distance measurement necessary for polarpoint determination, measurements must always begin with the reflector positioned at a point to which the absolute distance is known. The interferometer counting pulses are then added to or subtracted from this initial distance; this is called *quasiabsolute distance measurement* (source: Leica AG, Heerburgg, Switzerland).

5   EMERGING TECHNOLOGY AND CONCLUSIONS

5.1   Overview

The electronics industry is a fast-changing and significant industry that has evolved into the largest and most pervasive manufacturing industry in a relatively short period of time. It provides numerous essential products in our everyday lives, from lightbulbs to supercomputers. The electronics industry is a vital part of the world economy. It is "the largest

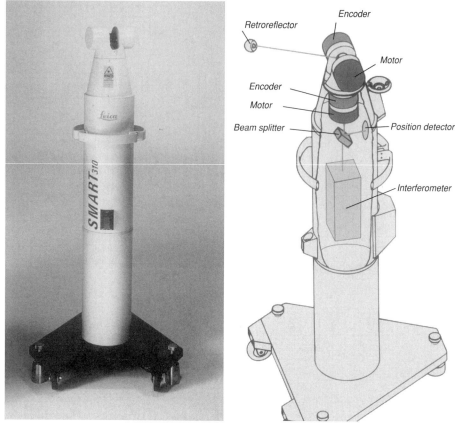

Figure 28 The Leica Smart310 (current model is LT500) is a transportable and climate-proof 3D laser tracking system with freely movable reflector sphere. It features a maximum targeting speed of 5 m/sec (6 m/sec) over 25 m (35 m) range with 0.7-in. (1.26-in.) angle resolution and 1.26-μm (1.26-μm) distance resolution. For one-man operation it features voice recognition and remote control (photo courtesy Leica AG).

manufacturing employer in the United States, accounting for nearly 11% of the U.S. gross domestic product. It is expected to grow at a rate of 4% per year throughout the remainder of the 1990s" (MCC, 1992).

A significant trend in the electronics industry today is toward making products more personal by making them more compact, multifunctional, and affordable than ever. As the trend towards miniaturization and high performance of products continues, the technologies that will dominate the electronics industry are *electronics packaging* and *assembly technology*.

Electronics packaging is a state-of-the-art technology based on the science of establishing interconnections ranging from zero-level packages (chip-level connections), first-level packages (either single-chip or multichip modules), second-level packages (e.g., printed circuit boards), and third-level packages (e.g., motherboards). There are at least three popular chip-level connections: face-up wire bonding, face-up tape-automated bonding, and flip-chip. Of these three technologies, the flip-chip provides the highest packaging density and performance and the lowest packaging profile. In order to achieve both lower cost and miniaturization in consumer products, much research has been done on the technology to accomplish assembly densities of 20 components/cm^2 based on 0.5-mm QFP pitch and passive component sizes of 1.0 × 0.5 mm. Assembly density in Japan is expected to reach 50 components/cm^2 by the year 2000. Passive components are expected to reach their size limitation at 0.8 mm × 0.4 mm before they are integrated into modules. Pin pitches will be as low as 0.15 mm. Low-cost resin board technologies will reach 50 micron lines and 50 micron vias with eight layers (Boulton, 1995).

Current plastic packages such as QFP that are surface-mounted onto PCB have effectively met various modern product requirements. It is quite predictable that future electronics products will require extremely thinner and lighter packaging methods. Existing precision robots are capable of satisfying present the assembly requirement of 0.4-mm leadframe pitch, and the trend is moving toward 0.15-mm leadframe pitch, giving rise to 800 pins in 30 sq mm and 1000 pins in 38 sq mm sizes. The late utilization of P-QFP beyond the current 0.4-mm pitch toward 0.15-mm pitch requires serious a enhancement of surface mount technology and electromigration resistance of both the plastic package and the PCB. Current assembly processes have achieved assembly defects of less than 20 PPM. Advanced precision robots have improved placement repeatability to ± 0.01 mm from ± 0.05 mm during the last six to seven years. In the next decade SMT machines will have more than 20 placement heads, placing more than 10 components per sec with better than 0.01-mm repeatability. Advanced TAB equipment will be able to place sub-0.1-mm pitch parts using machine vision, which has become essential as pitch size has reached the limits of human vision.

5.2 Emerging Robotic Technology

5.2.1 Intelligent Programming Interface

Anyone who has tried to program a robot finds that the task is not always as easy as it first seems to be. In fact, the same goes for many gadgets, from high-tech to mundane devices. The human endeavor of applying human-like features to robots has resulted in a number of complex mathematical models and algorithms to be added to robot control software that simulates human intelligence and behavior. By implementing the AI (Artificial Intelligence)-related algorithms and database technology, robots can be made to imitate, to a certain extent, the human behavior of learning, reasoning, and contingency reaction, which will be useful to automatic programming of robots. The early attempts to automate robot programming started with reasoning geometry information from CAD files so that robot manipulators could perform automatic mating of parts. The emergence of machine vision systems enabled real-time acquiring of geometry information for automatic operation of robots, which minimized the chore of robot programming. Machine vision has become an indispensable, integral part of most automated manufacturing and processing applications, and the use of 3D vision systems and force/torque sensors endows a robot with a greater degree of intelligence in dealing with its surroundings.

Enhancing robot performance in terms of "intelligence" is not the only issue to be considered for future robots. The widespread concept of computer-integrated manufacturing (CIM) and flexible manufacturing system (FMS) forced machine tool manufacturers to standardize, both in hardware and software structure. Computer numerical control (CNC) devices are more or less standardized in terms of programming method, since only a handful of manufacturers dominate the market, whereas current robot controllers are far from being standardized.

The great expectations of intelligent robots and intelligent manufacturing systems have as yet failed to materialize. Instead, companies are starting to concentrate on mechatronics, a more conservative mechanics and electronics combination, and on motion control technologies, in order to tune a variety of machines to special purposes. Microprocessors are seemingly ubiquitous. Virtually every numerically controlled machine or industrial

Table 5 Challenges in Current Robotic Core Technologies and a Forecast on the Time of Its Commercialization[a]

Field	Technology Challenge	Commercialization (Year)
Sensing/perception	Steroscopic image recognition	2002
	Color (human-like perception)	2003
	Voice recognition (any human voice)	2005
Robotic hardware	Muscle-like actuator	2010
	1 mm square actuator	2000
	Articulated robot with human walking speed	2004
	Redundant manipulator	2000
	High-Powered small lightweight actuator	2000

[a] Source: Japan Robot Assn.

robot is equipped with at least one, in most cases complete with its own display screen and keyboard. Microprocessors exert full digital control over the manufacturing process and in addition network with other computers in the factory (Harashima, 1993).

5.2.2 Microelectromechanical Systems

Microelectromechanical systems (MEMS) are batch-fabricated miniature devices that convert physical parameters to or from electrical signals and depend on mechanical structures or parameters in important ways for their operation. Thus this definition includes batch-fabricated monolithic devices such as accelerometers, pressure sensors, microvalves, and gyroscopes fabricated by micromachining or similar processes. Also included are microassembled structures based on batch-fabricated parts, especially when batch assembly operations are used. It is expected that electronic signal processing will exist in most future MEMS, which implies that they will be composed of sensors, actuators, and integrated electronics. Trends toward increasing levels of integration are driving to realization such devices as monolithic chips or multichip modules. These microsystems will be critically important as they extend microelectronics beyond its traditional functions of information processing and communications into the additional areas of information gathering (sensing) and control (actuation). Semiconductor Equipment and Materials International (SEMI) has estimated that the world market for MEMS devices could reach $8 billion by the turn of the century. This does not count the much larger markets for finished products that could be leveraged by the price/performance advantages of MEMS devices incorporated into such products. Microelectromechanical systems promise to lead microelectronics into important new areas that will be revolutionized by low-cost data acquisition, signal processing, and control. These microsystems are expected to have a profound impact on society, but their development will require synergy among many different disciplines that may be slow to develop. Global leadership and cooperation will be required to realize the benefits of MEMS in a timely way (Wise, 1994).

5.2.3 Open Architecture

Currently most CNC, robot, motion, and discrete control applications within any industry incorporate proprietary control technologies. Although these proprietary technologies have been proven to be reliable and adequate for various applications, there are difficulties associated with using them, including vendor-dictated pricing structures, nonstandard interfaces, higher integration costs, higher costs of extension and upgrading, and the requirement of individual training for operation and maintenance.

A controller with an open, modular architecture will provide benefits such as reduced initial system cost, simplified integration tasks, easier incorporation of diagnostic functions for the controller, machine, and process, and better integration of user proprietary knowledge. The concept of open, modular control also facilitates the math-based manufacturing strategy being implemented in the industry. Math-based manufacturing requires easily reconfigurable machining operations and the integration of low-cost, high-speed communications in machining lines for transferring large amounts of data. Flexible, modular controllers are key enablers for making math-based manufacturing easier to implement and cost-justifiable. Requirements for open controller architecture include unrestricted usage of commercially available hardware and software products and that it be built to industrial standard specifications. The controller infrastructure must have the ability to perform all tasks in a deterministic fashion and satisfy specific timing requirements of an application. Finally, the controller hardware bus structure must be a de facto standard in the marketplace such as VME, ISA, or PCI (Sorensen, 1995).

5.3 Conclusions

Equipment is the key to advanced manufacturing. It must be an integral part of technology development. The robotic equipment and the automatic manufacturing strategy described in this chapter have exhibited the possibility of utilizing industrial robots and other mechatronic devices to realize a modern manufacturing facility that relentlessly pursues cost-effectiveness and quality. Business owners in general will not invest in robots unless they are more productive than human workers. It is therefore necessary to lower robot costs while increasing their speed and performance.

The term *mechatronics,* widely used in the context of representing an engineering discipline, has been defined as follows: "The synergetic integration of mechanical engineering with electronic and intelligent computer control in the design and manufacture of industrial products and processes" (Harashima, 1993). Most current manufacturing equipment and the equipment mentioned in this chapter are based on this technology, utilizing some forms of actuators, sensors, and controllers.

Although much proven mechatronic technology has been used in industry, there is more to be done in the areas of mechanics, control, and sensors to exceed the current scope of robot usage. One of the most critical technologies is motion control technology, which regulates the movement of objects. The continuing trend toward miniaturizing mechanisms requires smaller actuators and smaller controllers. Continuing advances in digital servo-drive technology will put more control over electrical motors while decreasing the size and cost of an equipment considerably. Intelligent control algorithms and automatic tuning utility will greatly enhance the usability of electrical motors in such demanding situations. The usage of intelligent sensors such as machine vision will further add more advanced features to mechatronic equipment. Further the ever-progressing availability of powerful computers and intelligent software will make an immense contribution to robotic technology.

Successful progress in the electronics industry will depend on technological breakthroughs and the process of ongoing improvements in production technologies for products such as camcorders and digital cellular telephones, for which the emphasis has been on miniaturization, lighter weight, lower cost, lower power consumption, and portability. The same features are now conspicuous in notebooks, palmtop computers, cellular phones, and personal and wearable digital assistants.

REFERENCES

Bailo, C., G. Alderson, and J. Yen. "Manufacturing Center. Requirements of Open, Modular Architecture Controllers Applications in the Automotive Industry." Version 1.1.

Boulton, W., ed. 1995. "JTEC Panel Report on Electronic Manufacturing and Packaging in Japan." International Technology Research Institute PB95-188116.

Harashima, F. 1993. "EXPERT OPINION: Mechatronics on the Move." *IEEE Spectrum* (January).

Harris Corp. 1997. "Lexicon of Semiconductor Terms."

Kim, S. 1994. "Automation for Inspection/Adjustment in Electric Product Manufacturing." In *Korean Control Conference.* 13–20.

MCC/Sandia National Laboratory. 1992. "Industrial Competitiveness in the Balance: A Net Technical Assessment of North American vs. Offshore Electronics Packaging Technology." September.

Mills, P. "The Demands on Flexible Assembly Automation in the 1990s: Standard Application Software Is the Solution." Adept Technology.

Sorensen, P. 1995. "Overview of a Modular, Industry Standards Based, Open Architecture Machine Controller." Paper presented at International Robotics and Vision Conference, Detroit.

Stanley, W. D., and G. R. Dougherty. 1984. *Digital Signal Processing.* Reston: Reston.

Tsuchiya, K. 1993. "Assembly Robots: The Present and the Future." *Machine Design* (February 26), 263.

United Nations Economic Commission for Europe, Working Party on Engineering Industries and Automation. 1995. "World Industrial Robots 1995." New York and Geneva: United Nations.

Wise, K. D. 1994. "Microelectromechanical Development in Japan." WTEC Report, September.

ADDITIONAL READING

Aleksander, I. *Reinventing Man: The Robot Becomes Reality.* New York: Holt, Rinehart & Winston, 1984.

Asada, H., and J.-J. E. Slotine. *Robot Analysis and Control.* New York: John Wiley & Sons, 1986.

Ayres, R., and S. Miller. *Robotics, Applications and Social Implications.* Cambridge: Ballinger, 1983.

Braconi, F. *Factory of the Future.* Norwalk: Business Communications, 1986.

Brady, M., et al., eds. *Robot Motion: Planning and Control.* Cambridge: MIT Press, 1982.

Craig, J. *Introduction to Robotics: Mechanics and Control.* Reading: Addison-Wesley, 1986.

Cugy, A. *Industrial Robot Specifications.* London: Kogan Page, AMACOM, 1984.

Delson, E. B. *The Factory Automation Industry.* New York: Find/SVP, 1987.

Engelberger, J. *Robotics in Practice: Management and Applications of Industrial Robots.* New York: AMACOM, 1980.

Fu, K. S., R. C. Gonzalez, and C. S. G. Lee. *Robotics: Control, Sensing, Vision, and Intelligence.* New York: McGraw-Hill, 1987.

Groover, M., et al. *Industrial Robotics: Technology, Programming, and Applications.* New York: McGraw-Hill, 1986.

Hunt, V. D. *Industrial Robotics Handbook.* New York: Industrial Press, 1983.

Klafter, R., et al. *Robotic Engineering: An Integrated Approach.* Englewood Cliffs: Prentice-Hall, 1989.

Nof, S. Y., ed. *Handbook of Industrial Robotics.* New York: John Wiley & Sons, 1985.

Smith, D., ed. *Industrial Robots: A Delphi Forecast of Markets and Technology.* Dearborn: Society of Manufacturing Engineers, 1983.

Snyder, W. *Industrial Robots: Computer Interfacing and Control.* Englewood Cliffs: Prentice-Hall, 1985.

Staugaard, A., Jr. *Robotics and AI: An Introduction to Applied Machine Intelligence.* Englewood Cliffs: Prentice-Hall, 1987.

Tver, D. F., and R. W. Bolzz. *Robotics Sourcebook and Dictionary.* New York: Industrial Press, 1983.

CHAPTER 58
ROBOTICS IN SPACE

John G. Webster
University of Wisconsin
Madison, Wisconsin

1 INTRODUCTION

Various kinds of robots have been developed to work in space. In this chapter two space robots, the Viking lander robot and shuttle remote manipulator system, are first introduced. Those two robotic systems represent milestones of robotics in space. The harsh environment of space is then described in Section 3. Several features of space are also described from the perspective of robotic design and operation. Section 4 describes a spacelab with a robot working in the shuttle cargo bay. In Section 5 three aspects of the space station are discussed: flight telerobotic servicer, space station robotics, and space automation. In Section 6 advanced issues in Mars rover, space manufacturing, and self-replication robots are introduced. Since a Mars rover, Sojourner, successfully traversed miles on Mars in 1997, research into constructing robots and self-replication robots that work in space mining and manufacturing facilities is still awaiting future development.

2 HISTORICAL PERSPECTIVE

2.1 Viking Project

The first robot in space was a furlable boom on the Viking lander that operated on the surface of Mars (Holmberg, Faust, and Holt, 1980). Two spacecraft were launched in 1975 and orbited Mars almost a year later. After a damp landing site was selected, the lander separated, entered the atmosphere, deployed a parachute, fired retrorockets to slow the descent, and landed on three legs. Figure 1 shows that each lander had two identical lander camera systems. These were used to:

1. Geologically characterize the Martian landscape
2. Look for macroscopic evidence of life
3. Acquire surface samples
4. Support the magnetic properties and physical properties investigations
5. Support atmospheric characterization

The surface sampler acquisition assembly (SSAA) consisted of a three-axis nine-ft (three-m) furlable tube boom and collector head (Figure 1) capable of acquiring surface samples at any location within an approximate 130 ft² (12.1 m²) area in front of the lander and delivering 1000-μm sieved or raw samples to the required experiments. The extendable/retractable boom element, combined with the integrated azimuth/elevation gimbal system, allowed the collector head to be commanded to any location within the articulation limits of the boom unit. After digging samples, the collector head was positioned over hoppers in the X-ray fluorescence funnel, the biology processor, or the gas chromatograph mass spectrometer (GCMS), and the samples fell by gravity. The processors performed organic analyses to search for extraterrestrial life.

The ability to receive and interpret data and make changes in the planned strategy normally required about two weeks. This was the time needed to prepare the software, check it to prevent errors that could be disastrous, and send and verify commands (Soffen, 1977). However, the lander had extensive autonomy and could perform sequences of operations during the period between ground-commanded instructions.

Handbook of Industrial Robotics, Second Edition, Edited by Shimon Y. Nof
ISBN 0-471-17783-0   © 1999 John Wiley & Sons, Inc.

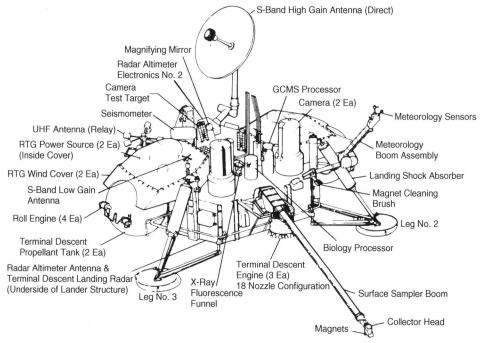

Figure 1   Viking lander robot. (From TRW Inc., 1988.)

2.2   Shuttle Remote Manipulator System

The second robot in space was the shuttle remote manipulator system (SRMS) (Ussher and Doetsch, 1985), a six-degree-of-freedom (d.o.f.) remotely controlled manipulator for cargo deployment and retrieval during on-orbit operations of the space shuttle orbiter vehicle. Figure 2 shows the arm in diagrammatic form. The shoulder joint has pitch and yaw, the elbow joint has pitch, and the wrist joint has pitch, yaw, and roll. The upper arm boom is made of 16 plies of tapered lightweight ultrahigh modulus GY/70/934 graphite epoxy composite 13 in. (33 cm) in diameter and 16 ft (4.9 m) long. The lower arm boom is 13 in. (33 cm) in diameter and 19 ft (5.8 m) long. Three manipulator positioning mechanisms and manipulator retention latches roll inboard to allow the cargo bay payload doors to close and hold the arm firmly in position during launch.

Two closed-circuit television cameras located on the arm—one on the elbow and one on the wrist—assist the payload specialist in maneuvering the end of the arm. A multilayer insulation thermal blanket system provides passive thermal control and keeps temperatures between −3 and +36°C. The SRMS can be switch-selected to operate in four standard modes.

1. In manual augmented mode the operator used two hand controllers to "fly" the end of the arm.
2. In single-joint control, using the control computer, a toggle switch can control a single joint.
3. In direct mode, without the computer, a toggle switch can control a single joint.
4. In automatic mode the computer uses prestored trajectories to obtain the required end-effector position.

Each SRMS joint is powered by an optically commutated, brushless DC motor providing a stall torque of 100 oz · in (0.7 N · m) and a no-load speed of approximately 90 radians per second. A pulse width modulated signal drives the three motor windings. A

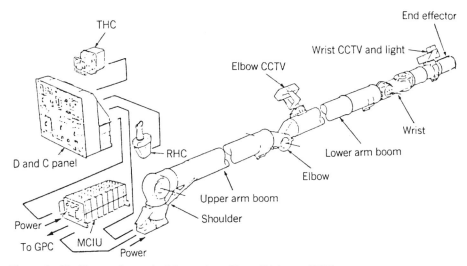

Figure 2 Shuttle remote manipulator system. (From Erickson, 1987.)

backdriveable reduction gearbox provides high torque. Dry-type lubricant provides low friction. A 16-bit optical encoder provides angular position data. The shoulder pitch joint has a gear ratio of 1842 and provides 1158 ft·lb (1570 N·m) of output torque. During dynamic braking, the motor acts as a generator and dumps electrical power onto the DC bus.

The end-effector provides an initial soft dock followed by rigidizing of the interface. It also has a large capture envelope. This is provided by closing three snare wires around a pin attached to a payload so that the pin is centralized within the end-effector.

3 SPACE ENVIRONMENT

When robotic systems are designed for operation in space, the harsh environment must be considered. One of the features of space is that a complete vacuum may exist. Thus liquids such as hydraulic fluid that might evaporate should be avoided. DC motors are normally used for actuators, but these should not have commutators, because the lack of moisture in space does not permit copper oxide to build so there is no commutator lubrication. To prevent rapid wear of carbon brushes, DC brushless motors are normally used. These are AC motors where the phases that drive the rotating field are provided by electric switching circuitry.

Another feature of space is that temperature is hard to control. Since there is no fluid surrounding the robot, heat can be dissipated only by conduction within the robot or its support or by radiation. Radiation is a problem because the large radiation from the sun heats up one side of the robot, while radiation into space cools the dark side.

Another feature of space is the lack of gravity. Thus structures can be made lighter without buckling. At the same time, items can float off into space since gravity does not keep them where they are placed.

Because of the harsh lighting conditions in space, solar glare and intense shadows, lighting systems must be carefully considered. Dark areas must be eliminated and glare reduced. Filters or other signal modulators can cut solar or reflected glares. Fill lighting can illuminate shadows.

4 SPACELAB EXPERIMENTS

Spacelab is a large laboratory that is carried to space in the space shuttle cargo bay and permits astronauts to enter to perform experiments. Figure 3 shows the ROTEX robot, a small 6-d.o.f. robot in the German Spacelab mission[1] (Hirzinger, 1987). Its gripper is

---

[1]ROTEX was flown on the space shuttle in 1993 (Oda, Kibe, and Yomagata, 1996).

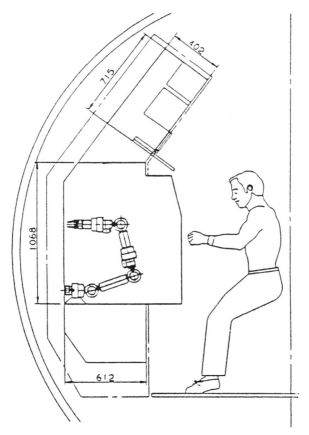

Figure 3   The German ROTEX robot. (From Hirzinger, 1987.)

provided with a number of sensors, especially a six-axis force-torque wrist sensor, grasp-ing force control, an array of nine laser range finders, and a pair of optical fibers to provide a stereo image out of the gripper. It is able to perform automatic, preprogrammed motions as well as teleoperated motions via an astronaut or an operator on the ground. It operates a biological experiment by pressing a piston that adds nutrient fluid; the experiment can be stopped by pressing another piston to add a fixative. It demonstrates servicing capabilities by assembling a mechanical grid structure, connecting an electrical plug, and grasping a floating a floating object.

5   SPACE STATION

In 1984 President Reagan committed the United States to a permanently manned, fully operational space station. The extended life of the program (15–20 years) and the wide range of planned capabilities present NASA with a major challenge in making effective use of all available resources, such as automation and robotics. The space station is designed to evolve from an initial modest capability (150 kW of power, 6–8 crew mem-bers) in the mid-1990s to a much higher capability system in the early twenty-first century (Pivirotto, 1986). A typical scenario involving the space station, various types of vehicles, robots, and ground stations is shown in Figure 4.

5.1   FLIGHT TELEROBOTIC SERVICER

5.1.1   Objectives

The flight telerobotic servicer (FTS) will use telerobotics for the assembly, checkout, and maintenance of the space station to minimize the amount of extravehicular activity re-

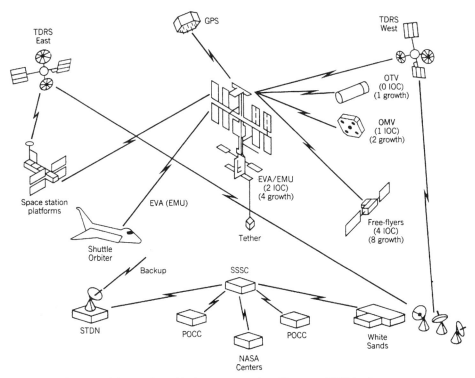

Figure 4   Spacecraft interacting with the space station. (Courtesy TRW Inc.)

quired of the crew (Goddard Space Flight Center, 1986). It should decrease the time necessary to construct and assemble the space station elements, release the crew to deal with contingencies, problem solving, and supervisory functions, and reduce the risk to personnel safety during the process.

5.1.2   Flight Demonstration

Prior to the start of the space station assembly, the space shuttle will carry a telerobotic unit to demonstrate operation and technical aspects of the FTS. Using the SRMS, an astronaut will be able to remove the robot from the pallet within the cargo bay where it is rigidly affixed for launch and position it next to the experiment and toolbox palette for assembly and serving demonstrations. The FTS will be powered through an umbilical cable and operated by an astronaut from the shuttle aft flight deck. The telerobotic unit will show, as a minimum, the capability to assemble space station thermal utility connections and remove and replace an orbital replacement unit in a space environment.

5.1.3   Space Station Flight System

The FTS will be a multipurpose tool that serves as an extension of the astronaut for performing extravehicular functions. It will be able to participate in the assembly of the space station and the maintenance of the space station during and after assembly. The orbiting maneuvering vehicle can transport the FTS to remote platforms and free-flyer spacecraft for in situ assembly and servicing operations, including refueling. The tasks that are being analyzed include servicing the Hubble space telescope and refueling the Gamma Ray observatory. A conceptual generic telerobot that could operate as an end-effector on the SRMS or a free-flying orbital maneuvering vehicle is shown in Figure 5.

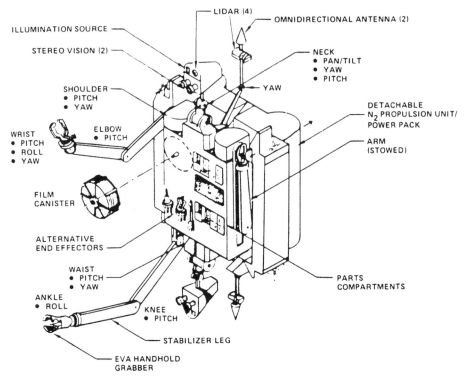

Figure 5 A conceptual generic telerobot. (From J. Webster, "Space Robots, Research," in *International Encyclopedia of Robotics: Applications and Automation,* vol. 3, ed. R. Dorf and S. Y. Nof. New York: John Wiley & Sons. 1988, 1635–1640.)

5.2 Space Station Robotics

5.2.1 Applications for Robotics

The requirement for robotic elements has been identified for numerous functions on the space station. This requirement includes both extravehicular functions and functions within the laboratory and habitation modules, as follows (Erickson, 1987):

1. Assembly of space station, attached payloads, and structures in space
2. Space servicing, maintenance, and repair of space station (external), attached payloads, free-flying satellites, and platforms
3. Support of deployment and retrieval and of docking and berthing
4. Transportation on space station
5. EVA support
6. Safe haven support
7. Laboratory functions—care of plants and animals, analysis of biological samples, and centrifuge access

5.2.2 Location of Robot

The NASA telerobotic plan is for a generic telerobot that may function in the following locations:

1. In the satellite servicing bay of the space station
2. On the SRMS as an end-effector
3. On the space station mobile servicing center as an end-effector
4. On an orbiting maneuvering vehicle

5. As a crawler on the space station truss

5.2.3   Characteristics

The characteristics of such a robot include the following:

1. Modular design exploiting orbit-replaceable units for servicing
2. Reach and strength at least equivalent to that of a suited astronaut
3. Controllability from the space shuttle, the space station, and the ground
4. Vision and force feedback for human operators
5. Compliance in the robot's manipulator joints
6. Adaptability to a pressurized environment
7. Accommodation to increased levels of autonomy
8. Accommodation to attachment to and operation on coorbiting and polar platforms

5.2.4   Extravehicular Activity Retriever

An extravehicular activity retriever, or extravehicular autonomous robot (EVAR) (Figure 6), is a typical space station robot. Basically an EVAR has two functions: rescuing an astronaut drifting away from the space station and retrieving parts and equipment in a space environment. Figure 6a shows an EVAR as an astronaut's helper. NASA Johnson Space Center (JSC) has developed such a voice-supervised, artificial-intelligent, free-flying robot (Rosheim, 1994). Technologies critical for developing EVAR include coordinated robot and satellite control technology and teleoperation technology (Oda, Kibe, and Yamagata, 1996). Those two technologies are accomplished based on the software architecture of EVAR (Figure 7). Five major subsystems, *perception, world model, reasoning, sensing,* and *acting,* consist the software architecture. Based on the continuous sensory perception, EVAR builds its own environmental knowledge.

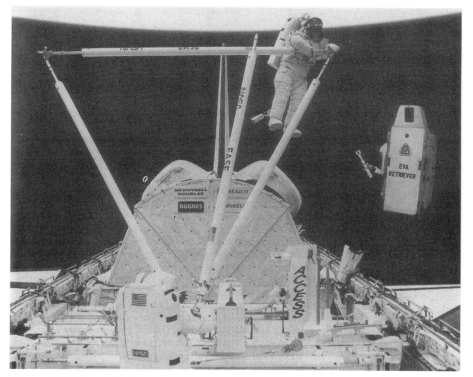

Figure 6   (a) Extravehicular activity retriever as an astronaut's helper (courtesy NASA Johnson Space Center; also in Rosheim, 1994).

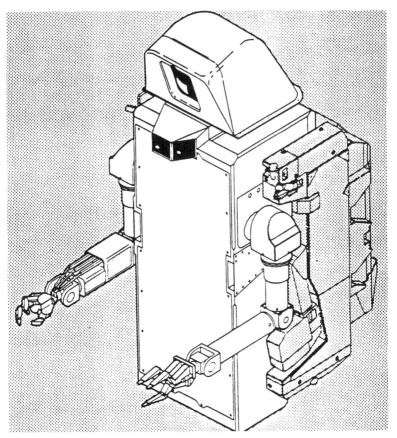

Figure 6 (continued)   (b) Body of EVAR. (From M. E. Rosheim, *Robot Evolution: The Development of Anthrobotics.* New York: John Wiley & Sons, 1994.)

5.3   Space Station Automation

There are three ascending levels of autonomy that build from simple machine concepts toward the cognitive processes more typical of humans. Each specific capability is named and then briefly described below (Heer and Bejezy, 1983; Firschein et al., 1985).

5.3.1   Teleoperation and Telepresence

Techniques of teleoperation and telepresence are discussed in Chapter 9. Basically, telepresence is a technique that provides operator-centered, intuitive interactions between the operators and a remote environment. This technique takes advantage of the natural cognitive and sensory-motor skills of an operator and effectively transfers them to the telerobots (Li et al., 1996). Figure 8 shows such a relationship between human operator space H and telerobot workspace W.

A testbed called FITT/DART[2] (Li et al., 1996) has been developed in NASA Johnson Space Center. FITT is a testbed in human operator space. It is a virtual reality environment. On the other hand, DART is a dual-arm dexterous robot equipped with a stereo camera platform, designed to provide an operator-centered perspective of the remote environment. In such an environment the operator is "immersed" virtually in the telerobot workspace (Figure 9). The resulting system provides a very flexible and efficient capability in many tasks.

---

[2]Full Immersion Telepresence Testbed/Dexterous Anthropomorphic Robotic Testbed (Li et al., 1996).

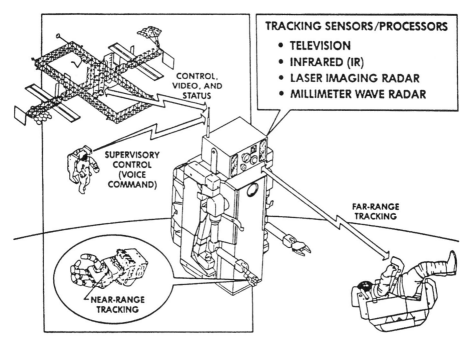

TRACKING SENSORS/PROCESSORS
- **TELEVISION**
- **INFRARED (IR)**
- **LASER IMAGING RADAR**
- **MILLIMETER WAVE RADAR**

CONTROL, VIDEO, AND STATUS

SUPERVISORY CONTROL (VOICE COMMAND)

FAR-RANGE TRACKING

NEAR-RANGE TRACKING

Figure 6 (continued) (c) Communication between EVAR and Space Station and Astronauts. (From Defigueiredo, R. J. P. and L. M. Jenkins, "Space Robots," in *International Encyclopedia of Robotics: Applications and Automation,* vol. 3, ed. R. Dorf and S. Y. Nof. New York: John Wiley & Sons. 1988, 1626–1634.)

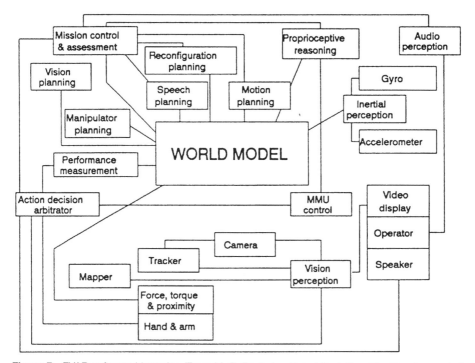

Figure 7   EVAR software hierarchy. (From M. E. Rosheim, *Robot Evolution: The Development of Anthrobotics.* New York: John Wiley & Sons, 1994.)

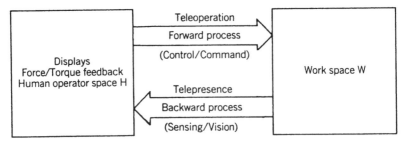

Figure 8   Teleoperation and telepresence. (From Defigueiredo, R. J. P., and L. M. Jenkins, "Space Robots," in *International Encyclopedia of Robotics: Applications and Automation,* vol. 3, ed. R. Dorf and S. Y. Nof. New York: John Wiley & Sons. 1988, 1626–1634.)

5.3.2   Supervisory Control and Adaptive Robotics

- *Objective location* can be accomplished by integrating information through vision, touch, or other senses.
- *Proprioception* can be accomplished by sensing the positions and motions of the robot's own articulated structures, such as arms, fingers, legs, feet, and "necks."
- *Effort sensing* can be determined by sensing external forces and torques exerted on the robot's body and limbs, especially on its grippers or other end-effectors.
- *Grasp sensing* may involve interpretation of high-resolution tactile array images or finger-joint torques.
- *Position determination* of a mobile robot can be accomplished by navigation satellites, vision, inertial guidance, or some combination of these and other means.
- *Task-level control* is a method of specifying and executing the procedure (program, algorithm) to accomplish a task.
- *Effector control* requires computing each joint actuator's positions, motions, and effectors in order to position or move an articulated structure such as an arm or leg in any specified way.
- *Coordinate conversion* requires transforming positions, velocities, forces, torques, and other spatial quantities from one reference frame to another fast enough to support real-time control of arms, legs, cameras, etc.
- *Adaptability* requires adjusting preprogrammed motions "on the fly" to match the actual positions of objects around the robot, usually on the basis of sensory information.
- *Effort control* requires the exertion of a controlled force and torque in arbitrary directions on an object with the robot's limbs.
- *Programmable compliance* requires adjustment of the effective mechanical compliance of the robot's limbs to suit specific task requirements.
- *Gripper control* requires the control of gripper action and grasping force to maintain a firm hold on an object without damaging it and without allowing it to slip.
- *Long-range navigation* requires position determination and course planning for traversing distances that are quite large relative to the size of the robot.
- *Short-range navigation* requires obstacle detection and avoidance for traversing distances that are comparable to the robot in size.
- *Locomotion* requires a propulsion system to move the robot over large distances or adjust its position with respect to objects in the work area.

5.3.3   Intelligent Robots

- *Multisensory integration* requires the combination of information from different kinds of sensors for external events and conditions.
- *Situational assessment* deduces from sensory observations and previous knowledge the important facts about the robot's surroundings.

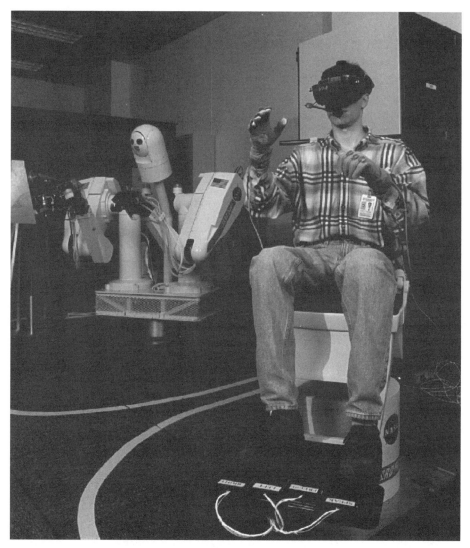

Figure 9 FITT interface and DART robot (courtesy NASA Johnson Space Center, ⟨http://tommy.jsc.nasa.gov/~li/dartfitt.html⟩).

- A *human–machine interface* is required for the best communication with the space station crew.
- *Automatic planning* devises a complex schedule of activities in order to accomplish a particular mission.
- *Plan execution and monitoring* compares the current situation to the situation anticipated in the plan, noting any problems and taking advantage of any unplanned-for advantages that occur.
- *Automatic replanning* generates a new plan to suit the present circumstances.

- *Knowledge representation* for intelligent robots is the representation in the computer of many different kinds of knowledge for implementing perception as well as manipulation and mobility.

6 ADVANCED ISSUES

6.1 Mars Rover

A Mars rover could traverse thousands of miles in a one-year mission over terrain too difficult for a manned vehicle to cover (Pivirotto, 1986). The Martian areas of greatest geological interest, such as escarpments and volcanoes, lie many kilometers from suitable (i.e., flat) landing sites. A rover could use a pair of video cameras returning intermittent 3D television frames to a terrestrial operator. The operator designates a ca 320–655-ft (100–200-m) path for the rover to follow that avoids obstacles and includes references to preselected landmarks. The planned course is transmitted to the rover and stored on board, and the vehicle autonomously follows the course. At the end of the designated path the rover sends another stereo panorama and the cycle repeats. The rover would need a sample handling and containment system, sensors, a control system, and a manipulator (with auxiliary devices such as core drills).

In July 1997 Sojourner (Figure 10), a Mars rover, successfully drove off the rear ramp of Mars Pathfinder Lander (now renamed the Dr. Carl Sagan Memorial Station) onto the Martian surface. The project provided detailed geological information of the Martian area for future research. A new Mars prototype, Rocky 7 (Figure 11), has been developed in the Jet Propulsion Laboratory (Volpe et al., 1996) of NASA. The major differences between Sojourner and Rocky 7 are in four areas: motor system, computer, software, and manipulation system. Those differences are listed in Table 1.

6.2 Space Manufacturing

A study group has suggested a permanent, growing, highly automated space manufacturing facility (SMF) based on the utilization of ever-increasing fractions of nonterrestrial

Figure 10 Sojourner Mars rover (courtesy NASA Jet Propulsion Laboratory, ⟨http://mpfwww.arc.nasa.gov/tasks/scirover/homepage.html⟩).

Figure 11   Rocky 7—A prototype of Mars rover (courtesy NASA Jet Propulsion Laboratory, ⟨http: //mpfwww.arc.nasa.gov/tasks/scirover/homepage.html⟩).

material (Freitas and Gilbreath, 1982). First, the original SMF would be highly dependent on Earth for its raw material inputs. This dependency would lessen as nonterrestrial sources of raw materials, especially the Moon and asteroids, are developed. Second, the SMF would be run almost entirely by teleoperation, but later these teleoperators might be largely replaced by autonomous robots. Finally, the SMF would originally manufacture solar power stations, communications satellites, and a number of other products difficult or impossible to make anywhere but in space (e.g., certain biomedical substances, and foamy metals), but should eventually also begin to produce some outputs for use in other NASA missions in space or back on Earth. Examples include pressure vessels, integrated circuits and other electronic components for robots and computers, laser communication links, gigantic antennas, lunar teletourism equipment, and solar sails.

6.3   Self-Replicating Robots

A study group has suggested that a lunar manufacturing facility (LMF) be designed as an automated, multiproduct, remotely controlled, reprogrammable facility capable of constructing duplicates of itself, which would themselves be capable of further replication (Freitas and Gilbreath, 1982). Successive new systems need not be exact copies of the original, but could, by remote design and control, be improved, recognized, or enlarged so as to reflect changing human requirements. Theoretical concepts of machine duplication and automation are well developed.

An engineering demonstration project could be initiated to begin with simple replication of robot assembler by robot assembler from supplied parts, and proceed in phased steps to full reproduction of a complete machine processing or factory system by another machine processing system, supplied ultimately only with raw materials. The raw materials of the lunar surface, and the material processing techniques available in a lunar environment, are probably sufficient to support an automated LMF capable of self-replication and growth. Such a LMF would incorporate paving robots, mining robots,

Table 1   A Comparison of Two Mars Rovers: Sojourner and Rocky 7[a]

System	Sojourner	Rocky 7
Motor control	Bang-bang on/off control of motor actuation is used based on actuator positions monitored by optical encoders. No variable rate motion is possible.	Accurate variable-rate motion is accomplished with PID motor serving, pulse width modulation (PWM) of motor currents, and optical encoder monitoring of output shaft position.
Computer	Custom boards with 80C85 CPU, 100 KPS	Commercial 3U VME, 68060 CPU, 100 MIPS
Software	C and assembly. Unix development environment. Silicon Graphics Inventor-based operator interface.	Wind River VxWorks real-time operating system. Real Time Innovations ControlShell, Network Data Delivery System.
		Unix development environment. C++, C, Lisp software. On-board stero correlation. HTML/Java-based operator interface.
Manipulation system	APXS deployment mechanism: 1-d.o.f. mechanism for deploying the one science instrument.	An arm with 4 d.o.f. for digging dump, grasping rocks, carrying one sample, and pointing an integrated optical spectrometer and its calibration target
		A mast with 3 d.o.f. is used to raise a stereo camera pair to visually inspect the entire vehicle, and deploy another 0.5-kg instrument canister to targets in a large area around the rover.

[a]Courtesy NASA Jet Propulsion Laboratory.

chemical processing sectors, fabrication sectors, assembly sectors, computer control and communication systems, a solar canopy, a seed mass, and sources of power and information.

ACKNOWLEDGMENTS

The author would like to thank Mr. Chin-Yin Huang from Purdue University for helping in the preparation of this chapter. Mr. Carlos Palacios from Purdue University also contributed.

REFERENCE

Defigueiredo, R. J. P., and L. M. Jenkins. 1988. "Space Robots." In *International Encyclopedia of Robotics: Applications and Automation.* Vol. 3. Ed. R. Dorf and S. Y. Nof. New York: John Wiley & Sons. 1626–1634.

Erickson, J. D. 1987. "Manned Spacecraft Automation and Robotics." *Proceedings of the IEEE* **75,** 417–426.

Firschein, O., et al. 1985. "NASA Space Station Automation: AI-Based Technology Review." SRI International Project 7268.

Freitas, R. A., and W. P. Gilbreath, eds. 1982. "Advanced Automation for Space Mission." NASA Conference Publication 2255.

Goddard Space Flight Center. 1986. *Preliminary Program Plan for the Space Station Telerobotic Service Program.* Office of Space Station, NASA.

Heer, E., and A. K. Bejczy. 1983. "Control of Robot Manipulators for Handling and Assembly in Space." *Mechanism and Machine Theory* **18,** 23–35.

Hirzinger, G. 1987. "The Space and Telerobotic Concept of DFVLR ROTEX." In *Proceedings of the 1987 International Conference on Robotics and Automation.* Vol. 1. 443–449.

Holmberg, N. A., R. P. Faust, and H. M. Holt. 1980. "Viking '75 Spacecraft Design and Test Summary, Volume 1—Lander Design." NASA Reference Publication 1027, November.

Li, L., et al. 1996. "Development of a Telepresence Controlled Ambidextrous Robot for Space Application." In *Proceedings of the 1996 IEEE International Conference on Robotics and Automation.* 58–63.

Oda, M., K. Kibe, and F. Yamagata. 1996. "ETS-VII Space Robot In-Orbit Experiment Satellite." In *Proceedings of the 1996 IEEE International Conference on Robotics and Automation.* 739–744.

Pivirotto, D. S. 1986. "Unmanned Space Systems." Paper presented at 13th Annual Symposium Association of Unmanned Vehicle Systems.

Rosheim, M. E. 1994. In *Robot Evolution: The Development of Anthrobotics.* New York: John Wiley & Sons.

Soffen, G. A. 1977. "The Viking Project." *Journal of Geophysical Research* **82,** 3959–3970

Ussher, T. H., and K. H. Doetsch. 1985. "An Overview of the Shuttle Remote Manipulator System." In *Space Shuttle Technical Conference.* NASA CP-2342-Pt-2. 892–904.

Volpe, R., et al. 1996. "The Rocky 7 Mars Rover Prototype." In *Proceedings of IROS '96.* 1558–1564.

Webster, J. 1988. "Space Robots, Research." In *International Encyclopedia of Robotics: Applications and Automation.* Vol. 3. Ed. R. Dorf and S. Y. Nof. New York: John Wiley & Sons. 1635–1640.

CHAPTER 59

APPLIANCE INDUSTRY

Erich Niedermayr
Joseph Pössinger
Norbert Roth
Siemens AG
Munich, Germany

1 INTRODUCTION, SCOPE, AND OBJECTIVES

The main objective of this chapter is to discuss important design and engineering issues based on representative and successful robot applications in the appliance industry. However, in the appliance industry, as well as in other industrial branches, dramatic changes have taken place that have of course had a significant impact on production automation and therefore on the installation and use of industrial robots. In the beginning we shall therefore outline significant aspects, constraints, and trends to support a better understanding of correlations and give a realistic estimation of the potential of robot applications.

To give a precise definition of the appliance industry is not that easy, and depending on the country and culture one may group the various tasks differently, e.g., electronics, automotive, etc.

In this chapter we give two examples of plants in Germany. One is taken from the production of household appliances, the other from the production of electrical switch gear for safeguarding electrical power supply. Aside from the fact that these parts are completely different in design, dimensions, and functionality, one can well compare them because of similarities in high production volumes, long product life cycle, large product variants based on regional standards and requirements, very narrow profit margins, etc. These specific constraints and the marketing conditions discussed below need to be considered at the beginning of each automation project in order to support an economic and successful robot installation.

2 THE CHANGING FACE OF PRODUCTION

Change, increasingly dynamic in nature, marks the industrial scene today and will undoubtedly continue to do so in the future. A company's economic success depends more and more on its ability to adapt as quickly and effectively as possible to changing conditions in a turbulent environment. Product innovation has to go hand in hand with the provision of efficient development and manufacturing processes. The main specific factors driving these changes are:

- Orientation to constantly changing customer requirements
- Integration of people into the production process
- Globalization of competition
- Technological advances in miniaturization, microelectronics, and software

2.1 Customer Focus

Customer focus, in the sense of delivering maximum customer satisfaction, is the key to success in the marketplace. If *production* is supposed to mean every step in the value chain that goes into creating the features a customer expects in a product, then the term has to be given a much broader definition than its previous strictly manufacturing sense. Once a company takes a business decision to launch a particular product in a particular

---

Handbook of Industrial Robotics, Second Edition, Edited by Shimon Y. Nof
ISBN 0-471-17783-0 © 1999 John Wiley & Sons, Inc.

market, then the production process starts right at the product definition stage, i.e., with the functional and economic evaluation of the product functions required from the customer's perspective. This process extends through the stages of product and process development to parts manufacture, assembly, and inspection. Nor does it stop there; customers increasingly want life-cycle support from the manufacturer of the product. In the service sector production-related services are becoming increasingly complex, mainly because of the growing software element in many products. Industry also has to take greater responsibility for the recycling and disposal of products at the end of their service life. This is a result of growing environmental awareness and the legal requirements. See also Chapter 54.

2.2   Integrating People

For years efforts to achieve greater productivity were inspired by Taylorist principles of splitting work down into tasks and then automating these functional units. Not until the late 1980s did engineers begin to question whether the concept of universal automation, and the high level of investments involved, really made economic sense in view of the complexity of the processes that needed to be managed. The answer was lean production. In its practical application this was a strategy for reducing costs that meant not only moving away from high levels of automation but, more importantly, putting people back at the center of the production process.

The flexibility required comes from a creative, motivated, and responsive workforce. Automation should not be rejected out of hand, however. Applied at the right level and in the right place, automation is often the best solution for problems related to quality, cost, and time. Innovative production strategies see future organizational structures as being made up of autonomous, self-optimizing, and evolving units that are market-led.

2.3   Globalization and Regional Competition

A company operating globally needs to have a presence in the main world markets. Being close to the market means having direct contact with the customer in order to identify and offer the best product variants for a particular market. In doing this companies often have to outsource part of the added value in the particular country because economic policy constraints require this regional input of added value; because democratization and liberalization is opening up new markets in regions with long-term requirements; or because low labor costs and good technological resources promise economic advantages. The resulting changes in the internal and external value chain lead to the creation of a production network. Companies concentrate on strategically important activities in order to gain competitive advantage, while activities that are not seen as part of the company's core competences are outsourced to regional partners. There are also benefits to be gained from collaborative ventures between partners with complementary roles in the value chain. Strategic alliances are sought with competitors in order to accomplish technologically specialized projects. Such alliances offer time and cost benefits while at the same time minimizing the risk of the individual company seeking the alliance. That company can then focus on systems integration, particularly mechanical, electronic, and software systems, and concentrate its efforts on key technologies.

2.4   The Role of Technologies

The value-adding elements in production are shifting more and more to the beginning and end of the value chain. One reason for this is that production has always been the focus of intense efforts to increase productivity, and much of the potential for improvement has already been achieved. Other important drivers of this change are microelectronics and software. Their effect is twofold. Increasing miniaturization and integration have meant that many product functions previously implemented as discrete components are now being integrated virtually from the outset and no longer show up as added value in the actual production process. Further productivity increases have come from the impact that electronics have had on the efficiency of the actual production tools. Other effects come from hardware being replaced by software or by new technologies like microsystems engineering and surface mounting. And once the product is in service, product complexity and recycling issues mean that the need for customer support is increasingly a feature of this phase.

3   PRODUCTION AUTOMATION ISSUES AND REQUIREMENTS

After the somewhat exaggerated expectations of the early 1980s, industrial robots (IR) have become an essential factor in planning and configuring automation systems. The ability to program freely positions and paths that can be reproduced precisely—the core function of IR—represents a central task in dynamic production automation. IRs can be used universally, are highly developed, and can be integrated well into existing information structures. Consequently they have become an important element in innovative automation engineering.

This positive picture should not, however, be allowed to disguise the general problems of industrial automation. The industrial robot has ensured for itself a secure position in automation engineering in all areas where high quality is demanded in conjunction with work that involves a large number of variants and is spatially complex and adaptable. Use is thus limited—mainly for economic reasons—to particular production processes that have varying significance in different areas of production.

In Figure 1 the facts for and against the use of IRs in the appliance industry (from the viewpoint of a large electrical company in Germany) are shown. The drawing visualizes the potential volume of possible robot applications as a basin fed or reduced by volume streams representing the different economic and technological facts.

New technological trends and processes do sometimes allow the area in which robots are used to be expanded, but complex production processes are sometimes considerably simplified thanks to new constructional developments, the widespread replacement of mechanical components by microelectronics, and new materials.

A good example is the application of spraypainting in manufacturing household appliances. This was formerly a typical application area for robots, but nowadays these applications have greatly diminished because of refined stamping processes with coated metal sheets.

The cost of peripherals such as devices for arranging, feeding, and transport is a significant economic factor for automation procedures using robots, as are the expense of design and engineering, setup and optimization services, and maintenance and service costs.

However, two important aspects should be emphasized here:

1. A cost comparison between conventional and robot-supported automation is worthwhile only if the entire life-cycle costs, including miscellaneous production adaptations, are taken into account. However, the higher engineering costs that the use of robots entails mostly lead to a better understanding of the process and thus to improvements in products and quality.

2. Compared to manual assembly and manufacturing, automation yields a more consistent and generally higher-quality product. However, this compares with the significant costs entailed in the accurate provision and transport of components required because of current inadequacies in sensory capabilities. In addition to other social aspects, this makes it more difficult to use robots on manual assembly lines.

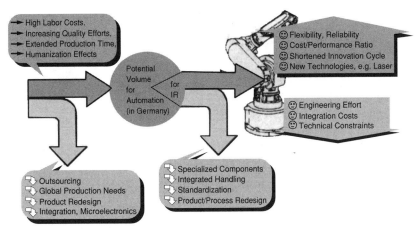

Figure 1

The global changes to which an international company is exposed naturally also affect the use of robots to a varying extent. Highly complex systems and increasing quality requirements, the desire to be able to respond fast to market changes with new products, and high labor costs are familiar arguments for using modern automation engineering.

However, important changes such as the commitment to national value creation for market access and international alliances for development and production should also be mentioned.

4 DESIGN AND ENGINEERING ISSUES

Using two representative and successful applications in the appliance industry, we shall outline important design and engineering issues and provide the reader with important instructions and realization knowhow.

4.1 Steps to Realization

In general the following issues have to be considered and made clear when robot applications are to be planned and realized in that sector:

- Formulation of the job, the actual state, and the application-specific conditions.
- Is an IR a suitable means of production for this application (task volume, required flexibility, number of degrees of freedom)?
- Is the environment prepared for an IR application (acceptance, maintenance, service, etc.)?
- Is the product suitable for a robot application?
- Are the new technologies safeguarded (production tests/simulations, etc.)?
- Is quality assurance of every step of the process guaranteed (use of sensors, malfunction routines, etc.)?
- Is the material flow optimized (safe insertion of the workpieces, provided with sufficient buffers, malfunction routines, removal of rejects, short idle times, etc.)?
- Is the plant working/used to capacity, in order to achieve a high efficiency (if possible, multiple shifts, high maneuverability, high availability, etc.)?
- A man–machine interface as simple and clear as possible has to be created (good integration possibilities of workplaces for manual operations, simple control panel, clear safety concept, etc.).
- Are the control concept and the information flow adapted to the existing production structure?

4.2 Example of Application: Insertion of Sealings

In the field of production of devices and instruments, especially in the field of household appliances, the assembly of so-called pliant hollow parts and profiled seals and tubes represents an essential market capacity. Industrial robots are well suited for application in this sector due to their kinematic flexibility and accuracy and the easy programmability of their sequences of motion. It is, however, necessary to adjust the periphery and the grippers/tools to the production requirements and to provide as many automation conformance means on the product as necessary.

In the following we show how door seals of a dishwasher can be inserted with the help of industrial robots.

4.2.1 Formulation of the Job

In dishwashers the door is sealed with the help of a profiled joint made of rubber that is embedded into the container. Figure 2 shows a container and the inserted profiled seal as well as a cross section of the sealing bed and two cross sections of profiles that have to be inserted.

The handling and mounting of the profiled seal into the container should be automated, but the different profile shapes have to be considered. The aim of the development was a fully automatic assembly line that would replace four human workers.

The problems with manual mounting of the profiled seals are:

1. Finding personnel for this stressful task ($\rightarrow$ strain on the wrists causes a high percentage of sickness)

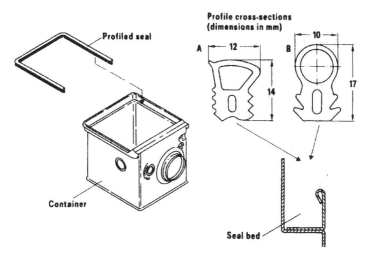

Figure 2 Dishwasher container with profiled seal.

2. Difficulty in guaranteeing a standard quality due to the different cutting lengths of the profiles seals when they are rolled in manually

4.2.2 Development of Technologies and Choice of Industrial Robots

As there was until now no suitable automatic production process on the market, a new technology for the mounting of profiled seals had to be developed. Therefore possible joining processes and placing-at-disposal processes were chosen to develop a laying tool for the most suitable principle. Due to various experimental tests the respective design can be ensured. Due to possible modifications of the container, for example when the model is modified, or possible changes to the shape of the profiled seals, an adjustable continuous insertion of the profile seals by use of the kinematic and control-technological flexibility of an IR was chosen. A six-axis articulated robot arm had to be chosen because of the spatial sequences of motion and the embedding process that has to advance in line with the orientation of the laying tool. For the working space to be used as efficiently as possible, the IR was mounted above the container-holding fixture.

4.2.3 Process Assurance by Performing Production Tests

The initial tests had shown that the embedding of profiled seals with the help of a rolling-in method is possible. The profiled seal has to be guided tangentially to the press-in roller. Because of expected tolerance problems and the changing behavior of the profiled seals of different charges from various producers, additional production tests were performed. These tests were supposed to be as similar to the real production process as possible.

In order to keep the expenditure for the production tests low, we started with profiled seals cut into suitable lengths and a container pick-up unit. The press-in roller of the laying tool was driven by a controlled DC motor so that no tensile forces can be performed on the profiled seals when it is mounted.

4.2.4 Results of the Production Test

Some thousands of profiled seals were embedded automatically and the required quality standard was kept. Various seals could be built in by changing the robot software. Furthermore, essential parameters were determined: the shape of the press-in roller and the possibility of manufacturing with continuous profiled seals by a driven guiding chamber. The operators as well as the workers become accustomed to their "steel colleague." The test plant and future use of the technology were also demonstrated to the factory committee. Even if there were some negative comments, the new technology was in general accepted positively. In order to complete the automation project positively, it is essential to integrate all the people involved at an early stage.

4.2.5 Integration of the Technology into the Production Process

It is very important that the developed production line be integrated optimally in the production flow. With the experience gained from the production tests a new laying tool could be designed. The flexible manufacturing cell was fully automated (Figure 3). The container was moved by a overhead conveyor from the previous process to the production line so that the profiled seal could be mounted automatically. Due to cycle time reasons, before the insertion station a station for applying a lubricant and after the insertion station a station for rolling in the end of the profiled seal were installed. These three stations, where the container is kept in a defined position, were connected by a fast linear transfer unit. A continous sealing profile was used at the automatic line.

4.2.6 Experiences with Application

In the plant the workers operate in three shifts. The reliability over several years is exeptionally high (>92%). The insertion quality remains excellent. Another advantage of this system is that essential modifications of the behavior of the profiled seals, for example when the supplier changes, can be adjusted quickly by adapting the insertion program.

4.3 Example of Application: Assembly of Electrical Switching Devices

Another important sector in the appliance industry is electronic switching devices. Especially in the field of assembly, there is great automation potential.

The following example demonstrates the installation of a transformer subassembly for grounding current-protection switches.

4.3.1 Formulation of the Job

Until now the transformer subassemblies were manufactured manually with few additional facilities. The following work had to be done:

- Winding a 2-mm-thick copper wire on a coil carrier
- Covering a magnetic ring
- Cutting the wire ends to an individual length
- Insulating the wire ends
- Shaping the wire ends
- Brazing supply terminals to the wire ends

Figure 4 demonstrates the respective work processes.

A reason for automation is that manual installation is very stressful labor due to the extreme strain on the muscles when the thick, solid wire has to be wound up and bent. However, the transformer subassembly as a safety-relevant component requires a very

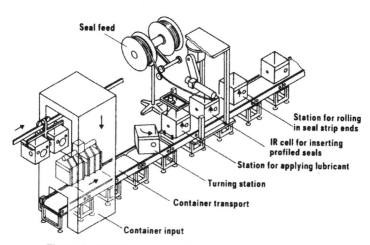

Figure 3 Industrial robot cell for insertion of profiled seals.

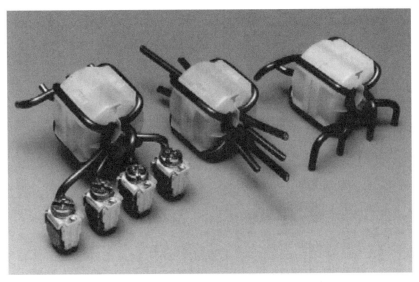

Figure 4

high quality standard. Besides the rationalization effects, further economic goals such as stock decrease and reduction of overall production time should be achieved.

4.3.2 Development of the Concept and Choice of the Industrial Robot

Because the product was to be manufactured fully automatically, different variations were considered. Facts that favored the use of industrial robots were:

1. The spatial complex shaping process could not be solved efficiently with the conventional automation components if the required flexibility was to be maintained.
2. Except for the wire, there were only small parts that had to be fed automatically, so the peripheral equipment could be designed to be fairly simple and less expensive. Furthermore, a division into suboperations with adequate labor content, those with longer cycle times, was possible.

Finally, the assembly line was divided into four flexible manufacturing cells loosely connected via a belt-driven transport system. A central component of every cell is an IR. For three of those cells a six-axis articulated-arm robot and for the winding of the thick solid wire a three-axis robot with two Cartesian axes and a gripper swivel axis were chosen.

4.3.3 Product Design

Before realization, the product has to be made suitable for automation. In this case the following modifications were performed on the product:

- Standardization of the wire used
- Optimization of the shaping process for the IR
- Reinforcement of the coil carrier in order to protect the magnetic core
- Use of pliant supply terminals
- Standardization of the outer geometry of the coil body
- Feedability of the supply terminals
- Technology assurance

Because the technologies for winding thick solid wire and for shaping the wire ends were not well developed yet, experimental tests in our lab, and finally production tests, were absolutely necessary. The results of these production tests, which were performed with minimal effort on peripheral equiment, turned out to be positive. They gave important hints for the development of the actual production line and confirmed the planned concept.

4.3.4 Quality Assurance

Since the product is a safety-relevant component, quality assurance of every single production step is very important. There especially has to be found a new way to braze the supply terminals to the wire ends in order to be able to guarantee 100% verification for the brazing joints. In order to do that, sensors were integrated that controlled the amount of material that had to be brazed as well as the energy supply. In most cases the absolutely necessary process assurances can be carried out with relatively simple sensors. In general, one can say that the simpler the chosen sensors, the easier the development of the required software and respectively the smaller the fault liability.

4.3.5 Installation of the Production Line and Material Flow

Figure 5 shows that the production line consists of four industrial robot modules.

In module 1 the thick solid wire is wound onto the carrier of the coils. The loading and unloading of the coils are handled by the robot of module 2.

Module 2 prepares the parts for the winding process (control of coils, maintenance of the winding module, and cutting and spreading of the wire ends).

Modules 3 and 4, for capacity reasons, are installed as duplicates and are equipped for shaping and insulating the wire ends as well as brazing the supply terminals to the wire ends. In spite of the higher peripheral requirements, the decision was made to divide the production line in this way. Because the technology used in the modules is highly complicated, it was considered very important to be able to produce with half the capacity in case one of the modules failed.

An important criterion when such a production line is planned is that the connections of the modules provide sufficient buffer capacity. The machine also works during break times and a fault clearance is also possible without an idle time for other modules. This feature is made possible by a new pallet concept (40 subassemblies can be transported on one pallet) and supply lifts/elevators for up to ten pallets that are connected via a transportation belt.

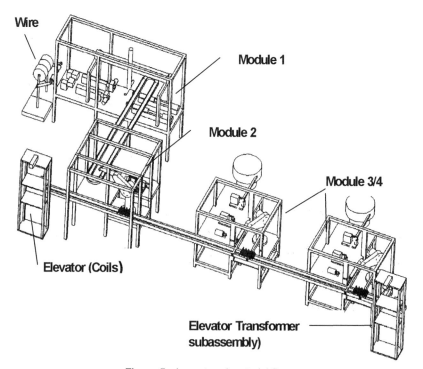

Figure 5 Layout and material flow.

Additional malfunction routines and ejection possibilities in every module guarantee operation with a high degree of reliability. Furthermore, little supervision of the process is required.

4.3.6 Training for Maintenance

To carry out an automation project successfully, it is important to collaborate with the involved personnel at a very early stage.

As soon as we had a concrete design and were sure that our plans were technically feasible and of economic use, we informed the factory committee and the managers of the shop floor. In this way our ideas became widely accepted.

The workers and personnel for setting up the production lines were not only informed at an early stage but also trained. In practice, however, it was proven that these steps though important, were not sufficient. In case of a malfunction of the system the maintenance specialist often realized that there was no mechanical problem. So an electrican was consulted, who sometimes needed to ask an electronics engineer for his opinion. Therefore we found that the maintenance technicians responsible for highly automated plants need special and profound training. Besides basic mechanical training, they must also acquire knowledge in the fields of electronics, software, and service.

The measures we took proved to be very effective. The reliablility of the whole system became much better and remained at a high level.

5 CONCLUSION

The appliance industry still offers a wide range of potential robot applications, due basically to higher quality requirements, shortened innovation cycles, and the large number of product variants. However, the fierce competition and global changes in production automation mandate careful analysis of the various aspects, such as product design, mastering of the automated process, factory infrastructure, and qualification of personnel.

Industrial robots have become a reliable, highly developed component of modern automation engineering and can be universally used and well integrated into existing information structures. The following aspects of robot applications should also be emphasized:

- The application of robots provides a more consistent and generally higher-quality product compared to manual assembly and manufacturing.
- The higher engineering effort for robotized automation leads to a better understanding of the different technological processes involved and thus to improvements in products and quality.
- This knowhow, together with adequate automation equipment, most often is of significant importance in case production has to be moved to or replicated in other countries.

ADDITIONAL READING

Feldmann, K., N. Roth, and H. Rottbauer. "Relevance of Assembly in Global Manufacturing." *Annals of the CIRP* **45**(2), 545–552, 1996.

Freund, B., H. König, and N. Roth. "Impact of Information Technologies on Manufacturing." *International Journal of Technology Management* **13**(3), 215–228, 1997.

Niedermayr, E. "Neue Aufgaben der Informationsverarbeitung in der Fertigungsautomatisierung, 6." Paper delivered at Kooperationssymposium TU Wien, Steyr, Austria, 1992.

———. "Potential and Constraints in Robot Assembly." UN/IFR World Industrial Robots, Statistics Report, 1996.

Niedermayr, E., K. Kempkens, and N. Roth. "Advanced Robot Control in Industrial and Non-traditional Applications." *Proceedings of IEEE International Workshop on Intelligent Robots and Systems, IROS 90,* 1990.

Roth, N., and C. Weyrich. "Produktion im Wandel." *Siemens Zeitschrift Special FuE* (Winter), 2–4, 1996/1997.

CHAPTER 60

FOOD AND AGRICULTURE ROBOTICS

Yael Edan
Ben-Gurion University of the Negev
Beer Sheva, Israel

1 INTRODUCTION

Robots can be applied to all food production phases, from field operations for a variety of agricultural tasks such as cultivating, transplanting, spraying, trimming, and selective harvesting; to inspection and quality sorting; to processing or packaging operations. Food and agricultural robots have the potential of raising the quality of the produce, lowering production costs, and reducing the drudgery of manual labor. Many techniques for automation and mechanization of farm machinery have been developed and commercialized, such as combine harvesters, tree shakers, and sprayers. However, these automatic machines are limited to once-over harvesting, usually of fruit directed to processing markets. Picking of fresh-market fruit is not possible using existing machines due to excessive mechanical damage and lack of selective handling capabilities.

Despite the tremendous number of robotic applications in industry, very few robots are operational in agriculture production, and only recently have robots been commercially introduced to food applications. In contrast to industrial applications, which are repetitive, well defined, and known a priori, in agricultural operations a robot must deal with an unstructured, uncertain, varying environment that cannot be predetermined. The fruit is randomly located in the field and is difficult to detect and reach, being hidden by leaves and positioned among branches. Environmental conditions are hostile (dust, dirt, and extreme temperature and humidity) and continually change (bumpy terrains, changing illumination conditions—clouds, sun direction—wind that moves leaves and fruit). Food products are natural objects that have a high degree of variability (shape, size, and quality) and are often presented at packaging sites by systems that cannot control their position or orientation. The objects are soft, fragile, and easily damaged and must be handled gently.

For fast and robust operation to be achieved in such a complex and dynamic environment, food and agricultural robots must be equipped with sensing, scheduling, and adaptive planning capabilities. An intelligent sensor-based control system that enables flexible, real-time multitasking must be structured for efficient planning, coordination, and control of the tasks. Special grippers capable of gently handling objects that vary in size and shape must be designed. Compared to precision requirements for industrial robots, those for agricultural robots are much lower. However, a greater degree of compliance is usually required. The unstructured nature of the external environment increases the chances of collisions. Moreover, the machines are operated by personnel who usually do not have an expert technical background. Therefore inherent safety and reliability of the robotic system is an important feature. Food safety is also an issue in several applications requiring the manipulator to be washable and reliable against leakage of contamination. Maintenance is another factor to be considered from different aspects: increase in machine complexity (to overcome the complex natural environment and provide inherent safety) requires an increased skill level from maintenance personnel. A spread-out service area increases maintenance costs, but minimum turnaround time for repair is sometimes essential—when the crop is ready for harvesting no delays are allowed.

Another factor limiting agricultural and food robotization is machine cost: the machine has to be very sophisticated to enable robust operation in the natural environment, and

Handbook of Industrial Robotics, Second Edition, Edited by Shimon Y. Nof
ISBN 0-471-17783-0 © 1999 John Wiley & Sons, Inc.

this increases its cost. However, the annual cost of a machine is difficult to allocate for the individual farmer since most agricultural machines cannot be used year-round due to the seasonal nature of agriculture. Furthermore, the market price of the agricultural product is usually low and hence the actual cost of the robot must be low to make it economically justified.

2 DESIGN

2.1 System Design

Designing a robotic system includes selecting the most appropriate manipulator (e.g., cylindrical, Cartesian) for the specific task and defining its motions; selecting the number of arms and their operational configuration; determining parameters such as speed and acceleration, and selecting the type of drive (i.e., hydraulic, pneumatic, or electric). The solution for similar problems in industry has been to simulate robot performance using commercial software packages that provide animated, graphical representation of the time-varying solutions. Engineers select robots by evaluating, in simulation, alternative manipulators integrated into a workcell, considering different workcell setups, material, tool flow, and control strategies. However, the agricultural environment is loosely structured, i.e., there is large variability between successive tasks and even between successive items even though they are nominally identical. When different machine designs are evaluated during the development process, performance differences can occur due to changes in the machine and the crop or interaction between the two. Experiments are difficult if not impossible to reproduce since crop conditions change with time and vary spatially in the field. Thus not all conditions can be predicted and performance must be derived based on a statistical analysis. Moreover, robotic performance must be evaluated for a variety of crop conditions. A systematic method to evaluate robot performance for an unstructured environment must be applied and includes the following steps:

- *Data modeling:* Horticultural practices such as spatial geometry of the fruit and fruit distribution highly affect machine performance and therefore should be considered throughout the design of the robot. Actual field coordinates of fruit in typical rows must be measured and serve as input for the simulation model. However, since there are different crop conditions that can be adapted, the machine's performance must be evaluated for a variety of practiced horticulture conditions to determine the one that yields the best results. New feasible datasets should be generated based on distribution curves of measured data by modifying all features that could affect the design. Since this is crop-dependent, it must be done according to the horticultural practices of the specific crop evaluated. Simulations should be performed for actual field coordinates (measured data) and generated datasets that fit common horticultural practices.

- *Graphic simulation* is important to initiate the design process. It permits assessment of alternative intuitive design concepts and designs in a timely manner. This step should include the general system configuration and should be conducted using a robotic simulation package. 3D color animation provides realism that helps confirm the simulation validity and provides insight into the systems operations. At this stage technical feasibility and predicted cycle times can be computed.

- *Numerical simulation and sensitivity analysis:* Since graphic simulation is time-consuming, exact design parameters should be evaluated by developing numerical simulation programs that model the systems dynamics. To simplify the simulation process, each dependent variable should be varied separately. Performance should be evaluated for a variety of possible test cases and analyzed statistically. The specific design parameters evaluated should include type of robot (Cartesian, cylindrical), number of arms, configuration (serial, tandem, parallel), actuator speeds, and horticultural practices (e.g., planting distance, fruit distribution).

- *Motion planning:* In selecting the robot the time required by the robot to complete the task is an important factor. An algorithm was developed for finding the near-minimum-time path through N given task points (N fruit) of a robot with known dynamic and kinematic constraints. The sequence of motions can be defined by the traveling salesman algorithm using the geodesic distance in inertia space as the cost function. Application of the algorithm to citrus and melon harvesting indicated that fruit distribution and crop parameters, in addition to robot characteristics, influence cycle times. Therefore it is important to use this algorithm to select the most time-efficient robot and crop design.

- *Powering of agricultural robot arms:* Most food automation and agricultural robots have electrical manipulators and pneumatic grippers. While most food applications utilize off-the-shelf industrial robots, most agricultural robots are custom-designed. Electrical manipulators are used since they are easy to design, their control is easy to develop and implement, and required torques and speeds are achievable. Pneumatic manipulators are advantageous since they are faster and lighter, but their control is more complicated. Accuracies achieved with pneumatics are acceptable for the agricultural domain. Since pneumatics provides a cheap solution, its application feasibility in agriculture has been proven, and its control is becoming easily implemented, it is estimated that more and more agricultural applications will shift towards pneumatics. In parallel, developments in hydraulic control and the availability of hydraulics on the farm might make hydraulics a common option for farm applications that require high torques when leakage of contamination is not critical. R&D in hydraulic agricultural robots has been directed to develop light mass systems.

Since each agricultural robot must be custom-designed and adapted to the specific crop, the design methodology is important for evaluating new applications before expensive prototypes are built and field-tested. It has been proven to decrease development costs tremendously. Based on simulations performed for a multitude of design and crop parameters, the best set of design parameters can be derived and guidelines provided for the kind of crop parameters to be adapted to improve performance of the robot. To optimize performance, the tradeoff between the number of systems and their performance must be determined systematically using statistical and sensitivity analysis with numerical simulation tools. Finally, optimum configuration should be determined by an economic analysis based on a cost/benefit ratio.

2.2   Gripper Design

The gripper must be designed for the specific product and with the necessary compliance due to the variability in the object's dimensions, shape, and position and should be capable of gently handling the delicate products. Due to the frequency of use of the mechanism and the hostile environment in which it operates, the gripper should be made as durable and simple as possible. The number of moving parts should be minimized, and they should be protected from dirt and moisture. All sensors and electrical components must be protected as well. The structural parts must be sturdy enough but also must be kept as lightweight as possible to reduce the load exerted on the robot. The feasibility of the design concept of the gripper can be evaluated only in the field under actual operating conditions. Therefore an initial prototype should be constructed as early as possible in the design of the robotic system and manually activated and tested in the field.

A procedure for minimizing the gripper's weight has been developed by combining (1) stress analysis using the finite element method to validate that the gripper is strong enough to withstand the load of the melon for infinite lifetime and (2) optimization of the gripper's dimensions. In addition, the influence of different grippers on the internal stresses and the deformations of fruit handled by different types of grippers should be evaluated. This can be achieved by measuring physical properties of the fruit (e.g., elasticity modulus, failure strength, ripeness, shape, size) and using these data as input to a finite element analysis that analyzes the compression of the fruit for different gripper designs.

Versatile grippers capable of imitating human hand attributes have been investigated by two different approaches (Figure 1): developing simple and inexpensive pneumatic solutions and utilizing complex grippers with electronic sensing and closed-loop control.

2.3   Sensors

The sensing system must consist of sensors to detect the fruit's location (position and orientation), guide the robot manipulator to the fruit, determine fruit ripeness/quality, and monitor the robot's location. For robust operation in the unpredictable and dynamic environment the robot must be equipped with a redundant multisensor system for each of these tasks.

Figure 1 Grippers for robotic harvesting: (a) melon; (b) lettuce (courtesy Professor Makoto Dohi, Shimane University); (c) transplanting; (d) apple (courtesy Dr. Ik.Joo Jang, Kyungpook National University).

2.3.1 Vehicle Location

Continuous update of the vehicle's location is essential for accurate steering of the vehicle along the row and for accurate target interception (since the vehicle advances in an uneven terrain, its velocity is unstable). The true distance must be derived from the path traveled to coordinate between the carrier position at time of target detection and the position at time of intercept. A wheeled mobile robot can estimate the current robot point and orientation by accumulating the rotation of the wheels using encoders, gyro compasses, etc. However, due to inherent sensor inaccuracies, the rough surface (which causes slip), and the fact that the estimation errors accumulate as the robot moves, an additional position sensor is essential. This sensor can be a global sensor, e.g., laser, GPS, which provides absolute position, or a sensor that provides relative location by continuous local sensing of the environment. In both cases algorithms must be developed to correct the inherent accumulated errors.

2.3.2 Guidance

Automatic guidance sensors for mobile agricultural systems include mechanical sensing, ultrasonic, radio frequency, gyroscopes, leader cables, and optical systems. Several autonomous guidance systems have been developed applying different sensing approaches: optical techniques for an automated plowing system; photo detectors to detect the furrow;

infrared sensors as indicators for rotary tilling; a vision-guided, battery-powered lawn tractor; and positioning sensing systems based on lasers, passive radar beacons, and geographic positioning systems. Microcomputers are used to calculate the tractor position error, and different control algorithms, e.g., on/off, proportional, fuzzy, are used to position and set the correct steering angle.

2.3.3 Detection

Vision sensors tend to be the most suitable technique for dealing with the wide range of size, shape, and color of random, partially occluded targets. Real-time dedicated imaging hardware is essential for real-time response. Usually two levels of sensors are used: the first for target detection for global path planning and the second for local guidance and accurate fine-positioning. The first level has been applied using different sensors, e.g., gray-level, color, infrared. Fine-positioning has been achieved using a second image sensor, tactile sensors, laser ranging, and ultrasonic sensors.

2.3.4 Ripeness

Fruit ripeness/quality must be determined for selective harvesting and many packaging and handling operations. Several nondestructive techniques are being explored to evaluate food quality and ripeness, including vision, near-infrared, acoustic, and sniffer. However, despite intensive R&D in this area, most of these systems, except for machine vision, are not yet available as commercial units. Examples of fruit ripeness sensors for selective harvesting include color sensing and electronic sniffers. Examples for quality sorting include near-infrared and machine vision.

2.4 Control System

An autonomous system must react to changing operating conditions, make real-time decisions based on sensory input, and respond to conflicting goals. This requires sensor-based control with a flexible, real-time multitasking and parallel programming environment that integrates perception, planning, and control. Different approaches have been employed for intelligent control of agricultural robots: distributed control, state network control, hierarchical blackboard, and behavior-based architecture.

3 APPLICATIONS

3.1 Food Applications

By application of vision guidance and development of grippers capable of handling delicate products, several applications of robots to food handling and packaging have been developed: poultry products, prawn handling, egg candling tasks, meal-ready-to-eat pouch inspection, pork grading, and automatic sorting of oysters and dried mushrooms.

Robotic applications in food processing, such as meat cutting, sheep and lamb carcass processing, fish processing, and meat pork grading, have been increasing due to the shortage of skilled labor and the unpleasant and hazardous working conditions. The main problems are the hostile environment and the large size and shape variation between individual objects. Successful application requires the capability of adapting to variations in the fish/carcass size/shape/orientation, which has been achieved by developing force feedback control and continuous path planning and image processing feedback.

3.2 Agriculture Processing Operations

Many labor-intensive processing operations have attracted robotic development, including micropropagation, pruning of container-grown ornamental plants, transplanting, grafting, and cutting. All operations are conducted in controlled environments, and several commercial applications exist. Work has concentrated on gripper design and image analysis techniques.

3.3 Greenhouse Robots

Implementation of robots in greenhouses can help reduce hazards of automatic spraying, improve work comfort and labor efficiency, and potentially increase accuracy of operations. Autonomous operation of vehicles inside the greenhouse is easier since the environment is relatively controllable and more structured than the external agricultural environment. Several prototype automatic four-wheel vehicles have been developed for greenhouse transport operations, guided by, e.g., infrared guidance and leader cable network routing. Guidance is fully autonomous, but all tasks are still performed manually. In initial research toward a complete autonomous robot in Japan, a multipurpose manipulator (tomato picker and selective sprayer) attached to an AGV advances along the rows. Tomato, cherry tomato, and cucumber harvesting robots have been developed.

3.4 Harvesting Robots

Extensive research has been conducted in applying robots to a variety of agricultural harvesting tasks: apples, asparagus, citrus, cucumbers, grapes, lettuce, melons, mushrooms, tomatoes, and watermelons. R&D focused on object detection in a natural environment, gripper and manipulator design, and motion control from a static point. Autonomous guidance of these systems has not been investigated, although several self-guided vehicles have been suggested: a fruit-picking manipulator mounted on a dedicated platform in either a between-rows configuration (Figure 2) or a self-gantry system that overhangs the trees. The best picking rates are 2–3 sec per fruit.

3.4.1 Fruit Location

Several sensing technologies have been investigated for fruit detection: vision (black and white, color); infrared; structured light. Only 75–85% of the total fruits are claimed to be identified regardless of the sensing technique or algorithm. An innovative sensor that emits red and infrared beams (Figure 3) provides an example of possible solutions to increasing fruit detection.

3.4.2 Gripper and Manipulator Design and Motion Control

An average 2-sec picking cycle is a common requirement for agricultural robots. Since the workspace is relatively large, this implies that the actuators must have high velocities. On the other hand, the actuators can be less precise than usually required in industrial operations (centimeters instead of millimeters). In addition, the manipulator must be capable of overcoming obstacles such as branches and limbs. Due to the variability of the dimension and position of the objects, the end-effectors must be adaptive enough to provide the robot with the necessary compliance.

3.5 Field Operations

Autonomous guidance of tractors and combines along the rows is commercially applied for plowing operations, rotary tilling, detasseling, and automatic spraying. An autonomous alfalfa harvester has been developed that consists of a GPS unit to determine harvester location, a compass for heading information, and a pair of video cameras for viewing the region in front of the harvester. Maximum speed was 4.5 mph. Feasibility of robotic transplanting in the field has also been demonstrated.

Autonomous operation of a speed sprayer in an orchard was achieved using fuzzy logic control of image processing and ultrasonic sensors and steered by two hydraulic cylinders. Current research directions include adaptive chemical and fertilizer application by different methods, such as combining global positioning in the field by GPS and characteristics of the soil–crop system using GIS systems, and site-specific spraying/robotic weed control by directing spray (or cutting) onto the weeds that are identified by machine vision.

Several robots for selective pruning of trees and vines exist. System computation efficiency and the degree of cordon bending limit the speed at which the robot arm can move along and still correctly position the end-effector. The system is guided by two cameras mounted as eye-on-hand followers. At least 80% of the position updates are placed within ± 1 cm for speeds between 7.5 and 12.3 cm/s.

3.6 Animal Operations

Milking cows two or three times a day for seven days a week is a burden to the farmer. Attaching the milking machine to the cow is the last remaining major repetitive task that the worker in the modern milking parlor must perform. Several milking robots have been developed for this purpose in Europe and are being commercially introduced into farms. The task for the robot of locating the cow's teats and attaching the machine is complicated by morphological differences between cows (teats vary in size, spacing, and angle of projection, and conformation of a specific cow changes between milkings) and the need to adapt to the cow's motions during the teat attachment. In most systems a two-stage teat location process is applied. First, the approximate teat positions are determined by

(a)

(b)

(c)

Figure 2   Platforms for fruit-picking robots: (a) Korean apple harvester (courtesy Dr. Ik.Joo Jang, Kyungpook National University); (b) VIP melon picker; (c) Japanese grape harvester (courtesy Professor Makoto Dohi, Shimane University).

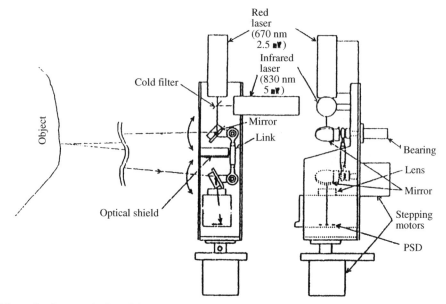

Figure 3   Custom-designed Japanese sensor for fruit detection (courtesy Professor Makoto Dohi, Shimane University).

dead reckoning using body-position sensors, ultrasonic proximity sensors, or vision systems. Final attachment is achieved by fine-position sensors using arrays of light beams mounted on the robot arm. Development of technology alone is not sufficient for full success of robotic milking. Several other factors, such as designing an efficient layout system and changing strategic management and herd management concepts, must be considered and integrated into the dairy routines.

Robot shearing operations have been developed and commercially applied in Australia. The sheep is constrained with straps on a movable platform. Position clippers using force feedback hydraulically control the actual shearing of the wool. Path computations are continuously updated during the shearing process.

4   SUMMARY

Dealing with natural objects requires a high level of sophistication from robots. Operating in unstructured and dynamic environments further complicates the problem. The decreasing costs of robots have enabled their wide introduction into commercial food applications and animal operations (automatic milking and shearing). The validity of the technical feasibility of agricultural robots has been widely demonstrated. Nevertheless, commercial application of robots in complex agriculture applications is still unavailable despite the tremendous amount of research in the last decade. The main limiting factors are production inefficiencies and lack of economic justification. The current unsatisfactory results of harvesting efficiency are due to problems in fruit identification (75–85%), low cycle times (3–4 sec per fruit), and the inability to autonomously deal with obstacles (branches, limbs). It is generally assumed that harvest efficiency will not be higher than 90%. Typical harvesting rates assumed are 2 sec per fruit for a two-arm configuration. Under these assumptions the cost of robot harvesting was found to be higher than manual labor cost (assuming a cost of $100,000 for a two-arm robot). It was concluded that even if robot harvesting tasks are equal to those of manual picking they will not justify the cost of development. Even if the cost of manual labor increases by 50%, the development cost will just break even. Future technological developments might lead to improved fruit detection (by multiple sensors or a new sensor capable of real-time distinction between fruit and background) and reduced cycle times. Obstacle avoidance is still at the research phase, and a possible available solution is to reduce the need for it by horticulture modifications. However, the most promising factor to enable economic feasibility of agriculture robots is to ensure they provide high fruit quality. Multiple uses of the robot (for additional tasks such as transplanting, spraying, and postharvest tasks such as sorting and sizing), combined with increasing labor costs and demand for high-quality produce on one hand and decreasing computer/sensor costs with increasing power on the other hand, will further enhance their feasibility.

ADDITIONAL READING

System Design

Chen, C., and R. G. Holmes. "Simulation and Optimal Design of a Robot Arm for Greenhouse." ASAE Paper No. 90-1537. St. Joseph: ASAE, 1990.

Edan, Y., and G. E. Miles. "Animated, Visual Simulation of Robotic Melon Harvesting." ASAE Paper No. 897612. St. Joseph: ASAE, 1989.

————. "Systems Engineering of Agricultural Robot Design." *IEEE Transactions on Systems, Man, and Cybernetics* **24**(8), 1259–1264, 1994.

Edan, Y., et al. "Near-Minimum-Time Task Planning for Fruit-Picking Robots." *IEEE Transactions on Robotics and Automation* **7**(1), 48–56, 1991.

Herman, T., B. Bonicelli, and F. Sevila. "On-Line Predictive Algorithm for High Speed and Power Hydraulic Robots." In *AGENG 92*, 1992.

Lang, Z., and S. Molnar. "A Pneumatic Driven Robot Harvester Model." AGENG Report No. 94-D-124, 1994.

Miles, G. E., and L. J. Kutz. "Robots in Transplanting." *Biotechnology in Agriculture and Forestry* **17**, 452-468.

Norman, D. W., J. A. Throop, and W. W. Gunkel. "Electrohydraulic Robot Design for Mechanized Grapevine Pruning." ASAE Paper No. 91-1598. St. Joseph: ASAE, 1991.

Papadopoulous, E., B. Mu, and R. Frenette. "Modelling and Identification of Electrohydraulic Articulated Forestry Machine." In *IEEE International Conference on Robotics and Automation*, 1997. 60–65.

Srivastava, A. K., C. E. Goering, and R. P. Rohrbach. *Engineering Principles of Agricultural Machines*. St. Joseph: ASAE, 1994.

Ting, K. C., and G. A. Giacomelli. "Automation-Culture-Environment Based Systems Analysis of Transplant Production." In *International Symposium on Transplant Production Systems*, Japan, 1992.

Ting, K. C., Y. Yang, and W. Fang. "Stochastic Modeling of Robotic Workcell for Seedling Plug Transplant." ASAE Paper No. 90-1539. St. Joseph: ASAE, 1990.

Umeda, M., K. Namikawa, and M. Iida. "Development of Agricultural Hydraulic Robot II." AGENG Report No. 94-D-132, 1994.

Gripper Design

Amaha, K., and T. Takakura. "Development of a Robotic Hand for Harvesting Cucumber Fruits." *Journal of Agricultural Meterology* **45**, 93–97, 1989.

Cardenas-Weber, M., et al. "Melon Material Properties and Finite Element Analysis of Melon Compression with Application to Robotic Gripping." *Transactions of the ASAE* **34**, 920–929, 1991.

Edan, Y., et al. "Robot Gripper Analysis: Finite Element Modeling and Optimization." *Applied Engineering in Agriculture* **8**(4), 563–570, 1992.

Kondo, N., et al. "Agricultural Robots: Manipulators and Fruit Harvesting Hands." ASAE Paper No. 92-3518. St. Joseph: ASAE, 1992.

Kondo, N., et al. "Robotic Harvesting Hands for Fruit Vegetables." ASAE Paper No. 94-3073. St. Joseph: ASAE, 1994.

Kondo, N., et al. "End-Effectors for Petty-Tomato Harvesting Robot." *Acta Horticulturae* **399**, 239–245, 1995.

Monta, M., et al. "End Effector for Tomato Harvesting Robot." ASAE Paper No. 96-3007. St. Joseph: ASAE, 1996.

Monta, M., et al. "Agricultural Robot: Grape Berry Thinning Hand." ASAE Paper No. 92-3519. St. Joseph: ASAE, 1992.

Tedford, J. D. "Development in Robot Gripper for Soft Fruit Packing in New Zealand." *Robotica* **8**(4), 270–278, 1990.

Ting, K. C., et al. "Robot Workcell for Transplanting of Seedlings Part II—End Effector Development." *Transactions of the ASAE* **33**(3), 1013–1017, 1990.

Simonton, W. "Robotic End Effector for Handling Greenhouse Plant Material." *Transactions of the ASAE* **34**(6), 2615–2621, 1991.

Wolf, I., et al. "Developing Grippers for a Melon Harvesting Robot." ASAE Paper No. 90-7504. St. Joseph: ASAE, 1990.

Vehicle Location and Guidance

Bonicelli, B., and M. O. Monod. "A Self-Propelled Plowing Robot." ASAE Paper No. 87-1064. St. Joseph: ASAE, 1987.

Brown, N. H., J. N. Wilson, and H. C. Wood. "Image Analysis for Vision-Based Agricultural Vehicle Guidance." ASAE Paper No. 90-1623. St. Joseph: ASAE, 1990.

Fehr, B. W., and J. B. Gerrish. "Vision Guided Off-Road Vehicle." ASAE Paper No. 89-7516. St. Joseph: ASAE, 1989.

Harries, G. O., and B. Ambler. "A Tractor Guidance System Using Optoelectronic Remote Sensing Techniques." *Journal of Agricultural Engineering Research* 26, 33–53, 1981.

Kazaz, I., and S. Gan-Mor. "Leader Cable Architecture for Guidance of Agricultural Vehicles." ASAE Paper No. 93-1056. St. Joseph: ASAE, 1993.

Mizrach, A., et al. "Evaluation of a Laser Method for Guidance of Field Machinery." *Computers and Electronics in Agriculture* 10, 135–149.

Ochs, E. S., J. A. Throop, and W. W. Gunkel. "Vision Control and Rough Terrain Compensation of a Robotic Grape Pruner." ASAE Paper No. 92-7043. St. Joseph: ASAE, 1992.

Reid, J. F., and S. W. Searcy. "An Algorithm for Separating Guidance Information from Row Crop Images." *Transactions of the ASAE* 31(6), 1624–1632, 1988.

Shmulevich, I., G. Zeltzer, and A. Brunfeld. "A Laser Scanning System for Guidance of Field Machinery." *Transactions of the ASAE* 32(2), 425–430, 1989.

Shropshire, G., C. Peterson, and K. Fisher. "Field Experience with Differential GPS." ASAE Paper No. 93-1073. St. Joseph: ASAE, 1993.

Throop, J. A., and E. Ochs. "Positioning with Changing Ground Speed of Mobile Robots." ASAE Paper No. 91-7031. St. Joseph: ASAE, 1991.

Tillet, N. D. "Automatic Guidance Sensors for Agricultural Field Operations: A Review." *Journal of Agricultural Engineering Research* 50, 167–187, 1991.

Toda, M., T. Okamoto, and T. Torou. "Studies on Autonomous Agricultural Robots." ASAE Paper No. 93-3091. St. Joseph: ASAE, 1993.

Toda, M. et al. "Studies on Autonomous Vehicles for Agricultural Robotics." ASAE Paper No. 93-3091. St. Joseph: ASAE, 1992.

Detection Sensors

Aoyama, J., et al. "Development of Missing Plant Detection System for Seedling Trays." ASAE Paper No. 92-7045. St. Joseph: ASAE, 1992.

Benady, M., and G. E. Miles. "Locating Melons for Robotic Harvesting Using Structured Light." ASAE Paper No. 92-7021. St. Joseph: ASAE, 1992.

Cardenas-Weber, M., A. Hetzroni, and G. E. Miles. "Machine Vision to Locate Melons and Guide Robotic Harvesting." ASAE Paper No. 91-7006. St. Joseph: ASAE, 1991.

Cerruto, E., and G. Schillaci. "Algorithms for Fruit Detection in the Citrus Canopy." Milano: AGENG Report No. 94-D-016, 1994.

Davis, P. F. "Orientation-Independent Recognition of Chrysanthemum Nodes by an Artificial Neural Network." *Computers and Electronics in Agriculture* 5, 305–314, 1991.

Dobrusin Y., et al. "Real-Time Image Processing for Robotic Melon Harvesting." ASAE Paper No. 92-3515. St. Joseph: ASAE, 1992.

Fujuira, T. "Agricultural Robots (1): Vision Sensing System." ASAE Paper No. 92-3517. St. Joseph: ASAE, 1992.

Harrel, R. C., D. C. Slaughter, and P. D. Adsit. "A Fruit Tracking System for Robotic Harvesting." *Machine Vision and Applications* 2, 69–80, 1989.

Gamage, L. B., R. G. Gosine, and C. W. de Silva. "Extraction of Rules from Natural Objects for Automated Mechanical Processing." *IEEE Transactions on Systems, Man, and Cybernetics* 26(1), 105–119, 1996.

Gao, W., et al. "Selective Harvesting Robot for Crisp Head Vegetable: 2D Recognition and Measurement of 3D Shape." *Journal of Japanese Society of Agricultural Machinery* 58(4), 35–43, 1996.

Kondo, N., and Y. Shibano. "Methods of Detecting Fruit by Visual Sensor Attached to Manipulator (Part 3)." *Journal of Japanese Society of Agricultural Machinery* 52(4), 75–82, 1990.

Kondo, N., et al. "Visual Sensor for Cucumber Harvesting Robot." In *Proceedings of the Food Processing Automation III Conference*, 462–471, 1994.

Kondo, N., et al. "Visual Sensing Algorithm for Chrysanthemum Cutting Sticking Robot." *Acta Horticulturae* (December), 383–388, 1996.

Li, Y. F., and M. H. Lee. "Applying Vision Guidance in Robotic Food Handling." *IEEE Robotics and Automation Magazine* (March), 4–12, 1996.

Maw, B. W., H. L. Brewer, and S. J. Thompson. "Photoelectric Transducer for Detecting Seedlings." *Transactions of the ASAE* 29(4), 912, 1986.

Nangle, J. A., G. E. Rehkugler, and J. A. Throop. "Grapevine Cordon Following Using Digital Image Processing." *Transactions of the ASAE* 31(1), 309–315, 1989.

Rabatel, G., A. Bourely, and F. Sevila. "Object Detection with Machine Vision in Outdoor Complex Scenes: The Case of Robotic Harvest of Apples." In *Proceedings of European Robotics and Intelligent Systems Conference*, Corfu, Greece, June 1991. 395–403.

Shutske, J. M., et al. "Collision Avoidance Sensing for Slow Moving Agricultural Vehicles." ASAE Paper No. 97-5008, St. Joseph: ASAE.

Tokuda, M., and K. Namikawa. "Discernment of Watermelon Fruits Using Image Processing." *Journal of Japanese Society of Agriculture Machinery* **57**(2), 13–20.

Whittaker, A. D., et al. "Fruit Location in a Partially Occluded Image." *Transactions of the ASAE* **30**(3) 591–596, 1987.

Yamashita, J., and N. Kondo. "Agricultural Robots (1): Vision Sensing System." ASAE Paper No. 92-3517. St. Joseph: ASAE, 1992.

Fruit Quality/Ripeness Sensors

Benady, M., et al. "Determining Melon Ripeness by Analyzing Headspace Gas Emissions." *Transactions of the ASAE* **38**(1), 251–257, 1995.

Gao, W., et al. "Selective Harvesting Robot for Crisp Head Vegetables." *Journal of Japanese Society of Agricultural Machinery* **58**(4), 35–43, 1996.

BARD. "Nondestructive Technologies for Quality Evaluation of Fruits and Vegetables." In *Proceedings of the International Workshop funded by BARD*. St. Joseph: ASAE, 1993.

Control Systems

Brandon, J. R., S. W. Searcy, and R. J. Babowicz. "Distributed Control for Vision Based Tractor Guidance." ASAE Paper No. 89-7517. St. Joseph: ASAE, 1989.

Edan, Y., B. E. Engel, and G. E. Miles. "Intelligent Control System Simulation of an Agricultural Robot." *Journal of Intelligent and Robotic Systems* **8**, 267–284, 1993.

Harrell, R. C., et al. "The Florida Robotic Grove Lab." *Transactions of the ASAE* **33**(2), 391–399, 1990.

Hoffman. "A Perception Architecture for Autonomous Harvesting." ASAE Paper No. 96-3068. St. Joseph: ASAE, 1996.

Tajima, K., J. Tatsuno, and K. Tamaki. "A Control of the Agricultural Robot with Flexible Boom Electrically Driven by Solar Light Power." ASAE Paper No. 97-3092. St. Joseph: ASAE, 1997.

Food Applications

Bourely, A. J., T. C. Hsia, and S. K. Upadhyaya. "Robotic Egg Candling System." *California Agriculture* **41**(1–2), 22–24, 1987.

Chen, S., and W.-H. Cjang. "Machine Vision Guided Robotic Sorting of Fruits." In *Proceedings of the Food Processing Automation III Conference,* 1994. 432–441.

Dreyer, J. "Palletizing and Packaging with Robotic Automation." In *Proceedings of the Food Processing Automation III Conference,* 1994. 132–138.

Jerney, T. D. "Replacing Touch Labor with Robots in Food Packaging Tasks." In *Proceedings of the Food Processing Automation III Conference,* 1994. 116–123.

Goldenberg, A. A., and A. Seshan. "An Approach to Automation of Pork Grading." *Food Research Journal* **27**, 191–193, 1994.

Goldenberg, A. A., and Z. Lu. "Automation of Meat Pork Grading Process." *Computers and Electronics in Agriculture* **16**, 125–135, 1997.

Kassler, M. "Robosorter: A System for Simultaneous Sorting of Food Products." In *Proceedings of the Food Processing Automation III Conference,* 1994. 414–418.

Kassler, M. "Robotics and Prawn Handling." *Robotica* **8**(4), 299–302, 1990.

Khodabandehloo, K. "Robotic Handling and Packaging of Poultry Products." *Robotica* **8**(4), 279–284, 1990.

Khodabandehloo, K., and P. N. Brett. "Intelligent Robot Systems for Automation in the Food Industry." In *Proceedings of the Institute of Mechanical Engineers,* 1990. 247–254.

Masinick, D. P. "Applications in Robotic Packaging and Palletization." In *Proceedings of the Food Processing Automation III Conference,* 1994. 124–131.

McMurray, G. V., et al. "Humane Level Performance Robotics for the Poultry Industry." In *Proceedings of the Food Processing Automation IV Conference,* 1995. 494–503.

Purnell, G., N. A. Maddock, and K. Khodabandehloo. "Robot Deboning for Beef Forequarters." *Robotica* **8**(4), 303–310, 1990.

Taylor, M. G., and R. G. Templer. "A Washable Robot Suitable for Meat Processing." *Computers and Electronics in Agriculture* **16**, 113–123, 1997.

Tojeiro, P., and F. W. Wheaton. "Computer Vision Applied to Oyster Orientation." ASAE Paper No. 87-3046. St. Joseph: ASAE, 1987.

Wang, D. L., and K. C. Ting. "Feasibility Study of Robotic Workcell for Meal-Ready-to-Eat Pouch Inspection." In *Proceedings of the Food Processing Automation III Conference,* 1994. 139–148.

Agricultural Processing Operations

Beam, S. M., et al. "Robotic Transplanting: Simulation, Design, Performance Tests." ASAE Paper No. 917027. St. Joseph: ASAE, 1996.

Hwang, H., and C. H. Lee. "System Integration for Automatic Sorting of Dried Oak Mushrooms." ASAE Paper No. 96-3006. St. Joseph: ASAE, 1996.

Hwang, H., and F. E. Sistler. "A Robotic Pepper Transplanter." *Applied Engineering in Agriculture* **2**(1), 2–5, 1986.

Kutz, L. J., and J. B. Craven. "Evaluation of Photoelectric Sensors for Robotic Transplanting." *Transactions of the ASAE* **10**(1), 295–301, 1994.

Kutz, L. J., et al. "Robotic Transplanting of Bedding Plants." *Transactions of the ASAE* **30**(3), 586–590, 1987.

Nishiura, N., et al. "High Frequency Vibrating Blade System for Vegetative Tissue Processing in Grafting Robot System." In *Proceedings of the Food Processing Automation III Conference,* 1994. 165–174.

Noguchi, N., et al. "Development of a Tillage Robot Using a Position Sensing System and a Geomagnetic Direction Sensor." ASAE Paper No. 97-3090. St. Joseph: ASAE, 1997.

Roux, P., et al. "Tree Pruning Mechanism and Automation in Forestry." AGENG Report No. 94-D-127, 1994.

Ryu, K. H., et al. "End-Effector for a Robotic Transplanter." ASAE Paper No. 97-3091. St. Joseph: ASAE, 1997.

Simonton, W. "Automatic Geranium Stock Processing in a Robotic Workcell." *Transactions of the ASAE* **33**, 2074–2080, 1990.

Tai, Y. W., P. P. Ling, and K. C. Ting. "Machine Vision Assisted Robotic Seedling Transplanting." ASAE Paper No. 92-7044. St Joseph: ASAE, 1992.

Tillett, R. D. "Vision-Guided Planting of Dissected Microplants." *Journal of Agricultural Engineering Research* **46**, 197–205, 1990.

———. "Robotic Manipulators in Horticulture: A Review." *Journal of Agricultural Engineering Research* **55**, 89–105, 1993.

Greenhouse Operations

Hayaski, S., and O. Sakue. "Tomato Harvesting by Robotic System." ASAE Paper No. 96-3067. St Joseph: ASAE, 1996.

Kawamura, N., et al. "Study on Agricultural Robot." Memoirs of the College of Agriculture: Kyoto University, No. 129, 1986. 29–46.

Kondo, N., et al. "Request to Cultivation Method for Tomato Harvesting Robot." *Acta Horticulturae* **319**, 567–572, 1992.

Kondo, N., et al. "Cucumber Harvesting Robot." In *Proceedings of the Food Processing Automation Conference III,* 1994. 451–460.

Subrata, I., et al. "Cherry Tomato Harvesting Robot Using 3D Vision Sensor." *Journal of the Japanese Society for Agricultural Machinery* **58**(4), 45–52, 1996.

Yamashita, J., et al. "Agricultural Robots (4): Automatic Guided Vehicle for Greenhouses." ASAE Paper No. 92-3544. St. Joseph: ASAE, 1992.

Harvesting Robots

Amaha, K., H. Shono, and T. Takakura. "A Harvesting Robot of Cucumber Fruits." ASAE Paper No. 89-7053. St. Joseph: ASAE, 1989.

Armdt, G., R. Rudziejewski, and V. A. Stewart. "On the Future of Automatic Asparagus Harvesting Technology." *Computers and Electronics in Agriculture* **16**, 137–145, 1997.

Dohi, M. "Multipurpose Robot for Vegetable Production II." *Journal of Japan Society for Agricultural Machinery* **56**(2), 101–108, 1994.

Edan, Y., et al. "Robotic Melon Harvesting." ASAE Paper No. 94-3073. St. Joseph: ASAE, 1994.

Gioco, M. "Robotics Applied to Selective Harvesting of Green Asparagus." AGENG Report No. 94-D-142, 1994.

Grand d'Esnon et al. "Magali: A Self-Propelled Robot to Pick Apples." ASAE Paper No. 87-1037. St. Joseph: ASAE, 1987.

Harrell, R. C., et al. "Robotic Picking of Citrus." *Robotica* **8**, 269–278, 1990.

Humburg, D. S., and J. F. Reid. "Field Performance for Machine Vision for Selective Harvesting of Asparagus." *Applied Engineering in Agriculture* **2**(1), 2–5, 1986.

Jang, I. J. "Studies on Development of Unmanned Robot for Mechanized Orchard Mode." College of Agriculture, Kyungpok National University, Taegu, Korea, 1996.

Juste, J., and I. Fornes. "Contributions to Robotic Harvesting in Citrus in Spain." In *Proceedings of the Ag-ENG 90 Conference,* 1990.

Kassay, L. "Hungarian Robotic Apple Harvester." ASAE Paper No. 92-7042. St. Joseph: ASAE, 1992.

———. "Results of Research of the AUFO 12/9 Armed Fruit Picking Robot." ASAE Paper No. 97-3094. St. Joseph: ASAE, 1997.

Kondo, N. "Harvesting Robot Based on Physical Properties of Grapevine." *Journal of Agricultural Research Quarterly* **29**(3), 171–177, 1995.

Sarig, Y. "Robotics of Fruit Harvesting: A State-of-the-Art Review." *Journal of Agricultural Engineering Research* **54**, 265–280, 1993.

Sittichareonchai, A., and F. Sevila. "A Robot to Harvest Grapes." ASAE Paper No. 89-7074. St. Joseph: ASAE, 1989.

Field Operations

Bar, A., Y. Edan, and Y. Alper. "Robotic Transplanting: Simulation and Adaptation." ASAE Paper No. 96-3008. St. Joseph: ASAE, 1996.

Benady, M., et al. "Design of a Field Crops Robotic Machine." ASAE Paper No. 91-7028. St. Joseph: ASAE, 1991.

Hoffman, R. "Demeter: An Autonomous Alfalfa Harvesting System." ASAE Paper No. 96-3005. St. Joseph, ASAE, 1996.

Hwang, H., and F. E. Sistlet. "A Robotic Pepper Transplanter." *Applied Engineering in Agriculture* **2**(1), 2–5, 1986.

Kawamuru, N., and K. Namikawa. "Automatic Steering of Tractor with Rotary Tilling." Research Report on Agricultural Machinery, Kyoto University, Kyoto, Japan, 1984.

Thompson, P., and F. Sevila. "Design of a Robotized Weed Controller." AGENG Report No. 94-D-130, 1994.

Yamashita, J., et al. "Agricultural Robots (5): Six Wheel Drive-Tractor for Orchards." ASAE Paper No. 92-1518. St. Joseph: ASAE, 1992.

Animal Operations

Artmann, R. "Sensor Systems for Milking Robots." *Computers and Electronics in Agriculture* **17**, 19–40, 1997.

Bottema, J. "Automatic Milking: Reality." In *Proceedings of the International Symposium on Prospects for Automatic Milking,* 1992. 63–65.

Devir, S., E. Maltz, and J. H. M. Metz. "Strategic Management Planning and Implementation at the Milking Robot Dairy Farm." *Computers and Electronics in Agriculture* **17**, 95–110, 1997.

Duck, M. "Evolution of Duvesldorf Milking Robot." In *Proceedings of the International Symposium on Prospects for Automatic Milking,* 1992. 49—54.

Frost, A. R. "Robotic Milking: A Review." *Robotica* **8**(4), 311–318, 1990.

Hillerton, J. E. "Milking Equipment for Robotic Milking." *Computers and Electronics in Agriculture* **17**, 41–51, 1997.

Ipema, A. H. "Integration of Robotic Milking in Dairy Housing Systems: Review of Cow Traffic and Milking Capacity Aspects." *Computers and Electronics in Agriculture* **17**, 79–94, 1997.

Rossing, W., and P. H. Hogewerf. "State of the Art of Automatic Milking Systems." *Computers and Electronics in Agriculture* **17**, 1–17, 1997.

Sphar, S. L., and E. Maltz. "Herd Management for Robot Milking." *Computers and Electronics in Agriculture* **17**, 53–62, 1997.

Toth., L., J. Bak, and T. Liptay. "The Hungarian Milking Robot." In *Proceedings of the International Symposium on Prospects for Automatic Milking,* 1992. 569–571.

Trevelyan, J. P. 1989. "Sensing and Control for Sheep-Shearing Robots." *IEEE Transactions on Robotics and Automation* **5**(6), 716–727, 1989.

van der Linde, R., and J. Lubbernik. "Robotic Milking System: Design and Performance." In *Proceedings of the International Symposium on Prospects for Automatic Milking,* 1992. 55–62.

Xu, B., and D. Aneshansley. "A Vision-Guided Automatic Dairy Cow's Teat-Cup Attachment System." ASAE Paper No. 91-3513. St. Joseph: ASAE, 1991.

Economics of Robot Operations

Dijkhuizen, A. A., et al. "Economics of Robot Application." *Computers and Electronics in Agriculture* **17**, 111–121, 1997.

Edan, Y., M. Benady, and G. E. Miles. "Economic Analysis of Robotic Melon Harvesting." ASAE Paper No. 92-1512. St. Joseph: ASAE, 1992.

Fang, W., K. C. Ting, and G. A. Giacomelli. "Engineering Economics of Scara Robot Based Plug Transplanting Workcell." ASAE Paper No. 91-7026. St. Joseph: ASAE, 1991.

Harrell, R. "Economic Analysis of Robotic Citrus Harvesting in Florida." *Transactions of the ASAE* **30**(2), 298–304.

Spharim, I., and R. Nakar. *A Robot for Picking Oranges—A Techno-economic Simulator.* Bet Dagan, Israel: Agricultural Research Organization, Volcani Center, 1987.

CHAPTER 61

APPAREL, WIRE, AND WOODWORKING INDUSTRIES

Manfred Schweizer
Thomas Hörz
Claus Scholpp
Fraunhofer Institute for Manufacturing Engineering and Automation (IPA)
Stuttgart, Germany

1 INTRODUCTION

The reasons for the necessity of automation in industrial production are well known. Approximately 80% of the total production time of nearly every production process is used for material handling, so it seems reasonable to use automated systems, e.g., industrial robots, for the handling tasks.

The handling of workpieces depends mainly on the behavior of the material. Generally materials are characterized as one of two different kinds: *limp* and *rigid*. The requirements of robot systems are completely different for these two kinds of materials.

To explain the handling of limp materials we look at the apparel industry and the wire industry. To explain the handling of rigid but soft materials we look at the woodworking industry.

2 ROBOT TECHNOLOGY HANDLING LIMP MATERIALS

The behavior of limp materials normally cannot be foreseen, especially if the material is three-dimensional. The automatic handling of textiles or similar materials is extremely complicated.

What causes problems is mainly that parts with nonconsistent configurations are exposed to forces of all kinds. The greatest difficulties are encountered when fabrics or materials of a similar nature are handled fully automatically. This is why manual production methods have until today dominated most of the branches of industry processing textiles. Because of this situation, and because of the high wage level in Germany, the transfer of production facilities is increasing constantly in these branches. If this trend is to be stopped or even reversed, concerted efforts must be made to develop suitable technologies that will enable manufacturers to produce competitive products even in a high-wage country such as Germany.

2.1 The Sewing Cell as an Integral Innovation in Clothing Robotics

In the sewing department there are possibilities for automation, especially before the workpieces become three-dimensional. The following problems must be solved:

- Gripping the workpieces from a pile
- Isolating the gripped pieces from the pile
- Transporting the parts
- Placing the parts in the desired position with the desired orientation under or in front of the needle

The greatest problem is that all these tasks have to be carried out with a reliability of more than 98%. Although a great variety of grippers for fabric with different working

Handbook of Industrial Robotics, Second Edition, Edited by Shimon Y. Nof
ISBN 0-471-17783-0   © 1999 John Wiley & Sons, Inc.

methods are available, none of these systems meets the requirements of all users. Nevertheless, a gripping system suitable for almost every kind of material can be found.

The kinematics mechanisms of the necessary handling system consist of several subsystems: kinematics, control systems (point-to-point or continuous-path), and drive systems (pneumatic, hydraulic, electric). For installation in the apparel industry an expensive six-axis industrial robot is not necessary when a smaller and cheaper handling system meets the requirements sufficiently. Because the workpieces to be handled by the same automation have very different dimensions, a modular handling system would be better.

We can now construct a handling system if we know:

- The dimensions of the workpieces
- The cycle times of the sewing operation
- The corresponding gripper and its weight

The premises are that the selected workplace is equipped with an automated sewing machine that automatically guides the workpieces during the sewing operation and can be started with a binary signal: the workpieces are delivered to the automated workplace in a box. We can integrate the handling system into the sewing operation.

A flexible sewing cell has been developed for workpieces collaboration through the Fraunhofer Institute IPA and the Hohenstein Clothing Physiological Institute in Bönnigheim. The entire system consists of the following components (Figure 1):

- Magazine unit for two magazines
- Positioning system for flat workpieces
- Dual-axis handling system
- Commercially available sewing machine

The prototype of this sewing cell is designed for the production of trousers. Depending on the type of sewing machine used, either single parts can be stitched or the side seams of trousers legs can be closed. At the beginning of a working cycle the handling system, using needle grippers, takes one trousers leg out of the magazine and puts it on the orientation platform. The positioning unit, which is the heart of the system, then arranges the trousers leg in the right orientation. The positioning unit (Figure 2) consists of rolls for the movement of the trousers leg and sensors that gather the information as to the actual orientation of the workpiece to the control system.

The positioning unit is also able to arrange and place, if necessary, workpieces of different sizes and materials one on top of the other without manual adaptation. That provides the possibility to process trousers legs, sleeves, coats, jacket parts, and more. The positioning accuracy is approximately ± 1 mm and therefore meets the requirements of the apparel industry.

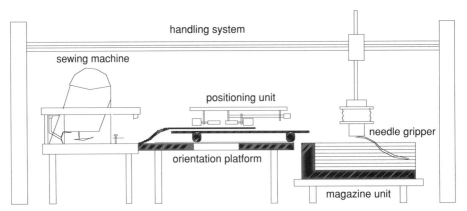

Figure 1 Automatic sewing cell.

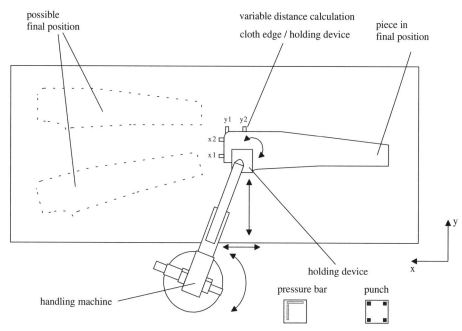

Figure 2   Positioning system.

The sewing cell operates on the basis of a user-friendly human–machine interface provided by a PC. Information about system conditions is constantly displayed on the screen, e.g., the current processing step, the next machining step, fault conditions, and instructions for their elimination are displayed to the operator together with general information such as the number of processed parts.

As long as it is impossible to obtain parameters such as material type and material thickness from a production host computer, the information is entered by the operator via PC.

More years of intensive development efforts will be necessary before direct linkage between cutting systems and flexible sewing cells is possible. Aside from the corresponding preparation of existing data, mainly machines and systems will need to be developed.

Task- and order-specific adjustments of the velocities of upper/lower conveyor, thread tension, pressure force of the pressing foot, or even the automated change of needle or thread, are not yet adequate. Peripheral components, e.g., positioning units and magazine filling stations, are not available either.

The goals of competitiveness, the idea of just-in-time production, and shorter fashion cycles, bringing increasing requirements for flexibility of a company, will accelerate development and attract the necessary resources and capital.

2.2   Automation of Lamp Wiring as an Example of Robot Technology in the Wiring Industry

Wire harnesses, cable sets, and single wires are used today in nearly every product to transmit current or signals for various applications. Industrial applications include automobile industry, the white goods industry (washing machines, kitchen equipment), the brown goods industry (stereos, televisions, video recorders), and the light industry. Every field has its own requirements in functionality, piece numbers, lots, and quality, but nearly all the named industrial fields are characterized by a very low degree of automation in the assembly of the wires and cables. The main reasons are:

- Difficulty of handling wires and cables, which are nonrigid parts
- Great range of variants and types

- Small lots and piece numbers
- Inadequate flexibility of existing machines on the market
- Missing assembly-oriented design

One of the decisive automation handicaps is the design of the product. A wire harness consists of different components such as wires, connectors, terminals, tapes, tubes for protection, and bellows. The main problem, especially in the automobile industry, is the basic structure of the product. Wire harnesses today are designed mainly for functionality, not assembly. This means that very complex structures exist that are difficult to assemble with a higher degree of automation, or even manually. The technique of termination is another problem. One of the most commonly used technique is by crimping, which involves a complicated assembly sequence while routing the wires. In many cases the economical realization of an assembly system suffers from the great range of types and variants. There are many different wire cross-crimping sections and colors. For the assembly sequences of cutting, displacement of the insulation, crimping, and block loading for very simple structures, machines with high productivity and acceptable flexibility are already available. The missing, and most expensive, steps are still routing the wires and completing the wire harness with additional features such as tubes, tapes, and bellows.

In the assembly of lamps, components such as chokes and sockets are fitted into lamp housings and then wired up. From the point of view of assembly techniques, conductors are items difficult to keep in position and therefore very difficult to manipulate and assemble without the aid of methods and tools that have to be specially developed. In industrial production today, conductor harnesses are predominantly manually assembled as preassembly groups and fitted into the end-product in a final assembly stage.

The demands placed on a system for the fully automatic wiring of lamps are:

- Complete cycle fully automated
- Process-integrated quality control
- Program flexibility
- CAD/CAM coupling
- High availability and productivity

Simplification, compared to the classic preassembly of conductor harnesses, can be achieved by directly assembling the individual conductors in the lamp housings. In this way it is possible to eliminate operations such as complete precutting, terminating and order picking of the individual conductors. For direct assembly to be carried out economically and with appropriate component variability, the shape of the components was to be redesigned. Previously, individual lamp components were designed for the use of screws or plugs for connection purposes. It was necessary to redesign the connection geometry for sockets, chokes, etc., since the technique of making connections by piercing the insulation proved to be the most assembly-friendly form of connection for the set task (see Figure 3). Designing the individual parts and component assemblies in a form suitable for assembly can consequently be seen as the key to possible greater automation, subject to economic and technical limiting conditions.

The requirements for program flexibility and consequently minimum tool change and shutdown times, with a greater multiplicity of types and variations, necessitate the integration of tolerance compensating systems in a fully automatic cycle. The analysis of the product range shows that production tolerances (of housings, components, etc.) are always on the order of several millimeters. The systems to be applied are therefore to be designed so that the maximum tolerances that occur can be dealt with in production. The system best suited to this purpose is a vision system, in which the exact position of the components before the assembly operation can be recorded. Knowledge of the exact position is obtained with the aid of markings provided specially for optical measurement of the components (three holes in each injection molding in the region of the connection zone). After the identification of the actual position, the deviation from the desired position can be entered in a sequential control program and the assembly cycle started.

The flange-mounted tool on the SCARA robot, with integrated slip control and synchronization of the tool path feed rate for feeding the conductors, is designed as a multifunction tool (see Figure 4). The vision system for determining the exact position of the

Before   **Afterwards**

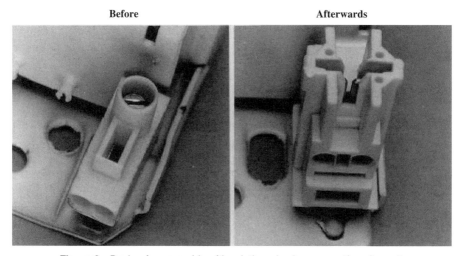

Figure 3   Design for assembly of insulation-piercing connection element.

components and the two light sources required for the measurement are fitted directly onto the tool. With the aid of a pressing stamp and the robot moving in the z-direction, the contact is formed. The conductor is cut off by the tool, directly after the contact, with an integrated cutting system. The force transducers to measure the insertion forces are integrated directly into the z-axis of the industrial robot.

With the single-position system shown in Figure 5, the lamp housings are fed in with the aid of transfer systems, each type being identified with the aid of the vision system and a bar code provided on the housings. After identification, the appropriate robot sequence program is loaded. In order to be able to compensate for tolerances, the industrial robot moves the vision system to the assumed desired position of the components and measures their exact position. For this operation it is possible to refer back directly to CAD data and consequently drastically reduce not only the material flow but also the flow of information. After the components are measured, the wiring process can be commenced. On the basis of extensive test runs and multiple optimization of the system, an average conductor laying speed of approximately 530mm/sec can now be achieved.

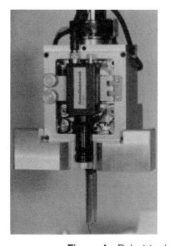
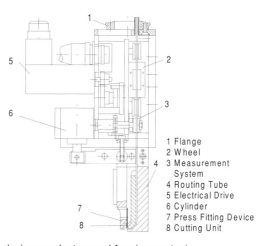

1 Flange
2 Wheel
3 Measurement
  System
4 Routing Tube
5 Electrical Drive
6 Cylinder
7 Press Fitting Device
8 Cutting Unit

Figure 4   Robot tool for laying conductors and forming contacts.

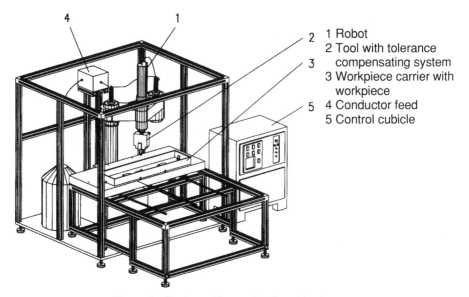

1 Robot
2 Tool with tolerance
   compensating system
3 Workpiece carrier with
   workpiece
4 Conductor feed
5 Control cubicle

Figure 5 Single-position system for wiring lamps.

3 ROBOT TECHNOLOGY FOR HANDLING WOOD PARTS

The application potential for robots in the woodworking industry is discussed below. Industrial robots have been available for flexible automation of the manufacturing process for more than 20 years now. But in the automotive and the household appliances industry, as well as in the electrical engineering sector, good growth rates could be achieved by using robots, while the number of industrial robots used in the woodworking industry is still relatively low.

The manufacture of customized products, such as in the woodworking industry is usually performed manually according to traditional handcrafting. For intermediate products with high volumes and many identical work cycles (for instance, office and kitchen furniture), state-of-the-art production lines and CNC processing centres are often utilized. These woodworking systems are customized for certain processing steps. But changes in consumer behavior are forcing enterprises to increase flexibility under similar efficiency to be able to stand up to international competition. Due to the change from a supplier market to a buyer market, enterprises must react more flexibly to the market and more quickly to customer desires. For the companies concerned, this means that they will have to cope with increasing product variety, higher quality expectations, and shorter product life cycles and delivery times.

Which path can and must an enterprise take to live up to these requirements of flexibility due to customer wishes? Surely continuous series manufacture is still justified, even though a strong trend towards complete stationary manufacture is clearly recognizable. Increasingly, in addition to existing processes such as drilling, milling, and grooving, integration of assembly tasks such as force-inserting hinge joints, dowelling, and nailing, is required of the stand alone machines.

Small, intermediate, and large companies are equally faced with this situation. This trend also confirms the great interest of many industrial enterprises in the application of flexible assembly and robot technology in combination with stationary CNC machines.

Industrial robots have been available for flexible automation in manufacturing for more than 20 years. In contrast to other industries, this technology found its way into the woodworking industry relatively slowly. This has also led to the fact that, in spite of overproportionate annual growth rates in industries like the automobile industry, the household appliances industry, and electrical engineering, the number of industrial robots utilized by the woodworking industry is still relatively small.

3.1   Robot Technology Within the Woodworking Industry

3.1.1   The Current Situation

Japan dominates the world market of the woodworking industry, with 2,365 installed industrial robots (Figure 6).

Specific examples of applications are found in the functions of *coating, handling,* and *separating.* Two applications developed within the scope of research work at the IPA are described below.

3.1.2   Flexibly Automated Processing and Preassembly of Rack Furniture

Especially in the upholstered furniture industry, considerable adjusting times for machining and preassembly of rack furniture parts are incurred. In connection with an industrial robot (Figure 7), the following process steps and system components have been integrated:

- Commissioning of furniture components into suitable assembly installations
- Handling of furniture components by the industrial robot
- Machining of furniture components using the industrial robot
- Assembly of individual components into a finished furniture frame under support of the industrial robot

The controller of the assembly cell identifies by a code-reader, which is mounted on the assembly installation, the particular furniture frame-variant to be processed next. The assembly installation is designed in such a way that all component models of a product range may be supplied by it. After they are gathered from the assembly installation, the workpieces are inserted into a highly flexible work-locating fixture, which holds securely, in a defined manner, all the individual parts needed for production. The fixture includes a vacuum fixing plate for plate-shaped workpieces, and a mechanical fixture for all other workpieces. Workpieces are handled by a suitable universal gripping device. After the

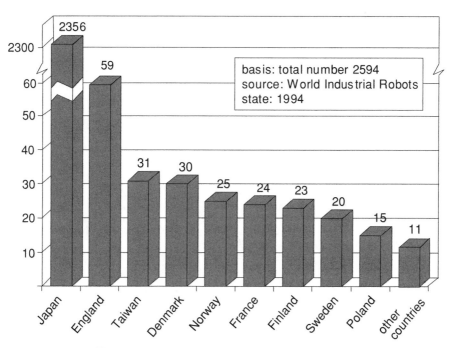

Figure 6   Number of industrial robots in woodworking.

Figure 7   Complete view of the robot cell for processing and preassembly of rack furniture.

individual workpiece has been gripped, processing by the robot follows. Processing may include drilling, milling, grinding, etc. The processing tool is designed for multiple steps so that tool changing is avoided.

After the cutting process, the workpiece is prepared for assembly. This may, for instance, be the assembly of wood dowels or pressing of threaded sleeves or connecting fittings. Next the workpieces are inserted into an assembly device according to the processing sequence, depending on the intended type of connection. Dowel connections are completed in the frame press while the robot takes a new workpiece from the assembly installation and places it in the work-locating fixture.

All parts of the system, including the robot control, are connected via CAN bus (controller area network) modules (Figure 8). This control guarantees high flexibility of the system without expensive retooling or programming work in case of model changes. For instance, when additional system components are needed they may be optionally integrated into the current control system. For control of program sequence and as a programming platform for the CAN bus modules, a PC can take over the functions of a cell computer.

3.1.3   Future Application Fields of Robot Technology in the Woodworking Industry

In a survey performed by the Fraunhofer Institute for Manufacturing Engineering and Automation (IPA) of woodworking operations of different sizes, the criteria shown in Figure 9 were found to be the main obstacles to automation.

Small volume, small lot sizes, and great range of variants and models as well as frequent product changes require highly flexible automation systems. The time- and cost-intensive assembly sector can be helped by the robotics option. Surely the woodworking industry in the future will not consist of "plants without employees." On the other hand, *hybrid production and assembly* is a suitable approach to rationalization in the woodworking industry. The work processes consist of manual work steps that are performed by the worker and, relative to flexible production and assembly systems, may be seen as a total integration for manufacturing wood products.

4   OUTLOOK

On the basis of many studies for woodworking robotics, a forecast for future application fields of robot technology was prepared. It shows that due to the technical possibilities

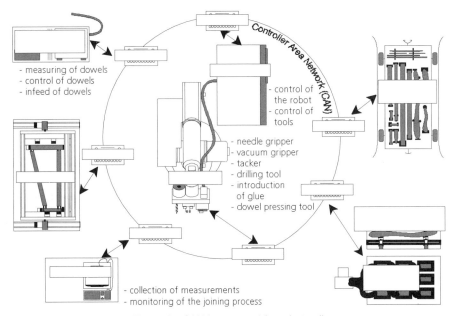

- measuring of dowels
- control of dowels
- infeed of dowels

Controller Area Network (CAN)

- control of the robot
- control of tools

- needle gripper
- vacuum gripper
- tacker
- drilling tool
- introduction of glue
- dowel pressing tool

- collection of measurements
- monitoring of the joining process

Figure 8   CAN bus control for robot cell.

and the current competitiveness in this industry, future implementation of industrial robots in the woodworking industry will mainly be in the areas of *assembly, handling, separating, coating,* and *shaping.* The main application fields and expected development of installed automation solutions with robots are shown in Figure 10.

production

volume too small		71%
small lot sizes		64%
too large variety of variants and models		79%
missing engineering solutions on the market		7%
flexibility of the solutions offered too small		21%
lack of experience		21%

product

frequent changes of products		64%
large workpiece dimensions		50%
large workpiece tolerance		29%
raw material is a natural product		36%

efficiency

efficiency too low		71%
investment volume too high		50%

0   20   40   60   80   100%
percentile frequency of statement

source: IPA survey;   basis: woodworking enterprises of different size;   state: 1/94

Figure 9   Obstacles to automation in the woodworking industry.

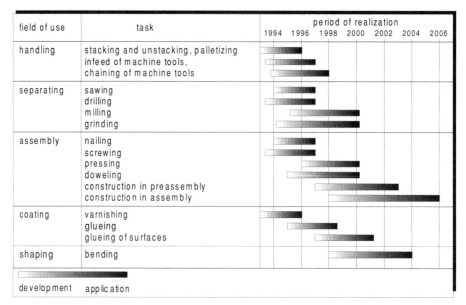

Figure 10 Expected development of automation solutions using robotics for the woodworking industry.

REFERENCES

Cramer, R. 1995. *Redesign of Electric Components for the Flexible Automated Assembly.* Stuttgart, Germany: Fraunhofer Institute for Manufacturing Engineering and Automation (IPA).

Hörz, T. 1996. *Flexible automatisierte Bearbeitung und Vormontage von Gestellmöbeln mit Industrieroboter.* Stuttgart, Germany: Fraunhofer Institute for Manufacturing Engineering and Automation.

Krockenberger, O., and H. Nollek, 1991. *Handling with an Automatic Sewing Cell for Trouser Legs.* Stuttgart, Germany: Fraunhofer Institute for Manufacturing Engineering and Automation.

Schweizer, M., and T. Hörz. 1996. "Robotertechnologie zur Standortsicherung." *HK Holz- und Kunststoffverarbeitung* (April), 60–65.

Schweizer, M., H. Dreher, and T. Hörz. 1995. "Robotertechnik 2, 3; Zukünftige Einsatzfelder der Robotertechnik in der Holzverarbeitung, Heutiger Stand in der Holzverarbeitung." *HK Holz- und Kunststoffverarbeitung* (February), 95–96; (March), 254–257.

Warnecke, H.-J., H. Dreher, and T. Hörz. 1995. "Robotertechnik 1; Einsatzpotentiale in der holzverarbeitenden Industrie." *HK Holz- und Kunststoffverarbeitung* (January), 76–79.

"World Industrial Robots 1994, Statistics 1983–1993 and Forecasts to 1997." GE.94-03808-October 1994–900, ISDN 92-1-100686-4, Geneva: United Nations, 1994.

ADDITIONAL READING

Boothroyd, G., and L. Alting. "Design for Assembly and Disassembly." Paper presented at 42nd General Assembly of CIRP, Aix-en-Provence, France, 1992.

Emmerich, H. *Flexible Montage von Leitungssätzen mit Industrierobotern.* Berlin: Springer-Verlag, 1992.

Emmerich, H., and S. Koller. "Neues Rationalisierungspotential ausschöfpen." *Schweizer Maschinenmarkt.* 1992.

Nof, S. Y., Wilhelm, W. E., and Warnecke, H. J. Industrial Assembly, Longon: Chapman and Hall, 1997.

Susnjara, K., "Robots in the Woodworking Industry," Chapter 46 in the Handbook of Industrial Robotics, (S. Y. Nof, Ed.), New York: John Wiley and Sons, 1985, pp. 879–886.

Warnecke, H. J., M. Schweizer, and H. Emmerich. "Hochflexible Montage von Verdrahtungssätzen." Paper presented at International Congress on Industrial Handling, Zürich, 1990.

CHAPTER 62

ROBOTICS IN CONSTRUCTION AND SHIPBUILDING

Kinya Tamaki
Aoyama Gakuin University
Tokyo, Japan

1 INTRODUCTION

Today more than 380,000 robots are used in Japanese industry. This is about 60% of the total world population. Most of them are used in the manufacturing industry (Hasegawa, 1996*b*). Many are used in the electronics, electric, automotive, and precision industries. The recent progress in robotics technology will expand the areas of robot applications from manufacturing factories to nonmanufacturing industry; for example, construction, shipbuilding, services, and housekeeping work. The number of robots in nonmanufacturing fields is not yet large, but it has been forecasted that the demand for these robots will become huge in the next century, and the robot manufacturing industry will support the economy of the country.

This chapter describes the historical progress of construction robots and the current status and future directions of building construction and shipbuilding robots in Japan. Research and development of construction robotics in Japan has continued for more than 20 years, since the beginning in 1978. Japanese contractors and machinery manufacturers are in the forefront of both development of construction robotics and the actual introduction of robots into sites. Nowadays Japan has the most robots in the world, not only in the manufacturing field but also in the construction field. In the last section of this chapter, the application of shipbuilding robotics is described as a related area of construction robotics, which also has significant importance worldwide (Reeve and Rongo, 1996).

2 THE HISTORICAL PATH OF CONSTRUCTION ROBOT RESEARCH IN THE WORLD

One representative international conference is the International Symposium on Automation and Robotics in Construction (ISARC), which was first organized in 1984. At that time construction robots were more an idea than a reality. At the 13th ISARC (1996), held in Tokyo, 960 articles had accumulated, representing diversified domains of application. These proceedings reflect how things have changed. The whole spectrum of topics is covered: from hard automation to autonomous robots; from component technologies to complete system implementation; from theory to practical application; from robot hardware to information technology; from lessons learned to prospects for the future. Construction robotics and automation is clearly no longer a fledgling discipline.

In 1991 the International Association for Automation and Robotics in Construction (IAARC) was founded to promote and organize international activities within the framework of automation and robotics in construction. Yasuyoshi Miyatake, the former fourth president of the IAARC, reported an intermediate analysis of articles presented at the past 13 ISARC conferences (Miyatake, 1997). The distribution of articles in the application domain was analyzed: tunneling and underground, 18%; assembly and finishing, 14%; inspection and repair and maintenance, 10%; earthwork and foundation, 8%; survey, 8%; masonry construction, 7%; concrete placing and finishing, 5%; material handling, 5%; road construction, 5%; others, 20%. Figure 1 illustrates the chronological trend over the development stages of *conceptual, prototypical,* and *commercial.* The number of conceptual articles remains more than half throughout the whole history. As time goes on, the

Handbook of Industrial Robotics, Second Edition, Edited by Shimon Y. Nof
ISBN 0-471-17783-0 © 1999 John Wiley & Sons, Inc.

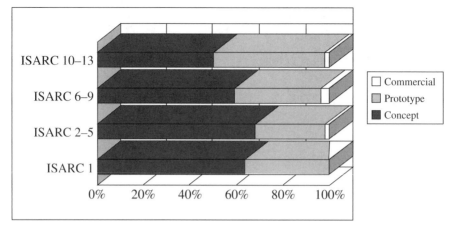

Figure 1 Chronological trend: proportion of development stages: concept, prototype, commercial. (From Y. Miyatake, "The Historical Path of ISARC," 14th ISARC Keynote Speech, June 1997.)

number of articles related to prototypes has gradually increased, while the articles on commercialization have remained few.

The general tendency is that U.S. and EC (such as U.K.) researchers cover the methodology, research survey, and computer application fields, while the Japanese cover the elemental technologies and development fields. Recently Japanese general contractors and construction machinery manufacturers have tried to develop and implement several unique construction methods for automated structural steel work and precast as a strategy towards a future production system using some computer control technology. Furthermore, they are beginning to develop the conception of a computer-integrated construction (CIC) system toward the next generation. Some examples of these construction systems are described in Section 5.

3 QUESTIONNAIRE SURVEYS ON ROBOTICS SYSTEM DEVELOPMENT IN JAPAN

The automation committee of the Japan Construction Mechanization Association (JCMA) has been conducting a questionnaire survey of construction automation and robotization every two years. Some topics of the new survey are as follows:

1. Purpose of construction automation and robotization (answered in Table 1).
2. Expense of R&D: maximum expense for one machine was $5M (a tunnel machine) and 42 items were more than $1M. The average expense per item was $.75M.
3. Diffusion level of automated construction machines and construction robots: most diffused items and actually operated numbers are as shown in Table 2. They reported that the number of machines working in more than 10 places was 38. In recent years R&D in the fields of civil engineering and building construction has gradually increased.

More than 100 types of construction robot have already been developed. However, as previously mentioned, very few have been commercialized or produced for practical use because of difficult conditions in development, application, and management. This is explained in detail in Section 4.

The Construction Robotics Committee, Building Contractors Society, in Japan has been conducting a study for the past seven years to support the development and promotion of construction robotics. The CRC has targeted users of robots such as construction companies, machine manufacturers and lease/rental companies, and designers, and has been presenting the results of these studies (Masao et al., 1996).

1. *Design of robots.* Responses to the question "What robot design issues do you have to solve, and how do you solve these problems?" are indicated in Figure 2.

Table 1 Purpose of Automation and Robotization[a]

Purpose	Share	Accomplishment Ratio
a. Working condition improvement	15%	58%
b. Ease of operation	12%	43%
c. Safety promotion	16%	62%
d. Cost savings	5%	20%
e. Performance improvement	14%	75%
f. Quality improvement	12%	49%
g. Labor savings	18%	50%
i. Unmanned operation	8% ·	69%

[a] JMCA.

Answer (e), "Assessment of currently replaceable tasks by robots," chosen by 65% of respondents, received the highest response ratio, followed by (d), 37%, and (h), 12%. As a solution to these problems, (a), "Survey or experiment to obtain quantitative specifications," received the highest response ratio at 62%, followed by (b), 39%, and (g), 31%.

2. *Demonstration experiments.* Responses to the question "How many construction sites are used to perform demonstration experiments before the robot is ready to be used?" are indicated in Figure 3. The answer that received the highest response ratio was (b), "3 sites," chosen by 35%, followed by (c), "5 sites," chosen by 29%, and (d), "10 sites," chosen by 22%. In reality, it is quite difficult to standardize robots due to the different conditions and environments at each site, and developers indicate that different adjustments are necessary depending on the conditions encountered at each individual site.

3. *Evaluation of robot's performance.* In response to the question "How do you evaluate the robot's performance during demonstration experiments?" the answers "Whether or not development objectives are attained" and "Evaluation by skilled laborers" both showed the highest response ratio, at 78%, followed by "Evaluation by general contractor employees," at 46%. Developers indicate that while it is

Table 2 Popular Robots and Their Working Numbers[a]

Item	Working Number
a. Wireless controlled steel beam assembly manipulator	460
b. Mortar spraying robot	265
c. Automated road surface cutting system	113
d. Remote control panel handling manipulator	105
e. Automated snow-removing grader	94
f. Concrete floor-finishing robot	30
g. Automated dam wall concrete form	26
h. Heavy duty remote control manipulator	20
i. Steel beam welding robot	10
j. Steel pillar uprightness measuring system	10

[a] JMCA.

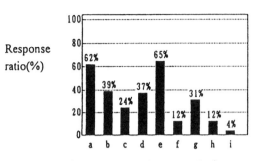

c. Experiment to establish proper
    conditions for robot control
d. The way tasks are performed by
    skilled laborers
e. Assessment of currently
    replaceable tasks by robots
f. Simulation
g. Inspection of robot's performance,
    speed, etc.

a. Survey or experiment to obtain
    quantitative specifications
b. Experiment on mobility, finish
    function, etc.

h. Advance evaluation of robot's
    capability such as ability to
    avoid obstacles, etc.
i. Other

Figure 2 Issues to be examined for robot design and countermeasures. (From K. Masao et al., "Study on Development and Utilization of Construction Robots," Paper presented at 13th ISARC, June 1996.)

important to evaluate whether the development objectives are attained by the completed robot, it is also necessary to consider evaluation made by skilled laborers.

4. *Reasons for discontinuing development.* Reasons given for the response "End development after prototype" are indicated in Figure 4. Among developers who discontinued development at the prototype creation stage, 28% selected (d), "No future development of prototype was possible," and 13% selected (e), "Development time and development budget, etc. was not sufficient and development was discontinued," showing an approximate total of 40%. Developers who actually developed a prototype to the point of practical use totaled 34%, including those answering (a), "Made the prototype work for practical use" (25%) and (c), "Made the prototype work for practical use with major modification" (9%).

4 CHARACTERISTICS OF AUTOMATION AND ROBOTICS IN CONSTRUCTION

In the construction industry, compared to other manufacturing industries, standardization of products and across-the-board mechanization and robotization have been difficult. The causes lie in the outdoor nature of production in the construction sites, the peculiarities of working conditions being different for each project, and the large number of on-site workers required. Automation and robotics in building construction can be characterized as follows (Ueno, 1994; Warszawski and Navon, 1996):

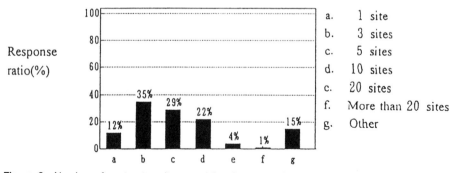

a. 1 site
b. 3 sites
c. 5 sites
d. 10 sites
e. 20 sites
f. More than 20 sites
g. Other

Figure 3 Number of contractor sites used for demonstration experiments. (From K. Masao et al., "Study on Development and Utilization of Construction Robots," Paper presented at 13th ISARC, June 1996.)

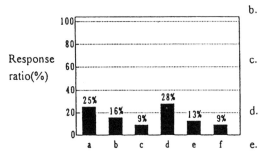

b. Attained anticipated objectives by gaining development know-how at the prototype creation stage

c. Made the prototype work for practical use with major modifications

d. No future development of the prototype was possible

e. Development time and budget was not sufficient and development was discontinued

a. Made the prototype work for practical use

f. Other

Figure 4   Reasons for discontinuing development after creation of prototype. (From K. Masao et al., "Study on Development and Utilization of Construction Robots," Paper presented at 13th ISARC, June 1996.)

1. Unlike the manufacturing industry, with its mass production, construction is a monoproduct industry. For example, construction work consists of less repetitive tasks than does factory work.

2. Building material elements are large, and high weight capacities of robots must be required.

3. Robots are required to be usable in outdoor weather conditions.

4. The operation of the robots is required to be easy for workers.

5. Robots require mobile function when being transferred to different working sections or other construction sites.

6. Robots must be light enough not to affect the load requirements of the floor, and robust enough to withstand the dirt and shocks of the construction work.

7. Robots need very advanced and thorough maintenance work in view of the mechanical and electronic systems they employ.

In the construction industry, in spite of much earnest effort, we have not yet reached the point of enjoying the results of automation and robotization R&D efforts. There are other problems that cannot be solved just by the design of the robotics system (Ueno, 1994; Hasegawa, 1996a).

1. For effective utilization of automation and robotics systems in construction, the design of building and material elements requires the use of standardization, which is not available at the present.

2. Because building design and construction methods are not designed for automation and robotics construction, their applicability is limited.

3. Lack of economic justification for heavy R&D funds includes the following two problems: First, ownership. Who should own their costs and maintain the automation and robotics systems? Second, very small market size compared with the manufacturing industry. We must therefore consider standardizing the robots and utilizing the R&D results as common properties and also coordinate the opportunities of robot applications.

4. Restructuring the industrial structure of construction to match the management organization of the automation and robotics in production. The construction industry consists of a multilayered structure of general contractors, subcontractors, construction machinery manufacturers, engineering design office, and so forth.

5 TOTALLY AUTOMATED BUILDING CONSTRUCTION SYSTEMS IN JAPAN

Several general contractors and robot machinery manufacturers in Japan have collaboratively developed the innovative automated building construction systems. By introducing these systems they aim to reduce the construction period, improve the working environment, decrease volume of construction waste, and increase productivity.

The Shimizu Manufacturing system by Advanced Robotics Technology (SMART) was developed and first applied on a substantial scale in 1992 (Maeda, 1994), and the improved SMART was used in 1997 in the construction of a building in Yokohama (Maeda, 1997). As shown in Figure 5, the main frame of the system is made up of a temporary steel framework roof called the *hat truss* and four jacking towers that support it. Vertical and horizontal conveying systems are located on the underside of the hat truss, and a lift-up mechanism for raising the hat truss is incorporated in the bottom of each tower.

5.1 Construction Procedures

Figure 6 shows the flow of the SMART construction procedures at the Yokohama site. The hat truss is assembled in advance into blocks as large as possible on the ground and then lifted and pieced together using a large crane. After the conveying equipment and other items are attached to the underside of the hat truss, it is raised and connected with the tops of the jacking towers. Together with this, the protective covering mesh sheet is attached and the assembly of the construction plant is completed.

5.2 Repeated Construction Based on Cyclic Process

Figure 7 illustrates the repeated construction process (cyclic process) for a standard floor. The cyclic process includes assembly of steel-frame, positioning of floor slabs (Figure 8), attachment of exterior wall panels, and positioning of the facilities and machinery unit and the interior partition materials unit. Finally, the lift-up mechanism automatically raises the construction plant as a whole unit. While synchronizing the operation of the lift-up mechanism, which consists of four jacking towers supporting the total plant weight of

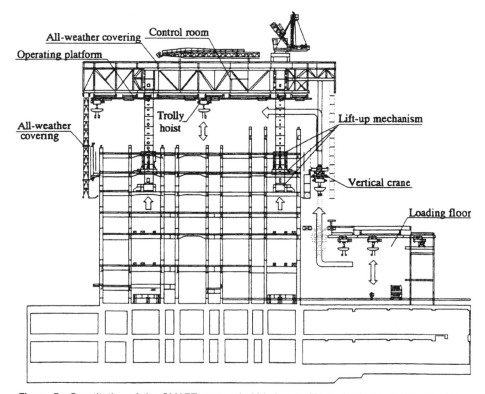

Figure 5 Constitution of the SMART system in Yokohama. (From J. Maeda and Y. Miyatake, "Improvement of a Computer Integrated and Automated Construction System for High-Rise Building and Its Application for RC (Rail City) Yokohama Building," Paper presented at 14th ISARC, June 1997.)

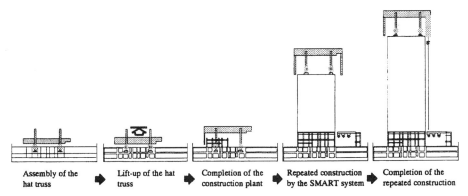

| Assembly of the hat truss | ➡ | Lift-up of the hat truss | ➡ | Completion of the construction plant | ➡ | Repeated construction by the SMART system | ➡ | Completion of the repeated construction |

Figure 6 Construction procedure of SMART system. (From J. Maeda and Y. Miyatake, "Improvement of a Computer Integrated and Automated Construction System for High-Rise Building and Its Application for RC (Rail City) Yokohama Building," Paper presented at 14th ISARC, June 1997.)

1,650 tons, the structure climbs the height of one floor, four meters, within about an hour and a half (Figure 9).

Figure 10 illustrates the control room, in which all equipment, such as the lift-up system and the material transportation system, is controlled by computers. The operations and motions of individual equipment are displayed in real time on monitor screens.

5.3 Assembling and Welding of Steel-Frame

The steel-frame columns and beams transported by the automated conveying system have self-guiding joints that allow them to be attached without being touched by a human hand (Figure 11). Removal of the sling is also done automatically. Adjustment of vertical accuracy of positioned columns is carried out by an automated laser measurement system. While reading the system's digital display, an operator uses the adjustment bolts built into the column to adjust its vertical alignment.

Figure 12 is a multilayer welding robot for welding column joints. The welding robot travels along rails and welds the joints of the mutual steel columns, including the corners. The robot is also able to check the welded portion with a laser sensor. Developments in welding robot technology for construction and shipbuilding are described in Shimizu et al. (1995), Okumoto et al. (1997), and Sorenti (1997).

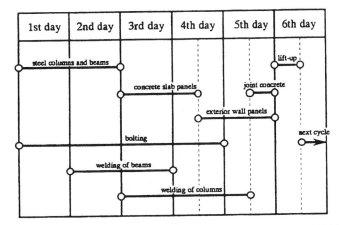

Figure 7 Cyclic process (5.5 days) for a standard floor. (From J. Maeda and Y. Miyatake, "Improvement of a Computer Integrated and Automated Construction System for High-Rise Building and Its Application for RC (Rail City) Yokohama Building," Paper presented at 14th ISARC, June 1997.)

Figure 8   Installation of floor slab pane (courtesy Shimizu Corp.).

6   ROBOT APPLICATION TECHNOLOGY IN CIM SYSTEM AT SHIPYARDS

6.1   Concept of CIM System for Shipbuilding by Hitachi Zosen Corp.

The significant factors in a computer-integrated manufacturing (CIM) system are integration of autonomous distributed databases and network technology, which ensure free access between multiple databases. The essential databases within the shipyard CIM system are CAD/CAM and drawings and specifications, material-management, and manufacturing, as shown in Figure 13. Nearly 70% of total database size is involved in the CAD/CAM and design database and the manufacturing database, compared to the database regarding management items. Therefore development of elemental technologies, such as CAD/CAM, NC/FA, and automated manufacturing simulation, is a key factor for completing the shipyard CIM system.

6.2   3D CAD System for Shipbuilding and Mobile Articulated NC Robot for Welding

The HICADEC was developed for the 3D CAD system of shipbuilding. The mobile welding robot HIROBO, with an articulated structure, is controlled not by teaching

Figure 9   Lift-up system (courtesy Shimizu Corp.).

Figure 10   Control room (courtesy Shimizu Corp.).

playback but by NC off-line programming. One operator can simultaneously operate more than three or four robots with small size and easy operation.

The simple parametric processing CAM system (SIMCAM) creates NC welding data for the purpose of controlling HIROBOs dispersed on individual huge blocks making up a ship structure. A significant feature of the SIMCAM is its ability to automatically generate the reliable NC data to complete collision avoidance motions against peripheral

Figure 11   Steel column erection (courtesy Shimizu Corp.).

Figure 12   Welding robot for column joints (courtesy Shimizu Corp.).

workpieces involved in the block, based on the results from 20 repeated motion simulations. Other CAD/CIM systems for shipyards are described in Bucher, Toftner, and Vaagenes (1994) and Sorenti (1997).

6.3   Integrated Information Technology Linked with CAD/CAM and Robot System

The CAMEX, as shown in Figure 14, is linked with the CAM system (SIMCAM) and a robot group system composed of multiple HIROBOs. It was developed to create NC data for the whole robot group system as an integrated CAD/CAM system, connecting as well with 3D ship structural data with the aid of the HICADEX. The NC data from the CAMEX are transferred to the station controller through the LAN and are downloaded

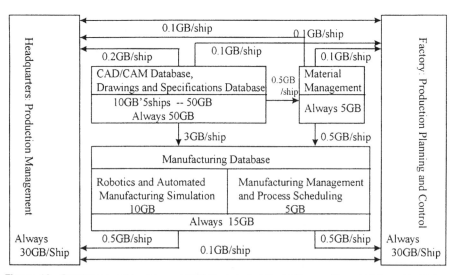

Figure 13   Databases within shipyard CIM System in Hitachi Zosen. (From T. Miyazaki, "Robot Application Technology for Shipyard CIMS Implementation," *Journal of Japan Welding Society* **63**(1). [In Japanese.])

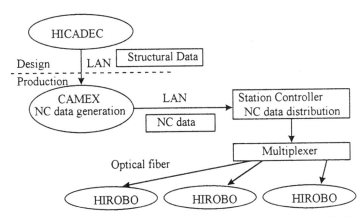

Figure 14 Linkage between CAD/CAM/robot system in Hitachi Zosen. (From T. Miyazaki, "Robot Application Technology for Shipyard CIMS Implementation," Journal of Japan Welding Society **63**(1). [In Japanese.])

through the optical fibers into the individual HIROBOs distributed at various shipyard shops.

6.4 Off-Line Teaching Technology of 3D Computer Graphic Simulator

Figure 15 illustrates a processing flow of the off-line teaching system for the welding robot HIROBO, which has applied the 3D computer graphic simulator ROBOTICS, connecting with the 3D CAD system CATIA. The CATIA generates 3D solid models of workpieces based on the 3D coordinates data included in CAD data files. The workpiece modeling data are incorporated into and remodeled within ROBOTICS.

The ROBOTICS models and defines the robot joint coordinates data of the HIROBO and the space coordinates data of the welding work environment. These modeling data are then stored into individual libraries. The initial task planning for the welding operations is interactively performed with reference to the mutual modeling data, such as the workpieces and robots within the virtual work environment. Additionally, the feasible task planning data are completed from the results of the repetitive 3D computer graphic simulation, considering collision avoidance. After the task data are converted into robot motion programs by the use of a robot language tool, the practical programming data for the welding robot are loaded into a robot controller for the HIROBO. Another off-line programming method is described in McGhee et al. (1997).

6.5 Integrated Applications of Robotics in Shipbuilding Worldwide

Advances in robotics applications in shipbuilding are described in Reeve, Rongo, and Blomquist (1997), Sagatun and Kjelstad (1996), and Blasko, Howser, and Moniak (1995) and illustrated in Figure 16.

7 CONCLUSIONS

R&D of construction and shipbuilding robotics has achieved remarkable progress during the last 20 years. A survey reveals that Japan and Germany are the only countries where private companies manufacture building robots (Warszawski and Navon, 1996). The Japanese construction industry is said to be in the forefront in these fields. However, more than half of the approximately 100 robots already developed are almost at trial-manufacturing level.

Therefore these robots must be continuously improved through actually utilized experiences, their capabilities repetitively refined until reaching a well-used level so as to be put in practical use at many general sites. To break through difficult obstructions to the diffusion of construction robots, countermeasures should be considered involving not only individual developers, but also cooperation between multiple organizations, such as mutual general contractors, subcontractors, machinery manufacturers, material manufacturers, engineering architecture offices, academic or public research institutes, and administrative management agencies.

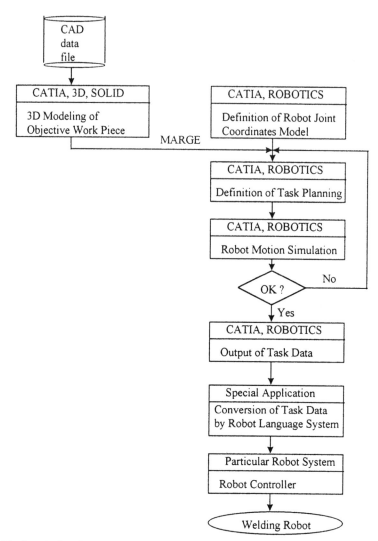

Figure 15 Processing flow of off-line teaching by using robot simulation system. (From Watanabe, "Current Situation and Problems of Application Technology for 3D Off-Line Robot in Teaching System Corresponded to CIM," Technical Report, Kansai Branch of Japan Welding Society, 1992. [In Japanese.])

Recently the trend in R&D has progressed from single-unit robots to total building construction automation, improving on the new construction methods and material design adaptable for robotization at the early phase of starting fundamental construction design. This chapter has discussed one automated building construction system, the SMART system by Shimizu Corp., but other general contractors have developed and implemented unique systems, such as AMURAD by Kajima Corp., T-Up by Taisei Corp., ABC and Big Canopy by Obayashi Corp., "Roof Push Up" by Takenaka Corp., and AKATUSKI by Fujita Corp.

Computer integration and production information management under the basic concept of the CIM and CIC system will be increasingly more important for future construction and shipbuilding automation. Moreover, autonomous distributed-cooperative network communication and database management technologies will be indispensable for establishing advanced control and management systems in future computer-integrated sites.

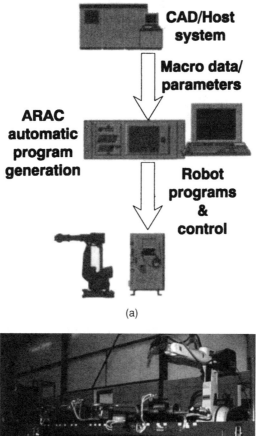

(a)

(b)

Figure 16 CIM production systems for the shipbuilding industry (courtesy Kranendonk Factory Automation BV). (a) Automatic program generation from CAD data by ARAC, Arithmetic Robot Application Control. (b) Welding of excentrical (exhaust) tube connections. Mathematical control of path synchronization of robot and external axes by the ARAC system. (c) Stud welding line for marine chains. Variable mix of different links in one chain. The work is automatically divided between eight robot stations. (d) FMS systems for welding of elevator parts. Programming of 13,500 variants within two months with ARAC macro programming technology. Integrated control of stacker crane system. (e) Dual robot profile cutting line for shipbuilding profiles. One-piece production with data from CAD system. (f) Robot welding systems for ship diesel engine frames. Programs are transferred from one robot cell to another by means of the ARAC system. New programs are made using ARAC macro technology. (g) Robot web line for welding of shipbuilding panels in one-piece production. (h) Arc welding maintenance of ship components. (i) Profile cutting line and CNC bending line marker.

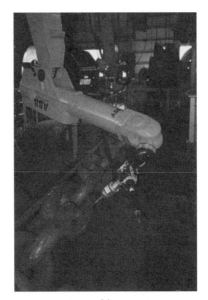

(c)

(d)

(e)

Figure 16   (Continued)

(f)

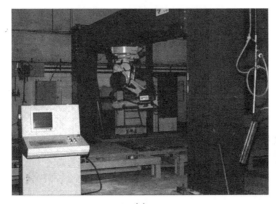

(g)

(h)

Figure 16   (Continued)

(i)

Figure 16   (Continued)

REFERENCES

Blasko, G., B. Howser, and D. Moniak. 1993. "Evaluation of the Hitachi Zosen Welding Robots for Shipbuilding." *Journal of Ship Production* **9**(1).

Bucher, H., P. Toftner, and P. Vaagenes. 1994. "Computer Integration of CAD Modelling and Robotized Shopfloor Production Systems in Shipyards." In *Proceedings of the 8th International Conference on Computer Applications in Shipbuilding,"* Bremen, Germany, September.

Hasegawa, Y. 1996a. "Current Status Key Issues for Construction Automation and Robotics in Japan." Paper presented at 13th ISARC, June.

———. 1996b. "A New Wave of Japanese Robotization." Paper presented at 4th ICARU, Singapore, December.

Katano, H., et al. 1996. "Questionnaire on Robotization in Building Work for Chartered Architects." Paper presented at 13th ISARC, June.

Maeda, J. 1994. "Development and Application of the SMART System." Paper presented at 11th ISARC, June.

Maeda, J., and Y. Miyatake. 1997. "Improvement of a Computer Integrated and Automated Construction System for High-Rise Building and Its Application for RC (Rail City) Yokohama Building." Paper presented at 14th ISARC, June.

Masao, K., et al. 1996. "Study on Development and Utilization of Construction Robots." Paper presented at 13th ISARC, June.

McGhee, S., et al. 1997. "Automatic Programming System for Shipyard Robots." *Journal of Ship Production* **13**(2).

Miyatake, Y. 1997. "The Historical Path of ISARC." 14th ISARC Keynote Speech, June.

Miyazaki, T. 1994. "Robot Application Technology for Shipyard CIMS Implementation." *Journal of Japan Welding Society* **63**(1). [In Japanese.]

Nakazima, H., and K. Miyawaki. 1993. "Welding Robot from Now on—Large and Small Robot." *Journal of Japan Welding Society* **62**(1). [In Japanese.]

Okumoto, Y. 1997. "Advanced Welding Robot System to Ship Hull Assembly." *Journal of Ship Production* **13**(2).

Reeve, R., and R. Rongo. 1996. "Shipbuilding Robotics and Economics." *Journal of Ship Production* **12**(1).

Reeve, R., R. Rongo, and P. Blomquist. 1997. "Flexible Robotics for Shipbuilding." *Journal of Ship Production* **13**(1).

Sagatun, S., and K. Kjelstad. 1996. "Robot Technology in the Shipyard Production Environment." *Journal of Ship Production* **12**(1).

Shimizu, I., et al. 1995. "Development of Welding Robot Technology for Civil Engineering and Construction." Nippon Steel Technical Report No. 65, April.

Sorenti, P. 1997. "Efficient Robotic Welding for Shipyards—Virtual Reality Simulation Holds the Key." *The Industrial Robot* **24**(4), 278–281.

Ueno, T. 1994. "A Japanese View on the Role of Automation and Robotics in Next Generation Construction." Paper presented at 11th ISARC, June.

Warszawski, A., and R. Navon. 1996. "Survey of Building Robots." Paper presented at 13th ISARC, June.

Watanabe. 1992. "Current Situation and Problems of Application Technology for 3D Off-Line Robot in Teaching System Corresponded to CIM." Technical Report, Kansai Branch of Japan Welding Society. [In Japanese.]

CHAPTER 63

PROCESS INDUSTRIES

Thomas D. Jerney
Adept Technology, Inc.
San Jose, California

1 BACKGROUND

Applications of robotics in the process industries serve a wide variety of products (Kano, 1994), including petroleum, plastics, steel and other metals (Fukushima, 1996; Kunisaki, 1994), paper and glass (Otsuka and Kubota, 1995; Jiang, Chen, and Wang, 1994), building materials, pharmaceuticals and detergents (Karathanassi, Iossifidis, and Rokos, 1996; Rovetta, Bucci, and Serio, 1988), food and beverage, and more. Within the process industries the packaging function is common and has a significant robotics market. With very high production rates on most items, automation has been widely used to improve quality and throughput and lower costs. Besides packaging, other robotics applications include load/unload and material handling (Seida and Franke, 1995), quality inspection, and laboratory operations (Berry and Giarrocco, 1994; Panussis et al., 1996; Tanaka, 1992; Courtney, Beck, and Martin, 1991). Robotics solutions for these applications are reviewed in this chapter.

Often activities in the process industries are considered by workers to be dangerous, dirty, and difficult. As a result there is a problem of labor shortage, and automation is inevitable. In a conventional automation system there is a cooperative interaction between human operators and machines. Frequently a machine is expected to perform a variety of tasks, and robotics is necessary. For instance, a robot-automated labeling system for steel rolls has been developed (Fukushima, 1996).

Virtually all "hard" packaged goods (bottles, cans, cartons, etc.) are produced on packaging machinery specially developed to enable fully automated filling, weighing, capping, sealing, labeling, cartoning, etc. Current automation can handle hundreds of items per minute, much faster than any manual process. Robots are rarely a viable contender to replace these dedicated machines; it would take many robots to equal the rates available on most packaging machines, and this alternative would be far from optimal.

Manual production and packaging functions, where "touch labor" (labor that touches the product in the process) is used, are still common in many packaging industries. Typical examples are processes with relatively low production rates, including flexible packages, pharmaceutical kits, bakery items, confectionery, and meat and poultry portions. Many of these processes have not been automated because there is no practical way to accomplish the tasks the touch laborer performs.

Robots have been used in packaging functions, replacing some touch labor tasks, since about 1980. Initial applications included creating fancy candy assortments and performing heavy lifting tasks such as palletizing full cases. Confectionery applications were popular for robotization because the handling tasks are repetitive, reprogrammability for different assortments is important, and the robot did not have to be installed in a "washdown" environment. Many of these products also yield larger profit margins, enabling their producers adequate capital investment.

The use of robots to replace touch labor in packaging has rapidly grown in the past 10 years as robot speeds have increased, sensory capabilities (e.g., vision guidance) have been introduced, robots with environmental tolerance (washdown capability) have become available, and economic justifications have improved. Applications have evolved in industries producing consumer goods, which will be described later in this chapter.

Handbook of Industrial Robotics, Second Edition, Edited by Shimon Y. Nof
ISBN 0-471-17783-0 © 1999 John Wiley & Sons, Inc.

The production of the typical consumer-disposable item includes *upstream, midstream,* and *downstream* functions. Upstream functions involve the conversion of raw materials into the product; for example, mixing ingredients, baking, coating. Midstream functions include placement of the product in a consumer package; for example, placing a chocolate truffle in an assortment tray. Downstream functions include case packing the completed cartons and palletizing the cases. Rates decrease as the products move downstream and are combined in multiples for handling. Robots are most commonly used in midstream and downstream functions.

Rates are always critical in packaging automation. A manual operator working with two hands can move in excess of 100 pieces/min (50–60 pieces/min is more typical in sustained production). Robot rates have been steadily improving, from 50 pieces/min in 1980 to over 130 pieces/min currently. Investments are most commonly based on labor savings, so the tradeoff between the robot and the manual operator is critical. A robot commonly has one arm vs. the human's two arms, which is a disadvantage. On the other hand, multiple part grippers on a single robot arm can greatly improve its throughput. It is now common for a single robot to handle as many products as two or more manual operators.

Investments in automation are rationalized on economic terms. Dedicated machines (upstream) that produce the product may have no practical manual alternative: high-speed enrobers for candy bars perform a function that humans could not duplicate. The typical robot application does have a manual alternative; indeed, most applications are replacements for existing touch labor operations. The investments in robot-based systems are then commonly compared with manual labor costs. Low-skilled handling jobs, especially in the production of food items, are usually not high-wage positions. This has impeded the rationale for investing in midstream and downstream food packaging automation.

This effect is shown in Figure 1, where labor costs in the automotive and food-related industries are compared with robot labor. As one would expect, the auto industry has used robots in much greater volume than the food industry, and began this conversion in the 1960s. The trends of increasing robot speeds (lower robot labor cost) and higher costs

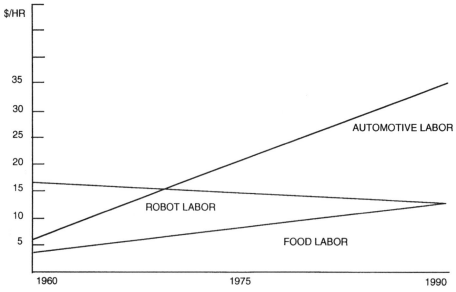

COST OF LABOR
VS.
COST OF ROBOT BASED AUTOMATION

Figure 1 Cost comparison between automotive, process, and robot labor (courtesy Adept Technology).

for food packaging workers are accelerating the crossover where robot-based automation investments can be rationalized in packaging functions.

Management commonly looks at "payback" values in deciding on investments, with the shorter payback programs getting priority. One-year payback hurdles are commonly sought. Most robot-based automation projects are not able to meet a one-year hurdle; two to five years is more common. The true costs of manual labor in handling tasks are being acknowledged as including much more than the operator's wages. Additional factors are:

- Recruitment costs, especially if the position has high turnover.
- Cost of supervision, benefits, production lost on breaks, etc.
- Costs related to injuries, and the related medical claims and lost work. Carpal tunnel syndrome (CTS) is a common affliction of workers performing highly repetitive manual tasks.
- Quality-related concerns, since the touch laborer cannot provide complete consistency in tasks performed.

When these factors are considered, the investment in automation can be more easily justified.

Technical issues have had a large effect on the evolution of robots in packaging. Early robots were very difficult to program; this increased the cost of implementation and also raised support issues with the end users. Programs for advanced applications involving line tracking or vision guidance could result in thousands of lines of code. New software packages have been developed that have imbedded, process-specific knowledge and a graphical interface; the robot can be programmed with "point-and-click" actions at the monitor.

Implementation of robot-based solutions has also been slowed by the custom nature of each project. Users familiar with dedicated machines having standardized descriptions and performance are uncomfortable with acquiring a seemingly custom solution employing a robot. The dedicated machine is easy to understand and buy; the robot-based system requires much more definition, carries more perceived risks, and is harder to acquire and implement.

The sensory capability of robot-based systems has been enhanced, especially in machine vision for guiding the robot to pick items appearing randomly on a moving belt. This capability is now commonplace (Figure 2) and is relatively easy to set up. Many robots can also interface with a moving line, so that items can be picked and placed on the fly as the robot tracks their motion.

High speed is usually accomplished with four-axis robots, either SCARA (selective compliance assembly robot arm) or Cartesian configuration. Five- and six-axis robots suffer from higher moving mass and less directly coupled drives; they have more difficulty reaching very high speeds. With most processes designed around the reach of the typical human, robot reach of 30 to 40 in. is common.

Grippers are designed to pick products at high rates without damaging them. Vacuum pick-ups are most often used. Delicate items may require gentle fingers to acquire the product by gripping on the sides. Multiple pick grippers are common; using them, the robot moves around and acquires items one at a time until a group is assembled on the gripper (Figure 3). As an alternative, products can be grouped in a staging area for simultaneous pick-up by the gripper. With multiple picks the robot throughput is dramatically increased, since the robot handles more products in each cycle.

2 CURRENT APPLICATION EXAMPLES

2.1 Bakeries

In bakeries a product typically emerges from the cooking process on a moving stainless conveyor and is manually assembled or transferred to the consumer package or wrapper in-feed. Robots in automated systems assemble and transfer cookies, doughnuts, biscuits, and bread items from the oven line to the packaging line and organize them for consumer presentation (Figure 4). The robots use machine vision to locate and inspect the individual product items. Multiple robot cells vacuum-pick baked items from the main conveyor to achieve required throughput. Machine precision is possible for the placements, although it is not usually required.

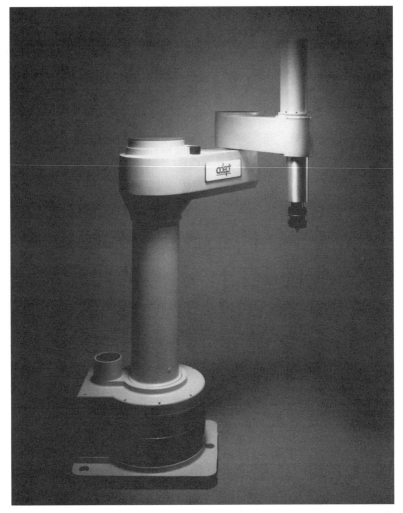

Figure 2 A four-axis, direct-drive SCARA robot is typically guided by machine vision for packaging applications (courtesy Adept Technology).

Cookie lines have been automated to lower labor costs, and they eliminate a source of CTS exposure. The automation investments are based on a two-and-a-half-year payback with experience from injury liabilities figured in. Actual rates exceed 80 cookies/min for each cell, with a typical line packing off 1000 cookies/min using up to 16 robot workcells. Gripping technology is primarily vacuum, with some mechanical grippers used for frosted items like doughnuts, muffins, and cupcakes (Figure 5).

2.2 Meat and Poultry

The meat and poultry industry has many demanding manual tasks and a high incidence of CTS injury claims. Robots have not been able to replace initial slaughter and cutting operations, but once the meat item is in consumer portions, they can perform the packaging functions. A good application is final packing ground beef patties or chicken and fish filets (see Figure 6). Here items appear randomly on a main conveyor exiting the freezer and must be packed organized in layers to a case or HFFS in-feed. The manual operation is both tedious and uncomfortable, as the packing room is cold and damp and the freezer line runs nonstop.

Robots used in meat handling may also have vacuum grippers (see Figure 7), but here an entire layer (six or more patties) may be picked before the robot moves to the case.

Figure 3 Picking multiple products with multiple fingers (courtesy SIG).

Handling rates of 70–80 pieces/min per robot are routinely achieved, exceeding the typical manual rate per operator. Several robot cells are placed downstream from the patty former and freezer to handle the volume. Machine vision locates the pieces, inspects them for size and shape, and directs the robots for picking.

Flexibility of a software-controlled system allows the producer to run multiple products and pack patterns on the same line with minimal changeover time. Automation cells have been justified on a two-year payback considering just labor cost savings. The cost equation has allowed for running the automated cells through breaks and on a longer work day when required.

Robots are also used to pack sausages from bulk to a horizontal form–fill–seal machine (HFFS) (see Figure 8). Vacuum grippers pick a four-sausage array from random positions,

Figure 4 Multiple robot arms inspect, assemble, and transfer baked goods from an oven line to a packaging line (courtesy TechniStar).

Figure 5 Mechanical fingers are used when picking delicate products at over 80 units per minute per cell (courtesy TechniStar).

alternating the curvature, and move them to the package cavity for placement. Robots can place the sausages even when the HFFS is indexing. Rates exceed 90 sausages/min per robot cell. Justification of this system was based on a three- to three-and-a-half-year payback, with actual experience resulting in repeat orders for the system integrator.

2.3 Pharmaceutical Production

Packing of pill assortments to blister packages has been automated with vision-guided robots. The user needed 100% assurance that the assortment was correct and the ability to change assortment content through software commands only. In the system shown in Figure 9 two robots are used in conjunction with guidance vision and seven cameras. Hopper and feeder systems deliver the pills to the first robot, which picks them up with a four-up vacuum gripper and places them into the blister packs. The second robot loads and unloads the packs onto a sealing machine. The vision system can detect subtle differences in shape between the pills. It is used at both stations to verify that the fill is accurate.

Robots are also used to create pharmaceutical kits and handle difficult items like IV bags. Robots are deployed in the production of disposable items such as blood test kits, sutures, needles, and tapes.

2.4 Case Packing of Bags

A common manual task is packing bags off a vertical form–fill–seal (VFFS) machine to cases. Several dedicated machines exist for automated case packing of bagged product, but they typically have bag staging, pack pattern, and case size flexibility limits that prevent covering the variety of needs of the specialty packer. A vision-guided robot, picking moving bags from their observed positions and packing them to software-controlled patterns, gives the flexibility required (Figure 10).

Machine vision inspection for orientation, labels, and print registration ensures good consumer presentation. Shingled, layered, staggered, and stacked pack patterns are available in one robotics cell, which may also handle 4–32-ounce bags. Multiple suction cups

Figure 6   Final packing of ground beef patties or chicken or fish filets (courtesy TechniStar).

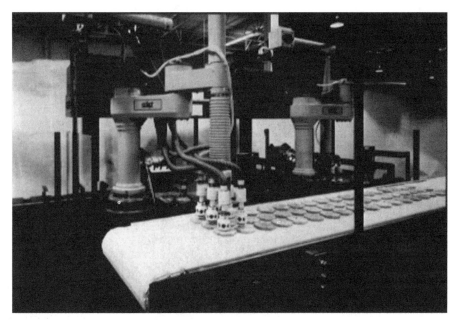

Figure 7   Robots picking meat products with multiple vacuum grippers (courtesy Dimension Industries).

Figure 8 Sausages are picked with vacuum grippers and placed on a horizontal form–fill–seal machine (courtesy Adec Robot AG).

minimize puckering on fragile bags. Pick-and-place pressure is regulated to reduce the potential for product damage, and bags are not dropped into the case.

Handling rates are slower in this application due to the larger moves involved. Sixty bags/min is commonly achieved. Here the user considered the real costs of worker injuries in justifying the investment in automation.

2.5 Airline Meals

Several tasks may be automated using robots in the preparation of airline meals. In preparation of the kits, silverware must be sorted out of the washer. Machine vision, used in conjunction with a robotics picker, can reassemble the knife–fork–spoon combinations.

Standard condiments and dishes are placed to the tray line using robots working from known feed locations. Salads, desserts, and entrees may also be placed; a "smart" gripper enables one robot to handle a wide variety of tray and dish sizes.

2.6 Candy

Robots pack fancy candies in their display cartons. A multiple robot line (see Figure 11) uses vision-guided robots to locate the individual chocolates appearing randomly on the incoming conveyor, then place them at specific locations on their tray, which moves on a continuous moving carton line. This line handles 600 pieces/min and can be changed from one assortment pack pattern to another in minutes. The customer, in Switzerland, automated this line primarily to relieve a labor shortage problem.

Wrapped candy pouches are handled by the robots shown in Figure 12. The pouches are filled at a vertical form–fill–seal machine and dropped with random spacing on a "smart" belt. The belt converts the random spacing to regular spacing so the robots can pick them in groups for transfer to the cartoner bucket in-feed. Each robot handles 90 pouches/min and performs a nonsynchronous transfer between two continuous moving belts. The buckets proceed through the horizontal cartoner, where they are placed in consumer packages. Four robots feed the cartoner, resulting in 60 cartons/min for the

Figure 9 A two-robot cell with seven-camera vision guidance system for packing variable pill assortments in blister packages (courtesy Adec Robot AG).

line. The robots replaced seven manual positions on each of three shifts, allowing a projected payback of less than two years.

2.7 Palletizing

Robots are available for palletizing cases after they are filled (Figure 13). Cases are typically presented on a conveyor single file, and need to be picked and stacked on standard 40 × 48-in. pallets. Stacking patterns are variable, based on case dimensions, and with multiple products (cases) running at one time the palletizing robot may need to serve several pallets. The work area is arranged to accommodate all of the pallets and allow forklift access for removing complete loads.

Software has been developed that enables creation of pack patterns based on optimal density and stacking strength. A 3D graphic presentation aids in visualizing the complete layers and pallet. The pattern is established off-line on a PC and then downloaded to the robot.

3 ROBOTS IN PROCESS LABORATORIES

Another typical robotics application in the process industries is associated with laboratories. Robots are used in process laboratories to automate physical operations such as data collection or task functions (Laboratory Automation Systems). It is estimated that

Figure 10 Vision-guided robotics packing of bagged products in cases can follow variable packing patterns (courtesy TechniStar).

75% of all laboratory costs are associated with paperwork and sample preparation activities. This market and its labor-intensive processing requirements for the most part have not changed in 50 years. Today, even with sophisticated and highly sensitive equipment for chemical analysis, no standard exists for hardware design, equipment interface or software protocol.

Of approximately 28,000 analytical chemistry laboratories in the United States, it is estimated that approximately 24,000 have a serious need for productivity tools. The average laboratory will spend between $20,000 and $65,000 per year for general automation hardware and software, extending over a three- to five-year period. Assuming $50,000 is spent per year, the U.S. market size is $1.2 billion. The corresponding market for the rest of the world is even larger, with 118,000 laboratories requiring automation for a market size of approximately $5.9 billion.

3.1 Clinical Lab Market

The predominant force driving the use of Laboratory Automation Systems in clinical laboratories is the urgency of making very low-cost, high-quality laboratory medical care widely available. Laboratory Automation Systems can accomplish numerous benefits, including enhancing lab productivity, maximizing routine testing, minimizing specimen misplacement and loss through automated specimen tracking, reducing turnaround time, improving documentation, and automating reflex and repeat testing. Laboratory Automation Systems can be modular, conveyor-based, or a complete fixed turnkey system.

Laboratory Automation Systems may be described as performing a conglomerate of functions, including transfer, loading, packaging, and dispensing operations. However, they are more than likely to perform a function very specific to a particular experiment or test. Specific laboratory operations performed may include weighing, pipetting, reagent dispensing, mixing, screw capping, liquid/liquid extraction, and solid phase extraction.

Laboratory Automation Systems have been used in hospital laboratories for the handling of blood samples and other specimens and in research laboratories for the development of bacterial and viral cultures. Laboratory Automation Systems are ideal for these environments because they both remove humans from possible infectious environments and relieve humans from the tedious and repetitive tasks involved with such operations. In laboratory operations repeatability, reliability, and an ability to decelerate gently are cited as important features.

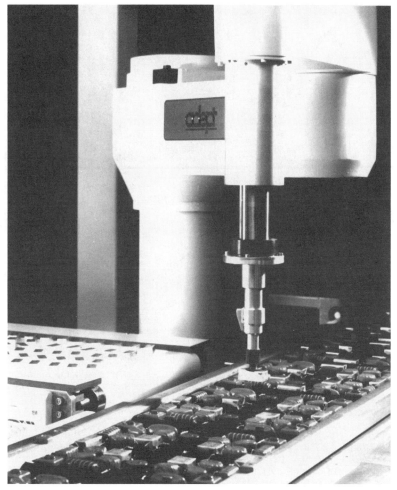

Figure 11 A multiple vision-guided robot line for candy assembly in variable packing patterns (courtesy SIG).

3.2   Pharmaceutical/Biotech Market

With the advent of competition in the pharmaceutical industry from generic drug makers, pharmaceutical companies are under intense pressure to commercialize new drugs. Consequently, pharmaceutical companies are doubling and tripling their drug screening efforts by making more compounds in order to find new drugs. Furthermore, in the biotechnology area some companies have begun to screen large libraries as a means of drug discovery. The only way life science companies can cope with the mounds of data needing to be analyzed in order to produce more "hits" in the drug screening process is to embrace automation.

In a study published by the *Financial Times* it was estimated that getting a new drug to the market may cost $271 million. A good portion of the cost is associated with discovering the right compounds that have a chance for success, but most of the cost is due to government regulation and data requirements. Clinical testing takes an average of seven years. Companies are constantly trying to improve productivity wherever they can. As an example, Zeneca used to administer drugs to human volunteers 30 months after synthesis; they now do it in less than 14 months.

In 1993 only 39 new chemical entities (NCEs) were introduced onto the market, and only 4 had the potential to become blockbusters (i.e., sales in excess of $750 million U.S.

Figure 12 Robotics handling of pouches: a conveyor belt regulates randomly spaced pouches so that they can be picked up by the robots and fed to a cartoner (courtesy Dimension Industries).

per annum). There were only 14 blockbusters between 1969 and 1990. To remain ahead of competition, pharmaceutical companies will need to increase the number of compounds synthesized in a frenzied attempt to discover elusive NCEs.

Laboratory Automation Systems are being used for a wide range of biotechnology and drug screening applications. A Laboratory Automation System has at least one robot arm for sample handling, a pipetting arm under the control of user-programmable system software installed on a personal computer, and analytical instrumentation and a high-level programming environment for programming the Laboratory Automation System. The end user loads the work surface with trays or stacks of plates or racks of tubes awaiting samples and/or reagents. The end user then programs the Laboratory Automation System to perform a series of specific tasks.

Laboratory Automation Systems are often used in conjunction with other instruments, such as plate readers, and a variety of robotics instrumentation that expands the capabilities of the workcell (see Figure 14). Pharmaceutical companies have been using Laboratory Automation Systems for some time to perform high-throughput drug screening programs, and now biotechnology companies are discovering the utility of these units for molecular biology experimentation.

In contrast to Laboratory Automation Systems used in clinical laboratories, research laboratories require an open architecture where scientists have the flexibility to customize the parameters or functions of the automated system or workstation for each type of experiment.

4 GUIDELINES FOR ROBOT APPLICATIONS

Applications of robotics in the process industries can be selected most effectively if they fit the capabilities of commercially available robots. Typical specifications for a commercial robot are:

- Robot speeds up to 80 pieces/min, single pick, single place. Higher rates may be achieved with multiple pick–place gripping.

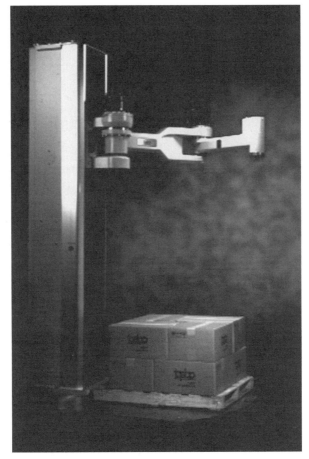

Figure 13 A case-palletizing articulated robot (courtesy Adept Technologies).

- Reach: 31 in. (similar to manual reach)
- Weight capacity: 20 lb
- Dexterity: four axes
- Versatility: all functions software-controlled and changeable; movable (portable) workstations available
- Sensory: line tracking to 15 in./sec; machine vision with recognition and inspection capability; integrated force sensing available
- Gripping: vacuum or mechanical
- Software: graphical programming interface with process knowledge imbedded
- Environment: fully capable of "washdown" and operation in a condensing environment

Cycle rates as listed above have been enabled by direct drive technology used in controlling manipulator axes and by advanced software for motion control. Direct drive allows motion transmission without gears and prevents motion degradation over time, even in a high-cycle environment. New control software optimizes robot motions and servo-parameters automatically to reach peak cycle rates.

Manipulators can be grouped to handle high volumes of products by sharing the work.

A new capability for computer-controlled robots is the ability to work in a washdown environment. The manipulator arm (see Figure 15) is sealed to prevent moisture from entering the mechanism and condensation from forming inside the arm. Materials and design details are selected to allow cleanup and sanitation and meet USDA requirements.

Figure 14 Laboratory Automation Systems are often used in conjunction with other instruments, such as plate readers, and a variety of robotics instrumentation that expands the capabilities of the workcell (courtesy CRS Robotics).

This washdown capability is required for handling meat and cheese and is desired in many bakery, catering, and candy operations.

Integrating vision with the robot gives it the ability to recognize, locate, and inspect product. The vision processor exists on the same computer bus with the robot controller and communicates directly with it for high-speed operation. Having the same programming language and calibration of the vision and robot computers makes the system easy to set up and train.

One vision system can provide data to more than one robot. Cameras are located above the incoming conveyor to detect the product location and orientation. The vision system can inspect the product for defects, determine the placement point, assemble these data, and tell the manipulator what to do on the fly. Placement can then be verified with a second camera and multiple component kits checked for presence of all items.

Typical machine vision capabilities are:

- Viewing area: 2D, with field of view as appropriate
- Number of cameras: 1–8
- Resolution: > one part in 500
- Gray levels: 128
- Recognition: 1–10 pieces/sec in full frame
- Presentation requirements: no orientation restraints, can handle touching and overlapping parts
- Inspection tools: full library, including rulers, object, point and line finders, and image manipulation to enhance areas of interest or defects

The operation of the vision system in cooperation with the robot can be transparent to the programmer and operator when it is integrated; this feature has greatly facilitated the use of vision in packaging and other process applications.

5 EMERGING TRENDS

The trend in robot design is toward a machine that is faster, cheaper, and easier to use. New robot arms are being introduced that can achieve over 120 picks/min, with a single

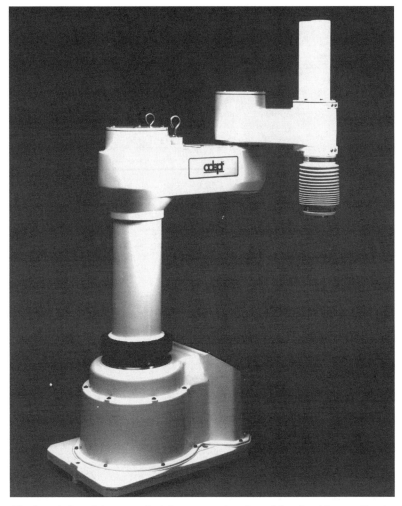

Figure 15 A sealed washdown manipulator arm can handle moist and wet items without moisture penetrating the mechanism and can be washed down to prevent contamination (courtesy Adept Technologies).

tool. Hardware prices decrease as controllers take advantage of new low-cost computing technology. System prices decrease as system integrators develop standardized solutions and new high-level software simplifies programming.

Manufacturers of dedicated machines (for example, case-packing machines) move closer to robotics capabilities in their machines as they incorporate the programmability and sensor (machine vision) features previously available only with robots. Hybrid machines, where the robot is imbedded in a more complex machine, are evolving.

As the installed base of robots grows, customers gain confidence in the viability of robotics solutions. The proliferation of computers into society improves the acceptance of computer controlled machinery such as robots. More staff has been exposed to operation and programming, so implementations carry a lower training contents. Support staff requirements are reduced for follow-on applications. Economic justification for automation improves as labor costs and injury liabilities for touch labor increase. This trend, coupled with lower robot system cost, shortens the payback time. In the long run it can be predicted that few touch labor functions will remain in the process industries for production and packaging operations.

ACKNOWLEDGMENT

Kim Paczay, CRS Robotics Corp., contributed to the section on laboratory robotics.

REFERENCES

Berry, W. F., and V. Giarrocco. 1994. "Automated Simulated Distillation Using an Articulated Laboratory Robot System." *Journal of Automatic Chemistry* **16**(6), 205–209.

Courtney, P., M. S. Beck, and W. J. Martin. 1991. "A Vision Guided Life-Science Laboratory Robot." *Measurement Science and Technology* **2**, 97–101

Fukushima, F. 1996. "Robot-Automated Labeling System." *Robot* (July), 46–52

Jiang P., H. Chen, and Y. Wang. 1994. "A Vision-Guided Curve Tracking and Input System for Glass Cutting Robot." In *Proceedings of 2nd Asian Conference on Robotics and Its Applications,* Beijing, October 13–15. Beijing: International Academic. 665–669.

Kano, Y. 1994. "Application of Robots in Food Manufacturers and Chemical Industries." *Robot* (July), 11–19.

Karathanassi, V., C. Iossifidis, and D. Rokos. 1996. "Application of Machine Vision Techniques in the Quality Control of Pharmaceutical Solutions." *Computers in Industry* **32**(2), 169–179.

Kunisaki, A. 1994. "Press-Formed Steel Sheet Palletizing System." In *Proceedings of 2nd Asian Conference on Robotics and Its Applications*, Beijing, October 13–15. Beijing: International Academic. 67–70.

Otsuka, K., and H. Kubota. 1995. "Application of Robot in the Glass Industry." *Robot* (May), 64–69.

Panussis, D. A., et al. 1996. "Automated Plaque Picking and Arraying on a Robotic System Equipped with a CCD Camera and a Sampling Device Using Intramedic Tubing." *Laboratory Robotics and Automation* **8**(4), 195–203.

Rovetta, A., M. Bucci, and O. Serio. 1988. "High Quality and High Speed Production: Integration of Vision Systems in a Detergent Manufacturing Factory." In *Proceedings of ISATA 19th International Symposium on Automotive Technology,* Monte Carlo, Monaco, October 24–28. Vol. 1. Croydon, U.K.: Allied Automotive. 269–279.

Seida, S. B., and E. A. Franke. 1995. "Unique Applications of Machine Vision for Container Inspection and Sorting." In *Proceedings of the Food Processing Automation IV,* Chicago, November 3–5. St. Joseph: ASAE. 102–108.

Tanaka, H. 1992. "An Instance of Using Laboratory Automation (LA) Robot for the Pharmaceutical Analysis." *Robot* (May), 55–59.

CHAPTER **64**

SERVICES

Gay Engelberger
HelpMate Robotics Inc.
Danbury, Connecticut

FOREWORD

A Handbook of *Industrial* Robotics must at least recognize that the factory floor is not the sole domain for robotics, albeit the $6 billion-dollar robotics market (as of 1997) is almost exclusively industrial. Human employment in manufacturing pales before the number of us providing services. Many of the roles for people in service are just as mind-numbing, repetitive, unrewarding, and even dangerous as the jobs that industrial robots are now claiming. Joseph F. Engelberger, founder of the industrial robot industry and now Chairman of HelpMate Robotics Inc., provides the following definition of a service robot: "A Service Robot utilizes any or all of the robotic capabilities of mobility, sensory perception, manipulation and artificial intelligence to perform tasks of a service nature, as differentiated from manufacturing, the domain of industrial robots."

The potential market for *Robotics in Service* (the title of J. F. Engelberger's 1989 book) promises to be much larger than that for manufacturing. However, and it's a big "however," service jobs do demand more capabilities than the industrial jobs that first succumbed to robotics. A blind industrial robot, fixed to one station and programmed in record–playback mode, is still adequate for the majority of factory tasks now being done by robots.

Service robots are likely to require mobility, sensory perception (both visual and tactile), sensory fusion algorithms, closed-loop control based upon interaction with the environment, a modicum of artificial intelligence, and user-friendliness unheard of in manufacturing. Moreover, that friendliness in close juxtaposition to fellow human workers implies a higher order of safety standards. Most service robots will not abide being caged in. Table 1 shows the fundamental distinctions between manufacturing robots and service robots.

Happily for industrial robotics, the attributes developed for service jobs can be continually migrated to manufacturing to permit robotizing ever more sophisticated jobs. Figure 1 in Chapter 1 is a current listing of a roboticist's "toolchest."

We can now consider some of the service tasks that are being addressed by robotics researchers. In some cases commercial success can already be reported.

1 HOSPITAL MATERIAL TRANSPORT

1.1 HelpMate Robotics

HelpMate Robotics Inc., located in Danbury, Connecticut, founded in 1984 and formerly known as Transitions Research Corp., manufactures HelpMate, a trackless robotic courier designed to perform material transport tasks throughout a hospital environment.

The HelpMate robotic courier is a flexible automation system designed specifically for hospital transport of medications, lab samples, supplies, meals, medical records, and equipment between departments and nursing stations. It currently performs these tasks on a round-the-clock basis in over 80 hospitals, most often in support of programs to improve service even while staff levels are being reduced. A front view of HelpMate is shown in Figure 1 and a back view is shown in Figure 2.

Handbook of Industrial Robotics, Second Edition, Edited by Shimon Y. Nof
ISBN 0-471-17783-0 © 1999 John Wiley & Sons, Inc.

Table 1   Fundamental Differences Between Manufacturing and Service Robots

Service Robots	Manufacturing Robots
Usually mobile	Usually fixed
Unstructured workplaces	Highly structured workplaces
Safety inherent in robot	Safety external to robot
Sensory feedback control	Articulation control
Speed not critical	Speed essential

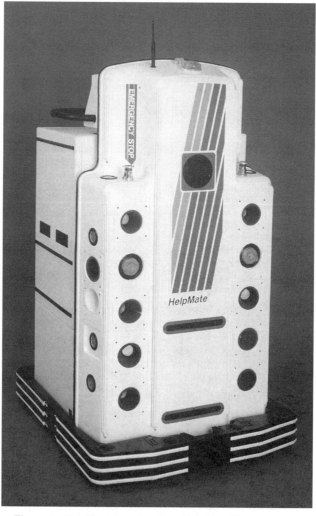

Figure 1   HelpMate front (courtesy HelpMate Robotics Inc.).

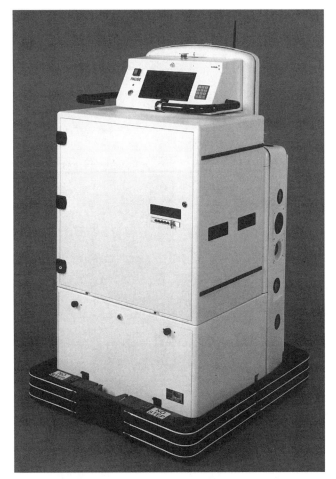

Figure 2   HelpMate back (courtesy HelpMate Robotics Inc.).

HelpMate navigates by following a map of the hospital corridors that has been programmed into its on-board memory along with the locations of desired stops. HelpMate
uses on-board sensors, such as infrared and ultrasound, to identify objects (and people)
in its way and to find clear paths around these obstacles. HelpMate travels from floor to
floor by use of one or more of the existing service elevators, communicating via spread
spectrum radio frequency

HelpMate comes equipped with a standard backpack configuration that provides six
cubic feet of cargo space and can handle up to 200 pounds of payload. HelpMate is
powered by batteries that provide over 12 hours of continuous operation.

2   SECURITY AND SURVEILLANCE

2.1   Cybermotion

Cybermotion, Inc., of Salem, Virginia, has been developing autonomous mobile robots
for indoor security applications since 1984.

Their SR-2 and SR-3 families of Security Robots (Figure 3) perform routine patrol of
offices, warehouses, and commercial space. They are equipped with a wide range of
sensors, all integrated into the operation of the vehicle. These include sensors for smoke,
gas, temperature, humidity, ambient lighting, flame, body heat, radar, inventory tagging,
puddles, etc. The vehicles provide threat assessment for air quality, fire, and intrusion.
They are used for prevention, investigation, and management tasks.

Figure 3 Security and surveillance robot (courtesy Cybermotion, Inc.).

Cybermotion uses a layered programming and control architecture designed to allow its robots maximum flexibility. At the bottom of this architecture, the robot uses dead reckoning, uncertainty modeling, fuzzy logic, state moderators, learning, and multiple navigation "agents." The vehicle interprets a control program that is transmitted to it and invokes the various navigation agents as it moves about. Most standard navigation agents are based on sonar, but Lidar and other agents can be used, depending upon the environment. Multiple agents can be active at the same time, and a fuzzy logic-based arbitrator determines which corrections are most likely to be correct and applies them in whole or in part.

Cybermotion has been working for the Department of Defense to develop an enhanced version of the SR Robot series, called *MDARS* (Mobile Detection Assessment and Response System). MDARS is being developed under the management of the U.S. Army PSEMO (Physical Security Equipment Management Office) and the technical direction of NRaD (Naval Research and Development). The vehicle will be used to patrol military warehouses and perform routine electronic inventory control via a short-range R.F. tag reader system. It is equipped with digital video processing and transmission and an automatic IR Illuminator, as well as a very high-speed pan-and-tilt system with continuous rotation capability. Developments under this program are being made commercially available as well.

3   FLOOR CLEANING

3.1   The Kent Company

The Kent Company of Elkhart, Indiana, has developed RoboKent (Figure 4), a fully robotic, self-teaching, self-operating floor cleaning robot. It comes in two models, a fully automated wet scrubber-vac used on hard floors and a sweeper-vac useful on all floor surfaces.

Both machines are successfully cleaning hallways and smaller conference and meeting rooms in hospitals, schools and colleges, and industrial locations. Also, both machines share important technology, utilizing a self-teaching system rather than manual programming. RoboKent uses a combination of sonar, infrared, and wheel revolution counting for determining the area to be cleaned and the machines' location and for obstacle avoidance. Manual programming, bar codes, and external reference targets are not required in order to use either RoboKent.

3.2   Cybermotion

Cybermotion has developed a fully autonomous vacuuming robot (Figure 5) under exclusive contract with Cyberclean Technologies of Richmond, Virginia. This vacuuming robot performs intricate cleaning maneuvers, including the operation of elevators, and then returns to its charging station automatically. The robot has a high-power vacuuming deck with a beater bar and elevator mechanism attached to the rear of its turret.

The system vacuums while adjusting the height of the deck to provide a groomed appearance to carpets. The robot carries enough batteries to provide continuous operation from between six and eight hours, depending on the floor surface.

Figure 4   Floor cleaning robots (courtesy The Kent Company).

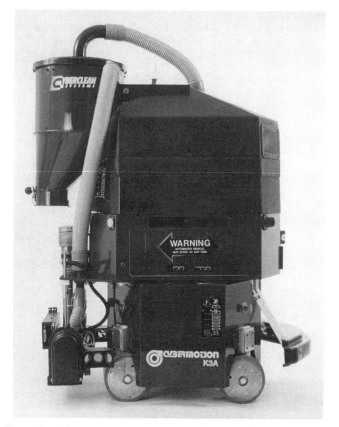

Figure 5   Autonomous vacuuming robot (courtesy Cybermotion, Inc.).

4   HAZARDOUS ENVIRONMENT APPLICATIONS

4.1   Cybermotion

Cybermotion robots have been used as the platforms for many inspection systems in the nuclear field. These applications include systems to detect contamination of floor surfaces, and visual inspection systems. Their most advanced system is ARIES (Autonomous Robotic Inspection Experimental System) (Figure 6). The ARIES project is a cooperative research and development program with Clemson University, the University of South Carolina, and Cybermotion, Inc. This platform provides an automated method for visually characterizing mixed waste storage drums which allows personnel to perform their jobs with reduced exposure to potentially dangerous environments. Based upon this analysis the integrity of the drum is determined. Drums can be marked for repackaging if considered suspect based upon site operating parameters.

4.2   RedZone Robotics

RedZone Robotics, Inc. of Pittsburgh, Pennsylvania, develops, manufactures, and integrates specialized robotic systems, inspection and control systems, and intelligent vehicle technologies for work in hazardous environments.

RedZone has developed Houdini (Figure 7), a work platform for tank waste retrieval applications. Houdini is tethered, hydraulically powered, and track-driven, with a folding frame chassis that allows it to fit through confined accessways as small as 22.5 inches in diameter. Modeled after bulldozers and backhoes from the construction industry, and hardened to work in nuclear and other hazards, Houdini provides unique work capabilities in a flexible configuration that facilitates in-tank deployment.

Figure 6 Robotic inspection in hazardous areas (courtesy Cybermotion, Inc.).

RedZone has also developed Rosie (Figure 8), a next-generation, construction-grade worksystem that enables reliable decommissioning and dismantling operations in hazardous environments. Rosie is a fully decontaminable, modular system that allows multiple configurations and tooling options to best fit the application.

4.3 Remotec

Remotec, Inc., founded in 1983, located in Oak Ridge, Tennessee, and a subsidiary of Northrop Grumman, has developed a family of remote mobile robots for dealing with the most hazardous environments. They are commonly used for explosives handling, nuclear surveillance and maintenance, HazMat response and SWAT operations.

The ANDROS MARK V-A (Figure 9), is the largest, strongest robot in the Remotec family. With articulated tracks, the MARK V-A can maneuver over rough terrain and obstacles, climb stairs, and cross ditches as wide as 24 inches. It is environmentally sealed to operate in any weather condition and in areas of high temperature/humidity. It can operate on any surface, including sand, mud, gravel, and grass. The ANDROS MARK

Figure 7   Waste retrieval by Houdini (courtesy RedZone Robotics, Inc.).

Figure 8   Robotics for decommissioning and dismantling installations in hazardous environments (courtesy RedZone Robotics, Inc.).

Figure 9 A mobile robot developed for highly hazardous environments (courtesy Remotec, Inc.).

V-A is equipped with two television cameras for remote viewing and a dexterous manipulator for hazardous tasks.

5 PHARMACY

5.1 Automated Healthcare

Automated Healthcare, Inc., a wholly owned subsidiary of McKesson Corp., is located in Pittsburgh, Pennsylvania, and manufactures and markets pharmacy automation systems. The RxBOT (Figure 10) is a centralized drug distribution system that automates the storage retrieval and dispensing of unit dose bar-coded patient medications and complements standalone decentralized drug dispensing systems. A robotic arm operates on vertical and horizontal rails and is programmed to retrieve medications and deposit them

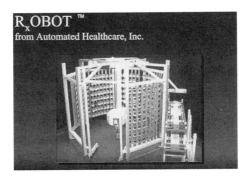

Figure 10 A robotic system for dispensing and distribution of drugs and medications (courtesy Automated Healthcare, Inc.).

into patient-specific cassettes. The system uses bar codes to verify, retrieve, and track medications from drug wholesaler to patient.

5.2   Automated Prescription Services

Autoscript II (Figure 11), developed by HelpMate Robotics, Inc. and marketed by Automated Prescription Services (APS), is a material-handling system that automatically fills prescriptions in high-volume mail order pharmacy operations. Autoscript II works with a pill-dispensing system, called *Autoscript,* already developed by APS. A technician uses a barcoder to transfer prescriptions into a computer, which then communicates the information to Autoscript II. Autoscript II then travels through the numerous pill storage aisles until it reaches the location of the matching bar code and communicates the prescription to Autoscript, which automatically dispenses the pills into a container held by Autoscript II. Autoscript II is capable of handling numerous prescriptions before returning to the technician. In an eight-hour shift the robot fills 2,500 prescriptions.

6   SURGERY

Additional information on medical robotics can be found in Chapter 65.

6.1   Computer Motion

Computer Motion of Goleta, California, has developed AESOP (Automated Endoscopic System for Optimal Positioning), a robotic arm that holds and moves the laparoscope for the surgeon during minimally invasive procedures (Figure 12), eliminating the need for a human scope-holding assistant. The surgeon controls the position of the scope with a foot pedal, hand control, or voice. There is no jiggling of the scope from hand movement. Movements of the laparoscope are smooth, and the video image remains steady throughout the procedure. A programmable memory allows the robot to remember operative sites and return to them. AESOP received clearance to be marketed for the U.S. Food and Drug Administration in December 1993.

6.2   Integrated Surgical Systems

Integrated Surgical Systems, Inc. (ISS), Sacramento, California, developed ROBODOC (Figure 13) to perform a critical function in total hip replacement surgery where the ball-and-socket joint and a portion of the femur are replaced by a prosthetic device consisting of an acetablar cup and a femoral stem. The surgical robot accurately machines a cavity in the femur for the femoral stem component.

ROBODOC consists of a Sankyo Seiki custom five-axis surgical robot mounted on a mobile base and equipped with a force sensor, bone motion monitor, custom surgical tools, and an operating room monitor that displays informational messages and surgical

Figure 11   High-volume robotic prescription filler (courtesy Automated Prescription Services).

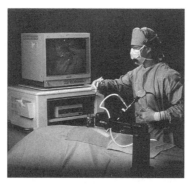

Figure 12   Robotic laparoscope as a surgeon aid (courtesy Computer Motion, Inc.).

simulation. The front-end to the system is the ORTHODOC Preoperative Planning Workstation, featuring an IBM RS-6000 Workstation and related equipment.

7   ENTERTAINMENT

7.1   Honeybee Robotics

Honeybee Robotics, a New York City-based automation company, has developed RoboTender (Figure 14), an automated, interactive bartending system that utilizes a programmable industrial robot to perform a wide variety of bartending and entertainment tasks and activities.

The configuration of the RobotTender system can be customized to a particular setting in terms of number and type of stations (i.e., how many different types and brands of liquor, beer, wine, etc.), lighting, voice interaction, etc. The RoboTender system is designed to work with another bartender or a waitress in what would normally be a multibartender setting. The attending bartender or waitress enters drink orders and delivers drinks from the RoboTender to the customer. Drinks can be ordered manually (via a computer keyboard or touchscreen) or verbally (via voice interactive software).

Once the drink order is entered, the RoboTender takes a plastic glass from the automated cup dispenser and proceeds to the first station. The circular work area includes

Figure 13   Robotic support for hip replacement (courtesy Integrated Surgical Systems, Inc.).

Figure 14 Robotic bartender (courtesy Honeybee Robotics, Inc.).

automatic dispensing stations for ice, soft drinks and mixers, wine, beer, and liquor. The ice and liquids from these stations go directly from the dispenser into the glass. Once the drink has been prepared, the RoboTender brings the glass to a delivery station, where it is collected by the bartender/server and then served to the customer. The RoboTender can be programmed to mix over 100 different drinks and has the capability of learning many more selections.

REFERENCES

Engelberger, J. F. 1989. *Robotics in Service.* London: Kogan Page.

ADDITIONAL READING

Everett, H. R. *Sensors for Mobile Robots: Theory and Application.* Wellesley: A. K. Peters, 1995.
Miller, R. K., *The Service Robot Market.* International Service Robot Assn, 1995.
Service Robot: An International Journal. Bradford, West Yorkshire: MCB Press.

CHAPTER 65

MEDICAL ROBOTICS AND COMPUTER-INTEGRATED SURGERY

Russell H. Taylor
The Johns Hopkins University
Baltimore, Maryland

1 INTRODUCTION

As in many other parts of our society, robots are increasingly used in healthcare. Applications include lab automation, hospital logistics support (Engelberger, 1989*b*, 1989*c*), rehabilitation (Engelberger, 1989*a*), and surgery. Many of these uses (e.g., lab automation, logistics) are remarkably similar to comparable applications in nonhealth areas, and we will exclude them from further discussion here. Surgical applications are, however, unique in both their potential to change medical practice and their technical challenges. In the subsequent discussion we consider *medical robots* to be programmable manipulation systems used in the execution of interventional medical procedures, principally surgery.

In the next 20 years medical robots and related computer-integrated systems will influence surgery and interventional medicine as profoundly as industrial robots and computer-integrated manufacturing have influenced industrial production over the past 20 years. These systems represent a novel partnership between human surgeons and machines, exploiting the complementary strengths of each (see Table 1) to overcome human limitations in traditional surgical practice. Their introduction will be largely driven by their potential to provide *better* and more *cost-effective* care than can be provided by an unaided physician.

In the discussion that follows we will first address the limitations of current practice and very briefly discuss the issue of cost-effectiveness, which is often raised in connection with advanced technology in medicine. Second, we will discuss the role played by robotic systems within the broader context of computer-integrated surgery. This context is vital to proper consideration of robotic surgery. The introduction of robots into surgery will change surgical processes just as profoundly as robots changed manufacturing processes. Just as with manufacturing, it is crucial to consider the entire *system* or workstation, rather than the single tool. Similarly, it is crucial to focus on surgical robot*ics*—i.e., the integration of programmable devices, sensors, and interfaces into surgery—rather than simply surgical robot*s*, in the sense of arm-like manipulators wielding surgical instruments. Finally, we will provide a brief discussion of several critical issues related to the design of medical robotic systems.

1.1 The Limitations of Traditional Surgery

For centuries surgery has been practiced in essentially the same way. The surgeon formulates a general diagnosis and surgical plan, makes an incision to get access to the target anatomy, performs the procedure using hand tools with visual or tactile feedback, and closes the opening. Modern anesthesia, sterility practices, and antibiotics have made this approach spectacularly successful. But human limitations have brought this classical approach to a point of diminishing returns:

- *Planning and feedback:* Surgeons cannot see through tissue, nor can they always see the difference between diseased and normal tissue. Advances in medical im-

Handbook of Industrial Robotics, Second Edition, Edited by Shimon Y. Nof
ISBN 0-471-17783-0 © 1999 John Wiley & Sons, Inc.

Table 1 Complementary Strengths of Human Surgeons and Robots

	Strengths	Limitations
Humans	• Excellent judgment • Excellent hand–eye coordination • Excellent dexterity (at natural "human" scale) • Able to integrate and act on multiple information sources • Easily instructable • Versatile and able to improvise	• Prone to fatigue and inattention • Tremor limits fine motion • Limited manipulation ability and dexterity outside natural scale • Bulky end-effectors (hands) • Limited geometric accuracy • Hard to keep sterile • Affected by radiation, infection
Robots	• Excellent geometric accuracy • Untiring and stable • Immune to ionizing radiation • Can be designed to operate at many different scales of motion and payload • Able to integrate multiple sources of numerical and sensor data	• Poor judgment • Difficulty adapting to new situations • Limited dexterity • Limited hand–eye coordination • Limited ability to integrate and interpret complex information

aging (X-rays, CT, MRI, ultrasound, etc.) provide much richer information for diagnosis and planning and to a more limited extent for intraoperative feedback. But it is still hard to couple such image information to the surgeon's natural hand–eye coordination.

- *Precision:* Many anatomical structures (e.g., retina, small nerves, vascular structures) are so small that natural hand tremor makes repairing them tedious or impossible. Delicate and vital structures can be destroyed by a small slip (e.g., in spine and brain surgery) or a misguided instrument. Even in less demanding tasks, such as tissue retraction, excessive force can cause severe damage.

- *Access and dexterity:* The dissection required to gain access to the target is often far more traumatic than the actual repair. Although "minimally invasive" methods such as endoscopic surgery and intravascular catheterization promote faster healing and shorter hospital stays, they severely limit the surgeon's dexterity, visual feedback, and manipulation precision. As in everyday life, many surgical tasks are done better with three or four hands, further complicating access.

- *Accuracy:* Even with perfect feedback and access, humans are very poor at carrying out tasks requiring geometric accuracy, such as precisely machining a bone to receive an orthopaedic implant or injecting a pattern of radiation seeds into a tumor. Existing "stereotactic" methods are cumbersome and do not work well in many parts of the body, especially in mobile soft tissue.

- *Specialization and access to care:* As surgery becomes ever more specialized, it becomes very difficult to make expert knowledge available for difficult cases performed outside of a few major medical centers.

1.2 The Challenge of Cost-Effective Care

Society is faced increasingly with hard choices between the cost of surgical treatments and the cost of leaving conditions untreated. Estimated expenditures for surgery and related costs exceed $150 billion each year in the United States. There is a widespread perception that advances in technology are the prime factor driving increased costs. Despite the undoubted capital costs associated with new technology, our thesis is that *this perception is wrong*. Advances in technology can *reduce* both the direct costs of treatment and the indirect costs to society of untreated disease and deformity by:

1. Reducing the morbidity and hospitalization associated with invasive procedures
2. Reducing surgical complications and errors
3. Improving consistency

4. Making possible cure or amelioration of hitherto untreatable conditions at an affordable price

2   COMPUTER-INTEGRATED SURGICAL SYSTEMS

Figure 1 illustrates the overall structure of computer-integrated surgery (CIS). CIS systems exploit a variety of technologies (robots, sensors, human–machine interfaces) to connect the "virtual reality" of computer models of the patient to the "actual reality" of the operating room. In the discussion that follows we provide one taxonomy for considering ways to exploit this synergy:

1. CASP/CASE, or surgical CAD/CAM
2. Surgical augmentation systems
3. Surgical simulation systems
4. Surgical assistant systems

For each of these areas we will provide a brief overview, scribe the relevance of "robotic" systems, and briefly mention current research challenges and opportunities.

2.1   "CASP/CASE" Systems

These systems are analogous to CAD/CAM systems for manufacturing. They integrate computer-assisted surgical planning (CASP) with robots or other computer-assisted surgical execution (CASE) systems to accurately execute optimized patient-specific treatment plans. We define two dominant modes for such systems: *stereotactic* CASP/CASE and *model interactive* CASP/CASE.

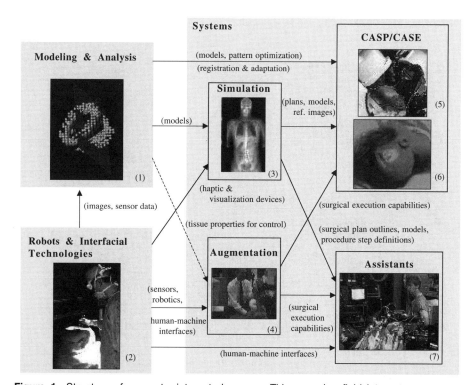

Figure 1   Structure of computer-integrated surgery. This emerging field integrates computer-based modeling and analysis of patient anatomy with robots, sensors, and other interfacial technologies into systems that significantly enhance the ability of human surgeons to plan and carry out surgical interventions. (Photo credits: 1. Elliot McVeigh, Johns Hopkins University; 2. Mike Blackwell, Carnegie-Mellon University; 3. James Anderson, Johns Hopkins University; 4, 5, 7. Russell Taylor, Johns Hopkins University; 6. Eric Grimson, Massachusetts Institute of Technology.)

Stereotactic CASP/CASE exploits the positional accuracy of machines to position surgical tools in space, in much the same way that numerically controlled machine tools are used in parts fabrication. Current examples include the widely cited ROBODOC system for hip replacement surgery (Mittelstadt et al., 1996; Taylor et al., 1994; Bauer, Borner, and Lahmer, 1997) (see Figure 2), robots for placing instruments into the brain (Kwoh et al., 1988; Lavallee et al., 1996b; Drake et al., 1991a, 1991b, 1994) (see Figure 3), and numerous radiation therapy systems (e.g., Adler et al., 1994; Troccaz et al., 1994; Schlegel, 1996; Schreiner et al., 1997). Model-interactive CASP/CASE systems, on the other hand, rely on the surgeon to control the motions of the surgical instruments, which may either be conventional hand tools or sophisticated robotic manipulation devices. The computer uses sophisticated visualizations, *augmented reality,* or other techniques to provide the surgeon with real-time feedback in executing the planned procedure. In the simplest variation, also called *surgical navigation* (e.g., Zinreich et al., 1993; Kosugi et al., 1988; Lavallee et al., 1994; Nolte et al., 1994; Maciunas, 1993; Maciunas et al., 1993), surgical instruments are tracked in 3D and preoperative CT or MRI images are "registered" to the patient by locating fiducial markers or, more recently, by various simple rigid-body 2D-to-3D and 3D-to-3D registration schemes (e.g., Lavallee, 1996). A simple display shows the position of the instruments superimposed on the preoperative images. With variations, these systems are now widely used in brain surgery and are beginning to be used in some orthopaedics and spine procedures (e.g., Lavallee et al., 1994; Nolte et al., 1994; Lavallee et al., 1996a; Simon et al., 1997). More sophisticated variations have been developed to combine simple manipulation aids and 3D graphics with navigation and optimized surgical plans in such areas as craniofacial surgery (e.g., Cutting et al., 1991, 1992, 1995; Cutting, Bookstein, and Taylor, 1996; Taylor et al., 1991), orthopaedics (e.g., Simon et al., 1997), and neurosurgery (Kall, Kelly, and Goerss, 1985; Ettinger et al., 1997).

The advantages of CASP/CASE are comparable to those of manufacturing CAD/CAM:

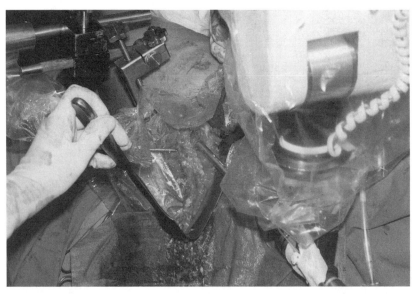

Figure 2 Surgical CAD/CAM. This photograph shows the ROBODOC system machining a socket into a patient's femur to receive a cementless total hip replacement prosthesis. The choice of implant and desired position in the femur is determined before surgery by the surgeon, based on CT images. In the operating room surgery proceeds normally until it is time to make the cavity to receive the implant. A combination of force-compliant guiding and autonomous tactile search by the robot is used to locate the femur relative to the robot's coordinate system, and the robot machines the cavity with roughly an order-of-magnitude better accuracy than is achievable free-hand. The surgeon then inserts the prosthesis manually and completes the procedure in the normal fashion. (Photo courtesy Integrated Surgical Systems, Inc. and B. G. Unfallklinik.)

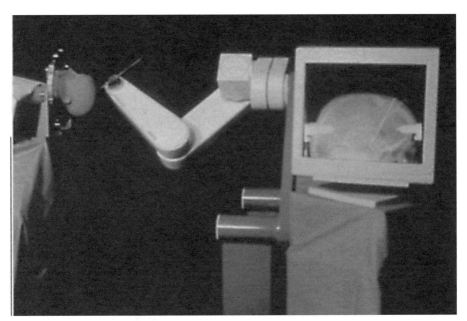

Figure 3 Surgical CAD/CAM—Stereotactic Brain Surgery. In this case the robot is accurately placing needles into the patient's brain at a target points determined before surgery from CT or MRI images. One advantage provided by the robot is greater speed and convenience compared to conventional mechanical stereotactic frames. (Photo courtesy Integrated Surgical Systems, Inc.)

- *Improved quality:* Greater accuracy and better control results in lower surgical complication rates and faster healing times. For example, robotic machining of implant cavities has essentially a zero fracture rate, compared to fracture rates of about 3% for manual hip surgery (18% for revision cases) (Schurman and Maloney, 1992). Further, robot-prepared implant cavities fit approximately 10 times better than manual cavities ($\cong 0.1$ mm vs. ≥ 1 mm), and recent controlled studies on laboratory animals (Bauer, Borner, and Lahmer, 1997) have shown that this improved fit and placement make a significant difference in histologically assessed healing and gait recovery.

- *Consistent execution:* The ability to execute a procedure the same way each time, and exactly as planned, promotes learning and process improvement. This is especially significant when combined with the inherent ability of computer-based systems to *record* relevant data for later analysis and analysis of surgical outcomes.

- *Design optimization:* The ability to execute a procedure as planned makes plan optimization worthwhile. This leverage was part of the original rationale for ROBODOC, and has been noted in other CASP/CASE areas, including craniofacial surgery (Cutting, Bookstein, and Taylor, 1996), orthopaedics (e.g., DiGioia, Jaramaz, and O'Toole, 1994), and in radiosurgery and brachytherapy (e.g., Adler et al., 1994; Vaillant et al., 1997).

- *New capabilities:* Stereotactic CASP/CASE interventions are often much less invasive than traditional surgery. This not only promotes faster recovery and saves costs, but also can enable treatment for otherwise inoperable conditions. For example, at most about 25–40% of liver cancer is operable by conventional surgery, while perhaps 60% of patients might be candidates for localized pattern therapy (e.g., implanted seed or drug injections) delivered by a robot.

- *Lower costs:* Improved quality translates directly into reduced cost. For example, simply eliminating femoral fractures could save in excess of $60M for the roughly 150,000 primary and 25,000 revision hip replacements in the United States each year. Similarly, CASP/CASE injection therapy for liver cancer might cost $7,500 vs. $22000 for surgical resection and could help more people.

Both CASP/CASE variations crucially depend upon the use of patient-specific anatomical models derived from medical images and on the ability of computer-based systems to use real-time sensing in the operating room to "register" plans and patient models to the actual patient. Today this is only achieved with bones (reasonably well), soft tissues like the brain that are constrained by bone (not as well), or targets for which a large safety margin is provided, such as is sometimes practiced in radiosurgery. Current research topics in this field include:

- Accurate and efficient methods for registering preoperative images and models to real-time images, especially for deformable soft tissue structures
- Methods for using this information to update preoperative plans based on new information
- Compact, adaptable robotic systems for therapy delivery that are usable with a variety of imaging modalities, including real-time fluoroscopy, ultrasound, CT, and MRI imaging
- Real-time imaging sensors and display devices for OR use
- Interactive optimization methods for surgical plans.

2.2 Surgical Augmentation Systems

These systems fundamentally extend human sensory-motor abilities to overcome many of the limitations of traditional surgery. They are essential building blocks both for CASP/CASE and more sophisticated surgical assistant systems. However, they can also be used directly by human surgeons in otherwise conventional surgical settings.

There is current research on telerobotic systems to augment human manipulation abilities in such areas as endoscopic surgery (e.g., Sackier and Wang, 1996; Neisius, Dautzenberg, and Trapp, 1994; Paggetti et al., 1997; Petelin, 1994; Taylor et al., 1995; Wang, 1994; Goradia, Taylor, and Auer, 1997; Cohn et al., 1994; Sastry, Cohn, and Tendick, 1997) and microsurgery (e.g., Schenck et al., 1995; Mitsuishi et al., 1995; Salcudean, Ku, and Bell, 1997). Although the motion scale involved is different, these systems present many of the same challenges and opportunities. Much microsurgery augmentation research has stressed motion and force scaling and the elimination of hand tremor. Limited dexterity robotic manipulators for eye surgery have been developed with reported motion precision ranging from 10 to 100 μm. However, significant advances are still needed in mechanisms, sensing, actuation, control, haptic interfaces, and ergonomic design for such devices to become practical and cost-effective microsurgical systems. Generally these systems must be made much smaller and less intrusive into the surgical field. Distal dexterity and force sensitivity must be greatly improved without reliance on unsafe actuators. Alternatives to conventional master–slave force-reflecting telemanipulation need to be developed in order to reduce system costs and to permit better integration into surgical environments. One approach combines a single manipulator with hands-on force-compliant guidance by the surgeon, integrated with higher-level interfaces to CASP/CASE systems for navigational guidance, enforcement of safety constraints, etc. This approach has been explored by Davies et al. (Troccaz, Peshkin, and Davies, 1997) for knee replacement surgery. A number of extensions are being pursued at Johns Hopkins University for neuroendoscopy (Goradia, Taylor, and Auer, 1997) (see Figure 4), microsurgery (Kumar et al., 1997), and spine applications.

Several groups (e.g., Mitsuishi et al., 1995; Satava, 1992; Green et al., 1992; Kavoussi et al., 1994) are addressing the goal of permitting expert surgeons to participate in surgical procedures without being physically present at the operating table. Such systems could improve health care delivery to remote areas or in special circumstances such as military conflicts or civil disasters. Even within a university hospital they can significantly improve surgeon productivity and improve "mentoring" of residents and fellows. Experience at Johns Hopkins University (Kavoussi et al., 1994) has shown that the effectiveness of such mentoring is greatly enhanced if the surgeon mentor actively assists in the procedure, and there is considerable interest in extending the capabilities provided for doing this. These systems need many of the same advances in robotics and human–machine communication required by other surgical augmentation systems. Although there is a large literature in remote teleoperation for space and undersea applications, coping with communication delays and bandwidth limitations will require significant advances in improved tissue property modeling and surgical simulation capabilities, as well as greatly improved systems for insuring safety. Advances in these areas are synergistic with the advances needed for CASP/CASE and surgical assistant systems.

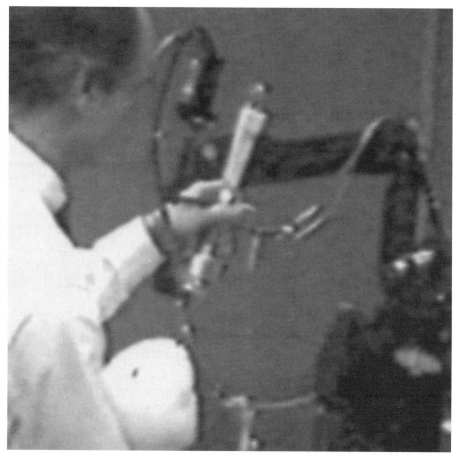

Figure 4 Surgical augmentation: experimental steady-hand manipulation system for neuroendoscopy being developed at Johns Hopkins University using the IBM/JHU LARS manipulator. The robot has a special mechanical structure that decouples translational and rotational motions at a remote motion center, here located at the point where the endoscope enters the skull. (Photo courtesy R. H. Taylor, Johns Hopkins University.)

2.3   Simulation Systems

Although they are currently being developed for training (e.g., Satava, 1993, 1995, 1996; Anderson et al., 1996; Dawson, Kaufman, and Meglin, 1996; Delp et al., 1996; Lasko et al., 1995; Higgins et al., 1990; Preminger et al., 1996; Bauer et al., 1995), surgical simulators are a vital part of the CASP/CASE paradigm. There has been some work on patient-specific simulation for pretreatment planning in areas such as radiation therapy (Schlegel, 1996; Morantz and Wara, 1995), orthopaedics (Raibert, 1997; Funda et al., 1994), brain surgery (Dow, Nowinski, and Serra, 1996; S. Fang et al., 1995a; A. Fang et al., 1995b; Goradia et al., 1997), and intravascular procedures (Dawson, Kaufman, and Meglin, 1996; Chui et al., 1996), typically using models based on medical images, with additional tissue properties inferred from anatomical atlases or standard models. Better methods are needed for constructing such models and combining them with careful experimental studies to predict the effects of surgical instruments and manipulations, for validating the predictions of simulators against reality, and for "scoring" simulations against treatment goals and constraints.

Robots and robotic devices are beginning to be used as part of simulators to provide realistic haptic feedback to surgeons (e.g., Delp et al., 1996; Brett et al., 1995; Raibert, 1997). In these applications the robot design requirements are similar to those for the "master" in force-reflecting master–slave telerobotic systems. The main difference is that the surgeon interacts with a numerical simulation of a physical system (e.g., tissue compliances) rather than with an actual "slave" system.

2.4 Surgical Assistant Systems

These systems work *cooperatively* with a surgeon to automate many of the tasks performed by surgical assistants. Existing systems are limited to very simple tasks such as laparoscopic camera aiming (Sackier and Wang, 1996; Neisius, Dautzenberg, and Trapp, 1994; Taylor et al., 1995; Funda et al., 1994; Hurteau et al., 1994), limb positioning (McEwen, 1993; "Andronics Orthopaedic Robotic Devices," 1989), tissue retraction (McEwen et al., 1989), and microscope control. These systems can save costs by reducing the number of people required to perform a surgical procedure and by performing assistive tasks such as retraction more consistently and with less trauma to the patient. However, achieving any of these advantages requires effective means for the surgeon to *supervise* the assistant system without having to continually *control* it. Further, replacing a human assistant with a robot requires that the robot be versatile enough to perform substantially *all* of the work done by the human being replaced.

Current systems typically require the surgeon to explicitly control the assistant with a "master" device such as a joystick or foot pedal or to invoke simple discrete actions through a simple voice recognition system (e.g., Uecker et al., 1994). The Johns Hopkins/IBM LARS system for laparoscopic surgery (Taylor et al., 1995; Funda et al., 1993, 1994a, 1994b, 1994c; Eldridge et al., 1996; Taylor et al., 1996) extended this paradigm somewhat by permitting a surgeon to designate target anatomy on the video display by using an instrument-mounted mouse and relying on image processing to generate control inputs to reposition the robot. More recently, there has been exploration of the use of this system to perform force-controlled retraction of internal organs, using control strategies based on experiments with force-sensing retraction instruments (Poulose et al., 1998).

Extending assistant systems beyond very simple applications will require fundamental advances in the ability of computer systems to interpret complex supervisory commands within the context of surgical situations. In turn, this will require significant advances in the modeling of anatomy and surgical procedures, real-time image-processing and registration techniques to enable computers to maintain these models and follow the progress of surgical procedures, and advanced interfaces to allow the surgeon and assistant to communicate.

3 MEDICAL ROBOT SYSTEMS DESIGN CONSIDERATIONS

Medical robots have the same basic components as any other robot system (manipulator, controller, end-effector, communications interfaces, etc.), and many of the design challenges are familiar to anyone who has developed an industrial system. However, the unique demands of the surgical environment, together with the emphasis on *cooperative* execution of surgical tasks, rather than unattended automation, do create some unusual challenges. A few of these follow:

- *Safety* is paramount in any surgical robot and must be given careful attention at all phases of system design. Each element of the hardware and software should be subjected to rigorous validation at all phases, ranging from design through implementation and manufacturing to actual deployment in the operating room. Redundant sensing and consistency checks are practically essential for all safety-critical functions. Reliability experience gained with a particular design or component adapted from industrial applications is useful but *not* sufficient or even always particularly relevant, since designs must often be adapted for operating room conditions. It is important both to guard against both the effects of electrical, electronic, or mechanical component failure *and* the more insidious effects of a perfectly functioning robot subsystem correctly executing an improper motion command caused by misregistration to the patient. Further excellent discussion may be found in Taylor (1996) and Davies (1996) and in a number of papers on specific systems.
- *Sterility* is also a crucial concern. Usually this is achieved by covering most of the robot with a sterile bag or drape and then separately sterilizing the instruments or end-effectors. Autoclaving is the most universal and popular sterilization method, but unfortunately can be very destructive for electromechanical components, force

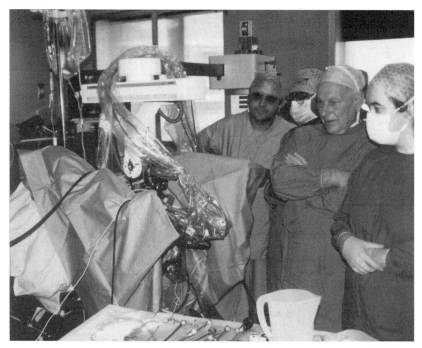

Figure 5   Surgical CAD/CAM: robot for prostate surgery. This robot is performing a transurethral prostatectomy, which requires the robot to remove a conically shaped volume of tissue from the inside of the patient's prostate. For safety reasons the robot has a specialized 4-d.o.f. structure that restricts motions of the cutting instrument to three kinematically decoupled reorientation motions and an insertion motion along the instrument's axis. (Photo courtesy Brian Davies, Imperial College.)

sensors, or the like. Other common methods include gas (slow, but usually kindest to equipment) and soaking.

- *Manipulator* design is very important in medical robots. Several early systems (e.g., Kwoh et al., 1988) used essentially unmodified industrial robots. Although this is perhaps marginally acceptable in a research system that will simply position a

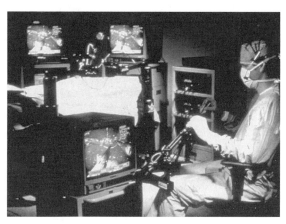

Figure 6   Surgical augmentation: force-reflecting teloperated robot, called Zeuss, intended for minimally invasive cardiac surgery. (Photo courtesy Computer Motion, Inc.)

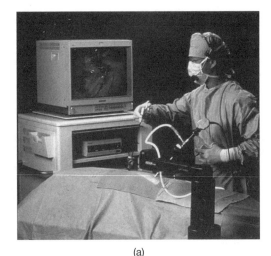

(a)

(b)

Figure 7 Two surgical assistant robots. (*a*) AESOP, a clinically applied system for holding a laparoscopic camera, operated by computer joystick or foot control. (*b*) IBM/JHU LARS, an experimental system combining the use of simple image processing and interactive on-screen pointing by the surgeon via a small mouse clipped to the surgeon's instruments. (Photos courtesy of Computer Motion, Inc., and Johns Hopkins University.)

guide and then be turned *off* before any contact is made with a patient, and use of an unmodified robot capable of high speeds is inherently suspect. Great care needs to be taken to protect both the patient and operating room personnel from runaway conditions. It is generally better to make several crucial modifications to any industrial robot that will be used in surgery. These include:

(**a**) installation of redundant position sensing;

(**b**) changes in gear ratios to slow down maximum end-effector speed; and

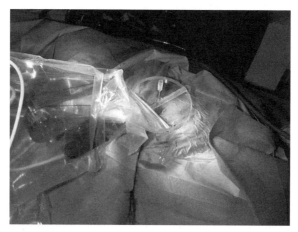

Figure 8   Simple 1-d.o.f. "robot" for driving a needle into soft tissue. This radiolucent device, called PAKY, permits the surgeon to insert a needle into an organ under real-time X-ray guidance without exposing his or her hands to radiation. It has been applied clinically in percutaneous kidney stone removal procedures, and it is suitable for use either stand alone with a passive lockable positioning device or as the end-effector of an active robot. The photograph shows the use of the PAKY in a percutaneous kidney stone removal procedure. The PAKY is held by a two degree-of-freedom remote-center-of-motion (RCM) robotic arm developed by the Johns Hopkins Urology Department. This subsystem is part of a modular family of medical robots being developed at Johns Hopkins University as a successor to the LARS system shown in Figure 7(b) and intended to be used for a variety of tasks, including image-guided implant therapy and microsurgery. (Photo courtesy D. Stoinovici, The Johns Hopkins University.)

    (c)   through evaluation and possible redesign for electrical safety and sterility.

Because the speed/work volume design points for industrial and surgical applications are very different, a more recent trend has emphasized design of custom manipulator kinematic structures for specific classes of applications. Some examples may be found in Taylor et al. (1996b), Nathan et al. (1994), Stoianovici et al., (1997), Erbse et al. (1997), Grace et al. (1993), and Brandt et al. (1997).

    Many surgical applications (e.g., in laparoscopy and neuroendoscopy) require surgical instruments to pass through a narrow opening into the patient's body. This constraint has led a number of groups to consider two rather different approaches in designing robots for such applications. The first approach (e.g., Figure, 4 Figure 5, Figure 7b, and Taylor et al. (1991), Neisius, Dautzenberg, and Trapp (1994), Eldridge et al. (1996), and Nathan et al. (1994)) uses goniometers, parallel five-bar linkages, or other means to decouple instrument motions about an "isocenter" that is placed at the entry portal. The second approach (e.g., Figure 6, Figure 7a, and Sackier and Wang (1996), Wang (1994), and Hurteau et al. (1994)) relies on passive compliance to cause the surgical instrument to comply to the entry portal constraint. In this case the robot's wrist typically has two unactuated, but encoded, rotary axes proximal to the surgical instrument holder. Both approaches have merit, and they can be combined fruitfully (Funda et al., 1994a). The first approach is generally more precise and provides a more stable platform for stereotactic procedures. The second approach has the advantages of being simple and of automatically accommodating patient motions. A fuller discussion of the tradeoff can be found in Funda et al. (1994a).

    Surgical manipulators are not always active devices. Often the human surgeon provides some or all of the motive power while the computer provides real-time navigational or other assistance (e.g., Taylor et al., 1991; Troccaz, Peshkin, and Davies, 1997; Nathan et al., 1994; Stoianovici et al., 1997; Ho, Hibberd, and Davies, 1995; and Harris et al., 1997).

    Because medical robots are often used together with imaging, materials are also an important concern in surgical manipulator design equipment (e.g., Stoianovici et al., 1997; Masamune et al., 1995). Figure 8 shows one example of a simple 1-d.o.f. radiolucent mechanisms that can be used to drive needles into soft tissue (Stoianovici et al., 1997).

This device is intended to be used with fluoroscopic X-rays or CT scanners, and it can be employed either with a simple support clamp or as the end-effector of an active robot.

4 CONCLUSION

Medical robots are like any other robots in the sense that they are machines designed to serve a particular purpose. Just like industrial robots, they are most effective when considered as part of an integrated system to exploit their unique capabilities, and many of the lessons learned from computer-integrated manufacturing may be applied to their introduction in surgery. At the same time, they differ from most industrial robots in several crucial respects, including their use *cooperatively* with human surgeons rather than as means of *replacing* surgeons, and in some of the unique safety and workspace requirements for surgery. These differences, together with the articulate and demanding nature of surgeon end users, make medical robotics an especially fruitful area for robotics technology research as well as an important end application for robotic systems.

ACKNOWLEDGMENT AND DISCLOSURE

While at IBM Research the author codeveloped the prototype of the ROBODOC™ system for orthopaedic surgery, which is currently manufactured and marketed by Integrated Surgical Systems (ISS) of Davis, California. The author currently receives research funding through ISS as part of NIST ATP Cooperative Agreement Number 94-01-0228. Also, the author is currently a member of the Scientific Advisory Board of ISS. Because of this service, the author has a small financial interest in ISS.

REFERENCES

Adler, J., et al. 1994. "Image-Guided Robotic Radiosurgery." In *Proceedings of First International Symposium on Medical Robotics and Computer Assisted Surgery (MRCAS '94)*, Pittsburgh. Vol. 2. 291–297.

Anderson, J. H., et al. 1996. "da Vinci: A Vascular Catheterization and Interventional Radiology-Based Training and Patient Pretreatment Planning Simulator." *JVIR* 7(1), Part 2:373.

"Andronics Orthopaedic Robotic Devices." 1989. Vancouver: Andronics Devices.

Bauer, A., M. Borner, and A. Lahmer. 1997. "Robodoc—Animal Evaluation and Clinical Evaluation." In *Proceedings of First Joint Conference of CVRMed and MRCAS*, Grenoble, France. Berlin: Springer-Verlag. 561–564.

Bauer, A., et al. 1995. "Virtual Reality in the Surgical Arthroscopical Training." In *Proceedings of 2nd International Symposium on Medical Robotics and Computer Assisted Surgery (MRCAS '95)*, Baltimore. New York: Wiley-Liss. 350–354.

Brandt, G., et al. 1997. "A Compact Robot for Image-Guided Orthopaedic Surgery: Concept and Preliminary Results." In *Proceedings of First Joint Conference of CVRMed and MRCAS*, Grenoble, France. Berlin: Springer-Verlag. 767–776.

Brett, P. N., et al. 1995. "A Computer Aided Tactile Simulator for Invasive Tools." In *Proceedings of Second International Symposium on Medical Robotics and Computer Assisted Surgery (MRCAS '95)*, New York: Wiley-Liss. 324–328.

Chui, C. K., et al. 1996. "Potential Field and Anatomy Vasculature for Realtime Computation in da Vinci." In *First Visible Human Conference*, Bethesda.

Cohn, M. B., et al. "Millirobotics for Telesurgery." In *Proceedings of First International Symposium on Medical Robotics and Computer Assisted Surgery (MRCAS '94)*, Pittsburgh.

Cutting, C. B., F. L. Bookstein, and R. H. Taylor. 1996. "Applications of Simulation, Morphometrics and Robotics in Craniofacial Surgery." In *Computer-Integrated Surgery*. Ed. R. H. Taylor et al. Cambridge: MIT Press. 641–662.

Cutting, C., et al. 1991. "Comprehensive Three-Dimensional Cephalometric System for the Planning and Execution of Craniofacial Surgical Procedures." In *Proceedings of Fourth Biannual Meeting of the International Society of Cranio-Maxillofacial Surgery*, Santiago de Compostela.

Cutting, C., et al. 1992. "Computer Aided Planning and Execution of Cranofacial Surgical Procedures." In *Proceedings of IEEE Engineering in Medicine and Biology Conference*, Paris.

Cutting, C., et al. 1995. "Optical Tracking of Bone Fragments During Craniofacial Surgery." In *Proceedings of Second International Symposium on Medical Robotics and Computer Assisted Surgery (MRCAS '95)*, Baltimore.

Davies, B. L. 1996. "A Discussion of Safety Issues for Medical Robots." In *Computer-Integrated Surgery*. Ed. R. H. Taylor et al. Cambridge: MIT Press. 287–298.

Dawson, S., J. Kaufman, and D. Meglin. 1996. "An Interactive Virtual Reality Trainer–Simulator for Interventional Radiology." *JVIR* 7(1), Part 2:374.

Delp, S. L., et al. 1996. "Surgical Simulation: Practicing Trauma Techniques for Emergency Medicine." In *Medical Simulation and Training*. 22–23, 30.

DiGioia, A. M., B. Jaramaz, and R. V. O'Toole. 1994. "An Integrated Approach to Medical Robotics and Computer Assisted Surgery in Orthopaedics." In *Proceedings of First International Symposium on Medical Robotics and Computer Assisted Surgery* (*MRCAS '94*), Pittsburgh.

Dow, D., W. L. Nowinski, and L. Serra. 1996. "Workbench Surface Editor of Brain Cortical Surface." In *Proceedings of SPIE Medical Imaging 1996: Image Display,* Newport Beach.

Drake, J. M., et al. 1991*a*. "Computer- and Robot-Assisted Resection of Thalamic Astrocytomas in Children." *Neurosurgery* **29**(1), 27–31.

Drake, J. M., et al. 1991*b*. "Robotic and Computer Assisted Resection of Brain Tumors." In *Fifth International Conference on Advanced Robotics,* Pisa.

Drake, J. M., et al. 1994. "Frameless Stereotaxy in Children." *Pediatric Neurosurgery* **20,** 152–159.

Eldridge, B., et al. 1996. "A Remote Center of Motion Robotic Arm for Computer Assisted Surgery." *Robotica* **14**(1), 103–109.

Engelberger, J. 1989*a*. "Aiding the Handicapped and Elderly." In *Robots in Service.* Cambridge: MIT Press. 210–217.

———. 1989*b*. "Paranurse." In *Robots in Service.* Cambridge: MIT Press. 134–138.

———. 1989*c*. "Parapharmacist." In *Robots in Service.* Cambridge: MIT Press. 139–140.

Erbse, S., et al. 1997. "Development of an Automatic Surgical Holding System Based on Ergonomic Analysis." In *Proceedings of First Joint Conference of CVRMed and MRCAS,* Grenoble, France. Berlin: Springer-Verlag. 737–746.

Ettinger, G. J., et al. "Experimentation with a Transcranial Magnetic Stimulation System for Functional Brain Imaging." In *Proceedings of First Joint Conference of CVRMed and MRCAS,* Grenoble, France. Berlin: Springer-Verlag. 477–486.

Fang, S., et al. 1995*a*. "Geometric Model Reconstruction from Cross Sections for Branched Structures in Medical Applications." In *Proceedings of SPIE Vision Geometry IV,* San Diego. 329–340.

Fang, A., et al. 1995*b*. "Three-Dimensional Talairach-Tournoux Atlas." In *SPIE Medical Imaging 1995: Image Display,* San Diego.

Funda, J., et al. 1993. "Optimal Motion Control for Teleoperative Surgical Robots." In *1993 SPIE International Symposium on Optical Tools for Manufacturing and Advanced Automation,* Boston.

Funda, J., et al. 1994*a*. "Comparison of Two Manipulator Designs for Laparoscopic Surgery." In *1994 SPIE International Symposium on Optical Tools for Manufacturing and Advanced Automation,* October, Boston. 172–183.

Funda, J., et al. 1994*b*. "An Experimental User Interface for an Interactive Surgical Robot." In *First Proceedings of International Symposium on Medical Robotics and Computer Assisted Surgery* (*MRCAS '94*), Pittsburgh.

Funda, J., et al. 1994*c*. "Image Guided Command and Control of a Surgical Robot." In *Proceedings of Medicine Meets Virtual Reality II,* San Diego.

Goradia, T. M., R. H. Taylor, and L. M. Auer. 1997. "Robot-Assisted Minimally Invasive Neurosurgical Procedures: First Experimental Experience." In *Proceedings of First Joint Conference of CVRMed and MRCAS,* Grenoble, France. Berlin: Springer-Verlag. 319–322.

Goradia, T. M., et al. 1997. "Dandy: A Virtual Reality Workbench for Skull-Based Neurosurgery." In *American Association of Neurological Surgeons,* Denver.

Grace, K. W., et al. 1993. "Six Degree of Freedom Micromanipulator for Ophthalmic Surgery." In *IEEE International Conference on Robotics and Automation,* Atlanta.

Green, P., et al. 1992. "Telepresence: Advanced Teleoperator Technology for Minimally Invasive Surgery (abstract)." *Surgical Endoscopy* **6**.

Harris, S. J., et al. 1997. "Experiences with Robotic Systems for Knee Surgery." In *Proceedings of First Joint Conference of CVRMed and MRCAS.* Grenoble, France. Berlin: Springer-Verlag. 757–766.

Higgins, G. A., et al. 1990. "Virtual Reality Surgery: Implementation of a Coronary Angioplasty Training Simulator." In *Surgical Technology International IV.*

Ho, S. C., R. D. Hibberd, and B. L. Davies. 1995. "Robot Assisted Knee Surgery." *IEEE Engineering in Medicine and Biology Magazine* **14**(3), 292–300.

Hurteau, R., et al. 1994. "Laparoscopic Surgery Assisted by a Robotic Cameraman: Concept and Experimental Results." In *IEEE Conference on Robotics and Automation,* San Diego.

Kall, B. A., P. J. Kelly, and S. J. Goerss. 1985. "Interactive Stereotactic Surgical System for the Removal of Intracranial Tumors Utilizing the CO_2 Laser and CT-Derived Database." *IEEE Transactions on Biomedical Engineering* **BME-32**(2), 112–116.

Kavoussi, L., et al. 1994. "Telerobotic-Assisted Laparoscopic Surgery: Initial Laboratory and Clinical Experience." *Urology* **44**(1), 15–19.

Kosugi, Y., et al. 1988. "An Articulated Neurosurgical Navigation System Using MRI and CT Images." *IEEE Transactions on Biomedical Engineering* **35**(2), 147–152.

Kumar, R., et al. 1997. "Robot-Assisted Microneurosurgical Procedures, Comparative Dexterity Experiments." In *Society for Minimally Invasive Therapy 9th Annual Meeting, Abstract Book Vol. 6, Supplement 1.* Tokyo.

Kwoh, Y. S., et al. 1988. "A Robot with Improved Absolute Positioning Accuracy for CT Guided Stereotactic Brain Surgery." *IEEE Transactions on Biomedical Engineering* **35**(2), 153–160.

Lasko, H. A., et al. 1995. "A Fully Immersive Cholecystectomy Simulator." In *Interactive Technology and the New Paradigm for Health Care.* Amsterdam: IOS Press.

Lavallee, S. 1996. "Registration for Computer-Integrated Surgery: Methodology, State of the Art." In *Computer-Integrated Surgery.* Ed. R. H. Taylor et al. Cambridge: MIT Press. 77–98.

Lavallee, S., et al. 1994. "Computer-Assisted Knee Anterior Cruciate Ligament Reconstruction First Clinical Trials." In *Proceedings of First International Symposium on Medical Robotics and Computer Assisted Surgery (MRCAS '94),* Pittsburgh.

Lavallee, S., et al. 1996*a*. "Computer-Assisted Spinal Surgery Using Anatomy-Based Registration." In *Computer-Integrated Surgery.* Ed. R. H. Taylor, et al. Cambridge: MIT Press. 425–449.

Lavallee, S., et al. 1996*b*. "Image-Guided Operating Robot: A Clinical Application in Stereotactic Neurosurgery." In *Computer-Integrated Surgery.* Ed. R. H. Taylor et al. Cambridge: MIT Press. 343–352.

Maciunas, R. J. 1993. *Interactive Image-Guided Neurosurgery.* American Association of Neurological Surgeons.

Maciunas, R. J., et al. 1993. "Beyond Stereotaxy: Extreme Levels of Application Accuracy Are Provided by Implantable Fiducial Markers for Interactive Image-Guided Neurosurgery." In *Interactive Image-Guided Neurosurgery.* Ed. R. J. Maciunas. American Association of Neurological Surgeons. Chapter 21.

Masamune, K., et al. 1995. "Development of a MRI Compatible Needle Insertion Manipulator for Stereotactic Neurosurgery." In *Proceedings of Second International Symposium on Medical Robotics and Computer Assisted Surgery (MRCAS '95),* Baltimore, November. New York: Wiley-Liss. 165–172.

McEwen, J. A. 1993. "Solo Surgery with Automated Positioning Platforms." In *Proceedings of NSF Workshop on Computer Assisted Surgery,* Washington, D.C.

McEwen, J. A., et al. 1989. "Development and Initial Clinical Evaluation of Pre-Robotic and Robotic Retraction Systems for Surgery." In *Proceedings of Second Workshop on Medical and Health Care Robotics,* Newcastle-on-Tyne.

Mitsuishi, M., et al. 1995. "A Tele-micro-surgery System with Co-located View and Operation Points and Rotational-Force-Feedback-Free Master Manipulator." In *Proceedings of Second International Symposium on Medical Robotics and Computer Assisted Surgery (MRCAS '95),* Baltimore.

Mittelstadt, B., et al. 1996. "The Evolution of a Surgical Robot from Prototype to Human Clinical Use." In *Computer-Integrated Surgery.* Ed. R. H. Taylor et al. Cambridge: MIT Press. 397–407.

Morantz, R. A., and W. M. Wara. 1995. "Gamma Knife Radiosurgery in the Treatment of Brain Tumors." *Cancer Control.* 300–308.

Nathan, M. S., et al. 1994. "Devices for Automated Resection of the Prostate." In *Proceedings of First International Symposium on Medical Robotics and Computer Assisted Surgery (MRCAS '94),* Pittsburgh.

Neisius, B., P. Dautzenberg, and R. Trapp. 1994. "Robotic Manipulator for Endoscopic Handling of Surgical Effectors and Cameras." In *Proceedings of First International Symposium on Medical Robotics and Computer Assisted Surgery (MRCAS '94),* Pittsburgh.

Nolte, L. P., et al. 1994. "A Novel Approach to Computer Assisted Spine Surgery." In *Proceedings of First International Symposium on Medical Robotics and Computer Assisted Surgery (MRCAS '94),* Pittsburgh.

Paggetti, C., et al. 1997. "A System for Computer Assisted Arthroscopy." In *Proceedings of First Joint Conference of CVRMed and MRCAS,* Grenoble, France. Berlin: Springer-Verlag. 653–662.

Petelin, J. B. 1994. "Computer Assisted Surgical Instrument Control." In *Proceedings of Medicine Meets Virtual Reality II,* San Diego.

Poulose, B. K., et al. 1998. "Human versus Robotic Organ Retraction During Laparoscopic Fnissen Fundoplication." In *Society of American Gastrointestinal Endoscopic Surgeons (SAGES),* Seattle.

Preminger, G. M., et al. 1996. "Virtual Reality Simulations in Endoscopic Urologic Surgery." In *Medicine Meets Virtual Reality 4: Health Care in the Information Age—Future Tools for Transforming Medicine.* Amsterdam: IOS Press.

Raibert, M. 1997. "Virtual Surgery." ⟨http://www.bdi.com/html/virtual_surgery.html⟩.

Sackier, J. M., and Y. Wang. "Robotically Assisted Laparoscopic Surgery: From Concept to Development." In *Computer-Integrated Surgery.* Ed. R. H. Taylor et al. Cambridge: MIT Press. 577–580.

Salcudean, S. E., S. Ku, and G. Bell. 1997. "Performance Measurement in Scaled Teleoperation for Microsurgery." In *Proceedings of First Joint Conference of CVRMed and MRCAS,* Grenoble, France. Berlin: Springer-Verlag. 789–798.

Sastry, S. S., M. Cohn, and F. Tendick. 1997. *Millirobotics for Minimally-Invasive Surgery.* University of California—Berkeley.

Satava, R. 1992. "Robotics, Telepresence, and Virtual Reality: A Critical Analysis of the Future of Surgery." *Minimally Invasive Therapy* **1**, 357–363.

————. 1993. "Virtual Reality Surgical Simulation: The First Steps." *Surgical Endoscopy* **7,** 203–205.

————. 1995. "Virtual Reality, Telesurgery, and the New World Order of Medicine." *Journal of Image-Guided Surgery* **1**(1), 12–16.

————. 1996. "CyberSurgeon: Advanced Simulation Technologies for Surgical Education." In *Medical Simulation and Training.* 6–9.

Schenck, J. F., et al. 1995. "Superconducting Open-Configuration MR Imaging System for Image-Guided Therapy." *Radiology* **195,** 805–814.

Schlegel, W. 1996. "Requirements in Computer-Assisted Radiotherapy." In *Computer-Integrated Surgery.* Ed. R. H. Taylor et al. Cambridge: MIT Press. 681–691.

Schreiner, S., et al. "A System for Percutaneous Delivery of Treatment with a Fluoroscopically-Guided Robot." In *Proceedings of First Joint Conference of CVRMed and MRCAS,* Grenoble, France. Berlin: Springer-Verlag. 747–756.

Schurman, D. J., and W. J. Maloney. 1992. "Segmental Cement Extraction at Revision Total Hip Arthroplasty." *Clinical Orthopedics and Related Research* (December).

Simon, D. A., et al. 1997. "Development and Validation of a Navigational Guidance System for Acetabular Implant Placement." In *Proceedings of First Joint Conference of CVRMed and MRCAS,* Grenoble, France. Berlin: Springer-Verlag. 583–592.

Stoianovici, D., et al. 1997. "An Efficient Needle Injection Technique and Radiological Guidance Method for Percutaneous Procedures." In *Proceedings of First Joint Conference of CRVMed and MRCAS,* Grenoble, France. Berlin: Springer-Verlag. 295–298.

Taylor, R. H. 1996. "Safety." In *Computer-Integrated Surgery.* Ed. R. H. Taylor et al. Cambridge: MIT Press. 283–286.

Taylor, R. H., et al. 1991. "A Model-Based Optimal Planning and Execution System with Active Sensing and Passive Manipulation for Augmentation of Human Precision in Computer-Integrated Surgery." In *Proceedings of 1991 International Symposium on Experimental Robotics,* Toulouse, France. Berlin: Springer-Verlag.

Taylor, R. H., et al. 1994. "An Image-Directed Robotic System for Precise Orthopaedic Surgery." *IEEE Transactions on Robotics and Automation* **10**(3), 261–275.

Taylor, R. H., et al. 1995. "A Telerobotic Assistant for Laparoscopic Surgery." *IEEE Engineering in Medicine and Biology Magazine* **14**(3), 279–291.

Taylor, R. H., et al. 1996*b.* "A Telerobotic Assistant for Laparoscopic Surgery." In *Computer-Integrated Surgery.* Ed. R. H. Taylor *et al.* Cambridge: MIT Press. 581–592.

Troccaz, J., M. Peshkin, and B. L. Davies. 1997. "The Use of Localizers, Robots, and Synergistic Devices in CAS." In *Proceedings of First Joint Conference of CVRMed and MRCAS,* Grenoble, France. Berlin: Springer-Verlag. 727–736.

Troccaz, J., et al. 1994. "Patient Set-Up Optimization for External Conformal Radiotherapy." In *Proceedings of First International Symposium on Medical Robotics and Computer Assisted Surgery* (*MRCAS '94*), Pittsburgh.

Uecker, D. R., et al. 1994. "A Speech-Directed Multi-Modal Man-Machine Interface for Robotically Enhanced Surgery." In *First International Symposium on Medical Robotics and Computer Assisted Surgery* (*MRCAS '94*), Pittsburgh.

Vaillant, M., et al. 1997. "A Path-Planning Algorithm for Image Guided Neurosurgery." In *Proceedings of First Joint Conference of CVRMed and MRCAS,* Grenoble, France. Berlin: Springer-Verlag. 467–476.

Wang, Y. 1994. "Robotically Enhanced Surgery." In *Medicine Meets Virtual Reality II,* San Diego.

Zinreich, S. J., et al. 1993. "Frameless Stereotaxic Integration of CT Imaging Data: Accuracy and Initial Applications." *Radiology* **188**(3), 735–742.

ADDITIONAL READING

Proceedings of Second International Symposium on Medical Robotics and Computer Assisted Surgery (*MRCAS '95*), Baltimore, June 1995. New York: Wiley-Liss.

Proceedings of First Joint Conference of CVRMed and MRCAS, Grenoble, France, 1997. Berlin: Springer-Verlag.

Taylor, R., and S. D. Stulberg. "Medical Robotics Working Group Report." In *Proceedings of NSF Workshop on Medical Robotics and Computer-Assisted Medical Interventions* (*RCAMI*), Bristol, England, 1997. Ed. A. Digioia et al. Excerpts reprinted in *Journal of Computer Aided Surgery* **2,** 69–101, 1997.

Taylor, R., G. Bekey, and J. Funda, eds. *Proceedings of NSF Workshop on Computer Assisted Surgery,* Washington, D.C., 1993.

Taylor, R. H., et al., eds. *Computer-Integrated Surgery.* Cambridge: MIT Press, 1996.

PART 11

ROBOTICS AROUND THE WORLD

CHAPTER **66**
ROBOTICS AROUND THE WORLD

José A. Ceroni
Chin-Yin Huang
Marco A. Lara
NaRaye P. Williams
Purdue University
West Lafayette, Indiana

Donald A. Vincent
Robotic Industries Association
Ann Arbor, Michigan

1 INTRODUCTION

Robotics is helping companies innovate, automate, and compete. Throughout the world there is a growing awareness among companies of all sizes, in virtually every industry, that robotics can help them stay globally competitive. Yet there are thousands of companies in North America alone who have not installed even one robot in either their operations or their products. The potential for robotics automation is enormous, and the benefits of using robot technology are becoming more widely understood. With 12,000 or more new robot orders annually received by U.S.-based robotics companies and many untapped opportunities still available throughout the world, the robotics industry will be one of the most important global industries of the twenty-first century.

Robotics is booming for several reasons. The technology is reliable, cost-justified, and suited for a wide range of tasks that cannot be performed effectively by manual labor or fixed automation. Also, many applications have emerged that cannot be performed successfully without robots. In the electronics industry miniaturization is driving the demand for robots. In the automotive industry, which accounts for approximately half of the 82,000 installed robots in the United States, robots are essential for certain tasks such as spot welding, sealing, and painting.

What does the future hold for robotics? We believe robotics will show a substantial increase in manufacturing applications, not just in the United States, but around the world. Some of the fastest growing robotics users are found in the Pacific Rim, while regions such as Latin America are just getting started. Material-handling applications, such as parts transfer, machine tending, packaging, and palletizing, cut across many industries. Manufacturers of consumer goods, electronics, food and beverages, and other nonautomotive products are now taking advantage of robots to become stronger global competitors. Small, medium, and large companies in just about every industry are taking a fresh look at robots to see how this powerful technology can help them solve manufacturing challenges.

We also believe that nonmanufacturing applications in tasks such as security, material transport, and commercial cleaning will rapidly accelerate. Prime potential growth areas include nuclear clean-up, hazardous materials handling, and undersea exploration. Other important robot applications of the future are space exploration, agriculture, assisting the elderly, surgery, police work, pumping gas, and household chores. As Joe Engelberger likes to say, many applications start with a simple question: "Do you think a robot can do this?" In the twenty-first century, more often than not the answer will be "yes." It is

Handbook of Industrial Robotics, Second Edition, Edited by Shimon Y. Nof
ISBN 0-471-17783-0   © 1999 John Wiley & Sons, Inc.

an exciting time for the worldwide robotics industry, as robotics plays a key role in making countries more productive.

2 STATISTICS

This chapter presents statistical data collected by the Robotic Industries Association (RIA), the UN/Economic Commission for Europe (UN/ECE), and the International Federation of Robotics (IFR).[1]

According to the RIA, the industry's trade group, the North American robotics industry had its best year ever in 1997. A total of 12,459 robots were shipped, an increase of 28% over 1996. With a value of nearly $1.1 billion, shipment revenue topped the billion-dollar mark for the first time ever. Robot shipments have been on a dramatic upswing for the past five years, rising 172% in units and 136% in dollars since 1992. New orders also topped the billion-dollar mark, as a total of 12,149 robots valued at $1.104 billion were ordered last year. Though new orders rose just 1% in 1997, they are up 131% in units and 122% in dollars in the past five years, with 1997 being even stronger than anticipated. Shipments continued to soar, and the industry set another record for new orders despite some expectations of a downturn after five straight years of double-digit growth (Robotics Industries Association, 1997).

The statistical data reported by the UN/ECE are almost exclusively based on information which national robot associations have supplied to the IFR. For several years the UN/ECE and the IFR have cooperated closely in the compilation, processing and analysis of worldwide statistics on industrial robots. The UN/ECE *World Industrial Robots 1997* reports the investment in robotics surged by 11% in 1996 and is forecasted to increase by 13% per year in the period 1996 to 2000. The total stock of operational industrial robots is estimated at 680,000 units, up 6% over 1995 (United Nations, 1997).

While the number of industrial robots is the key measure of the robotics industry, the numbers should actually be used only as a leading indicator of robotics. There are many robotics applications, as described throughout this handbook, that integrate robotics intelligence beyond the traditional industrial robot manipulator. Robotics devices in space, service, medical and health applications, quality inspection, transportation applications, and even bank ATMs may not be counted as industrial robots, but they serve industry and humanity on a daily basis, faithfully and effectively.

REFERENCES

Robotic Industries Association. 1997. *The Robotics Market, 1997 Results and Trends.*
Rudall, B. H. 1997. "World Industrial Robots—Statistical Data to 1998." *Robotica* **15,** 247–249.
United Nations Economic Commission for Europe and International Federation of Robotics. 1997. *World Industrial Robots 1997.* Geneva: United Nations.

---

[1] See Chapter 2 for additional statistics focusing on the Japanese robotics market.

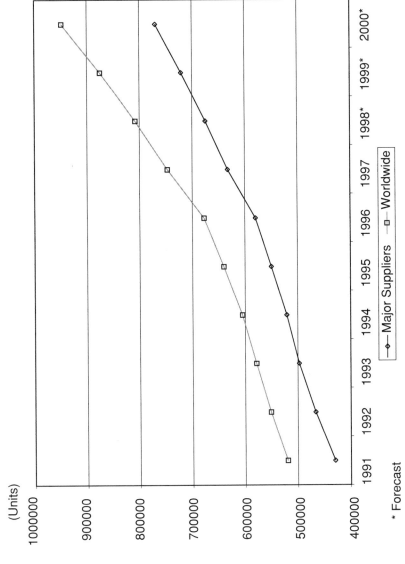

Figure 1 Operating robot population worldwide and by the major suppliers. (a) The "major robot suppliers" are considered to be France, Germany, Italy, Japan, the U.K., and the U.S., also known as the "Big Six." (b) Graphs were prepared based on data in United Nations Economic Commission for Europe and International Federation of Robotics, *World Industrial Robots 1997*. Geneva: United Nations.

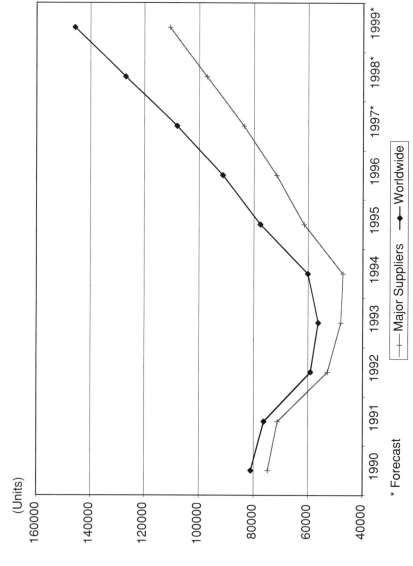

(Units)

160000

140000

120000

100000

80000

60000

40000

1990   1991   1992   1993   1994   1995   1996   1997*   1998*   1999*

* Forecast

—+— Major Suppliers   —◆— Worldwide

Figure 2 Shipment of robots worldwide and by the major suppliers. (a) Graphs were prepared based on data in B. H. Rudall, "World Industrial Robots—Statistical Data to 1998" *Robotics* **15**, 247–249, 1997.

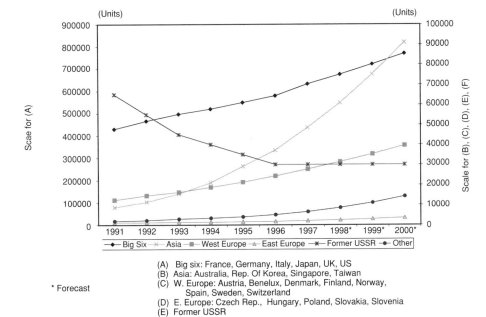

(A) Big six: France, Germany, Italy, Japan, UK, US
(B) Asia: Australia, Rep. Of Korea, Singapore, Taiwan
(C) W. Europe: Austria, Benelux, Denmark, Finland, Norway,
      Spain, Sweden, Switzerland
(D) E. Europe: Czech Rep., Hungary, Poland, Slovakia, Slovenia
(E) Former USSR
(F) Other

* Forecast

Figure 3 Operating robot population by economic region (est.). (a) Graphs were prepared based on data in United Nations Economic Commission for Europe and International Federation of Robotics, *World Industrial Robots 1997*. Geneva: United Nations.

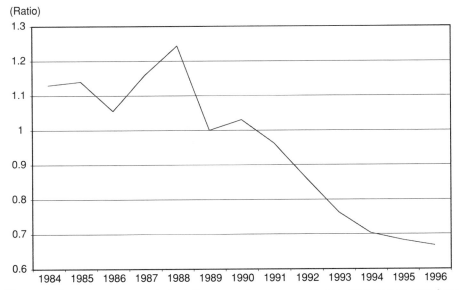

Figure 4 Average price per robot relative to employee compensation. (a) Based on data from France, Germany, U.K., and U.S. (b) Graph was prepared based on data in United Nations Economic Commission for Europe and International Federation of Robotics, *World Industrial Robots 1997*. Geneva: United Nations.

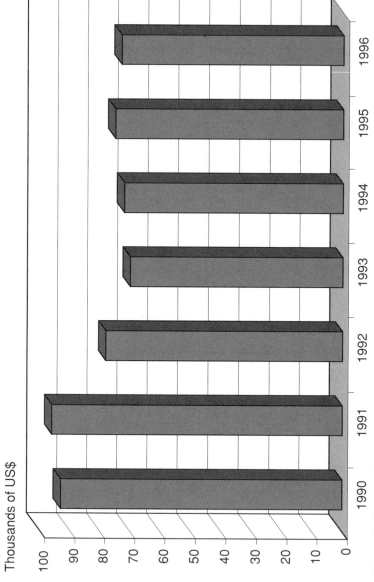

Figure 5 Average unit price of robots. (a) Based on data from France, Germany, Italy, Republic of Korea, U.K., and U.S. (b) Graph was prepared based on data in United Nations Economic Commission for Europe and International Federation of Robotics, *World Industrial Robots 1997*. Geneva: United Nations.

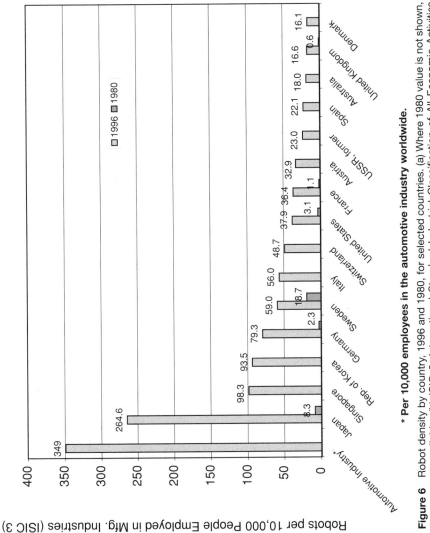

Figure 6 Robot density by country, 1996 and 1980, for selected countries. (a) Where 1980 value is not shown, data are not available. (b) ISIC 3: International Standard Industrial Classification of All Economic Activities revision 3. Graph was prepared based on data in United Nations Economic Commission for Europe and International Federation of Robotics, *World Industrial Robots 1997*. Geneva: United Nations.

* **Per 10,000 employees in the automotive industry worldwide.**

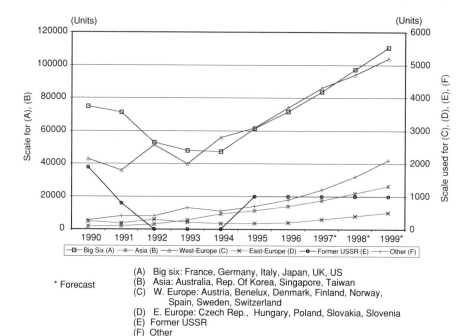

(A) Big six: France, Germany, Italy, Japan, UK, US
* Forecast
(B) Asia: Australia, Rep. Of Korea, Singapore, Taiwan
(C) W. Europe: Austria, Benelux, Denmark, Finland, Norway,
       Spain, Sweden, Switzerland
(D) E. Europe: Czech Rep., Hungary, Poland, Slovakia, Slovenia
(E) Former USSR
(F) Other

Figure 7   Worldwide shipment of robots by economic region. (a) Graphs were prepared based on data in B. H. Rudall, "World Industrial Robots—Statistical Data to 1998," *Robotics* **15,** 247–249, 1997.

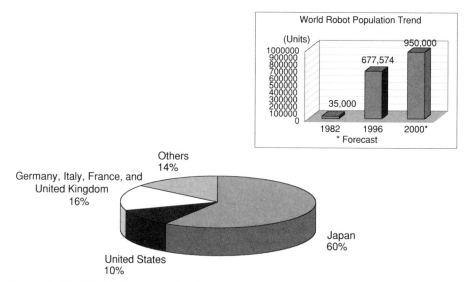

Figure 8   1996 World robot population. (a) Graphs were prepared based on data in United Nations Economic Commission for Europe and International Federation of Robotics, *World Industrial Robots 1997.* Geneva: United Nations.

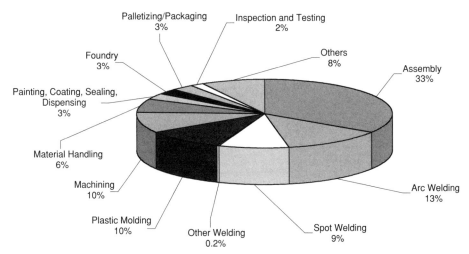

Figure 9   Worldwide operating robots by application, 1996. (a) Graph was prepared based on data in United Nations Economic Commission for Europe and International Federation of Robotics, *World Industrial Robots 1997*. Geneva: United Nations.

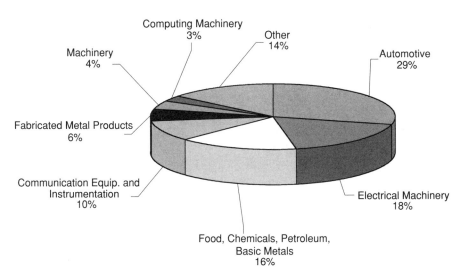

Figure 10   Worldwide operating robots by industry, 1996. (a) Graph was prepared based on data in From United Nations Economic Commission for Europe and International Federation of Robotics, *World Industrial Robots 1997*. Geneva: United Nations.

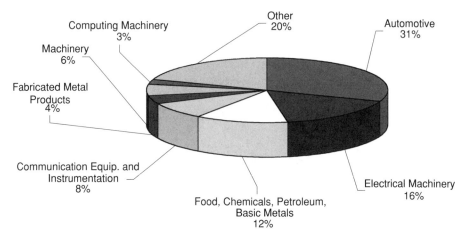

Figure 11 Worldwide shipment of robots by industry, 1996. (a) Graph was prepared based on data in United Nations Economic Commission for Europe and International Federation of Robotics, *World Industrial Robots 1997*. Geneva: United Nations. *Note:* The differences from data in Figure 10 are explained mainly by replacement of obsolete robots.

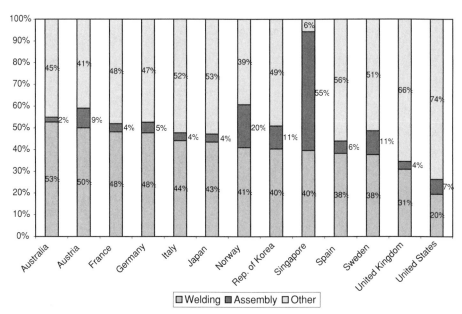

Figure 12 Share of welding, assembly, and other application purpose for the 1996 shipment of robots for selected countries. (a) Graph was prepared based on data in United Nations Economic Commission for Europe and International Federation of Robotics, *World Industrial Robots 1997*. Geneva: United Nations.

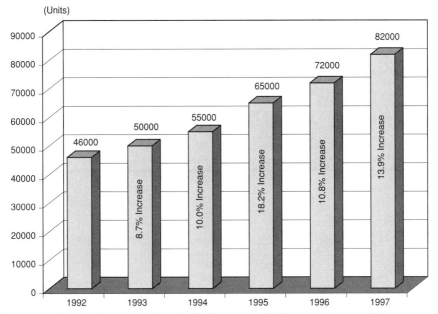

Figure 13   Growth of robot population in North America. (a) Graph was prepared based on data in Robotic Industries Association, *The Robotics Market, 1997 Results and Trends,* 1997.

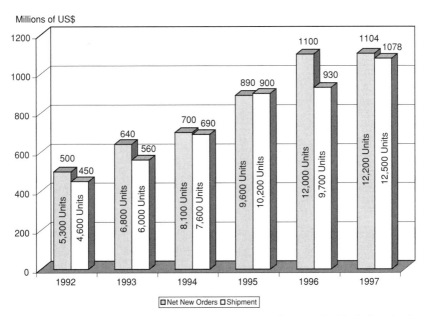

Figure 14   Units and value of new orders and shipments of robots for North America-based companies. (a) Graphs were prepared based on data in Robotic Industries Association, *The Robotics Market, 1997 Results and Trends,* 1997.

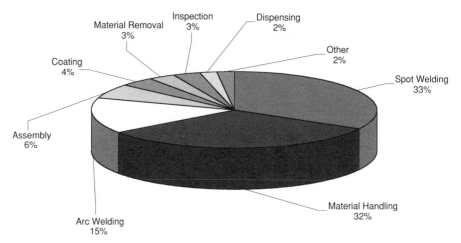

Figure 15 Leading applications in 1997 new orders of robots (North America). (a) Graph was prepared based on data in Robotic Industries Association, *The Robotics Market, 1997 Results and Trends,* 1997.

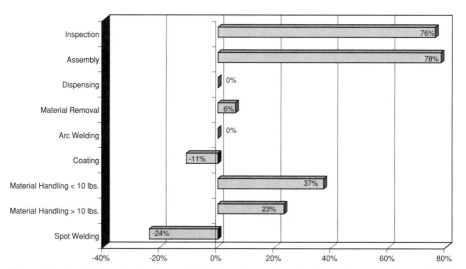

Figure 16 Application trends in new North American orders of robot units comparison of 1996 to 1997. (a) Graph was prepared based on data in Robotic Industries Association, *The Robotics Market, 1997 Results and Trends,* 1997.

3   SAMPLE OF ROBOTICS EQUIPMENT SPECIFICATIONS

Table 1   Articulated Robots

Nomenclature for axis (just this table):

Axis 1 = Rotation	Axis 2 = Upper arm	Axis 3 = Front arm
Axis 4 = Rotation wrist	Axis 5 = Bending wrist	Axis 6 = Twisting wrist

Payload (kgs/lbs)	Operation Range Axis	Operation Range Degrees	Reach (mm/in.)	Repeatability (mm/in.)	Speed Axis	Speed Degrees/sec
5/11	1	340	1444/56.9	Less than ±0.1/0.004	1	120
	2	140			2	120
	3	135			3	135
	4	300			4	280
	5	230			5	280
	6	600			6	280
10/22	1	±180	1613/63.5	Less than 0.1/0.004	1	150
	2	+150–90			2	150
	3	+150–110			3	150
	4	±200			4	260
	5	±135			5	260
	6	±200			6	400
30/66	1	±180	1785/69.5	Within ±0.15/0.006	1	120
	2	+150–75			2	120
	3	+135–110			3	120
	4	±240			4	200
	5	±190			5	200
	6	±350			6	300
60/132	1	±150	2080/82.0	Within ±0.20/0.008	1	120
	2	+90–75			2	120
	3	+90–130			3	120
	4	±240			4	140
	5	±190			5	160
	6	±190			6	240
100/220	1	±150	2420/95.4	Within ±0.4/0.016	1	112
	2	+65–60			2	112
	3	+30–115			3	112
	4	±240			4	140
	5	±190			5	140
	6	±350			6	240
120/264	1	360	2800/110	±0.5/0.02	1	100
	2	140			2	100
	3	133			3	100
	4	600			4	210
	5	240			5	150
	6	600			6	210

Table 2   SCARA Robots

Payload (kgs/lbs)	Operation Range Joint	Value	Repeatability Axis	Value	Maximum Speed Axes/Joint	Value
5/11	1	±90 deg	X–Y	±0.02 mm	X–Y	5100 mm/sec
	2	±135 deg	Z	±0.02 mm	Z	830 mm/sec
	3	150 mm	Roll	±0.02 deg	Roll	770 deg/sec
	4 (Roll)	±220 deg				
10/22	1	±117 deg	X–Y	±0.03 mm	X–Y	5200 mm/sec
	2	±135 deg	Z	±0.02 mm	Z	610 mm/sec
	3	250 mm	Roll	±0.02 deg	Roll	510 deg/sec
	4 (Roll)	±360 deg				
70/154[a]	1	220 deg	X–Y	±0.3 mm	1	120 deg/sec
	2	300 deg	Z	±0.3 mm	2	160 deg/sec
	3	2000 mm	Roll	±0.5 deg	3	1000 mm/sec
	4 (Roll)	±180 deg			4 (Roll)	225 deg/sec
150/330[a]	1	220 deg	X–Y	±0.3 mm	1	140 deg/sec
	2	300 deg	Z	±0.3 mm	2	140 deg/sec
	3	2000 mm	Roll	±0.5 deg	3	1500 mm/sec
	4 (Roll)	±360 deg			4 (Roll)	225 deg/sec

[a] SCARA Robots with vertical arm displacement.

Table 3   Gantry Robots

Payload (kgs/lbs)	Speed Axis	Value	Acceleration Axis	Value	Accuracy Axis	Value
80/176	X	1.6 m/sec	X	2.5 m/sec^2	X	±0.2 mm
	Y	2.0 m/sec	Y	4.0 m/sec^2	Y	±0.1 mm
	Z	1.6 m/sec	Z	3.2 m/sec^2	Z	±0.1 mm
	C, B	180 deg/sec	C, B	447 deg/sec^2	C, B	<±0.05 deg
300/660	X	1.25 m/sec	X	1.7 m/sec^2	X	±0.2 mm
	Y	1.60 m/sec	Y	2.5 m/sec^2	Y	±0.1 mm
	Z	1.25 m/sec	Z	2.5 m/sec^2	Z	±0.1 mm
	C, B	90 deg/sec	C, B	224 deg/sec^2	C, B	<±0.05 deg
700/1540	X	1.0 m/sec	X	1.0 m/sec^2	X	±0.4 mm
	Y	1.0 m/sec	Y	1.0 m/sec^2	Y	±0.2 mm
	Z	1.0 m/sec	Z	1.0 m/sec^2	Z	±0.2 mm
	C, B	70 deg/sec	C, B	70 deg/sec^2	C, B	<±0.05 deg
1200/2640	X	0.5 m/sec	X	0.5 m/sec^2	X	±0.4 mm
	Y	0.5 m/sec	Y	0.5 m/sec^2	Y	±0.2 mm
	Z	0.5 m/sec	Z	0.5 m/sec^2	Z	±0.2 mm
	C, B	35 deg/sec	C, B	35 deg/sec^2	C, B	<±0.05 deg

Table 4   Robotics-Related Vision Systems

Type	Key Features
Integrated vision system	• CPU: MC68030 @ 33 MHz, 32-bit data bus • Coprocessors: MC68882 floating point coprocessor. Dedicated vision coprocessors • Image digitizer: 8 bit (256 gray colors). Single-channel A/D converter (four optional) stores up to four images • Memory: 4–32 MB using standards SIMMs • Video input: Up to 4 cameras multiplexed, stationary or arm mounted. RS-170 or CCIR video formats. Supports internal or external synchronization • Video output: PC's VGA Monitor at 800 × 600 (noninterlaced)
Multifunctional vision system	• General processor Intel 80376, 16 MHz • Digital signal processor 50 MHz • Graphics processor • Camera inputs: 4 • Video rate: 50/60 Hz • Scan modes: interlaced • Image memory: 512 × 512 pixels • Video signal quantification: 256 levels of gray • Segmentation time: 40–120 ms • Process time per full image (typical): 500–3000 ms • Position accuracy (typical): 1/5000 of scene length • Orientation accuracy (typical): better than 0.1 deg • Graphical user interface
Intelligent vision system	• CPU: 33 MHz i960 RISC CPU • RAM: 1 Mbyte (72 pin SIMM). 4 Mbytes (IDS 3) • Frame store: 521 K bytes • Technology: CMOS • Image size: 768 × 574 pixel • Digital resolution: 512 × 512 × 8 bits • Pixel size: 10.5 μm square • Frame rate: 25/30 frames/sec • Lens compatibility: 2/3 inch format • Video format: CCIR/EIA/noninterlaced
Guidance vision system	• System packaging: single-slot PCI bus plug-in board • Processors: 80 MHz and advanced vision co-processors • Video input/output: RS-170 video input sources, 2 on 15 pin high-density-Dsub, 2 on header, 8-bit flash converter and programmable 256 × 8 LUT. Composite video input. Near-real-time video display over PCI bus • On-board I/O: Asynchronous trigger input (opto-isolated). Accept/reject output (opto-isolated). Strobe output. Integrated light controller

Table 5   Robotics-Related Sensors

Type	Key Features
Photoelectric sensor (Retroreflective laser sensor with polarization filter)	• Small-diameter laser beam for precise detection tasks • Polarization filter to detect shiny objects • Nominal range Sn/Actual range Sb: 5.0 m/4.0 m • Power indicator/Output indicator: green LED/yellow LED • Alignment aid/soiled lens indicator: yellow flashing LED • Light source/wave length: red laser diode/675 nm • Laser class: class 2 • Voltage supply range: 10–30 VDC • Supply current: less than or equal to 50 mA • Temperature range: –10 . . . 50°C • Housing material: zinc diecast • Protection class: IP 67
Ultrasonic proximity sensor	• Scanning range far limit Sde: 400 . . . 2500 mm • Sonic beam angle: 8° • Sonic frequency: 120 kHz • Hystersis: 4% Sde • Voltage supply range: 12–30 VDC • Residual ripple: less than 10% Vs • Supply current: less than 30 mA • Response time t on/t off: less than or equal to 250 ms • Repeatability: less than or equal to 2 mm or 1% (Sde–St) • Temperature range: 0 . . . 60°C • Housing material: brass nickel plated • Protection class: IP 67
Inductive proximity sensor	• Max. operating distance: 2.5 mm ± 20% • Secure sensing range: 0 . . . 1.8 mm • Hysteresis: 20% or less of operation distance • Repeatability: axial direction, perpendicular to axial: 0.04 mm or less • Supply voltage: 12–24 VDC ± 10%; Ripple P-P 10% or less • Current consumption: 0.8 mA or less • Max. response frequency: 1 kHz • Temperature range: –25 . . . 70°C • Protection class: IP 67
Pressure sensor	• Related pressure range: 0 to 1.00 MPa • Setting pressure range: –0.050 to 1.00 MPa • Pressure resistance: 1.47 MPa • Supply voltage: 12 to 24 VDC ±10%; Ripple P-P 10% or less • Current consumption: 50 mA or less • Repeatability: Within ±0.2% F.S. ±1 digit • Response time: 2.5 ms or less • Temperature range: –10 . . . 50°C • Protection class: IP 40, IP 67

Table 6   Grippers

Type	Total stroke	Closing force	Backdrive force (N)	Opening force	Repeatability (mm)	Self-weight (g)
2-Jaw, parallel with CAM	54 mm	700 N	1750 (to open) 625 (to close)	250 N	±0.05	3195
2-Jaw, parallel with CAM	50 mm	1150 N	3000 (to open) 625 (to close)	250 N	±0.05	3131
2-Jaw, parallel mini-size	7° wedge	6 Ncm	–	6 Ncm	±0.1	41
2-Jaw, parallel mini-size	7° wedge	20 Ncm	–	8 Ncm	±0.1	96
2-Jaw, parallel mini-size	7° wedge	40 Ncm	–	20 Ncm	±0.1	96
2-Jaw, parallel finger-type	0–8 mm (adjustable)	7.5 Ncm	–	11.5 Ncm	±0.1	35
2-Jaw, parallel finger-type	0–10 mm (adjustable)	55 Ncm	–	28 Ncm	±0.1	80
2-Jaw, parallel finger-type	0–10 mm (adjustable)	270 Ncm	–	170 Ncm	±0.1	265
3-Jaw	3 × 3 mm	43 N	124 (to open) 210 (to close)	72 N	±0.1	180
3-Jaw	3 × 6 mm	240 N	700 (to open) 1200 (to close)	400 N	±0.1	1670
3-Jaw	3 × 8 mm	355 N	1000 (to open) 1600 (to close)	540 N	±0.1	2800

Table 7   Robotics-Related Controllers

Performance	User Interfaces	Machine Interfaces	Program Features
• Controlled axes: 12 • Memory: up to 10 MB in RAM; up to 20,000 instructions • Main processor: 32 bit w/on-board floating-point calculation • Control principle: self-optimizing motion control—no trimming	• Operator's panel: in cabinet or external • Programming unit: portable; joystick and keypad; display 16 lines × 40 characters; Window style communication • Off-line: robot programming on a PC • Languages: choice among 10 national languages	• Digital inputs/ outputs: 96/96; 24 V or 110 V AC • Analog inputs/ outputs: 4/4; ±10 V and 20 mA • Remote I/O: PLC • Serial channels: RS232 three, 300-19, 200 bits/s; RS485 one, 300-38, 400 bits/s • VME: VME-bus	• Pull-down menus, dialogues and joystick for robot motion, function keys and windows • Cut, paste, copy, delete, search, change functions • Manager functions for 5 different user levels • Powerful user-friendly robot languages

Table 8   Robotics-Related Feeder Systems

Type	Key Features
Vibratory feeder bowls	• Available sizes (in inches): 2¾ to 42 • Multiple coil design for balanced feed on 6-in. and larger units • Peripheral toe clamp for bowl mounting • Standard voltage is 115 V 60 Hz. 230 V and 50 Hz also available • Vibratory frequency 120 cps on models with diameter of 2¾-in. through 24-in. low-profile style • 60 cps on 24-in. through 42-in. welded style • Delivery rates vary from moderate to high speeds depending on part characteristics
Centrifugal feeders	• Rigid disc design appropriate for feeding and orienting larger parts • Combined disc and rim drive system • Delivery rates vary from moderate to high speeds depending on part characteristics • Outer wall diameter: 71 in. • Overall height: 46.9 in. ±1 in. • Discharge height: 32.47 in. • Bowl outer diameter: 56 in. • Disc diameter: 47.10 in. • Rim width: 4.13 in. • Single 90 VDC motor drive system • Motor size: ¾ hp • Supply voltage: 115 VAC
Storage hopper	• Cubic foot capacity ranges from 1.0 to 10.0 • Adjustable speed control • 115 VAC, 60 Hz (230 VAC, 50 Hz optional) • Quick dump door optional • Noncontact limit switch optional
Vertical bucket elevator	• Cubic foot storage capacity: 1.0 to 10.0 • Low bulk load height: approx. 28 in. at 1 cu. ft capacity; approx. 42 in. at 5 cu. ft capacity • 115 VAC, 1 phase, 5 amp • Capable of handling most parts that can be fed in standard bulk supply hopper
Overhead prefeeder	• Design for handling small parts and engineered to interface directly with part feeders • Hopper width and length: 18 × 18 in. • Overall height: 15.25 in. • Discharge height: 2 in. ±1 • Belt width: 5 in. • Motor size ¹⁄₃₀ hp • Supply voltage/Motor voltage: 115 VAC/90 VDC

Table 9   Assembly Machinery

Type	Key Features
Synchronous carousel assembly	• Cam-driven, synchronous indexing, high-production assembly machine. Suitable for smaller parts assembly. High-speed assembly in a minimum of floor space • Basic machinery can be customized by tooling it to meet any particular assembly application • Besides special tooling, optional equipment can include a profile plate to check part configuration, a shot-pin device to locate the sprocket, a photoelectric scanner for identification, and any type of semiautomatic or fully automatic control system required to meet production rates • Some typical applications: 　• Circuit breaker: 972 per hr (3.7 sec cycle) 　• Sprinkler head: 1773 per hr (2.0 sec cycle) and 25 models 　• Washing machine agitator shaft: 750 per hr (4.8 sec cycle) 　• Windshield wiper motor: 1200 per hr (3.0 sec cycle)
Nonsynchronous carousel assembly	• System designed to operate efficiently with modular components, basic stations, and a wide variety of special engineered stations. This type of assembly system is well suited to both single products and product families with multiple models. • Advantages: well suited for larger or more complex assemblies and flexibility (mixture of automatic, semi-automatic, and manual stations; variable cycle time; incorporation of repair loops; addition or deletion of stations) • Some typical applications: 　• Washing mach. transmission: production rate 310 per hr; 26 stations 　• Wheel bearing: production rate 200 per hr; 22 stations 　• Air conditioning compressor: production rate 300 per hr; 50 stations 　• Disc brake: production rate 250 per hr; 29 stations 　• Rack & pinion steering gear: production rate 240 per hr; 53 stations
Assembly transport system	• Conveyor with a universal rotate/translate module that performs the functions of a corner, a lift-and-rotate and a lift-and-translate without lifting the pallet. • No sliding friction • Drive shafts and clutches are eliminated. • Power consumption reduced 25–75% • Roller run only when a pallet is moving in that zone • Wide range of pallet sizes (18–48 in.) • Can handle weights up to 1500 lb. • Individual conveyor section can run at different speeds • High speeds possible due to no-impact stops (150 fpm) • Rotate/translate module can rotate a pallet in either direction and translate (redirect) it forward, backward, left or right

4 ROBOTICS JOURNALS

The following table provides information on some journals dedicated to robotics and the robotics industry.

Table 10   Classification of Journals by Type and Subject

	Publication Type				Publication Subject Matter										
	Country of Publication	Language of Publication	Publication Type	Publication Frequency	Artificial Intelligence	Automation	Industrial Application	Modelling & Simulation	Product Developments	Prototyping & Demonstration	Robot Design/Specifications	Robotics Systems Theory	Technological Advances	Other	Not Available
AI Today	US	EN	AS	BM	•		•								
A F R I Liason	FR	FR		Q		•	•		•						
Advanced Manufacturing Technology	US	EN	N	M			•								
Advanced Robotics	NE	EN	AS	BM									•		
Advances in Automation and Robotics	US	EN		I		•			•						
Advances in Design and Manufacturing	NE	EN		I					•						
Archives of Control Sciences	PL	EN		Q											•
Artificial Intelligence	NE	EN	AS	18/	•								•		
Australian Robot Association Newsletter	AT	EN	N	Q											•
Automatica and Robotica	SP			12/											•
Automation	JA	JA		M	•	•	•								
Automation News	US	EN		SM		•	•								
Automatizacion de la Produccion	SP	SP		6/		•									
Automatizacion Integrada y Revista de Robotica	SP	SP		M											•
Automazione Oggi	IT	IT		20/		•									
Autonomous Robots	NE	EN	AS	Q			•					•			
Axes Robotique	FR	FR		10/											•
Bibliography of Robotic and Technical Resources	US	EN	B	A										•	
Business Ratio Report: Industrial Robots	UK	EN	TP	A										•	
Canadian Institute for Advanced Research in Artificial Intelligence and Robotics	CA				•							•	•		

	Publication Type				Publication Subject Matter										
	Country of Publication	Language of Publication	Publication Type	Publication Frequency	Artificial Intelligence	Automation	Industrial Application	Modelling & Simulation	Product Developments	Prototyping & Demonstration	Robot Design/Specifications	Robotics Systems Theory	Technological Advances	Other	Not Available
Chino Ido Robotto Shinpojumu Shiryo (Intelligent Robot Symposium Proceedings)	JA	JA	P	BY											•
Componentes, Equipos y Sistemas De Automatica y Robotica	SP	SP				•	•		•						
Conference on Remote Systems Technology, Proceedings	NE		P	A							•				
Engineering Design and Automation	US	EN	AS	Q		•									
Flexible Automation	GE	GE	TP	6/		•									
IEEE International Conference on Robotics and Automation Proceedings	US	EN	P	A		•									
IEEE Robotics and Automation Magazine	US	EN		Q		•						•	•		
IEEE Transactions on Robotics and Automation	US	EN	AS	BM		•	•			•					
IFR Robotics Newsletter	SW	EN	N	4/								•			
IIE Solutions	US	EN	NP	M		•		•							•
Industrial News	UK	EN	TP	BM							•				
Industrial Robot	UK	EN	TP	BM			•		•						
Integrated Manufacturing Systems	UK	EN		6/		•	•				•		•		
International Conference on Advanced Robotics Proceedings	JA	JA	P	I											•
International Journal of Robotics and Automation	CN	EN	AS	4/		•	•	•			•				
International Journal of Robotics Research	US	EN		BM			•								
International Robotics Product Database	US	EN	C	A							•				
International Symposium on Automation and Robotics in Construction Proceedings	JA	JA	P	I		•	•					•			
Internal Symposium on Industrial Robots, Proceedings	JA	JA	P	I		•									
JARA Robot News	JA	EN	N	BM			•								
Jiqiren/Robot	PC	CH		BM										•	
Journal de la Robotique et Informatique	FR	FR		11/			•								
Journal of Intelligent and Robotic Systems	NE	EN	AS	M	•		•						•		
Journal of Machinery Manufacture and Reliability	US	EN	AS	BM		•		•							
Journal of Robotic Systems	US	E/J	AS	M			•				•	•			
Journal of Robotics and Mechatronics	JA	EN		BM											

	Publication Type				Publication Subject Matter										
	Country of Publication	Language of Publication	Publication Type	Publication Frequency	Artificial Intelligence	Automation	Industrial Application	Modelling & Simulation	Product Developments	Prototyping & Demonstration	Robot Design/Specifications	Robotics Systems Theory	Technological Advances	Other	Not Available
Kensetsu Robotto Shinpojumu Ronbunshu (Symposium on Construction Robotics in Japan, Proceedings)	JA	J(E)	P	A			•								
Kompass Professional Machines	FR		D											•	
Laboratory Robotics and Automation	US	EN	AS	BM		•		•					•		
Military Robotics Newsletter	US	EN	N	SM			•								
Mobile Robots	US				•	•									
Nihon Robotto Gakkai Robotto Shinpojumu Yokoshu (Robotics Society of Japan, Preprints of Robotics Symposium)	JA	J/E	P	A											•
Nihon Robotto Gakkai Gakaujutsu Koenkai Yokoshu (Robotics Society of Japan, Preprints of the Meeting)	JA	J/E		A											•
Nihon Robotto Gakkaishi (Robotics Society of Japan Journal)	JA	J(E)		BM											•
Pascale 34. Robotique, Automatique et Automatisation des Processus Industriels	FR	F/E	BG	10/		•	•								
Precision Machinery	US	EN		4/											•
Problemi Na Tekhnicheskata Kibernetika I Robotika (Problems of Engineering Cybernetics and Robotics)	BU	V	AS	I			•		•			•			
Productique-Affaires	FR	F/E		18/									•		
Progress in Robotics and Intelligent Systems	US	EN	AS	A					•						
RIA Quarterly Statistics Report—Robotics	US	EN		Q			•								
Robot Explorer	US	EN	N	8/								•	•	•	
Robot Times	US	EN		Q									•		
Roboter	GR	GR	TP	4/											•
Robotic Age	AT	EN		Q											•
Robotica	UK	EN	AS	BM		•	•					•			
Robotics and Autonomous Systems	NE	EN	AS	8/				•							
Robotics and Computer-Integrated Manufacturing	UK	EN	AS	Q			•					•			
Robotics and Expert Systems	US	EN	P	I	•										
Robotics and Manufacturing	US	EN		BY										•	
Robotics Review	US	EN		A											•
Robotics Today	US	EN	TP	Q		•	•								

	Publication Type				Publication Subject Matter										
	Country of Publication	Language of Publication	Publication Type	Publication Frequency	Artificial Intelligence	Automation	Industrial Application	Modelling & Simulation	Product Developments	Prototyping & Demonstration	Robot Design/Specifications	Robotics Systems Theory	Technological Advances	Other	Not Available
Robotikusu Mekatoronikusu Koenkai Koen Ronbunshu (Annual Conference on Robotics and Mechatronics)	JA	EN	P	A											•
Robotronics Age Newsletter	CN	EN	N	M											•
Robotto (Robot)	JA	J(E)		BM											•
Robotto Sensa Shinpojumu Yokoshu (Robot Sensor Symposium, Preprints)	JA	J(E)		BY											•
Sangyoyo Robotto Riyo Gijutsu Koshukai Tekisuto (Text of Lectures on Utilization Techniques of Industrial Robots)	JA	JA		3/											•
Series in Robotics and Automated Systems	SI	EN	MS	I	•										
Service Robot	UK	EN	TP	3/											•
Specifications and Application of Industrial Robots in Japan: Manufacturing Fields	JA	E		BY			•				•				
Specifications and Applications of Industrial Robots in Japan: Non-Manufacturing Fields	JA	EN		BY			•				•				
Tecniche Dell'Automazione e Robotica	IT	IT		M			•								
World Robot Statistics	SW	EN		A											•

Country of Publication:		Language of Publication:		Publication Type:		Publication Frequency:	
AT	Australia	Ch	Chinese	AS	Academic / Scholarly Pub.	A	Annual
BU	Bulgaria	EN	English	B	Bibliography	BM	Bimonthly
CN	Canada	F / E	French / English	C	Catalog	BY	Biyearly
FR	France	FR	French	D	Directory	I	Irregular
GR	Germany	GR	German	MS	Monographic Series	M	Monthly
JA	Japan	IT	Italian	N	Newsletter	SM	Semimonthly
NE	Netherlands	JA	Japanese	P	Proceedings	Q	Quarterly
PC	People's Rep. of China	J / E	Japanese / English	TP	Trade Publication	x/	x / Year
PL	Poland	J(E)	Japan. Summaries /				
SI	Singapore		Eng. Text				
SP	Spain	SP	Spanish				
SW	Sweden						
UK	United Kingdom						
US	United States						

5 ROBOTICS AND RELATED JOURNALS

Although not dedicated to providing information on robotics, the following journals often carry information on the robotics development and research (revised from HBIR, 1985).

Australia
  Australian Machinery and Production
    Engineering
  Australian Welding Journal
  Electrical Engineer
  Journal of the Institution of Engineers
  Metals of Australia
Austria
  Schweisstechnik
  Diagramm
Belgium
  Manutention Mécanique et Automation
  Revue M (Mecanique)
Bulgaria
  Mashinostroene
  Teoretichna i Prilozhnaa Mekhanika
Canada
  Canadian Machinery and Metalworking
Czechoslovakia
  Slévárenstvi
  Strojinrenska Vyroba
  Techmeká Práce
Germany
  Biological Cybernetics
  Bleeh Rohre
  Der Plastverarbeiter mit Sonderdruck
  Die Computerzeitung
  Die Maschine
  Die Wirtschaft
  DVS-Berichte
  Elektronik
  Elektronikindustrie
  Elektrotechnische Zeitschrift
  Feingerätetechnik
  Feinwerktechnik und Messtechnik
  Fertigungtechnik und Berieb
  Fördern und Heben
  Hebezeuge und Fördermittel
  Industrie-Anzeiger
  Kunststoffe
  Lecture Notes in Computer Science
  Maschinen-Anlagen + Verfahren
  Maschinenbautechnik
  Maschinenmarkt + Europa Industrie
    Revue
  Maschine und Werkzeug
  Messen-Steuern-Regeln
  Metallverarbeitung
  Praktiker
  Produktion
  Regelungstechnik
  Regelungstechnische Praxis
  Schweissen und Schneiden
  Schweisstechnik
  Seewirtschaft
  Siemens Energietechnik

  Sozialistische Rationalisierung in der
    Elektrotechnik/Elektronik
  Technische Gemeinschaft
  Technisches Zentralblatt für die
    Gesamte Technik
  VEM-Elektro-Anlagenbahn
  VDI
  Werkstatt und Betrieb
  Werkstattstechnik
  Wissenschaftlich-Technische
    Informationen
  Wissenschaftliche Zeitschrift der
    Technischen Hochschule Ilmenau
  Zeitschrift für Angewandte Mathematik
    und Mechanik
  ZIS-Mitteilungen
  Zeitschrift für Wirtschaftliche Fertigung
Finland
  Konepajamies
France
  Energie fluide. hydraulique,
    pneumatique asservissements,
    lubrification
  Fondeur aujourd'hui
  Machine moderne
  Machine-outil
  Manutention
  Métaux déformation
  Nouvel automatisme
  Soudage et techniques connexes
  L'Usine nouvelle
Hungary
  Automatizálás
  Bányászati és Kohászati Lapok Öntöde
    Az Orzágos Magyar Bányászati és
    Koháztáti Egyesület Lapja
  Gépgyartástechnológia
  Ipargazdaság
  Mérés és Automatika
Italy
  La tecnica professionale
  Maccine
  Rivista de Meccanica
  Transport Industriali
Japan
  Chemical Engineering
  Hydraulics and Pneumatics
  Japan Economic Journal
  Japan Light Metal Welding
  Journal of the Instrumentation Control
    Association
  Journal of the Japan Welding Society
  Mechanical Automation
  Mechanical Design
  Mechanical Engineering
  Mitsubishi Denki Giho
  Promoting Machine Industry in Japan

Robotpia
Science of Machine
Journal of the Institute of Electrical
   Engineers of Japan
Toshiba Review
Transactions of the Institute of
   Electronics and Comm. Engineering
Transactions of the Society of
   Instruments and Control Engineers
Welding Technique
Netherlands
Ingenieur
International Journal of Production
   Economics
Iron Age Metalworking International
Metalbewerking
Polytechnisch Tijdschrift
New Zealand
Automation and Control
Poland
Automateka Kolejowa
Biuletyn Informacyjiny Institutu
   Maszyn Matematyeznyh
Mechanik
Przeglad Mechaniczny
Przeglad Spawalnietwa
Wiadomosci Elektrotechniezne
Zeszyty Naukowe
Rumania
Constructia de Masini
International Journal of Informatics
Russia and the Former Soviet
Republics
Avtomatika i Telemekhanika
Avtomatizatsiya
   Proizvodstvennykh Protssov v
   Mashinostroenii i
   Priborostroenii
Avtomatizatsiya
   Tekhnologicheskikh Protsessov
Avtomatizirovannyy Elekropivod
Vestnik Mashinostroeniya
Voprosy Dinamiki i Prochnosti
Izvestiya Akademii Nauk SSSR
Isvestiya Vysshikh Uchebnykh
   Zavedeniy
Izvestiya Leningradskogo
Kuznechno-stampovochnoye
   Proizvodstvo
Liteynoye Proizvodstvo
Mashinovedeniye
Mashinostroitel
Mekhanizatsiya i Avtomatizatsiya
   Proizvdstva
Mekhanizatsiya I Elektrifikatsiya
   Sel'skogo Khozyaystva
Priborostroeniye
Promyshlennyy Transport
Svarochnoye Proizvodstvo
Stanki i Instrument
Stroitel'nyye i Dorohnyye Mashiny
Sudostroeniye

Trudy Vsesoyuznogo
   Nauchnoissledovatel'skogo
   Instituta Ugol'nogo
   Mashinostroeniya
Trudy Leningradskogo
   Politekhnicheskogo Instituta
Trudy Moskovskogo Energeticheskogo
   Instituta
Elektrotekhnika
Elektrotekhnicheskaya Promyshlenost'
Spain
Regulación y Mando Automatico
Switzerland
CIRP Annals
Elektroniker
Management Zeitschrift Ind. Organis.
Schweisstechnik Soudure
Schweizerische Technische Zeitschrift
Technica
Technische Rundschau
Zeitschrift Schweisstechnik
United Kingdom
Assembly Automation
Assembly Engineering
Automotive Engineering
British Foundryman
Control and Instrumentation
Design Engg. Materials and
   Components
Electrical Review
Engineer
Foundry Trade Journal
Hydraulic Pneumatic Mechanical Power
International Journal of Man–Machine
   Studies
International Journal of Production
   Research
Machinery & Production Engineering
Manufacturing Engineer
Materials Handling News
Mechanism & Machine Theory
Metals and Materials
Metalworking Production
New Electronica New Scientist
Pattern Recognition
Plastics in Engineering
Robot News International
Sensor Review
Sheet Metals Industries
Welding and Metal Fabrication
United States
American Machinist
American Metalmarket
ASME Trans. on Dynamics,
   Measurement, and Control
Assembly Engineering
Automatic Machining
Compressed Air
Computer Graphic and Image
   Processing
Design News
Electronics

Futurist
Hydraulics and Pneumatics
IEEE Spectrum
IEEE Transactions on Automatic
  Control
IEEE Transactions on Industrial
  Electronics and Control
  Instrumentation
IEEE Transactions on Power Apparatus
  and Systems
IEEE Transactions on Systems, Man,
  and Cybernetics
IIE Transactions
Industrial Engineering
Industrial Robots International
Information and Control
Intl. Journal of Computer and
  Information Science

Iron Age
Journal of Manufacturing Systems
Machine and Tool Blue Book
Manufacturing Engineer
Material Handling Engineering
Mechanical Engineering
Modern Material Handling
Plating and Surface Finishing
Product Engineering
Production
Robotics World
Tooling and Production
Welding Design and Fabrication
Welding Engineer
Welding Journal
Yugoslavia
Automatika

6   CONTACT INFORMATION OF SOME ROBOTICS AND AUTOMATION ASSOCIATIONS

Australian Robotics and Automation
  Association Inc.
G.P.O. Box 1527
Sydney N.S.W. 2001
AUSTRALIA
  Web Site: http://www.araa.asn.au
  Tel: (02) 9959 3239
  Fax: (02) 9959 4632

Austria IFAC Beirat Österreich
Floragasse 7A
A-1040 VIENNA
Austria
  Tel: +43 (1) 504 18 35
  Fax: +43 (1) 504 18 359

Chinese Society of Automation
1001, No 293
Song Chiang Rd.
Taipei
China
  Tel: +886 (2) 505 4168
  Fax: +886 (2) 503 0554

DIRA
Vestre Skolevej 39
DK-8464 Galten
DENMARK
  Tel/Fax: +45 86 94 64 10

Robotics Society in Finland
c/o Suomen Automation Tuki Oy
Asemapäällikönkatu 12 C
FIN-00520 HELSINKI
Finland
  http://www.roboyhd.fi/indexeng.html
  Tel: +358 (9) 584 008 20
  Fax: +358 (9) 146 1650

SYMAP
Syndicat de la machine-outil, du soudage,
  de l'assemblage
et de la productique associée
Maison de la mecanique
45, rue Louis Blanc
F-92400 COURBEVOIE
France
  Tel: +33 (1) 4717 6700
  Fax: +33 (1) 4717 6725

Robotik + Automation im VDMA
P.O. Box 71 08 64
DE-640498 FRANKFURT (MAIN)
Germany
  Tel: +49 (69) 660 314 66
  Fax: +49 (69) 660 314 59

Hungarian Robotics Association
Népszínház utca 8
H-1081 BUDAPEST
Hungary
  Tel: +36 (1) 114 2620
  Fax: +36 (1) 113 9183

Israel: Robotics & FMS Club, Israel SME
  (#319)
Meirav Hadracha
4, Nevatim St.
Petach-Tikva 49561, Israel
  Tel. +972-3-922-8422
  Fax +972-3-922-8433

SIRI-Associazione Italiana di Robotica
Viale Fulvio Testi, 128
I-20092 CINISELLO BALSAMO (MI)
Italy
  Tel: +39 (2) 262 55 257
  Fax: +39 (2) 262 55 349

JARA-Japan Robot Association
c/o Kikaishinko Bldg.
3-5-8 Shibakoen, Minato-ku
Tokyo 105
Japan
  Tel: +81 (3) 3434 2919
  Fax: +81 (3) 3578 1404

Malaysia Centre for Robotics and
  Industrial Automation
c/o Universiti Sains Malaysia
(Perak Branch Campus)
Seri Iskandar, 31750 Tronoh
Perak
Malaysia
  Tel: +60 (5) 367 77 94
  Fax: +60 (5) 367 77 98

Federation of Norwegian Engineering
  Industries (TBL)
Box 7072-H
N-0306 OSLO 3
Norway
  Tel: +47 (22) 59 66 00
  Fax: +47 (22) 59 66 69

Polish Society for Measurement,
  Automatic Control and Robotics—
  POLSPAR
Czackiego Str. 3/5
PL-00950 WARSZAWA
Poland
  Tel: +48 (22) 26 87 31
  Fax: +48 (22) 27 29 49

Singapore Industrial Automation
  Association
151 Chin Swee Road
#03-13 Manhattan House
Singapore 169876
  http://www.asia-mfg.com/siaa
  Tel: (65) 734-6911
  Fax: (65) 235-5721

Asociación Española de Robótica (AER)
Rambla de Catalunya 70, 3o 2a
E-08007 BARCELONA
Spain
  Tel: +34 (3) 215 5760
  Fax: +34 (3) 215 2307

International Federation of Robotics (IFR)
P.O. Box 5510
SE-114 85 STOCKHOLM
Sweden
  http://www.ifr.org
  Tel: +46 8 782 08 00
  Fax: +46 8 660 33 78

Swedish Robot Association—SWIRA
P.O. Box 5510
SE-114 85 STOCKHOLM
Sweden
  Tel: +46 (8) 782 0800
  Fax: +46 (8) 660 3378

Schweizerische Gesellschaft für
  Automatik
c/o Automatic Control Lab.
ETH-Zentrum, ETL
CH-8092 ZÜRICH
Switzerland
  Tel: +46 (8) 782 0800
  Fax: +46 (8) 660 3378

The British Robot Association (BRA)
Aston Science Park
Love Lane
Birmingham B7 4BJ
United Kingdom
  http://www.bra-automation.co.uk/
  Tel: (0121) 628 1745
  Fax: (0121) 628 1746

Association for Robotics in Hazardous
  Environments (RHE)
P.O. Box 3724
Ann Arbor, MI 48106
United States of America
  Web Site: *http://www.robotics.org/*
    htdocs/meetria/rhe.html
  Tel: +1 (734) 994 6088
  Fax: +1 (734) 994 3338

Automated Imaging Association
P.O. Box 3724
Ann Arbor, MI 48106
United States of America
  Web Site: *http://www.robotics.org/*
    htdocs/meetria/aia.html
  Tel: +1 (734) 994 6088
  Fax: +1 (734) 994 3338

International Service Robot Association
  (ISRA)
P.O. Box 3724
Ann Arbor, MI 48106
United States of America
  Web Site: *http://www.robotics.org/*
    htdocs/meetria/isra.html
  Tel: +1 (734) 994 6088
  Fax: +1 (734) 994 3338

Robotic Industries Association (RIA)
900 Victors Way
P.O. Box 3724
Ann Arbor, MI 48106
United States of America
  http://www.robotics.org
  Tel: +1 (734) 994-6088
  Fax: +1 (734) 994-3338

USA: Robotic International, SME
One SME Drive
P.O. Box 930
Darborn, MI 48121-9300, USA
  http://www.SME.org/
  Tel. +1-313-271-1500
  Fax +1-313-271-2861

7 ROBOTICS RESEARCH LABORATORIES

Almost every university has one or more robotics laboratory. The following list of web sites is a sample.

Autonomous and Intelligent Machines: http://198.110.216.210/

Boston University—A/ME Robotics and Control Group: http://robotics.bu.edu/

Brown University—AI/Robotics Lab: http://www.cs.brown.edu/research/robotics/

Bucknell University—Bucknell Robotics Laboratory: http://sun.bucknell.edu/
~robotics/brl.html

California Institute of Technology—Robotics Group: http://robby.caltech.edu/

Carnegie-Mellon University—/http://www.cs.cmu.edu/scs/project-homes.html/
Robocog

Cornell University—Cornell Computer Science Robotics and Vision Laboratory: http:
//simon.cs.cornell.edu/Info/Projects/csrvl/csrvl.html

Georgia Tech—Intelligent Machine Dynamics: http://davinci.marc.gatech.edu/

Georgia Tech—Mobile Robotics: http://www.cc.gatech.edu/aimosaic/robot-lab/
MRLHome.html

Harvard University—hrl: http://hrl.harvard.edu/

Johns Hopkins University—Robot Kinematics and Motion Planning Laboratory: http:
//caesar.me.jhu.edu/

Massachusetts Institute of Technology—AI Lab: http://www.ai.mit.edu/

Stanford University—Robotics Laboratory: http://robotics.stanford.edu/

Stanford University—Dextrous Manipulation Laboratory: http://cdr.stanford.edu/
html/Touch/touchpage.html

UC Berkeley—Robotics & Intelligent Machines Laboratory: http://
robotics.eecs.berkeley.edu/

University of Massachusetts—Laboratory for Perceptual Robotics: http://www-
robotics.cs.umass.edu/lpr.html

University of Southern California—Modular Robotics Laboratory: http://
www.usc.edu/users/goldberg/mrl.html

German Aerospace Center—Institute of Robotics and System Dynamics (dlr): http://
www.op.dlr.de/FF-DR/

Politecnico di Milano Laboratory of Electronics and Information—Artificial
Intelligence and Robotics Project: http://www.elet.polimi.it/section/compeng/air/

University of Edinburgh—Mobile Robot Group: http://www.dai.ed.ac.uk/groups/
mrg/MRG.html

University of Western Australia—Telerobotics: http://telerobot.mech.uwa.edu.au/

8 ROBOTICS INTERNET RESOURCES

Medical Robotics and Computer Assisted Surgery Jumpstation: http://
www.ius.cs.cmu.edu/mrcas/mmenu.html

Mobile Robots Laboratory: Interesting Sites Around the Web: http://
www.cc.gatech.edu/ai/robot-lab/mrl-jump-points.html

Robotics FAQ: http://www.cs.cmu.edu/groups/ai/html/faps/ai/robotics/top.html

Robotics Internet Resources Compendium: http://sun.bucknell.edu/~robotics/
rirc.html

Robotics Internet Resources Page: http://www-robotics.cs.umass.edu/robotics.html

PART 12
ROBOTICS TERMINOLOGY

ROBOTICS TERMINOLOGY

José A. Ceroni
Shimon Y. Nof
Purdue University
West Lafayette, Indiana

INTRODUCTION

This section provides a comprehensive list of terms that are relevant to the field of industrial robotics. It augments and revises the robotics terminology from the first edition of this Handbook, which was originally prepared by Joseph Jablonowski and Jack W. Posey. The definitions of the terms reflect their meaning in the field of robotics.

Almost all the terms and definitions come from the material covered in the Handbook. Of the 530 terms compiled for the Handbook's first edition, 82 were deleted or revised, and 371 new terms were added for a total of 819 terms. This increase reflects the expansion and maturing of the robotics field. The contribution of new terms from each of the 10 areas of the Handbook was according to the list in Table 1.

Table 1

Area	New Terms Added
1 Development of Industrial Robotics	20
2 Mechanical Design	48
3 Control and Intelligence	41
4 Programming and Intelligence	38
5 Organizational and Economic Aspects	27
6 Applications: Planning Techniques	27
7 Applications: Design and Integration	36
8 Robotics in Processes	24
9 Robotics in Operations	26
10 Robotics in Various Applications	29
From multiple areas	55
Total new terms	371
Terms reviewed or updated from the first edition	448
Total terms in the Robotics Terminology	819

This terminology is not intended to be an exhaustive, final word on the vocabulary of the field of robotics; rather, it is a reflection of the importance the authors have placed on various theories, practices, and equipment in the field.

ROBOTIC TERMS

A

Accommodation, Active: A control technique that integrates information from sensors with the robot's motion in order to respond to felt forces. Used to stop a robot when forces reach set levels, in guiding tasks like edge tracing and insertion, and to provide the robot with a capability to compensate for errors in the positioning and orientation of a workpiece.

Accommodation, Passive: The capability of a manipulator to correct for residual positioning errors through the sensing of reaction forces and torques as sensed by a compliant wrist. No sensors, actuators, or controls are involved.

Accuracy:

1. The quality, state, or degree of conformance to a recognized standard or specification.
2. The ability of a robot to position its end-effector at a programmed location in space. Accuracy is characterized by the difference between the position to which the robot tool-point automatically goes and the originally taught position, particularly at nominal load and normal operating temperature.

Compare with Repeatability.

Actuator: A motor or transducer that converts electrical, hydraulic, or pneumatic energy into power for motion or reaction.

Adaptive Arc Welding System: An arc welding system consisting of an off-line programming system and a real-time controller. The off-line programming system allows the planning of the entire automated welding process, the real-time controller provides the system with the capability for implementing the welding plans and dealing with the anomalies that may arise during their execution.

Adaptive Control: A control method used to achieve near-optimum performance by continuously and automatically adjusting control parameters in response to measured process variables. Its operation is in the conventional manner of a machine tool or robot with two additional components:

1. At least one sensor which is able to measure working conditions; and
2. A computer algorithm which processes the sensor information and sends suitable signals to correct the operation of the conventional system.

Adaptive Network Based Fuzzy Inference System (ANFIS):   A fuzzy inference system tuned with a backpropagation algorithm based on some collection of input–output data. This setup allows fuzzy systems to learn. ANFIS are highly specialized for speed and cannot accept all the customization options that basic fuzzy inference systems allow.

Adaptive Robot: A robot equipped with one or more external sensors, interfaced with other machines and communicating with other computers. This type of robot could exhibit aspects of intelligent behavior, considerably beyond unaided human capabilities, by detecting, measuring, and analyzing data about its environment and using both passive and active means for interaction. Sensory data could include signals from electromagnetic spectrum, acoustic signals, measurements of temperature, pressure, and humidity, measurements of physical and chemical properties of materials, detection of contaminants, and electrical signals (see also Responsive Robots).

Adaptive System: A robotic system with adaptive control capability having:

1. At least one sensor which is able to measure the working conditions; and
2. A computer algorithm which processes the sensor information and sends suitable signals to correct the operations of the conventional system.

Agent: A computing hardware and/or software-based system that has the following properties:

1. Autonomy: Agents operate without the direct intervention of humans or others, and have certain control over their actions and internal state.
2. Social ability: Agents interact with other agents (and possibly humans) via some agent-communication language.
3. Reflexivity: Agents perceive their environment, which may be the physical world, a user via a graphical user interface, a collection of other agents, the Internet, or

perhaps all these combined, and react in a timely fashion to changes that occur in this environment.

4. Proactiveness: Agents do not simply act in response to their environment, but are able to exhibit goal-directed behavior by taking the initiative.

The importance of computer agents is their ability to perform complex intelligent activities in a highly distributed global environment with the advantages of synergy and in parallel to their human users.

Air Jet Sensor: A sensor, located on the dispenser nozzle of a dispensing system, that uses an air jet to detect any break in bead application.

Air Motor: A device that converts pneumatic pressure and flow into continuous rotary or reciprocating motion.

AL: A research-oriented motion language developed by the Stanford University Artificial Intelligence Laboratory, designed primarily to deal with robot positions and velocities. AL's features include joint interpolated motion, force sensing and application, changeable tools, and vision verification. It is based on the ALGOL language.

Algorithm: A set of specific rules for the solution of a problem in a finite number of steps.

AML (A Manufacturing Language): An interactive, structured robot programming language developed by IBM and capable of handling many operations, including interfacing and data processing, that go beyond the programming of robot motions. Command categories include robot motion and sensor commands. AML supports joint interpolated motion and force sensing.

Analog Control: Control signals that are processed through analog means. Analog control can be electronic, hydraulic, or pneumatic.

Android: A robot that resembles a human being.

Angular Heading Changes: Changes in the path direction of a mobile robot. In multiwheeled systems it is caused when two or more wheels are actively driven with the others idling.

Anthropomorphic Robot: Also known as a *jointed-arm robot*. A robot with all rotary joints and motions similar to a person's arm.

Arm: An interconnected set of of links and powered joints comprising a manipulator which support or move a wrist, hand, or end-effector.

Arm Joint Accuracy (AJA): One of the Robot Motion Economy measures. It measures the accuracy of the robot arm during motion. In point-to-point tasks the accuracy is important at the end of the motion, while in continuous path tasks accuracy is an important measure throughout the motion.

Arm Joint Load (AJL): One of the Robot Motion Economy measures. It estimates the load to which the arm is subjected during task motions.

Arm Joint Utilization (AJU): One of the Robot Motion Economy measures. It is calculated based on the total weighted movement of all joints during the task performance and is based on the kinematics of robot motions. Motions of joints closer to the robot base are weighted more heavily to reflect relatively higher wear.

Articulated Robot: A robot arm which contains at least two consecutive revolute joints acting around parallel axes resembling human arm motion. The work envelope is formed by partial cylinders or spheres. The two basic type of articulated robots, vertical and horizontal, are sometimes called anthropomorphic because of the resemblance to the motions of the human arm.

Artificial Intelligence (AI): The ability of a machine system to perceive anticipated or unanticipated new conditions, decide what actions must be performed under the conditions, and plan the actions accordingly. The main areas of application are knowledge-based systems, computer sensory systems, and machine learning.

Artificial Muscles: A chemical compound designed to emulate the behavior of biological muscles. The compounds contract and expand by the application of positive electrical charges. Expected applications range from noiseless and quiet propulsion systems to human implants reducing the rejection of organ transplants. Another variety are the muscle wires made from titanium alloy that contract when electricity is passed through them and can lift thousands of times their own weight.

Artificial Neural Networks (ANN): A complementing approach of neural networks and artificial intelligence which is based on serial processing through layers of symbol manipulation. It is also referred to as *connectionism* or *parallel distributed processing*. The main applications of ANN are pattern recognition, planning, and control.

Assembly (Robotic): Robot manipulation of components resulting in a finished assembled product. Presently available simple robots can be used for simple assembly operations, such as mating two parts together. However, for more complex assembly operations robots require better control and intelligence for achieving the required positioning accuracy and sensory feedback. Examples of typical applications include the insertion of light bulbs into instrument panels, the assembly of computer hard drives, the insertion or placement of components on printed circuit boards, the assembly of small electric motors, and furniture assembly.

Assembly Constraints: Logical conditions that determine the set of all feasible assembly sequences for a given product. Assembly constraints can be of two types: geometric precedence constraints (those arising from the part geometry) and process constraints (those arising from assembly process issues).

Assembly Planner: A system for automating robot programming. The assembly planner examines a computer-assisted design database to produce task-level program. A task planner then creates a robot-level program from the task-level program.

Assembly Sequencing: The determination of one or more sequences in which the parts can be put together to guarantee a feasible assembly.

Asynchronous Program Control: A type of robot program control structure that allows the execution of event-driven steps of a program. Typical events can be hardware errors, program function key interrupts, or sensors exceeding specified ranges.

Atlantis: A hybrid behavior-based robot control architecture developed at NASA's Jet Propulsion Laboratory and consisting of a three-layer architecture. The reactive layer is a behavior-based system. At the other end of this hybrid system is a deliberative layer which plans and maintains world models. The intervening interface layer, referred to as the *sequencer*, controls sequences of reactivity and deliberation. In addition, the sequencer is used to handle cognizant failures, i.e., self-recognition when deliberation is necessary due to a failure to achieve the system's goals by reactive methods alone.

Atomic Force Microscope (AFM): A device for mapping surface atomic structure by measuring the force acting on the tip of a sharply pointed wire or other object that is moved over the surface.

Attended Program Verification (APV): A special mode of robot operation designed to confirm that a robot's programmed path and process performance are consistent with expectations is defined by the new standard. This is a testing mode when the robot is allowed to move at full-programmed speed, which presumably exceeds the slow speed velocity limits.

Automated Storage/Retrieval System (AS/RS): Automated warehouses where parts stored in bins are retrieved by automated equipment (usually a stacker crane) and delivered to different collection and distribution points. A computer-based system keeps track of how many items are stored in which bins and controls the system that selects the desired bin.

Automatically Guided Vehicle (AGV): Also known as *robot cart*. Wire- or rail-guided carts used to transport raw materials, tools, finished parts, and in-process parts over relatively great distances and through variable, programmable routes. The parts are usually loaded on pallets.

Automation: Automatically controlled operation of an apparatus, process, or system by mechanical or electronic devices that replace human observation, effort, and decision.

Autonomous Robot: A robot with some degree of autonomy.

Autonomous Robot Architecture (AURA): The first robot control architecture to merge classical artificial intelligence planning techniques with behavior-based control for robot navigation. AURA partitions the navigational task into two phases, planning and execution.

Autonomy: The capability of an entity to create and control the execution of its own plans and/or strategies. For instance, in mobile robots the ability of the robot for determining the trajectory to reach a specific location or pose.

AUTOPASS: An experimental task-oriented robot programming language developed by IBM, by which a manufacturing task to be accomplished is described. In a task-oriented language such as AUTOPASS the robot must recognize various high-level, real-world terms. AUTOPASS supports parallel processing, can interface with complex sensory systems, and permits the use of multiple manipulators. It is a PL/I-based language.

Axis: A traveled path in space, usually referred to as a linear direction of travel in any three dimensions. In Cartesian coordinate systems, labels of X, Y, and Z are commonly used to depict axis directions relative to Earth. X refers to a directional plane or line parallel to Earth. Y refers to a directional plane or line that is parallel to Earth and perpendicular to X. Z refers to a directional plane or line that is vertical to and perpendicular to the Earth's surface.

Axis, Prismatic: Also known as *translational axis*. An assembly between two rigid members in a mechanism enabling one to have a linear motion relative to and in contact with the other.

Axis, Rotational: Also known as *rotatory axis*. An assembly connecting two rigid members in a mechanism which enables one to rotate in relation to the other around a fixed axis.

B

B-Splines: Curves defined by a set of control points. It is used to represent the robot's end-effector trajectories that account for system dynamics and actuator limitations.

Backlash: The free play in a power transmission system, such as a gear train, resulting in a characteristic form of hysteresis.

Backlighting: Position of the light source in a vision system. The light source is placed behind the object of which the image is about to be acquired.

Backward Reasoning: A strategy for searching the rule base in an expert system that acts like a problem solver by beginning with a hypothesis and seeking out more information until the hypothesis is either proved or disproved.

Bang-Bang Control: A binary control system which rapidly changes from one mode or state to the other. In motion systems this applies to direction only. Bang-bang control is often mechanically actuated, hence the name.

Bang-Bang Robot: A robot in which motions are controlled by driving each axis or degree of freedom against a mechanical limit stop.

Base: The platform or structure to which a robot arm is attached; the end of a kinematic chain of arm links and joints, opposite to that which grasps or processes external objects.

Batching: The operation of a robotic system such that tasks of one family of operations, called a *batch*, are performed together. Batching is required when a system cannot easily perform a variety of tasks in random order. Different setups may be required for different families of tasks. Batching reduces the number of setups by performing one family of tasks, followed by another batch from another family, and so on.

Baud: The unit of measure of signalling speed in data communications. Baud is equal to the number of bits of signal events per second.

Behavioral Controller: A reactive, behavior-based control system, emphasizing the importance of tightly coupling sensing and action without the use of intervening representations or world models. A simple behavioral controller may include three primitive behaviors: avoiding obstacles, moving towards a perceived goal, and remaining on a path.

The coordination process determines the aggregate overt action of the robot by selecting or combining the individual behavioral outputs in a meaningful way.

Bidding Mechanism: A real-time technique for allocating the resources of a Multi-Robot System to competing jobs arriving to the system.

Blackboard Systems: Domain-specific problem-solving systems that exploit the blackboard architecture and exhibit a characteristically incremental and opportunistic problem-solving style. The blackboard architecture has three defining features: a global database called the *blackboard*, independent knowledge sources that generate solution elements on the blackboard, and a scheduler to control knowledge source activity.

Blob: A cluster of adjacent pixels of the same nature (gray level, or color in binary image, etc.) which usually represents an object or region in the field of view.

Bus Interface: A network topology linking a number of computers by a single circuit with all messages broadcast to the entire network.

Business Process Reengineering (BPR):   See Process Reengineering

C

Cable Drive: Also known as *tendon drive*. The transmission of mechanical power from an actuator to a remote mechanism via a flexible cable and pulleys.

CAD Solid Modeler: A software interface part of the Computer-Aided Design system which allows the user to build a database, with a valid and complete geometric description of the robot and its environment. This CAD database includes the models of the robot links and all the objects in the cell environment: machines, fixtures, feeders, grippers, parts, etc.

CAD/CAM: An acronym for Computer-Aided Design and Computer-Aided Manufacturing.

CAD/CAM Coupling: The integration of manufacturing capabilities and constraints (CAM) into the design process (CAD).

Calibration: The procedures used in determining actual values which describe the geometric dimensions and mechanical characteristics of a robot or multibody structure. A robot calibration system must consist of appropriate robot modeling techniques, accurate measurement equipment, and reliable model parameter determination methods. For practical improvement of a robot's absolute accuracy, error compensation methods are required which use calibration results.

Camera Coordinate Frame: A coordinate system assigned to the camera in a robotics vision system.

Canny's Edge Operator: Operator for line feature extraction from vision stereo images.

Capacitive Proximity Switch: Proximity sensor utilizing the dielectric property of the material to be sensed to alter the capacitance field set up by the sensor. The sensor can be used to detect objects constructed of nonconductive material, allowing the robot to detect the presence or absence of objects made of wood, paper, or plastic.

Cartesian Path Tracking: The travel of the hand or end-effector of a manipulator in a path described by Cartesian coordinates. The manipulator joint displacements can be determined by means of an inverse Jacobian transformation.

Cartesian Robot: See Rectangular Robot.

Cartesian Space: See Rectangular Robot.

Cell Formation: See Group Technology.

Cellular Robotic System: A hybrid layout organization in which robots and machining centers are grouped into work centers called *cells*, that process parts with similar shapes or processing requirements.

Center: A manufacturing unit consisting of two or more cells and the material transport and storage buffers that interconnect them.

Centralized Control: Control exercised over an extensive and complex system from a single controller.

Chain Drive: The transmission of power via chain and mating-toothed sprocket wheels.

Chain Robot Arm: A robot arm designed especially for use as a monoarticulate chain manipulator. Special design considerations include a cross-sectional profile no larger than that of the chain links, a fail-safe return to a straightened position, high-powered DC or AC servo motors, and continuous gas cooling of the drives. Actual and planned applications include repair and inspection of nuclear reactors and active liquor tanks housed in concrete cells.

Chebyshev Travel: Motion of the robot arm simultaneously in the X and Y directions of the robot's workspace coordinate system.

Chip Mounter: A robotic device employed for the high-speed automatic mounting of surface-mount type electronic components into integrated circuit boards. Modern high-speed chip mounters utilize high-speed vision assistance achieving mounting cycles of 0.15 sec/chip.

CIM/Robotics: The integration of various technologies and functional areas of the organization to produce an entirely integrated manufacturing organization. The motivation for CIM/Robotics has been based on the perceived need for manufacturing industry to respond to changes more rapidly than in the past. CIM/robotics promises many benefits, including increased machine utilization, reduced work-in-process inventory, increased productivity of working capital, reduced number of machine tools, reduced labor costs, reduced lead times, more consistent product quality, less floorspace, and reduced set-up costs.

Circle Point Analysis: A class of robot calibration methods where each joint is identified as a screw, i.e., as a line in space. From knowledge of all of the joint screws, the robot's kinematic parameters can be extracted directly.

Circular Interpolation: A robot programming feature which allows a robot to follow a circular path through three points it is taught.

Class I Testing: Robot testing procedure requiring no specific parameter optimization in the robot's operation.

Class II Testing: Robot testing procedure in which the robot operates under optimum cycle time conditions.

Class III Testing: Robot testing procedure in which the robot operates under optimum repeatability conditions.

Class IV Testing: Robot testing procedure that allows the optimization of robot performance characteristics beyond those covered by testing classes II and III.

Closed-Loop Control: The use of a feedback loop to measure and compare actual system performance with desired performance. This strategy allows the robot control to make any necessary adjustments.

Coating (Robotic):   Robot manipulation of a coating tool, e.g., a spray gun, to apply some material, such as paint, varnish, stain, or plastic powder, to the surface of either a stationary or a moving part. These coatings are applied to a wide variety of parts, including automotive body panels, appliances, and furniture. Other uses include applying resin and chopped glass fiber to molds for producing glass-reinforced plastic parts and spraying epoxy resin between layers of graphite broadgoods in the production of advanced composites. The benefits of robotic coating are higher product quality through more uniform application of material, reduced costs by eliminating human labor and reducing waste coating material, and reduced exposure of humans to toxic materials.

Cobot: A robot that is designed specifically to collaborate with humans. It is typically a safe, mechanically passive robotic device that is intended for direct physical contact with a human operator. The operator supplies the power for motions, while the cobot responds with software-controlled movements.

Cognitive Functions: A group of mental processes that involve acquisition, coding, storing, manipulation, and recall of spatial and other information.

Collaboration: The active participation and work of the sub-systems towards accomplishing collaborative integration. It can be characterized as mandatory, optional, or concurrent, and it can occur internally (among sub-systems of the same system) or externally (among sub-systems of different systems). An important function of collaborative integration is to overcome conflicts among the sub-systems.

Collapsible Frame Structure: A mobile robot system that uses almost its entire structural system in an active fashion to reconfigure itself. An example of this type of robot is Houdini, a robot developed for the Department of Energy as an internal tank cleanup system that would remotely access hazardous (or petrochemical) waste-storage tanks through a 24-inch diameter opening in the collapsed position, and then open up to a full 8-foot by 6-foot footprint and carry a manipulator arm with exchangeable end-effector tooling and articulated plow.

Common Object Request Broker Architecture (CORBA): A distributed object technology defined by the Object Management Group (OMG). CORBA is the standard architecture for developing object-oriented applications that would run across a diversity of multivendor products and operating environments.

Complex Joint: An assembly between two closely related rigid members enabling one member to rotate in relation to the other around a mobile axis.

Compliance: A feature of a robot which allows for mechanical float in the tooling in relation to the robot tool mounting plate. This feature enables the correction of misalignment errors encountered when parts are mated during assembly operations or loaded into tight-fitting fixture or periphery equipment.

Compliant Assembly: The deliberate placement of a known, engineered, and relatively large compliance into tooling in order to avoid wedging and jamming during rigid part assembly.

Compliant Motion: The motion required of a robot when in contact with a surface because of uncertainty in the world model and the inherent inaccuracy of a robot.

Compliant Support: In rigid part assembly, compliant support provides both lateral and angular compliance for at least one of the mating parts.

Computer Integrated Manufacturing—Open Systems Architecture (CIMOSA): A European proposal for an open system architecture for Computer-Integrated Manufacturing. CIMOSA provides a reference architecture to help particular enterprises build their own architecture, describing four aspects of the enterprise (function, information, resource, and organization) at three modeling levels (requirements definition, design specification, and implementation description). The enterprise model provides semantic unification of concepts shared in the CIM system. Alternative reference architectures are the Purdue Enterprise Reference Architecture (PERA), developed at Purdue University in Indiana, and the GRAI-GIM architecture, developed at the University of Bordeaux, France.

Computer Numerical Control (CNC): A numerical control system with a dedicated mini- or micro-computer which performs the functions of data processing and control.

Computer Vision: Also known as *machine vision*. The use of computers or other electronic hardware to acquire, interpret, and process visual information. It involves the use of visual sensors to create an electronic or numerical analog of a visual scene, and computer processing to extract intelligence from this representation. Examples of applications have been in inspection, measurement of critical dimensions, parts sorting and presentation, visual servoing for manipulator motions, and automatic assembly.

Computer-Aided Design (CAD): The use of an interactive-terminal workstation, usually with graphics capability, to automate the design of products. CAD includes functions like drafting and parts fitup.

Computer-Aided Engineering (CAE): Engineering analysis performed at a computer terminal with information from a CAD database. It includes mass property analysis, finite element analysis, and kinematics.

Computer-Aided Manufacturing (CAM): Working from a product design likely to exist in a CAD database, CAM encompasses the computer-based technologies that physically produce the product, including part-program preparation, process planning, tool design, process analysis, and part processing by numerically controlled machines.

Computer-Assisted Surgical Execution (CASE): The use of robotics devices in the execution of the surgical plan outlined during the surgical planning stage. The execution control can be computer or surgeon based.

Computer-Assisted Surgical Planning (CASP): The use of an interactive computer interface for planning the surgery procedure over the patient. It provides the surgical team with feedback information through the simulation of the surgical procedures.

Computer-Integrated Manufacturing (CIM): The philosophy dictating that all functions within a manufacturing operation be database driven and that information from within any single database be shared by other functional groups. CIM includes major functions of operations management (purchasing, inventory management, order entry), design and manufacturing engineering (CAD, NC programming, CAM), manufacturing (scheduling, fabrication, robotics, assembly, inspection, materials handling), and storage and retrieval (inventories, incoming inspection, shipping, vendor parts). Modern CIM covers manufacturing and service enterprise integration.

Computer-Integrated Surgery (CIS): The integration of computer-based modeling and analysis of patient anatomy with robots, sensors, and other interface technologies into systems that enhance the human surgeon ability to plan and carry out surgical interventions.

Concurrent Engineering (CE): The integration of product design and process planning into a common activity. It helps improve the quality of early design decisions and thereby reduces the length and cost of the design process.

Confined Sluicing End-Effector (CSEE): A special robot's end-effector for radioactive waste management developed at the Oak Ridge National Laboratory in Tennessee. The end-effector sucks waste out of the tank, but it also has a series of high-pressure cutting jets for breaking up solid waste.

Connectivity Analysis: An image analysis technique used to determine whether adjacent pixels of the same color constitute a blob, for which various user-specified features can be calculated. Used to separate multiple objects in a scene from each other and from the background.

Contact Sensor: A grouping of sensors consisting of tactile, touch, and force/torque sensors. A contact sensor is used to detect contact of the robot hand with external objects.

Continuous Path Control: A type of robot control in which the robot moves according to a replay of closely spaced points programmed on a constant time base during teaching. The points are first recorded as the robot is guided along a desired path, and the position of each axis is recorded by the control unit on a constant time basis by scanning axis encoders during the robot motion. The replay algorithm attempts to duplicate that motion. Alternatively, a continuous path control can be accomplished by interpolation of a desired path curve between a few taught points.

Continuous Path System: A type of robot movement in which the tool performs the task while the axes of motion are moving. All axes of motion may move simultaneously, each at a different velocity, in order to trace a required path or trajectory.

Control: The process of making a variable or system of variables conform to what is desired.

1. A device to achieve such conformance automatically
2. A device by which a person may communicate commands to a machine

Control Hierarchy: A system in which higher-level control elements are used to control lower-level ones and the results of lower-level elements are utilized as inputs by higher-level elements. Embodied in the phrase "distributed intelligence," in which processing units (usually microprocessors) are dispersed to control individual axes of a robot.

Control Loops: Segments of control of a robot. Each control loop corresponds to a drive unit which actuates one axis of motion of the manipulator.

Control System: The system which implements the designed control scheme, including sensors, manual input and mode selection elements, interlocking and decision-making circuitry, and output elements to the operating mechanism.

Controlled Passive Devices: Operator interface using actuators that cannot impart power to the device, but can control the way the device responds to externally applied power. For teleoperator input devices the input power is supplied by the human operator. By controlled passive actuation the response of the device can provide to the human a haptic display of conditions present at the remote manipulator.

Controller: A hardware/software device that continuously measures the value of a variable quantity or condition and then automatically acts on the controlled equipment to correct any deviation from a desired preset value.

Controller Area Network (CAN): A bus network connection of the system controllers of the devices in the robot cell. It enhances the reprogrammability of each cell component, as well as the connectivity flexibility of further equipment in the cell.

Conveyor Tracking Robot: A robot synchronized with the movement of a conveyor. Frequent updating of the input signal of the desired position on the conveyor is required.

Cooperation: The willingness and readiness of subsystems to share or combine their tasks and resources as in "open systems."

Cooperative Behavior: Any agent behavior requiring negotiation and depending on directed communication in order to assign particular tasks to the participants. The cooperative behavior can be explicit: a set of interactions which involve exchanging information or performing actions in order to benefit another agent; or implicit: actions that are a part of the agent's own goal-achieving behavior repertoire, but have effects in the world that help other agents achieve their goals.

Coordinate Measuring Machine (CMM): A Cartesian robotic device armed with a probe as tool for measuring parts data (e.g., dimension, location of features, or cylindricity). The probe is a transducer, or sensor, that converts physical measurements into electrical signals, using various measuring systems within the probe structure.

Coordinate Transformation: In robotics, a 4×4 matrix used to describe the positions and orientations of coordinate frames in space. It is a suitable data structure for the description of the relative position and orientation between objects. Matrix multiplication of the transformations establishes the overall relationship between objects.

Coordinated Joint Motion: Also known as *coordinated axis control*. Control wherein the axes of the robot arrive at their respective endpoints simultaneously, giving a smooth appearance to the motion.

Coordination: The means for integrating or linking together different systems whose decisions and actions are interdependent in order to achieve common goals. Coordination can be classified into *programmed*, where the activities are dictated by organizational goals and objectives and plans are specified in advance by the organization; *nonprogrammed*, where activities usually occur as a result of some emergency; and *by feedback*, which combines both the programmed and nonprogrammed modes. Also, the additional information processing performed when multiple, connected agents pursue goals that a single agent pursuing the same goals would not perform.

Crisp Set: In Fuzzy Logic Control, a collection of distinct objects. It is defined in such a way as to dichotomize the elements of a given universe of discourse into two groups: members and nonmembers. This dichotomization is defined by a characteristic function.

Cybernetics: The theoretical study of control and information processes in complex electronic, mechanical, and biological systems, considered by some as the theory of robots.

Cycle Time: The period of time from starting to finishing an operation. Cycle time is used to determine the nominal production rate of a robotic system.

Cycle Time Chart: A graphical representation of the components of the cycle time of a robotics operation. It resembles the graphs utilized in project scheduling.

Cylindrical Robot: Also known as *cylindrical coordinate robot* or *columnar robot*. A robot, built around a column, that moves according to a cylindrical coordinate system in which the position of any point is defined in terms of an angular dimension, a radial dimension, and a height from a reference plane. The outline of a cylinder is formed by the work envelope. Motions usually include rotation and arm extension.

D

DC Servo Drives: Electric motors controlled using a feedback mechanism. A transducer feedback and a speed control form a servo loop. DC servo drives are controlled through a voltage change; the motor runs faster if a higher voltage is applied. It consists of stator, rotor, commutator, bearing, and housing.

Decision-Making: The process of drawing conclusions from limited information or conjecture.

Decision Support System: The means by which the improvements offered by operations research techniques can be translated into overall system performance.

Decoupler: A device for the containment of parts between machines in manufacturing cells for improving flow flexibility. It reduces the time and/or functional dependency of one machine on another within the cell. Moreover, it frees the worker or the robot from the machines so that the worker or robot can move to other machines in the cell.

Defuzzifier: Fuzzy Logic Controller whose function is the mapping back of fuzzy values into the actual values.

Degrees of Freedom: The number of independent ways the end-effector can move. It is defined by the number of rotational or translational axes through which motion can be obtained. Every variable representing a degree of freedom must be specified if the physical state of the manipulator is to be completely defined.

Deployed Robot: A manipulator that enters an otherwise sealed and prohibited zone through an authorized path of entry in order to employ an inspection device, tool, and so on in a predetermined task under direct or programmed control.

Depth Information and Perception: Positional data of parts extracted from vision system stereo images. It contributes to simplify and enhance tasks such as bin picking of parts by avoiding complex and expensive fixture and feeding mechanisms. Techniques such as the laser-based Structured Light Imaging are utilized to extract the depth information and generate the depth perception.

Derivative Control: A control scheme whereby the actuator drive signal is proportional to the time derivative of the difference between the input (desired output) and the measured actual output.

Design Flexibility: See Design for Flexible Manufacturing and Assembly (DFMA).

Design for Flexible Manufacturing and Assembly (DFMA):   The design of a product so that it can be produced easily and economically. DFMA identifies product-design characteristics that are inherently easy to manufacture, focuses on the design of component parts that are relatively easy to fabricate and assemble, and integrates product design with process planning. DFMA improves the quality of the product design and reduces the time and cost of both product design and manufacture.

Design for Reliability: Robotics systems design strategies for increasing their reliability. Strategies can be classified in:

1. Fault avoidance
2. Fault tolerance
3. Fault detection
4. Error recovery.

Design Process (of Robots):   A multistep process beginning with a description of the range of tasks to be performed. Several viable alternative configurations are then determined, followed by an evaluation of the configurations with respect to the sizing of components and dynamic system performance. Based on appropriate technical and economic criteria, a configuration is then selected. If no configuration meets the criteria, the process may be repeated in an iterative manner until a configuration is selected.

Dexterous Manipulators: Robotic arms with the capability to move their payload with consequently facility and quickness within the manipulator's workspace.

Dielectrophoresis: The ability of an uncharged material to move when subjected to an electric field.

Digital Signal Processing (DSP):   A signal analysis in which one or more analog inputs are sampled at regular intervals, converted to digital form, and fed to a memory.

Dimensional Testing: See Inspection (Robotic).

Direct Drive Motor: Electric motors employed for compact and lightweight design of robotic material handling transporters. Characteristics such as high torque (torque/weight ratio of 6.0 Nm/kg or more) for climbing slope, smooth acceleration and deceleration which will protect valuable payload in motion from vibration/physical shocks are fundamental. Direct drive motors are also maintenance free, as the need for lubrication and other maintenance activities is eliminated.

Direct Drive Torque Motors (DDM):   See Direct Drive Motor.

Direct Dynamics: Also known as *forward dynamics* or *integral dynamics*. The determination of the trajectory of the manipulator from known input torques at the joints.

Direct Kinematics: Also known as *forward kinematics*. The determination of the position of the end-effector from known joint displacements.

Direct Manipulation Interfaces: User interfaces designed based on the idea that users should be allowed to manipulate computer interfaces in a way that is analogous to the way they manipulate objects in space.

Direct Matching: A class of techniques in image analysis which matches new images or portions of images directly with other images. Images are usually matched with a model or template which was memorized earlier.

Direct Numerical Methods: Numerical methods for determining optimal robot trajectories. Direct methods are designed to minimize the functional cost. Compared to indirect methods, direct methods prove to be quite robust and globally convergent, but computationally more expensive.

Disassembly: The inverse of the assembly process, in which products are decomposed into parts and subassemblies. In product remanufacturing the disassembly path and the termination goal are not necessarily fixed, but rather are adapted according to the actual product condition.

Disparity, Convergent: Also known as *crossed disparity*. In stereo image depth analysis convergent disparity is a condition of an object point located in front of the fixation point. The optic axes of two cameras will have to cross or converge in order to fixate at the object point.

Disparity, Divergent: Also known as *uncrossed disparity*. In stereo image depth analysis divergent disparity is a condition of an object point located behind the fixation point. The optic axes of two cameras will have to diverge or uncross in order to fixate at the object point.

Dispensing System (Robotic):   Several components combined to apply a desired amount of adhesive, sealant, or other material in a consistent, uniform bead along accurate trajectories. The components are a five- or six-axis robot; containers, pumps, and regulators; a programmable controller; and a dispenser or gun. Examples of robotic dispensing include sealer to car underbody wheelhouse components, urethane bead on windshield periphery before installation, and sealant applications on appliances.

Distal: The direction away from a robot base toward the end-effector of the arm.

Distributed Artificial Intelligence (DAI):   A computational approach focusing on negotiation and coordination of multi-agent environments in which agents can vary from knowledge-based systems to sorting algorithms, and approaches can vary from heuristic search to decision theory. DAI can be divided into two subfields: Distributed Problem Solving (DPS) and Multi-agent Systems (MAS).

Distributed Control: A collection of modules, each with its own specific control function, interconnected tightly to carry out an integrated data acquisition and control application.

Distributed Numerical Control (DNC):   The use of a computer for inputting data to several physically remote numerically controlled machine tools.

Distributed Problem Solving (DPS):   Study of how a loosely coupled network of problem-solving nodes (processing elements) can solve problems that are beyond the capabilities of the nodes individually. Each node (sometimes called *agent*) is a sophisticated system that can modify its behavior as circumstances change and plan its own communication and cooperation strategies with other nodes.

Drift: The tendency of a system to gradually move away from a desired response.

Drive Power: The source or means of supplying energy to the robot actuators to produce motion.

Drum Sequencer: The mechanically programmed rotating device that uses limit switches to control a robot or other machine.

Duty Cycle: The time intervals devoted to starting, running, stopping, and idling when a device is used for intermittent duty.

Dynamic Accuracy

1. Degree of conformance to the true value when relevant variables are changing with time.
2. Degree to which actual motion corresponds to desired or commanded motion.

E

Edge Detection: An image analysis technique in which information about a scene is obtained without acquiring of an entire image. Locations of transition from black to white and white to black are recorded, stored, and connected through a process called *connectivity* to separate objects in the image into blobs. The blobs can then be analyzed and recognized for their respective features.

Edge Finding: A type of gray-scale image analysis used to locate boundaries between regions of an image. There are two steps in edge finding: locating pixels in the image that are likely to be an edge, and linking candidate edge points together into a coherent edge.

Elbow: The joint which connects the upper arm and forearm of a robot.

Electro-Reological Fluid (ERF):   A functional fluid whose mechanical property such as viscosity is changed considerably at high speed and reversibly by the applied voltage. The fluid can be classified roughly into *suspension* (Bingham fluid) and *homogeneous* (liquid crystal).

Electronics Data Interchange (EDI):   The direct computer-to-computer exchange between two organizations of standard business transaction documents such as invoices, bills of lading, and purchase orders. It differs from e-mail in that it transmits an actual structured transaction as opposed to an unstructured text message.

Emergency Stop: A method using hardware-based components that overrides all other robot controls and removes drive power from the robot actuators to bring all moving parts to a stop.

EMILY: A high-level, functional robot programming language developed by IBM using a relatively simple processor. Considered to be a primitive motion-level language that incorporates simple and straight-line motion but not continuous path motion. EMILY has gripper operation commands and is based on assembly-level language.

Encoder: A transducer used to convert linear or rotary position to digital data.

End-Effector: Also known as *end-of-arm tooling* or, more simply, *hand*. The subsystem of an industrial robot system that links the mechanical portion of the robot (manipulator) to the part being handled or worked on and gives the robot the ability to pick up and transfer parts and/or handle a multitude of differing tools to perform work on parts. It is commonly made up of four distinct elements: a method of attachment of the hand or tool to the robot tool mounting plate, power for actuation of tooling machines, mechanical linkages, and sensors integrated into the tooling. Examples include grippers, paint spraying nozzles, welding guns, and laser gauging devices.

End-Effector, Turret: A number of end-effectors, usually small, that are mounted on a turret for quick automatic change of end-effectors during operation.

End-of-Axis Control: Controlling the delivery of tooling through a path or to a point by driving each axis of a robot in sequence. The joints arrive at their preprogrammed positions on a given axis before the next joint sequence is actuated.

End-of-Life Value: The usefulness of the components of a product at the end of its service life. This value should compensate for the cost of the remanufacturing or disassembly process of the product.

Endpoint Control: Control wherein the motions of the axes are such that the endpoint moves along a prespecified type of path line (straight line, circle, etc.)

Endpoint Rigidity: The resistance of the hand, tool, or endpoint of a manipulator arm to motion under applied force.

Enterprise Integration: The integration of sophisticated automation equipment, computer hardware and software, planning, control, and data management methodologies into the enterprise functional areas with the objective of enhancement of the enterprise ability to compete in the global market.

Enterprise Reengineering: The radical redesign of business processes, combining steps to cut waste and eliminating repetitive, paper-intensive tasks in order to improve cost, quality, and service and maximize the benefits of information technology.

Enterprise Resource Planning (ERP): A business management system that integrates all facets of the business, including planning, manufacturing, sales, and marketing.

Ergonomics: The study of human capability and psychology in relation to the working environment and the equipment operated by the worker (see also Robot Ergonomics).

Error Recovery: An ability in intelligent robotic systems to detect a variety of errors and, through programming, take corrective action to resolve the problem and complete the desired process.

Error-Absorbing Tooling: A type of robot end-effector able to compensate for small variations in position and orientation. Especially suitable for assembly tasks, where the insertion of components demands tight tolerance positioning and orientation of the parts (see also Remote Center Compliance Device).

Exoskeleton: An articulated teleoperator mechanism whose joints correspond to those of a human arm. When attached to the arm of a human operator, it will move in correspondence to his or her arm. Exoskeletal devices are sometimes instrumented and used for master–slave control of manipulators.

Expert System (ES): A computer program, usually based on artificial intelligence techniques, that performs decision functions similar to those of a human expert and, on demand, can justify to the user its line of reasoning. Typical applications in the field of robotics are high-level robot programming, planning and control of assembly, and processing and recovery of errors.

External Sensor: A feedback device that is outside the inherent makeup of a robot system, or a device used to effect the actions of a robot system that are used to source a signal independent of the robot's internal design.

Eye-in-Hand System: A robot vision system in which the camera is mounted on or near the robot gripper. This arrangement eases the calculation of object location and orientation and eliminates blind-spot problems encountered in using a static overhead camera.

F

Facility Design Language (FDL): A language tool developed at Purdue University in Indiana to enhance the design capabilities of virtual robotic systems and permit concurrent and distributed engineering techniques to be used in the general area of facility design. FDL is based on TDL, the task description component of Robcad, and supports control and communication, material handling and sensor functions, flow and process geometry functions (such as collision avoidance), data integration, and conflict resolution functions.

Failure Modes and Effects Analysis (FMEA): A systematic procedure for documenting the effects of system malfunctions on reliability and safety

Fault Tolerance: The capability of a system to perform in accordance with design specifications even when undesired changes in the internal structure or external environment occur.

Fault Tree Analysis (FTA): A top-down approach for developing fault trees. The analysis starts with a malfunction or accident and works downwards to basic events (bottom of the tree).

Feature Extractor: A program used in image analysis to compute the values of attributes (features) considered by the user to be possibly useful in distinguishing between different shapes of interest.

Feature Set (Image): A group of standard feature values gleaned from a known image of a prototype and used to recognize an object based on its image. The feature set is computed for an unknown image and compared with the feature sets of prototypes. The unknown object is recognized as identical to the prototype whose feature values vary the least from its own.

Feature Values: Numerical values used to describe prototypes in a vision system. The feature values that are independent of position and orientation can be used as a basis of comparison between two images. Examples of single feature values are the maximum length and width, centroid, and area.

Feature Vector: A data representation for a part. The data can include area, number of holes, minimum and maximum diameter, perimeter, and so on. The vision system usually generates the feature vector for a part during teach-in. The feature vector is then used in image analysis.

Fiber Optics: The technique of transmitting light through long, thin, flexible fibers of glass, plastic, or other transparent materials; bundles of parallel fibers can be used to transmit complete images.

Fiducial Points: Reference points on the face of a printed circuit board. Fiducial points are used for high-precision location to perform robotic component insertion or surface mount placement.

Finishing (Robotic): A finishing robot is usually of the continuous-path type and capable of multiple program storage with random access. These two features allow for the smooth, intricate movements duplicating the human wrist which are needed to perform high-speed, efficient finishing tasks such as spraypainting or coating. Finishing robots usually require some additional features not normally required for other types of robots: noise filters to prevent interference of electrical noise from electrostatic spraying devices, an explosion-proof remote control operator's panel in the spray area, gun and cap cleaners, and a cleaning receptacle for cleaning the gun after prolonged use or color changes.

First-Generation Robot Systems: Robots with little, if any, computer power. Their only intelligent functions consist of learning a sequence of manipulative actions choreographed by a human operator using a teach box. The factory world around them has to be prearranged to accommodate their actions. Necessary constraints include precise workpiece positioning, care in specifying spatial relationships with other machines, and safety for nearby humans and equipment.

Fixed Assembly System: A group of equipment dedicated to assemble one product type only.

Fixture: A device used for holding and positioning a workpiece without guiding the tool.

Flexi-Arm: A manipulator arm designed to have as much freedom as the human wrist and especially suited to coating and other finishing applications. The design provides full arching without regard to the pitch-and-yaw axis, minimizes the arm size, and eliminates the need for electrical wiring, hydraulic hoses, and actuators on the end of the arm. The arm can be fitted with a seventh axis to further increase gun (tool) mobility so the arm has better reach and access than a human spray finisher.

Flexibility (Gripper):   The ability of a gripper to conform to parts that have irregular shapes and adapt to parts that are inaccurately oriented with respect to the gripper.

Flexibility, Mechanical: Pliable or capable of bending. In robot mechanisms this may be due to joints, links, or transmission elements. Flexibility allows the endpoint of the robot to sag or deflect under a load and vibrate as a result of acceleration or deceleration.

Flexibility, Operational: Multipurpose robots that are adaptable and capable of being redirected, trained, or used for new purposes. Refers to the reprogrammability or multitask capability of robots.

Flexibility–Efficiency Trade-off: The trade-off between retaining a capability for rapid redesign or reconfiguration of the product to produce a range of different products, and being efficient enough to produce a large number of products at high levels of production and low unit cost.

Flexible Arm: A robot arm with mechanical flexibility, e.g., inflatable links or links made of mechanically flexible materials (contrast with Rigidity).

Flexible Assembly System (FAS):   An arrangement of assembly machines or stations, and a connecting transport system under control of a central computer that allows the assembly of several, not necessarily identical, workpieces simultaneously (see also Programmable Assembly System).

Flexible Fixturing: Fixture systems with the ability of accommodating several part types for the same type of operation. The fixture can be robotic and change automatically according to sensor input detecting the part change.

Flexible Fixturing Robots: Robots working in parallel, designed to hold and position parts on which other robots or people or automation can work.

Flexible Manufacturing System (FMS):   An arrangement of machine tools that is capable of standing alone, interconnected by a workpiece transport system, and controlled by a central computer. The transport subsystem, possibly including one or more robots, carries work to the machines on pallets or other interface units so that accurate registration is rapid and automatic. FMS may have a variety of parts being processed at one time.

Flight Telerobotic Servicer (FTS):   Telerobotics based system for the assembly, check-out, and maintenance of the International Space Station to minimize the amount of extravehicular activity required of the crew.

Folding Arm Manipulator: A manipulator designed to enter an enclosed area, such as the interior of a nuclear reactor, through a narrow opening. A control console outside the enclosed area is connected to the manipulator by an umbilical cord which carries command signals, telemetry, and services. Once inside the enclosed area, the manipulator unfolds and/or extends into a working position. Visual feedback from the work zone employs closed-circuit television through a remotely controlled camera on the manipulator feeding a visual display unit in the control console. The manipulator is designed to allow all relevant motions to be recovered in an emergency situation.

Force Control: A method of error detection in which the force exerted on the end-effector is sensed and fed back to the controller, usually by mechanical, hydraulic, or electric transducers.

Force Reflection: Also known as *bilateral master–slave control*. A category of teleoperator control incorporating the features of simple master–slave control and also providing the operator with resistance to motions of the master unit which corresponds to the resistance experienced by the slave unit.

Force–Torque Sensors: The sensors that measure the amount of force and torque exerted by the mechanical hand along three hand-referenced orthogonal directions and applied around a point ahead and away from the sensors.

Forearm: That portion of a jointed arm which is connected to the wrist and elbow.

Forward Dynamics: The computation of a trajectory resulting from an applied torque.

Forward Kinematics: The computation of the position or motion of each link as a function of the joint variables.

Forward Reasoning: A strategy for searching the rule base in an expert or knowledge-based system that begins with the information entered by the user and searches the rule base to arrive at a conclusion.

FREDDY: A pioneering robot system developed at Edinburgh University, Scotland, in the mid-1970s that used television cameras, a touch-sensitive manipulator, and a motor-controlled mobile viewing platform to study the acquisition of perceptual descriptions.

Function Based Sharing Control (FBSC): A strategy for combining robotic and teleoperated commands. It constructs a path in terms of a position dependent path planner.

FUNKY: Robot software developed by IBM for advanced motion guiding that produces robot programs through the use of a function keyboard and manual guiding device. Considered a point-to-point level language that is inherently unstructured, FUNKY has support for gripper commands, tool operations, touch sensor commands, and interaction with external devices.

Fuzzifier: One of the four basic components of the Fuzzy Logic Controller. A fuzzifier performs the function of fuzzification, which is a subjective valuation to transform measurement data into a valuation of a subjective value. Hence it can be defined as a mapping from an observed input space to labels of fuzzy sets in a specified input universe of discourse. Since the data manipulation in a FLC is based on fuzzy set theory, fuzzification is necessary and desirable at an early stage.

Fuzzy Adaptive Learning Control Network (FALCON): Feedforward multilayer network which integrates the basic elements and functions of a traditional fuzzy logic controller into a connectionist structure which has distributed learning abilities. In this connectionist structure the input and output nodes represent the input states and output control/decision signals, respectively, and in the hidden layers there are nodes functioning as membership functions and rules.

Fuzzy Basis Function Network (FBFN): A Neuro-Fuzzy Control system where a fuzzy system is represented as a series expansion of fuzzy basis functions (FBFs) which are algebraic superpositions of membership functions. Each FBF corresponds to one fuzzy logic rule.

Fuzzy Logic: Rule-based artificial intelligence that tolerates imprecision by using non-specific terms called "membership functions" to solve problems.

Fuzzy Logic Control (FLC): Control mechanism that incorporates the "expert experience" of a human operator in the design of the controller in controlling a process whose input/output relationship is described by a collection of fuzzy control rules (e.g., IF–THEN rules) involving linguistic variables rather than a complicated dynamic model.

Fuzzy Logic Controller: Composed of four principal components: a fuzzifier, a fuzzy rule base, an inference engine, and a defuzzifier.

Fuzzy Relations: A generalization of classical relations to allow for various degrees of association between elements. An important operation of fuzzy relations is the composition of fuzzy relations. There are two types of composition operators: max–min and min–max compositions, and these compositions can be applied to both relation–relation compositions and set–relation compositions.

Fuzzy Rule Base: A collection of fuzzy IF–THEN rules in which the preconditions and consequences involve linguistic variables. This collection of fuzzy control rules characterizes the simple input–output relation of the system.

Fuzzy Set: A collection of vaguely distinct objects. A fuzzy set introduces vagueness by eliminating the sharp boundary that divides members from nonmembers in the group.

Thus, the transition between the full membership and nonmembership is gradual rather than abrupt, and this is realized by a membership function. Hence, fuzzy sets may be viewed as an extension and generalization of the basic concepts of crisp sets.

G

Game Theoretic Approach: The mathematical study of games or abstract models of conflict situations from the viewpoint of determining an optimal policy or strategy.

Gantry Robot: An overhead-mounted, rectilinear robot with a minimum of three degrees of freedom and normally not exceeding six. Bench-mounted assembly robots that have a gantry design are not included in this definition. A gantry robot can move along its x and y axes traveling over relatively greater distances than a pedestal-mounted robot at high traverse speeds while still providing a high degree of accuracy for positioning. Features of a gantry robot include large work envelopes, heavy payloads, mobile overhead mounting, and the capability and flexibility to operate over the work area of several pedestal-mounted robots.

Genetic Algorithms (GA):   Search algorithms based on the mechanics of natural selection and natural genetics. Genetic algorithms have been used primarily in optimization and machine learning problems. They have three basic operators: reproduction of solutions based on their fitness, crossover of genes, and mutation for random change of genes. Another operator associated with each of these three operators is the selection operator, which produces survival of the fittest.

Geometric Dexterity: The ability of the robot to achieve a wide range of orientations of the hand with the tool center point in a specified position.

Geometric Modeler: A component of an off-line programming system which generates a world model from geometric data. The world model allows objects to be referenced during programming.

Global Positioning System (GPS):   A positioning or navigation system designed to use 18 to 24 satellites, each carrying atomic clocks, to provide a receiver anywhere on earth with extremely accurate measurements of its three-dimensional position, velocity, and time.

Graphic Simulation: A three-dimensional model generating static or dynamic displays of objects and workcell layout. It enables a programmer to view workcell objects from different directions (using perspective transformations) with variable magnification (zooming), as if the programmer were a "flying eye" observing a real workcell. An off-line programmer can interactively use the simulation for writing and debugging the program steps for a given task, observe the animation, and, when satisfied with the results, download the program into a real workcell controller and run it.

Grasp Planning: A capability of a robot programming language to determine where to grasp objects in order to avoid collisions during grasping or moving. The grasp configuration is chosen so that objects are stable in the gripper.

Gray-Scale Picture: A digitized image in which the brightness of the pixels can have more than two values, typically, 128 or 256 values. A gray-scale picture requires more storage space and more sophisticated image processing than a binary image.

Gripper: The grasping hand of the robot which manipulates objects and tools to fulfill a given task.

Gripper Design Factors: Factors considered during the design of a gripper in order to prevent serious damage to the tool or facilitate quick repair and alignment. The factors include: parts' or tools' shape, dimension, weight, and material; adjustment for realignment in the x and y direction; easy-to-remove fingers; mechanical fusing (shear pins, etc.); locating surface at the gripper–arm interface; spring loading in the z (vertical) direction; and specification of spare gripper fingers.

Gripper, External: A type of mechanical gripper used to grasp the exterior surface of an object with closed fingers.

Gripper, Internal: A type of mechanical gripper used to grip the internal surface of an object with open fingers.

Gripper, Soft: A type of mechanical gripper which provides the capability of conforming to part of the periphery of an object of any shape.

Gripper, Swing Type: A type of mechanical gripper which can move its fingers in a swinging motion.

Gripper, Translational: A type of mechanical gripper which can move its own fingers, keeping them parallel.

Gripper, Universal: A gripper capable of handling and manipulating many different objects of varying weights, shapes, and materials.

Gripping Surfaces: The surfaces, such as the inside of the fingers, on the robot gripper or hand that are used for grasping.

Gross Volume of Work Envelope: The volume of the work envelope determined by shoulder and elbow joints.

Group Behavior: See Cooperative Behavior.

Group Technology: A technique for grouping parts to gain design and operational advantages. For example, in robotics group technology is used to ensure that different parts are of the same part family in planning part processing for a workcell or designing widely usable fixtures for part families. Part grouping may be based on geometric shapes, operation processes, or both.

Groupe de recherché en automatisation integree (GRAI): See Computer Integrated Manufacturing Open Systems Architecture (CIMOSA).

Growing (Image): Transformation from an input binary image to an output binary image. Growing increases the number of one type of pixel for purposes of smoothing, noise elimination, and detection of blobs based on approximate size.

Guarded Motions: The motion required of a robot when approaching a surface. This motion is required because of uncertainty in the world model and the inherent inaccuracy of a robot. The goal of the guarded motion is to achieve a desired manipulator configuration on an actual surface while avoiding excessive forces.

H

Hand (Robot's): A fingered gripper sometimes distinguished from a regular gripper by having more than three fingers and more dexterous finger motions resembling those of the human hand.

Hand Coordinate System: A robot coordinate system based on the last axis of the robot manipulator.

Handchanger: A mechanism analogous to a toolchanger on a machining center or other machine tool. It permits a single robot arm to equip itself with a series of task-specific hands or grippers.

Haptic Interfaces: Devices controlled by human hand (or body) physical contact forces or exerting such forces on the human body. These devices are necessary elements of virtual reality machines. They can be programmed to give the operator arm and hand the sensation of forces associated with various arbitrary maneuvers, for instance feeling the smoothness or vibrations of a virtual surface.

Hard Automation: Also known as *fixed automation* or *hard tooling*. A nonprogrammable, fixed tooling which is designed and dedicated for specific operations that are not easily changeable. It may be reconfigured mechanically and is cost-effective for a high production rate of the same product.

Hard Tooling: Traditional tooling where every part to be processed in the robotic cell has its own fixtures and tools. It results in increased changeover time and processing delays.

Harmonic Drive: A drive system that uses inner and outer gear bands to provide smooth motion.

Heat Distortion Zone: Volume of the parts being joined by a welding process that are affected by the increase in temperature resulting from the welding method. These zones have their properties altered and may become brittle.

HELP (High Level Procedural Language):   A robot programming language, based on PASCAL/Fortran, developed at the DEA Corporation in Turin, Italy. HELP supports structured program design for robot operation and features flexibility to multiple arms, support of continuous path motion, force feedback and touch sensor commands, interaction with external devices, and gripper operation commands.

Heuristics: A method of solving a problem in which several approaches or methods are tried and evaluated for progress toward an optimal solution after each attempt (but an optimal solution is not guaranteed).

Heuristic Problem Solving: In computing logic, the ability to plan and direct actions to steer toward higher-level goals. This approach is the opposite of algorithmic problem solving.

Hexapod: A robot that uses six leglike appendages to stride over a surface.

Hierarchical Control: A distributed control technique in which the controlling processes are arranged in a hierarchy and distributed physically.

High-Level Language: A programming language that generates machine code from function-oriented statements that approach English.

High-Level Robot Programming: The control of a robot with a high-level language that contains English-like (or another natural language) commands. These commands then perform computations of numerous elementary operations in order to simplify complicated robot operations.

Hold: A stopping of all movement of the robot during its sequence in which some power is maintained on the robot; for example, on hydraulically driven robots, power is shut off to the servo-valves but is present in the main electrical and hydraulic systems.

Holonic Manufacturing System (HMS):   A new paradigm for the next-generation manufacturing systems, comprising multiagent manufacturing systems in which the entities (the holons) are both autonomous and cooperative building blocks of a manufacturing system for transforming, transporting, storing, and/or validating information and physical objects. Holonic manufacturing systems are human-centered, and therefore they aim to enable the cooperation amongst humans and machines that preserves the ability to plan and act autonomously for both of them. The strength of the holonic organization (holarchy) is that it enables the construction of very complex systems that are nonetheless efficient in the use of resources, highly resilient to disturbances (both internal and external), and viable by adapting to changes in the environment in which they exist.

Holonomic Constraints: An integrable set of differential equations that describes the restrictions on the motion of a system; a function relating several variables, in the form $f(x_1, \ldots, x_n) = 0$, in optimization or physical problems.

Holonomic System: A system in which the constraints are such that the original coordinates can be expressed in terms of independent coordinates and possibly of the time.

Home Robots: Small mobile vehicles fitted with a relatively slow-moving arm and hand, and visual and force/tactile sensors, controlled by joysticks and speech, with a number of accessories specialized for carrying objects, cleaning, and other manipulative tasks.

Homogeneous Transform: A 4×4 matrix which represents the rotation and translation of vectors in the joint coordinate systems. It is used to compute the position and orientation of any coordinate system with respect to any other coordinate system.

Horizontal Integration: The integration of activities at the same hierarchical level in the enterprise by investigating the passing of material and information from one activity to another. The analysis can be supported through activity modeling.

Hostile Environments: Robot's work environments characterized by high temperatures, vibration, moisture, pollution, or electromagnetic or nuclear radiation.

HRL (High Robot Language): Robot motion software, based on LISP and Fortran, developed at the University of Tokyo. HRL is used to describe manipulator motions for mechanical assemblies and disassemblies. Its features include language extensions, world models, and orbit calculation commands.

Human Error Analysis (HEA): A system malfunction analysis technique which relates the tasks performed in the system to the malfunctions suffered.

Human Factors Engineering: The area of knowledge dealing with the capabilities and limitations of human performance in relation to design of machines, jobs, and other modifications of human's physical environment.

Human-Oriented Design: Design of a complex computerized system and interfaces for ease of learning, training, operating, and maintaining by human operators and users (see also User-Friendly).

Human–Machine Interface: See User Interface.

Human–Robot Interaction (HRI): The analysis and design of real and virtual interfaces with robots, including communication and off-line programming, adaptive and social behavior, anthropomorphic interfaces and robot devices, collaboration, and human-friendly interactions.

Humanoid: A robot designed to resemble human physical characteristics.

Humanoid Robotics: See Humanoid.

Hybrid Teleoperator/Robot Systems: A partially controlled robot for performing service tasks. Most of the intelligence is supplied by a human operator interfaced in a user-friendly manner to control switches, joysticks, and voice input devices to control the physical motion and manipulation of the robot.

Hydraulic Motor: An actuator consisting of interconnected valves and pistons or vanes which converts high-pressure hydraulic or pneumatic fluid into mechanical shaft translation or rotation. While hydraulic motors were popular for early robots, DC servo-drives have become more popular.

Hypermedia: Hypertext-based systems that combine data, text, graphics, video, and sound.

Hyperredundant Manipulators: See Underactuated Robotic System.

Hypertext: A data structure in which there are links between words, phrases, graphics, or other elements and associated information so that selection of a key object can activate a linkage and reveal the information.

I

ICRA: International Conference on Robotics and Automation.

Image Analysis: The interpretation of data received from an imaging device. For the three basic analysis approaches that exist, see Image Buffering, Edge Detection, and Windowing.

Image Buffering: An image analysis technique in which an entire image is digitized and stored in computer memory. Computer software uses the image data to detect features, such as an object's area, centroid location, orientation, perimeter, and others.

Imaging: The analysis of an image to derive the identity, position, orientation, or condition of objects in the scene. Dimensional measurements may also be performed.

Indexer: Device for transferring the partially completed assembly from one assembly station to the next. Indexers can be in-line (transference assembly lines, e.g., automobile assembly) or rotatory (rotary assembly tables).

Indirect Numerical Methods: Numerical methods for determining optimal robot trajectories. In indirect methods the solution which is an extremal is computed from the minimum principle (first-order necessary conditions) by solving a two-point boundary value problem.

Induction Motor: An alternating-current motor wherein torque is produced by the reaction between a varying or rotating magnetic field that is generated in stationary-field magnets and the current that is induced in the coils of the rotor.

Inductosyn: Trademark for Farrand Controls resolver, in which an output signal is produced by inductive coupling between metallic patterns, versus glass-scale position resolvers that use Moire-fringe patterns.

Industrial Robot: A mechanical device that can be programmed to perform a variety of tasks of manipulation and locomotion under automatic control. The attribute "industrial" distinguishes such production and service robots from science fiction robots and from purely software robots (such as search engines for the World Wide Web).

Inference Engine: The kernel of computational systems that infer, such as knowledge-based or expert systems or the Fuzzy Logic Controller. It is based on modeling human decision-making within the conceptual framework of fuzzy logic or approximate reasoning.

Information and Communication Technologies (ICT): The collection of technologies that deal specifically with processing, storing, and communicating information, including all types of computer and communication systems as well as reprographics methodologies.

Information View: A subset of data elements from a database. The view is supposed to satisfy the requirements of a user or of an enterprise function.

Infrared Sensor: A device that intercepts or demodulates infrared radiation that may carry intelligence.

Inspection (Robotic): Robot manipulation and sensory feedback to check the compliance of a part or assembly with specifications. In such applications robots are used in conjunction with sensors, such as a video camera, laser, or ultrasonic detector, to check part locations, identify defects, or recognize parts for sorting. Application examples include inspecting printed circuit boards and valve cover assemblies for automotive engines, sorting metal castings, and inspecting the dimensional accuracy of openings in automotive bodies.

Integral Control: A control scheme whereby the signal driving the actuator equals the time integral of the error signal.

Integrated Circuit (IC): An interconnected array of active and passive elements integrated with a single semiconductor substrate or deposited on the substrate by a continuous series of compatible processes, and capable of performing at least one complete electronic circuit function.

Integration: A process by which subsystems share or combine physical or logical tasks and resources so that the whole system can produce better (synergistic) results. Internal integration occurs among subsystems of the same system, and external integration is with subsystems of other systems. Integration depends on a cooperative behavior.

Intelligent Collision Avoidance and Warning Systems: Systems utilizing sophisticated ultrasound, capacitive, inferred, or microwave presence sensing system or computer vision to detect the presence of obstacles and then react appropriately by stopping, replanning their motions, or giving alarm signals.

Intelligent Programming Interface: Interface for automating the programming of robots, based on Artificial Intelligence-related algorithms and database technology, enabling robots to imitate, to a certain extent, the human behavior of learning, reasoning, and contingency reaction.

Intelligent Robot: A robot that can be programmed to execute performance choices contingent on sensory inputs.

Intelligent Transportation Systems (ITS): The combination of robotics and transportation, both in terms of robotic vehicles and in terms of knowledge-based driving and transportation management.

Interactive Manual–Automatic Control: A type of remote robot operation. Data from sensors integrated with the remote robot are used to adapt the real-time control actions to changes or variances in task conditions automatically through computer control algorithms.

Interactive Task Programming: Generation of task-based robot programs by the human operator with the assistance of a task instructions database. The main benefits of this programming mode are:

1. Automatic generation of motion program from task-level programs
2. Generation of a knowledge database for robot programming
3. Enablement operators to teach the robot behavior at the more natural task-instruction level

Interface: A shared boundary which might be a mechanical or electrical connection between two devices; a portion of computer storage accessed by two or more programs; or a device for communication with a human operator.

Interface Box (Input/Output): Additional equipment for the robot system's interface which is required for an application, but is not part of the robot system. For example, in spot-welding, an interface box can be used to control cooling water, shielding gas, a weld-gun servo-controller card, power supply, and AC input.

Interfacing (Robot with Vision): Calculating the relative orientation between the camera coordinate frame and robot coordinate system so that objects detected by the camera can be manipulated by the robot.

Interference Zone: Space contained in the work envelopes of more than one robot.

Interlock: To arrange the control of machines or devices so that their operation is interdependent in order to assure their proper coordination and synchronization.

Interpolator: A program in a system computer of a numerically controlled machine or robot that determines the calculated motion path (e.g., linear, circular, elliptic), between given end points.

Inverse Dynamics: The determination of torques to be exerted at the joints to move the manipulator along a desired trajectory and to exert the desired force at the end-effector.

Inverse Kinematics: The determination of joint displacements required to move the end-effector to a desired position and orientation.

Ironware: A nickname for a crew of industrial robotics equipment. Another version: Iron Collar.

ISIR: International Symposium on Industrial Robotics.

Islands of Automation: An approach used to introduce factory automation technology into manufacturing by selective application of automation. Examples include numerically controlled machine tools; robots for assembly, inspection, coating, or welding; automated assembly equipment; and flexible machining systems. Islands of automation should not be viewed as ends in themselves but as a means of forming integrated factory systems. They may range in size from an individual machine or workstation to entire departments. In nonintegrated or poorly integrated enterprises even fully automated manufacturing division can be considered an island.

ISRA: International Service Robot Association.

J

Jamming: In part assembly, jamming is a condition where forces applied to the part for part mating point in the wrong direction. As a result, the part to be inserted will not move.

JARA: Japan Robot Association

Job and Skill Analysis: A method for analyzing robot work methods. Job analysis focuses on what to do, while skills analysis focuses on how.

Job Shop: A discrete parts manufacturing facility characterized by a mix of products of relatively low volume production in batch lots.

Joint: A rotary or linear articulation or axis of rotational or translational (sliding) motion in a manipulator system.

Joint Acceleration: Rotational acceleration of the arm's joint, used in Forward Dynamics to obtain the joint's displacement and velocity.

Joint Coordinate: The position of a joint.

Joint Coordinate System: The set of all joint position values. In non-Cartesian robots, actually not a coordinate system.

Joint Elasticity: See Backlash.

Joint Encoder: A transducer whose output is a series of pulses (a binary number) representing the joint's shaft position.

Joint Geometry Information: Geometry data of the mating of the parts to be joined, assembled, or welded. In welding, the geometry of the welding joint is utilized by the robotic system for positioning of the welding gun/fingers and deposition/squeezing parameters.

Joint Level Calibration: First level of calibration procedures. It involves the determination of the relationship between the actual joint displacement and the signal generated by the joint displacement transducer.

Joint Level Control: A level of robot control which requires the programming of each individual joint of the robot structure to achieve the required overall positions.

Joint Position: The joint's pose as registered by the associated encoders.

Joint Rate Control: A category of teleoperator control which requires the operator to specify the velocity of each separate joint.

Joint Space: The space defined by a vector whose components are the angular or translational displacement of each joint of a multi-degree-of-freedom linkage relative to a reference displacement for each such joint.

Joint Torque: Momentum applied by the joint's actuator.

Joint Velocity: Angular speed at which the joint is rotating.

Just-in-Time (JIT): A philosophy and a collection of management methods and techniques whose focus is the execution of operations only when they are needed. The benefits derive from the integration into a smooth-running production. JIT was originated at Toyota as an effort to eliminate waste (particularly inventories). It has evolved into a system for the continuous improvement of all aspects of manufacturing operations.

K

Kinematic Chain: The combination of rotary and/or translational joints, or axes of motion.

Kinematic Constraints: Positions and orientations of the robot's end-effector for which it may not be possible to find an inverse kinematics map. These constraints derive from the kinematic structure of the manipulator, being especially critical in underactuated or redundant manipulators.

Kinematic Geometry: The representation of the robot's physical structure in the robot controller.

Kinematic Model: A mathematical model used to define the position, velocity, and acceleration of each link coordinate and the end-effector, excluding consideration of mass and force.

Kinematic Structure: The physical composition of the robot. It includes joints, links, actuators, power feeding mechanism, and tooling.

Kinematic-Loop Method: A class of robot calibration methods where the manipulator is placed into a number of poses providing joint and TCP (Tool Center Point) measurement information, and where all model parameters are identified simultaneously by exploitation of the differentiability of the system model as a function of its parameters.

Kinematics (of Robot, Manipulator):   The study of the mapping of joint coordinates to link coordinates in motion, and the inverse mapping of link coordinates to joint coordinates in motion.

Kitting Process: The organization of subassemblies, consumables, and tools in sets corresponding to one processed unit. The kits are fed into the robot workstation for their process.

Knowledge-Based Process Planning: A knowledge database-supported system for establishing the processes and parameters to be used in the manufacturing facility to convert a part from its initial state to its final form according to an engineering drawing.

Knowledge-Based System (KBS):   A computer system whose usefulness derives primarily from a database containing human knowledge in a computerized format.

Knowledge Supply Chains: Integrated processes that use knowledge resources in industry and academia to enhance an enterprise's competitiveness by providing it with timely talent and knowledge.

L

LAMA (LAnguage for Mechanical Assembly):   Robot programming software that is part of a system capable of transforming mechanical assembly descriptions into robot programs. LAMA allows a programmer to define assembly strategies. Force feedback is accomplished in the system by force sensors on the wrist that are capable of resolving X, Y, and Z components of the force and torque acting on the wrist.

Laser Interferometer: A laser-based instrument splits the laser light into two or more beams, which are subsequently reunited after traveling over different paths and display interference. Because of the monochromaticity and high intrinsic brilliance of laser light, it can operate with path differences in the interfering beams of hundreds of meters, in contrast to a maximum of about 20 centimeters for classical interferometers.

Laser Sensor: A range-measuring device which illuminates an object with a collimated beam. The backscattered light, approximately coaxial with the transmitted beam, is picked up by a receiver. The range, or distance, is determined by either:

1. Measuring the time delay for a pulse of light to travel from the sensor to the object and back
2. Modulating the beam and measuring the phase difference between the modulations on the backscattered and the transmitted signals

Laser Vision System: Laser-based sensing system to gather information for motion control, torch position and orientation, and welding parameters such as arc current, arc voltage, arc length, and torch weaving. Laser-based vision systems provide also the capability of performing 100 percent inspection.

Laser Welding (Robotic):   Welding systems utilizing a highly coherent beam of light focused to a very small spot, where the metal melts and leads to the formation of a keyhole (a cylindrical cavity filled with ionized metallic gas) that causes a pool of molten metal. When this metal hardens behind the keyhole, it forms a weld as the beam moves along.

Lateral Resolution: The ability of a sensor, such as an ultrasonic sensor, to distinguish between details in the direction of a scan. In simple ultrasonic sensors lateral resolution is poor but can be improved by using the concept of back-propagation.

Lead-Through Programming: Also known as *lead-through teaching*, *programming by guiding*, or *manual programming*. A technique for programming robot motion, usually following a continuous path motion, but sometimes referring also to teaching point-to-point motions by using a teach box. For continuous path motions the operator grasps a handle which is secured to the arm and guides the robot through the desired task or motions while the robot controller records movement information and any activation signals for external equipment. This programming approach contrasts with off-line programming, which can be accomplished away from the manipulator.

Lean Manufacturing System: The description of a manufacturing system operating under the Just-in-Time approach as depicted by Womack, Jones, and Roos in the book *The Machine That Changed the World*, which chronicles the automobile industry.

Learning Control: A control scheme whereby experience is automatically used to change control parameters or algorithms.

Learning Machines: Machines that are capable of improving their future actions as a result of analysis and appraisal of past actions.

Level of Automation: The degree to which a process has been made automatic. Relevant to the level of automation are questions of automatic failure recovery, the variety of situations which will be automatically handled, and the conditions under which manual intervention or action by human beings is required.

Light-Section Inspection: The use of a slit projector to project a slit of light on an object to be inspected and an image detector to interpret the slit image of the object. Depending on the specific application, the projector and image detector may be oriented to provide a direct reflection or diffused reflection. A feature of light-section inspection is that the detection process is essentially sequential, thereby allowing relatively easier image analysis compared to other three-dimensional inspection techniques.

Lights out Factory: A totally automated factory, equipped with intelligent control and manufacturing systems.

Limit-Detecting Hardware: A device for stopping robot motion independently from control logic.

Limit Switch: An electrical switch positioned to be switched when a motion limit occurs, thereby deactivating the actuator that causes the motion.

Limited Sequence Manipulator: A nonservo manipulator that operates between fixed mechanical stops. Such a manipulator can operate only on parts in a fixed location (position and orientation) relative to the arm.

Line Feature: A linear edge extracted from the analysis of a vision image of a part. Line features serve as references to extract additional part features and as orientational cues in the position of the object.

Linear Interpolation: A computer function automatically performed in the control that defines the continuum of points in a straight line based on only two taught coordinate positions. All calculated points are automatically inserted between the taught coordinate positions upon playback.

Link:

1. A robotic manipulator is built of joints and links; each link is a basic part of a mechanical limb;
2. A non-servo manipulator that operates between fixed mechanical stops. Such a manipulator can operate only on parts in a fixed location (position and orientation) relative to the arm.

Link Coordinate: A coordinate system attached to a link of a manipulator.

LISP: Acronym for list processing; a high-level computer language implemented at the Massachusetts Institute of Technology in 1958. LISP is useful in artificial intelligence applications.

Load Capacity: The maximum total weight that can be applied to the end of the robot arm without sacrifice of any of the applicable published specifications of the robot.

Load Deflection:

1. The difference in position of some point on a body between a nonloaded and an externally loaded condition.

2. The difference in position of a manipulator hand or tool, usually with the arm extended between a nonloaded condition (other than gravity) and an externally loaded condition. Either or both static and dynamic (inertial) loads may be considered.

Local Area Network (LAN):   Telecommunications network that requires its own dedicated channels and that encompasses a limited distance, usually one building or several buildings in close proximity.

Location Analysis: The first step of refining a robotic system design. The details of the placement of workstations, buffers, material-handling equipment, and so on, are worked out. Location analysis may also be applied to location of an automated facility with respect to existing facilities and location of workpieces, accessories, and tools within a workstation. Typical factors to be considered are transport time or costs, workstation dimensions, transporter reach, fixed costs, and capacity limits.

M

Machine Language: The lowest-level language used directly by a machine.

Machine Loading/Unloading (Robotic):   The use of the robot's manipulative and transport capabilities in ways generally more sophisticated than simple material handling. Robots can be used to grasp a workpiece from a conveyor belt, lift it to a machine, orient it correctly, and then insert or place it on the machine. After processing, the robot unloads the workpiece and transfers it to another machine or conveyor. Some applications include loading and unloading of hot billets into forging presses, machine tools such as lathes and machining centers, and stamping presses. Another application is the tending of plastic injection molding machines. The primary motivation for robotic machine loading/unloading is the reduction of direct labor cost. Overall productivity is also increased. The greatest efficiency is usually achieved when a single robot is used to service several machines. A single robot may also be used to perform other operations while the machines are performing their primary functions.

Machine Vision: See Vision System.

Machine Vision Inspection System: The combination of a control system, sensor system, image-processing system, and workpiece-handling system to automatically inspect, transport, and handle the disposition of objects. The system also can adapt to changing working environments to a certain degree. The workpiece handling system may consist of x-y-θ tables, limited sequence arms, robots, or other positioning devices to emulate the oculomotor or manual functions of human workers. The sensor system may consist of single or multiple visual sensors for partial/overall viewing of scenes and/or coarse/fine inspection (to emulate human peripheral/foveal or far/close vision).

Machine-Mounted Robot: A robot usually dedicated to the machine on which it is mounted. The robot is designed to swivel its arm to load a part for one machine operation and then set the part down and repick it at a different orientation for the next operation. Fixtures are used to hold and maintain part position during these motions.

Machining (Robotic):   Robot manipulation of a powered spindle to perform drilling, grinding, routing, or other similar operations on the workpiece. Sensory feedback may also be used. The workpiece can be placed in a fixture by a human, another robot, or a second arm of the same robot. In some operations, the robot moves the workpiece to a stationary powered spindle and tool, such as a buffing wheel. Because of accuracy requirements, expensive tool designs, and a lack of appropriate sensory feedback capabilities, robot applications in machining are limited at present and are likely to remain so until both improved sensing capabilities and better positioning accuracy can be justified.

Machining Center: A numerically controlled metal-cutting machine tool that uses tools like drills or milling cutters equipped with an automatic tool-changing device to exchange those tools for different and/or fresh ones. In some machining centers programmable pallets for part fixturing are also available.

Magnetic Pickup Devices: A type of end-of-arm tooling that can be used to handle parts with ferrous content. Either permanent magnets or electromagnets are used, with permanent magnets requiring a stripping device to separate the part from the magnet during part release.

Main Reference: A geometric reference which must be maintained throughout a production process; for example, spot welding. The compliance with the references of the component elements of a subassembly guarantees the geometry of the complete assembly.

Maintenance Program: A maintenance program will typically:

1. Specify, assign, and coordinate maintenance activities
2. Establish maintenance priorities and standards
3. Set maintenance schedules
4. Specify diagnosis and monitoring methods
5. Provide special tools, diagnostic equipment, and service kits
6. Provide control of spare parts inventory
7. Provide training to maintenance personnel and operators
8. Measure and document program effectiveness.

Major Axes (Motions):   These axes may be described as the independent directions an arm can move the attached wrist and end-effector relative to a point of origin of the manipulator such as the base. The number of robot arm axes required to reach world coordinate points is dependent on the robot configuration.

MAL (Multipurpose Assembly Language):   A Fortran-based robot programming software developed by the Milan Polytechnic Institute of Italy, primarily for the programming of assembly tasks. Multiple robot arms are supported by MAL.

Manipulation (Robotic):   The handling of objects, by moving, inserting, orienting, twisting, and so on, to be in the proper position for machining, assembling, or some other operation. In many cases it is the tool that is being manipulated rather than the object being processed.

Manipulator: A mechanism, usually consisting of a series of segments, or links, jointed or sliding relative to one another, for grasping and moving objects, usually in several degrees of freedom. It is remotely controlled by a human (manual manipulator) or a computer (programmable manipulator). The term refers mainly to the mechanical aspect of a robot.

Manipulator-Level Control: A level of robot control which involves specifying the robot movements in terms of world positions of the manipulator structure. Mathematical techniques are used to determine the individual joint values for these positions.

Manipulator Responsiveness: The ability of a manipulator to recreate a user's trajectories and impedance in time and space.

Manual Manipulator: A manipulator operated and controlled by a human operator. See Teleoperator.

Manufacturing Automation Protocol (MAP):   A collection of existing and emerging communication protocols, each developed by a standard-setting body. It adopts the ISO/OSI model and consists of seven layers: physical, data link, network, transport, session, presentation, and application.

Manufacturing Cell: A manufacturing unit consisting of two or more workstations or machines and the material transport mechanisms and storage buffers that interconnect them.

Manufacturing Flexibility: Ability of the manufacturing system to respond to changes in the product or process specifications. Changes may result from customer requirements on the product and/or process being utilized in its manufacturing, or from the incorporation of new technologies into the existing processes.

Manufacturing Message Specification (MMS):   ISO/IEC 9506 standard for specifying messages to be passed between various types of industrial processes control systems.

MAPLE: A PL/I-based robot language developed by IBM and used for computations with several extensions for directing a robot to execute complex tasks. MAPLE supports force feedback and proximity sensory commands, gripper commands, coordinate transformations, and simple and straight-line motion.

Mass Customization: See Programmable Automation.

Massively Parallel Assembly: The arrangement of microscopic parts on a reusable pallet and then pressing two pallets together, thereby assembling the entire array in parallel.

Master–Slave Control: Control strategy for teleoperated systems which allows the operator to specify the end position of the slave (remote) end-effector by specifying the position of a master unit. Commands are resolved into the separate joint actuators either by the kinematic similarity of the master and slave units or mathematically by a control unit performing a transformation of coordinates.

Material Handling (Robotic): The use of the robot's basic capability to transport objects. Typically, motion takes place in two or three dimensions, with the robot mounted stationary on the floor on slides or rails that enable it to move from one workstation to another, or overhead. Robots used in purely material-handling operations are typically nonservo or pick-and-place robots. Some application examples include transferring parts from one conveyor to another; transferring parts from a processing line to a conveyor; palletizing parts; and loading bins and fixtures for subsequent processing. The primary benefits of using robots for material handling are reduction of direct labor costs, removal of humans from tasks that may be hazardous, tedious, or exhausting, and less damage to parts during handling. It is common to find robots performing material-handling tasks and interfacing with other material-handling equipment such as containers, conveyors, guided vehicles, monorails, automated storage/retrieval systems, and carousels.

MCL (Manufacturing Control Language): A high-level programming language developed by the McDonnell-Douglas Aircraft Company and designed for off-line programming of workcells that may include a robot. MCL is structured with major and minor words that are combined to form a geometric entity or a description of desired motion. It supports more than one type of robot and peripheral devices, as well as simple and complex touch and vision sensors. Robot gripper commands are included.

Mechanical Grip Devices: The most widely used type of end-of-arm tooling in parts-handling applications. Pneumatic, hydraulic, or electrical actuators are used to generate a holding force which is transferred to the part via linkages and fingers. Some devices are able to sense and vary the grip force and grip opening.

Mechatronics: The synergetic integration of mechanical engineering with electronic and intelligent computer control in the design and manufacture of industrial products and processes.

Medical Robotics: The integration of robotics equipment in the planning and execution of medical procedures.

Memory Alloy: Metallic compounds with the ability to recover the original shape after its alteration by physical or other means such as electric charge applications. Often used in gripper-fingers to handle objects with unique shapes.

Metamorphic Robotics: A collection of independently controlled mechatronic modules, each of which has the ability to connect, disconnect, and climb over adjacent modules. Each module allows power and information to flow through itself and to its neighbors. A change in the manipulator morphology results from the locomotion of each module over its neighbors.

Micro Electromechanical Systems (MEMS): Mechanical components, whose size typically ranges from about 10 micrometers to a few hundred micrometers, with smallest feature size of less than a micron and overall size of up to a millimeter or more. These components are manufactured by efficient, highly automated fabrication processes using computer-aided design and analysis tools, lithographic pattern generation, and micromachining techniques such as thin film deposition and highly selective etching. MEMS applications are in accelerometers, oscillators, micro optical components, and microfluidic and biomedical devices. Interest is now shifting towards complex microsystems that combine sensors, actuators, computation, and communication in a single micro device.

Micromanipulation: Technology for the assembly, modification, and maintenance of three-dimensional microsystems, integrating sensing, actuating, and mechanical parts. It is classified into contact type and noncontact type.

Microrobot System: Robotics systems including micromachines, micromanipulators, and human–machine interfaces. Design of microrobot systems is closely related to issues of scaling problems and control methods. The main characteristics of a microrobot system are:

1. Simplicity
2. Preassembly
3. Functional integration
4. Multitude
5. Autonomous decentralization

Minimal Precedence Constraint (MPC) Method: A method for the generation of assembly sequences based on the identification of geometric precedence constraints that implicitly represent all geometrically feasible assembly sequences. The minimal precedence constraint for an assembly component is defined as the alternative assembly states that will prevent the assembly of this component.

Minor Axes (Motions): The independent attitudes relative to the mounting point of the wrist assembly on the arm by which the wrist can orient the attached end-effector.

Mobile Robot: A freely moving programmable industrial robot which can be automatically moved, in addition to its usual five or six axes, in another one, two, or three axes along a fixed or programmed path by means of a conveying unit. The additional degrees of freedom distinguish between linear mobility, area mobility, and space mobility. Mobile robots can be applied to tasks requiring workpiece handling, tool handling, or both.

Modal Analysis: A ground vibration test to determine experimentally the natural frequencies, mode shapes, and associated damping factors of a structure.

Model Interactive CASP/CASE: Computer-Integrated Surgery systems relying on the surgeon to control the motions of the surgical instruments (conventional or robotic manipulation devices).

Modular Robots: Robots that are built of standard independent building blocks, such as joints, arms, wrist grippers, controls, and utility lines, and are controlled by one general control system. Each modular mechanism has its own drive unit and power and communication links. Different modules can be combined by standard interface to provide a variety of kinematic structures designed to best solve a given application requirement. "Mobot," a contraction for modular robot, is a tradename of the Mobot Corporation of San Diego.

Modularity Concept: Design approach favoring the subassembly of components and their delivery to the final assembly operation. The modules are assembled without the constraints that the final assembly line may impose, allowing automated assembly operations, tighter controls and tolerances, and overlap of the assembly times.

Molded Interconnect Devices (MID): Molded plastic parts with a partially metal-plated surface forming an electric circuit pattern. MIDs unite the functions of conventional circuit boards, casings, connectors, and cables. The main technological advantage of MIDs is that they integrate electronic and mechanical elements onto circuit carriers with virtually any geometric shape. They have enabled entirely new functions and usually help to miniaturize electronic products.

Molecular Robotics: See Nano-Robotics.

Mono-Articulate Chain Manipulator: A manipulator at the end of a special type of chain used to enter an enclosed area through a narrow opening. The chain is constructed of box section links in such a manner as to allow the chain to articulate in one direction but not another. This design results in a chain which can be reeled or coiled but which forms a rigid element when extended.

Monorail Transporter: Material-handling devices designed to operate at much faster speed (up to 260 meters per minute) than the conventional flat conveyor systems or the mobile robots. Monorails are built for midair operation to connect assembly cells. A rail system should be attached on the ceiling or wall. Monorails improve facility utilization by increased storage density, and reduce facility size by less transport aisles.

Motion Economy Principles (MEPs): Principles that guide the development, trouble-shooting, and improvement of work methods and workplaces, adapted for robot work.

Motion Envelope: See Work Envelope.

Motion Instruction Level: Programming instruction set based on the specific motion instructions (see also Motion Planning, Fine, and Motion Planning, Gross).

Motion Planning, Fine: Dealing with uncertainty in the world model by using guarded motions when approaching a surface and compliant motions when in contact with a surface.

Motion Planning, Gross: Planning robot motions that are transfer movements for which the only constraint is that the robot and whatever it is carrying should not collide with objects in the environment.

Motion-Velocity Graphs: Graphs which show regions of maximum movement and velocity combinations for common arm and wrist motions. Such charts are used to ascertain the applicability of a robot for a particular task.

Mounting Plate: The means of attaching end-of-arm tooling to an industrial robot. It is located at the end of the last axis of motion on the robot. The mounting plate is sometimes used with an adapter plate to enable the use of a wide range of tools and tool power sources.

Multiagent Systems (MAS): One of the two subfields of Distributed Artificial Intelligence (DAI). It deals with heterogeneous, not necessarily centrally designed agents charged with the goal of utility-maximizing coexistence; for instance, fulfillment of production plans by a globally distributed enterprise.

Multi-gripper System: A robot system with several grippers mounted on a turret-like wrist, or capable of automatically exchanging its gripper with alternative grippers, or having a gripper for multiple parts. A type of mechanical gripper enabling effective simultaneous execution of two or more different jobs effectively.

Multi-hand Robot Systems: A class of robotic manipulators with more than one end-effector, enabling effective simultaneous execution of two or more different jobs. Design methods for each individual hand in a multi-hand system are similar to those of single hands, but must also consider the other hands.

Multiple Stage Joint: A linear motion joint consisting of sets of nested single-stage joints.

Multiplexed Teleoperation: The teleoperation by a single master station over several dissimilar slave robots. Conversely, several collaborating master operators can be multiplexed to control a single slave robot performing a given task.

Multi-robot System (MRS): A robotic system with two or more robots executing a set of tasks requiring the robots' collaboration.

Multi-sensor System: Arrangement of a series of sensors gathering information about the robot's task. Neural networks-based mechanisms can be utilized for performing the sensor data fusion.

Multi-spectral Color and Depth Sensing: The separation of the different colors of the spectrum in the captured image of the sensing device. The method records the energy levels of the red, green, blue, or infrared bands of the spectrum. Multi-spectral sensors are basic remote sensing data sources for quantitative thematic information.

N

Nanoassembly: Assembly of nanoscale components requiring a combination of precise positioning and chemical compliance (chemical affinity between atoms and molecules).

Nanolithography: A technique closely related to Nanomanipulation which involves removing or depositing small amounts of material by using Scanning Probe Microscopy (SPM).

Nanomanipulation: See Nanorobotics.

Nanorobotics: The science of designing, building, and applying robots capable of interaction with atomic- and molecular-sized objects.

Nanotechnology: Study of phenomena and structures with characteristic dimensions in the nanometer range. Some of its applications include:

1. Cell probes with dimensions ~1/1000 of the cell's size
2. Space applications, e.g., hardware to fly on satellites
3. Computer memory
4. Near-field optics, with characteristic dimensions ~20 nm
5. X-ray fabrication, systems that use X-ray photons
6. Genome applications, reading and manipulating DNA
7. Optical antennas

Natural Language: Any language spoken by humans. Any human or computer language whose rules reflect and describe current rather than prescribed usage; it is often loose and ambiguous in interpretation.

Navigation Controller: A control system that uses sensor information on two or more navigation factors, such as altitude, direction, and velocity, to compute course information for mobile robots or AGVs.

Nearest Neighbor Classifier: A method of object classification by statistical comparison of computed image features from an unknown object with the features known from prototype objects. The statistical distance between object and prototype is computed in the multidimensional feature space.

Negotiation Schemes: A multiple-way communication method for reaching a mutually agreed upon course of action, for instance, resource allocation and task assignment for distributed robots.

Neural Networks: An information-processing device modeled after biological networks of neurons, that utilizes a large number of simple interconnected modules, and in which information is stored by components that at the same time effect the connections between these modules. Often called *Artificial Neural Networks* (ANN), their applications cover four main areas:

1. The automatic generation of nonlinear mappings (e.g., robot control and noise removal from signals)
2. Decisions based on a massive amount of data (e.g., speech recognition and fault prediction)
3. Fast generation of near-optimal solution to a combinatorial optimization problem (e.g., airline scheduling and network routing)
4. Situations in which there exist more input variables than other approaches can consider.

Neuro-Fuzzy Control Systems (NFCS): Connectionist control models applying neural learning techniques to determine and tune the parameters and/or structure of the connectionist model. Neuro-fuzzy control systems require two major types of tuning: structural and parametric. Structural tuning concerns the tuning of the structure of fuzzy logic rules such as the number of variables to account for and for each input or output variable the partition of the universe of discourse; the number of rules and the conjunctions that constitute them.

Noncontact Inspection: The critical examination of a product, without touching it with mechanical probes, to determine its conformance to applicable quality standards or specifications.

Noncontact Proximity Sensors: See Proximity Sensor.

Noncontact Sensor: A type of sensor, including proximity and vision sensors, that functions without any direct contact with objects.

Nonkinematic calibration: Third level of calibration procedures. It involves the analysis of nonkinematic errors in positioning the robot's TCP that are due to effects such as joint compliance, friction, clearance, etc.

Nonmotion Instruction: Instruction controlling the I/O signals or the execution of the sequence of instructions in the robot program.

Nulling Time: The time required to reduce to zero, or close to zero, the difference between the actual and the programmed position of every joint.

Numerical Control: A method for the control of machine tool systems. A part program containing all the information, in symbolic "numerical" form, needed for processing a workpiece is stored on a medium such as paper or on magnetic media. The information is read into a computer controller, which translates the part program instructions to machine operations on the workpiece. See also Computerized Numerical Control.

O

Object-Level Control: A type of robot control where the task is specified in the most general form. A comprehensive database containing a world model and knowledge of application techniques is required. Knowledge-based algorithms are required to interpret instructions and apply them to the database to automatically produce optimized, collision-free robot programs.

Object-Oriented Methodologies (OO): Modeling, design, and programming methods that focus on organizing all relevant knowledge about objects, leading to benefits that include less errors, modularity of information, and the ability to inherit generic information for particular objects.

Object-Oriented Programming (OOP): A nonprocedural computing programming paradigm in which program elements are conceptualized as objects that can pass messages to each other. Each object has its own data and programming code and is internally self reliant; the program makes the object part of a larger whole by incorporating it into a hierarchy of layers. Object-oriented programming is an extension of the concept of modular programming. The modules are independent so that they can be copied into other programs. This capability raises the possibility of inheritance by copying and adding some new features to an existing object, and then moving the new object to a new program.

Off-Line Programming: Developing robot programs partially or completely without requiring the use of the robot itself. The program is loaded into the robot's controller for subsequent automatic action of the manipulator. An off-line programming system typically has three main components: geometric modeler, robot modeler, and programming method. Often it is associated with robot simulation to try to debug it before implementation. The advantages of off-line programming are reduction of robot downtime; removal of programmer from potentially hazardous environments; a single programming system for a variety of robots; integration with existing computer-aided design/computer-assisted manufacturing systems; simplification of complex tasks, and verification of robot programs prior to execution.

On-Line Programming: The use of a teach pendant for teach programming, which directs the controller in positioning the robot and interacting with auxiliary equipment. It is normally used for point-to-point motion and controlled path motion robots, and can be used in conjunction with off-line programming to provide accurate trajectory data.

On–Off Control: A type of teleoperator control in which joint actuators can be turned on or off in each direction at a fixed velocity.

One-Dimensional Scanning: The processing of an image one scan line at a time independent of all other scan lines. This technique simplifies processing but provides limited information. It is useful for inspection of products such as paper, textiles, and glass.

Open Architecture: A computer architecture whose specifications are made widely available to allow other parties to develop add-on peripherals for it.

Open Development Environment: The organization of the Computer-Aided Design system, allowing the researcher or the advanced user to access all the geometric and kinematic models in the system and develop new applications, new types of analysis, and new planning and synthesis algorithms.

Open-Loop Control: Control of a manipulator in which preprogrammed signals are delivered to the actuators without the actual response at the actuators being measured. This control is the opposite of closed-loop control.

Operating System: Software that controls the execution of computer programs and one that may provide scheduling, allocation, debugging, data management, and other management functions.

Optimal Control: A control scheme whereby the system response to a commanded input, given the dynamics of the process to be controlled and the constraints on measuring, is optimal according to a specified objective function or criterion of performance.

Orientation: Also known as *positioning*. The consistent movement or manipulation of an object into a controlled position and attitude in space.

Orientation Finding: The use of a vision system to locate objects so they can be grasped by the manipulator or mated with other parts.

Overlap:

1. Sharing of work space by several robots (space overlap)
2. Performance of some or all of an operation concurrently by more than one robot with one or more other operations.

Overshoot: The degree to which a system response to a step change in reference input goes beyond the desired value.

P

PAL: A robot motion software developed at Purdue University in Indiana by which robot tasks are represented in terms of structured Cartesian coordinates. PAL incorporates coordinate transformation, gripper, tool, and sensor-controlled operation commands.

Palletizing/Depalletizing: A term used for loading/unloading a carton, container, or pallet with parts in organized rows and possibly in multiple layers.

Pan:

1. Orientation of a view, as with a video camera, in azimuth.
2. Motion in the azimuth direction.

Parallel Manipulator: Robotic mechanisms containing two or more serial kinematic chains connecting the end-effector to the base. Generally, parallel manipulators can offer more accuracy in positioning and orienting objects than open-chain manipulators. They can also possess a high payload/weight ratio, and they are easily adaptable to position and force control.

Parallel Program Control: A robot program control structure which allows the parallel execution of independent programs.

Parallel Robots: See Parallel Manipulator.

Part Classification: A coding scheme, typically involving four or more digits, which specifies a discrete product as belonging to a part family according to group technology.

Part Mating: The action of assembling two parts together according to the assembly design specifications. It occurs in four stages: approach, chamfer crossing, one-point contact, and two-point contact.

Part Mating Theory: It predicts the success or failure mode of assembly of common geometry parts, such as round pegs and holes, screw threads, gears, and some simple prismatic part shapes. Two common failure modes during two-point contact are wedging and jamming.

Part Orientation: See orientation.

Part Program: A collection of instructions and data used in numerically controlled machine tool systems to produce a workpiece.

PASLA (Programmable ASsembly Robot Language): A robot programming language developed at Nippon Electric Company, Ltd. (NEC) of Japan. It incorporates coordinate guidance and sequencer functions. It is a motion-directed language that consists of 20 basic instructions.

Passive Manipulator: Robotic manipulator with one or more joints without actuators (passive joints) and at least one actuated joint (active joint). The design of this type of manipulator reduces weight, cost, and energy consumption. On the other hand, their control and dynamics are nonlinear and highly coupled. See also Underactuated Robotic System.

Path: The trajectory of the robot's end-effector, or the trajectory of a mobile robot, when performing a specific task.

Path Accuracy: For a path-controlled robot, this is the level of accuracy at which programmed path curves can be followed at nominal load.

Path Interpolation: At each interpolation interval, based on the information provided by the path planner, the path interpolator computes an intermediate position (between the start position and destination position). The eventual output of the path interpolator at each interpolation interval is a set of joint angles which forms the input to the servo-control loop to command the robot to move along the path.

Path Measuring System: A part of the mechanical construction of each axis which provides the position coordinate for the axis. Typically, for translational axes, potentiometers or ultrasound are used for path measuring systems. But for rotational axes, resolvers, absolute optical encoders, or incremental encoders are used. A path measuring system may be located directly on a robot axis or included with the drive system.

Pattern Comparison Inspection: The comparison of an image of an object with a reference pattern. The image and pattern are aligned with one another for detecting deviations of the object being inspected. The features extracted from the object image are compared with the features of the reference pattern. If both have the same features, the object is presumed to have no defects.

Pattern Recognition: A field of artificial intelligence in which image analysis is used to determine whether a particular object or data set corresponds to one of several alternatives or to none at all. The analysis system is provided in advance with the characteristics of several prototype objects so that it can classify an unknown object by comparing it with each of the different prototypes.

Payload: The maximum weight that a robot can handle satisfactorily during its normal operations and extensions.

Performance Specifications: The specification of various important parameters or capabilities in the robot design and operation. Performance is defined in terms of:

1. The quality of behavior
2. The degree to which a specified result is achieved
3. A quantitative index of behavior or achievement, such as speed, power, or accuracy

Peripheral Equipment: The equipment used in conjunction with the robot for a complete robotic system. This equipment includes grippers, conveyors, part positioners, and part or material feeders that are needed with the robot.

Perspective Transform: The mathematical relationship between the points in the object space and the corresponding points in a camera image. The perspective transform is a function of the location of the camera in a fixed coordinate system, its orientation as determined by its pan and tilt angles, and its focal length.

Photoelectric Sensors: A register control using a light source, one or more phototubes, a suitable optical system, an amplifier, and a relay to actuate control equipment when a change occurs in the amount of light reflected from a moving surface due to register marks, dark areas of a design, or surface defects.

Pick-and-Place: A grasp-and-release task, usually involving a positioning task.

Pick-and-Place Robot: Also known as *bang-bang robot*. A simple robot, often with only two or three degrees of freedom, that transfers items from a source to a destination via point-to-point moves.

Pin Through Hole Component (PTHC) Assembly: Assembly technique for electronic components. The component is positioned for aligning its pins with the holes, inserted, and then soldered or glued to fix its position and ensure the proper contact of the pins to the printed circuit board.

Pinch Point: Any point where it is possible for a part of the body to be injured between the moving or stationary parts of a robot and the moving or stationary parts of associated equipment, or between the material and moving parts of the robot or associated equipment.

Pipe-Crawling Robot: Mobile and legged robots designed for a specific task-set of sewer-line inspection. Due to the fact that mobile and legged systems work in a more unstructured and much more expansive world without controllable conditions, they are usually designed for a well-formulated task to reduce complexity and cost and increase reliability and overall performance.

Pitch: Also known as bend, particularly for the wrist. The angular rotation of a moving body about an axis that is perpendicular to its direction of motion and in the same plane as its top side.

Pixel: Also known as *photo-element* or *photosite*. It is an element of a digital picture or sensor. Pixel is short for picture-cell.

Playback Accuracy: The difference between a position command taught, programmed, or recorded in an automatic control system and the position actually produced at a later time when the recorded position is used to execute motion control.

PLC: See Programmable Logic Controller (PLC).

Pneumatic Pickup Device: The end-of-arm tooling such as vacuum cups, pressurized bladders, and pressurized fingers.

Point-to-Point Control: A robot motion control in which the robot can be programmed by a user to move from one position to the next. The intermediate paths between these points cannot be specified.

Point-to-Point Positioning: Open-loop operation mode of a robotic positioner. It utilizes multiple-stop mechanical or solid-state limit switches on all axes.

Point-to-Point System: The robot movement in which the robot moves to a numerically defined position and stops, performs an operation, moves to another numerically defined position and stops, and so on. The path and velocity while traveling from one point to the next generally have no significance.

Pooling of Facilities: The sharing of facilities (robot, machines, cells, etc.) for the manufacturing of a set of parts or products.

Pose: The robot's joints position for a particular end-effector position and orientation within the robot's workspace. Specific positions are named according to the tasks the robot is performing; for example, the home pose, which indicates the resting position of the robot's arm.

Position Control: A control by a system in which the input command is the desired position of a body.

Position Finding: The use of a vision system to locate objects so they can be grasped by a manipulator or mated with other parts.

Position Sensor: Device detecting the position of the rotor relative to the stator of the actuator. The rotor speed is derived from the position information by differentiation with respect to time. The servo-system uses this sensor data to control the position as well as the speed of the motor. In general, one of two sensor types is used: either the *pulse coder* type, also known as *digital encoder*, or the *resolver type*.

Positioners: Also known as *positioning table*, positioners are fixture devices for locating the parts to be processed in the required position and orientation. Positioners can be implemented as hard tooling devices or reprogrammable robotic devices which reduce the setup time and part changeover times. For instance, positioners are used in robotic arc welding to hold and position pieces to be welded. The movable axes of the positioner are sometimes considered additional robot axes. The robot controller controls all axes in order to present the seam to be welded to the robot's torch in the location and orientation taught or modified by adaptive feedback, or changes inserted by the operator, dynamically during execution.

Postprocessor: The programming software part of a robotic simulator/emulator, which converts the task program simulated into the specific robot's language motion instructions.

Precision (Robot): A general concept reflecting the robot's accuracy, repeatability, and resolution.

Presence Sensing: The use of a device designed, constructed, and installed to create a sensing field or space around a robot(s) and one that will detect an intrusion into such field or space by a person, another robot, and so on (constrast with Telepresence).

Pressurized Bladder: A pneumatic pickup device which is generally designed especially to conform to the shape of the part. The deflated bladder is placed in or around the part. Pressurized air causes the bladder to expand, contact the part, and conform to the surface of the part, applying equal pressure to all points of the contacted surface.

Pressurized Fingers: A pneumatic pickup device that has one straight half, which contacts the part to be handled, one ribbed half, and a cavity for pressurized air between the two halves. Air pressure filling the cavity causes the ribbed half to expand and "wrap" the straight side around a part.

Priority Capability: A feature of a robot program used to control a robot at a flexible machining center. If signals (service calls) are received from several machines simultaneously, the priority capability determines the order in which the machines will be served by the robot.

Prismatic Motion: The straight-line motion of an arm link relative to the link connected to it.

Problem Solving: A computational reasoning process based on three elements:

1. General knowledge common to all problems under solution
2. Specific knowledge about the selected specific problems to be solved
3. Reasoning knowledge appropriate for the selected problem domain

Procedural Reasoning System (PRS): A hybrid robot control architecture which continuously creates and executes plans but abandons them as the situation dictates. PRS plans are expanded dynamically and incrementally on an as-needed basis. Thus the system reacts to changing situations by changing plans as rapidly as possible should the need arise.

Process Control: The control of the product and associated variables of processes which are continuous in time (such as oil refining, chemical manufacture, water supply, and electrical power generation).

Process Reengineering: Focuses on the whole process but has a wider scope than the removal of waste. This approach questions whether the status quo is relevant to the future system. A reduced number of activities, organizational and job redesign, and new developments in Information Technology, such as document Image Processing (DIP) or expert systems, may be used. This type of change seeks to reduce the number of activities by up to 90 percent and addresses real strategic benefits.

Process Simplification: Analysis of processes within the frame of process reengineering, for locating and eliminating non-value-added activities as storage and inspection. The analysis is usually performed by teams seeking to remove these activities. The benefits of this type of change are related to the price of nonconformance in the process (the cost of scrap, rework, inspection, warranty departments, etc.)

Product Data Representation and Exchange (STEP): ISO 10303 standard which provides a means of describing product data throughout the life cycle of a product independent of any particular computer system. STEP standardizes the structure and meaning of items of product data in the form of data models to meet particular industrial needs but does not standardize the data values. The standard provides a series of data models for different engineering applications and also a methodology for creating the data models.

Product Description Exchange for STEP (PDES): A standard file format including all the information necessary to describe a product from its design to its production. It supports multiple-application domains (e.g., mechanical engineering and electronics).

Product-Centric Paradigm: Design and development of robotic manufacturing systems based on one or few products. Many automated systems developed under the old product-centric paradigm have failed economically because there was not enough demand for the product(s) the system was designed to produce. These product-specific systems have generally lacked the ability to be reconfigured at a cost that would allow them to meet the needs of additional products.

Program Simulation: Execution of the robot's program in a virtual environment that is different from the real application. The environment may be a computer representation of the robot and object interacting with it, a scale prototype system, or the same system with the omission of some of the elements of the operation (e.g., parts, conveyors, etc.).

Programmable: A robot capable of being instructed, under computer control, to operate in a specific manner or capable of accepting points or other commands from a remote source.

Programmable Assembly Line: A number of assembly stations interconnected with a buffered transfer system. A multilevel computer system controls the assembly line. The individual stations can be programmable, dedicated, or manual assembly.

Programmable Assembly System: A number of assembly stations or assembly centers in a stand-alone or interconnected configuration, possibly connected to one another or other equipment by a buffered transfer mechanism. Within the assembly station, equipment such as robots, dedicated equipment, programmable feeders, magazines, fixtures, and vision or other sensors is arranged as required by the operators assigned to the station. A suitable computer control system completes the configuration.

Programmable Automation: Automation for discrete parts manufacturing characterized by the features of flexibility to perform different actions for a variety of tasks, ease of programming to execute a desired task, and artificial intelligence to perceive new conditions, decide what actions must be performed under those conditions, and plan the actions accordingly. Although suitable primarily for increasing the productivity of batch manufacturing, programmable automation may also be applicable to mass production for the following reasons:

1. Reduction of the setup time for manufacturing short-lived products in a competitive world market (often called *Mass Customization*)
2. Lowering the cost of production equipment by reusing components, such as robots, sensors, and computers, that are commercially available and recyclable
3. Producing a sufficiently large sample of new products to enable testing of their technical and market performance

Programmable Controller: Short for Programmable Logic Controller.

Programmable Feeder: A part feeder that can deliver a wide range of product parts with known or desired orientation and without any replacement or retooling when switching to a different product part.

Programmable Fixture: A multipurpose, computer-controlled fixture that is capable of accepting and rigidly holding parts of different shapes which it then presents, upon request, to a manipulator or vision system in a specified orientation. This fixture might take the form of a rugged hand and three-degree-of-freedom wrist.

Programmable Logic Controller (PLC): A solid-state device which has been designated as a direct replacement for relays and "hard-wired" solid-state electronics. The logic can be altered without wiring changes. It has high reliability and fast response; operates in hostile industrial environments without fans, air conditioning, or electrical filtering; is programmed with a simple ladder diagram language; is easily reprogrammed with a portable panel if requirements change; is reusable if equipment is no longer required; has indicator lights provided at major diagnostic points to simplify troubleshooting.

Programmable Manipulator: A mechanism which is capable of manipulating objects by executing a program stored in its control computer (as opposed to manual manipulator, which is controlled by a human).

Programming (Robot):   The act of providing the control instructions required for a robot to perform its intended task.

Pronation: The orientation or motion toward a position with the back or protected side facing up or exposed. See supination.

Proportional Control: A control scheme whereby the signal which drives the actuator is monotonically related to the difference between the input command (desired output) and the measured actual output.

Proportional-Integral-Derivative (PID) Control: A control scheme whereby the signal which drives the actuator equals a weighted sum of the difference, time integral of the difference, and time derivative of the difference between the input (desired output) and the measured actual output. Used especially in process control.

Proprioception: The reception of stimuli produced by the robot itself. It can be accomplished by sensing the positions and motions of the robot's own articulated structures, such as arms, fingers, legs, feet, "necks," etc.

Prosthetic Robot: A programmable manipulator or device that substitutes for lost functions of human limbs.

Protocol: A defined means or process for receiving and transmitting data through communication channels. Often associated with operations that are influenced by the communicated information. For instance, a workflow protocol defines the information exchanged for distributed control logic in a production or service system.

Proximal: The area on a robot close to the base but away from the end-effector of the arm.

Proximity Sensor: A device which senses that an object is only a short distance (e.g., a few inches or feet) away and/or measures how far away it is. Proximity sensors typically work on the principles of triangulation of reflected light, elapsed time for reflected sound, intensity-induced eddy currents, magnetic fields, back pressure from air jets, and others.

Pseudo-Gantry Robot: A pedestal robot installed in the inverted position and mounted on slides which allows it to traverse over a work area.

PUMA (Programmable Universal Machine for Assembly):   Originated as a 1975 developmental project at General Motors Corporation, PUMA resulted in the specification for a human-arm-size, articulated, electrically driven robot that was later commercialized by Unimation, Inc.

Purdue Enterprise Reference Architecture (PERA):   See Computer-Integrated Manufacturing Open Systems Architecture (CIMOSA)

Pyloelectric Effect: An energy transformation from optical energy, supplied by a source external to the microrobotic device, to electric energy required for the device operation.

R

Radius Statistics: Statistics of shape features computed from the set of radius vectors of a blob. They are used in image analysis for determining orientation, distinguishing shapes, and counting corners.

RAIL: A generalized robot programming language based on PASCAL and developed by Automatix, Inc., Billerica, Massachusetts, for control of vision and arm manipulation with many constructs to support inspection.

Rapid Prototyping: The process of generating the trial version of a product. The goal is to generate inexpensively and quickly the fit, accessibility and other interrelated product aspects that will ensure an efficient manufacturing process.

RAPT (Robot APT): An APT-based task-level robot programming language developed at the University of Edinburgh in Scotland. RAPT describes objects in terms of their features (cylinders, holes, and faces, etc.) and has means for describing relationships between objects in a subassembly.

Rate Control: A control system where the input is the desired velocity of the controlled object.

RCCL (Robot Control "C" Library): A robot programming system based on the "C" language and developed at Purdue University in Indiana to specify tasks by a set of primitive system calls suitable for robot control. RCCL features include manipulator task description, sensor integration, updateable world representation, manipulator independence, tracking, force control, and Cartesian path programming.

RCL: An assembly-level, command-oriented language developed at Rensselaer Polytechnic Institute in New York to program a sequence of steps needed to accomplish a robot task. RCL supports simple and straight-line motion but has no continuous-path motion. The force feedback, gripper, and touch sensor commands are included in the language.

Reusability: The characteristic of hardware or software of being used by several tasks without significant modifications.

Reachability: A measure of motion economy, the number of useful positions or cells that a particular robot succeeds in reaching during its intended task execution.

Reactive Robots: See Responsive Robots.

Realistic Robot Simulation (RRS): A technique for the design and off-line programming of robotic workcells, based on the faithful virtual representation of the workcell elements (robots, AGVs, conveyor belts, and sensors) and the embedding of an industrial robot controller into the simulation system by means of a versatile message-based interface and a sophisticated environment model for intuitive simulation of grip and release operations, conveyor belts, and sensors. This representation is combined with a graphical 3D visualization of the workcell in real time. User interaction with the simulation system is based on a windows-oriented graphical user interface (GUI).

Reasoning: The computational process of deriving new facts from given ones. It includes elements for identifying the domain knowledge relevant to the problem at hand, resolving conflicts, activating rules, and interpreting the results.

Recognition: The act or process of identifying (or associating) an input with one of a set of possible known alternatives (see also Pattern Recognition).

Recognition Unit: A sensory device used to distinguish between random parts (e.g., by measuring their outer diameter).

Recovery Graph: A logical representation including all feasible disassembly sequences and their associated costs and revenues.

Rectangular Robot: Also known as *rectangular coordinate robot*, *Cartesian coordinate robot*, *Cartesian robot*, *rectilinear coordinate robot*, or *rectilinear robot*. A robot that moves in straight lines up and down and in and out. The degrees of freedom of the manipulator arm are defined primarily by a Cartesian coordinate axis system consisting of three intersecting perpendicular straight lines with origin at the intersection. This robot may lack control logic for coordinated joint motion. It is common for simple assembly tasks.

Redundant Manipulator: A robot manipulator with more degrees of freedom than the number required for a class of tasks or motions.

Reflexiveness: The ballistic mapping from an input (perception) to output (behavioral response) that can involve single or multiple actions. It allows robotic systems timely reaction to unexpected changes in the environment that could affect their autonomy and eventually the fulfillment of their goals.

Rehabilitation (Robotic): The application of robotics systems as substitutes for human or animal helpers or as replacements for lost human functions (robotic prosthetic/orthotics). Robotic helpers can serve as manipulators for disabled or guides for blind people, with their utilization being external to the human or animal body. On the other hand, robotic prosthetics involves the tight coupling of human and machine with the objective of working in cooperation (for instance, the implantation of artificial limbs). In prosthetic systems it is fundamental to establish resilient communication methods and media between the human body and its robotic counterpart, especially in the case of powered orthosies.

Relational Image Analysis Methods: Methods for image analysis that depend on the context of local features rather than global feature values.

Relative Coordinate System: Also known as *tool coordinate system*. A coordinate system whose origin moves relative to world or fixed coordinates.

Remanufacturing: A process in which worn-out products or parts are brought back to original specifications and conditions or converted into raw materials. Central to remanufacturing is the disassembly process for material and component isolation, since the objectives of product remanufacturing are to maximize the quantity and quality of parts obtained for repair and reuse and minimize the disposal quantities. Complementary processes to which products are subjected include refurbishment, replacement, repair, and testing.

Remote Center Compliance Device (RCC): In rigid part assembly, a passive support device that aids part insertion operations. The RCC can be used with assembly robots as well as with traditional assembly machines and fixed single-axis workstations. The main feature of the RCC is its ability to project its compliance center outside itself, hence the source of its particular abilities to aid assembly and the reason for the word ''remote'' in its name. Its major function is to act as a multiaxis ''float,'' allowing positional and angular misalignments between parts to be accommodated. The RCC allows a gripped part to rotate about its tip or to translate without rotating when pushed laterally at its tip. It enables successful mating between two parts, a tool and a part, a part and a fixture, a tool and a tool holder, and many other mating pairs. Instrumented versions (IRCC) add sensors to increase the abilities in complex assembly.

Remote Manipulation System (RMS): Space-based teleoperation system on the fleet of U.S. space shuttles. The manipulator arms, located in the main shuttle bay, are controlled directly by a human operator viewing the system through a window. Two three-axis variable-rate command joysticks provide velocity input commands from the operator to the robot's controller. Applications of the RMS include deployment and capture of satellite systems, space-based assembly, and construction of the International Space Station (ISS).

Remotely Operated Vehicle (ROV): An application of mobile robots controlled by teleoperation. Typical usage include underwater and space exploration and search and rescue as well as offshore oil-rig servicing and maintenance (see also Sojourner).

Remotely Piloted Vehicle (RPV): A robot aircraft controlled over a two-way radio link from a ground station or mother aircraft that can be hundreds of miles away; electronic guidance is generally supplemented by remote control television cameras feeding monitor receivers at the control station.

Repeatability: The envelope of variance of the robot tool point position for repeated cycles under the same conditions. It is obtained from the deviation between the positions and orientations reached at the end of several similar cycles. Contrast with Accuracy.

Replacement Flexibility: The ease with which a production line can continue to operate at a reduced rate in the event of failure of one of its components (robot or tooling). One way to attain replacement flexibility is by controlling other line robots to automatically compensate for the missing work of a faulty robot. Another way is to install robots at the end of a line to perform operations that were missed.

Replica Master: A control device which duplicates a manipulator arm in shape and serves for precise manipulator teaching. Control is achieved by servoing each joint of the manipulator to the corresponding joint of the replica master.

Resolution: The smallest incremental motion which can be produced by the manipulator. Serves as one indication of the manipulator accuracy. Three factors determine the resolution: mechanical resolution, control resolution, and programming resolution.

Resolved Motion Rate Control:

1. A control scheme whereby the desired velocity vector of the endpoint of a manipulator arm is commanded and from it the computer determines the joint's angular velocities to achieve the desired result.

2. Coordination of a robot's axes so that the velocity vector of the endpoint is under direct control. Motion in the coordinate system of the endpoint along specified directions or trajectories (line, circle, etc.) is possible. This technique is used in manual control of manipulators and as a computational method for achieving programmed coordinate axis control in robots.

Resolver: A rotary or linear feedback device that converts mechanical motion to analog electrical signals that represent motion or position.

Responsive Robots: Also known as adaptive robots or reactive robots. Robots equipped with a set of competences that they use in order to "survive" in unknown or partially unknown environments after completion of the designer's task. This type of robotics systems are especially suitable for conditions of unreliable or noisy sensors where the robot operation has to be responsive to unexpected events and conditions (see also Error Recovery).

Retroreflective Sensing: A photoelectric source consolidation method based on the aiming of the light beam into a white retro target feeding a photoelectric sensor.

Reversal Error: The deviation between the positions and orientations reached at the ends of several repeated paths.

RIA: Robotic Industries Association.

Rigidity: The property of a robot to retain its stiffness under loading and movement. Rigidity can be improved by features such as a cast-iron base, precision ball screws on all axial drives, ground and hardened spiral bevel gears in the wrist, brakes on the least stiff axes, and end-effector design that permits a workpiece or tool to be held snugly (Contrast with Flexi-Arm and with Flexibility, Mechanical).

ROBCAD: A three-dimensional graphical engineering software platform developed by Tecnomatix, Inc. for the design of robotics systems and robotic manufacturing processes. It facilitates concept analysis, layout design, and robot programming in the office environment.

ROBEX: An off-line programming system developed at the Machine Tool Laboratory in the Federal Republic of Germany for the control of a robotic arm. ROBEX supports collision detection and prevention.

ROBODOC System: A stereotactic CASP/CASE system for hip replacement surgery.

Robot: See Industrial Robot.

Robot Calibration (for Vision): The act of determining the relative orientation of the camera coordinate system with respect to the robot coordinate system.

Robot Dynamics: Mathematical models specifying the equations of manipulator motions subject to forces and relative to a chosen coordinate system.

Robot Ergonomics: The study and analysis of relevant aspects of robots in working environments, including interactions and collaboration between robots and with people. It is used to provide tools for the purpose of optimizing overall performance of the work system, including analysis of work characteristics, work methods analysis, robot selection, workplace design, performance measurement, and integrated human and robot ergonomics. Robot work should be optimized to:

1. Minimize the time/unit and cost of work produced
2. Minimize the amount of effort and energy expanded by the operator (robot and/or human)
3. Minimize the amount of waste, scrap, and rework
4. Maximize quality of work produced
5. Maximize safety

Robot Hand: See Hand (Robot's).

Robot Language: Set of instructions (keywords) providing the interface between the robot programmer and the robot controller. The robot controller interprets the compiled version of the constructed program and transforms it into robot motion and operation. The robot language can be motion- (robot displacements) or task- (robotics operations) based.

Robot Learning: The improvement of robot performance by experience. Robots learn by three different methods:

1. Being taught by an operator via a teach box
2. Being taught via off-line geometric database and programs
3. Learning from on-line experience

Robot learning is found in three main areas:

1. Robots learning about their operation
2. Human operators learning to accept and work with robots
3. Organizations learning to introduce, integrate, and effectively utilize robots

Robot Mobility: See Mobile Robot.

Robot Model Calibration: Second level of calibration procedures. It involves the basic kinematic geometry of the robot in addition to the joint-level calibration.

Robot Modeler: A component of an off-line programming system which describes the properties of jointed mechanisms. The joint structure, constraints, and velocity data are stored in the system to give a kinematic representation of the robot. Robot modelers are classified as kinematic, path control, or generalized types.

Robot Process Capability (RPC): Ability of a robotic system to consistently perform a job with a certain degree of accuracy, repeatability, reproducibility, and stability. It is a function of the task variables, such as speed of movement, spatial position, and load.

Robot Risk Assessment: A technique for economical evaluation of hazards in the robot's operation. Hazards are evaluated in terms of their severity and probability of occurrence, and countermeasures are identified for each hazard. Finally the method compares the control measures against the expected loss reduction.

Robot Selection Data: The following data are typically needed for the selection of a robot for a given task: work envelope, repeatability, accuracy, payload, speed, degrees of freedom, drive, control, and foundation type.

Robot Simulation: See Simulation; ROBCAD.

Robot System: A Robot System includes the robot(s) (hardware and software) consisting of the manipulator, power supply, and controller; the end-effector(s); any equipment, devices, and sensors required for the robot to perform its task; and any communications interface that is operating and monitoring the robot, equipment, and sensors. (This definition excludes the rest of the operating system hardware and software.)

Robot Task: Specification of the goals for the position of object being manipulated by the robot, ignoring the motions required by the robot to achieve these goals.

Robot Time and Motion (RTM) Method: A technique developed at Purdue University in Indiana to estimate the cycle time for given robot work methods without having to first implement the work method and measure its performance. RTM is analogous to the Methods Time Measurement technique for human work analysis. This system is made up of three major components: RTM elements, robot performance models, and an RTM analyzer.

Robot Workstation Design: The use of geometric requirements of a workstation, including gross size of moves as well as their directions, and design scenarios for carrying out operations in conjunction with an economic analysis to select robots, computers, and tooling for an operation.

Robot–Human Charts: Detailed relative lists of characteristics and skills of industrial robots and humans that were developed at Purdue University. These charts are used to aid engineers in determining whether a robot can perform a job, or as a guideline and reference for robot specifications. The charts contain three main types of characteristics: physical, mental and communicative, and energy.

Robot-Level Language: A robot system computer programming language with commands to access sensors and specify robot motions.

Robotic Assembly: The combination of robots, people, and other technologies for the purpose of assembly in a technologically and economically feasible manner. Robotic assembly offers an alternative with some of the flexibility of people and the uniform performance of fixed automation.

Robotic Autonomous Guidance: See Mobile Robot.

Robotic Cell: A robot-served cluster of workstations which contains no internal buffers for work-in-process and in which only a single family of parts is produced.

Robotic Fixturing: A programmable fixture system that can accommodate a set of parts for processing in the same workcell.

Robotic Welding System: Welding is currently the largest application area for industrial robots, including mainly spot and arc welding and the emerging laser welding. A robotic welding system comprises one or more robots, controls, suitable grippers for the work and the welding equipment, one or more compatible welding positioners with controls, and a suitable welding process with high-productivity filler. The correct safety barriers and screens, along with a suitable material-handling system, are also required.

Robotics: The science of designing, building, and applying robots.

Robotics Automation: The intelligent and interactive connection of perception to action through cognition and planning. It includes the following technologies: kinematics; dynamics; control and simulation of robots and automatic machines; sensing and perception—vision and other noncontact sensors; tactile and other contact sensing systems; systems control theory and applications as related to the modeling of robotics systems; robot mobility and navigation; robotics-related computer hardware and software components; architectures and systems; advanced command and programming languages for robots; linkages to computer-aided design; engineering and manufacturing information systems; electronic and manufacturing science and technology as related to robotics; human–machine interfaces as related to robotics and automation; and management of flexible automation.

Robotics System Design: Decisions concerning the operations that will be performed by the robotic system, selecting the equipment, and deciding how it is to be configured. This stage is critical to the eventual performance of the system since the choices made here limit the options that are available under the later system operation phase.

Robotics System Planning: The feasibility study and preliminary system design, usually with consideration of several alternatives, as well as preliminary economic evaluations.

Robust Processes: See Process Capability.

Roll: Also known as *twist*. The rotational displacement of a joint around the principal axis of its motion, particularly at the wrist.

ROSS: An object-oriented language suitable for use in a multi-robot distributed system. All processing in ROSS is done in terms of message passing among a collection of "actors" or "objects."

RPL (Robot Programming Language):   A robot motion software developed at SRI International in California. RPL was designed for the development, testing, and debugging of control algorithms for manufacturing systems consisting of manipulators, sensors, and auxiliary equipment. Continuous path motion is supported by this language. Features of RPL include a manual teach mode and commands for object recognition, gripper operation, feedback, touch sensors, and vision. The programming is based on LISP cast in a Fortran-like syntax.

RTM: See Robot Time and Motion Method.

Run-Length Encoding: A method of storing compressed data of a binary image. Each horizontal line of the image is represented by a run-length representation containing only the column numbers at which a transition from 0 to 1, or vice versa, takes place. Run-length encoding can provide considerable efficiency in image processing compared to pixel-by-pixel analyses.

S

Scanning Probe Microscopy (SPM): An instrument for removing small amounts of material from a silicon substrate that has been passivated with hydrogen by moving the tip of an SPM in a straight line over the substrate and applying a suitable voltage.

Scanning Tunneling Microscope (STM): An instrument for producing surface images with atomic-scale lateral resolution, in which a fine probe tip is raster-scanned over the surface at a distance of 0.5–1 nanometer and the resulting tunneling current, or the position of the tip required to maintain a constant tunneling current, is monitored.

Screw Theory: A geometric description of the robot kinematics, based on the fundamental fact that any arbitrary rigid body motion is equivalent to a rotation about a certain line combined with a translation parallel to that line.

Sealing (Robotic): The application by a robotic arm of the sealing compound in the precise sealing path and amount required by the product. Common applications are the application of adhesives and insulation material.

Seam Tracking: A control function to assure the quality of the weld seam. Various parameters of a robotic arc weld are monitored and the data are then used to correct both the cross-seam and stickout directions. Lateral welding torch motion is required and is provided by a weaving motion for seam tracking.

Search Routine: A robot function that searches for a precise location when it is not known exactly. An axis or axes move slowly in one direction until terminated by an external signal. It is used in stacking and unstacking of parts, locating workpieces, or inserting parts in holes.

Second-Generation Robot Systems: A robot with a computer processor added to the robot controller. This addition makes it possible to perform, in real time, the calculations required to control the motions of each degree of freedom in a cooperative manner to effect smooth motions of the end-effector along predetermined paths. It also becomes possible to integrate some simple sensors, such as force, torque, and proximity, into the robot system, providing some degree of adaptability to the robot's environment.

Secondary References: Geometric references used in assembling a main assembly component. The compliance with the references of the components guarantees the correct geometry of the completed assembly.

Security (Robotic): The application of a mobile robot as a security guard, patrolling a designated area while monitoring for intrusion and other unwanted conditions, such as fire, smoke, and flooding. The advantages of robotic systems in security are economical (with a robotic system replacing four to five people on a 24-hour-per-day basis), reduced training and supervision, the gathering and analysis of large amounts of information for decision-making, and better reactiveness than man-based security systems.

Selective Compliance Assembly Robotic Arm (SCARA): A horizontal-revolute configuration robot designed at Japan's Yamanachi University. The tabletop-size arm with permanently tilted, high-stiffness links sweeps across a fixtured area and is especially suited for small-parts insertion tasks in the vertical (z) direction.

Self-Assembly Approach: One of the two main approaches for building useful devices from nanoscale components and a natural evolution of traditional chemistry and bulk processing. Self-assembly has severe limitations because the structures produced tend to be highly symmetric, and the most versatile self-assembled systems are organic and therefore generally lack robustness.

Self-Organizing Systems: Systems that are able to affect or determine their own internal structure, for purposes of maintenance, repair, recovery, and adaptation.

Self-Replicating Robots: Robotic systems capable of constructing duplicates of themselves, which would then be capable of further self-replication.

Semiautonomous Control: A method for controlling a robot whereby a human operator sets up the robot system for some repetitive task and a computer subroutine then takes over to complete the assigned task.

Sensing: The feedback from the environment of the robot which enables the robot to react to its environment. Sensory inputs may come from a variety of sensor types, including proximity switches, force sensors, and machine vision systems.

Sensor: A device such as a transducer that detects a physical phenomenon and relays information about it to a control mechanism.

Sensor Coordinate System: A coordinate system mounted over the workspace of the robot and assigned to a sensor.

Sensor Fusion: Also known as *sensor integration*. The coordination and integration of data from diverse sources to produce a usable perspective for a robotics system. A large number of sensors can be applied, and the information they gather from the work environment or workpiece is analyzed and integrated in a unique meaningful stream of feedback data to the robotic manipulator(s).

Sensor Integration: See Sensor Fusion.

Sensor System: The components of a robot system which monitor and interpret events in the environment. Internal measurement devices, also considered sensors, are a part of closed axis-control loops and monitor joint position, velocity, acceleration, wrist force, and gripper force. External sensors update the robot model and are used for approximation, touch, geometry, vision, and safety. A data acquisition system uses data from sensors to generate patterns. A data processing system then identifies the patterns and generates frames for the dynamic world-model processor.

Sensor-Triggered Reflex: A preprogrammed emergency response of a manipulator, induced by a sensor upon detection of certain events, such as an intruder being detected in the workspace or an impending collision with unexpected obstacles.

SensorGlove: A robotics sensor capable of precision measurement of human gestures, with applications in surgery and telerobotics.

Sensory-Controlled Robot: Also known as *intelligent robot*. A robot whose program sequence can be modified as a function of information sensed from its environment. The robot can be servoed or non-servoed.

Sequencer: A controller which operates an application through a fixed sequence of events.

Sequential Program Control: A robot program control structure which allows the execution of a program in ordered steps. Standard structured programming language constructs for sequential control are branches, IF tests, and WHILE loops.

Service (Robotic): The application of robotics systems aiming to achieve a high level of flexibility, adaptability, safety, and efficiency in environments populated by humans. The learning cycle is continuous to ensure adaptability in steadily changing environments. The user of service robots is usually a nontechnical person. Applications include health and safety, cleaning and maintenance, security, office and food deliveries, entertainment, and other services.

Servo-Control Level: The lowest level of the control hierarchy. At this level, drive signals for the actuators are generated to move the joints. The input signals are joint trajectories in joint coordinates.

Servo-Controlled Robot: A robot driven by servo mechanisms, that is, motors or actuators whose driving signal is a function of the difference between command position and/or rate, and measured actual position and/or rate. Such a robot is capable of stopping at, or moving through, a practically unlimited number of points in executing motions through a programmed trajectory.

Servo-System: A control system for the robot in which the control computer issues motion commands to the actuators and internal measurement devices measure the motion and signal the results back to the computer. This process continues until the arm reaches the desired position.

Servo-mechanism: An automatic control mechanism consisting of a motor or actuator driven by a signal which is a function of the difference between commanded position and/or rate, and measured actual position and/or rate.

Servo-valve: A transducer whose input is a low-energy signal and whose output is a higher-energy fluid flow which is proportional to the low-energy signal.

Settling Time: The time a system requires to stay within 2 to 5 percent of its steady state value after an impulse has been issued to its controller. The impulse in a robotic system takes the form of a controller's command to move the robot arm or operate the end-effector.

Shake: The uncontrollable vibration of a robot's arm during, or at the end of, a movement.

SHAKEY: A pioneering man-sized mobile robot on wheels equipped with a range-finding device, camera, and other sensors. The SHAKEY project was developed in the late 1960s at Stanford Research Institute to study robot plan formation.

Shape Memory Alloy (SMA):   See Memory Alloy.

Shell Arm Structure: A robot arm structure designed to yield lower weight or higher strength to weight ratios. Such design is typically more expensive and generally more difficult to manufacture. Cast, extruded, or machined hollow beam-based structures are often applied, and though not as structurally efficient as pure monocoque designs, they can be more cost-effective.

Shipbuilding (Robotic):   The application of robotics systems for the welding, hull assembly, coating, and blasting of large hull structures of ships. This type of application requires special robotic systems for handling workpieces of typically 33 ft $\times$ 15 ft at subassembly stage and 65 ft $\times$ 49 ft (weighing as much as 700 tons) at assembly stage. Work piece fixtures must be designed for handling these dimensions and weight and allow the robots to perform the welding, coating, or abrasive operation. Production quantities also challenge the implementation of robotic systems. Ships are built in quantities of 10 to 20 units per design; however, a greater degree of standardization is present at the joints utilized for assembling the hull structures, allowing the robot implementation in repeated basic activities such as welding.

Shoulder: The manipulator arm linkage joint that is attached to the base.

Shrinking (Image):   The transformation from an input binary image to an output binary image to decrease the number of one type of pixel for purposes of smoothing, eliminating noise, and detecting blobs based on their approximate size.

SIGLA: An assembly-level manipulator model language developed by the Olivetti Corporation in Italy, in which the focus is on the end-effector's motion through space. SIGLA supports multiple arms, gripper operation, touch sensors, force feedback, parallel processing, tool operations, interaction with external devices, relative or absolute motion, and an anticollision command.

Simulation: A controlled statistical sampling technique for evaluating the performance of complex probabilistic, robotics manufacturing, and other systems. Robot simulation is helpful in the design of robotics mechanisms and control, design of robotics systems, and off-line programming and calibration.

Single Stage Joint: A linear motion joint made up of a moving surface which slides linearly along a fixed surface.

Single-Point Control of Motion: A safeguarding method for certain maintenance operations in which it is necessary to enter the restricted work envelope of the robot. A single-point control of the robot motion is used such that it cannot be overridden at any location, in a manner which would adversely affect the safety of the persons performing the maintenance function. Before the robot system can be returned to its regular operation, a deliberate separate action is required by the person responsible for it to release the single-point control.

Skeletonizing: The transformation from an input binary image to an output binary image. This transformation is similar to shrinking in that blobs are guaranteed to never shrink so far that they entirely disappear.

Skill Twist (Robot-Caused): The change in level of skill requirements of jobs eliminated and jobs created by industrial robotics. The performance requirement of jobs eliminated are mostly at the semiskilled or unskilled level, while those created are generally at a higher technical level.

Slew Rate: The maximum velocity at which a manipulator joint can move; a rate imposed by saturation in the servo-loop controlling that joint.

Slip Sensors: Sensors that measure the distribution and amount of contact area pressure between hand and objects positioned tangentially to the hand. They may be single-point, multiple-point (array), simple binary (yes–no), or proportional sensors.

Sojourner: A Mars rover that in 1997 successfully drove off the rear ramp of the Mars Pathfinder Lander onto the Martian surface. It was the first time an intelligent robot was reacting to unplanned events on the surface of another planet. Sojourner's dimensions were about 2 ft × 1.5 ft × 1 ft and it traveled at a speed of 0.4 inches per second. Remote commands from Earth took about 10 minutes to reach Sojourner.

Sonar Sensing: A system using underwater sound, at sonic or ultrasonic frequencies, to detect and locate objects in the sea or for communication; the commonest sensing type is echo-ranging sonar; other versions are passive sonar, scanning sonar, and searchlight sonar.

Sorting (Robotic): The integrated operation of a sensor system and a robot for the discrimination of two or more types of items, e.g., fruits, boxes, or workpieces. The sensor system can be as simple as dimensional-based (detection of several discrete gripper fingers apertures) or as complex as vision-based (recognition of distinctive features, including color). Once the item is identified and classified by the sensor system, the robot will sort it based on the output of the sensor system fed to the robot's controller.

Space Exploration (Robotic): The utilization of robotic systems in outer space, with five basic objectives:

1. Geologically characterize the landscape of the target location
2. Look for macroscopic evidence of life
3. Acquire surface samples
4. Support the magnetic and physical properties investigations
5. Support atmospheric characterization

Autonomy of the robotic system is a necessary characteristic for the exploration task given the robots' autonomy, which, based on the information gathered and its interpretation, allows the robotic system to perform sequences of operations during transmission-delayed ground commands. Endurance against the outer space harsh conditions, lift-off, and landing stresses is also a must for the robotic system. The most recent robotic explorer is the Sojourner (see Sojourner), the latest success in the program started by the Viking Lander sent for the exploration of Mars in 1975.

Space Manufacturing (Robotic): The concept of building a highly automated manufacturing facility (the Space Manufacturing Facility) which would initially be based on teleoperated robotic systems, later replaced by autonomous robotic systems. The objective is the manufacture of solar power stations, communication satellites, and products requiring outer space environmental conditions, using mainly nonterrestrial materials.

Space Robot: A robot used for manipulation or inspection in an earth orbit or deep space environment.

Spatial Resolution: A value describing the dimensions of an image by the number of available pixels; for example, 512 × 512. A relatively larger number of pixels implies a relatively higher image resolution.

Speech Recognition: The process of analyzing an acoustic speech signal to identify the linguistic message that was intended, so that a machine can correctly respond to spoken commands.

Speed Control Function: A feature of the robot control program used to adjust the velocity of the robot as it moves along a given path.

Spherical Robot: Also known as a *spherical coordinate robot* or *polar robot*. A robot operating in a spherical work envelope, or a robot arm capable of moving with rotation, arm inclination, and arm extension.

Splined Joint Trajectory: A technique for following a path described in Cartesian coordinates. Several points are selected from the Cartesian path and transformed into angular displacements of the robot's joints that are then controlled to move along straight-line segments in joint coordinates. These motions may or may not correspond precisely to the straight line specified in Cartesian coordinates, resulting in an error between the two paths that can be reduced by adding intermediate points.

Spot-Welding Robot: A robot used for spot-welding and consisting of three main parts: a mechanical structure composed of the body, arm and wrist; a welding tool; and a control unit. The mechanical structure serves to position the welding tool at any point within the working volume and orient the tool in any given direction so that it can perform the appropriate task.

Spraying (Robotic):   See Coating (Robotic).

Springback: The deflection of a manipulator arm when the external load is removed.

SRI Vision Module: A self-contained vision subsystem developed at SRI International to sense and process visual images in response to top-level commands from a supervisory computer.

Standard Data Exchange Format: Computer-Aided Design data formats for exporting and importing geometric data from one CAD system to another. Some standards are: IGES (Initial Graphics Exchange Standard), DXF (Drawing Transfer File), STEP (Product Data Representation and Exchange), and VDAFS (a European standard).

Static Deflection: Also known as *static behavior* or *droop*. Deformation of a robot structure considering only static loads and excluding inertial loads. Sometimes the term is used to include the effects of gravity loads.

Stepping Motor: A bidirectional, permanent-magnet motor which turns through one angular increment for each pulse applied to it.

Stereo Analysis, Area Based: A stereo image depth analysis in which a succession of windows in the left image are matched to corresponding windows in the right image by cross correlation.

Stereo Analysis, Edge-Based: A stereo image depth analysis characterized by candidate points for matching which represent changes in image intensity.

Stereo Imaging: The use of two or more cameras to pinpoint the location of an object point in a three-dimensional space. Also known as *stereo vision*.

Stereotactic CASP/CASE: Computer-Integrated Surgery systems exploiting the positional accuracy of machines to position surgical tools in space.

Sticking Effect: The resulting effect from adhesive forces in the manipulation of parts with size less than a millimeter and masses less than 10^{-6} kg where the gravitational and inertial forces may become insignificant.

Stiffness: The amount of applied force per unit of displacement of a compliant body.

Stop (Mechanical):   A mechanical constraint or limit on some motion. It can be set to stop the motion at a desired point.

Structured Light Imaging: Depth information extraction technique which consists of scanning a scene illuminated with a laser stripe, capturing the image of the stripe with an off-set camera, and then, through triangulation, calculating the 3D coordinates of each of the illuminated points in the scene so that three-dimensional patterns can be determined.

Super-Articulated Robot: Robotic systems having a large number of rotary joints in several places along the body. This type of robot usually resembles an elephant-trunk or a snake-like mechanism.

Superconductive Solenoids: A solenoid whose operation depends on superconductivity as produced by temperatures near the absolute zero.

Supervisory Control: The overall control and coordination of the robot system, including both the internal sections of the system and synchronization with the external environment comprising the operator, associated production devices, and possibly a higher-level control computer. This control strategy enables a higher level of robot decision-making. Lower-level controllers perform the control task continuously in real time and usually communicate back with the supervisory control level.

Supination: An orientation or motion toward a position with the front, or unprotected side, face up or exposed. See pronation.

Support and Containment Device: A type of end-of-arm tooling requiring no power, such as lifting forks, hooks, scoops, and ladles. The robot moves to a position beneath a part to be transferred, lifts to support and contain the part or material, and performs the transfer process.

Surface Mount Technology: The technique of mounting circuit components and their electrical connections on the surface of a printed board, rather than through holes.

Surgery (Robotic):   The application of robotic systems in the planning and execution of surgical procedures. The robotic system is aimed to overcome the limitations of traditional surgery by providing feedback information during the surgery (by X-ray, computer-assisted tomographic scanning (CAT), magnetic resonance scanning, or ultrasound), delivering precision required in surgical procedures where delicate and vital structures can be destroyed by a small hand slip (e.g., in spine, brain, or retina surgery) or a misguided instrument, increasing the access and dexterity and providing better accuracy and repeatability in implementing the planned surgical actions. The main stream of applications is stereotactic-based (the guiding of a probe through a planned trajectory to the damaged area); however, future applications will involve the robotic system in a more active role, performing surgical incisions with the assistance of an appropriate vision system. This type of applications will require multiple hand robots or the coordination of multiple single-hand robots.

Surgical Augmentation Systems: Computer-Integrated Surgery systems extending the sensory-motor abilities to overcome the limitations of traditional surgery. One of its applications involves the utilization of telerobotic systems to augment human manipulation abilities in microsurgery.

Surveillance Systems: Systems observing air, surface, or subsurface areas or volumes by visual, electronic, photographic, or other means, for intelligence gathering or other purposes.

Swarm Systems: Systems of a relatively large number of simple robotics members (hundreds or more). These organized groups of robots exhibit collectively intelligent behavior. Robots of different types, each executing different protocols, can be arranged in a variety of patterns to form structures such as a measuring swarm, a distributed sensor where each member of the swarm is sensitive to a specific value.

Swing Arm Robot: A robot arm typically with five- or six-axis, medium payload articulated arm mounted on a sixth or seventh axis swing-arm rotational base. It allows for greater reach and faster cycle times.

System Configuration: An iterative design process consisting of evaluation of the factors affecting the product and the production tasks, selection of a design concept based on these factors, and evaluation of the performance of the selected concept. Following the results of the evaluation, a system can be refined and reevaluated, or discarded.

System Planning: An early phase of robotic system selection during which the concern is to establish feasibility of the project and determine the initial estimates for system size, cost, return on investment, and other variables of interest to upper management.

System Reliability: The probability that a system will accurately perform its specified task under stated environmental conditions.

T

T3: The "Tomorrow Tool Today" was an industrial robot family and a robot programming language developed by Cincinnati Milacron, Inc. in the 1980s. The language was based on programmed motions to the required points. T3 supported straight-line motion taught with a joystick, continuous path motion, conveyor tracking, and interaction with external devices, sensors, and tool operations.

Tactile Sensing: The detection by a robot through contact by touch, force, pattern slip, and movement. Tactile sensing allows for the determination of local shape, orientation, and feedback forces of a grasped workpiece.

Task Planner: A part of a task-level programming language which transforms task-level specifications into robot-level specifications. The output of the task planner is a robot program for a specific robot to achieve a desired final state when executed from a specified initial state.

Task-Level Language: A computer programming language for robot programming, that requires specification of task goals for the positions of objects and operations. This goal specification is intended to be completely robot-independent; no positions or paths dependent on the robot geometry or kinematics are specified by the user.

TCP: See Tool Center Point.

TEACH: A robot programming software developed by the California Institute of Technology and the Jet Propulsion Laboratory in California to provide commands for vision, force, and other sensors and process synchronization commands which provide for concurrency in a systematic manner. This software supports multiple manipulators and other peripheral devices simultaneously.

Teach Pendant: Also known as *teach box*. A portable, hand-held programming device connected to the robot controller containing a number of buttons, switches, or programming keys used to direct the controller in positioning the robot and interfacing with auxiliary equipment. It is used for on-line programming.

Teach Pendant Programming: See On-Line Programming.

Teach Programming: Also known as *teaching* or *lead-through programming*. A method of entering a desired control program into the robot controller. The robot is manually moved by a teach pendant or led through a desired sequence of motions by an operator. The movement information as well as other necessary data are recorded by the robot controller as the robot is guided through the desired path.

Teach Restrict: A facility whereby the speed of movement of a robot, which during normal operation would be considered dangerous, is restricted to a safe speed during teaching.

Teleoperation: The use of robotic devices which have mobility and manipulative and some sensing capabilities and are remotely controlled by a human operator. Teleoperation is advantageous where human control is preferred but the environment is hazardous or undesirable, e.g., remote manipulation in space, nuclear reactors, explosive material handling, or clean-room assembly.

Telepresence: The mental state entered when sensory feedback has sufficient scope and fidelity to convince the user that he or she is physically present at the remote site.

Telescoping Joint: A linear motion joint consisting of sets of nested single-stage joints.

Template Matching: The comparison of sample object image against a stored pattern, or template. This technique is used for inspection by machine vision.

Textual Programming Systems: Programming system applying a text-based interface for inputting program instructions. It usually requires a longer time to master and provides less information than graphical programming systems.

TheoAgent: A hybrid reactive behavior-based robot control architecture developed at Carnegie-Mellon University in Pennsylvania. It embodies the philosophy: "Reacts when it can, plans when it must." Stimulus–response rules encode the reactivity options. The method for selecting the rules constitutes the planning component. TheoAgent is primarily concerned with learning new rules: in particular, learning to act correctly; learning to be more reactive (reduce time); and learning to perceive better (distinguish salient features).

Theodolite System: A noncontact measuring system for calibration employing a triangulation technique and utilized to determine the spatial location of the robot's end-effector.

Third-Generation Robot Systems: Robot systems characterized by the incorporation of multiple computer processors, each operating asynchronously to perform specific functions. A typical third-generation robot system includes a separate low-level processor for each degree of freedom and a master computer supervising and coordinating these processors as well as providing higher-level functions.

3D Laser Tracking System: A calibration system to accurately determine the robot's end-effector pose with a set of external measurement devices for a given set of joint displacements recorded by internal (joint-based) measurement devices. The system usually includes a tracker consisting of a mirror system with two motors and two encoders, a laser interferometer, and a precision distance sensor.

3D Orientational and Positional Errors: Errors generated in the three-dimensional analysis of stereo vision images. Errors result from the selection of features from both images and the three-dimensional object construction algorithm implemented.

Three-Roll Wrist: A wrist with three interference-free axes of rotational movement (pitch, yaw, and roll) that intersect at one point to permit a continuous or reversible tool rotation which simplifies the required end effector design by its extensive reachability. It was originally designed by Cincinnati Milacron, Inc.

Thresholding: A procedure of binarization of an image by segmenting it to black and white regions (represented by ones and zeroes). The gray level of each pixel is compared to a threshold value and then set to 0 or 1 so that binary image analysis can then be performed.

Tilt: The orientation of a view, as with a video camera, in elevation.

Time Effectiveness Ratio: A measure of performance for teleoperators which, when multiplied by the task time for the unencumbered hand, yields the task time for the teleoperator.

Time to Market: A measure of the time required by the production system to deliver a product to the market. Time to market is affected by the flexibility of the system to incorporate changes in product and process, as well as coordination and integration of the different production and management functions within the production system.

Tool Center Point (TCP): A tool-related reference point that lies along the last wrist axis at a user-specified distance from the wrist.

Tool Changing (Robotic): An alternative to dedicated, automatic tool changers that may be attractive because of an increased flexibility and a relatively lower cost. A robot equipped with special grippers can handle a large variety of tools which can be shared quickly and economically by several machines.

Tool Coordinate System: A coordinate system assigned to the end-effector.

Tool-Coordinate Programming: Programming the motion of each robot axis so that the tool held by the robot gripper is always held normal to the work surface.

Torque Control: A method to control the motions of a robot driven by electric motors. The torque produced by the motor is treated as an input to the robot joint. The torque value is controlled by the motor current.

Torque/Force Controller: A control system capable of sensing forces and torques encountered during assembly or movement of objects, and/or generating forces on joint torques by the manipulator, which are controlled to reach desired levels.

Touch Sensors: Sensors that measure the distribution and amount of contact area pressure between hand and objects perpendicular to the hand. Touch sensors may be single-point, multiple-point (array), simple binary (yes–no), or proportional sensors, or may appear in the form of artificial skin.

Track-Mounted Robot: A robot arm with typically four- to six-axis, medium payload articulated arm mounted on an auxiliary axis linear track that travels in the same direction as the flow of the production line.

Tracking: A continuous position-control response to continuously changing input requirements.

Tracking (Line): The ability of a robot to work with continuously moving production lines and conveyors. *Moving-base* line tracking and *stationary-base* line tracking are the two methods of line tracking.

Tracking Sensor: Sensors used by the robot to continuously adjust the robot path in real time while it is moving.

Training by Showing: The use of a vision system to view actual examples of prototype objects in order to acquire their visual characteristics. The vision system can then classify unknown objects by comparison with the stored prototype data.

Trajectory: A subelement of a cycle that defines lesser but integral elements of the cycle. A trajectory is made up of points at which the robot performs or passes through an operation, depending on the programming.

Transducer: A device that converts one form of energy into another; for instance, converting the movement along certain distance to a number of electrical pulses that can be counted.

Translation: A movement such that all axes remain parallel to what they were (i.e., without rotation).

Transport (Robotic): The acquisition, movement through space, and release of an object by a robot. Simple material-handling tasks requiring one- or two-dimensional movements are often performed by nonservo robots. More complicated operations, such as machine loading and unloading, palletizing, part sorting, and packaging, are typically performed by servo-controlled, point-to-point robots.

Transportation (Robotic): Similar to robotic transport, but over larger distances, involving intelligent robotic vehicles, automated roadways, and computer-supported transportation management systems.

Triangulation Ranging: Range-mapping techniques that combine direction calculations from a single camera and the previous known direction of projected light beams.

Tropism System Cognitive Architecture: A cognitive architecture for the individual behavior of robots in a colony, developed at the University of Southern California in California. Experimental investigation of the properties of the colony demonstrates the group's ability to achieve global goals, such as the gathering of objects, and to improve its performance as a result of learning, without explicit instructions for cooperation.

U

Ultrasonic Sensor: A range-measuring device which transmits a narrow-band pulse of sound towards an object. A receiver senses the reflected sound when it returns. The time it takes for the pulse to travel to the object and back is proportional to the range.

Underactuated Robotic System: Robotics system with fewer actuators than degrees of freedom. For this type of robotics system, traditional models of control do not work, since the linearization will always fail to be controllable.

Unidirectional Assembly: Assembly process performed in a single direction (usually top-down). This type of assembly process simplifies the planning of the robot motion during the assembly, as well as the design of the assembly and its components.

Unit Task Times: A method of predicting teleoperator performance in which the time required for a specific task is based on completion of unit tasks or component subtasks.

Universal Fixture: A fixture designed to handle a large variety of objects. See Programmable Fixture.

Universal Transfer Device (UTD):   A term first applied to a Versatran robot (Versatran was one of the pioneering robot manufacturers that was later acquired by Prab Company of Michigan) used for press loading at the Canton Forge Plant of the Ford Motor Company, and later to other robots at Ford plants. The use of the term was discontinued in 1980.

Unmanned Air-Vehicles (UAV):   An application of mobile flying robots under teleoperation control, mostly in military applications such as reconnaissance missions.

Unmanned Manufacturing Cells: Cells with the ability to operate autonomously, replacing the human's decision-making and sensory abilities while producing superior-quality parts (zero defects), without failures to disrupt the system. The cell must be able to react to changes in demand for the parts and therefore must be flexible. The cells must also be able to adapt to changes in the product mix and to accommodate changes in the design of existing parts for which the cell was initially designed planned.

Upper Arm: The portion of a jointed arm that is connected to the robot's shoulder.

User-Friendly: A common term implying ease of learning and operating a complex system by human users, especially via a computer interface. A more scientific term is *Human Oriented Design*.

User Interface: The interface between the robot and the operator through devices such as a teach pendant or PC. It provides the operator with the means to create programs, jog the robot, teach positions, and diagnose problems.

V

Vacuum Cups: A type of pneumatic pickup device which attaches to parts being transferred via a suction or vacuum pressure created by a venturi transducer or a vacuum pump. They are typically used on parts with a smooth surface finish, but can be used on some parts with nonsmooth surface finish by the adding of a ring of closed-cell foam rubber to the cup.

VAL (Versatile Assembly Language):   An assembly-level robot programming language developed by Unimation, Inc., in the late 1970s and an outgrowth of work done at California's Stanford University that provides the ability to define the task a robot is to perform. VAL's features include continuous path motion and matrix transformation.

VAL-II: An enhanced and expanded assembly-level robot control and programming system based on VAL and developed by Unimation, Inc. VAL-II includes the capabilities of VAL as well as a capability for communication with external computer systems at various levels, trajectory modifications in response to real-time data, standard interfaces to external sensors, computed or sensor-based trajectories, and facilities for making complex decisions.

Vehicle Navigation: A system incorporating control surfaces or other devices which adjusts and maintains the navigation course, and sometimes speed, of a robotic vehicle in accordance with signals received from a guidance system.

Velocity Control: A method to control the motions of a robot driven by electric motors. The robot arm is treated as a load disturbance acting on the motor's shaft. The velocity of the robot arm is controlled by manipulation of the motor voltage.

Vertical Integration: The integration of activities at different hierarchical levels in the enterprise by investigating the passing of information through a control system and an information system. The analysis can be supported through information modeling and an architecture which identifies where various actions and exchanges are taken.

Very Large Scale Integrated Circuit (VLSI):   A complex integrated circuit that contains at least 20,000 logic gates or 64,000 bits of memory.

Viability: The designed ability of Responsive Robots to adapt to changing conditions and unexpected events.

Vibratory Feeder: A feeding mechanism for small parts that causes piece parts to move upward as they vibrate up inclined ledges or tracks, spiraling around the inside of the bowl.

Virtual Fixture: Abstract sensorial data in a virtual reality environment which is overlaid on top of the remote space and used to guide the operator in the performance of tasks.

Virtual Manufacturing Components: Computer model representations which characterize properties of manufacturing components from viewpoints and perspectives of the various people concerned with a manufacturing system throughout its lifetime.

Virtual Reality (VR): A simulation of an environment that is experienced by a human operator provided with a combination of visual (computer-graphic), auditory, and tactile presentations generated by a computer program. Also known as *artificial reality, immersive simulation, virtual environment* and *virtual world*.

Virtual Robot: Virtual representation of a robot manipulator and part of a virtual reality system for robot programming developed at Fraunhofer-IPA, Stuttgart, Germany. The programming is done on a virtual robot and a virtual environment, with the programmer interaction being mediated by virtual reality input/output devices. Thus the programmer wears a sensing glove and a head mounted display and feels immersed in the application. He or she can navigate using a trackball, look at the scene from any angle, and see details that may not be visible in real life. Once the code is debugged, it is downloaded to a real robot controller connected to the same virtual reality engine, and the task is executed. Feedback from the sensors on the real robot is then used to fine-tune the program.

Vision System: A camera (or cameras) system interfaced to guide a robot to locate a part, identify it, direct the gripper to a suitable grasping position, pick up the part, and bring it to the work area. A coordinate transformation between the cameras and the robot must be carried out to enable proper operation of the system.

Vision, Three-Dimensional: The means of providing a robot with depth perception. With three-dimensional (stereo) vision, robots can avoid assembly errors, search for out-of-place parts, distinguish between similar parts, and correct positioning discrepancies (see also Stereo Imaging).

Vision, Two-Dimensional: The processing of two-dimensional images by a computer vision system to derive the identity, position, orientation, or condition of objects in the scene. It is useful in industrial applications, such as inspecting, locating, counting, measuring, and controlling industrial robots.

W

Warmup: A procedure used to stabilize the temperature of a robot's hydraulic components. A warmup usually consists of a limited period of movement and motion actions and is carried out to prevent program-positioning errors whenever the robot has been shut down for any length of time.

Weaving: In robotic arc welding, this is a motion pattern of the welding tool to provide a higher-quality weld. The robot controller produces a weaving pattern by controlling weave-width, left-and-right dwell, and crossing time.

Web Searching (Robotic): The traversing of the World Wide Web by a web robot (an autonomous computer program) retrieving recursively documents that are referenced. Web robots differ from web browsers in that browsers are operated by a human user and do not automatically retrieve referenced documents. Web robots can be used for selective indexing (deciding which sites to retrieve the documents from), HTML validation, link validation, "what's new" monitoring, and mirroring.

Wedging: In rigid part assembly, a condition where two-point contact occurs too early in part mating, leading to the part that is supposed to be inserted appearing to be stuck in the hole. Unlike jamming, wedging is caused by geometric rather than ill-proportioned applied forces.

Welding (Robotic):   Robot manipulation of a welding tool for spot or arc welding. Robots are used in welding applications to reduce costs by eliminating human labor, improve product quality through better welds, and, particularly in arc welding, minimize human exposure to harsh environments. Spot welding automotive bodies, normally performed by a point-to-point servo robot, is currently the largest single application for robots. In such applications robots make from about 40 to over 75 percent of the total spot welds on a given vehicle.

Windowing (Image):   An image analysis technique in which only selected areas of the image are analyzed. The area, or windows, may surround a hole or some other relevant aspect of a part in the field of view. Various techniques can be used to study features of the object in the window.

Windup: A colloquial term describing the twisting of a shaft under torsional load that may cause a positioning error; the twist usually unwinds when the load is removed.

Work Envelope: Also known as the *robot operating envelope* or *workspace*. The set of points representing the maximum extent or reach of the robot tool in all directions (see also Reachability).

Workflow: Automated controls of business that are "structured." It identifies the sequence of activities and the rules dictating the sequence.

Working Range:

1. The volume of space which can be reached by maximum extensions of the robot's axis.
2. The range of any variable within which the system normally operates.

Workspace: See Work Envelope.

Workstation: A location providing a stable, well-defined space for the implementation of related production tasks. Major components may include a station substructure or platform, tool and material storage, and locating devices to interface with other equipment. The workstation is traditionally defined as one segment of fixed technology, a portion of a fixed transfer machine, or a person working at one worksite. Now station can mean one or more robots at a single worksite, a robot dividing its time among several worksites, a robot serving one or several fixed workheads, or any other useful combination.

World Model: A model of the robot's environment containing geometric and physical descriptions of objects; kinematic descriptions of linkages; descriptions of the robot system characteristics; and explicit specification of the amount of uncertainty there is in model parameters. This model is useful in task-level programming.

World-Coordinate Programming: Programming the motion of each robot axis such that the tool center point is the center of the path with no regard to tool pose.

World-Coordinate System: A Cartesian coordinate system with the origin at the manipulator base. The X and Y axes are perpendicular and on a plane parallel to the ground, and the Z axis is perpendicular to both X and Y. It is used to reference a workpiece, jig, or fixture.

World Wide Web Technologies (WWW):   A set of standards for storing, retrieving, formatting, and displaying information using a client/server architecture, graphical user interfaces, and a hypertext language that enables dynamic links to other documents. Robotics applications include distributed programming and control and teleoperation.

Wrist: A set of joints, usually rotational, between the arm and the hand or end-effector, which allow the hand or end-effector to be oriented relative to the workpiece.

Wrist Force Sensor: A structure with some compliant sections and transducers that serve as force sensors by measuring the deflections of the compliant sections. The types of transducers used are strain-gauge, piezoelectric, magnetostrictive, and magnetic.

X

X-Y-θ **Table:**   A robotic mechanism used primarily for positioning parts by translational and rotational planar motions. It can be integrated into a vision system and serve as an intelligent workpiece conveyor/presenter which loads, transports, positions, and orients parts.

Y

Yaw: The angular displacement of a moving joint about an axis which is perpendicular to the line of motion and the top side of the body.

INDEX

ABOUT THE CD-ROM

INTRODUCTION

This CD-ROM presents visually some of the material included in the Handbook of Industrial Robotics, 2nd Edition. References to the corresponding Handbook chapters are indicated for each of the topics. The topics are organized in two main categories:

1) Design of robot manipulators
2) Robotics functions

The images and videos are not intended as an exhaustive compilation nor as a replacement for the Handbook. They complement the text by illustrating the robotics field with color and motion.

The CD application begins automatically, just place it in your CD-ROM drive. If it does not start, run AUTORUN.EXE from the CD. You may need to install Microsoft Media Player (included on the CD) to view the videos.

MINIMUM SYSTEM REQUIREMENTS

- IBM-compatible computer with a Pentium Processor or better, running Windows 95 or better.
- Color Monitor set to view 16-bit color or better.
- 16 MB RAM
- CD-ROM Drive
- Sound card not required, but is needed to hear MIDI files included on the CD.

USER ASSISTANCE

If you need assistance with installation or if you have a damaged disk, please contact Wiley Technical Support at:

Phone: (212) 850-6753
Fax: (212) 850-6800 (Attention: Wiley Technical Support)
Email: techhelp@wiley.com

CUSTOMER NOTE: IF THIS BOOK IS ACCOMPANIED BY SOFTWARE, PLEASE READ THE FOLLOWING BEFORE OPENING THE PACKAGE.

This software contains files to help you utilize the models described in the accompanying book. By opening the package, you are agreeing to be bound by the following agreement:

This software product is protected by copyright and all rights are reserved by the author, John Wiley & Sons, Inc., or their licensors. You are licensed to use this software on a single computer. Copying the software to another medium or format for use on a single computer does not violate the U.S. Copyright Law. Copying the software for any other purpose is a violation of the U.S. Copyright Law.

This software product is sold as is without warranty of any kind, either express or implied, including but not limited to the implied warranty of merchantability and fitness for a particular purpose. Neither Wiley nor its dealers or distributors assumes any liability for any alleged or actual damages arising from the use of or the inability to use this software. (Some states do not allow the exclusion of implied warranties, so the exclusion may not apply to you.)

WILEY